Engineering Mathematics

Volume 2

BABU RAM

Formerly Dean, Faculty of Physical Sciences,
Maharshi Dayanand University, Rohtak

Acquisitions Editor: Anita Yadav
Production Editor: Vamanan Namboodiri

ISBN: 978-81-317- 8503-4

First Impression, 2012

Published by Pearson India Education Services Pvt.Ltd,CIN:U72200TN2005PTC057128.
Formerly known as TutorVista Global Pvt Ltd, licensees of Pearson Education in South Asia

Head Office: 7th Floor, knowledge Boulevard, A-8(A) Sector-62, Noida (U.P) 201309, India

Registered Office: 4th floor, Software Block, Elnet Software City, TS 140 Block 2 & 9,
Rajiv Gandhi Salai, Taramani, Chennai - 600 113, Tamil Nadu, Fax: 080-30461003,
Phone: 080-30461060, www.pearson.co.in Email: companysecretary.india@pearson.com

Digitally printed in India by Repro India Ltd. in the year of 2017.

In the Memory of My Parents

Smt. Manohari Devi and Sri. Makhan Lal

Contents

Preface to the Revised Edition

The book Engineering Maths I caters to the requirements of the revised syllabi of various Indian universities. Eight more chapters have been incorporated in this book to enable the students cover all the topics in the syllabus. Accordingly, the contents of the present book has also been divided into two volumes. Volume I of the book consists of six new chapters, namely, Tangents and Normals, Rectification and Quadrature, Centre of Gravity and Moments of Inertia, Logic, Elements of Fuzzy Logic, and Graphs in addition to the eleven chapters of Part I of first edition of the book. The two chapters, new to this edition, Calculus of Variation and Dynamics are available on the Web site: www.pearsoned.co.in/baburam.

The contents of some of the previous chapters have been rearranged along with additions and refinements. A number of examples, selected generally from various university question papers, have been added in almost all chapters of this book.

I would like to thank Suresh Kumar Godara for the effort he has taken in preparing the figures of the book. My wife Meena Kumari and daughter-in-law Ritu provided the moral support during the revision of this book. My son Aman Kumar, working with Goldman-Sach, Bangalore, offered, as usual, valid comments on the contents of some chapters.

Special thanks are due to Thomas Mathew Rajesh, Anita Yadav and Vamanan Namboodiri at Pearson Education for their constructive support.

Suggestions and feedback on the contents of the book will be gratefully acknowledged.

Babu Ram

1 Preliminaries

In this chapter, we present some basic mathematical concepts and definitions which shall be frequently used in the forthcoming chapters. The knowledge of these concepts is essential due to their extensive use in the study of Fourier series and various integral transforms. For example, knowledge of real- and complex-valued functions—along with their continuity, differentiability, and integrability, and sequences, series, and special functions—is required time and again in this study.

1.1 SETS AND FUNCTIONS

According to Georg Cantor, a *set* may be viewed as a well-defined collection of objects, called the *elements* or *members* of the set.

The sets are denoted by capital letters such A, B, and C whereas its elements are denoted by lowercase letters such as *a*, *b*, and *c*. We write $a \in A$ if *a* is an element of the set A.

EXAMPLE 1.1

(a) $\mathbb{R} = \{x : x \text{ is a real number}\}$ is called the *set of real numbers*.

(b) $\mathbb{Z} = \{x : x \text{ is an integer}\}$ represents the set of *all integers*... $-4, -3, -2, -1, 0, 1, 2, 3, 4, \ldots$

(c) $\mathbb{N} = \{x : x \text{ is a positive integer or zero}\}$ represents the set consisting of 0, 1, 2, 3,...

(d) The set having no elements is represented by ϕ or { } and is called the *empty* (or *null* or *void*) set.

Definition 1.1. Let A and B be sets. Then A is called a subset of B, written as $A \subseteq B$ if and only if every element of A is also an element of B.

EXAMPLE 1.2

The set A = {6, 5, 8} is a subset of the set B = {2, 3, 6,7, 5, 8}.

Definition 1.2. Let A and B be sets. Then A is called a *proper* subset of B if and only if every element of A is in B, but there is at least one element of B that is not in A.

EXAMPLE 1.3

The set {5, 6, 8} is a proper subset of the set {1, 5, 6, 8, 4}.

Definition 1.3. Two sets A and B are called *equal* if every element of A is in B and every element of B is in A. Thus, A = B if and only if $A \subseteq B$ and $B \subseteq A$.

Definition 1.4. Suppose we are dealing with sets, all of which are subsets of a set U. Then, the set U is called a *universal set* or a *universe of discourse*.

Definition 1.5. Let A and B be subsets of a universal set U. Then the *union* of A and B, denoted by $A \cup B$, is the set of all elements *a* in *U* such that *a* is in A or *a* is in B.
Thus,

$$A \cup B = \{a \in U: a \in A \text{ or } a \in B\}.$$

Definition 1.6. Let A and B be subsets of a universal set U. Then, the *intersection* of A and B,

denoted by $A \cap B$, is the set of all elements a of U such that $a \in A$ and $a \in B$.
Thus,

$$A \cap B = \{a \in U\text{: } a \in A \text{ and } a \in B\}.$$

Definition 1.7. Let A and B be subsets of a universal set U. Then, the *difference* B−A (or the *relative complement of* A *in* B) is the set of all elements a in U such that $a \in B$ and $a \in A$.
Thus,

$$B - A = \{a \in U\text{: } a \in B \text{ and } a \in A\}.$$

Definition 1.8. Let A be a subset of a universal set U. Then *complement* of A, denoted by A^c, is the set of all elements a in U such that a is not in A.
Thus,

$$A^c = \{a \in U\text{: } a \in A\}.$$

Definition 1.9. Two sets A and B are called *disjoint* if and only if they have no element in common.

Definition 1.10. Let X and Y be arbitrary given sets. By a *function f* $f: X \to Y$ from the set X into Y, we mean a rule which assigns to each member x of X, a unique member $f(x)$ of Y.

The member $f(x)$ is called *image* of x under the function (mapping) f or the *value* of f at x. The set X is called the *domain* of f and the set Y is called the *codomain* of f. The set of elements $f(x)$, $x \in X$ is called the *range* of f. Thus, the range of f is a subset of Y.

If Y = $\mathbb{R}$, the set of real numbers, then f is called the *real-valued* function and if Y = $\mathbb{C}$, the set of complex numbers, then f is called the *complex-valued* function. If X = $\mathbb{C}$, then f is called a *function of complex variables*.

Definition 1.11. Let f: A → B be a mapping from the set A into the set B. If $f(x_1) = f(x_2) \Rightarrow x_1 = x_2$ for every $x_1, x_2 \in A$, then f is called *one-one mapping* or *injective mapping*.

Thus, a function $f: A \to B$ is injective if and only if the *images of distinct points of A are distinct*, that is, $x_1 \neq x_2 \Rightarrow f(x_1) \neq f(x_2)$.

EXAMPLE 1.4

Let $\mathbb{Z}^+$ be a set of positive integers and Y be a set of even positive integers. Then, the mapping f: $\mathbb{Z}^+ \to Y$ defined by $f(x) = 2x$ is injective. In fact, if $x, y \in \mathbb{Z}^+$, then $f(x) = 2x$, $f(y) = 2y$ and so, $f(x) = f(y)$ implies $2x = 2y$ and hence, $x = y$.

Definition 1.12. A function f which is not one-one is called a *many-to-one mapping*.

EXAMPLE 1.5

The function f defined by $f(x) = x^2$, $-\infty < x < \infty$ is not one-one, because 4 is the image of both −2 and 2.

Definition 1.13. Let $f: X \to Y$ be a map. If (X) = Y, that is, the range of f is the whole of Y, then f is called a *surjective* or *onto mapping*.
Thus, $f: X \to Y$ is onto if and only if for every point y in Y there exists at least one point x in X such that $f(x) = y$.

EXAMPLE 1.6

The linear function $f: \mathbb{R} \to \mathbb{R}$ defined by $f(x) = ax + b$, $x \in \mathbb{R}$ is surjective whereas the function $f: \mathbb{R} \to \mathbb{R}$ defined by $f(x) = \sin x$ is not surjective. In fact, there is no element in $\mathbb{R}$ for which $\sin x = 2$. Thus, the range of f is not equal to $\mathbb{R}$.

Consider f: $\mathbb{Z}^+ \to \mathbb{Z}^+$ defined by $f(x) = x^2$, $x \in \mathbb{Z}^+$. Then, if $x, y \in \mathbb{Z}^+$, we note that

$$f(x) = f(y) \Rightarrow x^2 = y^2$$

$$\Rightarrow (\pm x)^2 = (\pm y)^2$$

$$\Rightarrow \pm x = \pm y$$

$$\Rightarrow x = y\text{, since } x, y \in \mathbb{Z}^+.$$

Hence, f is one-one. Further

$$R(f) = \{1, 4, 9, \ldots\},$$

which is a proper subset of $\mathbb{Z}^+$ Thus, f is *not surjective.*

Definition 1.14. A mapping $f : \mathrm{X} \to \mathrm{Y}$ is called *bijective* if it is both injective and surjective.

For example, if X = $\{x \in \mathbb{R}, x \neq 0\}$ Then, the mapping $f : \mathrm{X} \to \mathrm{X}$ defined by $f(x) = \frac{1}{x}$ is one-one and onto and hence, bijective.

1.2 CONTINUOUS AND PIECEWISE CONTINUOUS FUNCTIONS

Definition 1.15. A function f: X → ℝ is said to be *continuous* at a point $x_0 \in \mathrm{X}$, if given $\varepsilon > 0$, there exists a $\delta > 0$ such that

$$| f(x) - f(x_0) | < \varepsilon \text{ whenever } | x - x_0 | < \delta.$$

Equivalently, we say that f is continuous at x_0 if $\lim_{x \to x_0} f(x) = f(x_0)$.

The *left-hand limit* of f at the point x_0 is defined by

$$\lim_{\substack{x \to x_0 \\ x < x_0}} f(x) = f(x_0 - 0),$$

provided the limit exists and is finite.

Similarly, the *right-hand limit* of f at the point x_0 is defined by

$$\lim_{\substack{x \to x_0 \\ x > x_0}} f(x) = f(x_0 - 0),$$

provided the limit exists and is finite.

At the point of continuity, both the left- and right-hand limit exist and

$$f(x_0 - 0) = f(x_0 + 0) = f(x_0).$$

At a point x_0 of discontinuity, if both $f(x_0 - 0)$ and $f(x_0 + 0)$ exist but are not equal, then x_0 is called a point of *discontinuity of the first kind* or a point of *jump discontinuity*. In such a case, $f(x_0 + 0) - f(x_0 - 0)$ is called the *jump* of the discontinuous function f at the point x_0.

EXAMPLE 1.7

Consider the function f defined by

$$f(x) = \begin{cases} x^2 & \text{for } x \leq 0 \\ 4x + 3 & \text{for } x > 0. \end{cases}$$

Then,

$$f(0-) = \lim_{x \to 0-} x^2 = 0 \text{ and}$$

$$f(0+) = \lim_{x \to 0+} 4x + 3 = 3.$$

Thus, $f(0-) \neq f(0+)$ and so, f has jump discontinuity at $x = 0$. The jump this discontinuity is

$$f(0+) - f(0-) = 3.$$

If at least one of left- and right-hand limit does not exist at a point x_0, then x_0 is called a point of *discontinuity of second kind.*

EXAMPLE 1.8

The function f defined by $f(x) = \frac{1}{t-3}$ has a discontinuity at $t = 3$. Since neither the left- nor the right-hand limit exists at $t = 3$, the function has a discontinuity of second kind.

Definition 1.16. A function f (real-valued or complex-valued) is called *piecewise* (or *sectionally*) *continuous* on the interval $[a, b]$ if there exists a partition $a = x_0 < x_1 < \ldots < x_n = b$ of $[a, b]$ such that f is continuous in each of the open interval (x_{i-1}, x_i), $i = 1, 2, \ldots, n$ and each of the limit $f(a+)$, $f(b-)$, $f(x_i+)$, and $f(x_i-)$ exists for $i = 1, 2, \ldots, n$.

The function f is called the *piecewise continuous on* ℝ if f is the piecewise continuous on each subinterval $[a, b]$ of ℝ.

Some of the properties of piecewise continuous functions are listed below:

(a) A function f continuous on $[a, b]$ is piecewise continuous there.

(b) The sum, difference, and product of two functions which are piecewise continuous on $[a, b]$ are piecewise continuous on $[a, b]$.

(c) A function piecewise continuous on $[a, b]$ is *bounded* on $[a, b]$, that is, there exists a positive constant M such that $| f(x) | \leq \mathrm{M}$ for all $x \in [a, b]$.

(d) The definite integral of a piecewise continuous function exists on [*a*, *b*] and

$$\int_a^b f(x)\,dx = \sum_{i=1}^{n} \int_{x_{i-1}}^{x_i} f(x)\,dx.$$

(e) The indefinite integral $\int_a^t f(x)\,dx,\ a \le t \le b$, exists and is continuous on [*a*, *b*].

EXAMPLE 1.9

The function *f* defined by

$$f(x) = \begin{cases} e^{-x^2/2} & \text{for } x > 0 \\ 0 & \text{for } x < 0 \end{cases}$$

has jump discontinuity at $x = 0$ (with jump 1) and is continuous elsewhere. Hence, *f* is piecewise continuous on [0, ∞] But the function *f* defined by

$$f(x) = \sin\frac{1}{x},\quad x \neq 0$$

$$f(0) = 0$$

is not piecewise continuous on [0, 1] because $f(0+)$ does not exist.

1.3 DERIVABILITY OF A FUNCTION AND PIECWISE SMOOTH FUNCTIONS

Definition 1.17. A function *f* defined on [*a*, *b*] is said to be derivable at *x* if

$$\lim_{h \to 0} \frac{f(x+h) - f(x)}{h} \text{ exists.}$$

In case the limit exists, then this limit is called the *derivative* of *f* at *x* and is denoted by $f'(x)$. The limits

$$\lim_{h \to 0-} \frac{f(x+h) - f(x)}{h} \text{ and } \lim_{h \to 0+} \frac{f(x+h) - f(x)}{h}$$

if exist are called, respectively, the *left*-and *right- hand derivative* and are again denoted by $f'_-(x)$ and $f'_+(x)$. Clearly, $f'(x)$ exists if both $f'_-(x)$ and $f'_+(x)$ exist and are equal.

Further, if a function is derivable at a point, then it is continuous at that point. But the converse need not be true. For example, the function f defined by $f(x) = |x|, x \in \mathbb{R}$ is continuous at $x = 0$ but it is not derivable at $x = 0$. In fact, $f'_-(x) = -1$ and $f'_+(x) = 1$ for this function.

Definition 1.18. A piecewise continuous function *f* on the interval [*a*, *b*] is called *piecewise smooth* if its derivative f' is piecewise continuous.

A function is called *piecewise smooth on* ℝ if it is piecewise smooth on each interval [*a*, *b*] of ℝ. Thus, the graph of a piecewise smooth function is either a continuous curve or a discontinuous curve, which can have a finite number of *corners* (points at which the curve has two distinct tangents).

EXAMPLE 1.10

The function with the graph (Figure 1.1).

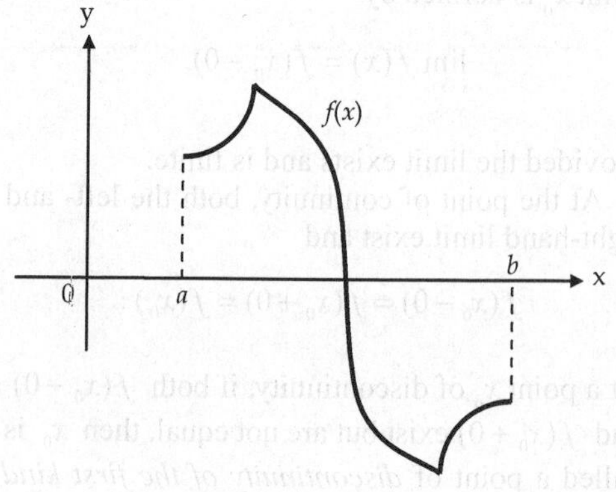

Figure 1.1

is continuous piecewise smooth, whereas the function with graph is discontinuous piecewise smooth (Figure 1.2)

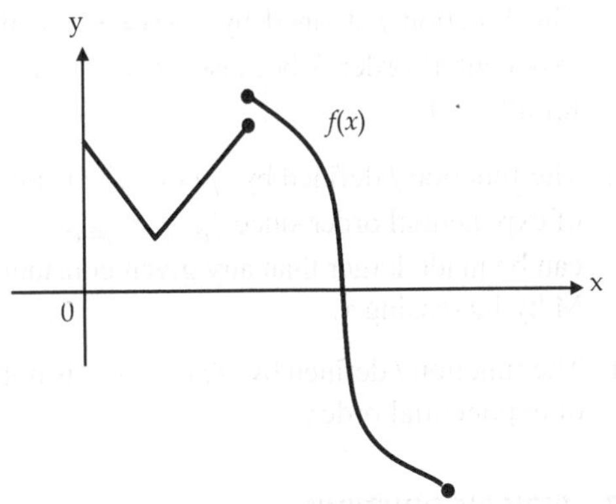

Figure 1.2

1.4 THE RIEMANN INTEGRAL

Let f be a *bounded real function* defined on $[a, b]$ and let

$$a = x_0 \le x_1 \le x_2 \le \ldots \le x_n = b$$

be a partition of $[a, b]$. Corresponding to each partition P of $[a, b]$, let

$$M_i = \text{lub}\, f(x), \ (x_{i-1} \le x \le x_i),$$

$$m_i = \text{glb}\, f(x), \ (x_{i-1} \le x \le x_i),$$

$$U(P, f) = \sum_{i=1}^{n} M_i\, (x_i - x_{i-1}), \text{ and}$$

$$L(P, f) = \sum_{i=1}^{n} m_i\, (x_i - x_{i-1}).$$

The U(P, f) and L(P, f) are called *upper Riemann sum* and *lower Riemann sum*, respectively. Further, let

$$\overline{\int_a^b} f(x)\, dx = \text{glb } U(P, f) \text{ and}$$

$$\underline{\int_a^b} f(x)\, dx = \text{lub } L(P, f),$$

where glb and lub are taken over all possible partitions P of $[a, b]$. Then, $\overline{\int_a^b} f$ is called the upper Riemann integral and $\underline{\int_a^b} f$ is called the lower Riemann integral of f over $[a, b]$.

A bounded real function f is said to be *Riemann-integrable* on $[a, b]$ if upper- and lower Riemann integral are equal and then, their common value, denoted by $\int_a^b f(x)dx$, is called the *Riemann-integral* of f.

If a or b is infinite or if f is unbounded for some $x \in [a,b]$, then the integral $\int_a^b f$ is called an *improper integral.*

Definition 1.19. A function f is called *absolutely integrable* on $\mathbb{R}$ *if*

$$\int_{-\infty}^{\infty} |\, f(t)\,|\, dt < \infty$$

is an improper Riemann integral.

EXAMPLE 1.11

The function $p_a(t)$, known as a *Block function* or *Rectangular pulse function of height* 1, defined by

$$p_a(t) = \begin{cases} 1 & \text{for } |\, t\,| \le \frac{a}{2}, \\ 0 & \text{otherwise} \end{cases}$$

is absolutely integrable.

1.5 THE CAUSAL AND NULL FUNCTION

Definition 1.20. A function f is called *causal* if $f(x) = 0$ for $x < 0$.

For example, f defined by

$$f(x) = \begin{cases} 0 & \text{for } x < 0 \\ \sin x & \text{for } x > 0 \end{cases}$$

is *a causal sine function.*

A delayed (by π/3) causal sine function is then

$$g(x) = \begin{cases} 0 & \text{for } x < \pi/3 \\ \sin x & \text{for } x > \pi/3 \end{cases}$$

Similarly, the function H defined by

$$H(t-a) = \begin{cases} 0 \text{ for } t < a \\ 1 \text{ for } t > a \end{cases}$$

is causal and is called *Heaviside's unit step function.*

Definition 1.21. A function f is said to be a *null function* if for all $t > 0$,

$$\int_0^t f(x)\,dx = 0.$$

EXAMPLE 1.12

The function f defined by

$$f(t) = \begin{cases} 1 & \text{for } t = 1/2 \\ -1 & \text{for } t = 1 \\ 0 & \text{otherwise} \end{cases}$$

is a null function.

The other examples are

(a) $N(t) = \begin{cases} 1 \text{ for } t = 0 \\ 0 \text{ for } t \neq 0 \end{cases}$

(b) a function which is identically zero (which is the only continuous null function).

1.6 FUNCTIONS OF EXPONENTIAL ORDER

Definition 1.22. A function f is said to be of *exponential order* or *exponential growth* α if there exist constants $\alpha \in \mathbb{R}$ and $M > 0$ such that for some $t_0 \geq 0$

$$|f(t)| \leq Me^{\alpha t}, \quad t \geq t_0$$

In such a case we write $f(t)$ is $O(e^{\alpha t})$.

EXAMPLE 1.13

(a) The function $f(x) = e^{3x} \sin x$ of exponential order $O(e^{3x})$ since with $M = 1$, $|f(x)| = |e^{3x} \sin x| \leq e^{3x}$.

(b) The function $\varepsilon(t)$ defined by

$$\varepsilon(t) = \begin{cases} 0 \text{ for } t < 0 \\ 1 \text{ for } t > 0 \end{cases}$$

is of exponential order with $M = 1$ and $\alpha = 0$ because $|\varepsilon(t)| \leq 1$.

(c) Every bounded function f is of exponential order with $M = 1$ and $\alpha = 0$ because $|f(t)| \leq M$ for some $M > 0$.

(d) The function f defined by $f(t) = i^2$ is of exponential order 3 because $|t^2| = t^2 < e^{3t}$ for all $t > 0$.

(e) The function f defined by $f(x) = e^{x^3}$ is not of exponential order since $|e^{-\alpha x} e^{x^3}| = e^{x^3 - \alpha x}$ can be made larger than any given constant M by increasing x.

(f) The function f defined by $f(x) = e^{3x^2}$ is not of exponential order.

1.7 PERIODIC FUNCTIONS

Definition 1.23. A function f is called *periodic* if there exists a constant $T > 0$ for which $f(x) = f(x+T)$ for any x in the domain of definition of f.

The smallest value of the constant T for which the above condition holds is called the *fundamental period* (or simply *period*) of the function f.

EXAMPLE 1.14

The functions $\sin t$ and $\cos t$ both have period $T = 2\pi$, whereas $\tan x$ has period $T = \pi$. Similarly, $A \sin(\omega x + \phi)$, where A, ω, and φ are constants is periodic with period $\frac{2\pi}{\omega}$. This function is called a *harmonic* of *amplitude* $|A|$, *angular frequency* ω, and *initial phase* φ. In fact,

$$A \sin\left[\omega\left(x + \frac{2\pi}{\omega}\right) + \phi\right] = A \sin[(\omega x + \phi) + 2\pi] = A \sin(\omega x + \phi).$$

If T is a period of the function f, then the numbers 2T, 3T, ... are also periods of f. In fact,

$$f(x) = f(x+T) = f(x+2T) = f(x+3T) = \ldots$$

and also

$$f(x) = f(x-T) = f(x-2T) = f(x-3T) = \ldots$$

Further, the sum, difference, product, or quotient of two functions period T is again a function of period T. Also, if periodic function f with period T is integrable on any intervals of length T, then it is integrable on any other interval of the same length and the value of the integrals is the same. Thus for any periodic function f with period T, we have

$$\int_{a}^{a+T} f(x)\,dx = \int_{b}^{b+T} f(x)\,dx$$

for any a and b.

1.8 EVEN AND ODD FUNCTIONS

Definition 1.24. Let the function f defined either on $\mathbb{R}$ or on some interval be symmetric with respect to the origin of co-ordinates. Then
(a) f is called *even* if

$$f(-t) = f(t) \text{ for all } t.$$

It follows from the definition that the graph of an even function f is symmetric with respect to the y-axis (Figure 1.3).

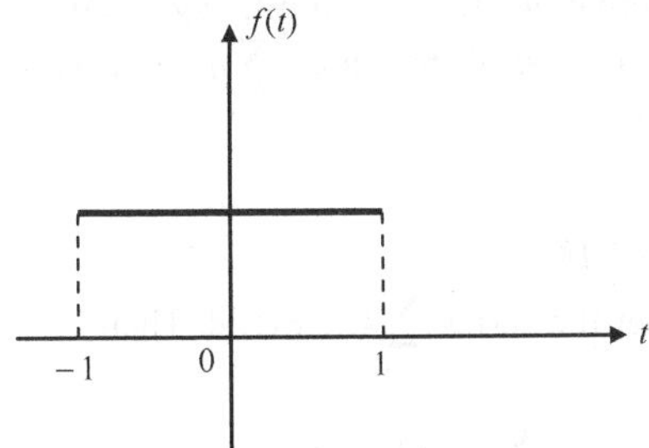

Figure 1.3

Thus interpreting the integral as an area, we have for an even function f,

$$\int_{-T}^{T} f(t)\,dt = 2\int_{0}^{T} f(t)\,dt \text{ for any T,}$$

provided that f is defined and integrable on [−T, T].

(b) f s called *odd* if

$$f(-t) = -f(t) \text{ for every } t.$$

The definition suggests that the graph of an odd function is symmetric with respect to the origin (Figure 1.4).

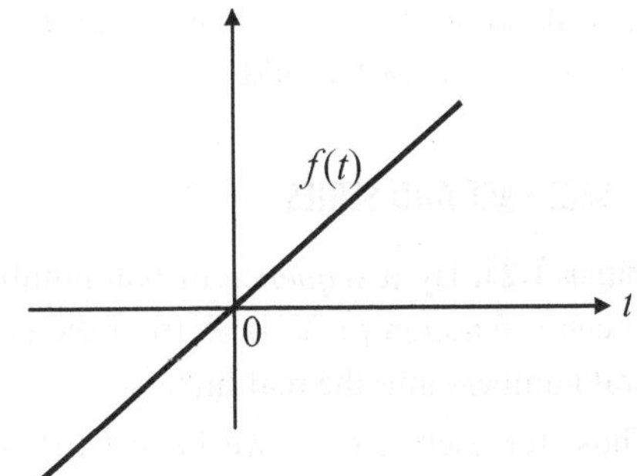

Figure 1.4

Thus, for odd function f, we have

$$\int_{-T}^{T} f(t)\,dt = 0 \text{ for any T,}$$

provided f is defined and integrable on [−T, T].

Let f and g be two even functions, then $h = fg$ is even. In fact,

$$h(-x) = (fg)(-x) = f(-x)g(-x)$$
$$= f(x)g(x) = h(x)\cdot$$

Similarly, if f and g both are odd functions, then $h = fg$ is even, since

$$h(-x) = f(-x)g\,(-x) = [-f(x)]\,[-g(x)]$$
$$= f(x)g(x) = h(x).$$

But if one of the functions f and g is even and the other is odd, then $h = fg$ is odd. In fact, let f be even and g odd. Then

$$h(-x) = f(-x)g(-x) = [f(x)][-g(x)]$$
$$= -f(x)g(x) = -h(x).$$

Thus, we have shown that

(a) The product of two even- or two odd functions is an even function.
(b) The product of an even function and an odd function is an odd function.

EXAMPLE 1.15

The function cos x is even, whereas sin x is odd. Similarly, $f(x) = x^2\,(-\pi \le x \le \pi)$ is even and so is $f(x)=|x|(-\pi \le x \le \pi)$. But $f(x)=x(-\pi < x < \pi)$ is odd.

1.9 SEQUENCE AND SERIES

Definition 1.25. By a *sequence* of real numbers, we mean a function $f: \mathbb{N} \to \mathbb{R}$ from the set of natural numbers into the real line.

Thus, for each $n \in \mathbb{N}$ we have $f(n) = x_n$, $x_n \in \mathbb{R}$ The term $x_n = f(n)$ is called the *n*th term of the sequence *f*. The sequence *f* is generally denoted by

$$f = \{x_1, x_2, \ldots, x_n, \ldots\}$$

or simply by $\{x_n\}$, (x_n), or $< x_n >$.

EXAMPLE 1.16

$\left\{\frac{1}{n}\right\}$ is the sequence $\left\{1, \frac{1}{2}, \frac{1}{3}, \ldots\right\}$ with *n*th term $x_n = \frac{1}{n}$.

Definition 1.26. A sequence $\{x_n\}$ is called *increasing* if $x_n \le x_{n+1}$ for all *n* and *decreasing* if $x_n \ge x_{n+1}$ for all *n*.

EXAMPLE 1.17

The sequence $\{x_n\}$, where $x_n = 2n$ is increasing.

Definition 1.27. A sequence $\{x_n\}$ is said to *converge* to the limit *x* if $\lim_{n\to\infty} |x_n - x| = 0$.

The limit of a convergent sequence of real number is unique.

A sequence $\{x_n\}$ which does not converge to a limit is called *divergent*.

Let $\{x_n\}$ and $\{y_n\}$ be two convergent sequences with limits *x* and *y*, respectively. Then

(a) $\lim_{n\to\infty} (\alpha x_n + \beta y_n) = \alpha x + \beta y, \alpha, \beta \in \mathbb{R}$ or $\mathbb{C}$

(b) $\lim_{n\to\infty} x_n y_n = xy$

(c) $\lim_{n\to\infty} \frac{x_n}{y_n} = \frac{x}{y}$ if $y \ne 0$.

Definition 1.28. A real-valued sequence $\{x_n\}$ is said to be a *Cauchy sequence* if for any $\varepsilon > 0$ there exists a positive integer n_0 such that $|x_n - x_m| < \varepsilon$ for all $n, m \ge n_0$.

Definition 1.29. Let $\{a_n\}$ be a sequence. An expression of the form

$$a_1 + a_2 + a_3 + \ldots = \sum_{n=1}^{\infty} a_n$$

is called an *infinite series* or just a *series*.

For the sequence $\{a_n\}$, let

$$S_n = \sum_{k=1}^{n} a_k = a_1 + a_2 + \ldots + a_n.$$

Then S_n is called the *partial sum* of the series $\sum_{n=1}^{\infty} a_n$.

Definition 1.30. A series $\sum_{n=1}^{\infty} a_n$ is called *convergent* if and only if the sequence of its partial sums $\{s_n\}$ converges to some limit, say *s*. Then, *s* is called the *sum* of the series $\sum_{n=1}^{\infty} a_n$ and we write $s = \sum_{n=1}^{\infty} a_n$.

EXAMPLE 1.18

Consider the series $\sum_{n=0}^{\infty} x^n$, $x \in \mathbb{R}$. Then

$$s_n = \sum_{k=0}^{n} x^k = 1 + x + x^2 + \ldots + x^n.$$

If $x = 1$, then

$$s_n = 1 + 1 + \ldots + 1 = n + 1$$

and so $\lim_{n\to\infty} s_n = \lim_{n\to\infty} n + 1 = \infty$. Thus the given series, known as *geometric series*, diverges for $x = 1$. If $x \ne 1$, then

$$s_n = \frac{1 - x^{n+1}}{1 - x}$$

If $|x| < 1$, then $\lim_{n\to\infty} x^n = 0$ and so the series converges to $\frac{1}{1-x}$, that is,

$$\sum_{n=0}^{\infty} x^n = \frac{1}{1-x} \text{ for } |x|<1.$$

For $|x|>1$, the partial sum s_n does not tend to a limit and so the series diverges for these values.

As an another example, the series $\sum \frac{1}{n^p}$ converges if $p > 1$ and diverges if $p \le 1$. This series is called *harmonic series.*

Suppose that the series $\sum a_n$ converges and has the sum s. Since $s_n - s_{n-1} = a_n$, we have

$$\lim_{n\to\infty} a_n = \lim_{n\to\infty}(s_n - s_{n-1}) = \lim_{n\to\infty} s_n - \lim_{n\to\infty} s_{n-1}$$
$$= s - s = 0.$$

Therefore it follows that if $\sum a_n$ converges, then $a_n \to 0$ as $n \to \infty$.

Definition 1.31. A series $\sum_{n=1}^{\infty} a_n$ is called *absolutely convergent* if $\sum_{n=1}^{\infty} |a_n|$ converges.

An *absolutely convergent series is convergent.* But the converse need not be true. For example, the series $\sum \frac{(-1)^n}{n}$ converges by Leibnitz's Rule but it is not absolutely convergent, because the series $1 + \frac{1}{2} + \frac{1}{3} + \ldots$ is divergent.

1.10 SERIES OF FUNCTIONS

Consider the series

$$\sum_{n=1}^{\infty} \frac{\sin nt}{n^2} \text{ and } \sum_{n=0}^{\infty} \frac{x^n}{n!}.$$

The partial sums of these series are polynomials. The terms of the sequences of partial sums are in fact functions.

Let $\{f_n\}$ be a sequence of functions defined on a set E and suppose that the sequence of numbers $\{f_n(x)\}$ converges for every $x \in \mathrm{E}$. Then the function f defined by

$$f(x) = \lim_{n\to\infty} f_n(x),\ x \in E$$

is called the *limit function* of $\{f_n\}$ and we say that $\{f_n\}$ *converges to f pointwise* on E.

Definition 1.32 A sequence of functions $\{f_n\}$ is said to *converge uniformly* on a set E to a function f if for every $\varepsilon > 0$ there exists an integer N such that $n \ge \mathrm{N}$ implies

$$|f_n(x) - f(x)| < \varepsilon \text{ for all } x \in \mathrm{E}$$

An infinite series of functions $\sum f_n(x)$ is said to *converge uniformly* on E if the sequence of its partial sums converges uniformly on E.

The following result on uniform convergence shall be used to derive *Fourier formulae.*

Theorem 1.1. (Term by Term Integration). If $f_1(x)$, $f_2(x)$, ..., $f_1(x)$, $f_2(x)$, ..., $f_n(x)$, are continuous functions of x in $[a, b]$ and if $\Sigma f_n(x)$ converges uniformly to $f(x)$ in $[a, b]$, then

$$\int_a^b f(x)\,dx = \int_a^b f_1(x)dx + \int_a^b f_2(x)dx + \ldots$$
$$+\int_a^b f_n(x)dx + \ldots$$

1.11 PARTIAL FRACTION EXPANSION OF A RATIONAL FUNCTION

Definition 1.33. A *rational function* F(s) is a function which has the form

$$F(s) = \frac{P(s)}{Q(s)},$$

where degree of the polynomial $Q(s)$ is greater than the degree of the polynomial $P(s)$, and $P(s)$ and $Q(s)$ have no common factor. The zeros of $Q(s)$ are called *poles* of $F(s)$.

Partial fraction expansion of a rational function will be required to determine inverse Laplace transform and inverse z-function of a rational function in the forthcoming chapters.

To find partial fraction expansion of a rational function $F(s)$, we recall that

(a) To each linear factor of the form $as + b$ of $Q(s)$, there corresponds a partial fraction of the form $\frac{A}{as+b}$, where A is a constant.

(b) To each repeated linear factor of the form $(as + b)^n$, there corresponds a partial fraction of the form

$$\frac{A_1}{as+b}+\frac{A_2}{(as+b)^2}+\ldots+\frac{A_n}{(as+b)^n},$$

where $A_1, A_2, \ldots, A_n$, are constants.

(c) To each quadratic factor of the form $as^2 + bs + c$ of $Q(s)$, there corresponds a partial fraction of the form $\frac{As+B}{as^2+bs+c}$, where A and B are constants.

(d) To each repeated quadratic factor of the form $(as^2+bs+c)^n$ of $Q(s)$, there corresponds a partial fraction of the form

$$\frac{A_1s+B_1}{as^2+bs+c}+\frac{A_2s+B_2}{(as^2+bs+c)^2}+\ldots$$

$$+\frac{A_ns+B_n}{(as^2+bs+c)^n},$$

where $A_1, A_2, \ldots, A_n$ are constants.

The constants are determined by clearing fractions and equating the like powers of both sides of the resulting equation.

EXAMPLE 1.19

Resolve into partial fractions:

$$\frac{3s+1}{(s-1)(s^2+1)}$$

Solution. According to the above discussed scheme, we have

$$\frac{3s+1}{(s-1)(s^2+1)}=\frac{A}{s-1}+\frac{Bs+C}{s^2+1}$$

and so

$$3s+1 = A(s^2+1)+(Bs+c)(s-1)$$

Taking $s = 1$ yields $A = 2$. Now taking $s = 0$, we have $1 = A - C = 2 - C$, which gives $C = 1$. Comparing the coefficients of s on both sides of the above equation, we have $3 = -B + C = -B + 1$ and so $B = -2$. Hence

$$\frac{3s+1}{(s-1)(s^2+1)}=\frac{2}{s-1}+\frac{-2s+1}{s^2+1}$$

EXAMPLE 1.20

Resolve into partial fraction:

$$\frac{4s+5}{(s-1)^2(s+2)}$$

Solution. Write

$$\frac{4s+5}{(s-1)^2(s+2)}=\frac{A}{s-1}+\frac{B}{(s-1)^2}+\frac{C}{s+2}$$

and so

$$4s+5 = A(s-1)(s+2)+B(s+2)+C(s-1)^2.$$

Taking $s = 1$ yields $B = 3$. Taking $s = -2$ yields $C = -\frac{1}{3}$ and equating the coefficients of s^2 on both sides, we get $A = \frac{1}{3}$. Hence

$$\frac{4s+5}{(s-1)^2(s+2)}=\frac{1}{3(s-1)}+\frac{3}{(s-1)^2}-\frac{1}{3(s+2)}$$

1.12 SPECIAL FUNCTIONS

In this section, we present some special functions having applications in science and engineering.

Definition 1.34. (The Gamma Function). The *gamma function* is defined by the improper integral

$$\Gamma(z)=\int_0^\infty e^{-u}u^{z-1}du, \quad \text{Re}(z)>0 \quad (1)$$

Substituting $z = 1$ in (1), we get

$$\Gamma(1)=\int_0^\infty e^{-u}du=\lim_{T\to 0}\int_0^T e^{-u}du$$

$$=\lim_{T\to 0}(1-e^{-T})=1.$$

Further, integration by parts yields

$$\Gamma(z+1)=\int_0^\infty e^{-u}u^z du$$

$$= 0 + z\int_0^\infty e^{-u}u^{z-1}du = z\Gamma(z), \qquad (2)$$

which is *recurrence formula* for $\Gamma(z)$. Using (2), we have

$$\Gamma(2) = 1,\ \Gamma(3) = 2,\ \Gamma(4) = 3!$$

and in general,

$$\Gamma(n+1) = n!,\ n = 0, 1, 2, \ldots$$

Thus gamma function is an extension of *factorial function.* The other important properties of gamma function are

$$\Gamma\left(\frac{1}{2}\right) = \sqrt{\pi} \qquad (3)$$

$$\Gamma(p)\Gamma(1-p) = \frac{\pi}{\sin p\pi},\quad 0 < p < 1 \qquad (4)$$

Definition 1.35. (The Beta Function). The *beta function* $\beta(m,n)$ is defined by

$$\beta(m,n) = \int_0^1 u^{m-1}(1-u)^{n-1}du,\ m > 0, n > 0.$$

The beta function has the following properties:

(*i*) $\beta(m,n) = \dfrac{\Gamma(m)\Gamma(n)}{\Gamma(m+n)}.$ (5)

(*ii*) $\displaystyle\int_0^{\pi/2} \sin^{2m-1}\theta\ \cos^{2n-1}\theta\ d\theta = \frac{1}{2}\beta(m, n)$

$$= \frac{(m)\ (n)}{2\ (m+n)} \qquad (6)$$

The relation (6) follows from (5). In fact,

$$\beta(m,n) = \int_0^1 u^{m-1}(1-u)^{n-1}\,du = \frac{\Gamma(m)\Gamma(n)}{\Gamma(m+n)}.$$

Substituting $u = \sin^2\theta$, we have

$$\beta(m,n) = 2\int_0^{\pi/2} \sin^{2m-1}\theta\ \cos^{2n-1}\theta\ d\theta$$

$$= \frac{\Gamma(m)\Gamma(n)}{\Gamma(m+n)},$$

and so (6) follows.

EXAMPLE 1.21

Show that

(a) $\displaystyle\int_0^{\pi/2} \sin^2\theta\ \cos^4\theta\ d\theta = \frac{\neq}{32}$

(b) $\displaystyle\int_0^{\pi/2} \sin^7\theta\ d\theta = \frac{16}{32}$

(c) $\displaystyle\int_0^{\pi/2} \frac{d\theta}{\sqrt{\tan\theta}} = \frac{\neq\sqrt{2}}{2}$

Solution. (a) Comparing the given integral with

$$\int_0^{\pi/2} \sin^{2m-1}\theta\ \cos^{2n-1}\theta d\theta,$$

we observe that $m = \frac{3}{2}$, $n = \frac{5}{2}$ and so using (6), we get

$$\int_0^{\pi/2} \sin^2\theta \cos^4\theta\ d\theta = \frac{\Gamma(3/2)\Gamma(5/2)}{2\Gamma[(3/2)+(5/2)]}$$

$$= \frac{3\sqrt{\pi}}{3!(16)}.\sqrt{\pi} = \frac{\pi}{32}.$$

(b) Comparing the given integral with

$$\int_0^{\pi/2} \sin^{2m-1}\theta\ \cos^{2n-1} d\theta,$$

we observe that $2m - 1 = 7$ and $2n - 1 = 0$ and so we get $m = 4$ and $n = \frac{1}{2}$. Hence, the application of (6) yields

$$\int_0^{\pi/2} \sin^7\theta\, d\theta = \frac{\Gamma(4)\Gamma(1/2)}{2\Gamma(9/2)} = \frac{3!(8)\sqrt{\pi}}{105\sqrt{\pi}} = \frac{16}{35}.$$

(c) We have

$$\int_0^{\pi/2} \frac{d\theta}{\sqrt{\tan\theta}} = \int_0^{\pi/2} \sin^{-1/2}\theta \cos^{1/2}\theta\ d\theta$$

Taking $2m - 1 = -\frac{1}{2}$ and $2n - 1 = \frac{1}{2}$, we get $m = \frac{1}{4}$, $n = \frac{3}{4}$. Hence, using (6) and the relation (4), we get

$$\int_0^{\pi/2} \frac{d\theta}{\sqrt{\tan\theta}} = \frac{\Gamma(1/4)\Gamma(3/4)}{2\Gamma(1)} = \frac{1}{2}.\frac{\pi}{\sin(\pi/4)}$$

$$= \frac{\pi\sqrt{2}}{2}.$$

Definition 1.36. (Bessel Function). A *Bessel function* of order n is defined by

$$J_n(t) = \frac{t^n}{2^n \Gamma(n+1)}$$

$$\times\left[1 - \frac{t^2}{2(n+2)} + \frac{t^4}{2.4(2n+2)(2n+4)} - \ldots\right] \quad (7)$$

The main properties of this function are

(a) $J_{-n}(t) = (-1)^n J_n(t)$, if n is *a positive integer* (8)

(b) $J_{n+1}(t) = \frac{2n}{t} J_n(t) - J_{n-1}(t)$ (9)

(c) $\frac{d}{dt}\{t^n J_n(t)\} = t^n J_{n-1}(t)$ (10)

Thus, if $n = 0$, we have

$$J_0'(t) = J_1(t) \text{ [using } (a)\text{]}.$$

Definition 1.37. (Error Function and Complementary Error Functions). The *error function* is defined by the integral

$$\text{erf}(z) = \frac{2}{\sqrt{\pi}} \int_0^z e^{-t^2}\, dt, \quad (11)$$

where z may be real or complex variable. This function appears in probability theory, heat conduction theory, and mathematical physics. When $z = 0$, erf $(0) = 0$, and

$$\text{erf}(\infty) = \frac{2}{\sqrt{\pi}} \int_0^\infty e^{-t^2}\, dt = \frac{\Gamma(1/2)}{\sqrt{\pi}} = 1.$$

The *complementary error function* is defined by the integral

$$\text{erfc}(z) = \frac{2}{\sqrt{\pi}} \int_z^\infty e^{-t^2}\, dt \quad (12)$$

Using the properties of the integrals, we note that

$$\text{erfc}(z) = \frac{2}{\sqrt{\pi}} \int_0^\infty e^{-t^2}\, dt - \frac{2}{\sqrt{\pi}} \int_0^z e^{-r^2}\, dt$$

$$= \frac{2}{\sqrt{\pi}}\left(\frac{\sqrt{\pi}}{2}\right) - \text{erf}(z)$$

$$= 1 - erf(z) \quad (13)$$

Definition 1.38. (Heaviside's Unit Step Function). The *Heaviside's unit step function* (also known as *delayed unit step function*) is defined by

$$H(t-a) = \begin{cases} 1 \text{ for } t > a \\ 0 \text{ for } t < a. \end{cases} \quad (14)$$

It delays the output until $t = a$ and then assumes a constant value of 1 unit.

If $a = 0$, then

$$\text{H}(t) = \begin{cases} 1 \text{ for } t > 0 \\ 0 \text{ for } t < 0, \end{cases} \quad (15)$$

which is generally called *unit step function*.

The graph of Heaviside's unit step function is shown in Figure 1.5.

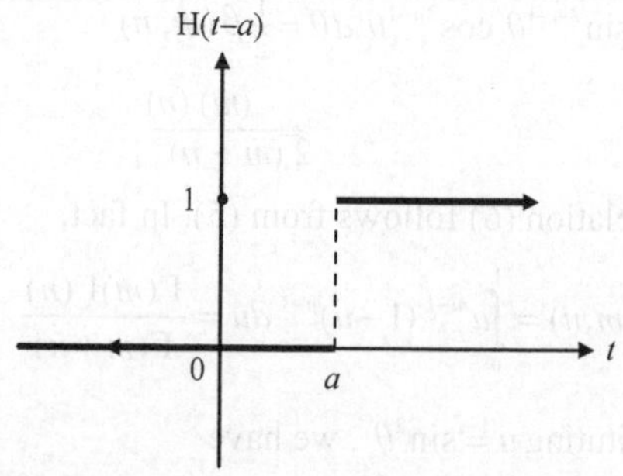

Figure 1.5 Graph of Heaviside's unit step function

This Function has jump discontinuity at $t = a$ with a jump of unit magnitude. The beauty of this function is that it acts like a switch to turn another function on or off at some time. For example, the function

$$g(t) = H(t-a)\cos 2\pi t$$

is zero for $t < a$ but assumes the graph of the cosine function for $t > a$ as shown in Figure 1.6.

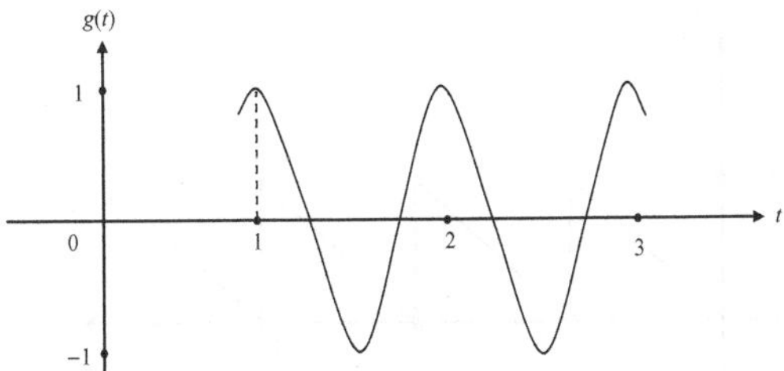

Figure 1.6 Graph of $g(t) = H(t-a)\cos 2\pi t$

Definition 1.39. (Pulse Unit Height and Duration T). The *pulse of unit height* and duration T is defined by

$$f(t) = \begin{cases} 1 \text{ for } 0 < t < T \\ 0 \text{ for } T < t \end{cases} \tag{16}$$

Definition 1.40. (Sinusoidal Pulse). The *sinusoidal pulse* is defined by

$$f(t) = \begin{cases} \sin at & \text{for } 0 < t < \pi/a \\ 0 & \text{for } \pi/a < t. \end{cases} \tag{17}$$

Definition 1.41. (Rectangle Function). A *rectangle function f* is defined by

$$f(t) = \begin{cases} 1 \text{ for } a < t < b \\ 0 \text{ otherwise.} \end{cases} \tag{18}$$

The graph of this function is shown in Figure 1.7.

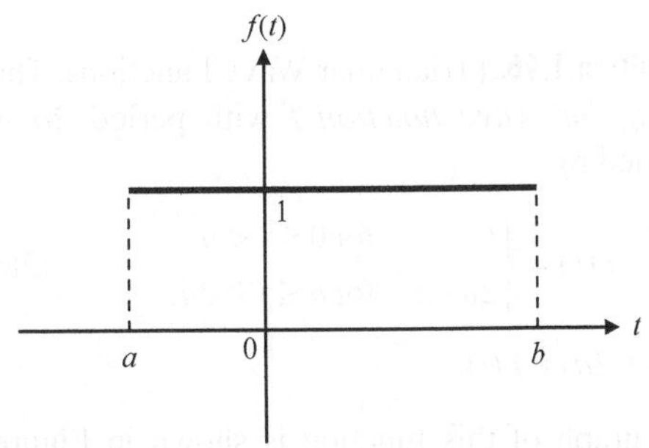

Figure 1.7 Graph of Rectangle Function

In term of Heaviside unit step function, we have

$$f(t) = H(t-a) - H(t-b)$$

If $a = 0$, then rectangle function reduces to pulse of unit height and duration b.

Definition 1.42. (Gate Function). The *gate function* is defined as

$$f_a(t) = \begin{cases} 1 \text{ for } |t| < a \\ 0 \text{ for } |t| > a \end{cases} \tag{19}$$

The graph of the gate function is shown in Figure 1.8.

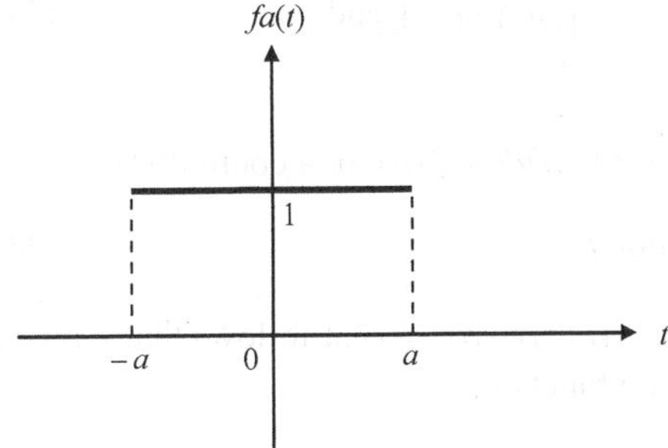

Figure 1.8 Graph of Gate Function

Definition 1.43. (Dirac Delta Function). Consider the function f_ε defined by

$$f_\varepsilon(t) = \begin{cases} \frac{1}{\varepsilon} & \text{for } 0 \le t \le \varepsilon \\ 0 & \text{for } t > \varepsilon, \end{cases} \tag{20}$$

where $\varepsilon > 0$. The graph of f_ε is shown in Figure 1.9.

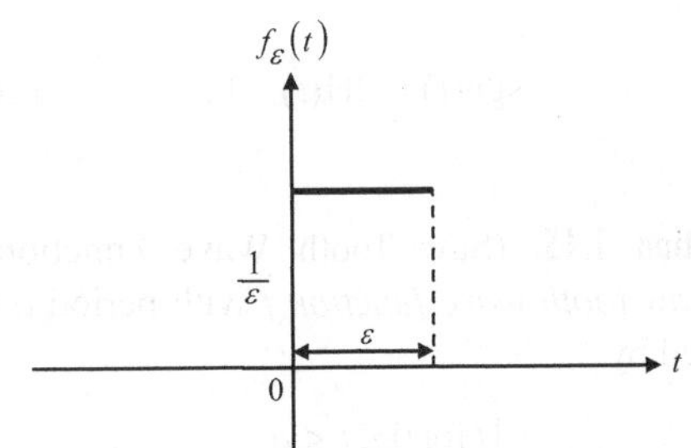

Figure 1.9 Graph of $f_\varepsilon(t)$

We note that as $\varepsilon \to 0$, the height of the rectangle increases indefinitely and width decreases in such a way that *its area is always equal to 1.*

We further note that $\lim_{\varepsilon \to 0} f_\varepsilon(t)$ does not exist. Even then we define a function δ as

$$\delta(t) = \lim_{\varepsilon \to 0} f_\varepsilon(t). \tag{21}$$

This "*generalized function* or *distribution*" $\delta(t)$ is called *Dirac delta function* and has the property

$$\delta(t) = 0 \text{ for } t \neq 0, \tag{22}$$

$$\int_{-\infty}^{\infty} \delta(t)\, dt = 1, \text{ and} \tag{23}$$

$$\int_{-\infty}^{\infty} \delta(t-a) f(t) dt = f(a) \text{ for a continuous function } f. \tag{24}$$

Since $\delta(t-a) = \delta(a-t)$, it follows that $\delta(t)$ in an *even* function.

Definition 1.44. (Signum Function). The *signum function*, denoted by sgn(t), is defined by

$$\text{sgn}(t) = \begin{cases} 1 & \text{for } t > 0 \\ -1 & \text{for } t < 0. \end{cases} \tag{25}$$

It H(t) is unit step function, then

$$\text{H}(t) = \frac{1}{2}[1 + \text{sgn}(t)]$$

and so

$$\text{sgn}(t) = 2\text{H}(t) - 1 \tag{26}$$

Definition 1.45. (Saw Tooth Wave Function). The *saw tooth wave function* f with period a is defined by

$$f(t) = \begin{cases} t \text{ for } 0 \le t < a \\ 0 \text{ for } t \le 0, \end{cases} \tag{27}$$

$$f(t+a) = f(t).$$

The graph of this function is shown in Figure 1.10.

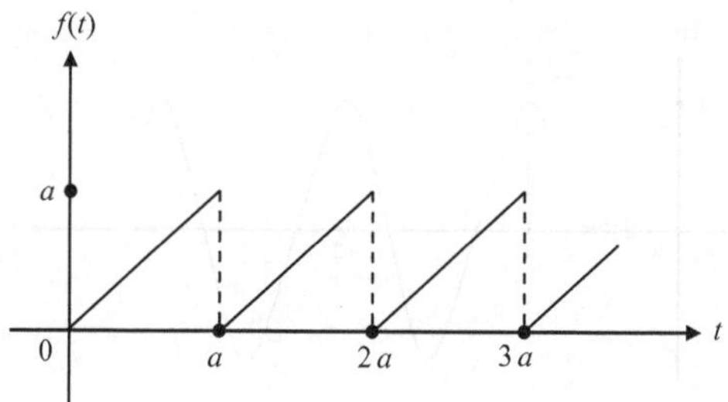

Figure 1.10 Graph of Saw-tooth Function with Period a

The *saw-tooth function with period* 2π is defined as

$$f(t) = \begin{cases} t \text{ for } -\pi < t < \pi \\ 0 \text{ elsewhere.} \end{cases}$$

The graph of this function is shown in Figure 1.11.

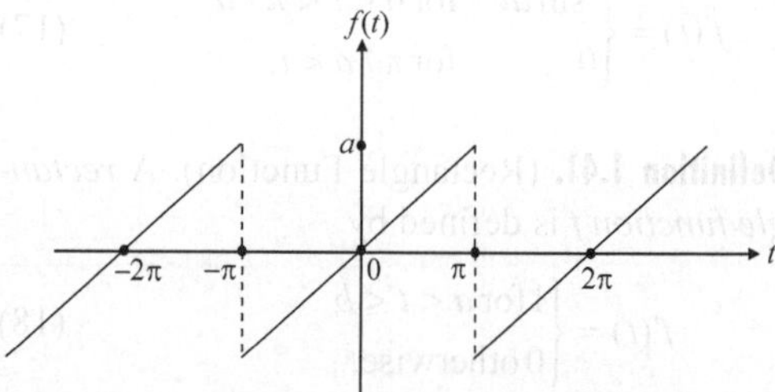

Figure 1.11 Graph of Saw Tooth Function with Period 2π

Definition 1.46. (Triangular Wave Function). The *triangular wave function* f with period $2a$ is defined by

$$f(t) = \begin{cases} t & \text{for } 0 \le t < a \\ 2a - t & \text{for } a \le t < 2a, \end{cases} \tag{28}$$

$$f(t + 2a) = f(t).$$

The graph of this function is shown in Figure 1.12.

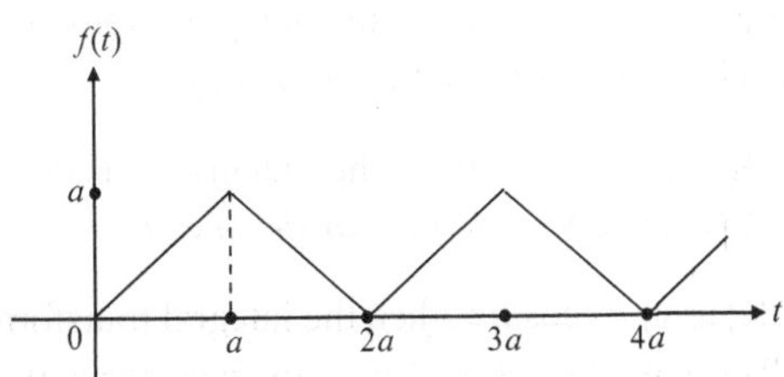

Figure 1.12 Graph of triangular wave function with period $2a$

Definition 1.47. (Half-wave Rectified Sinusoidal Function). The *half-wave rectified sinusoidal function f* with period 2π is defined by

$$f(t)=\begin{cases}\sin t & \text{for } 0<t<\pi\\ 0 & \text{for } \pi<t<2\pi,\end{cases} \tag{29}$$

$$f(t+2\pi)=f(t).$$

The graph of this function is shown in Figure 1.13.

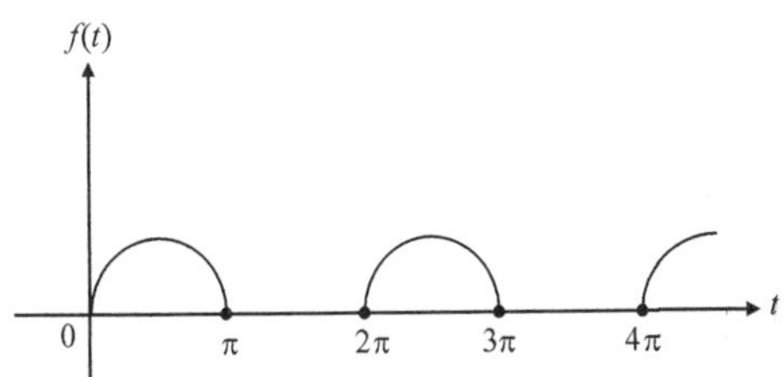

Figure 1.13 Graph of Half-wave Rectified Sine Function

Definition 1.48. (Full Rectified Sine Wave Function). The *full rectified sine wave function f* with period π is defined by

$$f(t)=\begin{cases}\sin t \text{ for } 0<t<\pi\\ -\sin t \text{ for } \pi<t<2\pi,\end{cases} \tag{30}$$

$$f(t+\pi)=f(t).$$

or by

$$f(t)=|\sin \omega t| \text{ with period } \pi/\omega.$$

The graph of this function is shown in Figure 1.14.

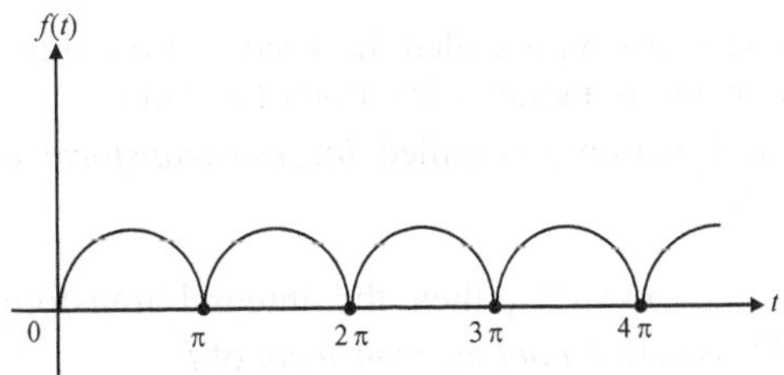

Figure 1.14 Graph of Full Rectifier with Period π

Definition 1.49. (Square Wave Function). The *square wave function f* with period $2a$ is defined by

$$f(t)=\begin{cases}1 \text{ for } 0<t<a\\ -1 \text{ for } a<t<2a\end{cases} \tag{31}$$

$$f(t+2a)=f(t).$$

In terms of Heaviside's unit step function, the definition converts to

$$f(t)=\mathrm{H}(t)-2\mathrm{H}(t-a)+2\mathrm{H}(t-2a)$$

$$-2\mathrm{H}(t-3a)+\ldots$$

The graph of this function is shown in Figure 1.15.

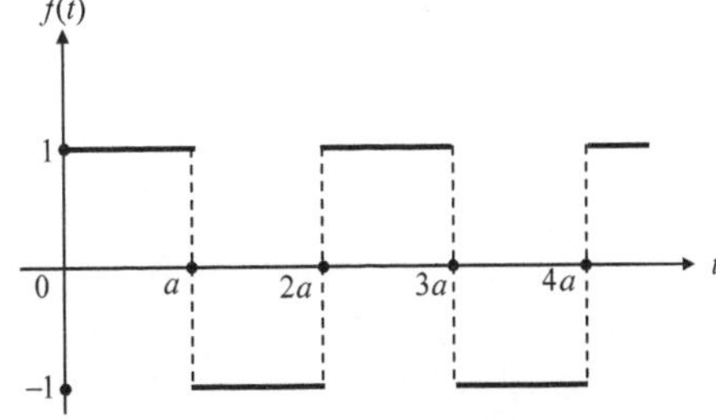

Figure 1.15 Graph of the Square Wave Function

1.13 THE INTEGRAL TRANSFORMS

Definition 1.50. The *integral transform* of a function *f* is defined by

$$I\{f(x)\} = \int_{x_1}^{x_2} f(x)K(w,x)dx, \qquad (32)$$

where K(*w, x*) is called the *kernel* of the transform and is a known function of *w* and *x*.
The function *f* is called *inverse transform* of $I\{f(x)\}$.

If $K(w,x) = e^{-wx}$, then the integral transform (32) is called *Laplace transform* of *f*.

If $K(w,x) = e^{-iwx}$, then the integral transform (32) is called *Fourier transform* of *f*.

If $K(w,x) = x^{w-1}$, then the integral transform (32) is called the *Mellin transform* of *f*.

If $K(w,x) = \cos nx$, then the integral transform (32) is called the *Fourier cosine transform* of *f*.

If $K(w,x) = \sin nx$, then the integral transform (32) is called the *Fourier sine transform* of *f*.

2 Linear Algebra

2.1 CONCEPTS OF GROUP, RING, AND FIELD

Definition 2.1. Let S be a non-empty set. Then a mapping $f : S \times S \to S$ is called a *binary operation* in S.

A non-empty set along with one or more binary operations defined on it is called an *algebraic structure*.

Definition 2.2. A non-empty set G together with a binary operation $f : G \times G \to G$ defined on it and denoted by $*$ is called a *group* if the following axioms are satisfied:

(G_1) **Associativity**: For $a, b, c \in G$,

$$(a * b) * c = a * (b * c)$$

(G_2) **Existence of Identity**: There exists an element e in G such that for all $a \in G$,

$$a * e = e * a = a$$

(G_3) **Existence of Inverse Element**: For each element $a \in G$, there exists an element $b \in G$, such that

$$a * b = b * a = e.$$

Definition 2.3. Let G be a group. If for every pair $a, b \in G$,

$$a * b = b * a,$$

then G is called a *commutative* (or *abelian*) group.

If $a * b \neq b * a$, then G will be called *non-abelian* or *non-commutative* group.

Definition 2.4. The number of elements in a group G is called the *order of the group* G and is denoted by $O(G)$. A group having a finite number of elements is called a *finite group*.

EXAMPLE 2.1

Let $\mathbb{Z}$ be the set of all integers and let $f : \mathbb{Z} \times \mathbb{Z} \to \mathbb{Z}$ defined by $f(a, b) = a * b = a + b$ be binary operation in $\mathbb{Z}$. Then

(i) $a + (b + c) = (a + b) + c$ for all $a, b, c \in \mathbb{Z}$

(ii) $a + 0 = 0 + a = a$ for all $a \in \mathbb{Z}$ and so 0 acts as an additive identity.

(iii) $a + (-a) = (-a) + a = 0$ for $a \in \mathbb{Z}$ and so $(-a)$ is the inverse of a.

(iv) $a + b = b + a$, $a, b \in \mathbb{Z}$ (Commutativity).

Hence, $(\mathbb{Z}, +)$ is an infinite additive abelian group.

EXAMPLE 2.2

The set of all integers $\mathbb{Z}$ cannot be a group under multiplication operation $f(a, b) = ab$. In fact, ± 1 are the only two elements in $\mathbb{Z}$ which have inverses.

EXAMPLE 2.3

The set of even integers $[0, \pm 2, \pm 4, \ldots]$ is an additive abelian group under addition.

EXAMPLE 2.4

The set of vectors V form an additive abelian group under addition.

EXAMPLE 2.5

We shall note in article 2.12 on matrices that the set of all $m \times n$ matrices form an additive abelian group.

EXAMPLE 2.6

The set $\{-1, 1\}$ is a multiplicative abelian group of order 2.

Definition 2.5. Let S be a set with binary operation $f(m,n) = mn$, then an element $a \in S$ is called

(i) **Left cancellative** if

$$ax = ay \Rightarrow x = y \text{ for all } x, y \in S,$$

(ii) **Right cancellative** if

$$xa = ya \Rightarrow x = y \text{ for all } x, y \in S.$$

If any element a is both left and right cancellative, then it is called *cancellative* (or *regular*). If every element of a set S is regular, then we say that *cancellation law* holds in S.

Theorem 2.1. If G is a group under the binary operation $f(a,b) = a * b = ab$, then *for* $a, b, c \in G$,

$$ab = ac \Rightarrow b = c \quad \text{(left cancellation law)}$$
$$ba = ca \Rightarrow b = c \quad \text{(right cancellation law)}$$

(Thus *cancellation law holds in a group*).

Proof. Since G is a group and $a \in G$, there exists an element $c \in G$ such that $ac = ca = e$. Therefore,

$$\begin{aligned} ab = ac &\Rightarrow c(ab) = c(ac) \\ &\Rightarrow (ca)b = c(ac) \\ &\Rightarrow eb = ce \\ &\Rightarrow b = c. \end{aligned}$$

Similarly, we can show that

$$ba = ca \Rightarrow b = c.$$

Theorem 2.2. Let G be a group. Then,

(a) The identity element of G is unique.
(b) Every $a \in G$ has a unique inverse.
(c) For every $a \in G$, $(a^{-1})^{-1} = a$
(d) For all $a, b \in G$

$$(ab)^{-1} = b^{-1}a^{-1}.$$

Proof. (a) Suppose that there are two identity elements e and e' in G. Then,

$$ee' = e \text{ since } e' \text{ is an identity element,}$$

and

$$ee' = e' \text{ since } e \text{ is an identity element.}$$

Hence $e = e'$.

(b) Suppose that an arbitrary element a in G has two inverses b and c. Then, $ab = ba = e$ and $ac = ca = e$.
Therefore,

$$(ba)c = ec = c$$

and

$$b(ac) = be = b.$$

But, by associativity in G,

$$(ba)c = b(ac).$$

Hence $b = c$.

(c) Since G is a group, every element $a \in G$ has its inverse a^{-1}. Then, $a^{-1}a = e$. Now

$$a^{-1}(a^{-1})^{-1} = e = a^{-1}a$$

By left cancellation law, it follows that $(a^{-1})^{-1} = a$.

(d) We have

$$\begin{aligned} (ab)(b^{-1}a^{-1}) &= a(bb^{-1})a^{-1} = aea^{-1} \\ &= aa^{-1} = e. \end{aligned}$$

Similarly

$$\begin{aligned} (b^{-1}a^{-1})(ab) &= b^{-1}(a^{-1}a)b \\ &= b^{-1}eb = b^{-1}b = e. \end{aligned}$$

Thus

$$(ab)(b^{-1}a^{-1}) = (b^{-1}a^{-1})(ab) = e.$$

Hence, by the definition of inverse,

$$(ab)^{-1} = b^{-1}a^{-1}.$$

Definition 2.6. A subset H of a group G is said to be a *subgroup* of G, if under the binary operation in G, H itself forms a group.

Every group G has two trivial subgroups, G itself and the identity group $\{e\}$.

The non-trivial subgroups of G are called *proper subgroups* of G.

EXAMPLE 2.7

The additive group $\mathbb{R}$ of real numbers is a subgroup of the additive group $\mathbb{C}$ of complex numbers.

Regarding subgroups, we have

Theorem 2.3. A non-empty subset H of a group G is a subgroup of G if and only if

(i) $a, b \in H \Rightarrow ab \in H$,
(ii) $a \in H \Rightarrow a^{-1} \in H$.

Conditions (i) and (ii) may be combined into a single one and we have "A non-empty subset H of a group G is a subgroup of G if and only if $a, b \in H \Rightarrow ab^{-1} \in H$."

Theorem 2.4. The intersection of two subgroups of a group is again a subgroup of that group.

Definition 2.7. Let G and H be two groups with binary operations $\phi : G \times G \to G$ and $\psi : H \times H \to H$, respectively, then a mapping $f : G \to H$ is said to be a *group homomorphism* if for all $a, b \in G$,

$$f(\phi(a,b)) = \psi(f(a), f(b)). \quad (1)$$

Thus if G is additive group and H is multiplicative group, then (1) becomes

$$f(a+b) = f(a).f(b).$$

If, in addition f is bijective, then f is called an *isomorphism*.

EXAMPLE 2.8

Let Z be additive group of integers. Then the mapping $f : Z \to H$, where H is the additive group of even integers defined by $f(a) = 2a$ for all $a \in Z$ is a group homomorphism. In fact, for $a, b \in Z$

$$f(a+b) = 2(a+b) = 2a + 2b = f(a) + f(b).$$

Also

$$f(a) = f(b) \Rightarrow 2a = 2b \Rightarrow a = b,$$

and so f is one-one homomorphism (called *monomorphism*).

Definition 2.8. Let G and H be two groups. If $f : G \to H$ is a homomorphism and e_H denotes the identity element of H, then the subset

$$K = \{x : x \in G, f(x) = e_H\}$$

of G is called the *kernel* of the homomorphism f.

Definition 2.9. A non-empty set R with two binary operation '+' and '.' is called a *ring* if the following conditions are satisfied.

(i) **Associativity of '+'**: if $a, b, c \in R$, then
$$a + (b + c) = (a + b) + c$$

(ii) **Existence of Identity for '+'**: There exists an element 0 in R such that
$$a + 0 = 0 + a = a \quad \text{for all } a \in R$$

(iii) **Existence of inverse with respect to '+'**: To each element $a \in R$, there exists an element $b \in R$ such that
$$a + b = b + a = 0$$

(iv) **Commutativity of '+'**: If $a, b \in R$, then
$$a + b = b + a$$

(v) **Associativity of '.'**: If $a, b, c \in R$, then
$$a \cdot (b \cdot c) = (a \cdot b) \cdot c$$

(vi) **Distributivity of '+' over '.'**: If $a, b, c \in R$, then
$$a \cdot (b + c) = a \cdot b + a \cdot c$$
(Left distributive law)

and

$$(a+b)\cdot c = a\cdot c + b\cdot c$$

(Right distributive law)

Let R be a ring. If there is an element 1 in R such that $a.1 = 1.a = a$ for every $a \in R$, then R is called a *ring with unit element.*

If R is a ring such that $a.b = b.a$ for every $a, b \in R$, then R is called *commutative ring.*

A ring R is said to be a *ring without zero divisors* if $a.b = 0 \Rightarrow a = 0$ or $b = 0$.

EXAMPLE 2.9

We have seen that $(\mathbb{Z}, +)$ is an abelian group. Further, if $a, b, c \in \mathbb{Z}$ then

$$a\cdot(b\cdot c) = (a\cdot b)\cdot c$$
$$a\cdot(b+c) = a\cdot b + a\cdot c$$
$$(a+b)\cdot c = a\cdot c + b\cdot c$$
$$a\cdot 1 = 1\cdot a = a$$
$$a\cdot b = b\cdot a$$
$$a\cdot b = 0 \Rightarrow a = 0 \text{ or } b = 0.$$

Hence $\mathbb{Z}$ is commutative ring with unity which is *without zero divisor.*

EXAMPLE 2.10

The set of even integers is a commutative ring but there does not exist any element b satisfying $b\,a = a\,b = a$ for $a \in R$. Hence, it is a ring without unity.

EXAMPLE 2.11

We shall see later on that the set of $n \times n$ matrices form a non-commutative ring with unity. This ring is a *ring with zero divisors*. For example, if $\begin{bmatrix} 1 & 0 \\ 0 & 0 \end{bmatrix}$ and $\begin{bmatrix} 0 & 0 \\ 1 & 0 \end{bmatrix}$, then their product is $\begin{bmatrix} 0 & 0 \\ 0 & 0 \end{bmatrix}$. But none of the given matrix is zero.

Definition 2.10. A commutative ring with unity is called an *integral domain* if it has no zero divisor.

For example, *ring of integers is an integral domain.*

Definition 2.11. A ring R with unity is said to be a *division ring* (or *skew field*) if every non-zero element of R has a multiplicative inverse.

Definition 2.12. A commutative division ring is called a *field.*

For example, the set of rational number $\mathbb{Q}$ under addition and multiplication forms a field. Similarly, $\mathbb{R}$ and $\mathbb{C}$ are also fields. Every field is an integral domain but the converse is not true. For example, the set of integers form an integral domain but is not a field. An important result is that

"Every finite integral domain is a field."

Definition 2.13. A subset S of ring R is called a *subring* of R if S is a ring under the binary operations in R. Thus, S will be a *subring* of R if

(i) $a, b \in S \Rightarrow a - b \in S$,

(ii) $a, b \in S \Rightarrow ab \in S$.

For example, the set of real numbers is a subring of the ring of complex numbers.

Definition 2.14. A mapping $f: R \to R'$ from the ring R into the ring R' is said to be a *ring homomorphism* if

(i) $f(a+b) = f(a) + f(b)$,

(ii) $f(a\cdot b) = f(a).f(b)$,
for all $a, b \in R$.

If, in addition, f is one-to-one and onto then f is called *ring isomorphism.*

2.2 VECTOR SPACE

Definition 2.15. A non-empty set V is said to be a *Vector Space* over the field F if

(i) V is an additive abelian group.

(ii) If for every $\alpha \in F$, $v \in V$, there is defined an element αv, called scalar multiple of α and v, in V subject to

$$\alpha(v+\omega) = \alpha v + \alpha\omega,$$
$$(\alpha+\beta)v = \alpha v + \beta v,$$
$$\alpha(\beta v) = (\alpha\beta)v,$$
$$1v = v,$$

for all $\alpha,\ \beta \in F,\ v,\ w \in V$, where 1 represents the unit elements of F under multiplication.

In the above definition, the elements of V are called *vectors* whereas the elements of F are called *scalars*.

EXAMPLE 2.12

Let $V_2 = \{(x, y) : x, y \in \mathbb{R}\}$ be a set of ordered pairs. Define addition and scalar multiplication on V_2 by

$$(x,y) + (x', y') = (x + x', y + y'),$$

and

$$\alpha(x,y) = (\alpha x, \alpha y).$$

Then V_2 is an abelian group under addition operation defined earlier and

$$\begin{aligned}\alpha\left[(x, y) + (x', y')\right] &= \alpha\left(x + x', y + y'\right)\\ &= \left(\alpha x + \alpha x', \alpha y + \alpha y'\right)\\ &= (\alpha x, \alpha y) + \left(\alpha x', \alpha y'\right)\\ &= \alpha(x, y) + \alpha(x', y'),\\ (\alpha + \beta)(x, y) &= ((\alpha + \beta)x, (\alpha + \beta)y)\\ &= (\alpha x + \beta x, \alpha y + \beta y)\\ &= (\alpha x, \alpha y) + (\beta x, \beta y)\\ &= \alpha(x, y) + \beta(x, y),\\ \alpha(\beta(x, y)) &= (\alpha\beta)(x, y),\\ 1.(x, y) &= (x, y).\end{aligned}$$

Hence, V_2 is a vector space over $\mathbb{R}$. It is generally denoted by $\mathbb{R}^2$.

Similarly, the set of n-tuples $(x_1, x_2, \ldots, x_n)$ form a vector space over $\mathbb{R}$ and is denoted by V_n or $\mathbb{R}^n$.

EXAMPLE 2.13

Let $f(x) = a_n x^n + a_{n-1}x^{n-1} + \ldots + a_1 x + a_0$ and $g(x) = b_n x^m + b_{m-1}x^{m-1} + \ldots + b_1 x + b_0$ be polynomials with coefficients from a field F. Suppose further that $m \leq n$, that is, $b_{m+1} = b_{m+2} = \ldots = b_n = 0$. Define addition of $f(x)$ and $g(x)$ as

$$\begin{aligned}f(x) + g(x) = (a_n + b_n)x^n + (a_{n-1} + b_{n-1})n^{n-1}\\ + \ldots (a_1 + b_1)x + (a_0 + b_0)\end{aligned}$$

Now define scalar multiplication of $f(x)$ by the scalar c as

$$\begin{aligned}cf(x) = ca_n x^n + ca_{n-1}x^n + ca_{n-1}x^{n-1} + \ldots\\ + ca_1 x + ca_0.\end{aligned}$$

With these operations of addition and scalar multiplication in the set $P_n(x)$ of all polynomials, it can be shown that $P_n(x)$ form a vector space.

EXAMPLE 2.14

The set of all matrices over a fields F form a vector space under the operation of addition of matrices and scalar multiplication of matrices.

Definition 2.16. Let V be a vector space over the field K and W be a subset of V. If W is a vector space under the operations of V, then it is called a *vector subspace* of V.

Thus, W will be a subspace of V if

(i) W is a subgroup of V,

(ii) $\lambda \in F,\ w \in W$ imply $\lambda\ w \in W$.

The conditions (i) and (ii) can be combined into a single condition, namely, $\lambda_1,\ \lambda_2 \in F$ and $w_1,\ w_2 \in W$ imply $\lambda_1 w_1 + \lambda_2 w_2 \in W$.

Definition 2.17. Let V be a vector space over a field F and let $v_1, v_2, \ldots, v_n \in V$. Then any element of the form $\alpha_1 v_1 + \alpha_2 v_2 + \ldots + \alpha_n v_n$, $\alpha_i \in F$ is called a *linear combination* over F of $v_1, v_2, \ldots v_n$.

Definition 2.18. Let S be a non-empty subset of the vector space V. Then the *linear span* of S, denoted by $L(S)$, is the set of all linear combinations of finite sets of the elements of S.

Definition 2.19. Let V be a vector space over a field F. Then $v_1, v_2, \ldots, v_n \in V$ are said to be *linearly independent* over F if for scalars $\lambda_1, \lambda_2, \ldots, \lambda_n \in F$, $\lambda_1 v_1 + \lambda_2 v_2 + \ldots + \lambda_n v_n = 0$ implies $\lambda_1 = \lambda_2 = \ldots = \lambda_n = 0$.

Definition 2.20. Let V be a vector space. Then $v_1, v_2, \ldots, v_n \in V$ are called *linearly dependent* if there exist $\lambda_1, \lambda_2, \ldots, \lambda_n \in F$, not all of them zero, such that $\lambda_1 v_1 + \lambda_2 v_2 + \ldots + \lambda_n v_n = 0$.

Thus, $v_1, v_2, \ldots, v_n$ are linearly dependent if they are not linearly independent.

Definition 2.21. An infinite subset S of a vector space V over a field F is said to be *linearly independent* if every finite subset of S is linearly independent.

Theorem 2.5. $L(S)$ is a subspace of V.

Proof. Let $v, w \in L(S)$. Then

$$\begin{aligned} v &= \lambda_1 s_1 + \lambda_2 s_2 + \ldots + \lambda_n s_n, \\ w &= \mu_1 t_1 + \mu_2 t_2 + \ldots + \mu_m t_m, \end{aligned}$$

where λ's and μ's are scalars and s_i and t_i are in S. Therefore, for $\alpha, \beta \in F$, we have

$$\begin{aligned} \alpha v + \beta w &= \alpha(\lambda_1 s_1 + \lambda_2 s_2 + \ldots + \lambda_n s_n) \\ &\quad + \beta(\mu_1 t_1 + \ldots + \mu_m t_m) \\ &= (\alpha\lambda_1)s_1 + (\alpha\lambda_2)s_2 + \ldots + (\alpha\lambda_n)s_n \\ &\quad + (\beta\mu_1)t_1 + \ldots + (\beta\mu_m)t_m \in L(S). \end{aligned}$$

Hence $L(S)$ is subspace of V.

Further, if S and T are subsets of a vector space V, then

(i) $S \subset T \Rightarrow L(S) \subset L(T)$

(ii) $L(S \cup T) = L(S) \cup L(T)$

(iii) $L(L(S)) = L(S)$

EXAMPLE 2.15

Let $v_1 = (1,\ 0)$ and $v_2 = (1,\ 0)$ be vectors in the vector space $R^2 = \{(x,\ y) : x, y \in R\}$. If $\lambda_1, \lambda_2 \in R$, then

$$\begin{aligned} \lambda_1 v_1 + \lambda_2 v_2 = 0 &\Rightarrow \lambda_1(1,\ 0) + \lambda_2(0,\ 1) = 0 \\ &\Rightarrow (\lambda_1,\ 0) + (0,\ \lambda_2) = 0 \\ &\Rightarrow (\lambda_1,\ \lambda_2) = 0 \\ &\Rightarrow \lambda_1 = \lambda_2 = 0. \end{aligned}$$

Hence, v_1, v_2 are linearly independent.

EXAMPLE 2.16

Let $v_1 = (1,0,1), v_2 = (0,1,0)$ and $v_3 = (1,1,1)$. Then we note that

$$\begin{aligned} v_1 + v_2 - v_3 &= (1,0,1) + (0,1,0) - (1,1,1) \\ &= (1+0-1, 0+1-1, 1+0-1) \\ &= (0,0,0). \end{aligned}$$

Hence, v_1, v_2 and v_3 are linearly dependent.

EXAMPLE 2.17

Determine whether the following set of vectors is linearly dependent or linearly independent.

$$\left\{ \begin{pmatrix} 1 & -3 \\ -2 & 4 \end{pmatrix}, \begin{pmatrix} -2 & 6 \\ 4 & -8 \end{pmatrix} \right\}$$

in the set of 2×2 matrices over the field of real numbers.

Solution. We check whether the only linear combination of vectors in the given set that equal to zero is the one in which all the coefficients are zero. So let λ_1 and λ_2 be scalars and

$$\lambda_1 \begin{pmatrix} 1 & -3 \\ -2 & 4 \end{pmatrix} + \lambda_2 \begin{pmatrix} -2 & 6 \\ 4 & -8 \end{pmatrix} = \begin{pmatrix} 0 & 0 \\ 0 & 0 \end{pmatrix}. \tag{2}$$

We have

$$\begin{pmatrix} \lambda_1 & -3\lambda_1 \\ -2\lambda_1 & 4\lambda_1 \end{pmatrix} + \begin{pmatrix} -2\lambda_2 & 6\lambda_2 \\ 4\lambda_2 & -8\lambda_2 \end{pmatrix} = \begin{pmatrix} 0 & 0 \\ 0 & 0 \end{pmatrix}$$

or

$$\begin{pmatrix} \lambda_1 - 2\lambda_2 & -3\lambda_1 + 6\lambda_2 \\ -2\lambda_1 + 4\lambda_2 & 4\lambda_1 - 8\lambda_2 \end{pmatrix} = \begin{pmatrix} 0 & 0 \\ 0 & 0 \end{pmatrix}.$$

Comparing the corresponding entries on both sides, we get

$$\left.\begin{aligned} \lambda_1 - 2\lambda_2 &= 0 \\ -3\lambda_1 + 6\lambda_2 &= 0 \\ -2\lambda_1 + 4\lambda_2 &= 0 \\ 4\lambda_1 - 8\lambda_2 &= 0 \end{aligned}\right\} \text{ all represent same equation.}$$

From any of these equations, we get $\lambda_2 = \frac{1}{2}\lambda_1$. So, let us take $\lambda_1 = 2$, then $\lambda_2 = 1$. Hence relation (2) is satisfied for non-zero scalars λ_1 and λ_2. Hence the given set is *linearly dependent*.

EXAMPLE 2.18

Find the span of the set $S = \{(1,0,0), (0,1,0)\}$ in $\mathfrak{R}^3$.

Solution. The span of the given set consists of all vectors in $\mathfrak{R}^3$ which are linear combination of $(1,0,0)$ and (0, 1, 0). Thus the span consists of vectors of the form $a(1,0,0) + b(0,1,0) = (a,b,0)$ for some scalars a and b. Hence the span of S is the xy-plane.

EXAMPLE 2.19

Show that the following set of vectors is linearly dependent.

$$\{(1,0,1), (1,1,0), (1,-1,1), (1,2,-3)\}$$

Solution. Let

$$\lambda_1(1,0,1) + \lambda_2(1,1,0) + \lambda_3(1,-1,1) + \lambda_4(1,2,-3) = (0,0,0) \quad (3)$$

or

$$(\lambda_1,0,\lambda_1) + (\lambda_2,\lambda_2,0) + (\lambda_3,-\lambda_3,\lambda_3) + (\lambda_4,2\lambda_4,-3\lambda_4) = (0,0,0)$$

or

$$(\lambda_1+\lambda_2+\lambda_3+\lambda_4,\ 0+\lambda_2-\lambda_3+2\lambda_4,\ \lambda_1+0+\lambda_3-3\lambda_4) = (0,0,0)$$

or

$$\lambda_1+\lambda_2+\lambda_3+\lambda_4 = 0,$$
$$\lambda_2-\lambda_3+2\lambda_4 = 0$$
$$\lambda_1+\lambda_3-3\lambda_4 = 0.$$

Solving these equations, we get

$$\lambda_1 = 5\lambda_4,\ \lambda_2 = -4\lambda_4,\ \lambda_3 = -2\lambda_4.$$

Taking $\lambda_4 = 1$, we get

$$\lambda_1 = 5,\ \lambda_2 = -4,\ \lambda_3 = -2,\ \lambda_4 = 1.$$

Hence (3) reduces to

$$5(1,0,1) - 4(1,1,0) - 2(1,-1,1) + 1(1,2,-3) = (0,0,0)$$

and so not all λ_1, λ_2, λ_3, λ_4 are zero. Hence the given set is linearly dependent.

EXAMPLE 2.20

Show that (i) the set $S = \{e^x, xe^x, x^2e^x\}$ in $C^2(-\infty,\infty)$ is linearly independent, (ii) the set $S = \{\sin(x+1), \sin x, \cos x\}$ in $C(0,\infty)$ is linearly independent.

Solution. (i) We have

$$S = \{e^x, xe^x, x^2e^x\}$$

Suppose that

$$\lambda_1 e^x + \lambda_2 xe^x + \lambda_3 x^2 e^x = 0(x) = 0 \quad \text{for all} \quad x \in (-\infty,\infty) \quad (4)$$

Differentiating, we get

$$\lambda_1 e^x + \lambda_2(e^x + xe^x) + \lambda_3(2xe^x + x^2e^x) = 0$$

or

$$(\lambda_1+\lambda_2)e^x + (\lambda_2+2\lambda_3)xe^x + \lambda_3 x^2 e^x = 0 \quad (5)$$

Differentiating (5), we get

$$(\lambda_1+\lambda_2)e^x + (\lambda_2+2\lambda_3)(e^x+xe^x) + \lambda_3(2xe^x + x^2e^x) = 0$$

or

$$(\lambda_1+2\lambda_2+2\lambda_3)e^x + (\lambda_2+4\lambda_3)xe^x + \lambda_3 x^2 e^x = 0 \quad (6)$$

Subtracting (4) from (5), we get

$$\lambda_2 e^x + 2\lambda_3 x e^x = 0 \quad (7)$$

Subtracting (5) from (6), we get

$$(\lambda_2 + 2\lambda_3)e^x + 2\lambda_3 x e^x = 0 \quad (8)$$

Subtracting (7) from (8), we have

$$2\lambda_3 e^x = 0$$
$$\Rightarrow \quad \lambda_3 = 0, \quad \text{since} \quad e^x \neq 0.$$

Putting $\lambda_3 = 0$ in (7), we have $\lambda_2 = 0$. Putting $\lambda_2 = \lambda_3 = 0$ in (4), we get $\lambda_1 = 0$. Hence the set S is linearly independent.

(ii) $S = \{\sin(x+1),\ \sin x,\ \cos x\}$. Let

$$\lambda_1 \sin(x+1) + \lambda_2 \sin x + \lambda_3 \cos x = 0(x) = 0 \quad (9)$$

Now (9) is possible only if $\lambda_1 = \lambda_2 = \lambda_3 = 0$. Hence the given system is linearly independent.

EXAMPLE 2.21

Determine whether the following polynomials span P_2:

$$P_1(x) = 2x^2 - x + 1,$$
$$P_2(x) = 4x^2 - x + 5,\ P_3(x) = 2x^2 - 2x - 2$$

Solution. Let

$$ax^2 + bx + c = \lambda_1(2x^2 - x + 1) + \lambda_2(4x^2 - x + 5) + \lambda_3(2x^2 - 2x - 2)$$
$$= (2\lambda_1 + 4\lambda_4 + 2\lambda_3)x^2 + (-\lambda_1 - \lambda_2 - 2\lambda_3)x + (\lambda_1 - 5\lambda_2 - 2\lambda_3).$$

Comparing coefficients of the powers of x on both sides, we get

$$2\lambda_1 + 4\lambda_2 + 2\lambda_3 = a$$
$$-\lambda_1 - \lambda_2 - 2\lambda_3 = b$$
$$\lambda_1 + 5\lambda_2 - 2\lambda_3 = c.$$

This system is inconsistent (the rank of coefficient matrix is 2 where as rank of the augmented matrix is 3). Therefore the given polynomials do not span P_2.

EXAMPLE 2.22

Show that the set $S = \{1 - t - t^3,\ -2 + 3t + t^2 + 2t^3,\ 1 + t^2 + 5t^3\}$ is linearly independent in P_3.

Solution. We have $S = \{1 - t - t^3,\ -2 + 3t + t^2 + 2t^3,\ 1 + t^2 + 5t^3\}$. Let

$$\lambda_1(1 - t - t^3) + \lambda_2(-2 + 3t + t^2 + 2t^3) + \lambda_3(1 + t^2 + 5t^3) = 0$$

or

$$(-\lambda_1 + 2\lambda_2 + 5\lambda_3)t^3 + (\lambda_2 + \lambda_3)t^2 + (-\lambda_1 + 3\lambda_2)t + (\lambda_1 - 2\lambda_2 + \lambda_3) = 0.$$

By equating coefficients of powers of t, we get

$$-\lambda_1 + 2\lambda_2 + 5\lambda_3 = 0$$
$$\lambda_2 + \lambda_3 = 0$$
$$-\lambda_1 + 3\lambda_2 = 0.$$

The coefficient matrix A of this system is non-singular, that is, $|A| \neq 0$. Therefore it has a trivial solution: $\lambda_1 = \lambda_2 = \lambda_3 = 0$. Hence S is linearly independent.

Theorem 2.6. Let V be a vector space over a field F. If $v_1,\ v_2,\ v_3, \ldots, v_n$ are linearly independent elements of V, then every element in their span has a unique representation in the form $\lambda_1 v_1 + \lambda_2 v_2 + \ldots \lambda_n v_n$ with $\lambda_i \in F$.

Proof. Every element in the linear span is of the form $\lambda_1 v_1 + \lambda_2 v_2 + \ldots + \lambda_n v_n$. Suppose that there are following two representations for an element: $\lambda_1 v_1 + \lambda_2 v_2 + \ldots + \lambda_n v_n$ and $\mu_1 v_1 + \mu_2 v_2 + \ldots + \mu_n v_n$

Then

$$\lambda_1 v_1 + \lambda_2 v_2 + \ldots + \lambda_n v_n = \mu_1 v_1 + \mu_2 v_2 + \ldots + \mu_n v_n$$

and so

$$(\lambda_1 - \mu_1)v_1 + (\lambda_2 - \mu_2)v_2 + \ldots + (\lambda_n - \mu_n)v_n = 0.$$

Since $v_1, v_2, v_3 \ldots v_n$ are linearly independent, we have

$$\lambda_1 - \mu_1 = 0, \lambda_2 - \mu_2 = 0, \ldots, \lambda_n - \mu_n = 0,$$

which yield

$$\lambda_1 = \mu_1, \ \lambda_2 = \mu_2, \ldots, \lambda_n = \mu_n.$$

Hence, representation of every element in the span is unique.

Theorem 2.7. If $v_1, v_2, v_3, \ldots, v_n$ are in V, then either they are linearly independent or some v_k is a linear combination of the preceding ones $v_1, v_2, v_3, \ldots, v_{k-1}$.

Proof. If $v_1, v_2, v_3, \ldots, v_n$ are linearly independent, we are done. So suppose that $v_1, v_2, v_3, \ldots, v_n$ are linearly dependent. Thus, $\alpha_1 v_1 + \alpha_2 v_2 + \ldots + \alpha_n v_n = 0$, where not all of $\alpha_1, \alpha_2, \ldots, \alpha_n$, are zero. Let k be the largest integer for which $\alpha_k \neq 0$. Since $\alpha_i = 0$ for $i > k$, $\alpha_1 v_1 + \ldots + \alpha_k v_k = 0$. Since $\alpha_k \neq 0$, we have

$$\begin{aligned} v_k &= \alpha_k^{-1}(-\alpha_1 v_1 - \alpha_2 v_2 - \ldots - \alpha_{k-1} v_{k-1}) \\ &= \left(-\alpha_k^{-1}\alpha_1\right) v_1 + \ldots + \left(-\alpha_k^{-1}\alpha_{k-1}\right) v_{k-1}. \end{aligned}$$

Hence, v_k is a linear combination of its predecessors.

Theorem 2.8. A system of vectors in a vector space is linearly dependent if and only if any one of the vectors in that system can be represented as a linear combination of the other vectors in the system.

Proof. Suppose that V is a vector space over the field F and let $v_1, v_2, v_3, \ldots, v_n \in V$ be linearly dependent. Then, by definition,

$$\lambda_1 v_1 + \lambda_2 v_2 + \ldots + \lambda_n v_n = 0,$$

where not all of the λ_i are zero. Suppose that $\lambda_1 \neq 0$, then

$$v_1 = -\frac{\lambda_2}{\lambda_1} v_2 - \frac{\lambda_3}{\lambda_3} v_3 - \ldots - \frac{\lambda_n}{\lambda_n} v_n.$$

Hence v_1 is linear combination of other vector in the system.

Conversely, suppose that v_1 is linear combination of $v_2, v_3, \ldots, v_n$, that is

$$v_1 = \lambda_2 v_2 + \lambda_3 v_3 + \ldots + \lambda_n v_n, \quad \lambda_i \in F$$

and so

$$(-1)v_1 + \lambda_2 v_2 + \lambda_3 v_3 + \ldots + \lambda_n v_n = 0.$$

Since the first coefficient is non-zero, it follows that $v_1, v_2, v_3, \ldots, v_n$ is a linearly dependent system.

Theorem 2.9. If a subsystem of a finite system of vectors in a vector space is linearly dependent, then the whole system is linearly dependent.

Proof. Let $v_1, v_2, v_3, \ldots, v_n \in V$ be a finite system of vectors in V. Suppose that $v_1, v_2, v_3, \ldots, v_k, k < n$ is linearly dependent. Therefore,

$$\begin{aligned} &\lambda_1 v_1 + \lambda_2 v_2 + \ldots + \lambda_k v_k + 0 v_{k+1} + 0 v_{k+2} + \ldots \\ &\quad + 0 v_n = 0, \end{aligned}$$

where not all of $\lambda_1, \lambda_2, \ldots, \lambda_k$ are zero. Hence, $v_1, v_2, v_3, \ldots, v_n$ is linearly dependent.

It follows from Theorem 2.9 that any superset of a linearly dependent set is also linearly dependent.

EXAMPLE 2.23

Show that the set

$$\{(1,1,0),(0,1,1),(1,0,-1),(1,1,1)\}$$

is linearly dependent.

Solution. We note that

$$1(1,1,0) - 1(0,1,1) - 1(1,0,-1) = (0,0,0)$$

Hence, the set

$$\{(1,1,0),(0,1,1),(1,0,-1)\}$$

is linearly dependent. Being superset of this linearly dependent set, the given set is also linearly dependent.

Definition 2.22. Let S be subset of a vector space V. If every element of V can be written as the linear combination of the elements of S, then S is called *generator* of V.

For example, let $V_2 = \{(x,y) : x,y \in \mathbb{R}\}$ be the vector space and let

$$v_1 = (1,0), v_2 = (0,1)$$

be vectors in V_2. If $(x,y) \in V_2$ be arbitrary, then

$$\begin{aligned}(x,y) &= x(1,0) + y(0,1)\\ &= xv_1 + yv_2\end{aligned}$$

Hence $S = \{v_1, v_2\}$ generates V_2.

Definition 2.23. Let S be subset of a vector space V. If

(i) S generates V, that is, $L(S) = V$, and
(ii) the elements of S are linearly independent, then S is called *basis* of the vector space V.

For example $\{(1,0),(0,1)\}$ is a basis of the vector space

$$V_2 = \{(x,y) : x,y \in \mathbb{R}\}.$$

Definition 2.24. The number of elements in the basis of a vector space is called the *dimension* of that vector space.

For example, the dimension of V_2 is 2.

If the number of elements in the basis of a vector space is finite, then the vector space is called *finite dimensional vector space*.

EXAMPLE 2.24

Let F be a field and

$$F^{(n)} = \{(x_1, x_2, \ldots, x_n) : x_i \in F\}$$

be a set of *n-tuples*. If we define addition and scalar multiplication in $F^{(n)}$ by

$$\begin{aligned}&(x_1, x_2, \ldots, x_n) + (y_1, y_2, \ldots, y_n)\\ &= (x_1 + y_1, x_2 + y_2, \ldots, x_n + y_n),\end{aligned}$$

and

$$\lambda(x_1, x_2, \ldots, x_n) = (\lambda x_1, \lambda x_2, \ldots, \lambda x_n).$$

Then $F^{(n)}$ becomes a vector space (linear space).

Let

$$\begin{aligned}&e_1 = (1,0,\ldots,0),\\ &e_2 = (0,1,0,\ldots,0), \ldots, e_n = (0,0,\ldots,1).\end{aligned}$$

Then,

$$\begin{aligned}(x_1, x_2, \ldots, x_n) &= x_1(1,0,\ldots,0) + x_2(0,1,0,\ldots,0)\\ &\quad + \ldots + x_n(0,0,\ldots,1)\\ &= x_1e_1 + x_2e_2 + \ldots + x_ne_n\end{aligned}$$

and so $\{e_1, e_2, \ldots, e_n\}$ generates $F^{(n)}$. Further

$$\begin{aligned}&\lambda_1e_1 + \lambda_2e_2 + \ldots + \lambda_ne_n = 0\\ &\Rightarrow \lambda(1,0,\ldots,0) + \lambda_2(0,1,\ldots,0)\\ &\quad + \ldots + \lambda_n(0,0,\ldots,1) = 0\\ &\Rightarrow (\lambda_1, \lambda_2, \ldots, \lambda_n) = 0\\ &\Rightarrow \lambda_1 = \lambda_2 = \ldots = \lambda_n = 0\end{aligned}$$

and so $\{e_1, e_2, \ldots, e_n\}$ is linearly independent. Hence, $\{e_1, e_2, \ldots, e_n\}$ is the *standard basis* of $F^{(n)}$ and so $F^{(n)}$ is n-dimensional.

EXAMPLE 2.25

Let $P_n(x)$ be the set of all polynomials of degree n over a field F. Then the set $\{1, x, x^2, \ldots, x^n\}$ is a basis, called the *standard basis* for $P_n(x)$.

EXAMPLE 2.26

Let M_n be the set of all matrices of order n and let E_{ij} be the matrix whose (i,j) th entry is 1 and all other entries are zero. Then (see Example 2.43) the set $\{E_{ij}\}, 1 \le i \le n, 1 \le j \le n$ is a basis for the vector space $M_n(F)$ over the field F.

Theorem 2.10. Let V be a vector space. Then $\beta = \{v_1, v_2, \ldots, v_n\}$ is a basis for V if and only if each $v \in V$ can be uniquely expressed as a linear combination of vectors of β.

Proof. Suppose first that $\beta = \{v_1, v_2, \ldots v_n\}$ is a basis for V. Then, by the definition of basis, each $v \in V$ can be expressed as a linear combination of vectors of β. To prove uniqueness of the expression, suppose that

$$v = a_1v_1 + a_2v_2 + \ldots + a_nv_n \tag{10}$$

and

$$v = b_1v_1 + b_2v_2 + \ldots + b_nv_n \tag{11}$$

are two representations of v with scalars $a_i, b_i, 1 \le i \le n$. Subtracting (11) from (10), we get

$$(a_1 - b_1)v_1 + (a_2 - b_2)v_2 + \ldots + (a_n - b_n)v_n = 0 \tag{12}$$

But, being a basis, β is linearly independent, Therefore (12) implies

$$a_1 - b_1 = 0,\ a_2 - b_2 = 0, \ldots, a_n - b_n = 0$$

or

$$a_1 = b_1,\ a_2 = b_2, \ldots, a_n = b_n.$$

Hence v is uniquely expressible as a linear combination of the vectors of β.

Conversely, suppose that each $v \in V$ is uniquely expressible as a linear combination.

$$v = a_1v_1 + a_2v_2 + \ldots + a_nv_n.$$

Thus $\beta = \{v_1, v_2, \ldots v_n\}$ generates V. Therefore to prove that β is a basis, it is sufficient to show that β is linearly independent. Suppose, on the contrary, that β is not linearly independent. Then there exist $b_1, b_2, \ldots, b_n$ not all zero such that

$$b_1v_1 + b_2v_2 + \ldots + b_nv_n = 0.$$

Also, 0 can be expressed as the linear combination

$$0v_1 + 0v_2 + \ldots + 0v_n = 0$$

Thus, we get two distinct expressions for the vector $0 \in v$. This contradicts the fact that each $v \in V$ is uniquely expressible as a linear combination of the vectors of β. Hence β is linearly independent and so is a basis.

Remark 2.1 The above theorem may also be stated as follows:

Theorem 2.11. If $\beta = \{v_1, v_2, \ldots, v_n\}$ span, a vector space V, then the following two statements are equivalent,

(i) $\{v_1, v_2, \ldots, v_n\}$ is a linearly independent set.
(ii) Each $v \in V$ can uniquely expressible as a linear combination of vectors in β.

Theorem 2.12. If $v_1, v_2, \ldots, v_n$ is a basis of a vector space V over a field F and if $w_1, w_2, \ldots, w_m$ in V are linearly independent over F, then $m \le n$.

Proof. Since $v_1, v_2, \ldots, v_n$ form a basis of V over F and $w_1 \in V$, we have

$$w_1 = \lambda_1v_1 + \lambda_2v_2 + \ldots + \lambda_nv_n, \quad \lambda_n \in F \tag{13}$$

and so $\{w_1, v_1, v_2, \ldots, v_n\}$ are linearly dependent. Being a member of linearly independent set, $w_1 \ne 0$. Therefore, there is at least one λ_i which is not zero. Suppose that $\lambda_1 \ne 0$. Then (13) implies

$$v_1 = \lambda_1^{-1}(w_1 - \lambda_2v_2 - \ldots - \lambda_nv_n).$$

Consider the set

$$S_1 = \{w_1, v_2, v_3, \ldots, v_n\}.$$

We claim that S_1 is a basis of V. To prove it, let $v \in V$.

Then

$$\begin{aligned} v &= \mu_1v_1 + \mu_2v_2 + \ldots + \mu_nv_n \\ &= \mu_1\{\lambda_1^{-1}(w_1 - \lambda_2v_2 - \ldots - \lambda_nv_n)\} \\ &\quad + \mu_2v_2 + \ldots + \mu_nv_n = 0 \\ &= (\mu_1\lambda_1^{-1})w_1 + \sum_{i=2}^{n}(\mu_i - \mu_1\lambda_1^{-1}\lambda_i)v_i. \end{aligned}$$

Hence, S_1 generates V. Further $\mu_1 w_1 + \mu_2 v_2 + \ldots + \mu_n v_n = 0$ implies

$$\mu_1(\lambda_1 v_1 + \lambda_2 v_2 + \ldots + \lambda_n v_n) + \mu_2 v_2 + \ldots + \mu_n v_n = 0$$

$$\Rightarrow (\lambda_1\mu_1)v_1 + (\mu_1\lambda_2 + \mu_2)v_2 + \ldots + (\mu_1\lambda_n + \mu_n)v_n = 0$$

$$\Rightarrow \lambda_1\mu_1 = \mu_1\lambda_2 + \mu_2 = \ldots = \mu_1\lambda_n + \mu_n = 0$$

(By linear independence of $v_1, \ldots, v_n$)

Since $\lambda_1 \neq 0$, we have $\mu_1 = 0$. Substituting $\mu_1 = 0$ in other equations, we get $\mu_2 = \mu_3 = \ldots = \mu_n = 0$. Hence, S_1 is linearly independent and so S_1 is a basis for V. As such

$$w_2 = a_1 w_1 + a_2 v_2 + \ldots + a_n v_n, \quad a_i \in F$$

Now, all $a_2, a_3, \ldots, a_n$ cannot be zero otherwise $a_1 w_1 - w_2 = 0$, which, due to linear independence of $\{w_1, w_2\}$ yields $a_1 = 0, -1 = 0$ (absurd). Suppose $a_2 \neq 0$. Then,

$$v_2 = a_2^{-1} w_2 - a_2^{-1}(a_1 w_1 + a_3 v_3 + \ldots + a_n v_n).$$

As in the case of S_1, we can prove that $S_2 = \{w_1, w_2, v_3, \ldots v_n\}$ generates V and is linearly independent. Hence, S_2 is a basis of V. Proceeding in this manner, we can show that $S_{m-1} = \{w_1, w_2, \ldots, w_{m-1}, v_m, \ldots v_n\}$ is a basis of V. Then,

$$w_m = c_1 w_1 + c_2 w_2 + \ldots + c_{m-1} w_{m-1} + c_m v_m + \ldots + c_n w_n.$$

Hence $n - (m-1) \geq 1$ or $m \leq n$.

Corollary 2.1. If V is a finite dimensional vector space over a field F, then any two bases of V have the same number of elements.

Proof. Let $v_1, v_2, \ldots, v_n$ and $w_1, w_2, \ldots w_m$ are two bases of V. Then $w_1, w_2, \ldots, w_m$ are linearly independent. Hence, by Theorem 2.10, $m \leq n$. Similarly, taking $w_1, w_2, \ldots, w_m$ as the basis and $v_1, v_2, \ldots, v_n$ as linearly independent set, we have $n \leq m$. Hence $m = n$, which proves the corollary.

Corollary 2.2. **(Extension to basis)** Let the set $\{v_1, v_2, \ldots, v_m\}$ be a linearly independent subset of an n-dimensional vector space V. Then there exist vectors $v_{m+1}, \ldots, v_n$ in V such that $\{v_1, v_2, \ldots, v_m, v_{m+1}, \ldots, v_n\}$ is a basis for V.

Proof. By Theorem 2.12, $m \leq n$. If $m = n$, then $\{v_1, v_2, \ldots, v_m\}$ is a basis for V. Further, if $m < n$, then $\{v_1, v_2, \ldots, v_m\}$ is not a basis of n-dimensional vector space V. Thus $\{v_1, v_2, \ldots, v_m\}$ does not generate V. Therefore there exists any non-zero vector v_{m+1} in V such that $v_{m+1} \notin$ span $\{v_1, v_2, \ldots, v_m\}$. Hence the set $\{v_1, v_2, \ldots, v_m, v_{m+1}\}$ is linearly independent. Continue with the process until we get n-linearly independent vectors $\{v_1, v_2, \ldots, v_m, v_{m+1}, \ldots, v_n\}$ which form a basis for V.

Corollary 2.3. If $\dim(V) = n$, then any linearly independent subset S of V consisting of n elements forms a basis of the vector space V.

Proof. Since $\dim(V) = n$, each basis of V has n elements. By Corollary, S can be extended to the basis of V. Since S has already n elements, therefore S is a basis of V.

EXAMPLE 2.27

Find a standard basis vector that can be added to the set

$$S = \{(-1, 2, 3), (1, -2, -2)\}$$

to produce a basis of $\Re^3$.

Solution. We have

$$S = \{(-1, 2, 3), (1, -2, -2)\}.$$

We note that

$$\lambda_1(-1, 2, 3) + \lambda_2(1, -2, -2) = (0, 0, 0)$$

$$\Rightarrow -\lambda_1 + \lambda_2 = 0,\ 2\lambda_1 - 2\lambda_2 = 0 \quad \text{and}$$

$$3\lambda_1 - 2\lambda_2 = 0.$$

Solving these equations, we have $\lambda_1 = \lambda_2 = 0$. Hence S is linearly independent.

The standard basis for $\mathfrak{R}^3$ is

$$\beta = \{(1,0,0),\ (0,1,0),\ (0,0,1)\}.$$

Now

$$\begin{aligned}\lambda_1(-1,2,3) + \lambda_2(1,-2,-2)& \\ +\lambda_3(1,0,0) = (0,0,0)& \\ \Rightarrow -\lambda_1 + \lambda_2 + \lambda_3 = 0& \\ 2\lambda_1 - 2\lambda_2 = 0& \\ 3\lambda_1 - 2\lambda_2 = 0&\end{aligned}$$

Solving these equations, we get $\lambda_1 = \lambda_2 = \lambda_3 = 0$. Hence the set

$$T = \{(-1,2,3),\ (1,-2,-2),\ (1,0,0)\}$$

is an extension of S to a basis of $\mathfrak{R}^3$.

EXAMPLE 2.28

Determine whether B is in the column space of the matrix A, and if so, express B as a linear combination of the column vectors of A if

$$A = \begin{bmatrix} 1 & -1 & 1 \\ 1 & 1 & -1 \\ -1 & -1 & 1 \end{bmatrix} \quad \text{and} \quad B = \begin{bmatrix} 2 \\ 0 \\ 0 \end{bmatrix}.$$

Solution. We are given that

$$A = \begin{bmatrix} 1 & -1 & 1 \\ 1 & 1 & -1 \\ -1 & -1 & 1 \end{bmatrix},\ B = \begin{bmatrix} 2 \\ 0 \\ 0 \end{bmatrix}$$

The column vectors of A are

$$\begin{bmatrix} 1 \\ 1 \\ -1 \end{bmatrix},\ \begin{bmatrix} -1 \\ 1 \\ -1 \end{bmatrix} \quad \text{and} \quad \begin{bmatrix} 1 \\ -1 \\ 1 \end{bmatrix}$$

We know that

$$\begin{bmatrix} 1 \\ -1 \\ 1 \end{bmatrix} = 0 \begin{bmatrix} 1 \\ 1 \\ -1 \end{bmatrix} - 1 \begin{bmatrix} -1 \\ 1 \\ -1 \end{bmatrix}$$

Also

$$\lambda_1 \begin{bmatrix} 1 \\ 1 \\ -1 \end{bmatrix} + \lambda_2 \begin{bmatrix} -1 \\ 1 \\ -1 \end{bmatrix} = \begin{bmatrix} 0 \\ 0 \\ 0 \end{bmatrix}$$

$$\Rightarrow \lambda_1 - \lambda_2 = 0,\ \lambda_1 + \lambda_2 = 0 \quad \text{and}$$
$$-\lambda_1 - \lambda_2 = 0.$$

Solving these equations, we get $\lambda_1 = \lambda_2 = 0$. Hence the columns $\begin{bmatrix} 1 \\ 1 \\ -1 \end{bmatrix}$ and $\begin{bmatrix} -1 \\ 1 \\ -1 \end{bmatrix}$ form the basis of the column space of A.

The vector B will be in the column space if it can be written as linear combination of the basis. Suppose

$$\begin{bmatrix} 2 \\ 0 \\ 0 \end{bmatrix} = a \begin{bmatrix} 1 \\ 1 \\ -1 \end{bmatrix} + b \begin{bmatrix} -1 \\ 1 \\ -1 \end{bmatrix}.$$

This yields

$$2 = a - b,\ 0 = a + b \quad \text{and} \quad 0 = -a - b.$$

Solving these equations, we get $a = 1,\ b = -1$. Thus

$$\begin{bmatrix} 2 \\ 0 \\ 0 \end{bmatrix} = 1 \begin{bmatrix} 1 \\ 1 \\ -1 \end{bmatrix} - 1 \begin{bmatrix} -1 \\ 1 \\ -1 \end{bmatrix}$$

Hence B is in the column space of A.

EXAMPLE 2.29

Find basis and dimension of

$$\begin{aligned} W = \{(a_1,a_2,a_3,a_4) \in \mathfrak{R}^4 : a_1 + a_2 = 0,& \\ a_2 + a_3 = 0, a_3 + a_4 = 0\}.& \end{aligned}$$

Solution. We have

$$W = \{(a_1,a_2,a_3,a_4) \in \mathfrak{R}^4\}$$

such that

$$a_1 + a_2 = 0$$
$$a_2 + a_3 = 0$$
$$a_3 + a_4 = 0.$$

Solving these equations, we get

$$a_1 = -a_2 = a_3 = -a_4.$$

Thus

$$W = \{(a, -a, a, -a) : a \in \Re\}$$

Clearly

$$(a, -a, a, -a) = a(1,0,0,0) - a(0,1,0,0) + a(0,0,1,0) - a(0,0,0,1).$$

Also $\{(1, 0, 0, 0), \ (0, 1, 0, 0), \ (0, 0, 1, 0), (0, 0, 0, 1)$ is linearly independent. Hence, basis for W is

$$\beta = \{(1,0,0,0), (0,1,0,0), (0,0,1,0), (0,0,0,1)\}$$

and

$$\dim(W) = 4.$$

EXAMPLE 2.30

Find a basis for the subspace of P_2 spanned by the vectors

$$1 + x, x^2, -2 + 2x^2, -3x.$$

Solution. The subspace is spanned by

$$1 + x, \ x^2, \ -2 + 2x^2, \ -3x.$$

It is easy show that the set of these vectors is linearly dependent. Let

$$-2 + 2x^2 = \lambda_1(1 + x) + \lambda_2 x^2 + \lambda_3(-3x)$$

Comparing coefficients of the powers of x, we get

$$-2 = \lambda_1, \ 0 = \lambda_1 - 3\lambda_3, \ 2 = \lambda_2.$$

Thus

$$\lambda_1 = -2, \ \lambda_3 = -\frac{2}{3}, \ \lambda_2 = 2. \quad \text{Hence}$$

$$-2 + 2x^2 = -2(1 + x) + 2x^2 - \frac{2}{3}(-3x).$$

Hence $-2 + 2x^2$ is discarded from the given set. The remaining set of vectors is $S = \{1 + x, -3x, x^2\}$ which is linearly independent. Hence S is a basis for the subspace.

EXAMPLE 2.31

Reduce $S = \{(1,0,0), \ (0,1,-1), \ (0,4,-3), (0,2,0)\}$ to obtain a basis of $\Re^3$.

Solution. We have

$$S = \{(1,0,0), (0,1,-1), (0,4,-3), (0,2,0)\}$$

We have to discard one vector from S. Let us try $(0,4,-3)$ to see whether it is a linear combinations of the remaining vectors. So, let

$$(0,4,-3) = \lambda_1(1,0,0) + \lambda_2(0,1,-1) + \lambda_3(0,2,0).$$

Then

$$0 = \lambda_1, \ 4 = \lambda_2 + 2\lambda_3 \quad \text{and} \quad -3 = -\lambda_2,$$

that is,

$$\lambda_1 = 0, \ \lambda_2 = 3, \ \lambda_3 = \frac{1}{2}.$$

Thus

$$(0,4,-3) = 0(1,0,0) + 3(0,1,-1) + \frac{1}{2}(0,2,0)$$

and so we discard $(0,4,-3)$. Thus the remaining set is

$$S_1 = \{(1, 0, 0), \ (0, 1, -1), \ (0, 2, 0)\},$$

The set S_1 is linearly independent and if $(a, b, c) \in \Re^3$, then

$$(a, b, c) = a(1, 0, 0) - c(0, 1, -1) + \frac{b + c}{2}(0, 2, 0).$$

Hence S_1 generates $\mathfrak{R}^3$. Hence S_1 is a basis of $\mathfrak{R}_3$.

EXAMPLE 2.32

Show that

$$S=\left\{\begin{bmatrix}1 & 2\\ 1 & -2\end{bmatrix},\begin{bmatrix}0 & -1\\ -1 & 0\end{bmatrix},\begin{bmatrix}0 & 2\\ 3 & 1\end{bmatrix},\begin{bmatrix}0 & 0\\ -1 & 2\end{bmatrix}\right\}$$

is a basis for M_{22}.

Solution. The given set is

$$S=\left\{\begin{bmatrix}1 & 2\\ 1 & -2\end{bmatrix},\begin{bmatrix}0 & -1\\ -1 & 0\end{bmatrix},\begin{bmatrix}0 & 2\\ 3 & 1\end{bmatrix},\begin{bmatrix}0 & 0\\ -1 & 2\end{bmatrix}\right\}.$$

We note that

$$\lambda_1\begin{bmatrix}1 & 2\\ 1 & -2\end{bmatrix}+\lambda_2\begin{bmatrix}0 & -1\\ -1 & 0\end{bmatrix}+\lambda_3\begin{bmatrix}0 & 2\\ 3 & 1\end{bmatrix}+\lambda_4\begin{bmatrix}0 & 0\\ -1 & 2\end{bmatrix}=\begin{bmatrix}0 & 0\\ 0 & 0\end{bmatrix}$$

implies

$$\begin{bmatrix}\lambda_1 & 2\lambda_1-\lambda_2+2\lambda_3\\ \lambda_1-\lambda_2+3\lambda_3-\lambda_4 & -2\lambda_1+\lambda_3+2\lambda_4\end{bmatrix}=\begin{bmatrix}0 & 0\\ 0 & 0\end{bmatrix}$$

Hence

$$\begin{aligned}\lambda_1&=0\\ 2\lambda_1-\lambda_2+2\lambda_3&=0\\ \lambda_1-\lambda_2+3\lambda_3-\lambda_4&=0\\ -2\lambda_1+\lambda_3+2\lambda_4&=0.\end{aligned}$$

Solving these equations, we get $\lambda_1=\lambda_2=\lambda_3=\lambda_4=0$. Hence S is linearly independent.

Further, let $\begin{bmatrix}a & b\\ c & d\end{bmatrix}\in M_{2\times 2}$. Then if

$$\begin{bmatrix}a & b\\ c & d\end{bmatrix}=\lambda_1\begin{bmatrix}1 & 2\\ 1 & -2\end{bmatrix}+\lambda_2\begin{bmatrix}0 & -1\\ -1 & 0\end{bmatrix}+\lambda_3\begin{bmatrix}0 & 2\\ 3 & 1\end{bmatrix}+\lambda_4\begin{bmatrix}0 & 0\\ -1 & 2\end{bmatrix}$$

or

$$\begin{bmatrix}a & b\\ c & d\end{bmatrix}=\begin{bmatrix}\lambda_1 & 2\lambda_1-\lambda_2+2\lambda_3\\ \lambda_1-\lambda_2+3\lambda_3-\lambda_4 & -2\lambda_3+\lambda_3+2\lambda_4\end{bmatrix}$$

Then

$$\begin{aligned}a&=\lambda_1\\ b&=2\lambda_1-\lambda_2+2\lambda_3\\ c&=\lambda_1+3\lambda_3-\lambda_4\\ d&=-2\lambda_1+\lambda_3+2\lambda_4.\end{aligned}$$

Solving these equations, we get

$$\lambda_1=a,\ \lambda_2=2d+4a-\frac{4}{7}(5a+b-c+3d),$$

$$\lambda_3=d+2a-\frac{2}{7}(5a+b-c+3d),$$

$$\lambda_4=\frac{1}{7}[5a+b-c+3d].$$

Thus the set S generates $M_{2\times 2}$. Hence S is a basis for $M_{2\times 2}$.

EXAMPLE 2.33

The set W of diagonal $n\times n$ matrices is a subspace of $M_{n\times n}(F)$. The basis for this subspace is the set consisting of matrices E_{ii}, where E_{ij} is the matrix having 1 as the (i,j) entry and 0 elsewhere. Clearly, $\dim(W)=n$.

EXAMPLE 2.34

The set of symmetric $n\times n$ matrix is a subspace W of $M_{n\times n}(F)$. The basis for the space W is the set of matrices E_{ij}, whose (i,j) th element is 1 and the other elements are zero. Also $E_{ij}=E_{ji}$, Since the matrices in W are symmetric. Thus

$$\begin{aligned}\dim W&=n+(n-1)+\cdots+1\\ &=\frac{1}{2}n(n+1).\end{aligned}$$

EXAMPLE 2.35

Extend the subset $S=\{(2,1,4,3),(2,1,2,0)\}$ to form a basis of the vector space $\mathfrak{R}^4$.

Solution. Let $e_1 = (1,0,0,0)$, $e_2 = (0,1,0,0)$, $e_3 = (0,0,1,0)$, $e_4 = (0,0,0,1)$, $v_1 = (2,1,4,3)$ and $v_2 = (2,1,2,0)$. Then $\{e_1, e_2, e_3, e_4\}$ form the standard basis of $\Re^4$. If α, β are real numbers, then

$$\alpha v_1 + \beta v_2 = \alpha(2,1,4,3) + \beta(2,1,2,0)$$
$$= (0,0,0,0) \quad \text{implies}$$
$$\alpha = \beta = 0.$$

Hence S is linearly independent. Consider the set $S_1 = \{v_1, v_2, e_1\}$. It is easy to show that it is linearly independent. Similarly, we can show that $S_2 = \{v_1, v_2, e_1, e_2\}$ is linearly independent. Since $\dim(\Re^4) = 4$, the corollary 2.3 implies that S_2 is a basis of $\Re^4$.

2.3 LINEAR TRANSFORMATION

Definition 2.25. Let V and W be two vector spaces over the same field F and let $T : V \to W$ be a mapping such that

(i) $T(v_1 + v_2) = T(v_1) + T(v_2), v_1, v_2, \in V$ and

(ii) $T(\lambda v) = \lambda T(v)$, $\lambda \in F$, $v \in V$.

Then, T is called a *linear transformation or vector space homomorphism* from V to W.

A one-to-one vector space homomorphism is called *vector space isomorphism.* Two vector spaces over the same field are said to be *isomorphic* if there exists an isomorphism from one onto another.

Theorem 2.13. Let V and W be vector spaces over F. Then, $T : V \to W$ is a linear transformation if and only if

$$T(\lambda_1 v_1 + \lambda_2 v_2) = \lambda_1 T(v_1) + \lambda_2 T(v_2),$$

where $\lambda_1, \lambda_2 \in F$ and $v_1, v_2 \in V$.

Proof. First suppose that $T : V \to W$ is a linear transformation. Then, by definition,

$$T(v_1 + v_2) = T(v_1) + T(v_2), \quad (14)$$
$$T(\lambda_1 v_1) = \lambda_1 T(v_1). \quad (15)$$

We have, by (15)

$$T(\lambda_2 v_2) = \lambda_2 T(v_2),$$

and then (14) yields

$$T(\lambda_1 v_1 + \lambda_2 v_2) = T(\lambda_1 v_1) + T(\lambda_2 v_2),$$
$$= \lambda_1 T(v_1) + \lambda_2 T(v_2).$$

Conversely, suppose that

$$T(\lambda_1 v_1 + \lambda_2 v_2 = \lambda_1 T(v_1) + \lambda_2 T(v_2).$$

Substituting $\lambda_1 = \lambda_2 = 1$, we have

$$T(v_1 + v_2) = T(v_1) + T(v_2),$$

and taking $\lambda_2 = 0$, we have

$$T(\lambda_1 v_1) = \lambda_1 T(v_1).$$

Hence, T is a linear transformation.

EXAMPLE 2.36

Define $T : V_2 \to V_2$ by

$$T(a_1, a_2) = (a_1, -a_2).$$

Then T is called the ***reflection*** about the x-axis. To prove that T is a linear transformation, let (a_1, a_2) and (a_3, a_4) be in V_2. Then

$$T((a_1,a_2)+(a_3,a_4)) = T(a_1+a_3, a_2+a_4)$$
$$= (a_1+a_3, -(a_2+a_4))$$
$$= (a_1,-a_2)+(a_3,-a_4)$$
$$= T(a_1,a_2)+T(a_3,a_4)$$

and

$$T(\lambda(a_1,a_2)) = T(\lambda a_1, \lambda a_2)$$
$$= (\lambda a_1, -\lambda a_2)$$
$$= \lambda(a_1, -a_2)$$
$$= \lambda T(a_1,a_2).$$

Hence T is a linear transformation.

EXAMPLE 2.37

Let V be a vector space and let u be a fixed non-zero vector in V. Define $T : V \to V$ by

$$T(a) = a + u.$$

Then T is called ***translation*** by the vector u.

Let a, b be in V. Then

$$T(a+b) = (a+b)+u$$

and

$$T(a)+T(b) = (a+u)+(b+u).$$

Thus

$$T(a+b) \neq T(a)+T(b).$$

Hence *translation* is *not* a linear transformation.

EXAMPLE 2.38

Let $T_\theta(a_1,a_2)$ be the vector obtained by rotating (a_1,a_2) counter clockwise by an angle θ. If $(a_1,a_2) \neq (0,0)$ and $T_\theta(0,0) = (0,0)$, then the map $T_\theta : V_2 \to V_2$ is called ***rotation*** of co-ordinates about the origin.

Let α be the angle that (a_1,a_2) makes with the positive x-axis. Then $a_1 = r\cos\alpha$ and $a_2 = r\sin\alpha$, where $r = \sqrt{a_1^2 + a_2^2}$. Now $T_\theta(a_1,a_2)$ has also length r and makes an angle $\alpha + \theta$ with the positive x-axis. Thus

$$\begin{aligned}
T_\theta(a_1,a_2) &= (r\cos(\alpha+\theta), r\sin(\alpha+\theta)) \\
&= (r\cos\alpha\cos\theta - r\sin\alpha\sin\theta, \\
&\qquad r\cos\alpha\sin\theta + r\sin\alpha\cos\theta) \\
&= (a_1\cos\theta - a_2\sin\theta, a_1\sin\theta + a_2\cos\theta)
\end{aligned}$$

is an *explicit expression* for the map T_θ.

We note that if (a_1,a_2), $(a_3,a_4) \in V_2$, then

$$\begin{aligned}
&T_\theta((a_1,a_2)+(a_3,a_4)) \\
&= T_\theta(a_1+a_3, a_2+a_4) \\
&= \Big((a_1+a_3)\cos\theta - (a_2+a_4)\sin\theta, \\
&\qquad (a_1+a_3)\sin\theta + (a_2+a_4)\cos\theta\Big) \\
&= (a_1\cos\theta - a_2\sin\theta, a_1\sin\theta + a_2\cos\theta) \\
&\quad + (a_3\cos\theta - a_4\sin\theta, a_3\sin\theta + a_4\cos\theta) \\
&= T_\theta(a_1,a_2) + T_\theta(a_3,a_4)
\end{aligned}$$

and

$$\begin{aligned}
&T_\theta(\lambda(a_1,a_2)) \\
&= T_\theta(\lambda a_1, \lambda a_2) \\
&= (\lambda a_1\cos\theta - \lambda a_2\sin\theta, \lambda a_1\sin\theta + \lambda a_2\cos\theta) \\
&= \lambda(a_1\cos\theta - a_2\sin\theta, a_1\sin\theta + a_2\cos\theta) \\
&= \lambda T_\theta(a_1,a_2).
\end{aligned}$$

Hence rotation of co-ordinates about the origin *is a linear transformation.*

EXAMPLE 2.39

The mapping $T : V_2 \to V_2$ defined by $T(a_1,a_2) = (a_1,0)$ is called the ***projection*** of V_2 on the x-axis. Similarly, the map $T : V_3 \to V_3$ defined by $T(a_1,a_2,a_3) = (a_1,a_2,0)$ is called the ***projection*** of V_3 on the xy-plane.

We note that if $a = (a_1,a_2,a_3)$, $b = (b_1, b_2,b_3)$ in V_3, then we have

$$\begin{aligned}
T(a+b) &= T(a_1+b_1, a_2+b_2, a_3+b_3) \\
&= (a_1+b_1, a_2+b_2, 0) \\
&= (a_1,a_2,0) + (b_1,b_2,0) \\
&= T(a) + T(b)
\end{aligned}$$

and

$$\begin{aligned}
T(\lambda a) &= T(\lambda a_1, \lambda a_2, \lambda a_3) \\
&= (\lambda a_1, \lambda a_2, 0) \\
&= \lambda(a_1,a_2,0) \\
&= \lambda T(a).
\end{aligned}$$

Hence *projection* is a linear transformation.

EXAMPLE 2.40

A mapping $T: \mathfrak{R} \to \mathfrak{R}$ is called a ***contraction mapping*** if there exists a number M, $0 < M < 1$ such that for distinct x and y in $\mathfrak{R}$,

$$|T(x) - T(y)| \leq M|x-y|.$$

The mapping T is called an ***expansion*** if for distinct x and y in $\mathfrak{R}$,

$$|T(x) - T(y)| > |x-y|.$$

In view of this definition, the map $T: \mathfrak{R} \to \mathfrak{R}$ defined by $T(x) = \lambda x$ is a contraction for real $0 < \lambda < 1$ and an expansion for real $\lambda > 1$.

In fact, for $0 < \lambda < 1$, we have

$$\begin{aligned} |T(x) - T(y)| &= |\lambda x - \lambda y| \\ &= \lambda|x-y| < |x-y|. \end{aligned}$$

Hence T is a contraction.

Similarly for $\lambda > 1$, we have

$$\begin{aligned} |T(x) - T(y)| &= |\lambda x - \lambda y| \\ &= \lambda|x-y| > |x-y| \end{aligned}$$

and so in that case T is an expansion

Further, we note that for this mapping,

$$\begin{aligned} T(x+y) &= \lambda(x+y) \\ &= \lambda x + \lambda y \\ &= T(x) + T(y) \end{aligned}$$

and for scalar μ,

$$\begin{aligned} T(\mu x) &= \lambda(\mu x) \\ &= \mu(\lambda x) \\ &= \mu T(x). \end{aligned}$$

Hence T is a linear transformation.

EXAMPLE 2.41

Let $P_n(\mathfrak{R})$ denote the vector space of all polynomials over $\mathfrak{R}$. Define $D: P_n(\mathfrak{R}) \to P_n(\mathfrak{R})$ by $D(f(x)) = f'(x)$, where $f'(x)$ denotes the derivative of $f(x)$. Then D is called the ***Differential operator***. We note that

$$\begin{aligned} D(f(x)+g(x)) &= (f(x)+g(x))' = f'(x)+g'(x) \\ &= D(f(x)+D(g(x)) \end{aligned}$$

and for real λ

$$D(\lambda f(x) = (\lambda f(x)' = \lambda f'(x) = \lambda D(f(x)).$$

Hence differential operator is a linear operator.

EXAMPLE 2.42

Let $M_{n\times n}(F)$ be the set of all $n \times n$ matrices over the field F. Then the map $T: .M_{n\times n}(F) \to F$ defined by $T(A) = tr(A)$ is a linear transformation. In fact, if A and B are two $n \times n$ matrices, then

$$\begin{aligned} T(A+B) &= tr(A+B) \\ &= tr(A) + tr(B), \text{ (by property of trace)} \\ &= T(A) + T(B) \end{aligned}$$

and

$$\begin{aligned} T(\lambda A) &= tr(\lambda A) \\ &= \lambda tr(A) \\ &= \lambda T(A). \end{aligned}$$

Hence T is a linear transformation.

Theorem 2.14. Let $T: V \to W$ be a linear transformation from a vector space V into another vector space W. If T is one-one and $v_1, v_2, \ldots, v_n$ are linearly independent vectors, then $Tv_1, Tv_2, \ldots Tv_n$ are linearly independent.

Proof. Suppose

$$\lambda_1 Tv_1 + \lambda_2 Tv_2 + \ldots + \lambda_n Tv_n = 0, \qquad (16)$$

that is,

$$T(\lambda_1 v_1 + \lambda_2 v_2 + \ldots + \lambda_n v_n) = 0$$
$$= T(0) \text{ (linearity of } T).$$

Since, T is one-one it implies that

$$\lambda_1 v_1 + \lambda_2 v_2 + \ldots + \lambda_n v_n = 0. \quad (17)$$

Since, $v_1, v_2, \ldots v_n$ are linearly independent, (17) yields $\lambda_1 = \lambda_2 = \ldots = \lambda_n = 0$. Thus, (16) implies $\lambda_1 \lambda_2 = \ldots = \lambda_n = 0$. Hence, $Tv_1, Tv_2 \ldots Tv_n$ are linearly independent.
We now state (without proof) the following result:

Theorem 2.15. Every n-dimensional vector space V over a field F is isomorphic to $F^{(n)}$.

Corollary 2.4. Two vector spaces of the same dimension over same field are isomorphic.

Proof. Suppose V and W are two n-dimensional vector spaces over F. Then, both are isomorphic to $F^{(n)}$ and so are isomorphic to each other.

Theorem 2.16. If U and W are two subspaces of a finite dimensional vector space V, then $\dim(U + W) = \dim U + \dim W - \dim(U \cap W)$.

Proof. Let $\dim V = n$, $\dim U = m$, $\dim W = p$, $\dim(U \cap W) = r$. Then

$$m \le n, \quad p \le n, \quad r \le n.$$

Let $\{v_1, v_2, \ldots, v_r\}$ be a basis for $U \cap W$. Therefore, it is linearly independent in $U \cap W$ and so in U and W. Thus, it can be extended to a basis for U, say

$$\{v_1, v_2, \ldots v_r, v_{r+1}, \ldots, u_m\}$$

and to basis of W, say

$$\{v_1, v_2, \ldots, v_r, w_r, w_{r+1}, w_{r+2}, \ldots, w_p\}.$$

We assert that the set

$$\{v_1, v_2, \ldots, v_r, u_{r+1}, u_{r+2}, \ldots, u_m, w_{r+1}, w_{r+2}, \ldots, w_p\}$$

is a basis for $U + W$. In this direction, let

$$\sum_{i=1}^{r} a_i v_i + \sum_{i=r+1}^{m} b_i u_i + \sum_{i=r+1}^{p} c_i w_i = 0 \quad (18)$$

or

$$\sum_{i=1}^{r} a_i v_i + \sum_{i=r+1}^{m} b_i u_i = - \sum_{i=r+1}^{p} c_i w_i = v, \text{ say} \quad (19)$$

The left hand side belong to U whereas the right hand side belongs to W. Hence, $v \in U \cap W = \{v_1, v_2, \ldots, v_r\}$. Therefore, $v = \sum_{i=1}^{r} \lambda_i v_i$ and so (19) transforms to

$$\sum_{i=1}^{r} \lambda_i v_i + \sum_{i=r+1}^{p} c_i w_i = 0.$$

But $\{v_1, v_2, \ldots, v_r, w_{r+1}, \ldots w_p\}$, being a basis of W, is linearly independent and so

$$\lambda_1 = \lambda_2 = \ldots = \lambda_r = c_{r+1} = \ldots = c_p = 0.$$

Thus, (18) transforms to

$$\sum_{i=1}^{r} a_i v_i + \sum_{i=r+1}^{m} b_i u_i = 0.$$

Again, $\{v_1, v_2, \ldots, v_r, u_{r+1}, u_{r+2}, \ldots, u_m\}$, being a basis of U, is linearly independent. Hence, $a_1 = a_2 = \ldots = a_r = b_{r+1} = \ldots = b_m = 0$. Hence, the set

$$\{v_1, v_2, \ldots, v_r, u_{r+1}, u_{r+2}, \ldots, u_m, w_{r+1}, \ldots, w_p\}$$

is linearly independent.

Further, let $z \in U + W$. Then $z = u + w$, where $u \in U$ and $w \in W$. Thus

$$z = \sum_{i=1}^{r} a_i v_i + \sum_{i=r+1}^{m} b_i u_i + \sum_{i=1}^{r} d_i v_i + \sum_{i=r+1}^{p} t_i w_i,$$

where a_i, b_i, d_i and t_i are suitable scalars from F. As such z is a linear combination of $v_1, v_2, \ldots, v_n, u_{n+1}, \ldots, u_m, w_{r+1}, \ldots, w_p$. Hence,

$$\{v_1, v_2 \ldots, v_r, u_{r+1}, u_{r+2}, \ldots, u_m, w_{r+1}, w_{r+2}, \ldots, w_p\}$$

is a basis for $U+W$. The number of elements in this basis is

$$r+(m-r)+(p-r)=m+p-r,$$

that is

$$\dim(U+W)=\dim U+\dim W-\dim(U\cap W).$$

This completes proof of the theorem.

Let V and W be the two vectors spaces over the same field. Then the set of all linear transformation $T\colon V\to W$ is denoted by Hom (V, W). Thus,

$$\text{Hom }(V,W)=\{T:T:V\to W \text{ is a linear transformation}\}.$$

If dim $V=m$, dim $W=n$, then, it can be shown that Hom (V, W) is a vector space of dimension mn over F. As a particular case, dimension of Hom (V, V) is m^2.

2.4 LINEAR ALGEBRA

Definition 2.26. A non-empty set A is said to be a *linear algebra* over the field F if

(i) A is a ring under the binary operation of addition and multiplication.

(ii) A is a vector space over F.

(iii) $\lambda(T_1 T_2)=(\lambda T_1)T_2=T_1(\lambda T_2)$, for T_1, $T_2\in A$, $\lambda\in F$.

If, in addition, A has multiplicative identity, then A is called *algebra with unity*.

If every non-zero element of an algebra A with unity has multiplicative inverse, then A is called a *division algebra*.

The set of linear transformations Hom (V, V) is an algebra. Another two important linear algebras are discussed in Examples 2.43 and 2.44.

EXAMPLE 2.43 (Matrix Algebra)

Let M_n be a set of matrices of order n over a field F. Then elements of M_n are

$$\begin{bmatrix} a_{11} & a_{12} & \dots & a_{1n}\\ a_{21} & a_{22} & \dots & a_{2n}\\ \dots & \dots & \dots & \dots\\ \dots & \dots & \dots & \dots\\ a_{n1} & a_{n2} & \dots & a_{nn}\end{bmatrix}, a_{ij}\in F.$$

If we define addition, multiplication, and scalar multiplication in M_n as

$$(a_{ij})+(b_{ij})=(a_{ij}+b_{ij}),$$

$$(a_{ij})\,(b_{ij})=\sum_{k=1}^{n}a_{ik}b_{kj},$$

$$c\,(a_{ij})=(ca_{ij}),\ c\in F.$$

Then, M_n is a linear algebra over F. Let E_{ij} be a matrix whose (i, j)th element is 1 and all other elements are zero. These n^2 matrices $E_{ij}(i, j=1,2,\dots n)$ are called unit matrices and form a basis for M_n since any matrix can be written as the linear combination of these matrices. Thus, the dimension of the matrix algebra formed by matrices of order n is n^2. This algebra is known as *Matrix Algebra*.

EXAMPLE 2.44 (Algebra of Quaternions)

Let $\mathbb{Q}$ be a four-dimensional vector space over the field $\mathbb{R}$ of real number whose basis is $\{1, i, j, k\}$. Here 1 acts as unit element. Multiplication of the remaining elements in the basis is taken as

$$i^2=-1,\ j^2=-1,\ k^2=-1,$$

$$ij=k,\ ji=-k,$$

$$jk=i,\ kj=-i,$$

$$ki=j,\ ik=-j.$$

If $a,b,c,d\in\mathbb{R}$, then every element $a+bi+cj+dk$ of $\mathbb{Q}$ is called *quaternion*.

It can be seen that the mapping

$$a+bi+cj+dk \to \begin{bmatrix} a+bi & c+di \\ -c+di & a-bi \end{bmatrix}$$

of $\mathbb{Q}$ into $\mathbb{C}_2$ is a ring isomorphism. Since, matrix multiplication is associative, multiplication in $\mathbb{Q}$ is also associative. We can show that $\mathbb{Q}$ is an algebra over F.

Since $ij \neq ji$, the elements in $\mathbb{Q}$ are not *commutative*.

If $q = a+bi+cj+dk \in \mathbb{Q}$, then the element $q' = a-bi-cj-dk$ of $\mathbb{Q}$ is called the *conjugate* of q. Further,

$$qq' = q'q = a^2+b^2+c^2+d^2.$$

If $q \neq 0$, then a, b, c, d cannot be zero and so qq' is non-zero. Therefore,

$$\begin{aligned} q^{-1} &= (qq')^{-1}q' \\ &= \frac{q'}{qq'} = \frac{a-bi-cj-dk}{a^2+b^2+c^2+d^2} \in \mathbb{Q}. \end{aligned}$$

Thus, every non-zero element of $\mathbb{Q}$ has multiplicative inverse. Hence, $\mathbb{Q}$ is *non-commutative division algebra.*

2.5 RANK AND NULLITY OF A LINEAR TRANSFORMATION

Definition 2.27. Let V and W be two linear spaces and $T: V \to W$ be a linear transformation. Then, the *range* of T is the set of all T-images. Thus

$$R(T) = \{T(v) : v \in V\}.$$

Definition 2.28. Let $T: V \to W$ be a linear transformation from a vector space V into another vector space W. Then the *kernel (null space)* of T is the set

$$N(T) = \{v \in V : T(v) = e_w\},$$

where e_w is the identity element of W.

The null space is also denoted by ker T.

Theorem 2.17. Let $T: V \to W$ be a linear transformation from a vector space V into the vector space W. Then

(i) $R(T)$ is a subspace of W

(ii) $N(T)$ is a subspace of V

(iii) T is one-one if and only if $N(T)$ consists of identity element of W

(iv) If $\{v_1, v_2, \dots v_n\}$ generates V, then $\{Tv_1, Tv_2, \dots Tv_n\}$ generates $R(T)$.

Proof. (i) Let $w_1, w_2, \in R(T)$. Then there are vectors v_1, v_2 in V such that $Tv_1 = w_1$ and $Tv_2 = w_2$. Since T is linear, we have

$$w_1 + w_2 = Tv_1 + Tv_2 = T(v_1+v_2).$$

But, V being a vector space, $v_1 + v_2 \in V$. Therefore, $T(v_1+v_2)$ belongs to $R(T)$ and so $w_1 + w_2 \in R(T)$.
Further if α in a scalar, then

$$\alpha w_1 = \alpha T(v_1) = T(\alpha v_1) \in R(T).$$

Thus, w_1, $w_2 \in R(T)$ and $\lambda \in F$ imply $w_1 + w_2 \in R(T)$ and $\lambda w_1 \in R(T)$. Hence, $R(T)$ is a subspace of W

(ii) Let v_1, $v_2 \in N(T)$, Then, by definition of null space, $Tv_1 = 0_w$ and $Tv_2 = 0_w$. Therefore,

$$T(v_1+v_2) = Tv_1 + Tv_2 = 0_w + 0_w = 0_w,$$

and so $v_1 + v_2 \in N(T)$. Further if α is a scalar, then

$$T(\lambda v_1) = \lambda T v_1 = \lambda 0_w = 0_w$$

and so $\lambda v_1 \in N(T)$. Hence, $N(T)$ is a subspace of V.

(iii) Suppose T is one-one. Then, $T(v_1) = T(v_2) \Rightarrow v_1 = v_2$. Now, let $v \in N(T)$. Then, $T(v) = 0_w = T(0_v)$ and so $v = 0_v$. Thus, every vector in $N(T)$ is 0_v. Hence, $N(T) = \{0_v\}$.
Conversely, assume that $N(T)$ consists of only zero element of V. Now let

$Tv_1 = Tv_2$. Then $T(v_1 - v_2) = Tv_1 - Tv_2 = 0_w$ and so $v_1 - v_2 \in N(T) = \{0_v\}$, which yields $v_1 = v_2$. Thus, $Tv_1 = Tv_2 \Rightarrow v_1 = v_2$ and so T is one-one.

(iv) Since $\{v_1, v_2, \ldots, v_n\}$ generates V, every vector v can be expressed as a linear combination of $v_1, v_2, \ldots, v_n$. The vector $Tv_1, Tv_2, \ldots, Tv_n \in R(T)$ and $R(T)$ is a subspace of W. Therefore, a linear combination of $Tv_1, Tv_2, \ldots, Tv_n$ is in $R(T)$.
On the other hand, let $w \in R(T)$. Then there exists a vector $v \in V$ such that $Tv = w$. But, $v = \alpha_1 v_1 + \alpha_2 v_2 + \ldots + \alpha_n v_n$. Therefore,

$$\begin{aligned} w = Tv &= T(\alpha_1 v_1 + \alpha_2 v_2 + \ldots + \alpha_n v_n) \\ &= \alpha_1 Tv_1 + \alpha_2 Tv_2 + \ldots + \alpha_n Tv_n. \end{aligned}$$

Hence, $R(T)$ is generated by $Tv_1, Tv_2, \ldots, Tv_n$.

Definition 2.29. Let $T: V \to W$ be a linear transformation from a vector space V into the vector space W. If $R(T)$ is finite dimensional, then the dimension of $R(T)$ is called the *rank* of T and is denoted by $r(T)$.

Definition 2.30. Let $T: V \to W$ be a linear transformation from a vector space V into the vector space W. If $N(T)$ is finite dimensional, then the dimension of $N(T)$ is called the *nullity* of T and is denoted by $n(T)$.

Theorem 2.18. Let $T: V \to W$ be a linear transformation from a vector space V into the vector space W. If $w_1, w_2, \ldots w_n$ are linearly independent vector of $R(T)$ and $v_1, v_2, \ldots, v_n$ are vectors in V such that $T(v_1) = w_1, T(v_2) = w_2, \ldots, T(v_n) = w_n$, then $v_1, v_2, \ldots, v_n$ are linearly independent.

Proof. Suppose

$$\lambda_1 v_1 + \lambda_2 v_2 + \ldots + \lambda_n v_n = 0. \tag{20}$$

Since T is linear,

$$\begin{aligned} 0_w = T(0_v) &= T(\lambda_1 v_1 + \lambda_2 v_2 + \ldots + \lambda_n v_n) \\ &= \lambda_1 T(v_1) + \lambda_2 T(v_2) + \ldots + \lambda_n T(v_n) \\ &= \lambda_1 w_1 + \lambda_2 w_2 + \ldots + \lambda_n w_n. \end{aligned} \tag{21}$$

But since $w_1, w_2, \ldots, w_n$ are linearly independent, (21) implies $\lambda_1 = \lambda_2 = \ldots = \lambda_n = 0$. Thus, we have proved that (20) implies that $\lambda_1 = \lambda_2 = \ldots = \lambda_n = 0$. Hence, $v_1, v_2, \ldots, v_n$ are linearly independent.

Theorem 2.19. Let $T: V \to W$ be a linear transformation from a finite dimensional vector space V into another vector space W. Then

$$\dim R(T) + \dim N(T) = \dim V$$

or equivalently,

$$r(T) + n(T) = \dim V.$$

(Thus, *rank + nullity = dimension of the domain vector space*).

Proof. Since the null space $N(T)$ is a subspace of finite dimensional space V, it is finite dimensional. So, let $n(T) = n$ and $\dim V = p (p \geq n)$. Suppose $\{v_1, v_2, \ldots, v_n\}$ is a basis for $N(T)$. Since $v_1, v_2, \ldots, v_n \in N(T)$, we have

$$Tv_1 = Tv_2 = \ldots = Tv_n = 0 \tag{22}$$

Now $\{v_1, v_2, \ldots, v_n\}$ is linearly independent in $N(T)$ and so in V. We extend this linearly independent set of vectors in V to a basis of V. So, let $\{v_1, v_2, \ldots, v_n, v_{n+1}, \ldots, v_p\}$ be a basis for V. We claim that the set

$$\{Tv_{n+1}, \quad Tv_{n+2}, \ldots, Tv_p\}$$

is a basis for $R(T)$. So $\{v_1, v_2, \ldots, v_n, v_{n+1}, \ldots v_p\}$ is a basis for V. By Theorem 2.17, $\{Tv_1, Tv_2, \ldots, Tv_{n+1}, \ldots, Tv_p\}$ generates $R(T)$. But, by (22), $Tv_i = 0$ for $i = 1, 2, 3, \ldots, n$. Hence, $\{Tv_{n+1}, Tv_{n+2}, \ldots, Tv_p\}$ generates $R(T)$. Further, let

$$\lambda_{n+1} Tv_{n+1} + \lambda_{n+2} Tv_{n+2} + \ldots + \lambda_p Tv_p = 0, \tag{23}$$

that is,

$$T(\lambda_{n+1} v_{n+1} + \lambda_{n+2} v_{n+2} + \ldots + \lambda_p Tv_p) = 0.$$

Then,

$$\lambda_{n+1}v_{n+1}+\lambda_{n+2}v_{n+2}+\ldots+\lambda_n v_p \in N(T).$$

But, $N(T)$ is subspace of V. Therefore,

$$\lambda_{n+1}v_{n+1}+\lambda_{n+2}v_{n+2}+\ldots+\lambda_n v_p$$
$$=\alpha_1 v_1+\alpha_2 v_2+\ldots+\alpha_n v_n$$

or

$\alpha_1 v_1 + \alpha_2 v_2 + \ldots + \alpha_n v_n - \lambda_{n+1}v_{n+1} - \ldots - \lambda_p v_p = 0$. Since $\{v_1, v_2, \ldots, v_n, v_{n+1}, \ldots, v_p\}$ is a basis of V, it implies that $\alpha_1 = \alpha_2 = \ldots = \alpha_n = \lambda_{n+1} = \ldots = \lambda_p = 0$. Hence, (23) implies that $\{Tv_{n+1}, Tv_{n+2}, \ldots, Tv_p\}$ is linearly independent. Hence, it is basis for $R(T)$ and so

$$\dim R(T) = r(T) = p-n = \dim V - \dim N(T)$$

Hence

$$r(T)+n(T)=\dim V.$$

Theorem 2.20. Let V and W be vector spaces of finite and equal dimension and let $T : V \to W$ be a linear transformation. Then the following statements are equivalent:

(a) T is one-one

(b) T is onto

(c) $r(T) = \dim(V)$

Proof. By Theorem 2.19, we have

$$r(T)+n(T)=\dim V. \tag{24}$$

Now using Theorem 2.17 (iii), we have

$$\begin{aligned} T \text{ is one-one} &\Leftrightarrow N(T)=\{0\} \\ &\Leftrightarrow n(T)=0 \\ &\Leftrightarrow r(T)=\dim V, \quad \text{using (24)} \\ &\Leftrightarrow r(T)=\dim W, \quad \text{by hypothesis} \\ &\Leftrightarrow \dim(R(T))=\dim W, \\ &\qquad \text{(by definition of rank)} \\ &\Leftrightarrow R(T)=W \\ &\Leftrightarrow T \quad \text{is onto.} \end{aligned}$$

Thus,

T is one-one $\Rightarrow r(T)=\dim V \Rightarrow T$ is onto.

EXAMPLE 2.45

Let $T : V_2 \to V_2$ be defined by

$$T(a_1, a_2) = (a_1, a_1+a_2).$$

Then T is a linear transformation and

$$T(a_1, a_2) = (0, 0) \Leftrightarrow a_1 = a_2 = 0.$$

Thus

$$N(T)=\{0\}.$$

Hence T is one-one by Theorem 2.17 Since V_2 is finite dimensional, Theorem 2.20 implies that T is also onto.

EXAMPLE 2.46

Let $T : V_3 \to V_2$ be a linear transformation defined by

$$T(a_1, a_2, a_3) = (a_1-a_2, a_1+a_3).$$

Then the null space of T is given by

$$\begin{aligned} N(T) &= \{(a_1, a_2, a_3) : T(a_1, a_2, a_3) = (0, 0)\} \\ &= \{(a_1,a_2,a_3) : (a_1-a_2, a_1+a_3) = (0,0)\} \\ &= \{(a_1,a_2,a_3) : a_1-a_2=0 \text{ and } a_1+a_3=0\} \\ &= \{(a_1, a_2, a_3) : a_1 = a_2 = -a_3\} \\ &= \{(a_1, a_1, -a_1)\} \\ &= \{a_1(1, 1, -1)\}. \end{aligned}$$

Thus $N(T)$ is the subspace generated by $(1, 1, -1)$.

To see whether T is onto, let (a, b) be arbitrary member of V_2. Then

$(a_1-a_2, a_1+a_3) = (a, b) \Rightarrow a_1-a_2 = a$ and $b = a_1+a_3$. Solving these equations, we get

$$a_2 = a_1 - a \text{ and } a_3 = b - a_1.$$

Therefore

$$T(a_1,\ a_1-a,\ b-a_1)=(a,\ b).$$

Thus to each vector $(a,\ b)$ in V_2, there exists a vector in V_3 such that $(a,\ b)$ is image of that vector under T. Hence $R(T)=V_2$ and so T is onto.

EXAMPLE 2.47

Consider the linear transformation $T\colon V_2\to V_3$ defined by $T(a_1,\ a_2)=(a_1+a_2,\ 0,\ 2a_1-a_2)$. Its null space is given by

$$\begin{aligned}N(T)&=\{(a_1,\ a_2):T(a_1,\ a_2)=(0,\ 0,\ 0)\}\\&=\{(a_1,\ a_2):(a_1+a_2,\ 0,\ 2a_1-a_2)\\&\qquad=(0,\ 0,\ 0)\}\\&=\{(a_1,\ a_2):a_1+a_2=0,\ 2a_1-a_2=0\}\\&=\{(0,\ 0)\}.\end{aligned}$$

Since empty set ϕ is a basis for the zero vector space, the dimension of the vector space $\{0\}$ is zero. Therefore

$$\text{nullity of } T=n(T)=\dim N(T)=0.$$

Since V_2 is generated by $e_1=(1,\ 0)$ and $e_2=(0,\ 1)$. Therefore $R(T)$ is generated by $T(e_1)$ and $T(e_2)$, that is, by $(1,\ 0,\ 2)$ and $(1,\ 0,\ -1)$. Also

$$\begin{aligned}&\lambda_1(1,\ 0,2)+\lambda_2(1,\ 0,\ -1)=(0,0)\\&\Rightarrow(\lambda_1,\ 0,\ 2\lambda_1)+(\lambda_2,\ 0,\ -\lambda_2)=(0,0)\\&\Rightarrow(\lambda_1+\lambda_2,\ 0,\ 2\lambda_1-\lambda_2)=(0,0)\\&\Rightarrow\lambda_1+\lambda_2=0,\quad 2\lambda_1-\lambda_2=0\\&\Rightarrow\lambda_1=\lambda_2=0.\end{aligned}$$

Consequently, $(1,\ 0,\ 2)$ and $(1,\ 0,\ -1)$ are also linearly independent. Hence $\dim(R(T))=2$. We note that

$$r(T)+n(T)=2+0=2=\dim V_2.$$

Hence Theorem 2.19 is verified.

EXAMPLE 2.48

Consider $T\colon V_3\to V_2$ defined by

$$T(a_1,\ a_2,\ a_3)=(a_1-a_2,\ 2a_3).$$

It is a linear transformation. Further,

$$\begin{aligned}N(T)&=\{(a_1,\ a_2,\ a_3):T(a_1,\ a_2,\ a_3)=(0,\ 0)\}\\&=\{(a_1,\ a_2,\ a_3):(a_1-a_2,\ 2a_3)=(0,\ 0)\}\\&=\{(a_1,\ a_2,\ a_3):a_1-a_2=0,\ 2a_3=0\}\\&=\{(a_1,\ a_2,\ a_3):a_1=a_2,\ a_3=0\}\\&=\{(a_1,\ a_1,\ 0)\}\\&=\{a_1(1,\ 1,\ 0)\}.\end{aligned}$$

Thus $N(T)$ is generated by $(1,\ 1,\ 0)$ and so

$$n(T)=\dim(N(T))=1.$$

Also, since $N(T)\neq\{0\}$, the map T is not one-one.

Further, $e_1=(1,\ 0,\ 0)$, $e_2=(0,\ 1,\ 0)$ and $e_3=(0,\ 0,\ 1)$ form a basis for V_3. Then $T(e_1)=(1,\ 0)$, $T(e_2)=(-1,\ 0)$ and $T(e_3)=(0,\ 2)$. Therefore

$$\begin{aligned}R(T)&=span\{T(e_1),\ T(e_2),\ T(e_3)\}\\&=span\{(1,0),\ (-1,0),\ (0,2)\}.\end{aligned}$$

Since $\dim V_2=2$, the set of these three vectors are linearly dependent. But $(-1,0)=-1(1,0)+0(0,2)$. Therefore, we may discard $(-1,0)$. Further,

$$\begin{aligned}&\lambda_1(1,0)+\lambda_2(0,2)=(0,0)\\&\qquad\Rightarrow(\lambda_1,0)+(0,2\lambda_2)=(0,0)\\&\qquad\Rightarrow(\lambda_1,2\lambda_2)=(0,0)\\&\qquad\Rightarrow\lambda_1=\lambda_2=0.\end{aligned}$$

Hence $(1,0)$ and $(0,2)$ are linearly independent.Thus,

$$R(T)=\{(1,0),\ (0,2)\}.$$

Hence $\dim(R(T)) = 2$. We note that

$$\begin{aligned} r(T) + n(T) &= \dim(R(T)) + \dim(N(T)) \\ &= 2 + 1 = 3 = \dim V_3. \end{aligned}$$

Hence Rank-Nullity Theorem is verified in this case.

EXAMPLE 2.49

Show that P_3 and M_{22} are isomorphic.

Solution. The given vector spaces are $P_3(x)$ and $M_{2\times2}$. Define

$$T : P_3 \to M_{2\times2} \text{ by}$$

$$T(f) = \begin{pmatrix} f(1) & f(2) \\ f(3) & f(4) \end{pmatrix}, \; f \in P_3.$$

Let $f_1 = a_1 + b_1x + c_1x^2 + d_1x^3$ and

$$f_2 = a_2 + b_2x + c_2x^2 + d_2x^3$$

be in P_3. Then,

$$\begin{aligned} f_1 + f_2 = &(a_1 + a_2) + (b_1 + b_2)x + (c_1 + c_2)x^2 \\ &+ (d_1 + d_2)x^3 \end{aligned}$$

and so

$$\begin{aligned} &T(f_1 + f_2) \\ &= \begin{pmatrix} a_1 + b_1 + c_1 + d_1 & a_1 + 2b_1 + 4c_1 + 8d_1 \\ a_1 + 3b_1 + 9c_1 + 27d_1 & a_1 + 4b_1 + 16c_1 + 64d_1 \end{pmatrix} \\ &+ \begin{pmatrix} a_2 + b_2 + c_2 + d_2 & a_2 + 2b_2 + 4c_2 + 8d_2 \\ a_2 + 3b_2 + 9c_2 + 27d_2 & a_2 + 4b_2 + 16c_2 + 64d_2 \end{pmatrix} \\ &= T(f_1) + T(f_2). \end{aligned}$$

Hence, T is a linear transformation.

Also $T(f) = 0$ only when f is a zero polynomial, that is if kernel (null space) consists of zero element only. Hence T is one-to-one. Also $\dim P_3 = \dim M_{2\times2} = 4$ (finite). Hence T is also onto. Hence T is an isomorphism and P_3 is isomorphic to $M_{2\times2}(F)$.

EXAMPLE 2.50

Is $T;\mathfrak{R}^3 \to \mathfrak{R}^3$ defined by $T(x,y,z) = (x + 3y, y, z + 2x)$ linear? Is it one-one, onto or both? Justify.

Solution. We have

$$T(x,\, y,\, z) = (x + 3y,\, y,\, z + 2x).$$

We note that T is linear.

The null space of T is given by

$$\begin{aligned} N(T) &= \{(x,\, y,\, z) : T(x,\, y,\, z) = (0,0,0)\} \\ &= \{(x,y,z) : (x + 3y,\, y,\, z + 2x) = (0,0,0)\} \\ &= \{(x,y,z) : x + 3y = 0,\, y = 0,\, z + 2x = 0\} \\ &= \{(x,y,z) : x = y = z = 0\} \\ &= \{(0,0,0)\} \end{aligned}$$

Thus, the null space consists only of $(0,0,0)$. Hence T is one-one. Further, let (a,b,c) be an arbitrary element of $\mathfrak{R}^3$. Then,

$$(x + 3y, y, z + 2x) = (a,b,c) \quad \text{implies that}$$

$$x + 3y = a,\; y = b,\; z + 2x = c.$$

Solving these, we get

$$x = a - 3b,\; y = b,\; z = c - 2a + 6b.$$

Therefore,

$$T(a - 3b, b, c - 2a + 6b) = (a,b,c).$$

Thus, to each vector (a,b,c) in $\mathfrak{R}^3$ there exists a vector in $\mathfrak{R}^3$ such that (a,b,c) is image of that vector under T. Hence T is onto. Thus T is both one-one and onto.

EXAMPLE 2.51

Find a basis for the row space of A and column space of A if

$$A = \begin{bmatrix} 1 & 4 & 5 & 2 \\ 2 & 1 & 3 & 0 \\ -1 & 3 & 2 & 0 \end{bmatrix}.$$

Also verify the dimension theorem for matrices.

Solution. The row vectors of matrix A are

$$(1, 4, 5, 2), (2, 1, 3, 0), (-1, 3, 2, 0),$$

which are linearly independent. Hence

$$\text{row rank of } A = 3.$$

The column vectors constituting the range of the matrix are

$$(1, 2, -1), (4, 1, 3), (5, 3, 2), (2, 0, 0).$$

These vectors are linearly dependent and we notice that

$$(5, 3, 2) = 1(1, 2, -1) + 1(4, 1, 3) + 0(2, 0, 0)$$

Therefore we discard (5, 3, 2) from the column vectors. The remaining vectors are linearly independent. Hence,

$$\text{column rank of } A = 3.$$

Thus

$$\text{rank of } A = 3.$$

Further, the given matrix A is a linear transformation from V_4 to V_3. For the kernel of A, we have

$$\begin{bmatrix} 1 & 4 & 5 & 2 \\ 2 & 1 & 3 & 0 \\ -1 & 3 & 2 & 0 \end{bmatrix} \begin{bmatrix} x_1 \\ x_2 \\ x_3 \\ x_4 \end{bmatrix} = \begin{bmatrix} 0 \\ 0 \\ 0 \end{bmatrix},$$

which implies

$$x_1 + 4x_2 + 5x_3 + 2x_4 = 0$$

$$2x_1 + x_2 + 3x_3 = 0$$

$$-x_1 + 3x_2 + 2x_3 = 0.$$

The last two equations yield $x_1 = x_2 = -x_3$. The first equation then yields $x_4 = 0$. Thus the null space is the set of all vectors of the form $(x_1, x_1, -x_1, 0)$. But

$$(x_1, x_1, -x_1, 0) = x_1(1, 1, -1, 0).$$

Thus, the kernel is generated by (1, 1, −1, 0). Hence,

$$\text{nullity of } A = 1.$$

We note that

$$\begin{aligned} \text{rank of } A + \text{nullity of } A &= 3 + 1 = 4 \\ &= \dim(V_4). \end{aligned}$$

Hence, the dimension theorem for matrices is verified.

EXAMPLE 2.52

Let $T_1: M_{22} \to \Re$ and $T_2: M_{22} \to M_{22}$ be linear transformations defined by

$$T_1(A) = tr(A) \quad \text{and} \quad T_2(A) = A^T.$$

Find $(T_1 \circ T_2)(A)$, where $A = \begin{bmatrix} a & b \\ c & d \end{bmatrix}$.

Solution. We have

$$T_1: M_{2\times 2} \to \Re$$

defined by

$$T_1(A) = tr(A)$$

and

$$T_2: M_{2\times 2} \to M_{2\times 2}$$

defined by

$$T_2(A) = A^T.$$

Then, for the given matrix A, we have

$$\begin{aligned} (T_1 o T_2)(A) &= T_1(T_2(A)) \\ &= T_1\left(A^T\right) \\ &= T_1\left(\begin{bmatrix} a & c \\ b & d \end{bmatrix}\right) \\ &= tr\left(\begin{bmatrix} a & c \\ b & d \end{bmatrix}\right) = a + d. \end{aligned}$$

EXAMPLE 2.53

Let $T : \mathfrak{R}^4 \to \mathfrak{R}^3$ be the linear transformation defined by

$$T(x_1, x_2, x_3, x_4) = (w_1, w_2, w_3),$$

where $w_1 = 4x_1 + x_2 - 2x_3 - 3x_4$, $w_2 = 2x_1 + x_2 + x_3 - 4x_4$, $w_3 = 6x_1 - 9x_3 + 9x_4$.
Find bases for the range and kernel of T.

Solution. The given linear transformation is

$$T(x_1,x_2,x_3,x_4) = (w_1,w_2,w_3),$$

where

$$w_1 = 4x_1 + x_2 - 2x_3 - 3x_4,$$
$$w_2 = 2x_1 + x_2 + x_3 - 4x_4,$$
$$w_3 = 6x_1 - 9x_3 + 9x_4.$$

Since $v_1 = (1, 0, 0, 0)$, $v_2 = (0, 1, 0, 0)$, $v_3 = (0, 0, 1, 0)$ and $v_4 = (0, 0, 0, 1)$ form a standard basis for $\mathfrak{R}^4$, we have

$$\begin{aligned} R(T) &= \text{span}\{(Tv_1, Tv_2, Tv_3, Tv_4)\} \\ &= \text{span}\{(4, 2, 6), (1, 1, 0), (-2, 1, -9), \\ &\qquad (-3, -4, 9)\}. \end{aligned}$$

Since

$$(2, 1, -9) = -\frac{3}{2}(4, 2, 6) + 4(1,1,0),$$

we have

$$R(T) = \text{span}\{(4, 2, 6), (1, 1, 0), (-3, 4, 9)\}.$$

Also, the set $\{(4, 2, 6), (1, 1, 0), (-3, 4, 9)\}$ is linearly independent. Hence, it forms a basis for $R(T)$.
Further,

$$T(x_1,x_2,x_3,x_4) = (w_1,w_2,w_3) = (0,0,0)$$

implies

$$4x_1 + x_2 - 2x_3 - x_4 = 0$$
$$2x_1 + x_2 + x_3 - 4x_4 = 0$$
$$6x_1 + 0 - 9x_3 + 9x_4 = 0.$$

Solving these equations, we get

$$x_2 = -\frac{8}{3}x_1,\ x_3 = \frac{2}{3}x_1,\ x_4 = 0.$$

Thus, the null space (kernel) consists vectors of the form $\left(x_1, -\frac{8}{3}x_1, \frac{2}{3}x_1, 0\right)$. Thus, kernel is generated by $(3, -8, 2, 0)$.

EXAMPLE 2.54

Let $T : \mathfrak{R}^2 \to \mathfrak{R}^2$ be defined by $T(x, y) = (x + y, x - y)$. Is T one-one? If so, find formula for $T^{-1}(x, y)$.

Solution. We have $T : \mathfrak{R}^2 \to \mathfrak{R}^2$ defined by

$$T(x,y) = (x+y, x-y).$$

We note that $T(x, y) = (0, 0)$ implies $x + y = 0$, $x - y = 0$, that is, $x = 0$, $y = 0$. Thus, the null space consists of only $(0,0)$ vector. Hence, T is one-one. Further, since $\mathfrak{R}^2$ is finite dimensional and T is one-one, it follows that T is also onto. Hence, T^{-1} exists. Let $(a, b) \in \mathfrak{R}^2$. Then,

$$\begin{aligned} T^{-1}(a,b) &= \left\{(x,y) \in \mathfrak{R}^2 : T(x,y) = (a,b)\right\} \\ &= \{(x,y) : (x+y, x-y) = (a,b)\} \\ &= \{(x,y) : x+y = a \quad \text{and} \quad x-y = b\} \\ &= \left\{(x,y) : x = \frac{a+b}{2},\ y = \frac{a-b}{2}\right\} \\ &= \left(\frac{a+b}{2}, \frac{a-b}{2}\right). \end{aligned}$$

Thus $T^{-1} : \mathfrak{R}^2 \to \mathfrak{R}^2$ is defined by

$$T^{-1}(a,b) = \left(\frac{a+b}{2}, \frac{a-b}{2}\right).$$

2.6 MATRIX OF A LINEAR TRANSFORMATION

Let V be an n-dimensional vector space over a field F. Let $T: V \to V$ be a linear transformation in V and let $b = \{v_1, v_2, \ldots, v_n\}$ be some fixed basis of V. Since T maps V into V, the elements $Tv_1, Tv_2, \ldots, Tv_n$ are all in V. Further, since $\{v_1, v_2, \ldots, v_n\}$ is a basis for V, we have

$$\begin{aligned} Tv_1 &= \alpha_{11}v_1 + \alpha_{12}v_2 + \ldots + \alpha_{1n}v_n \\ Tv_2 &= \alpha_{21}v_1 + \alpha_{22}v_2 + \ldots + \alpha_{2n}v_n \\ &\ldots \quad \ldots \quad \ldots \quad \ldots \\ &\ldots \quad \ldots \quad \ldots \quad \ldots \\ Tv_n &= \alpha_{n1}v_1 + \alpha_{n2}v_2 + \ldots + \alpha_{mn}v_n, \end{aligned}$$

where each scalar $\alpha_{ij} \in F$. Thus, we have

$$Tv_i = \sum_{j=1}^{n} \alpha_{ij}v_j \quad \text{for } i = 1, 2, \ldots, n.$$

The ordered set of n^2 number α_{ij} in F completely describes T and is used to represent the linear transformation T.

Definition 2.31. Let V be an n-dimensional vector space over F and let $\{v_1, v_2, \ldots, v_n\}$ be a basis of V over F. If $T: V \to V$ is a linear transformation, then the matrix

$$M(T) = \begin{bmatrix} \alpha_{11} & \alpha_{21} & \alpha_{31} & \ldots & \ldots & \alpha_{n1} \\ \alpha_{12} & \alpha_{22} & \alpha_{32} & \ldots & \ldots & \alpha_{n2} \\ \ldots & \ldots & \ldots & \ldots & \ldots & \ldots \\ \ldots & \ldots & \ldots & \ldots & \ldots & \ldots \\ \alpha_{1n} & \alpha_{2n} & \alpha_{3n} & \ldots & \ldots & \alpha_{nn} \end{bmatrix}$$

is called the *matrix of the linear transformation* T in basis $\{v_1, v_2, \ldots v_n\}$, where

$$Tv_i = \sum_{j=1}^{n} \alpha_{ij}v_j.$$

The matrix $m(T)$ is also denoted by $[T]$ or $[T]_\beta$.

EXAMPLE 2.55

Let $T: V_2(F) \to V_2(F)$ be a linear transformation defined by $T(a,b) = (a,0)$, called *projection* of V_2 on x-axis. The standard ordered basis of $V_2(F)$ is $\{(1, 0), (0, 1)\}$. We have

$$T(1, 0) = (1, 0) = 1(1, 0) + 0(0, 1)$$
$$T(0, 1) = (0, 0) = 0(1, 0) + 0(0, 1).$$

Therefore, the matrix of the linear transformation T is $M(T) = \begin{pmatrix} 1 & 0 \\ 0 & 0 \end{pmatrix}$.

EXAMPLE 2.56

Let $T: V_3(F) \to V_3(F)$ be defined by

$$T(x_1, x_2, x_3) = \left(x_1 - x_2 + x_3,\ 2x_1 + 3x_2 - \frac{1}{2}x_3,\ x_1 + x_2 - 2x_3\right).$$

The standard basis of $V_3(F)$ is $\{(1, 0, 0), (0, 1, 0), (0, 0, 1)\}$. Then

$$\begin{aligned} T(1, 0, 0) &= (1, 2, 1) = 1(1, 0, 0) \\ &\quad + 2(0, 1, 0) + 1(0, 0, 1) \\ T(0, 1, 0) &= (-1, 3, 1) = -1(1, 0, 0) \\ &\quad + 3(0, 1, 0) + 1(0, 0, 1) \\ T(0, 0, 1) &= \left(1, -\frac{1}{2}, -2\right) \\ &= 1(1, 0, 0) - \frac{1}{2}(0, 1, 0) \\ &\quad - 2(0, 0, 1) \end{aligned}$$

Hence, the matrix of the linear transformation T is

$$M(T) = \begin{bmatrix} 1 & -1 & 1 \\ 2 & 3 & -\frac{1}{2} \\ 1 & 1 & -2 \end{bmatrix}.$$

EXAMPLE 2.57

Let $T: V_2 \to V_3$ be the linear transformation defined by

$$T(x_1, x_2) = (x_1 + 3x_2,\ 0,\ 2x_1 - 4x_2).$$

The standard basis of V_2 is $\{(1, 0), (0, 1)\}$. Then

$$T(1, 0) = (1, 0, 2) = 1(1, 0, 0) + 0(0, 1, 0) + 2(0, 0, 1)$$

and

$$T(0,1) = (3, 0, -4) = 3(1, 0, 0) + 0(0, 1, 0) - 4(0, 0, 1).$$

Hence the matrix of linear transformation T is

$$M(T) = \begin{bmatrix} 1 & 3 \\ 0 & 0 \\ 2 & -4 \end{bmatrix}.$$

EXAMPLE 2.58

Let V be the set of all polynomials in x of degree $n-1$ or less, over a field F. Then V is a linear space. Let D be a linear transformation defined on V by

$$D(\beta_0 + \beta_1 x + \ldots + \beta_{n-1}x^{n-1}) = \beta_1 + 2\beta_2 x + \ldots + (n-1)\beta_{n-1}x^{n-2}.$$

We note that D is nothing but a *differential operator*.

Consider the base $\{v_1, v_2, \ldots, v_n\} = \{1, x, \ldots, x^{n-1}\}$ of V. Then

$$Dv_1 = D1 = 0 = 0v_1 + 0v_2 + \ldots + 0v_n$$

$$Dv_2 = Dx = 1 = 1v_1 + 0v_2 + \ldots + 0v_n$$

$$\ldots \quad \ldots \quad \ldots \quad \ldots$$

$$\ldots \quad \ldots \quad \ldots \quad \ldots$$

$$Dv_n = Dx^{n-1} = (n-1)x^{n-2} = 0v_1 + 0v_2 + \ldots + (n-1)v_{n-1} + 0v_n.$$

Therefore, the matrix D in the basis $\{1, x, \ldots, x^{n-1}\}$ is

$$M_1(D) = \begin{bmatrix} 0 & 1 & 0 & \ldots & \ldots & 0 \\ 0 & 0 & 2 & \ldots & \ldots & 0 \\ \ldots & \ldots & \ldots & \ldots & \ldots & \ldots \\ \ldots & \ldots & \ldots & \ldots & \ldots & n-1 \\ 0 & 0 & 0 & \ldots & \ldots & 0 \end{bmatrix}.$$

2.7 CHANGE-OF-BASIS MATRIX (TRANSFORMING MATRIX OR TRANSITION MATRIX)

Consider the real space $\mathfrak{R}^3$ with standard basis $\beta = \{e_1, e_2, e_3\}$.
Thus

$$e_1 = (1,0,0),\ e_2 = (0,1,0),\ e_3 = (0,0,1).$$

Let $\beta_1 = (e_1', e_2', e_3')$, where

$$e_1' = (5,-1,-2),\ e_2' = (2,3,0),\ e_3' = (-2,1,1).$$

We note that β_1 is also a basis for $\mathfrak{R}^3$. Further,

$$e_1' = 5e_1 - e_2 - 2e_3$$

$$e_2' = 2e_1 + 3e_2 + 0e_3$$

$$e_3' = -2e_1 + e_2 + e_3.$$

Thus

$$\begin{bmatrix} e_1' \\ e_2' \\ e_3' \end{bmatrix} = \begin{bmatrix} 5 & -1 & -2 \\ 2 & 3 & 0 \\ -2 & 1 & 1 \end{bmatrix} \begin{bmatrix} e_1 \\ e_2 \\ e_3 \end{bmatrix} = T\beta, \text{ say.}$$

Hence the matrix T changes the basis from β to β_1, and is so called *"change-of-basis matrix", "transforming matrix" or "transition matrix".*

2.8 RELATION BETWEEN MATRICES OF A LINEAR TRANSFORMATION IN DIFFERENT BASES

Let $S = \underset{\beta_1 \to \beta_2}{S}$ be the *change-of-basis matrix* that changes the basis β_1 into the basis β_2. Thus

$$\beta_2 = S\beta_1 \tag{25}$$

If M_1 and M_2 be the matrices of a linear transformation T in the basis β_1 and β_2 respectively, then

$$T\beta_1 = M_1\beta_1 \tag{26}$$

and

$$T\beta_2 = M_2\beta_2 \tag{27}$$

Using expression (25), the expression (27) reduces to

$$T(S\beta_1) = M_2(S\beta_1) \tag{28}$$

But

$$\begin{aligned} T(S\beta_1) &= S(T\beta_1) \\ &= S[M_1\beta_1], \quad \text{using (26)} \\ &= (SM_1)\beta_1 \end{aligned} \tag{29}$$

Hence expression (28) and (29) yield

$$(SM_1)\beta_1 = (M_2S)\beta_1.$$

If there is at least one i, $1 \le i \le n$, for which the ith row of the matrix SM_1 is different from the ith row of the matrix M_2S, then two district linear combinations of vectors $e_1, e_2, \ldots, e_n$ will be equal to each other and so contradicts the linear independence of the basis β_1. Hence

$$SM_1 = M_2S.$$

Since S is non-singular, we have

$$M_2 = SM_1S^{-1} \quad \text{and} \quad M_1 = S^{-1}M_2S \tag{30}$$

The second equation of (30) shows that M_1 and M_2 *are similar matrices.*

EXAMPLE 2.59

If in Example 2.58, we consider the basis $\{x^{n-1}, x^{n-2}, \ldots, x^2, x, 1\}$, then we get the matrix of D as

$$M_2(D) = \begin{bmatrix} 0 & 0 \ldots & \ldots & 0 & 0 \\ n-1 & 0 \ldots & \ldots & 0 & 0 \\ \ldots & n-2 \ldots & \ldots & 0 & 0 \\ \ldots & \ldots & \ldots & \ldots & \ldots \\ 0 & 0 & 0 \ldots & 1 & 0 \end{bmatrix}.$$

Thus, the *matrix of D depends completely on the basis. Although these matrices are different, they represent the same linear transformation.*

To verify the relation (30), let V be the vector space of all polynomials over F of degree 2 or less and let D be the differential operator defined by

$$D(\alpha_0 + \alpha_1 x + \alpha_2 x^2) = \alpha_1 + 2\alpha_2 x.$$

Consider the basis

$$B = \{v_1, v_2, \ldots, v_3\} = \{1, x, x^2\}.$$

Then

$$Dv_1 = D1 = 0 = 0v_1 + 0v_2 + 0v_3$$

$$Dv_2 = Dx = 1 = 1v_1 + 0v_2 + 0v_3$$

$$Dv_3 = Dx^2 = 2x = 0v_1 + 2v_2 + 0v_3.$$

Therefore, the matrix of D in this basis is

$$M_1(D) = \begin{bmatrix} 0 & 1 & 0 \\ 0 & 0 & 2 \\ 0 & 0 & 0 \end{bmatrix}.$$

Now, we consider the basis

$$\beta_1 = \{u_1, u_2, u_3\} = \{x^2, x, 1\}$$

Then

$$Du_1 = Dx^2 = 2x = 0u_1 + 2u_2 + 0u_3$$

$$Du_2 = Dx = 1 = 0u_1 + 0u_2 + 1u_3$$

$$Du_3 = D1 = 0 = 0u_1 + 0u_2 + 0u_3.$$

Therefore, the matrix of D in the basis β_1 is

$$M_2(D) = \begin{bmatrix} 0 & 0 & 0 \\ 2 & 0 & 0 \\ 0 & 1 & 0 \end{bmatrix}.$$

We note that

$$u_1 = x^2 = v_3 = 0v_1 + 0v_2 + 1v_3$$

$$u_2 = x = v_2 = 0v_1 + 1v_2 + 0v_3$$

$$u_3 = 1 = v_1 = 1v_1 + 0v_2 + 0v_3.$$

Therefore, the transforming matrix is

$$S = \underset{\beta\to\beta_1}{S} = \begin{bmatrix} 0 & 0 & 1 \\ 0 & 1 & 0 \\ 1 & 0 & 0 \end{bmatrix}.$$

Further, we note that

$$S^{-1} = \begin{bmatrix} 0 & 0 & 1 \\ 0 & 1 & 0 \\ 1 & 0 & 0 \end{bmatrix}.$$

Then

$$SM_1(D)S^{-1} = \begin{bmatrix} 0 & 0 & 1 \\ 0 & 1 & 0 \\ 1 & 0 & 0 \end{bmatrix}\begin{bmatrix} 0 & 1 & 0 \\ 0 & 0 & 2 \\ 0 & 0 & 0 \end{bmatrix} \times \begin{bmatrix} 0 & 0 & 1 \\ 0 & 1 & 0 \\ 1 & 0 & 0 \end{bmatrix}$$

$$= \begin{bmatrix} 0 & 0 & 1 \\ 0 & 1 & 0 \\ 1 & 0 & 0 \end{bmatrix}\begin{bmatrix} 0 & 1 & 0 \\ 2 & 0 & 0 \\ 0 & 0 & 0 \end{bmatrix}$$

$$= \begin{bmatrix} 0 & 0 & 0 \\ 2 & 0 & 0 \\ 0 & 1 & 0 \end{bmatrix} = M_2(D).$$

EXAMPLE 2.60

If

$$M_1(T) = \begin{bmatrix} 1 & 1 & 2 \\ -1 & 2 & 1 \\ 0 & 1 & 3 \end{bmatrix}$$

is the matrix of a linear transformation T: $V_3 \to V_3$ in the basis $v_1 = (1, 0, 0)$, $v_2 = (0, 1, 0)$, $v_3 = (0, 0, 1)$. Find the matrix of T in the basis $u_1 = (1, 1, 1)$, $u_2 = (0, 1, 1)$, $u_3 = (0, 0, 1)$

Solution. We note that

$$u_1 = (1, 1, 1) = v_1 + v_2 + v_3$$
$$= 1v_1 + 1v_2 + 1v_3$$
$$u_2 = (0, 1, 1) = v_2 + v_3$$
$$= 0v_1 + 1v_2 + 1v_3$$
$$u_3 = (0, 0, 1) = v_3$$
$$= 0v_1 + 0v_2 + 1v_3.$$

Therefore, the transforming matrix is

$$S = \underset{\beta\to\beta_1}{S} = \begin{bmatrix} 1 & 1 & 1 \\ 0 & 1 & 1 \\ 0 & 0 & 1 \end{bmatrix}.$$

Then

$$S^{-1} = \begin{bmatrix} 1 & -1 & 0 \\ 0 & 1 & -1 \\ 0 & 0 & 1 \end{bmatrix}$$

and so

$$SM_1(T)S^{-1} = \begin{bmatrix} 1 & 1 & 1 \\ 0 & 1 & 1 \\ 0 & 0 & 1 \end{bmatrix}\begin{bmatrix} 1 & 1 & 2 \\ -1 & 2 & 1 \\ 0 & 1 & 3 \end{bmatrix} \times \begin{bmatrix} 1 & -1 & 0 \\ 0 & 1 & -1 \\ 0 & 0 & 1 \end{bmatrix}$$

$$= \begin{bmatrix} 1 & 1 & 1 \\ 0 & 1 & 1 \\ 0 & 0 & 1 \end{bmatrix}\begin{bmatrix} 1 & 0 & 1 \\ -1 & 3 & -1 \\ 0 & 1 & 2 \end{bmatrix}$$

$$= \begin{bmatrix} 0 & 4 & 2 \\ -1 & 4 & 1 \\ 0 & 1 & 2 \end{bmatrix} = M_2(T)$$

is the required matrix of T in the basis $\{u_1, u_2, u_3\}$.

EXAMPLE 2.61

Find the transition matrix from basis $\beta = \{(1, 0), (0, 1)\}$ of $\mathfrak{R}^2$ to the basis $\beta_1 = \{(1, 1), (2, 1)\}$ of $\mathfrak{R}^2$.

Solution. The bases are

$$\beta = \{(1, 0), (0, 1)\} \quad \text{and}$$
$$\beta_1 = \{(1, 1), (2, 1)\}.$$

We have to find the transition matrix $\underset{\beta\to\beta_1}{S}$ form the basis β to the basis β_1. Since

$$(1,\ 1) = 1(1,\ 0) + 1(0,\ 1)$$
$$(2,\ 1) = 2(1,\ 0) + 1(0,\ 1).$$

Therefore the required transition matrix is

$$\underset{\beta \to \beta_1}{S} = \begin{bmatrix} 1 & 1 \\ 2 & 1 \end{bmatrix}.$$

EXAMPLE 2.62

For the basis $S = \{v_1,\ v_2,\ v_3\}$ of $\mathfrak{R}^3$, where $v_1 = (1,\ 1,\ 1)$, $v_2 = (1,\ 1,\ 0)$, $v_3 = (1,\ 0,\ 0)$, let $T : \mathfrak{R}^3 \to \mathfrak{R}^3$ be a linear transformation such that

$$T(v_1) = (2,\ -1,\ 4),\ T(v_2) = (3,\ 0,\ 1),$$
$$T(v_3) = (-1,\ 5,\ 1).$$

Find a formula for $T(x_1, x_2, x_3)$ and use it to find $T(2,\ 4,\ -1)$.

Solution. The given basis is

$$\beta_1 = \{v_1,\ v_2,\ v_3\}$$
$$= \{(1,\ 1,\ 1),\ (1,\ 1,\ 0),\ (1,\ 0,\ 0)\}.$$

Linear transformation $T : \mathfrak{R}^3 \to \mathfrak{R}^3$ is such that

$$T(v_1) = (2,\ -1,\ 4)$$
$$= 4(1,\ 1,\ 1) - 5(1,\ 1,\ 0) + 3(1,\ 0,\ 0)$$
$$T(v_2) = (3,\ 0,\ 1)$$
$$= 1(1,\ 1,\ 1) - 1(1,\ 1,\ 0) + 3(1,\ 0,\ 0)$$
$$T(v_3) = (-1,\ 5,\ 1)$$
$$= 1(1,\ 1,\ 1) + 4(1,\ 1,\ 0) - 6(1,\ 0,\ 0).$$

Therefore, the matrix of T with respect to the basis β_1 is

$$[T]_{\beta_1} = \begin{bmatrix} 4 & 1 & 1 \\ -5 & -1 & 4 \\ 3 & 3 & -6 \end{bmatrix}.$$

The standard basis for $\mathfrak{R}^3$ is

$$\beta = \{1,\ 0,\ 0),\ (0,\ 1,\ 0),\ (0,\ 0,\ 1)\}.$$

Since

$$(1,\ 0,\ 0) = 0(1,\ 1,\ 1) + 0(1,\ 1,\ 0) + 1(1,\ 0,\ 0)$$
$$(0,\ 1,\ 0) = 0(1,\ 1,\ 1) + 1(1,\ 1,\ 0) - 1(1,\ 0,\ 0)$$
$$(0,\ 0,\ 1) = 1(1,\ 1,\ 1) - 1(1,\ 1,\ 0) + 0(1,\ 0,\ 0).$$

Therefore the transition matrix is

$$S = \begin{bmatrix} 0 & 0 & 1 \\ 0 & 1 & -1 \\ 1 & -1 & 0 \end{bmatrix}$$

Now

$$S^{-1} = \frac{1}{|S|} adj\, S$$
$$= \frac{1}{-1} \begin{bmatrix} -1 & -1 & -1 \\ -1 & -1 & 0 \\ -1 & 0 & 0 \end{bmatrix}$$
$$= \begin{bmatrix} 1 & 1 & 1 \\ 1 & 1 & 0 \\ 1 & 0 & 0 \end{bmatrix}$$

Therefore matrix of the linear transformation T with respect to the basis β is

$$[T]_\beta = S[T]_{\beta_1} S^{-1}$$
$$= \begin{bmatrix} 0 & 0 & 1 \\ 0 & 1 & -1 \\ 1 & -1 & 0 \end{bmatrix} \begin{bmatrix} 4 & 1 & 1 \\ -5 & -1 & 4 \\ 3 & 3 & -6 \end{bmatrix} \begin{bmatrix} 1 & 1 & 1 \\ 1 & 1 & 0 \\ 1 & 0 & 0 \end{bmatrix}$$
$$= \begin{bmatrix} 0 & 0 & 1 \\ 0 & 1 & -1 \\ 1 & -1 & 0 \end{bmatrix} \begin{bmatrix} 6 & 5 & 4 \\ -2 & -6 & -5 \\ 0 & 6 & 3 \end{bmatrix}$$
$$= \begin{bmatrix} 0 & 6 & 3 \\ -2 & -12 & -8 \\ 8 & 11 & 9 \end{bmatrix}.$$

Thus

$$T\begin{pmatrix} x_1 \\ x_2 \\ x_3 \end{pmatrix} = \begin{bmatrix} 0 & 6 & 3 \\ -2 & -12 & -8 \\ 8 & 11 & 9 \end{bmatrix} \begin{bmatrix} x_1 \\ x_2 \\ x_3 \end{bmatrix}$$
$$= \begin{bmatrix} 6x_2 + 3x_3 \\ -2x_1 - 12x_2 - 8x_3 \\ 8x_1 + 11x_2 + 9x_3 \end{bmatrix},$$

and so

$$T(x_1, x_2, x_3) = (6x_2 + 3x_3,\ -2x_1 - 12x_2 - 8x_3,\ 8x_1 + 11x_2 + 9x_3).$$

Hence

$$T(2,4,-1) = (21,-44,51).$$

EXAMPLE 2.63

Let $T: \Re^2 \to \Re^3$ be the linear transformation defined by

$$T(x_1, x_2) = (x_2,\ -5x_1 + 13x_2,\ -7x_1 + 16x_2).$$

Find the matrix for the transformation T with respect to the bases

$$\beta = \{(3,\ 1)^T,\ (5,\ 2)^T\} \quad \text{for} \quad \Re^2$$

$$\text{and } \beta_1 = \{(1,\ 0,\ -1)^T,\ (-1,\ 2,\ 2)^T,\ (0,\ 1,\ 2)^T\} \text{ for } \Re^3.$$

Solution. We have

$$T\begin{bmatrix} x_1 \\ x_2 \end{bmatrix} = \begin{bmatrix} x_2 \\ -5x_1 + 13x_2 \\ -7x_1 + 16x_2 \end{bmatrix},$$

$$\beta = \left\{\begin{bmatrix} 3 \\ 1 \end{bmatrix}, \begin{bmatrix} 5 \\ 2 \end{bmatrix}\right\},$$

$$\beta_1 = \left\{\begin{bmatrix} 1 \\ 0 \\ -1 \end{bmatrix}, \begin{bmatrix} -1 \\ 2 \\ 2 \end{bmatrix}, \begin{bmatrix} 0 \\ 1 \\ 2 \end{bmatrix}\right\}$$

Then

$$T\left(\begin{bmatrix} 3 \\ 1 \end{bmatrix}\right) = \begin{bmatrix} 1 \\ -2 \\ -5 \end{bmatrix}$$

$$= 1\begin{bmatrix} 1 \\ 0 \\ -1 \end{bmatrix} + 0\begin{bmatrix} -1 \\ 2 \\ 2 \end{bmatrix} - 2\begin{bmatrix} 0 \\ 1 \\ 2 \end{bmatrix}$$

$$T\left(\begin{bmatrix} 5 \\ 2 \end{bmatrix}\right) = \begin{bmatrix} 2 \\ 1 \\ -3 \end{bmatrix}$$

$$= 3\begin{bmatrix} 1 \\ 0 \\ -1 \end{bmatrix} + 1\begin{bmatrix} -1 \\ 2 \\ 2 \end{bmatrix} - 1\begin{bmatrix} 0 \\ 1 \\ 2 \end{bmatrix}$$

Therefore, the matrix for T is

$$\underset{\beta,\beta_1}{T}\begin{bmatrix} 1 & 3 \\ 0 & 1 \\ -2 & -1 \end{bmatrix}$$

EXAMPLE 2.64

Find the reflection of $\Re^2$ about the line $y = mx$.

Solution. Let $T: \Re^2 \to \Re^2$ be the reflection. We know that the reflection of any point $(x,\ y)$ lying on the line $y = mx$ will be $(x,\ y)$ itself and the reflection of any point lying on the perpendicular to the line $y = mx$ is the point $-(x,\ y)$.

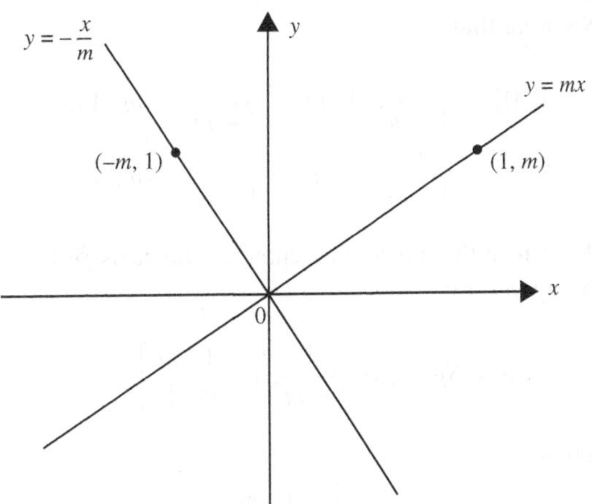

Therefore

$$T(1,\ m) = (1,\ m)$$

and

$$T(-m,\ 1) = -(-m,\ 1) = (m,\ -1).$$

Let

$$\beta_1 = \{(1,\ m),\ (-m,\ 1)\}.$$

Then

$$\lambda_1(1, m) + \lambda_2(-m, 1) = 0 \text{ implies}$$
$$\lambda_1 - m\lambda_2 = 0 \text{ and } m\lambda_1 + \lambda_2 = 0.$$

Solving these equations, we get $\lambda_1 = \lambda_2 = 0$. Hence the set β_1 is linearly independent. Also, β_1 generates $\mathfrak{R}^2$. Therefore it is a basis for $\mathfrak{R}^2$. Further,

$$T(1, m) = (1, m) = 1(1, m) + 0(-m, 1),$$
$$T(-m, 1) = (m, -1) = 0(1, m) - 1(-m, 1).$$

Hence the matrix of T with respect to the basis β_1 is

$$[T]\beta_1 = \begin{bmatrix} 1 & 0 \\ 0 & -1 \end{bmatrix}$$

The standard basis for $\mathfrak{R}^2$ is $\beta = \{(1, 0), (0, 1)\}$.
We note that

$$(1, 0) = \frac{1}{1+m^2}(1, m) - \frac{m}{1+m^2}(-m, 1),$$
$$(0, 1) = \frac{m}{1+m^2}(1, m) + \frac{1}{1+m^2}(-m, 1).$$

Therefore the matrix that changes the basis β_1 to β is given by

$$S = S_{\beta_1 \to \beta} = \frac{1}{1+m^2}\begin{bmatrix} 1 & m \\ -m & 1 \end{bmatrix}.$$

Then

$$S^{-1} = \begin{bmatrix} 1 & m \\ -m & 1 \end{bmatrix}$$

Therefore matrix of T with respect to β is

$$[T] = S[T]_{\beta_1} S^{-1}$$
$$= \frac{1}{1+m^2}\begin{bmatrix} 1 & -m \\ m & 1 \end{bmatrix}\begin{bmatrix} 1 & 0 \\ 0 & -1 \end{bmatrix}\begin{bmatrix} 1 & m \\ -m & 1 \end{bmatrix}$$
$$= \frac{1}{1+m^2}\begin{bmatrix} 1-m^2 & 2m \\ 2m & m^2-1 \end{bmatrix}. \qquad (31)$$

Thus for any $(x, y) \in \mathfrak{R}^2$, we have

$$T\begin{pmatrix} x \\ y \end{pmatrix} = \frac{1}{1+m^2}\begin{bmatrix} 1-m^2 & 2m \\ 2m & m^2-1 \end{bmatrix}\begin{bmatrix} x \\ y \end{bmatrix}$$
$$= \frac{1}{1+m^2}\begin{bmatrix} (1-m^2)x + 2my \\ 2mx + (m^2-1)y \end{bmatrix},$$

that is,

$$T(x, y) = \frac{1}{1+m^2}\Big((1-m^2)x + 2my,\ 2mx + (m^2-1)y\Big).$$

Deductions (i) Putting $m = 1$, the reflection operator about the line $y = x$ is given by

$$T(x, y) = \frac{1}{2}(2y, 2x) = (y, x).$$

Its matrix (putting $m = 1$ in (31)) with respect to the standard basis is

$$[T]_\beta = \frac{1}{2}\begin{bmatrix} 0 & 2 \\ 2 & 0 \end{bmatrix} = \begin{bmatrix} 0 & 1 \\ 1 & 0 \end{bmatrix}$$

(ii) Putting $m = 2$, the reflection operator T about the line $y = 2x$ is given by

$$T(x,y) = \frac{1}{5}(-3x + 4y,\ 4x + 3y)$$

and its matrix (putting $m = 2$ in (31)) with respect to the standard basis is

$$[T]_\beta = \frac{1}{5}\begin{bmatrix} -3 & 4 \\ 4 & 3 \end{bmatrix}.$$

EXAMPLE 2.65

Find the matrix of the *reflection operator* about the *yz*-plane on $\mathfrak{R}^3$.

Solution. The reflection relative to the *yz*-plane on $\mathfrak{R}^3$ is shown in the figure below:

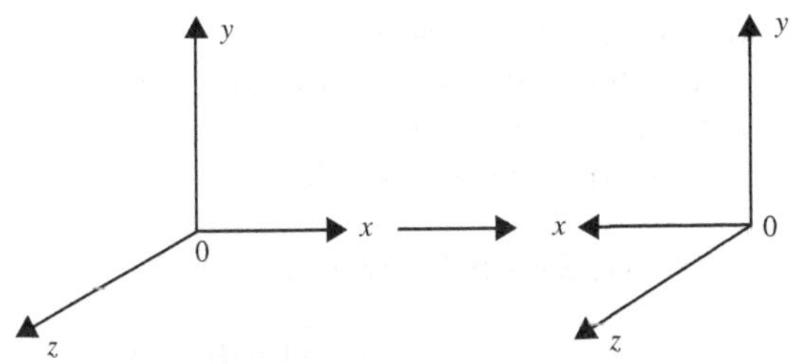

It changes the sign of the x-coordinate leaving the y and z-coordinate values unchanged. Thus the transformation (reflection operator) T is defined by

$$T(x, y, z) = (-x, y, z).$$

The standard basis for $\mathfrak{R}^3$ is

$$\beta = \{(1, 0, 0), (0, 1, 0), (0, 0, 1)\}.$$

Then

$$\begin{aligned} T(1, 0, 0) &= (-1, 0, 0) \\ &= -1(1, 0, 0) + 0(0, 1, 0) + 0(0, 0, 1) \\ T(0, 1, 0) &= (0, 1, 0) \\ &= 0(1, 0, 0) + 1(0, 1, 0) + 0(0, 0, 1) \\ T(0, 0, 1) &= (0, 0, 1) \\ &= 0(1, 0, 0) + 0(0, 1, 0) + 1(0, 0, 1). \end{aligned}$$

Hence the matrix of the reflection operator in the yz-plane is

$$[T]_\beta = \begin{bmatrix} -1 & 0 & 0 \\ 0 & 1 & 0 \\ 0 & 0 & 1 \end{bmatrix}.$$

EXAMPLE 2.66

Find the matrix of the *reflection operator* T about the x-axis with respect to the standard basis.

Solution. From Example 2.36, we have

$$T(x, y) = (x, -y).$$

The standard basis of $\mathfrak{R}^2$ is $\beta = \{(1, 0), (0, 1)\}$. Therefore,

$$\begin{aligned} T(1, 0) &= (1, 0) = 1(1, 0) + 0(0, 1) \\ T(0, 1) &= (0, -1) = 0(1, 0) - 1(0, 1). \end{aligned}$$

Hence, the matrix of T is

$$[T]_\beta = \begin{bmatrix} 1 & 0 \\ 0 & -1 \end{bmatrix}.$$

EXAMPLE 2.67

Using the concepts of rotation and reflection about the x-axis, find the matrix of the reflection operator about the line $y = x$.

Solution. The sequence of transformations to produce reflection about the line $y = x$ is

(i) Clockwise rotation by 45°
(ii) Reflection about the x-axis
(iii) Counterclockwise rotation by 45°.

The corresponding matrices with respect to the standard basis are

$$T_1 = \begin{bmatrix} \cos 45^\circ & \sin 45^\circ \\ -\sin 45^\circ & \cos 45^\circ \end{bmatrix} \text{(inverse of matrix } T_3\text{)},$$

$$T_2 = \begin{bmatrix} 1 & 0 \\ 0 & -1 \end{bmatrix}$$

and

$$T_3 = \begin{bmatrix} \cos 45^\circ & -\sin 45^\circ \\ \sin 45^\circ & \cos 45^\circ \end{bmatrix}.$$

Therefore matrix T of the reflection operator about the line $y = x$ is

$$\begin{aligned} T &= T_3 T_2 T_1 \\ &= \begin{bmatrix} \frac{1}{\sqrt{2}} & -\frac{1}{\sqrt{2}} \\ \frac{1}{\sqrt{2}} & \frac{1}{\sqrt{2}} \end{bmatrix} \begin{bmatrix} 1 & 0 \\ 0 & -1 \end{bmatrix} \begin{bmatrix} \frac{1}{\sqrt{2}} & \frac{1}{\sqrt{2}} \\ -\frac{1}{\sqrt{2}} & \frac{1}{\sqrt{2}} \end{bmatrix} \\ &= \begin{bmatrix} 1 & 0 \\ 0 & 1 \end{bmatrix} \end{aligned}$$

2.9 NORMED LINEAR SPACE

Definition 2.32. Let X be a linear space. A *norm* on X is a real-valued function $\|\cdot\|$ on X satisfying

(i) $\|x\| \geq 0$ for all $x \in X$

(ii) $\|x\| = 0$ if and only if $x = 0$

(iii) $\|x+y\| \leq \|x\| + \|y\|$ for all $x, y \in X$

(iv) $\|\alpha x\| = |\alpha|\ \|x\|$ for all $x \in X$ and all scalar α.

A linear space X along with a norm defined on it is called a *normed linear space* and is denoted by $(X, \|\cdot\|)$ or simply by X.

Definition 2.33. A normed linear space X is said to be *complete* if every Cauchy sequence in X converges to some point in X.

Definition 2.34. If $\int_{-\pi}^{\pi} |f(t)|^2\, dt < \infty$, then f is said to belong to the class $L^2[-\pi, \pi]$.

In $L^2[-\pi, \pi]$, the norm of f is defined by

$$\|f\| = \left[\int_{-\pi}^{\pi} |f(t)|^2 dt\right]^{1/2}$$

2.10 INNER PRODUCT SPACE

Let $\mathbb{C}$ be the field of complex numbers. Let the conjugate of a complex number c be denoted by $\bar{c}$. Thus, if $c = a+ib$; where a and b are real numbers, then $\bar{c} = a - ib$. The familiar properties of conjugation are: $(\bar{\bar{c}}) = c$, $\overline{(c+d)} = \bar{c}+\bar{d}$, $\overline{cd} = \bar{c}\,\bar{d}$, $|c| = (\bar{c}c)^{1/2}$ and $\bar{c} = c$ if and only if c is real.

Definition 2.35. *An inner product space* (or *pre-Hilbert space)* is a complex linear space X together with an inner product $(,) : X \times X \to \mathbb{C}$ such that

(i) $(x, y) = \overline{(y, x)}$

(ii) $(\lambda x + \mu y, z) = \lambda(x, z) + \mu(y, z)$

(iii) $(x, x) \geq 0$ and $(x, x) = 0 \Leftrightarrow x = 0$

for any $x, y, z \in X$ and $\lambda, \mu \in \mathbb{C}$.

Condition (i) clearly reduces to $(x, y) = (y, x)$ if X is a real vector space.

From (i) and (ii) we obtain

$$\begin{aligned}(x, \lambda y + \mu z) &= \overline{(\lambda y + \mu z, x)}\\ &= \overline{\lambda (y, x)} + \overline{\mu (z, x)}\\ &= \bar{\lambda}\overline{(y, x)} + \bar{\mu}\overline{(z, x)}\\ &= \bar{\lambda}(x, y) + \bar{\mu}(x, z).\end{aligned}$$

We also observe that if $(x, z) = (y, z)$ for all z, then $(x - y, z) = 0$ for all z. In particular, $(x - y, x - y) = 0$ and hence, by (iii), it follows that $x - y = 0$ or $x = y$.

In any inner product space, the following observations are immediate.

(a) $(x, y+z) = (x, y) + (x, z)$

(b) $(x, \lambda y) = \bar{\lambda}(x, y)$

(c) $(0, y) = (x, 0) = 0$

(d) $(x - y, z) = (x, z) - (y, z)$,
$(x, y - z) = (x, y) - (y, z)$.

Definition 2.36. In an inner product space, the *norm* (or *length*) of a vector x, denoted by $\|x\|$, is the non-negative real number defined by

$$\|x\| = (x, x)^{1/2}.$$

We observe that

$$\begin{aligned}\|\lambda x\|^2 &= (\lambda x, \lambda x) = \lambda(x, \lambda x)\\ &= \lambda\bar{\lambda}(x, x) = |\lambda|^2\ \|x\|^2\end{aligned}$$

and hence

$$\|\lambda x\| = |\lambda|\ \|x\|.$$

EXAMPLE 2.68

(a) Let $\mathbb{C}^n$ be the vector space of n-tuples. If $x = (\lambda_1, \lambda_2, \ldots, \lambda_n)$ and $y = (\mu_1, \ldots, \mu_n)$, define

$$(x, y) = \sum_{k=1}^{n} \lambda_k\, \bar{\mu}_k.$$

Then all the axioms for pre-Hilbert space are satisfied. Therefore, $\mathbb{C}^n$ is an inner product space. This space is known as *n-dimensional unitary space*. In this space, the norm of x is defined by

$$\|x\| = \left(\sum_{i=1}^{n} |\lambda_i|^2\right)^{1/2}$$

(b) Let $C[a, b]$ be the vector space of continuous functions defined on $[a, b]$, $a < b$. Define

$$(x,y) = \int_a^b x(t)\overline{y(t)}\, dt.$$

With respect to this inner product, $C[a, b]$ is a pre-Hilbert space. The norm of x in $C[a, b]$ is introduced by taking

$$\|x\| = \left(\int_a^b |x(t)|^2 dt\right)^{1/2}.$$

(c) Let $M_n(\mathfrak{R})$ be the vector space of matrices of order n over the field of real numbers. If A, B are in $M_n(\mathfrak{R})$, define inner product in $M_n(\mathfrak{R})$ by

$$(A, B) = tr(B^T A).$$

We note that

(i) $$(A,A) = tr(A^T A) = \sum_{i=1}^{n}\sum_{h=1}^{n} a_{ki}\, a_{hi} = \sum_{i=1}^{n}\sum_{h=1}^{n} a_{ki}^2.$$

If $A \neq 0$, then $a_{ki} \neq 0$ for some k and i and so $(A,A) > 0$ in that case.

(ii) $$\begin{aligned}(A+B, C) &= tr\left(C^T (A+B)\right)\\ &= tr\left(C^T A + C^T B\right)\\ &= tr\left(C^T A\right) + tr\left(C^T B\right)\\ &= (A, C) + (B, C)\end{aligned}$$

(iii) $$\begin{aligned}(A, B) &= tr\left(B^T A\right)\\ &= tr\left(AB^T\right)\\ &= (B, A) = \overline{(B, A)}.\end{aligned}$$

Hence $M_n(\mathfrak{R})$ is an inner product space.

EXAMPLE 2.69

In $\mathbb{C}^2$, define for $x = (\alpha_1, \alpha_2)$ and $y = (\beta_1, \beta_2)$,

$$(x,y) = 2\alpha_1\bar{\beta}_1 + \alpha_1\bar{\beta}_2 + \alpha_2\bar{\beta}_1 + \alpha_2\bar{\beta}_2$$

Then $\mathbb{C}^2$ is an inner product space.

Theorem 2.21. (Cauchy–Schwarz Inequality). If x and y are any two vectors in an inner product space, then

$$|(x, y)| \leq \|x\|\ \|y\|.$$

Proof. If $x = 0$ or $y = 0$, then $(x, y) = 0$ and the result holds. Otherwise, we have

$$\begin{aligned}&(x+\lambda y,\ x+\lambda y) \geq 0\\ \Rightarrow &(x,\ x+\lambda y) + \lambda(y,\ x+\lambda y) \geq 0\\ \Rightarrow &(x,\ x) + \bar{\lambda}(x,\ y) + \lambda\left[(y,\ x) + \bar{\lambda}(y,\ y)\right] \geq 0\\ \Rightarrow &(x,\ x) + \bar{\lambda}(x,\ y) + \lambda(y,\ x) + \lambda\bar{\lambda}(y,\ y) \geq 0\end{aligned}$$

Taking $\lambda = -\dfrac{(x,\ y)}{(y,\ y)}$, this implies

$$(x,\ x) - \frac{\overline{(x,\ y)}(x,\ y)}{(y,\ y)} - \frac{(x,\ y)\ (y,\ x)}{(y,\ y)} + \frac{(x,\ y)\ \overline{(x,\ y)}\ (y,\ y)}{(y,\ y)\ (y,\ y)} \geq 0$$

$$\Rightarrow (x,\ x) - \frac{|(x,\ y)|^2}{(y,\ y)} \geq 0$$

$$\Rightarrow |(x,\ y)|^2 \leq (x,\ x)\ (y,\ y) = \|x\|^2\ \|y\|^2$$

$$\Rightarrow |(x,\ y)| \leq \|x\|\ \|y\|,$$

which is the required inequality.

Applying Theorem 2.21 in the unitary space $\mathbb{C}^n$, we have

Corollary 2.5. If $\lambda_1, \lambda_2, \ldots, \lambda_n$ and $\mu_1, \mu_2, \ldots, \mu_n$ are complex numbers, then

$$\left|\sum_{k=1}^{n} \lambda_k \bar{\mu}_k\right| \leq \left(\sum_{k=1}^{n} |\lambda_k|^2\right)^{1/2} \left(\sum_{k=1}^{n} |\mu_k|^2\right)^{1/2}.$$

Corollary 2.6. If $f,\ g \in C[0,\ 1]$, then

$$\left|\int_0^1 f(t)\overline{g(t)}\,dt\right|^2 \leq \int_0^1 |f(t)|^2\,dt \int_0^1 |g(t)|^2\,dt.$$

Theorem 2.22. (Parallelogram Law). In a pre-Hilbert space

$$\|x+y\|^2 + \|x-y\|^2 = 2\|x\|^2 + 2\|y\|^2.$$

Proof. Writing out the expression on the left in terms of inner products, we have

$$\begin{aligned}\|x+y\|^2 + \|x-y\|^2 &= (x+y,\ x+y) \\ &\quad + (x-y,\ x-y) \\ &= (x,\ x) + (x,\ y) + (y,\ x) \\ &\quad + (y,\ y) + (x,\ x) \\ &\quad - (x,\ y) - (y,\ x) + (y,\ y) \\ &= 2(x,\ x) + 2(y,\ y) \\ &= 2\|x\|^2 + \|y\|^2.\end{aligned}$$

Theorem 2.23. (Polarization Identity). In a pre-Hilbert space,

$$(x,\ y) = \frac{1}{4}\Big[\|x+y\|^2 - \|x-y\|^2 + i\|x+iy\|^2 - i\|x-iy\|^2\Big]$$

Proof. We have

$$\begin{aligned}\|x+y\|^2 &= (x+y,\ x+y) = (x,\ x) + (y,\ y) \\ &\quad + (x,\ y) + (y,\ x) \\ &= \|x\|^2 + \|y\|^2 + (x,\ y) + (y,\ x) \quad (32)\end{aligned}$$

Replace y by $-y$, iy, and $-iy$ and obtain

$$\|x-y\|^2 = \|x\|^2 + \|y\|^2 - (x,\ y) - (y,\ x)$$

$$\|x+iy\|^2 = \|x\|^2 + \|y\|^2 - i(x,\ y) + i(y,\ x)$$

$$\|x-iy\|^2 = \|x\|^2 + \|y\|^2 + i(x,\ y) - i(y,\ x)$$

It follows that

$$-\|x-y\|^2 = -\|x\|^2 - \|y\|^2 + (x,\ y) + (y,\ x) \quad (33)$$

$$i\|x+iy\|^2 = i\|x\|^2 + i\|y\|^2 + (x,\ y) - (y,\ x) \quad (34)$$

$$-i\|x-iy\|^2 = -i\|x\|^2 - i\|y\|^2 + (x,\ y) - (y,\ x) \quad (35)$$

Adding (32), (33), (34), and (35), we get

$$\|x+y\|^2 - \|x-y\|^2 + i\|x+iy\|^2 - i\|x-iy\|^2$$

$= 4(x,\ y)$, which is the required equality.

Definition 2.37. In a pre-Hilbert space, $\|x-y\|$ is called the *distance* from x to y.

If we put $d(x,\ y) = \|x-y\|$, then we note that

(i) $d(x,\ y) \geq 0$

(ii) $d(x,\ y) = 0 \Leftrightarrow x = y$

(iii) $d(x,\ y) = d(y,\ x)$

(iv) $d(x,\ y)?\, d(x,\ z) + d(z,\ y)$ (Triangle inequality).

The function d is called *distance function* or a *metric*.

Theorem 2.24. In a pre-Hilbert space, every Cauchy Sequence is bounded.

Proof. Let $< x_n >$ be a Cauchy sequence and let N be an index such that $\|x_n - x_m\| \leq 1$, whenever $m,\ n \geq N$. If $n \geq N$ then

$$\begin{aligned}\|x_n\| &= \|(x_n - x_N) + x_N\| \\ &< \|x_n - x_N\| + \|x_N\| < 1 + \|x_N\|.\end{aligned}$$

Thus, if M is the largest of the numbers $1 + \|x_N\|,\ \|x_1\|, \ldots \|x_{N-1}\|$, we have $\|x_n\| \leq M$ for all n. Hence $< x_n >$ is bounded.

EXAMPLE 2.70

Let x and y be arbitrary elements in a pre-Hilbert space. Show that $\|x+y\| \leq \|x\| + \|y\|$.

Solution. Let x, y be in an inner product space. We have

$$\begin{aligned}\|x+y\|^2 &= (x+y,\ x+y)\\ &= (x,\ x)+(y,\ y)+(x,\ y)+(y,\ x)\\ &= (x,\ x)+(y,\ y)+(x,\ y)+(\overline{x,\ y})\\ &= (x,\ x)+(y,\ y)+2R(x,\ y)\\ &= \|x\|^2+\|y\|^2+2\|x\|\ \|y\|\\ &\quad \text{(Using Cauchy–Schwarz inequality)}\\ &= (\|x\|+\|y\|)^2,\end{aligned}$$

which yields

$$\|x+y\| \leq \|x\| + \|y\|.$$

Remark 2.2 We note that if $\|x\| = (x,\ x)^{1/2}$, then

(i) $\|x\| = (x,\ x)^{1/2} \geq 0$,

(ii) $\|x\| = 0 \Leftrightarrow x = 0$ since $\|x\| = 0 \Leftrightarrow (x,\ x) = 0 \Leftrightarrow x = 0$.

(iii) $\|\lambda x\| = |\lambda|\ \|x\|$ (already proved)

(iv) $\|x+y\| \leq \|x\| + \|y\|$.

It follows, therefore, that an *inner product space is a normed linear space.*

EXAMPLE 2.71

Let $f(t) = t$ and $g(t) = e^t$. Then f and g are continuous on $[0,\ 1]$. Define inner product $(f,\ g)$ by

$$\begin{aligned}(f,\ g) &= \int_0^1 f(t)\overline{g(t)}\ dt\\ &= \int_0^1 t\ e^t\ dt = 1.\end{aligned}$$

Also

$$\begin{aligned}\|f\| = (f,\ f)^{\frac{1}{2}} &= \left(\int_0^1 f^2(t)dt\right)^{\frac{1}{2}} = \left(\int_0^1 t^2 dt\right)^{\frac{1}{2}}\\ &= \left(\left[\frac{t^3}{3}\right]_0^1\right) = \left(\frac{1}{3}\right)^{\frac{1}{2}} = \frac{1}{\sqrt{3}},\end{aligned}$$

$$\begin{aligned}\|g\| = (g,\ g)^{\frac{1}{2}} &= \left(\int_0^1 g^2(t)dt\right)^{\frac{1}{2}}\\ &= \left(\int_0^1 e^{2t} dt\right)^{\frac{1}{2}} = \left(\left|\frac{e^{2t}}{2}\right|_0^1\right)^{\frac{1}{2}}\\ &= \sqrt{\frac{e^2-1}{2}}.\end{aligned}$$

Similarly, we can show that

$$\|f+g\| = \sqrt{\frac{11+3e^2}{6}}.$$

We note that

$$|(f,\ g)| = 1 \text{ and } \|f\| = \frac{1}{\sqrt{3}},\ \|g\| = \sqrt{\frac{e^2-1}{2}},$$

$$\|f\|\ \|g\| = \sqrt{\frac{e^2-1}{2}}\left(\frac{1}{\sqrt{3}}\right) = \frac{1}{\sqrt{6}}\sqrt{e^2-1}.$$

Thus

$$|(f,\ g)| \leq \|f\|\ \|g\|$$

and Cauchy–Schwarz inequality is satisfied. Also, we have

$$\|(x+y)\| \leq \|x\| + \|y\| \text{ (traingle inequality).}$$

Definition 2.38. Let x and y be vectors in a pre-Hilbert space H. Then x is said to be *orthogonal* to y (written as $x \perp y$) if $(x,\ y) = 0$. Since $(y,\ x) = \overline{(x,\ y)}$, it follows that $x \perp y \Leftrightarrow y \perp x$. It is also clear that $x \perp 0$ for every x. Moreover, since $(x,\ x) = \|x\|^2$, 0 is the only vector orthogonal to itself.

Definition 2.39. A vector x in an inner product space is said to be *orthogonal to a non-empty set* S, written as $x \perp S$, if $x \perp y$ for every $y \in S$.

Definition 2.40. If W is a subspace of an inner product space V, then the set of all vectors in V orthogonal to W is called the *orthogonal complement* of W and is denoted by $W^{\perp}$. Thus

$$W^{\perp} = \{x \in V : (x, w) = 0 \quad \text{for all} \quad w \in W\}.$$

If V is an inner product space, then it is clear that

(i) $V^{\perp} = \{0\}$ since, 0 is the only vector orthogonal to all vectors in V.

(ii) $\{0\}^{\perp} = V$.

Furthermore, we observe that if $w \in W \cap W^{\perp}$, then w must be self-orthogonal. But we know that 0 is the only vector orthogonal to itself. Hence, $w = 0$ and consequently $W \cap W^{\perp} = \{0\}$.

Definition 2.41. A set S of vectors is said to be *orthogonal set* if $x \perp y$ whenever x and y are distinct vectors in S.

Theorem 2.25. If x is orthogonal to each of $y_1, y_2, \ldots, y_n$, then x is orthogonal to every linear combination of the y_k.

Proof. If $x \perp y_k$ for $k = 1, 2, \ldots, n$ and $y = \sum_{k=1}^{n} \lambda_k y_k$,then

$$(x, y) = \sum_{k=1}^{n} \bar{\lambda}_k (x, y_k) = \sum_{k=1}^{n} \bar{\lambda}_k .0 = 0,$$

which proves the theorem.

Theorem 2.26. (Pythagorean Theorem). If $x \perp y$, then

$$\|x+y\|^2 = \|x-y\|^2 = \|x\|^2 + \|y\|^2.$$

Proof. If $x \perp y$, then $(x, y) = 0 = (y, x)$. Therefore,

$$\begin{aligned}\|x+y\|^2 &= (x+y, x+y) = (x, x) + (x, y) \\ &\quad + (y, x) + (y, y) \\ &= (x, x) + (y, y) = \|x\|^2 + \|y\|^2.\end{aligned}$$

Similarly,

$$\begin{aligned}\|x-y\|^2 &= (x-y, x-y) \\ &= (x, x) - (x, y) - (y, x) + (y, y) \\ &= (x, x) + (y, y) = \|x\|^2 + \|y\|^2.\end{aligned}$$

Hence the result.

Corollary 2.7. If $x_1, x_2, \ldots, x_n$ are orthogonal in an inner product space, then

$$\left\| \sum_{k=1}^{n} x_k \right\|^2 = \sum_{k=1}^{n} \|x_k\|^2.$$

Proof. We have seen in Theorem 2.26 that the result is true for $n = 2$. Assume inductively that

$$\left\| \sum_{k=1}^{n-1} x_k \right\|^2 = \sum_{k=1}^{n-1} \|x_k\|^2.$$

Setting $x = \sum_{k=1}^{n-1} x_k$ and $y = x_n$, we have $x \perp y$. Then

$$\begin{aligned}\left\| \sum_{k=1}^{n} x_k \right\|^2 &= \|x+y\|^2 = \|x\|^2 + \|y\|^2 \\ &= \sum_{k=1}^{n-1} \|x_k\|^2 + \|x_n\|^2 = \sum_{k=1}^{n} \|x_k\|^2.\end{aligned}$$

This completes the proof of the corollary.

Lemma 2.1. If W is a subspace of an inner product space V, then $W^{\perp}$ is subspace of V.

Proof. Let $x, y \in W^{\perp}$ and $\alpha, \beta \in \mathbb{C}$. To prove that $W^{\perp}$ is a subspace, it suffices to prove that $\alpha x + \beta y \in W^{\perp}$. Let $w \in W$. Then $(x, w) = (y, w) = 0$ and so

$$\begin{aligned}(\alpha x + \beta y, w) &= \alpha(x, w) + \beta(y, w) \\ &= \alpha.0 + \beta.0, \\ &= 0 + 0 = 0.\end{aligned}$$

Thus $\alpha x + \beta y \perp w$. Hence, $\alpha x + \beta y \in W^{\perp}$ and as such $W^{\perp}$ is a subspace of V.

Definition 2.42. The set of vectors $\{v_i\}$ in an inner product space is said to be *orthonormal* set if

(i) $(v_i, v_j) = 0,\ i \neq j$, that is $\{v_i\}$ is orthogonal.
(ii) $(v_i, v_i) = 1$, that is, each v_i is of length 1.

Lemma 2.2. If $\{v_i\}$ is an orthonormal set in V and if $w \in V$, then $u = w - (w, v_1)v_1 - (w, v_2)v_2 - \ldots - (w, v_n)v_n$ is orthogonal to each of $v_1, v_2, \ldots v_n$.

Proof. Computing (u, v_i) for any i, $1 \leq i \leq n$, we have

$$\begin{aligned}(u, v_i) &= (w - (w, v_1)v_1 - \ldots - (w, v_n)v_n, v_i)\\ &= (w, v_i) - (w, v_i)(v, v_i)\\ &\quad - (w, v_2)(v_2, v_i) - \ldots\\ &\quad - (w, v_i)(v_i, v_i) - \ldots - (w, v_n)(v_n, u_i)\\ &= (w, v_i) - 0 - 0 - \ldots\\ &\quad - (w, v_i)1 - \ldots - 0\\ &= (w, v_i) - (w, v_i) = 0.\end{aligned}$$

Hence the result follows.

Lemma 2.3. If $\{v_i\}$ is an orthonormal set, then the vectors in $\{v_i\}$ are linearly independent. If $w = \alpha_1 v_1 + \ldots + \alpha_n v_n$, then $\alpha_i = (w, v_i)$ for $i = 1, 2, \ldots, n$.

Proof. Suppose that

$$\alpha_1 v_1 + \alpha_2 v_2 + \ldots + \alpha_n v_n = 0. \qquad (36)$$

Therefore

$$\begin{aligned}0 &= (\alpha_1 v_1 + \alpha_2 v_2 + \ldots + \alpha_n v_n, v_i)\\ &= \alpha_1(v_1, v_i) + \alpha_2(v_2, v_i) + \ldots\\ &\quad + \alpha_i(v_i, v_i) + \ldots + \alpha_n(v_n, v_i)\\ &= \alpha_1.0 + \alpha_2.0 + \cdots + \alpha_i + \alpha_n.0 = \alpha_i.\end{aligned}$$

Thus (36) implies $\alpha_i = 0$ for $i = 1, 2, \ldots, n$. Hence, $\{v_i\}$ is linearly independent. Also, it is clear that if $w = \alpha_1 v_1 + \ldots + \alpha_n v_n$, then

$$(w, v_i) = \alpha_i \ldots \text{ for } i = 1, 2, \ldots, n.$$

Theorem 2.27. Let V be a finite dimensional inner product space. Then V has an orthonormal set as a basis.

Proof. Suppose that V is of dimension n over a field F and let $v_1, v_2, \ldots, v_n$ be a basis of V. From this basis, we construct an orthonormal set of n vectors $w_1, w_2, \ldots, w_n$ By Lemma (2.3), this set shall be linearly independent. Also every vector of V is also a linear combination of $\{w_1, w_2, \ldots, w_n\}$. Thus, the set $\{w_1, w_2, \ldots, w_n\}$ shall generate V and as such it will form a basis of V.

We first normalize v_1 by putting $w_1 = \frac{v_1}{\|v_1\|}$. Then

$$(w_1, w_1) = \left(\frac{v_1}{\|v_1\|}, \frac{v_1}{\|v_1\|}\right) = \frac{1}{\|v_1\|^2}(v_1, v_1) = 1,$$

and hence $\|w_1\| = 1$. Consider now $\alpha w_1 + v_2$ so that it may be orthogonal to w_1 for some value of α. For this we need $(\alpha w_1 + v_2, w_1) = 0$, that is, $\alpha(w_1, w_1) + (v_2, w_1) = 0$, that is, $\alpha.1 + (v_2, w_1) = 0$, which yields $\alpha = -(v_2, w_1)$. So we have $v_2 - (v_2, w_1)w_1$ orthogonal to w_1. We, therefore, substitute

$$w_2 = \frac{v_2 - (v_2, w_1)w_1}{\|v_2 - (v_2 - w_1)w_1\|}.$$

Then $\|w_2\| = 1$ and $(w_1, w_2) = 0$. Thus $\{w_1, w_2\}$ is orthogonal.

Adopting the above procedure consider $v_3 - (v_3, w_1)w_1 - (v_3, w_2)w_2$. It is orthogonal to w_1 and w_2. We normalize this by putting

$$w_3 = \frac{v_3 - (v_3, w_1)w_1 - (v_3, w_2)w_2}{\|v_3 - (v_3, w_1)w_1 - (v_3, w_2)w_2\|}.$$

We see that $(w_1, w_2) = 0, (w_2, w_3) = 0$, $(w_1, w_3) = 0$, and $\|w_1\| = \|w_2\| = \|w_3\| = 1$. Thus, $\{w_1, w_2, w_3\}$ is orthonormal. Assume inductively that orthonormal vectors $w_1, w_2, \ldots w_{n-1}$ have already been defined. The desired vector w_n must be a linear combination of $v_1, v_2, \ldots, v_n$ or equivalently of $w_1, w_2, \ldots w_{n-1}, v_n$. Moreover, it must be orthonormal to each of $w_1, w_2, \ldots w_{n-1}$. So we consider

$$v_n - \sum_{k=1}^{n-1} (v_k, w_k) w_k.$$

Then this vector is orthogonal to $w_1, w_2, \ldots w_{n-1}$. We then define

$$w_n \frac{v_n - \sum_{k=1}^{n-1} (v_n, w_k) w_k}{\left\| v_n - \sum_{k=1}^{n-1} (v_n, w_k) w_k \right\|}.$$

Then the set $[w_1, w_2, \ldots w_n]$ is an orthonormal set which serves as basis for V. This completes the proof of the theorem. The above theorem may also be stated as:

Theorem 2.28. (Gram-Schmidt Orthogonalization Process). If $\{v_1, v_2, \ldots, v_n\}$ is a linearly independent set in an inner product space V, then there exists an orthonormal set $\{w_1, w_2, \ldots w_n\}$ such that $\{w_1, w_2, \ldots w_n\}$ generates the same linear subspace as $\{v_1, v_2, \ldots, v_n\}$.

EXAMPLE 2.72

Let F be the real field and let V be the set of polynomials, in a variable x, over F of degree 2 or less. In V, we define inner product by

$$(p(x), q(x)) = \int_{-1}^{1} p(x) q(x) dx, \; p(x), q(x) \in V.$$

We start with the basis $\{v_1, v_2, v_3\} = \{1, x, x^2\}$ of V. In view of the above theorem, we have,

$$w_1 = \frac{v_1}{\|v_1\|} = \frac{1}{\left(\int_{-1}^{1} 1 dx \right)^{1/2}} = \frac{1}{\sqrt{2}},$$

and since $(v_2, w_1) = \left(v_2, \frac{1}{\sqrt{2}} \right) = \int_{-1}^{1} \frac{1}{\sqrt{2}} x dx = 0,$

$\|x\| = (x, x)^{1/2} = \left(\int_{-1}^{1} x^2 dx \right)^{1/2}$, we have

$$w_2 = \frac{v_2 - (v_2, w_1)}{\|v_2 - (v_2, w_1) w_1\|} = \frac{x}{\left(\int_{-1}^{1} x^2 dx \right)^{1/2}} = \frac{\sqrt{3}}{\sqrt{2}} x.$$

Similarly,

$$w_3 = \frac{v_3 - (v_3, w_1) w_1 - (v_3, w_2) w_2}{\|v_3 - (v_3, w_1) w_1 - (v_3, w_2) w_2\|}$$

$$= \frac{x^2 - \frac{1}{3}}{\left(\int_{-1}^{1} \left(x^2 - \frac{1}{3}\right)^2 dx \right)^{1/2}}$$

$$= \frac{\sqrt{10}}{4} \left(-1 + 3x^2 \right).$$

EXAMPLE 2.73

Consider the set of vectors $u = (1, 1, 0)$, $v = (1, -1, 1)$ and $w = (-1, 1, 2)$ in V_3. Then

$$(u, v) = 1(1) + 1(-1) + 0(1) = 0,$$

$$(v, w) = 1(-1) + (-1)(1) + 1(2) = 0,$$

$$(w, u) = 1(-1) + 1(1) + (2) = 0.$$

Thus, $\{u, v, w\}$ is an orthogonal set of non-zero vectors. Further,

$$\|u\| = (u, u)^{\frac{1}{2}} = \sqrt{2},$$

$$\|v\| = (v, v)^{\frac{1}{2}} = \sqrt{3},$$

$$\|w\| = (w, w)^{\frac{1}{2}} = \sqrt{6}.$$

Therefore u, v and w are not unit vector and so $\{u, v, w\}$ is not an orthonormal set. But $\left\{ \frac{u}{\sqrt{2}}, \frac{v}{\sqrt{3}}, \frac{w}{\sqrt{6}} \right\}$ is orthonormal.

EXAMPLE 2.74

Let

$$S = \{(1, 0, 0), (1, 1, 1), (1, 2, 3)\}.$$

We normalize S using Gram-Schmidt orthogonalization process. We have

$$w_1 = \frac{v_1}{\|v_1\|} = \frac{(1, 0, 0)}{\sqrt{1^2 + 0^2 + 0^2}} = (1, 0, 0),$$

$$w_2 = \frac{v_2 - (v_2,\ w_1)w_1}{\|v_2 - (v_2\ w_1)w_1\|}$$

$$= \frac{(1,\ 1,\ 1) - 1(1,\ 0,\ 0)}{\|(1,\ 1,\ 1) - 1(1,\ 0,\ 0)\|} = \frac{(0,\ 1,\ 1)}{\|(0,\ 1,\ 1)\|}$$

$$= \frac{(0,\ 1,\ 1)}{0^2+1^2+1^2} = \left(0,\ \frac{1}{\sqrt{2}},\ \frac{1}{\sqrt{2}}\right),$$

$$w_3 = \frac{v_3 - (v_3,\ w_1)w_1 - (v_3,\ w_2)w_2}{\|v_3 - (v_3,\ w_1)w_1 - (v_3,\ w_2)w_2\|}$$

$$= \frac{(1,\ 2,\ 3) - 1(1,\ 0,\ 0) - \frac{5}{\sqrt{2}}\left(0,\ \frac{1}{\sqrt{2}},\ \frac{1}{\sqrt{2}}\right)}{\left\|(1,\ 2,\ 3) - 1(1,\ 0,\ 0) - \frac{5}{\sqrt{2}}\left(0,\ \frac{1}{\sqrt{2}},\ \frac{1}{\sqrt{2}}\right)\right\|}$$

$$= \frac{\left(0,\ -\frac{1}{2},\ \frac{1}{2}\right)}{\frac{1}{\sqrt{2}}} = \left(0,\ -\frac{1}{\sqrt{2}},\frac{1}{\sqrt{2}}\right).$$

We note that

$$(w_1,\ w_2) = 0,\ (w_1,\ w_3) = 0,\ (w_2,\ w_3) = 0 \quad \text{and}$$

$$\|w_1\| = \|w_2\| = \|w_3\| = 1.$$

Hence $\{w_1,\ w_2,\ w_3\}$ is an orthonormal set.

Theorem 2.29. If V is finite dimensional inner product space and if W is a subspace of V, then $V = W + W^{\perp}$. More particularly, V is the direct sum of W and $W^{\perp}$.

Proof. As a subspace of the inner product space V, W is itself an inner product space (its inner product being that of V restricted W). Thus, we can find an orthonormal set $w_1,\ w_2,\ldots,w_r$ in V which is a basis of W. If $v \in V$, by Lemma 2.2, $v_0 = v - (v,\ w_1)w_1 - (v,\ w_2)w_2 - \ldots - (v,\ w_r)w_r$ is orthogonal to each $w_1,\ w_2,\ldots,w_r$ and so is orthogonal to W. Thus $v_0 \in W^{\perp}$. Since $v = v_0 + (v,\ w_1)w_1 + (v,\ w_2)w_2 + \ldots + (v,\ w_r)w_r$, it follows that $v \in W + W^{\perp}$. Therefore, $V = W + W^{\perp}$. Further, since $W \cap W^{\perp} = \{0\}$, it follows that $V = W \oplus W^{\perp}$.

Corollary 2.8. If V is a finite dimensional inner product space and W is a subspace of V, then $(W^{\perp})^{\perp} = W$.

Proof. Let $w \in W$. Then for any $u \in W^{\perp}$, we have $(w,\ u) = 0$ whence $w \in (W^{\perp})^{\perp}$. Hence $W \subset (W^{\perp})^{\perp}$. Also $V = W + W^{\perp}$ and $V = W^{\perp} + (W^{\perp})^{\perp}$. Since the sums are direct, it follows that

$$\dim W = \dim(W^{\perp})^{\perp}.$$

Since $W \subset (W^{\perp})^{\perp}$ and is of the same dimension as $(W^{\perp})^{\perp}$, it follows that $W = (W^{\perp})^{\perp}$.

EXAMPLE 2.75

Compute $d(f,g)$ *for* $f = \cos 2\pi x$ *and* $g = \sin 2\pi x$ in $C[0,1]$ with inner product $(f,g) = \int_0^1 f(x)g(x)dx$.

Solution. We know that

$$d(f,g) = \|f-g\|.$$

But

$$\|f-g\|^2 = (f-g,\ f-g)$$
$$= (f,\ f) + (g,\ g) - (f,\ g) - (g,\ f)$$
$$= \int_0^1 \cos^2 2\pi x dx + \int_0^1 \sin^2 2\pi x dx$$
$$- \int_0^1 \sin 2\pi x \cos 2\pi x dx$$
$$- \int_0^1 \cos 2\pi x \sin 2\pi x dx$$
$$= \frac{1}{2}\int_0^1 (1+\cos 4\pi x)dx$$
$$+ \int_0^1 (1-\cos 4\pi x)dx$$
$$-2\int_0^1 \frac{1}{2}(\sin 4\pi x)dx$$

$$= \frac{1}{2}\left[x + \frac{\sin 4\pi x}{4\pi}\right]_0^1$$

$$= +\frac{1}{2}\left[x + \frac{\sin 4\pi x}{4\pi}\right]_0^1$$

$$-\left[-\frac{\cos 4\pi x}{4\pi}\right]_0^1$$

$$= \frac{1}{2} + \frac{1}{2} + \frac{1}{4\pi} - \frac{1}{4\pi}$$

$$= 1.$$

Hence

$$d(f, g) = \|f - g\| = \sqrt{1} = 1.$$

EXAMPLE 2.76

Find a basis for the orthogonal complement of the subspace of $\Re^3$ spanned by the vectors $v_1 = (1, -1, 3), v_2 = (5, -4, -4)$ and $v_3 = (7, -6, 2)$.

Solution. Let

$$W = \text{Span} = \{(1, -1, 3), (5, -4, -4), (7, -6, 2)\}.$$

Then

Span $(W^{\perp}) = \{(a, b, c) : (a, b, c)$ is orthogonal to every vector in $W\}$

Now, if (a, b, c) is orthogonal to every vector in W, we have

$$a - b + 3c = 0$$

$$5a - 4b - 4c = 0$$

$$7a - 6b + 2c = 0.$$

Solving these equations, we get $a = b = c = 0$. Thus

$$\text{span } W^{\perp} = \{0, 0, 0\}.$$

But $(0, 0, 0) = 0(1, 0, 0) + 0(0, 1, 0) + 0(0, 0, 1)$. Hence, the basis of $W^{\perp}$ is $\{(1, 0, 0), (0, 1, 0), (0, 0, 1)\}$.

EXAMPLE 2.77

Let $W = \text{span}\left\{\left(\frac{4}{5}, 0, \frac{-3}{5}\right), (0, 1, 0)\right\}$. Express $w = (1, 2, 3)$ in the form of $w = w_1 + w_2$, where $w_1 \in W$ and $w_2 \in W^{\perp}$.

Solution. We have

$$W = span\left\{\left(\frac{4}{5}, 0, -\frac{3}{5}\right), (0, 1, 0)\right\}.$$

Let (a, b, c) be or orthogonal to $\left(\frac{4}{5}, 0, \frac{-3}{5}\right)$ and $(0, 1, 0)$. Then

$$\frac{4}{5}a - \frac{3}{5}c = 0, \; b = 0$$

or

$$c = \frac{4}{5}a, \; b = 0.$$

Thus $(a, b, c) = \left(a, 0, \frac{4a}{3}\right) = a\left(1, 0\frac{4}{3}\right)$ and so

$$W^{\perp} = span\left\{1, 0, \frac{4}{3}\right\}.$$

Let

$$(1, 2, 3) = \alpha\left(\frac{4}{5}, 0 - \frac{3}{5}\right) + \beta(0, 1, 0) + \lambda\left(1, 0\frac{4}{3}\right)$$

This yields $\alpha = -1$, $\beta = 2$ and $\lambda = \frac{9}{5}$. Now $(-1)\left(\frac{4}{5}, 0, -\frac{3}{5}\right) + 2(0, 1, 0) \in W$, that is, $\left(-\frac{4}{5}, 2, \frac{3}{5}\right) \in W$. Also $\frac{9}{5}\left(1, 0, \frac{4}{3}\right) \in W^{\perp}$. Then $(1, 2, 3) = \left(-\frac{4}{5}, 2, \frac{3}{5}\right) + \left(\frac{9}{5}, 0, \frac{12}{5}\right) = w_1 + w_2$,4 where $w_1 \in W$ and $w_2 \in W^{\perp}$.

EXAMPLE 2.78

Let $\Re^3$ have the Euclidean inner product. Transform the basis

$$S = \{(1, 0, 0), (3, 7, -2), (0, 4, 1)\}$$

in to an orthonormal basis using the Gram-Schmidt Process.

Solution. We have
$S = \{(1, 0, 0), (3, 7, -2), (0, 4, 1)\}$.
Let
$v_1 = (1, 0, 0)$, $v_2 = (3, 7, -2)$, $v_3 = (0, 4, 1)$.
Then

$$w_1 = \frac{v_1}{\|v_1\|} = \frac{(1,0,0)}{1} = (1,0,0),$$

$$\begin{aligned} w_2 &= \frac{v_2 - (v_2, w_1)w_1}{\|v_2 - (v_2, w_1)w_1\|} \\ &= \frac{(3,7,-2) - 3(1,0,0)}{\|(3,7,-2) - 3(1,0,0)\|} \\ &= \frac{(0,7,-2)}{\|(0,7,-2)\|} = \frac{(0,7,-2)}{\sqrt{53}}, \end{aligned}$$

$$\begin{aligned} w_3 &= \frac{v_3 - (v_3, w_1)w_1 - (v_3, w_2)w_2}{\|v_3 - (v_3, w_1)w_1 - (v_3, w_2)w_2\|} \\ &= \frac{(0,4,1) - 0 - \frac{26}{53}(0,7,-2)}{(0,4,1) - \frac{26}{53}(0,7,-2)} \\ &= \frac{(0, \frac{30}{53}, \frac{1}{53})}{\|(0, \frac{30}{53}, \frac{1}{53})\|} = \frac{53\,(0, \frac{30}{53}, \frac{1}{53})}{\sqrt{901}} \\ &= \frac{(0, 30, 1)}{\sqrt{901}}. \end{aligned}$$

Hence the orthonormal basis is

$$\{w_1, w_2, w_3\} = \left\{(1,0,0), \frac{(0,7,-2)}{\sqrt{53}}, \frac{(0,30,1)}{\sqrt{901}}\right\}.$$

EXAMPLE 2.79

For $U = \begin{bmatrix} u_1 & u_2 \\ u_3 & u_4 \end{bmatrix}$ and $V = \begin{bmatrix} v_1 & v_2 \\ v_3 & v_4 \end{bmatrix}$ in M_{22}, define inner product by

$$(U;V) = u_1v_1 + u_2v_2 + u_3v_3 + u_4v_4.$$

For the matrices $A = \begin{bmatrix} 2 & 6 \\ 1 & -6 \end{bmatrix}, B = \begin{bmatrix} 3 & 2 \\ 1 & 0 \end{bmatrix}$, verify Cauchy-Schwarz inequality and find the cosine of the angle between them.

Solution. We have

$$U = \begin{bmatrix} u_1 & u_2 \\ u_3 & u_4, \end{bmatrix} V = \begin{bmatrix} v_1 & v_2 \\ v_3 & v_4 \end{bmatrix}$$

$$(U, V) = u_1v_1 + u_2v_2 + u_3v_4 + u_4v_4.$$

Under this inner product

$$\begin{aligned} (A, B) &= 6 + 12 + 1 = 19 \\ \|A\| &= (A, A)^{\frac{1}{2}} = \sqrt{9+36+1+9} = \sqrt{50}, \\ \|B\| &= \sqrt{9+4+1} = \sqrt{14}. \end{aligned}$$

We note that

$$19 = |(A, B)|\sqrt{50}\sqrt{14}.$$

Hence Cauchy-Schwartz inequality is satisfied. If θ is the angle between A and B, then

$$\cos\theta = \frac{(A, B)}{\|A\| \ \|B\|} = \frac{19}{\sqrt{50}\sqrt{14}}$$

or

$$\theta = \cos^{-1}\left(\frac{19}{10\sqrt{7}}\right).$$

EXAMPLE 2.80

For the matrix $A = \begin{bmatrix} \frac{1+i}{2} & \frac{1+i}{2} \\ \frac{1-i}{2} & \frac{1-i}{2} \end{bmatrix}$, show that the row vectors form an orthonormal set in C^2. Also, find A^{-1}.

Solution. The give matrix is

$$A = \begin{bmatrix} \frac{1+i}{2} & \frac{1+i}{2} \\ \frac{1-i}{2} & \frac{-1+i}{2} \end{bmatrix}.$$

The row vectors are

$$\begin{aligned} v_1 &= [\ \tfrac{1+i}{2},\ \tfrac{1+i}{2}\ ] \text{ and} \\ v_2 &= (\ \tfrac{1-i}{2}, \tfrac{-1+i}{2}\ ) \end{aligned}$$

We note that

$$\begin{aligned} (v_1, v_2) &= \left(\frac{1+i}{2}\right)\left(\frac{1-i}{2}\right) + \left(\frac{1+i}{2}\right)\left(\frac{i-1}{2}\right) \\ &= \frac{1-i^2}{4} + \frac{i^2-1}{4} = 0 \end{aligned}$$

Thus the set $\{v_1, v_2\}$ is orthogonal. Further,

$$\|v_1\| = (v_1, v_1)^{\frac{1}{2}} = \left[\left|\frac{1+i}{2}\right|^2 + \left|\frac{1+i}{2}\right|^2\right]^{\frac{1}{2}} = 1$$

$$\|v_1\| = (v_2, v_2)^{\frac{1}{2}} = \left[|1-i|^{\frac{1}{2}} + |i-1|^2\right]^{\frac{1}{2}} = 1.$$

Hence $\{v_1, v_2\}$ is orthonormal.
Further,

$$|A| = \frac{i^2-1}{4} - \frac{1-i^2}{4} = \frac{-2}{4} - \frac{2}{4} = -1.$$

Therefore

$$A^{-1} = \frac{1}{|A|} adj\, A = \frac{1}{-1}\begin{bmatrix} \frac{-1+i}{2} & \frac{-1-i}{2} \\ \frac{i-1}{2} & \frac{i+1}{2} \end{bmatrix}$$

$$= \begin{bmatrix} \frac{1-i}{2} & \frac{1+i}{2} \\ \frac{1-i}{2} & \frac{-i-1}{2} \end{bmatrix}.$$

2.11 LEAST SQUARE LINE APPROXIMATION

Suppose that we have an empirical data in the form of n pairs of values (x_1, y_1), $(x_2, y_2), \ldots, (x_n, y_n)$, where the experimental errors are associated with the functional values $y_1, y_2, \ldots, y_n$ only. Then we seek a linear function

$$y = f(x) = a + bx \tag{37}$$

fitting the given points as well as possible. The equation (30) will not in general be satisfied by any of the n pairs. Substituting in (37) each of the n pairs of values in turn, we get

$$\left.\begin{aligned} e_1 &= y_1 - a - bx_1 \\ e_2 &= y_2 - a - bx_2 \\ &\cdots\cdots\cdots\cdots\cdots\cdots \\ &\cdots\cdots\cdots\cdots\cdots\cdots \\ e_n &= y_n - a - bx_n \end{aligned}\right\}, \tag{38}$$

where e_k, $k = 1, \ldots, n$ are measurement errors, called *residuals* or *deviations*. To know how far the curve $y = f(x)$ lies from the given data, the following errors are considered:

(i) Maximum error

$$e_\infty(f) = \max_{1\le k\le n} \{|y_k - a - bx_k|\}$$

(ii) Average error

$$e_A(f) = \frac{1}{n}\sum_{k=1}^{n} |y_k - a - bx_k|$$

(iii) Root mean square (RMS) error

$$e_{rms}(f) = \left[\frac{e_1^2 + \ldots + e_n^2}{n}\right]^{1/2}.$$

The least square line $y = f(x) = a + bx$ is the line that minimize the root mean square error $e_{rms}(f)$. But the quantity $e_{rms}(f)$ is minimum if and only if $\sum_{k=1}^{n} (y_k - a - bx_k)^2 = \sum_{k=1}^{n} e_k^2$ is minimum. Thus, in case of least-square line we are looking for a linear function $a + bx$ as an approximation to a function $y = f(x)$ when we are given the value of y at the points $x_1, \ldots, x_n$. We aim at minimizing the sum of the squared errors

$$e(a, b) = \sum_{i=1}^{n} (y_i - a - bx_i)^2. \tag{39}$$

Geometrically, if d_i is the vertical distance from the data point (x_i, y_i) to the point $(x_i, a + bx_i)$ on the line, then $d_i = y_i - a - bx_i$ (see Figure 2.1). We must minimize the sum of the squares of the vertical distances d_i, that is, the sum $\sum_{i=1}^{n} d_i^2$.

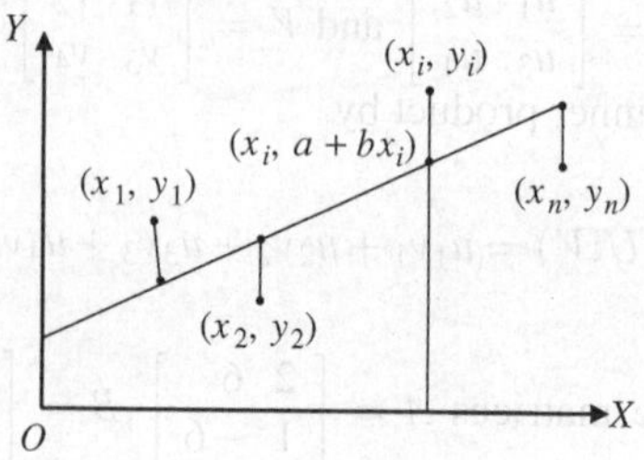

Figure 2.1.

To minimize $e(a,b)$, we equate to zero the partial derivative of (39) with respect to 'a' and with respect to 'b'. Thus

$$\frac{\partial e(a,b)}{\partial a} = \sum_{i=1}^{n} 2(y_i - a - bx_i) = 0$$

and

$$\frac{\partial e(a,b)}{\partial b} = \sum_{i=1}^{n} 2x_i(y_i - a - bx_i) = 0,$$

which are known as *normal equations*. We write these equations in the form

$$na + b\sum_{i=1}^{n} x_i = \sum_{i=1}^{n} y_i \tag{40}$$

and

$$a\sum_{i=1}^{n} x_i + b\sum_{i=1}^{n} x_i^2 = \sum_{i=1}^{n} x_i y_i. \tag{41}$$

The normal equations (40) and (41) can the solved for 'a' and 'b' using Cramer's rule or by some other method.

Using properties of adjoint of an operator (not in the course) it can be shown that matrix form of the normal equations is

$$A^T AX = A^T Y, \tag{42}$$

where

$$A = \begin{bmatrix} 1 & x_1 \\ 1 & x_2 \\ \vdots & \vdots \\ 1 & x_n \end{bmatrix}, X = \begin{bmatrix} a \\ b \end{bmatrix}, Y = \begin{bmatrix} y_1 \\ y_2 \\ \vdots \\ y_n \end{bmatrix}.$$

If rank $(A) = n$, then $A^T A$ is non-singular and so invertible. Therefore Pre-multiplying (42) by $(A^T A)^{-1}$, we get

$$X = (A^T A)^{-1} A^T Y.$$

Hence X is the least square solution.

EXAMPLE 2.81

Fit a straight line $y = a + bx$ to the following data by the method of least square:

$$\begin{matrix} x: & 0 & 1 & 3 & 6 & 8 \\ y: & 1 & 3 & 2 & 5 & 4 \end{matrix}$$

Solution. We have (as per our notations)

$$A = \begin{bmatrix} 1 & 0 \\ 1 & 1 \\ 1 & 3 \\ 1 & 6 \\ 1 & 8 \end{bmatrix}, X = \begin{bmatrix} a \\ b \end{bmatrix}, Y = \begin{bmatrix} 1 \\ 3 \\ 2 \\ 5 \\ 4 \end{bmatrix}.$$

Therefore,

$$A^T A = \begin{bmatrix} 1 & 1 & 1 & 1 & 1 \\ 0 & 1 & 3 & 6 & 8 \end{bmatrix} \begin{bmatrix} 1 & 0 \\ 1 & 1 \\ 1 & 3 \\ 1 & 6 \\ 1 & 8 \end{bmatrix}$$

$$= \begin{bmatrix} 5 & 18 \\ 18 & 110 \end{bmatrix}$$

and

$$(A^T A)^{-1} = \frac{1}{226} \begin{bmatrix} 110 & -18 \\ -18 & 5 \end{bmatrix}.$$

Hence,

$$X = (A^T A)^{-1} A^T Y$$

$$= \frac{1}{226} \begin{bmatrix} 110 & -18 \\ -18 & 5 \end{bmatrix} \begin{bmatrix} 1 & 1 & 1 & 1 & 1 \\ 0 & 1 & 3 & 6 & 8 \end{bmatrix} \begin{bmatrix} 1 \\ 3 \\ 2 \\ 5 \\ 4 \end{bmatrix}$$

$$= \frac{1}{226} \begin{bmatrix} 110 & -18 \\ -18 & 5 \end{bmatrix} \begin{bmatrix} 15 \\ 71 \end{bmatrix}$$

$$= \frac{1}{226} \begin{bmatrix} 372 \\ 85 \end{bmatrix} = \begin{bmatrix} 1.64 \\ 0.37 \end{bmatrix}.$$

Thus,

$$X = \begin{bmatrix} a \\ b \end{bmatrix} = \begin{bmatrix} 1.64 \\ 0.37 \end{bmatrix}$$

and so $a = 1.64$ and $b = 0.37$. Hence the line of best fit is

$$y = 1.64 + 0.37x.$$

2.12 MINIMAL SOLUTION TO A SYSTEM OF EQUATIONS

Let $AX = B$ be a system of equations represented in matrix form. Even if the system is consistent, it may not have a unique solution. In such cases, it is better to find a solution of minimal norm. A solution S is called a minimal Solution if $\|S\| \leq \|U\|$ for all other solutions U of the system. Using Operator Theory, it can be shown that if V is a solution of $(AA^*)V = B$, then $S = A^*V$ is the minimal solution to the system $AX = B$.

EXAMPLE 2.82

Find the minimal solution of the following system of equations:

$$x + y + z = 1$$
$$x + 2y + 4z = 2$$
$$x + 4y + 10z = 4.$$

Solution. The matrix form of the given system is $AX = B$, where

$$A = \begin{bmatrix} 1 & 1 & 1 \\ 1 & 2 & 4 \\ 1 & 4 & 10 \end{bmatrix}, X = \begin{bmatrix} x \\ y \\ z \end{bmatrix} \text{ and } B = \begin{bmatrix} 1 \\ 2 \\ 4 \end{bmatrix}.$$

This system (See Example 2.126) has infinite number of solutions. We have

$$A^* = \begin{bmatrix} 1 & 1 & 1 \\ 1 & 2 & 4 \\ 1 & 4 & 10 \end{bmatrix} \text{ and } AA^* = \begin{bmatrix} 3 & 7 & 15 \\ 7 & 21 & 49 \\ 15 & 49 & 117 \end{bmatrix}.$$

Then, consider $(AA^*)V = B$, that is, the system

$$3v_1 + 7v_2 + 15v_3 = 1$$
$$7v_1 + 21v_2 + 49v_3 = 2$$
$$15v_1 + 49v_2 + 117v_3 = 4.$$

We note that $V = \begin{bmatrix} \frac{1}{2} \\ \frac{-1}{14} \\ 0 \end{bmatrix}$ is a solution of this system. Thus, the minimal solution of the given system is $X = A^*V = \begin{bmatrix} \frac{3}{7} \\ \frac{5}{14} \\ \frac{3}{14} \end{bmatrix}$.

2.13 MATRICES

Definition 2.43. A rectangular array of mn real or complex numbers, arranged in m rows and n columns, is called an $m \times n$ matrix.

An $m \times n$ matrix A is represented by the symbol

$$A = \begin{bmatrix} a_{11} & a_{12} & .. & .. & a_{1n} \\ a_{21} & a_{22} & .. & .. & a_{2n} \\ .. & .. & .. & .. & .. \\ .. & .. & .. & .. & .. \\ a_{m1} & a_{m2} & .. & .. & a_{mn} \end{bmatrix} \text{ or}$$

$$A = \begin{pmatrix} a_{11} & a_{12} & .. & .. & a_{1n} \\ a_{21} & a_{22} & .. & .. & a_{2n} \\ .. & .. & .. & .. & .. \\ .. & .. & .. & .. & .. \\ a_{m1} & a_{m2} & .. & .. & a_{mn} \end{pmatrix},$$

where a_{ij} denotes the element in the *ith* row and *jth* column of the matrix. Each of the mn number constituting a $m \times n$ matrix is called *entry* (or *element*) of the matrix A. We generally abbreviate the symbol of the matrix A by A $= [a_{ij}]_{m\times n}$ or simply by $[a_{ij}]$. Further, if a matrix A $= [a_{ij}]$ has m rows and n columns, then it is said to be of order $m \times n$.

Definition 2.44. Two matrices A $= [a_{ij}]_{m\times n}$ and B $= [b_{ij}]_{m\times n}$ over a field F($\mathbb{R}$ or $\mathbb{C}$) are said to be *equal* if

(i) they are both of the same type, that is, have the same number of rows and columns.

(ii) the elements in the corresponding places of the two matrices are equal, that is $a_{ij} = b_{ij}$ for all pairs of i, j.

We observe that the relation of equality of two matrices is an equivalence relation. In fact,

(i) If A is any matrix, A = A (Reflexivity)

(ii) If A = B, then B = A (Symmetry)

(iii) If A = B and B = C, then A = C (Transitivity).

Definition 2.45. The elements a_{ii} of a matrix $A = [a_{ij}]$ are called the *diagonal elements* of A.

Definition 2.46. A matrix in which the number of rows is equal to the number of columns is called a *square matrix*.

If A is a square matrix having n rows and n columns then it is also called a *matrix of order n*.

Definition 2.47. If the matrix A is of order n, the elements $a_{11}, a_{22}, \ldots, a_{nn}$ are said to constitute the *main diagonal* of A and the elements $a_{n1}, a_{n-12}, \ldots, a_{1n}$ constitute its *secondary diagonal*.

Definition 2.48. A square matrix A = $[a_{ij}]$ is said to be a *diagonal matrix* if each of its non-diagonal element is zero, that is, if $a_{ij} = 0$ whenever $i \neq j$.

A diagonal matrix whose diagonal elements, in order, are $d_1, d_2, \ldots, d_n$ is denoted by Diag $[d_1, d_2, \ldots, d_n]$ or Diag $[a_{11}, a_{22}, \ldots, a_{nn}]$ if A = $[a_{ij}]$.

Definition 2.49. A diagonal matrix, whose diagonal elements are all equal is called a *scalar matrix*.

For example, the matrix

$$\begin{bmatrix} 2 & 0 & 0 \\ 0 & 2 & 0 \\ 0 & 0 & 2 \end{bmatrix}$$

is a scalar matrix of order 3.

Definition 2.50. A scalar matrix of order n, each of whose diagonal element is equal to 1 is called a *unit matrix* or *identity matrix* of order n and is denoted by I_n.

For example, the matrix

$$\begin{bmatrix} 1 & 0 & 0 & 0 \\ 0 & 1 & 0 & 0 \\ 0 & 0 & 1 & 0 \\ 0 & 0 & 0 & 1 \end{bmatrix}$$

is a unit matrix of order 4.

Definition 2.51. A matrix, rectangular or square, each of whose entry is zero is called a *zero matrix* or a *null matrix* and is denoted by **0**.

Definition 2.52. A matrix having 1 row and n column is called a *row matrix* (or a *row vector*). For example, the matrix

$$[2 \; 3 \; 5 \; 6 \; 2]$$

is a row matrix.

Definition 2.53. If a matrix has m rows and 1 column, it is called a *column matrix* (or a *column vector*).

For example, the matrix

$$\begin{bmatrix} 2 \\ 1 \\ 0 \\ -3 \end{bmatrix}$$

is a column matrix.

Definition 2.54. A *submatrix* of a given matrix A is defined to be either A or any array obtained on deleting some rows or columns or both of the matrix A.

Definition 2.55. A square submatrix of a square matrix A is called a *principal submatrix* if its diagonal elements are also the principal diagonal elements of the matrix A.

Thus to obtain principal submatrix, it is necessary to delete corresponding rows and columns. For example, the matrix

$$\begin{bmatrix} 3 & 1 \\ 4 & 3 \end{bmatrix}$$

is a principal submatrix of the matrix

$$\begin{bmatrix} 1 & 2 & 3 & 4 \\ 2 & 3 & 1 & 0 \\ 6 & 4 & 3 & 2 \\ 1 & 2 & 4 & 1 \end{bmatrix}.$$

Definition 2.56. A principal square submatrix is called *leading submatrix* if it is obtained by deleting only some of the last rows and the corresponding columns. For example,

$$\begin{bmatrix} 1 & 2 \\ 2 & 3 \end{bmatrix}$$

is the *leading principal submatrix* of the matrix

$$\begin{bmatrix} 1 & 2 & 3 & 4 \\ 2 & 3 & 1 & 0 \\ 6 & 4 & 3 & 2 \\ 1 & 2 & 4 & 1 \end{bmatrix}.$$

2.14 ALGEBRA OF MATRICES

Matrices allow the following basic operations:

(a) Multiplication of a matrix by a scalar.

(b) Addition and subtraction of two matrices.

(c) Product of two matrices.

However, the concept of dividing a matrix by another matrix is undefined.

Definition 2.57. Let α be a scalar (real or complex) and A be a given matrix. Then the *multiplication* of A = $[a_{ij}]$ by the scalar α is defined by

$$\alpha A = \alpha[a_{ij}] = [\alpha\, a_{ij}],$$

that is, each element of A is multiplied by the scalar α. The order of the matrix so obtained will be the same as that of the given matrix A.

For example

$$4\begin{bmatrix} 3 & 1 & 2 \\ 2 & 1 & 0 \end{bmatrix} = \begin{bmatrix} 12 & 4 & 8 \\ 8 & 4 & 0 \end{bmatrix}.$$

Definition 2.58. Two matrices A = $[a_{ij}]$ and B = $[b_{ij}]$ are said to be *comparable (conformable)* for addition/subtraction if they are of the same order.

Definition 2.59. Let A and B be two matrices of the same order, say $m \times n$. Then, the *sum of the matrices A* and *B* is defined by

$$C = [c_{ij}] = A + B = [a_{ij}] + [b_{ij}] = [a_{ij} + b_{ij}].$$

Thus,

$$c_{ij} = a_{ij} + b_{ij}, 1 \le i \le m, 1 \le j \le n,$$

The order of the new matrix C is same as that of A and B. Similarly,

$$C = A - B = [a_{ij}] - [b_{ij}] = [a_{ij} - b_{ij}]$$

Thus,

$$c_{ij} = a_{ij} - b_{ij} \quad for \quad 1 \le i \le m, \quad 1 \le j \le n$$

Definition 2.60. If $A_1, A_2, \ldots, A_n$ are n matrices which are conformable for addition and $\lambda_1, \lambda_2, \ldots \lambda_n$ are scalars, then $\lambda_1 A_1 + \lambda_2 A_2 + \ldots + \lambda_n A_n$ is called a *linear combination* of the matrices $A_1, A_2, \ldots, A_n$.

Let $A = [a_{ij}], B = [b_{ij}], C = [c_{ij}]$, be $m \times n$ matrices with entries from the complex numbers. Then the following properties hold:

(a) $A + B = B + A$ (Commutative law for addition)

(b) $(A + B) + C = A + (B + C)$ (Associative law for addition)

(c) $A + \mathbf{0} = \mathbf{0} + A = A$ (Existence of additive identity)

(d) $A + (-A) = (-A) + A = \mathbf{0}$ (Existence of inverse)

Thus the *set of matrices form an additive commutative group.*

2.15 MULTIPLICATION OF MATRICES

Definition 2.61. Two matrices $A = [a_{ij}]_{m\times n}$ and $B = [b_{ij}]_{p\times q}$ are said to *comparable* or *conformable* for the product AB if $n = p$, that is, if the number of columns in A is equal to the number of rows in B.

Definition 2.62. Let $A = [a_{ij}]_{m\times n}$ and $B = [b_{ij}]_{p\times q}$ be two matrices. Then, the product AB is the matrix $C = [c_{ij}]_{m\times q}$ such that

$$\begin{aligned} c_{ij} &= a_{i1}b_{1j} + a_{i2}b_{2j} + \ldots + a_{1n}b_{nj} \\ &= \sum_{k=1}^{n} a_{ik}b_{kj} \text{ for } 1 \le i \le m, 1 \le j \le n. \end{aligned}$$

Note that the c_{ij} [the (i, j)th element of AB] has been obtained by multiplying the ith row of A, namely $(a_{i1}, a_{i2}, \ldots, a_{in})$ with the jth column of B, namely

$$\begin{bmatrix} b_{1j} \\ b_{2j} \\ \cdots \\ \cdots \\ b_{nj} \end{bmatrix}$$

Remark 2.3 In the product AB, the matrix A is called *prefactor* and B is called *postfactor*.

EXAMPLE 2.83

Construct an example to show that product of two non-zero matrices may be a zero matrix.

Solution. Let

$$A = \begin{bmatrix} x & 0 \\ y & 0 \end{bmatrix}, \quad B = \begin{bmatrix} 0 & 0 \\ a & b \end{bmatrix}.$$

Then A and B are both 2×2 matrices. Hence, they are conformable for product. Now,

$$\begin{aligned} \mathrm{AB} &= \begin{bmatrix} x & 0 \\ y & 0 \end{bmatrix}\begin{bmatrix} 0 & 0 \\ a & b \end{bmatrix} \\ &= \begin{bmatrix} 0+0 & 0+0 \\ 0+0 & 0+0 \end{bmatrix} = \begin{bmatrix} 0 & 0 \\ 0 & 0 \end{bmatrix}. \end{aligned}$$

Definition 2.63. When a product AB = **0** such that neither A nor B is **0** then the factors A and B are called *divisors of zero*.

The above example shows that in the algebra of matrices, there exist divisors of zero, whereas in the algebra of complex numbers, there is no zero divisor.

EXAMPLE 2.84

Taking

$$A = \begin{bmatrix} 1 & 3 & 0 \\ -1 & 2 & 1 \\ 0 & 0 & 2 \end{bmatrix}, \quad B = \begin{bmatrix} 2 & 3 & 4 \\ 1 & 2 & 3 \\ -1 & 1 & 2 \end{bmatrix},$$

show that matrix multiplication is not, in general, commutative.

Solution. Both A and B are 3×3 matrices. Therefore, both AB and BA are defined. We have,

$$\begin{aligned} AB &= \begin{bmatrix} 1 & 3 & 0 \\ -1 & 2 & 1 \\ 0 & 0 & 2 \end{bmatrix}\begin{bmatrix} 2 & 3 & 4 \\ 1 & 2 & 3 \\ -1 & 1 & 2 \end{bmatrix} \\ &= \begin{bmatrix} 5 & 9 & 13 \\ -1 & 2 & 4 \\ -2 & 2 & 4 \end{bmatrix} \end{aligned}$$

and

$$\begin{aligned} BA &= \begin{bmatrix} 2 & 3 & 4 \\ 1 & 2 & 3 \\ -1 & 1 & 2 \end{bmatrix}\begin{bmatrix} 1 & 3 & 0 \\ -1 & 2 & 1 \\ 0 & 0 & 2 \end{bmatrix} \\ &= \begin{bmatrix} -1 & 12 & 11 \\ -1 & 7 & 8 \\ -2 & -1 & 5 \end{bmatrix}. \end{aligned}$$

Hence $AB \neq BA$.

EXAMPLE 2.85

Give an example to show that cancellation law does not hold, in general, in matrix multiplication.

Solution. Let

$$A=\begin{bmatrix}0&4\\0&5\end{bmatrix},\quad B=\begin{bmatrix}5&4\\0&0\end{bmatrix},$$
$$C=\begin{bmatrix}1&2\\0&0\end{bmatrix}.$$

Then A and B are conformable for multiplication. Similarly, A and C are also conformable for multiplication. Thus,

$$AB=\begin{bmatrix}0&4\\0&5\end{bmatrix}\begin{bmatrix}5&4\\0&0\end{bmatrix}=\begin{bmatrix}0&0\\0&0\end{bmatrix}$$

and

$$AC=\begin{bmatrix}0&4\\0&5\end{bmatrix}\begin{bmatrix}1&2\\0&0\end{bmatrix}=\begin{bmatrix}0&0\\0&0\end{bmatrix}.$$

Hence $AB=AC$, $A\neq \mathbf{0}$ without having $B=C$ and so we cannot ordinarily cancel A from $AB=AC$ even if $A\neq \mathbf{0}$.

Remark 2.4 The above examples show that in matrix algebra

(a) The commutative law $AB=BA$ does not hold true.

(b) There exist divisors of zero, that is, there exist matrices A and B such that $AB=\mathbf{0}$ but neither A nor B is zero.

(c) The cancellation law does not hold in general, that is, $AB=AC$, $A\neq 0$ does not imply in general that $B=C$.

2.16 ASSOCIATIVE LAW FOR MATRIX MULTIPLICATION

If $A=[a_{ij}]_{m\times n}$ $B=[b_{jk}]_{n\times p}$, and $C=[c_{kl}]_{p\times q}$ are three matrices with entries from the set of complex numbers, then

$(AB)C=A(BC)$.

(Associative Law for Matrix Multiplication).

2.17 DISTRIBUTIVE LAW FOR MATRIX MULTIPLICATION

If $A=[a_{ij}]_{m\times n}, B=[b_{jk}]_{n\times p}$, and $C=[c_{kl}]_{n\times p}$ are three matrices with elements from the set of complex numbers, then

$A(B+C)=AB+AC$

(Distributive Law for Matrix Multiplication).

Definition 2.64. The matrices A and B are said to be *anticommutative* or *anticommute* if $AB=-BA$.

For example, each of the Pauli Spin matrices (used in the study of electron spin in quantum mechanics)

$$\sigma_x=\begin{bmatrix}0&1\\1&0\end{bmatrix},\ \sigma_y=\begin{bmatrix}0&-i\\i&0\end{bmatrix},\ \sigma_z=\begin{bmatrix}1&0\\0&-1\end{bmatrix},$$

where $i^2=-1$ anticommute with the others. In fact,

$$\sigma_x\sigma_y=\begin{bmatrix}0&1\\1&0\end{bmatrix}\begin{bmatrix}0&-i\\i&0\end{bmatrix}=\begin{bmatrix}i&0\\0&-i\end{bmatrix},$$
$$\sigma_y\sigma_x=\begin{bmatrix}0&-i\\i&0\end{bmatrix}\begin{bmatrix}0&1\\1&0\end{bmatrix}=\begin{bmatrix}-i&0\\0&i\end{bmatrix}$$

and so $\sigma_x\sigma_y=-\sigma_y\sigma_x$.

Definition 2.65. If A and B are matrices of order n, then the matrix $AB-BA$ is called the *commutator* of A and B.

Definition 2.66. The sum of the main diagonal elements a_{ii}, $i=1,2,\ldots,n$ of a square matrix A is called the *trace* or *spur* of A.

Thus,

$$tr\,A=a_{11}+a_{22}+\ldots+a_{nn}.$$

Theorem 2.30. Let A and B be square matrices of order n and λ be a scalar. Then

(a) $tr\,(\lambda A)=\lambda\, tr\, A$,

(b) $tr\,(A+B)=tr\,A+tr\,B$,

(c) $tr\,(AB)=tr\,(BA)$.

Proof. Let

$$A=[a_{ij}]_{n\times n}\text{ and }B=[b_{ij}]_{n\times n}.$$

(a) We have

$$\lambda A=[\lambda\, a_{ij}]_{n\times n}$$

and so

$$tr\ (\lambda A) = \sum_{I=1}^{n} \lambda a_{ii} = \lambda \sum_{i=1}^{n} a_{ii} = \lambda\ tr\ A.$$

(b) We have

$$A + B = [a_{ij} + b_{ij}]_{n\times n}$$

and so

$$\begin{aligned} tr\ (A+B) &= \sum_{i=1}^{n} [a_{ii} + b_{ii}] \\ &= \sum_{i=1}^{n} a_{ii} + \sum_{i=1}^{n} b_{ii} \\ &= tr\ A + tr\ B. \end{aligned}$$

(c) We have

$$AB = [c_{ij}]_{n\times n}$$

where

$$c_{ij} = \sum_{k=1}^{n} a_{ik}\, b_{kj},$$

and

$$BA = [d_{ij}]_{n\times n},$$

where

$$d_{ij} = \sum_{k=1}^{n} b_{ik} a_{kj}.$$

Then

$$\begin{aligned} tr\ (AB) &= \sum_{i=1}^{n} c_{ii} = \sum_{i=1}^{n} \left(\sum_{k=1}^{n} a_{ik}\, b_{ki} \right) \\ &= \sum_{k=1}^{n} \left(\sum_{i=1}^{n} a_{ik}\, b_{ki} \right) = \sum_{k=1}^{n} \left(\sum_{i=1}^{n} b_{ki}\, a_{ik} \right) \\ &= \sum_{k=1}^{n} d_{kk} = tr\ (BA). \end{aligned}$$

EXAMPLE 2.86

If A and B are matrices of the same order say n, show that the relation $AB - BA = I_n$ does not hold good.

Solution. Suppose on the contrary that the relation $AB - BA = I_n$ holds true. Since A and B are of same order, AB and BA are also of order n. Therefore,

$$\begin{aligned} tr(AB - BA) &= tr\ I_n \\ \Rightarrow tr\ AB - tr\ BA &= tr\ I_n. \end{aligned}$$

Since $tr\ AB = tr\ BA$, we have

$0 = tr I_n = 1 + 1 + 1 \ldots + 1 = n,$

which is absurd. Hence, the given relation does not hold good.

Definition 2.67. An $n \times n$ matrix A is said to be *nilpotent* if $A^n = 0$ for some positive integer n.

The smallest positive integer n, for which $A^n = 0$, is called the *degree of nilpotence* of A.

For example, the matrix

$$\begin{bmatrix} 0 & 1 & 2 & -1 \\ 0 & 0 & 1 & 2 \\ 0 & 0 & 0 & 1 \\ 0 & 0 & 0 & 0 \end{bmatrix}$$

is nilpotent and the degree of nilpotence is 4. Similarly, the matrix

$$A = \begin{bmatrix} 6 & 9 \\ -4 & -6 \end{bmatrix}$$

is nilpotent with degree of nilpotence 2. In fact,

$$\begin{aligned} A^2 &= \begin{bmatrix} 6 & 9 \\ -4 & -6 \end{bmatrix} \begin{bmatrix} 6 & 9 \\ -4 & -6 \end{bmatrix} \\ &= \begin{bmatrix} 36-36 & 54-54 \\ -24+24 & -36+36 \end{bmatrix} = \begin{bmatrix} 0 & 0 \\ 0 & 0 \end{bmatrix}. \end{aligned}$$

It can be shown that every 2×2 nilpotent matrix A such that $A^2 = 0$ may be written in the form

$$\begin{bmatrix} \lambda\mu & \mu^2 \\ -\lambda^2 & -\lambda\mu \end{bmatrix},$$

where λ, μ are scalars. If A is real then λ, μ are also real.

Definition 2.68. A square matrix A is said to be *involutory* if $A^2 = I$

For example, the matrix

$$\begin{bmatrix} 1 & 0 \\ 0 & 1 \end{bmatrix}$$

is involutory.

Theorem 2.31. A matrix A is involutory if and only if $(I+A)(I-A) = 0$.

Proof. Suppose first that A is involutory, then

$$A^2 = I$$

or

$$I - A^2 = 0$$

or

$$I^2 - A^2 = 0 \text{ since } I^2 = I$$

or

$$(I+A)(I-A) = 0 \text{ since } AI = IA.$$

Conversely, let

$$(I+A)(I-A) = 0$$

Then,

$$I^2 - IA + AI - A^2 = 0$$

or

$$I^2 - A^2 + 0 = 0$$

or

$$I^2 - A^2 = 0$$

or

$$A^2 = I^2 = I.$$

Definition 2.69. A square matrix A is said to be *idempotent* if $A^2 = A$.

For example, I_n is idempotent.

2.18 TRANSPOSE OF A MATRIX

Definition 2.70. A matrix obtained by interchanging the corresponding rows and columns of a matrix A is called the *transpose matrix* of A.

The transpose of a matrix A is denoted by A^T (or by A'). Thus, if $A = [a_{ij}]_{m\times n}$, then $A^T = [a_{ji}]_{n\times m}$ is an $n \times m$ matrix. For example, the transpose of the matrix

$$\begin{bmatrix} 1 & 0 & 2 \\ 3 & 7 & 4 \\ 1 & 2 & 8 \end{bmatrix}$$

is

$$\begin{bmatrix} 1 & 3 & 1 \\ 0 & 7 & 2 \\ 2 & 4 & 8 \end{bmatrix}.$$

Further,

(i) The transpose of a row matrix is a column matrix. For example, if $A = [1\ 2\ 4\ 3]$, then

$$A^T = \begin{bmatrix} 1 \\ 2 \\ 4 \\ 3 \end{bmatrix}.$$

(ii) The transpose of a column matrix is a row matrix. For example, if

$$A = \begin{bmatrix} 3 \\ 8 \\ 3 \end{bmatrix},$$

then

$$A^T = [\ 3\ \ 8\ \ 3\ ]$$

(iii) If A is $m \times n$ matrix, then A^T is an $n \times m$ matrix. Therefore, the product AA^T, $A^T A$ are both defined and are of order $m \times m$ and $n \times n$, respectively.

If $A = [a_{ij}]_{m\times n}$ and $B = [b_{ij}]_{m\times n}$ are matrices of the same order and if λ is a scalar, then the transpose of matrix has the following properties:

(a) $(A^T)^T = A$

(b) $(\lambda A)^T = \lambda A^T$

(c) $(A+B)^T = A^T + B^T$

(d) $(AB)^T = B^T A^T$ (Reversal law).

2.19 SYMMETRIC, SKEW-SYMMETRIC, AND HERMITIAN MATRICES

Definition 2.71. A square matrix A is said to be *symmetric* if $A = A^T$.

Thus, $A = [a_{ij}]_{n\times n}$ is symmetric if $a_{ij} = a_{ji}$ for $i \le i \le n$, $1 \le j \le n$.

Definition 2.72. A square matrix $A = [a_{ij}]_{n\times n}$ is said to be *skew symmetric* if $a_{ij} = -a_{ji}$ for all i and j.

Thus square matrix is skew-symmetric if $A = -A^T$. For example,

$$\begin{bmatrix} a & h & g \\ h & b & f \\ g & f & c \end{bmatrix}$$

is symmetric matrix, whereas the matrix

$$\begin{bmatrix} 0 & 1 & 2 \\ -1 & 0 & 3 \\ -2 & -3 & 0 \end{bmatrix}$$

is a skew-symmetric matrix.

Properties of Symmetric and Skew-Symmetric Matrices

(a) In a skew-symmetric matrix A, all diagonal elements are zero. In fact, if A is skew-symmetric, then

$$a_{ij} = -a_{ji} \text{ for all } i \text{ and } j.$$
$$\Rightarrow a_{ii} = -a_{ii}$$
$$\Rightarrow a_{ii} = 0.$$

(b) The matrix which is both symmetric and skew-symmetric must be a null matrix. In fact, if $A = [a_{ij}]$ is symmetric, then

$$a_{ij} = a_{ji} \text{ for all } i \text{ and } j.$$

Further, if $A = [a_{ij}]$ is skew-symmetric, then $a_{ij} = -a_{ji}$ for all i and j. Adding, we get $2a_{ij} = 0$ for all i and j and so $a_{ij} = 0$ for all i and j. Hence, A is a *null matrix*. Thus, "Null matrix is the only matrix which is both symmetric and skew-symmetric."

(c) For any square matrix A, $A + A^T$ is a symmetric matrix and $A - A^T$ is a skew-symmetric matrix. In fact, we note that

(i) $(A+A^T)^T = A^T + (A^T)^T = A^T + A = A + A^T$ and so $A + A^T$ is symmetric.

(ii) $$(A-A^T)^T = A^T - (A^T)^T = A^T - A = -(A-A^T)$$

and so $A - A^T$ is skew-symmetric.

(d) Every square matrix A can be expressed uniquely as the sum of a symmetric and a skew-symmetric matrix. To show it, set

$$P = \frac{1}{2}(A+A^T) \text{ and } Q = \frac{1}{2}(A-A^T).$$

Then

$$\begin{aligned} P^T &= \left[\frac{1}{2}(A+A^T)\right]^T = \frac{1}{2}(A+A^T)^T \\ &= \frac{1}{2}\left(A^T + (A^T)^T\right) = \frac{1}{2}(A^T + A) \\ &= \frac{1}{2}(A+A^T) = P. \end{aligned}$$

and so P is symmetric. Further,

$$\begin{aligned} Q^T &= \left[\frac{1}{2}(A-A^T)\right]^T = \frac{1}{2}(A-A^T)^T \\ &= \frac{1}{2}\left(A^T - (A^T)^T\right) = \frac{1}{2}(A^T - A) \\ &= -\frac{1}{2}(A-A^T) = -Q, \end{aligned}$$

and so Q is skew-symmetric. Also $P + Q = A$. Thus A can be expressed as the sum of a symmetric and a skew-symmetric matrix.

To establish the uniqueness of the expression, let $A = P_1 + Q_1$, where P_1 is symmetric and Q_1

is skew-symmetric. It is sufficient to show that $P_1 = P$ and $Q_1 = Q$. We have,

$$A^T = (P_1 + Q_1)^T = P_1^T + Q_1^T = P_1 - Q_1$$

Thus

$$A + A^T = 2P_1 \text{ or } P_1 = \frac{1}{2}(A + A^T) = P.$$

Also,

$$Q_1 = A - P_1 = A - \frac{1}{2}(A + A^T)$$
$$= \frac{1}{2}(A - A^T) = Q.$$

Hence, the expression is unique.

(e) If A is a square matrix, then $A + A^T$ and AA^T are symmetric matrices. These facts follow from

(i) $$(A + A^T)^T = A^T + (A^T)^T = A^T + A = A + A^T$$

and

(ii) $(AA^T)^T = (A^T)^T A^T = AA^T$.

(f) If A and B are two symmetric matrices, then AB-BA is a skew-symmetric matrix.
In fact,

$$(AB - BA)^T = (AB)^T - (BA)^T$$
$$= B^T A^T - A^T B^T \text{ (Reversal Law)}$$
$$= BA - AB$$
$$= -(AB - BA).$$

(g) If A is a symmetric (skew-symmetric), then $B^T\ AB$ is a symmetric (skew-symmetric) matrix. In fact, if A is symmetric, $A^T = A$ and so

$$(B^T AB)^T = B^T A^T \left(B^T\right)^T = B^T A^T B$$
$$= B^T AB$$

and if A is skew-symmetric, then

$$\left(B^T AB\right)^T = B^T A^T \left(B^T\right)^T = B^T(-A)B$$
$$= -B^T AB.$$

EXAMPLE 2.87

Express the matrix

$$A = \begin{bmatrix} 1 & 2 & 4 \\ 3 & 0 & 2 \\ 7 & 2 & 5 \end{bmatrix}$$

as the sum of a symmetric matrix and a skew-symmetric matrix.

Solution. We know that every square matrix A can be expressed as the sum of symmetric matrix $\frac{1}{2}(A + A^T)$ and a skew-symmetric matrix $\frac{1}{2}(A - A^T)$. In the present case

$$\frac{1}{2}(A + A^T) = \frac{1}{2}\left[\begin{pmatrix} 1 & 2 & 4 \\ 3 & 0 & 2 \\ 7 & 2 & 5 \end{pmatrix} + \begin{pmatrix} 1 & 3 & 7 \\ 2 & 0 & 2 \\ 4 & 2 & 5 \end{pmatrix}\right]$$
$$= \frac{1}{2}\begin{pmatrix} 2 & 5 & 11 \\ 5 & 0 & 4 \\ 11 & 4 & 10 \end{pmatrix} = \begin{pmatrix} 1 & \frac{5}{2} & \frac{11}{2} \\ \frac{5}{2} & 0 & 2 \\ \frac{11}{2} & 2 & 5 \end{pmatrix}$$

and

$$\frac{1}{2}(A - A^T)$$
$$= \frac{1}{2}\left[\begin{pmatrix} 1 & 2 & 4 \\ 3 & 0 & 2 \\ 7 & 2 & 5 \end{pmatrix} - \begin{pmatrix} 1 & 3 & 7 \\ 2 & 0 & 2 \\ 4 & 2 & 5 \end{pmatrix}\right]$$
$$= \frac{1}{2}\begin{pmatrix} 0 & -1 & -3 \\ 1 & 0 & 0 \\ 3 & 0 & 0 \end{pmatrix} = \begin{pmatrix} 0 & -\frac{1}{2} & -\frac{3}{2} \\ \frac{1}{2} & 0 & 0 \\ \frac{3}{2} & 0 & 0 \end{pmatrix}$$

Hence

$$\begin{pmatrix} 1 & 2 & 4 \\ 3 & 0 & 2 \\ 7 & 2 & 5 \end{pmatrix} = \begin{pmatrix} 1 & \frac{5}{2} & \frac{11}{2} \\ \frac{5}{2} & 0 & 2 \\ \frac{11}{2} & 2 & 5 \end{pmatrix} + \begin{pmatrix} 0 & -\frac{1}{2} & -\frac{3}{2} \\ \frac{1}{2} & 0 & 0 \\ \frac{3}{2} & 0 & 0 \end{pmatrix}.$$

Definition 2.73. A matrix obtained from a given matrix A by replacing its elements by the corresponding conjugate complex numbers is called the *conjugate* of A and is denoted by $\bar{A}$.

Thus if $A = [a_{ij}]_{m\times n}$, then $\bar{A} = [\overline{a_{ij}}]_{m\times n}$, where $\bar{a}_{ij}$ denotes the complex conjugate of a_{ij}.

Definition 2.74. A matrix whose all elements are real is called a *real* matrix.

If A is a real matrix, then obviously $A = \bar{A}$ Further if A and B are two matrices, then

(a) $(\bar{\bar{A}}) = A$

(b) $\overline{(A+B)} = \bar{A} + \bar{B}$

(c) $\overline{(\lambda A)} = \bar{\lambda}\bar{A}$

(d) $\overline{(AB)} = \bar{A}\bar{B}$, where λ a complex number.

Definition 2.75. The transpose of the conjugate of a matrix A is called *transposed conjugate* or *tranjugate* of A and is denoted by A^θ or sometimes by A^*.

We observe that

$$(\bar{A})^T = \overline{(A^T)}.$$

For example, let

$$A = \begin{bmatrix} 2 & 1+2i & 3+4i \\ 1+i & 7 & 2+i \\ 3+2i & 4+i & 3+3i \end{bmatrix},$$

then

$$\bar{A} = \begin{bmatrix} 2 & 1-2i & 3-4i \\ 1-i & 7 & 2-i \\ 3-2i & 4-i & 3-3i \end{bmatrix}$$

and

$$A^\theta = \begin{bmatrix} 2 & 1-i & 3-2i \\ 1-2i & 7 & 4-i \\ 3-4i & 2-i & 3-3i \end{bmatrix}.$$

Let A and B the matrices, then the tranjugate of the matrix possesses the following properties:

(a) $(A^\theta)^\theta = A$

(b) $(A+B)^\theta = A^\theta + B^\theta$, A and B being of the same order.

(c) $(\lambda A)^\theta = \bar{\lambda} A^\theta$, λ being a complex number.

(d) $(AB)^\theta = B^\theta A^\theta$, A and B being conformable to multiplication.

Definition 2.76. A square matrix $A = [a_{ij}]$ is said to be *Hermitian* if $a_{ij} = \overline{a_{ji}}$ for all i and j.
Thus, a matrix is Hermitian if and only if $A = A^\theta$.
We note that

(a) A real Hermitian matrix is a real symmetric matrix.

(b) If A is Hermitian, then

$$a_{ii} = \overline{a_{ii}} \quad \text{for all } i,$$

and so a_{ii} is real for all i. Thus, every diagonal element of a Hermitian matrix must be real.

Definition 2.77. A square matrix $A = [a_{ij}]$ is said to be Skew-Hermitian if $a_{ij} = -\overline{a_{ji}}$ for all i and j. Thus, a matrix is Skew-Hermitian if $A = -A^\theta$.
We observe that

(a) A real Skew-Hermitian matrix is nothing but a real Skew-symmetric matrix.

(b) If A is Skew-Hermitian matrix, then $a_{ii} = -\bar{a}_{ii}$ or $a_{ii} + \bar{a}_{ii} = 0$ and so a_{ii} is either a pure imaginary number or must be zero. Thus the diagonal element of a Skew-Hermitian matrix must be a pure imaginary number or zero.

For example,

$$\begin{bmatrix} 2 & 3-4i & 2+3i \\ 3+4i & 0 & 7-5i \\ 2-3i & 7+5i & 4 \end{bmatrix}$$

is an Hermitian matrix, whereas, the matrix

$$\begin{bmatrix} 0 & 3+4i \\ -3+4i & i \end{bmatrix}$$

is Skew-Hermitian.
It can be shown easily that if A is any square matrix, then $A + A^\theta, AA^\theta, A^\theta A$ are Hermitian and $A - A^\theta$ is Skew-Hermitian.

EXAMPLE 2.88

Show that every square matrix can be uniquely expressible as the sum of a Hermitian matrix and Skew-Hermitian matrix.

Solution. As mentioned above, if A is any square matrix, then $A + A^\theta$ is Hermitian and $A - A^\theta$ is Skew-Hermitian. Therefore, $\frac{1}{2}(A + A^\theta)$ and

$\frac{1}{2}(A-A^\theta)$ are Hermitian and Skew-Hermitian, respectively, so that

$$A=\frac{1}{2}(A+A^\theta)+\frac{1}{2}(A-A^\theta),$$

which proves first part of our result. The uniqueness can be proved easily and is left to the reader.

EXAMPLE 2.89

Show that every square matrix A can be uniquely expressed as $P+iQ$ where P and Q, are Hermitian matrices.

Solution. We take

$$P=\frac{1}{2}(A+A^\theta) \text{ and } Q=\frac{1}{2i}(A-A^\theta).$$

Then $A=P+iQ$. Further,

$$\begin{aligned} P^\theta &= \left(\frac{1}{2}(A+A^\theta)\right)^\theta=\frac{1}{2}\left(A+A^\theta\right)^\theta \\ &= \frac{1}{2}A^\theta+\frac{1}{2}\left(A^\theta\right)^\theta=\frac{1}{2}(A^\theta+A) \\ &= \frac{1}{2}(A+A^\theta)=P, \end{aligned}$$

showing that P is Hermitian. Similarly,

$$\begin{aligned} Q^\theta &= \left[\frac{1}{2i}\left(A-A^\theta\right)\right]^\theta \\ &= \left(-\frac{1}{2i}\right)\left(A-A^\theta\right)^\theta \\ &= -\frac{1}{2i}\left\{A^\theta-\left(A^\theta\right)^\theta\right\} \\ &= -\frac{1}{2i}(A^\theta-A) \\ &= \frac{1}{2i}(A-A^\theta)=Q, \end{aligned}$$

showing that Q is also Hermitian. Thus $A=P+iQ$, where P and Q are Hermitian.

2.20 LOWER AND UPPER TRIANGULAR MATRICES

Definition 2.78. A square matrix $A=[a_{ij}]$, in which all elements above the main diagonal are zero, is called a *lower triangular matrix*.

Thus a matrix A is lower triangular if $a_{ij}=0$ for $i<j$.

Definition 2.79. A square matrix $A=[a_{ij}]$, in which all elements below the main diagonal are zero, is called an *upper triangular matrix*.

Thus a matrix A is upper triangular matrix if $a_{ij}=0$ for $i>j$.

For example,

$$\begin{bmatrix} 1 & 2 & -1 \\ 0 & 1 & 3 \\ 0 & 0 & 1 \end{bmatrix} \quad \textit{and} \quad \begin{bmatrix} 1 & 0 & 0 \\ 6 & -5 & 0 \\ 0 & 4 & 1 \end{bmatrix}$$

are, respectively, upper triangular and lower triangular matrices.

2.21 DETERMINANTS

The determinant plays a very important role in every branch of mathematics. It is a very powerful tool for solving systems of linear equations.

Definition 2.80. Let M_n be the set of all matrices of order n over the field $F(\Re \text{ or } C)$ and let $A=[a_{ij}]\in M_n$. Then the *determinant of order n* of the matrix A is defined as an *inductive mapping* $D_n: M_n \to F$ such that

(i) $D_n(a_{ij})=a_{ij}$, and

(ii) $D_n(A)=\sum_{i=1}^{n}(-1)^{i+1}a_{1i}D_{n-1}(A_{1i})$,

where A_{1i} is the matrix obtained by deleting the first row and ith column of the matrix A.

Obviously, the determinant of a matrix A is a *number* whereas the matrix A is itself a *system* of *numbers*. We denote the determinant of A by $|A|$ or by $\Delta(A)$. Thus, if $A=\begin{bmatrix} a_{11} & a_{12} \\ a_{21} & a_{22} \end{bmatrix}$, then

$$\begin{aligned} D_2(A) &= \sum_{i=1}^{2}(-1)^{i+1}a_{1i}D_1(A_{1i}) \\ &= a_{11}D_1(A_{11})-a_{12}D_1(A_{12}) \end{aligned}$$

$$= a_{11}D_1(a_{22}) - a_{12}D_1(a_{21})$$
$$= a_{11}a_{22} - a_{12}a_{21}.$$

Hence, in our notation,

$$\begin{vmatrix} a_{11} & a_{12} \\ a_{21} & a_{22} \end{vmatrix} = a_{11}a_{22} - a_{12}a_{21}$$

is a *second order determinant*.
Similarly, if

$$A = \begin{bmatrix} a_{11} & a_{12} & a_{13} \\ a_{21} & a_{22} & a_{23} \\ a_{31} & a_{32} & a_{33} \end{bmatrix},$$

then

$$|A| = \sum_{i=1}^{3}(-1)^{i+1}a_{1i}D_2(A_{1i})$$
$$= a_{11}D_2A_{11} - a_{12}D_2A_{12} + a_{13}D_2A_{13}$$
$$= a_{11}D_2\begin{bmatrix} a_{22} & a_{23} \\ a_{32} & a_{33} \end{bmatrix} - a_{12}D_2\begin{bmatrix} a_{21} & a_{23} \\ a_{31} & a_{33} \end{bmatrix}$$
$$+ a_{13}D_2\begin{bmatrix} a_{21} & a_{22} \\ a_{31} & a_{32} \end{bmatrix}$$
$$= a_{11}\begin{vmatrix} a_{22} & a_{23} \\ a_{32} & a_{33} \end{vmatrix} - a_{12}\begin{vmatrix} a_{21} & a_{23} \\ a_{31} & a_{33} \end{vmatrix}$$
$$+ a_{13}\begin{vmatrix} a_{21} & a_{22} \\ a_{31} & a_{32} \end{vmatrix}$$
$$= a_{11}(a_{22}a_{33} - a_{23}a_{32}) - a_{12}(a_{21}a_{33} - a_{23}a_{31})$$
$$+ a_{13}(a_{21}a_{32} - a_{22}a_{31}),$$

which is *third-order determinant of the matrix A*.

EXAMPLE 2.90

Compute the determinant

$$\begin{vmatrix} 6 & 8 & 9 \\ 2 & 4 & 5 \\ 0 & 5 & 7 \end{vmatrix}$$

Solution. We have

$$\begin{vmatrix} 6 & 8 & 9 \\ 2 & 4 & 5 \\ 0 & 5 & 7 \end{vmatrix} = 6(4\times 7 - 5\times 5) - 8(2\times 7 - 5\times 0)$$
$$+ 9(2\times 5 - 4\times 0)$$
$$= 18 - 112 + 90 = -4.$$

EXAMPLE 2.91

Show that the determinant of a triangular matrix is equal to the product of entries of the main diagonal of that matrix.

Solution. Consider first the upper triangular matrix

$$A = \begin{bmatrix} a_1 & b_1 & c_1 \\ 0 & b_2 & c_2 \\ 0 & 0 & c_3 \end{bmatrix} \quad \text{of third order.}$$

Then

$$|A| = a_1\begin{vmatrix} b_2 & c_2 \\ 0 & c_3 \end{vmatrix} - b_1\begin{vmatrix} 0 & c_2 \\ 0 & c_3 \end{vmatrix} + c_1\begin{vmatrix} 0 & b_2 \\ 0 & 0 \end{vmatrix}$$
$$= a_1(b_2c_3 - 0) - b_1(0-0) + c_1(0-0)$$
$$= a_1b_2c_3$$

Similarly, if

$$A = \begin{bmatrix} a_{11} & a_{12} & \dots & a_{1n} \\ 0 & a_{22} & \dots & a_{2n} \\ \dots & \dots & \dots & \dots \\ 0 & 0 & \dots & a_{nn} \end{bmatrix},$$

then

$$|A| = a_{11}a_{22}\dots a_{nn}.$$

Similar result holds for lower triangular matrix.

Properties of Determinant

1. Taking the *transpose does not alter the determinant.* To prove it, let

$$\Delta = \begin{vmatrix} a_{11} & a_{12} & a_{13} \\ a_{21} & a_{22} & a_{23} \\ a_{31} & a_{32} & a_{33} \end{vmatrix}$$

be a third order determinant. Taking transpose, we have

$$\Delta' = \begin{vmatrix} a_{11} & a_{21} & a_{31} \\ a_{12} & a_{22} & a_{32} \\ a_{13} & a_{23} & a_{33} \end{vmatrix}.$$

We note that

$$\Delta = a_{11}(a_{22}a_{33} - a_{23}a_{32}) - a_{12}(a_{21}a_{33} - a_{23}a_{31}) + a_{13}(a_{21}a_{32} - a_{22}a_{31}) \quad (43)$$

and

$$\Delta' = a_{11}(a_{22}a_{33} - a_{32}a_{23}) - a_{21}(a_{12}a_{33} - a_{32}a_{13}) + a_{31}(a_{12}a_{23} - a_{22}a_{13}). \quad (44)$$

The right hand sides of (43) and (44) are the same numbers. Hence $\Delta = \Delta'$. Thus the determinant remains unchanged under the reflection of its elements in the main diagonal. This property is also known as *Reflection Property*.

2. *Switching Property: Interchanging rows/ columns of a determinant changes its signature.*

   Let

$$\Delta = \begin{vmatrix} a_{11} & a_{12} & a_{13} \\ a_{21} & a_{22} & a_{23} \\ a_{31} & a_{32} & a_{33} \end{vmatrix}$$

and let Δ_1 be the determinant obtained by interchanging second and third row. Thus

$$\Delta_1 = \begin{vmatrix} a_{11} & a_{12} & a_{13} \\ a_{31} & a_{32} & a_{33} \\ a_{21} & a_{22} & a_{23} \end{vmatrix}$$

Then

$$\Delta_1 = a_{11}\begin{vmatrix} a_{32} & a_{33} \\ a_{22} & a_{23} \end{vmatrix} - a_{12}\begin{vmatrix} a_{31} & a_{33} \\ a_{21} & a_{23} \end{vmatrix} + a_{13}\begin{vmatrix} a_{31} & a_{32} \\ a_{21} & a_{22} \end{vmatrix}$$

$$= a_{11}(a_{32}a_{23} - a_{33}a_{22}) - a_{12}(a_{31}a_{23} - a_{33}a_{21}) + a_{13}(a_{31}a_{22} - a_{32}a_{21})$$

$$= -a_{11}(a_{22}a_{33} - a_{23}a_{32}) + a_{12}(a_{21}a_{33} - a_{23}a_{31}) - a_{13}(a_{21}a_{32} - a_{22}a_{31})$$

$$= -\Delta.$$

Similar result holds for interchange of any two rows/column.

3. *All zero Property: If all the elements of a row/ column of a determinant are zero, then the determinant is zero.*

   Let all the elements of the ith row of a determinant be zeros. Every term of the determinant must have, as a factor, one element of the ith row. Therefore all the terms of the determinant are zero and so $\Delta = 0$. Similar result holds for column having all its entries as 0.

4. *A determinant containing two identical rows is equal to zero.*

   Suppose that ith and jth row of the original determinant Δ are identical. If we interchange ith and jth row then the determinant becomes $\Delta_1 = -\Delta$. But since Δ and Δ_1 are the same determinant, we have

$$\Delta = -\Delta \quad \text{or} \quad 2\Delta = 0 \quad \text{or} \quad \Delta = 0.$$

5. *Scalar Multiple Property: If Δ_1 is the determinant obtained by multiplying a row/column of a determinant Δ by k, then $\Delta_1 = k\Delta$.*

   Suppose that Δ_1 is obtained from Δ by multiplying the ith row by a scalar k. Since each term of the determinant contains exactly one element of the ith row, therefore each term acquires the factor k and as such the determinant Δ_1 is k multiple of Δ. This property allows us to *factor out a common factor of all the elements of some row/column of that determinant.*

6. *Proportionality Property: A determinant with two proportional rows is equal to zero.*

In fact, if each element in ith row of a determinant is kth time the corresponding element in the jth row, then we can factor out k (by property (5)) and obtain a determinant with two identical rows. Hence the value of the determinant is zero by property (4).

7. *Sum Property: If all elements of any row/column consist of the sum of two terms, then the determinant can be expressed as the sum of two determinants.*

   To verify this, consider the third order determinant.

$$\Delta = \begin{vmatrix} a_1+\alpha_1 & b_1 & c_1 \\ a_2+\alpha_2 & b_2 & c_2 \\ a_3+\alpha_3 & b_3 & c_3 \end{vmatrix}$$

Then

$$\begin{aligned}\Delta &= (a_1+\alpha_1)(b_2c_3-c_2b_3)-b_1[(a_2+\alpha_2)c_3 \\ &\quad -c_2(a_3+\alpha_3)]+c_1[(a_2+\alpha_2)b_3 \\ &\quad -b_2(a_3+\alpha_3)] \\ &= [a_1(b_2c_3-c_2b_3)-b_1(a_2c_3-c_2a_3) \\ &\quad +c_1(a_2b_3-b_2a_3)]+[\alpha_1(b_2c_3-c_2b_3) \\ &\quad -b_1(\alpha_2c_3-c_2\alpha_3)+c_1(\alpha_2b_3-b_2a_3)] \\ &= \begin{vmatrix} a_1 & b_1 & c_1 \\ a_2 & b_2 & c_2 \\ a_3 & b_3 & c_3 \end{vmatrix} + \begin{vmatrix} \alpha_1 & b_1 & c_1 \\ \alpha_2 & b_2 & c_2 \\ \alpha_3 & b_3 & c_3 \end{vmatrix}\end{aligned}$$

8. *Invariance Property: If a determinant Δ_1 is obtained from a determinant Δ by adding to a row/ column k times some different row/column, then $\Delta_1 = \Delta$.*

   To verify it, let

$$\Delta_1 = \begin{vmatrix} a_1+ka_2 & b_1+kb_2 & c_1+kc_3 \\ a_2 & b_2 & c_2 \\ a_3 & b_3 & c_3 \end{vmatrix}$$

$$= \begin{vmatrix} a_1 & b_1 & c_1 \\ a_2 & b_2 & c_2 \\ a_3 & b_3 & c_3 \end{vmatrix} + \begin{vmatrix} ka_2 & kb_2 & kc_3 \\ a_2 & b_2 & c_2 \\ a_3 & b_3 & c_3 \end{vmatrix}$$

$$= \Delta + k\begin{vmatrix} a_2 & b_2 & c_3 \\ a_2 & b_2 & c_3 \\ a_3 & b_3 & c_3 \end{vmatrix}$$

$= \Delta + k(0)$, since two rows are identical

$= \Delta.$

9. Factor Property: *If the entries of a determinant Δ are polynomials in x and two different rows/ columns become identical for $x = a$, then $x = a$ is a factor of Δ.*

   Since the elements are polynomials in x, the determinant itself is a polynomial in x. Suppose that $\Delta = p(x)$. For $x = a$, two rows/columns become identical and so $\Delta = 0 = p(x)$ for $x = a$. Thus $p(a) = 0$. Therefore, by factor theorem, $x - a$ is a factor of $p(x)$ and hence of Δ.

Minors and Cofactors

Definition 2.81. Let Δ be a determinant of order n and let $1 \le k \le n-1$. Then the determinant obtained by deleting $n-k$ rows and $n-k$ columns in Δ is called the *minor of order k*.

Definition 2.82. Let Δ be a determinant of order n and let a_{ij} be an element of Δ. Then the *co-factor* of a_{ij}, denoted by A_{ij}, is defined as

$$A_{ij} = cof(a_{ij}) = (-1)^{i+j}|M_{ij}|,$$

where $|M_{ij}|$ is the determinant obtained by deleting ith row and jth column of the determinant Δ. The determinant $|M_{ij}|$ is called *minor of the element a_{ij}*. For example, consider the determinant

$$|A| = \begin{vmatrix} 1 & 5 & 7 \\ 3 & 2 & 1 \\ 0 & 4 & 8 \end{vmatrix}.$$

Then the determinants

$$\begin{vmatrix} 5 & 7 \\ 2 & 1 \end{vmatrix}, \begin{vmatrix} 2 & 1 \\ 4 & 8 \end{vmatrix}, \text{ and } \begin{vmatrix} 1 & 5 \\ 3 & 2 \end{vmatrix}$$

are minors of order 2, whereas

$$cof(2) = (-1)^{2+2}\begin{vmatrix} 1 & 7 \\ 0 & 8 \end{vmatrix} = \begin{vmatrix} 1 & 7 \\ 0 & 8 \end{vmatrix} = 8,$$

$$cof(4) = (1)^{3+2}\begin{vmatrix} 1 & 7 \\ 3 & 1 \end{vmatrix} = -(1-21) = 20.$$

Theorem 2.32. (Evaluation of Determinant): Let a_{ij} be an element of a determinant $|A|$ of order n. Then

$$|A| = \sum_{j=1}^{n} a_{ij}\, cof(a_{ij}), \quad 1 \le i \le n.$$

Proof. Let

$$|B| = \begin{vmatrix} a_{i1} & a_{i2} & a_{i3} & \dots & \dots & a_{in} \\ a_{21} & a_{22} & a_{23} & \dots & \dots & a_{2n} \\ \dots & \dots & \dots & \dots & \dots & \dots \\ \dots & \dots & \dots & \dots & \dots & \dots \\ a_{i-1,1} & a_{i-1,2} & a_{i-1,3} & \dots & \dots & a_{i-1,n} \\ a_{11} & a_{12} & a_{13} & \dots & \dots & a_{1n} \\ a_{i+1,1} & a_{i+2,2} & a_{i+1,2} & \dots & \dots & a_{i+1,n} \\ \dots & \dots & \dots & \dots & \dots & \dots \\ \dots & \dots & \dots & \dots & \dots & \dots \\ a_{n1} & a_{n2} & a_{n3} & \dots & \dots & a_{nn} \end{vmatrix} \tag{45}$$

be the determinant obtained from the determinant

$$|A| = \begin{vmatrix} a_{11} & a_{12} & a_{13} & \dots & \dots & a_{1n} \\ a_{21} & a_{22} & a_{23} & \dots & \dots & a_{2n} \\ \dots & \dots & \dots & \dots & \dots & \dots \\ \dots & \dots & \dots & \dots & \dots & \dots \\ a_{i-1,1} & a_{i-1,2} & a_{i-1,3} & \dots & \dots & a_{i-1,n} \\ a_{i1} & a_{i2} & a_{i3} & \dots & \dots & a_{in} \\ a_{i+1,1} & a_{i+2,2} & a_{i+1,3} & \dots & \dots & a_{i+1,n} \\ \dots & \dots & \dots & \dots & \dots & \dots \\ \dots & \dots & \dots & \dots & \dots & \dots \\ a_{n1} & a_{n2} & a_{n3} & \dots & \dots & a_{nn} \end{vmatrix} \tag{46}$$

by switching ith row of $|A|$ to the first row. To do this task $i-1$ permutations are needed. Therefore, by switching property, we have

$$|A| = (-1)^{i-1}|B|$$

$$= (-1)^{i-1}\sum_{j=1}^{n}(-1)^{j+1} b_{1j}|B_{1j}|,$$

using definition 2.80.

But we observe from (4) and (5) that

(i) $b_{1j} = a_{ij}, \quad 1 \le j \le n$
(ii) $B_{1j} = A_{ij}, \quad 1 \le j \le n.$

Hence

$$|A| = (-1)^{i-1}\sum_{j=1}^{n}(-1)^{j+1}(-1)^{j+1}a_{ij}|A_{ij}|$$

$$= \sum_{j=1}^{n}(-1)^{i+j}a_{ij}|A_{ij}|$$

$$= \sum_{j=1}^{n} a_{ij}\, cof(a_{ij}), \quad 1 \le i \le n.$$

Remark 2.5 Theorem 2.31 can be applied to expand a given determinant with the help of any row. If we switch jth column of $|A|$ to first column to get $|B|$, then the corresponding modifications in the proof of the theorem yields

$$|A| = \sum_{i=1}^{n} a_{ij}\, cof(a_{ij}), \quad 1 \le j \le n.$$

Hence, the determinant $|A|$ can also be evaluated with the help of any column.

If follows therefore that "*The determinant $|A|$ is equal to the sum of the elements of any row or column multiplied by their corresponding cofactors.*"

EXAMPLE 2.92

Expand the determinant

$$|A| = \begin{vmatrix} 5 & 8 & 7 \\ 3 & 4 & 1 \\ 2 & 1 & 8 \end{vmatrix}$$

(i) with the help of second row
(ii) with the help of third column.

Solution. (i) To evaluate with the help of ith row, we have

$$|A| = \sum_{j=1}^{n} a_{ij}\, cof(a_{ij}), \quad 1 \le i \le n.$$

In the present case, $i = 2$, $n = 3$. Hence

$$|A| = \sum_{j=1}^{n} a_{2j}\, cof(a_{2j})$$
$$= a_{21}\, cof(a_{21}) + a_{22}\, cof(a_{22}) + a_{23}\, cof(a_{23})$$
$$= 3\, cof(3) + 4\, cof(4) + 1\, cof(1)$$
$$= 3[(-1)^3(64-7)] + 4[(-1)^4(40-14)] + 1[(-1)^5(5-16)]$$
$$= -171 + 104 + 11 = -56.$$

(ii) The formula for expansion through column is.

$$|A| = \sum_{i=1}^{n} a_{ij}\, cof(a_{ij}), \quad 1 \le j \le n.$$

In the present case $j = 3$, $n = 3$. Thus

$$|A| = \sum_{i=1}^{3} a_{i3}\, cof(a_{i3})$$
$$= a_{13}\, cof(a_{13}) + a_{23}\, cof(a_{23}) + a_{33}\, cof(a_{33})$$
$$= 7\, cof(7) + 1\, cof(1) + 8\, cof(8)$$
$$= 7[(-1)^4(3-8)] + 1[(-1)^5(5-16)] + 8[(-1)^6(20-24)]$$
$$= -35 + 11 - 32 = -56.$$

Theorem 2.33. Let a_{ij} be an element of a determinant $|A|$. Then

$$\sum_{i=1}^{n} a_{ij}\, cof(a_{ik}) = 0, \quad j \ne k.$$

(In other words, *the sum of the products of the elements of any row or column with the cofactors of the elements of a different row or different column is zero*).

Proof. If we replace rth column of $|A|$ by its jth column, we get a determinant $|B|$, whose two columns, (kth and ith) are identical. Hence $|B| = 0$. Also $b_{ik} = a_{ij}$ and $B_{ir} = A_{ir}$. Therefore, by Theorem 2.31.

$$0 = |B| = \sum_{i=1}^{n} b_{ir}\, cof(b_{ik})$$
$$= \sum_{i=1}^{n} a_{ij}\, cof(a_{ik}), \quad j \ne k.$$

Definition 2.83. If A_{ij} is cofactor of a_{ij} in a determinant $|A|$ of order n, than the determinant

$$|A_{ij}| = \begin{vmatrix} A_{11} & A_{12} & \cdots & \cdots & A_{1n} \\ A_{21} & A_{22} & \cdots & \cdots & A_{2n} \\ \cdots & \cdots & \cdots & \cdots & \cdots \\ \cdots & \cdots & \cdots & \cdots & \cdots \\ A_{n1} & A_{n2} & \cdots & \cdots & A_{nn} \end{vmatrix}$$

is called the *cofactor determinant* of the determinant $|A|$. If is also called *adjugate* or *reciprocal determinant* of $|A|$.

Definition 2.84. Let $|A|$ and $|B|$ be two determinants of order n with corresponding matrices A and B. Then the product of $|A|$ and $|B|$ is defined to be the determinant of the matrix AB.

Thus , if

$$|A| = \begin{vmatrix} a_{11} & a_{12} & \cdots & \cdots & a_{1n} \\ a_{21} & a_{22} & \cdots & \cdots & a_{2n} \\ \cdots & \cdots & \cdots & \cdots & \cdots \\ \cdots & \cdots & \cdots & \cdots & \cdots \\ a_{n1} & a_{n2} & \cdots & \cdots & a_{nn} \end{vmatrix},$$

$$|B| = \begin{vmatrix} b_{11} & b_{12} & \cdots & \cdots & b_{1n} \\ b_{21} & b_{22} & \cdots & \cdots & b_{2n} \\ \cdots & \cdots & \cdots & \cdots & \cdots \\ \cdots & \cdots & \cdots & \cdots & \cdots \\ b_{n1} & b_{n2} & \cdots & \cdots & b_{nn} \end{vmatrix},$$

then

$$|A||B| = |AB| = |C|$$
$$= \begin{vmatrix} c_{11} & c_{12} & \cdots & \cdots & c_{1n} \\ c_{21} & c_{22} & \cdots & \cdots & c_{2n} \\ \cdots & \cdots & \cdots & \cdots & \cdots \\ \cdots & \cdots & \cdots & \cdots & \cdots \\ c_{n1} & c_{n2} & \cdots & \cdots & c_{nn} \end{vmatrix},$$

where

$$c_{ij} = \sum_{k=1}^{n} a_{ik}b_{kj}.$$

Thus , for example,

$$c_{11} = \sum_{k=1}^{n} a_{1k}b_{k1} = a_{11}b_{11} + a_{12}b_{21} + \cdots + a_{1n}b_{n1},$$

$$c_{12} = a_{11}b_{12} + a_{12}b_{22} + \cdots + a_{1n}b_{n2},$$

and so on. However, *due to reflection property, we can also imply the row by row multiplication rule or column by column multiplication rule.* For example, if we apply row by row multiplication, then

$$c_{11} = a_{11}b_{11} + a_{12}b_{12} + \cdots + a_{1n}b_{1n}$$
$$c_{12} = a_{11}b_{21} + a_{12}b_{22} + \cdots + a_{1n}b_{2n},$$

and so on.

Theorem 2.34. (Jacobi). If $|A|$ is a determinant of order n and $|A_{ij}|$ is the determinant of cofactors of the elements of $|A|$, then

$$|A_{ij}| = |A|^{n-1}$$

Proof. Let

$$A = \begin{vmatrix} a_{11} & a_{12} & \cdots & \cdots & a_{1n} \\ a_{21} & a_{22} & \cdots & \cdots & a_{2n} \\ \cdots & \cdots & \cdots & \cdots & \cdots \\ \cdots & \cdots & \cdots & \cdots & \cdots \\ a_{n1} & a_{n2} & \cdots & \cdots & a_{nn} \end{vmatrix}$$

and let cofactor determinant (or adjugate or reciprocal determinant of $|A|$) be

$$|A_{ij}| = \begin{vmatrix} A_{11} & A_{12} & \cdots & \cdots & A_{1n} \\ A_{21} & A_{22} & \cdots & \cdots & A_{2n} \\ \cdots & \cdots & \cdots & \cdots & \cdots \\ \cdots & \cdots & \cdots & \cdots & \cdots \\ A_{n1} & A_{n2} & \cdots & \cdots & A_{nn} \end{vmatrix}$$

Then (using row by row multiplication)

$$|A||A_{ij}| = |AA_{ij}| = \begin{vmatrix} |A| & 0 & \cdots & \cdots & 0 \\ 0 & |A| & 0 & \cdots & 0 \\ \cdots & \cdots & \cdots & \cdots & \cdots \\ \cdots & \cdots & \cdots & \cdots & \cdots \\ 0 & 0 & 0 & \cdots & |A| \end{vmatrix} = |A|^n,$$

since the value the determinant of a triangular matrix is the product of the diagonal elements. Hence

$$|A_{ij}| = |A|^{n-1}.$$

Theorem 2.35. (Laplace Expansion). Let $|A|$ be a determinant of order n and let $1 \le r \le n-1$. If r rows (or r columns) are arbitrarily chosen, then the sum of the products of all r-order minors contained in the chosen rows (or columns) by their cofactors is equal to the determinant $|A|$.

Thus the Laplace method gives an expansion of the determinant of order n in terms of the determinants of order r and $(n-r)$.

Working Rule for Laplace Expansion

Let $|A|$ be a determinant of order 3. Suppose that we want to expand $|A|$ along first two rows. We make all possible arrangements of three columns by taking two at a time. Let the arrangements be (1, 2), (1, 3), (2, 3). Observe that *in these arrangements, the first number is less than the second.* If $\left|A_{ij}^{kl}\right|$ denotes the minor obtained by deleting all rows and columns except rows i, j and columns k, l, then the expansion of $|A|$ along first two rows is

$$\begin{aligned}|A| &= (-1)^{1+2+1+2}\left|A_{12}^{12}\right|\left|A_{3}^{3}\right| \\ &+ (-1)^{1+2+1+3}\left|A_{12}^{13}\right|\left|A_{3}^{2}\right| \\ &+ (-1)^{1+2+2+3}\left|A_{12}^{23}\right|\left|A_{3}^{1}\right|.\end{aligned}$$

For example, consider the expansion of the determinant

$$|A| = \begin{vmatrix} 1 & 2 & 3 \\ 3 & 2 & 1 \\ 1 & 0 & 2 \end{vmatrix}$$

Then, using Laplace expansion along first two rows, we have

$$\begin{aligned}|A| &= (-1)^{1+2+1+2}\left|A_{12}^{12}\right|\left|A_{3}^{3}\right| \\ &+ (-1)^{1+2+1+3}\left|A_{12}^{13}\right|\left|A_{3}^{2}\right| \\ &+ (-1)^{1+2+2+3}\left|A_{12}^{23}\right|\left|A_{3}^{1}\right|\end{aligned}$$

$$= |2|\begin{vmatrix} 1 & 2 \\ 3 & 2 \end{vmatrix} - |0|\begin{vmatrix} 1 & 3 \\ 3 & 1 \end{vmatrix} + 1\begin{vmatrix} 2 & 3 \\ 2 & 1 \end{vmatrix}$$

$$= 2(2-6) - 0 + 1(2-6) = -8 - 4 = -12.$$

EXAMPLE 2.93

Show that

$$|A| = \begin{vmatrix} 1 & 1 & 1 \\ a & b & c \\ a^2 & b^2 & c^2 \end{vmatrix} = (a-b)(b-c)(c-a).$$

Solution. We have

$$\begin{vmatrix} 1 & 1 & 1 \\ a & b & c \\ a^2 & b^2 & c^2 \end{vmatrix} = \begin{vmatrix} 0 & 0 & 1 \\ a-b & b-c & c \\ a^2-b^2 & b^2-c^2 & c^2 \end{vmatrix}$$

$$C_1 \to C_1 - C_2$$
$$C_2 \to C_2 - C_3$$

$$= (a-b)(b-c)\begin{vmatrix} 0 & 0 & 1 \\ 1 & 1 & c \\ a+b & b+c & c^2 \end{vmatrix}$$

$$= (a-b)(b-c)[\{(b+c)-(a+b)\}]$$

$$= (a-b)(b-c)(c-a.)$$

Second Method. If we put $a = b$ in the given determinant that first and second columns become identical and therefore the determinant vanishes. Hence $(a-b)$ is a factor of the determinant.

Similarly, the determinant vanishes when we put $b = c$ and $c = a$. Hence $(b-c)$ and $(c-a)$ are also factors of the determinant. Further, since the given determinant is of order 3, therefore

$$|A| = k(a-b)(b-c)(c-a).$$

Comparing the coefficients of the leading term bc^2 in $|A|$ on both sides, we have $k = 1$. Hence

$$|A| = (a-b)(b-c)(c-a).$$

EXAMPLE 2.94

Evaluate the determinant

$$\triangle = \begin{vmatrix} 2bc-a^2 & c^2 & b^2 \\ c^2 & 2ca-b^2 & a^2 \\ b^2 & a^2 & 2ab-c^2 \end{vmatrix}$$

Solution. We have

$$\triangle = \begin{vmatrix} a & b & c \\ b & c & a \\ c & a & b \end{vmatrix}\begin{vmatrix} -a & c & b \\ -b & a & c \\ -c & b & a \end{vmatrix}$$

$$= \begin{vmatrix} a & b & c \\ b & c & a \\ c & a & b \end{vmatrix}\begin{vmatrix} a & b & c \\ b & c & a \\ c & a & b \end{vmatrix}, C_2 \leftrightarrow C_3$$

$$= \begin{vmatrix} a & b & c \\ b & c & a \\ c & a & b \end{vmatrix}^2$$

$$= [a(bc-a^2) - b(b^2-ac) + c(ab+c^2)]^2$$

$$= [3abc - a3 - b3 - c3]^2$$

$$= [a^3 + b^3 + c^3 - 3abc]^2.$$

EXAMPLE 2.95

Find the roots of the equation

$$\begin{vmatrix} x & -6 & -1 \\ 2 & -3x & x-3 \\ -3 & 2x & x+2 \end{vmatrix} = 0.$$

Solution. Putting $x = 2$ in the determinant, we observe that the first two rows are identical. Hence the determinant vanishes for $x = 2$. Thus $(x-2)$ is a factor of the determinant. Putting $x = -3$, first and third row become identical as so the determinant vanishes for $x = -3$ and so $(x+3)$ is a factor of the determinant. Substituting $x = 1$, the determinant becomes

$$\begin{vmatrix} 1 & -6 & -1 \\ 2 & -3 & -2 \\ -3 & 2 & 3 \end{vmatrix} = -\begin{vmatrix} 1 & -6 & 1 \\ 2 & -3 & 2 \\ -3 & 2 & -3 \end{vmatrix} = 0,$$

since first and third columns are identical. Hence $x-1$ is a factor of the determinant. Thus the given equation reduces to

$$(x-1)(x-2)(x+3) = 0,$$

which yields $x = 1$, $x = 2$, $x = -3$ as the solution.

EXAMPLE 2.96

Evaluate the determinant

$$\Delta = \begin{vmatrix} (b+c)^2 & a^2 & a^2 \\ b^2 & (c+a)^2 & b^2 \\ c^2 & c^2 & (a+b)^2 \end{vmatrix}$$

Solution. We have

$$\Delta = \begin{vmatrix} (b+c)^2 & a^2-(b+c)^2 & a^2-(b+c)^2 \\ b^2 & (c+a)^2-b^2 & 0 \\ c^2 & 0 & (a+b)^2-c^2 \end{vmatrix}$$

$$C_2 \to C_2 - C_1$$
$$C_3 \to C_3 - C_1$$

$$= \begin{vmatrix} (b+c)^2 & (a+b+c)(a-b-c) & (a+b+c)(a-b-c) \\ b^2 & (c+a-b)(a+b+c) & 0 \\ c^2 & 0 & (a+b+c)(a+b-c) \end{vmatrix}$$

$$= (a+b+c)^2 \begin{vmatrix} (b+c)^2 & a-b-c & a-b-c \\ b^2 & a-b+c & 0 \\ c^2 & 0 & a+b-c \end{vmatrix}$$

$$= (a+b+c)^2 \begin{vmatrix} 2bc & -2c & -2b \\ b^2 & a-b+c & 0 \\ c^2 & 0 & a+b-c \end{vmatrix}$$

$$R_1 \to R_1 - (R_2 + R_3)$$

$$= 2(a+b+c)^2 \begin{vmatrix} bc & -c & -b \\ b^2 & a-b+c & 0 \\ c^2 & 0 & a+b-c \end{vmatrix}$$

$$= 2(a+b+c)^2 \begin{vmatrix} 0 & -c & -b \\ b^2 & a-b+c & 0 \\ ac+bc & 0 & a+b-c \end{vmatrix}$$

$$C_1 \to C_1 + C_3)$$

$$= 2(a+b+c)^2[cb^2(a+b-c)$$
$$+ b(ac+bc)(a-b+c)]$$

$$= 2bc(a+b+c)^2[b(a+b-c)$$
$$+ (a+b+(a-b+c))]$$

$$= 2bc(a+b+c)^2[ab+b^2-bc+a^2-ab$$
$$+ ac+ab-b^2+bc]$$

$$= 2bc(a+b+c)^2[a(a+b+c)]$$

$$= 2abc(a+b+c)^3$$

EXAMPLE 2.97

If ω is a cube root of unity, show that

$$\Delta = \begin{vmatrix} 1 & \omega & \omega^2 \\ \omega & \omega^2 & 1 \\ \omega^2 & 1 & \omega \end{vmatrix} = 0.$$

Solution. We have

$$\Delta = \begin{vmatrix} 1+\omega+\omega^2 & \omega & \omega^2 \\ 1+\omega+\omega^2 & \omega^2 & 1 \\ 1+\omega+\omega^2 & 1 & \omega \end{vmatrix}, C_1 \to C_1 + C_2 + C_3$$

$$= \begin{vmatrix} 0 & \omega & \omega^2 \\ 0 & \omega^2 & 1 \\ 0 & 1 & \omega \end{vmatrix}, \text{since } 1+\omega+\omega^2 = 0$$

$= 0$, since first column consists of all zero entries

EXAMPLE 2.98

Evaluate

$$\Delta = \begin{vmatrix} 1 & 1 & 1 \\ a & b & c \\ bc & ca & ab \end{vmatrix}$$

Solution. We have

$$\Delta = \begin{vmatrix} 1 & 1 & 0 \\ a & b-a & c-b \\ bc & c(a-b) & a(b-c) \end{vmatrix}, \begin{matrix} C_2 \to C_2 - C_1 \\ C_3 \to C_3 - C_2 \end{matrix}$$

$$= 1[(b-a)a(b-c) - c(a-b)(c-b)]$$

$$= (b-a)[a(b-c) + c(c-b)]$$

$$= (b-a)(b-c)(a-c) = (a-b)(b-c)(c-a).$$

EXAMPLE 2.99

If A_i denotes the cofactor of a_i, show that

$$\begin{vmatrix} A_1 & B_1 & C_1 \\ A_2 & B_2 & C_2 \\ A_3 & B_3 & C_3 \end{vmatrix} = \begin{vmatrix} a_1 & b_1 & c_1 \\ a_2 & b_2 & c_2 \\ a_3 & b_3 & c_3 \end{vmatrix}^2$$

Solution. It is a particular case of Jacobi's Theorem. In fact, if

$$\Delta = \begin{vmatrix} a_1 & b_1 & c_1 \\ a_2 & b_2 & c_2 \\ a_3 & b_3 & c_3 \end{vmatrix},$$

$$\Delta' = \begin{vmatrix} A_1 & B_1 & C_1 \\ A_2 & B_2 & C_2 \\ A_3 & B_3 & C_3 \end{vmatrix},$$

then

$$\Delta\Delta' = \begin{vmatrix} \Delta & 0 & 0 \\ 0 & \Delta & 0 \\ 0 & 0 & \Delta \end{vmatrix} = \Delta^3$$

or

$$\Delta' = \Delta^2.$$

EXAMPLE 2.100

Expand

$$\Delta = \begin{vmatrix} 1+a^2-b^2 & 2ab & -2b \\ 2ab & 1-a^2+b^2 & 2a \\ 2b & -2a & 1-a^2-b^2 \end{vmatrix}$$

Solution. We have

$$\Delta = \begin{vmatrix} 1+a^2+b^2 & 0 & -2b \\ 0 & 1+a^2+b^2 & 2a \\ b(1+a^2+b^2) & -a(1+a^2+b^2) & 1-a^2-b^2 \end{vmatrix}$$

$$C_1 \to C_1 - bC_3$$
$$C_2 \to C_2 + aC_3$$

$$= \left(1+a^2+b^2\right)^2 \begin{vmatrix} 1 & 0 & -2b \\ 0 & 1 & 2a \\ b & -a & 1-a^2-b^2 \end{vmatrix}$$

$$= \left(1+a^2+b^2\right)^2 \begin{vmatrix} 1 & 0 & -2b \\ 0 & 1 & 2a \\ 0 & 0 & 1+a^2+b^2 \end{vmatrix}$$

$$R_3 \to R_3 - bR_1 + aR_2$$

$$= \left(1+a^2+b^2\right)^3, \text{ using triangle property.}$$

EXAMPLE 2.101

Expand the determinant

$$\Delta = \begin{vmatrix} 3 & 1 & -1 & 2 \\ -5 & 1 & 3 & -4 \\ 2 & 0 & 1 & -1 \\ 1 & -0 & 3 & -3 \end{vmatrix}$$

about the third row

Solution. We have

$$\Delta = 2\,cof(2) + 0\,cof(0) + 1\,cof(1) + (-1)cof(-1)$$

$$= 2(-1)^{3+1} \begin{vmatrix} 1 & -1 & 2 \\ 1 & 3 & -4 \\ -5 & 3 & -3 \end{vmatrix} + 1(-1)^{3+3} \begin{vmatrix} 3 & 1 & 2 \\ -5 & 1 & -4 \\ 1 & -5 & -3 \end{vmatrix} + (-1)(-1)^{3+4} \begin{vmatrix} 3 & 1 & -1 \\ -5 & 1 & 3 \\ 1 & -5 & 3 \end{vmatrix}$$

$$= 2[1(-9+12)+1(-3-20)+2(3+15)] + 1[3(-3-20)-1(15+4)+2(25-1)] + [3(3+15)-1(-15-3)-1(25-1)]$$

$$= 2(16) - 40 + 48 = 40$$

EXAMPLE 2.102

Using Laplace expansion about first and third row, evaluate the determinant of Example 2.101.

Solution. Using Laplace expansion, we have

$$\Delta = (1)^{1+3+1+2} \left|A_{13}^{12}\right| \left|A_{24}^{34}\right|$$

$$+(1)^{1+3+1+3}\left|A^{13}_{13}\right|\left|A^{24}_{24}\right|$$
$$+(-1)^{1+3+1+4}\left|A^{14}_{13}\right|\left|A^{23}_{24}\right|$$
$$+(-1)^{1+3+2+3}\left|A^{23}_{13}\right|\left|A^{14}_{24}\right|$$
$$+(-1)^{1+3+2+4}\left|A^{24}_{13}\right|\left|A^{13}_{24}\right|$$
$$+(-1)^{1+3+3+4}\left|A^{34}_{13}\right|\left|A^{12}_{24}\right|$$

$$= -\begin{vmatrix}3 & 1\\ 2 & 0\end{vmatrix}\begin{vmatrix}3 & -4\\ 3 & -3\end{vmatrix}+\begin{vmatrix}3 & -1\\ 2 & 1\end{vmatrix}\begin{vmatrix}1 & -4\\ -5 & -3\end{vmatrix}$$
$$-\begin{vmatrix}3 & 2\\ 2 & -1\end{vmatrix}\begin{vmatrix}1 & 3\\ -5 & 3\end{vmatrix}-\begin{vmatrix}1 & -1\\ 0 & 1\end{vmatrix}\begin{vmatrix}-5 & -4\\ 1 & -3\end{vmatrix}$$
$$+\begin{vmatrix}1 & 2\\ 0 & -1\end{vmatrix}\begin{vmatrix}-5 & 3\\ 1 & 3\end{vmatrix}$$
$$-\begin{vmatrix}-1 & 2\\ 1 & -1\end{vmatrix}\begin{vmatrix}-5 & 1\\ 1 & -5\end{vmatrix}$$
$$= -(-2)(-9+12)+(3+2)(-3-20)$$
$$-(-3-4)(3+15)-(1)(15+4)$$
$$+(-1)(-15-3)-(1-2)(25-1)$$
$$= 6-115+126-19+18+24 = 40.$$

2.22 ADJOINT OF A MATRIX

Definition 2.85. Let $A = [a_{ij}]$ be a square matrix of order n. Then the *cofactor* of a_{ij} is defined as

$$A_{ij} = cof(a_{ij}) = (-1)^{i+j}\left|M_{ij}\right|,$$

where M_{ij} is the matrix obtained by deleting ith row and jth column of the matrix A.

For example, if

$$A = \begin{bmatrix}1 & 3 & 2\\ 2 & 0 & 4\\ 3 & 6 & 5\end{bmatrix},$$

then

$$\text{cof}(a_{23}) = \text{cof}(4) = (-1)^5\begin{vmatrix}1 & 3\\ 3 & 6\end{vmatrix}$$
$$= -(6-9) = 3$$
$$\text{cof}(a_{32}) = \text{cof}(6) = (-1)^5\begin{vmatrix}1 & 2\\ 2 & 4\end{vmatrix} = 0.$$

Definition 2.86. Let $A = [a_{ij}]$ be a square matrix of order n. Then the *cofactor matrix* of A is defined to be the matrix $[A_{ij}]$, where A_{ij} denotes the cofactor of the entry a_{ij} in $|A|$.

For example,if

$$A = \begin{bmatrix}2 & 1 & 0\\ 0 & -3 & 1\\ -1 & -1 & 3\end{bmatrix},$$

then

$$A_{11} = (-1)^2(-9+1) = -8,$$
$$A_{12} = (-1)^3(0+1) = -1,$$
$$A_{13} = (-1)^4(0-3) = -3,$$
$$A_{21} = (-1)^3(3+0) = -3,$$
$$A_{22} = (-1)^4(6+0) = 6,$$
$$A_{23} = (-1)^5(-2+1) = 1,$$
$$A_{31} = (-1)^4(1+0) = 1,$$
$$A_{32} = (-1)^5(2+0) = -2,$$
$$A_{33} = (-1)^6(-6+0) = -6.$$

Hence, the cofactor matrix of A is given by

$$[A_{ij}] = \begin{bmatrix}-8 & -1 & -3\\ -3 & 6 & 1\\ 1 & -2 & -6\end{bmatrix}.$$

Definition 2.87. The *adjoint* of a square matrix $A = [a_{ij}]$ of order n is defined to the transpose of the cofactor matrix of A. Thus

$$\text{adj}\, A = \begin{bmatrix}A_{11} & A_{21} & \cdots & \cdots & A_{n1}\\ A_{12} & A_{22} & \cdots & \cdots & A_{n2}\\ A_{13} & A_{23} & \cdots & \cdots & A_{n3}\\ \cdots & \cdots & \cdots & \cdots & \cdots\\ A_{1n} & A_{2n} & \cdots & \cdots & A_{nn}\end{bmatrix}.$$

EXAMPLE 2.103

Find the adj A if

$$A = \begin{bmatrix}2 & 1 & 0\\ 0 & -3 & 1\\ -1 & -1 & 3\end{bmatrix}.$$

Solution. We have seen earlier that the cofactor matrix of A is

$$[A_{ij}] = \begin{bmatrix} -8 & -1 & -3 \\ -3 & 6 & 1 \\ 1 & -2 & -6 \end{bmatrix}.$$

Therefore,

$$\text{adj } A = [A_{ij}]^T = \begin{bmatrix} -8 & -1 & -3 \\ -3 & 6 & 1 \\ 1 & -2 & -6 \end{bmatrix}^T$$

$$= \begin{bmatrix} -8 & -3 & 1 \\ -1 & 6 & -2 \\ -3 & 1 & -6 \end{bmatrix}.$$

Theorem 2.36. Let A be an $n \times n$ matrix. Then

$$A(adj\,A) = |A|I_n = (adj\,A)\,A.$$

Proof. Since both A and $adj\,A$ are square matrices of order n, the product $A(\text{adj } A)$ and $(\text{adj } A)$ A are defined. Let

$$A = \begin{bmatrix} a_{11} & a_{12} & \dots & \dots & a_{1n} \\ a_{21} & a_{22} & \dots & \dots & a_{2n} \\ \dots & \dots & \dots & \dots & \dots \\ \dots & \dots & \dots & \dots & \dots \\ a_{n1} & a_{n2} & \dots & \dots & a_{nn} \end{bmatrix}$$

Then the cofactor matrix of A is

$$[A_{ij}] = \begin{bmatrix} A_{11} & A_{12} & \dots & \dots & A_{1n} \\ A_{21} & A_{22} & \dots & \dots & A_{2n} \\ \dots & \dots & \dots & \dots & \dots \\ \dots & \dots & \dots & \dots & \dots \\ A_{n1} & A_{n2} & \dots & \dots & A_{nn} \end{bmatrix}$$

and therefore,

$$adj\,A = \begin{bmatrix} A_{11} & A_{21} & \dots & \dots & A_{n1} \\ A_{12} & A_{22} & \dots & \dots & A_{n2} \\ \dots & \dots & \dots & \dots & \dots \\ \dots & \dots & \dots & \dots & \dots \\ A_{1n} & A_{2n} & \dots & \dots & A_{nn} \end{bmatrix}.$$

Thus

$$A\,(adj\,A) = \begin{bmatrix} a_{11} & a_{12} & \dots & \dots & a_{1n} \\ a_{21} & a_{22} & \dots & \dots & a_{2n} \\ \dots & \dots & \dots & \dots & \dots \\ \dots & \dots & \dots & \dots & \dots \\ a_{n1} & a_{n2} & \dots & \dots & a_{nn} \end{bmatrix}$$

$$\times \begin{bmatrix} A_{11} & A_{21} & \dots & \dots & A_{n1} \\ A_{12} & A_{22} & \dots & \dots & A_{n2} \\ \dots & \dots & \dots & \dots & \dots \\ \dots & \dots & \dots & \dots & \dots \\ A_{1n} & A_{2n} & \dots & \dots & A_{nn} \end{bmatrix}$$

But, we know that

$$a_{i1}A_{j1} + a_{i2}A_{j2} + \cdots + a_{in}A_{jn} = \begin{cases} |A| \; if\; i = j \\ 0 \; if\; i \neq j \end{cases}.$$

Therefore,

$$A\,(\text{adj } A) = \begin{bmatrix} |A| & 0 & \dots & \dots & 0 \\ 0 & |A| & \dots & \dots & 0 \\ \dots & \dots & \dots & \dots & \dots \\ \dots & \dots & \dots & \dots & \dots \\ 0 & 0 & \dots & \dots & |A| \end{bmatrix}$$

$$= |A| \begin{bmatrix} 1 & 0 & \dots & \dots & 0 \\ 0 & 1 & \dots & \dots & 0 \\ \dots & \dots & \dots & \dots & \dots \\ \dots & \dots & \dots & \dots & \dots \\ 0 & 0 & \dots & \dots & 1 \end{bmatrix}$$

$$= |A|I_n.$$

Similarly,

$$(\text{adj } A)A = |A|I_n.$$

Hence

$$A\,(\text{adj } A) = |A|\,I_n = (\text{adj } A)A.$$

Corollary 2.9. If $|A| \neq 0$, then

$$A\left(\frac{1}{|A|}\text{adj } A\right) = I_n = \left(\frac{1}{|A|}\text{adj } A\right)A.$$

2.23 THE INVERSE OF A MATRIX

Definition 2.88. A square matrix A of order n is said to be *invertible* if there exists another square matrix B of order n such that

$$AB = BA = I_n.$$

The matrix B is then called the *inverse of A*. If there exists no such matrix B, then A is called *non-invertible (singular)*. The inverse of A is denoted by A^{-1}. For example, if $A = \begin{bmatrix} 1 & 1 \\ 0 & 1 \end{bmatrix}$ and $B = \begin{bmatrix} 1 & -1 \\ 0 & 1 \end{bmatrix}$, then

$$AB = \begin{bmatrix} 1 & 1 \\ 0 & 1 \end{bmatrix}\begin{bmatrix} 1 & -1 \\ 0 & 1 \end{bmatrix} = \begin{bmatrix} 1 & 0 \\ 0 & 1 \end{bmatrix}$$

$$BA = \begin{bmatrix} 1 & -1 \\ 0 & 1 \end{bmatrix}\begin{bmatrix} 1 & 1 \\ 0 & 1 \end{bmatrix} = \begin{bmatrix} 1 & 0 \\ 0 & 1 \end{bmatrix}.$$

Thus $AB = BA = I_2$. Hence A is invertible and its inverse is B.

Theorem 2.37. The inverse of a square matrix is unique.

Proof. Suppose on the contrary that B and C are two inverse of a matrix A. Then

$$AB = BA = I_n \qquad (47)$$

and

$$AC = CA = I_n. \qquad (48)$$

Thus, we have

$$\begin{aligned} B &= B\,I_n \quad \text{(property of identity matrix)} \\ &= BAC \quad \text{[using (47)]} \\ &= (BA)\,C \quad \text{(Associative Law)} \\ &= I_nC \quad \text{[using (48)]} \\ &= C. \end{aligned}$$

Hence, inverse of A is unique.

Definition 2.89. A square matrix A is called *non-singular* if $|A| \neq 0$.

The square matrix A will be called *singular* if $|A| = 0$.

Theorem 2.38. A square matrix A is invertible if and only if it is non-singular.

Proof. *The condition is necessary.* Let A be invertible and let B be the inverse of A so that

$$AB = I = BA.$$

Therefore,

$$|A|\,|B| = |I| = 1.$$

Hence $|A| \neq 0$.

The condition is sufficient. Let A be non singular. Therefore, $|A| \neq 0$. Let

$$B = \frac{1}{|A|}(\text{adj}\,A).$$

Then

$$\begin{aligned} AB = A\left(\frac{1}{|A|}\text{adj}\,A\right) &= \frac{1}{|A|}(A\ \text{adj}\,A) \\ &= \frac{1}{|A|}[|A|I] = I. \end{aligned}$$

Similarly, $BA = I$. Hence, $AB = BA = I$ and so $B = \frac{1}{|A|}(\text{adj}\,A)$ is the inverse of A.

Theorem 2.39. Let A and B be two non-singular matrices of the same order. Then AB is non-singular and

$$(AB)^{-1} = B^{-1}A^{-1}.$$

Proof. Since

$$|AB| = |A|\,|B| \neq 0,$$

it follows that AB is non-singular and so invertible. Moreover,

$$\begin{aligned} (AB)\,(B^{-1}A^{-1}) &= A\left(B\,B^{-1}\right)A^{-1} \\ &= AIA^{-1} = AA^{-1} = I \end{aligned}$$

and

$$\begin{aligned} (B^{-1}A^{-1})\,(AB) &= B^{-1}\left(A^{-1}A\right)B = B^{-1}I\,B \\ &= B^{-1}B = I. \end{aligned}$$

Hence

$$(AB)\left(B^{-1}A^{-1}\right) = I = \left(B^{-1}A^{-1}\right)(AB),$$

which proves that $B^{-1}A^{-1}$ is the inverse of AB, that is,

$$(AB)^{-1} = B^{-1}A^{-1}$$

Theorem 2.40. If A is a non-singular matrix, then

$$\left(A^T\right)^{-1} = \left(A^{-1}\right)^T.$$

(Thus operations of transposing and inversion commute).

Proof. We note that

$$A^T\left(A^{-1}\right)^T = \left(A^{-1}A\right)^T = I^T = I$$

and

$$\left(A^{-1}\right)^T A^T = \left(AA^{-1}\right)^T = I^T = I.$$

Hence

$$A^T\left(A^{-1}\right)^T = I = \left(A^{-1}\right)^T A^T$$

and so

$$\left(A^T\right)^{-1} = \left(A^{-1}\right)^T.$$

Theorem 2.41. If a matrix A is invertible, then A^θ is invertible and

$$\left(A^\theta\right)^{-1} = \left(A^{-1}\right)^\theta.$$

Proof. We have

$$A^\theta\left(A^{-1}\right)^\theta = \left(A^{-1}A\right)^\theta = I^\theta = I$$

and

$$\left(A^{-1}\right)^\theta A^\theta = \left(AA^{-1}\right)^\theta = I^\theta = I.$$

Thus, $\left(A^{-1}\right)^\theta$ is the inverse of the A^θ.

2.24 METHODS OF COMPUTING INVERSE OF A MATRIX

1. Method of an Adjoint Matrix

If A is non-singular square matrix, then we have

$$A\left(\frac{1}{|A|}\operatorname{adj} A\right) = I = \left(\frac{1}{|A|}\operatorname{adj} A\right)A.$$

This relation yields

$$A^{-1} = \frac{1}{|A|}\operatorname{adj} A.$$

EXAMPLE 2.104

Find the inverse of the matrix

$$A = \begin{vmatrix} 3 & -3 & 4 \\ 2 & -3 & 4 \\ 0 & -1 & 1 \end{vmatrix}.$$

Solution. We have

$$|A| = \begin{vmatrix} 3 & -3 & 4 \\ 2 & -3 & 4 \\ 0 & -1 & 1 \end{vmatrix}$$

$$= 3(-3+4) + 3(2) + 4(-2) = 1.$$

Cofactors of the entries are

$$A_{11} = 1, \quad A_{12} = -2 \quad A_{13} = -2$$

$$A_{21} = -1, \quad A_{22} = 3 \quad A_{23} = 3$$

$$A_{31} = 0 \quad A_{32} = -4 \quad A_{33} = -3.$$

Therefore, the cofactor matrix is

$$[A_{ij}] = \begin{bmatrix} 1 & -2 & -2 \\ -1 & 3 & 3 \\ 0 & -4 & -3 \end{bmatrix}$$

and so

$$\operatorname{adj} A = \begin{bmatrix} 1 & -1 & 0 \\ -2 & 3 & -4 \\ -2 & 3 & -3 \end{bmatrix}.$$

Hence

$$A^{-1} = \frac{1}{|A|}\operatorname{adj} A = \begin{bmatrix} 1 & -1 & 0 \\ -2 & 3 & -4 \\ -2 & 3 & -3 \end{bmatrix}.$$

EXAMPLE 2.105

Find A^{-1} if

$$A = \begin{bmatrix} 1 & 2 & -1 \\ 3 & 0 & 2 \\ 4 & -2 & 5 \end{bmatrix}.$$

Solution. We have $|A| = -4$. The cofactor matrix is

$$[A_{ij}] = \begin{bmatrix} 4 & -7 & -6 \\ -8 & 9 & 10 \\ 4 & -5 & -6 \end{bmatrix}$$

and so

$$\text{adj } A = \begin{bmatrix} 4 & -8 & 4 \\ -7 & 9 & -5 \\ -6 & 10 & -6 \end{bmatrix}.$$

Hence

$$A^{-1} = -\frac{1}{4}\begin{bmatrix} 4 & -8 & 4 \\ -7 & 9 & -5 \\ -6 & 10 & -6 \end{bmatrix} = \begin{bmatrix} -1 & 2 & -1 \\ \frac{7}{4} & \frac{-9}{4} & \frac{5}{4} \\ \frac{3}{2} & \frac{-5}{2} & \frac{3}{2} \end{bmatrix}$$

2. Method Using Definition of Inverse

Let B be the inverse of matrix A, which is non-singular. Then, $AB = I$, that is,

$$\begin{bmatrix} a_{11} & a_{12} & \dots & \dots & a_{1n} \\ a_{21} & a_{22} & \dots & \dots & a_{2n} \\ \dots & \dots & \dots & \dots & \dots \\ \dots & \dots & \dots & \dots & \dots \\ a_{n1} & a_{n2} & \dots & \dots & a_{nn} \end{bmatrix}\begin{bmatrix} b_{11} & b_{12} & \dots & \dots & b_{1n} \\ b_{21} & b_{22} & \dots & \dots & b_{2n} \\ \dots & \dots & \dots & \dots & \dots \\ \dots & \dots & \dots & \dots & \dots \\ b_{n1} & b_{n2} & \dots & \dots & b_{nn} \end{bmatrix}$$

$$= \begin{bmatrix} 1 & 0 & \dots & \dots & 0 \\ 0 & 1 & \dots & \dots & 0 \\ \dots & \dots & \dots & \dots & \dots \\ \dots & \dots & \dots & \dots & \dots \\ 0 & 0 & \dots & \dots & 1 \end{bmatrix}$$

Multiplying the matrices on the left and then comparing the corresponding entries we can find $b_{11}, b_{12}, \dots, b_{nn}$. Then, B will be the inverse of A.

EXAMPLE 2.106

Find the inverse of

$$A = \begin{bmatrix} 1 & 2 & -1 \\ 0 & 1 & 3 \\ 0 & 0 & 1 \end{bmatrix}$$

Solution. The given matrix is upper triangular matrix. The readers may prove that inverse of an upper triangular matrix is also upper triangular matrix. Similarly, the inverse of a lower-triangular matrix is again a lower-triangular matrix. So, let

$$\begin{bmatrix} a & b & c \\ 0 & d & e \\ 0 & 0 & f \end{bmatrix}$$

be the inverse of the given matrix. Then, by definition of the inverse, we must have

$$\begin{bmatrix} 1 & 2 & -1 \\ 0 & 1 & 3 \\ 0 & 0 & 1 \end{bmatrix}\begin{bmatrix} a & b & c \\ 0 & d & e \\ 0 & 0 & f \end{bmatrix} = \begin{bmatrix} 1 & 0 & 0 \\ 0 & 1 & 0 \\ 0 & 0 & 1 \end{bmatrix}$$

or

$$\begin{bmatrix} a & b+2d & c+2e-f \\ 0 & d & e+3f \\ 0 & 0 & f \end{bmatrix} = \begin{bmatrix} 1 & 0 & 0 \\ 0 & 1 & 0 \\ 0 & 0 & 1 \end{bmatrix}$$

Equating corresponding entries, we get

$a = 1,\ d = 1,\ f = 1$

$b + 2d = 0$ so that $b = -2d = -2$

$e + 3f = 0$ so that $e = -3f = -3$

$c + 2e - f = 0$ so that $c = f - 2e = 1 + 6 = 7.$

Hence,

$$A^{-1} = \begin{bmatrix} 1 & -2 & 7 \\ 0 & 1 & -3 \\ 0 & 0 & 1 \end{bmatrix}.$$

Remark 2.6 We can also find the inverse of a lower triangular matrix by the above method.

3. Method of Matrix Equation

Let

$$a_{11}x_1 + a_{12}x_2 + \ldots + a_{1n}x_n = b_1$$
$$a_{21}x_1 + a_{22}x_2 + \ldots + a_{2n}x_n = b_2$$
$$\ldots \quad \ldots \quad \ldots \quad \ldots$$
$$\ldots \quad \ldots \quad \ldots \quad \ldots$$
$$a_{n1}x_1 + a_{n2}x_2 + \ldots + a_{nn}x_n = b_n$$

be a set of n equations in n variables $x_1, x_2, \ldots, x_n$ In matrix form, we can represent these equations by

$$AX = B,$$

where

$$A = \begin{bmatrix} a_{11} & a_{12} & \ldots & \ldots & a_{1n} \\ a_{21} & a_{22} & \ldots & \ldots & a_{2n} \\ \ldots & \ldots & \ldots & \ldots & \ldots \\ \ldots & \ldots & \ldots & \ldots & \ldots \\ a_{n1} & a_{n2} & \ldots & \ldots & a_{nn} \end{bmatrix},$$

$$X = \begin{bmatrix} x_1 \\ x_2 \\ \ldots \\ \ldots \\ x_n \end{bmatrix}, \quad B = \begin{bmatrix} b_1 \\ b_2 \\ \ldots \\ \ldots \\ b_n \end{bmatrix}$$

and A is called the *coefficient matrix*. If A is non-singular matrix, then A^{-1} exists. Premultiplying the matrix equation by A^{-1} we get

$$A^{-1}(AX) = A^{-1}B$$

or

$$(A^{-1}A)X = A^{-1}B$$

or

$$IX = A^{-1}B$$

or

$$X = A^{-1}B.$$

Hence, if we can represent $x_1, x_2, \ldots, x_n$ in terms of $b_1, b_2, \ldots, b_n$, then the coefficient matrix of this system will be the inverse of A.

EXAMPLE 2.107

Find the inverse of

$$A = \begin{bmatrix} 1 & 0 & -4 \\ 0 & -1 & 2 \\ -1 & 2 & 1 \end{bmatrix}.$$

Solution. We observe that $|A| = -1 \neq 0$. Thus, A is non-singular and so the inverse of A exists. We consider the matrix equation

$$\begin{bmatrix} 1 & 0 & -4 \\ 0 & -1 & 2 \\ -1 & 2 & 1 \end{bmatrix} \begin{bmatrix} x_1 \\ x_2 \\ x_3 \end{bmatrix} = \begin{bmatrix} b_1 \\ b_2 \\ b_3 \end{bmatrix},$$

which yields

$$x_1 - 4x_3 = b_1$$
$$-x_2 + 2x_3 = b_2$$
$$-x_1 + 2x_2 + x_3 = b_3.$$

Solving these equations for x_1, x_2, and x_3, we have

$$x_1 = 5b_1 + 8b_2 + 4b_3$$
$$x_2 = 2b_1 + 3b_2 + 2b_3$$
$$x_3 = b_1 + 2b_2 + b_3.$$

In matrix form, we have

$$\begin{bmatrix} x_1 \\ x_2 \\ x_3 \end{bmatrix} = \begin{bmatrix} 5 & 8 & 4 \\ 2 & 3 & 2 \\ 1 & 2 & 1 \end{bmatrix} \begin{bmatrix} b_1 \\ b_2 \\ b_3 \end{bmatrix},$$

that is

$$X = A^{-1}B.$$

Hence

$$A^{-1} = \begin{bmatrix} 5 & 8 & 4 \\ 2 & 3 & 2 \\ 1 & 2 & 1 \end{bmatrix}.$$

4. Method of Elementary Transformation (Gauss--Jordan Method)

The following transformations are called *elementary transformation* of a matrix:

(a) Interchanging of rows (columns).

(b) Multiplication of a row (column) by a non-zero scalar.

(c) Adding/subtracting k multiple of a row (column) to another row (column).

Definition 2.90. A matrix B is said to be *row* (column) *equivalent* to a matrix A if it is obtained from A by applying a finite number of elementary row (column) transformations. In such case, we write B $\sim$ A. In Gauss–Jordan Method, we perform the sequence of elementary row transformations on A and I simultaneously, keeping them side-by-side.

EXAMPLE 2.108

Using elementary row transformations, find A^{-1} if

$$A = \begin{bmatrix} 1 & 0 & 2 \\ 2 & -1 & 3 \\ 4 & 1 & 5 \end{bmatrix}.$$

Solution. Consider the augmented matrix

$$[A|I] = \left[\begin{array}{ccc|ccc} 1 & 0 & 2 & 1 & 0 & 0 \\ 2 & -1 & 3 & 0 & 1 & 0 \\ 4 & 1 & 5 & 0 & 0 & 1 \end{array}\right]$$

$$\sim \left[\begin{array}{ccc|ccc} 1 & 0 & 2 & 1 & 0 & 0 \\ 0 & -1 & -1 & -2 & 1 & 0 \\ 0 & 1 & -3 & -4 & 0 & 1 \end{array}\right] \quad \begin{array}{l} R_2 \to R_2 - 2R_1 \\ R_3 \to R_3 - 4R_1 \end{array}$$

$$\sim \left[\begin{array}{ccc|ccc} 1 & 0 & 2 & 1 & 0 & 0 \\ 0 & 1 & 1 & 2 & -1 & 0 \\ 0 & 1 & -3 & -4 & 0 & 1 \end{array}\right] \quad R_2 \to -R_2$$

$$\sim \left[\begin{array}{ccc|ccc} 1 & 0 & 2 & 1 & 0 & 0 \\ 0 & 1 & 1 & 2 & -1 & 0 \\ 0 & 0 & -4 & -6 & 1 & 1 \end{array}\right] \quad R_3 \to R_3 - R_2$$

$$\sim \left[\begin{array}{ccc|ccc} 1 & 0 & 2 & 1 & 0 & 0 \\ 0 & 1 & 1 & 2 & -1 & 0 \\ 0 & 0 & 1 & \frac{3}{2} & -\frac{1}{4} & -\frac{1}{4} \end{array}\right] \quad R_3 \to -\frac{1}{4}R_3$$

$$\sim \left[\begin{array}{ccc|ccc} 1 & 0 & 2 & 1 & 0 & 0 \\ 0 & 1 & 0 & \frac{1}{2} & -\frac{3}{4} & \frac{1}{4} \\ 0 & 0 & 1 & \frac{3}{2} & -\frac{1}{4} & -\frac{1}{4} \end{array}\right] \quad R_2 \to R_2 - R_3$$

$$\sim \left[\begin{array}{ccc|ccc} 1 & 0 & 0 & -2 & \frac{1}{2} & \frac{1}{2} \\ 0 & 1 & 0 & \frac{1}{2} & -\frac{3}{4} & \frac{1}{4} \\ 0 & 0 & 1 & \frac{3}{2} & -\frac{1}{4} & -\frac{1}{4} \end{array}\right] \quad R_1 \to R_1 - 2R_3$$

Hence

$$A^{-1} = \begin{bmatrix} -2 & \frac{1}{2} & \frac{1}{2} \\ \frac{1}{2} & -\frac{3}{4} & \frac{1}{4} \\ \frac{3}{2} & -\frac{1}{4} & -\frac{1}{4} \end{bmatrix}.$$

EXAMPLE 2.109

Using elementary row transformation, find the inverse of the matrix

$$A = \begin{bmatrix} 1 & 3 & 3 \\ 1 & 4 & 3 \\ 1 & 3 & 4 \end{bmatrix}.$$

Solution. Consider the augmented matrix

$$[A|I] = \left[\begin{array}{ccc|ccc} 1 & 3 & 3 & 1 & 0 & 0 \\ 1 & 4 & 3 & 0 & 1 & 0 \\ 1 & 3 & 4 & 0 & 0 & 1 \end{array}\right]$$

$$\sim \left[\begin{array}{ccc|ccc} 1 & 3 & 3 & 1 & 0 & 0 \\ 0 & 1 & -1 & 0 & 1 & -1 \\ 1 & 3 & 4 & 0 & 0 & 1 \end{array}\right] \quad R_2 \to R_2 - R_3$$

$$\sim \left[\begin{array}{ccc|ccc} 1 & 3 & 3 & 1 & 0 & 0 \\ 0 & 1 & -1 & 0 & 1 & -1 \\ 0 & 0 & 1 & -1 & 0 & 1 \end{array}\right] \quad R_3 \to R_3 - R_1$$

$$\sim \left[\begin{array}{ccc|ccc} 1 & 0 & 6 & 1 & -3 & 3 \\ 0 & 1 & -1 & 0 & 1 & -1 \\ 0 & 0 & 1 & -1 & 0 & 1 \end{array}\right] \quad R_1 \to R_1 - 3R_2$$

$$\sim \left[\begin{array}{ccc|ccc} 1 & 0 & 6 & 1 & -3 & 3 \\ 0 & 1 & 0 & -1 & 1 & 0 \\ 0 & 0 & 1 & -1 & 0 & 1 \end{array}\right] \quad R_2 \to R_2 + R_3$$

$$\sim \left[\begin{array}{ccc|ccc} 1 & 0 & 0 & 7 & -3 & -3 \\ 0 & 1 & 0 & -1 & 1 & 0 \\ 0 & 0 & 1 & -1 & 0 & 1 \end{array}\right] \quad R_1 \to R_1 - 6R_3.$$

Hence

$$A^{-1} = \begin{bmatrix} 7 & -3 & -3 \\ -1 & 1 & 0 \\ -1 & 0 & 1 \end{bmatrix}.$$

EXAMPLE 2.110

Find the inverse of the matrix

$$A = \begin{bmatrix} 1 & 1 & 3 \\ 1 & 3 & -3 \\ -2 & -4 & -4 \end{bmatrix},$$

by using elementary transformations.

Solution. Write $A = I_3A$, that is,

$$\begin{bmatrix} 1 & 1 & 3 \\ 1 & 3 & -3 \\ -2 & -4 & -4 \end{bmatrix} = \begin{bmatrix} 1 & 0 & 0 \\ 0 & 1 & 0 \\ 0 & 0 & 1 \end{bmatrix} A.$$

Now we reduce the matrix A to identity matrix I_3 by elementary row transformation keeping in mind that each such row transformation will apply to the prefactor I_3 on the right hand side.

Performing $R_2 \to R_2 - R_1$ and $R_3 \to R_3 + 2R_1$, we get

$$\begin{bmatrix} 1 & 1 & 3 \\ 0 & 2 & -6 \\ 0 & -2 & 2 \end{bmatrix} = \begin{bmatrix} 1 & 0 & 0 \\ -1 & 1 & 0 \\ 2 & 0 & 1 \end{bmatrix} A.$$

Performing $R_2 \to \frac{1}{2}R_2$, we get

$$\begin{bmatrix} 1 & 1 & 3 \\ 0 & 1 & -3 \\ 0 & -2 & 2 \end{bmatrix} = \begin{bmatrix} 1 & 0 & 0 \\ -\frac{1}{2} & \frac{1}{2} & 0 \\ 2 & 0 & 1 \end{bmatrix} A.$$

Performing $R_1 \to R_1 - R_2$ and $R_3 \to R_3 + R_2$, we get,

$$\begin{bmatrix} 1 & 0 & 6 \\ 0 & 1 & -3 \\ 0 & 0 & -4 \end{bmatrix} = \begin{bmatrix} \frac{3}{2} & -\frac{1}{2} & 0 \\ -\frac{1}{2} & \frac{1}{2} & 0 \\ 1 & 1 & 1 \end{bmatrix} A.$$

Performing $R_3 \to -\frac{1}{4}R_3$, we get

$$\begin{bmatrix} 1 & 0 & 6 \\ 0 & 1 & -3 \\ 0 & 0 & 1 \end{bmatrix} = \begin{bmatrix} \frac{3}{2} & -\frac{1}{2} & 0 \\ -\frac{1}{2} & \frac{1}{2} & 0 \\ -\frac{1}{4} & -\frac{1}{4} & -\frac{1}{4} \end{bmatrix} A.$$

Performing $R_1 \to R_1 - 6R_3$ and $R_2 \to R_2 + 3R_3$, we get

$$\begin{bmatrix} 1 & 0 & 0 \\ 0 & 1 & 0 \\ 0 & 0 & 1 \end{bmatrix} = \begin{bmatrix} 3 & 1 & \frac{3}{2} \\ -\frac{5}{4} & -\frac{1}{4} & -\frac{3}{4} \\ -\frac{1}{4} & -\frac{1}{4} & -\frac{1}{4} \end{bmatrix} A.$$

Thus,

$$I_3 = \begin{bmatrix} 3 & 1 & \frac{3}{2} \\ -\frac{5}{4} & -\frac{1}{4} & -\frac{3}{4} \\ -\frac{1}{4} & -\frac{1}{4} & -\frac{1}{4} \end{bmatrix} A.$$

Hence

$$A^{-1} = \begin{bmatrix} 3 & 1 & \frac{3}{2} \\ -\frac{5}{4} & -\frac{1}{4} & -\frac{3}{4} \\ -\frac{1}{4} & -\frac{1}{4} & -\frac{1}{4} \end{bmatrix}.$$

2.25 RANK OF A MATRIX

Definition 2.91. A matrix is said to be of *rank* r if it has at least one non-singular submatrix of order r but has no non-singular submatrix of order more than r.

Rank of a matrix A is denoted by $\rho(A)$.

A matrix is said to be of *rank zero* if and only if all its elements are zero.

EXAMPLE 2.111

Find the rank of the matrix

$$A = \begin{bmatrix} 1 & 3 & 4 & 2 \\ 2 & 4 & 6 & 2 \\ -1 & 5 & 4 & 6 \end{bmatrix}.$$

Solution. The matrix A is of order 3×4. Therefore, $\rho(A) \le 3$. We note that

$$|A_1| = \begin{vmatrix} 1 & 3 & 4 \\ 2 & 4 & 6 \\ -1 & 5 & 4 \end{vmatrix} = 0,$$

$$|A_2| = \begin{vmatrix} 1 & 3 & 2 \\ 2 & 4 & 2 \\ -1 & 5 & 6 \end{vmatrix} = 0,$$

$$|A_3| = \begin{vmatrix} 1 & 4 & 2 \\ 2 & 6 & 2 \\ -1 & 4 & 6 \end{vmatrix} = 0,$$

$$|A_4| = \begin{vmatrix} 3 & 4 & 2 \\ 4 & 6 & 2 \\ 5 & 4 & 6 \end{vmatrix} = 0.$$

Therefore, $\rho(A) \neq 3$. But, we have submatrix $B = \begin{bmatrix} 1 & 3 \\ 2 & 4 \end{bmatrix}$, whose determinant is equal to $-2 \neq 0$. Hence, by definition, $\rho(A) = 2$.

EXAMPLE 2.112

Find the rank of the matrix

$$A = \begin{bmatrix} 2 & 1 & -1 \\ 0 & 3 & -2 \\ 2 & 4 & -3 \end{bmatrix}.$$

Solution. Since $|A| = 0$, $\rho(A) \leq 2$. But, we note that $\begin{vmatrix} 2 & 1 \\ 0 & 3 \end{vmatrix} = 6 \neq 0$. Hence, $\rho(A) = 2$.

Remark 2.7 The rank of a matrix is, of course, uniquely defined when the elements are all explicitly given numbers, but not necessarily otherwise. For example, consider the matrix

$$A = \begin{bmatrix} 4-x & 2\sqrt{5} & 0 \\ 2\sqrt{5} & 4-x & \sqrt{5} \\ 0 & \sqrt{5} & 4-x \end{bmatrix}.$$

We have

$$|A| = (4-x)^3 - 25(4-x) = 0, \text{ if } x = 9, 4 \text{ or } -1.$$

When $x = 9$, we have the singular matrix

$$A = \begin{bmatrix} -5 & 2\sqrt{5} & 0 \\ 2\sqrt{5} & -5 & \sqrt{5} \\ 0 & \sqrt{5} & -5 \end{bmatrix},$$

which has non-singular submatrix

$$\begin{bmatrix} -5 & \sqrt{5} \\ \sqrt{5} & -5 \end{bmatrix}.$$

Thus, for $x = 9$ the rank of A is 2. Similarly, the rank is 2 when $x = 4$ or $x = -1$. For other values of x, $|A| \neq 0$ and so the rank of A is 3.

Theorem 2.42. Let A be an $m \times n$ matrix. Then $\rho(A) = \rho(A^T)$.

Proof. Suppose $\rho(A) = r$. Then, there is at least one square submatrix R of A of order r whose determinant is non-zero. If R^T is transpose of R, then it is submatrix of A^T. Since, the value of a determinant does not alter by interchanging the rows and columns, $|R^T| = |R| \neq 0$. Therefore, $\rho(A^T) \geq r$.

Now if A^T contains a square submatrix S of order $r+1$, then corresponding to S, S^T is a submatrix of A of order $r+1$. But $\rho(A) = r$. Therefore, $|S| = |S^T| = 0$. Thus, A^T cannot contain an $(r+1)$ rowed square submatrix with non-zero determinant. Thus, $\rho(A^T) \leq r$. Hence, $\rho(A^T) = r$.

Theorem 2.43. The rank of a matrix does not alter under elementary row (column) transformations.

Proof. Let $A = [a_{ij}]$ be an $m \times n$ matrix of rank r. We prove the theorem only for elementary row transformation. The proof for column transformation is similar.

Case I. Interchange of a pair of row does not alter the rank.

Let s be the rank of the matrix B obtained from the matrix A of rank r by elementary transformation $R_p \leftrightarrow R_q$. Let B_0 be any $(r+1)$ rowed square submatrix of B. The $(r+1)$ rows of B_0 are also the rows of some uniquely determined submatrix A_0 of A. The identical rows of A_0 and B_0 may occur in the same or in different relative positions. Since, the interchange of two rows of a determinant changes only the sign, we have

$$|B_0| = |A_0| \quad or \quad |B_0| = -|A_0|.$$

Since $\rho(A) = r$, every $(r+1)$-rowed minor of A vanishes, that is, $|A_0| = 0$. Hence, $|B_0| = 0$. Therefore, every $(r+1)$ rowed minor of B vanishes. Hence, $s = \rho(B) \leq r = \rho(A)$. But A can also be obtained from B by interchanging its rows. Therefore, $r \leq s$. Hence $r = s$.

Case II. Multiplication of the elements of a row by a non-zero number does not alter the rank.

Let s be the rank of the matrix B obtained from the matrix A of rank r by the elementary transformation $R_p \to kR_p (k \neq 0)$. If B_0 is any $(r+1)$-rowed submatrix of B, then there exists a uniquely determined submatrix A_0 of A such that $|B_0| = |A_0|$ (when pth row of B is one of those rows which are deleted to obtain B_0 from B) or $|B_0| = k|A_0|$ (when pth row of B is retained while obtaining B_0 from B). Since $\rho(A) = r$, every $(r+1)$-rowed submatrix has zero determinant, that is $|A_0| = 0$. Hence, $|B_0| = 0$. Thus every $(r+1)$-rowed submatrix of B vanishes. Hence $\rho(B) \leq r$, that is, $s \leq r$. On the other hand, A can be obtained from B by elementary transformation $R_p \to \frac{1}{k} R_p$. Therefore, we have $r \leq s$. Hence $r = s$.

Case III. Addition to the elements of a row, the product by any number k of the corresponding elements of any other row, does not alter the rank.

Let s be the rank of the matrix B obtained from the matrix A by elementary transformation $R_p \to R_p + kR_q$. Let B_0 be any $(r+1)$-rowed square submatrix of B and A_0 be the corresponding placed submatrix of A. The transformation $R_p \to R_p + kR_q$ has changed only the pth row of the matrix A. We know that the value of the determinant does not change if we add to the elements of any row the corresponding elements of any other row multiplied by some number. Therefore, if no row of the submatrix A_0 is a part of the pth row or if two rows of A_0 are parts of the pth and qth rows of A, then $|B_0| = |A_0|$. Since $\rho(A) = r$, we have $|A_0| = 0$ and consequently $|B_0| = 0$.

Again, if a row of A_0 is a part of the pth row of A, but no row is part of qth row, then

$$|B_0| = |A_0| + k|C_0|,$$

where C_0 is an $(r+1)$-rowed square matrix which can be obtained from A_0 by replacing the elements of A_0 in the row which corresponds to the pth row of A by the corresponding elements in the qth row of A. All the $(r+1)$ rows of the matrix C_0 are exactly the same as the rows of some $(r+1)$-rowed square submatrix of A, though arranged in some different order. Therefore, $|C_0| = \pm$ times some $(r+1)$-rowed minor of A. Since the rank of A is r, every $(r+1)$- rowed minor of A is also zero, so that $|A_0| = 0$, $|C_0| = 0$, and so in turn $|B_0| = 0$. Thus, every $(r+1)$- rowed square matrix of B has zero determinant. Hence, $s \leq r$. Also, since, A can be obtained from B by an elementary transformation, $R_p \to R_p + kR_p$ Therefore, as stated, $r \leq s$. Hence $r = s$.

EXAMPLE 2.113

Find the rank of the matrix

$$A = \begin{bmatrix} 3 & 2 & -1 \\ 4 & 2 & 6 \\ 7 & 4 & 5 \end{bmatrix}.$$

Solution. We have

$$A = \begin{bmatrix} 3 & 2 & -1 \\ 4 & 2 & 6 \\ 7 & 4 & 5 \end{bmatrix}$$

$$\sim \begin{bmatrix} -1 & 0 & -7 \\ 4 & 2 & 6 \\ 7 & 4 & 5 \end{bmatrix} R_1 \to R_1 - R_2$$

$$\sim \begin{bmatrix} 1 & 0 & 7 \\ 4 & 2 & 6 \\ 7 & 4 & 5 \end{bmatrix} R_1 \to -R_1$$

$$\sim \begin{bmatrix} 1 & 0 & 7 \\ 0 & 2 & -22 \\ 0 & 4 & -44 \end{bmatrix} \begin{matrix} R_2 \to R_2 - 4R_1 \\ R_3 \to R_3 - 7R_1 \end{matrix}$$

$$\sim \begin{bmatrix} 1 & 0 & 7 \\ 0 & 1 & -11 \\ 0 & 4 & -44 \end{bmatrix} R_2 \to \frac{1}{2} R_2$$

$$\sim \begin{bmatrix} 1 & 0 & 7 \\ 0 & 1 & -11 \\ 0 & 0 & 0 \end{bmatrix} R_3 \to R_3 - 4R_2.$$

Thus, $|A| = 0$. Therefore $\rho(A) \neq 3$. But, since

$$\begin{vmatrix} 1 & 0 \\ 0 & 1 \end{vmatrix} = 1 \neq 0,$$

it follows that $\rho(A) = 2$.

2.26 ELEMENTARY MATRICES

Definition 2.92. A matrix obtained from a unit matrix by a single elementary transformation is called an *elementary matrix*.

For example,

$$\begin{bmatrix} 0 & 0 & 1 \\ 0 & 1 & 0 \\ 1 & 0 & 0 \end{bmatrix}$$

is the elementary matrix obtained from I_3 by subjecting it to $C_1 \leftrightarrow C_3$. The matrix

$$\begin{bmatrix} 4 & 0 & 0 \\ 0 & 1 & 0 \\ 0 & 0 & 1 \end{bmatrix}$$

is the elementary matrix obtained from I_3 by subjecting it to $R_1 \to 4R_1$, whereas the matrix

$$\begin{bmatrix} 1 & 2 & 0 \\ 0 & 1 & 0 \\ 0 & 0 & 1 \end{bmatrix}$$

is the elementary matrix obtained from I_3 by subjecting it to $R_1 \to R_1 + 2R_1$. The elementary matrix obtained by interchanging the ith and jth row of a unit matrix I is denoted by E_{ij}. Since, we obtain same matrix by interchanging ith and jth row or ith and jth column, E_{ij} will also denote the elementary matrix obtained from A by interchanging ith and jth column. $E_i(k)$ denotes the elementary matrix obtained by multiplying the ith row or ith column of a unit matrix by k.

Similarly, $E_{ij}(m)$ denotes the elementary matrix obtained by adding to the elements of the ith row (column) of a unit matrix the m multiple of the corresponding elements of the jth row (column).

We note that $|E_{ij}| = -1$, $|E_i(k)| = k \neq 0$, $|E_{ij}(m)| = 1$. It follows, therefore, that *all the elementary matrices are non-singular and, hence, possess inverse.*

Theorem 2.44. Every elementary row (column) transformation of a matrix can be obtained by premultiplication (post-multiplication) with corresponding elementary matrix.

Proof. Let B be the matrix obtained from an $m \times n$ matrix A by row transformation. If E is elementary matrix obtained from I_m by the same row transformation, it is sufficient to show that $B = EA$.

Let

$$M = \begin{bmatrix} R_1 \\ R_2 \\ \dots \\ \dots \\ R_m \end{bmatrix}, \; N = [C_1 C_2 \dots C_n].$$

Then

$$MN = \begin{bmatrix} R_1C_1 & R_1C_2 & \dots & \dots & R_1C_n \\ R_2C_1 & R_2C_2 & \dots & \dots & R_2C_n \\ \dots & \dots & \dots & \dots & \dots \\ \dots & \dots & \dots & \dots & \dots \\ R_mC_1 & R_mC_2 & \dots & \dots & R_mC_n \end{bmatrix}$$

Clearly, a row transformation applied to M will be the row transformation applied to MN. Hence, elementary row transformation of a product MN of two matrices M and N can be obtained by subjecting the prefactor M to the same elementary row transformation.

Similarly, every elementary column transformation of a product MN can be obtained by subjecting the post-factor N to the same elementary column transformation.

Now, A is an $m \times n$ matrix and I_m is an identity matrix of order m. Therefore, $A = I_mA$. Hence, by the preceding arguments, if we apply a row transformation to A to get a matrix B, then this can be done by applying the same row transformation to I_m. Thus, if B is obtained from A by applying a row transformation and E is obtained from I_m by using the same row transformation, then $B = EA$.

Similarly, if B is obtained from A by subjecting it to a column transformation and E is obtained from I by subjecting it to the same column transformation then $B = AE$.

EXAMPLE 2.114

Let

$$A = \begin{bmatrix} 1 & 3 & 4 \\ 2 & 1 & 3 \\ 5 & 3 & 2 \end{bmatrix}$$

and

$$B = \begin{bmatrix} 1 & 3 & 4 \\ 4 & 2 & 6 \\ 5 & 3 & 2 \end{bmatrix}$$

Thus, B has been obtained from A by the row transformation $R_2 \to 2\,R_2$. Now, if, E is the elementary matrix obtained from I_3 by $R_2 \to 2\,R_2$ then

$$EA = \begin{bmatrix} 1 & 0 & 0 \\ 0 & 2 & 0 \\ 0 & 0 & 1 \end{bmatrix}\begin{bmatrix} 1 & 3 & 4 \\ 2 & 1 & 3 \\ 5 & 3 & 2 \end{bmatrix} = \begin{bmatrix} 1 & 3 & 4 \\ 4 & 2 & 6 \\ 0 & 0 & 1 \end{bmatrix} = B.$$

2.27 ROW REDUCED ECHELON FORM AND NORMAL FORM OF MATRICES

Definition 2.93. A matrix is said to be in *row-reduced echelon* form if

(i) The first non-zero entry in each non-zero row is 1.

(ii) The rows containing only zeros occur below all the non-zero rows.

(iii) The number of zeros before the first nonzero element in a row is less than the number of such zeros in the next row.

The rank of a matrix in row reduced echelon form is equal to the number of non-zero rows of the matrix. For example, the matrix,

$$\begin{bmatrix} 0 & 1 & 3 & 4 \\ 0 & 0 & 1 & 2 \\ 0 & 0 & 0 & 0 \end{bmatrix}$$

is in the row reduced echelon form and its rank is 2 (the number of non-zero rows).

Theorem 2.45. Every non-zero $m \times n$ matrix of rank r can be reduced, by a sequence of elementary transformation, to the form

$$\begin{bmatrix} I_r & 0 \\ 0 & 0 \end{bmatrix}$$

(*normal form* or *first canonical form*), where I_r is the identity matrix of order r.

Proof. Let $A = [a_{ij}]_{m\times n}$ be a matrix of rank r. Since A is non-zero, it has at least one element different from zero. Suppose $a_{ij} \neq 0$. Interchanging the first and ith row and then first and jth column we obtain a matrix B whose leading element is non-zero, say k.

Multiplying the elements of the first row of the matrix B by $\frac{1}{k}$, we obtain a matrix

$$C = \begin{bmatrix} 1 & c_{12} & c_{13} & \dots & \dots & c_{1n} \\ c_{21} & c_{22} & c_{23} & \dots & \dots & c_{2n} \\ \dots & \dots & \dots & \dots & \dots & \dots \\ \dots & \dots & \dots & \dots & \dots & \dots \\ c_{m1} & c_{m2} & c_{m3} & \dots & \dots & c_{mn} \end{bmatrix},$$

whose leading element is equal to 1. Subtracting suitable multiples of the first column of C from the remaining columns, and suitable multiples of first row from the remaining rows, we obtain a matrix

$$D = \begin{bmatrix} 1 & 0 & 0 & \dots & \dots & 0 \\ 0 & d_{22} & d_{23} & \dots & \dots & d_{2n} \\ \dots & \dots & \dots & \dots & \dots & \dots \\ \dots & \dots & \dots & \dots & \dots & \dots \\ 0 & d_{m2} & d_{m3} & \dots & \dots & d_{mn} \end{bmatrix},$$

in which all elements of the first row and first column except the leading element are equal to zero. If

$$\begin{bmatrix} d_{22} & d_{23} & \dots & \dots & d_{2n} \\ d_{32} & d_{34} & \dots & \dots & d_{3n} \\ \dots & \dots & \dots & \dots & \dots \\ \dots & \dots & \dots & \dots & \dots \\ d_{m2} & d_{m3} & \dots & \dots & d_{mn} \end{bmatrix} \neq 0,$$

we repeat the above process for this matrix and get a matrix

$$E = \begin{bmatrix} 1 & 0 & 0 & \cdots & \cdots & 0 \\ 0 & 1 & 0 & \cdots & \cdots & 0 \\ 0 & 0 & e_{33} & \cdots & \cdots & e_{3n} \\ \cdots & \cdots & \cdots & \cdots & \cdots & \cdots \\ 0 & 0 & 3_{m3} & \cdots & \cdots & e_{mn} \end{bmatrix}.$$

Continuing this process, we obtain a matrix

$$N = \begin{bmatrix} I_k & 0 \\ 0 & 0 \end{bmatrix}.$$

The rank of N is k. Since, the matrix N has been obtained from A by elementary transformations, $\rho(N) = \rho(A)$, that is, $k = r$. Hence, every non-zero matrix can be reduced to the form

$$\begin{bmatrix} I_k & 0 \\ 0 & 0 \end{bmatrix}$$

by a finite chain of elementary transformations.

Corollary 2.10. The rank of an $m \times n$ matrix A is r if and only if it can be reduced to the normal form by a sequence of elementary transformations.

Proof. If $\rho(A) = r$, then by the above theorem it can be reduced to normal form by a sequence of elementary transformations.

Conversely, let the matrix A has been reduced to normal form $\begin{bmatrix} I_r & 0 \\ 0 & 0 \end{bmatrix}$ by elementary transformations. Now the rank of $\begin{bmatrix} I_r & 0 \\ 0 & 0 \end{bmatrix}$ is r and we know that rank of a matrix is not altered by elementary transformation. Therefore, rank of A is also r.

Corollary 2.11. If A is an $m \times n$ matrix of rank r, there exist non-singular matrices P and Q such that

$$PAQ = \begin{bmatrix} I_r & 0 \\ 0 & 0 \end{bmatrix}.$$

Proof. Since A is an $m \times n$ matrix of rank r, it can be reduced to normal form $\begin{bmatrix} I_r & 0 \\ 0 & 0 \end{bmatrix}$ using a sequence of elementary transformations. Further, since the elementary row (column) transformations are equivalent to pre-(post) multiplication by the corresponding elementary matrices, we have

$$P_s P_{s-1} \ldots P_1 A Q_1 Q_2 \ldots Q_t = \begin{bmatrix} I_r & 0 \\ 0 & 0 \end{bmatrix}.$$

Now, since, each elementary matrix is non-singular and the product of non-singular matrices is again non-singular, it follows that $P_s P_{s-1} \ldots P_1$ and $Q_1 Q_2 \ldots Q_t$ are non-singular matrices, say P and Q. Hence

$$PAQ = \begin{bmatrix} I_r & 0 \\ 0 & 0 \end{bmatrix},$$

where P and Q are non-singular matrices.

2.28 EQUIVALENCE OF MATRICES

Definition 2.94. Two matrices whose elements are real or complex numbers are said to be *equivalent* if and only if each can be transformed into the other by means of elementary transformations.

If the matrix A is equivalent to the matrix B, then we write $A \sim B$

The relation of equivalence '$\sim$' in the set of all $m \times n$ matrices is an *equivalence relation*, that is, is reflexive, symmetric, and transitive.

Theorem 2.46. If A and B are equivalent matrices, then $\rho(A) = \rho(B)$.

Proof. If $A \sim B$, then B can be obtained from A by a finite number of elementary transformations. But elementary transformation do not alter the rank of a matrix. Hence $\rho(A) = \rho(B)$.

Theorem 2.47. If two matrices A and B have the same size and the same rank, they are equivalent.

Proof. Let A and B be two $m \times n$ matrices of the same rank r. Then they can be reduced to normal form by elementary transformations. Therefore,

$$A \sim \begin{bmatrix} I_r & 0 \\ 0 & 0 \end{bmatrix} \text{ and } B \sim \begin{bmatrix} I_r & 0 \\ 0 & 0 \end{bmatrix}$$

or, by symmetry of the relation of equivalence of matrices

$$A \sim \begin{bmatrix} I_r & 0 \\ 0 & 0 \end{bmatrix} \text{ and } \begin{bmatrix} I_r & 0 \\ 0 & 0 \end{bmatrix} \sim B.$$

Using transitivity of the relation, '$\sim$' we have $A \sim B$.

Theorem 2.48. If A and B are equivalent matrices, there exist non-singular matrices P and Q such that $B = PAQ$.

Proof. If $A \sim B$, then B can be obtained from A by a finite number of elementary transformations of A. But elementary row (column) transformations are equivalent to pre (post) multiplication by the corresponding elementary matrices. Therefore, there are elementary matrices $P_1, P_2, \ldots, P_s Q_1, Q_2, \ldots, Q_t$ such that

$$P_s\, P_{s-1} \ldots P_1 A\, Q_1\, Q_2 \ldots Q_t = B.$$

Since, each elementary matrix is non-singular and the product of non-singular matrices is non-singular, we have,

$$PAQ = B,$$

where $P = P_s\, P_{s-1} \ldots P_1$ and $Q = Q_1\, Q_2 \ldots Q_t$ are non-singular matrices.

Theorem 2.49. Any non-singular matrix of explicitly given numbers may be factored into the product of elementary matrices.

Proof. Any non-singular matrix A of order n and the identity matrix I_n have the same order and same rank. Hence $A \sim I_n$. Therefore, by the Theorem 2.48, there exist elementary matrices P_j and Q_j such that

$$A = P_s P_{s-1} \ldots P_1 I_n Q_1 Q_2 \ldots Q_t.$$

EXAMPLE 2.115

Reduce the matrix

$$A = \begin{bmatrix} 1 & -1 & 2 & -3 \\ 4 & 1 & 0 & 2 \\ 0 & 3 & 0 & 4 \\ 0 & 1 & 0 & 2 \end{bmatrix}$$

to normal form and, hence, find its rank.

Solution. We observe that

$$A \sim \begin{bmatrix} 1 & 0 & 0 & 0 \\ 4 & 5 & -8 & 14 \\ 0 & 3 & 0 & 4 \\ 0 & 1 & 0 & 2 \end{bmatrix} \begin{matrix} C_2 \to C_2 + C_1 \\ C_3 \to C_3 - 2C_1 \\ C_4 \to C_4 + 3C_1 \end{matrix}$$

$$\sim \begin{bmatrix} 1 & 0 & 0 & 0 \\ 0 & 5 & -8 & 14 \\ 0 & 3 & 0 & 4 \\ 0 & 1 & 0 & 2 \end{bmatrix} R_2 \to R_2 - 4R_1$$

$$\sim \begin{bmatrix} 1 & 0 & 0 & 0 \\ 0 & 1 & 0 & 2 \\ 0 & 3 & 0 & 4 \\ 0 & 5 & -8 & 14 \end{bmatrix} R_2 \leftrightarrow R_4$$

$$\sim \begin{bmatrix} 1 & 0 & 0 & 0 \\ 0 & 1 & 0 & 0 \\ 0 & 3 & 0 & -2 \\ 0 & 5 & -8 & 4 \end{bmatrix} C_4 \to C_4 - 2C_2$$

$$\sim \begin{bmatrix} 1 & 0 & 0 & 0 \\ 0 & 1 & 0 & 0 \\ 0 & 0 & 0 & -2 \\ 0 & 0 & -8 & 4 \end{bmatrix} \begin{matrix} R_3 \to R_3 - 3R_2 \\ R_4 \to R_4 - 5R_2 \end{matrix}$$

$$\sim \begin{bmatrix} 1 & 0 & 0 & 0 \\ 0 & 1 & 0 & 0 \\ 0 & 0 & -2 & 0 \\ 0 & 0 & 4 & -8 \end{bmatrix} C_3 \leftrightarrow C_4$$

$$\sim \begin{bmatrix} 1 & 0 & 0 & 0 \\ 0 & 1 & 0 & 0 \\ 0 & 0 & 1 & 0 \\ 0 & 0 & -2 & 1 \end{bmatrix} \begin{matrix} C_3 \to -\frac{1}{2}C_3 \\ C_4 \to -\frac{1}{8}C_4 \end{matrix}$$

$$\sim \begin{bmatrix} 1 & 0 & 0 & 0 \\ 0 & 1 & 0 & 0 \\ 0 & 0 & 1 & 0 \\ 0 & 0 & 0 & 1 \end{bmatrix} R_4 \to R_4 + 2R_3$$

$$= I_4.$$

Hence, $\rho(A) = 4$.

EXAMPLE 2.116

Reduce the matrix

$$\begin{bmatrix} 3 & 2 & -1 \\ 4 & 2 & 6 \\ 7 & 4 & 5 \end{bmatrix}$$

to the normal form and, hence, find its rank.

Solution. We note that

$$A = \begin{bmatrix} 3 & 2 & -1 \\ 4 & 2 & 6 \\ 7 & 4 & 5 \end{bmatrix}$$

$$\sim \begin{bmatrix} -1 & 0 & -7 \\ 4 & 2 & 6 \\ 7 & 4 & 5 \end{bmatrix} \quad R_1 \to R_1 - R_2$$

$$\sim \begin{bmatrix} 1 & 0 & 7 \\ 4 & 2 & 6 \\ 7 & 4 & 5 \end{bmatrix} \quad R_1 \to -R_1$$

$$\sim \begin{bmatrix} 0 & 0 & 7 \\ 0 & 2 & -22 \\ 0 & 4 & -44 \end{bmatrix} \quad \begin{matrix} R_2 \to R_2 - 4R_1 \\ R_3 \to R_3 - 7R_1 \end{matrix}$$

$$\sim \begin{bmatrix} 1 & 0 & 7 \\ 0 & 1 & -11 \\ 0 & 4 & -44 \end{bmatrix} \quad R_2 \to \frac{1}{2}R_2$$

$$\sim \begin{bmatrix} 1 & 0 & 7 \\ 0 & 1 & -11 \\ 0 & 0 & 0 \end{bmatrix} \quad R_3 \to R_3 - 4R_2$$

$$\sim \begin{bmatrix} 1 & 0 & 0 \\ 0 & 1 & -11 \\ 0 & 0 & 0 \end{bmatrix} \quad C_3 \to C_3 - 7C_1$$

$$\sim \begin{bmatrix} 1 & 0 & 0 \\ 0 & 1 & 0 \\ 0 & 0 & 0 \end{bmatrix} \quad C_3 \to C_3 + 11C_2$$

$$\sim \begin{bmatrix} I_2 & 0 \\ 0 & 0 \end{bmatrix}.$$

Hence $\rho(A) = 2$.

EXAMPLE 2.117

For the matrix

$$A = \begin{bmatrix} 1 & 1 & 1 \\ 1 & -1 & -1 \\ 3 & 1 & 1 \end{bmatrix},$$

find the non-singular matrices P and Q such that PAQ is in the normal form. Hence, find the rank of the matrix A.

Solution. We write

$$A = I_3 A I_3.$$

Thus,

$$\begin{bmatrix} 1 & 1 & 1 \\ 1 & -1 & -1 \\ 3 & 1 & 1 \end{bmatrix} = \begin{bmatrix} 1 & 0 & 0 \\ 0 & 1 & 0 \\ 0 & 0 & 1 \end{bmatrix} A \begin{bmatrix} 1 & 0 & 0 \\ 0 & 1 & 0 \\ 0 & 0 & 1 \end{bmatrix}.$$

We shall apply elementary transformations on A until it is reduced to normal form, keeping in mind that each row transformation will also be applied to the pre-factor I_3 of the product on the right and each column transformation will also be applied to the post-factor I_3 of the product on the right.

Performing $R_2 \to R_2 - R_1$, $R_3 \to R_3 - 3R_1$, we get,

$$\begin{bmatrix} 1 & 1 & 1 \\ 0 & -2 & -2 \\ 0 & -2 & -2 \end{bmatrix} = \begin{bmatrix} 1 & 0 & 0 \\ -1 & 1 & 0 \\ -3 & 0 & 1 \end{bmatrix} A \begin{bmatrix} 1 & 0 & 0 \\ 0 & 1 & 0 \\ 0 & 0 & 1 \end{bmatrix}$$

Performing $C_2 \to C_2 - C_1$, $C_3 \to C_3 - C_1$, we get

$$\begin{bmatrix} 1 & 0 & 0 \\ 0 & -2 & -2 \\ 0 & -2 & -2 \end{bmatrix} = \begin{bmatrix} 1 & 0 & 0 \\ -1 & 1 & 0 \\ -3 & 0 & 1 \end{bmatrix} A \begin{bmatrix} 1 & -1 & -1 \\ 0 & 1 & 0 \\ 0 & 0 & 1 \end{bmatrix}.$$

Performing $R_2 \to -\frac{1}{2}R_2$, we get

$$\begin{bmatrix} 1 & 0 & 0 \\ 0 & 1 & 1 \\ 0 & -2 & -2 \end{bmatrix} = \begin{bmatrix} 1 & 0 & 0 \\ \frac{1}{2} & -\frac{1}{2} & 0 \\ -3 & 0 & 1 \end{bmatrix} A \begin{bmatrix} 1 & -1 & -1 \\ 0 & 1 & 0 \\ 0 & 0 & 1 \end{bmatrix}.$$

Performing $R_3 \to R_3 + 2R_2$, we get

$$\begin{bmatrix} 1 & 0 & 0 \\ 0 & 1 & 1 \\ 0 & 0 & 0 \end{bmatrix} = \begin{bmatrix} 1 & 0 & 0 \\ \frac{1}{2} & -\frac{1}{2} & 0 \\ -2 & -1 & 1 \end{bmatrix} A \begin{bmatrix} 1 & -1 & -1 \\ 0 & 1 & 0 \\ 0 & 0 & 1 \end{bmatrix}.$$

Last, performing $C_3 \to C_3 - C_2$, we have

$$\begin{bmatrix} 1 & 0 & 0 \\ 0 & 1 & 0 \\ 0 & 0 & 0 \end{bmatrix} = \begin{bmatrix} 1 & 0 & 0 \\ \frac{1}{2} & -\frac{1}{2} & 0 \\ -2 & -1 & 1 \end{bmatrix} A \begin{bmatrix} 1 & -1 & 0 \\ 0 & 1 & -1 \\ 0 & 0 & 1 \end{bmatrix}$$

or

$$\begin{bmatrix} I_2 & 0 \\ 0 & 0 \end{bmatrix} = PAQ,$$

where

$$P = \begin{bmatrix} 1 & 0 & 0 \\ \frac{1}{2} & -\frac{1}{2} & 0 \\ -2 & -1 & 1 \end{bmatrix}, \quad Q = \begin{bmatrix} 1 & -1 & 0 \\ 0 & 1 & -1 \\ 0 & 0 & 1 \end{bmatrix}.$$

Since A is equivalent to $\begin{bmatrix} I_2 & 0 \\ 0 & 0 \end{bmatrix}$, we have $\rho(A) = 2$.

EXAMPLE 2.118

For the matrix

$$A = \begin{bmatrix} 1 & -1 & 1 \\ 1 & 1 & 1 \\ 3 & 1 & 1 \end{bmatrix},$$

find non-singular matrices P and Q such that PAQ is in the normal form. Hence find the rank of A.

Solution. Write $A = I_3\, A\, I_3$ that is,

$$A = \begin{bmatrix} 1 & 0 & 0 \\ 0 & 1 & 0 \\ 0 & 0 & 1 \end{bmatrix} A \begin{bmatrix} 1 & 0 & 0 \\ 0 & 1 & 0 \\ 0 & 0 & 1 \end{bmatrix}.$$

As, in the above example, we shall reduce A to normal form subjecting it to elementary transformations. Performing $R_2 \to R_2 - R_1,\ R_3 - 3R_1$, we have

$$\begin{bmatrix} 1 & -1 & 1 \\ 0 & 2 & 0 \\ 0 & 4 & -2 \end{bmatrix} = \begin{bmatrix} 1 & 0 & 0 \\ -1 & 1 & 0 \\ -3 & 0 & 1 \end{bmatrix} A \begin{bmatrix} 1 & 0 & 0 \\ 0 & 1 & 0 \\ 0 & 0 & 1 \end{bmatrix}.$$

Performing $C_2 \to C_2 + C_1,\ C_3 \to C_3 - C_1$ we have

$$\begin{bmatrix} 1 & 0 & 0 \\ 0 & 2 & 0 \\ 0 & 4 & -2 \end{bmatrix} = \begin{bmatrix} 1 & 0 & 0 \\ -1 & 1 & 0 \\ -3 & 0 & 1 \end{bmatrix} A \begin{bmatrix} 1 & 1 & -1 \\ 0 & 1 & 0 \\ 0 & 0 & 1 \end{bmatrix}.$$

Performing $R_2 \to \frac{1}{2}R_2$, we have

$$\begin{bmatrix} 1 & 0 & 0 \\ 0 & 1 & 0 \\ 0 & 4 & -2 \end{bmatrix} = \begin{bmatrix} 1 & 0 & 0 \\ -\frac{1}{2} & \frac{1}{2} & 0 \\ -3 & 0 & 1 \end{bmatrix} A \begin{bmatrix} 1 & 1 & -1 \\ 0 & 1 & 0 \\ 0 & 0 & 1 \end{bmatrix}.$$

Performing $R_3 \to R_3 - 4R_2$, we have

$$\begin{bmatrix} 1 & 0 & 0 \\ 0 & 1 & 0 \\ 0 & 0 & -2 \end{bmatrix} = \begin{bmatrix} 1 & 0 & 0 \\ -\frac{1}{2} & \frac{1}{2} & 0 \\ -1 & -2 & 1 \end{bmatrix} A \begin{bmatrix} 1 & 1 & -1 \\ 0 & 1 & 0 \\ 0 & 0 & 1 \end{bmatrix}$$

Performing $C_3 \to -\frac{1}{2}C_3$, we get

$$\begin{bmatrix} 1 & 0 & 0 \\ 0 & 1 & 0 \\ 0 & 0 & 1 \end{bmatrix} = \begin{bmatrix} 1 & 0 & 0 \\ -\frac{1}{2} & \frac{1}{2} & 0 \\ -1 & -2 & 1 \end{bmatrix} A \begin{bmatrix} 1 & 1 & \frac{1}{2} \\ 0 & 1 & 0 \\ 0 & 0 & -\frac{1}{2} \end{bmatrix}.$$

Hence,

$$I_3 = PAQ,$$

where

$$P = \begin{bmatrix} 1 & 0 & 0 \\ -\frac{1}{2} & \frac{1}{2} & 0 \\ -1 & -2 & 1 \end{bmatrix} \text{ and } Q = \begin{bmatrix} 1 & 1 & \frac{1}{2} \\ 0 & 1 & 0 \\ 0 & 0 & -\frac{1}{2} \end{bmatrix}$$

Since $A \sim I_3$, $\rho(A) = \rho(I_3) = 3$.

Theorem 2.50. The rank of the product of two matrices cannot exceed the rank of either matrix.

Proof. Let A be $m \times n$ and B be $n \times p$ matrices with rank r_A and r_B, respectively. Then, by Corollary 2.11, there exist non-singular matrices C and D of order m and n, respectively, such that

$$CAD = \begin{bmatrix} I_{rA} & 0 \\ 0 & 0 \end{bmatrix},$$

where $\begin{bmatrix} I_{rA} & 0 \\ 0 & 0 \end{bmatrix}$ denotes the normal form of A. Thus

$$A = C^{-1} \begin{bmatrix} I_{rA} & 0 \\ 0 & 0 \end{bmatrix} D^{-1}$$

so that

$$\begin{aligned} AB &= \left[C^{-1} \begin{bmatrix} I_{rA} & 0 \\ 0 & 0 \end{bmatrix} D^{-1} \right] B \\ &= C^{-1} \left(\begin{bmatrix} I_{rA} & 0 \\ 0 & 0 \end{bmatrix} (D^{-1}B) \right). \end{aligned}$$

Since C^{-1} is non-singular, AB has same rank as $\begin{bmatrix} I_{rA} & 0 \\ 0 & 0 \end{bmatrix}(D^{-1}B)$. But $\begin{bmatrix} I_{rA} & 0 \\ 0 & 0 \end{bmatrix}$ has zeros in the last $m - r_A$ rows and, hence, $\begin{bmatrix} I_{rA} & 0 \\ 0 & 0 \end{bmatrix}(D^{-1}B)$ also has only zeros in the last $m - r_A$ rows. Hence, the rank of $\begin{bmatrix} Ir_A & 0 \\ 0 & 0 \end{bmatrix}(D^{-1}B)$ is at most r_A. It follows, therefore, that

$$\rho(AB) \le r_A$$

Also

$$\begin{aligned} \rho(AB) &= \rho\left((AB)^T\right) \\ &= \rho[B^T A^T] \end{aligned}$$

But, as proved earlier in the above steps

$$\rho(B^T A^T) \le \rho(B^T) = \rho(B) = r_B.$$

Hence,

$$\rho(AB) \le r_B.$$

This completes the proof of the theorem.

EXAMPLE 2.119

Let A be any non-singular matrix and B a matrix such that AB exists. Show that

$$\rho(AB) = \rho(B).$$

Solution. Let $C = AB$. Since A is non-singular, therefore $B = A^{-1}C$. Since rank of the product of two matrices does not exceed the rank of either matrix, we have,

$$\rho(C) = \rho(AB) \le \rho(B)$$

and

$$\rho(B) = \rho(A^{-1}C) \le \rho(C).$$

Hence

$$\rho(C) = \rho(AB) \le \rho(B) \le \rho(C),$$

which yields

$$\rho(B) = \rho(C) = \rho(AB).$$

2.28.1 ROW AND COLUMN EQUIVALENCE OF MATRICES

Definition 2.95. A matrix A is said to be *row (column) equivalent* to B if B is obtainable from A by a finite number of elementary row (column) transformations of A.

Row equivalence of the matrices A and B is denoted by $A \overset{R}{\sim} B$ and column equivalence of A and B is denoted by $A \overset{C}{\sim} B$.

Theorem 2.51. Let A be an $m \times n$ matrix of rank r. Then there exists a non-singular matrix P such that

$$PA = \begin{bmatrix} G \\ \mathbf{0} \end{bmatrix},$$

where G is an $r \times n$ matrix of rank r and 0 is $(m - r) \times n$ matrix.

Proof. Since A is an $m \times n$ matrix of rank r, therefore there exist non-singular matrices P and Q such that

$$PAQ = \begin{bmatrix} I_r & 0 \\ 0 & 0 \end{bmatrix}.$$

But every non-singular matrix can be expressed as product of elementary matrices. So,

$$Q = Q_1 Q_2 \dots Q_t,$$

where $Q_1, Q_2 \dots Q_t$ are all elementary matrices. Thus,

$$PAQ_1\, Q_2 \dots Q_t = \begin{bmatrix} I_r & 0 \\ 0 & 0 \end{bmatrix}.$$

Since, elementary column transformation of a matrix is equivalent to post-multiplication with the corresponding elementary matrix, we post-multiply the left hand side of the above expression by the elementary matrices $Q_t^{-1} Q_{t-1}^{-1}, \dots, Q_2^{-1}, Q_1^{-1}$ successively and effect the corresponding column transformations in the right hand side, we get a relation of the form

$$PA \begin{bmatrix} G \\ \mathbf{0} \end{bmatrix}.$$

Since elementary transformations do not alter the rank

$$\rho(PA) = \rho(A) = r \text{ and so } \rho \begin{bmatrix} G \\ \mathbf{0} \end{bmatrix} = r,$$

which implies that $\rho(G) = r$ since G has r rows and last $m - r$ rows of $\begin{bmatrix} G \\ 0 \end{bmatrix}$ consist of zero elements only.

Theorem 2.52. Every non-singular matrix is row equivalent to a unit matrix.

Proof. Suppose that the matrix A is of order 1. Then $A = [a_{11}]$ which is clearly row equivalent to a unit matrix. We shall prove our result by induction on the order of the matrix. Let A be of order n. Since the result is true for non-singular matrix of order 1, we assume that the result is true for all matrices of order $n - 1$.

Let $A = [a_{ij}]$ be an $n \times n$ non-singular matrix. The first column of the matrix A has at least one non-zero element, otherwise $|A| = 0$, which contradicts the fact that A is non-singular. Let $a_{11} = k \neq 0$. By interchanging (if necessary) the pth row with the first row, we obtain a matrix B whose leading coefficient is $k \neq 0$. Multiplying the elements of the first row by $\frac{1}{k}$, we get the matrix.

$$C = \begin{bmatrix} 1 & c_{12} & c_{13} & \dots & \dots & c_{1n} \\ c_{21} & c_{22} & c_{23} & \dots & \dots & c_{2n} \\ \dots & \dots & \dots & \dots & \dots & \dots \\ \dots & \dots & \dots & \dots & \dots & \dots \\ c_{n1} & c_{n2} & c_{n3} & \dots & \dots & c_{nn} \end{bmatrix}$$

Using elementary row transformation, we get

$$D = \begin{bmatrix} 1 & d_{12} \ d_{13} \ \dots \ d_{1n} \\ 0 & \\ \dots & \mathbf{A_1} \\ \dots & \\ 0 & \end{bmatrix},$$

where A_1 is $(n-1) \times (n-1)$ matrix. The matrix A_1 is non-singular otherwise $|A_1| = 0$ and so $|D| = 0$. Since $A \sim D$, this will imply $|A| = 0$ contradicting the fact that A is non-singular. By induction hypothesis, A_1 can be transformed to I_{n-1} by elementary row transformations. Thus, we get a matrix M such that

$$M = \begin{bmatrix} 1 & d_{12} & d_{13} & \dots & \dots & d_{1n} \\ 0 & 1 & 0 & \dots & \dots & 0 \\ \dots & \dots & \dots & \dots & \dots & \dots \\ \dots & \dots & \dots & \dots & \dots & \dots \\ 0 & \dots & \dots & \dots & \dots & 1 \end{bmatrix}.$$

Further, use of elementary row transformation reduces M to the matrix.

$$I_n = \begin{bmatrix} 1 & 0 & 0 & \dots & \dots & 0 \\ 0 & 1 & 0 & \dots & \dots & 0 \\ \dots & \dots & \dots & \dots & \dots & \dots \\ \dots & \dots & \dots & \dots & \dots & \dots \\ 0 & \dots & 0 & \dots & \dots & 1 \end{bmatrix},$$

which completes the proof of the theorem.

Corollary 2.12. Let A be a non-singular matrix of order n. Then there exists elementary matrices $E_1, E_2, \dots, E_t$ such that

$$E_t E_{t-1} \dots E_2 E_1 A = I_n.$$

Proof. By the Theorem 2.52, non-singular matrix A can be reduced to I_n by finite number of elementary row transformations. Since elementary row transformation is equivalent to pre-multiplication by the elementary matrix, therefore, there exists elementary matrices $E_1, E_2, \ldots, E_t$ such that $E_t, E_{t-1} \ldots E_2 E_1 A = I_n$.

Corollary 2.13. Every non-singular matrix is a product of elementary matrices.

Proof. Let A be a non-singular matrix. Then, by Corollary 2.12, there exist elementary matrices $E_1, E_2, \ldots, E_t$ such that

$$E_t E_{t-1} \ldots E_2 E_1 A = I_n.$$

Pre-multiplying both sides by $(E_t E_{t-1} \ldots E_2 E_1)^{-1}$, we get

$$A = E_1^{-1} E_2^{-1} \ldots E_t^{-1}.$$

Since, inverse of an elementary matrix is also an elementary matrix, it follows that non-singular matrix can be expressed as a product of elementary matrices.

Corollary 2.14. The rank of a matrix does not alter by pre-multiplication or post-multiplication with a non-singular matrix.

Proof. Every non-singular matrix can be expressed as a product of elementary matrices. Also we know that elementary row (column) transformations are equivalent to pre-(post) multiplication with the corresponding elementary matrices. But elementary transformations do not alter the rank of a matrix. Hence, the rank of a matrix remains unchanged by pre-multiplication or post-multiplication with a non-singular matrix.

2.29 ROW RANK AND COLUMN RANK OF A MATRIX

Definition 2.96. Let A be any $m \times n$ matrix. Then the maximum number of linearly independent rows (columns) of A is called the *row rank (column rank)* of A.

The following theorem (stated without proof) shall be used in the sequel.

Theorem 2.53. The row rank, the column rank and the rank of a matrix are equal.

2.30 SOLUTION OF SYSTEM OF LINEAR EQUATIONS

Let

$$\left.\begin{array}{l} a_{11}x_1 + a_{12}x_2 + \ldots + a_{1n}x_n = b_1 \\ a_{21}x_1 + a_{22}x_2 + \ldots + a_{2n}x_n = b_2 \\ \ldots \quad \ldots \quad \ldots \quad \ldots \quad \ldots \\ \ldots \quad \ldots \quad \ldots \quad \ldots \quad \ldots \\ a_{m1}x_1 + a_{m2}x_2 + \ldots + a_{mn}x_n = b_m \end{array}\right\} \quad (49)$$

be a system of m linear equations in n unknown $x_1, x_2, \ldots, x_n$ The matrix form of this system is

$$AX = B,$$

where

$$A = \begin{bmatrix} a_{11} & a_{12} & \ldots & \ldots & a_{1n} \\ a_{21} & a_{22} & \ldots & \ldots & a_{2n} \\ \ldots & \ldots & \ldots & \ldots & \ldots \\ \ldots & \ldots & \ldots & \ldots & \ldots \\ a_{m1} & a_{m2} & \ldots & \ldots & a_{mn} \end{bmatrix}$$

is called *coefficient matrix* of the system,

$$X = \begin{bmatrix} x_1 \\ x_2 \\ \ldots \\ \ldots \\ x_n \end{bmatrix}$$

is the *column matrix of unknowns*, and

$$B = \begin{bmatrix} b_1 \\ b_2 \\ \ldots \\ \ldots \\ b_m \end{bmatrix}$$

is *column matrix of known numbers* or the *matrix of constants*. We call the system (49) as the system of *non-homogenous equations*.

Any set of values of $x_1, x_2, \ldots, x_n$ from a scalar field which simultaneously satisfy (49) is called a *solution*, over that field, of the system. When such a system has one or more solutions, it is said to be *consistent*, otherwise it is called *inconsistent*.

2.31 SOLUTION OF NON-HOMOGENOUS LINEAR SYSTEM OF EQUATIONS

(A) Matrix Inversion Method.

Consider the non-homogeneous system of linear equations $AX = B$, where A is non-singular $n \times n$ matrix. Since A is non-singular, A^{-1} exists. Pre-multiplication of $AX = B$ by A^{-1} yields

$$A^{-1}(AX) = A^{-1}B$$

or

$$(A^{-1}A)X = A^{-1}B$$

or

$$IX = A^{-1}B$$

or

$$X = A^{-1}B.$$

Thus if A is non-singular, then the given system of equation can be solved using inverse of A. This method is called the Matrix Inversion Method.

EXAMPLE 2.120

Solve

$$x + 2y - 3z = -4$$
$$2x + 3y + 2z = 2$$
$$3x - 3y - 4z = 11$$

by Matrix Inversion Method.

Solution. The matrix form of the system is $AX = B$, where

$$A = \begin{bmatrix} 1 & 2 & -3 \\ 2 & 3 & 2 \\ 3 & -3 & -4 \end{bmatrix}, X = \begin{bmatrix} x \\ y \\ z \end{bmatrix} \text{ and } B = \begin{bmatrix} -4 \\ 2 \\ 11 \end{bmatrix},$$

We note that

$$|A| = 1(-6) - 2(-14) - 3(-15) = 67 \neq 0.$$

Thus A is non-singular. Hence the required solution is given by

$$X = A^{-1}B. \tag{50}$$

The cofactor matrix of A is

$$[A_{ij}] = \begin{bmatrix} -6 & 14 & -15 \\ 17 & 5 & 9 \\ 13 & -8 & -1 \end{bmatrix}$$

and so

$$\text{adj } A = [A_{ij}]^T = \begin{bmatrix} -6 & 17 & 13 \\ 14 & 5 & -8 \\ -15 & 9 & -1 \end{bmatrix}.$$

Hence

$$A^{-1} = \frac{1}{|A|}\text{adj } A = \frac{1}{67}\begin{bmatrix} -6 & 17 & 13 \\ 14 & 5 & -8 \\ -15 & 9 & -1 \end{bmatrix}.$$

Substituting A^{-1} in (50), we get

$$\begin{bmatrix} x \\ y \\ z \end{bmatrix} = \frac{1}{67}\begin{bmatrix} -6 & 17 & 13 \\ 14 & 5 & -8 \\ -15 & 9 & -1 \end{bmatrix}\begin{bmatrix} -4 \\ 2 \\ 11 \end{bmatrix}$$
$$= \frac{1}{67}\begin{bmatrix} 201 \\ -134 \\ 67 \end{bmatrix} = \begin{bmatrix} 3 \\ -2 \\ 1 \end{bmatrix}.$$

Hence $x = 3$, $y = -2$, and $z = 1$.

B. Cramer's Rule.

If $|A| \neq 0$, then $AX = B$ has exactly one solution $x_j = \frac{|A_j|}{|A|}$, $j = 1, 2, \ldots, n$, where A_j is the matrix obtained from A by replacing the jth column of A by the column of b's.

Consider the matrix form $AX = B$ of the system of linear equations. Again Suppose that A is non-singular. Then, pre-multiplication of $AX = B$ by A^{-1} yields

$$X=A^{-1}B=\frac{1}{|A|}\begin{bmatrix} A_{11} & A_{21} & \dots & \dots & A_{n1} \\ A_{12} & A_{22} & \dots & \dots & A_{n2} \\ \dots & \dots & \dots & \dots & \dots \\ \dots & \dots & \dots & \dots & \dots \\ A_{1n} & A_{2n} & \dots & \dots & A_{nn} \end{bmatrix}\begin{bmatrix} b_1 \\ b_2 \\ \dots \\ \dots \\ b_n \end{bmatrix}$$

or

$$\begin{bmatrix} x_1 \\ x_2 \\ \dots \\ \dots \\ x_n \end{bmatrix} = \frac{1}{|A|}\begin{bmatrix} b_1A_{11}+b_2A_{21}+\dots+b_nA_{n1} \\ b_1A_{12}+b_2A_{22}+\dots+b_nA_{n2} \\ \dots \\ \dots \\ b_1A_{1n}+b_2A_{2n}+\dots+b_nA_{nn} \end{bmatrix}.$$

Therefore,

$$x_1 = \frac{1}{|A|}(b_1A_{11}+b_2A_{21}+\dots+b_nA_{n1}) = \frac{|A_1|}{|A|}$$

$$x_2 = \frac{1}{|A|}(b_1A_{12}+b_2A_{22}+\dots+b_nA_{n2}) = \frac{|A_2|}{|A|}$$

$$\dots \quad \dots \quad \dots \quad \dots$$

$$x_n = \frac{1}{|A|}(b_nA_{1n}+b_nA_{2n}+\dots+b_nA_{nn}) = \frac{|A_n|}{|A|},$$

where A_j is the matrix obtained from A by replacing the jth column of A by the column of b's.

EXAMPLE 2.121

Solve the system of linear equations

$$3x+y+2z=3$$
$$2x-3y-z=-3$$
$$x+2y+z=4$$

by Cramer's rule.

Solution. Let A be the coefficient matrix. Then

$$|A| = \begin{vmatrix} 3 & 1 & 2 \\ 2 & -3 & -1 \\ 1 & 2 & 1 \end{vmatrix} = 8$$

and so A is non-singular. Thus the Cramer's rule is applicable and we have

$$x = \frac{1}{|A|}\begin{vmatrix} 3 & 1 & 2 \\ -3 & -3 & -1 \\ 4 & 2 & 1 \end{vmatrix} = \frac{8}{8} = 1,$$

$$y = \frac{1}{|A|}\begin{vmatrix} 3 & 3 & 2 \\ 2 & -3 & -1 \\ 1 & 4 & 1 \end{vmatrix} = \frac{16}{8} = 2,$$

$$z = \frac{1}{|A|}\begin{vmatrix} 3 & 1 & 3 \\ 2 & -3 & -3 \\ 1 & 2 & 4 \end{vmatrix} = \frac{-8}{8} = -1.$$

Hence, the required solution is $x = 1$, $y = 2$, and $z = -1$.

Remark 2.8 The above two methods are applicable only when A is non-singular.

2.32 CONSISTENCY THEOREM

Definition 2.97. Let $AX = B$ be the matrix form of a given system of equations. Then the matrix

$$[A:B] = \begin{bmatrix} a_{11} & a_{12} & \dots & a_{1n} & b_1 \\ a_{21} & a_{22} & \dots & a_{2n} & b_2 \\ \dots & \dots & \dots & \dots & \dots \\ \dots & \dots & \dots & \dots & \dots \\ a_{m1} & a_{m2} & \dots & a_{mn} & b_m \end{bmatrix}$$

is called the *augmented matrix* of the given system of equations.

Definition 2.98. If a system of linear equations has one or more solution, it is said to be *consistent;* otherwise it is called *inconsistent.*

Theorem 2.54. (Consistency Theorem). The system of linear equations $AX = B$ is consistent if and only if the coefficient matrix A and the augmented matrix $[A:B]$ are of the same rank.

Proof. Let $C_1, C_2, \dots, C_n$ denote the column vectors of the matrix A. Then the equation $AX = B$ is equivalent to

$$[C_1C_2\dots C_n]\begin{bmatrix} x_1 \\ x_2 \\ \dots \\ \dots \\ x_n \end{bmatrix} = B$$

or

$$x_1C_1 + x_2C_2 + \ldots + x_nC_n = B. \qquad (51)$$

Let r be the rank of the matrix A. Then A has r linearly independent columns. Without loss of generality, we assume that $C_1, C_2, \ldots, C_r$ form a linearly independent set and so each of the remaining $n-r$ columns is a linear combination of these r columns $C_1, C_2, \ldots, C_r$.

Suppose the given system of linear equations is consistent. Therefore, there exist n scalar $k_1, k_2, \ldots, k_n$ such that

$$k_1C_1 + k_2C_2 + \ldots + k_nC_n = B. \qquad (52)$$

Now since each of $n - r$ columns $C_{r+1}, C_{r+2}, \ldots, C_n$ is a linear combination of first r columns $C_1, C_2, \ldots, C_r$, it follows from (52) that B is also a linear combination of $C_1, C_2, \ldots, C_r$ Thus, the maximum number of linearly independent columns of the matrix $[A : B]$ is also r. Therefore, the matrix $[A : B]$ is also of rank r. Hence, rank of A and the augmented matrix $[A : B]$ is the same.

Conversely, suppose that the matrices A and $[A : B]$ are of the same rank r. Then the maximum number of linearly independent columns of the matrix $[A : B]$ is r. But the first r columns $C_1, C_2, \ldots, C_r$ of the matrix $[A : B]$ had already formed a linearly independent set. Therefore, the column B should be expressed as a linear combination of $C_1, C_2 \ldots, C_r$ Hence, there are scalars $k_1, k_2, \ldots, k_r$, such that

$$k_1C_1 + k_2C_2 + \ldots + k_rC_r = B$$

or

$$k_1C_1 + k_2C_2 + \ldots + k_rC_r$$
$$+ 0C_{r+1}0C_{r+2} + \ldots + 0C_n = B \qquad (53)$$

Comparing (51) and (52), we get

$$x_1 = k_1,\ x_2 = k_2, \ldots, x_r = k_r,\ x_{r+1} = 0,$$
$$x_{r+2} = 0, \ldots, x_r = 0$$

as the solution of the equation $AX = B$. Hence, the given system of linear equations is consistent. This completes the proof of the theorem.

If the system of linear equations is consistent, then the following cases arise:

Case I. $m \geq n$, that is, number of equations is more than the number of unknowns. In such a case

(i) if $\rho(A) = \rho([A : B]) = n$, then the system of equations has a unique solution

(ii) if $\rho(A) = \rho([A : B]) = r < n$ then the $(n-r)$ unknowns are assigned arbitrary values and the remaining r unknowns can be determined in terms of these $(n-r)$ unknowns.

Case II. $m < n$, that is, the number of equations is less than the number of unknowns. In such a case

(i) if $\rho(A) = \rho([A : B]) = m$, then $n - m$ unknowns can be assigned arbitrary values and the values of the remaining m unknowns can be found in terms of these $n - m$ unknowns, which have already been assigned values

(ii) if $\rho(A) = \rho([A : B]) = r < m$, then the $(n - r)$ unknowns can be assigned arbitrary values and the values of remaining r unknowns can be found in terms of these $(n-r)$ unknowns, which have already been assigned values.

EXAMPLE 2.122

Show that the system

$$x + y + z = -3$$
$$3x + y - 2z = -2$$
$$2x + 4y + 7z = 7$$

of linear equations is not consistent.

Solution. The matrix form of the system is

$$AX = B$$

and the augmented matrix is

$$[A:B]=\begin{bmatrix}1 & 1 & 1 & -3\\ 3 & 1 & -2 & 2\\ 2 & 4 & 7 & 7\end{bmatrix}$$

$$\sim\begin{bmatrix}1 & 1 & 1 & -3\\ 0 & -2 & -5 & 7\\ 0 & 2 & 5 & 13\end{bmatrix}\quad \begin{matrix}R_2\to R_2-3R_1\\ R_3\to R_3-2R_1\end{matrix}$$

$$\sim\begin{bmatrix}1 & 1 & 1 & -3\\ 0 & -2 & -5 & 7\\ 0 & 0 & 0 & 20\end{bmatrix}\quad R_3\to R_3+R_2$$

Thus the number of non-zero rows in Echelon form of the matrix $[A:B]$ is 3. But

$$A\sim\begin{bmatrix}1 & 1 & 1\\ 0 & -2 & -5\\ 0 & 0 & 0\end{bmatrix}$$

and so $\rho(A)=2$.
Thus,

$$\rho(A)\neq\rho([A:B]).$$

Hence, the given system of equation is inconsistent.

EXAMPLE 2.123
Show that the equations

$$\begin{aligned}x+2y-z&=3\\ 3x-y+2z&=1\\ 2x-2y+3z&=2\\ x-y+z&=-1\end{aligned}$$

are consistent. Also solve them.

Solution. In matrix form, we have

$$AX=\begin{bmatrix}1 & 2 & -1\\ 3 & -1 & 2\\ 2 & -2 & 3\\ 1 & -1 & 1\end{bmatrix}\begin{bmatrix}x\\ y\\ z\end{bmatrix}=\begin{bmatrix}3\\ 1\\ 2\\ -1\end{bmatrix}.$$

The augmented matrix is

$$[A:B]=\begin{bmatrix}1 & 2 & -1 & 3\\ 3 & -1 & 2 & 1\\ 2 & -2 & 3 & 2\\ 1 & -1 & 1 & -1\end{bmatrix}$$

$$\sim\begin{bmatrix}1 & 2 & -1 & 3\\ 0 & -7 & 5 & -8\\ 0 & -6 & 5 & -4\\ 0 & -3 & 2 & -4\end{bmatrix}\quad \begin{matrix}R_2\to R_2-3R_1\\ R_3\to R_3-2R_1\\ R_4\to R_4-R_1\end{matrix}$$

$$\sim\begin{bmatrix}1 & 2 & -1 & 3\\ 0 & -1 & 0 & -4\\ 0 & -6 & 5 & -4\\ 0 & -3 & 2 & -4\end{bmatrix}\quad R_2\to R_2-R_3$$

$$\sim\begin{bmatrix}1 & 2 & -1 & 3\\ 0 & -1 & 0 & -4\\ 0 & 0 & 5 & 20\\ 0 & 0 & 2 & 8\end{bmatrix}\quad \begin{matrix}R_3\to R_3-6R_2\\ R_4\to R_4-3R_2\end{matrix}$$

$$\sim\begin{bmatrix}1 & 2 & -1 & 3\\ 0 & -1 & 0 & -4\\ 0 & 0 & 1 & 4\\ 0 & 0 & 1 & 4\end{bmatrix}\quad \begin{matrix}R_3\to \frac{1}{5}R_3\\ R_4\to \frac{1}{2}R_4\end{matrix}$$

$$\sim\begin{bmatrix}1 & 2 & -1 & 3\\ 0 & -1 & 0 & -4\\ 0 & 0 & 1 & 4\\ 0 & 0 & 0 & 0\end{bmatrix}\quad R_4\to R_4-R_3$$

$$\sim\begin{bmatrix}1 & 2 & -1 & 3\\ 0 & 1 & 0 & 4\\ 0 & 0 & 1 & 4\\ 0 & 0 & 0 & 0\end{bmatrix}\quad R_2\to -R_2$$

The number of non-zero rows in the echelon form is 3. Hence $\rho([A:B])=3$. Also

$$A\sim\begin{bmatrix}1 & 2 & -1\\ 0 & -1 & 0\\ 0 & 0 & 1\\ 0 & 0 & 0\end{bmatrix}.$$

Clearly, $\rho(A)=3$. Thus, $\rho(A)=\rho([A,B])$ and so the given system is consistent. Further, $r=n=3$. Therefore, the given system of equation has a unique solution. Rewriting the equation from the augmented matrix, we have

$$\begin{aligned} x+2y-z&=3\\ -y&=-4\\ z&=4 \end{aligned}$$

and so $x=-1$, $y=4$ and $z=4$ is the required solution.

EXAMPLE 2.124

For what values of λ and μ, the system of equations

$$\begin{aligned} x+y+z&=6\\ x+2y+3z&=10\\ x+2y+\lambda z&=\mu \end{aligned}$$

has (i) no solution (ii) a unique solution, and (iii) an infinite number of solutions.

Solution. The matrix form of the given system is

$$AX=\begin{bmatrix}1&1&1\\1&2&3\\1&2&\lambda\end{bmatrix}\begin{bmatrix}x\\y\\z\end{bmatrix}=\begin{bmatrix}6\\10\\\mu\end{bmatrix}=B.$$

Therefore, the augmented matrix is

$$[A:B]=\begin{bmatrix}1&1&1&6\\1&2&3&10\\1&2&\lambda&\mu\end{bmatrix}$$

$$\sim\begin{bmatrix}1&1&1&6\\0&1&2&4\\0&1&\lambda-1&\mu-6\end{bmatrix}\quad \begin{matrix}R_2\to R_2-R_1\\R_3\to R_3-R_1\end{matrix}$$

$$\sim\begin{bmatrix}1&1&1&6\\0&1&2&4\\0&0&\lambda-3&\mu-10\end{bmatrix}\quad R_3\to R_3-R_2.$$

If $\lambda\neq 3$, then $\rho(A)=3$ and $\rho([A:B])=3$. Hence, the given system of equations is consistent. Since $\rho(A)$ is equal to the number of unknowns, therefore, the given system of equations possesses a unique solution for any value of μ.

If $\lambda=3$ and $\mu\neq 10$, then $\rho(A)=2$ and $\rho([A:B])=3$. Therefore, the given system of equations is inconsistent and so has no solution.

If $\lambda=3$ and $\mu=10$ then $\rho(A)=\rho([A:B])=2$. Thus, the given system of equation is consistent. Further, $\rho(A)$ is less than the number of unknowns, therefore, in this case the given system of equations possesses an infinite number of solutions.

EXAMPLE 2.125

Determine the value of λ for which the system of equations

$$\begin{aligned} x_1+x_2+x_3&=2\\ x_1+2x_2+x_3&=-2\\ x_1+x_2+(\lambda-5)x_3&=\lambda \end{aligned}$$

(i) has no solution

(ii) has a unique solution.

Solution. The matrix form of the given system is

$$AX=\begin{bmatrix}1&1&1\\1&2&1\\1&2&\lambda-5\end{bmatrix}\begin{bmatrix}x_1\\x_2\\x_3\end{bmatrix}=\begin{bmatrix}2\\-2\\\lambda\end{bmatrix}=B.$$

Therefore, the augmented matrix is

$$[A:B]=\begin{bmatrix}1&1&1&2\\1&2&1&-2\\1&1&\lambda-5&\lambda\end{bmatrix}$$

$$\sim\begin{bmatrix}1&1&1&2\\0&1&0&-4\\0&0&\lambda-6&\lambda-2\end{bmatrix}\quad \begin{matrix}R_2\to R_2-R_1\\R_3\to R_3-R_1\end{matrix}$$

If $\lambda=6$, then $\rho(A)=2$ and $\rho([A:B])=3$. Therefore, the system is inconsistent and so possesses no solution.

If $\lambda \neq 6$, then $\rho(A) = \rho[A : B]) = 3$. Hence, the system is consistent in this case. Since $\rho(A)$ is equal to the number of unknowns, the system has a unique solution in this case.

EXAMPLE 2.126

Determine the value of λ for which the system of equations

$$x+y+z=1$$
$$x+2y+4z=\lambda$$
$$x+4y+10z=\lambda^2$$

possesses a solution and, hence, find its solution.

Solution. The given system of equations is expressed in the matrix form as

$$AX = \begin{bmatrix} 1 & 1 & 1 \\ 1 & 2 & 4 \\ 1 & 4 & 10 \end{bmatrix} \begin{bmatrix} x \\ y \\ z \end{bmatrix} = \begin{bmatrix} 1 \\ \lambda \\ \lambda^2 \end{bmatrix} = B.$$

Therefore, the augmented matrix is

$$[A:B] = \begin{bmatrix} 1 & 1 & 1 & 1 \\ 1 & 2 & 4 & \lambda \\ 1 & 4 & 10 & \lambda^2 \end{bmatrix} \sim \begin{bmatrix} 1 & 1 & 1 & 1 \\ 0 & 1 & 3 & \lambda-1 \\ 0 & 3 & 9 & \lambda^2-1 \end{bmatrix} \sim \begin{bmatrix} 1 & 1 & 1 & 1 \\ 0 & 1 & 3 & \lambda-1 \\ 0 & 0 & 0 & \lambda^2-3\lambda+2 \end{bmatrix}.$$

We note that

$$\rho(A) = \rho([A:B]) \text{ if } \lambda^2 - 3\lambda + 2 = 0.$$

Thus, the given equation is consistent if $\lambda^2 - 3\lambda + 2 = 0$, that is if $(\lambda-2)(\lambda-1) = 0$, that is, if $\lambda = 2$ or $\lambda = 1$. If $\lambda = 2$, then we have

$$[A:B] \sim \begin{bmatrix} 1 & 1 & 1 & 1 \\ 0 & 1 & 3 & 1 \\ 0 & 0 & 0 & 0 \end{bmatrix}$$

and so the given system of equations is equivalent to

$$x+y+z=1$$
$$y+3z=1.$$

These equations yields $y = 1 - 3z$, and $x = 2z$. Therefore, if $z = k$, an arbitrary constant, then $x = 2k, y = 1 - 3k$, and $z = k$ constitute the general solution of the given equation.

If $\lambda = 1$ then, we have

$$[A:B] \sim \begin{bmatrix} 1 & 1 & 1 & 1 \\ 0 & 1 & 3 & 0 \\ 0 & 0 & 0 & 0 \end{bmatrix}$$

and so the given system of equations is equivalent to

$$x+y+z=1$$
$$y+3z=0.$$

These equations yield $y = -3z$, $x = 1 + 2z$. Thus, if c is an arbitrary constant, then $x = 1 + 2c$, $y = -3c$, and $z = c$, constitute the general solution of the given system of equations.

EXAMPLE 2.127

Find the value of λ and μ for which the system of equations

$$3x+2y+z=6$$
$$3x+4y+3z=14$$
$$6x+10y+\lambda z=\mu$$

has (i) unique solution, (ii) no solution, and (iii) infinite number of solutions.

Solution. The given system of equations is expressed by the matrix equation

$$AX = \begin{bmatrix} 3 & 2 & 1 \\ 3 & 4 & 3 \\ 6 & 10 & \lambda \end{bmatrix} \begin{bmatrix} x \\ y \\ z \end{bmatrix} = \begin{bmatrix} 6 \\ 14 \\ \mu \end{bmatrix} = B.$$

Therefore, the augmented matrix is

$$[A:B] = \begin{bmatrix} 3 & 2 & 1 & 6 \\ 3 & 4 & 3 & 14 \\ 6 & 10 & \lambda & \mu \end{bmatrix}$$

$$\sim \begin{bmatrix} 3 & 2 & 1 & 6 \\ 0 & 2 & 2 & 8 \\ 0 & 6 & \lambda-2 & \mu-12 \end{bmatrix} \quad \begin{matrix} R_2 \to R_2 - R_1 \\ R_3 \to R_3 - 2R_1 \end{matrix}$$

$$\sim \begin{bmatrix} 3 & 2 & 1 & 6 \\ 0 & 2 & 2 & 8 \\ 0 & 0 & \lambda-8 & \mu-36 \end{bmatrix} \quad R_3 \to R_3 - 3R_2.$$

If $\lambda \neq 8$, then $\rho(A) = \rho([A:B]) = 3$ and so in this case the system is consistent. Further, since $\rho(A)$ is equal to number of unknowns, the given system has a unique solution.

If $\lambda = 8, \mu \neq 36$, then $\rho(A) = 2$ and $\rho([A:B]) = 3$. Hence, the system is inconsistent and has no solution.

If $\lambda = 8, \mu = 36$, then $\rho(A) = \rho([A:B]) = 2$. Therefore, the given system of equation is consistent. Since rank of A is less than the number of unknowns, the given system of equation has infinitely many solutions.

EXAMPLE 2.128

Using consistency theorem, solve the equation

$$x+y+z=9$$
$$2x+5y+7z=52$$
$$2x+y-z=0.$$

Solution. The matrix form of the given system of equations is

$$AX = \begin{bmatrix} 1 & 1 & 1 \\ 2 & 5 & 7 \\ 2 & 1 & -1 \end{bmatrix} \begin{bmatrix} x \\ y \\ z \end{bmatrix} = \begin{bmatrix} 9 \\ 52 \\ 0 \end{bmatrix} = B.$$

Therefore, the augmented matrix is

$$[A:B] = \begin{bmatrix} 1 & 1 & 1 & 9 \\ 2 & 5 & 7 & 52 \\ 2 & 1 & -1 & 0 \end{bmatrix}$$

$$\sim \begin{bmatrix} 1 & 1 & 1 & 9 \\ 0 & 3 & 5 & 34 \\ 0 & -1 & -3 & -18 \end{bmatrix} \quad \begin{matrix} R_2 \to R_2 - 2R_1 \\ R_3 \to R_3 - 2R_1 \end{matrix}$$

$$\sim \begin{bmatrix} 1 & 1 & 1 & 9 \\ 0 & -1 & -3 & -18 \\ 0 & 3 & 5 & 34 \end{bmatrix} \quad R_3 \leftrightarrow R_3$$

$$\sim \begin{bmatrix} 1 & 1 & 1 & 9 \\ 0 & -1 & -3 & -18 \\ 0 & 0 & -4 & 20 \end{bmatrix} \quad R_3 \to R_3 - 3R_2$$

$$\sim \begin{bmatrix} 1 & 1 & 1 & 9 \\ 0 & 1 & 3 & 18 \\ 0 & 0 & 1 & -5 \end{bmatrix} \quad \begin{matrix} R_2 \to -R_2 \\ R_3 \to R_3 - \frac{1}{4} R_3 \end{matrix}$$

Thus we get echelon form of the matrix $[A:B]$. The number of non-zero rows in this form is 3. Therefore $\rho([A:B]) = 3$. Further, since

$$A \sim \begin{bmatrix} 1 & 1 & 1 \\ 0 & 1 & 3 \\ 0 & 0 & 1 \end{bmatrix}.$$

Therefore $\rho(A) = 3$. Hence, $\rho(A) = \rho([A:B]) = 3$. This shows that the given system of equations is consistent. Also, since $\rho(A)$ is equal to the number of unknowns, the solution of the given system is unique. To find the solution, we note that the given system of equation is equivalent to

$$\begin{bmatrix} 1 & 1 & 1 \\ 0 & -1 & -3 \\ 0 & 0 & -4 \end{bmatrix} \begin{bmatrix} x \\ y \\ z \end{bmatrix} = \begin{bmatrix} 9 \\ -18 \\ -20 \end{bmatrix}$$

and so

$x+y+z=9,\ -y-3z=-18,\ -4z=-20$, which yields $z=5,\ y=3,$ and $x=1$ as the required solution

2.33 HOMOGENEOUS LINEAR EQUATIONS

Consider the following system of m homogeneous equations in n unknowns $x_1, x_2, \ldots, x_n$:

$$a_{11}x_1 + a_{12}x_2 + \ldots + a_{1n}x_n = 0$$
$$a_{21}x_1 + a_{22}x_2 + \ldots + a_{2n}x_n = 0$$
$$\cdots$$
$$\cdots$$
$$a_{m1}x_1 + a_{m2}x_2 + \ldots + a_{mn}x_n = 0.$$

The matrix form of this system is

$$AX = 0,$$

where

$$\begin{bmatrix} a_{11} & a_{12} & \cdots & \cdots & a_{1n} \\ a_{21} & a_{22} & \cdots & \cdots & a_{2n} \\ \cdots & \cdots & \cdots & \cdots & \cdots \\ \cdots & \cdots & \cdots & \cdots & \cdots \\ \cdots & \cdots & \cdots & \cdots & \cdots \\ a_{m1} & a_{m2} & \cdots & \cdots & a_{mn} \end{bmatrix}, X = \begin{bmatrix} x_1 \\ x_2 \\ \cdots \\ \cdots \\ x_n \end{bmatrix}, 0 = \begin{bmatrix} 0 \\ 0 \\ \cdots \\ \cdots \\ 0 \end{bmatrix}.$$

It is evident that $x_1 = 0$, $x_2 = 0, \ldots, x_n = 0$, that is, $X = \mathbf{0}$ is a solution of the given system of equations. This solution is called *trivial solution* of the given system.

Let X_1 and X_2 be two solution of $AX = \mathbf{0}$. Then $AX_1 = 0$ and $AX_2 = 0$ and so for arbitrary numbers k_1, k_2, we have

$$\begin{aligned} A(k_1X_1 + k2X_2) &= k_1(AX_1) + k_2(AX_2) \\ &= k_1 0 + k_2 0 = 0. \end{aligned}$$

If follows, therefore, that linear combination of two solutions of $AX = \mathbf{0}$ is also a solution. Hence, the collection of all solutions of the equation $AX = \mathbf{0}$ form a subspace of the vector space V_n.

Theorem 2.55. Let the rank of a matrix A be r. Then the number of linearly independent solutions of m homogeneous linear equations in n variables, $AX = \mathbf{0}$ is $(n - r)$.

Proof. Let

$$\begin{bmatrix} a_{11} & a_{12} & \cdots & \cdots & a_{1n} \\ a_{21} & a_{22} & \cdots & \cdots & a_{2n} \\ \cdots & \cdots & \cdots & \cdots & \cdots \\ \cdots & \cdots & \cdots & \cdots & \cdots \\ \cdots & \cdots & \cdots & \cdots & \cdots \\ a_{m1} & a_{m2} & \cdots & \cdots & a_{mn} \end{bmatrix} X = \begin{bmatrix} x_1 \\ x_2 \\ \cdots \\ \cdots \\ x_n \end{bmatrix}$$

Since $\rho(A) = r$, it has r linearly independent columns. Without loss of generality, suppose that the first r columns of the matrix A are linearly independent. We write

$$A = [C_1 C_2 \ldots C_n],$$

where $C_1,\ C_2, \ldots, Cn$ are column vectors of A. Therefore, $AX = \mathbf{0}$ can be written as vector equation.

$$\begin{aligned} x_1C_1 + x_2C_2 + \ldots + x_rC_r \\ + x_{r+1}C_{r+1} + \ldots + x_nc_n = 0. \end{aligned} \tag{54}$$

Since each of the vector $C_{r+1},\ C_{r+2}, \ldots, C_n$ is a linear combination of vectors $C_1, C_2, \ldots, C_r$, therefore

$$\begin{aligned} C_{r+1} &= p_{11}C_1 + p_{12}C_2 + \ldots + p_{1r}C_r, \quad (55) \\ C_{r+2} &= p_{21}C_1 + p_{22}C_2 + \ldots + p_{2r}C_r, \\ &\ldots \\ &\ldots \\ C_n &= p_{k1}C_1 + p_{k2}C2 + \ldots + p_{kr}C_r, \end{aligned}$$

where $k = n - r$. The expression (54) can be written as

$$\begin{aligned} p_{11}C_1 + p_{12}C_2 + \ldots + p_{1r}C_r - 1.C_{r+1} + \quad (56) \\ 0C_{r+2} + \ldots + 0C_n = 0 \\ p_{21}C_1 + p_{22}C_2 + \ldots + p_{2r}C_r + 0.C_{r+1} \\ -1C_{r+2} + \ldots + 0C_n = 0 \\ \ldots \\ \ldots \\ p_{k1}C_1 + p_{k2}C_2 + \ldots + p_{kr}C_r - 0.C_{r+1} \\ + 0C_{r+2} + \ldots - 1C_n = 0. \end{aligned} \tag{57}$$

Comparing (56) and (57), we note that

$$X_1 = \begin{bmatrix} p_{11} \\ p_{12} \\ \cdots \\ \cdots \\ p_{1r} \\ -1 \\ 0 \\ \cdots \\ \cdots \\ 0 \end{bmatrix}, X_2 = \begin{bmatrix} p_{11} \\ p_{12} \\ \cdots \\ \cdots \\ p_{2r} \\ 0 \\ -1 \\ 0 \\ \cdots \\ 0 \end{bmatrix}, \ldots, X_{n-r} = \begin{bmatrix} p_{k1} \\ p_{k2} \\ \cdots \\ \cdots \\ p_{kr} \\ 0 \\ 0 \\ \cdots \\ 0 \\ -1 \end{bmatrix}$$

are $(n - r)$ solutions of $AX = \mathbf{0}$. Suppose now that

$$c_1X_1 + c_2X_2 + \ldots + c_{n-r}X_{n-r} = \mathbf{0}.$$

Comparing $(r+1)$th, $(r+2)$th,..., nth component on both sides, we get

$$-c_1 = 0,\ -c_2 = 0, \ldots, c_{n-r} = 0.$$

Hence $X_1, X_2, \ldots, X_{n-r}$ are linearly independent. Suppose that the vector X, with components $x_1, x_2, \ldots, x_n$ is any solution of the equation $AX = \mathbf{0}$. We assert that X is linear combination of $x_1, x_2, \ldots, x_{n-r}$. To prove it, we note that the vector

$$X + x_{r+1}X_1 + x_{r+2}X_2 + \ldots + x_nX_{n-r} \qquad (58)$$

being linear combination of solutions is also a solution. Then the last $n-r$ components of the vector (58) are all zero. Let $z_1, z_2, \ldots, z_r$ be the first r components of the vector (58). Then the vector whose components are $(z_1, z_2, \ldots, z_r, 0, 0, \ldots, 0)$ is a solution of the equation $AX = \mathbf{0}$. Therefore from (54), we have

$$z_1C_1 + z_2C_2 + \ldots + z_rC_r = 0.$$

But the vector $C_1, C_2, \ldots, C_r$ are linearly independent. Hence $z_1 = z_2 = z_r = 0$. Hence (58) is a zero vector, that is,

$$X + x_{r+1}X_1 + x_{r+2}X_2 + \ldots + x_nX_{n-r} = 0$$

or

$$X = -x_{r+1}X_1 - x_{r+2}X_2 - \ldots - X_nX_{n-r}.$$

Thus, every solution is a linear combination of the $n-r$ linearly independent solution $X_1, X_2, \ldots, X_{n-r}$. It follows, therefore, that the set of solution $[X_1, X_2, \ldots, X_{n-r}]$ form a basis of vector space of all the solutions of the system of equations $AX = \mathbf{0}$.

Remark 2.9 Suppose we have a system of m linear equations in n unknowns. Thus, the coefficient matrix A is of order $m \times n$. Let r be the rank of A. Then, $r \le n$ (number of column of A).

If $r = n$, then $AX = \mathbf{0}$ possesses $n - n = 0$ number of independent solutions. In this case, we have simply the trivial solution (which forms a linearly dependent system).

If $r < n$, then there are $n - r$ linearly independent solutions. Further any linear combination of these solutions will also be a solution of $AX = \mathbf{0}$. Hence, in this case, the equation $AX = \mathbf{0}$ has infinite number of solutions.

If $m < n$, then since $r \le m$, we have $r < n$. Hence the system has a non-zero solution. The number of solutions of the equation $AX = \mathbf{0}$ will be infinite.

Theorem 2.56. A necessary and sufficient condition that a system of n homogeneous linear equations in n unknowns have non-trivial solutions is that coefficient matrix be singular.

Proof.
The condition is necessary. Suppose that the system of n homogeneous linear equations in n unknowns have a non-trivial solution. We want to show that $|A| = 0$. Suppose, on the contrary, $|A| \neq 0$. Then rank of A is n. Therefore, number of linearly independent solution is $n - n = 0$. Thus, the given system possesses no linearly independent solution. Thus, only trivial solution exists for the given system. This contradicts the fact that the given system of equation has non-trivial solution. Hence $|A| = 0$.
The condition is sufficient. Suppose $|A| = 0$. Therefore, $\rho(A) < n$. Let r be the rank of A. Then the given equation has $(n - r)$ linearly independent solutions. Since a linearly independent solution can never be zero, therefore, the given system must have a non-zero solution.

EXAMPLE 2.129

Solve

$$\begin{aligned} x + 3y - 2z &= 0 \\ 2x - y + 4z &= 0 \\ x - 11y + 14z &= 0. \end{aligned}$$

Solution. The matrix form of the given system of homogeneous equations is $AX = \mathbf{0}$, where

$$A=\begin{bmatrix}1&3&-2\\2&-1&4\\1&-11&14\end{bmatrix}, X=\begin{bmatrix}x\\y\\z\end{bmatrix}, 0=\begin{bmatrix}0\\0\\0\end{bmatrix}.$$

We note that

$$|A|=\begin{vmatrix}1&3&-2\\2&-1&4\\1&-11&14\end{vmatrix}=30-72+42=0.$$

Therefore A is singular, that is $\rho(A)<n$. Thus, the given system has a non-trivial solution and will have infinite number of solutions.
The given system is

$$\begin{bmatrix}1&3&-2\\2&-1&4\\1&-11&14\end{bmatrix}\begin{bmatrix}x\\y\\z\end{bmatrix}=0$$

$$\sim\begin{bmatrix}1&3&-2\\0&-7&8\\0&-14&16\end{bmatrix}\begin{bmatrix}x\\y\\z\end{bmatrix}=0,\quad \begin{matrix}R_2\to R_2-2R_1\\R_3\to R_3-R_1,\end{matrix}$$

$$\sim\begin{bmatrix}1&3&-2\\0&-7&8\\0&0&0\end{bmatrix}\begin{bmatrix}x\\y\\z\end{bmatrix}=0,\ R_3\to R_3-2R_2,$$

and so we have

$$x+3y-2z=0$$
$$-7y+8z=0.$$

These equations yield $y=\frac{8}{7}z$, $x=\frac{-10}{7}z$. Giving different values to z, we get infinite number of solutions.

EXAMPLE 2.130

Solve

$$x_1-x_2+x_3=0$$
$$x_1+2x_2-x_3=0$$
$$2x_1+x_2+3x_3=0.$$

Solution. In matrix form, we have $AX=\mathbf{0}$, where

$$A=\begin{bmatrix}1&-1&1\\1&2&-1\\2&1&3\end{bmatrix}, X=\begin{bmatrix}x_1\\x_2\\x_3\end{bmatrix}, 0=\begin{bmatrix}0\\0\\0\end{bmatrix}.$$

We note that $|A|=9\neq 0$. Thus A is non-singular. Hence, the given system of homogeneous equation has only trivial solution $x_1=x_2=x_3=0$.

EXAMPLE 2.131

Solve

$$2x-2y+5z+3w=0$$
$$4x-y+z+w=0$$
$$3x-2y+3z+4w=0$$
$$x-3y+7z+6w=0.$$

Solution. The matrix form of the given system of homogeneous equation is

$$AX=\begin{bmatrix}2&-2&5&3\\4&-1&1&1\\3&-2&3&4\\1&-3&7&6\end{bmatrix}\begin{bmatrix}x\\y\\z\\w\end{bmatrix}=0.$$

Performing row elementary transformations to get echelon form of A, we have

$$A=\begin{bmatrix}2&-2&5&3\\4&-1&1&1\\3&-2&3&4\\1&-3&7&6\end{bmatrix}\sim\begin{bmatrix}1&-3&7&6\\4&-1&1&1\\3&-2&3&4\\2&-2&5&3\end{bmatrix}\ R_1\leftrightarrow R_4$$

$$\sim\begin{bmatrix}1&-3&7&6\\0&11&-27&-23\\0&7&-18&-14\\0&4&-9&-9\end{bmatrix}\ \begin{matrix}R_2\to R_2-4R_1\\R_3\to R_3-3R_1\\R_4\to R_4-2R_1\end{matrix}$$

$$\sim\begin{bmatrix}1&-3&7&6\\0&4&-9&-9\\0&7&-18&-14\\0&4&-9&-9\end{bmatrix}\ R_2\to R_2-R_3$$

$$\sim\begin{bmatrix}1&-3&7&6\\0&4&-9&-9\\0&28&-72&-56\\0&4&-9&-9\end{bmatrix}\ R_3\to 4R_3$$

$$\sim\begin{bmatrix}1&-3&7&6\\0&4&-9&-9\\0&0&-9&7\\0&0&0&0\end{bmatrix}\ \begin{matrix}R_3\to R_3-7R_2\\R_4\to R_4-R_2.\end{matrix}$$

The above echelon form of A suggests that rank of A is equal to the number of non-zero rows. Thus $\rho(A) = 3$.
The number of unknowns is 4. Thus $\rho(A) < n$. Hence, the given system possesses non-trivial solution. The number of independent solution will be $(n-r) = 4-3 = 1$.

Further, the given system is equivalent to

$$\begin{bmatrix} 1 & -3 & 7 & 6 \\ 0 & 4 & -9 & -9 \\ 0 & 0 & -9 & 7 \\ 0 & 0 & 0 & 0 \end{bmatrix}\begin{bmatrix} x \\ y \\ z \\ w \end{bmatrix} = 0$$

and so, we have

$$x - 3y + 7z + 6w = 0$$
$$4y - 9z - 9w = 0$$
$$-9z + 7w = 0$$

These equations yield $z = \frac{7}{9}w,\ y = 4w,\ x = \frac{5}{9}w$. Thus taking $w = t$, we get $x = \frac{5}{9}t, y = 4t,\ z = \frac{7}{9}t,\ w = t$ as the general solution of the given equations.

EXAMPLE 2.132

Determine the value of λ for which the following equations have non-zero solutions:

$$x + 2y + 3z = \lambda x$$
$$3x + y + 2z = \lambda y$$
$$2x + 3y + z = \lambda z.$$

Solution. The matrix form of the given equation is

$$AX = \begin{bmatrix} 1-\lambda & 2 & 3 \\ 3 & 1-\lambda & 2 \\ 2 & 3 & 1-\lambda \end{bmatrix}\begin{bmatrix} x \\ y \\ z \end{bmatrix} = 0.$$

The given system will have non-zero solution only if $|A| = 0$, that is, if rank of A is less than 3. Thus for the existence of non-zero solution, we must have

$$\begin{vmatrix} 1-\lambda & 2 & 3 \\ 3 & 1-\lambda & 2 \\ 2 & 3 & 1-\lambda \end{vmatrix} = 0$$

or

$$\begin{vmatrix} 6-\lambda & 6-\lambda & 6-\lambda \\ 3 & 1-\lambda & 2 \\ 2 & 3 & 1-\lambda \end{vmatrix} = 0 \text{ using} R_1 \to R_1 + R_2 + R_3$$

or

$$(6-\lambda)\begin{vmatrix} 1 & 1 & 1 \\ 3 & 1-\lambda & 2 \\ 2 & 3 & 1-\lambda \end{vmatrix} = 0$$

or

$$(6-\lambda)\begin{vmatrix} 1 & 0 & 0 \\ 3 & -2-\lambda & -1 \\ 2 & 3 & -1-\lambda \end{vmatrix} = 0, \quad \begin{matrix} C_2 \to C_2 - C_1 \\ C_3 \to C_3 - C_1 \end{matrix}$$

or

$$(6-\lambda)[\lambda^2 + 3\lambda + 3] = 0,$$

which yields

$$\lambda = 6 \text{ and } \frac{-3 \pm \sqrt{9-12}}{2}.$$

Thus, the only real value of λ for which the given system of equation has a solution is 6.

2.34 CHARACTERISTIC ROOTS AND CHARACTERISTIC VECTORS

Let A be a matrix of order n, λ a scalar and

$$X = \begin{bmatrix} x_1 \\ x_2 \\ \dots \\ x_n \end{bmatrix} \text{ a column vector.}$$

Consider the equation

$$AX = \lambda X \tag{59}$$

Clearly $X = 0$ is a solution of (59) for any value of λ. The question arises whether there exist scalar λ and non-zero vector X, which simultaneously satisfy the equation (59). This problem is known as *characteristic value problem.* If I_n is unit matrix of order n, then (59) may be written in the form

$$(A - \lambda I_n)X = \mathbf{0}. \qquad (60)$$

The equation (60) is the matrix form of a system of n homogeneous linear equations in n unknowns. This system will have a non-trivial solution if and only if the determinant of the coefficient matrix $A - \lambda I_n$ vanishes, that is, if

$$|A-\lambda I_n| = \begin{vmatrix} a_{11}-\lambda & a_{12} & \cdots & \cdots & a_{1n} \\ a_{21} & a_{22}-\lambda & \cdots & \cdots & a_{2n} \\ \cdots & \cdots & \cdots & \cdots & \cdots \\ \cdots & \cdots & \cdots & \cdots & \cdots \\ a_{n1} & a_{n2} & \cdots & \cdots & a_{nn}-\lambda \end{vmatrix} = 0.$$

The expansion of this determinant yields a polynomial of degree n in λ, which is called the *characteristic polynomial* of the matrix A.

The equation $|A - \lambda I_n| = 0$ is called the *characteristic equation* or *secular equation* of A.

The n roots of the characteristic equation of a matrix A of an order n are called the *characteristic roots*, *characteristic values, proper values, eigenvalues,* or *latent roots* of the matrix A.

The set of the eigenvalues of a matrix A is called the *spectrum* of A.

If λ is an eigenvalue of a matrix A of order n, then a non-zero vector X such that $AX = \lambda X$ is called a *characteristic vector, eigen vector, proper vector*, or *latent vector* of A corresponding to the characteristic root λ.

Theorem 2.57. The equation $AX = \lambda X$ has a non-trivial solution if and only if λ is a characteristic root of A.

Proof. Suppose first that λ is a characteristic root of the matrix A. Then $|A - \lambda I_n| = 0$ and consequently the matrix $A - \lambda I$ is singular. Therefore, the matrix equation $(A - \lambda I)X = \mathbf{0}$ possesses a non-zero solution. Hence, there exists a non-zero vector X such that $(A - \lambda I)X = \mathbf{0}$ or $AX = \lambda X$.

Conversely, suppose that there exists a non-zero vector X such that $AX = \lambda X$ or $(A - \lambda I)X = \mathbf{0}$. Thus, the matrix equation $(A - \lambda I)X = \mathbf{0}$ has a non-zero solution. Hence $A - \lambda I$ is singular and so $|A - \lambda I| = \mathbf{0}$. Hence, λ is a characteristic root of the matrix A.

Theorem 2.58. Corresponding to a characteristics value λ, there correspond more than one characteristic vectors.

Proof. Let X be a characteristic vector corresponding to a characteristic root λ. Then, by definition, $X \neq 0$ and $AX = \lambda X$. If k is any non-zero scalar, then $kX \neq 0$. Further,

$$A(kX) = k(AX) = k(\lambda X) = \lambda(kX).$$

Therefore, kX is also a characteristic vector of A corresponding to the characteristic root λ.

Theorem 2.59. If X is a proper vector of a matrix A, then X cannot correspond to more than one characteristic root of A.

Proof. Suppose, on the contrary, X be a characteristic vector of a matrix A corresponding to two characteristic roots λ_1 and λ_2. Then, $AX = \lambda_1 X$ and $AX = \lambda_2 X$ and so $(\lambda_1 - \lambda_2)X = 0$. Since $X \neq 0$, this implies $\lambda_1 - \lambda_2 = 0$ or $\lambda_1 = \lambda_2$. Hence the result follows.

Theorem 2.60. Let $X_1, X_2, \ldots, X_n$ be non-zero characteristic vectors associated with distinct characteristic roots $\lambda_1, \lambda_2, \ldots \lambda_n$ of a matrix A. Then $X_1, X_2, \ldots, X_n$ are linearly independent.

Proof. Let $c_1, c_2, \ldots, c_n$ the constants such that

$$c_1X_1 + c_2X_2 + \ldots + c_nX_n = 0 \qquad (61)$$

Multiplying throughout by A and using the fact that $AX_i = \lambda_i X_i$, we get

$$c_1\lambda_1X_1 + c_2\lambda_2X_2 + \ldots + c_n\lambda_nX_n = 0 \qquad (62)$$

Repeating this process, we obtain successively

$$\left.\begin{array}{l} c_1\lambda_1^2X_1+c_2\lambda_2^2X_2+\ldots+c_n\lambda_n^2X_n=0\\ c_1\lambda_1^3X_1+c_2\lambda_2^3X_2+\ldots+c_n\lambda_n^3X_n=0\\ \ldots\\ \ldots\\ c_1\lambda_1^{k-1}X_1+c_2\lambda_2^{k-1}X_2+\ldots+c_n\lambda_n^{k-1}X_n=0 \end{array}\right\} \quad (63)$$

The k equations (61) through (63) in vector unknowns $X_1, X_2,\ldots,X_n$ can be written in the form

$$[c_1X_1\ c_2X_2\ldots\ldots c_nX_n]\begin{bmatrix} 1 & \lambda_1 & \lambda_1^2 & \cdots & \lambda_1^{k-1}\\ 1 & \lambda_2 & \lambda_2^2 & \cdots & \lambda_2^{k-1}\\ \cdots & \cdots & \cdots & \cdots & \cdots\\ \cdots & \cdots & \cdots & \cdots & \cdots\\ 1 & \lambda_n & \lambda_n^2 & \cdots & \lambda_n^{k-1} \end{bmatrix}=0.$$

Since $\lambda_1,\lambda_2,\ldots,\lambda_n$ are distinct, the right factor is a *non-singular Vander-monde matrix*. Since it is non-singular, its inverse exists. Post-multiplication by its inverse yields

$$[c_1X_1\ c_2X_2\ldots\ldots c_nX_n]=0$$

Since $X_1,X_2,\ldots,X_n$ are all non-zero, it follows that $c_1=c_2=c_n=0$. Thus, the relation (61) implies $c_1=c_2=c_n=0$. Hence, $X_1,X_2,\ldots,X_n$ are linearly independent.

Let $\phi(\lambda)=a_0\lambda^n+a_1\lambda^{n-1}+\ldots+a_{n-1}\lambda+a_n$ be the characteristic polynomial of a matrix A. Thus

$$\begin{aligned}|A-\lambda I| &= \phi(\lambda)\\ &= a_0\lambda^n+a_1\lambda^{n-1}+\ldots+a_{n-1}\lambda+a_n.\end{aligned}$$

If we put $\lambda=0$, then we get $|A|=a_n$. The diagonal term of $|A-\lambda I|$ is $(a_{11}-\lambda)(a_{22}-\lambda)\ldots(a_{nn}-\lambda)$ and this is the only product yielding λ^n and λ^{n-1}. Expanding the product, we obtain $(-1)^n\lambda^n$ and $(-1)^{n-1}\sum a_{ii}\lambda^{n-1}$ as the first two terms of $\phi(\lambda)$. Hence

$$a_0=(-1)^n \text{ and } a_1=(-1)^{n-1}\sum a_{ii}$$

In $\phi(\lambda)$, the coefficient of λ^{n-1}, namely $(-1)^{n-1}(a_{11}+a_{22}+\ldots+a_{nn})$ is of special interest. As we know, the term $a_{11}+a_{22}+\ldots+a_{nn}$ is called the *trace* or *spur* of the matrix A. It follows from the above discussion that the *sum of the eigenvalues of a matrix is equal to its trace* and the product of the eigenvalues of a matrix A is its determinant $|A|$.

Theorem 2.61. The characteristic roots of a Hermitian matrix are real.

Proof. Let λ be a characteristic root of a Hermitian matrix. Then there exists a non-zero vector X such that

$$AX=\lambda X. \quad (64)$$

Taking transpose conjugate, we get

$$X^\theta A^\theta=\bar{\lambda}X^\theta. \quad (65)$$

Pre-multiplying (64) by X^θ and post-multiplying (65) by X, we get

$$\begin{aligned}&X^\theta AX=\lambda X^\theta X, \text{ and}\\ &X^\theta A^\theta X=\bar{\lambda}X^\theta X.\end{aligned}$$

Since A is Hermitian, $A^\theta=A$ and, therefore, (64) and (65) imply

$$\lambda X^\theta X=\bar{\lambda}X^\theta X\Rightarrow(\lambda-\bar{\lambda})X^\theta X=0.$$

Since $X\neq 0$, we have $\lambda-\bar{\lambda}=0$ and so $\lambda=\bar{\lambda}$. Hence λ is real.

Corollary 2.15. The characteristic roots of a real symmetric matrix are all real.

Proof. Since a real symmetric matrix is Hermitian, it follows from Theorem 2.61 that the characteristic roots of a real symmetric matrix are all real.

Corollary 2.16. The characteristic roots of a Skew-Hermitian matrix are either pure imaginary or zero.

Proof. Let A be a Skew-Hermitian matrix. Then iA is Hermitian. Let λ be the characteristic root of A. Then $AX=\lambda X, X\neq 0$ or $(iA)X=(i\lambda)X$. Thus $i\lambda$ is a characteristic root of iA. But iA is Hermitian and characteristic roots of Hermitian matrix are real. Thus $i\lambda$ is real, which is possible only if λ is zero or pure imaginary. This proves the result.

Corollary 2.17. The characteristic roots of a skew symmetric matrix are either pure imaginary or zero.

Proof. Since a Skew-Symmetric matrix is Skew-Hermitian, the result follows from corollary 2.16.

EXAMPLE 2.133

Find the characteristic vectors of the matrix

$$A = \begin{bmatrix} 3 & 1 & 0 \\ 0 & 3 & 1 \\ 0 & 0 & 3 \end{bmatrix}$$

Solution. The characteristic equation of the given matrix is

$$[A - \lambda I] = \begin{vmatrix} 3-\lambda & 1 & 0 \\ 0 & 3-\lambda & 1 \\ 0 & 0 & 3-\lambda \end{vmatrix} = 0,$$

which yields $(3-\lambda)^3 = 0$. Thus 3 is the only distinct characteristic root of A. The characteristic vectors are given by non-zero solutions of the equation $(A-3I)X = 0$, that is,

$$\begin{bmatrix} 0 & 1 & 0 \\ 0 & 0 & 1 \\ 0 & 0 & 0 \end{bmatrix} \begin{bmatrix} x_1 \\ x_2 \\ x_3 \end{bmatrix} = \begin{bmatrix} 0 \\ 0 \\ 0 \end{bmatrix}.$$

The coefficient matrix of the equation is of rank 2. Therefore, number of linearly independent solution is $n - r = 1$. The above equation yields $x_2 = 0, x_3 = 0$. Therefore, $x_1 = 1, x_2 = 0, x_3 = 0$ is a non-zero solution of the above equation. Thus,

$$X = \begin{bmatrix} 1 \\ 0 \\ 0 \end{bmatrix}$$

is an eigenvector of A corresponding to the eigenvalue 3. Also any non-zero multiple of this vector shall be an eigenvector of A corresponding to $\lambda = 3$.

EXAMPLE 2.134

Find the eigenvalues and the corresponding eigenvectors of the matrix

$$A = \begin{bmatrix} 6 & -2 & 2 \\ -2 & 3 & -1 \\ 2 & -1 & 3 \end{bmatrix}.$$

Solution. The characteristic equation of the given matrix is

$$|A - \lambda I| = \begin{vmatrix} 6-\lambda & -2 & 2 \\ -2 & 3-\lambda & -1 \\ 2 & -1 & 3-\lambda \end{vmatrix} = 0$$

or

$$\begin{vmatrix} 6-\lambda & -2 & 0 \\ -2 & 3-\lambda & 2-\lambda \\ 2 & -1 & 2-\lambda \end{vmatrix} = 0, \quad C_3 \to C_3 - C_2$$

or

$$(2-\lambda)\begin{vmatrix} 6-\lambda & -2 & 0 \\ -2 & 3-\lambda & 1 \\ 2 & -1 & 1 \end{vmatrix} = 0$$

or

$$(2-\lambda)(\lambda-2)(\lambda-8) = 0.$$

Thus, the characteristic roots of A are $\lambda = 2, 2, 8$. The eigenvector of A corresponding to the eigenvalue 2 is given by $(A - 2I)X = \mathbf{0}$ or

$$\begin{bmatrix} 4 & -2 & 2 \\ -2 & 1 & -1 \\ 2 & -1 & 1 \end{bmatrix} \begin{bmatrix} x_1 \\ x_2 \\ x_3 \end{bmatrix} = \begin{bmatrix} 0 \\ 0 \\ 0 \end{bmatrix}$$

or

$$\begin{bmatrix} -2 & 1 & -1 \\ 4 & -2 & 2 \\ 2 & -1 & 1 \end{bmatrix} \begin{bmatrix} x_1 \\ x_2 \\ x_3 \end{bmatrix} = \begin{bmatrix} 0 \\ 0 \\ 0 \end{bmatrix}, R_1 \leftrightarrow R_2$$

or

$$\begin{bmatrix} -2 & 1 & -1 \\ 0 & 0 & 0 \\ 0 & 0 & 0 \end{bmatrix} \begin{bmatrix} x_1 \\ x_2 \\ x_3 \end{bmatrix} = \begin{bmatrix} 0 \\ 0 \\ 0 \end{bmatrix}, \quad \begin{matrix} R_2 \to R_2 + 2R_1 \\ R_3 \to R_3 + 2R_1 \end{matrix}$$

The coefficient matrix is of rank 1. Therefore, there are $n-r=3-1=2$ linearly independent solution. The above equation is

$$-2_{x1}+x_2-x_3=0.$$

Clearly,

$$X_1=\begin{bmatrix}-1\\0\\2\end{bmatrix} \text{ and } X_2=\begin{bmatrix}1\\2\\0\end{bmatrix}$$

are two linearly independent solutions of this equation. Then X_1 and X_2 are two linearly independent eigenvectors of A corresponding to eigenvalue 2. If k_1, k_2 are scalars not both equal to zero, then $k_1X_1+k_2X_2$ yields all the eigenvectors of A corresponding to the eigenvalue 2.

The characteristic vectors of A corresponding to the characteristic root 8 are given by $(A-8I)X=\mathbf{0}$ or by

$$=\begin{bmatrix}6-8 & -2 & 2\\ -2 & 3-8 & -1\\ 2 & -1 & 3-8\end{bmatrix}\begin{bmatrix}x_1\\x_2\\x_3\end{bmatrix}=\begin{bmatrix}0\\0\\0\end{bmatrix}$$

$$\sim\begin{bmatrix}-2 & -2 & 2\\ 0 & -3 & -3\\ 0 & -3 & -3\end{bmatrix}\begin{bmatrix}x_1\\x_2\\x_3\end{bmatrix}=\begin{bmatrix}0\\0\\0\end{bmatrix}, \begin{matrix}R_2\to R_2-R_1\\ R_3\to R_3+R_1\end{matrix}$$

$$\sim\begin{bmatrix}-2 & -2 & 2\\ 0 & -3 & -3\\ 0 & 0 & 0\end{bmatrix}\begin{bmatrix}x_1\\x_2\\x_3\end{bmatrix}=\begin{bmatrix}0\\0\\0\end{bmatrix}, R_3\to R_3-R_3$$

The coefficient matrix is of rank 2. Therefore, number of linearly independent solution is $n-r=3-2=1$. The above equations give

$$-2x_1-2x_2+2x_3=0$$
$$-3x_2-3x_3=0.$$

Hence $x_2=-x_3$. Taking $x_2=-1, x_3=1$, we get

$$x_1=2. \text{ Therefore } X_3=\begin{bmatrix}2\\-1\\1\end{bmatrix}$$

is an eigenvector of A corresponding to $\lambda=8$. Further, every non-zero multiple of X_3 is an eigenvector of A corresponding to the eigenvalue 8.

EXAMPLE 2.135

If A is non-singular, show that the eigenvalues of A^{-1} are the reciprocals of the eigenvalues of A.

Solution. Let λ be a characteristic root of the matrix A. Therefore, there exists non-zero vector X such that

$$AX=\lambda X$$
$$\Rightarrow A^{-1}AX=\lambda A^{-1}X$$
$$\Rightarrow \frac{1}{\lambda}X=A^{-1}X.$$

Hence $\frac{1}{\lambda}$ is a characteristic root of A^{-1} and X is the corresponding characteristic vector.

EXAMPLE 2.136

Show that the characteristic roots of a triangular matrix are just the diagonal elements of the matrix.

Solution. Let

$$\begin{bmatrix}a_{11} & a_{12} & \cdots & \cdots & a_{1n}\\ 0 & a_{22} & \cdots & \cdots & a_{2n}\\ \cdots & \cdots & \cdots & \cdots & \cdots\\ \cdots & \cdots & \cdots & \cdots & \cdots\\ 0 & 0 & \cdots & \cdots & a_{nn}\end{bmatrix}$$

be a triangular matrix of order n. Then

$$|A-\lambda I|=\begin{vmatrix}a_{11}-\lambda & a_{12} & \cdots & \cdots & a_{1n}\\ 0 & a_{22}-\lambda & \cdots & \cdots & a_{2n}\\ \cdots & \cdots & \cdots & \cdots & \cdots\\ \cdots & \cdots & \cdots & \cdots & \cdots\\ 0 & 0 & \cdots & \cdots & a_{nn}-\lambda\end{vmatrix}$$
$$=(a_{11}-\lambda)(a_{22}-\lambda)\cdots(a_{nn}-\lambda).$$

Hence, the roots of the characteristic equation $|A-\lambda I|=0$ are $a_{11}, a_{22},\ldots,a_{nn}$ which are the diagonal element of A.

EXAMPLE 2.137

Show that 0 is an eigenvalue of a matrix A if and only if A is singular.

Solution. If $\lambda = 0$ is an eigenvalue, it satisfies the characteristic equation $|A - \lambda I| = 0$ and so we have $|A| = 0$. Thus A is singular. Conversely if A is singular, then $|A| = 0$. Thus $\lambda = 0$ satisfy the equation $|A - \lambda I| = 0$ and so it is an eigenvalue.

2.35 THE CAYLEY-HAMILTON THEOREM

Let

$$\phi(\lambda) = a_0\lambda^n + a_1\lambda^{n-1} + \ldots + a_{n-1}\lambda + a_n$$

be the characteristic polynomial of a matrix A. Then

$$\phi(A) = a_0A^n + a_1A^{n-1} + \ldots + a_{n-1}A + a_nI_n,$$
$$a_0 = (-1)^n$$

is called the *characteristic function* of the matrix A. Concerning this function, we have the following famous theorem.

Theorem 2.62. (Cayley-Hamilton Theorem). Every square matrix A satisfies its characteristic equation $\phi(A) = 0$.

Proof. The characteristic matrix of A is $A - \lambda I_n$. Since the elements of $A - \lambda I_n$ are at most of the first degree in λ, the elements (cofactor) of the adjoint matrix of $A - \lambda I_n$ are of degree utmost $n - 1$ in λ. Therefore, we may represent adj $(A - \lambda I_n)$ as a matrix polynomial

$$adj(A - \lambda I_n) = B_0\lambda^{n-1} + B_1\lambda^{n-2} + \ldots + B_{n-2}\lambda + B_{n-1},$$

where B_k is the matrix whose elements are the coefficients of λ^k in the corresponding elements of adj $(A - \lambda I_n)$. But

$$(A - \lambda I_n) \text{ adj } (A - \lambda I_n) = |A - \lambda I_n| I_n,$$

that is,

$$A\, adj(A - \lambda I_n) - \lambda\, adj\,(A - \lambda I_n) = \phi(\lambda) I_n.$$

Substituting the expansion of adj $(A - \lambda I_n)$ from above and $\phi(\lambda) = a_0\lambda_n + a_1\lambda^{n-1} + \ldots + a_{n-1}\lambda + a_n$, we get

$$\begin{aligned} &A(B_0\lambda^{n-1} + B_1\lambda^{n-2} + \ldots + B_{n-2}\lambda + B_{n-1}) \\ &- \lambda(B_0\lambda^{n-1} + B_1\lambda^{n-2} + \ldots + B_{n-1}) \\ &= (a_0\lambda^n + a_1\lambda^{n-1} + \ldots + a_{n-1}\lambda + a_n)I_n. \end{aligned}$$

Comparing coefficients of like powers of λ on both sides, we get

$$\begin{aligned} -IB_0 &= a_0I_n \\ AB_0 - IB_1 &= a_1I_n \\ AB_1 - IB_2 &= a_2I_n \\ &\ldots\ldots\ldots \\ &\ldots\ldots\ldots \\ AB_{n-1} &= a_nI_n. \end{aligned}$$

Multiplying these successively by $A^n, A^{n-1}, \ldots, I_n$ and adding, we get,

$$0 = a_0A^n + a_1A^{n-1} + \ldots + a_{n-1}A + a_nI,$$

that is,

$$\phi(A) = \mathbf{0}.$$

This completes the proof of the theorem.

Corollary 2.18. If A is non-singular, then

$$\begin{aligned} A^{-1} &= \frac{-a_0}{a_n}A^{n-1} - \frac{a_1}{a_n}A^{n-2} - \ldots - \frac{a_{n-1}}{a_n}I \\ &= \frac{-1}{a_n}\left(a_0A^{n-1} + a_1A^{n-2} + \ldots + a_{n-1}I\right) \end{aligned}$$

Proof. By Cayley-Hamilton theorem, we have

$$a_0A^n + a_1A^{n-1} + \ldots + a_{n-1}A + a_nI = 0.$$

Pre-multiplication with A^{-1} yields

$$a_0A^{n-1} + a_1A^{n-2} + \ldots + a_{n-1} + a_nA^{-1} = 0.$$

or

$$A^{-1} = -\frac{a_0}{a_n}A^{n-1} - \frac{a_1}{a_n}A^{n-2} - \ldots - \frac{a_{n-1}}{a_n}I$$
$$= -\frac{1}{a_n}\left(a_0A^{n-1} + a_1A^{n-2} + \ldots + a_{n-1}I\right).$$

Remark 2.10 It follows from above that

$$A^n = -\frac{1}{a_0}\left(a_1A^{n-1} + a_2A^{n-2} + \ldots + a_nI\right).$$

Thus higher powers of a matrix can be obtained using lower powers of A.

EXAMPLE 2.138

Verify Cayley-Hamilton theorem for the matrix

$$A = \begin{bmatrix} 2 & -1 & 1 \\ -1 & 2 & -1 \\ 1 & -1 & 2 \end{bmatrix}$$

and hence find A^{-1}.

Solution. We have

$$|A - \lambda I| = \begin{vmatrix} 2-\lambda & -1 & 1 \\ -1 & 2-\lambda & -1 \\ 1 & -1 & 2-\lambda \end{vmatrix}$$
$$= -\lambda^3 + 6\lambda^2 - 9\lambda + 4.$$

Thus, the characteristic equation of the matrix$A$ is

$$\lambda^3 - 6\lambda^2 + 9\lambda - 4 = 0.$$

To verify Cayley-Hamilton theorem, we have to show that

$$A^3 - 6A^2 + 9A - 4I = 0. \qquad (66)$$

We have

$$A^2 = \begin{bmatrix} 6 & -5 & 5 \\ -5 & 6 & -5 \\ 5 & -5 & 6 \end{bmatrix},$$

$$A^3 = \begin{bmatrix} 22 & -21 & 21 \\ -21 & 22 & -21 \\ 21 & -21 & 22 \end{bmatrix}.$$

Then, we note that

$$A^3 - 6A^2 + 9A - 4I = \begin{bmatrix} 0 & 0 & 0 \\ 0 & 0 & 0 \\ 0 & 0 & 0 \end{bmatrix} = \mathbf{0}.$$

Further, pre-multiplying (66) by A^{-1}, we get

$$A^2 - 6A + 9I - 4A^{-1} = 0$$

and so

$$A^{-1} = \frac{1}{4}(A^2 - 6A + 9I)$$
$$= \frac{1}{4}\begin{bmatrix} 3 & 1 & -1 \\ 1 & 3 & 1 \\ -1 & 1 & 3 \end{bmatrix}.$$

2.36 ALGEBRAIC AND GEOMETRIC MULTIPLICITY OF AN EIGENVALUE

Definition 2.99. If λ is an eigenvalue of order m of matrix A, then m is called the *algebraic multiplicity* of λ.

Definition 2.100. If s is the number of linearly independent eigenvectors corresponding to the eigenvalue λ, then s is called the *geometric multiplicity* of λ.

If r is the rank of the coefficient matrix of $(A - \lambda I)X = \mathbf{0}$, then $s = n - r$, where n is the number of unknowns.

The geometric multiplicity of an eigenvalue cannot exceed its algebraic multiplicity.

2.37 MINIMAL POLYNOMIAL OF A MATRIX

Definition 2.101. A polynomial in x in which the coefficient of the highest power of x is unity is called a *monic polynomial.*

For example, $x^4 - x^3 2x^2 + x + 4$ is a monic polynomial of degree 4 over the field of real numbers.

Definition 2.102. The monic polynomial $m(x)$ of lowest degree such that $m(A) = 0$ is called the *minimal polynomial* of the matrix A.

If $m(x)$ is the minimal polynomial of a matrix A, then the equation $m(x) = 0$ is called the *minimal equation* of the matrix A.

Theorem 2.63. The minimal polynomial of a matrix is unique.

Proof. Suppose that the minimal polynomial of a matrix A is of degree r. Therefore, for no non-zero polynomial of degree less than r, we can have $m(A) = 0$. Let $m_1(x)$ and $m_2(x)$ be two minimal polynomial of A. Then

$$m_1(A) = A^r + a_1 A^{r-1} + \ldots + a_r I = \mathbf{0}$$
$$m_2(A) = A^r + b_1 A^{r-1} + \ldots + b_r I = \mathbf{0}.$$

Subtracting, we have

$$(b_1 - a_1)A^{r-1} + \ldots + (b_r - a_r)I = \mathbf{0}.$$

Thus, we have a polynomial $f(x)$ of degree $r-1$ such that $f(A) = \mathbf{0}$. Since its degree is less than r, this should be a zero polynomial. Hence

$$b_1 - a_1 = 0, \ldots, b_r - a_r = 0$$

and

$$a_1 = b_1, \ldots, b_r = a_r$$

proving that $m_1(A) = m_2(A)$. Hence, minimal polynomial of A is unique.

Theorem 2.64. Every polynomial $p(\lambda)$ such that $p(A) = \mathbf{0}$ is exactly divisible by the minimal polynomial $m(\lambda)$.

Proof. Let $q(\lambda)$ be the quotient when $p(\lambda)$ is divided by $m(\lambda)$ and let $r(\lambda)$ be the remainder, which is of degree less than the degree of $m(\lambda)$. Then, by division algorithm, we have,

$$p(\lambda) = m(\lambda)q(\lambda) + r(\lambda)$$

so

$\mathbf{0} = p(A) = m(A)q(A) + r(A) = \mathbf{0}.q(A)) + r(A)$, which yields $r(A) = \mathbf{0}$. Since $r(\lambda)$ is of degree less than the degree of $m(\lambda)$, it follows that $m(\lambda)$ is not a minimal polynomial unless $r(\lambda) = 0$. Thus,

$$p(\lambda) = m(\lambda)q(\lambda)$$

and hence $m(\lambda)$ divides $p(\lambda)$.

Corollary 2.19. The minimal polynomial of a matrix is a divisor of the characteristic polynomial of that matrix.

Proof. Let $\phi(\lambda)$ be the characteristic polynomial of a matrix A. Then by Cayley-Hamilton theorem, $\phi(A) = \mathbf{0}$. Let $m(\lambda)$ be the minimal polynomial of A. Then, by Theorem 2.64, $m(\lambda)$ divides $\phi(\lambda)$.

Corollary 2.20. Every root of the minimal equation of a matrix is also a characteristic root of the matrix.

Proof. Let $\phi(\lambda)$ be the characteristic polynomial of a matrix A and $m(\lambda)$ be its minimal polynomial. Then, by Corollary 2.19, $m(\lambda)$ divides $\phi(\lambda)$. Therefore, there exists a polynomial $q(\lambda)$ such that

$$\phi(\lambda) = m(\lambda)q(\lambda).$$

Now, suppose μ is a root of the equation $m(\lambda) = 0$. Therefore, $m(\mu) = 0$ and so

$$\phi(\mu) = m(\mu)q(\mu) = 0.$$

Hence μ is a root of the characteristic equation $\phi(\lambda) = 0$ and so μ is a characteristic root of A.

Theorem 2.65. If $\lambda_1, \lambda_2, \ldots, \lambda_n$ are the characteristic roots, distinct or not, of a matrix A of order n and if $g(A)$ is any polynomial function of A, then the characteristic roots of $g(A)$ are $g(\lambda_1), g(\lambda_2), \ldots, g(\lambda_n)$.

Proof. We have

$$|A - \lambda I_n| = (-1)^n(\lambda - \lambda_1)(\lambda - \lambda_2)\ldots(\lambda - \lambda_n).$$

We want to show that

$$|g(A) - \lambda I_n| = (-1)^n(\lambda - g(\lambda_1)) \times (\lambda - g(\lambda_2))\ldots(\lambda - g(\lambda_n)).$$

Suppose $g(x)$ is of degree r in x and that for a fixed value of λ, the roots of $g(x)-\lambda=0$ are $x_1,x_2,\ldots,x_r$. Then

$$g(x)-\lambda=\alpha(x-x_1)(x-x_2)\ldots(x-x_r),$$

where α is the coefficient of x^r in $g(x)$. Hence

$$g(A)-\lambda I_n=\alpha(A-x_1I_n)(A-x_2I_n)\ldots(A-x_rI_n).$$

Therefore if $\phi(\lambda)$ is the characteristic polynomial of A, then

$$\begin{aligned}
&|g(A)-\lambda I_n|\\
&=\alpha^n|(A-x_1I_n)|\;|A-x_2I_n|\ldots|A-x_rI_n|\\
&=\alpha^n\phi(x_1)\phi(x_2)\ldots\phi(x_r)\\
&=\alpha^n(-1)^n(x_1-\lambda_1)(x_1-\lambda_2)\ldots(x_1-\lambda_n)\\
&\quad\ldots(-1)^n(x_r-\lambda_1)(x_r-\lambda_2)\ldots(x_r-\lambda_n)\\
&=\alpha(\lambda_1-x_1)(\lambda_1-x_2)\ldots(\lambda_1-x_r)\\
&\quad\ldots\alpha(\lambda_n-x_1)(\lambda_n-x_2)\ldots(\lambda_n-x_r)\\
&=(g(\lambda_1)-\lambda)(g(\lambda_2)-\lambda)\ldots(g(\lambda_n)-\lambda)\\
&=(-1)^n(\lambda-g(\lambda_1))(\lambda-g(\lambda_2))\ldots(\lambda-g(\lambda_n)).
\end{aligned}$$

Hence, $g(\lambda_1),g(\lambda_2),\ldots,g(\lambda_n)$ are the characteristic roots of $g(A)$.

Theorem 2.66. Every root of the characteristic equation of a matrix is also a root of the minimal equation of the matrix.

Proof. Suppose $m(x)$ is the minimal polynomial of a matrix A. Then $m(A)=\mathbf{0}$. Let λ be a characteristic root of A. Then, by Theorem 2.65, $m(\lambda)$ is the characteristic root of $m(A)$. But $m(A)=0$ and so each of its characteristic root is zero. Hence $m(\lambda)=0$, which implies that λ is a root of the equation $m(x)=0$. This proves that every characteristic root of a matrix A is also a root of the minimal equation $m(x)=0$.

Corollary 2.20 and Theorem 2.66 combined together yield:

Theorem 2.67. A scalar λ is a characteristic root of a matrix if and only if it is a root of the minimal equation of that matrix.

Definition 2.103. An n-rowed matrix is said to be *derogatory* or *non-derogatory* according as the degree of its minimal equation is less than or equal to n.

It follows from the definition that a matrix is non-derogatory if the degree of its minimal polynomial is equal to the degree of its characteristic polynomial.

Theorem 2.68. If the roots of the characteristic equation of a matrix are all distinct, then the matrix is non-derogatory.

Proof. Let A be a matrix of order n having n distinct characteristic roots. By Theorem 2.67, each of these roots is also a root of the minimal polynomial of A. Therefore, the minimal polynomial of A is of degree n. Hence, by definition, A is non-derogatory.

EXAMPLE 2.139

Show that the matrix

$$A=\begin{bmatrix}7&4&-1\\4&7&-1\\-4&-4&4\end{bmatrix}$$

is derogatory.

Solution. We have

$$|A-\lambda I|=\begin{vmatrix}7-\lambda&4&-1\\4&7-\lambda&-1\\-4&-4&4-\lambda\end{vmatrix}$$
$$=-(\lambda-12)(3-\lambda)^2.$$

Therefore, roots of the characteristic equation $|A-\lambda I|=0$ are $\lambda=3,\ 3,\ 12$.

Since each characteristic root of a matrix is also a root of its minimal polynomial, therefore, $(x-3)$ and $(x-12)$ shall be factors of $m(x)$. Let. $g(x)=(x-3)(x-12)=x^2-15x+36$.

We have

$$A^2=\begin{bmatrix}69&60&-15\\60&69&-15\\-60&-60&24\end{bmatrix}$$

Then, we observe that

$$g(A) = A^2 - 15A + 36I = \begin{bmatrix} 0 & 0 & 0 \\ 0 & 0 & 0 \\ 0 & 0 & 0 \end{bmatrix}.$$

Thus $g(x)$ is the monomic polynomial of lowest degree such that $g(A) = 0$. Hence $g(x)$ is minimal polynomial of A. Since its degree is less than the order of the matrix A, the given matrix A is derogatory.

2.38 ORTHOGONAL, NORMAL AND UNITARY MATRICES

Definition 2.104. Let

$$X = \begin{bmatrix} x_1 \\ x_2 \\ \dots \\ \dots \\ x_n \end{bmatrix} \quad \text{and} \quad Y = \begin{bmatrix} y_1 \\ y_2 \\ \dots \\ \dots \\ y_n \end{bmatrix}$$

be two complex n-vectors. The inner product of X and Y denoted by (X, Y), is defined as

$$(X, Y) = X^{\theta} Y = [\bar{x}_1\ \bar{x}_2 \dots \bar{x}_n] \begin{bmatrix} y_1 \\ y_2 \\ \dots \\ \dots \\ y_n \end{bmatrix}$$

$$= \bar{x}_1 y_1 + \bar{x}_2 y_2 + \dots + \bar{x}_n y_n.$$

If X and Y are real, then their product becomes

$$(X, Y) = X^T Y = [x_1\ x_2 \dots x_n] \begin{bmatrix} y_1 \\ y_2 \\ \dots \\ \dots \\ y_n \end{bmatrix}$$

$$= x_1 y_1 + x_2 y_2 + \dots + x_n y_n.$$

Definition 2.105. Let X be a complex n-vector. Then the positive square root of the inner product of X with itself is called the *length* or *norm* of X. It is denoted by $\|X\|$.

For example, if

$$X = \begin{bmatrix} x_1 \\ x_2 \\ \dots \\ \dots \\ x_n \end{bmatrix}$$

then

$$\|X\| = \sqrt{(X, X)} = \sqrt{X^{\theta} X}$$

$$= \sqrt{|x_1|^2 + |x_2|^2 + \dots + |x_n|^2}.$$

Obviously, the length of a vector is zero if and only if the vector is a zero vector.

Definition 2.106. A vector X is called a *unit vector* if $\|X\| = 1$.

Definition 2.107. Two complex n-vectors X and Y are said to be *orthogonal* if

$$(X, Y) = X^{\theta} Y = 0.$$

Obviously, *zero is the only vector which is orthogonal to itself.*

Definition 2.108. A set S of complex n-vectors $X_1, X_2, \dots, X_n$ is said to be an orthogonal set if any two distinct vectors in S are orthogonal.

Definition 2.109. A set S of complex n-vectors $X_1, X_2, \dots, X_n$ is said to be an orthonormal set if

(i) S is an orthogonal set

(ii) each vector in S is a unit vector.

Thus the set $X_1, X_2, \dots, X_n$ is orthonormal if

$$(X_i, X_j) = \delta_{ij},\ i,\ j = 1, 2, \dots, n,$$

where δ_{ij} (called Kronecker delta) is defined as

$$\delta_{ij} = \begin{cases} 0 & \text{for } i \neq j \\ 1 & \text{for } i = j. \end{cases}$$

Theorem 2.69. An orthogonal set of non-zero vectors is linearly independent.

Proof. Let $S = [X_1, X_2, \ldots, X_n]$ be an orthogonal set of non-zero vectors. Let $c_1, c_2, \ldots, c_n$ be scalars such that

$$c_1X_1 + c_2X_2 + \ldots + c_nX_n = 0 \qquad (67)$$

Let $1 \le m \le n$. Then inner product of (67) with X_m is

$$(X_m, c_1X_1 + c_2X_2 + \ldots + c_nX_n) = (X_m, 0)$$

or

$$c_1(X_m, X_1) + c_2(X_m, X_2) + \ldots + c_m(X_m, X_m) + \ldots + c_n(X_m, X_n) = 0.$$

Since $(X_m, X_n) = 0$ for $m \neq n$, the above relation yields

$$c_m(X_m, X_m) = 0.$$

Since $X_m \neq 0$, the inner product $(X_m, X_m) \neq 0$. Hence $c_m = 0$, $m = 1, 2, \ldots, n$. Thus, (67) implies $c_1 = c_2 = \ldots = c_n = 0$. Hence, $X_1, X_2, \ldots, X_n$ are linearly independent.

Corollary 2.21. Every orthonormal set of vectors is linearly independent.

Proof. Since for every vector X_n, $(X_n, X_n) = 1$, the result follows from Theorem 2.69.

Definition 2.110. A square matrix U with complex element is said to be *unitary* if $U^\theta U = I$.

If U is unitary, then

$$\begin{aligned} & U^\theta U = I \\ \Rightarrow\ & |U^\theta U| = |I| \\ \Rightarrow\ & |U^\theta|\,|U| = |I| = 1 \\ \Rightarrow\ & |U| \neq 0. \end{aligned}$$

Hence, U is non-singular and so invertible. Thus, U^θ is the inverse of U and we have

$$U^\theta U = I = UU^\theta.$$

Hence, a matrix U is unitary if and only if

$$U^\theta U = UU^\theta = I.$$

If U is a unitary matrix, then the transformation $Y = UX$ is called a *unitary transformation.*

Theorem 2.70. The eigenvalues of a unitary matrix are of unit modulus.

Proof. Let λ be an eigenvalue of a unitary matrix. Therefore, there exists non-zero vector X such that

$$AX = \lambda X \qquad (68)$$

Therefore, taking transposed conjugate of (68), we get

$$\begin{aligned} & (AX)^\theta = (\lambda X)^\theta \\ \Rightarrow\ & X^\theta A^\theta = \bar{\lambda} X^\theta \end{aligned} \qquad (69)$$

By (68) and (69), we have

$$\begin{aligned} & X^\theta A^\theta AX = \bar{\lambda}\lambda\, X^\theta X \\ \Rightarrow\ & X^\theta X = \bar{\lambda}\lambda\, X^\theta X \\ \Rightarrow\ & (1 - \bar{\lambda}\lambda) X^\theta X = 0 \\ \Rightarrow\ & (1 - \bar{\lambda}\lambda) = 0 \text{ since } X^\theta X \neq 0 \\ \Rightarrow\ & \bar{\lambda}\lambda = 1 \\ \Rightarrow\ & |\lambda| = 1. \end{aligned}$$

Theorem 2.71. (i) If U is unitary matrix, then absolute value of $|U|$ is equal to 1.

(ii) Any two eigenvectors corresponding to the distinct eigenvalues of a unitary matrix are orthogonal.

Proof. (i) We have

$$|U^\theta| = |(\bar{U})^T| = |\bar{U}| = \overline{|U|}$$

Therefore

$$\begin{aligned} |U|^2 &= \overline{|(U)|} \cdot |U| = |U^\theta|\,|U| = |U^\theta\, U| \\ &= |I| = 1. \end{aligned}$$

Hence, absolute value of determinant of a unitary matrix is 1.

(ii) Let λ_1 and λ_2 be two distinct eigenvalues of a unitary matrix U and let X_1, X_2 be the corresponding eigenvectors. Then

$$UX_1 = \lambda_1 X_1 \quad (70)$$
$$UX_2 = \lambda_2 X_2 \quad (71)$$

Taking conjugate transpose of (71), we get

$$X_2^{\theta} U^{\theta} = \bar{\lambda}_2 X_2^{\theta} \quad (72)$$

Post-multiplying both sides of (72) by UX_1, we get

$$\begin{aligned} & X_2^{\theta} U^{\theta} U X_1 = \bar{\lambda}_2 X_2^{\theta} U X_1 \\ \Rightarrow\ & X_2^{\theta} X_1 = \bar{\lambda}_2 X_2^{\theta} \lambda_1 X_1 \text{ since } U^{\theta} U = I \\ & \text{and } UX_1 = \lambda_1 X_1 \\ \Rightarrow\ & X_2^{\theta} X_1 = \bar{\lambda}_2 \lambda_1 X_2^{\theta} X_1 \\ \Rightarrow\ & (1 - \bar{\lambda}_2 \lambda_1) X_2^{\theta} X_1 = 0 \quad (73) \end{aligned}$$

But eigenvalues of a unitary matrix are of unit modulus.

Therefore $\bar{\lambda}_2 \lambda_2 = 1$, that is, $\bar{\lambda}_2 = \dfrac{1}{\lambda_2}$. Thus (73) reduces to

$$\begin{aligned} & \left(1 - \frac{\lambda_1}{\lambda_2}\right) X_2^{\theta} X_1 = 0 \\ \Rightarrow\ & \left(\frac{\lambda_2 - \lambda_1}{\lambda_2}\right) X_2^{\theta} X_1 = 0 \\ \Rightarrow\ & X_2^{\theta} X_1 = 0 \text{ since } \lambda_1 \neq \lambda_2. \end{aligned}$$

Hence, X_1 and X_2 are orthogonal vectors.

Theorem 2.72. The product of two unitary matrices of the same order is unitary.

Proof. Let A and B be two unitary matrices of order n. Then

$$AA^{\theta} = A^{\theta} A = I \text{ and } BB^{\theta} = B^{\theta} B = I.$$

We have

$$\begin{aligned} (AB)^{\theta} (AB) &= \left(B^{\theta} A^{\theta}\right)(AB) \\ &= B^{\theta}(A^{\theta} A) B \\ &= B^{\theta} I B \\ &= B^{\theta} B = I. \end{aligned}$$

Hence, AB is an unitary matrix of order n. Similarly,

$$\begin{aligned} (BA)^{\theta} (BA) &= \left(A^{\theta} B^{\theta}\right)(BA) \\ &= A^{\theta} \left(B^{\theta} B\right) A \\ &= A^{\theta} I A \\ &= A^{\theta} A = I \end{aligned}$$

and so BA is unitary.

Theorem 2.73. The inverse of a unitary matrix of order n is an unitary matrix.

Proof. Let U be an unitary matrix. Then

$$\begin{aligned} & UU^{\theta} = I \\ \Rightarrow\ & \left(UU^{\theta}\right)^{-1} = I^{-1} = I \\ \Rightarrow\ & \left(U^{\theta}\right)^{-1} U^{-1} = I \\ \Rightarrow\ & \left(U^{-1}\right)^{\theta} U^{-1} = I \end{aligned}$$

Hence, U^{-1} is also a unitary matrix.

Remark 2.11 It follows from Theorem 2.73 that the set of unitary matrices is a group under the binary operation of multiplication. This group is called *unitary group*.

Definition 2.111. A square matrix P is said to be *orthogonal* if $P^T P = I$.

Thus, a real unitary matrix is called an orthogonal matrix.

If P is orthogonal, then

$$\begin{aligned} & P^T P = I \\ \Rightarrow\ & |P^T P| = |I| = 1 \\ \Rightarrow\ & |P^T|\,|P| = 1 \\ \Rightarrow\ & |P|^2 = 1 \\ \Rightarrow\ & |P| \neq 0. \end{aligned}$$

Thus P is invertible and have inverse as P^T. Hence $P^T P = I = PP^T$.

If P is an orthogonal matrix, then the transformation $Y = PX$ is called *orthogonal transformation*.

Theorem 2.74. The product of two orthogonal matrices of order n is an orthogonal matrix of order n.

Proof. Let A and B be orthogonal matrices of order n. Therefore, A and B are invertible. Further both AB and BA are matrices of order n. But

$$|AB| = |A|\,|B| \neq 0 \text{ and } |BA| = |B|\,|A| \neq 0$$

Therefore, AB and BA are invertible. Now

$$\begin{aligned} (AB)^T\,(AB) &= \left(B^T A^T\right)(AB) \\ &= B^T (A^T A) B \\ &= B^T I\, B \\ &= B^T B = I. \end{aligned}$$

Hence AB is orthogonal. Similarly BA is also orthogonal.

Theorem 2.75. If a matrix P is orthogonal, then P^{-1} is also orthogonal.

Proof. Since P is orthogonal, we have

$$\begin{aligned} & PP^T = I \\ \Rightarrow\ & (PP^T)^{-1} = I^{-1} = I \\ \Rightarrow\ & (P^T)^{-1} P^{-1} = I \\ \Rightarrow\ & (P^{-1})^T P^{-1} = I. \end{aligned}$$

Hence, P^{-1} is also orthogonal.

Remark 2.12 The above results show that the set of orthogonal matrices form a multiplication group called *orthogonal group*.

Theorem 2.76. Eigenvalues of an orthogonal matrix are of unit modulus.

Proof. Since an orthogonal matrix is a real unitary matrix, the result follows from Theorem 2.70.

Remark 2.13 It follows from Theorem 2.70 that ± 1 can be the only real characteristic roots of an orthogonal matrix.

Definition 2.112. A matrix A is said to be *normal* if and only if $A^\theta A = AA^\theta$.

For example, unitary, Hermitian, and Skew-Hermitian matrices are normal. Also, the diagonal matrices with arbitrary diagonal elements are normal.

Theorem 2.77. If U is unitary, then A is normal if and only if $U^\theta\, AU$ is normal.

Proof. We have

$$\begin{aligned} \left(U^\theta AU\right)^\theta \left(U^\theta AU\right) &= \left(U^\theta A^\theta U\right)\left(U^\theta AU\right) \\ &= U^\theta A^\theta \left(UU^\theta\right) AU \\ &= U^\theta A^\theta I\, AU \\ &= U^\theta A^\theta AU \end{aligned} \quad (74)$$

and similarly,

$$\left(U^\theta AU\right)\left(U^\theta AU\right)^\theta = U^\theta AA^\theta U \quad (75)$$

From (74) and (75), we note that $A^\theta A = AA^\theta$ if and only if

$$\left(U^\theta AU\right)^\theta \left(U^\theta AU\right) = \left(U^\theta AU\right)\left(U^\theta AU\right)^\theta.$$

Hence, A is normal if and only if $U^\theta AU$ is normal.

2.39 SIMILARITY OF MATRICES

Definition 2.113. Let A and B be matrices of order n. Then B is said to be *similar* to A if there exists a non-singular matrix P such that $B = P^{-1}AP$.

It can be seen easily that the relation of similarity of matrices is an equivalence relation.

If B is similar to A, then

$$\begin{aligned}|B| &= |P^{-1}AP| = |P^{-1}|\,|A|\,|P| \\ &= |P^{-1}|\,|P|\,|A| \\ &= |P^{-1}P|\,|A| \\ &= |I|\,|A| = |A|.\end{aligned}$$

Therefore it follows that similar matrices have the same determinant.

Theorem 2.78. Similar matrices have the same characteristic polynomial and hence the same characteristic roots.

Proof. Suppose A and B are similar matrices. Then there exists an invertible matrix P such that $B = P^{-1}AP$. Since

$$\begin{aligned}B - xI &= P^{-1}AP - xI \\ &= P^{-1}AP - P^{-1}(xI)P \\ &= P^{-1}(A - xI)P,\end{aligned}$$

we have

$$\begin{aligned}|B - xI| &= |P^{-1}(A - xI)P| \\ &= |P^{-1}|\,|P|\,|A - xI| \\ &= |P^{-1}P|\,|A - xI| \\ &= |A - xI|.\end{aligned}$$

Thus A and B have the same characteristic polynomial and so they have same characteristic roots.

Further if λ is characteristic root of A, then,

$$AX = \lambda X,\ X \neq 0$$

and so

$$\begin{aligned}B(P^{-1}X) &= (P^{-1}AP)P^{-1}X \\ &= P^{-1}AX = P^{-1}(\lambda X) \\ &= \lambda(P^{-1}X)\end{aligned}$$

This shows that $(P^{-1}X)$ is an eigenvector of B corresponding to its eigenvalue λ.

Corollary 2.22. If a matrix A is similar to a diagonal matrix D, the diagonal elements of D are the eigenvalues of A.

Proof. Since A and D are similar, they have same eigenvalues. But the eigenvalues of the diagonal matrix D are its diagonal elements. Hence the eigenvalues of A are the diagonal elements of D.

2.40 DIAGONALIZATION OF A MATRIX

Definition 2.114. A matrix A is said to be *diagonalizable* if it is similar to a diagonal matrix.

Theorem 2.79. *A matrix of order n is diagonalizable if and only if it possesses n linearly independent eigenvectors.*

Proof. Suppose first that A is diagonalizable. Then A is similar to a diagonal matrix

$$D = \text{diag}\ [\lambda_1\ \lambda_2 \ldots \lambda_n].$$

Therefore, there exists an invertible matrix $P = [X_1\ X_2 \ldots X_n]$ such that $P^{-1}AP = D$, that is, $AP = PD$ and so

$$A[X_1\ X_2 \ldots X_n] = [X_1\ X_2 \ldots X_n]\ \text{diag}\ [\lambda_1\ \lambda_2 \ldots \lambda_n]$$

or

$$[AX_1,\ AX_2 \ldots AX_n] = [\lambda_1 X_1\ \lambda_2 X_2 \ldots \lambda_n X_n].$$

Hence

$$AX_1 = \lambda_1 X_1,\ AX_2 = \lambda_2 X_2, \ldots, AX_n = \lambda_n X_n.$$

Thus, $X_1, X_2, \ldots, X_n$ are eigenvectors of A corresponding to the eigenvalues $\lambda_1, \lambda_2, \ldots, \lambda_n$, respectively. Since P is non-singular, its column vectors $X_1, X_2, \ldots, X_n$ are linearly independent. Hence A has n linearly independent eigenvectors.

Conversely suppose that A possesses n linearly independent eigenvectors $X_1, X_2, \ldots, X_n$ and let $\lambda_1, \lambda_2, \ldots, \lambda_n$ be the corresponding eigenvalues. Then

$$AX_1 = \lambda_1 X_1, \quad AX_2 = \lambda_2 X_2, \ldots, AX_n = \lambda_n X_n.$$

Let

$$P = [X_1, X_2, \ldots, X_n] \text{ and } D = diag[\lambda_1\ \lambda_2 \ldots \lambda_n].$$

Then

$$\begin{aligned} AP &= [AX_1, AX_2 \ldots AX_n] \\ &= [\lambda_1 X_1\ \lambda_2 X_2 \ldots \lambda_n X_n] \\ &= [X_1, X_2, \ldots, X_n]\ diag\ [\lambda_1\ \lambda_2 \ldots \lambda_n] = PD. \end{aligned}$$

Since the column vectors $X_1, X_2, \ldots, X_n$ of the matrix P are linearly independent, so P is invertible and P^{-1} exists.
Therefore,

$$\begin{aligned} AP = PD &\Rightarrow P^{-1}AP = P^{-1}PD \\ &\Rightarrow P^{-1}AP = D \\ &\Rightarrow \text{A is similar to D.} \Rightarrow A \text{ is diagonalizable.} \end{aligned}$$

Theorem 2.80. If the eigenvalues of a matrix of order n are all distinct, then it is similar to a diagonal matrix.

Proof. Suppose that a square matrix of order n has n distinct eigenvalues, $\lambda_1, \lambda_2, \ldots, \lambda_n$. As eigenvectors of a matrix corresponding to distinct eigenvalues are linearly independent, A has n linearly independent eigenvectors and so, by the above theorem, it is similar to a diagonal matrix.

The following result is very useful in diagonalization of a given matrix.

Theorem 2.81. The necessary and sufficient condition for a square matrix to be similar to a diagonal matrix is that geometric multiplicity of each of its eigenvalues coincide with the algebraic multiplicity.

EXAMPLE 2.140

Show that the matrix

$$A = \begin{bmatrix} 2 & 3 & 4 \\ 0 & 2 & -1 \\ 0 & 0 & 1 \end{bmatrix}$$

is not similar to diagonal matrix.

Solution. The characteristic equation of A is

$$|A - \lambda I| = \begin{vmatrix} 2-\lambda & 3 & 4 \\ 0 & 2-\lambda & -1 \\ 0 & 0 & 1-\lambda \end{vmatrix} = 0$$

and so

$$(2-\lambda)(2-\lambda)(1-\lambda) = 0.$$

Hence the eigenvalues of A are 2, 2, and 1. The eigenvector X of A corresponding to $\lambda = 2$ is given by $(A - 2I)X = 0$, that is, by

$$\begin{bmatrix} 0 & 3 & 4 \\ 0 & 0 & -1 \\ 0 & 0 & -1 \end{bmatrix} \begin{bmatrix} x_1 \\ x_2 \\ x_3 \end{bmatrix} = \begin{bmatrix} 0 \\ 0 \\ 0 \end{bmatrix}$$

$$\sim \begin{bmatrix} 0 & 3 & 4 \\ 0 & 0 & -1 \\ 0 & 0 & 0 \end{bmatrix} \begin{bmatrix} x_1 \\ x_2 \\ x_3 \end{bmatrix} = \begin{bmatrix} 0 \\ 0 \\ 0 \end{bmatrix}, \ R_3 \to R_3 - R_2.$$

The coefficient matrix is of rank 2. Hence number of linearly independent solution is $n - r = 1$. Thus geometric multiplicity of 2 is 1. But its algebraic multiplicity is 2. Therefore, geometric multiplicity is not equal to algebraic multiplicity. Hence A is not similar to a diagonal matrix.

EXAMPLE 2.141

Give an example to show that not every square matrix can be diagonalized by a non-singular transformation of coordinates.

Solution. Consider the matrix

$$A = \begin{bmatrix} 1 & 1 \\ 0 & 1 \end{bmatrix}.$$

The characteristic equation of A is

$$|A - \lambda I| = \begin{vmatrix} 1-\lambda & 1 \\ 0 & 1-\lambda \end{vmatrix} = 0$$

or

$$(1-\lambda)^2 = 0,$$

which yields the characteristic roots as $\lambda = 1, 1$.

The characteristic vector corresponding to $\lambda = 1$, is given by $(A-I)X = \mathbf{0}$, that is, by

$$\begin{bmatrix} 0 & 1 \\ 0 & 0 \end{bmatrix}\begin{bmatrix} x_1 \\ x_2 \end{bmatrix} = \begin{bmatrix} 0 \\ 0 \end{bmatrix}.$$

The rank of the coefficient matrix is 1 and so that number of linearly independent solution is $n-r = 2-1 = 1$. Thus the geometric multiplicity of characteristic root is 1, whereas algebraic multiplicity of the characteristic root is 2. Hence, the given matrix is not diagonalizable.

EXAMPLE 2.142

Show that the matrix

$$A = \begin{bmatrix} 8 & -8 & -2 \\ 4 & -3 & -2 \\ 3 & -4 & 1 \end{bmatrix}$$

is diagonalizable. Hence, find the transforming matrix and the diagonal matrix.

Solution. The roots of the characteristic equation

$$|A-\lambda I| = \begin{vmatrix} 8-\lambda & -8 & -2 \\ 4 & -3-\lambda & -2 \\ 3 & -4 & 1-\lambda \end{vmatrix} = 0$$

are 1, 2, 3. Since the eigenvalues are all distinct, A is similar to a diagonal matrix. Further, algebraic multiplicity of each eigenvalue is 1. So there is only one linearly independent eigenvector of A corresponding to each eigenvalue. Now the eigenvector corresponding to $\lambda = 1$ is given by $(A-I)X = 0$, that is, by

$$\begin{bmatrix} 7 & -8 & -2 \\ 4 & -4 & -2 \\ 3 & -4 & 0 \end{bmatrix}\begin{bmatrix} x_1 \\ x_2 \\ x_3 \end{bmatrix} = \begin{bmatrix} 0 \\ 0 \\ 0 \end{bmatrix},$$

$$\sim \begin{bmatrix} 7 & -8 & -2 \\ -3 & 4 & 0 \\ 3 & -4 & 0 \end{bmatrix}\begin{bmatrix} x_1 \\ x_2 \\ x_3 \end{bmatrix} = \begin{bmatrix} 0 \\ 0 \\ 0 \end{bmatrix}, \; R_2 \to R_2 - R_1$$

$$\sim \begin{bmatrix} 7 & -8 & -2 \\ -3 & 4 & 0 \\ 0 & 0 & 0 \end{bmatrix}\begin{bmatrix} x_1 \\ x_2 \\ x_3 \end{bmatrix} = \begin{bmatrix} 0 \\ 0 \\ 0 \end{bmatrix}, \; R_3 \to R_3 + R_2.$$

We note that rank of the coefficient matrix is 2. Therefore, there is only one linearly independent solution. Hence geometric multiplicity of the eigenvalues 1 is 1. The equation can be written as

$$7x_1 - 8x_2 - 2x_3 = 0$$
$$-3x_1 + 4x_2 = 0.$$

The last equation yields $x_1 = \frac{4}{3}x_2$. So taking $x_2 = 3$, we get $x_1 = 4$. Then the first equation yields $x_3 = 2$. Hence, the eigenvector corresponding to $\lambda = 1$ is

$$X_1 = \begin{bmatrix} 4 \\ 3 \\ 2 \end{bmatrix}.$$

Similarly, eigenvectors corresponding to $\lambda = 2$ and 3 are found to be

$$X_2 = \begin{bmatrix} 3 \\ 2 \\ 1 \end{bmatrix} \text{ and } X_3 = \begin{bmatrix} 2 \\ 1 \\ 1 \end{bmatrix}.$$

Therefore, the transforming matrix is

$$P = \begin{bmatrix} 4 & 3 & 2 \\ 3 & 2 & 1 \\ 2 & 1 & 1 \end{bmatrix},$$

and so the diagonal matrix is

$$P^{-1}AP = \begin{bmatrix} 1 & 0 & 0 \\ 0 & 2 & 0 \\ 0 & 0 & 3 \end{bmatrix}.$$

EXAMPLE 2.143

Diagonalize the matrix

$$A = \begin{bmatrix} 1 & 0 & -1 \\ 1 & 2 & 1 \\ 2 & 2 & 3 \end{bmatrix}.$$

Solution. The characteristic equation of the given matrix is

$$|A-\lambda I| = \begin{vmatrix} 1-\lambda & 0 & -1 \\ 1 & 2-\lambda & 1 \\ 2 & 2 & 3-\lambda \end{vmatrix} = 0$$

or

$$-\lambda^3+6\lambda^2-11\lambda+6=0.$$

The characteristic roots are $\lambda = 1, 2, 3$. Since the characteristic roots are distinct, the given matrix is diagonalizable and the diagonal elements shall be the characteristic roots 1, 2, 3.

The characteristic vectors corresponding to $\lambda = 1$ are given by $(A-I)X=0$, that is, by

$$\begin{bmatrix} 0 & 0 & -1 \\ 1 & 1 & 1 \\ 2 & 2 & 2 \end{bmatrix}\begin{bmatrix} x_1 \\ x_2 \\ x_3 \end{bmatrix} = \begin{bmatrix} 0 \\ 0 \\ 0 \end{bmatrix}$$

$$\sim \begin{bmatrix} 1 & 1 & 1 \\ 0 & 0 & -1 \\ 2 & 2 & 2 \end{bmatrix}\begin{bmatrix} x_1 \\ x_2 \\ x_3 \end{bmatrix} = \begin{bmatrix} 0 \\ 0 \\ 0 \end{bmatrix}, \ R_1 \leftrightarrow R_2$$

$$\sim \begin{bmatrix} 1 & 1 & 1 \\ 0 & 0 & -1 \\ 0 & 0 & 0 \end{bmatrix}\begin{bmatrix} x_1 \\ x_2 \\ x_3 \end{bmatrix} = \begin{bmatrix} 0 \\ 0 \\ 0 \end{bmatrix}, \ R_3 \to R_3 - 2R_1.$$

The rank of the coefficient matrix is 2. Therefore, there is only $3-2=1$ linearly independent solution. The above equation yields,

$$x_1+x_2+x_3=0$$
$$-x_3=0.$$

Hence, the characteristic vector corresponding to $\lambda = 1$ is

$$\begin{bmatrix} 1 \\ -1 \\ 0 \end{bmatrix}.$$

The characteristic vector corresponding to $\lambda = 2$ is given by $(A-2I)X=0$, that is, by

$$\begin{bmatrix} -1 & 0 & -1 \\ 1 & 0 & 1 \\ 2 & 2 & 1 \end{bmatrix}\begin{bmatrix} x_1 \\ x_2 \\ x_3 \end{bmatrix} = \begin{bmatrix} 0 \\ 0 \\ 0 \end{bmatrix}$$

$$\sim \begin{bmatrix} -1 & 0 & -1 \\ 0 & 0 & 0 \\ 1 & 2 & 0 \end{bmatrix}\begin{bmatrix} x_1 \\ x_2 \\ x_3 \end{bmatrix} = \begin{bmatrix} 0 \\ 0 \\ 0 \end{bmatrix}, \ \begin{matrix} R_2 \to R_2 + R_1 \\ R_3 \to R_3 - R_1 \end{matrix}$$

$$\sim \begin{bmatrix} -1 & 0 & -1 \\ 1 & 2 & 0 \\ 0 & 0 & 0 \end{bmatrix}\begin{bmatrix} x_1 \\ x_2 \\ x_3 \end{bmatrix} = \begin{bmatrix} 0 \\ 0 \\ 0 \end{bmatrix}, \ R_2 \leftrightarrow R_3.$$

The rank of the coefficient matrix is 2. Therefore, there is only $3-2=1$ linearly independent solution. The equation implies

$$-x_1-x_3=0$$
$$x_1+2x_2=0$$

which yields $x_1=2$, $x_2=-1$, $x_3=-2$. Therefore, the characteristic vector is

$$\begin{bmatrix} 2 \\ -1 \\ -2 \end{bmatrix}.$$

The characteristic vector corresponding to $\lambda = 3$ is given by

$$\begin{bmatrix} -2 & 0 & -1 \\ 1 & -1 & 1 \\ 2 & 2 & 0 \end{bmatrix}\begin{bmatrix} x_1 \\ x_2 \\ x_3 \end{bmatrix} = \begin{bmatrix} 0 \\ 0 \\ 0 \end{bmatrix}$$

$$\sim \begin{bmatrix} -2 & 0 & -1 \\ -1 & -1 & 0 \\ 2 & 2 & 0 \end{bmatrix}\begin{bmatrix} x_1 \\ x_2 \\ x_3 \end{bmatrix} = \begin{bmatrix} 0 \\ 0 \\ 0 \end{bmatrix}, \ R_2 \to R_2 + R_1$$

$$\sim \begin{bmatrix} -2 & 0 & -1 \\ -1 & -1 & 0 \\ 0 & 0 & 0 \end{bmatrix}\begin{bmatrix} x_1 \\ x_2 \\ x_3 \end{bmatrix} = \begin{bmatrix} 0 \\ 0 \\ 0 \end{bmatrix}, \ R_3 \to R_3 + 2R_2.$$

The rank of coefficient matrix is 2 and so there is $3-2=1$ independent solution. The equation yields,

$$-2x_1-x_3=0$$
$$-x_1-x_2=0.$$

and so the corresponding characteristic vector is

$$\begin{bmatrix} 1 \\ -1 \\ -2 \end{bmatrix}.$$

Thus, the transforming matrix is

$$P = \begin{bmatrix} 1 & 2 & 1 \\ -1 & -1 & -1 \\ 0 & -2 & -2 \end{bmatrix}.$$

We have $|P| = -2$ and the cofactors of P are

$$A_{11}=0, \quad A_{12}=-2, \quad A_{13}=2,$$
$$A_{21}=2, \quad A_{22}=-2, \quad A_{23}=2,$$
$$A_{31}=-1, \quad A_{32}=0, \quad A_{33}=1.$$

Therefore,

$$\text{adj } P = \begin{bmatrix} 0 & 2 & -1 \\ -2 & -2 & 0 \\ 2 & 2 & 1 \end{bmatrix},$$

and so

$$P^{-1} = \frac{1}{|P|}\text{adj } P = \begin{bmatrix} 0 & -1 & \frac{1}{2} \\ 1 & 1 & 0 \\ -1 & -1 & -\frac{1}{2} \end{bmatrix}.$$

Then we observe that

$$P^{-1}AP = \begin{bmatrix} 0 & -1 & \frac{1}{2} \\ 1 & 1 & 0 \\ -1 & -1 & -\frac{1}{2} \end{bmatrix} \begin{bmatrix} 1 & 0 & -1 \\ 1 & 2 & 1 \\ 2 & 2 & 3 \end{bmatrix}$$

$$\times \begin{bmatrix} 1 & 2 & 1 \\ -1 & -1 & -1 \\ 0 & -2 & -2 \end{bmatrix}$$

$$= \begin{bmatrix} 0 & -1 & \frac{1}{2} \\ 1 & 1 & 0 \\ -1 & -1 & -\frac{1}{2} \end{bmatrix} \begin{bmatrix} 1 & 4 & 3 \\ -1 & -2 & -3 \\ 0 & -4 & -6 \end{bmatrix}$$

$$= \begin{bmatrix} 1 & 0 & 0 \\ 0 & 2 & 0 \\ 0 & 0 & 3 \end{bmatrix} = \text{diag } [1\ 2\ 3].$$

Definition 2.115. Let A and B be square matrices of order n. Then B is said to be *unitarily similar* to A if there exists a unitary matrix U such that $B = U^{-1}AU$.

Theorem 2.82. (Existence Theorem). If A is Hermitian matrix, then there exists a unitary matrix U such that $U^{\theta}\ AU$ is a diagonal matrix whose diagonal elements are the characteristic roots of A, that is,

$$U^{\theta}\ AU = \text{diag}[\lambda_1\ \lambda_2 \ldots \lambda_n].$$

Proof. We shall prove Theorem 2.82 by induction on the order of A. If $n = 1$, then the theorem is obviously true. We assume that the theorem is true for all Hermitian matrices of order $n - 1$. We shall establish that the theorem holds for all Hermitian matrices of order n.

Let λ_1 be an eigenvalue of A. Thus λ_1 is real. Let X_1 be the eigenvector corresponding to the eigenvalues λ_1. Therefore $AX_1 = \lambda_1 X_1$. We choose an orthonormal basis of the complex vector space V_n having X_1 as a member. Therefore, there exists a unitary matrix S with X_1 as its first column. We now consider the matrix $S^{-1}AS$. Since X_1 is the first column of S, the first column of $S^{-1}AS$ is $S^{-1}AX_1 = S^{-1}\lambda_1 X_1 = \lambda_1\ S^{-1}X_1$. But $S^{-1}X_1$ is the first column of $S^{-1}S = I$. Therefore, the first column of $S^{-1}AS$ is $[\lambda_1\ 0 \ldots 0 \ldots 0]^T$. Since S is unitary, $S^{-1} = S^{\theta}$ and so

$$\left(S^{-1}AS\right)^{\theta} = S^{\theta}\ A^{\theta}\left(S^{-1}\right)^{\theta} = S^{\theta}\ A^{\theta} S = S^{-1}AS.$$

Hence $S^{-1}AS$ is Hermitian. Therefore, the first row of $S^{-1}AS$ is $[\lambda_1\ 0 \ldots 0 \ldots 0]$. Thus,

$$S^{-1}AS = \begin{bmatrix} \lambda_1 & 0 \\ 0 & B \end{bmatrix},$$

where B is a square matrix of order $n - 1$. Therefore, by induction hypothesis, there exists a unitary matrix V such that

$$V^{-1}BV = D_1,$$

where D_1 is a diagonal matrix of order $n - 1$.

Let $R = \begin{bmatrix} I & 0 \\ 0 & V \end{bmatrix}$ be a matrix of order n. Then R is invertible and $R^{-1} = \begin{bmatrix} I & 0 \\ 0 & V^{-1} \end{bmatrix}$. Now since V is unitary, $V^{\theta} = V^{-1}$ and so

$$R^{\theta} = \begin{bmatrix} I & 0 \\ 0 & V^{\theta} \end{bmatrix} = \begin{bmatrix} I & 0 \\ 0 & V^{-1} \end{bmatrix} = R^{-1}.$$

Hence, R is uniatary. Since R and S are unitary matrices of order n, SR is also unitary of order n.

Let $SR = U$. Then

$$\begin{aligned}
U^{-1}AU &= (SR)^{-1}A(SR) \\
&= \left(R^{-1}S^{-1}\right)A(SR) \\
&= R^{-1}(S^{-1}AS)R \\
&= \begin{bmatrix} I & 0 \\ 0 & V^{-1} \end{bmatrix}\begin{bmatrix} \lambda_1 & 0 \\ 0 & B \end{bmatrix}\begin{bmatrix} I & 0 \\ 0 & V \end{bmatrix} \\
&= \begin{bmatrix} \lambda_1 & 0 \\ 0 & V^{-1}B \end{bmatrix}\begin{bmatrix} I & 0 \\ 0 & V \end{bmatrix} = \begin{bmatrix} \lambda_1 & 0 \\ 0 & V^{-1}BV \end{bmatrix} \\
&= \begin{bmatrix} \lambda_1 & 0 \\ 0 & D_1 \end{bmatrix} = \text{diag}[\lambda_1\ \lambda_2 \dots \lambda_n].
\end{aligned}$$

As an immediate consequence of this theorem, we have

Corollary 2.23. If A is a real symmetric matrix, there exists an orthogonal matrix U such that U^TAU is a diagonal matrix, whose diagonal elements are the characteristic roots of A.

Theorem 2.83. If λ is an m-fold eigenvalue of Hermitian matrix A, then rank of $A - \lambda\ I_n$ is $n - m$.

Proof. By Theorem 2.82, there exists a unitary matrix U such that

$$U^*AU = \text{diag}[\lambda\lambda \dots \lambda\lambda_{m+1}\lambda_{m+2} \dots \lambda_n],$$

where λ occurs m times and $\lambda_{m+1}, \lambda_{m+2}, \dots, \lambda_n$ are all distinct from λ. Since U is unitary, subtracting λI_n from both sides of the above equation, we get

$$U^*[A - \lambda I_n]U = diag[00 \dots 0(\lambda_{m+1} - \lambda)\\ (\lambda_{m+2} - \lambda) \dots (\lambda_n - \lambda)].$$

Since U is non-singular, it follows that the rank of $A - \lambda I_n$ is same as that of the diagonal matrix on the right-hand side. But the rank of the matrix on the right-hand side is $n - m$ because $(\lambda_{m+1} - \lambda), (\lambda_{m+2} - \lambda), \dots, (\lambda_n - \lambda)$ are all non-zero.

Corollary 2.24. If λ is m-fold eigenvalues of a Hermitian matrix A, then there exists m linearly independent vectors of A associated with λ, that is, with λ_1 there is associated an m-dimensional space of characteristic vectors.

Theorem 2.84. With every Hermitian matrix A we can associate an orthonormal set of n characteristic vectors.

Proof. The eigenvectors associated with a given eigenvalue of A form a vector space for which we can construct an orthonormal basis by Gram-Schmidt process. For each A, there are n vectors in the basis so constructed. Also, the eigenvectors associated with distinct eigenvalues of a Hermitian matrix are orthogonal. It follows, therefore, that these n basis vectors constitute orthonormal set.

Theorem 2.84 indicates how the diagonalization process may be effected. In fact, we have the following theorem.

Theorem 2.85. If $U_1, U_2, \dots, U_n$ is an orthonormal system of eigenvectors associated respectively with the eigenvalues $\lambda_1 \dots \lambda_n$ of Hermitian matrix A and if U is the unitary matrix $[U_1\ U_2 \dots U_n]$, then

$$U^*AU = diag[\lambda_1\ \lambda_2 \dots \lambda_n].$$

(The vectors $U_1, U_2, \dots, U_n$ are often called a set of *principal axes* of A and the transformation with matrix U used to diagonalize A is called *principal axis transformation*).

Proof. We have $AU_j = \lambda_j\ U_j$, $j = 1, 2, \dots, n$, where λ_j is the eigenvalue associated with U_j. Thus, if $U = [U_1\ U_2 \dots U_n]$, then

$$[AU_1\ AU_2 \dots AU_n] = [\lambda_1 U_1\ \lambda_2 U_2 \dots \lambda_n U_n],$$

that is,

$$AU = U \text{ diag } [\lambda_1\ \lambda_2 \dots \lambda_n].$$

Since, U is unitary $U^{-1} = U^\theta$ and so premultiplication by U^θ yields

$$\begin{aligned}
U^\theta\ AU &= U^\theta U \text{ diag } [\lambda_1\ \lambda_2 \dots \lambda_n] \\
&= U^{-1}U \text{ diag } [\lambda_1\ \lambda_2 \dots \lambda_n] \\
&= \text{ diag } [\lambda_1\ \lambda_2 \dots \lambda_n].
\end{aligned}$$

EXAMPLE 2.144

Diagonalize the matrix

$$A=\begin{bmatrix} 2 & 1-2i \\ 1+2i & -2 \end{bmatrix}.$$

Solution. The characteristic equation of A is

$$|A-\lambda I|=\begin{vmatrix} 2-\lambda & 1-2i \\ 1+2i & -2-\lambda \end{vmatrix}=0$$

or

$$\lambda^2-9=0,$$

which yields the characteristic roots as $\lambda=-3$, 3. The eigenvectors corresponding to the eigenvalue -3 is given by $(A+3I)X=\mathbf{0}$, that is, by

$$\begin{bmatrix} 5 & 1-2i \\ 1+2i & 1 \end{bmatrix}\begin{bmatrix} x_1 \\ x_2 \end{bmatrix}=\begin{bmatrix} 0 \\ 0 \end{bmatrix},$$

which yields

$$5x_1-(1-2i)x_2=0$$
$$(1+2i)x_1+x_2=0.$$

Solving these equations, we get $x_1=1-2i$, $x_2=-5$. Hence, $X_1=\begin{bmatrix} 1-2i \\ -5 \end{bmatrix}$ is the eigenvector corresponding to $\lambda=-3$.

The eigenvector corresponding to $\lambda=3$ is given by $(A-3I)X=\mathbf{0}$ that is, by

$$\begin{bmatrix} -1 & 1-2i \\ 1+2i & -5 \end{bmatrix}\begin{bmatrix} x_1 \\ x_2 \end{bmatrix}=\begin{bmatrix} 0 \\ 0 \end{bmatrix},$$

which yields

$$-x_1+(1-2i)x_2=0$$
$$(1+2i)x_1-5x_2=0.$$

Solving these equations, we get $x_1=5$, $x_2=1+2i$. Thus the required eigenvector is $X_2=\begin{bmatrix} 5 \\ 1+2i \end{bmatrix}$. We note that

$$X_2^\theta X_1=[5 \quad 1-2i]\begin{bmatrix} 1-2i \\ -5 \end{bmatrix}$$
$$=5(1-2i)-5(1-2i)=0.$$

Thus $\{X_1, X_2\}$ is an orthogonal set. Now

$$\text{Norm of } X_1=\sqrt{|1-2i|^2+|-5|^2}=\sqrt{5+25}$$
$$=\sqrt{30}$$
$$\text{Norm of } X_2=\sqrt{|5|^2+|1+2i|^2}$$
$$=\sqrt{25+5}$$
$$=\sqrt{30}.$$

Therefore, normalized characteristic vectors are

$$U_1=\begin{bmatrix} \frac{1-2i}{\sqrt{30}} \\ \frac{-5}{\sqrt{30}} \end{bmatrix},\ U_2=\begin{bmatrix} \frac{5}{\sqrt{30}} \\ \frac{1+2i}{\sqrt{30}} \end{bmatrix}.$$

Hence the transforming unitary matrix is

$$U=\begin{bmatrix} \frac{1-2i}{\sqrt{30}} & \frac{5}{\sqrt{30}} \\ \frac{-5}{\sqrt{30}} & \frac{1+2i}{\sqrt{30}} \end{bmatrix}.$$

We note that

$$U^\theta AU=\begin{bmatrix} -3 & 0 \\ 0 & 3 \end{bmatrix}=\text{diag}\,[-3\ 3].$$

EXAMPLE 2.145

Diagonalize the Hermitian matrix

$$A=\begin{bmatrix} 5 & 2 & 0 & 0 \\ 2 & 2 & 0 & 0 \\ 0 & 0 & 5 & -2 \\ 0 & 0 & -2 & 2 \end{bmatrix}.$$

Solution. The characteristic equation of A is

$$|A-\lambda I| = \begin{vmatrix} 5-\lambda & 2 & 0 & 0 \\ 2 & 2-\lambda & 0 & 0 \\ 0 & 0 & 5-\lambda & -2 \\ 0 & 0 & -2 & 2-\lambda \end{vmatrix} = 0.$$

The characteristic roots are 1, 1, 6, 6. The characteristic vectors corresponding to $\lambda = 1$ are given by $(A-I)X = \mathbf{0}$, that is, by

$$\begin{bmatrix} 4 & 2 & 0 & 0 \\ 2 & 1 & 0 & 0 \\ 0 & 0 & 4 & -2 \\ 0 & 0 & -2 & 1 \end{bmatrix} \begin{bmatrix} x_1 \\ x_2 \\ x_3 \\ x_4 \end{bmatrix} = \begin{bmatrix} 0 \\ 0 \\ 0 \\ 0 \end{bmatrix},$$

which yields

$$\begin{aligned} 4x_1 + 2x_2 &= 0 \\ 2x_1 + x_2 &= 0 \\ 4x_3 - 2x_4 &= 0 \\ -2x_3 + x_4 &= 0 \end{aligned}$$

with the complete solution as

$$X_1 = \begin{bmatrix} 1 \\ -2 \\ 0 \\ 0 \end{bmatrix}, \; X_2 = \begin{bmatrix} 0 \\ 0 \\ 1 \\ 2 \end{bmatrix}.$$

These vectors are already orthogonal. The normalized vectors are

$$U_1 = \begin{bmatrix} \frac{1}{\sqrt{5}} \\ -\frac{2}{\sqrt{5}} \\ 0 \\ 0 \end{bmatrix} \quad \text{and} \quad U_2 = \begin{bmatrix} 0 \\ 0 \\ \frac{1}{\sqrt{5}} \\ \frac{2}{\sqrt{5}} \end{bmatrix}.$$

Similarly, the normalized vectors corresponding to $\lambda = 6$ are

$$U_3 = \begin{bmatrix} \frac{2}{\sqrt{5}} \\ \frac{1}{\sqrt{5}} \\ 0 \\ 0 \end{bmatrix} \text{ and } U_4 = \begin{bmatrix} 0 \\ 0 \\ -\frac{2}{\sqrt{5}} \\ \frac{1}{\sqrt{5}} \end{bmatrix}.$$

Hence, the transforming unitary matrix is

$$U = \begin{bmatrix} \frac{1}{\sqrt{5}} & 0 & \frac{2}{\sqrt{5}} & 0 \\ -\frac{2}{\sqrt{5}} & 0 & \frac{1}{\sqrt{5}} & 0 \\ 0 & \frac{1}{\sqrt{5}} & 0 & -\frac{2}{\sqrt{5}} \\ 0 & \frac{2}{\sqrt{5}} & 0 & \frac{1}{\sqrt{5}} \end{bmatrix}$$

and $U^\theta AU = \text{diag}[1\ 1\ 6\ 6]$.

2.41 TRIANGULARIZATION OF AN ARBITRARY MATRIX

Not every matrix can be reduced to diagonal form by a unitary transformation. But it is always possible to reduce a square matrix to a triangular form. In this direction, we have the following result.

Theorem 2.86. (Jacobi-Thoerem). Every square matrix A over the complex field can be reduced by a unitary transformation to upper triangular form with the characteristic roots on the diagonal.

Proof. We shall prove the theorem by induction on the order n of the matrix A. If $n = 1$, the theorem is obviously true. Suppose that the result holds for all matrices of order $n-1$. Let λ_1 be the characteristic root of A and U_1 denote the corresponding unit characteristic vector. Then $AU_1 = \lambda_1 U_1$. Let $\{U_1, U_2, \ldots, U_n\}$ be an orthonormal set, that is, $U = [U_1, U_2, \ldots, U_n]$. Then

$$U^\theta AU = \begin{bmatrix} U_1^\theta \\ U_2^\theta \\ \cdots \\ \cdots \\ U_n^\theta \end{bmatrix} \left[AU_1\ AU_2 \ldots AU_n \right]$$

$$= \begin{bmatrix} U_1^\theta \\ U_2^\theta \\ \cdots \\ \cdots \\ U_n^\theta \end{bmatrix} \left[\lambda_1 U_1\ AU_2 \ldots AU_n \right]$$

Since $U_1^\theta U_1 = I$ and $U_2^\theta U_1 = U_3^\theta U_1 = \ldots = U_n^\theta U_1 = 0$, we have

$$U^*AU = \begin{bmatrix} \lambda_1 & U_1^\theta AU_2 & \ldots & \ldots & U_1^\theta AU_n \\ 0 & U_2^\theta AU_2 & \ldots & \ldots & U_2^\theta AU_n \\ \ldots & \ldots & \ldots & \ldots & \ldots \\ \ldots & \ldots & \ldots & \ldots & \ldots \\ 0 & U_n^\theta AU_2 & \ldots & \ldots & U_n^\theta AU_n \end{bmatrix} = \begin{bmatrix} \lambda_1 & B \\ 0 & C \end{bmatrix}.$$

Now, by induction hypothesis, the matrix

$$C = \begin{bmatrix} U_2^\theta AU_2 & \ldots & \ldots & U_2^\theta AU_n \\ \ldots & \ldots & \ldots & \ldots \\ \ldots & \ldots & \ldots & \ldots \\ U_n^\theta AU_2 & \ldots & \ldots & U_n^\theta AU_n \end{bmatrix},$$

which is of order $n-1$, is triangularizable. Thus there exists a unitary matrix W of order $n-1$ which triangularize C, that is, $W^\theta CW$ is triangular. Let $V = \begin{bmatrix} I & 0 \\ 0 & W \end{bmatrix}$. Then $V^{-1} = \begin{bmatrix} I & 0 \\ 0 & W^{-1} \end{bmatrix}$ and

$$V^\theta = \begin{bmatrix} I & 0 \\ 0 & W^\theta \end{bmatrix} = \begin{bmatrix} I & 0 \\ 0 & W^{-1} \end{bmatrix} = V^{-1}.$$

Hence V is unitary and

$$V^\theta(U^\theta AU)V = \begin{bmatrix} I & 0 \\ 0 & W^\theta \end{bmatrix}\begin{bmatrix} \lambda_1 & B \\ 0 & C \end{bmatrix}\begin{bmatrix} I & 0 \\ 0 & W \end{bmatrix}$$

$$= \begin{bmatrix} \lambda_1 & B \\ 0 & W^\theta C \end{bmatrix}\begin{bmatrix} I & 0 \\ 0 & W \end{bmatrix} = \begin{bmatrix} \lambda_1 & BW \\ 0 & W^\theta CW \end{bmatrix},$$

where $W^\theta CW$ is upper triangular. Thus, we have

$$(UV)^\theta A(UV) = \begin{bmatrix} \lambda_1 & BW \\ 0 & W^\theta CW \end{bmatrix}$$

$$= \begin{bmatrix} \lambda_1 & b_{12} & \ldots & \ldots & b_{1n} \\ 0 & \lambda_2 & \ldots & \ldots & b_{2n} \\ \ldots & \ldots & \ldots & \ldots & \ldots \\ \ldots & \ldots & \ldots & \ldots & \ldots \\ 0 & 0 & \ldots & \ldots & \lambda_n \end{bmatrix}.$$

Since UV is unitary, the characteristic roots of the triangular matrix are the same as that of A. Thus, diagonal elements of triangular matrix are characteristic roots of A.

Theorem 2.87. A matrix A over the complex field can be diagonalized by a unitary transformation if and only if A is normal.

Proof. Suppose first that U is unitary and A can be diagonalized, that is, $U^\theta AU = \text{diag}[\lambda_1 \lambda_2 \ldots \lambda_n]$. Then $A = U\,\text{diag}[\lambda_1 \lambda_2 \ldots \lambda_n]U^\theta$ and so $A^\theta A = U(\text{diag}[\lambda_1 \lambda_2 \ldots \lambda_n])^\theta (\text{diag}[\lambda_1 \lambda_2 \ldots \lambda_n])\ U^\theta$ and $AA^\theta = U(\text{diag}[\lambda_1 \lambda_2 \ldots \lambda_n])(\text{diag}[\lambda_1 \lambda_2 \ldots \lambda_n])^\theta U^\theta$ But $\text{diag}[\lambda_1 \lambda_2 \ldots \lambda_n](\text{diag}[\lambda_1 \lambda_2 \ldots \lambda_n])^\theta = (\text{diag}[\lambda_1 \lambda_2 \ldots \lambda_n])^\theta\, \text{diag}[\lambda_1\ \lambda_2\ \ldots\ \lambda_n]$.
Hence $A^\theta A = AA^\theta$ and so A is normal.

Conversely, suppose A is normal. Then, by Theorem 2.77, there exists unitary matrix U such that $U^\theta AU = B$, where B is upper triangular. But $U^\theta AU$ is normal and so B is normal. Suppose that the upper triangular matrix B is

$$B = \begin{bmatrix} \lambda_1 & b_{12} & \ldots & \ldots & b_{1n} \\ 0 & \lambda_2 & \ldots & \ldots & b_{2n} \\ \ldots & \ldots & \ldots & \ldots & \ldots \\ \ldots & \ldots & \ldots & \ldots & \ldots \\ 0 & 0 & \ldots & \ldots & \lambda_n \end{bmatrix}$$

Since $B^\theta B = BB^\theta$, we have

$$\begin{bmatrix} \bar{\lambda}_1 & 0 & b_{13} & \ldots & \ldots & 0 \\ \bar{b}_{12} & \bar{\lambda}_2 & b_{23} & \ldots & \ldots & 0 \\ \ldots & \ldots & \ldots & \ldots & \ldots & \ldots \\ \ldots & \ldots & \ldots & \ldots & \ldots & \ldots \\ \ldots & \ldots & \ldots & \ldots & \ldots & \ldots \\ \bar{b}_{1n} & \bar{b}_{2n} & 0 & & & \bar{\lambda}_n \end{bmatrix}$$

$$\times \begin{bmatrix} \lambda_1 & b_{12} & b_{13} & \ldots & \ldots & b_{1n} \\ 0 & \lambda_2 & b_{23} & \ldots & \ldots & b_{2n} \\ \ldots & \ldots & \ldots & \ldots & \ldots & \ldots \\ \ldots & \ldots & \ldots & \ldots & \ldots & \ldots \\ \ldots & \ldots & \ldots & \ldots & \ldots & \ldots \\ 0 & 0 & 0 & \ldots & \ldots & \lambda_n \end{bmatrix}$$

$$= \begin{bmatrix} \lambda_1 & b_{12} & b_{13} & \dots & \dots & b_{1n} \\ 0 & \lambda_2 & b_{23} & \dots & \dots & b_{2n} \\ \dots & \dots & \dots & \dots & \dots & \dots \\ \dots & \dots & \dots & \dots & \dots & \dots \\ \dots & \dots & \dots & \dots & \dots & \dots \\ 0 & 0 & 0 & \dots & \dots & \lambda_n \end{bmatrix}$$

$$\times \begin{bmatrix} \bar{\lambda}_1 & 0 & b_{13} & \dots & \dots & 0 \\ \bar{b}_{12} & \bar{\lambda}_2 & b_{23} & \dots & \dots & 0 \\ \dots & \dots & \dots & \dots & \dots & \dots \\ \dots & \dots & \dots & \dots & \dots & \dots \\ \dots & \dots & \dots & \dots & \dots & \dots \\ \bar{b}_{1n} & \bar{b}_{2n} & 0 & \dots & \dots & \bar{\lambda}_n \end{bmatrix}$$

Comparison of 1-1 entries on both sides, we have

$$\bar{\lambda}_1\lambda_1 = \lambda_1\bar{\lambda}_1 + b_{12}\bar{b}_{12} + b_{13}\bar{b}_{13} + \ldots + b_{1n}\bar{b}_{1n}$$

or

$$0 = |b_{12}|^2 + |b_{13}|^2 + \ldots + |b_{1n}|^2,$$

which implies that $b_{12} = b_{13} = \ldots = b_{1n} = 0$. Similarly, comparison of 2-2 entries, we get

$$b_{23} = b_{24} = \ldots = b_{2n} = 0$$

and so on, Hence B is diagonal, that is, $U^\theta AU = \text{diag}[\lambda_1\ \lambda_2\ \ldots\ \lambda_n]$.

2.42 QUADRATIC FORMS

Definition 2.116. A homogeneous polynomial of the type

$$\sum_{i=1}^{n}\sum_{j=1}^{n} a_{ij}\, x_i\, x_j,$$

where a_{ij} are elements of a field F is called a *quadratic form* in n variables $x_1, x_2, \ldots, x_n$ over the field F.

If a_{ij} are real, then the quadratic form is called real *quadratic form*.

For example, $x_1^2 - 3x_1x_2 + x_2^2 + x_1x_3$ is a real quadratic form.

Theorem 2.88. Every quadratic form over a field F in n variables $x_1, x_2, \ldots, x_n$ can be expressed in the form of X^TBX, where B is a symmetric matrix of order n over F and X is a column vector $[x_1, x_2, \ldots, x_n]^T$.

Proof. Let

$$\sum_{i=1}^{n}\sum_{j=1}^{n} a_{ij}\, x_i\, x_j,$$

be a quadratic form over the field F in n variables $x_1, x_2, \ldots, x_n$. Since x_i, x_j are scalars, we have $x_i\,x_j = x_j\,x_i$. Therefore, the coefficient of $x_i\,x_j$ is $a_{ij} + a_{ji}$. Thus, we assign half of the coefficient to x_{ij} and half to x_{ji}. Let b_{ij} be another set of scalars such that $b_{ii} = a_{ii}$ and $b_{ij} = \frac{1}{2}(a_{ij} + a_{ji})$ for $i \neq j$. Then

$$\sum_{i=1}^{n}\sum_{j=1}^{n} a_{ij}\, x_i\, x_j = \sum_{i=1}^{n}\sum_{j=1}^{n} b_{ij}\, x_i\, x_j.$$

Since, $b_{ij} = b_{ji}$, the matrix $B = [b_{ij}]_{n\times n}$ is symmetric. We further note that if

$$X = \begin{bmatrix} x_1 \\ x_2 \\ \dots \\ \dots \\ x_n \end{bmatrix},$$

then

$$X^TBX = [x_1 x_2 \ldots x_n] \begin{bmatrix} b_{11} & b_{12} & \dots & \dots & b_{1n} \\ b_{21} & b_{22} & \dots & \dots & b_{2n} \\ \dots & \dots & \dots & \dots & \dots \\ \dots & \dots & \dots & \dots & \dots \\ b_{n1} & b_{n2} & \dots & \dots & b_{nn} \end{bmatrix} \begin{bmatrix} x_1 \\ x_2 \\ \dots \\ \dots \\ x_n \end{bmatrix}$$

$$= \sum_{i=1}^{n}\sum_{j=1}^{n} b_{ij}\, x_i\, x_j = \sum_{i=1}^{n}\sum_{j=1}^{n} a_{ij}\, x_i\, x_j.$$

The symmetric matrix B is called the *matrix of the quadratic form*

$$\sum_{i=1}^{n}\sum_{j=1}^{n} a_{ij}\, x_i\, x_j.$$

EXAMPLE 2.146

Find the matrix of the quadratic form $x_1^2 - 3x_1x_2 + x_2^2 + x_1x_3$.

Solution. The given quadratic form can be written as

$$x_1^2 - \frac{3}{2}x_1x_2 - \frac{3}{2}x_2x_1 + x_2^2 + \frac{1}{2}x_1x_3 + \frac{1}{2}x_3x_1.$$

Therefore, the matrix of the given quadratic form is

$$A = \begin{bmatrix} a_{11} & a_{12} & a_{13} \\ a_{21} & a_{22} & a_{23} \\ a_{31} & a_{32} & a_{33} \end{bmatrix},$$

where

$$a_{11} = 1, \quad a_{12} = -\frac{3}{2}\ a_{13} = \frac{1}{2}$$

$$a_{21} = -\frac{3}{2},\ a_{22} = 1\ a_{23} = 0$$

$$a_{31} = \frac{1}{2},\ a_{32} = 0\ a_{33} = 0.$$

Hence

$$A = \begin{bmatrix} 1 & -\frac{3}{2} & \frac{1}{2} \\ -\frac{3}{2} & 1 & 0 \\ \frac{1}{2} & 0 & 0 \end{bmatrix},$$

which is symmetric.

EXAMPLE 2.147

Find the quadratic form corresponding to the symmetric matrix.

$$A = \begin{bmatrix} 1 & 2 & 3 \\ 2 & 0 & 3 \\ 3 & 3 & 1 \end{bmatrix}$$

Solution. The required quadratic form is

$$X^T AX = [x_1x_2x_3] \begin{bmatrix} 1 & 2 & 3 \\ 2 & 0 & 3 \\ 3 & 3 & 1 \end{bmatrix} \begin{bmatrix} x_1 \\ x_2 \\ x_3 \end{bmatrix}$$

$$= [x_1x_2x_3] \begin{bmatrix} x_1 + 2x_2 + 3x3 \\ 2x_1 + 3x_2 \\ 3x_1 + 3x_2 + x_3 \end{bmatrix}$$

$$= x_1(x_1 + 2x_2 + 3x_3) + x_2(2x_1 + 3x_3)$$
$$+ x3(3x_1 + 3x_2 + x_3)$$
$$= x_1^2 + x_3^2 + 4x_1x_2 + 6x_1x_3 + 6x_2x_3.$$

2.43 DIAGONALIZATION OF QUADRATIC FORMS

We know that for every real symmetric matrix A there exists an orthogonal matrix U such that

$$U^T\ AU = \text{diag}[\lambda_1\,\lambda_2 \ldots \lambda_n],$$

where $\lambda_1, \lambda_2, \ldots, \lambda_n$ are characteristic roots of A.

Applying the orthogonal transformation $X = UY$ to the quadratic form X^TAX, we have

$$X^T AX = \lambda_1y_1^2 + \lambda_2y_2^2 + \ldots + \lambda_ny_n^2.$$

If the rank of A is r, then $n - r$ characteristic roots are zero and so

$$X^T AX = \lambda_1y_1^2 + \lambda_2y_2^2 + \ldots + \lambda_ry_r^2,$$

where $\lambda_1, \lambda_2, \ldots, \lambda_n$ are non-zero characteristic roots.

Definition 2.117. A square matrix B of order n over a field F is said to be *congruent* to another square matrix A of order n over F, if there exists a nonsingular matrix P over F such that $B = P^TAP$.

The relation of 'congruence of matrices' is an equivalence relation in the set of all $n \times n$ matrices over a field F. Further, let A be symmetric matrix and let B be congruent to A. Therefore, there exists a non-singular matrix P such that $B = P^TAP$. Then

$$B^T = (P^T\ AP)^T = P^TA^TP$$

$$= P^TAP, \text{ since } A \text{ is symmetric}$$

$$= B.$$

Hence, every *matrix congruent to a symmetric matrix is a symmetric matrix*.

Theorem 2.89. (Congruent reduction of a symmetric matrix). If A is any n rowed non-zero symmetric matrix of rank r over a field F, then there exists an n rowed non-singular matrix P over F such that

$$P^T AP = \begin{bmatrix} A_1 & \mathbf{0} \\ \mathbf{0} & \mathbf{0} \end{bmatrix},$$

where A_1 is a non-zero singular diagonal matrix of order r over F and each $\mathbf{0}$ is a null matrix of a suitable size.

Proof. We prove the theorem by induction. When $n = 1$, $r = 1$ also. The quadratic form is simply $a_{11}x_1^2, a_{11} \neq 0$ and the identity transformation $y_1 = x_1$ is the non-singular transformation. Suppose that the theorem is true for all symmetric matrices of order $n-1$, then we first show that there exists a matrix $B = [b_{ij}]_{n\times n}$ over F congruent to A such that $b_{11} \neq 0$. We take up the following cases.

Case I. If $a_{11} \neq 0$, then we take $B = A$.

Case II. If $a_{11} = 0$, but some diagonal element of A, say $a_{ii} \neq 0$. Then using $R_i \leftrightarrow R_1, C_i \leftrightarrow C_1$ to A, we obtain a matrix B congruent to A such that $b_{11} = a_{ii} \neq 0$.

Case III. Suppose that each diagonal element of A is zero. Since A is non-zero, there exists, non-zero element a_{ij} such that $a_{ij} = a_{ji} \neq 0$. Applying the congruent operation $R_i \to R_i + R_j$, $C_i \to C_i + C_j$ to A, we obtain a matrix $D = [d_{ij}]_{n\times n}$ congruent to A such that $d_{ii} = a_{ij} + a_{ji} = 2a_{ij} \neq 0$. Now, applying the congruent operation $R_i \to R_1$, $C_i\ C_1$ to D, we obtain a matrix $B = [b_{ij}]_{n\times n}$ congruent to D and therefore also congruent to A such that $b_{11} = d_{ii} \neq 0$. Hence, there exists a matrix $B = [b_{ij}]$ congruent to a symmetric matrix such that the leading element of B is non-zero. Since B is congruent to a symmetric matrix, therefore, B itself is symmetric. Since $b_{11} \neq 0$, all elements in the first row and first column except the leading element can be made zero by suitable congruent operation. Thus we have a matrix

$$C = \begin{bmatrix} a_{11} & 0 & \dots & 0 \\ 0 & & & \\ \dots & & B_1 & \\ 0 & & & \end{bmatrix}$$

congruent to B and, therefore, congruent to A such that B is a square matrix of order $n-1$. Further C is congruent to a symmetric matrix and so C is also symmetric. Consequently B_1 is also a symmetric matrix. By induction hypothesis, B_1 can be reduced to a diagonal matrix by congruent operation. So C can be reduced to a diagonal matrix by congruent operations. Thus, A is congruent to a diagonal matrix, say, diag $[\lambda_1\, \lambda_2 \dots \lambda_k \dots 0000]$. Thus there exists a non-singular matrix P such that

$$P^T AP = \text{diag}[\lambda_1\ \lambda_2\ \dots\ \lambda_k\ \dots\ 0\,0\,0\,0].$$

Since $\rho(A) = r$ and we know that rank does not alter by multiplying by a non-singular matrix, therefore, rank of $P^T AP = \text{diag}[\lambda_1\ \lambda_2\ \dots\ \lambda_k\ \dots\ 0\,0\,0\,0]$ is also r. So r elements of diag $[\lambda_1\ \lambda_2\ \dots\ \lambda_k\ \dots\ 0\ 0\ 0]$ are non-zero. Thus, $k = r$ and so

$$P^T AP = \text{diag}[\lambda_1\ \lambda_2\ \dots\ \lambda_r\ \dots\ 0\,0\,0\,0].$$

Corollary 2.25. Corresponding to every quadratic form $X^T AX$ over a field F, there exists a non-singular linear transformation $X = PY$ over F such that the form $X^T AX$ transforms to

$$\lambda_1 y_1^2 + \lambda_2 y_2^2 + \dots + \lambda_r y_r^2,$$

where $\lambda_1, \lambda_2, \dots, \lambda_r$ are scalars in F and r is the rank of the matrix A.

Definition 2.118. The rank of the symmetric matrix A is called the *rank of the quadratic form* $X^T AX$.

EXAMPLE 2.148

Find a non-singular matrix P such that $P^T AP$ is a diagonal matrix, where

$$A = \begin{bmatrix} 6 & -2 & 2 \\ -2 & 3 & -1 \\ 2 & -1 & 3 \end{bmatrix}.$$

Find the quadratic form and its rank.

Solution. Write $A = IAI$, that is,

$$\begin{bmatrix} 6 & -2 & 2 \\ -2 & 3 & -1 \\ 2 & -1 & 3 \end{bmatrix} = \begin{bmatrix} 1 & 0 & 0 \\ 0 & 1 & 0 \\ 0 & 0 & 1 \end{bmatrix} A \begin{bmatrix} 1 & 0 & 0 \\ 0 & 1 & 0 \\ 0 & 0 & 1 \end{bmatrix}$$

Using congruent operations, we shall reduce A to diagonal form. Performing congruent operations $R_2 \to R_2 + \frac{1}{3}R_1, C_2 \to C_2 + \frac{1}{3}C_1$ and $R_3 \to R_3 - \frac{1}{3}R_1, C_3 \to C_3 - \frac{1}{3}C_1$, we have

$$\begin{bmatrix} 6 & 0 & 0 \\ 0 & \frac{7}{3} & -\frac{1}{3} \\ 0 & -\frac{1}{3} & \frac{7}{3} \end{bmatrix} = \begin{bmatrix} 1 & 0 & 0 \\ \frac{1}{3} & 1 & 0 \\ -\frac{1}{3} & 0 & 1 \end{bmatrix} A \begin{bmatrix} 1 & \frac{1}{3} & -\frac{1}{3} \\ 0 & 1 & 0 \\ 0 & 0 & 1 \end{bmatrix}.$$

Now performing congruent operation $R_3 \to R_3 + \frac{1}{7}R_2, C_3 \to C_3 + \frac{1}{7}C_2$, we have

$$\begin{bmatrix} 6 & 0 & 0 \\ 0 & \frac{7}{3} & 0 \\ 0 & 0 & \frac{16}{7} \end{bmatrix} = \begin{bmatrix} 1 & 0 & 0 \\ \frac{1}{3} & 1 & 0 \\ \frac{2}{7} & \frac{1}{7} & 1 \end{bmatrix} A \begin{bmatrix} 1 & \frac{1}{3} & -\frac{2}{7} \\ 0 & 1 & \frac{1}{7} \\ 0 & 0 & 1 \end{bmatrix}.$$

Thus

$$diag \begin{bmatrix} 6 & \frac{7}{3} & \frac{16}{7} \end{bmatrix} = P^{-1} AP,$$

where

$$P = \begin{bmatrix} 1 & \frac{1}{3} & -\frac{2}{7} \\ 0 & 1 & \frac{1}{7} \\ 0 & 0 & 1 \end{bmatrix}$$

The quadratic form corresponding to the matrix A is

$$X^T AX = 6x_1^2 + 3x_2^2 + 3x_3^2 - 4x_1x_2 - 2x_2x_3 + 4x_3x_1. \quad (76)$$

The non-singular transformation $X = PY$ corresponding to the matrix P is

$$\begin{bmatrix} x_1 \\ x_2 \\ x_3 \end{bmatrix} = \begin{bmatrix} 1 & \frac{1}{3} & -\frac{2}{7} \\ 0 & 1 & \frac{1}{7} \\ 0 & 0 & 1 \end{bmatrix} \begin{bmatrix} y_1 \\ y_2 \\ y_3 \end{bmatrix},$$

which yields

$$x_1 = y_1 + \frac{1}{3}y_2 - \frac{2}{7}y_3$$
$$x_2 = y_2 + \frac{1}{7}y_3$$
$$x_3 = y_3.$$

Substituting these values in (76), we get

$$(PY)^T A(PY) = 6y_1^2 + \frac{7}{3}y_2^2 + \frac{16}{7}y_3^2.$$

It contains a sum of *three* squares. Thus, the rank of the quadratic form is 3.

Theorem 2.90. Let A be any n-rowed real symmetric matrix of rank r. Then there exists a real non-singular matrix P such that

$$P^T AP = \text{diag}[1\ 1\ \ldots\ 1 - 1 - 1 - 1 \ldots - 1\ 0\ 0\ 0 \ldots 0],$$

where 1 appears p times and -1 appears $r-p$ times.

Proof. Since A is a symmetric matrix of rank r, there exists a non-singular real matrix Q such that

$$Q^T AQ = \text{diag}[\lambda_1\ \lambda_2\ \ldots\ \lambda_r\ \ldots\ 0\ 0\ 0\ 0].$$

Suppose p of the non-zero diagonal elements are positive and $r-p$ are negative. Then by using congruence operations $R_i \leftrightarrow Rj, C_i \leftrightarrow C_j$, we can assume that first p elements $\lambda_1, \lambda_2, \ldots, \lambda_p$ are positive and $\lambda_{p+1}, \lambda_{p+2}, \ldots, \lambda_r$ are negative. Let

$$S = \text{diag}\left[\frac{1}{\sqrt{\lambda_1}} \frac{1}{\sqrt{\lambda_2}} \ldots \frac{1}{\sqrt{\lambda_p}} \frac{1}{\sqrt{-\lambda_{p+1}}} \ldots \frac{1}{\sqrt{-\lambda_r}} 1 1 1\right].$$

Then S is non-singular and $S^T = S$. Let $P = QS$. Then P is also real non-singular matrix and we have

$$P^T AP = (QS)^T A(QS) = S^T Q^T AQS$$

$$= S^T(\text{diag}[\lambda_1\,\lambda_2,\ldots\,\lambda_r 0\ldots 0])S$$
$$= S(\text{diag}[\lambda_1\,\lambda_2\ldots\lambda_r 0\ldots 0])S$$
$$= \text{diag}[1\;1\ldots 1\;-1\;-1\ldots -1\;0\ldots 0]$$

so that 1 appears p times and -1 appears $r-p$ times.

Corollary 2.26. If X^TAX is a real quadratic form of rank r in n variables, then there exists a real non-singular linear transformation $X = PY$ which transform X^TAX to the form

$$Y^TP^TAPY = y_1^2 + y_2^2 + \ldots + y_p^2 - y_{p+1}^2 - \cdots - y_r^2,$$

which is called *canonical form or normal form* of a real quadratic form.

The number of positive terms in the normal form of X^TAX is called the *index* of the quadratic form, whereas $p-(r-p) = 2p-r$ is called the *signature* of the quadratic form and is usually denoted by s.

A quadratic form X^TAX with a non-singular matrix A of order n is called *positive definite* if $n = r = p$, that is, if n = rank = index. A quadratic form is called *positive semi-definite* if $r < n$ and $r = p$. Similarly a quadratic form is called *negative definite* if its index is zero and $n = r$ and called *negative semi-definite* if $r < n$ and its index is zero.

EXAMPLE 2.149

Find the rank, index, and signature of the quadratic form $x^2 - 2y^2 + 3z^2 - 4yz + 6zx$.

Solution. The matrix of the given quadratic form is

$$A = \begin{bmatrix} 1 & 0 & 3 \\ 0 & -2 & -2 \\ 3 & -2 & 3 \end{bmatrix}.$$

Write $A = IAI$, that is,

$$\begin{bmatrix} 1 & 0 & 3 \\ 0 & -2 & -2 \\ 3 & -2 & -3 \end{bmatrix} = \begin{bmatrix} 1 & 0 & 0 \\ 0 & 1 & 0 \\ 0 & 0 & 1 \end{bmatrix} A \begin{bmatrix} 1 & 0 & 0 \\ 0 & 1 & 0 \\ 0 & 0 & 1 \end{bmatrix}.$$

Performing congruence operations $R_3 \to R_3 - 3R_1$, $C_3 \to C_3 - 3C_1$, we get

$$\begin{bmatrix} 1 & 0 & 0 \\ 0 & -2 & -2 \\ 0 & -2 & -6 \end{bmatrix} = \begin{bmatrix} 1 & 0 & 0 \\ 0 & 1 & 0 \\ -3 & 0 & 1 \end{bmatrix} A \begin{bmatrix} 1 & 0 & -3 \\ 0 & 1 & 0 \\ 0 & 0 & 1 \end{bmatrix}.$$

Performing congruence operations $R_3 \to R_3 - R_2$, $C_3 \to C_3 - C_2$, we have,

$$\begin{bmatrix} 1 & 0 & 0 \\ 0 & -2 & 0 \\ 0 & 0 & -4 \end{bmatrix} = \begin{bmatrix} 1 & 0 & 0 \\ 0 & 1 & 0 \\ -3 & -1 & 1 \end{bmatrix} A \begin{bmatrix} 1 & 0 & -3 \\ 0 & 1 & -1 \\ 0 & 0 & 1 \end{bmatrix}.$$

Performing $R_2 \to \frac{1}{\sqrt{2}}R_2$, $C_2 \to \frac{1}{\sqrt{2}}C_2$, and $R_3 \to \frac{1}{\sqrt{4}}R_3$, $C_3 \to \frac{1}{\sqrt{4}}C_3$, we get,

$$\begin{bmatrix} 1 & 0 & 0 \\ 0 & -1 & 0 \\ 0 & 0 & -1 \end{bmatrix} = \begin{bmatrix} 1 & 0 & 0 \\ 0 & \frac{1}{\sqrt{2}} & 0 \\ -\frac{3}{2} & -\frac{1}{2} & \frac{1}{2} \end{bmatrix} A \begin{bmatrix} 1 & 0 & -\frac{3}{2} \\ 0 & \frac{1}{\sqrt{2}} & -\frac{1}{2} \\ 0 & 0 & \frac{1}{2} \end{bmatrix}.$$

Hence $X = PY$ transforms the given quadratic form to $y_1^2 - y_2^2 - y_3^2$.

The rank of the quadratic form is 3 (the number of non-zero terms in the normal form.)

The number of positive terms is 1. Hence, the index of the quadratic form is 1.

We note that $2p - r = 2 - 3 = -1$. Therefore, signature of the quadratic form is -1.

2.44 MISCELLANEOUS EXAMPLES

EXAMPLE 2.150

Compute the inverse of $\begin{bmatrix} 2 & 1 & -1 \\ 0 & 2 & 1 \\ 5 & 2 & -3 \end{bmatrix}$ using elementary transformations.

Solution. Write $A = I_3A$, that is,

$$\begin{bmatrix} 2 & 1 & -1 \\ 0 & 2 & 1 \\ 5 & 2 & -3 \end{bmatrix} = \begin{bmatrix} 1 & 0 & 0 \\ 0 & 1 & 0 \\ 0 & 0 & 1 \end{bmatrix} A.$$

We reduce the matrix on the L.H.S of the equation to identity matrix by elementary row transformations, keeping in mind that each row transformation will apply to I_3 on the right hand side.

Interchanging R_1 and R_3, we get

$$\begin{bmatrix} 5 & 2 & -3 \\ 0 & 2 & 1 \\ 2 & 1 & -1 \end{bmatrix} = \begin{bmatrix} 0 & 0 & 1 \\ 0 & 1 & 0 \\ 1 & 0 & 0 \end{bmatrix} A.$$

Performing $R_1 \to R_1 - 2R_3$, we get

$$\begin{bmatrix} 1 & 0 & -1 \\ 0 & 2 & 1 \\ 2 & 1 & -1 \end{bmatrix} = \begin{bmatrix} -2 & 0 & 1 \\ 0 & 1 & 0 \\ 1 & 0 & 0 \end{bmatrix} A$$

Performing $R_3 \to R_3 - 2R_1$, we have

$$\begin{bmatrix} 1 & 0 & -1 \\ 0 & 2 & 1 \\ 0 & 1 & 1 \end{bmatrix} = \begin{bmatrix} -2 & 0 & 1 \\ 0 & 1 & 0 \\ 5 & 0 & -2 \end{bmatrix} A$$

Now performing $R_2 \to R_2 - R_3$, we get

$$\begin{bmatrix} 1 & 0 & -1 \\ 0 & 1 & 0 \\ 0 & 1 & 1 \end{bmatrix} = \begin{bmatrix} -2 & 0 & 1 \\ -5 & 1 & 2 \\ 5 & 0 & -2 \end{bmatrix} A$$

Now performing $R_3 \to R_3 - R_2$, we have

$$\begin{bmatrix} 1 & 0 & -1 \\ 0 & 1 & 0 \\ 0 & 0 & 1 \end{bmatrix} = \begin{bmatrix} -2 & 0 & 1 \\ -5 & 1 & 2 \\ 10 & -1 & -4 \end{bmatrix} A$$

Lastly, performing $R_1 \to R_1 + R_3$, we have

$$\begin{bmatrix} 1 & 0 & 0 \\ 0 & 1 & 0 \\ 0 & 0 & 1 \end{bmatrix} = \begin{bmatrix} 8 & -1 & -3 \\ -5 & 1 & 2 \\ 10 & -1 & -4 \end{bmatrix} A. = A^{-1}A.$$

Hence

$$A^{-1} = \begin{bmatrix} 8 & -1 & -3 \\ -5 & 1 & 2 \\ 10 & -1 & -4 \end{bmatrix}.$$

EXAMPLE 2.151

Using Cayley-Hamilton theorem, find A^{-1}, given the matrix

$$A = \begin{bmatrix} 13 & -3 & 5 \\ 0 & 4 & 0 \\ -15 & 9 & -7 \end{bmatrix}.$$

Solution. Proceeding as in Example 2.138, the characteristic equation is

$$|A - \lambda I| = \begin{vmatrix} 13-\lambda & -3 & 5 \\ 0 & 4-\lambda & 0 \\ -15 & 9 & -7-\lambda \end{vmatrix} = 0$$

or

$$(13-\lambda)(4-\lambda)(-7-\lambda) + 75(4-\lambda) = 0$$

or

$$\lambda^3 - 10\lambda^2 + 8\lambda + 64 = 0.$$

By Cayley's Hamilton theorem, we have

$$A^3 - 10A^2 + 8A64I = 0$$

or

$$A^{-1} = -\frac{1}{64}[A^2 - 10A + 8I]$$

$$= -\frac{1}{64}\left\{\begin{bmatrix} 94 & -6 & 30 \\ 0 & 16 & 0 \\ -90 & 18 & -26 \end{bmatrix} - 10\begin{bmatrix} 13 & -3 & 5 \\ 0 & 4 & 0 \\ -15 & 9 & -7 \end{bmatrix} + 8\begin{bmatrix} 1 & 0 & 0 \\ 0 & 1 & 0 \\ 0 & 0 & 1 \end{bmatrix}\right\}$$

$$= -\frac{1}{64}\begin{bmatrix} -28 & 24 & -20 \\ 0 & -16 & 0 \\ 60 & -72 & 52 \end{bmatrix}$$

$$= -\frac{1}{16}\begin{bmatrix} -7 & 6 & -5 \\ 0 & -4 & 0 \\ 15 & -18 & 13 \end{bmatrix}.$$

EXAMPLE 2.152

For the matrix:

$$A=\begin{bmatrix}2 & 1 & -3 & -6\\ 3 & -3 & 1 & 2\\ 1 & 1 & 1 & 2\end{bmatrix},$$

find non-singular matrices P and Q such that PAQ is in the normal form. Hence find the rank of A.

Solution. Write

$$A = I_3 A I_4,$$

that is,

$$\begin{bmatrix}2 & 1 & -3 & -6\\ 3 & -3 & 1 & 2\\ 1 & 1 & 1 & 2\end{bmatrix}=\begin{bmatrix}1 & 0 & 0\\ 0 & 1 & 0\\ 0 & 0 & 1\end{bmatrix}A\begin{bmatrix}1 & 0 & 0 & 0\\ 0 & 1 & 0 & 0\\ 0 & 0 & 1 & 0\\ 0 & 0 & 0 & 1\end{bmatrix}.$$

Performing elementary transformation $R_1 \leftrightarrow R_3$, we get

$$\begin{bmatrix}1 & 1 & 1 & 2\\ 3 & -3 & 1 & 2\\ 2 & 1 & -3 & -6\end{bmatrix}=\begin{bmatrix}0 & 0 & 1\\ 0 & 1 & 0\\ 1 & 0 & 0\end{bmatrix}A\begin{bmatrix}1 & 0 & 0 & 0\\ 0 & 1 & 0 & 0\\ 0 & 0 & 1 & 0\\ 0 & 0 & 0 & 1\end{bmatrix}$$

Performing $R_2 \to R_2 - 3R_1$ and $R_3 \to R_3 - 2R_1$, we get

$$\begin{bmatrix}1 & 0 & 0 & 0\\ 0 & -6 & -2 & -4\\ 0 & -1 & -5 & -10\end{bmatrix}$$

$$=\begin{bmatrix}0 & 0 & 1\\ 0 & 1 & -3\\ 1 & 0 & -2\end{bmatrix}A\begin{bmatrix}1 & 0 & 0 & 0\\ 0 & 1 & 0 & 0\\ 0 & 0 & 1 & 0\\ 0 & 0 & 0 & 1\end{bmatrix}.$$

Performing $C_2 \to C_2 - C_1, C_3 \to C_3 - C_1, C_4 \to C_4 - 2C_1$, we get

$$\begin{bmatrix}1 & 0 & 0 & 0\\ 0 & -6 & -2 & -4\\ 0 & -1 & -5 & -10\end{bmatrix}$$

$$=\begin{bmatrix}0 & 0 & 1\\ 0 & 1 & -3\\ 1 & 0 & -2\end{bmatrix}A\begin{bmatrix}1 & -1 & -1 & -2\\ 0 & 1 & 0 & 0\\ 0 & 0 & 1 & 0\\ 0 & 0 & 0 & 1\end{bmatrix}.$$

Now performing $R_3 \leftrightarrow R_2$, we have

$$\begin{bmatrix}1 & 0 & 0 & 0\\ 0 & -1 & -5 & -10\\ 0 & -6 & -2 & -4\end{bmatrix}$$

$$=\begin{bmatrix}0 & 0 & 1\\ 1 & 0 & -2\\ 0 & 1 & -3\end{bmatrix}A\begin{bmatrix}1 & -1 & -1 & -2\\ 0 & 1 & 0 & 0\\ 0 & 0 & 1 & 0\\ 0 & 0 & 0 & 1\end{bmatrix}.$$

Performing $R_2 \to -R_2$, we get

$$\begin{bmatrix}1 & 0 & 0 & 0\\ 0 & 1 & 5 & 10\\ 0 & -6 & -2 & -4\end{bmatrix}$$

$$=\begin{bmatrix}0 & 0 & 1\\ -1 & 0 & 2\\ 0 & 1 & -3\end{bmatrix}A\begin{bmatrix}1 & -1 & -1 & -2\\ 0 & 1 & 0 & 0\\ 0 & 0 & 1 & 0\\ 0 & 0 & 0 & 1\end{bmatrix}.$$

Now performing $R_3 \to R_3 + 6R_2$, we get

$$\begin{bmatrix}1 & 0 & 0 & 0\\ 0 & 1 & 5 & 10\\ 0 & 0 & 28 & 56\end{bmatrix}$$

$$=\begin{bmatrix}0 & 0 & 1\\ -1 & 0 & 2\\ -6 & 1 & 9\end{bmatrix}A\begin{bmatrix}1 & -1 & -1 & -2\\ 0 & 1 & 0 & 0\\ 0 & 0 & 1 & 0\\ 0 & 0 & 0 & 1\end{bmatrix}.$$

Performing $C_3 \to C_3 - 5C_2$ and $C_4 \to C_4 - 10C_2$, we get

$$\begin{bmatrix}1 & 0 & 0 & 0\\ 0 & 1 & 0 & 0\\ 0 & 0 & 28 & 56\end{bmatrix}$$

$$=\begin{bmatrix}0 & 0 & 1\\ -1 & 0 & 2\\ -6 & 1 & 9\end{bmatrix}A\begin{bmatrix}1 & -1 & 4 & 8\\ 0 & 1 & -5 & -10\\ 0 & 0 & 1 & 0\\ 0 & 0 & 0 & 1\end{bmatrix}.$$

Performing $R_4 \to \frac{1}{28}R_4$, we get

$$\begin{bmatrix} 1 & 0 & 0 & 0 \\ 0 & 1 & 0 & 0 \\ 0 & 0 & 1 & 2 \end{bmatrix}$$

$$= \begin{bmatrix} 0 & 0 & 1 \\ -1 & 0 & 2 \\ -\frac{3}{14} & \frac{1}{28} & \frac{9}{28} \end{bmatrix} A \begin{bmatrix} 1 & -1 & 4 & 8 \\ 0 & 1 & -5 & -10 \\ 0 & 0 & 1 & 0 \\ 0 & 0 & 0 & 1 \end{bmatrix}.$$

Performing $C_4 \to C_4 - 2C_3$, we get

$$\begin{bmatrix} 1 & 0 & 0 & 0 \\ 0 & 1 & 0 & 0 \\ 0 & 0 & 0 & 0 \end{bmatrix}$$

$$= \begin{bmatrix} 0 & 0 & 1 \\ -1 & 0 & 2 \\ -\frac{3}{14} & \frac{1}{28} & \frac{9}{28} \end{bmatrix} A \begin{bmatrix} 1 & -1 & 4 & 8 \\ 0 & 1 & -5 & 0 \\ 0 & 0 & 1 & -2 \\ 0 & 0 & 0 & 1 \end{bmatrix}.$$

or

$$\begin{bmatrix} I_3 & 0 \\ 0 & 0 \end{bmatrix} = PAQ,$$

where

$$P = \begin{bmatrix} 0 & 0 & 1 \\ -1 & 0 & 2 \\ -\frac{3}{14} & \frac{1}{28} & \frac{9}{28} \end{bmatrix} \text{ and}$$

$$Q = \begin{bmatrix} 1 & -1 & 4 & 0 \\ 0 & 1 & -5 & 0 \\ 0 & 0 & 1 & -2 \\ 0 & 0 & 0 & 1 \end{bmatrix}.$$

Also $\rho(A) = 3$.

EXAMPLE 2.153

(a) Find the rank of the matrix $\begin{bmatrix} 3 & -1 & 2 \\ -6 & 2 & 4 \\ -3 & 1 & 2 \end{bmatrix}$ by reducing it to the normal form

(b) For the matrix $A = \begin{bmatrix} 1 & 1 & 2 \\ 1 & 2 & 3 \\ 0 & -1 & -1 \end{bmatrix}$, find nonsingular matrices P and Q such that PAQ is in the normal form. Hence find the rank of A.

(c) Reduce the following matrix to column echelon and find its rank:

$$A = \begin{bmatrix} 1 & 1 & -1 & 1 \\ -1 & 1 & -3 & -3 \\ 1 & 0 & 1 & 2 \\ 1 & -1 & 3 & 3 \end{bmatrix}.$$

(d) Find all values of μ for which rank of the matrix

$$A = \begin{bmatrix} \mu & -1 & 0 & 0 \\ 0 & \mu & -1 & 0 \\ 0 & 0 & \mu & -1 \\ -6 & 11 & -6 & 1 \end{bmatrix}$$

is equal to 3.

Solution. (a) We have

$$A = \begin{bmatrix} 3 & -1 & 2 \\ -6 & 2 & 4 \\ -3 & 1 & 2 \end{bmatrix}$$

$$\sim \begin{bmatrix} 1 & -1 & 2 \\ -10 & 2 & 4 \\ -5 & 1 & 2 \end{bmatrix} \quad C_1 \to C_1 - C_3$$

$$\sim \begin{bmatrix} 1 & -1 & 2 \\ 0 & -8 & 24 \\ 0 & -4 & 12 \end{bmatrix} \quad \begin{matrix} R_2 \to R_2 + 10R_1 \\ R_3 \to R_3 + 5R_1 \end{matrix}$$

$$\sim \begin{bmatrix} 1 & -1 & 2 \\ 0 & 1 & -3 \\ 0 & -4 & 12 \end{bmatrix} \quad R_2 \to -\frac{1}{8}R_2$$

$$\sim \begin{bmatrix} 1 & -1 & 2 \\ 0 & 1 & -3 \\ 0 & 0 & 0 \end{bmatrix} \quad R_3 \to R_3 + 4R_2$$

$$\sim \begin{bmatrix} 1 & 0 & 0 \\ 0 & 1 & -3 \\ 0 & 0 & 0 \end{bmatrix} \quad \begin{matrix} C_2 \to C_2 + C_1 \\ C_3 \to C_3 - C_1 \end{matrix}$$

$$\sim \begin{bmatrix} 1 & 0 & 0 \\ 0 & 1 & 0 \\ 0 & 0 & 0 \end{bmatrix} \quad C_3 \to C_3 + 3C_2$$

$$= \begin{bmatrix} I_2 & 0 \\ 0 & 0 \end{bmatrix} \text{ (normal form)}$$

Hence $\rho(A) = 2$.
(b) Expressing the given matrix in the form $A = I_3 A I_3$, we have

$$\begin{bmatrix} 1 & 1 & 2 \\ 1 & 2 & 3 \\ 0 & -1 & -1 \end{bmatrix} = \begin{bmatrix} 1 & 0 & 0 \\ 0 & 1 & 0 \\ 0 & 0 & 1 \end{bmatrix} A \begin{bmatrix} 1 & 0 & 0 \\ 0 & 1 & 0 \\ 0 & 0 & 1 \end{bmatrix}.$$

Using the elementary transformation $R_2 \to R_2 - R_1$, we get

$$\begin{bmatrix} 0 & 1 & 2 \\ 0 & 1 & 1 \\ 0 & -1 & -1 \end{bmatrix} = \begin{bmatrix} 1 & 0 & 0 \\ -1 & 1 & 0 \\ 0 & 0 & 1 \end{bmatrix} A \begin{bmatrix} 1 & 0 & 0 \\ 0 & 1 & 0 \\ 0 & 0 & 1 \end{bmatrix}$$

Using the elementary column transformations $C_2 \to C_2 - C_1$ and $C_3 \to C_3 - 2C_1$, we have

$$\begin{bmatrix} 1 & 0 & 0 \\ 0 & 1 & 1 \\ 0 & -1 & 1 \end{bmatrix} = \begin{bmatrix} 1 & 0 & 0 \\ -1 & 0 & 0 \\ 0 & 0 & 1 \end{bmatrix} A \begin{bmatrix} 1 & -1 & -2 \\ 0 & 1 & 0 \\ 0 & 0 & 1 \end{bmatrix}$$

Operating $R_3 \to R_3 + R_2$, we get

$$\begin{bmatrix} 1 & 0 & 0 \\ 0 & 1 & 1 \\ 0 & 0 & 0 \end{bmatrix} = \begin{bmatrix} 1 & 0 & 0 \\ -1 & 0 & 0 \\ -1 & 0 & 1 \end{bmatrix} A \begin{bmatrix} 1 & -1 & -2 \\ 0 & 1 & 0 \\ 0 & 0 & 1 \end{bmatrix}$$

Now operating $C_3 \to C_3 - C_2$, we have

$$\begin{bmatrix} 1 & 0 & 0 \\ 0 & 1 & 0 \\ 0 & 0 & 0 \end{bmatrix} = \begin{bmatrix} 1 & 0 & 0 \\ -1 & 0 & 0 \\ -1 & 0 & 1 \end{bmatrix} A \begin{bmatrix} 1 & -1 & -1 \\ 0 & 1 & -1 \\ 0 & 0 & 1 \end{bmatrix}$$

or

$$\begin{bmatrix} I_2 & 0 \\ 0 & 0 \end{bmatrix} = PAQ,$$

where

$$P = \begin{bmatrix} 1 & 0 & 0 \\ -1 & 0 & 0 \\ -1 & 0 & 1 \end{bmatrix} \text{ and}$$

$$Q = \begin{bmatrix} 1 & -1 & -1 \\ 0 & 1 & -1 \\ 0 & 0 & 1 \end{bmatrix}.$$

Since elementary transformations do not alter the rank of a matrix,

$$\rho(A) = \rho \begin{bmatrix} I_2 & 0 \\ 0 & 0 \end{bmatrix} = 2.$$

(c) A Matrix is said to be in column echelon form if

(i) The first non-zero entry in each non-zero column is 1.

(ii) The column containing only zeros occurs next to all non-zero columns.

(iii) The number of zeros above the first nonzero entry in each column is less than the number of such zeros in the next column.

The given matrix is

$$A = \begin{bmatrix} 1 & 1 & -1 & 1 \\ -1 & 1 & -3 & -3 \\ 1 & 0 & 1 & 2 \\ 1 & -1 & 3 & 3 \end{bmatrix}$$

$$\sim \begin{bmatrix} 1 & 0 & 0 & 0 \\ -1 & 2 & -4 & -2 \\ 1 & -1 & 2 & 1 \\ 1 & -2 & 4 & 2 \end{bmatrix} \begin{matrix} C_2 \to C_2 - C_1 \\ C_3 \to C_3 + C_1 \\ C_4 \to C_4 - C_1 \end{matrix}$$

$$\sim \begin{bmatrix} 1 & 0 & 0 & 0 \\ -1 & 2 & -4 & 0 \\ 1 & -1 & 2 & 0 \\ 1 & -2 & 4 & 0 \end{bmatrix} C_4 \to C_4 + C_2$$

$$\sim \begin{bmatrix} 1 & 0 & 0 & 0 \\ -1 & 2 & 0 & 0 \\ 1 & -1 & 0 & 0 \\ 1 & -2 & 0 & 0 \end{bmatrix} C_3 \to C_3 + 2C_2$$

$$\sim \begin{bmatrix} 1 & 0 & 0 & 0 \\ -1 & 1 & 0 & 0 \\ 1 & -\frac{1}{2} & 0 & 0 \\ 1 & -1 & 0 & 0 \end{bmatrix} C_2 \to \frac{1}{2} C_2,$$

which is column echelon form. The number of nonzero column is two and therefore $\rho(A) = 2$.

(d) Similar to Remark 2.7

We are given that

$$A = \begin{bmatrix} \mu & -1 & 0 & 0 \\ 0 & \mu & -1 & 0 \\ 0 & 0 & \mu & -1 \\ -6 & 11 & -6 & 1 \end{bmatrix}$$

Therefore

$$|A| = \mu \begin{vmatrix} \mu & -1 & 0 \\ 0 & \mu & -1 \\ 11 & -6 & 1 \end{vmatrix} + 1 \begin{vmatrix} 0 & -1 & 0 \\ 0 & \mu & -1 \\ -6 & -6 & 1 \end{vmatrix}$$

$$= \mu^3 - 6\mu^2 + 11\mu - 6$$

$$= 0 \text{ if } \mu = 1,\ 2,\ 3.$$

For $\mu = 3$, we have the singular matrix

$$\begin{bmatrix} 3 & -1 & 0 & 0 \\ 0 & 3 & -1 & 0 \\ 0 & 0 & 3 & -1 \\ -6 & 11 & -6 & 1 \end{bmatrix},$$

which has non-singular sub-matrix

$$\begin{bmatrix} 3 & -1 & 0 \\ 0 & 3 & -1 \\ 0 & 0 & 3 \end{bmatrix}.$$

Thus for $\mu = 3$, the rank of the matrix A is 3. Similarly, the rank is 3 for $\mu = 2$ and $\mu = 1$. For other values of μ, we have $|A| \neq 0$ and so $\rho(A) = 4$ for other values of μ.

EXAMPLE 2.154

Solve the system of equations:

$$x + y + z = 6$$

$$x - y + 2z = 5$$

$$3x + y + z = 8$$

Solution. The augmented matrix is

$$[A : B] = \begin{bmatrix} 1 & 1 & 1 & 6 \\ 1 & -1 & 2 & 5 \\ 3 & 1 & 1 & 8 \end{bmatrix}$$

$$\sim \begin{bmatrix} 1 & 1 & 1 & 6 \\ 0 & -2 & 1 & -1 \\ 0 & -2 & -2 & -10 \end{bmatrix} \begin{matrix} R_2 \to R_2 - R_1 \\ R_3 \to R_3 - 3R_1 \end{matrix}$$

$$\sim \begin{bmatrix} 1 & 0 & 0 & 6 \\ 0 & -2 & 1 & -1 \\ 0 & -2 & -2 & -10 \end{bmatrix} \begin{matrix} C_2 \to C_2 - C_1 \\ C_3 \to C_3 - C_1 \end{matrix}$$

$$\sim \begin{bmatrix} 1 & 0 & 0 & 6 \\ 0 & 1 & -2 & -1 \\ 0 & -2 & -2 & -10 \end{bmatrix} C_2 \leftrightarrow C_3$$

$$\sim \begin{bmatrix} 1 & 0 & 0 & 0 \\ 0 & 1 & -2 & -1 \\ 0 & 0 & -6 & -12 \end{bmatrix} R_3 \to R_3 + 2R_1$$

$$\sim \begin{bmatrix} 1 & 0 & 0 & 6 \\ 0 & 1 & -2 & -1 \\ 0 & 0 & 0 & -9 \end{bmatrix} R_3 \to R_3 + 3R_2$$

It follows that $\rho(A) = 2$ and $\rho[A : B] = 3$. Hence the given equation is inconsistent.

EXAMPLE 2.155

Discuss the consistency of the system of equations:

$$2x - 3y + 6z - 5w = 3,\ y - 4z + w = 1,$$

$$4x - 5y + 8z - 9w = \lambda$$

for various values of λ. If consistent, find the solution.

Solution. The matrix equation is $AX = B$, where

$$A = \begin{bmatrix} 2 & -3 & 6 & -5 \\ 0 & 1 & -4 & 1 \\ 4 & -5 & 8 & -9 \end{bmatrix},\ X = \begin{bmatrix} x \\ y \\ z \\ w \end{bmatrix} \text{ and}$$

$$B = \begin{bmatrix} 3 \\ 1 \\ \lambda \end{bmatrix}$$

The augmented matrix is

$$[A:B]=\begin{bmatrix}2&-3&6&-5&3\\0&1&-4&1&1\\4&-5&8&-9&\lambda\end{bmatrix}$$

$$\sim\begin{bmatrix}2&-3&6&-5&3\\0&1&-4&1&1\\0&1&-4&1&\lambda-6\end{bmatrix}\quad R_3\to R_3-2R_1$$

$$\sim\begin{bmatrix}2&-3&6&-5&3\\0&1&-4&1&1\\0&0&0&0&\lambda-7\end{bmatrix}\quad R_3\to R_3-R_2$$

We note that $\rho(A)=\rho(A:B)$ if $\lambda-7=0$, that is, if $\lambda=7$. Thus the given equation is consistent if $\lambda=7$. Thus if $\lambda=7$, then we have

$$[A:B]=\begin{bmatrix}2&-3&6&-5&3\\0&1&-4&1&1\\0&0&0&0&0\end{bmatrix}$$

and so the given system of equations is equivalent to

$$2x-3y+6z-5\omega=3$$
$$y-4z+\omega=1.$$

Therefore if $w=k_1$, $z=k_2$, then $y=1+4k_2-k_1$ and $x=3+3k_2+k_1$. Hence the general solution of the system is $x=3+3k_2+k_1$, $y=1+4k_2-k_1$, $z=k_2$, $w=k_1$.

EXAMPLE 2.156

Test for consistency the following set of equations and solve if it is consistent: $5x+3y+7z=4$, $3x+26y+2z=9$, $7x+2y+10z=5$.

Solution. The augmented matrix is

$$[A:B]=\begin{bmatrix}5&3&7&4\\3&26&2&9\\7&2&10&5\end{bmatrix}$$

$$\sim\begin{bmatrix}15&9&21&12\\15&130&10&45\\7&2&10&5\end{bmatrix}\quad \begin{matrix}R_1\to R_1\\R_2\to 5R_2\end{matrix}$$

$$\sim\begin{bmatrix}15&9&21&12\\0&121&-11&33\\7&2&10&5\end{bmatrix}\quad R_2\to R_2-R_1$$

$$\sim\begin{bmatrix}33&21&49&28\\0&11&-1&3\\35&21&50&25\end{bmatrix}\quad \begin{matrix}R_1\to \frac{7}{3}R_1\\R_2\to \frac{1}{4}R_2\\R_3\to 5R_3\end{matrix}$$

$$\sim\begin{bmatrix}35&21&49&28\\0&11&-1&3\\0&-11&1&-3\end{bmatrix}\quad R_3\to R_3-R_1$$

$$\sim\begin{bmatrix}35&21&49&28\\0&11&-1&3\\0&0&0&0\end{bmatrix}\quad R_3\to R_3-R_2.$$

We observe that

$$\rho(A)=2,\ \rho([A:B])=2,$$

and so $\rho(A)=\rho([A:B])$. Hence the given system of equation is consistent. Further, the given system is equivalent to

$$35x+21y+49z=28$$
$$11y-z=3,$$

which yield $y=\frac{3+z}{11}$ and $x=\frac{7}{11}-\frac{16}{11}z$.

Taking $z=0$, we get a particular solution as

$$x=\frac{7}{11},\ y\frac{3}{11},\ z=0.$$

EXAMPLE 2.157

(a) Find the value of λ for which the equations

$$(\lambda-1)x+(3\lambda+1)y+2\lambda z=0,$$
$$(\lambda-1)x+(4\lambda-2)y+(\lambda+3)z=0,$$
$$2x+(3\lambda+1)y+3(\lambda-1)z=0.$$

are consistent, and find the ratios of $x:y:z$ when λ has the smallest of these values. What happens when λ has the greater of these values?

(b) Determine b such that the system of homogeneous equations

$$2x+y+2z=0, \quad x+y+3z=0 \text{ and}$$
$$4x+3y+bz=0$$

has (i) Trivial solution (ii) Non-trivial solution. Also find the non-trivial solution using matrix method.

Solution. (a) For consistency, the coefficient matrix A, in the matrix equation $AX=0$, should be singular.

Therefore, we must have

$$\begin{vmatrix} \lambda-1 & 3\lambda+1 & 2\lambda \\ \lambda-1 & 4\lambda-2 & \lambda+3 \\ 2 & 3\lambda+1 & 3(\lambda-1) \end{vmatrix}=0$$

$$\sim \begin{vmatrix} \lambda-1 & 3\lambda+1 & 2\lambda \\ 0 & \lambda-3 & 3-\lambda \\ 2 & 3\lambda+1 & 3(\lambda-1) \end{vmatrix}=0$$

$$\sim \begin{vmatrix} \lambda-1 & 3\lambda+1 & 5\lambda+1 \\ 0 & \lambda-3 & 0 \\ 2 & 3\lambda+1 & 6\lambda-2 \end{vmatrix}=0$$

$$\sim (\lambda-3)\begin{vmatrix} \lambda-1 & 5\lambda+1 \\ 2 & 6\lambda-2 \end{vmatrix}=0$$

$$\sim 2(\lambda-3)[(\lambda-1)(3\lambda-1)-(5\lambda+1)]=0$$

or

$$6\lambda(\lambda-3)^2=0, \text{ which yields } \lambda=0 \text{ or } \lambda=3.$$

When $\lambda=0$, the given system of equations reduces to

$$-x+y=0,$$
$$-x-2y+3z=0,$$
$$2x+y-3z=0.$$

The last two equations yield

$$\frac{x}{3}=\frac{y}{3}=\frac{z}{3} \text{ and so } x=y=z.$$

When $\lambda=3$, all the three equations become identical.

(b) The given system of equation is

$$x+y+3z=0,$$
$$2x+y+2z=0,$$
$$4x+3y+bz=0.$$

The system in matrix form is

$$\begin{bmatrix} 1 & 1 & 3 \\ 2 & 1 & 2 \\ 4 & 3 & b \end{bmatrix}\begin{bmatrix} x \\ y \\ z \end{bmatrix}=\begin{bmatrix} 0 \\ 0 \\ 0 \end{bmatrix}$$

This homogenous system will have a non-trivial solution only if $|A|=0$. Thus for non-trivial solution

$$\begin{vmatrix} 1 & 1 & 3 \\ 2 & 1 & 2 \\ 4 & 3 & b \end{vmatrix}=0$$

or

$$1(b-6)-1(2b-8)+3(6-4)=0$$

or

$$-b+8=0, \text{ which yields } b=8.$$

Thus for non-trivial solution, $b=8$. The coefficient matrix for non-trivial solution is

$$\begin{bmatrix} 1 & 1 & 3 \\ 2 & 1 & 2 \\ 4 & 3 & 8 \end{bmatrix} \sim \begin{bmatrix} 1 & 1 & 3 \\ 0 & -1 & -4 \\ 0 & -1 & -4 \end{bmatrix} \begin{matrix} R2 \to R_2-2R_1 \\ R3 \to R_3-4R_1 \end{matrix}$$

$$\sim \begin{bmatrix} 1 & 1 & 3 \\ 0 & -1 & -4 \\ 0 & 0 & 0 \end{bmatrix} \quad R3 \to R_3-R_2.$$

The last matrix is of rank 2. Thus the given system is equivalent to

$$x+y+3z=0$$
$$-y-4z=0.$$

Hence $y=-4z$ and then $x=z$. Taking $z=t$ the general solution is

$$x=t,\ y=-4t,\ z=t.$$

EXAMPLE 2.158

Prove that the sum of the eigenvalues of a matrix A is the sum of the elements of the principal diagonal.

Solution. If $A = [a_{ij}]$ be the matrix of order n, then the characteristic equation of the matrix A is

$$|A - \lambda I = \lambda^n - \lambda^{n-1}\left(\sum_{i=1}^{n} a_{ii}\right) + \ldots = 0.$$

Form the theory of equations, the sum of the roots $\lambda_1, \lambda_2, \ldots, \lambda_n$ is equal to negative of the coefficient of λ^{n-1}. Hence

$$\begin{aligned}\lambda_1 + \lambda_2 + \ldots + \lambda_n &= a_{11} + a_{22} + \ldots + a_{nn}\\ &= \text{Trace } A.\end{aligned}$$

EXAMPLE 2.159

(a) Find the eigen values of A^{-1} if the matrix A is

$$\begin{bmatrix} 2 & 5 & -1 \\ 0 & 3 & 2 \\ 0 & 0 & 4 \end{bmatrix}$$

(b) Find the eigenvalues and the corresponding vectors of the matrix

$$A = \begin{bmatrix} 1 & 0 & 0 \\ 0 & 2 & 1 \\ 2 & 0 & 3 \end{bmatrix}.$$

Solution. (a) By Example 2.136, the eigenvalues of triangular matrix are the diagonal elements. Hence the eigenvalues of A are 2, 3 and 4. Since the eigenvalues of A^{-1} are multiplicative inverses of the eigenvalues of the matrix A, the eigenvalues ofA^{-1} are $\frac{1}{2}, \frac{1}{3}$ and $\frac{1}{4}$.

(b) We have

$$A = \begin{bmatrix} 1 & 0 & 0 \\ 0 & 2 & 1 \\ 2 & 0 & 3 \end{bmatrix}$$

The characteristic equation of A is

$$|A - \lambda I| = \begin{vmatrix} 1-\lambda & 0 & 0 \\ 0 & 2-\lambda & 1 \\ 2 & 0 & 3-\lambda \end{vmatrix} = 0.$$

or

$$\lambda^3 - 6\lambda^2 + 11\lambda - 6 = 0,$$

which yields $\lambda = 1, 2, 3$. Hence the characteristic roots are 1, 2 and 3.

The eigenvector corresponding to $\lambda = 1$ is given by $(A - I)X = 0$, that is, by

$$\begin{bmatrix} 0 & 0 & 0 \\ 0 & 1 & 1 \\ 2 & 0 & 2 \end{bmatrix}\begin{bmatrix} x_1 \\ x_2 \\ x_3 \end{bmatrix} = \begin{bmatrix} 0 \\ 0 \\ 0 \end{bmatrix}.$$

Thus, we have

$$\begin{aligned} x_2 + x_3 &= 0,\\ 2x_1 + 2x_3 &= 0. \end{aligned}$$

Hence $x_1 = x_2 = -x_3$. Taking $x_3 = -1$, we get the vector

$$X_1 = \begin{bmatrix} 1 \\ 1 \\ -1 \end{bmatrix}$$

The eigenvector corresponding to the eigenvalue 2 is given by $(A - 2I)X = 0$, that is, by

$$\begin{bmatrix} -1 & 0 & 0 \\ 0 & 0 & 1 \\ 2 & 0 & 1 \end{bmatrix}\begin{bmatrix} x_1 \\ x_2 \\ x_3 \end{bmatrix} = \begin{bmatrix} 0 \\ 0 \\ 0 \end{bmatrix}.$$

This equation yields

$$X_2 = \begin{bmatrix} 0 \\ 1 \\ 0 \end{bmatrix} \text{ as one of the vector.}$$

Similarly, the eigenvector corresponding to $\lambda = 3$ is given by $(A - 3I)X = 0$ or by

$$\begin{bmatrix} -2 & 0 & 0 \\ 0 & -1 & 1 \\ 2 & 0 & 0 \end{bmatrix}\begin{bmatrix} x_1 \\ x_2 \\ x_3 \end{bmatrix} = \begin{bmatrix} 0 \\ 0 \\ 0 \end{bmatrix},$$

which yields $\begin{bmatrix} 0 \\ 1 \\ 1 \end{bmatrix}$ as one of the solution. Hence

$$X_3 = \begin{bmatrix} 0 \\ 1 \\ 1 \end{bmatrix}.$$

EXAMPLE 2.160

Find the sum and product of the eigen values of the matrix:

$$\begin{bmatrix} 1 & 2 & 3 & 4 \\ 2 & 1 & 5 & 6 \\ 7 & 4 & 3 & 2 \\ 4 & 3 & 0 & 5 \end{bmatrix}$$

Solution. The given matrix is

$$A = \begin{bmatrix} 1 & 2 & 3 & 4 \\ 2 & 1 & 5 & 6 \\ 7 & 4 & 3 & 2 \\ 4 & 3 & 0 & 5 \end{bmatrix}.$$

The sum of the eigenvalues is the trace (spur) of the matrix and so the sum is $1+1+3+5 = 10$. Product of the eigenvalues is equal to $|A|$. Expanding $|A|$, we get the product as 262.

EXAMPLE 2.161

One of the eigenvalues of $\begin{bmatrix} 7 & 4 & -4 \\ 4 & -8 & -1 \\ 4 & -1 & -8 \end{bmatrix}$ is -9. Find the other two eigenvalues.

Solution. The characteristic equation of the given matrix is

$$|A-\lambda I| = \begin{vmatrix} 7-\lambda & 4 & -4 \\ 4 & -8-\lambda & -1 \\ 4 & -1 & -8-\lambda \end{vmatrix} = 0.$$

or

$$(7-\lambda)[\lambda^2+16\lambda+63]-4[-28-4\lambda] \\ -4[28+4\lambda] = 0$$

or

$$\lambda^3+9\lambda^2-49\lambda-441 = 0.$$

Clearly $\lambda = -9$ satisfies this equation. Then, by synthetic division, the reduced equation is

$$\lambda^2 - 49 = 0,$$

which yields $\lambda = \pm 7$. Thus the eigenvalues of the given matrix are $-9, 7, -7$. The sum of the eigenvalues is $-9+7-7 = -9$, which is equal to the trace of the given matrix.

EXAMPLE 2.162

Verify that the following set of vectors in $\mathbb{R}^3$ is linearly dependent: (1, 0, 1), (1, 1, 1), (1, 1, 2) and (1, 2, 1): Also find the number of linearly independent vectors.

Solution. The vectors in $\mathbb{R}^3$ are given to be

$$v_1 = (1,\ 0,\ 1),\ v_2 = (1,\ 1,\ 1),$$
$$v_3 = (1,\ 1,\ 2),\ v_4 = (1,\ 2,\ 1).$$

Let

$$\lambda_1 v_1 + \lambda_2 v_2 + \lambda_3 v_3 + \lambda_4 v_4 = 0, \qquad (77)$$

This gives

$$\lambda_1(1,\ 0,\ 1) + \lambda_2(1,\ 1,\ 1) + \lambda_3(1,\ 1,\ 2) \\ + \lambda_4(1,\ 2,\ 1) = 0$$

or

$$\lambda_1 + \lambda_2 + \lambda_3 + \lambda_4 = 0$$
$$0\lambda_1 + \lambda_2 + \lambda_3 + 2\lambda_4 = 0$$
$$\lambda_1 + \lambda_2 + 2\lambda_3 + \lambda_4 = 0.$$

We have four variable and three equations. Thus there is one degree of freedom. We have

$$\frac{\lambda_1}{-1} = \frac{\lambda_2}{2} = \frac{\lambda_3}{0} = \frac{\lambda_4}{1} = k.$$

Therefore

$$\lambda_1 = -k,\ \lambda_2 = 2k,\ \lambda_3 = 0,\ \lambda_4 = -k.$$

Putting these values of λ_i is (77), we get

$$-kv_1 + 2kv_2 + 0v_3 - kv_4 = 0$$

or

$$v_1 - 2v_2 + 0v_3 + v_4 = 0.$$

Thus (77) is satisfied for $\lambda_1 = 1$, $\lambda_2 = -2$, $\lambda_3 = 0$ and $\lambda_4 = 1$. Since not all of λ_i are zeros, it follows that v_1, v_2, v_3, v_4 are linearly dependent.

EXAMPLE 2.163

What do you mean by an orthogonal matrix? Verify that the following matrix is orthogonal:

$$\begin{bmatrix} \cos\theta & 0 & \sin\theta \\ 0 & 1 & 0 \\ -\sin\theta & 0 & \cos\theta \end{bmatrix}$$

Solution. A square matrix P is said to be orthogonal if $P^T P = PP^T = I$. If

$$P = \begin{bmatrix} \cos\theta & 0 & \sin\theta \\ 0 & 1 & 0 \\ -\sin\theta & 0 & \cos\theta \end{bmatrix},$$

then

$$PP^T = \begin{bmatrix} \cos\theta & 0 & \sin\theta \\ 0 & 1 & 0 \\ -\sin\theta & 0 & \cos\theta \end{bmatrix} \begin{bmatrix} \cos\theta & 0 & -\sin\theta \\ 0 & 1 & 0 \\ \sin\theta & 0 & \cos\theta \end{bmatrix}$$

$$= \begin{bmatrix} \cos^2\theta + \sin^2\theta & 0 & \begin{matrix} -\sin\theta 0\cos\theta \\ +\sin\theta\cos\theta \end{matrix} \\ 0 & 1 & 0 \\ \begin{matrix} -\sin\theta\cos 2\theta \\ +\sin\theta\cos\theta \end{matrix} & & \cos^2\theta + \sin^2\theta \end{bmatrix}$$

$$= \begin{bmatrix} 1 & 0 & 0 \\ 0 & 1 & 0 \\ 0 & 0 & 1 \end{bmatrix} = I.$$

Hence P is orthogonal.

EXAMPLE 2.164

Show that the transformation

$$y_1 = \frac{1}{3}x_1 + \frac{2}{3}x_2 + \frac{2}{3}x_3,$$

$$y_2 = \frac{2}{3}x_1 + \frac{1}{3}x_2 - \frac{2}{3}x_3,$$

$$y_3 = \frac{2}{3}x_1 - \frac{2}{3}x_2 + \frac{1}{3}x_3 \text{ is orthogonal.}$$

Solution. In matrix form, we have

$$Y = PX.$$

where

$$Y = \begin{bmatrix} y_1 \\ y_2 \\ y_3 \end{bmatrix}, P = \begin{bmatrix} \frac{1}{3} & \frac{2}{3} & \frac{2}{3} \\ \frac{2}{3} & \frac{1}{3} & -\frac{2}{3} \\ \frac{2}{3} & -\frac{2}{3} & \frac{1}{3} \end{bmatrix} \text{ and}$$

$$X = \begin{bmatrix} x_1 \\ x_2 \\ x_3 \end{bmatrix}.$$

The transformation $Y = PX$ will be orthogonal if $P^T P = I$. To show it, we observe that

$$P^T P = \begin{bmatrix} \frac{1}{3} & \frac{2}{3} & \frac{2}{3} \\ \frac{2}{3} & \frac{1}{3} & -\frac{2}{3} \\ \frac{2}{3} & -\frac{2}{3} & \frac{1}{3} \end{bmatrix} \begin{bmatrix} \frac{1}{3} & \frac{2}{3} & \frac{2}{3} \\ \frac{2}{3} & \frac{1}{3} & -\frac{2}{3} \\ \frac{2}{3} & -\frac{2}{3} & \frac{1}{3} \end{bmatrix}$$

$$= \begin{bmatrix} 1 & 0 & 0 \\ 0 & 1 & 0 \\ 0 & 0 & 1 \end{bmatrix} = I.$$

Hence $Y = PX$ is orthogonal.

EXAMPLE 2.165

Diagonalise the matrix $A = \begin{pmatrix} 2 & 0 & 1 \\ 0 & 3 & 0 \\ 1 & 0 & 2 \end{pmatrix}$ through an orthogonal transformation.

Solution. We shall proceed as in Example 2.145. The characteristic equation of A is

$$|A-\lambda I| = \begin{vmatrix} 2-\lambda & 0 & 1 \\ 0 & 3-\lambda & 0 \\ 1 & 0 & 2-\lambda \end{vmatrix} = 0.$$

or

$$(2-\lambda)(3-\lambda)(2-\lambda)-(3-\lambda)=0$$

$$(3-\lambda)[(2-\lambda)^2-1]=0$$

$$(3-\lambda)[4+\lambda^2-4\lambda-1]=0$$

$$(3-\lambda)(\lambda^2-4\lambda+3]=0$$

$$\Rightarrow \lambda = 3,3,1.$$

The ch. vector corresponding to $\lambda = 3$ is given by

$$\left.\begin{aligned} -x_1+x_3 &= 0 \\ x_1-x_3 &= 0 \end{aligned}\right\} \quad (78)$$

Thus

$$X_1 = \begin{bmatrix} 1 \\ 0 \\ 1 \end{bmatrix}.$$

Let$X_2 = \begin{bmatrix} x \\ y \\ z \end{bmatrix}$ be another eigenvector of A corresponding to the eigenvalue 3 and orthogonal to X_1. Then

$x-y=0$, because it satisfies equation (78) and

$$x+z=0, \text{ using} X_2^{\theta} X_1 = 0.$$

Obviously $x=1$, $y=1$ and $z=-1$ is a solution. Therefore

$$X_2 = \begin{bmatrix} 1 \\ 1 \\ -1 \end{bmatrix}.$$

Further eigenvector corresponding to $\lambda = 1$ is given by $(A-I)X=0$, that is , by

$$\begin{bmatrix} 1 & 0 & 1 \\ 0 & 2 & 0 \\ 1 & 0 & 1 \end{bmatrix}\begin{bmatrix} x_1 \\ x_2 \\ x_3 \end{bmatrix} = 0.$$

This equation yields

$$x_1+x_3=0,\ 2x_2=0,\ x_1+x_3=0.$$

Thus $x_1 = 1$, $x_2 = 0$. $x_3 = -1$ and so Length (norm) of the vectors X_1, X_2, X_3 are respectively $\sqrt{2}$, $\sqrt{3}$, $\sqrt{2}$. Hence the orthogonal matrix is

$$P = \begin{bmatrix} \frac{1}{\sqrt{2}} & \frac{1}{\sqrt{3}} & \frac{1}{\sqrt{2}} \\ 0 & \frac{1}{\sqrt{3}} & 0 \\ \frac{1}{\sqrt{2}} & -\frac{1}{\sqrt{3}} & -\frac{1}{\sqrt{2}} \end{bmatrix}$$

and

$$P^T AP = dig\,[3\ 3\ 1].$$

EXAMPLE 2.166

(a) Show that the matrix $\begin{bmatrix} 3 & 1 & -1 \\ -2 & 1 & 2 \\ 0 & 1 & 2 \end{bmatrix}$, is diagonalizable. Hence, find P such that $P^{-1}AP$ is a diagonal matrix. Also obtain the matrix $B = A^2+5A+3I$.

(b) **Find a matrix** P which diagonalizes the matrix $A = \begin{bmatrix} 4 & 1 \\ 2 & 3 \end{bmatrix}$, verify $P^{-1}AP = D$, where D is the diagonal matrix.

(c) Show that the matrix $A = \begin{bmatrix} 3 & 1 & -1 \\ -2 & 1 & 2 \\ 0 & 1 & 2 \end{bmatrix}$, is diagonalizable. Hence, find P such that $P^{-1}AP$ is a diagonal matrix.

Solution. (a) The characteristic equation for the given matrix A is

$$|A-\lambda I| = \begin{vmatrix} 3-\lambda & 1 & -1 \\ -2 & 1-\lambda & 2 \\ 0 & 1 & 2-\lambda \end{vmatrix} = 0,$$

that is,

$$(3-\lambda)(\lambda^2-3\lambda)+4-2\lambda+2=0$$

or

$$\lambda^3-6\lambda^2+11\lambda-6=0:$$

By inspection, $\lambda = 1$ is a root. The reduced equation is $\lambda^2 - 5\lambda + 6 = 0$, which yields $\lambda = 2, 3$. Since all characteristic roots are distnict, the given matrix A is diagonalizable. To find the non-singular matrix P satisfying $P^{-1}AP = \text{diag}(1, 2, 3)$, we proceed as follows:

The characteristic vectors are given by $(A - \lambda I)X = 0$, that is, by

$$\begin{bmatrix} 3-\lambda & 1 & -1 \\ -2 & 1-\lambda & 2 \\ 0 & 1 & 2-\lambda \end{bmatrix} \begin{bmatrix} x_1 \\ x_2 \\ x_3 \end{bmatrix} = \begin{bmatrix} 0 \\ 0 \\ 0 \end{bmatrix}$$

and so

$$\left.\begin{aligned} (3-\lambda)x_1 + x_2 - x_3 &= 0 \\ -2x_1 + (1-\lambda)x_2 + 2x_3 &= 0 \\ 0x_1 + x_2 + (2-\lambda)x_3 &= 0 \end{aligned}\right\} \qquad (79)$$

For $\lambda = 1$, we get

$$2x_1 + x_2 - x_3 = 0$$

$$-2x_1 + 2x_3 = 0$$

$$x_2 + x_3 = 0$$

We note that $x_1 = 1,\ x_2 = -1,\ x_3 = 1$ satisfy these equations. Hence the eigenvector corresponding to $\lambda = 1$ is $[1 -1\ 1]^T$.

For $\lambda = 2$, we get from (79),

$$x_1 + x_2 - x_3 = 0$$

$$-2x_1 - x_2 + 2x_3 = 0$$

$$0x_1 + x_2 = 0$$

Clearly $x_1 = 1,\ x_2 = 0,\ x_3 = 1$ is a solution. Hence the eigenvector corresponding to $\lambda = 2$ is $[1\ 0\ 1]^T$.

For $\lambda = 3$, we have

$$x_2 - x_3 = 0$$

$$-2x_1 - 2x_2 + 2x_3 = 0$$

$$x_2 - x_3 = 0$$

Taking $x_3 = 1$, we get $x_2 = 1$ and $x_1 = 0$. Thus the eigenvector corresponding to $\lambda = 3$ is $[0\ 1\ 1]^T$.

Hence

$$P = \begin{bmatrix} 1 & 1 & 0 \\ -1 & 0 & 1 \\ 1 & 1 & 1 \end{bmatrix}$$

and

$$P^{-1}AP = \text{diag } [1\ 2\ 3] = D, \text{ say} \qquad (80)$$

Premultiplication by P and postmultiplication by P^{-1} reduces (80) to

$$A = P\,D\,P^{-1}.$$

Further,

$$A^n = P\,D^n\,P^{-1}.$$

Thus

$$A^2 = P\,D^2\,P^{-1}.$$

But

$$D = \begin{bmatrix} 1 & 0 & 0 \\ 0 & 2 & 0 \\ 0 & 0 & 3 \end{bmatrix}, \quad D^2 = \begin{bmatrix} 1 & 0 & 0 \\ 0 & 4 & 0 \\ 0 & 0 & 9 \end{bmatrix}$$

$$P = \begin{bmatrix} 1 & 1 & 0 \\ -1 & 0 & 1 \\ 1 & 1 & 1 \end{bmatrix} \text{ and } P^{-1} = \begin{bmatrix} -1 & -1 & 1 \\ 2 & 1 & -1 \\ -1 & 0 & 1 \end{bmatrix}.$$

Putting these values in $B = A^2 + 5A + 3I$, we get

$$B = A^2 + 5A + 3I = \begin{bmatrix} 25 & 8 & -8 \\ -18 & 9 & 18 \\ -2 & 8 & 19 \end{bmatrix}.$$

(b) We have

$$A = \begin{bmatrix} 4 & 1 \\ 2 & 3 \end{bmatrix}$$

The characteristic equation of A is

$$|A - \lambda I| = \begin{vmatrix} 4-\lambda & 1 \\ 2 & 3-\lambda \end{vmatrix} = 0$$

or

$$(4-\lambda)(3-\lambda)-2=0$$

or

$$\lambda^2-7\lambda+10=0$$

The characteristic roots are $\lambda = \frac{7+3}{2} = 2,\ 5$. Since the eigenvalues are distinct, the matrix A is diagonalizable. The eigenvector corresponding to $\lambda = 2$ is given by $(A-2I)X = 0$, that is by

$$\begin{bmatrix} 2 & 1 \\ 2 & 1 \end{bmatrix}\begin{bmatrix} x_1 \\ x_2 \end{bmatrix} = \begin{bmatrix} 0 \\ 0 \end{bmatrix}$$

or by

$$2x_1 + x_2 = 0 \quad \text{or} \quad x_1 = -\frac{x_2}{2}.$$

Putting $x_2 = 2$, we get

$$X_1 = \begin{bmatrix} -1 \\ 2 \end{bmatrix}.$$

Similarly, eigenvector corresponding to $\lambda = 5$ is given by $(A-5I)X = 0$ or by

$$\begin{bmatrix} -1 & 1 \\ 2 & -2 \end{bmatrix}\begin{bmatrix} x_1 \\ x_2 \end{bmatrix} = 0$$

or by

$$-x_1 + x_2 = 0$$
$$2x_1 - 2x_2 = 0$$

and so $x_1 = x_2$. Putting $x_2 = 1$, we get

$$X_2 = \begin{bmatrix} 1 \\ 1 \end{bmatrix}$$

Thus the transforming matrix is

$$P = \begin{bmatrix} -1 & 1 \\ 2 & 1 \end{bmatrix}$$

and

$$P^{-1} = \begin{bmatrix} -\frac{1}{3} & \frac{1}{3} \\ \frac{2}{3} & \frac{1}{3} \end{bmatrix}$$

Then

$$P^{-1}AP = \begin{bmatrix} -\frac{1}{3} & \frac{1}{3} \\ \frac{2}{3} & \frac{1}{3} \end{bmatrix}\begin{bmatrix} 4 & 1 \\ 2 & 3 \end{bmatrix}\begin{bmatrix} -1 & 1 \\ 2 & 1 \end{bmatrix}$$

$$= \begin{bmatrix} -\frac{2}{3} & \frac{2}{3} \\ \frac{10}{3} & \frac{5}{3} \end{bmatrix}\begin{bmatrix} -1 & 1 \\ 2 & 1 \end{bmatrix}$$

$$= \begin{bmatrix} 2 & 0 \\ 0 & 5 \end{bmatrix}.$$

(c) The characteristic matrix of the given matrix A is

$$|A-\lambda I| = \begin{vmatrix} 3-\lambda & 1 & -1 \\ -2 & 1-\lambda & 2 \\ 0 & 1 & 2-\lambda \end{vmatrix} = 0$$

or

$$(3-\lambda)[(1-\lambda)(2-\lambda)-2] - l[-4+2\lambda] - l(-2) = 0$$

or

$$(3-\lambda)(1-\lambda)(2-\lambda)-6+2\lambda+4 - 2\lambda+2 = 0$$

or

$$(3-\lambda)(1-\lambda)(2-\lambda) = 0.$$

Hence the given matrix A has distinct characteristic roots $\lambda = 1, 2, 3$. Consequently it is diagonalizable. Now the eigenvector corresponding to $\lambda = 1$ is given by $(A-I)X = 0$, that is, by

$$\begin{bmatrix} 2 & 1 & -1 \\ -2 & 0 & 2 \\ 0 & 1 & 1 \end{bmatrix}\begin{bmatrix} x_1 \\ x_2 \\ x_3 \end{bmatrix} = \begin{bmatrix} 0 \\ 0 \\ 0 \end{bmatrix}$$

Thus

$$\begin{aligned} 2x_1 + x_2 - x_3 &= 0 \\ -2x_1 + 0x_2 + 2x_3 &= 0 \\ 0x_1 + x_2 + x_3 &= 0 \end{aligned}$$

and so $x_1 = x_3 = -x_2$. Taking $x_2 = -1$, we get an eigenvector corresponding to $\lambda = 1$ as

$$X_1 = \begin{bmatrix} 1 \\ -1 \\ 1 \end{bmatrix}.$$

Now eigenvector corresponding to $\lambda = 2$ is given by $(A - 2I)X = 0$, that is, by

$$\begin{bmatrix} 1 & 1 & -1 \\ -2 & -1 & 2 \\ 0 & 1 & 0 \end{bmatrix} \begin{bmatrix} x_1 \\ x_2 \\ x_3 \end{bmatrix} = \begin{bmatrix} 0 \\ 0 \\ 0 \end{bmatrix}$$

Thus

$$\begin{aligned} x_1 + x_2 - x_3 &= 0 \\ -2x_1 - x_2 + 2x_3 &= 0 \\ x_2 &= 0 \end{aligned}$$

For this system $x_1 = 1,\ x_2 = 0,\ x_3 = 1$ is a solution. Therefore

$$X_2 = \begin{bmatrix} 1 \\ 0 \\ 1 \end{bmatrix}.$$

An eigenvector corresponding to $\lambda = 3$ is given by $(A - 3I)X = 0$, that is, by

$$\begin{bmatrix} 0 & 1 & -1 \\ -2 & -2 & 2 \\ 0 & 1 & -1 \end{bmatrix} \begin{bmatrix} x_1 \\ x_2 \\ x_3 \end{bmatrix} = \begin{bmatrix} 0 \\ 0 \\ 0 \end{bmatrix}$$

Thus

$$\begin{aligned} x_2 - x_3 &= 0 \\ -2x_1 + 2x_2 + 2x_3 &= 0 \\ x_2 - x_3 &= 0, \end{aligned}$$

which yields $x_1 = 0,\ x_2 = 1,\ x_3 = 1$ as one of the solution. Thus

$$X_3 = \begin{bmatrix} 0 \\ 1 \\ 1 \end{bmatrix}.$$

Therefore the transforming matrix is

$$P = \begin{bmatrix} 1 & 1 & 0 \\ -1 & 0 & 1 \\ 1 & 1 & 1 \end{bmatrix}$$

and so the diagonal matrix is

$$P^{-1}AP = \begin{bmatrix} -1 & -1 & 1 \\ 2 & 1 & -1 \\ -1 & 0 & 1 \end{bmatrix} \begin{bmatrix} 3 & 1 & -1 \\ -2 & 1 & 2 \\ 0 & 1 & 2 \end{bmatrix} \times \begin{bmatrix} 1 & 1 & 0 \\ -1 & 0 & 1 \\ 1 & 1 & 1 \end{bmatrix}$$

$$= \begin{bmatrix} 1 & 0 & 0 \\ 0 & 2 & 0 \\ 0 & 0 & 3 \end{bmatrix}.$$

EXAMPLE 2.167

Reduce the quadratic form $x^2 + 5y^2 + z^2 + 2xy + 6zx + 2yz$ to a canonical form through an Orthogonal transformation.

Solution. The given quadratic form can be written as

$$x^2 + xy + yx + 5y^2 + yz + yz + z^2 + 3zx + 3xz.$$

The matrix of the quadratic form is

$$A = \begin{bmatrix} 1 & 1 & 3 \\ 1 & 5 & 1 \\ 3 & 1 & 1 \end{bmatrix}.$$

Write $A = IAI$, that is,

$$\begin{bmatrix} 1 & 1 & 3 \\ 1 & 5 & 1 \\ 3 & 1 & 1 \end{bmatrix} = \begin{bmatrix} 1 & 0 & 0 \\ 0 & 1 & 0 \\ 0 & 0 & 1 \end{bmatrix} A \begin{bmatrix} 1 & 0 & 0 \\ 0 & 1 & 0 \\ 0 & 0 & 1 \end{bmatrix}.$$

Using congruent operations $R_2 \to R_2 - R_1$, $C_2 \to C_2 - C_1$ and $R_3 \to R_3 - 3R_1$, $C_3 \to C_3 - C_1$, we get

$$\begin{bmatrix} 1 & 0 & 0 \\ 0 & 4 & -2 \\ 0 & -2 & -8 \end{bmatrix} = \begin{bmatrix} 1 & 0 & 0 \\ -1 & 1 & 0 \\ -3 & 0 & 1 \end{bmatrix} A \begin{bmatrix} 1 & -1 & -3 \\ 0 & 1 & 0 \\ 0 & 0 & 1 \end{bmatrix}.$$

Now performing congruent operation $R_3 \to R_3 + \frac{1}{2}R_3$, $C_3 \to C_3 + \frac{1}{2}C_2$, we get

$$\begin{bmatrix} 1 & 0 & 0 \\ 0 & 4 & 0 \\ 0 & 0 & -9 \end{bmatrix} = \begin{bmatrix} 1 & 0 & 0 \\ -1 & 1 & 0 \\ -\frac{7}{2} & \frac{1}{2} & 1 \end{bmatrix} A \begin{bmatrix} 1 & -1 & -\frac{7}{2} \\ 0 & 1 & \frac{1}{2} \\ 0 & 0 & 1 \end{bmatrix}.$$

Thus

$$\text{diag}\,[1\ 4\ -9] = P^{-1}AP,$$

where

$$P = \begin{bmatrix} 1 & -1 & -\frac{7}{2} \\ 0 & 1 & \frac{1}{2} \\ 0 & 0 & 1 \end{bmatrix}.$$

Hence the required canonical form is

$$x^2 + 4y^2 - 9z^2.$$

EXAMPLE 2.168

Reduce the quadratic form $x^2 + y^2 + z^2 - 2xy - 2yz - 2zx$ to canonical form through an orthogonal transformation.

Solution. The matrix of the given quadratic form is

$$A = \begin{bmatrix} 1 & -1 & -1 \\ -1 & 1 & -1 \\ -1 & -1 & 1 \end{bmatrix}.$$

The characteristic equation of A is

$$|A - \lambda I| = \begin{vmatrix} 1-\lambda & -1 & -1 \\ -1 & 1-\lambda & -1 \\ -1 & -1 & 1-\lambda \end{vmatrix} = 0$$

or

$$(1-\lambda)[\lambda^2 - 2\lambda] + \lambda - 2 - 2 + \lambda = 0$$

or

$$\lambda^3 - 3\lambda^2 + 4 = 0,$$

which yields $\lambda = -1, 2, 2$.

The eigenvectors will be given by $(A - \lambda)X = \mathbf{0}$, which implies

$$\begin{bmatrix} 1-\lambda & -1 & -1 \\ -1 & 1-\lambda & -1 \\ -1 & -1 & 1-\lambda \end{bmatrix} \begin{bmatrix} x \\ y \\ z \end{bmatrix} = \begin{bmatrix} 0 \\ 0 \\ 0 \end{bmatrix}$$

or

$$(1-\lambda)x - y - z = 0,$$
$$-x + (1-\lambda)y - z = 0 \quad \text{and}$$
$$-x - y + (1-\lambda)z = 0.$$

For $\lambda = -1$, we have

$$2x - y - z = 0, \ -x + 2y - z = 0 \quad \text{and}$$
$$-x - y + 2z = 0.$$

Solving these equations, we get $x = y = z = 1$ and so the eigenvector is $[1\ 1\ 1]^T$. Its normalized form is $\left[\frac{1}{\sqrt{3}}\ \frac{1}{\sqrt{3}}\ \frac{1}{\sqrt{3}}\right]^T$.

Corresponding to $\lambda = 2$, we have $-x - y - z = 0$, $-x - y - z = 0$ and $-x - y - z = 0$. We note that $x = -2$, $y = 1$, $z = 1$ is a solution. Thus the eigenvector is $[-2\ 1\ 1]^T$ and its normalized form is $\left[\frac{-2}{\sqrt{6}}, \frac{1}{\sqrt{6}}, \frac{1}{\sqrt{6}}\right]^T$. To find the second vector, we have

$$-x - y - z = 0$$

and

$$-2x + y + z = 0 \quad \text{using} \quad X_2^{\theta} X_1 = 0.$$

We note that $x = 0$, $y = -1$ and $z = 1$ is a solution. The normalized vector is $\left[0, \frac{-1}{\sqrt{2}}\ \frac{1}{\sqrt{2}}\right]$.

Hence

$$P = \begin{bmatrix} \frac{1}{\sqrt{3}} & -\frac{2}{\sqrt{6}} & 0 \\ \frac{1}{\sqrt{3}} & \frac{1}{\sqrt{6}} & -\frac{1}{\sqrt{2}} \\ \frac{1}{\sqrt{3}} & \frac{1}{\sqrt{6}} & \frac{1}{\sqrt{2}} \end{bmatrix}$$

and

$$P^T AP = \text{diag } [-1\ 2\ 2].$$

EXERCISES

1. Let F be a field and let $V = \{(a_1, a_2)\} : a_1, a_2 \in F\}$, define addition in V by

$$(a_1, a_2) + (a_3, a_4) = (a_1 + a_3, a_2 + a_4)$$

and for $\lambda \in F$ and $(a_1, a_2) \in V$, define

$$\lambda(a_1, a_2) = (a_1, 0).$$

Show that V is not a vector space over F.
Hint. 1. $V = V$ property is not satisfied.

2. Let $V = \{0\}$. Define addition and scalar multiplication in V by $0+0=0$ and $C(0) = 0$ for $C \in F$. Show that V is a vector space (called *zero vector space*).

3. For what value of λ the set $\{(1+\lambda, 1, 1), (1, 1+\lambda, 1), (1, 1, 1+\lambda)\}$ is linearly independent.

Ans. $\lambda = -1$

4. Show that the subset $\{x^2 - 1, x+1, x-1\}$ of the vector space $P_3(x)$ of polynomials is linearly independent.

5. Show that the subset $\{(1,1,1,0), (3,2,2,1), (1,1,3,2), (1,2,6,5), (1,1,2,1)\}$ of V_u is linearly dependent.

6. Show that the following sets of vectors are linearly independent.

   (i) $S = \{(1,0,0,-1), (0,1,0,-1), (0,0,1,-1), (0,0,0,1)\}$ in V_4.

   (ii) $M = \left\{\begin{pmatrix} 1 & -2 \\ -1 & 4 \end{pmatrix}, \begin{pmatrix} -1 & 1 \\ 2 & -4 \end{pmatrix}\right\}$ in the set of 2×2 matrices over the field of real numbers.

   (iii) $S = \{1, x, x^2, \ldots, x^n\}$ in the set $P_n(x)$ of all polynomials of degree n.

7. Does the vector $(2,-1,1)$ is in the span of the set $s = \{(1,0,2), (-1,1,1)\}$?
Ans. No, because

$$(2,-1,1) = 1(1,0,2) - 1(-1,1,1).$$

8. Show that the set $S = \{(1,2,3), (3,1,0), (-2,1,3)\}$ does not form a basis for V_3.

9. Show that the subset $\{(0,0,1), (1,0,1), (1,-1,1), (3,0,1)\}$ is not a basis for V_3.

10. Show that the subset $\{1, x, (x-1)x, x(x-1)(x-2)\}$ from a basis for vector space of polynomials of degree 3.

11. Extend $S = \{(2,1,1), (1,3,0)\}$ to form a basis of R^3.

Ans. $\{(2,1,1), (1,3,0), (1,0,0)\}$.

12. Construct two subspaces M and N of V_4 such that $\dim(M) = 2$, $\dim(N) = 3$ and $\dim(M \cap N) = 1$.

Ans. $M = \{(1,0,0,0), (0,1,0,0)\}$
$N = \{(1,1,5,2), (1,2,3,0), (1,1,1,1)\}$

13. If u and v are distinct vectors in a vector space V, show that $\{u, v\}$ is linearly dependent if and only if u or v is a multiple of the other.

14. How many elements are there in the basis of vector space formed by the set of all matrices of order n.

Ans. n^2

15. Show that

$$e_1 = \begin{bmatrix} 1 \\ 0 \\ 0 \end{bmatrix} \text{ and } e_2 = \begin{bmatrix} 0 \\ 1 \\ 0 \end{bmatrix}$$

form a linearly independent set and describe its linear span geometrically.

Solution. We note that
$\alpha e_1 + \beta e_1 = 0$
implies

$$\alpha \begin{bmatrix} 1 \\ 0 \\ 0 \end{bmatrix} + \beta \begin{bmatrix} 0 \\ 1 \\ 0 \end{bmatrix} = \begin{bmatrix} 0 \\ 0 \\ 0 \end{bmatrix}$$

implies

$$\begin{bmatrix} \alpha \\ 0 \\ 0 \end{bmatrix} + \begin{bmatrix} 0 \\ \beta \\ 0 \end{bmatrix} = \begin{bmatrix} 0 \\ 0 \\ 0 \end{bmatrix}$$

implies

$$\begin{bmatrix} \alpha \\ \beta \\ 0 \end{bmatrix} = \begin{bmatrix} 0 \\ 0 \\ 0 \end{bmatrix}$$

Consequently, $\alpha = \beta = 0$. Hence $\{e_1 e_2\}$ form a linearly independent set.
Further, the linear span of $\{e_1 e_2\}$ is the set of all vectors of the form $\begin{bmatrix} \alpha \\ \beta \\ 0 \end{bmatrix}$ which is nothing but (x,y) plane and is a subset of the three dimensional Euclidean space.

16. Show that the vectors

$$v_1 = \begin{bmatrix} 1 \\ 1 \\ 1 \end{bmatrix}, \; v_2 = \begin{bmatrix} 0 \\ -1 \\ 2 \end{bmatrix}, \text{ and } v_3 = \begin{bmatrix} 2 \\ -1 \\ 8 \end{bmatrix}$$

are linearly dependent.

Solution. We note that the relation.

$$\alpha_1 \begin{bmatrix} 1 \\ 1 \\ 1 \end{bmatrix} + \alpha_2 \begin{bmatrix} 0 \\ -1 \\ 2 \end{bmatrix} + \alpha_3 \begin{bmatrix} 2 \\ -1 \\ 8 \end{bmatrix} = \begin{bmatrix} 0 \\ 0 \\ 0 \end{bmatrix}$$

is satisfied if we choose $\alpha_1 = 2$, $\alpha_2 = 3$ and $\alpha_3 = 1$. Thus $\alpha_1 v_1 + \alpha_2 v_2 + \alpha_3 v_3 = 0$ is satisfied by $\alpha_1, \alpha_2, \alpha_3$, where not all of these scalars are zero. Hence the set $\{v_1, v_2, v_3\}$ is linearly dependent.

17. Show that the mapping $T : V_3 \to V_1$ defined by $T(x_1, x_2, x_3) = x_1^2 + x_2^2 + x_3^2$ is not a linear transformation.

18. Show that the following maps are linear transformations:

    (i) $T: V_3 \to V_2$ defined by $T(a_1, a_2, a_3) = (a_1 - a_2, 2a_3)$.

    (ii) $T: M_{2\times 3}(F) \to M_{2\times 2}(F)$ defined by.

$$T\left(\begin{bmatrix} a_{11} & a_{12} & a_{13} \\ a_{21} & a_{22} & a_{23} \end{bmatrix}\right) = \begin{bmatrix} 2a_{11} - a_{12} & a_{13} + 2a_{12} \\ 0 & 0 \end{bmatrix},$$

where $M_{m\times n}(F)$ means the set of $m \times n$ matrices over the field F.

19. Let V be the vector space of continuous real valued functions over $\Re$. Show that the map $T: V \to \Re$ defined by

$$T(f) = \int_a^b f(t)dt, \; f \in V$$

is a linear transformation.

20. Find the matrix of the linear transformation $T: V_2 \to V_3$ defined by

$$T(a_1, a_2) = (2a_1 - a_2, 3a_2 + 4a_2, a_1)$$

Ans.

$$\begin{bmatrix} 2 & -1 \\ 3 & 4 \\ 1 & 0 \end{bmatrix}.$$

21. Let T: $V_2 \to V_3$ the defined by $T(x_1, x_2) = (x_1 - x_2, \; x_1, \; 2x_1 + x_2)$. Taking standard basis for V_2 and the basis (1, 1, 0), (0, 1, 1), (2, 2, 3) for V_3, find the matrix of the linear transformation T.

Ans.

$$\begin{bmatrix} -\frac{1}{3} & -1 \\ 0 & 1 \\ \frac{2}{3} & 0 \end{bmatrix}.$$

22. Find the matrix of anticlockwise rotation of the co-ordinates by an angle θ.

Hint : $T_\theta(a_1, a_2) = (a_1 \cos\theta - a_2 \sin\theta, a_1 \sin\theta + a_2 \cos\theta)$,
Basis $\beta = \{(1,0), (0,1)\}$.

Therefore

$$T(1,0) = (\cos\theta, \sin\theta) = \cos\theta(1,0) + \sin\theta(0,1),$$
$$T(0,1) = (-\sin\theta, \cos\theta) = -\sin\theta(1,0) + \cos\theta(0,1)$$

Hence

$$M(T_\theta) = \begin{bmatrix} \cos\theta & -\sin\theta \\ \sin\theta & \cos\theta \end{bmatrix}.$$

23. Find the matrix of the linear transformation $T: V_3 \to V_3$ defined by

$$T(a_1, a_2, a_3) = (a_1 - a_2 + a_3, 2a_1 + 3a_2 - \frac{1}{2}a_3, a_1 + a_2 - 2a_3).$$

Ans.

$$\begin{bmatrix} 2 & 6 & 0 \\ 0 & -\frac{3}{2} & -\frac{1}{4} \\ 1 & \frac{11}{2} & -\frac{3}{4} \end{bmatrix}$$

24. If $T: V_4 \times V_3$ is a linear map defined by

$$T(e_1) = (1, 1, 1), T(e_2) = (1, -1, 1),$$
$$T(e_3) = (1, 0, 0), T(e_4) = (1, 0, 1)$$

Show that $r(T) + n(T) = \dim(V_4)$.

25. Define a linear transformation T which is projection of the 3-axis along the xy-plane.

Ans. $T : \mathbb{R}^3 \to \mathbb{R}^3$ defined by $T(a_1, a_2, a_3) = (0, 0, a_3)$

26. Find (x,y), $\|x\|$, $\|y\|$, $\|x+y\|$ for the vectors $x = (2, 1+i, i)$ and $y = (2-i, 2, 1+2i)$ in $\not{C}^3$.

Ans. $(x,y) = 8 + 5i$
$\|x\| = \sqrt{7}$, $\|y\| = \sqrt{14}$,
$\|x+y\| = \sqrt{37}$.

27. Find the inner product of $v_1 = (1-i, 2+3i)$ and $v_2 = (2+i, 3-2i)$ in $\not{C}^2$.

Hint :

$$(v_1, v_2) = (1-i)(\overline{2+i}) + (2+3i)(\overline{3-2i})$$
$$= (1-i)(2-i) + (1+3i)(3+2i)$$
$$= (1-3i)(13i) = 39 + 13i$$

28. Show that $(A,B) = tr(A+B)$ on $M_{2\times 2}(\mathfrak{R})$ is not an inner product
Hint: $tr(A+B,C) \neq tr(A,C) + tr(B,C)$

29. Using Gram-Schmidt orthogonalization process normalize the following sets of linearly independent vectors:
   (i) $\{(1, 1, 0), (1, -1, 1), (-1, 1, 2)\}$ in V_3.
   (ii) $\{(1, 2, 1), (-1, 1, 0), (5, -1, 2)\}$ in V_3.
   (iii) $\{(1, 0, 1, 1), (-1, 0, -1, 1), (0, -1, 1, 1)\}$ in V_4.
   (iv) $\{(1, 0, 1, 0), (1, 1, 1, 1), (0, 1, 2, 1)\}$ in V_4.

Ans.
   (i) $\frac{1}{\sqrt{2}}(1,1,0)$, $\frac{1}{\sqrt{3}}(1,-1,1)$, $\frac{1}{\sqrt{6}}(-1,1,2)$
   (ii) $\frac{1}{\sqrt{6}}(1,2,1)$, $\frac{1}{\sqrt{66}}(-7,4,1)$, $\frac{1}{\sqrt{11}}(-1,1,3)$
   (iii) $\frac{1}{\sqrt{3}}(1,0,1,1)$, $\frac{1}{\sqrt{6}}(-1,0,-1,2)$, $\frac{1}{\sqrt{3}}(-1,-2,-1,0)$
   (iv) $\frac{1}{\sqrt{2}}(1,0,1,0)$, $\frac{1}{\sqrt{2}}(0,1,0,1)$, $\frac{1}{\sqrt{2}}(-1,0,1,0)$

30. Show that the system

$$s = \{e^{int} : 0 \le t \le 2\pi,\ n \text{ an integer}\}$$

is an orthonormal system with respect to the inner product defined by

$$(e^{m}, e^{int}) = \frac{1}{2\pi}\int_0^{2\pi} e^{m}\overline{e^{int}}\,dt$$

$$\textbf{Hint}: (e^{imt}, e^{int}) = \frac{1}{2\pi} e^{imt}..\overline{e^{int}}\,dt$$

$$= \frac{1}{2\pi}\int_0^{2\pi} e^{imt}.e^{-int}\,dt$$

$$= \frac{1}{2\pi}\int_0^{2\pi} e^{i(m-n)t}\,dt$$

$$= 0$$

and

$$(e^{int}, e^{int}) = \frac{1}{2\pi}\int_0^{2\pi} e^{int}.\overline{e^{int}}\,dt = \frac{1}{2\pi}\int_0^{2\pi} e^{i(n-n)t}\,dt$$

$$= \frac{1}{2\pi}\int_0^{2\pi} e^{0}\,dt$$

$$= \frac{1}{2\pi}\int_0^{2\pi} dt = 1.$$

31. Show that the sum of two inner products defined on a vector space is another inner product on that vector space.

32. Find the orthogonal complement of the set $\{(1,0,i),\ (1,2,1)\}$ in $\not{C}^3$.
Ans. $span\ (\{(i, -\frac{1}{2}(1+i), 1)\})$

33. If $w = span(\{e_1, e_2\})$ in V_3, find $w^{\perp}$.
Ans. $w^{\perp} = span(\{e_3\})$.

34. Fit a straight line $y = a + bx$ to the data: $(1,1),\ (2,3),\ (3,4),\ (4,6),\ (5,5)$
Ans. $y = 1.1 + 0.5x$

35. Fit a straight line $y = a + bx$ to the data $(2,2),\ (5,4),\ (6,6),\ (9,9),\ (11,10)$
Ans. $y = 0.9537x - 0.0944$.

36. Fit a least square line to the data $(1,2),\ (2,3),\ (3,5)$ and $(4,7)$.
Ans. $y = 1.7x$

37. Find the minimal solution for the following system:

$$x + 2y + z = 4$$
$$x - y + 2z = -11$$
$$x + 5y = 19.$$

Hint: Let A be the coefficient matrix. Then, the solution of $(AA^{*})U = B$, where $B = \begin{bmatrix} 4 \\ -11 \\ 19 \end{bmatrix}$ is $U = \begin{bmatrix} 1 \\ -2 \\ 0 \end{bmatrix}$. Then $X = A^{*}U = \begin{bmatrix} -1 \\ 4 \\ -3 \end{bmatrix}$ is the minimal solution.

38. Show that the subset $\{x^2 - 1, x + 1, x - 1\}$ of the vector space of polynomials is linearly independent.

39. Show that the subset $\{(1,\ 1,\ 1,\ 0),\ (3,\ 2,\ 2,\ 1), (1,\ 1,\ 3,\ -2), (1,\ 2,\ 6,\ -5),\ (1,\ 1,\ 2,\ 1)\}$ of V_4 is linearly dependent.

40. Show that the subset $\{(0,\ 0,\ 1),\ (1,\ 0,\ 1), (1,\ -1,\ 1), (3,\ 0,\ 1)\}$ is not a basis for V_3.

41. Use the principle of mathematical induction to show that if $A = \begin{bmatrix} 1 & 1 \\ 0 & 1 \end{bmatrix}$, then $A^n = \begin{bmatrix} 1 & n \\ 0 & 1 \end{bmatrix}$ for every positive integer n.

42. Express the matrix

$$A = \begin{bmatrix} 1 & 3 & 5 \\ 2 & -1 & 3 \\ 4 & 6 & 5 \end{bmatrix}$$

as the sum of a symmetric matrix and a Skew symmetric matrix.

Ans. $\begin{bmatrix} 1 & \frac{5}{2} & \frac{9}{2} \\ \frac{5}{2} & -1 & \frac{9}{2} \\ \frac{9}{2} & \frac{9}{2} & 5 \end{bmatrix} + \begin{bmatrix} 1 & \frac{1}{2} & \frac{1}{2} \\ -\frac{1}{2} & 0 & -\frac{3}{2} \\ -\frac{1}{2} & \frac{3}{2} & 0 \end{bmatrix}$

43. Show that

$$\begin{vmatrix} a & b & c \\ b+c & c+a & a+b \\ a^2 & b^2 & c^2 \end{vmatrix}$$

$$= -(a-b)(b-c)(c-a)(a+b+c)$$

44. If $a+b+c=0$, solve the equation

$$\begin{vmatrix} a-x & c & b \\ c & b-x & a \\ b & a & c-x \end{vmatrix} = 0$$

Ans. $0, \pm\sqrt{(a^2+b^2+c^2-ab-bc-ca)}$

45. If $\Delta = \begin{vmatrix} x & x^2 & 1+x^3 \\ y & y^2 & 1+y^3 \\ z & z^2 & 1+z^3 \end{vmatrix} = 0,$

show without expansion that $xyz = -1$

Hint:

$$\Delta = \begin{vmatrix} x & x^2 & 1 \\ y & y^2 & 1 \\ z & z^2 & 1 \end{vmatrix} + \begin{vmatrix} x & x^2 & x^3 \\ y & y^2 & y^3 \\ z & z^2 & z^3 \end{vmatrix} = 0$$

$$= \begin{vmatrix} x & x^2 & 1 \\ y & y^2 & 1 \\ z & z^2 & 1 \end{vmatrix} + xyz \begin{vmatrix} 1 & x & x^2 \\ 1 & y & y^2 \\ 1 & z & z^2 \end{vmatrix} = 0$$

$$= \begin{vmatrix} x & x^2 & 1 \\ y & y^2 & 1 \\ z & z^2 & 1 \end{vmatrix} + xyz \begin{vmatrix} x & x^2 & 1 \\ y & y^2 & 1 \\ z & z^2 & 1 \end{vmatrix} = 0$$

$$= \begin{vmatrix} x & x^2 & 1 \\ y & y^2 & 1 \\ z & z^2 & 1 \end{vmatrix} [1+xyz] = 0$$

Hence $1+xyz=0$ or $xyz=-1$

46. Using Laplace expansion, expand the determinant $\Delta = \begin{vmatrix} 0 & 1 & 2 & 3 \\ 1 & 0 & 3 & 0 \\ 2 & 3 & 0 & 1 \\ 3 & 0 & 1 & 2 \end{vmatrix}$ about first and third row:

Ans. 88

47. Find the adjoint of the matrix

$$A = \begin{bmatrix} 1 & 1 & 1 \\ 1 & 2 & -3 \\ 2 & -1 & 3 \end{bmatrix}$$

and verify the result $A(\text{adj } A) = (\text{adj } A)A = |A|I_n$.

Ans. $\text{adj } A = \begin{bmatrix} 3 & -4 & -5 \\ -9 & 1 & 4 \\ -5 & 3 & 1 \end{bmatrix}$

48. Find the inverse of the following matrices:

(i) $A = \begin{bmatrix} 0 & 1 & 2 \\ 1 & 2 & 3 \\ 3 & 1 & 1 \end{bmatrix}$

(ii) $B = \begin{bmatrix} \cos\alpha & -\sin\alpha & 0 \\ \sin\alpha & \cos\alpha & 0 \\ 0 & 0 & 1 \end{bmatrix}$

(iii) $C = \begin{bmatrix} 1 & 2 & 2 \\ 2 & 1 & -2 \\ -2 & 2 & -1 \end{bmatrix}$

(iv) $D = \begin{bmatrix} 1 & 3 & 3 \\ 1 & 4 & 3 \\ 1 & 3 & 4 \end{bmatrix}$

(v) $E = \begin{bmatrix} 1 & 1 & 3 \\ 1 & 3 & -3 \\ -2 & -4 & -4 \end{bmatrix}.$

Ans. (i) $\begin{bmatrix} \frac{1}{2} & -\frac{1}{2} & \frac{1}{2} \\ -4 & 3 & -1 \\ \frac{5}{2} & -\frac{3}{2} & \frac{1}{2} \end{bmatrix}$,

(ii) $\begin{bmatrix} \cos\alpha & \sin\alpha & 0 \\ -\sin\alpha & \cos\alpha & 0 \\ 0 & 0 & 1 \end{bmatrix}$,

(iii) $\begin{bmatrix} \frac{1}{3} & \frac{2}{3} & -\frac{2}{3} \\ \frac{2}{3} & \frac{1}{3} & \frac{2}{3} \\ \frac{2}{3} & -\frac{2}{3} & -\frac{1}{3} \end{bmatrix}$, (iv) $\begin{bmatrix} 7 & -3 & -3 \\ -1 & 1 & 0 \\ -1 & 0 & 1 \end{bmatrix}$,

(v) $-\frac{1}{8}\begin{bmatrix} -24 & -8 & -12 \\ 10 & 2 & 6 \\ 2 & 2 & 2 \end{bmatrix}$

49. Using Gauss Jordan method, find the inverse of the following matrices:

(i) $A = \begin{bmatrix} 3 & -3 & 4 \\ 2 & -3 & 4 \\ 0 & -1 & 1 \end{bmatrix}$

(ii) $B = \begin{bmatrix} 2 & -1 & 3 \\ 1 & 1 & 1 \\ 1 & -1 & 1 \end{bmatrix}$

(iii) $C = \begin{bmatrix} 3 & -2 & -1 \\ -4 & 1 & -1 \\ 2 & 0 & 1 \end{bmatrix}$

(iv) $D = \begin{bmatrix} 14 & 3 & -2 \\ 6 & 8 & -1 \\ 0 & 2 & -7 \end{bmatrix}$

Ans. (i) $\begin{bmatrix} 1 & -1 & 0 \\ -2 & 3 & -4 \\ -2 & 3 & -3 \end{bmatrix}$

(ii) $\begin{bmatrix} -1 & 1 & 2 \\ 0 & \frac{1}{2} & -\frac{1}{2} \\ 1 & -\frac{1}{2} & -\frac{3}{2} \end{bmatrix}$

(iii) $\begin{bmatrix} 1 & 2 & 3 \\ 2 & 5 & 7 \\ -2 & -4 & -5 \end{bmatrix}$

(iv) $-\frac{1}{654}\begin{bmatrix} -54 & 17 & 13 \\ 42 & -98 & 2 \\ 12 & -28 & 94 \end{bmatrix}$

50. Find the rank of the following matrices.

(i) $\begin{bmatrix} 0 & 1 & -3 & -1 \\ 0 & 0 & 1 & 1 \\ 3 & 1 & 0 & 2 \\ 1 & 1 & -2 & 0 \end{bmatrix}$ (ii) $\begin{bmatrix} 2 & 3 & -1 & -1 \\ 1 & -1 & -2 & -4 \\ 3 & 1 & 3 & -2 \\ 6 & 3 & 0 & -7 \end{bmatrix}$

(iii) $\begin{bmatrix} 2 & -1 & 3 & 4 \\ 0 & 3 & 4 & 1 \\ 2 & 3 & 7 & 5 \\ 2 & 5 & 11 & 6 \end{bmatrix}$ (iv) $\begin{bmatrix} 1 & 2 & 1 & 2 \\ 1 & 3 & 2 & 2 \\ 2 & 4 & 3 & 4 \\ 3 & 7 & 4 & 6 \end{bmatrix}$

(v) $\begin{bmatrix} 1 & 3 & 4 & 3 \\ 3 & 9 & 12 & 9 \\ 1 & 3 & 4 & 1 \end{bmatrix}$

Ans. (i) 3, (ii) 3, (iii) 3, (iv) 3, (v) 2

51. Show that no Skew-symmetric matrix can be of rank 1.

Hint: Diagonal elements are all zeros. If all non-diagonal positive elements are zero, then the corresponding negative elements are also zero and so rank shall be zero, If at least one of the elements is non-zero, then at least one 2-rowed minor is not equal to zero. Hence, rank is greater than or equal to 2.

52. Reduce the following matrices to normal form and, hence, find their ranks.

(i) $\begin{bmatrix} 5 & 3 & 14 & 4 \\ 0 & 1 & 2 & 1 \\ 1 & -1 & 2 & 0 \end{bmatrix}$

(ii) $\begin{bmatrix} 1 & 2 & 1 & 0 \\ -2 & 4 & 3 & 0 \\ 1 & 0 & 2 & -8 \end{bmatrix}$

(iii) $\begin{bmatrix} 1 & 1 & 2 \\ 1 & 2 & 3 \\ 0 & -1 & -1 \end{bmatrix}$

(iv) $\begin{bmatrix} 2 & -2 & 0 & 6 \\ 4 & 2 & 0 & 2 \\ 1 & -1 & 0 & 3 \\ 1 & -2 & 1 & 2 \end{bmatrix}$

Ans. (i) $\begin{bmatrix} 1 & 0 & 0 & 0 \\ 0 & 1 & 0 & 0 \\ 0 & 0 & 1 & 0 \end{bmatrix}$, Rank 3

(ii) $\begin{bmatrix} 1 & 0 & 0 & 0 \\ 0 & 1 & 0 & 0 \\ 0 & 0 & 1 & 0 \end{bmatrix}$, Rank 3

(iii) $\begin{bmatrix} 1 & 0 & 0 \\ 0 & 1 & 0 \\ 0 & 0 & 0 \end{bmatrix}$, Rank 2;

(iv) $\begin{bmatrix} I_3 & 0 \\ 0 & 0 \end{bmatrix}$, Rank 3

53. Find the inverse of the matrix

$$A = \begin{bmatrix} 2 & -1 & 3 \\ 1 & 1 & 1 \\ 1 & -1 & 1 \end{bmatrix}$$

using elementary operations

Ans. $\begin{bmatrix} -1 & 1 & 2 \\ 0 & \frac{1}{2} & -\frac{1}{2} \\ 1 & -\frac{1}{2} & -\frac{3}{2} \end{bmatrix}$

54. Using elementary transformation, find the inverse of the matrix

$$A = \begin{bmatrix} -1 & -3 & 3 & -1 \\ 1 & 1 & -1 & 0 \\ 2 & -5 & 2 & -3 \\ -1 & 1 & 0 & 1 \end{bmatrix}$$

Ans. $\begin{bmatrix} 0 & 2 & 1 & 3 \\ 1 & 1 & -1 & -2 \\ 1 & 2 & 0 & 1 \\ -1 & 1 & 2 & 6 \end{bmatrix}$

55. Test for consistency and solve the following system of equations.

(i) $x+y+z=6$

$x+2y+3z=14$

$x+4y+7z=30.$

Ans. Consistent; rank 2; $x=c-2, y=8-2c, z=c$ for arbitrary constant c

(ii) $2x+6y+11=0$

$6x+20y+6z+3=0$

$6y-18z+1=0$

Ans. Not consistent

(iii) $2x-y+3z=8$

$-x+2y+z=4$

$3x+y-4z=0$

Ans. Consistent, $x=2, y=2, z=2$

56. Find the values of λ for which the following system of linear equations will have no unique solution

$$3x-y+\lambda z=1$$
$$2x+y+z=2$$
$$x+2y-\lambda z=1$$

Will this system have any solution for these values of λ?

Ans. $\lambda = -\frac{7}{2}$ and the equations are inconsistent for this value. Hence no solution exists for $\lambda = -\frac{7}{2}$.

57. Discuss the existence and nature of solutions for all values of λ for the following system of equations.

$$x+y+4z=6$$
$$x+2y-2z=6$$
$$\lambda x+y+z=6$$

Ans. Unique Solution for $\lambda \neq \frac{7}{10}$. For $\lambda = \frac{7}{10}$, the equations are not consistent.

58. Solve the following equations using matrix

method

$$2x-y+3z=9$$
$$x+y+z=6$$
$$x-y+z=2$$

Ans. Coefficient matrix is non-singular. The unique solution is $x=1, y=2, z=3$

59. Determine the values of a and b for which the equations

$$x+2y+3z=4$$
$$x+3y+4z=5$$
$$x+3y+az=b$$

have (i) no solution, (ii) a unique solution, and (iii) an infinite number of solution.

Ans. (i) $a=4, b\neq 5$ (ii) $a\neq 4$ (iii) $a=4, b=5$

60. Solve completely the system of equations

$$x+y+z=0$$
$$2x-y-3z=0$$
$$3x-5y+4z=0$$
$$x+17y+4z=0$$

Ans. Trivial solution $x=y=z=0$.

61. Solve

$$4x+2y+z+3u=0$$
$$6x+3y+4z+7u=0$$
$$2x+y+u=0$$

Ans. $x=c_1,\ u=c_2, y=-2c_1-c_2, z=-c_2$

62. Find the eigenvalues of the matrix

$$\begin{bmatrix} a & h & g \\ 0 & b & 0 \\ 0 & 0 & c \end{bmatrix}$$ **Ans.** a, b, c

63. Find the eigenvalues and the corresponding eigenvectors for the given matrix.

$$\begin{bmatrix} 8 & -6 & 2 \\ -6 & 7 & -4 \\ 2 & -4 & 3 \end{bmatrix}$$

Ans. $0; 3; 15; c_1=\begin{bmatrix} \frac{1}{2} \\ 1 \\ 1 \end{bmatrix}, c_2=\begin{bmatrix} -4 \\ -2 \\ 4 \end{bmatrix},$

$$c_3=\begin{bmatrix} 2 \\ -2 \\ 1 \end{bmatrix}$$

64. If the characteristic roots of a matrix A are $\lambda_1, \lambda_2, \ldots, \lambda_n$, show that the characteristic roots of A^2 are $\lambda_1^2\ \lambda_2^2, \ldots, \lambda_n^2$.

Hint: $AX=\lambda X \Rightarrow A(AX)=\lambda(AX) \Rightarrow A^2X=\lambda(\lambda X)=\lambda^2 X$

65. Show that the matrices A and $B^{-1}AB$ have the same characteristic roots

Hint : $|B^{-1}AB-\lambda I| = |B^{-1}AB - B^{-1}\lambda IB|$
$= |B^{-1}(A-\lambda I)B|$
$= |B^{-1}||A-\lambda||B|$
$= |A-\lambda I||B^{-1}B|$
$= |A-\lambda I|$.

66. Verify Caley-Hamilton theorem for the matrix

$$A=\begin{bmatrix} 1 & 0 & 2 \\ 0 & 2 & 1 \\ 2 & 0 & 3 \end{bmatrix}$$

and, hence, find its inverse.

Ans. A satisfies $A^3-6A^2+7A+2I=0$

$$A^{-1}=\frac{1}{2}(A^2-6A+7I)=\begin{bmatrix} -3 & 0 & 2 \\ -1 & \frac{1}{2} & \frac{1}{2} \\ 2 & 0 & -1 \end{bmatrix}.$$

67. Find the minimal polynomial of the matrix.

$$\begin{bmatrix} 5 & -6 & -6 \\ -1 & 4 & 2 \\ 3 & -6 & -4 \end{bmatrix}$$

Ans. $x^2 - 3x + 2$.

68. Show that the matrix

$$\begin{bmatrix} a+ic & -b+id \\ b+id & -a-ic \end{bmatrix}$$

is unitary if and only if $a^2 + b^2 + c^2 + d^2 = 1$.

69. Show that the matrix

$$A = \begin{bmatrix} \frac{1}{3} & \frac{2}{3} & \frac{2}{3} \\ \frac{2}{3} & \frac{1}{3} & -\frac{2}{3} \\ \frac{2}{3} & -\frac{2}{3} & -\frac{1}{3} \end{bmatrix}$$

is orthogonal. **Hint:** Show that $A^T A = I$.

70. Show that the matrix

$$\begin{bmatrix} -9 & 4 & 4 \\ -8 & 3 & 4 \\ -16 & 8 & 7 \end{bmatrix}$$

is diagonalizable. Obtain the diagonalizing matrix *P*.

Ans. $P = \begin{bmatrix} 1 & 0 & 1 \\ 1 & 1 & 1 \\ 1 & -1 & 2 \end{bmatrix}$, diag$[-1, -1, 3]$

71. Diagonalize the matrix

$$\begin{bmatrix} 1 & -6 & -4 \\ 0 & 4 & 2 \\ 0 & -6 & -3 \end{bmatrix}$$

Ans. $\begin{bmatrix} 1 & 0 & 0 \\ 0 & 1 & 0 \\ 0 & 0 & 1 \end{bmatrix}$

72. Diagonalize the real-symmetric matrix

$$A = \begin{bmatrix} 3 & -1 & 1 \\ -1 & 5 & -1 \\ 1 & -1 & 3 \end{bmatrix}$$

Ans. $P = \begin{bmatrix} \frac{1}{\sqrt{2}} & \frac{1}{\sqrt{3}} & \frac{1}{\sqrt{6}} \\ 0 & \frac{1}{\sqrt{3}} & -\frac{2}{\sqrt{6}} \\ -\frac{1}{\sqrt{2}} & \frac{1}{\sqrt{3}} & \frac{1}{\sqrt{6}} \end{bmatrix}$, diag[2 3 6]

73. Find a non-singular matrix P such that $P^T AP$ is a diagonal matrix, where

$$A = \begin{bmatrix} 0 & 1 & 2 \\ 1 & 0 & 3 \\ 2 & 3 & 0 \end{bmatrix}$$

Ans. $\begin{bmatrix} 0 & -\frac{1}{2} & -3 \\ 1 & \frac{1}{2} & -2 \\ 0 & 0 & 1 \end{bmatrix}$

74. Reduce the quadratic form $x^2 + 4y^2 + 9z^2 + t^2 - 12yz + 6zx = 4xy - 2xt - 6zt$ to canonical form and find its rank and signature
Ans. $y_1^2 - y_2^2 + y_4^2$, Rank:3, Signature:1

75. Reduce the quadratic form $6x_1^2 + 3x_2^2 + 14x_3^2 + 4x_2x_3 + 18x_3x_1 + 4x_1x_2$ to canonical form and find its rank and signature.
Ans. $y_1^2 + y_2^2 + y_3^2$, Rank:3, Signature:3

3 Functions of Complex Variables

In this chapter, we deal with functions of complex variables which are useful in evaluating a large number of new definite integrals, the theory of differential equations, the study of electric fields, thermodynamics, and fluid mechanics.

3.1 BASIC CONCEPTS

Definition 3.1. A *complex number* z is an ordered pair (x, y) of real numbers x and y. If $z = (x, y)$ and $w = (u, v)$ are two complex numbers, then their addition and multiplication are defined as

$$z + w = (x, y) + (u, v) = (x + u, y + v)$$

$$zw = (x, y)\,(u, v) = (xu - yv, xv + yu).$$

With these operations of addition and multiplication, the complex numbers satisfy the same arithmetic properties as do the real numbers.

If we write the real number x as $(x, 0)$ and denote $i = (0, 1)$ (called imaginary number), then

$$z = (x, y) = (x, 0) + (0, y)$$

$$= (x, 0) + (y, 0)\,(0, 1)$$

$$= x + iy.$$

Thus, a complex number z can be expressed as $z = x + iy$, where x is called the real part of z and y is called the imaginary part of z. Thus $\text{Re}(z) = x$, $\text{Im}(z) = y$. Further,

$$i^2 = (0, 1)\,(0, 1) = (-1, 0) = -1$$

and so $i = \sqrt{-1}$.

The set of complex numbers is denoted by $\mathbb{C}$. Since a real number x can be written as $x = (x, 0) = x + i0$, the set $\mathbb{C}$ is an extension of $\mathbb{R}$. Further, since the complex number $z = x + iy$ is an ordered pair (x, y), we can represent such numbers by points in xy plane, called the *complex plane* or *Argand diagram* (Figure 3.1).

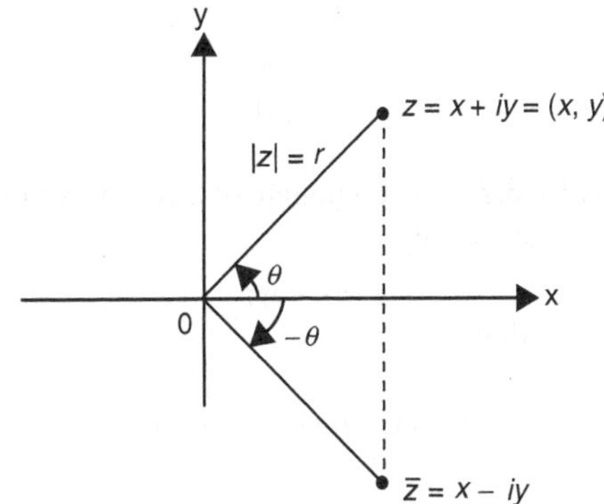

Figure 3.1 Argand diagram

The *modulus* (or *absolute value*) of z is

$$|z| = r = \sqrt{x^2 + y^2}$$

and

$$|zw| = |z|\,|w|.$$

Further,

$$|z + w| \le |z| + |w| \text{ (triangle inequality).}$$

Since $x = r\cos\theta$, $y = r\sin\theta$, we have

$$z = x + iy = r\cos\theta + i\,r\sin\theta$$

$$= r(\cos\theta + i\sin\theta),$$

which is called as the *polar form* of the complex number z. The angle θ is called the *amplitude* or *argument* of the complex number z and we have

$$\tan\theta = \frac{y}{x}.$$

Let $z = r(\cos\theta + i\sin\theta)$ and $w = R(\cos\phi + i\sin\phi)$ be two complex numbers. Then

$$\begin{aligned} zw &= rR[(\cos\theta\cos\phi - \sin\theta\sin\phi) \\ &\quad + i(\sin\theta\cos\phi + \cos\theta\sin\phi)] \\ &= rR[\cos(\theta+\phi) + i\sin(\theta+\phi)]. \end{aligned}$$

Hence, the *arguments are additive under multiplication*, that is,

$$\arg(zw) = \arg z + \arg w.$$

Similarly, we can show that

$$\arg\left(\frac{z}{w}\right) = \arg z - \arg w$$

and

$$\left|\frac{z}{w}\right| = \frac{|z|}{|w|}.$$

Definition 3.2. The conjugate of a complex number z is defined by

$$\bar{z} = x - iy.$$

We note that

$$z\bar{z} = (x+iy)(x-iy) = x^2 + y^2 = |z|^2,$$

$$\overline{z+w} = \bar{z} + \bar{w},$$

$$\overline{zw} = \bar{z}\,\bar{w}.$$

Consider complex numbers z with $|z| = 1$. All these numbers have distance 1 to the origin (0, 0) and so they form a circle with radius 1 and centre at the origin. This circle is called the *unit circle*.

Definition 3.3. For each $y \in \mathbb{R}$, the complex number e^{iy} is defined as

$$e^{iy} = \cos y + i\sin y,$$

which gives

$$e^{i\theta} = \cos\theta + i\sin\theta, 0 \le \theta < 2\pi,$$

known as Euler's formula. We note that

$$|e^{i\theta}| = \cos^2\theta + \sin^2\theta = 1,$$

$$\arg(e^{i\theta}) = \theta,\ e^{i\pi} = 1,\ \overline{e^{i\theta}} = e^{-i\theta},$$

$$e^{2\pi ik} = 1,\ k \in \mathbb{Z},\ e^{i(\theta+2k\pi)} = e^{i\theta},\ k \in \mathbb{Z}.$$

Since $e^{-i\theta} = \cos\theta - i\sin\theta$, we have

$$\cos\theta = \frac{e^{i\theta} + e^{-i\theta}}{2},\ \sin\theta = \frac{e^{i\theta} - e^{-i\theta}}{2i}.$$

For $z = x + iy$, we define e^z by

$$e^z = e^{x+iy} = e^x e^{iy} = e^x(\cos y + i\sin y)$$

and so

$$\mathrm{Re}(e^z) = e^x\cos y, \quad \mathrm{Im}(e^z) = e^x\sin y,$$

$$\arg(e^z) = \mathrm{Im}\,z, \quad |e^z| = e^x = e^{\mathrm{Re}(z)}.$$

Definition 3.4. The complex number $z = r(\cos\theta + i\sin\theta)$, with $r = |z|$ can be written as $z = r\,e^{i\theta} = |z|\,e^{i\theta}$, which is called exponential form of the complex number z.

For any non-zero complex number z, we define

$$z^0 = 1\ z^{n+1} = z^n \cdot z \text{ for } n \ge 0$$

and

$$z^{-n} = (z^{-1})^n \text{ if } z \ne 0, n > 0.$$

3.2 DE-MOIVRE'S THEOREM

Theorem 3.1. For any complex number $z = re^{i\theta}$ and $n = 0, \pm1, \pm2, \ldots$, we have

$$z^n = r^n e^{in\theta}.$$

Proof: For $n = 0$, the result is trivial since $z^0 = 1$. For $n = 1, 2, \ldots$ it can be proved by mathematical induction. For $n = -1, -2, \ldots$ let $n = -m$, where $m = 1, 2, \ldots$

Then

$$z^n = z^{-m} = (z^{-1})^m = \left(\frac{1}{r}e^{-i\theta}\right)^m = \left(\frac{1}{r}\right)^m e^{-im\theta}$$

$$= r^{-m}e^{i(-m\theta)} = r^n e^{in\theta}.$$

Substituting $r = 1$, we have

$$z^n = (e^{i\theta})^n = e^{in\theta}$$

$$\Rightarrow (\cos\theta + i\sin\theta)^n = \cos n\theta + i\sin n\theta,\ n \in I,$$

which is known as *De-Moivre's theorem for integral index.*

The De-Moivre's theorem also holds for rational index. To show it let $n = \frac{p}{q}$ be a rational number. Then,

$$\left(\cos\frac{\theta}{q} + i\sin\frac{\theta}{q}\right)^q = \cos q\frac{\theta}{q} + i\sin q\cdot\frac{\theta}{q}$$

$$= \cos\theta + i\sin\theta$$

and so taking qth root of both sides, we note that $\cos\frac{\theta}{q} + i\sin\frac{\theta}{q}$ is one of the values of $(\cos\theta + i\sin\theta)^{\frac{1}{q}}$. Therefore, $\left(\cos\frac{\theta}{q} + i\sin\frac{\theta}{q}\right)^p$ is one of the values of $(\cos\theta + i\sin\theta)^{\frac{p}{q}}$, that is, $\cos\frac{p}{q}\theta + i\sin\frac{p}{q}\theta$ is one of the values of $(\cos\theta + i\sin\theta)^{\frac{p}{q}}$. Hence, $\cos n\theta + i\sin n\theta$ is one of the values of $(\cos\theta + i\sin\theta)^n$. This proves De-Moivre's theorem for rational index $\frac{p}{q}$. However, in general, the restriction $-\pi < \theta \le \pi$ is necessary. For example, if $\theta = -\pi$, $n = \frac{1}{2}$, then the result is not valid.

If n is a positive integer, then De-Moivre's formula

$$(\cos\theta + i\sin\theta)^n = (\cos n\theta + i\sin n\theta)$$

implies

$$\cos n\theta = \sum_{k=0,\,even}^{n} (-1)^{\frac{k}{2}} \binom{n}{k} \cos^{n-k}\theta \sin^k\theta$$

and

$$\sin n\theta = \sum_{k=1,\,odd}^{n} (-1)^{\frac{k-1}{2}} \binom{n}{k} \cos^{n-k}\theta \sin^k\theta.$$

Thus, expansion of $\cos n\theta$ and $\sin n\theta$ can be obtained using the above formulas. For example,

$$\cos 7\theta = \sum_{k=0,\,even}^{7} (-1)^{\frac{k}{2}} \binom{7}{k} \cos^{7-k}\theta \sin^k\theta$$

$$= \cos^7\theta - \binom{7}{2}\cos^5\theta\sin^2\theta$$

$$+ \binom{7}{4}\cos^3\theta\sin^4\theta - \binom{7}{6}\cos\theta\sin^6\theta$$

$$= \cos^7\theta - 21\cos^5\theta\sin^2\theta$$

$$+ 35\cos^3\theta\sin^4\theta - 7\cos\theta\sin^6\theta$$

$$= \cos^7\theta - 21\cos^5\theta(1-\cos^2\theta) + 35\cos^3\theta$$

$$\times(1-\cos^2\theta)^2 - 7\cos\theta(1-\cos^2\theta)^3$$

$$= \cos^7\theta - 21\cos^5\theta + 21\cos^7\theta$$

$$+ 35\cos^3\theta(1 - 2\cos^2\theta + \cos^4\theta)$$

$$- 7\cos\theta(1 - 3\cos^2\theta + 3\cos^4\theta - \cos^6\theta$$

$$= 64\cos^7\theta - 112\cos^5\theta + 56\cos^2\theta - 7\cos\theta.$$

Similarly,

$$\sin 7\theta = \sum_{k=1,\,odd}^{7} (-1)^{\frac{k-1}{2}} \binom{7}{k}\cos^{7-k}\theta\sin^k\theta$$

$$= 7\cos^6\theta\sin\theta - 35\cos^4\theta\sin^3\theta$$

$$+ 21\cos^2\theta\sin^5\theta\ \sin^7\theta$$

$$= 7\sin\theta - 56\sin^3\theta + 112\sin^5\theta$$

$$- 64\sin^7\theta.$$

Substituting $z = e^{i\theta}$, we have

$$z^n = \cos n\theta + i\sin n\theta \text{ and}$$

$$z^{-n} = \cos\theta - i\sin n\theta.$$

Therefore,

$$z^n + \frac{1}{z^n} = 2\cos n\theta \text{ and } z^n - \frac{1}{z^n} = 2i\sin n\theta.$$

Thus,

$$z + \frac{1}{z} = 2\cos\theta \text{ and } z - \frac{1}{z} = 2i\sin\theta.$$

These expressions are useful in finding the expansion of $\cos^n \theta$ and $\sin^n \theta$. For example,

$$\left(z+\frac{1}{z}\right)^7 = (z^7+z^{-7}) + 7(z^5+z^{-5})$$

$$+ 21(z^3+z^{-3}) + 35(z+z^{-1})$$

or

$$(2\cos\theta)^7 = 2\cos 7\theta + 14\cos 5\theta + 42\cos 3\theta$$

$$+ 70\cos\theta$$

or

$$\cos^7\theta = 2^{-7}[2\cos 7\theta + 14\cos 5\theta$$

$$+ 42\cos 3\theta + 70\cos\theta]$$

$$= 2^{-6}[\cos 7\theta + 7\cos 5\theta + 21\cos 3\theta$$

$$+ 35\cos\theta].$$

Similarly, the expansion of $\left(z-\frac{1}{z}\right)^n$ gives $\sin^n\theta$. For example,

$$\left(z-\frac{1}{z}\right)^5 = \left(z^5-\frac{1}{z^5}\right) - 5\left(z^3-\frac{1}{z^3}\right)$$

$$+10\left(z-\frac{1}{z}\right)$$

or

$$(2i\sin\theta)^5 = 2i\sin 5\theta - 10i\sin 3\theta + 20i\sin\theta$$

or

$$\sin^5\theta = 2^{-4}[\sin 5\theta - 5\sin 3\theta + 10\sin\theta].$$

The following theorem is helpful to determine the nth root of a non-zero complex number.

Theorem 3.2. For $z_0 \neq 0$, there exist n values of z satisfying the equation $z^n = z_0$.

Proof: We have $z^n = z_0$, that is, $z = z_0^{\frac{1}{n}}$. Let $z_0 = r_0 e^{i\theta_0}$, $-\pi < \theta_0 \leq \pi$, and $z = re^{i\theta}$. Then

$$z^n = z_0 \Rightarrow (re^{i\theta})^n = r_0 e^{i\theta_0}$$

$$\Rightarrow r^n e^{in\theta} = r_0 e^{i\theta_0}$$

$$\Rightarrow r^n = r_0 \text{ and } n\theta = \theta_0 + 2k\pi,\ k \in I.$$

Therefore,

$$r = (r_0)^{\frac{1}{n}},\ \theta = \frac{\theta_0}{n} + \frac{2k\pi}{n},$$

where $(r_0)^{\frac{1}{n}}$ denotes the positive nth root of r_0. Hence, all values z given by

$$z = (r_0)^{\frac{1}{n}} e^{i\left(\frac{\theta_0+2k\pi}{n}\right)},\ k = 0, 1, 2, \ldots, n-1$$

satisfy the given equation $z^n = z_0$. These n values of z are called nth roots of z_0. The root corresponding to $k = 0$, that is, $c = (r_0)^{\frac{1}{n}} e^{i\frac{\theta_0}{n}}$ is called the *principal root*. In terms of the principal root, the nth roots of z_0 are

$$c,\ cw_n,\ cw_n^2, \ldots,\ cw_n^{n-1} \text{ where } w_n = e^{\frac{i2\pi}{n}}.$$

For the values of k other than 0, 1,..., $n-1$, the roots start repeating.

For example, to find the fifth roots of unity, we put $z_0 = 1$ so that $z_0 = 1(\cos 0 + i\sin 0)$. Thus, $c = (r_0)^{\frac{1}{5}} e^0 = 1$ and $w_n = e^{\frac{i2\pi}{5}}$. Hence, the fifth roots are

$$1,\ e^{\frac{i2\pi}{5}},\ e^{\frac{i4\pi}{5}},\ e^{\frac{i6\pi}{5}},\ e^{\frac{i8\pi}{5}}$$

or

$$\cos 0 + i\sin 0,\ \cos\frac{2\pi}{5} + i\sin\frac{2\pi}{5},$$

$$\cos\frac{4\pi}{5} + i\sin\frac{4\pi}{5},\ \cos\frac{6\pi}{5} + i\sin\frac{6\pi}{5},$$

$$\cos\frac{8\pi}{5} + i\sin\frac{8\pi}{5}.$$

As another example, we find fourth roots of the complex number $-8-8\sqrt{3}\,i$. We have

$$r^2 = 8^2 + (8\sqrt{3})^2 = 256 \text{ so } r = 16.$$

Therefore, we can write $z_0 = 16e^{\frac{-i2\pi}{3}}$. Then $c = (16)^{\frac{1}{4}} e^{\frac{-i2\pi}{12}} = 2e^{\frac{-i\pi}{6}}$ and $w_n = e^{\frac{i2\pi}{4}}$. Hence the roots are

$$2e^{\frac{-i\pi}{6}}, 2e^{\frac{-i\pi}{2}} \cdot e^{\frac{i\pi}{2}}, 2e^{\frac{-i\pi}{6}} \cdot e^{i\pi}, 2e^{-\frac{i\pi}{6}} \cdot e^{\frac{3i\pi}{4}}$$

or

$$2\left[\cos\left(-\frac{\pi}{6}\right) + i\sin\left(-\frac{\pi}{6}\right)\right], 2\left[\cos\frac{\pi}{3} + i\sin\frac{\pi}{3}\right]$$

$$2\left[\cos\frac{5\pi}{6} + i\sin\frac{5\pi}{6}\right], 2\left[\cos\frac{4\pi}{3} + i\sin\frac{4\pi}{3}\right].$$

De-Moivre's theorem can also be used to solve equations. For example, consider the equation $z^4 - z^3 + z^2 - z + 1 = 0$. Multiplying both sides by $(z + 1)$, we get $z^5 + 1 = 0$. Therefore,

$$z^5 = -1 = (\cos\pi + i\sin\pi)$$

$$= \cos(2n+1)\pi + i\sin(2n+1)\pi,$$

$$n = 0, 1, 2, \ldots$$

Therefore, the roots of the equation are given by

$$[\cos(2n+1)\pi + i\sin(2n+1)\pi]^{\frac{1}{5}}$$

$$= \left[\cos(2n+1)\frac{\pi}{5} + i\sin(2n+1)\frac{\pi}{5}\right].$$

Hence, the roots are

$$\cos\frac{\pi}{5} + i\sin\frac{\pi}{5}, \quad \cos\frac{3\pi}{5} + i\sin\frac{3\pi}{5}$$

$$\cos\pi - i\sin\pi = 1, \quad \cos\frac{7\pi}{5} + i\sin\frac{7\pi}{5}$$

$$\cos\frac{9\pi}{5} + i\sin\frac{9\pi}{5}.$$

But the root -1 corresponds to the factor $(z + 1)$. Therefore, the required roots are

$$\cos\frac{\pi}{5} + i\sin\frac{\pi}{5}, \cos\frac{3\pi}{5} + i\sin\frac{3\pi}{5}$$

$$\cos\frac{\pi}{5} - i\sin\frac{\pi}{5}, \cos\frac{3\pi}{5} - i\sin\frac{3\pi}{5}.$$

3.3 LOGARITHMS OF COMPLEX NUMBERS

Let z and w be complex numbers. *If* $w = e^z$, then z is called a logarithm of w to the base e. Thus $\log_e w = z$. If $w = e^z$, then

$$e^{z+2n\pi i} = e^z \cdot e^{2n\pi i} = e^z = w.$$

Therefore,

$$\log_e w = z + 2n\pi i, \quad n = 0, \pm 1, \pm 2, \ldots$$

Thus if z is logarithm of w, then $z + 2n\pi\, i$ is also logarithm of w. Hence, the *logarithm of a complex number has infinite values and so is a many-valued function.* The value $z + 2n\pi i$ is called the *general value* of $\log_e w$ and is denoted by $\mathrm{Log}_e\, w$. Thus

$$\mathrm{Log}_e w = z + 2n\pi i = 2n\pi i + \log_e w.$$

Substituting $n = 0$ in the general value, we get the *principal value* of z, that is, $\log_e w$.

3.3.1 Real and Imaginary Parts of Log $(x + iy)$

Let $x + iy = r(\cos\theta + i\sin\theta)$ so that $r = \sqrt{x^2 + y^2}$, $\theta = \tan^{-1}\frac{y}{x}$. Then

$$\mathrm{Log}(x + iy) = \log(x + iy) + 2n\pi i$$

$$= \log[r(\cos\theta + i\sin\theta)] + 2n\pi i$$

$$= \log(re^{i\theta}) + 2n\pi i$$

$$= \log r + \log e^{i\theta} + 2n\pi i$$

$$= \log r + 2n\pi i + i\theta$$

$$= \log(x^2 + y^2)^{\frac{1}{2}} + 2n\pi i + i\theta$$

$$= \frac{1}{2}\log(x^2 + y^2) + 2n\pi i + i\tan^{-1}\frac{y}{x}.$$

Hence

$$\mathrm{Re}[\mathrm{Log}(x + iy)] = \frac{1}{2}\log(x^2 + y^2)$$

and

$$\text{Im}\,[\text{Log}\,(x+iy)] = 2n\pi + \tan^{-1}\frac{y}{x}.$$

EXAMPLE 3.1

Separate the following into real and imaginary parts:

(i) Log $(1+i)$ (ii) Log $(4+3i)$.

Solution. (i) We have $x+iy = 1+i$ so that $r^2 = x^2 + y^2 = 1 + 1 = 2$. Therefore

$$\text{Re}[\text{Log}\,(1+i)] = \frac{1}{2}\log(x^2+y^2) = \frac{1}{2}\log 2$$

and

$$\text{Im}[\text{Log}(1+i)] = 2n\pi + \tan^{-1}\frac{y}{x}$$

$$= 2n\pi + \tan^{-1}\frac{1}{1}$$

$$= 2n\pi + \frac{\pi}{4} = (8n-1)\frac{\pi}{4}.$$

(ii) We have $x+iy = 4+3i$ so that $r^2 = x^2+y^2 = 25$.

Therefore

$$\text{Re}[\log(4+3i)] = \frac{1}{2}\log 5^2 = \log 5$$

and

$$\text{Im}[\log(4+3i)] = 2n\pi + \tan^{-1}\frac{3}{4}.$$

EXAMPLE 3.2

Express Log $[Log(\cos\theta + i\sin\theta)]$ in the form $A + iB$.

Solution. We have

$$\begin{aligned}\text{Log}(\cos\theta + i\sin\theta) &= \log(\cos\theta + i\sin\theta) + 2n\pi i\\ &= \log e^{i\theta} + 2n\pi i\\ &= i\theta + 2n\pi i\\ &= (2n\pi + \theta)i.\end{aligned}$$

Then,

$$\begin{aligned}&Log[Log(\cos\theta + i\sin\theta)]\\ &= Log[(2n\pi + \theta)i]\\ &= 2m\pi\, i + \log[(2n\pi+\theta)i]\,,\ m \in Z\\ &= 2m\pi\, i + \log[(2n\pi+\theta) + \log i]\\ &= 2m\pi i + \log(2n\pi+\theta) + \log e^{i\frac{\pi}{2}}\\ &= 2m\pi i + \log(2n\pi+\theta) + \frac{i\pi}{2}\\ &= \frac{i\pi}{2}(4m+1) + \log(2n\pi+\theta)\\ &= A + i\,B.\end{aligned}$$

where

$$A = \log(2n\pi+\theta),\ B = \frac{\pi}{2}(4m+1)$$

EXAMPLE 3.3

Find the general value of

(i) log (-3) (ii) log $(-i)$.

Solution. (i) Since

$$-3 = 3(-1) = 3(\cos\pi + i\sin\pi) = 3e^{i\pi}$$

therefore

$$\begin{aligned}\text{Log}(-3) &= \text{Log}\,(3e^{i\pi}) = 2n\pi i + \log(3e^{i\pi})\\ &= 2n\pi i + \log 3 + i\pi = \log 3 + i(2n+1)\pi.\end{aligned}$$

(ii) Since $-i = \cos\left(-\frac{\pi}{2}\right) + i\sin\left(-\frac{\pi}{2}\right) = e - \frac{i\pi}{2}$, therefore

$$\text{Log}(-i) = 2n\pi i + \log e^{-\frac{i\pi}{2}} = 2n\pi i - \frac{i\pi}{2}$$

$$= (4n-1)\frac{\pi i}{2}.$$

EXAMPLE 3.4

Show that

$$\tan\left(i\,log\,\frac{a-ib}{a+ib}\right) = \frac{2ab}{a^2-b^2}.$$

Solution. *Let $a + ib = r(\cos\theta + i\sin\theta)$.* Therefore $a = r\cos\theta$, $b = r\sin\theta$ and $\tan\theta = \frac{b}{a}$. Then

$$\tan\left(i\log\frac{a-ib}{a+ib}\right) = \tan\left[i\log\frac{r(\cos\theta - i\sin\theta)}{r(\cos\theta + i\sin\theta)}\right]$$

$$= \tan\left[i\log\frac{e^{-i\theta}}{e^{i\theta}}\right]$$

$$= \tan[i\log e^{-2i\theta}]$$

$$= \tan[i(-2i\theta)\log e)]$$

$$= \tan 2\theta = \frac{2\tan\theta}{1-\tan^2\theta}.$$

$$= \frac{\frac{2b}{a}}{1-\frac{b^2}{a^2}} = \frac{2ab}{a^2-b^2}.$$

EXAMPLE 3.5

Prove that

$$\tan^{-1}x = \frac{1}{2i}\log\left(\frac{1+ix}{1-ix}\right).$$

Solution. Taking

$$1 + ix = r(\cos\theta + i\sin\theta),$$

we have $r\cos\theta = 1$ and $r\sin\theta = x$. Therefore

$$\tan\theta = x, \text{ that is, } \theta = \tan^{-1}x.$$

Now

$$\frac{1}{2i}\log\frac{1+ix}{1-ix} = \frac{1}{2i}\log\frac{r(\cos\theta + \sin\theta)}{r(\cos\theta - i\sin\theta)}$$

$$= \frac{1}{2i}\log\frac{e^{i\theta}}{e^{-i\theta}}$$

$$= \frac{1}{2i}\log e^{2i\theta}$$

$$= \frac{1}{2i}.2i\theta$$

$$= \theta = \tan^{-1}x.$$

EXAMPLE 3.6

Show that

$$\cos\left[i\log\frac{a+ib}{a-ib}\right] = \frac{a^2-b^2}{a^2+b^2}.$$

Solution. Setting $a = r\cos\theta$, $b = r\sin\theta$, so that $\tan\theta = \frac{b}{a}$, we have

$$\log\left(\frac{a+ib}{a-ib}\right) = \log\frac{r(\cos\theta + i\sin\theta)}{r(\cos\theta - i\sin\theta)} = \log\frac{re^{i\theta}}{re^{-i\theta}}$$

$$= \log e^{2i\theta} = 2i\theta = 2i\tan^{-1}\frac{b}{a}.$$

Therefore

$$\cos\left[i\log\frac{a+ib}{a-ib}\right] = \cos[i(2i\theta)] = \cos 2\theta$$

$$= \frac{1-\tan^2\theta}{1+\tan^2\theta}$$

$$= \frac{1-\frac{b^2}{a^2}}{1+\frac{b^2}{a^2}} = \frac{a^2-b^2}{a^2+b^2}.$$

3.4 HYPERBOLIC FUNCTIONS

Let z be real or complex. Then

(i) $\frac{e^z - e^{-z}}{2}$ is called the hyperbolic sine of z and is denoted as $\sinh z$.

(ii) $\frac{e^z + e^{-z}}{2}$ is called the hyperbolic cosine of z and is denoted by $\cosh z$.

The other hyperbolic functions are defined in terms of hyperbolic sine and cosine as follows:

$$\tanh z = \frac{\sinh z}{\cosh z} = \frac{e^z - e^{-z}}{e^z + e^{-z}}$$

$$\coth z = \frac{\cosh z}{\sinh z} = \frac{e^z + e^{-z}}{e^z - e^{-z}}$$

$$\text{sech } z = \frac{1}{\cosh z} = \frac{2}{e^z + e^{-z}}$$

$$\text{cosech } z = \frac{1}{\sinh z} = \frac{2}{e^z - e^{-z}}.$$

If follows from the above definitions that

(i) $\sinh 0 = \frac{e^0 - e^{-0}}{2} = \frac{1-1}{2} = 0$

(ii) $\cosh 0 = \frac{e^0 + e^0}{2} = \frac{1+1}{2} = 1$

(iii) $\cosh z + \sinh z = \dfrac{e^z + e^{-z}}{2} + \dfrac{e^z - e^{-z}}{2} = e^z$

(iv) $\cosh z - \sinh z = \dfrac{e^z + e^{-z}}{2} - \dfrac{e^z - e^{-z}}{2} = e^{-z}.$

3.5 RELATIONS BETWEEN HYPERBOLIC AND CIRCULAR FUNCTIONS

(i) By definition

$$\sinh \theta = \frac{e^{\theta} - e^{-\theta}}{2}.$$

Substituting $\theta = iz$, we have

$$\sinh(iz) = \frac{e^{iz} - e^{-iz}}{2} = i\,\frac{e^{iz} - e^{-iz}}{2i} = i \sin z.$$

Similarly, by definition of circular function $\sin\theta$, we have

$$\sin \theta = \frac{e^{i\theta} - e^{-i\theta}}{2i}.$$

Substituting $\theta = iz$, we get

$$\sin(iz) = \frac{e^{-z} - e^{z}}{2i} = -\frac{(e^{-z} - e^{z})}{2i}$$

$$= \frac{i^2(e^{z} - e^{-z})}{2i} = i\,\frac{e^{z} - e^{-z}}{2} = i \sinh z$$

(ii) By definition

$$\cosh \theta = \frac{e^{\theta} + e^{-\theta}}{2}.$$

Substituting $\theta = iz$, we get

$$\cosh(iz) = \frac{e^{iz} + e^{-iz}}{2} = \cos z.$$

Similarly, by definition

$$\cos \theta = \frac{e^{i\theta} + e^{-i\theta}}{2}.$$

Substituting $\theta = iz$, we get

$$\cos(iz) = \frac{e^{-z} + e^{z}}{2} = \cosh z.$$

(iii) We note that

$$\tan(iz) = \frac{\sin(iz)}{\cos(iz)} = \frac{i \sinh z}{\cosh z} = i \tanh z$$

and

$$\tanh(iz) = \frac{e^{iz} - e^{-iz}}{e^{iz} + e^{-iz}} = \frac{i\,\dfrac{e^{iz} - e^{iz}}{2i}}{\dfrac{e^{iz} + e^{-iz}}{2}} = i\,\frac{\sin z}{\cos z}$$

$$= i \tan z.$$

(iv) We have

$$\cot(iz) = \frac{\cos(iz)}{\sin(iz)} = \frac{\cosh z}{i \sinh z} = \frac{i \cosh z}{i^2 \sinh z}$$

$$= -i \coth z$$

(v) We have

$$\sec(iz) = \frac{1}{\cos(iz)} = \frac{1}{\cosh z} = \operatorname{sech} z$$

(vi) Lastly

$$\operatorname{cosec}(iz) = \frac{1}{\sin(iz)} = \frac{1}{i \sinh z} = \frac{i}{i^2 \sinh z}$$

$$= -i \operatorname{cosech} z.$$

3.6 PERIODICITY OF HYPERBOLIC FUNCTIONS

We note that

(i) $\sinh(z + 2n\pi i) = \sinh z$. Therefore, $\sinh z$ is a periodic function with period $2\pi i$.

(ii) $\cosh(z + 2n\pi i) = \cosh z$ and so $\cosh z$ is also periodic with period $2\pi i$.

(iii) $\tanh(z + n\pi i) = \tanh z$ and so $\tanh z$ is periodic with period πi.

Further cosech z, sech z and coth z are reciprocals of sinh z, cosh z and tanh z, respectively, and are, therefore, periodic with period $2\pi i$, $2\pi i$, and πi, respectively.

EXAMPLE 3.7

Show that

(i) $\cosh^2 z - \sinh^2 z = 1$

(ii) $\operatorname{sech}^2 z + \tanh^2 z = 1$

(iii) $\coth^2 z - \operatorname{cosech}^2 z = 1.$

Solution. (i) Since $\cos^2 \theta + \sin^2 \theta = 1$, substituting $\theta = iz$, we get

$$\cos^2(iz) + \sin^2(iz) = 1$$

or

$$(\cosh z)^2 + (i \sinh z)^2 = 1$$

or

$$\cosh^2 z + i^2 \sinh^2 z = 1$$

or

$$\cosh^2 z - \sinh^2 z = 1.$$

(ii) Dividing both sides of the above expression by $\cosh^2 z$, we get

$$1 - \tanh^2 z = \text{sech}^2 z$$

or

$$\text{sech}^2 z + \tanh^2 z = 1.$$

(iii) Dividing both sides of (i) by $\sinh^2 z$, we get

$$\frac{\cosh^2 z}{\sinh^2 z} - 1 = \text{cosech}^{2z}$$

or

$$\coth^2 z - \text{cosech}^2 z = 1.$$

EXAMPLE 3.8

Show that

$$\sinh 2z = 2 \sinh z \cosh z = \frac{2 \tanh z}{1 - \tanh^2 z}.$$

Solution. Substitute $\theta = iz$ in trigonometric relation $\sin 2\theta = 2 \sin \theta \cos \theta$ to get

$$\sin(2iz) = 2 \sin(iz) \cos(iz)$$

or

$$i \sinh 2z = 2i \sinh z \cosh z$$

or

$$\sinh 2z = 2 \sinh z \cosh z.$$

Also, we know that

$$\sin 2\theta = \frac{2 \tan \theta}{1 + \tan^2 \theta}.$$

Substituting $\theta = iz$, we get

$$\sin(2iz) = \frac{2 \tan iz}{1 + \tan^2 iz} = \frac{2i \tanh z}{1 + (i \tanh z)^2}$$

or

$$i \sinh 2z = 2i \frac{\tanh z}{1 + i^2 \tanh^2 z} = \frac{2i \tanh z}{1 - \tanh^2 z}$$

or

$$\sinh 2z = \frac{2 \tanh z}{1 - \tanh^2 z}.$$

EXAMPLE 3.9

Show that

$$\tanh 3z = \frac{3 \tanh z + \tanh^3 z}{1 + 3 \tanh^2 z}.$$

Solution. Substituting $\theta = iz$ in the trigonometric relation $\tan 3\theta = \dfrac{3 \tan \theta - \tan 3\ \theta}{1 - 3 \tan^2 \theta}$, we get

$$\tan (3iz) = \frac{3 \tan(iz) - \tan^3(iz)}{1 - 3 \tan^2(iz)}$$

or

$$i \tanh 3z = \frac{3i \tanh z - (i \tanh z)^3}{1 - 3(i \tanh z)^2}$$

or

$$i \tanh 3z = \frac{3i \tanh z + i \tanh^3 z}{1 + 3 \tanh^2 z}$$

or

$$\tanh 3z = \frac{3 \tanh z + \tanh^3 z}{1 + 3 \tanh^2 z}.$$

Remark 3.1. Proceeding as in the above example, the following formulae of hyperbolic function can also be derived.

(i) $\sinh (x \pm y) = \sinh x \cosh y \pm \cosh x \sinh y$

(ii) $\cosh (x \pm y) = \cosh x \cosh y \pm \sinh x \sinh y$

(iii) $\tanh(x \pm y) = \dfrac{\tanh x \pm \tanh y}{1 \pm \tanh x \tanh y}$

(iv) $\cosh 2x = \cosh^2 x + \sinh^2 x = 2\cosh^2 x - 1$

$$= 1 + 2 \sinh^2 x = \frac{1 + \tanh^2 x}{1 - \tanh^2 x}$$

(v) $\tanh 2x = \dfrac{2 \tanh x}{1 + \tanh^2 x}$

(vi) $\sinh 3x = 3 \sinh x + 4 \sinh^3 x$

(vii) $\cosh 3x = 4 \cosh^2 x - 3 \cosh x$

(viii) $\sinh(A + B) + \sinh(A - B) = 2\sinh A \cosh B$

(ix) $\sinh(A + B) - \sinh(A - B) = 2\cosh A \sinh B$

(x) $\cosh(A+B)+\cosh(A-B)=2\cosh A\cosh B$

(xi) $\cosh(A+B)-\cosh(A-B)=2\sinh A\sinh B$

(xii) $\sinh C+\sinh D=2\sinh\frac{C+D}{2}\cosh\frac{C-D}{2}$

(xiii) $\sinh C-\sinh D=2\cosh\frac{C+D}{2}\sinh\frac{C-D}{2}$

(xiv) $\cosh C+\cosh D=2\cosh\frac{C+D}{2}\cosh\frac{C-D}{2}$

(xv) $\cosh C-\cosh D=2\sinh\frac{C+D}{2}\sinh\frac{C-D}{2}$.

EXAMPLE 3.10

Separate the following into real and imaginary parts.

(i) $\tan(x+iy)$ (ii) $\sec(x+iy)$ (iii) $\tan^{-1}(x+iy)$.

Solution. (i) We have

$$\tan(x+iy)=\frac{\sin(x+iy)}{\cos(x+iy)}$$

$$=\frac{2\sin(x+iy)}{2\cos(x+iy)}\cdot\frac{\cos(x-iy)}{\cos(x-iy)}$$

$$=\frac{\sin 2x+\sin 2iy}{\cos 2x+\cos 2iy}=\frac{\sin 2x+i\sinh 2y}{\cos 2x+\cosh 2y}$$

$$=\frac{\sin 2x}{\cos 2x+\cosh 2y}+i\frac{\sinh 2y}{\cos 2x+\cosh 2y}.$$

Hence,

$$\text{Re}(\tan(x+iy))=\frac{\sin 2x}{\cos 2x+\cosh 2y}$$

$$\text{Im}(\tan(x+iy))=\frac{\sinh 2y}{\cos 2x+\cosh 2y}.$$

(ii) $\sec(x+iy)=\frac{1}{\cos(x+iy)}=\frac{1}{2\cos(x+iy)}$

$$\times\frac{2\cos(x-iy)}{\cos(x-iy)}$$

$$=\frac{2(\cos x\cos iy+\sin x\sin iy)}{\cos 2x+\cos 2iy}$$

$$=\frac{2(\cos x\cosh y+i\sin x\sinh y)}{\cos 2x+\cosh 2y}$$

$$=\frac{2\cos x\cosh y}{\cos 2x+\cosh 2y}$$

$$+i\frac{2\sin x\sinh y}{\cos 2x+\cosh 2y}.$$

Therefore,

$$\text{Re}[\sec(x+iy)]=\frac{2\cos x\cosh y}{\cos 2x+\cosh 2y}$$

$$\text{Im}[\sec(x+iy)]=\frac{2\sin x\sinh y}{\cos 2x+\cosh 2y}$$

(iii) Suppose $\alpha+i\beta=\tan^{-1}(x+iy)$. Then $\alpha-i\beta=\tan^{-1}(x-iy)$.

Addition of these two expressions yields

$$2\alpha=\tan^{-1}(x+iy)+\tan^{-1}(x-iy)$$

$$=\tan^{-1}\frac{(x+iy)+(x-iy)}{1-(x+iy)(x-iy)}.$$

Therefore, $\alpha=\frac{1}{2}\tan^{-1}\frac{2x}{1x^2-y^2}$.

Similarly, subtracting $\alpha-i\beta$ from $\alpha+i\beta$, we get

$$2i\beta=\tan^{-1}(x+iy)-\tan^{-1}(x-iy)$$

$$=\tan^{-1}\frac{(x+iy)(x-iy)}{1+(x+iy)(x-iy)}$$

$$=\tan^{-1}i\frac{2y}{1+x^2+y^2}$$

$$=i\tanh^{-1}\frac{2y}{1+x^2+y^2}.$$

Hence,

$$\beta=\frac{1}{2}\tanh^{-1}\frac{2x}{1+x^2+y^2}.$$

EXAMPLE 3.11

Separate into real and imaginary parts:

(i) $\sin^{-1}(\cos\theta + i\sin\theta)$

(ii) $\tan^{-1}(e^{i\theta})$

Solution. (i) Let

$$\sin^{-1}(\cos\theta + i\sin\theta) = x + iy.$$

Then

$$\cos\theta + i\sin\theta = \sin(x+iy)$$

$$= \sin x\cos iy + \cos x\sin iy$$

$$= \sin x\cosh y + i\cos x\sinh y$$

Equating real and imaginary parts, we have

$$\cos\theta = \sin x\cosh y \qquad (1)$$

$$\sin\theta = \cos x\sinh y \qquad (2)$$

Squaring and adding (1) and (2), we get

$$1 = \sin^2 x\cosh^2 y + \cos^2 x\sin^2 hy$$

$$= \sin^2 x(1+\sinh^2 y) + (1-\sin^2 x)\sinh^2 y$$

$$= \sin^2 x + \sinh^2 y.$$

or

$$\sinh^2 y = 1 - \sin^2 x = \cos^2 x. \qquad (3)$$

From (2), we have

$$\sin^2\theta = \cos^2 x\sinh^2 y = \cos^4 x, \text{ using (3).}$$

or

$$\cos^2 x = \sin\theta$$

or

$$\cos x = \sqrt{\sin\theta}$$

or

$$x = \cos^{-1}\sqrt{\sin\theta} \text{ (Real Part)} \qquad (4)$$

Also, from (2),

$$\sinh y = \frac{\sin\theta}{\cos x} = \frac{\sin\theta}{\sqrt{\sin\theta}} = \sqrt{\sin\theta}$$

or

$$y = \sinh^{-1}(\sqrt{\sin\theta})$$

$$= \log\left[\sqrt{\sin\theta} + \sqrt{1+\sin\theta}\right] \text{ (Imaginary Part),}$$

since

$$\sinh^{-1}x = \log(x + \sqrt{x^2+1}).$$

(ii) Let

$$\tan^{-1}(e^{i\theta}) = \tan^{-1}(\cos\theta + i\sin\theta)$$

$$= x + iy. \qquad (5)$$

Then

$$\tan^{-1}(\cos\theta - \sin\theta) = x - iy \qquad (6)$$

Adding (i) and (ii), we have

$$2x = \tan^{-1}(\cos\theta + \sin\theta) + \tan^{-1}(\cos\theta - i\sin\theta)$$

$$= \tan^{-1}\frac{(\cos\theta + i\sin\theta) + (\cos\theta - i\sin\theta)}{1-(\cos\theta + i\sin\theta)(\cos\theta - i\sin\theta)}$$

$$= \tan^{-1}\frac{2\cos\theta}{1-1} = \tan^{-1}\infty.$$

Thus

$$\tan 2x = \infty = \tan\frac{\pi}{2},$$

and so

$$2x = n\pi + \frac{\pi}{2}$$

or

$$x = \frac{n\pi}{2} + \frac{\pi}{4} \text{ (Real Part).}$$

Now subtracting (6) from (5), we have

$$2iy = \tan^{-1}(\cos\theta + i\sin\theta)$$

$$- \tan^{-1}(\cos\theta - i\sin\theta)$$

$$= \tan^{-1}\frac{(\cos\theta + i\sin\theta) - (\cos\theta - i\sin\theta)}{1 + (\cos\theta + i\sin\theta)(\cos\theta - i\sin\theta)}$$

$$= \tan^{-1}\left(\frac{2i\sin\theta}{1+1}\right)$$

$$= \tan^{-1}(i\sin\theta)$$

and so

$$\tan(2iy) = i\sin\theta$$

or

$$i\tanh 2y = i\sin\theta$$

or

$$\frac{e^{2y}-e^{-2y}}{e^{2y}+e^{-2y}} = \sin\theta$$

or

$$\frac{e^{2y}+e^{-2y}}{e^{2y}-e^{-2y}} = \frac{1}{\sin\theta}.$$

Using compodendo and dividendo, we get

$$e^{4y} = \left[\frac{\cos\frac{\theta}{2}+\sin\frac{\theta}{2}}{\cos\frac{\theta}{2}-\sin\frac{\theta}{2}}\right]^2$$

or

$$e^{2y} = \frac{\cos\frac{\theta}{2}+\sin\frac{\theta}{2}}{\cos\frac{\theta}{2}-\sin\frac{\theta}{2}}$$

$$= \frac{1+\tan\frac{\theta}{2}}{1-\tan\frac{\theta}{2}}$$

$$= \tan\left(\frac{\pi}{4}+\frac{\theta}{2}\right)$$

or

$$2y = \log\tan\left(\frac{\pi}{4}+\frac{\theta}{2}\right)$$

or

$$y = \frac{1}{2}\log\tan\left(\frac{\pi}{4}+\frac{\theta}{2}\right) \text{ (Imaginary Part).}$$

EXAMPLE 3.12

Separate the following into real and imaginary parts.

(i) $\sinh(x+iy)$ (ii) $\coth(x+iy)$

Solution. (i) Since $\sin i\theta = i\sinh\theta$, we have

$$\sinh(x+iy) = \frac{1}{i}\sin i(x+iy) = \frac{1}{i}\sin(ix+i^2y)$$

$$= \frac{i}{i^2}\sin(ix-y)$$

$$= -i(\sin ix\cos y - \cos ix\sin y)$$

$$= -i(i\sinh x\cos y - \cosh x\sin y)$$

$$= \sinh x\cos y + i\cosh x\sin y.$$

Hence

$$\text{Re}[\sinh(x+iy)] = \sinh x\cos y$$

$$\text{Im}[\sinh(x+iy)] = \cosh x\sin y.$$

(ii) $\coth(x+iy)$

$$= \frac{\cosh(x+iy)}{\sinh(x+iy)} = \frac{\cos i(x+iy)}{\frac{1}{i}\sin i(x+iy)}$$

$$= i\,\frac{\cos(ix-y)}{\sin(ix-y)} = i\,\frac{2\sin(ix+y)\cos(ix-y)}{2\sin(ix+y)\sin(ix-y)}$$

$$= i\,\frac{\sin 2ix+\sin 2y}{\cos 2y-\cos 2ix} = i\,\frac{i\sinh 2x+\sin 2y}{\cos 2y-\cosh 2x}$$

$$= -\frac{\sinh 2x}{\cos 2y-\cosh 2x} + i\,\frac{\sin 2y}{\cos 2y-\cosh 2x}$$

$$= \frac{\sinh 2x}{\cosh 2x-\cos 2y} - i\,\frac{\sin 2y}{\cosh 2x-\cos 2y}.$$

EXAMPLE 3.13

Separate into real and imaginary parts:

$$\log\sin(x+iy).$$

Solution. We have

$\log(\sin(x+iy)$

$$= \log(\sin x\cos iy + \cos x\sin iy)$$

$$= \log(\sin x \cosh y + i \cos x \sinh y)$$

$$= \log(\alpha + i\beta).$$

where $\alpha = \sin x \cosh y$, $\beta = \cos x \sinh y$

$$= \frac{1}{2}\log(\alpha^2 + \beta^2) + i\tan^{-1}\frac{\beta}{\alpha}$$

$$= \frac{1}{2}\log(\sin^2 x\cosh^2 y + \cos^2 x\sinh^2 y)$$

$$+ i\tan^{-1}\left(\frac{\cos x \sinh y}{\sin x \cosh y}\right)$$

$$= \frac{1}{2}\log\left[\frac{1-\cos 2x}{2}\cdot\frac{\cosh 2y + 1}{2}\right.$$

$$\left. + \frac{1+\cos 2x}{2}\cdot\frac{\cosh 2y - 1}{2}\right]$$

$$+ i\tan^{-1}[\cot x \tanh y]$$

$$= \frac{1}{2}\log\left[\frac{1}{2}(\cosh 2y - \cos 2x)\right]$$

$$+ i\tan^{-1}[\cot x \tanh y].$$

EXAMPLE 3.14

Separate into real and imaginary parts:

$$\cosh^{-1}(x + iy)$$

Solution. Let

$$\cosh^{-1}(x + iy) = u + iv \qquad (7)$$

Then

$$\cosh^{-1}(x - iy) = u - iv \qquad (8)$$

Adding (8) and (9) we get

$$2u = \cosh^{-1}(x+iy) + \cosh^{-1}(x-iy)$$

$$= \cosh^{-1}\left[(x + iy)(x - iy)\right.$$

$$\left. + \sqrt{(x + iy)^2 - 1}\,.\,\sqrt{(x - iy)^2 - 1}\right]$$

$$= \cosh^{-1}\left[x^2 + y^2 + \sqrt{(x^2 - iy^2 - 1) + 2ixy}\right.$$

$$\left. \times \sqrt{(x^2 - y^2 - 1) - 2ixy}\right]$$

$$= \cosh^{-1}\left[x^2 + y^2 + \sqrt{(x^2 - y^2 - 1)^2 + 4x^2y^2}\right].$$

Therefore the real part is

$$u = \frac{1}{2}\cosh^{-1}\left[x^2 + y^2\right.$$

$$\left. + \sqrt{(x^2 - y^2 - 1)^2 + 4x^2y^2}\right] \text{ (Real Part)}$$

Subtracting (8) from (7), we get

$$2iv = \cosh^{-1}(x+iy) - \cosh^{-1}(x - iy)$$

$$= \cosh^{-1}\left[(x+iy)(x-iy)\right.$$

$$\left. - \sqrt{(x + iy)^2 - 1}\,.\,\sqrt{(x- iy)^2 - 1}\right]$$

$$= \cosh^{-1}\left[x^2 + y^2 - \sqrt{x^2 - y^2 - 1 + 2ixy}\right.$$

$$\left. \times \sqrt{(x^2 - y^2 - 1) - 2ixy}\right]$$

$$= \cosh^{-1}\left[x^2 + y^2 - \sqrt{(x^2 - y^2 - 1)^2 + 4x^2y^2}\right]$$

and so

$$\cos 2iv = x^2 + y^2 - \sqrt{(1 - x^2 + y^2)^2 + 4x^2y^2}$$

or

$$\cos 2v = x^2 + y^2 - \sqrt{(1 - x^2 + y^2)^2 + 4x^2y^2}$$

or

$$2v = \cos^{-1}x^2 + y^2 - \sqrt{(1 - x^2 + y^2)^2 + 4x^2y^2}$$

or

$$v = \frac{1}{2}\cos^{-1}x^2 + y^2$$

$$- \sqrt{(1 - x^2 + y^2)^2 + 4x^2y^2}\ .$$

(Imaginary Part).

EXAMPLE 3.15

If $u = \log\tan\left(\frac{\pi}{4} + \frac{\theta}{2}\right)$, show that

$$\tanh\frac{u}{2} = \tan\frac{\theta}{2}.$$

Deduce that $\tanh u = \sin\theta$.

Solution. We have

$$u = \log\tan\left(\frac{\pi}{4} + \frac{\theta}{2}\right)$$

Therefore

$$e^u = \tan\left(\frac{\pi}{4} + \frac{\theta}{2}\right)$$

or

$$e^{\frac{u}{2}}.e^{\frac{u}{2}} = \frac{1 + \tan\frac{\theta}{2}}{1 - \tan\frac{\theta}{2}}$$

or

$$\frac{e^{\frac{u}{2}}}{e^{-\frac{u}{2}}} = \frac{1 + \tan\frac{\theta}{2}}{1 - \tan\frac{\theta}{2}}.$$

Using compodendo and dividendo, we have

$$\frac{e^{\frac{u}{2}} - e^{-\frac{u}{2}}}{e^{\frac{u}{2}}} = e^{\frac{u}{2}} = \frac{\left(1 + \tan\frac{\theta}{2}\right) - \left(1 - \tan\frac{\theta}{2}\right)}{\left(1 + \tan\frac{\theta}{2}\right) + \left(1 - \tan\frac{\theta}{2}\right)}$$

or

$$\tanh\frac{u}{2} = \tan\frac{\theta}{2}. \qquad (9)$$

Further,

$$\tanh u = \frac{2\tanh\frac{u}{2}}{1 + \tanh^2\frac{u}{2}}$$

$$= \frac{2\tan\frac{u}{2}}{1 + \tan^2\frac{\theta}{2}}, \text{ using (9)}$$

$$= \sin\theta.$$

EXAMPLE 3.16

If $\sin(A + iB) = x + iy$, show that

(i) $x^2\operatorname{cosec}^2 A - y^2\sec^2 A = 1$

(ii) $\dfrac{x^2}{\cosh^2 B} + \dfrac{y^2}{\sinh^2 B} = 1.$

Solution. We have

$$x + iy = \sin(A + iB) = \sin A\cos iB + \cos A\sin iB$$
$$= \sin A\cosh B + i\cos A\sinh B.$$

Therefore, real and imaginary parts are

$$x = \sin A\cosh B \text{ and } y = \cos A\sinh B$$

(i) Form above, we have

$$\frac{x}{\sin A} = \cosh B \text{ and } \frac{y}{\cos A} = \sinh B.$$

Squaring and subtracting, we get

$$\frac{x^2}{\sin^2 A} - \frac{y^2}{\cos^2 A} = \cosh^2 B - \sinh^2 B = 1$$

or

$$x^2\operatorname{cosec}^2 A - y^2\sec^2 A = 1.$$

(ii) Again, from (i), we have

$$\frac{x}{\cosh B} = \sin A \text{ and } \frac{y}{\sinh B} = \cos A.$$

Squaring and adding we get

$$\frac{x^2}{\cosh^2 B} + \frac{y^2}{\sinh^2 B} = 1.$$

EXAMPLE 3.17

If $\sin(\theta + i\phi) = R(\cos\alpha + i\sin\alpha)$, show that

$$R^2 = \frac{1}{2}[\cosh 2\phi - \cos 2\theta]$$

and $\tan\alpha = \tanh\phi \cot\theta$.

Solution. We have

$$R(\cos\alpha + i\sin\alpha) = \sin(\theta + i\phi)$$

$$= \sin\theta \cos i\phi + \cos\theta \sin i\phi$$

$$= \sin\theta \cosh\phi + i\cos\theta \sinh\phi.$$

Equating real and imaginary parts, we get

$$R\cos\alpha = \sin\theta\cosh\phi \quad (10)$$

$$R\sin\alpha = \cos\theta\sinh\phi \quad (11)$$

Squaring (10) and (11) and adding, we have

$$R^2 = \sin^2\theta\cosh^2\phi + \cos^2\theta\sinh^2\phi$$

$$= \frac{1 - \cos 2\theta}{2}\cosh^2\phi + \frac{1 + \cos 2\theta}{2}\sinh^2\phi$$

$$= \frac{1}{2}[(\cosh^2\phi + \sinh^2\phi)$$

$$- \cos 2\theta(\cosh^2 - \sinh^2\phi)]$$

$$= \frac{1}{2}[\cosh 2\phi - \cos 2\theta].$$

Moreover, dividing (11) by (10), we have

$$\tan\alpha = \cot\theta\tanh\phi.$$

EXAMPLE 3.18

Show that

$$(\cosh x + \sinh x)^n = \cosh nx + \sinh nx,$$

where n is a positive integer.

Solution. We have

$$(\cosh x + \sin nhx)^n = \left(\frac{e^x + e^{-x}}{2} + \frac{e^x - e^{-x}}{2}\right)^n$$

$$= (e^x)^n = e^{xn}$$

$$= \cosh nx + \sinh nx.$$

EXAMPLE 3.19

If $x + iy = \cosh(u + iv)$, show that

(i) $x^2\sec^2 v - y^2\operatorname{cosec}^2 v = 1$

(ii) $\dfrac{x^2}{\cosh^2 u} + \dfrac{y^2}{\sinh^2 u} = 1.$

Solution. We are given that

$$x + iy = \cosh(u + iv)$$

$$= \cos i(u + iv) = \cos(iu - v)$$

$$= \cos iu\cos v - \sin iu\sin v$$

$$= \cosh u\cos v + i\sinh u\sin v.$$

Equating the real and imaginary parts, we get

$$x = \cosh u\cos v \text{ and } y = \sinh u\sin v$$

(i) From above, we have

$$\frac{x}{\cos v} = \cosh u \text{ and } \frac{y}{\sin v} = \sinh u.$$

Squaring and subtracting, we get

$$\frac{x^2}{\cos^2 v} - \frac{y^2}{\sin^2 v} = \cosh^2 u - \sinh^2 u = 1$$

or

$$x^2\sec^2 v - y^2\operatorname{cosec}^2 v = 1$$

(ii) From above, we also have

$$\frac{x}{\cosh u} = \cos v \text{ and } \frac{y}{\sinh u} = \sin v.$$

Squaring and adding, we get

$$\frac{x^2}{\cosh^2 u} + \frac{y^2}{\sinh^2 u} = \sin^2 v + \cos^2 v = 1.$$

3.7 SOME BASIC CONCEPTS

Definition 3.5. Let z_0 be a point in the complex plane and let ε be any positive number. Then the set of all points z such that $|z - z_0| < \varepsilon$ is called *ε-neighbourhood* of z_0.

A neighbourhood of a point z_0 from which z_0 is omitted is called a *deleted neighbourhood of* z_0. Thus $0 < |z - z_0| < \varepsilon$ is a deleted neighbourhood of z_0.

Definition 3.6. A point z_0 is called a *limit point, cluster point*, or *point of accumulation* of a point set S if every deleted neighbourhood of z_0 contains points of *S*.

We observe that if z_0 is a limit point of the point set *S*, then since ε is any positive number, *S* contains an infinite number of points. Hence, *a finite set has no limit point.*

Definition 3.7. The union of a set S and the set of its limit points is called the *closure* of S.

Definition 3.8. A set S is said to be *closed* if it contains all of its limit points.

Definition 3.9. A point z_0 is called an *interior point* of a point set S if there exists a neighbourhood of z_0 lying wholly in S.

Definition 3.10. A set S is said to be *open* if every point of S is an interior point.

Thus, a set S is open if for every $z \in S$, there exists a neighbourhood lying wholly in S.

Definition 3.11. An open set is said to be *connected* if any two points of the set can be joined by a polynomial arc (path) lying entirely in the set.

Definition 3.12. An open connected set is called a *domain* or *open region.*

Definition 3.13. The closure of an open region or domain is called *closed region.*

Definition 3.14. If to a domain we add some, all, or none of its limit points, then the set obtained is called the region.

Definition 3.15. A function $w = f(z)$, which assign a complex number w to each complex variable z is called a *complex-valued function of a complex variable z.*

If only one value of w corresponds to each value of z, we say that $w = f(z)$ is a *single-valued function* of z or that $f(z)$ is *single valued.*

If more than one value of w corresponds to a value of z, then $f(z)$ is called *multiple-valued* or *many-valued* function of z.

EXAMPLE 3.20

The function $f(z) = z^2$ is single-valued function whereas the function $f(z) = z^{1/2} = r^{1/2} e^{\frac{\theta+2k\pi}{2}}$, $k = 0, 1, \ldots, n - 1$ is multiple-valued having n *branches* (one for each value of k).

Consider

$$f(z) = z^2 = (x+iy)^2 = x^2 - y^2 + 2xiy.$$

This shows that a complex-valued function can be expressed as

$$f(z) = \phi(x, y) + i\psi(x, y),$$

where $\phi(x, y)$, $\psi(x, y)$ are real functions of the real variables x and y. The function ϕ is called *real part* and ψ is called *imaginary part* of $f(z)$.

Definition 3.16. The function $f(z)$ is said to have the *limit l* as z approaches z_0 if given $\varepsilon > 0$, there exists a $\delta > 0$ such that

$$|f(z) - l| < \varepsilon \quad \textit{whenever } 0 < |z - z_0| < \delta.$$

We then write $\lim_{z \to z_0} f(z) = l$, provided that the limit is *independent of the direction of approach of z to* z_0.

Definition 3.17. The function $f(z)$ is said to be *continuous* at z_0 if $\lim_{z \to z_0} f(z) = f(z_0)$, provided that the limit is *independent of the direction of approach* of z to z_0.

For example, let $f(z) = z^2$ for all z. Then, we note that $\lim_{z \to i} f(z) = f(i) = -1$. Hence f is continuous at $z = i$.

Definition 3.18. The single-valued function $f(z)$ defined on a domain (open connected set) D is said to be *differentiable* at z_0 if

$$\lim_{z \to z_0} \frac{f(z) - f(z_0)}{z - z_0}$$

exists and is *independent of the direction of approach* of z to z_0.

If this limit exists, then the same is called *derivative* of $f(z)$ at z_0 and is denoted by $f'(z_0)$.

3.8 ANALYTIC FUNCTIONS

Definition 3.19. If $f(z)$ is differentiable at all points of some neighbourhood $|z - z_0| < r$ of z_0, then $f(z)$ is said to be *analytic* (or *holomorphic*) *at* z_0.

If $f(z)$ is analytic at each point of a domain D, then $f(z)$ is called *analytic in that domain.*

EXAMPLE 3.21

Consider

$$f(z) = \frac{1+z}{1-z}.$$

We note that

$$f'(z) = \lim_{\Delta z \to 0} \frac{f(z+\Delta z) - f(z)}{\Delta z}$$

$$= \lim_{\Delta z \to 0} \frac{\frac{1+(z+\Delta z)}{1-(z+\Delta z)} - \frac{1+z}{1-z}}{\Delta z}$$

$$= \lim_{\Delta z \to 0} \frac{2}{(1-z-\Delta z)(1-z)} = \frac{2}{(1-z)^2},$$

independent of the direction of approach of Δz to 0, provided that $z \neq 1$. Thus $f(z)$ is analytic for all finite value of z except $z = 1$, where the derivative does not exist.

On the other hand, the function $f(z) = |z|^2$ is not analytic at any point since its derivative exists only at the point $z = 0$ and not throughout any neighbourhood.

Definition 3.20. A function which is analytic everywhere in the finite plane (that is everywhere except at ∞) is called an *entire function* or *integral function.*

For example, e^z, $\sin z$, and $\cos z$ are entire functions.

Definition 3.21. The point at which the function $f(z)$ is not analytic is called *singular point* of $f(z)$.

We notice that $z = 1$ is the singular point of $f(z)$ in Example 3.21.

Definition 3.22. The point z_0 is called an *isolated singularity* or *isolated singular point* of $f(z)$ if we can find $\delta > 0$ such that the circle $|z - z_0| = \delta$ encloses no singular point other than z_0.

If no such δ can be found, then z_0 is called *non-isolated singularity*.

Definition 3.23. The point z_0 is called a *pole of order n* of $f(z)$ if there exists a positive integer n such that $\lim\limits_{z \to z_0} (z - z_0)^n f(z) = A \neq 0$.

If $n = 1$, then z_0 is called a *simple pole.*

EXAMPLE 3.22

(i) $f(z) = \dfrac{1}{(z-1)(z-3)}$ has simple poles at $z = 1$ and $z = 3$.

(ii) $f(z) = \dfrac{1}{(z-2)^3}$ has a pole of order 3 at $z = 2$.

Regarding analyticity of a function $f(z)$, we have the following results.

Theorem 3.3. A necessary condition that $f(z) = u(x, y) + iv(x, y)$ be analytic in a domain D is that in D, the functions u and v satisfy the Cauchy-Riemann equations

$$\frac{\partial u}{\partial x} = \frac{\partial v}{\partial y}, \quad \frac{\partial u}{\partial y} = -\frac{\partial v}{\partial x}.$$

Proof: Let $f(z)$ be analytic in the domain D. Therefore, the limit

$$\lim_{\Delta z \to 0} \frac{f(z+\Delta z) - f(z)}{\Delta z}$$

must exist independent of the manner in which Δz approaches zero. Since $\Delta z = \Delta x + i\Delta y$,

$$\lim_{\Delta z \to 0} \frac{f(z+\Delta z) - f(z)}{\Delta z}$$

$$= \lim_{\substack{\Delta x \to 0 \\ \Delta y \to 0}} \frac{u(x+\Delta x, y+\Delta y)}{\Delta x + i\Delta y}$$

$$+\frac{iv(x+\Delta x, y+\Delta y)-u(x,y)+iv(x,y)}{\Delta x+i\Delta y} \tag{12}$$

must exist independent of the manner in which Δx and Δy approach zero.

Two cases arise:

(i) If $\Delta y = 0$, $\Delta x \to 0$, then (12) becomes

$$\lim_{\Delta x\to 0}\frac{u(x+\Delta x,\ y)+iv(x+\Delta x, y)}{\Delta x} - \frac{u(x,y)+iv(x,y)}{\Delta x}$$

$$= \lim_{\Delta x\to 0}\frac{u(x+\Delta x, y)-u(x,y)}{\Delta x} + \lim_{\Delta x\to 0}\frac{i[v(x+\Delta x, y)-v(x,y)]}{\Delta x}$$

$$= \frac{\partial u}{\partial x} + i\frac{\partial v}{\partial x}, \tag{13}$$

provided the partial derivatives exist.

(ii) If $\Delta x = 0$ and $\Delta y \to 0$, then (12) becomes

$$\lim_{\Delta y\to 0}\left[\frac{u(x,y+\Delta y-u(x,y)}{i\Delta y}+\frac{v(x,y+\Delta y)-v(x,y)}{\Delta y}\right]$$

$$= \frac{1}{i}\frac{\partial u}{\partial y}+\frac{\partial v}{\partial y} \tag{14}$$

For $f(z)$ to be analytic, these two limits in (13) and (14) should be identical. Hence a necessary condition for $f(z)$ to be analytic is

$$\frac{\partial u}{\partial x}+i\frac{\partial v}{\partial x} = -i\frac{\partial u}{\partial y}+\frac{\partial v}{\partial y}$$

and so

$$\frac{\partial u}{\partial x} = \frac{\partial v}{\partial y} \text{ and } \frac{\partial v}{\partial x} = -\frac{\partial u}{\partial y} \tag{15}$$

The equations given in (15) are called *Cauchy-Riemann Equations*.

Remark 3.2. The Cauchy-Riemann equations are not sufficient conditions for analyticity of a function. For example, we shall see that the function $f(z) = \sqrt{|xy|}$ is not analytic at the origin although Cauchy-Riemann equations are satisfied.

The following theorem provides us with sufficient conditions for a function to be analytic.

Theorem 3.4. If $f(z) = u(x, y) + iv(x, y)$ is defined in a domain D and the partial derivatives $\frac{\partial u}{\partial x}, \frac{\partial u}{\partial y}, \frac{\partial v}{\partial x}, \frac{\partial v}{\partial y}$ are continuous and satisfy Cauchy-Riemann equations, then $f(z)$ is analytic in D.

Proof: Since $\frac{\partial u}{\partial x}$ and $\frac{\partial u}{\partial y}$ are continuous, we have

$$\begin{aligned}
\Delta u &= u(x+\Delta x, y+\Delta y)-u(x,y)\\
&= [u(x+\Delta x, y+\Delta y)-u(x, y+\Delta y)]\\
&\quad +[u(x, y+\Delta y)-u(x,y)]\\
&= \left(\frac{\partial u}{\partial x}+\varepsilon_1\right)\Delta x+\left(\frac{\partial u}{\partial y}+\eta_1\right)\Delta y\\
&= \frac{\partial u}{\partial x}\Delta x+\frac{\partial u}{\partial y}\Delta y+\varepsilon_1\Delta x+\eta_1\Delta y,
\end{aligned}$$

where $\varepsilon_1 \to 0$ and $\eta_1 \to 0$ as $\Delta x \to 0$ and $\Delta y \to 0$, respectively.

Similarly, the continuity of $\frac{\partial v}{\partial x}$ and $\frac{\partial v}{\partial y}$ implies

$$\Delta v = \frac{\partial v}{\partial x}\Delta x+\frac{\partial v}{\partial y}\Delta y+\varepsilon_2\Delta x+\eta_2\Delta y,$$

where $\varepsilon_2 \to 0$ and $\eta_2 \to 0$ as $\Delta x \to 0$ and $\Delta y \to 0$, respectively. Hence

$$\begin{aligned}
\Delta f(z) = \Delta w = \Delta u + i\Delta u &= \left(\frac{\partial u}{\partial x}+\frac{i\partial u}{\partial x}\right)\Delta x\\
&\quad +\left(\frac{\partial u}{\partial y}+i\frac{\partial u}{\partial y}\right)\Delta y+\varepsilon\Delta x+\eta\Delta y,
\end{aligned}$$

where $\varepsilon = \varepsilon_1 + i\varepsilon_2 \to 0$ and $\eta = \eta_1 + i\eta_2 \to 0$ as $\Delta x \to 0$ and $\Delta y \to 0$. But, by Cauchy-Riemann equations

$$\frac{\partial u}{\partial x} = \frac{\partial v}{\partial y} \text{ and } \frac{\partial v}{\partial x} = -\frac{\partial u}{\partial y}.$$

Therefore,

$$\Delta w = \left(\frac{\partial u}{\partial x} + i\frac{\partial v}{\partial x}\right)\Delta x + \left(-\frac{\partial v}{\partial x} + i\frac{\partial u}{\partial x}\right)\Delta y + \varepsilon\Delta x + \eta\Delta y$$

$$= \left(\frac{\partial u}{\partial x} + i\frac{\partial v}{\partial x}\right)(\Delta x + i\Delta y) + \varepsilon\Delta x + \eta\Delta y.$$

Dividing by $\Delta z = \Delta x + i\Delta y$ and taking the limit as $\Delta z \to 0$, we get

$$\frac{dw}{dz} = f'(z) = \lim_{\Delta z \to 0} \frac{\Delta w}{\Delta z}$$

$$= \frac{\partial u}{\partial x} + i\frac{\partial v}{\partial x}.$$

Thus, the derivative exists and is unique. Hence $f(z)$ is analytic in D.

Remark 3.3. From above, we note that

$$f'(z) = \frac{\partial u}{\partial x} + i\frac{\partial v}{\partial x} = \frac{\partial u}{\partial x} - i\frac{\partial u}{\partial y}$$

using Cauchy-Riemann equations

and

$$f'(z) = \frac{\partial u}{\partial x} + i\frac{\partial v}{\partial x} = \frac{\partial v}{\partial x} + i\frac{\partial v}{\partial y}$$

using Cauchy-Riemann equations.

EXAMPLE 3.23

Show that the function $f(z) = \bar{z}$ is not analytic at any point.

Solution. We have

$$f'(z) = \lim_{\Delta z \to 0} \frac{\overline{z + \Delta z} - \bar{z}}{\Delta z}$$

$$= \lim_{\substack{\Delta x \to 0 \\ \Delta y \to 0}} \frac{\overline{x + iy + \Delta x + i\Delta y} - \overline{x + iy}}{\Delta x + i\Delta y}$$

$$= \lim_{\substack{\Delta x \to 0 \\ \Delta y \to 0}} \frac{x - iy + \Delta x - i\Delta y - (x - iy)}{\Delta x + i\Delta y}$$

$$= \lim_{\substack{\Delta x \to 0 \\ \Delta y \to 0}} \frac{\Delta x - i\Delta y}{\Delta x + i\Delta y}.$$

If we take $\Delta x = 0$, then the above limit is -1 and if we take $\Delta y = 0$, then this limit is 1. Since the limit depends on the manner in which $\Delta z \to 0$, the derivative does not exist and so $f(z)$ is not analytic.

Second Method: We have

$$f(z) = u + iv = \bar{z} = x - iy,$$

and so

$$u = x, v = -y,$$

$$\frac{\partial u}{\partial x} = 1, \frac{\partial v}{\partial y} = -1.$$

Thus Cauchy-Riemann equations are not satisfied. Hence the function is not analytic.

Theorem 3.5. If $f(z) = u(x, y) + iv(x, y)$ is analytic in a domain D, then u and v are harmonic, that is, they satisfy

$$\frac{\partial^2 u}{\partial x^2} + \frac{\partial^2 u}{\partial y^2} = 0 \quad \text{and} \quad \frac{\partial^2 v}{\partial x^2} + \frac{\partial^2 v}{\partial y^2} = 0.$$

Thus, for an analytic function $f(z)$, u, and v satisfy Laplace-equation

$$\frac{\partial^2 \phi}{\partial x^2} + \frac{\partial^2 \phi}{\partial y^2} = 0.$$

Proof: Since $f(z)$ is analytic in D, Cauchy-Riemann equations are satisfied and so

$$\frac{\partial u}{\partial x} = \frac{\partial v}{\partial y}, \tag{16}$$

$$\frac{\partial v}{\partial x} = -\frac{\partial u}{\partial y}. \tag{17}$$

Assuming that u and v have continuous second order partial derivatives, we differentiate both sides of (16) and (17) with respect to x and y, respectively, and get

$$\frac{\partial^2 u}{\partial x^2} = \frac{\partial^2 v}{\partial x \partial y} \tag{18}$$

and

$$\frac{\partial^2 v}{\partial y \partial x} = -\frac{\partial^2 u}{\partial y^2}. \tag{19}$$

The equations (18) and (19) imply

$$\frac{\partial^2 u}{\partial x^2} = -\frac{\partial^2 u}{\partial y^2}$$

and so

$$\frac{\partial^2 u}{\partial x^2}+\frac{\partial^2 u}{\partial y^2}=0.$$

Hence u is harmonic.

Similarly, differentiating (16) and (17) w.r.t. y and x respectively, we get

$$\frac{\partial^2 v}{\partial x^2}+\frac{\partial^2 v}{\partial y^2}=0.$$

Hence v is harmonic.

Definition 3.24. If $f(z) = u + iv$ is analytic and u and v both satisfy Laplace's equation, then u and v are called *conjugate harmonic functions* or simply *conjugate functions*.

EXAMPLE 3.24

Show that

$$u = e^{-x}(x\sin y - y\cos y)$$

is harmonic.

Solution. We are given that

$$u = e^{-x}(x\sin y - y\cos y).$$

Therefore,

$$\frac{\partial u}{\partial x} = e^{-x}\sin y - x\,e^{-x}\sin y + y\,e^{-x}\cos y,$$

$$\frac{\partial^2 u}{\partial x^2} = -2\,e^{-x}\sin y + x\,e^{-x}\sin y - y\,e^{-x}\cos y,$$

$$\frac{\partial u}{\partial y} = x\,e^{-x}\cos y + y\,e^{-x}\sin y - e^{-x}\cos y,$$

$$\frac{\partial^2 u}{\partial y^2} = -xe^{-x}\sin y + 2\,e^{-x}\sin y + y\,e^{-x}\cos y.$$

Thus, we have

$$\frac{\partial^2 u}{\partial x^2}+\frac{\partial^2 u}{\partial y^2}=0$$

and so u is harmonic.

EXAMPLE 3.25

If

$$u = e^{x}(x\cos y - y\sin y),$$

find v such that $f(z) = u + iv$ is analytic.

Solution. We want $f(z)$ to be analytic. So, by Cauchy-Riemann equations

$$\frac{\partial u}{\partial x}=\frac{\partial v}{\partial y} \quad \text{and} \quad \frac{\partial v}{\partial x}=-\frac{\partial u}{\partial y}.$$

Thus

$$\frac{\partial v}{\partial y}=\frac{\partial u}{\partial x}=e^{x}(x\cos y - y\sin y)+e^{x}\cos y$$

$$= x\,e^{x}\cos y - e^{x}y\sin y + e^{x}\cos y,$$

$$\frac{\partial v}{\partial x}=-\frac{\partial u}{\partial y}= x\,e^{x}\sin y + e^{x}\sin y + e^{x}y\cos y.$$

Now

$$dv=\frac{\partial v}{\partial x}dx+\frac{\partial v}{\partial y}dy$$

$$=(x\,e^{x}\sin y + e^{x}\sin y + e^{x}y\cos y)dx$$

$$+(x\,e^{x}\cos y - e^{x}y\sin y + e^{x}\cos y)dy$$

Therefore,

$$v=\int_{y\text{ constant}}[xe^{x}\sin y + e^{x}\sin y + e^{x}y\cos y]\,dx$$

$$+\int(xe^{x}\cos y - e^{x}y\sin y + e^{x}\cos y)\,dy$$

$$= e^{x}(x\sin y + \sin y + y\cos y) - e^{x}\sin y + C$$

$$= e^{x}(x\sin y + y\cos y) + C \text{ (constant)}.$$

Hence

$$f(z) = u + iv = e^{x}[x\cos y - y\sin y$$

$$+ ix\sin y + iy\cos y] + Ci$$

$$= e^{x}(x+iy)(\cos y + i\sin y) + Ci$$

$$= (x+iy)e^{x+iy} + Ci$$

$$= z\,e^{z} + Ci.$$

EXAMPLE 3.26

If $u_1(x, y) = \frac{\partial u}{\partial x}$ and $u_2(x, y) = \frac{\partial u}{\partial y}$, show that

$$f'(z) = u_1(z,0) - i\,u_2(z,0)$$

Solution. By Remark 3.3 we have

$$f'(z) = \frac{\partial u}{\partial x} - i\frac{\partial u}{\partial y}$$
$$= u_1(x,y) - iu_2(x,y)$$

Substituting $y = 0$, we get

$$f'(z) = u_1(x,0) - iu_2(x,0)$$

Replacing x by z, we have

$$f'(z) = u_1(z,0) - iu_2(z,0) \qquad (20)$$

Remark 3.4. (i) If $\frac{\partial v}{\partial y} = v_1(x, y)$ and $\frac{\partial v}{\partial x} = v_2(x, y)$, then as in Example 3.26, we have

$$f'(z) = v_1(z,0) + iv_2(z,0) \qquad (21)$$

(ii) Integrating (20) and (21), we get $f(z)$. This method of constructing an analytic function is called *Milne-Thomson's method.*

EXAMPLE 3.27

If $u = e^{-x}(x \sin y - y \cos y)$, determine the analytic function $u + iv$.

Solution. We have

$$u_1(x,y) = \frac{\partial u}{\partial x} = e^{-x}\sin y - x\,e^{-x}\sin y + y\,e^{-x}\cos y$$

$$u_2(x,y) = \frac{\partial u}{\partial y} = x\,e^{-x}\cos y + y\,e^{-x}\sin y - e^{-x}\cos y$$

so that, by Example 3.26, we get

$$f'(z) = u_1(z,0) - iu_2(z,0)$$
$$= 0 - i(z\,e^{-z} - e^{-z}) = i\,e^{-z} - i\,z\,e^{-z}.$$

Integrating, we get

$$f(z) = \int ie^{-z}dz + ize^{-z} - \int ie^{-z}dz + Ci$$
$$= i\,z\,e^{-z} + Ci.$$

Also, on separating real and imaginary parts, we get

$$v = e^{-x}(y \sin y + x \cos y) + C$$

EXAMPLE 3.28

Find the analytic function of which the imaginary part is $v = 3x^2y - y^3$.

Solution. We are given that

$$v = 3x^2y - y^3.$$

Therefore,

$$\frac{\partial v}{\partial x} = 6xy, \frac{\partial v}{\partial y} = 3x^2 - 3y^2.$$

Thus

$$v_1(x,y) = \frac{\partial v}{\partial y} = 3x^2 - 3y^2,$$

$$v_2(x,y) = \frac{\partial v}{\partial x} = 6xy.$$

Therefore,

$$f'(z) = v_1(z,0) + i\,v_2(z,0) = 3z^2.$$

Hence

$$f(z) = \int 3z^2 dz = 3\frac{z^2}{3} + C$$
$$= z^3 + C = (x+iy)^3 + C$$
$$= x^3 - 3xy^2 + 3ix^2y - iy^3 + C$$

Comparing real and imaginary parts, we have

$$u = x^3 - 3xy^2 + C$$
$$v = 3x^2y - y^3.$$

EXAMPLE 3.29

Show that the function $u = \frac{1}{2}\log(x^2 + y^2)$ is harmonic and find its harmonic conjugate and the analytic function.

Solution. We have

$$u = \frac{1}{2}\log(x^2 + y^2).$$

Therefore,

$$\frac{\partial u}{\partial x} = \frac{x}{x^2+y^2}, \quad \frac{\partial u}{\partial y} = \frac{y}{x^2+y^2},$$

$$\frac{\partial^2 u}{\partial x^2} = \frac{y^2-x^2}{x^2+y^2}, \quad \frac{\partial^2 u}{\partial y^2} = \frac{x^2-y^2}{x^2+y^2}.$$

Thus

$$\frac{\partial^2 u}{\partial x^2}+\frac{\partial^2 u}{\partial y^2}=0,$$

and so u is harmonic. Further,

$$u_1(x,y)=\frac{x}{x^2+y^2},\quad u_2(x,y)=\frac{y}{x^2+y^2}.$$

Therefore,

$$f'(z)=u_1(z,0)-iu_2(z,0)=\frac{1}{z}-i\,0=\frac{1}{z}.$$

Hence the integration yields

$$\begin{aligned}f(z)&=\int\frac{1}{z}dz=\log z+C\\&=\log(r\,e^{i\theta})+C\\&=\log r+i\theta+C\\&=\log(x^2+y^2)^{1/2}+i\tan^{-1}\frac{y}{x}+C\\&=\frac{1}{2}\log(x^2+y^2)+i\tan^{-1}\frac{y}{x}+C.\end{aligned}$$

Comparing real and imaginary parts, we get

$$u=\frac{1}{2}\log(x^2+y^2)\text{ and }v=\tan^{-1}\frac{y}{x}+C$$

EXAMPLE 3.30

Find analytic function whose real part is

$$u=\frac{\sin 2x}{\cosh 2y-\cos 2x}.$$

Solution. We have

$$u=\frac{\sin 2x}{\cosh 2y-\cos 2x}$$

So

$$\begin{aligned}\frac{\partial u}{\partial x}&=\frac{(\cosh 2y-\cos 2x)2\cos 2x-\sin 2x(2\sin 2x)}{(\cosh 2y-\cos 2x)^2}\\&=\frac{2\cos 2x\cosh 2y-2}{(\cosh 2y-\cos 2x)^2},\text{ and}\end{aligned}$$

$$\frac{\partial u}{\partial y}=\frac{-2\sin 2x\sinh 2y}{(\cosh 2y-\cos 2x)^2}.$$

Therefore,

$$\begin{aligned}f'(z)&=u_1(z,0)-iu_2(z,0)=\frac{2\cos 2z-2}{(1-\cos 2z)^2}+i(0)\\&=-\frac{2}{1-\cos 2z}=\frac{-2}{2\sin^2 z}=-\operatorname{cosec}^2 z.\end{aligned}$$

Integrating w.r.t. z, we get

$$f(z)=\int-\operatorname{cosec}^2 z\,dz=\cot z+Ci.$$

EXAMPLE 3.31

Find regular (analytic) function whose imaginary part is

$$v=\frac{x-y}{x^2+y^2}.$$

Solution. We are given that $v=\frac{x-y}{x^2+y^2}$. Therefore,

$$\frac{\partial v}{\partial x}=\frac{(x^2+y^2)-(x-y)2x}{(x^2+y^2)^2}=\frac{x^2+y^2-2x^2+2xy}{(x^2+y^2)^2},$$

$$\begin{aligned}\frac{\partial v}{\partial y}&=\frac{(x^2+y^2)(-1)-(x-y)(2y)}{(x^2+y^2)^2}\\&=\frac{-x^2-y^2-2xy+2y^2}{(x^2+y^2)^2}.\end{aligned}$$

Then

$$\begin{aligned}f'(z)&=v_1(z,0)+iv_2(z,0)\\&=-\frac{z^2}{z^4}+\frac{i(-z^2)}{z^4}\\&=\frac{-z^2(1+i)}{z^4}=\frac{-(1+i)}{z^2}.\end{aligned}$$

Hence, integration of $f'(z)$ yields

$$f(z)=\frac{1+i}{z}+C.$$

EXAMPLE 3.32

Find the regular function where imaginary part is

$$v=e^x\sin y.$$

Solution. We have

$$v_1(x,y)=\frac{\partial v}{\partial y}=e^x\cos y,\quad v_2(x,y)=\frac{\partial v}{\partial x}=e^x\sin y.$$

Therefore,

$$f'(z)=v_1(z,0)+iv_2(z,0)$$
$$=e^z+0.$$

Hence

$$f(z)=\int e^z dz=e^z+C.$$

EXAMPLE 3.33

In a two-dimensional fluid flow, the stream function ψ is given by $\psi = \tan^{-1}\frac{y}{x}$. Find the velocity potential.

Solution. The two-dimensional flow is represented by the function

$$f(z)=\phi+i\psi,$$

where ϕ is velocity potential and ψ is the stream potential. Thus, the imaginary part of the function is given as

$$\psi=\tan^{-1}\frac{y}{x}.$$

So

$$\frac{\partial\psi}{\partial x}=\frac{1}{1+\frac{y^2}{x^2}}\frac{d}{dx}\left(\frac{y}{x}\right)=-\frac{y}{x^2+y^2}$$

$$\frac{\partial\psi}{y}=\frac{1}{1+\frac{y^2}{x^2}}\frac{d}{dy}\left(\frac{y}{x}\right)=\frac{x}{x^2+y^2}.$$

Therefore,

$$f'(z)=\psi_1(z,0)+i\,\psi_2(z,0)$$
$$=\frac{z}{z^2}+i0=\frac{1}{z}.$$

Integrating, we get

$$f(z)=\log z+C$$
$$=\log(r\,e^{i\theta})+C$$
$$=\log r+i\theta.$$

Hence, real part = $\phi = \log r$

$$=\log(x^2+y^2)^{1/2}$$
$$=\frac{1}{2}\log(x^2+y^2).$$

EXAMPLE 3.34

If the potential function is log (x^2+y^2), find the flux function and the complex potential function.

Solution. The complex potential function is given by

$$f(z)=\phi+i\psi,$$

where ϕ is potential function and ψ is flux function. We are given that

$$\phi=\log(x^2+y^2).$$

To find $f(z)$ and ψ, we proceed as in Example 3.33 and get

$$f(z)=2\log z+C,$$
$$\psi=2\tan^{-1}\frac{y}{x}.$$

EXAMPLE 3.35

If $u - v = (x - y)(x^2 + 4xy + y^2)$ and $f(z) = u + iv$ is analytic function if $z = x + iy$, find $f(z)$ in terms of z.

Solution. We have

$$u+iv=f(z) \tag{22}$$

and so

$$iu-v=if(z) \tag{23}$$

Adding (22) and (23), we get

$$(u-v)+i(u+v)=(1+i)f(z)=F(z)$$
$$=U+iV,\text{ say.}$$

Then F(z) = U + iV is analytic function. We have

$$U=u-v=(x-y)(x^2+4xy+y^2).$$

Therefore,

$$\frac{\partial U}{\partial x} = 3x^2 + 6xy - 3y^2 = \phi_1(x, y),$$
$$\frac{\partial U}{\partial y} = 3x^2 - 6xy - 3y^2 = \phi_2(x, y).$$

Therefore, by Milne's method

$$F(z) = \int (\phi_1(z,0) - i\phi_2(z,0)]\, dz$$
$$= \int (3z^2 - i3z^2)\, dz$$
$$= (1-i)z^3 + C.$$

Thus,

$$(1+i)f(z) = (1-i)z^3 + C.$$

Hence

$$f(z) = \frac{1-i}{1+i}z^3 + C$$
$$= -iz^3 + C.$$

EXAMPLE 3.36

If $f(z) = u + iv$ is an analytic function of $z = x + iy$, show that the family of curves $u(x, y) = C_1$ and $v(x, y) = C_2$ form an orthogonal system.

Solution. Recall that two family of curves form an orthogonal system if they intersect at right angles at each of their points of intersection. Differentiating $u(x, y) = C_1$, we get

$$\frac{\partial u}{\partial x} + \frac{\partial u}{\partial y}.\frac{dy}{dx} = 0$$

or

$$\frac{dy}{dx} = \frac{-\frac{\partial u}{\partial x}}{\frac{\partial u}{\partial y}} = m_1, \text{ say.}$$

Similarly, differentiating $v(x, y) = C_2$, we get

$$\frac{dy}{dx} = -\frac{\frac{\partial v}{\partial x}}{\frac{\partial v}{\partial y}} = m_2, \text{ say.}$$

Using Cauchy-Riemann equations, we have

$$m_1 m_2 = \frac{-\frac{\partial u}{\partial x}}{\frac{\partial u}{\partial y}} . \frac{-\frac{\partial v}{\partial x}}{\frac{\partial v}{\partial y}} = \frac{-\frac{\partial v}{\partial y}}{\frac{\partial u}{\partial y}} . \frac{\frac{\partial u}{\partial y}}{\frac{\partial v}{\partial y}} = -1.$$

Hence, the two curves $u(x, y) = C_1$ and $u(x, y) = C_2$ are orthogonal.

Remark 3.5. If $f(z) = u + iv$ is an analytic function, then Example 3.29 implies that u = constant and v = constant intersect at right angle in the z-plane.

EXAMPLE 3.37

Obtain polar form of Cauchy-Riemann equations.

Solution. Since $x = r\cos\theta$, $y = r\sin\theta$, we have

$$x^2 + y^2 = r^2 \text{ and } \theta = \tan^{-1}\frac{y}{x}.$$

Therefore,

$$\frac{\partial r}{\partial x} = \frac{x}{r} = \cos\theta,$$
$$\frac{\partial r}{\partial y} = \frac{y}{r} = \sin\theta,$$

$$\frac{\partial \theta}{\partial x} = \frac{1}{1+\frac{y^2}{x^2}}\left(-\frac{y}{x^2}\right), = -\frac{y}{x^2+y^2} = -\frac{\sin\theta}{r},$$
$$\frac{\partial \theta}{\partial y} = \frac{x}{x^2+y^2} = \frac{\cos\theta}{r}.$$

Now

$$\frac{\partial u}{\partial x} = \frac{\partial u}{\partial r}\cdot\frac{\partial r}{\partial x} + \frac{\partial u}{\partial \theta}\frac{\partial \theta}{\partial x} = \frac{\partial u}{\partial r}\cos\theta - \frac{\partial u}{\partial \theta}\frac{\sin\theta}{r},$$
$$\frac{\partial u}{\partial y} = \frac{\partial u}{\partial r}\sin\theta + \frac{\partial u}{\partial \theta}\cdot\frac{\cos\theta}{r},$$
$$\frac{\partial v}{\partial x} = \frac{\partial v}{\partial r}\cos\theta - \frac{\partial v}{\partial \theta}\frac{\sin\theta}{r},$$
$$\frac{\partial v}{\partial y} = \frac{\partial v}{\partial r}\sin\theta + \frac{\partial v}{\partial \theta}\cdot\frac{\cos\theta}{r}.$$

But, by Cauchy-Riemann equations

$$\frac{\partial u}{\partial x} = \frac{\partial v}{\partial y}, \quad \frac{\partial u}{\partial y} = -\frac{\partial v}{\partial x}.$$

Hence

$$\frac{\partial u}{\partial r}\cos\theta - \frac{\partial u}{\partial \theta}\frac{\sin\theta}{r} = \frac{\partial v}{\partial r}\sin\theta + \frac{\partial v}{\partial \theta}\frac{\cos\theta}{r}, \quad (24)$$

and

$$\frac{\partial u}{\partial r}\sin\theta + \frac{\partial u}{\partial \theta}\frac{\cos\theta}{r} = -\frac{\partial v}{\partial r}\cos\theta + \frac{\partial v}{\partial \theta}\frac{\sin\theta}{r}. \quad (25)$$

Multiplying (24) by cos θ and (25) by sin θ and adding, we get

$$\frac{\partial u}{\partial r} = \frac{1}{r}\frac{\partial v}{\partial \theta} \quad (26)$$

Now multiplying (24) by – sin θ and (25) by cos θ and adding, we get

$$\frac{1}{r}\frac{\partial u}{\partial \theta} = -\frac{\partial v}{\partial r}. \quad (27)$$

The equations (26) and (27) are called *Cauchy-Riemann equations in polar form.*

EXAMPLE 3.38

Deduce from Example 3.37 that

$$\frac{\partial^2 u}{\partial r^2} + \frac{1}{r}\frac{\partial u}{\partial r} + \frac{1}{r^2}\frac{\partial^2 u}{\partial \theta^2} = 0.$$

Solution. The polar form of Cauchy-Riemann equations is

$$\frac{\partial u}{\partial r} = \frac{1}{r}\frac{\partial v}{\partial \theta}, \quad \frac{1}{r}\frac{\partial u}{\partial \theta} = -\frac{\partial v}{\partial r},$$

that is,

$$\frac{\partial u}{\partial r} = \frac{1}{r}\frac{\partial v}{\partial \theta} \quad (28)$$

and

$$\frac{\partial u}{\partial \theta} = -r\frac{\partial v}{\partial r} \quad (29)$$

Differentiating (28) with respect to r, we get

$$\frac{\partial^2 u}{\partial r^2} = -\frac{1}{r^2}\frac{\partial v}{\partial \theta} + \frac{1}{r}\frac{\partial^2 v}{\partial \theta\, \partial r} \quad (30)$$

Differentiating (30) with respect to θ, we have

$$\frac{\partial^2 u}{\partial \theta^2} = -r\frac{\partial^2 v}{\partial r\, \partial \theta} \quad (31)$$

Using (28), (30), and (31), we have

$$\frac{\partial^2 u}{\partial r^2} + \frac{1}{r}\frac{\partial u}{\partial r} + \frac{1}{r^2}\frac{\partial^2 u}{\partial \theta^2} = 0.$$

EXAMPLE 3.39

Find the analytic function $f(z) = u + iv$ if $u = a(1+\cos\theta)$.

Solution. By polar form of Cauchy-Riemann equations, we have

$$\frac{\partial u}{\partial r} = \frac{1}{r}\frac{\partial v}{\partial \theta}, \text{ and} \quad (32)$$

$$\frac{\partial u}{\partial \theta} = -r\frac{\partial v}{\partial r}. \quad (33)$$

From (33), we have

$$\frac{\partial v}{\partial r} = -\frac{1}{r}\frac{\partial u}{\partial \theta} = -\frac{1}{r}(-a\sin\theta) = \frac{a\sin\theta}{r}.$$

Integrating w.r.t r, we get

$$v = a\sin\theta\log r + \phi(\theta).$$

Hence, $f(z) = u + iv = a\,(1 + \cos\theta + i\sin\theta\log r) + \phi(\theta)$.

EXAMPLE 3.40

Show that the function $e^x(\cos y + i\sin y)$ is holomorphic and find its derivative.

Solution. Let

$$\begin{aligned} f(z) = u + iv &= e^x(\cos y + i\sin y) \\ &= e^x\cos y + i\,e^x\sin y. \end{aligned}$$

Thus

$$u = e^x\cos y, \quad v = e^x\sin y,$$

and so

$$\frac{\partial u}{\partial x} = e^x \cos y, \quad \frac{\partial u}{\partial y} = -e^x \sin y,$$

$$\frac{\partial v}{\partial x} = e^x \sin y, \quad \frac{\partial v}{\partial y} = e^x \cos y.$$

We note that

$$\frac{\partial u}{\partial x} = \frac{\partial v}{\partial y} \quad \text{and} \quad \frac{\partial u}{\partial y} = -\frac{\partial v}{\partial x}$$

and, hence, Cauchy-Riemann equations are satisfied. Since, partial derivative are continuous and Cauchy-Riemann equations are satisfied, it follows that $f(z)$ is analytic. Further,

$$\begin{aligned} f'(z) &= \frac{\partial u}{\partial x} + i\frac{\partial v}{\partial x} \\ &= e^x \cos y + ie^x \sin y \\ &= e^x(\cos y + i \sin y) = e^{x+iy} = e^z. \end{aligned}$$

We note that $f'(z) = f(z)$.

EXAMPLE 3.41

Show that the function

$$f(z) = \frac{x^3(1+i) - y^3(1-i)}{x^2+y^2} (z \neq 0), \ f(0) = 0$$

is continuous and satisfies Cauchy-Riemann equations at the origin, yet $f'(0)$ does not exist.

Solution. We observe that

$$\begin{aligned} \lim_{z\to 0} f(z) &= \lim_{\substack{x\to 0\\ y\to 0}} \frac{x^3(1+i) - y^3(1-i)}{x^2+y^2} \\ &= \lim_{y\to 0} \frac{-y^3(1-i)}{y^2} = \lim_{y\to 0}[-y(1-i)] = 0, \end{aligned}$$

$$\begin{aligned} \lim_{z\to 0} f(z) &= \lim_{\substack{y\to 0\\ x\to 0}} \frac{x^3(1+i) - y^3(1-i)}{x^2+y^2} \\ &= \lim_{x\to 0} \frac{x^3(1+i)}{x^2} = \lim_{x\to 0} x(1+i)] = 0. \end{aligned}$$

Also $f(0) = 0$. Now let both x and y tend to zero along the path $y = mx$. Then,

$$\begin{aligned} \lim_{z\to 0} f(z) &= \lim_{\substack{y\to mx\\ x\to 0}} \frac{x^3(1+i) - y^3(1-i)}{x^2+y^2} \\ &= \lim_{x\to 0} \frac{x^3(1+i) - m^3x^3(1-i)}{x^2+m^2x^2} \\ &= \lim_{x\to 0} \frac{x[1+i-m^3(1-i)]}{1+m^2} = 0. \end{aligned}$$

Thus $\lim_{z\to 0} f(z) = f(0)$, whatever may be the path of z tending to zero. Hence f is continuous at the origin.

Now let $\quad f(z) = u + iv,$

where $\quad u = \frac{x^3 - y^3}{x^2+y^2}, \quad v = \frac{x^3+y^3}{x^2+y^2}.$

Then $\quad u(0,0) = 0, \ v(0,0) = 0.$

Now, at the origin (0,0), we have

$$\begin{aligned} \frac{\partial u}{\partial x} &= \lim_{x\to 0} \frac{u(x,0) - u(0,0)}{x} = \lim_{x\to 0} \frac{x}{x} = 1, \\ \frac{\partial u}{\partial y} &= \lim_{y\to 0} \frac{u(0,y) - u(0,0)}{y} = \lim_{y\to 0} \frac{-y}{y} = -1, \\ \frac{\partial v}{\partial x} &= \lim_{x\to 0} \frac{v(x,0) - v(0,0)}{x} = \lim_{x\to 0} \frac{x}{x} = 1, \\ \frac{\partial v}{\partial y} &= \lim_{y\to 0} \frac{v(0,y) - v(0,0)}{y} = \lim_{y\to 0} \frac{y}{y} = 1. \end{aligned}$$

Hence at the origin

$$\frac{\partial u}{\partial x} = \frac{\partial v}{\partial y} \, and \, \frac{\partial u}{\partial y} = -\frac{\partial v}{\partial x}$$

and so the Cauchy-Riemann equations are satisfied.

But

$$\begin{aligned} f'(0) &= \lim_{z\to 0} \frac{f(z) - f(0)}{z} \\ &= \lim_{z\to 0} \frac{x^3 - y^3 + i(x^3+y^3)}{(x^2+y^2)(x+iy)}. \end{aligned}$$

If $z \to 0$ along $y = mx$, then

$$f'(0) = \lim_{x\to 0} \frac{x^3 - m^3x^3 + i(x^3 + m^3x^3)}{(x^2 + m^2x^2)(x + imx)}$$

$$= \frac{1 - m^3 + i(1 + m^3)}{(1 + m^2)(1 + im)}$$

and so that limit is not unique since it depends on m. Hence $f'(0)$ does not exist.

EXAMPLE 3.42

Show that function $f(z) = \sqrt{|xy|}$ is not regular at the origin, although the Cauchy-Riemann equations are satisfied at the origin.

Solution. We have

$$f(z) = u + iv = \sqrt{|xy|}.$$

Therefore,

$$u(x, y) = \sqrt{|xy|} \text{ and } v(x, y) = 0.$$

Then, at the origin

$$\frac{\partial u}{\partial x} = \lim_{x\to 0} \frac{u(x,0) - u(0,0)}{x} = \lim_{x\to 0} \frac{0-0}{x} = 0,$$

$$\frac{\partial u}{\partial y} = \lim_{y\to 0} \frac{u(0,y) - u(0,0)}{y} = \lim_{y\to 0} \frac{0-0}{y} = 0,$$

$$\frac{\partial v}{\partial x} = 0 \text{ and } \frac{\partial v}{\partial y} = 0.$$

Hence, Cauchy-Riemann equations are satisfied at the origin. But

$$f'(0) = \lim_{z\to 0} \frac{f(z) - f(0)}{z}$$

$$= \lim_{z\to 0} \frac{\sqrt{|xy|}}{(x + iy)}.$$

If $z \to 0$ along $y = mx$, then

$$f'(0) = \lim_{z\to 0} \frac{\sqrt{|mx^2|}}{x + imx} = \lim_{z\to 0} \frac{\sqrt{|m|}}{1 + im}.$$

The limit is not unique since it depends on m. Hence $f'(0)$ does not exist.

EXAMPLE 3.43

Show that the function

$$f(z) = e^{-z^{-4}} \; (z \neq 0),\; f(0) = 0$$

is not analytic at the origin, although Cauchy-Riemann equations are satisfied at that point.

Solution. We have

$$f(z) = e^{-z^{-4}} = e^{-\frac{1}{(x+iy)^4}} = e^{-\frac{(x-iy)^4}{(x^2+y^2)^4}}$$

$$= e^{-\frac{1}{r^8}(x^4+y^4-6x^2y^2)} . e^{4ixy(x^2-y^2)/r^8}$$

$$= e^{-\frac{1}{r^8}(x^4+y^4-6x^2y^2)} \left[\cos\frac{4xy(x^2-y^2)}{r^8} + i\sin\frac{4xy(x^2-y^2)}{r^8}\right].$$

Thus

$$u(x, y) = e^{-\frac{1}{r^8}(x^4+y^4-6x^2y^2)} \cos\frac{4xy(x^2-y^2)}{r^8},$$

$$v(x, y) = e^{-\frac{1}{r^8}(x^4+y^4-6x^2y^2)} \sin\frac{4xy(x^2-y^2)}{r^8}.$$

Hence, at the origin,

$$\frac{\partial u}{\partial x} = \lim_{x\to 0} \frac{u(x,0) - u(0,0)}{x} = \lim_{x\to 0} \frac{e^{-x^{-4}}}{x}$$

$$= \lim_{x\to 0} \frac{1}{x\left[1 + \frac{1}{x^4} + \frac{1}{2x^8} + \dots\right]} = \frac{1}{\infty} = 0.$$

Similarly,

$$\frac{\partial u}{\partial y} = 0, \quad \frac{\partial v}{\partial x} = 0, \quad \frac{\partial v}{\partial y} = 0.$$

Hence, Cauchy-Riemann equations are satisfied at the origin. But, taking $z = re^{i\pi/4}$, we have

$$f'(0) = \lim_{z\to 0} \frac{e^{-z^{-4}} - 0}{z} = \lim_{r\to 0} \frac{e^{r^{-4}}}{re^{i\pi/4}} = \infty.$$

Hence $f(z)$ is not analytic at $z = (0, 0)$.

EXAMPLE 3.44

Show that an analytic function with constant modulus is constant.

Solution. Let $f(z)$ be analytic with constant modulus. Thus

$$|f(z)|=|u+iv|=C\,(\text{constant})$$

and so

$$u^2+v^2=C^2.$$

Then, we have

$$2u\frac{\partial u}{\partial x}+2v\frac{\partial v}{\partial x}=0 \text{ and } 2u\frac{\partial u}{\partial y}+2v\frac{\partial v}{\partial y}=0.$$

Using Cauchy-Riemann equation, the above relations reduce to

$$u\frac{\partial u}{\partial x}-v\frac{\partial u}{\partial y}=0 \tag{34}$$

and

$$u\frac{\partial u}{\partial y}+v\frac{\partial u}{\partial x}=0. \tag{35}$$

Multipling (34) by u, (35) by v and adding, we get

$$(u^2+v^2)\frac{\partial u}{\partial x}=0.$$

Thus $\frac{\partial u}{\partial x}=0$ [if $f(z)\neq 0$]. Similarly, $\frac{\partial u}{\partial y}=0$, $\frac{\partial v}{\partial x}=0$, $\frac{\partial v}{\partial y}=0$. Since all the four partial derivatives are zero, the functions u and v are constant and consequently $u+iv$ is constant.

EXAMPLE 3.45

If $f(z)=u+iv$ is an analytic function of $z=x+iy$ and ψ any function of x and y with differential coefficient of first and second orders, then

$$\left(\frac{\partial\psi}{\partial x}\right)^2+\left(\frac{\partial\psi}{\partial y}\right)^2=\left\{\left(\frac{\partial\psi}{\partial u}\right)^2+\left(\frac{\partial\psi}{\partial v}\right)^2\right\}|f'(z)|^2$$

and

$$\frac{\partial^2\psi}{\partial x^2}+\frac{\partial^2\psi}{\partial y^2}=\left(\frac{\partial^2\psi}{\partial x^2}+\frac{\partial^2\psi}{\partial v^2}\right)|f'(z)|^2.$$

Solution. We have $\frac{\partial\psi}{\partial x}=\frac{\partial\psi}{\partial u}\cdot\frac{\partial u}{\partial x}+\frac{\partial\psi}{\partial v}\cdot\frac{\partial v}{\partial x}$

and

$$\frac{\partial\psi}{\partial y}=\frac{\partial\psi}{\partial u}\cdot\frac{\partial u}{\partial y}+\frac{\partial\psi}{\partial v}\cdot\frac{\partial v}{\partial y}=-\frac{\partial\psi}{\partial u}\cdot\frac{\partial v}{\partial x}+\frac{\partial\psi}{\partial v}\cdot\frac{\partial u}{\partial x}$$

by Cauchy-Riemann equations.

Therefore,

$$\left(\frac{\partial\psi}{\partial x}\right)^2+\left(\frac{\partial\psi}{\partial y}\right)^2=\left[\left(\frac{\partial\psi}{\partial u}\right)^2+\left(\frac{\partial\psi}{\partial v}\right)^2\right]\left[\left(\frac{\partial u}{\partial x}\right)^2+\left(\frac{\partial v}{\partial x}\right)^2\right]=\left[\left(\frac{\partial\psi}{\partial u}\right)^2+\left(\frac{\partial\psi}{\partial v}\right)^2\right]|f'(z)|^2,$$

since $f'(z)=\frac{\partial u}{\partial x}+i\frac{\partial v}{\partial x}$.

Now let us prove the second result. We have

$$f(z)=w=u+iv, \text{and } \bar{w}=u-iv$$

and so

$$u=\frac{1}{2}(w+\bar{w}),\ v=\frac{1}{2i}(w-\bar{w}).$$

Therefore

$$\frac{\partial}{\partial\omega}=\frac{\partial}{\partial u}\frac{\partial u}{\partial\omega}+\frac{\partial}{\partial v}\cdot\frac{\partial v}{\partial\omega}=\frac{1}{2}\left(\frac{\partial}{\partial u}-i\frac{\partial}{\partial v}\right),$$

$$\frac{\partial}{\partial\bar{w}}=\frac{\partial}{\partial u}\cdot\frac{\partial u}{\partial\bar{w}}+\frac{\partial}{\partial v}\cdot\frac{\partial v}{\partial\bar{w}}=\frac{1}{2}\left(\frac{\partial}{\partial u}+i\frac{\partial}{\partial v}\right).$$

Thus

$$\frac{\partial}{\partial w}\cdot\frac{\partial}{\partial\bar{w}}=\frac{1}{4}\left(\frac{\partial}{\partial u}-i\frac{\partial}{\partial v}\right)\left(\frac{\partial}{\partial u}+i\frac{\partial}{\partial v}\right),$$

that is,

$$4\frac{\partial}{\partial w}\cdot\frac{\partial}{\partial\bar{w}}=\frac{\partial^2}{\partial u^2}+\frac{\partial^2}{\partial v^2}.$$

Hence

$$\frac{\partial^2\psi}{\partial u^2}+\frac{\partial^2\psi}{\partial v^2}=4\frac{\partial^2\psi}{\partial w\,\partial\bar{w}} \tag{36}$$

But

$$4\frac{\partial^2}{\partial w\,\partial \bar{w}} = 4\left(\frac{\partial}{\partial z}\cdot\frac{\partial z}{\partial w}\right)\left(\frac{\partial}{\partial \bar{z}}\cdot\frac{\partial \bar{z}}{\partial \bar{w}}\right)$$

$$= 4\left(\frac{1}{f'(z)}\cdot\frac{\partial}{\partial z}\right)\left(\frac{1}{f'(\bar{z})}\cdot\frac{\partial}{\partial \bar{z}}\right)$$

$$= 4\left(\frac{1}{f'(z)\,f(\bar{z})}\right).\frac{\partial^2}{\partial z\,\partial \bar{z}}$$

$$= \frac{4}{|f'(z)|^2}\cdot\frac{\partial}{\partial z}\cdot\frac{\partial}{\partial \bar{z}}$$

$$= \frac{4}{|f'(z)|^2}\left[\frac{1}{2}\left(\frac{\partial}{\partial x}-i\frac{\partial}{\partial y}\right).\frac{1}{2}\left(\frac{\partial}{\partial x}+i\frac{\partial}{\partial y}\right)\right]$$

$$= \frac{1}{|f'(z)|^2}\left(\frac{\partial^2}{\partial x^2}+\frac{\partial^2}{\partial y^2}\right).$$

Hence (36) yields

$$\frac{\partial^2\psi}{\partial u^2}+\frac{\partial^2\psi}{\partial v^2} = \frac{1}{|f'(z)|^2}\left(\frac{\partial^2\psi}{\partial x^2}+\frac{\partial^2\psi}{\partial y^2}\right)$$

and so

$$\frac{\partial^2\psi}{\partial x^2}+\frac{\partial^2\psi}{\partial y^2} = \left(\frac{\partial^2\psi}{\partial u^2}+\frac{\partial^2\psi}{\partial v^2}\right)|f'(z)|^2.$$

EXAMPLE 3.46

If $f(z)$ is a regular function of z, show that

$$\left(\frac{\partial^2}{\partial x^2}+\frac{\partial^2}{\partial y^2}\right)|f(z)|^2 = 4|f'(z)|^2.$$

Solution. Since $z = x + iy$, we have

$$x = \frac{1}{2}(z+\bar{z}) \quad\text{and}\quad y = -\frac{i}{2}(z-\bar{z}).$$

Therefore,

$$\frac{\partial}{\partial z} = \frac{\partial}{\partial x}\cdot\frac{\partial x}{\partial z}+\frac{\partial}{\partial y}\cdot\frac{\partial y}{\partial z} = \frac{1}{2}\left(\frac{\partial}{\partial x}-i\frac{\partial}{\partial y}\right),$$

$$\frac{\partial}{\partial \bar{z}} = \frac{\partial}{\partial x}\cdot\frac{\partial x}{\partial \bar{z}}+\frac{\partial}{\partial y}\cdot\frac{\partial y}{\partial \bar{z}} = \frac{1}{2}\left(\frac{\partial}{\partial x}+i\frac{\partial}{\partial y}\right).$$

Thus

$$\frac{\partial}{\partial z}\cdot\frac{\partial}{\partial \bar{z}} = \frac{1}{4}\left(\frac{\partial}{\partial x}-i\frac{\partial}{\partial y}\right)\left(\frac{\partial}{\partial x}+i\frac{\partial}{\partial y}\right)$$

$$= \frac{1}{4}\left(\frac{\partial^2}{\partial x^2}+\frac{\partial^2}{\partial y^2}\right).$$

Hence

$$\left(\frac{\partial^2}{\partial x^2}+\frac{\partial^2}{\partial y^2}\right)|f(z)|^2 = 4\frac{\partial^2}{\partial z\,\partial \bar{z}}[f(z)f(\bar{z})]$$

$$= 4\frac{\partial}{\partial z}[f(z)\,f'(\bar{z})]$$

$$= 4\,f'(z)\;f'(\bar{z}) = 4\,|f'(z)|^2.$$

EXAMPLE 3.47

If $f(z)$ is a regular function of z such that $f'(z) \neq 0$, show that

$$\left(\frac{\partial^2}{\partial x^2}+\frac{\partial^2}{\partial y^2}\right)\log|f'(z)| = 0.$$

Solution. We have

$$\frac{\partial}{\partial z} = \frac{1}{2}\left(\frac{\partial}{\partial x}-i\frac{\partial}{\partial y}\right), \frac{\partial}{\partial \bar{z}} = \frac{1}{2}\left(\frac{\partial}{\partial x}+i\frac{\partial}{\partial y}\right).$$

Therefore

$$\frac{\partial^2}{\partial z\,\partial \bar{z}} = \frac{1}{4}\left(\frac{\partial^2}{\partial x^2}+\frac{\partial^2}{\partial y^2}\right),$$

that is,

$$\frac{\partial^2}{\partial x^2}+\frac{\partial^2}{\partial y^2} = 4\frac{\partial^2}{\partial z\,\partial \bar{z}} \tag{37}$$

But

$$\log|f'(z)| = \frac{1}{2}\log|f'(z)|^2 = \frac{1}{2}\log[f'(z)\;f'(\bar{z})\,]$$

$$= \frac{1}{2}\log f'(z) + \frac{1}{2}\log\; f'(\bar{z}).$$

Therefore (37) yields

$$\left(\frac{\partial^2}{\partial x^2}+\frac{\partial^2}{\partial y^2}\right)\log|f'(z)|$$

$$= 4\frac{\partial^2}{\partial z\,\partial \bar{z}}\left[\frac{1}{2}\log f'(z)+\frac{1}{2}\log f'(\bar{z})\right] = 0.$$

3.9 INTEGRATION OF COMPLEX-VALUED FUNCTIONS

The theory of Riemann-integrals can be extended to complex-valued functions. Integrals of complex-valued functions are calculated over certain types of curves in the complex plane. The following definitions are required for the complex integration.

Definition 3.25. A *continuous curve* or *arc* C in the complex plane joining the points $z(\alpha)$ and $z(\beta)$ are defined by the parametric representation

$$z(t) = x(t) + iy(t), \ \alpha \le t \le \beta,$$

where $x(t)$ and $y(t)$ are continuous real functions. The point $z(\alpha)$ is the *initial point* and $z(\beta)$ is the *terminal point* (Figure 3.2).

If $z(\alpha) = z(\beta)$, $\alpha \ne \beta$, then the endpoints coincide and the curve is called *closed curve*. A closed curve which does not intersect itself anywhere is called a *simple closed curve* (Figure 3.3). The curve is traversed counterclockwise.

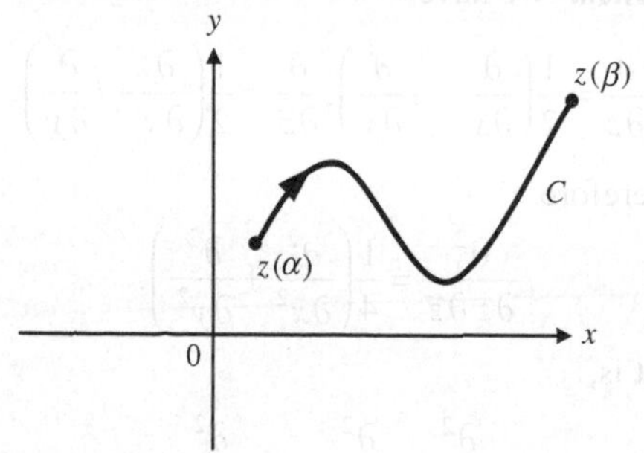

Figure 3.2 Curve C

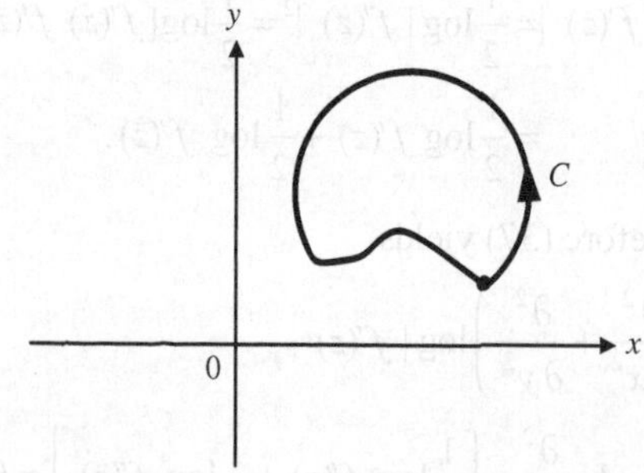

Figure 3.3 Simple Closed Curve

Definition 3.26. A continuous curve C: $z(t) = x(t) + iy(t)$, $\alpha \le t \le \beta$ is called *smooth curve* or *smooth arc* if $z'(t)$ is continuous in $[\alpha, \beta]$ and $z'(t) \ne 0$ in (α, β).

Definition 3.27. A piecewise smooth curve C is called a *contour*.

Thus, a curve C: $z(t) = x(t) + iy(t)$, $\alpha \le t \le \beta$ is a contour if there is a partition $\alpha = t_0 < t_1 < \ldots < t_n = \beta$ such that $z(t)$ is smooth on each subinterval $[t_{i-1}, t_i]$, $i = 1, 2, \ldots, n$.

Definition 3.28. A region in which every closed curve can be contracted to a point without passing out of the region is called a *simply connected region.*

A region which is not simply connected is called *multiply connected.*

Figure 3.4 illustrates the simply-connected and multiply-connected regions.

Thus, simply-connected region does not have any hole in it.

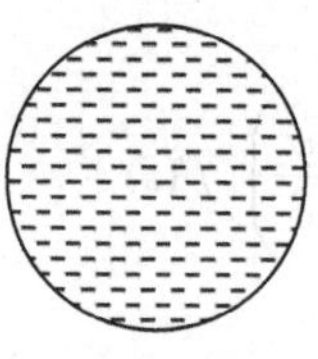

Simply-connected region

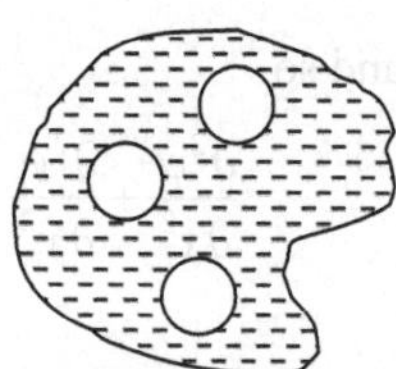

Multiply-connected region

Figure 3.4

Definition 3.29. The *Riemann-integral* of $f(z)$ over a contour C is defined as

$$\int_C f(z)\,dz = \int_\alpha^\beta f(z(t))z'(t)\,dt.$$

The integral on the right-hand side exists because the integrand is piecewise continuous.

We note that the following properties hold for the integral.

(i) $-\int_C f(z)\,dz = \int_{-C} f(z)\,dz$

(ii) if $C_1, C_2, \ldots, C_n$ are disjoint contours, then

$$\int_{C_1+C_2+\ldots+C_n} f(z)dz = \int_{C_1} f(z)dz + \int_{C_2} f(z)dz + \ldots + \int_{C_n} f(z)dz$$

(iii) if $f(z)$ is continuous on contour C, then

$$\left|\int_C f(z)\,dz\right| = \left|\int_\alpha^\beta f(z(t))\,z'(t)\,dt\right| \le \int_\alpha^\beta |f(z(t))|\,|z'(t)|\,dt = \int_C |f(z)|\,|dz|,$$

where

$$\int_C |dz| = \int_\alpha^\beta |z'(t)|\,dt = \int_\alpha^\beta \sqrt{[x'(t)]^2 + [y'(t)]^2}\,dt = L_c,\ \text{length of the curve } C.$$

Therefore, if $|f(z)| \le M$ on C, then

$$\left|\int_C f(z)\,dz\right| \le \int_C |f(z)|\,|dz| \le M\,L_c.$$

EXAMPLE 3.48

Evaluate $\int_C \bar{z}\,dz$ from $z = 0$ to $z = 4 + 2i$ along the curve C given by $z = t^2 + it$.

Solution. We have

$$\int_0^{4+2i} \bar{z}\,dz = \int_C \overline{(t^2+it)}\,dz = \int_C \overline{(t^2+it)}\,(2t+i)\,dt$$

The point $z = 0$ and $z = 4 + 2i$ correspond to $t = 0$ and $t = 2$, respectively. Hence the given integral is equal to

$$\int_0^2 (t^2 - it)(2t+i)\,dt = \int_0^2 (2t^3 - it^2 + t)\,dt = 10 - \frac{8i}{3}.$$

EXAMPLE 3.49

Evaluate $\int_0^{2+i} (\bar{z})^2\,dz$ along the line $y = \frac{x}{2}$.

Solution. Along the given line, we have $x = 2y$ and so $z = x + iy = 2y + iy = (2 + i)y$, $\bar{z} = (2 - i)y$, and $dz = (2 + i)dy$. Thus

$$\int_0^{2+i} (\bar{z})^2\,dz = \int_0^1 (2-i)^2\,y^2.(2+i)\,dy = 5(2-i)\left[\frac{y^3}{3}\right]_0^1 = \frac{5}{3}(2-i).$$

EXAMPLE 3.50

Evaluate $\int_0^{1+i} (x - y + ix^2)dz$ along the straight line from $z = 0$ to $z = 1 + i$.

Solution. As shown in Figure 3.5, the straight line from $z = 0$ to $z = 1 + i$ is OA. On this line, we have $y = x$ and so $z = x + iy$. Thus

$$dz = dx + i\,dy = dx + i\,dx = (1+i)dx.$$

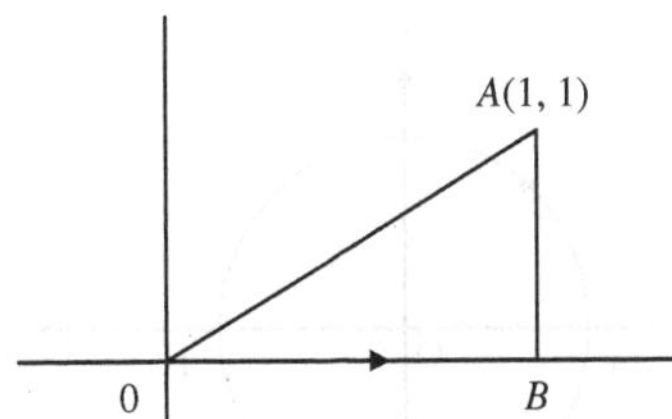

Figure 3.5

Hence

$$\int_{OA} (x - y + ix^2)\,dz = \int_0^1 (ix^2)(1+i)\,dx = (i-1)\left[\frac{x^3}{3}\right]_0^1 = \frac{i-1}{3}.$$

EXAMPLE 3.51

Evaluate $\int_0^{1+i} (x^2 + iy)\,dz$ along the path $y = x^2$.

Solution. We have $z = x + iy = x + ix^2$ and so $dz = dx + 2ixdx = (2ix + 1)dx$. Hence

$$\int_0^{1+i} (x^2 + iy)\, dz = \int_0^1 (x^2 + ix^2)(2ix+1)\, dx$$
$$= \left[(2i-2)\frac{x^4}{4} + (1+i)\frac{x^3}{3} \right]_0^1$$
$$= \frac{5i-1}{6}.$$

EXAMPLE 3.52

Show that $\int_C \frac{dz}{z} = -\pi i$ or πi according as C is the semi-circular arc of $|z| = 1$ above or below the x-axis.

Solution. Taking $z = re^{i\theta}$, we have (Figure 3.6) $dz = ir\, e^{i\theta}\, d\theta$. Therefore, for the semi-circular arc above the x-axis, we have

$$I_1 = \int_{C_1} \frac{dz}{z} = \int_{\pi}^{0} \frac{1}{re^{i\theta}}\, i\, r\, e^{i\theta} d\theta = i \int_{\pi}^{0} d\theta = -\pi i.$$

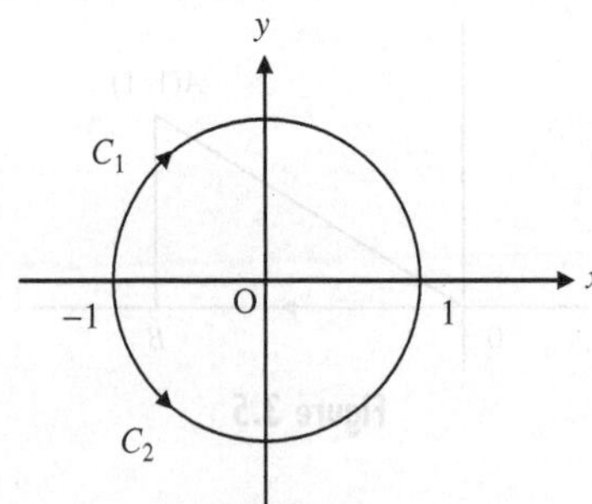

Figure 3.6

For the lower semi-circular arc, we have

$$I_2 = \int_{C_2} \frac{dz}{z} = i \int_{\pi}^{2\pi} d\theta = \pi i.$$

EXAMPLE 3.53

Evaluate $\int_{(0,3)}^{(2,4)} [(2y + x^2)dx + (3x - y)\, dy]$ along the parabola $x = 2t,\, y = t^2 + 3$.

Solution. The points (0, 3) and (2, 4) on the parabola correspond to $t = 0$ and $t = 1$, respectively. Thus, the given integral becomes

$$\int_0^1 [2(t^3 + 3) + 4t^2]\, 2\, dt + (6t - t^3 - 3)\, 2t\, dt$$
$$= \int_0^1 (24t^2 - 2t^3 - 6t + 12)\, dt = \frac{33}{2}.$$

EXAMPLE 3.54

Evaluate $\int_C (z - z^2)\, dz$, where C is the upper half of the unit circle $|z| = 1$.

Solution. The contour is $|z| = 1$. So let $z = e^{i\theta}$. Then $dz = i\, e^{i\theta}\, d\theta$. As shown in the Figure 3.7, the limits of integration become 0 to π. Hence

$$\int_C (z - z^2)\, dz = \int_0^{\pi} (e^{i\theta} - e^{2i\theta}) i\, e^{i\theta} d\theta$$

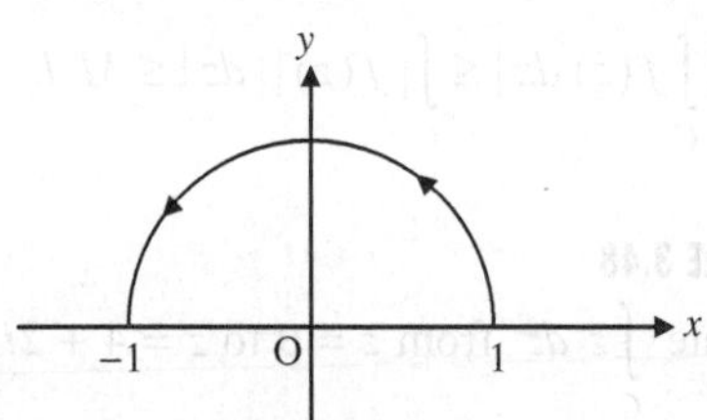

Figure 3.7

$$= i \int_0^{\pi} (e^{2i\theta} - e^{3i\theta})\, d\theta = i \left[\frac{e^{2i\theta}}{2i} - \frac{e^{3i\theta}}{3i} \right]_0^{\pi}$$
$$= \frac{1}{6}[3e^{2\pi i} - 2e^{3\pi i} - 3 + 2] = \frac{2}{3}.$$

EXAMPLE 3.55

Show that $\int_C (z-a)^n dz = 0$, where n is any integer not equal to -1 and C is the circle $|z - a| = r$ with radius r and centre at a.

Solution. Substituting $z - a = r\, e^{i\theta}$, we have $dz = ir\, e^{i\theta}$ and so the given integral reduces to

$$\int_0^{2\pi} r^n e^{ni\theta} .ir e^{i\theta} d\theta$$

$$= i r^{n+1} \int_0^{2\pi} e^{(n+1)i\theta} d\theta = i r^{n+1} \left[\frac{e^{(n+1)i\theta}}{i(n+1)} \right]_0^{2\pi}$$

$$= \frac{r^{n+1}}{n+1} [e^{2(n+1)\pi i} - 1] = 0, \quad n \neq -1.$$

Theorem 3.6. If $f(z)$ is continuous on a contour C of length L and $|f(z)| \le M$, then

$$\left| \int_C f(z)\, dz \right| \le ML.$$

Proof: Since

$$\left| \sum_{s=1}^{n} f(\xi_r)(z_r - z_{r-1}) \right| \le \sum_{r=1}^{n} | f(\xi_r) || z_r - z_{r-1} |,$$

taking limit as $n \to \infty$, we get

$$\left| \int_C f(z)\, dz \right| \le \int_C | f(z) || dz |$$

$$\le \mathrm{M} \int_C | dz |$$

$$\le \mathrm{ML}, \text{ since } \int_C | dz | = \mathrm{L}.$$

Theorem 3.7. (Cauchy's Integral Theorem). If $f(z)$ is an analytic function and if $f'(z)$ is continuous at each point within and on a closed contour C, then

$$\int_C f(z)\, dz = 0.$$

Proof: Since $z = x + iy$, we can write

$$\int_C f(z)\, dz = \int_C (u + iv)(dx + idy)$$

$$= \int_C [(u\, dx - v\, dy) + i(v\, dx + udy)]$$

$$= \int_C (u\, dx - v\, dy) + i \int_C (v\, dx + u\, dy)$$

Since $f'(z) = \frac{\partial u}{\partial x} + i\frac{\partial v}{\partial x} = \frac{\partial u}{\partial x} - i\frac{\partial u}{\partial y}$ and $f'(z)$ is continuous, it follows that u_x, u_y, v_x, and v_y are all continuous in the region D enclosed by the curve C. Hence, by Green's Theorem, we have

$$\int_C f(z)\, dz$$

$$= \int\int_D -\left(\frac{\partial u}{\partial y} + \frac{\partial v}{\partial x} \right) dx\, dy + i \int\int_D \left(\frac{\partial u}{\partial x} - \frac{\partial v}{\partial y} \right) dx\, dy$$

$$= \int\int_D \left(\frac{\partial v}{\partial x} - \frac{\partial v}{\partial x} \right) dx\, dy + i \int\int_D \left(\frac{\partial u}{\partial x} - \frac{\partial u}{\partial x} \right) dxdy$$

$$= 0,$$

the last but one step being the consequence of Cauchy-Riemann equations.

Theorem 3.7 was further generalized by Goursat in the form of the following theorem:

Theorem 3.8. (Cauchy-Goursat). Let $f(z)$ be analytic in a region R. Then for any closed contour C in R,

$$\int_C f(z)\, dz = 0.$$

(For proof, see E.C.Titchmarsh, *Theory of Functions*, Oxford University, Press).

Theorem 3.9. The function $F(z)$ defined by $F(z) = \int_a^z f(\xi)\, d\xi$, where z and a both are in domain D is an analytic function of z such that $F'(z) = f(z)$.

Proof: We have

$$F(z) = \int_a^z f(\xi) d\xi.$$

Therefore,

$$\frac{F(z) - F(z_0)}{z - z_0} - f(z_0)$$

$$= \frac{\int_a^z f(\xi) d\xi - \int_a^{z_0} f(\xi) d\xi}{z - z_0} - f(z_0)$$

$$= \frac{\int_{z_0}^{z} f(\xi)d\xi}{z - z_0} - f(z_0).$$

But,

$$f(z_0) = \frac{1}{z - z_0} \int_{z_0}^{z} f(z_0)d\xi.$$

Hence

$$\frac{F(z) - F(z_0)}{z - z_0} - f(z_0) = \frac{\left[\int_{z_0}^{z} \{f(\xi) - f(z_0)\}\right]}{z - z_0} d\xi.$$

Since f is continuous, we have

$$|f(\xi) - f(z_0)| < \varepsilon \text{ for } |\xi - z_0| < \delta.$$

Therefore

$$\left|\frac{F(z) - F(z_0)}{z - z_0} - f(z_0)\right| < \frac{\varepsilon}{z - z_0} \int_{z_0}^{z} d\xi = \varepsilon.$$

Thus, $F'(z_0) = f(z_0)$ and so $F(z)$ is differentiable and has $f(z_0)$ as its derivative. Hence F(z) is analytic.

Theorem 3.10. (Cauchy's Integral Formula). If $f(z)$ is analytic within and on any closed contour C and if a is a point within the contour C, then

$$f(a) = \frac{1}{2\pi i} \int_C \frac{f(z)}{z - a} dz.$$

Proof: Let $z = a$ be any point within the contour C. Describe a small circle γ about $z = a$, whose radius is r and which lies entirely within C. Consider the function

$$\phi(z) = \frac{f(z)}{z - a}.$$

This function is analytic at all points in the ring-shaped region between C and γ but it has a simple pole at $z = a$. Now, we take a cross cut by joining any point of C to any point of γ. Thus, we obtain a closed contour Γ as shown in Figure 3.8.

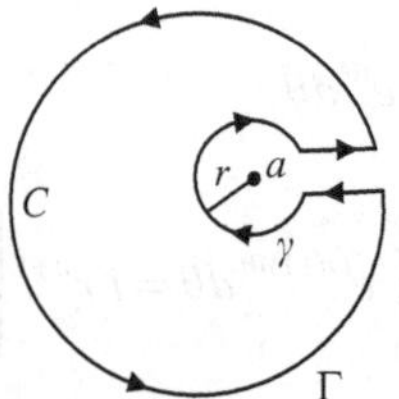

Figure 3.8

Hence, by Cauchy-Goursat theorem, we have

$$\int_\Gamma \phi(z)\, dz = 0,$$

which yields

$$\int_C \phi(z)\, dz - \int_\gamma \phi(z)\, dz = 0.$$

Thus,

$$\int_C \frac{f(z)}{z - a} dz = \int_\gamma \frac{f(z)}{z - a} dz = \int_\gamma \frac{f(z) - f(a) + f(a)}{z - a} dz$$

$$= \int_\gamma \frac{f(a)}{z - a} dz + \int_\gamma \frac{f(z) - f(a)}{z - a} dz \quad (38)$$

Since $f(z)$ is continuous at $z = a$, to each $\varepsilon > 0$ there exists a positive δ such that

$$|f(z) - f(a)| < \varepsilon \text{ whenever } |z - a| < \delta.$$

Moreover, by substituting $z - a = r\, e^{i\theta}$, we get

$$\int_\gamma \frac{f(a)}{z - a} dz = f(a) \int_\gamma \frac{dz}{z - a} = f(a) \int_0^{2\pi} \frac{ir\, e^{i\theta}}{re^{i\theta}} d\theta$$

$$= i\, f(a) \int_0^{2\pi} d\theta = 2\pi i\, f(a).$$

Hence (38) yields

$$\int_C \frac{f(z)}{z - a} dz - 2\pi i\, f(a) = \int_\gamma \frac{f(z) - f(a)}{z - a} dz$$

and so

$$\left|\int_C \frac{f(z)}{z - a} dz - 2\pi i\, f(a)\right| \le \int_\gamma \left|\frac{f(z) - f(a)}{z - a}\right| dz$$

$$< \varepsilon \int_0^{2\pi} d\theta, z - a = r\, e^{i\theta}$$

$$< 2\pi\varepsilon.$$

The left-hand side is independent of ε, and so vanishes. Consequently,

$$\int_C \frac{f(z)}{z-a}\,dz = 2\pi i\, f(a),$$

and, therefore,

$$f(a) = \frac{1}{2\pi i}\int_C \frac{f(z)}{z-a}\,dz.$$

Theorem 3.11. (Cauchy's Formula for Derivative of Analytic Function). If $f(z)$ is an analytic function in a region D, then its derivative at any point $z = a$ is represented by

$$f'(a) = \frac{1}{2\pi i}\int_C \frac{f(z)}{(z-a)^2}\,dz,$$

where C is any closed contour in D surrounding the point $z = a$.

Proof: Suppose that 2δ is the shortest distance from the point a to the contour C. Thus $|z-a| \geq 2\delta$ for every point z on C. If $|h| \leq \delta$, the point $a+h$ also lies within C, at a distance not less than δ from C (Figure 3.9). Therefore, by Cauchy's integral formula, we have

$$f(a) = \frac{1}{2\pi i}\int \frac{f(z)}{z-a}\,dz, \text{ and so}$$

$$f(a+h) = \frac{1}{2\pi i}\int_C \frac{f(z)}{z-a-h}\,dz.$$

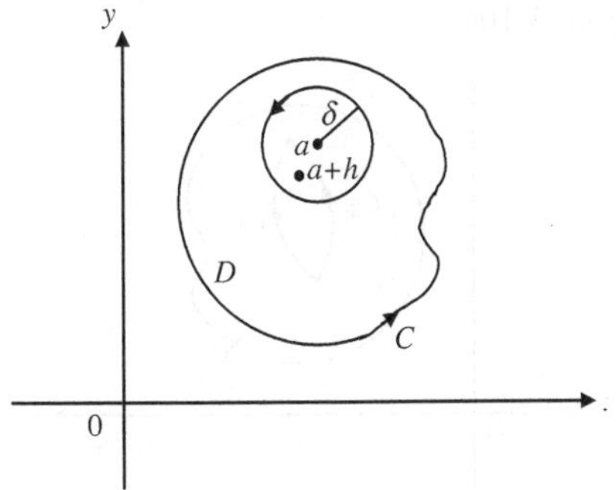

Figure 3.9

Thus

$$\frac{f(a+h)-f(a)}{h}$$

$$= \frac{1}{2\pi i h}\int_C \left(\frac{f(z)}{z-a-h} - \frac{f(z)}{z-a}\right) dz$$

$$= \frac{1}{2\pi i}\int \frac{f(z)\,dz}{(z-a)(z-a-h)}$$

$$= \frac{1}{2\pi i}\int \frac{z-a-h+h}{(z-a)^2(z-a-h)} f(z)dz$$

$$= \frac{1}{2\pi i}\int \frac{z-a-h}{(z-a)^2(z-a-h)} f(z)dz$$

$$+ \frac{1}{2\pi i}\int_C \frac{hf(z)}{(z-a)^2(z-a-h)}\,dz$$

$$= \frac{1}{2\pi i}\int_C \frac{f(z)}{(z-a)^2}\,dz$$

$$+ \frac{1}{2\pi i}\int_C \frac{hf(z)}{(z-a)^2(z-a-h)}\,dz$$

and so

$$\frac{f(a+h)-f(a)}{h} - \frac{1}{2\pi i}\int_C \frac{f(z)}{(z-a)^2}\,dz$$

$$= \frac{1}{2\pi i}\int_C \frac{hf(z)}{(z-a)^2(z-a-h)}\,dz.$$

Now

$$|z-a-h| \geq |z-a| - |h| \geq 2\delta - \delta = \delta.$$

Since $f(z)$ is analytic on C, it is continuous and so is bounded. Thus there exists a constant $M > 0$ such that $|f(z)| \leq M$. Therefore,

$$\left|\frac{f(a+h)-f(a)}{h} - \frac{1}{2\pi i}\int_C \frac{f(z)}{(z-a)^2}\,dz\right|$$

$$= \left|\frac{1}{2\pi i}\int_C \frac{hf(z)}{(z-a)^2(z-a-h)}\,dz\right|$$

$$\leq \frac{|h|}{2\pi}\int_a \frac{|f(z)|}{|z-a|^2|z-a-h|}|dz|$$

$$\leq \frac{M|h|}{2\pi}\int_C \frac{|dz|}{4\delta^2(\delta)} = \frac{|h|}{2\pi}\cdot\frac{ML}{4\delta^3},$$

since $\int_C |dz| = L$(length of C).

Letting $h \to 0$, we get

$$\lim_{h\to 0}\frac{f(a+h)-f(a)}{h} = \frac{1}{2\pi i}\int_C \frac{f(z)}{(z-a)^2}\,dz.$$

Hence

$$f'(a) = \frac{1}{2\pi i}\int_C \frac{f(z)}{(z-a)^2}\,dz.$$

Theorem 3.12. If $f(z)$ is analytic in a domain D, then $f(z)$ has, at any point $z = a$ *of* D, derivatives of all orders given by

$$f^{(n)}(a) = \frac{n!}{2\pi i}\int_C \frac{f(z)}{(z-a)^{n+1}}\,dz,$$

where C is any closed contour in D surrounding the point $z = a$.

Proof: By Cauchy's integral formulae, we have

$$f(a) = \frac{1}{2\pi i}\int_C \frac{f(z)}{(z-a)}\,dz,$$

$$f'(a) = \frac{1}{2\pi i}\int_C \frac{f(z)}{(z-a)^2}\,dz.$$

Thus the result is true for $n = 0$ and $n = 1$. We use mathematical induction on n. Suppose that the result is true for $n = m$. Thus

$$f^{(m)}(a) = \frac{m!}{2\pi i}\int_C \frac{f(z)}{(z-a)^{m+1}}\,dz.$$

Then

$$f^{(m+1)}(a)$$

$$= \lim_{h\to 0}\frac{f^{(m)}(a+h)-f^{(m)}(a)}{h}$$

$$= \lim_{h\to 0}\frac{m!}{2\pi ih}\left[\int_C \frac{f(z)dz}{(z-a-h)^{m+1}} - \int_C \frac{f(z)dz}{(z-a)^{m+1}}\right]$$

$$= \lim_{h\to 0}\frac{m!}{2\pi ih}\int_C\left[\frac{1}{(z-a)^{m+1}}\times\left\{\left(1-\frac{h}{z-a}\right)^{-m+1}-1\right\}\right]f(z)\,dz$$

$$= \lim_{h\to 0}\frac{m!}{2\pi ih}\int_C\left[\frac{1}{(z-a)^{m+1}}\times\left\{(m+1)\frac{h}{z-a}+O(h^2)\right\}\right]f(z)\,dz$$

$$= \frac{(m+1)!}{2\pi i}\int_C \frac{f(z)}{(z-a)^{m+1}}\,dz,$$

which shows that the theorem is also true for $n = m + 1$. Hence it is true for all values of n and we have

$$f^{(n)}(a) = \frac{n!}{2\pi i}\int \frac{f(z)\,dz}{(z-a)^{n+1}}.$$

Remark 3.6. Since each of $f'(a), f'(a) \ldots, f^{(n)}(a)$ have unique differential coefficient, it follows that *derivatives of an analytic function are also analytic functions.*

The following theorem is a sort of *converse of Cauchy's theorem.*

Theorem 3.13. (Morera's Theorem). If $f(z)$ is continuous in a region D and if the integral $\int f(z)dz$ taken round any closed contour in D is zero, then $f(z)$ is analytic inside D.

Proof: Let z_0 be any fixed and z any variable point of the domain D and let C_1, C_2 be any two continuous rectifiable curves in D joining z_0 to z (Figure 3.10).

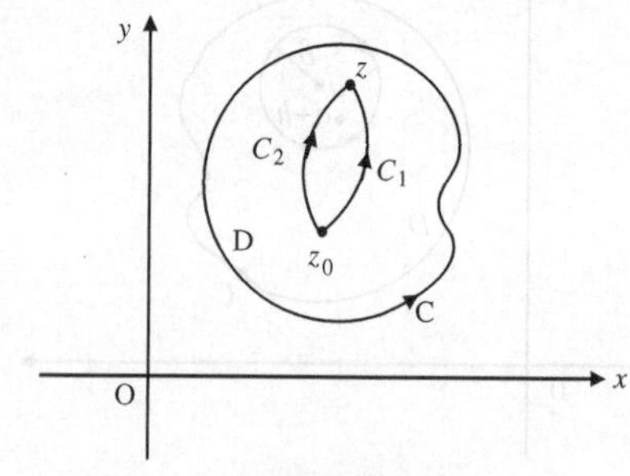

Figure 3.10

Then

$$\int_C f(z)\,dz = \int_{C_1} f(z)\,dz - \int_{C_2} f(z)\,dz = 0.$$

Thus the value of the integral is independent of the path. So, let

$$F(z) = \int_{z_0}^{z} f(\xi)\,d\xi.$$

Since $f(z) = \int_{z}^{z+h} \frac{f(z)}{h}\,d\xi$, we have

$$\frac{F(z+h)-F(z)}{h} - f(z)$$

$$= \frac{1}{h}\left[\int_{z_0}^{z+h} f(\xi)\,d\xi - \int_{z_0}^{z} f(\xi)d\xi\right] - \int_{z}^{z+h}\frac{f(z)}{h}\,d\xi$$

$$= \frac{1}{h}\left[\int_{z}^{z+h} [f(\xi)-f(z)]\,d\xi\right].$$

Since $f(z)$ is continuous, to every $\varepsilon > 0$ there exists $\eta > 0$ such that

$$|f(\xi)-f(z)| < \varepsilon \quad \text{whenever } |z-\xi| < \eta.$$

Thus

$$\left|\frac{F(z+h)-F(z)}{h} - f(z)\right|$$

$$\le \left|\frac{1}{h}\right| \int_{z}^{z+h} |f(\xi)-f(z)|\,|d\xi|$$

$$\le \frac{1}{|h|}\varepsilon\,|h| = \varepsilon.$$

Hence

$$F'(z) = f(z).$$

Since $F(z)$ is analytic, its derivative is also analytic. Therefore, $F'(z)$ is analytic and consequently $f(z)$ is analytic.

Theorem 3.14. (Cauchy's Inequality). If $f(z)$ is analytic within a circle $|z-a| = R$ *and if* $|f(z)| \le M$ *on C, then*

$$|f^{(n)}(a)| \le \frac{M.n!}{R^n}.$$

Proof: We know that

$$f^{(n)}(a) = \frac{n!}{2\pi i}\int_C \frac{f(z)}{(z-a)^{n+1}}\,dz.$$

Therefore,

$$|f^{(n)}(a)| = \left|\frac{n!}{2\pi i}\int_C \frac{f(z)}{(z-a)^{n+1}}\,dz\right|$$

$$\le \frac{n!}{2\pi}\int \frac{|f(z)|}{|(z-a)^{n+1}|}\,|dz|$$

$$= \frac{n!}{2\pi}\cdot\frac{M}{R^{n+1}}\int_C |dz|$$

$$= \frac{n!}{2\pi}\cdot\frac{M}{R^{n+1}}\,2\pi R$$

$$= \frac{Mn!}{R^n}.$$

Theorem 3.15. (Liouville's Theorem). A bounded entire function is constant.

Proof: Let $f(z)$ be bounded entire function. Then there exists a positive constant M such that $|f(z)| \le$ M. Let a be any point of the z-plane and C be the circumference of the circle $|z-a| = R$. Then, by Cauchy's integral formula, we have

$$|f'(a)| = \left|\frac{1}{2\pi i}\int_C \frac{f(z)}{(z-a)^2}\,dz\right|$$

$$\le \frac{1}{2\pi}\int \frac{|f(z)|}{|(z-a)^2|}\,|dz| \le \frac{M}{2\pi R^2}\int_C |dz|$$

$$= \frac{M}{2\pi R^2}.2\pi R$$

$$= \frac{M}{R}.$$

Since $f(z)$ is an entire function, R may be taken arbitrarily large and, therefore, M/R tends to zero as $R \to \infty$. Hence, $|f(z)| \le M$ leads us to $|f'(a)|$

= 0. Since a is arbitrary, we have $f'(a) = 0$ for all points in the z-plane. Hence $f(z)$ is constant.

Second Proof: By Cauchy inequality, we have

$$|f^{(n)}(a)| \leq \frac{n!M}{R^n}.$$

Thus, for $n = 1$, we get

$$|f'(a)| < \frac{M}{R} \to 0 \text{ as } R \to \infty.$$

Therefore, $|f(z)| \leq M$ implies $|f'(a)| = 0$. Since a is arbitrary, $f'(a) = 0$ for all points in the z-plane. Hence $f(z)$ is constant.

Remark 3.7. Since cos z and sin z are entire functions of complex variable z, it follows from Liouville's Theorem that *cos z and sin z are not bounded for complex z.*

Theorem 3.16. (Poisson's Integral Formula). Let $f(z)$ be analytic in the region $|z| \leq R$ and let $u(r, \theta)$ be the real part of $f(r e^{i\theta})$, $z = re^{i\theta}$. Then for $0 < r < R$,

$$u(r,\theta) = \frac{1}{2\pi}\int_0^{2\pi} \frac{(R^2 - r^2)u(R,\phi)}{R^2 - 2Rr\cos(\theta - \phi) + r^2}\, d\phi,$$

where ϕ is the value of θ on the circle $|z| = R$.

Proof: Let C be the circle $|z| = R$. Suppose $z = re^{i\theta}$ is a point within the domain $|z| < R$ and let $\xi = R e^{i\phi}$ be a point on the circle $|z| = R$ (Figure 3.11). Then, Cauchy's integral formula yields

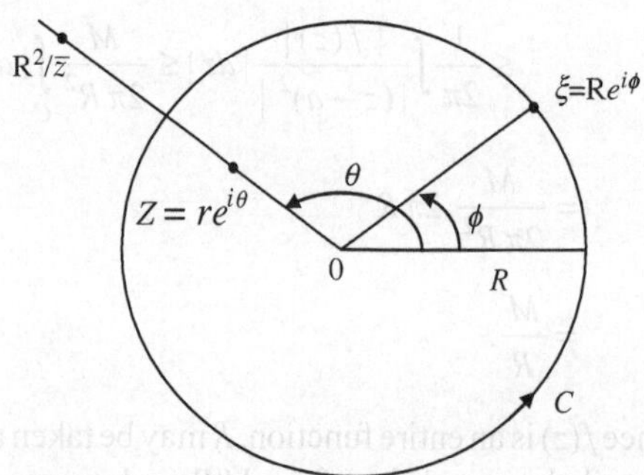

Figure 3.11

$$f(z) = u + iv = \frac{1}{2\pi i}\int_C \frac{f(\xi)}{(\xi - z)}\, d\xi \tag{39}$$

Now the point z being interior, the point $R^2/\bar{z}$ is the inverse point of z with respect to $|z| = R$ and, hence, lies outside the circle. Therefore, $\dfrac{f(\xi)}{\xi - \dfrac{R^2}{\bar{z}}}$ is analytic within C. Hence, by Cauchy's Goursat theorem, we have

$$0 = \frac{1}{2\pi i}\int_C \frac{f(\xi)}{\xi - \left(\dfrac{R^2}{\bar{z}}\right)}\, d\xi. \tag{40}$$

Subtracting (40) from (39), we have

$$f(z) = \frac{1}{2\pi i}\int \frac{f(\xi)\left(\dfrac{R^2}{\bar{z}} - z\right)}{(\xi - z)\left(\dfrac{R^2}{\bar{z}} - \xi\right)}\, d\xi$$

$$= \frac{1}{2\pi i}\int \frac{R^2 - z\bar{z}}{(\xi - z)(R^2 - \xi\bar{z})} f(\xi)\, d\xi.$$

Substituting $\xi = Re^{i\phi}$, $z = re^{i\theta}$, we get

$$f(re^{i\theta}) = \frac{1}{2\pi i}\int_0^{2\pi} \frac{f(Re^{i\phi})(R^2 - r^2)Re^{i\phi}}{(Re^{i\phi} - re^{i\theta})(R^2 - Re^{i\phi}.re^{-i\theta})}\, i d\phi$$

$$= \frac{1}{2\pi}\int_0^{2\pi} \frac{f(Re^{i\phi})(R^2 - r^2)}{(Re^{i\phi} - re^{i\theta})(-re^{-i\theta} + Re^{-i\phi})}\, d\phi$$

$$= \frac{1}{2\pi}\int_0^{2\pi} \frac{f(Re^{i\phi})(R^2 - r^2)}{R^2 + r^2 - 2Rr\cos(\theta - \phi)}\, d\phi.$$

Thus

$$u(r,\theta) + iv(r,\theta)$$

$$= \frac{1}{2\pi}\int_0^{2\pi} \frac{(R^2 - r^2)[u(R,\phi) + iv(R,\phi)]}{R^2 + r^2 - 2Rr\cos(\theta - \phi)}\, d\phi$$

Equating real and imaginary parts, we get

$$u(r,\theta)=\frac{1}{2\pi}\int_0^{2\pi}\frac{(R^2-r^2)\,u(R,\phi)}{R^2+r^2-2Rr\cos(\theta-\phi)}d\phi$$

and

$$v(r,\theta)=\frac{1}{2\pi}\int_0^{2\pi}\frac{(R^2-r^2)\,v(R,\phi)}{R^2+r^2-2Rr\cos(\theta-\phi)}d\phi.$$

EXAMPLE 3.56

If C is any simple closed curve, evaluate $\int_C \frac{dz}{z-a}$ if (*a*) a is outside C and (*b*) a is inside C.

Solution. Let

$$f(z)=\frac{1}{z-a}.$$

(i) If $z = a$ is outside C, then $f(z)$ is analytic everywhere inside C. Hence, by Cauchy's integral theorem $\int_C \frac{dz}{z-a}=0.$

(ii) If $z = a$ is inside C, let Γ be the circle of radius r with centre at a so that Γ is inside C. Then

$$\int_C \frac{dz}{z-a}=\int_\Gamma \frac{dz}{z-a}.$$

Substituting $z - a = r\,e^{i\theta}$, we get $dz = ir\,e^{i\theta}\,d\theta$ and so

$$\int_C \frac{dz}{z-a}=\int_\Gamma \frac{dz}{z-a}=\int_0^{2\pi}\frac{ir\,e^{i\theta}}{re^{i\theta}}d\theta=i\int_0^{2\pi}d\theta=2\pi i.$$

EXAMPLE 3.57

Evaluate $\int_C \frac{e^z}{z-2}dz$, where C is the circle

(i) $|z| = 3$ and (ii) $|z| = 1$.

Solution. (i) Let $f(z) = e^z$. Then $f(z)$ is analytic and $z = 2$ lies inside the circle $|z| = 3$. Therefore, by Cauchy's integral formula

$$f(2)=\frac{1}{2\pi i}\int_C \frac{f(z)}{z-2}dz=\frac{1}{2\pi i}\int_C \frac{e^z}{z-2}dz.$$

Thus

$$\int_C \frac{e^z}{z-2}dz=2\pi i\,f(2)=2\pi i\,e^2.$$

(ii) The point $z = 2$ lies outside the circle $|z| = 1$. Also the function $\frac{e^z}{z-2}$ is analytic within and on $|z| = 1$. Hence, by Cauchy's integral theorem

$$\int_C \frac{e^z}{z-2}dz=0.$$

EXAMPLE 3.58

Evaluate $\int_{|z|=1/2}\frac{e^z}{z^2+1}dz.$

Solution. The function $\frac{e^z}{z^2+1}=\frac{e^z}{(z+i)(z-i)}$ is analytic at all points except $z = \pm i$. Also the points $\pm i$ lie outside $|z| = 1/2$. Hence, by Cauchy-Goursat theorem, the given integral is equal to zero.

EXAMPLE 3.59

Using Cauchy's integral formula and Cauchy-Goursat theorem, evaluate the integral

$$\int_C \frac{z^2-z+1}{z-1}dz,$$

where C is the circle

(i) $|z| = 1$ and (ii) $= |z| = \frac{1}{2}$.

Solution. (i) Let $f(z) = z^2 - z + 1$. Then f is analytic in the circle $|z| = 1$ and $z = 1$ lies on C. Therefore, by Cauchy's integral formula,

$$f(1)=\frac{1}{2\pi i}\int_C \frac{f(z)}{z-1}dz.$$

But, $f(1) = 1$. Hence

$$\int_0 \frac{f(z)}{z-1}dz=2\pi i,$$

that is,

$$\int_C \frac{z^2-z+1}{z-1}dz=2\pi i.$$

(ii) The function $f(z) = z^2 - z + 1$ is analytic everywhere within $|z| = \frac{1}{2}$, Since $z = 1$ lies outside $|z| = \frac{1}{2}$, the function $\frac{z^2 - z + 1}{z - 1}$ is also analytic within $|z| = \frac{1}{2}$, Hence, by Cauchy-Goursat integral theorem

$$\int_{|z|=\frac{1}{2}} \frac{z^2 - z + 1}{z - 1}\, dz = 0.$$

EXAMPLE 3.60

Evaluate

$$\int_{|z|=3} \frac{\sin \pi z^2 + \cos \pi z^2}{(z-1)(z-2)}\, dz.$$

Solution. Since

$$\frac{1}{(z-1)(z-2)} = \frac{1}{z-2} - \frac{1}{z-1},$$

the given integral can be written as

$$\int_C \frac{\sin \pi z^2 + \cos \pi z^2}{(z-1)(z-2)}\, dz$$

$$= \int_C \frac{\sin \pi z^2 + \cos \pi z^2}{z-2}\, dz$$

$$- \int_C \frac{\sin \pi z^2 + \cos \pi z^2}{z-1}\, dz.$$

The points $z = 2$ and $z = 1$ lie within the circle $|z| = 3$ and the function $f(z) = \sin \pi z^2 + \cos \pi z^2$ is analytic within and on $|z| = 3$. Hence, by Cauchy's integral formula,

$$\int_C \frac{\sin \pi z^2 + \cos \pi z^2}{z-2}\, dz = 2\pi i\, f(2)$$

$$= 2\pi i[\sin 4\pi + \cos 4\pi]$$

$$= 2\pi i$$

and

$$\int_C \frac{\sin \pi z^2 + \cos \pi z^2}{z-2}\, dz = 2\pi i\, f(1)$$

$$= 2\pi i[\sin \pi + \cos \pi]$$

$$= -2\pi i,$$

Hence

$$\int_C \frac{\sin \pi z^2 + \cos \pi z^2}{(z-1)(z-2)}\, dz = 2\pi i - (-2\pi i) = 4\pi i.$$

EXAMPLE 3.61

Using Cauchy integral formula, evaluate the integral $\int_C \frac{e^{2z}}{(z-1)(z-2)}\, dz$, where C is the circle $|z| = 3$.

Solution. Let $f(z) = e^{2z}$. Then f is analytic within the circle $|z| = 3$. Also $z = 1$, $z = 2$ lie within $|z| = 3$. Hence, by Cauchy's integral formula, we have

$$\int_{|z|=3} \frac{e^{2z}}{(z-1)(z-2)}\, dz$$

$$= \int_{|z|=3} \frac{e^{2z}}{z-2}\, dz - \int_{|z|=3} \frac{e^{2z}}{z-1}\, dz$$

$$= 2\pi i\, f(2) - 2\pi i\, f(1)$$

$$= 2\pi i(e^4 - e^2).$$

EXAMPLE 3.62

Evaluate the integral $\int_C \frac{ze^z}{(z-a)^3}\, dz$, where the point a lies within the closed contour C.

Solution. Let $f(z) = ze^z$. Then f is analytic (rather entire). The point a lies within C. Therefore,

$$f''(a) = \frac{2!}{2\pi i} \int_C \frac{f(z)}{(z-a)^3}\, dz$$

But,

$$f'(a) = ze^z + e^z, \quad f''(z) = ze^z + 2e^z.$$

So

$$f''(a) = (a+2)e^a.$$

Hence

$$\int_C \frac{ze^z}{(z-a)^3}\,dz = \pi i(a+2)e^a.$$

EXAMPLE 3.63

Evaluate $\int_{|z|=3} \frac{e^{2z}}{(z+1)^4}\,dz$

Solution. The function $f(z) = e^{2z}$ is entire. The point $z = -1$ lies within the circle $|z| = 3$. Therefore, by Cauchy's integral formula, we have

$$f'''(-1) = \frac{3!}{2\pi i}\int_{|z|=3} \frac{e^{2z}}{(z+1)^4}\,dz.$$

But

$$f'(a) = 2e^{2z}, f''(z) = 4e^{2z}, \quad f'''(z) = 8e^{2z}$$

and so $f'''(-1) = 8\,e^{-2}$. Hence

$$\int_{|z|=3} \frac{e^{2z}}{(z+1)^4}\,dz = \frac{8\pi i\, e^{-2}}{3}.$$

EXAMPLE 3.64

Evaluate $I = \int_{|z+1-i|=2} \frac{z+4}{z^2+2z+5}\,dz$

Solution. We have

$$I = \int_{|z+1-i|=2} \frac{z+4}{z^2+2z+5}\,dz$$

$$= \int_{|z+1-i|=2} \frac{z+4}{(z+1+2i)\,(z+1-2i)}\,dz$$

$$= \int_{|z+1-i|=2} \frac{f(z)}{(z+1-2i)}\,dz,$$

where $f(z) = \frac{z+4}{(z+1+2i)}$. The point $-1-2i$ lies outside the contour $|z+1-i| = 2$, whereas the point $-1+2i$ lies within the contour (infact taking $z = -1+2i$ in $|z+1-i|$, the value is less than 2). Hence, by Cauchy's integral formula, we have

$$I = 2\pi i\, f(-1+2i)$$

$$= 2\pi i\left(\frac{-1+2i+4}{-1+2i+1+2i}\right) = \frac{\pi}{2}(3+2i).$$

EXAMPLE 3.65

Evaluate, $\int_C \frac{dz}{(z-a)^n}$, $n = 2, 3$, where C is closed curve containing a.

Solution. Here $f(z) = 1$ so that $f'(z) = f''(z) = f'''(z) = 0$. By Cauchy's integral formula

$$f^{(n)}(a) = \frac{n!}{2\pi i}\int_C \frac{dz}{(z-a)^{n+1}}.$$

Therefore,

$$\int_C \frac{dz}{(z-a)^2} = 2\pi i\, f'(a) = 0$$

$$\int_C \frac{dz}{(z-a)^3} = \frac{2\pi i}{2} f''(a) = 0.$$

EXAMPLE 3.66

Evaluate $\int_{|z|=\frac{1}{2}} \frac{3z^2+7z+1}{z+1}\,dz.$

Solution. Let $f(z) = 3z^2 + 7z + 1$. Then $f(z)$ is analytic within $|z| = \frac{1}{2}$. The point $z = -1$ lies outside the curve $|z| = \frac{1}{2}$. The function $\frac{3z^2+7z+1}{z+1}$ is analytic within and on $|z| = \frac{1}{2}$. Hence, by Cauchy-Goursat theorem

$$\int_{|z|=\frac{1}{2}} \frac{3z^2+7z+1}{z+1}\,dz = 0.$$

EXAMPLE 3.67

Evaluate $I = \int_{|z-1|=3} \frac{e^z}{(z+1)^2(z-2)} dz$

Solution. By partial fractions

$$\frac{1}{(z+1)^2(z-2)} = \frac{1}{9(z-2)} - \frac{1}{9(z+1)} - \frac{1}{3(z+1)^2}.$$

Hence

$$I = \frac{1}{9}\int_{|z-1|=3} \frac{e^z}{z-2} dz - \frac{1}{9}\int_{|z-1|=3} \frac{e^z}{z+1} dz$$

$$-\frac{1}{3}\int_{|z-1|=3} \frac{e^z}{(z+1)^2} dz.$$

The function $f(z) = e^z$ is an entire function and the points $z = -1$ and $z = 2$ lie in $|z-1| = 3$. Also $f'(z) = e^z$ and $f'(-1) = e^{-1}$. Hence, by Cauchy's integral formula,

$$I = \frac{1}{9}\cdot 2\pi i\, f(2) - \frac{1}{9}\cdot 2\pi\, i\, f(-1) - \frac{1}{3}\cdot 2\pi i\, f'(-1)$$

$$= \frac{2\pi i}{9}(e^2 - e^{-1} - 3e^{-1}) = \frac{2\pi i}{9}\left(e^2 - \frac{4}{e}\right).$$

3.10 POWER SERIES REPRESENTATION OF AN ANALYTIC FUNCTION

Definition 3.30. A series of the form $\sum_{n=0}^{\infty} a_n (z - z_0)^n$, where a_n and z_0 are fixed complex numbers and z is a complex variable, is called a *power series* in $(z - z_0)$.

The radius of the power series is given by

$$R = \frac{1}{\overline{\lim_{n\to\infty}}(|a_n|)^{1/n}}$$

or by

$$R = \lim_{n\to\infty}\left|\frac{a_n}{a_{n+1}}\right|,$$

provided that the limit exists. If $R = 0$, the series converges only for $z = z_0$. It converges absolutely if $|z - z_0| < R$ and uniformly if $|z - z_0| \le R_0 < R$. The series diverges for $|z - z_0| > R$.

The circle $|z - z_0| = R$, $0 < R < \infty$, is called the *circle of convergence*.

Theorem 3.17. A power series represents an analytic function inside its circle of convergence.

Proof: Suppose the power series $f(z) = \sum_{n=0}^{\infty} a_n z^n$ converges for $|z| < R$. Then if $\rho < R$, $a_n \rho^n$ is bounded and so $|a_n \rho^n| \le K$ for $K > 0$. Let $g(z) = \sum_{n=1}^{\infty} n a_n z^{n-1}$. Then if $|z| < \rho$ and $|z| + |h| < \rho$, we have

$$\frac{f(z+h) - f(z)}{h} - g(z)$$

$$= \sum_{n=0}^{\infty} a_n \left[\frac{(z+h)^n - z^n}{h} - n z^{n-1}\right].$$

But

$$\left|\frac{(z+h)^n - z^n}{h} - nz^{n-1}\right|$$

$$\le \frac{(|z| + |h|)^n - |z|^n}{|h|} - n|z|^{n-1}.$$

Hence

$$\left|\frac{f(z+h) - f(z)}{h} - g(z)\right|$$

$$\le \sum_{n=0}^{\infty} \frac{1}{\rho^n}\left[\frac{(|z| + |h|)^n - |z|^n}{|h|} - n|z|^{n-1}\right]$$

$$= K\left[\frac{1}{|h|}\left(\frac{\rho}{\rho - |z| - |h|} - \frac{\rho}{\rho - |z|}\right) - \frac{\rho}{(\rho - |z|)^2}\right]$$

$$= \frac{K\rho|h|}{(\rho - |z| - |h|)(\rho - |z|)^2} \to 0 \text{ as } h \to 0.$$

Hence $f(z)$ has the derivative $g(z)$ and so is analytic within the circle of convergence with radius R.

The converse of Theorem 3.17 is the following theorem due to Taylor.

Theorem 3.18. (Taylor). If $f(z)$ is analytic inside a disk $|z - z_0| < R$ with centre at z_0, then for all z in the disk

$$f(z) = \sum_{n=0}^{\infty} \frac{f^{(n)}(z_0)}{n!}(z - z_0)^n,$$

where $f^{(n)}(z_0)$ represents nth derivative of $f(z)$ at z_0.

The coefficients $\frac{f^{(n)}(z_0)}{n!}$ are called *Taylor's coefficients*. The infinite series is convergent if $|z - z_0| < \delta$, where δ is the distance from z_0 to the nearest point of C. If $\delta_1 < \delta$, then the series converges uniformly in the region $|z - z_0| \le \delta_1$.

Proof: Choose $\delta_2 = \frac{\delta + \delta_1}{2}$ so that $0 < \delta_1 < \delta_2 < \delta$ (Figure 3.12). Then, by the given hypothesis, $f(z)$ is analytic within and on a circle Γ defined by $|z - z_0| = \delta_2$.

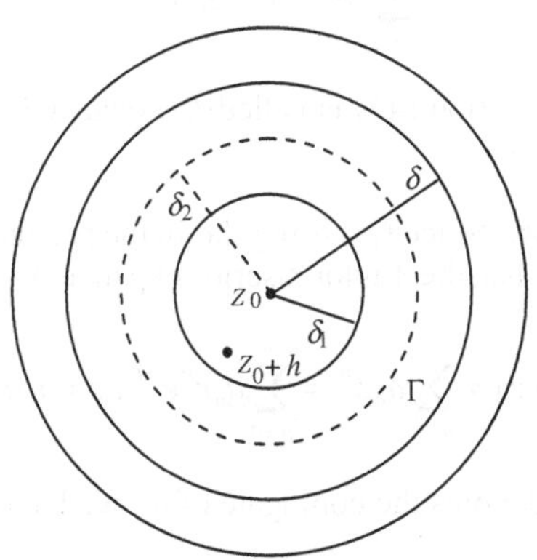

Figure 3.12

Let $z_0 + h$ be any point of the region defined by $|z - z_0| \le \delta_1$. Since $z_0 + h$ lies within the circle Γ, the Cauchy's integral formula yields

$$f(z_0 + h) = \frac{1}{2\pi i}\int_{\Gamma} \frac{f(z)}{z - z_0 - h}\,dz$$

$$= \frac{1}{2\pi i}\int_{\Gamma} \frac{f(z)}{(z - z_0)\left(1 - \frac{h}{z - z_0}\right)}\,dz$$

$$= \frac{1}{2\pi i}\int_{\Gamma} \frac{f(z)}{z - z_0}\left[1 + \frac{h}{z - z_0} + \frac{h^2}{(z - z_0)^2} + \ldots + \frac{h^n}{(z - z_0)^n} + \frac{h^{n+1}}{(z - z_0)^n (z - z_0 - h)}\right] dz$$

$$= \frac{1}{2\pi i}\int_{\Gamma} \frac{f(z)}{z - z_0}\,dz + \frac{h}{2\pi i}\int_{\Gamma} \frac{f(z)}{(z - z_0)^2}\,dz + \ldots + \frac{h^n}{2\pi i}\int_{\Gamma} \frac{f(z)}{(z - z_0)^{n+1}}\,dz + \frac{h^{n+1}}{2\pi i}\int_{\Gamma} \frac{f(z)}{(z - z_0)^{n+1}(z - z_0 - h)}\,dz$$

$$= f(z_0) + h f'(z_0) + \frac{h^2}{2!} f''(z_0) + \ldots + \frac{h^n}{n!} f^{(n)}(z_0) + \Delta_n$$

where

$$\Delta_n = \frac{h^{n+1}}{2\pi i}\int_{\Gamma} \frac{f(z)\,dz}{(z - z_0)^{n+1}(z - z_0 - h)}.$$

Hence

$$f(z_0 + h) = \sum_{n=0}^{\infty} \frac{f^{(n)}(z_0)}{n!}(z - z_0)^n + \Delta_n.$$

But $f(z)$ is bounded by virtue of its continuity. Thus, there exists a positive M such that $|f(z)| \le M$ on Γ. Further,

$$|z - z_0 - h| \ge |z - z_0| - |h| > \delta_2 - \delta_1$$

and since $|z - z_0| \le \delta_1$, we have

$$|h| = |z_0 + h - z_0| \le \delta_1.$$

Since $\delta_1 < \delta_2$, we get

$$|\Delta_n| \le \frac{|h|^{n+1}}{|2\pi i|}\int_{\Gamma} \frac{|f(z)|}{|z - z_0|^{n+1}|z - z_0 - h|}|dz|$$

$$\le \frac{M\,\delta_1^{n+1}}{2\pi\delta_2^{n+1}(\delta_2 - \delta_1)}\,2\pi\delta_2$$

$$\le M\left(\frac{\delta_1}{\delta_2}\right)^n \cdot \frac{\delta_1}{(\delta_2-\delta_1)} \to 0 \text{ as } n \to \infty.$$

Hence

$$f(z_0+h) = \sum_{n=0}^{\infty} \frac{f^{(n)}(z_0)}{n!}(z-z_0)^n.$$

Substituting $z_0 + h = z$, we get

$$f(z) = \sum_{n=0}^{\infty} \frac{f^{(n)}(z_0)}{n!}(z-z_0)^n. \quad (41)$$

Remark 3.8. (i) Theorem 18 was actually *invented by Cauchy* when he was in exile.

(ii) Substituting $z_0 = 0$, the Taylor's series reduces to

$$f(z) = \sum_{n=0}^{\infty} \frac{f^{(n)}(0)}{n!} z^n,$$

which is known as *Maclaurin's series*.

(iii) Using Cauchy's integral formula,

$$f^{(n)}(z_0) = \frac{n!}{2\pi i}\int_C \frac{f(z)}{(z-z_0)^{n+1}}\,dz,$$

the Taylor's series (41) reduces to

$$f(z) = \sum_{n=0}^{\infty} a_n (z-z_0)^n, \quad (42)$$

where

$$a_n = \frac{f^{(n)}(z_0)}{n!} = \frac{1}{2\pi i}\int \frac{f(z)}{(z-z_0)^{n+1}}\,dz.$$

Theorem 3.19. On the circumference of the circle of convergence of a power series, there must be at least one singular point of the function, represented by the series.

Proof: Suppose on the contrary that there is no singularity on the circumference $|z - z_0| = R$ of the circle of convergence of the power series

$$f(z) = \sum_{n=0}^{\infty} a_n (z-z_0)^n. \quad (43)$$

Then $f(z)$ is analytic in the disk $|z - z_0| < R + \varepsilon$, $\varepsilon > 0$. But, this implies that the series (43) converges in the disk $|z - z_0| < R + \varepsilon$. This contradicts the assumption that $|z - z_0| = R$ is the circle of convergence. Hence, there is at least one singular point of the function $f(z)$ on the circle of convergence of the power series $\sum_{n=0}^{\infty} a_n (z-z_0)^n$.

EXAMPLE 3.68

If the function $f(z)$ is regular for $|z| < R$ and has the Taylor's expansion $\sum_{n=0}^{\infty} a_n z^n$, show that for $r < R$

$$\frac{1}{2\pi}\int_0^{2\pi} |f(re^{i\theta})|^2\,d\theta = \sum_{n=0}^{\infty} |a_n|^2 r^{2n}.$$

Hence, show that if $|f(z)| \le M$, $|z| < R$, then

$$\sum_{n=0}^{\infty} |a_n|^2 r^{2n} \le M^2. \quad (44)$$

[The relation (44) is called *Parseval's inequality*].

Solution. Since $f(z)$ is regular in the region $|z| = r < R$, it has the Taylor's series expansion

$$f(z) = \sum_{n=0}^{\infty} a_n z^n = \sum_{n=0}^{\infty} a_n r^n e^{in\theta}, \ z = r e^{i\theta}.$$

If $\overline{a}_n$ denotes the conjugate of a_n, we have

$$f(\overline{z}) = \sum_{p=0}^{\infty} \overline{a}_p r^p e^{-ip\theta}.$$

Hence

$$|f(z)|^2 = f(z)f(\overline{z}) = \sum_{n=0}^{\infty} a_n r^n e^{in\theta} \sum_{p=0}^{\infty} \overline{a}_p r^p e^{-ip\theta}.$$

The two series for $f(z)$ and $f(\overline{z})$ are absolutely convergent and, hence, their product is uniformly convergent for the range $0 \le \theta \le 2\pi$. Thus, term-by-term integration is justified. On integration, all the terms for which $n \ne p$ vanish, for

$$\int_0^{2\pi} e^{il\theta}\, d\theta = 0, \quad l \neq 0$$

Hence, we have

$$\frac{1}{2\pi}\int_0^{2\pi} |f(z)|^2\, d\theta = \frac{1}{2\pi}\sum_{n=0}^{\infty} a_n \bar{a}_n\, r^{2n} \int_0^{2\pi} d\theta$$

$$= \frac{1}{2\pi}\sum_{n=0}^{\infty} |a_n|^2\, r^{2n} 2\pi$$

$$= \sum_{n=0}^{\infty} |a_n|^2\, r^{2n}. \tag{45}$$

If $|f(z)| \leq$ M, then (45) gives

$$\sum_{n=0}^{\infty} |a_n|^2\, r^{2n} = \frac{1}{2\pi}\int_0^{2\pi} |f(z)|^2\, d\theta$$

$$\leq \frac{1}{2\pi}\int_0^{2\pi} M^2\, d\theta = M^2.$$

EXAMPLE 3.69

Show that a function which has no singularities in the finite part of the plane or at ∞ is a constant.

Solution. Since the function $f(z)$ has no singularities in the finite part of the plane, it can be expanded in the Taylor's series in any circle $|z| = K$ (arbitrarily large). Thus

$$f(z) = \sum_{r=0}^{\infty} A_r z^r$$

and so

$$f\left(\frac{1}{z}\right) = \sum_{r=0}^{\infty} A_r\, z^{-r}. \tag{46}$$

Further, if $f(z)$ has no singularity at $z = \infty$, $f\left[\frac{1}{z}\right]$ has no singularity at $z = 0$. Since, $f(z)$ has no singularity in finite plane, $f\left[\frac{1}{z}\right]$ also has none in the finite plane. Hence, by Taylor's theorem,

$$f\left(\frac{1}{z}\right) = \sum_{r=0}^{\infty} B_r\, z^r \tag{47}$$

From (46) and (47), we have

$$\sum_{r=0}^{\infty} B_r\, z^r = \sum_{r=0}^{\infty} A_r\, z^{-r}.$$

But this is not possible unless $B_r = A_r = 0$ for all values of r except zero in which case $A_0 = B_0$ and then $f(z) = A_0 = B_0 =$ constant.

EXAMPLE 3.70

If a function is analytic for all finite value of z as $|z| \to \infty$, and $|f(z)| = \mathrm{A}|z|^k$, then show that $f(z)$ is a polynomial of degree less than or equal to k.

Solution. Let $f(z)$ be analytic in the finite part of z-plane and let $\lim_{|z|\to\infty} |f(z)| = \mathrm{A}|z|^k$ We assume that $f(z)$ is analytic inside a circle $|z| = R$, where R is large but finite. Then $f(z)$ has Taylor's series

$$f(z) = \sum_{n=0}^{\infty} a_n z^n, \tag{48}$$

where

$$a_n = \frac{1}{2\pi i}\int_C \frac{f(z)}{z^{n+1}}\, dz.$$

Hence

$$|a_n| \leq \frac{1}{2\pi}\int \frac{|f(z)|}{|z|^{n+1}} |dz|$$

$$= \frac{M}{2\pi\, R^{n+1}}\int_C |dz|, \quad M = \max|f(z)| \text{ on } C.$$

$$= \frac{M}{2\pi R^{n+1}}.2\pi R$$

$$= \frac{M}{R^n} = \frac{A|z|^k}{R^n} = \frac{AR^k}{R^n}$$

$$= \frac{A}{R^{n-k}} \to 0 \text{ as } R \to \infty \text{ for } n > k.$$

Thus, $a_n = 0$ for all $n > k$. Hence (37) implies that $f(z)$ is a polynomial of degree $\leq k$.

EXAMPLE 3.71

Expand $\sin z$ in Taylor series about $z = \frac{\pi}{4}$.

Solution. We have $f(z) = \sin z$. So,

$$f'(z) = \cos z, f''(z) = -\sin z,$$
$$f'''(z) = -\cos z, f^{(4)}(z) = \sin z, \ldots$$

and

$$f\left(\frac{\pi}{4}\right) = \frac{\sqrt{2}}{2}, f'\left(\frac{\pi}{4}\right) = \frac{\sqrt{2}}{2},$$

$$f''\left(\frac{\pi}{4}\right) = \frac{-\sqrt{2}}{2}, f'''\left(\frac{\pi}{4}\right) = \frac{-\sqrt{2}}{2},$$

$$f^{(4)}\left(\frac{\pi}{4}\right) = \frac{\sqrt{2}}{2}, \ldots$$

Hence,

$$f(z) = f\left(\frac{\pi}{4}\right) + f'\left(\frac{\pi}{4}\right)\left(z - \frac{\pi}{4}\right) + \frac{1}{2!}f''\left(\frac{\pi}{4}\right)\left(z - \frac{\pi}{4}\right)^2 + \frac{1}{3!}f'''\left(\frac{\pi}{4}\right)\left(z - \frac{\pi}{4}\right)^3 + \ldots$$

$$= \frac{\sqrt{2}}{2}\left[1 + [z - (\pi/4)] - \frac{[z-(\pi/4)]^2}{2!} - \frac{[z-(\pi/4)]^3}{3!} + \ldots\right],$$

which is the required expansion.

Now suppose that $f(z)$ is not analytic in a disk but only in an *annular region* (*ring-shaped region*) bounded by two concentric circles C_1 and C_2 and is also analytic on C_1 and C_2. The function $f(z)$ can then be expressed in terms of two series by the following theorem known as *Laurent theorem.*

Theorem 3.20. (Laurent). Let $f(z)$ be analytic in the annular region bounded by two concentric circles C_1 and C_2 with centre z_0 and radii R_1 and R_2, respectively, with $0 < R_1 < R_2$. If z is any point of the annulus, we have

$$f(z) = \sum_{n=0}^{\infty} a_n (z - z_0)^n + \sum_{n=1}^{\infty} \frac{b_n}{(z - z_0)^n},$$

where

$$a_n = \frac{1}{2\pi i} \int_{C_2} \frac{f(\xi)d\xi}{(\xi - z_0)^{n+1}}, \quad n = 0, 1, 2, \ldots$$

$$b_n = \frac{1}{2\pi i} \int_{C_1} \frac{f(\xi)d\xi}{(\xi - z_0)^{1-n}}, \quad n = 1, 2, 3, \ldots$$

and integration over C_1 and C_2 is taken in anti-clockwise direction.

Proof: Since $f(z)$ is analytic on the circles and within the annular region between the two circles, the Cauchy integral formula yields

$$f(z) = \frac{1}{2\pi i}\left[\int_{C_2} \frac{f(\xi)}{\xi - z} d\xi - \int_{C_1} \frac{f(\xi)}{\xi - z} d\xi\right], \quad (49)$$

where z is any point in the region D (Figure 3.13).

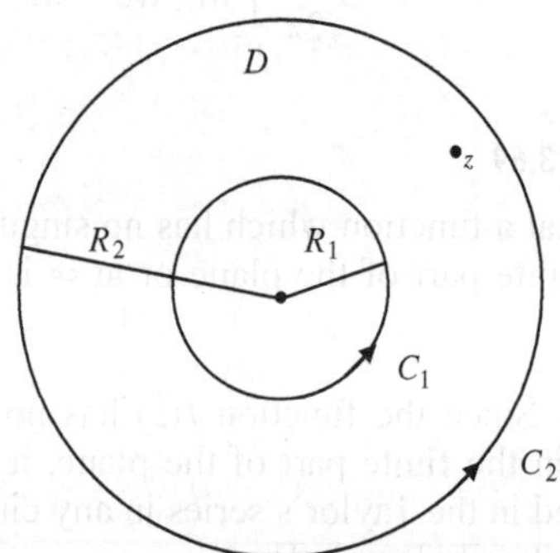

Figure 3.13

Now

$$\frac{1}{\xi - z} = \frac{1}{\xi - z_0 - (z - z_0)}$$

$$= \frac{1}{(\xi - z_0)\left(1 - \frac{z - z_0}{\xi - z_0}\right)}$$

$$= \frac{1}{\xi - z_0}\left(1 - \frac{z - z_0}{\xi - z_0}\right)^{-1}$$

$$= \frac{1}{\xi - z_0}\left[1 + \frac{z - z_0}{\xi - z_0} + \left(\frac{z - z_0}{\xi - z_0}\right)^2 + \ldots +\right.$$

$$+\ldots+\left(\frac{z-z_0}{\xi-z_0}\right)^n+\ldots\Bigg]$$

$$=\frac{1}{\xi-z_0}\sum_{n=0}^{\infty}\left(\frac{z-z_0}{\xi-z_0}\right)^n$$

$$=\sum_{n=0}^{\infty}\frac{(z-z_0)^n}{(\xi-z_0)^{n+1}}.$$

Therefore,

$$\frac{1}{2\pi i}\int_{C_2}\frac{f(\xi)}{\xi-z}d\xi=\frac{1}{2\pi i}\int_{C_2}f(\xi)\sum_{n=0}^{\infty}\frac{(z-z_0)^n}{(\xi-z_0)^{n+1}}d\xi$$

$$=\frac{1}{2\pi i}\sum_{n=0}^{\infty}\int_{C_2}\frac{f(\xi)(z-z_0)^n}{(\xi-z_0)^{n+1}}d\xi$$

$$=\sum_{n=0}^{\infty}a_n(z-z_0)^n,$$

where

$$a_n=\frac{1}{2\pi i}\int_{C_2}\frac{f(\xi)}{(\xi-z_0)^{n+1}}d\xi,\quad n=0,1,2,\ldots.$$

Further,

$$-\frac{1}{\xi-z}=\frac{1}{z-z_0-(\xi-z_0)}=\frac{1}{z-z_0}\left(1-\frac{\xi-z_0}{z-z_0}\right)^{-1}$$

$$=\frac{1}{z-z_0}\left[1+\frac{\xi-z_0}{z-z_0}+\left(\frac{\xi-z_0}{z-z_0}\right)^2+\ldots\right]$$

$$=\frac{1}{z-z_0}\sum_{n=0}^{\infty}\left(\frac{\xi-z_0}{z-z_0}\right)^n$$

$$=\sum_{n=0}^{\infty}\frac{(\xi-z_0)^n}{(z-z_0)^{n+1}}=\sum_{n=1}^{\infty}\frac{(\xi-z_0)^{n-1}}{(z-z_0)^n}.$$

Therefore,

$$-\frac{1}{2\pi i}\int_{C_1}\frac{f(\xi)}{\xi-z}d\xi=\frac{1}{2\pi i}\sum_{n=1}^{\infty}\int_{C_1}f(\xi)\frac{(\xi-z_0)^{n-1}}{(z-z_0)^n}d\xi$$

$$=\sum_{n=1}^{\infty}\frac{b_n}{(z-z_0)^n},$$

where

$$b_n=\frac{1}{2\pi i}\int_{C_1}\frac{f(\xi)}{(\xi-z_0)^{1-n}}d\xi.$$

Hence (49) becomes

$$f(z)=\sum_{n=0}^{\infty}a_n(z-z_0)^n+\sum_{n=1}^{\infty}b_n(z-z_0)^{-n},\quad (50)$$

where

$$a_n=\frac{1}{2\pi i}\int_{C_2}\frac{f(\xi)}{(\xi-z_0)^{n+1}}d\xi,$$

$$b_n=\frac{1}{2\pi i}\int_{C_1}\frac{f(\xi)}{(\xi-z_0)^{1-n}}d\xi.$$

Remark 3.9. Laurent's theorem is a generalization of Taylor's theorem. In fact, if $f(z)$ were analytic within and on C_2, then all the b_n are zero by Cauchy's theorem since the integrands are analytic within and on C_1. Also, then

$$a_n=\frac{f^{(n)}(z_0)}{n!},\quad n=0,1,2,\ldots$$

EXAMPLE 3.72

Expand $\frac{z-1}{z^2}$ in a Taylor series in powers of $z-1$ and determine the region of convergence.

Solution. We have

$$f(z)=\frac{z-1}{z^2}=\frac{1}{z}-\frac{1}{z^2},$$

$$f'(z)=-\frac{1}{z^2}+\frac{2}{z^3},$$

$$f''(z)=\frac{2}{z^3}-\frac{3.2}{z^4},$$

....

$$f^{(n)}(z)=\frac{(-1)^n n!}{z^{n+1}}+\frac{(-1)^{n+1}(n+1)!}{z^{n+2}}.$$

Hence

$$f(1)=0,\quad \frac{f^{(n)}(1)}{n!}=(-1)^{n+1}n.$$

Hence

$$f(z) = f(1) + \sum_{n=1}^{\infty} \frac{f^{(n)}(1)}{n!}(z-1)^n$$

$$= \sum_{n=1}^{\infty} (-1)^{n+1} n(z-1)^n.$$

The region of convergence is $|z-1| < 1$.

EXAMPLE 3.73

Determine the two Laurent series expansions in power of z of the function

$$f(z) = \frac{1}{z(1+z^2)}.$$

Solution. The function ceases to be regular at $z = 0$ and $z = \pm i$.

When $0 < |z| < 1$, then

$$f(z) = \frac{1}{z(1+z^2)} = \frac{1}{z}(1+z^2)^{-1}$$

$$= \frac{1}{z}[1 - z^2 + z^4 - z^6 + \ldots]$$

$$= \frac{1}{z} - z + z^3 - z^5 + \ldots$$

$$= \frac{1}{z} + \sum_{n=1}^{\infty} (-1)^n z^{2n-1}.$$

When $|z| > 1$, then $\left|\frac{1}{z}\right| < 1$ and so

$$f(z) = \frac{1}{z^3\left(1+\frac{1}{z^2}\right)} = \frac{1}{z^3}\left(1+\frac{1}{z^2}\right)^{-1}$$

$$= \frac{1}{z^3}\left[1 - \frac{1}{z^2} + \frac{1}{z^4} - \ldots\right]$$

$$= \frac{1}{z^3} - \frac{1}{z^5} + \frac{1}{z^7} - \ldots$$

$$= \sum_{n=1}^{\infty} (-1)^{n+1} \frac{1}{z^{2n+1}}.$$

EXAMPLE 3.74

Expand $f(z) = \dfrac{1}{(z+1)(z+3)}$ in Taylor's/Laurent's series valid for the region

(i) $|z| < 1$, (ii) $1 < |z| < 3$, (iii) $|z| > 3$, (iv) $0 < |z+1| < 2$.

Solution. We have

$$f(z) = \frac{1}{(z+1)(z+3)} = \frac{1}{2(z+1)} - \frac{1}{2(z+3)}.$$

(i) When $|z| < 1$, we have

$$f(z) = \frac{1}{2}(z+1)^{-1} - \frac{1}{2}(z+3)^{-1}$$

$$= \frac{1}{2}(1+z)^{-1} - \frac{1}{6}\left(1+\frac{z}{3}\right)^{-1}$$

$$= \frac{1}{2}[1 - z + z^2 - z^3 + \ldots] - \quad 1 - \quad +$$

$$-\frac{1}{6}\left[1 - \frac{z}{3} + \left(\frac{z}{3}\right)^2 - \left(\frac{z}{3}\right)^3 + \ldots\right]$$

$$= \frac{1}{3} - \frac{4}{9}z + \frac{13}{27}z^2 + \ldots$$

This is a Taylor's series valid for $|z| < 1$.

(ii) When $1 < |z| < 3$, we have for $|z| > 1$,

$$\frac{1}{2(z+1)} = \frac{1}{2z\left(1+\frac{1}{z}\right)} = \quad 2 \quad - \quad + \quad - \quad + \ldots$$

$$= \frac{1}{2z}\left(1 - \frac{1}{z} + \frac{1}{z^2} - \frac{1}{z^3} + \ldots\right)$$

$$= \frac{1}{2z} - \frac{1}{2z^2} + \frac{1}{2z^3} - \frac{1}{2z^4} + \ldots$$

and for $|z| < 3$,

$$\frac{1}{2(z+3)} = \frac{1}{6\left(1+\frac{z}{3}\right)} = \frac{1}{6}\left(1+\frac{z}{3}\right)^{-1}$$

$$= \frac{1}{6} - \frac{z}{18} + \frac{z^2}{54} - \frac{z^3}{162} + \ldots$$

Hence the Laurent series for $f(z)$ for the annulus $1 < |z| < 3$ is

$$f(z)=\ldots-\frac{1}{2z^4}+\frac{1}{2z^3}-\frac{1}{2z^2}+\frac{1}{2z}-\frac{1}{6}+\frac{z}{18}-\frac{z^2}{54}+\frac{z^3}{162}+\ldots$$

(iii) For $|z|>3$, we have

$$f(z)=\frac{1}{2(z+1)}-\frac{1}{2(z+3)}$$
$$=\frac{1}{2z}\left(1+\frac{1}{z}\right)-\frac{1}{2z}\left(1+\frac{3}{z}\right)^{-1}$$
$$=\frac{1}{z^2}-\frac{4}{z^3}+\frac{13}{z^4}-\ldots$$

(iv) When $0<|z+1|<2$, we substitute $z+1=u$, then we have $0<|u|<2$ and

$$f(z)=\frac{1}{u(u+2)}=\frac{1}{2u\left(1+\frac{u}{2}\right)}=\frac{1}{2u}\left(1+\frac{u}{2}\right)^{-1}$$
$$=\frac{1}{2u}-\frac{1}{4}-\frac{u}{8}-\frac{u^2}{16}+\ldots$$
$$=\frac{1}{2(z+1)}-\frac{1}{4}+\frac{z+1}{8}-\frac{(z+1)^2}{16}+\ldots.$$

EXAMPLE 3.75

Obtain Taylor's/Laurent's series expansion for $f(z)=\frac{(z-2)(z+2)}{(z+1)(z+4)}$ which are valid

(i) When $|z|<1$
(ii) When $1<|z|<4$
(iii) When $|z|>4$.

Solution. (i) We have

$$f(z)=\frac{(z-2)(z+2)}{(z+1)(z+4)}=1-\frac{1}{z+1}-\frac{4}{z+4}.$$

When $|z|<1$,

$$f(z)=1-(1+z)^{-1}-\frac{4}{4}\left(1-\frac{z}{4}\right)^{-1}$$
$$=1-[1-z+z^2-\ldots]-\left[1-\frac{z}{4}+\left(\frac{z}{4}\right)^2-\ldots\right]$$
$$=-1+(z-z^2+\ldots)+\left[\frac{z}{4}-\left(\frac{z}{4}\right)^2+\ldots\right]$$
$$=-1+\sum_{n=1}^{\infty}(-1)^{n+1}(1+4^{-n})z^n,$$

which is a Maclaurine's series.

(ii) When $1<|z|<4$, we have $\frac{1}{|z|}<1$ and $\frac{|z|}{4}<1$. Thus

$$f(z)=1-\frac{1}{z}\left(1+\frac{1}{z}\right)^{-1}-\frac{4}{4}\left(1+\frac{z}{4}\right)^{-1}$$
$$=\left[-\frac{1}{z}+\frac{1}{z^2}-\frac{1}{z^3}+\ldots\right]-\left[-\frac{z}{4}+\frac{z^2}{4^2}-\frac{z^3}{4^3}+\ldots\right]$$
$$=\sum_{n=1}^{\infty}(-1)^n\left[\frac{1}{z^n}-\left(\frac{z}{4}\right)^n\right],$$

which is a Laurent series.

(iii) When $|z|>4$, we have $\frac{4}{|z|}<1$. Hence

$$f(z)=1-\frac{1}{z}\left(1+\frac{1}{z}\right)^{-1}-\frac{4}{z}\left(1+\frac{4}{z}\right)^{-1}$$
$$=1-\frac{1}{z}\left[1-\frac{1}{z}+\frac{1}{z^2}-\ldots\right]-\frac{4}{z}\left[1-\frac{4}{z}+\frac{4^2}{z^2}-\ldots\right]$$
$$=1+\sum_{n=1}^{\infty}(-1)^n(1+4^n)\frac{1}{z^n},$$

which is again a Laurent's series.

EXAMPLE 3.76

Find series expansion of $f(z)=\frac{1}{(z-1)(z-2)}$ in the regions

(i) $1<|z|<2$, (ii) $|z|>2$, (iii) $0<|z-1|<1$.

Solution. (i) We have

$$f(z)=\frac{1}{(z-1)(z-2)}=\frac{1}{z-2}-\frac{1}{z-1}.$$

Now $|z| > 1$ implies $\frac{1}{|z|} < 1$ and $|z| < 2$ implies $\left|\frac{z}{2}\right| < 1$. Hence

$$f(z) = \frac{-1}{2\left(1-\frac{z}{2}\right)} - \frac{1}{z\left(1-\frac{1}{z}\right)}$$

$$= -\frac{1}{2}\left(1-\frac{z}{2}\right)^{-1} - \frac{1}{z}\left(1-\frac{1}{z}\right)^{-1}$$

$$= -\frac{1}{2}\left(1+\frac{z}{2}+\frac{z^2}{4}+\frac{z^3}{8}+\ldots\right) - \frac{1}{z}\left(1+\frac{1}{z}+\frac{1}{z^2}+\frac{1}{z^3}+\ldots\right)$$

$$= \ldots - \frac{1}{z^4} - \frac{1}{z^3} - \frac{1}{z^2} - \frac{1}{z} - \frac{1}{2} - \frac{z}{4} - \frac{z^2}{8} - \ldots,$$

which is a Laurent's series

(ii) When $|z| > 2$, then $\left|\frac{2}{z}\right| < 1$ and so

$$f(z) = \frac{1}{z\left(1-\frac{2}{z}\right)} - \frac{1}{z\left(1-\frac{1}{z}\right)}$$

$$= \frac{1}{z}\left(1-\frac{2}{z}\right)^{-1} - \frac{1}{z}\left(1-\frac{1}{z}\right)^{-1}$$

$$= \frac{1}{z}\left(1+\frac{2}{z}+\frac{4}{z^2}-\ldots\right) - \frac{1}{z}(1+z+z^2+\ldots)$$

$$= \ldots + \frac{4}{z^3} + \frac{2}{z^2} - 1 - z - z^2 - \ldots$$

(iii) When $0 < |z-1| < 1$, we substitute $z - 1 = u$ and get $0 < |u| < 1$. Then

$$f(z) = \frac{1}{u-1} - \frac{1}{u} = -\frac{1}{1-u} - \frac{1}{u}$$

$$= -(1-u)^{-1} - \frac{1}{u} = -(1+u+u^2+\ldots) - \frac{1}{u}$$

$$= -\frac{1}{u} - 1 - u - u^2 - \ldots$$

$$= -\frac{1}{z-1} - 1 - (z-1) - (z-1)^2 - \ldots,$$

which is a Laurent's series.

3.11 ZEROS AND POLES

Let $f(z)$ be analytic in a domain D. Then it can be expanded in Taylor's series about any point z_0 in D as

$$f(z) = \sum_{n=0}^{\infty} a_n (z-z_0)^n, \ldots$$

where

$$a_n = \frac{1}{2\pi i}\int \frac{f(z)}{(z-z_0)^{n+1}}\,dz.$$

If $a_0 = a_1 = a_2 = \ldots = a_{m-1} = 0$ but $a_m \neq 0$, then the first term in the above expansion is $a_m(z-z_0)^m$ and we say that $f(z)$ has a *zero of order m* at $z = z_0$.

If $f(z)$ satisfies the conditions of the Laurent's theorem, then

$$f(z) = \sum_{n=0}^{\infty} a_n (z-z_0)^n + \sum_{n=1}^{\infty} b_n (z-z_0)^{-n}$$

where

$$a_n = \frac{1}{2\pi i}\int_{C_2} \frac{f(\xi)d\xi}{(\xi-z_0)^{n+1}}, \quad n = 0,1,2,\ldots$$

$$b_n = \frac{1}{2\pi i}\int_{C_1} \frac{f(\xi)d\xi}{(\xi-z_0)^{1-n}}, \quad n = 1,2,3,\ldots$$

The term $\sum_{n=1}^{\infty} b_n (z-z_0)^{-n}$ is called the *principal part* of the function $f(z)$ at $z = z_0$.

Now there are the following three possibilities:

(i) If the principal part has only a *finite number of terms* given by

$$\frac{b_1}{z-z_0} + \frac{b_2}{(z-z_0)^2} + \ldots + \frac{b_n}{(z-a)^n}, b_n \neq 0,$$

then the point $z = z_0$ is called *a pole of order n*. If $n = 1$, then z_0 is called a *simple pole*.

(ii) If the principal part in Laurent expansion of $f(z)$ contains an *infinite number of terms*, then $z = z_0$ is called as *isolated essential singularity*.

(iii) If the principal part in Laurent expansion of $f(z)$ *does not contain any term*, that is, all b_n are zeros, then

$$f(z) = a_0 + a_1(z - z_0) + a_2(z - z_0^2) + \ldots + a_n(z - z_0)^2 + \ldots$$

and $z = z_0$ is called a *removable singularity*. Setting $f(z_0) = a_0$ makes $f(z)$ analytic at z_0.

From the Laurent expansion, it follows that a function $f(z)$ has a pole of order m at z_0 if and only if

$$f(z) = \frac{g(z)}{(z - z_0)^m},$$

where $g(z)$ is analytic at z_0 and $g(z_0) \neq 0$.

EXAMPLE 3.77

Find Laurent expansion of $\frac{z}{(z+1)(z+2)}$ about $z = -2$ and name the singularity.

Solution. We have $f(z) = \frac{z}{(z+1)(z+2)}$.

Substitute $z + 2 = u$. Then

$$\begin{aligned}\frac{z}{(z+1)(z+2)} &= \frac{u-2}{(u-1)u} = \frac{2-u}{u(1-u)}\\ &= \frac{2-u}{u}(1+u+u^2+u^3+\ldots)\\ &= \frac{2}{u}+1+u+u^2+u^3+\ldots\\ &= \frac{2}{z+2}+1+(z+2)+(z+2)^2+\ldots\end{aligned}$$

Thus, the Laurent expansion about $z = -2$ has only one term in the principal part. Hence $z = -2$ is a *simple pole*.

EXAMPLE 3.78

Find Laurent's expansion of $f(z) = \frac{e^{2z}}{(z-1)^3}$ about $z = 1$ and name the singularity.

Solution. We have $f(z) = \frac{e^{2z}}{(z-1)^3}$.

Substituting $z - 1 = u$, we get

$$\begin{aligned}\frac{e^{2z}}{(z-1)^3} &= \frac{e^{2(u+1)}}{u^3} = \frac{e^2}{u^3}.e^{2u}\\ &= \frac{e^2}{u^3}\left[1+2u+\frac{(2u)^2}{2!}+\frac{(2u)^3}{3!}+\ldots\right]\\ &= \frac{e^2}{u^3}+\frac{2e^2}{u^2}+\frac{2e^2}{u}+\frac{4e^2}{3}+\frac{2e^2u}{3}+\ldots\\ &= \frac{e^2}{(z-1)^3}+\frac{2e^2}{(z-1)^2}+\frac{2e^2}{z-1}+\frac{4e^2}{3}+\frac{2e^2(z-1)}{3}+\ldots.\end{aligned}$$

Thus, we obtain Laurent's series whose principal part consists of three terms. Hence, $f(z)$ has a pole of order 3 at $z = 1$. The function is analytic everywhere except for the pole of order 3 at $z = 1$. Hence, the series converges for all $z \neq 1$.

EXAMPLE 3.79

Find the Taylor and Laurent's series which represent the function $\frac{z^2-1}{z^2+5z+6}$ in the region

(i) $|z| < 2$ (ii) $2 < |z| < 3$ (iii) $|z| > 3$.

Solution. We have

$$\begin{aligned}f(z) &= \frac{z^2-1}{(z+3)(z+2)}\\ &= 1+\frac{3}{z+2}-\frac{8}{z+3}.\end{aligned}$$

(i) When $|z| < 2$, we have $\left|\frac{z}{2}\right| < 1$ and

$$\begin{aligned}f(z) &= 1+\frac{3}{2\left(1+\frac{z}{2}\right)}-\frac{8}{3\left(1+\frac{z}{3}\right)}\\ &= 1+\frac{3}{2}\left(1+\frac{z}{2}\right)^{-1}-\frac{8}{3}\left(1+\frac{z}{3}\right)^{-1}\\ &= 1+\frac{3}{2}\left(1-\frac{z}{2}+\frac{z^2}{4}-\ldots\right)-\frac{8}{3}\left(1-\frac{z}{3}+\frac{z^2}{9}-\ldots\right)\\ &= \left(1+\frac{3}{2}-\frac{8}{3}\right)-z\left(\frac{3}{4}-\frac{8}{9}\right)+z^2\left(\frac{3}{8}-\frac{8}{27}\right)-\ldots\\ &= -\frac{1}{6}+\sum_{n=1}^{\infty}(-1)^n\left(\frac{3}{2^{n+1}}-\frac{8}{3^{n+1}}\right)z^n.\end{aligned}$$

(ii) When $2 < |z| < 3$, we have $\left|\frac{2}{z}\right| < 1$ and $\left|\frac{z}{3}\right| < 1$. Hence

$$f(z) = 1 + \frac{3}{z\left(1+\frac{2}{z}\right)} - \frac{8}{3\left(1+\frac{z}{3}\right)}$$

$$= 1 + \frac{3}{z}\left(1+\frac{2}{z}\right)^{-1} - \frac{8}{3}\left(1+\frac{z}{3}\right)^{-1}$$

$$= 1 + \frac{3}{z}\left(1-\frac{2}{z}+\frac{4}{z^2}-\ldots\right) - \frac{8}{3}\left(1-\frac{z}{3}+\frac{z^2}{9}-\ldots\right)$$

$$= -\frac{5}{3} + \sum_{n=1}^{\infty}(-1)^{n+1}\frac{8}{3^{n+1}}z^n + \sum_{n=1}^{\infty}\frac{3(-2)^{n+1}}{z^n}$$

which is a Laurent's series

(iii) When $|z| > 3$, we have $\left|\frac{3}{z}\right| < 1$. Hence

$$f(z) = 1 + \frac{3}{z\left(1+\frac{2}{z}\right)} - \frac{8}{z\left(1+\frac{3}{z}\right)}$$

$$= 1 + \frac{3}{z}\left(1-\frac{2}{z}+\frac{4}{z^2}-\ldots\right) - \frac{8}{z}\left(1-\frac{3}{z}+\frac{9}{z^2}-\ldots\right)$$

$$= 1 + \sum_{n=1}^{\infty}\left[(-1)^n\left\{\frac{8(3)^{n-1}-3(2)^{n-1}}{z^n}\right\}\right],$$

which is a Laurent's series.

EXAMPLE 3.80

Find the singularities of $f(z) = \frac{z}{(z^2+4)^2}$ and indicate the character of the singularities.

Solution. We have

$$f(z) = \frac{z}{(z^2+4)^2} = \frac{z}{[(z+2i)(z-2i)]^2}$$

$$= \frac{z}{(z+2i)^2\,(z-2i)^2}.$$

Since $\lim_{z\to 2i}(z-2i)^2 f(z) = \lim_{z\to 2i}\frac{z}{(z+2i)^2} = \frac{1}{8i}$ $\neq 0$, it follows that $z = 2i$ is a pole of order 2. Similarly, $z = -2i$ is a pole of order 2. Further, we can find δ such that no other singularity other than $z = 2i$ lies inside the circle $|z - 2i| = \delta$ (for example, we may take $\delta = 1$). Hence $z = 2i$ is an isolated singularity. Similarly, $z = -2i$ is also an isolated singularity.

EXAMPLE 3.81

Find the nature and the location of the singularities of $f(z) = \frac{1}{z(e^z-1)}$. Show that if $0 < |z| < 2\pi$, the function can be expanded in Laurent's series.

Solution. We have

$$f(z) = \frac{1}{z(e^z-1)}.$$

The function ceases to be regular at $z = 0$ and $e^z - 1 = 0$, that is, $e^z = 1$ or for $e^z = e^{\pm 2n\pi i}$ or for $z = \pm 2n\pi i$, $n = 0, \pm 1, \pm 2, \ldots$ Thus, $z = 0$ is a *double pole* (pole of order 2). The other singularities are simple poles. Hence, the function can be expanded in Laurent's series in the annulus $0 < |z| < 2\pi$. We note that

$$f(z) = \frac{1}{z\left(1+z+\frac{z^2}{2!}+\frac{z^3}{3!}+\ldots-1\right)}$$

$$= \frac{1}{z^2\left(1+\frac{z}{2!}+\frac{z^2}{3!}+\ldots\right)}$$

$$= \frac{1}{z^2}\left[1+\frac{z}{2!}+\frac{z^2}{3!}+\ldots\right]^{-1}$$

$$= \frac{1}{z^2}\left[1-\left(\frac{z}{2!}+\frac{z^2}{3!}+\ldots\right)+\left(\frac{z}{2!}+\frac{z^2}{3!}+\ldots\right)^2+\ldots\right]$$

$$= \frac{1}{z^2}\left[1-\frac{z}{2!}+\left(\frac{1}{4}-\frac{1}{6}\right)z^2+z^3\left(\frac{1}{24}+\frac{1}{6}-\frac{1}{8}\right)+\ldots\right]$$

$$= \frac{1}{z^2}-\frac{1}{2z}+\frac{1}{12}-\frac{1}{120}z+\ldots$$

EXAMPLE 3.82

Show that $\frac{e^z}{z^3}$ has a pole of order 3 at $z = 0$.

Solution. We have

$$\frac{e^z}{z^3} = \frac{1}{z^3}\left(1 + z + \frac{z^2}{2!} + \frac{z^3}{3!} + \ldots\right)$$

$$= \frac{1}{z^3} + \frac{1}{z^2} + \frac{1}{2!z} + \frac{1}{3!} + \frac{1}{4!}z + \ldots$$

Thus, the principal part of the Laurent expansion consists of three terms and so $\frac{e^z}{z^3}$ has a pole of order 3 at $z = 0$.

EXAMPLE 3.83

Show that $z \sin \frac{1}{z}$ has essential singularity at $z = 0$.

Solution. We have

$$z \sin \frac{1}{z} = z\left\{\frac{1}{z} - \frac{1}{3!z^3} + \frac{1}{5!z^5} - \ldots\right\}$$

$$= 1 - \frac{1}{3!z^2} + \frac{1}{5!z^4} + \ldots$$

Since the series *does not terminate*, $z = 0$ is an *essential singularity*.

Definition 3.31. A function $f(z)$ is said to be *meromorphic* if it is analytic in the finite part of the plane except at a finite number of poles.

3.12 RESIDUES AND CAUCHY'S RESIDUE THEOREM

Definition 3.32. Let the Laurent series expansion of a function $f(z)$ at isolated singularity z_0 be

$$f(z) = \sum_{n=1}^{\infty} \frac{b_n}{(z - z_0)^n} + \sum_{n=0}^{\infty} a_n (z - z_0)^n \qquad (51)$$

The coefficient b_1, in the principal part of the expansion, given by

$$b_1 = \frac{1}{2\pi i} \int_C f(\xi)\, d\xi \qquad (52)$$

for the contour C: $|z - z_0| = r < R$ is called *residue* of $f(z)$ at z_0 and is denoted by $\text{Res}(z_0)$.

The residue of $f(z)$ at $z = \infty$ is defined by

$$-b_1 = \frac{-1}{2\pi i} \int_C f(\xi)\, d\xi.$$

It is the coefficient of $\frac{1}{z}$ with its sign changed in the expansion of $f(z)$ in the neighbourhood of $z = \infty$.

If $f(z)$ has a pole of order m at z_0, then the Laurent expansion of $f(z)$ is

$$f(z) = \frac{b_m}{(z - z_0)^m} + \frac{b_{m-1}}{(z - z_0)^{m-1}} + \ldots + \frac{b_1}{(z - z_0)}$$
$$+ a_0 + a_1(z - z_0) + a_2(z - z_0^2) + \ldots$$

Multiplying both sides by $(z - z_0)^m$, we have

$$(z - z_0)^m f(z) = b_m + b_{m-1}(z - z_0)$$
$$+ \ldots + b_1(z - z_0)^{m-1}$$
$$+ a_0(z - z_0)^m$$
$$+ a_1(z - z_0)^{m+1} + \ldots,$$

which is Taylor's series of the analytic function $(z - z_0)^m f(z)$. Differentiating both sides $m - 1$ times with respect to z, we have

$$\frac{d^{m-1}}{dz^{m-1}}[(z - z_0)^m f(z)] = b_1(m-1)!$$
$$+ m(m-1)\ldots 2a_0(z - z_0) + \ldots$$

Letting $z \to z_0$, we get

$$\lim_{z \to z_0} \frac{d^{m-1}}{dz^{m-1}}[(z - z_0)^m f(z)] = b_1(m-1)!$$

and so

$$b_1 = \text{Res}(z_0)$$
$$= \frac{1}{(m-1)!} \lim_{z \to z_0} \frac{d^{m-1}}{dz^{m-1}}[(z - z_0)^m f(z)]. \quad (53)$$

If z_0 is a simple pole, that is, $m = 1$, then

$$\text{Res}(z_0) = b_1 = \lim_{z \to z_0} (z - z_0) f(z). \qquad (54)$$

If

$$f(z) = \frac{p(z)}{q(z)},$$

where $p(z)$ and $q(z)$ are analytic at $z = z_0$, $p(z_0) \neq 0$ and $q(z)$ has a simple zero at z_0, that is, $f(z)$ has a simple pole at z_0. Then $q(z) = (z - z_0)\, g(z)$, $g(z_0) \neq 0$ and $q(z_0) = 0$. Hence (54) reduces to

$$\text{Res}(z_0) = \lim_{z \to z_0} (z - z_0)\, f(z) \qquad \lim_{z\ z} (\quad)$$

$$= \lim_{z \to z_0} (z - z_0) \frac{p(z)}{q(z)}$$

$$= \lim_{z \to z_0} \frac{p(z)}{\dfrac{q(z) - q(z_0)}{z - z_0}} = \frac{p(z_0)}{q'(z_0)} \qquad (55)$$

Thus, the residues at poles can be calculated using the formulas (53), (54), and (55).

EXAMPLE 3.84

Find residues of

(i) $f(z) = \dfrac{z^2 - 2z}{(z+1)^2\,(z^2+4)}$

(ii) $f(z) = \dfrac{e^{z^2}}{(z-i)^3}$

at all its poles.

Solution. (i) The function $f(z)$ has a pole of order 2 at $z = -1$ and simple poles at $z = \pm 2i$. Therefore,

$$\text{Res}(-1) = \lim_{z \to -1} \frac{d}{dz}\left[(z+1)^2 \frac{z^2 - 2z}{(z+1)^2 (z^2+4)}\right]$$

$$= \lim_{z \to -1} \frac{(z^2+4)(2z-2) - (z^2 - 2z)(2z)}{(z^2+4)^2}$$

$$= -\frac{14}{25}$$

$$\text{Res}(2i) = \lim_{z \to 2i} (z - 2i) \frac{z^2 - 2z}{(z+1)^2\,(z^2+4)}$$

$$= \lim_{z \to 2i} \frac{z^2 - 2z}{(z+1)^2 (z+2i)} = \frac{7+i}{25}$$

$$\text{Res}(-2i) = \lim_{z \to -2i} (z+2i) \frac{z^2 - 2z}{(z+1)^2\,(z-2i)\,(z+2i)}$$

$$= \frac{7-i}{25}$$

(ii) $f(z)$ has a pole of order 3 at $z = i$. Hence

$$\text{Res}(i) = \frac{1}{2!} \lim_{z \to i} \frac{d^2}{dz^2} [(z-i)^3 f(z)]$$

$$= \frac{1}{2} \lim_{z \to i} \frac{d^2}{dz^2} (e^{z^2})$$

$$= \lim_{z \to i} [2z^2 e^{z^2} + e^{z^2}]$$

$$= -\frac{1}{e}.$$

EXAMPLE 3.85

Find the residue of $f(z) = \cot z$ at its poles.

Solution. We have

$$f(z) = \cot z = \frac{\cos z}{\sin z}.$$

The poles of $f(z)$ are given by $\sin z = 0$. Thus $z = n\pi$, $n = 0, \pm 1, \pm 2, \ldots$ are the simple poles. Using formula (55), the residue at $z = n\pi$ is given by

$$\text{Res}(n\pi) = \lim_{z \to n\pi} \frac{\cos z}{\dfrac{d}{dz}(\sin z)} = \lim_{z \to n\pi} \frac{\cos z}{\cos z} = 1.$$

EXAMPLE 3.86

Find the residue at each pole of $f(z) = \dfrac{ze^{iz}}{z^2 + a^2}$.

Solution. We have

$$f(z) = \frac{ze^{iz}}{(z+ai)\,(z-ai)}$$

Therefore, $f(z)$ has simple poles at $z = \pm\, ai$. Now

$$\text{Res}(ai) = \lim_{z \to ai} (z - ai) f(z)$$

$$= \lim_{z \to ai} (z - ai) \frac{ze^{iz}}{(z+ai)\,(z-ai)}$$

$$= \lim_{z \to ai} \frac{ze^{iz}}{z + ai} = \frac{e^{-a}}{2},$$

$$\text{Res}(-ai) = \lim_{z \to -ai} (z + ai) f(z)$$

$$= \lim_{z \to -ai} \frac{ze^{iz}}{z - ai} = \frac{e^{a}}{2}.$$

EXAMPLE 3.87

Find the residue of $f(z) = \dfrac{1 - e^{2z}}{z^4}$ at its poles.

Solution. The function $f(z)$ has a pole of order 4 at $z = 0$. Therefore,

$$\text{Res}(0) = \frac{1}{3!} \lim_{z \to 0} \frac{d^3}{dz^3}[(z - 0)^4 f(z)]$$

$$= \frac{1}{6} \lim_{z \to 0} \frac{d^3}{dz^3}[1 - e^{2z}]$$

$$= \frac{1}{6} \lim_{z \to 0} \frac{d^3}{dz^3}\left[1 - \left\{1 + 2z + 4z^2 + 8z^3 + \frac{16z^4}{4!} + \ldots\right\}\right]$$

$$= \frac{1}{6} \lim_{z \to 0} \frac{d^3}{dz^3}\left[-2z - 2z^2 - \frac{8}{6}z^3 - \frac{16}{24}z^4 - \ldots\right]$$

$$= \frac{1}{6} \lim_{z \to 0}\left[-8 - \frac{294}{24}z - \ldots\right] = -\frac{4}{3}.$$

EXAMPLE 3.88

Find the residues of $f(z) = \dfrac{z^3}{(z-1)(z-2)(z-3)}$ at $z = 1, 2,$ and 3 and ∞ and show that their sum is zero.

Solution. The function

$$f(z) = \frac{z^3}{(z-1)(z-2)(z-3)}$$

has simple poles at $z = 1, 2,$ and 3. Now

$$\text{Res}(1) = \lim_{z \to 1}(z-1) f(z) = \lim_{z \to 1} \frac{z^3}{(z-2)(z-3)} = \frac{1}{2}$$

$$\text{Res}(2) = \lim_{z \to 2}(z-2) f(z) = \lim_{z \to 2} \frac{z^3}{(z-1)(z-3)} = -8$$

$$\text{Res}(3) = \lim_{z \to 3}(z-3) f(z) = \lim_{z \to 3} \frac{z^3}{(z-1)(z-2)} = \frac{27}{2}.$$

To find residue at ∞, we expand $f(z)$ in the neighbourhood of $z = \infty$ as follows:

$$f(z) = \frac{z^3}{z^3\left(1 - \frac{1}{z}\right)\left(1 - \frac{2}{z}\right)\left(1 - \frac{3}{z}\right)}$$

$$= \left(1 - \frac{1}{z}\right)^{-1}\left(1 - \frac{2}{z}\right)^{-1}\left(1 - \frac{3}{z}\right)^{-1}$$

$$= 1 + \frac{6}{z} + \textit{higher powers of } \frac{1}{z}.$$

Now residue at ∞ is coefficient of $\frac{1}{z}$ with sign changed. Thus $\text{Res}(\infty) = -6$. Hence, the sum of the residues equals $\frac{1}{2} - 8 + \frac{27}{2} - 6 = 0$.

To compute the values of integrals in our study, we shall require the following theorem.

Theorem 3.21. (Cauchy's Residue Theorem). If $f(z)$ is analytic within and on a closed contour C except at finitely many poles lying in C, then

$$\int_C f(z)\, dz = 2\pi i \,\Sigma R,$$

where ΣR denotes the sum of residues of $f(z)$ at the poles within C.

Proof: Let $z_1, z_2, \ldots, z_n$ be the n poles lying in C. Let $C_1, C_2, \ldots, C_n$ be the circles with centre $z_1, z_2, \ldots, z_n$ and radius ρ such that all these circles lie entirely within C and do not overlap (Figure 3.14)

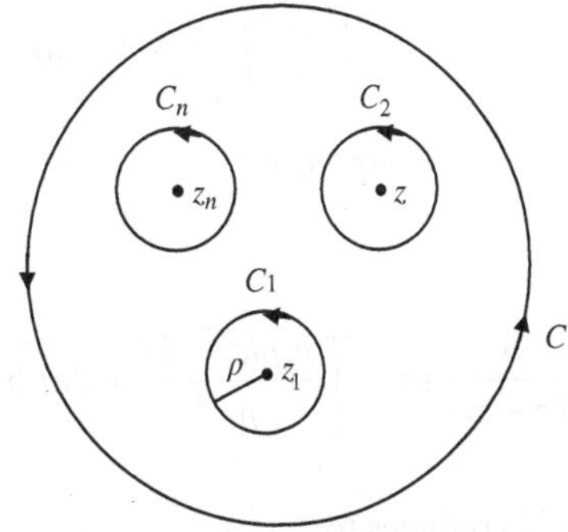

Figure 3.14

Then $f(z)$ is analytic in the region between C and the circles. Hence, by Cauchy-Goursat theorem

$$\int_C f(z)\,dz = \int_{C_1} f(z)\,dz + \int_{C_2} f(z)\,dz + \ldots + \quad f(z)\,dz$$

$$+ \int_{C_n} f(z)\,dz. \qquad (56)$$

If $f(z)$ has a pole of order m_1 at $z = z_1$, then

$$f(z) = \phi_1(z) + \sum_{r=1}^{m_1} \frac{b_r}{(z-z_1)^r},$$

where $\phi_1(z)$ is regular within and on C_1. Therefore,

$$\int_{C_1} f(z)\,dz = \int_{C_1} \phi_1(z)\,dz + \int_{C_1} \frac{b_1}{(z-z_1)}\,dz +$$

$$+ \int_{C_1} \frac{b_2}{(z-z_2)^2}\,dz + \ldots + \int_{C_1} \frac{bm_1}{(z-z_1)^r}\,dz. \qquad (57)$$

Since $\phi_1(z)$ is analytic within and on C_1, by Cauchy-Goursat theorem

$$\int_{C_1} \phi_1(z)\,dz = 0.$$

Moreover, substituting $z - z_1 = \rho\, e^{i\theta}$, we have

$$\int_{C_1} \frac{bm_1}{(z-z_1)^r}\,dz = \int_0^{2\pi} \frac{bm_1 . pie^{i\theta}}{e^{m_1}\, e^{m_1 i\theta}}\,d\theta$$

$$= \frac{ibm_1}{e^{m_1 - 1}} \int_0^{2\pi} e^{-(m_1-1)i\theta}\,d\theta = 0$$

for $m_1 \neq 1$

and

$$\int_{C_1} \frac{b_1}{(z-z_1)}\,dz = \int_0^{2\pi} \frac{b_1 \rho i e^{i\theta}\,d\theta}{\rho e^{i\theta}} = 2\pi i\, b_1.$$

Hence (57) reduces to

$$\int_{C_1} f(z)\,dz = 2\pi i R_1,$$

where R_1 is the residue of $f(z)$ at $z = z_1$. Similarly,

$$\int_{C_2} f(z)\,dz = 2\pi i\, R_2,$$

....

$$\int_{C_n} f(z)\,dz = 2\pi i\, R_n,$$

where R_i is the residue of $f(z)$ at $z = z_i$. Hence (56) becomes

$$\int_C f(z)\,dz = 2\pi i(R_1 + R_2 + \ldots + R_n) = 2\pi i\, \Sigma R,$$

b bm dz + ... + dz.

where $\Sigma R = R_1 + R_2 + \ldots + R_n$.

Remark 3.10. In the assumptions of Cauchy's integral formula, $f(z)$ is assumed to be analytic within and on a closed curve. Therefore, $\frac{f(z)}{z-a}$ has a simple pole at $z = a$. Then

$$\text{Res}(a) = \lim_{z\to a} (z-a)\frac{f(z)}{(z-a)} = f(a).$$

Hence, by Cauchy's Residue theorem, we have

$$\int_C \frac{f(z)}{z-a}\,dz = 2\pi i\, f(a),$$

that is,

$$f(a) = \frac{1}{2\pi i}\int_C \frac{f(z)}{z-a}\,dz.$$

It follows, therefore, that Cauchy's integral formula is a particular case of Cauchy's Residue theorem.

EXAMPLE 3.89

Evaluate

$$\int_C \frac{dz}{z^2(z+1)(z-1)}, \quad C : |z| = 3.$$

Solution. The integrand has simple poles at $z = 1$ and $z = -1$ and double poles at $z = 0$ lying in C. Therefore,

$$\text{Res}(1) = \lim_{z\to 1}(z-1)f(z) = \lim_{z\to 1}\frac{1}{z^2(z+1)} = \frac{1}{2},$$

$$\text{Res}(-1) = \lim_{z\to -1}(z+1)f(z) = \lim_{z\to -1}\frac{1}{z^2(z-1)} = -\frac{1}{2},$$

$$\text{Res}(0) = \lim_{z\to 0}\frac{d}{dz}[(z-0)^2 f(z)] = \lim_{z\to 0}\frac{d}{dz}\left[\frac{1}{(z+2)(z-1)}\right] = \lim_{z\to 0}\frac{-2z-1}{(z^2+z-2)^2} = \frac{-1}{4}.$$

Hence, by Cauchy-Residue theorem,

$$\int_C \frac{dz}{z^2(z+1)(z-1)} = 2\pi i\left[\frac{1}{2}-\frac{1}{2}-\frac{1}{4}\right] = -\frac{\pi i}{2}.$$

EXAMPLE 3.90

Evaluate the integral

$$\int_{|z|=1}\frac{4z^2-4z+1}{(z-2)(4+z^2)}dz.$$

Solution. Let

$$f(z) = \frac{4z^2-4z+1}{(z-2)(4+z^2)}.$$

The poles of $f(z)$ are $z = 2$, $z = \pm 2i$. We note that none of these poles lie in the curve $|z| = 1$. Thus the function is analytic within and on $|z| = 1$. Hence, by Cauchy-Goursat theorem,

$$\int_{|z|=1}\frac{4z^2-4z+1}{(z-2)(4+z^2)}dz = 0.$$

EXAMPLE 3.91

Evaluate $\int_C \frac{dz}{(z^2+4)^2}$, where C is the curve $|z-i| = 2$.

Solution. Let

$$f(z) = \frac{1}{(z^2+4)^2} = \frac{1}{(z+2i)^2(z-2i)^2}.$$

Thus $f(z)$ has two double poles at $z = 2i$ and $z = -2i$, out of which only $z = 2i$ lies within $|z-i| = 2$. Now

$$\text{Res}(2i) = \lim_{z\to 2i}\frac{d}{dz}[(z-2i)^2.f(z)] = \lim_{z\to 2i}\frac{d}{dz}\left[\frac{1}{(z+2i)^2}\right] = \lim_{z\to 2i}\left[\frac{-2z-4i}{(z+2i)^4}\right] = \frac{-i}{32}.$$

Hence, by Cauchy's Residue theorem, we have

$$\int_C \frac{dz}{(z^2+4)^2} = 2\pi i\left(-\frac{i}{32}\right) = \frac{\pi}{16}.$$

EXAMPLE 3.92

Evaluate

$$\int_C \frac{1-\cos 2(z-3)}{(z-3)^3}dz,$$

where C is the curve $|z-3| = 1$.

Solution. Expanding $\cos 2(z-3)$, we have

$$f(z) = \frac{1-\cos 2(z-3)}{(z-3)^3} = \frac{1}{(z-3)^3}\left[1-1+\frac{4(z-3)^2}{2!}-\frac{16(z-3)^4}{4!}+\dots\right] = \frac{2}{z-3}-\frac{16}{4!}(z-3)+\dots.$$

Thus $f(z)$ has a simple pole at $z = 3$. The Laurent's series is in the power of $z-3$. The coefficient of $\frac{1}{z-3}$ is 2. Hence, the residue of $f(z)$ at $z = 3$ is 2 and so by Cauchys Residue theorem

$$\int_C f(z)dz = 2\pi i(2) = 4\pi i.$$

EXAMPLE 3.93

Evaluate

$$\int_{|z|=3} \frac{\sin \pi z^2 + \cos \pi z^2}{(z-1)^2(z-2)} dz.$$

Solution. The integrand has simple pole at $z = 2$ and a pole of order 2 at $z = 1$. But these poles lie within $|z| = 3$. Now

$$\text{Res}(2) = \lim_{z\to 2} (z-2)\, f(z)$$

$$= \lim_{z\to 2} \frac{\sin \pi z^2 + \cos \pi z^2}{(z-1)^2}$$

$$= \sin 4\pi + \cos 4\pi = 1,$$

$$\text{Res}(1) = \lim_{z\to 1} \frac{d}{dz}[(z-1)^2 f(z)]$$

$$= \lim_{z\to 1} \frac{d}{dz}\left[\frac{\sin \pi^2 + \cos \pi z^2}{z-2}\right]$$

$$= \lim_{z\to 1}\left[\frac{(z-2)\,(2\pi z \cos \pi z^2 - 2\pi z \sin \pi z^2)}{(z-2)^2} - \frac{(\sin \pi z^2 + \cos \pi z^2)}{(z-2)^2}\right]$$

$$= 2\pi + 1.$$

Hence, by Cauchy's Residue theorem, we have

$$\int_{|z|=3} \frac{\sin \pi z^2 + \cos \pi z^2}{(z-1)^2\,(z-2)} dz = 2\pi i(2\pi + 2) = 4\pi i(\pi + 1).$$

EXAMPLE 3.94

Evaluate $I = \int_{|z|=3} \frac{z \sec z}{(1-z)^2} dz$

Solution. The integrand has a double pole at $z = 1$, which lies within the contour $|z| = 3$. Now

$$\text{Res}(1) = \lim_{z\to 1} \frac{d}{dz}\left[(z-1)^2 \frac{z \sec z}{(1-z)^2} dz\right]$$

$$= \lim_{z\to 1} \frac{d}{dz}[z \sec z]$$

$$= \lim_{z\to 1}[z \sec z \tan z + \sec z] = \sec 1[1 + \tan 1].$$

Hence, by Cauchy's Residue theorem

$$I = 2\pi i[\sec 1(1 + \tan 1)].$$

EXAMPLE 3.95

Evaluate $\int_{|z-1|=2} \frac{dz}{z^2 \sinh z}$.

Solution. Since

$$\sin hz = z + \frac{z^3}{3!} + \frac{z^5}{5!} + \frac{z^7}{7!} + \ldots,$$

$$f(z) = \frac{1}{z^2 \sinh z} = \frac{1}{z^2\left[z + \frac{z^3}{3!} + \frac{z^5}{5!} + \frac{z^7}{7!} + \ldots\right]}$$

$$= \frac{1}{z^3}\left[1 + \left(\frac{z^2}{3!} + \frac{z^4}{5!} + \ldots\right)\right]^{-1}$$

$$= \frac{1}{z^3}\left[1 - \left(\frac{z^2}{3!} + \frac{z^4}{5!} + \ldots\right) + \left(\frac{z^2}{3!} + \frac{z^4}{5!} + \ldots\right)^2 - \ldots\right]$$

$$= \frac{1}{z^3}\left[1 - \frac{z^2}{6} - \frac{z^4}{120} + \frac{z^4}{36} + \ldots\right]$$

$$= \frac{1}{z^3}\left[1 - \frac{z^2}{6} - \left(\frac{1}{36} - \frac{1}{120}\right) z^4 + \ldots\right]$$

$$= \frac{1}{z^3} - \frac{1}{6z} + \frac{7}{360} z^4 - \ldots$$

The coefficient of $\frac{1}{z}$ in this Laurent series in the powers of z is $-\frac{1}{6}$. Hence, residue at the pole $z = 0$ is

$$\text{Res}(0) = -\frac{1}{6}.$$

Hence, by Cauchy's Residue theorem,

$$I = 2\pi i\left(-\frac{1}{6}\right) = -\frac{\pi i}{3}.$$

3.13 EVALUATION OF REAL DEFINITE INTEGRALS

We shall now discuss the application of Cauchy's Residue theorem to evaluate real definite integrals.

(A) Integration Around the Unit Circle

We consider the integrals of the type

$$\int_0^{2\pi} f(\cos\theta, \sin\theta)\, d\theta, \qquad (58)$$

where the integrand is a rational function of $\sin\theta$ and $\cos\theta$. Substitutet $z = e^{i\theta}$. Then, $dz = i\,e^{i\theta}\,d\theta = iz\,d\theta$ and

$$\cos\theta = \frac{1}{2}\left(z + \frac{1}{z}\right),$$

$$\sin\theta = \frac{1}{2i}\left(z - \frac{1}{z}\right).$$

Thus (58) converts into the integral

$$\int_C \phi(z)dz, \qquad (59)$$

where $f(z)$ is a rational function of z and C is the unit circle $|z| = 1$. The integral (59) can be solved using Cauchy's Residue theorem.

EXAMPLE 3.96

Show that

$$I = \int_0^{2\pi} \frac{d\theta}{a + b\cos\theta} = \frac{2\pi}{\sqrt{a^2 - b^2}}, \quad a > b > 0.$$

Solution. Substituting $z = e^{i\theta}$, we get $d\theta = \dfrac{dz}{i\,z}$ and so

$$I = \frac{1}{i}\int_{|z|=1} \frac{dz}{z\left[a + \frac{b}{2}\left(z + \frac{1}{z}\right)\right]}$$

$$= \frac{2}{i}\int_{|z|=1} \frac{dz}{bz^2 + 2az + b}.$$

The poles of the integrand are given by

$$z = \frac{-2a \pm \sqrt{4a^2 - 4b^2}}{2b} = \frac{-a \pm \sqrt{a^2 - b^2}}{b}.$$

Thus the poles are

$$\alpha = \frac{-a + \sqrt{a^2 - b^2}}{b} \quad \text{and} \quad \beta = \frac{-a - \sqrt{a^2 - b^2}}{b}$$

Since $a > b > 0$, $|\beta| > 1$. But $|\alpha\beta| = 1$ (product of roots) so that $|\alpha| < 1$. Hence, $z = \alpha$ is the only simple pole lying within $|z| = 1$. Further

$$\text{Res}(\alpha) = \lim_{z\to\alpha} (z - \alpha).\frac{2}{bi(z - \alpha)(z - \beta)}$$

$$= \frac{2}{bi(\alpha - \beta)} = \frac{1}{i\sqrt{a^2 - b^2}}.$$

Hence

$$I = 2\pi i\left(\frac{1}{i\sqrt{a^2 - b^2}}\right) = \frac{2\pi}{\sqrt{a^2 - b^2}}.$$

EXAMPLE 3.97

Use calculus of residues to show that

$$\int_0^{2\pi} \frac{\cos 2\theta}{5 + 4\cos\theta}\, d\theta = \frac{\pi}{6}.$$

Solution. We have

$$\int_0^{2\pi} \frac{\cos 2\theta}{5 + 4\cos\theta}\, d\theta = \text{real part of} \int_0^{2\pi} \frac{e^{2i\theta}}{5 + 4\cos\theta}\, d\theta.$$

Now substituting $z = e^{i\theta}$, we get

$$\int_0^{2\pi} \frac{e^{2i\theta}}{5 + 4\cos\theta}\, d\theta$$

$$= \int_{|z|=1} \frac{z^2}{5 + 2\left(z + \frac{1}{z}\right)}.\frac{dz}{iz}$$

$$= \frac{1}{i}\int_{|z|=1} \frac{z^2}{2z^2 + 5z + 2}\, dz$$

$$= \frac{1}{i}\int_{|z|=1} \frac{z^2}{(2z + 1)(z + 2)}\, dz$$

$$= \frac{1}{2i}\int_{|z|=1} \frac{z^2}{\left(z + \frac{1}{2}\right)(z + 2)}\, dz.$$

The integrand has simple poles at $z = -\frac{1}{2}$ and $z = -2$ of which only $z = -\frac{1}{2}$ lies inside $|z| = 1$. Now

$$\text{Res}(-\tfrac{1}{2}) = \lim_{z\to -\frac{1}{2}} \left(z+\frac{1}{2}\right) f(z) = \lim_{z\to -\frac{1}{2}} \frac{z^2}{2i(z+2)}$$

$$= \frac{1}{12i}.$$

Hence

$$\int_0^{2\pi} \frac{e^{2i\theta}}{5+4\cos\theta} = 2\pi i . \frac{1}{12i} = \frac{\pi}{6}.$$

Equating real and imaginary parts, we have

$$\int_0^{2\pi} \frac{\cos 2\theta}{5+4\cos\theta}\, d\theta = \frac{\pi}{6} \quad \text{and} \quad \int_0^{2\pi} \frac{\sin 2\theta}{5+4\cos\theta}\, d\theta = 0.$$

EXAMPLE 3.98

Show that

$$\int_0^{2\pi} e^{\cos\theta} \cos(n\theta - \sin\theta) d\theta = \frac{2\pi}{n!}.$$

Solution. The given integral is the real part of the integral

$$\int_0^{2\pi} e^{\cos\theta} e^{-(n\theta-\sin\theta)i} d\theta = \int_0^{2\pi} e^{\cos\theta + i\sin\theta} . e^{-in\theta} d\theta$$

$$= \int_0^{2\pi} e^{e^{i\theta}} . e^{-in\theta} d\theta$$

$$= \frac{1}{i} \int_{|z|=1} \frac{e^z}{z^{n+1}} dz, \quad z = e^{i\theta}.$$

The integrand has a pole of order $n + 1$ at $z = 0$ which lies in $|z| = 1$. Then

$$\text{Res}(0) = \lim_{z\to 0} \frac{1}{n!} \frac{d^n}{dz^n} \left\{ z^{n+1} . \frac{e^z}{z^{n+1}} \right\}$$

$$= \lim_{z\to 0} \frac{1}{n!} \frac{d^n}{dz^n} \{e^z\} = \frac{1}{n!}.$$

Hence

$$I = 2\pi i \frac{1}{i} . \frac{1}{n!} = \frac{2\pi}{n!}.$$

Equating real and imaginary parts, we get

$$\int_0^{2\pi} e^{\cos\theta} \cos(n\theta - \sin\theta) d\theta = \frac{2\pi}{n!}$$

and

$$\int_0^{2\pi} e^{\cos\theta} \sin(n\theta - \sin\theta)\, d\theta = 0.$$

EXAMPLE 3.99

Show that

$$\int_0^{2\pi} \frac{\sin^2\theta}{a + b\cos\theta}\, d\theta = \frac{2\pi}{b^2}\{a - \sqrt{a^2 - b^2}\}, 0 < b < a.$$

Solution. Let

$$I = \int_0^{2\pi} \frac{\sin^2\theta}{a + b\cos\theta}\, d\theta.$$

Substitute $z = e^{i\theta}$ so that $\cos\theta = \frac{1}{2}\left(z + \frac{1}{z}\right)$, $\sin\theta = \frac{1}{2i}\left(z - \frac{1}{z}\right)$, and $dz = iz\, d\theta$. So

$$I = \frac{1}{i} \int_{|z|=1} \frac{\left[\frac{1}{2i}\left(z - \frac{1}{z}\right)\right]^2}{a + \frac{b}{2}\left(z + \frac{1}{z}\right)} . \frac{dz}{z}$$

$$= -\frac{1}{2i} \int_{|z|=1} \frac{(z^2-1)^2}{z^2(2az + bz^2 + b)} dz$$

$$= -\frac{1}{2ib} \int_{|z|=1} \frac{(z^2-1)^2}{z^2\left(z^2 + \frac{2a}{b} z + 1\right)} dz.$$

The integrand has a double pole at $z = 0$ and simple poles at $z = \alpha$ and $z = \beta$, where

$$\alpha = \frac{-a+\sqrt{a^2-b^2}}{b}, \quad \beta = \frac{-a-\sqrt{a^2-b^2}}{b}.$$

Since $a > b > 0$, $|\beta| > 1$. But $|\alpha\beta| = 1$ so that $|\alpha| < 1$. Thus, the pole inside $|z| = 1$ is a double pole at $z = 0$ and a simple pole at $z = \alpha$. Now

$$\text{Res}(0) = \text{coefficient of } \frac{1}{z} \text{ in } -\frac{(z^2-1)^2}{2ibz^2\left(z^2+\frac{2a}{b}z+1\right)}$$

$$= \text{coefficient of } \frac{1}{z} \text{ in } -\frac{1}{2ibz^2}(z^4+1-2z^2) \times\left(z^2+\frac{2az}{b}+1\right)^{-1}$$

$$= \text{coefficient of } \frac{1}{z} \text{ in } -\frac{1}{2ibz^2}(1-2z^2+z^4) \times\left(1-\frac{2a}{b}z-z^2-\dots\right)$$

$$= \frac{a}{i\,b^2} = -\frac{ai}{b^2},$$

$$\text{Res}(\alpha) = \lim_{z\to\alpha}(z-\alpha) \times\left[-\frac{1}{2ib}\frac{(z^2-1)^2}{z^2(z-2)(z-\beta)}\right]$$

$$= -\frac{(\alpha^2-1)^2}{2ib\alpha^2(\alpha-\beta)} = -\frac{1}{2ib}\frac{\left(\alpha-\frac{1}{\alpha}\right)^2}{(\alpha-\beta)}$$

$$= -\frac{1}{2ib}\frac{(\alpha-\beta)^2}{\alpha-\beta} \text{ since } \frac{1}{\alpha} = \beta$$

$$= -\frac{1}{2ib}(\alpha-\beta) = \frac{i}{b^2}\sqrt{a^2-b^2}.$$

Hence

$$I = 2\pi i\left[\frac{i}{b^2}\sqrt{a^2-b^2} - \frac{ai}{b^2}\right]$$

$$= \frac{2\pi}{b^2}[a-\sqrt{a^2-b^2}]$$

EXAMPLE 3.100

Evaluate

$$\int_0^{2\pi} \frac{\cos 3\theta}{5-4\cos\theta}\, d\theta$$

Solution. Let

$$I = \int_0^{2\pi} \frac{e^{i3\theta}}{5-4\cos\theta}\, d\theta$$

Putting $z = e^{i\theta}$, we get

$$I = \frac{1}{i}\int_{|z|=1} \frac{z^3}{5-2\left(z+\frac{1}{z}\right)}\frac{dz}{z} = \frac{1}{i}\int_{|z|=1} \frac{z^3}{5z-2z^2-2}\, dz$$

$$= -\frac{1}{i}\int_{|z|=1} \frac{z^3}{2\left(z^2-\frac{5}{2}z+1\right)}\, dz.$$

The poles of the integrand are given by $2z^2 - 5z + 2 = 0$ and so are $z = 2$ and $z = \frac{1}{2}$. Out of these poles, $z = \frac{1}{2}$ lies within $|z| = 1$. Then

$$\text{Res}\left(\frac{1}{2}\right) = -\frac{1}{i}\lim_{z\to 1/2}\left(z-\frac{1}{2}\right)\frac{z^3}{2\left(z-\frac{1}{2}\right)(z-2)}$$

$$= -\frac{1}{i}\lim_{z\to 1/2}\frac{z^3}{2(z-2)} = \frac{1}{24\,i}.$$

Hence

$$I = \frac{2\pi i}{24\,i} = \frac{\pi}{12}.$$

EXAMPLE 3.101

Evaluate

$$\int_0^{2\pi} \frac{d\theta}{(5-3\cos\theta)^2}.$$

Solution. Substituting $z = e^{i\theta}$, we have $dz = iz\, d\theta$ and so

$$I=\int_0^{2\pi}\frac{d\theta}{(5-3\cos\theta)^2}=\frac{1}{i}\int_{|z|=1}\frac{1}{\left[5-\frac{3}{2}\left(z+\frac{1}{z}\right)\right]^2}\frac{dz}{z}$$

$$=\frac{4}{i}\int_{|z|=1}\frac{z^2}{[10z-3z^2-3]^2}dz$$

$$=-\frac{4}{9i}\int_{|z|=1}\frac{z}{\left[z^2-\frac{10}{3}z+1\right]^2}dz.$$

The double poles of the integrand are given by $z^2-\frac{10}{3}+1=0$ and so the double poles are at $z=3$ and $z=\frac{1}{3}$. The double pole at $z=\frac{1}{3}$ lies in $|z|=1$. Now

$$\text{Res}\left(\frac{1}{3}\right)=-\frac{4}{9i}\lim_{z\to\frac{1}{3}}\frac{d}{dz}$$

$$\times\left[\left(z-\frac{1}{3}\right)^2\frac{z}{[z-(1/3)]^2(z-3)^2}\right]$$

$$=-\frac{4}{9i}\lim_{z\to 1/3}\frac{d}{dz}\left[\frac{z}{(z-3)^2}\right]$$

$$=-\frac{4}{9i}\lim_{z\to 1/3}\left[\frac{(z-3)^2-z[2(z-3)]}{(z-3)^4}\right]$$

$$=-\frac{4}{9i}\lim_{z\to\frac{1}{3}}\left[\frac{(z-3)-2z}{(z-3)^3}\right]$$

$$=-\frac{4}{9i}\lim_{z\to 1/3}\left[\frac{-z-3}{(z-3)^3}\right]$$

$$=-\frac{4}{9i}\left[\frac{-10/3}{(-8/3)^3}\right]=-\frac{40}{512\,i}.$$

Hence

$$I=2\pi i\left(\frac{-40}{512i}\right)=-\frac{5\pi}{32}.$$

EXAMPLE 3.102

Evaluate $\int_0^{\pi}\frac{a\,d\theta}{a^2+\sin^2\theta}, a>0.$

Solution. Let $I=\int_0^{\pi}\frac{a\,d\theta}{a^2+\sin^2\theta}=\int_0^{\pi}\frac{2a\,d\theta}{2a^2+2\sin^2\theta}$

$$=\int_0^{\pi}\frac{2a\,d\theta}{2a^2+(1-\cos 2\theta)}$$

$$=\int_0^{2\pi}\frac{a d\phi}{2a^2+1-\cos\phi},\quad 2\theta=\phi.$$

Substituting $z=e^{i\phi}$, we get

$$I=\frac{1}{i}\int_{|z|=1}\frac{adz}{z\left[2a^2+1-\frac{1}{2}\left(z+\frac{1}{z}\right)\right]}$$

$$=\frac{2a}{i}\int_{|z|=1}\frac{dz}{2z(2a^2+1)-z^2-1}$$

$$=2ai\int_{|z|=1}\frac{dz}{z^2-2z(2a^2+1)-1}.$$

The poles α and β of the integrand are $z=2a^2+1\pm 2a\sqrt{a^2+1}$. We note that $|\alpha|=|2a^2+1+2\sqrt{a^2+1}|>1$. Since $|\alpha\beta|=1$, we have $|\beta|=|2a^2+1-2\sqrt{a^2+1}|<1$. Hence the pole β lies in $|z|=1$. Therefore

$$\text{Res}(\beta)=2ai\lim_{z\to\beta}(z-\beta)\,f(z)$$

$$=2ai\lim_{z\to\beta}(z-\beta)\frac{1}{(z-\alpha)(z-\beta)}$$

$$=2ai\lim_{z\to\beta}\frac{1}{(z-\alpha)}=2ai\left[\frac{1}{(\beta-\alpha)}\right]$$

$$=2ai\left(\frac{1}{-4a\sqrt{a^2+1}}\right)=-\frac{i}{2\sqrt{a^2+1}}.$$

Hence, by Cauchy's residue theorem,

$$I=2\pi i\left(-\frac{i}{2\sqrt{a^2+1}}\right)=\frac{\pi}{\sqrt{a^2+1}}.$$

EXAMPLE 3.103

Evaluate

$$\int_0^{2\pi} \frac{d\theta}{1-2p\sin\theta+p^2} \quad (0<p<1).$$

Solution. We have

$$I=\int_0^{2\pi} \frac{d\theta}{1-2p\sin\theta+p^2}.$$

Substitute $z=e^{i\theta}$ so that $dz=iz\,d\theta$. Thus

$$I=\frac{1}{i}\int_{|z|=1} \frac{1}{1-\frac{2p}{2i}\left(z-\frac{1}{z}\right)+p^2}\cdot\frac{dz}{z}$$

$$=\frac{1}{i}\int_{|z|=1} \frac{1}{z\left(i-pz+\frac{p}{z}+ip^2\right)}dz$$

$$=\frac{1}{i}\int_{|z|=1} \frac{i\,dz}{-pz^2+p+z(p^2+1)\,i}$$

$$=-\int_{|z|=1} \frac{dz}{pz^2-z(p^2+1)i-p}$$

$$=-\int \frac{dz}{(pz-i)\,(z-pi)}.$$

The poles of the integrand are given by $z=pi$ and $z=\frac{i}{p}$. Out of these simple poles, the pole at $z=pi$ lies in $|z|=1$.

$$\text{Res}(pi)=\lim_{z\to pi}(z-pi)\frac{1}{(pz-i)\,(z-pi)}$$

$$=\lim_{z\to pi}\frac{1}{pz-i}=\frac{1}{p^2i-i}=\frac{1}{i(p^2-1)}.$$

Hence, by Cauchy Residue theorems, we have

$$I=-2\pi i\left(\frac{1}{i(p^2-1)}\right)=\frac{2\pi}{1-p^2}.$$

(B) Definite Integral of the Type $\int_{-\infty}^{\infty} F(x)\,dx$

We know that if $|F(z)|\le M$ on a contour C and if L is the length of the curve C, then

$$\left|\int_C F(z)\,dz\right|\le ML. \tag{60}$$

Now, suppose, that $|F(z)|\le M/R^k$ for $z=R\,e^{i\theta}$, $k>1$ and constant M. Then (60) implies

$$\left|\int_\Gamma F(z)\,dz\right|\le\frac{M}{R^k}(\pi R)=\frac{\pi M}{R^{k-1}},$$

where Γ is the semi-circular arc of radius R and length πR as shown in Figure 3.15.

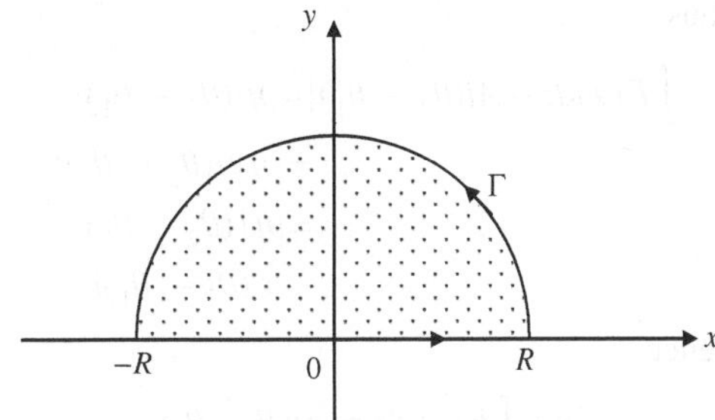

Figure 3.15

Then

$$\lim_{R\to\infty}\left|\int_\Gamma F(z)\,dz\right|=0$$

and so

$$\lim_{R\to\infty}\int_\Gamma F(z)\,dz=0.$$

We have thus proved the following result.

Theorem 3.22. If $|F(z)|\le M/R^k$ for $z=R\,e^{i\theta}$, $k>1$ and constant M, then $\lim_{R\to\infty}\int_\Gamma F(z)\,dz=0$, where Γ is the circular arc of radius R shown in Figure 3.15.

Equally important results are the following theorems:

Theorem 3.23. If C is an arc $\theta_1\le\theta\le\theta_2$ of the circle $|z|=R$ and if $\lim_{R\to\infty} zF(z)=A$, then

$$\lim_{R\to\infty}\int_C F(z)\,dz=i(\theta_2-\theta_1)A.$$

Proof: For sufficiently large value of R, we have

$$|zF(z)-A| < \varepsilon, \varepsilon > 0$$

or equivalently

$$z\,F(z) = A + \eta \text{ where } |\eta| < \varepsilon.$$

Therefore, substituting $z = R\,e^{i\theta}$

$$\int_C F(z)\,dz = \int_C \frac{A+\eta}{z}\,dz = \int_{\theta_1}^{\theta_2} \frac{(A+\eta)\,i\,R\,e^{i\theta}}{R\,e^{i\theta}}\,d\theta,$$
$$= Ai(\theta_2-\theta_1) + \eta i(\theta_2-\theta_1).$$

Thus

$$\left|\int F(z)dz - Ai(\theta_2-\theta_1)\right| = |\eta i(\theta_2-\theta_1)|$$
$$= |\eta|\,|i(\theta_2-\theta_1)|$$
$$= |\eta|\,(\theta_2-\theta_1)$$
$$< (\theta_2-\theta_1)\varepsilon.$$

Hence

$$\lim_{R\to\infty} \int_C F(z)\,dz = Ai(\theta_2-\theta_1).$$

Remark 3.11. (i) If $\lim_{R\to\infty} F(z) = 0$, then Theorem 3.23 implies that $\lim_{R\to\infty} \int_C F(z)\,dz = 0$.

(ii) The Theorem 3.23 shall be applied to integrals of the form $\int_{-\infty}^{\infty} \frac{P(x)}{Q(x)}\,dx$, where P($x$) and Q($x$) are polynomials such that

(i) The polynomial Q(x) has no real root

(ii) The degree of P(x) is at least two less than that of the degree of Q(x).

Theorem 3.24. (Jordan's Lemma). If $f(z) \to 0$ as $z \to \infty$ and $f(z)$ is meromorphic in the upper half-plane, then

$$\lim_{R\to\infty} \int_\Gamma e^{imz} f(z)\,dz = 0, \quad m > 0,$$

where Γ denotes the semi-circle $|z|$ = R, Im(z) > 0.

Proof: We shall use Jordan's inequality

$$\frac{2\theta}{\pi} \le \sin\theta \le \theta, \quad 0 \le \theta \le \frac{\pi}{2}$$

to prove our theorem. We assume that $f(z)$ has no singularities on Γ for sufficiently large value of R. Since $f(z) \to 0$ as $R \to \infty$, there exists $\varepsilon > 0$ such that $|f(z)| < \varepsilon$ when $|z| = R \le R_0$, $R_0 > 0$. Let Γ be any semi-circle with radius $R \ge R_0$. Substituting $z = \mathrm{R}\,e^{i\theta}$, we get

$$\int_\Gamma e^{imz} f(z)\,dz$$
$$= \int_0^\infty e^{im\,\mathrm{R}e^{i\theta}} f(R\,e^{i\theta}) R\,i\,e^{i\theta}\,d\theta$$
$$= \int_0^\pi e^{imR(\cos\theta + i\sin\theta)} f(R\,e^{i\theta})\,Rie^{i\theta}\,d\theta$$
$$= \int_0^\pi e^{i\,mR\cos\theta} . e^{-m\,R\sin\theta} f(R\,e^{i\theta})\,Rie^{i\theta}\,d\theta.$$

Thus, using Jordan's inequality, we have

$$\left|\int_\Gamma e^{imz} f(z)\,dz\right|$$
$$\le \int_0^\pi |e^{i\,mz\cos\theta}|\,|e^{-mR\sin\theta}|\,|f(\mathrm{R}e^{i\theta})|\,|Ri|\,|e^{i\theta}|\,d\theta$$
$$\le \int_0^\pi e^{-mR\sin\theta} \varepsilon R\,d\theta \quad \text{using } |f(R\,e^{i\theta})| < \varepsilon.$$
$$= 2\varepsilon R \int_0^{\pi/2} e^{-m\,R\sin\theta}\,d\theta = 2\varepsilon R \int_0^{\pi/2} e^{-2mR\theta/\pi}$$
$$= 2\varepsilon R \frac{1-e^{-mR}}{2mR/\pi} = \frac{\varepsilon\pi}{m}(1-e^{-mR}) < \frac{\varepsilon\pi}{m}.$$

Hence $\lim_{R\to\infty} \int_\Gamma e^{imz} f(z)\,dz = 0$.

Remark 3.12. Jordan's lemma should be used to evaluate integrals of the form

$$\int_{-\infty}^{\infty} \frac{P(x)}{Q(x)} \sin mx\,dx \quad \text{or} \quad \int_{-\infty}^{\infty} \frac{P(x)}{Q(x)} \cos mx\,dx, m > 0,$$

where P(x) and Q(x) are polynomials such that

(i) degree of Q(x) exceeds the degree of P(x)

(ii) the polynomial Q(x) has no real roots.

We shall make use of Theorem 3.22, 3.23, and 3.24 in evaluating the definite integrals of the form $\int_{-\infty}^{\infty} F(x)\,dx$.

EXAMPLE 3.104

Using contour integration, show that

$$\int_0^{\infty} \frac{dx}{(1+x^2)^2} = \frac{\pi}{4}.$$

Solution. Consider the integral

$$\int_C \frac{dz}{(1+z^2)^2},$$

where C is the contour consisting of a large semi-circle C of radius R together with the part of real axis from –R to R traversed in the counter-clockwise sense (Fig. 3.15).

The double poles of the integrand are $z = \pm i$, out of which the double pole $z = i$ lies within the contour C. Now

$$\text{Res}\,(i) = \lim_{z\to i} \frac{d}{dx}\left[(z-i)^2 \frac{1}{(z+i)^2(z-i)^2}\right]$$

$$= \lim_{z\to i} \frac{d}{dx}\left[\frac{1}{(z+i)^2}\right] = \lim_{z\to i}\left[\frac{2(z+i)}{(z+i)^4}\right]$$

$$= \lim_{z\to i} \frac{-2}{(z+i)^3} = \frac{-2}{(2i)^3} = \frac{1}{4i}.$$

Hence, by Cauchy's Residue theorem

$$\int_C \frac{dz}{(1+z^2)^2} = 2\pi i\left(\frac{1}{4i}\right) = \frac{\pi}{2},$$

that is,

$$\int_{\Gamma} \frac{dz}{(1+z^2)^2} + \int_{-R}^{R} \frac{dx}{(1+x^2)^2} = \frac{\pi}{2}. \qquad (61)$$

But, substituting $z = \text{R}\, e^{i\theta}$, we have

$$\frac{1}{(1+z^2)^2} = \frac{1}{(1+R^2 e^{2i\theta})^2} \le \frac{1}{(R^2-1)^2}$$

$$\le \frac{1}{R^4} \to 0 \text{ as } R \to \infty.$$

Hence, letting $R \to \infty$ in (61), we get

$$\int_{-\infty}^{\infty} \frac{dx}{(1+x^2)^2} = \frac{\pi}{2}$$

and so

$$\int_0^{\infty} \frac{dx}{(1+x^2)^2} = \frac{\pi}{4}.$$

EXAMPLE 3.105

Evaluate $\int_0^{\infty} \frac{dx}{x^6+1}$.

Solution. Consider $\int_C \frac{dz}{z^6+1}$, where C is the closed contour consisting of the line from $-R$ to R and the semi-circle Γ traversed in the positive sense. The simple poles of $\frac{1}{z^6+1}$ are

$$z = e^{\pi i/6}\ e^{3\pi i/6}, e^{5\pi i/6}, e^{7\pi i/6}, e^{9\pi i/6}, e^{11\pi i/6}.$$

But only three simple poles $e^{\pi i/6}, e^{3\pi i/6}$, and $e^{5\pi i/6}$ lie within C. Now

$$\text{Res}(e^{\pi i/6}) = \lim_{z\to e^{\pi i/6}} \frac{1}{\frac{d}{dz}(z^6+1)} = \lim_{z\to e^{\pi i/6}} \frac{1}{6z^5}$$

$$= \frac{1}{6} e^{-5\pi i/6}.$$

$$\text{Res}(e^{3\pi i/6}) = \lim_{z\to e^{3\pi i/6}} \frac{1}{6z^5} = \frac{1}{6} e^{-5\pi i/2},$$

$$\text{Res}(e^{5\pi i/6}) = \lim_{z\to e^{5\pi i/6}} \frac{1}{6z^5} = \frac{1}{6} e^{-25\pi i/6}.$$

Thus

$$\int_C \frac{1}{z^6+1}\,dz = 2\pi i \Sigma R = 2\pi/3,$$

that is,

$$\int_{\Gamma} \frac{1}{z^6+1}\,dz + \int_{-R}^{R} \frac{1}{x^6+1}\,dx = \frac{2\pi}{3}.$$

But, $\lim_{z\to\infty} zF(z) = \lim_{z\to\infty} \frac{z}{z^6+1} = 0$. Therefore by Theorem 3.23,

$$\int_\Gamma \frac{1}{z^6+1}\,dz \to 0 \text{ as } R \to \infty.$$

Hence, letting $R \to \infty$, we get

$$\int_{-\infty}^{\infty} \frac{1}{x^6+1}\,dx = \frac{2\pi}{3}$$

and so

$$\int_0^{\infty} \frac{1}{x^6+1}\,dx = \frac{\pi}{3}.$$

EXAMPLE 3.106

Show that

$$\int_{-\infty}^{\infty} \frac{x^2-x+2}{x^4+10x^2+9} = \frac{5\pi}{2}.$$

Solution. Consider the integral

$$\int_C \frac{z^2-z+2}{z^4+10z^2+9}\,dz$$

$$= \int_\Gamma \frac{z^2-z+2}{z^4+10z^2+9}\,dz + \int_{-R}^{R} \frac{z^2-z+2}{z^4+10z^2+9}\,dz.$$

The poles of the integrand are given by $z^4 + 10z^2 + 9 = 0$ which yields the simple poles at $z = \pm 3i, \pm i$. Out of these poles only $3i$ and i lie within semi-circle with radius R. Now

$$\text{Res}(3i) = \lim_{z\to 3i} (z-3i) \frac{1}{(z-3i)(z+3i)(z-i)(z+i)}$$

$$= \frac{7+3i}{48i}$$

$$\text{Res}(i) = \lim_{z\to i} (z-i) \frac{1}{(z-i)(z+i)(z-3i)(z+3i)}$$

$$= \frac{1-i}{16i}.$$

Thus

$$\int_\Gamma \frac{z^2-z+2}{z^4+10z^2+9}\,dz + \int_{-R}^{R} \frac{x^2-x+2}{x^4+10x^2+9}\,dx$$

$$= 2\pi\text{i}\left[\frac{7+3i}{48i} + \frac{1-i}{16i}\right] = \frac{5\pi}{12}.$$

Further, $z\,F(z) \to 0$ as $z \to \infty$. Therefore, by Theorem 3.23, we have

$$\int_\Gamma \frac{z^2-z+2}{z^4+10z^2+9}\,dz \to 0 \text{ as } R \to \infty.$$

Hence, letting $R \to \infty$, we get

$$\int_{-\infty}^{\infty} \frac{x^2-x+2}{x^4+10x^2+9}\,dx = \frac{5\pi}{12}.$$

EXAMPLE 3.107

Evaluate $\int_0^{\infty} \frac{\cos ax}{x^2+1}\,dx.$

Solution. Consider the integral

$$\int_C \frac{e^{iaz}}{z^2+1}\,dz = \int_\Gamma \frac{e^{iaz}}{z^2+1}\,dz + \int_{-R}^{R} \frac{e^{iaz}}{z^2+1}\,dz.$$

The poles of the integrand are $z = \pm i$ of which $z = i$ lies in C. Hence

$$\int_\Gamma \frac{e^{iaz}}{z^2+1}\,dz + \int_{-R}^{R} \frac{\cos ax}{x^2+1}\,dz = 2\pi i.\,(\text{residue at } i)$$

But

$$\text{Res(i)} = \lim_{z\to i} (z-i) \frac{e^{iaz}}{(z-i)(z+i)} = \frac{e^{-a}}{2\,i}.$$

Since $f(z) \to 0$ as $z \to \infty$, by Jordan's lemma, the integral $\int_\Gamma \frac{e^{iaz}}{z^2+1}\,dz \to 0$ as $R \to \infty$. Hence, in the limit as $R \to \infty$, we get

$$\int_{-\infty}^{\infty} \frac{\cos ax}{x^2+1}\,dx = 2\pi i \frac{e^{-a}}{2i} = \pi\, e^{-a}$$

and so

$$\int_0^{\infty} \frac{\cos ax}{x^2+1}\,dx = \frac{\pi e^{-a}}{2}.$$

EXAMPLE 3.108

Evaluate $\displaystyle\int_{-\infty}^{\infty} \frac{x^2}{(x^2+a^2)(x^2+b^2)}\,dx, \quad a, b > 0.$

Solution. Consider

$$\int_C \frac{z^2}{(z^2+a^2)(z^2+b^2)}\,dz,$$

where C is the contour consisting of the line from $-R$ to R and semi-circle Γ of radius R traversed in the positive sense. Then

$$\int_C \frac{z^2}{(z^2+a^2)(z^2+b^2)}\,dz \qquad (\qquad)(\qquad)$$

$$= \int_{-R}^{R} \frac{x^2}{(x^2+a^2)(x^2+b^2)}\,dx$$

$$+ \int_\Gamma \frac{z^2}{(z^2+a^2)(z^2+b^2)}\,dz.$$

But since $z\,f(z) \to \infty$ as $z \to \infty$, the second integral on the right tends to zero. Thus

$$\int_C \frac{z^2}{(z^2+a^2)(z^2+b^2)}\,dz$$

$$= \int_{-R}^{R} \frac{x^2}{(x^2+a^2)(x^2+b^2)}\,dx.$$

But the poles of the integrand of the integral on the left are $z = \pm\, ai$ and $z = \pm\, bi$ out of which $z = ai$ and $z = bi$ lie within C. Now

$$\text{Res}(ai) = \lim_{z\to ai} (z-ai)\frac{z^2}{(z-ai)(z+ai)(z^2+b^2)}$$

$$= \frac{a}{2i(a^2-b^2)},$$

$$\text{Res}(bi) = \lim_{z\to bi} (z-bi)\frac{z^2}{(z^2+a^2)(z+bi)(z-bi)}$$

$$= \frac{-b}{2i(a^2-b^2)}.$$

Hence

$$\int_C \frac{z^2}{(z^2+a^2)(z^2+b^2)}\,dz$$

$$= 2\pi i\left[\frac{a}{2i(a^2-b^2)} - \frac{b}{2i(a^2-b^2)}\right]$$

$$= \frac{\pi(a-b)}{a^2-b^2} = \frac{\pi}{a+b}.$$

Hence, as $R \to \infty$,

$$\int_{-\infty}^{\infty} \frac{x^2}{(x^2+a^2)(x^2+b^2)}\,dx = \frac{\pi}{a+b}.$$

EXAMPLE 3.109

Evaluate $\displaystyle\int_0^{\infty} \frac{x\sin x}{x^2+a^2}\,dx, \; a > 0.$

Solution. Consider

$$\int_C \frac{z\,e^{iz}}{z^2+a^2}\,dz,$$

where C is contour consisting of line from $-R$ to R and semi-circle with radius R traversed in positive sense. Then

$$\int_C \frac{z\,e^{iz}}{z^2+a^2}\,dz = \int_{-R}^{R} \frac{x\sin x}{x^2+a^2}\,dx + \int_\Gamma \frac{z\,e^{iz}}{z^2+a^2}\,dz.$$

Since $\displaystyle\lim_{z\to\infty} \frac{z}{z^2+a^2} = 0$, we have, by Jordan's lemma,

$$\int_\Gamma f(z)\,dz = 0.$$

The integrand has simple poles at $z = \pm\, ai$ of which $z = ai$ lies within C. Further

$$\text{Res}(ai) = \lim_{z\to ai} (z-ai)\frac{z\,e^{iz}}{(z-ai)(z+ai)}$$

$$= \lim_{z\to ai} \frac{z\,e^{iz}}{z+ai} = \frac{e^{-a}}{2}.$$

Hence

$$\int_C \frac{z\,e^{iz}}{z^2+a^2}\,dz = 2\pi i\left(\frac{e^{-a}}{2}\right) = \pi i\,e^{-a}.$$

Equating imaginary parts, we have

$$\int_{-\infty}^{\infty} \frac{x\sin x}{x^2+a^2} = \pi e^{-a}$$

and so

$$\int_0^{\infty} \frac{x\sin x}{x^2+a^2}\,dx = \frac{\pi}{2}e^{-a}.$$

EXAMPLE 3.110

Use calculus of residue to show that

$$\int_{-\infty}^{\infty} \frac{\cos x}{(x^2+a^2)(x^2+b^2)}\,dx$$

$$= \frac{\pi}{a^2-b^2}\left(\frac{e^{-b}}{b} - \frac{e^{-a}}{a}\right), \quad a > b > 0.$$

Solution. The integrand is of the form $\dfrac{P(x)}{Q(x)}$. So let us consider

$$\int_C f(z)\,dz = \int_C \frac{e^{iz}}{(z^2+a^2)(z^2+b^2)}\,dz,$$

where C is semi-circle Γ with radius R and the line from $-R$ to R. We have

$$\int_{-R}^{R} f(x)\,dx + \int_\Gamma f(z)\,dz = 2\pi i\,\Sigma\text{R}.$$

But, by Jordan's lemma

$$\int_\Gamma f(z)\,dz \to 0 \text{ as } R\to\infty.$$

Further, the poles of $f(z)$ are $z = \pm ai, \pm bi$, out of which $z = ai$ and $z = bi$ lie in the upper half-plane. Now

$$\text{Res}(ai) = \lim_{z\to ai} (z-ai)\frac{e^{iz}}{(z+ai)(z^2+b^2)(z-ai)}$$

$$= \frac{e^{-a}}{2ai(b^2-a^2)},$$

$$\text{Res}(bi) = \lim_{z\to bi} (z-bi)\frac{e^{iz}}{(z+bi)(z^2+a^2)(z-bi)}$$

$$= \frac{e^{-b}}{2bi(a^2-b^2)}.$$

Hence

$$\int_{-\infty}^{\infty} \frac{\cos x}{(x^2+a^2)(x^2+b^2)}\,dx$$

$$= 2\pi i\left[\frac{e^{-a}}{2ai(b^2-a^2)} + \frac{e^{-b}}{2bi(a^2-b^2)}\right]$$

$$= \frac{\pi}{a^2-b^2}\left[\frac{e^{-b}}{b} - \frac{e^{-a}}{a}\right].$$

EXAMPLE 3.111

Show that

$$\int_{-\infty}^{\infty} \frac{\sin x}{x^2+4x+5}\,dx = -\frac{\pi\sin 2}{e}.$$

Solution. Consider

$$\int_C f(z)\,dz = \int_C \frac{e^{iz}\,dz}{z^2+4z+5},$$

where C is the contour as in the above examples. Then

$$\int_C f(z)\,dz = \int_{-R}^{R} f(x)\,dx + \int_\Gamma f(z)\,dz = 2\pi i\Sigma\text{R}.$$

Since $\dfrac{1}{z^2+4z+5} \to 0$ as $z\to\infty$, by Jordan's lemma, $\displaystyle\int_\Gamma f(z)\,dz \to 0$ as $R\to\infty$. Further, the poles of $f(z)$ are $-2\pm i$. The pole $z=-2+i$ lie in the upper half-plane. Then

$$\text{Res}(2+i) = \lim_{z\to -2+i} (z+2-i)\frac{e^{iz}}{z^2+4z+5} = \frac{e^{-(1+2i)}}{2i}.$$

Hence

$$\int_C f(z)\,dz = \pi\, e^{-(1+2i)} = \frac{\pi}{e}(\cos 2 - i\sin 2).$$

Equating the imaginary parts, we get

$$\int_{-\infty}^{\infty} \frac{\sin x}{x^2+4x+5}\,dx = -\frac{\pi}{e}\sin 2.$$

EXAMPLE 3.112

Evaluate

$$\int_{-\infty}^{\infty} \frac{\sin x}{x(x^2-2x+2)}\,dx \text{ and } \int_{-\infty}^{\infty} \frac{1-\cos x}{(x^2-2x+2)}\,dx.$$

Solution. Consider

$$\int_C f(z)\,dz = \int_C \frac{1-e^{iz}}{z(z^2-2z+2)}\,dz,$$

where C is the contour consisting of a large semi-circle Γ of radius R in the upper half-plane and the real axis from $-R$ to R. We have

$$\int_C f(z)\,dz = \int_{-R}^{R} f(x)\,dx + \int_\Gamma f(z)\,dz = 2\pi i\Sigma R.$$

We observe that

$$\left|\int_\Gamma f(z)\,dz\right| \le \int_0^\pi \left|\frac{1-e^{iRe^{i\theta}}}{Re^{i\theta}(R^2e^{2i\theta}-2Re^{i\theta}+2)} R ie^{i\theta}d\theta\right|$$

$$\le \int_0^\pi \frac{1-e^{-R\sin\theta}}{R^2-2R+2}\,d\theta \to 0 \text{ as } R\to\infty,$$

since $\sin\theta$ is positive

Hence, when $R\to\infty$, we have

$$\int_C f(z)\,dz = \int_{-\infty}^{\infty} f(x)\,dx = 2\pi i\Sigma R.$$

The function $f(z)$ has simple poles at $z = 1\pm i$ of which $z=1+i$ lies in the upper half-plane. However, $z=0$ is not a pole because expanding $1-e^{iz}$ we see that z is a common factor of numerator and denominator. Let $\alpha = 1+i$ and $\beta = 1-i$. Then

$$\text{Res}(\alpha) = \lim_{z\to\alpha}(z-\alpha)\frac{1-e^{iz}}{z(z-\beta)(z-\alpha)}$$

$$= \lim_{z\to\alpha}\frac{1-e^{iz}}{z(z-\beta)} = \frac{1-e^{i\alpha}}{\alpha(\alpha-\beta)}$$

$$= \frac{1-e^{i-1}}{(1+i)(2i)} = \frac{(1-i)(1-e^{i-1})}{4i}$$

$$= \frac{1-i}{4i}\left[1-\frac{1}{e}e^{i}\right]$$

$$= \frac{1-i}{4i}\left[1-\frac{1}{e}(\cos 1 + i\sin 1)\right].$$

Thus

$$\int_{-\infty}^{\infty}\frac{1-e^{ix}}{x(x^2-2x+2)}\,dx$$

$$= 2\pi i\left[\frac{1-i}{4i}\left\{1-\frac{1}{e}(\cos 1+i\sin 1)\right\}\right]$$

$$= \frac{\pi}{2e}(1-i)\,[e-\cos 1 - i\sin 1].$$

Equating real and imaginary parts, we get

$$\int_{-\infty}^{\infty}\frac{1-\cos x}{x(x^2-2x+2)}\,dx = \frac{\pi}{2e}[e-\cos 1 - \sin 1]$$

and

$$\int_{-\infty}^{\infty}\frac{\sin x}{x(x^2-2x+2)}\,dx = \frac{\pi}{2e}[e-\cos 1+\sin 1].$$

EXAMPLE 3.113

Show that

$$\int_0^{\infty}\frac{\log(1+x^2)}{1+x^2}\,dx = \pi\log 2.$$

Solution. Consider

$$\int_C f(z)\,dz = \int_C \frac{\log(z+i)}{1+z^2}\,dz,$$

where C is the contour as in the above examples. We have

$$\int_C f(z)\,dz = \int_{-R}^{R} f(x)\,dx + \int_\Gamma f(z)\,dz = 2\pi i \Sigma \text{R}.$$

Substituting $z = Re^{i\theta}$, we can show that $\int_\Gamma f(z)\,dz \to 0$ as $R \to \infty$. Hence when $R \to \infty$, we get

$$\int_{-\infty}^{\infty} f(x)\,dx = 2\pi i \Sigma \text{R}.$$

But $f(z)$ has simple pole at $z=+i$ and a logarithmic singularity at $z=-i$, out of which $z=i$ lies inside C. Now

$$\text{Res}(i) = \lim_{z\to i}(z-i)\frac{\log(z+i)}{(z-i)(z+i)}$$

$$= \lim_{z\to i}\frac{\log(z+i)}{z+i} = \frac{\log 2i}{2i} = \frac{1}{2i}\left[\log 2 + i\frac{\pi}{2}\right].$$

Thus

$$\int_{-\infty}^{\infty}\frac{\log(x+i)}{x^2+1}\,dx = 2\pi i\left[\frac{1}{2i}\left(\log 2 + i\frac{\pi}{2}\right)\right]$$

$$= \pi\left[\log 2 + \frac{i\pi}{2}\right].$$

Comparing real parts

$$\int_{-\infty}^{\infty}\frac{\frac{1}{2}\log(1+x^2)}{1+x^2}\,dx = \pi\log 2$$

and so

$$\int_0^{\infty}\frac{\log(1+x^2)}{1+x^2}\,dx = \pi\log 2.$$

(C) Poles on the Real Axis

When the integrand has a simple pole on a real axis, we delete it from the region by indenting the contour. Indenting is done by drawing a small semi-circle having the pole as the centre. The procedure followed is called "indenting at a point."

EXAMPLE 3.114

Show that

$$\int_0^{\infty}\frac{\sin x}{x}\,dx = \frac{\pi}{2} \quad\text{and}\quad \int_0^{\infty}\frac{\cos x}{x}\,dx = 0.$$

Solution. Consider the integral

$$\int_C f(z)\,dz = \int_C \frac{e^{iz}}{z}\,dz,$$

where C is the contour (shown in Figure 3.16) consisting of

(i) real axis from ρ to R, where ρ is small and R is large

(ii) the upper half of the circle $|z| = R$

(iii) the real axis from $-R$ to $-\rho$

(iv) the upper half of the circle $|z| = \rho$.

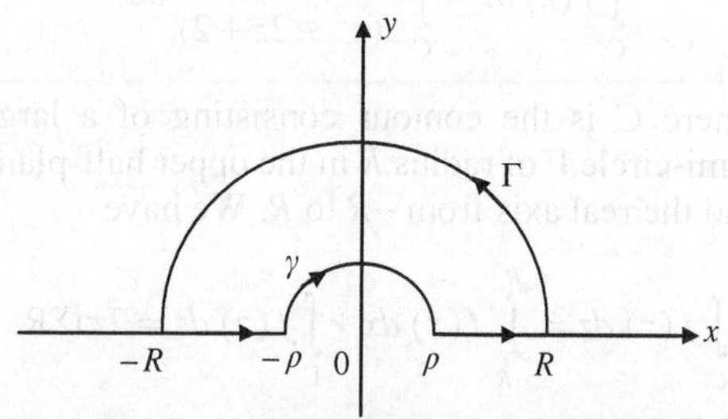

Figure 3.16

Since there is no singularity inside C, the Cauchy-Goursat theorem implies

$$\int_C f(z)\,dz = \int_\rho^R f(x)\,dx + \int_\Gamma f(z)\,dz$$

$$+ \int_{-R}^{\rho} f(x)\,dx + \int_\gamma f(z)\,dz = 0.$$

By Jordan's lemma, we have

$$\lim_{R\to\infty}\int_{\Gamma} f(z)\,dz = 0.$$

Further, since, $\lim_{e\to 0} zf(z) = 1$, we have

$$\lim_{\rho\to 0}\int_{\gamma} f(z)\,dz = i(0-\pi).1 = -\pi i.$$

Hence as $\rho \to 0$ and $R \to \infty$, we get

$$\int_0^{\infty} f(x)\,dx + \int_{-\infty}^{0} f(x)\,dx - \pi i = 0$$

and so

$$\int_{-\infty}^{\infty} f(x)\,dx = \pi i,$$

that is,

$$\int_{-\infty}^{\infty} \frac{e^{ix}}{x}\,dx = \pi \text{i}.$$

Equating real and imaginary parts, we get

$$\int_{-\infty}^{\infty} \frac{\cos x}{x}\,dx = 0 \quad\text{and}\quad \int_{-\infty}^{\infty} \frac{\sin x}{x}\,dx = \pi.$$

Hence $\displaystyle\int_0^{\infty} \frac{\cos x}{x}\,dx = 0$ and $\displaystyle\int_0^{\infty} \frac{\sin x}{x}\,dx = \frac{\pi}{2}.$

EXAMPLE 3.115

Evaluate

$$\int_0^{\infty} \frac{\sin x}{x(x^2+a^2)}\,dx \quad\text{and}\quad \int_0^{\infty} \frac{\cos x}{x(x^2+a^2)}\,dx, \quad a>0$$

Solution. Consider the integral

$$\int_C f(z)\,dz = \int_C \frac{e^{iz}}{z(z^2+a^2)}\,dz,$$

where C is the contour as shown in Figure 3.16.

Now $f(z)$ has simple poles at $z = 0, \pm\, ai$. Out of these, $z = 0$ lie on x-axis and $z = ai$ lies in the upper half-plane. Residue at $z = ai$ is

$$\text{Res}(ai) = \lim_{z\to ai} (z-ai)\frac{e^{iz}}{z(z-ai)\,(z+ai)}$$

$$= \lim_{z\to ai} \frac{e^{iz}}{z(z+ai)} = \frac{e^{-a}}{-2a^2}.$$

Hence

$$\int_C f(z)\,dz = \int_{\rho}^{R} f(x)dx + \int_{\Gamma} f(z)\,dz$$

$$+ \int_{-R}^{-\rho} f(x)dx + \int_{\gamma} f(z)\,dz$$

$$= 2\pi i\left[\frac{e^{-a}}{-2a^2}\right] = -\frac{\pi i}{a^2}\,e^{-a}.$$

Now

$$\int_{\Gamma} f(z)\,dz \to 0 \text{ as R} \to \infty.$$

Also substituting $z = \rho e^{i\theta}$, we note that

$$\int_{\gamma} f(z)\,dz = -\frac{\pi i}{a^2} \text{ as } \rho \to 0.$$

Hence as $\rho \to 0$ and $R \to \infty$, we have

$$\int_0^{\infty} f(x)\,dx + \int_{-\infty}^{0} f(x)\,dx - \frac{\pi i}{a^2} = -\frac{\pi i}{a^2}\,e^{-a}$$

and so

$$\int_{-\infty}^{\infty} f(x)\,dx = \frac{\pi i}{a^2}(1-e^{-a}),$$

that is,

$$\int_{-\infty}^{\infty} \frac{e^{ix}}{x(x^2+a^2)}\,dx = \frac{\pi i}{a^2}(1-e^{-a}).$$

Equating real and imaginary parts, we get

$$\int_{-\infty}^{\infty} \frac{\cos x}{x(x^2+a^2)}\,dx = 0 \quad\text{and}$$

$$\int_{-\infty}^{\infty} \frac{\sin x}{x(x^2+a^2)}\,dx = \frac{\pi}{a^2}(1-\ e^{-\,a}).$$

Thus

$$\int_0^\infty \frac{\cos x}{x(x^2+a^2)}\,dx = 0 \quad \text{and}$$

$$\int_0^\infty \frac{\sin x}{x(x^2+a^2)}\,dx = \frac{\pi}{2a^2}(1 - e^{-a}).$$

EXAMPLE 3.116

Show that

$$\int_0^\infty \frac{x^{p-1}}{1+x}\,dx = \frac{\pi}{\sin px},\ 0 \le p \le 1.$$

Solution. Consider the integral $\int_C \frac{z^{p-1}}{1+z}\,dz$, where C is the contour shown in Figure 3.17 and where AB and GH are actually coincident with the x-axis. Here $z = 0$ is a branch point and the real axis is the branch line. The integrand has a simple pole at $z = -1 = e^{\pi i}$ lying on the x-axis and inside C. Now

$$\operatorname{Res}(e^{\pi i}) = \lim_{z\to -1} (z+1)\frac{z^{p-1}}{z+1}$$

$$= \lim_{z\to -1} z^{p-1} = (e^{\pi i})^{p-1} = e^{(p-1)\pi i}.$$

Hence

$$\int_C \frac{z^{p-1}}{1+z}\,dz = 2\pi i e^{(p-1)\pi i}.$$

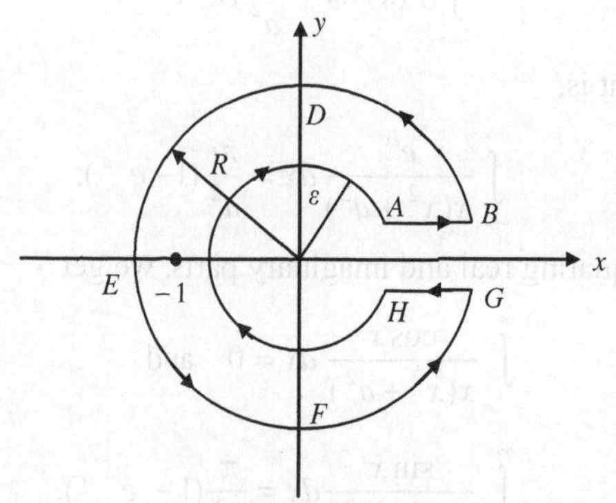

Figure 3.17

Thus

$$\int_\varepsilon^R \frac{x^{p-1}}{1+x}\,dx + \int_0^{2\pi} \frac{(Re^{i\theta})^{p-1} i R e^{i\theta}\,d\theta}{1+R e^{i\theta}}$$

$$+\int_R^\varepsilon \frac{(x e^{2\pi i})^{p-1}}{1+xe^{2\pi i}}\,dx + \int_{2\pi}^0 \frac{(\varepsilon e^{i\theta})^{p-1} i\varepsilon e^{i\theta}}{1+\varepsilon e^{i\theta}}\,d\theta$$

$$= 2\pi i e^{(p-1)}\pi i.$$

Now taking the limit as $R \to \infty$ and $\varepsilon \to 0$, the second and fourth integral approaches zero. Therefore,

$$\int_0^\infty \frac{x^{p-1}}{1+x}\,dx + \int_\infty^0 \frac{xe^{2\pi i(p-1)}x^{p-1}}{1+x}\,dx = 2\pi i e^{(p-1)\pi i},$$

that is,

$$[1 - e^{2\pi i(p-1)}]\int_0^\infty \frac{x^{p-1}}{1+x}\,dx = 2\pi i e^{(p-1)\pi i},$$

which yields

$$\int_0^\infty \frac{x^{p-1}}{1+x}\,dx = \frac{2\pi i}{e^{p\pi i} - e^{-p\pi i}} = \frac{\pi}{\sin p\pi}.$$

EXAMPLE 3.117

Using calculus of residue, evaluate

$$\int_0^\infty \sin x^2\,dx \quad \text{and} \quad \int_0^\infty \cos x^2\,dx.$$

Solution. Consider the integral $\int_C e^{iz^2}\,dz$, where C is the contour as shown in Figure 3.18. Here, AP is the arc of a circle with centre at the origin O and radius R.

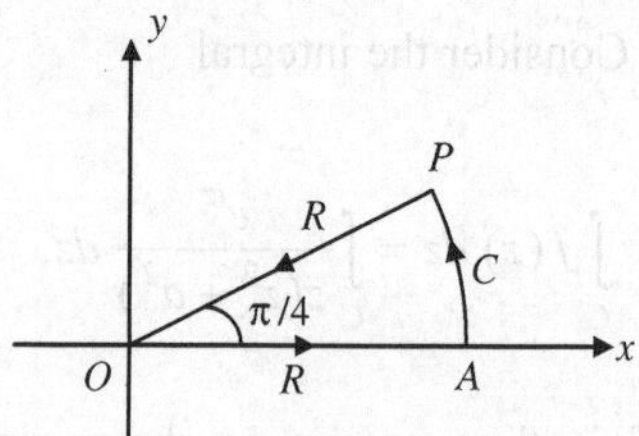

Figure 3.18

The function e^{iz^2} has no singularities within and on C. Hence by Cauchy-Goursat theorem

$$\int_C e^{iz^2}\,dz = 0.$$

Thus

$$\int_{OA} e^{iz^2}\,dz + \int_{AP} e^{iz^2}\,dz + \int_{PO} e^{iz^2}\,dz = 0,$$

that is,

$$\int_0^R e^{ix^2 dx} + \int_0^{\pi/4} e^{iR^2e^{2i\theta}}\, iR\,e^{i\theta}\,d\theta + \int_R^0 e^{ir^2} e^{\pi i/2} e^{\pi i/4}\,dr = 0,$$

or

$$\int_0^R (\cos x^2 + i\sin x^2)\,dx = e^{\pi i/4}\int_0^R e^{-r^2}\,dr$$

$$-\int_0^{\pi} e^{iR^2} e^{2i\theta}\, i\,\mathrm{R}e^{i\theta}\,d\theta$$

As $R \to \infty$

$$e^{\pi i/4}\int_0^{\infty} e^{-r^2}\,dr = \frac{\sqrt{\pi}}{2}\,e^{\pi i/4},$$

and

$$\left|\int_0^{\pi/4} e^{iR^2} e^{2i\theta}\, i\,R\,e^{i\theta}\,d\theta\right|$$

$$\le \int_0^{\pi/4} e^{-R^2\sin 2\theta}\,R\,d\theta$$

$$= \frac{R}{2}\int_0^{\pi/2} e^{-R^2\sin\phi}\,d\phi, \quad \phi = 2\theta$$

$$\le \frac{R}{2}\int_0^{\pi/2} e^{-2R^2\phi/\pi}\,d\phi, \quad 0 \le \phi \le \frac{\pi}{2}$$

$$= \frac{\pi}{4R}(1 - e^{-R^2}) \to 0 \ \text{ as } \ \mathrm{R} \to \infty.$$

Hence

$$\int_0^{\infty} (\cos x^2 + i\sin x^2)\,dx = \frac{\sqrt{\pi}}{2}e^{\pi i/4} = \frac{1}{2}\sqrt{\frac{\pi}{2}} + \frac{i}{2}\sqrt{\frac{\pi}{2}}.$$

Equating real and imaginary parts, we get

$$\int_0^{\infty} \cos x^2\,dx = \frac{1}{2}\sqrt{\frac{\pi}{2}} \quad \text{and} \quad \int_0^{\infty} \sin x^2\,dx = \frac{1}{2}\sqrt{\frac{\pi}{2}}.$$

3.14 CONFORMAL MAPPING

We know that a real-valued function $y = f(x)$ of a real variable x determines a curve in the xy-plane if x and y are interpreted as rectangular co-ordinates. But in case of analytic function $w = f(z)$ of a complex variable z, no such simple geometric interpretation is possible. In fact in this case, both z and w are complex numbers and, therefore, geometric representation of the function requires four real co-ordinates. But our geometry fails in a space of more than three dimensions. Thus, no geometric interpretation is possible of $w = f(z)$.

Suppose that we regard the points z and w as points in two different planes—the z-plane and the w-plane. Then we can interpret the functional relationship $w = f(z)$ as a mapping of points in the z-plane onto the points in the w-plane. Thus if $f(z)$ is regular on some set S in z-plane, there exists a set of points S' in the w-plane. The set S' is called the image of the set S under the function $w = f(x)$.

Let $f(z)$ be regular and single-valued in a domain D. If $z = x + iy$ and $w = u(x, y) + iv(x, y)$, the image of the continuous arc $x = x(t)$, $y = y(t)$, $(t_1 \le t \le t_2)$ is the arc $u = u(x(t), y(t))$, $v = v(x(t), y(t))$ under the mapping $w = f(z)$. Further, u and v are continuous in t if $x(t)$ and $y(t)$ are continuous. Therefore, $w = f(z)$ maps a continuous arc into a continuous arc.

Let the two curves C_1 and C_2 in the z-plane intersect at the point $P(x_0, y_0)$ at an angle and let C_1 and C_2 be mapped under $w = f(z)$ into the curves Γ_1 and Γ_2, respectively, in the w-plane. If Γ_1 and Γ_2 intersect at (u_0, v_0) at the same angle a such that the sense of angle is same in both cases, then $w = f(z)$ is called *conformal mapping. Thus, a mapping which preserves both the magnitude and the sense of the angles is called conformal.*

But, if a mapping preserves only the magnitude of angles but not necessarily the sense, then it is called *isogonal mapping.*

Theorem 3.25. The mapping $w = f(z)$ is conformal at every point z of a domain where $f(z)$ is analytic and $f'(z) \neq 0$.

Proof: Consider a smooth arc $z = z(t)$, which terminates at a point $z_0 = z(t_0)$ at which $f(z)$ is analytic. Let $w_0 = f(z_0)$ and $w = w(t) = f(z(t))$. Then

$$w - w_0 = \frac{f(z) - f(z_0)}{z - z_0}(z - z_0)$$

and so

$$\arg(w - w_0) = \arg\left[\frac{f(x) - f(z_0)}{z - z_0}\right]$$
$$+ \arg(z - z_0) \quad (62)$$

where $\arg(z - z_0)$ is the angle between the positive axis and the vector pointing from z_0 to z. If $z \to z_0$ along the smooth arc $z(t)$, then $\lim\limits_{z \to z_0} \arg(z - z_0)$ is the angle θ between the positive axis and the tangent to the arc at z_0. Similarly, $\arg(w - w_0)$ tends to the angle ϕ between the positive axis and the tangent to $w(t)$ at w_0. Hence, taking limit as $z \to z_0$, (62) reduces to (See Figure 3.19).

$$\phi = \arg f'(z) + \theta, \text{ provided } f'(z) \neq 0$$

Thus, the difference $\phi - \theta$ depends only on the point z_0 and not on the smooth arc $z = f(t)$ for which the angle θ and ϕ were computed. If $z_1(t)$ is another smooth arc terminating at z_0 and if the corresponding tangential directions are given by the angles θ_1 and ϕ_1, then

$$\phi_1 - \theta_1 = \arg f'(z), \text{ provided } f'(z) \neq 0.$$

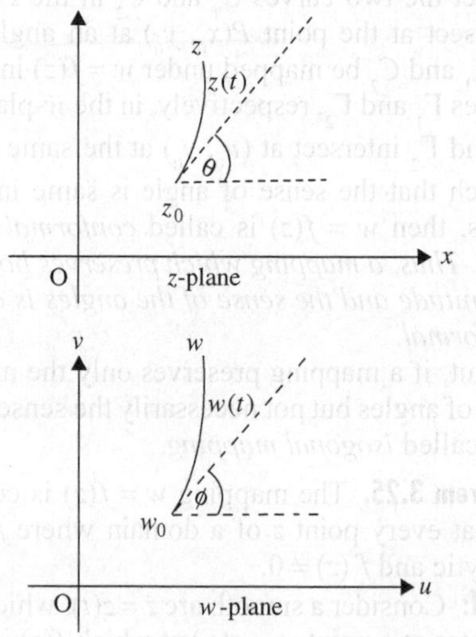

Figure 3.19

Hence

$$\phi_1 - \theta_1 = \phi - \theta$$

or

$$\phi_1 - \phi = \theta_1 - \theta, \quad (63)$$

where $\theta_1 - \theta$ is the angle between the arcs $z_1(t)$ and $z(t)$ and $\phi_1 - \phi$ is the angle between the images of these arcs. The expression (63) shows that the angle between the arcs is not changed by the mapping $w = f(z)$, provided $f' \neq 0$ at the point of intersection. Also (63) shows that the sense of angles is also preserved. Hence the mapping is conformal.

Bilinear (Mobius or Fractional) Transformation

Consider the transformation

$$w = \frac{az + b}{cz + d}, \quad ad - bc \neq 0, \quad (64)$$

where a, b, and c are complex constants. This can be written as

$$cwz + dw - az - b = 0, \quad (65)$$

which is linear in both w and z. Therefore, the mapping (64) is called *Bilinear* or *Mobius Transformation*. Also (64) can be written as

$$w = \frac{a}{c} + \frac{bc - ad}{c(cz + d)}, \quad ad - bc \neq 0. \quad (66)$$

The condition $ad - bc \neq 0$, called the *determinant of the transformation*, prevents (64) from degenerating into a constant.

A transformation $w = f(z)$ is said to be *univalent* if $z_1 \neq z_2$ implies $f(z_1) \neq f(z_2)$.

EXAMPLE 3.118

Show that the linear transformation $w = \dfrac{az + b}{cz + d}$ is a univalent transformation.

Solution. We have

$$w(z_1) = \frac{az_1 + b}{cz_1 + d}, \quad w(z_2) = \frac{az_2 + b}{cz_2 + d}.$$

Therefore,

$$w(z_1) - w(z_2) = \frac{(z_1 - z_2)(ad - bc)}{(cz_2 + d)(cz_1 + d)}.$$

Since $ad - bc \neq 0$, we note that $z_1 \neq z_2$ implies $w(z_1) \neq w(z_2)$. Hence $w = \dfrac{az+b}{cz+d}$ is univalent.

Particular cases of w $= \dfrac{a\,z+b}{c\,z+d}$, $ad - bc \neq 0$

(i) Substituting $c = 0$, $d = 1$, we get the transformation

$$w = az + b. \tag{67}$$

To find the effect of this transformation on a point in the z-plane, let us assume that $b = 0$. Thus $w = az$. Introducing polar co-ordinates we have $z = re^{i\theta}$. If $a = |a|\, e^{ia}$, then

$$w = r|a|e^{i(\theta+a)}$$

and so

$$|w| = r|a| \quad \text{and} \quad \arg w = \theta + a.$$

Thus, under the mapping $w = az$, all distances from the origin are multiplied by the same factor $|a|$ and the argument of all numbers z are increased by the same amount a. Hence the transformation $z \to az$ results in a magnification or contraction according as $|a| > 1$ or $|a| < 1$ and rotation of any geometric figure in the z-plane. In particular, the mapping $z \to az$ maps a circle into a circle. The addition of b to $w = az$ amounts only to a translation. If b is real, all points are translated horizontally by the same amount and if b is complex, then we will also have vertical translation. *Hence $w = az + b$ will always transform a circle into circle.*

(ii) Substituting $a = d = 0$, $b = c$ in $w = \dfrac{az+b}{cz+d}$, $ad - bc \neq 0$, we get

$$w = \frac{1}{z}, \tag{68}$$

which is the translation, called *inversion*. Setting $z = re^{i\theta}$, (68) reduces to

$$w = \frac{e^{-i\theta}}{r}. \tag{69}$$

Thus

$$|w| = \frac{1}{r} = \frac{1}{|z|} \text{ or } |w||z| = 1 \text{ and } \arg w = -\theta.$$

This means that the points of the w-plane corresponding to z has a modulus which is the reciprocal of the modulus of z. *Thus the mapping $w = \frac{1}{z}$ transforms points in the interior of the unit circle into the points in its exterior and vice- versa. The circumference of the unit circle is transformed into itself. But since* $\arg w = -\arg z$, *the circumference $|w| = 1$ is described in the negative sense if $|z| = 1$ is described in the positive sense.*

We note that $z = 0$ is mapped by $w = \frac{1}{z}$ to ∞ in the w-plane and $w = 0$ is mapped to ∞ in the z-plane. If we apply the mapping twice, we get the identity mapping. For any point z_0 in the z-plane, $\frac{1}{z_0}$ *is called the inverse of z_0 with respect to the circle $|z| = 1$.* That is why, the mapping $w = \frac{1}{z}$ is called *inversion.* The fixed points of the mapping are given by $z = \frac{1}{z}$, that is, by $z^2 = 1$. Hence ± 1 are the fixed points of the inversion.

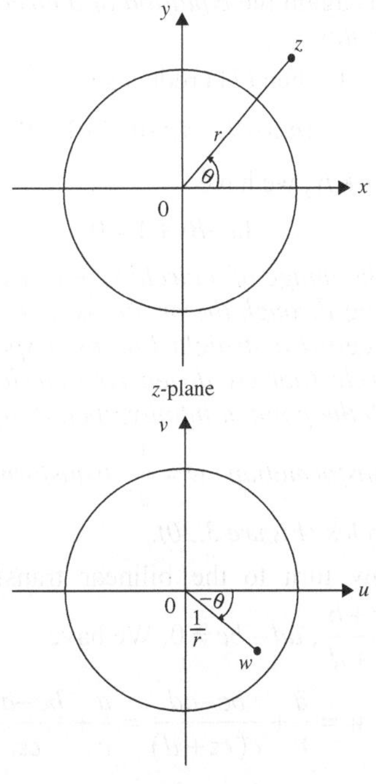

Figure 3.20

The mapping $w = \frac{1}{z}$ *transforms circles into circles*. To prove it, let the equation of circle in xy-plane be

$$x^2 + y^2 + Ax + By + C = 0,$$

where A, B, and C are real constants. Changing to polar co-ordinates, we have

$$r^2 + r(A\cos\theta + B\sin\theta) + C = 0. \quad (70)$$

If ρ, ϕ are polar coordinates in w-plane, then $w = \frac{1}{z} = \frac{e^{-i\theta}}{r}$ implies that $\rho = \frac{1}{r}$ and $\phi = -\theta$. Therefore, under the transformation $w = \frac{1}{z}$, the circle's equation (70) transforms to

$$\frac{1}{\rho^2} + \frac{1}{\rho}(A\cos\phi - B\sin\phi) + C = 0. \quad (71)$$

If $C \neq 0$, then

$$\rho^2 + \rho\left(\frac{A}{C}\cos\phi - \frac{B}{C}\sin\phi\right) + \frac{1}{C} = 0,$$

which is again the *equation of a circle in polar coordinates*.

If $C = 0$, then (71) reduces to

$$A\rho\cos\theta - B\rho\sin\theta + 1 = 0$$

If $w = u + iv$, we have

$$Au - Bv + 1 = 0.$$

Thus, the image of a circle $x^2 + y^2 + Ax + By = 0$ *passing through the origin, is a straight line. If we regard a straight line as a special case of a circle (namely degenerate circle) passing through the point at infinity, then it follows that the transformation* $w = \frac{1}{z}$ *transforms circles into circles (Figure 3.20).*

We now turn to the bilinear transformation $w = \frac{az+b}{cz+d}$, $ad - bc \neq 0$. We have

$$w = \frac{a}{c} + \frac{bc - ad}{c(cz+d)} = \frac{a}{c} + \frac{bc-ad}{cz_1}$$

$$= \frac{a}{c} + \frac{bc-ad}{c} z_2,$$

where

$$z_1 = cz + d \text{ and } z_2 = \frac{1}{z_1}.$$

Thus, the bilinear transformation $w = \frac{az+b}{cz+d}$, $ad - bc \neq 0$ splits into three successive transformations

$$z_1 = cz + d \quad (72)$$

$$z_2 = \frac{1}{z_1} \quad (73)$$

$$w = \frac{a}{c} + \frac{bc - cd}{c} z_2. \quad (74)$$

The transformations (72) and (74) are of the form $w = az + b$, whereas (73) is inversion. Hence, by the above discussion it follows that "The linear transformation $w = \frac{az+b}{cz+d}$, $ad - bc \neq 0$ maps circles in the z-plane onto circles in the w-plane. The point $z = -\frac{d}{c}$ is transformed by $w = \frac{az+b}{cz+d}$ into the point $w = \infty$, accordingly circles passing through the point $z = -\frac{d}{c}$ will be transformed into straight lines."

EXAMPLE 3.119

Find the condition where the transformation $w = \frac{az+b}{cz+d}$ transforms the unit circle in the w-plane into a straight line.

Solution. The given transformation is $w = \frac{az+b}{cz+d}$. Therefore,

$$|w| = 1 \Rightarrow w\bar{w} = 1$$

$$\Rightarrow \left(\frac{az+b}{cz+d}\right)\left(\frac{\bar{a}\bar{z}+\bar{b}}{\bar{c}\bar{z}+\bar{d}}\right) = 1$$

$$\Rightarrow (a\bar{a} - c\bar{c})z\bar{z} + (a\bar{b} - c\bar{d})z$$

$$+ (\bar{a}b - \bar{c}d)\bar{z} + b\bar{b} - d\bar{d} = 0.$$

In order that this equation represents a straight line, the coefficient of $z\bar{z}$ must vanish, that is,

$$a\bar{a} - c\bar{c} = 0 \text{ or } a\bar{a} = c\bar{c} \text{ or } |a| = |c|,$$

which is the required condition. If a and c are reals, then the condition becomes $a = c$.

EXAMPLE 3.120

Investigate the mapping $w = z^2$.

Solution. The given mapping is $w = z^2$. The derivative $\frac{dw}{dz} = 2z$ vanishes at the origin. Hence the mapping is not conformal at the origin. Taking $z = x + iy$ and $w = u + iv$, we have

$$u + iv = (x+iy)^2 = x^2 - y^2 + 2ixy.$$

Separating the real and imaginary parts, we get

$$u = x^2 - y^2 \text{ and } v = 2xy.$$

Therefore, the straight lines $u = a$ and $v = b$ in the w-plane correspond to the *rectangular hyperbolas*

$$x^2 - y^2 = a \text{ and } 2xy = b.$$

These hyperbolas cut at right angles except in the case $a = 0$, $b = 0$, when they intersect at the angle $\frac{\pi}{4}$.

Now, let $x = a$ be a straight line in the z-plane parallel to the y-axis. Then

$$u = a^2 - y^2 \text{ and } v = 2ay.$$

Elimination of y yields

$$v^2 = 4a^2\left(a^2 - u\right),$$

which is a parabola in w-plane having its vertex at $u = a^2$ on the positive real axis in the w-plane. This parabola open towards the negative side of the u-axis. The line $y = b$ corresponds to the curve

$$u = x^2 - b^2, \quad v = 2bx.$$

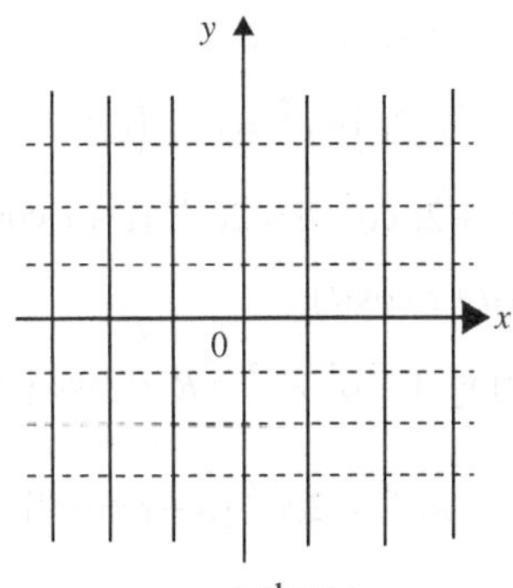

Figure 3.21

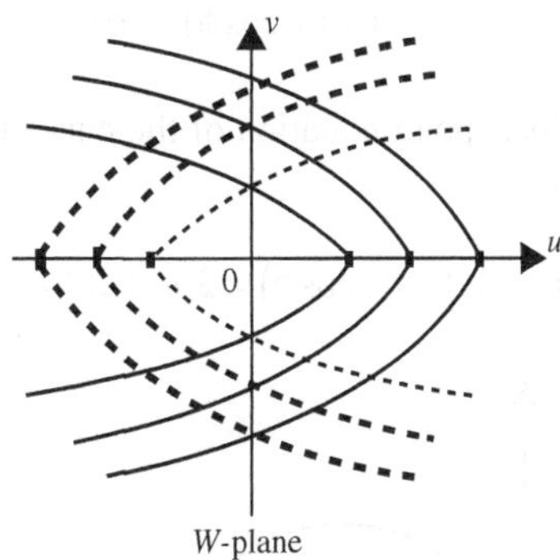

Figure 3.21 (continued)

Elimination of x yields

$$v^2 = 4b^2\left(u + b^2\right),$$

which is again a parabola, but pointing in the opposite direction and having vertex at $u = -b^2$, $v = 0$, focus on the origin and opening towards the positive side of u-axis (See Figure 3.21).

Hence the straight lines x = constant and y = constant correspond to the system of co-focal parabolas.

EXAMPLE 3.121

If a and c are reals, show that the transformation $w = z^2$ transforms the circle $|z - a| = c$ in the z-plane to a limacon in the w-plane.

Solution. We have $z - a = ce^{i\theta}$ so that

$$w - a^2 + c^2$$
$$= \left(a + ce^{i\theta}\right)^2 - a^2 + c^2$$

$$= a^2 + c^2 e^{2i\theta} + 2ace^{i\theta} - a^2 + c^2$$
$$= ce^{i\theta}\left(ce^{i\theta} + 2a\right) + c^2 = ce^{i\theta}\left(ce^{i\theta} + 2a + ce^{-i\theta}\right)$$
$$= ce^{i\theta}\left[2a + 2c\cos\theta\right] = 2ce^{i\theta}\left(a + c\cos\theta\right)$$
$$= 2ce^{i\theta}\left(a + c\cos\theta\right).$$

Substituting $w - a^2 + c^2 = R\,e^{i\theta}$, we get

$$Re^{i\phi} = 2ce^{i\theta}\left(a + c\cos\theta\right)$$

or

$$R = 2c\left(a + c\cos\phi\right), \quad \phi = \theta.$$

Therefore, polar equation of the curve in the w-plane is

$$R = 2c\left(a + c\cos\phi\right) = 2ac + 2c^2\cos\phi.$$

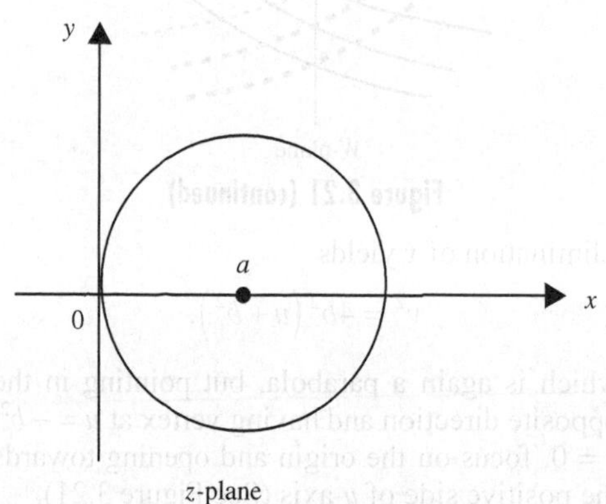

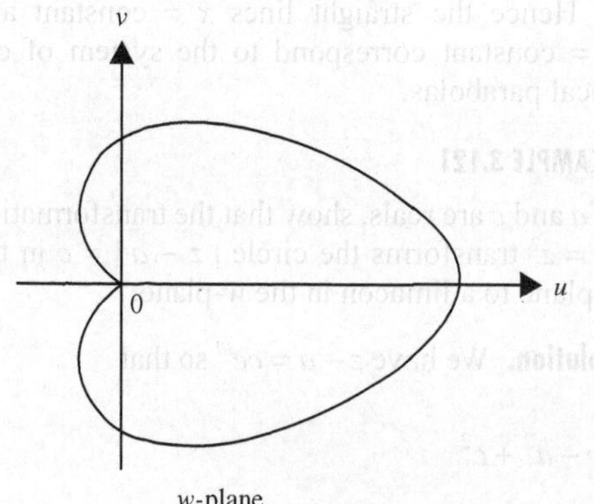

Figure 3.22

If we take $a = c$, that is, if the circle in z-plane touches the axis of y at (0, 0) and its centre is at $x = a$ and radius a, then the limacon degenerates into a cardiod $R = 2a^2(1 + \cos\phi)$ (see Figure 3.22).

EXAMPLE 3.122

Find the image in the w-plane of the circle $|z - 3| = 2$ in the z-plane under the inverse mapping $w = \frac{1}{z}$.

Solution. The image in the w-plane of the given circle $|z - 3| = 2$ in the z-plane under the inverse mapping $w = \frac{1}{z}$ is given by

$$\left|\frac{1}{w} - 3\right| = 2$$

or

$$\left|\frac{1}{u + iv} - 3\right| = 2$$

or

$$\left|\frac{u - iv}{u^2 + v^2} - 3\right| = 2$$

or

$$\left|\frac{u - iv}{u^2 + v^2} - 3\right|^2 = 4$$

or

$$\left[\left(\frac{u}{u^2 + v^2} - 3\right) - \frac{iv}{u^2 + v^2}\right] \times \left[\left(\frac{u}{u^2 + v^2} - 3\right) + \frac{iv}{u^2 + v^2}\right] = 4$$

or

$$\left(\frac{u}{u^2 + v^2} - 3\right)^2 + \frac{v^2}{\left(u^2 + v^2\right)^2} = 4$$

or

$$\frac{u^2 + v^2}{\left(u^2 + v^2\right)^2} - \frac{6u}{u^2 + v^2} + 5 = 0$$

or

$$1 - 6u + 5\left(u^2 + v^2\right) = 0$$

or

$$\left(u-\frac{3}{5}\right)^2+v^2=\frac{4}{25}=\left(\frac{2}{5}\right)^2.$$

If follows that image of $|z-3|=2$ is a circle with centre $\left(\frac{3}{5},0\right)$ and radius $\frac{2}{5}$.

On the other hand, $w=\frac{1}{z}$ implies

$$u+iv=\frac{1}{x+iy}=\frac{x-iy}{x^2+y^2}=\frac{x}{x^2+y^2}-\frac{iy}{x^2+y^2}.$$

Equating real and imaginary parts, we get

$$u=\frac{x}{x^2+y^2},\quad v=\frac{-y}{x^2+y^2}.$$

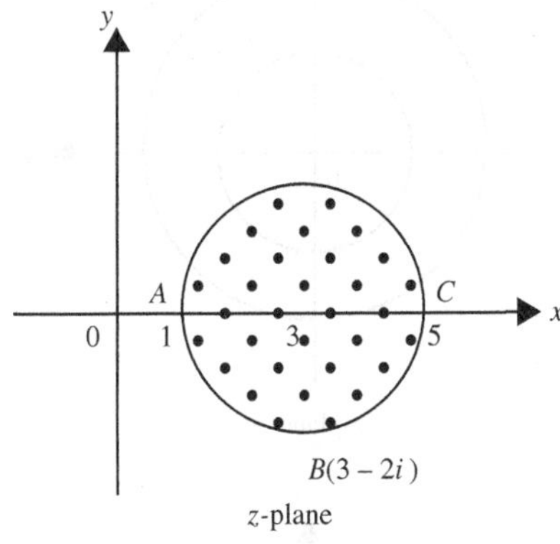

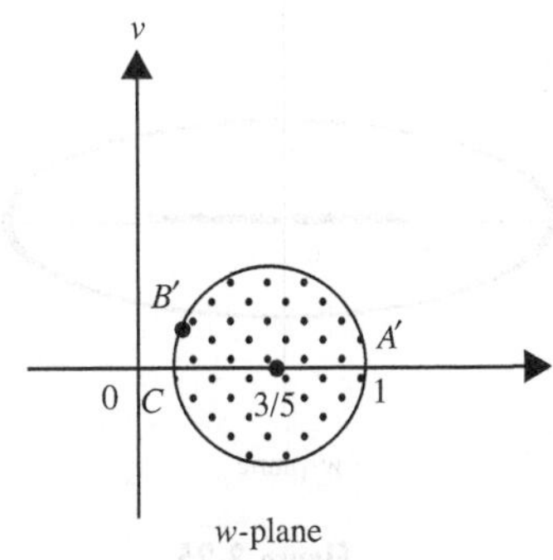

Figure 3.23

The centre (3, 0) of the circle in z-plane is mapped into $(u,v)=\left(\frac{1}{3},0\right)$ in the w-plane which is inside the mapped circle (See Figure 3.23). Therefore, under $w=\frac{1}{z}$, the region under the circle $|z-3|=2$ is mapped onto the region inside the circle in the w-plane.

We note that the point $A(1+i0)$ is mapped into (1, 0), $B(3-2i)$ into $B'\left(\frac{3}{13},\frac{2}{13}\right)$, and $C(5+i0)$ is mapped into the point $C'\left(\frac{1}{5},0\right)$. Thus as the point z traverse the circle in the z-plane in an anti-clockwise direction, the corresponding point w in the w-plane will also traverse the mapped circle is an anticlockwise direction.

EXAMPLE 3.123

Discuss the transformation $w=z+\frac{1}{z}$.

Solution. At $z=0$, w becomes infinite. Further $\frac{dw}{dz}=1-\frac{1}{z^2}$ vanishes at $z=\pm 1$. Thus, $z=\pm 1$ are the critical points and the function $w=z+\frac{1}{z}$ is not conformal at 0, 1, and −1. Substituting $z=re^{i\theta}$ and $w=u+iv$, we have

$$\begin{aligned}u+iv&=re^{i\theta}+\frac{1}{re^{i\theta}}\\&=r(\cos\theta+i\sin\theta)+\frac{1}{r}(\cos\theta-i\sin\theta)\\&=\left(r+\frac{1}{r}\right)\cos\theta+i\left(r-\frac{1}{r}\right)\sin\theta.\end{aligned}$$

Therefore,

$$u=\left(r+\frac{1}{r}\right)\cos\theta \tag{75}$$

$$v=\left(r+\frac{1}{r}\right)\sin\theta \tag{76}$$

If $r=1$, that is, if the radius of the circle in z-plane is unity, then we get $u=2\cos\theta$, $v=0$. Therefore, as θ varies from 0 to 2π in describing the unit circle in the z-plane, the domain described in the w-plane is the segment of the real axis between the points 2 and –2 twice, that is, the ellipse of minor axis 0 and major axis equal to 1. Moreover, (75) and (76) yield

$$\frac{u}{r+\frac{1}{r}} = \cos\theta, \quad \frac{v}{r-\frac{1}{r}} = \sin\theta.$$

Squaring and adding, we get

$$\frac{u^2}{\left(r+\frac{1}{r}\right)^2} + \frac{v^2}{\left(r-\frac{1}{r}\right)^2} = 1,$$

which is an ellipse in the w-plane and it corresponds to each of the two circles $|z| = r$ and $|z| = \frac{1}{r}$, since the equation of the ellipse does not change on changing r to $\frac{1}{r}$. Thus the major and minor axis of the ellipse in w-plane are $r+\frac{1}{r}$ and $r-\frac{1}{r}$. As $r \to 0$ or $r \to \infty$, both semi-axis tend to infinity. Thus, the inside and the outside of the unit circle in the z-plane both correspond to the whole w-plane, cut along the real axis from –1 to 1.

The fixed points of the given transformation are given by $z = z + \frac{1}{z}$. Therefore, $z = \infty$ in the fixed point (see Figures 3.24 and 3.25).

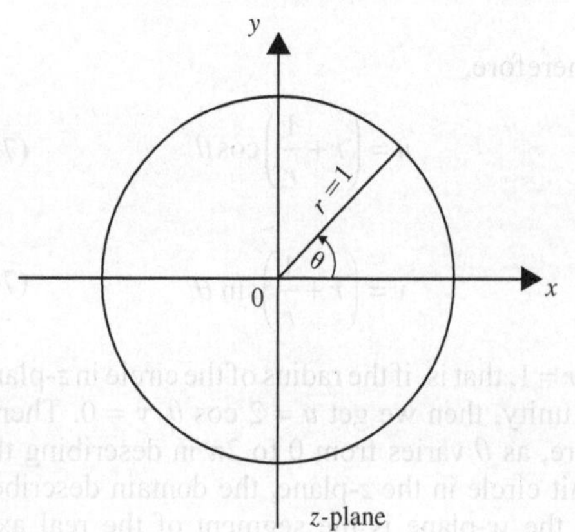

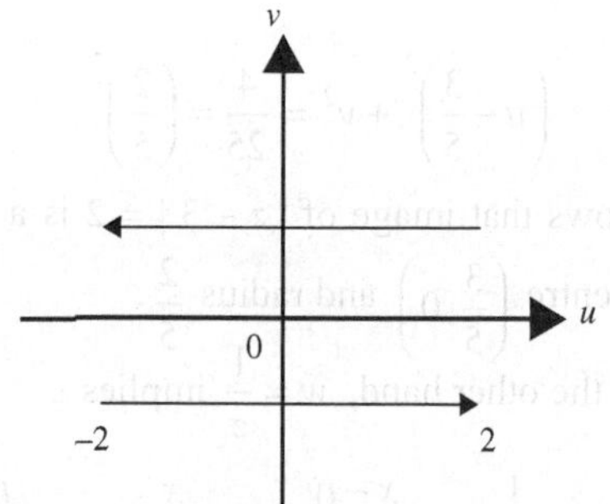

Figure 3.24

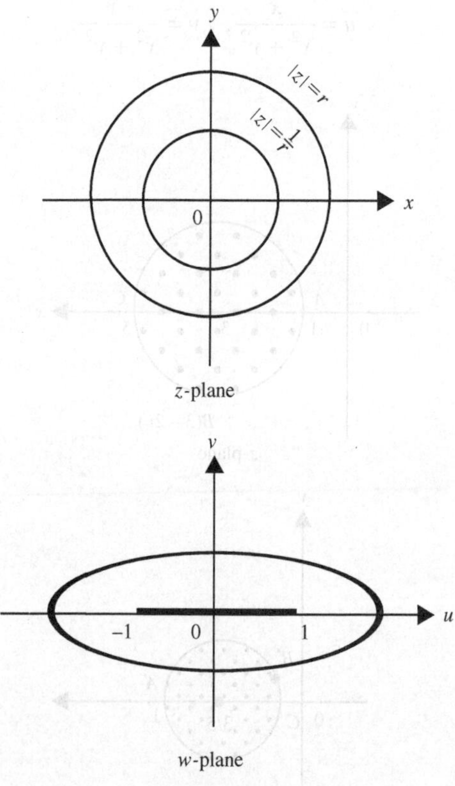

Figure 3.25

EXAMPLE 3.124

Examine the exponential transformation $w = e^z$.

Solution. Substituting $z = x + iy$ and $w = u + iv$, the exponential transformation $w = e^z$ yields

$$u+iv=e^{x+iy}=e^{x}(\cos y+i\sin y).$$

Equating real and imaginary parts, we have

$$u=e^{x}\cos y \quad \text{and} \quad v=e^{x}\sin y$$

or

$$\cos y=\frac{u}{e^{x}} \text{ and } \sin y=\frac{v}{e^{x}}.$$

Squaring and adding, we get

$$u^2+v^2=e^{2x} \tag{77}$$

Also

$$\frac{v}{u}=\tan y. \tag{78}$$

Let $x=a$ be a line parallel to the imaginary axis. Then (77) yields $u^2+v^2=e^{2a}$. Thus, the line parallel to y-axis is transformed into circles with centre at the origin (see Figure 3.26).

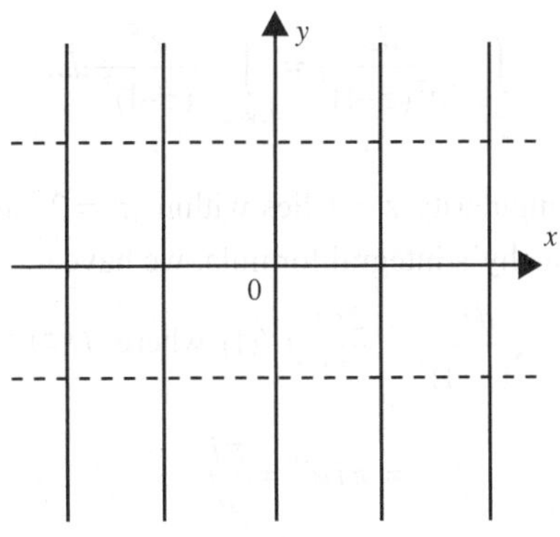

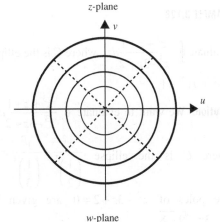

Figure 3.26

On the other hand, let $y=b$ be a line parallel to the x-axis. Then (78) yields

$$v=u\tan b.$$

Thus, the lines parallel to the x-axis are mapped by the transformation into rays emanating from the origin.

If $x=0$, then we have $u^2+v^2=1$. Hence the imaginary axis is mapped into unit circle $u^2+v^2=1$ in the w-plane (Please see Figure 3.27).

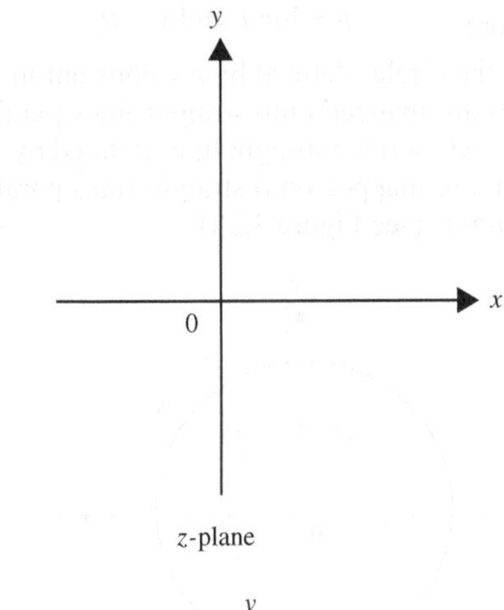

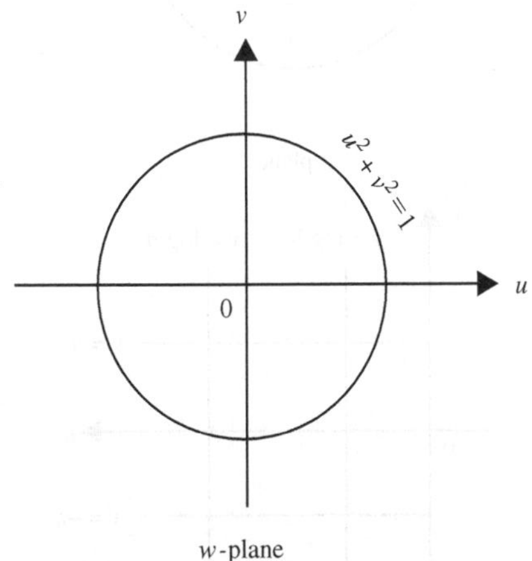

Figure 3.27

Moreover,

$$\frac{dw}{dz}=\frac{\partial u}{\partial x}+i\frac{\partial v}{\partial x}=e^{x}\cos y+ie^{x}\sin y$$
$$=e^{x}(\cos y+i\sin y)=e^{x}e^{iy}=e^{x+iy}=e^{z}\neq 0.$$

Therefore, the mapping $w = e^z$ is conformal everywhere in the complex plane.

EXAMPLE 3.125

Discuss logarithmic transformation $w = \log z$.

Solution. Substituting $z = re^{i\theta}$ and $w = u + iv$, we have

$$u + iv = \log\left(re^{i\theta}\right) = \log r + i\theta.$$

Therefore, $\quad u = \log r$ and $v = \theta$.

Hence, the circles defined by r = constant in the z-planes are mapped onto straight lines parallel to the v-axis and the straight lines defined by θ = constant are mapped onto straight lines parallel to the u-axis (see Figure 3.28)

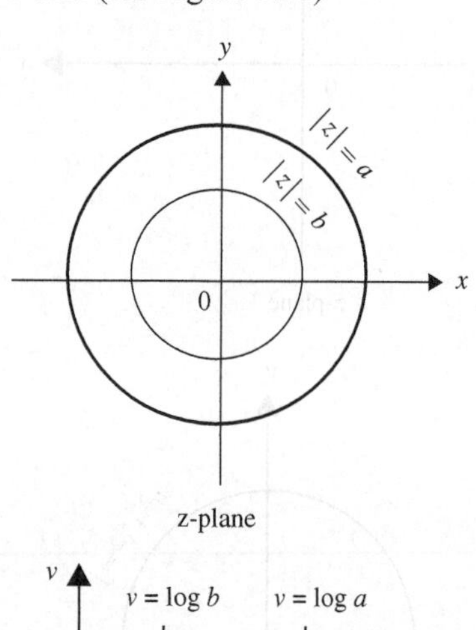

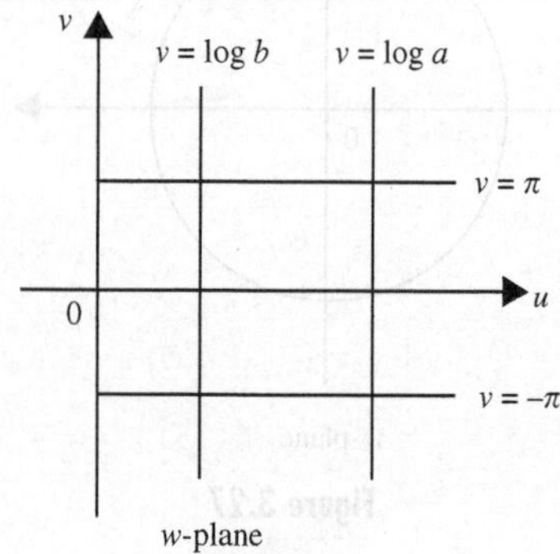

Figure 3.28

Since the derivative $\frac{dw}{dz} = \frac{1}{z}$ is infinite at the origin, the mapping is not conformal at the origin.

3.15 MISCELLANEOUS EXAMPLES

EXAMPLE 3.126

Evaluate $\int_c dz/(z-3)^2$ where C is the circle $|z| = 1$.

Solution. We have $\int_{|z|=1} \frac{dz}{(z-3)^2}$. The integrand is analytic except at $z=3$. But $z=3$ lies outside $|z| = 1$. Hence, by Cauchy-Gaursat theorem, the given integral is equal to zero.

EXAMPLE 3.127

Using Cauchy's integral formula, evaluate $\int_C \frac{dz}{e^z(z-1)^3}$, where C is $|z| = 2$.

Solution. We have

$$\int_{|z|=2} \frac{dz}{e^z(z-1)^3} = \int_{|z|=2} \frac{e^{-z}}{(z-1)^3} dz.$$

The singularity z = 1 lies within $|z| = 2$ and so by Cauchy's integral formula, we have

$$\int_{|z|=2} \frac{dz}{e^2(z-1)^3} = \frac{2\pi i}{2!} f''(1), \text{ where } f(z) = e^{-z}$$

$$= \pi i e^{-1} = \frac{\pi i}{e}.$$

EXAMPLE 3.128

Evaluate $\oint_C \frac{z^3 + z + 1}{z^3 - 3z + 2} dz$, where C is the ellipse $4x^2 + 9y^2 = 1$.

Solution. We want to evaluate $\oint_C \frac{z^3 + z + 1}{z^3 - 3z + 2} dz$, where C is the ellipse $\frac{x^2}{\left(\frac{1}{2}\right)^2} + \frac{y^2}{\left(\frac{1}{3}\right)^2} = 1$.

The poles of $z^2 - 3z + 2 = 0$ are given by $z = \frac{3 \pm \sqrt{9-8}}{2} = 2, 1$.

Both singularities $z = 2$ and $z = 1$ lie outside the contour C. Hence, by Cauchy–Gorsat Theorem,

$$\oint_C \frac{z^3+z+1}{z^3-3z+2}dz = 0.$$

EXAMPLE 3.129

Using Residue Theorem, show that

$$\int_0^{2\pi} \frac{d\theta}{a+b\sin\theta} = \frac{2\pi}{\sqrt{a^2-b^2}}, \quad a > b > 0.$$

Solution. Putting $z = e^{i\theta}$ so that $d\theta = \frac{dz}{iz}$ and $\sin\theta = \frac{1}{2i}\left(z - \frac{1}{z}\right)$, we get

$$I = \int_0^{2\pi} \frac{d\theta}{a+b\sin\theta} = \frac{2}{b}\int_{|z|=1} \frac{dz}{z^2+\frac{2iaz}{b}-1}.$$

Suppose that the poles are α and β. Than $\alpha+\beta = -\frac{2ia}{b}$ and $|\alpha\beta| = 1$. Then α, which is less than β lies inside $|z| = 1$ and

$$\text{Res}(\alpha) = \lim_{z\to\alpha}[(z-1)(f(z)]$$

$$= \frac{1}{\alpha-\beta} = \frac{1}{\sqrt{(\alpha+\beta)^2-4\alpha\beta}}$$

$$= \frac{1}{\sqrt{\left(-\frac{2ia}{b}\right)^2-4(-1)}} = \frac{b}{2i\sqrt{a^2-b^2}}.$$

Hence, by Cauchy's Residue Theorem,

$$I = 2\pi i\left[\frac{2}{b}\left(\frac{b}{2i\sqrt{a^2-b^2}}\right)\right] = \frac{2\pi}{\sqrt{a^2-b^2}}.$$

EXAMPLE 3.130

Show that the image of the hyperbole $x^2-y^2 = 1$, under the transformation $\omega = \frac{1}{z}$, is $r^2 = \cos 2\theta$.

Solution. Let $z = r\,e^{i\theta}$ so that $x = r\cos\theta$, $y =$ sin $r\sin\theta$. Let $\omega = \text{R}e^{i\phi}$. Then the inversion $\omega = \frac{1}{z}$ gives

$$\text{R}e^{i\phi} = \frac{1}{re^{i\theta}} \text{ and so } R = \frac{1}{r} \text{ and } \phi = -\theta.$$

The hyperbole $x^2 - y^2 = 1$, under this transformation becomes

$$r^2\cos^2\theta - r^2\sin^2\theta = 1$$

or

$$r^2(\cos^2\theta - \sin^2\theta) = 1$$

or

$$r^2\cos 2\theta = 1$$

or

$$\frac{1}{R^2}\cos(-2\phi) = 1$$

or

$$R^2 = \cos 2\phi.$$

Hence the hyperbole $x^2 - y^2 = 1$ transformation to the lemniscate $R^2 = \cos 2\phi$.

EXAMPLE 3.131

Show that the transformation $\omega = \frac{2z+3}{z-4}$ transform the circle $x^2+y^2-4x = 0$ into a straight line.

Solution. Let $z = x+iy$ and $\omega = u+iv$. Then

$$x = \frac{z+\bar{z}}{2},\ x^2+y^2 = z\bar{z} \text{ and } u = \frac{\omega+\bar{\omega}}{2}.$$

Therefore the equation of the given circle in z-plane reduces to

$$z\bar{z} - 2(z+\bar{z}) = 0 \quad (79)$$

The given transformation yields

$$\omega(z-4) = 2z+3$$

or

$$z(\omega-2) = 4\omega+3$$

or

$$z = \frac{4\omega+3}{\omega-2}.$$

Therefore

$$\bar{z} = \frac{4\bar{\omega}+3}{\bar{\omega}-2}.$$

Therefore $\omega = \dfrac{2z+3}{z-4}$ transforms the circle (79) into

$$\left(\frac{4\omega+3}{\omega-2}\right)\left(\frac{4\bar{\omega}+3}{\bar{\omega}-2}\right) - 2\left(\frac{4\omega+3}{\omega-2} + \frac{4\bar{\omega}+3}{\bar{\omega}-2}\right) = 0$$

or

$$12(\omega+\bar{\omega}) + 16\omega\bar{\omega} + 9$$
$$-2(8\omega\bar{\omega} - 5\omega - 5\bar{\omega} - 12) = 0$$

or

$$22(\omega+\bar{\omega}) + 33 = 0$$

or

$$2(\omega+\bar{\omega}) + 3 = 0$$

or

$$4u + 3 = 0,$$

which is a straight line in ω -plane

EXAMPLE 3.132

By the transformation $\omega = z^2$, show that the circle $|z-a| = c$ (a and c being real) in the z-plane corresponds to the binacon in the w-plane.

Solution. The equation of the given circle is

$$|z-a| = c$$

or

$$z - a = c\,e^{i\theta}$$

or

$$z = a + c\,e^{i\theta}.$$

Also then

$\omega = z^2$ implies

$$\omega - a^2 = z^2 - a^2 = (z-a)(z+a)$$
$$= c\,e^{i\theta}(2a + ce^{i\theta}).$$

Therefore

$$\omega - (a^2 - c^2) = c\,e^{i\theta}(2a + c\,e^{i\theta}) + c^2$$
$$= c\,e^{i\theta}[2a + c\,e^{i\theta} + c\,e^{-i\theta}]$$
$$= c\,e^{i\theta}[2a + c(e^{i\theta} + e^{-i\theta})]$$
$$= c\,e^{i\theta}[2a + 2c\cos\theta]$$
$$= 2c\,e^{i\theta}[a + c\cos\theta].$$

If we take the pole (origin) at $a^2 - c^2$, than we can take $\omega - (a^2 - c^2) = R\,e^{i\phi}$ and so

$$R\,e^{i\phi} = 2c\,e^{i\theta}(a + c\cos\theta),$$

which yields

$$R = 2c(a + c\cos\phi) \text{ and } \phi = \theta$$

or

$$R = 2c(a + c\cos\phi) \text{ (binacon in } w\text{-plane)}$$

Hence the circle $|z-a| = c$ is transformed into a binacon in the w-plane by the mapping $\omega = z^2$.

EXAMPLE 3.133

Find the bilinear transformation which maps the point $(-1, 0, 1)$ into the point $(0, i, 3i)$.

Solution. Let the required transformation be

$$\frac{(z-z_1)(z_2-z_3)}{(z-z_3)(z_2-z_1)} = \frac{(\omega-\omega_1)(\omega_2-\omega_3)}{(\omega-\omega_3)(\omega_2-\omega_1)}$$

We have $z_1 = -1$, $z_2 = 0$, $z_3 = 1$, $\omega_1 = 0$, $\omega_2 = i$ and $\omega_3 = 3i$.

Therefore

$$\frac{(z+1)(-1)}{(z-1)(1)} = \frac{\omega(i-3i)}{(\omega-3i)(i)}$$

or

$$\frac{z+1}{z-1} = \frac{2\omega}{\omega-3i}$$

or

$$z = \frac{3\omega-3i}{\omega+3i}$$

or

$$\omega = \frac{-3i(z+1)}{z-3},$$

which is the required bilinear transformation.

EXAMPLE 3.134

Find the image of infinite strip $0 < y < \frac{1}{2}$ under the transformation $\omega = \frac{1}{z}$.

Solution. We have

$$\omega = \frac{1}{z} \quad \text{or} \quad z = \frac{1}{\omega} = \frac{\bar{\omega}}{\omega\bar{\omega}} = \frac{\bar{\omega}}{|\omega|^2} = \frac{u-iv}{u^2+v^2}.$$

or

$$x+iy = \frac{u-iv}{u^2+v^2} = \frac{u}{u^2+v^2} - i\frac{v}{u^2+v^2}$$

Equating real and imaginary parts, we have

$$x = \frac{u}{u^2+v^2} \quad \text{and} \quad y = \frac{-v}{u^2+v^2}.$$

Now $y=0$ implies $v=0$ and $y=\frac{1}{2}$ yields

$$u^2+v^2 = -2v \text{ or } u^2+v^2+2v=0$$

or

$$u^2+(v+1)^2 = 1,$$

which is a circle with centre at $(0,-1)$ and radius 1 in the *w*-plane. It follows therefore that the line $y=0$ (x-axis) is mapped into $v=0$ (*u*-axis) and the line $y=\frac{1}{2}$ is transformed into the circle $u^2+(v+1)^2=1$. Thus the strip $0<y<\frac{1}{2}$ in the *z*-plane is mapped into the region between the u-axis and the circle $u^2+(v+1)^2=1$ under the given inversion $\omega=\frac{1}{z}$.

EXERCISES

1. Solve the equation $e^{2z-1} = 1+i$

   Hint: $1+i = \sqrt{2}\, e^{\frac{i\pi}{4}}$ (in exponential form). Therefore, $e^{2z-1} = \sqrt{2}\, e^{\frac{i\pi}{4}}$ and so $2z-1 = \log \sqrt{2} + i\left(2n\pi + \frac{\pi}{4}\right)$.

   Hence $z = \frac{1}{2} + \frac{1}{4}\log 2 + i\left(n+\frac{1}{8}\right)\pi$.

2. Find the values of $(1)^{\frac{1}{4}}$

   Ans. $\pm 1, \pm i$

3. Determine$(32)^{\frac{1}{5}}$

   Ans: $2,\ 2\left(\cos\frac{2\pi}{5} \pm i\sin\frac{2\pi}{5}\right)$, $2\left(\cos\frac{4\pi}{5} \pm i\sin\frac{4\pi}{5}\right)$.

4. Express $\cos^8\theta$ in a series ofcosines of multiples of θ.

   Ans: $2^{-7}[\cos 8\theta + 8\cos 6\theta + 28\cos 4\theta + 56\cos 2\theta + 35]$

5. Express $\sin^6\theta$ in a series of multiples of θ

   Ans. $-\frac{1}{2^5}[\cos 6\theta - 6\cos 4\theta + 15\cos 2\theta - 10]$

6. Show that $\sin 6\theta = 6\cos^5\theta \sin\theta - 20\cos^3\theta \sin 3\theta + 6\cos\theta \sin^5\theta$

7. Show that

   $$i\log\frac{x-i}{x+i} = \pi - 2\tan^{-1}x$$

8. Prove that

   $$\tan\theta\ 3x = \frac{3\tanh x + \tanh^3 x}{1+3\tanh^2 x}$$

9. If $x+iy = \tan(A+iB)$, show that $x^2+y^2-2y\coth 2B+1=0$

10. If $\tan^{-1}x + \tan^{-1}y + \tan^{-1}z = \pi$, show that $x + y + z = xyz$.

Hint: $\tan^{-1}\left[\dfrac{x+y+z-xyz}{1-xy-yz-zx}\right] = \pi$

or

$$\frac{x+y+z-xyz}{1-xy-yz-zx} = \tan\pi = 0.$$

Hence,

$$x+y+z-xyz = 0.$$

11. Separate $\tanh^{-1}(x + iy)$ into real and imaginary parts.

Ans. $u = \dfrac{1}{2}\tanh^{-1}\dfrac{2x}{1+x^2+y^2}\; v = \dfrac{1}{2}\tan^{-1}\dfrac{2y}{1-x^2-y^2}$

12. Separate into real and imaginary parts: $\tan^{-1}(x + iy)$

Ans. $u = \dfrac{1}{2}\tan^{-1}\dfrac{2x}{1-x^2-y^2}$

$$v = \frac{1}{2}\tanh^{-1}\frac{2y}{1+x^2+y^2}$$

13. Solve the equation $e^{2z-1} = 1 + i$

Hint: $1+i = \sqrt{2}\,e^{\frac{i\pi}{4}}$ (in exponential form).

Therefore, $e^{2z-1} = \sqrt{2}\,e^{\frac{i\pi}{4}}$ and so

$$2z-1 = \log\sqrt{2} + i\left(2n\pi + \frac{\pi}{4}\right).$$

Hence $z = \dfrac{1}{2} + \dfrac{1}{4}\log 2 + i\left(n + \dfrac{1}{8}\right)\pi.$

14. Find the values of $(1)^{\frac{1}{4}}$

Ans. $\pm 1, \pm i$

15. Determine $(3z)^{\frac{1}{5}}$

Ans. $2, 2\left(\cos\dfrac{2\pi}{5} \pm i\sin\dfrac{2\pi}{5}\right),$

$$2\left(\cos\frac{4\pi}{5} \pm i\sin\frac{4\pi}{5}\right).$$

16. Show that

$$\tan^{-1}x = \frac{1}{2i}\log\left(\frac{1+ix}{1-ix}\right).$$

Hint: Substituting $1+ix = r(\cos\theta + i\sin\theta)$, we have $r\cos\theta = 1$ and $r\sin\theta = x$ so that $\tan\theta = x$ or $\theta = \tan^{-1}x =$ L.H.S. Under the same substitution, we have R.H.S $= \dfrac{1}{2i}\cos\left(\dfrac{re^{i\theta}}{re^{-i\theta}}\right)$

$= \dfrac{1}{2i}\cos(e^{2i\theta}) = \dfrac{1}{2i}(2i\theta) = \theta.$ Hence the result.

17. If $x + iy = \tan(A + iB)$, show that

$$x^2 + y^2 - 2y\coth 2B + 1 = 0$$

18. If $\sin(\theta + i\phi) = R(\cos\alpha + i\sin\alpha)$, show that

(i) $R^2 = \dfrac{1}{2}\{\cosh 2\phi - \cos 2\theta\}$

(ii) $\tan\alpha = \tanh\phi\cot\theta$

Hint: $R(\cos\alpha + i\sin\alpha) = \sin\theta\cos i\phi + \cos\theta\sin i\phi = \sin\theta\cosh\phi + i\cos\theta\sinh\phi$. Therefore, equating real and imaginary parts, we get $R(\cos\alpha = \sin\theta\cosh\phi$ and $R\sin\alpha = \cos\theta\sinh\phi$. Squaring and adding, we get the result. Also dividing $R\sin\alpha = \sin\theta\sinh\alpha$ by $R\cos\alpha = \sin\theta\cosh\phi$, we get the second result.

19. Separate $\log\sin(x + iy)$ into real and imaginary parts.

Ans. $\text{Re}[\log\sin(x+iy)]$

$$= \frac{1}{2}\log\left[\frac{1}{2}(\cosh 2y - \cos 2x)\right]$$

$$\text{Im}[\log\sin(x+iy)] = \tan^{-1}(\cot x\tanh y).$$

20. Show that $u = y^3 - 3x^2y$ is a harmonic function. Find its harmonic conjugate and the corresponding analytic function $f(z)$ in terms of z.

Ans. $v = -3xy^2 + x^3 + C, f(z) = iz^3 + Ci$

21. Show that the function $u = x^3 - 3xy^2$ is harmonic and find the corresponding analytic function.

Ans. $f(z) = z^3 + C$

22. If $f(z)$ an analytic function of z, prove that

$$\left(\frac{\partial^2}{\partial x^2} + \frac{\partial^2}{\partial y^2}\right)|\text{Re}f(z)|^2 = 2|f'(z)|^2.$$

Hint: As in Example 3.51,

$$\frac{\partial^2}{\partial x^2}+\frac{\partial^2}{\partial y^2}=4\frac{\partial^2}{\partial z\,\partial \bar{z}}$$

$$|\text{Re}\, f(z)|^2=|u|^2=|\frac{1}{2}[f(z)+f(\bar{z})]|^2$$

Therefore,

$$\left(\frac{\partial^2}{\partial x^2}+\frac{\partial^2}{\partial y^2}\right)|\text{Re}\, f(z)|^2$$

$$=\frac{\partial^2}{\partial z\,\partial \bar{z}}|f(z)+f(\bar{z})|^2$$

$$=\frac{\partial^2}{\partial z\,\partial \bar{z}}[(f(z)+f(\bar{z})(f(\bar{z})+f(z))]$$

$$=\frac{\partial^2}{\partial z\,\partial \bar{z}}[f(z)+f(\bar{z})]^2$$

$$=\frac{\partial}{\partial z}.2[f(z)+f(\bar{z})]\,f'(z)$$

$$=2\,f'(z)\,f'(\bar{z})=2\,|f'(z)|^2.$$

23. Solve $\frac{\partial^2\phi}{\partial x^2}+\frac{\partial^2\phi}{\partial y^2}=x^2-y^2.$

Hint: $\frac{\partial^2}{\partial x^2}+\frac{\partial^2}{\partial y^2}=4\frac{\partial^2}{\partial z\,\partial \bar{z}},$

$$x^2-y^2=\frac{1}{2}(z^2+\bar{z}^2).$$

Therefore,

$$\frac{\partial^2\phi}{\partial x^2}+\frac{\partial^2\phi}{\partial y^2}=x^2-y^2 \text{ implies} \frac{1}{2}(z^2+\bar{z}^2)=4\frac{\partial}{\partial z}\frac{\partial\phi}{\partial \bar{z}}$$

$$=4\frac{\partial}{\partial z}\left(\frac{\partial\phi}{\partial \bar{z}}\right).$$

Integrating w.r.t z, we get

$$\frac{\partial\phi}{\partial \bar{z}}=\frac{z^3}{24}+\frac{z\,\bar{z}^2}{8}+\phi_1(\bar{z})$$

Integrating w.r.t. $\bar{z}$ now yields

$$\phi=\frac{z^3\bar{z}}{24}+\frac{z\,\bar{z}^3}{24}+\phi_1(\bar{z})+\phi_1(z)$$

Replacing z by $x+iy$ and $\bar{z}$ by $x-iy$, we get

$$\phi=\frac{1}{12}[(x^4-y^4)+\phi_1(x-iy)+\phi_2(x+iy)].$$

24. Find the analytic function $f(iz)=u+iv$, if $v=\left(r-\frac{1}{r}\right)\sin\theta,\ r\neq 0.$

Hint: By polar form of Cauchy-Riemann equation, $\frac{\partial u}{\partial r}=\frac{1}{r}\frac{\partial v}{\partial \theta},\ \frac{\partial u}{\partial \theta}=-r\frac{\partial v}{\partial r}$ (*).

Thus $\frac{\partial u}{\partial r}=\frac{1}{r}\left[r-\frac{1}{r}\right]\cos\theta=\left(1-\frac{1}{r^2}\right)\cos\theta.$

Integrating we get $u=\cos\theta\left(r+\frac{1}{r}\right)+\phi(\theta)$.

Then $\frac{\partial u}{\partial \theta}=-\sin\theta\left(r+\frac{1}{r}\right)+\phi'(0)$. But by (*) $\frac{\partial u}{\partial \theta}=-r\frac{\partial v}{\partial r}=-\left(r+\frac{1}{2}\sin\theta\right)$. Hence $\phi'(\theta)=0$, which implies that $\phi(\theta)$ is constant. Hence $u=\cos\theta\left(r+\frac{1}{r}\right)+C$ and

$$f(z)=u+iv=cos\theta\left(r+\frac{1}{r}\right)$$

$$+i\left(r-\frac{1}{r}\right)\sin\theta=C.$$

25. If $u=x^2-y^2$, find a function $f(z)=u+iv$ which is analytic.

Hint: $\frac{\partial u}{\partial x}=2x,\ \frac{\partial u}{\partial y}=-2y$ and so by Milne theorem, we have

$$f(z)=\int[u_1(z,0)-iu_2(z,0)]\,dz=\int 2z$$
$$=z^2+C=x^2-y^2+i(2xy+C).$$

26. If $f(z)=u+iv$ is an analytic function of z and $u-v=e^x(\cos y-\sin y)$, find $f(z)$.

Ans. e^z+C

27. Show that $f(z) = z + 2\bar{z}$ is not analytic anywhere in complex plane.

Hint: Cauchy-Riemann equations are not satisfied.

28. Show that $\int \frac{dz}{z-a} = 2\pi i$, where C is the circle $|z - a| = r$.

29. Evaluate $\int_0^{3+i} z^2\, dz$ along $x = 3y^2$.

Ans. 4 + 3i

Hint: $z = x + iy = 3y^2 + iy$, $dz = (6y + i)dy$ and so the integral is $\int_0^1 (3y^2 + iy)(6y + i)\, dy$.

30. Evaluate $\int_{|z-1|=2} \frac{e^{2z}}{(z+1)^4}\, dz$.

Ans. $\frac{8\pi i}{3e^2}$

31. Evaluate $\int_{|z-1|=3} \frac{e^z}{(z+1)^4\,(z-2)}\, dz$

Ans. $\frac{2\pi i}{81}\left(e^2 - \frac{13}{e}\right)$

32. Evaluate $\int_{|z-i|=1} \frac{e^z}{z^2+1}\, dz$.

Hint: $\frac{e^z}{z^2+1}$ is analytic at all points except $\pm i$. The point $z = i$ lies inside $|z - i| = 1$. So, let $f(z) = \frac{e^z}{z+i}$. Then, by Cauchy integral formula, the given integral $= \pi e^i$.

33. Evaluate $I = \int_{|z|=2} \frac{z^3 - 2z + 1}{(z-i)^2}\, dz$.

Hint: By Cauchy integral formula, $I = 2if'(i) = 2\pi i\,[3z^2 - 2]_{z=i} = 2i(-3-2) = -10\pi i$.

34. Expand log $(1 + z)$ in a Taylor series about the point $z = 0$ and find the region of convergence of the series.

Ans. $f(z) = z - \frac{z^2}{2} + \frac{z^3}{3} + \ldots + (-1)^{n-1}\frac{z^n}{n} + \ldots$

This series converges for $|z| < 1$.

35. Expand $f(z) = \frac{z}{(z^2-1)(z^2+4)}$ as a Laurent series about $1 < |z| < 2$.

Hint: Use partial fraction and take cases of $\left|\frac{1}{z}\right| < 1$ and $\left|\frac{z}{2}\right| < 1$.

Ans. $\frac{1}{3}\left(\frac{1}{z^5} - \frac{1}{z^3} - \frac{z}{4} + \frac{z^3}{16} - \frac{z^5}{64} + \ldots\right)$

36. If $0 < |z| < 4$, show that $\frac{1}{4z - z^2} = \sum_{n=0}^{\infty} \frac{z^{n-1}}{4^{n+1}}$.

37. Find the singularities with their nature of the function $\frac{e^{\frac{c}{z-a}}}{e^{z/a} - 1}$.

Ans. Simple poles at $= 2\pi\, nia$, $n = 0, \pm 1, \pm 2, \ldots$

38. Find residues at each poles of $f(z) = \frac{z^2 - 2z}{(z+1)^2(z^2+4)}$.

Ans. $\text{Res}(1) = -\frac{14}{25}$, $\text{Res}(2i) = \frac{7+i}{25}$, $\text{Res}(-2i) = \frac{7-i}{25}$

39. Evaluate $I = \int_{|z|=4} \frac{e^z}{(z^2+\pi^2)^2}\, dz$.

Ans. $-\frac{i}{\pi}$

40. Evaluate $\int_{|z-2|=2} \frac{3z^2+2}{(z-1)(z^2+9)}\, dz$.

Ans. πi

41. Evaluate $\int_{|z|=2} \tan z\, dz$. **Ans.** $-4\pi i$

42. Show that $\int_0^{2\pi} \frac{d\theta}{a + b\sin\theta} = \frac{2\pi}{\sqrt{a^2 - b^2}}$, $a > b > 0$.

43. Evaluate $\int_0^{2\pi} \frac{d\theta}{17-8\cos\theta}$. **Ans.** $\frac{\pi}{15}$

44. Show that $\int_0^{2\pi} \frac{d\theta}{(5-3\sin\theta)^2} = \frac{5\pi}{32}$.

45. Show that $\int_0^{\pi} \frac{\cos 2\theta}{1-2a\cos\theta+a^2}\, d\theta = \frac{\pi a^2}{1-a^2}$ $(a^2 < 1)$.

46. Show that $\int_0^{2\pi} \frac{d\theta}{2+\cos\theta} = \frac{2\pi}{\sqrt{3}}$.

47. Show that $\int_{-\infty}^{\infty} \frac{dx}{x^4+1} = \frac{\pi}{\sqrt{2}}$.

48. Show that $\int_{-\infty}^{\infty} \frac{x^3 \sin mx}{x^4+a^4} = \frac{\pi}{2} e^{-ma/\sqrt{2}} \times \cos^{ma}$ $\times \cos\left(\frac{ma}{\sqrt{2}}\right)$, $m > 0,\ a > 0$.

Hint: Use Jordan lemma, the poles are $ae^{(2n+1)\pi i/4}$, poles $ae^{\pi i/4}$ and $ae^{i3\pi/4}$ lie in the upper half-plane.

49. Show that

$$\int_0^{\infty} \frac{x dx}{(x^2+1)(x^2+2x+2)} = \frac{-\pi}{5}.$$

50. Show that

$$\int_{-\infty}^{\infty} \frac{x^2 dx}{(x^2+1)(x^2+4)} = \frac{\pi}{3}.$$

51. Show that

$$\int_0^{\infty} \frac{\sin^2 x}{x^2}\, dx = \frac{\pi}{2}.$$

52. Evaluate

$$\int_0^{\infty} \frac{dx}{x^4+x^2+1}$$

Ans. $\pi(\sqrt{3}/6)$

53. Show that

$$\int_0^{\infty} \frac{d\theta}{1-2r\cos\theta+r^2} = \frac{\pi}{1-r^2}.$$

54. Show that

$$\int_0^{\infty} \frac{\sin \pi x}{x(1-x^2)}\, dx = \pi.$$

55. Discuss the transformation $w = \sqrt{z}$.

Hint: Letting $z = x + iy$, $w = u + iv$, we have $u^2 - v^2 = x$ and $2uv = y$. The lines $x = a$ and $y = b$ correspond to the rectangular hyperbolas $u^2 - v^2 = a$ and $2uv = b$, which are orthogonal to each other. (see figure 3.29).

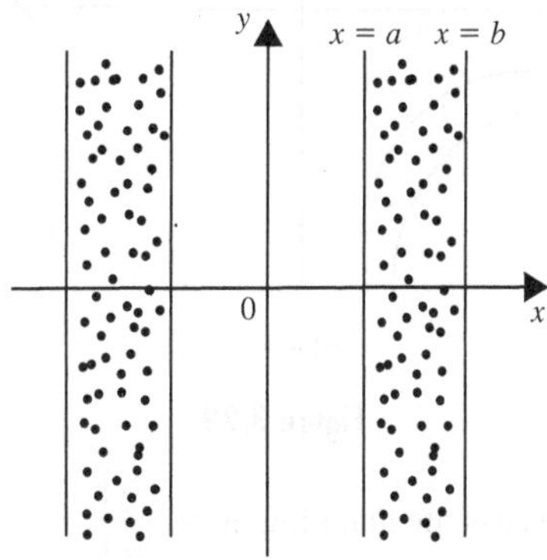

z-plane

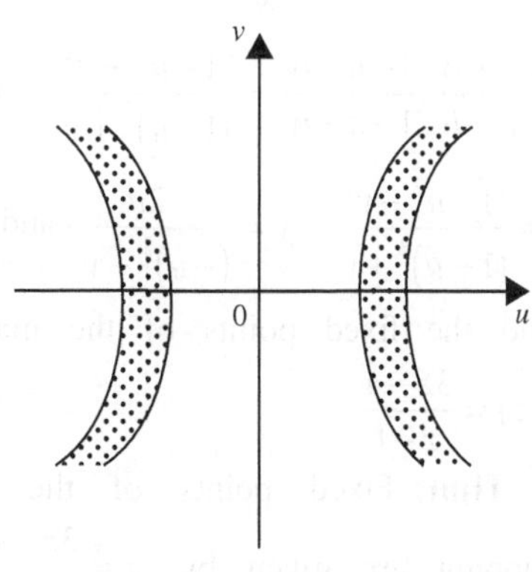

w-plane

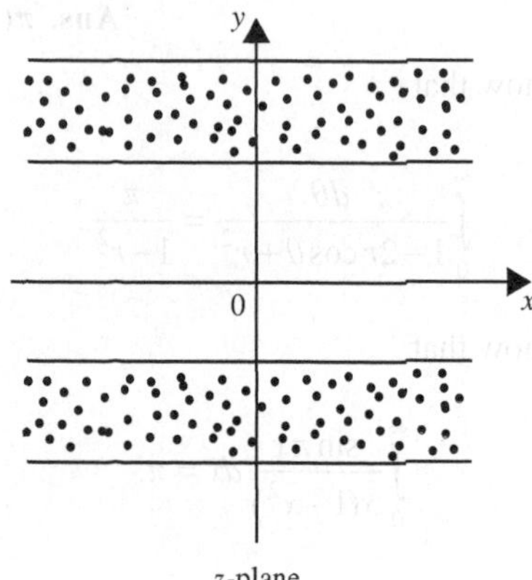

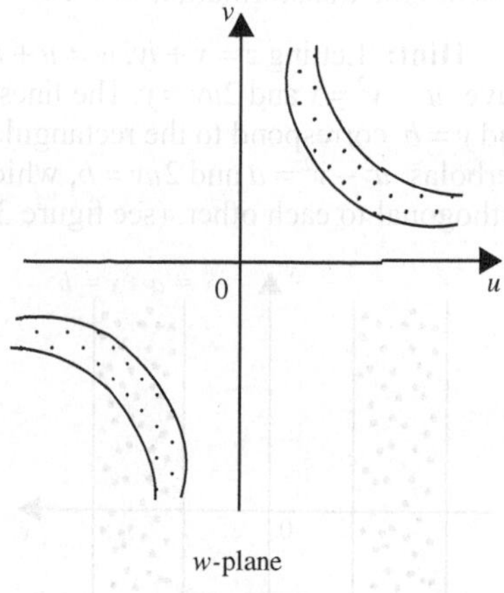

Figure 3.29

56. Discuss the mapping $w = \dfrac{z-1}{z+1}$.

Hint: $z = \dfrac{1+w}{1-w}$. Therefore, $x+iy = \dfrac{1+u+iv}{1-u-iv}\cdot\dfrac{1-u+iv}{1-u+iv} = \dfrac{1-u^2-v^2}{(1-u)^2+v^2}$ Hence

$x = \dfrac{1-u^2-v^2}{(1-u)^2+v^2}$, $y = \dfrac{2u}{(-u)^2+v^2}$ and so on.

57. Find the fixed points of the mapping $\omega(z) = \dfrac{3z-4}{z-1}$.

Hint: Fixed points of the given mapping are given by $z = \dfrac{3z-4}{z-1}$, $z^2 - z = 3z - 4$, or $z^2 - 4z + 4 = 0$. Hence $z = \dfrac{4 \pm \sqrt{16-16}}{2} = 2$ is the fixed point of the mapping.

58. Find the bilinear transformation that maps the points $z = -1, 0, 1$ in the z-plane on to the points $\omega = 0, i, 3i$ in the ω - plane.

Hint: The bilinear transformation is givenby $\omega = \dfrac{az+b}{cz+d}$, $ad - bc \neq 0$. Therefore, we have

$$0 = \frac{a(-1)+b}{c(-1)+d},\ i = \frac{a(0)+b}{c(0)+d},\ \text{and}$$

$$3i = \frac{a(1)+b}{c(1)+d}.$$

From the first equation we have $a = b$. Then second and the third equations yield $d = -ai$ and $c = \dfrac{ai}{3}$. Hence substituting these values in $\omega = \dfrac{az+b}{cz+d}$, we get $\omega = \dfrac{3(z+1)}{i(z-3)}$.

59. Find the bilinear transformation which maps the points $z = 1, i, -1$ onto the points $w = i, 0, -i$.

Hint: Proceeding as in Example 3.133, we have

$$\frac{(z-1)(i+1)}{(z+1)(i-1)} = \frac{(w-i)(0+i)}{(w+i)(0-i)}$$

which on cross multiplication yields

$$w = \frac{i-z}{i+z} = \frac{1+iz}{1-iz}.$$

60. Find the bilinear transformation which maps the points $i, -i, 1$ of the z-plane into $0, 1, \infty$ of the w-plane respectively.

Hint: We have $z_1 = i, z_2 = -i, z_3 = 1, w = 0, w = 1, w = \infty$. Therefore $w_1 = 0, w_2 = 1, w_3 = \infty$. Since $w_3 = 0$, we have

$$\frac{(z-z_1)(z_2-z_3)}{(z-z_3)(z_2-z_1)} = \frac{(w-w_1)(w_2-w_3)}{(w-w_3)(w_2-w_1)}$$

$$= \frac{(w-w_1)\left(\frac{w_2}{w_3}-1\right)}{\left(\frac{w}{w_3}-1\right)(w_2-w_1)} = \frac{w-w_1}{w_2-w_1} \text{ as } w_3 \to \infty.$$

$$\frac{(z-i)(-i-1)}{(z-1)(-i-i)} = \frac{w-0}{1-0}$$

or

$$w = \frac{1}{2}\left[\frac{iz+z+1+i}{iz-i}\right].$$

4 Ordinary Differential Equations

Differential equations play an important role in engineering and science. Many physical laws and relations appear in the form of differential equations. For example, the current I in an LCR circuit is described by the differential equation $LI'' + RI + \frac{1}{C}Q = E$, which is derived from Kirchhoff's laws. Similarly, the displacement y of a vibrating mass m on a spring is described by the equation $my'' + ky = 0$. The study of differential equations involves *formation of differential equations*, the *solutions of differential equations*, and the *physical interpretation* of the solution in terms of the given problem.

4.1 DEFINITIONS AND EXAMPLES

Definition 4.1 A *differential equation* is an equation which involves derivatives.

For example,

(a) $\frac{d^2x}{dt^2} + n^2 x = 0,$

(b) $\frac{d^2y}{dx^2} + \left(\frac{dy}{dx}\right)^2 = 0$

(c) $\frac{\partial^2}{\partial x^2} + \frac{\partial^2}{\partial y^2} = 0$ and

(d) $\frac{dy}{dx} = x + 1$

are differential equations.

We note that a differential equation may have the variables present only in the derivatives. For example in (b), the variables are present only in derivatives. Moreover, a differential equation may have more than one dependent variable. For example,

$$\frac{d\phi}{dt} + \frac{d\psi}{dt} = \phi + \psi$$

has two dependent variables ϕ and ψ and one independent variable t.

Definition 4.2 A differential equation involving derivatives with respect to a single independent variable is called an *ordinary differential equation.* For example,

$$\frac{d^2y}{dx^2} + 3y = 0$$

is an ordinary differential equation.

Definition 4.3 A differential equation involving partial derivatives with respect to two or more independent variables is called a *partial differential equation.*

For example,

$$\frac{\partial^2 u}{\partial t^2} = a^2\frac{\partial^2 u}{\partial x^2} \quad \text{and} \quad \frac{\partial u}{\partial t} = k\frac{\partial^2 u}{\partial x^2}$$

are partial differential equations.

Definition 4.4 The order of the highest derivative appearing in a differential equation free from radicals is called the *order* of that differential equation. For example the, order of the differential equation $y'' + 4y = 0$ is two, the order of the differential equation $\frac{\partial^2 u}{\partial t^2} = a^2\frac{\partial^2 u}{\partial x^2}$ is two, and order of the differential equation $y = x\dfrac{dy}{dx} + \dfrac{x}{dy/dx}$ is one.

Definition 4.5 The degree or exponent of the highest derivative appearing in a differential equation free from radicals and fractions is called the *degree* of the differential equation. For example, the degree of the differential equation

$$y\frac{dy}{dx} = x\left(\frac{dy}{dx}\right)^2 + 1$$

is two. Similarly, the degree of the differential equation

$$\left(\frac{d^3y}{dx^3}\right)^{\frac{2}{3}} = 1 + 2\frac{dy}{dx}$$

is two because the given equation can be written as

$$\left(\frac{d^3y}{dx^3}\right)^2 = \left(1 + 2\frac{dy}{dx}\right)^3.$$

4.2 FORMULATION OF DIFFERENTIAL EQUATION

The derivation of differential equations from physical or other problems is called *modelling.* The modelling involves the successive differentiations and elimination of parameters present in the given sytem.

EXAMPLE 4.1

Form the differential equation for *"free fall"* of a stone dropped from the height y under the action of force due to gravity g.

Solution. We know that equation of motion of the free fall is

$$y = \frac{1}{2}gt^2$$

Differentiating with respect to t, we get

$$\frac{dy}{dt} = gt.$$

Differentiating once more, we get

$$\frac{d^2y}{dt^2} = g,$$

which is the desired differential equation representing the free fall of a stone.

EXAMPLE 4.2

Form the differential equation of simple harmonic motion given by $x = A\cos(\omega t + \phi)$, where A and ϕ are constants.

Solution. To get the required differential equation, we have to differentiate the given relation and eliminate the constants A and ϕ. Differentiating twice, we get

$$\frac{dx}{dt} = -A\omega\sin(\omega t + \phi),$$

$$\frac{d^2x}{dt^2} = -A\omega^2\cos(\omega t + \phi) = -\omega^2 x. \quad (1)$$

Hence

$$\frac{d^2x}{dt^2} + \omega^2 x = 0$$

is the differential equation governing simple harmonic motion. Equation (1) shows that acceleration varies as the distance from the origin.

EXAMPLE 4.3

Find the differential equation governing the motion of a particle of mass m sliding down a frictionless curve.

Solution. Velocity of the particle at the starting point $P(x, y)$ is zero since it starts from rest. Let $\phi(x, u)$ be some intermediate point during the motion (see Figure 4.1). Let the origin O be the lowest point of the curve and let the length of the arc OQ be s.

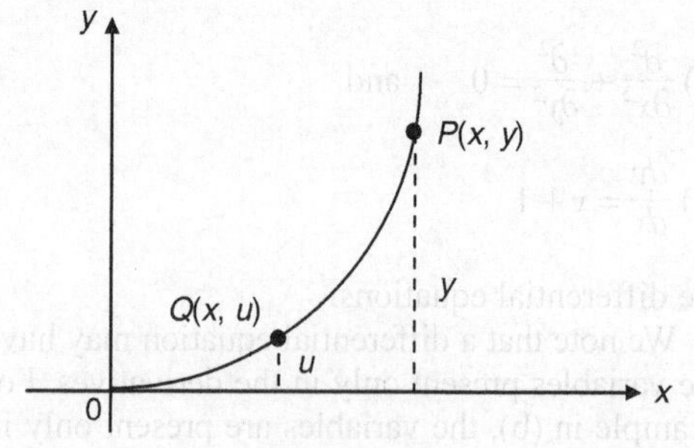

Figure 4.1

By Law of Conservation of Energy, we have potential energy at P + kinetic energy at P

= potential energy at Q + kinetic energy at Q,

and so

$$mgy + 0 = mgu + \frac{1}{2} m \left(\frac{ds}{dt}\right)^2.$$

Hence

$$\left(\frac{ds}{dt}\right)^2 = 2g\,(y - u)$$

is the required differential equation.

If the duration T_0 of descent is independent of the starting point, the solution of this differential equation comes out to be the equation of a *cycloid*. Thus the shape of the curve is a cycloid. This curve is called *Tautochrone Curve*.

EXAMPLE 4.4

Derive the differential equation governing a mass-spring system.

Solution. Let m be the mass suspended by a spring that is rigidly supported from one end (see Figure 4.2). Let

(i) rest position be denoted by $x = 0$, downward displacement by $x > 0$ and upward displacement be denoted by $x < 0$.

(ii) $k > 0$ be spring constant and $a > 0$ be damping constant.

(iii) $a\frac{dx}{dt}$ be the damping force due to medium (damping force is proportional to the velocity).

(iv) $f(t)$ be the external impressed force on m.

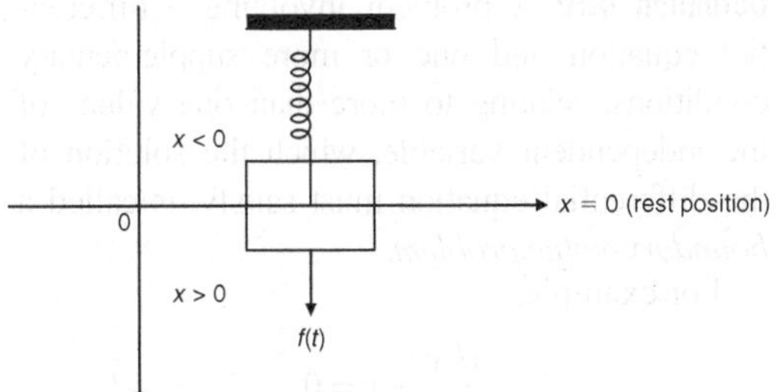

Figure 4.2

Then, by Newton's second law of motion, the sum of force acting on m is $m\frac{d^2x}{dt^2}$ and so

$$m\frac{d^2x}{dt^2} = -kx - a\frac{dx}{dt} + f(t),$$

that is,

$$\frac{d^2x}{dt^2} + \frac{a}{m}\frac{dx}{dt} + \frac{k}{m}x = f(t),$$

which is the required differential equation governing the system.

EXAMPLE 4.5

Find the differential equation of all circles of radius a and centre (h, k).

Solution. We know that the equation of the circle with radius a and centre (h, k) is

$$(x - h)^2 + (y - x)^2 = a^2 \tag{2}$$

Differentiating twice, we get

$$x - h + (y - k)\frac{dy}{dx} = 0, \tag{3}$$

$$1 + (y - k)\frac{d^2y}{dx^2} + \left(\frac{dy}{dx}\right)^2 = 0 \tag{4}$$

From (4), we have

$$y - k = -\frac{1 + \left(\frac{dy}{dx}\right)^2}{\frac{d^2y}{dx^2}}$$

and then (3) yields

$$x - h = -(y - k)\frac{dy}{dx} = \frac{\frac{dy}{dx}\left[1 + \left(\frac{dy}{dx}\right)^2\right]}{\frac{d^2y}{dx^2}}.$$

Substituting the values of $x - h$ and $y - k$ in (2), we get

$$\left[1 + \left(\frac{dy}{dx}\right)^2\right]^3 = a^2\left(\frac{d^2y}{dx^2}\right)^2,$$

which is the required differential equation. It follows that

$$\frac{\left[1 + \left(\frac{dy}{dx}\right)^2\right]^{\frac{3}{2}}}{\frac{d^2y}{dx^2}} = a,$$

and so the radius of curvature of a circle at any point is constant.

EXAMPLE 4.6

Form the differential equation from the equation $xy = Ae^x + B e^{-x}$.

Solution. Differentiating twice, we get

$$x\frac{dy}{dx} + y = Ae^x - B e^{-x}$$

$$x\frac{d^2y}{dx^2} + \frac{dy}{dx} + \frac{dy}{dx} = Ae^x + B e^{-x} = xy.$$

Hence

$$x\frac{d^2y}{dx^2} + 2\frac{dy}{dx} - xy = 0$$

is the required differential equation.

4.3 SOLUTION OF DIFFERENTIAL EQUATION

Definition 4.6 A solution of a differential equation is a functional relation between the variables involved such that this relation and the derivatives obtained from it satisfy the given differential equation.

For example, $x^2 + 4y = 0$ is a solution of the differential equation

$$\left(\frac{dy}{dx}\right)^2 + x\frac{dy}{dx} - y = 0. \quad (5)$$

In fact, differentiating $x^2 + 4y = 0$, we get

$$2x + 4\frac{dy}{dx} = 0$$

and so $\frac{dy}{dx} = -\frac{x}{2}$. Hence

$$\left(\frac{dy}{dx}\right)^2 + x\frac{dy}{dx} - y = \frac{x^2}{4} + x\left(-\frac{x}{2}\right) - y$$

$$= \frac{x^2}{4} - \frac{x^2}{2} - \left(-\frac{x^2}{4}\right) = 0.$$

Hence $x^2 + 4y = 0$ is a solution of (5).

Definition 4.7 A solution of a differential equation in which the number of arbitrary constants is equal to the order of the differential equation is called the *general* (or *complete*) *solution* of the differential equation.

For example $xy = Ae^x + Be^{-x}$ is the general solution of

$$x\frac{d^2y}{dx^2} + 2\frac{dy}{dx} - xy = 0.$$

Definition 4.8 A solution obtained from the general solution of a differential equation by giving particular values to the arbitrary constants is called a *particular solution* of that differential equation.

Definition 4.9 A problem involving a differential equation and one or more supplementary conditions, relating to one value of the independent variable, which the solution of the given differential equation must satisfy is called an *initial-value problem.*

For example,

$$\frac{d^2y}{dt^2} + y = \cos 2t,$$

$$y(0) = 1, y'(0) = -2$$

is an initial-value problem. Similarly, the problem

$$\frac{d^2y}{dx^2} + y = 0,$$

$$y(1) = 3, y'(1) = 4$$

is also an initial-value problem.

Definition 4.10 A problem involving a differential equation and one or more supplementary conditions, relating to more than one values of the independent variable, which the solution of the differential equation must satisfy, is called a *boundary-value problem.*

For example,

$$\frac{d^2y}{dx^2} + y = 0$$

$$y(0) = 2, y\left(\frac{\pi}{2}\right) = 4$$

is a boundary value problem.

4.4 DIFFERENTIAL EQUATIONS OF FIRST ORDER

We consider first the differential equations of first order. Let D be a open connected set in $\mathbb{R}^2$ and let $f: D \rightarrow \mathbb{R}$ be continuous. We discuss the problem of determining solution in D of the first order differential equation

$$\frac{dy}{dx} = f(x, y).$$

Definition 4.11 A real-valued function $f: D \rightarrow \mathbb{R}$ defined on the connected open set D in $\mathbb{R}^2$ is said to satisfy a *Lipschitz condition* in y on D with Lipschitz constant M if and only if

$$|f(x, y_2) - f(x, y_1)| \leq M |y_2 - y_1|$$

for all (x, y_1) and $(x, y_2) \in D$.

Regarding existence of solutions of first order differential equation, we have the following theorems.

Theorem 4.1 (Picard's Existence and Uniqueness Theorem). Let $f: D \rightarrow \mathbb{R}$ be continuous on open connected set D in $\mathbb{R}^2$ and satisfy Lipschitz condition in y on D. Then for every $(x_0, y_0) \in D$, the initial-value problem $\frac{dy}{dx} = f(x, y)$ has a solution passing through (x_0, y_0).

The solution obtained in Theorem 4.1 is unique. The Lipschitz condition in the hypothesis of Picard's theorem cannot be dropped because continuity of f without this condition will not yield unique solution. For example, consider the equation

$$\frac{dy}{dx} = y^{\frac{2}{3}}, \; y(0) = 0.$$

Clearly $\phi_1(x) = 0$ is a solution to this equation. Further substituting $y = \sin^3\theta$, we have $dy = 3\sin^2\theta \cos\theta \, d\theta$ and so

$$\frac{3\sin^2\theta \cos\theta \, d\theta}{dx} = \sin^2\theta$$

which yields $3\cos\theta \, d\theta = dx$ and $x = 3\sin\theta$. Thus $x^3 = 27\sin^3\theta = 27y$ and hence $y = \frac{x^3}{27}$ is also a solution. Therefore, the given initial-value problem does not have unique solution. The reason is that it does not satisfy Lipschitz condition.

Theorem 4.2 (Peano's Existence Theorem). Let f be a continuous real-valued function on a non-empty open subset D of the Euclidean space $\mathbb{R}^2$ and let $(x_0, y_0) \in D$. Then there is a positive real number α such that the first order differential equations $\frac{dy}{dx} = f(x, y)$ has a solution ϕ in the interval $[x_0, x_0 + \alpha]$ which satisfies the boundary condition $\phi(x_0) = y_0$.

Clearly, Peano's theorem is merely an existence theorem and not a uniqueness theorem.

We now consider certain basic types of first order differential equations for which an exact solution may be obtained by definite procedures. The most important of these types are separable equations, homogeneous equations, exact equations, and linear equations. The corresponding methods of solution involve various devices. We take these types of differential equations one by one.

4.5 SEPARABLE EQUATIONS

The first order differential equation

$$\frac{dy}{dx} = f(x, y) \tag{6}$$

is separable if f may be expressed as

$$f(x, y) = \frac{M(x)}{N(y)}, \quad (x, y) \in D \tag{7}$$

The function $M(x)$ and $N(y)$ are real-valued functions of x and y, respectively. Thus (6) becomes

$$N(y)\frac{dy}{dx} = M(x) \tag{8}$$

The equation (8) is solved by integrating with respect to x. Thus the solution is

$$\int N(y)dy = \int M(x)dx + C.$$

EXAMPLE 4.7

Solve

$$\frac{dy}{dx} = e^{x+y}, \; y(1) = 1. \text{ Find } y(-1).$$

Solution. We have

$$\frac{dy}{dx} = \frac{e^x}{e^{-y}}$$

and so

$$e^{-y}dy = e^x\,dx.$$

Integrating both sides

$$\int e^{-y}\,dy = \int e^x dx + C \quad \text{or} \quad -e^{-y} = e^x + C.$$

Using initial condition $y(1) = 1$, we get

$$-e^{-1} = e + C$$

and so $C = -\left(\frac{1+e^2}{e}\right)$.

Thus the solution is

$$-e^{-y} = e^x - \left(\frac{1+e^2}{e}\right).$$

Hence $y(-1)$ is given by

$$y(-1) = e^{-1} - \left(\frac{1+e^2}{e}\right)$$

$$= \frac{1}{e} - \frac{1+e^2}{e} = -e,$$

that is

$$-e^{-y} = -e$$

or

$$y = -1.$$

EXAMPLE 4.8

Solve $\frac{dy}{dx} = (4x + y + 1)^2, y(0) = 1.$

Solution. Substituting $4x + y + 1 = t$, we get

$$\frac{dy}{dx} = \frac{dt}{dx} - 4$$

Hence, the given equation reduces to

$$\frac{dt}{dx} - 4 = t^2$$

or

$$\frac{dt}{dx} = t^2 + 4$$

or

$$\frac{dt}{t^2 + 4} = dx.$$

Integrating both sides, we have

$$\int \frac{dt}{t^2 + 4}\,dt = \int dx + C$$

or

$$\frac{1}{2}\tan^{-1}\frac{t}{2} = x + C$$

or

$$\frac{1}{2}\tan^{-1}\frac{4x + y + 1}{2} = x + C$$

or

$$4x + y + 1 = 2\tan 2(x + C)$$

Using given initial conditions $x = 0$, $y = 1$, we get

$$\frac{1}{2}\tan^{-1} 1 = C$$

which gives $C = \frac{\pi}{8}$. Hence the solution is

$$4x + y + 1 = 2\tan(2x + \pi/4).$$

EXAMPLE 4.9

Solve

$$y - x\frac{dy}{dx} = a\left(y^2 + \frac{dy}{dx}\right).$$

Solution. The given equation can be written as

$$\frac{dy}{dx}(a + x) = y(1 - ay)$$

or

$$\frac{dy}{y(1 - ay)} = \frac{dx}{a + x}.$$

Integrating both sides, we have

$$\log y - \log(1 - ay) = \log(a + x) + C$$

or

$$\log\frac{y}{(a + x)(1 - ay)} = C.$$

Hence

$$y = K(a + x)(1 - ay), K \text{ constant}$$

is the general solution.

EXAMPLE 4.10

Solve

$$x(1+y^2)\,dx + y(1+x^2)dy = 0.$$

Solution. Dividing the given equation throughout by $(1+x^2)(1+y^2)$, we get

$$\frac{x}{1+x^2}dx + \frac{y}{1+y^2}dy = 0.$$

Integrating both sides, we have

$$\int \frac{x}{1+x^2}dx + \int \frac{y}{1+y^2}dy = C$$

or

$$\frac{1}{2}\int \frac{2x}{1+x^2}dx + \frac{1}{2}\int \frac{2y}{1+y^2}dy = C$$

or

$$\frac{1}{2}\log(1+x^2) + \frac{1}{2}\log(1+y^2) = C$$

or

$$\frac{1}{2}\log(1+x^2)(1+y^2) = C$$

or

$$\log(1+x^2)(1+y^2) = 2C = \log C$$

Hence

$$(1+x^2)(1+y^2) = K\text{ (constant)}$$

is the required solution.

EXAMPLE 4.11

Solve

$$\frac{dy}{dx} = e^{x-y} + x^2 e^{-y}.$$

Solution. We have

$$\frac{dy}{dx} = \frac{e^x}{e^y} + \frac{x^2}{e^y}$$

or

$$e^y \frac{dy}{dx} = e^x + x^2$$

or

$$e^y dy = (e^x + x^2)dx = e^x dx + x^2 dx.$$

Integrating both sides, we get

$$e^y = e^x + \frac{x^3}{3} + C\text{ (constant)}.$$

EXAMPLE 4.12

Solve

$$16y\frac{dy}{dx} + 9x = 0.$$

Solution. We are given that

$$16y\frac{dy}{dx} + 9x = 0$$

or

$$16y\,dy = -9x\,dx.$$

Integrating both sides, we have

$$16\frac{y^2}{2} = -9\frac{x^2}{2} + C$$

or

$$\frac{x^2}{16} + \frac{y^2}{9} = K\text{ (constant)},$$

which is the required solution and represents a family of ellipses.

EXAMPLE 4.13

Solve

$$\frac{dy}{dx} = (x. + y)^2.$$

Solution. Substituting $z = x + y$, we get $\frac{dz}{dx} = 1 + \frac{dy}{dx}$. Therefore, the given equation reduces to

$$\frac{dz}{dx} - 1 = z^2$$

or

$$\frac{dz}{dx} = 1 + z^2$$

or

$$\frac{dz}{1+z^2} = dx.$$

Integrating, we get

$$\tan^{-1} z = x + C.$$

Putting back the value of z, we get

$$\tan^{-1}(x + y) = x + C.$$

Hence

$$x + y = \tan(x + C).$$

EXAMPLE 4.14

Solve

$$(2x - 4y + 5)\frac{dy}{dx} = x - 2y + 3.$$

Solution. We have

$$2\left(x - 2y + \frac{5}{2}\right)\frac{dy}{dx} = x - 2y + 3.$$

Substituting $z = x - 2y$, we get $\frac{dz}{dx} = 1 - 2\frac{dy}{dx}$.

The equation becomes

$$(2z + 5)\frac{dz}{dx} = 4z + 11$$

or

$$(4z + 10)\frac{dz}{dx} = 2(4z + 11)$$

or

$$\frac{4z + 10}{4z + 11}\,dz = 2dx$$

or

$$\frac{4z + 11 - 1}{4z + 11}\,dz = 2dx$$

or

$$\left(1 - \frac{1}{4z + 11}\right)dz = 2dx$$

Integrating both sides, we have

$$z - \frac{1}{4}\log|4z + 11| = 2x + C.$$

Putting back the value of z, we have

$$4x + 8y + \log|4x - 8y + 11| = C.$$

EXAMPLE 4.15

Solve $(x + 2y)(dx - dy) = dx + dy$.

Solution. We have

$$(x + 2y)(dx - dy) = dx + dy$$

or

$$(x + 2y - 1)dx = (x + 2y + 1)dy$$

or

$$\frac{dy}{dx} = \frac{x + 2y - 1}{x + 2y + 1}.$$

Substituting $x + 2y = z$, we have $1 + 2\frac{dy}{dx} = \frac{dz}{dx}$ and so

$$\frac{dy}{dx} = \frac{1}{2}\left(\frac{dz}{dx} - 1\right).$$

Hence the above equation becomes

$$\frac{1}{2}\left(\frac{dz}{dx} - 1\right) = \frac{z - 1}{z + 1}$$

or

$$\frac{dz}{dx} = \frac{3z - 1}{z + 1}.$$

Separating the variables, we have

$$\frac{(z + 1)\,dz}{3z - 1} = dx$$

or

$$\frac{1}{3}\left(\frac{3z - 1 + 4}{3z - 1}\right)dz = dx$$

or

$$\frac{1}{3}\left(1 + \frac{4}{3z - 1}\right)dz = dx.$$

Integrating both sides, we get

$$\frac{1}{3}\int\left(1 + \frac{4}{3z - 1}\right)dz = \int dx + C$$

or

$$\frac{1}{3}\left(z + \frac{4}{3}\log(3z - 1)\right) = x + C$$

or

$$3z + 4\log(3z - 1) = 9x + k \text{ (constant)}$$

or

$$3(y - x) + 2\log(3x + 6y - 1) = \frac{k}{2} = K \text{ (constant)}.$$

4.6 HOMOGENEOUS EQUATIONS

Definition 4.12 An expression of the form

$$a_0x^n + a_1x^{n-1}y + a_2x^{n-2}y^2 + \ldots + a_ny^n,$$

in which every term is of the nth degree, is called a *homogeneous function* of degree n.

Definition 4.13 A differential equation of the form

$$\frac{dy}{dx} = \frac{f(x, y)}{\phi(x, y)},$$

where $f(x, y)$ and $\phi(x, y)$ are homogeneous functions of the same degree in x and y is called an *homogeneous equation.*

A homogeneous differential equation can be solved by substituting $y = vx$. Then $\frac{dy}{dx} = v + x\frac{dv}{dx}$. Putting the value of y and $\frac{dy}{dx}$ in the given equation, we get a differential equation in which variables can be separated. Integration then yields the solution in terms of v, which we replace with $\frac{y}{x}$.

EXAMPLE 4.16

Solve

$$(x^2 + y^2)dx - 2xy\,dy = 0.$$

Solution. We have

$$(x^2 + y^2)dx - 2xy\,dy = 0$$

and so

$$\frac{dy}{dx} = \frac{x^2 + y^2}{2xy} \tag{9}$$

This equation is homogeneous in x and y. So, put $y = vx$. We have $\frac{dy}{dx} = v + x\frac{dv}{dx}$. Hence (9) becomes

$$v + x\frac{dv}{dx} = \frac{x^2 + v^2x^2}{2vx^2} = \frac{1 + v^2}{2v}$$

or

$$x\frac{dv}{dx} = \frac{1 + v^2}{2v} - v = \frac{1 - v^2}{2v}$$

or

$$\frac{2v\,dv}{1 - v^2} = \frac{dx}{x}$$

or

$$\left(\frac{1}{1 - v} - \frac{1}{1 + v}\right)dv = \frac{dx}{x}.$$

Integrating both sides, we get

$$-\log(1 - v) - \log(1 + v) = \log x + C$$

or

$$-\log(1 - v^2) = \log x + C$$

or

$$\log x + \log(1 - v^2) = C$$

or

$$\log x(1 - v^2) = C$$

or

$$\log x\left(1 - \left(\frac{y}{x}\right)^2\right) = C$$

or

$$x\left(1 - \frac{y^2}{x^2}\right) = C.$$

Hence

$$x^2 - y^2 = Cx$$

is the general solution of the given differential equation.

EXAMPLE 4.17

Solve

$$x\frac{dy}{dx} = y + \sqrt{x^2 + y^2}.$$

Solution. We have

$$\frac{dy}{dx} = \frac{y + \sqrt{x^2 + y^2}}{x},$$

which is homogeneous in x and y. Put $y = vx$ so that $\frac{dy}{dx} = v + x\frac{dv}{dx}$. Hence the equation takes the form

$$v + x\frac{dv}{dx} = \frac{vx + \sqrt{x^2 + v^2x^2}}{x}$$

$$= v + \sqrt{1 + v^2}.$$

Thus

$$x\frac{dv}{dx} = \sqrt{1+v^2}$$

and so

$$\frac{dv}{\sqrt{1+v^2}} = \frac{dx}{x}.$$

Integrating both sides, we get

$$\int\frac{dv}{\sqrt{1+v^2}} = \int\frac{dx}{x} + C$$

or

$$\log(v+\sqrt{1+v^2}) = \log x + \log C = \log x\,C.$$

Hence

$$v+\sqrt{1+x^2} = x\,C.$$

Substituting the value of v, we get

$$\frac{y}{x}+\sqrt{1+\frac{y^2}{x^2}} = x\,C,$$

which is the required solution.

EXAMPLE 4.18

Solve

$$(y^2-2xy)\,dx = (x^2-2xy)dy.$$

Solution. The given differential equation can be expressed by

$$\frac{dy}{dx} = \frac{y^2-2xy}{x^2-2xy}.$$

Clearly it is a homogeneous equation. Put $y = vx$ so that $\frac{dy}{dx} = v + x\frac{dv}{dx}$. Hence

$$v+x\frac{dy}{dx} = \frac{v^2x^2-2x^2v}{x^2-2x^2v} = \frac{v^2-2v}{1-2v}.$$

or

$$x\frac{dv}{dx} = \frac{v^2-2v}{1-2v} - v.$$

$$= \frac{v^2-2v-v+2v^2}{(1-2v)}$$

$$= \frac{3v^2-3v}{1-2v}.$$

Thus

$$\frac{1-2v}{3v^2-3v}\,dv = \frac{dx}{x}$$

or

$$\frac{1-2v}{3(v^2-v)} = \frac{dx}{x}.$$

Integrating, we get

$$-\frac{1}{3}\log(v^2-v) = \log x + C$$

$$-\frac{1}{3}\log\left(\frac{y^2}{x^2}-\frac{y}{x}\right) = \log x + C$$

$$-\frac{1}{3}\log\left(\frac{y^2-xy}{x^2}\right) = \log x + C$$

$$\log x^3\left(\frac{y^2-xy}{x^2}\right) = \log C.$$

Hence

$$x(y^2-xy) = C.$$

EXAMPLE 4.19

Solve

$$x\frac{dy}{dx}+\frac{y^2}{x} = y.$$

Solution. We have

$$x\frac{dy}{dx} = y - \frac{y^2}{x} = \frac{xy-y^2}{x}$$

or

$$\frac{dy}{dx} = \frac{xy-y^2}{x^2},$$

which is homogeneous in x and y. Putting $y = vx$, we have $\frac{dy}{dx} = v + x\frac{dv}{dx}$. Hence the equation becomes

$$v+x\frac{dy}{dx} = \frac{xy-y^2}{x^2} = \frac{vx^2-v^2x^2}{x^2} = v - v^2$$

or

$$x\frac{dv}{dx} = -v^2$$

or

$$\frac{dv}{v^2} = -\frac{dx}{x}.$$

Integrating, we get

$$\frac{1}{v} = -\log x + \log C.$$

Putting $v = \frac{y}{x}$, we get

$$\frac{x}{y} = -\log x\, C.$$

EXAMPLE 4.20

Solve

$$(x^2 - y^2)dx = 2xy\, dy.$$

Solution. We have

$$\frac{dy}{dx} = \frac{x^2 - y^2}{2xy}$$

so that the given equation is homogeneous in x and y. Putting $y = vx$, the equation takes the form

$$v + x\frac{dv}{dx} = \frac{x^2 - v^2x^2}{2vx^2} = \frac{1 - v^2}{2v}.$$

Therefore,

$$x\frac{dv}{dx} = \frac{1 - v^2}{2v} - v = \frac{1 - v^2 - 2v^2}{2v} = \frac{1 - 3v^2}{2v}.$$

Now, variables separation yields

$$\frac{2v}{1 - 3v^2}\, dv = \frac{dx}{x}.$$

Integrating both sides, we get

$$-\frac{1}{3}\int \frac{-6v}{1 - 3v^2}\, dv = \int \frac{dx}{x} + C$$

or

$$-\frac{1}{3}\log(1 - 3v^2) = \log x + C$$

or

$$-\log(1 - 3v^2) = \log x^3 + 3C$$

or

$$\log x^3(1 - 3v^2) = -3C$$

or

$$x^3\left(1 - \frac{3y^2}{x^2}\right) = K$$

or

$$x(x^2 - 3y^2) = K.$$

EXAMPLE 4.21

Solve

$$(1 + e^{\frac{x}{y}}dx + e^{\frac{x}{y}}\left(1 - \frac{x}{y}\right) dy = 0.$$

Solution. We have

$$\frac{dx}{dy} = -\frac{e^{\frac{x}{y}}\left(1 - \frac{x}{y}\right)}{1 + e^{\frac{x}{y}}}.$$

Putting $x = vy$, we have $\frac{dx}{dy} = v + y\frac{dv}{dy}$ and so the above equation reduces to

$$v + y\frac{dv}{dy} = -\frac{e^v(1 - v)}{1 + e^v} = \frac{e^v(v - 1)}{1 + e^v}.$$

Hence

$$y\frac{dv}{dy} = \frac{e^v(v - 1)}{1 + e^v} - v = -\frac{v + e^v}{1 + e^v}$$

and so separation of variable gives

$$\frac{1 + e^v}{v + e^v}\, dv = -\frac{dy}{y}.$$

Integrating both sides, we get

$$\int \frac{1 + e^v}{v + e^v} = -\int \frac{dy}{y} + C$$

or

$$\log(v + e^v) = -\log y + \log k$$

or

$$\log(v + e^v) = \log\frac{k}{y}.$$

Thus

$$v + e^v = \frac{k}{y}.$$

But $v = \frac{x}{y}$. Hence

$$x + ye^{\frac{x}{y}} = k \text{ (constant)}.$$

4.7 EQUATIONS REDUCIBLE TO HOMOGENEOUS FORM

Equations of the form

$$\frac{dy}{dx} = \frac{ax + by + C}{a'x + b'y + C'}$$

can be reduced, by substitution, to the homogeneous form and then solved. Two cases arise:

(i) If $\frac{a}{a'} \neq \frac{b}{b'}$, then the substitution $x = X + h$ and $y = Y + k$, where h and k are suitable constants, makes the given equation homogeneous in X and Y.

(ii) If $\frac{a}{a'} = \frac{b}{b'}$, then the substitution $ax + by = z$ serves our purpose.

EXAMPLE 4.22

Solve

$$\frac{dy}{dx} = \frac{2x - y + 1}{x + 2y - 3}.$$

Solution. We observe that the condition $\frac{a}{a'} \neq \frac{b}{b'}$ is satisfied in the present case. So, we put

$$x = X + h \quad \text{and} \quad y = Y + k.$$

Therefore, $dx = dX$ and $dy = dY$ and given equation reduces to

$$\frac{dY}{dX} = \frac{2(X+h) - (Y+k) + 1}{X + h + 2(Y+k) - 3}$$

$$= \frac{2X - Y + 2h - k + 1}{X + 2Y + h + 2k - 3}.$$

We choose h and k such that

$$2h - k + 1 = 0, \qquad \text{and}$$

$$h + 2k - 3 = 0.$$

Solving these two equations, we get $h = \frac{1}{5}$, $k = \frac{7}{5}$.

Hence

$$\frac{dY}{dX} = \frac{2X - Y}{X + 2Y},$$

which is homogeneous in X and Y. So put $Y = vX$. Then

$$v + X\frac{dv}{dx} = \frac{2X - vX}{X + 2vX} = \frac{2 - v}{1 + 2v}$$

and so

$$X\frac{dv}{dx} = \frac{2 - v}{1 + 2v} - v = \frac{-2v^2 - 2v + 2}{1 + 2v}.$$

Now separation of variables yields

$$\frac{1 + 2v}{-2(v^2 + v - 1)}\, dv = \frac{dX}{X}.$$

Integrating both sides, we get

$$-\frac{1}{2}\int \frac{1 + 2v}{v^2 + v - 1}\, dv = \int \frac{dX}{X} + C$$

or

$$-\frac{1}{2}\log(v^2 + v - 1) = \log X + C$$

or

$$-\log(v^2 + v - 1) = 2\log X + C = \log X^2 + C$$

or

$$\log X^2(v^2 + v - 1) = \log k$$

or

$$X^2(v^2 + v - 1) = k$$

or

$$X^2\left(\frac{Y^2}{X^2} + \frac{Y}{X} - 1\right) = k$$

or

$$Y^2 + YX - X^2 = k$$

or

$$\left(y + \frac{7}{5}\right)^2 + \left(y + \frac{7}{5}\right)\left(x + \frac{1}{5}\right) - \left(x + \frac{1}{5}\right)^2 = k$$

or

$$y^2 + xy - x^2 + \left(\frac{14}{5} + \frac{1}{5}\right)y + \left(-\frac{2}{5} + \frac{7}{5}\right)x = k$$

or

$$y^2 + xy - x^2 + 3y + x = k.$$

EXAMPLE 4.23

Solve

$$\frac{dy}{dx} = \frac{2x + 3y + 4}{4x + 6y + 5}.$$

Solution. We note that the condition $\frac{a}{a'} = \frac{b}{b'}$ is satisfied in the present case. Hence put $2x + 3y = z$ so that $2 + 3\frac{dy}{dx} = \frac{dz}{dx}$. Hence the given equation reduces to

$$\frac{dz}{dx} = \frac{7z + 22}{2z + 5}$$

or

$$\frac{2z + 5}{7z + 22}\, dz = dx$$

Integrating both sides, we get

$$\int \frac{2z + 5}{7z + 22}\, dz = \int dx + C$$

or

$$\frac{2}{7}z - \frac{9}{49}\log(7z+22) = x + C.$$

Substituting $z = 2x + 3y$, we get

$$14(2x+3y) - 9\log(14x+21y+22) = 49x + C$$

or

$$21x - 42y + 9\log(14x+21y+22) = C$$

EXAMPLE 4.24

Solve

$$(x + 2y + 1)\,dx = (2x + 4y + 3)\,dy.$$

Solution. We have

$$\frac{dy}{dx} = \frac{x+2y+1}{2x+4y+3} = \frac{ax+by+C}{a'x+b'y+C}.$$

We observe that $\frac{a}{a'} = \frac{b}{b'} = \frac{1}{2}$. So we put $x + 2y = z$ and have $1 + 2\frac{dy}{dx} = \frac{dz}{dx}$. Then $\frac{dy}{dx} = \frac{\frac{dz}{dx} - 1}{2}$.

The given equation now reduces to

$$\frac{\left(\frac{dz}{dx}\right) - 1}{2} = \frac{z+1}{2z+3}$$

or

$$\frac{dz}{dx} = \frac{2z+2}{2z+3} + 1 = \frac{4z+5}{2z+3}$$

or

$$\left(\frac{2z+3}{4z+5}\right) dz = dx.$$

Integrating both sides, we have

$$\int \frac{2z+3}{4z+5}\,dz = \int dx + C$$

or

$$\int \left[\frac{1}{2} + \frac{1}{2(4z+5)}\right] dz = \int dx + C$$

or

$$\frac{1}{2}z + \frac{1}{8}\log(4z+5) = x + C$$

or

$$4z + \log(4z+5) = 4x + k \text{ (constant)}$$

or

$$4(x+2y) + \log(4x+8y+5) = 8x + k$$

or

$$4(2y - x) + \log(4x+8y+5) = k.$$

4.8 LINEAR DIFFERENTIAL EQUATIONS OF FIRST ORDER AND FIRST DEGREE

Definition 4.14 A differential equation is said to be *linear* if the dependent variable and its derivatives occur in the first degree and are not multiplied together.

Thus a linear differential equation of first order and first degree is of the form $\frac{dy}{dx} + Py = Q$, where P and Q are functions of x only

or

$\frac{dx}{dy} + Px = Q$, where P and Q are funtions of y only.

A linear differential equation of the first order is called *Leibnitz's linear equation.*

To solve the linear differential equation

$$\frac{dy}{dx} + Py = Q, \tag{10}$$

we multiply both sides by $e^{\int P dx}$ and get

$$\frac{dy}{dx}e^{\int P dx} + Pye^{\int P dx} = Qe^{\int P dx} \tag{11}$$

But

$$\frac{d}{dx}(y\,e^{\int P dx}) = \frac{dy}{dx}e^{\int P dx} + P\,ye^{\int P dx}.$$

Hence (11) reduces to

$$\frac{d}{dx}(y\,e^{\int P dx}) = Qe^{\int P dx},$$

which on integration yields

$$y\,e^{\int P dx} = \int Q\,e^{\int P dx}dx + C,$$

as the required solution.

The factor $e^{\int P dx}$ is called an *integrating factor* (I.F.) of the differential equation.

EXAMPLE 4.25

Solve

$$(x + 2y^3)\frac{dy}{dx} = y.$$

Solution. The given differential equation can be written as

$$y\frac{dx}{dy} = x + 2y^3$$

or

$$y\frac{dx}{dy} - x = 2y^3$$

or

$$\frac{dx}{dy} - \frac{1}{y}x = 2y^2$$

Comparing with $\frac{dx}{dy} + Px = Q$, we have $P = -\frac{1}{y}$ and $Q = 2y^2$. The integrating factor is

$$\text{I.F.} = e^{\int P dy} = e^{\int -\frac{1}{y} dy} = e^{-\log y} = e^{\log y^{-1}} = y^{-1}.$$

Therefore, the solution of the differential equation is

$$xy^{-1} = \int (2y^2) y^{-1}\, dy + C$$

$$= \int 2y\, dy + C = y^2 + C.$$

EXAMPLE 4.26

Solve

$$(1 + y^2)\frac{dx}{dy} = \tan^{-1} y - x.$$

Solution. The given equation can be expressed as

$$\frac{dx}{dy} + \frac{x}{1+y^2} = \frac{\tan^{-1} y}{1+y^2}$$

and so is Leibnitz's linear equation in x. Comparing with $\frac{dx}{dy} + Px = Q$, we get $P = \frac{1}{1+y^2}$ and $Q = \frac{\tan^{-1} y}{1+y^2}$. Therefore

$$\text{I.F.} = e^{\int P dy} = e^{\int \frac{1}{1+y^2} dy} = e^{\tan^{-1} y}.$$

Hence the solution of the differential equation is

$$xe^{\tan^{-1} y} = \int \frac{\tan^{-1} y}{1+y^2} e^{\tan^{-1} y}\, dy + C$$

$$= \int t\, e^t\, dt + C, \quad \tan^{-1} y = t$$

$$= t\, e^t - e^t + C \text{ (integration by parts)}$$

$$= (\tan^{-1} y - 1)\, e^{\tan^{-1} y} + C.$$

Hence

$$x = \tan^{-1} y - 1 + C\, e^{-\tan^{-1} y}.$$

EXAMPLE 4.27

Solve

$$\sin 2x \frac{dy}{dx} = y + \tan x.$$

Solution. We have

$$\frac{dy}{dx} - \frac{1}{\sin 2x} y = \frac{\tan x}{\sin 2x} = \frac{\sin x}{2\cos^2 x \sin x} = \frac{1}{2}\sec^2 x.$$

Thus, the given equation is linear in y. Now

$$\text{I.F.} = e^{\int -\operatorname{cosec} 2x\, dx} = e^{-\frac{1}{2}\log \tan x}$$

$$= e^{\log(\tan x)^{-\frac{1}{2}}} = (\tan x)^{-\frac{1}{2}}.$$

Hence, the solution of the given differential equation is

$$y(\tan x)^{-\frac{1}{2}} = \int \frac{1}{2}\sec^2 x (\tan x)^{-\frac{1}{2}}\, dx + C$$

$$= \frac{1}{2}\int (\sec^2 x)(\tan x)^{-\frac{1}{2}}\, dx + C$$

$$= \frac{1}{2}\frac{(\tan x)^{-\frac{1}{2}+1}}{-\frac{1}{2}+1} + C$$

$$= (\tan x)^{\frac{1}{2}} + C,$$

which can be expressed as

$$y = \tan x + C\sqrt{\tan x}.$$

4.9 EQUATIONS REDUCIBLE TO LINEAR DIFFERENTIAL EQUATIONS

Definition 4.15 An equation of the form

$$\frac{dy}{dx} + Py = Qy^n, \tag{12}$$

where P and Q are functions of x is called *Bernoulli's equation.*

The Bernoulli's equation can be reduced to Leibnitz's differential equation in the following way:

Divide both sides of (12) by y^n to get

$$y^{-n}\frac{dy}{dx}+P\,y^{1-n}=Q \tag{13}$$

Put $y^{1-n}=z$ to give

$$(1-n)\,y^{-n}\frac{dy}{dx}=\frac{dz}{dx}$$

or

$$y^{-n}\frac{dy}{dx}=\frac{1}{1-n}\frac{dz}{dx}.$$

Hence (13) reduces to

$$\frac{1}{1-n}\frac{dz}{dx}+Pz=Q$$

or

$$\frac{dz}{dx}+P(1-n)\,z=Q(1-n),$$

which is Leibnitz's linear equation in z and can be solved by finding the appropriate integrating factor.

EXAMPLE 4.28

Solve

$$\frac{dy}{dx}+x\sin 2y=x^2\cos^2 y.$$

Solution. Dividing throughout by $\cos^2 y$, we have

$$\sec^2 y\frac{dy}{dx}+\frac{2x\sin y\cos y}{\cos^2 y}=x^2$$

or

$$\sec^2 y\frac{dy}{dx}+2x\tan y=x^3.$$

Putting $\tan y=z$, we have $\sec^2 y\dfrac{dy}{dx}=\dfrac{dz}{dx}\cdot$ Hence, the given equation reduces to

$$\frac{dz}{dx}+2xz=x^3, \tag{14}$$

which is Leibnitz-equation in z and x. The integrating factor is given by

$$\text{I.F.}=e^{\int P dx}=e^{\int 2x\,dx}=e^{x^2}.$$

Hence solution of the equation (14) is

$$ze^{x^2}=\int x^3\cdot e^{x^2}dx+C=\int x(x^2e^{x^2})dx+C$$
$$=\frac{1}{2}\int 2x(x^2e^{x^2})dx+C=\frac{1}{2}\int te^t dt+C,\ x^2=t$$
$$=\frac{1}{2}(x^2-1)e^{x^2}+C$$

Putting back the value of z, we get

$$\tan y\,e^{x^2}=\frac{1}{2}(x^2-1)\,e^{x^2}+C$$

or

$$\tan y=\frac{1}{2}(x^2-1)+C\,e^{-x^2}.$$

EXAMPLE 4.29

Solve

$$\frac{dy}{dx}+y=xy^3.$$

Solution. Dividing throughout by y^3, we get

$$y^{-3}\frac{dy}{dx}+y^{-2}=x.$$

Put $y^{-2}=z$. Then $\dfrac{dz}{dx}=-2y^{-3}\dfrac{dy}{dx}$ and, therefore, the above differential equation reduces to

$$-\frac{1}{2}\frac{dz}{dx}+z=x$$

or

$$\frac{dz}{dx}-2z=-2x,$$

which is Leibnitz's equation in z. We have

$$\text{I.F.}=e^{-\int 2\,dx}=e^{-2x}.$$

Therefore, the solution of the equation in z is

$$z\,e^{-2x}=\int -2x\cdot e^{-2x}dx+C=\frac{1}{2}e^{-2x}(2x+1)+C$$

or

$$z=x+\frac{1}{2}+C\,e^{2x}$$

or

$$y^{-2}=x+\frac{1}{2}+C\,e^{2x}.$$

EXAMPLE 4.30

Solve

$$x\frac{dy}{dx}+y=x^3y^6.$$

Solution. Dividing throughout by y^6, we have

$$y^{-6}\frac{dy}{dx}+\frac{y^{-5}}{x}=x^2.$$

Putting $y^{-5}=z$, we have $-5y^{-6}\frac{dy}{dx}=\frac{dz}{dx}$ and so the preceeding equation transfers to

$$-\frac{1}{5}\frac{dz}{dx}+\frac{z}{x}=x^2$$

or

$$\frac{dz}{dx}-\frac{5z}{x}=-\,5x^2.$$

Now

$$\text{I.F.}=e^{-5\int\frac{1}{x}dx}=e^{-5\log x}=x^{-5}.$$

Therefore, the solution the above differential equation in z is

$$z\cdot x^{-5}=\int -5x^2x^{-5}\,dx+C$$

or

$$y^{-5}\cdot x^{-5}=-\,5\int x^{-3}\,dx+C=\frac{-5}{-2}x^{-2}+C$$

or

$$1=(10+Cx^2)x^3\,y^5.$$

4.10 EXACT DIFFERENTIAL EQUATION

Definition 4.16 A differential equation of the form

$$M(x,y)dx+N(x,y)dy=0$$

is called an *exact differential equation* if there exists a function $U\equiv U(x,y)$ of x and y such that $M(x,y)\,dx+N(x,y)\,dy=d\,U$.

Thus, a differential eqution of the form $M(x,y)\,dx+N(x,y)dy=0$ is an exact differential equation if $Mdx+Ndy$ is an exact differential.

For example, the differential equation $y^2\,dx+2xy\,dy=0$ is an exact equation because it is the total differential of $U(x,y)=xy^2$. In fact the coefficient of dy is $\frac{\partial F}{\partial y}(xy^2)=2xy$.

The following theorem tells us whether a given differential equation is exact or not.

Theorem 4.3 A necessary and sufficient condition for the differential equation $Mdx+Ndy=0$ to be exact is

$$\frac{\partial M}{\partial y}=\frac{\partial N}{\partial x},$$

where M and N are functions of x and y having continuous first order derivative at all points in the rectangular domain.

Proof: (1) *Condition is necessary*. Suppose that the differential equation $Mdx+Ndy=0$ is exact. Then there exists a function $U(x,y)$ such that

$$M\,dx+N\,dy=dU.$$

But, in term of partial derivatives, we have

$$dU=\frac{\partial U}{\partial x}\,dx+\frac{\partial U}{\partial y}\,dy.$$

Therefore,

$$M\,dx+N\,dy=\frac{\partial U}{\partial x}\,dx+\frac{\partial U}{\partial y}\,dy.$$

Equating coefficients of dx and dy, we get

$$M=\frac{\partial U}{\partial x}\text{ and }N=\frac{\partial U}{\partial y}.$$

Now

$$\frac{\partial M}{\partial y}=\frac{\partial^2 U}{\partial y\,\partial x}\text{ and }\frac{\partial N}{\partial x}=\frac{\partial^2 U}{\partial x\,\partial y}.$$

Since partial derivatives of M and N are continuous, we have

$$\frac{\partial^2 U}{\partial y\,\partial x}=\frac{\partial^2 U}{\partial x\,\partial y}.$$

Hence

$$\frac{\partial M}{\partial y}=\frac{\partial N}{\partial x}.$$

(2) *Condition is sufficient*. Suppose that M and N satisfy

$$\frac{\partial M}{\partial y}=\frac{\partial N}{\partial x}.$$

Let

$$\int M\,dx=U,$$

where y is treated as a constant while integrating M.

Then

$$\frac{\partial}{\partial x}\left(\int M\,dx\right)=\frac{\partial U}{\partial x}$$

or

$$M=\frac{\partial U}{\partial x}.$$

Therefore,

$$\frac{\partial M}{\partial y}=\frac{\partial^2 U}{\partial y\,\partial x}$$

and so

$$\frac{\partial M}{\partial y} = \frac{\partial^2 U}{\partial y\, \partial x} = \frac{\partial N}{\partial x}$$

Also, by continuity of partial derivatives,

$$\frac{\partial^2 U}{\partial y\, \partial x} = \frac{\partial^2 U}{\partial x\, \partial y}.$$

Thus

$$\frac{\partial M}{\partial y} = \frac{\partial^2 U}{\partial y\, \partial x} \text{ and } \frac{\partial N}{\partial x} = \frac{\partial^2 U}{\partial x\, \partial y}.$$

Integrating both sides of $\frac{\partial N}{\partial x} = \frac{\partial^2 U}{\partial x\, \partial y}$ with respect to x, we get

$$N = \frac{\partial U}{\partial y} + f(y).$$

Thus,

$$Mdx + Ndy = \frac{\partial U}{\partial x}\, dx + \left[\frac{\partial U}{\partial y} + f(y)\right] dy$$

$$= \frac{\partial U}{\partial x}\, dx + \frac{\partial U}{\partial y}\, dy + f(y)dy$$

$$= d\,U + f(y)dy$$

$$= d\left[U + \int f(y)dy\right]$$

Thus, $M\,dx + N\,dy$ is the exact differential of $U + \int f(y)\, dy$ and, hence, the differential equation $M\,dx + N\,dy = 0$ is exact.

4.11 THE SOLUTION OF EXACT DIFFERENTIAL EQUATION

In the proof of Theorem 4.3, we note that if $M\,dx + N\,dy = 0$ is exact, then

$$M\,dx + N\,dy = d\left(U + \int f(y)\, dy\right).$$

Therefore,

$$d\left(U + \int f(y)\, dy\right) = 0$$

or

$$dU + f(y)\, dy = 0$$

Integrating, we get the required solution as

$$U + \int f(y)\, dy = C \text{ (constant)}$$

or

$$\int_{y \text{ constant}} M\,dx + \int f(y)\, dy = C$$

or

$$\int_{y \text{ constant}} M\,dx + \int (\text{terms of } N \text{ not containing } x)\, dy = C.$$

EXAMPLE 4.31

Solve

$$(2x \cos y + 3x^2 y)\, dx + (x^3 - x^2 \sin y - y)\, dy = 0$$

Solution. Comparing with $M\,dx + N\,dy = 0$, we get

$$M = 2x \cos y + 3x^2 y,$$
$$N = x^3 - x^2 \sin y - y.$$

Then

$$\frac{\partial M}{\partial y} = -\,2x \sin y + 3x^2 = \frac{\partial N}{\partial x}.$$

Hence the given equation is exact. Therefore, the solution of the equation is given by

$$\int_{y \text{ constant}} Mdx + \int (\text{terms of } N \text{ not containing } x)\, dy = C$$

or

$$\int_{y \text{ constant}} (2x \cos y + 3x^2 y)\, dx + \int -y dy = C$$

or

$$x^2 \cos y + x^3 y - \frac{y^2}{2} = C.$$

EXAMPLE 4.32

Solve

$$(2xy + y - \tan y)\, dx + (x^2 - x \tan^2 y + \sec^2 y)\, dy = 0.$$

Solution. Comparing with $Mdx + Ndy = 0$, we observe that

$$M = 2xy + y - \tan y, \text{ and}$$
$$N = x^2 - x \tan^2 y + \sec^2 y.$$

Therefore,

$$\frac{\partial M}{\partial y} = 2x - \sec^2 y + 1,$$

$$\frac{\partial N}{\partial x} = 2x - \tan^2 y.$$

Hence, $\frac{\partial M}{\partial y} = \frac{\partial N}{\partial x}$ and so the equation is exact. Its solution is given by

$$\int_{y \text{ constant}} (2xy + y - \tan y)\, dx + \int \sec^2 y\, dy = C$$

or

$$x^2y + xy - x\tan y + \tan y = C.$$

EXAMPLE 4.33

Solve

$$(2xy\cos x^2 - 2xy + 1)\, dx + (\sin x^2 - x^2)\, dy = 0.$$

Solution. Comparing with $M\, dx + N\, dy = 0$, we note that

$$M = 2xy\cos x^2 - 2xy + 1,\ N = \sin x^2 - x^2.$$

Then

$$\frac{\partial M}{\partial y} = 2x\cos x^2 - 2x,\ \frac{\partial N}{\partial y} = 2x\cos x^2 - 2x.$$

Thus $\frac{\partial M}{\partial y} = \frac{\partial N}{\partial x}$ and, therefore, the given equation is exact. The solution of the given equation is

$$\int_{y \text{ constant}} (2xy\cos x^2 - 2xy + 1)\, dx + \int 0\, dy = C$$

or

$$y\int 2x\cos x^2 dx - 2y\int x\, dx + \int dx = 0$$

or

$$y\int \cos t\, dt - x^2y + x = C, \quad x^2 = t$$

or

$$y\sin x^2 - x^2y + x = C.$$

EXAMPLE 4.34

Solve

$$\frac{dy}{dx} + \frac{y\cos x + \sin y + y}{\sin x + x\cos y + x} = 0.$$

Solution. The given differential equation is

$$(y\cos x + \sin y + y)\, dx + (\sin x + x\cos y + x)\, dy = 0$$

Comparing with $M\, dx + N\, dy = 0$, we get

$$M = y\cos x + \sin y + y,$$
$$N = \sin x + x\cos y + x,$$

and so

$$\frac{\partial M}{\partial y} = \cos x + \cos y + 1,$$

$$\frac{\partial N}{\partial x} = \cos x + \cos y + 1.$$

Therefore, the given equation is exact and its solution is

$$\int_{y \text{ constant}} (y\cos x + \sin y + y)\, dx + \int 0\, dy = C$$

or

$$y\sin x + x\sin y + xy = C.$$

EXAMPLE 4.35

Solve

$$(\sec x\tan x\tan y - e^x)\, dx + \sec x\sec^2 y\, dy = 0.$$

Solution. Comparing with $M\, dx + N\, dy = 0$, we have

$$M = \sec x\tan x\tan y - e^x,$$
$$N = \sec x\sec^2 y.$$

Therefore,

$$\frac{\partial M}{\partial y} = \sec x\tan x\sec^2 y,$$

$$\frac{\partial N}{\partial x} = \sec x\tan x\sec^2 y.$$

Hence the equation is exact and its solution is given by

$$\int_{y \text{ constant}} (\sec x\tan x\tan y - e^x)\, dx + \int 0\, dx = C$$

or

$$\tan y\int \sec x\tan x\, dx - e^x = C$$

or

$$\tan y \sec x - e^x = 0$$

EXAMPLE 4.36

Solve

$$\left(1 + e^{\frac{x}{y}}\right) dx + e^{\frac{x}{y}}\left(1 - \frac{x}{y}\right) dy = 0.$$

Solution. Comparing with $M\,dx + N\,dy = 0$, we get

$$M = 1 + e^{\frac{x}{y}}, \qquad N = e^{\frac{x}{y}}\left(1 - \frac{x}{y}\right),$$

$$\frac{\partial M}{\partial y} = e^{\frac{x}{y}}\left(\frac{-x}{y^2}\right) = -\frac{x}{y^2}e^{\frac{x}{y}},$$

$$\frac{\partial N}{\partial x} = e^{\frac{x}{y}}\left(\frac{1}{y}\right)\left(1 - \frac{x}{y}\right) + e^{\frac{x}{y}}\left(\frac{-1}{y}\right) = -\frac{x}{y^2}e^{\frac{x}{y}}.$$

Hence the given differential equation is exact and its solution is

$$\int_{y \text{ constant}} \left(1 + e^{\frac{x}{y}}\right) dx + \int 0\, dx = C$$

or

$$x + y\, e^{\frac{x}{y}} = C.$$

EXAMPLE 4.37

Show that the differential equation

$$(ax + hy + g)\, dx + (hx + by + f)\, dt = 0$$

is the differential equation of a family of conics.

Solution. Comparing the given differential equation with $Mdx + Ndy = 0$, we get

$$M = ax + hy + g,\ N = hx + by + f,$$

$$\frac{\partial M}{\partial y} = h, \frac{\partial N}{\partial x} = h.$$

Hence the given differential equation is exact and its solution is

$$\int_{y \text{ constant}} (ax + hy + g)\, dx + \int (by + f)\, dy = C$$

or

$$a\frac{x^2}{2} + hyx + gx + b\frac{y^2}{2} + fy = C$$

or

$$ax^2 + bx^2 + 2hxy + 2gx + 2fy + k = 0,$$

which represents a family of conics.

4.12 EQUATIONS REDUCIBLE TO EXACT EQUATION

Differential equations which are not exact can sometimes be made exact on multiplying by a suitable factor called an *integrating factor*.

The integrating factor for $Mdx + Ndy = 0$ can be found by the following rules:

1. If $Mdx + Ndy = 0$ is a homogeneous equation in x and y, then $\frac{1}{Mx + Ny}$ is an integrating factor, provided $Mx + Ny \neq 0$.
2. If the equation $Mdx + Ndy = 0$ is of the form $f(xy)y\, dx + \phi(xy)x\, dy = 0$, then $\frac{1}{Mx - Ny}$ is an integrating factor, provided $Mx - Ny \neq 0$.
3. Let $M\,dx + N\,dy = 0$ be a differential equation. If $\frac{\frac{\partial M}{\partial y} - \frac{\partial N}{\partial x}}{N}$ is a function of x only, say $f(x)$, then $e^{\int f(x)\,dx}$ is an integrating factor.
4. Let $M\,dx + N\,dy = 0$. If $\frac{\frac{\partial N}{\partial x} - \frac{\partial M}{\partial y}}{M}$ is a function of y only, say $f(y)$, then $e^{\int f(y)\,dy}$ is an integrating factor.
5. For the equation

$$x^a y^b(mydx + nxdy) + x^{a'}y^{b'}(m'ydx + n'xdy) = 0$$

the integrating factor is $x^h y^k$, where h and k are such that

$$\frac{a + h + 1}{m} = \frac{b + k + 1}{n},$$

$$\frac{a' + h + 1}{m'} = \frac{b' + k + 1}{n'}.$$

EXAMPLE 4.38

Solve

$$y\, dx - x\, dy + \log x\, dx = 0.$$

Solution. The given equation is not exact. Dividing by x^2, we get

$$\frac{y}{x^2}\,dx - \frac{x\,dy}{x^2} + \frac{1}{x^2}\log x\,dx = 0$$

or

$$\frac{ydx - xdy}{x^2} + \frac{1}{x^2}\log x\,dx = 0$$

or

$$-d\left(\frac{y}{x}\right) + d\left(\int \frac{1}{x^2}\log x\,dx\right) = 0$$

or

$$d\left(\int \frac{1}{x^2}\log x dx - \frac{y}{x}\right) = 0$$

Thus $\frac{1}{x^2}$ is an integrating factor and on integration, we get the solution as

$$\int \frac{1}{x^2}\log x\,dx - \frac{y}{x} = C$$

On integration by parts, we have

$$-\frac{\log x}{x} + \int \frac{1}{x^2}\,dx = C + \frac{y}{x}.$$

or

$$-\frac{1}{x}\log x - \frac{1}{x} = C + \frac{y}{x}.$$

EXAMPLE 4.39

Solve

$$xdx + ydy = \frac{a^2(x\,dy - y\,dx)}{x^2 + y^2}.$$

Solution. We know that

$$d\left(\tan^{-1}\frac{y}{x}\right) = \frac{x\,dy - y\,dx}{x^2 + y^2}.$$

Therefore, the given differential equation is

$$xdx + ydy - a^2 d\left(\tan^{-1}\frac{y}{x}\right)$$

Integrating, we get

$$\frac{x^2}{2} + \frac{y^2}{2} - a^2\tan^{-1}\frac{y}{x} = C$$

or

$$x^2 + y^2 - 2a^2\tan^{-1}\frac{y}{x} = k \text{ (constant)}.$$

EXAMPLE 4.40

Solve

$$x\,dy - y\,dx + a(x^2 + y^2)\,dx = 0.$$

Solution. Dividing throughout by $x^2 + y^2$, we get

$$\frac{x\,dy}{x^2 + y^2} - \frac{y\,dx}{x^2 + y^2} + adx = 0$$

or

$$\frac{x\,dy - y\,dx}{x^2 + y^2} + adx = 0$$

or

$$d\left(\tan^{-1}\frac{y}{x}\right) + adx = 0.$$

Integrating, we get

$$\tan^{-1}\frac{y}{x} + ax = C.$$

EXAMPLE 4.41

Solve

$$(x^2 y - 2xy^2)\,dx = (x^3 - 3x^2y)\,dy = 0.$$

Solution. The given equation is homogeneous. Comparing it with $M\,dx + N\,dy = 0$, we have

$$M = x^2y - 2xy^2,\ N = -x^3 + 3x^2y,$$

$$\frac{\partial M}{\partial y} = x^2 - 4xy,\ \frac{\partial N}{\partial x} = -3x^2 + 6xy.$$

Therefore, $\frac{\partial M}{\partial y} \neq \frac{\partial N}{\partial x}$ and so the given equation is not exact. Further

$$Mx + Ny = x^3y - 2x^2y^2 - x^3y + 3x^2y^2 = x^2y^2 \neq 0.$$

Hence the integrating factor is $\frac{1}{Mx + Ny} = \frac{1}{x^2y^2}$.

Multiplying the given equation by $\frac{1}{x^2y^2}$ we have

$$\left(\frac{1}{y} - \frac{2}{x}\right)dx - \left(\frac{x}{y^2} - \frac{3}{y}\right)dy = 0 \quad (15)$$

Since

$$\frac{\partial}{\partial y}\left(\frac{1}{y} - \frac{2}{x}\right) = \frac{\partial}{\partial x}\left(\frac{x}{y^2} - \frac{3}{y}\right) = -\frac{1}{y^2},$$

the equation (15) is exact and so its solution is

$$\int_{y\text{ constant}} \left(\frac{1}{y} - \frac{2}{x}\right)dx - \int\left(-\frac{3}{y}\right)dy = C$$

or

$$\frac{x}{y} - 2\log x + 3\log y = C.$$

EXAMPLE 4.42

Solve

$$y\,dx + 2x\,dy = 0.$$

Solution. The given equation is of type $Mdx + Ndy = 0$ and is homogeneous. Further

$$M = y,\ N = 2x,$$

$$\frac{\partial M}{\partial y} = 1,\ \frac{\partial N}{\partial x} = 2,\ \frac{\partial M}{\partial y} \neq \frac{\partial N}{\partial x}.$$

Thus the equation is not exact. But $Mx + Ny = xy + 2xy = 3xy \neq 0$. Therefore, $\frac{1}{3xy}$ is the integrating factor. Multiplying the given equation throughout by $\frac{1}{3xy}$, we get

$$\frac{1}{3xy}y\,dx + \frac{2x}{3xy}\,dy = 0$$

or

$$\frac{1}{3x}\,dx + \frac{2}{3y}\,dy = 0.$$

The solution is

$$\frac{1}{3}\int \frac{1}{x}\,dx + \frac{2}{3}\int \frac{1}{y}\,dy = C$$

or

$$\frac{1}{3}\log x + \frac{2}{3}\log y = C$$

or

$$\log xy^2 = k = \log p$$

or

$$xy^2 = p \text{ (constant)}.$$

EXAMPLE 4.43

Solve

$$x^2y\,dx - (x^3 + y^3)\,dy = 0.$$

Solution. The given equation is homogeneous and comparing with $Mdx + N\,dy = 0$, we get

$$M = x^2y,\ N = -x^3 - y^3.$$

Then

$$Mx + Ny = x^3y - x^3y - y^4 = -y^4 \neq 0.$$

Thus the integrating factor is $\frac{1}{-y^4}$. Multiplying the given differential equation throughout by $-\frac{1}{y^4}$, we get

$$-\frac{1}{y^4}x^2y\,dx + \frac{1}{y^4}(x^3 + y^3)\,dy = 0$$

or

$$-\frac{x^2}{y^3}\,dx + \left(\frac{x^3}{y^4} + \frac{1}{y}\right)dy = 0$$

which is exact. Hence the required solution is

$$-\frac{1}{y^3}\int x^2\,dx + \int \frac{1}{y}\,dy = C$$

$$-\frac{x^3}{3y^3} + \log y = C.$$

EXAMPLE 4.44

Solve

$$y(xy + 2x^2y^2)dx + x(xy - x^2y^2)dy = 0.$$

Solution. The given differential equation is of the form

$$f(xy)\,y\,dx + \phi(xy)\,x\,dy = 0.$$

Also comparing with $M\,dx + Ndy = 0$, we get

$$M = y(xy + 2x^2y^2),\ N = x(xy - x^2y^2).$$

Therefore,

$$Mx - Ny = 3x^3y^3 \neq 0.$$

Thus, $\frac{1}{3x^3y^3}$ is the integrating factor. Multiplying throughout by $\frac{1}{3x^3y^3}$, we have

$$\frac{1}{3x^3y^3}(y(xy + 2x^2y^2))\,dx + \frac{1}{3x^3y^3}[x(xy - x^2y^2)]\,dy = 0.$$

or

$$\left(\frac{1}{3x^2y} + \frac{2}{3x}\right)dx + \left(\frac{1}{3xy^2} - \frac{1}{3y}\right)x\,dy = 0.$$

Since

$$\frac{\partial}{\partial y}\left(\frac{1}{3x^2y}+\frac{2}{3x}\right)=\frac{\partial}{\partial x}\left(\frac{1}{3xy^2}-\frac{1}{3y}\right),$$

the above equation is exact and its solution is

$$\int\limits_{y\text{ constant}}\left(\frac{1}{3x^2y}+\frac{2}{3x}\right)dx+\int\left(\frac{-1}{3y}\right)dy=C$$

or

$$-\frac{1}{3xy}+\frac{2}{3}\log x-\frac{1}{3}\log y=C$$

or

$$-\frac{1}{xy}+2\log x-\log y=k\text{ (constant)}.$$

EXAMPLE 4.45

Solve

$$(1+xy)y\,dx+(1-xy)x\,dy=0$$

Solution. The given differential equation is of the form

$$f(xy)ydx+\phi(xy)x\,dy=0.$$

Comparing with $M\,dx+Ndy=0$, we have

$$M=y+xy^2,\ N=x-x^2y.$$

Therefore,

$$Mx-Ny=2x^2y^2\neq 0.$$

Therefore, the integrating factor is $\frac{1}{2x^2y^2}$. Multiplying the given differential equation throughout by $\frac{1}{2x^2y^2}$, we get

$$\frac{1}{2x^2y^2}(y+xy^2)\,dx+\frac{1}{2x^2y^2}(x-x^2y)=0$$

or

$$\left(\frac{1}{2x^2y}+\frac{1}{2x}\right)dx+\left(\frac{1}{2xy^2}-\frac{1}{2y}\right)dy=0.$$

We note that this equation is exact. Hence its solution is

$$\int\limits_{y\text{ constant}}\left(\frac{1}{2x^2y}+\frac{1}{2x}\right)dx+\int\left(-\frac{1}{2y}\right)dy=C$$

or

$$\frac{1}{2y}\left(-\frac{1}{x}\right)+\frac{1}{2}\log x-\frac{1}{2}\log y=C$$

or

$$\log\frac{x}{y}-\frac{1}{xy}=k\text{ (constant)}.$$

EXAMPLE 4.46

Solve

$$(x^2y^2+xy+1)\,y\,dx+(x^2\,y^2-xy+1)\,x\,dy=0.$$

Solution. The given differential equation is of the form

$$f(xy)\,ydx+\phi(xy)\,xdx=0.$$

Moreover, comparing the given equation with $M\,dx+N\,dy=0$, we get

$$M=x^2y^3+xy^2+y,\ N=x^3y^2-x^2y+x.$$

Therefore

$$\begin{aligned}Mx-Ny&=x^3y^3+x^2y^2+xy-x^3y^3+x^2y^2-xy\\&=2x^2y^2\neq 0.\end{aligned}$$

Therefore, the integrating factor is $\frac{1}{2x^2y^2}\cdot$ Multiplying the given differential equation throughout by $\frac{1}{2x^2y^2}$,

we get

$$\frac{1}{2x^2y^2}(x^2y^3+xy^2+y)\,dx+\frac{1}{2x^2y^2}(x^3y^2-x^2y+x)\,dy=0$$

or

$$\left(\frac{y}{2}+\frac{1}{2x}+\frac{1}{2x^2y}\right)dx+\frac{1}{2}\left(x-\frac{1}{y}+\frac{1}{xy^2}\right)dy=0,$$

which is exact. Hence the solution of the equation is

$$\int\limits_{y\text{ constant}}\left(\frac{y}{2}+\frac{1}{2x}+\frac{1}{2x^2y}\right)dx+\frac{1}{2}\int\left(\frac{-1}{y}\right)dy=C$$

or

$$\left(\frac{xy}{2}+\frac{1}{2}\log x-\frac{1}{2yx}\right)-\frac{1}{2}\log y=k$$

or

$$xy-\frac{1}{xy}+\log\frac{x}{y}=k\text{ (constant)}.$$

EXAMPLE 4.47

Solve

$$y(2xy+1)\,dx + x(1+2xy-x^3y^3)\,dy = 0.$$

Solution. The differential equation in question is of the form

$$f(xy)\,ydx + \phi(xy)\,xdy = 0.$$

Further comparing the given equation with $M\,dx + N\,dy = 0$, we get

$$M = 2xy^2 + y,\ N = x + 2x^2y - x^4y^3.$$

Therefore,

$$Mx - Ny = 2x^2y^2 + xy - xy - 2x^2y^2 - x^4y^4$$

$$= -x^4y^4.$$

Thus the integrating factor is $\frac{1}{-x^4y^4}\cdot$ Multiplying the given differential equation throughout by $\frac{-1}{x^4y^4}$, we get

$$-\frac{1}{x^4y^4}(2xy^2+y)\,dx - \frac{1}{x^4y^4}(x+2x^2y-x^4y^3)\,dy$$

$$= 0.$$

or

$$\left(-\frac{2}{x^3y^2} - \frac{1}{x^4y^3}\right)dx + \left(-\frac{1}{x^3y^4} - \frac{2}{x^2y^3} + \frac{1}{y}\right)dy = 0,$$

which is exact. Hence the solution of the equation is

$$\int_{y\text{ constant}} \left(\frac{-2}{x^3y^2} - \frac{1}{x^4y^3}\right)dx + \int \frac{1}{y}\,dy = C$$

or

$$\frac{1}{x^2y^2} + \frac{1}{3x^3y^3} + \log y = C.$$

EXAMPLE 4.48

Solve

$$(x^2+y^2+2x)\,dx + 2y\,dy = 0.$$

Solution. Comparing the given equation with $Mdx + Ndy = 0$, we get

$$M = x^2+y^2+2x,\ N = 2y$$

which gives

$$\frac{\partial M}{\partial y} = 2y,\ \frac{\partial N}{\partial x} = 0.$$

Thus the equation is not exact. We have

$$\frac{\frac{\partial M}{\partial y} - \frac{\partial N}{\partial x}}{N} = \frac{2y}{2y} = 1 = x^0 \text{ (function of } x).$$

Therefore, $e^{\int 1\,dx} = e^x$ is the integrating factor. Multiplying the given differential equation throughout by e^x, we get

$$(x^2+y^2+2x)\,e^x dx + 2ye^x dy = 0,$$

which is exact. Hence the required solution is

$$\int_{y\text{ constant}} (x^2+y^2+2x)\,e^x\,dx + \int 0\,dy = C,$$

which yield

$$(x^2+y^2)\,e^x = C.$$

EXAMPLE 4.49

Solve

$$(xy^2 - e^{\frac{x}{3}})\,dx - x^2y\,dy = 0.$$

Solution. Comparing the given equation with $Mdx + N\,dy = 0$, we get

$$M = xy^2 - e^{\frac{x}{3}},\ N = -x^2y,$$

$$\frac{\partial M}{\partial y} = 2xy, \qquad \frac{\partial N}{\partial x} = -2xy.$$

Therefore,

$$\frac{\frac{\partial M}{\partial y} - \frac{\partial N}{\partial x}}{N} = \frac{2xy-(-2xy)}{-x^2y} = -\frac{4}{x},$$

which is a function of x only. Hence $e^{\int -\frac{4}{x}dx} = e^{-4\log x} = \frac{1}{x^4}$ is the integrating factor. Multiplying the equation throughout by $\frac{1}{x^4}$, we get

$$\left(\frac{y^2}{x^3} - \frac{1}{x^4}e^{\frac{x}{3}}\right) dx - \frac{y}{x^2}\, dy = 0,$$

which is exact. The required solution is, therefore,

$$\int_{y \text{ constant}} \left(\frac{y^2}{x^3} - \frac{1}{x^4}e^{\frac{x}{3}}\right) dx = C$$

or

$$-\frac{y^2}{2x^2} + \frac{1}{3}\int -\frac{3}{x^4}e^{\frac{x}{3}}\, dx = C$$

or

$$-\frac{y^2}{2x^2} + \frac{1}{3}\int e^t\, dt = C,\ t = \frac{x}{3}.$$

or

$$-\frac{y^2}{2x^2} + \frac{1}{3}e^t = C$$

or

$$-\frac{3y^2}{2x^2} + 2e^{\frac{x}{3}} = k \text{ (constant)}.$$

EXAMPLE 4.50

Solve

$$(xy^3 + y)\, dx + 2(x^2y^2 + x + y^4)\, dy = 0.$$

Solution. Comparing with $M\, dx + N\, dy = 0$, we get

$$M = xy^3 + y, \qquad N = 2x^2y^2 + 2x + 2y^4,$$

$$\frac{\partial M}{\partial y} = 3xy^2 + 1, \quad \frac{\partial N}{\partial x} = 4xy^2 + 2.$$

Since $\frac{\partial M}{\partial y} \neq \frac{\partial N}{\partial x}$, the equation is not exact. It is also not homogeneous. It is also not of the form $f(x\,y)\, y\, dx + \phi\,(x\,y) x\, dy = 0$. We note that

$$\frac{\frac{\partial N}{\partial x} - \frac{\partial M}{\partial y}}{M} = \frac{xy^2 + 1}{y(xy^2 + 1)} = \frac{1}{y} \text{ (function of } y \text{ alone).}$$

Hence, the integrating factor is $e^{\int \frac{1}{y} dy} = e^{\log y} = y$. Multiplying the given equation throughout by y, we get

$$y(xy^3 + y)\, dx + 2y(x^2y^2 + x + y^4)\, dy = 0$$

$$(xy^4 + y^2)\, dx + (2x^2y^3 + 2xy + y^5)\, dy = 0,$$

which is exact. Therefore, its solution is

$$\int_{y \text{ constant}} (xy^4 + y^2)\, dx + \int y^5 dy = C$$

or

$$\frac{x^2y^4}{2} + xy^2 + \frac{y^6}{6} = C.$$

EXAMPLE 4.51

Solve

$$(y^4 + 2y)\, dx + (xy^3 + 2y^4 - 4x)dy = 0.$$

Solution. The given differential equation is neither homogeneous nor is of the type $f(x\,y)\, y\, dx + \phi(x\,y)\, x\, dy = 0$. Comparing with $Mdx + Ndy = 0$, we get

$$M = y^4 + 2y,\ N = xy^3 + 2y^4 - 4x.$$

Therefore,

$$\frac{\partial M}{\partial y} = 4y^3 + 2,\ \frac{\partial N}{\partial x} = y^3 - 4,$$

which shows that the given equation is not exact. Further,

$$\frac{\frac{\partial N}{\partial x} - \frac{\partial M}{\partial y}}{M} = \frac{-3y^3 - 6}{y^4 + 2y} = \frac{-3(y^3 + 2)}{y(y^3 + 2)} = -\frac{3}{y},$$

which is a function of y alone. Therefore, the integrating factor is $e^{\int -\frac{3}{y} dy} = e^{-3\log y} = y^{-3} = \frac{1}{y^3}$. Multiplying the given differential equation throughout by $\frac{1}{y^3}$, we get

$$\frac{1}{y^3}(y^4 + 2y)\, dx + \frac{1}{y^3}(xy^3 + 2y^4 - 4x)\, dy = 0$$

or

$$\left(y + \frac{2}{y^2}\right) dx + \left(x + 2y - \frac{4x}{y^3}\right) dy = 0,$$

which is exact. The solution of the given differential equation is, therefore,

$$\int_{y\text{ constant}} \left(y+\frac{2}{y^2}\right) dx + \int 2y\, dy = 0$$

or

$$\left(y+\frac{2}{y^2}\right) x + y^2 = C.$$

EXAMPLE 4.52

Solve

$$y(xy+2x^2y^2)\, dx + x(xy - x^2y^2)\, dy = 0.$$

Solution. The given equation can be written as

$$xy(ydx+xdy) + x^2y^2(2ydx - xdy) = 0.$$

Comparing it with

$$x^a y^b (mydx + nxdy) + x^{a'} y^{b'} (m'y\, dx + n'x\, dy) = 0,$$

we note that $a = b = 1$, $a' = b' = 2$, $m = n = 1$ and $m' = 2$, $n' = -1$. Then the integrating factor is $x^h y^k$, where

$$\frac{a+h+1}{m} = \frac{b+k+1}{n}, \frac{a'+h+1}{m'} = \frac{b'+k+1}{n'},$$

that is,

$$\frac{h+2}{1} = \frac{k+2}{1}, \frac{3+h}{2} = \frac{3+k}{-1}$$

or

$$h+2 = k+2, \quad -3-k = 6+2k$$

or

$$h-k=0, \quad h+2k=-4.$$

Solving for h and k, we get $h = k = -3$. Hence, the integrating factor is $\frac{1}{x^3y^3}$. Multiplying throughout by $\frac{1}{x^3y^3}$, we get

$$\left(\frac{1}{x^2y}+\frac{2}{x}\right) dx + \left(\frac{1}{xy^2}-\frac{1}{y}\right) dy = 0,$$

which is exact. Therefore, the solution is

$$\int_{y\text{ constant}} \left(\frac{1}{x^2y}+\frac{2}{x}\right) dx + \int\left(-\frac{1}{y}\right) dy = C$$

or

$$\frac{1}{y}\left(-\frac{1}{x}\right) + 2\log x - \log y = C$$

or

$$-\frac{1}{xy} + 2\log x - \log y = C.$$

EXAMPLE 4.53

Solve

$$(y^2 + 2x^2y)\, dx + (2x^3 - xy)\, dy = 0.$$

Solution. The given equation can be written as

$$x^0y(y\, dx - x\, dy) + x^2y^0(2y\, dx + 2x\, dy) = 0.$$

Comparing this with

$$x^a y^b (mydx + nxdy + {} + x^{a'} y^{b'} (m'ydx + n'xdy) = 0,$$

we get

$$a = 0,\ b = 1,\ a' = 2,\ b' = 0,$$
$$m = 1,\ n = -1,\ m' = 2,\ n' = 2.$$

Then the integrating factor is $x^h y^k$, where

$$\frac{a+h+1}{m} = \frac{b+k+1}{n}, \frac{a'+h+1}{m'} = \frac{b'+k+1}{n'},$$

or

$$\frac{h+1}{1} = \frac{2+k}{.-1}, \frac{3+h}{2} = \frac{k+2}{2}$$

or

$$h+k=-3 \quad \text{and} \quad 2h-2k=-4.$$

Solving for h and k, we get $h = -\frac{5}{2}$ $k = -\frac{1}{2}$. Hence, $\frac{1}{x^{\frac{5}{2}}y^{\frac{1}{2}}}$ is the integrating factor. Multiplying the given differential equation by $\frac{1}{x^{\frac{5}{2}}y^{\frac{1}{2}}}$, we get

$$(x^{-\frac{5}{2}}y^{\frac{3}{2}} + 2x^{-\frac{1}{2}}y^{\frac{1}{2}})\, dx$$
$$+ (2x^{\frac{1}{2}}y^{\frac{1}{2}} + 2x^{-\frac{3}{2}}y^{\frac{1}{2}} + dy) = 0,$$

which is exact. Therefore, the required solution is

$$\int_{y\text{ constant}} (x^{-\frac{5}{2}}y^{\frac{3}{2}} + 2x^{-\frac{1}{2}}y^{\frac{1}{2}})\, dx + \int 0\, dy = 0$$

or

$$-\frac{2}{3}y^{\frac{3}{2}}x^{-\frac{3}{2}} + 4y^{\frac{1}{2}}x^{\frac{1}{2}} = C$$

or

$$-y^{\frac{3}{2}}x^{-\frac{3}{2}} + 6x^{\frac{1}{2}}y^{\frac{1}{2}} = C$$

or

$$6\sqrt{xy} - \left(\frac{y}{x}\right)^{\frac{3}{2}} = C.$$

EXAMPLE 4.54

Solve

$$(3xy - 2ay^2) + (x^2 - 2axy)\,dy = 0.$$

Solution. The given equation can be written as

$$xy^0(3ydx + xdy) + yx^0(-2aydx - 2axdy) = 0.$$

Comparing this with

$$x^a y^b(mydx + nxdy) + x^{a'}y^{b'}(m'ydx + n'xdy) = 0,$$

we get

$$a = 1, b = 0, a' = 0, b' = 1,$$
$$m = 3, n = 1, m' = -2a, n' = -2a.$$

Then the integrating factor is $x^h y^k$, where

$$\frac{a+h+1}{m} = \frac{b+k+1}{m}, \frac{a'+h+1}{m'} = \frac{b'+k+1}{n'}$$

or

$$\frac{2+h}{3} = \frac{k+1}{1}, \frac{h+1}{-2a} = \frac{k+2}{-2a}$$

or

$$h - 3k = 1, h - k = 1.$$

Thus $h = 1$, $k = 0$. Therefore, the integrating factor is x. Multiplying the given differential equation throughout by x, we get

$$(3x^2y - 2axy^2)\,dx + (x^3 - 2ax^2y)\,dy = 0.$$

Let

$$M = 3x^2y - 2axy^2, N = x^3 - 2ax^2y.$$

Then

$$\frac{\partial M}{\partial y} = 3x^2 - 4axy, \frac{\partial N}{\partial x} = 3x^2 - 4axy.$$

Hence $\frac{\partial M}{\partial y} = \frac{\partial N}{\partial x}$ and so the transformed equation is exact.

The required solution is

$$\int_{y\text{ constant}} (3x^2y - 2axy^2)\,dx + \int 0\,dy = 0$$

or

$$yx^3 - ax^2y^2 = C.$$

EXAMPLE 4.55

Solve

$$(2x^2y^2 + y)\,dx + (-x^3y + 3x)\,dy = 0.$$

Solution. Comparing the given differential equation with $Mdx + Ndy = 0$, we get

$$M = 2x^2y^2 + y, \quad N = -x^3y + 3x,$$

$$\frac{\partial M}{\partial y} = 4x^2y + 1, \quad \frac{\partial N}{\partial x} = 3 - 3x^2y.$$

Since $\frac{\partial M}{\partial y} \neq \frac{\partial N}{\partial x}$, the given equation is not exact. However, the given equation can be written in the form

$$x^2y\,(2ydx - xdy) + x^0y^0\,(ydx + 3xdy) = 0.$$

Comparing it with

$$x^a y^b(my\,dx + nx\,dy + x^{a'}y^{b'}(m'ydx + n'xdy) = 0,$$

we get

$$a = 2, b = 1, a' = 0, b' = 0,$$

$$m = 2, n = -1, m' = 1, n' = 3.$$

The integrating factor is $x^h y^k$, where

$$\frac{a+h+1}{m} = \frac{b+k+1}{n}, \frac{a'+h+1}{m'} = \frac{b'+k+1}{n'}$$

or

$$h + 2k = -7, 3h - k = -2.$$

Solving these equations for h and k, we get $h = -\frac{11}{7}$ and $k = -\frac{19}{7}$. Thus the integrating factor is $x^{-\frac{11}{7}}y^{-\frac{19}{7}}$. Multiplying the given equation throughout by $x^{-\frac{11}{7}}y^{-\frac{19}{7}}$, we get

$$(2x^{\frac{3}{7}}y^{-\frac{5}{7}} + x^{-\frac{11}{7}}y^{-\frac{12}{7}})\,dx$$
$$-(x^{\frac{10}{7}}y^{-\frac{12}{7}} - 3x^{-\frac{4}{7}}y^{-\frac{10}{7}})\,dy = 0.$$

This transformed equation is exact and its solution is

$$\int_{y\text{ constant}} (2x^{\frac{3}{7}}y^{-\frac{5}{7}} + x^{-\frac{11}{7}}y^{-\frac{12}{7}})\, dx = C$$

or

$$\frac{7}{5}x^{\frac{10}{7}}y^{-\frac{5}{7}} - \frac{7}{4}x^{-\frac{4}{7}}y^{-\frac{12}{7}} = C.$$

4.13 APPLICATIONS OF FIRST ORDER AND FIRST DEGREE EQUATIONS

The aim of this section is to form differential equations for physical problems like flow of current in electric circuits, Newton law of cooling, heat flow and orthogonal trajectories, and to find their solutions.

(A) Problems Related to Electric Circuits

Consider the *RCL* circuit shown in the Figure 4.3 and consisting of resistance, capacitor, and inductor connected to a battery.

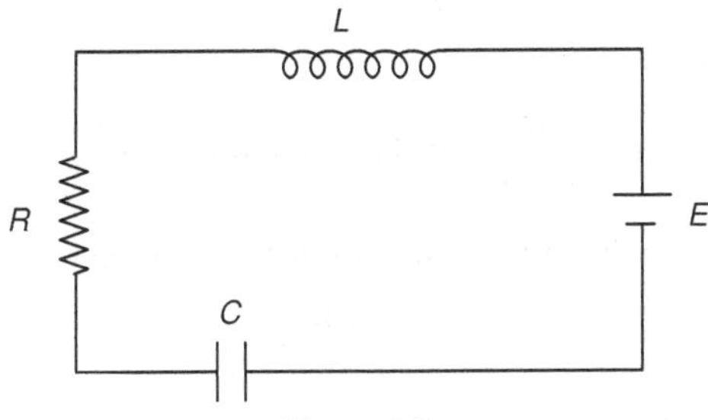

Figure 4.3

We know that the resistance is measured in ohms, capacitance in farads, and inductance in henrys. Let I denote the current flowing through the circuit and Q denote the charge. Since the current is rate of flow of charge, we have $I = \frac{dQ}{dt}$. Also, by Ohm's law, $\frac{V}{I} = R$ (resistance). Therefore, the voltage drop across a resistor R is RI. The voltage drop across the inductor L is $L\frac{dI}{dt}$ and the voltage drop across a capacitor is Q/C. If E is the voltage (e.m.f.) of the battery, then by Kirchhoff's law, we have

$$L\frac{dI}{dt} + RI + \frac{Q}{C} = E(t), \qquad (16)$$

where L, C, and R are constants. Since $I = \frac{dQ}{dt}$, we have $Q = \int_0^1 I(u)\, du$ and so (16) reduces to

$$L\frac{dI}{dt} + RI + \frac{1}{C}\int_0^t I(u)\, du = E(t).$$

The forcing function (input function), $E(t)$, is supplied by the battery (voltage source). The system described by the above differential equation is known as *harmonic oscillator*.

The equation (16) can be written as

$$\frac{dI}{dt} + \frac{R}{L}I + \frac{Q}{LC} = \frac{E}{L}(t) \qquad (17)$$

EXAMPLE 4.56

Given that $I = 0$ at $t = 0$, find an expression for the current in the LR circuit shown in the Figure 4.4.

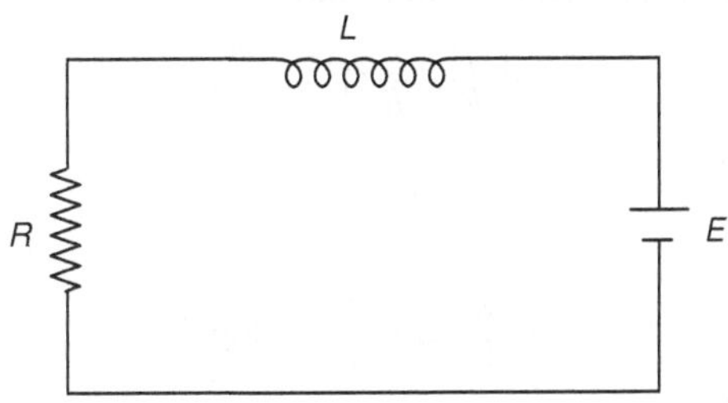

Figure 4.4

Solution. By Kirchhoff's law, we have

$$\frac{dI}{dt} + \frac{R}{L}I = \frac{E}{L}, \qquad (18)$$

which is Leibnitz's linear equation. Its integrating factor is $e^{\int\frac{R}{L}dt} = e^{\frac{Rt}{L}}$. Hence the solution of (18) is

$$Ie^{\frac{Rt}{L}} = \int\frac{E}{L}e^{\frac{Rt}{L}}\,dt + C = \frac{E}{L}\int e^{\frac{Rt}{L}}\,dt + C$$

$$= \frac{E}{L}\frac{e^{\frac{Rt}{L}}}{\frac{R}{L}} + C = \frac{E}{R}e^{\frac{Rt}{L}} + C.$$

Thus

$$I = \frac{E}{R} + Ce^{-\frac{Rt}{L}}. \qquad (19)$$

But $I(0) = 0$, therefore, (19) yields $0 = \frac{E}{R} + C$ and so $C = -\frac{E}{R}$. Hence

$$I=\frac{E}{R}-\frac{E}{R}e^{-\frac{Rt}{L}}=\frac{E}{R}(1-e^{-\frac{Rt}{L}}).$$

Clearly I increases with time t and attains its maximum value, $\frac{E}{R}$.

EXAMPLE 4.57

Find the time t when the current reaches half of its theoretical maximum in the circuit of Example 4.56.

Solution. From Example 4.56, we have

$$I=\frac{E}{R}(1-e^{-\frac{Rt}{L}}).$$

The maximum current is $\frac{E}{R}$. By the requirement of the problem, we must have

$$\frac{1}{2}\frac{E}{R}=\frac{E}{R}(1-e^{-\frac{Rt}{L}})$$

or

$$\frac{1}{2}=e^{-\frac{Rt}{L}}$$

or

$$-\frac{Rt}{L}=\log\frac{1}{2}=-\log 2$$

or

$$t=\frac{L}{R}\log 2.$$

EXAMPLE 4.58

In an LR circuit, an e.m.f. of 10 sin t volts is applied.If $I(0)=0$, find the current in the circuit.

Solution. In an LR circuit the current is governed by the differential equation

$$\frac{dI}{dt}+\frac{R}{L}I=\frac{E}{L}.$$

We are given that $E=10\sin t$. Therefore,

$$\frac{dI}{dt}+\frac{R}{L}I=\frac{10}{L}\sin t.$$

This is Leibnitz's linear equation with integrating factor as $e^{\int\frac{R}{L}dt}=e^{\frac{Rt}{L}}$. Therefore, its solution is

$$Ie^{\frac{Rt}{L}}=\int\frac{10}{L}\sin t\, e^{\frac{Rt}{L}}\,dt+C$$

$$=\frac{10}{L}\int e^{\frac{Rt}{L}}\sin t\,dt+C \qquad (20)$$

But we know (using integration by parts) that

$$\int e^{at}\sin bt\,dt=\frac{e^{at}}{a^2+b^2}(a\sin bt-b\cos bt)$$

$$=\frac{e^{at}}{a^2+b^2}\sin\left\{bt-\tan^{-1}\frac{b}{a}\right\}.$$

Therefore (20) reduces to

$$Ie^{Rt/L}=\frac{10}{L}\left[\frac{e^{\frac{Rt}{L}}}{\frac{R^2}{L^2}+1}\left(\frac{R}{L}\sin t-\cos t\right)\right]+C$$

$$=\frac{10L^2}{L^2}\left[\frac{e^{\frac{Rt}{L}}}{R^2+L^2}(R\sin t-L\cos t)\right]+C$$

$$=10\frac{e^{\frac{Rt}{L}}}{R^2+L^2}(R\sin t-L\cos t)+C.$$

Hence

$$I=\frac{10}{R^2+L^2}(R\sin t-L\cos t+Ce^{-\frac{Rt}{L}}.$$

Using the initial condition $I(0)=0$, we get

$$C=\frac{10L}{R^2+L^2}.$$

Hence

$$I=\frac{10}{R^2+L^2}(R\sin t-L\cos t+Le^{-\frac{Rt}{L}}).$$

EXAMPLE 4.59

If voltage of a battery in an LR circuit is $E_0\sin t$, find the current I in the circuit under the initial condition $I(0)=0$.

Solution. Proceeding as in Example 4.58, we get

$$I=\frac{E_0}{R^2+L^2}e^{-\frac{Rt}{L}}+\frac{E_0L}{R^2+L^2}\cos t+\frac{E_0R}{R^2+L^2}\sin t.$$

EXAMPLE 4.60

Find the current in the following electric circuit containing capacitor C, resistance R, and a battery of e.m.f. $E_0\sin\omega t$ with the initial condition $I(0)=0$.

Solution. The RC circuit of the problem is shown in Figure 4.5.

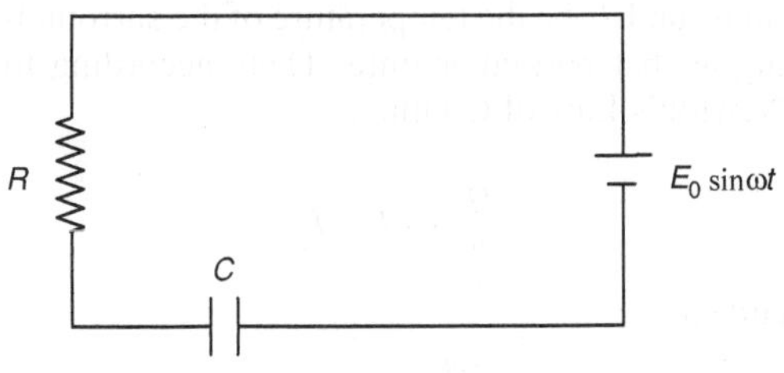

Figure 4.5

Using Kirchhoff's law, we have

$$RI + \frac{1}{C}\int_0^t I(u)\, du = E_0 \sin \omega t.$$

Differentiating both sides with respect to t, we have

$$R\frac{dI}{dt} + \frac{I}{C} = \omega E_0 \sin \omega t$$

or

$$\frac{dI}{dt} + \frac{I}{RC} = \frac{\omega E_0}{R} \cos \omega t,$$

which is Leibnitz's linear equation. The integrating factor is $e^{\int \frac{1}{RC} dt} = e^{\frac{t}{RC}}$. Therefore, the solution of the above first order equation is

$$I\, e^{\frac{t}{RC}} = \int \frac{\omega E_0}{R} \cos \omega t \cdot e^{\frac{t}{RC}} dt + C$$

$$= \frac{\omega E_0}{R} \int \cos \omega t\, e^{\frac{t}{RC}}\, dt + C$$

Using $\int e^{ax} \cos bx\, dx = \frac{e^{ax}}{a^2 + b^2}(a \cos bx + b \sin bx)$, we have

$$Ie^{\frac{t}{RC}} = \frac{\omega E_0}{R}\left[\frac{e^{\frac{t}{RC}}}{\left(\frac{1}{RC}\right)^2 + \omega^2}\left(\frac{1}{RC}\cos\omega t + \omega \sin\omega t\right)\right] + C$$

$$= \frac{\omega R^2 C^2 E_0}{R^2 C}\left[\frac{e^{\frac{t}{RC}}}{1 + R^2\omega^2 C^2}(\cos\omega t + RC\omega \sin \omega t)\right] + C$$

$$= \frac{\omega\, CE_0\, e^{\frac{t}{RC}}}{1 + R^2\omega^2 C^2}(\cos \omega t + RC\omega \sin \omega t) + C.$$

But $I(0) = 0$ implies $C = -\frac{\omega CE_0}{1 + R^2\omega^2 C^2}$.

Hence

$$I = \frac{\omega CE_0}{1 + R^2\omega^2 C^2}(\cos \omega t + RC\omega \sin \omega t - e^{-\frac{t}{RC}}).$$

EXAMPLE 4.61

A voltage $E\, e^{-at}$ is applied at $t = 0$ to a LR circuit. Find the current at any time t.

Solution. The differential equation governing the LR circuit is

$$\frac{dI}{dt} + \frac{R}{L} I = \frac{e.m.f}{L} = \frac{E\, e^{-at}}{L}.$$

As in Example 4.58, the integrating factor is $e^{\frac{Rt}{L}}$. Therefore, the solution of the above equation is

$$I\, e^{\frac{Rt}{L}} = \int \frac{E\, e^{-at}}{L} e^{\frac{Rt}{L}}\, dt + C = \frac{E}{L}\int e^{\frac{Rt}{L} - at} dt + C$$

$$= \frac{E}{L}\frac{e^{\frac{Rt}{L} - at}}{\frac{R}{L} - a} + C = \frac{E}{R - aL}\left[e^{\frac{Rt}{L} - at}\right] + C$$

and so

$$I = \frac{E}{R - aL} e^{-at} + C\, e^{-\frac{Rt}{L}}.$$

Using the initial condition $I(0) = 0$, we get

$$C = -\frac{R}{R - aL}.$$

Hence

$$I = \frac{E}{R - aL}\left[e^{-at} - e^{-\frac{Rt}{L}}\right].$$

EXAMPLE 4.62

An *RC* circuit has an e.m.f. given in volt by 400 cos 2*t*, a resistance of 100 ohms, and a capacitance of 10^{-2} farad. Initially there is no charge on the capacitor. Find the current in the circuit at any time *t*.

Solution. The equation governing the circuit is

$$RI + \frac{Q}{C} = E.$$

We are given that $R = 100$ ohms, $C = 10^{-2}$ farad and $E = 400 \cos 2t$. Thus, we have

$$I + Q = 4 \cos 2t$$

or

$$\frac{dQ}{dt} + Q = 4 \cos 2t, \text{ since } I = \frac{dQ}{dt}.$$

The integrating factor is $e^{\int 1 dt} = e^t$. Therefore, the solution is

$$Q \cdot e^t = 4 \int \cos 2t \, e^t \, dt$$

$$= 4\left[\frac{e^t}{5}[\cos 2t + 2 \sin 2t]\right] + C$$

$$= \frac{4}{5} e^t \cos 2t + \frac{8}{5} e^t \sin 2t + C.$$

Thus

$$Q = \frac{4}{5} \cos 2t + \frac{8}{5} \sin 2t + C\, e^{-t}.$$

Using the initial condition $Q(0) = 0$, we get $C = -\frac{4}{5}$. Therefore,

$$Q = \frac{4}{5} \cos 2t + \frac{8}{5} \sin 2t - \frac{4}{5} e^{-t}.$$

Hence

$$I = \frac{dQ}{dt} = -\frac{8}{5} \sin 2t + \frac{16}{5} \cos 2t + \frac{4}{5} e^{-t}.$$

(B) Problems Related to Newton's Law of Cooling

Newton's Law of Cooling states that *the time rate of change of the temperature of a body is proportional to the temperature difference between the body and its surrounding medium.*

Let *T* be the temperature of the body at any time *t* and T_0 be the temperature of the surrounding at that particular time. Then, according to Newton's Law of Cooling,

$$\frac{dT}{dt} \propto (T - T_0)$$

and so

$$\frac{dT}{dt} = -k(T - T_0), \tag{21}$$

the negative sign with the constant of proportionality is required to make $\frac{dT}{dt}$ negative in cooling process when *T* is greater than T_0, and positive in a heating process when *T* is less than T_0.

The equation (21) is first order differential equation and can be solved for *T*.

EXAMPLE 4.63

A body at a temperature of 50° F is placed outdoors, where the temperature is 100° F. If after 5 minutes, the temperature of the body is 60° F, find the time required by the body to reach a temperature of 75° F.

Solution. By Newton's Law of Cooling, we have

$$\frac{dT}{dt} = -k(T - T_0)$$

or

$$\frac{dT}{dt} + k\,T = k\,T_0.$$

But $T_0 = 100$ F. Therefore,

$$\frac{dT}{dt} + kT = 100k,$$

which is linear. The integrating factor is $e^{\int k\,dt} = e^{kt}$. Hence the solution is

$$T \cdot e^{kt} = 100 \int k\, e^{kt} dt + C = 100 e^{kt} + C$$

or

$$T = C\, e^{-kt} + 100.$$

When $t = 0$, $T = 50$, therefore, $C = -50$. Hence

$$T = -50e^{-kt} + 100.$$

Now it is given that $T = 60$ at $t = 5$. Hence

$$60 = -50e^{-5k} + 100 \text{ or } e^{-5k} = \frac{4}{5}.$$

Taking log, we get

$$k = -\frac{1}{5}\log\frac{4}{5} = -\frac{1}{5}(-0.223) = 0.045.$$

Hence $T = -50e^{0.045t} + 100$. When $T = 75$, we get $e^{-0.045t} = \frac{1}{2}$, which yields $-0.045t = \log\frac{1}{2}$ and so $t = 2.4$ minutes.

EXAMPLE 4.64

A body originally at 80° C cools down to 60° C in 20 minutes, the temperature of the air being 40° C. Find the temperature of the body after 40 minutes from the original.

Solution. By Newton's Law of Cooling, we have

$$\frac{dT}{dt} = -k(T - T_0)$$

and so variable separation gives

$$\frac{dT}{T - T_0} = -kdt.$$

Integrating, we have

$$\log(T - T_0) = -kt + \log C$$

or

$$T - T_0 = Ce^{-kt} \quad \text{or} \quad T - 40 = Ce^{-kt}.$$

But when $t = 0$, $T = 80°$C. Therefore, C = 80 – 40 = 40 and we have

$$T - 40 = 40e^{-kt}.$$

Now when $t = 20$, $T = 60°$. Therefore

$$20 = 40e^{-20k} \text{ or } e^{-20k} = \frac{1}{2}$$

or

$$-.20k = \log\frac{1}{2}, \text{ which yields } k = \frac{1}{20}\log 2.$$

Hence

$$T - 40 = 40e^{-\left(\frac{1}{20}\log 2\right)t}$$

When $t = 40$, we have

$$T - 40 = 40e^{-2\log 2}$$

or

$$T = 40 + 40e^{\log\frac{1}{4}} = 40 + \frac{40}{4} = 50° \text{ C}.$$

(C) Problems Relating to Heat Flow

The fundamental principles of heat conduction are:

(i) Heat always flow from a higher temperature to a lower temperature.
(ii) The quantity of heat in a body is proportional to its mass and temperature ($Q = mst$), where m is the mass, s is the specific heat, and t is the time.
(iii) The rate of heat flow across an area is proportional to the area and to the rate of change of temperature with respect to its distance normal to the area.

Let Q be the quantity of heat flow per second across a slab of area A and thickness δx and whose faces are kept at temperature T and $T + \delta T$. Then, by the above principles

$$Q \propto A\frac{dT}{dx}.$$

or

$$Q = -kA\frac{dT}{dx} \tag{22}$$

where k is a constant, called the *coefficient of thermal conductivity* and depends upon the material of the body. Negative sign has been taken since T decreases as x increases. The relation (22) is called the *Fourier's law of conductivity.*

EXAMPLE 4.65

The inner and outer surfaces of a spherical shell are maintained at temperature T_0 and T_1, respectively. If the inner and outer radii of the shell are

r_0 and r_1, respectively and k is the thermal conductivity, find the amount of heat lost from the shell per unit time. Also, find the temperature distribution through the shell.

Solution. We have

$$Q = -k(4\pi x^2)\frac{dT}{dx},$$

where x is the radius. Thus

$$dT = -\frac{Q}{4\pi k}\cdot\frac{dx}{x^2}.$$

Integrating, we get

$$T = -\frac{Q}{4\pi k}\left(-\frac{1}{x}\right) + C = \frac{Q}{4\pi xk} + C \qquad (23)$$

Now $T = T_0$ when $x = r_0$ and $T = T_1$ when $x = r_1$. Therefore, we have

$$T_0 = \frac{Q}{4\pi r_0 k} + C \text{ and } T_1 = \frac{Q}{4\pi r_1 k} + C.$$

Subtracting, we get

$$T_0 - T_1 = \frac{Q}{4\pi k}\left(\frac{1}{r_0} - \frac{1}{r_1}\right) = \frac{Q}{4\pi k}\left(\frac{r_1 - r_0}{r_0 r_1}\right)$$

or

$$Q = \frac{4\pi k\, r_0 r_1 (T_0 - T_1)}{r_1 - r_0}.$$

When $x = r_0$, $T = T_0$, then (23) gives

$$T_0 = \frac{Q}{4\pi r_0 k} + C$$

$$= \frac{4\pi k r_0 r_1 (T_0 - T_1)}{4\pi r_0 k(r_1 - r_0)} + C$$

$$= \frac{r_1(T_0 - T_1)}{r_1 - r_0} + C.$$

Thus

$$C = T_0 - \frac{r_1(T_0 - T_1)}{r_1 - r_0}$$

$$= \frac{T_0(r_1 - r_0) - r_1(T_0 - T_1)}{r_1 - r_0} = \frac{r_1 T_1 - r_0 T_0}{r_1 - r_0}$$

Hence (23) transforms to

$$T = \frac{4\pi k r_0 r_1 (T_0 - T_1)}{4\pi x k(r_1 - r_0)} + \frac{r_1 T_1 - r_0 T_0}{r_1 - r_0}$$

$$= \frac{1}{r_1 - r_0}\left[\frac{(T_0 - T_1) r_0 r_1}{x} + r_1 T_1 - r_0 T_0\right].$$

EXAMPLE 4.66

A spherical shell of inner and outer radii 10 cm and 15 cm, respectively, contains steam at 150° C. If the temperature of the outer surface of the shell is 40° C and thermal conductivity $k = 0.0025$, find the temperature half-way through the thickness of the shell under steady state conditions.

Solution. With the notation of Example 4.65, we have

$$r_0 = 10 \text{ cm}, r_1 = 15 \text{ cm}, x = 12.5 \text{ cm},$$
$$T_0 = 150°C, T_1 = 40° \text{ C}.$$

Hence

$$T = \frac{1}{r_1 - r_0}\left[\frac{(T_0 - T_1) r_0 r_1}{x} + r_1 T_1 - r_0 T_0\right]$$

$$= \frac{1}{5}\left[\frac{(150 - 40)150}{12.5} + 600 - 1500\right]$$

$$= \frac{1}{5}\left[\frac{16500}{12.5} - 900\right] = \frac{1}{5}[1320 - 900]$$

$= 84°$ C.

EXAMPLE 4.67

A long hollow pipe has a inner radius of r_0 cm and outer radius of r_1 cm. The inner surface is kept at a temperature T_0 and the outer surface at the temperature T_1. If thermal conductivity is k, find the heat lost per second of 1 cm length of the pipe. Also find the temperature distribution through the thickness of the pipe.

Solution. Let Q cal/sec be the constant quantity of heat flowing out radially through the surface of the pipe having radius x cm and length 1 cm. Then the area of the lateral surface is $2\pi x$. Therefore, by Fourier law,

$$Q = -kA\frac{dT}{dx} = -k(2\pi x)\frac{dT}{dx}$$

and so

$$dT = -\frac{Q}{2\pi k}\cdot\frac{dx}{x}.$$

Integrating, we get

$$T = -\frac{Q}{2\pi k}\log x + C. \tag{24}$$

When $x = r_0$, $T = T_0$, and so

$$T_0 = -\frac{Q}{2\pi k}\log r_0 + C \tag{25}$$

When $x = r_1$, $T = T_1$ and we have

$$T_1 = -\frac{Q}{2\pi k}\log r_1 + C. \tag{26}$$

Subtracting (26) from (25), we have

$$T_0 - T_1 = \frac{Q}{2\pi k}[\log r_0 - \log r_1] = \frac{Q}{2\pi k}\log\frac{r_1}{r_0}.$$

Thus

$$Q = \frac{2\pi k(T_0 - T_1)}{\log\frac{r_1}{r_0}}, \tag{27}$$

which gives the heat lost per second in 1 cm of the pipe. Further subtracting (25) from (24) and using (27), we get

$$T - T_0 = -\frac{Q}{2\pi k}[\log x - \log r_0] = -\frac{Q}{2\pi k}\log\frac{x}{r_0}$$

$$= -\frac{2\pi k(T_0 - T_1)}{2\pi k\log\frac{r_1}{r_0}}\log\frac{x}{r_0}$$

$$= \frac{(T_1 - T_0)}{\log\frac{r_1}{r_0}}\log\frac{x}{r_0}.$$

Hence,

$$T = T_0 + \frac{(T_1 - T_0)}{\log\frac{r_1}{r_0}}\log\frac{x}{r_0} \tag{28}$$

which gives the required temperature distribution through the thickness of the pipe.

EXAMPLE 4.68

A pipe 20 cm in diameter contains steam at 150°C and is protected with a covering 5 cm thick. If thermal conductivity is 0.0025 and the temperature of the outer surface of the covering is 40° C, find the temperature half-way through the covering under steady-state conditions.

Solution. Using the notation of Example 4.67, we have

$$r_0 = 10cm,\ r_1 = 15\ cm,\ x = 12.5,$$
$$T_0 = 150°\ \text{C},\ T_1 = 40°\ \text{C},\ k = 0.0025.$$

Hence using (28), we have

$$T = 150 + \frac{40 - 150}{\log\frac{15}{10}}\log\frac{12.5}{10}$$

$$= 150 - 110\log\frac{1.25}{1.5} = 84.5°\ \text{C}.$$

EXAMPLE 4.69

A long hollow pipe has an inner diameter of 10 cm and outer diameter of 20 cm. The inner surface is kept at 200° C and the outer surface at 50° C. The thermal conductivity is 0.12. How much heat will be lost per second from a portion of 1 cm of the pipe and what is the temeprature at a distance of 7.5 cm from the centre of the pipe?

Solution. From Example 4.67, we have

$$Q = \frac{2\pi k(T_0 - T_1)}{\log\frac{r_1}{r_0}}$$

$$= \frac{2\pi k(200 - 50)}{\log\frac{10}{5}} = \frac{300\pi k}{\log 2}$$

$$=\frac{300\,\pi\,(0.12)}{\log 2}=163\text{cal/sec.}$$

Also

$$T=T_0+\frac{(T_1-T_0)}{\log\frac{r_1}{r_0}}\log\frac{x}{r_0}=200+\frac{(50-200)}{\log 2}\log\frac{7.5}{5}$$

$$=200-150\log\frac{1.5}{2}=200-150(.58)=113°\text{ C.}$$

(D) Rate Problems

In some problems, the rate at which a quantity changes is a known function of the amount present and/or the time and it is desired to find the quantity itself. Radioactive nuclei decay, population growth, and chemical reactions are some of the phenomenon of this kind.

Let x be the amount of radioactive nuclei present after t years. Then $\frac{dx}{dt}$ represents the rate of decay. Since the nuclei decay at a rate proportional to the amount present, we have

$$\frac{dx}{dt}=kx, \qquad (29)$$

where k is constant of proportionality.

The law of chemical reaction states that the *rate of change of chemical reaction is proportional to the amount of substance present at that instant*. Thus, the differential equation (29) governs the chemical reaction of first order.

Moreover, if the rate at which amount of a substance increases or decreases is found to be jointly proportional to two factors, each factor being a linear function of x, then the chemical reaction is said to be of *second order*. For example, if a solution contains two substances whose amounts at the beginning are a and b respectively and if equal amount x of each substance changes in time t, then the amounts of the substance left in the solution at time t are $(a-x)$ and $(b-x)$ and, therefore, we have

$$\frac{dx}{dt}=k(a-x)\,(b-x).$$

Taking the case of population growth, we assume that the population is a continuous and differentiable function of time. Let x be the number of individuals in a population at time t. Then *rate of change of population is proporitional to the number of individuals in it at any time*. Thus equation (29) is valid for population growth also.

EXAMPLE 4.70

If 10% of 50 mg of a radioactive material decays in 2 hours, find the mass of the material left at any time t and the time at which the material has decayed to one-half of its initial mass.

Solution. Let x denote the amount of material present at time t. Then the equation governing the decay is

$$\frac{dx}{dt}=kx.$$

Variable separation gives

$$\frac{dx}{x}=k\,dt.$$

Integrating, we get

$$\log x=kt+\log C \quad\text{or}\quad x=Ce^{kt}.$$

At $t=0$, $x=50$. Therefore, $C=50$ and so $x=50\,e^{kt}$. At $t=2$, 10% of the mass present is decayed. Thus 5 mg of the substance has been decayed and 45 gm still remains. Therefore, $45=50\,e^{2k}$. Thus

$$k=\frac{1}{2}\log\frac{45}{50}=-0.053.$$

Hence mass of the material left at any time t is

$$x=50\,e^{-0.053\,t}.$$

Further, when half of the material is decayed, we have $x=25$ mg and so

$$25=50\,e^{-0.053\,t}$$

or

$$-0.053t=\log\frac{1}{2}$$

or

$$t=13\text{ hours.}$$

EXAMPLE 4.71

A tank contains 1000 litres of fresh water. Salt water which contains 150 gm of salt per litre runs into it at the rate of 5 litres/min and well stirred mixture runs out of it at the same rate. When will the tank contain 5000 gm of salt?

Solution. Let x denote the amount of salt in the tank at time t. Then

$$\frac{dx}{dt} = \text{IN} - \text{OUT}.$$

The brine flows at the rate of 5 litres/min and each litre contains 150 gm of salt. Thus

$$\text{IN} = 5 \times 150 = 750 \text{ gm/min}$$

Since the rate of outflow equals the rate of inflow, the tank contains 1000 litres of mixture at any time t. This 1000 litres contains x gm of salt at time t and so the concentration of the salt at time t is $\frac{x}{1000}$ gm/litres. Since mixture flows out at the rate of 5 litres/min, we have

$$\text{OUT} = \frac{x}{1000} \times 5 = \frac{x}{200} \text{ gm/litres}.$$

Thus the differential equation for x becomes

$$\frac{dx}{dt} = 750 - \frac{x}{200}. \qquad (30)$$

Since initially there was no salt in the tank, we have the initial condition $x(0) = 0$. The equation (30) is linear and separable. We have in fact

$$\frac{dx}{150000 - x} = \frac{dt}{200}.$$

Integrating, we get

$$-\log(150000 - x) = \frac{t}{200} + C. \qquad (31)$$

Using the initial condition $x(0) = 0$, we have

$$C = -\log 150000.$$

Hence (31) yields

$$\frac{t}{200} = \log 150000 - \log(150000 - x)$$

or

$$t = 200 \log \frac{150000}{150000 - x}.$$

If $x = 5000$ gm, then

$$t = 200 \log \frac{150000}{145000} = 200 \log \frac{30}{29} = 6.77\text{min}.$$

EXAMPLE 4.72

If the population of a city gets doubled in 2 years and after 3 years the population is 15,000, find the initial population of the city.

Solution. Let x denote the population at any time t and let x_0 be the initial population of the city. Then

$$\frac{dx}{dt} = kx,$$

which has the solution as

$$x = C\,e^{kt}.$$

At $t = 0$, $x = x_0$. Hence $C = x_0$. Thus

$$x = x_0 e^{kt} \qquad (32)$$

But at $t = 2$, $x = 2x_0$. Therefore

$$2x_0 = x_0 e^{2k} \text{ or } e^{2k} = 2$$

or

$$k = \frac{1}{2} \log 2 = 0.347$$

Hence (32) reduces to

$$x = x_0\, e^{0.347\,t}.$$

At $t = 3$, $x = 15,000$ and so

$$15,000 = x_0\, e^{(0.347)(3)} = x_0(2.832)$$

Hence

$$x_0 = \frac{15000}{2.832} = 5297.$$

(E) Falling Body Problems

Consider a body of mass m falling under the influence of gravity g and an air resistance, which is proportional to the velocity of the falling body. Newton's second law of motion states that the *net force acting on a body is equal to the time rate of change of the momentum of the body*.
Thus

$$F = \frac{d}{dt}(mv).$$

If m is assumed to be constant, then

$$F = m\frac{dv}{dt} \tag{33}$$

where F is the net force on the body and v is the velocity of the body at time t.

The falling body is under the action of two forces: (i) Force due to gravity which is given by the weight mg of the body (ii) the force due to resistance of air and that is $-kv$, where $k \geq 0$ is a constant of proportionality. Thus (33) yields

$$mg - kv = m\frac{dv}{dt}$$

or

$$\frac{dv}{dt} + \frac{k}{m}v = g, \tag{34}$$

which is *equation of motion for the falling body*.

If air resistance is negligible, then $k = 0$ and we have

$$\frac{dv}{dt} = g. \tag{35}$$

The differential equation (35) is separable and we have

$$dv = g\,dt.$$

Integrating, we get

$$v = gt + C.$$

But when $t = 0$, $v = 0$ and so $C = 0$. Hence

$$v = gt. \tag{36}$$

Also velocity is time rate of change of displacement x and so

$$\frac{dx}{dt} = gt \quad \text{or} \quad dx = gt\,dt.$$

Integrating, we get

$$x = \frac{1}{2}gt^2 + k \text{ (constant)}.$$

But at $t = 0$, the displacement is 0. Therefore $k = 0$. Hence

$$x = \frac{1}{2}gt^2. \tag{37}$$

EXAMPLE 4.73

A body of mass 16 kg is dropped from a height of 625 ft. Assuming that there is no air resistance, find the time required by the body to reach the ground.

Solution. By (37), we have (with $g = 32$ ft/sec^2)

$$x = \frac{1}{2}gt^2 = 16t^2,$$

therefore,

$$t^2 = \sqrt{\frac{x}{16}} = \sqrt{\frac{625}{16}} = \frac{25}{4} = 6.25 \text{ sec.}$$

(F) Orthogonal Trajectories

Recall that a curve which cuts every member of a given family of curves according to some definite law is called a *trajectory* of the family.

A curve which cuts every member of a given family of curves at right angles is called *orthogonal trajectories*. Further, two families of curves are said to be orthogonal if every member of either family cuts each member of the other family at right angles.

Consider a one-parameter family of curves in the xy-plane defined by

$$f(x, y, c) = 0, \tag{38}$$

where c denotes the parameter. Differentiating (38), with respect to x and eliminating c between (38) and the resulting equation, we get the differential equation of the family in question. Let the differential equation be

$$F\left(x, y, \frac{dy}{dx}\right) = 0. \tag{39}$$

To obtain the equation of the orthogonal trajectory, we replace $\frac{dy}{dx}$ by $-\frac{dx}{dy}$ and get the differential equation of orthogonal trajectory as $F\left(x, y, -\frac{dx}{dy}\right)$. Solution of this differential equation will yield the equation of the orthogonal trajectory.

In case of polar curves

$$f(r, \theta, c) = 0. \tag{40}$$

Differentiating (40) and eliminating c between (40) and the resulting equations, we get the differential equation of the family represented by (40). Let the differential equation be

$$F\left(r, \theta, \frac{dr}{d\theta}\right) = 0. \qquad (41)$$

Replacing $dr/d\theta$ by $-r^2\, d\theta/dr$ in (41), we get the differential equation of the orthogonal trajectory as

$$F\left(r, \theta - r^2 \frac{d\theta}{dr}\right) = 0. \qquad (42)$$

Solution of (42) will then yield the equation of the required orthogonal trajectory.

EXAMPLE 4.74

Find the orthogonal trajectories of the family of curves $x^2 + y^2 = cx$.

Solution. We have

$$x^2 + y^2 - cx = 0. \qquad (43)$$

Differentiating, we get

$$2x + 2y\frac{dy}{dx} = c. \qquad (44)$$

Eliminating c between (43) and (44) yields

$$2x + 2y\frac{dy}{dx} = \frac{x^2 + y^2}{x}$$

or

$$\frac{dy}{dx} = \frac{y^2 - x^2}{2xy}. \qquad (45)$$

The equation (45) is the differential equation of the family represented by (43). Therefore, the differential equation of the orthogonal trajectory is

$$\frac{dy}{dx} = \frac{2xy}{x^2 - y^2}. \qquad (46)$$

This is an homogeneous equation. Substituting $y = vx$, and separating variables, we get

$$\frac{dx}{x} + \left(-\frac{1}{v} + \frac{2v}{v^2 + 1}\right) dv = 0.$$

Integrating, we get

$$\log x - \log v + \log (v^2 + 1) = C$$

or

$$x(v^2 + 1) = kv, \; (C = \log k).$$

Substituting $v = \frac{y}{x}$, we get

$$x^2 + y^2 = ky.$$

EXAMPLE 4.75

Find the orthogonal trajectory of the family of the curves $xy = C$.

Solution. The equation of the given family of curves is

$$xy = C. \qquad (47)$$

Differentiating, we get

$$x\frac{dy}{dx} + y = 0$$

or

$$\frac{dy}{dx} = -\frac{y}{x}. \qquad (48)$$

Therefore, the differential equation of the family of orthogonal trajectory is

$$\frac{dy}{dx} = \frac{x}{y} \qquad (49)$$

or

$$x\, dx - y\, dy = 0.$$

Integrating, we get

$$x^2 - y^2 = k,$$

which is the equation of orthogonal trajectories called *equipotential lines* (shown in Figure 4.6).

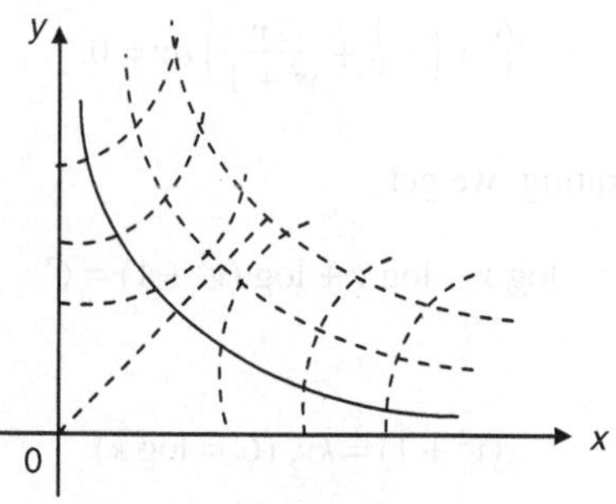

Figure 4.6

EXAMPLE 4.76

Find the orthogonal trajectories of the family of curves $y = ax^2$.

Solution. The given family represented by the equation

$$y = ax^2 \tag{50}$$

is a family of parabolas symmetric about y–axis with vertices at (0, 0). Differentiating with respect to x, we get

$$\frac{dy}{dx} = 2ax \tag{51}$$

Eliminating a between (50) and (51), we get

$$\frac{dy}{dx} = \frac{2xy}{x^2} = \frac{2y}{x}.$$

Therefore, differential equation of the orthogonal trajectory is

$$\frac{dy}{dx} = -\frac{x}{2y} \tag{52}$$

or

$$2y\,dy + x\,dx = 0.$$

Integrating, we get

$$2\frac{y^2}{2} + \frac{x^2}{2} = C$$

or

$$\frac{x^2}{2} + \frac{y^2}{1} = C. \tag{53}$$

The orthogonal trajectories represented by (53) are ellipses (shown in the Figure 4.7)

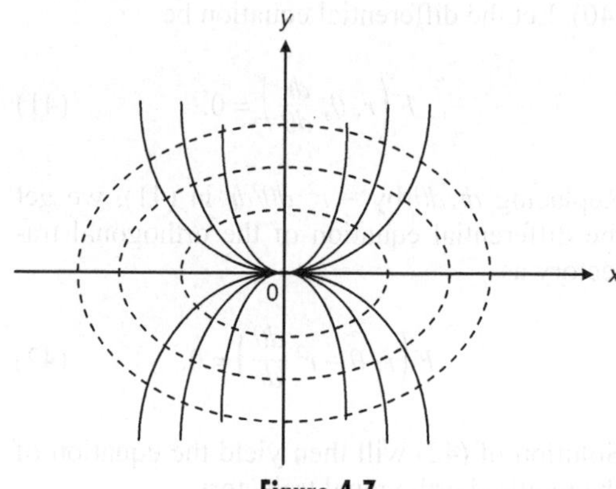

Figure 4.7

EXAMPLE 4.77

Show that the system of confocal and coaxial parabolas is self-orthogonal.

Solution. The equation of the family of confocal parabolas having x-axis as their axis is of the form

$$y^2 = 4a(x + a). \tag{54}$$

Differentiating, we get

$$y\frac{dy}{dx} = 2a$$

or

$$\frac{dy}{dx} = \frac{2a}{y}. \tag{55}$$

From (55), we have $a = \frac{y}{2}\frac{dy}{dx}$. Substituting this value in (54), we get

$$y^2 = 2y\frac{dy}{dx} + \left(x + \frac{1}{2}y\frac{dy}{dx}\right)$$

or

$$y\left(\frac{dy}{dx}\right)^2 + 2x\frac{dy}{dx} - y = 0, \tag{56}$$

which is the differential equation of the given family. Replacing $\frac{dy}{dx}$ by $-\frac{dx}{dy}$ in (56) we obtain (56) again. Hence, each member of family (54) cuts every other member of the same family orthogonally.

EXAMPLE 4.78

Find the orthogonal trajectories of the family of curves

$$\frac{x^2}{a^2}+\frac{y^2}{b^2+\lambda}=1.$$

Solution. The equation of the family of the given curve is

$$\frac{x^2}{a^2}+\frac{y^2}{b^2+\lambda}=1. \tag{57}$$

Differentiating with respect to x, we get

$$\frac{x}{a^2}+\frac{y}{b^2+\lambda}\frac{dy}{dx}=0. \tag{58}$$

From (57) and (58) we have respectively

$$b^2+\lambda=\frac{a^2y^2}{a^2-x^2} \quad \text{and} \quad b^2+\lambda=-\frac{a^2y}{x}\frac{dy}{dx}.$$

Hence

$$\frac{a^2y^2}{a^2-x^2}=-\frac{a^2y}{x}\frac{dy}{dx}$$

or

$$\frac{dy}{dx}=-\frac{xy}{a^2-x^2}.$$

Therefore, differential equation of the orthogonal trajectory is

$$\frac{dy}{dx}=\frac{a^2-x^2}{xy}$$

or

$$y\,dy=\frac{a^2-x^2}{x}\,dx=\frac{a^2}{x}\,dx-x\,dx.$$

Integrating, we get

$$\frac{y^2}{2}=a^2\log x-\frac{x^2}{2}+C$$

or

$$x^2+y^2=2a^2\log x+k \text{ (constant)},$$

which is the equation of the required orthogonal trajectories.

EXAMPLE 4.79

Find the orthogonal trajectory of the cardioid $r = a(1-\cos\theta)$.

Solution. The equation of the family of given cardioide is

$$r=a(1-\cos\theta). \tag{59}$$

Differentiating with respect to θ, we have

$$\frac{dr}{d\theta}=a\sin\theta. \tag{60}$$

Dividing (60) by (59), we get the differential equation of the given family as

$$\frac{1}{r}\frac{dr}{d\theta}=\frac{\sin\theta}{1-\cos\theta}=\frac{2\sin\frac{\theta}{2}\cos\frac{\theta}{2}}{2\sin^2\frac{\theta}{2}}$$

$$=\cot\frac{\theta}{2} \tag{61}$$

Replacing $\frac{dr}{d\theta}$ by $-r^2\frac{d\theta}{dr}$ in (61), we get

$$\frac{1}{r}\left(-r^2\frac{d\theta}{dr}\right)=\cot\frac{\theta}{2}$$

or

$$\frac{dr}{r}+\tan\frac{\theta}{2}\,d\theta=0, \tag{62}$$

which is the equation of the family of orthogonal trajectories. Integrating (62), we get

$$\log r - 2\log\cos\frac{\theta}{2} = \log C$$

or

$$\log r = \log C + \log\cos^2\frac{\theta}{2} = \log C\cos^2\frac{\theta}{2}$$

or

$$r = C\cos^2\frac{\theta}{2} = C(1+\cos\theta),$$

which is the equation of orthogonal trajectory of the given family.

EXAMPLE 4.80

Find the orthogonal trajectories of the family of curves $x^2 + y^2 = c^2$

Solution. The equation of the given family of curves is

$$x^2 + y^2 - c^2 = 0 \qquad (63)$$

Differentiating (63), we get

$$2x + 2y\frac{dy}{dx} = 0$$

or

$$\frac{dy}{dx} = -\frac{x}{y}, \qquad (64)$$

which is the differential equation representing the given curves. Therefore the differential equation of the required family of orthogonal trajectories is

$$\frac{dy}{dx} = \frac{y}{x}$$

or

$$\frac{dy}{y} = \frac{dx}{x} \qquad (65)$$

Integrating, we get

$$\log y = \log x + \log k$$

or

$$y = kx,$$

which is the equation of the orthogonal trajectories, which are straight lines through the origin as shown in the Figure 4.8.

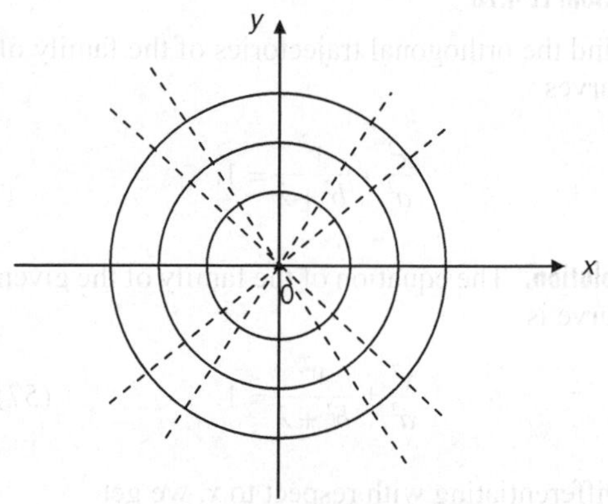

Figure 4.8

EXAMPLE 4.81

Find the orthogonal trajectories of the curves $r^2 = a^2\cos 2\theta$.

Solution. We are given that

$$r^2 = a^2\cos 2\theta = a^2(1 - 2\sin^2\theta). \qquad (66)$$

Differentiating (66) w.r.t. θ, we have

$$2r\frac{dr}{d\theta} = -2a^2\sin 2\theta. \qquad (67)$$

Dividing (67) by (66), we get

$$\frac{2}{r}\frac{dr}{d\theta} = \frac{-2\sin 2\theta}{\cos 2\theta}.$$

Replacing $\frac{dr}{d\theta}$ with $-r^2\frac{d\theta}{dr}$, we get

$$-2r\frac{d\theta}{dr} = -\frac{\sin 2\theta}{\cos 2\theta} = -2\tan 2\theta$$

or

$$\frac{dr}{r} = \cot 2\theta.$$

Integrating, we get

$$\log r = \frac{1}{2}\log \sin 2\theta + \log C$$

or

$$2\log r = 2\log C + \log \sin 2\theta$$

or

$$\log r^2 = \log C^2 + \log \sin 2\theta$$

or

$$r^2 = C^2 \sin 2\theta.$$

EXAMPLE 4.82

Find the orthogonal trajectories of the family

$$\frac{2a}{r} = 1 - \cos\theta.$$

Solution. We have

$$\frac{2a}{r} = 1 - \cos\theta,$$

or

$$r = \frac{2a}{1-\cos\theta}. \tag{68}$$

Differentiating (68) with respect to θ, we get

$$\frac{dr}{d\theta} = \frac{(1-\cos\theta)(0) - 2a(\sin\theta)}{(1-\cos\theta)^2}$$

$$= -\frac{2a\sin\theta}{(1-\cos\theta)^2}. \tag{69}$$

Dividing (68) by (69), we get

$$\frac{1}{r}\frac{dr}{d\theta} = -\frac{\sin\theta}{1-\cos\theta} = -\frac{2\sin\frac{\theta}{2}\cos\frac{\theta}{2}}{2\sin^2\frac{\theta}{2}}$$

$$= -\cot\frac{\theta}{2}. \tag{70}$$

Replacing $\frac{dr}{d\theta}$ by $-r^2\frac{d\theta}{dr}$ in (70), we get

$$\frac{1}{r}\left(-r^2\frac{d\theta}{dr}\right) = -\cot\frac{\theta}{2}$$

or

$$r\frac{d\theta}{dr} = \cot\frac{\theta}{2}$$

or

$$\frac{dr}{r} = \tan\frac{\theta}{2}\,d\theta = \frac{\sin\frac{\theta}{2}}{\cos\frac{\theta}{2}}\,d\theta.$$

Integrating, we get

$$\log r = -2\log\cos\frac{\theta}{2} + \log c$$

or

$$\log r + 2\log\cos\frac{\theta}{2} = \log c$$

or

$$\log r + \log\cos^2\frac{\theta}{2} = \log c$$

or

$$r\cos^2\frac{\theta}{2} = c$$

or

$$\frac{r}{2}(1+\cos\theta) = c$$

or

$$r = \frac{2c}{1+\cos\theta},$$

which is the required orthogonal trajectory.

4.14 EQUATIONS OF FIRST ORDER AND HIGHER DEGREE

Let $\frac{dy}{dx} = p$. Then the general form of the equations of first order and higher degree is

$$p^n + P_1 p^{n-1} + P_2 p^{n-2} + \dots + P_n = 0, \quad (71)$$

where $P_1, P_2, \dots, P_n$ are functions of x and y.

These equations can be divided into two subclasses:

(a) When left hand side of (71) can be resolved into rational factors of first degree.

(b) When left hand side of (71) cannot be factorized into rational factors of first degree.

We discuss several methods to solve equations of these subclasses.

4.15 EQUATIONS WHICH CAN BE FACTORIZED INTO FACTORS OF FIRST DEGREE

Let

$$p^n + P_1 p^{n-1} + P_2 p^{n-2} + \dots + P_n = 0 \quad (72)$$

be equation of first order and degree n. Let the factors of left hand side be

$$(p - R_1)(p - R_2)\dots(p - R_n),$$

where $R_1, R_2, \dots, R_n$ are function of x and y. Then (72) reduces to

$$(p - R_1)(p - R_2)\dots(p - R_n) = 0.$$

Equating each factor to zero, we have

$$p = R_1, p = R_2, \dots, p = R_n,$$

which are equations of first order and first degree. Suppose that solutions of these equation are

$$f_1(x,y,c_1) = 0, f_2(x,y,c_2) = 0, \dots, f_n(x,y,c_n) = 0.$$

Since each constant c_i take infinite number of values, these solution will even be general if all $c_i = c$.

Hence n solutions of the given equation are $f_1(x, y, c) = 0, f_2(x, y, c) = 0, \dots, f_n(x, y, c) = 0$. Combining all these solution into one, we have $f_1(x, y, c) f_2(x, y, c), \dots, f_n(x, y, c) = 0$ as a general solution.

EXAMPLE 4.83

Solve

$$p^2 + 2py \cot x = y^2.$$

Solution. We have

$$p^2 + 2py \cot x - y^2 = 0. \quad (73)$$

Solving for p, we have

$$p = \frac{-2y \cot x \pm \sqrt{4y^2 \cot^2 x + 4y^2}}{2}$$

$$= -y \cot x \pm y \operatorname{cosec} x$$

Hence the factors form of (73) is

$$(p + y\cot x - y\operatorname{cosec} x)(p + y\cot x + y\operatorname{cosec} x) = 0.$$

Thus we obtain two differential equations

$$p + y(\cot x - \operatorname{cosec} x) = 0 \quad (74)$$

and

$$p + y(\cot x + \operatorname{cosec} x) = 0 \quad (75)$$

Both of these equations are Leibnitz's linear equations. Integrating factor for (74) is

$$I.F. = e^{\int (\cot x - \operatorname{cosec} x)\,dx} = e^{[\log \sin x - \log(\operatorname{cosec} x - \cot x)]}$$

$$= e^{\log \frac{\sin x}{\operatorname{cosec} x - \cot x}} = e^{\log\left[\frac{\sin^2 x}{1 - \cos x}\right]}$$

$$= e^{\log(1 + \cos x)} = 1 + \cos x.$$

Therefore solution of (72) is

$$y\,(1+\cos x)=c$$

or

$$y\,(1+\cos x)-c=0 \tag{76}$$

Similarly, the solution of (75) is

$$y\,(1-\cos x)=c$$

or

$$y\,(1-\cos x)-c=0 \tag{77}$$

Hence, combining (76) and (77), the required solution is

$$[y\,(1+\cos x)-c]\,[y\,(1-\cos x)-c]=0.$$

EXAMPLE 4.84

Solve $xy\,p^2-(x^2+y^2)\,p+xy=0$

Solution. The given differential equation is

$$xy\,p^2-(x^2+y^2)\,p+xy=0.$$

Solving for p, we have

$$p=\frac{x}{y},\frac{y}{x}.$$

Hence the factorized form of the given equation is

$$\left(p-\frac{x}{y}\right)\left(p-\frac{y}{x}\right)=0.$$

Thus we have two equations

$$p-\frac{x}{y}=0 \tag{78}$$

and

$$p-\frac{y}{x}=0. \tag{79}$$

From (78), we have

$$y\,dy-x\,dx=0$$

Integrating, we get

$$y^2-x^2=c. \tag{80}$$

From (79), we have

$$\frac{dy}{dx}-\frac{y}{x}=0$$

The integrating factor is

$$I.F.=e^{-\int\frac{1}{x}dx}$$

$$=e^{-\log x}=\frac{1}{x}.$$

Therefore the solution of (79) is

$$y\left(\frac{1}{x}\right)=c$$

or

$$(y-cx)=0. \tag{81}$$

Hence the required solution is

$$(y-cx)\,(y^2-x^2-c)=0.$$

EXAMPLE 4.85

Solve $\dfrac{dy}{dx}-\dfrac{dx}{dy}=\dfrac{x}{y}-\dfrac{y}{x}.$

Solution. The given differential equation is

$$p-\frac{1}{p}=\frac{x}{y}-\frac{y}{x}$$

or

$$p^2-p\left(\frac{x}{y}-\frac{y}{x}\right)-1=0. \tag{82}$$

Factorizing the left hand side of (82), the equation reduces to

$$\left(p+\frac{y}{x}\right)\left(p-\frac{x}{y}\right)=0.$$

Thus we get two equations

$$p+\frac{y}{x}=0 \tag{83}$$

and

$$p-\frac{x}{y}=0 \tag{84}$$

From (83), we have

$$x\,dy+y\,dx=0$$

or

$$d\,(xy)=0.$$

Integrating, we get

$$xy = c. \qquad (85)$$

From (84), we have

$$y\,dy - x\,dx = 0.$$

Integrating, we get

$$y^2 - x^2 = c \qquad (86)$$

Combining (85) and (86), the required solution is

$$(xy - c)(y^2 - x^2 - c) = 0.$$

EXAMPLE 4.86

Solve $p^2 - 5p + 6 = 0$.

Solution. Factorizing the left hand side, the given differential equation reduces to

$$(p-3)(p-2) = 0.$$

Thus, we get two equations

$$\frac{dy}{dx} - 3 = 0 \qquad (87)$$

and

$$\frac{dy}{dx} - 2 = 0 \qquad (88)$$

From (87), we get

$$dy - 3dx = 0.$$

Integrating, we have

$$y - 3x = c. \qquad (89)$$

From equation (88), we have

$$dy - 2dx = 0.$$

Integrating we get

$$y - 2x = 0. \qquad (90)$$

Combining the solutions (89) and (90), the required solution is

$$(y - 3x - c)(y - 2x - c) = 0.$$

EXAMPLE 4.87

Solve $yp^2 + (x - y)p - x = 0$.

Solution. We have

$$yp^2 + (x - y)p - x = 0.$$

Resolving into factors, we get

$$(p - 1)(py + x) = 0.$$

Thus we get two equations

$$p - 1 = 0 \qquad (91)$$

and

$$py + x = 0 \qquad (92)$$

From equation (91), we have

$$\frac{dy}{dx} - 1 = 0$$

or

$$dy - dx = 0,$$

which, on integration, yields

$$y - x = c. \qquad (93)$$

From equation (92), we have

$$y\frac{dy}{dx} + x = 0$$

or

$$y\,dy + x\,dx = 0.$$

which, on integration, yields

$$y^2 + x^2 = c. \qquad (94)$$

Combining (93) and (94), the general solution of the given differential equation is

$$(y - x - c)(x^2 + y^2 - c) = 0.$$

4.16 EQUATIONS WHICH CANNOT BE FACTORIZED INTO FACTORS OF FIRST DEGREE

In case of the differential equation of first order and higher degree which cannot be factorized into factors of first degree, we deal with the following cases:

(a) Equation solvable for y

Let the given differential equation be

$$f(x, y, p) = 0 \tag{95}$$

Suppose that solving (95) for y ,we get

$$y = F(x, p). \tag{96}$$

On differentiating (96) with respect to x, we get

$$p = G\left(x, p, \frac{dy}{dx}\right) \tag{97}$$

Eliminating p from (96) and (97), we get the required solution of (95).

In case elimination of p from (96) and (97) is not possible, then solve (96) and (97) for x and y in term of p. These solutions for x and y give the required solution.

EXAMPLE 4.88

Solve the differential equation

$$y + px = x^4 p^2.$$

Solution. We have

$$y = -px + x^4 p^2 \tag{98}$$

Differentiating both sides with respect to x, we get

$$p = -p - x\frac{dp}{dx} + 4x^3p^2 + 2x^4 p\frac{dp}{dx}$$

or

$$2p + x\frac{dp}{dx} - 4x^3 p^2 - 2x^4 p\frac{dp}{dx} = 0$$

or

$$2p\frac{dx}{dp} + x - 4x^3p^2\frac{dx}{dp} - 2x^4 p = 0$$

or

$$2p\frac{dx}{dp} + x(1 - 2x^3 p) + x(1 - 2x^3 p) = 0$$

or

$$(1 - 2x^3 p)\left(x + 2p\frac{dx}{dp}\right) = 0.$$

Therefore

$$2p\frac{dx}{dp} + x = 0$$

or

$$\frac{dx}{dp} + \frac{x}{2p} = 0$$

or

$$\frac{dp}{p} = -2\frac{dx}{x}.$$

Integrating, we get

$$\log p + 2\log x = \log c$$

or

$$\log px^2 = \log c$$

or

$$px^2 = c$$

or

$$p = \frac{c}{x^2}.$$

Putting this value of p in (98), we get

$$y = -\frac{cx}{x^2} + x^4\frac{c^2}{x^4}$$

$$= -\frac{c}{x} + c^2$$

or

$$xy = c^2x - c.$$

EXAMPLE 4.89

Solve $y = xp^2 + p$.

Solution. The given differential equation is

$$y = xp^2 + p.$$

Differentiating both sides with respect to x, we get

$$p = p^2 + 2xp\frac{dp}{dx} + \frac{dp}{dx}$$

or

$$p^2\frac{dx}{dp} - p\frac{dx}{dp} + 2xp + 1 = 0$$

or

$$\frac{dx}{dp} + \frac{2x}{p-1} = -\frac{1}{p(p-1)} \text{ (Linear equation)} \quad (99)$$

The integrating factor is

$$I.F. = e^{\int \frac{2}{p-1}dp} = e^{2\log(p-1)} = (p-1)^2.$$

Thus the solution of (99) is

$$x(p-1)^2 = -\int \frac{1}{p(p-1)}(p-1)^2\,dp + c$$

or

$$x(p-1)^2 = -\int \frac{p-1}{p}\,dp + c$$

$$= -\int\left(1 - \frac{1}{p}\right)dp + c$$

$$= -[p - \log p] + c$$

$$= \log p - p + c$$

or

$$x = (\log p - p + c)(p-1)^{-2}.$$

Hence

$$x = (\log p - p + c)(p-1)^{-2}$$

and

$$y = xp^2 + p$$

taken together is the required solution of the given equation.

EXAMPLE 4.90

Solve

$$x^2p^4 + 2xp - y = 0.$$

Solution. We have $y = x^2p^4 + 2xp$. Differentiating both sides of the given differential equation with respect to x, we get

$$2xp^4 + 4x^2p^3\frac{dp}{dx} + 2p + 2x\frac{dp}{dx} - p = 0$$

or

$$p + 2xp^4 + 2x\frac{dp}{dx} + 4x^2p^3\frac{dp}{dx} = 0$$

or

$$p(1 + 2xp^3) + 2x\frac{dp}{dx}(1 + 2xp^3) = 0$$

or

$$\left(p + 2x\frac{dp}{dx}\right)(1 + 2xp^3) = 0.$$

Therefore

$$p + 2\frac{dp}{dx} = 0 \text{ (for non-singular solution)}$$

or

$$2\frac{dp}{dx} = -\frac{dx}{x}.$$

Integrating we get

$$2\log p + \log x = \log c$$

or

$$\log p^2 + \log x = \log c$$

or

$$\log xp^2 = \log c$$

or

$$xp^2 = c$$

or

$$p = \sqrt{\frac{c}{x}}.$$

Putting this value in the given equation, the required solution is $y = 2\sqrt{cx} + c^2$.

EXAMPLE 4.91

Solve

$$p^3 + p = e^y.$$

Solution. The given differential equation is

$$p^3 + p = e^y \quad \text{or} \quad p(p^2 + 1) = e^y.$$

Taking log of both sides, we get

$$y = \log p + \log (p^2 + 1) \tag{100}$$

Differentiating both sides of (100) with respect to x, we get

$$p = \frac{1}{p}\frac{dp}{dx} + \frac{2p}{1+p^2}\frac{dp}{dx}$$

or

$$p = \frac{dp}{dx}\left[\frac{1}{p} + \frac{2p}{1+p^2}\right]$$

or

$$dx = \left[\frac{1}{p^2} + \frac{2}{1+p^2}\right] dp.$$

Integrating, we get

$$x = -\frac{1}{p} + 2\tan^{-1} p + c \tag{101}$$

Therefore (100) and (101) taken together constitute the solution of the given differential equation.

EXAMPLE 4.92

Solve

$$y = x + a\tan^{-1} p.$$

Solution. The given differential equation is

$$y = x + a\tan^{-1} p. \tag{102}$$

Differentiating both sides of (102) with respect to x, we get

$$p = 1 + \frac{a}{1+p^2}\frac{dp}{dx}$$

or

$$(p-1) = \frac{a}{1+p^2}\frac{dp}{dx}$$

or

$$\frac{dx}{dp}(p-1) = \frac{a}{1+p^2}$$

or

$$\frac{dx}{dp} = \frac{a}{(p-1)(1+p^2)}$$

or

$$dx = \frac{a\,dp}{(p-1)(1+p^2)}$$

$$= a\left[\frac{1}{2(p-1)} - \frac{p}{2(1+p^2)} - \frac{1}{2(1+p^2)}\right] dp$$

$$= \frac{a}{2}\left[\frac{1}{p-1} - \frac{p}{1+p^2} - \frac{1}{1+p^2}\right] dp$$

Integrating, we get

$$x = \frac{a}{2}\left[\log\frac{p-1}{\sqrt{1+p^2}} - \tan^{-1} p\right] + c \tag{103}$$

Hence (102) and (103) taken together constitute the solution of the given differential equation.

(b) Equation Solvable for x.
Let the given differential equation be

$$f(x, y, p) = 0 \tag{104}$$

Let on solving (104) for x, we have

$$x = F(y, p). \tag{105}$$

Differentiating (105) with respect to y yields an equation of the type

$$\frac{1}{p} = G\left(y, p, \frac{dp}{dy}\right) \qquad (106)$$

The equation (106) being in two variables y and p, may possibly be solved. Let the solution be

$$\phi\,(y, p, c) = 0. \qquad (107)$$

Eliminating p from (105) and (107) yields the required solution.

In case the elimination is not possible, then (105) and (107) expressed in term of p is regarded as solution.

EXAMPLE 4.93

Solve

$$p = \tan\left(x - \frac{p}{1+p^2}\right).$$

Solution. The given differential equation yields

$$\tan^{-1} p = x - \frac{p}{1+p^2}$$

or

$$x = \frac{p}{1+p^2} + \tan^{-1} p \qquad (108)$$

Differentiating (108) with respect to y, we have

$$\frac{1}{p} = \left[\frac{1}{1+p^2} - \frac{2p^2}{(1+p^2)^2} + \frac{1}{1+p^2}\right]\frac{dp}{dy}$$

$$= \frac{2}{(1+p^2)^2}\frac{dp}{dy}$$

or

$$dy = \frac{2p}{(1+p^2)^2}\,dp$$

Integrating we get

$$y = -\frac{1}{1+p^2} + c \qquad (109)$$

Hence (108) and (109) constitute the solution of the given differential equation.

EXAMPLE 4.94

Solve

$$x - yp = ap^2$$

Solution. The given differential equation yields

$$x = yp + ap^2 \qquad (110)$$

Differentiating both sides of (110) with respect to y, we get

$$\frac{1}{p} = p + y\frac{dp}{dy} + 2ap\frac{dp}{dy}$$

$$= p + [y + 2ap]\frac{dp}{dy}$$

or

$$\frac{1}{p} - p = (y + 2ap)\frac{dp}{dy}$$

or

$$\frac{dy}{dp}\left(\frac{1-p^2}{p}\right) - y = 2ap$$

or

$$\frac{dy}{dp} + y\left(\frac{p}{p^2-1}\right) = \frac{-2ap^2}{p^2-1} \qquad (111)$$

The integrating factor for the linear equation (111) is

$$I.F. = e^{\int \frac{p}{p^2-1}dp} = e^{\frac{1}{2}\log(p^2-1)}$$

$$= e^{\log\sqrt{p^2-1}} = \sqrt{p^2-1}.$$

Hence the solution is

$$(y + ap)\sqrt{p^2 - 1} + a\cosh^{-1} p = c \qquad (112)$$

The expressions (110) and (112) taken together constitute the solution of the given equation.

EXAMPLE 4.95

Solve

$$y = x - a \log p$$

Solution. We have

$$x = y + a \log p \qquad (113)$$

Differentiating (113) with respect to y, we get

$$\frac{1}{p} = 1 + \frac{a}{p}\frac{dp}{dy}$$

or

$$\frac{1}{p} - 1 = \frac{a}{p}\frac{dp}{dy}$$

or

$$1 - p = a\frac{dp}{dy}$$

or

$$dy = \frac{a}{1-p}\,dp.$$

Integrating we get

$$y = -\,a \log(1 - p) + c \qquad (114)$$

Hence (113) and (114) constitute the required solution of the given equation.

EXAMPLE 4.96

Solve

$$y = 2px + y^2 p^3.$$

Solution. From the given equation, we have

$$x = \frac{y - y^2 p^3}{2p}. \qquad (115)$$

Differentiating (115) with respect to y, we get

$$\frac{1}{p} = \frac{1}{2}\left[\frac{\left(1 - 2yp^3 - 3y^2 p^2 \frac{dp}{dy}\right)p - (y - y^2p^3)\frac{dp}{dy}}{p^2}\right]$$

or

$$p + 2yp^4 + 2y^2 p^3 \frac{dp}{dy} + y\frac{dp}{dy} = 0$$

or

$$p(1 + 2yp^3) + y\frac{dp}{dy}(1 + 2yp^3) = 0$$

or

$$(1 + 2yp^3)\left(p + y\frac{dp}{dy}\right) = 0.$$

Therefore

$$p + y\frac{dp}{dy} = 0$$

or

$$\frac{d}{dy}(py) = 0$$

Integrating, we get

$$py = c \quad \text{or} \quad p = \frac{c}{y}.$$

Putting $p = \frac{c}{y}$ in the given equation, we get the required solution

$$y = \frac{2cx}{y} + \frac{c^3y^2}{y^3}$$

or

$$y^2 = 2cx + c^3.$$

4.17 CLAIRAUT'S EQUATION

Clairaut's equation is of the form

$$y = px + f(p). \qquad (116)$$

It is a first degree equation in x and y. Differentiating (116) with respect to x, we get

$$p = p + x\frac{dp}{dx} + f'(p)\frac{dp}{dx}$$

or

$$(x + f'(p))\frac{dp}{dx} = 0, \qquad (117)$$

where $f'(p) = \frac{df}{dx}\cdot$ Therefore either $(x + f'(p)) = 0$ or $\frac{dp}{dx} = 0$.

If we consider $\frac{dp}{dx} = 0$, then integration yields $p = c$. Putting $p = c$ in (116), the solution of Clairaut's equation is

$$y = cx + f(c). \qquad (118)$$

Thus solution of Clairaut's equation is obtained by replacing p by c in the given equation.

If we consider the factor $x + f'(p) = 0$ and eliminate p using (116), then we get a solution, called *singular solution* of the Clairaut's equation. This solution contains no constant.

EXAMPLE 4.97

Find the general and singular solution of $y = xp + \frac{a}{p}\cdot$

Solution. We have

$$y = xp + \frac{a}{p} \qquad (119)$$

It is of the form

$$y = px + f(p)$$

and so is a Clairaut's equation. Hence its general solution is obtained by replacing p by c in (119). Hence

$$y = xc + \frac{a}{c} \qquad (120)$$

is the required general solution.

Differentiating (120) with respect to c, we get

$$0 = x - \frac{a}{c^2}$$

or

$$c^2 = \frac{a}{x} \qquad (121)$$

From (120) and (121), we have

$$y^2 = x^2 c^2 + \frac{a^2}{c^2} + 2ax$$

$$= x^2\left(\frac{a}{x}\right) + \frac{xa^2}{a} + 2ax$$

$$= 4ax.$$

Hence the singular solution of the given equation is

$$y^2 = 4ax.$$

EXAMPLE 4.98

Solve

$$y = 2xp + y^2 p^3.$$

Solution. Putting $y^2 = u$, we have $2y\, dy = du$. Then

$$p = \frac{dy}{dx} = \frac{\frac{du}{2\sqrt{u}}}{dx} = \left(\frac{1}{2\sqrt{u}}\right)\left(\frac{du}{dx}\right)$$

$$= \frac{1}{2\sqrt{u}}P, \text{ where } P = \frac{du}{dx}\cdot$$

Therefore the given equation reduces to

$$\sqrt{u} = 2x\left(\frac{1}{2\sqrt{u}}\right)P + u\left(\frac{1}{8u\sqrt{u}}\right)P^3$$

or

$$u = xP + \frac{1}{8}P^3, \qquad (122)$$

which is Clairaut's equation. The general solution of (122) is

$$u = cx + \frac{1}{8}c^3.$$

Hence the solution of the given equation is

$$y^2 = cx + \frac{1}{8}c^3.$$

EXAMPLE 4.99

Solve $p = \log(px - y)$. Also obtain the singular solution.

Solution. The given equation is equivalent to

$$e^p = px - y$$

or

$$y = px - e^p, \quad (123)$$

which is Clairaut's equation. Its general solution is

$$y = cx - e^c. \quad (124)$$

From (124), we have

$$e^c = cx - y$$

and so

$$c = \log(cx - y), \quad (125)$$

which is another form of the general solution. Differentiating (125) with respect to c, we get

$$1 = \frac{x}{cx - y}$$

or

$$c = \frac{x+y}{x}.$$

Putting this value in (125), we get

$$\frac{x+y}{x} = \log\left[\frac{x+y}{x}\cdot x - y\right]$$

$$= \log x.$$

Thus

$$x + y = x\log x$$

or

$$y = x\log x - x = x(\log x - 1),$$

which is singular solution of the given equation.

EXAMPLE 4.100

Solve the differential equation

$$(py + x)(px - y) = a^2 p.$$

Solution. Putting $x^2 = u$ and $y^2 = v$, we get

$$2x\,dx = du \quad \text{and} \quad 2y\,dy = dv.$$

Therefore

$$p = \frac{dy}{dx} = \frac{\frac{dv}{(2y)}}{\frac{du}{(2x)}} = \frac{x}{y}\left(\frac{dv}{du}\right)$$

$$= \frac{x}{y}P, \text{ say.}$$

Putting this value of p in the given equation, we have

$$(Px + x)\left(\frac{x^2}{y}P - y\right) = \frac{a^2 x}{y}P$$

or

$$(P+1)(uP - v) = a^2 P$$

or

$$uP - v = \frac{a^2 P}{P+1}$$

or

$$v = uP - \frac{a^2 P}{P+1},$$

which is Clairaut's equation. Its solution is

$$v = uc - \frac{a^2 c}{c+1}.$$

Hence the general solution of the given equation is

$$y^2 = cx^2 - \frac{a^2 c}{c+1}.$$

4.18 HIGHER ORDER LINEAR DIFFERENTIAL EQUATIONS

Definition 4.17 A differential equation in which the dependent variable and its derivatives occur only in the first degree and are not multiplied together is called a *linear differential equation*.

Thus, a linear differential equation of nth order is of the form

$$a_0 \frac{d^n y}{dx^n} + a_1 \frac{d^{n-1}y}{dx^{n-1}} + a_2 \frac{d^{n-2}y}{dx^{n-2}} + ... + a_n y = F(x) \quad (126)$$

where $a_0, a_1, ..., a_n$ and $F(x)$ are functions of x alone.

If $a_0, a_1, ..., a_n$ are constants, then the above equation is called a *linear differential equation with constant* coefficients.

If F is identically zero, then the equation (126) reduces to

$$a_0 \frac{d^n y}{dx^n} + a_1 \frac{d^{n-1}y}{dx^{n-1}} + a_2 \frac{d^{n-2}y}{dx^{n-2}} + \cdots + a_n y = 0 \quad (127)$$

and is called a *homogeneous linear differential equation* of order n.

Definition 4.18 If $f_1, f_2, ..., f_n$ are n given functions and $c_1, c_2, ..., c_n$ are n constants, then the expression

$$c_1 f_1 + c_2 f_2 + \ldots + c_n f_n$$

is called a linear combination of $f_1, f_2, ..., f_n$.

Definition 4.19 The set of functions $\{f_1, f_2, ..., f_n\}$ is said to be linearly independent on $[a, b]$ if the relation

$$c_1 f_1 + c_2 f_2 + \ldots + c_n f_n = 0$$

for all $x \in [a, b]$ implies that

$$c_1 = c_2 = \ldots = c_n = 0.$$

Definition 4.20 The symbol $D = \frac{d}{dx}$ is called a differential operator. Similarly, $D^2 = \frac{d^2}{dx^2}$, $D^3 = \frac{d^3}{dx^3}$, ..., $D^n = \frac{d^n}{dx^n}$ are also regarded as operators. In terms of these symbols, the equation (126) takes the form

$$(a_0 D^n + a_1 D^{n-1} + \ldots + a_{n-1} D + a_n) y = F(x)$$

or

$$f(D)\, y = F(x),$$

where

$$f(D) = a_0 D^n + a_1 D^{n-1} + \ldots + a_{n-1} D + a_n.$$

We note that

$$D(u + v) = Du + Dv$$

$$D(\lambda u) = \lambda D(u)$$

$$(D^m D^n)(u) = D^{m+n}(u)$$

$$(D^m D^n)(u) = (D^n D^m)(u).$$

Theorem 4.4 Any linear combination of linearly independent solutions of the homogeneous linear differential equation is also a solution (in fact, complete solution) of that equation.

Proof: Let $y_1, y_2, ..., y_n$ be the solutions of the homogeneous linear differential equation

$$(D^n + a_1 D^{n-1} + a_2 D^{n-2} + \cdots + a_n) y = 0 \quad (128)$$

Therefore,

$$\left.\begin{array}{l} D^n y_1 + a_1 D^{n-1} y_1 + a_2 D^{n-2} y_1 + \cdots + a_n y_1 = 0 \\ D^n y_2 + a_1 D^{n-1} y_2 + a_2 D^{n-2} y_2 + \cdots + a_n y_2 = 0 \\ \cdots\cdots\cdots\cdots \\ D^n y_n + a_1 D^{n-1} y_n + a_2 D^{n-2} y_n + \cdots + a_n y_n = 0 \end{array}\right\} \quad (129)$$

Let

$$u = c_1 y_1 + c_2 y_2 + \ldots + c_n y_n.$$

Then

$$D^n u + a_1 D^{n-1} u + a_2 D^{n-2} u + \cdots + a_n u$$
$$= D^n (c_1 y_1 + c_2 y_2 + \cdots + c_n y_n)$$
$$+ a_1 D^{n-1} (c_1 y_1 + c_2 y_2 + \cdots + c_n y_n)$$
$$+ a_2 D^{n-2} (c_1 y_1 + c_2 y_2 + \cdots + c_n y_n)$$
$$+ \cdots + a_n (c_1 y_1 + c_2 y_2 + \cdots + c_n y_n)$$
$$= c_1 (D^n y_1 + a_1 D^{n-1} y_1 + \cdots + a_n y_1)$$
$$+ c_2 (D^n y_2 + a_1 D^{n-1} y_2 + \cdots + a_n y_2)$$
$$+ \cdots + c_n (D^n y_n + a_1 D^{n-1} y_n + \cdots + a_n y_n)$$
$$= 0 + 0 + \cdots + 0 = 0 \text{ using (129).}$$

Hence, $u = c_1 y_1 + c_2 y_2 + \ldots + c_n y_n$ is also a solution of the homogeneous linear differential equation (128). Since this solution contains n arbitrary constants, it is a general or a complete solution of (128).

EXAMPLE 4.101

Show that $c_1 \sin x + c_2 \cos x$ is a solution of $\frac{d^2y}{dx^2} + y = 0$.

Solution. Let

$$y_1 = \sin x, \quad y_2 = \cos x.$$

Then

$$\frac{dy_1}{dx} = \cos x, \quad \frac{dy_2}{dx} = -\sin x$$

$$\frac{d^2y_1}{dx^2} = -\sin x, \qquad \frac{d^2y_2}{dx^2} = -\cos x.$$

We note that

$$\frac{d^2y_1}{dx^2} + y_1 = -\sin x + \sin x = 0$$

and

$$\frac{d^2y_2}{dx^2} + y_2 = -\cos x + \cos x = 0.$$

Hence, $\sin x$ and $\cos x$ are solutions of the given equation. These two solutions are linearly independent. Therefore, their linear combination $c_1 \sin x + c_2 \cos x$ is also a solution of the given equation.

Theorem 4.5 If y_1 is a complete solution of the homogeneous equation $f(D)y = 0$ and y_2 is a particular solution containing no arbitrary constants of the differential equation $f(D)y = F(x)$, then $y_1 + y_2$ is the complete solution of the equation $f(D)y = F(x)$.

Proof: Since y_1 is a complete solution of the homogeneous differential equation $f(D)y = 0$, we have

$$f(D)\, y_1 = 0 \tag{130}$$

Further, since y_2 is a particular solution of linear differential equation $f(D)y = F(x)$, we have

$$f(D)\, y_2 = F(x). \tag{131}$$

Adding (130) and (131), we get

$$f(D)\, y_1 + f(D)\, y_2 = F(x)$$

or

$$f(D)\,(y_1 + y_2) = F(x).$$

Hence, $y_1 + y_2$ satisfies the equation $f(D)y = F(x)$ and so is the complete solution since it contains n arbitrary constants.

Definition 4.21 Let $f(\mathrm{D})y = F(x)$ be a linear differential equation with constant coefficients. If y_1 is a complete solution of $f(\mathrm{D})y = 0$ and y_2 is a particular solution of $f(\mathrm{D})y = F(x)$, then $y_1 + y_2$ is a complete solution of $f(\mathrm{D})y = F(x)$ and then y_1 is called the *complementary function* and y_2 is called the *particular integral* of the differential equation $f(D)\, y = F(x)$.

Consider the homogeneous differential equation $f(D)y = 0$. Then

$$(D^n + a_1 D^{n-1} + a_2^{n-2} + \ldots + a_n)\, y = 0. \tag{132}$$

Let $y = e^{mx}$ be a solution of (132). Then

$$Dy = me^{mx},\ D^2y = m^2 e^{mx},\ \ldots,\ D^n y = m^n e^{mx}$$

and so (132) transforms to

$$(m^n + a_1 m^{n-1} + \cdots + a^n)\, e^{mx} = 0.$$

Since $e^{mx} \neq 0$, we have

$$m^n + a_1 m^{n-1} + a_2 m^{n-2} + \cdots + a_n = 0. \tag{133}$$

It follows, therefore, that if e^{mx} is a solution of $f(D)y = 0$, then equation (133) is satisfied.

The equation (133) is called *auxiliary equation* for the differential equation $f(D)y = 0$.

4.19 SOLUTION OF HOMOGENEOUS LINEAR DIFFERENTIAL EQUATION WITH CONSTANT COEFFICIENTS

Consider the homogeneous linear differential equation

$$\frac{d^n y}{dx^n} + a_1 \frac{d^{n-1} y}{dx^{n-1}} + \ldots + a_{n-1}\frac{dy}{dx} + a_n y = 0.$$

The symbolic form of this equation is

$$(D^n + a_1 D^{n-1} + \ldots + a_{n-1} D + a_n)\, y = 0, \tag{134}$$

where $a_1, a_2, \ldots, a_n$ are constants. If $y = e^{mx}$ is a solution of (134), then

$$m^n + a_1 m^{n-1} + \ldots + a_{n-1} m + a_n = 0. \quad (135)$$

Three cases arise, according as the roots of (135) are *real and distinct, real and repeated or complex.*

Case I. Distinct Real Roots

Suppose that the auxiliary equation (135) has n distinct roots $m_1, m_2, \ldots, m_n$. Therefore, (135) reduces to

$$(m - m_1)(m - m_2)\ldots(m - m_n) = 0 \quad (136)$$

Equation (136) will be satisfied by the solutions of the equations

$$(D - m_1)y = 0,\ (D - m_2)y = 0,\ \ldots,$$
$$(D - m_n)y = 0.$$

We consider $(D - m_1)y = 0$. This can be written as

$$\frac{dy}{dx} - m_1 y = 0,$$

which is linear differential equation with integrating factor as $e^{-m_1 x}$. Therefore, its solution is

$$y.e^{-m_1 x} = \int 0.e^{-m_1 x} dx + c_1$$

or

$$y = c_1\, e^{m_1 x}.$$

Similarly,

the solution of $(D - m_2)\, y = 0$ is $c_2\, e^{m_2 x}$,

the solution of $(D - m_3)\, y = 0$ is $c_3\, e^{m_3 x}$,

. . .

. . .

the solution of $(D - m_n)\, y = 0$ is $c_n e^{m_n x}$.

Hence, the complete solution of homogeneous differential equation (134) is

$$y = c_1 e^{m_1 x} + c_2 e^{m_2 x} + \ldots + c_n e^{m_n x}. \quad (137)$$

Case II. Repeated Real Roots

Suppose that the roots m_1 and m_2 of the auxiliary equation are equal. Then the solution (137) becomes

$$y = c_1 e^{m_1 x} + c_2\, e^{m_2 x} + c_3\, e^{m_3 x} + \ldots + c_n\, e^{m_n x}$$
$$= (c_1 + c_2)\, e^{m_1 x} + c_3\, e^{m_2 x} + \ldots + c_n\, e^{m_n x}.$$

Since it contains $n - 1$ arbitrary constants, it is not a complete solution of the given differential equation. We shall show that the part of the solution corresponding to equal roots m_1 and m_2 is $(c_1x + c_2)\, e^{m_1x}$.

To prove it, consider the equation

$$(D - m_1)^2 y = 0,$$

that is,

$$(D - m_1)(D - m_1)\, y = 0.$$

Substituting $(D - m_1)y = U$, the above equation becomes

$$(D - m_1)\, U = 0$$

or

$$\frac{dU}{dx} - m_1 U = 0$$

or

$$\frac{dU}{U} = m_1 dx.$$

Integrating, we get

$$\log U = m_1 x + \log c_1$$

or

$$\log \frac{U}{c_1} = m_1 x \quad \text{or} \quad \frac{U}{c_1} = e^{m_1x}$$

and so

$$U = c_1\, e^{m_1x}.$$

Hence

$$(D - m_1)y = c_1 e^{m_1x}$$

or

$$\frac{dy}{dx} - m_1 y = c_1\, e^{m_1x},$$

which is again a linear equation with integrating factor e^{-m_1x}. Hence, the solution is

$$y.e^{-m_1x} = \int c_1 e^{m_1x}.e^{-m_1x} + c^2$$

$$= c_1 x + c_2,$$

which yields

$$y = (c_1x + c_2)e^{m_1x}.$$

Hence, the complete solution of the given differential equation is

$$y = (c_1x + c_2)\, e^{m_1x} + c_3 e^{m_3x} + \ldots + c_n e^{m_nx}.$$

Remark 4.1 If three roots of the auxiliary equation are equal, that is, $m_1 = m_2 = m_3$, then the complete solution will come out to be

$$y = (c_1x^2 + c_2x + c_3)\, e^{m_1x} + c_4 e^{m_4x} + \ldots + c_n e^{m_nx}.$$

In general, if $m_1 = m_2 = \ldots = m_k$, then the complete solution of the differential equation shall be

$$y = (c_1x^{k-1} + c_2x^{k-2} + \ldots + c_k)e^{m_kx} + c_{k+1}e^{m_{k+1}x} + \cdots + c_n\, e^{m_nx}.$$

Case III. Conjugate Complex Roots

(a) Suppose that the auxiliary equation has a non–repeated complex root $\alpha + i\beta$. Then, since the coefficients are real, the conjugate complex number $\alpha - i\beta$ is also a non–repeated root. Thus, the solution given in (137) becomes

$$y = c_1 e^{(\alpha+i\beta)x} + c_2 e^{(\alpha-i\beta)x} + c_3 e^{m_3x} + \ldots + c_n e^{m_nx}$$

$$= e^{\alpha x}(c_1 e^{i\beta x} + c_2 e^{-i\beta x}) + c_3 e^{m_3x} + \ldots + c_n e^{m_nx}$$

$$= e^{\alpha x}[c_1(\cos\beta x + i\sin\beta x) + c_2(\cos\beta x - i\sin\beta x)] + c_3 e^{m_3x} + \ldots + c_n e^{m_nx}$$

$$= e^{\alpha x}[(c_1 + c_2)\cos\beta x + i(c_1 - c_2)\sin\beta x] + c_3 e^{m_3x} + \ldots + c_n e^{m_nx}$$

$$= e^{\alpha x}[k_1\cos\beta x + k_2 \sin\beta x] + c_3 e^{m_3x} + \ldots + c_n e^{m_nx},$$

where $k_1 = c_1 + c_2$, $k_2 = i(c_1 - c_2)$.

(b) If two pairs of imaginary roots are equal, then

$$m_1 = m_2 = \alpha + i\beta \quad \text{and} \quad m_3 = m_4 = \alpha - i\beta.$$

Using Case II, the complete solution is

$$= e^{ax}[(c_1x + c_2)\cos\beta x + (c_3x + c_4)\sin\beta x] + c_5 e^{m_5x} + \ldots + c_n e^{m_nx}.$$

EXAMPLE 4.102

Solve

$$\frac{d^3y}{dx^3} + \frac{d^2y}{dx^2} + 4\frac{dy}{dx} + 4y = 0.$$

Solution. The symbolic form of the given equation is

$$(D^3 + D^2 + 4D + 4)\,y = 0.$$

Therefore its auxiliary equation is

$$m^3 + m^2 + 4m + 4 = 0.$$

By inspection −1 is a root. Therefore, $(m + 1)$ is a factor of $m^3 + m^2 + 4m + 4$. The synthetic division by $m + 1$ gives

−1	1	1	4	4
		−1	0	−4
	1	0	4	0

Therefore, the auxiliary equation is

$$(m + 1)\,(m^2 + 4) = 0$$

and so

$$m = -1 \text{ and } m \pm 2i.$$

Hence, the complementary solution is

$$y = c_1e^{-x} + e^{0x}(c_2\cos 2x + c_3\sin 2x)$$

$$= c_1e^{-x} + c_2\cos 2x + c_3\sin 2x.$$

EXAMPLE 4.103

Solve

$$\frac{d^3y}{dx^3} - 3\frac{d^2y}{dx^2} + 3\frac{dy}{dx} - y = 0.$$

Solution The symbolic form of the given equation is

$$(D^3 - 3D^2 + 3D - 1)\,y = 0.$$

Therefore, the auxiliary equation is

$$m^3 - 3m^2 + 3m - 1 = 0.$$

By inspection 1 is a root. Then synthetic division yields

1	1	−3	3	−1
		1	−2	1
	1	−2	1	0

Therefore, the auxiliary equation is

$$(m - 1)\,(m^2 - 2m + 1) = 0 \text{ or } (m - 1)^3 = 0.$$

Hence the roots are 1, 1, 1 and so the solution of the given equation is

$$y = (c_1 + c_2x + c_3x^2)e^x.$$

EXAMPLE 4.104

Find the general solution of

$$\frac{d^4y}{dx^4} - 5\frac{d^3y}{dx^3} + 6\frac{d^2y}{dx^2} + 4\frac{dy}{dx} - 8y = 0.$$

Solution The auxiliary equation is

$$m^4 - 5m^3 + 6m^2 + 4m - 8 = 0,$$

whose roots are 2, 2, 2, and − 1. Hence the general solution is

$$y = (c_1 + c_2x + c_3x^2)\,e^{2x} + c_4e^{-x}.$$

EXAMPLE 4.105

Solve $\frac{d^2y}{dx^2} - 2\frac{dy}{dx} + 10y = 0$ subject to the conditions $y(0) = 4$, $y'(0) = 1$.

Solution. The symbolic form of the given differential equation is

$$(D^2 - 2D + 10)\,y = 0.$$

Therefore, the auxiliary equation is

$$m^2 - 2m + 10 = 0,$$

which yields

$$m = 1 \pm 3i.$$

Therefore, the solution is

$$y = e^x(c_1 \cos 3x + c_2 \sin 3x).$$

Now,

$$y' = c_1[e^x \cos 3x - 3e^x \sin 3x]$$

$$+ c_2[e^x \sin 3x + 3e^x \cos 3x]$$

$$= e^x \cos 3x(c_1 + 3c_2) + e^x \sin x(c_2 - 3c_1).$$

The initial conditions $y(0) = 4$ and $y'(0) = 1$ yield

$$4 = c_1 \quad \text{and} \quad 1 = c_1 + 3c_2$$

and so $c_1 = 4$ and $c_2 = -1$. Hence the solution is

$$y = e^x(4 \cos 3x - \sin 3x).$$

EXAMPLE 4.106

Solve

$$\frac{d^3y}{dx^3} + y = 0.$$

Solution. The auxiliary equation is

$$m^3 + 1 = 0$$

or

$$(m + 1)\,(m^2 - m + 1) = 0.$$

Thus the roots are -1, $\frac{1 \pm \sqrt{3}i}{2}$. Hence, the general solution of the equation is

$$y = c_1 e^{-x} + e^{\frac{1}{2}x}\left(c_2 \cos \frac{\sqrt{3}}{2}x + c_3 \sin \frac{\sqrt{3}}{2}x\right).$$

EXAMPLE 4.107

Solve

$$\frac{d^4y}{dx^4} = m^4\, y.$$

Solution. The auxiliary equation for the given differential equation is

$$s^4 - m^4 = 0$$

or

$$(s + m)\,(s - m)\,(s^2 + m^2) = 0$$

and so $s = m, -m, \pm mi$. Hence the solution is

$$y = c_1\, e^{mx} + c_2\, e^{-mx} + c_3 \cos mx + c_4 \sin mx$$

$$= c_1[\cosh mx + \sinh mx] + c_2[\cosh mx - \sinh mx\,]$$

$$+ c_3 \cos mx + c_4 \sin mx$$

$$= (c_1 + c_2) \cosh mx + (c_1 - c_2) \sinh mx$$

$$+ c_3 \cos mx + c_4 \sin mx$$

$$= C_1 \cosh mx + C_2 \sinh mx + c_3 \cos mx$$

$$+ c_4 \sin mx,$$

where

$$C_1 = c_1 + c_2 \text{ and } C_2 = c_1 - c_2.$$

EXAMPLE 4.108

Solve $\quad 4\frac{d^3y}{dx^3} + 4\frac{d^2y}{dx^2} + \frac{dy}{dx} = 0.$

Solution. The auxiliary equation for the given differential equation is

$$4m^3 + 4m^2 + m = 0$$

or

$$m(4m^2 + 4m + 1) = 0.$$

Thus the roots are $m = 0$, $\frac{-1}{2}$, and $\frac{-1}{2}$. Hence the solution is

$$y = c_1 e^{0x} + (c_2 + c_3 x)\, e^{-\frac{x}{2}}$$
$$= c_1 + (c_2 + c_3 x)\, e^{-\frac{x}{2}}.$$

4.20 COMPLETE SOLUTION OF LINEAR DIFFERENTIAL EQUATION WITH CONSTANT COEFFICIENTS

We now discuss the methods of finding particular integral of a linear differential equation with constant coefficients so that complete solution of the equation may be found.

Definition 4.22 $\frac{1}{D} F(x)$ is that function of x which when operated upon by D yields $F(x)$.

Similarly, $\frac{1}{f(D)} F(x)$ is that function of x, free fromarbitraryconstant,whichwhenoperateduponby $f(D)$ yields $F(x)$.

Thus, $\frac{1}{D}$ is the inverse operator of D and $\frac{1}{f(D)}$ is the inverse operator of $f(D)$.

Theorem 4.6 $\frac{1}{D} F(x) = \int F(x)\, dx.$

Proof: Let

$$\frac{1}{D} F(x) = v. \qquad (138)$$

Operating both sides of (141) by D, we get

$$D \cdot \frac{1}{D} F(x) = Dv$$

or

$$F(x) = Dv = \frac{dv}{dx}$$

or

$$dv = F(x)\, dx.$$

Integrating, we get

$$v = \int F(x)\, dx,$$

where no constant of integration is added since $\frac{1}{D} F(x)$ contains no constant. Thus,

$$\frac{1}{D} F(x) = \int F(x)\, dx.$$

Hence, $\frac{1}{D}$ stands for integration.

Theorem 4.7 $\frac{1}{D-a} F(x) = e^{ax} \int F(x)\, e^{-ax}\, dx.$

Proof: Let

$$\frac{1}{D-a} F(x) = y. \qquad (139)$$

Operating both sides of (139) by D – a, we have

$$F(x) = (D - a)\, y = Dy - ay$$

$$= \frac{dy}{dx} - ay.$$

Therefore,

$$\frac{dy}{dx} - ay = F(x),$$

which is a linear differential equation with integrating factor e^{-ax}. Therefore, its solution is

$$y \,.\, e^{-ax} = \int F(x)\, e^{-ax}\, dx$$

or

$$y = e^{ax} \int F(x)\, e^{-ax} dx$$

or

$$\frac{1}{D-a} F(x) = e^{ax} \int F(x)\, e^{-ax} dx.$$

Theorem 4.8 $\frac{1}{f(D)} F(x)$ is the particular integral of $f(D)y = F(x)$.

Proof: The given linear differential equation is

$$f(D)\, y = F(x) \qquad (140)$$

Substituting $y = \frac{1}{f(D)} F(x)$ in (140), we have $F(x) = F(x)$, which is true. Hence $y = \frac{1}{f(D)} F(x)$ is a solution of (140).

4.20.1 Standard Cases of Particular Integrals

Consider the linear differential equation

$$f(D)y = F(x).$$

By Theorem 4.5, its particular integral is

$$\text{P.I.} = \frac{1}{f(D)} F(x).$$

Case I. When $F(x) = e^{ax}$

We have

$$f(D) = D^n + a_1 D^{n-1} + a_2 D^{n-2} + \ldots + a_n.$$

Therefore,

$$f(D)\, e^{ax} = (D^n + a_1 D^{n-1} \cdots + a_n) e^{ax}$$

$$= D^n e^{ax} + a_1 D^{n-1} e^{ax} + \cdots + a_{n-1} D e^{ax} + a_n e^{ax}$$

$$= a^n e^{ax} + a_1 a^{n-1} e^{ax} + \cdots + a_{n-1} a e^{ax} + a_n e^{ax}$$

$$= (a^n + a_1 a^{n-1} + \cdots + a_{n-1} a + a_n) e^{ax}$$

$$= f(a) e^{ax}.$$

Operating both sides by $\frac{1}{f(D)}$ yields

$$e^{ax} = \frac{1}{f(D)} [f(a)\, e^{ax}]$$

$$= f(a) \frac{1}{f(D)} e^{ax}.$$

Hence

$$\frac{1}{f(D)} e^{ax} = \frac{1}{f(a)} e^{ax}, \text{ provided } f(a) \neq 0. \quad (141)$$

If $f(a) = 0$, then $D - a$ is a factor of $f(D)$. So, let

$$f(D) = (D - a)\, \phi(D). \quad (142)$$

Then

$$\frac{1}{f(D)} e^{ax} = \frac{1}{(D-a)\phi(D)} e^{ax} = \frac{1}{(D-a)} \left[\frac{1}{\phi(D)} e^{ax}\right]$$

$$= \frac{1}{D-a} \left[\frac{1}{\phi(a)} e^{ax}\right] \text{ using (141) since } \phi(a) \neq 0$$

$$= \frac{1}{\phi(a)} \left[\frac{1}{D-a} e^{ax}\right]$$

$$= \frac{1}{\phi(a)} e^{ax} \int e^{ax} \cdot e^{-ax} dx, \text{ by Theorem 4.7}$$

$$= x \frac{e^{ax}}{\phi(a)} \quad (143)$$

Differentiating (142) with respect to D gives

$$f'(D) = \phi(D) + (D - a)\, \phi'(D).$$

Putting $D = a$, we get $f'(a) = \phi(a)$. Therefore, (143) reduces to

$$\frac{1}{f(D)} e^{ax} = x \frac{e^{ax}}{\phi(a)} = x \frac{e^{ax}}{\left[\frac{d}{dD} f(D)\right]_{D=a}}$$

$$= x \frac{e^{ax}}{f'(a)}, \text{ provided } f'(a) \neq 0. \quad (144)$$

If $f'(a) = 0$, then the rule can be repeated to give

$$\frac{1}{f(D)} e^{ax} = x^2 \frac{e^{ax}}{f''(a)} e^{ax},$$

provided $f''(a) \neq 0$ and so on.

Case II. When F(x) = sin (ax + b) or cos (ax + b)

We have

$$D \sin(ax + b) = a \cos(ax + b)$$
$$D^2 \sin(ax + b) = -a^2 \sin(ax + b)$$
$$D^3 \sin(ax + b) = -a^3 \cos(ax + b)$$
$$D^4 \sin(ax + b) = a^4 \sin(ax + b)$$
$$\cdots$$
$$\cdots$$

We note in general that

$$(D^2)^n \sin(ax + b) = (-a^2)^n \sin(ax + b).$$

Hence

$$f(D^2) \sin(ax + b) = f(-a^2) \sin(ax + b).$$

Operating on both sides by $\frac{1}{f(D^2)}$, we get

$$\sin(ax+b) = f(-a^2)\frac{1}{f(D^2)}\sin(ax+b).$$

Dividing both sides by $f(-a^2)$, we have

$$\frac{1}{f(-a^2)}\sin(ax+b) = \frac{1}{f(D^2)}\sin(ax+b).$$

Hence

$$\frac{1}{f(D^2)}\sin(ax+b) = \frac{1}{f(-a^2)}\sin(ax+b), \quad (145)$$

provided $f(-a^2) \neq 0$.
Similarly,

$$\frac{1}{f(D^2)}\cos(ax+b) = \frac{1}{f(-a^2)}\cos(ax+b), \quad (146)$$

provided $f(-a^2) \neq 0$.

If $f(-a^2) = 0$, then (145) and (146) are not valid. In such a situation, we proceed as follows:

By Euler's formula

$$e^{i(ax+b)} = \cos(ax+b) + i\sin(ax+b).$$

Thus

$$\frac{1}{f(D^2)}e^{i(ax+b)} = \frac{1}{f(D^2)}[\cos(ax+b) + i\sin(ax+b)]$$

or, by (144),

$$x\cdot\frac{1}{f'(D^2)}e^{i(ax+b)} = \frac{1}{f(D^2)}[\cos(ax+b) + i\sin(ax+b)]$$

or

$$x\cdot\frac{1}{f'(D^2)}[\cos(ax+b) + i\sin(ax+b)] = \frac{1}{f(D^2)}[\cos(ax+b) + i\sin(ax+b)]$$

Equating real and imaginary parts, we have

$$\frac{1}{f(D^2)}\cos(ax+b) = x\cdot\frac{1}{[f'(D^2)]_{D^2=-a^2}}\cos(ax+b) \quad (147)$$

provided that $f'(-a^2) \neq 0$, and

$$\frac{1}{f(D^2)}\sin(ax+b) = x\frac{1}{[f'(D^2)]_{D^2=-a^2}}\sin(ax+b), \quad (148)$$

provided $f'(-a^2) \neq 0$.

If $f'(-a^2) = 0$, then repeating the above process, we have

$$\frac{1}{f(D^2)}\cos(ax+b) = x^2\frac{1}{[f''(D^2)]_{D^2=-a^2}}\cos(ax+b)$$

provided $f''(-a^2) \neq 0$

and

$$\frac{1}{f(D^2)}\sin(ax+b) = x^2\frac{1}{[f''(D^2)]_{D^2=-a^2}}\sin(ax+b)$$

provided $f''(-a^2) \neq 0$.

Case III. When $F(x) = x^n$, n being positive integer

Since in this case

$$\text{P.I.} = \frac{1}{f(D)}F(x) = \frac{1}{f(D)}x^n,$$

we make the coefficient of the leading term of $f(D)$ unity, take the denominator in numerator and then expand by Binomial theorem. Operate the resulting expansion on x^n.

Case IV. When $F(x) = e^{ax}Q(x)$, where $Q(x)$ is some function of x

If G is a function of x, we have

$$D[e^{ax}G] = e^{ax}DG + ae^{ax}G = e^{ax}(D+a)G$$

$$D^2[e^{ax}G] = e^{ax}(D+a)^2G$$

$$\dots$$

$$\dots$$

$$D^n[e^{ax}G] = e^{ax}(D+a)^nG$$

Hence

$$f(D)\,[e^{ax}G] = e^{ax} f(D+a)G$$

Operating both sides by $\frac{1}{f(D)}$, we get

$$e^{ax}G = \frac{1}{f(D)}\,[e^{ax}f(D+a)]\,G$$

Putting $f(D+a)G = Q$, we have $G = \frac{Q}{f(D+a)}$ and so we have

$$e^{ax}\frac{1}{f(D+a)}\,Q = \frac{1}{f(D)}\,(e^{ax}Q)$$

or

$$\frac{1}{f(D)}\,(e^{ax}Q(x) = e^{ax}\frac{1}{f(D+a)}\,Q(x) \quad (149)$$

Case V. When F (x) = x Q (x)

Resolving $f(D)$ into linear factors, we have

$$f(D) = (D-m_1)\,(D-m_2)\,.\,.\,.\,(D-m_n).$$

Then, using partial fractions and Theorem 4.4, we have

$$\text{P.I.} = \frac{1}{f(D)}\,F(x)$$

$$= \frac{1}{(D-m_1)\,(D-m_2)\ldots(D-m_n)}\,F(x)$$

$$= \left[\frac{A_1}{D-m_1} + \frac{A_2}{D-m_2} + \cdots + \frac{A_n}{D-m_n}\right]F(x)$$

$$= \mathrm{A}_1 e^{m_1x}\int F(x)e^{-m_1x} + \mathrm{A}_2 e^{m_2x}\int F(x)e^{-m_2x}$$

$$+\cdots+ \mathrm{A}_n e^{m_nx}\int F(x)\,e^{-m_nx}.$$

EXAMPLE 4.109

Solve $4\frac{d^2y}{dx^2} + 4\frac{dy}{dx} - 3y = e^{2x}$.

Solution. The symbolic form of the given differential equation is

$$(4D^2 + 4D - 3)\,y = e^{2x}$$

and so the auxiliary equation is

$$4m^2 + 4m - 3 = 0.$$

The roots of A.E. are $m = \frac{1}{2}, -\frac{3}{2}$. Therefore, the complementary function is

$$\text{C.F.} = c_1 e^{\frac{x}{2}} + c_2 e^{-\frac{3x}{2}}.$$

Since 2 is not a root of the auxiliary equation, by (141), we have

$$\text{P.I.} = \frac{1}{f(2)}e^{2x} = \frac{1}{4\,(4) + 4\,(2) - 3}\,e^{2x} = \frac{e^{2x}}{21}.$$

Hence the complete solution of the given equation is

$$y = \text{C.F.} + \text{P.I.} = c_1 e^{\frac{x}{2}} + c_2\,e^{-\frac{3x}{2}} + \frac{e^{2x}}{21}.$$

EXAMPLE 4.110

Solve $\frac{d^2y}{dx^2} - 5\frac{dy}{dx} + 6y = e^{3x}$.

Solution. The symbolic form of the equation is

$$(D^2 - 5\mathrm{D} + 6)\,y = e^{3x}.$$

The auxiliary equation is

$$m^2 - 5m + 6 = 0$$

or

$$(m-3)\,(m-2) - 0.$$

Therefore $m = 2, 3$. Then

$$\text{C.F.} = c_1 e^{3x} + c_2\,e^{2x}.$$

Since 3 is a root of auxiliary equation, we use (143) and get

$$\text{P.I.} = x\,\frac{e^{3x}}{\phi(3)} = x\frac{e^{3x}}{(3-2)} = xe^{3x}.$$

Hence, the complete solution of the given equation is

$$y = \text{C.F.} + \text{P.I.} = c_1 e^{3x} + c_2 e^{2x} + x\,e^{3x}.$$

Remark 4.2 (a) In the above example, if we use (144), then

$$\text{P.I.} = x\frac{e^{ax}}{\left(\frac{d}{dD}f(D)\right)_{D=a}}$$

$$= x\frac{e^{3x}}{[2D-5]_{D=3}} = x\frac{e^{3x}}{6-5} = xe^{3x}.$$

(b) We can also find the particular integral in the above case by using Theorem 4.4. In fact, we have

$$\text{P.I.} = \frac{1}{f(D)}F(x) = \frac{1}{(D-3)(D-2)}e^{3x}$$

$$= \frac{1}{D-3}e^{2x}\int F(x)\,e^{-2x}\,dx$$

$$= \frac{1}{D-3}e^{2x}\int e^{3x}e^{-2x}\,dx = \frac{1}{D-3}e^{3x}$$

$$= e^{3x}\int e^{3x}\cdot e^{-3x}\,dx = e^{3x}\int e^{0}dx = x\,e^{3x}.$$

EXAMPLE 4.111

Solve

$$\frac{d^2y}{dx^2} - 3\frac{dy}{dx} + 2y = \cosh x.$$

Solution. The auxiliary equation is

$$m^2 - 3m + 2 = 0,$$

which yields $m = 1, 2$. Therefore,

$$\text{C.F.} = c_1e^x + c_2e^{2x}.$$

Now

$$\cosh x = \frac{e^x + e^{-x}}{2}.$$

Therefore,

$$\text{P.I.} = \frac{1}{f(D)}F(x) = \frac{1}{(D-1)(D-2)}\left[\frac{e^x+e^{-x}}{2}\right]$$

$$= \frac{1}{2}\cdot\frac{1}{D^2-3D+2}e^x + \frac{1}{2}\cdot\frac{1}{D^2-3D+2}e^{-x}$$

$$= \frac{1}{2}\frac{1}{D^2-3D+2}e^x + \frac{1}{2}\frac{1}{1+3+2}e^{-x}$$

$$= \frac{1}{2}x\frac{1}{\frac{d}{dD}[D^2-3D+2]_{D=1}}e^x + \frac{1}{12}e^{-x}$$

since $f(1) = 0$

$$= \frac{x}{2}\frac{1}{[2D-3]_{D=1}}e^x + \frac{1}{12}e^{-x}$$

$$= -\frac{x}{2}e^x + \frac{1}{12}e^{-x}.$$

Hence, the complete solution of the equation is

$$y = \text{C.F.} + \text{P.I.} = c_1e^x + c_2e^{2x} - \frac{x}{2}e^x + \frac{1}{12}e^{-x}.$$

EXAMPLE 4.112

Solve $\frac{d^2y}{dx^2} - 4y = e^x + \sin 2x.$

Solution. The auxiliary equation for the given differential equation is

$$m^2 - 4 = 0,$$

which yields $m = \pm 2$. Hence

$$\text{C.F.} = c_1e^{2x} + c_2e^{-2x}.$$

Now, using (141) and (145), we have

$$\text{P.I.} = \frac{1}{D^2-4}(e^x + \sin 2x) = \frac{1}{D^2-4}e^x + \frac{1}{D^2-4}\sin 2x$$

$$= \frac{e^x}{1^2-4} + \frac{1}{-4-4}\sin 2x = -\frac{1}{3}e^x - \frac{1}{8}\sin 2x.$$

Hence, the complete solution of the given differential equation is

$$y = \text{C.F.} + \text{P.I.} = c_1e^{2x} + c_2e^{-2x} - \frac{1}{3}e^x - \frac{1}{8}\sin 2x$$

EXAMPLE 4.113

Solve $\frac{d^3y}{dx^3} + \frac{d^2y}{dx^2} - \frac{dy}{dx} - y = \cos 2x.$

Solution. The auxiliary equation is

$$m^3 + m^2 - m - 1 = 0,$$

whose roots are 1, – 1, – 1. Therefore,

$$\text{C.F.} = c_1 e^x + (c_2 + c_3 x)\, e^{-x}$$

Further,

$$\text{P.I.} = \frac{1}{f(D)} F(x) = \frac{1}{D^3+D^2-D-1} \cos 2x$$

$$= \frac{1}{DD^2-D^2-D-1} \cos 2x$$

$$= \frac{1}{D(-4)+(-4)-D-1} \cos 2x$$

$$= \frac{1}{-5D-5} \cos 2x = -\frac{1}{5(D+1)} \cos 2x$$

$$= -\frac{(D-1)}{5(D^2-1)} \cos 2x = -\frac{1}{5}(D-1)\frac{1}{-4-1} \cos 2x$$

$$= \frac{1}{25}(D-1)\cos 2x = \frac{1}{25}(D\cos 2x - \cos 2x)$$

$$= \frac{1}{25}(-2\sin 2x - \cos 2x) = -\frac{1}{25}(2\sin 2x + \cos 2x).$$

Hence the complete solution is

$$y = c_1 e^x + (c_2 + c_2 x)\, e^{-x} - \frac{1}{25}(2\sin 2x + \cos 2x).$$

EXAMPLE 4.114

Solve $\dfrac{d^2y}{dx^2} - 4y = x \sinh x$.

Solution. The auxiliary equation is $m^2 - 4 = 0$ and $m = \pm 2$. Therefore,

$$\text{C.F.} = c_1 e^{2x} + c_2 e^{-2x}.$$

Further,

$$\text{P.I.} = \frac{1}{D^2-4} x \sinh x = \frac{1}{D^2-4} x\left(\frac{e^x - e^{-x}}{2}\right)$$

$$= \frac{1}{2}\left[\frac{1}{D^2-4} x e^x - \frac{1}{D^2-4} x e^{-x}\right]$$

$$= \frac{1}{2}\left[e^x \frac{1}{(D+1)^2-4} x - e^{-x}\frac{1}{(D-1)^2-4} x\right]$$

$$= \frac{1}{2}\left[e^x \frac{1}{D^2+2D-3} x - e^{-x}\frac{1}{D^2-2D-3} x\right]$$

$$= \frac{1}{2}\left[e^x \frac{1}{-3\left(1-\frac{2D}{3}-\frac{D^2}{3}\right)} x - e^{-x}\frac{1}{-3\left(1+\frac{2D}{3}-\frac{D^2}{3}\right)} x\right]$$

$$= -\frac{1}{6}\left[e^x\left\{1-\left(\frac{2D}{3}+\frac{D^2}{3}\right)\right\}^{-1} x\right.$$

$$\left. - e^{-x}\left\{1+\left(\frac{2D}{3}-\frac{D^2}{3}\right)\right\}^{-1} x\right]$$

$$= -\frac{1}{6}\left[e^x\left(1+\frac{2D}{3}+\ldots\right)x - e^{-x}\left(1-\frac{2D}{3}+\ldots\right)x\right]$$

$$= -\frac{1}{6}\left[e^x\left(x+\frac{2}{3}\right) - e^{-x}\left(x-\frac{2}{3}\right)\right]$$

$$= -\frac{x}{3}\left(\frac{e^x - e^{-x}}{2}\right) - \frac{2}{9}\left(\frac{e^x + e^{-x}}{2}\right)$$

$$= -\frac{x}{3}\sinh x - \frac{2}{9}\cosh x.$$

Hence the solution is

$$y = \text{C.F.} + \text{P.I.}$$

$$= c_1 e^{2x} + c_2 e^{-2x} - \frac{x}{3}\sinh x - \frac{2}{9}\cosh x.$$

EXAMPLE 4.115

Solve $\dfrac{d^2y}{dx^2} - 4y = x^2$.

Solution. The auxiliary equation is $m^4 - 4 = 0$ and so $m = \pm 2$. Therefore,

$$\text{C.F.} = c_1 e^{2x} + c_2 e^{-2x}.$$

Further,

$$\text{P.I.} = \frac{1}{f(D)} F(x) = \frac{1}{D^2-4} x^2$$

$$= -\frac{1}{4-D^2} x^2 = -\frac{1}{4\left(1-\frac{D^2}{4}\right)} x^2$$

$$= -\frac{1}{4}\left(1-\frac{D^2}{4}\right)^{-1} x^2$$

$$= -\frac{1}{4}\left[1+\frac{D^2}{4}+\ldots\right] x^2$$

$$= -\frac{1}{4} x^2 - \frac{1}{16} D^2(x^2)$$

$$= -\frac{1}{4} x^2 - \frac{1}{16}(2) = -\frac{1}{4} x^2 - \frac{1}{8}.$$

Hence the complete solution is

$$y = \text{C.F.} + \text{P.I.} = c_1 e^{2x} + c_2 e^{-2x} - \frac{1}{4} x^2 - \frac{1}{8}.$$

EXAMPLE 4.116

Solve $\frac{d^2y}{dx^2} + 4y = e^x + \sin 3x + x^2$.

Solution. The auxiliary equation is $m^2 + 4 = 0$ and so $m = \pm 2i$. Therefore,

$$\text{C.F.} = c_1 \cos 2x + c_2 \sin 2x.$$

Further,

$$\text{P.I.} = \frac{1}{f(D)} F(x) = \frac{1}{D^2+4}(e^x + \sin 3x + x^2)$$

$$= \frac{1}{D^2+4} e^x + \frac{1}{D^2+4} \sin 3x + \frac{1}{D^2+4} x^2$$

$$= \frac{1}{5} e^x + \frac{1}{-9+4} \sin 3x + \frac{1}{4}\left(1+\frac{D^2}{4}\right)^{-1} x^2$$

$$= \frac{1}{5} e^x - \frac{1}{5} \sin 3x + \frac{1}{4}\left(1-\frac{D^2}{4}+\ldots\right) x^2$$

$$= \frac{1}{5} e^x - \frac{1}{5} \sin 3x + \frac{x^2}{4} - \frac{1}{16}\cdot 2$$

$$= \frac{1}{5} e^x - \frac{1}{5} \sin 3x + \frac{1}{4} x^2 - \frac{1}{8}.$$

Hence the complete solution is
$y = \text{C.F.} + \text{P.I.}$

$$= c_1 \cos 2x + c_2 \sin 2x + \frac{1}{5} e^x - \frac{1}{5} \sin 3x + \frac{1}{4} x^2 - \frac{1}{8}.$$

EXAMPLE 4.117

Solve $\frac{d^2y}{dx^2} - 2\frac{dy}{dx} + y = xe^x \sin x$.

Solution. The auxiliary equation of the given differential equation is

$$m^2 - 2m + 1 = 0,$$

which yields $m = 1, 1$. Hence

$$\text{C.F.} = (c_1 + c_2 x)\, e^x.$$

The particular integral is

$$\text{P.I.} = \frac{1}{f(D)} F(x) = \frac{1}{(D-1)^2} xe^x \sin x$$

$$= e^x \frac{1}{(D+1-1)^2} x \sin x = e^x \frac{1}{D^2} x \sin x$$

$$= e^x \frac{1}{D} \int x \sin x\, dx = e^x \frac{1}{D}(-x \cos x + \sin x)$$

$$= e^x \int (-x \cos x + \sin x)\, dx$$

$$= e^x[-x \sin x - \cos x - \cos x]$$

$$= -e^x(x \sin x + 2 \cos x).$$

Hence the complete solution is

$y = \text{C.F.} + \text{P.I.}$

$$= (c_1 + c_2 x)\, e^x - e^x(x \sin x + 2 \cos x).$$

EXAMPLE 4.118

Solve $(D^2 - 4D + 3)y = \sin 3x \cos 2x$.

Solution. The auxiliary equation is

$$m^2 - 4m + 3 = 0,$$

which yields $m = 3, 1$. Therefore,

$$\text{C.F.} = c_1 e^{3x} + c_2 e^{x}.$$

Further

$$\text{P.I.} = \frac{1}{D^2-4D+3}[\sin 3x \cos 2x]$$

$$= \frac{1}{D^2-4D+3}\left[\frac{1}{2} 2\sin 3x \cos 2x\right]$$

$$= \frac{1}{D^2-4D+3}\left[\frac{1}{2}(\sin 5x + \sin x)\right]$$

$$= \frac{1}{2}\cdot\frac{1}{D^2-4D+3}\sin 5x + \frac{1}{2}\cdot\frac{1}{D^2-4D+3}\sin x$$

$$= \frac{1}{2}\left[\frac{1}{-25-4D+3}\sin 5x + \frac{1}{-1-4D+3}\sin x\right]$$

$$= \frac{1}{2}\left[\frac{1}{-22-4D}\sin 5x + \frac{1}{2-4D}\sin x\right]$$

$$= \frac{1}{2}\left[-\frac{1}{2(11+2D)}\sin 5x + \frac{1}{2(1-2D)}\sin x\right]$$

$$= \frac{1}{4}\left[-\frac{11-2D}{121-4D^2}\sin 5x + \frac{1+2D}{1-4D^2}\sin x\right]$$

$$= \frac{1}{4}\left[-\frac{11-2D}{121-4(-25)}\sin 5x + \frac{1+2D}{1-4(-1)}\sin x\right]$$

$$= \frac{1}{4}\left[-\frac{11-2D}{221}\sin 5x + \frac{1+2D}{5}\sin x\right]$$

$$= \frac{1}{4}\left[-\frac{1}{221}[11\sin 5x - 2D\sin 5x] + \frac{1}{5}(\sin x + 2D\sin x)\right]$$

$$= \frac{1}{4}\left[-\frac{11}{221}\sin 5x + \frac{10}{221}\cos 5x + \frac{1}{5}\sin x + \frac{2}{5}\cos x\right]$$

$$= -\frac{11}{884}\sin 5x + \frac{10}{884}\cos 5x + \frac{1}{20}\sin x + \frac{1}{10}\cos x.$$

Hence the complete solution is

$$y = c_1 e^{3x} + c_2 e^{x} - \frac{11}{884}\sin 5x + \frac{10}{884}\cos 5x + \frac{1}{20}\sin x + \frac{1}{10}\cos x.$$

EXAMPLE 4.119

Solve $(D^2 + 1)y = \operatorname{cosec} x$.

Solution. The auxiliary equation is $m^2 + 1 = 0$, which yields $m = \pm i$. Thus,

$$\text{C.F.} = c_1 \cos x + c_2 \sin x.$$

Now

$$\text{P.I.} = \frac{1}{f(D)}F(x) = \frac{1}{D^2+1}\operatorname{cosec} x$$

$$= \frac{1}{(D+i)(D-i)}\operatorname{cosec} x$$

$$= \frac{1}{2i}\left[\frac{1}{D-i} - \frac{1}{D+i}\right]\operatorname{cosec} x$$

$$= \frac{1}{2i}\left[\frac{1}{D-i}\operatorname{cosec} x - \frac{1}{D+i}\operatorname{cosec} x.\right]$$

But, by Theorem 4.4,

$$\frac{1}{D-i}\operatorname{cosec} x = e^{ix}\int \operatorname{cosec} x\, e^{-ix}\,dx$$

$$= e^{ix}\int \operatorname{cosec} x(\cos x - i\sin x)\,dx$$

$$= e^{ix}\int (\cot x - i)\,dx$$

$$= e^{ix}(\log \sin x - ix).$$

Similarly,

$$\frac{1}{D+i}\operatorname{cosec} x = e^{-ix}(\log \sin x + ix).$$

Therefore,

$$\text{P.I.} = \frac{1}{2i}[e^{ix}(\log \sin x - ix) - e^{-ix}(\log \sin x + ix)]$$

$$= \log \sin x\left(\frac{e^{ix} - e^{-ix}}{2i}\right) - x\left(\frac{e^{ix} + e^{-ix}}{2}\right)$$

$$= (\log \sin x)\sin x - x\cos x.$$

Hence the complete solution is

$$y = c_1 \cos x + c_2 \sin x + \sin x \log \sin x - x \cos x.$$

EXAMPLE 4.120

Solve $\dfrac{d^2y}{dx^2} + 5\dfrac{dy}{dx} + 6y = e^{-2x} \sin 2x.$

Solution. The auxiliary equation is $m^2 + 5m + 6 = 0$, which yields

$$m = \frac{-5 \pm \sqrt{25-24}}{2} = -2, -3.$$

Therefore,

$$\text{C.F.} = c_1 e^{-2x} + c_2 e^{-3x}.$$

Further

$$\text{P.I.} = \frac{1}{f(D)} F(x)$$

$$= \frac{1}{D^2 + 5D + 6} e^{-2x} \sin 2x$$

$$= e^{-2x} \frac{1}{(D-2)^2 + 5(D-2) + 6} \sin 2x$$

$$= e^{-2x} \frac{1}{D^2 + D} \sin 2x$$

$$= e^{-2x} \frac{1}{D-4} \sin 2x$$

$$= e^{-2x} \frac{D+4}{D^2-4} \sin 2x$$

$$= e^{-2x} \frac{D+4}{-8} \sin 2x$$

$$= \frac{e^{-2x}}{-8} [D \sin 2x + 4 \sin 2x]$$

$$= \frac{e^{-2x}}{-8} [2 \cos 2x + 4 \sin 2x].$$

Hence the complete solution is

$$y = \text{C.F.} + \text{P.I.}$$

$$= c_1 e^{-2x} + c_2 e^{-3x} - \frac{1}{8} e^{-2x} [2 \cos 2x + 4 \sin 2x].$$

EXAMPLE 4.121

Solve $\dfrac{d^3y}{dx^2} + y = \sin 3x - \cos^2 \dfrac{x}{2}.$

Solution. The auxiliary equation of the given differential equation is

$$m^3 + 1 = 0$$

or

$$(m + 1)(m^2 - m + 1) = 0$$

and so $m = -1, \frac{1}{2} \pm \frac{\sqrt{3}}{2} i$. Therefore,

$$\text{C.F.} = c_1 e^{-x} + e^{\frac{x}{2}} \left[c_2 \cos \frac{\sqrt{3}}{2} x + c_3 \sin \frac{\sqrt{3}}{2} x \right].$$

On the other hand,

$$\text{P.I.} = \frac{1}{D^3 + 1} \left[\sin 3x - \frac{1}{2}(1 + \cos x) \right]$$

$$= \frac{1}{D^3 + 1} \sin 3x - \frac{1}{2(D^3 + 1)} (1 + \cos x)$$

$$= \frac{1}{DD^2 + 1} \sin 3x - \frac{1}{2(1 + D^3)} [1 + \cos x]$$

$$= \frac{1}{-9D + 1} \sin 3x - \frac{1}{2(1 + D^3)} [1 + \cos x]$$

$$= \frac{1}{1 - 9D} \sin 3x - \frac{1}{2}(1 + D^3)^{-1}(1) - \frac{1}{2(1 + D^3)} \cos x$$

$$= \frac{1 + 9D}{1 - 81D^2} \sin 3x - \frac{1}{2}(1 - D^3 + \cdots)(1) - \frac{1}{2 + 2DD^2} \cos x$$

$$= \frac{1 + 9D}{1 - 81(-9)} \sin 3x - \frac{1}{2} - \frac{1}{2 - 2D} \cos x$$

$$= \frac{1 + 9D}{730} \sin 3x - \frac{1}{2} - \frac{2 + 2D}{4 - 4D^2} \cos x$$

$$= \frac{1}{730} (\sin 3x + 27 \cos 3x) - \frac{1}{2} - \frac{2 + 2D}{8} \cos x$$

$$=\frac{1}{730}(\sin 3x+27\cos 3x)-\frac{1}{2}-\frac{1}{4}\cos x+\frac{1}{4}\sin x$$

$$=\frac{1}{730}(\sin 3x+27\cos 3x)-\frac{1}{2}-\frac{1}{4}(\cos x-\sin x).$$

Hence the complete solution is

$$y=c_1e^{-x}+e^{\frac{x}{2}}(c_2\cos\frac{\sqrt{3}}{2}x+c_3\sin\frac{\sqrt{3}}{2}x)$$

$$+\frac{1}{730}(\sin 3x+27\cos 3x)-\frac{1}{2}-\frac{1}{4}(\cos x-\sin x).$$

EXAMPLE 4.122

Solve $\frac{d^3y}{dx^3}-3\frac{dy}{dx}+2y=x^2e^x$.

Solution. The symbolic form of the given equation is

$$(D^3-3D+2)\,y=x^2e^x.$$

Its auxiliary equation is

$$m^3-3m+2=0,$$

which yields $m = 1, 1, -2$. Therefore,

$$\text{C.F.}=(c_1+c_2x)\,e^x+c_3\,e^{-2x}.$$

Further

$$\text{P.I.}=\frac{1}{D^3-3D+2}x^2e^x$$

$$=e^x\frac{1}{(D+1)^3-3(D+1)+2}x^2$$

$$=e^x\frac{1}{D^3+3D^2}x^2=e^x\frac{1}{3D^2\,(1+\frac{D}{3})}x^2$$

$$=e^x\frac{1}{3D^2}\left[1-\frac{D}{3}+\left(\frac{D}{3}\right)^2-\ldots\right]x^2$$

$$=\frac{e^x}{3D^2}\left[1-\frac{D}{3}+\frac{D^2}{9}-\ldots\right]x^2$$

$$=\frac{e^x}{3D^2}\left(x^2-\frac{Dx^2}{3}+\frac{D^2x^2}{9}\right)$$

$$=\frac{e^x}{3D^2}\left(x^2-\frac{2x}{3}+\frac{2}{9}\right)$$

$$=\frac{e^x}{3D}\left[\int x^2dx-\frac{2}{3}\int x\,dx+\frac{2}{9}\int dx\right]$$

$$=\frac{e^x}{3D}\left[\frac{x^3}{3}-\frac{x^2}{3}+\frac{2}{9}x\right]$$

$$=\frac{e^x}{3}\left[\int\frac{x^3}{3}dx-\int\frac{x^2}{3}dx+\frac{2}{9}\int x\,dx\right]$$

$$=\frac{e^x}{3}\left[\frac{x^4}{12}-\frac{x^3}{9}+\frac{x^2}{9}\right]=\frac{e^x}{108}[3x^4-4x^3+4x^2]$$

$$=\frac{x^2e^x}{108}[3x^2-4x+4].$$

Hence complete solution is

$$y=\text{C.F.}+\text{P.I.}$$

$$=(c_1+c_2x)\,e^x+c_3e^{-2x}+\frac{x^2e^x}{108}[3x^2-4x+4].$$

EXAMPLE 4.123

Solve $(D^2-1)\,y=x\sin 3x+\cos x$.

Solution. The auxiliary equation for the given differential equation is $m^2-1=0$, and so $m=\pm1$. Therefore,

$$\text{C.F.}=c_1e^x+c_2e^{-x}.$$

Further,

$$\text{P.I.}=\frac{1}{D^2-1}(x\sin 3x+\cos x)$$

$$=\frac{1}{D^2-1}x(I.P.\ of\ e^{3ix})+\frac{1}{D^2-1}\cos x$$

$$=\text{I.P. of}\frac{1}{D^2-1}x\,e^{3ix}+\frac{1}{(-1^2)-1}\cos x$$

$$=\text{I.P. of}\left[e^{3ix}\frac{1}{(D+3i)^2-1}x\right]-\frac{1}{2}\cos x$$

$$=\text{I.P. of}\left[e^{3ix}\frac{1}{D^2+6iD-10}x\right]-\frac{\cos x}{2}$$

$$=\text{I.P. of}\left[e^{3ix}\frac{1}{-10\left(1-\frac{6}{10}iD-\frac{D^2}{10}\right)}x\right]-\frac{1}{2}\cos x$$

$$= -\frac{1}{10}\text{ I.P. of}\left[e^{3ix}\left(1-\left(\frac{3}{5}iD+\frac{D^2}{10}\right)\right)^{-1}x\right]-\frac{1}{2}\cos x$$

$$= -\frac{1}{10}\text{ I.P. of}\left[e^{3ix}\left(1+\frac{3}{5}iD+\frac{D^2}{10}+\ldots\right)x\right]-\frac{1}{2}\cos x$$

$$= -\frac{1}{10}\text{ I.P. of}\left[e^{3ix}\left(x+\frac{3}{5}i\right)\right]-\frac{1}{2}\cos x$$

$$= -\frac{1}{10}\text{ I.P. of}\left[(\cos 3x+i\sin 3x)\left(x+\frac{3}{5}i\right)\right]-\frac{1}{2}\cos x$$

$$= -\frac{1}{10}\text{ I.P. of}\left[x\cos 3x+ix\sin 3x+\frac{3i}{5}\cos 3x\right.$$

$$\left.-\frac{3}{5}\sin 3x\right]-\frac{1}{2}\cos x$$

$$= -\frac{1}{10}\text{ I.P. of}\left[x\cos 3x-\frac{3}{5}\sin 3x\right.$$

$$\left.+i\left(\frac{3}{5}\cos 3x+x\sin 3x\right)\right]-\frac{1}{2}\cos x$$

$$= -\frac{1}{10}\left(\frac{3}{5}\cos 3x+x\sin 3x\right)-\frac{1}{2}\cos x.$$

Hence the complete solution of the given differential equation is

$$y = \text{C.F.} + \text{P.I.} = c_1e^{-x}+c_2e^{-x}$$

$$-\frac{1}{10}\left(\frac{3}{5}\cos 3x+x\sin 3x\right)-\frac{1}{2}\cos x.$$

EXAMPLE 4.124

Solve $(D^2+1)\,y = \sin x \sin 2x + e^x x^2$.

Solution. The auxiliary equation for the given differential equation is

$$m^2+1=0, \text{ which yields } m = \pm 2.$$

Therefore

$$\text{C.F.} = c_1\cos x + c_2 \sin x.$$

The particular integral is given by

$$\text{P.I.} = \frac{1}{D^2+1}\sin x\sin 2x + \frac{1}{D^2+1}e^x x^2$$

$$= \frac{1}{D^2+1}\left[\frac{1}{2}(\cos x-\cos 3x)\right]+\frac{1}{D^2+1}e^x x^2$$

$$= \frac{1}{2(D^2+1)}\cos x - \frac{1}{2(D^2+1)}\cos 3x$$

$$+\,e^x\frac{1}{(D+!)^2+1}x^2$$

$$= \frac{x}{4}\sin x - \frac{1}{2(-9+1)}\cos 3x$$

$$+\frac{e^x}{2}\left(1+D+\frac{D^2}{2}\right)^{-1}x^2$$

$$= \frac{x}{4}\sin x + \frac{1}{16}\cos 3x + \frac{1}{2}e^x(x^2-2x+1).$$

Hence the complete solution is

$$y = \text{C.F.} + \text{P.I.}$$

$$= c_1\cos x + c_2\sin x + \frac{1}{4}x\sin x + \frac{1}{16}\cos 3x$$

$$+\frac{1}{2}e^x(x^2-2x+1).$$

4.21 APPLICATION OF LINEAR DIFFERENTIAL EQUATION

Linear differential equation play an important role in the analysis of electrical, machanical and other linear systems. Some of the applications of these equation are discussed below.

(A) Electrcial circuits

Consider an LCR circuit consisting of induction L, capacitor C, and resistance R. Then, using Kirchhoff's law, the equation governing the flow of current in the circuit is

$$L\frac{dI}{dt}+RI+\frac{Q}{C}=E(t),$$

where I is the current flowing in the circuit, Q is the charge, and E is the e.m.f. of the battery. Since $I=\frac{dQ}{dt}$, the above equation reduce to

$$L\frac{d^2Q}{dt}+R\frac{dQ}{dt}+\frac{Q}{C}=E(t) \qquad (150)$$

If we consider LC circuit without having e.m.f. source, then the differential equation describing the circuit is

$$L\frac{d^2Q}{dt^2}+\frac{Q}{C}=0$$

or

$$\frac{d^2Q}{dt^2}+\frac{Q}{LC}=0$$

or

$$\frac{d^2Q}{dt^2}+\mu^2Q=0,\ \mu^2=\frac{1}{LC}.$$

This equation represents *free electrical oscillations* of the current having period

$$T=\frac{2\pi}{\mu}=2\pi\sqrt{LC}.$$

EXAMPLE 4.125

In an LCR circuit, an inductance L of one hendry, resistance of 6 ohm, and a condeser of 1/9 farad have been connected through a battery of e.m.f. $E = \sin t$. If $I = Q = 0$ at $t = 0$, find charge Q and current I.

Solution. The differential equation for the given circuit is

$$L\frac{d^2Q}{dt^2}+R\frac{dQ}{dt}+\frac{Q}{C}=E(t).$$

Here L = 1, R = 6, C = $\frac{1}{9}$, $E(t) = \sin t$. Thus, we have

$$\frac{d^2Q}{dt^2}+6\frac{dQ}{dt}+9Q=\sin t$$

subject to $Q(0) = 0$, $Q'(0) = 0 = I(0)$. The auxiliary equation for this differential equation is $m^2 + 6m + 9 = 0$ and so $m = -3, -3$. Thus

$$\text{C.F.}=(c_1+c_2t)\,e^{-3t}.$$

Now

$$\text{P.I.}=\frac{1}{D^2+6D+9}\sin t=\frac{1}{6D+8}\sin t$$

$$=\frac{6D-8}{36D^2-64}\sin t=\frac{6D-8}{-100}\sin t$$

$$=-\frac{1}{100}[6D\sin t-8\sin t]$$

$$=-\frac{6}{100}\cos t+\frac{8}{100}\sin t.$$

Hence the complete solution is

$$Q=(c_1+c_2t)\,e^{-3t}-\frac{6}{100}\cos t+\frac{8}{100}\sin t.$$

Now $Q(0) = 0$ gives $0 = c_1 - \frac{6}{100}$ and so $c_1 = \frac{6}{100}$. Also

$$\frac{dQ}{dt}=-3c_1e^{-3t}+c_2(e^{-3t}-3te^{-3t})$$

$$+\frac{6}{100}\sin t+\frac{8}{100}\cos t.$$

Therefore $\frac{dQ}{dt}=0$ *at* $t = 0$ yields

$$0=-3c_1+c_2+\frac{8}{100}=-\frac{18}{100}+c_2+\frac{8}{100}$$

and so $c_2=\frac{1}{10}$. Hence

$$Q=\frac{6}{100}+\frac{t}{10}e^{-3t}-\frac{6}{100}\cos t+\frac{8}{100}\sin t$$

$$=\frac{5e^{-3t}}{50}(5t+3)-\frac{3}{50}\cos t+\frac{2}{25}\sin t.$$

Since $I=\frac{dQ}{dt}$, we have

$$I=\frac{5e^{-3t}}{50}-\frac{3}{50}(15t+3)e^{-3t}+\frac{3}{50}\sin t+\frac{2}{25}\cos t$$

$$=\frac{e^{-3t}}{50}(15t+4)+\frac{3}{50}\sin t+\frac{2}{25}\cos t.$$

EXAMPLE 4.126

Find the frequency of *free vibrations* in a closed electrical circuit with inductance L and capacity C in series.

Solution. Since there is no applied e.m.f., the differential equation governing this LC circuit is

$$L\frac{d^2Q}{dt^2}+\frac{Q}{C}=0$$

or

$$\frac{d^2Q}{dt^2} = -\frac{Q}{LC} = -\omega^2 Q,$$

where $\omega^2 = \frac{1}{LC}$. Thus the equation represents oscillatory current with period

$$T = \frac{2\pi}{\omega} = 2\pi\sqrt{LC}.$$

Then

$$\text{Frequency} = \frac{1}{T} = \frac{1}{2\pi\sqrt{LC}} \text{ per secound}$$

$$= \frac{60}{2\pi\sqrt{LC}} = \frac{30}{\pi\sqrt{LC}} \text{ per minute.}$$

EXAMPLE 4.127

The differential equation for a circuit in which self-inductane and capacitance neutralize each other is

$$L\frac{d^2i}{dt^2} + \frac{i}{C} = 0.$$

Find the current i as a function of t, given that I is maximum current and $i = 0$ when $t = 0$.

Solution. We have

$$\frac{d^2i}{dt^2} + \frac{i}{LC} = 0.$$

The auxiliary equation is $m^2 + \frac{1}{LC} = 0$ and so $m = \pm\frac{i}{\sqrt{LC}}$. Hence the solution is

$$i = c_1 \cos\frac{1}{\sqrt{LC}}t + c_2 \sin\frac{1}{\sqrt{LC}}t.$$

Since $i = 0$ at $t = 0$, we have $c_1 = 0$ and so

$$i = c_2 \sin\frac{t}{\sqrt{LC}}.$$

For maximum current I, we have

$I = c_2 \max \sin\frac{t}{\sqrt{LC}} = c_2$. Hence $i = I\sin\frac{t}{\sqrt{LC}}$.

EXAMPLE 4.128

An LCR circuit with batary e.m.f. E sin *pt* is tuned to resonance so that $p^2 = \frac{1}{LC}$. *Show* that for small value of $\frac{R}{L}$ the current in the circuit at time t is given by $\frac{Et}{2L}\sin pt$.

Solution. The differential equation governing the LCR circuit is $L\frac{d^2Q}{dt^2} + R\frac{dQ}{dt} + \frac{Q}{C} = E(t)$ = E sin *pt*.

The auxiliary equation is

$$m^2 + \frac{R}{L}m + \frac{1}{LC} = 0,$$

which yields

$$m = \frac{-\frac{R}{L} \pm \sqrt{\frac{R^2}{L^2} - \frac{4}{LC}}}{2}$$

$$= \frac{-\frac{R}{L} \pm \sqrt{-\frac{4}{LC}}}{2} \text{ since } \frac{R}{L} \text{ is small}$$

$$= -\frac{R}{2L} \pm \frac{1}{\sqrt{LC}}i$$

$$= -\frac{R}{L} \pm pi \text{ since } p^2 = \frac{1}{LC}.$$

Therefore

$$\text{C.F.} = e^{-\frac{R}{L}t}(c_1 \cos pt + c_2 \sin pt)$$

$$= \left(1 - \frac{Rt}{2L}\right)(c_1 \cos pt + c_2 \sin pt)$$

rejecting higher power of $\frac{R}{L}$.

Further

$$\text{P.I.} = \frac{1}{LD^2 + RD + \frac{1}{C}}(E \sin pt)$$

$$= \frac{E}{-Lp^2 + RD + \frac{1}{C}} \sin pt$$

$$= \frac{E}{-\frac{L}{LC} + \frac{1}{C} + RD} \sin pt$$

$$= \frac{E}{RD} \sin pt = \frac{E}{R}\frac{D}{D^2} \sin pt$$

$$= -\frac{E}{Rp^2} D \sin pt$$

$$= -\frac{E}{Rp^2} p \cos pt = \frac{E}{Rp} \cos pt$$

Thus, the complete solution is

$$q = \left(1 - \frac{Rt}{2L}\right)(c_1 \cos pt + c_2 \sin pt) - \frac{E}{Rp} \cos pt$$

Using the initial condition $q = 0$ for $t = 0$, we get

$$0 = c_1 - \frac{E}{Rp} \text{ or } c_1 = \frac{E}{Rp}.$$

Also

$$i = \frac{dq}{dt} = \left(1 - \frac{Rt}{2L}\right)(-c_1 \sin pt + c_2 \cos pt)p$$

$$- \frac{R}{2L}(c_1 \cos pt + c_2 \sin pt) + \frac{E}{R} \sin pt.$$

Using the initial condition $i = 0$ for $t = 0$, we get

$$0 = pc_2 - \frac{Rc_1}{2L}$$

$$= pc_2 - \frac{R}{2L}\left(\frac{E}{Rp}\right)$$

$$= pc_2 - \frac{RE}{2Lp}$$

and so $c_2 = \frac{E}{2Lp^2}$. Hence, the solution is

$$i = \left(1 - \frac{Rt}{2L}\right) - \left(\frac{E}{Rp} \sin pt + \frac{2}{2Lp^2} \cos pt\right) p$$

$$- \frac{R}{2L}\left(\frac{E}{Rp} \cos pt + \frac{2}{2Lp^2} \sin pt\right) + \frac{E}{R} \sin pt$$

$$= \frac{Et}{2L} \sin pt$$

4.22 MASS-SPRING SYSTEM

In examplc 4.4, we have seen that the differential equation governing a Mass-spring system is

$$m\frac{d^2x}{dt^2} + a\frac{dx}{dt} + kx = f(t)$$

where m is the mass, a $\frac{dx}{dt}$ the damping force due to the medium, k is spring constant, and x represents the displacement of the mass.

This is exactly the same differential equation which occurs in LCR electric circuits. When $a = 0$, the motion is called damped. If $f(t) = 0$, then the motion is called *forced*.

EXAMPLE 4.129

A mass of 10 kg is attached to a spring having spring constant 140 N/m. The mass is started in motion from the equilibrium position with a velocity of 1 m/sec in the upward direction and with an applied external force $f(t) = 5 \sin t$. Find the subsequent motion of the mass if the force due to air resistance is $-90\, x'$ N.

Solution. The describing differential equation is

$$\frac{d^2x}{dt^2} + \frac{a}{m}\frac{dx}{dt} + \frac{k}{m}x = \frac{1}{m}f(t).$$

Here $a = 90$, $k = 140$, and $m = 10$. Therefore we have

$$\frac{d^2x}{dt^2} + 9\frac{dx}{dt} + 14x = \frac{1}{2}\sin t.$$

The auxiliary equation is $m^2 + 9m + 14 = 0$, and so $m = -2, -7$. Therefore

$$\text{C.F.} = c_1e^{-2t} + c_2e^{-7t}.$$

Further

$$\text{P.I.}=\frac{1}{D^2+9D+14}\left(\frac{1}{2}\sin t\right)=\frac{1}{2(-1+9D+14)}\sin t$$

$$=\frac{1}{2}\frac{1}{(9D=13)}\sin t=\frac{1}{2}\frac{9D-13}{2(-81-169)}\sin t$$

$$=-\frac{1}{500}(-13\sin t+9\cos t)$$

$$=\frac{13}{500}\sin t-\frac{9}{500}\cos t.$$

Hence, the complete solution is

$$x=c_1e^{-2t}+c_2e^{-7t}+\frac{13}{500}\sin t-\frac{9}{500}\cos t.$$

Using the initial condition $x(0) = 0$, we get $0 = c_1 + c_2 - \frac{9}{500}$ and so $c_1 + c_2 = \frac{9}{500}$. Since $\frac{dx}{dt}(0) = -1$ (initial velocity in upper direction), we get

$$-1=-2c_1-7c_2+\frac{13}{500}\text{ or }2c_1+7c_2=\frac{513}{500}.$$

Solving for c_1 and c_2, we get $c_1=-\frac{90}{500}$, $c_2=\frac{99}{500}$. Hence

$$x=\frac{1}{500}(-90e^{-2t}+99e^{-7t}+13\sin t-9\cos t).$$

EXAMPLE 4.130

If in a mass spring system mass = 4kg, spring consistance and initial velocity, then find the subsequent motion of the weight. Show that the resonance occurce in the case.

Solution. The governing equation is

$$\frac{d^2x}{dt^2}+\frac{a}{m}\frac{dx}{dt}+\frac{k}{m}x=\frac{1}{m}f(t).$$

Therefore, we have

$$\frac{d^2x}{dt^2}+16x=2\sin 4t.$$

The auxiliary equation is $m^2 + 16 = 0$ and so $m = \pm 4i$. Therefore,

$$\text{C.F.}=c_1\cos 4t+c_2\sin 4t.$$

Now

$$\text{P.I.}=\frac{1}{D^2+16}(2\sin 4t)=\frac{2t}{2D}(\sin 4t)$$

$$=\frac{t\,D(\sin 4t)}{-16}=-\frac{t}{16}(4\cos 4t)$$

$$=-\frac{t}{4}\cos 4t.$$

Hence the complete solution is

$$x=c_1\cos 4t+c_2\sin 4t-\frac{t}{4}\cos 4t.$$

Using initial condition $x(0) = 0$, we have $0 = c_1$. Differentiating w.r.t, we get

$$\frac{dx}{dt}=-4c_1\sin 4t+4c_2\cos 4t$$

$$-\frac{1}{4}[\cos 4t-4t\sin 4t]$$

Now $\frac{dx}{dt} = 0$ at $t = 0$. Therefore, $0 = 4c_2 - \frac{1}{4}$ which gives $c_2=\frac{1}{16}$. Hence

$$x=\frac{1}{16}\sin 4t-\frac{t}{4}\cos 4t.$$

We observe that $x(t) \to \infty$ as $t \to \infty$ due to the presence of the term $t\cos 4t$. This term is called a *secular term*, The presence of secular term causes resonce because the solution becomes unbounded.

4.23 SIMPLE PENDULUM

The system in which a heavy particle (bob) is attached to one end of light inextensible string, the other end of which is fixed and oscillates under the action of gravity force in a verticle plane is called a simple pendulum. To describe its motion, let m be the mass of the particles, l be the length of the string, and O be the fixed point of the string (Figure 4.9)

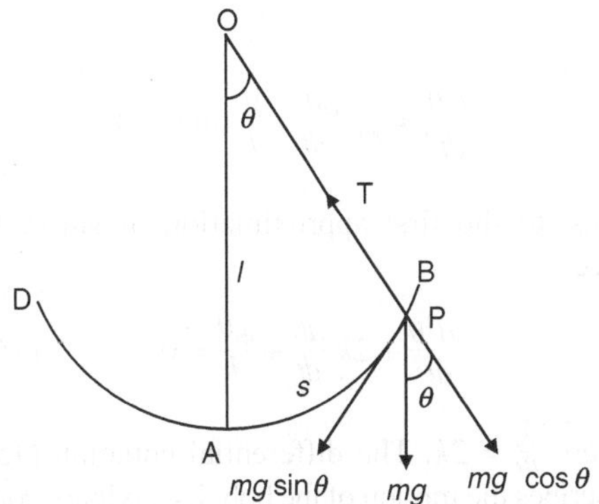

Figure 4.9

Let P be the position of the heavy particle at any time t and let $\angle AOP = \theta$, where OA is verticle line through O. Then the force acting on the bob are
(a) weight mg acting
(b) tension T in the string.
Resolving mg, we note that tension is balanced by mg cosθ. The equation of motion along the tangent is

$$m\frac{d^2s}{dt^2} = -mg\sin\theta$$

or

$$\frac{d^2}{dt^2}(l\theta) = -g\sin\theta$$

or

$$\frac{d^2\theta}{dt^2} = -\frac{g}{l}\sin\theta$$

$$= -\frac{g}{l}\left[\theta - \frac{\theta^3}{3!} + ..\right] \text{ (using expansion for } \sin\theta)$$

$$= -\frac{g\theta}{l} \text{ to the first approximation.}$$

Thus the differential equation describing the motion of bob is

$$\frac{d^2\theta}{dt^2} + \omega^2\theta = 0,$$

where $\omega^2 = \frac{g}{l}$ The auxiliary equation of this equation is

$$m^2 + \omega^2 = 0$$

and so $m = \pm\,\omega i$. Therefore, the solution is

$$\theta = c_1\cos\omega t + c_2\sin\omega t$$

$$= c_1\cos\sqrt{\frac{g}{l}}t + c_2\sin\sqrt{\frac{g}{l}}t.$$

The motion in case of simple pendulum is *simple harmonic motion* where time period

$$T = \frac{2\pi}{\omega} = 2\pi\sqrt{\frac{l}{g}}.$$

The motion of the bob from one extreme position to the extreme position on the other on the side of A is called *a beat or a swing*. Threrefore,

$$\text{Time for one swing} = \frac{1}{2}T = \pi\sqrt{\frac{l}{g}}$$

The pendulum which beats once every second is called a *second's pendulum*. Therefore, the number of beats in a second pendulum in one day is equal to the number of seconds in a day. Thus, a second pendulum beats 86,400 times a day.

Now, since, the time for 1 beat in a second's pendulum is 1 sec, we have

$$1 = \pi\sqrt{\frac{l}{g}} = \pi\sqrt{\frac{l}{981}}$$

and so l = 99.4 cm is the length of second's pendulum.

EXAMPLE 4.131

The differential equation of a simple pendulum is $\frac{d^2x}{dt^2} + \omega_0^2 x = F_0 \sin nt$, where ω_0 and F_0 are constants. If initially $\frac{dx}{dt} = 0$, determine the motion when $\omega_0 = n$

Solution. We have

$$\frac{d^2x}{dt^2} + n^2x = F_0\sin nt \text{ since, } \omega_0 = n.$$

The auxiliary equation is $m^2 + n^2 = 0$ and so $m = \pm\, ni$. Therefore,

$$\text{C.F.} = c_1\cos nt + c_2\sin nt.$$

Now

$$\text{P.I.} = \frac{F_0}{D^2 + n^2}\sin nt = \frac{F_0 t}{2D}\sin nt.$$

$$= \frac{F_0 t}{2}\int \sin nt\,dt = -\frac{F_0 t}{2n}\cos nt$$

Thus the complete solution is

$$x = c_1 \cos nt + c_2 \sin nt - \frac{F_0 t}{2n} \cos nt.$$

Initially $x = 0$ for $t = 0$, so $c_1 = 0$.
Also

$$\frac{dx}{dt} = -nc_1 \sin nt + nc_2 \cos nt$$

$$-\frac{F_0}{2n} [\cos nt - tn \sin nt].$$

But for $t = 0$, $\frac{dx}{dt} = 0$. Therefore, $0 = nc_2 - \frac{F_0}{2n}$ and so $c_2 = \frac{F_0}{2n^2}$.
Hence

$$x = \frac{F_0}{2n^2} \sin nt - \frac{F_0 t}{2n} \cos nt$$

$$= \frac{F_0}{2n^2} (\sin nt - nt \cos nt).$$

EXAMPLE 4.132

A simple pendulum of length l is oscillating through a small angle θ in a medium in which the resistance is proportional to velocity. Find the differential equation of its motion. Discuss the motion and find the period of oscillation.

Solution Please refer to Figure 4.9 of article 4.23. we have $\underline{|AOP} = \theta$, $AP = s = l\theta$, $OA = l$. The resistance is proportional to the velocity and so it is $\lambda \frac{ds}{dt}$, where λ is a constant. Therefore the equation of motion along the tangent is

$$m \frac{d^2 s}{dt^2} = -mg \sin\theta - \lambda \frac{ds}{dt}.$$

or

$$\frac{d^2(l\theta)}{dt^2} + \frac{\lambda}{m} \frac{d(l\theta)}{dt} + g \sin\theta = 0$$

or

$$\frac{d^2\theta}{dt^2} + \frac{\lambda}{m} \frac{d\theta}{dt} + \frac{g}{l} \sin\theta = 0$$

Thus, to the first approximation of sin θ, we have

$$\frac{d^2\theta}{dt^2} + 2k \frac{d\theta}{dt} + \frac{g\theta}{l} = 0, \quad (151)$$

where $\frac{\lambda}{m} = 2k$. The differential equation (151) describes the motion of the bob. Its auxiliary equation is

$$m^2 - 2km + \frac{g}{l} = 0,$$

whose roots are $m = -k \pm \sqrt{k^2 - \omega^2}$,

where $\omega^2 = \frac{g}{l}$. Since the oscillatory motion of the bob is possible only if $k < \omega$, the roots are $-k \pm i\sqrt{\omega^2 - k^2}$. Therefore the solution of differential equation (151) is

$$\theta = e^{-kt}\left[C_1 \cos\left(\sqrt{\omega^2 - k^2}\right) t + C_2 \sin\left(\sqrt{\omega^2 - k^2}\right) t\right],$$

which gives a vibratory motion of period

$$\frac{2\pi}{\sqrt{\omega^2 - k^2}}.$$

4.24 DIFFERENTIAL EQUATION WITH VARIABLE COEFFICIENTS

We shall consider differential equations with variable coefficients which can be reduced to linear differential equations with constant coefficients.

The general form of a linear equation of second order is

$$\frac{d^2y}{dx^2} + P\frac{dy}{dx} + Qy = R,$$

where P, Q and R are functions of x only. For example,

$$\frac{d^2y}{dx^2} + x\frac{dy}{dx} + x^2y = \cos x$$

is a linear equation of second order.

If the coefficients P and Q are constants, then such type of equation can be solved by finding complementary function and particular integral as discussed earlier. But, if P and Q are not constants but variable, then there is no general method to solve such problem. We shall discuss seven special methods to handle such problems.

4.25 METHOD OF SOLUTION BY CHANGING THE INDEPENDENT VARIABLE

Let

$$\frac{d^2y}{dx^2} + P\frac{dy}{dx} + Qy = R, \qquad (152)$$

be the given linear equation of second order. We change the independent variable x to z by taking

$$z = f(x).$$

then

$$\frac{dy}{dx} = \frac{dy}{dz}\cdot\frac{dz}{dx},$$

and

$$\frac{d^2y}{dx^2} = \frac{dy}{dz}\left(\frac{d^2z}{dx^2}\right) + \frac{d^2y}{dz^2}\left[\frac{dz}{dx}\right]^2$$

Substituting these values in (152), we get

$$\frac{d^2y}{dz^2}\left[\frac{dz}{dx}\right]^2 + \frac{dy}{dz}\left(\frac{d^2z}{dx^2}\right) + P\frac{dy}{dz}\cdot\frac{dz}{dx} + Q\,y = R$$

or

$$\frac{d^2y}{dz^2}\left[\frac{dz}{dx}\right]^2 + \frac{dy}{dz}\left[\frac{d^2z}{dx^2} + p\,\frac{dz}{dx}\right] + Q\,y = R$$

or

$$\frac{d^2y}{dz^2} + \frac{dy}{dz}\left[\frac{\frac{d^2z}{dx^2} + P\frac{dz}{dx}}{\left(\frac{dz}{dx}\right)^2}\right] + \frac{Q}{\left[\frac{dz}{dx}\right]^2}y = \frac{R}{\left[\frac{dz}{dx}\right]^2}$$

or

$$\frac{d^2y}{dz^2} + P_1\frac{dy}{dz} + Q_1y = R_1,$$

where

$$P_1 = \frac{\frac{d^2z}{dx^2} + P\frac{dz}{dx}}{\left(\frac{dz}{dx}\right)^2}, \quad Q_1 = \frac{Q}{\left[\frac{dz}{dx}\right]^2},$$

and

$$R_1 = \frac{R}{\left[\frac{dz}{dx}\right]^2}.$$

Using the functional relation between z and x, it follows that P_1, Q_1 and R_1 are functions of x. Choose z so that $P_1 = 0$, that is,

$$\frac{d^2z}{dx^2} + P\frac{dz}{dz} = 0,$$

which yields

$$\frac{dz}{dx} = e^{-\int P dx}$$

or

$$z = \int e^{-\int P dx}\,dx.$$

If for this value of z, Q_1 becomes constant or a constant divided by z^2, then the equation (152) can be integrated to find its solution. *We note that Q_1 becomes constant if we choose* $\left(\frac{dz}{dx}\right)^2 = Q$.

EXAMPLE 4.133

Solve $\frac{d^2y}{dx^2} + \cot x\frac{dy}{dx} + 4y\cos ec^2x = 0.$

Solution. Comparing with the standard form, we get

$$P = \cot x,\ Q = 4 \operatorname{cosec}^2 x \text{ and } R = 0.$$

If we choose z such that $P_1 = 0$, then z is given by

$$z = \int e^{-\int P dx} dx = \int e^{-\int \cot x\, dx} dx$$

$$= \int e^{\log \operatorname{cosec} x} dx = \int \operatorname{cosec} x\, dx$$

$$= \log \tan \frac{x}{2}.$$

Then

$$Q_1 = \frac{Q}{\left(\frac{dz}{dx}\right)^2} = \frac{4 \operatorname{cosec}^2 x}{\operatorname{cosec}^2 x} = 4,$$

and

$$R_1 = \frac{R}{\left(\frac{dz}{dx}\right)^2} = \frac{0}{\operatorname{cosec}^2 x} = 0.$$

Therefore the given equation reduces to

$$\frac{d^2y}{dz^2} + 4y = 0,$$

or, in symbolic form,

$$(D^2 + 4)\, y = 0.$$

The auxiliary equation for this differential equation is $m^2 + 4 = 0$, which yields $m \pm 2i$. Therefore the solution of the given equation is

$$y = c_1 \cos 2z + c_2 \sin 2z$$

$$= c_1 \cos\left(2 \log \tan \frac{x}{2}\right) + c_2 \sin\left(2 \log \tan \frac{x}{2}\right).$$

EXAMPLE 4.134

Solve

$$x\frac{d^2y}{dx^2} - \frac{dy}{dx} - 4x^3 y = x^5.$$

Solution. Dividing throughout by x, the given differential equation reduces to

$$\frac{d^2y}{dx^2} - \frac{1}{x}\frac{dy}{dx} - 4x^2 y = x^4.$$

Comparing with the standard form, we have

$$P = -\frac{1}{x},\ Q = -4x^2, \text{ and } R = x^4.$$

Choose z so that $P_1 = 0$. Then

$$z = \int e^{-\int P dx} dx = \int e^{\int \frac{1}{x} dx} dx$$

$$= \int e^{\log x} dx = \int x\, dx = \frac{x^2}{2}.$$

Further,

$$Q_1 = \frac{Q}{\left(\frac{dz}{dx}\right)^2} = \frac{-4x^2}{x^2} = -4,$$

and

$$R_1 = \frac{R}{\left(\frac{dz}{dx}\right)^2} = \frac{x^4}{x^2} = x^2.$$

Hence the given equation reduces to

$$\frac{d^2y}{dz^2} - 4y = x^2 = 2z$$

or

$$(D^2 - 4)\, y = 2z.$$

The auxiliary equation for this differential equation is $m^2 - 4 = 0$ and so $m = \pm 2$. Therefore

$$\text{C.F.} = c_1 e^{2z} + c_2 e^{-2z} = c_1 e^{x^2} + c_2 e^{-x^2}.$$

Now

$$\text{P.I.} = \frac{1}{D^2 - 4}(2z) = \frac{-2}{4}\left(1 - \frac{D^2}{4}\right)^{-1}(z)$$

$$= \frac{-1}{2}\left[1 + \frac{D^2}{4} + \cdots\right](z)$$

$$= -\frac{z}{2} = -\frac{x^2}{4}.$$

Hence the complete solution of the given differential equation is

$$y = \text{C.F.} + \text{P.I.} = c_1 e^{x^2} + c_2 e^{-x^2} - \frac{1}{4}x^2$$

EXAMPLE 4.135

Solve

$$x^2 \frac{d^2y}{dx^2} - 2x\frac{dy}{dx} + 2y = x^3.$$

(This equation is Cauchy-Euler equation and has also been solved in Example 4.160 by taking $t = \log x$).

Solution. Dividing the given equation throughout by x^2, we get

$$\frac{d^2y}{dx^2} - \frac{2}{x}\frac{dy}{dx} + \frac{2}{x^2}y = x.$$

Comparing with the standard form, we have

$$P = -\frac{2}{x},\ Q = \frac{2}{x^2} \text{ and } R = x.$$

Choosing z such that $P_1 = 0$, we have

$$z = \int e^{-\int P dx} dx = \int e^{2\int \frac{1}{x}dx} dx$$

$$= \int e^{2\log x} dx = \int x^2\, dx = \frac{x^3}{3}.$$

Therefore the given equation reduces to

$$\frac{d^2y}{dz^2} + Q_1 y = R_1,$$

where

$$Q_1 = \frac{Q}{\left[\frac{dz}{dx}\right]^2} = \frac{\frac{2}{x^2}}{(x^2)^2} = \frac{2}{x^6} = \frac{2}{9z^2}$$

and

$$R_1 = \frac{R}{\left[\frac{dz}{dx}\right]^2} = \frac{x}{x^4} = \frac{1}{x^3} = \frac{1}{3z}.$$

Hence the equation reduces to

$$\frac{d^2y}{dx} + \frac{2y}{9z^2} = \frac{1}{3z}$$

or

$$9z^2\frac{d^2y}{dz^2} + 2y = 3z.$$

Let $X = \log z$ so that $z = e^X$ and (see Example 4.158), $z^2\frac{d^2y}{dz^2} = D(D-1)y$. Thus the equation now reduces to

$$[9D(D-1) + 2]\,y = 3e^X$$

or

$$(9D^2 - 9D + 2)\,y = 3e^X$$

The auxiliary equation for this symbolic equation is

$$m^2 - 9m + 2 = 0,$$

which yields $m = \frac{2}{3}, \frac{1}{3}$. Hence the complementary function is given by

$$\text{C.F.} = c_1 e^{\frac{2}{3}X} + c_2 e^{\frac{1}{3}X}$$

$$= c_1 e^{\frac{2}{3}\log z} + c_2 e^{\frac{1}{3}\log z}$$

$$= c_1 z^{\frac{2}{3}} + c_2 z^{\frac{1}{3}} = c_3 x^2 + c_4 x.$$

Further,

$$\text{P.I.} = 3\,\frac{1}{9D^2 - 9D + 2}\,e^X$$

$$= 3\,\frac{1}{9 - 9 + 2}\,e^X = \frac{3}{2}\,e^X = \frac{3}{2}\,e^{\log z}$$

$$= \frac{3}{2}\,z = \frac{3}{2}\left(\frac{x^3}{3}\right) = \frac{1}{2}\,x^3.$$

Hence the complete solution of the given differential equation is

$$y = \text{C.F.} + \text{P.I.} = c_3 x^2 + c_4 x + \frac{1}{2}\,x^3.$$

EXAMPLE 4.136

Solve $\dfrac{d^2y}{dx^2} + \dfrac{2}{x}\dfrac{dy}{dx} + \dfrac{a^2}{x^4}\cdot y = 0$

Solution. Choose z such that

$$\left(\frac{dz}{dx}\right)^2 = Q = \frac{a^2}{x^4}$$

we have

$$\frac{dz}{dx} = \pm\frac{a}{x^2},\ z = \mp\frac{a}{x}$$

Changing the independent variable from x to z by the relation $z = \frac{a}{x}$, we have

$$\frac{d^2y}{dz^2} + P_1\frac{dy}{dz} + Q_1 y = R_1,$$

$$\text{where } P_1 = \frac{\dfrac{d^2z}{dx^2} + P\dfrac{dz}{dx}}{\left(\dfrac{dz}{dx}\right)^2} = \frac{\dfrac{2a}{x^3} + \dfrac{2}{x}\left(-\dfrac{a}{x^2}\right)}{\left(-\dfrac{a}{x^2}\right)^2} = 0$$

$$Q_1 = \frac{Q}{\left(\dfrac{dz}{dx}\right)^2} = 1 \text{ and } R_1 = \frac{R}{\left(\dfrac{dz}{dx}\right)^2} = 0$$

Thus, the transformed equation is

$$\frac{d^2y}{dz^2} + y = 0$$

Hence, the solution is

$$y = c_1 \cos z + c_2 \sin z$$

$$y = c_1 \cos\frac{a}{x} + c_2 \sin\frac{a}{z}$$

EXAMPLE 4.137

Solve $x\dfrac{d^2y}{dx^2} - \dfrac{dy}{dx} - 4x^3y = x^5$.

Solution. The given equation can be written as

$$\frac{d^2y}{dx^2} - \frac{1}{x}\frac{dy}{dx} + 4x^2y = x^4$$

Choosing z, such that

$$\left(\frac{dz}{dx}\right)^2 = Q = 4x^2$$

$$\text{or} \quad \frac{dz}{dx} = 2x,$$

Therefore,

$$z = x^2$$

Now changing the independent variable from x to z by the relation $z = x^2$, we have

$$\frac{d^2y}{dz^2} + P_1\frac{dy}{dz} + Q_1 y = R_1,$$

$$\text{where } P_1 = \frac{\dfrac{d^2z}{dx^2} + P\dfrac{dz}{dx}}{\left(\dfrac{dz}{dx}\right)^2} = \frac{2 + \left(\frac{-1}{x}\right)2x}{(2x)^2} = 0,$$

$$Q_1 = \frac{Q}{\left(\dfrac{dy}{dz}\right)^2} = 1$$

$$\text{and} \quad R_1 = \frac{R}{\left(\dfrac{dz}{dx}\right)^2} = \frac{x^4}{(2x)^2} = \frac{x^2}{4} = \frac{z}{2}$$

The given equation is transformed to

$$\frac{d^2y}{dz^2} + y = \frac{z}{4},$$

whose $C.F. = c_1 \cos z + c_2 \sin z$. Further,

$$P.I. = \frac{1}{4} \cdot \frac{1}{D^2+1} \cdot z = \frac{1}{4}(1+D^2)^{-1}z$$

$$= \frac{1}{4}\{1 - D^2 + D^4 - \ldots\}z$$

$$= \frac{1}{4}\{z - 0 + 0\} = \frac{z}{4}$$

Hence,

$$y = c_1 \cos z + c_2 \sin z + \frac{z}{4}$$

$$= c_1 \cos x^2 + c_2 \sin x^2 + \frac{x^2}{4}$$

EXAMPLE 4.138

Solve $(1+x^2)^2 \dfrac{d^2y}{dx^2} + 2x(1+x^2)\dfrac{dy}{dx} + 4y = 0$

Solution. The given equation can be written as

$$\frac{d^2y}{dx^2} + \frac{2x}{1+x^2}\frac{dy}{dx} + \frac{4}{(1+x^2)^2}y = 0.$$

Taking z, such that

$$\left(\frac{dz}{dx}\right)^2 = Q = \frac{4}{(1+x^2)^2},$$

we have

$$\frac{dz}{dx} = \frac{2}{1+x^2} \text{ or } z = 2\tan^{-1}x.$$

Changing the independent variable from x to z by the relation $z = 2\tan^{-1}x$, we have

$$\frac{d^2y}{dz^2} + P_1\frac{dy}{dz} + Q_1 y = R_1$$

where $$P_1 = \frac{\dfrac{d^2z}{dx^2} + P\dfrac{dz}{dx}}{\left(\dfrac{dz}{dx}\right)^2}$$

$$= \frac{-\dfrac{4x}{(1+x^2)^2} + \dfrac{2x}{1+x^2}\cdot\dfrac{2}{1+x^2}}{\dfrac{4}{(1+x^2)^2}}$$

$$= 0$$

$$Q_1 = \frac{Q}{\left(\dfrac{dz}{dx}\right)^2} = 1,$$

and

$$R_1 = \frac{R}{\left(\dfrac{dz}{dx}\right)^2} = 0$$

Therefore the transformed equation is

$$\frac{d^2y}{dz^2} + y = 0,$$

whose solution is

$$y = c_1 \cos z + c_2 \sin z$$

$$= c_1 \cos(2\tan^{-1}x) + c_2 \sin(2\tan^{-1}x)$$

$$= c_1 \cos\left(\tan^{-1}\frac{2x}{1-x^2}\right) + c_2 \sin\left(\tan^{-1}\frac{2x}{1-x^2}\right)$$

$$= c_1 \frac{1-x^2}{1+x^2} + c_2 \frac{2x}{1+x^2}$$

EXAMPLE 4.139

Solve $x^6 \dfrac{d^2y}{dx^2} + 3x^5 \dfrac{dy}{dx} + a^2 y = \dfrac{1}{x^2}$.

Solution. The given equation can be written as

$$\frac{d^2y}{dx^2} + \frac{3}{x}\frac{dy}{dx} + \frac{a^2}{x^6}y = \frac{1}{x^8}.$$

Choosing z, such that

$$\left(\frac{dy}{dx}\right)^2 = Q = \frac{a^2}{x^6},$$

we have

$$\frac{dy}{dx} = \frac{a}{x^3} \quad \text{or} \quad z = -\frac{a}{2x^2}.$$

Changing the independent variable from x to z by the relation $z = -\dfrac{a}{2x^2}$, we have

$$\frac{d^2y}{dz^2} + P_1\frac{dy}{dz} + Q_1 = R_1$$

where $P_1 = \dfrac{\dfrac{d^2z}{dx^2} + P\dfrac{dz}{dx}}{\left(\dfrac{dz}{dx}\right)^2} = 0, Q_1 = \dfrac{Q}{\left(\dfrac{dz}{dx}\right)^2} = 1$

and $R_1 = \dfrac{R}{\left(\dfrac{dz}{dx}\right)^2} = \dfrac{1}{a^2x^2} = -\dfrac{2z}{a^3}$

Thererfore, the transformed eqation is

$$\frac{d^2y}{dz^2} + y = -\frac{2z}{a^3},$$

whose $C.F. = c_1 \cos z + c_2 \sin z$

$$= c_1 \cos\left(-\frac{a}{2x^2}\right) + c_2 \sin\left(-\frac{a}{2x^2}\right)$$

$$= c_1 \cos\frac{a}{2x^2} + c_2 \sin\frac{a}{2x^2}$$

and $P.I. = \dfrac{1}{D^2+1}\cdot\left(-\dfrac{2z}{a^3}\right)$

$$= -\frac{2}{a^3}.(1+D^2)^{-1}.z$$

$$= -\frac{2}{a^3}(1 - D^2 + D^4 \ldots)\, z$$

$$= -\frac{2z}{a^3} = \frac{1}{a^2x^2}.$$

Hence

$$y = c_1 \cos\frac{a}{2x^2} + c^2 \sin\frac{a}{2x^2} + \frac{1}{a^2x^2}.$$

EXAMPLE 4.140

Solve $(x^3 - x)\dfrac{d^2y}{dx^2} + \dfrac{dy}{dx} + n^2x^3y = 0.$

Solution. The given equation can be written as

$$\frac{d^2y}{dx^2} + \frac{1}{x\,(x^2-1)}\frac{dy}{dx} + \frac{n^2x^2}{x^2-1}\cdot y = 0$$

Choosing z, such that

$$\left(\frac{dz}{dx}\right)^2 = Q = \frac{n^2x^2}{x^2-1}$$

we have

$$\frac{dy}{dx} = \frac{nx}{\sqrt{(x^2-1)}}$$

or $\quad z = n\sqrt{(x^2-1)}$

Changing the independent variable from x to z by the relation $z = n\sqrt{(x^2-1)}$. we have

$$\frac{d^2y}{dz^2} + P_1\frac{dy}{dz} + Q_1 y = R_1$$

where $P_1 = \dfrac{\dfrac{d^2z}{dx^2} + P\dfrac{dz}{dx}}{\left(\dfrac{dz}{dx}\right)^2}$

$$= \frac{\dfrac{-n}{(x^2-1)^{\frac{3}{2}}} + \dfrac{1}{x(x^2-1)}\cdot\dfrac{nx}{\sqrt{x^2-1}}}{\dfrac{n^2x^2}{(x^2-1)}}$$

$$= 0.$$

$$Q_1 = \frac{Q}{\left(\dfrac{dz}{dx}\right)^2} = 1 \text{ and } R_1 = \frac{R}{\left(\dfrac{dz}{dx}\right)^2} = 0.$$

Therefore the transformed equation is

$$\frac{d^2y}{dz^2} + y = 0,$$

whose solution is

$$y = c_1 \sin(z + c_2)$$

$$= c_1 \sin\left[n\sqrt{(x^2-1)} + c_2\right]$$

EXAMPLE 4.141

Solve $\dfrac{d^2y}{dx^2} - \dfrac{dy}{dx} - 4x^3y = 8x^3 \sin x^2.$

Solution. The given equation can be written as

$$\frac{d^2y}{dx^2} - \frac{1}{x}\frac{dy}{dx} - 4x^3y = 8x^3 \sin x^2.$$

Choosing z, such that

$$\left(\frac{dz}{dx}\right)^2 = 4x^2$$

we have

$$\frac{dy}{dx} = 2x \quad \text{or} \quad z = x^2.$$

Changing the independent variable from x to z by the relation $z = x^2$, we have

$$\frac{d^2y}{dz^2} + P_1\frac{dy}{dx} + Q_1 y = R_1, \text{ where}$$

$$P_1 = \frac{\frac{d^2z}{dx^2} + P\frac{dz}{dx}}{\left(\frac{dz}{dx}\right)^2} = 0,\ Q_1 = \frac{Q}{\left(\frac{dz}{dx}\right)^2} = -1$$

and $R_1 = \dfrac{R}{\left(\frac{dz}{dx}\right)^2} = 2\sin x^2 = 2\sin z.$

Thus, the transformed equation is

$$\frac{d^2y}{dz^2} - y = 2\sin z,$$

whose $C.F. = c_1e^z + c^2e^{-z}$ and

$$P.I. = \frac{1}{D^2 - 1}\cdot(2\sin z)$$

$$= \frac{1}{-1^2 - 1}\cdot \sin z = -\sin z.$$

Hence,

$$y = c_1e^{x^2} + c_2e^{-x^2} - \sin x^2.$$

4.26 METHOD OF SOLUTION BY CHANGING THE DEPENDENT VARIABLE

This method is also called the *method of removing first derivative*. Let

$$\frac{d^2y}{dx^2} + P\frac{dy}{dx} + Qy = R \tag{153}$$

be the given linear equation of second order. We change the dependent variable y by taking $y = vz$.

Then

$$\frac{dy}{dx} = v\frac{dz}{dx} + z\frac{dv}{dx},$$

$$\frac{d^2y}{dx^2} = v\frac{d^2z}{dx^2} + 2\frac{dv}{dx}\cdot\frac{dz}{dx} + z\frac{d^2v}{dx^2}.$$

Therefore equation (153) reduces to

$$z\frac{d^2v}{dx^2} + \left(2\frac{dz}{dx} + Pz\right)\frac{dv}{dx} + \left(\frac{d^2z}{dx^2} + P\frac{dz}{dx} + Qz\right)v = R$$

or

$$\frac{d^2v}{dx^2} + \left(P + \frac{2}{z}\frac{dz}{dx}\right)\frac{dv}{dx} + \frac{1}{z}\left(\frac{d^2z}{dx^2} + P\frac{dz}{dx} + Qz\right)v = \frac{R}{Z}$$

or

$$\frac{d^2v}{dx^2} + P_1\frac{dv}{dx} + Q_1v = R_1, \tag{154}$$

where

$$P_1 = P + \frac{2}{z}\frac{dz}{dx},$$

$$Q_1 = \frac{1}{z}\left(\frac{d^2z}{dx^2} + P\frac{dz}{dx} + Qz\right),\ R_1 = \frac{R}{z}.$$

By making proper choice of z, any desired value can be assigned to P_1 or Q_1. In particular, if $P_1 = 0$, then

$$P + \frac{2}{z}\cdot\frac{dz}{dx} = 0$$

or

$$\frac{dz}{dx} = -\frac{Pz}{2}$$

or

$$\frac{dz}{z} = -\frac{1}{2}Pdx.$$

Integrating, we get

$$\log z = -\frac{1}{2}\int Pdx$$

or

$$z = e^{-\frac{1}{2}\int P dx}.$$

Therefore

$$\frac{dz}{dx} = -\frac{P}{2} e^{-\frac{1}{2}\int P dx}$$

and

$$\frac{d^2z}{dx^2} = e^{-\frac{1}{2}\int P dx}\left[\frac{P^2}{4} - \frac{1}{2}\frac{dP}{dx}\right].$$

Putting these values in Q_1, we get

$$Q_1 = Q - \frac{1}{2}\frac{dP}{dx} - \frac{P^2}{4}.$$

Hence the equation (154) reduces to

$$\frac{d^2v}{dx^2} + \left(Q - \frac{1}{2}\frac{dP}{dx} - \frac{P^2}{4}\right) v = R_1. \quad (155)$$

The equation (155) is called the **normal form** of the given differential equation (153) and we observe that this equation *does not have first derivative*. That is why, the present method is called the **method of removing first derivative**. The normal form (155) can be solved by using already discussed methods.

EXAMPLE 4.142

Solve $\frac{d^2y}{dx^2} - 2\tan x\frac{dy}{dx} - 5y = 0$ by the method of removing first derivative.

Solution. Comparing the given equation with standard form, we have

$$P = -2\tan x, \ Q = -5, \ R = 0.$$

Therefore

$$Q_1 = Q - \frac{1}{2}\frac{dP}{dx} - \frac{P^2}{4} = -5 + \sec^2 x - \tan^2 x$$

$$= -5 + (1 + \tan^2 x) - \tan^2 x = -4,$$

and

$$z = e^{-\frac{1}{2}\int P dx} = e^{-\frac{1}{2}\int \tan x \, dx} = \sec x.$$

The normal form of the given equation is

$$\frac{d^2v}{dx^2} + Q_1 v = R_1$$

or

$$\frac{d^2v}{dx^2} - 4v = 0$$

or

$$(D^2 - 4)v = 0.$$

The auxiliary equation for this symbolic form is $m^2 - 4 = 0$. Therefore $m = \pm 2$ and so

$$v = c_1 e^{2x} + c_2 e^{-2x}$$

Hence the required solution is

$$y = vz = (c_1 e^{2x} + c_2 e^{-2x}) \sec x.$$

EXAMPLE 4.143

Solve $\frac{d^2y}{dx^2} - 4x\frac{dy}{dx} + (4x^2 - 3)y = e^{x^2}$.

Solution. We have

$$Q_1 = Q - \frac{1}{2}\frac{dP}{dx} - \frac{P^2}{4}$$

$$= 4x^2 - 3 - \frac{1}{2}(-4) - 4x^2 = -1$$

and

$$z = e^{-\frac{1}{2}\int P dx} = e^{2\int x \, dx} = e^{x^2}.$$

The normal form of the given equation is

$$\frac{d^2v}{dx^2} - v = \frac{e^{x^2}}{z} = \frac{e^{x^2}}{e^{x^2}} = 1$$

or

$$(D^2 - 1)v = 1.$$

Therefore A.E. is $m^2 - 1 = 0$, which yields $m = \pm 1$.

Thus

$$\text{C.F.} = c_1 e^x + c_2 e^{-x}.$$

Moreover

$$\text{P.I.} = -1.$$

Hence

$$v = \text{C.F.} + \text{P.I.} = c_1 e^x + c_2 e^{-x} - 1.$$

But $y = vz$. Therefore

$$y = (c_1 e^x + c_2 e^{-x} - 1)\, e^{x^2}.$$

EXAMPLE 4.144

Solve $x^2 \dfrac{d^2y}{dx^2} - 2(x^2 + x)\dfrac{dy}{dx} + (x^2 + 2x + 2)\, y = 0$

Solution.

The given equation can be written as

$$\frac{d^2y}{dx^2} - 2\left(1 + \frac{1}{x}\right)\frac{dy}{dx} + \left(1 + \frac{2}{x} + \frac{2}{x^2}\right) y = 0$$

Comparison with the standard form yields

$$P = -2\left(1 + \frac{1}{x}\right)$$

$$Q = 1 + \frac{2}{x} + \frac{2}{x^2} \text{ and } R = 0$$

Putting $y = uv$,the normal form is $\dfrac{d^2v}{dx^2} + Xv = Y$

where $u = e^{-\frac{1}{2}\int P\,dx} = e^{-\frac{1}{2}\int\left\{1+\frac{1}{x}\right\}dx}$

$$= e^x + \log u = xe^x$$

$$X = Q - \frac{1}{2}\frac{dp}{dx} - \frac{1}{4}P^2$$

$$= 1 + \frac{2}{x} + \frac{2}{x^2} - \frac{1}{2}\frac{2}{x^2} - \frac{1}{4}\cdot 4\left(1 + \frac{1}{x}\right)^2$$

$$= 0$$

and $\quad Y = Re^{\frac{1}{2}\int p\,dx} = 0$

Therefore, the normal form is

$$\frac{d^2v}{dx^2} = 0$$

On integration, we get

$$v = c_1 x + c_2$$

Therefore, the complete solution of given equation is

$$y = uv = xe^x \{c_1 x + c_2\}$$

EXAMPLE 4.145

Solve $\dfrac{d^2y}{dx^2} - 2\tan\dfrac{dy}{dx} + 5y = \sec x \cdot e^x$

Solution.

Here, $P = -2\tan x$, $Q = 5$ and $R = \sec x \cdot e^X$. Putting $y = uv$ in the given equation, we get

$$\frac{d^2y}{dx^2} + Xv = Y,$$

where $u = e^{-\frac{1}{2}\int P\,dx} = e^{\int \tan x\,dx}$

$$= e^{\log \sec x} = \sec x,$$

$$X = Q - \frac{1}{2}\frac{dP}{dx} - \frac{1}{4}P^2$$

$$= 5 + \frac{1}{2}\cdot 2\sec^2 x - \frac{1}{4}\cdot 4\tan^2 x = 6,$$

$$Y = Re^{\frac{1}{2}\int P\,dx} = e^x$$

Therefore, the equation reduces to the form:

$$\frac{d^2v}{dx^2} + 6v = e^x,$$

whose $C.F. = c_1 \cos\sqrt{6x} + c_2 \sin\sqrt{6}\,x$

and $P.I. = \dfrac{1}{D^2 + 6}e^x = \dfrac{e^x}{7}$

Therefore,

$$v = c_1 \cos\sqrt{6}x + c_2 \sin\sqrt{6}x + \frac{e^x}{7}$$

Hence, the solution of the given equation is

$$y = uv = \sec x\,\{c_1 \cos\sqrt{6}x + c_2 \sin\sqrt{6}x + \frac{e^x}{7}\}$$

EXAMPLE 4.146

Solve $\dfrac{d^2y}{dx^2} - 4x\dfrac{dy}{dx} + (4x^2 - 3)\,y = e^{x^2}$.

Solution. Here, $P = -4x$, $Q = 4x^2 - 3$, $R = e^{x^2}$ Putting $y = uv$, the normal form is

$$\frac{d^2v}{dx^2} + Xv = Y$$

where $u = e^{-\frac{1}{2}\int P dx} = e^{x^2}$

$$X = Q - \frac{1}{2}\frac{dP}{dx} - \frac{1}{4}P^2$$

$$= 4x^2 - 3 - \frac{1}{2}(-4) - \frac{1}{4}(16x^2) = -1$$

and $\quad Y = Re^{-\frac{1}{2}\int P dx} = 1.$

Therefore, the normal form is $\dfrac{d^2v}{dx^2} - v = 1$,

whose $C.F. = c_1e^x + c^2e^{-x}$ and $P.I. = \dfrac{1}{D^2 - 1}\cdot 1$ $= -(1 - D^2)^{-1}.1 = -1$

Therefore,

$$v = c_1e^x + c_2e^{-x} - 1$$

Hence, the solution of the giving equation is

$$y = uv = e^{x^2}\,(c_1e^x + c_2e^{-x} - 1)$$

EXAMPLE 4.147

Solve $\dfrac{d^2y}{dx^2} + \dfrac{2}{x}\dfrac{dy}{dx} + \left(1 + \dfrac{2}{x^2}\right)y = xe^x$.

Solution. Here, $P = -\frac{2}{x}$, $Q = 1 + \dfrac{2}{x^2}$ and $R = xe^x$.

Putting $y = uv$, the normal form is

$$\frac{d^2v}{dx^2} + Xv = Y,$$

where $u = e^{-\frac{1}{2}\int P dx} = e^{-\frac{1}{2}\int \frac{1}{x}dx} = e^{\log x} = x$

$$X = Q - \frac{1}{2}\frac{dP}{dx} - \frac{1}{4}P^2 = 1$$

and $\quad Y = Re^{\frac{1}{2}\int P dx}$

$$= xe^x.\,e^{-\int \frac{dx}{x}} = xe^x.\,e^{-\log x} = e^x$$

Thus, the normal form of the given equation is

$$\frac{d^2v}{dx^2} + v = e^x$$

whose $C.F. = c_1\cos x + c_2\sin x$ and $P.I. = \dfrac{1}{D^2+1}\cdot e^x = \dfrac{e^x}{2}$

Therefore,

$$v = c_1\cos x + c_2\sin x + \frac{1}{2}e^x.$$

Hence, the solution of the given equation is

$$y = uv = x\left(c_1\cos x + c_2\sin x + \frac{1}{2}e^x\right)$$

EXAMPLE 4.148

1. Solve $\dfrac{d^2y}{dx^2} - 4x\dfrac{dy}{dx} + (4x^2 - 1)y = -3e^{x^2}\sin 2x$

2. Solve $\dfrac{d^2y}{dx^2} + 2x\dfrac{dy}{dx} + (x^2 + 5)\,y = xe^{-\frac{1}{2}x^3}$

Solutions. 1. Here, $P = -4x$, $Q = 4x^2 - 1$ and $R = -3e^{x^2}\sin 2x$.

Putting $y = uv$; the given equation reduces to

$$\frac{d^2y}{dx^2} + Xv = Y$$

where $u = e^{-\frac{1}{2}\int P dx} = e^{x^2}$,

$$X = Q - \frac{1}{2}\frac{dP}{dx} - \frac{1}{4}P^2$$

$$= 4x^2 - 1 - \frac{1}{2}(-4) - \frac{1}{4}\cdot 16x^2 = 1,$$

$$Y = Re^{\frac{1}{2}\int P dx} = -3\sin 2x$$

Therefore, the reduced equation is

$$\frac{d^2v}{dx^2} + v = -3 \sin 2x,$$

whose $C.F. = c_1 \cos x + c_2 \sin x$ and

$$P.I. = \frac{1}{D^2+1}(-3 \sin 2x) = \frac{-3}{-2^2+1} \sin 2x$$

$$= \sin 2x$$

$$v = c_1 \cos x + c_2 \sin x + \sin 2x$$

Therefore, the solution of the given equation is

$$y = uv = e^{x^2}(c_1 \cos x + c_2 \sin x + \sin 2x)$$

2. Here, $P = 2x$, $Q = x^2 + 5$, $R = xe^{-\frac{1}{2}x^3}$

Putting $y = uv$; the given equation is transformed to

$$\frac{d^2y}{dx^2} + Xv = Y$$

where $u = e^{-\frac{1}{2}\int P dx} = e^{-\frac{1}{2}\int 2x\, dx} = e^{\frac{x^2}{2}}$

$$X = Q - \frac{1}{2}\frac{dP}{dx} - \frac{1}{4}P^2$$

$$= x^2 + 5 - 1 - x^2 = 4.$$

and $\qquad Y = Re^{\frac{1}{2}\int P dx} = x$

The transformed equation is

$$\frac{d^2v}{dx^2} + 4v = x.$$

The A.E. is $m^2 + 4 = 0$. Therefore $m = \pm 2i$, and so

$$C.F. = c_1 \cos(2x + c_2)$$

Further,

$$P.I. = \frac{x}{D^2+4} = \frac{1}{4}\left(1+\frac{D^2}{4}\right)^{-1} x = \frac{x}{4}.$$

Hence, the solution is

$$y = uv = e^{\frac{x^2}{2}}\left[c_1 \cos(2x + c^2) + \frac{1}{4}x\right]$$

4.27 METHOD OF UNDETERMINED COEFFICIENTS

This method is used to find Particular integral of the differential equation $F(D) = X$, where the input (forcing) function X consists of the sum of the terms, each of which possesses a finite number of essentially different derivatives. A trial solution consisting of terms in X and their finite derivatives is considered. Putting the values of the derivatives of the trial solution in $f(D)$ and comparing the coefficient on both sides of $f(D) = X$, the P.I. can be found. Obviously *the method fails if X consists of terms like sec x and tan x having infinite number of different derivatives.* Further, *if any term in the trial solution is a part of the complimentary function, then that terms should be multiplied by x and then tried.*

EXAMPLE 4.149

Solve $\frac{d^2y}{dx^2} + y = e^x + \sin x.$

Solution. The symbolic form of the given differential equation is

$$(D^2 + 1)y = e^x + \sin x.$$

The auxiliary equation is $m^2 + 1 = 0$. Therefore $m = \pm i$ and so

$$\text{C.F.} = c_1 \cos x + c_2 \sin x.$$

The forcing function consists of terms e^x and $\sin x$. Their derivatives are e^x and $\cos x$. So consider the trial solution.

$$y = a\, e^x + bx \sin x + cx \cos x.$$

Then

$$\frac{dy}{dx} = ae^x + b(x \cos x + \sin x)$$
$$+ c(\cos x - x \sin x)$$

$$\frac{d^2y}{dx} = ae^x + b\,[\cos x - x \sin x + \cos x]$$
$$+ c\,[- \sin x - \sin x - x \cos x]$$

$$= ae^x + 2b \cos x - bx \sin x - 2c \sin x$$
$$- cx \cos x$$

$$= ae^x + (2b - cx) \cos x - (bx + 2c) \sin x.$$

Substituting the values of $\frac{d^2y}{dx^2}$ and y in the given equation, we get

$$2ae^x + 2b \cos x - 2c \sin x = e^x + \sin x.$$

Comparing corresponding coefficients, we have

$$2a = 1,\ 2b = 0 \text{ and } 2c = -1.$$

Thus $a = \frac{1}{2}$, $b = 0$, $c = -\frac{1}{2}$ and so

$$\text{P.I.} = \frac{1}{2} e^x - \frac{1}{2} x \cos x.$$

Hence the solution is

$$y = c_1 \cos x + c_2 \sin x + \frac{1}{2} e^x - \frac{1}{2} x \cos x.$$

EXAMPLE 4.150

Solve $\frac{d^2y}{dx^2} - 4\frac{dy}{dx} + 4y = e^x \sin x.$

Solution. The symbolic form of the given differential equation is

$$m^2 - 4m + 4 = 0$$

which yields $m = 2, 2$. Hence

$$C.F. = (c_1 + c_2 x)\, e^{2x}.$$

The forcing function is $e^x \sin x$. Its derivative is $e^x \cos x + e^x \sin x$. Therefore, we consider

$$y = ae^x \sin x + be^x \cos x$$

as the trial solution. Then

$$\frac{dy}{dx} = a\,(e^x \cos x + e^x \sin x)$$
$$+ b\,(e^x \cos x - e^x \sin x)$$

$$= (a + b)\, e^x \cos x + (a - b)\, e^x \sin x,$$

$$\frac{d^2y}{dx^2} = (a + b)\,[e^x \cos x - e^x \sin x]$$
$$+ (a - b)\,[e^x \cos x + e^x \sin x]$$

$$= 2a\, e^x \cos x - 2be^x \sin x.$$

Substituting the values of $\frac{d^2y}{dx^2}, \frac{dy}{dx}$ and y in the given differential equation, we get

$$2ae^x \cos x - 2be^x \sin x - 4\,(a + b)\, e^x \cos x$$
$$- 4\,(a - b)\, e^x \sin x$$
$$+ 4\,ae^x \sin x + 4\,be^x \cos x = e^z \sin x$$

or

$$- 2ae^x \cos x + 2be^x \sin x = e^x \sin x.$$

Comparing coefficients, we get $2b = 1$ or $b = \frac{1}{2}$.

Hence

$$\text{P.I.} = \frac{1}{2} e^x \cos x$$

and so the complete solution of the given differential equation is

$$y = \text{C.F.} + \text{P.I.} = (c_1 + c_2 x)\, e^{2x} + \frac{1}{2} e^x \cos x.$$

4.28 METHOD OF REDUCTION OF ORDER

This method is used to find the complete solution of $\frac{d^2y}{dx^2} + P\frac{dy}{dx} + Qy = R$, where P, Q and R are function of x only, and when part of complementary function is known. So, let u, a function of x, be a part of the complementary function of the above differential equation. Then

$$\frac{d^2u}{dx^2} + P\frac{du}{dx} + Qu = 0 \qquad (156)$$

Let $y = uv$ be the complete solution of the given differential equation, where v is also a function of x. Then

$$\frac{dy}{dx} = u\frac{dv}{dx} + v\frac{du}{dx},$$

$$\frac{d^2y}{dx^2} = u\frac{d^2v}{dx^2} + 2\frac{du}{dx}\cdot\frac{dv}{dx} + v\frac{d^2v}{dx^2}.$$

Substituting these values of $\frac{d^2y}{dx^2}, \frac{dy}{dx}$ and y in the given equation, we get

$$u\frac{d^2v}{dx^2} + 2\frac{du}{dx}\cdot\frac{dv}{dx} + v\frac{d^2v}{dx^2} + P\left(u\frac{dv}{dx} + \frac{du}{dx}\right)$$

$$+ Quv = R$$

or

$$u\frac{d^2v}{dx^2} + \left(2\frac{du}{dx} + Pu\right)\cdot\frac{dv}{dx}$$

$$+\left(\frac{d^2u}{dx^2} + P\frac{du}{dx} + Qu\right)v = R$$

or, using (156)

$$u\frac{d^2v}{dx^2} + \left(2\frac{du}{dx} + Pu\right)\frac{dv}{dx} = R$$

or, division by u yields,

$$\frac{d^2v}{dx^2} + \left(\frac{2}{u}\frac{du}{dx} + P\right)\frac{dv}{dx} = \frac{R}{u}$$

or, taking $\frac{dv}{dx} = z$, we get

$$\frac{dz}{dx} + \left(\frac{2}{u}\frac{du}{dx} + P\right)z = \frac{R}{u} \qquad (157)$$

which is a first order differential equation in z and x. The integration factor for (157) is

$$\text{I.F.} = e^{\left(\int \frac{2}{u}dx + \int pdx\right)} = u^2 e^{\int pdx}.$$

Therefore, the solution of (157) is

$$zu^2 e^{\int Pdx} = \int \frac{R}{u}\left(u^2 e^{\int pdx}\right)dx + c_1$$

or

$$z = \frac{1}{u^2 e^{\int Pdx}}\left[\int \frac{R}{u}\left(u^2 e^{\int Pdx}\right)dx + c_1\right]$$

or

$$\frac{dv}{dx} = \frac{1}{u^2}e^{-\int Pdx}\left[\int Ru\, e^{\int Pdx}\,dx + c_1\right] \qquad (158)$$

Integrating (158) with respect to x, we get

$$v = \int \frac{1}{u^2}e^{-\int Pdx}\left[\int Ru\, e^{\int Pdx}\,dx + c_1\right] + c_2,$$

where c_1 and c_2 are constants of integration. Hence the complete solution of the given differential equation is

$$y = uv$$

$$= u\left\{\int \frac{1}{u^2}e^{-\int Pdx}\left[\int Rue^{\int Pdx}\,dx + c_1\right]\right\} + c_2u.$$

4.28.1 Method to Find the Particular Integral or the First Part of the Solution for

$$\frac{d^2y}{dx^2} + P\frac{dy}{dx} + Qy = 0 \qquad (159)$$

I. $y = e^{mx}$ is a solution

If $y = e^{mx}$, then $\frac{dy}{dx} = me^{mx}$ and $\frac{d^2y}{dx^2} = m^2e^{mx}$

Thus if $y = e^{mx}$ is a solution of equation (159), then

$$(m^2 + Pm + Q)\, e^{mx} = 0$$

or

$$m^2 + Pm + Q = 0$$

Deductions:

(i) Put $m = 1$, *i, e.,* $y = e^X$ is solution of (159) if $1 + P + Q = 0$.

(ii) Put $m = 2$, *i, e.,* $y = e^{2x}$ is solution of (159) if $1 + \frac{P}{2} + \frac{Q}{2^2} = 0$

(iii) Put $m = a$, *i, e.,* $y = e^{ax}$ is solution of (159) if $1 + \frac{P}{a} + \frac{Q}{a^2} = 0.$

II. $y = x^m$ is a solution

If $y = x^m$ then $\frac{dy}{dx} = mx^{m-1}$ and

$$\frac{d^2y}{dx^2} = m(m-1)x^{m-2}$$

If $y = x^m$ is a solution of (159), then

$$m(m-1)x^{m-2} + Pmx^{m-1} + Qx^m = 0$$

or $$m(m-1)x^{m-2} + Pmx^{m-1} + Qx^m = 0$$

or $$m(m-1) + Pmx + Qx^2 = 0$$

Deductions:

(i) Put $m = 1$,i,e., $y = x$ is solution of (159) if $P + Qx = 0$.

(ii) Put $m = 2$,i,e., $y = x^2$ is solution of (159) if $2 + 2Px + Qx^2 = 0$.

(iii) Put $m = 3$,i,e., $y = x^3$ is solution of (159) if $6 + 3Px^2 + Qx^3 = 0$.

The above deduction helps in finding the integral belonging to complementary function of the given equation.

EXAMPLE 4.151

Find the complete solution of

$$\frac{d^2y}{dx^2} - 4x\frac{dy}{dx} + (4x^2 - 2)y = 0,$$

if $y = e^{x^2}$ is an integral included in the complementary solution.

Solution. Let $y = uv$, where $u = e^{x^2}$ be the complete solution of the given equation. Then

$$\frac{d^2v}{dx^2} + \left[P + \frac{2}{u}\frac{du}{dx}\right]\frac{dv}{dx} = 0,$$

where

$$P = -4x, Q = 4x^2 - 2 \text{ and } R = 0.$$

Thus

$$\frac{d^2v}{dx^2} + \left[-4x + \frac{2}{e^{x^2}}(2xe^{x^2})\right]\frac{dv}{dx} = 0$$

or

$$\frac{d^2v}{dx^2} + [4x - 4x]\frac{dv}{dx} = 0$$

or

$$\frac{d^2v}{dx^2} = 0.$$

Integrating, we get

$$\frac{dv}{dx} = c_1.$$

Integrating once more, we get

$$v = c_1x + c_2.$$

Hence the complete solution is

$$y = uv = e^{x^2}[c_1x + c_2].$$

EXAMPLE 4.152

Solve $\sin^2 x\frac{d^2y}{dx^2} = 2y$, given that $y = \cot x$ is an integral included in the complementary function.

Solution. The given equation is

$$\frac{d^2y}{dx^2} - (2\cos ec^2 x)y = 0.$$

Comparing with

$$\frac{d^2y}{dx^2} + P\frac{dy}{dx} + Qy = R,$$

we get

$$P = 0, Q = -2\cos ec^2 x, R = 0.$$

Therefore, putting $y = uv = (\cot x)v$, the reduced equation is

$$\frac{d^2v}{dx^2} + \left[P + \frac{2}{u}\frac{du}{dx}\right]\frac{dv}{dx} = 0,$$

that is,

$$\frac{d^2v}{dx^2} + \left[\frac{2}{\cot x}(-\cos ec^2 x)\right]\frac{dv}{dx} = 0$$

or

$$\cot x\frac{d^2v}{dx^2} - 2\cos ec^2 x\frac{dv}{dx} = 0. \quad (160)$$

Let $\frac{dv}{dx} = z$. Then (160) reduces to

$$\cot x \frac{dz}{dx} - (2 \cos ec^2 x) z = 0$$

or

$$\frac{dz}{z} = 2 \frac{\cos ec^2 x}{\cot x} dx. \qquad (161)$$

Integrating (161), we get

$$\log z = -2 \log \cot x + \log c_1$$

or

$$\log z + \log \cot^2 x = \log c_1$$

or

$$z = c_1 \tan^2 x$$

or

$$\frac{dv}{dx} = c_1 \tan^2 x = c_1 (\sec^2 x - 1) \qquad (162)$$

Integrating (162), we get

$$v = c_1 (\tan x - x) + c_2,$$

where c_1 and c_2 are constants of integration. Hence the complete solution is

$$y = uv = \cot x [c_1 (\tan x - x) + c_2]$$
$$= c_1 (1 - x \cot x) + c_2 \cot x.$$

EXAMPLE 4.153

Solve $x^2 \frac{d^2y}{dx^2} - 2x (1 + x) \frac{dy}{dx} + 2 (1 + x) y = x^3$

Solution. The given equation can be written as

$$\frac{d^2y}{dx^2} - 2 \left\{\frac{1}{x} + 1\right\} \frac{dy}{dx} + 2 \left\{\frac{1}{x^2} + \frac{1}{x}\right\} y = x \quad (163)$$

Therefore comparing with the standard form we have

$$P = -2 \left(\frac{1}{x} + 1\right),$$

$$Q = 2 \left\{\frac{1}{x^2} + \frac{1}{x}\right\}, \; R = x$$

we note that

$$P + Qx = -2 \left\{\frac{1}{x} + 1\right\} + 2 \left\{\frac{1}{x^2} + \frac{1}{x}\right\} = 0.$$

Therefore, $y = x$ is a part of C.F. Putting $y = vx$, we have

$$\frac{dy}{dx} = v + x \frac{dv}{dx}$$

and $$\frac{d^2y}{dx^2} = \frac{d^2v}{dx^2} \cdot x + 2 \frac{dv}{dx}$$

We have from (163)

$$x \frac{d^2v}{dx^2} + 2 \frac{dv}{dx} - 2 \left(\frac{1}{x} + 1\right) \left(x \frac{dv}{dx} + v\right)$$
$$+ 2 \left(\frac{1}{x^2} + \frac{1}{x}\right) vx = x$$

or

$$x \frac{d^2v}{dx^2} + \{2 - 2 - 2x\} \frac{dv}{dx}$$
$$+ \left\{\frac{-2}{x} - 2 + \frac{2}{x} + 2\right\} v = x$$

or

$$\frac{d^2v}{dx^2} - 2 \frac{dv}{dx} = 1$$

or $$\frac{dp}{dx} - 2p = 1 \; [\text{where } p = \frac{dv}{dx}]$$

which is a linear equation. For this equation

$$I.F. = e^{-2\int dx} = e^{-2x}$$

Therefore

$$p. e^{-2x} = \int 1 \cdot e^{-2x} dx + c_1 = -\frac{1}{2} e^{-2x} + c_1$$

or $$p = \frac{dv}{dx} = -\frac{1}{2} + c_1 e^{2x}$$

Interegrating we get

$$v = -\frac{1}{2} x^2 + c_1 \frac{e^{2x}}{2} + c_2$$

Therefore, the complete solution is

$$y = vx = -\frac{1}{2} x^2 + \frac{c_1}{2} x e^{2x} + c_2 x$$

EXAMPLE 4.154

Solve $x^2 \frac{d^2y}{dx^2} - (x^2 + 2x) \frac{dy}{dx} + (x + 2) y = x^3 e^x$

Solution. The given equation can be written as

$$\frac{d^2y}{dx^2}-\left(1+\frac{2}{x}\right)\frac{dy}{dx}+\left(\frac{1}{x}+\frac{2}{x^2}\right)y=xe^x \quad (164)$$

Hence, $P=-\left(1+\frac{2}{x}\right)$, $Q=\frac{1}{x}+\frac{2}{x^2}$ and $R=xe^x$
Since $P+Qx=0$, therefore, $y=x$ is a part of C. F. Putting $y=vx$, we have

$$\frac{dy}{dx}=v+x\cdot\frac{dv}{dx}\Rightarrow\frac{d^2y}{dx^2}=x\frac{d^2v}{dx^2}+2\frac{dv}{dx}$$

Therefore (164) yields

$$\frac{d^2v}{dx^2}-\frac{dv}{dx}=e^x$$

or $$\frac{dv}{dx}-p=e^x, \qquad \text{where } p=\frac{dv}{dx},$$

which is a linear eqaution. For this equation

$$I.F.=e^{-\int dx}=e^{-x}$$

Therefore

$$p.e^{-x}=\int e^{-x}\cdot e^x+c_1=x+c_1$$

or

$$p=\frac{dv}{dx}=xe^x+c_1e^x$$

On integration, we have

$$v=xe^x-e^x+c_1e^x+c_2$$

Therefore, the complete solution is

$$y=vx=x^2e^x-xe^x+c_1xe^x+c_2x$$

EXAMPLE 4.155

Solve $\frac{d^2y}{dx^2}-\cot x\frac{dy}{dx}-(1-\cot x)y=e^x\sin x$

Solution. We note that, $P+Q+1=0$, therefore $y=e^x$ is a part of C.F. Put $y=ve^x$ so that

$$\frac{dy}{dx}=e^x\cdot\frac{dv}{dx}+v\cdot e^x \text{ and}$$

$$\frac{d^2y}{dx^2}=e^x\cdot\frac{d^2v}{dx^2}+2\frac{dv}{dx}\cdot e^x+v\cdot e^x$$

Putting these values in the given equation, we get

$$\frac{d^2y}{dx^2}+(2-\cot x)\frac{dv}{dx}=\sin x$$

or $$\frac{dv}{dx}+(2-\cot x)p=\sin x \quad \left[\text{where } p=\frac{dv}{dx}\right]$$

which is a linear equation. Now

$$I.F.=e^{\int(2-\cot x)dx}=e^{2x-\log\sin x}=\frac{e^{2x}}{\sin x}.$$

Therefore

$$p\cdot\frac{e^{2x}}{\sin x}=\int\frac{e^{2x}}{\sin x}\cdot\sin x\,dx+c_1=\frac{1}{2}e^{2x}+c_1$$

or

$$p=\frac{dv}{dx}=\frac{1}{2}\sin x+c_1e^{-2x}\sin x$$

On integration, we have

$y=ve^x$

$$=-\frac{1}{2}e^x\cos x-\frac{c_1}{5}e^{-x}(2\sin x+\cos x)+c_2e^x$$

EXAMPLE 4.156

Solve $x\frac{dv}{dx}-y=(x-1)\left(\frac{d^2y}{dx^2}-x+1\right)$

Solution. The given equation can be written as

$$\frac{d^2y}{dx^2}-\frac{x}{x-1}\frac{dy}{dx}+\frac{y}{x-1}=x-1.$$

Here $P+Qx=0$, therefore $y=x$ is a part of C.F. Put $y=vx$, so that

$$\frac{dy}{dx}=\frac{dv}{dx}\cdot x+v \text{ and } \frac{d^2y}{dx^2}=\frac{d^2v}{dx^2}+2\frac{dv}{dx}.$$

Thus, the given equation reduces to

$$\frac{d^2v}{dx^2}+\left(-\frac{x}{x-1}\frac{2}{x}\right)\frac{dv}{dx}=\frac{x-1}{x}$$

or $$\frac{dp}{dx}+\left(-\frac{x}{x-1}+\frac{2}{x}\right)p=\frac{x-1}{x} \text{ where } x=\frac{dv}{dx}$$

which is a linear equation. Now

$$I.F. = e^{-\int \frac{x}{x-1}dx + \int \frac{2}{x}dx}$$

$$= e^{-\int \left(1+\frac{1}{x-1}\right)dx + \int \frac{2}{x}dx}$$

$$= e^{-x-\log(x-1)+2\log x} = \frac{x^2}{(x-1)}e^{-x}.$$

Therefore,

$$p.\frac{x^2e^{-x}}{x-1} = \int \frac{x-1}{x}\frac{x^2}{x-1}e^{-x}\,dx + c_1$$

$$= \int xe^{-x}\,dx + c_1$$

$$= -xe^{-x} - e^{-x} + c_1.$$

or

$$p = \frac{dv}{dx} = -\frac{x-1}{x} - \frac{(x-1)}{x^2} + \frac{c_1(x-1)e^x}{x^2}$$

$$= -1 + \frac{1}{x^2} + c_1\left(\frac{1}{x} - \frac{1}{x^2}\right)e^x.$$

Integrating, we get

$$v = -x - \frac{1}{x} + c_1 \cdot \frac{1}{x}e^x + c_2$$

Therefore, the complete solution is

$$y = vx = -x^2 - 1 + c_1e^x + c_2x$$

$$= c_1e^x + c_2x - (1 - x^2).$$

EXAMPLE 4.157

Solve $(1-x^2)\frac{d^2y}{dx^2} + x\frac{dy}{dx} - y = x(1-x^2)^{\frac{3}{2}}$

Solution. The given equation can be written as

$$\frac{d^2y}{dx^2} + \frac{x}{1-x^2}\frac{dy}{dx} - \frac{x}{1-x^2}y = x(1-x^2)^{\frac{1}{2}}$$

Here, $P + Qx = 0$. Therefore $y = x$ is a part of C.F. Putting $y = vx$, the given equation reduce to

$$\frac{d^2v}{dx^2} + \left(\frac{x}{1-x^2} + 2\right)\frac{dy}{dx} = x(1-x^2)^{\frac{1}{2}}$$

or $\quad \frac{dp}{dx} + \left(\frac{x}{1-x^2} + \frac{2}{x}\right)p = \sqrt{(1-x^2)},$

where $\quad p = \frac{dv}{dx},$

which is a linear equation. Now

$$I.F. = e^{\int\left(\frac{x}{1-x^2}+\frac{2}{x}\right)dx}$$

$$= e^{-\frac{1}{2}\log(1-x^2)+2\log x}$$

$$= \frac{x^2}{\sqrt{1-x^2}}$$

Therefore,

$$p\cdot\frac{x^2}{\sqrt{(1-x^2)}}\int x^2\,dx + c_1 + \frac{x^3}{3} + c_1$$

or

$$p = \frac{dv}{dx}$$

$$= \frac{1}{3}x\sqrt{(1-x^2)} + \frac{c_1}{x^2}\sqrt{(1-x^2)}$$

$$= \frac{1}{3}x\sqrt{(1-x^2)} + c_1(1-x^2)^{\frac{1}{2}}\cdot\frac{1}{x^2}.$$

Integrating, we have

$$v = -\frac{1}{9}(1-x^2)^{\frac{3}{2}} + c_1(1-x^2)^{\frac{1}{2}}\left(-\frac{1}{x}\right) - c_1\int\frac{dx}{\sqrt{(1-x^2)}} + c_2$$

$$= -\frac{1}{9}(1-x^2)^{\frac{3}{2}} - \frac{c_1}{x}(1-x^2)^{\frac{1}{2}} - c_1\sin^{-1}x + c_2.$$

Therefore, the complete solution is

$$y = vx = \frac{1}{9}x(1-x^2)^{\frac{3}{2}} - c_1\{x\sin^{-1}x + \sqrt{(1+x^2)}\} + c_2x.$$

EXAMPLE 4.158

Solve $(x+2)\frac{d^2y}{dx^2}+(2x+5)\frac{dy}{dx}+2y=(x-1)e^x$.

Solution. The given equation reduces to

$$\frac{d^2y}{dx^2}-\frac{2x+5}{x+2}\frac{dy}{dx}+\frac{2}{x+2}y=\frac{x+1}{x+2}e^x.$$

Here, $\frac{P}{2}+\frac{Q}{2^2}+1=0$. Therefore $y=e^{2x}$ is a solution of this equation.

Thus Putting $y=ve^{2x}$, the equation reduces to

$$(x+2)\frac{d^2v}{dx^2}+(2x+3)\frac{dv}{dx}=(x+1)e^{-x}$$

or $$\frac{d^2v}{dx^2}+\frac{2x+3}{x+2}\frac{dv}{dx}=\frac{x+1}{x+2}e^{-x}$$

or $$\frac{dp}{dx}+\frac{2x+3}{x+2}p=\frac{x+1}{x+2}e^{-x}, \text{ where } p=\frac{dv}{dx},$$

which is a linear equation. For this equation

$$I.F.=e^{\int\frac{2x+3}{x+2}dx}$$

$$=e^{\int\left(2-\frac{1}{x+2}\right)dx}$$

$$=e^{2x-\log(x+2)}=\frac{e^{2x}}{x+2}.$$

Therefore

$$p\cdot\frac{e^{2x}}{x+2}=\int\frac{(x+1)}{(x+2)^2}e^x\,dx+c_1$$

$$=\int\left\{\frac{1}{x+2}+\frac{1}{(x+2)^2}\right\}e^xdx+c_1$$

$$=\frac{e^x}{x+2}+c_1$$

or

$$p=\frac{dv}{dx}=e^{-x}+c_1e^{-2x}(x+2).$$

Integrating, we get

$$v=-e^{-x}-\frac{1}{2}c_1e^{-2x}(x+2)-\frac{1}{4}c_1e^{-2x}+c_2$$

$$=-e^{-x}-\frac{1}{4}c_1(2x+5)e^{-2x}+c_2.$$

Hence, the complete solution is

$$y=ve^{2x}=-e^x-\frac{1}{4}c_1(2x+5)+c_2e^{2x}.$$

EXAMPLE 4.159

Solve $x^2\frac{d^2y}{dx^2}+1+\frac{2}{x}\cot x-\frac{2}{x^2}y=x\cos x$ given that $\frac{\sin x}{x}$ is a C.F.

Solution.

Put $y=v\frac{\sin x}{x}$ so that

$$\frac{dy}{dx}=\frac{dv}{dx}\frac{\sin x}{x}+v\left(\frac{x\cos x-\sin x}{x^2}\right)$$

$$=\frac{dv}{dx}\frac{\sin x}{x}+v\left(\frac{\cos x}{x}-\frac{\sin x}{x^2}\right)$$

and

$$\frac{d^2y}{dx^2}=\frac{d^2v}{dx^2}\frac{\sin x}{x}+2\frac{dv}{dx}\cdot\left(\frac{\cos x}{x}-\frac{\sin x}{x^2}\right)$$
$$+v\left(-\frac{\sin x}{x}-2\frac{\cos x}{x}+\frac{\sin x}{x^3}\right)$$

The given equation reduces to

$$\frac{d^2v}{dx^2}\frac{\sin x}{x}+2\left(\frac{\cos x}{x}-\frac{\sin x}{x^2}\right)\frac{dv}{dx}$$
$$+\left(-\frac{\sin x}{x}-\frac{2\cos x}{x}+\frac{2\sin x}{x^3}\right)$$
$$+\left(1+\frac{2\cot x}{x}\frac{2}{x^2}\right)v\frac{\sin x}{x}=x\cos x$$

or $$\frac{d^2v}{dx^2}+2\left(\cot x-\frac{1}{x}\right)\frac{dv}{dx}=x^2\cot x$$

or $$\frac{dp}{dx}+2\left(\cot x-\frac{1}{x}\right)p=x^2\cot x$$

where $p=\frac{dv}{dx}$ which is a linear equation.

For this equation,

$$I.F.=e^{\int 2\left(\cot x-\frac{1}{x}\right)dx}$$

$$=e^{2(\log\sin x-\log x)}=\frac{\sin^2 x}{x^2}.$$

Therefore,

$$p.\frac{\sin^2 x}{x^2} = \int x^2 \cot x \cdot \frac{\sin^2 x}{x^2}\,dx + c_1$$

$$= \frac{1}{2}\int \sin 2x\,dx + c_1$$

$$= \frac{1}{4}\cos 2x + c_1.$$

or,

$$p = \frac{dv}{dx} = -\frac{1}{x^2}\,\mathrm{cosec}^2 x$$

$$+ \frac{1}{2}x^2 + c_1 x^2\,\mathrm{cosec}^2 x$$

Integrating, we get

$$v = \left(c_1 - \frac{1}{4}\right)[-x^2 \cot x + 2x \log \sin x$$

$$- 2\int \log \sin x\,dx] + \frac{x^3}{6} + c_2.$$

Therefore complete solution is

$$y = v \cdot \frac{\sin x}{x}$$

$$= \left(c_1 - \frac{1}{4}\right)\left[-x \cos x + 2 \sin x \log \sin x\right.$$

$$\left. - \frac{2\sin x}{x} \cdot \int \log \sin x\,dx\right] + \frac{x^2 \sin x}{6} + c_2 \frac{\sin x}{x}$$

4.29 THE CAUCHY-EULER HOMOGENEOUS LINEAR EQUATION

Consider the following differential equation with variable coefficients:

$$x^n \frac{d^n y}{dx^n} + a_1 x^{n-1} \frac{d^{n-1} y}{dx^{n-1}} + \cdots + a_{n-1} x \frac{dy}{dx}$$

$$+ a_n = F(x), \qquad (165)$$

where a_i are constant and F is a function of x. This equation is known as *Cauchy-Euler homogeneous linear equation* (or *equidimensional equation*). The Cauchy-Euler homogeneous linear equation can be reduced to linear differential equation with constant coefficients by putting $x = e^t$ or $t = \log x$. Then

$$\frac{dy}{dx} = \frac{dy}{dt} \cdot \frac{dt}{dx} = \frac{dy}{dt} \cdot \frac{1}{x}$$

or

$$x\frac{dy}{dx} = \frac{dy}{dt} = Dy.$$

Now

$$\frac{d^2y}{dx^2} = \frac{d}{dx}\left(\frac{1}{x}\frac{dy}{dt}\right) = -\frac{1}{x^2}\frac{dy}{dt} + \frac{1}{x}\frac{d^2y}{dt^2} \cdot \frac{dt}{dx}$$

$$= -\frac{1}{x^2}\frac{dy}{dt} + \frac{1}{x^2}\frac{d^2y}{dt^2}$$

and so

$$x^2\frac{d^2y}{dx^2} = \frac{d^2y}{dt^2} - \frac{dy}{dt} = D^2 y - Dy = D(D-1)y.$$

Similarly,

$$x^3\frac{d^3y}{dx^3} = D(D-1)(D-2)y$$

and so on. Putting these values in (165), we obtain a linear differential equation with constant coefficients which can be solved by using the methods discussed already.

EXAMPLE 4.160

Solve $x^2\frac{d^2y}{dx^2} - 2x\frac{dy}{dx} + 2y = x^3.$

Solution. This is a Cauchy-Euler equation. Putting $x = e^t$ or $t = \log x$, we have $x\frac{dy}{dx} = Dy$, $x^2\frac{d^2y}{dx^2} = D(D-1)y$. Hence the given equation transforms to

$$(D(D-1) - 2D + 2)y = e^{3t}$$

or

$$(D^2 - 3D + 2)y = e^{3t},$$

which is a linear differential equation with constant coefficient. The auxiliary equation is $m^2 - 3m + 2 = 0$ and so $m = 1, 2$. Therefore

$$\text{C.F.} = c_1 e^t + c_2 e^{2t}.$$

The particular integral is

$$\text{P.I.} = \frac{1}{f(D)} F(x) = \frac{1}{D^2 - 3D + 2} e^{3t}$$

$$= \frac{1}{9 - 9 + 2} e^{3t} = \frac{1}{2} e^{3t}.$$

Hence the complete solution is

$$y = c_1 e^t + c_2 e^{2t} + \frac{1}{2} e^{3t}$$

Returning back to the variable x, we have

$$y = c_1 x + c_2 x^2 + \frac{1}{2} x^3.$$

EXAMPLE 4.161

Solve the Cauchy-Euler equation

$$x^2 \frac{d^2y}{dx^2} - x\frac{dy}{dx} + y = \log x.$$

Solution. Putting $x = e^t$, we have

$$x\frac{dy}{dx} = Dy \text{ and } x^2 \frac{d^2y}{dx^2} = D(D-1)\,y$$

and so the equation transforms to

$$(D(D-1) - D + 1)\,y = t$$

or

$$(D-1)^2 y = t.$$

The complementary function is

$$\text{C.F.} = (c_1 + c_2 t)\, e^t.$$

Now

$$\text{P.I.} = \frac{1}{(D-1)^2} t = (1-D)^{-2} t$$

$$= (1 + 2D + \cdots)\, t = t + 2.$$

Hence the complete solution is

$$y = \text{C.F.} + \text{P.I.} = (c_1 + c_2 t)\, e^t + t + 2.$$

Returning back to x, we get

$$y = (c_1 + c_2 \log x)\, x + \log x + 2.$$

EXAMPLE 4.162

Solve Cauchy-Euler equation

$$x^2 \frac{d^2y}{dx^2} + x\frac{dy}{dx} + y = \log x \sin(\log x).$$

Solution. Putting $x = e^t$, this equation transforms to

$$(D^2 + 1)\, y = t \sin t.$$

The complementary function is

$$\text{C.F.} = c_1 \cos t + c_2 \sin t.$$

Further

$$\text{P.I.} = \frac{1}{D^2 + 1} t \sin t = \text{I.P. of } \frac{1}{D^2 + 1} t\, e^{it}$$

$$= \text{I.P. of } e^{it} \frac{1}{(D + i)^2 + 1}$$

$$= \text{I.P. of } e^{it} \frac{1}{2iD\left(1 + \frac{D}{2i}\right)} t$$

$$= \text{I.P. of } e^{it} \cdot \frac{1}{2i} \cdot \frac{1}{D}\left(1 - \frac{iD}{2}\right)^{-1} t$$

$$= \text{I.P. of } e^{it} \left(\frac{-it^2}{4} + \frac{t}{4}\right)$$

$$= \text{I.P. of } (\cos t + i \sin t)\left(\frac{-it^2}{4} + \frac{t}{4}\right)$$

$$= -\frac{t^2}{4} \cos t + \frac{t}{4} \sin t.$$

Therefore, the complete solution is

$$y = c_1 \cos t + c_2 \sin t - \frac{t^2}{4} \cos t + \frac{t}{4} \sin t.$$

Returning back to x, we get

$$y = c_1 \cos(\log x) + c_2 \sin(\log x) - \frac{1}{4} (\log x)^2 \cos(\log x) + \frac{1}{4} \log x \sin(\log x).$$

EXAMPLE 4.163

Solve the Cauchy-Euler equation

$$x^2 \frac{d^2y}{dx^2} - x\frac{dy}{dx} + 2y = x \log x.$$

Solution. Putting $x = e^t$, the equation reduces to

$$(D(D-1) - D + 2)\, y = t\, e^t$$

or

$$(D^2 - 2D + 2)\, y = t\, e^t$$

The C.F. for this equation is

$$\text{C.F.} = e^t (c_1 \cos t + c_2 \sin t).$$

The particular integral is

$$\text{P.I.} = \frac{1}{D^2 - 2D + 2}\, t\, e^t$$

$$= e^t \frac{1}{(D+1)^2 - 2(D+1) + 2}\, t$$

$$= e^t \frac{1}{D^2 + 1}\, t = e^t (1 - D^2)^{-1}\, t$$

$$= e^t (t - 0) = t\, e^t.$$

Therefore, the complete solution is

$$y = \text{C.F.} + \text{P.I.} = e^t (c_1 \cos t + c_2 \sin t) + t\, e^t$$

$$= x\,(c_1 \cos(\log x))$$

$$+ c_2 \sin(\log x) + x \log x.$$

EXAMPLE 4.164

Solve $x^2 \dfrac{d^2y}{dx^2} - 2y = x^2 + \frac{1}{x}\cdot$

Solution. Putting $x = e^t$, the given equation reduces to

$$(D(D-1) - 2)\, y = e^{2t} + \frac{1}{e^t}\cdot$$

The auxiliary equation is $m^2 - m - 2 = 0$ and so $m =' 2, -1$. Therefore,

$$\text{C.F.} = c_1 e^{2t} + c_2 e^{-t}.$$

Moreover,

$$\text{P.I.} = \frac{1}{D^2 - D - 2}\left(e^{2t} + \frac{1}{e^t}\right)$$

$$= \frac{1}{D^2 - D - 2}\, e^{2t} + \frac{1}{D^2 - D - 2}\,(e^{-t})$$

$$= t \frac{1}{[2D - 1]_{D=2}}\, e^{2t} + t \frac{1}{(2D-1)_{D=-1}}\, e^{-t}$$

$$= \frac{t}{3} e^{2t} - \frac{1}{3} t e^{-t}.$$

Thus, the complete solution is

$$y = \text{C.F.} + \text{P.I.} = c_1 e^{2t} + c_2 e^{-t} + \frac{t}{3} e^{2t} - \frac{t}{3} e^{-t}$$

$$= c_1 x^2 + \frac{c_2}{x} + \frac{1}{3}\left(x^2 - \frac{1}{x}\right) \log x.$$

EXAMPLE 4.165

Solve the Cauchy-Euler equation

$$x^2 \frac{d^2y}{dx^2} + 2x \frac{dy}{dx} - 20y = (x+1)^2.$$

Solution. Putting $x = e^t$, the given equation transforms into

$$(D(D-1) + 2D - 20)\, y = e^{2t} + 2e^t + 1$$

or

$$(D^2 + D - 20)\, y = e^{2t} + 2e^t + 1.$$

The auxiliary equation is $m^2 + m - 20 = 0$ and so $m' =' -5, 4$. Therefore,

$$\text{C.F.} = c_1 e^{-5t} + c_2 e^{4t}.$$

Now

$$\text{P.I.} = \frac{1}{D^2 + D - 20}\, e^{2t} + \frac{2}{D^2 - D - 20}\, e^t$$

$$+ \frac{1}{D^2 - D - 20}\, x^{0t}$$

$$= -\frac{1}{14} e^{2t} + \frac{2}{-20} e^t - \frac{1}{20}\cdot$$

Thus the complete solution is

$$y = \text{C.F.} + \text{P.I.}$$

$$= c_1 e^{-5t} + c_2 e^{4t} - \frac{1}{14} e^{2t} - \frac{1}{10} e^t - \frac{1}{20}$$

$$= c_1 x^{-5} + c_2 x^4 - \frac{1}{14} x^2 - \frac{x}{10} - \frac{1}{20}$$

EXAMPLE 4.166

Solve the differential equations:

$$(x^2D^2 - xD - 3)y = x^2 (\log x)^2.$$

Solution. The given equation

$$x^2 \frac{d^2y}{dx^2} - x\frac{dy}{dx} - 3y = x^2 (\log x)^2$$

is a Cauchy-Eular homogeneous linear equation. So put $x = e^t$ so that $t = \log x$. Then

$$x\frac{dy}{dx} = Dy \text{ and } x^2\frac{d^2y}{dx^2} = D(D-1)y,$$

and the given equation transforms to

$$[D(D-1) - D - 3]\, y = t^2 e^{2t}$$

or

$$(D^2 - 2D - 3)\, y = t^2 e^{2t}.$$

The roots of the auxiliary equation $m^2 - 2m - 3 = 0$ are $m = 3$ and $m = -1$. Therefore

$$C.F. = c_1 e^{2t} + c_2 e^{-t}$$

Now

$$P.I. = \frac{1}{D^2 - 2D - 3}(t^2 e^{2t})$$

$$= e^{2t}\frac{1}{[(D+2)^2 - 2(D+2) - 3]}t^2$$

$$= e^{2t}\frac{1}{D^2 - 2D - 3}t^2$$

$$= -\frac{e^{2t}}{3}\frac{1}{\left(1 - \frac{2}{3}D - \frac{D^2}{3}\right)}t^2$$

$$= -\frac{1}{3}e^{2t}\left[1 - \left(\frac{2D}{3} + \frac{D^2}{3}\right)\right]^{-1} t^2$$

$$= -\frac{1}{3}e^{2t}\left[1 + \frac{2D}{3} + \frac{D^2}{3} + \frac{4D^2}{9} + \ldots\right]t^2$$

$$= -\frac{1}{3}e^{2t}\left[1 + \frac{2D}{3} + \frac{7D^2}{9} + \ldots\right]t^2$$

$$= -\frac{1}{3}e^{2t}\left[t^2 + \frac{2}{3}D(t^2) + \frac{7}{9}D^2(t^2)\right]$$

$$= -\frac{1}{3}e^{2t}\left[t^2 + \frac{4}{3}t + \frac{14}{9}\right]$$

Therefore the complete solution is

$$y = \text{C.F.} + \text{P.I.}$$

$$= c_1 e^{3t} + c_2 e^{-t} - \frac{1}{3}e^{2t}\left[t^2 + \frac{4}{3}t + \frac{14}{9}\right]$$

$$= c_1 x^3 + \frac{c_2}{x} - \frac{x^2}{3}\left[(\log x)^2 + \frac{4}{3}\log x + \frac{14}{9}\right].$$

EXAMPLE 4.167

Solve the differential equation

$$\frac{x^2d^2y}{dx^2} - 2x\frac{dy}{dx} - 4y = x^2 + 2\log x.$$

Solution. The given Cauchy-Euler homogeneous linear equation is

$$x^2\frac{d^2y}{dx^2} - 2x\frac{dy}{dx} - 4y = x^2 + 2\log x.$$

Substitute $x = e^t$ so that $t = \log x$. Then

$$x\frac{dy}{dx} = Dy \text{ and } \frac{d^2y}{dx^2} = D(D-1)\,y.$$

Therefore the given equation reduces to

$$D(D-1)\,y - 2Dy - 4y = e^{2t} + 2t$$

or

$$(D^2 - 3D - 4)\,y = e^{2t} + 2t.$$

The characteristic equation for the equation is $m^2 - 3m - 4 = 0$, which yields $m = 4, -1$. Thus

$$C.F. = c_1 e^{4t} + c_2 e^{-t}$$

The particular integral is given by

$$P.I. = \frac{1}{D^2 - 3D - 4}e^{2t} + \frac{2^t}{D^2 - 3D - 4}e^{-t}.$$

$$= \frac{e^{2t}}{4 - 6 - 4} - \frac{2}{4\left(1 - \frac{D - 3D}{4}\right)}$$

$$= -\frac{e^{2t}}{6} - \frac{1}{2}\left(1 - \frac{D^2 - 3D}{4}\right)^{-1} t$$

$$= -\frac{1}{6}e^{2t} - \frac{1}{2}\left(1 + \frac{D^2 - 3D}{4} + \dots\right) t$$

$$= -\frac{1}{6}e^{2t} - \frac{1}{2}t - \frac{1}{2}\left[\frac{D^2}{4} - \frac{3D}{4}\right] t$$

$$= -\frac{1}{6}e^{2t} - \frac{1}{2}t + \frac{3}{8}.$$

Hence the complete solution of the given differential equation is

$$y = C.F + P.I. = c_1 e^{4t} + c_2 e^{-t} - \frac{1}{6}e^{2t} - \frac{1}{2}t + \frac{3}{8}$$

$$= c_1 x^4 + \frac{c_2}{x} - \frac{x^2}{6} - \frac{1}{2}\log x + \frac{3}{8}.$$

4.30 LEGENDRE'S LINEAR EQUATION

An equation of the form

$$(ax+b)^n \frac{d^n y}{dx^n} + a_1 (ax+b)^{n-1} \frac{d^{n-1} y}{dx^{n-1}} + \dots a_n y = F(x) \qquad (166)$$

where a_n are constants and F is a function of x, is called *Legendre's linear equation.*

To reduce the Legendre's equation to a linear differential equation with constant coefficient, we put $ax + b = e^t$ or $t = \log(ax + b)$. Then

$$\frac{dy}{dx} = \frac{dy}{dt} \cdot \frac{dt}{dx} = \frac{a}{ax+b}\frac{dy}{dt}$$

or

$$(ax+b)\frac{dy}{dx} = aDy.$$

Further

$$\frac{d^2y}{dx^2} = \frac{d}{dx}\left(\frac{a}{ax+b}\frac{dy}{dt}\right) = \frac{a^2}{(ax+b)^2}\left(\frac{d^2y}{dt^2} - \frac{dy}{dt}\right)$$

and so

$$(ax+b)^2 \frac{d^2y}{dx^2} = a^2(D^2 - D)\,y = a^2D\,(D-1)y.$$

Similarly,

$$(ax+b)^3 \frac{d^3y}{dx^3} = a^3D\,(D-1)\,(D-2)\,y$$

and so on.

Putting these values in (166), we get a linear differential equation with constant coefficients which can be solved by usual methods.

EXAMPLE 4.168

Solve $(2x+3)^2 \dfrac{d^2y}{dx^2} - 2\,(2x+3)\dfrac{dy}{dx} - 12y = 6x.$

Solution. Putting $2x + 3 = e^t$ or $t = \log(2x + 3)$, the given equation reduces to

$$(4\,(D^2 - D) - 4D - 12)\,y = 6\,\frac{e^t - 3}{2} = 3e^t - 9$$

or

$$(4D^2 - 8D - 12)\,y = 3e^t - 9.$$

The auxiliary equation is

$$4m^2 - 8m - 12 = 0,$$

which yields $m = 3, -1$. Therefore,

$$\text{C.F.} = c_1 e^{-t} + c_2 e^{3t}.$$

Now

$$\text{P.I.} = \frac{1}{4D^2 - 8D - 12}(3e^t - 9) = \frac{3}{-16}e^t + \frac{3}{4}.$$

Hence the complete solution is

$$y = c_1 e^{-t} + c_2 e^{3t} - \frac{3}{16}e^t + \frac{3}{4}$$

$$= \frac{c_1}{(2x+3)} + c_2(2x+3)^3 - \frac{3}{16}(2x+3) + \frac{3}{4}.$$

EXAMPLE 4.169

Solve Legendre' equation

$$(1+x)^2 \frac{d^2y}{dx^2} + (1+x)\frac{dy}{dx} + y = 4\cos\log(1+x).$$

Solution. Putting $x + 1 = e^t$ or $t = \log(x + 1)$, the given equations transforms to $(1^2(D^2 - D) + 1D + 1)\,y = 4\cos t$

or

$$(D^2 + 1)\, y = 4 \cos t.$$

The auxiliary equation is $m^2 + 1 = 0$ and so $m = \pm i$ Therefore,

$$\text{C.F.} = c_1 \cos t + c_2 \sin t.$$

The particular integral is

$$\text{P.I.} = \frac{4}{D^2+1} \cos t = t \frac{4\,(2D)}{[4D^2]_{D^2=-1}} \cos t$$

$$= \frac{8t}{-4} D \cos t = -\,2t\,(-\sin t) = 2t \sin t.$$

Thus, the complete solution is

$$y = c_1 \cos t + c_2 \sin t + 2t \sin t$$

$$= c_1 \cos (\log (x+1)) + c_2 \sin (\log (x+1))$$

$$+\, 2 \log (x+1) \sin (\log (x+1)).$$

EXAMPLE 4.170

Solve the Legendre's equation

$$(1+x)^2 \frac{d^2y}{dx^2} + (1+x) \frac{dy}{dx} + y = \sin (2 \log (1+x)).$$

Solution. Putting $x + 1 = e^t$, the given equation reduces to

$$(D^2 + 1) + y = \sin (2t).$$

The complementary function is

$$\text{C.F.} = c_1 \cos t + c_2 \sin t$$

Further,

$$\text{P.I.} = \frac{1}{D^2+1} \sin [2t] = \frac{1}{-4+1} \sin 2t$$

$$= -\frac{1}{3} \sin 2t.$$

Therefore, the complete solution is

$$y = c_1 \cos t + c_2 \sin t - \frac{1}{3} \sin 2t$$

$$= c_1 \cos (\log (x+1)) + c_2 \sin (\log (x+1))$$

$$- \frac{1}{3} \sin (2 \log (x+1)).$$

EXAMPLE 4.171

Solve $(1+x)^2 \frac{d^2y}{dx^2} + (1+x) \frac{dy}{dx} + y = 2 \sin (\log (x+1))$.

Solution. Putting $x + 1 = e^t$, the given equation transforms to

$$(D^2 + 1)\, y = 2 \sin t.$$

Its complementary function is given by

$$\text{C.F.} = c_1 \cos t + c_2 \sin t.$$

Its particular integral is given by

$$\text{P.I.} = \frac{1}{D^2+1} (2 \sin t) = \frac{2}{D^2+1} \sin t$$

$$= \frac{2t}{[2D]_{D^2=-1}} \sin t = \frac{D \sin t}{-1} = -\, t \cos t.$$

Therefore, the complete solution is

$$y = c_1 \cos t + c_2 \sin t - t \cos t$$

$$= c_1 \cos (\log (x+1)) + c_2 \sin (\log (x+1))$$

$$-\, \log (x+1) \cos (\log (x+1)).$$

EXAMPLE 4.172

Solve

$$(2x+3)^2 \frac{d^2y}{dx^2} + 5\,(2x+3) \frac{dy}{dx} + y = 4x.$$

Solution. We have

$$(2x+3)^2 \frac{d^2y}{dx^2} + 5\,(2x+3) \frac{dy}{dx} + y = 4x.$$

which is Legendre's linear equation. So putting $2x + 3 = e^t$ or $t = \log (2x+3)$, the given equation reduces to

$$[4\,(D^2 - D) + 10D + 1]\, y = 4 \frac{e^t - 3}{2}$$

or

$$(4D^2 + 6D + 1)\, y = 2e^t - 6.$$

The auxiliary equation is

$$4m^2 + 6m + 1 = 0,$$

which yields

$$m = \frac{-6 \pm \sqrt{36-16}}{8} = -\frac{6 \pm 2\sqrt{5}}{8}$$

$$= \frac{-3 \pm \sqrt{5}}{4}$$

Therefore

$$C.F. = c_1 e^{\left(\frac{-3+\sqrt{5}}{4}\right)} t + c_2 e^{\left(\frac{-3+\sqrt{5}}{4}\right)} t$$

Further,

$$\text{P.I.} = \frac{1}{4D^2+6D+1}(2e^t - 6) = \frac{2}{11}e^t - 6$$

$$= \frac{2}{11}(2x+3) - 6.$$

Hence the solution is

$$y = \text{C.F.} + \text{P.I.}$$

$$= c_1(2x+3)^{\frac{-3+\sqrt{5}}{4}} + c_2(2x+3)^{\frac{-3-\sqrt{5}}{4}}$$

$$+ \frac{2}{11}(2x+3) - 6.$$

4.31 METHOD OF VARIATION OF PARAMETERS TO FIND PARTICULAR INTEGRAL

Let $y_1(x), y_2(x), \ldots, y_n(x)$ be the functions defined on $n-1$ derivaties on $[a, b]$. Then the determinant

$$W(y_1, y_3 \cdots y_n) = \begin{vmatrix} y_1 & y_2 & \cdots & y_n \\ y'_1 & y'_2 & \cdots & y'_n \\ \cdots & \cdots & \cdots & \cdots \\ \cdots & \cdots & \cdots & \cdots \\ y_1^{(n-1)} & y_2^{(n-1)} & \cdots & y_n^{(n-1)} \end{vmatrix}$$

is called the Wronskian of the set $\{y_1, y_2, \ldots, y_n\}$.

If the Wronskian of a set of n function on $[a, b]$ is non-zero for atleast one point in $[a, b]$, then the set of n functions is linear independent.

If the Wronskian is identically zero on $[a, b]$ and each of the functions is a solution of the same linear differential equation, then the set of functions is linearly dependent.

The method of variation of parameters is applicable to the differential equations of the form

$$\frac{d^2y}{dx^2} + p\frac{dy}{dx} + qy = F(x), \qquad (167)$$

where p, q and F are functions of x.

Let the complementancy functions of (167) be $y = c_1y_1 + c_2y_2$. Then y_1 and y_2 satisfy the equation

$$\frac{d^2y}{dx^2} + p\frac{dy}{dx} + qy = 0, \qquad (168)$$

Replacing c_1 and c_2 (regarded as parameters) by unknown functions $u(x)$ and $v(x)$, we assume that particular integral of (167) is

$$y = uy_1 + vy_2. \qquad (169)$$

Differentiating (169) with respect to x, we get

$$y' = uy'_1 + xy'_2 + u'y_1 + x'y_2 \qquad (170)$$

$$= uy'_1 + xy'_2,$$

under the assumption that

$$u'y_1 + v'y_2 = 0 \qquad (171)$$

Differentiating (170) with respect to x, we get

$$y'' = uy''_1 + vy''_2 + u'y_1 + v'y'_2. \qquad (172)$$

Substituting the values of y, y', and y'' from (169), (170), and (172) in (167), we have

$$uy''_1 + vy''_2 + u'y'_1 + v'y'_2 + p(uy'_1 + vy'_2)$$

$$+ q(uy_1 + vy_2) = F(x)$$

or

$$u(y''_1 + py'_1 + qy_1) + v(y''_2 + py'_2 + qy_2)$$

$$+ u'y'_1 + v'y'_2 = F(x).$$

Since y_1, y_2 satisfy (168), the above expression reduces to

$$u'y'_1 + v'y'_2 = F(x). \qquad (173)$$

Solving (171) and (173), we get

$$u' = -\frac{y_2F(x)}{W} \text{ and } v' = \frac{y_1F(x)}{W},$$

where

$$W = \begin{vmatrix} y_1 & y_2 \\ y'_1 & y'_2 \end{vmatrix} = y_1 y'_2 - y'_1 y_2.$$

Integrating, we have

$$u = -\int \frac{y_2 F(x)}{W}\, dx,\ v = \int \frac{y_1 F(x)}{W}\, dx.$$

Substituting the valueof u and v in (169), we get

$$P.I. = y = -y_1 \int \frac{y_2 F(x)}{W}\, dx + y_2 \int \frac{y_1 F(x)}{W}\, dx.$$

EXAMPLE 4.173

Using method of variation of parameters, solve

$$\frac{d^2y}{dx^2} - 6\frac{dy}{dx} + 9y = \frac{e^{3x}}{x^2}.$$

Solution. The auxiliary equation for the given differential equation is

$$m^2 - 6m + 9 = 0,$$

which yields $m = 3, 3$. Therefore,

$$C.F. = (c_1 + c_2 x)\, e^{3x}.$$

Thus, we get

$$y_1 = e^{3x} \text{ and } y^2 = xe^{3x}.$$

The Wronskian of y_1, y_2 is

$$W = \begin{vmatrix} y_1 & y_2 \\ y'_1 & y'_2 \end{vmatrix} = \begin{vmatrix} e^{3x} & xe^{3x} \\ 3\,e^{3x} & (3x+1)e^{3x} \end{vmatrix} = e^{6x}$$

Therefore,

$$\text{P.I.} = -y_1 \int \frac{y_2 F(x)}{W}\, dx + y_2 \int \frac{y_1 F(x)}{W}\, dx.$$

$$= -e^{3x} \int \frac{xe^{3x}.e^{3x}}{x^2 e^{6x}}\, dx + xe^{3x} \int \frac{e^{3x}.e^{3x}}{x^2 e^{6x}}$$

$$= -e^{3x} \int \frac{dx}{x} + xe^{3x} \int \frac{1}{x^2}\, dx$$

$$= -\, e^{3x} \log x + xe^{3x}\left(-\frac{1}{x}\right) = -e^{3x}\,(\log x + 1).$$

Hence the complete solution is

$$y = \text{C.F.} + \text{P.I.} = (c_1 + c_2 x)\, e^{3x} - e^{3x}\,(\log x + 1)$$

$$= [k + c_2 x - \log x]\, e^{3x} \text{ where } k = c_1 - 1.$$

EXAMPLE 4.174

Using method of variation of parameters, solve,

$$\frac{d^2y}{dx^2} + y = \sec x.$$

Solution. The auxiliary equation for the given differential equation is $m^2 + 1 = 0$ and so $m = \pm i$. Thus

$$\text{C.F.} = c_1 \cos x + c_2 \sin x.$$

To find P.I., let

$$y_1 = \cos x \text{ and } y_2 = \sin x.$$

Then

$$W = \begin{vmatrix} \cos x & \sin x \\ -\sin x & \cos x \end{vmatrix} = \cos^2 x + \sin^2 x = 1.$$

Therefore,

$$\text{P.I.} = -y_1 \int \frac{y_2 F(x)}{W}\, dx + y_2 \int \frac{y_1 F(x)}{W}\, dx.$$

$$= -\cos x \int \frac{\sin x \sec x}{1}\, dx$$

$$+ \sin x \int \frac{\cos x \sec x dx}{1}$$

$$= \cos x \log \cos x + x \sin x.$$

Hence the complete solution is

$$y = C.F. + P.I. = c_1 \cos x + c_2 \sin x$$

$$+ \cos x \log \cos x + x \sin x.$$

EXAMPLE 4.175

Solve the given equation using method of variation of parameters

$$\frac{d^2y}{dx^2} + y = \operatorname{cosec} x$$

Solution. The symbolic form of the differential equation is

$$(D^2 + 1)\, y = \operatorname{cosec} x.$$

Its auxiliary equation is $m^2 + 1 = 0$ and so $m = \pm i$. Therefore,

$$\text{C.F.} = c_1 \cos x + c_2 \sin x.$$

To find P.I., let

$$y_1 = \cos x \text{ and } y_2 = \sin x.$$

Then Wronskian

$$W = \begin{vmatrix} y_1 & y_2 \\ y'_1 & y'_2 \end{vmatrix} = \begin{vmatrix} \cos x & \sin x \\ -\sin x & \cos x \end{vmatrix} = 1.$$

Therefore,

$$\text{P.I.} = -y_1 \int \frac{y_2 F(x)}{W} dx + y_2 \int \frac{y_1 F(x)}{W} dx$$

$$= -\cos x \int \sin x \operatorname{cosec} x dx$$

$$+ \sin x \int \cos x \operatorname{cosec} x dx$$

$$= -\cos x \int dx + \sin x \int \frac{\cos x}{\sin x} dx$$

$$= -x \cos x + \sin x \log \sin x.$$

Hence the complete solution is

$$y = \text{C.F.} + \text{P.I.} = c_1 \cos x + c_2 \sin x$$

$$- x \cos x + \sin x \log \sin x.$$

EXAMPLE 4.176

Solve the given equation using method of variation of parameters

$$\frac{d^2y}{dx^2} - 2\frac{dy}{dx} = e^x \sin x.$$

Solution. The auxiliary equation is $m^2 - 2m = 0$ or $m(m-2) = 0$ and so $m = 0, 2$. Hence

$$\text{C.F.} = c_1 + c_2 e^{2x}.$$

Now let

$$y_1 = 1 \text{ and } y_2 = e^{2x}.$$

Then the Wronskian of y_1, y_2 is

$$W = \begin{vmatrix} 1 & e^{2x} \\ 0 & 2e^{2x} \end{vmatrix} = 2e^{2x}.$$

Therefore,

$$\text{P.I.} = -y_1 \int \frac{y_2 F(x)}{W} dx + y_2 \int \frac{y_1 F(x)}{W} dx$$

$$= -\int \frac{e^{2x}.e^x}{xe^{2x}} dx + e^{2x} \int \frac{e^x \sin x}{2e^{2x}} dx$$

$$= -\frac{1}{2}\int e^x \sin x dx + \frac{e^{2x}}{2}\int e^{-x} \sin x dx$$

$$= -\frac{1}{2} e^x \sin x.$$

Hence the complete solution is

$$y = \text{C.F.} + \text{P.I.}$$

$$= c_1 + c_2 e^{2x} - \frac{1}{2} e^x \sin x.$$

EXAMPLE 4.177

Solve $y'' - 2y' + 2y = e^x \tan x$

Solution. The auxiliary equation is $m_2 - 2m + 2 = 0$ and so $m = \dfrac{2 \pm \sqrt{4-8}}{2} = 1 \pm i$. Hence

$$\text{C.F.} = e^x (c_1 \cos x + c_2 \sin x).$$

Let

$$y_1 = e^x \cos x \text{ and } y_2 = e^x \sin x.$$

Then the Wronskian of y_1, y_2 is

$$W = \begin{vmatrix} e^x \cos x & e^x \sin x \\ e^x (\cos x - \sin x) & e^x (\cos x - \sin x) \end{vmatrix} = e^{2x}$$

Therefore,

$$\text{P.I.} = -y_1 \int \frac{y_2 F(x)}{W} dx + y_2 \int \frac{y_1 F(x)}{W} dx$$

$$= -e^x \cos x \int \frac{e^x \sin x e^x \tan x}{e^{2x}} dx$$

$$+ e^x \sin x \int \frac{e^x \cos x e^x \tan x}{e^{2x}} dx$$

$$= -e^x \cos x \int (\sec x - \cos x)\, dx$$

$$+ e^x \sin x \int \sin x dx$$

$$= -e^{-x}\cos x\,[\log(\sec x + \tan x) - \sin x]$$

$$-e^{x}\sin x\cos x$$

$$= -e^{-x}\cos x\log(\sec x + \tan x).$$

Hence, the complete solution is

$$y = \text{C.F.} + \text{P.I.} = e^{x}(c_1\cos x + c_2\sin x)$$

$$-\,e^{-x}\cos x\log(\sec x + \tan x).$$

EXAMPLE 4.178

Using method of variation of parameters, solve the differential equation

$$\frac{d^2y}{dx^2} + 4y = \tan 2x.$$

Solution. The symbolic form of the given differential equation is

$$(D^2 + 4)\,y = \tan 2x.$$

Its auxiliary equation is $m^2 + 4 = 0$, which yields $m = \pm 2i$. Thus

$$\text{C.F.} = c_1\cos 2x + c_2\sin 2x.$$

To find P.I., let

$$y_1 = \cos 2x \text{ and } y_2 = \sin 2x.$$

Then Wronskian W is

$$W = \begin{vmatrix} \cos 2x & \sin 2x \\ -2\sin x & 2\cos 2x \end{vmatrix} = 2.$$

Hence

$$\text{P.I.} = -y_1\int \frac{y_2F(x)}{W}\,dx + y_2\int\frac{y_1F(x)}{W}\,dx$$

$$= -\frac{\cos 2x}{2}\int \sin 2x\tan 2x\,dx$$

$$+\frac{\sin 2x}{2}\int \cos 2x\tan 2x\,dx$$

$$= -\frac{\cos 2x}{2}\int\frac{\sin^2 2x}{\cos 2x}\,dx$$

$$+\frac{\sin 2x}{2}\int \cos 2x\tan 2x\,dx$$

$$= -\frac{\cos 2x}{2}\int\frac{1-\cos^2 2x}{\cos 2x}\,dx$$

$$+\frac{\sin 2x}{2}\int \cos 2x\tan 2x\,dx$$

$$= -\frac{1}{2}\cos 2x\int(\sec 2x - \cos 2x)\,dx$$

$$+\frac{1}{2}\sin 2x\int \sin 2x dx$$

$$= -\frac{1}{4}\cos 2x\,[\log(\sec 2x + \tan 2x) - \sin 2x]$$

$$-\frac{1}{4}\sin 2x\cos 2x$$

$$= -\frac{1}{4}\cos 2x\log(\sec 2x + \tan 2x).$$

Hence the complete solution is

$$y = \text{C.F.} + \text{P.I.} = c_1\cos 2x + c_2\sin 2x$$

$$-\frac{1}{4}\cos 2x\log(\sec 2x + \tan 2x).$$

EXAMPLE 4.179

Solve the equation using the method of variation of parameters

$$\frac{d^2y}{dx^2} - 2\frac{dy}{dx} + y = e^{x}\log x$$

Solution. The auxiliary equation is $m^2 - 2m + 1 = 0$ and so $m = 1, 1$. Thus

$$\text{C.F.} = (c_1 + c_2x)\,e^{x}.$$

To find P.I., let $y_1 = e^x$, $y_2 = xe^x$. Then

$$W = \begin{vmatrix} e^x & xe^x \\ e^x & (x+1)e^x \end{vmatrix} = e^{2x}$$

Therefore,

$$\text{P.I} = -y_1\int\frac{y_2F(x)}{W}\,dx + y_2\int\frac{y_1F(x)}{W}\,dx$$

$$= -e^x \int \frac{xe^x.e^x}{e^{2x}} \log x\, dx + xe^x \int \frac{e^x.e^x \log x}{e^{2x}} dx$$

$$= -e^x \int x \log x dx + xe^x \int \log x dx$$

$$= -e^x \left[\frac{x^2}{2} \log x - \int \frac{x^2}{2x} dx\right] + xe^x \left[x \log x \int \frac{x}{x} dx\right]$$

$$= -e^x \left(\frac{x^2}{2} \log x - \frac{x^2}{4}\right) + x^2 e^x \log x - x^2 e^x$$

$$= -e^x \frac{x^2}{2} \log x + \frac{x^2}{4} e^x + x^2 e^x \log x - x^2 e^x$$

$$= \frac{1}{2} e^x x^2 \log x - \frac{3}{4} x^2 e^x = \frac{1}{4} x^2 e^x (2 \log x - 3)$$

Hence the complete solution is

$$y = \text{C.F.} + \text{P.I.}$$
$$= (c_1 + c_2 x)\, e^x + \frac{1}{4} x^2 e^x (2 \log x - 3).$$

EXAMPLE 4.180

Solve by using the method of variation of parameters $(D^2 + 9)y = \cot 3x$.

Solution. The auxiliary equation is $m^2 + 9 = 0$, which yields $m = 3i$. Therefore

$$\text{C.F.} = c_1 \cos 3x + c_2 \sin 3x.$$

To find P.I., let $F(x) = \cot 3x$, $y_1 = \cos 3x$ and $y_2 = \sin 3x$.

Then Wronskian

$$W = \begin{vmatrix} y_1 & y_2 \\ y'_1 & y'_2 \end{vmatrix} = \begin{vmatrix} \cos 3x & \sin 3x \\ -3 \sin 3x & 3 \cos 3x \end{vmatrix} = 3.$$

Therefore

$$\text{P.I.} = -y_1 \int \frac{y_2 F(x)}{W} dx + y_2 \int \frac{y_1 F(x)}{W} dx$$

$$= \frac{-1}{3} \cos 3x \int \sin 3x \cot 3x\, dx$$

$$+ \frac{1}{3} \sin 3x \int \cos 3x \cot 3x\, dx$$

$$= \frac{-1}{3} \cos 3x \int \cos 3x\, dx + \frac{1}{3} \sin 3x \int \frac{\cos^2 3x}{\sin 3x} dx$$

$$= \frac{-1}{9} \cos 3x \sin 3x + \frac{1}{3} \sin 3x \left[\int \frac{1 - \sin^2 3x}{\sin 3x} dx\right]$$

$$= -\frac{1}{9} \cos 3x \sin 3x$$

$$+ \frac{1}{3} \sin 3x \int (\operatorname{cosec} 3x - \sin 3x) dx$$

$$= -\frac{1}{9} \cos 3x \sin 3x$$

$$+ \frac{1}{3} \sin 3x \left[\frac{1}{3} \log (\operatorname{cosec} x - \cot 3x) \frac{\cos 3x}{3}\right]$$

$$= \frac{1}{9} \sin 3x \left[\log (\operatorname{cosec} 3x - \cot 3x)\right]$$

EXAMPLE 4.181

Solve $(D^2 + a^2)y = \tan ax$ by the method of variation of parameters.

Solution. We have $(D^2 + a^2)y = \tan ax$. It's auxiliary equation is $m^2 + a^2 = 0$, which yields $m = \pm ai$.

Therefore

$$\text{C.F.} = c_1 \cos ax + c_2 \sin ax.$$

To find particular integral, let

$$y_1 = \cos ax \text{ and } y_2 = \sin ax.$$

Then

$$y'_1 = -a \sin ax \text{ and } y'_2 = a \cos ax.$$

Thus the Wronskian W is given by

$$W = \begin{vmatrix} y_1 & y_2 \\ y'_1 & y'_2 \end{vmatrix} = \begin{vmatrix} \cos ax & \sin ax \\ -a \sin ax & a \cos ax \end{vmatrix} = a.$$

Hence P.I.

$$= -y_1 \int \frac{y_2 F(x)}{W} dx + y_2 \int \frac{y_1 F(x)}{W} dx$$

$$= \frac{-\cos ax}{a} \int \sin ax \tan ax dx + \frac{\sin ax}{a} \int \cos ax$$

$$\tan ax\, dx$$

$$= \frac{-\cos ax}{a}\int \frac{\sin^2 ax}{\cos ax}\,dx + \frac{\sin ax}{a}\int \cos ax \tan ax\,dx$$

$$= \frac{-\cos ax}{a}\int \frac{1-\cos^2 ax}{\cos ax}dx + \frac{\sin ax}{a}\int \cos ax \tan ax\,dx$$

$$= \frac{-\cos ax}{a}\int (\sec ax - \cos ax)\,dx + \frac{\sin ax}{a}\int \sin ax\,dx$$

$$= \frac{-\cos ax}{a^2}\int [\log(\sec ax + \tan ax) - \sin ax] - \frac{1}{a^2}\sin ax \cos ax$$

$$= \frac{-\cos ax}{a^2}[\log(\sec ax + \tan ax)].$$

Hence the complete solution of the given differential equation is

$$y = \text{C.F.} + \text{P.I.}$$

$$= c_1 \cos ax + c_2 \sin ax - \frac{\cos ax}{a^2} \times [\log(\sec ax + \tan ax)].$$

EXAMPLE 4.182

Solve the differential equation by the method of variations of parameters:

$$y'' + y = \sec^2 x$$

Solution. The given differential equation is

$$y'' + y = \sec^2 x.$$

The symbolic form of the equation is

$$(D^2 + 1)\,y = \sec^2 x.$$

The auxiliary equation is $m^2 + 1 = 0$, which yields $m = \pm i$. Therefore

$$\text{C.F.} = c_1 \cos x + c_2 \sin x.$$

To find particular integral, let

$$y_1 = \cos x \text{ and } y_2 = \sin x.$$

Then

$$y'_1 = -\sin x \text{ and } y'_2 = \cos x.$$

The Wronskian W is given by

$$W = \begin{vmatrix} y_1 & y_2 \\ y'_1 & y'_2 \end{vmatrix} = \begin{vmatrix} \cos x & \sin x \\ -\sin x & \cos x \end{vmatrix} = 1.$$

Therefore

$$\text{P.I.} = -y_1 \int \frac{y_2 F(x)}{W}\,dx + y_2 \int \frac{y_1 F(x)}{W}\,dx$$

$$= -\cos x \int \sin x \sec^2 x\,dx + \sin x \int \cos x \sec^2 x\,dx$$

$$= -\cos x \int \sec x \tan x\,dx + \sin x \int \sec x \cos x\,dx$$

$$= -\cos x \sec x + \sin x \log(\sec x + \tan x)$$

$$= -1 + \sin x \log(\sec x + \tan x).$$

Hence the complete solution is

$$y = \text{C.F.} + \text{P.I.}$$

$$= c_1 \cos x + c_2 \sin x - 1 + \sin x \log(\sec x + \tan x).$$

EXAMPLE 4.183

Find the general solution of the equation $y'' + 16y = 32 \sec 2x$, using method of variation of parameters.

Solution. The symbolic form of the given differential equation is

$$(D^2 + 16)\,y = 32 \sec 2x.$$

The roots of the auxiliary equation $m^2 + 16 = 0$ are $m = \pm 4i$. Therefore the complementary function is

$$\text{C.F.} = c_1 \cos 4x + c_2 \sin 4x.$$

To find the particular integral, let $F(x) = 32 \sec 2x$, $y_1 = \cos 4x$, $y_2 = \sin 4x$. Then the wronskian W is given by

$$W = \begin{vmatrix} y_1 & y_2 \\ y_1^1 & y_2^1 \end{vmatrix} = \begin{vmatrix} \cos 4x & \sin 4x \\ -4\sin x & 4\cos 4x \end{vmatrix} = 4.$$

By the method of variation of parameters, we have

$$\text{P.I.} = -y_1 \int \frac{y_2 F(x)}{W} dx + y_2 \int \frac{y_1 F(x)}{W} dx$$

$$= -\cos 4x \int \frac{\sin 4x\,(32 \sec 2x)}{4} dx$$

$$+ \sin 4x \int \frac{\cos 4x\,(32 \sec 2x)}{4} dx$$

$$= -8 \cos 4x \int \sec 2x\,(2 \sin 2x \cos 2x)\, dx$$

$$+ 8 \sin 4x \int (2 \cos^2 2x - 1)\sec 2x\, dx$$

$$= -16 \cos 4x \int \sin 2x\, dx + 8 \sin 4x$$

$$\times \int (\cos 2x - \sec 2x)\, dx$$

$$= -16 \cos 4x \left(\frac{-\cos 2x}{2}\right)$$

$$+ 8 \sin 4x \left[\sin 2x - \frac{1}{2} \log(\sec 2x + \tan 2x)\right]$$

$$= 8 \cos 4x \cos 2x + 8 \sin 4x \sin 2x$$

$$- 4 \sin 4x \log(\sec 2x + \tan 2x)$$

$$= 8 \cos(4x - 2x) - 4 \sin 4x \log(\sec 2x + \tan 2x)$$

$$= 8 \cos 2x\ \ 4 \sin 4x \log(\sec 2x + \tan 2x).$$

Hence the required solution is

$$y = \text{C.F.} + \text{P.I.}$$

$$= c_1 \cos 4x + c_2 \sin 4x + 8 \cos 2x$$

$$- 4 \sin 4x \log(\sec 2x + \tan 2x)$$

EXAMPLE 4.184

Solve $\frac{d^2y}{dx^2} + \frac{dy}{dx} - 2y = \frac{1}{1-e^x}$ by using the method of variation of parameters.

Solution. The symbolic form of the given differential equation is

$$(D^2 + D - 2)\, y = \frac{1}{1-e^x}.$$

The corresponding auxiliary equation is

$$m^2 + m - 2 = 0,$$

which yields $m = 1, -2$. Hence

$$\text{C.F.} = c_1 e^x + c_2 e^{-2x}.$$

Let

$$y_1 = e^x,\ y_2 = e^{-2x} \text{ so that } y'_1 = e^x \text{ and}$$

$$y'_2 = -2e^{-2x}$$

Let $F(x) = \frac{1}{1-e^x}$. Then Wronskian W is given by

$$W = \begin{vmatrix} y_1 & y_2 \\ y'_1 & y'_2 \end{vmatrix} = \begin{vmatrix} e^x & e^{-3x} \\ e^x & -2e^{-2ex} \end{vmatrix} = -3\, e^x e^{-2x}$$

Therefore

$$\text{P.I.} = -y_1 \int \frac{y_2 F(x)}{W} dx + y_2 \int \frac{y_1 F(x)}{W} dx$$

$$= -e^x \int \frac{e^{-2x}}{-3e^x e^{-2x}(1-e^x)} dx$$

$$+ e^{-2x} \int \frac{e^x}{-3\, e^x e^{-2x}(1-e^x)} dx$$

$$= \frac{1}{3} e^x \int \frac{e^{-x}}{1-e^x} dx - \frac{1}{3} e^{-2x} \frac{e^{2x}}{1-e^x} dx$$

But

$$\int \frac{e^{-x}}{1-e^x} dx = \int \frac{1}{z^2(1-z)} dz,\ e^x = z$$

$$= \int \left[\frac{1}{z} + \frac{1}{z^2} + \frac{1}{1-z}\right] dz$$

$$= \log z - \frac{1}{z} - \log (1-z)$$

$$= \log \frac{z}{1-z} - \frac{1}{z} = \log \frac{e^x}{1-e^x} - e^{-x}$$

and

$$\int \frac{e^{2x}}{1-e^x}\, dx = \frac{1}{2} \int \frac{dz}{1-z} = -\frac{1}{2} \log (1-z)$$

$$= -\frac{1}{2} \log (1-e^x).$$

Therefore,

$$\text{P.I.} = \frac{1}{3} e^x \left[\log \frac{e^x}{1-e^x} - e^{-x}\right] + \frac{1}{6} e^{-2x} \log (1-e^x).$$

Hence the complete solution is

$$y = c_1 e^x + c_2 e^{-2x} + \frac{1}{3} e^x \left[\log \frac{e^x}{1-e^x} - e^{-x}\right]$$

$$+ \frac{1}{6} e^{-2x} \log (1-e^x).$$

EXAMPLE 4.185

Solve the following differential equation by the method of variation of parameters:

$$x^2 \frac{d^2y}{dx^2} + 2x \frac{dy}{dx} - 12y = x^3 \log x.$$

Solution. The given differential equation with variable coefficients is

$$x^2 \frac{d^2y}{dx^2} + 2x \frac{dy}{dx} - 12y = x^3 \log x. \quad (174)$$

Putting $x = e^z$ so that $z = \log x$, equation (174) reduces to

$$[D (D-1) + 2D - 12]\, y = z\, e^{3z}$$

or,

$$(D^2 + D - 12) y = z\, e^{3z} \quad (175)$$

The auxiliary equation for the differential Equation (175) is

$$m^2 + m - 12 = 0,$$

which yields $m = 3, -4$. Therefore,

$$\text{C.F.} = c_1 e^3 z + c_2 e^{-4z}$$

$$= c_1 x^3 + c_2 x^{-4}.$$

Also, the given equation can be written as

$$\frac{d^2y}{dx^2} + \frac{2}{x} \cdot \frac{dy}{dx} - \frac{12}{x^2} = x \log x.$$

Therefore, we set $F(x) = x \log x$. Then, the Wronskian W is given by

$$W = \begin{vmatrix} y_1 & y_2 \\ y'_1 & y'_2 \end{vmatrix} = \begin{vmatrix} x^3 & x^{-4} \\ 3x^2 & -4x^{-5} \end{vmatrix} = -7x^{-2}.$$

Hence,

$$\text{P.I.} = -y_1 \int \frac{y_2 F(x)}{W}\, dx + y_2 \int \frac{y_1 F(x)}{W}\, dx$$

$$= -x^3 \int \frac{x^{-4}.x \log x}{-7x^{-2}}\, dx + x^{-4} \int \frac{x^3.x \log x}{-7x^{-2}}\, dx$$

$$= \frac{1}{7} x^3 \int \frac{\log x}{x}\, dx - \frac{1}{7} x^{-4} \int x^6 \log x\, dx$$

$$= \frac{1}{14} x^3 \int \frac{2 \log x}{x}\, dx - \frac{1}{7} x^{-4} \int x^6 \log x\, dx$$

$$= \frac{1}{14} x^3 (\log x)^2 - \frac{1}{7} x^{-4} \left[\log x . \frac{x^7}{7} - \int \frac{1}{x} . \frac{x^7}{7}\, dx\right]$$

$$= \frac{1}{14} x^3 (\log x)^2 - \frac{1}{49} x^{-4} \left[\frac{1}{7} x^7 \log x - \frac{1}{x} \cdot \frac{x^7}{7}\right]$$

$$= \frac{1}{14} x^3 (\log x)^2 - \frac{1}{49} x^3 \log x + \frac{1}{343} x^3.$$

Hence, the complete solution of the equation is

$$y = \text{C.F.} + \text{P.I.}$$

$$= c_1 x^3 + c_2 x^{-4} + \frac{1}{14} x^3 (\log x)^2$$

$$- \frac{1}{49} x^3 \log x + \frac{1}{343} x^3.$$

EXAMPLE 4.186

Solve the following differential equation using the method of variation of parameters:

$$x^2 \frac{d^2y}{dx^2} + x\frac{dy}{dx} - y = x^2 e^x.$$

Solution. The given equation is

$$x^2 \frac{d^2y}{dx^2} + x\frac{dy}{dx} - y = x^2 e^x. \qquad (176)$$

or,

$$\frac{d^2y}{dx^2} + \frac{1}{x}\frac{dy}{dx} - \frac{1}{x^2} y = e^x.$$

Thus, we set $F(x) = e^x$. Further, putting $x = e^z$ so that $z = \log x$, Equation (176) reduces to

$$[D(D-1) + D - 1]\, y = e^{2z} \cdot e^{e^z} \quad (177)$$

The auxiliary equation for Equation (177) is

$$m^2 - 1 = 0,$$

which yields $m = \pm 1$. Therefore,

$$\text{C.F} = c_1 e^z + c_2 e^{-z}$$

$$= c_1 x + \frac{c_2}{x}.$$

Therefore, the parts of the C.F. are x and $\frac{1}{x}$. So, let

$$y_1 = x \quad \text{and} \quad y_2 = \frac{1}{x}.$$

Then, the Wronskian is

$$W = \begin{vmatrix} y_1 & y_2 \\ y'_1 & y'_2 \end{vmatrix} = \begin{vmatrix} x & \frac{1}{x} \\ 1 & 1 - \frac{1}{x^2} \end{vmatrix} = \frac{-2}{x}$$

Therefore,

$$\text{P.I.} = -y_1 \int \frac{y_2 F(x)}{W} dx + y_2 \int \frac{y_1 F(x)}{W} dx$$

$$= -x \int \frac{e^x \cdot \frac{1}{x}}{-\frac{2}{x}\, dx} + \frac{1}{x} \int \frac{x e^x}{-\frac{2}{x}\, dx}$$

$$= \frac{1}{2} x \int e^x\, dx - \frac{1}{2x} \int x^2 e^x\, dx$$

$$= \frac{1}{2} x e^x - \frac{1}{2x}\left[x^2 e^x - \int 2x e^x dx\right]$$

$$= \frac{1}{2} x e^x - \frac{1}{2x}\left[x^2 e^x - 2\left\{x e^x - \int e^x dx\right\}\right]$$

$$= \frac{1}{2} x e^x - \frac{1}{2} x e^x + \frac{e^x}{x} = e^x + \frac{e^x}{x}.$$

Hence, the complete solution of the given equation is

$$y = \text{C.F} + \text{P.I.}$$

$$= c_1 x + \frac{c_2}{x} + e^x + \frac{e^x}{x}.$$

4.32 SOLUTION IN SERIES

The method of series solution of differential equations is applied to obtain solutions of *linear differential equations with variable coefficients.* Consider the differential equation

$$P_0(x)\frac{d^2y}{dx^2}+P_1(x)\frac{dy}{dx}+P_2(x)y=0, \quad (178)$$

where $P_0(x)$, $P_1(x)$, and $P_2(x)$ are polynomials in x. This equation can be written as

$$\frac{d^2y}{dx^2}+\frac{P_1(x)}{P_0(x)}\frac{dy}{dx}+\frac{P_2(x)}{P_0(x)}y=0. \quad (179)$$

The point $x = a$ is called an *ordinary point* of the equation (178) or (179) if the functions $\frac{P_1(x)}{P_0(x)}$ and $\frac{P_2(x)}{P_0(x)}$ are analytic at $x = a$. In other words, $x = a$ is an ordinary point of (178) if $P_0(a) \neq 0$.

If either (or both) of $\frac{P_1(x)}{P_0(x)}$ or/and $\frac{P_2(x)}{P_0(x)}$ is (are) not analytic at $x = a$, then $x = a$ is called a *singular point* of the equation (178) or (179). Thus $x = a$ is a singular point of (178) if $P_0(a) = 0$.

Further, let

$$Q_1(x)=(x-a)\frac{P_1(x)}{P_0(x)},$$

$$Q_2(x)=(x-a)^2\frac{P_2(x)}{P_0(x)}.$$

If Q_1 and Q_2 are both analytic at $x = a$, then $x = a$ is called a *regular singular point* of (178) otherwise it is called *irregular point* of (178).

For example, consider the equation

$$9x(1-x)\frac{d^2y}{dx^2}-12\frac{dy}{dx}+4y=0.$$

We have

$$\frac{P_1(x)}{P_0(x)}=\frac{-12}{9x(1-x)} \text{ and } \frac{P_2(x)}{P_0(x)}=\frac{4}{9x(1-x)},$$

which are not analytic at $x = 0$ and $x = 1$. Hence $x = 0$ and $x = 1$ are singular points of the given equation. Further, at $x = 0$,

$$Q_1(x)=\frac{-12}{9(1-x)},\ Q_2(x)=\frac{4}{-(1-x)},$$

which are analytic at $x = 0$. Thus $x = 0$ is a regular singular point.

For $x = 1$, we have

$$Q_1(x)=\frac{12}{9x},\ Q_2(x)=\frac{-4}{9x},$$

which are analytic if $x = 1$. Hence $x = 1$ is also regular.

4.32.1 Solution About Ordinary Point

If $x = a$ is an ordinary point of the differential equation (178), then its every solution can be expressed in the form

$$y=a_0+a_1(x-a)+a_2(x-a)^2+a_3(x-a)^3+\ldots, \quad (180)$$

where the power series converges in some interval $|x - a| < R$ about a. Thus the series may be differentiated term by term on this interval and we have

$$\frac{dy}{dx}=a_1+2a_2(x-a)+3a_3(x-a)^2+\ldots,$$

$$\frac{d^2y}{dx^2}=2a_2+6a_3(x-a)+\ldots.$$

Substituting the values of y, $\frac{dy}{dx}$, and $\frac{d^2y}{dx^2}$ is (178), we get an equation of the type

$$c_0+c_1(x-a)+c_2(x-a)^2+\ldots=0, \quad (181)$$

where the coefficients c_0, c_1, and c_2 are functions of a. Then (181) will be valid for all x in $|x - a| < R$ if all $c_1, c_2, \ldots$ are zero. Thus

$$c_0=c_1=c_2=\ldots.0. \quad (182)$$

The coefficients a_i of (180) are obtained from (182). In case (181) is expressed in powers of x, then equating to zero the coefficients of the various powers of x will determine $a_2, a_3, a_4, \ldots$ In terms of a_0 and a_1. The relation obtained by equating to zero the coefficient of x^n is called the *recurrence relation.*

4.32.2 Solution About Singular Point (Forbenious Method)

If $x = a$ is a regular singular point of (178), then the equation has at least one non–trivial solution of the form

$$y = (x - a)^m [a_0 + a_1(x - a) + a_2(x - a)^2 + \dots], \tag{183}$$

where m is a definite constant (real or complex) and the series on the right converges at every point of the interval of convergence with centre a. Differentiating (183) twice, we get

$$\frac{dy}{dx} = ma_0(x - a)^{m-1} + (m + 1)a_1(x - a)^m + \dots$$

$$\frac{d^2y}{dx^2} = m(m - 1)a_0(x - a)^{m-2} + m(m + 1)a_1(x - a)^{m-1} + \dots.$$

Substituting the values of y, $\frac{dy}{dx}$ and $\frac{d^2y}{dx^2}$ in (178), we get an equation of the form

$$c_0(x - a)^{m+k} + c_1(x - a)^{m+k+1} + c_2(x - a)^{m+k+2} + \dots = 0, \tag{184}$$

where k is an integer and the coefficients c_i are functions of m and a_i. In order that (184) be valid in $|x - a| < R$, we must have

$$c_0 = c_1 = c_2 = \dots = 0. \tag{185}$$

On equating to zero the coefficient c_0 in (184), we get a quadratic equation in m,called *indicial equation*, which gives the value of m. The two roots m_1 and m_2 of the indicial equation are called *exponents of the differential equation* (178). The coefficients $a_1, a_2, a_3, \dots$ are obtained in terms of a_0 from $c_1 = c_2 = .. = 0$. Putting the values of $a_1, a_2, \dots$ in (183), the solution of (178) is obtained.

If $m_1 - m_2 \neq 0$ or a positive integer, then the complete solution of equation (178) is

$$y = c_1(y)_{m_1} + c_2(y)_{m_2}$$

If $m_1 - m_2 = 0$, that is, the roots of indicial equation are equal, then the two independent solutions are obtained by substituting the value of m in y and $\frac{\partial y}{\partial m}$. Thus, in this case,

$$y = c_1(y)_{m_1} + c_2\left(\frac{\partial y}{\partial m}\right)_{m_1}$$

If $m_1 - m_2$ is a positive integers making a coefficient of y infinite when $m = m_2$, then the form of y is modified by replacing a_0 by $k(m - m_2)$. Two independent solutions of the differential equation (178) are then

$$y = c_1(y)_{m_2} + c_2\left(\frac{\partial y}{\partial m}\right)_{m_2}$$

If $m_1 - m_2$ is a positive integer making a coefficient sof y indeterminate when $m = m_2$, then the complete solution of (178) is

$$y = c_1(y)_{m_2}.$$

EXAMPLE 4.187

Find the power series solution of the equation

$$(1 - x^2)\frac{d^2y}{dx^2} - 2x\frac{dy}{dx} + 2y = 0$$

in powers of x, that is, about $x = 0$.

Solution. Let

$$y = a_0 + a_1x + a_2x^2 + a_3x^3 + a_4x^4 + \dots.$$

Differentiating twice we get

$$\frac{dy}{dx} = a_1 + 2a_2x + 3a_3x^2 + 4a_4x^3 + \dots + na_nx^{n-1} + \dots$$

$$\frac{d^2y}{dx^2} = 2a_2 + 6a_3x + \dots + n(n - 1)a_nx^{n-2} + \dots.$$

Substituting the values of y, $\frac{dy}{dx}$ and $\frac{d^2y}{dx^2}$ in the given differential equation, we get

$$(1 - x^2)[2a_2 + 6a_3x + \dots + n(n - 1)a_nx^{n-2} + \dots]$$
$$-2x[a_1 + 2a_2x + 3a_3x^2 + \dots + na_nx^{n-1} + \dots]$$
$$+ 2(a_0 + a_1x + a_2x^2 + \dots + a_nx^n + \dots) = 0$$

or

$$2(a_2+a_0)+(6a_3-2a_1)x + (12a_4-2a_2-4a_2+2a_2)x^2 + \ldots + [(n+2)(n+1)a_{n+2} - n(n-1)a_n - 2na_n + 2a_n]x^n + \ldots = 0$$

or

$$2(a_2+a_0)+6a_3x+(12a_4-4a_2)x^2 + \ldots + [(n^2+3n+2)a_{n+2} - (n^2-n)a_n + 2a_n(1-n)]x^n + \ldots = 0.$$

Equating to zero the coefficients of the various powers of x, we get

$$a_2 = -a_0,\ a_3 = 0,\ a_4 = -\tfrac{1}{3}a_0, \ldots$$

$$a_{n+2}[n^2+3n+2] + a_n(-n^2-n+2) = 0.$$

Taking $n = 3, 4, \ldots$

$$20a_5 - 10a_3 = 0 \Rightarrow a_5 = 0$$

$$30a_6 - 18a_4 = 0 \Rightarrow a_6 = \frac{18}{30}a_4 = -\frac{1}{5}a_0$$

$$\ldots$$

$$\ldots$$

Therefore,

$$y = a_0 + a_1x + a_2x^2 + a_3x^3 + a_4x^4 + a_5x^5 + a_6x^6 + \ldots$$

$$= a_0 + a_1x + a_0x^2 - \tfrac{1}{3}a_0x^4 - \tfrac{1}{5}a_0x^6 + \ldots$$

$$= a_0\left(1 - x^2 - \tfrac{1}{3}x^4 - \tfrac{1}{5}x^6 - \ldots\right) + a_1x.$$

EXAMPLE 4.188

Find the solution in series of the equation

$$\frac{d^2y}{dx^2} + x\frac{dy}{dx} + x^2y = 0$$

about $x = 0$.

Solution. The point $x = 0$ is a regular point of the given differential equation. So, let the required solution be

$$y = a_0 + a_1x + a_2x^2 + a_3x^3 + a_4x^4 + \ldots + a_nx^n + \ldots.$$

Then differentiating twice, we get

$$\frac{dy}{dx} = a_1 + 2a_2x + 3a_3x^2 + 4a_4x^3 + \ldots + na_nx^{n-1} + \ldots$$

$$\frac{d^2y}{dx^2} = 2a_2 + 6a_3x + 12a_4x^2 + 20a_5x^3 + \ldots + n(n-1)a_nx^{n-2} + \ldots.$$

Substituting the values of y, $\frac{dy}{dx}$, and $\frac{d^2y}{dx^2}$ in the given equation, we get

$$[2a_2 + 6a_3x + 12a_4x^2 + 20a_5x^3 + \ldots + n(n-1)a_nx^{n-2} + \ldots] + x[a_1 + 2a_2x + 3a_3x^2 + 4a_4x^3 + \ldots + na_nx^{n-1} + \ldots] + x^2[a_0 + a_1x + a_2x^2 + a_3x^3 + a_4x^4 + \ldots + a_nx^n + \ldots] = 0$$

or

$$2a_2 + (6a_3 + a_1)x + (12a_4 + 2a_2 + a_0)x^2 + (20a_5 + 3a_3 + a_1)x^3 + (30a_6 + 4a_4 + a_2)x^4 + (42a_7 + 5a_5 + a_3)x^5 + \ldots + \ldots$$
$$[(n+2)(n+1)a_n + 2 + na_n + a_{n-2}]x^n + \ldots = 0.$$

Equating to zero the coefficients of the various powers of x, we get

$$a_2 = 0,\ a_3 = -\frac{1}{6}a_1,$$

$$12a_4 + 2a_2 + a_0 = 0, \text{ which yields } a_4 = -\frac{1}{12}a_0$$

$$20a_5 + 3a_3 + a_1 = 0 \text{ which yields } a_5 = -\frac{1}{40}a_1$$

$$30a_6 + 4a_4 + a_2 = 0 \text{ which yields } a_6 = \frac{1}{90}a_0,$$

and so on. Hence

$$y = a_0 + a_1x - \frac{1}{6}a_1x^3 - \frac{1}{12}a_0x^4 - \frac{1}{40}a_1x^5 + \frac{1}{90}a_0x^6 \ldots$$

$$= a_0\left(1 - \frac{1}{12}x^4 + \frac{1}{90}x^6 - \ldots\right) + a_1\left(x - \frac{1}{6}x^3 - \frac{1}{40}x^5 - \ldots\right)$$

EXAMPLE 4.189

Find power series solution of the equation

$$\frac{d^2y}{dx}+xy=0$$

in powers of x, that is, about $x = 0$.

Solution. We note that $x = 0$ is a regular point of the given equation. Therefore, its solution is of the form

$$y=a_0+a_1x+a_2x^2+a_3x^3+a_4x^4+a_5x^5$$
$$+a_6x^6+\ldots+a_nx^n+\ldots .$$

Differentiating twice successively, we get

$$\frac{dy}{dx}=a_1+2a_2x+3a_3x^2+4a_4x^3+5a_5x^4$$
$$+6a_6x^5+\ldots+na_nx^{n-1}+\ldots$$
$$\frac{d^2y}{dx^2}=2a_2+6a_3x+12a_4x^2+20a_5x^3$$
$$+30a_6x^4+\ldots+n(n-1)x^{n-2}+\ldots$$

Putting the values of y, $\frac{dy}{dx}$ and $\frac{d^2y}{dx^2}$ in the given equation, we get

$$2a_2+6a_3x+12a_4x^2+20a_5x^3+30a_6x^4$$
$$+\ldots+n(n-1)a_nx^{n-2}+\ldots$$
$$+x[a_0+a_1x+a_2x^2+a_3x^3+a_4x^4+a_5x^5+\ldots]=0$$

or

$$2a_2+(6a_3+a_0)x+(12a_4+a_1)x^2+(20a_5+a_2)x^3$$
$$+\ldots+[(n+2)(n+1)a_{n+2}+a_{n+1}]x^n+\ldots .$$

Equating to zero the coefficients of various powers of x, we get

$a_2=0$, $6a_3+a_0=0$ which yields $a_3=-\frac{1}{6}a_0$

$12a_4+a_1=0$ which yields $a_4=\frac{-1}{12}a_1$

$20a_5+a_2=0$ which yields $a_5=0$

$(n+2)(n+1)a_{n+2}+a_{n-1}=0.$

Putting $n = 4, 5, 6, 7, \ldots$, we get

$30a_6+a_3=0$ which yields $a_6=-\frac{a_3}{30}=\frac{1}{180}a_0$

$42a_7+a_4=0$ which yields $a_7=-\frac{a_4}{42}=\frac{1}{504}a_1$,

and so on.

Hence

$$y=a_0\left(1-\frac{1}{6}x^3+\frac{1}{180}x^6-\ldots\right)$$
$$+a_1\left(x-\frac{1}{12}x^4+\frac{1}{504}x^7-\ldots\right).$$

EXAMPLE 4.190

Find series solution of the differential equation

$$x\frac{d^2y}{dx^2}+\frac{dy}{dx}-y=0$$

about $x = 0$.

Solution. The point $x = 0$ is a regular singular point of the given equation. So, let

$$y=a_0x^m+a_1x^{m+1}+a_2x^{m+2}+\ldots,\ a_0\neq 0.$$

Differentiating twice in succession, we get

$$\frac{dy}{dx}=\sum_{n=0}^{\infty}(n+m)a_nx^{n+m-1}$$

and

$$\frac{d^2y}{dx^2}=\sum_{n=0}^{\infty}(n+m)(n+m-1)a_nx^{n+m-2}.$$

Substituting the values of y, $\frac{dy}{dx}$ and $\frac{d^2y}{dx^2}$ in the given equation, we get

$$x\sum_{n=0}^{\infty}(n+m)(n+m-1)a_nx^{n+m-2}$$
$$+\sum_{n=0}^{\infty}(n+m)a_nx^{n+m-1}-\sum_{n=0}^{\infty}a_nx^{n+m}=0$$

or

$$x\sum_{n=0}^{\infty}(n+m)(n+m-1)a_nx^{n+m-1}$$
$$+\sum_{n=0}^{\infty}(n+m)a_nx^{n+m-1}-\sum_{n=0}^{\infty}a_nx^{n+m}=0$$

or

$$\sum_{n=0}^{\infty}(n+m)^2a_nx^{n+m-1}-\sum_{n=0}^{\infty}a_nx^{n+m}=0$$

or

$$\sum_{n=-1}^{\infty}(n+m+1)^2a_{n+1}x^{n+m}-\sum_{n=0}^{\infty}a_nx^{n+m}=0$$

or

$$m^2 a_0 x^{m-1} + \sum_{n=0}^{\infty} (n+m+1)^2 a_{n+1} x^{n+m} - \sum_{n=0}^{\infty} a_n x^{n+m} = 0$$

or

$$m^2 a_0 x^{m-1} + \sum_{n=0}^{\infty} x^{n+m} \times [(n+m+1)^2 a_{n+1} - a_n] = 0 \qquad (186)$$

Therefore, the indicial equation is $m^2 = 0$, which yields $m = 0, 0$. Equating to zero other coefficients in (186), we get

$$(n+m+1)^2 a_{n+1} = a_n,\ n \geq 0.$$

Therefore,

$$a_1 = \frac{1}{(m+1)^2} a_0,$$

$$a_2 = \frac{1}{(m+2)^2} a_1 = \frac{1}{(m+1)^2(m+2)^2} a_0,$$

$$a_3 = \frac{1}{(m+3)^2} a_2 = \frac{1}{(m+1)^2(m+2)^2(m+3)^2} a_0,$$

and so on. Thus

$$y = a_0 x^m + \frac{1}{(m+1)^2} a_0 x^{m+1} + \frac{1}{(m+1)^2(m+2)^2} a_0 x^{m+2} + \frac{1}{(m+1)^2(m+2)^2(m+3)^2} a_0 x^{m+3} + \ldots$$

$$= a_0 x^m \left[1 + \frac{1}{(m+1)^2} x + \frac{1}{(m+1)^2(m+2)^2} x^2 + \frac{1}{(m+1)^2(m+2)^2(m+3)^2} x^3\right].$$

Putting $m = 0$, we get one solution of the given differential equation as

$$y_1 = c_1 [1 + x + \tfrac{1}{4} x^2 + \tfrac{1}{36} x^3 + \ldots]$$

Further,

$$\frac{\partial y}{\partial m} = a_0 x^m \log x \left[1 + \frac{1}{(m+1)^2} x + \frac{1}{(m+1)^2(m+2)^2} x^2 + \frac{1}{(m+1)^2(m+2)^2(m+3)^2} x^3 + \ldots\right]$$

$$+ a_0 x^m - \frac{2x}{(m+1)^3} - \frac{1}{(m+1)^2 (m+2)^2} \times \left(\frac{2}{(m+1)} + \frac{2}{(m+2)^2}\right) x^2 + \ldots].$$

Therefore,

$$\left(\frac{\partial y}{\partial m}\right)_{m=0} = a_0 \log \left[1 + x + \tfrac{1}{4} x^2 + \tfrac{1}{36} x^3 + \ldots\right] + a_0 - 2x - \tfrac{1}{4} (2+1) x^2 + \ldots].$$

Hence the solution of the given equations is

$$y = c_1 (y)_{m=0} + c_2 \left(\frac{\partial y}{\partial m}\right)_{m=0}$$

$$= (c_1 + c_2 \log x) \left[1 + x + \tfrac{1}{4} x^2 + \tfrac{1}{36} x^3 + \ldots\right] - 2c_2 \left[x + \tfrac{1}{4}\left(1 + \tfrac{1}{2}\right) x^2 + \ldots\right].$$

EXAMPLE 4.191

Find the series solution near $x = 0$ of the differential equation

$$x^2 \frac{d^2 y}{dx^2} + x \frac{dy}{dx} + x^2 y = 0.$$

(This equation can also the written as $x^2 \frac{d^2 y}{dx^2} + x \frac{dy}{dx} + xy = 0$. and is known as *Bessel's Equation of order zero.*)

Solution. The point $x = 0$ is a regular singular point of the given equation. So, let

$$y = \sum_{n=0}^{\infty} a_n x^{n+m}.$$

Differentiating twice in succession, we get

$$\frac{dy}{dx} = \sum_{n=0}^{\infty} (n+m)a_n x^{n+m-1}$$

$$\frac{d^2y}{dx^2} = \sum_{n=0}^{\infty} (n+m)(n+m-1)a_n x^{n-m-2}.$$

Putting the values of y, $\frac{dy}{dx}$ and $\frac{d^2y}{dx^2}$ in the given equation, we get

$$x\sum_{n=0}^{\infty} (n+m)(n+m-1)a_n x^{n+m-2}$$

$$+\sum_{n=0}^{\infty} (n+m)a_n x^{n+m-1} + x\sum_{n=0}^{\infty} a_n x^{n+m} = 0$$

or

$$\sum_{n=0}^{\infty} (n+m)(n+m-1)a_n x^{n+m-1}$$

$$+\sum_{n=0}^{\infty} (n+m)a_n x^{n+m-1} + \sum_{n=0}^{\infty} a_n x^{n+m+1} = 0$$

or

$$\sum_{n=0}^{\infty} (n+m)^2 a_n x^{n+m-1} + \sum_{n=0}^{\infty} a_n x^{n+m+1} = 0$$

or

$$\sum_{n=-1}^{\infty} (n+m+1)^2 a_{n+1} x^{n+m} + \sum_{n=0}^{\infty} a_n x^{n+m+1} = 0$$

or

$$m^2 a_0 x^{m-1} + \sum_{n=0}^{\infty} (n+m+1)^2 a_{n+1} x^{n+m}$$

$$+\sum_{n=0}^{\infty} a_n x^{n+m+1} = 0$$

Therefore, the indicial equation is $m^2 = 0$, which yields $m = 0, 0$. Equating to zero the coefficients of powers of x^m, x^{m+1} x^{m+2}, ..., we get

$(n+m+1)^2\, a_1 = 0$ which yields $a_1 = 0$,

$(m+2)^2\, a_2 + a_0 = 0$ and so $a_2 = -\dfrac{a_0}{(m+2)^2}$,

$(m+3)^2 a_3 + a_1 = 0$ and so $a_3 = 0$,

$(m+4)^2 a_4 + a_2 = 0$ and so $a_4 = \dfrac{-a_2}{(m+4)^2}$

$$= \frac{a_0}{(m+2)^2(m+4)^2},$$

and so on. Hence,

$$y = a_0 x^m - \frac{a_0}{(m+2)^2} x^{m+2} + \frac{a_0}{(m+2)^2(m+4)^2} x^{m+4} - \ldots$$

$$= a_0 x^m \left[1 - \frac{1}{(m+2)^2} x^2 + \frac{1}{(m+2)^2} x^4 - \ldots\right].$$

Further,

$$\frac{\partial y}{\partial m} = a_0 x^m \log x \left[1 - \frac{1}{(m+2)^2} x^2 + \frac{1}{(m+2)^2(m+4)^2} x^4 - \ldots\right]$$

$$+ a_0 x^m \left[\frac{2x^2}{(m+2)^3} - \frac{x^4}{(m+2)^2(m+4)^2} \times \left(\frac{2}{m+2} + \frac{2}{m+4}\right) + \ldots\right].$$

Therefore,

$$\left(\frac{\partial y}{\partial m}\right)_{m=0} = a_0 \log x\left[1 - \frac{1}{4}x^2 + \frac{1}{64}x^4 - \ldots\right] + a_0\left[\frac{x^2}{4} - \frac{3x^4}{2.4.16} + \ldots\right].$$

Hence the solution of the given equation is

$$y = c_1(y)_{m=0} + c_2\left(\frac{\partial y}{\partial m}\right)_{m=0}$$

$$= c_1\left[1 - \frac{1}{2^2}x^2 + \frac{1}{2^2.4^2} x^4 - \frac{1}{2^2.4^2.6^2} x^6 + \ldots\right]$$

$$+ c_2 \log x\left[1 - \frac{1}{2^2}x^2 + \frac{1}{2^2.4^2}x^4 - \frac{1}{2^2.4^2.6^2}x^6 + \ldots\right]$$

$$+ a_0\left[\frac{1}{2^2}x^2 - \frac{1}{2^2.4^2}\left(1 + \frac{1}{2}\right)x^4 + \frac{1}{2^2.4^2.6^2}\left(1 + \frac{1}{2} + \frac{1}{3}\right)x^6 - \ldots\right]$$

$$= (c_1 + c_2 \log x)\left[1 - \frac{1}{2^2} x^2 + \frac{1}{2^2.4^2} x^4 - \frac{1}{2^2.4^2.6} x^6 + \ldots\right] + a_0\left[\frac{1}{2^2}x^2 - \frac{1}{2^2.4^2}\left(1 + \frac{1}{2}\right)x^4 + \frac{1}{2^2.4^2.6^2}\left(1 + \frac{1}{2} + \frac{1}{3}\right)x^6 + \ldots\right].$$

The solution

$$c_1(y)_{m=0} = c_1\left[1 - \frac{1}{2^2}x^2 + \frac{1}{2^2.4^2}x^4 - \frac{1}{2^2.4^2.6^2}x^6 + \ldots\right]$$

is called *Bessel function of the first kind of order zero* and is represented by $J_0(x)$, where as the solution.

$$c_2\left(\frac{\partial y}{\partial m}\right)_{m=0} = c_2 \log x\left[1 - \frac{1}{2^2}x^2 + \frac{1}{2^2.4^2}x^4 - \frac{1}{2^2.4^2.6^2}x^6 - \ldots\right]$$
$$+ a_0\left[\frac{1}{2^2}x^2 - \frac{1}{2^2.4^2}\left(1+\frac{1}{2}\right)x^4 + \frac{1}{2^2.4^2.6^2}\left(1+\frac{1}{2}+\frac{1}{5}\right)x^6 - \ldots\right]$$

is called *Neumann function or Bessel function of second kind of order zero* and is denoted by $Y_0(x)$.

EXAMPLE 4.192

Find series solution about $x = 0$ of the differential equations

$$2x(1-x)\frac{d^2y}{dx^2} + (1-x)\frac{dy}{dx} + 3y = 0.$$

Solution. Evidently $x = 0$ is a regular singular point of the given equation. So, let

$$y = \sum_{n=0}^{\infty} a_n x^{n+m}, \qquad a_n \neq 0.$$

Differentiating twice in succession, we get

$$\frac{dy}{dx} = \sum_{n=0}^{\infty}(n+m)a_n x^{n+m-1},$$

$$\frac{d^2y}{dx^2} = \sum_{n=0}^{\infty}(n+m)(n+m-1)a_n x^{n+m-2}.$$

Substituting the values of y, $\frac{dy}{dx}$, and $\frac{d^2y}{dx^2}$ in the given equation, we get

$$2x(1-x)\sum_{n=0}^{\infty}(n+m)(n+m-1)a_n x^{n+m-2} + (1-x)\sum_{n=0}^{\infty}(n+m)a_n x^{n+m-1} + 3\sum_{n=0}^{\infty}a_n x^{n+m} = 0.$$

or

$$2\sum_{n=0}^{\infty}(n+m)(n+m-1)a_n x^{n+m-1} - 2\sum_{n=0}^{\infty}(n+m)(n+m-1)a_n x^{n+m} + \sum_{n=0}^{\infty}(n+m)a_n x^{n+m-1} - \sum_{n=0}^{\infty}(n+m)a_n x^{n+m} + 3\sum_{n=0}^{\infty}a_n x^{n+m} = 0$$

or

$$\sum_{n=0}^{\infty}(n+m)(2n+2m-1)a_n x^{n+m-1} - \sum_{n=0}^{\infty}[(n+m)(2n+2m-1) - 3]a_n x^{n+m} = 0$$

or

$$\sum_{n=-1}^{\infty}(n+m+1)(2n+2m+1)a_{n+1}x^{n+m} - \sum_{n=0}^{\infty}[(n+m)(2n+2m-1) - 3]a_n x^{n+m} = 0$$

or

$$m(2m-1)a_0 x^{m-1} + \sum_{n=0}^{\infty}(n+m+1)(2n+2m+1)a_{n+1}x^{n+m} - \sum_{n=0}^{\infty}(n+m)(2n+2m-1) - 3a_n x^{n+m} = 0.$$

Therefore, the indicial equation is

$$m(2m-1) = 0, \text{ which yields } m = 0, \tfrac{1}{2}.$$

Equating to zero the other coefficients, we get

$$(n+m+1)(2n+2m+1)a_{n+1}$$
$$=[(n+m)(2n+2m-1)-3]a_n$$
$$=[2n^2+2m^2+4nm-m-n-3]\,a_n$$

or

$$a_{n+1}=\frac{2n^2+2m^2+4nm-m-n-3}{(n+m+1)(2n+2m+1)}\,a_n.$$

For $m=0$, we get

$$a_{n+1}=\frac{2n^2-n-3}{(n+1)(2n+1)}\,a_n=\frac{(n+1)(2n-3)}{(n+1)(2n+1)}\,a_n$$
$$=\frac{2n-3}{2n+1}\,a_n,\ n\geq 0.$$

Putting $n=0,1,2,\ldots$, we have

$a_1=-3a_0,\ a_2=-\frac{1}{3}a_1=a_0,$

$a_3=\frac{1}{5}a_2=\frac{1}{5}a_0,\ a_4=\frac{3}{7}a_3=\frac{3}{35}a_0,$ and so on.

Therefore, the solution for $m=0$ is

$$y_1=a_0x^0+a_1x+a_2x^2+a_3x^3+\ldots$$
$$=a_0-3a_0x+a_0x^2+\frac{1}{5}a_0x^3+\frac{3}{35}a_0x^4+\ldots$$
$$=a_0[\,1-3x+x^2+\tfrac{1}{5}x^3+\tfrac{3}{35}x^4+\ldots]$$

For $m=\frac{1}{2}$, we have

$$a_{n+1}=\frac{2n^2+2m^2+4mn-m-n-3}{(n+m+1)(2n+2m+1)}\,a_n$$
$$=\frac{2n^2+\frac{1}{2}+2n-\frac{1}{2}-n-3}{\left(n+\frac{3}{2}\right)(2n+2)}\,a_n$$
$$=\frac{2n^2+n-3}{(2n+3)(n+1)}\,a_n$$
$$=\frac{(2n+3)(n-1)}{(2n+3)(n+1)}\,a_n=\frac{n-1}{n+1}\,a_n.$$

Putting $n=0,1,2,\ldots$, we get

$a_1=-a_0,\qquad a_2=0,$

$a_3=\frac{1}{4}a_2=0,\qquad a_4=\frac{1}{2}a_3=0,$

and so on. Therefore, the solution corresponding to $m=\frac{1}{2}$ is

$$y_2=a_0x^{\frac{1}{2}}+a_1x^{\frac{3}{2}}+a_2x^{\frac{5}{2}}+a_3x^{\frac{7}{2}}+\ldots$$
$$=a_0x^{\frac{1}{2}}-a_0x^{\frac{3}{2}}=a_0x^{\frac{1}{2}}(1-x).$$

Hence, the general solution of the given equation is

$$y=c_1y_1+c_2y_2$$
$$=c_1\left[1-3x+x^2+\tfrac{1}{5}x^3+\tfrac{3}{35}x^4+\ldots\right]$$
$$+c_2x^{\frac{1}{2}}(1-x).$$

EXAMPLE 4.193

Find series solution about $x=0$ of the differential equation

$$x^2\frac{d^2y}{dx^2}-x\frac{dy}{dx}-\left(x^2+\frac{5}{4}\right)y=0.$$

Solution. The point $x=0$ is a regular singular pointof the given equation. So, let

$$y=\sum_{n=0}^{\infty}a_nx^{n+m},\ a_0\neq 0.$$

Differentiating twice in succession, we have

$$\frac{dy}{dx}=\sum_{n=0}^{\infty}(n+m)a_nx^{n+m-1},$$
$$\frac{d^2y}{dx}=\sum_{n=0}^{\infty}(n+m)(n+m-1)a_nx^{n+m-2}.$$

Putting the values of y, $\frac{dy}{dx}$, and $\frac{d^2y}{dx^2}$ in the given differential equation, we have

$$\sum_{n=0}^{\infty}(n+m)(n+m-1)a_nx^{n+m}$$
$$-\sum_{n=0}^{\infty}(n+m)a_nx^{n+m}-\sum_{n=0}^{\infty}a_nx^{n+m+2}$$
$$-\frac{5}{4}\sum_{n=0}^{\infty}a_nx^{n+m}=0$$

or

$$\sum_{n=0}^{\infty}\left[(n+m)(n+m-1)-(n+m)-\frac{5}{4}\right]a_nx^{n+m}$$
$$-\sum_{n=0}^{\infty}a_nx^{n+m+2}=0$$

or

$$\sum_{n=0}^{\infty}\left[(n+m)(n+m-2)-\frac{5}{4}\right]a_n x^{n+m}$$

$$-\sum_{n=2}^{\infty} a_{n-2}x^{n+m}=0$$

or

$$\left[m(m-2)-\frac{5}{4}\right]a_0x^m+\left[(m+1)(m-1)-\frac{5}{4}\right]a_1x^{m+1}$$

$$-\sum_{n=2}^{\infty}\left[\left((n+m)(n+m-2)-\frac{5}{4}\right)a_n-a_{n-2}\right]x^{n+m}=0.$$

Therefore, the indicial equation is

$m(m-2)-\frac{5}{4}=0$, which yields $m=\frac{5}{2},-\frac{1}{2}$.

Equating to zero the other coefficients, we get

$\left[(m+1)(m-1)-\frac{5}{4}\right]a_1=0$ and so $a_1=0$

and

$\left[(n+m)(n+m-2)-\frac{5}{4}\right]a_n=a_{n-2}$ for $n\geq 2$.

For $m=-\frac{1}{2}$, we get

$$\left[\left(n-\frac{1}{2}\right)\left(n-\frac{5}{2}\right)-\frac{5}{4}\right]a_n=a_{n-2};\ n\geq 2 \qquad (187)$$

or

$$a_n=\frac{1}{n(n-3)}a_{n-2},\ n\geq 2, n\neq 3 \qquad (188)$$

Hence, for $m=-\frac{1}{2}$, we have, from (188),

$a_2=-\frac{1}{2}a_0$. Putting $n=3$ in (187), we have

$$\left[\left(3-\frac{1}{2}\right)\left(3-\frac{5}{2}\right)-\frac{5}{4}\right]a_3=a_1 \text{ or } 0.\ a_3=a_1$$

and so a_3 may be any constant. Further, from(188), we have

$$a_4=-\frac{1}{8}a_0,\ a_5=\frac{a_3}{10},\ a_6=\frac{-a_0}{144},$$

$$a_7=\frac{a_3}{280}, \text{ and so on.}$$

Hence for $m=-\frac{1}{2}$, the required solution is

$$y=a_0x^{-\frac{1}{2}}\left(1-\frac{x^2}{2}-\frac{x^4}{8}-\frac{x^6}{144}-\ldots\right)$$

$$+a_3x^{-\frac{1}{2}}\left(x^3+\frac{x^5}{10}+\frac{x^7}{280}+\ldots\right). \qquad (189)$$

Since this solution contains two constants a_0 and a_3 and $a_0\neq 0$, this is general solution of the given differential equation. By taking $m=\frac{5}{2}$, the solution is

$$y=x^{\frac{5}{2}}\left(1+\frac{x^2}{10}+\frac{x^4}{280}+\ldots\right)$$

Hence, (189) is general solution

EXAMPLE 4.194

Find series solution about $x=0$ of the differential equation

$$9x(1-x)\frac{d^2y}{dx^2}-12\frac{dy}{dx}+4y=0.$$

Solution. The point $x=0$ is a regular singular point of the given equation. Let

$$y=\sum_{n=0}^{\infty}a_nx^{n+m},\ a_0\neq 0.$$

Differentiation of y with respect to x yields

$$\frac{dy}{dx}=\sum_{n=0}^{\infty}(n+m)a_nx^{n+m-1},$$

$$\frac{d^2y}{dx^2}=\sum_{n=0}^{\infty}(n+m)(n+m-1)a_nx^{n+m-2}.$$

Substituting the values of y, $\frac{dy}{dx}$, and $\frac{d^2y}{dx^2}$ in the given equation, we get

$$9\sum_{n=0}^{\infty}(n+m)(n+m-1)a_nx^{n+m-1}$$

$$-9\sum_{n=0}^{\infty}(n+m)(n+m-1)a_nx^{n+m}$$

$$-12\sum_{n=0}^{\infty}(n+m)a_nx^{n+m-1}+4\sum_{n=0}^{\infty}a_nx^{n+m}=0$$

or

$$\sum_{n=0}^{\infty}(n+m)(9n+9m-21)a_n x^{n+m-1}$$

$$-\sum_{n=0}^{\infty}[9(n+m)(n+m-1)-4]a_n x^{n+m}=0$$

or

$$\sum_{n=0}^{\infty}(n+m+1)(9n+9m-12)a_{n+1}x^{n+m}$$

$$-\sum_{n=0}^{\infty}[9(n+m)(n+m-1)-4]a_n x^{n+m}=0$$

or

$$m(9m-21)a_0x^{m-1}+\sum_{n=0}^{\infty}[(n+m-1)$$

$$\times(9n+9m-12)a_{n+1}$$

$$-(9(n+m)(n+m-1)-4)a_n]x^{n+m}=0.$$

Therefore, the indicial equation is

$m(9m-21)=0$, which yields $m=0,\frac{7}{3}$.

We note that the roots are distinct and their difference is not an integer. Equating other coefficients of the powers of x to zero, we get

$$(n+m+1)(9n+9m-12)a_{n+1}$$

$$=[9(n+m)(n+m-1)-4]a_n$$

or

$$a_{n+1}=\frac{9(n+m)(n+m-1)-4}{(n+m+1)(9n+9m-12)}a_n,\ n\geq 0.$$

For $m=0$, we get

$$a_{n+1}=\frac{3m+1}{3m+3}a_n,\ n\geq 0.$$

Putting $n=0,1,2,3,\ldots$, we obtain

$$a_1=\frac{1}{3}a_0,\ a_2=\frac{2}{3}a_1=\frac{2}{9}a_0,$$

$$a_3=\frac{7}{9}a_2=\frac{14}{81}a_0,\ a_4=\frac{5}{6}a_3=\frac{35}{243}a_0,$$

and so on. Thus, the solution corresponding to $m=0$ is

$$y_1=a_0\left[1+\frac{x}{3}+\frac{2}{9}x^2+\frac{14}{81}x^3+\frac{35}{243}x^4+\ldots\right].$$

For $m=-\frac{7}{3}$, we have

$$a_{n+1}=\frac{9\left(n-\frac{7}{3}\right)\left(n-\frac{7}{3}-1\right)}{3\left(n-\frac{7}{3}+1\right)(3n-7-4)}$$

$$=\frac{3n-6}{3n-4}a_n,\ n\geq 0.$$

Putting $n=0,1,2,3,\ldots$, we get

$$a_1=\frac{3}{2}a_0,\ a_2=3a_1=\frac{9}{2}a_0,$$

$$a_3=0,\ a_4=0,\ a_5=0,\ \ldots.$$

Thus, the solution corresponding to $m=-\frac{7}{3}$ is

$$y_2=a_0x^{\frac{7}{3}}\left(1+\frac{3}{2}x+9x^2\right)$$

Hence, the general solution of the given differential equation is

$$y=c_1y_1+c_2y_2$$

$$=c_1\left[1+\frac{x}{3}+\frac{2}{9}x^2+\frac{14}{81}x^3+\frac{35}{243}x^4+\ldots\right]$$

$$+c_2x^{\frac{7}{3}}\left(1+\frac{3}{2}x+9x^2\right).$$

EXAMPLE 4.195

Find series solution about $x=0$ of the differential equation

$$(1+x^2)\frac{d^2y}{dx^2}+x\frac{dy}{dx}-y=0.$$

Solution. Since $x=0$ is a regular point of the given equation, so let

$$y=a_0+a_1x+a_2x^2+\ldots+a_nx^n+\ldots.$$

Then,

$$\frac{dy}{dx}=a_1+2a_2x+3a_3x^2+4a_4x^3+\ldots$$

$$+na_nx^{n-1}+\ldots$$

$$\frac{d^2y}{dx^2}=2a_2+6a_3x+12a_4x^2+20a_5x^3$$

$$+\ldots+n(n-1)a_nx^{n-2}+\ldots.$$

Substituting the values of y, $\frac{dy}{dx}$, and $\frac{d^2y}{dx^2}$ in the given equation, we get

$$(1+x^2)[2a_2+6a_3x+12a_4x^2+20a_5x^3$$

$$+\ldots+n(n-1)a_nx^{n-2}]+x[a_1+2a_2x$$

$$+3a_3x^2+4a_4x^3+\ldots+a_nx^{n-1}+\ldots]$$

$$-[a_0+a_1x+a_2x^2+\ldots+a_nx^n+\ldots]=0.$$

Equating to zero the coefficients of various powers of x, we get

$$2a_2 - a_0 = 0 \text{ and so } a_2 = \frac{a_0}{2},$$

$$6a_3 + a_1 - a_1 = 0 \text{ and so } a_3 = 0,$$

$$2a_2 + 12a_4 + 2a_2 - a_2 = 0 \text{ and so}$$

$$a_4 = -\frac{1}{4}a_2 = -\frac{1}{8}a_0,$$

$$6a_3 + 20a_5 + 3a_3 + a_3 = 0 \text{ and so } a_5 = 0,$$

and in general,

$$n(n-1)a_n + (n+2)(n+1)a_{n+2} + na_n - na_n = 0$$

or

$$a_{n+2} = \frac{n(n+1)}{(n+1)(n+2)}a_n.$$

Putting $n = 4, 5, \ldots$, we get

$$a_6 = \frac{12}{30}a_4 = \frac{2}{5}a_4 = -\frac{1}{20}a_0$$

and so on. Thus, the required solution is

$$y = a_0\left(1 - \frac{1}{2}x^2 - \frac{1}{8}x^4 - \frac{1}{20}x^6 - \ldots\right) + a_1x.$$

4.33 BESSEL'S EQUATION AND BESSEL'S FUNCTION

The equation

$$x^2\frac{d^2y}{dx^2} + x\frac{dy}{dx} + (x^2 - n^2)y = 0,$$

where n is a non–negative real number, is called Bessel's equation of order n. It occurs in problems related to vibrations, electric fields, heat conduction, etc. For $n = 0$, we have already found its solution in Example 4.191. Now we find series solution of Bessel's equation of order n. This equation can be written as

$$\frac{d^2y}{dx^2} + \frac{1}{x}\frac{dy}{dx} + \left(1 - \frac{n^2}{x^2}\right) = 0.$$

Since $\frac{1}{x}$ and $\left(1 - \frac{n^2}{x^2}\right)$ are not analytic at 0, it follows that 0 is a singular point of the given equation. But $x\left(\frac{1}{x}\right)$ and $x^2\left(1 - \frac{n^2}{x^2}\right)$ are analytic at 0. Therefore, $x = 0$ is a regular singular point of the equation. So, let

$$y = \sum_{m=0}^{\infty} a_m x^{m+r} \neq 0$$

be the series solution of the equation about $x = 0$. Differentiating twice in succession, we get

$$\frac{dy}{dx} = \sum_{m=0}^{\infty}(m+r)\,a_m x^{m+r-1}$$

$$\frac{d^2y}{dx^2} = \sum_{m=0}^{\infty}(m+r)(m+r-1)\,a_m x^{m+r-2}.$$

Putting the values of y, $\frac{dy}{dx}$, and $\frac{d^2y}{dx^2}$ in the given differential equation, we get

$$\sum_{m=0}^{\infty}(m+r)(m+r-1)\,a_m x^{m+r} + \sum_{m=0}^{\infty}(m+r)a_m x^{m+r} + \sum_{m=0}^{\infty}a_m x^{m+r+2} - n^2\sum_{m=0}^{\infty}a_m x^{m+r}$$

or

$$\sum_{m=0}^{\infty}[(m+r)^2 - n^2]a_m x^{m+r} + \sum_{m=0}^{\infty}a_m x^{m+r+2} = 0$$

or

$$\sum_{m=0}^{\infty}[(m+r)^2 - n^2]a_m x^{m+r} + \sum_{m=2}^{\infty}a_{m-2}x^{m+r} = 0$$

or

$$(r^2 - n^2)a_0x^r + [(r+1)^2 - n^2]\,a_1x^{r+1} + \sum_{m=2}^{\infty}[((m+r)^2 - n^2)a_m + a_{m-2}]x^{m+r} = 0.$$

Equating to zero the coefficient of lower power of x, we get the indicial equation as

$$r^2 - n^2 = 0, \text{ which yields } r = n, -n.$$

Equating other coefficients to zero, we get

$$[(r+1)^2 - n^2]a_1 = 0 \text{ or } a_1 = 0$$

and

$$[(m+r)^2 - n^2]a_m + a_{m-2} = 0,$$

which yields

$$a_m = \frac{a_{m-2}}{(m+r)^2 - n^2},\ m \geq 2.$$

Putting $m = 2, 3, 4, 5, 6, \ldots$, we get

$$a_3 = a_5 = a_7 = \ldots = 0$$

and

$$a_2 = -\frac{a_0}{(r+2)^2 - n^2}$$

$$a_4 = \frac{(-1)^2 a_0}{\{(r+2)^2 - n^2\}\{(r+4)^2 - n^2\}}$$

$$a_6 = \frac{(-1)^3 a_0}{\{(r+2)^2 - n^2\{(r+4)^2 - n^2\}\{(r+6)^2 - n^2\}}$$

and so on. For $r = n$, we have

$$a_1 = 0,\ a_2 = -\frac{a_0}{(n+2)^2 - n^2} = -\frac{a_0}{4(n+1)},$$

$$a_4 = \frac{a_0}{4^2 2!(n+1)(n+2)},$$

$$a_6 = \frac{-a_0}{4^3 3!(n+1)(n+2)(n+3)}, \text{ and so on,}$$

where as $a_3 = a_5 = a_7 = \ldots = 0$. Thus the solution corresponding to $r = n$ is

$$y_1 = a_0 x^n\, 1 - \left[\frac{1}{4(n+1)}x^2 + \frac{1}{4^2.2!(n+1)(n+2)}x^4 - \frac{1}{4^3.3!(n+1)(n+2)(n+3)}x^6 + \ldots\right] \quad (190)$$

Similarly, for $r = -n$, the solution is

$$y_2 = a_0 x^{-n}\left[1 - \frac{1}{4(1-n)}x^2 + \frac{1}{4^2.2!(1-n)(2-n)}x^4 - \frac{1}{4^3.3!(1-n)(2-n)(3-n)}x^6 + \ldots\right]. \quad (191)$$

We observe that $y_1 = y_2$ for $n = 0$. Further, y_1 is meaningless if n is a negative integer and y_2 is meaningless if n is a positive integer. Hence if n is non–zero and non–integer, then the general solution of the Bessel's equation of order n is

$$y = c_1 y_1 + c_2 y_2.$$

But y_1 can be expressed as

$$y_1 = a_0 x^n \Gamma(n+1)\left[\frac{1}{\Gamma(n+1)} - \frac{1}{2^2(n+1)\Gamma(n+1)}x^2 + \frac{1}{2^4.2!(n+1)(n+2)(n+1)}x^4 + \frac{1}{2^6.3!(n+1)(n+2)(n+3)(n+1)}x^6 + \ldots\right]$$

$$= a_0 x^n \Gamma(n+1)\left[\frac{1}{\Gamma(n+1)} - \frac{1}{2^2\Gamma(n+2)}x^2 + \frac{1}{2^4.2!\,\Gamma(n+3)}x^4 + \frac{1}{2^6.3!\,\Gamma(n+4)}x^6 + \ldots\right]$$

$$= a_0 x^n \Gamma(n+1)\sum_{m=0}^{\infty}\frac{(-1)^m x^{2m}}{2^{2m}.m!\,\Gamma(n+m+1)}$$

$$= a_0 2^n \Gamma(n+1)\sum_{m=0}^{\infty}\frac{(-1)^m}{m!\,\Gamma(n+m+1)}\left(\frac{x}{2}\right)^{2m+n}$$

$$= \sum_{m=0}^{\infty}\frac{(-1)^m}{m!\,\Gamma(n+m+1)}\left(\frac{x}{2}\right)^{2m+n}, \quad (192)$$

where $a_0 = \frac{1}{2^n\Gamma(n+1)}$. The solution (192) is called the *Bessel's function of the first kind of order n* and is denoted by $J_n(x)$. Thus,

$$J_n(x) = \sum_{m=0}^{\infty}\frac{(-1)^m}{m!\,\Gamma(n+m+1)}\left(\frac{x}{2}\right)^{2m+n}. \quad (193)$$

Replacing n by $-n$ in $J_n(x)$, we get

$$J_{-n}(x) = \sum_{m=0}^{\infty}\frac{(-1)^m}{m!\,\Gamma(-n+m+1)}\left(\frac{x}{2}\right)^{2m-n}, \quad (194)$$

which is called *Bessel's function of the first kind* of order $-n$. Thus, the complete solution of the Bessel's equation of order n may be expressed as

$$y = c_1 J_n(x) + c_2 J_{-n}(x), \quad (195)$$

whose n is not an integer.

When n is an integer, let $y = u(x)J_n(x)$ be a solution of the Bessel's equation of order n. Then

$$\frac{dy}{dx} = u'(x)J_n(x) + u(x)J'_n(x)$$

$$\frac{d^2y}{dx^2} = u''(x)J_n(x) + J'_n(x)u'(x) + u'(x)J'_n(x) + u(x)J''_n(x)$$

$$= u''(x)J_n(x) + 2u'(x)J'_n(x) + u(x)J''_n(x).$$

Putting the values of y, $\frac{dy}{dx}$, and $\frac{d^2y}{dx^2}$ in the given equation, we get

$$u(x)[x^2J''_n(x) + xJ'_n(x) + (x^2 - n^2)J_n(x)] + x^2u''(x)J_n(x) + 2x^2u'(x)J'_n(x) + xu'(x)J_n(x) = 0.$$

Since $J_n(x)$ is a solution of the given equation,we have

$$x^2J''_n(x) + xJ'_n(x) + (x^2 - n^2)J_n(x) = 0.$$

Therefore, the above expression reduces to

$$x^2u''(x)J_n(x) + 2x^2u'(x)J'_n(x) + xu'(x)J_n(x) = 0$$

or

$$\frac{u''(x)}{u'(x)} + 2\frac{J'_n(x)}{J(x)} + \frac{1}{x} = 0$$

or

$$\frac{d}{dx}(\log u'(x)) + 2\frac{d}{dx}(\log J_n(x)) + \frac{d}{dx}(\log x) = 0$$

or

$$\frac{d}{dx}[\log(xu'(x)J_n^2(x))] = 0.$$

Integrating this expression, we get

$$\log(xu'(x)J_n^2(x)) = \log B \text{ so that } xu'(x)J_n^2(x) = B.$$

Thus

$$u'(x) = \frac{B}{xJ_n^2(x)} \text{ and so } u(x) = B\int\frac{dx}{xJ_n^2(x)} + A.$$

Hence, the complete solution is

$$y = u(x)J_n(x) = \left[A + B\int\frac{dx}{xJ_n^2(x)}\right]J_n(x)$$

$$= AJ_n(x) + BJ_n(x)\int\frac{dx}{xJ_n^2(x)}$$

$$= AJ_n(x) + BY_n(x),$$

where

$$Y_n(x) = J_n(x)\int\frac{dx}{xJ_n^2(x)}$$

is called the Bessel function of the second kind of order n or the Neumann function.

EXAMPLE 4.196

Show that

$$J_{-n}(x) = (-1)^n J_n(x).$$

Solution. We have

$$J_n(x) = \sum_{m=0}^{\infty}\frac{(-1)^m}{m!\,\Gamma(n+m+1)}\left(\frac{x}{2}\right)^{2m+n}$$

and

$$J_{-n}(x) = \sum_{m=0}^{\infty}\frac{(-1)^m}{m!\,\Gamma(-n+m+1)}\left(\frac{x}{2}\right)^{2m-n}.$$

But for positive integer n,

$$\Gamma(n+m+1) = (n+m)! \text{ and}$$

$$\Gamma(-n+m+1) = (m-n)!$$

Therefore,

$$J_n(x) = \sum_{m=0}^{\infty}\frac{(-1)^m}{m!\,(n+m)!}\left(\frac{x}{2}\right)^{2m+n}$$

and

$$J_{-n}(x) = \sum_{m=0}^{\infty}\frac{(-1)^m}{m!\,(m-1)}\left(\frac{x}{2}\right)^{2m-n}.$$

Since, $(-n)!$ is infinite for $n > 0$, we have $J_{-n}(x)$

$$= \sum_{m=0}^{\infty}\frac{(-1)^m}{m!\,(m-1)}\left(\frac{x}{2}\right)^{2m-n}$$

$$= \sum_{m=0}^{\infty}\frac{(-1)^{m+n}}{m!\,(n+m)!}\left(\frac{x}{2}\right)^{2m+n}, \text{y changing } m \text{ to } m+n$$

$$= (-1)^n\sum_{m=0}^{\infty}\frac{(-1)^m}{m!(n+m)!}\left(\frac{x}{2}\right)^{2m+n} = (-1)^nJ_n(x).$$

EXAMPLE 4.197

Show that

(i) $\frac{d}{dx}[x^nJ_n(x)] = x^nJ_{n-1}(x)$

(ii) $\frac{d}{dx}[x^{-n}J_n(x)] = -x^{-n}J_{n+1}(x)$

(iii) $J_n(x) = \frac{x}{2n}[J_{n-1}(x) + J_{n+1}(x)]$

(iv) $J'_n(x) = \frac{1}{2}[J_{n-1}(x) - J_{n+1}(x)]$.

These results are known as *recurrence formulae for the Bessel's function $J_n(x)$*.

Solution. (i) We know that

$$J_n(x) = \sum_{m=0}^{\infty} \frac{(-1)^m}{m!\,\Gamma(n+m+1)}\left(\frac{x}{2}\right)^{2m+n}$$

Therefore,

$$x^n J_n(x) = \sum_{m=0}^{\infty} \frac{(-1)^m}{m!\,\Gamma(n+m+1)}\frac{x^{2(m+n)}}{2^{2m+n}}$$

and so

$$\frac{d}{dx}(x^n J_n(x)) = \sum_{m=0}^{\infty} \frac{(-1)^m}{m!\,\Gamma(n+m+1)}\cdot\frac{x^{2(m+n)-1}}{2^{2m+n}}$$

$$= x^n \sum_{m=0}^{\infty} \frac{(-1)^m}{m!\Gamma(n-1+m+1)}\cdot\left(\frac{x}{2}\right)^{n-1+2m}$$

$$= x^n J_{n-1}(x).$$

(ii) Multiplying the expression for $J_{-n}(x)$ by x^{-n} throughout and differentiating, we get (ii)

(iii) Part (i) implies

$$x^n J'_n(x) + nx^{n-1}J_n(x) = x^n J_{n-1}(x)$$

and so

$$J'_n(x) + \frac{n}{x}J_n(x) = J_{n+1}(x). \qquad (196)$$

Similarly, part (ii) yields

$$-J'_n(x) + \frac{n}{x}J_n(x) = J_{n+1}(x). \qquad (197)$$

Adding (196) and (197), we obtain

$$\frac{2n}{x}J_n(x) = J_{n-1}(x) + J_{n+1}(x)$$

or

$$J_n(x) = \frac{x}{2n}[J_{n-1}(x) + J_{n+1}(x)]$$

or

$$J_{n+1}(x) = \frac{2n}{x}J_n(x) - J_{n-1}(x).$$

(iv) Subtracting (196) from (197) yields

$$2J'_n(x) = J_{n-1}(x) + J_{n+1}(x)$$

or

$$J'_n(x) = \frac{1}{2}[J_{n-1}(x) + J_{n+1}(x)].$$

EXAMPLE 4.198

Show that $e^{\frac{1}{2}x\left(z-\frac{1}{z}\right)}$ is the generating function ofthe Bessel's functions.

Solution. We have

$$e^{\frac{1}{2}x\left(z-\frac{1}{z}\right)} = e^{\frac{1}{2}xz}\,.\,e^{-\frac{1}{2}xz^{-1}}$$

$$= \left[1 + \left(\frac{1}{2}x\right)z + \left(\frac{1}{2}x\right)^2\frac{z^2}{2!} + \ldots + \frac{\left(\frac{1}{2}x\right)^r z^r}{r!} + \ldots\right]$$

$$\times\left[1 + \left(-\frac{1}{2}x\right)(z^{-1}) + \left(-\frac{1}{2}x\right)^2\frac{z^{-2}}{2!} + \ldots + \frac{\left(-\frac{1}{2}x\right)^r z^{-r}}{r!}\right].$$

The coefficient of z^n in this expansion is

$$\frac{\left(\frac{1}{2}x\right)^n}{n!} + \frac{\left(\frac{1}{2}x\right)^{n+1}\left(-\frac{1}{2}x\right)^2}{(n+1)!} + \frac{\left(\frac{1}{2}x\right)^{n+2}\left(-\frac{1}{2}x\right)^2}{(n+2)!2!} + \ldots$$

$$= \sum_{m=0}^{\infty}\frac{(-1)^m}{m!(n+m)!}\left(\frac{1}{2}x\right)^{2m+n}$$

$$= \sum_{m=0}^{\infty}\frac{(-1)^m}{m!\,\Gamma(n+m+1)}\left(\frac{x}{2}\right)^{2m+n} = J_n(x)$$

Thus

$$e^{\frac{1}{2}x\left(z-\frac{1}{z}\right)} = \sum_{n=-\infty}^{\infty} z^n J_n(x).$$

Hence, $e^{\frac{1}{2}x\left(z-\frac{1}{z}\right)}$ is the *generating function of the Bessel's function.*

EXAMPLE 4.199

Show that

(i) $J_{\frac{1}{2}}(x) = \sqrt{\frac{2}{\pi x}}\sin x$

(ii) (i) $J_{-\frac{1}{2}}(x) = \sqrt{\frac{2}{\pi x}} \cos x$

(iii) $\int x J_0^2(x)dx = \frac{1}{2}x^2[J_0^2(x)dx + J_1^2(x)dx].$

Solution. (i) Putting n =1/2 in the expression for $J_n(x)$, we have

$$J_{\frac{1}{2}}(x) = \sum_{m=0}^{\infty} \frac{(-1)^m}{m!\Gamma\left(\frac{1}{2}+m+1\right)} \left(\frac{x}{2}\right)^{2m+\frac{1}{2}}$$

$$= \left(\frac{x}{2}\right)^{\frac{1}{2}} \sum_{m=0}^{\infty} \frac{(-1)^m}{m!\Gamma\left(m+\frac{3}{2}\right)} \left(\frac{x}{2}\right)^{2m}$$

$$= \left(\frac{x}{2}\right)^{\frac{1}{2}} \left[\frac{1}{\Gamma\left(\frac{3}{2}\right)} - \frac{1}{\Gamma\left(\frac{5}{2}\right)}\left(\frac{x}{2}\right)^2 + \frac{1}{2!\Gamma\left(\frac{7}{2}\right)}\left(\frac{x}{2}\right)^4 - \dots\right]$$

$$= \left(\frac{x}{2}\right)^{\frac{1}{2}} \left[\frac{1}{\frac{1}{2}\Gamma\left(\frac{1}{2}\right)} - \frac{1}{\frac{3}{2}\frac{1}{2}\Gamma\left(\frac{1}{2}\right)}\left(\frac{x}{2}\right)^2 + \frac{1}{2.\frac{5}{3}.\frac{3}{2}.\frac{1}{2}\Gamma\left(\frac{1}{2}\right)}\left(\frac{x}{2}\right)^4 - \dots\right]$$

$$= \left(\frac{x}{2}\right)^{\frac{1}{2}} \frac{2}{\Gamma\left(\frac{1}{2}\right)}\left[1 - \frac{x^2}{3!} + \frac{x^4}{5!} - \dots\right]$$

$$= \frac{\sqrt{2}}{\sqrt{x}\sqrt{\pi}}\left[x - \frac{x^3}{3!} + \frac{x5}{5!}\dots\right] = \sqrt{\frac{2}{\pi x}} \sin x.$$

(ii) Putting $n = \frac{1}{2}$ in the expression for $J_{-n}(x)$ and proceeding as in part (i), we obtain part (ii).

(iii) We have

$$\int x J_0^2(x)dx$$

$$= \frac{x^2}{2}J_0^2(x) - \int \frac{x^2}{2} 2\, J_0(x)\, J'_0(x)$$

(integration by parts)

$$= \frac{x^2}{2}J_0^2(x) + \equiv x^2\, J_0(x)\, J_1(x)\, dx$$

since $J'_0(x) = -J_1(x)$

$$= \frac{x^2}{2}J_0^2(x) + \equiv x\, J_1(x).\ \frac{d}{dx}(x J_1(x))\, dx$$

since $\frac{d}{dx}(x J_1(x)) = x\, J_0(x)$

$$= \frac{x^2}{2}J_0^2(x) + \frac{1}{2}(x J_1(x))^2 = \frac{x^2}{2}[J_0^2(x) + J_1^2(x)].$$

EXAMPLE 4.200

Show that $J_n(\lambda_i x)$ is a solution of

$$x^2\frac{d^2y}{dx} + x\frac{dy}{dx} + (\lambda_i^2 x^2 - n^2)y = 0.$$

Solution. We know that $J_n(z)$ is a solution of the Bessel's equation

$$z^2\frac{d^2y}{dx} + z\frac{dy}{dx} + (z^2 - n^2)y = 0. \qquad (198)$$

Put $z = \lambda_i x$. Then

$$\frac{dy}{dz} = \frac{dy}{dx}.\frac{dx}{dz} = \frac{1}{\lambda_i}\frac{dy}{dx}$$

and

$$\frac{d^2y}{dz^2} = \frac{1}{\lambda_i^2}\frac{d^2y}{dx^2}.$$

Putting the values of z, $\frac{dy}{dz}$, and $\frac{d^2y}{dz^2}$ in (198), we get

$$\lambda_i^2 x^2\left(\frac{1}{\lambda_i^2}\frac{d^2y}{dx^2}\right) + \lambda_i x\left(\frac{1}{\lambda_i}\frac{dy}{dx}\right) + (\lambda_i^2 x^2 - n^2)y = 0.$$

or

$$x^2\frac{d^2y}{dx} + x\frac{dy}{dx} + (\lambda_i^2 x^2 - n^2)y = 0. \qquad (199)$$

Hence $J_n(\lambda_i x)$ is a solution of (199).

EXAMPLE 4.201

If $\lambda_1, \lambda_1, \dots$ are the roots of $J_n(ax) = 0$, $a \in \Re$, show that the system

$$J_n(\lambda_1 x), J_n(\lambda_2 x), J_n(\lambda_3 x), \dots$$

of Bessel's functions is an orthogonal system, with weight function x, on $[0, a]$, $a \in \Re$.

Solution. Let

$$u_i = J_n(\lambda_i x) \text{ and } u_j = J_n(\lambda_j x), \quad i \neq j.$$

Then, by Example 4.200, we have

$$x^2 u_i'' + x u_i' + (\lambda_i^2 x^2 - n^2)u_i = 0 \qquad (200)$$

and

$$x^2 u_j'' + x u_j' + (\lambda_j^2 x^2 - n^2)u_j = 0. \qquad (201)$$

Multiplying (200) by u_j and (201) by u_i and subtracting, we get

$$x^2(u_ju_i''-u_iu_j'')+x(u_ju_i'-u_iu_j')$$
$$+x^2u_iu_j(\lambda_i^2-\lambda_j^2)=0.$$

or

$$x(u_ju_i''-u_iu_j'')+(u_ju_i'-u_iu_j')=x(\lambda_i^2-\lambda_j^2)u_iu_j$$

or

$$x\frac{d}{dx}(u_ju_i'-u_iu_j')+(u_ju_i'-u_iu_j')=x(\lambda_i^2-\lambda_j^2)u_iu_j$$

or

$$\frac{d}{dx}\left[x\left(u_ju_i'-u_iu_j'\right)\right]=x(\lambda_j^2-\lambda_i^2)u_iu_j. \quad (202)$$

Integrating (202) with respect to x in $[0, a]$, we get

$$\left[x(u_ju_i'-u_iu_j')\right]_0^a=(\lambda_j^2-\lambda_i^2)\int_0^a xu_iu_jdx$$

or

$$a\left[\lambda_iJ_n(\lambda_ja)J_n'(\lambda_ia)-\lambda_jJ_n(\lambda_ia)J_n'(\lambda_ja)\right]$$
$$=(\lambda_j^2-\lambda_i^2)\int_0^a xu_iu_jdx. \quad (203)$$

Since $J_n(\lambda_ja)=J_n(\lambda_ia)=0$, it follows that

$$(\lambda_j^2-\lambda_i^2)\int_0^a xu_iu_j=0, \qquad \lambda_i\neq\lambda_j$$

or

$$\int_0^a xJ_n(\lambda_ix)\,J_n(\lambda_jx)\,dx=0, \quad i\neq j.$$

Hence the system $\{J_n(\lambda_ix)\}$ forms an orthogonal system with weight x.

Remark 4.3. From (203), we have

$$\int_0^a xJ_n(\lambda_ix)\,J_n(\lambda_jx)\,dx$$
$$=\frac{a\left[\lambda_iJ_n(\lambda_ja)J_n'(\lambda_ia)-\lambda_jJ_n(\lambda_ia)J_n'(\lambda_ja)\right]}{\lambda_j^2-\lambda_i^2}$$

If $\lambda_i=\lambda_j$, then right hand side of the above expression is of $\frac{0}{0}$ form. We assume that λ_i is a root of $J_n(ax)=0$ and $\lambda_j\to\lambda_i$. Then we have

$$\lim_{\lambda_j\to\lambda_i}\int_0^a xJ_n(\lambda_ix)\,J_n(\lambda_jx)\,dx$$
$$=\lim_{\lambda_j\to\lambda_i}\frac{a\lambda_iJ_n(\lambda_ja)J_n'(\lambda_ia)}{(\lambda_j^2-\lambda_i^2)}$$

Or, using L'Hospital's Rule,

$$\int_0^a xJ_n^2(\lambda_ix)\,dx=\lim_{\lambda_j\to\lambda_i}\frac{a^2\lambda_iJ_n'(\lambda_ja)J_n'(\lambda_ia)}{2\lambda_j}$$
$$=\frac{a^2}{2}[J_n'(\lambda_ia)]^2$$
$$=\frac{a^2}{2}J_{n+1}^2(\lambda_ia).$$

Hence

$$\frac{2}{a^2J_{n+1}^2(\lambda_ia)}\int_0^a xJ_n^2(\lambda_ix)\,dx=1.$$

EXAMPLE 4.202

Show that $x^nJ_n(x)$ is a solution of the differential equation

$$x^2\frac{d^2y}{dx}+(1-2n)\frac{dy}{dx}+xy=0.$$

Solution. Let $y=x^nJ_n(x)$. Then

$$\frac{dy}{dx}=\frac{d}{dx}(x^nJ_n(x))=x^nJ_{n-1}, \text{ and}$$

$$\frac{d^2y}{dx^2}=\frac{d}{dx}(x^nJ_{n-1})=x^nJ_{n-1}'+nx^{n-1}J_{n-1}.$$

Putting the values of y, $\frac{dy}{dx}$, and $\frac{d^2y}{dx^2}$ in the given equation, we get

$$x^2\frac{d^2y}{dx}+(1-2n)\frac{dy}{dx}+xy$$
$$=x^{n+1}\left[J_{n-1}'-\frac{n-1}{x}J_{n-1}\right]+x^{n+1}J_n$$
$$=x^{n+1}(-J_n)+x^{n+1}J_n=0.$$

EXAMPLE 4.203

Express $J_4(x)$ is term of $J_0(x)$ and $J_1(x)$.

Solution. We know that

$$J_n(x) = \frac{x}{2n}\,[J_{n-1}(x) + J_{n+1}(x)] \quad \text{(recurrence formula)}$$

or

$$J_{n+1}(x) = \frac{2n}{x}J_n(x) - J_{n-1}(x).$$

Putting $n = 1, 2$, and 3, we get

$$J_2(x) = \frac{2}{x}J_1(x) - J_0(x)$$

$$J_3(x) = \frac{4}{x}J_2(x) - J_1(x)$$

$$= \frac{4}{x}\left[\frac{2}{x}J_1(x) - J_0(x)\right] - J_1(x)$$

$$= \frac{8}{x^2}J_1(x) - \frac{4}{x}J_0(x) - J_1(x)$$

$$= \left(\frac{8}{x} - 1\right)J_1(x) - \frac{4}{x}J_0(x).$$

$$J_4(x) = \frac{6}{x}J_3(x) - J_2(x)$$

$$= \frac{6}{x}\left[\left(\frac{8}{x} - 1\right)J_1(x) - \frac{4}{x}J_0(x)\right] - \left[\frac{2}{x}J_1(x) - J_0(x)\right]$$

$$= \frac{48}{x^3}J_1(x) - \frac{6}{x}J_1(x) - \frac{24}{x^2}J_0(x) - \frac{2}{x}J_1(x) + J_0(x)$$

$$= \left(\frac{48}{x^3} - \frac{8}{x}\right)J_1(x) + \left(1 - \frac{24}{x^2}\right)J_0(x)$$

EXAMPLE 4.204

Solve

$$x^2\frac{d^2z}{dx^2} + x\frac{dz}{dx} + (x^2 - 64)z = 0.$$

Solution. The given equation is evidently Bessel's equation of order 8, which is an integer. Therefore, its general solution is

$$z = c_1J_8(x) + c_2Y_8(x).$$

where c_1 and c_2 are arbitrary constants.

EXAMPLE 4.205

Solve the differential equation

$$x\frac{d^2y}{dx^2} + \frac{dy}{dx} + \frac{1}{4}y = 0.$$

Solution. Let $z = \sqrt{x}$. Then

$$\frac{dy}{dx} = \frac{dy}{dz}\cdot\frac{dz}{dx} = \frac{dy}{dz}\cdot\frac{1}{2\sqrt{x}}, \text{ and}$$

$$\frac{d^2y}{dx^2} = \frac{dy}{dz}\cdot\frac{1}{2}\left(-\frac{1}{2}x^{-\frac{3}{2}}\right) + \frac{1}{2\sqrt{x}}\left(\frac{d^2y}{dz^2}\cdot\frac{1}{2\sqrt{x}}\right)$$

$$= \frac{1}{4x}\frac{d^2y}{dz^2} - \frac{1}{4x^{\frac{3}{2}}}\frac{dy}{dz} = \frac{1}{4z^2}\frac{d^2y}{dz^2} - \frac{1}{4z^3}\frac{dy}{dz}.$$

Putting these values of the partial derivatives in the given equation, we get

$$z^2\left[\frac{1}{4z^2}\frac{d^2y}{dz^2} - \frac{1}{4z^3}\frac{dy}{dz}\right] + \frac{1}{2z}\frac{dy}{dz} + \frac{1}{4}y = 0$$

or

$$\frac{d^2y}{dz^2} - \frac{1}{z}\frac{dy}{dz} + \frac{2}{z}\frac{dy}{dz} + y = 0$$

or

$$z\frac{d^2y}{dz^2} + \frac{dy}{dz} + zy = 0$$

or

$$z^2\frac{d^2y}{dz^2} + z\frac{dy}{dz} + (z^2 - 0^2)y = 0,$$

which is Bessel's equation of order 0. Therefore, its general solution is

$$y = c_1J_0(z) + c_2y_0(z) = c_1J_0(\sqrt{x}) + c_2Y_0(\sqrt{x}).$$

EXAMPLE 4.206

Show that

$$e^{\frac{x}{2}\left(z-\frac{1}{z}\right)} = J_0(x) + \left(z - \frac{1}{z}\right) J_1(x).$$

$$+ \left(z^2 + \frac{1}{z^2}\right) J_2(x) + \left(z^3 - \frac{1}{z^3}\right) J_3(x) + \ldots$$

Hence or otherwise show that

$$J_0(x) = \frac{1}{\pi} \int_0^{\pi} \cos(x \cos\theta)\, d\theta$$

Solution. We know that

$$e^{\frac{x}{2}\left(z-\frac{1}{z}\right)} = \sum_{n-\infty}^{\infty} z^n J_n(x) \quad \text{(gerenating functions)}$$

$$= J_0 + \sum_{n+1}^{\infty} (J_n z^n + J_n z^{-n})$$

$$= J_0 + \sum_{n+1}^{\infty} (J_n z^n + (-1)^n J_n z^{-n})$$

$$= J_0 + \sum_{n+1}^{\infty} [z_n + (-1)^n z^{-n}]\, J_n$$

$$= J_0 + \left(z - \frac{1}{z}\right) J_1 + (z^2 + z^{-2})\, J_2$$

$$+ (z^3 - z^{-3})\, J_3 + \ldots \qquad (204)$$

Putting $z = e^i$ we get $z - \frac{1}{z} = 2i \sin\phi$, $z^2 + \frac{1}{z^2} +$ $2\cos 2\phi$, and so on. Therefore, (204) reduces to

$$e^{i(x \sin\phi)} = J_0 + (2i \sin\phi)\, J_1 + (2 \cos 2\phi) J_2 + \ldots$$

or

$$\cos(x \sin\phi) + i \sin(x \sin\phi) = J_0 + (2i \sin\phi)\, J_1$$

$$+ (2\cos 2\phi)\, J_2 + \ldots$$

Separating real imaginary parts, we get

$$\cos(x \sin\phi) = J_0 + (2\cos 2\phi)\, J_2$$

$$+ (2\cos 4\phi) J_4 + \ldots \qquad (205)$$

and

$$\sin(x \sin\phi) = (2\sin\phi)\, J_1 + (2\sin 3\phi)\, J_3$$

$$+ (2\sin 5\phi)\, J_5 + \ldots, \qquad (206)$$

which are called *Jacobi's series.*

Puting $\phi = \frac{\pi}{2} - \theta$ in (205), we get

$$\cos(x \cos\theta) = J_0 - (2\cos 2\theta)\, J_2 + \ldots$$

Therefore,

$$\int_0^{\pi} \cos(x \cos\theta)\, d\theta$$

$$= \int_0^{\pi} J_0 d\theta - \int_0^{\pi} (2\cos 2\theta)\, J_2 d\theta + \ldots = \pi J_0.$$

Hence

$$J_0 = \frac{1}{\pi} \int_0^{\pi} \cos(x \cos\theta)\, d\theta.$$

EXAMPLE 4.207

Solve the differential equation

$$x\frac{d^2y}{dx^2} + \frac{dy}{dx} - ixy = 0.$$

Solution. The given equation can be written as

$$x^2 \frac{d^2y}{dx^2} + x\frac{dy}{dx} - ix^2 y = 0.$$

Putting $z = \sqrt{-i}x$, we have

$$\frac{dy}{dx} = \frac{dy}{dz} \cdot \frac{dz}{dx} = \sqrt{-i}\, \frac{dy}{dz},$$

$$\frac{d^2y}{dx^2} = -i\, \frac{d^2y}{dz^2}.$$

Thus, the given equation reduce to

$$z^2 \frac{d^2y}{dz^2} + z\frac{dy}{dz} + (z^2 - 0)y + 0,$$

which is Bessl's equation of order zero. Its solution is

$$y = J_0(z) = J_0(\sqrt{-i}x) = J_0\left(i^{\frac{3}{2}}x\right)$$

$$= 1 - \frac{i^3x^2}{2^2} + \frac{i^6x^4}{2^4(2!)^2} - \frac{i^9x^6}{2^6(3!)^2} + \frac{i^{12}x^8}{2^8(4!)^2} - \ldots$$

$$= \left[1 - \frac{x^4}{2^2.4^2} + \frac{x^8}{2^2.4^2.6^2.8^2} - \ldots\right]$$

$$+ i\left[\frac{x^2}{2^2} - \frac{x^6}{2^2.4^2.6^2} + \frac{x^{10}}{2^2.4^2.6^2.8^2.10^2} - \ldots\right]$$

$$= ber\, x + i\, bei\, x,$$

where

$$ber\, x = 1 + \sum_{m=1}^{\infty} (-1)^m \frac{x^{4m}}{2^2.4^2.6^2\ldots(4m)^2}$$

is called *Bessel's real (or ber) function* and

$$bei\, x = -\sum_{m=1}^{\infty} (-1)^m \frac{x^{4m-2}}{2^2.4^2.6^2\ldots(4^{m-2})^2}$$

called *Bessel's imaginary* (or *bei*) *function.*

4.34 FOURIER-BESSEL EXPANSION OF A CONTINUOUS FUNCTION

We have proved in Example 4.201 that Bessel's function $J_n(\lambda x)$ form an orthogonal set with weight x. This property allows us to expand a continous function f in Fourier-Bessel series in a range 0 to a. So, let

$$f(x) = c_1 J_n(\lambda_1 x) + c_2 J_n(\lambda_2 x) + \ldots + c_n J_n(\lambda_n x) \quad (207),$$

where λ_i are the roots of the equation $J_n(\lambda_a) = 0$. To determine the coefficients c_i, we multiply both sides of the equation (207) by $xJ_n(\lambda_i x)$ and integrate with respect to x between the limits 0 to a. Using orthogonal property and the remark 4.3,

we have

$$\int_0^a xf(x)\, J_n(\lambda_i x)\, dx = c_i \int_0^a x J_n(\lambda_i x)\, dx$$

$$= c_i\left[\frac{a^2}{2} J^2_{n+1}(\lambda_i a)\right]$$

Hence

$$c_i = \frac{2}{a^2 J^2_{n+1}(\lambda_i a)} \int_0^a x\, f(x)\, J_n(\lambda_i x)\, dx.$$

Substituting the values of c_i in the expression (207), we get the Fourier-Bessel series of f.

EXAMPLE 4.208

Expand $f(x) = x^2 (0 < x < a)$ in Fourier-Bessel series in $J_2(\lambda_i x)$ if $\lambda_n a$ are the positive roots of $J_2(x) = 0$.

Solution. Let the Fourier-Bessel series of the function $f(x) = x^2$ $(0 < x < a)$ be

$$x^2 = \sum_{n=1}^{\infty} c_n J_2(\lambda_n x).$$

We know that

$$c_i = \frac{2}{a^2 J^2_{n+1}(\lambda_i a)} \int_0^a xf(x)\, J_n(\lambda_i x)\, dx$$

$$= \frac{2}{a^2 J^2_3(\lambda_i a)} \int_0^a x^2 J_2(\lambda_i x)\, dx$$

$$= \frac{2}{a^2 J^2_3(\lambda_i a)} \left[\frac{x^3 J_3(\lambda_i x)}{\lambda_i}\right]_0^a$$

$$= \frac{2a}{\lambda_i J_3(\lambda_i a)}$$

Hence the Fourier-Besel series for $f(x) = x^2$ is

$$x^2 = 2a \sum_{n=1}^{\infty} \frac{J_2(\lambda_n x)}{\lambda_n J_3(\lambda_n a)}.$$

EXAMPLE 4.209

Expand $f(x) = 1$ in Fourier-Bessel series in $J_0(\lambda_n x)$ if $\lambda_1, \lambda_2, \ldots, \lambda_n$ are the roots of $J_0(x) = 0$.

Solution. Let the Fourier-Bessel series of $f(x) = 1$ be

$$1 = \sum_{\infty}^{1} c_n J_0(\lambda_n x)$$

Since $f(x) = 1$ and $a = 1$, the coefficients c_i are given by

$$c_i = \frac{2}{J_1^2(\lambda_i)} \int_0^1 x J_0(\lambda_i x)\, dx$$

$$= \frac{2}{J_1^2(\lambda_i)} \left[\frac{xJ_i(\lambda_i x)}{\lambda_i} \right]_0^1$$

$$= \frac{2}{\lambda_i J_1(\lambda_i)}$$

Hence the required Fourier-Bessel series is

$$1 = 2 \sum_{n=1}^{\infty} \frac{J_0(\lambda_n x)}{\lambda_n J_1(\lambda_n)}$$

4.35 LEGENDRE'S EQUATION AND LEGENDRE'S POLYNOMIAL

The differential equation

$$(1 - x^2) \frac{d^2y}{dx^2} - 2x \frac{dy}{dx} + n(n+1)y = 0 \quad (208)$$

is called Legendre's equation of order n, where n is a real number. The equation can be written is the form

$$\frac{d^2y}{dx^2} - \frac{2x}{1 - x^2} \frac{dy}{dx} + \frac{n(n+1)}{1 - x^2} y = 0.$$

Since $-\frac{2x}{1 - x^2}$ and $\frac{n(n+1)}{1 - x^2}$ are analytic at 0, the point $x = 0$ is a regular point of the Legendre's equation. So let

$$y = \sum_{m=0}^{\infty} a_m x^m$$

be the power series solution of the given Legendre's equation about $x = 0$. Then

$$\frac{dy}{dx} = \sum_{m=1}^{\infty} m a_m x^{m-1}, \quad \text{and}$$

$$\frac{d^2y}{dx^2} = \sum_{m=2}^{\infty} m(m-1)\, a_m x^{m-2}.$$

Putting the values of y, $\frac{dy}{dx}$, and $\frac{d^2y}{dx^2}$ in the given equation, we get

$$(1 - x^2) \sum_{m=2}^{\infty} m(m-1)\, a_m x^{m-2}$$

$$- 2x \sum_{m=1}^{\infty} m a_m x^{m-1} + n(n+1) \sum_{m=0}^{\infty} a_m x^m = 0$$

or

$$\sum_{m=2}^{\infty} m(m-1) a_m x^{m-2} - \sum_{m=2}^{\infty} m(m-1) a_m x^m$$

$$- 2 \sum_{m-1}^{\infty} m a_m x^m + n(n+1) \sum_{m-0}^{\infty} a_m x^m = 0$$

or

$$\sum_{m=0}^{\infty} (m+2)(m+1) a_{m+2} x^m - \sum_{m=z}^{\infty} m(m-1) a_m x^n$$

$$- 2 \sum_{m=1}^{\infty} m a_m x^m + n(n+1) \sum_{m=0}^{\infty} a_m x^m = 0$$

or

$$[2a^2 + n(n+1) a_0] + [6a_3 - 2a_1 + n(n+1)a_1]x$$

$$+ \sum_{m=2}^{\infty} [(m+2)(m+1)\, a_{m+2}$$

$$+ (n-m)(n+m+1) a_m] x^m = 0.$$

Equating to zero the coefficient of powers of x, we get

$2a_2 + n(n+1) a_0 = 0$, which yields

$$a_2 = -\frac{n(n+1)}{2!} a_0$$

$6a_3 - 2a_1 + n(n+1)a_1 = 0$, which gives

$$a_3 = -\frac{(n-1)(n+2)}{3!}a_1,$$

and in general

$$a_{m+2} = -\frac{(n-m)(n+m+1)}{(m+1)(m+2)}a_m,\ m \geq 2.$$

Putting $m = 0, 1$, we observe that a_2 and a_3 are the same as found above. Therefore,

$$a_{m+2} = -\frac{(n-m)(n+m+1)}{(m+1)(m+2)}a_m,\ m \geq 0,$$

which is called the *recurrence solution of the Legendre's equation.* Putting $m = 2, 3, \ldots$ we have

$$a_4 = \frac{(n-2)\,n(n+1)(n+3)}{4!}a_0,$$

$$a_5 = \frac{(n-3)(n-1)(n+2)(n+4)}{5!}a_1$$

and so on. Hence

$$y = a_0 + a_1 x - \frac{n(n+1)}{2!}a_0 x^2 - \frac{(n-1)(n+2)}{3!}x^3$$

$$+ \frac{(n-2)n\,(n+1)(n+3)}{4!}a_0 x^4$$

$$+ \frac{(n-3)(n-1)(n+2)(n+4)}{5!}a_1 x^5 + \ldots$$

$$= a_0\left[1 - \frac{n(n+1)}{2!}x^2 + \frac{(n-2)n(n+1)(n+3)}{4!}x^4 + \ldots\right]$$

$$+ a_1\left[x - \frac{(n-1)(n+2)}{3!}x^3 + \frac{(n-3)(n-1)(n+2)(n+4)}{4!}x^5 + \ldots\right]$$

$$= a_0 y_1(x) + a_1 y_2(x), \text{ say.}$$

Thus, y_1 contains only even powers of x while y_2 contains odd powers of x. Also y_1 and y_2 are linearly independent.

If n is even, then $y_1(x)$ is a polynomial of degree n and if n is odd, then $y_2(x)$ is a polynomial of degree n. Moreover if $a_n = \frac{(2n)!}{2^n(n!)^2}$, then

$$P_n(x) = \begin{cases} a_0 + a_2 x^2 + \ldots + a_n x^n \text{ if } n \text{ is even} \\ a_1 x + a_3 x^3 + \ldots + a_n x^n \text{ if } n \text{ is odd} \end{cases}$$

is called the *Legendre polynomial of degree n.* We note that

$$P_0(x) = 1,\ P_1(x) = x,$$

$$P_2(x) = \frac{1}{2}(3x^2 - 1),\ P_3(x) = \frac{1}{2}(5x^2 - 3x),$$

$$P_4(x) = \frac{1}{8}(35x^4 - 30x^2 + 3), \text{ and so on.}$$

Evidently, the value of the polynomial $P_n(x)$, $n = 0, 1, 2, \ldots$ is 1 for $x = 1$.

EXAMPLE 4.210

Show that

$$P_n(x) = \frac{1}{n!\,2n}\frac{d^n}{dx^n}(x^2 - 1)^n$$

(Rodrigue's formula):

Solution. Putting $u = (x^2 - 1)^n$, we have

$$\frac{du}{dx} = 2nx(x^2 - 1)^{n-1} = \frac{2nxu}{x^2 - 1}$$

and so

$$(1 - x^2)\frac{du}{dx} + 2nxu = 0. \quad (209)$$

Differentiating (209), $(n + 1)$ times by Leibnitz's theorem, we get

$$(1 - x^2)\,u_{n+2} - 2(n+1)\,xu_{n+1} - n(n+1)u_n$$

$$+ 2n\,[xu_{n+1} + (n+1)u_n] = 0$$

or

$$(1 - x^2)u_{n+2} - 2xu_{n+1} + n(n+1)u_n = 0$$

or

$$(1 - x^2)\frac{d^2u_n}{dx^2} - 2x\frac{du_n}{dx} + n(n+1)u_n = 0.$$

Thus, $u_n = \frac{d^n}{dx^n}(x^2 - 1)^n$ satisfies the Legendre's equation

$$(1 - x^2)\frac{d^2y}{dx^2} - 2x\frac{dy}{dx} + n(n+1)y = 0.$$

Since the solution of Legendre's equation is $P_n(x)$, we have

$$P_n(x) = cu_n = c\frac{d^n}{dx^n}(x^2 - 1)^n. \qquad (210)$$

Putting $x = 1$ in (210), we get

$$1 = c\left[\frac{d^n}{dx^n}(x-1)^n(x+1)^n\right]_{x=1}$$

$$= c\,[n!\,(x+1)^n + \text{term containing } x - 1 \text{ along with its powers}]_{x=1}$$

$$= c\,n!\,2^n,$$

and so $c = \frac{1}{n!\,2^n}$. Thus,

$$P_n(x) = \frac{1}{n!\,2^n}\frac{d^n}{dx^n}(x^2 - 1)^n.$$

EXAMPLE 4.211

Show that

$$(1 - 2xt + t^2)^{-\frac{1}{2}} = \sum_{n=0}^{\infty} t^n P_n(x),$$

$$|x| < 1,\ |t| < \tfrac{1}{3}.$$

[Thus $(1 - 2xt + t^2)^{-\frac{1}{2}}$ is a generating function of the Legendre's polynomials.]

Solution. We have, by binomial expansion,

$$[1 - (2xt - t^2)]^{\frac{1}{2}}$$

$$= 1 + \frac{1}{2}(2xt - t^2) + \frac{1.3}{2^2.2!}(2xt - t^2)^2 + \ldots$$

$$+ \frac{1.3\ldots(2n-5)}{2^{n-2}(n-2)!}(2xt - t^2)^{n-2}$$

$$+ \frac{1.3\ldots(2n-3)}{2^{n-1}(n-1)!}(2xt - t^2)^{n-1}$$

$$+ \frac{1.3\ldots(2n-1)}{2^n n!}(2xt - t^2)^n + \ldots$$

$$= 1 + xt + \left(-\frac{1}{2} + \frac{3}{8}\cdot 4x^2\right)t^2 + \ldots$$

$$+ \left[\frac{1.3\ldots(2n-1)}{2^n.n!}(2x)^n\right.$$

$$- \frac{1.3\ldots(2n-3)}{2^{n-1}(n-1)!}(n-1)(2x)^{n-2}$$

$$\left. + \frac{1.3\ldots(2n-5)(n-2)(n-3)}{2^{n-2}(n-2)!\,1.2}(2x)^{n-4} + \ldots\right]t^n$$

$$+ \ldots$$

$$= 1 + xt + \frac{1}{2}(3x^2 - 1)t^2 + \ldots$$

$$+ \frac{1.3\ldots(2n-1)}{n!}\left[x^n - \frac{n(n-1)}{2(2n-1)}x^{n-2}\right.$$

$$\left. + \frac{n(n-1)(n-2)(n-3)}{2.4(2n-1)(2n-3)}x^{n-4} + \ldots\right]t^n + \ldots$$

$$= P_0(x) + tP_1(x) + t^2P_2(x) + \ldots + t^n P_n(x) + \ldots$$

$$= \sum_{n=0}^{\infty} t^n P_n(x).$$

Hence, $(1-2xt+t^2)^{-\frac{1}{2}}$ generates Legendre polynomials $P_n(x)$.

EXAMPLE 4.212

Express the polynomial x^3+2x^2-x-3 in terms of Legendre's polynomials.

Solution. We know that

$$P_0(x)=1,\ P_1(x)=x,$$

$$P_2(x)=\frac{1}{2}(3x^2-1) \text{ and so } x^2=\frac{2}{3}P_2(x)+\frac{1}{3},$$

$$P_3(x)=\frac{1}{2}(5x^3-3x) \text{ and so } x^3=\frac{2}{5}P_3(x)+\frac{3}{5}P_1(x).$$

Hence,

$$x^3+2x^2-x-3$$

$$=\left[\frac{2}{5}P_3(x)+\frac{3}{5}P_1(x)\right]+2\left[\frac{2}{3}P_2(x)+\frac{1}{3}\right]$$

$$-P_1(x)-3P_0(x)$$

$$=\frac{2}{5}P_3(x)+\frac{4}{3}P_2(x)-\frac{2}{5}P_1(x)-\frac{7}{3}P_0(x).$$

EXAMPLE 4.213

Show that Legendre polynomials are *orthogonal* on the interval $[-1, 1]$.

Solution. Let P_m and P_n be Legendre polynomials for $m \neq n$. We want to show that

$$\int_{-1}^{1} P_m(x)\,P_n(x)\,dx=0,\ m\neq n.$$

Since P_m and P_n satisfy Legendre's equation, we have

$$(1-x^2)\,P_m''(x)-\ 2xP_m'(x)$$

$$+\,m(m+1)\,P_m(x)=0 \qquad (211)$$

$$(1-x^2)\,P_n''(x)+2x\,P_n'(x)$$

$$+\,n(n+1)\,P_n(x)=0. \qquad (212)$$

Multiplying (211) by $P_n(x)$ and (212) by $P_m(x)$ and subtracting, we get

$$(1-x^2)\,[P_n(x)\ P_m''(x)-P_m(x)\ P_n''(x)]$$

$$-\,2x\,[P_n(x)\,P_n''(x)+P_m(x)\,P_n''(x)]$$

$$=(n^2-m^2+n-m)\,P_m(x)\,P_n(x)$$

or

$$\frac{d}{dx}\,[(1-x^2)\,(P_n(x)\,P_m'(x)-P_m(x)\,P_n'(x))]$$

$$=(n-m)\,(n+m+1)\,P_m(x)\,P_n(x).$$

Integration on the interval $[-1, 1]$ yields

$$[(1-x^2)\,(P_n(x)\,P_m'(x)-P_m(x)\,P_n'(x)]_{-1}^{1}$$

$$=(n-m)\,(n+m+1)\int_{-1}^{1} P_m(x)\,P_n(x)dx$$

or

$$0=(n-m)\,(n+m+1)\int_{-1}^{1} P_m(x)\,P_n(x)dx$$

or

$$\int_{-1}^{1} P_m(x)\,P_n(x)\,dx=0,\ m\neq n.$$

Hence, $\{P_n(x)\}$ forms an orthogonal system.

EXAMPLE 4.214

Show that $\left\{\sqrt{\frac{2n+1}{2}}\,P_n(x)\right\}$ forms an *orthonormal* system.

Solution. The system of polynomials $\left\{\sqrt{\frac{2n+1}{2}}P_n(x)\right\}$ will form an orthonormal system if

$$\frac{2n+1}{2}\int_{-1}^{1} P_m(x)\,P_n(x)\,dx = \begin{cases} 0 \text{ for } m \neq n \\ 1 \text{ for } m = n. \end{cases}$$

By Example 4.213, it follows that

$$\int_{-1}^{1} P_m(x)\,P_n(x)\,dx = 0,\ m \neq n$$

If remains to show that $\int_{-1}^{1} P_n^2(x)dx = \frac{2}{2n+1}$.

In this direction, we have by generating functions,

$$(1 - 2xt + t^2)^{-\frac{1}{2}} = \sum_{m=0}^{\infty} t^m\,P_m(x)$$

and

$$(1 - 2xt + t^2)^{-\frac{1}{2}} = \sum_{m=0}^{\infty} t^n P_n(x).$$

Multiplying these two expressions, we get

$$(1 - 2xt + t^2)^{-1} = \sum_{m=0}^{\infty}\sum_{n=0}^{\infty} t^{m+n} P_m(x) P_n(x).$$

Integrating w.r.t. x on $[-1, 1]$, we have

$$\int_{-1}^{1} (1 - 2xt + t^2)^{-1}\,dx$$

$$= \sum_{m=0}^{\infty}\sum_{n=0}^{\infty}\int_{-1}^{1} t^{m+n} P_m(x)P_n(x)\,dx.$$

Since $\int_{-1}^{1} P_m(x)\,P_n(x)\,dx = 0$ for $m \neq n$, the above expression reduces to

$$\int_{-1}^{1} (1 - 2xt + t^2)^{-1}\,dx = \sum_{n=0}^{\infty}\int_{-1}^{1} t^{2n}\,P_n^2(x)\,dx.$$

Thus

$$\sum_{n=0}^{\infty}\int_{-1}^{1} t^{2n}\,P_n^2(x)\,dx = \left[-\frac{1}{2t}\log\,(1 + t^2 - 2xt)\right]_{x=-1}^{1}$$

$$= -\frac{1}{2t}\log\frac{(1-t)^2}{(1+t)^2} = \frac{1}{t}\log\frac{1+t}{1-t}$$

$$= \frac{1}{t}\left[2\left(t + \frac{t^3}{3} + \frac{t^5}{5} + \ldots\right)\right]$$

$$= 2\left(1 + \frac{t^2}{3} + \frac{t^4}{5} + \ldots + \frac{t^{2n}}{2n+1} + \ldots\right)$$

Comparing the coefficients of t^{2n}, we have

$$\int_{-1}^{1} P_n^2(x)\,dx = \frac{2}{2n+1}$$

or

$$\frac{2n+1}{2}\int_{-1}^{1} P_n^2(x)\,dx = 1$$

EXAMPLE 4.215

Using Rodrigue's formula, find the polynomial expression for $P_4(x)$.

Solution. The Rodrigue's formula is

$$P_n(x) = \frac{1}{2^n n!}\frac{d^n}{dx^n}(x^2 - 1)^n.$$

Therefore,

$$P_4(x) = \frac{1}{2^n 4!}\frac{d^4}{dx^4}(x^2 - 1)^4 = \frac{1}{384}\frac{d^4}{dx^4}(x^2 - 1)^4.$$

So let

$$y = (x^2 - 1)^4 = x^8 - 4x^6 + 6x^4 - 4x^2 + 1.$$

Then,

$$\frac{dy}{dx} = 8x^7 - 24x^5 + 24x^3 - 8x$$

$$\frac{d^2y}{dx^2} = 56x^6 - 120x^4 + 72x^2 - 8$$

$$\frac{d^3y}{dx^3} = 336x^5 - 480x^3 + 144x$$

$$\frac{d^4y}{dx^4} = 1680x^4 - 1440x^2 + 144.$$

Therefore,

$$P_4(x) = \frac{1}{384}[1680x^4 - 1440x^2 + 144]$$

$$= \frac{1}{8}[35x^4 - 30x^2 + 3].$$

EXAMPLE 4.216

Establish the following recurrence relations for Legendre polynomial.

(i) $(n+1)P_{n+1}(x) = (2n+1)xP_n(x) - nP_{n-1}(x)$

(ii) $nP_n(x) = xP'_n(x) - P'_{n-1}(x)$

(iii) $(2n+1)P_n(x) = P'_{n+1}(x) - P'_{n-1}(x)$

(iv) $(n+1)P_n(x) = P'_{n+1}(x) - xP'_n(x)$

(v) $(1-x^2)P'_n(x) = n(P_{n-1}(x) - xP_n(x))$

(vi) $(1-x^2)P'_n(x) = (n+1)(xP_n(x) - P_{n-1})$.

Solution. (i) The generating formula for $P_n(x)$ is

$$(1-2xt+t^2)^{-\frac{1}{2}} = \sum_{n=0}^{\infty} t^n P_n(x).$$

Differentiating w.r.t. t, we get

$$-\frac{1}{2}(1-2xt+t^2)^{-\frac{3}{2}}(-2x+2t)$$

$$= \sum_{n=0}^{\infty} nP_n(x)t^{n-1}$$

or

$$(x-t)(1-2xt+t^2)^{-\frac{1}{2}} = (1-2xt+t^2) \times \sum_{n=0}^{\infty} nP_n(x)\,t^{n-1}$$

or

$$(x-t)\sum_{n=0}^{\infty} t^n P_n(x) = (1-2xt+t^2)\sum_{n=0}^{\infty} nt^{n-1}P_n(x).$$

Comparing coefficients of t^n, we get

$$xP_n(x) - P_{n-1}(x) = (n+1)P_{n+1}(x) - 2nxP_n(x) + (n-1)P_{n-1}(x).$$

or

$$(n+1)P_{n+1}(x) = (2n+1)xP_n(x) - nP_{n-1}(x)$$

(ii) Differentiating the generating formula w.r.t. x, we get

$$-\frac{1}{2}(1-2xt+t^2)^{-\frac{3}{2}}(-2t) = \sum_{n=0}^{\infty} t^n P'_n(x)$$

or

$$t(1-2xt+t^2)^{-\frac{3}{2}} = \sum_{n=0}^{\infty} t^n P'_n(x) \quad (213)$$

On the other hand, differentiation of the generating formula w.r.t. t yields

$$(x-t)(1-2xt+t^2)^{-\frac{3}{2}} = \sum_{n=0}^{\infty} nt^{n-1}P_n(x). \quad (214)$$

Dividing (214) by (213), we get

$$\frac{(x-t)}{t} = \frac{\sum_{n=0}^{\infty} nt^{n-1}P_n(x)}{\sum_{n=0}^{\infty} t^n P'_n(x)}$$

or

$$\sum_{n=0}^{\infty} nt^n P_n(x) = (x-t)\sum_{n=0}^{\infty} t^n P'_n(x).$$

Comparing the coefficients of t^n on both sides, we have

$$nP_n(x) = xP'_n(x) - P'_{n-1}(x).$$

(iii) From (i), we have

$$(n+1)P_{n+1}(x) = (2n+1)xP_n(x) - nP_{n-1}(x).$$

Differentiating w.r.t. x, we get

$$(n+1)\,P'_{n+1}(x) = (2n+1)\,P_n(x)$$

$$+\,(2n+1)\,xP'_n(x) - nP'_{n-1}(x) \quad (215)$$

But from part (ii)

$$x\,P'_n(x) = n\,P_n(x) + P'_{n-1}(x)$$

Therefore, (215) reduces to

$$(n+1)\,P'_{n+1}(x) = (2n+1)\,P_n(x)$$

$$+\,(2n+1)\,x[nP_n(x) + P'_{n-1}(x)] - n\,P'_{n-1}(x)$$

or

$$(2n+1)\,P_n(x) = P'_{n+1} - P'_{n-1}$$

(iv) From part (ii), we have

$$(2n+1)\,P_n(x) = P'_{n+1}(x) - P'_{n-1}(x)$$

and from part (ii), we have

$$nP_n(x) = x\,P'_n(x) - P'_{n-1}(x)$$

Subtracting, we obtain

$$(n+1)\,P_n(x) = P'_{n+1}(x) - x\,P'_n(x)$$

(v) From part (iv), we have

$$(n+1)\,P_n\,(x) = P'_{n+1}(x) - x\,P'_n(x)$$

Replacing n by $n-1$, we get

$$nP_{n-1}(x) = P'_n(x) - x\,P'_n(x). \quad (216)$$

Also, from part (ii), we have

$$nP_n(x) = xP'_n(x) - P'_{n-1}(x)$$

or

$$nxP_n(x) = x^2\,P'_n(x) - xP'_{n-1}(x) \quad (217)$$

Subtracting (217) from (216), we get

$$nP_{n-1}(x) - nxP_n(x) = (1-x^2)\,P'_n(x)$$

or

$$(1-x^2)\,P'_n(x) = n[P_{n-1}\,(x) - xP_n(x)].$$

(vi) From (216), we have

$$nxP_{n-1}(x) = xP'_n(x) - x^2\,P'_{n-1}(x) \quad (218)$$

and from part (ii), we have

$$nP_n(x) = xP'_n(x) - P'_{n-1}(x) \quad (219)$$

Subtracting (219) from (218), we obtain

$$n[xP_{n-1}(x) - P_n(x)] = (1-x^2)\,P'_{n-1}(x).$$

Replacing n by $n+1$, we get

$$(1-x^2)\,P'_n(x) = (n+1)\,[xP_n(x) - P_{n+1}\,(x)]$$

EXAMPLE 4.217

Show that if n is odd, then

$$P'_n(x) = (2n-1)P_{n-1}(x) + (2n-5)\,P_{n-3}(x)$$

$$+\,(2n-9)\,P_{n-5}(x) + \ldots + 3P_1(x).$$

Solution. We have

$$(2n+1)\,P_n(x) = P'_{n+1}(x) - P'_{n-1}(x)$$

(recurrence formula).

Changing n to $(n-1)$, $(n-3)$, $(n-5)$,…, we get

$$(2n-1)P_{n-1}(x) = P'_n(x) - P'_{n-2}(x)$$

$$(2n-5)P_{n-3}(x) = P'_{n-2}(x) - P'_{n-4}(x)$$

$$(2n-9)P_{n-5}(x) = P'_{n-4}(x) - P'_{n-6}(x)$$

$$\cdots$$

$$\cdots$$

$$3P_1(x) = P'_2(x) - P'_0(x) \text{ if n is odd:}$$

Adding these equations, we get

$$(2n-1)P_{n-1}(x) = (2n-5)P_{n-3}(x)$$

$$+ (2n-9)P'_{n-5}(x) + \ldots + 3P_1(x)$$

$$= P'_n(x) - P'_0(x) = P'_n(x) - 0$$

$$= P'_n(x), \text{ since } P_0(x) = 1.$$

EXAMPLE 4.218

Show that

$$\int_{-1}^{1} xP_n(x)\,P_{n-1}(x) = \frac{2n}{4n^2-1}$$

Solution. We know that

$$(n+1)P_{n+1}(x) = (2n+1)xP_n(x) - nP_{n-1}(x).$$

Changing n to $n-1$, we get

$$nP_n(x) = (2n-1)xP_{n-1}(x) - (n-1)P_{n-2}(x)$$

or

$$xP_{n-1}(x) = \frac{1}{2n-1}[nP_n(x) + (n-1)P_{n-2}(x)]$$

Therefore,

$$\int_{-1}^{1} xP_n(x)\,P_{n-1}(x)\,dx$$

$$= \frac{1}{2n-1}\int_{-1}^{1}[nP_n(x) + (n-1)P_{n-2}(x)]\,P_n(x)\,dx$$

$$= \frac{n}{2n-1}\int_{-1}^{1}P_n^2(x)dx + \frac{n-1}{2n-1}\int_{-1}^{1}P_n(x)\,P_{n-2}(x)dx$$

$$= \frac{1}{2n-1}\int_{-1}^{1}P_n^2(x)\,dx + 0$$

using orthogonal property

$$= \frac{n}{2n-1}\left[\frac{2}{2n+1}\right] = \frac{2n}{4n^2-1}$$

EXAMPLE 4.219

Show that

(i) $P_n(-x) = (-1)^n P_n(x)$

(ii) $P_n(1) = 1$

(iii) $P_n(-1) = (-1)^n$.

Solution. (i) Generating function formula is

$$(1-2xt+t^2)^{-\frac{1}{2}} = \sum_{n=0}^{\infty} t^n P_n(x)$$

Therefore,

$$\sum_{n=0}^{\infty} t^n P_n(-x) = (1+2xt+t^2)^{-\frac{1}{2}}$$

$$= [1-2x(-t)+t^2]^{-\frac{1}{2}}$$

$$= \sum_{n=0}^{\infty}(-t)^n P_n(x) = \sum_{n=0}^{\infty}(-1)^n(-t)^n P_n(x)$$

and so $P_n(-x) = (-1)^n P_n(x)$.

(ii) We have

$$\sum_{n=0}^{\infty} P_n(1)\, t^n = (1-2t+t^2)^{-\frac{1}{2}} = (1-t)^{-1}$$

$$= 1 + t + t^2 + \ldots + t^n + \ldots$$

$$= \sum_{n=0}^{\infty} t^n$$

Hence $P_n(1) = 1$.

(iii) We note that

$$\sum_{n=0}^{\infty} t^n P_n(-1) = (1-2t+t^2)^{-\frac{1}{2}}$$

$$(1-t)^{-1} = \sum_{n=0}^{\infty} (-1)^n t^n$$

Hence $P_n(-1) = (-1)^n$.

Second Method: From part (i)

$$P_n(-x) = (-1)^n P_n(x).$$

Therefore,

$$P_n(-1) = (-1)^n P_n(1) = (-1)^n \text{ using (ii)}.$$

4.36 FOURIER–LEGENDRE EXPANSION OF A FUNCTION

We have established in Examples 4.213 and 4.214 that $\sqrt{\frac{2n+1}{2}}\ \{P_n(x)\}$ constitutes an orthonormal system. Therefore, as in the case of Fourier expansion, the expansion in terms of $P_n(x)$ of a given function $f(x)$ is possible. Let

$$f(x) = \sum_{n=0}^{\infty} c_n P_n(x). \qquad (220)$$

Multiplying both sides by $P_n(x)$ and integrating both sides over $[-1, 1]$, we get

$$\int_{-1}^{1} f(x) P_n(x) dx = c_n \int_{-1}^{1} P_n^2(x)\, dx = c_n \left(\frac{2}{2n+1}\right)$$

and so

$$c_n = \frac{2n+1}{2} \sum_{-1}^{1} f(x) P_n(x)\, dx$$

The coefficients c_n are called Fourier–Legendre coefficients and the expansion (220) is called Fourier–Legendre expansion of $f(x)$.

EXAMPLE 4.220

Expand

$$f(x) = \begin{cases} 0, & -1 < x < 0 \\ 1, & 0 < x < 1. \end{cases}$$

in Fourier–Legendre series.

Solution. We have

$$c_n = \frac{2n+1}{2} \int_{-1}^{1} f(x) P_n(x)\, dx = \frac{2n+1}{2} \int_{-1}^{1} P_n(x) dx$$

Therefore,

$$c_0 = \frac{1}{2} \int_{-1}^{1} P_0(x)\, dx = \frac{1}{2} \int_{-1}^{1} dx = \frac{1}{2}$$

$$c_1 = \frac{3}{2} \int_{-1}^{1} P_1(x)\, dx = \frac{3}{2} \int_{-1}^{1} x dx = \frac{3}{2}\left(\frac{1}{2}\right) = \frac{3}{4},$$

$$c_2 = \frac{5}{2} \int_{-1}^{1} P_2(x)\, dx = \frac{5}{2} \int_{-1}^{1} \frac{3x^2-1}{2}\, dx$$

$$= \frac{5}{4} [x^3 - x]_0^1 = 0,$$

$$c_3 = \frac{7}{2} \int_{-1}^{1} P_3(x)\, dx = \frac{7}{2} \int_{-1}^{1} \frac{5x^2-3x}{2}\, dx$$

$$= \frac{7}{4} \left[5\frac{x^4}{4} - 3\frac{x^2}{2}\right]_0^1 = -\frac{7}{16}.$$

$$c_4 = \frac{9}{2} \int_{-1}^{1} P_4(x) dx = \frac{9}{2} \int_{-1}^{1} \frac{35x^4 - 30x^2 + 3}{8}\, dx = \frac{27}{16},$$

and so on. Hence, Fourier–Legendre expansion of the given function is

$$f(x) = \sum_{n=0}^{\infty} c_n P_n(x)$$

$$= \frac{1}{2} P_0(x) + \frac{3}{4} P_1(x) - \frac{7}{16} P_3(x) + \frac{27}{16} P_4(x) + \ldots$$

4.37 MISCELLANEOUS EXAMPLES

EXAMPLE 4.221

(a) Show that $\int_{-1}^{1} (1 - x^2) [P'_n(x)]^2 dx = \frac{2n(n+1)}{2n+1}$.

(b) Show that $\int_{-1}^{1} f(x) P_n(x) dx = \frac{(-1)^n}{2^n n!} \int_{-1}^{1} (x^2 - 1)^n f^{(n)}(x) dx$.

(c) Show that $J_3(x) + 3J_0'(x) + 4J_0'''(x) = 0$

(d) $P_{2n}(0) = \frac{(-1)^2 (2n)!}{2^{2n} (n!)^2}$.

Solution. (a) We have (see solved Exercise 52, chapter 4)

$$\frac{d}{dx} [(1 - x^2) P'_n(x)] P_n(x) + n(n+1) P_n(x) = 0$$

or

$$n(n+1) P_n(x) = -\frac{d}{dx} [(1 - x^2) P'_n(x)] P_n(x). \quad (221)$$

Multiplying both sides of (221) by $P_n(x)$ and integrating between the limits -1 and 1, we get

$$n(n+1) \int_{-1}^{1} P_n^2(x)\, dx$$

$$= \int_{-1}^{1} [(1 - x^2) P'_n(x)]' P_n(x)\, dx$$

$$= \{-P_n(x) [(1 - x^2) P'_n(x)]\}_{-1}^{1}$$

$$+ \int_{-1}^{1} P'_n(x) (1 - x^2) P'_n(x)\, dx$$

$$= \int_{-1}^{1} (1 - x^2) [P'_n(x)]^2\, dx \quad (222)$$

But

$$\int_{-1}^{1} P_n^2(x)\, dx = \frac{2}{2n+1}.$$

Therefore (222) implies

$$\int_{-1}^{1} (1 - x^2) [P'_n(x)]^2\, dx = \frac{2n(n+1)}{2x+1}.$$

(b) By Rodrigue's formula, we have

$$P_n(x) = \frac{1}{2^n n!} \frac{d^n}{dx^n} (x^2 - 1)^n.$$

Therefore integrating by parts, we have

$$\int_{-1}^{1} f(x) P_n(x) dx = \frac{1}{2^n n!} \int_{-1}^{1} f(x) \frac{d^n}{dx^n} (x^2 - 1)^n dx$$

$$= \frac{1}{2^n n!} \left[\left\{ f(x) \frac{d^{n-1}}{dx^{n-1}} (x^2 - 1)^n \right\}_{-1}^{1} - \int_{-1}^{1} f'(x) \frac{d^{n-1}}{dx^{n-1}} (x^2 - 1)^n dx \right]$$

$$= \frac{(-1)^n}{2^n n!} \int_{-1}^{1} f'(x) \frac{d^{n-1}}{dx^{n-1}} (x^2 - 1)\, dx$$

$$= \frac{(-1)^2}{2^n n!} \int_{-1}^{1} f''(x) \frac{d^{n-1}}{dx^{n-1}} (x^2 - 1)^n dx$$

(using integration by parts second time)

....

....

....

$$= \frac{(-1)^n}{2^n n!} \int_{-1}^{1} f^{(n)}(x) (x^2 - 1)^n dx$$

(using integration by parts nth time)

(c) We know that

$$d/dx\, [x^{-n} J_n(x)] = -x^{-n} J_{n+1}(x)$$

Putting $n = 0$, we have

$$J'_0(x) = -J_1(x).$$

Therefore

$$J''_0(x) = -J'_1(x)$$
$$= -\frac{1}{2}[J_0 - J_2]$$

and

$$J'''_0(x) = -\frac{1}{2}[J'_0(x) - J'_2(x)]$$
$$= -\frac{1}{2}J'_0(x) + \frac{1}{2}\left[\frac{1}{2}\{J_1(x) - J_3(x)\}\right]$$
$$= -\frac{1}{2}J'_0(x) + \frac{1}{4}J_1(x) - \frac{1}{4}J_3(x)$$
$$= -\frac{1}{2}J'_0(x) + \frac{1}{4}[-J'_0(x)] - \frac{1}{4}J_3(x)$$
$$= -\frac{3}{4}J'_0(x) - \frac{1}{4}J_3(x).$$

Hence

$$4J'''_0(x) + 3J'_0(x) + J_3(x) = 0:$$

(d) Since $(1 - 2xt + t^2)^{-\frac{1}{2}}$ is generating function of the Legendre's polynomial, we have

$$(1 - 2xt + t^2)^{-\frac{1}{2}} = \Sigma\, t^n P_n(x)$$

or

$$\Sigma\, t^n P_n(0) = 1 - \frac{1}{2}t^2 + \frac{1.3}{2.4}t^4$$
$$- \ldots (-1)^r \frac{1.3.5...(2r-1)}{2.4.6..(2r)} t^{2r} + \ldots$$

Equating coefficients of t^{2n} on both sides, we get

$$P_{2n}(0) = (-1)^n \frac{1.3.5...(2n-1)}{2.4.6...(2n)}$$
$$= (-1)^n \frac{1.2.3.4...(2n-1)2^n}{[2.4.6...(2n)]^2}$$
$$= (-1)^n \frac{(2n)!}{[2^n(1.2...n)]^2} = (-1)^n \frac{(2n)!}{2^{2n}(n!)^2}.$$

EXAMPLE 4.222

Solve the following differential equation in series:

$$2x^2\frac{d^2y}{dx^2} - x\frac{dy}{dx} + (x-5)y = 0.$$

Solution. Comparing the given equation with

$$\frac{d^2y}{dx^2} + P(x)\frac{dy}{dx} + Q(x)y = 0,$$

we get

$$P(x) = -\frac{1}{2x},\quad Q(x) = \frac{x-5}{2x^2}.$$

At $x = 0$, both $P(x)$ and $Q(x)$ are not analytic. Thus $x = 0$ is a singular point. Further

$$xP(x) = -\frac{1}{2} \text{ and } x^2Q(x) = \frac{x-5}{2},$$

which are analytic at $x = 0$. Hence $x = 0$ is a regular singular point. So, let

$$y = \sum_{n=0}^{\infty} a_n x^{n+m},\ a_n \neq 0.$$

Differentiating twice, we get

$$\frac{dy}{dx} = \sum_{n=0}^{\infty}(n+m)\,a_n x^{n+m-1}$$
$$\frac{d^2y}{dx^2} = \sum_{n=0}^{\infty}(n+m)(n+m-1)\,a_n x^{n+m-2}.$$

Substituting the values of y, $\frac{dy}{dx}$ and $\frac{d^2y}{dx^2}$ in the given equation, we have

$$\sum_{n=0}^{\infty}2(n+m)(n+m-1)\,a_n x^{n+m}$$
$$-\sum_{n=0}^{\infty}(n+m)\,a_n x^{n+m}$$
$$+\sum_{n=0}^{\infty}a_n x^{n+m+1} - \sum_{n=0}^{\infty}5a_n x^{n+m}$$

or

$$\sum_{n=0}^{\infty}2(n+m)(n+m-1) - (n+m) - 5\,]\,a_n x^{n+m}$$
$$+\sum_{n=0}^{\infty}\ a_n x^{n+m+1} = 0$$

or

$$\sum_{n=0}^{\infty}[2n^2 + 2m^2 + 4mn - 3n - 3m - 5]\,a_n x^{n+m}$$
$$+\sum_{n=0}^{\infty}a_n x^{n+m+1} = 0$$

or

$$\sum_{n=0}^{\infty}[2n^2+2m^2+4mn-3n-3m-5]\,a_n x^{n+m} + \sum_{n=0}^{\infty} a_n x^{n+m+1} = 0$$

or

$$\sum_{n=0}^{\infty}[2n^2+2m^2+4mn-3n-3m-5]\,a_n x^{n+m} + \sum_{n=0}^{\infty} a_{n-1} x^{n+m} = 0$$

or

$$(2m^2-3m-5)a_0x^m \times \sum_{n=0}^{\infty}[2n^2+2m^2+4mn-3n-3m-5]\,a_n + a_{n-1}]x^{m+n} = 0.$$

Therefore the indicial equation is

$$2m^2-3m-5=0, \text{ which yields } m=\frac{5}{2}, -1.$$

Equating to zero the other coefficients, we have

$$(2n^2+2m^2+4mn-3n-3m-5)\,a_n = -a_{n-1} \text{ for } n \geq 1 \qquad (223)$$

For $m=\frac{5}{2}$, we have from (223),

$$(2n^2+7n)a_n = -a_{n-1},\ a \geq 1$$

or

$$a_n = -\frac{a_{n-1}}{2n^2+7n},\ n \geq 1.$$

Thus, putting $n = 1, 2, 3, \ldots$ we get

$$a_1 = -\frac{a_0}{9}$$

$$a_2 = \frac{a_0}{198}$$

$$a_3 = -\frac{a_0}{7722}$$

and so on. Therefore

$$y_1 = a_0 x^{\frac{5}{2}}\left[1-\frac{x}{9}+\frac{x^2}{198}-\frac{x^3}{7722}+\ldots\right].$$

For $m=-1$, the relation (223) yields

$$(2n^2-7n)a_n = -a_{n-1},\ n \geq 1$$

or

$$a_n = \frac{-a_{n-1}}{2n^2-7n},\ n \geq 1.$$

Therefore

$$a_1 = -\frac{a_0}{-5} = \frac{a_0}{5}$$

$$a_2 = -\frac{a_1}{-6} = \frac{a_0}{30}$$

$$a_3 = -\frac{a_2}{-3} = \frac{a_0}{90}$$

and so on. Hence

$$y_2 = a_0 x^{-1}\left[1+\frac{x}{5}+\frac{x^2}{30}+\frac{x^3}{90}+\ldots\right]$$

Therefore the general solution is

$$y = c_1y_1 + c_2y_2$$
$$= Ax^{\frac{5}{2}}\left[1+\frac{x}{9}+\frac{x^2}{198}-\frac{x^3}{7722}+\ldots\right] + Bx^{-1}\left[1+\frac{x}{5}+\frac{x^2}{30}+\frac{x^3}{90}+\ldots\right]$$

EXAMPLE 4.223

For Bessel's polynomial $J_n(x)$ of degree n, show that

(a) $J_2'(x) = (1-\frac{4}{x^2})\,J_1(x) + \frac{2}{x}J_0(x)$

(b) $J_0^2 + 2(J_1^2 + J_2^2 + J_3^2 + \ldots) = 1.$

Solution. (a) We know that

$$J_n'(x) = \frac{1}{2}\,[J_{n-1}(x) - J_{n+1}(x)].$$

Therefore

$$J_2'(x) = \frac{1}{2}\,[J_1(x) - J_3(x)]. \qquad (224)$$

Also

$$J_{n+1}(x) = \frac{2n}{x}J_n(x) - J_{n-1}(x).$$

Therefore

$$J_3(x) = \frac{4}{x} J_2(x) - J_1(x)$$

$$= \frac{4}{x}\left[\frac{2}{x} J_1(x) - J_0(x)\right] - J_1(x)$$

$$= \left(\frac{8}{x^2 - 1}\right) J_1(x) - \frac{4}{x} J_0(x).$$

Putting this value in (224), we get

$$J_2'(x) = \left(1 - \frac{4}{x^2}\right) J_1(x) + \frac{2}{x} J_0(x).$$

(b) From example 4.206, the Jacobi's series are

$$J_0 + 2J_2 \cos 2\phi + 2J_4 \cos 4\phi + \ldots$$

$$= \cos(x \sin \phi)$$

and

$$2J_1 \sin \phi + 2J_3 \sin 3\phi + 2J_5 \sin 5\phi + \ldots$$

$$= \sin(x \sin \phi).$$

Squaring these equations and integrating with respect to ϕ in the interval $(0, \pi)$, we get

$$\pi J_0^2 + 2\pi J_2^2 + 2\pi J_4^2 + \ldots$$

$$= \int_0^{\pi} \cos^2(x \sin \phi)\, d\phi \qquad (225)$$

and

$$2\pi J_1^2 + 2\pi J_3^2 + 2\pi J_5^2 + \ldots$$

$$= \int_0^{\pi} \sin^2(x \sin \phi)\, d\phi \qquad (226)$$

Adding (225) and (226) we get

$$\pi J_0^2 + 2\pi J_1^2 + 2\pi J_2^2 + \ldots = \int_0^{\pi} d\phi = \pi.$$

Hence

$$J_0^2 + 2J_1^2 + 2J_2^2 + 2J_3^2 + \ldots = 1.$$

4.38 SIMULTANEOUS LINEAR DIFFERENTIAL EQUATIONS WITH CONSTANT COEFFICIENT

Differential equations, in which there are two or more dependent variables and a single independent variable are called *simultaneous linear equations.* the aim of this section is to solve a system of linear differential equations with constant coefficients. The solution is obtained by eliminating all but one of the dependent variables and then solving the resultant equations by usual methods.

EXAMPLE 4.224

Solve the simultaneous equations

$$\frac{dx}{dt} = 7x - y, \quad \frac{dy}{dt} = 2x + 5y.$$

Solution. In symbolic form, we have

$$(D - 7)x + y = 0 \qquad (227)$$

$$(D - 5)y - 2x = 0 \qquad (228)$$

Operating (227) by $(D - 5)$ and subtracting (228) from it, we get

$$(D - 5)(D - 7)x + 2x = 0$$

or

$$(D^2 - 12D + 35 + 2)x = 0$$

or

$$(D^2 - 12D + 37)x = 0.$$

Its auxiliary equation is $m^2 - 12m + 37 = 0$, which yields $m = 6 \pm i$. Therefore, its complete solution is $x = e^{6t}(c_1 \cos t + c_2 \sin t)$.
Now

$$\frac{dx}{dt} = \frac{d}{dt}\left(c_1 e^{6t} \cos t\right) + \frac{d}{dt}\left(c_2 e^{6t} \sin t\right)$$

$$= c_1[6e^{6t} \cos t - e^{6t} \sin t]$$

$$+ c_2[6e^{6t} \sin t + e^{6t} \cos t]$$

$$= e^{6t}[(6c_1 + c_2)\cos t + (6c_2 - c_1)\sin t]$$

Putting the values of x and $\frac{dx}{dt}$ in (227), we get

$$e^{6t}[(6c_1 + c_2)\cos t + (6c_2 - c_1)\sin t] - 7e^{6t}[c_1\cos t + c_2\sin t] + y = 0.$$

Therefore,

$$y = 7e^{6t}(c_1\cos t + c_2\sin t) - e^{6t}[(6c_1 + c_2)\cos t + (6c_2 - c_1)\sin t]$$

$$= e^{6t}[(c_1 - c_2)\cos t + (c_1 + c_2)\sin t].$$

Hence the solution is

$$x = e^{6t}(c_1\cos t + c_2\sin t),$$

$$y = e^{6t}[(c_1 - c_2)\cos t + (c_1 + c_2)\sin t].$$

EXAMPLE 4.225

Solve

$$\frac{d^2y}{dt^2} + \frac{dy}{dt} - 2y = \sin t,\ \frac{dx}{dy} + x - 3y = 0.$$

Solution. *The* given system of equations are

$$(D^2 + D - 2)\,y = \sin t \qquad (229)$$

$$(D + 1)\,x - 3y = 0 \qquad (230)$$

The auxiliary equation for (229) is

$$m^2 + m - 2 = 0$$

and so $m = -2, 1$. Therefore,

$$\text{C.F.} = c_1\,e^t + c_2\,e^{-2t}.$$

$$\text{P.I.} = \frac{1}{D^2 + D - 2}\sin t = \frac{1}{D - 3}\sin t$$

$$= \frac{D + 3}{D^2 - 9}\sin t = -\frac{1}{10}(D + 3\sin t)$$

$$= -\frac{3}{10}\sin t - \frac{1}{10}\cos t.$$

Therefore, the complete solution of (229) is

$$y = c_1\,e^t + c_2 e^{-2t} - \frac{1}{10}(\cos t + 3\sin t).$$

Putting this value of y in (230), we get

$$\frac{dx}{dt} + x = 3[c_1 e^t + c_2 e^{-2t} - \frac{1}{10}(\cos t + 3\sin t)]$$

The integrating factor is $e^{\int 1dt} = e^t$. Therefore,

$$x.e^t = 3\int e^t\left[c_1 e^t + c_2 e^{-2t} - \frac{1}{10}(\cos t + 3\sin t)\right]dt,$$

which yields

$$x = \frac{3}{2}c_1 e^{2t} - 3c_2 e^{-t} - \frac{3}{10}e^t(\cos t - 2\sin t).$$

Therefore, the solution is

$$x = \frac{3}{2}c_1 e^{2t} - 3c_2 e^{-t} - \frac{3}{10}e^t(\cos t - 2\sin t)$$

$$y = c_1 e^t + c_2 e^{-2t} - \frac{1}{10}(\cos t + 3\sin t).$$

EXAMPLE 4.226

A mechanical system with two degrees of freedom satisfies the equation

$$2\frac{d^2x}{dt^2} + 3\frac{dy}{dt} = 4,\ 2\frac{d^2y}{dt^2} - 3\frac{dx}{dt} = 0$$

under the condition that $x, y, \frac{dx}{dt}, \frac{dy}{dt}$ all vanish at $t = 0$. Find x and y.

Solution. The given equations are

$$2D^2x + 3Dy = 4$$

$$2D^2y - 3Dx = 0 \qquad (231)$$

or

$$4D^3x + 6D^2y = 2D(4) = 0$$

$$-9Dx + 6D^2y = 0.$$

Subtracting, we get

$$4D^3x + 9Dx = 0$$

or

$$(4D^3 + 9D)\,x = 0.$$

Auxiliary equation is

$$4m^3 + 9m = 0, \text{ and so } m = 0, \pm\frac{3}{2}i.$$

Hence

$$x = c_1 + c_2 \cos\frac{3}{2}t + c^3 \sin\frac{3}{2}t.$$

At $t = 0$, $x = 0$. therefore, $0 = c_1 + c_2$ or $c_1 = -c_2$.

Also

$$\frac{dx}{dt} = -\frac{3}{2}c_2 \sin\frac{3}{2}t + \frac{3}{2}c_3 \cos\frac{3}{2}t$$

At $t = 0$, $\frac{dx}{dt} = 0$ and so $0 = \frac{3}{2}c_3$ or $c_3 = 0$. thus

$$x = c_1 - c_1 \cos\frac{3}{2}t.$$

Therefore,

$$\frac{dx}{dt} = \frac{3}{2}c_1 \sin\frac{3}{2}t \text{ and } \frac{d^2x}{dt^2} = \frac{9}{4}c_1 \cos\frac{3}{2}t.$$

Putting this value of $\frac{dx}{dt}$ in (231), we get

$$2D^2y - \frac{9}{2}c_1 \sin\frac{3}{2}t = 0 \text{ or } D^2y = \frac{9}{4}c_1 \sin\frac{3}{2}t.$$

Integrating, we get

$$Dy = -\frac{3}{2}c_1 \cos\frac{3}{2}t + k.$$

Using initial condition $\frac{dy}{dt} = 0$ at $t = 0$, we get $k = \frac{3}{2}c_1$.

Therefore,

$$Dy = -\frac{3}{2}c_1 \cos\frac{3}{2}t + \frac{3}{2}c_1.$$

Integrating again, we have

$$y = -c_1 \sin\frac{3}{2}t + \frac{3}{2}c_1 t + k.$$

When $t = 0$, $y = 0$. So we have $k = 0$. Hence

$$y = -c_1 \sin\frac{3t}{2} + \frac{3}{2}c_1 t.$$

Further, putting the value of $\frac{d^2x}{dt^2}$ and $\frac{dy}{dt}$ in the first of the given equations, we get $c_1 = \frac{8}{9}$.

Hence

$$x = \frac{8}{9}\left(1 - \cos\frac{3t}{2}\right), y = \frac{4}{3}t - \frac{8}{9}\sin\frac{3t}{2}.$$

EXAMPLE 4.227

Solve the following simultaneous differential equations:

$$\frac{dx}{dt} + 2x + 3y = 0$$

$$3x + \frac{dy}{dt} + 2y = 2e^{2t}.$$

Solution. We are given that

$$\frac{dx}{dt} + 2x + 3y = 0 \quad (232)$$

$$3x + \frac{dy}{dt} + 2y = 2e^{2t} \quad (233)$$

Putting $\frac{d}{dt} = D$, these equations reduce to

$$(D + 2)x + 3y = 0 \quad (234)$$

$$3x + (D + 2)y = 2e^{2t}. \quad (235)$$

Multiplying (234) by 3 and operating (235) by $D + 2$, we get

$$3(D + 2)x + 9y = 0 \quad (236)$$

$$3(D + 2)x + [(D + 2)(D + 2)]y = 2De^{2t} = 4e^{2t} \quad (237)$$

Subtracting (236) from (237), we have

$$(D^2 + 4D - 5)y = 4e^{2t} \quad (238)$$

The auxiliary equation for (238) is

$$m^2 + 4m - 5 = 0,$$

which yields $m = -5, 1$. Hence

$$\text{C.F.} = c_1e^{-5t} + c_2e^1.$$

Further,

$$\text{P.I.} = \frac{1}{D^2 + 4D - 5}(4e^{2t})$$

$$= \frac{4}{7}e^{2t}.$$

Hence solution of (238) is

$$y = \text{C.F.} + \text{P.I.} = c_1e^{-5t} + c_2e^t + \frac{4}{7}e^{2t} \quad (239)$$

Therefore

$$\frac{dy}{dt} = -5c_1e^{-5t} + c_2e^t + \frac{8}{7}e^{2t}$$

Putting this value of $\frac{dy}{dt}$ in (233), we get

$$x = \frac{1}{3}\left[2e^{2t} - \frac{dy}{dt} - 2y\right]$$

$$= \frac{1}{3}\left[2e^{2t} + 5c_1e^{-5t} - c_2e^t - \frac{8}{7}e^{2t} - 2c_1e^{-5t} - 2c_2e^t - \frac{8}{7}e^{2t}\right]$$

$$= \frac{1}{3}\left[-\frac{2}{7}e^{2t} + 3c_1e^{-5t} - 3c_2e^t\right]$$

$$= c_1e^{-5t} - c_2e^t - \frac{2}{21}e^{2t} \qquad (240)$$

Hence (240) and (239) constitute the solution of the given simultaneous differential equations.

EXAMPLE 4.228

Solve the following simultaneous differential equations:

$$\frac{d^2x}{dt^2} + y = \sin t$$

$$\frac{d^2y}{dt^2} + x = \cos t.$$

Solution. Putting $\frac{d}{dt} = D$, the given equations reduce to

$$D^2x + y = \sin t \qquad (241)$$

$$x + D^2y = \cos t. \qquad (242)$$

Operating (241) with D^2, we get

$$D^4x + D^2y = D\sin t = -\sin t. \qquad (243)$$

Subtracting (242) from (243), we get

$$(D^4 - 1)\,x = -\sin t - \cos t. \qquad (244)$$

Auxiliary equation for (244) is

$$m^4 - 1 = 0 \text{ or } (m^2 - 1)(m^2 + 1) = 0.$$

and so $m = 1, -1, i, -i$. Therefore,

$$\text{C.F.} = c_1e^t + c_2e^{-t} + c_3\cos t + c_4\sin t.$$

Further,

$$\text{P.I.} = \frac{1}{D^4 - 1}\,[-\sin t - \cos t]$$

$$= -\frac{t}{4}\cos t + \frac{t}{4}\sin t.$$

Hence

$$x = c_1e^t + c_2e^{-t} + c_3\cos t + c_4\sin t - \frac{t}{4}\cos t + \frac{t}{4}\sin t. \qquad (245)$$

Now

$$\frac{dx}{dt} = c_1e^t - c_2e^{-t} - c_3\sin t + c_4\cos t + \frac{t}{4}\sin t + \frac{t}{4}\cos t - \frac{1}{4}\cos t + \frac{1}{4}\sin t$$

and so

$$\frac{d^2x}{dt^2} = c_1e^t + c_2e^{-t} - c_3\cos t + c_4\sin t + \frac{t}{4}\cos t - \frac{t}{4}\sin t + \frac{1}{2}\sin t + \frac{1}{2}\cos t \qquad (246)$$

Putting this value of $\frac{d^2x}{dt^2}$ from (246) in equation (241), we have

$$y = \sin t - D^2x$$

$$= -c_1e^t - c_2e^{-t} + c_3\cos t + c_4\sin t + \frac{1}{4}(2 + t)(\sin t - \cos t) \qquad (247)$$

Hence equations (245) and (247) constitute the solution.

EXAMPLE 4.229

Solve the simultaneous equations

$$\frac{dx}{dt} + 2y = e^t,$$

$$\frac{dy}{dt} - 2x = e^{-t}$$

Solution. The given equations are

$$Dx + 2y = e^t \qquad (248)$$

$$-2x + Dy = e^{-t} \qquad (249)$$

Operating D on (248) and multiplying (249) by 2, we get

$$D^2x + 2Dy = De^t = e^t, \quad (250)$$

$$-4x + 2Dy = 2e^{-t} \quad (251)$$

Subtracting (251) from (250), we get

$$(D^2 + 4)x = e^t - 2e^{-t} \quad (252)$$

The auxiliary equation for (252) is

$$m^2 + 4 = 0$$

and so $m = \pm 2i$. Therefore

$$\text{C.F.} = c_1 \cos 2x + c_2 \sin 2x.$$

Further,

$$\text{P.I.} = \frac{1}{D^2+4}(e^t - 2e^{-t})$$

$$= \frac{1}{D^2+4}e^t - \frac{2}{D^2+4}e^{-t}$$

$$= \frac{1}{5}e^t - \frac{2}{5}e^{-t}.$$

Therefore solution of (252) is

$$x = \text{C.F.} + \text{P.I.}$$

$$= c_1 \cos 2x + c_2 \sin 2x + \frac{1}{5}e^t - \frac{2}{5}e^{-t}. \quad (253)$$

Differentiating (252) with respect to t, we have

$$\frac{dx}{dt} = -2c_1 \sin 2x + 2c_2 \cos 2x + \frac{1}{5}e^t + \frac{2}{5}e^{-t}.$$

Putting this value of $\frac{dx}{dt}$ in (248), we get

$$y = \frac{1}{2}[e^t - Dx]$$

$$= \frac{1}{2}[e^t + 2c_1 \sin 2x - 2c_2 \cos 2x$$

$$-\frac{1}{5}e^t - \frac{2}{5}e^{-t}]$$

$$= c_1 \sin 2x - c_2 \cos 2x + \frac{2}{5}e^t - \frac{1}{5}e^{-t} \quad (254)$$

Hence (253) and (254) constitute the solution of the given system of equations.

EXERCISES

1. Form differential equation from the following equations

   (a) $y = ae^{2x} + be^{-3x} + ce^x$

   Ans. $x\frac{d^2y}{dx^2} + 2\frac{dy}{dx} - xy = 0$

   (b) $x = A\cos(nt + \alpha)$

   Ans. $\frac{d^2x}{dt^2} + n^2x = 0$

2. Solve the separable equations

   (a) $(x-4)y^4\,dx - x^3(y^3-3)\,dy = 0$

   Ans. $-\frac{1}{x} + \frac{2}{x^2} + \frac{1}{y} - \frac{1}{y^3} = c$

   (b) $\frac{dy}{dx} = e^{2x-3y} + 4x^2e^{-3y}$

   Ans. $3e^x - 2e^{3y} + 8x^3 = c$

   (c) $\frac{dy}{dx} - x\tan(y-x) = 1$

   Ans. $\log\sin(y-x) = \frac{1}{2}x^2 + c$

   (d) $(x-y)^2\frac{dy}{dx} = a^2$

   Ans. $a\log\left(\frac{x-y-a}{x-y+a}\right) = 2y + k$

   (e) $\frac{dx}{dt} = x^2 - 2x + 2$

   Ans. $x = 1 + \tan(t + c)$

   (f) $x\frac{dy}{dx} + \cot y = 0$ if $y(\sqrt{2}) = \frac{\pi}{4}$

   Ans. $x = 2\cos y$

3. Solve the homogeneous equations :

   (a) $\frac{dy}{dx} = \frac{y}{x} + \sin\frac{y}{x}$

   Ans. $2x\tan^{-1}(cx)$

   (b) $ydx - xdy = \sqrt{x^2+y^2}\,dx$

   Ans. $y + \sqrt{x^2+y^2} = c$

   (c) $2xy\frac{dy}{dx} = 3y^2 + x^2$

   Ans. $x^2 + y^2 = cx^3$

   (d) $y' = \frac{y+x}{x}$

   Ans. $y = x\log|cx|$

4. Reduce the following equations to homogeneous form and solve them

(a) $(2x + y - 3)\,dy = (x + 2y - 3)\,dx$

Ans. $(y - x)^3 = k(y + x - 2)$

(b) $(x + 2y)(dx - dy) = dx + dy$

Ans. $\log\left(x + y + \frac{1}{3}\right) + \frac{3}{2}(y - x) + k$

(c) $\frac{dy}{dx} + \frac{ax + hy + g}{hx + by + f} = 0$

Ans. $ax^2 + 2hxy + by^2 + 2gx + 2fy + c = 0$

5. Solve the following linear equations:

(a) $x^2 \frac{dy}{dx} = 3x^2 - 2yx + 1$

Ans. $y = x + \frac{1}{x} + \frac{c}{x^2}$

(b) $(x + 1)\frac{dy}{dx} - y = e^{3x}(x + 1)^2$

Ans. $y = (x + 1)\left(\frac{1}{3}e^{3x} + c\right)$

(c) $\frac{dy}{dx} = -\frac{x + y\cos x}{1 + \sin x}$

Ans. $y(1 + \sin x) = c - \frac{x^2}{2}$

(d) $y\log y\,dx + (x - \log y)\,dy = 0$

Ans. $x = \frac{1}{2}\log y + \frac{c}{\log y}$

(e) $x\log x\frac{dy}{dx} + y = \log x^2$

Ans. $y = \log x + \frac{c}{\log x}$

6. Solve the following equations

(a) $(x^3y^2 + xy)\,dx = dy$

Ans. $\frac{1}{y} = x^2 - 2 + ce^{-\frac{x2}{2}}$

(b) $\frac{dy}{dx} - \frac{\tan y}{1 + x} = (1 + x)\,e^x \sec y$

Ans. $\sin y = (1 + x)(e^x + c)$

(c) $\frac{dz}{dx} + \frac{z}{x}\log z = \frac{z}{x}(\log z)^2$

Ans. $\frac{1}{\log z} = 1 + cx$

(d) $\frac{dy}{dx} + y\tan x = y^3\sec x$

Ans. $\cos^2 x = y^2(c + 2\sin x)$

7. Solve the exact equations

(a) $\left[y\left(1 + \frac{1}{x}\right) + \cos y\right]dx + (x + \log x - x\sin y)\,dy = 0$

Ans. $y(x + \log x) + x\cos y = 0$

(b) $[5x^4 + 3x^2y^2 - 2xy^3]dx + (2x^3y - 3x^2y^2 - 5y^4)\,dy = 0$

Ans. $x^5 + x^3y^2 - x^2y^3 - y^5 = 0$

(c) $\frac{2x}{y^3}dx + \frac{y^2 - 3x^2}{y^4}dy = 0$

Ans. $x^2 - y^2 = cy^3$

(d) $(x^2 + y^2 - a^2)x\,dx + (x^2 - y^2 - b^2)\,y\,dy = 0$

Ans. $x^4 + 2x^2y^2 - y^4 - 2a^2x^2 - 2b^2y^2 = c$

8. Solve the following equations which are reducible to exact equations:

(a) $(y^2 + 2x^2y)\,dx + (2x^3 - xy)\,dy = 0$

Ans. $6\sqrt{xy} - \left(\frac{y}{x}\right)^{\frac{3}{2}} = c$

(b) $(x^2 + y^2 + 1)\,dx - 2xy\,dy = 0$

Ans. $x^2 - \frac{1}{x} - \frac{y^2}{x} = c$

(c) $(xy\,e^{\frac{x}{y}} + y^2)\,dx - x^2 e^{\frac{x}{y}}\,dy = 0$

Ans. $3\log x - 2\log y + \frac{y}{x} = c$

(d) $x\,dy - y\,dx + a(x^2 + y^2)\,dx = 0$

Ans. $ax + \tan^{-1}\frac{y}{x} = c$

(e) $x\,dy - y\,dx = xy^2\,dx$

Ans. $\frac{x^2}{2} + \frac{x}{y} = c$

(f) $\frac{dy}{dx} = \frac{x^3 + y^3}{xy^2}$

Ans. $3\log x - \left(\frac{y}{x}\right)^3 = c$

9. In RC circuit containing steady voltage V, find the change at any time t. Also derive expression for the current flowing through the circuit.

Ans. $Q = CV(1 - e^{-\frac{t}{RC}}$ and $I = \frac{V}{R}e^{-\frac{t}{RC}}$

10. An RL circuit has an e.m.f. of 3 sin 2t volts, a resistance of 10 ohms, an inductance of

0.5 henry, and an initial current of 6 amp. Find the current in the circuit at any time t.

Ans. $I = \frac{609}{101} e^{-20t} + \frac{30}{101} \sin 2t - \frac{3}{101} \cos 2t$

11. If the air is maintained at 30° C and the temperature of the body cools from 80° C to 60° C in 12 min, find the temperature of the body after 24 min.

Ans. 48° C

12. A pipe 20 cm in diameter contains steam at 200° C. It is covered by a layer of insulation 6 cm thick and thermal conductivity 0.0003. If the temperature of the outer surface is 30° C, find the heat loss per hour from 2 metre length of the pipe.

Ans. 490,000 cal.

13. A steam pipe 20 cm in diameter is covered with an insulating sheath 5cm thick, the conductivity of which is 0.00018. If the pipe has the constant temperature 100° C, and outer surface of the sheath is kept at 30° C, find the temperature of the sheath as a function of the distance x from the axis of the pipe. How much heat is lost per hour through a section 1 metre long.

Ans. $T = 497.5 - 172.6 \log x$ and $Q = 70300$ cal.

14. A tank contains 5000 litres of fresh water. Salt water which contains 100 gm of salt per litre flows into it at the rate of 10 litre/min and the mixture kept uniform by stirring, runs out at the same rate. When will the tank contain 2,00,000 gm of salt? How long will it take for the quantity of salt in the tank to increase from 1,50,000 gm to 2,50,000 gm?

Ans. 4 h 2.52 min, 2 h 48.23 min

15. The amount x of a substance present in a certain chemical reaction at time t is given by $\frac{dx}{dt} + \frac{x}{10} = 2 - 1.5\, e^{-\frac{t}{10}}$. If at $t = 0, x = 0.5$, find x at $t = 10$.

Ans. $20 - 69/2e$

16. Radium decomposes at a rate proportional to the amount present. If a fraction p of the original amount disappears in 1 year, how much will it remain at the end of 21 years?

Hint: use $x = x_0\, e^{kt}$. Using given data $k = \log \frac{1}{p}$]

Ans. $\left(1 - \frac{1}{p}\right)^{21}$ times the original amount

17. Find the orthogonal trajectories of the family of curves $r = a(1 + \cos \theta)$.

Ans. $r = c\,(1 - \cos \theta)$

18. Find the orthogonal trajectories of the family of parabolas $y^2 = 4ax$.

Ans. $2x^2 + y^2 = c$

19. Find the orthogonal trajectories of the family of curves $ay^2 = x^3$.

Ans. $3y^2 + 2x^2 = c^2$

20. Solve the following differential equations:

(i) $p^2 - 7p + 12 = 0$

(ii) $p\,(p + y) = x\,(x + y)$

(iii) $x + yp^2 = p\,(1 + xy)$

(iv) $xp^2 + (y - x)\,p - y = 0$

Ans.

(i) $(y - 4x - c)\,(y - 3x - c) = 0$

(ii) $(2y - x^2 + c)\,(y + x + ce^{-x} - 1) = 0$

(iii) $(2y - x^2 - c)\,(2x - y^2 - c) = 0$

(iv) $(y - x - c)\,(xy - c) = 0$

21. Solve the following differential equation.

(i) $xp^2 + x = 2yp$

(ii) $y = 2px + p^n$

(iii) $xp^2 - yp - y = 0$

(iv) $y = p \sin p + \cos p$

(v) $4y = x^2 + p^2$

Ans.

(i) $2cy = c^2x^2 + 1$

(ii) $x = cp^{-2} - \frac{np^{n-1}}{n+1}$, $y = \frac{1-n}{1+n} p^n + \frac{2c}{p}$

(iii) $y = cp^2 e^p$

$x = c(p+1)e^p$

(iv) $x = c + \sin p$

$y = p \sin p + \cos p$

(v) $\log(p - x) = \frac{x}{p-x} + c$

$4y = x^2 + p^2$

22. Solve the following differential equations.

(i) $p^3 y + 2px = y$

(ii) $x = y + p^2$

(iii) $x + py = p^3$

(iv) $y = 2px + p^2 y$

Ans.

(i) $y^2 = 2cx + c^3$

(ii) $x = c - [2p + 2\log(p-1)]$,

$y = c - [p^2 + 2p + 2\log(p-1)]$

(iii) $x + py = p^3$,

$y = p^2 - 2 + \frac{c}{\sqrt{1+p^2}}$

(iv) $y^2 = 2cx + c^2$

23. Solve the differential equation.

(i) $p = \sin(y - xp)$

Hint: $\sin^{-1} p = y - xp$ (Clairaut's equation)

(ii) $xy(y - px) = x + py$

Hint: Put $x^2 = u$, $y^2 = v$ to convert to Clairaut's equation.

(iii) $x^2(y - px) = yp^2$

(iv) $y + px = x^4 p^2$

Hint: Put $\frac{1}{x} = u$.

Ans.

(i) $y = cx + \sin^{-1} c$

(ii) $y^2 = cx^2 + (1 + c)$

(iii) $y^2 = cx^2 + c^2$

(iv) $y = \frac{c}{x} + c^2$

24. Solve the following differential equation

(a) $\frac{d^4y}{dx^4} - 4\frac{d^3y}{dx^3} + 14\frac{d^2y}{dx^2} - 20\frac{dy}{dx} + 25y = 0$

Hint: Roots of A.E. are $1 + 2i$, $1 - 2i$, $1 + 2i$, $1 - 2i$.

Ans. $y = e^x [(c_1 + c_2x)\sin 2x + (c_3 + c_4x)\cos 2x$

(b) $\frac{d^2y}{dx^2} - 6\frac{dy}{dx} + 25y = 0, y(0) = -3, y'(0) = -1.$

Ans. $y = e^{3x}(2\sin 4x - 3\cos 4x)$

(c) $\frac{d^4y}{dx^4} + 8\frac{d^2y}{dx^2} + 16y = 0.$

Ans. $y = (c_1 + c_2x)\cos 2x + (c_3 + c_4x)\sin 2x$

(d) $\frac{d^2x}{dt^2} + 6\frac{dx}{dt} + 9x = 0$

Ans. $x = (c_1 + c_2t)\, e^{-3t}$

(e) $\frac{d^2y}{dx^2} + (a + b)\frac{dy}{dx} + aby = 0$

Ans. $y = c_1 e^{-ax} + c_2 e^{-bx}$

(f) $(D^3 + D^2 + 4D + 4)\, y = 0$

Ans. $y = c_1 e^{-x} + c_2 \cos 2x + c_3 \sin 2x$

(g) $\frac{d^3y}{dx^2}\ 2\frac{dy}{dx} + 2y = 0$. Are the solutions linearly independent?

Ans. $y = c_1 e^x + c_2 e^{2x}$, Wronskian

$$W(e^x, e^2x) = \begin{vmatrix} e^x & e^{2x} \\ e^x & 2e^{2x} \end{vmatrix} = e^{3x} \neq 0$$

Therefore, e^x, e^{2x} are linearly independent.

25. Solve the following differential equations:

(a) $\frac{d^2y}{dx^2} + a^2 y = \tan ax$

Ans. $y = c_1 \cos ax + c_2 \sin ax$

$-\frac{1}{a^2}\cos ax \log(\sec ax + \tan ax)$

(b) $(D^2 - 3D + 2)\, y = 6\, e^{-3x} + \sin 2x.$

Ans. $y = c_1 e^x + c_2 e^{2x}$

$+ \frac{3}{10} e^{-3x} + \frac{1}{20}(3\cos 2x - \sin 2x)$

(c) $\frac{d^2y}{dx^2} - 4y = x^2 + 2x.$

Ans. $y = c_1e^{2x} + c_2e^{-2x} - \frac{1}{4}\left(x^2 + \frac{1}{2}\right)$

(d) $(D^2 - 4D + 4)\, y = 8x^2\, e^{2x} \sin 2x.$

Ans. $y = (c_1 + c_2x)\, e^{2x}$
$- e^{2x} [4x \cos 2x + (2x^2 - 3) \sin 2x]$

(e) $(D^2 - 1)y = x \sin x + (1 + x^2)e^x.$

Ans. $y = c_1e^x + c_2e^{-x} - \frac{1}{2}(x \sin x + \cos x)$
$+ \frac{1}{12} xe^x (2x^2 - 3x + 9)$

(f) $(D - 1)^2 (D + 1)^2\, y = \sin^2 \frac{x}{2} + e^x + x.$

Ans. $y = (c_1 + c_2x)\, e^x + (c^3 + c^4x)\, e^{-x}$
$-\frac{1}{2} - \frac{1}{8} \cos x + e^x + x$

(g) $(D^2 + 4)\, y = e^x + \sin 2x.$

Ans. $y = c_1 \cos 2x + c_2 \sin 2x + \frac{1}{5} e^x - \frac{x}{4} \cos 2x$

(h) $\frac{d^3y}{dx^2} + 3\frac{dy}{dx} + 2y = 4 \cos^2 x.$

Ans. $y = c_1e^{-x} + c_2e^{-2x} + 1$
$+ \frac{1}{10}(3 \sin 2x - \cos 2x)$

(i) $\frac{d^3y}{dx^2} - 3\frac{dy}{dx} + 2y = xe^{3x} + \sin 2x.$

Ans. $y = c_1e^x + c_2e^{2x} + \frac{1}{4} e^{3x} (2x - 3)$
$+ \frac{1}{20}(3 \cos 2x - \sin 2x)$

(j) $\frac{d^3y}{dx^3} + 2\frac{d^2y}{dx^2} + \frac{dy}{dx} = x^2e^{2x} + \sin^2 x.$

Ans. $y = c_1 + (c_2 + c_3x)e^{-x} + \frac{e^{2x}}{18}\left(x^2 - \frac{7}{8}x + \frac{11}{6}\right)$
$+ \frac{1}{100}(3 \sin 2x + 4 \cos 2x)$

(k) $\frac{d^2y}{dx^2} + 3\frac{dy}{dx} + 2y = e^{ex}.$

Ans. $y = c_1e^{-x} + c_2e^{-2x} + e^{-2x}e^{ex}$

(l) $(D^2 - \text{D})\, y = 2x + 1 + 4 \cos x + 2e^x.$

Ans. $y = c_1 + c_2\, e^x + c_3e^{-x}$
$+ x\, e^x - (x^2 + x) - 2 \sin x$

26. An uncharged condenser of capacity C is charged by applying an e.m.f. E sin $\frac{t}{\sqrt{LC}}$ through leads of self inductance L and negligible resistance. Then find the charge at the condenser plate at time t.

Ans. $\frac{EC}{2} \sin\left[\frac{t}{\sqrt{LC}} - \frac{t}{\sqrt{LC}} \cos \frac{t}{\sqrt{LC}}\right]$

27. A circuit consists of an inductance L and a capacitor of capacity C in series. An alternating e.m.f. E sin nt is applied to the circuit at time $t = 0$, the initial current and charge on the condenser being zero. Prove that the current at time t is $\frac{nE}{L(n^2 - \omega^2)}$ (cos ωt – cos ωt), where CL $\omega^2 = 1$. If $n = \omega$, show that current at time t is $\frac{Et \sin \omega t}{2L}$.

Hint: Solve $L\frac{d^2Q}{dt^2} + \frac{Q}{C} = E$ sin nt under the conditions $I(0) = 0$, $Q(0) = 0$. Then use $I = \frac{dQ}{dt}$.

28. A pendulum of length l hangs against a wall at an angle θ to the horizontal. Show that the time of complete oscillation is $2\pi \sqrt{\frac{l}{g \sin \theta}}$.

29. A simple pendulum has a period T. When the length of the string is increased by a small fraction $\frac{1}{n}$ of its original length, the period is T'. Show that (approximately) $\frac{1}{n} = \frac{2(T' - T)}{T}$.

30. How many seconds a clock would lose per day if the length of its pendulum were increased in the ratio 900:901.

Ans. 48 s/day

31. Solve the following differential equations of second order by changing the independent variable.

(i) $x^4\frac{d^2y}{dx^2} + 2x^3\frac{dy}{dx} + 4y = 0$

Ans. $y = c_1 \cos \frac{2}{x} + c_2 \sin \left(\frac{2}{x}\right)$

(ii) $x^6 \frac{d^2y}{dx^2} + 3x^5 \frac{dy}{dx} + a^2 y = \frac{1}{x^2}$

Ans. $y = c_1 \cos \frac{a}{2x^2} - c_2 \sin \frac{a}{2x^2} + \frac{1}{a^2 x^2}$

(iii) $x^2 \frac{d^2y}{dx^2} - x \frac{dy}{dx} + y = \log x$

Ans. $y = (c_1 + c_2 \log x)\, x + \log x + 2$

(iv) $\cos x \frac{d^2y}{dx^2} + \sin x \frac{dy}{dx} - (2\cos^3 x) y = 2\cos^5 x$

Ans. $y = c_1 e^{\sqrt{2}\sin x} + c_2 e^{-\sqrt{2}\sin x} + \sin^2 x$

32. Solve the following differential equation of second order by changing the dependent variable.

(i) $\frac{d^2y}{dx^2} + 2n \cot nx \frac{dy}{dx} + (m^2 - n^2)\, y = 0$

Ans. $y = \frac{1}{\sin nx} [c_1 \cos mx + c_2 \sin mx]$

(ii) $x^2 \frac{d^2y}{dx^2} - 2y = x^2 + \frac{1}{x}$

Ans. $y = c_1 x^2 + \frac{c_2}{x} + \frac{1}{3}\left(x^2 - \frac{1}{x}\right) \log x$

(iii) $\cos^2 x \frac{d^2y}{dx^2} - (2 \sin x \cos x) \frac{dy}{dx} + (\cos^2 x) y = 0$

Ans. $y = [c_1 \cos(\sqrt{2}x) + c_2 \sin(\sqrt{2}x)]$

$- \sec x$

(iv) $\left[\frac{d^2y}{dx^2} + y\right] \cot x + 2\left[y \tan x + \frac{dy}{dx}\right] = 0$

Ans. $y = (c_1 + c_2 x) \cos x$

33. Solve by method of undetermined coefficients.

(i) $\frac{d^2y}{dx^2} + y = 2e^x + \cos x$

Ans. $y = c_1 \cos x + c_2 \sin x + e^x + \frac{1}{2} + x \sin x$

(ii) $\frac{d^2y}{dx^2} - 2\frac{dy}{dx} + y = x^2 e^x$

Ans. $y = (c_1 + c_2 x)\, e^x + \frac{x^4}{12} e^x$

(iii) $\frac{d^2y}{dx^2} + 2\frac{dy}{dx} + y = x - e^x$

Ans. $y = (c_1 + c_2 x)\, e^{-x} + x - 2 - \frac{e^x}{4}$

(iv) $y'' + 4y = 2 \sin 3x$

Hint: C.F. $= c_1 \cos 2x + c_2 \sin 2x$. Trial solution is $y = a \sin 3x + b \cos 3x$.

Ans. $c_1 \cos 2x + c_2 \sin 2x - \frac{2}{5} \sin 3x$

34. Solved by method of reduction of order.

(i) $x^2 \frac{d^2y}{dx^2} - 2x(1 + x) \frac{dy}{dx} + 2(1 + x)\, y = x^3$ if $y = x$ is a solution of the C.F.

Ans. $y = x\, [-\frac{x}{2} + \frac{c_1}{2} e^{2x} + c_2]$

(ii) $(1 - x^2) \frac{d^2y}{dx^2} + x \frac{dy}{dx} - y = x\,(1 - x^2)^{\frac{3}{2}}$ if $y = x$ is a part of C.F.

Ans. $y = \dfrac{-x(1 - x^2)^{\frac{3}{2}}}{9} - c_1 [\sqrt{1 - x^2} + x \sin^{-1} x]$

$+ c_2 x$

(iii) $\frac{d^2y}{dx^2} + 2\frac{dy}{dx} + y = x - e^x$

Ans. $y = (c_1 + c_2 x)\, e^{-x} + x - 2 - \frac{e^x}{4}$

35. Solve the following equations using the method of variation of parameters:

(a) $\frac{d^2y}{dx^2} + a^2 y = \sec ax.$

Ans. $y = c_1 \cos ax + c_2 \sin ax + \frac{1}{a^2} \cos ax$

$\log(\cos ax) + \frac{1}{a} x \sin ax$

(b) $\frac{d^2y}{dx^2} + y = x \sin x.$

Ans. $y = c_1 \cos x + c_2 \sin x + \frac{x}{4} \sin x$

$- \frac{x^2}{4} \cos x$

(c) $\frac{d^2y}{dx^2} + 4y = \tan 2x.$

Ans. $y = c_1 \cos 2x + c_2 \sin 2x$

$- \frac{1}{4} \cos 2x \log(\sec 2x + \tan 2x)]$

(d) $\frac{d^2y}{dx^2} - 2\frac{dy}{dx} + 2y = e^x \tan x.$

Ans. $y = e^x (c_1 \cos x + c_2 \sin x)$
$-e^x \cos x \log (\sec x + \tan x)$

(e) $(D^2 + 1)\, y = \frac{1}{1 + \sin x}$

Ans. $y = c_1 \cos x + c_2 \sin x$
$+ \sin x \log(1 + \sin x) - x \cos x - 1$

(f) $(D^2 + 1)y = x \sin x.$

Ans. $y = c_1 \cos x + c_2 \sin x + \frac{x}{4} \sin x$
$-\frac{x^2}{4} \cos x$

36. Solve the following Cauchy-Euler equations:

(a) $x^3 \frac{d^3y}{dx^3} + 2x^2 \frac{d^2y}{dx^2} + 2y = 10\left(x + \frac{1}{x}\right).$

Ans. $y = \frac{c_1}{x} + x\,[c_2 \cos (\log x) + c_3 \sin (\log x)]$

(b) $x^2 \frac{d^2y}{dx^2} - 3x \frac{dy}{dx} + 5y = x^2 \sin (\log x)$

Ans. $y = x^2\, [c_1 \cos (\log x + c_2 \sin (\log x)]$
$-\frac{1}{2} x^2 \log x \cos (\log x)$

(c) $x^2 \frac{d^2y}{dx^2} - 2x \frac{dy}{dx} - 4y = x^2 + 2 \log x$

Ans. $y = \frac{c_1}{x} + c_2 x^4 - \frac{x^6}{6} - \frac{1}{2} \log x + \frac{3}{8}$

(d) $x^2 \frac{d^3y}{dx^3} - 4x \frac{d^2y}{dx^2} + 6 \frac{dy}{dx} = 4$

Hint: Write the given equation in the form

$x^3 \frac{d^3y}{dx^3} - 4x^2 \frac{d^2y}{dx^2} + 6x \frac{dy}{dx} = 4x$

Ans. $y = c_1 + c_2 x^3 + c_3 x^4 + \frac{2}{3} x$

(e) $x^2 \frac{d^2y}{dx^2} - 4x \frac{dy}{dx} + 6y = x^2$

Ans. $y = \frac{1}{x} (c_1 + c_2 \log x) + \frac{1}{x} \log \frac{x}{1-x}$

(f) $x^2 \frac{d^2y}{dx^2} - 3x \frac{dy}{dx} + y = \log \frac{\sin (\log x) + 1}{x}$

Ans. $y = c_1 x^{\sqrt{3}+2} + c_2 x^{2-\sqrt{3}} + \frac{1}{61x} \log x$
$[5 \sin (\log x + 6 \cos (\log x)] + \frac{2}{61} [21 \sin$
$(\log x) + 191 \cos (\log x)] + \frac{1}{6x} (1 + \log x)$

(g) $x^3 \frac{d^3y}{dx^3} + 3x^2 \frac{d^2y}{dx^2} + x \frac{dy}{dx} + 8y$
$= 65 \cos (\log x)$

Ans. $y = c_1 x^{-2} + x\,[c_2 \cos \sqrt{3}\,(\log x) + c^3 \sin$
$(\sqrt{3} \log x)] + 8 \cos (\log x) - \sin (\log x)$

37. Solve the following Legendre's linear equations:

(a) $(3x + 2)\frac{d^2y}{dx^2} + 3\,(3x + 2)\frac{dy}{dx} - 36y = 3x^2$
$+ 4x + 1.$

Ans. $y = c_1(3x + 2)^2 + c_2\,(3x + 2)^{-2}$
$+ \frac{1}{108} [(3x + 2)^2 \log (3x + 2) + 1$

(b) $(x + 1)^2 \frac{d^2y}{dx^2} + (x + 1) \frac{dy}{dx}$
$= (2x + 3)(2x + 4).$

Ans. $y = c_1 + c_2 \log (x + 1) + [\log (x + 1)]^2$
$+ x^2 + 8x$

(c) $(1 + 2x)^2 \frac{d^2y}{dx^2} - 6\,(1 + 2x) \frac{dy}{dx} + 16y$
$= 8(1 + 2x)^2$

Ans. $(1 + 2x)^2\, [c_1 + c_2 \log (1 + 2x)$
$+ \log (1 + 2x)]$

Solution about regular points

38. Solve the following differential equations about $x = 0$.

(i) $\frac{d^2y}{dx^2} + x^2 y = 0$

Ans. $y = a_0[x - \frac{x^4}{4.3} + \frac{x^8}{8.7.4.3} - \frac{x^{12}}{12.11.8.7.4.3} + ...]$

$+ a_1[x - \frac{x^5}{5.4} + \frac{x^9}{9.8.5.4} - \frac{x^{13}}{13.12.9.8.5.4} + ...]$

(ii) $(x^2 + 1)\frac{d^2y}{dx^2} + xy\frac{dy}{dx} - xy = 0$

Ans. $y = a_0\left(1 - \frac{1}{6}x^3 - \frac{3}{40}x^5 + ...\right) + a_1\left(x - \frac{1}{6}x^3 + \frac{1}{12}x^4 + \frac{3}{40}x^5 + ...\right)$

(iii) $\frac{d^2y}{dx^2} + x\frac{dy}{dx} + (2x^2 + 1)\,y = 0$

Ans. $y = a_0\left(1 - \frac{1}{2}x^2 - \frac{1}{24}x^4 + ...\right) + a_1\left(x - \frac{x^3}{3} - \frac{x^5}{30} - ...\right)$

(iv) $\frac{d^2y}{dx^2} + x\frac{dy}{dx} + (3x + 2)y = 0$

Ans. $y = a_0(1 - x^2 - \frac{x^3}{2} + \frac{x^4}{3} + \frac{11x^5}{40} + ...) + a_1\left(x - \frac{x^3}{2} - \frac{x^4}{4} + \frac{x^5}{8} + ...\right)$

39. Find the power series solution in the powers of $(x - 1)$ of the initial value problem.

$$\frac{xd^2y}{dx^2} + \frac{dy}{dx} + 2y = 0, y(1) = 1, y'(1) = 2.$$

Hint: $x = 1$ is the ordinary point. So consider $y = \sum_{n=0}^{\infty} a_n(x - 1)^n$, find $\frac{dy}{dx}$ and $\frac{d^2y}{dx^2}$, put in the given equation. Use the initial conditions in finding a_0 and a_1 by putting $x = 1$ in y and $\frac{dy}{dx}$. Other coefficients to be found by equating to zero the powers of $(x - 1)$.

Ans. $y = 1 + 2(x - 1) - 2(x - 1)^2 + \frac{2}{3}(x - 1)^3 - \frac{1}{6}(x - 1)^4 + ...$

Solution about regular singular point

40. Solve the following differential equations about $x = 0$

(i) $x(1 - x)\frac{d^2y}{dx^2} + (1 - x)\frac{dy}{dx} - y = 0$

Ans. $y = a_0(1 + \log x)(1 + x + \frac{2x^2}{4} + \frac{2.5}{4.9}x^3 + ...) + a_0(-2x - x^2 - ...)$

(ii) $x(1 - x)\frac{d^2y}{dx^2} - (1 + 3x)\frac{dy}{dx} - y = 0$

Ans. $y = (A + B\log x)(1.2x^2 + 2.3x^3 + 3.4x^4 + ...) + B(-1 + x + 2x^2 + 11x^3 + ...)$

(iii) $2x^2\frac{d^2y}{dx^2} + \frac{xdy}{dx} + (x^2 - 1)\,y = 0$

Ans. $y = c_1x\left(1 - \frac{x^2}{14} + \frac{x^4}{616} + ...\right) + c_2\sqrt{x}\left(1 - \frac{x^2}{2} + \frac{x^4}{40} + ...\right)$

(iv) $(2x + x^3)\frac{d^2y}{dx^2}\frac{dy}{dx} - 6xy = 0$

Ans. $y = c_1\left(1 + 3x^2 + \frac{3x^4}{5} + ...\right) + c_2x^{\frac{3}{2}}\left(1 + \frac{3x^2}{8} - \frac{3x^4}{128} + ...\right)$

(v) $4x\frac{d^2y}{dx^2} + 2(1 - x)\frac{dy}{dx} - y = 0$

Ans. $y = c_1\left(1 + \frac{x}{2!} + \frac{x^2}{2^2.2!} + \frac{x^3}{2^3.3!} + ...\right) + c_2x^{\frac{1}{2}}\left(1 + \frac{x}{1.3} + \frac{x^2}{1.3.5} + \frac{x^3}{1.3.5.7} + ...\right)$

(vi) $x\frac{d^2y}{dx^2} + \frac{dy}{dx} + 2y = 0$

Ans. $y = (c_1 + c_2\log x)\left(1 - 2x + \frac{2^2x^2}{(2!)^2}...\right) + c_2(4x - 3x^2 + ...)$

Bessel's Equation and Bessel Function

41. Write the general solution of the differential equation

$$x^2\frac{d^2y}{dx^2} + x\frac{dy}{dx} + \left(x^2 - \frac{9}{16}\right)y = 0$$ in terms of $J_n(x)$.

Ans. $y = c_1 J_{-\frac{3}{4}}(x) + c_2 J_{-\frac{3}{4}}(x)$.

42. Express $J_5(x)$ in terms of $J_0(x)$ and $J_1(x)$.

Ans. $J_5(x) = \left(\frac{384}{x^4} - \frac{72}{x^2} - 1\right) J_1(x) + \left(\frac{12}{x} - \frac{192}{x^3}\right) J_0(x)$

43. Solve $x \frac{d^2y}{dx^2} + 2\frac{dy}{dx} + \frac{1}{2} xy = 0$ in terms of Bessel's function.

Ans. $y = \sqrt{x}\left[c_1 J_{\frac{1}{2}}\left(\frac{x}{\sqrt{2}}\right) + + c_2 J_{-\frac{1}{2}}\left(\frac{x}{\sqrt{2}}\right)\right]$

44. Show that $\int J_3(x)dx = -J_2(x) - \frac{2}{x} J_0(x) + c$.

45. Show that $J_0^2(x) + 2J_1^2(x) + 2J_2^2(x) + 2J_3^2(x) + \ldots = 1$

Hint: Squaring and integrating the Jocobi series.

$\cos(x\sin\phi) = J_0 + 2J_2\cos 2\phi + 2J_4\cos 4\phi + \ldots;$

$\sin(x\sin\phi) = 2J_1 \sin\phi + 2J_3 \sin 3\phi + 2J_5 \sin 5\phi + \ldots;$

we get

$$\int_0^{\pi} \cos^2(x \sin\phi)d\phi = \pi(J_0^2 + 2J_2^2 + 2J_4^2 + \ldots)$$

$$\int_0^{\pi} \sin^2(x \sin\phi)d\phi = \pi(2J_1^2 + 2J_3^2 + 2J_5^2 + \ldots)$$

Adding these expressions, we get the required result.

46. Show that

(i) $\cos x = J_0 - 2J_1 + 2J_4 + \ldots$

$$= J_0 + 2\sum_{n=1}^{\infty} (-1)^n J_{2n},$$

(ii) $\sin x = 2J_1 - 2J_3 + 2J_5 + \ldots$

$$= 2\sum_{n=1}^{\infty} (-1)^n J_{2n} + 1$$

Hint: Put $\phi = \frac{\pi}{2}$ in the Jocobi series

$\cos(x\sin\phi) = J_0 + 2J_2 \cos 2\phi + 2J_4\cos 4\phi + \ldots;$

and

$\sin(x\sin\phi) = 2J_1 \sin\phi + 2J_3 \sin 3\phi + 2J_5 \sin 5\phi + \ldots$

47. Show that

$$\frac{d}{dx}[J_n^2(x)] = \frac{x}{2n}[J_{n-1}^2(x) - J_{n+1}^2(x)].$$

Legendre Equation and Legendre Polynomial

48. Express in the terms of Legendre polynomial

(i) $4x^3 - 2x^2 - 3x + 8$

Ans. $\frac{8}{5} P_3(x) - \frac{4}{3} P_2(x) - \frac{3}{5} P_1(x) + \frac{22}{3} P_0(x)$

(ii) $x^4 + 3x^3 - x^2 + 5x - 2$

Ans. $\frac{8}{35} P_4(x) + \frac{6}{5} P_3(x) - \frac{2}{21} P_2(x) + \frac{34}{5} P_1(x) - \frac{224}{105} P_0(x)$

(iii) $4x^2 - 3x + 2$

Ans. $\frac{8}{3} P_2(x) - 3P_1(x) + \frac{10}{3} P_0(x)$

49. Show that

$$P_n(x) = P'_{n+1}(x) - 2xP'_n(x) + P'_{n-1}(x)$$

Hint: Use recurrence formulae

$$(2n + 1)P_n(x) = P'_{n+1}(x) - P'_{n-1}(x) \quad \text{(i)}$$

and

$$n P_n(x) = xP'_n(x) - P'_{n-1}(x) \quad \text{(ii)}$$

From (*i*), $P_n(x) = P'_{n+1}(x) - P'_{n-1}(x) - 2n P_n(x)$

$= P'_{n+1}(x) - P'_{n-1}(x) - 2x P'_n(x) + 2 P'_{n-1}(x)$

$= P'_{n+1}(x) - 2x P'_n(x) + P'_{n-1}(x)$

50. Show that $\int_{-1}^{1} P_n(x)dx = \begin{cases} 2 \text{ if } n = 0 \\ 0 \text{ if } n \geq 1 \end{cases}$

51. Show that

$$\int_{-1}^{1} xP_n(x)P'_n(x)dx = \frac{2n}{2n+1}.$$

Hint: Use integration by parts and

$$\int_{-1}^{1} P_n^2(x)dx = \frac{2}{2n+1}$$

52. Using Rodrigue's formula, show that $P_n(x)$ satisfies the differential equation

$$\frac{d}{dx}\left[(1-x^2)\frac{d}{dx}P_n(x)\right] + n(n+1)P_n(x) = 0.$$

Hint: Let $y = (x^2 - 1)^n$. Find $\frac{dy}{dx}$ and differentiate once more to get $(x^2 - 1)\frac{d^2y}{dx^2} + 2xy_1 = 2n(xy_1 + n)$.
Then use Leibnitz's theorem to get

$$(1-x^2)\,y_{n+2} - 2xy_{n+1} + n(n+1)y_n = 0$$

or

$$[(1-x^2)\,y_{n+1}] + n(n+1)y_n = 0$$

or

$$\frac{d}{dx}\left[(1-x^2)\frac{d}{dx}\left(\frac{d^n}{dx^n}(x^2-1)^n\right)\right] + n(n+1)\frac{d^n}{dx^n}(x^2-1)^n = 0$$

53. Expand

$$f(x) = \begin{cases} 0, & -1 < x < 0 \\ x, & 0 < x < 1 \end{cases}$$

in Fourier–Legendre series

Ans. $f(x) = \frac{1}{4}P_0(x) + \frac{1}{2}P_1(x) + \frac{5}{16}P_2(x) - \frac{3}{32}P_3(x) + \ldots$

54. Solve the simultaneous equations:

$$\frac{dx}{dt} + 2y + \sin t = 0,$$

$$\frac{dy}{dt} - 2x - \cos t = 0$$

under the condition that $x = 0$, $y = 1$ when $t = 0$.

Ans. $x = \cos 2t - \cos t$
$y = \sin 2t + \cos 2t - \sin t$

55. Solve

$$\frac{dx}{dt} + y = \sin t, \quad \frac{dy}{dt} + x = \cos t$$

Ans. $x = c_1e^t + c_2e^{-t}$
$y = \sin t - c_3e^t + c_4e^{-t}$

56. Solve

$$\frac{dx}{dt} + \omega y = 0, \quad \frac{dy}{dt} - \omega x = 0.$$

Ans. $x = \text{A}\cos\omega t + \text{B}\sin\omega t$
$y = \text{A}\sin\omega t - \text{B}\cos\omega t$

57. Solve

$$\frac{dx}{dt} - y = t, \quad \frac{dy}{dt} + x = t^2$$

Ans. $x = \frac{t^2}{2} + \frac{t^4}{12} + c_1t + c_2$
$y = \frac{t^3}{3} + c_3$

58. Solve

$$\frac{dx}{dt} + 2x + 3y = 0$$

$$\frac{dy}{dt} + 3x + 2y = 2e^{2t}.$$

Ans. $x = c_1e^t + c_2e^{-5t} + \frac{6}{7}e^{2t}$
$y = c_2e^{-5t} - c_1e^t + \frac{8}{7}e^{2t}$

5 Partial Differential Equations

A differential equation which involves partial derivatives with respect to two or more independent variables is called a *partial differential equation.* Such equations appear in the description of physical processes in applied sciences and engineering. The equations

$$\frac{\partial^2 u}{\partial x^2}+\frac{\partial^2 u}{\partial y^2}=0$$

(two-dimensional Laplace equation),

$$\frac{\partial^2 u}{\partial t^2}=c^2\frac{\partial^2 u}{\partial x^2} \text{ (wave equation)},$$

and

$$\frac{\partial^2 u}{\partial x^2}=k\frac{\partial u}{\partial t} \text{ (heat conduction equation)}$$

are all partial differential equations.

The order of the highest partial derivative occurring in a partial differential equation is called the *order* of that partial differential equation. The degree of the highest partial derivative appearing in the partial differential equation free from radicals and fractions is called the *degree* of that partial differential equation. For example,

$$\frac{\partial^2 u}{\partial x^2}+\frac{\partial^2 u}{\partial y^2}=x^2+y^2$$

is a partial differential equation of order two and degree one.

5.1 FORMULATION OF PARTIAL DIFFERENTIAL EQUATION

Partial differential equation can be formulated in two ways:

(i) By the elimination of arbitrary constants from a given relation involving dependent and independent variables: If the number of arbitrary constants equals the number of independent variables, then the partial differential equation formed is of first order. On the other hand, if the number of arbitrary constants is more than the number of independent variables, then the partial differential equation formed will be second or higher order.

(ii) By the elimination of arbitrary functions from the given functional relations: In this case, the order of the differential equation is the same as the number of arbitrary functions involved in the equation.

The following examples will illustrate the procedure of formation of partial differential equation. We shall use the following notation for the derivatives:

$$p=\frac{\partial z}{\partial x},\ q=\frac{\partial z}{\partial y},\ r=\frac{\partial^2 z}{\partial x^2},$$

$$s=\frac{\partial^2 z}{\partial x\partial y},\ t=\frac{\partial^2 z}{\partial y^2}.$$

EXAMPLE 5.1

Form the partial differential equations from the relations

(i) $z=ax+a^2y^2+b,$

(ii) $2z=(ax+y)^2+b.$

Solution. (i) Differentiating z with respect to x and y, we get

$$\frac{\partial z}{\partial x}=a,\ \frac{\partial z}{\partial y}=2a^2y.$$

Eliminating a between these relations, we have

$$\frac{\partial z}{\partial y} = 2\left(\frac{\partial z}{\partial x}\right)^2 y,$$

or

$$q = 2p^2 y,$$

which is a partial differential equation of first order.

(ii) Differentiation with respect to x, yields

$$\frac{\partial z}{\partial x} = a(ax + y).$$

Differentiating with respect to y, we get

$$\frac{\partial z}{\partial y} = ax + y.$$

Therefore, we have

$$\frac{\partial z}{\partial x} = a\frac{\partial z}{\partial y} = \frac{1}{x}\left(\frac{\partial z}{\partial y} - y\right)\frac{\partial z}{\partial y}$$

or

$$x\frac{\partial z}{\partial x} = \left(\frac{\partial z}{\partial y} - y\right)\frac{\partial z}{\partial y},$$

or, in notations

$$px = (q - y)q$$

or

$$px + qy = q^2,$$

which is the required partial differential equation.

EXAMPLE 5.2

Eliminate the arbitrary function f from the relation

$$z = y^2 + 2f\left(\frac{1}{x} + \log y\right).$$

Solution. The given relation is

$$z = y^2 + 2f\left(\frac{1}{x} + \log y\right).$$

Differentiating z partially with respect to x, we get

$$\frac{\partial z}{\partial x} = 2f'\left(\frac{1}{x} + \log y\right)\left(-\frac{1}{x^2}\right)$$

or

$$-px^2 = 2f'\left(\frac{1}{x} + \log y\right). \qquad (1)$$

Differentiating z with respect to y, we get

$$\frac{\partial z}{\partial y} = 2y + 2f'\left(\frac{1}{x} + \log y\right)\left(\frac{1}{y}\right)$$

or

$$q = 2y + 2f'\left(\frac{1}{x} + \log y\right)\frac{1}{y}$$

or

$$qy - 2y^2 = 2f'\left(\frac{1}{x} + \log y\right). \qquad (2)$$

The relations (1) and (2) yield

$$-px^2 = qy - 2y^2$$

or

$$px^2 + qy = 2y^2,$$

which is a partial differential equation of first order.

EXAMPLE 5.3

Eliminate arbitrary function f from the relation

$$f(x^2 + y^2, z - xy) = 0.$$

Solution. We are given that

$$f(x^2 + y^2, z - xy) = 0.$$

Let

$$x^2 + y^2 = u, \; z - xy = v.$$

Then the given relation becomes

$$f(u, v) = 0.$$

Differentiating with respect to x, we get

$$\frac{\partial f}{\partial u}\cdot\frac{\partial u}{\partial x} + \frac{\partial f}{\partial v}\cdot\frac{\partial v}{\partial x} = 0$$

or

$$f_u(2x) + f_v\left(\frac{\partial z}{\partial x} - y\right). \quad (3)$$

Differentiating $f(u, v) = 0$ with respect to y, we get

$$\frac{\partial f}{\partial u} \cdot \frac{\partial u}{\partial y} + \frac{\partial f}{\partial v} \cdot \frac{\partial v}{\partial y} = 0$$

or

$$f_u(2y) + f_v\left(\frac{\partial z}{\partial y} - x\right) = 0. \quad (4)$$

Eliminating f_u and f_v from (3) and (4), we get

$$\begin{vmatrix} 2x & p-y \\ 2y & q-x \end{vmatrix} = 0$$

or

$$2x(q - x) - 2y(p - y) = 0$$

or

$$2xq - 2x^2 - 2yp + 2y^2 = 0$$

or

$$py - qx = y^2 - x^2.$$

EXAMPLE 5.4

Eliminate the function f from the relation

$$f(xy + z^2, x + y + z) = 0.$$

Solution. The given relation is

$$f(xy + z^2, x + y + z) = 0.$$

We put

$$xy + z^2 = u, \; x + y + z = v,$$

so that the given equation becomes

$$f(u, v) = 0.$$

Differentiating with respect to x, we get

$$\frac{\partial f}{\partial u} \cdot \frac{\partial u}{\partial x} + \frac{\partial f}{\partial v} \cdot \frac{\partial v}{\partial x} = 0$$

or

$$f_u\left(y + 2z\frac{\partial z}{\partial x}\right) + f_v\left(1 + \frac{\partial z}{\partial x}\right) = 0 \quad (5)$$

Differentiating f with respect to y, we have

$$\frac{\partial f}{\partial u}\frac{\partial u}{\partial y} + \frac{\partial f}{\partial v} \cdot \frac{\partial v}{\partial y} = 0$$

or

$$f_u\left(x + 2z\frac{\partial z}{\partial y}\right) + f_v\left(1 + \frac{\partial z}{\partial y}\right) = 0 \quad (6)$$

Eliminating f_u and f_v from (5) and (6), we have

$$\begin{vmatrix} y + 2pz & 1 + p \\ x + 2qz & 1 + q \end{vmatrix} = 0$$

or

$$(y + 2pz)(1 + q) - (1 + p)(x + 2qz) = 0$$

or

$$y + 2pz + yq + 2pqz - [x + 2qz + px + 2pqz] = 0$$

or

$$p(2z + 2qz - 2qz - x) + q(y - 2z) = x - y$$

or

$$p(x - 2z) + q(2z - y) = y - x.$$

EXAMPLE 5.5

Form partial differential equation from the relation

(i) $xyz = \phi(x + y + z)$

(ii) $z = f(x^2 - y^2)$.

Solution. (i) Put $x + y + z = u$, $xyz = v$. Then the relation becomes

$$\phi(u) - v = 0.$$

Differentiating with respect to x, we get

$$\frac{\partial \phi}{\partial u}\frac{\partial u}{\partial x}=\frac{\partial v}{\partial x}$$

or

$$\frac{\partial \phi}{\partial u}\left(1+\frac{\partial z}{\partial x}\right)=y\left(z+x\frac{\partial z}{\partial x}\right),$$

that is

$$\frac{\partial \phi}{\partial u}(1+p)=y(z+xp). \tag{7}$$

Differentiating with respect to y, we get

$$\frac{\partial \phi}{\partial u}(1+q)=x(z+yq). \tag{8}$$

From (7) and (8), we have

$$\frac{1+p}{1+q}=\frac{y(z+xp)}{x(z+yq)}$$

or

$$x(y-z)p+y(z-x)q=z(x-y).$$

(ii) Differentiating z with respect to x, we have

$$\frac{\partial z}{\partial x}=f'(x^2-y^2)2x$$

or

$$p=f'(x^2-y^2)2x. \tag{9}$$

Differentiating z with respect to y, we have

$$\frac{\partial z}{\partial y}=f'(x^2-y^2)\,(-2y)$$

or

$$q=f'(x^2-y^2)\,(-2y). \tag{10}$$

Dividing (9) by (10), we get

$$\frac{p}{q}=-\frac{x}{y} \text{ or } py+qx=0.$$

EXAMPLE 5.6

Form the partial differential equations from the relations

(i) $z=f(x+it)+g(x-it)$

(ii) $z=f_1(y+2x)+f_2(y-3x)$.

Solution. (i) Differentiating z with respect to x, we have

$$p=\frac{\partial z}{\partial x}=f'(x+it)+g'(x-it).$$

$$r=\frac{\partial^2 z}{\partial x^2}=f''(x+it)+g''(x-it). \tag{11}$$

Differentiating z with respect to t, we have

$$q=\frac{\partial z}{\partial t}=if'(x+it)-ig'(x-it)$$

$$t=\frac{\partial^2 z}{\partial t^2}=i^2f''(x+it)+i^2g''(x-it). \tag{12}$$

Adding (11) amd (12), we get

$$\frac{\partial^2 z}{\partial x^2}+\frac{\partial^2 z}{\partial y^2}=0 \quad \text{or} \quad r+t=0,$$

which is partial differential equation of second order and whose solution is $z=f(x+it)+g(x-it)$.

(ii) Differentiating z with respect to x, we get

$$\frac{\partial z}{\partial x}=2f_1'(y+2x)-3f_2'(y-3x),$$

$$\frac{\partial^2 z}{\partial x^2}=4f_1''(y+2x)+9f_2''(y-3x).$$

Differentiating z with respect to y, we get

$$\frac{\partial z}{\partial y}=f_1'(y+2x)+f_2'(y-3x),$$

$$\frac{\partial^2 z}{\partial y^2}=f_1''(y+2x)+f_2''(y-3x).$$

Further,

$$\frac{\partial^2 z}{\partial x\partial y}=2f_1''(y+2x)-3f_2''(y-3x).$$

Therefore,

$$\frac{\partial^2 z}{\partial x^2}+\frac{\partial^2 z}{\partial x\partial y}-6\frac{\partial^2 z}{\partial y^2}=0$$

EXAMPLE 5.7

Form partial differential equation from the relation

$$z=yf(x)+xg(y).$$

Solution. We have

$$\frac{\partial z}{\partial x} = yf'(x) + xg(y),$$

$$\frac{\partial z}{\partial y} = f(x) + xg'(y),$$

$$\frac{\partial^2 z}{\partial x \partial y} = f'(x) + g'(y).$$

Then

$$x\frac{\partial z}{\partial x} + y\frac{\partial z}{\partial y} - z = xyf'(x) + xyg'(y) = xy\frac{\partial^2 z}{\partial x \partial y}.$$

Hence

$$xy\frac{\partial^2 z}{\partial x \partial y} = x\frac{\partial z}{\partial x} + y\frac{\partial z}{\partial y} - z,$$

which is the required partial differential equation.

EXAMPLE 5.8

Form partial differential equation from the relation

$$2z = \frac{x^2}{a^2} + \frac{y^2}{b^2}.$$

Solution. Differentiating z partially with respect to x and y, respectively, we get

$$2\frac{\partial z}{\partial x} = \frac{2x}{a^2} \text{ which yields } \frac{1}{a^2} = \frac{p}{x}$$

and

$$2\frac{\partial z}{\partial y} = \frac{2y}{b^2} \text{ which yields } \frac{1}{b^2} = \frac{q}{y}.$$

Substituting for $\frac{1}{a^2}$ and $\frac{1}{b^2}$ in the given equation, we get

$$2z = px + qy.$$

5.2 SOLUTIONS OF A PARTIAL DIFFERENTIAL EQUATION

A functional relation between x, y and z is called a *solution* of the first order partial differential equation $f(x, y, z, p, q) = 0$ if this relation along with the derivatives $p = \frac{\partial z}{\partial x}$ and $q = \frac{\partial z}{\partial y}$ satisfies that partial differential equation.

A solution such as $\phi(x, y, z, a, b) = 0$ of the equation $f(x, y, z, p, q) = 0$ which contains two arbitrary constants is called a *complete solution* or *complete integral* of that equation.

A solution obtained from the complete solution $\phi(x, y, z, a, b) = 0$ by giving particular values to the constants a and b is called a *particular integral* or a *particular solution* of the partial differential equation $f(x, y, z, p, q) = 0$.

We will now discuss the methods to find solutions of partial differential equation of first order.

(A) Direct Integration Method

This method is applicable to the partial differential equations which contain only one partial derivative. We should keep in mind that *the constant of integration will be arbitrary function of the variable kept constant.*

EXAMPLE 5.9

Solve

$$\frac{\partial^3 z}{\partial x^2 \partial y} = \cos(2x + 3y).$$

Solution. Integrating the given equation twice with respect to x, we get

$$\frac{\partial^2 z}{\partial x \partial y} = \frac{1}{2}\sin(2x + 3y) + f(y), \text{ and}$$

$$\frac{\partial z}{\partial y} = -\frac{1}{4}\cos(2x + 3y) + xf(y) + \phi(y).$$

Integrating now with respect to y, we get the required solution as

$$z = -\frac{1}{12}\sin(2x + 3y) + xf_1(y) + \phi_1(y)\psi(x),$$

where $f_1(y)$ and $\phi_1(y)$ are integral of $f(y)$ and $\phi(y)$ with respect to y.

EXAMPLE 5.10

Solve

$$\frac{\partial^2 z}{\partial x^2} + z = 0$$

subject to the condition that when $x = 0$, $z = e^y$ and $\frac{\partial z}{\partial x} = 1$.

Solution. If z was a function of x alone, then since the roots of the auxiliary equation would have been imaginary, the solution would have been

$$z = c_1 \cos x + c_2 \sin x,$$

where c_1, c_2 are constants. But here z is a function of x and y, therefore c_1 and c_2 are functions of y. Therefore, the solution of the given equation is

$$z = f(y) \cos x + \phi(y) \sin x.$$

Then

$$\frac{\partial z}{\partial x} = -f(y) \sin x + \phi(y) \cos x.$$

As $x = 0$ implies $z = e^y$ and also since we have $\frac{\partial z}{\partial x} = 1$ for $x = 0$, then,

$$e^y = z = f(y) \text{ and } 1 = \phi(y).$$

Hence

$$z = \sin x + e^y \cos x.$$

EXAMPLE 5.11

Solve

$$\frac{\partial^2 z}{\partial x^2} = a^2 z,$$

given that when $x = 0$, $\frac{\partial z}{\partial x} = a \sin y$ and $\frac{\partial z}{\partial y} = 0$.

Solution. If z were a function of x only, then the solution would have been $z = c_1 e^{ax} + c_2 e^{-ax}$. But z is a function of x and y. Therefore, c_1 and c_2 should be a function of y. Thus, the solution is

$$z = f_1(y)e^{ax} + f_2(y)e^{-ax}.$$

Therefore, using the condition that $\frac{\partial z}{\partial x} = a \sin y$ for $x = 0$, we have

$$a \sin y = a[f_1(y) - f_2(y)]$$

and so

$$f_1(y) = \sin y + f_2(y).$$

Thus

$$z = (\sin y + f_2(y))e^{ax} + f_2(y)e^{-ax}.$$

Now

$$\frac{\partial z}{\partial y} = [\cos y + f_2'(y)]e^{ax} + f_2'(y)e^{-ax}.$$

Now using the condition that $\frac{\partial z}{\partial y} = 0$ for $x = 0$, we have

$$0 = \cos y + f_2'(y) + f_2'(y)$$

and so

$$f_2'(y) = -\frac{\cos y}{2}.$$

Hence integrating, we get

$$f_2(y) = -\frac{1}{2} \sin y.$$

Thus

$$f_1(y) = \sin y - \frac{1}{2} \sin y = \frac{1}{2} \sin y$$

and so

$$z = \frac{1}{2} \sin y e^{ax} - \frac{1}{2} \sin y e^{-ax}$$

or

$$z = \sin y \sinh ax.$$

EXAMPLE 5.12

Solve

$$\frac{\partial^2 z}{\partial x \partial y} = \sin x \sin y,$$

given that $\frac{\partial z}{\partial y} = -2 \sin y$ when $x = 0$, *and* $z = 0$ when y is an odd multiple of $\pi/2$.

Solution. We are given that

$$\frac{\partial^2 z}{\partial x \partial y} = \sin x \sin y.$$

Therefore, integrating with respect to x, keeping y constant, we have

$$\frac{\partial z}{\partial y} = -\cos x \sin y + f(y).$$

When $x = 0$, $\frac{\partial z}{\partial y} = -2 \sin y$. Therefore, $-2 \sin y = -\sin y + f(y)$ and so $f(y) = \sin y$. Thus

$$\frac{\partial z}{\partial y} = -\cos x \sin y - \sin y.$$

Integrating with respect to y, keeping x constant, we get

$$z = \cos x \cos y + \cos y + \phi(x).$$

Now when y in an odd multiple of $\pi/2$, $z = 0$. So $\phi(x) = 0$. Hence

$$z = \cos x \cos y + \cos y,$$

which is the required solution.

EXAMPLE 5.13

Solve

$$\frac{\partial^2 z}{\partial y^2} = z,$$

given that when $y = 0$, $z = e^x$ and $\frac{\partial z}{\partial y} = e^{-x}$.

Solution. The solution is

$$z = \phi_1(x)e^y + \phi_2(x)e^{-y}.$$

Now when $y = 0$, $z = e^x$. Therefore,

$$e^x = \phi_1(x) + \phi_2(x).$$

Further

$$\frac{\partial z}{\partial y} = \phi_1(x)e^y - \phi_2(x)e^{-y}.$$

When $y = 0$, $\frac{\partial z}{\partial y} = e^{-x}$ and so

$$e^{-x} = \phi_1(x) - \phi_2(x).$$

Adding the above expressions for e^x and e^{-x}, we get

$$2\phi_1(x) = e^x + e^{-x} \text{ or } \phi_1(x) = \frac{e^x + e^{-x}}{2}$$
$$= \cosh x$$

Substituting this value of $\phi_1(x)$ in $e^x = \phi_1(x) + \phi_2(x)$, we get $\phi_1(x) = \sinh x$. Hence

$$z = \cosh x e^y + \sinh x e^{-y}.$$

(B) Lagrange's Method

The partial differential equation of the form $Pp + Qq = R$, where P, Q, and R are functions of x, y, z and $p = \frac{\partial z}{\partial x}$, $q = \frac{\partial z}{\partial y}$, is known as *Lagrange's linear equation.* Regarding the solution of this equation, we have the following theorem.

Theorem 5.1. The general solution of a linear partial differential equation

$$Pp + Qq = R \tag{13}$$

is of the form

$$f(u, v) = 0, \tag{14}$$

where $f(u, v)$ is an arbitrary function of

$$\left.\begin{aligned} u(x, y, z) &= c_1 \\ \text{and} \quad v(x, y, z) &= c_2 \end{aligned}\right\} \tag{15}$$

which form a solution of

$$\frac{dx}{P} = \frac{dy}{Q} = \frac{dz}{R}. \tag{16}$$

Proof: Since $u(x, y, z) = c_1$ is a solution of (16),

$$\frac{\partial u}{\partial x} dx + \frac{\partial u}{\partial y} dy + \frac{\partial u}{\partial z} dz = 0$$

and

$$\frac{dx}{P} = \frac{dy}{Q} = \frac{dz}{R}$$

must be compatible. Therefore,

$$Pu_x + Qu_y + Ru_z = 0. \tag{17}$$

Similarly

$$Pv_x + Qv_y + Rv_z = 0. \tag{18}$$

Solving (17) and (18) for P, Q, and R, we get

$$\frac{P}{u_y v_z - u_z v_y} = \frac{Q}{u_z v_x - u_x v_z} = \frac{R}{u_x v_y - v_x u_y}$$

or

$$\frac{P}{\dfrac{\partial(u, v)}{\partial(y, z)}} = \frac{Q}{\dfrac{\partial(u, v)}{\partial(x, z)}} = \frac{R}{\dfrac{\partial(u, v)}{\partial(x, y)}}. \quad (19)$$

Differentiating (14) w.r.t x and y, we get

$$\frac{\partial f}{\partial u}(u_x + pu_z) + \frac{\partial f}{\partial v}(v_x + pv_z) = 0, \quad (20)$$

and

$$\frac{\partial f}{\partial u}(u_y + qu_z) + \frac{\partial f}{\partial v}(v_y + qv_z) = 0. \quad (21)$$

Eliminating $\frac{\partial f}{\partial u}$ and $\frac{\partial f}{\partial v}$ from (20) and (21), we get

$$\begin{vmatrix} u_x + pu_z & v_x + pv_z \\ u_y + qu_z & v_y + qv_z \end{vmatrix} = 0.$$

Simplifying we get

$$p\frac{\partial(u, v)}{\partial(y, z)} + q\frac{\partial(u, v)}{\partial(z, x)} = \frac{\partial(u, v)}{\partial(x, y)}. \quad (22)$$

Now equation (19) and (22) yield

$$Pp + Qq = R.$$

The equations (16) are called *auxiliary equations* or *subsidiary equations.*

EXAMPLE 5.14

Solve

$$(x^2 - yz)p + (y^2 - zx)q = z^2 - xy.$$

Solution. The given Lagranges' differential equation is

$$(x^2 - yz)p + (y^2 - zx)q = z^2 - xy.$$

The corresponding auxiliary equations are

$$\frac{dx}{x^2 - yz} = \frac{dy}{y^2 - zx} = \frac{dz}{z^2 - xy}. \quad (23)$$

Therefore,

$$\frac{dx - dy}{(x - y)(x + y + z)} = \frac{dy - dz}{(y - z)(x + y + z)}$$

or

$$\frac{dx - dy}{x - y} = \frac{dy - dz}{y - z}.$$

Integrating we get

$$\log(x - y) = \log(y - z) + \log a$$

or

$$\log\left(\frac{x - y}{y - z}\right) = \log a$$

or

$$\frac{x - y}{b - z} = a \quad (24)$$

Using x, y, z as multiplier, each fraction of (23) is equal to

$$\frac{xdx + ydy + zdz}{x^3 + y^3 + z^3 - 3xyz}$$

$$= \frac{xdx + ydy + zdz}{(x + y + z)(x^2 + y^2 + z^2 - xy - yz - zx)}. \quad (25)$$

Also each fraction of (23) is equal to

$$\frac{dx + dy + dz}{x^2 + y^2 + z^2 - xy - yz - zx}. \quad (26)$$

From (25) and (26), we have

$$\frac{xdx + ydy + zdz}{x + y + z} = dx + dy + dz$$

or

$$xdx + ydy + zdz = (x + y + z)d(x + y + z).$$

Integrating, we get

$$\frac{x^2}{2} + \frac{y^2}{2} + \frac{z^2}{2} = \frac{(x + y + z)^2}{2} + c$$

or

$$x^2 + y^2 + z^2 = (x + y + z)^2 + b$$

or

$$xy + yz + zx = b. \quad (27)$$

From (24) and (27), the general solution is

$$f\left(\frac{x-y}{y-z}, xy+yz+zx\right)=0,$$

where f is an arbitrary function.

EXAMPLE 5.15

Solve

$$(mz-ny)p+(nx-lz)q=ly-mx.$$

Solution. The auxiliary equations are

$$\frac{dx}{mz-ny}=\frac{dy}{nx-lz}=\frac{dz}{ly-mx}. \tag{28}$$

Using multipliers x, y, and z, each fraction of (28) is equal to

$$\frac{xdx+ydy+zdz}{0}.$$

Therefore, $xdx + ydy + zdz = 0$, which on integration gives

$$x^2+y^2+z^2=a. \tag{29}$$

Now using multipliers l, m, and n, each fraction of (28) is equal to

$$\frac{ldx+mdy+ndz}{0}$$

and so $l\,dx + m\,dy + n\,dz = 0$, which on integration gives

$$lx+my+nz=b. \tag{30}$$

Hence, from (29) and (30), the required solution is

$$x^2+y^2+z^2=f(lx+my+nz),$$

where f is an arbitrary function.

EXAMPLE 5.16

Solve

$$(x^2-y^2-z^2)p+2xyq=2xz.$$

Solution. The auxiliary equations are

$$\frac{dx}{x^2-y^2-z^2}=\frac{dy}{2xy}=\frac{dz}{2xz}. \tag{31}$$

The last two members yield

$$\frac{dy}{y}=\frac{dz}{z},$$

which on integration yields

$$\log y=\log z+\log a$$

or

$$\log\frac{y}{z}=\log a \text{ or } \frac{y}{z}=a. \tag{32}$$

Using x, y, z as multipliers, we get each fraction of (31) equal to

$$\frac{xdx+ydy+zdz}{x(x^2+y^2+z^2)}.$$

Therefore,

$$\frac{xdx+ydy+zdz}{x(x^2+y^2+z^2)}=\frac{dz}{2xz}$$

or

$$\frac{2xdx+2ydy+2zdz}{x^2+y^2+z^2}=\frac{dz}{z}.$$

Integrating this equation, we get

$$\log(x^2+y^2+z^2)=\log z+\log b$$

or

$$\frac{x^2+y^2+z^2}{z}=b. \tag{33}$$

Therefore, by (32) and (33), the general solution is

$$\frac{x^2+y^2+z^2}{z}=f\left(\frac{y}{z}\right)$$

or

$$x^2+y^2+z^2=zf\left(\frac{y}{z}\right),$$

where f is an arbitrary function.

EXAMPLE 5.17

Solve

$$z(xp-yq)=y^2-x^2.$$

Solution. The given equation is

$$xzp - yzq = y^2 - x^2.$$

The auxiliary equations are

$$\frac{dx}{xz} = \frac{dy}{-yz} = \frac{dz}{y^2 - x^2} = \frac{xdx + ydy + zdz}{0}. \quad (34)$$

The first and second member of (34) imply

$$\frac{dx}{x} = \frac{dy}{-y},$$

which on integration yields

$$\log x + \log y = \log c$$

or

$xy = c$, where c is a constant of integration. (35)
The last member of (34) implies

$$xdx + ydy + zdz = 0,$$

which on integration yields

$$\frac{x^2}{2} + \frac{y^2}{2} + \frac{z^2}{2} = a$$

or

$$x^2 + y^2 + z^2 = 2a = b. \quad (36)$$

Therefore, the general solution of the given partial differential equation is

$$f(xy, x^2 + y^2 + z^2) = 0,$$

where f is an arbitrary function.

EXAMPLE 5.18

Solve

$$(z^2 - 2yz - y^2)p + (xy + zx)q = xy - zx.$$

Solution. The corresponding auxiliary equations are

$$\frac{dx}{z^2 - 2yz - y^2} = \frac{dy}{xy + zx} = \frac{dz}{xy - zx}$$

$$= \frac{xdx + ydy + zdz}{0} \quad (37)$$

The second and third members of (37) gives

$$\frac{dy}{y + z} = \frac{dz}{y - z}$$

or

$$(y - z)dy = (y + z)dz$$

or

$$ydy - (zdy + ydz) - zdz = 0$$

or

$$ydy - d(yz) - zdz = 0.$$

Integration we get

$$y^2 - 2yz - z^2 = a. \quad (38)$$

The last member of (37) gives

$$xdx + ydy + zdz = 0$$

Integrating we get

$$x^2 + y^2 + z^2 = b \quad (39)$$

Therefore (38) and (39) yield the general solution of the differential equation as

$$f(x^2 + y^2 + z^2, y^2 - 2yz - z^2) = 0.$$

EXAMPLE 5.19

Solve

$$p \tan x + q \tan y = \tan z.$$

Solution. The corresponding auxiliary equations are

$$\frac{dx}{\tan x} = \frac{dy}{\tan y} = \frac{dz}{\tan z}. \quad (40)$$

Integrating the first two members of (40), we have

$$\log \sin x = \log \sin y = \log a$$

or

$$\frac{\sin x}{\sin y} = a. \tag{41}$$

Similarly, the last two members of (40) yield

$$\frac{\sin y}{\sin z} = b. \tag{42}$$

Therefore, (41) and (42) yield the general solution of the given equation as

$$f\left(\frac{\sin x}{\sin y}, \frac{\sin y}{\sin z}\right) = 0,$$

where f is an arbitrary function.

EXAMPLE 5.20

Solve

$$x(y - z)p + y(z - x)q = z(x - y).$$

Solution. The auxiliary equations for the given equation are

$$\frac{dx}{xy - xz} = \frac{dy}{yz - yx} = \frac{dz}{zx - zy}$$

$$= \frac{dx + dy + dz}{0}. \tag{43}$$

The last member of (43) yields

$$dx + dy + dz = 0,$$

which on integration gives

$$x + y + z = a. \tag{44}$$

Also, using $\frac{1}{x}$, $\frac{1}{y}$, and $\frac{1}{z}$ as the multipliers, we get

$$\frac{dx}{x} + \frac{dy}{y} + \frac{dz}{z} = 0,$$

which on integration yields

$$\log x + \log y + \log z = \log b$$

or

$xyz = b$, where b is a constant of integration. (45)

From (44) and (45), the general solution of the given partial differential equation is

$$f(x + y + z, xyz) = 0,$$

where f is an arbitrary function.

EXAMPLE 5.21

Solve

$$\left(\frac{1}{z} - \frac{1}{y}\right)p + \left(\frac{1}{x} - \frac{1}{z}\right)q = \frac{1}{y} - \frac{1}{x}.$$

Solution. The corresponding auxiliary equations are

$$\frac{dx}{\frac{1}{z} - \frac{1}{y}} = \frac{dy}{\frac{1}{x} - \frac{1}{z}} = \frac{dz}{\frac{1}{y} - \frac{1}{x}}. \tag{46}$$

We note that each member of (46) is equal to

$$\frac{dx + dy + dz}{0}.$$

Thus

$$dx + dy + dz = 0,$$

which on integration gives

$$x + y + z = a. \tag{47}$$

Further using $\frac{1}{x}$, $\frac{1}{y}$, $\frac{1}{z}$ as the multipliers, each member of (46) is equal to

$$\frac{\frac{1}{x}\,dx + \frac{1}{y}\,dy + \frac{1}{z}\,dz}{0}$$

and so

$$\frac{1}{x}\,dx + \frac{1}{y}\,dy + \frac{1}{z}\,dz = 0.$$

Integrating, we get

$$\log x + \log y + \log z = \log b$$

or

$xyz = b$, where b is a constant of integration. (48)

Combining (47) and (48), the general solution of the equation is

$$f(x + y + z, xyz) = 0,$$

where f is an arbitrary function.

EXAMPLE 5.22

Solve

$$p\sqrt{x} + q\sqrt{y} = \sqrt{z}.$$

Solution. The auxiliary equations are

$$\frac{dx}{\sqrt{x}} = \frac{dy}{\sqrt{y}} = \frac{dz}{\sqrt{z}}.$$

On integration, the first two members of auxiliary equation give

$$\sqrt{x} - \sqrt{y} = c_1.$$

Similarly, on integration, the last two members give

$$\sqrt{y} - \sqrt{z} = c_2.$$

Therefore, the required solution is

$$f(\sqrt{x} - \sqrt{y}, \sqrt{y} - \sqrt{z}) = 0,$$

where f is an arbitrary function.

EXAMPLE 5.23

Solve $xp + yq = 3z$.

Solution. The auxiliary equations are

$$\frac{dx}{x} = \frac{dy}{y} = \frac{dz}{3z}$$

The first two members, on integration, yield

$$\log x = \log y + \log c \text{ or } \frac{x}{y} = c.$$

The first and third members on integration yield

$$\log x = \frac{1}{3}\log z + \log b$$

or

$$\log x^3 = \log z + \log b^3$$

or

$$\frac{x^3}{z} = a.$$

Hence, the required general solution is

$$f\left(\frac{x}{y}, \frac{x^3}{z}\right) = 0,$$

where f is an arbitrary function.

EXAMPLE 5.24

Solve

$$(z - y)p + (x - z)q = y - x.$$

Solution. The corresponding auxiliary equations are

$$\frac{dx}{z-y} = \frac{dy}{x-z} = \frac{dz}{y-x} = \frac{dx + dy + dz}{0}. \quad (49)$$

The last member of (49) yields

$$dx + dy + dz = 0,$$

which on integration, gives

$$x + y + z = a. \quad (50)$$

Further using x, y, z as the multipliers, each fraction of (49) is equal to

$$\frac{xdx + ydx + zdz}{0}.$$

Thus,

$$xdx + ydx + zdz = 0,$$

which on integration gives

$$x^2 + b^2 + z^2 = b. \quad (51)$$

From (50) and (51), the general solution of the given equation is

$$x^2 + y^2 + z^2 = f(x + y + z),$$

where f is an arbitrary function.

EXAMPLE 5.25

Solve

$$x(y^2 - z^2)p + y(z^2 - x^2)q = z(x^2 - y^2).$$

Solution. The corresponding auxiliary equations are

$$\frac{dx}{x(y^2 - z^2)} = \frac{dy}{y(z^2 - x^2)} = \frac{dz}{z(x^2 - y^2)}$$

$$= \frac{xdx + ydy + zdz}{0} \quad (52)$$

The last member of (52) implies

$$xdx + ydy + zdz = 0,$$

which on integration gives

$$x^2 + y^2 + z^2 = a. \tag{53}$$

On the other hand, using $\frac{1}{x}, \frac{1}{y}$, and $\frac{1}{z}$ as the multipliers, each member of (52) is equal to

$$\frac{\frac{1}{x}\,dx + \frac{1}{y}\,dy + \frac{1}{z}\,dz}{0},$$

and so

$$\frac{1}{x}\,dx + \frac{1}{y}\,dy + \frac{1}{z}\,dz = 0.$$

Integrating, we have

$$\log x + \log y + \log z = \log b$$

or

$$xyz = b. \tag{54}$$

Hence (53) and (54) provide the general solution as

$$f(x^2 + y^2 + z^2, xyz) = 0,$$

where f is an arbitrary function.

EXAMPLE 5.26

Solve the partial differential equation:

$$x^2(y - z)p + y^2(z - x)q = z^2(x - y)$$

Solution. The given partial differential equation is

$$x^2(y - 3)p + y^2(3 - x)q = z^2(x - y).$$

It is of the form $Pp + Qq = R$. The auxiliary equation for the Lagrange's method is

$$\frac{dx}{P} = \frac{dy}{Q} = \frac{dz}{R},$$

that is,

$$\frac{dx}{x^2(y - z)} = \frac{dy}{y^2(z - x)} = \frac{dz}{z^2(x - y)} \tag{55}$$

Using multipliers $\frac{1}{x}, \frac{1}{y}$ and $\frac{1}{z}$, each fraction of (55) is equivalent to

$$\frac{\frac{dx}{x} + \frac{dy}{y} + \frac{dz}{z}}{0}$$

and so

$$\frac{dx}{x} + \frac{dy}{y} + \frac{dz}{z} = 0 \tag{56}$$

Integrating (56), we get

$$\log x + \log y + \log z = \log a$$

or

$$\log xyz = \log a$$

or

$$xyz = a \tag{57}$$

Now using multipliers $\frac{1}{x^2}, \frac{1}{y^2}, \frac{1}{z^2}$, each fraction of (55) is equivalent to

$$\frac{\frac{dx}{x^2} + \frac{dy}{y^2} + \frac{dz}{z^2}}{0} \tag{58}$$

Integrating (58), we get

$$\frac{1}{x} + \frac{1}{y} + \frac{1}{z} = b \tag{59}$$

Hence, from (57) and (59), the required solution is

$$f\left(\frac{1}{x} + \frac{1}{y} + \frac{1}{z}, xyz\right) = 0.$$

EXAMPLE 5.27

Solve the partial differential equation

$$x^2p + y^2q = (x + y)z$$

Solution. We have

$$x^2p + y^2q = (x + y)z.$$

The auxiliary equation for Lagrange's method are

$$\frac{dx}{x^2} = \frac{dy}{y^2} = \frac{dz}{(x+y)z}. \quad (60)$$

Form (60), we get

$$\frac{dx - dy}{x^2 - y^2} = \frac{dz}{(x+y)z},$$

or

$$\frac{dx - dy}{x - y} = \frac{dz}{z}. \quad (61)$$

Integrating (61), we get

$$\log(x - y) = \log z + \log c. \quad (62)$$

Again, from (62), we have

$$\frac{dx}{x^2} = \frac{dy}{y^2}.$$

Integrating both sides, we get

$$\int \frac{dx}{x^2} = \int \frac{dy}{y^2}$$

or

$$\frac{1}{x} = \frac{1}{y} + D$$

or

$$\frac{1}{x} - \frac{1}{y} = D$$

or

$$\frac{x - y}{xy} = -D. \quad (63)$$

From (62) and (63), the required solution of the given partial differential equation is

$$f\left(\frac{x-y}{z}, \frac{x-y}{xy}\right)$$

EXAMPLE 5.28

Solve $(2z - y)p + (x + z)q + 2x + y = 0$.

Solution. We have

$$(2z - y)p + (x + z)\, q + 2x + y = 0.$$

The auxiliary equations are

$$\frac{dx}{2z - y} = \frac{dy}{x + z} = -\frac{dz}{2x + y} \quad (64)$$

Using multipliers x, y and z, each fraction of (64) is equal to $\frac{xdx + ydy + zdz}{0}$. Therefore $xdx + ydy + zdz = 0$. Integrating, we get

$$x^2 + y^2 + z^2 = a \quad (65)$$

Using multipliers 1, −2 and −1, each fraction of (64) is equal to $\frac{dx - 2dy - dz}{0}$, and so $dx - 2dy - dz = 0$, which on integration yields

$$x - 2y - z = b \quad (66)$$

Hence, from (65) and (66), the required solution is

$$x^2 + y^2 + z^2 = f(x - 2y - z),$$

where f is an arbitrary function.

5.3 NON-LINEAR PARTIAL DIFFERENTIAL EQUATIONS OF THE FIRST ORDER

A partial differential equation, which involves first order partial derivatives p and q with degree higher than one and the products of p and q, is called as a *non-linear partial differential equation of the first order*. A general method for solving non-linear partial differential equation is due to Charpit and is known, after his name, as Charpit's method.

5.4 CHARPIT'S METHOD

Consider the partial differential equation

$$f(x, y, z, p, q) = 0, \quad (67)$$

which may be non-linear in p and q. The fundamental idea in this method is to find another *first order* partial differential equation.

$$g(x, y, z, p, q, a) = 0, \quad (68)$$

where a is a constant, so that

(i) equations (67) and (68) can be solved for p and q to give

$$p = p(x, y, z, a) \text{ and } q = q(x, y, z, a)$$

(ii) the equation

$$dz = pdx + qdy \qquad (69)$$

is integrable.

The main step is to find the function g. After this, we can solve it for p and q along with (67). Substituting the values of p and q in (69) and then integrating, we get the required solution. Since this integral will contain two arbitrary constants a and b, it will be a complete integral of the partial differential equation (67). The condition (i) will be satisfied if

$$J = \frac{\partial(f, g)}{\partial(p, q)} = \begin{vmatrix} \frac{\partial f}{\partial p} & \frac{\partial g}{\partial p} \\ \frac{\partial f}{\partial q} & \frac{\partial g}{\partial q} \end{vmatrix} \neq 0 \qquad (70)$$

Assuming that (70) holds, the equation (69) exists and the condition (ii) will be satisfied if

$$p\left(\frac{\partial q}{\partial z}\right) + q\left(-\frac{\partial p}{\partial z}\right) - \left(\frac{\partial p}{\partial y} - \frac{\partial q}{\partial x}\right) = 0$$

or

$$p\frac{\partial q}{\partial z} + \frac{\partial q}{\partial x} = q\frac{\partial p}{\partial z} + \frac{\partial p}{\partial y}. \qquad (71)$$

Putting the values of p and q as functions of x, y, and z in (67) and (68) differentiating with respect to x, we get

$$\frac{\partial f}{\partial x} + \frac{\partial f}{\partial p}\frac{\partial p}{\partial x} + \frac{\partial f}{\partial q}\frac{\partial q}{\partial x} = 0$$

and

$$\frac{\partial g}{\partial x} + \frac{\partial g}{\partial p}\frac{\partial p}{\partial x} + \frac{\partial g}{\partial q}\frac{\partial q}{\partial x} = 0.$$

Therefore,

$$\left(\frac{\partial f}{\partial x}\frac{\partial g}{\partial q} - \frac{\partial f}{\partial q}\frac{\partial g}{\partial p}\right)\frac{\partial q}{\partial x} = \frac{\partial f}{\partial x}\frac{\partial g}{\partial p} - \frac{\partial f}{\partial p}\frac{\partial g}{\partial x}$$

or

$$\frac{\partial q}{\partial x} = \frac{1}{J}\left\{\frac{\partial f}{\partial x}\frac{\partial g}{\partial p} - \frac{\partial f}{\partial p}\frac{\partial g}{\partial x}\right\}$$

Similarly,

$$\frac{\partial p}{\partial y} = \frac{1}{J}\left\{-\frac{\partial f}{\partial y}\frac{\partial g}{\partial q} + \frac{\partial f}{\partial q}\frac{\partial g}{\partial y}\right\}$$

$$\frac{\partial p}{\partial z} = \frac{1}{J}\left\{-\frac{\partial f}{\partial z}\frac{\partial g}{\partial q} + \frac{\partial f}{\partial q}\frac{\partial g}{\partial z}\right\}$$

$$\frac{\partial p}{\partial z} = \frac{1}{J}\left\{-\frac{\partial f}{\partial z}\frac{\partial g}{\partial q} + \frac{\partial f}{\partial q}\frac{\partial g}{\partial z}\right\} \qquad (72)$$

Substituting from (72) into the equation (71), we have

$$\frac{1}{J}\left[p\left(\frac{\partial f}{\partial z}\frac{\partial g}{\partial q} - \frac{\partial f}{\partial p}\frac{\partial g}{\partial z}\right) + \left(\frac{\partial f}{\partial x}\frac{\partial f}{\partial p} - \frac{\partial f}{\partial p}\frac{\partial g}{\partial x}\right)\right]$$

$$= \frac{1}{J}\left[q\left(-\frac{\partial f}{\partial z}\frac{\partial g}{\partial q} + \frac{\partial f}{\partial q}\frac{\partial g}{\partial z}\right) + \left(-\frac{\partial f}{\partial y}\frac{\partial g}{\partial q} - \frac{\partial f}{\partial q}\frac{\partial g}{\partial y}\right)\right]$$

or

$$\left(-\frac{\partial f}{\partial p}\right)\frac{\partial g}{\partial x} + \left(-\frac{\partial f}{\partial q}\right)\frac{\partial g}{\partial y} + \left(-p\frac{\partial f}{\partial p} - q\frac{\partial f}{\partial q}\right)\frac{\partial g}{\partial z}$$

$$+ \left(p\frac{\partial f}{\partial z} + \frac{\partial f}{\partial x}\right) + \left(q\frac{\partial f}{\partial z} + \frac{\partial f}{\partial y}\right)\frac{\partial f}{\partial q} = 0. \qquad (73)$$

The equation (73) is linear in the variables x, y, z, p, q and has the auxiliary equations

$$\frac{dx}{-\frac{\partial f}{\partial p}} = \frac{dy}{-\frac{\partial f}{\partial q}} = \frac{dz}{-p\frac{\partial f}{\partial p} - q\frac{\partial f}{\partial q}} = \frac{dp}{\frac{\partial f}{\partial x} + p\frac{\partial f}{\partial z}}$$

$$= \frac{dq}{\frac{\partial f}{\partial y} + q\frac{\partial f}{\partial z}} \qquad (74)$$

Any of the solutions of equations (74) satisfies equation (73). If such a solution involves p or q it can be taken for (68). Thus, once equation (68) has been found, the problem reduces to solve (67) and (68) for p and q and then to integrate the equation (69).

EXAMPLE 5.29

Solve by Charpit's method

$$px + qy = pq.$$

Solution. We have

$$f = px + qy - pq = 0. \tag{75}$$

Since

$$\frac{\partial f}{\partial x} = p, \frac{\partial f}{\partial y} = q, \frac{\partial f}{\partial p} = x - q, \frac{\partial f}{\partial q} = y - p,$$

the Charpits's auxiliary equations are

$$\frac{dx}{-\frac{\partial f}{\partial p}} = \frac{dy}{-\frac{\partial f}{\partial q}} = \frac{dp}{\frac{\partial f}{\partial x}} = \frac{dq}{\frac{\partial f}{\partial y}}$$

or

$$\frac{dx}{q - x} = \frac{dy}{p - y} = \frac{dp}{p} = \frac{dq}{q}.$$

Taking the last pair, we have

$$\frac{dp}{p} = \frac{dq}{q}.$$

Integrating this equation, we get

$$\log p = \log q + \log a$$

or

$$\frac{p}{q} = a. \tag{76}$$

Substituting $p = aq$ is (75), we get

$$aqx + qy - aq^2 = 0$$

or

$$ax + y - aq = 0$$

or

$$q = \frac{ax + y}{a}.$$

Therefore,

$$p = ax + y.$$

Now

$$dz = pdx + qdy$$

$$= (ax + y)dx + \frac{ax + y}{a} dy$$

or

$$adz = a(ax + y)dx + (ax + y)dy$$

$$= (ax + y)\,(adx + dy).$$

Integrating, we get

$$az = \frac{(ax + y)^2}{2} + c,$$

which is the required solution of the equation.

EXAMPLE 5.30

Solve by Charpit's method

$$z^2 = pqxy.$$

Solution. We have

$$z^2 - pqxy = 0. \tag{77}$$

Therefore,

$$\frac{\partial f}{\partial x} = -pqy, \frac{\partial f}{\partial y} = -pqx,$$

$$\frac{\partial f}{\partial z} = 2z, \frac{\partial f}{\partial p} = -qxy, \frac{\partial f}{\partial q} = -pxy$$

and, thus, the Charpit's auxiliary equations are

$$\frac{dx}{qxy} = \frac{dy}{pxy} = \frac{dz}{pqxy + pqxy} = \frac{dp}{-pqy + 2zp}$$

$$= \frac{dq}{-pqx + 2zq}$$

or

$$\frac{dx}{qxy} = \frac{dy}{pxy} = \frac{dz}{2pqxy} = \frac{dp}{p(2z - qy)}$$

$$= \frac{dq}{q(2z - px)}. \tag{78}$$

From (78), we get

$$\frac{\frac{dp}{p} - \frac{dq}{q}}{px - qy} = \frac{\frac{dy}{y} - \frac{dx}{x}}{px - qy}$$

and so

$$\frac{dp}{p} + \frac{dx}{x} = \frac{dy}{y} + \frac{dq}{q}.$$

On Integrating we get

$$\text{lop } p + \log x = \log y + \log q + \log a^2$$

or

$$px = a^2qy. \qquad (79)$$

Substituting this value of px in (77), we get

$$z^2 = a^2 q^2 y^2 \text{ or } q = \frac{z}{ay}$$

and then (79) implies $p = \frac{az}{x}$. Therefore, $dz = p\,dx + q\,dy$ becomes

$$dz = \frac{az}{x}\,dx + \frac{z}{ay}\,dy$$

or

$$\frac{dz}{z} = a\frac{dx}{x} + \frac{1}{a}\frac{dy}{y}.$$

On integrating we get

$$\log z = a \log x + \frac{1}{a}\log y + \log b$$

$$= \log x^a + \log y^{1/a} + \log b = \log bx^a y^{1/a}.$$

or

$$z = bx^a y^{1/a}.$$

EXAMPLE 5.31

Solve by Charpit's method

$$(p^2 + q^2)y = qz.$$

Solution. The given equation is

$$f \equiv (p^2 + q^2)y - qz = 0.$$

We have

$$\frac{\partial f}{\partial x} = 0, \frac{\partial f}{\partial y} = p^2 + q^2, \frac{\partial f}{\partial z} = -q,$$

$$\frac{\partial f}{\partial p} = 2py, \frac{\partial f}{\partial q} = 2qy - z.$$

Therefore, Charpits's auxiliary equations are

$$\frac{dx}{-2py} = \frac{dy}{z - 2qy} = \frac{dz}{-qz} = \frac{dp}{-pq} = \frac{dq}{p^2}.$$

From last two members, we have

$$p^2\,dp = -\,pq\,dq$$

or

$$pdp = -qdq$$

and so integration implies

$$p^2 + q^2 = a^2. \qquad (80)$$

Substituting this value of $p^2 + q^2$ in the given equation, we have

$$a^2y = qz$$

and so

$$q = \frac{a^2y}{z},$$

which with the help of (80) implies

$$p = \sqrt{a^2 - q^2} = \sqrt{a^2 - \frac{a^4y^2}{z^2}}$$

$$= \frac{a}{z}\sqrt{z^2 - a^2y^2}.$$

Therefore,

$$dz = pdx + qdy$$

$$= \frac{a}{z}\sqrt{z^2 - a^2y^2}\,dx + \frac{a^2y}{z}\,dy.$$

or

$$zdz = a\sqrt{z^2 - a^2y^2}\,dx + a^2y\,dy$$

or

$$\frac{zdz - a^2ydy}{\sqrt{z^2 - a^2y^2}} = a\,dx$$

or

$$\frac{\frac{1}{2}d(z^2 - a^2y^2)}{\sqrt{z^2 - a^2y^2}} = a\,dx.$$

Integrating, we get

$$\sqrt{z^2 - a^2y^2} = ax + b$$

or

$$z^2 = (ax + b)^2 + a^2y^2.$$

EXAMPLE 5.32

Solve by Charpit's method

$$(p^2 + q^2)x = pz$$

Solution. The given equation is

$$f \equiv (p^2 + q^2)x - pz = 0.$$

We have

$$\frac{\partial f}{\partial x} = p^2 + q^2, \frac{\partial f}{\partial y} = 0, \frac{\partial f}{\partial z} = -p,$$

$$\frac{\partial f}{\partial p} = 2px - z, \frac{\partial f}{\partial q} = 2qx.$$

Therefore, Charpit's auxiliary equations are

$$\frac{dx}{z - 2px} = \frac{dy}{-2qx} = \frac{dz}{p(z - 2px) - 2q^2x}$$

$$= \frac{dp}{p^2 + q^2 - p^2} = \frac{dq}{-pq}$$

or

$$\frac{dx}{z - 2px} = \frac{dy}{-2qx} = \frac{dz}{-pz} = \frac{dp}{q^2} = \frac{dq}{-pq}.$$

The last two members give

$$-pq\,dp = q^2dq$$

or

$$-pdp = qdp,$$

which on integration gives

$$p^2 + q^2 = a^2. \tag{81}$$

Substituting this in the given equation, we get

$$a^2x = pz \quad \text{or} \quad p = \frac{a^2x}{z}.$$

Subsituting this value of p in (81), we have

$$q = \sqrt{a^2 - p^2} = \sqrt{a^2 - \frac{a^4x^2}{z^2}} = \frac{a}{z}\sqrt{z^2 - a^2x^2}.$$

Therefore,

$$dz = pdx + qdy = \frac{a^2x}{z}\,dx + \frac{a}{z}\sqrt{z^2 - a^2x^2}\,dy$$

or

$$zdz = a^2\,xdx + a\sqrt{z^2 - a^2x^2}\,dy$$

or

$$\frac{zdz - a^2x\,dx}{\sqrt{z^2 - a^2x^2}} = ady$$

or

$$\frac{\frac{1}{2}d(z^2 - a^2x^2)}{\sqrt{z^2 - a^2x^2}} = ady.$$

On integrating, we get

$$\sqrt{z^2 - a^2x^2} = ay + b$$

or

$$z^2 = a^2x^2 + (ay + b)^2.$$

EXAMPLE 5.33

Solve by Charpit's method

$$P = (z + qy)^2.$$

Solution. Given

$$f \equiv p - (z + qy)^2 = 0$$

or

$$p - z^2 - q^2y^2 - 2yzq = 0.$$

Thus

$$\frac{\partial f}{\partial x} = 0, \frac{\partial f}{\partial y} = -2q^2y - 2zq, \frac{\partial f}{\partial z} = -2z - 2yq$$

$$\frac{\partial f}{\partial p} = 1, \frac{\partial f}{\partial q} = -2qy^2 - 2yz.$$

Therefore, Charpit's auxiliary equations are

$$\frac{dx}{-1} = \frac{dy}{2y(z+qy)} = \frac{dz}{-p+2qy(z+qy)}$$

$$= \frac{dp}{-2p(z+qy)} = \frac{dq}{-4q(z+qy)}.$$

From second and fifth members, we get

$$\frac{dy}{y} = \frac{dq}{-2q} \text{ and so } q = \frac{a}{y^2}.$$

Putting this value of q in the given equation, we get

$$p = \left(z + \frac{a}{y}\right)^2.$$

Then $dz = pdx + qdy$ becomes

$$dz = \left(z + \frac{a}{y}\right)^2 dx + \frac{a}{y^2} dy$$

or

$$dx = \frac{d\left(z + \frac{1}{y}\right)}{\left(z + \frac{a}{y}\right)^2}.$$

Hence,

$$x + \frac{1}{z + \frac{a}{y}} = b$$

or

$$x + \frac{y}{a + yz} = b,$$

is the required solution.

EXAMPLE 5.34

Solve by Charpit's method $z = p^2x + q^2y$.

Solution. The given equation is

$$f \equiv z - p^2x - q^2y = 0.$$

Thus

$$\frac{\partial f}{\partial x} = -p^2, \frac{\partial f}{\partial y} = -q^2, \frac{\partial f}{\partial z} = 1$$

$$\frac{\partial f}{\partial p} = -2px, \frac{\partial f}{\partial q} = -2qy.$$

Therefore, Charpit's auxiliary equations are

$$\frac{dx}{2px} = \frac{dy}{2qy} = \frac{dz}{2p^2x + 2q^2y} = \frac{dp}{-p^2 + p}$$

$$= \frac{dq}{-q^2 + q}$$

or

$$\frac{dx}{2px} = \frac{dy}{2qy} = \frac{dz}{2(p^2x + q^2y)} = \frac{dp}{p(1-p)}$$

$$= \frac{dq}{q(1-q)}.$$

It follows that

$$\frac{p^2dx + 2pxdp}{p^2x} = \frac{q^2dy + 2qy\,dq}{q^2y}$$

or

$$\frac{dx}{x} + \frac{2dp}{p} = \frac{dy}{y} + \frac{2dq}{q},$$

which yields

$$\log x + 2\log p = \log y + 2\log q + \log a$$

or

$xp^2 = aq^2y$, where a is constant. (82)

Putting the value of xp^2 in the given equation, we get

$$z = aq^2y + q^2y = q^2(y + ay)$$

or

$$q = \sqrt{\frac{z}{y + ay}}$$

and then (82) yields $p = \sqrt{\frac{az}{x+ax}}$ · Then $dz = p\,dx + q\,dy$ becomes

$$dz = \sqrt{\frac{az}{x+ax}}\,dx + \sqrt{\frac{z}{y+ay}}\,dy$$

or

$$\sqrt{\frac{1+a}{z}}\,dz = \sqrt{\frac{a}{x}}\,dx + \sqrt{\frac{1}{y}}\,dy.$$

On integrating, we get

$$\sqrt{(1+a)z} = \sqrt{ax} + \sqrt{y} + b.$$

EXAMPLE 5.35

Solve by Charpit's method

$$q + xp = p^2.$$

Solution. We are given that

$$f \equiv q + xp - p^2 = 0.$$

Thus,

$$\frac{\partial f}{\partial x} = p, \frac{\partial f}{\partial y} = 0, \frac{\partial f}{\partial z} = 0,$$

$$\frac{\partial f}{\partial p} = x - 2p, \frac{\partial f}{\partial q} = 1.$$

Therefore, Charpit's auxiliary equations are

$$\frac{dx}{-x+2p} = \frac{dy}{-1}$$

$$= \frac{dz}{2p^2 - px - q} = \frac{dp}{p} = \frac{dq}{0}. \tag{83}$$

From second and fourth members of (83), we have

$$-\frac{dy}{1} = \frac{dp}{p}$$

which yields

$$-y = \log p + \log c = \log pc$$

and so $p = ae^{-y}$. Putting this value of p in the given equation, we get $q = a^2e^{-2y} - axe^{-y}$. Then $dz = p\,dx + q\,dy$ becomes

$$dz = ae^{-y}\,dx + (a^2\,e^{-2y} - axe^{-y})dy.$$

On integrating, we get

$$z = axe^{-y} + \frac{a^2e^{-2y}}{-2} - \frac{axe^{-y}}{-1} + b$$

$$= 2axe^{-y} - \frac{a^2}{2}\,e^{-2y} + b.$$

EXAMPLE 5.36

Using Charpit's method, find the complete integral of the euation $xp + 3yq = 2(z - x^2q^2)$.

Solution. We have

$$f \equiv xp + 3yq - 2z + 2x^2q^2 = 0.$$

Therefore

$$\frac{\partial f}{\partial x} = p + 4xq^2, \frac{\partial f}{\partial y} = 3q, \frac{\partial f}{\partial z} = -2,$$

$$\frac{\partial f}{\partial p} = x, \frac{\partial f}{\partial q} = 3y + 4x^2q.$$

Thus, the Charpit's auxiliary equations are

$$\frac{dx}{-x} = \frac{dy}{-3y - 4x^2q} = \frac{dz}{-px - q(3y + 4x^2q)}$$

$$= \frac{dp}{p + 4xq^2 - 2p} = \frac{dq}{3q - 2q}$$

or

$$\frac{dx}{-x} = \frac{dy}{-3y - 4x^2q} = \frac{dz}{-px - q(3y + 4x^2q)}$$

$$= \frac{dp}{4xq^2 - q} = \frac{dq}{q}.$$

Thus

$$\frac{dq}{q} = \frac{dx}{-x},$$

which yields $qx = a$ (constant). Putting this value of q in the given equation, we have

$$p = \frac{2(z - x^2q^2)}{x} - \frac{3yq}{x} = \frac{2(z - a^2)}{x} - \frac{3ya}{x^2}.$$

Therefore, $dz = p\,dx + q\,dy$ gives

$$dz = \left[\frac{2(z - a^2)}{x} - \frac{3ya}{x^2}\right]dx + \frac{a}{x}\,dy$$

or

$$x^2dz = [2x(z - a^2) - 3ya]dx + ax\,dy$$

$$= 2x(z - a^2)dy - 3ya\,dx + ax\,dy$$

or

$$x^2dz - 2x(z - a^2)dx = -3ya\,dx + ax\,dy$$

or

$$x^4 d\left(\frac{z-a^2}{x^2}\right) - 3ya\,dx + ax\,dy$$

or

$$d\left(\frac{z-a^2}{x^2}\right) - \frac{3ay}{x^4}\,dx + \frac{a}{x^3}\,dy = d\left(\frac{ay}{x^3}\right).$$

On integrating, we have the required solution as

$$\frac{z-a^2}{x^2} = \frac{ay}{x^3} + b$$

or

$$z = a\left(a + \frac{y}{x}\right) + bx^2,$$

where a and b are arbitrary constants.

EXAMPLE 5.37

Solve by Charpit's method $pxy + pq + qy = yz$.

Solution. The given equation is

$$f \equiv pxy + pq + qy - yz = 0.$$

Then

$$\frac{\partial f}{\partial x} = py, \frac{\partial f}{\partial y} = px + q - z, \frac{\partial f}{\partial z} = -y$$

$$\frac{\partial f}{\partial p} = xy + q, \frac{\partial f}{\partial q} = p + y.$$

Therefore, Charpit's auxiliary equations are

$$\frac{dx}{-xy-q} = \frac{dy}{-y-p} = \frac{dz}{p((xy+q)-(p+y)q}$$

$$= \frac{dp}{py-py} = \frac{dq}{(px+q-z)-yq}$$

From the fourth member of the auxiliary equations, we have

$$dp = 0 \text{ or } p = a \text{ (constant)}.$$

Putting this value in the given equation, we have

$$axy + aq + qy = yz,$$

or

$$q(a+y) = yz - axy$$

or

$$q = \frac{yz - axy}{a+y}.$$

Then $dz = p\,dx + q\,dy$ becomes

$$dz = adx + \frac{yz - axy}{a+y}\,dy = adx + \frac{y(z-ax)}{a+y}\,dy$$

or

$$dz - adx = \frac{y(z-ax)}{a+y}\,dy$$

or

$$\frac{dz - adx}{z - ax} = \frac{y}{a+y}\,dy = \frac{y-a+a}{a+y}\,dy$$

$$= dy - \frac{a}{a+y}\,dy.$$

On integrating, we get

$$\log(z - ax) = y - a\log(a+y) + b.$$

5.5 SOME STANDARD FORMS OF NON-LINEAR EQUATIONS

We have observed that the general method of finding solution is usually much longer. However, there are a few standard forms of the partial differential equation $f(x, y, z, p, q) = 0$ which can be solved by very short methods. We discuss these standard forms one-by-one.

(A) Equations of the Form $f(p, q) = 0$

Equation of this standard form involves only p and q and not x, y, and z explicitly. The Charpit's auxiliary equations for such equations are

$$\frac{dp}{0} = \frac{dq}{0}.$$

Therefore, $dp = 0$ and so $p = a$ (constant). Putting this value of p in $f(p, q) = 0$ yields $f(a, q) =$

0, which gives $q = b$. Now $f(p, q) = 0$ implies $f(a, b) = 0$ and so $b = \phi(a)$. Thus, $dz = p\,dx + q\,dy$ reduces to

$$dz = a dx + \phi(a)\, dy.$$

Integrating, we get

$$z = ax + \phi(a)y = c,$$

where a, b, and c are constants and a and b are connected by the relation $f(a, b) = 0$.

EXAMPLE 5.38

Solve

(i) $p^2 + q^2 = 1$

(ii) $p + q = pq$.

Solution. (i) The equation is of the form $f(p, q) = 0$, Therefore, its solution is

$$z = ax + by + c,$$

where $a^2 + b^2 = 1$. Thus $b^2 = 1 - a^2$ or $b = \sqrt{1 - a^2}$. Hence the solution is

$$z = ax + \sqrt{1 - a^2}\, y + c.$$

(ii) The given equation is of the form $f(p, q) = 0$ and so its solution is $z = ax + by + c$, where $a + b = ab$, that is, $b(a - 1) = a$ or $b = \frac{a}{a-1}$. Thus

$$z = ax + \frac{a}{a-1} y + c.$$

EXAMPLE 5.39

Solve $(x + y)(p + q)^2 + (x - y)(p - q)^2 = 1$.

Solution. First we reduce the given equation to a standard form $f(p, q) = 0$. Put

$$x + y = U^2, (x - y) = V^2. \quad (84)$$

Then

$$p = \frac{\partial z}{\partial x} = \frac{\partial z}{\partial U}\cdot\frac{\partial U}{\partial x} + \frac{\partial z}{\partial V}\cdot\frac{\partial V}{\partial x}$$

$$= \frac{\partial z}{\partial U}\left(\frac{1}{2U}\right) + \frac{\partial z}{\partial V}\left(\frac{1}{2V}\right),$$

$$q = \frac{\partial z}{\partial y} = \frac{\partial z}{\partial U}\cdot\frac{\partial U}{\partial y} + \frac{\partial z}{\partial V}\cdot\frac{\partial V}{\partial y}$$

$$= \frac{\partial z}{\partial U}\left(\frac{1}{2U}\right) + \frac{\partial z}{\partial V}\left(-\frac{1}{2V}\right).$$

Therefore,

$$\left.\begin{aligned} P + q &= \frac{1}{U}\frac{\partial z}{\partial U}, \\ p - q &= \frac{1}{V}\frac{\partial z}{\partial V} \end{aligned}\right\} \quad (85)$$

Putting the values of $x + y$, $x - y$, $P + q$ and $p - q$ from (84) and (85) in the given equation, we get

$$\left(\frac{\partial z}{\partial U}\right)^2 + \left(\frac{\partial z}{\partial V}\right)^2 = 1.$$

Therefore, we get equation of the form $f(p, q) = 0$. Hence the solution is

$$z = aU + bV + c,$$

where $a^2 + b^2 = 1$. Thus $b = \sqrt{1 - a^2}$. Hence the solution is

$$z = aU + \sqrt{1 - a^2}V + c$$

$$= a\sqrt{x + y} + \sqrt{1 - a^2}\,\sqrt{x - y} + c.$$

EXAMPLE 5.40

Solve $x^2p^2 + y^2q^2 = z^2$.

Solution. The given equation can be written as

$$\left(\frac{px}{z}\right)^2 + \left(\frac{qy}{z}\right)^2 = 1.$$

We convert it to the form $f(p, q) = 0$. To do so, put

$$\frac{dx}{x} = dX, \frac{dy}{y} = dY, \frac{dz}{z} = dZ,$$

that is,

$$X = \log x, Y = \log y, \text{ and } Z = \log z.$$

Then

$$\frac{\partial Z}{\partial X}=\frac{\partial Z}{\partial z}\cdot\frac{\partial z}{\partial x}\cdot\frac{\partial x}{\partial X}=\frac{1}{z}\cdot\frac{x}{1}\cdot\frac{\partial z}{\partial x}=\frac{x}{z}\frac{\partial z}{\partial x}=\frac{px}{z},$$

$$\frac{\partial Z}{\partial Y}=\frac{\partial Z}{\partial z}\cdot\frac{\partial z}{\partial y}\cdot\frac{\partial y}{\partial Y}=\frac{1}{z}\cdot\frac{y}{1}\cdot\frac{\partial z}{\partial y}=\frac{y}{z}\frac{\partial z}{\partial y}=\frac{qy}{z}.$$

Hence the given equation reduces to

$$\left(\frac{\partial Z}{\partial X}\right)^2+\left(\frac{\partial Z}{\partial Y}\right)^2=1.$$

Therefore, its solution is

$$Z=aX+bY+c,$$

where $a^2+b^2=1$ or $b=\sqrt{1-a^2}$. Thus

$$Z=aX+\sqrt{1-a^2}Y+c.$$

Returning back to x, y, z, we get

$$\log z=a\log x+\sqrt{1-a^2}\log y+c.$$

EXAMPLE 5.41

Solve $(x-y)(px-qy)=(p-q)^2$.

Solution. We convert first the given equation to the form $f(p, q)=0$. To this end, let us put

$$x+y=U,\ xy=V.$$

Then

$$p=\frac{\partial z}{\partial x}=\frac{\partial z}{\partial U}\frac{\partial U}{\partial x}+\frac{\partial z}{\partial V}\frac{\partial V}{\partial x}=\frac{\partial z}{\partial U}+y\frac{\partial z}{\partial V}$$

and

$$q=\frac{\partial z}{\partial y}=\frac{\partial z}{\partial U}\cdot\frac{\partial U}{\partial y}+\frac{\partial z}{\partial V}\cdot\frac{\partial V}{\partial y}=\frac{\partial z}{\partial U}+x\frac{\partial z}{\partial V}.$$

Then

$$px-qy=\left[x\frac{\partial z}{\partial U}+xy\frac{\partial z}{\partial V}\right]-\left[y\frac{\partial z}{\partial U}+xy\frac{\partial z}{\partial V}\right]$$

$$=(x-y)\frac{\partial z}{\partial U}$$

and

$$p-q=(y-x)\frac{\partial z}{\partial V}$$

or

$$(p-q)^2=(x-y)^2\left(\frac{\partial z}{\partial V}\right)^2.$$

Thus the given equation reduces to

$$\frac{\partial z}{\partial U}=\left(\frac{\partial z}{\partial V}\right)^2.$$

Hence the solution is

$$z=aU+bV+c,\text{ where } a=b^2$$

$$=aU+\sqrt{a}\,V+c$$

$$=a(x+y)+\sqrt{a}\,xy+c.$$

(B) Equation of the form $f(z, p, q)=0$

These types of equations do not contain x and y. In this case, the Charpit's auxiliary equations are

$$\frac{dx}{f_p}=\frac{dy}{f_p}=\frac{dx}{pf_p+qf_q}=\frac{dp}{-(f_x+pf_z)}$$

$$=\frac{dq}{-(f_y+qf_z)}.$$

Since $f_x=f_y=0$, we have

$$\frac{dp}{p}=\frac{dq}{q},$$

and so $p=aq$ (or $q=ap$). Putting this value in the given equation, we find p and q. Putting these values of p and q in $dz=p\,dx+q\,dy$ and integrating, the required solution is obtained.

EXAMPLE 5.42

Solve $p(1+q)=qz$.

Solution. The given equation is of the form $f(z, p, q)=0$. Therefore, Charpit's auxiliary equations give $p=aq$. Putting this value of p in the given equation, we get

$$aq(1+q)=qz \text{ or } a(1+q)=z$$

and so $q = \frac{z-a}{a}$. Hence $p = z - a$. Therefore, $dz = p\,dx + q\,dy$ becomes

$$dz = (z-a)dx + \left(\frac{z-a}{a}\right)dy$$

or

$$adz = a(z-a)dx + (z-a)dy$$

or

$$\frac{adz}{z-a} = adx + dy.$$

On integrating, we get

$$a\log(z-a) = ax + y + \log b$$

or

$$\log(z-a)^a = be^{ax+y}.$$

EXAMPLE 5.43

Solve $z = p^2 + q^2$.

Solution. The equation is of the form $f(z, p, q) = 0$. So taking $p = aq$, we have $z = a^2q^2 + q^2 = q^2(a^2 + 1)$ and so $q = \sqrt{\frac{z}{a^2+1}}$. Then $p = \frac{a\sqrt{z}}{\sqrt{a^2+1}}$. Therefore,

$$dz = pdx + qdy = \frac{a\sqrt{z}}{\sqrt{a^2+1}}dx + \frac{\sqrt{z}}{\sqrt{a^2+1}}dy$$

or

$$\sqrt{a^2+1}\,\frac{dz}{\sqrt{z}} = adx + dy$$

or

$$2\sqrt{a^2+1}\,\sqrt{z} = ax + y + b$$

or

$$4(a^2+1)z = (ax + y + b)^2.$$

EXAMPLE 5.44

Solve the equation $q^2 = z^2p^2(1-p^2)$.

Solution. The given equation is of the type $f(z, p, q) = 0$. So, by Charpit's auxiliary equations, we have $q = ap$. Putting this value in the given equation, we have

$$a^2p^2 = z^2p^2(1-p^2)$$

or

$$p^2 = \frac{z^2-a^2}{z^2}$$

or

$$p = \frac{\sqrt{z^2-a^2}}{z}$$

and so $q = \frac{a\sqrt{z^2-a^2}}{z}$. Thus

$$dz = pdx + qdy = \frac{\sqrt{z^2-a^2}}{z}dx + \frac{a\sqrt{z^2-a^2}}{z}dy$$

or

$$\frac{z\,dz}{\sqrt{z^2-a^2}} = dx + ady$$

or

$$d(\sqrt{z^2-a^2}) = dx + ady.$$

On integrating, we get

$$\sqrt{z^2-a^2} = x + ay + c$$

or

$$z^2 - a^2 = (x + ay + b)^2.$$

EXAMPLE 5.45

Solve $z^2(p^2 + q^2 + 1) = a^2$.

Solution. The given equation is of the form $f(z, p, q) = 0$. So, Charpit's auxiliary equations yield $q = ap$. Substituting this value in the given equation, we have

$$z^2(p^2 + a^2p^2 + 1) = a^2$$

or

$$p^2(1+a^2) = \frac{a^2}{z^2} - 1 = \frac{a^2-z^2}{z^2}$$

or

$$p = \sqrt{\frac{a^2 - z^2}{z^2(1 + a^2)}} = \frac{1}{z}\sqrt{\frac{a^2 - z^2}{1 + a^2}}$$

and so $q = \frac{a}{z}\sqrt{\frac{a^2 - z^2}{1 + a^2}}$. Thus

$$dz = pdx + qdy$$

$$= \frac{1}{z}\sqrt{\frac{a^2 - z^2}{1 + a^2}}\, dx + \frac{a}{z}\sqrt{\frac{a^2 - z^2}{1 + a^2}}\, dy$$

or

$$\frac{z\, dz}{\sqrt{a^2 - z^2}} = \frac{1}{\sqrt{1 + a^2}}\, dx + \frac{a}{\sqrt{1 + a^2}}\, dy$$

or

$$-d(\sqrt{a^2 - z^2}) = \frac{1}{\sqrt{1 + a^2}}\, dx + \frac{a}{\sqrt{1 + a^2}}\, dy.$$

On integrating, we get

$$-\sqrt{a^2 - z^2} = \frac{1}{\sqrt{1 + a^2}}\, x + \frac{a}{\sqrt{1 + a^2}}\, y + b$$

or

$$(-\sqrt{a^2 - z^2})(\sqrt{1 + a^2}) = x + ay + c$$

or

$$(a^2 - z^2)(1 + a^2) = (x + ay + c)^2.$$

EXAMPLE 5.46

Solve $z^2(p^2 + q^2 + 1) = 1$.

Solution. The given equation is of the type $f(z, p, q) = 0$ and so $q = ap$. Thus the given equation implies

$$z^2(p^2 + a^2 p^2 + 1) = 1$$

or

$$p^2(1 + a^2) = \frac{1}{z^2} - 1 = \frac{1 - z^2}{z^2}$$

or

$$p = \frac{\sqrt{(1 - z^2)}}{z\sqrt{1 + a^2}}.$$

Then $q = \frac{a}{z}\sqrt{\frac{1 - z^2}{1 + a^2}}$. Therefore, $dz = pdx + q\, dy$ becomes

$$dz = \frac{1}{z}\sqrt{\frac{1 - z^2}{1 + a^2}}\, dx + \frac{a}{z}\sqrt{\frac{1 - z^2}{1 + a^2}}\, dy$$

or

$$\sqrt{1 + a^2}\, \frac{z\, dz}{\sqrt{1 - z^2}} = dx + ady$$

or

$$\sqrt{1 + a^2}\, d\left(\sqrt{1 - z^2}\right) = dx + ady.$$

On integrating, we get

$$\sqrt{1 + a^2}\, (\sqrt{1 - z^2}) = x + ay + b$$

or

$$(1 + a^2)(1 - z^2) = (x + ay + b)^2.$$

EXAMPLE 5.47

Solve $zpq = p + q$.

Solution. The given equation is of the type $f(z, p, q) = 0$. So $q = ap$ and putting this value of q in the given equation, we get

$$zap^2 = ap + p$$

and so $p = \frac{a + 1}{az}$. Then $q = \frac{a + 1}{z}$. Hence

$$dz = \frac{a + 1}{az}\, dx + \frac{a + 1}{z}\, dy$$

or

$$zdz = \frac{a + 1}{a}\, dx + (a + 1)\, dy.$$

On integrating, we get

$$\frac{z^2}{2} = \frac{a + 1}{a}\, x + (a + 1)y + c$$

or

$$z^2 = \frac{2}{a}(a + 1)x + 2(a + 1)y + b.$$

EXAMPLE 5.48

Solve $z^2(p^2 x^2 + q^2) = 1$.

Solution. We reduce the given equation to the form $f(z, p, q) = 0$ putting

$$\frac{dx}{x} = dX,$$

so that $X = \log x$. Then

$$\frac{\partial z}{\partial x} = \frac{\partial z}{\partial X} \cdot \frac{\partial X}{dx} = \frac{\partial z}{\partial X} \cdot \frac{1}{x},$$

which yields

$$px = \frac{\partial z}{\partial X}.$$

Therefore, the given equation becomes

$$z^2\left[\left(\frac{\partial z}{\partial X}\right)^2 + q^2\right] = 1, \qquad (86)$$

which is of the form $f(z, P, q) = 0$, where $P = \frac{\partial z}{\partial X}$. Therefore, we have $q = aP$. Putting in (86), we get

$$z^2[P^2 + a^2P^2] = 1$$

or

$$P^2(1 + a^2) = \frac{1}{z^2}$$

or

$$P = \frac{1}{z\sqrt{1 + a^2}}.$$

Then $q = \frac{a}{z\sqrt{1 + a^2}}$ and so the equation $dz = P\,dX + q dy$ becomes

$$dz = \frac{1}{z\sqrt{1 + a^2}}\,dX + \frac{a}{z\sqrt{1 + a^2}}\,dy$$

or

$$z(\sqrt{1 + a^2})dz = dX + a\,dy.$$

Integrating, we get the required solution as

$$\sqrt{1 + a^2}\,\frac{z^2}{2} = X + ay + b$$

or

$$\sqrt{1 + a^2}\,z^2 = 2X + 2ay + 2b$$

$$= 2(\log x + ab) + c.$$

(c) Separable Equations

The equations of the form $f(x, p) = g(y, q)$, in which z is absent and the terms containing x and p can be separated from those containing y and q. The Carpit's auxiliary equations in this case become

$$\frac{dx}{f_p} = \frac{dy}{-g_q} = \frac{dz}{pf_p - qf_q} = \frac{dp}{-f_x} = \frac{dq}{g_y}.$$

Therefore,

$$f_x\,dx + f_p\,dp = 0,$$

which implies that f is a constant, that is, $f(x, p) = a$. Therefore, $g(y, q) = a$. Hence solving for p, we get $p = \phi(x)$ and solving for q, we have $q = \psi(y)$. Therefore, $dz = p\,dx + q\,dy$ gives

$$dz = \phi(x)dx + \psi(y)dy$$

and so

$$z = \int \phi(x)dx + \int \psi(y)dy + c.$$

EXAMPLE 5.49

Solve $p^2 - q^2 = x - y$.

Solution. The given equation can be separated as $p^2 - x = q^2 - y$ and, therefore, by Charpit's auxiliary equations,

$$p^2 - x = a = q^2 - y.$$

Thus

$$p = \sqrt{a + x} \text{ and } q = \sqrt{a + y}.$$

Thus $dz = p\,dx + q\,dy$ reduces to

$$dz = \sqrt{a + x}\,dx + \sqrt{a + y}\,dy$$

On integrating, we get the solution as

$$z = \frac{2}{3}(a + x)^{3/2} + \frac{2}{3}(a + y)^{3/2} + b.$$

EXAMPLE 5.50

Solve $z(p^2 - q^2) = x - y$

Solution. The given equation can be written as

$$(\sqrt{z}p)^2 - (\sqrt{z}q)^2 = x - y.$$

Substitute

$$\sqrt{z}\, dz = dZ, \text{ so that } Z = \frac{2}{3} z^{3/2}.$$

Then

$$\frac{\partial Z}{\partial x} = \frac{\partial Z}{\partial z} \cdot \frac{\partial z}{\partial x} = \sqrt{z}\, \frac{\partial z}{\partial x} = \sqrt{z}\, p,$$

$$\frac{\partial Z}{\partial y} = \frac{\partial Z}{\partial z} \cdot \frac{\partial z}{\partial y} = \sqrt{z}\, \frac{\partial z}{\partial y} = \sqrt{z}\, q.$$

Therefore, the given equation transforms to

$$\left(\frac{\partial Z}{\partial x}\right)^2 - \left(\frac{\partial Z}{\partial y}\right)^2 = x - y$$

or

$$P^2 - Q^2 = x - y,$$

where $P = \frac{\partial Z}{\partial x}$, $Q = \frac{\partial Z}{\partial y}$. Thus

$$P^2 - x = Q^2 - y = a,$$

which yields

$$P = \sqrt{a + x},\ Q = \sqrt{a + y}.$$

Thus $dz = p\, dx + q\, dy$ transforms to

$$dZ = P\, dx + Q\, dy = \sqrt{a + x}\, dx + \sqrt{a + y}\, dy.$$

On integrating, we get

$$Z = \frac{2}{3}(a + x)^{3/2} + \frac{2}{3}(a + y)^{3/2} + c$$

or

$$\frac{2}{3} z^{3/2} = \frac{2}{3}(a + x)^{3/2} + \frac{2}{3}(a + y)^{3/2} + c$$

or

$$z^{3/2} = (a + x)^{3/2} + (a + y)^{3/2} + b.$$

EXAMPLE 5.51

Solve $p^2 + q^2 = x + y$.

Solution. The given equation is separable and we have

$$p^2 - x = y - q^2 = a.$$

Thus

$$p = \sqrt{a + x} \text{ and then } q = \sqrt{y - a}.$$

Therefore, $dz = p\, dx + q\, dy$ takes the form

$$dz = \sqrt{a + x}\, dx + \sqrt{y - a}\, dy.$$

Integrating, we get the required solution as

$$z = \frac{2}{3}(a + x)^{3/2} + \frac{2}{3}(y - a)^{3/2} + b.$$

EXAMPLE 5.52

Solve $p + q = \sin x + \sin y$.

Solution. The given equation is separable, that is, of the form $f(x, p) = g(y, q)$. Thus

$$p - \sin x = \sin y - q = a$$

and so

$$p = a + \sin x$$
$$q = \sin y - a$$

Thus, $dz = pdx + q\, dy$ takes the form

$$dz = (a + \sin x)\, dx + (\sin y - a)\, dy.$$

On integrating, we get

$$z = ax - \cos x - \cos y - ay + b$$
$$= a(x - y) - (\cos x + \cos y) + b.$$

EXAMPLE 5.53

Solve $z^2 (p^2 + q^2) = x^2 + y^2$.

Solution. The given equation can be written as

$$(zp)^2 + (zq)^2 = x^2 + y^2.$$

Put $Z = \frac{1}{2} z^2$. Then

$$\frac{\partial Z}{\partial x} = \frac{\partial Z}{\partial z} \cdot \frac{\partial z}{\partial x} = z\frac{\partial z}{\partial x} = zp,$$

$$\frac{\partial Z}{\partial y} = \frac{\partial Z}{\partial z} \cdot \frac{\partial z}{\partial y} = z\frac{\partial z}{\partial y} = zq.$$

Hence the given equation becomes

$$\left(\frac{\partial Z}{\partial x}\right)^2 + \left(\frac{\partial Z}{\partial y}\right)^2 = x^2 + y^2$$

or

$$P^2 + Q^2 = x^2 + y^2,$$

where $P = \frac{\partial Z}{\partial x}$ and $Q = \frac{\partial Z}{\partial y}$. This equation is of the form $f(x, P) = g(y, Q)$. Therefore,

$$P^2 - x^2 = y^2 - Q^2 = a,$$

which yields

$$P = \sqrt{a + x^2},\ Q = \sqrt{y^2 - a}.$$

Thus the equation $dZ = P\,dx + Q\,dy$ becomes

$$dZ = \sqrt{a + x^2}\,dx + \sqrt{y^2 - a}\,dy.$$

On integrating, we get

$$Z = \frac{1}{2} x\sqrt{x^2 + a} + \frac{a}{2}\log\left(x + \sqrt{x^2 + a}\right)$$
$$+ \frac{1}{2} y\sqrt{y^2 - a} - \frac{a}{2}\log\left(y + \sqrt{y^2 - a}\right) + b$$

or

$$z^2 = x\sqrt{x^2 + a} + a\log\left(x + \sqrt{x^2 + a}\right)$$
$$+ y\sqrt{y^2 - a} - a\log\left(y + \sqrt{y^2 + a}\right) + 2b$$
$$= x\sqrt{x^2 + a} + y\sqrt{y^2 - a}$$
$$+ a\log\frac{x + \sqrt{x^2 + a}}{y + \sqrt{y^2 - a}} + 2b.$$

(D) Clairut's Equation

Any first order partial differential equation of the form

$$z = px + qy + f(p, q)$$

is called a *Clairut's equation.*

For such equations, the Charpit's auxiliary equations are

$$\frac{dx}{p + f_p} = \frac{dy}{y + f_q} = \frac{dz}{px + qy + pf_p + qf_q} = \frac{dp}{0}$$
$$= \frac{dq}{0}.$$

The last two members yield

$$p = a \text{ and } q = b.$$

Therefore $dz = pdx + qdy$ implies

$$dz = a\,dx + b\,dy,$$

which on integration gives the required solution as

$$z = ax + by + c, \text{ where } c = f(a, b).$$

EXAMPLE 5.54

Solve $(pq - p - q)(z - px - qy) = pq$.

Solution. The given equation can be written as

$$pqz - p^2xq - pq^2y - pz + p^2x + pqy - qz$$
$$+ pqx + q^2y = pq$$

or

$$z(pq - p - q) = -px(p + q - pq)$$
$$-qy(p + q - pq) + pq$$

or

$$z = px + qy + \frac{pq}{pq - p - q},$$

which is Clairut's equation. Therefpre, replacing p by a and q by b, we get the required solution as

$$z = ax + by + \frac{ab}{ab - a - b}.$$

EXAMPLE 5.55

Solve $pqz = p^2(xq + p^2) + q^2(yp + q^2)$.

Solution. Dividing throughout by pq, we get

$$z = \frac{p}{q}(xq + p^2) + \frac{q}{p}(yp + q^2)$$

$$= px + qy + \frac{p^3}{q} + \frac{q^3}{p}.$$

Thus, the given equation is Clairut's equation. So, replacing p by a and q by b, we get the solution

$$z = ax + by + \frac{a^3}{b} + \frac{b^3}{a}$$

or

$$z = ax + by + \frac{a^4 + b^4}{ab}.$$

EXAMPLE 5.56

Solve $z = px + qy + \sqrt{1 + p^2 + q^2}$.

Solution. It is a Clairut's equation and, therefore, replacing p by a and q by b, we get the solution

$$z = ax + by + \sqrt{1 + a^2 + b^2}.$$

EXAMPLE 5.57

Solve $4xyz = pq + 2px^2y + 2qxy^2$.

Solution. Dividing throughout by xy, we get

$$4z = 2px + 2qy + \frac{pq}{xy}.$$

Putting $x^2 = X$ and $y^2 = Y$, we get

$$p = \frac{\partial z}{\partial x} = \frac{\partial z}{\partial X} \cdot \frac{\partial X}{\partial x} = 2x \frac{\partial z}{\partial X}$$

$$q = \frac{\partial z}{\partial y} = \frac{\partial z}{\partial Y} \cdot \frac{\partial Y}{\partial y} = 2y \frac{\partial z}{\partial Y}.$$

Therefore, the equation transforms to

$$4z = 4x^2 \frac{\partial z}{\partial Y} + 4y^2 \frac{\partial z}{\partial Y} + 4 \frac{\partial z}{\partial X} \cdot \frac{\partial z}{\partial Y}$$

or

$$z = XP + YQ + PQ,$$

where

$$P = \frac{\partial z}{\partial X}, \; Q = \frac{\partial z}{\partial Y}.$$

This equation is Clairut's equation and so the solution is

$$z = aX + bY + ab = ax^2 + by^2 + ab.$$

EXAMPLE 5.58

Solve $pq(px + qy - z)^3 = 1$.

Solution. The given equation is

$$pq(px + qy - z)^3 = 1.$$

Therefore,

$$(px + qy - z)^3 = \frac{1}{pq}$$

or

$$px + qy - z = \frac{1}{(pq)^{1/3}}$$

or

$$z = px + qy - \frac{1}{(pq)^{1/3}},$$

which is Clairut's equation. Hence its solution is

$$z = ax + by - \frac{1}{(ab)^{1/3}},$$

EXAMPLE 5.59

Solve $(px + qy - z)^2 = 1 + p^2 + q^2$.

Solution. We have

$$(px + qy - z)^2 = 1 + p^2 + q^2.$$

Therefore

$$px + qy - z = \pm\sqrt{1 + p^2 + q^2}.$$

or

$$z = px + qy \pm \sqrt{1 + p^2 + q^2},$$

which is Clariut's equations. Hence its solution is

$$z = ax + by \pm \sqrt{1 + a^2 + b^2}.$$

EXAMPLE 5.60

Solve $z = px + qy + p^2q^2$ and find the complete and singular solutions.

Solution. We have $z = px + qy + p^2q^2$, which is a Clairut's equation. Hence its complete integral is

$$z = ax + by + a^2b^2 \tag{87}$$

The singular integral is obtained by eliminating a and b between (87) and the equations

$$0 = x - \frac{\partial f}{\partial a} = x - 2ab^2, \tag{88}$$

$$0 = y - \frac{\partial f}{\partial b} = y - 2a^2b. \tag{89}$$

From (87), (88) and (89), we have

$$\frac{x}{z} = \frac{2\,ab^2}{z} = \frac{2ab^2}{2a^2b^2 + 2a^2b^2 + a^2b^2} = \frac{2ab^2}{5a^2b^2} = \frac{2}{5a}$$

and

$$\frac{y}{z} = \frac{2a^2b}{5a^2b^2} = \frac{2}{5b}.$$

Therefore $a = \frac{2z}{5x}$ and $b = \frac{2z}{5y}$, Putting these values in (87), we get

$$z = \frac{2z}{5} + \frac{2z}{5} + \left(\frac{4z^2}{25xy}\right)^2 = \frac{4z}{5} + \frac{16z^4}{625x^2y^2}$$

or

$$1 = \frac{4}{5} + \frac{16z^3}{625x^2y^2}$$

or

$$16z^3 = 105x^2y^2.$$

5.6 LINEAR PARTIAL DIFFERENTIAL EQUATIONS WITH CONSTANT COEFFICIENTS

A linear partial differential equation of order n is an equation of the form

$$\left(a_0\frac{\partial^n z}{\partial x^n} + a_1\frac{\partial^n z}{\partial x^{n-1}\partial y} + a_2\frac{\partial^n z}{\partial x^{n-2}\partial y^2} + \ldots + a_n\frac{\partial^n z}{\partial y^n}\right)$$

$$+\left(b_0\frac{\partial^{n-1} z}{\partial x^{n-1}} + b_1\frac{\partial^{n-1} z}{\partial x^{n-2}\partial y} + \ldots + b_{n-1}\frac{\partial^{n-1} z}{\partial y^{n-1}}\right)$$

$$+ \ldots + \left(m_0\frac{\partial z}{\partial x} + m_1\frac{\partial z}{\partial y}\right) + \ldots + nz = f(xy), \tag{90}$$

where the coefficients $a_0, a_1, \ldots, b_0, b_1, \ldots, m_0, m_1, \ldots$, are constants or functions of x and y. In case these coefficients are all constants, then such a partial differential equation is called a *linear partial differential equation with constant coefficients*. On the other hand, if these coefficients are function of x or y or both, then the above equation turned out to be *linear partial differential equation with variable coefficients*. In case, the orders of all partial derivatives involved in (90) are same, then it is called a *homogeneous linear partial differential equation*. Similarly, if the orders of the partial derivatives are distinct, then (90) is called *non-homogeneous linear partial differential equation*. For Example

(i) the equations

$$2\frac{\partial^2 z}{\partial x^2} + 7\frac{\partial^2 z}{\partial x\partial y} + 5\frac{\partial^2 x}{\partial y^2} = \sin x$$

and

$$\frac{\partial^3 z}{\partial x^3} - 3\frac{\partial^3 z}{\partial x^2\partial y} + 4\frac{\partial^3 z}{\partial y^3} = e^{x+2y}$$

are homogeneous linear partial differential equations with constant coefficients, whereas the equations

$$\frac{\partial z}{\partial x}\frac{\partial^2 z}{\partial y^2} = e^x$$

and

$$\frac{\partial^2 z}{\partial x^2} + 2\frac{\partial^2 z}{\partial x\partial y} + \frac{\partial^2 z}{\partial y^2} + 4\frac{\partial z}{\partial y} + 3\frac{\partial z}{\partial y} + z$$

$$= \sin(x + y)$$

are non-homogeneous linear partial differential equations with constant coefficients.

Similarly, the equation

$$x^2\frac{\partial^2 z}{\partial x^2}+y\frac{\partial^2 z}{\partial x\partial y}+x\frac{\partial^2 z}{\partial y^2}+x=e^x$$

is linear partial differential equation with variable coefficients.

(A) Homogeneous Linear Partial Differential Equations with Constant Coefficients

Consider the homogeneous linear partial differential equation

$$a_0\frac{\partial^n z}{\partial x^n}+a_1\frac{\partial^n z}{\partial x^{n-1}\partial y}+a_2\frac{\partial^n z}{\partial x^{n-2}\partial y^2}+\dots$$

$$+a_n\frac{\partial^n z}{\partial y^n}=f(x,y)$$

with constant coefficients $a_0, a_1, \dots, a_n$. Let

(i) D stands for $\frac{\partial}{\partial x}$

(ii) D' stands for $\frac{\partial}{\partial y}$

(iii) DD' stands for $\frac{\partial^2}{\partial x\partial y}$

(iv) D^2 stands for $\frac{\partial^2}{\partial x^2}$

(v) D'^2 stands for $\frac{\partial^2}{\partial y^2}$

and so on. Then the above equation transforms to

$$(a_0D^n+a_1D^{n-1}D'+\dots+a_nD'^n)z=f(x,y).$$

or

$$\phi(D, D')z=f(x,y). \qquad (91)$$

As, in the case of ordinary linear differential equation with constant coefficients, the complete solution of (91) consists of two parts:

(a) Complementary function (C.F), which is the solution of $\phi(D, D')z=0$ and shall have n arbitrary functions for partial differential equation of order n.

(b) Particular integral (P.I), which is the particular solution of $\phi(D, D')z=f(x,y)$.

The complete solution of (90) is then given by $z=$ C.F. + P.I.

Theorem 5.2. Let $\phi(D, D')z=0$ be a linear partial differential equation with constant coefficients. If $u_1, u_2, \dots, u_m$ are m solutions of $\phi(D, D')z=0$, then their linear combination $\sum_{i=1}^{m} e_iu_i$ is also a solution of $\phi(D, D')z=0$.

Proof: Since u_i, $1\le i\le m$ is a solution of $\phi(D, D')$ $z=0$, we have $\phi(D, D')u_i=0$ for $1\le i\le m$. Therefore

$$\phi(D, D')\left(\sum_{i=1}^{m}c_iu_i\right)=\sum_{i=1}^{m}\phi(D, D')\,(c_iu_i)$$

$$=\sum_{i=1}^{m}c_i\phi(D, D')u_i$$

$$=\sum_{i=1}^{m}c_i(0)=0$$

and hence $\sum_{i=1}^{m}c_iu_i$ is also a solution of $\phi(D, D')$ $z=0$.

Theorem 5.3. Let $\phi(D, D')z=f(x,y)$ be a linear partial differential equation with constant coefficients. If u is a solution of $\phi(D, D')z=0$ and v is a solution of $\phi(D, D')z=f(x,y)$, then $u+v$ is a solution of $\phi(D, D')z=f(x,y)$.

Proof: Since u is a solution of $\phi(D, D')z=0$, we have

$$\phi(D, D')u=0. \qquad (92)$$

Similarly, since v is a solution of $\phi(D, D')z=f(x,y)$, we have

$$\phi(D, D')v=f(x,y) \qquad (93)$$

Then, by (92) and (93), we get

$$\phi(D, D')(u+v)=\phi(D, D')u+\phi(D, D')v$$

$$=0+f(x,y)=f(x,y).$$

Hence $u+v$ is a solution of $\phi(D, D')z=f(x,y)$.

(B) Complementary Function of Second Order Linear Homogeneous Partial Differential Equation

Consider the second order linear homogeneous partial differential equation

$$a_0\frac{\partial^2 z}{\partial x^2}+a_1\frac{\partial^2 z}{\partial x\partial y}+a_2\frac{\partial^2 z}{\partial y^2}=0. \qquad (94)$$

The symbolic form of equation (94) is

$$(a_0D^2+a_1DD'+a_2D'^2)z=0. \qquad (95)$$

The auxiliary equation (A.E.) for equation (95) is

$$a_0m^2+a_1m+a_2=0, \qquad (96)$$

which is obtained by putting $D=m$ and $D'=1$ in the symbolic form (95). The auxiliary equation (96), being a quadratic equation, has two roots, say m_1 and m_2, which may be *real and distinct* or *real and repeated.*

Case I. Let the roots of the auxiliary equation (96) be real and distinct. Then equation (94) may be written as

$$a_0(D-m_1D')(D-m_2D')z=0. \qquad (97)$$

Now, for $i=1, 2$, the solution of $(D-m_iD')z=0$ is also a solution of (97) and hence of (94). This yields

$$\frac{\partial z}{\partial x}-m_i\frac{\partial z}{\partial y}=0$$

or

$$p-m_iq=0, \qquad (98)$$

which is Lagrange's linear equation of first order. The auxiliary equations of (98) are

$$\frac{dx}{1}=\frac{dy}{-m_i}=\frac{dz}{0},$$

which yield

$$dy+m_idx=0 \text{ and } dz=0.$$

Then, integration yields

$$y+m_ix=c_1 \text{ and } x=c_2.$$

Hence

$$z=f_i\,(y+m_ix),$$

where f_i are arbitrary functions. Thus

$$z=f_1(y+m_1x) \text{ and } z=f_2(y+m_2x)$$

are solutions of the partial differential equation (94). Since linear combination of solutions of partial differential equation is also a solution the solution of (94) is

$$z=f_1(y+m_1x)+f_2\,(y+m_2x).$$

Proceeding in a similar way, the solution for the homogeneous linear partial differential equation of order n with constant coefficients for distinct roots of A.E.) is

$$z=f_1(y+m_1x)+f_2(y+m_2x)+\ldots$$
$$+f_n(y+m_nx).$$

Case II. Let the roots of the auxiliary equation be real and repeated. Thus for partial differential equation (94), we have $m_1=m_2$. Then equation (95) can be written as

$$a_0(D-m_1D')^2z=0$$

or

$$(D-m_1D')^2z=0. \qquad (99)$$

Put $(D-m_1D')z=u$, then (99) becomes

$$(D-m_1D')u=0.$$

Therefore, as in case 1, its solution is

$$u=f_1(y+m_1x)$$

so that

$$(D-m_1D')z=f_1(y+m_1x)$$

or

$$\frac{\partial z}{\partial x} - m_1 \frac{\partial z}{\partial y} = f_1(y + m_1 x),$$

or

$$p - m_1 q = f_1(y + m_1 x),$$

which is a Lagrange's linear equation with auxiliary equation

$$\frac{dx}{1} = \frac{dy}{-m_1} = \frac{dz}{f_1(y + m_1 x)}.$$

Thus

$$dy + m_1 dx = 0 \text{ and } dx = \frac{dz}{f_1(y + m_1 x)}$$

or

$$dy + m_1 dx = 0 \text{ and } dz = f_1(y + m_1 x)dx. \quad (100)$$

Integration of first member of (100) yields

$$y + m_1 x = c_1. \quad (101)$$

Using (101) in the second member of (100), we have

$$dz = f_1(c_1)dx,$$

which, on integration, yields

$$\begin{aligned} z &= f_1(c_1)x + c_2 \\ &= f_1(c_1)x + f_2(c_1), \text{ putting } c_2 = f_2(c_1) \\ &= x f_1(y + m_1 x) + f_2(y + m_1 x). \end{aligned}$$

Thus, the solution of (94), in this case, is

$$z = f_2(y + m_1 x) + x f_1(y + m_1 x).$$

Proceeding similarly, it can be shown that if the root m_1 of homogeneous linear partial differential equation of order n is repeated r times, then the complementary function of that partial differential equation is given by

$$\begin{aligned} z = f_1(y + m_1 x) + x f_2(y + m_1 x) + x^2 f_3\,(y + m_1 x) \\ + \ldots + x^{n-1} f_r\,(y + m_1 x), \end{aligned}$$

where $f_1, f_2, \ldots, f_r$ are arbitrary functions.

Remark 5.1. If $a_0 = 0$ in (94), then we have

$$(a_1 DD' + a_2 D'^2)\, z = 0$$

or

$$D'(a_1 D + a_2 D')\, z = 0$$

Therefore the solution of $D'z = 0$ is also a solution of (94). The solution of $D'z = 0$, that is, solution of $\frac{\partial z}{\partial y} = 0$ is $z = \phi(x)$, where ϕ is an arbitrary function.

EXAMPLE 5.61

Solve $\frac{\partial^2 z}{\partial x^2} - 4\frac{\partial^2 z}{\partial x \partial y} + 4\frac{\partial^2 z}{\partial y^2} = 0.$

Solution. The symbolic form of the given homogeneous linear partial differential equation of order 2 is

$$r - 4s + 4t = 0,$$

or

$$(D^2 - 4DD' + 4D'^2)z = 0.$$

Putting D = m, and D′ = 1, we get the auxiliary equation for the given equation as

$$m^2 - 4m + 4 = 0 \text{ or } (m - 2)^2 = 0,$$

which yields $m = 2, 2$. Thus the general solution of the given equation is

$$z = f_1(y + 2x) + x f_2(y + 2x).$$

EXAMPLE 5.62

Solve $(D^3D'^2 + D^2D'^3)z = 0$.

Solution. We have

$$(D^3D'^2 + D^2D'^3)z = 0.$$

or

$$[D'^2 D^2\,(D + D')]z = 0.$$

The part of the general solution (see Remark 5.1) corresponding to the factor D'^2 is $f_1(x) + yf_2(x)$. Further, the auxiliary equation of $D^2(D + D')z = 0$ is

$$m^2(m+1) = 0,$$

which yields $m = 0, 0, -1$. Therefore part of the general solution corresponding to the factor $D^2(D + D')$ is

$$f_3(y + 0x) + xf_4(y + 0x) + f_5(y - x).$$

Hence the general solution of the given partial differential equation is

$$z = f_1(x) + yf_2(y) + f_3(y) + xf_4(y) + f_5(y - x),$$

where f_1, f_2, f_3, f_4, f_5 are arbitrary functions.

(C) Particular Integral (P.I.) of $\phi(D,D')z = f(x, y)$

Since

$$\phi(D, D')\left[\frac{1}{\phi(D, D')} f(x, y)\right] = f(x, y),$$

the operator $\frac{1}{\phi(D, D')}$ is called the *inverse operator* of the operator $\phi(D, D')$. Further, $\frac{1}{\phi(D, D')} f(x, y)$ is a particular solution of the equation $\phi(D, D')z = f(x, y)$, because

$$\phi(D, D')\left[\frac{1}{\phi(D, D')} f(x, y)\right] = f(x, y).$$

We now discuss four different cases for finding particular integral according to different forms of $f(x, y)$.

Case I. *When $f(x, y) = x^m y^n$, where m and n are constants.* In this case, the particular integral $\frac{1}{\phi(D, D')} f(x, y)$ is obtained by expanding $\frac{1}{\phi(D, D')}$ into an infinite series in ascending powers of either D or D'. The particular integral obtained by using D and D' may not be identical. So, any of the two particular integral may be used. However, if $m < n$, then it is advised to expand $\frac{1}{\phi(D, D')}$ in ascending powers of D while for $n < m$, we should expand $\frac{1}{\phi(D, D')}$ in ascending power of D'

Case II. *When $f(x, y) = \sin(ax + by)$ or $\cos(ax + by)$.* In this case

$$\text{P.I.} = \frac{1}{\phi(D^2, DD', D'^2)} [\sin(ax + by) \text{ or } \cos(ax + by)].$$

Replace D^2 by $-a^2$, DD' by $-ab$ and D'^2 by $-b^2$ so that

$$\text{P.I.} = \frac{1}{\phi(-a^2, -ab, -b^2)} [\sin(ax + by)] \text{ or } \cos(ax + by).$$

In case $\phi(a, b) = 0$ then $bD - aD'$ must be a factor of $\phi(D, D')$. If $bD - aD'$ is repeated r times, then the value of $\frac{1}{(bD - aD')^r} f(ax + by) = \frac{x^r}{b^r\, r!} f(ax + by)$.

Case III. When $f(x, y) = e^{ax+by}$. We note that

$$D[e^{ax+by}] = a\, e^{ax+by},$$

$$D'[e^{ax+by}] = b\, e^{ax+by},$$

$$D^2[e^{ax+by}] = a^2\, e^{ax+by},$$

$$DD'[e^{ax+by}] = ab\, e^{ax+by},$$

$$D'^2[e^{ax+by}] = b^2\, e^{ax+by}.$$

Therefore

$$(D^2 + a_1 DD' + a_2 D'^2)\, e^{ax+by}$$

$$= (a^2 + a_1 ab + a_2 b^2)\, e^{ax+by}$$

or

$$\phi(D, D')e^{ax+by} = \phi(a, b)\, e^{ax+by},$$

or

$$\frac{\phi(D, D')e^{ax+by}}{\phi(a, b)} = f(x, y).$$

Operating $\frac{1}{\phi(D, D')}$ on both sides, we get

$$\frac{1}{\phi(a, b)} e^{ax+by} = \frac{1}{\phi(D, D')} f(x, y) = \text{P.I.}$$

Hence, for $f(x, y) = e^{ax+by}$, we have

$$\text{P.I.} = \frac{1}{\phi(a, b)} e^{ax+by}$$

Case IV. When $f(x, y)$ is any function of x and y, specially other that the above discussed three case, then resolve $\frac{1}{\phi(D, D')}$ into partial fractions, treating $\phi(D, D')$ as a function of D alone and operate each partial fraction on $f(x, y)$ as

$$\frac{1}{(D - m D')} f(x, y) = \int f(x, c - mx) dx,$$

that is, replacing y by $c - mx$ in $f(x, y)$ and integrate it with respect to x. After integration replace c by $y + mx$. Repeat the process with the second factor and so on.

EXAMPLE 5.63

Solve $\frac{\partial^2 z}{\partial x^2} - 4\frac{\partial z}{\partial x \partial y} + 4\frac{\partial^2 z}{\partial y^2} = e^{2x+y}$

Solution. From Example 5.61, the complementary function for this partial differential equation is

$$\text{C.F.} = f_1(y + 2x) + x f_2(y + 2x).$$

Further,

$$\text{P.I.} = \frac{1}{\phi(D, D')} f(x, y) = \frac{1}{(D - 2D')^2} e^{2x+y}.$$

Now if we put $D = 2$ and $D' = 1$, the rule III fails. Therefore we use rule IV and have (taking $y = c - mx = c - 2x$) from $(D - 2D')u = e^{2x+y}$, the solution

$$u = \int f(x, c - mx) dx = \int e^{2x+c-2x} dx$$

$$= \int e^c dx = x e^c = x e^{2x+y}.$$

Now from $(D - 2D')z = u = x e^{2x+y}$, the solution is

$$z = \int x e^{2x+c-2x} dx = \frac{1}{2} x^2 e^c = \frac{1}{2} x^2 e^{2x+y}.$$

Hence the complete solution of the given equation is

$$z = f_1(y + 2x) + x f_2(y + 2x) + \frac{1}{2} x^2 e^{2x+y}.$$

EXAMPLE 5.64

Solve $\frac{\partial^2 z}{\partial x^2} - 2\frac{\partial^2 z}{\partial x \partial y} = \sin x \cos 2y.$

Solution. The symbolic form of the given linear homogeneous partial differential equation of order 2 is

$$(D^2 - 2\,DD')z = \sin x \cos 2y.$$

The auxiliary equation for this equation is

$$m^2 - 2m = 0,$$

which yields $m = 0, 2$. Therefore

$$\text{C.F.} = f_1(y + 0) + f_2(y + 2x)$$

$$= f_1(y) + f_2(y + 2x).$$

Further,

$$\text{P.I.} = \frac{1}{\phi(D, D')} f(x, y)$$

$$= \frac{1}{(D^2 - 2DD')} [\sin x \cos 2y]$$

$$= \frac{1}{(D^2 - 2DD')} \left[\frac{1}{2}\{\sin(x + 2y) + \sin(x - 2y)\}\right]$$

$$= \frac{1}{2} \cdot \frac{1}{D^2 - 2DD'} [\sin(x + 2y)]$$

$$+ \frac{1}{2} \cdot \frac{1}{D^2 - 2DD'} [\sin(x - 2y)]$$

$$= \frac{1}{2}\left(\frac{1}{-1 + 4}\right) \sin(x + 2y) + \frac{1}{2}\left(\frac{1}{-1 - 4}\right)$$

$$\times x \sin(x - 2y)$$

$$= \frac{1}{6}\sin(x+2y) - \frac{1}{10}\sin(x-2y).$$

Hence the complete solution is

$$z = \text{C.F.} + \text{P.I.} = f_1(y) + f_2(y+2x)$$
$$+ \frac{1}{6}\sin(x+2y) - \frac{1}{10}\sin(x-2y).$$

EXAMPLE 5.65

Solve the partial differential equation

$$\frac{\partial^2 z}{\partial x^2} + \frac{\partial^2 z}{\partial x \partial y} - 6\frac{\partial^2 z}{\partial y^2} = \cos(2x+y).$$

Solution. The symbolic form of the given homogeneous linear partial differential equation of order 2 is

$$(D^2 + DD' - 6D'^2)z = \cos(2x+y).$$

The auxiliary equation for the equation is

$$m^2 + m - 6 = 0,$$

which yields $m = 2, -3$. Therefore

$$\text{C.F.} = f_1(y+2x) f_2(y-3x).$$

To find particular integral, we have

$$\text{P.I.} = \frac{1}{D^2 + DD' - 6D'^2}\cos(2x+y).$$

But putting $D^2 = -4$, $DD' = -2$ and $D'^2 = -1$, we get

$$\frac{1}{D^2 + DD' - 6D'^2} = \frac{1}{-4-2+6} = \frac{1}{0} = \infty$$

and so particular integral cannot be found by the rule for $\sin(ax+by)$ and $\cos(ax+by)$. Now

$$\frac{1}{D^2 + DD' - 6D'^2} = \frac{1}{(D-2D')(D+3D')}$$
$$= \frac{1}{5(D-2D')} - \frac{1}{5(D+3D')}$$

and so (by case IV),

$$\text{P.I.} = \frac{1}{5(D-2D')}\cos(2x+y)$$
$$- \frac{1}{5(D+3D')}\cos(2x+y)$$
$$= \frac{1}{5}\int \cos(2x+c_1-2x)dx$$
$$- \frac{1}{5}\int \cos(2x+c_2+3x)dx$$
$$= \frac{1}{5}\int (\cos c_1)dx - \frac{1}{5}\int \cos(5x+c_2)dx$$
$$= \frac{1}{5}x\cos c_1 - \frac{1}{25}\sin(5x+c_2)$$
$$= \frac{1}{5}x\cos(2x+y) - \frac{1}{25}\sin(y+2x).$$

Hence the complete solution of the given equation is

$$z = \text{C.F.} + \text{P.I.}$$
$$= f_1(y+2x) + f_2(y-3x) + \frac{1}{5}x\cos(2x+y)$$
$$- \frac{1}{25}\sin(y+2x).$$

EXAMPLE 5.66

Solve the partial differential equation

$$\frac{\partial^2 z}{\partial x^2} - \frac{\partial^2 z}{\partial y^2} = x^2 y.$$

Solution. The symbolic form of the given equation is

$$(D^2 - D'^2)^z = x^2 y.$$

Its auxiliary equation is

$$m^2 - 1 = 0,$$

which yields $m = 1, -1$. Therefore

$$\text{C.F.} = f_1(y+x) + f_2(y-x).$$

Further,

$$\text{P.I.} = \frac{1}{D^2 - D'^2}x^2 y = \frac{1}{D^2}\left(1 - \frac{D'^2}{D^2}\right)^{-1}(x^2 y)$$

$$= \frac{1}{D^2}\left(1 + \frac{D'^2}{D^2} + \ldots\right)(x^2 y)$$

$$= \frac{1}{D^2}\left[x^2 y + \frac{1}{D^2}\{D'^2(x^2 y)\} + \ldots\right]$$

$$= \frac{1}{D^2}\left[x^2 y + \frac{1}{D^2}\{x^2(0)\}\right] = \frac{1}{D^2}(x^2 y)$$

$$= \frac{1}{D}\left(\frac{x^3}{3}y\right) = \frac{x^4 y}{12}.$$

Hence the complete solution of the given partial differential equation is

$$z = f_1(y + x) + f_2(y - x) + \frac{x^4 y}{12}.$$

EXAMPLE 5.67

Solve the partial differential equation

$$\frac{\partial^2 z}{\partial x^2} + \frac{\partial^2 z}{\partial x \partial y} - 6\frac{\partial^2 z}{\partial y^2} = y \cos x.$$

Solution. As in Example 5.70, we have

$$\text{C.F.} = f_1(y + 2x) + f_2(y - 3x)$$

Further,

$$\text{P.I.} = \frac{1}{(D - 2D')(D + 3D')}(y \cos x).$$

Let

$$(D + 3D')u = y \cos x.$$

Then

$$u = \int (c + 3x) \cos x, \text{ since } y = c - mx = c + 3x$$

$$= (c + 3x) \sin x - \int 3 \sin x dx$$

$$= (c + 3x)\sin x + 3 \cos x = y \sin x + 3 \cos x.$$

Now $(D - 2D')z = u = y \sin x + 3 \cos x$ yields

$$z = \int [(c - 2x) \sin x + 3 \cos x]dx, \text{ because}$$

$$y = c - mx = c - 2x$$

$$= (c - 2x)(-\cos x) - (-2)(-\sin x) + 3 \sin x$$

$$= (y + 2x - x)(-\cos x) - 2 \sin x + 3 \sin x$$

$$= -y \cos x + \sin x.$$

Hence the complete solution is

$$z = f_1(y + 2x) + f_2(y - 3x) + \sin x - y \cos x.$$

EXAMPLE 5.68

Solve the partial differential equation

$$\frac{\partial^3 z}{\partial x^3} - 3\frac{\partial^3 z}{\partial x^2 y} + 4\frac{\partial^3 z}{\partial y^3} = e^{x+2y}.$$

Solution. The symbolic form of the given equation is

$$(D^3 - 3D^2D' + 4D'^3)z = e^{x+2y}.$$

Its auxiliary equation is

$$m^3 - 3m^2 + 4 = 0,$$

which yields $m = -1, 2, 2$. Therefore

$$\text{C.F.} = f_1(y - x) + f_2(y + 2x) + xf_3(y + 2x).$$

Also,

$$\text{P.I.} = \frac{1}{D^3 - 3D^2 D' + 4D'^3}(e^{x+2y})$$

$$= \frac{1}{1 - 6 + 32} e^{x+2y} = \frac{1}{27} e^{x+2y}.$$

Hence the complete solution of the given linear homogeneous partial differential equation of order 2 is

$$z = f_1(y - x) + f_2(y + 2x) + xf_3(y + 2x) + \frac{1}{27} e^{x+2y}.$$

EXAMPLE 5.69

Solve the partial differentialequation

$$(D^3 - 4D^2D' + 4DD'^2)z = \cos(2x + y).$$

Solution. The auxiliary equation for the given partial differential equation is

$$m^3 - 4m^2 + 4m = 0 \text{ or } m(m^2 - 4m + 4) = 0,$$

which yields $m = 0, 2, 2$. Hence

$$\text{C.F.} = f_1(y) + f_2(y+2x) + xf_3(y+2x).$$

Further,

$$\text{P.I.} = \frac{1}{\phi(D, D')}\cos(2x+y).$$

But, taking $D^2 = -4$, $DD' = -2$, $D'2 = -1$, we have

$$\frac{1}{\phi(D, D')} = \frac{1}{D^3 - 4D^2D' + 4DD'^2} = \frac{1}{0}.$$

Therefore the rule fails. But then

$$\text{P.I.} = \frac{1}{D(D-2D')^2}\cos(2x+y)$$

$$= \frac{1}{(D-2D')^2}\left[\frac{1}{D}\cos(2x+y)\right]$$

$$= \frac{1}{(D-2D')^2}\int \cos(2x+y)dx$$

$$= 2\frac{1}{(D-2D')^2}\sin(2x+y).$$

Now since $\frac{1}{D-2D'} = \frac{1}{0}$ for $D = 2$, $D' = 1$ and the factor $D - 2D'$ is repeated twice, we have

$$\text{P.I.} = \frac{1}{2}\left[\frac{x^2}{1^2 2!}\text{Sin}(2x+y)\right] = \frac{x^2}{4}\sin(2x+y).$$

Hence the complete solution of the given partial differential equation is

$$z = f_1(y) + f_2(y+2x) + xf_3(y+2x) + \frac{x^2}{4}\sin(2x+y).$$

EXAMPLE 5.70

Solve the partial Differential equations

(i) $\frac{\partial^3 z}{\partial x^2} - 4\frac{\partial^3 z}{\partial x^2 \partial y} + 4\frac{\partial^2 z}{\partial x \partial y^2} = 2\sin(3x+2y).$

(ii) $\frac{\partial^2 z}{\partial x^2} + \frac{\partial^2 z}{\partial y^2} = \cos mx.\ \cos ny.$

(iii) $\frac{\partial^2 z}{\partial x^2} - 2\frac{\partial^2 z}{\partial x \partial y} = \sin x \cdot \cos 2y.$

Solution. (i) As in the above Example, the complimentary function of the given equation is

$$\text{C.F.} = f_1(y) + f_2(y+2x) + xf_3(y+2x).$$

Further,

$$f(x, y) = 2\sin(3x+2y),$$

$$\phi(D, D') = D^3 - 4D^2D' + 4DD'^2,$$

$$\phi(a, b) = 27 - 4.9.2 + 4.3.4 = 3 \neq 0.$$

Therefore

$$\text{P.I.} = \frac{1}{D^3 - 4D^2D' + 4DD'^2}[2\sin(3x+2y)]$$

$$= \frac{1}{D(D^2 - 4DD' + 4D'^2)}[2\sin(3x+2y)]$$

$$= \frac{1}{D(-9+24-16)}[2\sin(3x+2y)]$$

$$= -\frac{2}{D}\sin(3x+2y)dx$$

$$= \frac{2}{3}\cos(3x+2y).$$

Hence the complete solution of the given partial differential equation is

$$z = f_1(y) + f_2(y+2x) + xf_3(y+2x) + \frac{2}{3}\cos(3x+2y)$$

(ii) The symbolic form of the equation is

$$(D^2 + D'^2)z = \cos mx \cos ny.$$

The auxiliary equation is

$m^2 + 1 = 0$, which yields $m = \pm i$.

Therefore

$$\text{C.F.} = f_1(y+ix) + f_2(y-ix).$$

Further,

$$\text{P.I.} = \frac{1}{D^2 + D'^2}\cos mx \cos ny$$

$$= \frac{1}{D^2 + D'^2} \times \left[\frac{1}{2}\{\cos(mx+ny) + \cos(mx-ny)\}\right]$$

$$= \frac{1}{2}\left[\frac{1}{D^2 + D'^2} \cos(mx + ny) + \frac{1}{D^2 + D'^2} \cos(mx - ny) \right]$$

$$= \frac{1}{2}\left[\frac{1}{m^2 + n^2} \int\int \cos u du du + \frac{1}{m^2 + n^2} \int\int \cos v \, dv \, dv \right],$$

where $mx + ny = u$ and $mx - ny = v$

$$= \frac{1}{2}\left[\frac{1}{m^2 + n^2} (-\cos u) + \frac{1}{m^2 + n^2} (-\cos v) \right]$$

$$= -\frac{1}{2(m^2 + n^2)} [\cos u + \cos v]$$

$$= \frac{-1}{2(m^2 + n^2)} [\cos(mx + ny) + \cos(mx - ny)]$$

$$= \frac{-1}{m^2 + n^2} \cos mx \cos ny.$$

Hence the solution is

$$z = \text{C.F.} + \text{P.I.}$$

$$= f_1(y + ix) + f_2(y - ix) - \frac{1}{m^2 + n^2} \cos mx \cos ny.$$

(iii) The symbolic form of the given equation is

$$(D^2 - 2DD')z = \sin x \cos 2y.$$

The auxiliary equation is

$$m^2 - 2m = 0, \text{ which yields } m = 0, 2.$$

Therefore

$$\text{C.F.} = f_1(y) + f_2(y + 2x),$$

where f_1 and f_2 are arbitrary functions. Further,

$$\text{P.I.} = \frac{1}{D^2 - 2DD'} \cdot \sin x \cos 2y$$

$$= \frac{1}{D^2 - 2DD'} \cdot \frac{1}{2} [\sin(x + 2y) + \sin(x - 2y)]$$

$$= \frac{1}{2}\left[\frac{1}{D^2 - 2DD'} \sin(x + 2y) + \frac{1}{D^2 - 2DD'} \sin(x - 2y) \right]$$

$$= \frac{1}{2}\left[\frac{1}{1^2 - 2(1)(2)} \int\int \sin u \, du \, du + \frac{1}{1^2 - 2(1)(2)} \int\int \sin v \, dv \, dv \right],$$

where $x + 2y = u$, $x - 2y = v$

$$= \frac{1}{2}\left[\frac{1}{3} \sin u - \frac{1}{5} \sin v \right]$$

$$= \frac{1}{6} \sin(x + 2y) - \frac{1}{10} \sin(x - 2y).$$

Hence the complete solution of the given partial differential equation is

$$z = \text{C.F.} + \text{P.I.} = f_1(y) + f_2(y + x) + \frac{1}{6} \sin(x + 2y) - \frac{1}{10} \sin(x - 2y).$$

(D) Solution of Non-homogeneous Linear Partial Differential Equation

As pointed out earlier, a partial differential equation $\phi(D, D')z = f(x, y)$ is called *non-homogeneous* if all the terms of $\phi(D, D')$ are not of the same order.

The complete solution of such equation also consists of two parts-complimentary function (C.F.) and particular integral (P.I). *The particular integral of such equation can be found using the rules of homogeneous linear partial differential equations.* The complementary function of such equation can be found by the following two methods:

Method I. Let $\phi(D, D')$ be factorized as

$$\phi(D, D') = (a_1 D + b_1 D' + c_1)(a_2 D + b_2 D' + c_2) \times \ldots (a_n D + b_n D') + c_n),$$

where a_i, b_i and c_i are constants. The solution of the equation

$$a_i D + b_i D' + c_i = 0 \qquad (102)$$

will be a solution of $\phi(D,D')=0$. The Lagrage's auxiliary equations for (102) are

$$\frac{dx}{a_i}=\frac{dy}{b_i}=\frac{dz}{-c_i z}.$$

Thus

$$b_i dx - a_i dy = 0$$

and

$$\frac{dz}{z} = -\frac{c_i}{a_i}dx \text{ or } \frac{dz}{z} = -\frac{c_i}{b_i}dy.$$

Integrating these equations, we get

$$b_i x - a_i y = k_1 \qquad \text{(constant)}$$

and

$$\log z = -\frac{c_i}{a_i}x + \log k_2 \text{ if } a_i \neq 0$$

or

$$\log z = -\frac{c_i}{b_i}y + \log k_3 \text{ if } b_i \neq 0,$$

that is,

$$b_i x - a_i y = k_1$$

and

$$z = k_2 e^{-\frac{c_i}{a_i}x} \text{ if } a_i \neq 0 \text{ or } z = k_3 e^{-\frac{c_i}{b_i}y} \text{ if } b_i \neq 0.$$

Hence the general solution of (102) is

$$z = e^{-\frac{c_i}{a_i}x} f_i(b_i x - a_i y) \text{ if } a_i \neq 0$$

or

$$z = e^{-\frac{c_i}{b_i}y} f_i(b_i x - a_i y) \text{ if } b_i \neq 0,$$

where f_i is an arbitrary function.

The solution corresponding to the other factors can be found similarly. The complementary function of the given equation is the sum of all solutions obtained as above.

In case of repeated factors, the solution of $(a_i D + b_i D' + c_i)^n$ shall be

$$z = e^{-\frac{c_i}{a_i}x} \sum_{i=1}^{n} x^{i-1} f_i(b_i x - a_i y) \text{ if } a_i \neq 0$$

or

$$z = e^{-\frac{c_i}{b_i}y} \sum_{i=1}^{n} y^{i-1} f_i(b_i x - a_i y) \text{ if } b_i \neq 0.$$

Method II. Suppose that $z = e^{ax+by}$ is a trial solution of $\phi(D, D')z = 0$. Then

$$D(e^{ax+by}) = \frac{\partial}{\partial x}(e^{ax+by}) = a\, e^{ax+by}$$

and

$$D'(e^{ax+by}) = \frac{\partial}{\partial y}(e^{ax+by}) = b\, e^{ax+by}.$$

Therefore $\phi(D, D')z = 0$ implies

$$\phi(a, b)\, e^{ax+by} = 0.$$

Since $e^{ax+by} \neq 0$, we have

$$\phi(a, b) = 0. \tag{103}$$

Suppose that solving (103) for b in terms of a, we get

$$b = f_1(a), f_2(a), \ldots$$

Then

$$\text{C.F.} = \sum c_i\, e^{ax + f_1(a)y}.$$

EXAMPLE 5.71

Solve $(D - D' - 1)(D - D' - 2)z = e^{2x-y}$.

Solution. The factors of $\phi(D, D')$ in the given non homogeneous partial differential are already of the type $a_i D + b_i D' + c_i$. Therefore we use method I in this case. In this case $a_1 = b_1 = -1$, $c_1 = -1$ and $a_2 = 1$, $b_2 = -1$, $c_2 = -2$.

Therefore

$$\text{C.F.} = e^{-\frac{c_1}{a_1}x} f_1(x+y) + e^{-\frac{c_2}{a_2}x} f_2(x+y)$$

$$= e^x f_1(y+x) + e^{2x} f_2(y+x).$$

Moreover,

$$\text{P.I.} = \frac{1}{(D - D' - 1)(D - D' - 2)} e^{2x-y}$$

$$= \frac{1}{(2)(1)} e^{2x-y}, \text{ replacing } D \text{ by 2 and } D' \text{ by } -1$$

$$= \frac{1}{2} e^{2x-y}.$$

Hence the complete solution is

$$z = \text{C.F.} + \text{P.I.}$$

$$= e^x f_1(y+x) + e^{2x} f_2(y+x) + \frac{1}{2} e^{2x-y}.$$

EXAMPLE 5.72

Solve the non-homogeneous partial differential equation $(D^2 - DD' - D' - 1)z = \cos(x + 2y)$.

Solution. We have

$$\phi(D, D') = D^2 - DD' + D' - 1$$

$$= (D-1)(D - D' + 1).$$

Thus $a_1 = 1$, $b_1 = 0$, $c_1 = -1$ and $a_2 = 1$, $b_2 = -1$, $c_2 = 1$. Therefore

$$\text{C.F.} = e^x f_1(y) + e^{-x} f_2(y+x).$$

Moreover,

$$\text{P.I.} = \frac{1}{D^2 - DD' + D' - 1} \cos(x+2y)$$

$$= \frac{1}{-1 + 2 + D' - 1} \cos(x + 2y)$$

$$= \frac{1}{D'} \cos(x+2y) = \int \cos(x+2y)\, dy$$

$$= \frac{1}{2} \sin(x + 2y)$$

Hence the complete equation is

$$z = e^x f_1(y) + e^{-x} f_2(y+x) + \frac{1}{2} \sin(x+2y).$$

EXAMPLE 5.73

Solve the partial differential equation

$$(D^3 - 3DD' + D' + 4)z = e^{2x+y}.$$

Solution. We note that $\phi(D, D') = D^3 - 3DD' + D' + 4$ is irreducible (does not have linear factor). Therefore we use method II in this case. We have

$$\phi(a, b) = a^3 - 3ab + 4 = 0$$

Being a cubic in a, it has three values of a in terms of b. Let these values be $f_1(b)$, $f_2(b)$ and $f_3(b)$. Then

$$\text{C.F.} = c_1 e^{f_1(b)x + by} + c_2 e^{f_2(b)x + by} + c_3 e^{f_3(b)x + by}.$$

Further,

$$\text{P.I.} = \frac{1}{D^3 - 3DD' + D' + 4} e^{2x+y}$$

$$= \frac{1}{8 - 6 + 1 + 4} e^{2x+y} = \frac{1}{7} e^{2x+y}.$$

Hence the complete solution is

$$z = \text{C.F.} + \text{P.I.}$$

$$= c_1 e^{f_1(b)x+by} + c_2 e^{f_2(b)x+by} + c_3 e^{f_3(b)x+by} + \frac{1}{7} e^{2x+y}.$$

EXAMPLE 5.74

Solve $(2D^4 - 3D^2D' + D'^2)z = 0$.

Solution. We have

$$2D^4 - 3D^2D' + D'^2 = (2D^2 - D')(D^2 - D')$$

$$= \phi_1(D, D')\, \phi_2(D, D').$$

Since $2D^2 - D'$ is irreducible, we have

$$\phi_1(a, b) = 2a^2 - b = 0$$

or

$$b = 2a^2.$$

Therefore contribution to C.F. due to $\phi_1(D, D')z = 0$ is

$$c_1 e^{ax+2a^2y}.$$

Again, since $D^2 - D'$ is irreducible, we have

$$\phi_2(a', b') = a'^2 - b' = 0 \text{ or } b' = a'^2.$$

Therefore contribution to C.F. due to $\phi_2(D, D')z = 0$ is

$$c_2 e^{a'x + a'^2 y}.$$

Hence the complete solution is

$$z = c_1 e^{ax + 2a^2 y} + c_2 e^{a'x + a'^2 y}.$$

EXAMPLE 5.75

$(D^2 - D'^2 - 3D + 3D')z = xy + e^{x+2y}$.

Solution. The given equation can be written as

$$(D - D')(D + D' - 3)z = xy + e^{x+2y}$$

Therefore, using method I, we have

$$C.F. = f_1(y + x) + e^{3x} f_2(y - x),$$

P.I. corresponding to xy is given by

$$\frac{1}{(D-D')(D+D'-3)}\, xy$$

$$= -\frac{1}{3D}\left(1 - \frac{D'}{D}\right)^{-1}\left(1 - \frac{D}{3} - \frac{D'}{3}\right)^{-1} \cdot xy$$

$$= -\frac{1}{3D}\left(1 + \frac{D'}{D} \ldots\right)\left(1 + \frac{D}{3} + \frac{D'}{3} + \frac{2DD'}{9} \ldots\right) xy$$

$$= -\frac{1}{3D}\left(1 + \frac{D}{3} + \frac{D'}{3} + \frac{D'}{D} + \frac{D'}{3} + \frac{2DD'}{9} \ldots\right) xy$$

$$= -\frac{1}{3D}\left(xy + \frac{y}{3} + \frac{2}{3}x + \frac{1}{D}x + \frac{2}{9}\right)$$

$$= -\frac{1}{3}\left(\frac{x^2}{2} \cdot y + \frac{xy}{3} + \frac{x^2}{3} + \frac{x^2}{6} + \frac{2}{9} \cdot x\right)$$

P.I. corresponding to e^{x+2y} is given by

$$\frac{1}{(D + D' - 3)(D - D')} \cdot e^{x+2y}$$

$$= \frac{1}{(D + D' - 3)} \cdot \frac{1}{1 - 2}\, e^{x+2y}$$

$$= -\frac{1}{(D + D' - 3)} \cdot e^{x+2y}$$

$$= -e^x \frac{1}{(D' - 2)}\, e^{2y}$$

$$= -e^x e^{2y} \frac{1}{D' + 2 - 2} \cdot 1$$

$$= -e^{x+2y} \frac{1}{D'} \cdot 1$$

$$= -y\, e^{x+2y}.$$

$\therefore$ The solution is $z = f_1(y + x) + e^{2x} f_2(y - x)$

$$-\frac{1}{3}\left(\frac{x^2}{2}y + \frac{xy}{3} + \frac{x^3}{3} + \frac{2}{9}x\right) - ye^{x+2y}.$$

5.7 EQUATIONS REDUCIBLE TO HOMOGENEOUS LINEAR FORM

An equation in which the coefficient of derivative of any order is a multiple of the variables of the same degree may be transformed into the partial differential equations with constants coefficients.

For this, we substitute

$$x = e^X, y = e^Y$$

so that, $X = \log x$ and $Y = \log y$,

$$\frac{\partial z}{\partial x} = \frac{\partial z}{\partial X} \cdot \frac{\partial X}{\partial x} = \frac{1}{x}\frac{\partial z}{\partial X}$$

or $$x\frac{\partial z}{\partial x} = \frac{\partial z}{\partial X}.$$

$$\therefore \quad x\frac{\partial}{\partial x} \equiv \frac{\partial}{\partial X} \equiv D(\text{say}).$$

Now $$x\frac{\partial}{\partial x}\left(x^{n-1}\frac{\partial^{n-1} z}{\partial x^{n-1}}\right)$$

$$= x^n\frac{\partial^n z}{\partial x^n} + (n-1)\,x^{n-1}\frac{\partial^{n-1} z}{\partial x^{n-1}}$$

$$x^n\frac{\partial^n z}{\partial x^n} = \left(x\frac{\partial}{\partial x} - n + 1\right)x^{n-1}\frac{\partial^{n-1} z}{\partial x^{n-1}}.$$

Putting $n = 2, 3, \ldots$, we have

$$x^2\frac{\partial^2 z}{\partial x^2} = (D-1)\,x\frac{\partial z}{\partial x} = D(D-1)z$$

$$x^3\frac{\partial^3 z}{\partial x^3} = (D-2)\,x^2\frac{\partial^2 z}{\partial x^2}$$

$$= (D-2)\,(D-1)\,Dz \text{ etc.}$$

Similarly, $y\dfrac{\partial z}{\partial y} = \dfrac{\partial z}{\partial Y} = D'z$

$$y^2\frac{\partial^2 z}{\partial y^2} = D'\,(D'-1)z \text{ etc.}$$

and $xy\dfrac{\partial^2 z}{\partial x\,\partial y} = DD'z$ etc.

Substituting in the given equation, it reduces to the form

$$F(D, D')z = V$$

which is an equation having constant coefficients and can easily be solved by the methods discussed earlier.

EXAMPLE 5.76

Solve $x^2\dfrac{\partial^2 z}{\partial x^2} + 2xy\dfrac{\partial^2 z}{\partial x\,\partial y} + y^2\dfrac{\partial^2 z}{\partial y^2} = 0.$

Solution. Substituting $x = e^X$, $y = e^Y$ so that $X = \log x$, $Y = \log y$ and denoting $\dfrac{\partial}{\partial X}$ and $\dfrac{\partial}{\partial Y}$ by D and D', respectively, the given equation reduces to

$$[D(D-1) + 2DD' + D'\,(D'-1)]z = 0$$

or $$(D + D')\,(D + D' - 1)z = 0$$

Hence, the solution is (using method I)

$$z = f_1\,(Y - X) + e^X f_2\,(Y - X)$$

$$= f_1\,(\log y - \log x) + x f_2\,(\log y - \log x)$$

$$= f_1\left(\log\frac{y}{x}\right) + x f_2\left(\log\frac{y}{x}\right) = \phi_1\left(\frac{y}{x}\right) + x\phi_2\left(\frac{y}{x}\right).$$

EXAMPLE 5.77

Solve $x^2\dfrac{\partial^2 z}{\partial x^2} - y^2\dfrac{\partial^2 z}{\partial y^2} - y\dfrac{\partial z}{\partial y} + x\dfrac{\partial z}{\partial x} = 0.$

Solution. Substituting $x = e^X$, $y = e^Y$ so that $X = \log x$, $Y = \log y$, and denoting $\dfrac{\partial}{\partial X}$ and $\dfrac{\partial}{\partial Y}$ by D and D', respectively, the given equation reduces to

$$[D(D-1) - D'\,(D'-1) - D' + D]z = 0$$

or $$(D^2 - D'^2)z = 0$$

$$\therefore \quad z = f_1\,(Y + X) + f_2\,(Y - X)$$

$$= f_1\,(\log y + \log x) + f_2\,(\log y - \log x)$$

$$= f_1\,(\log xy) + f_2\left\{\log\left(\frac{y}{x}\right)\right\}$$

$$\phi_1\,(xy) + \phi_2\,(y/x)$$

EXAMPLE 5.78

Solve $x^2\dfrac{\partial^2 z}{\partial x^2}+2xy\dfrac{\partial^2 z}{\partial x\partial y}+y^2\dfrac{\partial^2 z}{\partial y^2}-nx\dfrac{\partial z}{\partial x}$

$$-ny\frac{\partial z}{\partial y}+nz=x^2+y^2.$$

Solution. Substituting $x=e^X$, $y=e^Y$, if D and D' denote $\dfrac{\partial}{\partial X}$ and $\dfrac{\partial}{\partial Y}$, the given equation reduces to

$$[D(D-1)+2DD'+D'(D'-1)$$

$$-nD-nD'+n]z=e^{2X}+e^{2Y}$$

or $\quad (D+D'-1)(D+D'-n)z=e^{2X}+e^{2Y}$

$\therefore \quad C.F.=e^X f_1(Y-X)+e^{nX}f_2(Y-X)$

$$=xf_1(\log y-\log x)+x^n f_2(\log y-\log x)$$

$$=xf_1\left(\log\frac{y}{x}\right)+x^n f_2\left(\log\frac{y}{x}\right)$$

$$=x\phi_1\left(\frac{y}{x}\right)+x^n\phi_2\left(\frac{y}{x}\right).$$

Further,

$$P.I.=\frac{1}{(D+D')(D+D'-n)}\cdot e^{2X}$$

$$+\frac{1}{(D+D'-1)(D+D'-n)}\cdot e^{2Y}$$

$$=\frac{e^{2X}}{(2+0-1)(2+0-n)}$$

$$+\frac{e^{2Y}}{(0+2-1)(0+2-n)}$$

$$=\frac{e^{2X}+e^{2Y}}{-n}=\frac{x^2+y^2}{2-n}$$

$\therefore$ The solution is

$$z=x\phi_1\left(\frac{y}{x}\right)+x^n\phi_2\left(\frac{y}{x}\right)+\frac{x^2+y^2}{2-n}.$$

EXAMPLE 5.79

Solve $x^2\dfrac{\partial^2 z}{\partial x^2}-4xy\dfrac{\partial^2 z}{\partial x\partial y}+4y^2\dfrac{\partial^2 z}{\partial y^2}+6y\dfrac{\partial z}{\partial y}=x^3y^4.$

Solution. Substituting $x=e^X$, $y=e^Y$ and denoting $\dfrac{\partial}{\partial X}$ and $\dfrac{\partial}{\partial Y}$ by D and D', the equation reduces to

$$[D(D-1)-4DD'+4D'(D'-1)+6D']z$$

$$=e^{3X+4Y}$$

or $\quad (D-2D')(D-2D'-1)z=e^{3X+4Y}$

$$\text{C.F.}=f_1(Y+2X)+e^X f_2(Y+2X)$$

$$=f_1(\log y+2\log x)+xf_2(\log y+2\log x)$$

$$=f_1(\log yx^2)+xf_2(\log yx^2)$$

$$=\phi_1(yx^2)+x\phi_2(yx^2).$$

$$P.I.=\frac{1}{(D-2D')(D-2D'-1)}\cdot e^{3X+4Y}$$

$$=\frac{e^{3X+4Y}}{(5-8)(3-8-1)}=\frac{1}{30}x^3y^4$$

$\therefore$ The solution is

$$z=\phi_1(yx^2)+x\phi_2(yx^2)+\frac{1}{30}x^3y^4.$$

5.8 CLASSIFICATION OF SECOND ORDER LINEAR PARTIAL DIFFERENTIAL EQUATIONS

Consider the second order linear partial differential equation

$$A\frac{\partial^2 u}{\partial x^2}+B\frac{\partial^2 u}{\partial x\partial y}+C\frac{\partial^2 u}{\partial y^2}+D\frac{\partial u}{\partial x}+E\frac{\partial u}{\partial y}+F(u)=0 \qquad (103)$$

where A, B, C, D, E, and F are real constants. The equation (103) is said to be

(i) hyperbolic if $B^2-4AC>0$

(ii) parabolic if $B^2-4AC=0$

(iii) elliptic if $B^2-4AC<0$.

For example, the equation

$$\frac{\partial^2 u}{\partial x^2}-\frac{\partial^2 u}{\partial y^2}=0 \text{ (special case of wave equation)}$$

is hyperbolic, since A = 1, B = 0, C = –1, and $B^2-4AC=4>0$. This equation is satisfied by small transverse displacement of the points of a vibrating string.

On the other hand, the equation

$$\frac{\partial^2 u}{\partial x^2}+\frac{\partial^2 u}{\partial y^2}=0$$

(two-dimensional Laplace equation)

is *elliptic* since A = 1, B = 0, C = 1, and $B^2-4AC=-4<0$. This equation is satisfied by the steady temperature at points of a thin rectangular plate. Similarly, the *heat equation*

$$\frac{\partial u}{\partial t}=c\frac{\partial^2 u}{\partial x^2}$$

is *parabolic.*

5.9 THE METHOD OF SEPARATION OF VARIABLES

The method of separation of variables is a very powerful method for obtaining solutions for certain problems involving partial differential equations. Problems those are of great physical interest can be solved by this method. For example, wave equation, heat equation, and Laplace equation can be solved by this method.

This method involves a solution which breaks up into a product of functions, each of which contains only one of the variables. If the partial differential equation involves n independent variables $x_1, x_2, \ldots, x_n$, we first assume that the equation possesses product solution of the form $X_1 X_2 \ldots X_n$, where X_i is a function of only x_i ($i = 1, 2, \ldots, n$). This basic assumption will produce ordinary differential equations, one in each of the unknown functions X_i($i = 1, 2, \ldots, n$). We solve these n ordinary differential equations, which may also involve initial or boundary conditions. The solution of these n equations will produce particular solutions of the form $X_1 X_2 \ldots X_n$ satisfying some supplementary conditions of the original problem. Then these particular solutions are combined by superposition rule to produce a solution of the problem.

EXAMPLE 5.80

Solve $\dfrac{\partial^2 z}{\partial x^2}-2\dfrac{\partial z}{\partial x}+\dfrac{\partial z}{\partial t}=0$

Solution. Let

$$z = X(x)\,Y(y), \qquad (104)$$

be a trial solution of the given partial differential equation. Then the given equation reduces to

$X''(x)\,Y(y) - 2X'(x)Y(y) + X(x)\,Y'(y) = 0$

Separating the variables, we get

$$\frac{X''(x)-2X'(x)}{X(x)}=-\frac{Y'(y)}{Y(y)} \qquad (105)$$

The left-hand side is a function of x only, whereas the right-hand side is a function of y only. Since x and y are different variables, equality in (105) can occur only if the left-hand side and right-hand side are both equal to a constant, say a. Thus, we get two ordinary differential equations

$$X''(x)-2X'(x)-aX(x)=0 \qquad (106)$$

and

$$Y'(y)+aY(y)=0 \qquad (107)$$

The auxiliary equation for (106) is

$$m^2 - 2m - a = 0,$$

which yields $m = 1 \pm \sqrt{1 + a}$. Therefore, the solution of (106) is

$$X(x) = c_1 e^{(1 + \sqrt{1+a})x} + c_2 e^{(1 + \sqrt{1+a})x}.$$

The auxiliary equation for (107) is

$$m + a = 0$$

and so $m = -a$. Therefore, the solution of (107) is

$$Y(y) = c_3 e^{-ay}.$$

Substituting the value of $X(x)$ and $Y(y)$ into (104), we get

$$z = X(x)\, Y(y) = [c_1 e^{(1 + \sqrt{1+a})x}$$

$$+ c_2 e^{(1 - \sqrt{1+a})x}]\, c_3 e^{-ay}$$

$$= [c_4 e^{(1 + \sqrt{1+a})x} + c_5 e^{(1 - \sqrt{1+a})x}]\, e^{-ay},$$

which is the required solution of the given differential equation.

EXAMPLE 5.81

Solve $\frac{\partial u}{\partial x} = 2\frac{\partial u}{\partial t} + u$, subject to the condition $u(x, 0) = 6e^{-3x}$.

Solution. The given equation involves two independent variables x and t. So let

$$u(x, t) = X(x)T(t). \quad (108)$$

Then the given equation transforms to

$$X'(x)\, T(t) = 2X(x)\, T'(t) + X(x)\, T(t)$$

or

$$X'(x)\, T(t) = (2T'(t) + T(t))X(x).$$

Separating variables, we obtain

$$\frac{X'(x)}{X(x)} = \frac{2T'(t) + T(t)}{T(t)}. \quad (109)$$

Since x and t are different variables, the left-hand side and right-hand side are both equal to a constant, say a. Therefore, we get two ordinary differential equations

$$\frac{X'(x)}{X(x)} = a \quad (110)$$

and

$$2T'(t) + T(t) - aT(t) = 0. \quad (111)$$

The solution of (110) is

$$\log X(x) = ax + \log c_1$$

or

$$\log \frac{X(x)}{c_1} = ax$$

or

$$X(x) = c_1 e^{ax} \quad (112)$$

Equation (111) can be written as

$$2T'(t) = T(t)\,(a - 1)$$

or

$$\frac{T'(t)}{T(t)} = \frac{1}{2}\,(a - 1)$$

and so its solution is

$$\log T(t) = \frac{1}{2}\,(a - 1)t + \log c_2$$

or

$$\log \frac{T(t)}{c_2} = \frac{1}{2}\,(a - 1)t$$

or

$$T(t) = c_2\, e^{\frac{1}{2}(a-1)t} \quad (113)$$

Putting the values of $X(x)$ and $T(t)$ obtained from (112) and (113) in (108), the solution is

$$u(x, t) = X(x)\, T(t) = c_1 c_2 e^{ax} \cdot e^{\frac{1}{2}(a-1)t}.$$

Using the initial condition $u(x, 0) = 6e^{-3x}$, we get $6e^{-3x} = c_1 c_2 e^{ax}$.

So

$$c_1 c_2 = 6 \text{ and } a = -3.$$

Hence, the solution of the given partial differential equation is

$$u(x, t) = 6e^{-3x}\, e^{-2t} = 6e^{-(3x + 2t)}.$$

EXAMPLE 5.82

Solve $4\dfrac{\partial u}{\partial x} + \dfrac{\partial u}{\partial y} = 3u$ subject to the condition that $u(0, y) = 3e^{-y} - 5e^{-5y}$.

Solution. The given equation involves two variables x and y. So, let

$$u(x, y) = X(x)\, Y(y). \tag{114}$$

Then the given equation reduces to

$$4X'(x)\ Y(y) + X(x)\ Y'(y) = 3X(x)\ Y(y)$$

or equivalently,

$$4\frac{X'(x)}{X(x)} = \frac{3Y(y) - Y'(y)}{Y(y)}$$

Since x and y are independent variables, the lefthand side and right-hand side are both equal to some constant, say a. Thus, we get two differential equations

$$4\frac{X'(x)}{X(x)} = a, \tag{115}$$

and

$$\frac{3Y(y) - Y'(y)}{Y(y)} = a \tag{116}$$

The solution of (115) is

$$\log X(x) = \frac{a}{4}x + \log c_1$$

or

$$\frac{X(x)}{c_1} = e^{\frac{a}{4}x}$$

or

$$X(x) = c_1 e^{\frac{a}{4}x}. \tag{117}$$

The equation (116) can be written as

$$\frac{Y'(y)}{Y(y)} = 3 - a$$

and so its solution is

$$\log Y(y) = (3 - a)y + \log c_2$$

or

$$Y(y) = c_2\, e^{(3-a)y}. \tag{118}$$

Using (117) and (118), the solution (114) reduce to

$$u(x, y) = c_1 c_2\, e^{\frac{a}{4}x} \,.\, e^{(3-a)y}.$$

Therefore, using the initial condition $u(0, y) = 3e^{-y} - e^{-5y}$, we get

$$3e^{-y} - e^{-5y} = c_1 c_2\, e^{(3-a)y}.$$

Now

$$3e^{-y} - e^{-5y} = c_1 c_2\, e^{-(a-3)y} + 0\ e^{-5y}$$

gives one value of set as

$$c_1\, c_2 = 3 \text{ and } a - 3 = 1,$$

that is,

$$c_1\, c_2 = 3 \text{ and } a = 4.$$

The other set of values is given by

$$3e^{-y} - e^{-5y} = 0.e^{-y} + c_1 c_2 e^{-(a-3)y}$$

and that set is

$$c_1\, c_2 = -1 \text{ and } a - 3 = 5,$$

that is,

$$c_1\, c_2 = -1 \text{ and } a = 8,$$

Thus we get two solutions

$$u_1(x, y) = 3e^{\frac{4}{4}x}.e^{(3-4)y} = 3e^{x}.\, e^{-y} = 3e^{x-y}$$

and

$$u_2(x, y) = -1\ e^{\frac{8}{4}x}.\, e^{(3-8)y} = -e^{2x}.\, e^{-5y}$$

$$= -e^{2x-5y}.$$

Hence the required solution is

$$u(x, y) = u_1(x, y) + u_2(x, y)$$

$$= 3e^{x-y} - e^{2x-5y}.$$

5.10 CLASSICAL PARTIAL DIFFERENTIAL EQUATIONS

(A) One-Dimensional Heat Equation

Consider the flow of heat through a homogeneous metallic thin bar with area of cross-section A and with insulated sides so that the heat flows only in one direction perpendicular to an end of the bar. We take this end as origin and direction of heat flow as positive x-axis. The temperature u at any point of the bar depends on the distance x of the point from the fixed end, taken as the origin, and time t. We further assume that the temperature at all points of same cross-section is same.

The quantity of heat flow per second across any area of cross-section A is proportional to the area A and the rate of change of temperature with respect to distance x, normal to A. Thus, if, Q_1 is the quantity of heat that flows across the cross-section A, then

$$Q_1 = -kA\left(\frac{\partial u}{\partial x}\right)_x \text{ per second,}$$

where k is the coefficient of conductivity. We have taken negative sign on the right side because as x increases, u decreases.

Similarly, the quantity Q_2 of heat that flows per second across the cross-section at a distance $x + \delta x$ is given by

$$Q_2 = -kA\left(\frac{\partial u}{\partial x}\right)_{x+\delta x} \text{ per second.}$$

Hence the amount of heat retained per second by the slab of thickness x is

$$Q_1 - Q_2 = kA\left[\left(\frac{\partial u}{\partial x}\right)_{x+\delta x} - \left(\frac{\partial u}{\partial x}\right)_x\right].$$

But the rate of change of heat in a solid is $s\rho A \delta x \frac{\partial u}{\partial t}$, where s is specific heat of the material, ρ is the density, $A\delta x$ is the volume, and $\frac{\partial u}{\partial t}$ is rate of change of temperature with time. Hence

$$s\rho A\delta x \frac{\partial u}{\partial t} = kA\left[\left(\frac{\partial u}{\partial x}\right)_{x+\delta x} - \left(\frac{\partial u}{\partial x}\right)_x\right]$$

or

$$\frac{\partial u}{\partial t} = \frac{k}{s\rho}\left[\left(\frac{\partial u}{\partial x}\right)_{x+\delta x} - \left(\frac{\partial u}{\partial x}\right)_x\right] / \delta x$$

Taking limit as $\delta x \to 0$, we get

$$\frac{\partial u}{\partial t} = c^2 \frac{\partial^2 u}{\partial x^2}, \qquad (119)$$

where $c^2 = \frac{k}{s\rho}$ is called *diffusivity* of the bar material. Equation (119) is called *heat flow equation or heat conduction equation or onedimensional heat equation.*

We now solve *one-dimensional heat equation* using method of separation of variables.

EXAMPLE 5.83

Solve one dimensional *heat equation*

$$\frac{\partial u}{\partial t} = k\frac{\partial^2 u}{\partial x^2},\ x \in (0, L),$$

with initial condition $u(x, 0) = f(x)$ and the boundary conditions $u(0,t) = 0$, $u(L, t) = 0$, $t \geq 0$.

Solution. The boundary conditions show that the temperature at both ends of the rod is kept at zero. We shall use the method of separation of variables. So let,

$$u(x, t) = T(t)\, X(x).$$

Then, the given heat equation reduces to

$$T'(t)\, X(x) = k\, X''(x)\, T(t),$$

which gives

$$\frac{T'(t)}{T(t)} = k\frac{X''(x)}{X(x)}. \qquad (120)$$

The left-hand side of (120) is a function of t only whereas the right-hand side is a function of x only. Since t and x are independent variables, equality in (120) can occur only if the left-hand side and right-hand side are both equal to a constant, say C. Thus, we have

$$\frac{T'(t)}{T(t)} = k\frac{X''(x)}{X(x)} = C \text{ (constant of separation).}$$

The boundary conditions imply that

$$u(0, t) = T(t)\,X(0) = 0 \tag{121}$$

and

$$u(L, t) = T(t)\,X(L) = 0 \tag{122}$$

Since we do not want trivial solution $T(t) = 0$, the relations (121) and (122) yield $X(0) = 0$ and $X(L) = 0$.

If $C = 0$, then $X''(x) = 0$ and so $X(x) = ax + b$. Therefore, $X(0) = 0$ implies $b = 0$ and $X(L) = 0$ implies $aL = 0$. Thus $a = b = 0$ and as such $X(x)$ is a trivial solution. Hence $C = 0$ is discarded.

Let $C \neq 0$, then the characteristic equation of $X''(x) - \frac{C}{k}X(x) = 0$ is $s^2 - \frac{C}{k} = 0$ which yields two roots s_1 and s_2 such that $s_2 = -s_1$. The general solution is, therefore,

$$X(x) = \alpha\, e^{s_1 x} + \beta\, e^{-s_1 x} \tag{123}$$

The boundary condition $X(0) = 0$ implies $\alpha + \beta = 0$ and so $\beta = -\alpha$.

The boundary condition $X(L) = 0$ implies $\alpha(e^{s_1 L} - e^{-s_1 L}) = 0$ (since $\beta = -\alpha$).

Now α cannot be zero, because $\alpha = 0$ gives $\beta = 0$ and so (123) has trivial solution. Hence $e^{s_1 L} - e^{-s_1 L} = 0$, which yields $e^{2s_1 L} = 1$ and so $s_1 = \frac{in\pi}{L}$, where $n \neq 0$ is an integer. Thus $\frac{C}{k} = s_1^2$ yields $C = \frac{-kn^2\pi^2}{L^2}$. Hence (123) reduces to

$$X_n(x) = \alpha\left[e^{\frac{in\pi x}{L}} - e^{-\frac{in\pi x}{L}}\right] = 2\alpha\left[\sin\frac{n\pi x}{L}\right]$$

Also the equation $T'(t) - CT(t) = 0$ has characteristic equation as $s - C = 0$ and so $s = C$. So the fundamental solution is

$$T_n(t) = \alpha_0\, e^{-Ct} = \alpha_0 e^{-\frac{kn^2\pi^2}{L^2}t} = \alpha_0 e^{-\left(\frac{n\pi}{L}\right)^2 kt}.$$

Hence the fundamental solution of the heat equation is given by

$$\begin{aligned} u_n(x,t) &= T_n(t)X_n(x) \\ &= A_n e^{-\left(\frac{n\pi}{L}\right)^2 kt}\sin\frac{n\pi x}{L} \end{aligned}$$

for $n = 1, 2, \ldots;\ A_n = 2\alpha\alpha_0$.

By the principle of superposition (principle of adding all solutions for $n = 1, 2, 3, \ldots,$), we have

$$u(x,t) = \sum_{n=1}^{\infty} A_n e^{-\left(\frac{n\pi}{L}\right)^2 kt}\sin\left(\frac{n\pi x}{L}\right).$$

Using initial condition $u(x, 0) = f(x)$, we have

$$f(x) = \sum_{n=1}^{\infty} A_n \sin\left(\frac{n\pi x}{L}\right), \quad 0 \le x \le L. \tag{124}$$

Thus the coefficients A_n are nothing but the Fourier sine coefficients of the function $f(x)$ with respect to the system $\sin\frac{n\pi x}{L}$, that is,

$$A_n = \frac{2}{L}\int_0^L f(x)\sin\left(\frac{n\pi x}{L}\right)dx.$$

Substituting the value of A_n in $u(x, t)$ shall yield the solution of the given heat equation. For example, if $u(x, 0) = x$, $0 < x < 2\pi$, then

$$A_n = \frac{1}{\pi}\int_0^{2\pi} x\sin\frac{nx}{2}\,dx = -\frac{4}{n}(-1)^n,$$

and so in that case the solution of the heat equation becomes

$$\begin{aligned} u(x,t) &= \sum_{n=1}^{\infty} -\frac{4}{n}(-1)^n e^{-n^2\pi^2 kt/L^2}\sin\left(\frac{n\pi x}{L}\right), \\ &= \sum_{n=1}^{\infty} -\frac{4}{n}(-1)^n e^{-n^2 kt/4}\sin\left(\frac{nx}{2}\right) \end{aligned}$$

since $L = 2\pi$

EXAMPLE 5.84

Solve

$$\frac{\partial u}{\partial t} = k\frac{\partial^2 u}{\partial x^2},\ 0 < x < 2\pi$$

with the condition

$$u(x, 0) = x^2,\ u(0, t) = u(2\pi, t) = 0$$

Solution. From Example 5.83, we have

$$u(x,t) = \sum_{n=1}^{\infty} A_n\, e^{-n^2\pi^2 kt/L^2} \sin\frac{n\pi x}{L},\quad L = 2\pi$$

$$= \sum_{n=1}^{\infty} A_n e^{-n^2 kt/L^2} \sin\frac{nx}{2},$$

where

$$A_n = \frac{1}{\pi}\int_0^{2\pi} x^2 \sin\frac{nx}{2}\, dx$$

$$= \frac{-2}{\pi n}\left[x^2 \frac{nx}{2}\right]_0^{2\pi} + \frac{4}{\pi n}\int_0^{2\pi} x\cos\frac{nx}{2}\, dx$$

$$= \frac{8\pi}{n}(-1)^n + \frac{8}{\pi n^2}\left[x\sin\frac{nx}{2}\right]_0^{2\pi} - \frac{8}{\pi n^2}\int_0^{2\pi} \sin\frac{nx}{2}\, dx$$

$$= \frac{8}{n}(-1)^n\left[\frac{2}{\pi n^2} - \pi\right].$$

Hence

$$u(x,t) = \sum_{n=1}^{\infty} \frac{8}{n}(-1)^n\left[\frac{2}{\pi n^2} - \pi\right] e^{-n^2 kt/4} \sin\frac{nx}{2}.$$

EXAMPLE 5.85

Solve the heat conduction equation

$$\frac{\partial u}{\partial t} = k\frac{\partial^2 u}{\partial x^2},\ 0 < x < L,\ t > 0$$

with the boundary condition $ux(0, t) = 0$, $u_x(L, t) = 0$, $t \ge 0$ and the initial condition $u(x, 0) = f(x)$, $0 \le x \le L$.

Solution. The boundary conditions show that the ends of the rod are insulated. As in Example 5.83,

we have

$$\frac{T'(t)}{T(t)} = k\frac{X''(x)}{X(x)} = C \tag{125}$$

Differentiating $u(x, t) = T(t)\, X(x)$ with respect to x, we have

$$u_x(x, t) = T(t)X'(x).$$

Therefore, the boundary value conditions yield

$$u_x(0, t) = T(t)\, X'(0) = 0.$$

and

$$u_x(L, t) = T(t)\, X'(L) = 0,$$

and so

$$X'(0) = X'(L) = 0$$

Now if $C = 0$, then (125) implies that $X''(x) = 0$ and so $X(x) = ax + b$. Therefore, $X'(x) = a$ and so $X'(0) = a = 0$ and so $X(x) = b$. Therefore, $C = 0$ is an eigenvalue with eigen function a constant.

If $C \ne 0$, then the characteristic equation of $X''(x) - \frac{C}{k}X(x) = 0$ is $s^2 - \frac{C}{k} = 0$ which yield, two roots s_1 and s_2 with $s_2 = -s_1$. Therefore, the general solution is

$$X(x) = a\, e^{s_1 x} + \beta\, e^{-s_1 x} \tag{126}$$

We have now from (126),

$$X'(x) = a\, s_1 e^{s_1 x} - \beta\, e^{-s_1 x}$$

So the boundary condition $X'(0) = 0$ and $X'(L) = 0$ implies

$$a - \beta = 0 \text{ yielding } a = \beta,$$

and

$$a\,s_1(e^{s_1L} - e^{-s_1L}) = 0.$$

If $\alpha = 0$, then $\beta = 0$, and so we will have a trivial solution. Therefore, $e^{s_1L} - e^{-s_1L} = 0$, which gives $e^{2s_1L} = 1$ and so $s_1 = \frac{in\pi}{L}$. Then $\frac{C}{k} = s_1^2$ implies $C = -\frac{kn^2\pi^2}{L^2}$. Hence (126) becomes

$$\begin{aligned} X_n(x) &= \alpha[e^{in\pi x/L} + e^{-in\pi x/L}] \\ &= 2\alpha\cos\frac{n\pi x}{L}. \end{aligned}$$

Moreover (see Example 5.83), $T'(t) - CT(t) = 0$ has general solution as

$$T_n(t) = \alpha_0\, e^{-n^2\pi^2kt/L^2}.$$

Thus, the general solution of the heat equation becomes

$$\begin{aligned} u_n(x,t) &= T_n(t)X_n(x) \\ &= \frac{A_0}{2} + A_n e^{-\frac{n^2\pi^2kt}{L^2}}\cos\frac{n\pi x}{L}. \end{aligned}$$

Now superposition of general solutions yield

$$u(x,t) = \frac{A_0}{2} + \sum_{n=1}^{\infty} A_n e^{-\frac{n^2\pi^2kt}{L^2}}\cos\frac{n\pi x}{L}. \tag{127}$$

The initial condition $u(x, 0) = f(x)$, $0 \le x \le L$ gives

$$f(x) = \frac{A_0}{2} + \sum_{n=1}^{\infty} A_n\cos\frac{n\pi x}{L},\quad 0 \le x \le L.$$

Thus, the Fourier coefficients A_n of $f(x)$ with respect to the system $\cos\frac{n\pi x}{L}$ are given by

$$A_n = \frac{2}{L}\int_0^L f(x)\cos\frac{n\pi x}{L}\,dx, n = 0, 1, 2, \ldots$$

substituting the value of A_n in (127) will yield the solution of the given heat conduction equation.

EXAMPLE 5.86

An insulated rod of length l has its ends A and B maintained at 0°C and 100°C, respectively, until steady state condition prevails. If B is suddenly reduced to 0°C and maintained at 0°C, find the temperature at a distance x from A at time t.

Solution The heat flow equation is

$$\frac{\partial u}{\partial t} = c^2\frac{\partial^2 u}{\partial x^2}.$$

Prior to the sudden change of temperature at the end B, the temperature u depends only upon x and not on t. Hence the equation of heat flow is $\frac{\partial^2 u}{\partial x^2} = 0$, whose solution is $u = ax + b$. Since $u = 0$ for $x = 0$ and $u = 100$ for $x = l$, we get $0 = b$ and $100 = al$ or $a = \frac{100}{l}$. Thus $u(x) = \frac{100}{l}x$. Hence $u(x) = \frac{100}{l}x$ gives the temperature at $t = 0$, that is, $u(x, 0) = \frac{100}{l}x$ is the initial condition. The boundary conditions for the subsequent flow are $u(0, t) = 0$ and $u(l, t) = 0$ for all values of t. Therefore, by Example 5.83, the solution of the problem is

$$u(x,t) = \sum_{n=1}^{\infty} A_n\, e^{-\left(\frac{n\pi}{l}\right)^2 c^2 t}\sin\frac{n\pi x}{l},$$

where

$$A_n = \frac{2}{l}\int_0^l f(x)\,\sin\frac{n\pi x}{l}\,dx.$$

Here $f(x) = \frac{100x}{l}$. Therefore,

$$\begin{aligned} A_n &= \frac{2}{l}\int_0^l \frac{100}{l}x\sin\frac{n\pi x}{l}\,dx \\ &= \frac{200}{l^2}\int_0^l x\sin\frac{n\pi x}{l}\,dx \\ &= \frac{200}{l^2}\left(-\frac{l^2}{n\pi}\cos n\pi\right), \text{ using integration by parts} \\ &= \frac{200}{n\pi}(-1)^{n+1}, \text{ since } \cos n\pi = (-1)^n. \end{aligned}$$

Hence

$$u(x,t) = \frac{200}{\pi}\sum_{n=1}^{\infty}\frac{(-1)^{n+1}}{n}\sin\frac{n\pi x}{l}\,e^{-\frac{c^2n^2\pi^2t}{l^2}}.$$

EXAMPLE 5.87

A bar 10 cm long with insulated sides has its ends A and B maintained at temperature 50°C and 100° C, respectively, until steady state condition prevails. The temperature at A is suddenly raised to 90°C and at the same time lowered to 60°C at B. Find the temperature distribution in the bar at time t.

Solution. Prior to sudden change, the temperature distribution is described by $\frac{\partial^2 u}{\partial x^2} = 0$, whose solution is

$$u = ax + b.$$

Since $u = 50°C$ for $x = 0$ and $u = 100°C$ for $x = 10$ cm, we get $50 = b$ and $100 = 10a + b = 10a + 50$ and so $a = 5$. Hence the initial temperature distribution in the rod is $u(x, 0) = 5x + 50$. Similarly, taking $u = 90°C$ for $x = 0$ and $u = 60°C$ for $x = 10$ the final temperature distribution in the rod is $u_s(x, 0) = 3x + 90$. We want to find the temperature distribution during the intermediate period, measuring time from the instant when the end temperature was changed. Let

$$u(x, t) = u_s(x) + u_t(x, t),$$

where $u_s(x)$ is the steady state temperature and $u_t(x, t)$ is the transient temperature distribution which decreases as time increases. Thus

$$u_s(x) = -3x + 90,$$

and $u_t(x, t)$ satisfies the heat equation

$$\frac{\partial u}{\partial t} = c^2 \frac{\partial^2 u}{\partial x^2}.$$

Hence the solution is

$$u(x, t) = -3x + 90 + \sum_{n=1}^{\infty} A_n \sin \frac{n\pi x}{l} e^{\frac{-c^2 n^2 \pi^2 t}{l^2}}$$

Here $l = 10$ and A_n are determined by

$$A_n = \frac{2}{10} \int_0^{10} f(x) \sin \frac{n\pi x}{l} dx,$$

where $f(x) = u_t(x, 0) = u(x, 0) - u_s(x, 0) = 5x + 50 - (-3x + 90) = 8x - 40$. Thus

$$A_n = \frac{1}{5} \int_0^{10} (8x - 40) \sin \frac{n\pi x}{10} dx$$

$$= \frac{1}{5} \left[(8x - 40) \left(-\frac{10}{n\pi} \cos \frac{n\pi x}{10} \right) - 8 \left(-\frac{100}{n^2\pi^2} \sin \frac{n\pi x}{10} \right) \right]_0^{10}$$

$$= \frac{1}{5} \left[-\frac{400}{n\pi} \cos n\pi - \frac{400}{n\pi} \right]$$

$$= -\frac{80}{n\pi} (\cos n\pi + 1)$$

$$= \begin{cases} 0 & \text{if } n \text{ is odd} \\ -\frac{160}{n\pi} & \text{if } n \text{ is even} \end{cases}$$

Hence

$$u(x,t) = -3x + 90 - \frac{160}{\pi} \sum_{\text{even}} \frac{1}{n} e^{-\frac{c^2 n^2 \pi^2 t}{100}} . \sin \frac{n\pi x}{10}$$

$$= -3x + 90 - \frac{160}{\pi} \sum_{m=1}^{\infty} \frac{1}{2m} e^{-\frac{c^2 m^2 \pi^2 t}{25}} . \sin \frac{m\pi x}{5}$$

$$= -3x + 90 - \frac{80}{\pi} \sum_{m=1}^{\infty} \frac{1}{m} e^{-\frac{c^2 m^2 \pi^2 t}{25}} . \sin \frac{m\pi x}{5}.$$

EXAMPLE 5.88

A rod of length l with insulated side is initially at a uniform temperature u_0. Its ends are suddenly cooled at 0°C and kept at that temperature. Find the temperature function $u(x, t)$.

Solution. We have to solve the heat equation

$$\frac{\partial u}{\partial t} = c^2 \frac{\partial^2 u}{\partial x^2}$$

under the initial condition $u(x, 0) = u_0$, $u(0, t) = 0$, $u(l, t) = 0$ because both ends are kept at zero temperature. Therefore, by Example 5.83 the solution is

$$u(x, t) = \sum_{n=1}^{\infty} A_n e^{\frac{-c^2 n^2 \pi^2 t}{l^2}} \cdot \sin \frac{n\pi x}{l},$$

where

$$A_n = \frac{2}{l} \int_0^l u(x, o) \sin \frac{n\pi x}{l} dx$$

$$= \frac{2}{l} \int_0^l u_0 \sin \frac{n\pi x}{l} dx$$

$$= \begin{cases} 0 & \text{for even } n \\ \frac{4u_0}{n\pi} & \text{For odd } n. \end{cases}$$

Hence the temperature distribution is given by

$$u(x, t) = \frac{4u_0}{\pi} \sum_{odd\ n} \frac{1}{n} e^{\frac{-c^2 n^2 \pi^2 t^2}{l^2}} \cdot \sin \frac{n\pi x}{l}$$

$$= \frac{4u_0}{\pi} \sum_{n=1}^{\infty} \frac{1}{n} e^{\frac{-c^2 n^2 \pi^2 t^2}{l^2}} \cdot \sin \frac{n\pi x}{l} \sin \frac{(2n-1)\pi x}{l}.$$

EXAMPLE 5.89

Find the temperature $u(x, t)$ in a bar which is perfectly insulated laterally, whose ends are kept at temperature 0°C and whose initial temperature is $f(x) = x(100 - x)$, given that its length is 10 cm, constant cross-section of area 1 cm^2, density 10.6 gm/cm^3, thermal conductivity 1.04 cal/cm deg sec, and specific heat 0.056 cal/gm deg.

Solution. As in Example 5.83, the temperature distribution is given by

$$u(x, t) = \sum_{n=1}^{\infty} A_n e^{\frac{-c^2 n^2 \pi^2 t}{l^2}} \cdot \sin \frac{n\pi x}{l}.$$

Here $l = 10$ cm, $c^2 = \frac{k}{s} = 1.75$ and A_n are given below

$$A_n = \frac{2}{l} \int_0^l u(x, 0) \sin \frac{n\pi x}{l} dx$$

$$= \frac{1}{5} \int_0^{100} x(100 - x) \sin \frac{n\pi x}{l} dx$$

$$= \frac{400}{n^3 \pi^3} [1 - (-1)^n]$$

$$= \begin{cases} 0 & \text{for even } n \\ \frac{800}{(n^3\pi^3)} & \text{for odd } n. \end{cases}$$

Hence

$$u(x, y) = \frac{800}{\pi^3} \sum_{odd\ n} \frac{1}{n^3} e^{-0.0175 n^2 \pi^2 t} \cdot \sin \frac{n\pi x}{10}$$

$$= \frac{800}{\pi^3} \sum_{n=1}^{\infty} \frac{1}{(2n-1)^3} e^{-0.0175(2n-1)^2 \pi^2 t} \times \left(\sin \frac{(2n-1)\pi x}{10} \right).$$

(B) One-dimensional Wave Equation

Consider a uniform elastic string of length L stretched tightly with its ends fixed on the x-axis at $x = 0$ and $x = L$. Let for each x in the interval $0 < x < L$, the string is displaced into the xy-plane and let for each x, the displacement from the x-axis be given by $f(x)$, where f is a function of x (Fig. 5.1).

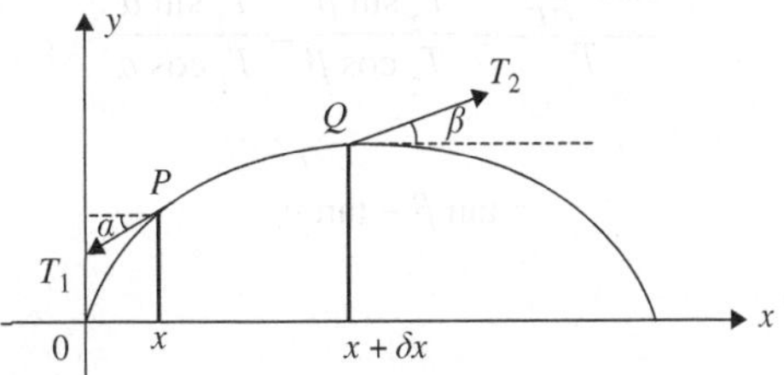

Figure 5.1

We assume that the string perfectly flexible, is of constant linear density ρ and of constant tension T at all times. We also assume that the motion takes place entirely in the *xy*-plane and that each point on the string moves on a straight line perpendicular to the *x*-axis as the string vibrates. The displacement *y* at each point of the string and the slope $\frac{\partial y}{\partial x}$ are small compared to the length L. Further, no external force acts upon the string during motion and angle between the string and the *x*-axis at each point is sufficiently small.

Let *m* be the mass per unit length of the string and let *PQ* be an element of length δs of the string. Then mass of *PQ* is $m\ \delta s$. The tensions T_1 and T_2 at *P* and *Q* are respectively tangential to the curve (Figure 5.1) making angles α and β with the horizontal direction.

Since there is no motion in the horizontal direction, we have

$$T_1 \cos \alpha = T_2 \cos \beta = \text{constant} = T, \text{ say. } (128)$$

By Newton's Second Law of Motion, the equation of motion in the vertical direction is

$$m\delta s \frac{\partial^2 y}{\partial t^2} = T_2 \sin \beta - T_1 \sin \alpha$$

or

$$\frac{m\delta s \frac{\partial^2 y}{\partial t^2}}{T} = \frac{T_2 \sin \beta}{T} - \frac{T_1 \sin \alpha}{T}$$

Using (128), we get

$$\frac{m\delta s \frac{\partial^2 y}{\partial t^2}}{T} = \frac{T_2 \sin \beta}{T_2 \cos \beta} - \frac{T_1 \sin \alpha}{T_1 \cos \alpha}$$

$$= \tan \beta - \tan \alpha$$

or

$$\frac{\partial^2 y}{\partial t^2} = \frac{T}{m\delta s} [\tan \beta - \tan \alpha]$$

$$= \frac{T}{m\delta s} [\tan \beta - \tan \alpha],\ \delta s \approx \delta x$$

$$= \frac{T}{m} \left[\frac{\left(\frac{\partial y}{\partial x}\right)_{x+\delta x} - \left(\frac{\partial y}{\partial x}\right)_x}{\delta x} \right]$$

$$= \frac{T}{m} \frac{\partial^2 y}{\partial x^2} \text{ in the limit as } \delta x \to 0$$

Hence

$$\frac{\partial^2 y}{\partial t^2} = a^2 \frac{\partial^2 y}{\partial x^2},\ 0 < x < \text{L}, t > 0. \qquad (129)$$

where $a^2 = T/m$ is a constant related to tension in the vibrating string of length L having fixed ends. The boundary conditions are

$$y(0, t) = y(\text{L}, t) = 0,\ t \geq 0,$$

and initial conditions are

$$y(x, 0) = f(x),\ 0 \leq x \leq \text{L}$$

$$y_t(x, 0) = 0,\ 0 \leq x \leq \text{L}.$$

The equation (129) is called one-dimensional wave equation. We find its solution in the following Example 5.90.

EXAMPLE 5.90

Solve the wave equation

$$\frac{\partial^2 u}{\partial t^2} = a^2 \frac{\partial^2 u}{\partial x^2},\ 0 < x < \text{L}, t > 0.$$

where a is a constant related to tension in the vibrating string of length L having fixed ends. The boundary conditions and initial conditions are

$$u(0, t) = u(L, t) = 0, t \geq 0$$

$$u(x, 0) = f(x), 0 \leq x \leq L$$

$$u_t(x, 0) = 0, 0 \leq x \leq L.$$

Solution. Let

$$u(x, t) = T(t)X(x)$$

Then the wave equation takes the form

$$T''(t)X(x) = a^2 X''(x)T(t)$$

or

$$\frac{T''(t)}{T(t)} = a^2 \frac{X''(t)}{X(t)} = C$$

(constant of separation).

The boundary conditions $u(0, t) = u(L, t) = 0$, $t \geq 0$ imply $X(0) = 0$ and $X(L) = 0$. Further, the condition $u_t(x, 0) = 0$ gives $T'(0)\, X(x) = 0$, $0 \leq x \leq L$. Therefore, $T'(0) = 0$.

The auxiliary equation for $X(x)$ is $s^2 - \frac{C}{a^2} = 0$ which yields $s_2 = -s_1$ as the two roots. Therefore the fundamental (general) solution is

$$X(x) = \alpha e^{s_1 x} + \beta e^{-s_1 x}. \tag{130}$$

The boundary condition $X(0) = 0$ and $X(L) = 0$ gives $\alpha + \beta = 0$, that is, $\beta = -\alpha$ and

$$0 = \alpha\, e^{s_1 L} + \beta e^{-s_1 L} = \alpha(e^{s_1 L} - e^{-s_1 L}).$$

Since $\alpha = 0$ implies $\beta = 0$ and the solution then becomes trivial, so $\alpha \neq 0$ and thus $e^{s_1 L} = e^{-s_1 L}$ and so $e^{2s_1 L} = 1$. Thus $s_1 = \frac{in\pi}{L}$, $n \neq 0$ being an integer. Therefore, $s_1^2 = \frac{C}{a^2}$ yields $C = -\frac{a^2 n^2 \pi^2}{L^2}$. The fundamental solution (130) now takes the form

$$X_n(x) = \alpha[e^{in\pi x/L} - e^{\ in\pi x/L}]$$
$$= 2i\ \alpha \sin \frac{n\pi x}{L}.$$

Further, characteristic equation for $T''(t) - CT(t) = 0$ is $s^2 - C = 0$, that is, $s^2 + \frac{a^2 n^2 \pi^2}{L^2} = 0$. Thus the fundamental solution is

$$T_n(t) = \alpha \cos\left(\frac{n\pi a t}{L}\right) + \beta \sin\left(\frac{n\pi a t}{L}\right).$$

We have

$$T'(t) = \frac{n\pi\alpha}{L}\left(-\alpha \sin \frac{n\pi a t}{L} + \beta \cos \frac{n\pi a t}{L}\right).$$

The condition $T'(0) = 0$ yields $\beta = 0$ and so

$$T_n(t) = \alpha \cos\left(\frac{n\pi\alpha t}{L}\right), \quad n = 1, 2, 3, \ldots$$

Hence the fundamental solution for the wave equation is

$$u_n(x, t) = T_n(t)\, X_n(x)$$

$$= A_n \cos \frac{n\pi a t}{L} \sin \frac{n\pi x}{L}, n = 1, 2, \ldots; A_n = 2a\alpha_0.$$

Superposition of the fundamental solutions implies

$$u(x,t) = \sum_{n=1}^{\infty} A_n \cos\frac{n\pi at}{L}\sin\frac{n\pi x}{L}. \tag{131}$$

Using initial condition $u(x, 0) = f(x)$, we have

$$f(x) = \sum_{n=1}^{\infty} A_n \sin\frac{n\pi x}{L}, \quad 0 \le x \le \mathrm{L}.$$

The Fourier coefficients A_n of $f(x)$ with respect to the system $\sin\frac{n\pi x}{L}$ is given by

$$\mathrm{A}_n = \frac{2}{L}\int_0^L f(x)\sin\frac{n\pi x}{L}\,dx.$$

Putting the values of A_n in (131), we get the solution of the wave equation.

EXAMPLE 5.91

A tightly stretched flexible string has its ends fixed at $x = 0$ and $x = 1$. At time $t = 0$, the string is given a shape defined by $f(x) = \mu x(l - x)$, where m is a constant and then released. Find the displacement of any point x of the string at any time $t > 0$.

Solution. By Example 5.90, we have

$$y(x,t) = \sum A_n \cos\frac{n\pi at}{l}\sin\frac{n\pi x}{l},$$

where

$$A_n = \frac{2}{l}\int_0^1 f(x)\sin\frac{n\pi x}{l}\,dx.$$

Here $f(x) = y(x, 0) = \mu x(1 - x)$. Hence integrating by part, we have

$$\mathrm{A}_n = \frac{2}{l}\int_0^l \mu x(l-x)\sin\frac{n\pi x}{l}\,dx$$

$$= \frac{2\mu}{l}\left[-\frac{2l^3}{n^3\pi^3}\cos n\pi + \frac{2l^3}{n^3\pi^3}\right]$$

$$= \frac{4\mu\, l^2}{n^3\pi^3}[1-(-1)^n]$$

$$= \begin{cases} \frac{8\mu\, l^2}{n^3\pi^3} & \text{for odd } n \\ 0 & \text{for even } n \end{cases}$$

Hence

$$y(x,t) = \frac{8\mu l^2}{\pi^3}\sum_{\text{odd } n}\frac{1}{n^3}\cos\frac{n\pi at}{l}\sin\frac{n\pi x}{l}$$

$$= \frac{8\mu\, l^2}{\pi^3}\sum_{n=1}^{\infty}\frac{1}{(2n-1)^3}\cos\frac{(2n-1)\pi at}{l} \times \sin\frac{(2n-1)\pi x}{l}.$$

EXAMPLE 5.92

A string is stretched and fastened to two points, l distant apart. Motion is started by displacing the string in the form $y = a\sin\frac{\pi x}{l}$ from which it is released at time $t = 0$. Show that the displacement of any point of the string at a distance x from one end at any time t is given by

$$y(x, t) = a \sin \frac{\pi x}{l} \cos \frac{\pi ct}{l}.$$

Solution. By Example 5.90,

$$y(x, t) = \sum_{n=1}^{\infty} A_n \cos \frac{\pi nct}{l} \sin \frac{\pi nx}{l},$$

where

$$A_n = \frac{2}{l} \int_0^l f(x) \sin \frac{n\pi x}{l}\, dx,$$

and $f(x) = u(x, 0)$. Here

$$u(x, 0) = a \sin \frac{\pi x}{l}.$$

Therefore,

$$A_n = \frac{2}{l} \int_0^l a \sin \frac{\pi x}{l} \sin \frac{n\pi x}{l}\, dx$$

$$= \frac{2a}{l} \int_0^l \sin \frac{\pi x}{l} \sin \frac{n\pi x}{l}\, dx,$$

which vanishes for all values of n except $n = 1$. Therefore,

$$A_1 = \frac{2a}{l} \int_0^l \sin^2 \frac{\pi x}{l}\, dx = \frac{a}{l} \int_0^l \left(1 - \cos \frac{2\pi x}{l}\right) dx$$

$$= \frac{a}{l} \left[x - \frac{l}{2\pi} \sin \frac{2\pi x}{l}\right]_0^l = \frac{a}{l}(l) = a.$$

Hence

$$y(x, t) = A_1 \cos \frac{\pi ct}{l} \sin \frac{\pi x}{l} = a \cos \frac{\pi ct}{l} \sin \frac{\pi x}{l}.$$

EXAMPLE 5.93

Show that the solution of the wave equation $\frac{\partial^2 y}{\partial t^2} = c^2 \frac{\partial^2 y}{\partial x^2}$ can be expressed in the form

$$y(x, t) = \phi(x + ct) + \psi(x - ct). \qquad (132)$$

If $y(x, 0) = f(x)$ and $\frac{\partial y}{\partial t}(x, 0) = 0$, show that $y(x, t) = \frac{1}{2}[f(x + ct) + f(x - ct)]$ [The solution (132) is called the *D'Alembert's solution of the wave equation*].

Solution. Put

$$u = x + ct \text{ and } v = x - ct$$

so that $\frac{\partial u}{\partial x} = 1$ and $\frac{\partial v}{\partial x} = 1$, $\frac{\partial u}{\partial t} = c$, $\frac{\partial v}{\partial t} = -c$.

Then y becomes a function of u and v and we have

$$\frac{\partial y}{\partial x} = \frac{\partial y}{\partial u} \cdot \frac{\partial u}{\partial x} + \frac{\partial y}{\partial v} \cdot \frac{\partial v}{\partial x} = \frac{\partial y}{\partial u} + \frac{\partial y}{\partial v}$$

and

$$\frac{\partial^2 y}{\partial x^2} = \frac{\partial}{\partial x}\left(\frac{\partial y}{\partial u} + \frac{\partial y}{\partial v}\right)$$

$$= \frac{\partial}{\partial u}\left(\frac{\partial y}{\partial u} + \frac{\partial y}{\partial v}\right) + \frac{\partial}{\partial v}\left(\frac{\partial y}{\partial u} + \frac{\partial y}{\partial v}\right)$$

$$= \frac{\partial^2 y}{\partial u^2} + 2\frac{\partial^2 y}{\partial u \partial v} + \frac{\partial^2 y}{\partial v^2},$$

supposing $\frac{\partial^2 y}{\partial u \partial v} = \frac{\partial^2 y}{\partial v \partial u}$.

Similarly,

$$\frac{\partial^2 y}{\partial t^2} = c^2 \left[\frac{\partial^2 y}{\partial u^2} - 2\frac{\partial^2 y}{\partial u \partial v} + \frac{\partial^2 y}{\partial v^2}\right].$$

Substituting these values of $\frac{\partial^2 y}{\partial x^2}$ and $\frac{\partial^2 y}{\partial t^2}$ in the wave equation, we get

$$\frac{\partial^2 y}{\partial u \partial v} = 0.$$

Integrating with respect to v, we get

$\frac{\partial y}{\partial u} = f(u)$, where f is an arbitrary function of u.

Now integrating w.r.t u, we have

$$y = \int f(u)\, du + \psi(v), \text{ where } \psi(v)$$

is an arbitrary function of v

$$= \phi(u) + \psi(v), \text{ say,}$$

since $\int f(u)\,du$ is a function of u only. Thus

$$y = \phi(x + ct) + \psi(x - ct). \qquad (133)$$

Now let initially $y(x, 0) = f(x)$ and $\frac{\partial y}{\partial t}(x, 0) = 0$. Differentiating (133) w.r.t. t, we get

$$\frac{\partial y}{\partial t} = c\phi'(x + ct) - c\psi'(x - ct).$$

But at $t = 0$, $\phi'(x) = \psi'(x)$, and $y(x, 0) = \phi(x) + \psi(x) = f(x)$. Thus, we have $\phi(x) = \psi(x) + k$. Therefore,

$$2\psi(x) + k = f(x)$$

or

$$\psi(x) = \frac{1}{2}[f(x) - k]$$

and then

$$\phi(x) = \frac{1}{2}[f(x) + k].$$

Hence the solution becomes

$$y(x, t) = \frac{1}{2}[f(x + ct) + k] + \frac{1}{2}[f(x - ct) - k]$$

$$= \frac{1}{2}[f(x + ct) + f(x - ct)].$$

EXAMPLE 5.94

Find the deflection of a vibrating string of unit length having fixed ends with initial velocity zero and initial deflection $f(x) = k(\sin x - \sin 2x)$.

Solution. We are given that

$$y(x, 0) = k(\sin x - \sin 2x) = f(x).$$

By D'Alembert method, the solution is

$$y(x, t) = \frac{1}{2}[f(x + ct) + f(x - ct)]$$

$$= \frac{1}{2}[k\{\sin(x + ct) - \sin 2(x + ct)\}]$$

$$+ k\{\sin(x - ct) - \sin 2(x - ct)g\}]$$

$$= k[\sin x \cos ct - \sin 2x \cos 2ct]$$

Since $y(x, 0) = k(\sin x - \sin 2x)$, we have

$$\frac{\partial y}{\partial t}(x, 0) = k[c \sin x \sin ct + 2c \sin 2x \sin 2ct]_{t=0}$$

$$= 0.$$

Thus the given boundary conditions are satisfied.

(C) Two-dimensional Heat Equation

Consider the flow of heat in a metal plate in the XOY plane. Suppose that the temperature at any point of the plate depends on x, y, and t (time) and not on the z-coordinate. Then this type of flow of heat is known as *two-dimensional heat flow*. It lies entirely in XOY plane and is zero along the normal to the plane XOY. The equation governing this type of flow is

$$\frac{\partial u}{\partial t} = c^2\left(\frac{\partial^2 u}{\partial x^2} + \frac{\partial^2 u}{\partial y^2}\right)$$

where $c^2 = \frac{k}{s\rho}$, s is specific heat and ρ is the density of the metal plate.

In steady state, u is independent of time and so $\frac{\partial u}{\partial t} = 0$. Therefore, the above equation transforms to

$$\frac{\partial^2 u}{\partial x^2} + \frac{\partial^2 u}{\partial y^2} = 0, \qquad (134)$$

which is known as *two-dimensional Laplace equation*.

We now find the solution of two dimensional Laplace equation by separation of variable method. Since there are two independent variables, let

$$u(x, y) = X(x)Y(y). \qquad (135)$$

Then using (135), the equation (134) reduces to

$$X''(x)Y(y) + X(x)Y''(y) = 0.$$

Therefore, separation of variables yields

$$\frac{X''(x)}{X(x)} = -\frac{Y''(y)}{Y(y)} = k \text{ (separation parameter)}.$$

Hence we get two differential equations

$$\frac{X''(x)}{X(x)} - kX(x) = 0 \quad (136)$$

and

$$Y''(y) + kY(y) = 0. \quad (137)$$

Now the following three cases are to be considered.

Case I. If $k > 0$, then $k = p^2$, where p is real. In this case the equations (136) and (137) take the form

$$\frac{d^2X}{dx^2} - p^2X = 0 \quad (138)$$

and

$$\frac{d^2Y}{dy^2} + p^2Y = 0. \quad (139)$$

The auxiliary equation for (138) is $m^2 - p^2 = 0$ and so $m = \pm p$. Thus the solution of (138) is

$$X(x) = c_1e^{px} + c_2e^{-px}.$$

The auxiliary equation for (139) is $m^2 + p^2 = 0$ and so $m = \pm pi$. Hence the solution of (139) is

$$Y(y) = c_3 \cos py + c_4 \sin py.$$

Thus, the solution of the Laplace equations in this case is

$$u(x, y) = X(x)Y(y)$$

$$= (c_1e^{px} + c_2e^{-px})$$

$$\times (c_3 \cos py + c_4 \sin py). \quad (140)$$

Case II. If $k = 0$, then equations (136) and (137) reduce to

$$\frac{d^2X}{dx^2} = 0 \quad \text{and} \quad \frac{d^2Y}{dy^2} = 0,$$

whose solutions are

$$X(x) = c_5x + c_6 \text{ and } Y(y) = c_7y + c_8.$$

Thus the solution of the Laplace equation is of the form

$$u(x, y) = (c_5\, x + c_6)\,(c_7\, y + c_8). \quad (141)$$

Case III. Let $k < 0$. Then $k = -p^2$. Then equations (136) and (137) reduce to

$$\frac{d^2X}{dx^2} + p^2X = 0 \text{ and } \frac{d^2Y}{dx^2} - p^2Y = 0,$$

whose solutions are, respectively

$$X(x) = c_9 \cos px + c_{10} \sin px$$

and

$$Y(y) = c_{11}\, e^{py} + c_{12}\, e^{-py}.$$

Thus the solution of the Laplace equation in this case is

$$u(x, y) = (c_9 \cos px + c_{10} \sin px)\,(c_{11}\, e^{py}$$

$$+ c_{12}\, e^{-py}). \quad (142)$$

In all these cases c_i, $i = 1, 2, \ldots, 12$ are constant of integration and are calculated using the boundary conditions.

EXAMPLE 5.95

Solve $\frac{\partial^2 u}{\partial x^2} + \frac{\partial^2 u}{\partial y^2} = 0$, subject to the conditions $u(x, 0) = 0$, $u(x, a) = 0$, $u(x, y) \to 0$ as $x \to \infty$ when $x \ge 0$ and $0 \le y \le a$.

Solution. The suitable method of separation of variables in this case is

$$u(x, y) = (c_1e^{px} + c_2e^{-px})(c_3 \cos py$$

$$+ c_4 \sin py).$$

Since $u(x, y) \to 0$ as $x \to \infty$, we have $c_1 = 0$ for all y. Therefore,

$$u(x, y) = c_2e^{-px}\,[c_3 \cos py + c_4 \sin py].$$

Since $u(x, 0) = 0$, we obtain $0 = c_2\, c_3\, e^{-px}$ and so $c_3 = 0$ since $c_2 \ne 0 \ne e^{-px}$ for all x. Therefore,

$$u(x, y) = c_2\, c_4\, e^{-px} \sin py.$$

Now using $u(x, a) = 0$, we get $0 = c_2 c_4 e^{-px} \sin pa$ and so $\sin pa = 0$, that is, $\sin pa = \sin n\pi$. Hence $pa = n\pi$ or $p = \frac{n\pi}{a}$, $n = 0, \pm 1, \pm 2, \ldots$. Therefore,

$$u_n(x, y) = A_n e^{-\frac{n\pi x}{a}} \sin \frac{n\pi y}{a},$$

$$n = 0, 1, 2, \ldots$$

Using principles of superposition, the solution of the wave equation is

$$u(x, y) = \sum A_n e^{-\frac{n\pi x}{a}} \sin \frac{n\pi y}{a}, \quad A_n \text{ constant.}$$

EXAMPLE 5.96

A rectangular plate with insulated surface is 10 cm wide and so long compared to its width that it may be considered in length without introducing an appreciable error. If the temperature of the short edge $y = 0$ is given by $u = 20$ for $0 \le x \le 5$ and $u = 20(10 - x)$ for $5 \le x \le 10$ and the two long edges $x = 0$, $x = 10$ as well as the other short edge are kept at 0°C, show that the temperature u at any point (x, y) is given by

$$u = \frac{800}{\pi^2} \sum_{n=1}^{\infty} \frac{(-1)^{n+1}}{(2n-1)^2} e^{-\frac{(2n-1)\pi y}{10}} . \sin \frac{(2n-1)\pi x}{10}.$$

Solution. To find the temperature $u(x, t)$, we have to solve the Laplace equation

$$\frac{\partial^2 u}{\partial x^2} + \frac{\partial^2 u}{\partial y^2} = 0$$

subject to the conditions

$$u(0, y) = 0, \quad y \ge 0$$

$$u(10, y) = 0, \quad y \ge 0$$

$$u(x, \infty) = 0, \quad 0 \le x \le 10$$

$$u(x, 0) = \begin{cases} 20x & 0 \le x \le 5 \\ 20(10 - x) & 5 \le x \le 10. \end{cases}$$

We know that three possible solutions of Laplace equations are

(i) $u = (c_1 e^{px} + c_2 e^{-px})(c_3 \cos py + c_4 \sin py)$

(ii) $u = (c_1 \cos px + c_2 \sin px)(c_3 e^{py} + c_4 e^{-py})$

(iii) $u = (c_1 x + c_2)(c_3 x + c_4)$.

Now solution (i) is not suitable because for $x = 0$, $u \ne 0$. The solution (iii) is also not suitable because it does not satisfy $u(x, \infty) = 0$ in $0 \le x \le 10$. Thus only (ii) is possible. Since $u(0, y) = 0$, (ii) yields

$$0 = c_1(c_3 e^{py} + c_4 e^{-py})$$

and so $c_1 = 0$. Thus

$$u(x, y) = c_2 \sin px (c_3 e^{py} + c_4 e^{-py}).$$

Since $u(10, y) = 0$, we have

$$0 = c_2 \sin 10\, p(c_3 e^{py} + c_4 e^{-py})$$

and $\sin 10\, p = 0 = \sin n\pi$. Thus $p = \frac{n\pi}{10}$, $n = 0, \pm 1, \pm 2, \ldots$ Also $u(x, \infty) = 0$. Therefore $c_3 = 0$. Hence the solution is

$$u_n(x, y) = c_2 c_4 e^{-\frac{n\pi y}{10}} . \sin \frac{n\pi x}{10}$$

$$= A_n e^{-\frac{n\pi y}{10}} . \sin \frac{n\pi x}{10}.$$

Hence the general solution is

$$u(x, y) = \sum_{n=1}^{\infty} A_n e^{-\frac{n\pi y}{10}} . \sin \frac{n\pi x}{10}$$

Now

$$u(x, 0) = \sum_{n=1}^{\infty} A_n \sin \frac{n\pi x}{10}$$

Therefore, A_n is Fourier sine coefficient of $u(x, 0)$. Thus

$$A_n = \frac{2}{l} \int_0^l u(x, 0) \sin \frac{n\pi x}{10}\, dx$$

$$= \frac{1}{5} \left(\int_0^5 20 \sin \frac{n\pi x}{10}\, dx + \int_5^{10} 20(10 - x) \sin \frac{n\pi x}{10}\, dx \right)$$

$$= \frac{800}{n^2\pi^2} \sin \frac{n\pi}{2}$$

$$= \begin{cases} 0 \text{ for even } n \\ \frac{800}{n^2\pi^2} (-1)^{\frac{n-1}{2}} \text{ for odd } n. \end{cases}$$

Hence the required solution is

$$u(x, y) = \frac{800}{\pi^2} \sum_{odd\, n} \frac{(-1)^{\frac{n-1}{2}}}{n^2} e^{-\frac{n\pi y}{10}} . \sin \frac{n\pi x}{10}$$

$$= \frac{800}{\pi^2} \sum_{1}^{\infty} \frac{(-1)^{n+1}}{(2n-1)^2} \times e^{-\frac{(2n-1)\pi y}{10}} . \sin \frac{(2n-1)\pi x}{10}.$$

EXAMPLE 5.97

Solve

$$\frac{\partial^2 u}{\partial x^2} + \frac{\partial^2 u}{\partial y^2} = 0$$

subject to the conditions $u(0, y) = u(l, y) = u(x, 0) = u(x, a) = \sin \frac{n\pi x}{l}$.

Solution. As per the given conditions, out of the three solutions of the Laplace equation, the solution

$$u(x, y) = (c_1 \cos px + c_2 \sin px)\,(c_3 e^{py} + c_4 e^{-py})$$

is suitable. The boundary condition $u(0, y) = 0$ implies $c_1 = 0$. Therefore,

$$u(x, y) = c_2 \sin px(c_3 e^{py} + c_4 e^{-py}).$$

Now using the condition $u(l, y) = 0$ gives

$$0 = c_2 \sin pl(c_3 e^{py} + c_4 e^{-py})$$

and so

$$\sin pl = 0 = \sin n\pi, \quad n = 0, 1, 2, \ldots$$

giving $p = \frac{n\pi}{l}$. Thus the solution becomes

$$u(x, y) = c_2 \sin \frac{n\pi x}{l} (c_3 e^{n\pi y/l} + c_4 e^{-n\pi y/l}).$$

Now the use of condition $u(x, 0) = 0$ gives

$$0 = c_2 \sin \frac{n\pi x}{l} (c_3 + c_4)$$

and so $c_3 + c_4 = 0$ or $c_4 = -c_3$. Thus the solution reduces to

$$u(x, y) = c_2\, c_3 \sin \frac{n\pi x}{l} (e^{n\pi y/l} - e^{-n\pi y/l}).$$

$$= A_n \sin \frac{n\pi x}{l} \sinh^{n\pi y/l}.$$

Now the last condition $u(x, a) = \sin \frac{n\pi x}{l}$ yields

$$\sin \frac{n\pi x}{l} = A_n \sin \frac{n\pi x}{l} \sinh \frac{n\pi x}{l}$$

or

$$A_n = \frac{1}{\sinh \frac{n\pi x}{l}}.$$

Hence

$$u(x, y) = \frac{\sin n\pi x}{l} \cdot \frac{\sin n\pi y/l}{\sin n\pi\, a/l}.$$

EXAMPLE 5.98

Solve

$$\frac{\partial^2 u}{\partial x^2} + \frac{\partial^2 u}{\partial y^2} = 0,\ 0 \le x \le a,\ 0 \le y \le b$$

subject to the conditions

$$u(0, y) = u(a, y) = u(x, b) = 0,\ u(x, 0) = x(a - x).$$

Solution. Three possible solutions to the given equation are

(i) $u(x, y) = (c_1 e^{px} + c_2 e^{-px})(c_3 \cos py + c_4 \sin py)$

(ii) $u(x, y) = (c_1 \cos px + c_2 \sin px)\,(c_3 e^{py} + c_4 e^{-py})$

(iii) $(c_1 x + c_2)\,(c_3 y + c_4)$.

Since u ≠ 0 for $x = 0$, the solution (i) does not satisfy $u(0, y) = 0$. The solution (iii) does not satisfy $u(x, 0) = a(a - x)$. Hence solution (ii) is suitable. Now $u(0, y) = 0$ implies $0 = c_1(c_3 e^{py} + c_4 e^{-py})$ and so $c_1 = 0$. Thus

$$u(x, y) = c_2 \sin px(c_3 e^{py} + c_4 e^{-py}).$$

Now $u(a, y) = 0$ implies $0 = c_2 \sin pa\, (c_3 e^{py} + c_4 e^{-py})$ and so $\sin pa = 0 = \sin n\pi$, $n = 0, \pm 1, \pm 2, \ldots$. Thus $p = \frac{n\pi}{a}$, $n = 0, \pm 1, \pm 2, \ldots$ Further $u(x, b) = 0$ implies

$$0 = c_2 \sin\frac{n\pi x}{a}\left(c_3 e^{\frac{n\pi b}{a}} + c_4 e^{-\frac{n\pi b}{a}}\right)$$

and so

$$\text{either } c_4 = \frac{c_3 e^{\frac{n\pi b}{a}}}{e^{-\frac{n\pi b}{a}}} \text{ or } c_3 = -\frac{c_4 e^{-\frac{n\pi b}{a}}}{e^{\frac{n\pi b}{a}}}.$$

For the first case,

$$u_1(x, y) = c_2 \sin\frac{n\pi x}{a}\left(c_3 e^{\frac{n\pi y}{a}} - \frac{c_3 e^{\frac{n\pi b}{a}}}{e^{-\frac{n\pi b}{a}}} e^{-\frac{n\pi y}{a}}\right)$$

$$= c_2 \sin\frac{n\pi x}{a}\left(e^{\frac{n\pi y}{a}} - \frac{e^{\frac{n\pi b}{a}}}{e^{-\frac{n\pi b}{a}}} e^{-\frac{n\pi y}{a}}\right)$$

$$= 2c_2 c_3 \sin\frac{n\pi x}{a}\left[\frac{e^{\frac{n\pi y}{a}} e^{-\frac{n\pi b}{a}} - e^{\frac{n\pi b}{a}} e^{-\frac{n\pi y}{a}}}{2e^{-\frac{n\pi b}{a}}}\right]$$

$$= 2c_2 c_3 \sin\frac{n\pi x}{a} \frac{\sinh\frac{n\pi}{a}(b - y)}{e^{-\frac{n\pi b}{a}}}$$

Similarly, for the second case,

$$u_2(x, y) = 2c_2 c_4 \sin\frac{n\pi x}{a}\left[\frac{\sinh\frac{n\pi}{a}(b - y)}{e^{-\frac{n\pi b}{a}}}\right]$$

Therefore, the solution is

$$u(x, y) = 2c_2 c_3 u_1(x, y) - 2c_2 c_4 u_2(x, y)$$

$$= A_n \sin\frac{n\pi x}{a} \frac{\sinh\frac{n\pi}{a}(b - y)}{\sinh\frac{n\pi b}{a}}.$$

But $u(x, 0) = a - x$. Thus

$$x(a - x) = \sum A_n \frac{\sin n\pi x}{a}.$$

Therefore,

$$A_n = \frac{2}{a}\int_0^a x(a - x) \sin\frac{n\pi x}{a}\, dx$$

$$= \frac{2}{a}\left[a(ax - x^2)\left(-\frac{\cos\frac{n\pi x}{a}}{n\pi}\right) - (a - 2x)a^2\left(-\frac{\sin\frac{n\pi x}{a}}{n^2\pi^2}\right) + (-2)a^3\left(\frac{\cos\frac{n\pi x}{a}}{n^3\pi^3}\right)\right]_0^a$$

$$= \frac{2}{a}\left[0 - 0 - \frac{2a^3}{n^3\pi^3}\cos\frac{n\pi a}{a} + \frac{2a^3}{n^3\pi^3}\right]$$

$$= \frac{4a^2}{n^3\pi^3}[1 - (-1)^n]$$

$$= \begin{cases} 0 & \text{for even } n \\ \frac{800}{n^2\pi^2} & \text{for odd } n. \end{cases}$$

Hence the solution becomes

$$u(x, y) = \frac{8a^2}{\pi^3}\sum_{odd\, n} \frac{1}{n^3} \sin\frac{n\pi x}{a} \cdot \frac{\sinh n\pi(b - y)/a}{\sinh n\pi b/a}$$

$$= \frac{8a^2}{\pi^3}\sum_{n=0}^{\infty} \frac{1}{(2n + 1)^3} \sin\frac{(2n + 1)\pi x}{a} \times \frac{\sinh\frac{(2n + 1)\pi}{a}(b - y)}{\sinh\frac{(2n + 1)\pi b}{a}}.$$

EXAMPLE 5.99

A rectangular plate with insulated surface is 8 cm wide and so long compared to its width that it may the considered infinite in length. If the temperature along one short edge $y = 0$ is given by $u(x, 0) = 100 \sin\frac{\pi x}{8}$, $0 < x < 8$ while two long edges $x = 0$ and $x = 8$ as well as the other short edges are kept at 0°C. Find the steady state temperature.

Solution. The partial differential equation governing the problem is

$$\frac{\partial^2 u}{\partial x^2} + \frac{\partial^2 u}{\partial y^2} = 0$$

along with the boundary conditions

$$u(0, y) = 0,\ u(8, y) = 0,\ u(x, y) = 0 \text{ as } y \to \infty,$$

$$u(x, 0) = 100 \sin\frac{\pi x}{8}.$$

The solution

$$u(x, y) = (c_1 \cos px + c_2 \sin px)(c_3 e^{py} + c_4 e^{-py})$$

is the suitable solution under the given conditions. The use of initial condition $u(0, y) = 0$ yields

$$0 = c_1(c_3e^{py} + c_4e^{-py})$$

and so c_1 =. Thus the solution becomes

$$u(x, y) = c_2 \sin px(c_3e^{py} + c_4e^{-py}).$$

The condition $u(8, y) = 0$ implies

$$0 = c_2 \sin 8p(c_3e^{py} + c_4e^{-py})$$

and so $\sin 8p = 0$. Therefore, $\sin 8p = \sin n\pi$ and so $p = \frac{n\pi}{8}$, $n = 0, \pm 1, \pm 2, \ldots$. The solution becomes

$$u(x, y) = c_2 \sin \frac{n\pi x}{8}(c_3e^{n\pi y/8} + c_4e^{-n\pi y/8}).$$

The condition $u(x, y) = 0$ as $y \to \infty$ implies $c_3 = 0$. Hence

$$u(x, y) = c_2\, c_4\, e^{-\frac{n\pi y}{8}} \cdot \sin \frac{n\pi x}{8}$$

$$= A_n\, e^{-\frac{n\pi y}{8}} . \sin^{\frac{n\pi x}{8}}$$

But $u(x, 0) = 100 \sin \frac{\pi x}{8}$. Therefore,

$$100 \sin \frac{\pi x}{8} = \text{An} \sin \frac{n\pi x}{8}$$

and so $A_n = 100$, $n = 1$. Hence the solution is

$$u(x, y) = 100\, e^{-\frac{\pi y}{8}} \sin \frac{\pi x}{8}.$$

(D) Two-dimensional Wave Equation

Consider the vibrations of a tightly stretched membrane, such as membrane of drum. We assume that the membrane is uniform and that the tension per unit length in the membrane is same at every point of it in all directions. Let m be the mass per unit area and T be the tension per unit length of the membrane. Let u be the displacement of an element $\delta x\, \delta y$ of the membrane perpendicular to the xy-plane. Then the forces acting on the membrane are.

(i) The forces $T\,\delta y$ tangential to the membrane on its opposite edges of length δy acting respectively at angles α and β to the horizontal.

(ii) The forces $T\delta x$ acting on the opposite edges of length δx (Figure 5.2).

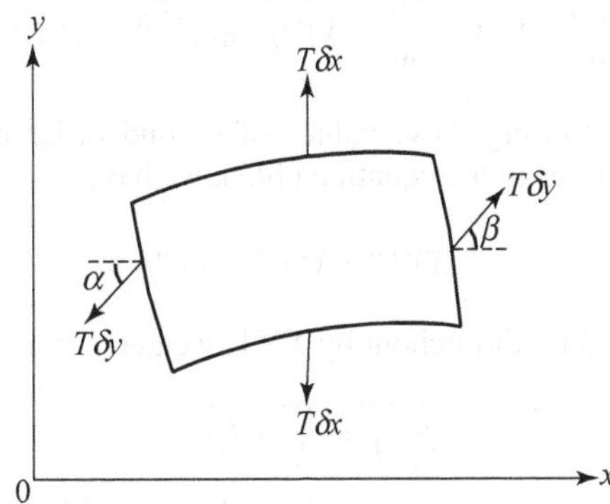

Figure 5.2

The angles α and β, being small, the vertical component of forces corresponding to (i) is

$$(T\,\delta y)\sin\beta - (T\,\delta y)\sin\alpha = T\,\delta y\,(\tan\beta - \tan\alpha)$$

$$= T\,\delta y\left[\left(\frac{\partial u}{\partial x}\right)_{x+\delta x} - \left(\frac{\partial u}{\partial x}\right)_x\right]$$

$$= T\,\delta y\delta x\left[\frac{\left(\frac{\partial u}{\partial x}\right)_{x+\delta x} - \left(\frac{\partial u}{\partial x}\right)_x}{\delta x}\right]$$

$$= T\,\delta x\delta y\,\frac{\partial^2 u}{\partial x^2} \text{ in the limit as } \delta x \to 0.$$

Similarly, the vertical component of forces corresponding to (ii) is $T\,\delta x\delta y\,\frac{\partial^2 u}{\partial y^2}$. Hence the equation of motion of the element $\delta x\delta y$ is

$$m\delta x\delta y\,\frac{\partial^2 u}{\partial t^2} = T\,\delta x\delta y\left(\frac{\partial^2 u}{\partial x^2} + \frac{\partial^2 u}{\partial y^2}\right)$$

or

$$\frac{\partial^2 u}{\partial t^2} = c_2\left(\frac{\partial^2 u}{\partial x^2} + \frac{\partial^2 u}{\partial y^2}\right) \qquad (143)$$

where $c_2 = \frac{T}{m}$. The equation (143) is called **Two-Dimensional Wave Equation.**

We now find the solution of the equation (143) by separation of variables method. So let

$$u\,(x, y, t) = X(x)Y(y)T(t)$$

be a trial solution of (143). Then

$$\frac{\partial^2 u}{\partial x^2} = X''YT, \frac{\partial^2 u}{\partial y^2} = XY''T, \text{ and } \frac{\partial^2 u}{\partial t^2} = XYT''.$$

Substituting these values of second order derivatives in the equation (143), we have

$$\frac{1}{c_2} XYT'' = X''YT + XY''T$$

Dividing throughout by XYT, we get

$$\frac{1}{c_2}\frac{T''}{T} = \frac{X''}{X} + \frac{Y''}{Y}.$$

Since x, y, and t are independent variables, the above equation is valid if each term in this is constant. Choosing the constants as $-k^2$, $-l^2$, and $-(k^2 + l^2)$, we get

$$\frac{d^2X}{dx^2} + k^2X = 0, \tag{144}$$

$$\frac{d^2Y}{dy^2} + l^2Y = 0, \tag{145}$$

$$\frac{d^2T}{dt^2} + (k^2 + l^2)\, c^2T = 0, \tag{146}$$

The solutions to the homogenous equations (144) (145) (146) are respectively

$$X = c_1 \cos kx + c_2 \sin kx,$$

$$Y = c_3 \cos ly + c_4 \sin ly,$$

$$T = c_5 \cos \sqrt{k^2 + l^2}\; ct + c_6 \sin \sqrt{k^2 + l^2}\; ct.$$

Hence the solution of the two-dimensional wave equation (143) is

$$u(x,y,t) = (c_1 \cos kx + c_2 \sin kx) \times (c_3 \cos ly + c_4 \sin ly) \times \left(c_5 \cos \sqrt{k^2 + l^2}ct + c_6 \sin \sqrt{k^2 + l^2}ct\right)$$

EXAMPLE 5.100

Find the deflection $u(x, y, t)$ of the square membrane with $a = b = c = 1$ if the initial velocity is zero and the initial deflection

$$f(x) = A \sin \pi x \sin 2\pi y.$$

Solution. We know that the vibrations of a membrane are governed by two-dimensional wave equation

$$\frac{\partial^2 u}{\partial t^2} = c^2\left(\frac{\partial^2 u}{\partial x^2} + \frac{\partial^2 u}{\partial y^2}\right).$$

The membrane is stretched between the lines

$$x = 0, x = 1, y = 0, \text{ and } y = 1.$$

Therefore the boundary conditions are

$$u(0, y, t) = 0, u(1, y, t) = 0,$$
$$u(x, 0, t) = 0, \text{ and } u(x, 1, t) = 0.$$

The initial conditions are

$$u(x, y, 0) = f(x, y)$$

$$= A \sin \pi x \sin 2\pi y \text{ and } \left(\frac{\partial u}{\partial t}\right)_{t=0} = 0.$$

We know that the solution of two-dimensional wave equation is

$$u(x,y,t) = (c_1 \cos kx + c_2 \sin kx) \times (c_3 \cos ly + c_4 \sin ly) \times \left(c_5 \cos \sqrt{k^2 + l^2}ct + c_6 \sin \sqrt{k^2 + l^2}ct\right)$$

It is given that $c = 1$. Therefore

$$u(x,y,t) = (c_1 \cos kx + c_2 \sin kx) \times (c_3 \cos ly + c_4 \sin ly) \times \left(c_5 \cos \sqrt{k^2 + l^2}\,t + c_6 \sin \sqrt{k^2 + l^2}\,t\right).$$

Using the boundary condition $u(0, y, t) = 0$, we get

$$0 = c_1(c_3 \cos ly + c_4 \sin ly) \times \left(c_5 \cos \sqrt{k^2 + l^2}\,t + c_6 \sin \sqrt{k^2 + l^2}\,t\right).$$

Thus $c_1 = 0$ and the solution reduces to

$$u(x,y,t) = c_2 \sin kx(c_3 \cos ly + c_4 \sin ly) \times \left(c_5 \cos\sqrt{k^2+l^2}\,t + c_6 \sin\sqrt{k^2+l^2}\,t\right) \quad (147)$$

Now the boundary condition $u(1, y, t) = 0$ gives $0 = c_2 \sin k(c_3 \cos ly + c_4 \sin ly)$

$$0 = c_2 \sin k(c_3 \cos ly + c_4 \sin ly) \times \left(c_5 \cos\sqrt{k^2+l^2}\,t + c_6 \sin\sqrt{k^2+l^2}\,t\right),$$

and so $\sin k = 0$, which implies $k = m\pi$. Hence (147) reduces to

$$u(x,y,t) = c_2 \sin m\pi x(c_3 \cos ly + c_4 \sin ly) \times \left(c_5 \cos\sqrt{k^2+l^2}\,t + c_6 \sin\sqrt{k^2+l^2}\,t\right). \quad (148)$$

The boundary condition $u(x, 0, t) = 0$ implies

$$0 = c_2 \sin m\pi x\left[c_3\left(c_5 \cos\sqrt{k^2+l^2}\,t + c_6 \sin\sqrt{k^2+l^2}\,t\right)\right]$$

and so $c_3 = 0$. Thus (148) reduces to

$$u(x,y,t) = c_2 c_4 \sin m\pi x \sin ly \times \left(c_5 \cos\sqrt{k^2+l^2}\,t + c_6 \sin\sqrt{k^2+l^2}\,t\right). \quad (149)$$

Lastly, the boundary condition $u(x, 1, t) = 0$ yields

$$0 = c_2 c_4 \sin m\pi x \sin l \times \left(c_5 \cos\sqrt{k^2+l^2}\,t + c_6 \sin\sqrt{k^2+l^2}\,t\right)$$

and so $\sin l = 0$ or $l = n\pi$. Therefore (149) reduces to

$$u(x,y,t) = c_2 c_4 \sin m\pi x \sin n\pi y \times \left(c_5 \cos\sqrt{k^2+l^2}\,t + c_6 \sin\sqrt{k^2+l^2}\,t\right).$$
$$= \sin m\pi x \sin n\pi y(A_{mn} \cos pt + B_{mn} \sin pt),$$

where $p = \sqrt{k^2+l^2}$ (150)

Therefore superposition of general solutions yields

$$u(x, y, t) = \sum_{m=1}^{\infty}\sum_{n=1}^{\infty} \sin m\pi x \sin n\pi y \times (A_{mn} \cos pt + B_{mn} \sin pt). \quad (151)$$

The application of initial condition $\left(\frac{\partial u}{\partial t}\right)_{t=0} = 0$ implies $B_{mn} = 0$ and use of the initial condition $f(x, y) = A \sin \pi x \sin 2\pi y$ at $t = 0$ yields

$$A \sin\pi x \sin 2\pi y = \sum_{m=1}^{\infty}\sum_{n=1}^{\infty} A_{mn} \sin m\pi x \sin n\pi y. \quad (152)$$

The right hand side of (152) is a Double Fourier Series of $A \sin \pi x \sin 2\pi y$. The Fourier coefficients are given by

$$A_{mn} = \frac{2}{1}\cdot\frac{2}{1}\int_0^1\int_0^1 A \sin \pi x \sin 2\pi y \times [\sin mx \sin n\pi y]\, dxdy.$$

We note that

$$A_{m1} = A_{m3} = A_{m4} = \ldots = 0,$$
$$A_{22} = A_{32} = A_{42} = \ldots = 0,$$

$$A_{m2} = 4A \int_0^1\int_0^1 \sin \pi x \sin m\pi x \sin^2(2\pi y)\, dxdy$$
$$= 2A \int_0^1\int_0^1 \sin \pi x \sin m\pi x\, [1 - \cos 4y]\, dxdy$$
$$= 2A \int_0^1\int_0^1 \sin \pi x \sin m\pi x\, dx = 0, \text{ and}$$

$$A_{12} = 2A \int_0^1 \sin^2\pi x\, dx = A\int_0^1 (1 - \cos 2\pi x)\, dx = A.$$

Hence the solution (151) becomes

$$u(x,y,t) = A \sin \pi x \sin 2\pi y \cos pt, \quad \text{since } m = 1 \text{ and } n = 2$$
$$= A \cos\left(\pi\sqrt{5}t\right) \sin \pi x \sin 2\pi y,$$

since

$$p = \sqrt{k^2 + l^2} = \sqrt{m^2\pi^2 + n^2\pi^2} = \pi\sqrt{m^2 + n^2}$$
$$= \pi\sqrt{1^2 + 2^2} = \pi\sqrt{5}.$$

5.11 SOLUTIONS OF LAPLACE EQUATION

(A) Laplace Equation in Two Dimensions

(1) **Cartesion Form**. Please see Article 5.10 (C).

(2) **Polar Form**. Putting $x = r\cos\theta$, $y = r\sin\theta$ in $\frac{\partial^2 u}{\partial x^2} + \frac{\partial^2 u}{\partial y^2} = 0$, the Laplace equation in polar coordinates is given by

$$\frac{\partial^2 u}{\partial r^2} + \frac{1}{r}\frac{\partial u}{\partial r} + \frac{1}{r^2}\frac{\partial^2 u}{\partial \theta^2} = 0. \qquad (153)$$

Let $u(r, \theta) = R(r)F(\theta)$ be trial solution of (153). Then

$$\frac{\partial u}{\partial r} = F\frac{dR}{dr}, \quad \frac{\partial^2 u}{\partial r^2} = F\frac{d^2R}{dr^2}, \quad \frac{\partial^2 u}{\partial \theta^2} = R\frac{d^2F}{d\theta^2}.$$

Therefore equation (153) reduces to

$$r^2F\frac{d^2R}{dr^2} + rF\frac{dR}{dr} + R\frac{d^2F}{d\theta^2} = 0.$$

Separation of variables yield

$$\frac{r^2\frac{d^2R}{dr^2} + r\frac{dR}{dr}}{R} = -\frac{\frac{d^2F}{d\theta^2}}{F} = k \text{ (constant)}$$

Thus, we get two equations

$$r^2\frac{d^2R}{dr^2} + r\frac{dR}{dr} - kR = 0 \qquad (154)$$

and

$$\frac{d^2F}{d\theta^2} + kF = 0. \qquad (155)$$

The equation (154) is a homogeneous linear differential equation. So, substituting $r = e^z$ the equation (154) reduces to

$$\frac{d^2R}{dz^2} - kR = 0. \qquad (156)$$

Case I. $k = p^2$ (positive)

The auxiliary equation of (156) is $m^2 - p^2 = 0$ and so $m = \pm p$. Thus the solution of (156) is

$$R = c_1e^{pz} + c_2e^{-pz} \qquad (157)$$

The auxiliary equation of (155) is $m^2 + p^2 = 0$ and so $m = ip$. Thus the solution of (155) is

$$F = c_3\cos p\theta + c_4\sin p\theta \qquad (158)$$

Case II. $k = -p^2$ (negative)

In this case the auxiliary equation for (156) is $m^2 + p^2 = 0$ and so $m = \pm ip$. Thus the solution of (156) is

$$R = c_1\cos pz + c_2\sin pz$$
$$= c_1\cos(p\log r) + c_2\sin(p\log r). \qquad (159)$$

The auxiliary equation of (155) is $m^2 - p^2 = 0$, which yields $m = \pm p$. Thus solution of (155) is

$$F = c_3e^{p\theta} + c_4e^{-p\theta} \qquad (160)$$

Case III. $k = 0$

In this case the solutions of (156) and (155) are respectively

$$R = c_1z + c_2 = c_1\log r + c_2 \qquad (161)$$

and

$$F = c_3\theta + c_4. \qquad (162)$$

Hence the three possible solution of the Laplace equation (153) are

$$u = (c_1 r^p + c_2 r^{-p})(c_3 \cos p\theta + c_4 \sin p\theta) \quad (163)$$

$$u = [c_1 \cos(p \log r) + c_2 \sin(p \log r)]\left(c_3 e^{p\theta} + c_4 e^{-p\theta}\right), \quad (164)$$

$$u = (c_1 \log r + c_2)(c_3\theta + c_4). \quad (165)$$

Out of these three solutions, the solution consistent with the boundary conditions is considered. Generally, the solution (163) is considered to be most suitable.

(B) Laplace Equation in Three Dimensions

(1) **Cartesian Form.** The Laplace equation in three dimensions is

$$\frac{\partial^2 u}{\partial x^2} + \frac{\partial^2 u}{\partial y^2} + \frac{\partial^2 u}{\partial z^2} = 0. \quad (166)$$

Let $u(x, y, z) = X(x)\, Y(y)\, Z(z)$ be the trial solution. Then (166) reduces to

$$X''YZ + XY''Z + XYZ'' = 0$$

Dividing throughout by XYZ, we get

$$\frac{X''}{X} + \frac{Y''}{Y} + \frac{Z''}{Z} = 0$$

that is,

$$\frac{1}{X}\frac{d^2X}{dX^2} + \frac{1}{Y}\frac{d^2Y}{dY^2} + \frac{1}{Z}\frac{d^2Z}{dZ^2} = 0 \quad (167)$$

Since x, y, z are independent, (167) is possible only if

$$\frac{1}{X}\frac{d^2X}{dX^2}, \frac{1}{Y}\frac{d^2Y}{dY^2} \text{ and } \frac{1}{Z}\frac{d^2Z}{dZ^2} \text{ are all constant.}$$

Let

$$\frac{1}{X}\frac{d^2X}{dX^2} = k^2, \frac{1}{Y}\frac{d^2Y}{dY^2} = l^2 \text{ and } \frac{1}{Z}\frac{d^2Z}{dZ^2} = -(k^2 + l^2).$$

Thus, we get three equations

$$\frac{d^2X}{dX^2} - k^2X = 0 \quad (168)$$

$$\frac{d^2Y}{dY^2} - l^2Y = 0 \quad (169)$$

and

$$\frac{d^2Z}{dZ^2} + (k^2 + l^2) = 0 \quad (170)$$

The solution of (168), (169), and (170) are respectively

$$X = c_1 e^{kx} + c_2 e^{-kx}$$

$$Y = c_3 e^{ly} + c_4 e^{-ly} \text{ and}$$

$$Z = c_5 \cos\sqrt{k^2 + l^2}z + c_6 \sin\sqrt{k^2 + l^2}z.$$

Hence the solution of the three dimensional Laplace equation in Cartesian form is

$$u = \left(c_1 e^{kx} + c_2 e^{-kx}\right)\left(c_3 e^{ly} + c_4 e^{-ly}\right) \times \left(c_5 z \cos\sqrt{k^2 + l^2} + c_6 z \sin\sqrt{k^2 + l^2}\right).$$

The choice of the constant depends on the initial and boundary conditions.

(C) Laplace Equation in Cylindrical Co-ordinates

Substituting $x = r\cos\theta$, $y = r\sin\theta$ and $z = z$ in the Cartesian form of three dimensional Laplace equation, we get

$$\frac{\partial^2 u}{\partial r^2} + \frac{1}{r}\frac{\partial u}{\partial r} + \frac{1}{r^2}\frac{\partial^2 u}{\partial \theta^2} + \frac{\partial^2 u}{\partial z^2} = 0. \quad (171)$$

Let $u(r, \theta, z) = R(r)\, F(\theta)\, Z(z)$ be the trial solution of (171). Then (171) takes the form

$$R''FZ + \frac{1}{r}R'FZ + \frac{1}{r^2}RF''Z + RFZ'' = 0.$$

Dividing throughout by R F Z, we get

$$\frac{R''}{R} + \frac{1}{rR}R' + \frac{F''}{r^2F} + \frac{Z''}{Z} = 0$$

or

$$\frac{1}{R}\left[\frac{d^2R}{dr^2} + \frac{1}{r}\frac{dR}{dr}\right] + \frac{1}{r^2F}\frac{d^2F}{d\theta^2} + \frac{1}{Z}\frac{d^2Z}{dz^2} = 0. \quad (172)$$

Taking

$$\frac{d^2F}{d\theta^2} = -n^2F \text{ and } \frac{d^2Z}{dz^2} = k^2Z, \quad (173)$$

we have

$$\frac{1}{R}\left[\frac{d^2R}{dr^2} + \frac{1}{r}\frac{dR}{dr}\right] - \frac{n^2}{r^2} + k^2 = 0$$

or

$$r^2\frac{d^2R}{dr^2}+r\frac{dR}{dr}+(k^2r^2-n^2)=0. \quad (174)$$

The solution of the Bessel's equation (174) is

$$R=c_1J_n(kr)+c_2Y_n(kr).$$

The solution of (173) are

$$F=c_3\cos n\theta+c_4\sin n\theta,$$
$$Z=c_5e^{kz}+c_6e^{-kz}.$$

Thus the solution of the Laplace equation (171) is

$$u=[c_1J_n(kr)+c_2Y_n(kr)](c_3\cos n\theta+c_4\sin n\theta)\times\left(c_5e^{kz}+c_6e^{-kz}\right),$$

which is called *cylindrical harmonic.*

(D) Laplace Equation in Spherical Co-ordinates

Substituting $x=r\sin\theta\cos\phi$, $y=r\sin\theta\sin\phi$ and $z=r\cos\theta$ in the three dimensional Cartesian form of the Laplace equation, we get

$$\frac{\partial^2u}{\partial r^2}+\frac{2}{r}\frac{\partial u}{\partial r}+\frac{1}{r^2}\frac{\partial^2u}{\partial\theta^2}+\frac{\cot\theta}{r^2}\frac{\partial u}{\partial\theta}+\frac{1}{r^2\sin^2\theta}\frac{\partial^2u}{\partial\phi^2}=0. \quad (175)$$

Let $u(r,\theta,\phi)=R(r)\,G(\theta)\,H(\phi)$ be a trial solution of (175). Then (175) takes the form

$$R''GH+\frac{2}{r}R'GH+\frac{1}{r^2}RG''H+\frac{\cot\theta}{r^2}RG'H+\frac{1}{r^2\sin^2\theta}RGH''=0.$$

Dividing this equation throughout by R G H, we have

$$\frac{1}{R}\left[r^2\frac{d^2R}{dr^2}+2r\frac{dR}{dr}\right]+\frac{1}{G}\left[\frac{d^2G}{d\theta^2}+\cot\theta\frac{dG}{d\theta}\right]+\frac{1}{H\sin^2\theta}\frac{d^2H}{d\phi^2}=0. \quad (176)$$

Taking

$$\frac{1}{R}\left[r^2\frac{d^2R}{dr^2}+2r\frac{dR}{dr}\right]=n(n+1) \quad (177)$$

and

$$\frac{1}{H}\frac{d^2H}{d^2}=-m^2, \quad (178)$$

the equation (176) reduces to

$$\frac{d^2G}{d\theta^2}+\cot\theta\frac{dG}{d\theta}+[n(n+1)-m^2\operatorname{cosec}^2\theta]\,G=0.$$

The solution of this Legendre's Equation is

$$G=c_1P_n^m(\cos\theta)+c_2Q_n^m(\cos\theta).$$

Further, the solution of equation (178) is

$$H=c_3\cos m\phi+c_4\sin m\phi.$$

Taking $R=r^k$ in (177), we get

$$k(k-1)+2k=n(n+1)$$

or

$$(k^2-n^2)+(k-n)=0$$

or

$$(k-n)(k+n+1)=0$$

and so $k=n,-n-1$. Therefore

$$R=c_5r^n+c_6r^{-n-1}.$$

Hence the general solution of the Laplace Equation (175) is

$$u=\sum_{n=0}^{\infty}\sum_{m=0}^{\infty}c_1P_n^m(\cos\theta)+c_2Q_n^m(\cos\theta)\times(c_3\cos m\phi+c_4\sin m\phi)(c_5r^n+c_6r^{-n-1}),$$

which is called *spherical harmonic.*

5.13 TELEPHONE EQUATIONS OF A TRANSMISSION LINE

Let O be the transmitting end and R the receiving end of a transmission cable carrying an electric current with resistance R ohms/km, inductance L henries/km, capacitance C farads/km and leakance G mhos/km. Let P be a point on the cable at a distance x km from the transmitting end O and let $v(x, t)$and $i(x, t)$ be respectively the instantaneous voltage and current at the point P at any time t. Consider a small segment PQ of length δx of the cables.

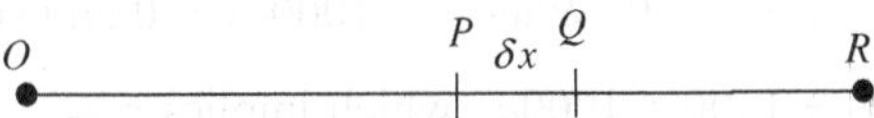

The voltage drop across PQ is the sum of the voltage drop due to the resistance R and the voltage drop due to the inductance L. Thus the voltage drop across the section PQ is given by

$$-\delta v = R\, i\,\delta x + L\,\delta x\,\frac{\partial i}{\partial t}$$

or

$$-\frac{\delta v}{\delta x} = R\, i + L\frac{\partial i}{\partial t}.$$

Taking limit as $\delta x \to 0$, we have

$$-\frac{\partial v}{\partial x} = Ri + L\frac{\partial i}{\partial t}$$

or

$$\frac{\partial v}{\partial x} + \left(R + L\frac{\partial}{\partial t}\right) i = 0. \qquad (179)$$

On the other hand, the current loss across the segment PQ is the sum of current loss due to capacitance and the current loss due to leakance. Therefore

$$-\delta i = C\frac{\partial v}{\partial t}\,\delta x + Gv\,\delta x$$

or

$$-\frac{\delta i}{\delta x} = C\frac{\partial v}{\partial t} + Gv.$$

Taking limit as $\delta x \to 0$, we have

$$-\frac{\partial i}{\partial x} = C\frac{\partial v}{\partial t} + Gv$$

or

$$\frac{\partial i}{\partial x} + \left(C\frac{\partial}{\partial t} + G\right) v = 0. \qquad (180)$$

Differentiating (179) partially with respect to x, we get

$$\frac{\partial^2 v}{\partial x^2} + R\frac{\partial i}{\partial x} + L\frac{\partial^2 i}{\partial t\partial x} = 0. \qquad (181)$$

Operating (180) by $R + L\frac{\partial}{\partial t}$, we have

$$R\frac{\partial i}{\partial x} + L\frac{\partial^2 i}{\partial t\partial x} + R\left(C\frac{\partial}{\partial t} + G\right)v$$

$$+ LC\frac{\partial^2 v}{\partial t^2} + LG\frac{\partial v}{\partial t} = 0. \quad (182)$$

Subtracting (182) from (181), we have

$$\frac{\partial^2 v}{\partial x^2} = LC\frac{\partial^2 v}{\partial t^2} + (LG + RC)\frac{\partial v}{\partial t} + RGV. \quad (183)$$

Similarly, operating (179) by $\left(C\frac{\partial}{\partial t} + G\right)$, we get

$$C\frac{\partial^2 v}{\partial t\partial x} + RC\frac{\partial i}{\partial t} + LC\frac{\partial^2 i}{\partial t^2} + G\frac{\partial v}{\partial x} + RGi$$

$$+ GL\frac{\partial i}{\partial t} = 0. \qquad (184)$$

Differentiating (180) partially with respect to x, we have

$$\frac{\partial^2 i}{\partial x^2} + C\frac{\partial^2 v}{\partial t\partial x} + G\frac{\partial v}{\partial x} = 0. \quad (185)$$

Subtracting (185) from (184), we get

$$\frac{\partial^2 i}{\partial x^2} = LC\frac{\partial^2 i}{\partial t^2} + (LG + RC)\frac{\partial i}{\partial t} + RGi = 0.(186)$$

We observe that the equation (186) can be obtained by replacing v in equation (183) by i. The equations (183) and (186) are called *telephone equations*.

If $R = G = 0$, then the telephone equations reduce to

$$\frac{\partial^2 v}{\partial x^2} = LC\frac{\partial^2 v}{\partial t^2} \text{ and } \frac{\partial^2 i}{\partial x^2} = LC\frac{\partial^2 i}{\partial t^2},$$

which are called *radio equations*.

If $L = G = 0$, then the telephone equations reduce to

$$\frac{\partial^2 v}{\partial x^2} = RC\frac{\partial v}{\partial t},$$

and

$$\frac{\partial^2 i}{\partial x^2} = RC\frac{\partial i}{\partial t},$$

which are called *telegraph equations*. Setting $c_2 = \frac{1}{RC}$, we note that the telegraph equations are similar to the heat equation $\frac{\partial v}{\partial t} = c_2\frac{\partial^2 v}{\partial x^2}$.

EXAMPLE 5.101

A transmission line 1000 kilometers long is initially under steady state condition with potential 1300 volts at the sending edge ($x = 0$) and 1200 volts at the receiving end ($x = 1000$). The terminal end of the line is suddenly grounded, but the potential at the source is kept at 1300 volts. If inductance and leakage are negligible, determine the potential $v(x, t)$.

Solution. When inductance L and leakage G are negligible, then the partial differential equation governing the given problem is

$$\frac{\partial^2 v}{\partial x^2} = RC\frac{\partial v}{\partial t}$$

or

$$\frac{\partial v}{\partial t} = \frac{1}{RC}\frac{\partial^2 v}{\partial x^2}, \tag{187}$$

where R and C represent resistance and capacitance. The equation (187) is of *one dimensional heat equation* type and is known as *telegraph equation*.

Since the transmission line is initially under steady state, therefore $\frac{\partial v}{\partial t} = 0$ and so (187) reduces to

$$\frac{\partial^2 v}{\partial x^2} = 0,$$

which yields (on integration)

$$v = c_1 x + c_2. \tag{188}$$

Now

(i) When $x = 0$, $v = 1300$ volts, then (187) yields $c_2 = 1300$. When $x = 1000$, $v = 1200$ volts and so

$$1200 = 1000\,c_1 + c_2 = 1000c_1 + 1300$$

and so $c_1 = \frac{1}{10}$. Hence (187) implies

$$v = 1300 - \frac{x}{10} \text{ at } t = 0. \tag{189}$$

(ii) *When the terminal end is grounded*, we have $v = 1300$ for $x = 0$ and so (188) yields $c_2 = 1300$. When $x = 1000$, $v = 0$ and so $0 = 1300 + 1000c_1$, which implies $c_1 = -\frac{13}{10}$.

Thus $v_g = 1300 - \frac{13}{10}x$ (when terminal is grounded).

If $v_t(x, t)$ is the transient part of the solution, then

$$v(x, t) = v_g + v_t(x, t), \tag{190}$$

where (see Example 5.83)

$$v_t(x, t) = \sum_{n=1}^{\infty} B_n \sin\frac{n\pi x}{1000}\, e^{-\frac{n^2\pi^2 t}{RC(1000)^2}}.$$

Thus (190) reduces to

$$v(x, t) = 1300 - \frac{13x}{10} + \sum_{n=1}^{\infty} B_n \sin\left(\frac{n\pi x}{1000}\right) e^{-\frac{n^2\pi^2 t}{RC(1000)2}}.$$

When $t = 0$, we have

$$v(x, 0) = 1300 - \frac{13}{10}x + \sum_{n=1}^{\infty} B_n \sin\left(\frac{n\pi x}{1000}\right).$$

or, using (189), we have

$$1300 - \frac{x}{10} = 1300 - \frac{13}{10}x + \sum_{n=1}^{\infty} B_n \sin\left(\frac{n\pi x}{1000}\right)$$

or

$$\frac{6x}{5} = \sum_{n=1}^{\infty} B_n \sin\left(\frac{n\pi x}{1000}\right).$$

Thus B_n is Fourier (sine) coefficient of the function $\frac{6x}{5}$ and as such

$$B_n = \frac{2}{L}\int_0^L f(x)\sin\frac{n\pi x}{1000}\,dx$$

$$= \frac{2}{1000}\int_0^{1000} \frac{6x}{5}\sin\frac{n\pi x}{1000}\,dx$$

$$= -\frac{6}{2500}\left[-\frac{(1000)^2}{n\pi}\cos n\pi\right] = \frac{2400}{n\pi}(-1)^n.$$

Hence the required solution is

$$v(x, t) = 1300 - \frac{13x}{10}$$

$$+ \frac{2400}{\pi}\sum_{n=1}^{\infty}\frac{(-1)^n}{n}\sin\frac{n\pi x}{1000}\, e^{-\frac{n^2\pi^2 t}{RC(1000)^2}}.$$

EXAMPLE 5.102

Find a solution of radio equation $\frac{\partial^2 v}{\partial x^2} = LC\frac{\partial^2 v}{\partial t^2}$, when a periodic e.m.f $V_0\cos pt$ is applied at the and $x = 0$ of the line.

Solution. The given radio equation is

$$\frac{\partial^2 v}{\partial x^2} = LC\frac{\partial^2 v}{\partial t^2}.$$

We are given that

$$v(0, t) = V_0\cos pt = f(t).$$

By D'Alembert method, the solution is

$$v(x,t) = \frac{1}{2}\left[f(t + x\sqrt{LC}) + f(t - x\sqrt{LC})\right]$$

$$= \frac{1}{2}\left[V_0\cos p(t + x\sqrt{LC}) + V_0\cos p(t - x\sqrt{LC})\right]$$

$$= \frac{V_0}{2}\left[\cos pt\cos px\sqrt{LC} - \sin pt\sin px\sqrt{LC} + \cos pt\cos px\sqrt{LC} + \sin pt\sin px\sqrt{LC}\right]$$

$$= V_0\cos pt\cos px\sqrt{LC}.$$

Since the solution implies $v(0, t) = V_0\cos pt$ and

$$\frac{\partial v}{\partial x}(0, t) = \left[V_0\cos pt(-p\sqrt{LC}\sin px\sqrt{LC})\right]_{x=0}$$

$$= 0,$$

the boundary conditions for the D'Alembert's method are satisfied.

EXAMPLE 5.103

Neglecting the resistance R and the leakance G in a transmission line 1 *km*. long, obtain the voltage drop $v(x, t)$, t seconds after the ends are suddenly grounded taking the initial conditions

$$i(x, 0) = i_0 \text{ and } v(x, 0) = v_0\sin\frac{\pi x}{l}.$$

Solution. The problem is governed by the radio equation

$$\frac{\partial^2 v}{\partial x^2} = LC\frac{\partial^2 v}{\partial t^2}.$$

Since the ends are suddenly earthed, we have

$$v(0, t) = v(l, t) = 0.$$

Also, when $t = 0$, we have $i = i_0$ (constant). Therefore $\frac{\partial i}{\partial t} = 0$, which implies $C\frac{\partial v}{\partial t} = 0$ and so $\left(\frac{\partial v}{\partial t}\right)_{t=0} = 0$.

Let $v = X(x)\ T(t)$ be a trial solution. Then the radio equation reduces to

$$T(t)\frac{d^2X(x)}{dx^2} = LC\,X(x)\frac{d^2T(t)}{dt^2}$$

or

$$\frac{1}{X(x)}\frac{d^2X(x)}{dx^2} = \frac{LC}{T(t)}\frac{d^2T(t)}{dt^2} = -p^2, \text{ say}.$$

Thus we get two equations

$$(D^2 + p^2)X(x) = 0 \text{ and } \left(D^2 + \frac{p^2}{LC}\right)T(t) = 0.$$

The solution of these equations are

$$X(x) = c_1\cos px + c_2\sin px$$

and

$$T(t) = c_3\cos\frac{p}{\sqrt{LC}}t + c_4\sin\frac{p}{\sqrt{LC}}t.$$

Hence solution of the radio equation is

$$v(x,t) = (c_1 \cos px + c_2 \sin px)\times\left(c_3 \cos\frac{p}{\sqrt{LC}}t + c_4 \sin\frac{p}{\sqrt{LC}}t\right).$$

The initial condition $v(0, t) = 0$ implies

$$0 = c_1\left(c_3 \cos\frac{p}{\sqrt{LC}}t + c_4 \sin\frac{p}{\sqrt{LC}}t\right)$$

and so $c_1 = 0$. Therefore

$$v(x,t) = c_2 \sin px\left(c_3 \cos\frac{p}{\sqrt{LC}}t + c_4 \sin\frac{p}{\sqrt{LC}}t\right)$$

The initial condition $v(l, t) = 0$ implies

$$0 = c_2 \sin pl\left(c_3 \cos\frac{p}{\sqrt{LC}}t + c_4 \sin\frac{p}{\sqrt{LC}}t\right)$$

and so $\sin pl = 0$ or $p = \frac{n\varpi}{l}$. Hence the solution takes the form

$$v(x,t) = c_2 \sin\frac{n\pi x}{l}\left(c_3 \cos\frac{n\pi t}{l\sqrt{LC}} + c_4 \sin\frac{n\pi t}{l\sqrt{LC}}\right).$$

Differentiating partially with respect to t, we get

$$\frac{\partial v}{\partial t} = c_2 \sin\frac{n\pi x}{l}\left[\frac{-n\pi c_3}{l\sqrt{LC}}\sin\frac{n\pi t}{l\sqrt{LC}} + \frac{n\pi c_4}{l\sqrt{LC}}\cos\frac{n\pi t}{l\sqrt{LC}}\right].$$

Now the initial condition $\left(\frac{\partial v}{\partial t}\right)_{t=0} = 0$ implies $c_4 = 0$. Hence the solution reduces to

$$v(x,t) = -\frac{n\pi}{l\sqrt{LC}}c_2 c_3 \sin\frac{n\pi x}{l}\cos\frac{n\pi t}{l\sqrt{LC}}$$

$$= A_n \sin\frac{n\pi x}{l}\cos\frac{n\pi t}{l\sqrt{LC}},$$

$$A_n = -\frac{n\pi}{l\sqrt{LC}}c_2 c_3.$$

Therefore the general solution is

$$v(x,t) = \sum_{n=1}^{\infty} A_n \sin\frac{n\pi x}{l}\cos\frac{n\pi t}{l\sqrt{LC}}.$$

But $v(x, 0) = v_0 \sin\frac{\varpi x}{l}$. Therefore

$$v_0 \sin\frac{\pi x}{l} = \sum_{n=1}^{\infty} A_n \sin\frac{n\pi x}{l}.$$

Hence $A_1 = v_0$, $A_2 = A_3 = \ldots = 0$ and the required solution is

$$v(x,t) = V_0 \sin\frac{\pi x}{l}\cos\frac{\pi t}{l\sqrt{LC}}$$

5.14 MISCELLANEOUS EXAMPLE

EXAMPLE 5.104

Find the temperature in a thin metal rod of length L, with both the ends insulated (so that there is no passage of heat through the ends) and with initial temperature in the rod $\sin\frac{\varpi x}{L}$.

Solution. As shown is Example 5.85, since the ends of the rod of length L are insulated, the solution is

$$U(x,t) = \frac{A_0}{2} + \sum_{n=1}^{\infty} A_n e^{-\frac{n^2\pi^2 kt}{L^2}}\cos\frac{n\pi x}{L},$$

where

$$A_n = \frac{2}{L}\int_0^L f(x)\cos\frac{n\pi x}{L}dx,\ n = 0,\ 1,\ 2, \ldots.$$

Here $f(x) = \sin\frac{\varpi x}{L}$. Therefore

$$A_n = \frac{2}{L}\int_0^L \sin\frac{\pi x}{L}\cos\frac{n\pi x}{L}dx$$

$$= \frac{2}{L}\int_0^L \frac{1}{2}\left[\sin\frac{(n+1)\pi x}{L} - \frac{\sin(n-1)\pi x}{L}\right]dx$$

$$= \frac{1}{L}\left[-\frac{\cos(n+1)\pi\frac{x}{L}}{(n+1)\pi} + \frac{\cos(n-1)\pi\frac{x}{L}}{(n-1)\pi}\right]_0^L$$

$$= \frac{1}{L}\left[-\frac{\cos(n+1)\pi}{(n+1)\pi}\right] + \frac{\cos(n-1)\pi}{(n-1)\pi} + \frac{1}{(n+1)\pi} - \frac{1}{(n-1)\pi}$$

$$= \frac{1}{L}\left[\frac{(-1)^{n+2}}{(n+1)\pi} + \frac{(-1)^{n-1}}{(n-1)\pi} + \frac{1}{(n+1)\pi} - \frac{1}{(n-1)\pi}\right]$$

$$=\frac{1}{\pi L}\left[\frac{(-1)^n}{(n+1)}-\frac{(-1)^n}{(n-1)}+\frac{1}{(n+1)}-\frac{1}{(n-1)}\right]$$

$$=\frac{-2}{\pi L}[1+(-1)^n].$$

Putting $n = 0$. we have $A_0 = -\frac{4}{\pi L}$. Hence

$$u(x,t) = -\frac{2}{\pi L}$$

$$-\frac{2}{\pi L}\sum_{n=1}^{\infty}(1+(-1)^n)e^{-\frac{n^2\pi^2 kt}{L^2}}\cos\frac{n\pi x}{L}.$$

EXAMPLE 5.105

A string is stretched between two fixed points at a distant $2l$ apart and the points of the string are given initial velocities

$$v=\begin{cases}\frac{cx}{l}, 0<x<1\\ \frac{c(2l-x)}{l}, l<x<2l\end{cases},$$

where x being the distance from an end point, Find the displacement of the string at any time.

Solution. The given problem is governed by one-dimensional wave equation $\frac{\partial^2 y}{\partial x^2} = c_2\frac{\partial^2 y}{\partial t}$, under the conditions

$$y(0,t)=y(2l,t)=0, t\ge 0$$
$$y(x,0)=0$$
$$\left(\frac{\partial y}{\partial t}\right)_{t=0}=\begin{cases}\frac{cx}{l}, 0\le x<l\\ \frac{c(2l-x)}{l}, l<x<2l\end{cases}$$

Let

$$y(x, t) = T(t)\,X(x)$$

be a trial solution of the given partial differential equation. Then the wave equation takes the form

$$T''(t)X(x) = c_2X''(x)\,T(t).$$

Separation of variables yields $\frac{T''(x)}{T(t)} = c_2\frac{X''(x)}{X(x)} = k$ (constant of separation). The boundary conditions $y(0, t) = y(2l, t) = 0$, $t \ge 0$ imply $X(0) = 0$ and $X(2l) = 0$. The auxiliary equation for $X''(x)\ \frac{k}{c_2}X(x) = 0$ is $s^2 - \frac{k}{c_2} = 0$, which yields s_1 and $-s_1$ as the two roots. Therefore the fundamental solution is

$$X(x) = \alpha\, e^{s_1 x} + \beta\, e^{-s_1 x}. \tag{191}$$

The boundary conditions $X(0) = 0 = X(2l)$ imply $= \alpha\,(e^{2s_1 l} - e^{-2s_1 l})$. $0 = \alpha\, e^{s_1(2l)} + \beta\, e^{-s_1(2l)} = \alpha\,(e^{2s_1 l}$ $e^{2s_1 l})$. Since $\alpha = 0$ implies $\beta = 0$ and since the solution in that case becomes trivial, it follows that $\alpha \ne 0$ and hence $e^{2s_1 l} = e^{-2s_1 l}$ or $e^{4s_1 l} = 1$ or $s_1 = \frac{in\pi}{2l}$, when $n \ne 0$ is an integer. Also, s_1 being the root, we have $s_1^2 = \frac{k}{c^2}$. Therefore $k = \frac{c^2 n^2\pi^2}{4l^2}$. The solution (191) now reduces to

$$X_n(x)=\alpha\left[e^{\frac{in\pi x}{2l}}-e^{\frac{-in\pi x}{2l}}\right]$$
$$=2\alpha\sin\frac{n\,\pi x}{2l} \tag{192}$$

On the other hand, the characteristic equation for $T''(t) - kT(t) = 0$ is $s^2 - k = 0$, that is $s^2 + \frac{c^2n^2\pi^2}{4l^2} = 0$. Therefore the fundamental solution in this case is

$$T_n(t)=\gamma\cos\left(\frac{n\pi ct}{2l}\right)+\delta\sin\left(\frac{n\pi ct}{2l}\right). \tag{193}$$

From (192) and (193), the fundamental solution of the given wave equation is

$$y_n(x,t)=T_n(t)X_n(x)$$
$$=2\alpha\left[\gamma\cos\left(\frac{n\pi ct}{2l}\right)+\delta\sin\left(\frac{n\pi ct}{2l}\right)\right]$$
$$\times\sin\frac{n\pi x}{2l}.$$

Since the string was initially at rest, $y\,(x, 0) = 0$. This implies $y_n(x, t) = B_n \sin\left(\frac{n\pi ct}{2l}\right)\sin\left(\frac{n\pi x}{2l}\right)$.

Now the superposition of the fundamental solutions implies

$$y(x, t) = \sum_{n=1}^{\infty} B_n \sin\left(\frac{n\pi ct}{2l}\right) \sin\left(\frac{n\pi x}{2l}\right) \quad (194)$$

Differentiating (194) partially with respect to t, we get

$$\frac{\partial y}{\partial t} = \sum_{n=1}^{\infty} B_n \sin\left(\frac{n\pi x}{2l}\right)\left[\cos\left(\frac{n\pi ct}{2l}\right)\left(\frac{n\pi c}{2l}\right)\right]$$

$$= \frac{\pi c}{2l}\sum_{n=1}^{\infty} n\, B_n \cos\left(\frac{n\pi ct}{2l}\right) \sin\left(\frac{n\pi x}{2l}\right) \quad (195)$$

But, at $t = 0$, we are given that

$$\frac{\partial y}{\partial t} = \begin{cases} \frac{cx}{l}, 0 < x < l \\ \frac{c(2l-x)}{l}, \; l < x < 2l. \end{cases}$$

Therefore

$$\frac{\pi c}{2l}\sum_{n=1}^{\infty} n\, B_n \sin\left(\frac{n\pi x}{2l}\right) = \begin{cases} \frac{cx}{l}, & 0 < x < l \\ \frac{c(2l-x)}{l}, & l < x < 2l. \end{cases}$$

The series on the left hand side is a Fourier sine series for $0 < x < 2l$. Therefore

$$\frac{\pi c}{2l} n\, B_n = \frac{2}{2l}\int_0^{2l} f(x) \sin\left(\frac{n\pi x}{2l}\right) dx$$

$$= \frac{c}{l^2}\left[\int_0^l x \sin\left(\frac{n\pi x}{2l}\right) dx. + \int_l^{2l} (2l - x) \sin\left(\frac{n\pi x}{2l}\right) dx\right]$$

$$= \frac{c}{l^2}\left\{\left[x\left(\frac{-\cos\left(\frac{n\pi x}{2l}\right)}{\pi^x/2l}\right) + \frac{\sin\left(\frac{n\pi x}{2l}\right)}{n^2\pi^2/(4l^2)}\right]_0^l + \left[(2l-x)\left(\frac{-\cos\left(\frac{n\pi x}{2l}\right)}{n\pi/2l}\right) - \frac{\sin\left(\frac{n\pi x}{2l}\right)}{\frac{n^2\pi^2}{4l^2}}\right]_l^{2l}\right\}$$

$$= \frac{c}{l^2}\left[\frac{8l^2 \sin\left(\frac{n\pi}{2}\right)}{n^2\pi^2} - 4l^2\ \frac{\sin\ n\pi}{n^2\pi^2}\right]$$

$$= \frac{4c}{n^2\pi^2}\left[2 \sin\frac{n\pi}{2} - \sin\ n\pi\right]$$

$$= \frac{8c}{n^2\pi^2}\sin\frac{n\pi}{2} = \pm\frac{8c}{n^2\pi^2}.$$

Therefore

$$B_n = \frac{16lc}{cn^3\pi^3},$$

And the required solution is

$$y\ (x, t) = \frac{16\ lc}{c\pi^3}\sum\frac{1}{n^3} \sin\left(\frac{n\pi ct}{2l}\right) \sin\left(\frac{n\pi x}{2l}\right).$$

EXAMPLE 5.106

A semi circular plate of radius a centimeter has insulated faces and heat flows in plane curves. The bounding diameter is kept at 0° C and the semi-circumference is maintained at a temperature given by

$$u(a, \theta) = \begin{cases} \frac{k\theta}{\pi}, 0 < \theta \le \pi l2 \\ \frac{k(\pi-\theta)}{\pi}, \pi l2 \le \theta < \theta \end{cases}$$

Find the steady-state temperature distribution $u(r, \theta)$.

Solution. We want to solve the Laplace Equation

$$r^2\frac{\partial^2 u}{\partial r^2} + r\frac{\partial u}{\partial r} + \frac{\partial^2 u}{\partial \theta^2} = 0, \quad (196)$$

subject to the conditions

$$u(r, 0) = 0,\ u(r, \pi) = 0,\ u(0, \theta) = 0,$$

and

$$u(a,\ \theta) = \begin{cases} \frac{k\theta}{\pi},\ 0 < \theta \le \frac{\pi}{2} \\ \frac{k(\pi-\theta)}{\pi},\ \frac{\pi}{2} \le \theta < \pi. \end{cases}$$

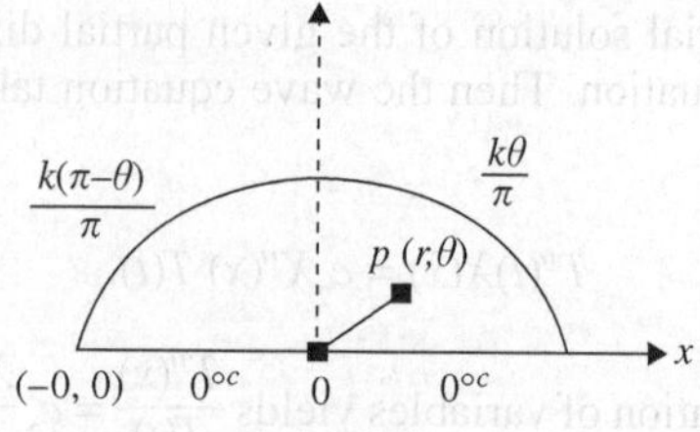

Figure 5.3

The solution to the equation (196) is

$$u(r, \theta) = (c_1 r^p + c_2 r^{-p})(c_3 \cos p\theta + c_4 \sin p\theta) \quad (197)$$

Using the boundary condition u(r, 0)= 0, we get $c_3 = 0$. Thus

$$u(r, \theta) = (c_1 r^{-p} + c_2 r^{-p}) c_4 \sin p\theta. \quad (198)$$

Now the condition $u(0, \theta) = 0$ implies

$$0 = c_2 \text{ since } r \to 0 \textit{ implies } u = 0.$$

Thus (198) transforms to

$$u(r, \theta) = c_1\, c_4\, r^p \sin p\theta = b_n r^p \sin p\theta, \textit{ say}.$$

Now $u(r, \pi) = 0$ implies

$$0 = b_n r^p \sin p\pi,$$

which yields $p\pi = n\pi$, that is, $p = n$ (integer). Hence the solution is

$$u_n(r, \theta) = b_n r^n \sin n\theta. \quad (199)$$

The superposition of the solutions implies

$$u(r, \theta) = \sum_{n=1}^{\infty} b_n r^n \sin n\theta. \quad (200)$$

Thus

$$u(a, \theta) = \sum_{n=1}^{\infty} b_n a^n \sin n\theta \quad (201)$$

The series on the right hand side of (201) is the half-range sine series of $u(a, \theta)$. Therefore

$$a^n b_n = \frac{2}{\pi}\left[\int_0^{\pi/2} \frac{k\theta}{\pi} \sin n\,\theta\, d\theta + \int_{\pi/2}^{\pi} \frac{k(\pi-\theta)}{\pi} \sin n\theta\, d\theta\right]$$

$$= \frac{2k}{\pi}\left[0\left(\frac{-\cos n\theta}{n}\right) + \frac{\sin n\theta}{n^2}\right]_0^{\pi/2} + \left[(\pi-\theta)\left(\frac{-\cos n\theta}{n}\right) - \frac{\sin n\theta}{n^2}\right]_{\pi/2}^{\pi}$$

$$= \frac{2k}{n}\left[\frac{-\pi}{2n}\cos\frac{n\pi}{2} + \frac{\sin\frac{n\pi}{2}}{n^2} + \frac{\pi}{2n}\cos\frac{n\pi}{2n} + \frac{\sin\frac{n\pi}{2}}{n^2}\right]$$

$$= \frac{4k}{\pi n^2} \sin\frac{n\pi}{2}.$$

Thus

$$b_n = \frac{4k}{\pi a^n n^2} \sin\frac{n\pi}{2}.$$

Hence the solution of the Laplace equation is $u(r, \theta) = \frac{4k}{\pi}\sum \frac{1}{n^2}\left(\frac{r}{a}\right)^n \sin\frac{n\pi}{2} \sin n\theta.$

EXAMPLE 5.107

Use the method of separation of variables to solve the equation $\frac{\partial^2 V}{\partial x^2} = \frac{\partial V}{\partial t}$.

Solution. Let $V(x, t) = T(t)X(x)$ be the possible solution of $\frac{\partial V}{\partial t} = \frac{\partial^2 V}{\partial x^2}$. Then the given heat equation reduces to

$$T'(t)\, X(x) = X''(x)\, T(t).$$

Separation of variable yields

$$\frac{T'(t)}{T(t)} = \frac{X''(x)}{X(x)}. \quad (202)$$

The left hand side of (100) is a function of t only, where as the right hand side is a function of x only. Since x and t are independent variables, (202) is valid only if each side is equal to some constant C, called the constant of separation. Thus

$$\frac{T'(t)}{T(t)} = \frac{X''(x)}{X(x)} = C.$$

Three cases arise:

Case I. When C is positive, Let $c = k^2$. Then

For $\dfrac{T'(t)}{T(t)} = c$ $= k^2$	For $\dfrac{X''(x)}{X(x)} = c$ $= k^2$
We have $$T'(t) - k^2T(t) = 0.$$ The auxiliary equation is $$m - k^2 = 0$$ and so $m = k^2$ and $$\text{C.F.} = c_1 e^{k^2 t}.$$ Hence $$T(t) = c_1 e^{k^2 t}.$$	We have $$X''(x) - k^2X(x) = 0.$$ The auxiliary equation is $$m^2 - k^2 = 0$$ and so $m = \pm k$. Hence $$\text{C.F.} = c_2 e^{kx} + c_3 e^{-kx}$$ and thus the solution is $$X(x) = c_2 e^{kx} + c_3 e^{-kx}.$$

Case II. Where C is negative, let $c = -k^2$. Then

For $\dfrac{T'(t)}{T(t)} = -k^2$	For $\dfrac{X''(x)}{X(x)} = -k^2$
We have $$T'(t) + k^2T(t) = 0.$$ The auxiliary equation is $$m + k^2 = 0,$$ Which yields $m = -k^2$. Thus the solution is $$T(t) = c_1 e^{-k^2 t}$$	We have $$X''(x) + k^2X(x) = 0.$$ The auxiliary equation is $$m^2 + k^2 = 0,$$ which yields $m = \pm ki$. Hence the solution is $$X(x) = c_2 \cos kx + c_3 \sin kx.$$

Case III. Where $C = 0$. Then

$\frac{T'(t)}{T(t)} = 0$ and so $T'(t) = 0$, which implies $T(t) = c_1$.	$$\frac{X''(x)}{X(x)} = 0$$ and so $X''(x) = 0$, which implies $X(x) = c_2 x + c_3$.

The above three cases suggest that the possible solutions of the given heat equation are:

(i) $V(x,t) = c_1 e^{k^2 t}(c_2 e^{kx} + c_3 e^{-kx})$

(ii) $V(x,t) = c_1 e^{-k^2 t}(c_2 \cos kx + c_3 \sin kx)$

(iii) $V(x,t) = c_1 (c_2 x + c_3)$.

EXAMPLE 5.108

Solve the following equations by method of separation of variables.

(a) $\dfrac{\partial^2 u}{\partial x \partial t} = e^{-t} \cos x$,
given that $u=0$ when $t=0$ and $\dfrac{\partial u}{\partial t}=0$ when $x=0$.

(b) $\dfrac{\partial u}{\partial x} = 4\dfrac{\partial u}{\partial y}$, $u(0, y) = 8\, e^{-3y}$.

Solution. (a) Let

$$u = X(x)\, T(t)$$

be a trial solution of the given differential equation. Then

$$\frac{\partial u}{\partial t} = X(x)T'(t),$$

$$\frac{\partial^2 u}{\partial x\,\partial t} = T'(t)X'(x).$$

Therefore the given equation transforms to

$$T'(t)X'(x) = e^{-t}\cos x.$$

The separation of variable yields

$$e^{t}T'(t) = \frac{\cos x}{X'(x)} = a \text{ (constant)}$$

Then

$$e^{t}T'(t) = a$$

or

$$dT = ae^{-t}dt \qquad (203)$$

Integration of (203) yields

$$T = -ae^{-t} + c_1 \qquad (204)$$

Further,

$$X'(x) = \frac{1}{a}\cos x\, dx \qquad (205)$$

Integrating (205), we get

$$X = \frac{1}{a}\sin x + c_2 \qquad (206)$$

From (204) and (206), we get

$$u(x,t) = \left(\frac{1}{a}\sin x + c_2\right)(-ae^{-t} + c_1) \qquad (207)$$

Since u is zero when $t = 0$, we have

$$0 = \left(\frac{1}{a}\sin x + c_2\right)(-a + c_1)$$

This is possible if $(-a + c_1) = 0$ or if $c_1 = a$. Further, the solution (206) implies

$$\frac{\partial u}{\partial t} = \left(\frac{1}{a}\sin x + c_2\right)(ae^{-t}). \qquad (208)$$

Using the condition that $\frac{\partial u}{\partial t} = 0$ when, $x = 0$, we get from (208) that $0 = c_2 a\, e^{-t}$, which implies $c_2 = 0$.

Substituting the values c_1 and c_2 in the solution (207), we have

$$u(x,t) = \frac{1}{a}\sin x[-ae^{-t} + a] = \sin x(1 - e^{-t}).$$

(b) The given partial differential equation is

$$\frac{\partial u}{\partial x} = 4\frac{\partial u}{\partial y},\quad u(0,y) = 8\,e^{-3y} \qquad (209)$$

Let

$$u = X(x)Y(y)$$

be a trial solution of the given partial differential equation Then (209) implies

$$X'(x)Y(y) = 4X(x)Y'(y)$$

Separating the variables, we get

$$\frac{X'(x)}{X(x)} = 4\frac{Y'(y)}{Y(y)} = a \text{ (constant)}$$

Thus we get two ordinary differential equations

$$\frac{X'(x)}{X(x)} = a \qquad (210)$$

and

$$4\frac{Y'(x)}{X(x)} = a \qquad (211)$$

Integrating (211), we get

$$\log X = ax + \log c_1$$

or

$$X = c_1 e^{ax} \qquad (212)$$

Integrating (211), we have

$$\log Y = \frac{a}{4}y + \log c_2$$

or

$$Y = c_2 e^{\frac{ay}{4}} \qquad (213)$$

Therefore

$$u = X(x)Y(y) = c_1c_2\ e^{ax}\ e^{\frac{ay}{4}}$$

$$= c_1c_2e^{ax+\frac{ay}{4}} = c_1c_2\, e^{\frac{4ax+ay}{4}}$$

But $u(0, y) = 8\,e^{-3y}$. Therefore

$$8e^{-3y} = c_1c_2e^{\frac{ay}{4}},$$

which implies $c_1c_2 = 8$ and $\frac{a}{4} = -3$. Thus $a = -12$. Hence the solution is

$$u(x,y) = 8e^{\frac{-48x - 12y}{4}} = 8e^{-12x - 3y}.$$

EXERCISES

1. Form the partial differential equation from the following:

(a) $z = ax + by + a^2 + b^2$

Ans. $z = px + qy + p^2 + q^2$

(b) $z = (a + x^2)(b + y^2)$

Ans. $pq = 4xyz$

(c) $(x - a)^2 + (y - b)^2 + z^2 = c^2$

Ans. $z^2(p^2 + q^2 + 1) = c^2$

2. Find the differential equation of all planes which are at a constant distance 'a' from the origin.

Ans. $z = px + qy + a\sqrt{1 + p^2 + q^2}$

3. Form partial differential equation by eliminating the arbitrary functions from the following:

(a) $z = f_1(x) f_2(y)$

Ans. $z\dfrac{\partial^2 z}{\partial x \partial y} - \dfrac{\partial z}{\partial x} \cdot \dfrac{\partial z}{\partial y} = 0$

(b) $z = f\left(\frac{y}{x}\right)$

Ans. $px + qy = 0$

(c) $z = xf_1(x + t) + f_2(x + c)$

Ans. $\dfrac{\partial^2 z}{\partial x^2} - 2\dfrac{\partial^2 z}{\partial x \partial t} + \dfrac{\partial^2 z}{\partial t^2} = 0$

(d) $z = f(x) + e^y g(x)$

Ans. $\dfrac{\partial^2 z}{\partial y^2} = \dfrac{\partial z}{\partial y}$.

4. Solve:

(a) $\dfrac{\partial^2 z}{\partial x^2} + z = 0$ given that when $x = 0, z = e^y$, and $\dfrac{\partial z}{\partial x} = 1$.

Ans. $z = e^y \cos x + \sin x$

(b) $\dfrac{\partial^2 z}{\partial y^2} = \sin(xy)$

Ans. $z = -\dfrac{1}{x^2}\sin(xy) + y f(x) + \phi(x)$

(c) $\dfrac{\partial^2 u}{\partial x \partial t} = e^{-t}\cos x$

Ans. $u = -e^{-t}\sin x + f(x) + g(t)$

5. Solve the Lagrange's linear equations":

(a) $\dfrac{y^2 z}{x} p + xzq = y^2$

Ans. $\phi(x^3 - y^3, x^2 - z^2) = 0$

(b) $x^2(y - z)p + y^2(z - x)q = z^2(x - y)$

Ans. $f\left(\frac{1}{x} + \frac{1}{y} + \frac{1}{z}, xyz\right) = 0$

(c) $x^2 p + y^2 q = (x + y)z$

Ans. $f\left(\frac{xy}{z}, \frac{1}{y} - \frac{1}{x}\right) = 0$

(d) $y^2 p - xyq = x(z - 2y)$

Ans. $f(x^2 + y^2, yz - y^2) = 0$

(e) $xp + yq = 3z$

Ans. $f\left(\frac{x}{y}, \frac{x^3}{z}\right) = 0$

(f) $(y - z)p + (x - y)q = z - x$

Ans. $f\left(x + y + z, \dfrac{x^2}{2} + yz\right) = 0$

(g) $p - q = \log(x+y)$

Ans. $x\log(x + y) - z = f(x + y)$

(h) $px(z - 2y^2) = (z - qy)(z - y^2 - 2x^3)$

Ans. $f\left(\frac{y}{z}, \frac{z}{x} - \dfrac{a^2 z^2}{x} + x^2\right) = 0$

6. Solve the following equations by Charpit's method:

(a) $2z + p^2 + qy + 2y^2 = 0$

Ans. $y^2[(x - a)^2 + y^2 + 2z] = b$.

(b) $2(z + xp + yp) = yp^2$

Ans. $z = \dfrac{ax}{y^2} - \dfrac{a^2}{4y^3} + \dfrac{b}{y}$

(c) $2zx - px^2 - 2pxy + pq = 0$

Ans. $z = ay + b(x^2 - a)$

(d) $1 + p^2 = qz$

Ans. $\dfrac{z^2}{2} \pm \left[\dfrac{z}{2}\sqrt{z^2 - 4a^2} - 2a^2\log\left(z + \sqrt{z^2 - 4a^2}\right)\right] = 2ax + 2y + b]$

(e) $p(q^2 + 1) + (b - z)q = 0$

Ans. $q\sqrt{z - a - b} = \sqrt{ax} + \dfrac{1}{\sqrt{a}}y + c$

7. Solve:

(a) $(y - x)(qx - px) = (p - q)^2$

Hint: Convert to $f(P, Q) = 0$ form

Ans. $z = a(x + y) + \sqrt{a}\, xy + c$

(b) $p^2 + p = q^2$

Ans. $z = ax + \sqrt{a^2 + ay + c}$

(c) $z^2\left(\frac{p^2}{x^2} + \frac{q^2}{y^2}\right) = 1$

Ans. $z = ax^2 + \sqrt{1 - a^2y^2} + c$

(d) $p^2x + q^2y = z$

Hint: Put $\frac{dx}{\sqrt{x}} = dX, \frac{dy}{\sqrt{y}} = dY, \frac{dz}{\sqrt{z}} = dZ$

Ans. $2\sqrt{z} = 2a\sqrt{x} + 2\sqrt{y}\,\sqrt{1-a^2} + c$

(e) $\sqrt{p} + \sqrt{q} = 1$

Ans. $z = ax + (1 - \sqrt{a})^2 y + c.$

(f) $pq = x^m y^n z^{2l}$

Hint: Put $x^m dx = dX$, $y^n\, dy = dY$ and $z^{-l}dz = dZ$

Ans. $\frac{z^{l-1}}{1-l} = a\frac{x^{m+1}}{m+1} + \frac{y^{n+1}}{a(n+1)} + c$

8. Solve:

(a) $p + q = z.$

Ans. $(1 + a)\log z = z + ay + b$

(b) $z^2(p^2z^2 + q^2) = 1$

Ans. $(z^2 + a^2)^3 = 9(x + ay + b)^2$

(c) $z^2 = 1 + p^2 + q^2$

Ans. $z = \cosh\frac{x+ay+b}{\sqrt{a^2+1}}$

(d) $p^2 = qz$

Ans. $z = b\, e^{ax + a^2y}$

(e) $z = pq + p + q = 0$

Ans. $z = ax - ay/(1 + a) + b$

9. Solve

(a) $p - x^2 = q + y^2$

Ans. $z = \frac{1}{3}(x^3 - y^3) + a(x + y) + b$

(b) $q = xyp^2$

Ans. $z = 2\sqrt{ax} + \frac{1}{2}ay^2 + b$

(c) $pq = x^4y^3z^4$

Ans. $z = \dfrac{-1}{c\frac{x^5}{5} + \frac{y^4}{4c} + b}$

(d) $\sqrt{p} + \sqrt{q} = 2x$

Ans. $z = \frac{1}{6}(2x + a)^3 + a^2y + b$

(e) $p^2 + q^2 = z^2(x + y)$

Ans. $\log z = \frac{2}{3}(a + x)^{3/2} + \frac{2}{3}(y - a)^{3/2} + c$

10. Solve:

(a) $z = px + qy + \sin(p + q)$

Ans. $z = ax + by + \sin(a + b)$

(b) $z = px + qy - 2\sqrt{pq}$

Ans. $z = ax + by - 2\sqrt{ab}$

Solve:

11. $\frac{\partial^2 z}{\partial x^2} - \frac{\partial^2 z}{\partial y^2} = 0.$

Ans. $z = f_1(y + x) + f_2(y - x).$

12. $r + t + 2s = 0.$

Ans. $z = f_1(y - x) + xf_2(y - x).$

13. $(4D^2 + 12D.D' + 9D'^2)z = 0.$

Ans. $z = f_1(2y - 3x) + xf_2(2y - 3x).$

14. $2\frac{\partial^2 z}{\partial x^2} - 3\frac{\partial^2 z}{\partial x\partial y} - 2\frac{\partial^2 z}{\partial y^2} = 0.$

Ans. $z = f_1(2y - x) + f_2(y + 2x).$

15. $\frac{\partial^4 z}{\partial x^4} + \frac{\partial^4 z}{\partial y^4} = 2\frac{\partial^4 z}{\partial x^2\partial y^2}$

Ans. $z = f_1(y + x) + xf_2(y + x) + f_3(y - x) + xf_4(y - x).$

16. $(D^2 - a^2D'^2)z = x^2.$

Ans. $z = f_1(y + ax) + xf_2(p - ax) + \frac{1}{12}x^4.$

17. $\frac{\partial^2 z}{\partial x^2} + 3\frac{\partial^2 z}{\partial x\partial y} + 2\frac{\partial^2 z}{\partial y^2} = 2x + 3y.$

Ans. $z = f_1(y - x) + f_2(y - 2x) - \frac{7}{6}x^3 + \frac{3}{2}x^2y.$

18. $\frac{\partial^2 z}{\partial x^2} + 3\frac{\partial^2 z}{\partial x \partial y} + 2\frac{\partial^2 z}{\partial y^2} = 6(x+y)$

Ans. $z = f_1(y-x) + f_2(y-2x) - \frac{1}{3}x^3 + \frac{1}{2}x^2 y.$

19. $(D^2 + 3DD' + 2D'^2)z = 12xy.$

Ans. $z = f_1(y-x) + f_2(y-2x) + 2x^2 y - 6x^3.$

20. $\frac{\partial z}{\partial x} + \frac{\partial z}{\partial y} = \sin x.$

Ans. $z = f(y-x) - \cos x.$

21. $r - 2s + t = \sin(2x+3y).$

Ans. $z = f_1(x+y) + xf_2(x+y) - \sin(2x+3y).$

22. $(D^3 - 4D^2 D' + 4DD'^2)z = \cos(2x+y).$

Ans. $z = f_1(y) + f_2(y+2x) + xf_3(y+2x) + \frac{1}{4}x^2 \sin(2x+y).$

23. $\frac{\partial^3 z}{\partial x^3} - 4\frac{\partial^2 z}{\partial x^2 \partial y} + 4\frac{\partial^2 z}{\partial x \partial y^2} = \sin(2x+y)$

Ans. $z = f_1(y) + f_2(y+2x) + xf_3(y+2x) - \frac{x^2}{4}\cos(2x+y).$

24. $(D^2 + D'^2)z = 12(x+y).$

Ans. $z = f_1(y+ix) + f_2(y-ix) + (x+y)^3.$

25. $(2D^2 - 5DD' + 2D'^2) = 5\sin(2x+y).$

Ans. $z = f_1(2y+x) + f_2(y+2x) - (5x/3)\cos(2x+y).$

26. $\frac{\partial^2 z}{\partial x^2} + \frac{\partial^2 z}{\partial x \partial y} - 6\frac{\partial^2 z}{\partial y^2} = y\sin x.$

Ans. $z = f_1(y-3x) + f_2(y+2x) - (y\sin x + \cos x).$

27. $t + s + q = 0$

Ans. $z = f_1(x) + e^{-x} f_2(y-x).$

28. $(D + D' - 1)(D + 2D' - 2)z = 0.$

Ans. $z = e^x f_1(y-x) + e^{2x} f_2(y-2x).$

29. $(D^2 + DD' + D' - 1)z = \sin(x+2y).$

Ans. $z = e^{-x} f_1(y) + e^x f_2(y-x) - \frac{1}{20}[2\cos(x+2y) + 4\sin(x+2y)]$

30. $(D^2 - DD' - 2D)z = \sin(3x+4y) + x^2 y.$

Ans. $z = f_1(y) + e^{2x} f_2(y+x) + \frac{1}{15}\sin(3x+4y) + \frac{2}{15}\cos(3x+4y) - \frac{1}{6}x^3 y - \frac{1}{4}x^2 y + \frac{1}{12}x^3 - \frac{1}{4}xy + \frac{1}{4}x^2 - \frac{1}{8}y + \frac{3}{8}x.$

31. $(D^2 - DD' - 2D)z = \sin(3x+4y) - e^{2x+y}.$

Ans. $z = f_1(y) + e^{2x} f_2(y+x) + \frac{1}{15}\sin(3x + 4y) + \frac{2}{15}\cos(3x+3y) + \frac{1}{2}e^{2x+y}$

32. $x^2\frac{\partial^2 z}{\partial x^2} - y^2\frac{\partial^2 z}{\partial y^2} = xy.$

Ans. $z = x\phi_1(y/x) + \phi_2(xy) + xy\log x$

33. $(D^2 - D')z = 2y - x^2.$

Ans. $z = \sum Ae^{hx+h^2 y} - \left(y^2 + \frac{x^4}{12}\right).$

34. $x^2\frac{\partial^2 z}{\partial x^2} - y^2\frac{\partial^2 z}{\partial y^2} = x^2 y.$

Ans. $z = x\phi_1(y/x) + \phi_2(xy) + \frac{1}{2}x^2 y.$

35. $\frac{\partial^2 z}{\partial x^2} - 4\frac{\partial^2 z}{\partial x \partial y} + 4\frac{\partial^2 z}{\partial y^2} + \frac{\partial z}{\partial x} - 2\frac{\partial z}{\partial y} = e^{x+y}.$

Ans. $z = f_1(y+2x) + e^{-x} f_2(y+2x) + \frac{1}{2}ye^{x+y}.$

36. $\frac{\partial^2 y}{\partial t^2} - a^2\frac{\partial^2 y}{\partial x^2} = E\sin pt.$

Hint: Roots of A.E are $m = \pm a$. Therefore

$$\text{P.I.} = \frac{1}{D^2 - 2D'^2} E\sin pt = -\frac{E}{p^2}\sin pt$$

$$\text{P.I.} = \frac{1}{D^2 - 2D'^2} E\sin pt = -\frac{E}{p^2}\sin pt$$

Ans. $y = f_1(x-at) + f_2(x+at) = \frac{E}{p^2}\sin pt.$

37. $(D^2 - DD' - 2D'^2)z = e^x (y - 1)$

Ans. $z = f_1(y - x) + f_2(y + 2x) + y\,e^x$

38. $(D^2 - DD')z = \cos x \cos 2y$

Hint: Convert $\cos x \cos 2y$ into $\frac{1}{2}[\cos(x + 2y) + \cos(x - 2y)]$ and then use standard method to find P.I.

Ans. $z = f_1(y) + f_2(y + x) - \frac{1}{6}\cos(x - 2y) + \frac{1}{2}\cos(x + 2y)$

39. $(D^3 - 7DD'^2 - 6D'^3)z = x^2 + xy^2 + y^3 + \cos(x-y)$

Ans. $f_1(y - x) + f_2(y - 2x) + f_3(y + 3x) + \frac{5}{72}x^6 + \frac{1}{60}x^5 + \frac{7}{50}x^5y + \frac{1}{24}x^4x^2 + \frac{1}{6}x^3y^3 + \frac{x}{4}\cos(x - y).$

40. $(D^2 - 2DD' + D'^2)z = \sin x$

Ans. $z = f_1(y + x) + xf_2(y + x) - \sin x.$

41. $(D^2 - D'^2)z = \tan^3 x \tan y - \tan^3 y \tan x$

Ans. $z = f_1(y + x) + f_2(y - x) + \frac{1}{2}\tan x \tan y.$

42. $\frac{\partial^2 z}{\partial x^2} - \frac{\partial^2 z}{\partial y^2} = e^{x+2y}$

Ans. $z = f_1(y + x) + f_2(y - x) - \frac{1}{3}e^{x+y}$

Solve the following non-homogeneous partial differential equations:

43. $(D^2 - DD' + D')z = x^2 + y^2$

Ans. $z = f_1(y) + e^{-x} f_2(y + x) + \frac{x^3}{3} - x^2 + xy^2 + 6x$

44. $(D^2 + D + D')z = 0$

Ans. $z = c\,e^{ax-(a^2+a)y}$, where a and c are constants.

45. $(2DD' + D'2 - 3D')z = 3\cos(3x - 2y)$

Ans. $x = f_1(x) + e^{3y}f_2(2y - x) + \frac{3}{50}[3\sin(3x - 2y) + 4\cos(3x - 2y)]$

46. $r - s + p = 1$

Ans. $z = f_1(y) + e^{-x} f_2(y + x) + x$

47. $(D^2 - D' - 1)z = x^2y$

Ans. $z = c_1\, e^{ax+(a^2-1)y} - x^2y + x^2 - 2y + 8,$

where a and c are constants.

48. Solve the following by the method of separation of variables

(a) $\frac{\partial^2 u}{\partial x \partial t} = e^{-t}\cos x,\ u(0) = 0,$

$\frac{\partial u}{\partial t} = 0$ at $t = 0$.

Ans. $u = (1 - e^{-t})\sin x$

(b) $3\frac{\partial u}{\partial x} + 2\frac{\partial u}{\partial y} = 0,\ u(x, 0) = 4e^{-x}.$

Ans. $3e^{-5x-3y} + 2e^{-3x-2y}$

(c) $\frac{\partial^2 u}{\partial x^2} - \frac{\partial u}{\partial y} - 2u = 0,\ u(0, y) = 0,$

$\frac{\partial u}{\partial x}(0, y) = 1 + e^{-3y}.$

Ans. $u = \frac{1}{\sqrt{2}}\sin\sqrt{2x} + e^{-3y}\sin x$

(d) $\frac{\partial^2 x^2}{\partial x^2} = h^2\frac{\partial u}{\partial t},\ u(0, t) = u(l, t) = 0,$

$u(x, 0) = \sin\frac{\pi x}{l}$

Ans. $u(x, t) = e^{-\frac{\pi 2t2}{l2}}\sin\frac{\pi x}{l}.$

49. Solve the equation $\frac{\partial u}{\partial t} = \frac{\partial^2 u}{\partial x^2}$ with boundary $u(x, 0) = 3\sin n x,\ u(0, t) = 0,\ u(l, t) = 0,$ where $0 < x < 1,\ t > 0.$

Ans. $u(x, t) = 3\sum_{n=1}^{\infty} e^{-n^2\pi^2 t}.\ \sin(n\pi x).$

50. Solve $\frac{\partial u}{\partial t} = k\frac{\partial^2 u}{\partial x^2},\ u_x(0, l) = 0,\ u(x, y) \rightarrow$ finite number as $t \rightarrow \infty,\ u(0, l) = lx - x^2.$

Ans. $u(x, t) = \frac{1}{6}l^2 - \frac{l^2}{\pi^2}\sum_{n=1}^{\infty}\frac{1}{n^2}\cos\frac{2n\pi x}{l}e^{-\frac{4\pi 2n2\,kt}{l2}}$

51. A tightly stretched string with fixed end points $x = 0$ and $x = 1$ is initially at rest in its equilibrium position. If it is set vibrating by giving to each of its points a velocity of $\lambda x(l - x)$, find the displacement of the string at any distance x from one end at any time t.

$$\textbf{Ans. } y(x, t) = \frac{8\lambda l^3}{c\pi^4} \sum_{m=1}^{\infty} \frac{1}{(2m-1)^4}$$

$$\sin\frac{(2m-1)\pi e^t}{l} \times \sin\frac{(2m-1)\pi x}{l}.$$

52. Solve the wave equation $\frac{\partial^2 y}{\partial t^2} = 4\frac{\partial^2 y}{\partial x^2}$ subject to the conditions

$y(0, t) = 0,$

$$\left(\frac{\partial y}{\partial t}\right)_{t=0} = 3\sin(2\pi x) - 2\sin(5\pi x)$$

$$\textbf{Ans. } y(x, t) = \frac{1}{20\pi}[15\sin(4\pi t)\sin(2\pi x) - 4\sin(5\pi x)\sin(10\pi t)]$$

53. A tightly stretched string with fixed points $x = 0$ and $x = l$ is initially in a position given by $y = y_0 \sin^3\left(\frac{\pi x}{l}\right)$. If it is released from rest from this position, find the displacement $y(x, t)$ at any time t.

Ans. $y(x, t)$

$$= \frac{y_0}{4}\left(3\sin\frac{\pi x}{l}\cos\frac{\pi x c}{l} - \sin\frac{3\pi x}{l}\cos\frac{3\pi ct}{l}\right)$$

54. An infinitely long plate uniform plate is bounded by two parallel edges and an end at right angle to them. The breadth is π. This end is maintained at a temperature u_0 at all points and the other edges are at zero temperature. Determine the temperature at any point of the plate in a steady state.

$$\textbf{Ans. } u(x, y) = \frac{440}{\pi}[e^{-y}\sin x + \frac{1}{3}e^{-3y}\sin 3x + \frac{1}{5}e^{-5y}\sin 4x + \ldots]$$

55. A homogeneous rod of conducting material of length 100 cm has its ends kept at zero temperature and the initial temperature is

$$u(x, 0) = \begin{cases} x, & 0 \le x \le 50 \\ 100 - x, & 50 \le x \le 100. \end{cases}$$

Find the temperature $u(x, t)$ at any time t.

$$\textbf{Ans. } u(x, t) = \frac{400}{\pi^2}\sum_{n=1}^{\infty}\frac{(-1)^n}{(2n+1)^2}e^{-\frac{(2n+1)^2c^2\pi^2 t}{100^2}}$$

56. A tightly stretched unit square membrane starts vibrating from rest and its initial displacement $k \sin 2\pi x \sin y$. Show that the deflection at any instant is $k\sin 2\pi x \sin \pi y \cos \sqrt{5}\pi ct$.

57. Show that the deflection of a rectangular membrane ($0 \le x \le a$, $0 \le y \le b$) with fixed boundary and starting from rest satisfying $u(x, y, 0) = xy(a - x)(b - y)$ is given by

$$u(x, y, t) = \sum_{m=1}^{\infty}\sum_{n=1}^{\infty} A_{mn}\cos ckt \times \sin\frac{m\pi x}{a}\sin\frac{n\pi y}{b},$$

where

$$A_{mn} = \frac{16a^2b^2}{m^3n^3\pi^6}(1 - \cos m\pi)(1 - \sin n\pi) \text{ and}$$

$$k^2 = \pi^2\left(\frac{m^2}{a^2} + \frac{n^2}{b^2}\right).$$

6 Fourier Series

In the early 18th century, the work of C. Maclaurin and B. Taylor led to the representation of functions like sin x, cos x, e^x, and arc tan x as power series expansions. By the middle of the eighteenth century it became important to study the possibility of representation of the given function by infinite series other than the power series. Since many phenomena like vibration of string, the voltages and currents in electrical networks, electromagnetic signals, and movement of pendulum are periodic in nature, physicist, and mathematicians discussed the possibility of representing a periodic function as an infinite series involving sinusoidal (sin x and cos x) functions. In his classic *Theorie analytique de la chaleur*, published in 1822, the French physicist Jean Baptiste Joseph Fourier announced in his work on *heat conduction* that an arbitrary periodic function could be expanded in a series of sinusoidal functions. Thus, the aim of the theory of *Fourier series* is to determine the conditions under which the periodic functions can be represented as linear combinations of sine and cosine functions. These combinations are called Fourier series and the coefficients that occur in the combinations are called *Fourier-coefficients*.

6.1 TRIGONOMETRIC SERIES

Let T denote the period of the periodic functions. Our aim is to approximate arbitrary periodic function as linear combination of sine and cosine functions. Therefore in that situation the sine and cosine functions must also have period T. Obviously, the functions $\sin \frac{2\pi t}{T}$, $\cos \frac{2\pi t}{T}$, $\sin \frac{4\pi t}{T}$, $\cos \frac{4\pi t}{T}$, and so on have period T. The constant function also has period T. Thus the functions $\sin \frac{2\pi nt}{T}$, and $\cos \frac{2\pi nt}{T}$, $n \in N$ have period T. Put $\frac{2\pi}{T} = \omega_0$. Then the functions $\sin n\omega_0 t$ and $\cos n\omega_0 t$, $n \in N$ have period T. The constant $\omega_0 = \frac{2\pi}{T}$ is called the fundamental frequency. The functions $\sin \omega_0 t$ and $\cos \omega_0 t$ will complete exactly one cycle on an interval of length T whereas functions $\sin n\omega_0 t$ and $\cos n\omega_0 t$ with $n > 1$ will complete several cycles (Figure 6.1). Thus, frequencies of $\sin n\omega_0 t$ and $\cos n\omega_0 t$ are integer multiples of ω_0. The linear combinations, called superpositions, are again periodic with period 2π.

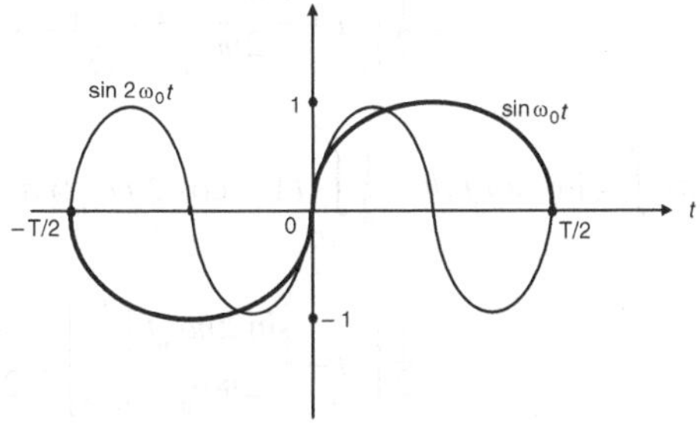

Figure 6.1 Sin $n\omega_0 t$ for $n = 1, 2$

Definition 6.1. An expression of the form

$$f(t) = a_0 + \sum_{k=1}^{n} (a_k \cos k\,\omega_0 t + b_k \sin k\,\omega_0 t),$$

$$\omega_0 = \frac{2\pi}{T},$$

where $a_0, a_1, a_2, \ldots$ are constants, is called a *trigonometric polynomial* with period T.

If $|a_n| + |b_n| > 0$, then the number n is called the order of the trigonometric polynomial.

Definition 6.2. An expression of the form

$$\frac{a_0}{2} + \sum_{n=1}^{\infty} (a_n \cos n\omega_0 t + b_n \sin n\omega_0 t), \qquad (1)$$

where $a_0, a_1, a_2, \ldots$ are constants, is called a *trigonometric series*. The constants a_n, b_n are called *coefficients of the series*.

The free (first) term has been taken as $\frac{a_0}{2}$ so that it can be found directly from the formula for a_n by taking $n = 0$, while finding Fourier coefficients.

The following trigonometric identities shall be required in the forthcoming discussion.

(a) $$\int_{-T/2}^{T/2} \cos n\omega_0 t \, dt = \left|\frac{\sin n\omega_0 t}{n\omega_0}\right|_{-T/2}^{T/2} = 0,$$

(b) $$\int_{-T/2}^{T/2} \sin n\omega_0 t \, dt = \left|\frac{-\cos n\omega_0 t}{n\omega_0}\right|_{-T/2}^{T/2} = 0,$$

(c) $$\int_{-T/2}^{T/2} \cos^2 n\omega_0 t \, dt = \frac{1}{2}\int_{-T/2}^{T/2} (1 + \cos 2\, n\omega_0 t)\, dt$$

$$= \frac{1}{2}\left[\left| t + \frac{\sin 2n\omega_0 t}{2n\omega_0}\right|_{-T/2}^{T/2}\right] = \frac{T}{2},$$

(d) $$\int_{-T/2}^{T/2} \sin^2 n\omega_0 t \, dt = \frac{1}{2}\int_{-T/2}^{T/2} (1 - \cos 2\, n\omega_0 t)\, dt$$

$$= \frac{1}{2}\left[\left| t - \frac{\sin 2n\omega_0 t}{2n\omega_0}\right|_{-T/2}^{T/2}\right] = \frac{T}{2},$$

(e) $$\int_{-T/2}^{T/2} \sin n\omega_0 t \cos m\omega_0 t \, dt$$

$$= \frac{1}{2}\int_{-T/2}^{T/2} [\sin (n+m)\, \omega_0 t + \sin (n-m)\, \omega_0 t]\, dt$$

$= 0$ using (b),

(f) If $n, m \in$ N; $n \neq m$; then

$$\int_{-T/2}^{T/2} \cos n\omega_0 t \cos m\omega_0 t \, dt$$

$$= \frac{1}{2}\int_{-T/2}^{T/2} [\cos (n+m)\, \omega_0 t + \cos (n-m)\, \omega_0 t]\, dt$$

$= 0$ using (a),

(g) If $n, m \in$ N, $n \neq m$, then

$$\int_{-T/2}^{T/2} \sin n\omega_0 t \sin m\omega_0 t \, dt$$

$$= \frac{1}{2}\int_{-T/2}^{T/2} [\cos (n-m)\, \omega_0 t - \cos (n+m)\, \omega_0 t]\, dt$$

$= 0$ using (a).

6.2 FOURIER (OR EULER) FORMULAE

Let $f(t)$, defined on $\left[-\frac{T}{2}, \frac{T}{2}\right]$, be the sum of the trigonometric series (1). Thus

$$f(t) = \frac{a_0}{2} + \sum_{n=1}^{\infty} (a_n \cos n\omega_0 t + b_n \sin n\omega_0 t). \qquad (2)$$

Suppose that this trigonometric series converges uniformly in $\left[-\frac{T}{2}, \frac{T}{2}\right]$. Then term-by-term integration of the series is valid. Therefore, integration of (2) term-by-term yields

$$\int_{-T/2}^{T/2} f(t)\, dt = \int_{-T/2}^{T/2} \frac{a_0}{2}\, dt + \sum_{n=1}^{\infty} a_n \int_{-T/2}^{T/2} \cos n\omega_0 t \, dt$$

$$+ \sum_{n=1}^{\infty} b_n \int_{-T/2}^{T/2} \sin n\omega_0 t \, dt$$

$$= \frac{a_0}{2} T + 0 + 0 = \frac{a_0}{2} T,$$

and so

$$a_0 = \frac{2}{T}\int_{-T/2}^{T/2} f(t)\, dt \qquad (3)$$

Now multiplying (2) by cos $m\omega_0 t$ and integrating, we obtain

$$\int_{-T/2}^{T/2} f(t)\cos m\omega_0 t\,dt = \frac{a_0}{2}\int_{-T/2}^{T/2} \cos m\omega_0 t\,dt$$

$$+\sum_{n=1}^{\infty} a_n \int_{-T/2}^{T/2} \cos m\omega_0 t \cos n\omega_0 t\,dt$$

$$+\sum_{n=1}^{\infty} b_n \int_{-T/2}^{T/2} \cos m\omega_0 t \sin n\omega_0 t\,dt$$

$$= 0 + \sum_{n=1}^{\infty} a_n \int_{-T/2}^{T/2} \cos m\omega_0 t \cos n\omega_0 t\,dt + 0$$

But

$$\int_{-T/2}^{T/2} \cos m\omega_0 t \cos n\omega_0 t\,dt = \begin{cases} 0 & \text{for } m \neq n \\ T/2 & \text{for } m = n \end{cases}$$

Hence,

$$\int_{-T/2}^{T/2} f(t)\cos m\omega_0 t = \frac{a_n}{2}\,T,$$

which yields

$$a_n = \frac{2}{T}\int_{-T/2}^{T/2} f(t)\cos m\omega_0 t\,dt \tag{4}$$

Similarly, multiplying (2) by sin $m\omega_0 t$ and integrating, we get

$$b_n = \frac{2}{T}\int_{-T/2}^{T/2} f(t)\sin n\omega_0 t\,dt \tag{5}$$

Note that if we put $n = 0$ in (4), we obtain (3). That is why, we take $\frac{a_0}{2}$ in (1) instead of taking a constant a_0.

In the above discussion, the interval of integration has been $\left(-\frac{T}{2}, \frac{T}{2}\right)$, whose length is precisely *one period* T. However, to determine the coefficients a_n and b_n, we can integrate over any other interval of length T. Sometimes (0, T) is also taken as interval of integration.

The formulae (4) and (5) are called Fourier or *Euler formulae*, the numbers a_n and b_n are called *Fourier coefficients* and the series $\frac{a_0}{2} + \sum_{n=1}^{\infty}(a_n \cos n\omega_0 t + b_n \sin n\omega_0 t)$ *is called Fourier series* of the function f where a_n and b_n are Fourier coefficients determined by (4) and (5).

When the periodic function f is real, then a_n and b_n are real and the *n*th term $a_n \cos n\omega_0 t + b_n \sin n\omega_0 t$, in the Fourier series, is called the *nth harmonic*. This term can also be written as a single cosine term in the following form:

$$a_n \cos n\omega_0 t + b_n \sin n\omega_0 t = \sqrt{a_n^2 + b_n^2}\cos(n\omega_0 t + \phi_n),$$

where

$$\tan \phi_n = \frac{b_n}{a_n} \text{ if } a \neq 0,$$

$$\phi_n = -\frac{\pi}{2} \text{ if } a_n = 0.$$

The factor $\sqrt{a_n^2 + b_n^2}$ is the *amplitude* of the *n*th harmonic and ϕ_n is the *initial phase*. The initial phase tells us how far the nth harmonic is shifted relative to cos $n\omega_0 t$.

If a_n and b_n are Fourier coefficients for f, then we write

$$f \sim \frac{a_0}{2} + \sum_{n=1}^{\infty}(a_n \cos n\omega_0 t + b_n \sin n\omega_0 t),$$

until and unless we know that the series converges to f. Thus we can replace '~' by '=' only if the Fourier series converges to f.

Deductions. (a) If f is even in $\left[-\frac{T}{2}, \frac{T}{2}\right]$, then $f(t)\cos n\omega_0 t$ is also even and so

$$a_n = \frac{4}{\mathrm{T}}\int_0^{\mathrm{T}/2} f(t)\cos n\omega_0 t\,dt$$

Further, since product of an even function f with odd function sin $n\omega_0 t$ is odd, we have

$$b_n = \frac{2}{\mathrm{T}}\int_0^{\mathrm{T}/2} f(t)\sin n\omega_0 t\,dt = 0.$$

Thus, if f is an even function, then its Fourier series will consist of cosine terms only.

(b) If f is an odd function in $\left[-\frac{T}{2}, \frac{T}{2}\right]$, then $f(t)\cos n\omega_0 t$ will be odd and $f(t)\sin n\omega_0 t$ will be even. Therefore, in this case,

$$a_n = 0 \text{ and } b_n = \frac{4}{T}\int_0^{T/2} f(t)\sin n\omega_0 t\, dt.$$

Thus, if f is an odd function, then its Fourier series will consist of sine terms only.

(c) As discussed above, to determine the Fourier coefficients a_n and b_n, we can, in general, integrate the integrand over any interval of length T (period). For example, if we take $T = 2l$ and the interval as $(-l, l)$, then $\omega_0 = \frac{2\pi}{T} = \frac{\pi}{l}$. Since $\sin\frac{\pi t}{l}$ and $\cos\frac{\pi t}{l}$ have period $2l$, the Fourier series valid in $(-l, l)$ takes the form

$$f(t) \sim \frac{a_0}{2} + \sum_{n=1}^{\infty}\left(a_n \cos\frac{n\pi t}{l} + b_n \sin\frac{n\pi t}{l}\right),$$

where

$$a_n = \frac{2}{T}\int_{-T/2}^{T/2} f(t)\cos m\omega_0 t\, dt$$

$$= \frac{2}{2l}\int_{-l}^{l} f(t)\cos\frac{n\pi t}{l}\, dt$$

$$= \frac{1}{l}\int_{-l}^{l} f(t)\cos\frac{n\pi t}{l}\, dt \qquad (6)$$

and similarly

$$b_n = \frac{1}{l}\int_{-l}^{l} f(t)\sin\frac{n\pi t}{l}\, dt \qquad (7)$$

In particular, if $T = 2\pi$ and interval of integration is $(-\pi, \pi)$, then

$$a_n = \frac{1}{\pi}\int_{-\pi}^{\pi} f(t)\cos nt\, dt, \qquad (8)$$

$$b_n = \frac{1}{\pi}\int_{-\pi}^{\pi} f(t)\sin nt\, dt. \qquad (9)$$

Similarly, if we carry out integration over $(c, c + T)$, where T is the period of the function f, then we have

$$a_n = \frac{2}{T}\int_c^{c+T} f(t)\cos m\omega_0 t\, dt, \qquad (10)$$

$$b_n = \frac{2}{T}\int_c^{c+T} f(t)\sin m\omega_0 t\, dt. \qquad (11)$$

Taking $c = 0$, we get the interval of integration as $(0, T)$ and

$$a_n = \frac{2}{T}\int_0^{T} f(t)\cos m\omega_0 t\, dt, \qquad (12)$$

$$b_n = \frac{2}{T}\int_0^{T} f(t)\sin m\omega_0 t\, dt. \qquad (13)$$

6.3 PERIODIC EXTENSION OF A FUNCTION

Let f be a function defined on the interval $\left(-\frac{T}{2}, \frac{T}{2}\right)$. By *periodic extension* of f we mean that f is defined by $f(t + kT) = f(t)$ for all $k \in \mathbb{Z}$. The extended function is then a periodic function of period T.

EXAMPLE 6.1

Consider the sawtooth function f defined on the interval $\left(-\frac{T}{2}, \frac{T}{2}\right) = (-\pi, \pi)$ by $f(t) = t$. The graph of f is shown in the Figure 6.2.

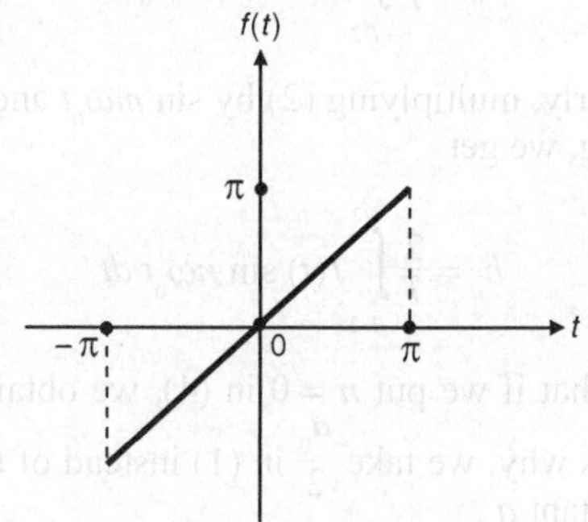

Figure 6.2 Graph of $f(t) = t$

The graph of the extended periodic function with period $T = 2\pi$ is then as shown in the Figure 6.3.

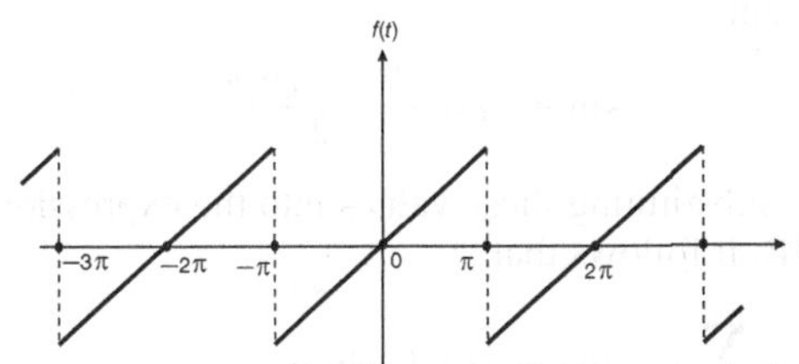

Figure 6.3 Extended Periodic Function $f(t) = t$ with Period 2π

It is a piecewise smooth function discontinuous at the points $t = (2k + 1)\pi$, $k = 0, \pm 1, \pm 2, \ldots$

EXAMPLE 6.2

Consider the function f defined by $f(t) = t^2$ on $(-\pi, \pi)$. The graph of f is a parabola, shown in the Figure 6.4.

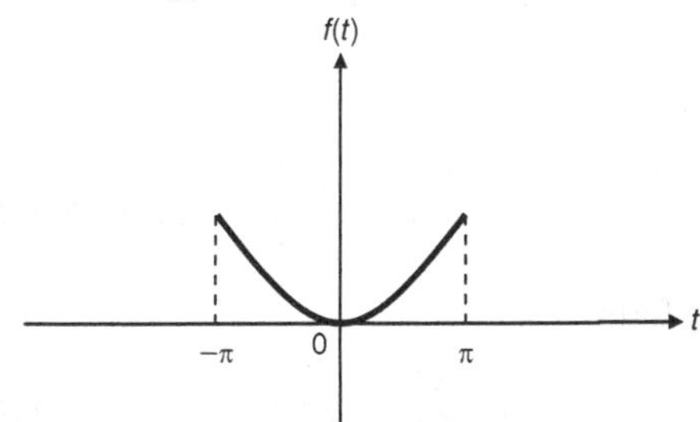

Figure 6.4 Graph of $f(t) = t^2$

The periodic extension of f is then a function of period 2π shown in the Figure 6.5.

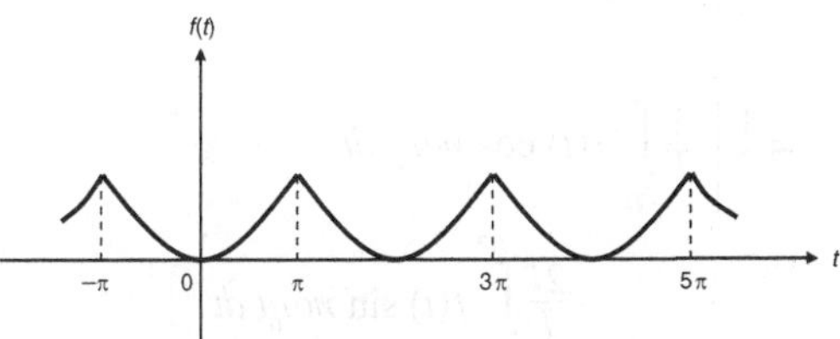

Figure 6.5 Periodically Extended Function $f(t) = t^2$

6.4 FOURIER COSINE AND SINE SERIES

We have seen that if f is an even function, then its Fourier series consists of cosine terms only, whereas for an odd function, the Fourier series consists of sine terms only. Sometimes one would like to obtain a Fourier series containing only cosine terms or sine terms for an *arbitrary function* on the interval $\left(0, \frac{T}{2}\right)$. Such series are called *Fourier cosine series* and *Fourier sine series*, respectively.

To obtain Fourier cosine series for an arbitrary function f defined on $\left(0, \frac{T}{2}\right)$, make the *even extension* of f from the interval $\left(0, \frac{T}{2}\right)$ onto the interval $\left(-\frac{T}{2}, 0\right)$ by defining $f(-t) = f(t)$ for $-\frac{T}{2} < t < 0$ and, subsequently, extend the function periodically with period T. The function, thus, created is now an even function on the interval $\left(-\frac{T}{2}, \frac{T}{2}\right)$ and so its Fourier series will consists of cosine terms only. This function is equal to the original function on the interval $\left(0, \frac{T}{2}\right)$.

To obtain Fourier sine series, we first make *odd extension* of f from the interval $\left(0, \frac{T}{2}\right)$ on the interval $\left(-\frac{T}{2}, 0\right)$ by defining $f(-t) = -f(t)$ for $-\frac{T}{2} < t < 0$ and, subsequently, extend the function periodically with period T. The function thus created is an odd function on the interval $\left(-\frac{T}{2}, \frac{T}{2}\right)$ and so its Fourier series shall consists of only sine terms. This newly created function is equal to the original function on the interval $\left(0, \frac{T}{2}\right)$.

The process of obtaining Fourier cosine series or Fourier sine series of an arbitrary function by making even or odd extension is called *forced series development*. The Fourier series so obtained is called *half-range series*.

EXAMPLE 6.3

Consider the function $f(t) = 1 (0 < t < \pi)$ and suppose that we want to have odd extension of f to $(-\pi, 0)$. So we have to define $f(-t) = -f(t)$. Thus $f(-t) = -1$ in $(-\pi, 0)$. Thus, the graph of the extended function is as shown in Figure 6.6.

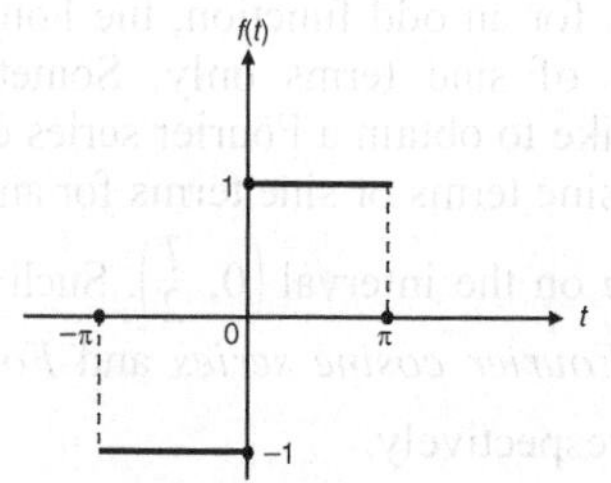

Figure 6.6 Odd Extension of f

The period of this extended function is 2. We then extend this periodically to obtain the following graph (Figure 6.7):

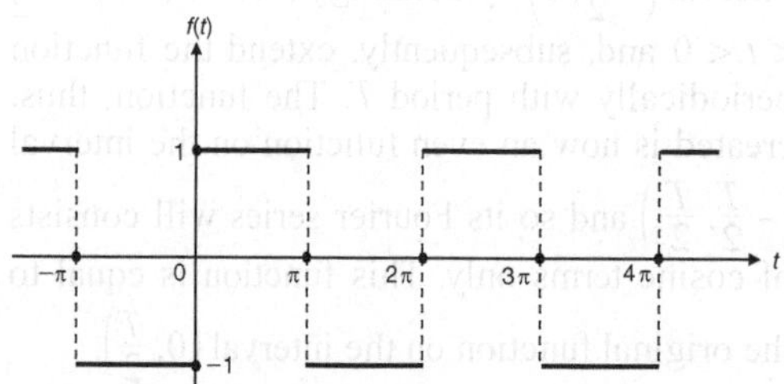

Figure 6.7 Periodic Extension of f

6.5 COMPLEX FOURIER SERIES

Let f be a an integrable function on the interval $\left(-\frac{T}{2}, \frac{T}{2}\right)$. Then the Fourier series of f is

$$f(t) \sim \frac{a_0}{2} + \sum_{n=1}^{\infty} (a_n \cos n\omega_0 t + b_n \sin n\omega_0 t), \quad (14)$$

where $\omega_0 = \frac{2\pi}{T}$ and a_n, b_n are determined by Euler's formulae

$$a_n = \frac{2}{T}\int_{-T/2}^{T/2} f(t)\cos n\omega_0 t\, dt,$$

$$b_n = \frac{2}{T}\int_{-T/2}^{T/2} f(t)\sin n\omega_0 t\, dt.$$

By Euler's formula, relating trigonometric and exponential functions, we have

$$\cos n\omega_0 t = \frac{e^{in\omega_0 t} + e^{in\omega_0 t}}{2}$$

and

$$\sin n\omega_0 t = \frac{e^{in\omega_0 t} - e^{in\omega_0 t}}{2}.$$

Substituting these values into the expression (14), it follows that

$$\frac{a_0}{2} + \sum_{n=1}^{\infty} (a_n \cos n\omega_0 t + b_n \sin n\omega_0 t)$$

$$= \frac{a_0}{2} + \sum_{n=1}^{\infty} \left(a_n \frac{e^{in\omega_0 t} + e^{-in\omega_0 t}}{2} + b_n \frac{e^{in\omega_0 t} - e^{-in\omega_0 t}}{2i}\right)$$

$$= \frac{a_0}{2} + \sum_{n=1}^{\infty} \left[\frac{1}{2}(a_n - ib_n)e^{in\omega_0 t} + \frac{1}{2}(a_n + ib_n)e^{-in\omega_0 t}\right]$$

$$= c_0 + \sum_{n=1}^{\infty} \left(c_n e^{in\omega_0 t} + c_{-n} e^{-in\omega_0 t}\right) = \sum_{n=-\infty}^{\infty} c_n e^{in\omega_0 t},$$

where

$$c_0 = \frac{a_0}{2},\ c_n = \frac{a_n - ib_n}{2},\ c_{-n} = \frac{a_n + ib_n}{2}, \text{ for } n \in \mathrm{N} \quad (15)$$

The form

$$f(t) \sim \sum_{n=-\infty}^{\infty} c_n e^{in\omega_0 t}, \quad (16)$$

is called *complex form of the Fourier series* of f. The coefficients c_n are *complex Fourier coefficients* of the function f. We note that

$$c_n = \frac{1}{2}(a_n - ib_n)$$

$$= \frac{1}{2}\left[\frac{2}{T}\int_{-T/2}^{T/2} f(t)\cos n\omega_0 t\, dt - \frac{2i}{T}\int_{-T/2}^{T/2} f(t)\sin n\omega_0 t\, dt\right]$$

$$= \frac{1}{T}\int_{-T/2}^{T/2} f(t)\left[\frac{e^{in\omega_0 t} + e^{-in\omega_0 t}}{2} - i\,\frac{e^{in\omega_0 t} - e^{-in\omega_0 t}}{2i}\right]$$

$$= \frac{1}{T}\int_{-T/2}^{T/2} f(t)\, e^{-in\omega_0 t} dt. \quad (17)$$

Similarly,

$$c_{-n} = \frac{a_n + ib_n}{2} = \frac{1}{T}\int_{-T/2}^{T/2} f(t)\, e^{in\omega_0 t}\, dt. \qquad (18)$$

If f is real, then c_n are c_{-n} are complex conjugates. Thus, the complex Fourier coefficients c_n are defined by:

$$c_n = \frac{1}{T}\int_{-T/2}^{T/2} f(t)\, e^{-in\omega_0 t}\, dt,\ n \in \mathbb{Z} \qquad (19)$$

The term $e^{in\omega_0 t}$ in Fourier series $\sum_{n=-\infty}^{\infty} c_n e^{in\omega_0 t}$ is called *time-harmonic function.*

6.6 SPECTRUM OF PERIODIC FUNCTIONS

Let f be a periodic function defined for $t \in \mathbb{R}$. If t is time variable, then we say that the periodic function f is defined in the *time domain.* Further, each Fourier coefficient in the Fourier expansion of f is associated with a specific frequency $n\omega_0$. Also if the series converges to f, then the function f is completely determined by their Fourier coefficients. Therefore, we say that f is described by the Fourier coefficients in the *frequency domain.*

The signals are generally interpreted in terms of frequencies. For example, sound is expressed in terms of frequency as *pitch* whereas light is expressed in term of frequency as *colour*.

Definition 6.3. The sequence of Fourier coefficients c_n with $n \in \mathbb{Z}$, which describe a function in the frequency domain, is called the *spectrum of the funtion.* Since n assumes only integer values, the spectrum is called a *discrete spectrum or line spectrum.*

Definition 6.4. The sequence of absolute values of c_n, that is, $\{|c_n|\}$ is called amplitude spectrum, whereas,the sequence $\{\arg(c_n)\}$ is called phase spectrum of the function.

6.7 PROPERTIES OF FOURIER COEFFCIENTS

Following theorems describe the properties of Fourier coefficients:

Theorem 6.1. (Linearity). Let c_n and d_n be, respectively. The Fourier coefficients of f and g. Then the Fourier coefficients of $af + bg$, $a, b \in \mathbb{C}$ shall be $ac_n + bd_n$.(Thus Fourier coefficients of linear combinations of functions is equal to the same linear combinations of the Fourier coefficients of the individual functions).

Proof: Let $\hat{C}_n$ be the Fourier coefficients of $af + bg$. Then

$$\hat{C}_n = \frac{1}{T}\int_{-T/2}^{T/2} [af(t) + bg(t)]\, e^{-in\omega_0 t}\, dt$$

$$= \frac{a}{T}\int_{-T/2}^{T/2} f(t)\, e^{-in\omega_0 t}\, dt$$

$$+ \frac{b}{T}\int_{-T/2}^{T/2} g(t)\, e^{-in\omega_0 t}\, dt$$

$$= ac_n + bd_n.$$

Theorem 6.2. (Conjugation). If Fourier coefficients of f are c_n, then Fourier coefficients of $\overline{f(t)}$ are $\overline{c}_{-n}$.

Proof: Since $\overline{e^{in\omega_0 t}} = e^{-in\omega_0 t}$, the Fourier coefficients of $\overline{f(t)}$ are given by

$$\frac{1}{T}\int_{-T/2}^{T/2} \overline{f(t)}\, e^{-in\omega_0 t}\, dt$$

$$= \frac{1}{T}\int_{-T/2}^{T/2} \overline{f(t)\, e^{in\omega_0 t}}\, dt$$

$$= \frac{1}{T}\int_{-T/2}^{T/2} \overline{f(t)\, e^{-i(-n)\omega_0 t}}\, dt = \overline{c}_{-n}.$$

Theorem 6.3. (Shift in Time). If c_n are Fourier coefficients of f, then the Fourier coefficient of $f(t - t_0)$ is $e^{-in\omega_0 t_0}\, c_n$.

Proof: We have

$$\frac{1}{T}\int_{-T/2}^{T/2} f(t - t_0)\, e^{-in\omega_0 t}\, dt$$

$$= e^{-in\omega_0 t_0} \cdot \frac{1}{T}\int_{-T/2}^{T/2} f(t - t_0)\, e^{-in\omega_0 (t - t_0)}\, dt$$

$$= e^{-in\omega_0 t_0}\cdot \frac{1}{T}\int_{-T/2}^{T/2} f(u)\, e^{-in\omega_0 u}\,du,\quad t - t_0 = u$$

$$= e^{-in\omega_0 t_0}\, c_n.$$

Theorem 6.4. (Time reversal). If c_n are the Fourier coefficients of $f(t)$, then c_{-n} will be the Fourier coefficients of $f(-t)$.

Proof: Putting $-t = u$, we get

$$\frac{1}{T}\int_{-T/2}^{T/2} f(-t)\, e^{-in\omega_0 t}\,dt$$

$$= \frac{1}{T}\int_{T/2}^{-T/2} f(u)\, e^{-in\omega_0(-u)}\,d(-u)$$

$$= \frac{1}{T}\int_{-T/2}^{T/2} f(u)\, e^{-i(-n)\omega_0 u}\,du = c_{-n}.$$

Theorem 6.5. (Bessel's inequality). Let c_n be the Fourier coefficients of piecewise continuous periodic function f with period T. Then

$$\sum_{n=-\infty}^{\infty} |c_n|^2 \le \frac{1}{T}\int_{-T/2}^{T/2} |f(t)|^2\,dt.$$

Proof: Let

$$S_n(t) = \sum_{k=-n}^{n} c_k e^{ik\omega_0 t} \tag{20}$$

be the partial sum of the Fourier series $\sum_{n=-\infty}^{\infty} c_n e^{ik\omega_0 t}$. Let $-n \le k \le n$. Then

$$\frac{1}{T}\int_{-T/2}^{T/2} |f(t) - S_n(t)|\, e^{-ik\omega_0 t}$$

$$= \frac{1}{T}\int_{-T/2}^{T/2} f(t)\, e^{-ik\omega_0 t} - \frac{1}{T}\int_{-T/2}^{T/2} S_n(t)\, e^{-ik\omega_0 t}$$

$$= c_k - \frac{1}{T}\sum_{l=-n}^{n} c_l \int_{-T/2}^{T/2} e^{i(l-k)\omega_0 t}\,dt, \text{ using (20)}$$

But

$$\int_{-T/2}^{T/2} e^{i(l-k)\omega_0 t}\,dt = \begin{cases} 0 & \text{for } l \ne k \\ T & \text{for } l = k. \end{cases}$$

Hence

$$\frac{1}{T}\int_{-T/2}^{T/2} [f(t) - S_n(t)]\, e^{-ik\omega_0 t}$$

$$= c_k - \frac{1}{T} T c_k = c_k - c_k = 0 \tag{21}$$

Using (21), it follows that

$$\frac{1}{T}\int_{-T/2}^{T/2} [f(t) - S_n(t)]\,\overline{S}_n(t)\,dt$$

$$= \sum_{k=-n}^{n} \overline{c}_k \int_{-T/2}^{T/2} [f(t) - S_n(t)]\, e^{-ik\omega_0 t}\,dt = 0. \tag{22}$$

But (22) implies

$$\frac{1}{T}\int_{-T/2}^{T/2} [f(t) - S_n(t)]\,\overline{[f(t) - S_n(t)]}\,dt$$

$$= \int_{-T/2}^{T/2} [f(t) - S_n(t)]\,\overline{f(t)}\,dt$$

$$= \int_{-T/2}^{T/2} f(t)\overline{f(t)}\,dt - \int_{-T/2}^{T/2} S_n(t)\,\overline{f(t)}\,dt$$

$$= \int_{-T/2}^{T/2} |f(t)|^2\,dt - \sum_{k=-n}^{n} c_k \int_{-T/2}^{T/2} e^{ik\omega_0 t}\,\overline{f(t)}\,dt$$

$$= \int_{-T/2}^{T/2} |f(t)|^2\,dt - \sum_{k=-n}^{n} c_k \int_{-T/2}^{T/2} \overline{e^{ik\omega_0 t}}\, f(t)\,dt$$

$$= \int_{-T/2}^{T/2} |f(t)|^2\,dt - \sum_{k=-n}^{n} c_k\, T\, \overline{c}_k \text{ (by definition of } c_k)$$

$$= \int_{-T/2}^{T/2} |f(t)|^2\,dt - T\sum_{k=-n}^{n} |c_k|^2.$$

The integrand on the left-hand side is equal to $|f(t) - S_n(t)|^2 \geq 0$. Hence the left-hand side is also greater than or equal to zero. Hence

$$T\sum_{k=-n}^{n} |c_k|^2 \leq \int_{-T/2}^{T/2} |f(t)|^2\, dt. \tag{23}$$

Since (23) holds for any $n \in$ N, letting $n \to \infty$, we have

$$\sum_{n=\infty}^{\infty} |c_k|^2 = \lim_{n\to\infty} \sum_{k=-n}^{n} |c_k|^2 \leq \frac{1}{T}\int_{-T/2}^{T/2} |f(t)|^2\, dt,$$

which proves the theorem.

Remark 6.1. For real form of the Fourier series, we have $c_n = \dfrac{a_n - ib_n}{2}$ so that Bessel's inequality takes the form

$$\frac{a_0^2}{2} + \sum_{n=1}^{\infty} \left(a_0^2 + b_0^2\right) \leq \frac{1}{T}\int_{-T/2}^{T/2} |f(t)|^2\, dt. \tag{24}$$

Theorem 6.6. (Riemann-Lebesgue Lemma). Let f be a piecewise continuous periodic function with Fourier coefficients c_n. Then $\lim_{n\to\infty} c_n = \lim_{n\to-\infty} c_n = 0$.

Proof: Since f is piecewise continuous, so is $|f(t)|^2$. Hence the integral $\frac{1}{T}\int_{-T/2}^{T/2} |f(t)|^2$ dt is finite. But, by Bessel's inequality, we have

$$\sum_{n=\infty}^{\infty} |c_n|^2 \leq \frac{1}{T}\int_{-T/2}^{T/2} |f(t)|^2\, dt.$$

Since right-hand side is finite, the series $\sum_{n=\infty}^{\infty} |c_k|^2$ of positive terms is convergent. Hence $c_n \to 0$ as $n \to \pm\infty$.

Remark 6.2. Since $c_n = \dfrac{a_n - ib_n}{2}$, it follows from the Riemann–Lebesgue lemma that a_n, b_n tend to zero as $n \to \infty$.

Theorem 6.7. (Mean Convergence of Fourier Series in $L^2[-\pi, \pi]$). Let $f \in L^2[-\pi, \pi]$. Then the Fourier series of f converges in mean to f in $L^2[-\pi, \pi]$.

Proof: A function f is said to belong to the class $L^p[-\pi, \pi]$, $0 < p \leq \infty$, if $\int_{-\pi}^{\pi} |f(t)|^p\, dt < \infty$. Further, a sequence of functions $f_n \in L^p[-\pi, \pi]$, $0 < p \leq \infty$ is said to *converge in norm or converge in mean with index* p to $f \in L^p[-\pi, \pi]$ if

$$\lim_{n\to\infty} \left[\int_{-\pi}^{\pi} |f_n(t) - f(t)|^p\, dt\right]^{\frac{1}{p}} = 0.$$

A necessary and sufficient condition for convergence of the sequence of functions $f_n \in L^p$ is that given $\varepsilon > 0$ there exists N such that

$$\|f_n - f_m\|_{L^p} < \varepsilon \text{ for } n, m \geq N,$$

that is,

$$\left[\int |f_n - f_m|^p\right]^{\frac{1}{p}} < \varepsilon \text{ for } n, m \geq N.$$

Let $\{S_n(x)\}$ be the sequence of partial sums of the Fourier series $\sum_{k=0}^{\infty} (a_k \cos k\omega_0 t + b_k \sin k\omega_0 t)$ for $f \in L^2[-\pi, \pi]$. Then if $m < n$, we have

$$\frac{1}{\pi}\int_{-\pi}^{\pi} |S_n(x) - S_m(x)|^2\, dx$$

$$= \frac{1}{\pi}\int_{-\pi}^{\pi} \left|\sum_{k=m+1}^{n} (a_k \cos k\omega_0 t + b_k \sin k\omega_0 t)\right|^2 dt.$$

$$\leq \sum_{k=m+1}^{n} \left|a_k^2 + b_k^2\right|.$$

But, by Bessels inequality, $\sum_{k=1}^{\infty} \left|a_k^2 + b_k^2\right| < \infty$ for $f \in L^2[-\pi, \pi]$. Therefore there exists a number N such that $\sum_{k=m+1}^{n} \left|a_k^2 + b_k^2\right| < \varepsilon$ for $m > N$. Hence

$$\| S_n(x) - S_m(x)\|_{L^2} < \varepsilon \text{ for } n, m \geq N,$$

and so the Fourier series of f converges in mean to f.

Theorem 6.8. (Riesz–Fischer Theorem). If $\{a_k\}$ and $\{b_k\}$ are the sequences of real numbers such that

$$\frac{a_0^2}{2}+\sum_{k=1}^{\infty}\left(a_k^2+b_k^2\right)<\infty,$$

then there exists a function f such that $\int_{-\pi}^{\pi}|f(t)|^2\,dt<\infty$ and whose Fourier coefficients are precisely a_k and b_k.

Proof: Let

$$S_n(t)=\frac{a_0}{2}+\sum_{k=1}^{n}(a_k\cos k\omega_0 t+b_k\sin k\omega_0 t),$$

where $-\frac{T}{2}\le t\le\frac{T}{2}$. If $m<n$, then

$$\frac{2}{T}\int_{-T/2}^{T/2}(S_n-S_m)^2$$

$$=\frac{2}{T}\int_{-T/2}^{T/2}\sum_{k=m+1}^{n}(a_k\cos k\omega_0 t+b_k\sin k\omega_0 t)^2\,dt$$

$$=\sum_{k=m+1}^{n}\left(a_k^2+b_k^2\right). \tag{25}$$

But, by the given hypothesis, the series $\sum_{k=1}^{n}\left(a_k^2+b_k^2\right)$ is convergent. Hence the tail $\sum_{k=m+1}^{n}\left(a_k^2+b_k^2\right)$ tends to zero as $n\to\infty$. The left-hand side of (25) is nothing but $\frac{2}{T}\|S_n-S_m\|$. Therefore, $\|S_n-S_m\|\to 0$ as $m,n\to\infty$. Thus $\{S_n\}$ is a Cauchy sequence. But f is in $L^2\left[-\frac{T}{2},\frac{T}{2}\right]$ space, which is complete. Hence this sequence of partial sum converges to some function f in $L^2\left[-\frac{T}{2},\frac{T}{2}\right]$. Thus for any $p=0,1,2,\ldots$, we have

$$\lim_{n\to\infty}\frac{2}{T}\int_{-T/2}^{T/2}S_n(t)\cos p\omega_0 t\,dt$$

$$=\frac{2}{T}\int_{-T/2}^{T/2}f(t)\cos p\omega_0 t\,dt.$$

But if $n\ge p$, we have

$$\frac{2}{T}\int_{-T/2}^{T/2}S_n(t)\cos p\omega_0 t\,dt=a_p.$$

Hence

$$\frac{2}{T}\int_{-T/2}^{T/2}f(t)\cos p\omega_0 t\,dt=a_p\ (p=0,1,2,\ldots).$$

Thus a_p are the Fourier cosine coefficients of f. Similarly, it can be proved that b_p are the Fourier sine coefficients of f.

6.8 DIRICHLET'S KERNEL

Definition 6.5. The Dirichlet's kernel $D_n(t)$ is defined by

$$D_n(t)=\frac{1}{2}+\cos t+\ldots+\cos nt. \tag{26}$$

This is a periodic function and plays an important role in the convergence of trigonometric series.Multiplying both sides of (26) with $2\sin\frac{t}{2}$, we get

$$2\sin\frac{t}{2}D_n(t)=\sin\frac{t}{2}+2\cos t\sin\frac{t}{2}$$

$$+2\cos 2t\sin\frac{t}{2}+\ldots+2\cos nt\sin\frac{t}{2} \tag{27}$$

Applying the formula

$$2\cos\alpha\sin\beta=\sin(\alpha+\beta)-\sin(\alpha-\beta),$$

the equation (27) reduces to

$$2\sin\frac{t}{2}D_n(t)$$

$$=\sin\frac{t}{2}+\left(\sin\frac{3}{2}t-\sin\frac{t}{2}\right)+\left(\sin\frac{5}{2}t-\sin\frac{3}{2}t\right)$$

$$+\ldots+\left(\sin\left(n+\frac{1}{2}\right)t-\sin\left(n-\frac{1}{2}\right)t\right)$$

$$= \sin\left(n+\frac{1}{2}\right)t.$$

Hence

$$D_n(t) = \frac{\sin\left(n+\frac{1}{2}\right)t}{2\sin\frac{t}{2}}. \tag{28}$$

We note that

(a) Dirichlet's kernel is *an even periodic function.*

(b) the expression (28) for Dirichlet's kernel implies that

$$D_n(t) \leq \frac{1}{2\sin\frac{t}{2}}. \tag{29}$$

Simple differentiation shows that $\frac{\sin t}{t}$ decreaes in the interval $\left(0, \frac{\pi}{2}\right)$. Therefore,

$$\frac{\sin t}{t} \geq \frac{\sin\frac{\pi}{2}}{\frac{\pi}{2}} = \frac{2}{\pi},$$

which means

$$\frac{\sin t}{t} \geq \frac{2}{\pi} \text{ for } 0 \leq t \leq \frac{\pi}{2},$$

that is,

$$\frac{\sin\frac{t}{2}}{\frac{t}{2}} \geq \frac{2}{\pi} \text{ for } 0 \leq \frac{t}{2} \leq \frac{\pi}{2},$$

that is,

$$\sin\frac{t}{2} \geq \frac{t}{\pi} \text{ for } 0 \leq t \leq \pi.$$

Hence (29) yields

$$|D_n(t)| \leq \frac{\pi}{2t} \text{ for } 0 < |t| \leq \pi,$$

and so

$$|D_n(t)| = O\left(\frac{1}{t}\right). \tag{30}$$

Sometimes it will be required that if $\delta \leq |t| \leq \pi$, then

$$|D_n(t)| \leq \frac{\pi}{2\delta}. \tag{31}$$

(c) The expression (26) shows that

$$|D_n(t)| \leq \frac{1}{2} + \underbrace{1 + 1 + \ldots + 1}_{n \text{ times}} = n + \frac{1}{2}. \tag{32}$$

(d) Integrating the expression (26) over the interval $\left[-\frac{T}{2}, \frac{T}{2}\right]$, we have

$$\int_{-T/2}^{T/2} D_n(t)dt = \frac{T}{2}. \tag{33}$$

(e) Since $D_n(t)$ is an even function, we have

$$L_n = \frac{1}{\pi}\int_{-\pi}^{\pi} |D_n(t)|\, dt = \frac{2}{\pi}\int_0^{\pi} |D_n(t)|\, dt.$$

Expanding $\frac{\sin\left(n+\frac{1}{2}\right)t}{2\sin\frac{t}{2}}$ we have $D_n(t) = \frac{\sin nt}{t} +$ O(1). Therefore,

$$L_n = \frac{2}{\pi}\int_0^{\pi} \left|\frac{\sin nt}{t}\right| dt + O(1).$$

Since $\int_0^{\pi} \left|\frac{\sin nt}{t}\right| dt \approx \frac{2}{\pi}.\log n$, it follows that

$$L_n \approx \frac{4}{\pi^2}\log n.$$

Thus, we have shown that

$$L_n = \frac{1}{\pi}\int_{-\pi}^{\pi} |D_n(t)|\, dt \approx \frac{4}{\pi^2}\log n. \tag{34}$$

The constant L_n in the expression (34) is called *Lebesgue constant.*

Remarks 6.3. (a) The expression

$$\bar{D}_n(t) = \sin t + \sin 2t + \ldots + \sin nt$$

is called *conjugate* Dirichlet's *kernel.* It takes the form

$$\bar{D}_n(t) = \frac{\cos\frac{t}{2} - \cos\left(n+\frac{1}{2}\right)t}{2\sin\frac{t}{2}}$$

and so

$$\bar{D}_n(t) \leq \frac{1}{\sin\frac{t}{2}}$$

(b) The expressions

$$K_n(t) = \frac{1}{n+1}\sum_{p=0}^{n} D_p(x)$$

and

$$\overline{K}_n(t) = \frac{1}{n+1}\sum_{p=0}^{n} \overline{D}_p(x)$$

are called the *Feres's kernel* and *conjugate Fejer's kernel*, respectively.

6.9 INTEGRAL EXPRESSION FOR PARTIAL SUMS OF A FOURIER SERIES

Let

$$S_n(t) = \frac{a_0}{2} + \sum_{k=1}^{n} (a_k \cos k\omega_0 t + b_k \sin k\omega_0 t)$$

be the partial sum of the Fourier series. Replacing a_k and b_k by the defining integrals, we obtain

$$S_n(t) = \frac{1}{T}\int_{-T/2}^{T/2} f(u)\, du$$

$$+\frac{2}{T}\sum_{k=1}^{n}\left[\left(\int_{-T/2}^{T/2} f(u)\cos k\omega_0 u\, du\right)\cos k\omega_0 t\right.$$

$$\left.+\left(\int_{-T/2}^{T/2} f(u)\sin k\omega_0 u\, du\right)\sin k\omega_0 t\right]$$

$$= \frac{2}{T}\int_{-T/2}^{T/2} f(u)\left[\frac{1}{2} + \sum_{k=1}^{n}(\cos k\omega_0 u \cos k\omega_0 t\right.$$

$$\left. + \sin k\omega_0 u \sin mk\omega_0 t)\right] du$$

$$= \frac{2}{T}\int_{-T/2}^{T/2} f(u)\left[\frac{1}{2} + \sum_{k=1}^{n}\cos k\omega_0(u-t)\right] du$$

$$= \frac{2}{T}\int_{-T/2}^{T/2} f(u)\, D_n(u-t)\, du$$

Substituting $u - t = x$, we have

$$S_n(t) = \frac{2}{T}\int_{(-T/2)-t}^{(T/2)-t} f(x+t)\, D_n(x)\, dx$$

$$= \frac{2}{T}\int_{-T/2}^{T/2} f(x+t)\, D_n(x)\, dx,$$

using the fact that the functions $f(x + t)$ and $D_n(x)$ are periodic in the variable x with period T and the length of the interval $\left[-\frac{T}{2} - t, \frac{T}{2} - t\right]$ is T and so the integral over $\left[-\frac{T}{2} - t, \frac{T}{2} - t\right]$ is same as the integral over $\left[-\frac{T}{2}, \frac{T}{2}\right]$.

Since $D_n(x)$ is even, that is, $D_n(-x) = D_n(x)$, the partial sum becomes

$$S_n(t) = \frac{2}{T}\int_{-T/2}^{0} f(x+t)\, D_n(x)dx$$

$$+ \frac{2}{T}\int_{0}^{T/2} f(x+t)\, D_n(x)dx$$

$$= \frac{2}{T}\int_{0}^{T/2} f(t-x)\, D_n(-x)dx$$

$$+ \frac{2}{T}\int_{0}^{T/2} f(x+t)\, D_n(x)dx$$

$$= \frac{2}{T}\int_{0}^{T/2} [f(x+t) + f(t-x)\, D_n(x)dx$$

$$= \frac{2}{T}\int_{0}^{T/2} [f(t+x) - f(t+) + f(t-x) - f(t-)]\, D_n(x)dx$$

$$+ \frac{2}{T}\int_{0}^{T/2} [f(t+) + f(t-)]\, D_n(x)dx \qquad (35)$$

Since $\int_0^{\pi/2} D_n(x)\, dx = \frac{T}{4}$ by (33), the second term in (35) is $\frac{f(t+) + f(t-)}{2}$. Thus, we get

$$S_n(t) = \frac{f(t+) + f(t-)}{2} + \frac{2}{T}\int_0^{T/2} [f(t+x) -$$

$$f(t+) + f(t-x) - f(t-)]\, D_n(x)dx.$$

6.10 FUNDAMENTAL THEOREM (CONVERGENCE THEOREM) OF FOURIER SERIES

The following theorem shows that the Fourier series of a piecewise smooth function converges to that function at each point of continuity.

Theorem 6.9. (Fundamental Theorem of Fourier Series). Let f be a piecewise smooth periodic function with period T defined on R with Fourier coefficients a_n and b_n. Then for any $t \in$ R,

$$\frac{a_0}{2} + \sum_{n=1}^{\infty} (a_n \cos n\omega_0 t + b_n \sin n\omega_0 t)$$

$$= \frac{1}{2}[f(t+) + f(t-)].$$

[At the point of continuity $f(t+) = f(t-) = f(t)$ and so, in that case, the right-hand side becomes $f(t)$].

Proof: If $S_n(t)$ is the partial sum of the Fourier series, then we have established (above) that

$$S_n(t) = \frac{f(t+) + f(t-)}{2} + \frac{2}{T}\int_0^{T/2} [f(t+x) - f(t+) + f(t-x) - f(t-)]\, D_n(x)dx,$$

where $D_n(x)$ is the Dirichlet's kernel. To prove the theorem, it is sufficient to show that the term

$$I_n(t) = \frac{2}{T}\int_0^{T/2} [f(t+x) - f(t+) + f(t-x) - f(t-)]\, D_n(x)dx$$

tends to zero as $n \to \infty$. To this end, we have

$$I_n(t) = \frac{2}{T}\int_0^{T/2} \frac{[f(t+x) - f(t+) + f(t-x) - f(t-)]}{x} \times \frac{x \sin\left(n + \frac{1}{2}\right)\omega_0 x}{2 \sin \omega_0 \frac{x}{2}}\, dx$$

$$= \frac{2}{T}\int_0^{T/2} Q(x) \sin\left(n + \frac{1}{2}\right)\omega_0 x\, dx, \qquad (36)$$

where

$$Q(x) = \frac{[f(t+x) - f(t+) + f(t-x) - f(t-)]}{x} \times \frac{x}{2 \sin \omega_0 \frac{x}{2}}.$$

For $x = 0$, the denominator of $Q(x)$ equals 0 and so integral $I_n(t)$ is not defined for $x = 0$. But since f is piecewise smooth,

$$\lim_{x \to 0} \frac{f(t+x) - f(t+)}{x} = f'(t+)$$

(right-hand derivative)

and

$$\lim_{x \to 0} \frac{f(t-x) - f(t-)}{x} = f'(t-)$$

(left-hand derivative)

exist. Also $\lim_{x \to 0} \frac{x}{\sin \omega_0 \frac{x}{2}}$ exists. Hence, $Q(x)$ is piecewise continuous function. Further, both $Q(x)$ and $\sin\left(n + \frac{1}{2}\right)\omega_0 x$ being odd, the integrand in (36) is an even function. Therefore, (36) can be written as

$$I_n(t) = \frac{2}{2T}\int_{-T/2}^{T/2} Q(x) \sin\left(n + \frac{1}{2}\right)\omega_0 x\, dx$$

$$= \frac{1}{2}\cdot\frac{2}{T}\int_{-T/2}^{T/2} \left[Q(x) \sin n\,\omega_0 x \cos \omega_0 \frac{x}{2} + \cos n\omega_0 tx \sin \omega_0 \frac{x}{2}\, dx\right]$$

$$= \frac{1}{2}\cdot\frac{2}{T}\int_{-T/2}^{T/2} Q(x) \cos \omega_0 \frac{x}{2} \sin n\omega_0 x\, dx$$

$$+ \frac{1}{2}\cdot\frac{2}{T}\int_{-T/2}^{T/2} Q(x) \sin \omega_0 \frac{x}{2} \cos n\omega_0 x\, dx.$$

But $\frac{2}{T}\int_{-T/2}^{T/2} Q(x) \cos \omega_0 \frac{x}{2} \sin n\omega_0 x \, dx$ is the Fourier coefficient b_n for the function $Q(x) \cos \omega_0 \frac{x}{2}$ whereas $\frac{2}{T}\int_{-T/2}^{T/2} Q(x) \sin \omega_0 \frac{x}{2} \cos n\omega_0 x \, dx$ is Fourier coefficient a_n for the function $Q(x) \sin \omega_0 \frac{x}{2}$. By Riemann–Lebesgue lemma, both of these coefficients tend to zero as $n \to \infty$. Hence $I_n(t) \to 0$ as $n \to \infty$. It follows, therefore, that the Fourier series converges to $\frac{1}{2}[f(t+) + f(t-)]$.

Remark 6.4. It follows from Theorem 6.9 that if two periodic piecewise smooth functions have the same Fourier series, that is, if their Fourier coefficients are equal, then these functions must be equal at all points of continuity. This assertion is known as *Uniqueness theorem.*

Remark 6.5. The assumption in the convergence theorem may be written as:

(a) the function f is periodic and single-valued.
(b) f is piecewise continuous.
(c) f has finite number of maxima and minima in a period.

These three conditions are called *Dirichlet's conditions*.

6.11 APPLICATIONS OF FUNDAMENTAL THEOREM OF FOURIER SERIES

As consequences of fundamental theorem of Fourier series, we have the following results:

Theorem 6.10. (Fourier Series of a Product of-Functions). Let f and g be piecewise smooth periodic functions with Fourier coefficients c_n and d_n respectively. Then $h = fg$ has a convergent Fourier series with Fourier coefficients p_n given by

$$p_n = \sum_{k=-\infty}^{\infty} c_k d_{n-k}.$$

Proof: Since f and g are piecewise smooth periodic functions, so is $h = fg$. Therefore, by fundamental theorem of Fourier series, h has a convergent Fourier series. The Fourier coefficients p_n of h are given by

$$p_n = \frac{1}{T}\int_{-T/2}^{T/2} f(t)\, g(t)\, e^{-in\omega_0 t}\, dt.$$

Since f is piecewise smooth periodic function, by fundamental theorem of Fourier series, it can be replaced by its Fourier series at the points of continuity. But in integration, the values at the points of discontinuity are of no importance. Therefore,

$$p_n = \frac{1}{T}\int_{-T/2}^{T/2} \sum_{k=-\infty}^{\infty} c_k\, e^{ik\omega_0 t} g(t) e^{-in\omega_0 t}\, dt.$$

Changing the order of integration and summation,we have

$$p_n = \sum_{k=-\infty}^{\infty} c_k \frac{1}{T}\int_{-T/2}^{T/2} g(t)\, e^{-i(n-k)\omega_0 t}\, dt = \sum_{k=-\infty}^{\infty} c_k d_{n-k}.$$

Theorem 6.11. (Parseval's Identity). Let f and g be piecewise smooth periodic function with Fourier coefficients c_n and d_n, respectively. Then

$$\frac{1}{T}\int_{-T/2}^{T/2} f(t)\, \overline{g(t)}\, dt = \sum_{k=-\infty}^{\infty} c_k \overline{d_k}.$$

Proof: Since d_n is Fourier coefficient of $g(t)$, by Theorem 6.2, the Fourier coefficient of $\overline{g(t)}$ shall be $\overline{d}_{-n}$. Now if p_n be the Fourier coefficients of the product $f\overline{g}$, Theorem 6.10 implies that

$$p_n = \sum_{k=-\infty}^{\infty} c_k \overline{d}_{-(n-k)}.$$

In particular,

$$p_0 = \sum_{k=-\infty}^{\infty} c_k \overline{d}_k.$$

But, by definition,

$$p_0 = \frac{1}{T}\int_{-T/2}^{T/2} f(t)\, \overline{g(t)}\, dt.$$

Hence

$$\frac{1}{T}\int_{-T/2}^{T/2} f(t)\, \overline{g(t)}\, dt = \sum_{k=-\infty}^{\infty} c_k \overline{d}_k.$$

Theorem 6.12. (Parseval's equality). Let f be a piecewise smooth periodic function with Fourier coefficient c_n. Then

$$\frac{1}{T}\int_{-T/2}^{T/2} |f(t)|^2 dt = \sum_{k=-\infty}^{\infty} |c_k|^2$$

Proof: Taking $f(t) = g(t)$ in Theorem 6.11, we have

$$\frac{1}{T}\int_{-T/2}^{T/2} f(t)\overline{f(t)}\, dt = \sum_{k=-\infty}^{\infty} c_k \bar{c}_k$$

and so

$$\frac{1}{T}\int_{-T/2}^{T/2} |f(t)|^2\, dt = \sum_{k=-\infty}^{\infty} |c_k|^2.$$

Definition 6.6. The integral $\frac{1}{T}\int_{-T/2}^{T/2} |f(t)|^2\, dt$ is called the *power* of periodic time continuous signal f. Thus, if f is piecewise smooth periodic function, then by Theorem 6.11, its power can be calculated using Fourier coefficients. In fact

$$P = \sum_{k=-\infty}^{\infty} |c_k|^2.$$

6.12 CONVOLUTION THEOREM FOR FOURIER SERIES

Definition 6.7. The *convolution product* of two piecewise smooth periodic functions f and g with period T is defined by

$$(f * g)(t) = \frac{1}{T}\int_{-T/2}^{T/2} f(u)\, g(t-u)\, du.$$

The convolution product is very useful in system analysis. We note that for $k \in Z$,

$$(f * g)(t + kT) = \frac{1}{T}\int_{-T/2}^{T/2} f(u)\, g(t + kT - u)\, du$$

$$= \frac{1}{T}\int_{-T/2}^{T/2} f(u)\, g(t-u)\, du$$

$$= (f * g)(t),$$

since g being periodic, $g(t - u + kT) = g(t - u)$. It follows, therefore, that convolution product of periodic function is also *periodic* with the *same period.*

Theorem 6.13. (Convolution Theorem for Fourier Series). If f and g are piecewise smooth periodic functions with Fourier coefficients c_n and d_n, then $f * g$ has a convergent Fourier series with Fourier coefficients $c_n\, d_n$ [denoted by $(f * g)_n$].

Proof: Since f and g are piecewise smooth periodic function, $f * g$ is also piecewise smooth periodic function. Hence, by Fundamental theorem of Fourier series, it has a convergent Fourier series. Further

$$(f * g)_n = \frac{1}{T}\int_{-T/2}^{T/2} (f * g)(t)\, e^{-in\omega_0 t}\, dt$$

$$= \frac{1}{T^2}\int_{-T/2}^{T/2}\left(\int_{-T/2}^{T/2} f(u)\, g(t-u)\, du\right) e^{-in\omega_0 t}\, dt.$$

Changing the order of integration, we get

$$(f * g)_n = \frac{1}{T}\int_{-T/2}^{T/2}\left(\frac{1}{T}\int_{-T/2}^{T/2} g(t-u)\, du\, e^{-in\omega_0(t-u)}\, dt\right) \times f(u)\, e^{-in\omega_0 u}\, du$$

$$= \frac{1}{T} d_n \int_{-T/2}^{T/2} f(u)\, e^{-in\omega_0 t}\, du = c_n d_n.$$

6.13 INTEGRATION OF FOURIER SERIES

Sometimes, the Fourier series of a function is known but not the function itself. In such cases, the following problems arise:

(a) If Fourier series of the function f of period 2π is given, can we calculate $\int_a^b f(x)\, dx$ over arbitrary interval $[a, b]$?

(b) If Fourier series of the function f is known, can we find the Fourier series of the function $F(x) = \int_0^x f(t)\, dt$?

The following theorem provides the answer to the above-posed problems.

Theorem 6.14. Let

$$f(t) \sim \frac{a_0}{2} + \sum_{n=1}^{\infty} (a_n \cos nt + b_n \sin nt)$$

be the Fourier series of an absolutely integrable function of period 2π. Then $\int_a^b f(t)\, dt$ can be found by term-by-term integration of the Fourier series (irrespective of the convergence), that is,

$$\int_a^b f(t)\, dt = \frac{a_0}{2}(b-a)$$

$$+ \sum_{n=1}^{\infty} \frac{a_n(\sin nb - \sin na) - b_n(\cos nb - \cos na)}{n}.$$

Moreover, the integral of f has the Fourier series expansion in $(-\pi, \pi)$ given by

$$\int_0^x f(t)\, dt = \sum \frac{b_n}{n}$$

$$+ \sum_{n=1}^{\infty} \frac{-b_n \cos nx + [a_n + (-1)^{n+1} a_0] \sin nx}{n}.$$

Proof: Let

$$F(x) = \int_0^x f(t)\, dt - \frac{a_0}{2} x.$$

Then F is continuous and has absolutely integrable derivative (except for a finite number of points). Moreover,

$$F(x+2\pi) = \int_0^{x+2\pi} f(t)\, dt - \frac{a_0}{2}(x+2\pi)$$

$$= \int_0^x f(t)\, dt + \int_x^{x+2\pi} f(t)\, dt - \frac{a_0}{2} x - \pi a_0$$

$$= F(x) + \int_{-\pi}^{\pi} f(t)\, dt - \pi a_0$$

$$= F(x) + \pi a_0 - \pi a_0, \text{ since } a_0 = \frac{1}{\pi}\int_{-\pi}^{\pi} f(t)\, dt$$

$$= F(x).$$

Therefore, F is periodic with period 2π. Hence F can be expressed as a Fourier series

$$F(x) = \frac{A_0}{2} + \sum (A_n \cos nx + B_n \sin nx) \tag{37}$$

where

$$A_n = \frac{1}{\pi}\int_{-\pi}^{\pi} F(x) \cos nx\, dx$$

$$= \frac{1}{\pi}\left| F(x) \frac{\sin nx}{n} \right|_{-\pi}^{\pi} - \frac{1}{n\pi}\int_{-\pi}^{\pi} F'(x) \sin nx\, dx$$

$$= 0 - \frac{1}{n\pi}\int_{-\pi}^{\pi} \left(f(x) - \frac{a_0}{2}\right) \sin nx\, dx$$

$$= -\frac{1}{n\pi}\int_{-\pi}^{\pi} f(x) \sin nx\, dx + \frac{a_0}{2n\pi}\int_{-\pi}^{\pi} \sin nx\, dx$$

$$= -\frac{1}{n\pi} \cdot \pi b_n + 0 = -\frac{b_n}{n},$$

and Similarly

$$B_n = \frac{a_n}{n}$$

Thus (37) reduces to

$$F(x) = \frac{A_0}{2} + \sum_{n=1}^{\infty} \frac{a_n \sin nx - b_n \cos nx}{n}$$

and so

$$\int_0^x f(t)\, dt = \frac{A_0}{2} + \frac{a_0}{2} x$$

$$+ \sum_{n=1}^{\infty} \frac{a_n \sin nx - b_n \cos nx}{n} \tag{38}$$

Putting $x = b$ and $x = a$ in (38) and subtracting, we get

$$\int_a^b f(t)\, dt = \frac{a_0}{2}(b-a)$$

$$+ \sum_{n=1}^{\infty} \frac{a_n(\sin nb - \sin na) - b_n(\cos nb - \cos na)}{n}.$$

It follows, therefore, that the *Fourier series (even divergent) can be integrated term-by-term in any interval.*

Now if we put $x = 0$ in (38), we get

$$\frac{A_0}{2} = \sum_{n=1}^{\infty} \frac{b_n}{n} \tag{39}$$

But (see Example 6.7)

$$\frac{x}{2}=\sum_{n=1}^{\infty}\frac{(-1)^{n+1}}{n}\sin nx.$$

Hence (38) reduces to

$$\int_0^x f(t)\,dt=\sum_{n=1}^{\infty}\frac{b_n}{n}$$

$$+\sum_{n=1}^{\infty}\frac{-b_n\cos nx+[a_n+(-1)^{n+1}a_0]\sin nx}{n}.$$

Remark 6.6. The expression (39) shows that for any Fourier series, the series $\sum\frac{b_n}{n}$ converges. This fact helps us to differentiate the Fourier series of absolutely integrable functions from other trigonometric series. For example, the series

$$\sum_{n=2}^{\infty}\frac{\sin nx}{\log n},$$

converges everywhere but cannot be a Fourier series since the series

$$\sum\frac{b_n}{n}=\sum\frac{1}{n\log n}$$

is divergent.

6.14 DIFFERENTIATION OF FOURIER SERIES

Regarding differentiation of Fourier series, we have the following theorem:

Theorem 6.15. Let f be a continuous function of period 2π having an absolutely integrable derivative (except at certain points). Then the Fourier series of f' can be obtained from the Fourier series of the function f by term-by-term differentiation.

Proof: By Convergence theorem, the Fourier series of f converges to f. So let

$$f(x)=\frac{a_0}{2}+\sum_{n=1}^{\infty}(a_n\cos nx+b_n\sin nx). \qquad (40)$$

If a'_n and b'_n denote Fourier coefficients of f', then

$$a'_0=\frac{1}{\pi}\int_{-\pi}^{\pi}f'(x)\,dx=f(\pi)-f(-\pi)=0,$$

$$a'_n=\frac{1}{\pi}\int_{-\pi}^{\pi}f'(x)\cos nx\,dx$$

$$=\frac{1}{\pi}[\cos nxf(x)]_{-\pi}^{\pi}+\frac{n}{\pi}\int_{-\pi}^{\pi}\sin nxf(x)\,dx$$

$$=0+\frac{n}{\pi}\int_{-\pi}^{\pi}f(x)\sin nx\,dx=nb_n,$$

$$b'_n=\frac{1}{\pi}\int_{-\pi}^{\pi}f'(x)\sin nx\,dx$$

$$=\frac{1}{\pi}[f(x)\sin nx]_{-\pi}^{\pi}-\frac{n}{\pi}\int_{-\pi}^{\pi}f(x)\cos nx\,dx$$

$$=0-na_n=-na_n.$$

Hence the Fourier series of f' is given by

$$f'(x)\sim\sum_{n=1}^{\infty}n(b_n\cos nx-a_n\sin nx),$$

which is nothing but the series obtained from (40) by term-by-term differentiation.

6.15 EXAMPLES OF EXPANSIONS OF FUNCTIONS IN FOURIER SERIES

EXAMPLE 6.4

Expand in Fourier series the function f defined by

$$f(x)=\begin{cases}0 & \text{for } -\pi\le x<0\\ 1 & \text{for } 0\le x\le\pi.\end{cases}$$

Deduce that sum of the Gregory series $1-\frac{1}{3}+\frac{1}{5}-\frac{1}{7}+\frac{1}{9}-\ldots$ is $\frac{\pi}{4}$

Solution. Taking periodic extension of the functcion, the graph of f is shown in Figure 6.8. The extended function is of period 2π. So, we have

$$a_n=\frac{1}{\pi}\int_{-\pi}^{\pi}f(x)\cos nx\,dx=\frac{1}{\pi}\int_0^{\pi}\cos nx\,dx$$

$$=\frac{1}{\pi}\left[\frac{\sin nx}{n}\right]_0^{\pi}$$

$$= \begin{cases} 0 \text{ for } n = 1, 2, .., \\ 1 \text{ for } n = 0 \end{cases}$$

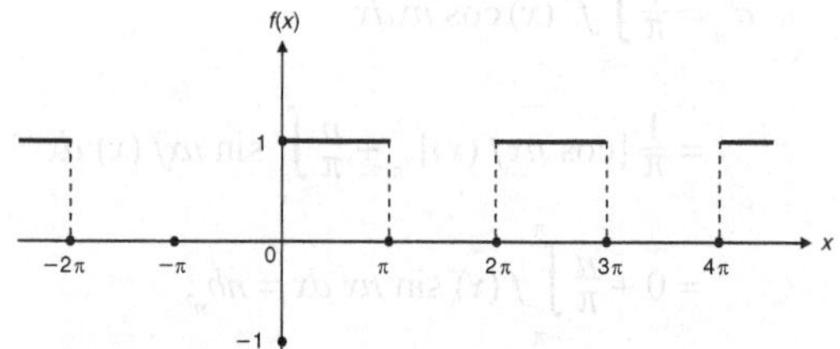

Figure 6.8 Graph of Periodically Extended f

$$b_n = \frac{1}{\pi}\int_{-\pi}^{\pi} f(x) \sin nx \, dx$$

$$= \frac{1}{\pi}\int_{0}^{\pi} \sin nx \, dx = \frac{1 - \cos n\pi}{n\pi}$$

$$= \begin{cases} \frac{2}{n\pi} \text{ for } n = 1, 3, 5, \ldots \\ 0 \text{ for } n = 2, 4, 6, \ldots \end{cases}$$

Hence the Fourier series of f is given by

$$f \sim \frac{1}{2} + \frac{2}{\pi}\left[\frac{\sin x}{1} + \frac{\sin 3x}{3} + \frac{\sin 5x}{5} + \ldots\right].$$

We have used the symbol ~ because the series does not converge to f. In fact, we note that $f(0) = 1$ by definition of f. But $x = 0$ in the series yields the sum as $\frac{1}{2}$.

At $x = \frac{\pi}{2}$, we have

$$1 = \frac{1}{2} + \frac{2}{\pi}\left[1 - \frac{1}{3} + \frac{1}{5} - \frac{1}{7} + \frac{1}{9} - \ldots\right],$$

which yields

$$1 - \frac{1}{3} + \frac{1}{5} - \frac{1}{7} + \frac{1}{9} - \ldots = \frac{\pi}{4}.$$

EXAMPLE 6.5

Determine Fourier series of the function f defined by

$$f(x) = \begin{cases} -\pi & \text{for } -\pi < x < 0 \\ x & \text{for } 0 < x < \pi. \end{cases}$$

Prove that $\frac{1}{1^2} + \frac{1}{3^2} + \frac{1}{5^2} + \ldots = \frac{\pi^2}{8}$.

Solution. Taking periodic extension of the given function, we have a function of period 2π. Then

$$a_0 = \frac{1}{\pi}\int_{-\pi}^{\pi} f(x)dx = \frac{1}{\pi}\left[\int_{-\pi}^{0} f(x)dx + \int_{0}^{\pi} f(x)dx\right]$$

$$= \frac{1}{\pi}\int_{-\pi}^{0} (-\pi)dx + \frac{1}{\pi}\int_{0}^{\pi} x\,dx = \frac{1}{\pi}\left[-\pi^2 + \frac{\pi^2}{2}\right] = -\frac{\pi}{2},$$

$$a_n = \frac{1}{\pi}\int_{-\pi}^{\pi} f(x)\cos nx \, dx = \frac{1}{\pi}\left[\int_{-\pi}^{0} -\pi \cos nx dx + \int_{0}^{\pi} x \cos nx \, dx\right]$$

$$= \frac{1}{\pi}\left[\frac{\cos n\pi}{n^2} - \frac{1}{n^2}\right] = \begin{cases} -\frac{2}{\pi n^2} & \text{for odd } n \\ 0 & \text{for even } n, \end{cases}$$

$$b_n = \frac{1}{\pi}\int_{-\pi}^{\pi} f(x) \sin nx \, dx$$

$$= \frac{1}{\pi}\left[\int_{-\pi}^{0} -\pi \sin nx dx + \int_{0}^{\pi} x \sin nx \, dx\right]$$

$$= \frac{1}{n}(1 - 2\cos n\pi).$$

Hence Fourier series expansion of f is

$$f(x) = \frac{a_0}{2} + \sum_{n=1}^{\infty} (a_n \cos nx + b_n \sin nx)$$

$$= -\frac{\pi}{4} - \frac{2}{\pi}\left(\frac{\cos x}{1^2} + \frac{\cos 3x}{3^2} + \frac{\cos 5x}{5^2} + \ldots\right)$$

$$+ 3\sin x - \frac{\sin 2x}{2} + \frac{3\sin 3x}{3} - \frac{\sin 4x}{4} + \ldots$$

Taking $x = 0$, we get

$$f(0) = -\frac{\pi}{4} - \frac{2}{\pi}\left(\frac{1}{1^2} + \frac{1}{3^2} + \frac{1}{5^2} + \ldots\right).$$

But, by Convergence Theorem, we have

$$f(0) = \frac{1}{2}[f(0+) + f(0-)] = \frac{1}{2}(-\pi + 0) = -\frac{\pi}{2}.$$

Hence

$$-\frac{\pi}{2} = -\frac{\pi}{4} - \frac{2}{\pi}\left(\frac{1}{1^2} + \frac{1}{3^2} + \frac{1}{5^2} + \ldots\right),$$

which yields

$$\frac{1}{1^2}+\frac{1}{3^2}+\frac{1}{5^2}+\ldots=\frac{\pi^2}{8}.$$

EXAMPLE 6.6

Expand $f(x) = x^2$, $-\pi < x < \pi$ in Fourier series and show that

(a) $\sum_{n=1}^{\infty}\frac{1}{n^2}=\frac{\pi^2}{6}$

(b) $\sum_{n=1}^{\infty}\frac{1}{(2n-1)^2}=\frac{\pi^2}{8}$

(c) $\sum_{n=1}^{\infty}\frac{1}{n^4}=\frac{\pi^2}{90}$.

Solution. The function is defined in the interval $(-\pi, \pi)$. The periodic extension of f is continuous and smooth (see Example 6.2). Since $f(x) = f(-x)$, the function is even. Hence, the Fourier coefficients $b_n = 0$ for $n = 1, 2, \ldots$. To calculate a_n, we use integration by parts and get

$$a_0=\frac{2}{\pi}\int_0^{\pi}x^2\,dx=\frac{2\pi^2}{3},$$

$$a_n=\frac{1}{\pi}\int_{-\pi}^{\pi}f(x)\cos nx dx=\frac{2}{\pi}\int_0^{\pi}x^2\cos nx dx$$

$$=\frac{2}{\pi}\left[x^2\frac{\sin nx}{n}-2x\frac{-\cos nx}{n^2}+2\left(-\frac{\sin nx}{n^3}\right)\right]_0^{\pi}$$

$$=\frac{2}{\pi}\left[2\pi\frac{\cos n\pi}{n^2}\right]=\frac{4}{n^2}(-1)^n,\text{ since }\cos n\pi=(-1)^n.$$

Since f is continuous and smooth, the Fourier series of f converges to f and so the Fourier series is

$$f(x)=\frac{\pi^2}{3}+4\sum_{n=1}^{\infty}\frac{(-1)^n}{n^2}\cos nx. \quad (41)$$

Derivations. (a) Substituting $x = \pi$ in (41), we get

$$\pi^2=\frac{\pi^2}{3}+4\sum_{n=1}^{\infty}\frac{(-1)^n}{n^2}(-1)^n=\frac{\pi^2}{3}+4\sum_{n=1}^{\infty}\frac{1}{n^2},$$

which yields

$$\frac{\pi^2}{6}=\sum_{n=1}^{\infty}\frac{1}{n^2}. \quad (*)$$

(b) Now putting $x = 0$ in (41), we get

$$0=\frac{\pi^2}{3}+4\sum_{n=1}^{\infty}\frac{(-1)^n}{n^2}.$$

Thus

$$4\left[-\frac{1}{1^2}+\frac{1}{2^2}-\frac{1}{3^2}+\ldots\right]=-\frac{\pi^2}{3} \quad (42)$$

that is,

$$\frac{1}{1^2}-\frac{1}{2^2}+\frac{1}{3^2}-\ldots=\frac{\pi^2}{12}. \quad (43)$$

Adding (*) and (43), we get

$$\frac{1}{1^2}+\frac{1}{2^2}+\frac{1}{3^2}+\ldots=\frac{\pi^2}{8}$$

or

$$\sum_{n=1}^{\infty}\frac{1}{(2n-1)^2}=\frac{\pi^2}{8}.$$

(c) Applying Parseval's equality, we get

$$\int_{-\pi}^{\pi}[f(x)]^2\,dx=2\pi a_0^2+\pi\sum_{n=1}^{\infty}\left(a_n^2+b_n^2\right),$$

that is

$$\int_{-\pi}^{\pi}(x^2)^2\,dx=2\pi\left(\frac{\pi^2}{3}\right)^2+\pi\sum_{n=1}^{\infty}\frac{16}{n^4},$$

that is,

$$\frac{2}{5}\pi^5=\frac{2}{9}\pi^5+\pi\sum_{n=1}^{\infty}\frac{16}{n^4},$$

which yields

$$\sum_{n=1}^{\infty}\frac{1}{n^4}=\frac{\pi^4}{90}.$$

EXAMPLE 6.7

Obtain the Fourier series for the function $f(x) = 2x + 1$, $-\pi < x < \pi$. Hence deduce Fourier series for x and the line $y = mx + c$.

Solution. Using Fourier formulae, we have

$$a_0=\frac{1}{\pi}\int_{-\pi}^{\pi}(2x+1)\,dx=\frac{1}{\pi}[x^2+x]_{-\pi}^{\pi}=2,$$

$$a_n=\frac{1}{\pi}\int_{-\pi}^{\pi}(2x+1)\cos nx\,dx$$

$$= \frac{1}{\pi}\left[(2x+1)\left(\frac{\sin nx}{n}\right) - 2\left(-\frac{\cos nx}{n^2}\right)\right]_{-\pi}^{\pi}$$

$$= \frac{1}{\pi}\left[\frac{2\cos n\pi}{n^2} - \frac{2\cos n\pi}{n^2}\right] = 0,$$

$$b_n = \frac{1}{\pi}\int_{-\pi}^{\pi} (2x+1)\sin nx\, dx$$

$$= \frac{1}{\pi}\left[(2x+1)\left(-\frac{\cos nx}{n}\right) - 2\left(-\frac{\sin nx}{n^2}\right)\right]_{-\pi}^{\pi}$$

$$= \frac{1}{\pi}\left[\frac{(2\pi+1)(-\cos n\pi)}{n} + \frac{(-2\pi+1)(-\cos n\pi)}{n}\right]$$

$$= \frac{1}{\pi}\left[-\frac{4\pi\cos n\pi}{n}\right] = -\frac{4}{n}(-1)^n.$$

Since the function is continuous and smooth, by Fundamental theorem, Fourier series converges to f and we have

$$f(x) = 2x + 1 = 1 - 4\sum_{n=1}^{\infty}\frac{(-1)^n}{n^2}\sin nx,\ -\pi < x < \pi,$$

that is,

$$x = 2\sum_{n=1}^{\infty}\frac{(-1)^{n+1}}{n}\sin nx,$$

which is the Fourier series for x in $[-\pi, \pi]$. Comparing $mx + c =$ with $2x + 1$, we get

$$mx + c = c - 2m\sum_{n=1}^{\infty}\frac{(-1)^n}{n}\sin nx.$$

EXAMPLE 6.8

Find the Fourier series for the function f defined by $f(x) = x - x^2$, $-\pi < x < \pi$. Deduce that $\frac{1}{1^2} - \frac{1}{2^2} + \frac{1}{3^2} - \frac{1}{4^2} + \dots = \frac{\pi^2}{12}$.

Solution. The periodic extension of f is of period 2π. Using Euler's formulae, we have

$$a_0 = \frac{1}{\pi}\int_{-\pi}^{\pi}(x - x^2)\,dx = \frac{-2\pi^2}{3},$$

$$a_n = \frac{1}{\pi}\int_{-\pi}^{\pi}(x - x^2)\cos nx dx$$

$$= \frac{1}{\pi}\left[(x - x^2)\left(\frac{\sin nx}{n}\right) - (1 - 2x)\left(-\frac{\cos nx}{n^2}\right) + (-2)\left(-\frac{\sin nx}{n^3}\right)\right]_{-\pi}^{\pi}$$

$$= \frac{1}{\pi}\left[\frac{(1-2\pi)\cos n\pi}{n^2} - \frac{(1+2\pi)\cos n\pi}{n^2}\right]$$

$$= \frac{4}{n^2}(-1)^n.$$

Similarly, one can show that

$$b_n = \frac{1}{\pi}\int_{-\pi}^{\pi}(x - x^2)\sin nx\,dx = -\frac{2(-1)^n}{n}.$$

Hence the Fourier series is

$$f(x) = -\cdot\frac{\pi^2}{3} + 4\left[\frac{\cos x}{1^2} - \frac{\cos 2x}{2^2} + \frac{\cos 3x}{3^2} - \dots\right] + 2\left[\frac{\sin x}{1} - \frac{\sin 2x}{2} + \frac{\sin 3x}{3} - \dots\right].$$

Putting $x = 0$, we get

$$0 = -\frac{\pi^2}{3} + 4\left(\frac{1}{1^2} - \frac{1}{2^2} + \frac{1}{3^2} - \dots\right)$$

and so

$$\frac{1}{1^2} - \frac{1}{2^2} + \frac{1}{3^2} - \dots = \frac{\pi^2}{12}.$$

EXAMPLE 6.9

If a is a real number, find the Fourier series of the function f defined by

$$f(x) = e^{ax}, -\pi < x < \pi$$

$$f(x + 2\pi) = f(x),\ x \in R,$$

Deduce the value of the series $\sum_{n=1}^{\infty}\frac{(-1)^n}{a^2 + n^2}$.

Solution. Using Euler's formulae, we have

$$a_n + ib_n = \frac{1}{\pi}\int_{-\pi}^{\pi} e^{ax+inx}\,dx = \frac{1}{\pi(a+in)}\left[e^{ax+inx}\right]_{-\pi}^{\pi}$$

$$= \frac{1}{\pi(a+in)}e^{in\pi}\left[e^{a\pi} - e^{-a\pi}\right]$$

$$= \frac{2(-1)^n}{\pi(a+in)}\sinh a\pi$$

$$= \frac{2(-1)^n (a - in)}{\pi(a^2 + n^2)} \sinh a\pi.$$

Equating real and imaginary parts, we have

$$a_n = \frac{2a(-1)^n \sinh a\pi}{\pi(a^2 + n^2)},\ b_n = \frac{2n \sinh a\pi}{\pi(a^2 + n^2)},$$

$$a_0 = \frac{2a \sinh a\pi}{\pi a^2} = \frac{2 \sinh a\pi}{\pi a}.$$

The series shall also converge to f due to piecewise continuity and smoothness. Hence

$$f(x) = \frac{\sinh a\pi}{\pi}\left[\frac{1}{a} + 2\sum_{n=1}^{\infty} \frac{(-1)^n}{a^2 + n^2}(a \cos nx - n \sin nx)\right].$$

Putting $x = 0$, we get

$$e^0 = 1 = \frac{\sinh a\pi}{\pi}\left[\frac{1}{a} + 2\sum_{n=1}^{\infty} \frac{(-1)^n a}{a^2 + n^2}\right]$$

and so

$$\sum_{n=1}^{\infty} \frac{(-1)^n}{a^2 + n^2} = \frac{1}{2a^2}(a\pi \operatorname{cosec} a\pi - 1).$$

EXAMPLE 6.10

Find the Fourier series of the function

$$f(x) = |x|, -2 \le x \le 2$$
$$f(x) = f(x + 4).$$

Solution. The period of the given function is 4. Therefore, the Fourier series shall be

$$\frac{a_0}{2} + \sum_{n=1}^{\infty}\left(a_n \cos\frac{n\pi x}{2} + b_n \sin\frac{n\pi x}{2}\right),$$

where

$$a_0 = \frac{1}{2}\int_{-2}^{2} |x| dx = 2,$$

$$a_n = \frac{1}{2}\int_{-2}^{2} |x| \cos\frac{n\pi x}{2}\, dx$$

$$= \int_0^2 x \cos\frac{n\pi x}{2}\, dx \text{ (integrand is even)}$$

$$= \left[\frac{2x}{n\pi}\sin\frac{n\pi x}{2}\right]_0^2 - \int_0^2 \frac{2}{n\pi}\sin\frac{n\pi x}{2}\, dx$$

$$= 0 + \frac{4}{n^2\pi^2}[-1 + (-1)^n] = \begin{cases} 0 & \text{for even } n \\ -\frac{8}{n^2\pi^2} & \text{for odd } n. \end{cases}$$

Since the given function is even, $b_n = 0$ for $n = 1, 2, \ldots$ Also, the function is continuous on $\left[0, \frac{1}{2}\right]$. Hence, by Convergence theorem,

$$f(x) = 1 - \frac{8}{\pi^2}\left[\cos x + \frac{1}{9}\cos 3x + \frac{1}{25}\cos 5x + \ldots\right].$$

EXAMPLE 6.11

Expand $f(t) = 1 - t^2, -1 \le t \le 1$ in Fourier series.

Solution. Periodically extended function of the given function is of period 2. Therefore, its Fourier shall be

$$\frac{a_0}{2} + \sum_{n=1}^{\infty} a_n \cos n\pi t.$$

By Euler's formula, we have

$$a_0 = \frac{1}{1}\int_{-1}^{1}(1 - t^2)dt = \frac{4}{3},$$

$$a_n = \frac{1}{1}\int_{-1}^{1}(1 - t^2)\cos n\pi t\, dt$$

$$= \left[(1 - t^2)\frac{\sin n\pi t}{n\pi}\right]_{-1}^{1} - \int_{-1}^{1}(-2t)\frac{\sin n\pi t}{n\pi}\, dt$$

$$= 0 + \frac{2}{n\pi}\int_{-1}^{1} t \sin n\pi t\, dt$$

$$= \frac{2}{n\pi}\left[\frac{t(-\cos n\pi t)}{n\pi}\right]_{-1}^{1} - \frac{2}{n\pi}\int_{-1}^{1}\frac{-\cos n\pi t}{n\pi}\, dt$$

$$= \frac{-4}{n^2\pi^2}(-1)^n + \frac{2}{n^2\pi^2}\left[\frac{\sin n\pi t}{n\pi}\right]_{-1}^{1} = -\frac{4}{n^2\pi^2}(-1)^n.$$

Hence the Fourier series is

$$f(x) \sim \frac{2}{3} + \sum_{n=1}^{\infty}\frac{-4}{n^2\pi^2}(-1)^n \cos n\pi t$$

$$= \frac{2}{3} + \frac{4}{\pi^2}\left[\cos \pi t - \frac{\cos 2\pi t}{2^2} + \frac{\cos 3\pi t}{3^2} - \ldots\right].$$

EXAMPLE 6.12

Determine the Fourier series for sawtooth function f defined by

$f(t) = t; t \in (-\pi, \pi)$
$f(t) = f(t + 2)$

Solution. The periodic extension of the function is piecewise smooth and its graph is shown in Figure 6.9. This extended function is an odd function with period T = 2π. Therefore, the *Fourier series will have only sine terms*. Using Euler formulae, we have

$$a_0 = \frac{2}{T}\int_{-T/2}^{T/2} f(t)\,dt = \frac{1}{\pi}\int_{-\pi}^{\pi} t\,dt = 0,$$

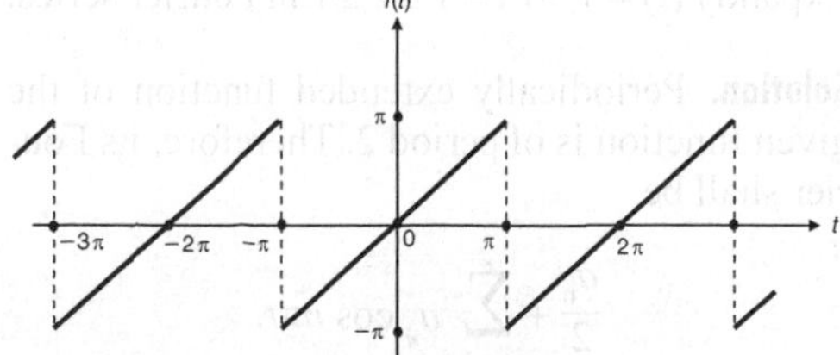

Figure 6.9

$$b_n = \frac{1}{\pi}\int_{-\pi}^{\pi} t \sin nt\,dt$$

$$= -\frac{1}{n\pi}\left[t\cos nt\right]_{-\pi}^{\pi} + \frac{1}{n\pi}\int_{-\pi}^{\pi} \cos nt\,dt$$

$$= -\frac{2}{n}\cos n\pi + 0 = -\frac{2}{n}(-1)^n, n \in \mathrm{N}.$$

Since the function is piecewise continuous and smooth, we have

$$f(t) = \sum_{n=1}^{\infty} (-1)^{n+1}\frac{2}{n}\sin nt$$

$$= 2\left(\sin t - \frac{\sin 2t}{2} + \frac{\sin 3t}{3} - \ldots\right).$$

EXAMPLE 6.13

Determine the Fourier series of the square wave function f defined by

$$f(x) = \begin{cases} -k & \text{for } -\pi < x < 0 \\ k & \text{for } 0 < x < \pi, \end{cases}$$

$$f(x) = f(x + 2\pi).$$

Deduce that $1 - \frac{1}{3} + \frac{1}{5} - \frac{1}{7} + \cdots = \frac{\pi}{4}$.

Solution. The graph of periodically extended square wave function f is shown in the Figure 6.10. The function is of period $T = 2\pi$. Further,

$$f(-x) = \begin{cases} -k & \text{for } -\pi < x < 0, \text{ that is, } 0 < x < -\pi \\ k & \text{for } 0 < -x < \pi, \text{ that is, } -\pi < x < 0 \end{cases}$$

$$= -\begin{cases} k & \text{for } 0 < x < \pi \\ -k & \text{for } -\pi < x < 0 \end{cases}$$

$$= -f(x).$$

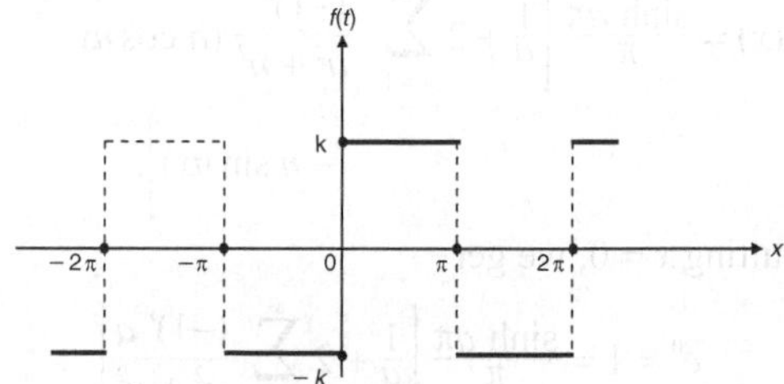

Figure 6.10

Thus, f is an odd function and so its Fourier series consists of sine terms only. We have

$$b_n = \frac{1}{\pi}\int_{-\pi}^{\pi} f(x)\sin nx\,dx = \frac{2}{\pi}\int_{0}^{\pi} k \sin nx\,dx$$

$$= \frac{2k}{\pi}\left[-\frac{\cos nx}{n}\right]_0^{\pi} = \begin{cases} 0 & \text{for even } n \\ \frac{4k}{n\pi} & \text{for odd } n. \end{cases}$$

Hence, the function being piecewise continuous and smooth, the Fourier series for f is given by

$$f(x) = \sum_{n=1}^{\infty} \frac{4k}{n\pi}\sin nx, \ n \text{ odd}$$

$$= \frac{4k}{\pi}\left[\sin x + \frac{1}{3}\sin 3x + \frac{1}{5}\sin 5x + \ldots\right].$$

Taking $x = \frac{\pi}{2}$, we get

$$k = \frac{4k}{\pi}\left[1 - \frac{1}{3} + \frac{1}{5} - \frac{1}{7} + \ldots\right]$$

and so

$$1 - \frac{1}{3} + \frac{1}{5} - \frac{1}{7} + \ldots = \frac{\pi}{4}.$$

EXAMPLE 6.14

Determine the Fourier series for the periodic triangle function f with period T defined for $0 < a \le \frac{T}{2}$ on $\left(-\frac{T}{2}, \frac{T}{2}\right)$ by

$$= \begin{cases} 1 - \frac{|x|}{a} & \text{for } |x| \le a \\ 0 & \text{for } a < |x| \le T/2. \end{cases}$$

Solution. The graph of the periodically extended triangle function is shown in the Figure 6.11.

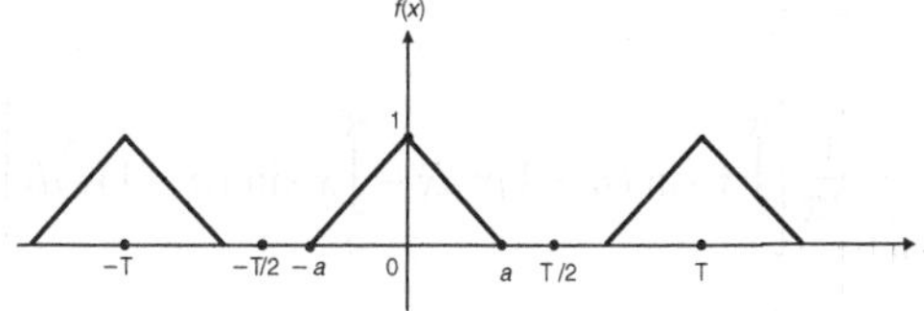

Figure 6.11

The function is periodic with period T and is also even. So the Fourier coefficients b_n are zero. For a_n, we have

$$a_n = \frac{4}{T}\int_0^{T/2} f(x)\cos n\omega_0 x\, dx,\ \omega_0 = \frac{2\pi}{T}$$

$$= \frac{4}{T}\int_0^{a} \left(1 - \frac{x}{a}\right) \cos n\omega_0 x\, dx$$

$$= \frac{4}{Tn\omega_0}\left[\left\{\left(1 - \frac{x}{a}\right) \sin n\omega_0 x\right\}_0^a + \frac{1}{a}\int_0^a \sin n\omega_0 x\, dx\right]$$

$$= \frac{4}{aTn\omega_0}\int_0^a \sin n\omega_0 x\, dx = \frac{4}{aTn\omega_0}\left[-\frac{\cos n\omega_0 x}{n\omega_0}\right]_0^a$$

$$= \frac{4}{aTn^2\omega_0^2}[1 - \cos n\omega_0 a] = \frac{8}{aTn^2\omega_0^2}\sin^2\left(\frac{n\omega_0 a}{2}\right),$$

and

$$a_0 = \frac{4}{T}\int_0^a \left(1 - \frac{x}{a}\right) dx = \frac{2a}{T}.$$

Hence

$$f(x) = \frac{a}{T} + \frac{8}{aTn^2\omega_0^2}\sum_{n=1}^{\infty} \frac{1}{n^2}\sin^2\left(n\omega_0 \frac{a}{2}\right)\cos nx.$$

EXAMPLE 6.15

Determine the Fourier series for periodic block function f with period T > 0 and $0 \le a \le$ T and defined by

$$f(x) = \begin{cases} 1 & \text{for } |x| \le a/2 \le T/2 \\ 0 & \text{for } a/2 < |x| \le T/2. \end{cases}$$

Solution. The graph of periodically extended function with period T is shown in the Figure 6.12. This function is also even.

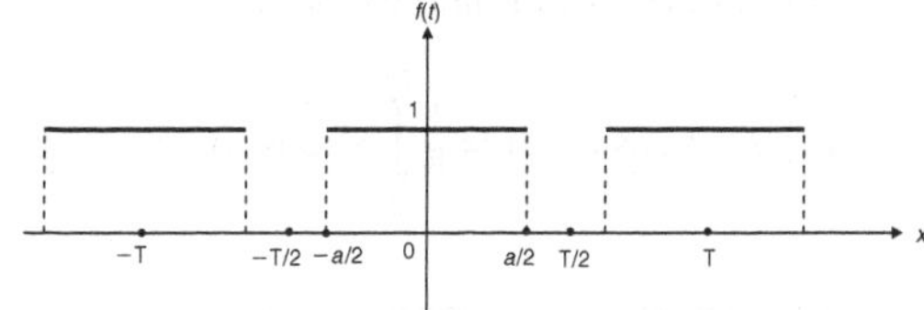

Figure 6.12

We have

$$a_n = \frac{2}{T}\int_{-T/2}^{T/2} f(x)\cos n\omega_0 x\, dx, \quad \omega_0 = \frac{2\pi}{T}$$

$$= \frac{4}{T}\int_0^{a/2} \cos n\omega_0 x\, dx = \frac{4}{T}\left[\frac{\sin n\omega_0 x}{n\omega_0}\right]_0^{a/2}$$

$$= \frac{4}{Tn\omega_0}\sin\left(n\omega_0 \frac{a}{2}\right),\ n \ne 0,$$

whereas

$$a_0 = \frac{4}{T}\int_0^{a/2} dx = \frac{2a}{T}.$$

Since f is piecewise continuous and smooth, the Convergence theorem of Fourier series yields

$$f(x) = \frac{a}{T} + \frac{4}{T\omega_0}\sum_{n=1}^{\infty} \frac{1}{n}\sin\left(n\omega_0 \frac{a}{2}\right)\cos nx$$

EXAMPLE 6.16

Expand $f(x) = x^2$, $0 < x < 2\pi$ in a Fourier series assuming that the function is of period 2π.

Solution. We are given that $T = 2\pi$. The graph of the periodically extended function is shown in the Figure 6.13.

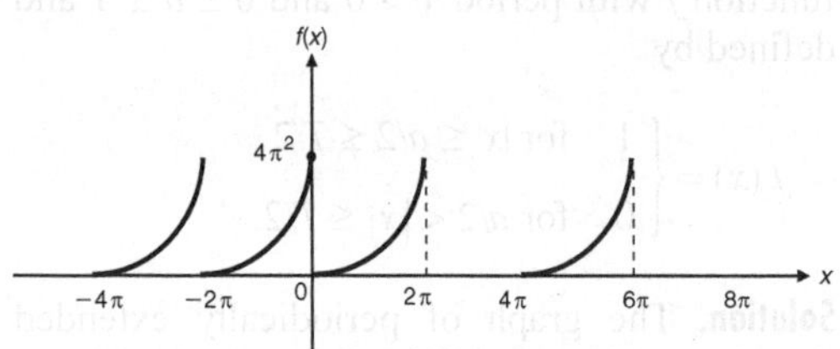

Figure 6.13

Using Fourier formulae, we have

$$a_n = \frac{1}{\pi}\int_0^{2\pi} f(x)\cos nx\, dx = \frac{1}{\pi}\int_0^{2\pi} x^2 \cos nx\, dx$$

$$= \frac{1}{\pi}\left[x^2\frac{\sin nx}{n} - 2x\left(-\frac{\cos nx}{n^2}\right) + 2\left(-\frac{\sin nx}{n^3}\right)\right]_0^{2\pi}$$

$$= \frac{4}{n^2},\ n \neq 0.$$

For $n = 0$, we get

$$a_0 = \frac{1}{\pi}\int_0^{2\pi} x^2\, dx = \frac{8\pi^2}{3}.$$

Similarly,

$$b_n = \frac{1}{\pi}\int_0^{2\pi} x^2 \sin nx\, dx = -\frac{4\pi}{n}.$$

Thus the required Fourier series is

$$\frac{4\pi^2}{3} + \sum_{n=1}^{\infty}\left(\frac{4}{n^2}\cos nx - \frac{4\pi}{n}\sin nx\right).$$

At $x = 0$ and $x = 2\pi$, the series converges to $2\pi^2$.

EXAMPLE 6.17

Develop Fourier series for $f(x) = x \sin x$, $0 < x < 2\pi$.

Solution. Using Euler's formulae, we have

$$a_0 = \frac{1}{\pi}\int_0^{2\pi} x \sin x\, dx$$

$$= \frac{1}{\pi}[x(-\cos x)]_0^{2\pi} - \frac{1}{\pi}\int_0^{2\pi}(-\cos x)\, dx = -2,$$

$$a_n = \frac{1}{\pi}\int_0^{2\pi} x \sin x \cos nx\, dx$$

$$= \frac{1}{2\pi}\int_0^{2\pi} x[\sin(n+1)x - \sin(n-1)x]\, dx$$

$$= \frac{1}{2\pi}\left[\int_0^{2\pi} x\sin(n+1)x\, dx - \int_0^{2\pi} x\sin(n-1)x\, dx\right]$$

$$= \frac{1}{2\pi}\left\{\left[\frac{x(-\cos(n+1)x)}{n+1} + \frac{\sin(n+1)x}{(n+1)^2}\right]_0^{2\pi} - \left[\frac{x(-\cos(n-1)x)}{n-1} + \frac{\sin(n-1)x}{(n-1)^2}\right]_0^{2\pi}\right\}$$

$$= \frac{1}{2\pi}\left[\frac{-2\pi}{n+1} + \frac{2\pi}{n-1}\right] = \frac{2}{n^2-1},\ n \neq 1,$$

$$a_1 = \frac{1}{\pi}\int_0^{2\pi} x \sin x \cos x\, dx = \frac{1}{2\pi}\int_0^{2\pi} x \sin 2x\, dx = -\frac{1}{2},$$

while

$$b_n = \frac{1}{\pi}\int_0^{2\pi} x \sin x \sin nx\, dx$$

$$= \frac{1}{2\pi}\int_0^{2\pi} x[\cos(n-1)x - \cos(n+1)x]\, dx$$

$$= \frac{1}{2\pi}\left[\frac{1}{(n-1)^2} - \frac{1}{(n-1)^2} - \frac{1}{(n+1)^2} + \frac{1}{(n+1)^2}\right]$$

$$= 0,\ n \neq 1$$

and

$$b_1 = \frac{1}{\pi}\int_0^{2\pi} x \sin^2 x\, dx = \frac{1}{2\pi}\int_0^{2\pi} x(1 - \cos 2x)\, dx$$

$$= \frac{1}{2\pi}[2\pi^2] = \pi.$$

Hence the Fourier expansion of f is

$$\frac{a_0}{2}+\sum_{n=1}^{\infty}(a_n \cos nx + b_n \sin nx)$$

$$= -1-\frac{1}{2}\cos x+\pi \sin x+2\sum_{n=1}^{\infty}\frac{\cos nx}{n^2-1}.$$

EXAMPLE 6.18

Find the Fourier series of $f(x) = x$, $0 < x < 2\pi$.

Solution. The periodic extension of the given function f is shown in the Figure 6.14.

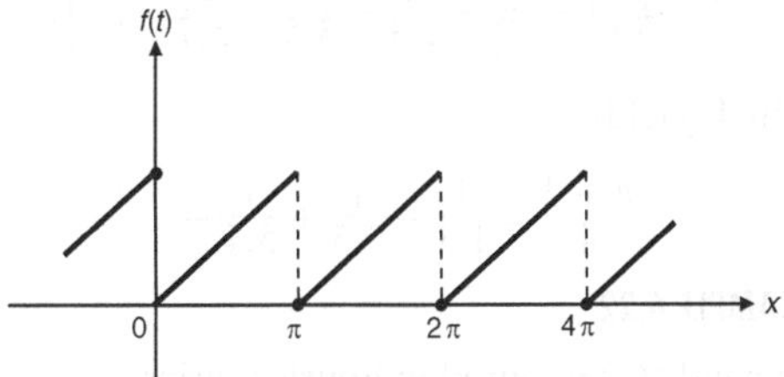

Figure 6.14

We have

$$a_0 = \frac{1}{\pi}\int_0^{2\pi} x\, dx = 2\pi,$$

$$a_n = \frac{1}{\pi}\int_0^{2\pi} x \cos nx dx = 0 - \frac{1}{\pi n}\int_0^{2\pi} \sin nx dx = 0,$$

$$b_n = \frac{1}{\pi}\int_0^{2\pi} x \sin nx dx = -\frac{2}{n}.$$

Therefore

$$x = \pi - 2\left(\sin x + \frac{\sin 2x}{2} + \frac{\sin 3x}{3} + \ldots\right),\ 0 < x < 2\pi.$$

EXAMPLE 6.19

Expand $f(x) = e^{-x}$, $0 < x < 2\pi$.

Solution. Using integration by parts, we have

$$a_n = \frac{1}{\pi}\int_0^{2\pi} f(x) \cos nx dx = \frac{1}{\pi}\int_0^{2\pi} e^{-x} \cos nx\, dx$$

$$= \left(\frac{1-e^{-2\pi}}{\pi}\right)\cdot\frac{1}{n^2+1},$$

$$b_n = \frac{1}{\pi}\int_0^{2\pi} f(x) \sin nx dx = \frac{1}{\pi}\int_0^{2\pi} e^{-x} \sin nx\, dx$$

$$= \frac{1-e^{-2\pi}}{\pi}\left(\frac{n}{n^2+1}\right).$$

Hence

$$f(x) \sim \frac{1-e^{-2\pi}}{\pi}\left[\frac{1}{2}+\left(\frac{1}{2}\cos x+\frac{1}{5}\cos 2x+\ldots\right)\right.$$
$$\left.+\left(\frac{1}{2}\sin x+\frac{2}{5}\sin 2x+\ldots\right)\right].$$

EXAMPLE 6.20

Find the Fourier series for $f(x) = \pi x$, $0 \le x \le 2$.

Solution. Comparing the interval with (0, T), we have T = 2. The function is odd and, therefore, the Fourier series shall consists of only sine terms. We have

$$b_n = \frac{2}{T}\int_0^{T} f(x) \sin n\pi x\, dx = \int_0^{2} \pi x \sin n\pi x\, dx$$

$$= \pi\left[x\frac{(-\cos n\pi x)}{n\pi}\right]_0^2 + \frac{1}{n}\int_0^2 (\cos n\pi x)\, dx = -\frac{2}{n}.$$

Hence, the required Fourier series is

$$-2\sum_{n=1}^{\infty}\frac{\sin n\pi x}{n}.$$

EXAMPLE 6.21

Determine the Fourier series of the *half-wave rectified sinusoidal* defined by

$$f(t) = \begin{cases} \sin t & \text{for } 0 < t < \pi \\ 0 & \text{for } \pi < t < 2\pi, \end{cases}$$
$$f(t) = f(t+2\pi).$$

Deduce that

(a) $\frac{1}{1.3}+\frac{1}{3.5}+\frac{1}{5.7}+\ldots = \frac{1}{2}$

(b) $\frac{1}{1.3}-\frac{1}{3.5}+\frac{1}{5.7}-\ldots = \frac{\pi-2}{4}.$

Solution. The graph of the function is shown in the Figure 6.15.

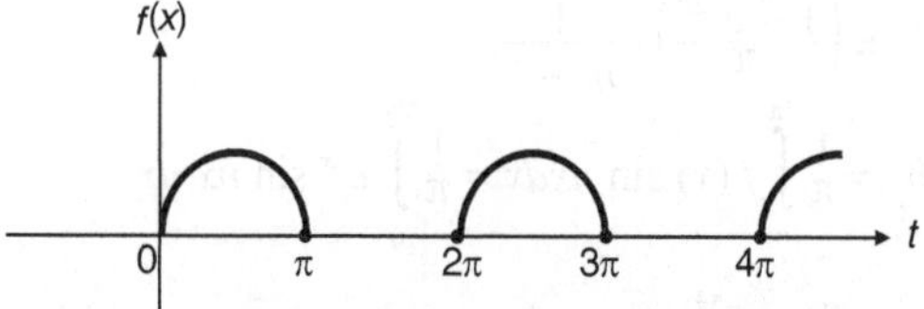

Figure 6.15

Using ther Euler formulae, we have

$$a_0 = \frac{1}{\pi}\int_0^{2\pi} f(t)\,dt = \frac{1}{\pi}\int_0^{\pi} \sin t\,dt = \frac{2}{\pi},$$

$$a_n = \frac{1}{\pi}\int_0^{2\pi} f(t)\cos nt\,dt = \frac{1}{\pi}\int_0^{\pi} \sin t\cos nt\,dt$$

$$= \frac{1}{2\pi}\int_0^{\pi} [\sin(n+1)t - \sin(n-1)t]dt$$

$$= \frac{1}{2\pi}\left[-\frac{\cos(n+1)t}{n+1} + \frac{\cos(n-1)}{n+1}\right]_0^{\pi}$$

$$= \begin{cases} -\dfrac{2}{\pi(n^2-1)} & \text{for even } n \\ 0 & \text{for odd } n. \end{cases}$$

$$b_n = \frac{1}{\pi}\int_0^{2\pi} f(t)\sin nt\,dt = \frac{1}{\pi}\int_0^{\pi} \sin t\sin nt\,dt$$

$$= \frac{1}{2\pi}\int_0^{2\pi} [\cos(n-1)t - \cos(n+1)t]\,dt$$

$$= \frac{1}{2\pi}\left[\frac{\sin(n-1)t}{n-1} - \frac{\sin(n+1)t}{n+1}\right]_0^{\pi} = 0,\ n \neq 1$$

For $n = 1$, we get

$$b_1 = \frac{1}{\pi}\int_0^{\pi} \sin t\sin t\,dt = \frac{1}{\pi}\int_0^{\pi} \sin^2 t\,dt = \frac{1}{2}.$$

Hence the required Fourier series is

$$f(t) \sim \frac{1}{\pi} - \frac{2}{3\pi}\cos 2t - \frac{2}{15\pi}\cos 4t$$

$$- \frac{2}{35\pi}\cos 6t + \ldots + \frac{1}{2}\sin t$$

$$= \frac{1}{\pi}\left[1 + \frac{\pi}{2}\sin t - \frac{2}{3}\cos 2t - \frac{2}{3.5}\cos 4t\right.$$

$$\left. - \frac{2}{3.5.7}\cos 6t - \ldots\right]$$

Deductions. (a) Putting $t = \pi$ in this Fourier series, we get $f(\pi) = 0$, and therefore,

$$0 = \frac{1}{\pi}\left[1 - \frac{2}{1.3} - \frac{2}{3.5} - \frac{2}{5.7} - \ldots\right]$$

or

$$\frac{1}{1.3} + \frac{1}{3.5} + \frac{1}{5.7} + \ldots = \frac{1}{2}.$$

(b) Putting $t = \frac{\pi}{2}$, we have $f\left(\frac{\pi}{2}\right) = \sin\frac{\pi}{2} = 1.$ Therefore, the Fourier series reduces to

$$1 = \frac{1}{\pi}\left[1 + \frac{\pi}{2} + \frac{2}{1.3} - \frac{2}{3.5} + \frac{2}{5.7} - \ldots\right],$$

which yields

$$\frac{\pi - 2}{4} = \frac{1}{1.3} - \frac{1}{3.5} + \frac{1}{5.7} - \ldots$$

EXAMPLE 6.22

Expand $f(x) = |\sin x|$ in Fourier series.

Solution. The given function is defined for all x and is continous, piecewise smooth, and even. Its graph is shown in the Figure 6.16.

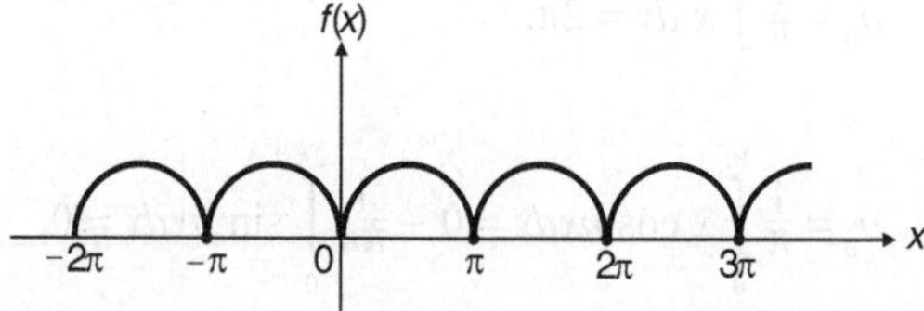

Figure 6.16

By Fundamental theorem of Fourier series, the function $f(x) = |\sin x|$ is everywhere equal to its convergent Fourier series. Since f is even, the series shall consists of only cosine terms. We have

$$a_0 = \frac{1}{\pi}\int_{-\pi}^{\pi} f(x)\,dx = \frac{2}{\pi}\int_0^{\pi} \sin x\,dx = \frac{4}{\pi}$$

and

$$a_n = \frac{2}{\pi}\int_0^{\pi} \sin x\cos nx\,dx$$

$$= \frac{1}{\pi}\int_0^{\pi} [\sin(n+1)x - \sin(n-1)x]\,dx$$

$$= -\frac{1}{\pi}\left[\frac{\cos(n+1)x}{n+1} - \frac{\cos(n-1)}{n-1}\right]_0^{\pi}$$

$$= -\frac{1}{\pi}\left[\frac{(-1)^{n+1}-1}{n+1} - \frac{(-1)^{n+1}-1}{n-1}\right]$$

$$= \frac{-2}{\pi(n^2-1)}[(-1)^n + 1], \text{ for } n \neq 1,$$

while for $n = 1$, we get

$$a_1 = \frac{2}{\pi}\int_0^{\pi} \sin x \cos x\, dx = \frac{1}{\pi}\int_0^{\pi} \sin 2x\, dx = 0.$$

Hence

$$f(x) = |\sin x|$$

$$= \frac{2}{\pi} - \frac{4}{\pi}\left(\frac{\cos 2x}{3} + \frac{\cos 4x}{15} + \frac{\cos 6x}{35} + \ldots\right)$$

EXAMPLE 6.23

Determine *sine series* expansion of the function f defined by $f(x) = 1$, $0 < x < \pi$.

Solution. The graph of the given function is shown in the Figure 6.17. We wish to obtain a Fourier sine series for this function. Making the odd extention of f onto the interval $(-\pi, 0)$

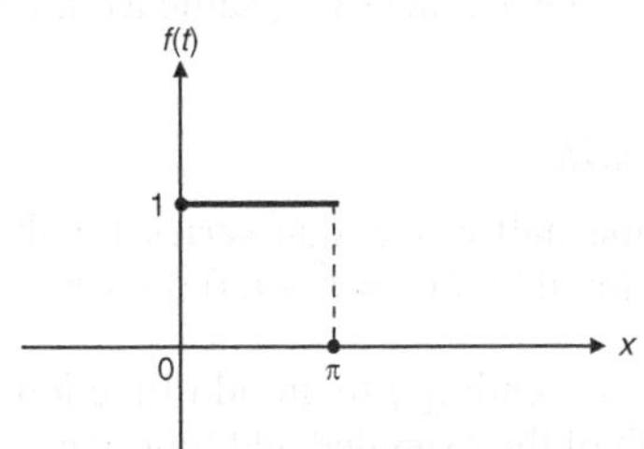

Figure 6.17

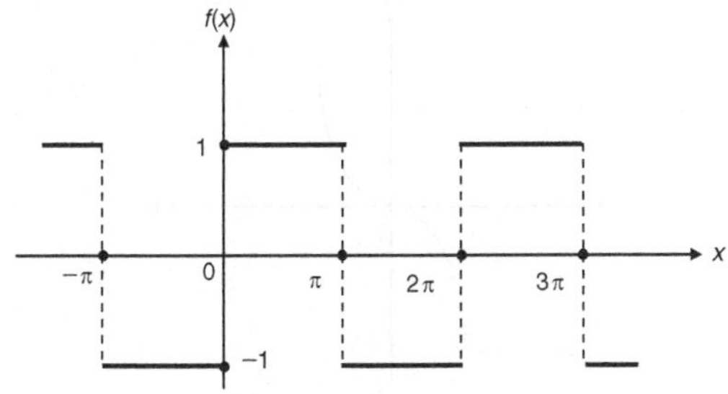

Figure 6.18

produce a discontinuity at $x = 0$. We, thus, get an odd function on the interval $(-\pi, \pi)$. We then extend it periodically with period 2π over the whole x-axis. The odd extension and subsequent periodic extensions are shown in the Figure 6.18.

The Fundamental theorem of Fourier series is applicable. The Fourier series will converge to 1 for $0 < x < \pi$. Outside the interval $0 < x < \pi$, it converges to the function as shown in the Figure 6.18 with the sum of the series being equal to zero at the points $0, \pm, \pm 2, \pm 3, \ldots$. Since the function, thus, created is odd, $a_n = 0$ for $n = 0, 1, 2, \ldots$ For b_n, we have

$$b_n = \frac{1}{\pi}\int_{-\pi}^{\pi} f(t)\sin nt\, dt = \frac{1}{\pi}\int_{-\pi}^{0} (-1)\sin nt\, dt + \frac{1}{\pi}\int_0^{\pi} \sin nt\, dt$$

$$= \frac{2}{\pi}\int_0^{\pi} \sin nt\, dt = \frac{2}{\pi}\left[\frac{-\cos nt}{n}\right]_0^{\pi}$$

$$= \frac{2}{\pi n}[1 - (-1)^n].$$

Hence,

$$f(x) = 1 = \sum_{n=1}^{\infty} \frac{2}{n\pi}[1 - (-1)^n]\sin nx$$

$$= \frac{4}{\pi}\left[\sin x + \frac{\sin 3x}{3} + \frac{\sin 5x}{5} + \ldots\right] \text{ for } 0 < x < \pi.$$

EXAMPLE 6.24

Find cosine series for the function f defined by

$$f(x) = \begin{cases} x & \text{for } 0 \le x \le L/2 \\ L - x & \text{for } L/2 \le x \le L, \end{cases}$$

Solution. The even extension of f in $(-L, L)$ and subsequent periodic extension are shown in the Figure 6.19.

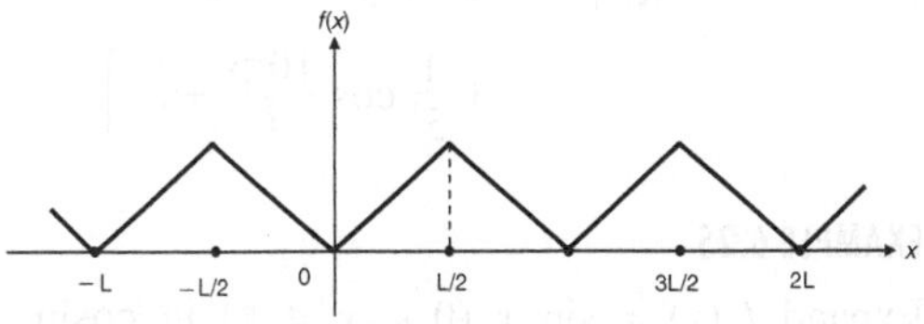

Figure 6.19

We have

$$a_0 = \frac{2}{L}\int_0^L f(x)\,dx = \frac{2}{L}\left[\int_0^{L/2} x\,dx + \int_{L/2}^{L} (L-x)\,dx\right]$$

$$= \frac{2}{L}\left[\frac{2L^2}{8}\right] = \frac{L}{2},$$

$$a_n = \frac{2}{L}\int_0^L f(x)\cos\frac{n\pi x}{L}\,dx$$

$$= \frac{2}{L}\left[\int_0^{L/2} x\frac{\cos n\pi x}{L}\,dx + \int_{L/2}^{L}(L-x)\,dx\cos\frac{n\pi x}{L}\,dx\right]$$

$$= \frac{L}{n\pi}\sin\frac{n\pi}{2} + \frac{2L}{n^2\pi^2}\cos\frac{n\pi}{2} - \frac{2L}{n^2\pi^2}$$

$$- \frac{L}{n\pi}\sin\frac{n\pi}{2} - \frac{2L}{n^2\pi^2}\cos n\pi + \frac{2L}{n^2\pi^2}\cos\frac{n\pi}{2}$$

$$= \frac{4L}{n^2\pi^2}\cos\frac{n\pi}{2} - \frac{2L}{n^2\pi^2}\cos n\pi - \frac{2L}{n^2\pi^2}$$

$$= \frac{4L}{n^2\pi^2}\cos\frac{n\pi}{2} - \frac{2L}{n^2\pi^2}[1+\cos n\pi]$$

$$= \frac{4L}{n^2\pi^2}\cos\frac{n\pi}{2} - \frac{2L}{n^2\pi^2}\left(2\cos^2\frac{n\pi}{2}\right)$$

$$= \frac{4L}{n^2\pi^2}\cos\frac{n\pi}{2}\left[1-\cos\frac{n\pi}{2}\right]$$

$$= \frac{4L}{n^2\pi^2}\cos\frac{n\pi}{2}\cdot 2\sin^2\frac{n\pi}{4}.$$

Thus

$$a_1 = 0,\ a_2 = -\frac{2L}{\pi^2},\ a_3 = 0,\ a_4 = 0,\ a_5 = 0,$$

$$a_6 = \frac{2L}{9\pi^2},\ a_7 = a_8 = a_9 = a_{10} = -\frac{2L}{25\pi^2},\ldots$$

Therefore,

$$f(x) = \frac{L}{4} - \frac{2L}{\pi^2}\left[\cos\frac{2\pi x}{L} + \frac{1}{3^2}\cos\frac{6\pi x}{L} + \frac{1}{5^2}\cos\frac{10\pi x}{L} + \ldots\right].$$

EXAMPLE 6.25

Expand $f(x) = \sin x$ $(0 < x < \pi)$ in cosine series.

Solution. The graph of the given function is shown in the Figure 6.20.

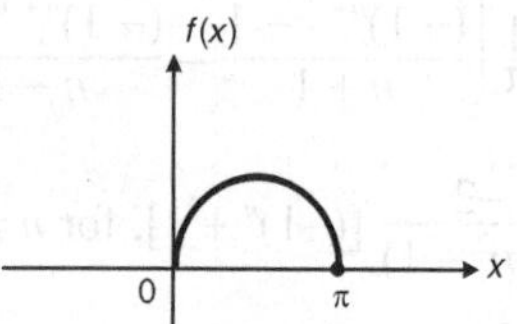

Figure 6.20

We extend sin x to an even function on the interval $(-\pi, \pi)$ and then extend it periodically with period 2. The graph of the extended function then becomes as shown in Figure 6.21.

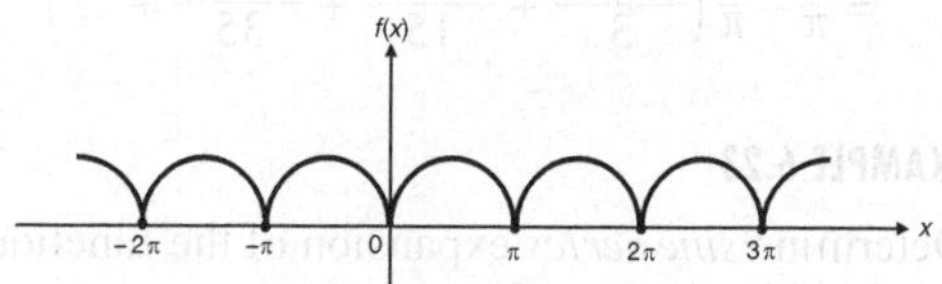

Figure 6.21

Since the function so created is even, $b_n = 0$. For this extended function, we have already calculated a_n, $n \neq 1$, a_1 and a_0 in Example 6.22. The Fourier series is, therefore, same as in Example 6.22.

EXAMPLE 6.26

Determine half-range sine series for the function f defined by $f(t) = t^2 + t$, $0 \le t \le \pi$.

Solution. Extending f to an odd function we get the graph of the extended odd function as shown in the Figure 6.22.

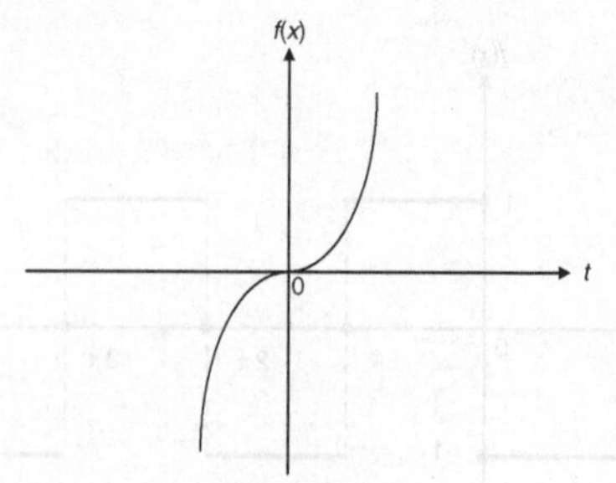

Figure 6.22

Since the extended function is odd, $a_n = 0$. For b_n, we have

$$b_n = \frac{1}{\pi}\int_{-\pi}^{\pi} f(t)\sin nt\,dt = \frac{1}{\pi}\int_{-\pi}^{0} f(t)\sin nt\,dt + \frac{1}{\pi}\int_{0}^{\pi} f(t)\sin nt\,dt$$

$$= \frac{2}{\pi}\int_0^{\pi}(t^2+t)\sin nt\,dt$$

$$= \frac{2}{\pi}\int_0^{\pi}\left[(t^2+t)\left(-\frac{\cos nt}{n}\right)\right]_0^{\pi} - \frac{2}{\pi}\int_0^{\pi}(2t+t)\left(-\frac{\cos nt}{n}\right)dt$$

$$= \frac{2}{\pi}\left[-\frac{(\pi^2-\pi)}{n}(-1)^n + \frac{2}{n^3}((-1)^n - 1)\right].$$

Thus, the required Fourier series is

$$f(t) = \frac{2}{\pi}\sum_{n=1}^{\infty}\left[\frac{(\pi^2-\pi)}{n}(-1)^{n+1} + \frac{2}{n^3}((-1)^n - 1)\right]\sin nt,\ t\in(0,\pi)$$

EXAMPLE 6.27

Find the *half-range sine series* for the function f defined by

$$f(x) = \begin{cases} x & \text{for } 0 < x < \pi/2. \\ \pi - x & \text{for } \pi/2 < x < \pi. \end{cases}$$

Solution. Extending f as an odd function in the interval $(-\pi, \pi)$, we have $a_n = 0$ for $n = 0, 1, 2, \ldots$. For b_n, we have

$$b_n = \frac{2}{\pi}\int_0^{\pi} f(x)\sin nx\,dx$$

$$= \frac{2}{\pi}\left[\int_0^{\pi/2} x\sin nx\,dx + \int_{\pi/2}^{\pi}(\pi - x)\sin nx\,dx\right]$$

$$= \frac{2}{\pi}\left[x\left(\frac{-\cos nx}{n}\right) + \frac{\sin nx}{n^2}\right]_0^{\pi/2}$$

$$+ \frac{2}{\pi}\left[(\pi - x)\left(-\frac{\cos nx}{n}\right) - \frac{\sin nx}{n^2}\right]_{\pi/2}^{\pi} = \frac{4}{\pi n^2}\sin\frac{n\pi}{2}.$$

Therefore, the *forced series development* of f is

$$f(x) = \sum_{n=1}^{\infty}\frac{4}{\pi n^2}\sin\frac{n\pi}{2}\sin nx$$

$$= \frac{4}{\pi}\left[\frac{\sin x}{1^2} - \frac{\sin 3x}{3^2} + \frac{\sin 5x}{5^2} - \ldots\right].$$

EXAMPLE 6.28

Find the Fourier series of the following function:

$$f(x) = \begin{cases} x^2 & \text{for } 0 \le x \le \pi \\ -x^2 & \text{for } -\pi \le x \le 0. \end{cases}$$

Solution. The given function is an odd extension of the function $f(x) = x^2$, $0 \le x \le \pi$ to the interval $-\pi \le x \le \pi$. Since the extended function is odd, the Fourier series shall consists of only sine terms. We have

$$b_n = \frac{1}{\pi}\int_{-\pi}^{\pi} f(x)\sin nx\,dx = \frac{2}{\pi}\int_0^{\pi} x^2\sin x\,dx$$

$$= \frac{2}{\pi}\left\{\left[x^2\left(-\frac{\cos nx}{n}\right)\right]_0^{\pi} - \int_0^{\pi} 2x\left(-\frac{\cos nx}{n}\right)dx\right\}$$

$$= \frac{-2}{n\pi}[(-1)^n\pi^2] + \frac{4}{n\pi}\left[x\,\frac{\sin nx}{n}\right]_0^{\pi} - \frac{4}{n^2\pi}\int_0^{\pi}\sin nx\,dx$$

$$= \frac{-2}{n\pi}[(-1)^n\pi^2] + \frac{4}{n\pi}\left[-\frac{\cos nx}{n}\right]_0^{\pi}$$

$$= -\frac{2}{n\pi}[(-1)^n\pi^2] + \frac{4}{n^3\pi}[(-1)^n - 1].$$

Hence, the Fourier sine series is given by

$$f(x) \sim 2\left(\pi - \frac{4}{\pi}\right)\sin x - \pi\sin 2x + \frac{2}{3}\left(\pi - \frac{4}{9\pi}\right)\sin 3x - \frac{\pi}{2}\sin 4x + \ldots.$$

EXAMPLE 6.29

Sketch the amplitude spectrum of the function f defined by

$$f(t) = t, \quad t \in (-\pi, \pi).$$

Solution. The fourier coeffiecients of this *saw tooth function*, as derived in Example 6.12, are

$$a_n = 0,\ b_n = -\frac{2}{n}(-1)^n.$$

Therefore

$$c_n = \frac{a_n - ib_n}{2} = \frac{(-1)^n i}{n}.$$

So the line spectrum is $\left\{\frac{(-1)^n i}{n}\right\}$. The amplitude spectrum is then given by

$$|c_n| = \frac{1}{|n|},\ n \neq 0, \quad |c_0| = 0.$$

The sketch of the amplitude spectrum is thus as shown in the Figure 6.23.

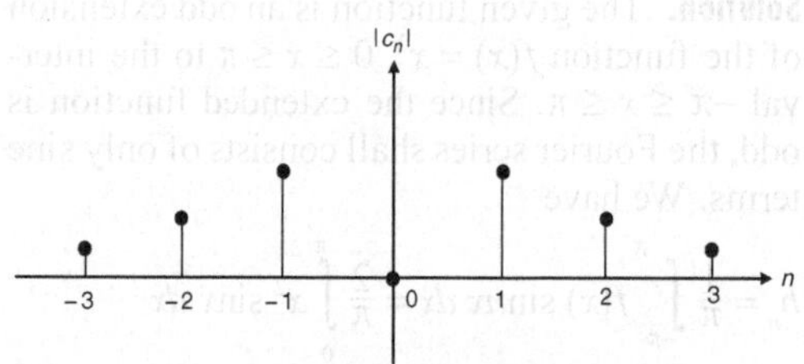

Figure 6.23

EXAMPLE 6.30

Find complex Fourier series for the function f defined by

$$f(x) = e^x,\ -\pi < x < \pi$$
$$f(x + 2\pi) = f(x).$$

Solution. By definition of complex Fourier coefficients, we have

$$c_n = \frac{1}{T}\int_{-T/2}^{T/2} f(t)e^{-in\omega_0 t}\, dt$$

Here T = 2π. Therefore, $\omega_0 = \frac{2\pi}{T} = 1$. Thus

$$c_n = \frac{1}{2\pi}\int_{-\pi}^{\pi} e^x . e^{-inx}\, dx = \frac{1}{2\pi}\int_{-\pi}^{\pi} e^{x-inx}\, dx$$

$$= \frac{1}{2\pi(1-in)}\left[e^{x-inx}\right]_{-\pi}^{\pi}$$

$$= \frac{1}{2\pi(1-in)}\left[e^{(1-in)\pi} - e^{-(1-in)\pi}\right]$$

$$= \frac{(-1)^n}{2(1-in)}\left[e^{\pi} - e^{-\pi}\right], \quad \text{since } e^{-inx} = (-1)^n$$

$$= \frac{2(-1)^n(1+in)}{2\pi(1+n^2)}\sinh\pi$$

$$= \frac{(-1)^n(1+in)}{\pi(1+n^2)}\sinh\pi.$$

Hence, the complex Fourier series is given by

$$f(x) \sim \sum_{n=-\infty}^{\infty} \frac{(-1)^n(1+in)}{\pi(1+n^2)}\sinh\pi\, e^{inx}.$$

EXAMPLE 6.31

Using shift property, derive the Fourier coefficients of $g(t) = t - \pi,\ 0 < t < 2\pi$ from the Fourier coefficients of $f(t) = t,\ -\pi < t < \pi,\ f(t) = f(t + 2\pi)$.

Solution. In Example 6.12, we have seen that Fouier coefficients of $f(t) = t,\ -\pi < t < \pi$ are $a_n = 0$ and $b_n = \frac{-2}{n}(-1)^n$.

Therefore,

$$c_n = \frac{a_n - ib^n}{2} = \frac{i}{n}(-1)^n.$$

The periodically extended graph of $g(t)$ is shown in the Figure 6.24.

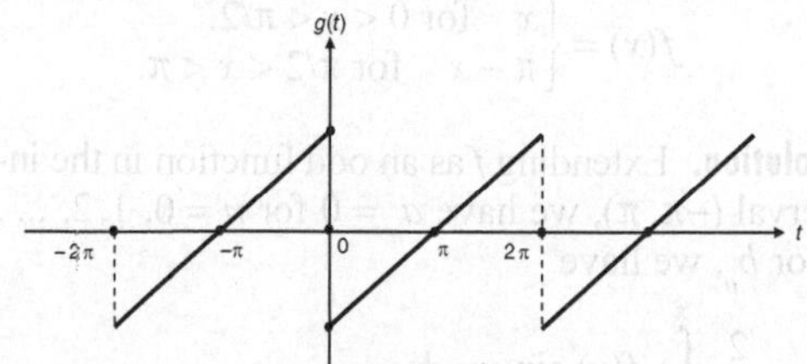

Figure 6.24

Clearly, $g(t) = f(t - \pi)$. Hence, by shift property, the coefficients of g are

$$d_n = c_n e^{-in\omega_0\pi}, \quad \omega_0 = \frac{2\pi}{2\pi} = 1$$

$$= c_n e^{-i\pi n} = \frac{i}{n}(-1)^n.e^{-i\pi n} = \frac{i}{n}(-1)^{2n} = \frac{i}{n}.$$

Verification. For the function $g(t)$, we have $a_n = 0$ and

$$b_n = \frac{2}{2\pi}\int_0^{2\pi} (t-\pi)\sin nt\, dt$$

$$= \frac{-1}{n\pi}[(t-\pi)\cos nt]_0^{2\pi}$$

$$+ \frac{1}{n\pi}\int_0^{2\pi} (1-\pi)\cos nt\, dt = -\frac{2}{n},$$

and so

$$c_n = \frac{a_n - ib_n}{2} = \frac{0 + 2\frac{i}{n}}{2} = \frac{i}{n}.$$

EXAMPLE 6.32

Find complex Fourier coefficients for the functions f defined by

$$f(t) = \begin{cases} 0 & \text{for } -\pi \le t \le 0. \\ 1 & \text{for } \quad 0 \le t \le \pi. \end{cases}$$

Using time reversal property of Fourier coefficients deduce the Fourier coeficients of $f(-t)$.

Solution. The Fourier coefficients of f (see Example 6.4) are $a_n = 0$ for $n = 1, 2, \ldots$, and $b_n = \frac{1-(-1)^n}{n\pi}$. Therefore,

$$c_n = \frac{a_n - ib_n}{2} = \left[\frac{(-1)^n - 1}{2n\pi}\right] i.$$

Let d_n be the Fourier coefficients of $f(-t)$. Then, by time reversal property,

$$d_n = c_{-n} = \frac{a_n + ib_n}{2} = \left(\frac{1-(-1)^n}{2n\pi}\right) i.$$

Verification. We have

$$f(-t) = \begin{cases} 0 & \text{for } -\pi < -t < 0 \\ 1 & \text{for } \quad 0 \le -t \le \pi. \end{cases}$$

that is,

$$f(-t) = \begin{cases} 0 & \text{for } \quad 0 < t < \pi \\ 1 & \text{for } \; -\pi < t < 0. \end{cases}$$

For this function $a_n = 0, n = 1, 2, \ldots$ and

$$b_n = \frac{1}{\pi}\int_{-\pi}^{\pi} f(t)\sin nt\, dx = \frac{1}{\pi}\left[-\frac{\cos nt}{n}\right]_{-\pi}^{0}$$

$$= \frac{\cos n\pi - 1}{n\pi} = \frac{(-1)^n - 1}{n\pi}$$

So

$$d_n = \frac{a_n - ib_n}{2} = \left(\frac{1-(-1)^n}{2n\pi}\right) i.$$

EXAMPLE 6.33

Given the series $x = 2\sum_{n=1}^{\infty} \frac{(-1)^{n+1}}{n} \sin nx$, show that

$$\sum_{n=1}^{\infty} \frac{1}{n^2} = \frac{\pi^2}{6}.$$

Solution. The Parsevel's equality states that

$$\int_{-\pi}^{\pi} |f(x)|^2 = 2\pi\, a_0^2 + \pi \sum_{n=1}^{\infty}\left(a_n^2 + b_n^2\right).$$

For this example, the left-hand side of this equality is

$$\int_{-\pi}^{\pi} x^2\, dx = \left[\frac{x^3}{3}\right]_{-\pi}^{\pi} = \frac{2}{3}\pi^3.$$

Since $a_n = 0$, $b_n = \frac{2(-1)^{n+1}}{n}$, the right-hand side of the equality is $\pi\sum_{n=1}^{\infty}\frac{4}{n^2}$. Hence, $\sum_{n=1}^{\infty}\frac{1}{n^2} = \frac{\pi^2}{6}$.

EXAMPLE 6.34

Expand $f(x) = x (0 < x < 2\pi)$ in Fourier series and deduse that $\sum_{n=1}^{\infty} \frac{4-(-1)^n}{4n^2} = \frac{3\pi^2}{16}$.

Solution. In Example 6.18, we have seen that the Fourier expansion of the given function is

$$x = \pi - 2\sum_{n=1}^{\infty}\frac{1}{n}\sin nx.$$

Using integration of Fourier series with $t = \frac{\pi}{2}$, we get

$$\int_0^{\pi/2} x\, dx = \pi\left(\frac{\pi}{2} - 0\right) - 2\sum_{n=1}^{\infty} -\frac{1}{n}\left[\frac{1}{n}\cos\frac{nx}{2}\right]_0^{\pi/2},$$

that is,

$$\frac{\pi^2}{8} = \frac{\pi^2}{2} + 2\sum_{n=1}^{\infty} \frac{1}{n^2}\left[\cos\frac{n\pi}{2} - \cos 0\right]$$

$$= \frac{\pi^2}{2} + 2\sum_{n=1}^{\infty} \frac{(-1)^n}{(2n)^2} - 2\sum_{n=1}^{\infty} \frac{1}{n^2}.$$

Therefore,

$$\frac{3\pi^2}{16} = \sum_{n=1}^{\infty} \frac{4-(-1)^2}{4n^2}.$$

EXAMPLE 6.35

Expand $f(x) = x^2$ $(-\pi < x < \pi)$ in Fourier series and deduse the value of the series

$$\sum_{k=1}^{\infty} \frac{(-1)^{k-1}}{(2k-1)^3}.$$

Solution. As per Example 6.6, the Fourier series expansion of $f(x) = x^2$ $(-\pi < x < \pi)$ is

$$x^2 = \frac{\pi^2}{3} + 4\sum_{n=1}^{\infty} \frac{(-1)^n}{n^2} \cos nx.$$

Using integration of Fourier series with $t = \pi/2$, we have

$$\int_0^{\pi/2} x^2\, dx = \frac{\pi^2}{3}\left(\frac{\pi}{2}\right) + 4\sum \frac{(-1)^n}{n^3} \sin(n\pi/2),$$

or

$$\frac{\pi^3}{24} = \frac{\pi^3}{6} + 4\sum_{n=1}^{\infty} \frac{(-1)^n}{n^3} \sin(n\pi/2)$$

$$= \frac{\pi^3}{6} + 4\left[-\sum_{k=1}^{\infty} \frac{(-1)^{k-1}}{(2k-1)^3}\right]$$

which yields

$$\sum_{k=1}^{\infty} \frac{(-1)^{k-1}}{(2k-1)^3} = \frac{1}{4}\left[\frac{\pi^3}{6} - \frac{\pi^3}{24}\right] = \frac{\pi^3}{32}.$$

EXAMPLE 6.36

Expand $f(x) = x(-\pi < x < \pi)$ in Fourier series and deduse the Fourier series for $f(x) = x^2$ $(-\pi < x < \pi)$.

Solution. As per Example 6.12, we have

$$x = \sum_{n=1}^{\infty} \frac{2}{n} (-1)^n \sin nx.$$

Term-by-term integration yields

$$\frac{x^2}{2} = \sum_{n=1}^{\infty} \frac{2}{n^2} (-1)^n \cos nx + A \text{ (constant)}. \quad (44)$$

Integration both sides with respect to x between the limits $-\pi$ and π, we have

$$\int_{-\pi}^{\pi} \frac{x^2}{2}\, dx = \sum_{n=1}^{\infty} \frac{2}{n^2} (-1)^n \int_{-\pi}^{\pi} \cos nx\, dx + \int_{-\pi}^{\pi} A\, dx,$$

which yields $A = \frac{\pi^2}{6}$. Hence, (44) reduces to

$$x^2 = \frac{\pi^2}{3} + \sum_{n=1}^{\infty} \frac{4(-1)^n}{n^2} \cos nx,$$

which is the required Fourier series for $f(x) = x^2$ $(-\pi < x < \pi)$.

EXAMPLE 6.37

Verify Riemann-Lebesgue lemma for the function $f(x) = 2x + 1, -\pi < x < \pi$.

Solution. The Fourier coefficients of this function (see Example 3.7) are

$$a_n = 0, n = 1, 2, \ldots \text{ and } b_n = -\frac{4}{n}(-1)^n.$$

Clearly $b_n \to 0$ as $n \to \infty$. Thus Riemann-Lebesgue lemma is valid.

EXAMPLE 6.38

Let f and g be periodic function with period 2π defined on $(-\pi, \pi)$ by $f(x) = g(x) = x$. Find the Fourier series of fg over $(-\pi, \pi)$.

Solution. In Example 6.12, we have seen that the Fourier coefficients of f are $a_n = 0$ and $b_n = -\frac{2}{n}(-1)^n$. Therefore, $c_n = d_n = \frac{a_n - ib_n}{2}$ $= \frac{(-1)^n i}{n}$. Thus

$$c_n = d_n = \begin{cases} \frac{(-1)^n i}{n} & \text{for } n \neq 0 \\ 0 & \text{for } n = 0. \end{cases}$$

If f_n is complex Fourier coefficients of fg, then we know that

$$f_n = \sum_{k=-\infty}^{\infty} c_k d_{n-k}.$$

Therefore,

$$f_0 = \sum_{k=-\infty}^{\infty} c_k d_{-k} = \sum_{k=-\infty}^{\infty} c_k c_{-k}.$$

But $c_{-k} = \frac{a_n + ib_n}{2} = \frac{-(-1)^n i}{n}$. Hence $f_0 = 2\sum_{k=1}^{\infty}\frac{1}{k^2}$ $= \frac{\pi^2}{3}$, since $\sum_{k=1}^{\infty}\frac{1}{k^2} = \frac{\pi^2}{6}$ (by Examples 6.6, 6.33).

Further,

$$f_n = \sum_{\substack{k=-\infty \\ k\neq 0,\, k\neq n}}^{\infty} c_k d_{n-k} = \sum_{\substack{k=-\infty \\ k\neq 0,\, k\neq n}}^{\infty} c_k c_{n-k}$$

$$= \sum_{\substack{k=-\infty \\ k\neq 0,\, k\neq n}}^{\infty} \frac{i}{k}(-1)^k . \frac{i}{(n-k)}(-1)^{n-k}$$

$$= -(-1)^n \sum_{\substack{k=-\infty \\ k\neq 0,\, k\neq n}}^{\infty} \frac{1}{k(n-k)}$$

$$= -(-1)^n \frac{1}{n} \sum_{\substack{k=-\infty \\ k\neq 0,\, k\neq n}}^{\infty} \left(\frac{1}{k} - \frac{1}{n-k}\right)$$

$$= \frac{-(-1)^n}{n}\left[\left(-\frac{1}{n} + \sum_{\substack{k=-\infty \\ k\neq 0}}^{\infty} \frac{1}{k}\right) + \left(-\frac{1}{n} + \sum_{\substack{k=-\infty \\ k\neq n}}^{\infty} \frac{1}{n-k}\right)\right]$$

$$= \frac{-(-1)^n}{n}\left[-\frac{2}{n}\right] = \frac{2}{n^2}(-1)^n.$$

Hence

$$f_n = \begin{cases} \frac{2}{n^2}(-1)^n & \text{for } n \neq 0 \\ \frac{\pi^2}{3} & \text{for } n = 0. \end{cases}$$

Therefore,

$$x^2 = \frac{\pi^2}{3} + 2\sum_{n=-\infty}^{\infty} \frac{1}{n^2}(-1)^n e^{inx}.$$

Verification. We have seen in Example 6.6 that Fouier coefficients of $f(x) = x^2$, $-\pi < x < \pi$ are $a_0 = \frac{2\pi^2}{3}$ and $a_n = \frac{4}{n^2}(-1)^n$. Then complex coefficients are

$$f_0 = \frac{\pi^2}{3} \text{ and } f_n = \frac{a_n - ib_n}{2} = \frac{2}{n^2}(-1)^n,\ n = 1, 2, \ldots$$

6.16 METHOD TO FIND HARMONICS OF FOURIER SERIES OF A FUNCTION FROM TABULAR VALUES

The harmonics of a Fourier series of a function can even be found if the function is not defined explicitly but the tabular values of the function are given. This is possible using mean value of a function

The mean value a function f over (a, b) is defined by

$$\frac{1}{b-a}\int_a^b f(x)\,dx.$$

By the definition of Fourier coefficients, we have

$$a_0 = \frac{1}{\pi}\int_0^{2\pi} f(x)dx = 2\left[\frac{1}{2\pi}\int_0^{2\pi} f(x)dx\right]$$

$= 2 \times$ mean value of f over $(0, 2\pi)$,

$$a_n = \frac{1}{\pi}\int_0^{2\pi} f(x)\cos nx\, dx$$

$$= 2\left[\frac{1}{2\pi}\int_0^{2\pi} f(x)\cos nx\, dx\right]$$

$= 2 \times$ mean value of $f\cos nx$ over $(0, 2\pi)$,

$$b_n = \frac{1}{\pi}\int_0^{2\pi} f(x)\sin nx\, dx$$

$$= 2\left[\frac{1}{2\pi}\int_0^{2\pi} f(x)\sin nx\, dx\right]$$

$= 2\times$mean value of $f(x)\sin nx$ over $(0, 2\pi)$,

Thus the Fourier coefficient can be determined by the mean values of $f(x)$, $f(x)\cos nx$ and $f(x)\sin nx$. Hence the nth harmonic $a_n \cos nx + b_n \sin nx$ can be determined using the tabular values of f, $f(x)\cos nx$ and $f(x)\sin nx$.

The following example illustrates the method discussed above.

EXAMPLE 6.39.

Find the first two harmonics of the Fourier Series of $y = f(x)$ from the data:

x^0:	0	30	60	90	120	150
y:	298	356	373	337	254	155

x^0:	180	210	240	270	300	330
y:	80	51	60	93	147	221

Solution. The *Mean value* of a function f over (a, b) is defined by $\frac{1}{b-a}\int_a^b f(x)\,dx$. The Fourier coefficients in terms of mean values can be expressed as

$$a_0 = 2\left(\frac{1}{2\pi}\right)\int_0^{2\pi} f(x)\,dx$$

$$= 2\,[\text{mean value of } f \text{ in } (0, 2\pi)],$$

$$a_n = 2\left(\frac{1}{2\pi}\right)\int_0^{2\pi} f(x)\cos nx\,dx$$

$$= 2[\text{ mean value of } f(x)\cos nx \text{ in } (0, 2\pi)],$$

$$b_n = 2\left(\frac{1}{2\pi}\right)\int_0^{2\pi} f(x)\sin nx dx$$

$$= 2[\text{ mean value of } f(x)\sin nx \text{ in } (0, 2\pi).$$

We want to find the first two harmonics, that is,$a_1 \cos x + b_1 \sin x$ and $a_2 \cos 2x + b_2 \sin 2x$.

The values of sin x, sin $2x$, cos x, cos $2x$, $f(x)$ sin x, $f(x)$ sin $2x$, $f(x)$ cos x and $f(x)$ cos $2x$ are given in the following table:

x	0^0	30^0	60^0	90^0	120^1	150^0	180^0	210^0	240^0	270^0	300^0	330^0
$f(x)$	298	356	373	337	254	155	80	51	60	93	147	221
sin x	0	0.5	0.87	1	0.87	0.5	0	−0.5	−0.87	−1.0	−0.87	−0.5
sin $2x$	0	0.87	0.87	0	−0.87	−0.87	0	0.87	0.87	0	−0.87	−0.87
cos x	1	0.87	0.5	0	−0.5	−0.87	−1	−0.87	−0.5	0	0.5	0.87
cos $2x$	1	0.5	0.5	−1	−0.5	−0.5	1	0.5	−0.5	−1	−0.5	0.5
$f(x)$ sin x	0	178	324.51	337	220.98	75	0	−25.5	−52.2	−93	−127.89	−110.5
$f(x)$ sin $2x$	0	309.72	324.51	0	−220.98	−134.85	0	44.37	52.2	0	−127.89	−192.27
$f(x)$ cos x	298	309.72	186.5	0	−127	−134.85	−80	−44.37	−30	0	73.5	192.27
$f(x)$ cos $2x$	298	178	186.5	−337	−127	−75	80	25.5	−30	−93	−73.5	110.5

We note that

$$\sum f(x) = 2425,$$

$$\sum f(x)\cos x = 643.77,$$

$$\sum f(x)\sin x = 726.40$$

$$\sum f(x)\cos 2x = 143.0 \text{ and}$$

$$\sum f(x)\sin 2x = 54.81.$$

Therefore

$$a_1 = 2\left[\frac{\sum f(x)\cos x}{12}\right] = \frac{643.77}{6} = 107.295$$

$$a_2 = 2\left[\frac{\sum f(x)\cos 2x}{12}\right] = \frac{143.0}{6} = 23.83$$

$$b_1 = 2\left[\frac{\sum f(x)\sin x}{12}\right] = \frac{726.4}{6} = 121.07$$

$$b_2 = 2\left[\frac{\sum f(x)\sin 2x}{12}\right] = \frac{54.81}{6} = 9.135.$$

Hence the first two harmonics are

$$a_1 \cos x + b_1 \sin x = 107.295 \cos x + 121.07\sin x$$

and

$$a_2 \cos 2x + b_2 \sin 2x = 23.83 \cos 2x + 9.135\sin 2x.$$

6.17 SIGNALS AND SYSTEMS

Definition 6.8. A *signal* is a function of one or more independent variable(s) which convey information.The independent variable may be *time, space*, etc.

For example, in electrical network, the voltage E(t) is a signal, which is defined as a function of time.

Definition 6.9. A *system* is a mapping F which assigns a unique output to an input.

Definition 6.10. Let y be a uniquely determined output corresponding to an input x under the system F, theny is called *response* of the system to the input x and we write $y = \mathrm{F}x$ or $x \rightarrow y$.

6.18 CLASSIFICATION OF SIGNALS

Definition 6.11. If the signal f, as a mapping, is real valued, then it is called a *real signal*.

Definition 6.12. If the signal f, as a mapping, is complex-valued, then it is called *complex signal*. It is of the form $f = f_1 + if_2$, where f_1 is called real part of the complex signal and f_2 is called the imaginary part of f. If $f_1 = f_2 = 0$, then the signal is called the *null signal*.

Definition 6.13. A signal which is a function of time variable t, $t \in \mathrm{R}$, is called *continuous time signal*.

For example, in electrical networks and mechanical systems, the signals are functions of the time variable. Similarly, temperature of a room, and speech signals are continuous time signals.

A continuous time signal f is said to be *bounded* if there exists a positive constant K such that $|f(t)| \leq \mathrm{K}$, $t \in \mathrm{R}$.

Definition 6.14. Signals which are defined at discrete time are called *discrete signals*. Thus discrete time signals can be considered as a function defined on Z or a part of Z (the set of integers).

For example, energy consumption in a state in the years 2001, 2002, . . ., 2006 is a discrete time signal.

A discrete time signal $f[n]$ is called *bounded* if there exists a positive constant K such that $|f[n]| \leq \mathrm{K}$, $n \in \mathrm{N}$.

Definition 6.15. A continuous time signal f is called periodic with period $\mathrm{T} > 0$ if $f(t + \mathrm{T}) = f(t)$, $t \in \mathrm{R}$.

For example, sinusoidal (sin t, cos t) are periodic signals. The sinusoidal are real signals, which in the continuous time case can be written as $f(t) = \mathrm{A}\cos(\omega t + \phi_0)$, $t \in \mathrm{R}$, where A is the *amplitude*, ω is the radial *frequency*,and ϕ_0 the *initial phase* of the signal. The frequency ω equals $\frac{2\pi}{\mathrm{T}}$, where T is *period.*

Definition 6.16. A discrete time signal $f[n]$ is called periodic with period $\mathrm{N} \in \mathrm{N}$ if $f[n + \mathrm{N}] = f[n]$, $n \in \mathrm{Z}$.

In discrete-time case, the sinusoidal signals have the form $f[n] = \mathrm{A}\cos(\omega n + \phi_0)$, $n \in \mathrm{N}$,where A is amplitude, ω is frequency, ϕ_0 is initial phase, and period $\mathrm{N} = \frac{2\pi}{\omega}$.

Definition 6.17. If, both, dependent and independent variables of a signal are continuous in nature, then it is called an *analog signal*. These signals arise when a physical wave form is converted into an electrical signal.

For example, telephone speech signals and TVsignals are analog signals.

Definition 6.18. If both dependent and independent variables of a signal are *discrete in nature*, then it is called a *digital signal*. These signal comprise of pulses occurring at discrete interval of time.

For example, telegraph and teleprinter signals are digital signals.

Definition 6.19. A complex signal $f(t)$ is called a *time-harmonic continuous time signal* if $f(t) = ce^{i\omega t}$, $t \in R$, where c is a complex variable.

A time-harmonic continuous time signal is *bounded since*

$$|ce^{i\omega t}| = |c|\, |e^{i\omega t}| = |c| \text{ for } t \in R.$$

If we set $c = Ae^{i\phi_0}$, where $A = |c|$ and ϕ_0 is the argument, then

$$f(t) = Ae^{i\phi_0} . e^{i\omega t} = Ae^{i(\omega t + \phi_0)}.$$

Thus, $f(t)$ can be represented in the complex plane by a point on the circle with origin as the centre and A as the radius. At $t = 0$, the argument is equal to ϕ_0, the initial phase (Figure 6.25).

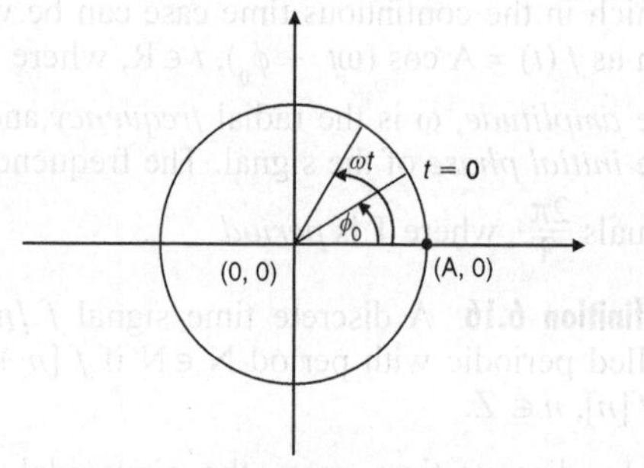

Figure 6.25

In the complex plane, the signal $f(t)$ corresponds to a circular movement with constant angular velocity $|\omega|$. Further, time-harmonic signal $f(t)$ is periodic with period $\frac{2\pi}{|\omega|}$. The real number ω is called frequency of the time harmonic signal, A the amplitude and ϕ_0 the initial phase. Further, by Euler's formula

$$f(t) = A\, e^{i(\omega t + \phi_0)}$$

$$= A[\cos(\omega t + \phi_0) + i \sin(\omega t + \phi_0)].$$

Thus *sinusoidal signal is the real part of a time-harmonic signal.* Also

$$A \cos(\omega t + \phi_0) = A\left[\frac{e^{i(\omega t+\phi_0)} + e^{-i(\omega t+\phi_0)}}{2}\right]$$

$$= \frac{ce^{i\omega t} + ce^{-i\omega t}}{2}, \; c = Ae^{i\phi_0}.$$

Definition 6.20. The *power* P of a continuous-time signal $f(t)$ is defined by

$$P = \lim_{A\to\infty} \frac{1}{2A} \int_{-A}^{A} |f(t)|^2\, dt.$$

Definition 6.21. The *power of a periodic continuous time signal* $f(t)$ with period T is defined by

$$P = \frac{1}{T} \int_{-T/2}^{T/2} |f(t)|^2\, dt.$$

Definition 6.22. A continuous time signal whose power is *finite* is called a *power signal.*

For example, a periodic signal is a power signal.In particular, sinusoidal waves are power signals.

Definition 6.23. The energy-content (total energy) of a continuous-time signal is defined by

$$E = \int_{-\infty}^{\infty} |f(t)|^2\, dt.$$

Definition 6.24. A continuous time signal with a finite energy content is called an *energy signal.*

For example, *rectangular pulse is an energy signal.*

Definition 6.25. The *power* P of a discrete-time signal $f[n]$, is defined by

$$P = \lim_{M\to\infty} \frac{1}{2M}\sum_{n=-M}^{M} |f[n]|^2.$$

Definition 6.26. If the power of a discrete-time signal is finite, then the signal is called a *discrete time power-signal*.

Definition 6.27. The *power of a periodic discrete-time signal* $f[n]$ with period N is defined by

$$P = \frac{1}{N}\sum_{n=0}^{N-1} |f[n]|^2.$$

Definition 6.28. The energy content E of a discrete time signal $f[n]$ is defined by

$$E = \sum_{n=-\infty}^{\infty} |f[n]|^2.$$

Definition 6.29. If the energy content of a discrete time signal is finite, then the signal is called an *energy-signal*.

Definition 6.30. A continuous time signal $f(t)$ is called *causal* if $f(t) = 0$ for $t < 0$.

Definition 6.31. A discrete-time signal $f[n]$ is called *causal* if $f[n] = 0$ for $n < 0$.

It follows from the above definitions that *periodic signals (except the null sequence) are not causal.*

Definition 6.32. Let $f(t)$ be a signal. If there exists t_0 such that $f(t) = 0$ for $t < t_0$, then t_0 is called the *switch on time* of the signal $f(t)$.

6.19 CLASSIFICATION OF SYSTEMS

Definition 6.33. A system F(t) is called linear if for two inputs x_1 and x_2 and arbitrary complex numbers a and b,

$$F(ax_1 + bx_2) = aF(x_1) + bF(x_2).$$

Thus for continuous time system, we can write

$$ax_1(t) + bx_2(t) \to a(Fx_1)(t) + b(Fx_2)(t),$$

whereas for discrete time signal

$$ax_1[n] + bx_2[n] \to a(Fx_1)[n] + b(Fx_2)[n].$$

For example, system F defined by F(t) = 3t is linear. In fact,

$$\mathrm{F}(t_1 + t_2) = 3(t_1 + t_2) = 3t_1 + 3t_2 = \mathrm{F}(t_1) + \mathrm{F}(t_2),$$

$$\mathrm{F}(\alpha t) = 3(\alpha t) = \alpha(3t) = \alpha\mathrm{F}(t).$$

Definition 6.34. A system for which power of the output equals the power of the input is called *all-pass system*.

Definition 6.35. A continuous time system is called *time-invariant* if for each input $u(t)$ and each $t_0 \in \mathrm{R}$,

$$u(t) \to y(t) \text{ implies } u(t - t_0) \to y(t - t_0).$$

Similarly, a discrete time system is called time invariant if for each input $u[n]$ and each $n_0 \in \mathrm{Z}$,

$$u[n] \to y[n] \text{ implies } u[n - n_0] \to y[n - n_0].$$

Definition 6.36. A system which is both linear and time-invariant is called a *linear time-invariant system* (or LTI *system*).

We now show that for linear time-invariant system F(t), the response (whenever exists) to a timeharmonic signal is again a time-independent signal with the same frequency.

Theorem 6.16. Let F be a linear time invariant system, u a time-harmonic input with frequency ω for which response exist. Then the output y is also a time harmonic signal with the same frequency ω.

Proof: Let $u(t)$ be the time-harmonic input with frequency ω and $y(t)$ the corresponding output. Thus, $u(t) = c\,e^{i\omega t}$, where c is a complex number and $\omega \in \mathrm{R}$. Since the system is time invariant, we have

$$F(u(t - t_0)) = y(t - t_0).$$

But

$$u(t - t_0) = c\, e^{i\omega(t - t_0)} = c\, e^{-i\omega t_0} \cdot e^{i\omega t} = e^{-i\omega t_0} u(t).$$

Since F is linear, we have

$$F(u(t - t_0)) = F(e^{-i\omega t_0} u(t)) = e^{-i\omega t_0} F(u(t))$$

$$= e^{-i\omega t_0} y(t).$$

Thus

$$F(u(t - t_0)) = y(t - t_0) = e^{-i\omega t_0} y(t).$$

Putting $t = 0$, we have

$$y(-t_0) = e^{-i\omega t_0} y(0).$$

Replacing $-t_0$ by t, we have

$$y(t) = y(0)e^{-i\omega t} = c\, e^{i\omega t},\ c \text{ a complex constant},$$

which shows that the response is again a time harmonic signal with frequency ω.

Remark 6.7. The complex constant c is a function of the frequency ω. This function is called *frequency response* (*system function or transfer function*) of the system. For continuous time systems, the transfer function is denoted by $H(\omega)$ and for discrete time systems by $H(e^{i\omega})$. Thus

$$e^{i\omega t} \to H(\omega)\, e^{i\omega t} \text{ for continuous time system},$$

and

$$e^{i\omega n} \to H(e^{i\omega})e^{i\omega n} \text{ for a discrete time system},$$

Since $H(\omega)$ is complex, we can write it as

$$H(\omega) = |H(\omega)|e^{i\Phi(\omega)},$$

where $|H(\omega)|$ and $\Phi(\omega)$ are, respectively, the*modulus* and *argument* of $H(\omega)$. The function $|H(\omega)|$ is called the *amplitude response* and $\Phi(\omega)$ is called *phase response.*

Definition 6.37. A system F is said to be *stable* if the response of each bounded signal is again bounded. For example, let time-harmonic signal be the input signal, then response $ce^{i\omega t}$ to this input signal exists and is bounded.

Definition 6.38. A system F is called real if the response to every real input is again real.

Consider the sinusoidal input $u(t) = A\cos(\omega t + \phi)$. We can consider it as the real part of the time harmonic signal $ce^{i\omega t}$ with $c = Ae^{i\phi_0}$. The response to this harmonic signal is $c\, H(\omega)\, ce^{i\omega t}$. If the system is real, the response of sinusoidal input $u(t)$ is equal the real part of $c\, H(\omega)\, ce^{i\omega t}$. But $H(\omega) = |H(\omega)|\, e^{i\Phi(\omega)}$.

Hence

$$y(t) = \mathrm{Re}[A\, e^{i\phi_0}\, |H(\omega)|e^{i\Phi(\omega)}e^{i\omega t}]$$

$$= A\,|H(\omega)|\, \cos(\omega t + \phi_0 + \Phi(\omega)).$$

Definition 6.39 A continuous-time system F is called causal if for each two inputs $u(t)$ and $v(t)$ and for each $t_0 \in R$,

$$u(t) = v(t) \Rightarrow (Fu)(t) = (Fv)(t) \text{ for } t < t_0.$$

Similarly, a discrete time system F is called causal if for each two inputs $u[n]$ and $v[n]$ and for each $n_0 \in Z$,

$$u[n] = v[n] \Rightarrow (Fu)[n] = (Fv)[n] \text{ for } n < n_0.$$

Regarding causal systems we have the following theorem:

Theorem 6.17. A linear time-invariant system F is causal if and only if the response to each causal input is again causal.

6.20 RESPONSE OF A STABLE LINEAR TIMEINVARIANT CONTINUOUS TIME SYSTEM (LTC SYSTEM) TO A PIECEWISE SMOOTH AND PERIODIC INPUT

We know that the response to the time-harmonic signal $e^{i\omega t}$ of frequency ω is equal to $H(\omega)\, e^{i\omega t}$, where $H(\omega)$ in the transfer function. Let the Fourier expansion of the periodic input $u(t)$ be

$$u(t) = \sum_{n=-\infty}^{\infty} u_n\, e^{in\omega_0 t},$$

where $\omega_0 = \frac{2\pi}{T}$ and u_n is the line spectrum of $u(t)$.

Since $u(t)$ is piecewise smooth, Convergence Theorem of Fourier series is applicable.

The following theorem gives the line spectrum y_n of the response $y(t)$.

Theorem 6.18. Let $y(t)$ be the response of a stable LTC system to a piecewise smooth and periodic input $u(t)$ with period T, fundamental frequency ω_0 and line spectrum u_n. Let $H(\omega)$ be the transfer function of the system. Then $y(t)$ is again periodic with period T and the line spectrum y_n of $y(t)$ is given by

$$y_n = H(n\omega_0)\, u_n,\ n = 0, \pm 1, \pm 2, \ldots$$

and so

$$y(t) = \sum_{n=-\infty}^{\infty} u_n H(n\omega_0)\, e^{in\omega_0 t}.$$

Proof: Since u_n is the line spectrum of the periodic input $u(t)$, we have

$$u(t) = \sum_{n=-\infty}^{\infty} u_n e^{in\omega_0 t},\ \omega_0 = \frac{2\pi}{T}.$$

Since response to time-harmonic signal $e^{i\omega t}$ of frequency ω is $H(\omega)\, e^{i\omega t}$, the response of $e^{in\omega_0 t}$ is $H(n\omega_0)\, e^{in\omega_0 t}$. Therefore, by the linearity of the system, we have

$$u_n e^{in\omega_0 t} \to u_n H(n\omega_0)\, e^{in\omega_0 t}.$$

Therefore, the line spectrum y_n of $y(t)$ is

$$y_n = H(n\omega_0)\, u_n.$$

Now, by superposition rule, we have

$$y(t) = \sum_{n=-\infty}^{\infty} H(n\omega_0) u_n\, e^{in\omega_0 t}$$

where $y(t)$ is clearly periodic with period T.

6.21 APPLICATION TO DIFFERENTIAL EQUATIONS

Consider the following differential equation of order m:

$$a_m \frac{d^m y}{dt^m} + a_{m-1} \frac{d^{m-1} y}{dt^{m-1}} + \ldots + a_1 \frac{dy}{dt} + a_0 y$$

$$= b_n \frac{d^n u}{dt^n} + b_{n-1} \frac{d^{n-1} u}{dt^{n-1}} + \ldots + b_1 \frac{du}{dt} + b_0 u, \quad (45)$$

with $n \le m$, where $a_0, a_1, \ldots, a_m$ and $b_0, b_1, \ldots, b_n$ are constants with $a_m \neq 0$ and $b_n \neq 0$. This equation describes the relation between an input $u(t)$ and the corresponding output $y(t)$. Let

$$P(s) = a_m s^m + a_{m-1} s^{m-1} + \ldots + a_1 s + a_0$$

and

$$Q(s) = b_n s^n + b_{n-1} s^{n-1} + \ldots + b_1 s + b_0.$$

The polynomial P(s) is called the *characteristics polynomial* of the differential equation (45). The transfer function $H(\omega)$ can be found by the following theorem:

Theorem 6.19. Let the differential equation (45) describes an LTC-system and have $P(s)$ as the characteristic polynomial. If for all ω, $P(i\omega) \neq 0$, then

$$H(\omega) = \frac{Q(i\omega)}{P(i\omega)}.$$

Proof: To get frequency response, we substitute the input $u(t) = e^{i\omega t}$ in the given differential equation. Then the response $y(t)$ is of the form $H(\omega)\, e^{i\omega t}$. Since the derivative of $e^{i\omega t}$ is $i\omega\, e^{i\omega t}$, substitution into the differential equation yields

$$P(i\omega)\, H(\omega)\, e^{i\omega t} = Q(i\omega)\, e^{i\omega t}$$

and so

$$H(\omega) = \frac{Q(i\omega)}{P(i\omega)}.$$

EXAMPLE 6.40

Solve the differential equation

$$\frac{dy}{dt} + 3y = \cos 3t,\ y(0) = 0.$$

Solution. The characteristic equation is

$$s + 3 = 0 \text{ so that } s = 3.$$

Therefore, the *homogeneous solution* (*eigen function*) is $c_1 e^{-3t}$.

Further, taking $P(s) = s + 3$, $Q(s) = 1$, we have

$$H(\omega) = \frac{Q(i\omega)}{P(i\omega)} = \frac{1}{i\omega + 3}.$$

Since

$$u(t) = \cos 3t = \frac{e^{i3t} + e^{-i3t}}{2}$$

and

$$e^{i3t} \to H(3)\, e^{3it}$$

$$e^{-i3t} \to H(-3)e^{-3it},$$

therefore,

$$\cos 3t \to \frac{H(3)e^{3it} + H(-3)e^{-3it}}{2}$$

$$= \frac{e^{3it}}{2(3+3i)} + \frac{e^{-3it}}{2(3-3i)}$$

$$= \frac{1}{6}\left[\frac{(\cos 3t + i\sin 3t)(1-i) + (\cos 3t - i\sin 3t)(1+i)}{(1+i)(1-i)}\right]$$

$$= \frac{1}{6}[\cos 3t + \sin 3t].$$

Hence the complete solution is

$$y(t) = \frac{1}{6}[\cos 3t + \sin 3t] + c_1 e^{-3t}.$$

Putting $t = 0$ and using the initial condition $y(0) = 0$, we have $0 = \frac{1}{6} + c_1$ and so $c_1 = -\frac{1}{6}$. Thus, the complete solution is

$$y(t) = \frac{1}{6}[\cos 3t + \sin 3t] - \frac{1}{6}e^{-3t}.$$

EXAMPLE 6.41

Solve $\frac{d^2y}{dt^2} + 5\frac{dy}{dt} + 6y = 2\sin t$, $t \geq 0$ subject to the conditions $y'(0) = 0$, $y(0) = 0$.

Solution. The characteristic equation $s^2 + 5s + 6 = 0$ yields $s = -3, -2$. Thus the homogeneous solution is

$$c_1 e^{-3t} + c_2 e^{-2t}.$$

Now taking

$$P(s) = s^2 + 5s + 6 \text{ and } Q(s) = 2,$$

we have

$$H(\omega) = \frac{Q(i\omega)}{P(i\omega)} = \frac{2}{6 + 5i\omega - \omega^2}.$$

Now

$$u(t) = \frac{2(e^{it} - e^{-it})}{2i} = \frac{e^{it} - e^{-it}}{i}$$

and

$$e^{it} \to H(1)\, e^{it}$$

$$e^{-it} \to H(1)\, e^{-it}.$$

Therefore,

$$\frac{e^{it} - e^{-it}}{i} \to \frac{1}{i}[H(1)e^{it} - H(-1)e^{-it}]$$

$$= \frac{1}{i}\left[\frac{2}{5+5i}e^{it} - \frac{2}{5-5i}e^{-it}\right]$$

$$= \frac{2}{5i}\left[\frac{e^{it}}{1+i} - \frac{e^{it}}{1-i}\right]$$

$$= \frac{2}{5i}\left[\frac{(\cos t + i\sin t)(1-i) - (\cos t - i\sin t)(1+i)}{2}\right]$$

$$= \frac{2}{5}[\sin t - \cos t].$$

Hence the complete solution is

$$y(t) = c_1 e^{-3t} + c_2 e^{-2t} + \frac{2}{5}(\sin t - \cos t).$$

Putting $t = 0$, we get

$$c_1 + c_2 = \frac{2}{5}.$$

Putting $t = 0$ in $y'(t)$, we get

$$3c_1 + 2c_2 = \frac{2}{5}.$$

Solving these equations, we get $c_1 = -\frac{2}{5}$, $c_2 = \frac{4}{5}$. Hence

$$y(t) = \frac{2}{5}\,[2e^{-2t} - e^{-3t} + \sin t - \cos t].$$

EXAMPLE 6.42

Find the power of the output for the following electric network (Figure 6.26).

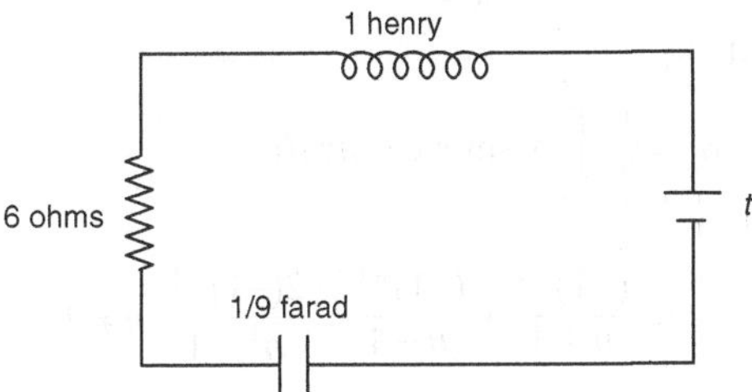

Figure 6.26

Solution By Kirchoff's law, the differential equation for the given circuit is

$$L\frac{d^2Q}{dt^2} + R\frac{dQ}{dt} + \frac{Q}{C} = E(t).$$

Here L = 1 henry, R = 6 ohms, $C = \frac{1}{9}$ farad, and $E(t) = t$. Hence the differential equation for the system is

$$\frac{d^2Q}{dt^2} + 6\frac{dQ}{dt} + 9Q = t,\ -\pi < t < \pi.$$

The input $u(t) = t$ is periodic with period 2π. The characteristic polynomial is $P(s) = s^2 + 6s + 9$, and $Q(s) = 1$.

Therefore, the transfer function is given by

$$H(\omega) = \frac{Q(i\omega)}{P(i\omega)} = \frac{1}{9 + 6i\omega - \omega^2}\cdot$$

The line spectrum (see Example 6.29) of the input function is $\left\{\frac{(-1)^n}{n}i\right\}$. The amplitude spectrum is

$$|c_n| = \frac{1}{|n|},\ n \neq 0,\ |c_0| = 0.$$

Therefore, by Theorem 6.17, the amplitude spectrum $|y_n|$ of the output is equal to

$$|y_n| = |H(n\omega_0)c_n| = \left|\frac{1}{-(n^2\omega_0^2 + 9)}c_n\right|$$

$$= \frac{1}{|n|\,(n^2\omega_0^2 + 9)} \text{ for } n \neq 0$$

and

$$|y_n| = 0 \text{ for } n = 0.$$

Therefore, power of $y(t)$ is

$$P = \frac{1}{2\pi}\int_0^{2\pi} |y(t)|^2\,dt = \sum_{n=-\infty}^{\infty} |y_n|^2$$

$$= \sum_{n=-\infty}^{\infty} \frac{1}{n^2(n^2\omega_0^2 + 9)^2}.$$

EXAMPLE 6.43

Show that the following LTC system representing an electric network consisting of L, C, R, and with $\alpha = \frac{1}{RC}$ is an all-pass system:

$$\frac{d^2y}{dt^2} - 2\alpha\frac{dy}{dt} + \alpha^2 y = \frac{d^2u}{dt^2} - \alpha^2 u,$$

with $u(t)$ a periodic input.

Solution. Let

$P(s) = s^2 - 2\alpha + \alpha^2$ (characteristic polynomial),

$Q(s) = s^2 - \alpha^2$.

Then the transfer function (frequency response) is

$$H(\omega) = \frac{Q(i\omega)}{P(i\omega)} = \frac{(i\omega)^2 - \alpha^2}{(i\alpha)^2 - 2i\omega + \alpha^2} = \frac{i\omega + \alpha}{i\omega - \alpha}\cdot$$

Now $u(t)$ is a periodic input with period T. Let u_n be the line spectrum of $u(t)$. Then the line spectrum of the response is

$$y_n = H(n\omega_0)\,u_n = \frac{in\omega_0 + \alpha}{in\omega_0 - \alpha}u_n.$$

Therefore, the amplitude spectrum $|y_n|$ of the response (output) is

$$|y_n| = \left|\frac{in\omega_0 + \alpha}{in\omega_0 - \alpha}u_n\right| = \left|\frac{in\omega_0 + \alpha}{in\omega_0 - \alpha}\right||u_n| = |u_n|.$$

It follows, therefore, that the amplitude spectrum is invariant under the given system. Using Parseval's identity for periodic function, we have

$$P=\frac{1}{T}\int_0^T |y(t)|^2 dt=\sum_{n=-\infty}^{\infty}|y_n|^2=\sum_{n=-\infty}^{\infty}|u_n|^2$$

$$=\frac{1}{T}\int_0^T |u(t)|^2\,dt.$$

Hence, the power of the output is equal to the power of the input and so the system is all-pass system.

6.22 APPLICATION TO PARTIAL DIFFERENTIAL EQUATIONS

Please see Examples 5.83 to 5.92 of chapter 5.

6.23 MISCELLANEOUS EXAMPLES

EXAMPLE 6.44

Expand $f(x)=\pi x$ from $x=-c$ to $x=c$ as a Fourier series.

Solution. The given function is an odd function of period $2c$. Therefore $a_n=0$ and

$$b_n=\frac{\pi}{c}\int_{-c}^{c} x\sin\frac{n\pi x}{c}\,dx$$

$$=\frac{\pi}{c}\left\{\left[x\left(\frac{-\cos\frac{n\pi x}{c}}{\frac{n\pi}{c}}\right)\right]_{-c}^{c}+\frac{c}{n\pi}\int_{-c}^{c}\cos\frac{n\pi x}{c}\,dx\right\}$$

$$=\frac{\pi}{c}\left\{\left[\frac{xc}{n\pi}\left(-\cos\frac{n\pi x}{c}\right)\right]_{-c}^{c}+\frac{c}{n\pi}\left[\frac{\sin\frac{n\pi x}{c}}{\frac{n\pi}{c}}\right]_{-c}^{c}\right\}$$

$$=-\frac{1}{n}\left[x\cos\frac{n\pi x}{c}\right]_{-c}^{c}+\frac{c}{n^2\pi}\left[\sin\frac{n\pi x}{c}\right]_{-c}^{c}$$

$$=-\frac{1}{n}[c\cos n\pi+c\cos n\pi]+0$$

$$=-\frac{2c}{n}\cos n\pi=-\frac{2c}{n}(-1)^n.$$

Hence

$$f(x)=-2c\sum_{n=1}^{\infty}\frac{1}{n}(-1)^n\sin\frac{n\pi x}{c}.$$

EXAMPLE 6.45

Express $f(x)=x\sin x$, $0<x<\pi$ as a Fourier cosine series. Hence, deduce the sum of the series $1+\frac{2}{1.3}-\frac{2}{3.5}+\frac{2}{5.7}-\ldots\infty=\frac{\pi}{2}.$

Solution. Similar to Example 6.17 except the limits. We note that $x\sin x$ is an even function and so the Fourier coefficients $b_n=0$. Further

$$a_0=\frac{1}{\pi}\int_{-\pi}^{\pi}x\sin x\,dx=2$$

and

$$a_n=\frac{1}{\pi}\int_{-\pi}^{\pi}x\sin x\cos nx\,dx$$

$$=\frac{(-1)^{n+1}}{n+1}+\frac{(-1)^{n-1}}{n-1}=\frac{2(-1)^{n-1}}{n^2-1},\ n\neq 1.$$

Also,

$$a_1=\frac{1}{\pi}\int_{-\pi}^{\pi}x\sin x\cos x\,dx=-\frac{1}{2}$$

Hence,

$$x\sin x=1-\frac{1}{2}\cos x+\sum_{n=2}^{\infty}\frac{2(-1)^{n-1}}{n^2-1}\cos nx$$

$$=1-\frac{1}{2}\cos x$$

$$-2\left[\frac{\cos 2x}{1.3}-\frac{\cos 3x}{2.4}+\frac{\cos 4x}{3.5}-\ldots\right]$$

Putting $x=\frac{\pi}{2}$, we get

$$\frac{\pi}{2}=1+\frac{2}{1.3}-\frac{2}{3.5}+\frac{2}{5.7}-\ldots$$

EXAMPLE 6.46

Find the Fourier series for $f(x)=\frac{(\pi-x)^2}{4}$ in the interval $(0,2\pi)$ and hence deduce

$$\frac{1}{1^2}+\frac{1}{2^2}+\frac{1}{3^2}+\ldots=\frac{\pi^2}{6}$$

Solution. We have $f(x)=\frac{(\pi-x)^2}{4}$ and the interval is $(0,2\pi)$. Therefore the Fourier coefficients are

$$a_0 = \frac{1}{\pi}\int_0^{2\pi} f(x)dx = \frac{1}{\pi}\int_0^{2\pi} \frac{(\pi - x)^2}{4}\, dx$$

$$= \frac{1}{4\pi}\left[\frac{(\pi - x)^3}{-3}\right]_0^{2\pi} = \frac{\pi^2}{6}$$

$$a_n = \frac{1}{\pi}\int_0^{2\pi} f(x)\cos nx\, dx$$

$$= \frac{1}{\pi}\int_0^{2\pi} \frac{(\pi - x)^2}{4}\cos nx\, dx$$

$$= \frac{1}{4\pi}\left[\left\{(\pi - x)^2 \frac{\sin nx}{n}\right\}_0^{2\pi} + \int_0^{2\pi} 2(\pi - x)\frac{\sin nx}{n}\, dx\right]$$

$$= \frac{-1}{2\pi n^2}(-2\pi) = \frac{1}{n^2},$$

$$b_n = \frac{1}{\pi}\int_0^{2\pi} f(x)\sin nx\, dx = \frac{1}{\pi}\int_0^{2\pi} \frac{(\pi - x)^2}{4}\sin nx\, dx$$

$$= \frac{1}{4\pi}\left[\left\{(\pi - x)^2 \frac{\cos nx}{-n}\right\}_0^{2\pi} - \int_0^{2\pi} 2(\pi - x)\frac{\cos nx}{n}\, dx\right]$$

$$= \frac{1}{2\pi n^2}\left[\frac{\cos nx}{n}\right]_0^{2\pi} = 0.$$

Therefore the Fourier series is

$$f(x) \sim \frac{a_0}{2} + \Sigma(\cos nx + \sin nx)$$

$$= \frac{\pi^2}{12} + \sum_{n=1}^{\infty} \frac{\cos nx}{n^2}$$

$$= \frac{\pi^2}{12} + \frac{\cos x}{1^2} + \frac{\cos 2x}{2^2} + \frac{\cos 3x}{3^2} + \frac{\cos 4x}{4^2} + \ldots$$

Putting $n = 0$ in the above equation, we get

$$\frac{\pi^2}{6} = \frac{1}{12} + \frac{1}{2^2} + \frac{1}{3^2} + \frac{1}{4^2} + \ldots$$

EXAMPLE 6.47

Examine whether the function $f(x) = x \cos x$ is even or odd and find its Fourier series in $(-\pi, \pi)$.

Solution. We have $f(x) = x \cos x$ as a product of odd and even functions x and $\cos x$ respectively. But the product of odd and even function is odd. Hence f is odd. Since f is odd, $a_n = 0$. Further,

$$b_n = \frac{2}{\pi}\int_0^{\pi} x\cos x \sin nx\, dx$$

$$= \frac{1}{\pi}\int_0^{\pi} 2x\cos x\sin nx\, dx$$

$$= \frac{1}{\pi}\int_0^{\pi} x\,[\sin(n+1)x + \sin(n-1)x]\, dx$$

$$= \frac{1}{\pi}\left[x\left\{-\frac{\cos(n+1)x}{n+1} - \frac{\cos(n-1)x}{n-1}\right\}\right.$$

$$\left.-\left\{-\frac{\sin(n+1)x}{(n+1)^2} - \frac{\sin(n-1)x}{(n-1)^2}\right\}\right]_0^{\pi}$$

$$= \frac{1}{\pi}\left[\pi\left\{-\frac{\cos(n+1)\pi}{n+1} - \frac{\cos(n-1)\pi}{n-1}\right\}\right],\ n \neq 1$$

$$= -\left[\frac{\cos(n+1)\pi}{n+1} + \frac{\cos(n-1)\pi}{n-1}\right],\ n \neq 1.$$

If $n \neq 1$ and n is odd, then both $n - 1$ and $n + 1$ are even. Therefore

$$b_n = -\left[\frac{-1}{n+1} - \frac{1}{n-1}\right] = \frac{2n}{n^2 - 1}.$$

If $n \neq 1$ and n is even, than both $n - 1$ and $n + 1$ are odd. Therefore

$$b_n = -\left[\frac{-1}{n+1} + \frac{1}{n-1}\right] = -\frac{2n}{n^2 - 1}.$$

When $n = 1$, we have

$$b_1 = \frac{2}{\pi}\int_0^{\pi} x\cos x\sin x\, dx$$

$$= \frac{1}{\pi}\int_0^{\pi} x\sin 2x\, dx = -\frac{1}{2}.$$

Hence

$$f(x) = x\cos x$$

$$= -\frac{1}{2}\sin x - \frac{4}{3}\sin 2x + \frac{3}{4}\sin 3x - \ldots$$

EXAMPLE 6.48

Obtain a half-range cosine series for

$$f(x)=\begin{cases} kx & 0 \le x \le \frac{L}{2} \\ k(L-x) & \frac{L}{2} \le x \le L. \end{cases}$$

Deduce the sum of the series

$$\frac{1}{1^2}+\frac{1}{3^2}+\frac{1}{5^2}+\ldots$$

Solution. As in Example 6.24, we have

$$a_0=\frac{kL}{2} \text{ and } a_n=\frac{8kL}{n^2\pi^2}\cos\frac{n\pi}{2}\sin^2\frac{n\pi}{4}.$$

Therefore

$$f(x)=\frac{kL}{4}-\frac{2kL}{\pi^2}$$

$$\times\left[\cos\frac{2\pi x}{L}+\frac{1}{3^2}\cos\frac{6\pi x}{L}+\frac{1}{5^2}\cos\frac{10\pi x}{L}+\ldots\right]$$

Putting $x = 0$, we get

$$0=\frac{kL}{4}-\frac{2kL}{\pi^2}\left[1+\frac{1}{3^2}+\frac{1}{5^2}+\ldots\right]$$

or

$$\frac{-kL}{4}=-\frac{2kL}{\pi^2}\left[1+\frac{1}{3^2}+\frac{1}{5^2}+\ldots\right]$$

or

$$\frac{\pi^2}{8}=\frac{1}{1^2}+\frac{1}{3^2}+\frac{1}{5^2}+\ldots.$$

EXAMPLE 6.49

Find the Fourier series of periodicity 2 for

$$f(x)=\begin{cases} x, -1 < x \le 0 \\ x+2\; 0 < x \le 1. \end{cases}$$

Hence, show that the sum of the series $1-\frac{1}{3}+\frac{1}{5}-\frac{1}{7}+\ldots=\frac{\pi}{4}$.

Solution. The period of the function is 2. Therefore the Fourier series shall be

$$\frac{a_0}{2}+\sum_{n=1}^{\infty}(a_n\cos n\,\omega_0 t+b_n\sin n\,\omega_0 t),$$

$$\omega_0=\frac{2\pi}{T}=\pi.$$

or

$$\frac{a_0}{2}+\sum_{n-1}^{\infty}(a_n\cos n\,\pi\, t+b_n\sin x\pi\, t),$$

where

$$a_0=\frac{2}{T}\int_{-1}^{1} f(x)\,dx,$$

$$a_n=\frac{2}{T}\int_{-1}^{1} f(x)\cos n\pi x\,dx,$$

$$b_n=\frac{2}{T}\int_{-1}^{1} f(x)\sin n\pi x\,dx.$$

With $T = 2$. we are given that

$$f(x)=\begin{cases} x, -1 < x < 0 \\ x+2,\; 0 < x \le 1. \end{cases}$$

Therefore

$$a_0=\int_{-1}^{1} f(x)dx=\int_{-1}^{0} f(x)\,dx+\int_{0}^{1} f(x)\,dx$$

$$=\int_{-1}^{0} x\,dx+\int_{0}^{1}(x+2)\,dx$$

$$=\left[\frac{x^2}{2}\right]_{-1}^{0}+\left[\frac{x^2}{2}+2x\right]_{0}^{1}=2,$$

$$a_n=\int_{-1}^{1} f(x)\cos n\pi x\,dx$$

$$=\int_{-1}^{0} x\cos n\pi x\,dx+\int_{0}^{1}(x+2)\cos n\pi x\,dx$$

$$=\left[x\,\frac{\sin n\pi x}{n\pi}-\left(-\frac{\cos n\pi x}{n^2\pi^2}\right)\right]_{-1}^{0}$$

$$+\left[(x+2)\,\frac{\sin n\,\pi x}{n\pi}\left(-\frac{\cos n\pi x}{n^2\pi^2}\right)\right]_{1}^{0}$$

$$=\left[\frac{1}{n^2\pi^2}-\frac{\cos n\pi}{n^2\pi^2}\right]+\left[\frac{\cos n\,\pi}{n^2\pi^2}-\frac{1}{n^2\pi^2}\right]=0,$$

$$b_n=\int_{-1}^{1} f(x)\sin n\pi x\,dx$$

$$=\int_{-1}^{0} x\sin n\pi x\,dx+\int_{0}^{1}(x+2)\sin n\pi x\,dx$$

$$=\left[x\left(\frac{-\cos n\pi x}{n\pi}\right)-\left(\frac{-\sin n\pi x}{n^2\pi^2}\right)\right]_{-1}^{0}$$

$$+\left[(x+2)\left(\frac{-\cos n\pi x}{n\pi}\right)-\left(-\frac{\sin n\pi x}{n^2\pi^2}\right)\right]_{0}^{1}$$

$$= -\frac{\cos n\pi}{n\pi} + \left[-\frac{3\cos n\pi}{n\pi} + \frac{2}{n\pi}\right]$$

$$= -4\frac{\cos n\pi}{n\pi} + \frac{2}{n\pi} = \frac{2}{n\pi}[1 - 2(-1)^n].$$

Hence

$$f(x) = 1 + \frac{2}{\pi}\sum_{n=1}^{\infty}\frac{1}{n}[1 - 2(-1)^n]\sin n\pi x$$

$$= 1 + \frac{2}{\pi}\left[3\sin\pi x + \frac{3}{3}\sin 3\pi x + \frac{3}{5}\sin 5\pi x + \ldots\right]$$

Putting $x = \frac{1}{2}$, we get

$$\frac{1}{2} + 2 = 1 + \frac{2}{\pi}\left[3 - \frac{3}{3} + \frac{3}{5} - \frac{3}{7} + \ldots\right]$$

or

$$\frac{3}{2} = \frac{2}{\pi}\left[3 - \frac{3}{3} + \frac{3}{5} - \frac{3}{7} + \ldots\right]$$

or

$$\frac{\pi}{4} = 1 - \frac{1}{3} + \frac{1}{5} - \frac{1}{7} + \ldots$$

EXERCISES

1. Find the Fourier series to represent x^2 in the interval $(-l, l)$.

   Hint: see Example 6.6

2. Find the Fourier series of the function

   $$f(x) = \begin{cases} \sin\frac{x}{2} & \text{for } 0 \le x \le \pi \\ -\sin\frac{x}{2} & \text{for } \pi \le x \le 2\pi \end{cases}$$

   $$f(x) = f(x + 2\pi)$$

   Ans. $-\frac{8}{\pi}\sum_{n=1}^{\infty}\frac{n\sin(2nx)}{(2n+1)(2n-1)}$

3. Derive Fourier series for e^{-ax}, $-\pi < x < \pi$ and deduce series for $\frac{\pi}{\sinh\pi}$.

   Hint: Similar to Example 6.9

   Ans: $\frac{2\sinh a\pi}{\pi}\left(\frac{1}{2a} - \frac{a\cos a}{1^2 + a^2} + \frac{a\cos 2a}{2^2 + a^2} - \ldots\right)$

   $$+\left(\frac{\sin x}{1^2 + a^2} - \frac{2\sin 2x}{2^2 + a^2} + \ldots\right)$$

   $$\frac{\pi}{\sinh\pi} = 2\left(\frac{1}{2^2 + 1} - \frac{1}{3^2 + 1} + \frac{1}{4^2 + 1} - \ldots\right)$$

4. Show that for $-\pi < x < \pi$,

   $$\cosh ax = \frac{2a^2}{\pi}\sinh a\pi \times \left[\frac{1}{2a^2} + \sum_{n=1}^{\infty}\frac{(-1)^n}{a^2 + n^2}\cos nx\right].$$

   Hint: $\cosh ax = \frac{1}{2}(e^{ax} + e^{-ax})$, so add the series of Example 6.9 and Exercise 6 (given above).

5. An alternating current, after passing through a rectifier, has the form

   $$i = \begin{cases} I_0\sin x & \text{for } 0 \le x \le \pi \\ 0 & \text{for } \pi \le x \le 2\pi, \end{cases}$$

   where I_0 is maximum current and the period is 2π. Express i as a Fourier series.

   Hint: See Example 6.21.

   Ans. $\frac{I_0}{p}\left(1 + \frac{\pi}{2}\sin\theta - \frac{2}{1.3}\cos 2\theta - \frac{2}{3.5}\cos 4\theta - \frac{2}{5.7}\cos 6\theta + \ldots\right)$

6. Determine Fourier Series expansion of the function

   $$f(x) = \begin{cases} 2 & \text{for } 0 < x < \frac{2\pi}{3} \\ 1 & \text{for } \frac{2\pi}{3} < x < \frac{4\pi}{3} \\ 0 & \text{for } \frac{4\pi}{3} < x < 2\pi. \end{cases}$$

   Ans. $1 + \frac{3}{\pi} \times \left(\sin x + \frac{\sin 2x}{2} + \frac{\sin 4x}{4} + \frac{\sin 5x}{5} + \ldots\right)$

7. Expand $f(t) = t^2$ in Fourier sine series on the interval (0, 1).

   Ans. $\sum_{n=1}^{\infty}\frac{2}{n\pi}\left[\frac{2((-1)^n - 1)}{n^2\pi^2}\right] - (-1)^n\sin n\pi t$

8. Find Fourier series expansion of the function f defined by

$$f(t) = \begin{cases} \pi^2 & \text{for } -\pi < t < 0 \\ (t-\pi)^2 & \text{for } 0 \le t \le \pi, \end{cases}$$

$f(t) = f(t + 2\pi)$.

Determine the value of $\sum_{n=1}^{\infty} \frac{1}{n^2}$.

Ans. $f(t) = \frac{2\pi^2}{3} + \sum_{n=1}^{\infty}$

$$\left[\frac{2}{n^2}\cos nt + \frac{(-1)^n}{n}\pi \sin nt\right] - \frac{4}{\pi}\sum_{n=1}^{\infty}\frac{\sin(2n-1)t}{(2n-1)^2}$$

Putting $t = 0$, we get $\sum_{n=1}^{\infty} \frac{1}{n^2} = \frac{\pi^2}{6}$

9. Find half range cosine series for the function

$$f(x) = \begin{cases} x & \text{for } 0 < x < \frac{\pi}{2} \\ \pi - x & \text{for } \frac{\pi}{2} < x < \pi. \end{cases}$$

Ans. $\frac{\pi}{4} - \frac{2}{\pi}\left(\cos 2x + \frac{\cos 6x}{32} + \frac{\cos 10x}{52} + \ldots\right)$

10. Find the Fourier sine series for

$$f(x) = \begin{cases} \frac{1}{4} - x & \text{for } 0 < x < \frac{1}{2} \\ x - \frac{3}{4} & \text{for } \frac{1}{2} < x < 1. \end{cases}$$

Hint: Taking odd extension, the extended function is odd over (–1, 1) and so find b_n using $b_n = \frac{2}{1}\int_0^1 f(x)\sin n\pi x\, dx$.

Ans. $\left(\frac{1}{\pi} - \frac{4}{\pi^2}\right)\sin \pi x + \left(\frac{1}{3\pi} + \frac{4}{3^2\pi^2}\right)\sin 3\pi x$

$$+\left(\frac{1}{5\pi} - \frac{4}{5^2\pi^2}\right)\sin 5\pi x + \ldots$$

11. Find Fourier cosine series for $f(x) = (x - 1)^2, 0 < x < 1$.

Ans. $\frac{1}{3} + \frac{4}{\pi^2}\sum_{n=1}^{\infty}\frac{1}{n^2}\cos n\pi x$

12. Verify Riemann-Lebesgue lemma for the function $f(x) = e^{ax}, -\pi < x < \pi$.

Hint: see Example 6.9, show that $a_n \to 0$, $b_n \to 0$ as $n \to \infty$).

13. Find half-range sine series for e^x, $0 < x < 1$.

Ans. $2\pi \sum_{n=1}^{\infty} \frac{n[1 - e(-1)^n]}{(1 + n^2\pi^2)} \sin n\pi x$

14. Given the series

$$t(\pi - t) = \frac{8}{\pi}\sum_{n=1}^{\infty}\frac{\sin(2n-1)}{(2n-1)^3}, 0 \le t \le \pi,$$

show that $\sum_{n=1}^{\infty}\frac{1}{(2n-1)^6} = \frac{\pi^6}{960}$

Hint: Use Parseval's equality.

15. Find Fourier series e^{-x} in $(0, \pi)$

Hint: $b_n = \frac{2}{\pi}\int_{n=1}^{\infty} e^{-x}\sin nx\, dx$

$= \frac{2n}{\pi}\frac{[(-1)^n e^{-x} - 1]}{1 + n^2}$ and so the series is

$$\frac{2}{\pi}\sum_{n=1}^{\infty}\frac{n[e^{-\pi}(-1)^{n-1}]}{1+n^2}\sin nx.$$

16. Find Fourier sine series for the function f defined by

$$f(x) = \begin{cases} x & \text{for } 0 \le x \le L/2 \\ L - x & \text{for } L/2 \le x \le L. \end{cases}$$

Hint: Extend f as odd function in $(-L, L)$. Then

$$b_n = \frac{2}{L}\int_0^L f(x)\sin\frac{n\pi x}{L}dx = \frac{2}{L}\left[\int_0^{L/2} + \int_{L/2}^{L}\right]$$

$$= -\frac{L}{n\pi}\cos\frac{n\pi}{2} + \frac{2L}{n^2\pi^2}\sin\frac{n\pi}{2}$$

$$+ \frac{L}{n\pi}\cos\frac{n\pi}{2} + \frac{2L}{n^2\pi^2}\sin\frac{n\pi}{2}$$

$$= \frac{4L}{n^2\pi^2}\sin\frac{n\pi}{2}$$

Hence

$$f(x) = \sum_{n=1}^{\infty}\frac{4L}{n^2p^2}\sin\frac{n\pi}{2}\sin\frac{n\pi x}{L}$$

$$= \frac{4L}{\pi^2}\int_{n=1}^{\infty}\frac{1}{n^2}\sin\frac{n\pi}{2}\sin\frac{n\pi x}{L}.$$

17. Solve the heat equation $\frac{\partial u}{\partial t} = \frac{\partial^2 u}{\partial x^2}$, $0 < x < 2$, $t > 0$

$$u_x(0, t) = 0,\ u(2, t)$$

$$u_x(x, 0) = \begin{cases} 1 & \text{for } 0 < x < 1 \\ 2 - x & \text{for } 1 \le x \le 2. \end{cases}$$

Hint: $u(x, t) = \frac{A_0}{2} + \sum_{n=1}^{\infty} A_n e^{-n^2\pi^2 t/L^2} \cos\frac{n\pi x}{L}$, $L = 2$

$$= \frac{A_0}{2} + \sum_{n=1}^{\infty} A_n e^{-n^2\pi^2 t/4} \cos\frac{n\pi x}{2},$$

$$A_n = \int_0^1 x \cos\frac{n\pi x}{2} + \int_1^2 (2 - x) \cos\frac{np\, x}{2}\, dx$$

$$= \frac{-4}{n^2\pi^2}\left[\cos\frac{n\pi}{2} - \cos n\pi\right].$$

Thus, $u(x, t)$ is equal to

$$\frac{3}{4} - \frac{4}{\pi^2} \sum_{n=1}^{\infty} \left[\cos\frac{n\pi}{2} - \cos n\pi\right]$$

$$\times e^{-n^2\pi^2 t/4} \cos\frac{n\pi x}{2}.$$

18. Solve $\frac{\partial u}{\partial t} = k\frac{\partial^2 u}{\partial x^2}$, $0 < x < \pi$ subject to

$$u(x, 0) = e^{-x},\ 0 < x < \pi$$

$$u(0, t) = 0,\ u(\pi, t) = 0,\ t \ge 0.$$

Ans. $u(x, t) = \frac{2}{\pi} \sum_{n=1}^{\infty} \frac{n[(-1)^n e^{-p} - 1]}{1 + n^2}$

$$\times e^{-n^2 kt} \sin nx$$

7 Fourier Transform

During the study of Fourier series, we confined ourselves to periodic functions. To a periodic function f we assigned Fourier coefficients c_n, $n \in \mathbb{Z}$ and then defined the Fourier series as a trigonometric series with coefficients taken as Fourier coefficients. We then discussed the convergence and some other properties of Fourier series. But we generally encounter non-periodic functions in many applications. Our aim in this chapter is to develop a concept, called *Fourier transform*, in which to a non-periodic function f, we shall assign for each $\omega \in \mathbb{R}$ a function F defined on $\mathbb{R}$ such that $F(\omega) \in \mathbb{C}$. This function F will be called Fourier transform of the non-periodic function f. The difference, we note, in a Fourier series and Fourier transform is that here series shall be replaced by an integral, called Fourier integral. The benefit is obvious because the present study will also be helpful in solving differential and partial differential equations of non-periodic functions.

7.1 FOURIER INTEGRAL THEOREM

A function $f: \mathbb{R} \to C$ is called *absolutely integrable* on $\mathbb{R}$ if $\int_{-\infty}^{\infty} |f(t)|dt$ exists as an improper Riemann integral. Further, the value of $\lim_{T\to\infty} \int_{-T}^{T} f(t)\, dt$ is called the *Cauchy principal value* of the integral $\int_{-\infty}^{\infty} f(t)\, dt$, provided that the limit exists.

The ultimate aim of this section is to establish Fourier integral theorem which is crucial for the study of Fourier and Laplace transforms. The following two fundamental theorems (7.1 and 7.2) together with Theorem 7.3 shall be required to prove the Fourier Integral Theorem 7.4.

Theorem 7.1. Let f be piecewise continuous on $[a, b]$. Then

$$\lim_{\omega\to\infty} \int_a^b f(t) \sin \omega t\, dt = \lim_{\omega\to\infty} \int_a^b f(t) \cos \omega t\, dt = 0.$$

Proof: We may assume, without loss of generality, that f is continuous on $[a, b]$ since we can prove the theorem for a finite number of sub-intervals on which it is continuous. Let $t = u + \frac{\pi}{\omega}$. Then

$$\int_a^b f(t) \sin \omega t\, dt = -\int_{a-\pi/\omega}^{b-\pi/\omega} f(u + \pi/\omega) \sin \omega u\, du,$$

and so

$$2\int_a^b f(t) \sin \omega t\, dt = \int_a^b f(u) \sin \omega u\, du$$

$$-\int_{a-\frac{\pi}{\omega}}^{b-\frac{\pi}{\omega}} f\left(u+\frac{\pi}{\omega}\right) \sin \omega u\, du$$

$$= -\int_{a-\frac{\pi}{\omega}}^{a} f\left(u+\frac{\pi}{\omega}\right) \sin \omega u\, du$$

$$-\int_{a}^{b-\frac{\pi}{\omega}} f\left(u+\frac{\pi}{\omega}\right) \sin \omega u\, du$$

$$+\int_{a}^{b-\frac{\pi}{\omega}} f(u)\sin\omega u\,du+\int_{b-\frac{\pi}{\omega}}^{b} f(u)\sin\omega u\,du$$

$$=-\int_{a-\frac{\pi}{\omega}}^{a} f\left(u+\tfrac{\pi}{\omega}\right)\sin\omega u\,du$$

$$+\int_{b-\frac{\pi}{\omega}}^{b} f(u)\sin\omega u\,du$$

$$-\int_{a}^{b-\frac{\pi}{\omega}} \left[f\left(u+\tfrac{\pi}{\omega}\right)-f(u)\right]\sin\omega u\,du.$$

Since f is continuous on $[a, b]$, it is uniformly continuous. Therefore, taking ω large enough, we have

$$|f(u+\tfrac{\pi}{\omega})-f(u)|<\frac{\varepsilon}{b-a},\ \varepsilon>0,\ u\in[a,b].$$

Also for large ω, $\frac{\pi}{\omega}<\frac{\varepsilon}{2M}$, where $|f(t)|\le M$ for $t\in[a, b]$. Therefore,

$$2\left|\int_a^b f(t)\sin\omega t\,dt\right|<M\,\frac{\varepsilon}{2M}+M\,\frac{\varepsilon}{2M}$$
$$+\frac{\varepsilon}{b-a}(b-a)=2\varepsilon.$$

Since $\varepsilon>0$ is arbitrary, it follows that

$$\lim_{\omega\to\infty}\int_a^b f(t)\sin\omega t\,dt=0.$$

Similarly, we can show that

$$\lim_{\omega\to\infty}\int_a^b f(t)\cos\omega t\,dt=0.$$

Remark 7.1. Since

$$\int_a^b f(t)\,e^{-i\omega t}\,dt=\int_a^b f(t)\cos\omega t\,dt$$
$$-i\int_a^b f(t)\sin\omega t\,dt,$$

it follows from Theorem 7.1 that

$$\lim_{\omega\to\pm\infty}\int_a^b f(t)\,e^{-i\omega t}\,dt=0.$$

Theorem 7.2. (Cantor-Lebesgue Lemma). If f is absolutely integrable and piecewise continuous on $\mathbb{R}$, then

$$\lim_{\omega\to\pm\infty}F(\omega)=\lim_{\omega\to\pm\infty}\int_{-\infty}^{\infty} f(t)\,e^{-i\omega t}\,dt=0.$$

Proof: Let $\varepsilon>0$. Since f is absolutely integrable, there exist $a, b\in\mathbb{R}$ such that

$$\int_{-\infty}^{a}|f(t)|dt+\int_{b}^{\infty}|f(t)|\,dt<\frac{\varepsilon}{2}.\qquad(1)$$

Also, by Theorem 7.1, we have

$$\left|\int_a^b f(t)\,e^{-i\omega t}\,dt\right|<\frac{\varepsilon}{2}\text{ for large }|\omega|.\qquad(2)$$

Hence, the triangle inequality, the fact that $|e^{-i\omega t}|=1$ and the relations (1) and (2) yield

$$\left|\int_{-\infty}^{\infty} f(t)\,e^{-i\omega t}\,dt\right|\le\left|\int_{-\infty}^{a} f(t)\,e^{-i\omega t}\,dt\right|$$
$$+\left|\int_{b}^{\infty} f(t)\,e^{-i\omega t}\,dt\right|+\left|\int_{a}^{b} f(t)\,e^{-i\omega t}\,dt\right|$$
$$\le\int_{-\infty}^{a}|f(t)|\,dt+\int_{b}^{\infty}|f(t)|\,dt+\left|\int_{a}^{b} f(t)\,e^{-i\omega t}\,dt\right|$$
$$<\frac{\varepsilon}{2}+\frac{\varepsilon}{2}=\varepsilon\text{ for large}|\omega|,$$

which proves Theorem 7.2.

Theorem 7.3. Let f be an absolutely integrable and piecewise smooth function on $\mathbb{R}$. Then

$$\lim_{\omega\to\infty}\frac{1}{\pi}\int_{-\infty}^{\infty} f(t-u)\,\frac{\sin\omega u}{u}\,du$$
$$=\frac{1}{2}[f(t+)+f(t-)],\ t\in\mathbb{R}.$$

Proof: Splitting the integral in the left-hand side of the assertion and changing u to $-u$, we have

$$\lim_{\omega\to\infty}\frac{1}{\pi}\int_{-\infty}^{\infty} f(t-u)\,\frac{\sin\omega u}{u}\,du$$

$$=\lim_{\omega\to\infty}\frac{1}{\pi}\int_{-\infty}^{0} f(t-u)\,\frac{\sin\omega u}{u}\,du$$

$$+ \lim_{\omega\to\infty} \frac{1}{\pi} \int_0^{\infty} f(t+u) \frac{\sin \omega u}{u}\, du$$

$$= \lim_{\omega\to\infty} \frac{1}{\pi} \int_0^{\infty} [f(t-u)+f(t+u)] \frac{\sin \omega u}{u} du \quad (3)$$

Also putting $\omega u = v$, we note that

$$\lim_{\omega\to\infty} \int_0^1 \frac{\sin \omega u}{u}\, du = \lim_{\omega\to\infty} \int_0^{\omega} \frac{\sin v}{v} dv$$

$$= \int_0^{\infty} \frac{\sin v}{v}\, dv = \frac{\pi}{2}. \qquad (4)$$

Multiplying (4) throughout by $\frac{f(t+)+f(t-)}{\pi}$, we get

$$\frac{1}{2}[f(t+)+f(t-)]$$

$$= \lim_{\omega\to\infty} \frac{1}{\pi} \int_0^{1} [f(t+)+f(t-)] \frac{\sin \omega u}{u}\, du. \quad (5)$$

Hence, from (3) and (5), we have

$$\lim_{\omega\to\infty} \frac{1}{\pi} \int_{-\infty}^{\infty} f(t-u) \frac{\sin \omega u}{u}\, du - \frac{1}{2}[f(t+)+f(t-)]$$

$$= \lim_{\omega\to\infty} \frac{1}{\pi} \int_0^{\infty} [f(t-u)+f(t+u)] \frac{\sin \omega u}{u}\, du$$

$$- \lim_{\omega\to\infty} \frac{1}{\pi} \int_0^{1} [f(t+)+f(t-)] \frac{\sin \omega u}{u}\, du$$

$$= \lim_{\omega\to\infty} \frac{1}{\pi} \int_0^{1} [f(t-u)+f(t+u)] - f(t+)$$

$$- f(t-)] \frac{\sin \omega u}{u}\, du$$

$$+ \lim_{\omega\to\infty} \frac{1}{\pi} \int_1^{\infty} [f(t-u)+f(t+u)] \frac{\sin \omega u}{u}\, du$$

$$= \lim_{\omega\to\infty} I_1 + \lim_{\omega\to\infty} I_2, \text{ say} \qquad (6)$$

Thus we have

$$I_1 = \frac{1}{\pi} \int_0^{1} f(t-u) - f(t-)] \frac{\sin \omega u}{u}\, du$$

$$+ \frac{1}{\pi} \int_0^{1} [f(t+u) - f(t+) \frac{\sin \omega u}{u}\, du.$$

Since f is piecewise smooth,

$$\lim_{u\to 0} \frac{f(t-u)-f(t-)}{u} = -f'(t-)$$

and

$$\lim_{u\to 0} \frac{f(t+u)-f(t+)}{u} = f'(t+).$$

Thus

$$\frac{f(t-u)-f(t-)}{u} \text{ and } \frac{f(t+u)-f(t+)}{u}$$

are piecewise continuous on $\mathbb{R}$ and also absolutely integrable. Hence, by Theorem 7.1, $I_1 \to 0$ as $\omega \to \infty$. For I_2, we define auxiliary function $g(u)$ by

$$g(u) = \begin{cases} \dfrac{f(t-u)+f(t+u)}{u} & \text{for } u \geq 1 \\ 0 & \text{otherwise,} \end{cases}$$

and so

$$I_2 = \frac{1}{\pi} \int_{-\infty}^{\infty} g(u) \sin \omega u\, du.$$

The function g is again piecewise continuous on $\mathbb{R}$. Since $\frac{1}{u} < 1$ if $u > 1$ and f is absolutely integrable, it follows that g is absolutely integrable. Hence by Theorem 7.1, $\lim_{\omega\to\infty} I_2 = 0$. Then (6) yields

$$\lim_{\omega\to\infty} \frac{1}{\pi} \int_{-\infty}^{\infty} f(t-u) \frac{\sin \omega u}{u}\, du = \frac{1}{2}[f(t+)+f(t-)].$$

Remark 7.2. If, in Theorem 7.3, f is assumed continuous in place of piecewise continuous, then

$\frac{1}{2}[f(t+)+f(t-)] = f(t)$ and so

$$\lim_{\omega\to\infty} \frac{1}{\pi} \int_{-\infty}^{\infty} f(t-u) \frac{\sin \omega u}{u}\, du = f(t).$$

Now we are in a position to prove the Fourier integral theorem.

Theorem 7.4. (Fourier Integrable Theorem). Let f be an absolutely integrable and piecewise smooth function on $\mathbb{R}$. Then the integral

$$\int_{-\infty}^{\infty} f(t)\, e^{-i\omega t}\, dt$$

converges absolutely and uniformly for ω in $[-T, T]$, $t \in \mathbb{R}$ and

$$\frac{1}{2\pi}\int_{-\infty}^{\infty}\int_{-\infty}^{\infty} f(u)\, e^{i\omega(t-u)}\, du\, d\omega = \frac{1}{2}[f(t+) + f(t-)],$$

where the integration with respect to ω is in Cauchy principal value sense.

Proof: Since $|e^{i\omega t}| = 1$, we have

$$\int_{-\infty}^{\infty} |f(t)e^{-i\omega t}|\, dt = \int_{-\infty}^{\infty} |f(t)|\, dt < \infty,\ t \in \mathbb{R}.$$

Hence, the integral $\int_{-\infty}^{\infty} f(t)\, e^{-i\omega t}\, dt$ converges absolutely and uniformly for $\omega \in [-T, T]$.

The hypothesis of the theorem allows us to interchange the order of integration and so, we have

$$\frac{1}{2\pi}\int_{-\infty}^{\infty}\int_{-\infty}^{\infty} f(u)\, e^{i\omega(t-u)}\, du\, d\omega$$

$$= \lim_{T\to\infty}\frac{1}{2\pi}\int_{-T}^{T}\int_{-\infty}^{\infty} f(u)\, e^{i\omega(t-u)}\, du\, d\omega$$

$$= \lim_{T\to\infty}\frac{1}{2\pi}\int_{-\infty}^{\infty} f(u)\int_{-T}^{T} e^{i\omega(t-u)}\, d\omega\, du$$

$$= \lim_{T\to\infty}\frac{1}{\pi}\int_{-\infty}^{\infty} f(u)\,\frac{\sin T(t-u)}{t-u}\, du$$

$$= \lim_{T\to\infty}\frac{1}{\pi}\int_{-\infty}^{\infty} f(t-u)\,\frac{\sin Tu}{u}\, du$$

$= \frac{1}{2}[f(t+) - f(t-)]$, by Theorem 7.3.

7.2 FOURIER TRANSFORMS

Let f be an absolutely integrable and piecewise smooth function on $\mathbb{R}$. If we put

$$F(\omega) = \int_{-\infty}^{\infty} f(u)\, e^{-i\omega u}\, du, \tag{7}$$

then Fourier integral theorem asserts that

$$\frac{1}{2\pi}\int_{-\infty}^{\infty} F(\omega)\, e^{i\omega u}\, d\omega = \frac{1}{2}[f(t+) - f(t-)]. \tag{8}$$

The function F defined by (7) is called *Fourier transform* of f. Thus, we define Fourier transform of a function as follows:

Definition 7.1 If f is absolutely integrable, then the function F defined by

$$F(\omega) = \int_{-\infty}^{\infty} f(t)\, e^{-i\omega t}\, dt$$

is called the *Fourier transform* (*spectrum* or *spectral density*) of f.

$|F(\omega)|$ is called *amplitude spectrum*, arg. $F(\omega)$ is called *phase spectrum* and $|F(\omega)|^2$ is called *energy spectrum* of f.

The condition that f is absolutely integrable is sufficient for the existence of Fourier transform of a function f. Under this condition the integral on the right-hand side converges. To show it, we note that

$$e^{-i\omega t} = \cos \omega t - i \sin \omega t,$$

$$|e^{-i\omega t}| = \sqrt{\cos^2 \omega t + \sin^2 \omega t} = 1,$$

$$|f(t)\, e^{-i\omega t}| = |f(t)|\,|e^{-i\omega t}| = |f(t)|.$$

Since f is absolutely integrable, we have

$$|F(\omega)| = \int_{-\infty}^{\infty} |f(t)\, e^{-i\omega t}|\, dt = \int_{-\infty}^{\infty} |f(t)|\, dt < \infty$$

and so the Fourier transform of f exists.

Remark 7.3. The condition of absolute integrability of a function is not a necessary condition for the existence of its Fourier transform. In fact, there are functions like sin ωu, cos ωu, $\frac{\sin u}{u}$, and unit step function which are not absolutely integrable but have Fourier transform.

If f is continuous, then (8) reduces to

$$f(t) = \frac{1}{2\pi}\int_{-\infty}^{\infty} F(\omega)\, e^{i\omega t}\, d\omega \qquad (9)$$

Formula (9) is called the *Inversion Formula* and $f(t)$ is then called *Inverse Fourier Transform* of $F(\omega)$.

We note that

$$F(\omega) = \int_{-\infty}^{\infty} f(t)\, e^{-i\omega t}\, dt = \int_{-\infty}^{\infty} f(t)\cos \omega t\, dt$$

$$- i\int_{-\infty}^{\infty} f(t)\sin \omega t\, dt = \mathrm{F_R}(\omega) + - i\mathrm{F_I}(\omega)$$

Equating real and imaginary parts, we have

$$\mathrm{F_R}(\omega) = \int_{-\infty}^{\infty} f(t)\cos \omega t\, dt \qquad (10)$$

$$\mathrm{F_I}(\omega) = -\int_{-\infty}^{\infty} f(t)\sin \omega t\, dt \qquad (11)$$

Further, since f is real, we note that

$$\mathrm{F_R}(-\omega) = \int_{-\infty}^{\infty} f(t)\cos(-\omega t)\, dt = \int_{-\infty}^{\infty} f(t)\cos \omega t\, dt$$

$$= \mathrm{F_R}(\omega), \qquad (12)$$

proving that $\mathrm{F_R}(\omega)$ is *even*. Similarly, we can show that

$$\mathrm{F_I}(-\omega) = -F_I(\omega), \qquad (13)$$

which implies that $\mathrm{F_I}(\omega)$ is an odd function. Now, since

$$\mathrm{F}(\omega) = \mathrm{F_R}(\omega) + i\mathrm{F_I}(\omega),$$

the expression (12) and (13) imply

$$\mathrm{F}(-\omega) = \mathrm{F_R}(\omega) + i\,\mathrm{F_I}(-\omega) = \mathrm{F_R}(\omega) - i\mathrm{F_I}(\omega)$$

$$= \mathrm{F}^*(\omega) \text{ (conjugate of } \mathrm{F}(\omega)).$$

Theorem 7.5. (Uniqueness Theorem). Let f and g be absolutely integrable and piecewise smooth functions on $\mathbb{R}$ with Fourier transforms $F(\omega)$ and $G(\omega)$, respectively. If $F(\omega) = G(\omega)$, then $f =$ g for all t at which f and g are continuous.

Proof: Since f and g are continuous, Fourier integral theorem yields

$$f(t) = \frac{1}{2\pi}\int_{-\infty}^{\infty} F(\omega)\, e^{i\omega t}\, d\omega$$

and

$$g(t) = \frac{1}{2\pi}\int_{-\infty}^{\infty} G(\omega)\, e^{i\omega t}\, d\omega.$$

Since F $(\omega) = \mathrm{G}(\omega)$, it follows that $f = g$ for t at which f and g are continuous.

7.3 FOURIER COSINE AND SINE TRANSFORMS

Definition 7.2. Let f be an absolutely integrable function on $\mathbb{R}$. Then the function

$$\mathrm{F_c}(\omega) = \int_{0}^{\infty} f(t)\cos \omega t\, dt$$

is called *Fourier cosine transform* of f, while the function

$$\mathrm{F_s}(\omega) = \int_{0}^{\infty} f(t)\sin \omega t\, dt$$

is called *Fourier sine transform* of f.

If f is even, then $f(t)$ cos ωt is even and $f(t)$ sin ωt is odd. Therefore,

$$\mathrm{F}(\omega) = \int_{-\infty}^{\infty} f(t)\, e^{-i\omega t}\, dt$$

$$= \int_{-\infty}^{\infty} f(t)\,(\cos \omega t - i \sin \omega t)\,dt$$

$$= \int_{-\infty}^{\infty} f(t) \cos \omega t\,dt - \int_{-\infty}^{\infty} f(t) \sin \omega t\,dt$$

$$= 2\int_{0}^{\infty} f(t) \cos \omega t\,dt - 0 = 2F_c(\omega).$$

Similarly, if f is odd, then $F(\omega) = -2iF_s(\omega)$.

We now obtain versions of Fourier integral theorem for even and odd functions.

Theorem 7.6. (Fourier Integral Theorem for Even Functions). Let f be an even absolutely integrable piecewise smooth function on ℝ. Then

$$\frac{2}{\pi}\int_{0}^{\infty} F_c(\omega) \cos \omega t\,d\omega = \frac{1}{2}[f(t+) + f(t-)],$$

where

$$F_c(\omega) = \int_{0}^{\infty} f(t) \cos \omega t\,dt$$

is the Fourier cosine transform of f.

Proof: If f is even, then by scaling property (see Remark 7.4) $F(\omega)$ is also even. Thus, $F(\omega)\cos \omega t$ is even and we have for $T > 0$,

$$\int_{-T}^{T} F(\omega) \cos \omega t\,d\omega = 2\int_{0}^{T} F(\omega) \cos \omega t\,d\omega$$

and

$$\int_{-T}^{T} F(\omega) \sin \omega t\,d\omega = 0.$$

Hence, as a Cauchy principal value, we have

$$\int_{-\infty}^{\infty} F(\omega)e^{i\omega t}\,d\omega = 2\int_{0}^{\infty} F(\omega) \cos \omega t\,d\omega.$$

But, for even function,

$$F(\omega) = 2F_c(\omega).$$

Therefore,

$$\int_{-\infty}^{\infty} F(\omega)e^{i\omega t}\,d\omega = 4\int_{0}^{\infty} F_c(\omega) \cos \omega t\,d\omega.$$

Hence Fourier integral theorem for even functions takes the form

$$\frac{1}{2}[f(t+) + f(t-)] = \frac{1}{2\pi}\int_{-\infty}^{\infty} F(\omega)e^{i\omega t}d\omega$$

$$= \frac{4}{2\pi}\int_{0}^{\infty} F_c(\omega) \cos \omega t\,d\omega$$

$$= \frac{2}{\pi}\int_{0}^{\infty} F_c(\omega) \cos \omega t\,d\omega.$$

Theorem 7.7. (Fourier Integral Theorem for Odd Functions). Let f be an odd absolutely integrable piecewise smooth function on ℝ. Then

$$\frac{2}{\pi}\int_{0}^{\infty} F_s(\omega) \sin \omega t\,d\omega = \frac{1}{2}[f(t+) + f(t-)],$$

where $F_s(\omega)$ is Fourier sine transform of f.

Proof: If f is odd, $f(t) \sin \omega t$ is even and $f(t) \cos \omega t$ is odd. Therefore, as in Theorem 7.6, we have

$$\int_{-\infty}^{\infty} F(\omega) \sin \omega t\,d\omega = 2\int_{0}^{\infty} F(\omega) \sin \omega t\,dt,$$

$$\int_{-\infty}^{\infty} F(\omega) \cos \omega t\,d\omega = 0,$$

and so

$$\int_{-\infty}^{\infty} F(\omega)\,e^{i\omega t}d\omega = 2i\int_{0}^{\infty} F(\omega) \sin \omega t\,dt.$$

But for odd function,

$$F(\omega) = -2i\,F_s(\omega).$$

Therefore,

$$\int_{-\infty}^{\infty} F(\omega)\,e^{i\omega t}d\omega = 4\int_{0}^{\infty} F_s(\omega) \sin \omega t\,d\omega.$$

Hence the Fourier integral theorem for odd functions, takes the form

$$\frac{2}{\pi}\int_0^{\infty} F_s(\omega)\sin\omega t\, d\omega = \frac{1}{2}[f(t+)+f(t-)].$$

7.4 PROPERTIES OF FOURIER TRANSFORMS

The Fourier transform F{ $f(t)$} of a function f satisfies a large number of properties that are satisfied by Fourier series and Laplace transforms. We establish these properties in the form of the following theorems.

Theorem 7.8. (Linearity of Fourier Transform). Let $F(\omega)$ and $G(\omega)$ be Fourier transforms of functions f and g, respectively. Then $aF(\omega) + bG(\omega)$ is the Fourier transform of $af + bg$, $a, b \in \mathbb{C}$. (Thus Fourier transform of a linear combination of functions is a linear combination of Fourier transforms of those functions).

Proof: By linearity of integration, we have

$$F\{af(t)+bg(t)\} = \int_{-\infty}^{\infty} [af(t) + bg(t)]\, e^{-i\omega t}\, dt$$

$$= a\int_{-\infty}^{\infty} f(t)\, e^{-i\omega t}\, dt + b\int_{-\infty}^{\infty} g(t)\, e^{-i\omega t}\, dt$$

$$= aF(\omega) + bG(\omega).$$

Thus, Fourier transform is a linear transformation.

Theorem 7.9. (Scaling Property). Let $F(\omega)$ be the Fourier transform of a function f and $a \in \mathbb{R}$ with $a \neq 0$ Then

$$F\{f(at)\} = \frac{1}{|a|} F\left(\frac{\omega}{a}\right).$$

Proof: Assume first that $a > 0$. Substituting $at = u$, we have

$$F\{f(at)\} = \frac{1}{a}\int_{-\infty}^{\infty} f(u)\, e^{-i\frac{\omega}{a}u}\, du = \frac{1}{a} F\left(\frac{\omega}{a}\right).$$

Now let $a < 0$. Then the substitution $at = u$ yields

$$F\{f(at)\} = \frac{1}{a}\int_{\infty}^{-\infty} f(u)\, e^{-i\left(\frac{\omega}{a}\right)u}\, du$$

$$= \left(-\frac{1}{a}\right)\int_{-\infty}^{\infty} f(u)\, e^{-i\left(\frac{\omega}{a}\right)u}\, du = \frac{1}{|a|} F\left(\frac{\omega}{a}\right).$$

Remark 7.4. Taking $a = -1$, the scaling property of Fourier transform yields

$$F\{f(-t)\} = F\left(\frac{\omega}{-1}\right) = F(-\omega),$$

which is known as *time reversal.*

Further, if f is even, then $f(-t) = f(t)$ and so, from the time reversal property, we have

$$F(\omega) = F\{f(t)\} = F\{f(-t)\} = F(-\omega).$$

Hence, for *even function f, Fourier transforms of f is also even.*

Theorem 7.10. (Shifting in Time Domain). Let $F(\omega)$ be the Fourier transform of a function f Then for a fixed $a \in \mathbb{R}$, one has

$$F\{f(t-a)\} = e^{-i\omega a} F(\omega).$$

Proof: Substituting $t - a = u$, we get

$$F\{f(t-a)\} = \int_{-\infty}^{\infty} f(t-a)\, e^{-i\omega t}\, dt$$

$$= \int_{-\infty}^{\infty} f(u)\, e^{-i\omega(a+u)}\, du$$

$$= e^{-i\omega a}\int_{-\infty}^{\infty} f(u)\, e^{-i\omega u}\, du = e^{-i\omega a} F(\omega).$$

Thus, when a function is shifted in time domain through quantity a, then its spectrum (Fourier transform) is multiplied by the factor $e^{-i\omega a}$. Since $|e^{-i\omega a}| = 1$, this property *does not change the amplitude spectrum but only changes the phase spectrum*. That is why, $e^{-i\omega a}$ is called a *phase factor*.

Theorem 7.11. (Shifting in the Frequency Domain). Let $F(\omega)$ be the Fourier transform of a function f Then for $a \in \mathbb{R}$, one has

$$F\{e^{iat} f(t)\} = F(\omega - a).$$

Proof: We have

$$F\{e^{iat} f(t)\} = \int_{-\infty}^{\infty} [f(t)e^{iat}]\, e^{-i\omega t}\, dt$$

$$= \int_{-\infty}^{\infty} f(t)\, e^{-i(\omega - a)t}\, dt = \mathrm{F}(\omega - a).$$

Theorem 7.12. (Symmetry or Duality). Let $F(\omega)$ be the Fourier transform of f. Then

$$F\{F(t)\} = 2\pi f(-\omega).$$

Proof: By Fourier integral theorem, we have

$$f(t) = \frac{1}{2\pi}\int_{-\infty}^{\infty} F(\omega)\, e^{i\omega t}\, d\omega.$$

Changing t to $-t$ yields

$$f(-t) = \frac{1}{2\pi}\int_{-\infty}^{\infty} F(\omega)\, e^{-i\omega t}\, d\omega.$$

Now interchanging t and ω, we have

$$f(-\omega) = \frac{1}{2\pi}\int_{-\infty}^{\infty} F(t)\, e^{-i\omega t}\, dt = \frac{1}{2\pi}\mathrm{F}\{\mathrm{F}(t)\}$$

and so

$$\mathrm{F}\{\mathrm{F}(t)\} = 2\pi f(-\omega).$$

Theorem 7.13. (Self Duality). Let f and g be piecewise smooth and absolutely integrable functions with Fourier transforms $F(\omega)$ and $G(\omega)$, respectively.
Then

$$\int_{-\infty}^{\infty} f(x)\, G(x)\, dx = \int_{-\infty}^{\infty} F(x)\, g(x)\, dx.$$

Proof: Changing the order of integration (permissible by the hypothesis of the theorem), we have

$$\int_{-\infty}^{\infty} f(x)\, G(x)\, dx = \int_{-\infty}^{\infty} f(x)\left(\int_{-\infty}^{\infty} g(y)\, e^{-ixy}\, dy\right)dx$$

$$= \int_{-\infty}^{\infty}\int_{-\infty}^{\infty} f(x)\, g(y)\, e^{-ixy}\, dy\, dx$$

$$= \int_{-\infty}^{\infty} g(y)\left(\int_{-\infty}^{\infty} f(x)\, e^{-ixy}\, dx\right) dy$$

$$= \int_{-\infty}^{\infty} g(y)\, F(y)\, dy$$

$$= \int_{-\infty}^{\infty} g(x)\, F(x)\, dx,$$

replacing the dummy variable y by x.

Theorem 7.14. (Differentiation in Time Domain). Let f be a continuously differentiable function with Fourier transform $F(\omega)$ and let $f(t) \to 0$ as $t \to \pm\infty$.

Then the Fourier transform of f' exists and

$$F\{f'(t)\} = i\omega\, F(\omega).$$

In general, if f is n times continuously differentiable and $\lim_{t\to\pm\infty} f^{(k)}(t) = 0$ for each $k = 0, 1, 2, \ldots, n-1$, then

$$F\{f^{(n)}(t) = (i\omega)^n\, F(\omega).$$

Proof: Since f' is continuous, integration by parts yields

$$\lim_{\substack{A\to-\infty\\ B\to\infty}} \int_A^B f'(t)\, e^{-i\omega t}\, dt = \lim_{\substack{A\to-\infty\\ B\to\infty}} \left[f(t)\, e^{-i\omega t}\right]_A^B$$

$$+ \lim_{\substack{A\to-\infty\\ B\to\infty}} i\omega \int_A^B f(t)\, e^{-i\omega t}\, dt$$

$$= \lim_{B\to\infty} f(B)\, e^{-i\omega B} - \lim_{A\to-\infty} f(A)\, e^{-i\omega A}$$

$$+ i\omega \int_{-\infty}^{\infty} f(t)\, e^{-i\omega t}\, dt$$

$$= \lim_{B\to\infty} f(B)\, e^{-i\omega B} - \lim_{A\to-\infty} f(A)\, e^{-i\omega A} + i\omega F(\omega),$$

provided that F(ω) exists. Since $f(t) \to 0$ as $t \to \pm\infty$,

$$\lim_{B\to\infty} f(B)\, e^{-i\omega B} = 0, \quad \lim_{A\to-\infty} f(A)\, e^{-i\omega A} = 0.$$

Hence

$$F\{f'(t)\} = \int_{-\infty}^{\infty} f'(t)\, e^{-i\omega t}\, dt = i\omega\, F(\omega).$$

Thus, differentiation in the time domain corresponds to multiplication of the Fourier transform by $i\omega$, provided that $\lim_{t\to\pm\infty} f(t) = 0$.

Applying the above-derived result repeatedly n times, we have

$$\begin{aligned} F\{f^{(n)}(t)\} &= i\omega F\{f^{(n-1)}(t)\} = (i\omega)^2 F\{f^{(n-2)}(t)\} \\ &= \ldots = (i\omega)^n F\{f(t)\}. \end{aligned}$$

Remark 7.5. The above expression does not guarantee the existence of the Fourier transform of $f^{(n)}(t)$, it only indicates that if the Fourier transform exists, then it is given by $(i\omega)^n$ F(ω).

Theorem 7.15. (Differentiation in the Frequency Domain). If f, tf, t^2f,.., t^nf are absolutely integrable and $F(\omega)$ is Fourier transform of f, then

$$\frac{d^n}{d\omega^n}(F(\omega)) = (-i)^n F\{t^n f(t)\},\ n = 1, 2, \ldots,$$

or equivalently,

$$F\{t^n f(t)\} = (-i)^n \frac{d^n}{d\omega^n} F(\omega).$$

Proof: From the definition of spectrum, we have

$$F(\omega) = \int_{-\infty}^{\infty} f(t)\, e^{-i\omega t}\, dt.$$

Differentiating under the integral sign, we obtain

$$\frac{d}{d\omega}(F(\omega)) = \int_{-\infty}^{\infty} f(t) \frac{\partial}{\partial\omega}\{e^{-i\omega t}\} dt$$

$$= \int_{-\infty}^{\infty} (-it) f(t)\, e^{-i\omega t}\, dt = (-i)\, F\{tf(t)\},$$

$$\frac{d^2}{d\omega^2}(F(\omega)) = \int_{-\infty}^{\infty} (-it)^2 f(t) e^{-i\omega t} dt = (-i)^2 F\{t^2 f(t)\},$$

and so on. In general,

$$\frac{d^n}{d\omega^n}(F(\omega)) = (-i)^n F\{t^n f(t)\},\ n = 1, 2, \ldots$$

Theorem 7.16. (Integration in Time Domain). Let f be a continuous and absolutely integrable function with Fourier transform $F(\omega)$. If

$\lim_{t\to\infty} \int_{-\infty}^{t} f(u)du = 0$ (or we may say $F(0) = 0$), *then for* $\omega \neq 0$, one has

$$F\left\{\int_{-\infty}^{t} f(u)\, du\right\} = \frac{F(\omega)}{i\omega}.$$

Proof: Consider the function

$$\Phi(t) = \int_{-\infty}^{t} f(u)\, du.$$

Since f is continuous, Φ is continuously differentiable function and $\Phi'(t) = f(t)$ by fundamental theorem of integral calculus. Since $\lim_{t\to\pm\infty} \Phi(t) = 0$, Theorem 7.14 implies

$$F\{f(t)\} = F\{\Phi'(t)\} = i\omega F\{\Phi(t)\},$$

and so

$$F\{\Phi(t)\} = \frac{1}{i\omega} F\{f(t)\} = \frac{1}{i\omega} F(\omega),$$

that is,

$$\mathrm{F}\left\{\int_{-\infty}^{t} f(u)\, du\right\} = \frac{F(\omega)}{i\omega}.$$

7.5 SOLVED EXAMPLES

EXAMPLE 7.1

Show that the *Heaviside's unit step function* H defined by

$$H(t) = \begin{cases} 1 & \text{for } t \geq 0 \\ 0 & \text{otherwise} \end{cases}$$

does not have Fourier transform.

Solution. We note that

$$\int_{-\infty}^{\infty} H(t)\, e^{-i\omega t}\, dt = \int_{0}^{\infty} e^{-i\omega t}\, dt = \lim_{T\to\infty} \int_{0}^{T} e^{-i\omega t}\, dt$$

$$= \lim_{T\to\infty} \frac{1}{-i\omega} [e^{-i\omega t}]_0^{T} = \frac{i}{\omega}(\lim_{T\to\infty} e^{-i\omega T} - 1).$$

Since $\lim_{T\to\infty} \sin T\omega$ and $\lim_{T\to\infty} \cos T\omega$ do not exist, the $\lim_{T\to\infty} e^{-i\omega T}$ does not exist. Hence, the given function *does not have the Fourier transform* (in terms of the definition of the transform, however (see Remark 7.3), its transform exists as a generalized function as shown in Example 7.12).

EXAMPLE 7.2.

Find the Fourier transform of the *gate function* f defined by

$$f(t) = \begin{cases} 1 & \text{for } |t| < a \\ 0 & \text{for } |t| > a \end{cases}$$

and hence evaluate

(a) $\displaystyle\int_{-\infty}^{\infty} \frac{\sin a\omega \cos \omega t}{\omega}\, d\omega$

(b) $\displaystyle\int_{0}^{\infty} \frac{\sin \omega}{\omega}\, d\omega.$

Solution. By the definition of spectrum, we have

$$\mathrm{F}(\omega) = \int_{-\infty}^{\infty} f(t)\, e^{-i\omega t}\, dt$$

$$= \int_{-\infty}^{-a} f(t)\, e^{-i\omega t}\, dt + \int_{-a}^{a} f(t)\, e^{-i\omega t}\, dt + \int_{a}^{\infty} f(t)\, e^{-i\omega t}\, dt$$

$$= \int_{\infty}^{a} -f(-u)\, e^{i\omega u}\, du + \int_{-a}^{a} f(t)\, e^{-i\omega t}\, dt + \int_{a}^{\infty} f(t)\, e^{-i\omega t}\, dt$$

$$= \int_{a}^{\infty} f(-u)\, e^{-i\omega u}\, du + \left[\frac{e^{-i\omega t}}{-i\omega}\right]_{-a}^{a} + 0$$

$$= 0 + \left[\frac{e^{i\omega a} - e^{-i\omega a}}{i\omega}\right] + 0 = \frac{2\sin \omega a}{\omega}, \omega \neq 0.$$

For $\omega = 0$, we have

$$\mathrm{F}(\omega) = \int_{-\infty}^{\infty} f(t)\, dt = \int_{-a}^{a} dt = 2a.$$

Deductions: (a) By Inversion formula (9),

$$\frac{1}{2\pi}\int_{-\infty}^{\infty} \mathrm{F}(\omega)\, e^{i\omega t}\, d\omega = f(t)$$

and so

$$\frac{1}{2\pi}\int_{-\infty}^{\infty} e^{i\omega t}\left(\frac{2\sin \omega a}{\omega}\right) d\omega = \begin{cases} 1 & \text{for } |t| < a \\ 0 & \text{for } |t| > a, \end{cases}$$

which gives

$$\int_{-\infty}^{\infty} e^{i\omega t}\left(\frac{\sin \omega a}{\omega}\right) d\omega = \begin{cases} \pi & \text{for} |t| < a \\ 0 & \text{for } |t| > a. \end{cases} \quad (14)$$

Since $e^{i\omega t} = \cos \omega t + i \sin \omega t$, equating real part in (14), we get

$$\int_{-\infty}^{\infty} \frac{\cos \omega t \sin \omega a}{\omega} d\omega = \begin{cases} \pi & \text{for } |t| < a \\ 0 & \text{for } |t| > a. \end{cases}$$

(b) Putting $t = 0$ and $a = 1$ in the deduction (a), we get

$$\int_{-\infty}^{\infty} \frac{\sin \omega}{\omega} d\omega = \pi$$

and so

$$\int_{0}^{\infty} \frac{\sin \omega}{\omega} d\omega = \frac{\pi}{2}.$$

EXAMPLE 7.3

Find the Fourier transform of f defined by

$$f(t) = \begin{cases} e^{-at} & \text{for } t > 0 \\ 0 & \text{for } t < 0, \end{cases} \quad a > 0.$$

(In terms of Heaviside's unit step function we can write $f(t) = e^{-at}\text{H}(t)$)

Solution. By definition,

$$\text{F}(\omega) = \int_{-\infty}^{\infty} f(t)\, e^{-i\omega t}\, dt = \int_{0}^{\infty} e^{-(a+i\omega)t}\, dt$$

$$= \left[\frac{1}{-(a+i\omega)}\, e^{-(a+i\omega)t}\right]_0^{\infty} = \frac{1}{a+i\omega}.$$

EXAMPLE 7.4

Find the Fourier transform of

$$f_a(t) = e^{-at}\text{H}(t) - e^{at}\text{H}(-t),\ a > 0,$$

and hence find the Fourier transform of *signum function* defined by

$$\text{sgn}(t) = \begin{cases} 1 & \text{for } t > 0 \\ -1 & \text{for } t < 0. \end{cases}$$

Solution. The graph of the function $f_a(t)$ is shown in the Figure 7.1.

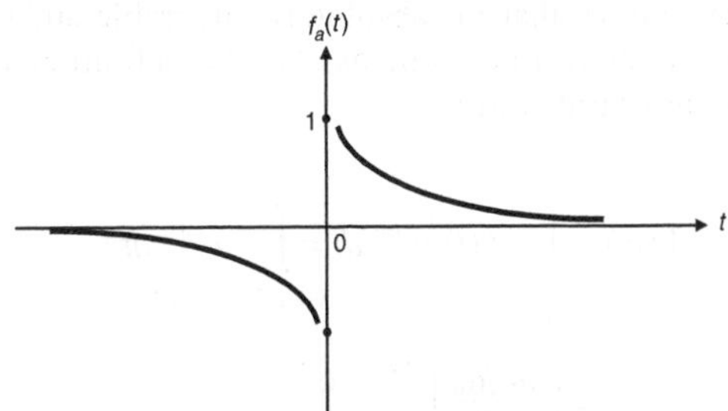

Figure 7.1

We have

$$\text{F}(\omega) = \int_{-\infty}^{\infty} f_a(t) e^{-i\omega t}\, dt$$

$$= -\int_{-\infty}^{0} e^{(a-i\omega)t}\, dt + \int_{0}^{\infty} e^{-(a+i\omega)t}\, dt$$

$$= \left[\frac{1}{a+i\omega} - \frac{1}{a-i\omega}\right] = -\frac{2i\omega}{a^2+\omega^2}.$$

Letting $a \to 0$, we have $f_a(t) \to \text{sgn}\, t$ and so

$$\text{F}\{\text{sgn}(t)\} = \lim_{a\to 0} -\left(\frac{2i\omega}{a^2+\omega^2}\right) = \frac{2}{i\omega},$$

or equivalently,

$$\text{F}\{i\, \text{sgn}(t)\} = \frac{2}{\omega}.$$

EXAMPLE 7.5

Find the Fourier transform of *Block function (rectangular pulse function)* $f(t)$ of height 1 and duration a defined by

$$f(t) = \begin{cases} 1 & \text{for} |t| \le \frac{a}{2} \\ 0 & \text{otherwise.} \end{cases}$$

Solution. The graph of the function f is shown in the Figure 7.2.

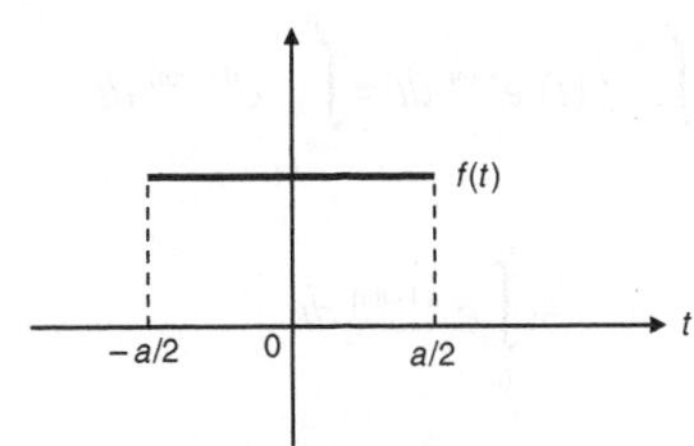

Figure 7.2

We notice that f is absolutely integrable and so its Fourier transform exists. By definition of Fourier transform,

$$F(\omega) = \int_{-\infty}^{\infty} f(t)\, e^{-i\omega t}\, dt = \int_{-a/2}^{a/2} e^{-i\omega t}\, dt$$

$$= \left[\frac{e^{-i\omega t}}{-i\omega}\right]_{-a/2}^{a/2}$$

$$= \frac{e^{ia\omega/2} - e^{-ia\omega/2}}{i\omega} = \frac{2\sin(a\omega/2)}{\omega},\ \omega \neq 0,$$

whereas, for $\omega = 0$, we have

$$F(0) = \int_{-\infty}^{\infty} f(t)\, dt = \int_{-a/2}^{a/2} dt = a.$$

EXAMPLE 7.6

Find the Fourier transform of f defined by $f(t) = e^{-|t|}$.

Solution. The graph of the function is shown in the Figure 7.3.

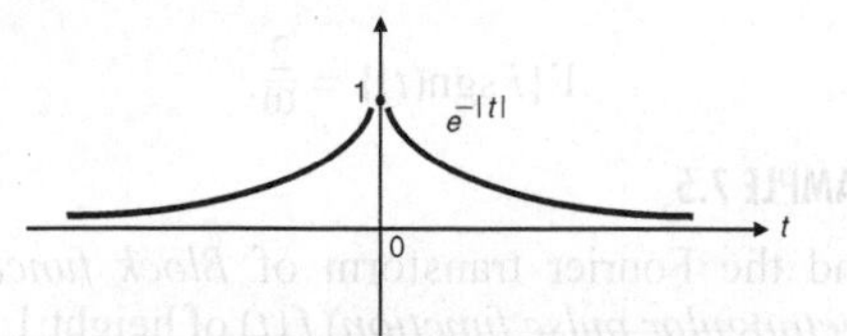

Figure 7.3

Since

$$f(t) = \begin{cases} e^{t} & \text{for } t < 0 \\ e^{-t} & \text{for } t \geq 0, \end{cases}$$

the definition of Fourier transform yields

$$F(\omega) = \int_{-\infty}^{\infty} f(t)\, e^{-i\omega t}\, dt = \int_{-\infty}^{0} e^{t(1-i\omega)}\, dt$$

$$+ \int_{0}^{\infty} e^{-t(1+i\omega)}\, dt$$

$$= \frac{1}{1-i\omega} + \frac{1}{1+i\omega} = \frac{2}{1+\omega^2}.$$

EXAMPLE 7.7

Find the Fourier transform of the *triangle function* defined for $a > 0$ by

$$f(t) = \begin{cases} 1 - \frac{|t|}{a} & \text{for} |t| \leq a \\ 0 & \text{otherwise.} \end{cases}$$

Solution. The graph of the function is shown in the Figure 7.4

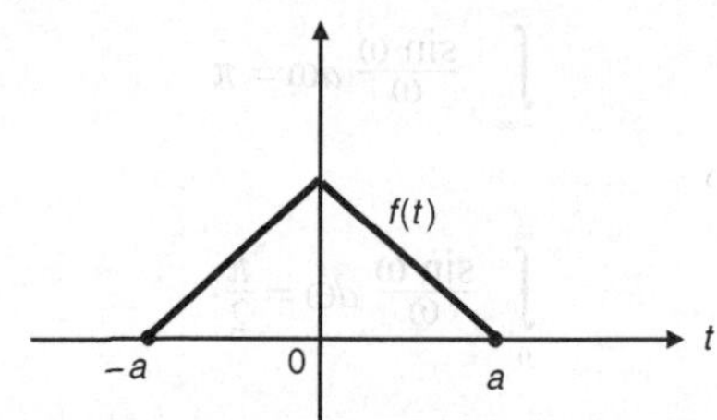

Figure 7.4

We note that f is absolutely integrable and

$$f(-t) = f(t). \tag{15}$$

By definition of Fourier transform, we have

$$F(\omega) = \int_{-\infty}^{\infty} f(t)\, e^{-i\omega t}\, dt$$

$$= \int_{0}^{\infty} f(t)\, e^{-i\omega t}\, dt + \int_{-\infty}^{0} f(t)\, e^{-i\omega t}\, dt$$

$$= \int_{0}^{\infty} f(t)\, e^{-i\omega t}\, dt + \int_{0}^{\infty} f(t)\, e^{i\omega t}\, dt \quad \text{by (15)}$$

$$= \int_{0}^{\infty} f(t)\, [e^{-i\omega t} + e^{i\omega t}] = 2\int_{0}^{\infty} f(t) \cos \omega t\, dt$$

$$= 2\int_{0}^{a} \left(1 - \frac{t}{a}\right) \cos \omega t\, dt$$

$$= \left[2\left(1 - \frac{t}{a}\right)\frac{\sin \omega t}{\omega}\right]_{0}^{a} + \frac{2}{a}\int_{0}^{a} \frac{\sin \omega t}{\omega}\, dt$$

$$= \frac{2}{a\omega}\left[-\frac{\cos \omega t}{\omega}\right]_0^a = \frac{2}{a\omega^2}[1 - \cos a\omega]$$

$$= \frac{4\sin^2(a\omega/2)}{a\omega^2}, \ \omega \neq 0.$$

For $\omega = 0$, $\cos \omega t = 1$, so we have

$$F(0) = 2\int_0^a \left(1 - \frac{t}{a}\right) dt = a.$$

EXAMPLE 7.8

Find the Fourier transform of *Gauss-function* f defined for $a > 0$ by

$$f(t) = e^{-at^2}.$$

Solution. The graph of the Gauss function is shown in the Figure 7.5

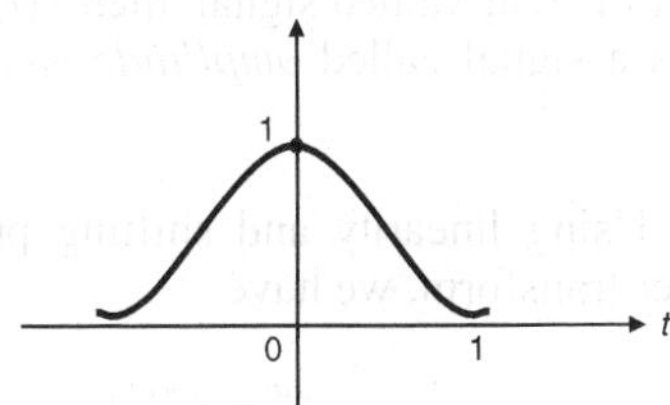

Figure 7.5

Since $f(t) = f(-t)$, we have

$$F(\omega) = \int_{-\infty}^{\infty} e^{-at^2} e^{-i\omega t} dt = 2\int_0^{\infty} e^{-at^2} \cos \omega t \, dt. \quad (16)$$

Differentiating (16) with respect to ω, we have

$$F'(\omega) = -2\int_0^{\infty} t \sin \omega t \, e^{-at^2} dt$$

$$= \frac{1}{a}\int_0^{\infty} (e^{-at^2})' \sin \omega t \, dt,$$

$$\text{since } (e^{-at^2})' = -2a\, e^{-at^2}$$

$$= \frac{1}{a}[e^{-at^2} \sin \omega t]_0^{\infty}$$

$$- \frac{\omega}{a}\int_0^{\infty} e^{-at^2} \cos \omega t \, dt = -\frac{\omega}{2a} F(\omega).$$

Dividing throughout by $F(\omega)$, we have

$$\frac{F'(\omega)}{F(\omega)} = -\frac{\omega}{2a}.$$

Integrating, we get

$$\log F(\omega) = -\frac{\omega^2}{4a} + C \text{ (constantof integration)}$$

and so

$$F(\omega) = e^C e^{-\frac{\omega^2}{4a}} = A\, e^{-\frac{\omega^2}{4a}}$$

Also, if we put $\omega = 0$, then $A = F(0)$. But

$$F(0) = \int_{-\infty}^{\infty} e^{-at^2} dt = \frac{1}{\sqrt{a}}\int_{-\infty}^{\infty} e^{-u^2} du, \ t = \frac{u}{\sqrt{a}}$$

$$= \frac{2}{\sqrt{a}}\int_0^{\infty} e^{-u^2} du = \frac{2}{\sqrt{a}} \cdot \frac{\sqrt{\pi}}{2} = \frac{\sqrt{\pi}}{a}.$$

Hence

$$F(\omega) = \frac{\sqrt{\pi}}{a} \cdot e^{-\frac{\omega^2}{4a}}.$$

EXAMPLE 7.9

Find Fourier transform of $f(t) = e^{-t^2/2}$.

Solution. This is a particular case of Example 7.8 for $a = \frac{1}{2}$. Therefore,

$$F(\omega) = \sqrt{2\pi}\, e^{-\frac{\omega^2}{2}}.$$

EXAMPLE 7.10

Find the Fourier transform of Dirac delta function $\delta(x)$ defined by

$$\delta(x) = 0, \ x \neq 0, \text{ and } \int_{-\infty}^{\infty} \delta(x)\, dx = 1.$$

Solution. The Dirac delta function is not a function in the classical sense but a function in the generalized sense. Thus, it is a *generalized* *function* or a *distribution*. Since for $t \neq 0$, $\delta(t) = 0$ and for $t = 0$, $e^{-i\omega t} = 1$, we have

$$F(\omega) = \int_{-\infty}^{\infty} \delta(t)\, e^{-i\omega t}\, dt = \int_{-\infty}^{\infty} \delta(t)\, dt = 1.$$

EXAMPLE 7.11

Show the Fourier transform of 1 is $2\pi\delta(\omega)$.

Solution. Since (by Example 7.10), $F\{\delta(t)\} = 1$, by inversion formula, we have

$$\delta(t) = \frac{1}{2\pi}\int_{-\infty}^{\infty} e^{i\omega t}\, d\omega$$

(called integral representation of delta function). Interchanging t and ω, we have

$$\delta(\omega) = \frac{1}{2\pi}\int_{-\infty}^{\infty} e^{i\omega t}\, dt.$$

Since delta function is even, we have

$$\delta(\omega) = \delta(-\omega) = \frac{1}{2\pi}\int_{-\infty}^{\infty} e^{-i\omega t}\, dt = \frac{1}{2\pi} F\{1\}.$$

Hence

$$F\{1\} = 2\pi\delta(\omega).$$

EXAMPLE 7.12

Find the Fourier transform of Heaviside's unit step function $H(t)$.

(As pointed out in Example 7.1, the Fourier transform of this function can be found only by using generalized function Dirac delta)

Solution. We take help of signum function, defined by

$$\text{sgn}(t) = \begin{cases} -1 & \text{for } t < 0 \\ 1 & \text{for } t > 0. \end{cases}$$

Then,

$$H(t) = \frac{1}{2}[1 + \text{sgn}(t)].$$

Therefore, by linearity, Examples 7.4 and 7.11, we have

$$F\{H(t)\} = \frac{1}{2} F\{1\} + \frac{1}{2} F\{\text{sgn}(t)\} = \frac{1}{2} 2\pi\delta(\omega) + \frac{2}{i\omega}$$

$$= \pi\delta(\omega) + \frac{1}{i\omega} = \pi\left[\delta(\omega) + \frac{1}{i\pi\omega}\right].$$

EXAMPLE 7.13

If $F(\omega)$ is the Fourier transform of a function f, then show that

$$F\{f(t)\cos at\} = \frac{F(\omega - a)}{2} + \frac{F(\omega + a)}{2}.$$

(This result is known as *Modulation theorem*. In fact, if f is real-valued signal, then $f(t)\cos at$ describes a signal called *amplitude modulated signal*).

Solution. Using linearity and shifting property of Fourier transform, we have

$$F\{f(t)\cos at\} = F\left\{f(t)\,\frac{e^{iat} + e^{-iat}}{2}\right\}$$

$$= \frac{1}{2} F\{f(t)\, e^{iat}\} + \frac{1}{2} F\{f(t)\, e^{-iat}\}$$

$$= \frac{1}{2} F(\omega - a) + \frac{1}{2} F(\omega + a)$$

$$= \frac{1}{2}[F(\omega - a) + F(\omega + a)].$$

EXAMPLE 7.14

Using Modulation theorem, find the Fourier transform of $f(t)\cos bt$, where f is defined by

$$f(t) = \begin{cases} 1 & \text{for } |t| < a \\ 0 & \text{for } |t| > a. \end{cases}$$

Solution. If $F(\omega)$ is the Fourier transform of f, then Modulation theorem asserts that

$$F\{f(t)\cos bt\} = \frac{F(\omega - b)}{2} + \frac{F(\omega + b)}{2} \qquad (17)$$

We know (see Example 7.5) that the Fourier transform of the function f is

$$F(\omega) = \begin{cases} \frac{2\sin \omega a}{\omega} & \text{for } \omega \neq 0 \\ 2a & \text{for } \omega = 0. \end{cases}$$

Therefore, (17) yields

$$F\{f(t)\cos bt\} = \begin{cases} \frac{\sin(\omega-b)\,a}{\omega-b} + \frac{\sin(\omega+b)a}{\omega+b}. \\ 2a \end{cases}$$

EXAMPLE 7.15

Find the Fourier transform of $e^{-a|t|}$, $a > 0$.

Solution. By definition

$$F(\omega) = \int_{-\infty}^{\infty} e^{-a|t|}\, e^{-i\omega t}\, dt.$$

But

$$e^{-a|t|} = \begin{cases} e^{at} & \text{for } t < 0 \\ e^{-at} & \text{for } t \geq 0. \end{cases}$$

Therefore,

$$F(\omega) = \int_{-\infty}^{0} e^{at}.e^{-i\omega t}\, dt + \int_{0}^{\infty} e^{-at}\, e^{-i\omega t}\, dt$$

$$= \int_{-\infty}^{0} e^{(a-i\omega)t}\, dt + \int_{0}^{\infty} e^{-(a+i\omega)t}\, dt$$

$$= \left(\frac{1}{a-i\omega} + \frac{1}{a+i\omega}\right) = \frac{2a}{a^2+\omega^2}.$$

Second Method: Since $e^{-a|t|}$ is even, we have

$$F(\omega) = 2F_c(\omega) = 2\int_{0}^{\infty} f(t)\cos \omega t\, dt$$

$$= 2\int_{0}^{\infty} e^{-at}\cos \omega t\, dt$$

$$= \frac{2a}{a^2+\omega^2} \text{ (on integrating by parts).}$$

EXAMPLE 7.16

Find Fourier transform of $f(t) = t\,e^{-a|t|}$, $a > 0$.

Solution. By Example 7.15,

$$F\{e^{-a|t|}\} = \frac{2a}{a^2+\omega^2}.$$

Therefore, by Theorem 18.15,

$$F\{t\,e^{-a|t|}\} = (-i)\frac{d}{d\omega}\left(\frac{2a}{a^2+\omega^2}\right) = \frac{4ia\omega}{(a^2+\omega^2)^2}.$$

EXAMPLE 7.17

Find Fourier sine transform of $f(t) = t\,e^{-at}$, $a > 0$.

Solution. We have

$$F_s(\omega) = \int t\,e^{-at}\sin \omega t\, dt.$$

But we know that

$$\int_{0}^{\infty} e^{-at}\cos \omega t\, dt = \frac{a}{a^2+\omega^2}.$$

Differentiating both sides with respect to ω, we get

$$\int_{0}^{\infty} t\,e^{-at}\sin \omega t\, dt = \frac{2a\omega}{(a^2+\omega^2)^2}$$

and so

$$F_s(\omega) = \frac{2a\omega}{(a^2+\omega^2)^2}.$$

EXAMPLE 7.18

Find the Fourier sine transform of $f(t) = 1/t\; e^{-at}$. Deduce that $F\left\{\frac{1}{t}\right\} = -i\pi$.

Solution. Integration by parts yields

$$\int_{0}^{\infty} e^{-at}\sin \omega t\, dt = \frac{\omega}{a^2+\omega^2}.$$

Integrating both sides with respect to a, we have

$$\int_0^\infty \frac{1}{t} e^{-at} \sin \omega t \, dt = \int_a^\infty \frac{\omega}{a^2+\omega^2} \, da$$

$$= \frac{\pi}{2} - \tan^{-1}\frac{a}{\omega} = \tan^{-1}\frac{\omega}{a}.$$

Thus

$$F_s(\omega) = \tan^{-1}\frac{\omega}{a}.$$

If $a \to 0$, then

$$F_s\left\{\frac{1}{t}\right\} = \frac{\pi}{2},$$

which implies that

$$F\left\{\frac{1}{t}\right\} = -2i\, F_s\left\{\frac{1}{t}\right\} = -i\pi.$$

EXAMPLE 7.19

Find Fourier cosine and Fourier sine transforms of the function f defined by

$$f(t) = e^{-at}, \ a \text{ is a constant.}$$

Deduce the value of

$$\int_0^\infty \frac{\cos nx}{a^2+x^2} dx \text{ and } \int_0^\infty \frac{x \sin nx}{a^2+x^2} dx.$$

Solution. We have

$$F(\omega) = F_C(\omega) + i\, F_s(\omega) = \int_0^\infty e^{-at} e^{-i\omega t} dt$$

$$= \int_0^\infty e^{-(a+i\omega)t} dt = \left[\frac{e^{-(a+i\omega)t}}{-(a+i\omega)}\right]_0^\infty$$

$$= \frac{1}{(a+i\omega)} = \frac{a-i\omega}{a^2+\omega^2}.$$

Hence

$$F_C(\omega) = \frac{a}{a^2+\omega^2} \text{ and } F_s(\omega) = \frac{-\omega}{a^2+\omega^2}.$$

Using Fourier integral theorem for cosine and sine transforms, we have

$$\frac{2}{\pi}\int_0^\infty F_C(\omega)\cos \omega t \, d\omega = e^{-at}$$

and

$$\frac{2}{\pi}\int_0^\infty F_s(\omega)\sin \omega t \, d\omega = e^{-at},$$

which, respectively, yield

$$\int_0^\infty \frac{1}{a^2+\omega^2}\cos \omega t \, d\omega = \frac{\pi}{2a} e^{-at}$$

and

$$\int_0^\infty \frac{\omega \sin \omega t}{a^2+\omega^2} d\omega = -\frac{\pi}{2} e^{-at}.$$

Changing ω to x and t to n, we have

$$\int_0^\infty \frac{\cos nx}{a^2+x^2} dx = \frac{\pi}{2a} e^{-na}$$

and

$$\int_0^\infty \frac{x \sin nx}{a^2+x^2} dx = -\frac{\pi}{2} e^{-na}.$$

EXAMPLE 7.20

Solve the integral equation

$$\int_0^\infty f(t)\cos at \, dt = e^{-a}.$$

Solution. By definition of Fourier cosine transform,

$$F_C(\omega) = \int_0^\infty f(t)\cos \omega t \, dt.$$

Therefore, the given equation reduces to

$$F_C(\omega) = e^{-\omega}.$$

Using Fourier integral theorem for Fourier cosine transform, we have

$$f(t) = \frac{2}{\pi}\int_0^\infty F_c(\omega)\cos\omega t\, d\omega$$

$$= \frac{2}{\pi}\int_0^\infty e^{-\omega}\cos\omega t\, d\omega$$

$$= \frac{2}{\pi}\left(\frac{1}{1+t^2}\right),$$

$$\text{since} \int_0^\infty e^{-ax}\cos bx\, dx = \frac{a}{a^2+b^2}.$$

EXAMPLE 7.21

Solve the integral equation

$$\int_0^\infty f(t)\sin\omega t\, dt = \begin{cases} 1 & \text{for } 0 \le t < 1 \\ 2 & \text{for } 1 \le t < 2 \\ 0 & \text{for } t \ge 2. \end{cases}$$

Solution. The given integral equation can be written in terms of Fourier sine transform as

$$F_s(\omega) = \begin{cases} 1 & \text{for } 0 \le t < 1 \\ 2 & \text{for } 1 \le t < 2 \\ 0 & \text{for } t \ge 2. \end{cases}$$

Using Fourier integral theorem for Fourier sine transform, we get

$$f(t) = \frac{2}{\pi}\int_0^\infty F_s(\omega)\sin\omega t\, d\omega$$

$$= \frac{2}{\pi}\left[\int_0^1 \sin\omega t\, d\omega + \int_1^2 \sin\omega t\, d\omega\right]$$

$$= \frac{2}{\pi}\left[-\frac{\cos\omega t}{t}\right]_0^1 + \frac{4}{\pi}\left[-\frac{\cos\omega t}{t}\right]_1^2$$

$$= \frac{2}{\pi t}[-\cos t + 1] + \frac{4}{\pi t}[-\cos 2t + \cos t]$$

$$= \frac{2}{\pi t}(1 - 2\cos 2t + \cos t).$$

EXAMPLE 7.22

Solve the integral equation

$$\int_0^\infty f(t)\cos\omega t\, d\omega = \begin{cases} 1-\omega & \text{for } 0 \le \omega \le 1 \\ 0 & \text{for } \omega > 1. \end{cases}$$

Hence evaluate $\int_0^\infty \frac{\sin^2 t}{t^2}\, dt.$

Solution. We have

$$\int_0^\infty f(t)\cos\omega t\, d\omega = F_c(\omega)$$

$$= \begin{cases} 1-\omega & \text{for } 0 \le \omega \le 1 \\ 0 & \text{for } \omega > 1. \end{cases}$$

By Inversion theorem for Fourier cosine transforms, we have

$$f(t) = \frac{2}{\pi}\int_0^\infty F_c(\omega)\cos\omega t\, d\omega$$

$$= \frac{2}{\pi}\int_0^1 (1-\omega)\cos\omega t\, d\omega$$

$$= \frac{2}{\pi}\left[(1-\omega)\frac{\sin\omega t}{t}\right]_0^1$$

$$-\frac{2}{\pi}\int_0^1 (-1)\frac{\sin\omega t}{t}\, d\omega$$

$$= \frac{2}{t\pi}\left[-\frac{\cos\omega t}{t}\right]_0^1 = \frac{2}{t^2\pi}[1-\cos t].$$

Then

$$F_c(\omega) = \frac{2}{\pi}\int_0^\infty \frac{1-\cos t}{t^2}\cos\omega t\, dt$$

$$= \begin{cases} 1-\omega & \text{for} \quad 0 \le \omega \le 1 \\ 0 & \text{for} \quad \omega > 1. \end{cases}$$

If $\omega \to 0$, then

$$\frac{2}{\pi}\int_0^\infty \frac{1-\cos t}{t^2}\, dt = 1,$$

that is,

$$\int_0^\infty \frac{2\sin^2 t/2}{t^2}\, dt = \frac{\pi}{2}.$$

Substituting $\frac{t}{2} = u$, we get

$$\int_0^\infty \frac{\sin^2 u}{u^2}\, du = \frac{\pi}{2}.$$

Changing the dummy variable, we get

$$\int_0^\infty \frac{\sin^2 t}{t^2}\, dt = \frac{\pi}{2}.$$

EXAMPLE 7.23

Find $f(t)$ if $F_s(\omega) = \frac{\omega}{1+\omega^2}$.

Solution. By Fourier integral theorem for Fourier sine transform,

$$f(t) = \frac{2}{\pi}\int_0^\infty F_s(\omega)\sin\omega t\, d\omega$$

$$= \frac{2}{\pi}\int_0^\infty \frac{\omega}{1+\omega^2}\sin\omega t\, d\omega$$

$$= \frac{2}{\pi}\int_0^\infty \frac{\omega^2}{\omega(1+\omega^2)}\sin\omega t\, d\omega$$

$$= \frac{2}{\pi}\int_0^\infty \frac{(1+\omega^2)-1}{\omega(1+\omega^2)}\sin\omega t\, d\omega$$

$$= \frac{2}{\pi}\int_0^\infty \frac{\sin\omega t}{\omega}\, d\omega - \frac{2}{\pi}\int_0^\infty \frac{\sin\omega t}{\omega(1+\omega^2)}\, d\omega$$

$$= \frac{2}{\pi}\cdot\frac{\pi}{2} - \frac{2}{\pi}\int_0^\infty \frac{\sin\omega t}{\omega(1+\omega^2)}\, d\omega$$

$$= 1 - \frac{2}{\pi}\int_0^\infty \frac{\sin\omega t}{\omega(1+\omega^2)}\, d\omega. \tag{18}$$

Therefore,

$$\frac{df}{dt} = -\frac{2}{\pi}\int_0^\infty \frac{\partial}{\partial t}\left(\frac{\sin\omega t}{\omega(1+\omega^2)}\right) d\omega$$

$$= -\frac{2}{\pi}\int_0^\infty \frac{\cos\omega t}{(1+\omega^2)}\, d\omega \tag{19}$$

and

$$\frac{d^2 f}{dt^2} = -\frac{2}{\pi}\int_0^\infty \left[-\frac{\omega\sin\omega t}{1+\omega^2}\right] d\omega$$

$$= \frac{2}{\pi}\int_0^\infty \frac{\omega\sin\omega t}{1+\omega^2}\, d\omega = f(t).$$

Thus, we get a differential equation

$$\frac{d^2 f}{dt^2} - f(t) = 0.$$

The roots of the characteristic equation $s^2 - 1 = 0$ of this equation are $s = \pm 1$. Therefore, the fundamental solution is

$$f(t) = c_1 e^t + c_2 e^{-t} \tag{20}$$

Then

$$f'(t) = c_1 e^t - c_2 e^{-t} \tag{21}$$

Putting $t = 0$ in (18), we get $f(0) = 1$. Therefore, (20) yields

$$1 = c_1 + c_2 \tag{22}$$

From (19), we have

$$f'(0) = -\frac{2}{\pi}\int_0^\infty \frac{d\omega}{1+\omega^2} = -\frac{2}{\pi}[\tan^{-1}\omega]_0^\infty$$

$$= -\frac{2}{\pi}\cdot\frac{\pi}{2} = -1.$$

From (21), we have

$$f'(0) = c_1 - c_2 = -1 \tag{23}$$

Now (22) and (23) yields

$$c_1 = 0,\ c_2 = 1.$$

Hence (20) yields

$$f(t) = e^{-t}.$$

7.6 COMPLEX FOURIER TRANSFORMS

Let a complex valued function $f(t)$ be continuous and have a piecewise continuous derivative in any finite interval. Suppose, further, that $g(t) = e^{yt} f(t)$ is absolutely integrable for some y. Then by Fourier inversion theorem, we have

$$g(t) = \frac{1}{2\pi}\int_{-\infty}^{\infty} e^{ixt}\int_{-\infty}^{\infty} g(u)\, e^{-ixu}\, du\, dx,$$

that is,

$$e^{yt} f(t) = \frac{1}{2\pi}\int_{-\infty}^{\infty} e^{ixt}\int_{-\infty}^{\infty} f(u)\, e^{-iu\,(x+iy)}\, du\, dx.$$

and so

$$f(t) = \frac{1}{2\pi}\int_{-\infty}^{\infty} e^{it(x+iy)}\int_{-\infty}^{\infty} f(u)\, e^{-iu(x+iy)}\, du\, dx$$

$$= \frac{1}{2\pi}\int_{-\infty+iy}^{\infty+iy} e^{itz}\int_{-\infty}^{\infty} f(u)\, e^{-iuz}\, du\, dz,$$

where the integral in the z-plane is taken along a straight line $x + iy$, y fixed and $-\infty < x < \infty$ such that $f(t)\, e^{yt}$ is absolutely integrable. More generally, we have the following theorem.

Theorem 7.17. (Complex form of Fourier Integral Theorem). Let $f(t)$ be a complex valued function which is piecewise smooth in any finite interval. Let $f(t)\, e^{\gamma t}$ be absolutely integrable for some real γ. Then

$$\frac{1}{2\pi}\int_{-\infty+iy}^{\infty+iy} e^{itz}\int_{-\infty}^{\infty} f(u)\, e^{-iuz}\, du\, dz = \frac{1}{2}\,[f(t+) + f(t-)],$$

where the integration in the z-plane is along the line $x + iy$, $-\infty < x < \infty$, y being fixed.

The expression

$$\mathrm{F}(z) = \int_{-\infty}^{\infty} f(u)\, e^{-iuz}\, du.$$

is called the *complex Fourier transform* of f.

If f satisfies the hypothesis of Theorem 7.17, then the inverse Fourier transform is given by

$$\frac{1}{2}\,[f(t+) + f(t-)] = \frac{1}{2\pi}\int_{-\infty+iy}^{\infty+iy} e^{itz}\,\mathrm{F}(z)\, dz \text{ for real } y.$$

We state (without proof) a result from complex analysis which asserts the analyticity of the Fourier transform.

Theorem 7.18. Let $f(t)$ be a piecewise continuous function such that $|f(t)| \le K\, e^{-bt}$, $0 \le t < \infty$ and $|f(t)| \le M\, e^{-at}$, $-\infty < t \le 0$, $a < b$. Then the Fourier transform of $f(t)$ exists and is analytic function of z for $a < Im(z) < b$. Also

$$F'(z) = \int_{-\infty}^{\infty} [-iu\, f(u)]\, e^{-iuz}\, du.$$

EXAMPLE 7.24

Find the complex Fourier transform of $f(t) = \sin \omega t$, $0 \le t < \infty$, $\omega > 0$, $f(t) = 0$, $-\infty < t \le 0$ and verify the inverse transform theorem.

Solution. Since $|f(t)| \le 1$ for $0 \le t < \infty$ and $|f(t)| = 0$ for $-\infty < t \le 0$, the conditions of Theorem 7.18 are satisfied. Therefore, the transform of the function is analytic for $-\infty < \mathrm{Im}(z) < 0$. We have

$$F(z) = \int_{-\infty}^{\infty} f(t)\, e^{-izt}\, dt = \int_{0}^{\infty} \sin \omega t\, e^{-izt}\, dt$$

$$= \int_{0}^{\infty} \frac{e^{i\omega t} - e^{-i\omega t}}{2i}\, e^{-i\,zt}\, dt$$

$$= \frac{1}{2i}\left[\int_{0}^{\infty} e^{i\,(\omega - z)t}\, dt - \int_{0}^{\infty} e^{-i\,(\omega + z)t}\, dt\right]$$

$$= \frac{1}{2}\left[\frac{1}{\omega - z} + \frac{1}{\omega + z}\right] = \frac{\omega}{\omega^2 - z^2}.$$

To verify the inverse transform theorem, we note that $\mathrm{F}(z) = \frac{\omega}{\omega^2 - z^2}$ is analytic in the extended z-plane except at $z = \pm\omega$. Therefore, this function is the analytic continuation of the Fourier transform to the rest of the plane. We wish to evaluate the integral

$$\frac{1}{2\pi}\int_{-\infty+iy}^{\infty+iy} \frac{\omega}{\omega^2 - z^2}\, e^{i\,zt}\, dz = \lim_{R\to\infty} \frac{1}{2\pi}\int_{C} \frac{\omega\, e^{i\,zt}}{\omega^2 - z^2}\, dz,$$

where C is the contour shown in Figure 7.6.

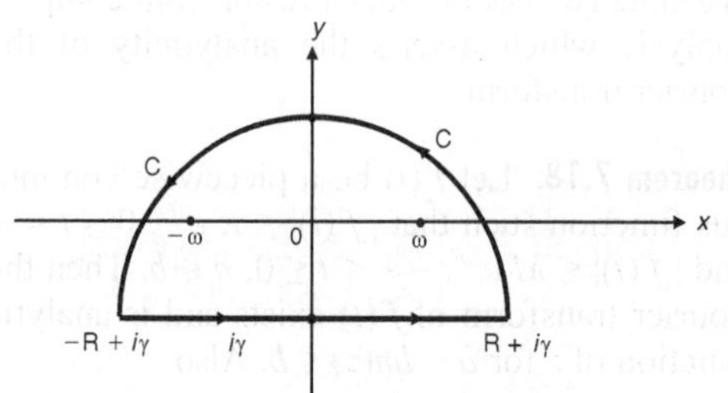

Figure 7.6

Two simple poles ω and $-\omega$ are inside the contour. By Cauchy residue theorem, we have

$$\frac{1}{2\pi}\int_{-\infty+i\gamma}^{\infty+i\gamma}\frac{\omega e^{izt}}{\omega^2-z^2}\,dz = -i\left(\frac{\omega e^{i\omega t}}{2\omega}-\frac{\omega e^{-i\omega t}}{2\omega}\right) = \sin\omega t,$$

provided that we can show that contribution of the semicircular arc goes to zero as $R\to\infty$. On the semi-circular contour $z = i\gamma + Re^{i\theta}$, $0 \le \theta \le \pi$ and $|e^{izt}| = |e^{-\gamma t}\, e^{itR}\,(\cos\theta + i\sin\theta)| = e^{-\gamma t}\, e^{-tR\sin\theta} \le e^{-\gamma t}$.

Therefore, on this part of the contour, we have

$$\left|\frac{1}{2\pi}\int\frac{\omega e^{izt}}{\omega^2 - z^2}\,dz\right| \le \frac{\omega e^{-\gamma t}}{2\pi}\,\frac{\pi R}{(R-\gamma)^2-\omega^2}$$

$$\to 0 \text{ as } R\to\infty.$$

If $t \le 0$, we close the contour with a semicircle below the x-axis, since there are no poles inside the contour the result is zero. Also, on the semi-circular arc, $z = i\gamma + Re^{i\theta}$, $\pi \le \theta \le 2\pi$ and

$$\left|\frac{1}{2\pi}\int\frac{\omega e^{izt}}{\omega^2-z^2}\,dz\right| \le \frac{\omega e^{-\gamma t}(\pi R)}{(R-\gamma)^2-\omega^2}$$

$$\to 0 \text{ as } R\to\infty.$$

EXAMPLE 7.25

Find Fourier transform of cos at, $t > 0$.

Solution. By Example 7.24,

$$F\{\sin at\} = \frac{a}{a^2-z^2}.$$

Also

$$\sin at = a\int_0^t \cos at\,dt.$$

Therefore, by Theorem 7.16, we have

$$F\{\sin at\} = aF\left\{\int_0^t \cos at\,dt\right\} = \frac{a}{iz}F\{\cos at\}.$$

Therefore,

$$F\{\cos at\} = \frac{iz}{a}F\{\sin at\} = \frac{iz}{a}\left(\frac{a}{a^2-z^2}\right) = \frac{iz}{a^2-z^2}.$$

7.7 CONVOLUTION THEOREM

Let f and g be absolutely integrable functions with $F(\omega)$ and $G(\omega)$ as their Fourier transforms, respectively. The question arises "which function has its Fourier transform as $F(\omega)\,G(\omega)$? Is it fg or something else?" The answer is "it is not fg". For example, consider the rectangular pulse function

$$f(t) = \begin{cases} 1 & \text{for } |t| \le \frac{a}{2} \\ 0 & \text{otherwise.}\end{cases}$$

Its Fourier transform is $\dfrac{2\sin(a\omega/2)}{\omega}$. The product of the function with itself is $f(t)$ itself while the product of its Fourier transform with itself is $\dfrac{4\sin^2(a\omega/2)}{\omega^2}$ which is not equal to $\dfrac{2\sin(a\omega/2)}{\omega}$.

The concept of convolution helps us in finding the function corresponding to the known product of two Fourier transforms.

Definition 7.3 Let f and g be two given functions. The convolution of f and g, denoted by $f * g$, is defined by

$$(f * g)(t) = \int_{-\infty}^{\infty} f(u)\,g(t-u)\,du;\; t\in\mathbb{R};$$

provided the integral exists.

An important case is that in which

$$f(t) = 0 \text{ for } t < 0$$

$$g(t) = 0 \text{ for } t < 0.$$

Then

$$(f * g)(t) = \int_0^t f(u)\,g(t-u)\,du.$$

Theorem 7.19. (Convolution Theorem). Let f and g be piecewise continuous, absolutely integrable and bounded functions with Fourier transforms $F(\omega)$ and $G(\omega)$, respectively. Then $f * g$ is absolutely integrable and

$$F\{f * g\} = F(\omega)G(\omega).$$

Proof: First we show that, under the given hypothesis, the convolution exists and is bounded. Since g is bounded, $|g(t-u)| \leq M$ (constant). Therefore, absolute integrability of f yields

$$|(f * g)(t)| = \left|\int_{-\infty}^{\infty} f(u)\, g(t-u)\, du\right|$$

$$\leq \int_{-\infty}^{\infty} |f(u)||g(t-u)|\, du \leq M \int_{-\infty}^{\infty} |f(u)|\, du < \infty,$$

which shows that $f * g$ is defined and bounded.

We now show that $f * g$ is absolutely integrable. To this end, we note that the absolute integrability of f and g imply

$$\int_{-\infty}^{\infty} \left|\int_{-\infty}^{\infty} f(u)\, g(t-u)\, dt\right| du$$

$$\leq \int_{-\infty}^{\infty} \left(\int_{-\infty}^{\infty} |f(u)\, g(t-u)|\, dt\right) du$$

$$= \int_{-\infty}^{\infty} |f(u)| \left(\int_{-\infty}^{\infty} |g(t-u)|\, dt\right) du$$

$$= \int_{-\infty}^{\infty} |f(u)|\, du \int_{-\infty}^{\infty} |g(v)|\, dv < \infty,\ v = t-u.$$

Thus $f * g$ absolutely integrable and so Fourier transform $f * g$ exists. By definition of Fourier transform and by changing the order of integration (permissible under the given condition), we have

$$F\{(f * g)(t)\} = \int_{-\infty}^{\infty} (f * g)(t)\, e^{-i\omega t}\, dt$$

$$= \int_{-\infty}^{\infty} \left(\int_{-\infty}^{\infty} f(u)\, g(t-u)\, e^{-i\omega(t-u)}\, e^{-i\omega u}\, du\right) dt$$

$$= \int_{-\infty}^{\infty} f(u)\, e^{-i\omega u} \left(\int_{-\infty}^{\infty} g(t-u)\, e^{-i\omega(t-u)}\, dt\right) du$$

$$= \int_{-\infty}^{\infty} f(u)\, e^{-i\omega u} \left(\int_{-\infty}^{\infty} g(v)\, e^{-i\omega v}\, dv\right) du,\ v = t-u,$$

$$= \int_{-\infty}^{\infty} f(u)\, e^{-i\omega u}\, (G(\omega))\, du = G(\omega) \int_{-\infty}^{\infty} f(u)\, e^{-i\omega u}\, du$$

$$= G(\omega)\, F(\omega) = F(\omega)\, G(\omega).$$

EXAMPLE 7.26

If $F\{f(t)\} = \dfrac{1}{(1+i\omega)^2}$ and $f(t) = 0$ for $t < 0$, find $f(t)$ using Convolution Theorem.

Solution. We are given that

$$F\{f(t)\} = \frac{1}{(1+i\omega)^2} = \frac{1}{1+i\omega} \cdot \frac{1}{1+i\omega}$$

Let $F(\omega) = G(\omega) = \dfrac{1}{1+i\omega}$. Then, by Convolution theorem,

$$F\{f * g\} = F(\omega)\, G(\omega)$$

or

$$F^{-1}[F(\omega)\, G(\omega)] = f * g.$$

But, by Example 7.3, $\dfrac{1}{1+i\omega}$ is the Fourier transform of e^{-t}, $t > 0$. Therefore

$$F^{-1}\{F(\omega)\, G(\omega)\} = \int_{-\infty}^{\infty} f(u)\, g(t-u)\, du$$

$$= \int_0^t e^{-u}\, e^{-(t-u)}\, du = \int_0^t e^{-t}\, du = e^{-t} \int_0^t du = t\, e^{-t}.$$

EXAMPLE 7.27

If $F\{f(t)\} = \dfrac{1}{\omega(\omega^2 - 1)}$, where $f(t) = 0$ for $t < 0$, find $f(t)$ using Convolution theorem.

Solution. Let

$$F_1(\omega) = \frac{1}{\omega} \text{ and } F_2(\omega) = \frac{1}{\omega^2 - 1} = -\frac{1}{1-\omega^2}.$$

Then

$$f_1(t) = i\, H(t) \text{ and } f_2(t) = -H(t) \sin t.$$

Therefore, by Convolution theorem

$$F\{f_1 * f_2\} = F_1(\omega)\, F_2(\omega),$$

that is,

$$f(t) = -\int_0^t i\, H(t)\, H(t-u) \sin(t-u)\, du$$

$$= -iH(t)\int_0^t \sin(t-u)\,du = iH(t)\,[\cos(t-u)]_0^t$$

$$= iH(t)\,[1-\cos t].$$

7.8 PARSEVAL'S IDENTITIES

Theorem 7.20. (Parseval's Identities). Let $F(\omega)$ and $G(\omega)$ be Fourier transforms of absolutely integrable piecewise smooth functions f and g respectively. Then

(a) $$\frac{1}{2\pi}\int_{-\infty}^{\infty} F(\omega)\,\overline{G(\omega)}\,dv = \int_{-\infty}^{\infty} f(t)\,\overline{g(t)}\,dt$$

(b) $$\frac{1}{2\pi}\int_{-\infty}^{\infty} |F(\omega)|^2\,d\omega = \int_{-\infty}^{\infty} |f(t)|^2\,dt.$$

[Identity (b) is also known as Plancherel's identity].

Proof: (a) By Fourier integral theorem,

$$g(t) = \frac{1}{2\pi}\int_{-\infty}^{\infty} G(\omega)\,e^{i\omega t}\,dt$$

Taking complex conjugate, we get

$$\overline{g(t)} = \frac{1}{2\pi}\int_{-\infty}^{\infty} \overline{G(\omega)}\,e^{-i\omega t}\,dt.$$

Therefore

$$\int_{-\infty}^{\infty} f(t)\,\overline{g(t)}\,dt = \frac{1}{2\pi}\int_{-\infty}^{\infty} f(t)\,dt\int_{-\infty}^{\infty} \overline{G(\omega)}\,e^{-i\omega t}\,d\omega$$

$$= \frac{1}{2\pi}\int_{-\infty}^{\infty}\int_{-\infty}^{\infty} f(t)\,\overline{G(\omega)}\,e^{-i\omega t}\,dt\,d\omega$$

$$= \frac{1}{2\pi}\int_{-\infty}^{\infty}\int_{-\infty}^{\infty} f(t)\,\overline{G(\omega)}\,e^{-i\omega t}\,d\omega\,dt$$

$$= \frac{1}{2\pi}\int_{-\infty}^{\infty} \overline{G(\omega)}\,d\omega\int_{-\infty}^{\infty} f(t)\,e^{-i\omega t}\,dt$$

$$= \frac{1}{2\pi}\int_{-\infty}^{\infty} \overline{G(\omega)}\,F(\omega)\,d\omega$$

$$= \frac{1}{2\pi}\int_{-\infty}^{\infty} F(\omega)\,\overline{G(\omega)}\,d\omega.$$

(b) Taking $f(t) = g(t)$ in identity (*a*) we have

$$\int_{-\infty}^{\infty} f(t)\overline{f(t)}\,dt = \frac{1}{2\pi}\int_{-\infty}^{\infty} F(\omega)\,\overline{F(\omega)}\,d\omega,$$

and so

$$\int_{-\infty}^{\infty} |f(t)|^2 = \frac{1}{2\pi}\int_{-\infty}^{\infty} |F(\omega)|^2\,d\omega.$$

Remark 7.6(i): Identity (b) *asserts that the Fourier transform of an absolutely integrable function is also absolutely integrable.*

Since a signal f is said to have *finite energy content* if f is absolutely integrable, the identity (b) shows that *Fourier transform of an energy signal is also an energy signal.*

(ii) The Parseval's identity for Fourier cosine transform and Fourier sine transform takes the form

$$\frac{2}{\pi}\int_0^{\infty}[F_c(\omega)]^2\,d\omega = \int_0^{\infty}|f(t)|^2\,dt$$

and

$$\frac{2}{\pi}\int_0^{\infty}[F_s(\omega)]^2\,d\omega = \int_0^{\infty}|f(t)|^2\,dt.$$

EXAMPLE 7.28

Using Parseval's identity, show that

(a) $$\int_0^{\infty}\frac{\sin^2 t}{t^2}\,dt = \frac{\pi}{2}$$

(b) $$\int_0^{\infty}\frac{t^2}{(t^2+1)^2}\,dt = \frac{\pi}{4}$$

(c) $$\int_0^{\infty}\frac{dt}{(t^2+1)^2} = \frac{\pi}{4}.$$

Solution. (a) We know (see Example 7.5) that the Fourier transform of rectangular pulse function f of height 1 and duration a is $\dfrac{2\sin(a\omega/2)}{\omega}$.

Therefore, by Parseval's identity,

$$\int_{-\infty}^{\infty}|f(t)|^2\,dt = \frac{1}{2\pi}\int_{-\infty}^{\infty}\frac{4\sin^2(a\omega/2)}{\omega^2}\,d\omega.$$

The left-hand side is

$$\int_{-\infty}^{\infty}|f(t)|^2\,dt = \int_{-a/2}^{a/2}dt = a.$$

Putting $u = \dfrac{a\omega}{2}$ in the right-hand side, we have

$$\frac{a}{\pi}\int_{-\infty}^{\infty}\frac{\sin^2 u}{u^2}\,du.$$

Therefore,

$$a=\frac{a}{\pi}\int_{-\infty}^{\infty}\frac{\sin^2 u}{u^2}\,du,$$

which yields

$$\int_0^{\infty}\frac{\sin^2 u}{u^2}\,du=\frac{\pi}{2}.$$

(b) By Example 7.19, the Fourier sine transform of e^{-t} is

$$F_s\{e^{-t}\}=\int_0^{\infty}e^{-t}\sin\omega t\,dt=\frac{-\omega}{1+\omega^2}$$

Therefore, by Parseval's identity, we get

$$\frac{2}{\pi}\int_0^{\infty}[F_s(\omega)]^2\,d\omega=\int_0^{\infty}|f(t)|^2\,dt,$$

that is,

$$\frac{2}{\pi}\int_0^{\infty}\frac{\omega^2\,d\omega}{(1+\omega^2)^2}=\int_0^{\infty}e^{-2t}\,dt$$

and so

$$\int_0^{\infty}\frac{\omega^2}{(1+\omega^2)^2}\,d\omega=\frac{\pi}{2}\int_0^{\infty}e^{-2t}\,dt=\frac{\pi}{4}.$$

Changing the dummy variable,we get the required result.

(c) The Fourier cosine transform (Example 7.19) of e^{-t} is

$$F_c\{e^{-t}\}=\int_0^{\infty}e^{-t}\cos\omega t\,dt=\frac{1}{1+\omega^2}$$

Therefore, by Parseval's identity

$$\frac{2}{\pi}\int_0^{\infty}[F_c(\omega)]^2\,d\omega=\int_0^{\infty}|f(t)|^2\,dt,$$

that is,

$$\frac{2}{\pi}\int_0^{\infty}\frac{d\omega}{(1+\omega^2)^2}=\int_0^{\infty}e^{-2t}\,dt=\frac{1}{2}$$

and so

$$\int_0^{\infty}\frac{d\omega}{(1+\omega^2)^2}=\frac{\pi}{4}.$$

EXAMPLE 7.29

Using Parseval's identity, find the value of the integral

$$\int_0^{\infty}\frac{dt}{(a^2+t^2)(b^2+t^2)}.$$

Solution. Let $f(t)=e^{-at}$, $g(t)=e^{-bt}$. Then

$$F_c(\omega)=\int_0^{\infty}f(t)\cos\omega t\,dt=\int_0^{\infty}e^{-at}\cos\omega t\,dt$$

$$=\frac{a}{a^2+\omega^2},$$

$$G_c(\omega)=\int_0^{\infty}e^{-bt}\cos\omega t\,dt=\frac{b}{b^2+\omega^2}.$$

Therefore, by Parseval's identity,

$$\frac{2}{\pi}\int_0^{\infty}F_c(\omega)\,G_c(\omega)\,d\omega=\int_0^{\infty}f(t)\,g(t)\,dt,$$

that is,

$$\frac{2ab}{\pi}\int_0^{\infty}\frac{1}{(a^2+\omega^2)(b^2+\omega^2)}d\omega=\left[\frac{e-(a+b)\,\omega}{-(a+b)}\right]_0^{\infty}$$

$$=\frac{1}{a+b}.$$

Hence

$$\int_0^{\infty}\frac{1}{(a^2+\omega^2)(b^2+\omega^2)}\,d\omega=\frac{\pi}{2ab(a+b)}.$$

EXAMPLE 7.30

Find energy spectrum of the function

$$f(t)=\begin{cases}e^{-at} & \text{for } t\ge 0\\ 0 & \text{for } t<0.\end{cases}$$

Solution. We have

$$F(\omega)=\int_{-\infty}^{\infty}f(t)\,e^{-i\omega t}\,dt=\int_0^{\infty}e^{-at}\,e^{-i\omega t}\,dt$$

$$=\int_0^{\infty}e^{-(a+i\omega)t}\,dt=\left[\frac{e^{-(a+i\omega)t}}{-(a+i\omega)}\right]_0^{\infty}$$

$$=\frac{1}{a+i\omega}=\frac{a-i\omega}{a^2+\omega^2}.$$

Hence

Energy spectrum $=|F(\omega)|^2=F(\omega)\,\overline{F(\omega)}$

$$\frac{a-i\omega}{a^2+\omega^2}\cdot\frac{a+i\omega}{a^2+\omega^2}=\frac{1}{a^2+\omega^2}.$$

7.9 FOURIER INTEGRAL REPRESENTATION OF A FUNCTION

We know that if a function f is piecewise smooth on $(-l, l)$ and periodic with period $2l$, then it has the Fourier series representation

$$f(x) = \frac{a_0}{2} + \sum_{n=1}^{\infty}\left(a_n \cos\frac{n\pi x}{l} + b_n \sin\frac{n\pi x}{l}\right), \quad (24)$$

where

$$a_n = \frac{1}{l}\int_{-l}^{l} f(t)\cos\frac{n\pi t}{l}\,dt \text{ and } b_n = \frac{1}{l}\int_{-l}^{l} f(t)\sin\frac{n\pi t}{l}\,dt.$$

Putting values of a_n and b_n in (24), changing the order of summation and integration and using the identity cos A cos B + sin A sin B = cos (A – B), we get

$$f(x) = \frac{1}{2l}\int_{-l}^{l} f(t)\,dt + \frac{1}{l}\int_{-l}^{l} f(t)\sum_{n=1}^{\infty}\cos\frac{n\pi(t-x)}{l}\,dt. \quad (25)$$

If we further assume that f is absolutely integrable, then $\int_{-\infty}^{\infty} |f(x)|\,dx < \infty$ and so

$$\lim_{l\to\infty}\frac{1}{2l}\int_{-l}^{l} f(t)\,dt = 0.$$

For the remaining part of (25) put $\Delta s = \frac{\pi}{l}$. Then $\Delta s \to 0$ as $l \to \infty$. Thus

$$f(x) = \lim_{\Delta s\to 0}\frac{1}{\pi}\int_{-\pi/\Delta s}^{\pi/\Delta s} f(t)\sum_{n=1}^{\infty}\cos[n\Delta s(t-x)]\,\Delta s\,dt.$$

When Δs is small, the points $n\Delta s$ are equally spaced along the x-axis, so let $n\Delta s = \lambda$. Then

$$f(x) = \lim_{\Delta s\to 0}\frac{1}{\pi}\int_{-\pi/\Delta s}^{\pi/\Delta s} f(t)\int_{0}^{\infty}\cos\lambda(t-x)\,d\lambda\,dt$$

$$= \frac{1}{\pi}\int_{0}^{\infty}\int_{-\infty}^{\infty} f(t)\cos\lambda\,(t-x)dt\,d\lambda$$

$$= \frac{1}{\pi}\int_{0}^{\infty}\int_{-\infty}^{\infty} f(t)\,(\cos\lambda t\cos\lambda x + \sin\lambda t\sin\lambda x)dt\,d\lambda$$

Equivalently, we can write

$$f(x) = \int_{0}^{\infty}[A(\lambda)\cos\lambda x + B(\lambda)\sin\lambda x]\,d\lambda, \quad (26)$$

where

$$A(\lambda) = \frac{1}{\pi}\int_{-\infty}^{\infty} f(t)\cos\lambda t\,dt \text{ and}$$

$$B(\lambda) = \frac{1}{\pi}\int_{-\infty}^{\infty} f(t)\sin\lambda t\,dt.$$

Then representation (26) is called the *Fourier integral representation of the function f.*

If f is an *even function,* then $f(-x) = f(x)$, and so

$$A(\lambda) = \frac{1}{\pi}\int_{-\infty}^{\infty} f(t)\cos\lambda t\,dt = \frac{2}{\pi}\int_{0}^{\infty} f(t)\cos\lambda t\,dt$$

and

$$B(\lambda) = \frac{1}{\lambda}\int_{-\infty}^{\infty} f(t)\sin\lambda t\,dt = 0.$$

Hence for an even function f,

$$f(x) = \int_{0}^{\infty} A(\lambda)\cos\lambda x\,d\lambda. \quad (27)$$

The expression (27) is called *Fourier cosine integral representation* of the function f.

Similarly for an *odd function f,*

$$B(\lambda) = \frac{2}{\pi}\int_{0}^{\infty} f(t)\sin\lambda t\,dt \text{ and } A(\lambda) = 0,$$

and so

$$f(x) = \int_{0}^{\infty} B(\lambda)\sin\lambda x\,dt. \quad (28)$$

The expression (28) is called *Fourier sine integral representation* of the function f.

EXAMPLE 7.31

Find a Fourier sine integral representation for

$$f(x) = \begin{cases} 1 & \text{for } 0 \le x \le \pi \\ 0 & \text{for } x > \pi. \end{cases}$$

Hence evaluate the integral

$$\int_{0}^{1}\frac{1-\cos(\pi\lambda)}{\lambda}\sin(\lambda t)\,dt.$$

Solution. By definition, the Fourier sine representation of f is

$$f(x) = \int_0^\infty B(\lambda) \sin \lambda t \, d\lambda,$$

where

$$B(\lambda) = \frac{2}{\pi}\int_0^\infty f(t) \sin \lambda t \, dt = \frac{2}{\pi}\int_0^\pi \sin \lambda t \, dt$$

$$= \frac{2}{\pi}\left[\frac{\cos \lambda t}{-\lambda}\right]_0^\pi = \frac{2}{\pi}\left[\frac{1-\cos \lambda\pi}{\lambda}\right].$$

Thus

$$f(x) = \frac{2}{\pi}\int_0^\infty \left[\frac{1-\cos \lambda\pi}{\lambda}\right] \sin \lambda t \, d\lambda.$$

Deduction: We have

$$\int_0^\infty \left(\frac{1-\cos \lambda\pi}{\lambda}\right) \sin \lambda t \, d\lambda = \frac{\pi}{2} f(x)$$

$$= \begin{cases} \frac{\pi}{2} \text{ for } 0 \le x < \pi \\ 0 \text{ for } x > \pi. \end{cases}$$

Since f is discontinuous at $x = \pi$, Theorem 7.7 yields

$$\int_0^\infty \left(\frac{1-\cos\lambda\pi}{\lambda}\right) \sin \lambda t \, dt = \frac{\pi}{2}\left[\frac{f(\pi+0)+f(\pi-0)}{2}\right]$$

$$= \frac{\pi}{2}\left[\frac{1+0}{2}\right] = \pi/4.$$

EXAMPLE 7.32

Find Fourier cosine integral representation of

$$f(x) = \begin{cases} \cos x & \text{for } 0 < x < \frac{\pi}{2} \\ 0 & \text{for } x > \frac{\pi}{2}. \end{cases}$$

Solution. We have

$$A(\lambda) = \frac{2}{\pi}\int_0^{\pi/2} \cos \lambda x \cos x \, dx = \frac{2\cos(\pi\lambda/2)}{\pi(1-\lambda^2)},$$

and so

$$f(x) = \frac{2}{\pi}\int_0^\infty \frac{\cos(\pi\lambda/2)}{(1-\lambda^2)} \cos \lambda x \, d\lambda.$$

7.10 FINITE FOURIER TRANSFORMS

The *Finite cosine transform* of a function f is defined by

$$F_c(n) = \int_0^l f(x) \cos \frac{n\pi x}{l} \, dx, \; n \text{ an integer}$$

and the *inverse finite Fourier cosine transform* of $F_c(n)$ is then given by

$$f(x) = \frac{1}{l} F_c(0) + \frac{2}{l}\sum_{n=1}^\infty F_c(n) \cos \frac{n\pi x}{l}.$$

Similarly, the transform

$$F_s(n) = \int_0^l f(x) \sin \frac{n\pi x}{l} \, dx, \; n \text{ an integer}$$

is called *finite Fourier sine transform of f*.

The inverse of this transform is defined as

$$f(x) = \frac{2}{l}\sum_{n=1}^\infty F_s(n) \sin \frac{n\pi x}{l}.$$

We note that *finite Fourier transforms are actually sequences of numbers* rather than continuous functions. If we compare these transforms with Fourier coefficients, we note that

$$a_n = \frac{2}{\pi} F_c(n), \; n = 0, 1, 2, \ldots$$

$$b_n = \frac{2}{\pi} F_s(n), \; n = 1, 2, \ldots.$$

In complex form $2\pi c_n$ is the finite Fourier transform of f over $[-\pi, \pi]$, where

$$c_n = \frac{1}{2\pi}\int_{-\pi}^{\pi} f(x)\, e^{-inx} \, dx.$$

EXAMPLE 7.33

Find finite Fourier sine transform of sin *at*.

Solution. We have

$$F_s(n) = \int_0^\pi \sin at \sin nt \, dt = \frac{1}{2}\int_0^\pi 2 \sin at \sin nt \, dt$$

$$= \frac{1}{2}\int_0^\pi [\cos (a-n)t - \cos(a+n)] \, dt$$

$$= \frac{1}{2}\left[\frac{\sin (a-n)t}{a-n} - \frac{\sin(a+n)t}{a+n}\right]_0^\pi = 0, a \ne n.$$

For $a = n$, we have

$$F_s(n) = \frac{1}{2}\int_0^\pi 2\sin^2 nt \, dt = \frac{1}{2}\int_0^\pi (1-\cos 2nt)dt = \frac{\pi}{2}.$$

Therefore,

$$F_s(n) = \begin{cases} 0 & \text{for } n \ne a \\ \pi/2 & \text{for } n = a. \end{cases}$$

EXAMPLE 7.34

Find finite Fourier transform and finite Fourier sine transform of rectangular pulse function [*top hat function*] defined by

$$f(x) = \begin{cases} 1 & \text{for } x \in [0, \pi] \\ 0 & \text{otherwise.} \end{cases}$$

Solution. The finite Fourier transform of this function is

$$\int_0^\pi e^{-inx}\,dx = \left[-\frac{e^{-inx}}{in}\right]_0^\pi = \frac{1}{in}(1 - e^{-in\pi})$$

$$= \frac{1}{in}(1 - (-1)^n).$$

The finite Fourier sine transform is

$$\int_0^\pi \sin nx\,dx = \frac{1}{n}(1 - (-1)^n),\ n \text{ an integer}$$

$$= \frac{2}{2k+1},\ k \text{ an integer.}$$

7.11 APPLICATIONS OF FOURIER TRANSFORMS

We have already discussed in the present chapter, the use of Fourier transform to solve certain integral equations and to evaluate certain integrals. The Fourier transform plays an important role in the *study of transfer of signals in communication system*. The signals, given in the form of ordinary differential equation with constant coefficients, can be analysed by using Fourier transforms. Also, the partial differential equations can be solved by using Fourier transform methods, thereby exploring physical phenomenon like heat conductions. Therefore, in the next two sections, we take up the applications of Fourier transform to solve ordinary differential equations and partial differential equations.

7.12 APPLICATION TO DIFFERENTIAL EQUATIONS

Consider the linear *n*th order differential equation with constant coefficients

$$a_n \frac{d^n y}{dt^n} + a^{n-1}\frac{d^{n-1}y}{dt^{n-1}} + \ldots + a_1\frac{dy}{dt} + a_0 y = f(t).$$

Assume that f has a Fourier transform F{f}. Then taking Fourier transform of both sides of the equation, we have

$$[a_n (i\omega)^n + a_{n-1}(i\omega)^{n-1} + \ldots + a_n (i\omega) + a0]\mathrm{F}\{y(t)\}$$

$$= \mathrm{F}\{f(t)\}.$$

Hence, if F{$y(t)$} exists, then

$$\mathrm{F}\{y(t)\} = \frac{\mathrm{F}\{f(t)\}}{P(i\omega)},$$

where

$$\mathrm{P(D)} = a_n \mathrm{D}^n + a_{n-1}\mathrm{D}^{n-1} + \ldots + a_1 D + a_0$$

is the operator on the left hand side of the differential equation.

Assume that $\frac{1}{P(i\omega)}$ has an inverse transform

$$g(t) = \frac{1}{2\pi}\int_{-\infty+i\gamma}^{\infty+i\gamma} \frac{e^{i\omega t}}{P(i\omega)}\,d\omega.$$

Then using the Convolution theorem, we obtain

$$y(t) = \frac{1}{2\pi}\int_{-\infty}^{\infty} f(u)\,g(t-u)\,du.$$

The solution $y(t)$ so obtained can be verified by putting it in the differential equation.

The following results using complex Fourier transform can also be used to find the solution of a given differential equation.

Theorem 7.21. Let f be a continuous function for which the Fourier integral theorem holds and let $F(z)$ be the Fourier transform of f. If $F(z)$ is analytic in some strip $a < Im(z) < b$, $P(iz)$ has no zeroes in this strip and

$$f(t) = \frac{1}{2\pi}\int_{-\infty+i\gamma}^{\infty+i\gamma} F(z)\,e^{izt}\,dz$$

converges uniformly in t for some λ satisfying $a < \gamma < b$ Then

$$y(t) = \frac{1}{2\pi}\int_{-\infty+i\gamma}^{\infty+i\gamma} \frac{F(z)e^{izt}}{P(iz)}\,dz$$

is a *solution of the differential equation $P(D)\,y = f(t)$.*

Proof: The hypothesis of the theorem implies that the integral

$$y(t)=\frac{1}{2\pi}\int_{-\infty+iy}^{\infty+iy}\frac{F(z)e^{izt}}{P(iz)}\,dz \tag{29}$$

converges uniformly in t. Therefore, we can differentiate it under the integral sign with respect to t giving thereby

$$Dy=\frac{1}{2\pi}\int_{-\infty+iy}^{\infty+iy}\frac{(iz)\,F(z)\,e^{izt}}{P(iz)}\,dz \tag{30}$$

Multiplying (30) by a_1 and (29) by a_0 and adding, we get

$$a_1Dy+a_0y=\frac{1}{2\pi}\int_{-\infty+iy}^{\infty+iy}\frac{[a_1(iz)+a_0]\,F(z)e^{izt}}{P(iz)}\,dz.$$

Similarly, the higher derivatives of (30) exist and we get

$$P(D)y=\frac{1}{2\pi}\int_{-\infty+iy}^{\infty+iy}\frac{P(iz)\,F(z)\,e^{izt}}{P(iz)}\,dz$$

$$=\frac{1}{2\pi}\int_{-\infty+iy}^{\infty+iy}F(z)\,e^{izt}\,dz=f(t),$$

proving that (29) is a solution of the differential equation $P(D)y=f(t)$.

EXAMPLE 7.35

Find a solution of

$$\frac{d^2y}{dt^2}+3\frac{dy}{dt}+2y=e^{-|t|},$$

using Fourier transform.

Solution. Taking Fourier transform of both sides, we have

$$[(iz)^2+3iz+2]\,F\{y(t)\}=F\{e^{-|t|}\}=\frac{2}{1+z^2}\cdot$$

Let

$$P(iz)=(iz)^2+3iz+2 \text{ and } F\{e^{-|t|}\}=F(z).$$

Then the solution is given by

$$y(t)=\frac{1}{2\pi}\int_{-\infty+iy}^{\infty+iy}\frac{F(z)\,e^{izt}}{P(iz)}\,dz.$$

Thus

$$y(t)=\frac{1}{2\pi}\int_{-\infty+iy}^{\infty+iy}\frac{2}{1+z^2}\cdot\frac{e^{izt}}{[(iz)^2+3iz+2]}\,dz$$

$$=\frac{1}{\pi}\int_{-\infty+iy}^{\infty+iy}\frac{e^{izt}}{(1+z^2)\,(-z^2+3iz+2)}\,dz$$

$$=-\frac{1}{\pi}\int_{-\infty+iy}^{\infty+iy}\frac{e^{izt}}{(z-i)(z+i)(z^2-3iz-2)}\,dz.$$

Case I. $t>0$.

The singularities within the contour are

(a) simple pole $z=2i$

(b) double pole at $z=i$.

Therefore for $t>0$, we have by Cauchy's residue theorem,

$$y(t)=-\frac{1}{\pi}\,2\pi i\,\Sigma R,$$

where ΣR is the sum of the residues of the integrand at the singularities.

But

$$\text{Res}(2i)=\lim_{z\to 2i}(z-2i)\frac{e^{izt}}{(z-i)\,(z^2+1)\,(z-2i)}$$

$$=\frac{ie^{-2t}}{3}.$$

Residue at the double pole $z=i$ is

$$\text{Res}(i)=\lim_{z\to i}\frac{d}{dz}\left[(z-i)^2\frac{e^{izt}}{(z-i)^2(z-2i)\,(z+i)}\right]$$

$$=\lim_{z\to i}\left[\frac{d}{dz}\left(\frac{e^{izt}}{z^2-iz+2}\right)\right]$$

$$=\lim_{z\to i}\left[\frac{it(z^2-iz+2)\,e^{izt}-2ze^{izt}+ie^{izt}}{(z^2-iz+2)^2}\right]$$

$$=\frac{2it\,e^{-t}-ie^{-t}}{4}.$$

Hence

$$y(t)=-\frac{1}{\pi}.2\pi i\left[\frac{ie^{-2t}}{3}+\frac{ite^{-t}}{2}-\frac{ie^{-t}}{4}\right]$$

$$=\frac{2}{3}e^{-2t}+te^{-t}-\frac{1}{2}e^{-t}.$$

Verification: We have $y' = -\frac{4}{3}e^{-2t} + \frac{3}{2}e^{-t} - te^{-t}$, $y'' = \frac{8}{3}e^{-2t} - \frac{5}{2}e^{-t} + te^{-t}$. Therefore, $y'' + 3y' + 2y = e^{-t}$.

Case II. $t < 0$.

If $t < 0$, then we close the contour in the lower half-plane and, hence, the simple pole $z = -i$ is the only singularity. Therefore,

$$y(t) = -\left(-\frac{1}{\pi}\right)2\pi i \text{ (residue of integrand at } z = -i)$$
$$= 2i \text{ (residue at } z = -i)$$

But residue at $z = -i$ is

$$\text{Res}(-i) = \lim_{z\to -i}(z+i)\frac{e^{izt}}{(z-i)^2(z+i)(z-2i)} = \frac{e^t}{12i}.$$

Hence

$$y(t) = \frac{2i.e^t}{12i} = \frac{e^t}{6}$$

Verification. We have $y' = \frac{e^t}{6}$, $y'' = \frac{e^t}{6}$ and, therefore,

$$y'' + 3y' + 2y = e^t.$$

EXAMPLE 7.36

Find a solution of

$$\frac{d^2y}{dt^2} + 3\frac{dy}{dt} + 2y = \text{H}(t)\sin \omega t$$

for $t > 0$ satisfying $\lim_{t\to 0+} y(t) = 0$ and $\lim_{y\to 0-} y'(t) = 1$.

Solution. Taking Fourier transform of both sides of the given equation, we get

$$[(iz)^2 + 3iz + 2]\, F\{y(t)\} = F\{\text{H}(t)\sin \omega t\}$$
$$= \frac{\omega}{\omega^2 - z^2} \text{ for } t > 0.$$

Let

$$\text{P}(iz) = (iz)^2 + 3iz + 2 \text{ and } \text{F}(z) = \frac{\omega}{\omega^2 - z^2}$$

Then the solution is given by

$$y(t) = \frac{1}{2\pi}\int_{-\infty+i\gamma}^{\infty+i\gamma} \frac{\text{F}(z)\, e^{izt}}{\text{P}(iz)}\, dz$$

Thus

$$y(t) = \frac{1}{2\pi}\int_{-\infty+i\gamma}^{\infty+i\gamma} \frac{\omega}{\omega^2 - z^2}\cdot\frac{e^{izt}}{(iz)^2 + 3iz + 2}\, dz$$

$$= \frac{\omega}{2\pi}\int_{-\infty+i\gamma}^{\infty+i\gamma} \frac{e^{izt}}{(z^2 - \omega^2)(z^2 - 3iz - 2)}\, dz$$

The integral on the right-hand side can be evaluated by contour integration. The singularities within the contour are poles $z = \pm\omega$, $z = 2i$, and $z = i$ (Figure 7.7)

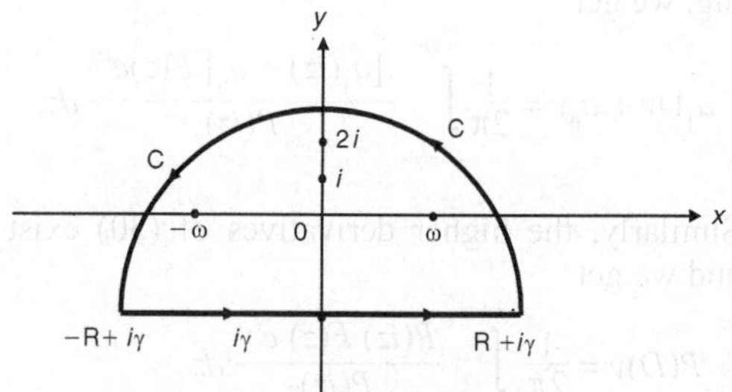

Figure 7.7

Further

$$\text{Res}(i) = \lim_{z\to i}(z-i)\frac{e^{izt}}{(z-i)(z^2-\omega^2)(z-2i)}$$
$$= \frac{e^{-t}}{i(\omega^2+1)},$$

$$\text{Res}(2i) = \lim_{z\to 2i}(z-2i)\frac{e^{izt}}{(z-i)(z-2i)(z^2-\omega^2)}$$
$$= \frac{ie^{-2t}}{\omega^2+4},$$

$$\text{Res}(\omega) = \lim_{z\to\omega}(z-\omega)\frac{e^{izt}}{(z-i)(z-2i)(z^2-\omega^2)}$$
$$= \frac{e^{i\omega t}}{2\omega(\omega^2 - 3i\omega - 2)},$$

$$\text{Res}(-\omega) = \lim_{z\to-\omega}(z+\omega)\frac{e^{izt}}{(z^2-3iz-2)(z+\omega)(z-\omega)}$$
$$= \frac{-e^{-i\omega t}}{2\omega(\omega^2 + 3i\omega - 2)}.$$

Therfore, by Cauchy-residue theorem,

$$y(t) = \frac{\omega}{2\pi}.2\pi i\left[\frac{ie^{-2t}}{\omega^2+4} + \frac{e^{-t}}{i(\omega^2+1)} + \frac{e^{i\omega t}}{2\omega(\omega^2-3i\omega-2)} - \frac{e^{-i\omega t}}{2\omega(\omega^2+3i\omega-2)}\right]$$

$$= \frac{-\omega e^{-2t}}{\omega^2+4} + \frac{\omega e^{-t}}{\omega^2+1} + \frac{ie^{i\omega t}}{2(\omega^2-3i\omega-2)} - \frac{ie^{-i\omega t}}{2(\omega^2+3i\omega-2)}.$$

EXAMPLE 7.37

Using Convolution theorem for Fourier transforms, solve

$$\frac{d^2y}{dx^2} - y = -H(1-|x|), \quad -\infty < x < \infty$$

$$y(x) \to 0 \text{ and } y'(x) \to 0 \text{ as } |x| \to \infty.$$

Solution. The Heaviside's unit step function H (Figure 7.8) is defined by

$$H(1-|x|) = \begin{cases} 1 & \text{for } |x| < 1 \\ 0 & \text{for } |x| > 1. \end{cases}$$

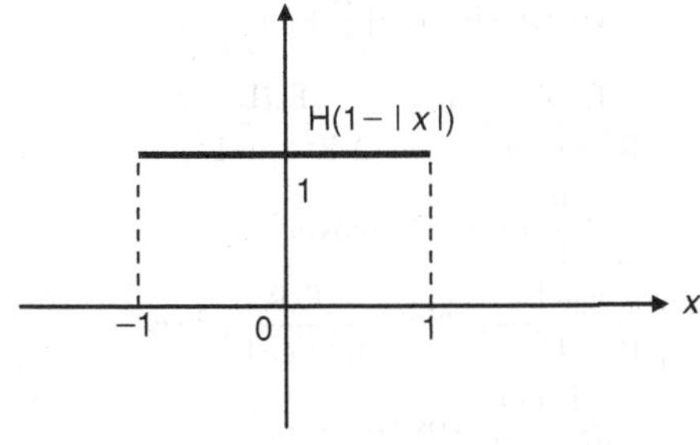

Figure 7.8

Taking Fourier transform of both sides of the given equation, we get

$$\{(i\omega)^2 - 1\}\, F\{y(t)\} = -F\{H(1-|x|)\}$$

$$= -\int_{-1}^{1} e^{-i\omega x}\,dx = -\frac{2\sin\omega}{\omega}.$$

Therefore

$$(\omega^2+1)\, F\{y|t|\} = \frac{2\sin\omega}{\omega},$$

and so

$$F\{y|t|\} = \frac{2\sin\omega}{\omega(\omega^2+1)}.$$

But

$$F^{-1}\left\{\frac{2\sin\omega}{\omega}\right\} = H(1-|x|)$$

and

$$F^{-1}\left\{\frac{1}{\omega^2+1}\right\} = \frac{1}{2}e^{-|x|}.$$

Therefore, by Convolution theorem

$$y(x) = \frac{1}{2}\int_{-\infty}^{\infty} e^{-|u|}\, H(1-|x-u|)\,du = \frac{1}{2}\int_{x-1}^{x+1} e^{-|u|}\,du$$

$$= \begin{cases} \sinh(1)e^{x} & \text{for } -\infty < x < -1 \\ 1 - e^{-1}\cosh x & \text{for } -1 \le x \le 1 \\ \sinh(1)\,e^{-x} & \text{for } 1 < x < \infty, \end{cases}$$

because

$$\int_{x-1}^{x+1} e^{u-|u|}\,du = \begin{cases} \int_{x-1}^{x+1} e^{u}\,du & \text{for } -\infty < x < -1 \\ \int_{x-1}^{0} e^{u}du + \int_{0}^{x+1} e^{-u}\,du & \text{for } -1 \le x \le 1 \\ \int_{x-1}^{x+1} e^{-u}\,du & \text{for } 1 < x < \infty. \end{cases}$$

EXAMPLE 7.38

Given that current I = 0 at $t = 0$, find I in the following LR circuit (Figure 7.9) for $t > 0$.

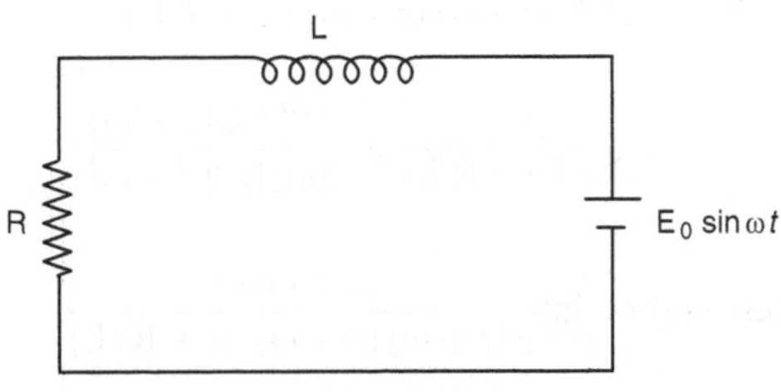

Figure 7.9

Solution. Since the voltage drop across a resistance R is RI and voltage drop across the inductor L is $L\frac{dI}{dt}$, by Kirchhoff's law, the differential equation governing the given circuit is

$$L\frac{dI}{dt} + RI = E_0 \sin \omega t,\ I(0) = 0,$$

where L, R, E_0 and ω are constants.
Write the equation in the form

$$\frac{dI}{dt} + \frac{R}{L}I = \frac{E_0}{L}\sin \omega t,\ I(0) = 0 \qquad (31)$$

Taking Fourier transform of both sides of (31), we get

$$\left(iz + \frac{R}{L}\right)F\{I(t)\} = \frac{E_0}{L}F\{\sin \omega t\} = \frac{E_0}{L}\cdot\frac{\omega}{\omega^2 - z^2}.$$

Let

$$P(iz) = iz + \frac{R}{L}$$

and

$$G(z) = \frac{E_0\omega}{L(\omega^2 - z^2)}.$$

Then the solution of (31) is given by

$$I(t) = \frac{1}{2\pi}\int_{-\infty+iy}^{\infty+iy} \frac{G(z)\,e^{izt}}{P(iz)}\,dz$$

$$= -\frac{E_0\omega}{2\pi L}\int_{-\infty+iy}^{\infty+iy} \frac{e^{izt}}{(z^2-\omega^2)(iz+R/L)}\,dz.$$

We shall evaluate the integral using Contour integration. The singularities of the integrand are simple poles at $z = \pm\,\omega$ and $z = i\frac{R}{L}$. Further,

$$\text{Res}(\omega) = \lim_{z\to\omega} \frac{(z-\omega)\,e^{izt}}{(z-\omega)(z+\omega)(iz+R/L)}$$

$$= \frac{e^{i\omega t}}{2\omega(i\omega + R/L)} = \frac{e^{i\omega t}(R/L - i\omega)}{2\omega(R^2/L^2 + \omega^2)}$$

$$\text{Res}(-\omega) = \lim_{z\to-\omega} \frac{(z+\omega)\,e^{izt}}{(z+\omega)(z-\omega)(iz+R/L)}$$

$$= \frac{e^{-i\omega t}}{(-2\omega)(R/L - i\omega)}$$

$$= \frac{e^{-i\omega t}(R/L + i\omega)}{(-2\omega)(R^2/L^2 + \omega^2)}$$

$$\text{Res}(iR/L) = \lim_{z\to iR/L} \frac{(z - iR/L)\,e^{izt}}{(z+\omega)(z-\omega)(iz+R/L)}$$

$$= \frac{1}{i}\lim_{z\to iR/L}\frac{e^{izt}}{z^2-\omega^2}$$

$$= -\frac{1}{i}\frac{e^{-Rt/L}}{R^2/L^2+\omega^2}$$

$$= \frac{i\,L^2 e^{-Rt/L}}{R^2 + L^2\omega^2}.$$

Hence, by Cauchy's residue theorem, we get

$$I(t) = -\frac{E_0\omega}{2\pi L}.2\pi i\sum R$$

$$= \frac{E_0\omega L}{R^2+L^2\omega^2}e^{-Rt/L} - \frac{E_0\omega i}{L}\left[\frac{e^{i\omega t}(R/L - i\omega)L^2}{2\omega(R^2+\omega^2L^2)} + \frac{e^{-i\omega t}(R/L+i\omega)L^2}{(-2\omega)(R^2+\omega^2L^2)}\right]$$

$$= \frac{E_0\omega L}{R^2+L^2\omega^2}e^{-Rt/L} - \frac{E_0 iL}{2}\left[\frac{e^{i\omega t}(R/L - i\omega)}{R^2+\omega^2L^2} - \frac{e^{-i\omega t}(R/L+i\omega)}{R^2+\omega^2L^2}\right]$$

$$= \frac{E_0\omega L}{R^2+L^2\omega^2}e^{-Rt/L} - \frac{E_0 iL}{2(R^2+\omega^2L^2)}\left[(\cos\omega t + i\sin\omega t)\left(\frac{R}{L} - i\omega\right) - (\cos\omega t - i\sin\omega t)\left(\frac{R}{L} + i\omega\right)\right]$$

$$= \frac{E_0\omega L}{R^2+L^2\omega^2}e^{-Rt/L} - \frac{E_0 iL}{2(R^2+\omega^2L^2)} \times [2i\frac{R}{L}\sin\omega t - 2i\omega\cos\omega t]$$

$$= \frac{E_0\omega L}{R^2+L^2\omega^2}e^{-Rt/L} + \frac{E_0R}{R^2+\omega^2L^2}\sin\omega t - \frac{E_0L\omega}{R^2+\omega^2L^2}\cos\omega t.$$

7.13 APPLICATION TO PARTIAL DIFFERENTIAL EQUATIONS

Fourier transform can also be applied to solve some boundary-value and initial-value problems for partial differential equations with constant coefficients. Consider the heat equation

$$\frac{\partial u}{\partial t} = k\frac{\partial^2 u}{\partial x^2}, x \in \mathbb{R}; t > 0.$$

with $u(x, 0) = f(x)$, $x \in \mathbb{R}$ and $t > 0$ and bounded $u(x, t)$. This equation represents heat conduction in a rod of infinite length. We shall use method of separation of variables. So let

$$u(x, t) = \mathrm{T}(t)\,\mathrm{X}(x) \tag{32}$$

Then the given heat equation becomes

$$\mathrm{T}'(t)\,\mathrm{X}(x) = k\mathrm{X}''(x)\,\mathrm{T}(t)$$

or

$$\frac{\mathrm{T}'(t)}{k\mathrm{T}(t)} = \frac{\mathrm{X}''(x)}{\mathrm{X}(x)} = C \text{ (constant of separation)}$$

Thus

$$\mathrm{X}''(x) - \mathrm{C}\mathrm{X}(x) = 0, \tag{33}$$

$$\mathrm{T}'(t) - \mathrm{C}k\mathrm{T}(t) = 0. \tag{34}$$

The characteristic equation of (34) is $s - Ck = 0$ and so $\mathrm{T}(t) = \alpha e^{Ckt}$. Since $\mathrm{T}(t)$ should be bounded, we have $\mathrm{C} \le 0$. We may thus choose $\mathrm{C} = -\omega^2$ for real ω. If we take $\omega = 0$, then (33) implies that $\mathrm{X}(x) = \alpha x + \beta$. Since $\mathrm{X}(x)$ should be bounded, we must have $\alpha = 0$. For $\omega \neq 0$, the equation (33) has general solution $\mathrm{X}(x) = \alpha e^{i\omega x} + \beta e^{-i\omega x}$. This function is bounded for all α and β since

$$|\mathrm{X}(x)| \le |\alpha||e^{i\omega x}| + |\beta||e^{-i\omega x}| = |\alpha| + |\beta|$$

Thus, we have

$$\mathrm{X}(x)\,\mathrm{T}(t) = e^{i\omega x}\,e^{-k\omega 2t},\ \omega \in \mathbb{R}$$

By superposition rule, the linearity property not only holds for finite sum of inputs but also for infinite sum of inputs and even for "continuous sums" that is for integrals. Thus

$$u(x, t) = \int_{-\infty}^{\infty} \mathrm{F}(\omega)\,e^{-k\omega^2 t}\,e^{i\omega x}\,d\omega$$

for some function $\mathrm{F}(\omega)$.

Putting $t = 0$, we have by initial condition,

$$f(x) = \int_{-\infty}^{\infty} \mathrm{F}(\omega)\,e^{i\omega x}\,d\omega.$$

Thus, $\mathrm{F}(\omega)$ is the Fourier transform of $\frac{f(x)}{2\pi}$. Therefore,

$$\mathrm{F}(\omega) = \frac{1}{2\pi}\int_{-\infty}^{\infty} f(x)\,e^{-i\omega x}\,dx.$$

Remark 7.7. Instead of using variable separation method, we could have started by taking Fourier transform of both sides of the equation. By doing so, we would have

$$\int_{-\infty}^{\infty} \frac{\partial u}{\partial t}\,e^{-i\omega x}\,dx = k\int_{-\infty}^{\infty} \frac{\partial^2 u}{\partial x^2}\,e^{-i\omega x}\,dx,$$

that is,

$$\frac{\partial}{\partial t}\mathrm{F}(\omega) = k(-i\omega)^2\,\mathrm{F}(\omega),$$

that is,

$$\frac{\partial}{\partial t}\mathrm{F}(\omega) + k\omega^2\mathrm{F}(\omega) = 0$$

The solution of this differential equation is

$$\mathrm{F}(\omega) = A\,e^{-k\omega^2 t}$$

The initial condition $u(x, 0) = f(x)$ implies

$$\mathrm{F}\{u(x, 0)\} = \mathrm{F}\{f\}.$$

Therefore substituting $t = 0$ in the above solution we get $\mathrm{A} = \mathrm{F}\{f\}$. Thus,

$$\mathrm{F}(\omega) = \mathrm{F}\{f\}e^{-k\omega^2 t}.$$

Taking inverse transform, we get

$$u(x, t) = \frac{1}{2\pi}\int_{-\infty}^{\infty} \mathrm{F}\{f\}e^{-k\omega^2 t}\,e^{i\omega x}\,d\omega.$$

EXAMPLE 7.39

Solve the heat equation

$$\frac{\partial u}{\partial t} = k\frac{\partial^2 u}{\partial x^2}, x \in \mathbb{R}, t > 0$$

with $u(x, 0) = f(x)$ and bounded (x, t), where

$$f(x) = \begin{cases} 1 \text{ for } |x| < a \\ 0 \text{ for } |x| > a. \end{cases}$$

Solution. The solution of the given equation is

$$u(x, t) = \int_{-\infty}^{\infty} \mathrm{F}(\omega)\,e^{-k\omega^2 t}\,e^{i\omega x}\,d\omega,$$

where

$$F(\omega) = \frac{1}{2\pi}\int_{-\infty}^{\infty} f(x)\, e^{-i\omega x}\, dx = \frac{1}{2\pi}\int_{-a}^{a} e^{-i\omega x}\, dx$$

$$= \frac{1}{2\pi}\left[\frac{e^{-i\omega x}}{-i\omega}\right]_{-a}^{a} = \frac{1}{\pi}\frac{\sin a\omega}{\omega}.$$

Thus,

$$u(x, t) = \frac{1}{\pi}\int_{-\infty}^{\infty} \frac{\sin a\omega}{\omega}\, e^{-k\omega^2 t} e^{\,i\omega x}\, d\omega$$

$$= \frac{1}{\pi}\int_{-\infty}^{\infty} \frac{\sin a\omega}{\omega}\, e{-}k\omega^2 t\,(\cos \omega x + i \sin \omega x)\, d\omega$$

$$= \frac{2}{\pi}\int_{0}^{\infty} \frac{\sin a\omega}{\omega}\, e-^{k\omega^2 t} \cos \omega x d\omega$$

$$= \frac{1}{\pi}\int_{0}^{\infty} e^{-k\omega^2 t}\, \frac{\sin (a + x)\omega + \sin(a - x)\omega}{\omega}\, d\omega.$$

EXAMPLE 7.40

Solve Example 7.39 without using the method of separation of variables.

Solution. Taking Fourier transform with respect to x on both sides of the given heat equation, we get

$$\int_{-\infty}^{\infty} \frac{\partial u}{\partial t}\, e^{-i\omega x}\, dx = k\int_{-\infty}^{\infty} \frac{\partial^2 u}{\partial x^2}\, e^{-i\omega x}\, dx$$

or

$$\frac{\partial}{\partial t} F\{u\} = k(i\omega)^2\, F\{u\} = -k\omega^2\, F\{u\}$$

or

$$\frac{\partial}{\partial t} F\{u\} + k\omega^2\, F\{u\} = 0$$

Hence

$$F\{u\} = A\, e^{-k\omega^2 t} \qquad (35)$$

Since $u(x, 0) = f(x)$, the Fourier transform of $u(x, 0)$ is

$$F\{u(x, 0)\} = \int_{-\infty}^{\infty} f(x)\, e^{-i\omega x}\, dx$$

$$= \int_{-a}^{a} e^{-i\omega x}\, dx = \frac{2 \sin \omega a}{\omega}.$$

Hence putting $t = 0$ in (35) yields

$$A = \frac{2 \sin \omega a}{\omega}.$$

Thus

$$F\{u\} = \frac{2 \sin \omega a}{\omega}\, e^{-k\omega^2 t},$$

The application of Fourier integral theorem now yields

$$u(x, t) = \frac{1}{2\pi}\int_{-\infty}^{\infty} \frac{2 \sin \omega a}{\omega}\, e^{-k\omega^2 t}\, e^{i\omega x}\, d\omega$$

$$= \frac{1}{\pi}\int_{-\infty}^{\infty} \frac{\sin \omega a}{\omega}\, e^{-k\omega^2 t}\,(\cos \omega x + i \sin \omega x)\, d\omega$$

$$= \frac{2}{\pi}\int_{0}^{\infty} \frac{\sin \omega a}{\omega}\, e^{-k\omega^2 t} \cos \omega x d\omega$$

$$= \frac{1}{\pi}\int_{0}^{\infty} e^{-k\omega^2 t}\, \frac{\sin (a + x)\,\omega + \sin (a - x)\omega}{\omega}\, d\omega.$$

EXAMPLE 7.41

Solve

$$\frac{\partial u}{\partial t} = \frac{\partial^2 u}{\partial x^2},\ t > 0$$

subject to $u(x, 0) = e^{-x^2}$.

Solution. Taking Fourier transform of the given heat equation, we get

$$\frac{\partial}{\partial t} F\{u\} + \omega^2 F\{u\} = 0.$$

The solution of this equation is

$$F\{u\} = A\, e^{-\omega^2 t} \qquad (36)$$

The initial condition is $u(x, 0) = -e^{x^2}$. Taking Fourier transform of this condition, we have

$$F\{u(x, 0)\} = F\{e^{-x^2}\}$$

$$= \int_{-\infty}^{\infty} e^{-x^2}\, e^{-i\omega x}\, dx = \int_{-\infty}^{\infty} e^{-(x^2 + i\omega x)}\, dx$$

$$= \int_{-\infty}^{\infty} e\left[\left(x + \frac{i\omega}{2}\right)^2 + \frac{\omega^2}{4}\right] dx$$

$$= e^{-\omega^2/4}\int_{-\infty}^{\infty} e^{-\left(\frac{x + i\omega}{2}\right)^2}\, dx$$

$$= e^{-\omega^2/4}\int_{-\infty}^{\infty} e^{-u^2}\, du = \sqrt{\pi}\, e^{-\omega^2/4}, \qquad (37)$$

since Gauss integral $\int_{-\infty}^{\infty} e^{-u^2}\,du = \sqrt{\pi}$. The equations (36) and (37) yield $A = \sqrt{\pi}\, e^{-\omega^2/4}$.

Hence

$$F\{u\} = \sqrt{\pi}\, e^{-\omega^2\left(t+\frac{1}{4}\right)}.$$

Taking inverse transform,we have by Example 7.8,

$$\begin{aligned} u(x,t) &= \frac{1}{2\pi}\int_{-\infty}^{\infty} \sqrt{\pi}\, e^{-\omega^2\left(t+\frac{1}{4}\right)} e^{i\omega x}\, d\omega \\ &= \frac{\sqrt{\pi}}{2\pi}\int_{-\infty}^{\infty} e^{-\omega^2(1+4t)/4}\, e^{i\omega x}\, d\omega \\ &= \frac{\sqrt{\pi}}{2\pi}\cdot 2\int_{-\infty}^{\infty} e^{-\left(\frac{1+4t}{4}\right)\omega^2} e^{i\omega x}\, d\omega \\ &= \frac{1}{\sqrt{\pi}}\cdot \frac{\sqrt{\pi}}{\left(\frac{1+4t}{4}\right)^{1/2}} e^{-\frac{x^2}{1+4t}} \\ &= \frac{2}{(1+4t)^{1/2}} e^{-\frac{x^2}{1+4t}}. \end{aligned}$$

EXAMPLE 7.42

Solve

$$\frac{\partial u}{\partial t} = k\frac{\partial^2 u}{\partial x^2},\ x > 0$$

subject to the conditions

$u(x, 0) = e^{-x}$,
$u(0, t) = 0,\ u(x, t) = 0,\ t \ge 0.$

Solution. This is a case of semi-infinite bar whose ends are kept at zero temperature. Since the boundary conditions do not involve derivative, we use sine transform. Taking Fourier sine transform, we get

$$\int_{-\infty}^{\infty} \frac{\partial u}{\partial t}\sin \omega x\, dx = k\int_{-\infty}^{\infty} \frac{\partial^2 u}{\partial x^2}\sin \omega x\, dx$$

or

$$\frac{\partial}{\partial t} F_s(\omega) = k(i\omega)^2 F_s(\omega)$$

or

$$\frac{\partial}{\partial t} F_s(\omega) + \omega^2 k\, F_s(\omega) = 0.$$

The solution of this equation is

$$F_s(\omega) = A\, e^{-k\omega^2 t}. \tag{38}$$

Taking Fourier sine transform of $u(x, 0) = e^{-x}$, we have

$$F_s\{e^{-x}\} = \frac{\omega}{1+\omega^2}.$$

Therefore, taking $t = 0$ in (38), we have $A = \frac{\omega}{1+\omega^2}$.
Hence

$$F_s(\omega) = \frac{\omega}{1+\omega^2} e^{-k\omega^2 t}.$$

Therefore, Fourier inversion formula for sine transform yields

$$u(x,t) = \frac{2}{\pi}\int_0^{\infty} \frac{\omega}{1+\omega^2} e^{-k\omega^2 t}\sin \omega x\, d\omega.$$

EXAMPLE 7.43

Solve

$$\frac{\partial u}{\partial t} = k\frac{\partial^2 u}{\partial x^2},\ t > 0,\ 0 \le x \le \pi$$

subject to the conditions

$$u(x, 0) = 2x,\ u(0, t) = u(\pi, t) = 0.$$

Solution. The solution of the given heat equation is

$$u(x,t) = \frac{2}{\pi}\int_0^{\infty} F_s(\omega)\, e^{-\omega^2 kt}\sin \omega x\, d\omega.$$

But integration by parts yield

$$F_s(\omega) = \int_0^{\infty} 2x \sin \omega x\, dx = -\frac{2\pi}{\omega}\cos \omega\pi.$$

Therefore,

$$u(x,t) = -4\int_0^{\infty} \frac{\cos \omega\pi}{\omega} e^{-\omega^2 kt}\sin \omega x\, d\omega.$$

EXAMPLE 7.44

Solve

$$\frac{\partial u}{\partial t} = k\frac{\partial^2 u}{\partial x^2},$$

subject to the conditions

$$u(x, 0) = 0, x \geq 0$$
$$u_x(0, t) = -\mu(\text{constant}), t > 0.$$

Solution. The problem concerns infinite half-plane and involves derivative of $u(x, t)$. Therefore, we use Fourier cosine transform to solve the problem. So taking Fourier cosine transform of the given heat equation, we get

$$\int_0^\infty \frac{\partial u}{\partial t} \cos \omega x \, dx = k \int_0^\infty \frac{\partial^2 u}{\partial x^2} \cos \omega x \, dx$$

or

$$\frac{\partial}{\partial t} F_c\{u\} = k\left[\frac{\partial u}{\partial x} \cos \omega x\right]_0^\infty + k\omega \int_0^\infty \frac{\partial u}{\partial x} \sin \omega x \, dx$$

$$= -k\frac{\partial}{\partial x} u(0, t) + k\omega\,[u \sin \omega x]_0^\infty$$

$$- k\omega^2 \int_0^\infty u \cos \omega x \, dx$$

$$= k\mu - k\omega^2 F_c\{u\}$$

or

$$\frac{\partial}{\partial t} F_c\{\mu\} + k\omega^2 F_c\{u\} = k\mu.$$

The integration factor for this equation is

$$e^{\int k\omega^2 \, dt} = e^{k\omega^2 t}$$

and so the solution is

$$e^{k\omega^2 t} F_c(u) = A + k\mu \int e^{k\omega^2 t} \, dt$$

$$= A + \frac{\mu}{\omega^2} e^{k\omega^2 t}. \tag{39}$$

The initial condition $u(x, 0) = 0$ implies $F_c\{u(x, 0)\} = 0$. Therefore, (39) yields $A = -\frac{\mu}{\omega^2}$. Hence the solution is

$$F_c\{u\} = \frac{\left(-\frac{\mu}{\omega^2} + \frac{\mu}{\omega^2} e^{k\omega^2 t}\right)}{e^{k\omega^2 t}} = \frac{\mu}{\omega^2}(1 - e^{-k\omega^2 t}).$$

Taking inverse Fourier cosine transform, we get

$$u(x, t) = \frac{2\mu}{\pi} \int_0^\infty \frac{\cos \omega x}{\omega^2} (1 - e^{-k\omega^2 t}) \, d\omega.$$

EXAMPLE 7.45

Solve the heat equation

$$\frac{\partial u}{\partial t} = k \frac{\partial^2 u}{\partial x^2}, \; 0 < x < \infty$$

subject to the conditions

$$u(x, 0) = e^{-ax}, a > 0$$
$$u_x(0, t) = 0, u_x(x, t) = 0, t > 0.$$

Solution. In this problem, the ends of the bar have been insulated and kept at zero temperature. The boundary conditions involve derivative and so we use Fourier cosine transform. Taking Fourier cosine transform, we have

$$\frac{\partial}{\partial u} F_c(\omega) + k\omega^2 F_c(\omega) = 0.$$

The fundamental solution to this equation is

$$F_c(\omega) = A e^{-k\omega^2 t}. \tag{40}$$

Taking Fourier transform of the initial condition $u(x, 0) = e^{-ax}$, we get

$$F_c\{e^{-ax}\} = \int_0^\infty e^{-ax} \cos \omega x \, dx = \frac{a}{a^2 + \omega^2}.$$

Therefore, for $t = 0$, the solution (40) implies $A = \frac{a}{a^2 + \omega^2}$. Hence

$$F_c(\omega) = \frac{a}{a^2 + \omega^2} e^{-k\omega^2 t}.$$

Now using Fourier integral theorem for Fourier cosine transforms, we have

$$u(x, t) = \frac{2a}{\pi} \int_{-\infty}^\infty \frac{1}{a^2 + \omega^2} e^{-k\omega^2 t} \cos \omega x \, d\omega.$$

EXAMPLE 7.46

Solve

$$\frac{\partial u}{\partial t} = k \frac{\partial^2 u}{\partial x^2}, \; 0 < x < \pi$$

subject to the conditions

$$u(x, 0) = e^{-x}, 0 < x < \pi$$
$$u(0, t) = 0, u(\pi, t) = 0, t \geq 0.$$

Solution. The problem concerns *finite half-space* and, therefore, we should use *finite Fourier sine transform*. So taking finite Fourier sine transform, we have

$$\frac{\partial}{\partial t} F_s(n) = k\int_0^{\pi} \frac{\partial^2 u}{\partial x^2} \sin nx\, dx$$

or

$$\frac{\partial}{\partial t} F_s(n) = k\left\{\left[\frac{\partial u}{\partial x}\sin nx\right]_0^{\pi} - n\int_0^{\pi}\frac{\partial u}{\partial x}\cos nx\, dx\right\}$$

$$= -nk\left\{\left[u\cos nx\right]_0^{\pi} + n\int_0^{\pi} u\sin nx\, dx\right\}$$

$$= -n^2 k F_s(n)$$

or

$$\frac{\partial^2}{\partial t} F_s(n) + n^2 k\, F_s(n) = 0.$$

The solution of this equation is

$$F_s(n) = Ae^{-kn^2 t}. \tag{41}$$

The initial condition is $u(x, 0) = e^{-x}$. Taking its finite Fourier sine transform, we have

$$F_s(n, 0) = \int_0^{\pi} e^{-x}\sin nx\, dx$$

$$= \left[\frac{e^{-x}\cos nx}{-n}\right]_0^{\pi} - \frac{1}{n}\int_0^{\pi} e^{-x}\cos nx\, dx$$

$$= \left[e^{-x}\frac{\cos nx}{-n}\right]_0^{\pi} - \frac{1}{n}\left\{\left[e^{-x}\frac{\sin nx}{n}\right]_0^{\pi} - \int_0^{\pi} -e^{-x}\frac{\sin nx}{n}\, dx\right\}$$

$$= \left[e^{-x}\frac{\cos nx}{-n}\right]_0^{\pi} - \frac{1}{n^2}\left[e^{-x}\sin nx\right]_0^{\pi} - \frac{1}{n^2}\int_0^{\pi} e^{-x}\sin nx\, dx.$$

Thus,

$$\left(1+\frac{1}{n^2}\right)F_s(n,0) = \left[e^{-x}\frac{\cos nx}{-n}\right]_0^{\pi} = \frac{1-(-1)^n e^{-\pi}}{n}$$

or

$$F_s(n)\, \frac{n[1-(-1)^h e^{-\pi}]}{1+n^2}. \tag{42}$$

Putting $t = 0$ in (41), the expression (42) implies

$$A = \frac{n\,[1-(-1)^n e^{-\pi}]}{1+n^2}.$$

Thus (41) reduces to

$$F_s(n) = \frac{n[1-(-1)^n e{-}\pi]}{1+n^2}.e^{-n2kt}.$$

Hence, by inverse finite Fourier sine transform, we have

$$u(x, t) = \frac{2}{\pi}\int_{n=1}^{\infty} \frac{n\,[1-(-1)^n e^{-\pi}]}{1+n^2}\, e^{-n^2 kt}\sin nx.$$

EXAMPLE 7.47

Solve *one-dimensional wave equation*

$$\frac{\partial^2\phi}{\partial x^2} = \frac{1}{c^2}\frac{\partial^2\phi}{\partial t^2}$$

with conditions $\phi(x, 0) = f(x)$ and $\phi_t(x, 0) = 0$.

Solution. Taking Fourier transform of both sides, we have

$$\int_{-\infty}^{\infty} \frac{\partial^2\phi}{\partial x^2} e^{-i\omega x}\, dx = \frac{1}{c^2}\int_{-\infty}^{\infty}\frac{\partial^2\phi}{\partial t^2} e^{-i\omega x}\, dx$$

or

$$-\omega^2 F\{\phi\} = \frac{1}{c^2}\frac{\partial^2}{\partial t^2}\int_{-\infty}^{\infty} \phi e^{-i\omega x}\, dx = \frac{1}{c^2}\frac{\partial^2}{\partial t^2} F\{\phi\}.$$

Thus,

$$\frac{\partial^2}{\partial t^2} F\{\phi\} + \omega^2 c^2 F\{\phi\} = 0$$

and so

$$F\{\phi\} = A\cos c\omega t + B\sin c\omega t. \tag{43}$$

The initial condition $\phi(x, 0) = f(x)$ implies that

$$\int_{-\infty}^{\infty} \phi e^{-i\omega x}\, dx = \int_{-\infty}^{\infty} f(x)\, e^{-i\omega x}\, dx,$$

which means that

$$F\{\phi\} = F\{f\} \text{ at } t = 0. \tag{44}$$

The condition $\frac{\partial\phi}{\partial t} = 0$ at $t = 0$ implies

$$\frac{\partial}{\partial t}\left\{\int_{-\infty}^{\infty} \phi e^{-i\omega t}\, dx\right\} = 0,$$

which means that,

$$\frac{\partial}{\partial t}[\mathrm{F}\{\phi\}] = 0 \text{ at } t = 0. \tag{45}$$

Using (43) and (44), we get A = F(f). Further, differentiating (43) w.r.t. t, we get

$$\frac{\partial}{\partial t}[\mathrm{F}\{\phi\}] = -\mathrm{A}\, c\omega \sin c\omega t + \mathrm{B}\, c\omega \cos c\omega t$$

Therefore by (45), we have $0 = \mathrm{B}\, c\omega$ which yields B = 0. Hence (43) reduces to

$$\mathrm{F}\{\phi\} = \mathrm{F}\{f\} \cos c\omega t.$$

Now Fourier inversion theorem yields

$$\phi(x, t) = \frac{1}{2\pi}\int_{-\infty}^{\infty} \mathrm{F}(\omega) \cos c\omega t\, e^{i\omega x} d\omega.$$

The solution can further be simplified to give

$$\phi(x, t) = \frac{1}{2\pi}\int_{-\infty}^{\infty} \mathrm{F}(\omega)\, e^{i\omega x}\left[\frac{e^{i\omega t} + e^{-i\omega t}}{2}\right] d\omega$$

$$= \frac{1}{2}\left[\frac{1}{2\pi}\int_{-\infty}^{\infty} \mathrm{F}(\omega)\, e^{i\omega(x+ct)}\, d\omega + \frac{1}{2\pi}\int_{-\infty}^{\infty} \mathrm{F}(\omega)\, e^{i\omega(x-ct)}\, d\omega\right]$$

$$= \frac{1}{2}[f(x+ct) + f(x-ct)].$$

EXAMPLE 7.48

Solve *two-dimensional Laplace's equation*

$$\frac{\partial^2\phi}{\partial x^2} + \frac{\partial^2\phi}{\partial y^2} = 0, \ y > 0$$

subject to the conditions

$$\phi_x \text{ and } \phi \to 0 \text{ as } x^2 + y^2 \to \infty$$

$$\phi(x, 0) = \begin{cases} 1 \text{ for } |x| \le 1 \\ 0 \text{ for } |x| > 1. \end{cases}$$

Solution. The problemis known as *Dirichlet's problem or boundary value problem of the first kind.* Taking Fourier transform with respect to x, we have

$$(i\omega)^2 \mathrm{F}\{\phi\} + \int_{-\infty}^{\infty} \frac{\partial^2\phi}{\partial y^2} e^{-i\omega x} dx = 0$$

or

$$-\omega^2 \mathrm{F}\{\phi\} + \frac{\partial^2}{\partial y^2}\left(\int_{-\infty}^{\infty} \phi\, e^{-i\omega x} dx\right) = 0$$

or

$$-\omega^2 \mathrm{F}\{\phi\} + \frac{\partial^2}{\partial y^2}[\mathrm{F}\{\phi\}] = 0$$

or

$$\frac{\partial^2 \mathrm{F}\{\phi\}}{\partial y^2} - \omega^2 \mathrm{F}\{\phi\} = 0. \tag{46}$$

The conditions that ϕ and its derivative vanish identically as $x \to \pm\infty$ have been used since we have applied derivative theorem to calculate Fourier transform of $\frac{\partial^2\phi}{\partial x^2}$.

Since $\mathrm{F}\{\phi\} \to 0$ for large y, the solution to equation (46) is

$$\mathrm{F}\{\phi\} = \mathrm{A}e^{-|\omega| y} \tag{47}$$

The initial condition is

$$\phi(x, 0) = \begin{cases} 1 & \text{for } |x| \le 1 \\ 0 & \text{for } |x| > 1. \end{cases}$$

Therefore,

$$\mathrm{F}\{\phi(x, 0)\} = \int_{-\infty}^{\infty} \phi(x, 0) e^{-i\omega x} dx = \int_{-1}^{1} e^{-i\omega x} dx$$

$$= \frac{e^{i\omega} - e^{-i\omega}}{i\omega} = \frac{2\sin\omega}{\omega}.$$

Thus for $y = 0$, (47) yields $\mathrm{A} = \frac{2\sin\omega}{\omega}$ and so

$$\mathrm{F}\{\phi\} = \frac{2\sin\omega}{\omega} e^{-|\omega| y}.$$

To find ϕ, we use convolution theorem. So, let

$$\frac{2\sin\omega}{\omega} = \mathrm{F}_1(\omega), \text{ and}$$

$$e^{-|\omega| y} = \mathrm{F}_2(\omega).$$

(Fourier transform with respect to x)

Then

$$F^{-1}\left\{\frac{2\sin\omega}{\omega}\right\} = \phi(x, 0)$$

and

$$F^{-1}\{e^{-|\omega|y}\} = \frac{1}{2\pi}\int_{-\infty}^{\infty} e^{-|\omega|y}e^{i\omega x}\,d\omega.$$

Since $e^{-|\omega|y}$ is even, we have

$$F^{-1}\{e^{-|\omega|y}\} = \frac{2}{2\pi}\int_{0}^{\infty} e^{-\omega y}\cos\omega x\,d\omega$$

$$= \frac{1}{\pi}\left[\frac{y}{x^2+y^2}\right], \text{ since}$$

$$\int_{0}^{\infty} e^{-at}\cos st\,dt = \frac{a}{s^2+a^2},\ a > 0.$$

Hence Convolution theorem yields

$$\phi(x, y) = F^{-1}\{F_1(\omega)F_2(\omega)\}$$

$$= \frac{1}{\pi}\int_{-\infty}^{\infty} \phi(x, 0)\frac{y}{(x-y)^2+y^2}\,dt$$

$$= \frac{y}{\pi}\int_{-1}^{1}\frac{1}{(x-t)^2+y^2}\,dt,$$

by definition of $\phi(x, 0)$,

$$= \frac{y}{x}\left[\tan^{-1}\frac{x-1}{y} + \tan^{-1}\frac{x+1}{y}\right].$$

Remark 7.8. The solution formula

$$\phi(x, y) = \frac{y}{\pi}\int_{-\infty}^{\infty}\frac{f(t)}{(x-t)^2+y^2}\,dt,\ y > 0,$$

is known as *Poisson integral formula* for the half-plane.

EXAMPLE 7.49

Solve two-dimensional Laplace equation

$$\frac{\partial^2\phi}{\partial x^2} + \frac{\partial^2\phi}{\partial y^2} = 0$$

subject to the conditions

$$\phi(x, 0) = f(x)$$

$$\frac{\partial\phi}{\partial y} = 0 \text{ at } y = 0.$$

Solution. Taking Fourier transform of the equation with respect to x, we get

$$\frac{\partial^2}{\partial y^2}F\{\phi\} - \omega^2 F\{\phi\} = 0.$$

The solution of this equation is

$$F\{\phi\} = A\,e^{\omega y} + B\,e^{-\omega y}. \tag{48}$$

But $\phi(x, 0) = f(x)$. Thus

$$F\{\phi(x, 0)\} = \int_{-\infty}^{\infty} f(x)e^{-i\omega x}\,dx = F\{f\}.$$

Thus (48) gives

$$F\{f\} = A + B. \tag{49}$$

Differentiating (48) w.r.t. y, we get

$$F'\{\phi\} = A\,\omega e^{\omega y} - B\,\omega e^{-\omega y}.$$

But $\frac{\partial\phi}{\partial y} = 0$ at $y = 0$. Therefore,

$$0 = A\omega - B\omega, \tag{50}$$

which gives A = B. Hence (49) yields

$$A = B = \frac{1}{2}F\{f\}.$$

Thus (48) reduces to

$$F\{\phi\} = \frac{1}{2}F\{f\}\,[e^{\omega y} + e^{-\omega y}].$$

Taking inverse transform, we have

$$\phi(x, y) = \frac{1}{2\pi}\int\frac{1}{2}F\{f\}\,[e^{\omega y} + e^{-\omega y}]\,e^{i\omega x}\,d\omega$$

$$= \frac{1}{2}\frac{1}{2\pi}\int_{-\infty}^{\infty} F\{f\}\,e^{\omega y}\,e^{i\omega x}\,d\omega$$

$$+ \frac{1}{2\pi}\int_{-\infty}^{\infty} F\{f\}\,e^{-\omega y}e^{i\omega x}\,d\omega]$$

$$= \frac{1}{2}\left[\frac{1}{2\pi}\int_{-\infty}^{\infty} F\{f\}\,e^{i(x-iy)\omega}\,d\omega\right.$$

$$\left.+ \frac{1}{2\pi}\int_{-\infty}^{\infty} F\{f\}\,e^{i(x+iy)\omega}\,d\omega\right]$$

$$= \frac{1}{2}[f(x-iy) + f(x+iy)].$$

EXERCISES

1. Find Fourier transform of $f(t) = 1 - t^2$, $-1 < t < 1$ and zero otherwise. Also evaluate $\int_0^\infty \frac{(t\cos t - \sin t)^2}{t^6}\, dt$.

Hint: Integrating by parts, we have

$$F(\omega) = \int_{-1}^{1} (1 - t^2)\, e^{-i\omega t}\, dt$$

$$= \frac{4}{\omega^3}(\omega \cos \omega - \sin \omega).$$

Use Parseval's theorem to get the value $\frac{\pi}{15}$ of the integral in question.

2. Find energy spectrum of the function

$$f(t) = \begin{cases} a & \text{for } |t| < T \\ 0 & \text{otherwise}. \end{cases}$$

Hint: $F(\omega) = \frac{2a \sin \omega t}{\omega}$ and so

$$|F(\omega)|^2 = \frac{4a^2 \sin^2 \omega t}{\omega^2}$$

3. Find Fourier cosine transform of $f(t) = te^{-at}$, $a > 0$.

Ans. $\frac{a^2-\omega^2}{(a^2+\omega^2)^2}$

4. Find the Fourier sine transform of $f(t) = e^{-|t|}$.

Ans. $\frac{\omega}{\omega^2+1}$

5. Find Fourier cosine transform of $2e^{-5x}+5e^{-2x}$.

Ans. $10\left(\frac{1}{\omega^2+4} + \frac{1}{\omega^2+25}\right)$

6. Find Fourier cosine transform of

$$f(x) = \frac{1}{a^2 + x^2}.$$

Ans. $\frac{\pi e^{-a\omega}}{2a}$

7. Find the function whose cosine transform is $\frac{\sin a\omega}{\omega}$, $a > 0$.

Hint: Use Fourier integral theorem for cosine transform and the fact that $\int_0^\infty \frac{\sin ax}{x}\, dx$

$$= \begin{cases} \pi/2 & \text{for } a > 0 \\ -\pi/2 & \text{for } a < 0. \end{cases}$$

Ans. $\begin{cases} 0 & \text{for } x > a \\ 1 & \text{for } x < a. \end{cases}$

8. Determine $f(t)$ if its Fourier cosine transform is $\frac{1}{1+\omega^2}$.

Ans. e^{-t}

9. Find finite Fourier cosine transform of $(1 - \frac{x}{\pi})^2$.

Hint: $F_c(\omega) = \int_0^\pi f(x) \cos \frac{\omega\pi x}{\pi}\, dx$

$$= \int_0^\pi \left(1 - \frac{x}{\pi}\right)^2 \cos \omega x\, dx$$

$$= \begin{cases} 2/(\pi\omega^2) & \text{for } \omega > 0 \\ \pi/3 & \text{for } \omega = 0. \end{cases}$$

10. Let $F\{f(t)\} = \frac{1}{\omega(\omega^2-1)}$ where $f(t) = 0$ for $t < 0$. Find $f(t)$ without using Convolution theorem.

Hint: By partial fractions $\frac{1}{\omega(\omega^2-1)} = \frac{-1}{\omega} + \frac{1}{2}\cdot\frac{1}{\omega-1} + \frac{1}{2}\cdot\frac{1}{\omega+1}$ and so $f(t) = -i\mathrm{H}(t) - \frac{i}{2}\mathrm{H}(t)(e^{it} + e^{-it}) = -i\,\mathrm{H}(t) + \frac{1}{2}\mathrm{H}(t)\cos t = -i\mathrm{H}(t)\,[1 - \cos t]$

11. Evaluate the integral $\int_0^\infty \frac{\sin at}{t(a^2+t^2)}\, dt$.

Hint: Take

$$f(t) = e^{-at},\ g(t) = \begin{cases} 1 & \text{for } 0 < t < a \\ 0 & \text{for } t > a. \end{cases}$$

Then $F_c(\omega) = \frac{a}{a^2+\omega h2}$, $G_c(\omega) = \frac{\sin a\omega}{\omega}$. Use Parseval's identity to get $\frac{2}{\pi}\int_0^\infty \frac{a \sin a\omega}{\omega(a^2+\omega^2)}\, d\omega = \int_0^a e^{-at}\, dt = \frac{1-e^{-a2}}{a}$ and so value of the integral is $\frac{\pi}{2a^2}(1 - e^{-a2})$

12. Find $f(x)$ if its finite Fourier sine transform is

$$F_s(n) = \frac{1 - \cos n\pi}{n^2\pi^2},\ n = 1, 2, \ldots, 0 < x < \pi.$$

Ans. $\frac{2}{\pi^3}\sum_{n=1}^{\infty}\frac{1-\cos n\pi}{n^2}\sin nx$

13. Find $f(x)$ if its finite sine transform is given by
$F_s(n) = \frac{2\pi(-1)^{n-1}}{n^3}$, $n = 1, 2, \ldots, 2 < x < \pi$.

Hint:

$$f(x) = \frac{2}{l}\sum_{n=1}^{\infty} F_s(n)\frac{\sin n\pi x}{l}$$
$$= 4\sum_{n=1}^{\infty}\frac{(-1)^{n-1}}{n^3}\sin nx.$$

14. Find Fourier sine integral representation of

$$f(x) = \begin{cases} \cos x & \text{for } 0 < x < \pi/2 \\ 0 & \text{for } x > \pi/2. \end{cases}$$

Ans. $\frac{2}{\pi}\int_0^{\infty}\left(\frac{\lambda - \sin \pi\lambda/2}{\lambda^2 - 1}\right)\sin \lambda x\, d\lambda$

15. Find Fourier integral representation of the rectangular pulse function

$$f(x) = \begin{cases} 1 & \text{for } |t| \le 1 \\ 0 & \text{otherwise.} \end{cases}$$

Hint: $A\lambda = \frac{1}{\pi}\int_{-\infty}^{\infty} f(t)\cos \lambda t\, dt = \frac{1}{\pi}\int_{-1}^{1}\cos \lambda t\, dt = \frac{2\sin\lambda}{\pi\lambda}$ and $B(\lambda) = 0$. Hence $f(x) = \frac{2}{\pi}\int_0^{\infty}\frac{\sin\lambda}{\lambda}\cos \lambda x\, d\lambda$.

16. Solve $\frac{\partial u}{\partial t} = k\frac{\partial^2 u}{\partial x^2}$, $x \in \mathbb{R}$, $t > 0$, subject to $u(x, 0) = f(x) = \frac{1}{1+x^2}$ for $x \in \mathbb{R}$ and $t > 0$
$u(x, t)$ is bound.

Hint: $u(x,t) = \int_{-\infty}^{\infty} F(\omega)\, e^{-k\omega^2 t} e^{i\omega x} dx$, where

$$F(\omega) = \frac{1}{2\pi}\int_{-\infty}^{\infty} f(x)\, e^{-i\omega x} dx$$
$$= \frac{1}{2\pi}F\{f(x)\} = \frac{1}{2\pi}\{\pi e^{-|\omega|}\}.$$

Therefore, $u(x,t) = \frac{1}{2}\int_{-\infty}^{\infty} e^{-|\omega|}\, e^{-k\omega^2 t} e^{i\omega x} d\omega$.

17. Solve $\frac{\partial u}{\partial t} = k\frac{\partial^2 u}{\partial x^2}$, $0 < x < \pi$ subject to

$u(x, 0) = 2x \qquad 0 < x < \pi$
$u(0, t) = u(\pi, t) = 0, \qquad t \ge 0.$

Hint: Proceed as in Example 7.43. $F_s(n) = \frac{2\pi}{n}(-1)^{n+1}e^{-\omega^2 kt}$. Using inverse formula for Fourier sine transform we have $u(x,t) = 4\sum_{n=1}^{\infty}\frac{(-1)^{n+1}}{n}e^{-n^2 kt}\sin nx$.

18. Solve $\frac{\partial^2 u}{\partial x^2} + \frac{\partial^2 u}{\partial y^2} = 0$, $0 < x < \pi$, $0 < y < \pi$ under the conditions $u(x, 0) = \sin^2 x$, $u(0, y) = u(\pi, y) = u(x, \pi) = 0$.

Ans. $u(x,y) = -\frac{4}{\pi}\sum_{n=1}^{\infty}\frac{\sin nx \sinh n(\pi - y)}{n(n^2-4)\sinh n\pi}$, n odd.

8 Discrete Fourier Transform

Discrete Fourier transform (DFT) is a powerful tool for frequency analysis of discrete time signals. It transforms discrete-time signal of finite length into discrete-frequency sequence of finite length. As we shall observe, the DFT can also be used to calculate Fourier coefficients.

8.1 APPROXIMATION OF FOURIER COEFFICIENTS OF A PERIODIC FUNCTION

Let f be a real-valued periodic function with period T over the interval [0, T]. We divide [0, T] into N subintervals I_n of equal length T/N. Thus

$$I_n = \left[\frac{(n-1)T}{N}, \frac{nT}{N}\right], n = 1, 2, \dots, N.$$

The linear interpolation $l_n(t)$ of $f(t)$ is given by

$$l_n(t) = \frac{N}{T}\left[\left(\frac{nT}{N} - t\right)f(n-1) + \left(t - \frac{(n-1)T}{N}\right)f(n)\right].$$

Thus the graph (Figure 8.1) of $l_n(t)$ consists of the line segment connecting the points $\left(\frac{(n-1)T}{N}, f(n-1)\right)$ and $\left(\frac{nT}{N}, f(n)\right)$.

Then the integral of f over the interval I_n can be approximated as

$$\int_{(n-1)T/N}^{nT/N} f(t)\,dt \approx \int_{(n-1)T/N}^{nT/N} l_n(t)\,dt = \frac{T}{2N}[f[n-1] + f[n]],$$

where $f[n] = f\left(n\frac{T}{N}\right)$, $n = 1, 2, \dots, N$.

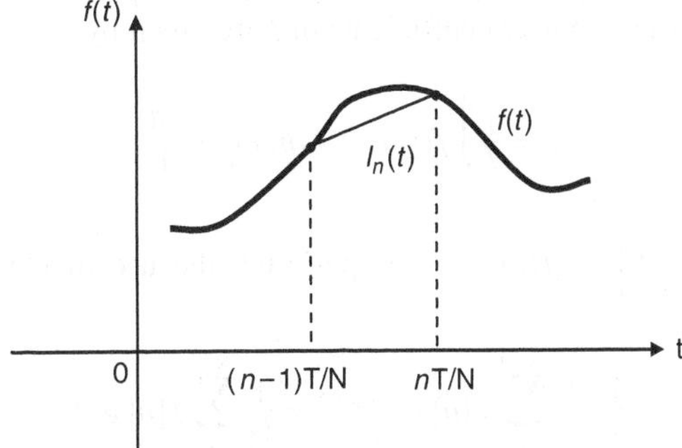

Figure 8.1

Summing all the approximations over [0, T], we have

$$\int_0^T f(t)\,dt = \sum_{n=1}^{N} \int_{(n-1)t/N}^{nT/N} f(t)\,dt \approx \sum_{n=1}^{N} \frac{T}{2N}[f[n-1] + f(n)]$$

$$= \frac{T}{2N}[f[0] + 2f[1] + \cdots + 2f[N-1] + f[N]].$$

Since f is periodic of period N, $f[0] = f[N]$ and so

$$\int_0^T f(t)\,dt \approx \frac{T}{N}[f[0] + f[1] + \cdots + f[N-1]] \quad (1)$$

gives approximation of the integral over one period starting from the sampling $f[n]$ of the fuunction f.

The right-hand side is nothing but Riemannsum and we know that Riemann-sum tends to the integral when length of the subintervals tends to zero. Since piecewise smooth functions are Riemann-integrable, for such functions, we have

$$\frac{T}{N}[f[0]+f[1]+\cdots+f[N-1]] \to \int_0^T f(t)dt \text{ as } N \to \infty. \qquad (2)$$

Expression (1) is called *trapezoidal rule*. Using this rule, we now approximate Fourier coefficients of a periodic function f.

Let f be a periodic function with period T. Then Fourier coefficients of f are given by

$$c_k = \frac{1}{T}\int_0^T f(t)\, e^{-ik\omega_0 t}\, dt,\ \omega_0 = \frac{2\pi}{T}.$$

Since $f(t)\, e^{-ik\omega_0 t}$ is periodic, the use of (1) yields

$$c_k \approx \frac{1}{T}\cdot\frac{T}{N}\sum_{n=0}^{N-1} f[n]\, e^{-ik\omega_0 nT/N} = \frac{1}{N}\sum_{n=0}^{N-1} f[n]\, e^{-2\pi ink/N}$$

Thus

$$c_k \approx \frac{1}{N}F[k], \qquad (3)$$

where

$$F[k] = \sum_{n=0}^{N-1} f[n]\, e^{-2\pi ink/N}.$$

It follows from (3) that $F[k]/N \to c_k$ as $N \to \infty$.

It may be mentioned that (3) cannot be a good approximation to c_k because for piecewise smooth signals, the Fourier coefficients c_k tends to 0 as $|k| \to \infty$ (Riemann-Lebesgue lemma).

We further note that

$$F[k+N] = \sum_{n=0}^{N-1} f[n]\, e^{-2\pi in(k+N)/N}$$

$$= \sum_{n=0}^{N-1} f[n]\, e^{-2\pi ink/N}\cdot e^{-2\pi in} = F[k].$$

Hence F[k] is *periodic* with *period* N.

8.2 DEFINITION AND EXAMPLES OF DFT

Definition 8.1. Let $f[n]$ be a periodic discrete-time signal with period N. Then the sequence F[k] defined by

$$F[k] = \sum_{n=0}^{N-1} f[n]\, e^{-2\pi ink/N},\ k \in \mathbb{Z}$$

is called the *N-point discrete Fourier transform* (*N-point DFT*) of $f[n]$.

Since F[k] is periodic with period N, it follows that DFT converts a periodic discrete signal into a periodic discrete signal, again having the same period. The sequence F[k] is called the *description of the signal in the k-domain* whereas $f[n]$ is called the *description in the n-domain* or *time domain*. Because of the close relationship of F[k] with Fourier coefficients (as seen above), the k-domain is also called *frequency domain* and F[k] is called *discrete spectrum* of $f[n]$.

Theorem 8.1. Let $f[n]$ be a periodic discrete signal with period N. Then for any integer p

$$\sum_{n=p}^{p+N-1} f[n] = \sum_{n=0}^{N-1} f[n].$$

Proof: Set $p = mN + l$, $0 \le l \le N-1$. Since f is periodic, we have

$$\sum_{n=p}^{p+N-1} f[n] = f[p]+f[p+1]+\cdots+f[p+N-1]$$

$$= f[p]+f[p+1]+\cdots+f[N-1] + f[N]+\cdots+f[N+l-1]$$

$$= f[p]+f[p+1] + \cdots + f[N-1]+f[0]+\cdots+f[l-1]$$

$$= f[0]+f[1]+\cdots+f[N-1]$$

$$= \sum_{n=0}^{N-1} f[n].$$

Remark 8.1. Since F[k] is periodic with period N, the Theorem 8.1 can be applied to F[k] also.

Further DFT for N = 2M + 1 can also be written as

$$F[k] = \sum_{n=-M}^{M} f[n]\, e^{-2\pi ink/N}.$$

EXAMPLE 8.1

Calculate the 4-point DFT of F[n] = {1, 1, 0, 0}.

Solution. DFT of the signal $f[n]$ is given by

$$F[k] = \sum_{n=0}^{3} f[n]\, e^{-2\pi ink/4}$$

$$= f[0] + f[1]e^{-2\pi ink/4} + f[2]\, e^{-4\pi ink/4} + f[3]\, e^{-6\pi ink/4}$$

$$= 1 + e^{-\pi ik/2} + 0 + 0$$

$$= 1 + e^{-\pi ik/2}.$$

Therefore,

$$F[0] = 1 + 1 = 2$$

$$F[1] = 1 - i$$

$$F[2] = 0$$

$$F[3] = 1 + i.$$

Thus

$$F[k] = (F[0], F[1], F[2], F[3]) = (2, 1 - i, 0, 1 + i).$$

The *amplitude spectrum* (*magnitude*) of F[k] (Figure 8.2) is

$$|F[k]| = (2, \sqrt{1^2 + (-1)^2}, 0, \sqrt{1^2 + 1^2})$$

$$= (2, 1.4142, 0, 1.4142).$$

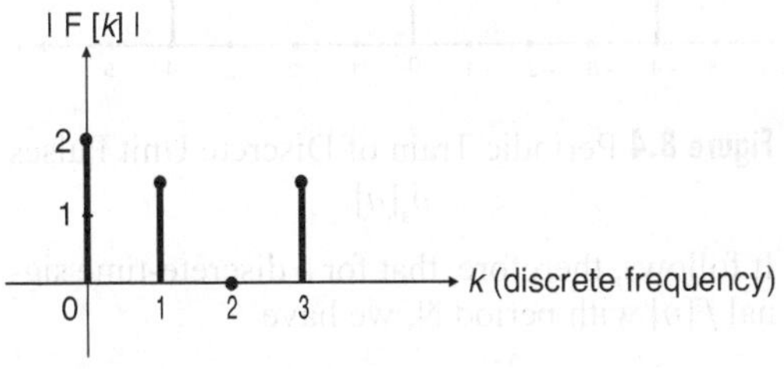

Figure 8.2 Amplitude Spectrum of F[k]

EXAMPLE 8.2

Calculate the 4-point DFT of $f[n]$ with period 4 and given by $f[-2] = 1, f[-1] = 0, f[0] = 2, f[1] = 0$.

Solution. The 4-point DFT of the given signal is

$$F[k] = \sum_{n=-2}^{1} f[n]\, e^{-2\pi ink/4}$$

$$= f[-2]\, e^{4\pi ik/4} + f[-1]\, e^{2\pi ik/4}$$

$$+ f[0] + f[1]\, e^{-2\pi ik/4}$$

$$= e^{\pi ik} + 2$$

Thus

$$F[-2] = 1 + 2 = 3$$
$$F[-1] = -1 + 2 = 1$$
$$F[0] = 3$$
$$F[1] = -1 + 2 = 1$$

and so

$$F[k] = (3, 1, 3, 1).$$

The amplitude spectrum of F[k] (Figure 8.3) is

$$|F[k]| = (3, 1, 3, 1).$$

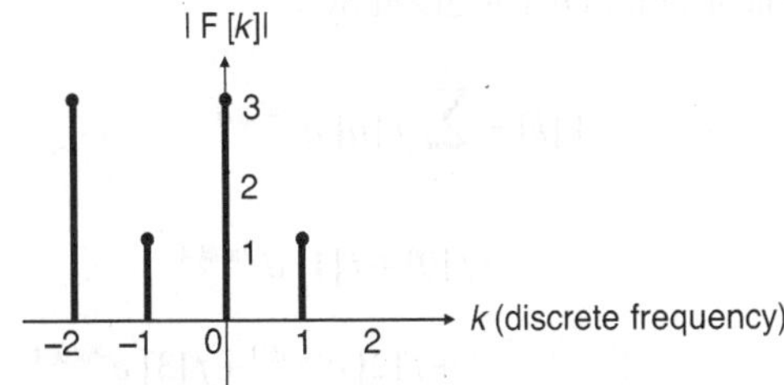

Figure 8.3 Amplitude Spectrum of $F[k]$

EXAMPLE 8.3

Calculate the N-point DFT of a finite-time sequence $f[n]$ defined by

$$f[n] = \begin{cases} 1 & \text{for } 0 \le n \le L-1 \\ 0 & \text{otherwise.} \end{cases}$$

Solution. The N-point DFT of the given signal is

$$F[k] = \sum_{n=0}^{N-1} f[n]\, e^{-2\pi ikn/N}$$

$$= \sum_{n=0}^{L-1} e^{-2\pi ikn/N}$$

$$= 1 + e^{-2\pi ik/N} + e^{-4\pi ik/N} + \dots + e^{-2(L-1)ik/N},$$

which is G.P. with first term 1 and common ratio $e^{-2\pi ik/N}$. Therefore,

$$F[k] = \frac{1[1-(e^{-2\pi ik/N})^{L-1+1}]}{1-e^{-2\pi ik/N}}$$

$$= \frac{1-e^{-2\pi ikL/N}}{1-e^{-2\pi ik/N}},\ k = 0, 1, 2, \dots, N-1.$$

EXAMPLE 8.4

Find 4-point DFT of discrete-time signal $f[n] = (-1)^n$ for $n \in \mathbb{Z}$.

Solution. The given signal is

$$f[n] = (-1)^n, \quad n \in \mathbb{Z}$$

The 4-point DFT is given by

$$F[k] = \sum_{n=0}^{3} f[n]\, e^{-2\pi ink/4}$$

$$= f[0] + f[1]\, e^{-2\pi ink/4} + f[2]\, e^{-4\pi ik/4} + f[3]\, e^{-6\pi ik/4}$$

$$= 1 - e^{-\frac{\pi ik}{2}} + e^{-\pi ik} - e^{-3\pi ik/2}$$

Thus

$$F[0] = 1 - 1 + 1 - 1 = 0$$

$$F[1] = 1 - e^{-i\pi/2} + e^{-i\pi} - e^{-i\,3\pi/2}$$

$$= 1 + i - 1 - i = 0$$

$$F[2] = 1 + 1 + 1 + 1 = 4$$

$$F[3] = 1 + i - 1 - i = 0.$$

Hence

$$F[k] = (0, 0, 4, 0).$$

8.3 INVERSE DFT

Consider the equation $z^N = 1$ in the complex plane. Set $\omega = e^{2\pi i/N}$. Then the N distinct roots of this equation are

$$z_j = e^{2\pi ij/N} = \omega^j,$$

$$j = 0, 1, 2, \dots, N-1$$

Therefore, for arbitrary integer n, we have

$$\frac{1}{N}\sum_{k=0}^{N-1} e^{2\pi ink/N} = \frac{1}{N}\sum_{k=0}^{N-1} \omega^{nk}$$

$$= \begin{cases} 1 & \text{if } n \text{ is an integral multiple of N} \\ \frac{1}{N}\cdot\frac{1-\omega^{nN}}{1-\omega^n} = 0 & \text{otherwise} \end{cases}$$

Now let N be a positive integer. Then *periodic train of discrete unit pulses* $\delta_N[n]$ with period N is defined by

$$\delta_N[n] = \begin{cases} 1 & \text{if } n \text{ is an integer multiple of N} \\ 0 & \text{otherwise.} \end{cases}$$

For example, the graph $\delta_4[n]$ is shown in Figure 8.8.

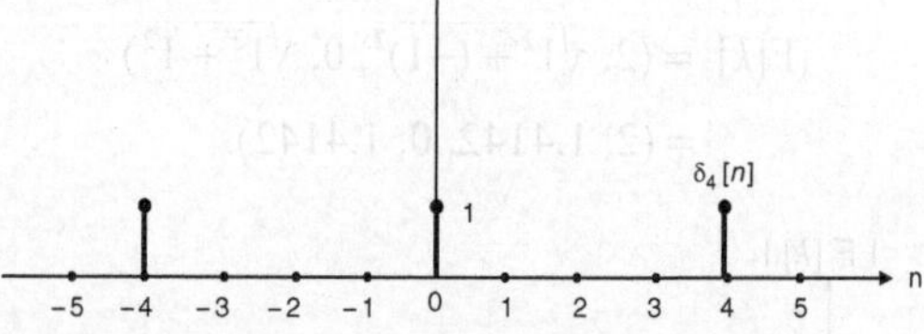

Figure 8.4 Periodic Train of Discrete Unit Pulses $\delta_4[n]$

It follows, therefore, that for a discrete-time signal $f[n]$ with period N, we have

$$f[n] = \sum_{k=0}^{N-1} f[k]\, \delta_N[n-k], n \in \mathbb{Z} \qquad (4)$$

Also, in terms of periodic train of discrete unit pulses $\delta_N[n]$, we have

$$\frac{1}{N}\sum_{k=0}^{N} e^{2\pi ink/\omega} = \delta_N[n], n \in \mathbb{Z}. \qquad (5)$$

Theorem 8.2. (Fundamental Theorem of the DFT). Let $f[n]$ be a periodic discrete-time signal with period N and let its DFT be given by

$$F[k] = \sum_{n=0}^{N-1} f[n]\, e^{-2\pi ink/N}.$$

Then

$$f[n] = \frac{1}{N}\sum_{k=0}^{N-1} F[k]e^{2\pi ink/N}.$$

Proof: Putting the value of F[k], the change in order of summation yields

$$\frac{1}{N}\sum_{k=0}^{N-1} F[k]e^{2\pi ink/N} = \frac{1}{N}\sum_{k=0}^{N-1}\sum_{l=0}^{N-1} f[l]2^{2\pi ik(n-l)/N}$$

$$= \frac{1}{N}\sum_{l=0}^{N-1} f[l]\sum_{k=0}^{N-1} e^{2\pi ik(n-l)/N}$$

$$= \sum_{l=0}^{N-1} f[l]\cdot\frac{1}{N}\sum_{k=0}^{N-1} e^{2\pi ik(n-l)/N}$$

$$= \sum_{l=0}^{N-1} f[l]\delta_N[n-l] \; [using\,(5)]$$

$$= f[n] \; [using\,(4)].$$

Thus, an arbitrary periodic discrete-time signal $f[n]$ with period N can be expressed as a *linear combination of the time harmonic signals* $e^{2\pi ink/N}$, $k = 0, 1, 2. \ldots, N-1$, which are themselves periodic with period N.

The quantity $2\pi/N$ is called the *fundamental frequency.*

Definition 8.2. The transformation assigning the signal $f[n]$ to DFT F[k] is called the *inverse discrete Fourier transform.*

EXAMPLE 8.5

Find DFT of $\delta_N[n]$, the periodic train of discrete unit pulses with period N and verify fundamental theorem of DFT.

Solution. We have

$$\delta_N[n] = \begin{cases} 1 & \text{if } n \text{ integer multiple of N} \\ 0 & \text{otherwise.} \end{cases}$$

Therefore,

$$F[k] = \sum_{n=0}^{N-1} \delta_N[n]e^{-2\pi ink/N} = \delta_N[0] = 1.$$

On the other hand, by Fundamental theorem of DFT, we have

$$f[n] = \frac{1}{N}\sum_{k=0}^{N-1} F[k]\, e^{2\pi ik/N}$$

$$= \frac{1}{N}\sum_{k=0}^{N-1} e^{2\pi ink/N} = \delta_N[n] \qquad \text{(by (5)).}$$

EXAMPLE 8.6

Given that F[k] = (2, 1 − i, 0, 1 + i), find $f[n]$.

Solution. By Fundamental theorem,

$$f[n] = \frac{1}{N}\sum_{k=0}^{N-1} F[k]\, e^{2\pi ink/N}$$

Here N = 4 and

$$F[0] = 2, F[1] = 1-i, F[2] = 0, F[3] = 1+i.$$

Therefore

$$f[n] = \frac{1}{4}[F[0] + F[1]\, e^{2\pi in/4} + F[2]\, e^{4\pi in/4} + F[3]\, e^{6\pi in/4}]$$

Thus

$$f[0] = \frac{1}{4}[2 + 1 - i + 1 + i] = 1,$$

$$f[1] = 1, f[2] = 0, f[3] = 0.$$

Hence

$$f[n] = (1, 1, 0, 0).$$

EXAMPLE 8.7

Find the DFT of the periodic discrete-time signal $f_1[n] = e^{2\pi inl/N}$ with period N. Also, verify Fundamental theorem of DFT.

Solution. We have

$$F[k] = \sum_{n=0}^{N-1} f_1[n]\, 2^{-2\pi ink/N} = \sum_{n=0}^{N-1} e^{2\pi inl/N} \cdot e^{-2\pi ink/N}$$

$$= \sum_{n=0}^{N-1} e^{2\pi in(l-k)/N} = N\delta_N(l-k).$$

Now, using fundamental theorem of DFT,

$$f_1[n] = \frac{1}{N}\sum_{k=0}^{N-1} N\delta_N(l-k)\, e^{2\pi ink/N}$$

$$= \sum_{k=0}^{N-1} \delta_N(l-k)\, e^{2\pi ink/N} = e^{2\pi inl/N}.$$

EXAMPLE 8.8

Find the 4-point inverse DFT of the discrete signal F[k] with period 4 given by F[0] = 1, F[1] = 0, F[2] = 0, F[3] = 1.

Solution. We have

$$F[k] = (1, 0, 0, 1).$$

Using Fundamental theorem for the DFT, we have

$$f[n] = \frac{1}{N}\sum_{k=0}^{N-1} F[k]\, e^{2\pi ink/N} = \frac{1}{4}\sum_{k=0}^{3} F[k]\, e^{2\pi ink/4}$$

$$= \frac{1}{4}[F[0] + F[1]e^{2\pi in/4} + F[2]\, e^{4\pi in/4} + F[3]\, e^{6\pi in/4}]$$

$$= \frac{1}{4}[F[0] + F[1]\, e^{i\pi n/2} + F[2]\, e^{i\pi n} + F[3]\, e^{3i\pi n/2}]$$

$$= \frac{1}{4}[F[0] + F[3]\, e^{3i\pi n/2}]$$

Thus

$$f[0] = \frac{1}{4}[F[0] + F[3]] = \frac{1}{2},$$

$$f[1] = \frac{1}{4}[F[0] + F[3]e^{i3\pi/2}] = \frac{1}{4}[1 - i],$$

$$f[2] = \frac{1}{4}[F[0] + F[3]e^{3i\pi}] = \frac{1}{4}[1 + 1(-1)] = 0,$$

$$f[3] = \frac{1}{4}[F[0] + F[3]e^{9i\pi/2}] = \frac{1}{4}[1 + i] = \frac{1+i}{4}.$$

Hence

$$f[n] = \left(\frac{1}{2}, \frac{1-i}{4}, 0, \frac{1+i}{4}\right).$$

8.4 PROPERTIES OF DFT

Theorem 8.3. (Linearity Property). Let F[k] and $G[k]$ be the DFT of discrete-time signals $f[n]$ and $g[n]$, respectively. Then DFT of $af[n] + bg[n]$ is a $F[k] + b\, G[k]$ for a, b $\in \mathbb{C}$

Proof: We note that

$$\sum_{n=0}^{N-1} [af[n] + bg[n]]\, e^{-2\pi ink/N}$$

$$= \sum_{n=0}^{N-1} (af[n])e^{-2\pi ink/N} + \sum_{n=0}^{N-1} (bg[n]e^{-2\pi ink/N})$$

$$= a\, F[k] + b\, G[k],$$

which proves our assertion.

Theorem 8.4. (Reciprocity). Let $F[k]$ be the DFT of discrete-time signal $f[n]$. Then the DFT of F[n] is $Nf[-k]$.

Proof: By Fundamental theorem of DFT, we have

$$f[n] = \frac{1}{N}\sum_{k=0}^{N-1} F[k]\, e^{2\pi ink/N}.$$

Interchanging the variables n and k, we get

$$f[k] = \frac{1}{N}\sum_{n=0}^{N-1} F[n]\, e^{2\pi ink/N},$$

and so

$$f[-k] = \frac{1}{N}\sum_{n=0}^{N-1} F[n]\, e^{-2\pi ink/N}.$$

Hence

$$\text{DFT}\{F[n]\} = \sum_{n=0}^{N-1} F[n]\, e^{-2\pi ink/N} = Nf[-k].$$

Theorem 8.5. (Time Reversal). Let $F[k]$ be the DFT of periodic discrete-time signal $f[n]$. Then DFT of $f[-n]$ is $F[-k]$.

Proof: We have, by periodicity of $f[n]$,

$$\text{DFT}\ \{f[-n]\} = \sum_{n=0}^{N-1} f[-n]\, e^{-2\pi ink/N}$$

$$= \sum_{n=0}^{N-1} f[N-n]\, e^{2\pi i(N-n)k/N}$$

$$= \sum_{n=1}^{N} f[n]\, e^{2\pi ink/N}$$

$$= \sum_{n=0}^{N-1} f[n]\, e^{2\pi ink/N}, \text{ by Theorem 8.1}$$

$$= F[-k], \text{ by definition of } F[k]$$

Remark 8.2. As a consequence of Theorem 8.5, we have

if $f[n]$ is even, then $F[k]$ is even

if $f[n]$ is odd, then $F[k]$ is odd.

Theorem 8.6. [Shift in n-domain (Time-Domain)]. Let $F[k]$ be the DFT of periodic discrete-time signal $f[n]$. Then the DFT of $f[n-l]$ is $e^{-2\pi ilk/N} F[k]$.

Proof: If $f[n]$ is a discrete-time signal, then $f[n-l]$ is also a discrete-time signal and we have

$$\text{DFT}\ \{f(n-l)\} = \sum_{n=0}^{N-1} f[n-l] e^{-2\pi ink/N}$$

$$= \sum_{m=-l}^{N-1-l} f[m]\, e^{-2\pi i(m+l)k/N},\ m = n-l,$$

$$= e^{-2\pi ilk/N} \sum_{m=-1}^{N-1-l} f[m]\, e^{-2\pi imk/N}$$

$$= e^{-2\pi ilk/N} \sum_{m=0}^{N-1} f[m]\, e^{-2\pi imk/N},$$

by Theorem 8.1

$$= e^{-2\pi ilk/N} F[k]$$

Remark 8.3. (a) Since

$$|e^{-2\pi ilk/N} F[k]| = |F[k]|,$$

it follows that *amplitude spectrum does not change under a shift in the n-domain.*

(b) Since

$$\arg[e^{-2\pi ilk/N} f[k]] = \arg[F[k]] - 2\pi lk/N$$

and since change $-2\pi lk/N$ is linear in k, therefore, a *shift in the n-domain results in a linear phase shift.*

Theorem 8.7. (Shift in the k-Domain). Let $F[k]$ be the DFT of a periodic discrete-time signal $f[n]$. Then DFT of $e^{2\pi inl/N} f[n]$ is $F[k-l]$.

Proof: We note that

$$\text{DFT}\{e^{2\pi inl/N} f[n]\} = \sum_{n=0}^{N-1} e^{2\pi inl/N} f[n]\, e^{-2\pi ink/N}$$

$$= \sum_{n=0}^{N-1} f[n]\, e^{-2\pi in(k-l)/N}$$

$$= F[k-l].$$

EXAMPLE 8.9

Show that

$$\sum_{n=0}^{N-1} e^{-2\pi ink/N} = N\delta_N[k]$$

and hence derive

$$\delta_N[n] = \frac{1}{N}\sum_{k=0}^{N-1} e^{2\pi ink/N},\ n \in \mathbb{Z}.$$

Solution. By Example 8.5, the DFT of $\delta_N[n]$ is the constant signal 1. But, by reciprocity rule (Theorem 8.4),

$$\text{DFT}\{F[n]\} = N f[-k].$$

Therefore,

$$\text{DFT}\{1\} = N\delta_N[-k]$$

$$= N\,\delta_N[k], \text{ since } \delta_N[n] \text{ is even.}$$

Therefore, by definition of DFT, we have

$$\text{DFT}\{1\} = \sum_{n=0}^{N-1} e^{-2\pi ink/N} = N\,\delta_N[k]. \qquad (6)$$

Interchanging n and k in (6), we get

$$\frac{1}{N}\sum_{k=0}^{N-1} e^{-2\pi ink/N} = \delta_N[n].$$

EXAMPLE 8.10

Find the DFT of

$$f[n] = \sum_{l=-m}^{m} \delta_N[n-l],\ 2m < N.$$

Solution. By linearity property

$$\text{DFT}\{f[n]\} = \sum_{l=-m}^{m} \text{DFT}\{\delta_N[n-l]\}$$

Since DFT $\{\delta_N[n]\}$ = 1, by shift property in n-domain, we have

$$\text{DFT}\{\delta_N[n-l]\} = e^{-2\pi ilk/N}\,.\ \text{DFT}\{\delta_N[n]\}$$

$$= e^{-2\pi ilk/N}.$$

Hence

$$\text{DFT}\{f[n]\} = \sum_{l=-m}^{m} e^{-2\pi ilk/N},$$

which is a geometric series with ratio $e^{-2\pi ik/N}$. Hence $F[k] = \text{DFT}\{f[n]\}$

$$= \begin{cases} 2m+1 & \text{if } k \text{ is a multiple of N} \\ \dfrac{e^{2\pi\, i\, mk/N} - e^{-2\pi\, i\,(m+1)k/N}}{1 - e^{-2\pi\, i\, k/N}} & \text{otherwise} \end{cases}$$

$$= \begin{cases} 2m+1 & \text{if } k \text{ is a multiple of N} \\ \dfrac{e^{\pi i(2m+1)k\ N} - e^{-\pi i(2m+1)k\ N}}{e^{\pi ik/N} - e^{-\pi ik/N}} & \text{otherwise} \end{cases}$$

$$= \begin{cases} 2m+1 & \text{if } k \text{ is a multiple of N} \\ \dfrac{\sin(\pi(2m+1)k/N)}{\sin(\pi k/N)} & \text{otherwise.} \end{cases}$$

EXAMPLE 8.11

Find the DFT of the periodic discrete time signal

$$f[n] = \begin{cases} 2m+1 & \text{if } k \text{ is a multiple of N} \\ \dfrac{\sin(2m+1)\pi n\ N}{\sin(n\pi/N)} & \text{otherwise.} \end{cases}$$

Solution. From Example 8.10, we note that $f[n]$ is the DFT of $g[n] = \sum_{l=-m}^{m} \delta_N[n-l]$. Therefore, by reciprocity, the DFT of $f[n]$ is $Ng[-k]$. Thus

$$\text{DFT}\{f[n]\} = N\sum_{l=-m}^{m} \delta_N(-k-l)$$

$$= N\sum_{l=-m}^{m} \delta_N(k+l),$$

since $\delta_N[n]$ is even.

EXAMPLE 8.12

If $f[n]$ is a real periodic discrete time signal with period 4 and $F[0] = 2$, $F[1] = 1 - i$, $F[2] = 0$, find $f[n]$ for $n = 0, 1, 2, 3$.

Solution. Since $F[k]$ is periodic with period N, we have for real signal $f[n]$

$$F[N-k] = F[-k] = \overline{F[k]}.$$

Thus

$$F[3] = F[4-1] = F[-1] = \overline{F[1]} = \overline{1-i} = 1+i.$$

Thus

$$F[k] = (2, 1-i, 0, 1+i).$$

Then, by Example 8.6, we have

$$f[0] = 1, f[1] = 1, f[2] = 0, f[3] = 0.$$

8.5 CYCLICAL CONVOLUTION AND CONVOLUTION THEOREM FOR DFT

Definition 8.3. Let $f[n]$ and $g[n]$ be periodic discrete-time singals with period N. Then the *cyclical convolution* $(f * g)[n]$ of these two signals is defined by

$$(f * g)[n] = \sum_{l=0}^{N-1} f[l]\, g[n-l].$$

EXAMPLE 8.13

Find cyclical convolution of the signals

$$f[n] = \{1, 2, 3, 4\},\ g[n] = \{5, 6, 7, 8\}.$$

Solution. To evaluate the convolution in a simple fashion, we display the two sequences around two concentric circles as shown in Figure 8.5.

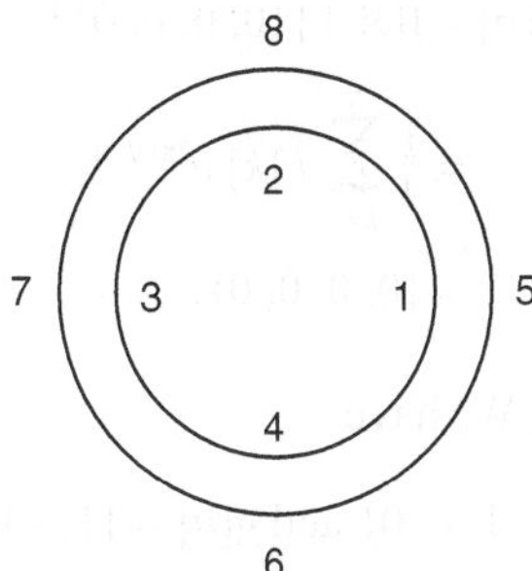

Figure 8.5 ($l = 0$)

Then

$$(f * g)[0] = (1)(5) + (2)(8) + (3)(7) + (4)(6) = 66.$$

For $l = 1$, rotate the outer circle counterclockwise by one position to get the Figure 8.6.

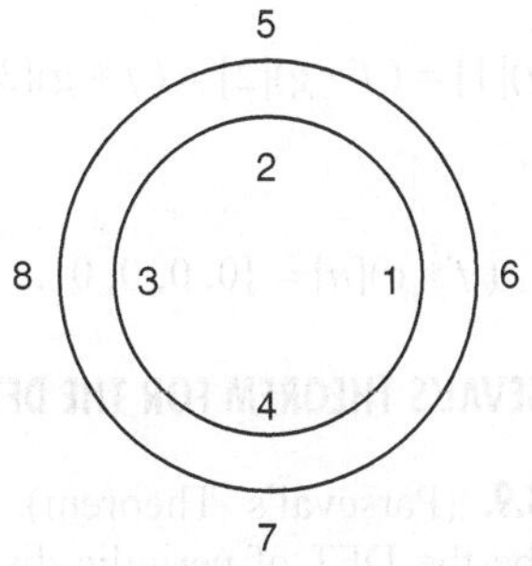

Figure 8.6 ($l = 1$)

Then

$$(f * g)[1] = (1)(6) + (2)(5) + (3)(8) + (4)(7) = 68.$$

For $l = 2$, rotate further outer circle counterclockwise by one position to get the Figure 8.7.

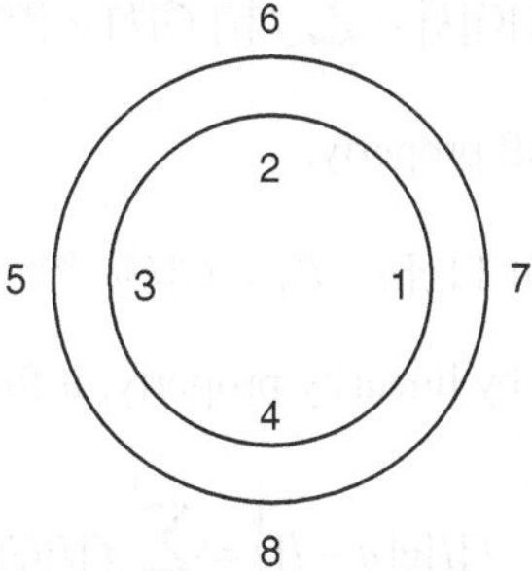

Figure 8.7 ($l = 2$)

Then

$$(f * g)[2] = (1)(7) + (2)(6) + (3)(5) + (4)(8) = 66.$$

For $l = 3$, rotate further the outer circle counterclockwise by one position to get the Figure 8.8.

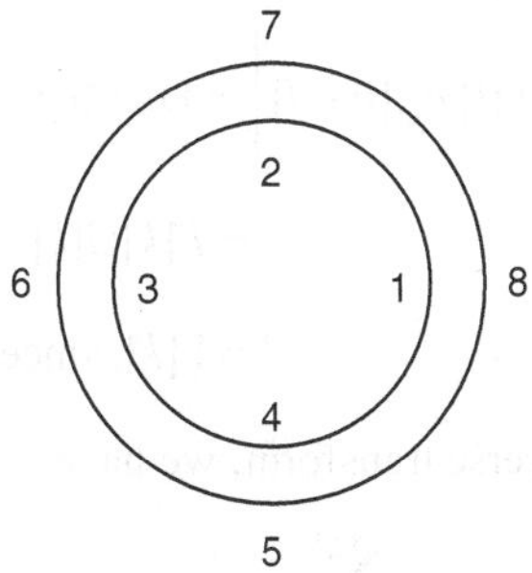

Figure 8.8 ($l = 3$)

Then

$$(f * g)[3] = (1)(8) + (2)(7) + (3)(6) + (4)(5) = 60.$$

Hence

$$(f * g)[n] = (66, 68, 66, 60).$$

Theorem 8.8. (Convolution Theorem). Let $F[k]$ and $G[k]$ be the DFT of periodic discrete-time signals $f[n]$ and $g[n]$, respectively. Then

$$DFT\{(f * g)[n]\} = F[k]G[k].$$

Proof: Since

$$F[k]=\sum_{l=0}^{N-1} f[l]\, e^{-2\pi ilk/N},$$

we have

$$F[k]G[k]=\sum_{l=0}^{N-1} f[l]\, G[k]\, e^{-2\pi ilk/N}.$$

But, by shift property,

$$\text{DFT}\{g[n-l]\}=G[k]e^{-2\pi ilk/N}.$$

Therefore, by linearity property, it follows that

$$\text{DFT}\left\{\sum_{l=0}^{N-1} f[l]g[n-l]\right\}=\sum_{l=0}^{N-1} f[l]G[k]e^{-2\pi ilk/N}$$

$$=F[k]\,G[k].$$

Hence

$$\text{DFT}\,\{(f*g)[n]\}=F[k]G[k].$$

Corollary 8.1. We have

$$\text{DFT}\left\{\sum_{l=0}^{N-1} f[l]\,\delta_N\,[n-l]\right\}=\text{DFT}\{(f*\delta_N)[n]\}$$

$$=F[k]G[k]$$

$$=F[k],\ \text{since } G[k]=1.$$

Taking inverse transform, we have

$$f[n]=\sum_{l=0}^{N-1} f[l]\,\delta_N\,[n-l].$$

Remark 8.4. Since

$$\text{DFT}\{(f*g)[n]\}=F[k]G[k]=G[k]F[k]$$

$$=\text{DFT}\{(g*f)[n]\},$$

it follows that

$$(f*g)[n]=(g*f)[n],\ n\in\mathbb{Z}.$$

Hence *cyclic convolution is commutative.*

EXAMPLE 8.14

Determine circular convolution of the discrete-time signals $f[n]=\{1,1,0,0\}$ and $g[n]=(-1)^n$, $n=0,1,2,3$ using Convolution theorem.

Solution. By Examples 8.1 and 8.4, we have

$$F[k]=\{2,\,1-i,\,0,\,1+i\}\ \text{and}$$

$$G[k]=\{0,0,4,0\}.$$

By Convolution theorem,

$$\text{DFT}\{(f*g)[n]\}=F[k]G[k]$$

$$=\{2,\,1-\text{i},\,0,\,1+\text{i}\}\ \{0,0,4,0\}$$

$$=\{0,0,0,0\}.$$

By Fundamental theorem, we have

$$(f*g)[n]=\text{IDFT}\{\{0,0,0,0\}\}$$

$$=\frac{1}{4}\sum_{k=0}^{3} F[k]\,e^{2\pi ink/4}$$

$$=\{0,0,0,0\}.$$

Verification. We have

$$f[n]=\{1,1,0,0\}\ \text{and}\ g[n]=\{1,-1,1,-1\}.$$

By concentric circles method, we note that

$$(f*g)[0]=(1)(1)+(1)(-1)+(0)(1)+(0)(-1)$$

$$=1-1=0$$

Similarly, we observe that

$$(f*g)[1]=(f*g)[2]=(f*g)[3]=0.$$

Hence

$$(f*g)[n]=\{0,0,0,0\}.$$

8.6 PARSEVAL'S THEOREM FOR THE DFT

Theorem 8.9. (Parseval's Theorem). Let $F[k]$ and $G[k]$ be the DFT of periodic discrete-time signals $f[n]$ and $g[n]$, respectively. Then

$$\sum_{n=0}^{N-1} f[n]\,\overline{g[n]} = \frac{1}{N}\sum_{k=0}^{N-1} F[k]\,\overline{G[k]}.$$

Proof: By Fundamental theorem of DFT, we have

$$\sum_{n=0}^{N-1} f[n]\,\overline{g[n]} = \frac{1}{N}\sum_{n=0}^{N-1} \overline{g[n]}\sum_{k=0}^{N-1} F[k]\,e^{2\pi ink/N}$$

Changing the order of summation, we get

$$\sum_{n=0}^{N-1} f[n]\,\overline{g[n]} = \frac{1}{N}\sum_{k=0}^{N-1} F[k]\sum_{n=0}^{N-1} \overline{g[n]}\,e^{2\pi ink/N}$$

$$= \frac{1}{N}\sum_{k=0}^{N-1} \mathrm{F}[k]\,\overline{\sum_{n=0}^{N-1} g[n]\,e^{-2\pi ink/N}}$$

$$= \frac{1}{N}\sum_{k=0}^{N-1} \mathrm{F}[k]\,\overline{\mathrm{G}[k]}.$$

Corollary 8.2. If $f[n] = g[n]$, then Parseval's theorem yields

$$\sum_{n=0}^{N-1} |f[n]|^2 = \frac{1}{N}\sum_{k=0}^{N-1} |\mathrm{F}[k]|^2.$$

Thus power of the signal $f[n]$ is given by

$$\mathrm{P} = \frac{1}{\mathrm{N}}\sum_{n=0}^{N-1} |f[n]|^2 = \frac{1}{\mathrm{N}^2}\sum_{n=0}^{N-1} |F[k]|^2.$$

EXAMPLE 8.15

Calculate the power of the signal $f[n] = \{1, 1, 0, 0\}$.

Solution. From Example 8.1, we have

$$\mathrm{F}[0] = 2,\ \mathrm{F}[1] = 1 - i,\ \mathrm{F}[2] = 0,$$

$$\mathrm{F}[3] = 1 + i.$$

Therefore,

$$\mathrm{P} = \frac{1}{16}\sum_{k=0}^{3} |F[k]|^2 = \frac{1}{16}[4 + 2 + 2] = \frac{1}{2}.$$

8.7 MATRIX FORM OF THE DFT

Let $f[n]$ be periodic discrete-time signal with period N. Then the DFT of $f[n]$ is given by

$$\mathrm{F}[k] = \sum_{n=0}^{N-1} f[n]\,e^{-2\pi ink/N}.$$

Letting $w_{\mathrm{N}} = e^{-2\pi i/N}$, we have

$$\mathrm{F}[k] = \sum_{n=0}^{N-1} f[n]w_N^{nk} = f[0]w_N^{0} + f[1]w_N^{k} + f[2]w_N^{2k}$$

$$+ \ldots f[N-1]w_N^{(N-1)k}.$$

Thus to calculate F[k] for certain value of k, we have to perform N – 1 multiplications and N – 1 additions. Thus 2N – 2 elementary operations are needed to find F[k] for certain value of k. Hence, to find F[0], F[1], …, F[N – 1] we need $\mathrm{N}(2\mathrm{N} - 2) = 2\mathrm{N}^2 - 2\mathrm{N}$ elementary operations. It follows, therefore, that to find N-point DFT, the number of elementary operations needed is of the order N^2.

From the above expression, we have

$$\mathrm{F}[0] = f[0]w_N^{0} + f[1]w_N^{0} + \ldots + f[N-1]w_N^{0}$$

$$\mathrm{F}[1] = f[0]w_N^{0} + f[1]w_N^{1} + \ldots + f[N-1]w_N^{(N-1)}$$

$$\mathrm{F}[2] = f[0]w_N^{0} + f[1]w_N^{2}\,\mathrm{N}$$

$$+ \ldots + f[N-1]w_N^{2(N-1)}$$

. . . .

. . . .

$$\mathrm{F}[N-1] = f[0]w_N^{0} + f[1]w_N^{N-1}$$

$$+ \ldots + f[N-1]w_N^{(N-1)^2}.$$

Thus, in matrix form, we have

$$\begin{bmatrix} F[0] \\ F[1] \\ F[2] \\ \cdots \\ \cdots \\ F[N-1] \end{bmatrix} = \begin{bmatrix} w_N^0 & w_N^0 & \cdots\cdots & w_N^0 \\ w_N^0 & w_N & \cdots\cdots & w_N^{N-1} \\ w_N^0 & w_N^2 & \cdots\cdots & w_N^{2(N-1)} \\ \cdots & \cdots & \cdots\cdots & \cdots \\ \cdots & \cdots & \cdots\cdots & \cdots \\ w_N^0 & w_N^{N-1} & \cdots\cdots & w_N^{(N-1)^2} \end{bmatrix} \begin{bmatrix} f[0] \\ f[1] \\ f[2] \\ \cdots \\ \cdots \\ f[N-1] \end{bmatrix}$$

$$= \begin{bmatrix} 1 & 1 & \cdots\cdots & 1 \\ 1 & w_N & \cdots\cdots & w_N^{N-1} \\ 1 & w_N^2 & \cdots\cdots & w_N^{2(N-1)} \\ \cdots & \cdots & \cdots\cdots & \cdots \\ \cdots & \cdots & \cdots\cdots & \cdots \\ 1 & w_N^{N-1} & \cdots\cdots & w_N^{(N-1)^2} \end{bmatrix} \begin{bmatrix} f[0] \\ f[1] \\ f[2] \\ \cdots \\ \cdots \\ f[N-1] \end{bmatrix}$$

where $w_N = e^{-2\pi i/N}$ is called the *phase rotation factor* or *twiddle factor* for the DFT. The N × N matrix on the right hand side is called the *DFT transformation matrix*.

EXAMPLE 8.16

Compute 3-point DFT of the periodic discrete-time signal $f[n] = \{1, 0, 1\}$.

Solution. The 3-point DFT of the given signal is given by

$$\begin{bmatrix} 1 & 1 & 1 \\ 1 & w_3^1 & w_3^2 \\ 1 & w_3^2 & w_3^4 \end{bmatrix} \begin{bmatrix} 1 \\ 0 \\ 1 \end{bmatrix}.$$

But, by periodicity,

$$w_3^4 = w_3^{(1+3)} = w_3^1.$$

Further

$$w_3^1 = e^{-2\pi i/3} = \cos\frac{2\pi}{3} - i\sin\frac{2\pi}{3} = \frac{-1-\sqrt{3}\,i}{2},$$

$$w_2^3 = (e^{-2\pi i/3})^2 = \frac{-1+\sqrt{3}\,i}{2}.$$

Thus the DFT is

$$\begin{bmatrix} F[0] \\ F[1] \\ F[2] \end{bmatrix} = \begin{bmatrix} 1 & 1 & 1 \\ 1 & \frac{-1-\sqrt{3}\,i}{2} & \frac{-1+\sqrt{3}\,i}{2} \\ 1 & \frac{-1+\sqrt{3}\,i}{2} & \frac{-1-\sqrt{3}\,i}{2} \end{bmatrix} \begin{bmatrix} 1 \\ 0 \\ 1 \end{bmatrix}$$

$$= \begin{bmatrix} 2 \\ \frac{1}{2}(1+\sqrt{3}\,i) \\ \frac{1}{2}(1-\sqrt{3}\,i) \end{bmatrix}.$$

Hence

$$F[0] = 2,\ F[1] = \frac{1}{2}(1+\sqrt{3}i),$$

$$F[2] = \frac{1}{2}(1-\sqrt{3}i).$$

EXAMPLE 8.17

Calculate 4-point DFT of the periodic discrete-time signal $f[n] = \{1,2,3,4\}$ with period 8.

Solution. The required DFT in matrix form is.

$$\begin{bmatrix} F[0] \\ F[1] \\ F[2] \\ F[3] \end{bmatrix} = \begin{bmatrix} 1 & 1 & 1 & 1 \\ 1 & w_4^1 & w_4^2 & w_4^3 \\ 1 & w_4^2 & w_4^4 & w_4^6 \\ 1 & w_4^3 & w_4^6 & w_4^9 \end{bmatrix} \begin{bmatrix} f[0] \\ f[1] \\ f[2] \\ f[3] \end{bmatrix}$$

Using periodicity, it reduces to

$$\begin{bmatrix} F[0] \\ F[1] \\ F[2] \\ F[3] \end{bmatrix} = \begin{bmatrix} 1 & 1 & 1 & 1 \\ 1 & w_4^1 & w_4^2 & w_4^3 \\ 1 & w_4^2 & w_4^0 & w_4^2 \\ 1 & w_4^3 & w_4^2 & w_4^1 \end{bmatrix} \begin{bmatrix} 1 \\ 2 \\ 3 \\ 4 \end{bmatrix}$$

But

$$w_4^0 = 1,\ w_4^1 = e^{-2\pi i/4} = -i,$$

$$w_4^2 = (-i)^2 = -1,$$

$$w_4^3 = (e^{-2\pi i/4})^3 = i.$$

Hence

$$\begin{bmatrix} F[0] \\ F[1] \\ F[2] \\ F[3] \end{bmatrix} = \begin{bmatrix} 1 & 1 & 1 & 1 \\ 1 & -i & -1 & i \\ 1 & -1 & 1 & -1 \\ 1 & i & -1 & -i \end{bmatrix} \begin{bmatrix} 1 \\ 2 \\ 3 \\ 4 \end{bmatrix}$$

$$= \begin{bmatrix} 1+2+3+4 \\ 1-2i-3+4i \\ 1-2+3-4 \\ 1+2i-3-4i \end{bmatrix} = \begin{bmatrix} 10 \\ -2+2i \\ -2 \\ -2-2i \end{bmatrix}$$

Hence

$$F[0] = 10,\ F[1] = -2+2i,\ F[2] = -2$$

$$F[3] = -2-2i$$

8.8 N-POINT INVERSE DFT

By Fundamental theorem for DFT, we have

$$f(n) = \frac{1}{N}\sum_{k=0}^{N-1} F[k]e^{2\pi ink/N} = \frac{1}{N}\sum_{k=0}^{N-1} F[k]w_N^{-Kn}, \quad (7)$$

where $w_N = e^{-2\pi i/N}$.

The matrix form of (7) comes out to be

$$\begin{bmatrix} f[0] \\ f[1] \\ f[2] \\ \cdots \\ \cdots \\ f[N-1] \end{bmatrix} = \frac{1}{N} \begin{bmatrix} 1 & 1 & 1 & \cdots & 1 \\ 1 & w_N & w_N^2 & \cdots & w_N^{N-1} \\ 1 & w_N^2 & w_N^4 & \cdots & w_N^{2(N-1)} \\ \cdots & \cdots & \cdots & \cdots & \cdots \\ \cdots & \cdots & \cdots & \cdots & \cdots \\ 1 & w_N^{N-1} & w_N^{2(N-1)} & \cdots & w_N^{(N-1)^2} \end{bmatrix}^* \times \begin{bmatrix} F[0] \\ F[1] \\ F[2] \\ \cdots \\ \cdots \\ F[N-1] \end{bmatrix}, \quad (8)$$

Where $[\cdot]^*$ denotes complex conjugate of the matrix $[\cdot]$.

EXAMPLE 8.18

Find 3-point inverse DFT of

$$F(k) = \left(2, \frac{1}{2}(1+\sqrt{3}\,i), \frac{1}{2}(1-\sqrt{3}\,i)\right).$$

Solution. As in Example 8.16 the matrix for the DFT is

$$\begin{bmatrix} 1 & 1 & 1 \\ 1 & \frac{-1-\sqrt{3}\,i}{2} & \frac{-1+\sqrt{3}\,i}{2} \\ 1 & \frac{-1+\sqrt{3}\,i}{2} & \frac{-1-\sqrt{3}\,i}{2} \end{bmatrix}$$

Therefore, by 8.8 we have

$$\begin{bmatrix} f[0] \\ f[1] \\ f[2] \end{bmatrix} = \frac{1}{3} \begin{bmatrix} 1 & 1 & 1 \\ 1 & \frac{-1+\sqrt{3}i}{2} & \frac{-1-\sqrt{3}i}{2} \\ 1 & \frac{-1-\sqrt{3}i}{2} & \frac{-1+\sqrt{3}i}{2} \end{bmatrix} \begin{bmatrix} 2 \\ \frac{1}{2}(1+\sqrt{3}i) \\ \frac{1}{2}(1-\sqrt{3}i) \end{bmatrix}$$

$$= \frac{1}{3}\begin{bmatrix} 3 \\ 0 \\ 3 \end{bmatrix} = \begin{bmatrix} 1 \\ 0 \\ 1 \end{bmatrix}.$$

Hence,

$$f[0] = 1, f[1] = 0, f[2] = 1.$$

8.9 FAST FOURIER TRANSFORM (FFT)

We have seen that the calculation of the DFT involve elementary operations of order N^2. Therefore, for large N the method becomes more cumbersome.

Cooley and Tukey (*Math. Com.* 19 (1965), 297– 301) published an algorithm which, under certain conditions, reduces the number of computations for computing DFT. This algorithm is known as *fast Fourier transform (FFT). A FFT is thus not a transform but is simply an efficient numerical implementation of the DFT.*

The FFT requires only N $\log_2$ N operations to compute DFT, where N is a power of 2. So assume that $N = 2^m$, where m is a positive integer. The N-point DFT of the discrete-time signal $f[n]$ is given by

$$F(k) = \sum_{n=0}^{N-1} f[n]\, e^{-2\pi ink/N},\ k = 0, 1, \ldots, N-1.$$

$$= \sum_{n=0}^{N-1} f[n]\, w_N^{nk},$$

where $w_N = e^{-2\pi i/N}$ is the *phase factor* Separation of $f[n]$ into even and odd-numbered points yields

$$F(k) = \sum_{n\ \text{even}} f[n] w_N^{nk} + \sum_{n\ \text{odd}} f[n] w_N^{nk}$$

$$= \sum_{m=0}^{\frac{N}{2}-1} f[2m] w_N^{2mk} + \sum_{m=0}^{\frac{N}{2}-1} f[2m+1]\, w_N^{(2m+1)k}$$

$$= \sum_{m=0}^{\frac{N}{2}-1} f[2m] w_N^{2mk} + w_N^k \sum_{m=0}^{\frac{N}{2}-1} f[2m+1] w_N^{2mk}$$

But

$$w_N^2 = (e^{-2\pi i/N})^2 = e^{-2\pi i/(N/2)} = w_{N/2}.$$

Therefore

$$F(k) = \sum_{m=0}^{\frac{N}{2}-1} f[2m] w_{N/2}^{mk} + w_N^k \sum_{m=0}^{\frac{N}{2}-1} f[2m+1] w_{N/2}^{mk}$$

$$= G[k] + w_N^k H[k],$$

where $G[k]$ is the DFT of even numbered points of the signal $f[n]$ and $H[k]$ is the DFT of odd numbered points of the signal $f[n]$. Since $f[n]$ and $F[k]$ are periodic with period N, G[k] and H[k] are also periodic with period N/2.

EXAMPLE 8.19

Use 4-point FFT to computer Fourier transform of

$$f[n] = \{1,2,3,4\}.$$

Solution. We have

$$F(k) = \sum_{m=0}^{\frac{N}{2}-1} f[2m] w_{N/2}^{mk} + w_N^k \sum_{m=0}^{\frac{N}{2}-1} f[2m+1] w_{N/2}^{mk}$$

$$= G[k] + w_N^k H(k).$$

We have N = 4, so

$$G(k) = \sum_{m=0}^{1} f[2m] w_2^{mk} = f[0] w_2^0 + f[2] w_2^k,$$

$$H(k) = \sum_{m=0}^{1} [2m+1] w_2^{mk} = f[1] w_2^0 + f[3] w_2^k.$$

Further $G[k]$ and $H[k]$ are periodic with period $\frac{N}{2} = 2$. Therefore,

$$G[3] = G[1] \text{ and } H[3] = H[1].$$

Also

$$w_2^0 = 1,\quad w_2^1 = e^{-2\pi i/2} = -1,$$

$$w_2^2 = (-1)^2 = 1, w_2^3 = -1,$$

$$w_4^0 = -1, w_4^1 = -i,$$

$$w_4^2 = -1, w_4^3 = i.$$

Thus

$$F[0] = G[0] + w_4^0 H[0]$$

$$= [f[0]w_2^0 + f[2]w_2^0] + w_4^0[f[1]w_2^0 + f[3]w_2^0]$$

$$= (1+3) + 1(2+4) = 10,$$

$$F[1] = G[1] + \omega_4^1 H[1]$$

$$= [f[0]w_2^0 + f[2]w_2^1] + w_4^1[f[1]w_2^0 + f[3]w_2^1]$$

$$= [1(1) + 3(-1)] + (-i)[2(1) + 4(-1)]$$

$$= -2 + 2i,$$

$$F[2] = G[2] + w_4^2 H[2]$$
$$= [f[0]w_2^0 + f[2]w_2^2] + w_4^2[f[1]w_2^0 + f[3]w_2^2]$$
$$= [1(1) + 3(1)^2] + (-1)[2(1) + 4(1)] = -2,$$
$$F[3] = G[3] + w_4^3 H[3]$$
$$= G[1] + w_4^3 H[1]$$
$$= -2 + i[-2] = -2 - 2i.$$

Hence

$$F[k] = \{10, -2 + 2i, -2, -2 - 2i\}.$$

EXAMPLE 8.20

Use 4-point FFT to compute DFT of the periodic discrete-time signal with period 4 given by

$$f[-1] = 2,\ f[0] = i,\ f[1] = 1,\ f[2] = i.$$

Solution: We have

$$F[k] = G[k] + w_N^k H(k),$$

where

$$G(k) = f[0]w_2^0 + f[2]w_2^k \text{ and}$$

$$H(k) = f[-1]w_2^0 + f[1]w_2^k.$$

Thus

$$F[0] = G[0] + w_4^0 + H[0]$$
$$= [f[0]w_2^0 + f[2]w_2^0]$$
$$+ w_4^0[f[-1]w_2^0 + f[1]w_2^0]$$
$$= 2i + 1(2 + 1) = 2i + 3,$$
$$F[1] = G[1] + w_4^1 H[1]$$
$$= [f[0]w_2^0 + f[1]w_2^1]$$
$$+ w_4^1[f[-1]w_2^0 + f[1]w_2^1]$$
$$= [i(1) + i(-1)] + (-i)[2(1)$$
$$+ 1(-1)] = -i,$$
$$F[2] = G[2] + w_4^2 H[2]$$
$$= [f[0]w_2^0 + f[2]w_2^2] + w_4^2[f[-1]w_2^0 + f[1]w_2^2]$$
$$= 2i + (-1)[2(1) + 1(1)] = 2i - 3,$$
$$F[3] = G[3] + w_4^3 H[3]$$
$$= G[1] + w_4^3 H[1],$$

by periodicity of G and H

$$= 0 + i(1) = i.$$

Hence

$$F[k] = \{2i + 3, -i, 2i - 3, i\}.$$

EXERCISES

1. Determine 10-piont DFT of the sequence
$$s[n] = \{1 \text{ for } 2 \le n \le 6$$
$$0 \text{ for } n = 0, 1, 7, 8, 9.$$
Ans. $e^{-4\pi ik/5} \dfrac{\sin \dfrac{nk}{2}}{\sin \dfrac{nk}{10}}$

2. Determine the 5-point DFT of the periodic-time signal $f[n]$ with period 5 and defined by
$$f[-2] = -1, f[\ 1] = -2, f[0] = 0, f[1] = 2,$$
$$f[2] = 1$$
Ans. $-2i \sin \dfrac{4\pi k}{5} - 4i \sin \dfrac{2\pi k}{5}$

3. Find 2-point DFT of the discrete-time signal $f[n] = (-1)^n$ for $n \in \mathbb{Z}$.
Ans. $F[0] = 0, F[1] = 2.$

4. Show that $\delta_N[n]$ is an even signal.

5. If $f[n]$ is a real periodic discrete-time signal with period 4 and $F[0] = 1, F[1] = i, F[2] = 0$, determine $f[n]$ for $n = 0, 1, 2, 3$.
Hint: Since $f[n]$ is real, the periodicity of $F[k]$ implies
$$F[3] = F[4 - 1] = F[-1] = \overline{F[1]} = -i.$$
Using Fundamental theorem for the DFT,
$$f[n] = \frac{1}{4}\sum_{k=0}^{N-1} F[k]e^{2\pi ink/4}$$
$$= \frac{1}{4}[1 + i\,e^{\pi in/2} - ie^{3\pi in/2}]$$
which yields $f[0] = 1, f[1] = -1, f[2] = 1,$ $f[3] = 3.$

6. Calculate the power of the periodic discrete time signal with period N defined by $f(n) = \sin\frac{2\pi n}{N}$.

   Hint: $f[n] = \frac{e^{2\pi in\ N} - e^{-2\pi\ in\ N}}{2i}$. Find DFT to get $F[-1] - \frac{N}{2i}$, $F[1] = \frac{N}{2i}$, $F[0] = F[2] = F[3] = \ldots = F[N-2] = 0$. Thus $P = \frac{1}{N^2}\left(\frac{N^2}{4} + \frac{N^2}{4}\right) = \frac{1}{2}$.

7. Using DFT transformation matrix, find inverse DFT of the DFT $F[k] = \{10, -2 + 2i, -2, -2 - 2i\}$.

   Ans. $f[n] = \{1, 2, 3, 4\}$

8. Find circular convolution of the two discrete time sequences $f[n] = \{1, 2, 1, 2\}$ and $g[n] = \{3, 2, 1, 4\}$.

   Ans. $\{16, 14, 17, 14\}$

9. Using Convolution theorem, find circularconvolution of $f[n] = \{2, 1, 2, 1\}$ and $g[n] = \{1, 2, 3, 4\}$.

   Ans. $\{14, 16, 14, 16\}$.

10. Use FFT to evaluate DFT of the 4-point sequence $\{1, 1, 0, 0\}$.

    Hint: $F[0] = G[0] + w_4^0 H[0] = 1 + 1(1) = 2$, $F[1] = G[1] + w_4^1 H[1] = [1(1) + 0] + (i)[1] = 1 - i$

    $F[2] = 0$, $F[3] = 1 + i$.

9 Laplace Transform

The study of Laplace transform is essential for engineers and scientists because these transforms provide easy and powerful means of solving differential and integral equations. The Laplace transforms directly provides the solution of differential equations with given boundary values without finding the general solution first. A Laplace transform is an extension of the continuous-time Fourier transform motivated by the fact that this transform can be used to a wider class of signals than the Fourier transform can. In fact, Fourier transform does not converge for many signals whereas the Laplace transform does. Fourier transform is not applicable to initial-value problems whereas Laplace transform is applicable. Also some functions like sinusoidal functions and polynomials do not have Fourier transform in the usual sense without the introduction of generalized functions. Causal functions (which assume zero value for $t < 0$) are best handled by Laplace transforms.

9.1 DEFINITION AND EXAMPLES OF LAPLACE TRANSFORM

Definition 9.1. Let $f(t)$ be a function of t defined for $t > 0$. Then the *one-sided Laplace transform* (or merely *Laplace transform*) of $f(t)$, denoted by $L\{f(t)\}$ or $F(s)$, is defined by

$$L\{f(t)\} = F(s) = \int_0^\infty e^{-st} f(t)dt,\ s \in \mathbb{R} \text{ or } \mathbb{C},$$

provided that the integral converges for some value of s.

The defining integral is called the *Laplace integral.*

Definition 9.2. The *two-sided Laplace transform* of a function $f(t)$ is defined by

$$L\{f(t)\} = \int_{-\infty}^{\infty} e^{-st} f(t)\, dt$$

for all values of s, real or complex for which the integral converges.

The defining integral in this case is called *two-sided Laplace integral.*

The symbol L, which transforms $f(t)$ *into* $F(s)$ is called *Laplace operator.* Thus the Laplace transform of a function exists only if the Laplace integral converges. The following theorem provides sufficient conditions for the existence of Laplace transform.

Theorem 9.1. Let $f(t)$ be piecewise continuous in every finite interval $0 \le t \le N$ and be of exponential order γ for $t > N$. Then Laplace transform of $f(t)$ exists for all $s > \gamma$.

Proof: Since $f(t)$ is piecewise continuous on every finite interval $[0, N]$ and e^{-st} is also piecewise continuous on $[0, N]$ for $N > 0$, it follows that $e^{-st} f(t)$ is integrable on $[0, N]$. For any positive number N, we have

$$\int_0^\infty e^{-st} f(t)\, dt = \int_0^N e^{-st} f(t)\, dt + \int_N^\infty e^{-st} f(t)\, dt.$$

By the above arguments, the first integral on the right exists. Further, since $f(t)$ is of exponential order γ for $t > N$, there exists constant M such that $|f(t)| \le M e^{\gamma t}$ for $t \ge 0$ and so

$$\left|\int_N^\infty e^{-st} f(t)\, dt\right| \le \int_N^\infty |e^{-st} f(t)|\, dt \le \int_0^\infty |e^{-st} f(t)|\, dt$$

$$\leq \int_0^\infty e^{-st}\, M e^{-\gamma t}\, dt = \frac{M}{s-\gamma}.$$

Thus the Laplace transform L{ $f(t)$} exists for $s > \gamma$.

Remark 9.1. The conditions of the Theorem 9.1 are *sufficient but not necessary* for the existence of Laplace transform of a function. Thus Laplace transform may exist even if these conditions are not satisfied. For example, $f(t) = t^{-\frac{1}{2}}$ does not satisfy these conditions but its Laplace transform does exist (see Example 9.8).

EXAMPLE 9.1

Find Laplace transform of *unit step function f* defined by $f(t) = 1,\ t \geq 0$.

Solution. By definition of Laplace transform, we have

$$\text{L}\{f(t)\} = \int_0^\infty e^{-st}\, dt = \lim_{T\to\infty} \int_0^T e^{-st}\, dt$$

$$= \lim_{T\to\infty} \left[\frac{e^{-st}}{-s}\right]_0^T = \lim_{T\to\infty} \frac{1-e^{-st}}{s} = \frac{1}{s} \text{ if } s>0.$$

EXAMPLE 9.2

Find the Laplace transform of the *unit ramp function f* defined by $f(t) = t,\ t \geq 0$.

Solution. Using integration by parts, we get

$$\text{L}\{f(t)\} = \int_0^\infty te^{-st}\, dt = \lim_{T\to\infty} \int_0^T te^{-st}\, dt$$

$$= \lim_{T\to\infty} \left\{\left[t\left(\frac{e^{-st}}{-s}\right)\right]_0^T - \left[\frac{e^{-st}}{s^2}\right]_0^T\right\}$$

$$= \lim_{T\to\infty} \left(\frac{1}{s^2} - \frac{e^{-sT}}{s^2} - \frac{Te^{-sT}}{s}\right) = \frac{1}{s^2} \text{ if } s > 0.$$

EXAMPLE 9.3

Find L{$f(t)$}, where $f(t) = [t],\ t > 0$.

Solution. We have

$$\text{L}\{f(t)\} = \int_0^\infty e^{-st} f(t)\, dt$$

$$= \int_0^1 e^{-st}(0)dt + \int_1^2 e^{-st}\, dt + \int_2^3 e^{-st}\, 2dt + \ldots$$

$$= \left|\frac{e^{-st}}{-s}\right|_1^2 + 2\left|\frac{e^{-st}}{-s}\right|_2^3 + 3\left|\frac{e^{-st}}{-s}\right|_3^4 + \ldots.$$

$$= \frac{e^{-s}}{s}(1-e^{-s}) + 2\,\frac{e^{-2s}}{s}(1-e^{-s}) + 3\,\frac{e^{-3s}}{s}(1-e^{-s}) + \ldots$$

$$= \frac{e^{-s}}{s}(1-e^{-s})\,[1 + 2\,e^{-s} + 3\,e^{-2s} + \ldots]$$

$$= \frac{e^{-s}}{s}(1-e^{-s})\frac{1}{(1-e^{-s})^2}$$

$$= \frac{e^{-s}}{s(1-e^{-s})} = \frac{1}{s(e^{s}-1)}.$$

EXAMPLE 9.4

Find Laplace transform of $f(t) = e^{at},\ t \geq 0$.

Solution. By definition of Laplace transform, we have

$$\text{L}\{f(t)\} = \int_0^\infty e^{-st} e^{at}\, dt = \lim_{T\to\infty} \int_0^T e^{-(s-a)t}\, dt$$

$$= \lim_{T\to\infty} \left[\frac{e^{-(s-a)}}{-(s-a)}\right]_0^T = \lim_{T\to\infty} \frac{1-e^{-(s-a)T}}{s-a}$$

$$= \frac{1}{s-a}, \text{ if } s > a.$$

The result of this example holds for complex numbers also.

EXAMPLE 9.5

Find Laplace transforms of $f(t) = \sin at$ and $g(t) = \cos at$.

Solution. Since

$$\int e^{at} \sin bt\, dt = \frac{e^{at}(a \sin bt - b \cos bt)}{a^2+b^2},$$

and

$$\int e^{at} \cos bt\, dt = \frac{e^{at}(a \cos bt + b \sin bt)}{a^2+b^2},$$

We have

$$L\{\sin at\} = \int_0^{\infty} e^{-st} \sin at \, dt$$

$$= \lim_{T\to\infty} \int_0^{T} e^{-st} \sin at \, dt$$

$$= \lim_{T\to\infty} \left[\frac{e^{-st}(-s \sin at - a \cos at)}{s^2 + a^2} \right]_0^T$$

$$= \lim_{T\to\infty} \left[\frac{a}{s^2 + a^2} - \frac{e^{-st}(s \sin aT + a \cos aT)}{s^2 + a^2} \right]$$

$$= \frac{a}{s^2 + a^2} \text{ if } s > 0,$$

and

$$L\{\cos at\} = \int_0^{\infty} e^{-st} \cos at \, dt$$

$$= \lim_{T\to\infty} \int_0^{T} e^{-st} \cos at \, dt$$

$$= \lim_{T\to\infty} \left[\frac{e^{-st}(-s \cos at + a \sin at)}{s^2 + a^2} \right]_0^T$$

$$= \lim_{T\to\infty} \left[\frac{s}{s^2 + a^2} - \frac{e^{-st}(s \cos aT - a \sin aT)}{s^2 + a^2} \right]$$

$$= \frac{s}{s^2 + a^2} \text{ if } s > 0.$$

EXAMPLE 9.6

Find the Laplace transforms of $f(t) = \sinh at$ and $g(t) = \cosh at$.

Solution. Since $\sinh at = \frac{e^{at} - e^{-at}}{2}$, we have

$$L\{\sinh at\} = L\left\{ \frac{e^{at} - e^{-at}}{2} \right\} = \int_0^{\infty} e^{-st} \left(\frac{e^{at} - e^{-at}}{2} \right) dt$$

$$= \frac{1}{2} \int_0^{\infty} e^{-st} e^{at} \, dt - \frac{1}{2} \int_0^{\infty} e^{-st} e^{-at} \, dt$$

$$= \frac{1}{2} L\{e^{at}\} - \frac{1}{2} L\{e^{-at}\}$$

$$= \frac{1}{2} \left[\frac{1}{s-a} - \frac{1}{s+a} \right] = \frac{a}{s^2 - a^2}, \; s > |a|.$$

Again, since $\cosh at = \frac{e^{at} + e^{-at}}{2}$, proceeding as above, we have

$$L\{\cosh at\} = \frac{1}{2} L\{e^{at}\} + \frac{1}{2} L\{e^{-at}\}$$

$$= \frac{1}{2} \left[\frac{1}{s-a} + \frac{1}{s+a} \right]$$

$$= \frac{s}{s^2 - a^2}, \; s > |a|.$$

EXAMPLE 9.7

Find Laplace transform of $f(t) = t^n$, where n is a positive integer.

Solution. Putting $st = u$, we have

$$L\{f(t)\} = \int_0^{\infty} e^{-st} t^n \, dt = \int_0^{\infty} e^{-u} \left(\frac{u}{s}\right)^n \cdot \frac{du}{s}$$

$$= \frac{1}{s^{n+1}} \int e^{-u} u^{(n+1)-1} \, du = \frac{\Gamma(n+1)}{s^{n+1}}$$

$$\text{for } s > 0 \text{ and } n + 1 \geq 0,$$

by the definition of gamma function. Since n is positive integer, $\Gamma(n+1) = n!$ and so

$$L\{t^n\} = \frac{n!}{s^{n+1}}$$

Remark 9.2. Integrating the defining formula for Laplace transform of t^n by parts, we have

$$L\{t^n\} = \int_0^{\infty} t^n e^{-st} \, dt$$

$$= \left[\frac{-t^n e^{-st}}{s} \right]_0^{\infty} + \frac{n}{s} \int_0^{\infty} t^{n-1} e^{-st} \, dt.$$

The integral on the right exists and the lower limit can be used in the first term if $n \geq 1$. Since $s > 0$, the exponent in the first term goes to zero as t tends to infinity. Thus, we obtain a general *recurrence formula,*

$$L\{t^n\} = \frac{n}{s} L\{t^{n-1}\}, \; n \geq 1.$$

Hence, by induction, we get the sequence

$$L\{t^0\} = L\{1\} = \frac{1}{s} \text{ (by Example 9.1)}$$

$$L\{t\} = \frac{1}{s} L\{t^0\} = \frac{1}{s^2}$$

$$L\{t^2\} = \frac{2}{s^3}$$

......

......

$$L\{t^n\} = \frac{n!}{s^{n+1}}.$$

EXAMPLE 9.8

Find the Laplace transform of $f(t) = t^{-\frac{1}{2}}$.

Solution. The condition $n + 1 > 0$ of Example 9.7 is satisfied and so

$$L\{t^{-\frac{1}{2}}\} = \frac{\Gamma(1/2)}{s^{\frac{1}{2}}} = \frac{\sqrt{\pi}}{s^{\frac{1}{2}}} = \sqrt{\frac{\pi}{s}}.$$

It may be mentioned here that the function $f(t) = t^{-\frac{1}{2}}$ does not satisfy the conditions of Theorem 9.1, even then the Laplace transform of this function exists. Thus the conditions of Theorem 9.1 are sufficient but not necessary for the existence of Laplace transform of a given function.

EXAMPLE 9.9

Find the Laplace transform of the function f defined by

$$f(t) = \begin{cases} t & \text{for } 0 \le t \le 3 \\ 0 & \text{for } t > 3. \end{cases}$$

Solution. The graph of the function f is shown in the Figure 9.1.

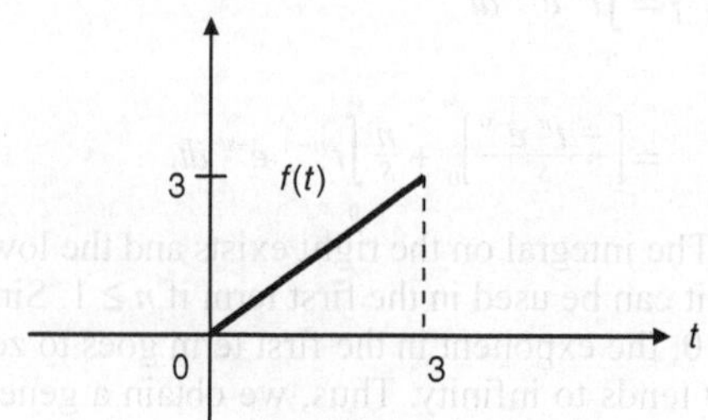

Figure 9.1

Using integration by parts, we have

$$L\{f(t)\} = \int_0^\infty e^{-st} f(t)\,dt = \int_0^3 e^{-st}\,t\,dt$$

$$= \left[t \cdot \frac{e^{-st}}{-s}\right]_0^3 - \int_0^3 \frac{e^{-st}}{-s}\,dt$$

$$= -\frac{3}{s} e^{-3s} + \frac{1}{s}\left[\frac{e^{-st}}{-s}\right]_0^3$$

$$= -\frac{3}{s} e^{-3s} - \frac{1}{s^2}[e^{-3s} - 1]$$

$$= \frac{1}{s^2}[1 - e^{-3s}] - \frac{3}{s} e^{-3s} \quad \text{for } s > 0.$$

EXAMPLE 9.10

Find the Laplace transform of the function f defined by $f(t) = \sqrt{t^n}$, $n \ge 1$ and odd integer.

Solution. Integration by parts yields

$$L\{f(t)\} = \int_0^\infty \sqrt{t^n}\, e^{-st}\,dt$$

$$= -\left[\frac{\sqrt{t^n}\, e^{-st}}{s}\right]_0^\infty + \frac{n}{2s}\int_0^\infty \sqrt{t^{n-2}}\, e^{-st}\,dt.$$

If $n \ge 1$, the lower limit can be used in the first term on the right and thus the integral exists. Thus

$$L\{\sqrt{t^n}\} = \frac{n}{2s} L\{\sqrt{t^{n-2}} - 2\},\ n \ge 1 \text{ and odd.}$$

Thus we obtain a sequence of formulas given below:

$$L\left\{\frac{1}{\sqrt{t}}\right\} = \frac{\sqrt{\pi}}{\sqrt{s}} \quad \text{(Example 9.8)}$$

$$L\{\sqrt{t}\} = \frac{\sqrt{\pi}}{2\sqrt{s^3}}$$

$$L\{\sqrt{t^3}\} = \frac{3\sqrt{\pi}}{4\sqrt{s^5}}$$

....

....

$$L\{\sqrt{t^n}\} = \frac{(n+1)!\,\sqrt{\pi}}{2^{(n+1)}\left(\frac{n+1}{2}\right)!\,\sqrt{s^{(n+2)}}}.$$

EXAMPLE 9.11

Find the Laplace transform of $\operatorname{erf}(\sqrt{z})$ and $\operatorname{erf}(z)$.

Solution. Recall that the *error function* is defined by the integral

$$\text{erf}(z) = \frac{2}{\sqrt{\pi}}\int_0^z e^{-t^2}\,dt,$$

Where the variable z may be real or complex.

The graph of erf(z), where z is real is shown in the Figure 9.2.

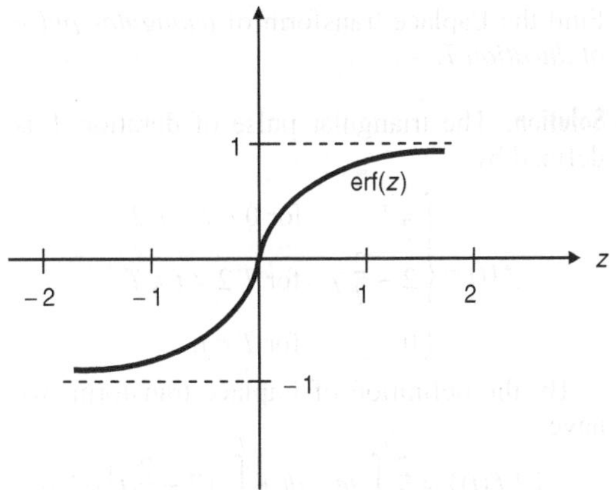

Figure 9.2 The Error Function

Let us find Laplace transform of erf($\sqrt{z}$). Using series expansion of e^{-t^2}, we have

L{erf($\sqrt{z}$)}

$$= L\left\{\frac{2}{\sqrt{\pi}}\int_0^{\sqrt{t}}\left[1 - t^2 + \frac{t^4}{2!} - \frac{t^6}{3!} + \ldots\right]dt\right\}$$

$$= \frac{2}{\sqrt{\pi}} L\left\{t^{\frac{1}{2}} - \frac{t^{\frac{3}{2}}}{3} + \frac{t^{\frac{5}{2}}}{5.2!} - \frac{t^{\frac{7}{2}}}{7.3!} + \ldots\right\}$$

$$= \frac{2}{\sqrt{\pi}}\left\{\frac{\Gamma(3/2)}{s^{\frac{3}{2}}} - \frac{\Gamma(5/2)}{3s^{\frac{5}{2}}} + \frac{\Gamma(7/2)}{5.2!s^{\frac{7}{2}}} - \frac{\Gamma(9/2)}{7.3!s^{\frac{9}{2}}} + \ldots\right\}$$

$$= \frac{1}{s^{\frac{3}{2}}} - \frac{1}{2s^{\frac{5}{2}}} + \frac{1.3}{2.4s^{\frac{7}{2}}} - \frac{1.3.5}{2.4.6s^{\frac{9}{2}}} + \ldots$$

$$= \frac{1}{s^{\frac{3}{2}}}\left[1 - \frac{1}{2s} + \frac{1.3}{2.4}\frac{1}{s^2} - \frac{1.3.5}{2.4.6}\frac{1}{s^3} + \ldots\right]$$

$$= \frac{1}{s^{\frac{3}{2}}}\left(1 + \frac{1}{s}\right)^{-\frac{1}{2}}$$

$$= \frac{1}{s\sqrt{(s+1)}},\ s > -1.$$

We now find the Laplace transform of erf (z). we have

$$L\{\text{erf}(z)\} = \int_0^\infty e^{-st}\,\text{erf}(z)\,dz$$

$$= \int_0^\infty e^{-st}\frac{2}{\sqrt{\pi}}\int_0^z e^{-x^2}\,dx\,dz$$

Changing the order of integration (Figure 9.3), we get

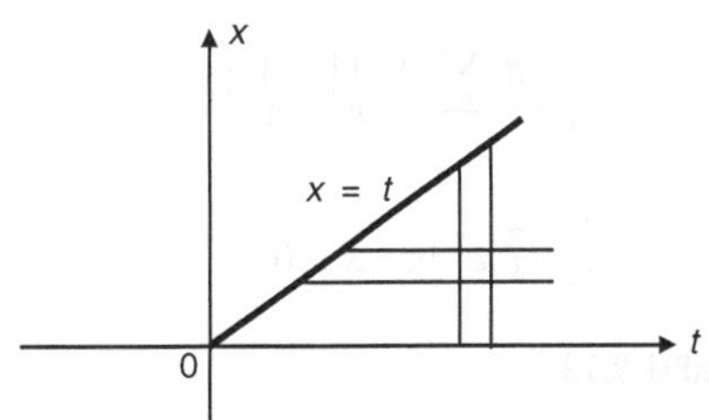

Figure 9.3

$$L\{\text{erf}(z)\} = \frac{2}{\sqrt{\pi}}\int_0^\infty e^{-x^2}\int_x^\infty e^{-st}\,dt\,dx = \frac{2}{s\sqrt{\pi}}\int_0^\infty e^{-(x^2+sx)}dx$$

$$= \frac{2}{s\sqrt{\pi}}e^{\frac{s^2}{4}}\int_0^\infty e^{-\left(x+\frac{s}{2}\right)^2}dx,$$

because $x^2 + sx = \left(x + \frac{s}{2}\right)^2 - \frac{s^2}{4}$.

Taking $u = x + \frac{s}{2}$, we have

$$L\{\text{erf}(z)\} = \frac{2}{s\sqrt{\pi}}e^{\frac{s^2}{4}}\int_{s/2}^\infty e^{-u^2}\,du$$

and so

$$L\{\text{erf}(z)\} = \frac{2}{s\sqrt{\pi}}e^{\frac{s^2}{4}}\,\text{erf}\,c\left(\frac{s}{2}\right),\ s > 0.$$

EXAMPLE 9.12

Find the Laplace transform of $f(t) = \sin\sqrt{t}$.

Solution. The series expansion for $\sin\sqrt{t}$ is

$$\sin\sqrt{t} = t^{\frac{1}{2}} - \frac{t^{\frac{3}{2}}}{3!} + \frac{t^{\frac{5}{2}}}{5!} - \frac{t^{\frac{7}{2}}}{7!} + \ldots$$

$$= \sum_{n=0}^{\infty} \frac{(-1)^n \, t^{n+\frac{1}{2}}}{(2n+1)!}$$

Therefore

$$L\{\sin\sqrt{t}\} = \sum_{n=0}^{\infty} \frac{(-1)^n}{(2n+1)!} L\left\{t^{n+\frac{1}{2}}\right\}$$

$$= \sum_{n=0}^{\infty} \frac{(-1)^n}{(2n+1)} \frac{\Gamma\left(n+\frac{3}{2}\right)}{s^{n+\frac{3}{2}}}$$

$$= \sum_{n=0}^{\infty} \frac{(-1)^n}{(2n+1)!} \cdot \frac{(2n+2)!}{2^{2n+2}(n+1)!} \sqrt{\pi} \cdot \frac{1}{s^{n+\frac{3}{2}}}$$

$$= \sum_{n=0}^{\infty} \frac{(-1)^n}{(2n+1)!} \cdot \frac{(2n+2)!}{2^{2n+1}\, n!} \cdot \frac{\sqrt{\pi}}{s^{n+\frac{3}{2}}}$$

$$= \frac{1}{2s}\sqrt{\frac{\pi}{s}} \sum_{n=0}^{\infty} \frac{(-1)^n}{n!}\left(\frac{1}{4s}\right)^n$$

$$= \frac{1}{2s}\sqrt{\frac{\pi}{s}}\, e^{-\frac{1}{4s}}, \quad s > 0.$$

EXAMPLE 9.13

Find the Laplace transform of the *pulse of unit height and duration* T.

Solution. The pulse of unit height and duration T is defined by

$$f(t) = \begin{cases} 1 & \text{for } 0 < t < T \\ 0 & \text{for } T < t. \end{cases}$$

therefore,

$$L\{f(t)\} = \int_0^T e^{-st}\, dt = \frac{1 - e^{-sT}}{s}.$$

EXAMPLE 9.14

Find the Laplace transform of sinusoidal (sine) pulse.

Solution. Sinusoidal pulse is defined by

$$f(t) = \begin{cases} \sin at & \text{for } 0 < t < \pi/a \\ 0 & \text{for } \pi/a < t. \end{cases}$$

Therefore, by the definition of Laplace transform we have

$$L\{f(t)\} = \int_0^{\infty} \sin at\, e^{-st}\, dt = \frac{a\left(1 + e^{-\frac{s\pi}{a}}\right)}{s^2 + a^2}.$$

The denominator of $L\{f(t)\}$ here is zero at $s = \pm\, ia$. But, since $e^{\pm ia} = -1$, the numerator also becomes zero. Thus $L\{f(t)\}$ have no poles and is an entire function.

EXAMPLE 9.15

Find the Laplace transform of *triangular pulse of duration T.*

Solution. The triangular pulse of duration T is defined by

$$f(t) = \begin{cases} \frac{2}{T}t & \text{for } 0 < t < T/2 \\ 2 - \frac{2}{T}t & \text{for } T/2 < t < T \\ 0 & \text{for } T < t. \end{cases}$$

By the definition of Laplace transform, we have

$$L\{f(t)\} = \frac{2}{T}\int_0^{T/2} te^{-st}\, dt + \int_{T/2}^{T} \left(2 - \frac{2}{T}t\right) e^{-st}\, dt$$

$$= \frac{2}{T}\left(\frac{1 - 2e^{-sT/2} + e^{-sT}}{s^2}\right).$$

EXAMPLE 9.16

Find the Laplace transform of a function defined by

$$f(t) = \begin{cases} \frac{t}{a} & \text{for } 0 < t < a \\ 1 & \text{for } t < a. \end{cases}$$

Solution. Integrating by parts,

$$L\{f(t)\} = \int_0^a \frac{t}{a} e^{-st}\, dt + \int_a^{\infty} e^{-st}\, dt$$

$$= \frac{-e^{-sa}}{s} - \frac{e^{-sa} - 1}{as^2} + \frac{e^{-sa}}{s} = \frac{1 - e^{-sa}}{as^2}.$$

EXAMPLE 9.17

Find the Laplace transform of f defined by

$$f(t) = \begin{cases} e^t & \text{for } 0 < t < 1 \\ 0 & \text{for } t > 1. \end{cases}$$

Solution. By definition of Laplace transform, we have

$$L\{f(t)\} = \int_0^1 e^t \cdot s^{-st}\, dt = \int_0^1 e^{(-s+1)t}\, dt = \frac{e^{1-s} - 1}{1 - s}.$$

EXAMPLE 9.18

Find the Laplace transform of a function f defined by

$$f(t) = \begin{cases} \sin t & \text{for } 0 < t < \pi \\ 0 & \text{for } t > \pi. \end{cases}$$

Solution. The integration by parts yields

$$\begin{aligned} L\{f(t)\} &= \int_0^{\pi} e^{-st} \sin t\, dt = 0 + \frac{1}{s}\int_0^{\pi} e^{-st} \cos t\, dt \\ &= \frac{1}{s}\left[\frac{e^{-st}}{-s}\cos t\right]_0^{\pi} + \frac{1}{s}\int_0^{\pi} \frac{e^{-st}}{-s} \sin t dt \\ &= \frac{1}{s^2}[1 + e^{-s\pi}] - \frac{1}{s^2}\int_0^{\pi} e^{-st} \sin t dt. \end{aligned}$$

Thus,

$$\left(\frac{s^2+1}{s^2}\right) L\{f(t)\} = \frac{1}{s^2}[1 + e^{-s\pi}],$$

which yields

$$L\{f(t)\} = \frac{1+e^{-s\pi}}{s^2+1}.$$

EXAMPLE 9.19

Find the Laplace transform of the function f_ε defined by

$$f_\varepsilon(t) = \begin{cases} \frac{1}{\varepsilon} & \text{for } 0 < t < \varepsilon \\ 0 & \text{for } t > \varepsilon, \end{cases}$$

where $\varepsilon > 0$. Deduce the Laplace transform of the *Dirac delta function.*

Solution. The graph of the function f_ε is shown in the Figure 9.4.

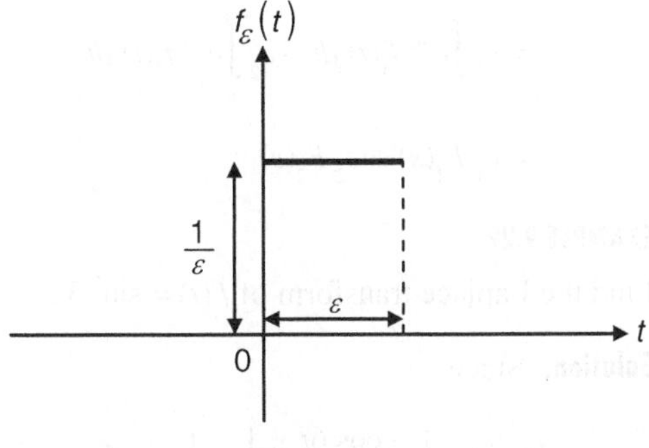

Figure 9.4

We observe that as $\varepsilon \to 0$, the height of the rectangle increases indefinitely and the width decreases in such a way that its area is always equals to 1.

The Laplace transform of f_ε is given by

$$\begin{aligned} L\{f_\varepsilon(t)\} &= \int_0^{\infty} e^{-st} f_\varepsilon(t)\, dt = \int_0^{\varepsilon} e^{-st} f_\varepsilon(t)\, dt \\ &\quad + \int_e^{\delta} e^{-st} f_\varepsilon(t)\, dt \\ &= \frac{1}{\varepsilon}\int_0^{\varepsilon} e^{-st}\, dt = \frac{1-e^{-s\varepsilon}}{s\varepsilon}. \end{aligned}$$

Further we note that

$$\begin{aligned} \lim_{\varepsilon\to 0} L\{f_\varepsilon(t)\} &= \lim_{\varepsilon\to 0} \frac{1-e^{-s\varepsilon}}{s\varepsilon} \\ &= \lim_{\varepsilon\to 0} \frac{1-\left(1 - s\varepsilon + \frac{s^2\varepsilon^2}{2!} - \ldots\right)}{s\varepsilon} \\ &= \lim_{\varepsilon\to 0}\left(1 - \frac{s\varepsilon}{2} + \ldots\right) = 1. \end{aligned}$$

Also, we observe from the definition of f_ε that $\lim_{\varepsilon\to 0} f_\varepsilon(t)$ does not exist and so $L\{\lim_{\varepsilon\to 0} f_\varepsilon(t)\}$ is not defined. Even then it is useful to define a function δ as $\delta(t) = \{\lim_{\varepsilon\to 0} f_\varepsilon(t)\}$ such that

$$\begin{aligned} L\{\delta(t)\} &= L\left\{\lim_{\varepsilon\to 0}\{f_\varepsilon(t)\}\right\} \\ &= \lim_{\varepsilon\to 0}[L\{f_\varepsilon(t)\}] = 1 \end{aligned}$$

The function $\delta(t)$ is called the *Dirac delta function or unit impulse function* having the properties

$$\delta(t) = 0,\ t \neq 0, \text{ and } \int_0^{\infty} \delta(t)\, dt = 1.$$

EXAMPLE 9.20

Find the Laplace transform of *Heaviside's unit step function* defined by

$$H(t-a) = \begin{cases} 1 & \text{for } t > a \\ 0 & \text{for } t < a. \end{cases}$$

Solution. The Heavisid's unit step function is also known as *delayed unit step function* and occurs in the electrical systems. It delays the output until $t = a$ and then assumes a constant value of one unit. Its graph is shown in Figure 9.5

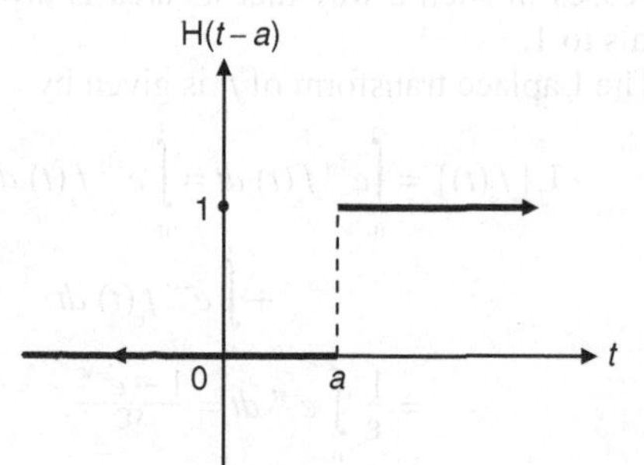

Figure 9.5

The Laplace transform of Heaviside's unit function is given by

$$L\{H(t-a)\} = \int_0^\infty e^{-st} H(t-a)\, dt = \int_a^\infty e^{-st}\, dt$$

$$= \lim_{T\to\infty} \int_a^T e^{-st}\, dt = \lim_{T\to\infty} \left[\frac{e^{-st}}{-s}\right]_a^T = \frac{e^{-sa}}{s}.$$

EXAMPLE 9.21

Find the Laplace transform of *rectangle function* defined by

$$g(t) = \begin{cases} 1 & \text{for } a < t < b \\ 0 & \text{otherwise.} \end{cases}$$

Solution. The graph of this function is shown in the Figure 9.6. Clearly, this function can be expressed in terms of Heaviside's unit function as

$$g(t) = H(t-a) - H(t-b).$$

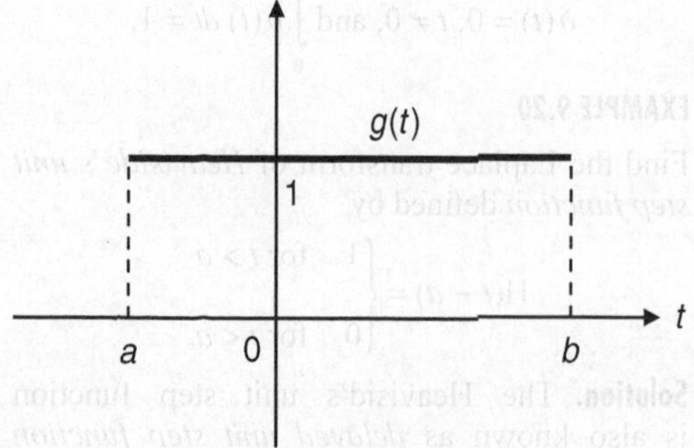

Figure 9.6

Further if $a = 0$, then g becomes pulse of unit height and duration b (Example 9.13). The Laplace transform of rectangle function g is given by

$$L\{g(t)\} = L\{H(t-a) - L\{H(t-b)\}$$

$$= \frac{e^{-sa}}{s} - \frac{e^{-sb}}{s}$$

$$= \frac{e^{-sa} - e^{-sb}}{s}.$$

9.2 PROPERTIES OF LAPLACE TRANSFORMS

While studying the following properties of Laplace transforms, we assume that the Laplace transforms of the given functions exist.

Theorem 9.2. (Linearity of the Laplace Transform). If c_1 and c_2 are arbitrary constants (real or complex) and $f_1(t)$ and $f_2(t)$ are functions with Laplace transforms $F_1(s)$ and $F_2(s)$, respectively, then

$$L\{c_1 f_1(t) + c_2 f_2(t)\} = c_1\, L\{f_1(t)\} + c_2\, L\{f_2(t)\}$$

$$= c_1\, F_1(s) + c_2\, F_2(s).$$

Thus L is a linear operator.

Proof: Using the definition of Laplace transform and the linearity property of integral, we have

$$L\{c_1 f_1(t) + c_2 f_2(t)\}$$

$$= \int_0^\infty e^{-st}\, [c_1 f_1(t) + c_2 f_2(t)]\, dt$$

$$= c_1 \int_0^\infty e^{-st} f_1(t)\, dt + c_2 \int_0^\infty e^{-st} f_2(t)\, dt$$

$$= c_1\, F_1(s) + c_2\, F_2(s).$$

EXAMPLE 9.22

Find the Laplace transform of $f(t) = \sin^2 3t$.

Solution. Since

$$\sin^2 3t = \frac{1 - \cos 6t}{2} = \frac{1}{2} - \frac{1}{2}\cos 6t,$$

We have

$$L\{\sin^2 3t\} = L\left\{\frac{1}{2} - \frac{1}{2}\cos 6t\right\} = \frac{1}{2}L\{1\}$$

$$-\frac{1}{2}L\{\cos 6t\}$$

$$= \frac{1}{2}\left(\frac{1}{s}\right) - \frac{1}{2}\left(\frac{s}{s^2+6^2}\right), s > 0$$

$$= \frac{1}{2}\left[\frac{1}{s} - \frac{s}{s^2+36}\right] = \frac{18}{s(s^2+36)}, s > 0.$$

EXAMPLE 9.23

Find the Laplace transform of $f(t) = e^{4t} + e^{2t} + t^3 + \sin^2 t$.

Solution. Since $\sin^2 t = \frac{1-\cos 2t}{2}$, by theorem 9.2, we have

$$L\{f(t)\} = L\{e^{4t} + e^{2t} + t^3 + \sin^2 t\}$$

$$= L\{e^{4t}\} + L\{e^{2t}\} + L\{t^3\} + \frac{1}{2}L\{1\} - \frac{1}{2}L\{\cos 2t\}$$

$$= \frac{1}{s-4} + \frac{1}{s-2} + \frac{6}{s^4} + \frac{1}{2s} - \frac{s}{2(s^2+4)}$$

$$= \frac{1}{s-4} + \frac{1}{s-2} + \frac{6}{s^4} + \frac{s}{2(s^2+4)}, s > 0.$$

EXAMPLE 9.24

Find Laplace transform of $f(t) = \sin^3 2t$.

Solution. Since $\sin 3t = 3\sin t - 4\sin^3 t$, we have

$$\sin^3 t = \frac{3}{4}\sin t - \frac{1}{4}\sin 3t$$

and so

$$\sin^3 2t = \frac{3}{4}\sin 2t - \frac{1}{4}\sin 6t.$$

hence, by linearity of L, we get

$$L\{f(t)\} = \frac{3}{4}L\{\sin 2t\} - \frac{1}{4}L\{\sin 6t\}$$

$$= \frac{3}{4}\left[\frac{2}{s^2+4}\right] - \frac{1}{4}\left[\frac{6}{s^2+36}\right], s > 0$$

$$= \frac{3}{2}\left[\frac{1}{s^2+4} - \frac{1}{s^2+36}\right]$$

$$= \frac{48}{(s^2+4)(s^2+36)}.$$

EXAMPLE 9.25

Find the Laplace transform of

$$f(t) = \sin at \sin bt.$$

Solution. We have

$$f(t) = \frac{1}{2}(2\sin at \sin bt)$$

$$= \frac{1}{2}[\cos(at - bt) - \cos(at + bt)]$$

$$= \frac{1}{2}\cos(a-b)t - \frac{1}{2}\cos(a+b)t.$$

Therefore, using linearity, we have

$$L\{f(t)\} = \frac{1}{2}L\{\cos(a-b)t\} - \frac{1}{2}L\{\cos(a+b)t\}$$

$$= \frac{1}{2}\left[\frac{s}{s^2+(a-b)^2}\right] - \frac{1}{2}\left[\frac{s}{s^2+(a+b)^2}\right], s > 0$$

$$= \frac{2abs}{(s^2+(a-b)^2)(s^2+(a+b)^2)}, s > 0.$$

EXAMPLE 9.26

Find the Laplace transform of

$$f(t) = \sin(\omega t + \phi), t \geq 0.$$

Solution. Since

$$\sin(\omega t + \phi) = \sin \omega t \cos\phi + \cos \omega t \sin \phi,$$

we have by linearity of the operator L,

$$L\{f(t)\} = \cos\phi\, L\{\sin \omega t\} + \sin\phi\, L\{\cos \omega t\}$$

$$= \cos\phi\left(\frac{\omega}{s^2+\omega^2}\right) + \sin\phi\left(\frac{s}{s^2+\omega^2}\right), s > 0$$

$$= \frac{1}{s^2+\omega^2}(\omega\cos\phi + s\sin\phi), s > 0.$$

EXAMPLE 9.27

Determine Laplace ttransform of the *square wave function f* defined by

$$f(t) = \mathrm{H}(t) - 2\mathrm{H}(t-a) + 2\mathrm{H}(t-2a)$$
$$- 2\mathrm{H}(t-3a) + \ldots$$

Solution. We note that

$$f(t) = \mathrm{H}(t) - 2\mathrm{H}(t-a) = 1 - 2(0) = 1,\ 0 < t < a$$

$$f(t) = \mathrm{H}(t) - 2\mathrm{H}(t-a) + 2\mathrm{H}(t-2a)$$
$$= 1 - 2(1) + 2(0) = -1,\ 0 < a < t < 2a$$

and so on. Thus the graph of the function is as shown in the Figure 9.7.

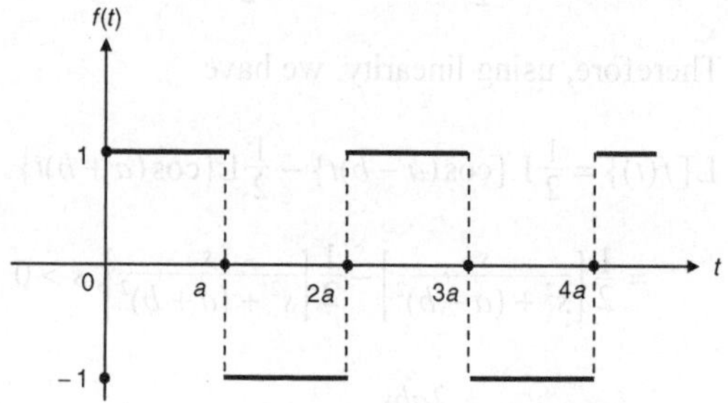

Figure 9.7

By linearity of Laplace operator, we have

$$\mathrm{L}\{f(t)\} = \mathrm{L}\{\mathrm{H}(t) - 2\mathrm{L}\{\mathrm{H}(t-a)\}$$
$$+ 2\mathrm{L}\{\mathrm{H}(t-2a)\} - 2\mathrm{L}\{\mathrm{H}(t-3a)\} + \ldots$$
$$= \frac{1}{s} - 2\frac{e^{-sa}}{s} + 2\frac{e^{-2sa}}{s} - 2\frac{e^{-3sa}}{s} + \ldots$$
$$= \frac{1}{s}[1 - 2e^{-sa}\{1 - e^{-sa} + e^{-2sa} - \ldots\}]$$
$$= \frac{1}{s}\left[1 - 2e^{-sa}\left(\frac{1}{1+e^{-sa}}\right)\right]$$
$$= \frac{1}{s}\left(\frac{1-e^{-sa}}{1+e^{-sa}}\right) = \frac{1}{s}\left(\frac{e^{\frac{sa}{2}} - e^{-\frac{sa}{2}}}{e^{\frac{sa}{2}} + e^{-\frac{sa}{2}}}\right)$$
$$= \frac{1}{s}\tanh\left(\frac{sa}{2}\right).$$

EXAMPLE 9.28

Find the Laplace transform of

$$f(t) = (\sin t - \cos t)^2,\ t \geq 0.$$

Solution. Since

$$(\sin t - \cos t)^2 = \sin^2 t + \cos^2 t - 2\sin t\cos t$$
$$= 1 - \sin 2t,$$

we have

$$\mathrm{L}\{f(t)\} = \mathrm{L}\{1\} - \mathrm{L}\{\sin 2t\} = \frac{1}{s} - \frac{2}{s^2+4}$$
$$= \frac{s^2 - 2s + 4}{s(s^2+4)},\ s > 0.$$

EXAMPLE 9.29

Find Laplace transform of $f(t) = 2 + \sqrt{t} + \frac{1}{\sqrt{t}}$, $t > 0$.

Solution. By linear of L, we have

$$\mathrm{L}\{f(t)\} = 2\mathrm{L}\{1\} + \mathrm{L}\{\sqrt{t}\} + \left\{\frac{1}{\sqrt{t}}\right\}$$
$$= \frac{2}{s} + \frac{\Gamma\left(\frac{1}{2}+1\right)}{s^{\frac{1}{2}+1}} + \frac{\sqrt{\pi}}{s}$$
$$= \frac{2}{s} + \frac{\sqrt{\pi}}{2s^{\frac{3}{2}}} + \sqrt{\frac{\pi}{s}},\ s > 0.$$

EXAMPLE 9.30

Find Laplace transform of $e^{at}\cos bt$ and $e^{at}\sin bt$, where a and b are real.

Solution. Let $f(t) = e^{(a+ib)t}$. Then (see Example 9.4)

$$\mathrm{L}\{f(t)\} = \frac{1}{s-(a+ib)} = \frac{1}{s-a-ib}$$
$$= \frac{1}{(s-a)-ib} \cdot \frac{(s-a)+ib}{(s-a)+ib}$$
$$= \frac{(s-a)+ib}{(s-a)^2+b^2} \qquad (1)$$

Also

$$e^{(a+ib)t} = e^{at}[\cos bt + i\sin bt]$$
$$= e^{at}\cos bt + i\,e^{at}\sin bt$$

Hence

$$L\{f(t)\} = L\{e^{at}\cos bt + i\, e^{at}\sin bt\}$$

$$= L\{e^{at}\cos bt\} + i\, L\{e^{at}\sin bt\}$$

$$\text{(by linearity of L)} \qquad (2)$$

Thus, by (1) and (2), we have

$$L\{e^{at}\cos bt\} + iL\{e^{at}\sin bt\} = \frac{(s-a)+ib}{(s-a)^2+b^2}.$$

Comparing real and imaginary parts, we have

$$L\{e^{at}\cos bt\} = \frac{s-a}{(s-a)^2+b^2}$$

and

$$L\{e^{at}\sin bt\} = \frac{b}{(s-a)^2+b^2}.$$

EXAMPLE 9.31

Find the Laplace transform of $f(t) = \int_0^t \frac{\sin u}{u}\,du.$

Solution. Using series expansion of sin u, we have

$$\int_0^t \frac{\sin u}{u}\,du = \int_0^t \frac{1}{u}\left[u - \frac{u^3}{3!} + \frac{u^5}{5!} - \frac{u^7}{7!} + \ldots\right]du$$

$$= \int_0^t \left[1 - \frac{u^2}{3!} + \frac{u^4}{5!} - \frac{u^6}{7!} + \ldots\right]du$$

$$= t - \frac{t^3}{3.3!} + \frac{t^5}{5.5!} - \frac{t^7}{7.7!} + \ldots$$

Therefore,

$$L\{f(t)\} = L\left\{t - \frac{t^3}{3.3!} + \frac{t^5}{5.5!} - \frac{t^7}{7.7!} + \ldots\right\}$$

$$= \frac{1}{s^2} - \frac{1}{3.3!}\cdot\frac{3!}{s^4} + \frac{1}{5.5!}\cdot\frac{5!}{s^6} - \frac{1}{7.7!}\cdot\frac{7!}{s^8}$$

$$+ \ldots \text{ (by linearity of L)}$$

$$= \frac{1}{s^2} - \frac{1}{3s^4} + \frac{1}{5s^6} - \frac{1}{7s^8} + \ldots$$

$$= \frac{1}{s}\left[\frac{1}{s} - \frac{(1/s)^3}{3} + \frac{(1/s)^5}{5} - \frac{(1/s)^7}{7} + \ldots\right]$$

$$= \frac{1}{s}\tan^{-1}\frac{1}{s}.$$

EXAMPLE 9.32

Find Laplace transform $f(t) = \cosh at - \cos at.$

Solution. By linearity of the Laplace operator, we have

$$L\{f(t)\} = L\{\cosh at - \cos at\}$$

$$= L\{\cosh at\} - L\{\cos at\}$$

$$= \frac{s}{s^2-a^2} - \frac{s}{s^2+a^2},\ s > |a|$$

$$= \frac{s^3 + a^2 s - s^3 + a^2 s}{s^4 - a^4} = \frac{2a^2 s}{s^4 - a^4}.$$

EXAMPLE 9.33

Find the Laplace transform of Bessel's *function of order zero.*

Solution. Recall that Bessel's function of order zero is defined by

$$J_0(t) = 1 - \frac{t^2}{2} + \frac{t^4}{2^2.4^2} - \frac{t^6}{2^2.4^2.6^2} + \ldots$$

Therefore,

$$L\{J_0(t)\} = L\left\{1 - \frac{t^2}{2^2} + \frac{t^4}{2^2.4^2} - \frac{t^6}{2^2.4^2.6^2} + \ldots\right\}$$

$$= L\{1\} - \frac{1}{2^2}L\{t^2\} + \frac{1}{2^2.4^2}L\{t^4\}$$

$$- \frac{1}{2^2.4^2.6^2}L\{t^6\} + \ldots$$

$$= \frac{1}{s} - \frac{1}{2^2}\frac{2!}{s^3} + \frac{1}{2^2 4^2}\frac{4!}{s^5} - \frac{1}{2^2.4^2 6^2}\frac{6!}{s^7} + \ldots$$

$$= \frac{1}{s}\left[1 - \frac{1}{2}\left(\frac{1}{s^2}\right) + \frac{1.3}{2.4}\left(\frac{1}{s^4}\right) - \frac{1.3.5}{2.4.6}\left(\frac{1}{s^6}\right) + \ldots\right]$$

$$= \frac{1}{s}\left[\left(1 + \frac{1}{s^2}\right)^{-\frac{1}{2}}\right] \text{ (using binomial theorem)}$$

$$= \frac{1}{\sqrt{(s^2+1)}}$$

Theorem 9.3. [First Shifting (Translation) Property]. If $f(t)$ is a function of t for $t > 0$ and $L\{f(t)\} = F(s)$, then

$$L\{e^{at} f(t)\} = F(s-a).$$

Proof: We are given that

$$L\{f(t)\} = F(s) = \int_0^\infty e^{-st} f(t)\, dt.$$

By the definition of Laplace transform, we have

$$L\{e^{at} f(t)\} = \int_0^\infty e^{-st} (e^{at} f(t))\, dt = \int_0^\infty e^{-(s-a)} f(t)\, dt$$

$$= F(s - a).$$

EXAMPLE 9.34

Find the Laplace transform of $g(t) = e^{-t} \sin^2 t$.

Solution. We have (see Example 9.23)

$$L\{\sin^2(t)\} = F(s) = \frac{2}{s(s^2+4)}.$$

Therefore, using first shifting property, we get

$$L\{g(t)\} = F(s - a)$$

$$= \frac{2}{(s+1)[(s+1)^2+4]}, \text{ since } a = -1.$$

$$= \frac{2}{(s+1)(s^2+2s+5)}.$$

EXAMPLE 9.35

Find Laplace transform of $g(t) = t^3 e^{-3t}$.

Solution. Since

$$L\{t^3\} = F(s) = \frac{3!}{s^4} = \frac{6}{s^4},$$

We have by shifting property,

$$L\{g(t)\} = L\{e^{-3t} \cdot t^3\} = F(s-a) = \frac{6}{(s+3)^4},$$

since $a = -3$.

EXAMPLE 9.36

Using first-shifting property, find Laplace transforms of $t \sin at$ and $t \cos at$.

Solution. Since $L\{t\} = \frac{1}{s^2}$, we have

$$L\{t\, e^{iat}\} = L\{t \cos at\} + iL\{t \sin at\}$$

$$= F(s-a) = \frac{1}{(s-ia)^2} = \frac{(s+ia)^2}{[(s-ia)(s+ia)]^2}$$

$$= \frac{(s^2-a^2)+i(2as)}{(s^2+a^2)^2}$$

Equating real and imaginary parts, we have

$$L\{t \cos at\} = \frac{s^2-a^2}{(s^2+a^2)^2}$$

and

$$L\{t \sin at\} = \frac{(2as)}{(s^2+a^2)^2}.$$

EXAMPLE 9.37

Find the Laplace transform of $f(t) = e^{at} \cosh bt$.

Solution. Since

$$L\{\cosh bt\} = \frac{s}{s^2-b^2}, \; s > |b|,$$

The shifting property yields

$$L\{e^{at} \cosh bt\} = F(s-a)$$

$$= \frac{s-a}{(s-a)^2-b^2}, \; s > |b| + a.$$

EXAMPLE 9.38

Find Laplace transform of

$$f(t) = e^{-3t}(2\cos 5t + 3 \sin 5t).$$

Solution. Since

$$L\{2 \cos 5t - 3 \sin 5t\}$$

$$= 2L\{\cos 5t\} - 3L\{\sin 5t\}$$

$$= \frac{2s}{s^2+25} - \frac{3 \times 5}{s^2+25} = \frac{2s-15}{s^2+25} = F(s),$$

therefore, shifting property yields

$$L\{f(t)\} = F(s-a) \text{ with } a = -3$$

$$= \frac{2(s+3)-15}{(s+3)^2+25} = \frac{2s-9}{s^2+6s+34}.$$

EXAMPLE 9.39

Find Laplace transform of $f(t) = \sinh 3t \cos^2 t$.

Solution. We know that $\cos^2 t = \frac{1+\cos 2t}{2}$. Therefore,

$$L\{\cos^2 t\} = \frac{1}{2}L\{1\} + \frac{1}{2}L\{\cos 2t\}$$

$$= \frac{1}{2}\left[\frac{1}{s} + \frac{s}{s^2+4}\right] = \frac{s^2+2}{s(s^2+4)},\ s > 0.$$

Therefore, by first shifting theorem, we have

$$L\{f(t)\} = L\{\sinh 3t \cos^2 t\} = L\left\{\frac{e^{3t}-e^{-3t}}{2}\cos^2 t\right\}$$

$$= \frac{1}{2}L\{e^{3t}\cos^2 t\} - \frac{1}{2}L\{e^{-3t}\cos^2 t\}$$

(by linearity of L)

$$= \frac{1}{2}\left[\frac{(s-3)^2+2}{(s-3)[(s-3)^2+4]}\right]$$

$$-\frac{1}{2}\left[\frac{(s+3)^2+2}{(s+3)[(s+3)^2+4]}\right]$$

$$= \frac{1}{2}\left[\frac{s^2-6s+11}{(s-3)(s^2-6s+13)}\right.$$

$$\left.-\frac{s^2+6s+11}{(s+3)(s^2+6s+13)}\right].$$

EXAMPLE 9.40

Find the Laplace transform of cosh *at* sin *bt*.

Solution. Let F(*s*) be Laplace transform of $f(t)$, $t > 0$ and let

$$g(t) = f(t)\cosh at.$$

Then

$$L\{g(t)\} = L[f(t)\cosh at] = L\,\frac{e^{at}+e^{-at}}{2}f(t)$$

$$= \frac{1}{2}L(e^{at}f(t)) + \frac{1}{2}L(e^{-at}f(t))$$

$$= \frac{1}{2}[F(s-a) + F(s+a)]$$

(use of first shifting theorem).

We take $f(t) = \sin bt$. Then $F(s) = \frac{b}{s^2+b^2}$ and, therefore, using above result, we have

$$L\{(\cosh at)\sin bt\}$$

$$= \frac{1}{2}\left[\frac{b}{(s-a)^2+b^2} + \frac{b}{(s+a)^2+b^2}\right].$$

EXAMPLE 9.41

Find the Laplace transform of

$$f(t) = \cosh 4t \sin 6t.$$

Solution. Taking $a = 4$, $b = 6$ in Example 9.40, we get

$$L\{\cosh 4t \sin 6t\} = \frac{6(s^2+52)}{s^4+40s^2+2704}.$$

Theorem 9.4. (Second Shifting Property). Let F(*s*) be the Laplace transform of $f(t)$, $t > 0$ and let *g* be a function defined by

$$g(t) = \begin{cases} f(t-a) & \text{for } t > a \\ 0 & \text{for } t < a. \end{cases}$$

Then

$$L\{g(t)\} = e^{-as}F(s).$$

Proof: Using the substitution $t - a = u$, we have

$$L\{g(t)\} = \int_0^\infty e^{-st}g(t)\,dt$$

$$= \int_0^a e^{-st}g(t)\,dt + \int_a^\infty e^{-st}g(t)\,dt$$

$$= 0 + \int_a^\infty e^{-st}f(t-a)\,dt$$

$$= \int_0^\infty e^{-s(u+a)}f(u)\,du$$

$$= e^{-as}\int_0^\infty e^{-su}f(u)\,du = e^{-as}F(s).$$

EXAMPLE 9.42

Find the Laplace transform of the function *f* defined by

$$f(t)=\begin{cases}\cos\left(t-\frac{2\pi}{3}\right) & \text{for } t>\frac{2\pi}{3}\\ 0 & \text{for } t<\frac{2\pi}{3}.\end{cases}$$

Solution. We know that $L\{\cos t\}=\frac{s}{s^2+1}$, $s>0$. Therefore, by second shifting property,

$$L\{f(t)\}=e^{-\frac{2\pi s}{3}}L\{\cos t\}=\frac{se^{-\frac{2\pi s}{3}}}{s^2+1},\ s>0.$$

EXAMPLE 9.43

Find the Laplace transform of the *sine function switched on at time t* = 3.

Solution. The given function is defined by

$$f(t)=\begin{cases}\sin t & \text{for } t\ge 3\\ 0 & \text{for } t<3.\end{cases}$$

Using Heaviside's unit step function H, this function can be expressed as

$$f(t)=H(t-3)\sin t.$$

To use second-shift theorem, we first write $\sin t$ as

$$\sin t=\sin(t-3+3)$$

$$=\sin(t-3)\cos 3+\cos(t-3)\sin 3.$$

Then

$$L\{f(t)\}=L\{H(t-3)\sin(t-3)\cos 3\}$$
$$+\{H(t-3)\cos(t-3)\sin 3\}$$

$$=\cos 3e^{-3s}L\{\sin t\}+\sin 3e^{-3s}L\{\cos t\}$$

$$=\cos 3e^{-3s}\frac{1}{s^2+1}+\sin 3\, e^{-3s}\frac{s}{s^2+1}$$

$$=\frac{e^{-3s}}{s^2+1}(\cos 3+s\sin 3).$$

EXAMPLE 9.44

Find the Laplace transform of the function *f* defined by

$$f(t)=\begin{cases}(t-1)^2 & \text{for } t\ge 1\\ 0 & \text{for } 0\le t<1.\end{cases}$$

Solution. This function is just the function $g(t)=t^2$ delayed by 1 unit of time and its graph is shown in Figure 9.8.

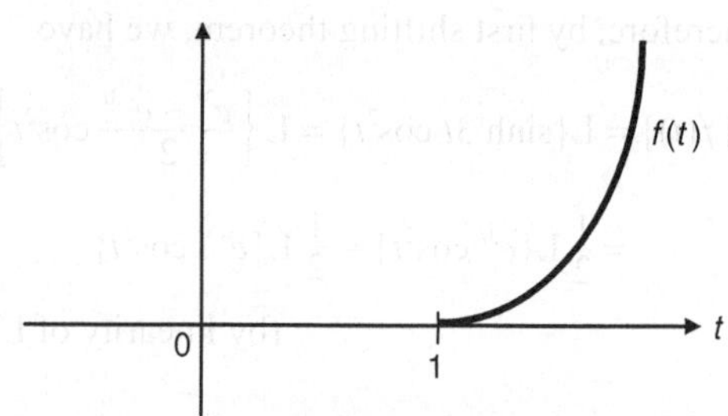

Figure 9.8

Therefore, by second shift property, we have

$$L\{f(t)\}=e^{-s}L\{t^2\}=\frac{2e^{-s}}{s^3},\quad \text{Re}(s)>0.$$

EXAMPLE 9.45

Find Laplace transform of the function *f* defined by

$$f(t)=\begin{cases}(t-4)^5 & \text{for } t>4\\ 0 & \text{for } t<4.\end{cases}$$

Solution. Using second shift property, we have

$$L\{f(t)\}=e^{-4s}[L\{t^5\}]=e^{-4s}\cdot\frac{5!}{s^6}=120\frac{e^{-4s}}{s^6}.$$

Theorem 9.5. (Change of Scale Property). If $F(s)$ is the Laplace transform of $f(t)$ for $t>0$, then for any positive constant a,

$$L\{f(at)\}=\frac{1}{a}F\left(\frac{s}{a}\right).$$

Proof: We are given that

$$F(s)=L\{f(t)\}=\int_0^\infty e^{-st}f(t)\,dt.$$

Taking $u = at$, we have

$$L\{f(at)\} = \int_0^\infty e^{-st} f(at)\,dt = \int_0^\infty e^{-\frac{su}{a}} f(u)\,\frac{du}{a}$$

$$= \frac{1}{a}\int_0^\infty e^{-\frac{su}{a}} f(u)\,du = \frac{1}{a}\,F\left(\frac{s}{a}\right).$$

EXAMPLE 9.46

Find the Laplace of $f(t) = \cos 6t$.

Solution. Since $L\{\cos t\} = \frac{s}{s^2+1}$, the change of scale property implies

$$L\{\cos 6t\} = \frac{1}{6}\left(\frac{s/6}{(s/6)^2+1}\right)$$

$$= \frac{1}{6}\left(\frac{s}{6[(s^2/36)+1]}\right) = \frac{s}{s^2+36}.$$

EXAMPLE 9.47

Using change of scale property, find the Laplace transform of $J_0(at)$.

Solution. Let $J_0(t)$ be Bessel's function of order zero. By Example 9.33, $L\{J_0(t)\} = \frac{1}{\sqrt{s^2+1}}$. Therefore, by change of scale property,

$$L\{J_0(at)\} = \frac{1}{a}\,F\left(\frac{s}{a}\right) = \frac{1}{a}\cdot\frac{1}{\sqrt{(s/a)^2+1}} = \frac{1}{\sqrt{s^2+a^2}}.$$

Theorem 9.6. (Laplace Transform of Derivatives).

Let f be a function such that

(a) f is continuous for all t, $0 \le t \le N$

(b) f is of exponential order γ for $t > N$

(c) f' is sectionally continuous for $0 \le t \le N$.

Then the Laplace transform of f' exists and is given by

$$L\{f'(t)\} = sF(s) - f(0),$$

where $F(s)$ is the Laplace transform of f.

Proof: The existence of the Laplace transform is established by Theorem 9.1. Further, integrating by parts, we have

$$L\{f'(t)\} = \int_0^\infty e^{-st} f'(t)\,dt = \lim_{T\to\infty}\int_0^T e^{-st} f'(t)\,dt$$

$$= \lim_{T\to\infty}\left\{[e^{-st} f(t)]_0^T + s\int_0^T e^{-st} f(t)dt\right\}$$

$$= \lim_{T\to\infty}\left\{[e^{-st} f(T) - f(0)] + s\int_0^T e^{-st} f(t)dt\right\}$$

$$= s\int_0^\infty e^{-st} f(t)\,dt - f(0)$$

$$= sF(s) - f(0),$$

the last but one step being the consequence of the fact that f is of exponential order and so $\lim_{T\to\infty} e^{-sT} f(T) = 0$ for $s > \gamma$.

EXAMPLE 9.48

Find Laplace transform of $g(t) = \sin at \cos at$.

Solution. Let $f(t) = \sin^2 at$. Then

$$f'(t) = 2a \sin at \cos at.$$

Since

$$L\{f'(t)\} = sF(s) - f(0),$$

We have

$L\{2a \sin at \cos at\}$

$$= sL\{\sin^2 at\} - 0 = sL\{\sin^2 at\}$$

$$= \frac{2a^2}{s(s^2+4a^2)} \qquad \text{(see Example 9.22).}$$

Hence

$$L\{\sin at \cos at\} = \frac{a}{(s^2+4a^2)}.$$

EXAMPLE 9.49

Using Laplace transform of $\cos bt$, find the Laplace transform of $\sin bt$.

Solution. We want to find $L\{\sin bt\}$ from $L\{\cos bt\}$. So, let $f(t) = \cos bt$. Then $f'(t) = -b \sin bt$ and so

$$L\{f'(t)\} = sF(s) - f(0) = sL\{\cos bt\} - 1$$

$$= s\left(\frac{s}{s^2+b^2}\right) - 1 = \frac{s^2}{s^2+b^2} - 1.$$

Thus

$$L\{-b \sin bt\} = \frac{-b^2}{s^2+b^2}.$$

Hence

$$L\{\sin bt\} = \frac{b}{s^2+b^2}.$$

EXAMPLE 9.50

Find Laplace transform of *Bessel's function of order* 1.

Solution. Let $J_1(t)$ be Bessel's function of order 1. We know that

$$\frac{d}{dt}\{t^n J_n(t)\} = t^n J_{n-1}(t).$$

If $n = 0$, we have

$$J'_0(t) = J_{-1}(t) = -J_1(t).$$

Hence

$$L\{J_1(t)\} = -L\{J'_0(t)\} = -[sL\{J_0(t) - J_0(0)]\}$$

$$= -\left[\frac{s}{\sqrt{s^2+1}} - 1\right]$$

$$= 1 - \frac{s}{\sqrt{s^2+1}} \quad \text{(see Example 9.33)}$$

$$= \frac{\sqrt{s^2+1} - s}{\sqrt{s^2+1}}.$$

Theorem 9.7. If $L\{f(t)\} = F(s)$, then

$$L\{f''(t)\} = s^2F(s) - sf(0) - f'(0)$$

if $f(t)$ and $f'(t)$ are continuous for $0 \le t \le N$ and of exponential order for $t >$ N whereas $f''(t)$ is sectionally continuous for $0 \le t \le N$.

Proof: By theorem 9.6, we have

$$L\{g'(t)\} = sG(s) - g(0).$$

Taking $g(t) = f'(t)$, we have

$$L\{f''(t)\} = sL\{f'(t)\} - f'(0)$$

$$= s[sF(s) - f(0)] - f'(0)$$

$$= s^2F(s) - sf(0) - f'(0).$$

EXAMPLE 9.51

Using Theorem 9.7, find $L\{\sin at\}$, $t \ge 0$.

Solution. Let $f(t) = \sin at$. Then

$$f'(t) = a\cos at, \; f''(t) = -a^2 \sin at.$$

By Theorem 9.7,

$$L\{f''(t)\} = s^2 F(s) - sf(0) - f'(0)$$

and so

$$L\{-a^2 \sin at\} = s^2 L\{\sin at\} - a$$

which yields

$$(s^2 + a^2)L\{\sin at\} = a$$

and hence

$$L\{\sin at\} = \frac{a}{s^2+a^2}, \; s > 0.$$

EXAMPLE 9.52

Using Laplace transform of derivatives, find $L\{t\cos at\}$.

Solution. Let $f(t) = t\cos at$. Then

$$f'(t) = \cos at - at \sin at$$

$$f''(t) = -2a \sin at - a^2 t \cos at.$$

But

$$L\{f''(t)\} = s^2 L\{f(t)\} - sf(0) - f'(0)$$

$$= s^2 L\{f(t)\} - 1$$

and so

$$L\{-2a \sin at - a^2 t \cos t\} = s^2 L\{t \cos at\} - 1,$$

that is,

$$(s^2 + a^2)L\{t\cos at\} = 2aL\{\sin at\} + 1$$

$$= -2a\left(\frac{a}{(s^2+a^2)}\right) + 1$$

$$= \frac{s^2 - a^2}{s^2 + a^2},$$

and so

$$L\{t\cos at\} = \frac{s^2 - a^2}{(s^2 + a^2)^2}.$$

Theorem 9.7 can be generalized to higher order derivatives in the form of the following result:

Theorem 9.8. Let $L\{f(t)\} = F(s)$. Then

$$L\{f^{(n)}(t)\} = s^n F(s) - s^{n-1}f(0) - s^{n-2}f'(0) - \ldots - sf^{(n-2)}(0) - f^{(n-1)}(0),$$

if $f(t), f'(t), \ldots f^{(n-1)}(t)$ are continuous for $0 \le t \le N$ and of exponential order for $t > N$ whereas $f^{(n)}(t)$ is piecewise continuous for $0 \le t \le N$.

Proof. We shall prove our result using mathematical induction. By theorem 9.6 and 9.7, we have

$$L\{f'(t)\} = sF(s) - f(0),$$

$$L\{f''(t)\} = s^2 F(s) - sf(0) - f'(0).$$

Thus the theorem is true for $f'(t)$ and $f''(t)$. Suppose that the result is true for $f^{(n)}(t)$. Then

$$L\{f^{(n)}(t) = s^n F(s) - s^{n-1}f(0) - \ldots - f^{(n-1)}(0)$$

Then application of Theorem 9.6 yields $L\{f^{(n+1)}(t)\}$

$$= s\{s^n F(s) - s^{n-1}f(0) - \ldots - f^{(n-1)}(0)] - f^{(n)}(0)$$

$$= s^{n+1}F(s) - s^n f(0) - \ldots - sf^{(n-1)}(0) - f^{(n)}(0),$$

which shows that the result holds for $(n+1)^{\text{th}}$ derivative also. Hence by mathematical induction, the result holds.

EXAMPLE 9.53

Using Theorem 9.8, fiind $L\{t^n\}$.

Solution. We have $f(t) = t^n$. Therefore,

$$f'(t) = nt^{n-1}, f''(t) = n(n-1)t^{n-2}, \ldots, f^{(n)}(t) = n!$$

Now use of Theorem 9.9 yields

$$L\{f^{(n)}(t)\} = L\{n!\} = s^n L\{t^n\} - s^{n-1}f(0) - \ldots - f^{(n-1)}(0)$$

But

$$f(0) = f'(0) = f''(0) = \ldots = f^{(n-1)}(0) = 0.$$

Therefore

$$L\{n!\} = s^n L\{t^n\},$$

which gives

$$L\{t^n\} = \frac{L\{n!\}}{s^n} = \frac{n!\, L\{1\}}{s^n} = \frac{n!}{s^{n+1}}.$$

Theorem 9.9. (Multiplication by t^n). If $L\{f(t)\} = F(s)$, then

$$L\{tf(t)\} = -\frac{d}{ds}F(s),$$

and in general

$$L\{t^n f(t)\} = (-1)^n \frac{d^n}{ds^n}F(s).$$

Proof: By definition of Laplace transform,

$$F(s) = \int_0^\infty e^{-st} f(t)\, dt.$$

Then, by Leibnitz-rule for differentiating under the integral sign, we have

$$\frac{dF}{ds} = \frac{d}{ds}\int_0^\infty e^{-st} f(t)\, dt = \int_0^\infty \frac{d}{ds}(e^{-st} f(t)\, dt)$$

$$= \int_0^\infty -t\, e^{-st} f(t)\, dt = -\int_0^\infty e^{-st}(tf(t))\, dt$$

$$= -L\{tf(t)\}$$

and so

$$L\{tf(t)\} = -\frac{d}{ds}F(s).$$

Thus the theorem is true for $n = 1$. To obtain the general form, we use mathematical induction. So, assume that the result is true for $n = m$. Thus

$$L\{t^m f(t)\} = (-1)^m \frac{d^m}{ds^m}F(s) = (-1)^m F^{(m)}(s).$$

Therefore,

$$\frac{d}{ds}[L(t^m f(t)] = (-1)^m F^{(m+1)}(s),$$

that is,

$$\frac{d}{ds}\int_0^\infty e^{-st}\, t^m f(t)\, dt = (-1)^m F^{(m+1)}(s),$$

which, on using Leibnitz rule, yields

$$-\int_0^\infty e^{-st}\, t^{m+1} f(t)\, dt = (-1)^m F^{(m+1)}(s),$$

and so

$$L\{t^{m+1} f(t)\} = (-1)^{m+1} F^{(m+1)}(s).$$

Hence, the result follows by mathematical induction.

EXAMPLE 9.54

Find Laplace transform of $f(t) = t\sin^2 t$.

Solution. Let $f(t) = t\sin^2 t$. We know that $L\{\sin^2 t\} = \frac{2}{s(s^2+4)}$. Therefore

$$L\{t\sin^2 t\} = -\frac{d}{ds}\left\{\frac{2}{s(s^2+4)}\right\} = 2\left[\frac{3s^2+4}{s^2(s^2+4)^2}\right].$$

EXAMPLE 9.55

Find Laplace transform of $f(t) = te^{-t}\cosh t$.

Solution. We know that

$$L\{\cosh t\} = \frac{s}{s^2-1}.$$

Therefore,

$$L\{t\cosh t\} = -\frac{d}{ds}\left(\frac{s}{s^2-1}\right) = \frac{s^2+1}{(s^2-1)^2}.$$

Then, by Theorem 9.3, we have

$$L\{e^{-t}\cosh t\} = \frac{(s+1)^2+1}{((s+1)^2-1)^2} = \frac{s^2+2s+2}{(s^2+2s)^2}.$$

EXAMPLE 9.56

Using Theorem 9.9, evaluate $\int_0^\infty te^{-2t}\sin t\, dt$ and $\int_0^\infty te^{-2t}\cos t\, dt$.

Solution. We know that $L\{\sin t\} = \frac{1}{s^2+1}$. Therefore, by Theorem 9.9, we have

$$L\{t\sin t\} = -\frac{d}{ds}\left(\frac{1}{s^2+1}\right) = \frac{2s}{(s^2+1)^2}.$$

But $\int_0^\infty e^{-2t}(t\sin t)dt$ is the Laplace transform of $t\sin t$ with $s = 2$. Hence

$$\int_0^\infty e^{-2t}(t\sin t)dt = L\{t\sin t\} = \left[\frac{2s}{(s^2+1)^2}\right]_{s=2} = \frac{4}{25}.$$

In a similar way, we can show that

$$\int_0^\infty te^{-2t}\cos t\, dt = L\{t\cos t\} \text{ with } s = 2$$

$$= \left[\frac{s^2-1}{(s^2+1)^2}\right]_{s=2} = \frac{3}{25}.$$

EXAMPLE 9.57

Evaluate the integral $I = \int_0^\infty e^{-x^2} dx$.

Solution. Putting $t = x^2$, we get

$$I = \frac{1}{2}\int_0^\infty e^{-t}\, t^{-\frac{1}{2}}\, dt = \frac{1}{2} L\{t^{-\frac{1}{2}}\} \text{ with } s = 1$$

$$= \frac{1}{2}\cdot\frac{\Gamma\left(-\frac{1}{2}+1\right)}{s^{\frac{1}{2}}} \text{ with } s = 1$$

$$= \frac{1}{2}\Gamma\left(\frac{1}{2}\right) = \frac{\sqrt{\pi}}{2}.$$

EXAMPLE 9.58

Find Laplace transform of $f(t) = e^{-2t}\, t\cos t$.

Solution. We know that $L\{\cos t\} = \frac{s}{s^2+1}$. Therefore, by Theorem 9.9, we have

$$L\{t\cos t\} = -\frac{d}{ds}\left(\frac{s}{s^2+1}\right) = \frac{s^2-1}{(s^2+1)^2}.$$

Now using first shifting property, we have

$$L\{e^{-2t}\, t\cos t\} = \frac{(s+2)^2+1}{((s+2)^2+1)^2} = \frac{s^2+4s+3}{(s^2+4s+5)^2}.$$

EXAMPLE 9.59

Find the Laplace transform of $f(t) = t^2 e^{-2t}\cos t$.

Solution. As in Example 9.58, $L\{t\cos t\} = \frac{s^2-1}{(s^2+1)^2}$.

Therefore,

$$L\{t^2 \cos t\} = -\frac{d}{ds}\left(\frac{s^2-1}{(s^2+1)^2}\right).$$

Then using first-shifting property, we have

$$L\{t^2 e^{-2t} \cos t\} = 2\left(\frac{s^3+10s^2+25s+22}{(s^2+4s+5)^3}\right).$$

EXAMPLE 9.60

Find Laplace transform of $f(t) = t^n e^{-at}$.

Solution. Since $L\{e^{-at}\} = \frac{1}{s+a}$, we have

$$L\{t^n e^{-at}\} = (-1)^n \frac{d^n}{ds^n}\left(\frac{1}{s+a}\right) = (-1)^{2n} \frac{n!}{(s+a)^{n+1}}.$$

Theorem 9.10. (Division by t). If $L\{f(t)\} = F(s)$, then

$$L\left\{\frac{f(t)}{t}\right\} = \int_s^\infty F(u)\,du,$$

provided $\lim_{t\to 0} \frac{f(t)}{t}$ exists.

Proof: Put $g(t) = \frac{f(t)}{t}$. So, $f(t) = tg(t)$ and

$$L\{f(t)\} = L\{tg(t)\}$$

$$= -\frac{d}{ds} L\{g(t)\}, \text{ by Theorem 9.9}$$

$$= -\frac{dG}{ds}.$$

Then integration yields

$$G(s) = -\int_\infty^s F(u)\,du = \int_s^\infty F(u)\,du,$$

that is,

$$L\left\{\frac{f(t)}{t}\right\} = \int_s^\infty F(u)\,du.$$

Remark 9.3. By Theorem 9.10, we have

$$L\left\{\frac{f(t)}{t}\right\} = \int_0^\infty e^{-st}\frac{f(t)}{t}\,dt = \int_s^\infty F(u)\,du.$$

Letting $s \to 0+$ and assuming that the integral converges, it follows that

$$\int_0^\infty \frac{f(t)}{t}\,dt = \int_0^\infty F(u)du.$$

For example, if $f(t) = \sin t$, then $F(s) = \frac{1}{s^2+1}$ and so

$$\int_0^\infty \frac{\sin t}{t}\,dt = \int_0^\infty \frac{du}{u^2+1} = [\tan^{-1} u]_0^\infty = \frac{\pi}{2}.$$

EXAMPLE 9.61

Find the Laplace transform of

$$f(t) = \frac{\cos 2t - \cos 3t}{t}.$$

Solution. By linearity of L, we have

$$L\{\cos 2t - \cos 3t\} = L\{\cos 2t\} - L\{\cos 3t\}$$

$$= \frac{s}{s^2+4} - \frac{s}{s^2+9}.$$

Therefore, by theorem 9.10, we get

$$L\left\{\frac{\cos 2t - \cos 3t}{t}\right\}$$

$$= \int_s^\infty \frac{u}{u^2+4}\,du - \int_s^\infty \frac{u}{u^2+9}\,du$$

$$= \frac{1}{2}[\log(u^2+4)]_s^\infty - \frac{1}{2}[\log(u^2+9)]_s^\infty$$

$$= \frac{1}{2}\left[\log\frac{u^2+4}{u^2+9}\right]_s^\infty$$

$$= \frac{1}{2}\lim_{u\to\infty}\left(\log\frac{u^2+4}{u^2+9}\right) - \frac{1}{2}\log\frac{s^2+4}{s^2+9}$$

$$= \frac{1}{2}\log\left(\lim_{u\to\infty}\frac{1+(4/u^2)}{1+(9/u^2)}\right) - \frac{1}{2}\log\frac{s^2+4}{s^2+9}$$

$$= 0 + \frac{1}{2}\log\frac{s^2+9}{s^2+4} = \frac{1}{2}\log\frac{s^2+9}{s^2+4}.$$

EXAMPLE 9.62

Find the Laplace transform of $f(t) = \frac{e^{-at} - e^{-bt}}{t}$.

Solution. We have

$$L\{e^{-at} - e^{-bt}\} = L\{e^{-at}\} - L\{e^{-bt}\}$$

$$= \frac{1}{s+a} - \frac{1}{s+b}.$$

Therefore, proceeding as in Example 9.61, we have

$$L\{f(t)\} = \int_s^\infty \left[\frac{1}{u+a} - \frac{1}{u+b}\right] du = \left[\log\frac{u+a}{u+b}\right]_s^\infty.$$

$$= \lim_{u\to\infty} \log\frac{u+a}{u+b} - \log\frac{s+a}{s+b}$$

$$= 0 - \log\frac{s+a}{s+b} = \log\frac{s+b}{s+a}.$$

EXAMPLE 9.63

Find the Laplace transform of $f(t) = \frac{1-\cos 2t}{t}$.

Solution. We have

$$L\{1 - \cos 2t\} = L\{1\} - L\{\cos 2t\} = \frac{1}{s} - \frac{s}{s^2+4}.$$

Therefore, by Theorem 9.11, we get

$$L\left\{\frac{1-\cos 2t}{t}\right\} = \int_s^\infty \left[\frac{1}{u} - \frac{u}{u^2+4}\right] du$$

$$= \left[\log u - \left(\frac{1}{2}\log(u^2+4)\right)\right]_s^\infty$$

$$= \left[\frac{1}{2}\log u^2 - \frac{1}{2}\log(u^2+4)\right]_s^\infty$$

$$= \frac{1}{2}\log\left(\frac{s^2+4}{s^2}\right).$$

EXAMPLE 9.64

Using Remark 9.3, evaluate the integral

$$\int_0^\infty \frac{e^{-t} - e^{-3t}}{t}\, dt.$$

Solution. By Remark 9.3, we have

$$\int_0^\infty \frac{e^{-t} - e^{-3t}}{t}\, dt$$

$$= \int_0^\infty L\{e^{-t} - e^{-3t}\}\, ds = \int_0^\infty \left[\frac{1}{u+1} - \frac{1}{u+3}\right] du$$

$$= [\log(u+1) - \log(u+3)]_0^\infty = \left[\log\frac{u+1}{u+3}\right]_0^\infty$$

$$= \log\left[\lim_{u\to\infty}\frac{u+1}{u+3}\right] - \log\frac{1}{3}$$

$$= \log\left[\lim_{u\to\infty}\frac{1+(1/u)}{1+(3/u)}\right] - \log\frac{1}{3}$$

$$= \log 1 - \log\frac{1}{3} = \log 3.$$

EXAMPLE 9.65

Find the Laplace transform of $f(t) = \frac{1-e^t}{t}$.

Solution. Since $L\{1 - e^t\} = L\{1\} - L\{e^t\} = \frac{1}{s} - \frac{1}{s-1}$, we have

$$L\{f(t)\} = \int_s^\infty \left(\frac{1}{u} - \frac{1}{u-1}\right) du$$

$$= [\log u - \log(u-1)]_s^\infty$$

$$= \left[\log\frac{u}{u-1}\right]_s^\infty = -\log\left[\frac{1}{1-(1/s)}\right]$$

$$= \log\left(\frac{s-1}{s}\right).$$

EXAMPLE 9.66

Find Laplace transform of $f(t) = \frac{\sin at}{t}$.

Solution. We know that

$$L\{\sin at\} = \frac{a}{s^2+a^2}.$$

Therefore, by Theorem 9.10,

$$L\left\{\frac{\sin at}{t}\right\} = a\int_s^\infty \frac{du}{u^2+a^2} = \tan^{-1}\left(\frac{a}{s}\right).$$

Theorem 9.11. (Laplace Transform of Integrals). If $L\{f(t)\} = F(s)$, then

$$L\left\{\int_0^t f(u)\, du\right\} = \frac{F(s)}{s}.$$

Proof: The function $f(t)$ should be integrable in such a way that

$$g(t) = \int_0^t f(u)\, du$$

is of exponential order. Then $g(0) = 0$ and $g'(t) = f(t)$. Therefore,

$$L\{g'(t)\} = sL\{g(t)\} - g(0) = sL\{g(t)\}$$

and so

$$L\left\{\int_0^t f(u)\, du\right\} = L\{g(t)\} = \frac{L\{g'(t)\}}{s}$$

$$= \frac{L\{f(t)\}}{s} = \frac{F(s)}{s}.$$

EXAMPLE 9.67

Find Laplace transform of $\int_0^t \frac{\sin u}{u}\, du$.

Solution. From Example 9.66, we have

$$L\left\{\frac{\sin t}{t}\right\} = \tan^{-1}\left(\frac{1}{s}\right).$$

Therefore, by Theorem 9.11, we have

$$L\left\{\int_0^t \frac{\sin u}{u}\, du\right\} = \frac{1}{s}\tan^{-1}\left(\frac{1}{s}\right).$$

(The function $\int_0^t \frac{\sin u}{u}\, du$ is called *sine integral function,* denoted by $Si(t)$, which is used in optics).

EXAMPLE 9.68

Find the Laplace transform of $\int_0^t \cos 2u\, du$.

Solution. We know that $L\{\cos 2t\} = \frac{s}{s^2+4}$ and so

$$L\left\{\int_0^t \cos 2u\, du\right\} = \frac{1}{s}\cdot\frac{s}{s^2+4} = \frac{1}{s^2+4}.$$

EXAMPLE 9.69

Find Laplace transform of *cosine integral function*

$$Ci(t) = \int_\infty^t \frac{\cos u}{u}\, du,\ t > 0$$

Solution. We know that $L\{\cos t\} = \frac{s}{s^2+1}$. Therefore, by Theorem 9.10

$$L\left\{\frac{\cos t}{t}\right\} = \int_s^\infty \frac{u}{u^2+1}\, du = \frac{1}{2}\int_s^\infty \frac{2u}{u^2+1}\, du$$

$$= \frac{1}{2}\log(1+s^2).$$

Now Theorem 9.11 yields the Laplace transform of cosine integral function $Ci(t)$ as given below:

$$L\{Ci(t)\} = L\left\{\int_\infty^t \frac{\cos u}{u}\, du\right\} = L\left\{-\int_t^\infty \frac{\cos u}{u}\, du\right\}$$

$$= -L\left\{\int_t^\infty \frac{\cos u}{u}\, du\right\}$$

$$= -\left[\frac{1}{s}\left(\frac{1}{2}\log(1+s^2)\right)\right]$$

$$= -\frac{1}{2s}\log(1+s^2)$$

EXAMPLE 9.70

Find the Laplace transform of the *exponential integral* defined by

$$Ei(t) = \int_t^\infty \frac{e^{-u}}{u}\, du,\ t > 0.$$

Solution. We know that $L\{e^{-u}\} = \frac{1}{s+1}$. Therefore, by Theorem 9.10, we have

$$L\left\{\frac{e^{-u}}{u}\right\} = \int_s^\infty \frac{1}{u+1}\, du = \log(1+s).$$

Now application of Theorem 9.11 yields

$$L\{Ei(t)\} = \frac{1}{s}\log(1+s).$$

EXAMPLE 9.71

Find Laplace transform of $\int_0^t e^t \frac{\sin t}{t}\, dt$.

Solution. We know that $L\{\sin t\} = \frac{1}{s^2+1}$. Therefore,

$$L\left\{\frac{\sin t}{t}\right\} = \int_s^\infty \frac{1}{u^2+1}\, du = [\tan^{-1} u]_s^\infty.$$

$$= \frac{\pi}{2} - \tan^{-1} s = \cot^{-1} s.$$

Therefore, by first-shifting property, we have

$$L\left\{e^t \frac{\sin t}{t}\right\} = \cot^{-1}(s-1).$$

Hence

$$L\left\{\int_0^t e^t \frac{\sin t}{t}\,dt\right\} = \frac{1}{s}\cot^{-1}(s-1).$$

Theorem 9.12. (Laplace Transform of a Periodic Function). Let f be a periodic function with period T so that $f(t) = f(t+T)$. Then

$$L\{f(t)\} = \frac{1}{1-e^{-sT}}\int_0^T e^{-st} f(t)\,dt.$$

Proof: We begin with the definition of Laplace transform and evaluate the integral using periodicity of $f(t)$. Thus we have

$$L\{f(t)\} = \int_0^\infty e^{-st} f(t)\,dt$$

$$= \int_0^T e^{-st} f(t)\,dt + \int_T^{2T} e^{-st} f(t)dt + \int_{2T}^{3T} e^{-st} f(t)\,dt + \ldots + \int_{(n-1)T}^{nT} e^{-st} f(t)dt + \ldots$$

provided that the series on the right-hand side converges. This is true since the function f satisfies the condition for the existence of its Laplace transform. Consider the integral

$$\int_{(n-1)T}^{nT} e^{-st} f(t)\,dt$$

and substitute $u = t-(n-1)T$. Since f has period T, we have

$$\int_{(n-1)T}^{nT} e^{-st} f(t)\,dt = e^{-s(n-1)T}\int_0^T e^{-su} f(u)\,du,$$

$$n = 1, 2, \ldots$$

Thus, we have

$$L\{f(t)\} = (1 + e^{-sT} + e^{-2sT} + \ldots)\int_0^T e^{-st} f(t)\,dt$$

$$= \frac{1}{1-e^{-sT}}\int_0^T e^{-st} f(t)\,dt \quad \text{(summing the G.P.)},$$

since $1 + r + r^2 + r^3 + \ldots = \frac{1}{1-r}, |r| < 1.$

EXAMPLE 9.72

Find Laplace transform of the *half-wave rectified sinusoidal*

$$f(t) = \begin{cases} \sin t & \text{for } 2n\pi < t < (2n+1)\pi \\ 0 & \text{for } (2n+1)\pi < t < (2n+2)\pi, \end{cases}$$

for $n = 0, 1, 2, \ldots$

Solution. The graph of the half-wave rectified sinusoidal function f is shown in the Figure 9.9

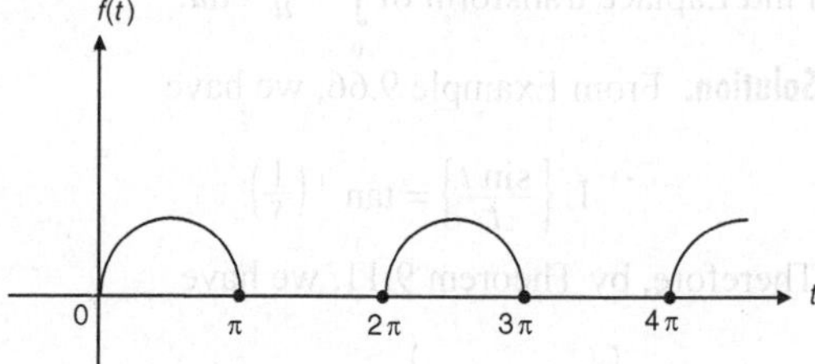

Figure 9.9

The function is of period 2π. So, Theorem 9.12 yields

$$L\{f(t)\} = \frac{1}{1-e^{-sT}}\int_0^T e^{-sT} f(t)\,dt$$

$$= \frac{1}{1-e^{-2\pi s}}\int_0^{2\pi} e^{-st} f(t)\,dt$$

$$= \frac{1}{1-e^{-2\pi s}}\int_0^{\pi} e^{-st}\sin t\,dt$$

$$= \frac{1}{1-e^{-2\pi s}}\left[\frac{e^{-st}(-s\sin t - \cos t)}{s^2+1}\right]_0^\pi$$

$$= \frac{1}{1-e^{-2\pi s}}\left(\frac{1+e^{-\pi s}}{s^2+1}\right)$$

$$= \frac{1}{(1-e^{-2\pi s})(s^2+1)}$$

EXAMPLE 9.73

Find the Laplace transform of *full rectified sine wave* defined by the expression

$$f(t) = \begin{cases} \sin t & \text{for } 0 < t < \pi \\ -\sin t & \text{for } \pi < t < 2\pi \end{cases}$$

$$f(t) = f(t + \pi).$$

Solution. The graph of the rectified sine wave is shown in the Figure 9.10.

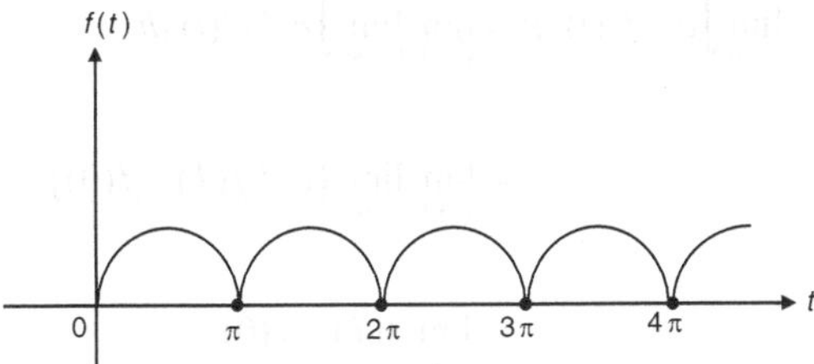

Figure 9.10

The function f has period π and, therefore, application of Theorem 9.12 yields

$$L\{f(t)\} = \frac{1}{1 - e^{-\pi s}} \int_0^{\pi} e^{-st} \sin t \, dt$$

$$= \frac{1}{1 - e^{-\pi s}} \left[\frac{1 + e^{-\pi s}}{s^2 + 1} \right]$$

$$= \frac{1 + e^{-\pi s}}{(1 - e^{-\pi s})(s^2 + 1)}.$$

EXAMPLE 9.74

Find Laplace transform of the *triangular wave function* defined by

$$f(t) = \begin{cases} t & \text{for } 0 \le t < a \\ 2a - t & \text{for } a \le t < 2a, \end{cases}$$

$$f(2a + t) = f(t).$$

Solution. The graph of triangular wave function is shown in the Figure 9.11.

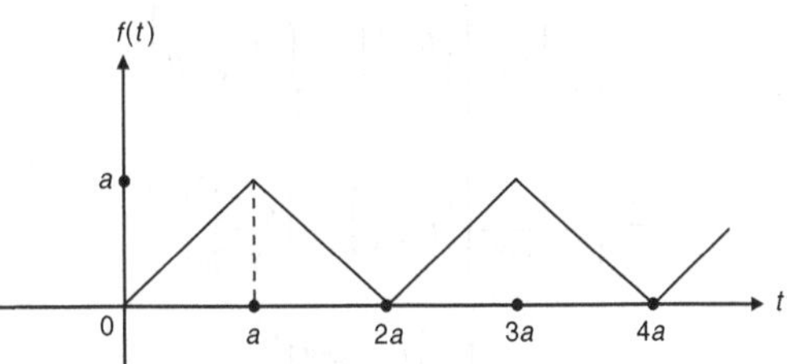

Figure 9.11

Using the formula for the Laplace transform of a periodic function, we have

$$L\{f(t)\} = \frac{1}{1 - e^{-2as}} \int_0^{2a} e^{-st} f(t) \, dt$$

$$= \frac{1}{1 - e^{-2as}} \left[\int_0^{a} t e^{-st} \, dt + \int_a^{2a} (2a - t) e^{-st} \, dt \right].$$

Now integration by parts gives

$$L\{f(s)\} = \frac{1}{1 - e^{-2as}} (e^{-as} - 1)^2 \cdot \frac{1}{s^2}$$

$$= \frac{1}{s^2} \left(\frac{1 - e^{-as}}{1 + e^{as}} \right) = \frac{1}{s^2} \tanh \left(\frac{as}{2} \right).$$

EXAMPLE 9.75

Find the Laplace transform of the *saw tooth wave function,* whose graph is shown in Figure 9.12.

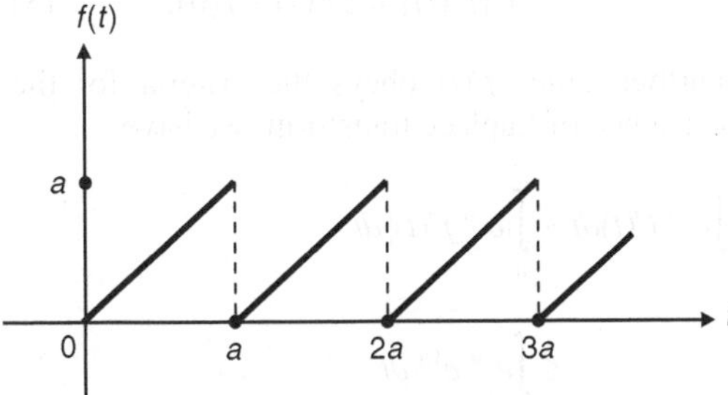

Figure 9.12

Solution. It is a periodic function with period a. Therefore,

$$L\{f(t)\} = \frac{1}{1 - e^{-as}} \int_0^{a} e^{-st} f(t) \, dt = \frac{1}{1 - e^{-as}} \int_0^{\infty} t e^{-st} \, dt$$

$$= \frac{1}{1-e^{-as}}\left\{\left[\frac{te^{-st}}{-s}\right]_0^a - \int_0^a \frac{e^{-st}}{-s}\,dt\right\}$$

$$= \frac{1}{1-e^{-as}}\left\{\left[t\,\frac{e^{-st}}{-s}\right]_0^t - \left[\frac{e^{-st}}{s^2}\right]_0^a\right\}$$

$$= \frac{1}{1-e^{-as}}\left[t\,\frac{e^{st}}{-s} - \frac{e^{-st}}{s^2}\right]_0^a$$

$$= \frac{1}{1-e^{-as}}\left[\frac{ae^{-sa}}{-s} - \frac{e^{-sa}}{s^2} + \frac{1}{s^2}\right]$$

$$= \frac{1}{1-e^{-as}}\left[-\frac{ae^{-sa}}{s} + \frac{1-e^{-sa}}{s^2}\right]$$

$$= \frac{1}{s^2} - \frac{ae^{-sa}}{s(1-e^{-as})}.$$

9.3 LIMITING THEOREMS

Theorem 9.13. (Initial Value). Let $f(t)$ and its derivative $f'(t)$ be piecewise continuous and of exponential order, then

$$\lim_{t\to 0} f(t) = \lim_{s\to\infty} sF(s),$$

provided the limits exist.

Proof: We know that

$$L\{f'(t)\} = sF(s) - f(0). \tag{3}$$

Further, since $f'(t)$ obeys the criteria for the existence of Laplace transform, we have

$$\int_0^\infty e^{-st} f'(t)dt \le \int_0^\infty |e^{-st} f'(t)|\,dt$$

$$\le \int_0^\infty e^{-st}\,e^{Mt}\,dt$$

$$= -\frac{1}{M-s} \to 0 \text{ as } s \to \infty.$$

Hence (3) implies $sF(s) - f(0) \to 0$ as $s \to \infty$, and so

$$\lim_{s\to 0} sF(s) = f(0) = \lim_{t\to 0} f(t).$$

Theorem 9.14. (Final Value). If the limits indicated exist, then

$$\lim_{t\to\infty} f(t) = \lim_{s\to 0} sF(s).$$

Proof: Using Laplace transform of the derivative, we have

$$L\{f'(t)\} = \int_0^\infty e^{-st} f'(t)dt = sF(s) - f(0) \tag{4}$$

Then

$$\lim_{s\to 0}\int_0^\infty e^{-st} f'(t)\,dt = \lim_{s\to 0}\lim_{T\to\infty}\int_0^T e^{-st} f'(t)\,dt$$

$$= \lim_{s\to 0}\lim_{T\to\infty}\{e^{-sT} f(T) - f(0)\}$$

$$= \lim_{T\to\infty} f(T) - f(0)$$

$$= \lim_{t\to\infty} f(t) - f(0).$$

Hence relation (4) yields

$$\lim_{t\to\infty} f(t) - f(0) = \lim_{s\to 0} sF(s) - f(0)$$

and so

$$\lim_{t\to\infty} f(t) = \lim_{s\to 0} sF(s).$$

EXAMPLE 9.76

Verify the validity of limiting theorems by considering $f(t) = e^t$.

Solution. We observe that

$$L\{e^t\} = F(s) = \frac{1}{s-1}.$$

$$\lim_{t\to 0} f(t) = f(0) = 1,$$

$$\lim_{s\to\infty} sF(s) = \lim_{s\to\infty}\frac{s}{s-1} = 1.$$

Thus, initial value theorem is verified in this case.

On the other hand,

$$\lim_{t\to\infty} f(t) = \lim_{t\to\infty} e^{-t} = 0,$$

$$\lim_{s\to 0} sF(s) = \lim_{s\to 0} \frac{s}{s+1} = 0,$$

and so the final value theorem also holds for e^{-t}.

EXAMPLE 9.77

Suppose Laplace transform of a function f is given by

$$F(s) = \frac{18}{s(s^2+36)}.$$

Find $f(0)$ and $f'(0)$.

Solution. By initial value theorem, we have

$$f(0) = \lim_{t\to 0} f(t) = \lim_{s\to\infty} s\,F(s)$$

$$= \lim_{s\to\infty} \frac{18s}{s(s^2+36)} = 0,$$

and

$$f'(0) = \lim_{s\to\infty} [s^2\,F(s) - s f(0)]$$

$$= \lim_{s\to\infty} \left[\frac{18s^2}{s(s^2+36)} - 0\right] = 0.$$

EXAMPLE 9.78

Let $F(s) = \frac{1}{s-1}$. Can we find $f(\infty)$ using final value theorem?

Solution. In this case, the final value theorem does not apply because

$$\lim_{t\to\infty} f(t) = \lim_{t\to\infty} e^{t} \to \infty \text{ as } t \to \infty.$$

9.4 MISCELLANEOUS EXAMPLES

EXAMPLE 9.79

Find the Laplace transforms of

(i) $e^{4t}\sin 2t\cos t$,

(ii) $\sinh t\cos^2 t$

Solution. (i) We have

$$\sin 2t\cos t = \frac{1}{2}[\sin 3t + \sin t].$$

Therefore

$$L\{\sin 2t\cos t\} = \frac{1}{2}L\{\sin 3t\} + \frac{1}{2}L\{\sin t\}$$

$$= \frac{1}{2}\left[\frac{3}{s^2+9} + \frac{1}{s^2+1}\right].$$

Now using shifting property, we get

$$L\{e^{4t}\sin 2t\cos t\}$$

$$= \frac{1}{2}\left[\frac{3}{(s-4)^2+9} + \frac{1}{(s-4)^2+1}\right]$$

$$= \frac{1}{2}\left[\frac{3}{s^2-8s+25} + \frac{1}{s^2-8s+17}\right].$$

(ii) Since $\cos^2 t = \frac{1+\cos 2t}{2}$. Therefore

$$L\{\cos^2 t\} = \frac{1}{2}L\{1\} + \frac{1}{2}L\{\cos 2t\}$$

$$= \frac{1}{2}\left[\frac{1}{s} + \frac{s}{s^2+4}\right] = \frac{s^2+2}{s(s^2+4)},\ s > 0.$$

Therefore, by shifting property,

$$L\{\sinh t\cos^2 t\} = L\left\{\frac{e^t - e^{-t}}{2}\cos^2 t\right\}$$

$$= \frac{1}{2}L\{e^t\cos^2 t\} - \frac{1}{2}L\{e^{-t}\cos^2 t\}$$

$$= \frac{1}{2}\left\{\frac{(s-1)^2+2}{(s-1)\,[(s-1)^2+4]}\right\}$$

$$- \frac{1}{2}\left\{\frac{(s+1)^2+2}{(s+1)[(s+1)^2+4]}\right\}$$

$$= \frac{1}{2}\left[\frac{s^2-2s+3}{(s-1)(s^2-2s+5)}\right.$$

$$\left. - \frac{s^2+2s+3}{(s+1)\,(s^2+2s+5)}\right].$$

EXAMPLE 9.80

Find the Laplace transforms of the following:

$$\frac{1}{t}(\cos at - \cos bt).$$

Solution. As in Example 9.61, we have

$$L\left\{\frac{\cos at - \cos bt}{t}\right\}$$

$$= \int_s^\infty \left[\frac{u}{u^2+a^2} - \frac{u}{u^2+b^2}\right] du$$

$$= \left[\frac{1}{2}\log(u^2 + a^2) - \frac{1}{2}\log(u^2 + b^2)\right]_s^\infty$$

$$= \frac{1}{2}\left[\log\frac{u^2 + a^2}{u^2 + b^2}\right]_s^\infty$$

$$= \frac{1}{2}\left[\log\left\{\lim_{s\to\infty}\frac{1+\frac{a^2}{s^2}}{1+\frac{b^2}{s^2}}\right\}\right] - \frac{1}{2}\log\frac{a^2+s^2}{b^2+s^2}$$

$$= -\frac{1}{2}\log\frac{a^2+s^2}{b^2+s^2} = \frac{1}{2}\log\frac{b^2+s^2}{a^2+s^2}.$$

EXAMPLE 9.81

Express $f(t) = \begin{cases} 1 & if \quad 0 < t \le 1 \\ t & if \quad 1 < t \le 2 \\ t^2 & if \quad t > 2 \end{cases}$ in terms of unit step function and hence find $Lf(t)$.

Solution. In terms of Heavyside s unit step function, we have

$$f(t) = H(t) + tH(t-1) + t^2 H(t-2).$$

Therefore

$$L\{f(t)\} = L\{H(t)\} + L\{tH(t-1)\} + L\{t^2 H(t-2)\}$$

$$= \frac{1}{s} - \frac{d}{ds}\left\{\frac{e^{-s}}{s}\right\} + (-1)^2\frac{d}{ds}\left\{\frac{e^{-2s}}{s}\right\}$$

$$= \frac{1}{s} - \frac{e^{-s}}{s} - \frac{e^{-s}}{s^2} - \frac{e^{-2s}}{s^2} - \frac{2e^{-2s}}{s}$$

$$= \frac{1}{s}[1 - e^{-s} - 2e^{-2s}] - \frac{1}{s^2}[e^{-s} + e^{-2s}].$$

EXAMPLE 9.82

Using Laplace transform evaluate $\int_0^\infty e^{-at}\frac{\sin^2 t}{t}\,dt$.

Solution. We have

$$L\{\sin^2 t\} = L\left\{\frac{1}{2}(1 - \cos 2t)\right\} = \frac{1}{2}\left[\frac{1}{s} - \frac{s}{s^2+4}\right].$$

Therefore

$$L\{e^{-at}\sin^2 t\} = \frac{1}{2}\left[\frac{1}{s+a} - \frac{s+a}{(s+a)^2+4}\right].$$

Further,

$$\int_0^\infty \frac{e^{-at}\sin^2 t}{t}\,dt = \int_0^\infty L\{e^{-at}\sin^2 t\}\,ds$$

$$= \frac{1}{2}\left[\int_0^\infty \frac{ds}{s+a} - \int_0^\infty \frac{s+a}{(s+a)^2+4}\,ds\right]$$

$$= \frac{1}{2}\left[\log\frac{s+a}{\sqrt{(s+a)^2+4}}\right]_0^\infty$$

$$= \frac{1}{2}\left[0 - \frac{1}{2}\log\frac{a}{\sqrt{(a^2+4)}}\right]$$

$$= \frac{1}{2}\log\frac{\sqrt{a^2+4}}{a}$$

EXCERCISES

1. Find the Laplace transforms of

(a) $4t + 6e^{4t}$ **Ans.** $\left(\frac{4}{s^2} + \frac{6}{s-4}\right)$

(b) $e^{-4t}\sin 5t$ **Ans.** $\left(\frac{5}{s^2+8s+41}\right)$

(c) $\frac{1}{t}(\sin at - at\cos at)$

Ans. $\tan^{-1}\left(\frac{a}{s}\right) - \frac{as}{s^2+a^2}$

(d) $\frac{t}{2a}\sin at$ **Ans.** $\left(\frac{s}{(s^2+a^2)^2}\right)$

(e) $\frac{t^{x-1}e^{at}}{\Gamma(x)}$ **Ans.** $\frac{1}{(s-a)^x}, x > 0$

(f) $f(t) = \begin{cases} t+1 & \text{for } 0 \le t \le 2 \\ 3 & \text{for } t > 2 \end{cases}$

Ans. $\frac{1}{s}+\frac{1}{s^2}(e^{-2s}-1)$

(g) $f(t)=t\mathrm{H}(t-a)$ **Ans.** $\frac{(1+as)e^{-as}}{s^2}$

(h) Null function defined by $\int_0^t n(t)\,dt=0$ for all t. **Ans.** 0

2. Show that the Laplace transforms of the following functions do not exist:

(a) e^{t^2} **Hint:** Not of exponential order

(b) $e^{\frac{1}{t}}$ **Hint:** Not defined at $t=0$

(c) $f(t)=\begin{cases}1 & \text{for even } t\\ 0 & \text{for odd } t\end{cases}$

Hint: has infinite number of finite jumps and so condition of piecewise continuity is not satisfied

(d) t^{-n}, n is positive integer.

3. Find $\mathrm{L}\left\{\frac{\sin h\,\omega t}{t}\right\}$ **Ans.** $\frac{1}{2}\log\frac{s+\omega}{s-\omega}$, $s>|\omega|$

4. Find the Laplace transform of a function whose graph is shown in the Figure 9.13.

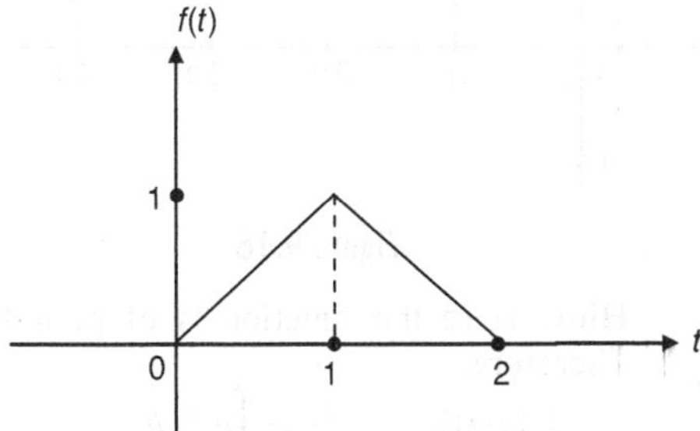

Figure 9.13

Hint: The function is defined by

$$f(t)=\begin{cases}t & \text{for } 0\le t<1\\ 2-t & \text{for } 1\le 2<2.\end{cases}$$

Therefore,

$$L\{f(t)\}=\int_0^1 te^{-st}\,dt+\int_1^2(2-t)\,e^{-st}\,dt$$

$$=\frac{1}{s^2}(1-e^{-s})^2.$$

5. Find the Laplace transform of step function f defined by $f(t)=n,\ n\le t<n+1,\ n=0,1,2,\ldots$.

Hint: The graph of step function is shown in the Figure 9.14.

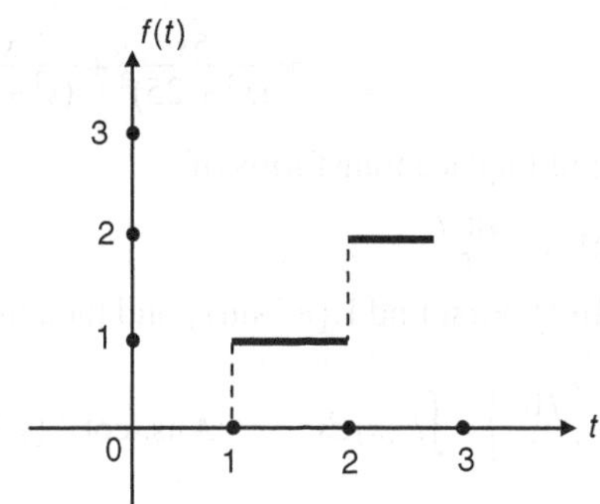

Figure 9.14

$$\mathrm{L}\{f(t)\}=\int_0^\infty e^{-st}f(t)\,dt$$

$$=\int_1^2 e^{-st}\,dt+2\int_2^3 e^{-st}\,dt+3\int_3^4 e^{-st}\,dt+\ldots$$

$$=\left[\frac{e^{-st}}{-s}\right]_1^2+2\left[\frac{e^{-st}}{-s}\right]_2^3+3\left[\frac{e^{-st}}{-s}\right]_3^4+\ldots$$

$$=\frac{e^{-s}}{s}(1+e^{-s}+e^{-2s}+\ldots)$$

$$=\frac{e^{-s}}{s(e^{s}-1)}.$$

6. Find the Laplace transforms of

(a) $te^{-2t}\sin 2t$ **Ans.** $\frac{4(s-2)}{(s^2-4s+8)^2}$

(b) $t^2\cos at$ **Ans.** $\frac{2s^3-6a^2s}{(s^2+a^2)^3}$

(c) $t\sin 3t\cos 2t$.

Hint: $\sin 3t\cos 2t=\frac{1}{2}(2\sin 3t\cos 2t)$

$$=\frac{1}{2}\sin(3t+2t)+\sin(3t-2t)$$

$$=\frac{1}{2}\sin 5t+\frac{1}{2}\sin t.$$

Therefore,

$$L\{\sin 3t \cos 2t\} = \frac{3s^2 + 15}{(s^2 + 1)(s^2 + 25)},$$

$s > 0$ and so

$$L\{s \sin 3t \cos 2t\} = -\frac{d}{ds}\frac{(3s^2 + 15)}{(s^2 + 1)(s^2 + 25)}$$

$$= \frac{5s}{(s^2 + 25)^2} + \frac{s}{(s^2 + 1)^2}.$$

7. Find Laplace transforms of

(a) $e^{-t}\dfrac{\sin t}{t}$

Hint: First find $L\{e^{-t} \sin t\}$ and then use

$L\left\{\dfrac{f(t)}{t}\right\} = \displaystyle\int_s^{\infty} F(s)\,ds.$ **Ans.** $\cot^{-1}(s + 1)$.

(b) $\dfrac{e^{at} - \cos 6t}{t}$ **Ans.** $\log \dfrac{s^2 - 36}{s - a}$

8. Evaluate $I = \displaystyle\int_0^{\infty} e^{-t}\frac{\sin^2 t}{t}\,dt.$

Hint: $I = \displaystyle\int_0^{\infty} L\{f(t)\}\,du$, where $f(t) = e^{-t}$ $\sin^2 t$. But $L\{e^{-t}\sin^2 t\} = \dfrac{2}{(s + 1)(s^2 + 2s + 5)}$. Therefore,

$I = \displaystyle\int_0^{\infty} \frac{2}{(s + 1)(s^2 + 2s + 5)ds}$ **Ans.** $\dfrac{1}{2}\log 5$

9. Show that $L\{tf'(t)\} = -\left\{s\dfrac{d}{ds}F(s) + F(s)\right\}$.

10. Show that $L\left\{\dfrac{1}{t}(\sin at - at\cos at)\right\} = \tan^{-1}\left(\dfrac{a}{s}\right) - \dfrac{as}{s^2 + a^2}$.

11. Show that $L\{t^2 f''(t)\} = s^2\dfrac{d^2}{ds^2}F(s) + 4s\dfrac{d}{ds}F(s) + 2F(s)$.

12. Use initial value theorem to find $f(0)$ and $f'(0)$ for the function f for which $F(s) = \dfrac{s}{s^2 - 5s + 12}$ **Ans.** $f(0) = 1, f'(0) = 5$

13. Find the Laplace transform of the square wave function with graph shown in the Figure 9.15.

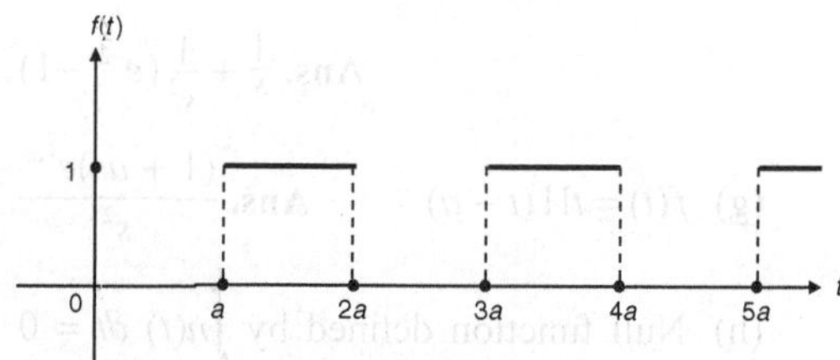

Figure 9.15

Hint: The function is of period $2a$. therefore,

$$L\{f(t)\} = \frac{1}{1 - e^{-2as}}\int_a^{2a} e^{-st}\,dt = \frac{e^{-as}}{s(1 + e^{-as})}$$

14. Express the function $f(t)$ in Exercise 13 in terms of Heaviside's unit step function and then find its Laplace transform.

Hint: $f(t) = H(t - a) - H(t - 2a) + H(t + 3a) - H(t - 4a) + \ldots$

15. Find the Laplace transform the of square wave function with graph given in the Figure 9.16.

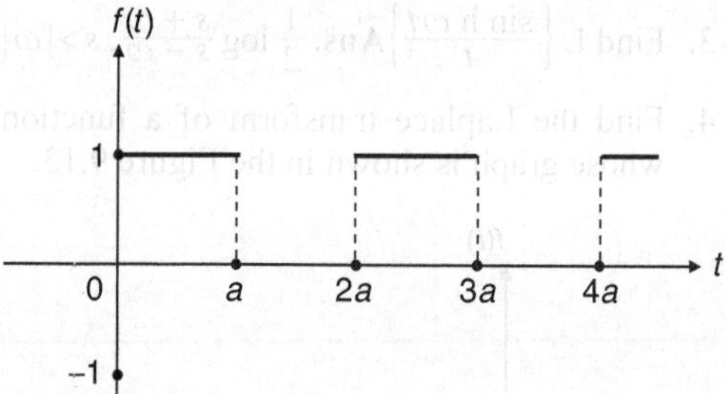

Figure 9.16

Hint: Here the function is of period $2a$. Therefore,

$$L\{f(t)\} = \frac{1}{1 - e^{-2as}}\int_0^{a} e^{-st}\,dt$$

$$= \frac{1}{1 - e^{-2as}}\left[\frac{e^{-st}}{-s}\right]_0^a = \frac{1}{s(1 + e^{-as})}.$$

16. Solve Example 9.27 using periodicity

Hint: $L\{f(t)\} = \dfrac{1}{e^{-2as}}\displaystyle\int_0^{2a} e^{-st} f(t)\,dt = \frac{1}{1 - e^{-2as}}$ $\times\left[\displaystyle\int_0^{a} e^{-st}\,dt + \int_a^{2a} -e^{-st}\,dt\right]$ which on using

Examples 9.13 and 9.15 yields the transform as

$$\frac{1}{s}\tanh\frac{as}{2}.$$

17. Find Laplace transform of the half-wave rectified sine function f defined by

$$f(t) = \begin{cases} \sin \omega t & \text{for } \dfrac{2n\pi}{\omega} < t < \dfrac{(2n+1)\pi}{\omega} \\ & \text{for } \dfrac{(2n+1)\pi}{\omega} < t < \dfrac{(2n+2)\pi}{\omega}. \end{cases}$$

Hint: The function is periodic with period $\frac{2\pi}{\omega}$. Therefore,

$$L\{f(t)\} = \frac{1}{1-e^{-2\pi s/\omega}}\int_0^{\pi/\omega} e^{-st}\sin\omega t\,dt$$

$$= \frac{1}{1-e^{-2\pi s/\omega}}\int_0^{\pi/\omega} e^{-st}\sin\omega t\,dt$$

$$= \frac{\omega}{s^2+\omega^2}(1+e^{-\pi s/\omega}).$$

18. Establish relation between Laplace transform and Fourier transform of a function.

Hint: Let ϕ be a function defined by

$$\phi(t) = \begin{cases} e^{-xt} f(t) & \text{for } t > 0 \\ 0 & \text{for } t < 0. \end{cases}$$

Then the Fourier transform of ϕ is

$$F(y) = F\{\phi(t)\} = \int_0^{\infty} e^{-iyt}\cdot e^{-xt} f(t)\,dt$$

$$= \int_0^{\infty} e^{-(x+iy)t} f(t)\,dt$$

$$= \int_0^{\infty} e^{-st} f(t)\,dt = L\{f(t)\}.$$

Thus Laplace transform of $f(t)$ is equal to the Fourier transform of $e^{-st} f(t)$.

10 Inverse Laplace Transform

Like the operations of addition, multiplication, and differentiation, the Laplace transform has also its inverse. During the process of solving physical problems like differential equations, it is necessary to invoke the inverse transform of the Laplace transform. Thus given a Laplace transform F(s) of a function f, we would like to know what f is. Hence, we are concerned with the solution of the integral equation,

$$\int_0^\infty e^{-st} f(t)\, dt = \text{F}(s).$$

10.1 DEFINITION AND EXAMPLES OF INVERSE LAPLACE TRANSFORM

Definition 10.1. Let f have Laplace transform F(s), that is, L$\{f(t)\}$ = F(s), then $f(t)$ is called an *inverse Laplace transform* of F(s) and we write

$$L^{-1}\{\text{F}(s)\} = f(t),\ t \geq 0.$$

The transformation L^{-1} is called *inverse Laplace operator* and it maps the Laplace transform of a function back to the original function.

We know that Laplace transform F(s) of a function $f(t)$ is uniquely determined due to the properties of integrals. However, this in not true for the inverse transform. For example, if $f(t)$ and $g(t)$ are two functions that are identical except for a finite number of points, they have the same transform F(s) since their integrals are identified. Therefore, either $f(t)$ or $g(t)$ is the inverse transform of F(s). Thus inverse transform of a given function F(s) is uniquely determined only upto an additive null function [a function $n(t)$ for which $\int_0^t n(u)\, du = 0$ for all t].

The following examples show that $L^{-1}\{\text{F}(s)\}$ can be more than one function.

(a) Let $f(t)$ = sin ωt, $t \geq 0$. Then L$\{f(t)\}$ $= \dfrac{\omega}{s^2 + w^2}$. Thus

$$L^{-1}\left\{\frac{\omega}{s^2+\omega^2}\right\} = \sin \omega t.$$

Now let

$$g(t) = \begin{cases} \sin \omega t & \text{for } t > 0 \\ 1 & \text{for } t = 0. \end{cases}$$

Then

$$\text{L}\{g(t)\} = \frac{\omega}{s^2+\omega^2},$$

and so

$$L^{-1}\frac{\omega}{s^2+\omega^2} = g(t).$$

Hence there are two inverse transforms of $\dfrac{\omega}{s^2+\omega^2}$.

(b) Let $f(t) = e^{-3t}$ and

$$g(t) = \begin{cases} 0 & \text{for } t = 0 \\ e^{-3t} & \text{for otherwise.} \end{cases}$$

Then both $f(t)$ and $g(t)$ have same Laplace transform $\dfrac{1}{s+3}$. Thus $\dfrac{1}{s+3}$ has two inverse Laplace transforms $f(t)$ and $g(t)$.

But the following theorem shows that the Laplace transform is one-one mapping.

Theorem 10.1. (Lerch's Theorem). Distinct continuous functions on $[0, \infty)$ have distinct Laplace transforms.

Thus, if we restrict ourselves to continuous functions on $[0, \infty)$, then the inverse transform $L^{-1}\{F(s)\} = f(t)$ is uniquely defined. Since many of the functions, we generally deal with, are solutions to the differential equations and hence continuous, the assumption of the theorem is satisfied.

EXAMPLE 10.1

Find inverse Laplace transform of $\frac{a}{s^2+a^2}, \frac{s}{s^2+a^2}, \frac{1}{s-a}, \frac{e^{-as}}{s}, \frac{s}{s^2-a^2}$ and $\frac{n!}{s^{n+1}}$.

Solution. We know that

$$L\{\sin at\} = \frac{a}{s^2+a^2}, \; L\{\cos at\} = \frac{s}{s^2+a^2},$$

$$L\{e^{at}\} = \frac{1}{s-a}, \; L\{H(t-a)\} = \frac{e^{-sa}}{s},$$

$$L\{\cosh at\} = \frac{s}{s^2-a^2}, \text{ and}$$

$$L\{t^n\} = \frac{n!}{s^{n+1}}, \; n \text{ being non-negative integer.}$$

Therefore,

$$L^{-1}\left\{\frac{a}{s^2+a^2}\right\} = \sin at,$$

$$L^{-1}\left\{\frac{s}{s^2+a^2}\right\} = \cos at,$$

$$L^{-1}\left\{\frac{1}{s-a}\right\} = e^{at},$$

$$L^{-1}\left\{\frac{e^{-sa}}{s}\right\} = H(t-a)$$

(Heavyside's unit step function),

$$L^{-1}\left\{\frac{s}{s^2-a^2}\right\} = \cosh at,$$

and

$$L^{-1}\left\{\frac{n!}{s^{n+1}}\right\} = t^n, \; n \text{ being non-negative integer,}$$

EXAMPLE 10.2

Find $L^{-1}\left(\frac{1}{\sqrt{s}}\right)$.

Solution. Since $L\left\{\frac{1}{t^{\frac{1}{2}}}\right\} = \frac{\Gamma(1/2)}{s^{\frac{1}{2}}} = \sqrt{\frac{\pi}{s}}$, it follows that

$$L^{-1}\left\{\frac{1}{\sqrt{s}}\right\} = \frac{1}{\sqrt{\pi t}}.$$

10.2 PROPERTIES OF INVERSE LAPLACE TRANSFORM

The operational properties used in finding the Laplace transform of a function are also used in constructing the inverse transform. We, thus, have the following properties of inverse transform

Theorem 10.2. (Linearity Property). If $F_1(s)$ and $F_2(s)$ are Laplace transforms of $f_1(t)$ and $f_2(t)$, respectively, and a_1 and a_2 are arbitrary constants, then

$$L^{-1}\{a_1 F_1(s) + a_2 F_2(s)\} = a_1 L^{-1}\{F_1(s)\} + a_2 L^{-1}\{F_2(s)\} = a_1 f_1(t) + a_2 f_2(t).$$

Proof: Since

$$L\{a_1 f_1(t) + a_2 f_2(t)\} = a_1 L\{f_1(t)\} + a_2 L\{f_2(t)\} = a_1 F_1(s) + a_2 F_2(s),$$

We have

$$L^{-1}\{a_1 F_1(s) + a_2 F_2(s)\} = a_1 f_1(t) + a_2 f_2(t) = a_1 L^{-1}\{F_1(s)\} + a_2 L^{-1}\{F_2(s)\}.$$

EXAMPLE 10.3

Find the inverse Laplace transform of

$$\frac{1}{2s}+\frac{4}{3(s-a)}+\frac{s}{s^2+16}.$$

Solution. By linearity of inverse Laplace transform, we have

$$L^{-1}\left\{\frac{1}{2s}+\frac{4}{3(s-a)}+\frac{s}{s^2+16}\right\}$$

$$=\frac{1}{2}L^{-1}\left\{\frac{1}{s}\right\}+\frac{4}{3}L^{-1}\left\{\frac{1}{s-a}\right\}$$

$$+L^{-1}\left\{\frac{s}{s^2+16}\right\}=\frac{1}{2}+\frac{4}{3}e^{at}+\cos 4t.$$

EXAMPLE 10.4

Find inverse Laplace transform of

$$\frac{5}{s-3}+\frac{s}{s^2+4}+\frac{3}{s-7}.$$

Solution. By linearity, we have

$$L^{-1}\left\{\frac{5}{s-3}+\frac{s}{s^2+4}+\frac{3}{s-7}\right\}$$

$$=5L^{-1}\left\{\frac{1}{s-3}\right\}+L^{-1}\left\{\frac{s}{s^2+4}\right\}$$

$$+3L^{-1}\left\{\frac{1}{s-7}\right\}=5e^{3t}+\cos 2t+3e^{7t}.$$

EXAMPLE 10.5

Find inverse Laplace transform of

$$-\frac{3}{2s}+\frac{4}{3(s-1)}+\frac{1}{6(s+2)}.$$

Solution. By linearity of inverse Laplace transform, we have

$$L^{-1}\left\{-\frac{3}{2s}+\frac{4}{3(s-1)}+\frac{1}{6(s+2)}\right\}$$

$$=-\frac{3}{2}L^{-1}\left\{\frac{1}{s}\right\}+\frac{4}{3}L^{-1}\left\{\frac{1}{s-1}\right\}+\frac{1}{6}L^{-1}\left\{\frac{1}{s+2}\right\}$$

$$=-\frac{3}{2}+\frac{4}{3}e^{t}+\frac{1}{6}e^{-2t}.$$

EXAMPLE 10.6

Find $L^{-1}\left\{\frac{1}{\sqrt{s}+a}\right\}$.

Solution. Since

$$\frac{1}{\sqrt{s}+a}=\frac{1}{\sqrt{s}}=-\frac{a}{\sqrt{s}(\sqrt{s}+a)}=\frac{1}{\sqrt{s}}-\frac{a(\sqrt{s}-a)}{\sqrt{s}(s-a^2)},$$

therefore,

$$L^{-1}\left\{\frac{1}{\sqrt{s}+a}\right\}=L^{-1}\left\{\frac{1}{\sqrt{s}}\right\}-aL^{-1}\frac{\sqrt{s}-a}{\sqrt{s}(s-a^2)}$$

$$=L^{-1}\left\{\frac{1}{\sqrt{s}}\right\}-aL^{-1}\left\{\frac{1}{s-a^2}\right\}+a^2L^{-1}\left\{\frac{1}{\sqrt{s}(s-a^2)}\right\}$$

$$=\frac{1}{\sqrt{\pi t}}-ae^{a^2t}+ae^{a^2t}\,erf(a\sqrt{t})$$

$$=\frac{1}{\sqrt{\pi t}}-ae^{a^2t}(1-erf(a\sqrt{t})$$

$$=\frac{1}{\sqrt{\pi t}}-ae^{a^2t}\,erf(a\sqrt{t})$$

Theorem 10.3. (First Shifting Property). If F(s) is Laplace transform of $f(t)$, then

$$L^{-1}\{F(s-a)\}=e^{at}f(t).$$

Proof: We know that

$$L\{e^{at}f(t)\}=F(s-a).$$

Therefore,

$$L^{-1}\{F(s-a)\}=e^{at}f(t).$$

EXAMPLE 10.7

Find inverse Laplace transform of $\frac{s-5}{s^2+6s+13}$.

Solution. Since

$$\frac{s-5}{s^2+6s+13}=\frac{(s+3)-8}{(s+3)^2+4},$$

we have

$$L^{-1}\left\{\frac{s-5}{s^2+6s+13}\right\}=L^{-1}\left\{\frac{s+3}{(s+3)^2+4}\right\}$$

$$-4L^{-1}\left\{\frac{2}{(s+3)^2+4}\right\}$$

$$=e^{-3t}\cos 2t-4e^{-3t}\sin 2t$$

$$=e^{-3t}(\cos 2t-4\sin 2t).$$

EXAMPLE 10.8

Find inverse Laplace transform of $\dfrac{2s-3}{s^2+4s+13}$.

Solution. Since

$$\frac{2s-3}{s^2+4s+3}=\frac{2s+4-7}{(s+2)^2+9}=\frac{2(s+2)-7}{(s+2)^2+9},$$

we have

$$L^{-1}\left\{\frac{2s-3}{s^2+4s+13}\right\}=L^{-1}\left\{\frac{2(s+2)}{(s+2)^2+9}\right\}$$

$$-7L^{-1}\left\{\frac{1}{(s+2)^2+9}\right\}$$

$$=2e^{-2t}\cos 3t-\frac{7}{3}e^{-2t}\sin 3t.$$

EXAMPLE 10.9

Find inverse Laplace transform of $\dfrac{s}{(s+1)^2}$.

Solution. We note that

$$\frac{s}{(s+1)^2}=\frac{s+1-1}{(s+1)^2}=\frac{1}{s+1}-\frac{1}{(s+1)^2}.$$

Therefore,

$$L^{-1}\left\{\frac{s}{(s+1)^2}\right\}=L^{-1}\left\{\frac{1}{s+1}\right\}-L^{-1}\left\{\frac{1}{(s+1)^2}\right\}$$

$$=e^{-t}\cdot 1-e^{-t}\cdot t=e^{-t}(1-t).$$

EXAMPLE 10.10

Find inverse Laplace transform of

$$\frac{1}{s^2+4s+13}-\frac{s+4}{s^2+8s+97}+\frac{s+2}{s^2-4s+29}.$$

Solution. We have

$$F(s)=\frac{1}{s^2+4s+13}-\frac{s+4}{s^2+8s+97}+\frac{s+2}{s^2-4s+29}$$

$$=\frac{1}{(s+2)^2+9}-\frac{s+4}{(s+4)^2+81}+\frac{s-2+4}{(s-2)^2+25}.$$

Therefore,

$$L^{-1}\{F(s)\}=\frac{1}{3}L^{-1}\left\{\frac{3}{(s+2)^2+9}\right\}$$

$$-L^{-1}\left\{\frac{s+4}{(s+4)^2+81}\right\}$$

$$+L^{-1}\left\{\frac{s-2}{(s-2)^2+25}\right\}$$

$$+\frac{4}{5}L^{-1}\left\{\frac{5}{(s-2)^2+25}\right\}$$

$$=\frac{1}{3}e^{-2t}\sin 3t-e^{-4t}\cos 9t$$

$$+e^{2t}\cos 5t+\frac{4}{5}e^{2t}\sin 5t.$$

EXAMPLE 10.11

Find the inverse Laplace transform of $\dfrac{s+2}{s^2-4s+13}$.

Solution. We have

$$\frac{s+2}{s^2-4s+13}=\frac{(s-2)+4}{(s-2)^2+9}$$

$$=\frac{(s-2)}{(s-2)^2+9}+\frac{4}{(s-2)^2+9}$$

$$=\frac{s-2}{(s-2)^2+9}+\frac{4}{3}\cdot\frac{3}{(s-2)^2+9}.$$

Therefore, by linearity of inverse Laplace transform and shifting property, we get

$$L^{-1}\left\{\frac{s+2}{s^2-4s+13}\right\}=L^{-1}\left\{\frac{s-2}{(s-2)^2+9}\right\}$$

$$+\frac{4}{3}L^{-1}\left\{\frac{3}{(s-2)^2+9}\right\}$$

$$= e^{2t}\cos 3t + \frac{4}{3}e^{2t}\sin 3t$$

$$= e^{2t}\left(\cos 3t + \frac{4}{3}\sin 3t\right).$$

Theorem 10.4. (Second Shifting Property). If $L^{-1}\{F(s)\} = f(t)$, then $L^{-1}\{e^{-sa}F(s)\} = g(t)$, where

$$g(t) = \begin{cases} f(t-a) & \text{for } t > a \\ 0 & \text{for } t < a \end{cases}$$

Proof: Since $L\{g(t)\} = e^{-sa}F(s)$, it follows that $L^{-1}\{e^{-sa}F(s)\} = g(t)$.

Second Proof: By definition of Laplace transform, we have

$$F(s) = \int_0^\infty e^{-st} f(t)\, dt.$$

Therefore,

$$e^{-sa}F(s) = \int_0^\infty e^{-sa}e^{-st} f(t)\, dt = \int_0^\infty e^{-s(t+a)} f(t)\, dt$$

$$= \int_0^\infty e^{-su} f(u-a)\, du, \quad t + a = u.$$

$$= \int_0^\infty e^{-su}(0)\, du + \int_a^\infty e^{-su} f(u-a)\, du$$

$$= L\{g(t)\}.$$

Hence

$$L^{-1}\{e^{-sa}F(s)\} = g(t).$$

EXAMPLE 10.12

Find inverse Laplace transform of $-\dfrac{e^{-\pi s/2}}{s^2+1}$.

Solution. We have

$$-\frac{e^{-\pi s/2}}{s^2+1} = e^{-\pi s/2}\left(-\frac{1}{s^2+1}\right)$$

$$= e^{-\pi s/2}F(s), \text{ where } F(s) = -\frac{1}{s^2+1}.$$

But

$$L^{-1}\{F(s)\} = L^{-1}\left\{-\frac{1}{s^2+1}\right\} = -L^{-1}\left\{\frac{1}{s^2+1}\right\}.$$

Therefore, by second-shifting property,

$$L^{-1}\left\{-\frac{e^{-\pi s/2}}{s^2+1}\right\} = g(t),$$

Where

$$g(t) = \begin{cases} -\sin(t-\pi/2) & \text{for } t > \pi/2 \\ 0 & \text{for } t < \pi/2 \end{cases}$$

$$= -\sin\left(t - \frac{\pi}{2}\right)\left[H\left(t - \frac{\pi}{2}\right)\right]$$

$$= \cos t\left[H\left(t - \frac{\pi}{2}\right)\right],$$

where $H(t)$ denotes Heavyside's unit step function.

EXAMPLE 10.13

Find inverse Laplace transform of $\dfrac{\omega e^{-sa}}{s^2+\omega^2}$.

Solution. We have

$$L^{-1}\left\{\frac{\omega}{s^2+\omega^2}\right\} = \sin \omega t.$$

Therefore, by second-shifting property,

$$L^{-1}\left\{e^{-sa}\frac{\omega}{s^2+\omega^2}\right\} = g(t),$$

where

$$g(t) = \begin{cases} \sin \omega(t-a) & \text{for } t > a \\ 0 & \text{for } t < a. \end{cases}$$

EXAMPLE 10.14

Find inverse transform of $\dfrac{2e^{-s}}{s^3}$, $\text{Re}(s) > 0$.

Solution. We have

$$\frac{2e^{-s}}{s^3} = 2e^{-s}\cdot\frac{1}{s^3}.$$

Since $L^{-1}\left\{\dfrac{1}{s^3}\right\} = t^2$, therefore, by second-shifting property,

$$L^{-1}\left\{\frac{2e^{-s}}{s^3}\right\} = g(t),$$

where

$$g(t) = \begin{cases} 2(t-1)^2 & \text{for } t \ge 1 \\ 0 & \text{for } 0 \le t \le 1. \end{cases}$$

EXAMPLE 10.15

Find $L^{-1}\left\{\frac{A}{s} - \frac{s}{s^2+1}e^{-sa}\right\}$, where A is a constant.

Solution. We have

$$L^{-1}\left\{\frac{1}{s}\right\} = 1 \quad \text{and} \quad L^{-1}\left\{\frac{s}{s^2+1}\right\} = \cos t.$$

Therefore, using linearity property and second–shifting property, we have

$$L^{-1}\left\{\frac{A}{s} - \frac{s}{s^2+1}e^{-sa}\right\}$$

$$= AL^{-1}\left\{\frac{1}{s}\right\} - L\left\{e^{-sa}\frac{s}{s^2+1}\right\} = A - g(t),$$

where

$$g(t) = \begin{cases} \cos(t-a) & \text{for } t > a \\ 0 & \text{for } t < a. \end{cases}$$

$$= H(t-a)\cos(t-a).$$

Hence

$$L^{-1}\left\{\frac{A}{s} - \frac{s}{s^2+1}e^{-sa}\right\}$$

$$= A - H(t-a)\cos(t-a).$$

EXAMPLE 10.16

Find the inverse Laplace transform of $\frac{e^{-7s}}{(s-3)^3}$.

Solution. Since

$$L^{-1}\left(\frac{1}{(s-3)^3}\right) = \frac{1}{2}t^2 e^{3t},$$

By second-shifting property, we have

$$L^{-1}\left\{e^{-7s}\frac{1}{(s-3)^3}\right\} = g(t),$$

Where

$$g(t) = \begin{cases} \frac{1}{2}(t-7)^2 e^3(t-7) & \text{for } t > 7 \\ 0 & \text{for } 0 \le t \le 7 \end{cases}$$

$$= \frac{1}{2}H(t-7)(t-7)^2 e^3(t-7).$$

Theorem 10.5. (Change of Scale Property). If $L^{-1}\{F(s)\} = f(t)$, then

$$L^{-1}\{F(as)\} = \frac{1}{a}f\left(\frac{t}{a}\right).$$

Proof: By definition of Laplace transform,

$$F(s) = \int_0^\infty e^{-st} f(t)\,dt.$$

Therefore,

$$F(as) = \int_0^\infty e^{-ast} f(t)\,dt$$

$$= \frac{1}{a}\int_0^\infty e^{-su} f\left(\frac{u}{a}\right) du,\ u = at$$

$$= \frac{1}{a}L\left\{f\left(\frac{t}{a}\right)\right\}.$$

Hence

$$L^{-1}\{F(as)\} = \frac{1}{a}f\left(\frac{t}{a}\right).$$

Remark 10.1. It follows from Theorem 10.5 that if $L^{-1}\{F(s)\} = f(t)$, then $L^{-1}\left\{F\left(\frac{s}{a}\right)\right\} = af(at)$ for $a > 0$.

EXAMPLE 10.17

Find the inverse transform of $\frac{s}{(s/2)^2+4}$.

Solution. Since

$$L^{-1}\left\{\frac{s}{s^2+4}\right\} = \cos 2t,$$

We have

$$L^{-1}\left\{\frac{s}{(s/2)^2+4}\right\} = 2L^{-1}\left\{\frac{s/2}{(s/2)^2+4}\right\}$$

$$= 2.2\cos(2(2t)) = 4\cos 4t.$$

Theorem 10.6. (Inverse Laplace Transform of Derivatives). If $L^{-1}\{F(s)\} = f(t)$, then

$$L^{-1}\{F^{(n)}(s)\} = (-1)^n t^n f(t).$$

Proof: Since

$$L\{t^n f(t)\} = (-1)^n \frac{d^n}{ds^n} F(s),$$

We have

$$L^{-1}\{F^{(n)}(s)\} = (-1)^n t^n f(t)$$

EXAMPLE 10.18

Find $L^{-1}\left\{\frac{1-s^2}{(s^2+1)^2}\right\}$.

Solution. We know that

$$L^{-1}\left\{\frac{s}{s^2+1}\right\} = \cos t.$$

Further

$$\frac{d}{ds}\left(\frac{s}{s^2+1}\right) = \frac{1-s^2}{(s^2+1)^2}.$$

Therefore, by Theorem 10.6, we have

$$L^{-1}\left\{\frac{1-s^2}{(s^2+1)^2}\right\} = -t\cos t.$$

EXAMPLE 10.19

Find $L^{-1}\left\{\log\frac{s+a}{s+b}\right\}$.

Solution. We note that

$$\frac{d}{ds}\log\frac{s+a}{s+b} = \frac{d}{ds}[\log(s+a) - \log(s+b)]$$

$$= \frac{1}{s+a} - \frac{1}{s+b}.$$

Therefore, the use of Theorem 10.6 yields

$$L^{-1}\left\{\frac{1}{s+a} - \frac{1}{s+b}\right\} = -t f(t)$$

and so

$$L^{-1}\left\{\frac{1}{s+a}\right\} - L^{-1}\left\{\frac{1}{s+b}\right\} = -t f(t),$$

that is,

$$e^{-at} - e^{-bt} = -t f(t).$$

Hence

$$f(t) = \frac{1}{t}(e^{-bt} - e^{-at}).$$

EXAMPLE 10.20

Find $L^{-1}\left\{\log\frac{s^2+a^2}{s^2+b^2}\right\}$.

Solution. Since

$$\frac{d}{ds}\log\frac{s^2+a^2}{s^2+b^2} = \frac{d}{ds}[\log(s^2+a^2) - \log(s^2+b^2)]$$

$$= \frac{2s}{s^2+a^2} - \frac{2s}{s^2+b^2},$$

Theorem 10.6 yields

$$L^{-1}\left\{\frac{2s}{s^2+a^2} - \frac{2s}{s^2+b^2}\right\} = -t f(t),$$

or

$$2\cos at - 2\cos bt = -t f(t),$$

or

$$f(t) = \frac{2}{t}(\cos bt - \cos at).$$

EXAMPLE 10.21

Find $L^{-1}\left\{\log\frac{1+s}{s}\right\}$.

Solution. Since

$$\frac{d}{ds}\log\frac{1+s}{s} = \frac{d}{ds}[\log(1+s) - \log s]$$

$$= \frac{1}{s+1} - \frac{1}{s},$$

Therefore,

$$L^{-1}\left\{\frac{1}{s+1} - \frac{1}{s}\right\} = -t f(t)$$

or

$$e^{-t} - 1 = -t f(t)$$

or

$$f(t) = \frac{1-e^{-t}}{t}.$$

EXAMPLE 10.22

Find $L^{-1}\left\{\log\left(1+\frac{1}{s^2}\right)\right\}$.

Solution. Since

$$\frac{d}{ds}\left\{\log\frac{s^2+1}{s^2}\right\}=\frac{d}{ds}[\log(s^2+1)-\log s^2]$$

$$=\frac{2s}{s^2+1}-\frac{2s}{s^2}=2\left[\frac{s}{s^2+1}-\frac{1}{s}\right],$$

we have

$$L^{-1}\left\{2\left(\frac{s}{s^2+1}-\frac{1}{s}\right)\right\}=-t\,f(t)$$

and so

$$2\cos t-2=-t\,f(t)$$

or

$$f(t)=\frac{2(1-\cos t)}{t}.$$

EXAMPLE 10.23

Find $L^{-1}\left\{\tan^{-1}\frac{1}{s}\right\}, s>0,$

Solution. Since

$$\frac{d}{ds}\left(\tan^{-1}\frac{1}{s}\right)=\frac{1}{1+(1/s)^2}\left(-\frac{1}{s^2}\right)$$

$$=\frac{-1}{1+(1/s^2)}\left(\frac{1}{s^2}\right)=-\frac{1}{s^2+1},$$

it follows that

$$L^{-1}\left\{-\frac{1}{s^2+1}\right\}=-t\,f(t),$$

that is,

$$-1(\sin t)=-t\,f(t).$$

Hence

$$f(t)=\frac{\sin t}{t}.$$

EXAMPLE 10.24

Find $L^{-1}\left\{\log\frac{s^2+1}{(s-1)^2}\right\}.$

Solution. Since

$$\frac{d}{ds}\left(\log\frac{s^2+1}{(s-1)^2}\right)=\frac{d}{ds}[\log(s^2+1)-2\log(s-1)]$$

$$=\frac{2s}{s^2+1}-\frac{2}{s-1},$$

we have

$$L^{-1}\left\{\frac{2s}{s^2+1}-\frac{2}{s-1}\right\}=-t\,f(t),$$

which yields

$$2[\cos t-e^t]=-t\,f(t)$$

or

$$f(t)=\frac{2(e^t-\cos t)}{t}.$$

EXAMPLE 10.25

Find $L^{-1}\left\{\frac{s+2}{(s^2+4s+5)^2}\right\}.$

Solution. We have

$$\frac{s+2}{(s^2+4s+5)^2}=\frac{s+2}{((s+2)^2+1)^2}$$

Therefore,

$$L^{-1}\left\{\frac{s+2}{(s^2+4s+5)^2}\right\}=L^{-1}\left\{\frac{s+2}{((s+2)^2+1)^2}\right\}$$

$$=e^{-2t}\cdot L^{-1}\left\{\frac{s}{(s^2+1)^2}\right\}$$

We now find $L^{-1}\left\{\frac{s}{(s^2+1)^2}\right\}$. We note that

$$\frac{d}{ds}\left(\frac{1}{(s^2+1)}\right)=-\frac{2s}{(s^2+1)^2}.$$

Therefore, by Theorem 10.6

$$L^{-1}\left\{-\frac{2s}{(s^2+1)^2}\right\}=-t\,L^{-1}\left\{\frac{1}{s^2+1}\right\}=-t\sin t$$

and so

$$L^{-1}\left\{\frac{s}{(s^2+1)^2}\right\}=\frac{1}{2}t\sin t.$$

Hence

$$L^{-1}\left\{\frac{s+2}{(s^2+4s+5)^2}\right\}=\frac{1}{2}te^{-2t}\sin t.$$

EXAMPLE 10.26

Find $L^{-1}\left\{\frac{s+3}{(s^2+6s+13)^2}\right\}.$

Solution. We have

$$\frac{s+3}{(s^2+6s+13)^2} = \frac{s+3}{((s+3)^2+4)^2}.$$

Therefore,

$$L^{-1}\left\{\frac{s+3}{(s^2+6s+13)^2}\right\} = L^{-1}\left\{\frac{s+3}{((s+3)^2+4)^2}\right\}$$

$$= e^{-3t}\, L^{-1}\left\{\frac{s}{(s^2+4)^2}\right\}.$$

To find $L^{-1}\left\{\frac{s}{(s^2+4)^2}\right\}$, we note that

$$\frac{d}{ds}\left(\frac{1}{s^2+4}\right) = \frac{-2s}{(s^2+4)^2}.$$

Therefore,

$$L^{-1}\left\{-\frac{2s}{(s^2+4)^2}\right\} = -t\, L^{-1}\left\{\frac{1}{s^2+4}\right\}$$

$$= -\frac{t}{2}\sin 2t$$

and so

$$L^{-1}\left\{\frac{s}{(s^2+4)^2}\right\} = \frac{1}{4}\, t \sin 2t.$$

Consequently, we get

$$L^{-1}\left\{\frac{s+3}{(s^2+6s+13)^2}\right\} = \frac{1}{4}\, t\, e^{-3t}\sin 2t.$$

Theorem 10.7. (Inverse Laplace Transform of Integrals). If $L^{-1}\{F(s)\} = f(t)$, then

$$L^{-1}\left\{\int_s^{\infty} F(u)\, du\right\} = \frac{f(t)}{t}.$$

Proof: Since,

$$L\left\{\frac{f(t)}{t}\right\} = \int_s^{\infty} F(u)\, du,$$

we have

$$L^{-1}\left\{\int_s^{\infty} F(u)\, du\right\} = \frac{f(t)}{t}.$$

EXAMPLE 10.27

Find

$$L^{-1}\left\{\int_s^{\infty}\left(\frac{1}{u-1}-\frac{1}{u+1}\right) du\right\}$$

Solution. BY Theorem 10.7, we have

$$L^{-1}\left\{\int_s^{\infty}\left(\frac{1}{u-1}-\frac{1}{u+1}\right) du\right\} = \frac{L^{-1}\left\{\frac{1}{u-1}-\frac{1}{u+1}\right\}}{t}.$$

But

$$L^{-1}\left\{\frac{1}{u-1}-\frac{1}{u+1}\right\} = e^{t} - e^{-t}.$$

Therefore,

$$L^{-1}\left\{\int_s^{\infty}\left(\frac{1}{u-1}-\frac{1}{u+1}\right) du\right\} = \frac{e^t - et}{t}.$$

Theorem 10.8. (Multiplication by s^n). If $L^{-1}\{F(s)\} = f(t)$ and $f(0) = 0$, then

$$L^{-1}\{s\,F(s)\} = f'(t).$$

Proof: We know that

$$L\{f'(t)\} = s\,F(s) - f(0)$$

and so

$$L^{-1}\{s\,F(s) - f(0)\} = f'(t),$$

that is,

$$L^{-1}\{s\,F(s)\} = f'(t).$$

Remark 10.2. If $f(0) \neq 0$, then

$$L^{-1}\{s\,F(s) - f(0)\} = f'(t)$$

and so

$$L^{-1}\{s\,F(s)\} = f'(t) + f(0)\,\delta(t),$$

where $\delta(t)$ is the Dirac delta function.

EXAMPLE 10.28

Find $L^{-1}\left\{\frac{s^2}{s^2+1}\right\}$.

Solution. We know that

$$L^{-1}\left\{\frac{s}{s^2+1}\right\} = \cos t = f(t)$$

and $f(0) = 1 \neq 0$ for $t = 0$. Therefore,

$$L^{-1}\left\{\frac{s^2}{s^2+1}\right\} = L^{-1}\{s\,F(s)\}$$

$$= f'(t) + f(0)\,\delta(t) = \delta(t) - \sin t.$$

EXAMPLE 10.29

Find $L^{-1}\left\{\frac{s^3}{s^2+1}\right\}$.

Solution. From the Example 10.28,

$$L^{-1}\left\{\frac{s^2}{s^2+1}\right\} = \delta(t) - \sin t = f(t).$$

Since $f(0) = 0$, we have

$$L^{-1}\left\{\frac{s^3}{s^2+1}\right\} = L^{-1}\{s\,F(s)\} = f'(t)$$

$$= \frac{d}{dt}\{\delta(t) - \sin t\} = \delta'(t) - \cos t.$$

Theorem 10.9. (Division by s). If $L^{-1}\{F(s)\} = f(t)$, then

$$L^{-1}\left\{\frac{F(s)}{s}\right\} = \int_0^t f(u)\,du.$$

Proof: We know that

$$\mathrm{L}\left\{\int_0^t f(u)\,du\right\} = \frac{F(s)}{s}.$$

Therefore,

$$L^{-1}\left\{\frac{F(s)}{s}\right\} = \int_0^t f(u)\,du.$$

Remark 10.3. Consider

$$g(t) = \int_0^t\int_0^v f(u)\,du\,dv.$$

Then

$$g'(t) = \int_0^t f(u)\,du \text{ and } g''(t) = f(t).$$

also $g(0) = g'(0) = 0$. Thus

$$\mathrm{L}\{g''(t)\} = s^2\,\mathrm{L}\{g(t)\} - s\,g(0) - g'(0)$$
$$= s^2\,\mathrm{L}\{g(t)\},$$

that is,

$$\mathrm{L}\{f(t)\} = s^2\,\mathrm{L}\{g(t)\}$$

or

$$F(s) = s^2\,\mathrm{L}\{g(t)\}$$

or

$$\mathrm{L}\{g(t)\} = \frac{F(s)}{s^2}$$

Hence

$$\mathrm{L}^{-1}\left\{\frac{F(s)}{s^2}\right\} = g(t) = \int_0^t\int_0^v f(u)\,du\,dv$$

$$= \int_0^t\int_0^t f(t)\,dt^2.$$

In general, we have

$$\mathrm{L}^{-1}\left\{\frac{F(s)}{s^n}\right\} = \underbrace{\int_0^t\int_0^t \ldots \int_0^t}_{n \text{ times}} f(t)\,dt^n.$$

EXAMPLE 10.30

Find $L^{-1}\left\{\frac{1}{s(s+1)}\right\}$.

Solution. Since $L^{-1}\left\{\frac{1}{s+1}\right\} = e^{-t}$, we have

$$L^{-1}\left\{\frac{1}{s(s+1)}\right\} = \int_0^t e^{-u}\,du = [-e^{-u}]_0^t$$

$$= -[e^{-t} - 1] = 1 - e^{-t}.$$

EXAMPLE 10.31

Find $L^{-1}\left\{\frac{1}{s^2(s+1)}\right\}$.

Solution. From Example 10.30, we have

$$L^{-1}\left\{\frac{1}{s(s+1)}\right\} = 1 - e^{-t}$$

Therefore, by Theorem 10.9, we have

$$L^{-1}\left\{\frac{1}{s^2(s+1)}\right\} = \int_0^t (1-e^{-u})\,du = [u+e^{u}]_0^t$$

$$= t + e^t - 1.$$

EXAMPLE 10.32

Find $L^{-1}\left\{\frac{s}{(s^2+a^2)^2}\right\}$ and deduce $L^{-1}\left\{\frac{1}{(s^2+a^2)^2}\right\}$ from it.

Solution. Since

$$\frac{d}{ds}\left(\frac{a}{s^2+a^2}\right) = -\frac{2as}{(s^2+a^2)^2},$$

we have

$$L^{-1}\left\{-\frac{2as}{(s^2+a^2)^2}\right\} = -t\,f(t),$$

Where $f(t) = L^{-1}\left\{\frac{a}{s^2+a^2}\right\} = \sin at$ and so

$$-2a\,L^{-1}\left\{\frac{s}{(s^2+a^2)^2}\right\} = -t\sin at$$

Hence

$$L^{-1}\left\{\frac{s}{(s^2+a^2)^2}\right\} = \frac{t\sin at}{2a}$$

Now Theorem 10.9 yields

$$L^{-1}\left\{\frac{1}{(s^2+a^2)^2}\right\}$$

$$= \frac{1}{2a}\int_0^t u\sin au\,du$$

$$= \frac{1}{2a}\left\{\left[u\,\frac{-\cos au}{a}\right]_0^t - \int_0^t -\frac{\cos au}{a}\,du\right\}$$

$$= \frac{1}{2a}\left[-\frac{t\cos at}{a} + \frac{\sin at}{a^2}\right]$$

$$= \frac{1}{2a^3}(\sin at - at\cos at).$$

10.3 PARTIAL FRACTIONS METHOD TO FIND INVERSE LAPLACE TRANSFORM

While working with physical problems, we come across some functions that are not immediately recognizable as the Laplace transform of some elementary functions. In such cases, we decompose the given function into partial fractions and then write down the inverse of each fraction. This method is used when inverse transform of a rational function is required.

We can also use the concept of *simple poles* of the rational function F(s) to find inverse Laplace transform. Let

$$F(s) = \frac{P(s)}{Q(s)} = \frac{P(s)}{(s-\alpha_1)(s-\alpha_2)\ldots(s-\alpha_n)},$$

$$\alpha_i \neq \alpha_j,$$

where P(s) is a polynomial of degree less than n. In terms of complex analysis, we call $\alpha_1, \alpha_2, \ldots \alpha_n$ the simple poles of F(s). The Partial fraction decomposition shall be

$$F(s) = \frac{A_1}{s-\alpha_1} + \frac{A_2}{s-\alpha_2} + \ldots + \frac{A_n}{s-\alpha_n}.$$

Multiplying both sides of this equation by $s - \alpha_i$, and letting $s \to \alpha_i$, we have

$$A_i = \lim_{s\to\alpha_i} (s-\alpha_i)\,F(s).$$

Again, in terms of complex analysis, A_i is called the *residue of F(s) at the poles* α_i.

Now, let L{$f(t)$} = F(s). Then

$$f(t) = L^{-1}\{F(s)\} = \sum_{i=1}^{n} L^{-1}\left(\frac{A_i}{s-\alpha_i}\right) = \sum_{i=1}^{n} A_i\,e^{\alpha_i t}$$

$$= \sum_{i=1}^{n} \lim_{s\to\alpha_i} (s-\alpha_i)F(s)\,e^{\alpha_i t}.$$

EXAMPLE 10.33

Find $L^{-1}\left\{\frac{1}{s^3(s^2+1)}\right\}$.

Solution. Write

$$\frac{1}{s^3(s^2+1)} = \frac{A}{s} + \frac{B}{s^2} + \frac{C}{s^3} + \frac{Ds+E}{s^2+1},$$

which yields

$$1 = As^2(s^2+1) + Bs(s^2+1) + C(s^2+1) + (Ds+E)s^3.$$

Putting $s = 0$, we get $C = 1$. Comparing coefficients of s on both sides, we get $B = 0$. Comparing coefficients of s^2 on both sides, we get $A + C = 0$ which yields $A = -C = -1$. Comparing coefficients of s^3, we get $B + E = 0$ and so $E = -B = 0$.Comparing coefficients of x^4, we have $A + D = 0$ and so $D = -A = 1$. Thus

$$\frac{1}{s^3(s^2+1)} = -\frac{1}{s} + \frac{1}{s^3} + \frac{s}{s^2+1}.$$

Hence, by Example 10.1,

$$L^{-1}\left\{\frac{1}{s^3(s^2+1)}\right\} = L^{-1}\left\{-\frac{1}{s} + \frac{1}{s^3} + \frac{s}{s^2+1}\right\}$$

$$= -1 + \frac{1}{2}t^2 + \cos t$$

$$= \frac{1}{2}(t^2 + 2\cos t - 2).$$

EXAMPLE 10.34

Find $L^{-1}\left\{\frac{2}{(s+1)(s^2+1)}\right\}$.

Solution. Write

$$\frac{2}{(s+1)(s^2+1)} = \frac{A}{s+1} + \frac{Bs+C}{s^2+1}.$$

Clearing fractions, we have

$$2 = A(s^2+1) + (Bs+C)(s+1).$$

Putting $s = -1$, we have A = 1. Comparing the coefficient of s^2 and that of s^0, we get

$$0 = A + B, \text{ which yields } B = -1$$

and

$$2 = A + C, \text{ which yields } C = 1.$$

Hence

$$\frac{2}{(s+1)(s^2+1)} = \frac{1}{s+1} + \frac{-s+1}{s^2+1}$$

$$= \frac{1}{s+1} - \frac{s}{s^2+1} + \frac{1}{s^2+1}.$$

Therefore,

$$L^{-1}\left\{\frac{2}{(s+1)(s^2+1)}\right\} = e^{-t} - \cos t + \sin t.$$

EXAMPLE 10.35

Find $L^{-1}\frac{s^2}{(s-1)(s+2)(s-3)}$.

Solution. The simple poles of the rational function F(s) are $s = 1, -2$, and 3. Therefore,

$$f(t) = L^{-1}\{F(s)\} = \lim_{s\to 1}(s-1)F(s)\,e^t$$

$$+ \lim_{s\to -2}(s+2)\,F(s)\,e^{-2t} + \lim_{s\to 3}(s-3)\,F(s)\,e^{3t}$$

$$= -\frac{1}{6}e^t + \frac{4}{(-3)(-5)}e^{-2t} + \frac{9}{(2)(5)}e^{3t}$$

$$= -\frac{1}{6}e^t + \frac{4}{15}e^{-2t} + \frac{9}{10}e^{3t}.$$

EXAMPLE 10.36

Find $L^{-1}\left\{\frac{4s+5}{(s-1)^2\,(s+2)}\right\}$.

Solution. Write

$$\frac{4s+5}{(s-1)^2(s+2)} = \frac{A}{s-1} + \frac{B}{(s-1)^2} + \frac{C}{s+2},$$

which gives

$$4s + 5 = A(s+2)\,(s-1) + B(s+2) + C(s-1)^2.$$

Putting $s = 1$, we get 9 = 3 B and so B = 3. Equating the coefficients of s and s^2, we get A + B – 2C = 4 and A + C = 0. These equations give $C = -\frac{1}{3}$ and $A = \frac{1}{3}$. Thus

$$\frac{4x+5}{(s-1)^2(s+2)} = \frac{1}{3(s-1)} + \frac{3}{(s-1)^2} - \frac{1}{3(s+2)}.$$

Hence

$$L^{-1}\left\{\frac{4s+5}{(s-1)^2(s+2)}\right\} = \frac{1}{3}e^t + 3t\,e^t - \frac{1}{3}e^{-2t}.$$

EXAMPLE 10.37

Find $L^{-1}\left\{\dfrac{s^2}{s^2+6s^2+11s+6}\right\}$.

Solution. We have

$$F(s)=\frac{s^2}{s^3+6s^2+11s+6}.$$

Clearly $s=-1,-2$, and -3 are roots of the polynomial $s^3+6s^2+11s+10$. Hence the simple poles of $F(s)$ are $-1,-2$, and -3. Therefore,

$$f(t)=L^{-1}\{F(s)\}=\lim_{s\to-1}(s+1)\,F(s)\,e^{-t}$$

$$+\lim_{s\to-2}(s+2)\,F(s)\,e^{-2t}+\lim_{s\to-3}(s+3)\,F(s)\,e^{-3t}$$

$$=\frac{1}{2}e^{-t}-4e^{-2t}+\frac{9}{2}e^{-3t}.$$

EXAMPLE 10.38

Find $L^{-1}\left\{\dfrac{s^3}{s^4-a^4}\right\}$.

Solution. We have

$$\frac{s^3}{s^4-a^4}=\frac{s^3}{(s-a)(s+a)(s^2+a^2)}$$

$$=\frac{A}{s-a}+\frac{B}{s+a}+\frac{Cs+D}{s^2+a^2},$$

which yields

$$s^3=A(s+a)(s^2+a^2)+B(s-a)(s^2+a^2)$$
$$+(Cs+D)(s^2-a^2)$$

$$=A(s^3+as^2+a^2s+a^3)+B(s^3-as^2+a^2s-a^3)$$
$$+C(s^3-sa^2)+d(s^2-a^2).$$

Putting $s=a,-a$, we get $A=B=\frac{1}{4}$. Comparing coefficients of s, and s^2, we get $C=\frac{1}{2}$, $D=0$. Thus

$$\frac{s^3}{s^4-a^4}=\frac{1}{4(s-a)}+\frac{1}{4(s+a)}+\frac{s}{2(s^2+a^2)}.$$

Hence

$$L^{-1}\left\{\frac{s^3}{s^4-a^4}\right\}=\frac{1}{4}e^{at}+\frac{1}{4}e^{-at}+\frac{1}{2}\cos at$$

$$=\frac{1}{2}\cos at+\frac{1}{2}\left(\frac{e^{at}+e^{-at}}{2}\right)$$

$$=\frac{1}{2}[\cos at+\cosh at].$$

EXAMPLE 10.39

Find $L^{-1}\left\{\dfrac{s^2}{(s^2-a^2)(s^2-b^2)(s^2-c^2)}\right\}$.

Solution. Let

$$F(s)=\frac{s^2}{(s^2-a^2)(s^2-b^2)(s^2-c^2)}$$

$$=\frac{s^2}{(s-a)(s+a)(s-b)(s+b)(s-c)(s+c)}.$$

The simple poles of $F(s)$ are $a,-a,b,-b,c$, and $-c$.
Therefore,

$$L^{-1}\{F(s)\}=\lim_{s\to a}(s-a)\,F(s)\,e^{at}$$

$$+\lim_{s\to -a}(s+a)\,F(s)e^{-at}+\lim_{s\to b}(s-b)F(s)e^{bt}$$

$$+\lim_{s\to -b}(s+b)\,F(s)e^{-bt}+\lim_{s\to c}(s-c)F(s)e^{ct}$$

$$+\lim_{s\to -c}(s+c)F(s)\,e^{-ct}$$

$$=\frac{a^2}{2a(a^2-b^2)(a^2-c^2)}e^{at}$$
$$-\frac{a^2}{2a(a^2-b^2)(a^2-c^2)}e^{-at}$$
$$+\frac{b^2}{2b(b^2-a^2)(b^2-c^2)}e^{bt}$$
$$-\frac{b^2}{2b(b^2-a^2)(b^2-c^2)}e^{-bt}$$
$$+\frac{c^2}{2c(c^2-a^2)(c^2-b^2)}e^{ct}$$
$$-\frac{c^2}{2c(c^2-a^2)(c^2-b^2)}e^{-ct}.$$

$$= \frac{a(e^{at} - e^{-at})}{2(a^2 - b^2)(a^2 - c^2)} + \frac{b(e^{bt} - e^{-bt})}{2(b^2 - a^2)(b^2 - c^2)} + \frac{c(e^{ct} - e^{-ct})}{2(c^2 - a^2)(c^2 - b^2)}$$

$$= \frac{a \sinh at}{(a^2 - b^2)(a^2 - c^2)} + \frac{b \sinh bt}{(b^2 - a^2)(b^2 - c^2)} + \frac{c \sinh ct}{(c^2 - a^2)(c^2 - b^2)}.$$

EXAMPLE 10.40

Find $L^{-1}\left\{\frac{2s^2 - 6s + 5}{s^3 - 6s^2 + 11s - 6}\right\}$.

Solution. Let

$$F(s) = \frac{2s^2 - 6s + 5}{s^3 - 6s^2 + 11s - 6}.$$

We observe that $s = 1$, $s = 2$, and $s = 3$ are the roots of the equation $s^3 - 6s^2 + 11s - 6 = 0$. Thus

$$F(s) = \frac{2s^2 - 6s + 5}{(s - 1)(s - 2)(s - 3)}.$$

Therefore,

$$L^{-1}\{F(s)\} = \lim_{s\to 1} (s - 1) F(s) e^t + \lim_{s\to 2} (s - 2) F(s) e^{2t} + \lim_{s\to 3} (s - 3) F(s) e^{3t} = \frac{1}{2} e^t - e^{2t} + \frac{5}{2} e^{3t}.$$

EXAMPLE 10.41

Find $L^{-1}\left\{\frac{1}{s^3 - a^3}\right\}$.

Solution. We have

$$F(s) = \frac{1}{s^3 - a^3} = \frac{1}{(s - a)(s^2 + as + a^2)} = \frac{A}{s - a} + \frac{Bs + C}{s^2 + as + a^2}.$$

Therefore,

$$1 = A(s^2 + as + a^2) + (Bs + C)(s - a).$$

Putting $s = a$, we have $A = \frac{1}{3a^2}$. Comparing the coefficients of constant terms and of s^2, we get respectively, $a^2A - aC = 1$ and $A + B = 0$, which in turn imply $B = -A = -\frac{1}{3a^2}$ and $C = -\frac{2}{3a}$. Thus

$$F(s) = \frac{1}{3a^2(s - a)} - \frac{1}{3a^2}\left(\frac{s + 2a}{s^2 + as + a^2}\right).$$

Hence

$$L^{-1}\{F(s)\} = \frac{1}{3a^2} L^{-1}\left\{\frac{1}{s - a}\right\} - \frac{1}{3a^2} L^{-1}\left(\frac{\left(s + \frac{a}{2}\right) + \frac{3a}{2}}{\left(s + \frac{a}{2}\right)^2 + \frac{3a^2}{4}}\right)$$

$$= \frac{1}{3a^2} e^{at} - \frac{1}{3a^2}\left[L^{-1}\left(\frac{s + \frac{a}{2}}{\left(s + \frac{a}{2}\right)^2 + \frac{3a^2}{4}}\right) + L^{-1}\left(\frac{\frac{3a}{2}}{\left(s + \frac{a}{2}\right)^2 + \frac{3a^2}{4}}\right)\right]$$

$$= \frac{1}{3a^2} e^{at} - \frac{1}{3a^2}\left[e^{-\frac{at}{2}}\left(\cos \frac{\sqrt{3}a}{2} t + \sqrt{3} \sin \frac{\sqrt{3}\,a}{2} t\right)\right]$$

$$= \frac{1}{3a^2}\left[e^{at} - e^{-at}\left(\cos \frac{\sqrt{3}a}{2} t + \sqrt{3} \sin \frac{\sqrt{3}\,a}{2} t\right)\right]$$

EXAMPLE 10.42

Find $L^{-1}\left\{\frac{4s + 5}{(s - 1)^2(s + 2)}\right\}$

Solution. Write

$$F(s) = \frac{4s + 5}{(s - 1)^2(s + 2)} = \frac{A}{s - 1} + \frac{B}{(s - 1)^2} + \frac{C}{s + 2},$$

and so

$$4s + 5 = A(s - 1)(s + 2) + B(s + 2) + C(s - 1)^2.$$

Taking $s = 1$, we get $B = 3$. Similarly, taking $s = -2$, we get $C = -\frac{1}{3}$. Now equating coefficients of s^2, we have $A + C = 0$ and so $A = -C = \frac{1}{3}$. Hence

$$F(s) = \frac{1}{3(s - 1)} + \frac{3}{(s - 1)^2} - \frac{1}{3(s + 2)}.$$

Hence

$$L^{-1}\{F(s)\} = \frac{1}{3}e^{t} + 3t\,e^{t} - \frac{1}{3}e^{-2t}.$$

EXAMPLE 10.43

Find $L^{-1}\left\{\frac{2s+1}{(s+2)^2(s-1)^2}\right\}$.

Solution. Write

$$F(s) = \frac{2s+1}{(s+2)^2(s-1)^2}$$

$$= \frac{A}{s+2} + \frac{B}{(s+2)^2} + \frac{C}{s-1} + \frac{D}{(s-1)^2}$$

and so

$$2s + 1 = A(s+2)(s-1)^2 + B(s-1)^2 + C(s-1)(s+2)^2 + D(s+2)^2.$$

Putting $s = 1$ yields $D = \frac{1}{3}$. Putting $s = -2$, we have $B = -\frac{1}{3}$. Comparing coefficients of s^3, we have $A + C = 0$ and comparing coefficient of s^2, we have $B + D - 3C = 0$. Thus $C = A = 0$ and so

$$F(s) = \frac{2s+1}{(s+2)^2(s-1)^2} = -\frac{1}{3(s+2)^2} + \frac{1}{3(s-1)^2}.$$

Hence

$$L^{-1}\{F(s)\} = -\frac{1}{3}t\,e^{-2t} + \frac{1}{3}t\,e^{t} = \frac{1}{3}t(e^{t} - e^{-2t}).$$

10.4 HEAVISIDE'S EXPANSION THEOREM

Let $F(s) = \frac{P(s)}{Q(s)}$, where $P(s)$ and $Q(s)$ have no common factors and the degree of $Q(s)$ is greater than the degree of $P(s)$. if $Q(s)$ has distinct zeros, then the following theorem gives us the inverse Laplace transform of $F(s)$.

Theorem 10.10. (Heaviside's Expansion Formula). Let $P(s)$ and $Q(s)$ be polynomials in s, where degree of $Q(s)$ is greater than that of $P(s)$. if $Q(s)$ has n distinct zeros $a_1, a_2, \ldots, a_n$, then

$$L^{-1}\left\{\frac{P(s)}{Q(s)}\right\} = \sum_{i=1}^{n} \frac{P(a_i)}{Q'(a_i)} e^{a_i t}.$$

Proof: Since $Q(s)$ has n distinct zeros, it factorizes into n linear factors and so by partial fractions, we have

$$\frac{P(s)}{Q(s)} = \frac{A_1}{s-a_1} + \frac{A_2}{s-a_2} + \ldots + \frac{A_i}{s-a_i} + \ldots + \frac{A_n}{s-a_n}.$$

Multiplying throughout by $s - a_i$ and lettings $s \to a_i$, we obtain

$$A_i = \lim_{s\to a_i} \frac{P(s)}{Q(s)}(s-a_i) = \lim_{s\to a_i} P(s)\left(\frac{s-a_i}{Q(s)}\right)$$

$$= \lim_{s\to a_i} P(s) \lim_{s\to a_i}\left(\frac{s-a_i}{Q(s)}\right)$$

$$= P(a_i)\lim_{s\to a_i}\frac{1}{Q'(s)} \text{ (using L'Hospital's rule)}$$

$$= \frac{P(a_i)}{Q'(a_i)}.$$

Hence

$$\frac{P(s)}{Q(s)} = \sum_{i=1}^{n} \frac{P(Q_i)}{Q'(a_i)} \frac{1}{(s-a_i)}$$

and so

$$L^{-1}\left\{\frac{P(s)}{Q(s)}\right\} = \sum_{i=1}^{n} \frac{P(a_i)}{Q'(a_i)} L^{-1}\left\{\frac{1}{s-a_i}\right\}$$

$$= \sum_{i=1}^{n} \frac{P(a_i)}{Q'(a_i)} e^{a_i t}.$$

EXAMPLE 10.44

Find $L^{-1}\left\{\frac{s}{(s-3)(s^2+4)}\right\}$ using Heaviside's expansion formula.

Solution. Let

$$F(s) = \frac{P(s)}{Q(s)} = \frac{s}{(s-3)(s^2+4)}.$$

Thus

$$P(s) = s,\ Q(s) = (s-3)(s^2+4)$$

$$= s^3 - 3s^2 + 4s - 12$$

and

$$Q'(s) = 3s^2 - 6s + 4.$$

Also roots of $Q(s) = 0$ are 3, $2i$, and $-2i$. Therefore, by Heaviside's expansion formula

$$L^{-1}\left\{\frac{P(s)}{Q(s)}\right\} = \sum_{i=1}^{3} \frac{P(a_i)}{Q'(a_i)} e^{a_i t} = \frac{P(3)}{Q'(3)} e^{3t} + \frac{P(2i)}{Q'(2i)} e^{2it} + \frac{P(-2i)}{Q'(-2i)} e^{-2it}$$

$$= \frac{3}{13} e^{3t} + \frac{2i}{-8-12i} e^{2it} - \frac{2i}{-8+12i} e^{-2it}$$

$$= \frac{3}{13} e^{3t} - \frac{(3+2i)}{26} (\cos 2t + i \sin 2t) + \frac{2i-3}{26} (\cos 2t - i \sin 2t)$$

$$= \frac{3}{13} e^{3t} - \frac{3}{13} \cos 2t + \frac{2}{13} \sin 2t.$$

EXAMPLE 10.45

Find $L^{-1}\left\{\frac{3s+7}{(s-3)(s+1)}\right\}$ using Heaviside's expansion formula.

Solution. We have

$$F(s) = \frac{P(s)}{Q(s)} = \frac{3s+7}{(s-3)(s+1)} = \frac{3s+7}{s^2-2s-3}.$$

Here $P(s) = 3s + 7$, $Q(s) = s^2 - 2s - 3$, $Q'(s) = 2s - 2$, and zeros of $Q(s)$ are 3 and -1. Hence, by Heaviside's expansion formula,

$$L^{-1}\{F(s)\} = \sum_{i=1}^{2} \frac{P(a_i)}{Q'(a_i)} e^{a_i t} = \frac{P(3)}{Q'(3)} e^{3t} + \frac{P(-1)}{Q'(-1)} e^{-t}$$

$$= \frac{16}{4} e^{3t} + \frac{4}{-4} e^{-t} = 4e^{3t} - e^{-t}.$$

10.5 SERIES METHOD TO DETERMINE INVERSE LAPLACE TRANSFORM

Let $L\{f(t)\} = F(s)$. In certain situation, we observed that Laplace transform of $f(t)$ can be obtained by expressing $f(t)$ in terms of power series and then by taking transform term-by-term. The same technique proves useful in finding inverse Laplace transforms. Thus, if

$$F(s) = \frac{a_0}{s} + \frac{a_1}{s^2} + \frac{a_2}{s^3} + \ldots,$$

Then

$$f(t) = a_0 + a_1 t + a_2 \frac{t^2}{2!} + \ldots,$$

where the series on the right-hand side may be summable to a known function.

EXAMPLE 10.46

Find $L^{-1}\left\{\frac{e^{-1/s}}{s}\right\}$.

Solution. Expansion in series yields

$$\frac{1}{s} e^{-1/s} = \frac{1}{s}\left[1 - \frac{1}{s} + \frac{1}{2!s^2} - \frac{1}{3!s^3} + \frac{1}{4!s^4} - \ldots\right]$$

$$= \frac{1}{s} - \frac{1}{s^2} + \frac{1}{2!s^3} - \frac{1}{3!s^4} + \ldots$$

$$= \sum_{n=0}^{\infty} \frac{(-1)^n}{n!\, s^{n+1}}.$$

Inversion of the series term-by-term yields

$$L^{-1}\{F(s)\} = \sum_{n=0}^{\infty} \frac{(-1)^n}{n!} L^{-1}\left\{\frac{1}{s^{n+1}}\right\}$$

$$= \sum_{n=0}^{\infty} \frac{(-1)^n t^n}{(n!)^2} = J_0(2\sqrt{t}),$$

where $J_0(t)$ is the Bessel's function of order zero.

EXAMPLE 10.47

Find $L^{-1}\left\{\frac{1}{s} \sin \frac{1}{s}\right\}$.

Solution. We have

$$\frac{1}{s} \sin \frac{1}{s} = \frac{1}{s}\left[\frac{1}{s} - \frac{(1/s)^3}{3!} + \frac{(1/s)^5}{5!} - \frac{(1/s)^7}{7!} + \ldots\right]$$

$$= \frac{1}{s^2} - \frac{1}{s^4 .3!} + \frac{1}{s^6 .5!} - \frac{1}{s^8\, 7!} + \ldots$$

Therefore,

$$L^{-1}\left\{\frac{1}{s} \sin \frac{1}{s}\right\} = L^{-1}\left\{\frac{1}{s^2}\right\} - \frac{1}{3!} L^{-1}\left\{\frac{1}{s^4}\right\} + \frac{1}{5!} L^{-1}\left\{\frac{1}{s^6}\right\} - \frac{1}{7!} L^{-1}\left\{\frac{1}{s^8}\right\} + \ldots$$

$$= t - \frac{1}{(3!)^2} t^3 + \frac{1}{(5!)^2} t^5 - \frac{1}{(7!)^2} t^7 + \ldots$$

EXAMPLE 10.48

Find $L^{-1}\left\{\tan^{-1}\frac{1}{s}\right\}$.

Solution. We have

$$\tan^{-1}\left(\frac{1}{s}\right)=\frac{1/s}{1}-\frac{(1/s)^3}{3}+\frac{(1/s)^5}{5}-\frac{(1/s)^7}{7}+\ldots$$

$$=\frac{1}{s}-\frac{1}{3s^3}+\frac{1}{5s^5}-\frac{1}{7s^7}+\ldots$$

Therefore, inversion of the series term-by-term yields

$$L^{-1}\left\{\tan^{-1}\left(\frac{1}{s}\right)\right\}=1-\frac{1}{3}\cdot\frac{t^2}{2!}+\frac{1}{5}\cdot\frac{t^4}{4!}-\frac{1}{7}\cdot\frac{t^6}{6!}+\ldots$$

$$=1-\frac{t^2}{3!}+\frac{t^4}{5!}-\frac{t^6}{7!}+\ldots$$

$$=\frac{1}{t}\left(t-\frac{t^3}{3!}+\frac{t^5}{5!}-\frac{t^7}{7!}+\ldots\right)=\frac{\sin t}{t}.$$

EXAMPLE 10.49

Find $L^{-1}\left\{\frac{1}{\sqrt{s+a}}\right\}$

Solution. Let

$$F(s)=\frac{1}{\sqrt{s+a}}=\frac{1}{\sqrt{s}}\left(1+\frac{a}{s}\right)^{-1/2}$$

$$=\frac{1}{\sqrt{s}}\left[1-\frac{1}{2}\left(\frac{a}{s}\right)+\frac{\frac{1}{2}\cdot\frac{3}{2}}{2!}\left(\frac{a}{s}\right)^2-\frac{\frac{1}{2}\cdot\frac{3}{2}\cdot\frac{5}{2}}{3!}\left(\frac{a}{s}\right)^3+\ldots\right]$$

$$=\sum_{n=0}^{\infty}\frac{(-1)^n\,1.3.5\ldots(2n-1)a^n}{2^n\,n!\,s^{n+(1\,2)}}\quad|s|>|a|.$$

Hence

$$L^{-1}\{F(s)\}=\sum_{n=0}^{\infty}\frac{(-1)^n\,1.3.5\ldots(2n-1)a^n\,t^{n-(1/2)}}{2^n\,n!\,\Gamma(n+1/2)}$$

$$=\frac{1}{\sqrt{t}}\sum_{n=0}^{\infty}\frac{(-1)^n\,1.3.5\ldots(2n-1)a^n\,t^n}{2^n\,n!\,\Gamma(n+1/2)}.$$

But, using $\Gamma(v+1)=v\Gamma(v)$, we have

$$\Gamma\left(n+\frac{1}{2}\right)=\Gamma\left(\frac{1}{2}\right)\left(\frac{1.3.5\ldots(2n-1)}{2^n}\right)$$

$$=\sqrt{\pi}\left(\frac{1.3.5\ldots(2n-1)}{2^n}\right).$$

Hence

$$L^{-1}\{F(s)\}=\frac{1}{\sqrt{t}}\sum_{n=0}^{\infty}\frac{(-1)^n\,a^n\,t^n}{\sqrt{\pi}\,n!}=\frac{1}{\sqrt{\pi t}}e^{-at}.$$

10.6 CONVOLUTION THEOREM

The convolution of two functions plays an important role in a number of physical applications. It is generally expedient to resolve a Laplace transform into the product of two transforms when the inverse transform of both transforms are known.

Definition 10.2. The *convolution* (*or Faltung*) of two given functions $f(t)$ and $g(t)$, $t > 0$ is defined by the integral

$$(f*g)(t)=\int_0^t f(u)\,g(t-u)\,du,$$

which, of course, exists if f and g are piecewise continuous.

The integral $\int_0^t f(u)\,g(t-u)\,du$ represents a *superposition* of effects of magnitude $f(u)$ occurring at time $t = u$ for which $g(t-u)$ is the *influence function or response* of a system to a unit impulse defined at time $t = u$.

If we substitute $v = t - u$, then

$$(f*g)(t)=\int_0^t g(v)f(t-v)\,dv=(g*f)(t).$$

Hence, the convolution is *commutative.* Also, it follows from the definition that

$$a(f*g)=af*g=f*ag,\ \text{a constant}$$

$$f*(g+h)=(f*g)+(f*h)$$

(distributive property).

Further, we note that

$$[f*(g*h)](t)$$

$$=\int_0^t f(u)(g*h)(t-u)du$$

$$= \int_0^t f(u) \left(\int_0^{t-u} g(v)\, h(t-u-v) dv \right) du$$

$$= \int_0^t f(u) \int_u^t g(z-u) h(t-z) dz\, du,\ v = z - u$$

$$= \int_0^t \left(\int_0^z f(u)\, g(z-u)\, du \right) h(t-z)\, dz$$

$= [(f * g) * h](t).$

Hence

$$f * (g * h) = (f * g) * h,$$

and so *convolution is associative.*

EXAMPLE 10.50

Evaluate $\cos t * \sin t$.

Solution. By definition,

$$\cos t * \sin t = \int_0^t \cos u \sin (t-u)\, du$$

$$= \int_0^t \frac{1}{2} [\sin t + \sin (t - 2u)]\, du$$

$$= \frac{1}{2} \sin t\, [u]_0^t + \frac{1}{4} [\cos (t - 2u)]_0^t$$

$$= \frac{1}{2} t \sin t + \frac{1}{4} [\cos(-t) - \cos t] = \frac{1}{2} t \sin t.$$

EXAMPLE 10.51

Evaluate $e^t * t$.

Solution. By definition of convolution,

$$e^t * t = \int_0^t e^u (t-u)\, du$$

$$= [t\, e^u]_0^t - [(ue^u - e^u)]_0^t = e^t - t - 1.$$

Convolution is very significant in the sense that the Laplace transform of the convolution of two functions is the product of their respective Laplace transforms. We prove this assertion in the form of the following theorem.

Theorem 10.11. (Convolution Theorem). If f and g are piecewise continuous on $[0, \infty)$ and of exponential order γ, then

$$L\{(f * g)(t)\} = L\{f(t)\} . L\{g(t)\} = F(s) G(s),$$

or equivalently,

$$L^{-1}\{F(s)\, G(s)\} = (f * g)(t).$$

Proof: Using the definition of the Laplace transform, we have

$$L\{(f * g)(t)\} = \int_0^\infty e^{-st} \int_0^t f(u)\, g(t-u)\, du\, dt$$

$$= \int_0^\infty \int_0^t e^{-st} f(u)\, g(t-u)\, du\, dt.$$

The domain of this double integral is an infinite wedge bounded by $t = 0$, $t = \infty$, $u = 0$, and $u = t$ as displayed in Figure 10.1. Due to hypotheses on f and g, the double integral on the right hand side converges absolutely and so we can perform change of order of integration.

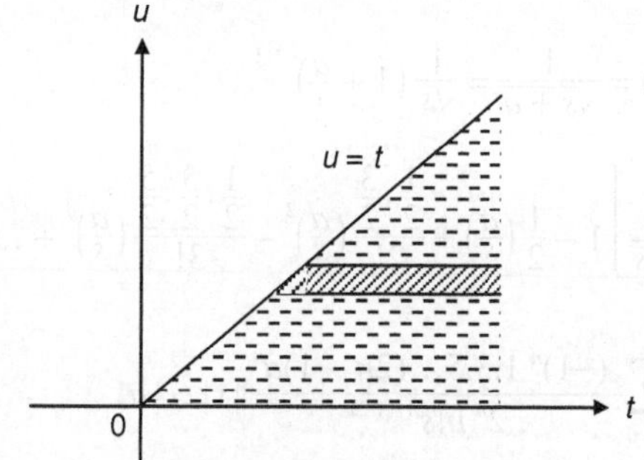

Figure 10.1

We take a strip parallel to the axis of t. Then t varies from u to ∞ and u varies from 0 to ∞. Hence

$$L\{f(t) * g(t)\} = \int_0^\infty \int_u^\infty e^{-st} f(u)\, g(t-u)\, dt\, du$$

$$= \int_0^\infty \int_u^\infty e^{-su} . e^{-s(t-u)} f(u)\, g(t-u)\, dt\, du$$

$$= \int_0^\infty e^{-su} f(u) \int_u^\infty e^{-s(t-u)} g(t-u)\, dt\, du.$$

Implementing the change of variable $t-u=v$ in the inner integral, we get

$$L\{f(t)*g(t)\} = \int_0^\infty e^{-su} f(u) \int_0^\infty e^{-sv} g(v)\,dv\,du$$

$$= \int_0^\infty e^{-su} f(u)\, G(s)\,du$$

$$= G(s)\int_0^\infty e^{-su} f(u)\,du$$

$$= G(s)F(s) = F(s)G(s).$$

Consequently,

$$L^{-1}\{F(s)G(s)\} = f(t)*g(t).$$

EXAMPLE 10.52

Find $L\{e^{at} * e^{bt}\}$.

Solution. By Convolution Theorem,

$$L\{e^{at} * e^{bt}\} = L\{e^{at}\}.\, L\{e^{bt}\} = \frac{1}{(s-a)(s-b)}.$$

EXAMPLE 10.53

Find $L^{-1}\left\{\dfrac{1}{(s-a)(s-b)}\right\}$.

Solution. By Convolution theorem

$$L^{-1}\left\{\frac{1}{(s-a)(s-b)}\right\} = e^{at} * e^{bt}$$

$$= \int_0^t e^{at} e^{b(t-u)}\,du = \frac{e^{at}-e^{bt}}{a-b}, \quad a \neq b.$$

EXAMPLE 10.54

Using Convolution Theorem, find $L^{-1}\left\{\dfrac{1}{s^3(s^2+1)}\right\}$.

Solution. Let $F(s) = \frac{1}{s^3}$, $G(s) = \frac{1}{s^2+1}$ and so $f(t) = \frac{1}{2}t^2$ and $g(t) = \sin t$.

By Convolution Theorem and integration by parts, we have

$$L^{-1}\{F(s)G(s)\}$$

$$= f*g = g*f = \frac{1}{2}\int_0^t \sin u\,(t-u)^2\,du$$

$$= \frac{1}{2}\left\{[-(t-u)^2 \cos u]_0^t - \int_0^t 2(t-u)\cos u\,du\right\}$$

$$= \frac{1}{2}\left[t^2 - 2\left\{[(t-u)\sin u]_0^t + \int_0^t \sin u\,du\right\}\right]$$

$$= \frac{1}{2}[t^2 + 2\cos t - 2].$$

EXAMPLE 10.55

Using Convolution theorem, find $L^{-1}\left\{\dfrac{1}{s^2(s-1)}\right\}$.

Solution. Let $F(s) = \frac{1}{s^2}$ and $G(s) = \frac{1}{s-1}$. Then, $f(t) = t$ and $g(t) = e^t$. Therefore, by Convolution Theorem,

$$L^{-1}\{F(s)G(s)\} = f*g = t*e^t$$

$$= \int_0^t e^u (t-u)\,du = e^t - t - 1.$$

EXAMPLE 10.56

Use Convolution Theorem to find $L^{-1}\left\{\dfrac{s}{(s^2+a^2)^2}\right\}$. (see also Example 10.32).

Solution. Let

$$\frac{s}{(s^2+a^2)^2} = \frac{s}{s^2+a^2}\cdot\frac{1}{s^2+a^2} = F(s)\,G(s).$$

Then

$$f(t) = L^{-1}\{F(s)\} = \cos at \text{ and}$$

$$g(t) = L^{-1}\{G(s)\} = \frac{\sin at}{a}.$$

By Convolution Theorem,

$$L^{-1}\{F(s)G(s)\} = f*g = \int_0^t \cos au\,\frac{\sin a(t-u)}{a}\,du$$

$$= \frac{1}{2a}\int_0^t [\sin at + \sin a(t-2u)]\,du$$

$$= \frac{1}{2a}\sin at\,[u]_0^t + \frac{1}{4a^2}[\cos a(t-2u)]_0^t$$

$$= \frac{1}{2a} t \sin at + \frac{1}{4a^2} [\cos a(-t) - \cos at]$$

$$= \frac{1}{2a} t \sin at.$$

EXAMPLE 10.57

Using Convolution Theorem, find $L^{-1}\left\{\frac{1}{s^2(s^2+a^2)}\right\}$.

Solution. Let

$$\frac{1}{s^2(s^2+a^2)} = \frac{1}{s^2} \cdot \frac{1}{(s^2+a^2)} = F(s)\,G(s).$$

Then

$$f(t) = L^{-1}\{F(s)\} = t \text{ and}$$

$$g(t) = L^{-1}\{G(s)\} = \frac{\sin at}{a}.$$

By Convolution Theorem,

$L^{-1}\{F(s)G(s)\}$

$$= \frac{1}{a}\sin at * t = \frac{1}{a}\int_0^t (t-u)\sin au\, du$$

$$= \frac{1}{a}\left\{-\left[\frac{1}{a}(t-u)\cos au\right]_0^t - \int_0^t \frac{\cos au}{a}\,du\right\}$$

$$= \frac{1}{a}\left\{\frac{t}{a} - \left[\frac{\sin au}{a^2}\right]_0^t\right\} = \frac{1}{a}\left[\frac{t}{a} - \frac{\sin at}{a^2}\right]$$

$$= \frac{1}{a^2}\left[t - \frac{1}{a}\sin at\right].$$

EXAMPLE 10.58

Using Convolution Theorem, find $L^{-1}\left\{\frac{1}{s^2(s+1)^2}\right\}$.

Solution. Let

$$\frac{1}{s^2(s+1)^2} = \frac{1}{s^2} \cdot \frac{1}{(s+1)^2} = F(s)G(s).$$

Then

$$f(t) = L^{-1}\{F(s)\} = t \text{ and}$$

$$g(t) = L^{-1}\{G(s)\} = te^{-t}$$

So, by Convolution Theorem, we have

$$L^{-1}\{F(s)G(s)\} = g * f$$

$$= \int_0^t (ue^{-u})(t-u)\,du = \int_0^t (tu-u^2)e^{-u}\,du$$

$$= -[(tu-u^2)e^{-u}]_0^t - \int_0^t -(t-2u)e^{-u}\,du$$

$$= \int_0^t (t-2u)\,e^{-u}\,du = -[(t-2u)e^{-u}]_0^t - 2\int_0^t e^{-u}\,du$$

$$= te^{-t} + t - 2\left[\frac{e^{-u}}{-1}\right]_0^t = te^{-t} + t + 2e^{-t} - 2$$

$$= (t+2)e^{-t} + t - 2.$$

EXAMPLE 10.59

Using Convolution Theorem, find $L^{-1}\left\{\frac{1}{(s^2+a^2)^2}\right\}$.

Solution. Let

$$\frac{1}{(s^2+a^2)^2} = \frac{1}{(s^2+a^2)} \cdot \frac{1}{(s^2+a^2)} = F(s)\,G(s).$$

Then

$$f(t) = g(t) = \frac{\sin at}{a}.$$

Therefore, by Convolution Theorem,

$L^{-1}\{F(s)G(s)\}$

$$= f * g = \frac{1}{a^2}\int_0^t \sin au \sin a(t-u)\,du$$

$$= \frac{1}{a^2}\int_0^t \frac{1}{2}[\cos a(2u-t) - \cos at]\,du$$

$$= \frac{1}{2a^2}\left[\int_0^t \cos a(2u-t)\,du - \int_0^t \cos at\,du\right]$$

$$= \frac{1}{2a^2}\left\{\left[\frac{\sin a(2u-t)}{2a}\right]_0^t - \cos at[u]_0^t\right\}$$

$$= \frac{1}{2a^2}\left[\frac{\sin at}{2a} - \frac{\sin a(-t)}{2a} - t\cos at\right]$$

$$= \frac{1}{2a^3}[\sin at - at\cos at].$$

EXAMPLE 10.60

Use Convolution Theorem to find $L^{-1}\left\{\frac{a^2}{s(s+a)^2}\right\}$.

Solution. Let

$$\frac{a^2}{s(s+a)^2} = \frac{1}{s}\cdot\frac{a^2}{(s+a)^2} = F(s)G(s).$$

Then

$$f(t) = L^{-1}\left\{\frac{1}{s}\right\} = 1.$$

To find $g(t) = L^{-1}\left\{\frac{a^2}{s(s+a)^2}\right\}$, we note that

$$\frac{d}{ds}\left(\frac{1}{s+a}\right) = \frac{-1}{(s+a)^2}.$$

Therefore,

$$L^{-1}\left\{\frac{-1}{(s+a)^2}\right\} = -t\,L^{-1}\left\{\frac{1}{s+a}\right\} = -t\,e^{-at}$$

and so

$$L^{-1}\left\{\frac{a^2}{(s+a)^2}\right\} = a^2\,t\,e^{-at}.$$

Application of Convolution Theorem yields

$$L^{-1}\{F(s)G(s)\} = \int_0^t a^2\,ue^{-au}\,du = a^2\int_0^t ue^{-au}\,du$$

$$= a^2\left\{\left[\frac{ue^{-au}}{-a}\right]_0^t - \int_0^t \frac{e^{-au}}{-a}du\right\}$$

$$= a^2\left\{\frac{te^{-at}}{-a} - \left[\frac{e^{-au}}{a^2}\right]_0^t\right\}$$

$$= -\,at\,e^{-at} - e^{-at} + 1$$

$$= 1 - e^{-at}(at+1).$$

EXAMPLE 10.61

Find $L^{-1}\left\{\frac{1}{\sqrt{s}(s-1)}\right\}$ using Convolution Theorem.

Solution. Let

$$\frac{1}{\sqrt{s}(s-1)} = \frac{1}{\sqrt{s}}\cdot\frac{1}{s-1} = F(s)G(s).$$

Then

$$f(t) = L^{-1}\{F(s)\} = \frac{1}{\sqrt{\pi t}} \text{ and}$$

$$g(t) = L^{-1}\{G(s)\} = e^t.$$

Hence Convolution Theorem yields

$$L^{-1}\left\{\frac{1}{\sqrt{s}(s+1)}\right\} = L^{-1}\{F(s)G(s)\} = f * g$$

$$= \int_0^t \frac{1}{\sqrt{\pi u}}\,e^{(t-u)}\,du = \frac{e^t}{\sqrt{\pi}}\int_0^t \frac{e^{-u}}{\sqrt{u}}\,du$$

$$= \frac{2e^t}{\sqrt{\pi}}\int_0^{\sqrt{t}} e^{-v^2}\,dv,\ u = v^2$$

$$= e^t\,\frac{2}{\sqrt{\pi}}\int_0^{\sqrt{t}} e^{-v^2}\,dv = e^t\,\text{erf}\,\sqrt{t},$$

where erf $\sqrt{t}$ is the *error function.*

EXAMPLE 10.62

Using Convolution Theorem, establish Euler's formula for Beta function:

$$\beta(a,b) = \frac{\Gamma(a)\,\Gamma(b)}{\Gamma(a+b)}.$$

Solution. We know that Beta function is defined by

$$\beta(a,b) = \int_0^1 x^{a-1}(1-x)^{b-1}\,dx.$$

Let

$$f(t) = t^{a-1},\ g(t) = t^{b-1},\ a, b > 0.$$

Then

$$(f * g)(t) = \int_0^t u^{a-1}(t-u)^{b-1}\,du.$$

Putting $u = vt$, we get

$$(f * g)(t) = t^{a+b-1}\int_0^1 v^{a-1}(1-v)^{b-1}\,dv$$

$$= t^{a+b-1}\,\beta(a, b).$$

But by Convolution Theorem,

$$L\{f * g\} = L\{f(t)\}L\{g(t)\}.$$

Therefore,

$$L\{t^{a+b-1}\beta(a, b)\} = L\{t^{a-1}\}\,L\{t^{b-1}\}$$

$$= \frac{\Gamma(a)}{s^a}\cdot\frac{\Gamma(b)}{s^b} = \frac{\Gamma(a)\,\Gamma(b)}{s^{a+b}}$$

and so

$$t^{a+b-1}\beta(a, b) = L^{-1}\left\{\frac{\Gamma(a)\,\Gamma(b)}{s^{a+b}}\right\}$$

$$= \Gamma(a)\,\Gamma(b)L^{-1}\left\{\frac{1}{s^{a+b}}\right\}$$

$$= \Gamma(a)\,\Gamma(b)\,\frac{t^{a+b-1}}{\Gamma(a+b)},$$

which implies

$$\beta(a, b) = \frac{\Gamma(a)\,\Gamma(b)}{\Gamma(a+b)},$$

the required Euler's Formula for Beta function.

Remark 10.4. It also follows from Example 10.62 that

$$L\left\{\int_0^t u^{a-1}(t-u)^{b-1}\,du\right\} = \frac{\Gamma(a)\,\Gamma(b)}{s^{a+b}}.$$

EXAMPLE 10.63

Using Convolution Theorem, show that

$$J_0(t) = \frac{2}{\pi}\int_0^1 \frac{\cos tx}{\sqrt{1-x^2}}\,dx,$$

where J_0 represents Bessel's function of order zero.

Solution. We know that

$$L\{J_0(t)\} = \frac{1}{\sqrt{s^2+1}} = \frac{1}{\sqrt{s+i}}\cdot\frac{1}{\sqrt{s-i}}$$

and so

$$J_0(t) = L^{-1}\left\{\frac{1}{\sqrt{s+i}}\cdot\frac{1}{\sqrt{s-i}}\right\} = L^{-1}\{F(s)G(s)\},$$

where

$$F(s) = \frac{1}{\sqrt{s+i}},\ G(s) = \frac{1}{\sqrt{s-i}}.$$

Since

$$L^{-1}\left\{\frac{1}{\sqrt{s+a}}\right\} = \frac{1}{\sqrt{\pi t}}\,e^{-at},$$

we have

$$f(t) = \frac{1}{\sqrt{\pi t}}\,e^{-it} \text{ and } g(t) = \frac{1}{\sqrt{\pi t}}\,e^{it}.$$

Hence, by Convolution Theorem

$$J_0(t) = L^{-1}\{F(s)G(s)\} = f * g$$

$$= \int_0^t \frac{1}{\sqrt{\pi u}}\,e^{-iu}\left(\frac{1}{\sqrt{\pi}(t-u)^{1/2}}\,e^{i(t-u)}\right)du$$

$$= \frac{1}{\pi}\int_0^t \frac{e^{i(t-2u)}}{\sqrt{u(t-u)}}\,du.$$

Putting $u = tv$, we get

$$J_0(t) = \frac{1}{\pi}\int_0^1 \frac{e^{it}(1-2v)}{\sqrt{v(1-v)}}\,dv.$$

Now putting $x = 1 - 2v$, we have

$$J_0(t) = \frac{1}{\pi}\int_{-1}^1 \frac{e^{itx}}{\sqrt{1-x^2}}\,dx$$

$$= \frac{1}{\pi}\int_{-1}^1 \frac{\cos tx}{\sqrt{1-x^2}}\,dx + \frac{i}{\pi}\int_{-1}^1 \frac{\sin tx}{\sqrt{1-x^2}}\,dx$$

$$= \frac{2}{\pi}\int_0^1 \frac{\cos tx}{\sqrt{1-x^2}}\,dx,$$

The second integral vanishes because $\sin x$ is odd.

10.7 COMPLEX INVERSION FORMULA

The complex inversion formula is a technique for computing directly the inverse of a Laplace transform. For this technique, we require the parameter s to be complex variable. The contour integration is the main tool applied in this technique.

Let f be a continuous function possessing a Laplace transform and defined in $(-\infty, \infty)$ such that $f(t) = 0$ for $t < 0$. Let $s = x + iy$ be a complex variable. Then

$$\begin{aligned} L\{f(t)\} = F(s) &= \int_0^{\infty} e^{-st} f(t)\, dt \\ &= \int_{-\infty}^{\infty} e^{-(x+iy)t} f(t)\, dt \\ &= \int_{-\infty}^{\infty} e^{-iyt} (e^{-xt} f(t))\, dt. \end{aligned}$$

The integral on the right is the Fourier transform of the function $e^{-xt} f(t)$. Thus the *Laplace transform of $f(t)$ is the Fourier transform of the function $e^{-xt} f(t)$.*

Suppose that f and its derivative f' are piecewise continuous on $(-\infty,\infty)$, that is, both f and f' are continuous in any finite interval except possibly for a finite number of jump discontinuities. Suppose, further, that f is absolutely integrable, that is, $\int_{-\infty}^{\infty} |f(t)| dt < \infty$. Then Fourier integral theorem asserts that

$$\frac{f(t+)+f(t-)}{2} = \frac{1}{2\pi} \int_{-\infty}^{\infty} \int_{-\infty}^{\infty} f(u)\, e^{is(t-u)}\, du\, ds,$$

where the integration with respect to s is in the Cauchy principal value sense.

If f were continuous on $(-\infty,\infty)$, then $\frac{f(t+)+f(t-)}{2}$ converts to $f(t)$ in the above expansion. Further, in terms of Fourier transform of $f(t)$, the above expression becomes

$$\frac{f(t+)+f(t-)}{2} = \frac{1}{2\pi} \int_{-\infty}^{\infty} F(s) e^{-ist}\, ds,$$

where $F(s) = \int_{-\infty}^{\infty} e^{-isu} f(u) du$ is Fourier transform of $f(t)$.

Theorem 10.12. (Complex Inversion Formula). Let f be continuous on $[0, \infty]$, $f(t) = 0$ for $t < 0$, f be of exponential order γ and f' be piecewise continuous on $[0,\infty)$. If $L\{f(t)\} = F(s)$ and the real number γ is so chosen that all the singularities of $F(s)$ lie in the left of the line Re $(s) > \gamma$, then

$$f(t) = L^{-1}\{F(s)\} = \frac{1}{2\pi i} \int_{\gamma-i\infty}^{\gamma+i\infty} e^{st} F(s)\, ds, \; t > 0.$$

Proof: By definition of Laplace transform,

$$F(s) = \int_0^{\infty} e^{-su} f(u)\, du.$$

We have then

$$\begin{aligned} &\lim_{T\to\infty} \frac{1}{2\pi i} \int_{\gamma-iT}^{\gamma+iT} e^{st} F(s) ds \\ &= \lim_{T\to\infty} \frac{1}{2\pi i} \int_{\gamma-iT}^{\gamma+iT} \int_0^{\infty} e^{st} e^{-su} f(u)\, du\, ds. \end{aligned}$$

Let $s = \gamma + iy$ so that $ds = idy$. Therefore,

$$\begin{aligned} &\lim_{T\to\infty} \frac{1}{2\pi i} \int_{\gamma-iT}^{\gamma+iT} e^{st} F(s) ds \\ &= \lim_{T\to\infty} \frac{1}{2\pi} e^{\gamma t} \int_{-T}^{T} e^{iyt}\, dy \int_0^{\infty} e^{-iyu}\, e^{-\gamma u} f(u)\, du \\ &= \frac{1}{2\pi} e^{\gamma t} \int_{-\infty}^{\infty} \int_0^{\infty} e^{iy(t-u)} (e^{-\gamma u} f(u))\, du\, dy. \end{aligned}$$

The given hypothesis and Theorem 10.1 imply that $L\{f(t)\}$ converges absolutely for $\text{Re}(s) > \gamma$, that is,

$$\int_0^{\infty} |e^{-st} f(t)| dt = \int_{-\infty}^{\infty} e^{-xt} |f(t)| dt < \infty, \; x > \gamma.$$

Thus $e^{-xt} f(t)$ is absolutely integrable and so Fourier integral theorem asserts that

$$\begin{aligned} &\frac{1}{2\pi} \int_{-\infty}^{\infty} \int_{-\infty}^{\infty} (e^{-\gamma u} f(u))\, e^{iy(t-u)} du\, dy \\ &= e^{-\gamma t} f(t) \text{ for } t > 0. \end{aligned}$$

Hence

$$\lim_{T\to\infty}\frac{1}{2\pi i}\int_{\gamma-iT}^{\gamma+iT} e^{st}F(s)\,ds = \frac{1}{2\pi}\cdot 2\pi e^{-\gamma t}f(t)\,e^{\gamma t}$$

$$= f(t),\ t>0,$$

and so

$$f(t)=\frac{1}{2\pi i}\int_{\gamma-i\infty}^{\gamma+i\infty} e^{st}F(s)\,ds,\ t>0.$$

Remark 10.5. The expression

$$f(t)=\frac{1}{2\pi i}\int_{\gamma-i\infty}^{\gamma+i\infty} e^{st}F(s)\,ds$$

is called *complex inversion formula, Bromwich integral formula, Fourier – Mell in inversion formula, or fundamental theorem of Laplace transform.* In practice, the integral in the above expression is evaluated by considering the contour integration along the contour Γ_R, ABCDEA, known as *Bromwich contour* shown in the Figure 10.2.

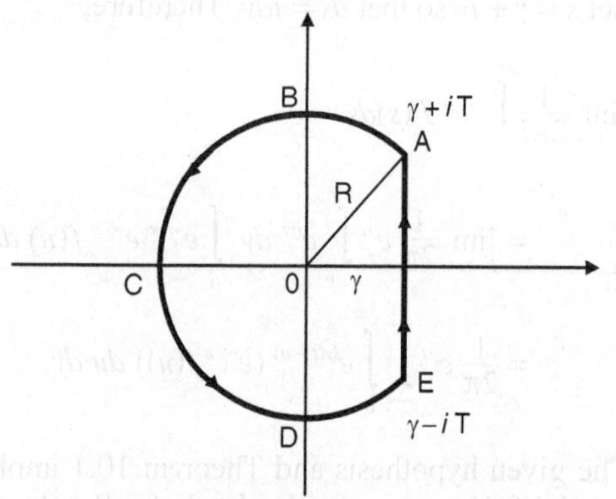

Figure 10.2 Bromwich Contour

The vertical line at γ is known as *Bromwich line.* Thus the contour Γ_R consists of arc C_R (ABCDE) of radius R and Centre at the origin and the Bromwich line EA. Thus,

$$\frac{1}{2\pi i}\int_{\Gamma_R} e^{st}F(s)\,ds = \frac{1}{2\pi i}\int_{C_R} e^{st}F(s)\,ds$$

$$+\frac{1}{2\pi i}\int_{EA} e^{st}F(s)\,ds.$$

Since F(s) is analytic for Re(s) = $x > \gamma$, all singularities of F(s) must lie to the left of the Bromwich line. Thus, by Cauchy residue theorem, we have

$$\frac{1}{2\pi i}\int_{\Gamma R} e^{st}F(s)\,ds = \sum_{k=1}^{n}\text{Res}\,(z_k),$$

where Res(z_k) is the residue of the function at the poles $s = z_k$. Since $e^{st} \neq 0$, multiplying F(s) by e^{st} does not affect the status of the poles z_k of F(s).

If we can show that

$$\lim_{R\to\infty}\int_{C_R} e^{st}F(s)\,ds = 0,$$

Then letting $R\to\infty$, we get

$$f(t)=\lim_{T\to\infty}\frac{1}{2\pi i}\int_{\gamma-iT}^{\gamma+iT} e^{ts}F(s)\,ds = \sum_{k=1}^{n}\text{Res}(z_k),$$

and so inverse function f can be determined.

The following theorem shows that *Laplace transform is one–to–one.*

Theorem 10.13. Let $f(t)$ and $g(t)$ be two piecewise smooth functions of exponential order and let $F(s)$ and $G(s)$ be the Laplace transforms of $f(t)$ and $g(t)$ respectively. If $F(s) = G(s)$ in a half-place $Re(s) > \gamma$, then $f(t) = g(t)$ at all points where f and g are continuous.

Proof: Suppose f and g are continuous at $t \in$ R. By complex inversion formula, we have

$$f(t)=\frac{1}{2\pi i}\int_{\gamma-i\infty}^{\gamma+i\infty} e^{st}F(s)\,ds,$$

$$g(t)=\frac{1}{2\pi i}\int_{\gamma-i\infty}^{\gamma+i\infty} e^{st}G(s)\,ds.$$

Since F(s) = G(s), it follows that $f(t) = g(t)$. Thus L{$f(t)$} = L{$g(t)$} implies that $f(t) = g(t)$ and so Laplace operator is one–to–one.

Generally, we see that most of the Laplace transforms satisfy the growth restriction

$$|F(s)| \le \frac{M}{|s|^p}$$

for all sufficiently large values of $|s|$ and some $p > 0$. Obviously, $F(s) \to 0$ as $|s| \to \infty$. Therefore, the following result (stated without proof) is helpful.

Theorem 10.14. Let for s on C_R, $F(s)$ satisfies the growth restriction

$$|F(s)| \le \frac{M}{|s|^p} \text{ for } p > 0, \text{ all } R > R_0.$$

Then

$$\lim_{R\to\infty} \int_{C_R} e^{st} F(s)\, ds = 0, t > 0.$$

EXAMPLE 10.64

Show that $F(s) = \frac{s}{s^2 - a^2}$ satisfies growth restriction condition.

Solution. We have

$$F(s) = \frac{s}{s^2 - a^2},$$

and so

$$|F(s)| \le \frac{|s|}{|s^2 - a^2|} \le \frac{|s|}{|s|^2 - |a|^2}.$$

If $|s| \ge 2|a|$, then $|a|^2 \le \frac{|s|^2}{4}$ and so $|s|^2 - |a|^2 \ge \frac{3}{4}|s|^2$ and we have

$$|F(s)| \le \frac{4/3}{|s|}.$$

EXAMPLE 10.65

Find the Laplace transform of $f(t) = \cosh at$ and verify the inversion formula.

Solution. By Example 10.6, we have $F(s) = \frac{s}{s^2 - a^2}$. The function $F(s)$ is analytic except at poles $s = a$ and $s = -a$. the Bromwich contour is shown in the Figure 10.3.

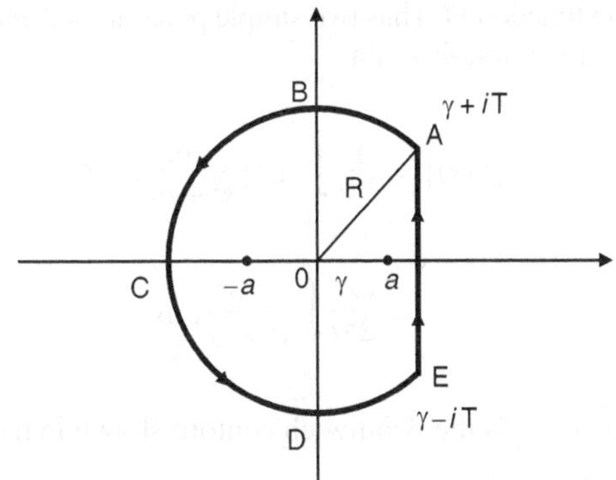

Figure 10.3

By inversion formula, we have

$$L^{-1}\left\{\frac{s}{s^2 + a^2}\right\} = \frac{1}{2\pi i}\int_{\gamma - i\infty}^{\gamma + i\infty} e^{st}\left(\frac{s}{s^2 - a^2}\right) ds$$

$$= \frac{1}{2\pi i}\int_{\Gamma_R} \frac{se^{st}}{s^2 - a^2}\, ds.$$

Further, $F(s)$ satisfies growth restriction condition. Therefore, integral over contour C_R (arc ABCDE) tends to zero as $R\to\infty$. Now

$$\text{Res}(a) = \lim_{s\to a}(s - a)e^{st} F(s) = \lim_{s\to a} \frac{se^{st}}{s+a} = \frac{ae^{at}}{2a},$$

$$\text{Res}(-a) = \lim_{s\to(-a)}(s + a)\, e^{st} F(s) = \lim_{s\to(-a)} \frac{se^{st}}{s - a}$$

$$= \frac{e^{-at}}{-2a}(-a).$$

Hence

$$L^{-1}\{F(s)\} = \frac{e^{at} + e^{-at}}{2} = \cosh at,$$

and so inversion formula is verified.

EXAMPLE 10.66

Find $L^{-1}\left\{\frac{\omega}{s^2 + \omega^2}\right\}$, $s > 0$ using inversion formula.

Solution. We have

$$F(s) = \frac{\omega}{s^2 + \omega^2}, s > 0$$

The function F(s) has two simple poles at $s = \pm i\omega$. By inversion formula

$$L^{-1}\{F(s)\} = \frac{1}{2\pi i}\int_{\gamma - i\infty}^{\gamma + i\infty} e^{st}\left(\frac{\omega}{s^2+\omega^2}\right) ds$$

$$= \frac{\omega}{2\pi i}\int_{\Gamma_R} \frac{e^{st}}{s^2+\omega^2}\, ds,$$

Where Γ_R is the Bromwich contour shown in the Figure 10.4:

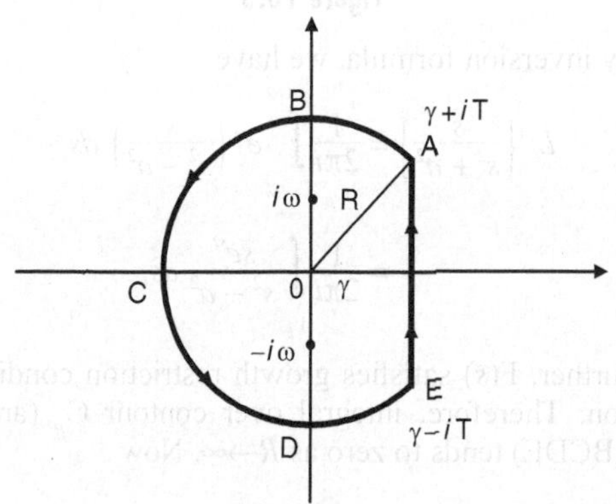

Figure 10.4

We have

$$\text{Res}(i\omega) = \lim_{s\to i\omega}(s - i\omega)e^{st}F(s) = \lim_{s\to i\omega}\frac{e^{st}}{s + i\omega}$$

$$= \frac{e^{i\omega t}}{2i\omega},$$

$$\text{Res}(-i\omega) = \lim_{s\to -i\omega}(s + i\omega)e^{st}F(s)$$

$$= \lim_{s\to -i\omega}\frac{e^{st}}{s - i\omega} = \frac{e^{-i\omega t}}{-2i\omega}.$$

Further, let $s = \gamma + R\,e^{i}$, $\frac{\pi}{2} \le \ \le \frac{3\pi}{2}$. Then the integral over the contour C_R yields

$$\left|\int_{\pi/2}^{3\pi/2}\frac{e^{\gamma t}e^{tR(\cos\ +i\sin\ )}R\,e^{i\theta}}{(\gamma + \text{Re}^{i}\ )^2+\omega^2}\right| \le \frac{\pi e^{\gamma t}R}{R^2-\gamma^2-\omega^2}$$

$$\to 0 \text{ as } R \to \infty.$$

Hence

$$L^{-1}\{F(s)\} = \frac{\omega}{2\pi i}2\pi i\,(\text{sum of residue at} \pm i\omega)$$

$$= \frac{e^{i\omega t} - e^{-i\omega t}}{2i} = \sin\omega t.$$

EXAMPLE 10.67

Find $L^{-1}\left\{\frac{1}{s(s^2+a^2)}\right\}$ using inversion formula.

Solution. We have

$$F(s) = \frac{1}{s(s^2+a^2)^2} = \frac{1}{s(s-ia)^2\,(s+ia)^2}.$$

Thus F(s) has a simple pole at $s = 0$ and a pole of order 2 at $s = \pm\, ia$. Further, F(s) satisfies growth restriction condition. Therefore, integral over the contour C_R goes to zero as $R\to\infty$. Further,

$$\text{Res}(0) = \lim_{s\to 0} se^{st}F(s) = \lim_{s\to 0}\frac{e^{st}}{(s^2+a^2)^2} = \frac{1}{a^4}$$

$$\text{Res}(ia) = \lim_{s\to ai}\frac{d}{ds}(s-ia)^2e^{ts}F(s)$$

$$= \lim_{s\to ai}\frac{d}{ds}\left(\frac{e^{ts}}{s(s+ia)^2}\right) = \frac{it}{4a^3}e^{iat} - \frac{e^{iat}}{2a^4},$$

$$\text{Res}(-ia) = \lim_{s\to -ia}\frac{d}{ds}(s+ia)^2e^{ts}F(s)$$

$$= \lim_{s\to -ia}\frac{d}{ds}\left(\frac{e^{ts}}{s(s-ia)^2}\right) = \frac{-it}{4a^3}e^{-iat} - \frac{e^{-iat}}{2a^4}.$$

Hence

$$f(t) = \frac{1}{a^4} + \frac{it}{4a^3}(e^{iat} - e^{-iat}) - \frac{1}{2a^4}(e^{iat}+e^{-iat})$$

$$= \frac{1}{a^4}\left(1 - \frac{a}{2}t\sin at - \cos at\right).$$

EXAMPLE 10.68

Find $L^{-1}\left\{\frac{s}{(s+1)^3\,(s-1)^2}\right\}$ using inversion formula.

Solution. We have

$$F(s) = \frac{s}{(s+1)^3(s-1)^2}.$$

The function F(s) has poles of multiplicity 2 at $s = 1$ and poles of multiplicity 3 at $s = -1$. Residue at $s = 1$ is given by

$$\text{Res}(1) = \lim_{s\to 1} \frac{d}{ds}\left[\frac{(s-1)^2 se^{st}}{(s+1)^3 (s-1)^2}\right]$$

$$= \lim_{s\to 1} \frac{d}{ds}\left(\frac{se^{st}}{(s+1)^3}\right)$$

$$= \frac{1}{16} e^t (2t-1),$$

$$\text{Res}(-1) = \lim_{s\to -1} \frac{1}{2!}\frac{d^2}{ds^2}\left[\frac{(s+1)^3 se^{st}}{(s+1)^3(s-1)^2}\right]$$

$$= \lim_{s\to -1} \frac{1}{2}\frac{d^2}{ds^2}\left(\frac{se^{st}}{(s-1)^2}\right)$$

$$= \frac{1}{16} e^{-t} (1-2t^2).$$

The value of the integral over the contour C_R tends to Zero as $R \to \infty$. Hence

$$f(t) = \text{sum of residues at the poles}$$

$$= \frac{1}{16} e^{-t} (1-2t^2) + \frac{1}{16} e^t (2t-1).$$

EXAMPLE 10.69

Derive Heaviside's expansion formula using complex inversion formula.

Solution. Let $F(s) = \frac{P(s)}{Q(s)}$, where P(s) and Q(s) are polynomials having no common factors (roots) and degree of Q(s) is greater than the degree of P(s). Suppose Q(s) has simple zeros at $z_1, z_2, \ldots, z_m$. If degree of P(s) and Q(s) are n and m, respectively, then for $a_0 \neq 0$, $b_m \neq 0$,

$$\frac{P(s)}{Q(s)} = \frac{a_n s^n + a_{n-1} s^{n-1} + \ldots + a_0}{b_m s^m + b_{m-1} s^{m-1} + \ldots + b_0}$$

$$= \frac{a_n + \frac{a_{n+1}}{s} + \ldots + \frac{a_0}{s^n}}{s^{m-n}\left(b_m + \frac{b_{m-1}}{s} + \ldots + \frac{b_0}{s^m}\right)}.$$

For sufficiently lage $|s|$, we have

$$|a_n + \frac{a_{n-1}}{s} + \ldots + \frac{a_0}{s^n}| \leq |a_n| + |a_{n-1}| + \ldots + |a_0|$$
$$= \mathbf{C_1} \textbf{ say,}$$

$$|b_m + \frac{a_{m-1}}{s} + \ldots + \frac{b_0}{s^n}| \geq |b_m| - \frac{|b_{m-1}|}{|s|} - \ldots - \frac{|b_0|}{|s|^m}$$
$$\geq \frac{|b_m|}{2} = \mathbf{C_2} \textbf{ say,}$$

and so

$$|F(s)| = \frac{|P(s)}{Q(s)|} \leq \frac{C_1/C_2}{|s|^{m-n}}.$$

Thus, F(s) satisfies growth restriction condition. Further,

$$\text{Res}(z_n) = \lim_{s\to z_n} (s-z_n)\, e^{st} F(s)$$

$$= \lim_{s\to z_n} \frac{e^{st} P(s)}{\frac{Q(s)-Q(z_n)}{s-z_n}}, \text{ since } Q(Z_n) = 0$$

$$= \frac{e^{z_n t} P(z_n)}{Q'(z_n)}.$$

Hence, by inversion formula, we have

$$f(z) = \sum_{n=1}^{m} \frac{P(z_n)}{Q'(z_n)} e^{tz_n},$$

which is the required Heaviside's expansion formula.

EXAMPLE 10.70

Find $L^{-1}\left(\frac{1}{s(1+e^{as})}\right)$, using complex inversion formula.

Solution. Let $F(s) = \frac{1}{s(1+e^{as})}$. Then F($s$) has a simple pole at $s = 0$. Further, $1 + e^{as} = 0$ yields

$$e^{as} = -1 = e^{(2n-1)\pi i}, n = 0, \pm 1, \pm 2, \ldots.$$

and so $s_n = \left(\frac{2n-1}{a}\right) \pi i$, $n = 0, \pm 1, \pm 2, \ldots$ are also poles of F(s). Also, $\frac{d}{ds}(1+e^{as})_{s=s_n} = -a \neq 0$. Therefore, s_n are simple poles. Now

$$\text{Res}(0) = \lim_{s\to 0} s\, e^{st} F(s) = \frac{1}{2},$$

$$\text{Res}\left[\left(\frac{2n-1}{a}\right)\pi i\right] = \lim_{s\to s_n} (s - s_n)\, e^{st} F(s)$$

$$= \lim_{s\to s_n} \frac{(s - s_n)e^{st}}{s(1+e^{as})} \left(\frac{0}{0} \text{ form}\right)$$

$$= \lim_{s\to s_n} \frac{e^{st} + t\, e^{st}(s - s_n)}{a\, s e^{as} + 1 + e^{as}}, \text{ by L'Hospital rule}$$

$$= \frac{e^{tsn}}{a\, s_n\, e^{asn}} = -\frac{e^{t}\left(\frac{2n-1}{a}\right)\pi i}{(2n-1)\pi i}.$$

Also, it can be shown that F(s) satisfies growth restriction condition. Hence, by inversion formula, at the points of continuity of f, we have

$$f(t) = \text{sum of residues at the poles}$$

$$= \frac{1}{2} - \sum_{n=-\infty}^{\infty} \frac{1}{(2n-1)\pi i}\, e^{t\left(\frac{2n-1}{a}\right)\pi i}$$

$$= \frac{1}{2} - \frac{2}{\pi} \sum_{n=1}^{\infty} \frac{1}{(2n-1)} \sin\left(\frac{2n-1}{a}\right)\pi t.$$

EXAMPLE 10.71

Find $L^{-1}\{e^{-a\sqrt{s}}\}$, $a > 0$.

Solution. We know (Exercise 12) that

$$L^{-1}\left\{\frac{e^{-a\sqrt{s}}}{s}\right\} = erfc\left(\frac{a}{2\sqrt{t}}\right),$$

that is,

$$L\left\{erfc\left(\frac{a}{2\sqrt{t}}\right)\right\} = e^{-\frac{a\sqrt{s}}{s}}.$$

Therefore,

$$L\left\{\frac{d}{dt}\, erfc\left(\frac{a}{2\sqrt{t}}\right)\right\} = s\, F(s) - f(0)$$

$$= s\, \frac{e^{-a\sqrt{s}}}{s}$$

because $erfc\left(\frac{a}{2\sqrt{t}}\right) \to 0$ as $t \to 0$.

Thus

$$L\left\{\frac{a}{2\sqrt{\pi t^3}}\, e^{-a^2/(4t)}\right\} = e^{-a\sqrt{s}}.$$

Hence

$$L^{-1}\{e^{-a\sqrt{s}}\} = \frac{a}{2\sqrt{\pi t^3}}\, e^{-a^2/(4t)}.$$

10.8 MISCELLANEOUS EXAMPLES

EXAMPLE 10.72

Find inverse Laplace transform of

$$\frac{(se^{-s/2} + \pi e^{-s})}{(s^2 + \pi^2)}.$$

Solution. Using linearity and shifting properties, we have

$$L^{-1}\left\{\frac{se^{-\frac{s}{2}} + \pi e^{-s}}{s^2 + \pi^2}\right\}$$

$$= L^{-1}\left\{\frac{se^{-\frac{s}{2}}}{s^2 + \pi^2}\right\} + L^{-1}\left\{\frac{\pi e^{-s}}{s^2 + \pi^2}\right\}$$

$$= \cos \pi\left(t - \frac{1}{2}\right) H\left(t - \frac{1}{2}\right) + \sin \pi(t-1)\, H(t-1)$$

$$= \sin \pi t\left[H\left(t - \frac{1}{2}\right)\right] - \sin \pi\, t[H(t-1)]$$

$$= \sin \pi t\left[H\left(t - \frac{1}{2}\right) - H(t-1)\right],$$

where $H(t)$ denotes Heavyside's unit step function.

EXAMPLE 10.73

Find the inverse Laplace transform of

$$\frac{1}{s^2(s^2 + a^2)}.$$

Solution. We know that

$$L\{\sin at\} = \frac{a}{s^2 + a^2}.$$

Therefore

$$L^{-1}\left\{\frac{1}{s^2 + a^2}\right\} = \frac{1}{a} \sin at.$$

Then

$$L^{-1}\left\{\frac{a}{s(s^2 + a^2)}\right\} = \frac{1}{a}\int_0^t \sin at\, dt = \frac{1}{a}\left[-\frac{\cos at}{a}\right]_0^t$$

$$= -\frac{1}{a^2}[\cos at - 1]$$

and

$$L^{-1}\left\{\frac{a}{s^2(s^2+a^2)}\right\} = -\frac{1}{a^2}\int_0^t (\cos at - 1)dt$$

$$= -\frac{1}{a^2}\left[\frac{\sin at}{a} - t\right]_0^t$$

$$= -\frac{1}{a^3}\sin at + \frac{t}{a^2}$$

$$= \frac{1}{a^2}\left[t - \frac{1}{a}\sin at\right].$$

Note: This question can also be solved using Convolution Theorem (see Example 10.57)

EXAMPLE 10.74

Find $L^{-1}\left[\cot^{-1}\left(\frac{2}{s+1}\right)\right]$.

Solution. Since

$$\frac{d}{ds}\left[\cot^{-1}\left(\frac{2}{s+1}\right)\right] = \frac{-1}{1+\left(\frac{2}{s+1}\right)^2}\left(-\frac{2}{(s+1)^2}\right)$$

$$= \frac{2}{s^2+2s+5},$$

We have

$$L^{-1}\left\{\frac{2}{s^2+2s+5}\right\} = -t f(t)$$

or

$$L^{-1}\left\{\frac{2}{(s+1)^2+4}\right\} = -t f(t)$$

or

$$e^{-t}\sin 2t = -t f(t).$$

Hence

$$f(t) = -\frac{1}{t}e^{-t}\sin 2t.$$

EXAMPLE 10.75

Apply Convolution Theorem to evaluate

$$L^{-1}\left(\frac{1}{s^2+a^2}\right)^2.$$

Solution. Proceeding as in Example 10.56, we have

$$F(s) = G(s) = \frac{1}{s^2+a^2}.$$

Then

$$f(t) = L^{-1}\{F(s)\} = \frac{\sin at}{a}$$

$$g(t) = L^{-1}\{G(s)\} = \frac{\sin at}{a}.$$

Therefore, by Convolution Theorem,

$$L^{-1}\{F(s)\,G(s)\}$$

$$= f * g = \int_0^t \frac{\sin au}{a}\cdot\frac{\sin a(t-u)}{a}\,du.$$

$$= \frac{1}{a^2}\int_0^t \frac{1}{2}[\cos a(2u-t) - \cos at]\,du$$

$$= \frac{1}{2a^2}\left[\int_0^t \cos a(2u-t)du - \int_0^t \cos at\,du\right]$$

$$= \frac{1}{2a^2}\left\{\left[\frac{\sin a(2u-t)}{2a}\right]_0^t - t\cos at\right\}$$

$$= \frac{1}{2a^2}\left[\frac{\sin at}{2a} + \frac{\sin at}{2a} - t\cos at\right]$$

$$= \frac{1}{2a^2}\left[\frac{\sin at}{a} - t\cos at\right]$$

$$= \frac{1}{2a^3}[\sin at - at\cos at].$$

EXERCISES

1. Find inverse Laplace transform of

(a) $\frac{2s+6}{(s^2+6s+10)^2}$ **Ans.** $te^{-3t}\sin t$

(b) $\frac{s}{(s^2+4)^2}$ **Ans.** $\frac{1}{4}t\sin 2t$

(c) $\log\frac{s+1}{s-1}$ **Ans.** $\frac{2\sinh t}{t}$

(d) $\cot^{-1}\frac{s}{\pi}$

Hint: $\frac{d}{ds}\left\{\cot^{-1}\frac{s}{\pi}\right\} = -\frac{\pi}{s^2+\pi^2}$ implies

$L^{-1}\left\{-\frac{\pi}{s^2+\pi^2}\right\} = -tf(t)$ implies

$\pi[\frac{1}{\pi}\sin \pi t] = tf(t)$

Ans. $f(t) = \frac{\sin \pi t}{t}$

(e) $\frac{1}{s}\log\left(1+\frac{1}{s^2}\right)$

Hint: Use Example 10.22 and Theorem 10.9

Ans. $2\int_0^t \frac{1-\cos u}{u}\,du$

(f) $\frac{2s-3}{s^2+4s+13}$

Ans. $2e^{-2t}\cos 3t - \frac{7}{3}e^{-2t}\sin 3t$

(g) $\frac{s}{s^4+4a^4}$

Hint: Use partial fraction method

Ans. $\frac{1}{2a^2}\sin at \sinh at$

(h) $\frac{a(s^2-2a^2)}{s^4+4a^4}$

Ans. $\cos at \sinh at$

(i) $\frac{s^2+6}{(s^2+1)(s^2+4)}$

Ans. $f(t) = \frac{1}{3}(5\sin t - \sin 2t)$

(j) $\frac{s}{s^4+s^2+1}$ (partial fraction method)

Ans. $\frac{2}{\sqrt{3}}\sin h\,\frac{t}{2}\sin\frac{\sqrt{3}}{2}t$

(k) $\frac{2s^2+5s-4}{s^3+s^2-2s}$

Hint: Has simple poles, so use residue method

Ans. $2+e^t-e^{-2t}$

(l) $\frac{2s^2-4}{(s+1)(s-2)(s-3)}$

Ans. $-\frac{1}{6}e^{-t}-\frac{4}{3}e^{2t}+\frac{7}{2}e^{3t}$

2. Use first shift property to find $L^{-1}\left\{\frac{1}{\sqrt{s+a}}\right\}$.

Hint: $L^{-1}\left\{\frac{1}{\sqrt{s+a}}\right\}=e^{-at}L^{-1}\left\{\frac{1}{\sqrt{s}}\right\}=e^{-at}\cdot\frac{t^{\frac{1}{2}-1}}{\Gamma(1/2)}$

Ans. $\frac{e^{-at}}{\sqrt{\pi t}}$

3. Solve Exercise 1(k) using Heaviside's expansion formula.

4. Use Heaviside's expansion formula to fiind

$L^{-1}\left\{\frac{27-12s}{(s+4)(s^2+9)}\right\}$

Ans. $3e^{-4t}-3\cos 3t$

5. Use series method to find $L^{-1}\{e^{-\sqrt{s}}\}$.

Ans. $\frac{1}{2\sqrt{\pi}\,t^{3/2}}e^{-\frac{1}{4t}}$

6. Show that $L^{-1}\left\{\frac{1}{s}\cos\frac{1}{s}\right\} = 1-\frac{t^2}{(2!)^2}+\frac{t4}{(4!)^2}-\frac{t^6}{(6!)^2}+\ldots$

7. Evaluate $\sin t * t^2$.

Ans. $t^2+2\cos t-2$

8. Find $L\{\sin t * t^2\}$.

Ans. $\frac{2}{(s^2+1)s^3}$

9. Use convolution Theorem to find the inverse Laplace transforms of the following:

(a) $\frac{1}{s(s-a)}$ **Ans.** $\frac{e^{at}-1}{a}$.

(b) $\frac{a^2}{(s^2+a^2)^2}$

Ans. $\frac{1}{2a}(\sin at - at\cos at)$

(c) $\frac{4}{s^3+s^2+s+1}$

Ans. $2(e^{-t}-\cos t+\sin t)$

(d) $\frac{s+2}{(s^2+4s+5)^2}$

Ans. $\frac{1}{2}te^{-2t}\sin t$

10. Verify complex inversion formula for $F(s) = \frac{1}{s(s-a)}$.

Hint: Simple poles at 0 and a, satisfies growth restriction condition, Res(0) = $-1/a$, Res(a) = e^{at}/a

Ans. $f(t) = \frac{1}{a}(e^{at} - 1)$

11. Using complex inversion formula, find the inverse Laplace transform of the following:

(a) $\frac{s}{s^2+a^2}$ **Ans.** $\cos at$

(b) $\frac{1}{(s+1)(s-2)^2}$

Ans. $\frac{1}{9}e^{-t} + \frac{1}{3}te^{2t} - \frac{1}{9}e^{2t}$

(c) $\frac{1}{(s^2+1)^2}$ **Ans.** $\frac{1}{2}(\sin t - t\cos t)$

12. Find $L^{-1}\left\{\frac{e^{-a\sqrt{s}}}{s}\right\}$, $a > 0$.

Ans. $1 - erf\left(\frac{a}{2\sqrt{t}}\right)$ or $erfc\left(\frac{a}{2\sqrt{t}}\right)$.

11 Applications of Laplace Transform

Laplace transform is utilized as a tool for solving linear differential equations, integral equations, and partial differential equations. It is also used to evaluate the integrals. The aim of this chapter is to discuss these applications.

11.1 ORDINARY DIFFERENTIAL EQUATIONS

Recall that a *differential equation* is an equation where the unknown is in the form of a derivative. The *order* of an ordinary differential equation is the highest derivative attained by the unknown. Thus the equation

$$\frac{d^2y}{dt^2} + \alpha\frac{dy}{dt} + \beta y = f(t)$$

is of *second order*, whereas the equation.

$$\left(\frac{dy}{dx}\right)^3 + y = \sin x$$

is a *first order* differential equation.

Theorem 9.8, opens up the possibility of using Laplace transform as a tool for solving *ordinary differential equations*. Laplace transforms, being linear, are useful only for solving linear differential equations. Differential equations containing powers of the unknown or expression such as tan x, e^x cannot be solved using Laplace transforms.

The results

$$L\{f'(t)\} = sF(s) - f(0)$$

and

$$L\{f''(t)\} = s^2F(s) - sf(0) - f'(0)$$

will be used frequently for solving ordinary differential equations. To solve linear ordinary differential equation by the Laplace transform method, we first convert the equation in the unknown function $f(t)$ into an equation in F(s) and find F(s). The inversion of F(s) then yields $f(t)$.

Since $f(0)$, $f'(0)$, and $f''(0)$ appear in Laplace transform of derivatives of f, the Laplace transform method is best suited to *initial value problems* (where auxiliary conditions are *all* imposed at $t = 0$). The solution by Laplace method with initial conditions automatically built into it. We need not add particular integral to complementary function and then apply the auxiliary conditions.

(a) Ordinary Differential Equations with Constant Solution

In case of an ordinary differential equation with constant coefficients, the transformed equation for F(s) turns out to be an algebraic one and, therefore, the Laplace transform method is powerful tool for solving this type of ordinary differential equations. If

$$a_n\frac{d^ny}{dt^n} + a^{n-1}\frac{d^{n-1}y}{dt^{n-1}} + \ldots + a_0y = f(t)$$

with $y(0) = y_0$, $y'(0) = y_1, \ldots, y^{(n-1)}(0) = y_{n-1}$, then $f(t)$ is called *input*, *excitation*, or *forcing function* and $y(t)$ is called the *output* or *response*. Further, the following results suggests that if $f(t)$ is continuous and of exponential order, then $y(t)$ is also continuous and of exponential order.

Theorem 11.1. If $a_ny^{(n)} + a_{n-1}y^{(n-1)} + \ldots + a_0y = f(t)$ is n^{th} order linear non-homogeneous equation with constant coefficients and f is continuous on $[0, \infty)$ and of exponential order, then $y(t)$ is also continuous and of exponential order.

EXAMPLE 11.1

Find the general solution of the differential equation

$$y''(t) + k^2 y(t) = 0.$$

Solution. Assume that the value of the unknown function at $t = 0$ be denoted by the constant A, and the value of its first derivative at $t = 0$ by the constant B. Thus

$$y(0) = \text{A and } y'(0) = \text{B}.$$

Taking Laplace transform of both sides of the given differential equations, we have

$$\text{L}\{y''(t)\} + k^2 \text{L}\{y(t)\} = 0$$

But

$$\text{L}\{y''(t)\} = s^2\text{Y}(s) - sy(0) - y'(0)$$

$$= s^2\text{Y}(s) - \text{A}s - \text{B}.$$

Therefore,

$$s^2\text{Y}(s) - \text{A}s - \text{B} + k^2\text{Y}(s) = 0.$$

The solution of this algebraic equation in Y(s) is

$$\text{Y}(s) = A\frac{s}{s^2 + k^2} + \frac{\text{B}}{k}\frac{k}{s^2 + k^2}.$$

Taking inverse Laplace transform, we get

$$y(t) = \text{A}\cos kt + \frac{\text{B}}{k}\sin kt,$$

where A and B are constants since the initial conditions were not given.

EXAMPLE 11.2

Solve

$$\frac{d^2x}{dt^2} + x = \text{A}\sin t,\ x(0) = x_0,\ x'(0) = v_0.$$

Show that the phenomenon of *resonance* occurs in this case.

Solution. Taking Laplace transform, we get

$$s^2\text{X}(s) - sx(0) - x'(0) + \text{X}(s) = \frac{A}{s^2 + 1}$$

or

$$(s^2 + 1)\,\text{X}(s) = \frac{A}{s^2 + 1} + sx_0 + v_0$$

or

$$\text{X}(s) = \frac{A}{s^2 + 1} + \frac{s}{s^2 + 1}x_0 + \frac{v_0}{s^2 + 1}.$$

Taking inverse Laplace transform, we have

$$x(t) = \frac{A}{2}(\sin t - t\cos t) + x_0\cos t + v_0\sin t.$$

We note that $x(t) \to \infty$ as $t \to \infty$ due to the term $t\cos t$. This term is called a *secular term. The presence of secular term causes resonance,* because the solution becomes unbounded.

Remark 11.1. If we consider the equation $\frac{d^2x}{dt^2} + k^2x = \text{A}\sin t,\ k \neq 1$, then there will be no secular term in the solution and so the system will be *purely oscillatory.*

EXAMPLE 11.3

Solve the initial value problem

$$y'(t) + 3y(t) = 0,\ y(0) = 1.$$

Solution. Taking Laplace transform, we get

$$\text{L}\{y'(t)\} + 3\text{L}\{y(t)\} = 0,$$

which yields

$$s\text{Y}(s) - y(0) + 3\text{Y}(s) = 0.$$

Since $y(0) = 1$, we have

$$s\text{Y}(s) - 3\text{Y}(s) = 1,$$

an algebraic equation whose solution is

$$\text{Y}(s) = \frac{1}{s + 3}.$$

Taking inverse Laplace transform leads to

$$y(t) = e^{-3t}.$$

EXAMPLE 11.4

Solve the initial value problem

$$\frac{d^2y}{dt^2} - 2\frac{dy}{dt} - 8y = 0,\ y(0) = 3,\ y'(0) = 6.$$

Solution. The given equation is

$$y''(t) - 2y'(t) - 8y = 0, y(0) = 3, y'(0) = 6.$$

Laplace transform leads to

$$L\{y''(t)\} - 2L\{y'(t)\} - 8L\{y\} = 0,$$

that is,

$$s^2Y(s) - sy(0) - y'(0) - 2\{sY(s) - y(0)\} - 8Y(s) = 0$$

and so using initial conditions, we have

$$(s^2 - 2s - 8)\, Y(s) - 3s = 0.$$

Hence

$$Y(s) = \frac{3s}{s^2 - 2s - 8} = 3\left[\frac{s - 1 + 1}{(s - 1)^2 - 9}\right]$$

$$= 3\left[\frac{s - 1}{(s - 1)^2 - 9} + \frac{1}{(s - 1)^2 - 9}\right].$$

Taking inverse Laplace transform, we get

$$y(t) = 3L^{-1}\left\{\frac{s - 1}{(s - 1)^2 - 9}\right\} + 3L^{-1}\left\{\frac{1}{(s - 1)^2 - 9}\right\}$$

$$= 3e^t \cosh 3t + e^t \sinh 3t.$$

EXAMPLE 11.5

Solve the initial value problem

$$y''' + y'' = e^t + t + 1, y(0) = y'(0) = y''(0) = 0.$$

Solution. Taking Laplace transform of both sides of the given equation, we have

$$L\{y'''(t)\} + L\{y''(t)\} = L\{e^t\} + L\{t\} + L\{1\},$$

that is,

$$s^3Y(s) - s^2y(0) - sy'(0) - y''(0)$$

$$+ s^2Y(s) - sy(0) - y'(0) = \frac{1}{s - 1} + \frac{1}{s^2} + \frac{1}{s}$$

Since $y(0) = y'(0) = y''(0) = 0$, we have

$$s^3Y(s) + s^2Y(s) = \frac{1}{s - 1} + \frac{1}{s^2} + \frac{1}{s},$$

and so

$$Y(s) = \frac{2s^2 - 1}{s^4(s - 1)(s + 1)}.$$

Using partial fraction decomposition, we have

$$Y(s) = -\frac{1}{s^2} + \frac{1}{s^4} - \frac{1}{2(s + 1)} - \frac{1}{2(s - 1)}.$$

Taking inverse transform yields

$$y(t) = L^{-1}\left\{-\frac{1}{s^2} + \frac{1}{s^4} - \frac{1}{2(s + 1)} - \frac{1}{2(s - 1)}\right\}$$

$$= -t + \frac{1}{6}t^3 - \frac{1}{2}e^{-t} + \frac{1}{2}e^t.$$

Verification: We have

$$y' = -1 + \frac{1}{2}t^2 + \frac{1}{2}e^{-t} + \frac{1}{2}e^t,$$

$$y'' = t - \frac{1}{2}e^{-t} + \frac{1}{2}e^t,$$

$$y''' = 1 + \frac{1}{2}e^{-t} + \frac{1}{2}e^t.$$

Adding y'' and y''', we get

$$y'' + y''' = t + e^t + 1 \text{ (the given equation)}.$$

EXAMPLE 11.6

Solve

$$\frac{d^2y}{dt^2} + 2\frac{dy}{dt} - 3y = \sin t, y(0) = y'(0) = 0.$$

Solution. Taking Laplace transform of both sides of the given equation, we take

$$L\{y'''(t)\} + 2L\{y'(t)\} - 3L\{y(t)\} = L\{\sin t\},$$

which yields

$$s^2Y(s) - sy(0) - y'(0) + 2\{sY(s) - y(0)\} - 3Y(s)$$

$$= \frac{1}{s^2 + 1}.$$

Using the given initial conditions, we have

$$s^2Y(s) + 2sY(s) - 3Y(s) = \frac{1}{s^2+1}$$

and so

$$Y(s) = \frac{1}{(s^2+1)(s^2+2s-3)} = \frac{s-1}{2(s^2+1)}$$

$$-\frac{s+1}{2(s^2+2s-3)}$$

$$= \frac{s}{2(s^2+1)} - \frac{1}{2(s^2+1)} - \frac{1}{2}\left[\frac{s+1}{(s+1)^2-4}\right].$$

Taking inverse Laplace transform, we have

$$y(t) = \frac{1}{2}\cos t - \frac{1}{2}\sin t - \frac{1}{2}e^{-t}\sinh 2t.$$

EXAMPLE 11.7

Solve

$$\frac{d^2y}{dt^2} - 6\frac{dy}{dt} + 9y = t^2e^{3t}, y(0) = 2, y'(0) = 6.$$

Solution. Taking Laplace transform, we get

$$s^2Y(s) - sy(0) - y'(0) - 6(sY(s) - y(0)) + 9Y(s)$$

$$= \frac{2}{(s-3)^3}.$$

Using initial conditions, we have

$$s^2Y(s) - 2s - 6 - 6sY(s) + 12 + 9Y(s)$$

$$= \frac{2}{(s-3)^3}$$

or

$$(s^2 - 6s + 9)Y(s) = 2(s-3) + \frac{2}{(s-3)^3}$$

or

$$Y(s) = \frac{2}{s-3} + \frac{2}{(s-3)^5}.$$

Taking inverse Laplace transform yields

$$y(t) = 2e^{3t} + \frac{1}{12}t^4e^{-3t}.$$

EXAMPLE 11.8

Solve

$$y'' - 3y' + 2y = t, y(0) = 0 \text{ and } y'(0) = 0.$$

Solution. Taking Laplace transform yields

$$s^2Y(s) - sy(0) - y'(0) - 3[sY(s) - y(0)]$$

$$+ 2Y(s) = \frac{1}{s^2}.$$

Making use of initial value conditions, we have

$$(s^2 - 3s + 2)\, Y(s) = \frac{1}{s^2}$$

and so

$$Y(s) = \frac{1}{s^2(s^2 - 3s + 2)}$$

$$= \frac{1}{4(s-2)} - \frac{1}{(s-1)} + \frac{3}{4s} + \frac{1}{2s^2}.$$

Taking inverse Laplace transform, we get

$$y(t) = \frac{1}{4}e^{2t} - e^t + \frac{3}{4} + \frac{t}{2}.$$

EXAMPLE 11.9

Solve

$$y' + 2y = 1 - H(t-1), y(0) = 2,$$

where H(t) is Heaviside's unit step function.

Solution. Taking Laplace transform leads to

$$sY(s) + 2 + 2Y(s) = \frac{1}{s} - \frac{e^{-s}}{s}$$

or

$$(s+2)\,Y(s) = \frac{1}{s} - \frac{e^{-s}}{s} - 2$$

or

$$Y(s) = \frac{1}{s(s+2)} - \frac{e^{-s}}{s(s+2)} - \frac{2}{s+2}.$$

But, by partial fraction, we have

$$\frac{1}{s(s+2)} = \frac{1}{2s} - \frac{1}{2(s+2)}.$$

Therefore,

$$Y(s) = \frac{1}{2s} - \frac{1}{2(s+2)} - \left(\frac{e^{-s}}{2s} - \frac{e^{-s}}{2(s+2)}\right) - \frac{2}{s+2}.$$

Taking inverse transform, we get

$$y(t) = \frac{1}{2} - \frac{1}{2}e^{-2t} - \frac{1}{2}H(t-1)$$

$$+ \frac{1}{2}e^{-2(t-1)}H(t-1) + 2e^{-2t}$$

$$= \begin{cases} \frac{1}{2} + \frac{3}{2} e^{-2t} \text{ for } 0 \le t < 1 \\ \frac{3}{2} e^{-2t} + \frac{1}{2} e^{-2(t-1)} \text{ for } t \ge 1. \end{cases}$$

EXAMPLE 11.10

Solve

$$\frac{d^2y}{dt^2} + 2\frac{dy}{dt} + 5y = e^{-t} \sin t, y(0) = 0, y'(0) = 1.$$

Solution. Taking Laplace transform yields

$$s^2Y(s) - sy(0) - y'(0) + 2(sY(s) - y(0)) + 5Y(s) = \frac{1}{(s+1)^2 + 1}.$$

Using initial conditions, we get

$$s^2Y(s) + 2sY(s) + 5Y(s) = 1 + \frac{1}{(s+1)^2 + 1}$$

or

$$(s^2 + 2s + 5)\, Y(s) = 1 + \frac{1}{(s+1)^2 + 1}$$

or

$$Y(s) = \frac{1}{s^2 + 2s + 5} + \frac{1}{(s^2 + 2s + 5)(s^2 + 2s + 2)}$$

$$= \frac{s^2 + 2s + 3}{(s^2 + 2s + 5)(s^2 + 2s + 2)}.$$

Using partial fractions, we have

$$Y(s) = \frac{2}{3(s^2 + 2s + 5)} + \frac{1}{3(s^2 + 2s + 2)}$$

$$= \frac{2}{3((s+1)^2 + 1)}.$$

Taking inverse transform, we get

$$y(t) = \frac{2}{3} \cdot \frac{1}{2} e^{-t} \sin 2t + \frac{1}{3} e^{-t} \sin t$$

$$= \frac{1}{3} e^{-t} (\sin t + \sin 2t).$$

EXAMPLE 11.11

Solve the equation of motion

$$m\frac{d^2x}{dt^2} + k\frac{dx}{dt} = mv_0 \delta(t), x(0) = x'(0) = 0,$$

which represents the motion of a pellet of mass m fired into a viscous gas from a gun at time $t = 0$ with a muzzle velocity v_0 and where $\delta(t)$ is Dirac delta function, $x(t)$ is displacement at time $t \ge 0$ and $k > 0$ is a constant.

Solution. The condition $x'(0) = 0$ implies that the pellet is initially at rest for $t < 0$. Taking the Laplace transform of both sides, we have

$$m[s^2X(s) - sx(0) - x'(0)] + k[sX(s) - x(0)] = 1 . mv_0.$$

Using the given conditions, this expression reduces to

$$(ms^2 + ks)\, X(s) = mv_0$$

or

$$X(s) = \frac{mv_0}{ms^2 + ks} = \frac{v_0}{s\left(s + \frac{k}{m}\right)}$$

Use of partial fractions yields

$$X(s) = \frac{v_0}{s\left(s + \frac{k}{m}\right)} = \frac{A}{s} + \frac{B}{s + \frac{k}{m}}$$

or

$$v_0 = A\left(s + \frac{k}{m}\right) + Bs.$$

Comparing coefficients, we get

$$v_0 = A\frac{k}{m}, \text{ which yields } A = \frac{mv_0}{k}$$

and

$$0 = A + B, \text{ which gives } B = -\frac{mv_0}{k}.$$

Hence

$$X(s) = \frac{mv_0}{ks} - \frac{mv_0}{k\left(s + \frac{k}{m}\right)}.$$

Then, application of inverse Laplace transform yields

$$x(t) = \frac{mv_0}{k}\left(1 - e^{-\frac{k}{m}t}\right).$$

The graph of $x(t)$ is shown in the Figure 11.1

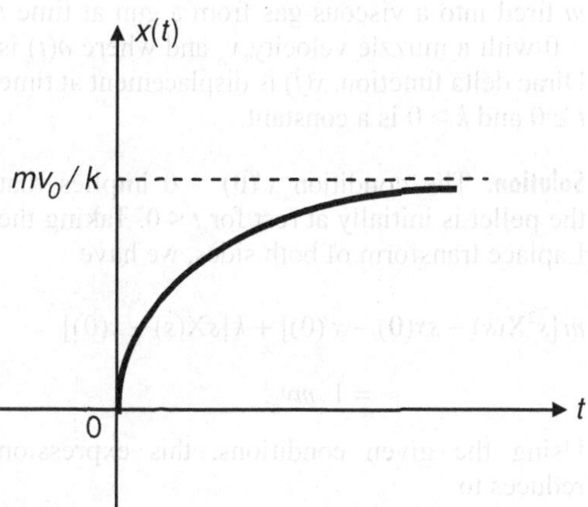

Figure 11.1

The velocity is given by

$$\frac{dx}{dt} = x'(t) = v_0 e^{-\frac{k}{m}t}.$$

We observe that $\lim_{t\to 0+} x'(t) = v_0$ and $\lim_{t\to 0-} x'(t) = 0.$ This indicates instantaneous jump in velocity at $t = 0$ from a rest state to the value v_0. The graph of $x'(t)$ is shown in the Figure 11.2.

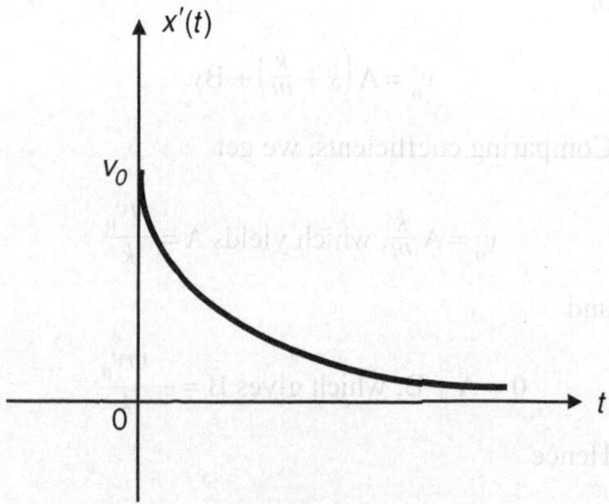

Figure 11.2

EXAMPLE 11.12

Solve boundary value problem

$$\frac{d^2y}{dt^2} + 9y = \cos 2t, y(0) = 1, y\left(\frac{\pi}{2}\right) = -1.$$

Solution. Suppose that $y'(0) = A$. Then taking Laplace transform, we have

$$s^2Y(s) - sy(0) - y'(0) + 9\,Y(s) = \frac{s}{s^2+4}$$

or

$$(s^2+9)Y(s) = s + A + \frac{s}{s^2+4},$$

and so

$$Y(s) = \frac{s+A}{s^2+9} + \frac{s}{(s^2+9)(s^2+4)}$$

$$= \frac{4s}{5(s^2+9)} + \frac{A}{s^2+9} + \frac{s}{5(s^2+4)}$$

(partial fractions).

Taking inverse Laplace transform yields

$$y(t) = \frac{4}{5}\cos 3t + \frac{A}{3}\sin 3t + \frac{1}{5}\cos 2t.$$

Since $y\left(\frac{\pi}{2}\right) = 1$, putting $t = \frac{\pi}{2}$, we get $A = \frac{12}{5}$. Hence

$$y(t) = \frac{4}{5}\cos 3t + \frac{4}{5}\sin 3t + \frac{1}{5}\cos 2t.$$

EXAMPLE 11.13

Solve

$$\frac{d^2x}{dt^2} + 6\frac{dx}{dt} + 9x = \sin t \; (t \geq 0)$$

subject to the conditions $x(0) = x'(0) = 0$.

Solution. Taking Laplace transform of both sides of the given equations yields

$$s^2X(s) - sx(0) - x'(0) + 6(sX(s) - x(0)) + 9X(s) = \frac{1}{s^2+1}.$$

Using the initial conditions, we have

$$(s^2+6s+9)\,X(s) = \frac{1}{s^2+1}$$

or

$$X(s) = \frac{1}{(s^2+1)(s+3)^2}$$

$$= \frac{A}{s+3} + \frac{B}{(s+3)^2} + \frac{Cs+D}{s^2+1}.$$

Comparing coefficients of the powers of s, we get

$$A = \frac{3}{50}, \qquad B = \frac{1}{10},$$

$$C = -\frac{3}{50}, \qquad D = \frac{2}{25}.$$

Hence

$$X(s) = \frac{3}{50(s+3)} + \frac{1}{10(s+3)^2} - \frac{3s}{50(s^2+1)} + \frac{2}{25(s^2+1)}.$$

Application of inverse Laplace transform gives

$$x(t) = \frac{3}{50}e^{-3t} + \frac{e^{-3t}t}{10} - \frac{3}{50}\cos t + \frac{2}{25}\sin t$$

$$= \frac{e^{-3t}}{50}(5t+3) - \frac{3}{50}\cos t + \frac{2}{25}\sin t.$$

The term $\frac{e^{-3t}}{50}(5t+3)$ is the particular solution, called the transient response since it dies away for large time, whereas the terms $-\frac{3}{50}\cos t + \frac{2}{25}\sin t$ is called the complementary function (sometimes called *steady state response* by engineers since it persists). However, there is nothing steady about it.

(b) Problems Related to Electrical Circuits

Consider the RCL circuit, shown in the Figure 11.3, consisting of *resistance*, *capacitor*, and *inductor* connected to a battery.

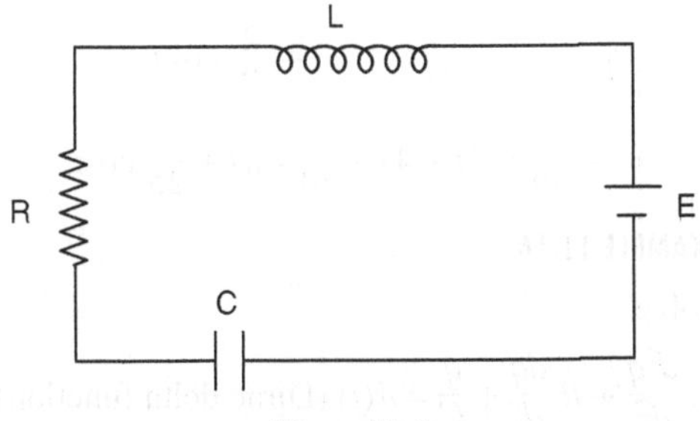

Figure 11.3

We know that resistance R is measured in *ohms*, capacitance C is measured in *farads*, and inductance is measured in *henrys*.

Let I denote the current flowing through the circuit and Q denote the charge. Then current I is related to Q by the relation $I = \frac{dQ}{dt}$. Also

(a) By Ohm's law, $\frac{V}{I} = R$ (resistance). Therefore, the voltage drop V across a resistor R is RI.

(b) The voltage drop across the inductor L is $L\frac{dI}{dt}$.

(c) The voltage drop across a capacitor is $\frac{Q}{C}$.

Thus, if E is the voltage (potential difference) of the battery, then by Kirchhoff's law, we have

$$L\frac{dI}{dt} + RI + \frac{Q}{C} = E(t),$$

where L, C, and R are constants. In terms of current, this equation becomes

$$L\frac{dI}{dt} + RI + \frac{1}{C}\int_0^t I(u)\,du = E(t),$$

because $I = \frac{dQ}{dt}$ implies $Q = \int_0^t I(u)\,du$.

In terms of charge, this differential equation takes the form

$$L\frac{d^2Q}{dt^2} + R\frac{dQ}{dt} + \frac{Q}{C} = E(t),$$

which is a differential equation of second order with constant coefficients L, R, and 1/C. The forcing function (input function) E(t) is supplied by the battery (voltage source). The system described by the above differential equation is known as *harmonic oscillator*.

EXAMPLE 11.14

Given that I = Q = 0 at $t = 0$, find I in the LR circuit (Figure 11.4) for $t > 0$.

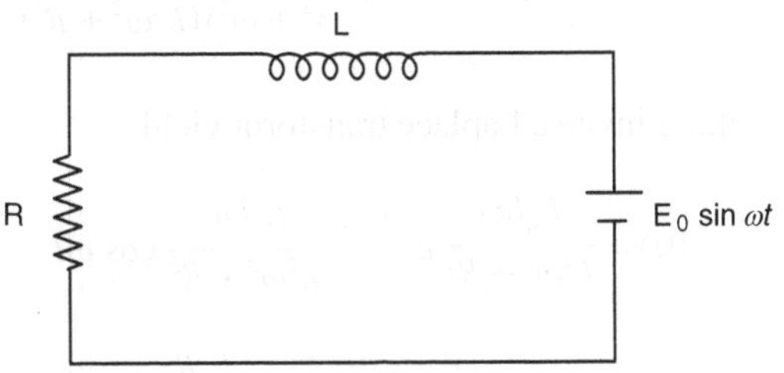

Figure 11.4

Solution. By Kirchhoff's law, the differential equation governing the given circuit is

$$L\frac{dI}{dt} + RI = E_0 \sin \omega t, \ I(0) = 0,$$

where L, R, E_0, and ω are constants. Taking Laplace transform of both sides, we have

$$L[sF(s) - I(0)] + RF(s) = \frac{E_0\omega}{s^2+\omega^2},$$

where F(*s*) denotes the Laplace transform of I. Using the given initial condition, we have

$$(Ls + R)\,F(s) = \frac{E_0\omega}{s^2+\omega^2}$$

which yields

$$F(s) = \frac{E_0\omega}{(Ls+R)(s^2+\omega^2)} = \frac{E_0\frac{\omega}{L}}{\left(s+\frac{R}{L}\right)(s^2+\omega^2)}$$

$$= \frac{A}{\left(s+\frac{R}{L}\right)} + \frac{Bs+C}{s^2+\omega^2}.$$

Comparison of coefficients of different powers of *s* yields

$$A = \frac{E_0L\omega}{L^2\omega^2+R^2},\ B = \frac{-E_0L\omega}{L^2\omega^2+R^2},\ C = \frac{E_0R\omega}{L^2\omega^2+R^2}.$$

Hence

$$F(s) = \frac{E_0L\omega}{\left(s+\frac{R}{L}\right)(L^2\omega^2+R^2)} - \frac{sE_0L\omega}{(s^2+\omega^2)(L^2\omega^2+R^2)}$$

$$+ \frac{E_0R\omega}{(s^2+\omega^2)(L^2\omega^2+R^2)}.$$

Taking inverse Laplace transform yields

$$I(t) = \frac{E_0L\omega}{L^2\omega^2+R^2}e^{-\frac{R}{L}t} - \frac{E_0L\omega}{L^2\omega^2+R^2}\cos\omega t$$

$$+ \frac{E_0R}{L^2\omega^2+R^2}\sin\omega t.$$

EXAMPLE 11.15

Given that I = Q = 0 at $t = 0$, find charge Q and current I in the following circuit (Figure 11.5) for $t > 0$.

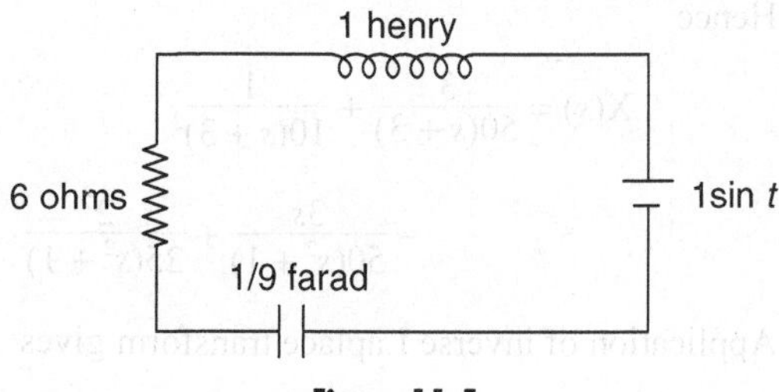

Figure 11.5

Solution. By Kirchhoff's law, the differential equation for the given circuit is

$$L\frac{d^2Q}{dt^2} + R\frac{dQ}{dt} + \frac{Q}{C} = E(t).$$

Here L = 1, R = 6, C = $\frac{1}{9}$, E(t) = sin t. Thus we have

$$\frac{d^2Q}{dt^2} + 6\frac{dQ}{dt} + 9Q = \sin t\,(t>0),$$

subject to Q(0) = 0, Q′(0) = I(0) = 0. By Example 11.13, the solution of this equation is

$$Q(t) = \frac{e^{-3t}}{50}(5t+3) - \frac{3}{50}\cos t + \frac{2}{25}\sin t.$$

Then

$$I(t) = \frac{dQ}{dt} = \frac{5e^{-3t}}{50} + \frac{3}{50}(5t+3)e^{-3t} + \frac{3}{50}\sin t$$

$$+ \frac{2}{25}\cos t$$

$$= -\frac{e^{-3t}}{50}(15t+4) + \frac{3}{50}\sin t + \frac{2}{25}\cos t.$$

EXAMPLE 11.16

Solve

$$L\frac{d^2q}{dt^2} + R\frac{dq}{dt} + \frac{q}{C} = \delta(t) \text{ (Dirac delta function)}$$

under conditions $q(0) = q'(0) = 0$.

Solution. Applying Laplace transform to both sides of the given equation, we find

$$\left(Ls^2 + Rs + \frac{1}{C}\right) Q(s) = 1$$

or

$$Q(s) = \frac{1}{Ls^2 + Rs + \frac{1}{C}} = \frac{1}{L\left(s^2 + \frac{R}{L}s + \frac{1}{LC}\right)}.$$

Suppose the roots of $s^2 + \frac{R}{L}s + \frac{1}{LC}$ are s_1 and s_2. Then

$$s_1 = \frac{-R + \sqrt{R^2 - (4L/C)}}{2L} \text{ and}$$

$$s_2 = \frac{-R - \sqrt{R^2 - (4L/C)}}{2L}.$$

Let us suppose R > 0. Then three cases arise:

(a) If $R^2 - \frac{4L}{C} < 0$, then s_1 and s_2 are complex and $s_1 = \bar{s}_2$.

(b) If $R^2 - \frac{4L}{C} = 0$, then s_1 and s_2 are real and $s_1 = s_2$.

(c) If $R^2 - \frac{4L}{C} > 0$, then s_1 and s_2 are real and $s_1 \neq s_2$.

Case (a). Using partial fractions, we have

$$Q(s) = \frac{1}{L(s - s_1)(s - s_2)}$$

$$= \frac{1}{L(s_1 - s_2)} \times \left[\frac{1}{s - s_1} - \frac{1}{s - s_2}\right].$$

Taking inverse Laplace transform yields

$$q(t) = \frac{1}{L(s_1 - s_2)[e^{s_1 t} - e^{s_2 t}]}.$$

If we put

$$\omega_0 = \frac{1}{2L}\sqrt{\frac{4L}{C} - R^2} \text{ and } \sigma = -\frac{R}{L},$$

then $s_1 = \bar{s}_2 = \sigma + i\omega_0$ and so $s_1 - s_2 = 2i\omega_0$. Therefore,

$$q(t) = \frac{1}{2Li\omega_0}\left(e^{(\sigma + i\omega_0)t} - e^{(\sigma - i\omega_0)t}\right)$$

$$= \frac{1}{L\omega_0} e^{\sigma t} \frac{e^{i\omega_0 t} - e^{-i\omega_0 t}}{2i}$$

$$= \frac{1}{L\omega_0} e^{\sigma t} \sin \omega_0 t, \ \sigma < 0.$$

Thus, the impulse response $q(t)$ is a *damped sinusoidal* with frequency ω_0. That is why, this case is called *damped vibration* or *undercritical damping* (Figure 11.6).

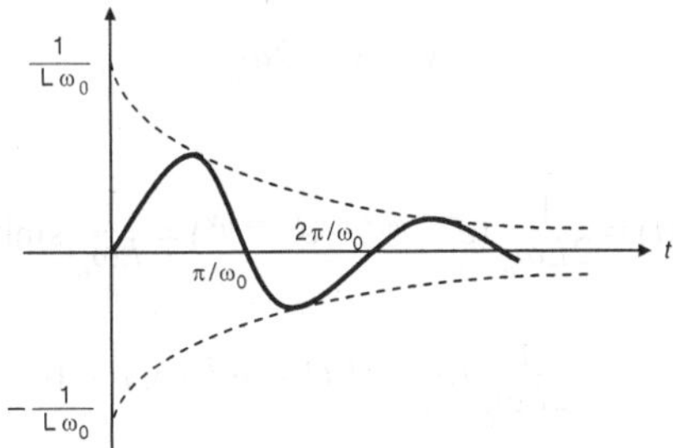

Figure 11.6

Case (b) In this case $s_1 = s_2 = -\frac{R}{2L}$ and so

$$Q(s) = \frac{1}{L(s - \sigma)^2}.$$

Taking inverse transform, we get

$$q(t) = \frac{te^{\sigma t}}{L}, \ \sigma < 0.$$

This case is called *critical damping* (Figure 11.7)

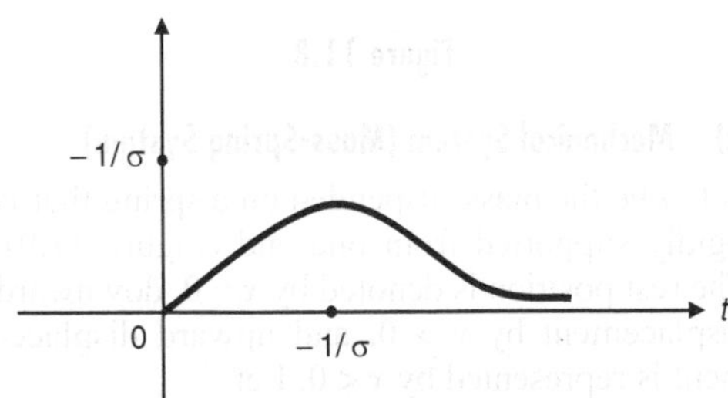

Figure 11.7

Case (c) As in case (a), we have

$$q(t) = \frac{1}{L(s_1 - s_2)}(e^{s_1 t} - e^{s_2 t}).$$

Since L > 0 and C > 0, we have $R > \sqrt{R^2 - \frac{4L}{C}}$ and so $s_2 < s_1 < 0$. Thus $q(t)$ is the sum of two exponentially damped functions. Put

$$\omega_0 = \frac{1}{2L}\sqrt{R^2 - \frac{4L}{C}} \text{ and } \sigma = -\frac{R}{2L}.$$

Then, we have

$$s_1 = \sigma + \omega_0,\ s_2 = \sigma - \omega_0.$$

Therefore,

$$s_1 - s_2 = 2\omega_0$$

and

$$q(t) = \frac{1}{2L\omega_0}\left(e^{(\sigma+\omega_0)t} - e^{(\sigma-\omega_0)t}\right) = \frac{1}{L\omega_0}\sinh \omega_0 t$$

$$= \frac{1}{2L\omega_0}\left(e^{(\sigma+\omega_0)t}\left(1 - e^{-2\omega_0 t}\right)\right),\ \sigma < 0.$$

Since $\sigma + \omega_0 < 0$, the impulse response $q(t)$ is damped hyperbolic sine. This case is called overdamped or overcritical damping (Figure 11.8).

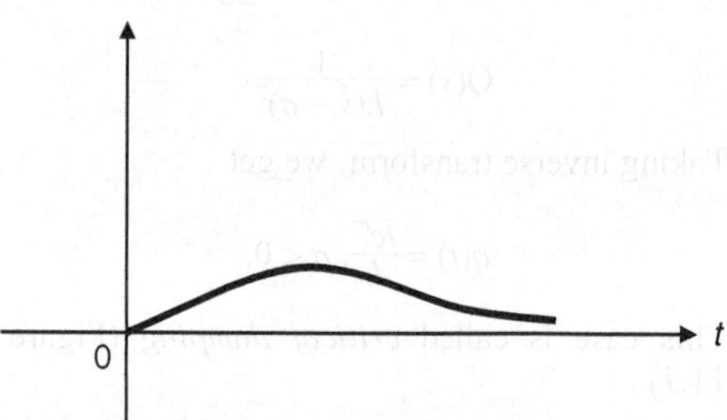

Figure 11.8

(c) Mechanical System (Mass-Spring System)

Let m be the mass suspended on a spring that is rigidly supported from one end (Figure 11.9). The rest position is denoted by $x = 0$, downward displacement by $x > 0$, and upward displacement is represented by $x < 0$. Let

(i) $k > 0$ be the *spring constant* (or *stiffness*) and $a > 0$ be the *damping constant*.

(ii) $a\dfrac{dx}{dt}$ be the damping force due to medium (air, etc.). Thus, damping force is proportional to the velocity.

(iii) $f(t)$ represents all external impressed forces on m. It is also called *forcing or excitation*.

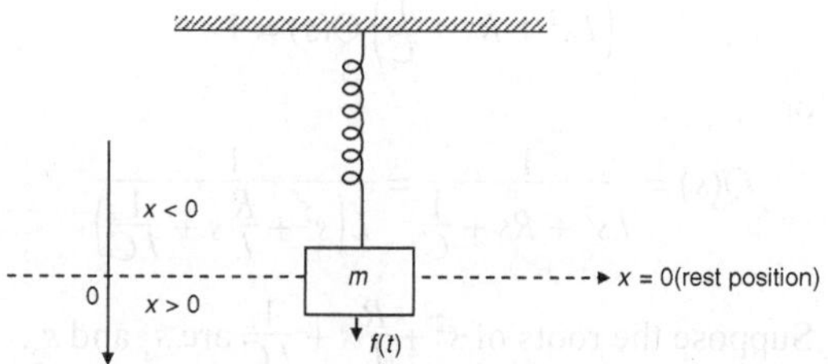

Figure 11.9

By Newton's second law of motion, the sum of forces acting on m equals $m\dfrac{d^2x}{dt^2}$ and so

$$m\frac{d^2x}{dt^2} = -kx - a\frac{dx}{dt} + f(t).$$

Thus the equation of motion is

$$m\frac{d^2x}{dt^2} + a\frac{dx}{dt} + kx = f(t) \qquad (1)$$

This is exactly the same differential equation which occurs in harmonic oscillator.

If $a = 0$, the motion is called *undamped* whereas if $a \neq 0$, the motion is called damped. Moreover, if $f(t) = 0$, that is, if there is no impressed forces, then the motion is called *forced.*

The equation (1) can be written as

$$\frac{d^2x}{dt^2} + \frac{a}{m}\frac{dx}{dt} + \frac{k}{m} = f(t)/m, \qquad (2)$$

where $f(t)/m$ is now the external impressed force (or excitation force) per unit mass.

EXAMPLE 11.17

Solve the equation of motion

$$\frac{d^2x}{dt^2} + 2b\frac{dx}{dt} + \lambda^2 x = \delta(t),\ x(0) = x'(0) = 0$$

for $0 < b < \lambda$.

[Clearly this is equation (2) with $\frac{a}{m} = 2b, \frac{k}{m} = \lambda^2$]

Solution. We want to find the response of the given mechanical system to a unit impulse. Taking Laplace transform, we get

$$\{s^2X(s) - sx(0) - x'(0)\}$$

$$+ 2b\{sX(s) - x(0)\} + {}^2X(s) = 1.$$

Taking note of the given conditions, we have

$$(s^2 + 2bs + \lambda^2)\, X(s) = 1$$

or

$$X(s) = \frac{1}{s^2 + 2bs + \lambda^2} = \frac{1}{(s+b)^2 + (\lambda^2 - b^2)}.$$

Taking inverse Laplace transform yields

$$x(t) = e^{-bt}\left(\frac{1}{\sqrt{\lambda^2 - b^2}} \sin \sqrt{\lambda^2 - b^2}\, t\right),$$

which is clearly a case of damped oscillation (Figure 11.10).

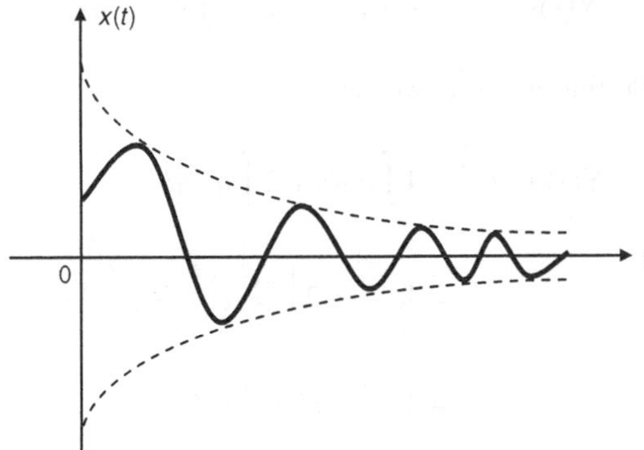

Figure 11.10

Also we note that

$$X(s) = \frac{1}{s^2 + 2bs + \lambda^2}\, L\{\delta(t)\}.$$

Thus we conclude that

Response = Transfer function × Input:

(d) Ordinary Differential Equations with Polynomial (Variable) Coefficients

We know that

$$L\{t^n f(t)\} = (-1)^n \frac{d^n}{ds^n} F(s),$$

where $F(s) = L\,\{f(t)\}$. Thus for $n = 1$, we have

$$L\{tf(t)\} = -F'(s).$$

Hence, if $f'(t)$ satisfies the sufficient condition for the existence of Laplace transform, then

$$L\{tf'(t)\} = -\frac{d}{ds} L\{f'(t)\} = -\frac{d}{ds}(sF(s) - f(0))$$

$$= -sF'(s) - F(s).$$

Similarly for $f''(t)$,

$$L\{tf''(t)\} = -\frac{d}{ds} L\{f''(t)\}$$

$$= -\frac{d}{ds}\{s^2F(s) - sf(0) - f'(0)\}$$

$$= -s^2F'(s) - 2sF(s) + f(0).$$

The above-mentioned derivations are used to solve linear differential equations whose coefficients are first degree polynomials.

EXAMPLE 11.18

Solve

$$ty'' + y' + ty = 0,\ y(0) = 1,\ y'(0) = 0.$$

Solution. Taking Laplace transform, we have

$$L\{ty''\} + L\{y'\} + L\{ty\} = 0$$

or

$$-\frac{d}{ds} L\{y''(t)\} + \{sY(s) - y(0)\} - \frac{d}{ds}\{Y(s)\} = 0$$

or

$$\frac{d}{ds}\{s^2Y(s) - sy(0) - y'(0)\} + \{sY(s) - y(0)\} - \frac{d}{ds} Y(s) = 0$$

which on using initial conditions yields

$$-\left[s^2 \frac{dY(s)}{ds} + 2sY(s)\right] + sY(s) - \frac{dY(s)}{ds} = 0$$

or

$$(s^2 + 1)\frac{dY(s)}{ds} + sY(s) = 0$$

or

$$\frac{dY(s)}{Y(s)} + \frac{s\,ds}{s^2 + 1} = 0.$$

Integrating, we have

$$\log Y(s) + \frac{1}{2}\log(s^2+1) = A \text{ (constant)}$$

and so

$$Y(s) = \frac{A}{\sqrt{s^2+1}}.$$

Taking inverse Laplace transform, we get

$$y(t) = A\,J_0(t),$$

where $J_0(t)$ is Bessel function of order zero.

Putting $t = 0$ and using initial condition $y(0) = 1$, we have

$$1 = AJ_0(0) = A.$$

Hence the required solution is

$$y(t) = J_0(t).$$

EXAMPLE 11.19

Solve

$$y'' + ty' - 2y = 4,\ y(0) = -1,\ y'(0) = 0.$$

Solution. Taking Laplace transform yields

$$L\{y''(t)\} + L\{ty'(t)\} - 2L\{y(t)\} = 4L\{1\}$$

or

$$s^2Y(s) - sy(0) - y'(0) - \frac{d}{ds}L\{y'(t)\} - 2Y(s) = \frac{4}{s}$$

or

$$s^2Y(s) - sy(0) - y'(0) - \frac{d}{ds}(sY(s) - y(0)) - 2Y(s) = \frac{4}{s}.$$

On using the initial values, we have

$$s^2Y(s) + s - \left(s\frac{dY(s)}{ds} + Y(s)\right) - 2Y(s) = \frac{4}{s}$$

or

$$\frac{s\,dY(s)}{ds} - (s^2-3)Y(s) = -\frac{4}{s} + s$$

or

$$\frac{dY(s)}{ds} + \left(\frac{3}{s} - s\right)Y(s) = -\frac{4}{s^2} + 1.$$

The integrating factor is

$$e^{\int\left(\frac{3}{s}-s\right)ds} = s^3e^{-\frac{s^2}{2}}.$$

Therefore,

$$\frac{d}{ds}\left[Y(s)\cdot s^3e^{-\frac{s^2}{2}}\right] = -\frac{4}{s^2}s^3e^{-\frac{s^2}{2}} + s^3e^{-\frac{s^2}{2}},$$

and so integration yields

$$Y(s)s^3e^{-\frac{s^2}{2}} = -4\int se^{-\frac{s^2}{2}}ds + \int s^3e^{-\frac{s^2}{2}}ds.$$

Putting $u = -\frac{s^2}{2}$, we get

$$Y(s)\,s^3\,e^{-\frac{s^2}{2}} = 4\int e^u du + 2\int u\,e^u du$$

$$= 4e^{-\frac{s^2}{2}} + 2\left(\frac{-s^2}{2}e^{-\frac{s^2}{2}} - e^{-\frac{s^2}{2}}\right) + A$$

$$= 2e^{-\frac{s^2}{2}} - s^2e^{-\frac{s^2}{2}} + C.$$

Thus,

$$Y(s) = \frac{2}{s^3} - \frac{1}{s} + \frac{C}{s^3}e^{\frac{s^2}{2}}.$$

Since $Y(s) \to 0$ as $s \to \infty$, we must have $C = 0$ and so

$$Y(s) = \frac{2}{s^3} - \frac{1}{s}.$$

Taking inverse Laplace transform, we get

$$y(t) = t^2 - 1.$$

EXAMPLE 11.20

Solve

$$ty'' + 2y' + ty = 0,\quad y(0) = 1,\quad y(\pi) = 0.$$

Solution. Let $y'(0) = A$ (constant). Taking Laplace transform of both sides, we obtain

$$-\frac{d}{ds}\{s^2Y(s) - sy(0) - y'(0)\} + 2\{sY(s) - y(0)\} - \frac{d}{ds}\{Y(s)\} = 0$$

and so

$$-s^2Y'(s) - 2sY(s) + y(0) + 2sY(s) - 2y(0) - Y'(s) = 0.$$

Using boundary conditions, we get

$$-(s^2+1)Y'(s) - 1 = 0$$

or

$$Y'(s) = \frac{-1}{s^2+1}.$$

Integration yields

$$Y(s) = -\tan^{-1} s + B \text{ (constant)}.$$

Since $Y(s)$ tends to zero as $s \to \infty$, we must have $B = \pi/2$. Hence,

$$Y(s) = \frac{\pi}{2} - \tan^{-1} s = \tan^{-1}\left(\frac{1}{s}\right).$$

Taking inverse Laplace transform, we have

$$y(t) = L^{-1}\left\{\tan^{-1}\left(\frac{1}{s}\right)\right\} = \frac{\sin t}{t}.$$

This solution clearly satisfies $y(\pi) = 0$.

EXAMPLE 11.21

Solve

$$ty'' + y' + 2y = 0, \qquad y(0) = 1.$$

Solution. Taking Laplace transform gives

$$-\frac{d}{ds}(s^2Y(s) - sy(0) - y'(0)) + (sY(s) - y(0)) + 2Y(s) = 0$$

or

$$-s^2Y'(s) - 2sY(s) + y(0) + sY(s) - y(0) + 2Y(s) = 0$$

or

$$-s^2Y'(s) - sY(s) + 2Y(s) = 0$$

or

$$Y'(s) + \left(\frac{1}{s} - \frac{2}{s^2}\right)Y(s) = 0.$$

The integrating factor is

$$e^{\int\left(\frac{1}{s}-\frac{2}{s^2}\right)ds} = e^{\log s + \frac{2}{s}} = se^{\frac{2}{s}}.$$

Therefore,

$$\frac{d}{ds}\{Y(s)se^{2/s}\} = 0.$$

Integrating, we have

$$Y(s)se^{2/s} = A \text{ (constant)}$$

or

$$Y(s) = \frac{Ae^{-\frac{2}{s}}}{s}.$$

Since $e^x = \sum_{n=0}^{\infty} \frac{x^n}{n!}$, taking $x = -\frac{2}{s}$, we have

$$Y(s) = A\sum_{n=0}^{\infty} \frac{(-1)^n 2^n}{n!\, s^{n+1}}.$$

Taking inverse Laplace transform,we get

$$y(t) = A\sum_{n=0}^{\infty} \frac{(-1)^n 2^n t^n}{(n!)^2}.$$

The condition $y(0) = 1$ now yields $A = 1$. Hence

$$y(t) = \sum_{n=0}^{\infty} \frac{(-1)^n 2^n t^n}{(n!)^2} = J_0(2\sqrt{2t}),$$

where J_0 is Bessel's function of order zero.

EXAMPLE 11.22

Solve

$$ty'' - y' = -1, \qquad y(0) = 0.$$

Solution. Taking Laplace transform of both sides of the given equation,

$$-\frac{d}{ds}\{s^2Y'(s) - sy(0) - y'(0)\} - \{sY(s) - y(0)\} = -\frac{1}{s}$$

or

$$-s^2Y'(s) - 2sY(s) + y(0) - sY(s) + y(0) = -\frac{1}{s}$$

or

$$-s^2Y'(s) - 3sY(s) = -\frac{1}{s}$$

or

$$Y'(s) + \frac{3}{s}Y(s) = \frac{1}{s^3}.$$

The integrating factor is

$$e^{\int \frac{3}{s} ds} = e^{3 \log s} = s^3.$$

Therefore,

$$\frac{d}{ds}(Y(s)s^3) = \frac{1}{s^3}s^3 = 1.$$

Integrating

$$Y(s)s^3 = s + A \text{ (constant)}$$

and so

$$Y(s) = \frac{1}{s^2} + \frac{A}{s^3}.$$

Taking inverse Laplace transform, we get

$$y(t) = t + Bt^2,$$

where B is constant. Obviously, the solution satisfies $y(0) = 0$.

EXAMPLE 11.23

Solve

$$ty'' + (t+1)y' + 2y = e^{-t}, y(0) = 0.$$

Solution. Taking Laplace transform of both sides gives

$$-\frac{d}{ds}\{s^2Y(s) - sy(0) - y'(0)\} - \frac{d}{ds}\{sY(s) - y(0)\}$$

$$+ \{sY(s) - y(0)\} + 2Y(s) = \frac{1}{s+1}$$

or

$$-s^2Y'(s) - 2sY(s) - \{sY'(s) + Y(s)\} + sY(s)$$

$$-y(0) + 2Y(s) = \frac{1}{s+1}$$

or

$$-s^2Y'(s) - sY'(s) - 2sY(s) + sY(s) + Y(s) = \frac{1}{s+1}$$

or

$$-(s^2+s)Y'(s) - (s-1)Y(s) = \frac{1}{s+1}$$

or

$$Y'(s) + \frac{s-1}{s^2+s} = \frac{-1}{s(s+1)^2}.$$

The integration factor is

$$e^{\int \frac{s-1}{s^2+s} ds} = e^{\int \left(-\frac{1}{s} + \frac{2}{s+1}\right) ds} = \frac{(s+1)^2}{s}.$$

Therefore,

$$\frac{d}{ds}\left(Y(s)\frac{(s+1)^2}{s}\right) = \frac{-1}{s(s+1)^2} \cdot \frac{(s+1)^2}{s} = -\frac{1}{s^2}.$$

Integrating, we get

$$Y(s)\frac{(s+1)^2}{s} = \int -\frac{1}{s^2} ds = \frac{1}{s} + C$$

and so

$$Y(s) = \frac{1}{(s+1)^2} + \frac{Cs}{(s+1)^2}.$$

By initial value theorem $y(0) = \lim_{s \to 0} sY(s) = 0$ and so $C = 0$. Hence

$$Y(s) = \frac{1}{(s+1)^2}.$$

Taking inverse Laplace transform, we get

$$y(t) = t\, e^{-t}.$$

11.2 SIMULTANEOUS DIFFERENTIAL EQUATIONS

The Laplace transforms convert a pair of differential equations into simultaneous algebraic equations in parameters. After that we solve these equations for Laplace transforms of the variables and then apply inverse Laplace operators to get the required solution.

EXAMPLE 11.24

Solve the simultaneous differential equations

$$3x' + y' + 2x = 1, \qquad x' + 4y' + 3y = 0$$

subject to the conditions $x(0) = 0$, $y(0) = 0$.

Solution. Taking Laplace transform, we get

$$3\{sX(x) - x(0)\} + \{sY(s) - y(0)\} + 2X(x) = \frac{1}{s}$$

and

$$sX(s) - x(0) + 4\{sY(s) - y(0)\} + 3Y(s) = 0.$$

Using the initial conditions, these equations reduce to

$$(3s + 2)\,X(s) + sY(s) = \frac{1}{s} \qquad (3)$$

and

$$sX(s) + (4s + 3)Y(s) = 0. \qquad (4)$$

Multiplying (3) and (4) by s and $(3s + 2)$ respectively and then subtracting, we get

$$Y(s) = -\frac{1}{11s^2 + 17s + 6} = \frac{-1}{(11s + 6)(s + 1)},$$

and then using (4), we have

$$X(s) = \frac{4s + 3}{s(11s + 6)(s + 1)}.$$

We deal with $X(s)$ first. Using partial fraction, we have

$$X(s) = \frac{1}{2s} - \frac{3}{10}\left(\frac{1}{s + (6/11)}\right) - \frac{1}{5(s + 1)}.$$

Taking inverse transform, we have

$$x(t) = \frac{1}{2} - \frac{3}{10}\,e^{-\frac{6}{11}t} - \frac{1}{5}\,e^{-t}$$

$$= \frac{1}{10}\left((5 - 3e^{-\frac{6}{11}t} - 2e^{-t}\right).$$

Further, poles of $Y(s)$ are $-\frac{6}{11}$ and -1. Hence

$$y(t) = \lim_{s \to \frac{-6}{11}} \left(s + \frac{6}{11}\right) Y(s)\, e^{-\frac{6}{11}t}$$

$$+ \lim_{s \to -1} (s + 1)Y(s)e^{-t} = \frac{1}{5}\left(e^{-t} - e^{-\frac{6}{11}t}\right).$$

EXAMPLE 11.25

Solve the simultaneous differential equations

$$\frac{dx}{dt} = 2x - 3y,\ \frac{dy}{dt} = y - 2x$$

subject to the conditions $x(0) = 8, y(0) = 3$.

Solution. Taking Laplace transform and using the given conditions, we have

$$sX(s) = 2X(s) - 3Y(s) + 8$$

and

$$sY(s) = Y(s) - 2X(s) + 3.$$

Thus

$$(s - 2)\,X(s) + 3Y(s) = 8,$$

and

$$2X(s) + (s - 1)Y(s) = 3.$$

Solving these algebraic equations, we get

$$X(s) = \frac{8s - 17}{s^2 - 3s - 4} = \frac{8s - 17}{(s + 1)(s - 4)},$$

and

$$Y(s) = \frac{3s - 22}{s^2 - 3s - 4} = \frac{3s - 22}{(s + 1)(s - 4)}.$$

Using partial fractions, these yields

$$X(s) = \frac{5}{s + 1} + \frac{3}{s - 4},\quad Y(s) = \frac{5}{s + 1} - \frac{2}{s - 4}.$$

Hence taking inverse Laplace transform, we get

$$x(t) = 5e^{-t} + 3e^{4t},\ y(t) = 5e^{-t} - 2e^{4t}.$$

EXAMPLE 11.26

Solve

$$\frac{dx}{dt} - y = e^t, \qquad \frac{dy}{dt} + x = \sin t$$

subject to the conditions $x(0) = 1, y(0) = 0$.

Solution. Taking Laplace transform and using the given conditions, we have

$$sX(s) - Y(s) = \frac{1}{s - 1} + 1 = \frac{s}{s - 1},$$

$$sY(s) + X(s) = \frac{1}{s^2 + 1}.$$

Solving these equations, we get

$$X(s) = \frac{s^4 + s^2 + s - 1}{(s - 1)(s^2 + 1)^2},$$

$$Y(s) = \frac{-s^4 + s^3 - 2s^2}{s(s-1)(s^2+1)^2} = \frac{-s^3 + s^2 - 2s}{(s-1)(s^2+1)^2}.$$

Now

$$X(s) = \frac{s^4 + s^2 + s - 1}{(s-1)(s^2+1)^2}$$

$$= \frac{A}{s-1} + \frac{Bs + C}{(s^2+1)} + \frac{Ds + E}{(s^2+1)^2}.$$

Comparison of coefficients yields

$$A = B = C = \frac{1}{2} \quad \text{and } E = 1.$$

Thus

$$X(s) = \frac{1}{2(s-1)} + \frac{s}{2(s^2+1)} + \frac{1}{2(s^2+1)} + \frac{1}{(s^2+1)^2}.$$

Hence

$$x(t) = \frac{1}{2}e^t + \frac{1}{2}\cos t + \frac{1}{2}\sin t + \frac{1}{2}(\sin t - t\cos t)$$

$$= \frac{1}{2}[e^t + \cos t + 2\sin t - t\cos t].$$

Now consider Y(s). We have

$$Y(s) = \frac{-s^3 + s^2 - 2s}{(s-1)(s^2+1)^2}$$

$$= \frac{A}{s-1} + \frac{Bs + C}{(s^2+1)} + \frac{Ds + E}{(s^2+1)^2}.$$

Comparing coefficients, we get

$$A = -\frac{1}{2}, B = \frac{1}{2}, C = -\frac{1}{2}, D = 2, E = 0,$$

and so

$$Y(s) = \frac{-1}{2(s-1)} + \frac{s}{2(s^2+1)}$$

$$-\frac{1}{2(s^2+1)} + \frac{2s}{(s^2+1)^2}.$$

Hence

$$y(t) = -\frac{1}{2}e^t + \frac{1}{2}\cos t - \frac{1}{2}\sin t + t\sin t.$$

EXAMPLE 11.27

The co-ordinates (x, y) of a particle moving along a plane curve at any time t are given by

$$\frac{dy}{dt} + 2x = \sin 2t, \; \frac{dx}{dt} - 2y = \cos 2t, \; t > 0.$$

If at $t = 0$, $x = 1$ and $y = 0$, show by using transforms, that the particle moves along the curve $4x^2 + 4xy + 5y^2 = 4$.

Solution. Using Laplace transform, we get

$$sY(s) - y(0) + 2X(s) = \frac{2}{s^2+4}$$

and

$$sX(s) - x(0) - 2Y(s) = \frac{s}{s^2+4}.$$

Using the given conditions, we have

$$sY(s) + 2X(s) = \frac{2}{s^2+4}$$

and

$$sX(s) - 2Y(s) = 1 + \frac{s}{s^2+4} = \frac{s^2+s+4}{s^2+4}.$$

Solving for X(s) and Y(s), and using partial fractions, we have

$$X(s) = \frac{s^3 + s^2 + 4s + 4}{(s^2+4)^2} = \frac{s}{s^2+4} + \frac{1}{s^2+4}$$

$$= \frac{s}{s^2+4} + \frac{1}{2}\frac{2}{s^2+4},$$

$$Y(s) = \frac{-2s^2 - 8}{(s^2+4)^2} = -\frac{2}{s^2+4}.$$

Hence taking inverse transform, we get

$$x(t) = \cos 2t + \frac{1}{2}\sin 2t,$$

$$y(t) = -\sin 2t.$$

We observe that

$$4x^2 + 4xy + 5y^2 = 4(\cos^2 2t + \sin^2 2t) = 4,$$

and hence the particle moves along the curve $4x^2 + 4xy + 5y^2 = 4$.

EXAMPLE 11.28

Solve the following system of equations:

$$x(t) - y''(t) + y(t) = e^{-t} - 1,$$

$$x'(t) + y'(t) - y(t) = -3e^{-t} + t,$$

subject to $x(0) = 0, y(0) = 1, y'(0) = -2$.

Solution. Taking Laplace transform yields

$$X(s) - \{s^2Y(s) - sy(0) - y'(0)\} + Y(s) = \frac{1}{s+1} - \frac{1}{s}$$

and

$$sX(s) - x(0) + sY(s) - y(0) - Y(s) = \frac{-3}{s+1} + \frac{1}{s^2}.$$

Using the given conditions, we have

$$X(s) - s^2Y(s) + s - 2 + Y(s) = \frac{-1}{s(s+1)}$$

and

$$sX(s) + sY(s) - 1 - Y(s) = \frac{-3s^2 + s + 1}{(s+1)s^2}$$

or

$$X(s) - (s^2 - 1)Y(s) = 2 - s - \frac{1}{s(s+1)}$$

$$= \frac{-s^3 + s^2 + 2s - 1}{s(s+1)}$$

and

$$sX(s) + (s-1)Y(s) = 1 - \frac{3s^2 - s - 1}{(s+1)s^2}$$

$$= \frac{s^3 - 2s^2 + s + 1}{(s+1)s^2}.$$

Solving for X(s) and Y(s), we have

$$X(s) = \frac{1}{s^2(s+1)} = \frac{1}{s^2} + \frac{1}{s} + 1 - \frac{1}{s},$$

$$Y(s) = \frac{s^2 - s - 1}{s^2(s+1)} = \frac{1}{s+1} - \frac{1}{s^2}.$$

Hence, taking inverse Laplace transform, we get

$$x(t) = t + e^{-t} - 1, y(t) = e^{-t} - t.$$

EXAMPLE 11.29

Given that I(0) = 0, find the current I in RL-network shown in the Figure 11.11.

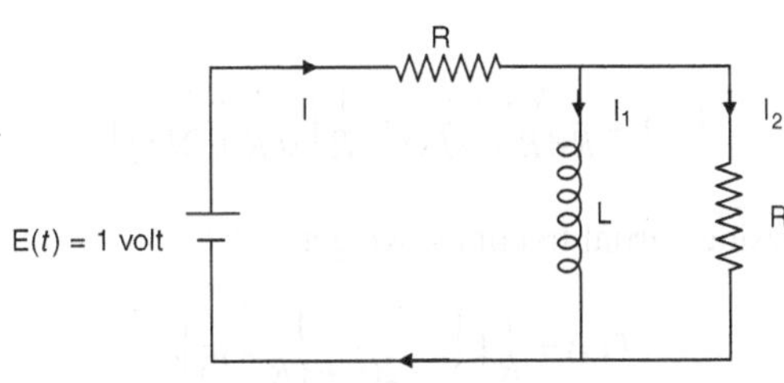

Figure 11.11

Solution. We note that $I = I_1 + I_2$ and so $RI = RI_1 + RI_2$, or equivalently, $RI_2 = RI - RI_1$. By Kirchhoff's law, we have

(a) In the closed loop containing R and L,

$$RI + L\frac{dI_1}{dt} = E = 1 \qquad (5)$$

(b) In the closed loop containing two resistances R,

$$RI + RI_2 = E = 1$$

or

$$RI + RI - RI_1 = 1$$

or

$$2RI - RI_1 = 1. \qquad (6)$$

We want to solve (5) and (6) under the conditions $I(0) = I_1(0) = 0$. Taking Laplace transform yields

$$RF(s) + L\{sG(s)I_1(0)\} = \frac{1}{s}$$

and

$$2R\,F(s) - RG(s) = \frac{1}{s}.$$

Using $I_1(0) = 0$, we have

$$RF(s) + LsG(s) = \frac{1}{s}. \qquad (7)$$

and

$$2RF(s) - RG(s) = \frac{1}{s}. \qquad (8)$$

Multiplying (7) by R and (8) by Ls and adding, we get

$$(R^2 + 2RLs)F(s) = \frac{R}{s} + L = \frac{R + Ls}{s}$$

or

$$F(s) = \frac{R + Ls}{Rs(R + 2Ls)} = \frac{1}{R}\left(\frac{R + Ls}{s(R + 2Ls)}\right).$$

Using partial fractions, we get

$$F(s) = \frac{1}{R}\left(\frac{1}{s} - \frac{1}{2(s + (R/2L))}\right).$$

Taking inverse Laplace transform yields

$$I(t) = \frac{1}{R}\left[1 - \frac{1}{2}e^{-\frac{R}{2L}t}\right] = \frac{1}{2R}\left[2 - e^{-\frac{R}{2L}t}\right].$$

11.3 DIFFERENCE EQUATIONS

A relationship between the values of a function $y(t)$ and the values of the function at different arguments $y(t + h)$, h constant, is called a *difference equation.* For example,

$$y(n+2) - y(n+1) + y(n) = 2$$

and

$$y(n+2) - 2y(n) + y(n-1) = 1$$

are difference equations.

A relation between the terms of a sequence $\{x_n\}$ is also a difference equation. For example,

$$x_{n+1} + 2x_n = 8$$

is a difference equation.

Difference equations (also called *recurrence relations*) are closely related to differential equations and their theory is basically the same as that of differential equations.

Order of a difference equation is the difference between the largest and smallest arguments occurring in the difference equation divided by the unit of increment.

For example, the order of the difference equation $a_{n+2} - 3a_{n+1} + 2a_n = 5^n$ is $\frac{n+2-n}{1} = 2.$

Solution of a difference equation is an expression for y_n which satisfies the given difference equation.

The aim of this section is to solve difference equations using Laplace transform.

We first make the following *observations*:

(A) Let $f(t) = a^{[t]}$, where $[t]$ is the greatest integer less than or equal to t and $a > 0$. Then $f(t)$ is of exponential order and by definition,

$$L\{f(t)\} = \int_0^\infty e^{-st} f(t)\,dt = \int_0^\infty e^{-st} a^{[t]} dt$$

$$= \int_0^1 e^{-st} a^0 dt + \int_1^2 e^{-st} a^1 dt + \int_2^3 e^{-st} a^2 dt + \ldots$$

$$= \frac{1 - e^{-s}}{s} + \frac{a(e^{-s} - e^{-2s})}{s} + \frac{a^2(e^{-2s} - e^{-3s})}{s} + \ldots$$

$$= \frac{1 - e^{-s}}{s}[1 + ae^{-s} + a^2e^{-s} + \ldots]$$

$$= \frac{1 - e^{-s}}{s(1 - ae^{-s})}\ (\text{Re}(s) > \max(0, \log a)).$$

(B) If $L^{-1}F\{(s)\} = f(t)$, then we know that

$$L^{-1}\{e^{-s}F(s)\} = \begin{cases} f(t-1) & \text{for } t > 1 \\ 0 & \text{for } t < 1 \end{cases}$$

Also, by observation (1) above, we have

$$L^{-1}\left\{\frac{1 - e^{-s}}{s(1 - ae^{-s})}\right\} = a^n \text{ for } n = 0, 1, 2, \ldots,$$

$$n \le t < n + 1.$$

Therefore,

$$L^{-1}\left\{\frac{(1 - e^{-s})e^{-s}}{s(1 - ae^{-s})}\right\}$$

$$= f(t-1)$$

$$= a^n \text{ for } n \le t - 1 < n + 1, n = 0, 1, 2,$$

$$= a^n \text{ for } n \le t < n + 1, n = 1, 2, 3, \ldots$$

(C) If $f(t) = na^{n-1}$ for $n \le t < n + 1$, $n = 0, 1, 2, \ldots$, then

$$L\{f(t)\} = \int_0^\infty e^{-st} f(t)\,dt$$

$$= \int_1^2 e^{-st}\,dt + 2a\int_2^3 e^{-st}\,dt + 4a\int_3^4 e^{-st}\,dt + \ldots$$

$$= \frac{e^{-s} - e^{-2s}}{s} + 2a\left[\frac{e^{-2} - e^{-3s}}{s}\right] + 4a^2\left[\frac{e^{-3s} - e^{4s}}{s}\right] + \ldots$$

$$= \frac{e^{-s}(1 - e^{-s})}{s}\,[1 + 2ae^{-s} + 4a^2 e^{-2s} + \ldots]$$

$$= \frac{e^{-s}(1 - e^{-s})}{s} \cdot \frac{1}{(1 - ae^{-s})^2} = \frac{e^{-s}(1 - e^{-s})}{s(1 - ae^{-s})^2}.$$

Hence

$$L^{-1}\frac{e^{-s}(1 + e^{-s})}{s(1 - ae^{-s})^2} = f(t) = na^{n-1},\ n = 0, 1, 2, \ldots$$

EXAMPLE 11.30

Solve

$$a_{n+2} - 4a_{n+1} + 3a_n = 0,\ a_0 = 0,\ a_1 = 1.$$

Solution. Let us define

$$y(t) = a_n,\ n \le t < n + 1,\ n = 0, 1, 2, \ldots$$

Then the given difference equation reduces to

$$y(t + 2) - 4y(t + 1) + 3y(t) = 0.$$

Thus

$$L\{y(t + 2)\} - 4L\{y(t + 1)\} + 3L\{y(t)\} = 0 \quad (9)$$

Now

$$L\{y(t + 2)\} = \int_0^\infty e^{-st} y(t + 2)\,dt$$

$$= \int_2^\infty e^{-s(u-2)} y(u)\,du,\ u = t + 2$$

$$= e^{2s}\int_0^\infty e^{-su} y(u)\,du - e^{2s}\int_0^2 e^{-su} y(u) du.$$

$$= e^{2s} L\{y(t)\} - e^{2s}\int_0^1 e^{-su}\,a_0 du - e^{2s}\int_1^2 e^{-su} a_1 du$$

$$= e^{2s} L\{y(t)\} - e^{2s}\left(\frac{e^{-s} - e^{-2s}}{s}\right) \text{ since } a_0 = 0,\ a_1 = 1$$

$$= e^{2s} L\{y(t)\} - \frac{e^s}{s}(1 - e^{-s}),$$

$$L\{y(t + 1)\} = \int_0^\infty e^{-st} y(t + 1)\,dt$$

$$= \int_1^\infty e^{-s(u-1)} y(u)\,du,\ u = t + 1$$

$$= \int_0^\infty e^{-s(u-1)} y(u)\,du - \int_0^1 e^{-s(u-1)} y(u)\,du$$

$$= e^s\int_0^\infty e^{-su} y(u)\,du - e^s\int_0^1 e^{-su} y(u)\,du$$

$$= e^s L\{y(t)\} - e^s\int_0^1 e^{-su} a_0 du$$

$$= e^s L\{y(t)\} \text{ since } a_0 = 0.$$

Hence (9) becomes

$$e^{2s} L\{y(t)\} - \frac{e^s}{s}(1 - e^{-s}) - 4e^s L\{y(t)\} + 3L\{y(t)\} = 0,$$

which yields

$$L\{y(t)\} = \frac{e^s(1 - e^{-s})}{s(e^{2s} - 4e^s + 3)}$$

$$= \frac{e^s(1 - e^{-s})}{2s}\left(\frac{1}{e^s - 3} - \frac{1}{e^s - 1}\right)$$

$$= \frac{1 - e^{-s}}{2s}\left(\frac{1}{1 - 3e^{-s}} - \frac{1}{1 - e^{-s}}\right)$$

$$= \frac{1 - e^{-s}}{2s(1 - 3e^{-s})} - \frac{1 - e^{-s}}{2s(1 - e^{-s})}$$

$$= \frac{1}{2} L\{3^{[t]}\} - \frac{1}{2} L\{1\}, \text{ by observation (1).}$$

Hence inversion yields

$$a_n = \frac{1}{2}[3^n - 1],\ n = 0, 1, 2, \ldots$$

EXAMPLE 11.31

Solve the difference equation

$$y(t + 1) - y(t) = 1,\ y(t) = 0,\ t < 1.$$

Solution. Taking Laplace transformation of both sides, we get

$$L\{y(t+1)\} - L\{y(t)\} = L\{1\}.$$

But, as in Example 11.30, we have

$$L\{y(t+1)\} = e^s L\{y(t)\},$$

and so

$$e^s L\{y(t)\} - L\{y(t)\} = \tfrac{1}{s}$$

or

$$L\{y(t)\} = \frac{1}{s(e^s - 1)}.$$

Taking inverse Laplace transform, we have

$$y(t) = L^{-1}\left\{\frac{1}{s(e^s - 1)}\right\}$$

$$= [t],\ t > 0$$

EXAMPLE 11.32

Solve

$$a_{n+2} - 4a_{n+1} + 3a_n = 5^n,\ a_0 = 0,\ a_1 = 1$$

Solution. We define

$$y(t) = a_n,\ n \le t < n+1,\ n = 0, 1, 2, \ldots$$

Then the difference equation becomes

$$y(t+2) - 4y(t+1) + 3y(t) = 5^n. \quad (10)$$

By observation (B) and Example 11.30, we have

$$L[5^n] = \frac{1 - e^{-s}}{s(1 - 5e^{-s})},$$

$$L\{y(t+2)\} = e^{2s} L\{y(t)\} - \tfrac{e^s}{s}(1 - e^{-s}),$$

$$L\{y(t+1)\} = e^s L\{y(t)\}.$$

Taking Laplace transform of both sides of (10), we have

$$L\{y(t+2)\} - 4L\{y(t+1)\} + 3L\{y(t)\} = L\{5^{[t]}\}$$

or

$$e^{2s} L\{y(t)\} - \tfrac{e^s}{s}(1 - e^{-s}) - 4e^s L\{y(t)\} + 3L\{y(t)\} = L\{5[t]\}$$

or

$$\{e^{2s} - 4e^s + 3\} L\{y(t)\} = \frac{e^s(1 - e^{-s})}{s} + L\{5^{[t]}\}.$$

Hence

$$L\{y(t)\} = \frac{e^s(1 - e^{-s})}{s(e^{2s} - 4e^s + 3)} + \frac{L\{5^{[t]}\}}{e^{2s} - 4e^s + 3}$$

$$= \tfrac{1}{2} L\{3^{[t]}\} - \tfrac{1}{2} L\{1\} + \frac{L\{5^{[t]}\}}{e^{2s} - 4e^s + 3}.$$

But

$$\frac{L\{5^{[t]}\}}{e^{2s} - 4e^s + 3} = \frac{1 - e^{-s}}{s(1 - 5e^{-s})} \cdot \frac{1}{e^{2s} - 4e^s + 3}$$

$$= \frac{e^s - 1}{s(e^s - 5)(e^s - 3)(e^s - 1)}$$

$$= \frac{e^s - 1}{s}\left(\frac{1/8}{e^s - 5} - \frac{1/4}{e^s - 3} + \frac{1/8}{e^s - 1}\right)$$

$$= \frac{1 - e^{-s}}{s}\left(\frac{1/8}{1 - 5e^{-s}} - \frac{1/4}{1 - 3e^{-s}} + \frac{1/8}{1 - e^{-s}}\right)$$

$$= \tfrac{1}{8} L\{1\} + \tfrac{1}{8} L\{5^{[t]}\} - \tfrac{1}{4} L\{3^{[t]}\}.$$

Hence

$$L\{y(t)\} = \tfrac{-3}{8} L\{1\} + \tfrac{1}{4} L\{3^{[t]}\} + \tfrac{1}{8} L\{5^{[t]}\}$$

and so

$$a_n = \frac{-3}{8} + \tfrac{1}{4} 3^n + \tfrac{1}{8} 5^n.$$

EXAMPLE 11.33

Find explicit formula (solution) for *Fibonacci sequence*:

$$a_{n+2} = a_{n+1} + a_n,\quad a_0 = 0,\ a_1 = 1.$$

Solution. Define

$$y(t) = a_n,\quad n \le t < n+1,\quad n = 0, 1, 2, \ldots$$

Then the given difference equation reduces to

$$y(t+2) - y(t+1) - y(t) = 0.$$

Taking Laplace transform, we have

$$L\{y(t+2)\} - L\{y(t+1)\} - L\{y(t)\} = 0.$$

But

$$L\{y(t+2)\} = e^{2s} L\{y(t)\} - \frac{e^s(1-e^{-s})}{s}$$

$$L\{y(t+1)\} = e^s L\{y(t)\}.$$

Therefore, we get

$$(e^{2s} - e^s - 1) L\{y(t)\} = \frac{e^s(1-e^{-s})}{s}$$

or

$$L\{y(t)\} = \frac{e^s(1-e^{-s})}{s(e^{2s} - e^s - 1)}$$

$$= \frac{e^s(1-e^{-s})}{s}\left[\frac{\frac{1}{\sqrt{5}}}{e^s - \frac{1+\sqrt{5}}{2}} - \frac{\frac{1}{\sqrt{5}}}{e^s - \frac{1-\sqrt{5}}{2}}\right]$$

$$= \frac{1-e^{-s}}{s}\left[\frac{1}{\sqrt{5}}\left(\frac{1}{1-\frac{1+\sqrt{5}}{2}e^{-s}} - \frac{1}{1-\frac{1-\sqrt{5}}{2}e^{-s}}\right)\right]$$

$$= \frac{1}{\sqrt{5}}\left[L\left\{\left(\frac{1+\sqrt{5}}{2}\right)^{[t]}\right\} - L\left\{\left(\frac{1-\sqrt{5}}{2}\right)^{[t]}\right\}\right].$$

Hence

$$a_n = \frac{1}{\sqrt{5}}\left[\left(\frac{1+\sqrt{5}}{2}\right)^n - \left(\frac{1-\sqrt{5}}{2}\right)^n\right], n \geq 0.$$

EXAMPLE 11.34

Solve the *differential-difference equation*

$$y''(t) - y(t-1) = f(t), y(t) = 0, y'(t) = 0 \text{ for } t \leq 0,$$

$$f(t) = \begin{cases} 0 \text{ for } t \leq 0 \\ 2t \text{ for } t > 0. \end{cases}$$

Solution. Taking Laplace transform of both sides, we get

$$L\{y''(t)\} - L\{y(t-1)\} = L\{f(t)\}$$

or

$$s^2 L\{y(t)\} - sy(0) - y(0) - e^{-s} L\{y(t)\} = \frac{2}{s^2}$$

or

$$(s^2 - e^{-s}) L\{y(t)\} = \frac{2}{s^2}$$

or

$$L\{y(t)\} = \frac{2}{s^2(s^2 - e^{-s})} = \frac{2}{s^4\left(1 - \frac{e^{-s}}{s^2}\right)}$$

$$= \frac{2}{s^4}\left(1 + \frac{e^{-s}}{s^2} + \frac{e^{-2s}}{s^4} + \frac{e^{-3s}}{s^6} + \ldots\right)$$

$$= 2\left(\frac{1}{s^4} + \frac{e^{-s}}{s^6} + \frac{e^{-2s}}{s^8} + \frac{e^{-3s}}{s^{10}} + \ldots\right)$$

$$= 2\sum_{n=0}^{\infty} \frac{e^{-ns}}{s^{2n+4}}.$$

But

$$L^{-1}\left\{\frac{e^{-ns}}{s^{n+4}}\right\} = \frac{(t-n)^{2n+3}}{(2n+3)!}.$$

Hence

$$y(t) = 2\sum_{n=0}^{[t]} \frac{(t-n)^{2n+3}}{(2n+3)!}.$$

11.4 INTEGRAL EQUATIONS

Equations of the form

$$f(t) = g(t) + \int_a^b K(t, u) f(u)\, du$$

and

$$g(t) = \int_a^b K(t, u) f(u)\, du,$$

where the function $f(t)$ to be determined appears under the integral sign are called *integral equations.*

In an integral equation, $K(t, u)$ is called the kernel. If a and b are constants, the equation is called a *Fredholm integral equation.* If a is a constant and $b = t$, then the equation is called a *Volterra integral equation.*

If the kernel $K(t, u)$ is of the form $K(t - u)$, then the integral $\int_0^t K(t-u) f(u)\, du$ represents convolution. Thus, we have

$$f(t) = g(t) + \int_0^t K(t-u)f(u)du = g(t) + K(t)*f(t).$$

Such equations are called *convolution-type integral equations*. Taking Laplace transform of convolution-type integral equation, we have

$$L\{f(t)\} = L\{g(t)\} + L\{K(t)*f(t)\}$$
$$= L\{g(t)\} + L\{K(t)\}\, L\{f(t)\},$$

by using Convolution theorem. Hence

$$(1 - L\{K(t)\}\, L\{f(t)\}) = L\{g(t)\},$$

which implies

$$L\{f(t)\} = \frac{L\{g(t)\}}{1 - L\{K(t)\}}.$$

Taking inverse Laplace transform yields the solution $f(t)$.

EXAMPLE 11.35

Solve the integral equation

$$f(t) = e^{-t} + \int_0^t \sin(t-u)\, f(u)\, du.$$

Solution. Taking Laplace transform of both sides of the given equation, we get

$$L\{f(t)\} = L\{e^{-t}\} + L\{\sin t\}\ L\{f(t)\},$$

which yields

$$L\{f(t)\} = \frac{L\{e^{-t}\}}{1 - L\{\sin t\}} = \frac{s^2+1}{s^2(s+1)}.$$

Using partial fractions, we obtain

$$L\{f(t)\} = \frac{2}{s+1} + \frac{1}{s^2} - \frac{1}{s}.$$

Taking inverse Laplace transform yields

$$f(t) = 2e^{-t} + t - 1.$$

EXAMPLE 11.36

Solve

$$\int_0^t f(u)\, f(t-u)\, du = 16 \sin 4t.$$

Solution. The given equation in convolution form is

$$f(t) * f(t) = 16 \sin 4t.$$

Taking Laplace transform, we get

$$L\{f(t) * f(t)\} = 16\, L\{\sin 4t\}$$

or

$$L\{f(t)\}L\{f(t)\} = 16\, L\{\sin 4t\}$$

(using convolution theorem).

or

$$[L\{f(t)\}]^2 = \frac{16(4)}{s^2+16} = \frac{64}{s^2+16}.$$

or

$$L\{f(t)\} = \frac{\pm 8}{\sqrt{s^2+16}}.$$

Taking inverse Laplace transform yields

$$f(t) = \pm 8\, J_0(4t),$$

where J_0 is Bessel's function of order zero.

Definition 11.1. The convolution-type integral equation of the form

$$\int_0^t \frac{f(u)}{(t-u)^n} du = g(t),\ 0 < n < 1$$

is called *Abel's integral equation*.

We consider below examples of this type of integral equations.

EXAMPLE 11.37

Solve the integral equation

$$1 + 2t - t^2 = \int_0^t f(u) \frac{1}{\sqrt{t-u}}\, du.$$

Solution. The given equation is a special case of Abel's integral equation. The convolution form of this equation is

$$1 + 2t - t^2 = f(t) * \frac{1}{\sqrt{t}}.$$

Taking Laplace transform yields

$$L\{f(t)\}\, L\left\{\frac{1}{\sqrt{t}}\right\} = L\{1\} + 2L\{t\} - L\{t^2\}$$

or

$$L\{f(t)\}\sqrt{\frac{\pi}{s}} = \frac{1}{s} + \frac{2}{s^2} - \frac{2}{s^3}$$

or

$$L\{f(t)\} = \frac{1}{\sqrt{\pi}}\left[\frac{1}{s^{1/2}} + \frac{2}{s^{3/2}} - \frac{2}{s^{5/2}}\right].$$

Taking inverse transform, we get

$$f(t) = \frac{1}{\sqrt{\pi}}\left[\frac{t^{-1/2}}{\Gamma(1/2)} + \frac{2t^{1/2}}{\Gamma(3/2)} - \frac{2t^{3/2}}{\Gamma(5/2)}\right]$$

$$= \frac{1}{\sqrt{\pi}}\frac{t^{-1/2}}{\sqrt{\pi}} + \frac{2t^{1/2}}{(1/2)\sqrt{\pi}} - \frac{2t^{3/2}}{(3/2)(1/2)\sqrt{\pi}}$$

$$= \frac{1}{\pi}\left[t^{-1/2} + 4t^{1/2} - 8t^{3/2}\right].$$

EXAMPLE 11.38 (Tautochrone Curve)

A particle (bead) of mass m is to slide down a frictionless curve such that the duration T_0 of descent due to gravity is independent of the starting point. Find the shape of such curve (known as *Tautochrone curve*).

Solution. Velocity of the bead at the starting point is zero since it starts from rest at that point, say P with co-ordinates (x, y). Let $Q = (x, u)$ be some intermediate point during the motion. Let the origin O be the lowest point of the curve (Figure 11.12). Let the length of the arc OQ be s.

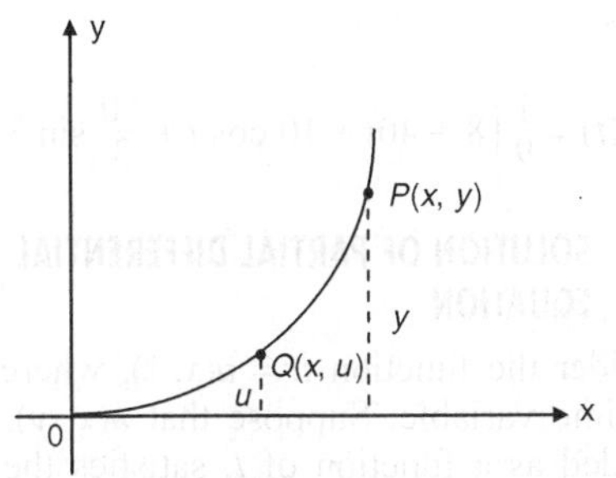

Figure 11.12

By law of conservation of energy, potential energy at P + kinetic energy at P = potential energy at Q + Kinetic energy at Q, that is,

$$mgy + 0 = mgu + \frac{1}{2}m\left(\frac{ds}{dt}\right)^2,$$

where $\frac{ds}{dt}$ is the instantaneous velocity of the particle at Q.

Thus

$$\left(\frac{ds}{dt}\right)^2 = 2g(y-u)$$

and so

$$\frac{ds}{dt} = -\sqrt{2g(y-u)},$$

negative sign since s decreases with time. The total time T_0 taken by the particle to go from P to Q is

$$T_0 = \int_0^{T_0} dt = \int_y^0 \frac{-ds}{\sqrt{2g(y-u)}} = \int_0^y \frac{ds}{\sqrt{2g(y-u)}}$$

If $\frac{ds}{du} = f(u)$, then $ds = f(u)\,du$ and so

$$T_0 = \frac{1}{\sqrt{2g}}\int_0^y \frac{f(u)}{\sqrt{y-u}}\,du$$

The convolution form of this integral equation is

$$T_0 = \frac{1}{\sqrt{2g}} f(y) * \frac{1}{\sqrt{y}}.$$

Taking Laplace transform of both sides and using Convolution theorem, we have

$$L\{T_0\} = \frac{1}{\sqrt{2g}} L\{f(y)\}\, L\left\{\frac{1}{\sqrt{y}}\right\}$$

or

$$L\{f(y)\} = \frac{\sqrt{2g}\, T_0/s}{\sqrt{T/s}} = \frac{\sqrt{2g/\pi}}{s^{1/2}}T_0 = \frac{C_0}{s^{1/2}},$$

where C_0 is a constant. Inverse Laplace transform then yields

$$f(y) = \frac{C}{\sqrt{y}}.$$

Since $f(y) = \frac{ds}{dy} = \sqrt{1 + \left(\frac{dx}{dy}\right)^2}$, we get

$$1 + \left(\frac{dx}{dy}\right)^2 = \frac{C^2}{y}$$

or

$$\left(\frac{dx}{dy}\right)^2 = \frac{C^2}{y} - 1 = \frac{C^2 - y}{y}$$

or

$$\frac{dx}{dy} = \sqrt{\frac{C^2 - y}{y}}$$

or

$$x = \int \sqrt{\frac{C^2 - y}{y}}\, dy.$$

Putting $y = C^2 \sin^2 \frac{\theta}{2}$, we get

$$x = \frac{C^2}{2}(\theta + \sin\theta),\ y = \frac{C^2}{2}(1 - \cos\theta),$$

which are the parametric equations of a *cycloid.*

11.5 INTEGRO-DIFFERENTIAL EQUATIONS

An integral equation in which various derivatives of the unknown function $f(t)$ are also present is called an *integro-differential equation.* These types of equations can also be solved by the method of Laplace transform.

EXAMPLE 11.39

Solve the following integro-differential equation:

$$y'(t) = \int_0^t y(u)\cos(t - u)\, du,\ y(0) = 1.$$

Solution. We write the given equation in convolution form as

$$y'(t) = y(t) * \cos t.$$

Taking Laplace transform and using Convolution theorem yields

$$L\{y'(t)\} = L\{y(t)\}\, L\{\cos t\}$$

or

$$sL\{y(t)\} - y(0) = L\{y(t)\}\,\frac{s}{s^2 + 1}$$

or

$$\left(s - \frac{s}{s^2 + 1}\right) L\{y(t)\} = 1, \text{ since } y(0) = 1$$

or

$$L\{y(t)\} = \frac{s^2 + 1}{s^3} = \frac{1}{s} + \frac{1}{s^3}.$$

Taking inverse Laplace transform, we get

$$y(t) = 1 + \frac{1}{2}t^2.$$

EXAMPLE 11.40

Solve

$$y'(t) + 5\int_0^t y(u)\cos 2(t - u)\, dy = 10,\ y(0) = 2.$$

Solution. Convolution form of the equation is

$$y'(t) + 5\cos t * y(t) = 10.$$

Taking Laplace transform and using Convolution theorem, we have

$$sL\{y(t)\} - y(0) + \frac{5sL\{y(t)\}}{s^2 + 4} = \frac{10}{s}$$

or

$$L\{y(t)\} = \frac{2s^3 + 10s^2 + 8s + 40}{s^2(s^2 + 9)}$$

$$= \frac{1}{9}\left\{\frac{8}{s} + \frac{40}{s^2} + \frac{10s}{s^2 + 9} + \frac{9 + 50}{s^2 + 9}\right\}.$$

Hence

$$y(t) = \frac{1}{9}\left(8 + 40t + 10\cos t + \frac{50}{3}\sin 3t\right).$$

11.6 SOLUTION OF PARTIAL DIFFERENTIAL EQUATION

Consider the function $u = u(x, t)$, where $t \geq 0$ is a time variable. Suppose that $u(x, y)$, when regarded as a function of t, satisfies the sufficient conditions for the existence of its Laplace transform.

Denoting the Laplace transform of $u(x, t)$ with respect to t by $U(x, s)$, we see that

$$U(x, s) = L\{u(x, t)\} = \int_0^\infty e^{-st}\, u(x, t)\, dt.$$

The variable x is the untransformed variable. For example,

$$L\{e^{a(x-t)}\} = e^{ax}\, L\{e^{-at}\} = e^{ax}\,\frac{1}{s+a}.$$

Theorem 11.2. Let $u(x, t)$ be defined for $t \geq 0$. Then

(a) $L\left\{\dfrac{\partial u}{\partial x}\right\} = \dfrac{d}{dx}(U(x, s))$

(b) $L\left\{\dfrac{\partial u}{\partial t}\right\} = sU(x, s) - u(x, 0)$

(c) $L\left\{\dfrac{\partial^2 u}{\partial x^2}\right\} = \dfrac{d^2}{dx^2}(U(x, s))$

(d) $L\left\{\dfrac{\partial^2 u}{\partial t^2}\right\} = s^2 U(x, s) - s\, u(x, 0) - \dfrac{\partial u}{\partial t}(x, 0).$

Proof: (a) We have, by Leibnitz?s rule for differentiating under the integration,

$$L\left\{\frac{\partial u}{\partial x}\right\} = \int_0^\infty e^{-st}\frac{\partial u}{\partial x}\, dt = \frac{d}{dx}\int_0^\infty e^{-st}\, u(x, t)\, dt$$

$$= \frac{d}{dx}(U(x, s)).$$

(b) Integrating by parts, we get

$$L\frac{\partial u}{\partial t} = \int_0^\infty e^{-st}\,\frac{\partial u(x, t)}{\partial t}\, dt$$

$$= \lim_{T\to 0}\int_0^T e^{-st}\,\frac{\partial u(x, t)}{\partial t}\, dt$$

$$= \lim_{T\to 0}\left\{[e^{-st}\, u(x, t)]_0^T + s\int_0^T e^{-st}\, u(x, t)\, dt\right\}$$

$$= s\int_0^T e^{-st}\, u(x, t) dt - u(x, 0)$$

$$= sU(x, s) - u(x, 0).$$

(c) Taking $V = \dfrac{\partial u}{\partial x}$, we have by (a),

$$L\left\{\frac{\partial^2 u}{\partial x^2}\right\} = L\left\{\frac{\partial V}{\partial x}\right\} = \frac{d}{dx}(V(x, s))$$

$$= \frac{d}{dx}\left(\frac{d}{dx}(U(x, s))\right) = \frac{d^2}{dx^2}(U(x, s)).$$

(d) Let $v = \dfrac{\partial u}{\partial t}$. Then

$$L\left\{\frac{\partial^2 u}{\partial t^2}\right\} = L\left\{\frac{\partial v}{\partial t}\right\} = sV(x, s) - v(x, 0)$$

$$= s[sU(x, s) - u(x, 0) - \frac{\partial u}{\partial t}(x, 0)$$

$$= s^2 U(x, s) - u(x, 0) - \frac{\partial u}{\partial t}(x, 0).$$

Theorem 11.2. suggest that if we apply Laplace transform to both sides of the given partial differential equation, we shall get an ordinary differential equation in U as a function of single variable x. This ordinary differential equation is then solved by the usual methods.

EXAMPLE 11.41

Solve

$$\frac{\partial u}{\partial x} = \frac{\partial u}{\partial t},\; u(x, 0) = x,\; u(0, t) = t,$$

Solution. Taking Laplace transform, we get

$$L\left\{\frac{\partial u}{\partial x}\right\} = L\left\{\frac{\partial u}{\partial t}\right\}.$$

Using Theorem 11.2, we get

$$\frac{d}{dx}[U(x, s)] = s\, U(x, s) - u(x, 0) = s\, U(x, s) - x.$$

Thus, we have first order differential equation

$$\frac{d}{dx}[U(x, s)] - sU(x, s) = -x$$

The integrating factor is

$$e^{\int -s\, dx} = e^{-sx}.$$

Therefore,

$$U(x, s)\, e^{-sx} = \int -xe^{-sx}\, dx$$

$$= -\left[x\frac{e^{-sx}}{-s} - \int \frac{e^{-sx}}{-s}\, dx\right] + C$$

$$= \frac{x e^{-sx}}{s} + \frac{e^{-sx}}{s^2} + C$$

(constant of integration).

This yields

$$U(x, s) = \frac{x}{s} + \frac{1}{s^2} + Ce^{sx}. \qquad (11)$$

Now the boundary condition $u(0, t)$ is a function of t. Taking Laplace transform of this function, we have

$$U(0, s) = L\{u(0, t)\} = L\{t\} = \frac{1}{s^2}.$$

Then taking $x = 0$ in (11), we have

$$\frac{1}{s^2} = \frac{1}{s^2} + C$$

and so C = 0. Thus, we have

$$U(x, s) = \frac{x}{s} + \frac{1}{s^2}.$$

Taking inverse Laplace transform, we have

$$u(x, t) = x + t.$$

EXAMPLE 11.42

Solve the partial differential equation

$$\frac{\partial u}{\partial t} + x\frac{\partial u}{\partial x} = x, x > 0, t > 0$$

with the initial and boundary conditions $u(x, 0) = 0, x > 0$ and $u(0, t) = 0$ for $t > 0$.

Solution. Taking Laplace transform with respect to t, we get

$$L\left\{\frac{\partial u}{\partial t}\right\} + L\left\{x\frac{\partial u}{\partial x}\right\} = L\{x\}$$

which yields

$$sU(x, s) - u(x, 0) + x\frac{d}{dx}U(x, s) = \frac{x}{s}.$$

Since $u(x, 0) = 0$, this reduces to

$$\frac{d}{dx}U(x, s) + \frac{s}{x}U(x, s) = \frac{1}{s}. \qquad (12)$$

The integrating factor is

$$e^{\int \frac{s}{x}dx} = e^{s\log x} = x^s.$$

Therefore solution of (12) is

$$U(x, s)x^s = \frac{1}{s}\int x^s\, dx + C$$

$$= \frac{1}{s}\,\frac{x^{s+1}}{s+1} + C = \frac{x^{s+1}}{s(s+1)} + C$$

and so

$$U(x, s) = \frac{x}{s(s+1)} + C \text{ (constant of integration).} \qquad (13)$$

Now since $U(0, t) = 0$, its Laplace transform is 0, that is, $U(0, s) = 0$. Therefore, (13) implies $C = 0$.

Hence

$$U(x, s) = \frac{x}{s(s+1)} = x\left(\frac{1}{s} - \frac{1}{s+1}\right).$$

Taking inverse Laplace transform, we get the solution as

$$u(x, t) = x(1 - e^{-t}).$$

EXAMPLE 11.43

Solve

$$\frac{\partial u}{\partial t} = \frac{\partial^2 u}{\partial x^2}, x > 0, t > 0$$

under the conditions

$$u(x, 0) = 1,\ u(0, t) = 0 \text{ and } \lim_{x\to\infty} u(x, t) = 1.$$

Solution. The given equation is heat conduction equation in a solid,where $u(x, t)$ is the temperature at position x at any time t and diffusivity is 1. The boundary condition $u(0, t) = 0$ indicates that temperature at $x = 0$ is 0 and $\lim_{x\to\infty} u(x, t) = 1$ indicates that the temperature for large values of x is 1 whereas $u(x, 0) = 1$ represents the initial temperature 1 in the semi-infinite medium $(x > 0)$ (Figure 11.13).

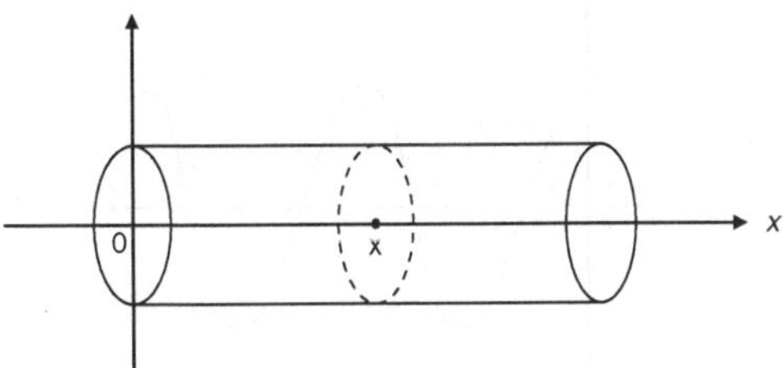

Figure 11.13

Taking Laplace transform, yields

$$sU(x, s) - u(x, 0) = \frac{d^2}{dx^2}U(x, s).$$

Since $u(x, 0) = 1$, we have

$$\frac{d^2}{dx^2}U(x, s) - sU(x, s) = -1.$$

The general solution of this equation is

$$U(x, s) = \text{C.F.} + \text{P.I.}$$

$$= [c_1 e^{\sqrt{s}x} + c_2 e^{-\sqrt{s}x}] + \frac{1}{s}. \qquad (14)$$

The conditions $u(0, t) = 0$ yields

$$U(0, s) = L\{u(0, t)\} = 0, \qquad (15)$$

whereas $\lim_{x\to\infty} u(x, t) = 1$ yields

$$\lim_{x\to\infty} U(x, s) = \lim_{x\to\infty} L\{u(x, t)\} = L\left\{\lim_{x\to\infty} u(x, t)\right\}$$

$$= L\{1\} = \frac{1}{s}. \qquad (16)$$

Now (14) and (16) imply $c_1 = 0$. Then (15) implies $c_2 = -\frac{1}{s}$. Hence

$$U(x, s) = \frac{1}{s} - \frac{e^{-\sqrt{s}x}}{s}.$$

Taking inverse Laplace transform, we get

$$u(x, t) = 1 - L^{-1}\left\{\frac{e^{-\sqrt{s}x}}{s}\right\}$$

$$= 1 - \left(1 - \text{erf}\left(\frac{x}{2\sqrt{t}}\right)\right) = erf\left(\frac{x}{2\sqrt{t}}\right).$$

EXAMPLE 11.44

Solve

$$\frac{\partial u}{\partial t} = \frac{\partial^2 u}{\partial x^2}, x > 0, t > 0$$

subject to the conditions

$$u(x, 0) = 0, x > 0,$$

$$u(0, t) = t, t > 0 \text{ and } \lim_{x\to\infty} u(x, t) = 0.$$

Solution. Taking Laplace transform, we have

$$\frac{d^2}{dx^2}U(x, s) - sU(x, s) = 0.$$

The solution of this equation is

$$U(x, s) = c_1 e^{\sqrt{s}x} + c{+} e^{-\sqrt{s}x} \qquad (17)$$

Since $\lim_{x\to\infty} u(x, t) = 0$, we have

$$\lim_{x\to\infty} U(x, s) = \lim_{x\to\infty} L\{u(x, t)\}$$

$$= L\left\{\lim_{x\to\infty} u(x, t)\right\} = L\{0\} = 0(\text{finite}).$$

Therefore, $c_1 = 0$ and (17) reduces to

$$U(x, s) = c_2 e^{-\sqrt{s}x}. \qquad (18)$$

Also, since $u(0, t) = t$, we have

$$U(0, s) = \frac{1}{s^2}.$$

Hence, (18) yields $c_2 = \frac{1}{s^2}$. Thus

$$U(x, s) = \frac{1}{s^2} e^{-\sqrt{s}x}.$$

Since $L^{-1}\{e^{-\sqrt{s}}x\} = \frac{x}{2\sqrt{\pi t^3}} e^{-\frac{x^2}{4t}}$, by Convolution theorem, we have

$$u(x, t) = \int_0^t (t-u)\frac{x}{2\sqrt{\pi u^3}} e^{-\frac{x^2}{4u}}\, du.$$

Putting $\lambda = \frac{x^2}{4u}$, we get

$$u(x, t) = \frac{2}{\sqrt{\pi}} \int_{x/2\sqrt{t}}^{\infty} e^{-\lambda^2}\left(t - \frac{x^2}{4\lambda^2}\right) d\lambda.$$

EXAMPLE 11.45

Solve

$$\frac{\partial u}{\partial t} = 2\frac{\partial^2 u}{\partial x^2}$$

subject to the conditions $u(0, t) = 0$, $u(5, t) = 0$, $u(x, 0) = \sin \pi x$.

Solution. Taking Laplace transform and using $u(x, 0) = \sin \pi x$, we get

$$\frac{d^2}{dx^2} U(x, s) - \frac{s}{2} U(x, s) = -\frac{1}{2} \sin \pi x.$$

Complementary function for this equation is $c_1 e^{\sqrt{\frac{s}{2}}x} + c_2 e^{-\sqrt{\frac{s}{2}}x}$ and particular integral is $\frac{1}{2(\pi^2 + (s/2))} \sin \pi x$. Thus the complete solution is

$$U(x, s) = c_1 e^{\sqrt{\frac{s}{2}}x} + c_2 e^{-\sqrt{\frac{s}{2}}x} + \frac{1}{2(\pi^2 + (s/2))} \sin \pi x \qquad (19)$$

Since $u(0, t) = 0$, we have $U(0, t) = 0$ and since $u(5, t) = 0$, $U(5, t) = 0$. Therefore, (19) gives

$$c_1 + c_2 = 0 \text{ and } c_1 e^{5\sqrt{s/2}} + c_2 e^{5(-\sqrt{s/2})} = 0.$$

These relations imply $c_1 = c_2 = 0$. Hence

$$U(x, s) = \frac{1}{2(\pi^2 + s/2)} \sin \pi x = \frac{1}{s + 2\pi^2} \sin \pi x.$$

Taking inverse Laplace transform, we get

$$U(x, t) = e^{-2\pi^2 t} \sin \pi s.$$

EXAMPLE 11.46

Solve one-dimensional wave equation

$$\frac{\partial^2 u}{\partial t^2} = a^2 \frac{\partial^2 y}{\partial x^2}, x > 0, t > 0$$

subject to the condition $y(x, 0) = 0$, $x > 0$, $y_t(x, 0) = 0$, $x > 0$, $y(0, t) = \sin \omega t$ and $\lim\limits_{x\to\infty} y(x, t) = 0$.

Solution. The displacement is only in the vertical direction and is given by $y(x, t)$ at position x and time t. For a vibrating string, the constant a equals $\sqrt{\frac{T}{\rho}}$, where T is tension in the string and ρ is mass per unit length of the vibrating string (Figure 11.14).

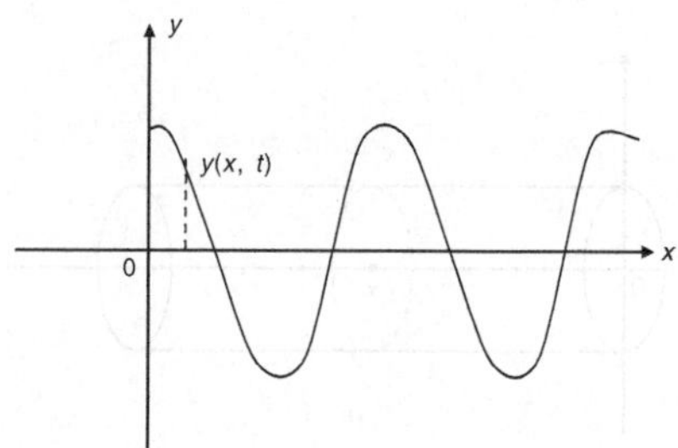

Figure 11.14

Taking Laplace transform, we get

$$s^2 Y(x, s) - sy(x, 0) - y_t(x, 0) - a^2 \frac{d^2}{dx^2} Y(x, s) = 0$$

or

$$\frac{d^2}{ds^2} Y(x, s) - \frac{s^2}{a^2} = 0 \qquad (20)$$

The general solution of (20) is

$$Y(x, s) = c_1 e^{\frac{s}{a}x} + c_2 e^{-\frac{s}{a}x} \qquad (21)$$

The condition $\lim\limits_{x\to\infty} y(x, t)$ implies $c_1 = 0$. Since $y(0, t) = \sin \omega t$, we have

$$Y(0, s) = \{y(0,t)\} = \frac{\omega}{s^2 + \omega^2}.$$

Therefore, (21) implies $c_2 = \frac{\omega}{s^2 + \omega^2}$ and so

$$Y(x, s) = \frac{\omega}{s^2 + \omega^2} e^{-\frac{s}{a}x}.$$

Taking inverse Laplace transform, we have

$$y(x, t) = \begin{cases} \sin \omega \left(t - \frac{x}{a}\right) & \text{for } t > \frac{x}{a} \\ 0 & \text{for } t < \frac{x}{a}. \end{cases}$$

$$= \sin \omega \left(t - \frac{x}{a}\right) H\left(t - \frac{x}{a}\right).$$

EXAMPLE 11.47

Solve

$$\frac{\partial^2 y}{\partial t^2} = \frac{\partial^2 y}{\partial x^2}, \text{ for } 0 < x < 1, t > 0$$

subject to $y(x, 0) = 0$, $0 < x < 1$; $y(0, t) = 0$, $t > 0$, $y(1, t) = 0$, $t > 0$ and $y_t(x, 0) = x$, $0 < x < 1$.

Solution. Taking Laplace transform and using $y(x, 0) = 0$ and $y_t(x, 0) = x$, we get

$$\frac{d^2Y(x, s)}{dt^2} - s^2Y(x, s) = x,$$

whose solution is given by

$$Y(x, s) = c_1 \cosh sx + c_2 \sinh sx - \frac{x}{s^2}.$$

Now $y(0, t) = 0$ implies that $Y(0, s) = 0$ and so $c_1 = 0$. Similarly, $y(1, t) = 0$ implies $Y(1, s) = 0$ and so $c_2 \sinh s - \frac{1}{s^2} = 0$. Thus $c_2 = \frac{1}{s^2 \sinh x}$. Hence

$$Y(x, s) = \frac{1}{s^2 \sinh x} \cdot \sinh sx - \frac{x}{s^2}.$$

This function has simple poles at $n\pi i$, $n = \pm 1, \pm 2, \ldots$, and a pole of order 2 at $s = 0$. Now

$$\text{Res}(n\pi i) = \lim_{s \to n\pi i} (s - n\pi i)e^{ts} \cdot \frac{\sinh sx}{s^2 \sinh s}$$

$$= \lim_{s \to n\pi i} \frac{(s - n\pi i)}{\sinh s} \lim_{s \to n\pi i} e^{ts} \frac{\sinh sx}{s^2}$$

$$= \frac{1}{\cosh n\pi i} \cdot e^{n\pi i t} \frac{\sinh n\pi i x}{n^2\pi^2}$$

$$= \frac{(-1)^{n+1}}{n^2\pi^2} e^{n\pi i t} \sin n\pi x,$$

$\text{Res}(0) = xt.$

Hence, by Complex inversion formula,

$$y(x, t) = xt + \sum_{n=1}^{-\infty} \frac{(-1)^{n+1}}{n^2\pi^2} e^{n\pi i t} \sin n\pi x - xt$$

$$= \frac{2}{\pi^2} \sum_{n=1}^{\infty} \frac{(-1)^{n+1}}{n^2} \sin n\pi x \sin n\pi t.$$

11.7 EVALUATION OF INTEGRALS

Laplace transforms can be used to evaluate certain integrals. In some cases the given integral is a special case of a Laplace transform for a particular value of the transform variable s. To evaluate an integral containing a free parameter, we first take Laplace transform of the integrand with respect to the free parameter. The resulting integral is then easily evaluated. Then we apply inverse Laplace transform to get the value of the given integral. In some cases, Theorem 9.9, regarding Laplace transform is used to evaluate the given integral.

EXAMPLE 11.48

Evaluate the integral

$$I = \int_0^\infty e^{-t} \frac{\sin t}{t} dt,$$

and show that $\int_0^\infty \frac{\sin t}{t} dt = \frac{\pi}{2}$.

Solution. We know that

$$L\left\{\frac{\sin t}{t}\right\} = \int_0^\infty e^{-st} \frac{\sin t}{t} dt = \tan^{-1} \frac{1}{s}. \quad (22)$$

Setting $s = 1$, we get

$$\int_0^\infty e^{-t} \frac{\sin t}{t} dt = \tan^{-1} 1 = \frac{\pi}{4}.$$

Further, letting $s \to 0$ in (22), we get

$$\int_0^\infty \frac{\sin t}{t} dt = \tan^{-1} \infty = \frac{\pi}{2}.$$

EXAMPLE 11.49

Evaluate the integral

$$\int_0^\infty \frac{\sin tx}{x(1 + x^2)} dx.$$

Solution. Let

$$f(t) = \int_0^\infty \frac{\sin tx}{x(1 + x^2)} dx.$$

Taking Laplace transform with respect to t, we have

$$F(s) = \int_0^\infty \frac{1}{x(1 + x^2)} dx \int_0^\infty e^{-st} \sin tx \, dx$$

$$= \int_0^\infty \frac{dx}{x(1 + x^2)(s^2 + x^2)}$$

$$= \frac{1}{s^2 - 1} \int_0^\infty \left(\frac{1}{1 + x^2} - \frac{1}{s^2 + x^2}\right) dx$$

$$= \frac{1}{s^2-1}\left(\frac{\pi}{2}-\frac{\pi}{2s}\right) = \frac{\pi}{2}\frac{1}{s^2-1}\left(\frac{s-1}{s}\right)$$

$$= \frac{\pi}{2}\left(\frac{1}{s(s+1)}\right) = \frac{\pi}{2}\left(\frac{1}{s}-\frac{1}{s+1}\right).$$

Taking inverse Laplace transforms, we get

$$f(t) = \frac{\pi}{2}(1-e^{-t}).$$

EXAMPLE 11.50

Evaluate

$$\int_0^\infty \frac{\sin^2 tx}{x^2}\,dx.$$

Solution. We have

$$f(t) = \int_0^\infty \frac{\sin^2 tx}{x^2}\,dx = \int_0^\infty \frac{1-\cos(2tx)}{2x^2}\,dx.$$

Taking Laplace transform with respect to t, we have

$$F(s) = \frac{1}{2}\int_0^\infty \frac{1}{x^2}\left(\frac{1}{s}-\frac{s}{4x^2+s^2}\right)dx = \frac{2}{s}\int_0^\infty \frac{dx}{4x^2+s^2}$$

$$= \frac{1}{s}\int_0^\infty \frac{dy}{y^2+s^2} = \frac{1}{s^2}\left[\tan^{-1}\frac{y}{s}\right]_0^\infty = \pm\frac{\pi}{2s^2}.$$

Thus, taking inverse Laplace transformation, we get

$$f(t) = \pm\frac{\pi t}{2} = \frac{\pi t}{2}\operatorname{sgn} t.$$

EXAMPLE 11.51

Evaluate

$$\int_{-\infty}^\infty \sin x^2\,dx.$$

Solution. Let

$$f(t) = \int_0^\infty \sin t\,x^2\,dx$$

Taking Laplace transform, we get

$$L\{f(t)\} = \int_0^\infty e^{-st}\,dt\int_0^\infty \sin t\,x^2\,dx$$

$$= \int_0^\infty dx\int_0^\infty e^{-st}\sin tx^2 dt$$

$$= \int_0^\infty L\{\sin tx^2\}\,dx = \int_0^\infty \frac{x^2}{s^2+x^4}dx.$$

Put $x^2 = s\tan\theta$, that is, $x = \sqrt{s}\,\sqrt{\tan\theta}$. Then

$$dx = \sqrt{s}\cdot\frac{1}{2}(\tan\theta)^{-1/2}\sec^2\theta\,d\theta.$$

Therefore,

$$L\{f(t)\} = \frac{1}{2}\int_0^{\pi/2}\frac{s^{3/2}\tan\theta(\tan\theta)^{-1/2}\sec^2\theta d\theta}{s^2(1+\tan^2\theta)}$$

$$= \frac{1}{2\sqrt{s}}\int_0^{\pi/2}\sqrt{\tan\theta}\,d\theta = \frac{1}{2\sqrt{s}}\int_0^{\pi/2}\sin^{1/2}\theta\cos^{-1/2}d\theta.$$

But we know that

$$\int_0^{\pi/2}\sin^{2m-1}\theta\cos^{2n-1}\theta d\theta = \frac{1}{2}\beta(m,n) = \frac{\Gamma(m)\Gamma(n)}{2\Gamma(m+n)}.$$

Taking $2m-1 = \frac{1}{2}$ and $2n-1 = -\frac{1}{2}$, we get $m = \frac{3}{4}$ and $n = \frac{1}{4}$. Hence using the relation $\Gamma(p)\Gamma(1-p) = \frac{\pi}{\sin p\pi}$, $0 < p < 1$, we have

$$L\{f(t)\} = \frac{1}{4\sqrt{s}}\frac{\Gamma(3/4)\Gamma(1/4)}{\Gamma(1)} = \frac{1}{4\sqrt{s}}\pi\sqrt{2} = \frac{\pi\sqrt{2}}{4\sqrt{s}}.$$

Taking inverse Laplace transform yields

$$f(t) = \frac{\pi\sqrt{2}}{4}\left(\frac{t^{-1/2}}{\sqrt{\pi}}\right) = \frac{\sqrt{2\pi}}{4}t^{-1/2}.$$

Putting $t = 1$, we get

$$\int_0^\infty \sin x^2\,dx = \frac{1}{2}\sqrt{\frac{\pi}{2}}$$

and so

$$\int_{-\infty}^\infty \sin x^2\,dx = \sqrt{\frac{\pi}{2}}.$$

EXAMPLE 11.52

Evaluate the integral

$$\int_0^\infty \frac{\cos tx}{x^2+1}\,dx,\ t > 0.$$

Solution. Let

$$f(t) = \int_0^\infty \frac{\cos tx}{x^2+1}\, dx.$$

Taking Laplace transform with respect to t, we get

$$L\{f(t)\} = L\left\{\int_0^\infty \frac{\cos tx}{x^2+1}\, dx\right\} = \left\{\int_0^\infty L\,\frac{\cos tx}{x^2+1}\right\} dx$$

$$= \int_0^\infty \frac{s}{(x^2+1)(s^2+x^2)}\, dx$$

$$= s\int_0^\infty \frac{dx}{(x^2+1)(s^2+x^2)}$$

$$= \frac{s}{s^2-1}\int_0^\infty \left(\frac{1}{x^2+1} - \frac{1}{s^2+x^2}\right) dx$$

$$= \frac{s}{s^2-1}\left[\tan^{-1} x - \tan^{-1}\frac{x}{s}\right]_0^\infty$$

$$= \frac{s}{s^2-1}\left(\frac{\pi}{2} - \frac{\pi}{2s}\right) = \frac{\pi/2}{s+1}.$$

Now, taking inverse Laplace transform, we have

$$f(t) = \frac{\pi}{2} e^{-t}, t > 0.$$

EXAMPLE 11.53

Evaluate

$$\int_0^\infty t\, J_0(t)\, dt,$$

where J_0 is Bessel?s function of order zero.

Solution. We know that

$$L\{J_0(t)\} = \frac{1}{\sqrt{s^2+1}}.$$

Therefore,

$$L\{tJ_0(t)\} = -\frac{d}{ds}\{L\{J_0(t)\}\} = -\frac{d}{ds}\left\{\frac{1}{\sqrt{s^2+1}}\right\}$$

$$= -\frac{2s}{(s^2+1)^{3/2}}.$$

But, by definition

$$L\{tJ_0(t)\} = \int_0^\infty e^{-st}\, tJ_0(t)\, dt.$$

Hence

$$\int_0^\infty e^{-st}\, t\, J_0(t)\, dt = -\frac{2s}{(s^2+1)^{3/2}}.$$

Taking $s = 0$, we get

$$\int_0^\infty t\, J_0(t)\, dt = 0.$$

EXAMPLE 11.54

Evaluate

$$\int_0^\infty e^{-2t} \operatorname{erf} \sqrt{t} dt.$$

Solution. We have

$$L\{\operatorname{erf}\sqrt{t}\} = \int_0^\infty e^{-st} \operatorname{erf} \sqrt{t} dt = \frac{1}{s\sqrt{s+1}}.$$

Taking $s = 2$, we get

$$\int_0^\infty e^{-2t} \operatorname{erf}\sqrt{t} dt = \frac{1}{2\sqrt{3}}.$$

EXAMPLE 11.55

Evaluate

$$\int_0^t \operatorname{erf}\sqrt{u}\, \operatorname{erf}\sqrt{(t-u)}\, du.$$

Solution. Let

$$f(t) = \int_0^t \operatorname{erf}\sqrt{u}\, \operatorname{erf}\sqrt{t-u}\, du.$$

Then, by Convolution theorem, we have

$$F(t) = L\{\operatorname{erf}\sqrt{t}\}\, L\{\operatorname{erf}\sqrt{t}\}$$

$$= \frac{1}{s\sqrt{s+1}} \cdot \frac{1}{s\sqrt{s+1}} = \frac{1}{s^2\sqrt{s+1}}$$

$$= \frac{-1}{s} + \frac{1}{s^2} + \frac{1}{s+1}.$$

Taking inverse transform, we get

$$f(t) = -1 + t + e^{-t}.$$

EXERCISES

1. Solve the following initial value problems:

(a) $y'(t) + 3y(t) = 0, x(1) = 1.$

Ans. $y(t) = e^{3(1-t)}$

(b) $\frac{d^2y}{dt^2} + y = 1, y(0) = y'(0) = 0.$

Ans. $1 - \cos t$

(c) $y'' + y = e^{-t}, y(0) = A, y'(0) = B.$

Ans. $y(t) = \frac{1}{2}e^{-t} + \left(A - \frac{1}{2}\right)\cos t + \left(B + \frac{1}{2}\right)\sin t$

(d) $\frac{d^2y}{dt^2} + y = 0, y(0) = 1, y'(0) = 0.$

Ans. $y(t) = \cos t.$

(e) $\frac{d^2y}{dt^2} + a^2y = f(t), y(0) = 1, y'(0) = -2.$

Hint: $Y(s) = \frac{s-2}{s^2+a^2} + \frac{F(s)}{s^2+a^2}$. But by Convolution theorem

$$L^{-1}\left\{\frac{F(s)}{s^2+a^2}\right\} = f(t) * \frac{\sin at}{a} \text{ and so}$$

$$y(t) = L^{-1}\left\{\frac{s}{s^2+a^2}\right\} - L^{-1}\left\{\frac{2}{s^2+a^2}\right\} + f(t) * \frac{\sin at}{a}$$

Ans. $\cos at - 2\frac{\sin at}{a} + \frac{1}{a}\int_0^t f(u)\sin a(t-u)\,du$

(f) $\frac{d^2y}{dt^2} + y = 3\sin 2t, y(0) = 3, y'(0) = 1.$

Ans. $-\sin 2t + 3\cos t + 3\sin t$

(g) $\frac{d^2x}{dt^2} + 5\frac{dx}{dt} + 6x = 2e^{-t}\ (t \geq 0), x(0) = 1$ and $x'(0) = 0$

Ans. $e^{-t} + e^{-2t} - e^{-3t}, t \geq 0$

(h) $\frac{d^2x}{dt^2} + 6\frac{dx}{dt} + 9x = 0, x(0) = x(0) = 0.$

Ans. $x(t) = 0$

(i) $\frac{d^2x}{dt^2} + 2b\frac{dx}{dt^2} + \lambda^2x = 0, x(0) = x'(0) = 0.$

Ans. $x(t) = e^{-bt}(c_1 \sin\sqrt{\lambda^2 - b^2}t + c_2 \cos\sqrt{\lambda^2 - b^2}t$

2. Solve $y' - 2ty = 0, y(0) = 1$ and show that its solution does not have Laplace transform.

Ans. $y(t) = e^{t^2}$ (not of exponential order)

3. Solve $ty'' + y = 0, y(0) = 0.$

Hint: Proceed as in Example 11.20

Ans. $C\sum_{n=0}^{\infty} \frac{(-1)^n t^{n+1}}{(n+1)!\,n!}$

4. Given that $I = Q = 0$ at $t = 0$, find current I in the LC circuit given for $t > 0$ in Figure 11.15.

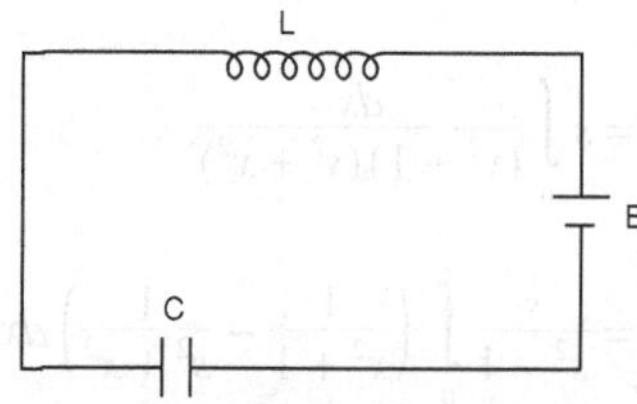

Figure 11.15

Hint: The differential equation governing the circuit is $L\frac{dI}{dt} + \frac{1}{C}\int_0^t I(u)\,du = E.$

Application of Laplace transform yields

$$Ls\,F(s) + \frac{F(s)}{Cs} = \frac{E}{s}, \text{ that is,}$$

$$F(s) = \frac{EC}{LCs^2 + 1} = \frac{E}{L\left(s^2 + \frac{1}{LC}\right)}$$

Ans. $I(t) = E\sqrt{\frac{C}{L}}\sin\frac{1}{\sqrt{LC}}t$

5. Given that $I = Q = 0$ at $t = 0$, find charge and current in the circuit shown in Figure 11.16.

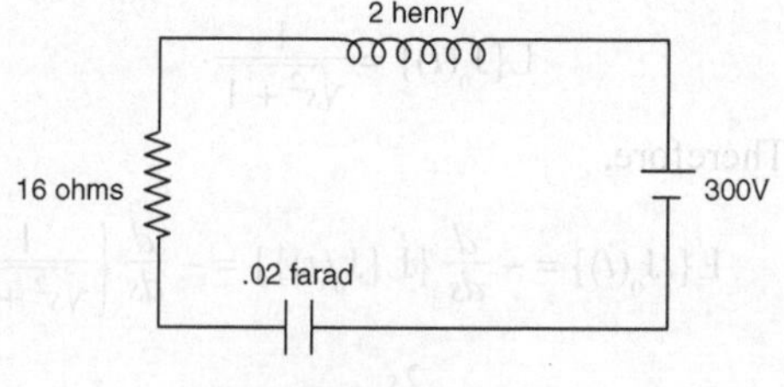

Figure 11.16

Hint: The governing equation is $\frac{d^2Q}{dt^2} + 8\frac{dQ}{dt} + 25Q = 150$, $F(s) = 150/[s(s^2+8s+25)]$ and so inversion gives $Q(t) = 6 - 6e^{-4t}\cos 3t - 8e^{-4t}\sin 3t$. Then $I(t) = 50\, e^{-4t}\sin 3t$.

6. Solve the following systems of differential equations:

(a) $\frac{dx}{dy} + x - y = 1 + \sin t$, $\frac{dy}{dt} - \frac{dx}{dt} + y = t - \sin t$, with $x(0) = 0$, $y(0) = 1$

Ans. $x(t) = t + \sin t$, $y(t) = t + \cos t$

(b) $\frac{dy}{dt} = -z$, $\frac{dz}{dt} = y$ with $y(0) = 1$, $z(0) = 0$.

Ans. $y(t) = \cos t$, $z(t) = \sin t$

7. Solve $y'' + 4y = 4\cos 2t$, $y(0) = y'(0) = 0$. Does resonance occur in this case?

Hint: $Y(s) = \frac{4s}{(s^2+4)^2}$ and so $y(t) = 4\left[\frac{t}{4}\sin 2t\right] = t\sin 2t$.

Note that $y(t) \to \infty$ as $t \to \infty$. Hence, there shall be resonance.

8. Solve the following difference equations:

(a) $3y(t) - 4y(t-1) + y(t-2) = t$, $y(t) = 0$ for $t < 0$.

Ans. $y(t) = \frac{t}{3} + \frac{1}{2}\sum_{n=1}^{[t]}\left(1 - \frac{1}{3^n}\right)(t-n)$

(b) $a_{n+2} - 2a_{n+1} + a_n = 0$, $a_0 = 0$, $a_1 = 1$.

Ans. $a_n = n$

(c) $a_n = a_{n-1} + 2a_{n-2}$, $a_0 = 1$, $a_1 = 8$.

Ans. $a_n = 3(2^n) - 2(-1)^n$, $n \ge 0$

(d) $a_n = 2a_{n-1} - a_{n-2}$, $a_1 = 1.5$, $a_2 = 3$

Ans. $1.5\, n$

(e) $y(t) - y(t-1) = t^2$

Ans. $y(t) = 2\sum_{n=0}^{[t]}\frac{(t-n)^{n+3}}{(n+3)!}$

(f) $y''(t) - y(t-1) = \delta(t)$, $y(t) = y'(t) = 0$, $t \le 0$.

Hint: $s^2L\{y(t)\} - e^{-s}L\{y(t)\} = L\{\ (t)\}$ and so $L\{y(t)\} = \frac{1}{s^2}\left(\frac{1-e^{-s}}{s^2}\right)$. But

$$L^{-1}\left\{\frac{e^{-ns}}{s^{2n+2}}\right\} = \begin{cases}\frac{(t-n)^{2n+1}}{(2n+1)!} & \text{for } t \ge n\\ 0 & \text{otherwise}\end{cases}$$

Hence $y(t) = \sum_{n=0}^{[t]}\frac{(t-n)^{2n+1}}{(2n+1)!}$

9. Solve the integral equations:

(a) $f(t) = 1 + \int_0^t \cos(t-u) f(u)\, du$

Ans. $f(t) = 1 + \frac{2}{\sqrt{3}}\sin\left(\frac{\sqrt{3}}{2}t\right)e^{t/2}$

(b) $y(t) = \sin t + 2\int_0^t y(u)\cos(t-u)\, du$

Ans. $y(t) = te^t$

(c) $y(t) = t + \frac{1}{6}\int_0^t y(u)(t-u)^3\, du$

Ans. $y(t) = \frac{1}{2}(\sinh t + \sin t)$

(d) $f(t) = \int_0^t \sin u\,(t-u)\, du$

Ans. 0

(e) $y'(t) + 3y(t) + 2\int_0^t y(u)\, du = t$, $y(0) = 1$.

Hint: $L\{y(t)\} = \frac{s^2+1}{s(s^2+3s+2)}$. Use partial fractions and then use inversion to give $y(t) = \frac{1}{2} - 2e^{-t} + \frac{5}{2}e^{-2t}$

10. Solve the partial differential equation $x\frac{\partial u}{\partial t} + \frac{\partial u}{\partial x} = x$, $x > 0$, $t > 0$ subject to the conditions $u(x, 0) = 0$ for $x > 0$ and $u(0, t) = 0$ for $t > 0$.

Hint: Using Laplace transform, $\frac{d}{dx}U(x, s) + x\, sU(x, s) = \frac{x}{s}$, integrating factor is $e^{\frac{1}{2}x^2}$

and so $U(x, s) = \frac{1}{s^2}\left[1 - e^{-\frac{1}{2}sx^2}\right]$. Inversion yields $u(x, t) = \begin{cases} t & \text{for } t < x^2/2 \\ x^2/2 & \text{for } 2t > x^2. \end{cases}$

11. Find the bounded solution of $\frac{\partial u}{\partial t} = \frac{\partial^2 u}{\partial x^2}$, $x > 0$, $t > 0$ for $u(0, t) = 1$, $u(x, 0) = 0$.

Hint: Application of Laplace transform yields $U(x, s) = c_1 e^{\sqrt{s}x} + c_2 e^{-\sqrt{s}x}$ for bounded $u(x, t)$, $U(x, s)$ must be bounded and so $c_1 = 0$. Further $U(0, s) = L\{u(0, t)\} = L\{1\} = \frac{1}{s}$. Therefore, $\frac{1}{s} = c_2$ and so $U(x, s) = \text{erfc}\,(x/2\sqrt{t})$.

12. Solve $\frac{\partial^2 u}{\partial t^2} = a^2 \frac{\partial^2 y}{\partial x^2}$, $x > 0$, $t > 0$ for $u(x, 0) = 0$, $y_t(x, 0) = 0$, $x > 0$, $y(0, t) = f(t)$ with $f(0) = 0$ and $\lim_{x\to\infty} y(x, t) = 0$.

Hint: see Example 11.54

Ans. $y(x, t) = f\left(t - \frac{x}{a}\right) H\left(t - \frac{x}{a}\right)$

13. Solve $\frac{\partial^2 u}{\partial t^2} = \frac{\partial^2 u}{\partial x^2}$ $x>0$, $t>0$ for $u(0, t) = 10 \sin 2t$, $u(x, 0) = 0$, $u_t(x, 0) = 0$ and $\lim_{x\to\infty} u(x, t) = 0$

Ans. $u(x, t) = \begin{cases} 10\sin 2(t - x) & \text{for } t > x \\ 0 & \text{for } t < x. \end{cases}$

14. Evaluate the integrals:

(a) $\int_0^\infty J_0(t)\, dt$

Ans. 1

(b) $\int_0^\infty \frac{e^{-t} - e^{-2t}}{t}\, dt$

Hint: $L\{e^{-t} - e^{-2t}\} = \frac{1}{s+1} - \frac{1}{s+2}$ and so

$$L\left\{\frac{e^{-t} - e^{-2t}}{t}\right\} = \int_s^\infty \left(\frac{1}{t+1} - \frac{1}{t+2}\right) dt$$

$$= \log\left(\frac{s+2}{s+1}\right),$$

that is,

$$\int_0^\infty e^{-st}\left(\frac{e^{-t} - e^{-2t}}{t}\right) dt = \log\left(\frac{s+2}{s+1}\right).$$ Taking $s=0$,

we get the value of the given integral equal to log 2

(c) $\int_0^\infty \cos x^2 dx$ (Proceed as in Example 11.59)

Ans. $\frac{1}{2}\sqrt{\frac{\pi}{2}}$

(d) $\int_0^\infty \frac{x \sin at}{x^2 + a^2}\, dx$; $a, t > 0$.

Ans. $f(t) = \frac{\pi}{2} e^{-t}$

12 The z-transform

In Chapters 9–11, we studied the Laplace transform for continuous-time signals (input functions) The z-transform is the finite or discrete-time version of the Laplace transform. This transform is useful for solving initial-value problems whose continuous analogs are treated by Laplace transform. It has many properties in common with the Laplace transform. We know that continuous-time systems are described by differential equations whereas discrete-time systems are described by difference equations. So we use z-transform to solve difference equations that are approximations to the differential equations of the initial-value problems treated by Laplace transform. So, we shall consider discrete-time signals $f[n]$ (also denoted by $s[n]$) that are non-periodic.

12.1 SOME ELEMENTARY CONCEPTS

Before defining z-transform, let us learn some elementary concepts which shall be helpful in the study of z-transforms. We recall that the signals defined at discrete time are called *discrete-time signals*. It is thus a function defined on $\mathbb{Z}$, the set of integers. A discrete-time signal can be represented by a graph, a functional relation or in the form of a sequence. For example, the expression

$$f[n] = \begin{cases} 2 \text{ for } n = 0 \\ 0 \text{ for } n \neq 0 \end{cases}$$

is a functional representation of the signal $f[n]$. The graphical representation of $f[n]$ is shown in the Figure 12.1.

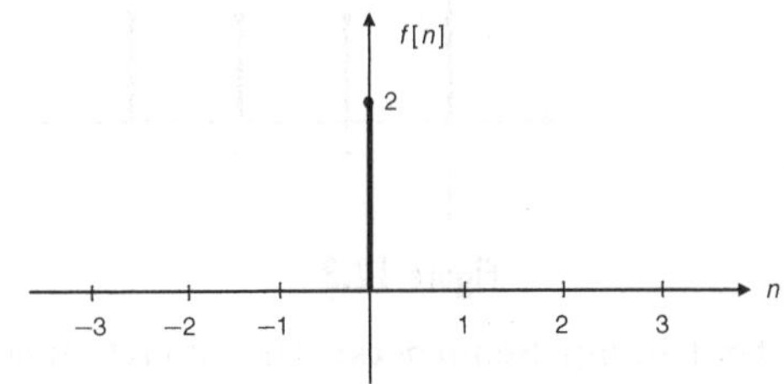

Figure 12.1

The sequence representation of the above signal is

$$f[n] = \ldots 0, 0, 0, \underset{\uparrow}{2}, 0, 0, 0, \ldots$$

where ↑ represents the value at the origin ($n = 0$) Some basic discrete-time signals are described below:

1. **Unit Impulse Sequence:** The discrete-time signal defined by

$$\delta[n] = \begin{cases} 1 \text{ for } n = 0 \\ 0 \text{ for } n \neq 0 \end{cases}$$

is called *unit impulse sequence.* Its graphical representation is shown in the Figure 12.2.

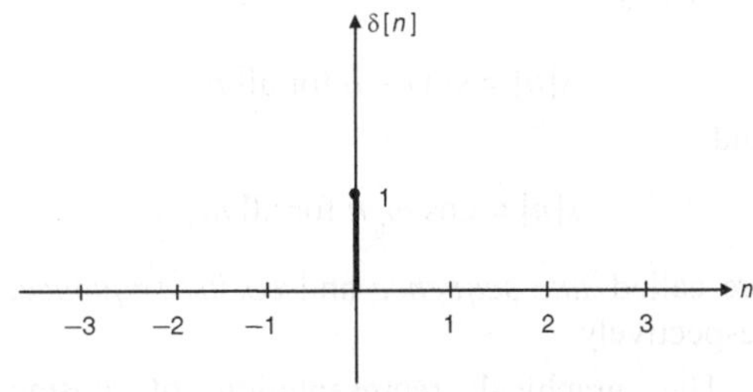

Figure 12.2

2. Unit-step Sequence: The discrete-time signal defined by

$$u[n] = \begin{cases} 1 \text{ for } n \geq 0 \\ 0 \text{ for } n < 0 \end{cases}$$

is called *unit-step sequence.* Its graphical representation is shown in the Figure 12.3.

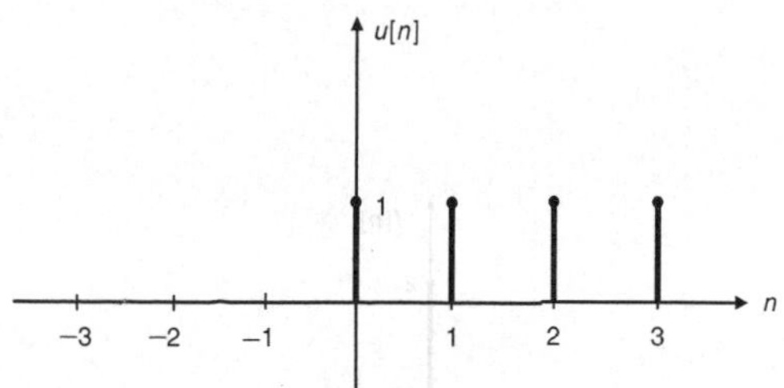

Figure 12.3

3. Unit-Ramp Sequences: The discrete-time signal defined by

$$r[n] = \begin{cases} n \text{ for } n \geq 0 \\ 0 \text{ for } n < 0 \end{cases}$$

is called *unit-ramp sequence.* Its graphical representation is shown in the Figure 12.4.

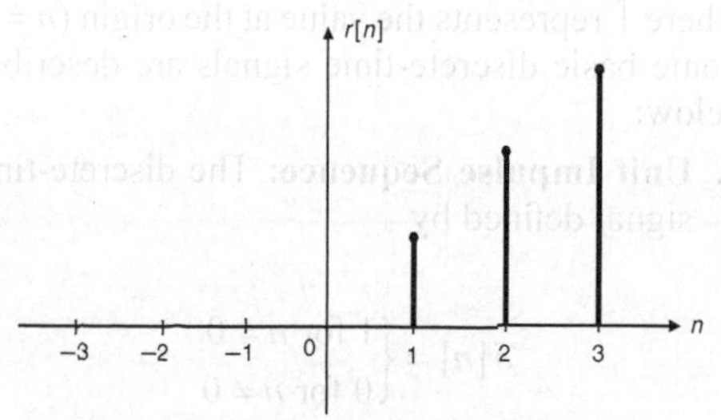

Figure 12.4

4. Sinusoidal Sequences: The sequences defined by

$$s[n] = \sin \omega_0 n \text{ for all } n$$

and

$$s[n] = \cos \omega_0 n \text{ for all } n,$$

are called *sine sequence* and *cosine sequence*, respectively.

The graphical representation of cosine sequence is shown in the Figure 12.5.

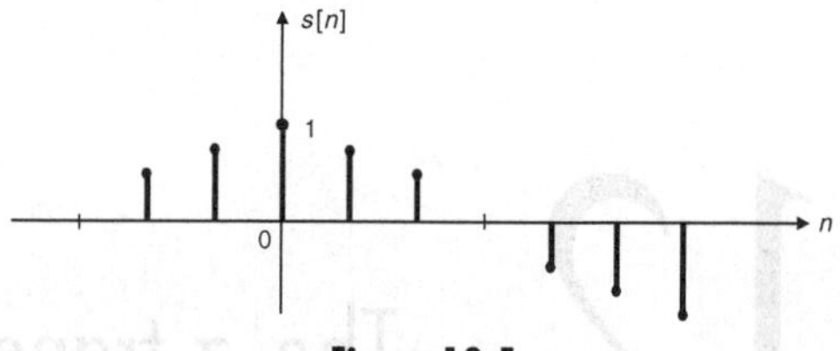

Figure 12.5

5. Exponential Sequence: The sequence defined by

$$e[n] = A^n \text{ for all values of } n,$$

is called *exponential sequence.* If A is real, then $e[n]$ is called *real sequence.* Its graphical representation for A < 1 is shown in Figure 12.6

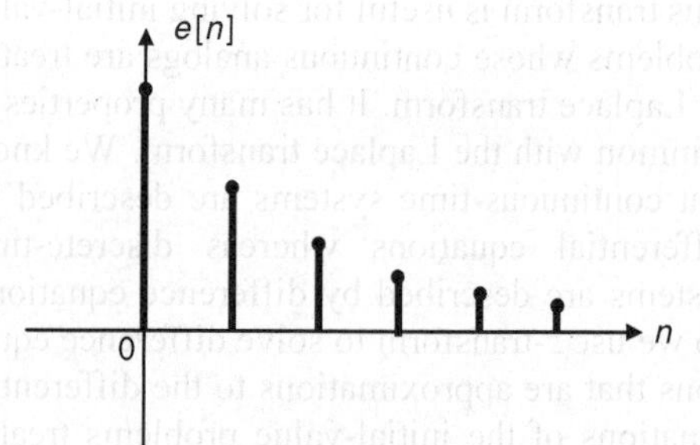

Figure 12.6

The change in the dependent and independent variable (time n) in a discrete-time signal can be made in the following ways:

1. Time shifting: Let the independent variable (time) n be replaced by $n - k$, where k is an integer. If k is a positive integer, then this shifting will cause *delay of signal by k units* of time. If k is a negative integer, then the shifting will *advance the signal by* $|k|$ units of time.

For example, let graph of a sequence $s[n]$ be as shown in the Figure 12.7.

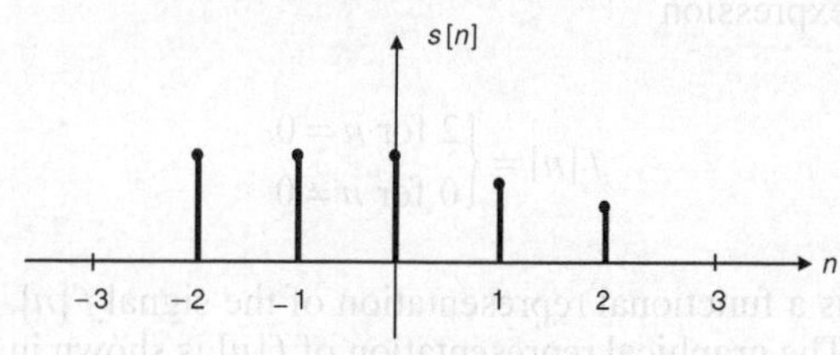

Figure 12.7

Then, $s[n-1]$ will be delayed by one unit and its graph shall be governed by the Figure 12.8.

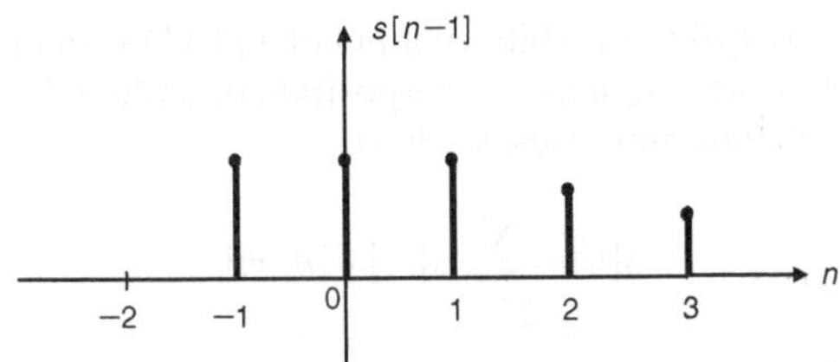

Figure 12.8

Similarly, the graph of $s[n+1]$ shall be as shown in the Figure 12.9.

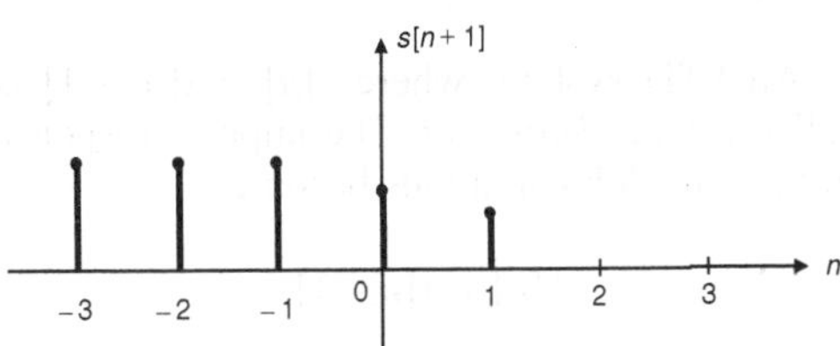

Figure 12.9

2. Folding: If the independent variable (time) n is replaced by $-n$, then the signal gets folded, that is, becomes *mirror image* of the original signal about the time origin ($n = 0$).

For example, folding of the signal $s[n]$ represented by Figure 12.7 is the sequence shown in the Figure 12.10.

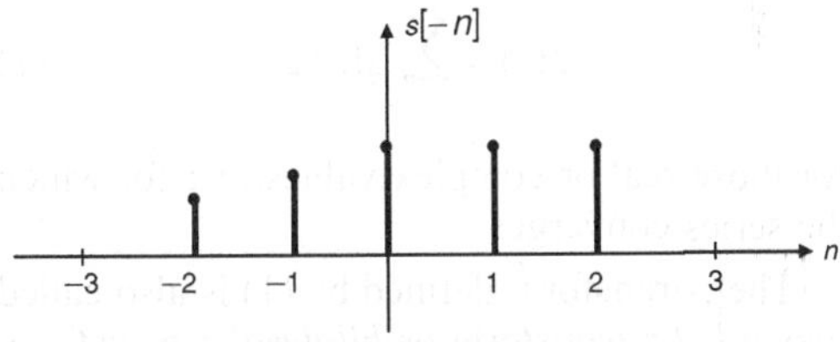

Figure 12.10

3. Time-scaling (down sampling): If we multiply the independent variable n by m, where m is an integer, then $s[mn]$ is the sequence whose terms are m times the terms of the original sequence $s[n]$.

For example, let graph of $s[n]$ be as shown in the Figure 12.11.

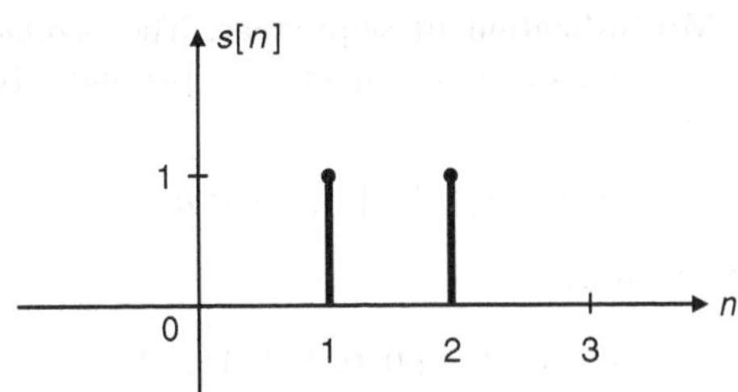

Figure 12.11

Then the graph of $s[2n]$ takes the form as shown in the Figure 12.12.

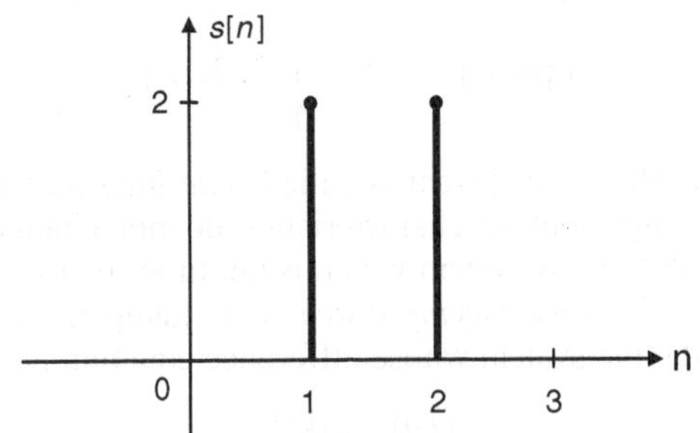

Figure 12.12

The *transformation* in *the dependent variable* (signal amplitude) can be done as follows:

1. Addition of Sequences: If $s_1[n]$ and $s_2[n]$ are two discrete-time sequences, then their sum is defined by

$$y[n] = s_1[n] + s_2[n], -\infty < n < \infty.$$

For example, if

$$s_1[n] = \{\ldots, 0, 0, \underset{\uparrow}{1}, 1, 1, \ldots\}$$

and

$$s_2[n] = \{\ldots, 0, 0, 1, \underset{\uparrow}{2}, 3, \ldots\},$$

then their sum is

$$y[n] = \{\ldots, 0, 0, 2, \underset{\uparrow}{3}, 4, \ldots\}$$

2. Multiplication of sequences: The *product* of two discrete-time sequences $s_1[n]$ and $s_2[n]$ is defined by

$$y[n] = s_1[n]\, s_2[n],\ -\infty < n < \infty.$$

For example, if

$$s_1[n] = \{\ldots, 0, 0, \underset{\uparrow}{1}, 1, 1, \ldots\}$$

and

$$s_2[n] = \{\ldots, 0, 0, \underset{\uparrow}{1}, 2, 3, \ldots\},$$

then their product is the sequence given by

$$y[n] = \{\ldots, 0, 0, \underset{\uparrow}{1}, 2, 3, \ldots\}$$

Recall that a system is called *time invariant* if its input-output characteristics do not change with time. A system which is not time-invariant is called *time varying system*. For example, consider the system whose difference equation is

$$y(n) = ns[n].$$

The response to delayed input is

$$y(n, k) = ns[n-k],$$

while the delayed response is

$$y(n-k) = (n-k)\, s[n-k].$$

Thus

$$y(n, k) \neq y(n-k)$$

and so the given discrete-time system is not time invariant.

On the other hand, a system described by the difference equation

$$y(n) = s[n] - s[n-1]$$

is time-invariant. In fact,

$$y(n, k) = s[n-k] - s[n-k-1],$$

$$y(n-k) = s[n-k] - s[n-k-1]$$

and so

$$y(n, k) = y(n-k).$$

The discrete-time systems that are linear and time invariant are called *LTD systems*.

The response to the discrete unit pulse $\delta[n]$ is called the *impulse response* of an *LTD* system and is denoted by $h[n]$.

If $s[n]$ is an arbitrary input of an LTD system, then, writing it as a superposition of shifted discrete unit impulses, we have

$$s[n] = \sum_{m=-\infty}^{\infty} s[m]\, \delta[n-m].$$

Since the system is time-invariant, the response to $\delta[n-m]$ is $h[n-m]$. So, using superposition rule, we have response $y[n]$ as

$$y[n] = \sum_{m=-\infty}^{\infty} s[m]\, h[n-m].$$

An LTD system, where $y[n] = s[n-1]$ is called a *time-delay unit*. The impulse response of the time delay unit thus becomes

$$h[n] = \delta[n-1].$$

Thus for a time-delay unit, the response becomes

$$y[n] = \sum_{m=-\infty}^{\infty} s[m]\, \delta[n-m-1].$$

12.2 DEFINITION OF Z-TRANSFORM

Definition 12.1. Let $s[n]$ be a discrete-time signal. The z-transform of $s[n]$ is defined by

$$Z(z) = \sum_{n=-\infty}^{\infty} s[n]\, z^{-n} \tag{1}$$

for those real or complex values of z for which the series converges.

The z-transform defined by (1) is also called *two-sided z-transform* or *bilateral z-transform*. The defining series is a two-sided power series having not only positive integer powers of z but also negative integer powers of z. This series has convergence properties similar to those of a power series.

The part

$$\sum_{n=0}^{\infty} s[n]\, z^{-n} \tag{2}$$

of the series in (1) is called *causal part* whereas the part

$$\sum_{n=-\infty}^{-1} s[n]\, z^{-n} \qquad (3)$$

is called *anti-causal part*. The anti-causal part can be rewritten as

$$\sum_{n=-\infty}^{-1} s[n]\, z^{-n} = \sum_{n=1}^{\infty} s[-n]\, z^{n},$$

which is a power series in z with coefficients $s[-n]$.

12.3 CONVERGENCE OF Z-TRANSFORM

The z-transform $Z(z)$ is *said to converge* if $\sum_{n=-M}^{N} s[n]\, z^{-n}$ *converges* for $M \to \infty$ and $N \to \infty$ independently from each other.

Thus the z-transform converges if and only if both the causal part and anti-causal part converge. Let $Z_+(z)$ and $Z_-(z)$ denote the sums of the causal part and anti-causal part, respectively. Then in case of convergence, we have

$$Z_+(z) = \sum_{n=0}^{\infty} s[n]\, z^{-n},\ Z_-(z) = \sum_{n=-\infty}^{-1} s[n]\, z^{-n},$$

and

$$Z(z) = Z_+(z) + Z_-(z).$$

We note that the anti-causal part is a power series in z. Also if we put $w = \frac{1}{z}$, then causal part is also a power series in w. Further, recall that if R is the radius of convergence of complex power series $\sum_{n=0}^{\infty} a_n z^n$,then

(a) if R = 0, then power series converges only for $z = 0$

(b) if R = ∞, the power series converges absolutely for all complex z

(c) if R > 0, the power series converges absolutely for $|z| < R$ and diverges for $|z| > R$.

In the light of this result, we discuss convergence of z-transform.

Let R_2 be the radius of convergence of anticausal part and R_1^{-1} the radius of convergence of the power series $\sum_{n=0}^{\infty} s[n]w^n$. The anti-causal part converges absolutely for $|z| < R_2$ and the causal part converges for $|z| > R_1$. It follows, therefore, that if $R_1 < R_2$ then the *z-transform converges in the ring* $R_1 < |z| < R_2$ (Figure 12.13).

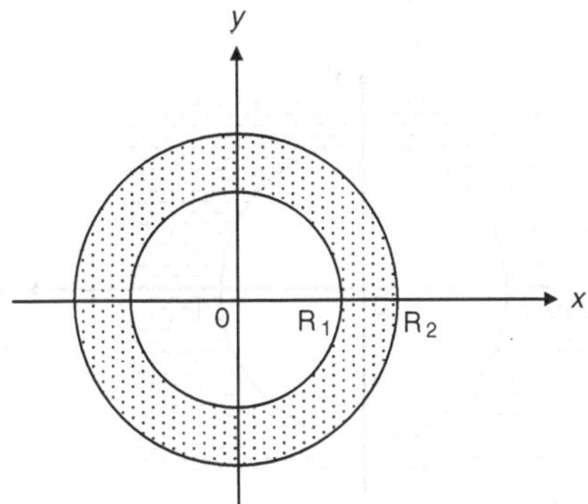

Figure 12.13 Shaded Area Shows the Region of Convergence of z-transform

The ring-shaped region $R_1 < |z| < R_2$ is called the *region of convergence* of the z-transform. For $|z| < R_1$, the anti-causal part converges while the causal part diverges and so z-transform diverges. Similarly, for $|z| > R_2$, the causal part converges while the anti-causal part diverges and so again the z-transform diverges.

If $R_1 > R_2$, then the z-transform diverges for every complex z and so the region of convergence in this case is empty.

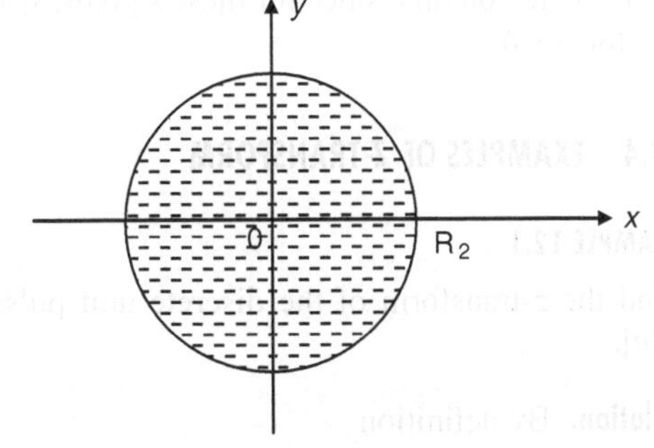

Figure 12.14

If $R_1 = 0$, then causal part converges for every $z \neq 0$, and the region of convergence is the interior of the circle with radius R_2 with the exception of $z = 0$ (Figure 12.14). If all the terms of the causal part are zero, with the possible exception of the terms with $n = 0$, then $z = 0$ also falls in the region of convergence.

If $R_2 = \infty$, (Figure 12.15), then the region of convergence is the exterior of the circle with radius R_1.

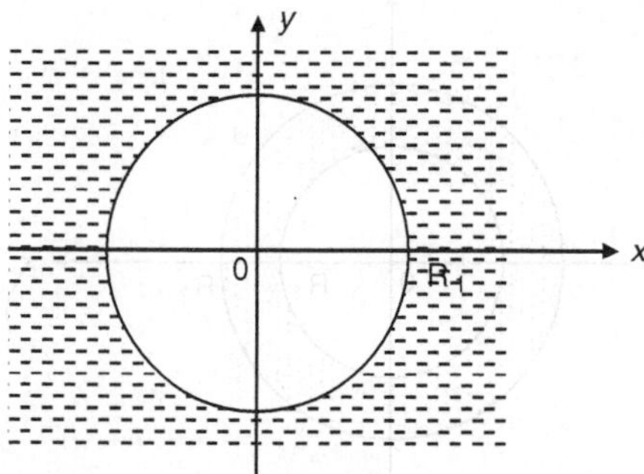

Figure 12.15

Generally we come across the signals that have been switched on a certain moment of time. Such signals are called *signals with a finite switch on time*. Thus for these signals there exists N such that $s[n] = 0$ for all $n < N$ and so anti-causal part consists of only a finite number of non-zero terms. As such anticausal part converges for all z implying $R_2 = \infty$. Thus for signals with a finite switch-on time, the region of convergence is exterior of a circle with radius R_1. For example, the causal signals are with a finite switch on time since for these signals, $s[n] = 0$ for $n < 0$.

12.4 EXAMPLES OF Z-TRANSFORM

EXAMPLE 12.1

Find the z-transform of the discrete unit pulse $\delta[n]$.

Solution. By definition

$$\delta[n] = \begin{cases} 1 & \text{for } n = 0 \\ 0 & \text{for } n \neq 0 \end{cases}$$

Therefore,

$$Z(z) = \sum_{n=-\infty}^{\infty} \delta[n]\, z^{-n}$$

$$= \ldots + \delta[-2]z^2 + \delta[-1]z + \delta[0]z^0 + \delta[1]z^{-1} + \delta[2]z^{-2} + \ldots$$

$$= \ldots + 0.z^2 + 0.z + 1.z^0 + 0.z^{-1} + 0.z^{-2} + \ldots = 1$$

Thus, the series converges for every complex z. Hence, the region of convergence is the whole complex plane.

EXAMPLE 12.2

Find the z-transform of the signal $s[n]$ defined by

$$s[n] = \begin{cases} \frac{1}{n} & \text{for } n > 0 \\ 0 & \text{for } n = 0 \\ 2^n & \text{for } n < 0. \end{cases}$$

Solution. We have

$$Z(z) = \sum_{n=-\infty}^{\infty} s[n]\, z^{-n}.$$

$$= \sum_{n=-\infty}^{-1} s[n]\, z^{-n} + \sum_{n=0}^{\infty} s[n]\, z^{-n}$$

$$= \sum_{n=1}^{\infty} s[-n]\, z^{n} + \sum_{n=0}^{\infty} s[n]\, z^{-n}$$

$$= \sum_{n=1}^{\infty} \left(\frac{z}{2}\right)^n + 0 + \sum_{n=1}^{\infty} \frac{z^{-n}}{n}$$

$$= \sum_{n=1}^{\infty} \left(\frac{z}{2}\right)^n + \sum_{n=1}^{\infty} \left(\frac{1}{z}\right)^n.$$

Thus the anti-causal part is the power series $\sum_{n=1}^{\infty} \left(\frac{z}{2}\right)^n$ which is a geometric series, convergent for $|z| < 2$ (radius of convergence). The causal part is a power series in $\frac{1}{z}$ which converges for $\left|\frac{1}{z}\right| < 1$. Thus, the causal part converges for $|z| > 1$ and diverges for $|z| < 1$. Hence the region of convergence of the z-transform of $s[n]$ is the ring-shaped region $1 < |z| < 2$.

EXAMPLE 12.3

Find the z-transform of

$$s[n] = \{1, 2, \underset{\uparrow}{3}, 4, 5\}.$$

Solution. We have

$$s[n] = \{1, 2, \underset{\uparrow}{3}, 4, 5\}.$$

Therefore

$$Z(z) = \sum_{n=-\infty}^{\infty} s[n]\, z^{-n} = \sum_{n=-2}^{2} s[n]\, z^{-n}$$

$$= s[-2]z^2 + s[-1]z^1 + s[0]z^0 + s[1]\, z^{-1} + s[2]z^{-2}$$

$$= z^2 + 2z + 3 + 4z^{-1} + 5z^{-2}.$$

The region of convergence of Z(z) is the entire z-plane except $z = 0$ and $z = \infty$.

EXAMPLE 12.4

Find the z-transform of

$$s[n] = \delta(n-k),\ k > 0$$

Solution. By definition

$$Z(z) = \sum_{n=-\infty}^{\infty} \delta(n-k)\, z^{-n}.$$

Putting $n - k = m$, we get

$$Z(z) = \sum_{m=-\infty}^{\infty} \delta(m)\, z^{(-k+m)}$$

$$= z^{-k} \sum_{m=-\infty}^{\infty} \delta(m) z^{-m} = z^{-k},$$

since $\sum_{m=-\infty}^{\infty} \delta(m)\, z^{-m} = 1$ by Example 12.1.

Obviously, the region of convergence of Z(z) is entire plane except $z = 0$.

EXAMPLE 12.5

Find z-transform of the unit step function $u[n]$.

Solution. Since

$$u[n] = \begin{cases} 1 & \text{for } n \geq 0 \\ 0 & \text{for } n < 0, \end{cases}$$

we have

$$Z(z) = \sum_{n=-\infty}^{\infty} u[n] z^{-n} = \sum_{n=0}^{\infty} z^{-n}$$

$$= \frac{1}{1 - \frac{1}{z}} = \frac{z}{z-1},\ |z| > 1.$$

The region of convergence is $|z| > 1$.

EXAMPLE 12.6

Find z-transform of the unit ramp sequence $r[n] = nu[n]$.

Solution. We have

$$Z(z) = \sum_{n=-\infty}^{\infty} r[n] z^{-n} = \sum_{n=0}^{\infty} nz^{-n}$$

$$= z \sum_{n=0}^{\infty} nz^{-(n+1)} = -z \frac{d}{dz} \sum_{n=0}^{\infty} nz^{-n}$$

$$= -z \frac{d}{dz}\left\{\frac{z}{z-1}\right\}, \text{ by Example 12.5}$$

$$= \frac{z}{(z-1)^2},\ |z| > 1.$$

EXAMPLE 12.7

Find z-transform of $s[n] = A^n\, u[n]$, $A \neq 0$.

Solution. Since

$$u[n] = \begin{cases} 1 & \text{for } n \geq 0 \\ 0 & \text{for } n < 0, \end{cases}$$

the sequence $s[n]$ is a causal sequence. We have

$$Z(z) = \sum_{n=-\infty}^{\infty} s[n] z^{-n} = \sum_{n=-\infty}^{\infty} A^n u[n] z^{-n}$$

$$= \sum_{n=0}^{\infty} A^n z^{-n} = \sum_{n=0}^{\infty} \left(\frac{A}{z}\right)^n$$

$$= 1 + \frac{A}{z} + \left(\frac{A}{z}\right)^2 + \ldots = \frac{1}{1 - \frac{A}{z}}$$

$$= \frac{z}{z - A}, \; |z| > |A|.$$

The region of convergence is $\left|\frac{A}{z}\right| < 1$ or $|z| > |A|$.

EXAMPLE 12.8

Find one sided z-transform of $s[n] = \cosh nx$.

Solution. We have

$$Z\{\cosh nx\} = \frac{1}{2} Z\{e^{nx} + e^{-nx}\}.$$

But, by Example 12.7,

$$Z\{e^{nx}\} = \frac{z}{z - e^x}, \; Z\{e^{-nx}\} = \frac{z}{z - e^{-x}}.$$

Therefore,

$$Z\{\cosh nx\} = \frac{1}{2}\left[\frac{z}{z - e^x} + \frac{z}{z - e^{-x}}\right]$$

$$= \frac{z(z - \cosh x)}{z^2 - 2z \cosh x + 1}.$$

EXAMPLE 12.9

Find z-transform of $s[n] = \frac{1}{\Gamma(n+1)}$, $n = 0, 1, \ldots$

Solution. We have

$$s[n] = \frac{1}{\Gamma(n+1)} = \frac{1}{n!}$$

and so

$$Z(z) = \sum_{n=0}^{\infty} \frac{z^{-n}}{n!} = e^{\frac{1}{z}} \text{ for all } z$$

EXAMPLE 12.10

Find z-transform of $s[n] = u[n] \cos w_0 n$.

Solution. We have

$$s[n] = u[n] \cos w_0 n.$$

$$= u[n]\left(\frac{e^{iw_0 n} + e^{-iw_0 n}}{2}\right).$$

Therefore,

$$Z(z) = \sum_{n=-\infty}^{\infty} u[n]\left(\frac{e^{iw_0 n} + e^{-iw_0 n}}{2}\right) z^{-n}$$

$$= \frac{1}{2}\sum_{n=0}^{\infty} e^{iw_0 n} z^{-n} + \frac{1}{2}\sum_{n=0}^{\infty} e^{-iw_0 n} z^{-n}$$

$$= \frac{1}{2}\sum_{n=0}^{\infty}\left(\frac{e^{iw_0}}{z}\right)^n + \frac{1}{2}\sum_{n=0}^{\infty}\left(\frac{e^{-iw_0}}{z}\right)^n$$

$$= \frac{1}{2}\left[1 + \frac{e^{iw_0}}{z} + \frac{e^{2iw_0}}{z^2} + \ldots\right]$$

$$+ \frac{1}{2}\left[1 + \frac{e^{-iw_0}}{z} + \frac{e^{-2iw_0}}{z^2} + \ldots\right]$$

$$= \frac{1}{2}\left[\frac{1}{1 - \left(\frac{e^{iw_0}}{z}\right)}\right] + \frac{1}{2}\left[\frac{1}{1 - \left(\frac{e^{-iw_0}}{z}\right)}\right]$$

$$= \frac{1}{2}\left[\frac{z}{z - e^{iw_0}}\right] + \frac{1}{2}\left[\frac{z}{z - e^{-iw_0}}\right]$$

$$= \frac{z}{2}\left[\frac{2z - 2\cos w_0}{z^2 - 2z \cos w_0 + 1}\right]$$

$$= \frac{z^2 - z \cos w_0}{z^2 - 2z \cos w_0 + 1}, \; \left|\frac{e^{iw_0}}{z}\right| < 1,$$

$$= \frac{1 - z^{-1} \cos w_0}{1 - 2z^{-1} \cos w_0 + z^{-2}}.$$

The region of convergence is $\left|\frac{e^{iw_0}}{z}\right| < 1$ or $|z| > |e^{iw_0}| = 1$ and $|z| > |e^{-iw_0}| = 1$. Thus, the region of convergence is the exterior of the unit circle $|z| = 1$ as shown in the Figure 12.16.

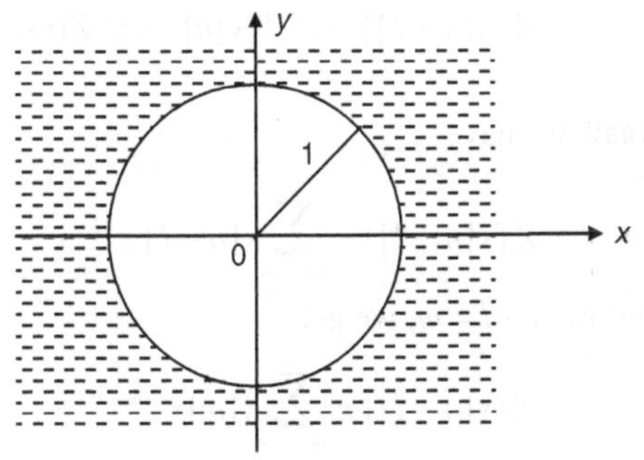

Figure 12.16

EXAMPLE 12.11

Find z-transform of

$$s[n] = \begin{cases} 0 & \text{for } n \geq 0 \\ -A^n & \text{for } n < 0, \end{cases}$$

where A is a non-zero complex number.

Solution. We have

$$Z(z) = \sum_{n=-\infty}^{-1} s[n]z^{-n} + \sum_{n=0}^{\infty} s[n]z^{-n}.$$

By the definition of $s[n]$, the causal part $\sum_{n=0}^{\infty} s[n]z^{-n}$ consists of all zero terms and so converges for every complex z to the sum 0. The anti-causal part is

$$\sum_{n=-\infty}^{-1} -A^n z^{-n} = -\sum_{n=1}^{\infty} \left(\frac{z}{A}\right)^n$$

$$= -\left[\frac{z}{A} + \left(\frac{z}{A}\right)^2 + \left(\frac{z}{A}\right)^3 + \ldots\right],$$

(geometric series)

$$= -\left[\frac{\frac{z}{A}}{1-\left(\frac{z}{A}\right)}\right] = \frac{z}{z-A}, \quad |z| < |A|.$$

The region of convergence of the z-transform is $\left|\frac{z}{A}\right| < 1$ or $|z| < |A|$, that is, the interior of the circle in the complex plane with radius $|A|$ as shown in the Figure 12.17.

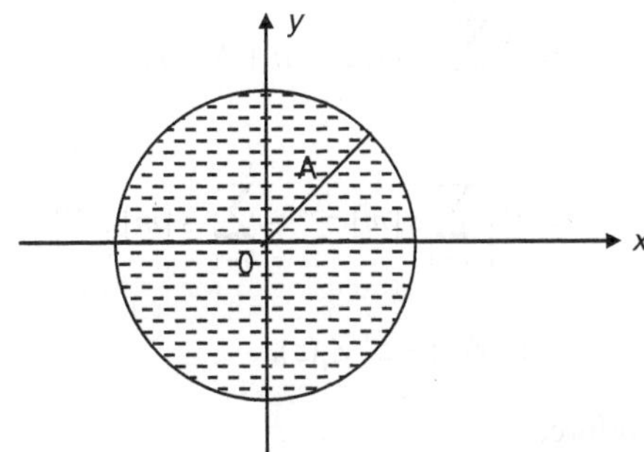

Figure 12.17

EXAMPLE 12.12

Find z-transform of $s[n] = 2^{n-1} u[n-1]$.

Solution. We have

$$Z(z) = \sum_{n=-\infty}^{\infty} 2^{n-1} u[n-1] z^{-n}.$$

Putting $m = n - 1$, we get

$$Z(z) = \sum_{m=-\infty}^{\infty} 2^m u[m] z^{-(m+1)}$$

$$= z^{-1} \sum_{m=-\infty}^{\infty} 2^m u[m] z^{-m} = z^{-1} \sum_{m=0}^{\infty} 2^m z^{-m}$$

$$= z^{-1}\left[1 + \frac{2}{z} + \frac{4}{z^2} + \frac{8}{z^3} \ldots\right]$$

$$= \left[\frac{1}{z} + \frac{2}{z^2} + \frac{4}{z^3} + \frac{8}{z^4} + \ldots\right] = \frac{1}{z-2}, \left|\frac{2}{z}\right| < 1.$$

EXAMPLE 12.13

If $s[n]$ is a periodic causal sequence of integral period N, show that

$$Z\{s[n]\} = \frac{z^N}{z^N - 1} Z_1(z),$$

where

$$Z_1(z) = \sum_{m=0}^{N-1} s[m]z^{-m}.$$

Solution. We have

$$Z(z) = \sum_{n=0}^{\infty} s[n]z^{-n} = z^N \sum_{n=0}^{\infty} s[n+N] z^{-(n+N)}$$

$$= z^N \sum_{m=N}^{\infty} s[m]\, z^{-m},\ n+N=m$$

$$= z^N \left[\sum_{m=0}^{\infty} s[m]\, z^{-m} - \sum_{m=0}^{N-1} s[m]\, z^{-m} \right]$$

$$= z^N Z(z) - z^N Z_1(z).$$

Therefore,

$$(z^N - 1)\, Z(z) = z^N Z_1(z),$$

and so

$$Z(z) = \frac{z^N}{z^N - 1} Z_1(z).$$

Remark 12.1. We observe that the signals in Examples 12.7 and 12.11 are different but their z-transforms are equal. Of course, their region of convergence also differs. It follows from these examples that $s[n]$ is not uniquely determined when only $Z(z)$ is given. Region of convergence, therefore, plays an important role. Thus, one should be very careful while finding inverse z-transform.

12.5 PROPERTIES OF THE Z-TRANSFORM

Most of the properties of the z-transform are analogous to the properties of Laplace transform. We present these properties in the form of the following theorems:

Theorem 12.1. (Linearity Property). Let $Z_1(z)$ and $Z_2(z)$ be z-transform of $s_1[n]$ and $s_2[n]$, respectively. Then

$$Z\{a_1 s_1[n] \pm a_2 s_2[n]\} = a_1 Z\{s_1[n]\} \pm a_2 Z\{s_2[n]\}$$

$$= a_1 Z_1(z) \pm a_2 Z_2(z).$$

Proof: The proof follows directly from the definition of the z-transform.

Theorem 12.2. (Time Shifting or Shifting in the n-domain).

Let l be an integer and $Z\{z\}$ the z-transform of $s[n]$. Then

$$Z\{s[n-l]\} = z^{-l} Z\{s[n]\} = z^{-l} Z(z).$$

Proof: We have

$$Z\{s[n-l]\} = \sum_{n=-\infty}^{\infty} s[n-l]\, z^{-n}.$$

Putting $n - l = m$, we get

$$Z\{s[n-l]\} = \sum_{m=-\infty}^{\infty} s[m] z^{-(m+l)}$$

$$= z^{-l} \sum_{m=-\infty}^{\infty} s[m] = z^{-l} Z(z).$$

Thus if the signal is delayed by l units of time, then z-transform is z^{-l} times of the original z-transform.

Similarly, for advanced signal, we have

$$Z\{s[n+l] = z^l Z(z).$$

Remark 12.2. For causal system, we have

$$Z\{s[n+l]\} = z^l Z(z) - z^l s[0] - z^{l-1} s[1] + \ldots + zs[l-1].$$

In fact,

$$Z\{s[n+l]\} = \sum_{n=0}^{\infty} s[n+l] z^{-n} = \sum_{N=l}^{\infty} s[N] z^{-(N-l)}$$

$$= \sum_{N=l}^{\infty} s[N] z^{l-N} = z^l \sum_{N=l}^{\infty} s[N] z^{-N}$$

$$= z^l \sum_{N=0}^{\infty} s[N] z^{-N} - z^l [s[0] + s[1] z^{-1} + \ldots + s[l-1] z^{-(l-1)}]$$

$$= z^l Z(z) - z^l s[0] - z^{l-1} s[1] + \ldots + zs[l-1].$$

Theorem 12.3. (Scaling in the z-domain). Let $Z(z)$ be the z-transform of $s[n]$. Then

$$Z\{a^n s[n]\} = Z\left(\frac{z}{a}\right).$$

Proof: We have

$$Z\left(\frac{z}{a}\right) = \sum_{n=-\infty}^{\infty} s[n]\left(\frac{z}{a}\right)^{-n} = \sum_{n=-\infty}^{\infty} a^n s[n] z^{-n}$$

$$= Z\{a^n s[n]\}.$$

Theorem 12.4. (Time Reversal). Let $Z(z)$ be the z-transform of $s[n]$. Then

$$Z\{s[-n]\} = Z\left(\frac{1}{z}\right),$$

where $s[-n]$ represents mirror image of the signal $s[n]$.

Proof: We have

$$Z\{s[-n]\} = \sum_{n=-\infty}^{\infty} s[-n]z^{-n}.$$

Putting $n = -m$, we get

$$Z\{s[-n]\} = \sum_{m=-\infty}^{\infty} s[m]z^{m} = \sum_{m=-\infty}^{\infty} s[m]\left(\frac{1}{z}\right)^{-m}$$

$$= Z\left(\frac{1}{z}\right).$$

Theorem 12.5. (Conjugation). If $Z(z)$ is the z-transform of $s[n]$, then

$$Z\{\overline{s[n]}\} = \overline{Z(\overline{z})}.$$

Proof: We have

$$Z\{\overline{s[n]}\} = \sum_{n-\infty}^{\infty} \overline{s[n]}z^{-n} = \overline{\sum_{n=-\infty}^{\infty} s[n]\,(\overline{z})^{-n}} = \overline{Z(\overline{z})}.$$

Remark 12.3. In case the signal $s[n]$ is real, then $s[n] = \overline{s[n]}$ and we have $Z(z) = \overline{Z(\overline{z})}$. Thus if a is zero of $Z(z)$, then $Z(a) = \overline{Z(\overline{a})}$ implies $Z(\overline{a}) = 0$ and so $\overline{a}$ is also a zero of the z-transform. It follows, therefore, that zeros of the z-transform of a real signal lie symmetrically with respect to the real axis (Figure 12.18).

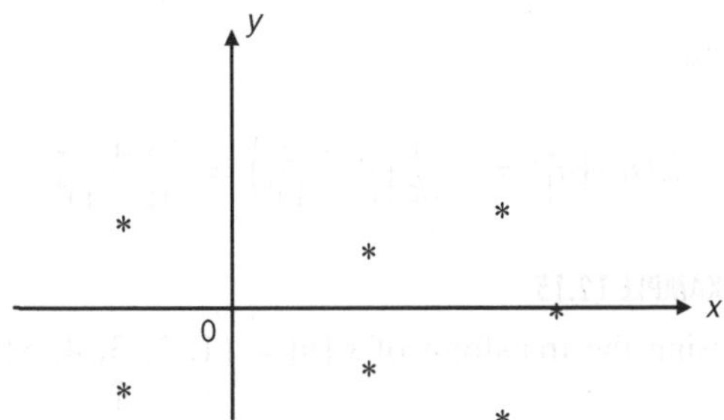

Figure 12.18

Theorem 12.6. (Differentiation in the z-domain). If $Z(z)$ is the z-transform of $s[n]$, then

$$Z\{ns[n]\} = -z\frac{d}{dz}Z(z).$$

Proof: Since power series can be differentiated term by term in its region of convergence, we have

$$\frac{d}{dz}Z(z) = \frac{d}{dz}\sum_{n=-\infty}^{\infty} s[n]\, z^{-n}$$

$$= -\sum_{n=-\infty}^{\infty} n\, s[n]\, z^{-n-1}$$

$$= -z^{-1}\sum_{n=-\infty}^{\infty} ns[n]z^{-n} = \frac{1}{z}Z\{ns[n]\}.$$

Hence

$$Z\{ns[n]\} = -z\frac{d}{dz}Z(z).$$

EXAMPLE 12.14

Using the z-transform of the unit step function $u[n]$, deduce the z-transforms of $nu[n]$, $n^2u[n]$, and $n^3u[n]$.

Solution. By Example 12.5, we have

$$Z\{u[n]\} = \frac{z}{z-1}, |z| > 1.$$

Therefore, by differentiation property, we get

$$Z\{nu[n]\} = -z\frac{d}{dz}Z\{u[n]\}$$

$$= -z\frac{d}{dz}\left(\frac{z}{z-1}\right) = \frac{z}{(z-1)^2},$$

$$Z\{n^2u[n]\} = -z\frac{d}{dz}\left(\frac{z}{(z-1)^2}\right) = \frac{z^2+z}{(z-1)^3},$$

and

$$Z\{n^3u[n]\} = -z\frac{d}{dz}\left(\frac{z^2+z}{(z-1)^3}\right) = \frac{z^3+4z^2+z}{(z-1)^4}.$$

EXAMPLE 12.15

Using the transform of $s_1[n] = \{1, 2, \underset{\uparrow}{3}, 4, 5\}$ from Example 12.3, find the z-transform of $s_2[n] = \{1, \underset{\uparrow}{2}, 3, 4, 5\}$.

Solution. We have

$$s_2[n] = s_1[n+1].$$

Therefore, by using time-shifting property, we get

$$Z\{s_2[n]\} = Z\{s_1[n+1]\} = zZ\{s_1[n]\}$$

$$= z\,[z^2 + 2z + 3 + 4z^{-1} + 5z^{-2}]$$

$$= z^3 + 2z^2 + 3z + 4 + 5z^{-1}.$$

EXAMPLE 12.16

Using differentiation in the z-domain and shift property, show that

$$Z\left\{\binom{n}{k} a^n u[n]\right\} = \frac{a^k z}{(z-a)^{k+1}},$$

$$K = 0, 1, 2, \ldots \text{ and } |z| > |a|.$$

(This result is useful in finding inverse z-transform of the expressions of the type $\frac{a^k z}{(z-a)^{k+1}}$, $k = 0, 1, 2, \ldots$ and $|z| > |a|$.)

Solution. We know that

$$u(n) = \begin{cases} 1 & \text{for } n \ge 0 \\ 0 & \text{for } n < 0. \end{cases}$$

By Example 12.5, we have

$$Z\{u[n]\} = \frac{z}{z-1}, \; |z| > 1.$$

Differentiation in the z-plane implies

$$Z\{nu[n]\} = -z\frac{d}{dz}\left\{\frac{z}{z-1}\right\} = \frac{z}{(z-1)^2}, \; |z| > 1.$$

Now using shift property, we get

$$Z\{(n-1)u[n-1]\} = \frac{1}{z}\cdot\frac{z}{(z-1)^2} = \frac{1}{(z-1)^2}.$$

Again using differentiation in the z-plane, we have

$$Z\{n(n-1)\,u[n-1]\} = -z\frac{d}{dz}\left\{\frac{1}{(z-1)^2}\right\}$$

$$= \frac{2z}{(z-1)^3}, \; |z| > 1.$$

Also

$$n(n-1)\,u[n-1] = n(n-1)\,u[n].$$

Repeating the above process k time, we get

$$Z\{n(n-1)(n-2)\ldots(n-k+1)\,u[n]\}$$

$$= \frac{k!\,z}{(z-1)^{k+1}}, \; |z| > 1$$

or equivalently

$$Z\left\{\binom{n}{k} u[n]\right\} = \frac{z}{(z-1)^{k+1}}, \; |z| > 1. \quad (4)$$

Since $\binom{n}{k} = 0$ for $k > n$, the relation (4) holds for $k \le n$. We now use scaling property to obtain

$$Z\left\{\binom{n}{k} a^n u[n]\right\} = \frac{z/a}{((z/a)-1)^{k+1}}, \; \left|\frac{z}{a}\right| > 1$$

$$= \frac{a^k z}{(z-a)^{k+1}}, \; |z| > |a|.$$

Remark 12.4. Since $Z\{u[n]\} = \frac{z}{z-1}$, we have by time reversal property,

$$Z\{u[-n]\} = \frac{1/z}{(1/z)-1} = \frac{1}{1-z}.$$

Then, by differentiation rule, we get

$$Z\{nu[-n]\} = -z\frac{d}{dz}\left(\frac{1}{1-z}\right) = -\frac{z}{(1-z)^2}$$

Using shifting property, we have

$$Z\{(n-1)\,u[-n-1]\} = \frac{-1}{(1-z)^2}$$

Again using differentiation rule,

$$Z\{n(n-1)u[-n-1]\} = -z\frac{d}{dz}\left\{\frac{-1}{(1-z)^2}\right\} = \frac{2z}{(1-z)^3}.$$

Proceeding in this way, we get

$$Z\left\{\binom{n}{k}u[-n-1]\right\} = -\frac{z}{(z-1)^{k+1}}$$

Then scaling property implies

$$Z\left\{\binom{n}{k}a^n\,u[-n-1]\right\} = \frac{a^k z}{(z-a)^{k+1}},\ |z| < |a|.$$

12.5.1. Table of z-transforms

In the light of properties and examples of z-transform, discussed so far, we now list the z-transforms of various sequences in the form of the following table. This table will help in finding the inverse z-transforms.

Sequence	z–transforms	R.O.C
$\delta[n]$	1	Entire complex plane
$\delta[n+1]$	z	all z
$u[n]$	$\frac{z}{z-1}$	$\lvert z\rvert > 1$
$u[-n]$	$\frac{1}{1-z}$	$\lvert z\rvert > 1$
$u[n-1]$	$\frac{1}{z-1}$	$\lvert z\rvert > 1$
$u[n+1]$	$\frac{z^2}{z-1}$	$\lvert z\rvert > 1$
$a^n u[n],\ a \neq 0$	$\frac{z}{z-a}$	$\lvert z\rvert > \lvert a\rvert$
$(-1)^n u[n]$	$\frac{z}{z+1}$	$\lvert z\rvert > 1$
$2^{n-1}u[n-1]$	$\frac{1}{z-2}$	$\lvert z\rvert > 2$
$nu[n]$	$\frac{z}{(z-1)^2}$	$\lvert z\rvert > 1$
$(n-1)u[n-1]$	$\frac{1}{(z-1)^2}$	$\lvert z\rvert > 1$
$n(n-1)u[n-1]$	$\frac{2z}{(z-1)^3}$	$\lvert z\rvert > 1$
$n(n-1)u[n]$	$\frac{2z}{(z-1)^3}$	$\lvert z\rvert > 1$
$\frac{1}{n!}$	$e^{\frac{1}{z}}$	all z
$u[n]\cos\omega_0 n$	$\frac{z^2 - z\cos\omega_0}{z^2 - 2z\cos\omega_0 + 1}$	$\lvert z\rvert > 1$

12.6 INVERSE Z-TRANSFORM

To find inverse z-transform of $Z(z)$, we mean to obtain the discrete-time sequence $s[n]$ from $Z(z)$. The inverse z-transform can be obtained in the following ways:

(A) Contour Integration Method

By definition

$$Z(z) = \sum_{n=-\infty}^{\infty} s[n]z^{-n}.$$

Multiplying both sides by z^{m-1} and integrating over a closed contour Γ within the region of convergence and enclosing the origin, we have

$$\oint_\Gamma Z(z)z^{m-1}\,dz = \oint_\Gamma \sum_{n=\infty}^{\infty} s[n]z^{-n+m-1}\,dz,$$

where Γ is taken exactly once in anti-clockwise direction. Since the series is convergent on this contour, changing the order of integration and summation, we have

$$\oint_\Gamma Z(z)\,z^{m-1}\,dz = \sum_{n=\infty}^{\infty} s[n]\oint_\Gamma z^{m-n-1}\,dz = 2\pi i\,s[m],$$

because, by Cauchy's integral theorem,

$$\frac{1}{2\pi i}\oint_\Gamma z^{m-n-1}\,dz = \begin{cases}1 & \text{for } m = n\\ 0 & \text{for } m \neq n.\end{cases}$$

Hence

$$s[n] = \frac{1}{2\pi i}\oint_\Gamma Z(z)\, z^{n-1}\, dz$$

$$= \frac{1}{2\pi i}\cdot 2\pi i \text{ [sum of the residues at poles of } Z(z)\, z^{n-1}]$$

$$= \text{sum of the residues at poles of } Z(z)\, z^{n-1}.$$

In case there is no poles of $Z(z)\, z^{n-1}$ inside Γ for one or more values of n, then $s[n] = 0$ for these values.

EXAMPLE 12.17

Find the inverse z-transform of

$$Z(z) = \frac{z}{z-1},\ |z| > 1.$$

Solution. The contour integration formula for finding inverse z-transform is

$$s[n] = \frac{1}{2\pi i}\oint_\Gamma Z(z)\, z^{n-1}\, dz = \frac{1}{2\pi i}\oint_\Gamma \frac{z}{z-1}\, z^{n-1}\, dz$$

$$= \frac{1}{2\pi i}\oint_\Gamma \frac{z^n}{z-1}\, dz = \frac{1}{2\pi i}\oint_{|z|=R>1} \frac{z^n}{z-1}\, dz.$$

For $n \geq 0$, the pole $z = 1$ of the integrand lies within the contour. Residue at $z = 1$ is

$$\text{Res}(1) = \lim_{z\to 1}(z-1)\frac{z^n}{z-1} = 1.$$

Thus, for $n \geq 0$,

$$s[n] = \frac{2\pi i}{2\pi i} = 1.$$

For $n < 0$, the poles are $z = 0$ and $z = 1$. Then for $n = -1$, we have

$$\text{Res}(0) = \lim_{z\to 0}(z-0)\frac{1}{z(z-1)} = \lim_{z\to 0}\frac{1}{z-1} = -1,$$

$$\text{Res}(1) = \lim_{z\to 1}(z-1)\frac{1}{z(z-1)} = \lim_{z\to 1}\frac{1}{z} = 1.$$

Hence

$$s[n] = \frac{1}{2\pi i}\cdot 2\pi i(-1+1) = 0.$$

Similarly for $n = -2, -3, \ldots$, we will have $s[n] = 0$.
Thus

$$s[n] = 0 \text{ for } n < 0.$$

Hence

$$s[n] = \begin{cases} 1 & \text{for } n \geq 0 \\ 0 & \text{for } n < 0. \end{cases}$$

and so the inverse z-transform is the unit step function $u[n]$.

EXAMPLE 12.18

Find the inverse z-transform of

$$Z(z) = \frac{z^3}{z^2-1}.$$

Solution. By contour integral formula, we have

$$s[n] = \frac{1}{2\pi i}\oint_\Gamma Z(z)\, z^{n-1}\, dz = \frac{1}{2\pi i}\oint_\Gamma \frac{z^{n+1}}{z^2-1}\, dz.$$

For $n \geq -2$, the poles of the integrand are $z = 1, -1$. There is no region of convergence containing the unit circle. So we take $|z| = R > 1$. Now

$$\text{Res}(1) = \lim_{z\to 1}(z-1)\frac{z^{n+2}}{(z-1)(z+1)} = \frac{1}{2},$$

$$\text{Res}(-1) = \lim_{z\to -1}(z+1)\frac{z^{n+2}}{(z-1)(z+1)} = \frac{(-1)^n}{-2}.$$

Thus for $n \geq -2$, we have

$$s[n] = \frac{1}{2} - \frac{(-1)^n}{2} = [1-(-1)^n].$$

For $n = -3$, the poles of the integrand are $z = 0, z = 1, z = -1$ and so

$$\text{Res}(0) = \lim_{z\to 0}(z-0)\frac{1}{(z-0)(z-1)(z+1)} = -1,$$

$$\text{Res}(1) = \lim_{z\to 1}(z-1)\frac{1}{z(z+1)(z-1)} = \frac{1}{2},$$

$$\text{Res}(-1) = \lim_{z\to -1}(z+1)\frac{1}{z(z+1)(z-1)} = \frac{1}{2},$$

Thus for $n = -3$, we have $s[n] = 0$. The value of $s[n]$ for $n = -4, -5, \ldots$ can be found similarly.

EXAMPLE 12.19

Find inverse z-transform of

$$Z(z) = \frac{2z}{(z-1)(z-2)}.$$

Solution. By contour integration method, we have

$$s[n] = \frac{1}{2\pi i}\oint_{\Gamma} Z(z)\, z^{n-1}\, dz$$

$$= \frac{1}{2\pi i}\oint_{\Gamma} \frac{2z}{(z-1)(z-2)}\, z^{n-1}\, dz$$

$$= \frac{1}{2\pi i}\oint_{\Gamma} \frac{2z^n}{(z-1)(z-2)}\, dz.$$

For $n \geq 0$, the poles of the integrand are $z = 1$, $z = 2$. Then

$$\text{Res }(1) = \lim_{z\to 1}(z-1)\frac{2z^n}{(z-1)(z-2)} = \frac{2}{-1} = -2,$$

$$\text{Res }(2) = \lim_{z\to 2}(z-2)\frac{2z^n}{(z-1)(z-2)} = 2^{n+1}.$$

Hence for $n \geq 0$,

$$s[n] = \frac{1}{2\pi i}\cdot 2\pi i[-2 + 2^{n+1}] = 2[-1 + 2^n].$$

We now take up the case where $n < 0$. For $n = -1$, the integrand becomes

$$\frac{2}{z(z-1)(z-2)}$$

and so its poles are $z = 0$, $z = 1$ and $z = 2$. In this case, we have

$$\text{Res }(0) = \lim_{z\to 0}(z-0)\frac{2}{z(z-1)(z-2)} = 1,$$

$$\text{Res }(1) = \lim_{z\to 1}(z-1)\frac{2}{z(z-1)(z-2)} = -2,$$

$$\text{Res }(2) = \lim_{z\to 2}(z-2)\frac{2}{z(z-1)(z-2)} = 1.$$

Thus the sum of residues is zero and so

$$s[n] = 0 \text{ for } n = -1.$$

For $n = -2$, there will be 0 as the poles of order 2 and $z = 1, 2$ as the simple poles. The sum of residues will again be zero. Proceeding in this way we see that $s[n] = 0$ for $n < 0$. Hence

$$s[n] = 2(-1 + 2^n)\, u[n].$$

EXAMPLE 12.20

Find the signal with a finite switch on time having

$$Z(z) = \frac{z}{\left(z-\frac{1}{2}\right)(z-2)}.$$

Solution. By contour integration method, we have

$$s[n] = \frac{1}{2\pi i}\oint_{\Gamma} Z(z)\, z^{n-1}\, dz$$

$$= \frac{1}{2\pi i}\oint_{\Gamma} \frac{z}{\left(z-\frac{1}{2}\right)(z-2)}\, z^{n-1}\, dz$$

$$= \frac{1}{2\pi i}\oint_{\Gamma} \frac{z^n}{\left(z-\frac{1}{2}\right)(z-2)}\, dz$$

$$= \text{sum of residues at poles within } \Gamma.$$

For $n \geq 0$, the poles of the integrand are $z = \frac{1}{2}$ and $z = 2$ lying in $|z| = R > 2$. Then

$$\text{Res}\left(\frac{1}{2}\right) = \lim_{z\to(1/2)}\left(z-\frac{1}{2}\right)\frac{z^n}{\left(z-\frac{1}{2}\right)(z-2)} = -\frac{2^{1-n}}{3}.$$

$$\text{Res }(2) = \lim_{z\to 2}\frac{(z-2)z^n}{\left(z-\frac{1}{2}\right)(z-2)} = \frac{2^{n+1}}{3}.$$

Hence

$$s[n] = \frac{2^{n+1}}{3} - \frac{2^{1-n}}{3} = \frac{1}{3}(2^{n+1} - 2^{1-n}),\ n \geq 0.$$

Now we take the case when $n < 0$. Let $n = -1$. Then the integrand is $\dfrac{1}{z\left(z-\frac{1}{2}\right)(z-2)}$ and so it has simple poles at $z = 0, \frac{1}{2}$ and 2. We have

$$\text{Res }(0) = \lim_{z\to 0}(z-0)\frac{1}{(z-0)\left(z-\frac{1}{2}\right)(z-2)} = 1,$$

$$\text{Res}\left(\frac{1}{2}\right) = \lim_{z \to \frac{1}{2}} \left(z - \frac{1}{2}\right) \frac{1}{(z-0)\left(z-\frac{1}{2}\right)(z-2)} = -\frac{4}{3}.$$

$$\text{Res}\,(2) = \lim_{z \to 2} (z-2) \frac{1}{(z-0)\,(z-2)\left(z-\frac{1}{2}\right)} = \frac{1}{3}.$$

Thus sum of residues is zero. Similarly, for other negative values of n, the sum of the residues will be zero. Hence

$$s[n] = \frac{1}{3}(2^{n+1} - 2^{1-n})\, u[n].$$

(B) Partial Fractions Method

In practice, the z-transforms are rational functions. So, let

$$Z(z) = \frac{P(z)}{Q(z)},$$

where P(z) and Q(z) have no factor in common and degree of P(z) is less than that of Q(z). We express $Z(z)$ into partial fractions and then find the inverse z-transform of each term.

EXAMPLE 12.21

Find inverse z-transform of

$$Z(z) = \frac{z^3}{z^2 - 1}.$$

Solution. Since the degree of numerator is greater than that of dominator, dividing the numerator by the denominator, we have

$$Z(z) = z + \frac{z}{z^2 - 1} = z + \frac{z}{(z-1)\,(z+1)}$$

$$= z + \frac{1}{2}\left[\frac{z}{z-1} - \frac{z}{z+1}\right] \text{ (partial fractions).}$$

The region of convergence for $Z(z)$ is $|z| > 1$ and its poles are $z = 1$ and $z = -1$. But, using table of z-transform given in 12.5.1, we have

$$Z^{-1}\{z\} = \delta[n+1],$$

$$Z^{-1}\left\{\frac{z}{z-1}\right\} = u[n], \text{ and}$$

$$Z^{-1}\left\{\frac{z}{z+1}\right\} = (-1)^n\, u[n].$$

Hence

$$s[n] = \delta[n+1] + \frac{1}{2}u[n] - \frac{1}{2}(-1)^n\, u[n]$$

$$= \delta[n+1] + \frac{1}{2}u[n]\,[1 - (-1)^n].$$

EXAMPLE 12.22

Find the inverse z-transform of

$$Z(z) = \frac{z+3}{z-2}.$$

Solution. Dividing the numerator by the denominator, we get

$$Z(z) = 1 + \frac{5}{z-2}.$$

The region of convergence of $Z(z)$ is $|z| > 2$. But

$$Z^{-1}\{1\} = s[n]$$

$$Z^{-1}\left\{\frac{1}{z-2}\right\} = 2^{n-1}u[n-1].$$

Hence, the required inverse z-transform is

$$s[n] = \delta[n] + 5\,[2^{n-1}u[n-1]].$$

EXAMPLE 12.23

Find the inverse z-transform of

$$Z(z) = \frac{2z^2 + 3z}{(z+2)\,(z-4)}.$$

Solution. We have

$$Z(z) = \frac{2z^2 + 3z}{(z+2)\,(z-4)}. \quad (5)$$

Since the degree of numerator and that of denominator is same, we write (5) as

$$\frac{Z(z)}{z} = \frac{2z+3}{(z+2)(z-4)}$$

$$= \frac{1}{6(z+2)} + \frac{11}{6(z-4)} \text{ (by partial fractions.)}$$

and so

$$Z(z) = \frac{z}{6(z+2)} + \frac{11z}{6(z-4)}.$$

Since $Z\{a^n u[n]\} = \frac{z}{z-a}$, $a \neq 0$, we have

$$Z^{-1}\left\{\frac{z}{z+2}\right\} = (-2)^n u[n], \text{ and}$$

$$Z^{-1}\left\{\frac{z}{z-4}\right\} = 4^n u[n].$$

Hence, the required inverse z-transform is

$$s[n] = \frac{1}{6}(-2)^n u[n] + \frac{11}{6}(4^n u[n]).$$

EXAMPLE 12.24

Find the inverse z-transform of

$$Z(z) = \frac{1}{z(z-2)^2}.$$

Solution. We have

$$\frac{Z(z)}{z} = \frac{1}{z^2(z-2)^2}$$

$$= \frac{1}{4}\left(\frac{1}{z} + \frac{1}{z^2} - \frac{1}{z-2} + \frac{1}{(z-2)^2}\right)$$

(partial fractions)

and so

$$Z(z) = \frac{1}{4}\left(1 + \frac{1}{z} - \frac{z}{z-2} + \frac{z}{(z-2)^2}\right).$$

Since $Z\{2^n u[n]\} = \frac{z}{z-2}$, differentiation in z-plane implies

$$Z\{2^n n\, u[n]\} = -z\frac{d}{dz}\left\{\frac{z}{z-2}\right\} = \left\{\frac{2z}{(z-2)^2}\right\}.$$

Therefore

$$Z\{2^{n-1} n\, u[n]\} = \frac{z}{(z-2)^2}.$$

Hence, the required inverse z-transform is

$$s[n] = \frac{1}{4}[\delta[n] + \delta[n-1] - 2^n u[n] + n2^{n-1} u[n]].$$

EXAMPLE 12.25

Find the inverse z-transform of

$$Z(z) = \frac{z}{\left(z-\frac{1}{2}\right)\left(z-\frac{1}{4}\right)}.$$

Solution. We have

$$Z(z) = \frac{z}{\left(z-\frac{1}{2}\right)\left(z-\frac{1}{4}\right)}$$

$$= \frac{2}{z-\frac{1}{2}} - \frac{1}{z-\frac{1}{4}} \quad \text{(partial fractions).}$$

Since $Z\{a^n u[n]\} = \frac{z}{z-a}$, the shifting property implies

$$Z\{an\text{–}1\, u[n-1]\} = \frac{1}{z} \cdot \frac{z}{z-a} = \frac{z}{z-a},\ a \neq 0.$$

Taking $a = \frac{1}{2}$ and $\frac{1}{4}$, the required inverse z-transform is

$$s[n] = 2\left(\frac{1}{2}\right)^{n-1} u[n-1] - \left(\frac{1}{4}\right)^{n-1} u[n-1]$$

$$= 4\left[\left(\frac{1}{2}\right)^n - \left(\frac{1}{4}\right)^n\right] u[n-1].$$

EXAMPLE 12.26

Find inverse z-transform of

$$Z(z) = \frac{z^4 + z^3 - z^2 - z + 1}{z^2 + 2z + 1}.$$

Solution. Here the degree of numerator is greater than the degree of the denominator. So, we start from $\frac{Z(z)}{z}$. Thus

$$\frac{Z(z)}{z} = \frac{z^4 + z^3 - z^2 - z + 1}{z(z^2 + 2z + 1)}$$

$$= \frac{(z-1)(z^3 + 2z^2 + z) + 1}{z(z^2 + 2z + 1)}$$

$$= \frac{z(z-1)(z^2 + 2z + z) + 1}{z(z^2 + 2z + 1)}$$

$$= z - 1 + \frac{1}{z(z^2 + 2z + 1)}$$

$$= z - 1 + \frac{1}{z} - \frac{1}{z+1} - \frac{1}{(z+1)^2}$$

(partial fractions).

Therefore,

$$Z(z) = z^2 - z + 1 - \frac{z}{z+1} - \frac{z}{(z+1)^2}.$$

Hence, for $|z| > 1$, the required inverse z-transform is

$$s[n] = \delta[n+2] - \delta[n+1] + \delta[n] - (-1)^n u[n] - n(-1)^{n-1} u[n].$$

EXAMPLE 12.27

Find inverse z-transform of

$$Z(z) = \frac{z^3 + 3z^2 + 3z - 4}{z^3 + 3z^2 - 4}.$$

Solution. The numerator and denominator are of the same degree. We first divide and get

$$Z(z) = 1 + \frac{3z}{z^3 + 3z^2 - 4}$$

$$= 1 + \frac{1}{3(z-1)} - \frac{1}{6(z+2)} + \frac{2}{(z+2)^2}$$

(partial fractions).

Therefore,

$$s[n] = \delta[n] + \frac{1}{3} u[n-1] - \frac{1}{6} (-2)^{n-1} u[n-1] + 2(n-1)(-2)^{n-2} u[n-1].$$

(C) Power Series Method for Finding Inverse z-transform

If $Z(z)$ can be written as

$$Z(z) = \sum_{n=-\infty}^{\infty} c_n z^{-n},$$

with a given region of convergence, then

$$s[n] = c_n \text{ for all } n.$$

EXAMPLE 12.28

Find inverse z-transform of

$$Z(z) = \frac{1}{1-z}, \; |z| < 1.$$

Solution. We have

$$Z(z) = \frac{1}{1-z} = -\left(\frac{1}{z-1}\right) = -\left[\frac{1}{z} + \frac{1}{z^2} + \frac{1}{z^3} + \ldots\right]$$

$$= -\frac{1}{z} - \frac{1}{z^2} - \frac{1}{z^3} - \ldots = \sum_{n=1}^{\infty} (-1) z^{-n},$$

which yields $s[n] = -1$, $n > 0$. (6)

Second Method:

$$Z(z) = \sum_{n=-\infty}^{\infty} c_n z^{-n} = \ldots\ldots + c_{-2} z^2 + c_{-1} z^1 + c_0 z^0 + c_1 z^{-1} + c_2 z^{-2} + \ldots \quad (7)$$

Comparing (6) and (7), we have

$$s[0] = c_0 = 0, \quad s[1] = c_1 = -1,$$

$$s[2] = c_2 = -1, \quad s[3] = c_3 = -1,$$

and so on.

EXAMPLE 12.29

Find inverse z-transform of

$$Z(z) = \frac{z^2 - 1}{z^3 + 2z + 4}.$$

Solution. We have

$$Z(z) = \frac{z^2 - 1}{z^3 + 2z + 4} = \frac{1}{z} - \frac{3}{z^3} - \frac{4}{z^4} + \ldots \quad (8)$$

But

$$Z(z) = \sum_{n=-\infty}^{\infty} c_n z^{-n} = \ldots + c_{-3} z^3 + c_{-2} z^3 + c_{-1} z^3 + c_0 z^0 + c_1 z^{-1} + c_2 z^{-2} + c_3 z^{-3} + c_4 z^{-4} + \ldots \quad (9)$$

Comparing (8) and (9), we have

$$s[0] = c_0 = 0, \quad s[1] = c_1 = 1,$$

$$s[2] = c_2 = 0, \quad s[3] = c_3 = -3,$$

$$s[4] = c_4 = -4, \text{ and so on.}$$

12.7 CONVOLUTION THEOREM

Let $s_1[n]$ and $s_2[n]$ be two discrete-time signals. Their *convolution* is defined by

$$(s_1 * s_2)[n] = \sum_{m=-\infty}^{\infty} s_1[m]\, s_2[n-m].$$

Then *Convolution theorem* of the z-transform reads as:

Theorem 12.7. (Convolution Theorem). Let $s_1[n]$ and $s_2[n]$ be discrete-time signals having z-transforms $Z_1(z)$ and $Z_2(z)$, respectively, and let V be the intersection of the regions of convergence of these transforms. Then

$$Z\{(s_1 * s_2)[n]\} = Z_1(z)\, Z_2(z),\ z \in V.$$

EXAMPLE 12.30

Using Convolution theorem, find the inverse z-transform of $\frac{z^2}{(z-1)(z-2)}$.

Solution. Let

$$Z_1(z) = \frac{z}{z-1} \quad |z| > 1,$$

$$Z_2(z) = \frac{z}{z-2}, \quad |z| > 2.$$

Then, the intersection of the two region of convergence is $|z| > 2$ and so the convolution product is defined. We have

$$s_1[n] = u[n],$$

$$s_2[n] = 2nu[n].$$

Therefore, by Convolution theorem,

$$Z^{-1}\left\{\frac{z^2}{(z-1)(z-2)}\right\} = \sum_{k=0}^{n} 1.2^{n-k}$$

$$= 2^n \sum_{k=0}^{n} \left(\frac{1}{2}\right)^k$$

(geometric series)

$$= 2^{n+1}\left[1 - \left(\frac{1}{2}\right)^{n+1}\right].$$

EXAMPLE 12.31

Using Convolution theorem find the inverse z-transform of $\frac{z(z+1)}{(z-1)^3}$.

Solution. We are given that

$$Z(z) = \frac{z(z+1)}{(z-1)^3} = \frac{z}{(z-1)^2} \cdot \frac{z+1}{z-1}$$

$$= \frac{z}{(z-1)^2}\left[\frac{z}{z-1} + \frac{1}{z-1}\right].$$

Let

$$Z_1(z) = \frac{z}{(z-1)^2} \text{ and } Z_2(z) = \frac{z}{z-1} + \frac{1}{z-1}.$$

Then

$$s_1[n] = n.1^n u[n],$$

$$s_2[n] = u[n] + u[n-1].$$

Hence, by Convolution theorem, we get

$$Z^{-1}\left\{\frac{z(z+1)}{(z-1)}\right\} = s_1[n] * s_2[n]$$

$$= \sum_{m=0}^{n} m\{u[n-m] + u[n-m-1]\}$$

$$= n^2.$$

EXAMPLE 13.32

Let $s_1[n] = \delta[n]$, the discrete unit pulse and $s_2[n] = 2^n u[n]$, where $u[n]$ is the unit step function. Verify Convolution theorem for inverse z-transform in this case.

Solution. We have

$$\delta[n] = \begin{cases} 1 & \text{for } n = 0 \\ 0 & \text{otherwise,} \end{cases}$$

and

$$u[n] = \begin{cases} 1 & \text{for } n \geq 0 \\ 0 & \text{for } n < 0. \end{cases}$$

If $Z_1(z)$ and $Z_2(z)$ are the z-transforms of $s_1[n]$ and $s_2[n]$, respectively, then $Z_1(z) = 1$, with region of convergence as whole complex plane and (see Example 12.7) $Z_2(z) = \frac{z}{z-2}$, with region of convergence $|z| > 2$.

The intersection of these two regions is $|z| > 2$. Thus the region is non-empty and the convolution product is well defined. Hence, by Convolution theorem, we have

$$Z^{-1}\{Z_1(z)Z_2(z)\} = Z^{-1}\left\{\frac{z}{z-2}\right\}$$

$$= \sum_{m=-\infty}^{\infty} \delta[m]2^{n-m}\, u[n-m]$$

$$= 2^n[n], \text{using definiton of } \delta[n].$$

EXAMPLE 13.33

Use Convolution theorem to find inverse z-transform of

$$Z(z) = \frac{z^2}{\left(z-\frac{1}{2}\right)\left(z-\frac{1}{3}\right)}.$$

Solution. Let

$$Z_1(z) = \frac{z}{z-\frac{1}{2}}, \quad |z| > \frac{1}{2}$$

$$Z_2(z) = \frac{z}{z-\frac{1}{3}}, \quad |z| > \frac{1}{3}.$$

Then

$$s_1[n] = \left(\frac{1}{2}\right)^n u[n] = 2^{-n}\, u[n],$$

$$s_2[n] = \left(\frac{1}{3}\right)^n u[n] = 3^{-n}\, u[n].$$

The common region of convergence is $|z| > \frac{1}{2}$. Therefore, by Convolution theorem, we have

$$Z^{-1}(z) = Z^{-1}\{Z_1(z)\, Z_2(z)\}$$

$$= \sum_{m=0}^{n} 2^{-m}\, u[m]3^{-n+m}\, u[n-m]$$

$$= 3^{-n} \sum_{m=0}^{n} 2^{-m}\, u[n] = 3^{-n} \sum_{m=0}^{n} \left(\frac{3}{2}\right)^m u[n]$$

$$= 3^{-n}\, u[n]\left[1 + \frac{3}{2} + \left(\frac{3}{2}\right)^2 + \ldots\right]$$

$$= 3^{-n}\left[2\left(\frac{3^{n+1}}{2^{n+1}} - 1\right)\right] u[n]$$

$$= [3.2^{-n} - 2.3^{-n}]\, u[n].$$

12.8 THE TRANSFER FUNCTION (OR SYSTEM FUNCTION)

The *transfer function* (*system function*) H(z) of a system is defined by

$$\text{H}(z) = \sum_{n=-\infty}^{\infty} h[n]z^{-n} \tag{10}$$

Thus transfer function is z-transform of the impulse response $h[n]$ to the unit impulse $\delta[n]$.

Further, we know that response $y[n]$ to the input $s[n]$ is

$$y[n] = \sum_{n=-\infty}^{\infty} s[m]\, h[n-m] = (s * h)[n].$$

Taking z-transform of both sides, we have by convolution theorem,

$$Z\{y[n]\} = Z\{(s * h)[n]\} = Z\{s[n]\}\, Z\{h[n]\}$$

$$= \text{H}(z)\, Z\{s[n]\} \text{ using(10)}.$$

Hence z-transform of the output is ordinary multiplication of the transfer function H(z). It follows that

$$\text{H}(z) = \frac{Z\{y[n]\}}{Z\{s[n]\}}.$$

EXAMPLE 12.34

Find transfer function H(z) and the unit pulse response $h[n]$ of a system described by the difference equation

$$y[n] = \frac{1}{2}y[n-1] + 2s[n],$$

where $y[n]$ is the response to the input $s[n]$.

Solution. Taking z-transform of both sides of the difference equation, we get

$$Z\{y[n]\} = \frac{1}{2}z^{-1}\, Z\{y[n]\} + 2z\{s[n]\}$$

or

$$\left(1-\frac{1}{2}z^{-1}\right)Z\{y[n]\}=2Z\{s[n]\}$$

or

$$\mathrm{H}(z)=\frac{Z\{y[n]\}}{Z\{s[n]\}}=\frac{2}{1-\frac{1}{2z}}=\frac{4z}{2z-1},$$

where as the unit pulse response is given by

$$h[n]=Z^{-1}\{\mathrm{H}[z]\}$$

$$=Z^{-1}\left\{\frac{2}{1-(1/2)z^{-1}}\right\}=2\left(\frac{1}{2}\right)^{n}u[n].$$

12.9 SYSTEMS DESCRIBED BY DIFFERENCE EQUATIONS

Consider the difference equation

$$y[n]+b_1y[n-1]+\ldots+b_My[n-\mathrm{M}]$$

$$=a_0u[n]+a_1u[n-1]+\ldots+a_N[n-\mathrm{N}]\quad(11)$$

describing a system with input $u[n]$ and response $y[n]$. Let Y(z) and U(z) denote the z-transforms of $y[n]$ and $u[n]$, respectively. Applying z-transform to both sides of (11), we get

$$(1+b_1z^{-1}+\ldots+b_{\mathrm{M}}z^{-M})\,\mathrm{Y}(z)$$

$$=(a_0+a_1z^{-1}+\ldots+a_{\mathrm{N}}z^{-\mathrm{N}})\,\mathrm{U}(z).$$

Therefore, the transfer function is

$$\mathrm{H}(z)=\frac{Y(z)}{U(z)}=\frac{a_0+a_1z^{-1}+\ldots+a_Nz^{-N}}{1+b_1z^{-1}+\ldots+b_Mz^{-M}}.$$

Thus the transfer function is a rational function given by

$$\mathrm{H}(z)=z^{M-N}\frac{a_N+a_{N-1}z+\ldots+a_0z^N}{b_M+b_{M-1}z+\ldots+z^M}.$$

In what follows, we shall consider causal system. To solve difference equations we shall apply z-transform to it and then shall take inverse z-transform to get the response of the given signal.

EXAMPLE 12.35

Find transfer function and response of the system described by the difference equation

$$y[n+2]-3y[n+1]+2y[n]=\delta[n].$$

Solution. The transfer function is given by

$$\mathrm{H}(z)=\frac{1}{z^2-3z+2}=\frac{Y(z)}{U(z)}=\frac{Y(z)}{1}=Y(z).$$

Thus

$$\mathrm{H}(z)=Y(z)=\frac{1}{z^2-3z+2}=\frac{1}{(z-2)\,(z-1)}$$

$$=\frac{1}{z-2}-\frac{1}{z-1}\qquad\text{(partial fractions)}.$$

Taking inverse z-transform, we get

$$h[n]=y[n]=2^{n-1}u[n-1]-u[n-1].$$

EXAMPLE 12.36

Find the transfer function and impulse response of the causal system described by the difference equation

$$y[n]-3y[n-1]+2y[n-2]=u[n]+u[n-1].$$

Solution. The transfer function is

$$\mathrm{H}(z)=\frac{1+z^{-1}}{1-3z^{-1}+2z^{-2}}=\frac{z+z^2}{z^2-3z+2}$$

$$=1+\frac{4z-2}{z^2-3z+2}$$

$$=1-\frac{2}{z-1}+\frac{6}{z-1}\quad\text{(partial fractions)}.$$

The impulse response is therefore given by

$$h[n]=\delta[n]-2(1)^{n-1}u[n-1]+6(2)^{n-1}u[n-1].$$

EXAMPLE 12.37

Solve the difference equation

$$y[n+2]+3y[n+1]+2y[n]=0,$$

$$y[0]=1,y[1]=2,$$

describing a causal system.

Solution. Taking z-transform, we get (see Remark 12.2)

$$z^2 Y(z) - z^2 y[0] - zy[1] + 3zY(z) - 3zy[0] + 2Y(z) = 0.$$

Using given conditions, we have

$$(z^2 + 3z + 2)\, Y(z) = z^2 + 5z$$

or

$$Y(z) = \frac{z^2 + 5z}{z^2 + 3z + 2} = \frac{4z}{z+1} - \frac{3z}{z+2}.$$

Taking inverse z-transform, we get

$$y[n] = 4(-1)^n u[n] - 3(-2)^n u[n].$$

EXAMPLE 12.38

Solve the difference equation

$$y[n] - 3y[n-1] + 3y[n-2] - y[n-3] = u[n-2],$$

describing a causal system, with $y[0] = y[-1] = y[-2] = 0$.

Solution. Taking the z-transform of both sides of the given difference equation, we have

$$[1 - 3z^{-1} + 3z^{-2} - z^{-3}]\, Y(z) = z^{-2} U(z).$$

Thus for $|z| > 1$, we have

$$Y(z) = \frac{z^{-2}}{1 - 3z^{-1} + 3z^{-2} - z^{-3}}\, U(z)$$

$$= \frac{z}{(z-1)^3}\, U(z) = \frac{z}{(z-1)^3} \cdot \frac{z}{z-1}$$

$$= \frac{z^2}{(z-1)^4}.$$

Now $Y(z) z^{n-1} = \dfrac{z^{n+1}}{(z-1)^4}$ has a pole of order 4 at $z = 1$, which lies within the contour $|z| = R > 1$. Moreover,

$$\operatorname{Res}(1) = \lim_{z \to 1} \frac{1}{3!} \frac{d^3}{dz^3} \left[(z-1)^4 \frac{z^{n+1}}{(z-1)^4} \right]$$

$$= \frac{1}{6} \lim_{z \to 1} [(n+1)\, n(n-1) z^{n-1}]$$

$$= \frac{1}{6} [(n+1)\, n(n-1)]$$

Hence, by contour integration method, the inverse z-transform is given by

$$y[n] = \frac{1}{6} [n(n-1)\,(n+1)].$$

EXAMPLE 12.39

Solve the difference equation

$$y[n+1] + 2y[n] = n,\ y[0] = 1.$$

Solution. Application of z-transform gives

$$Z\{Y(z) - y[0]\} + 2Y(z) = \frac{z}{(z-1)^2}$$

or

$$(z+2)\, Y(z) = z + \frac{z}{(z-1)^2}$$

or

$$Y(z) = \frac{z}{z+2} + \frac{z}{(z+2)\,(z-1)^2}.$$

Using partial fractions, we have

$$Y(z) = \frac{z}{z+2} + z \left[\frac{1}{9(z+2)} - \frac{1}{9(z-1)} + \frac{3}{9(z-1)^2} \right]$$

$$= \frac{10}{9} \frac{z}{(z+2)} - \frac{1}{9} \cdot \frac{z}{(z-1)} + \frac{3}{9} \cdot \frac{z}{(z-1)^2}.$$

Taking inverse z-transform, we have

$$y[n] = \frac{10}{9} (-2)^n u[n] - \frac{1}{9} u[n] + \frac{3n}{9} u[n].$$

EXAMPLE 12.40

Solve the difference equation

$$y[n+2] - 2y[n+1] + y[n] = 3n + 5,$$

subject to the condition $y[0] = y[1] = 0$.

Solution. Application of z-transform, along with the given conditions, yields

$$z^2 Y(z) - 2zY(z) + Y(z) = 3 \frac{z}{(z-1)^2} + \frac{5z}{z-1}$$

or

$$Y(z) = \frac{3z}{(z-1)^2(z^2-2z+1)}$$

$$+\frac{5z}{(z-1)(z^2-2z+1)}$$

$$= \frac{3z}{(z-1)^4} + \frac{5z}{(z-1)^3} \text{ (partial fractions).}$$

Taking inverse z-transform, we have

$$y[n] = 3\binom{n}{3}u[n] + 5\binom{n}{2}u[n]$$

$$= \left[\frac{1}{2}n(n-1)(n-2) + \frac{5}{2}n(n-1)\right]u[n]$$

$$= \left[\frac{1}{2}n(n-1)[n-2+5]\right]u[n]$$

$$= \frac{1}{2}n(n-1)(n+3)\,u[n].$$

EXAMPLE 12.41

Solve the difference equation

$$y[n+1] = y[n] + y[n-1], y[0] = y[1] = 1.$$

Solution. The given difference equation represents *Fibonacci sequence.* Taking z-transform of both sides, we get

$$z[Y(z) - y(0)] = Y(z) + z^{-1}[Y(z)] + y[-1]$$

and so

$$\left(z - 1 - \frac{1}{z}\right)Y(z) = z.$$

Hence

$$Y(z) = \frac{z^2}{z^2 - z - 1} = z\left[\frac{z}{(z-a)(z-b)}\right],$$

where

$$a = \frac{1}{2}(1+\sqrt{5}), b\frac{1}{2}(1+\sqrt{5}).$$

But

$$\frac{z}{(z-a)(z-b)} = \frac{a}{(a-b)(z-a)} + \frac{b}{(b-a)(z-b)}.$$

(partial fractions).

Hence

$$Y(z) = \left(\frac{a}{a-b}\right)\frac{z}{z-a} + \frac{b}{(b-a)}\cdot\frac{z}{z-b}.$$

Taking inverse z-transform, we have

$$y[n] = \left[\frac{a}{a-b}a^n + \frac{b}{b-a}b^n\right]u[n]$$

$$= \frac{a^{n+1} - b^{n+1}}{a-b}u[n]$$

$$= \frac{a^{n+1} - b^{n+1}}{a-b}, n = 0, 1, 2, \ldots$$

EXERCISES

1. Find z-transform of following sequences:

   (a) $s[n] = \{0, 0, 1, 2, 3\}$ (arrow ↑ under the first 1)

   Ans. $z^{-2} + 2z^{-3} + 3z^{-4}$, ROC: entire z-plane except $z = 0$

   (b) $s[n] = u[n] \sin w_0 n$

   Ans. $\frac{z^{-1}\sin\omega_0}{1 - 2z^{-1}\cos\omega_0 + z^{-2}}$, ROC: $|z| > 1$

   (c) $s[n] = e^{inx}$

   Hint: similar to $Z\{A^n\}$.

   Ans. $\frac{z}{z - e^{ix}}$

   (d) $s[n] = \cos nx$

   Ans. $\frac{z(z - \cos x)}{z^2 - 2z\cos x + 1}$

   (e) $s[n] = \sin nx$

   Ans. $\frac{3\sin x}{z^2 - 2z\cos x + 1}$

   Hint: For (*d*) and (*e*) use $e^{inx} = \cos nx + i \sin nx$ and part (c).

2. Using partial fraction method, find inverse z-transform of

$$Z(z) = \frac{4z^2 - 2z}{z^3 - 5z^2 + 8z - 4}.$$

   Hint: $Z(z) = \frac{2}{z-1} + \frac{2}{z-2} + \frac{12}{(z-2)^2}$

Ans. $2(1)^{n-1}u[n-1]+2(2)^{n-1}u[n-1]$
$+12(n-1)(2)^{n-2}u[n-1]$

3. Using Convolution theorem find inverse z-transform of

$$Z(z)=\frac{z^2}{(z-2)(z-3)}.$$

Hint: $Z_1(z)=\frac{z}{z-2}$, $|z|>2$; $Z_2(z)=\frac{z}{z-3}$, $|z|>3$, so $s_1[n]=2^n u[n]$, $s_2[n]=3^n u[n]$.

Therefore, $Z^{-1}\{Z_1(z), Z_2(z)\}$

$$=\sum_{m=0}^{n}2^m 3^{n-m}u[n-m]$$

$$=3^{n+1}\left[1-\left(\frac{2}{3}\right)^{n+1}\right]u[n],$$

$$n=0,1,2....$$

4. Find the inverse z-transform of $Z(z)=\frac{z}{z-a}$.

Hint: $\frac{z}{z-a}=\left(\frac{1-a}{z}\right)^{-1}=\frac{1+a}{z}+\frac{a^2}{z^2}+\ldots+\frac{a^n}{z^n}+\ldots$ and so $s[0]=1$, $s[1]=a$, $s[2]=a^2,\ldots, s[n]=a^n,\ldots.$

5. Find inverse z-transform of
$Z(z)=\frac{z}{z^2-5z+6}$.

Hint: Use partial fraction method

Ans. 3^n-2^n

6. Find the inverse z-transform of $Z(z)=\frac{1}{(z-1)^2(z-2)}$.

Ans. $2^{n-1}-n$

7. Using z-transform solve the following difference equations:

(a) $y[n+2]+4y[n+1]+3y[n]=2^n$, $y[0]=0$, $y[1]=1$.

Ans. $\frac{1}{3}(-1)^n+\frac{1}{15}2^n-\frac{2}{5}(-3)^n$

(b) $y[n+2]-5y[n+1]+6y[n]=2^n$, $y[0]=0$, $y[1]=0$.

Ans. $2^{n+1}-3^n-n\,2^{n-1}$

8. Show that $Z\{na^{n-1}\}=\frac{z}{(z-a)^2}$.

9. Find transfer function and the impulse response of a causal system described by the difference equation

$$y[n]+\frac{1}{2}y[n-1]=u[n].$$

Ans. $H(z)=\frac{z}{z+\frac{1}{2}}$ and $h[n]=\left(-\frac{1}{2}\right)^n$

13 Elements of Statistics and Probability

13.1 INTRODUCTION

Statistics is the science of assembling, analysing, characterizing, and interpreting the collection of data (information expressed numerically). The methods used for this purpose are called *statistical methods*. The general characteristics of data are:

1. Data shows a tendency to concentrate at certain values, usually somewhere in the centre of the distribution. Measures of this tendency are called *measures of central tendency* or *averages*.
2. The data varies about a measure of central tendency and the measures of deviation are called *measures of variability* or *dispersion*.
3. The data in a frequency distribution may fall into symmetrical or asymmetrical patterns. The measures of the degree of asymmetry are called the *measures of skewness*.
4. The measures of peakedness or flatness of the frequency curves are called *measures of kurtosis*.

If the figures in the original data are put into groups, then those groups are called *classes*. The difference between the upper and lower limits of a class is called the *width of the class* or simply the *class interval*. The number of observations in a class interval is called the *frequency*. The mid-point or the mid-value of the class is called the *class mark*. The table showing the classes and the corresponding frequencies is called a *frequency table*. The set of ungrouped data summarized by distributing it into a number of classes along with their frequencies is known as *frequency distribution*. The *cumulative frequency* (written as cum *f*) of the *n*th class in a frequency distribution is the sum of the frequencies beginning with the first and ending with the *n*th frequency. Thus

$$\text{Cum}\, f_n = \sum_{i=1}^{n} f_i$$

For example, consider the following table:

Marks in Physics (class)	Number of Students (f)	Cum (f)
50–60	5	5
60–70	16	21
70–80	24	45
80–90	25	70
90–100	20	90
Total	90	

In this frequency table, the marks obtained by 90 students in physics have been divided into classes with class interval 10. The frequency for the interval 50–60 is 5 whereas it is 16 for the class interval 60–70. The cumulative frequency of the class interval 70–80 is 45.

13.2 MEASURES OF CENTRAL TENDENCY

The commonly used measures of central tendency are mean, median, and mode. We define these concepts one by one.

1. **The Mean:** The *arithmetic mean* $\bar{x}$ of a set of n values $x_1, x_2, \ldots, x_n$ of a variate is defined by the formula

$$\bar{x} = \frac{1}{n}\sum_{i=1}^{n} x_i.$$

The *weight* of a value of variate is a numerical multiplier assigned to indicate its relative importance. The *weighted arithmetic mean* of set of variates $x_1, x_2, \ldots, x_n$ with weights $w_1, w_2, \ldots, w_n$, respectively, is defined by

$$\bar{x} = \frac{\sum_{i=1}^{n} w_i x_i}{\sum_{i=1}^{n} w_i}$$

Thus, in a frequency distribution, if $x_1, x_2, \ldots, x_n$ are the mid-values of the class intervals having frequencies $f_1, f_2, \ldots, f_n$, respectively, then

$$\bar{x} = \frac{\sum_{i=1}^{n} f_i x_i}{\sum_{i=1}^{n} f_i}.$$

Let $d_i = x_i - A$. Then

$$\sum_{i=1}^{n} f_i d_i = \sum_{i=1}^{n} f_i x_i - A \sum_{i=1}^{n} f_i.$$

Therefore,

$$\frac{\sum_{i=1}^{n} f_i d_i}{\sum_{i=1}^{n} f_i} = \frac{\sum_{i=1}^{n} f_i x_i}{\sum_{i=1}^{n} f_i} - A = \bar{x} - A$$

or

$$\bar{x} = A + \frac{\sum_{i=1}^{n} f_i d_i}{\sum_{i=1}^{n} f_i}.$$

This formula, obtained by shifting the origin, is more convenient to find the mean.

2. **The Median:** Suppose that n values $x_1, x_2, \ldots, x_n$ of a variate have been arranged in the following order of magnitudes,

$$x_1 \le x_2 \le x_3 \le \ldots \le x_n.$$

Then the *median* of this ordered set of values is the value $x_{\frac{n+1}{2}}$ when n is odd, and the value $\frac{1}{2}\left(x_{\frac{n}{2}} + x_{\frac{n}{2}+1}\right)$ when n is even.

The median for the discrete frequency distribution is obtained as follows:

(i) Determine $\frac{1}{2}\sum f_i$

(ii) Note the cumulative frequency just greater than $\frac{1}{2}\sum f_i$.

(iii) Find the value of x corresponding to the cumulative frequency obtained in step (ii). This value will be the median.

The median for the continuous frequency distribution is obtained as follows:

(a) Note the class corresponding to the cumulative frequency just greater than $\frac{1}{2}\sum f_i$. This class is known as *median class*.

(b) Compute the value of median by the formula.

$$\text{Median} = L + \frac{h}{f}\left(\frac{1}{2}\sum f_i - c\right),$$

where
L is the lower limit of the median class
f is the frequency of the median class
h is the width of the median class
c is the cumulative frequency of the class preceeding the median class.

3. **The Mode:** The *mode* is defined as that value of a variate which occurs most frequently.

For example, in the frequency distribution

x:	1	2	3	4	5	6
f:	3	7	28	10	9	5

the value of x corresponding to the maximum frequency, namely, 28 is 3. Hence mode is 3.

For a grouped distribution, mode is given by

$$\text{Mode} = L + \frac{\Delta_1}{\Delta_1 + \Delta_2} h,$$

where
L = lower limit of the class containing the mode
Δ_1 = excess of modal frequency (maximum) over frequency of preceeding class
Δ_2 = excess of modal frequency over frequency of succeeding class
h = width of modal class.

The empirical relationship between mean, median, and mode of a frequency distribution is

$$\text{mean} - \text{mode} = 3\,(\text{mean}-\text{median}).$$

However, for a symmetrical distribution, the mean, median, and mode coincide. For example, consider the following distribution

Class-interval:	0–10	10–20	20–30	30–40	40–80	50–60
Frequency:	6	8	14	26	17	10

The maximum frequency is 26 and $h = 10$. Further,

$L=30, \Delta_1 = 26 - 14 = 12, \Delta_2 = 26 - 17 = 9.$

Therefore

$$\text{Mode} = 30 + \frac{12}{12+9}(10) = 30 + \frac{120}{21}$$

$$= 35.714$$

Apart from the above measures of central tendency, we consider now the following partition values of the frequency:

The *partition values* are those values which divide the series of frequencies into a number of equal parts.

The three values, which divide the series of the given frequencies into four equal parts are called *quartiles.* The *lower (first) quartile*, Q_1, is the value which exceeds 25% of the observations and is exceeded by 75% of the observations. The *second quartile*, Q_2, coincides with the mean whereas the *third quartile*, Q_3, is the value which exceeds 75% observations and has 25% observations after it. In fact, if $N = \sum_{i=1}^{n} f_i$, L the lower limit of the median class, h the magnitude of the median class, and f the frequency of the median class, then

$$Q_1 = L + \frac{\frac{N}{4} - Cum(f)}{f} \cdot h, \text{ and}$$

$$Q_3 = L + \frac{\frac{3N}{4} - Cum(f)}{f} \cdot h.$$

Similarly, the 9 values which divide the frequency series into 10 equal parts are called *deciles* whereas the 99 values which divide the frequency series into 100 equal parts are called *percentites.*

EXAMPLE 13.1

Determine the mean, median, and mode for the following data:

Mid value:	15	20	25	30	35	40	45	50	55
Frequency:	2	22	19	14	3	4	6	1	1
Cum f:	2	24	43	57	60	64	70	71	72

Solution. For the given frequency distribution, we have

$$\text{Mean } \bar{x} = \frac{\sum_{i=1}^{n} f_i x_i}{\sum_{i=1}^{n} f_i}$$

$$= \frac{\begin{array}{c}2(15) + 22(20) + 19(25) + 14(30) + 3(35) \\ + 4(40) + 6(45) + 1(50) + 1(55)\end{array}}{+ 22 + 19 + 14 + 3 + 4 + 6 + 1 + 1}$$

$$= \frac{2005}{72} = 27.85.$$

To compute the median, we note that $\frac{1}{2}\sum f_i = 36$. The median class (corresponding to cum frequency 43) is (20–30). Width of the median class is 10. Frequency of the median class is 19. The cumulative frequency of the class preceding to median class is 24. Therefore,

$$\text{Median} = L + \frac{h}{f}\left(\frac{1}{2}\sum f_i - c\right)$$

$$= 20 + \frac{10}{19}(36\text{–}24)$$

$$= 23.32$$

To calculate the mode, we note that the maximum frequency is 22, that is, the modal frequency is 22. Then the modal class is (15–25). Therefore,

$$\text{Mode} = L + \frac{\Delta_1}{\Delta_1 + \Delta_2} h,$$

$$= 15 + \frac{22 - 2}{(22 - 2) + (22 - 19)} \cdot 10$$

$$= 15 + \frac{200}{3} = 23.69$$

EXAMPLE 13.2

Obtain the median for the following distribution:

x:	1	2	3	4	5	6	7	8	9
f:	8	10	11	16	20	25	15	9	6

Solution. For the given discrete frequency distribution, we have $\frac{1}{2}\sum f_i = \frac{120}{2} = 60$. The cumulative frequencies are

8, 18, 29, 45, 65, 90, 105, 114 ,120

The cumulative frequency just greater then 60 is 65. The value of x corresponding to 65 is 5. Hence the median is 5.

EXAMPLE 13.3

Given that the median value is 46, find the missing frequencies for the following incomplete frequency distribution:

Class:	10–20	20–30	30–40	40–50	50–60	60–70	70–80	Total
f:	12	30	–	65	–	25	18	229

Solution. Suppose that the frequency of the class 30–40 be f_1 and that for the class 50–60 be f_2. Also $\sum f_i = 229$. Therefore,

$$f_1 + f_2 + (12 + 30 + 65 + 25 + 18) = 229$$

and so $f_1 + f_2 = 79$. Since the median is 46, the median class is 40–50. Therefore, using the formula

$$\text{Mode} = L + \frac{h}{f}\left(\frac{1}{2}\sum f_i - c\right),$$

we have,

$$46 = 40 + \frac{10}{65}\left(\frac{229}{2} - c\right),$$

where c is the cumulative frequency of the class preceding the median class. Since the cumulative frequency are

12, 42, 42 $+f_1$, 107 $+f_1$, 132 $+f_1+f_2$, 157 $+f_1+f_2$, 175 $+f_1+f_2$,

the value of c is 42 $+f_1$. Hence

$$46 = 4 + \frac{10}{65}\left(\frac{229}{2} - (42 + f_1)\right)$$

which yields $f_1 = 33.5 \approx 34$. Then $f_2 = 79 - 34 = 45$. Hence the missing frequencies are 34 and 45.

13.3 MEASURES OF VARIABILITY (DISPERSION)

The measures of central tendency give us idea of the concentration of the observation about the central part of the distribution. They fail to give information whether the values are closely packed about the central value or widely scattered away from it. The two different distributions may have the same mean and same total frequency, yet they may differ in the sense that the individual values spread about the average differently. Thus, the measures of central tendency must be supplemented by some other measures to have the complete idea of distribution. One such measure is *dispersion*.

The degree to which numerical data tends to spread about an average value is called *variability* or *dispersion* of the data.

We now define some of the important measures of dispersion.

1. **Range:** The range is the difference of the greatest and the least values in the distribution. This is the simplest but a crude measure of dispersion.

2. **The Mean Deviation**: The mean deviation of a set of n values $x_1, x_2, \ldots, x_n$ of a variate is defined as the arithmetic mean of their absolute deviations from their average A (usually mean, median, or mode).

Thus if we consider the average as arithmetic mean of $x_1, x_2, \ldots, x_n$ then

$$\text{Mean deviation (M.D.)} = \frac{1}{n}\sum_{i=1}^{n} |x_i - \bar{x}|.$$

If $x_i | f_i$, $i = 1, 2, \ldots, n$ is the frequency distribution, the

$$\text{M.D} = \frac{1}{N}\sum_{i=1}^{n} f_i |x_i - \bar{x}|,\ N = \sum_{i=1}^{n} f_i$$

3. **The Variance**: Since mean deviation is based on all the observations, it is a better measure of dispersion than the range. But, in the definition, we have converted all minus signs to plus before averaging the deviations. Another method ofeliminating minus sign is to square the deviations and then average these squares. This step gives rise to a most powerful measure of dispersion, called variance, defined as follows:

The *variance*, S^2, of a sample of n values $x_1, x_2, \ldots, x_n$ of a variate with arithmetic mean x is defined as the $\frac{1}{n}$th of the sum of squares of their deviations from the mean. Thus

$$S^2 = \frac{1}{n}\sum_{i=1}^{n} (x_i - \bar{x})^2.$$

If $x_i|f_i$; i = 1,2,..., n is the frequency distribution then

$$S^2 = \frac{1}{N}\sum_{i=1}^{n} f_i(x_i - \bar{x})^2,\ N = \sum_{i=1}^{n} f_i$$

4. **The Standard Deviation**: It is defined as the positive square root of the variance. If is denoted by σ.Thus

$$\sigma = \left[\frac{1}{N}\sum_{i=1}^{n}(x_i - \bar{x})^2\right]^{\frac{1}{2}}.$$

In case of frequency distribution $x_i|f_i$, $i = 1, 2, . . ., n$, we have

$$\sigma = \left[\frac{1}{N}\sum_{i=1}^{n} f_i(x_i - \bar{x})^2\right]^{\frac{1}{2}},\ N = \sum_{i=1}^{n} f_i$$

5. **Quartile Deviation**: The *quartile deviation* Q is defined as

$$Q = \frac{1}{2}(Q_3 - Q_1),$$

where Q_1 and Q_3 are the first and third quartiles of the distribution, respectively.

Theorem 13.1. For the frequency distribution $x_i|f_i$, i = 1, 2, . . ., n

$$S^2 = \frac{1}{\sum f_i}\sum f_i x_i^2 - (\bar{x})^2.$$

Proof: We have

$$S^2 = \frac{1}{\sum f_i}\sum f_i(x_i - \bar{x})^2$$

$$= \frac{1}{\sum f_i}\sum f_i\left[x_i^2 + (\bar{x})^2 - 2x_i\bar{x}\right]$$

$$= \frac{1}{\sum f_i}\sum f_i x_i^2 + \frac{1}{\sum f_i}\sum f_i(\bar{x})^2 - \frac{2\bar{x}}{\sum f_i}\sum f_i x_i$$

$$= \frac{1}{\sum f_i}\sum f_i x_i^2 + (\bar{x})^2 - \frac{2\bar{x}}{\sum f_i}\sum f_i x_i$$

$$= \frac{1}{\sum f_i}\sum f_i x_i^2 + (\bar{x})^2 - 2(\bar{x})^2$$

$$= \frac{1}{\sum f_i}\sum f_i x_i^2 - (\bar{x})^2.$$

The ratio of the standard deviation to the mean is known as the *coefficient of variation*. Thus

$$\text{Coefficient of variation} = \frac{\sigma}{\bar{x}}.$$

Theorem 13.2. Variance and, hence, the standard is independent of the change of origin.

Proof: From above, the variance is given by

$$S^2 = \frac{1}{\sum f_i}\sum f_i(x_i - (\bar{x})^2.$$

Let $d_i = x_i - A$. Then

$$x_i - \bar{x} = (x_i - A) - (\bar{x} - A) = d_i - (\bar{x} - A)$$

and so

$$\sum f_i(x_i - (\bar{x})^2 = \sum f_i\left[d_i - (\bar{x} - A)\right]^2$$

$$= \sum f_i d_i^2 + (\bar{x} - A)^2 \times \sum f_i - 2(\bar{x} - A)\sum f_i d_i$$

$$= \sum f_i d_i^2 - \frac{(\sum f_i d_i)^2}{\sum f_i}, \text{ since}$$

$$\bar{x} = A + \frac{\sum f_i d_i}{\sum f_i}.$$

Therefore,

$$S^2 = \frac{\sum f_i d_i^2}{\sum f_i} - \left(\frac{\sum f_i d_i}{\sum f_i}\right)^2$$

and

$$\sigma = \sqrt{\frac{\sum f_i d_i^2}{\sum f_i} - \left(\frac{\sum f_i d_i}{\sum f_i^2}\right)^2}.$$

Moments: *The rth moment about the mean* $\bar{x}$ of a distribution, denoted by μ_r, is defined by

$$\propto_r = \frac{1}{N}\sum_{i=1}^{n} f_i(x_i - (\bar{x})^r,\ N = \sum_{i=1}^{n} f_i.$$

The moment about any point a is defined by

$$\propto'_r = \frac{1}{N}\sum_{i=1}^{n} f_i(x_i - a)^r,\ N = \sum_{i=1}^{n} f_i.$$

We note that

$$\alpha_0 = 1 = \alpha_0',$$

$$\alpha_1 = \frac{1}{N}\sum_{i=1}^{n} f_i(x_i - (\bar{x}) = \frac{1}{N}\sum f x_i - \frac{1}{N}\sum f_i \bar{x}$$

$$= \bar{x} - \bar{x}\cdot\left(\frac{1}{N}\sum f_i\right) = \bar{x} - \bar{x} = 0,$$

$$\alpha_1' = \frac{1}{N}\sum_{i=1}^{n} f_i(x_i - a)$$

$$= \frac{1}{N}\sum_{i=1}^{n} f x_i - \frac{1}{N}\sum_{i=1}^{n} f_i a = \bar{x} - a \text{ and}$$

$$\alpha_2 = \frac{1}{N}\sum_{i=1}^{n} f x_i (x_i - \bar{x})^2 = \sigma^2$$

where σ is the standard deviation.
If can be shown that

$$\alpha_2 = \alpha_2' - \alpha_1'$$

$$\alpha_3 = \alpha_3' - 3\alpha_2'\alpha_1' + 2\alpha_1'^3$$

$$\alpha_4 = \alpha_4' - 4\alpha_3'\alpha_1' + 6\alpha_2'\alpha_1' - 3\alpha_1'^4$$

13.4 MEASURE OF SKEWNESS

As pointed out earlier, the measure of skewness is the degree of asymmetry or the departure from the symmetry. Regarding skewness, we have

(a) *Pearson's coefficient of skewness, which equals* $\frac{mean - mode}{\sigma}$

(b) Coefficient of skewness based on third moment is given by

$$\gamma_i = \sqrt{\beta_1},$$

where

$$\beta_1 = \frac{\alpha_3^2}{\alpha_2^2}.$$

Therefore, *the simplest measure of skewness* is $\gamma_i = \sqrt{\beta_1}$.

13.5 MEASURES OF KURTOSIS

The measures of peakness or flatness of the frequency curve, called the *measure of kurtosis*, is defined by

$$\beta_2 = \frac{\alpha_4}{\alpha_2^2}.$$

Further, $\gamma_2 = \beta_2 - 3$ yields the excess of *kurtosis*. The curves with $\gamma_2 > 0$, that is, $\beta_2 > 3$ are called *leptokurtic* and the curves with $\gamma_2 < 0$, that is, $\beta_2 < 3$ are called *platykurtic*. The curve (normal curve) for which $\gamma_2 = 0$, that is, $\beta_2 = 3$ is called *mesokurtic*. Thus, *the normal curve is symmetrical about its mean.*

EXAMPLE 13.4

The following table shows the marks obtained by 100 candidates in an examination. Calculate the mean, median, and standard deviation:

Marks Obtained:	1–10	11–20	21–30	31–40	41–50	51–60
No.of candidates	3	16	26	16	16	8

Solution. We form the table shown below:

Class	Mid-Value x	Frequency f	Cum. frequency	fx	fx^2
1–10	5.5	3	3	16.5	90.75
11–20	15.5	16	19	248	3844
21–30	25.5	26	45	663	16906.5
31–40	35.5	31	76	1100.5	39067.75
41–50	45.5	16	92	728	33124
51–60	55.5	8	100	444	24642
		100		3200	117675

Then

$$\text{Mean } (\bar{x}) = \frac{\sum f x_i}{\sum f_i} = \frac{3200}{100} = 32.$$

Since $\frac{1}{2}\sum f_i = 50$, the median class is corresponding to the cum frequency 76. Thus the median class is 31–40. Therefore,

$$\text{Median} = L + \frac{h}{f}\left(\frac{1}{2}\sum f_i - c\right)$$

$$= 31 + \frac{10}{31}(50 - 45) = 32.6.$$

Further, variance is given by

$$S^2 = \frac{1}{\sum f_i}\sum f_i x_1^2 - (\bar{x})^2$$

$$= \frac{117675}{100} - (32)^2 = 152.75$$

and so the standard deviation is

$$\sigma = \sqrt{S^2} = 12.36 \approx 12.4.$$

EXAMPLE 13.5

The score obtained by two batsmen A and B in 10 matches are follows:

A	30	44	66	62	60	34	80	46	20	38
B	34	46	70	38	55	48	60	34	45	30

Determine who is more efficient and consistent.

Solution. The mean $\bar{x}_A$ for the batsman A is

$$\bar{x}_A = \frac{1}{10}\sum x_i = \frac{480}{10} = 48.$$

The variance for the batsman A is

$$S_A^2 = \frac{1}{10}\sum (x_i - \bar{x}_A)^2$$

$$= \frac{1}{10}[(48-30)^2 + (48-44)^2 + (48-66)^2 + (48-62)^2 + (48-60)^2 + (48-34)^2 + (48-80)^2 + (48-46)^2 + (48-20)^2 + (48-38)^2]$$

$$= \frac{1}{10}[324 + 16 + 324 + 196 + 144 + 196 + 1024 + 4 + 784 + 100]$$

$$= 314.8.$$

The coefficient of variation $= \frac{\sigma_A}{\bar{x}_A} = \frac{\sqrt{314.8}}{48} = 0.37.$

On the other hand, the mean $\bar{y}_B$ for the batsman B is

$$\bar{y}_B = \frac{1}{10}\sum y_i = \frac{460}{10} = 46.$$

The variance for the batsman B is

$$S_B^2 = \frac{1}{10}\sum (y_i - \bar{y}_B)^2$$

$$= \frac{1}{10}[(46-34)^2 + (46-46)^2 + (46-70)^2 + (46-38)^2 + (46-55)^2 + (46-48)^2 + (46-60)^2 + (46-34)^2 + (46-45)^2 + (46-30)^2]$$

$$= \frac{1}{10}(144 + 0 + 576 + 64 + 81 + 4 + 196 + 144 + 1 + 256)$$

$$= 146.6\,.$$

The coefficient of variation $= \frac{\sigma_B}{\bar{y}_B} = \frac{\sqrt{146.6}}{46} = 0.26.$

Since the average of batsman A is greater than the average of B, we conclude that A is a better scorer and hence is more efficient. But the coefficient of variance of B is less than the coefficient of variance of A, therefore, it follows that B is more consistent than A.

EXAMPLE 13.6

The first three moments of a distribution about the value 2 of the variable are 1, 16, and –40. Find the mean, variance, and third moment of the distribution about the value 2.

Solution. We are given that

$$\propto_1' = 1, \propto_2' = 16, \propto_3' = -40.$$

Since $N = \sum f_i$, we have

$$1 = \propto_1' = \frac{1}{N}\sum f_i(x_i - a) = \frac{1}{N}\sum f_i(x_i - 2),$$

$$= \frac{1}{N}\sum f_i x_i - \frac{2}{N}\sum f_i = \bar{x} - 2,$$

and so

$$\text{Mean}(\bar{x}) = 3.$$

The variance is

$$S^2 = \propto_2 = \propto_2' - \propto_1'^2 = 16 - 1 = 15.$$

The third moment $\propto_3$ is given by

$$\propto_3 = \propto_3' - 3\propto_2'\propto_1' + 2\propto_1'^3$$

$$= -40 - 3(16)(1) + 2(1)^3$$

$$= -40 - 48 + 2 = -86.$$

EXAMPLE 13.7

Determine Pearson's coefficient of skewness for the data given below:

Class:	10–19	20–29	30–39	40–49
f:	5	9	14	20
Class:	50–59	60–69	70–79	80–89
f:	25	15	8	4

Solution. We form the table given below:

Class	Mid-value (x):	Frequency (f)	Cum. frequency	fx	fx^2
10–19	14.5	5	5	72.5	1051.25
20–29	24.5	9	14	220.5	5402.25
30–39	34.5	14	28	483	16663.5
40–49	44.5	20	48	890	39605
50–59	54.5	25	73	1362.5	74256.25
60–69	64.5	15	88	967.5	62403.75
70–79	74.5	8	96	596	44402
80–89	84.5	4	100	338	28561
		100		4930	272345

Then

$$\text{Mean } (\bar{x}) = \frac{4930}{100} = 49.3.$$

The maximum frequency is 25, that is, the modal frequency is 25. Therefore, the modal class is 50–59. Hence

$$\text{Mode} = L + \frac{\Delta_1}{\Delta_1 + \Delta_2} h = 50 + \frac{(25-20)9}{(25-20)+(25-15)}$$

$$= 50 + \frac{45}{15} = 53.0$$

Also

$$\sigma^2 = \frac{1}{\sum f_i} \sum f_i x_i^2 - (\bar{x})^2 = 2723.45 - 2430.49$$

$$= 292.96,$$

and so $\sigma = 17.12$.

$$\text{Pearson's coefficient of skewness} = \frac{\text{mean} - \text{mode}}{\sigma}$$

$$= \frac{49.3 - 53.0}{17.12} = -0.22.$$

13.6 COVARIANCE

Suppose that the pair of random variable X and Y take n pairs of observations as follows:

$$(x_1, y_1), (x_2, y_2), \ldots, (x_n, y_n).$$

The arithmetic means of the observed values of X and Y are, respectively

$$\bar{x} = \frac{1}{n}\sum_{i=1}^{n} x_i \text{ and } \bar{y} = \frac{1}{n}\sum_{i=1}^{n} y_i \qquad (1)$$

The deviations of the observed values of X and Y from their respective means are

$$x_1 - \bar{x}, x_2 - \bar{x}, \ldots, x_n - \bar{x}$$

and

$$y_1 - \bar{y}, y_2 - \bar{y}, \ldots, y_n - \bar{y}$$

respectively. The *covariance* of X and Y, denoted by Cov (X,Y) is defined by

$$\text{Cov}(X,Y) = \frac{1}{n}\sum_{i=1}^{n} (x_i - \bar{x})(y_i - \bar{y}).$$

However, if $\bar{x}$ and $\bar{y}$ are not whole numbers, then the task of calculating Cov(X,Y) by this formula is time-consuming and cumbersome. A simplified expression for Cov (X,Y) can be derived as follows:

Using (1) we get

$$\sum_{i=1}^{n} (x_i - \bar{x})(y_i - \bar{y}) = \sum_{i=1}^{n} (x_i y_i - \bar{x} y_i - x_i \bar{y} - \bar{x}\bar{y})$$

$$= \sum_{i=1}^{n} x_i y_i - \bar{x}\sum_{i=1}^{n} y_i - \bar{y}\sum_{i=1}^{n} x_i + \bar{x}\bar{y}\sum_{i=1}^{n} 1$$

$$= \sum_{i=1}^{n} x_i y - \bar{x}(n\bar{y}) - \bar{y}(n\bar{x}) + n\bar{x}\,\bar{y}$$

$$= \sum_{i=1}^{n} x_i y_i - n\bar{x}\,\bar{y}$$

Hence

$$\text{Cov}(X,Y) = \frac{1}{n}\left[\sum_{i=1}^{n} x_i y_i - n\bar{x}\bar{y}\right]$$

$$= \frac{1}{n}\sum_{i=1}^{n} x_i y_i - \bar{x}\,\bar{y}.$$

It may be proved that *covariance is not affected by the change of origin but is affected by the change of scale.*

EXAMPLE 13.8

Find the covariance between x and y for the following data:

x:	3	4	5	8	7	9	6	2	1
y:	4	3	4	7	8	7	6	3	2

Solution. We have $n = 9$, $\sum x_i = 45$, $\sum y_i = 44$,

$$\bar{x} = \frac{1}{n}\sum x_i = \frac{45}{9} = 5, \bar{y} = \frac{1}{n}\sum y_i = \frac{44}{9} \text{ and}$$

$$\sum x_i y_i = 263.$$

Therefore,

$$\text{Cov}(X,Y) = \frac{1}{n}\sum_{i=1}^{n} x_i y_i - \bar{x}\,\bar{y}$$

$$= \frac{263}{9} - \frac{5(44)}{9}$$

$$= \frac{43}{9} = 4.78.$$

EXAMPLE 13.9

Calculate the covariance between height and weight of the following five persons:

Height in cm:	150	148	148	152	154
Weight in kg:	65	64	63	65	67

Solution. Since covariance is not affected by change of origin, we take $u_i = x_i - 148$ and $v_i = y_i - 65$ and get the following table:

x_i	y_i	$u_i = x_i - 148$	$v_i = y_i - 65$	$u_i v_i$
150	65	2	0	0
148	64	0	−1	0
148	63	0	−2	0
152	65	4	0	0
154	67	6	2	12
		12	**−1**	**12**

Therefore,

$$\text{Cov}(X,Y) = \frac{1}{n}\sum u_i v_i - \bar{u}\,\bar{v}$$

$$= \frac{12}{5} - \frac{12}{5}\left(\frac{-1}{5}\right)$$

$$= \frac{72}{25} = 2.88 \text{ cm kg.}$$

13.7 CORRELATION AND COEFFICIENT OF CORRELATION

The relation in which changes in one variable are associated or followed by changes in the other variable is called *correlation.* The data connecting such two variables is called *bivariate population.* For example, there is a correlation between the yield of a crop and the amount of rainfall.

The correlation between the variables x and y are of the following types:

(1) Positive correlation: If the variables behave likely, that is, increase in the value of x results in a corresponding increase in the values of y or decrease in the values of x results in a corresponding decrease in the values of y, then the correlation between x and y is said to be *positive.*

For example, in the following data the correlation between x and y is positive.

x	50	42	30	60	20	70
y	70	60	45	80	30	90

(2) Negative Correlation: If the increase in the values of x results in a corresponding decrease in the values of y or decrease in the values of y results in a corresponding increase in the values of x, then the correlation between the variables x and y is said to be *negative.*

For example, the correlation between x and y in the following table is negative.

x	18	20	30	42	60
y	60	55	45	35	28

(3) Linear Correlation: If the change in the values of x bears a constants ratio with the corresponding change in the values of y, then the correlation between x and y is called a *linear correlation.* Thus the graph of the points $\{(x_i, y_i)\}$, in this case, will be straight line.

(4) Non-linear or Curvilinear Correlation: If the graphs of the points $\{(x_i, y_i)\}$ is not a straight line, then the correlation between x and y is called *non-linear* or *curvilinear correlation.*

(5) Perfect Correlation: If the percentage change in the variable x is followed by same percentage change in the variable y, then the correlation between x and y is called a *perfect correlation.* In this case also, the graph of the points $\{(x_i, y_i)\}$ will be a straight line.

(6) Simple and Multiple Correlations: Relationship between two variables is called *simple correlation,* whereas the relationship among three or more variable is called *multiple correlation.*

A scale-free (numerical) measure for a relation between a pair of variables is called the *coefficient of correlation* or *correlation coefficient.*

The coefficient of correlation between two quantitative variables X and Y is defined by

$$\rho(X,Y) = \frac{\text{Cov}(X,Y)}{\sigma_x \sigma_y},$$

where

$\sigma_x = \sqrt{\frac{1}{n}\sum_{i=1}^{n}(x_i - \bar{x})^2}$ is the standard deviation for X-series.

$\sigma_y = \sqrt{\frac{1}{n}\sum_{i=1}^{n}(y_i - \bar{y})^2}$ s the standard deviation for Y-series.

Since the dimensions of the numerator and denominator in the definition of $\rho(X,Y)$ are same, it follows that $\rho(X,Y)$ in non-dimensional quantity. *$\rho(X,Y)$ measures the degree of linear association* between the two variates. If two variates are not related, then $\rho(X,Y) = 0$. However, if $\rho(X,Y) = 0$ we cannot say that the two variables are not related.

We note that

$$\rho(X,Y) = \frac{Cov(X,Y)}{\sigma_x \sigma_y}$$

$$= \frac{\sum_{i=1}^{n}(x_i - \bar{x})(y_i - \bar{y})}{\sqrt{\sum_{i=1}^{n}(x_i - \bar{x})^2}\sqrt{\sum_{i=1}^{n}(y_i - \bar{y})^2}}$$

But

$$\sum_{i=1}^{n}(x_i - \bar{x})(y_i - \bar{y}) = n\,\text{Cov}(X,Y)$$

$$= n\left[\frac{1}{n}\sum x_i y_i - \bar{x}\bar{y}\right]$$

$$= \sum x_i y_i - \frac{1}{n}\sum x_i \sum y_i,$$

$$\sum_{i=1}^{n}(x_i - \bar{x})^2 = \sum_{i=1}^{n} x_i^2 - \frac{1}{n}\left(\sum_{i=1}^{n} x_i\right)^2, \text{ and}$$

$$\sum_{i=1}^{n}(y_i - \bar{y})^2 = \sum_{i=1}^{n} y_i^2 - \frac{1}{n}\left(\sum_{i=1}^{n} y_i\right)^2.$$

Therefore,

$$\rho(X,Y) = \frac{\sum x_i y_i - \frac{1}{n}\sum x_i \sum y_i}{\sqrt{\sum x_i^2 - \frac{1}{n}\left(\sum x_i\right)^2}\sqrt{\sum y_i^2 - \frac{1}{n}\left(\sum y_i\right)^2}}$$

$$= \frac{n\sum x_i y_i - \sum x_i \sum y_i}{\sqrt{n\sum x_i^2 - \left(\sum x_i\right)^2}\sqrt{n\sum y_i^2 - \left(\sum y_i\right)^2}}$$

which is called *Karl-Pearson's coefficient of correlation* or *product moment correlation coefficient.*

In case of *grouped data* the coefficient of correlation between X and Y is defined by

$$\rho(X,Y) = \frac{n\sum f_i u_i v_i - \left(\sum f_i u_i\right)\left(\sum f_i v_i\right)}{\sqrt{n\sum f_i u_i^2 - \left(\sum f_i u_i\right)^2}\sqrt{n\sum f_i v_i^2 - \left(\sum f_i v_i\right)^2}},$$

where

u_i = deviation of the central values from the assumed mean say a, of x series

$= \frac{x_i - a}{h}$, h being positive factor of all deviation,

f_i = frequency corresponding to pair (xi, yi),

$n = \sum f_i$

A measure to deal with qualitative characteristic like character, morality, intelligence, beauty etc, is called *Spearman's Rank Correlation Coefficient* which and is defined by

$$r = 1 - \frac{6\sum d^2}{n(n^2 - 1)}$$

where

d^2 = sum of squares of the difference of two ranks,

n = number of paried observations.

Remark 13.1

(i) Since $\rho(X,Y) = \frac{\text{Cov}(X,Y)}{\sigma_x \sigma_y}$ and denominator contains positive square roots, it follows that.the sign of $\rho(X, Y)$ is the same as that of Cov (X, Y)

(ii) $-1 \leq \rho(X, Y) \leq 1$.

(iii) If $\rho(X, Y) = 1$, then the variables X and Y are not only statistically related but also functionally related. There exists a linear relationship of the form

$$Y = a + bX, b \geq 0$$

or

$$X = c + dY, d \geq 0,$$

which are straight lines with positive slopes.In this case, the variables have *perfect positive correlation.*

(iv) If $\rho(X,Y) = -1$, then there exists a linear relationship of the form

$$Y = a - bX, b \geq 0$$

or

$$X = c - dY, d \geq 0$$

which are straight lines with negative slopes.In this case, the variables have *perfect negative correlations.*

(v) If $\rho(X, Y)$ is close to 1, there is a high degree of positive correlation and if it is close to −1,then there is a high degree of negative correlation.

(vi) If $\rho(X, Y)$ is close to 0 in magnitude, we cannot draw any conclusion about the existence of relation between the variables. To reach at some conclusion, in such a case, we have to draw scatter diagram.

EXAMPLE 13.10

Calculate the covariance and the coefficient of correlation between X and Y if

$$n = 10, \sum x = 60, \sum y = 60,$$

$$\sum x^2 = 400, \sum y^2 = 580 \text{ and}$$

$$\sum xy = 305.$$

Solution. For the given data

$$\text{Cov}(X, Y) = \frac{1}{n} \sum xy - \overline{xy}$$

$$= \frac{1}{10}(305) - \left(\frac{60}{10}\right)\left(\frac{60}{10}\right) = -5.5,$$

$$\rho(X, Y) = \frac{n\sum x_i y_i - \sum x_i \sum y_i}{\sqrt{n\sum x_i^2 - \left(\sum x_i\right)^2}\sqrt{n\sum y_i^2 - \left(\sum y_i\right)^2}}$$

$$= \frac{3050 - 3600}{\sqrt{4000 - 3600}\sqrt{5800 - 3600}}$$

$$= \frac{-550}{20\sqrt{2200}} = -\frac{11}{4\sqrt{22}} = 0.586$$

EXAMPLE 13.11

Find the Karl Pearson coefficient of correlatio- between the industrial production and export using the following data:

Production

(in crore tons): 55 56 58 59 60 60 62

Export

(in crore tons): 35 38 38 39 44 43 45

Solution. Here $n = 7$. Put $u_i = x_i - 60$, $v_i = y_i - 38$. Then we have the followig table:

x	y	u	v	u^2	v^2	uv
55	35	−5	−3	25	9	15
56	38	−4	0	16	0	0
58	38	−2	0	4	0	0
59	39	−1	1	1	1	-1
60	44	0	6	0	36	0
60	43	0	5	0	25	0
62	45	2	7	4	49	14
		−10	16	50	120	28

Therefore,

$$\rho(X, Y) = \frac{n\sum u_i v_i - \sum u_i \sum v_i}{\sqrt{n\sum u_i^2 - \left(\sum u_i\right)^2}\sqrt{n\sum v_i^2 - \left(\sum v_i\right)^2}}$$

$$= \frac{196 + 160}{\sqrt{350 - 100}\sqrt{840 - 256}}$$

$$= \frac{356}{\sqrt{250}\sqrt{584}} = \frac{356}{382.08} = 0.93.$$

Since $\rho(X, Y)$ is close to 1, there is high degree of positive correlation.

EXAMPLE 13.12

Calculate the coefficient of correlation between the marks obtained by 8 students in Mathematics and Statistics.

Students	A	B	C	D	E	F	G	H
Mathematics	25	30	32	35	37	40	42	45
Statistics	08	10	15	17	20	23	24	25

Solution. Setting $u = x - 35$, $v = y - 17$, we get the following table:

Math(x)	Stat.(y)	u	v	u^2	v^2	uv
25	8	−10	−9	100	81	90
30	10	−5	−7	25	49	35
32	15	−3	−2	9	4	6
35	17	0	0	0	0	0
37	20	2	3	4	9	6
40	23	5	6	25	36	30
42	24	7	7	49	49	49
45	25	10	8	100	64	80
Total		6	6	312	292	296

Now Karl Pearson coefficient of correlation is defined by

$$\rho(X, Y) = \frac{Cov\,(X, Y)}{\sigma_x \sigma_y}$$

$$= \frac{n\sum u_i v_i - \sum u_i \sum v_i}{\sqrt{n\sum u_i^2 - \left(\sum u_i\right)^2}\sqrt{n\sum v_i^2 - \left(\sum vi\right)^2}}$$

$$= \frac{8(296) - 6(6)}{\sqrt{2496 - 36}\,\sqrt{2336 - 36}}$$

$$= \frac{2332}{\sqrt{2460}\,\sqrt{2300}}$$

$$= 0.938.$$

Since $\rho(X, Y)$ is close to 1, there is *high degree of positive correlation.*

EXAMPLE 13.13

Calculate the coefficient of correlation between x (Marks in Mathematics) and y (Marks in Statistics) for the following data:

y \ x	10-40	40-70	70-100	Total
0-30	5	20	–	25
30-60	–	28	2	30
60-90	–	32	13	45
Total	5	80	15	100

Solution: We have grouped data. Therefore

$\rho(X, Y)$

$$= \frac{n\sum f_i u_i v_i - \left(\sum f_i u_i\right)\left(\sum f_i v_i\right)}{\sqrt{n\sum f_i u_i^2 - \left(\sum f_i u_i\right)^2}\sqrt{n\sum f_i v_i^2 - \left(\sum f_i v_i\right)^2}}$$

In term of mid-values, the given table ist

y \ x	25	55	85
15	5	20	–
45	–	28	2
75	–	32	13

We have

$$u_i = \frac{x_i - a}{h},\ a = 55,\ h = 30,$$

$$v_i = \frac{y_i - b}{h},\ b = 45,\ h = 30.$$

Thus

$$u_1 = \frac{25 - 55}{30} = -1, \quad u_i^2 = 1$$

$$u_2 = \frac{55 - 55}{30} = 0, \quad u_2^2 = 0$$

$$u_3 = \frac{85 - 55}{30} = 1,\ u_2^3 = 1.$$

$$v_1 = \frac{15 - 45}{30} = 1,\ v_2^1 = 1$$

$$v_2 = \frac{45 - 45}{30} = 0,\ v_2^2 = 0$$

$$v_3 = \frac{75 - 45}{30} = 1,\ v_2^3 = 1.$$

In term of u_i and v_i, we have the following table:

v_i \ u_i	−1	0	1	Total
−1	5	20	–	25
0	–	28	2	30
1	–	32	13	45
Total	5	80	15	100

Then

$$n = \sum f_x = 5 + 80 + 15 = 100,$$

$$\sum f_x u_i = 5(-1) + 80(0) + 15(1) = 10,$$

$$\sum f_y v_i = 25(-1) + 30(0) + 45(1) = 20,$$

$$\sum f_x u_i^2 = 5(1) + 80(0) + 15(1) = 20,$$

$$\sum f_y v_i^2 = 25(1) + 30(0) + 45(1) = 70,$$

$$\sum f_x u_i v_i = [5(-1)(-1) + (0)(-1)(0) + 0(-1)(1)]$$

$$+ [20(0)(-1) + 28(0)(0) + 32(0)(1)]$$

$$+ 0(1)(-1) + 2(1)(0) + 13(1)(1)]$$

$$= 5 + 0 + 13 = 18$$

$$\sum f_y u_i v_i = [5(-1)(1) + 20(0)(-1) + 0(1)(-1)]$$

$$+ [0(-1)(-1) + 28(0)(0) + 2(1)(0)]$$

$$+ [0(-1)(1) + 32(0)(1) + 13(1)(1)]$$

$$= 5 + 13 = 18.$$

Hence

$$\rho(X, Y) = \frac{100(18) - 10(20)}{\sqrt{100(20) - (10)^2}\ \sqrt{100\,(70) - (20)^2}}$$

$$= \frac{1600}{\sqrt{1900}\ \sqrt{6600}}$$

$$= \frac{16}{\sqrt{1254}} = 0.452.$$

EXAMPLE 13.14

The correlation table given below shows the ages (y) of husband and (x) of wife of 53 married couples living together on the census night of 1991. Calculate the coefficient ofcorrelation between the age of the husband and that of the wife.

y \ x	15–25	25–35	35–45	45–55	55–65	65–75	Total
15-25	1	1	–	–	–	–	2
25-35	2	12	1	–	–	–	15
35-45	–	4	10	1	–	–	15
45-55	–	–	3	6	1	–	10
55-65	–	–	–	2	4	2	8
65-75	–	–	–	–	1	2	3
Total	3	17	14	9	6	4	53

Solution. We have grouped data. Therefore

$\rho(X,Y)$

$$= \frac{n\sum f_i u_i v_i - \sum f_i u_i \sum f_i v_i}{\sqrt{n\sum f_i u_i^2 - \left(\sum f_i u_i\right)^2}\ \sqrt{n\sum f_i v_i^2 - \left(\sum f_i v_i\right)^2}}$$

In term of mid values, the given table reduces to

y \ x	20	30	40	50	60	70	Total
20	1	1	–	–	–	–	**2**
30	2	12	1	–	–	–	**15**
40	–	4	10	1	–	–	**15**
50	–	–	3	6	1	–	**10**
60	–	–	–	2	4	2	**8**
70	–	–	–	–	1	2	**3**
Total	**3**	**17**	**14**	**9**	**6**	**4**	**53**

We have

$$u_i = \frac{x_i - a}{h},\ a = 40,\ h = 10,$$

$$v_i = \frac{y_i - b}{h},\ b=40,\ h=10.$$

Thus

$$u_1 = \frac{20-40}{10} = 2, u_1^2 = 4$$

$$u_2 = \frac{30-40}{10} = 1, u_2^2 = 1$$

$$u3 = \frac{40-40}{10} = 0, u_3^2 = 0$$

$$u_4 = \frac{50-40}{10} = 1, u_4^2 = 1$$

$$u_5 = \frac{60-40}{10} = 2, u_5^2 = 4$$

$$u_6 = \frac{70-40}{10} = 3, u_6^2 = 9$$

and similarly

$$v_1 = 2, v_1^2 = 4$$

$$v_2 = 1, v_2^2 = 1$$

$$v_3 = 0, v_3^2 = 0$$

$$v_4 = 1, v_4^2 = 1$$

$$v_5 = 2, v_5^2 = 4$$

$$v_6 = 3, v_6^2 = 9.$$

Now, in term of u_i and v_i, we have the following table:

vi \ ui	−2	−1	0	1	2	3	Total
−2	1	1	–	–	–	–	**2**
−1	2	12	1	–	–	–	**15**
0	–	4	10	1	–	–	**15**
1	–	–	3	6	1	–	**10**
2	–	–	–	2	4	2	**8**
3	–	–	–	–	1	2	**3**
Total	**3**	**17**	**14**	**9**	**6**	**4**	**53**

Then

$$n = \sum f_x = \sum f_y = 53,$$

$$\sum f_x u_i = 3(-2) + 17(-1) + 14(0) + 9(1) + 6(2) + 4(3) = 10,$$

$$\sum f_y v_i = 2(-2) + 15(-1) + 15(0) + 10(1) +8(2) + 3(3) = 16,$$

$$\sum f_x u_i^2 = 3(4) + 17(1) + 14(0) + 9(1) + 6(4) + 4(9) = 98,$$

$$\sum f_y v_i^2 = 2(4) + 15(1) + 15(0) + 10(1) +8(4) + 3(9) = 92,$$

$$\sum f_x u_i v_i = [1(-2)(-2) + 2(-2)(-1)] + [1(1)(-2) + 12(-1)(-1)] + [6(1)(1) + 2(1)(2)] + [1(1)(2) + 4(2)(2) + 1(2)(3)] + [2(3)(2) + 2(3)(3)]. = 86$$

Similarly $\sum f_y u_i v_i = 86.$

Hence

$$\rho(X, Y) = \frac{53(86) - (10)(16)}{\sqrt{53(98)-(10)^2}\ \sqrt{53(92)-(16)^2}} = \frac{4358}{\sqrt{5094}\ \sqrt{4620}}$$

Since ρ(X, Y) is close to 1, there is high degree of positive correlation.

EXAMPLE 13.15

The marks obtained by ten students in Mathematics and statistics are as follows. Compute the coefficient of correlation of ranks.

Marks in Mathematics	Marks in Statistics
66	71
78	69
51	54
76	68
80	72
55	65
72	41
56	59
42	64
43	56

Solution. The table of ranks is

Student	Marks in Math(x)	Marks in Stat. (y)	Ranks in $x(R_1)$	Ranks in $y(R_1)$	$d = R_1 - R_2$	d^2
1	66	71	5	2	3	9
2	78	69	2	3	−1	1
3	51	54	8	9	−1	1
4	76	68	3	4	−1	1
5	80	72	1	1	0	0
6	55	65	7	5	2	4
7	72	41	4	10	−6	36
8	56	59	6	7	−1	1
9	42	64	10	6	4	16
10	43	56	9	8	1	1

We have $n = 10, \sum d^2 = 70$.

Hence Spearman's Rank Correlation Coefficient is given by

$$r = 1 - \frac{6\sum d^2}{n(n^2 - 1)}$$

$$=1 - \frac{6(70)}{10(100 - 1)}$$

$$=1 - \frac{420}{490}$$

$$=0.576.$$

EXAMPLE 13.16

Six dancers in a function were ranked by three judges in the following order:

First Judge:	2	3	1	4	6	5
Second Judge:	4	2	3	5	1	6
Third Judge:	1	5	2	6	3	4

Which pair of Judges has the nearest evaluation?

Solution: Let R_1, R_2 and R_3 denote the ranks given by the first, second and third judges respectively. Then the rank table is given by:

Dancer	R_1	R_2	R_3	$d_{12} = R_1 - R_2$	$d_{13} = R_1 - R_3$	$d_{23} = R_2 - R_3$
1	2	4	1	−2	1	5
2	3	2	5	1	−2	−3
3	1	3	2	−2	−1	1
4	4	5	6	−1	−2	−1
5	6	1	3	5	3	−2
6	5	6	4	−1	1	2

d_{12}^2	d_{13}^2	d_{23}^2
4	1	25
1	4	9
4	1	1
1	4	1
25	9	4
1	1	4

Hence rank correlation coefficients are

$$r_{12} = 1 - \frac{6\sum d_{12}^2}{n(n^2-1)} = 1 - \frac{6(36)}{6(35)} = -0.03$$

$$r_{13} = 1 - \frac{6\sum d_{13}^2}{n(n^2-1)} = 1 - \frac{20}{35} = 0.572$$

$$r_{23} = 1 - \frac{6\sum d_{23}^2}{n(n^2-1)} = 1 - \frac{44}{35} = 0.257.$$

Since r_{12} is negative, the opinions of first and second judge regarding dancing are opposite to each other. Similarly opinions of second and third judges are different. But r_{13} is positive. Therefore opinion of first and third judges regarding dance are similar.

13.8 REGRESSION

The value of the coefficient of correlation indicates whether statistical relationship exists between the variables X and Y. However, it does not give any expression for this statistical relationship. Regression analysis gives us a method for finding such expression.

Suppose that for a given value of x, we wish to determine the value of y. Thus we want to have an equation of the form

$$y = f(x). \tag{2}$$

The function f is called a *regression function* while equation (2) is called the *regression equation* of Y on X.

On the other hand, if for a given value of y we wish to find value of x, then we want to establish an equation of the form

$$x = g(y). \tag{3}$$

The function g is called *regression function* and equation (3) is called *regression equation* of X on Y.

We consider equation (2). Let (x_i, y_i), $i = 1, 2, \ldots, n$ be observed values in a given data. Then the estimate at x_i is $f(x_i)$, while the actual value is y_i. Thus, the error in the observed values are

$$y_1 - f(x_1), y_2 - f(x_2), \ldots, y_n - f(x_n).$$

The regression equation is good if these errors are small. Here we consider the case of linear regression only. Thus we wish to express $f(x)$ and $g(y)$ in the form of linear polynomials of the form

$$f(x) = a + bx \text{ and } g(y) = c + dy.$$

We shall obtain these expressions using *least square approximation.*

Suppose we want to find the regression of Y on X. Let the approximation be

$$y = a + bx. \tag{4}$$

Let (x_i, y_i), $i = 1, 2, \ldots, n$ be observed values. Then the errors of estimation are

$$y_1 - (a + bx_1), y_2 - (a + bx_2), \ldots, y_n - (a + bx_n):$$

Our aim is to find a and b such that the sum of squares of the errors is minimum. Thus we want to minimize $\sum_{i=1}^{n} [y_i - (a + bx_i)]^2$. With these values of a and b, $y = a + bx$ is called the *best approximation in the least square sense.* For minimizing $\sum_{i=1}^{n} [y_i - (a + bx_i)]^2$, its first derivatives with respect to a and b should be equal to zero. Thus we have

$$\sum_{i=1}^{n} [y_i - (a + bx_i)] = 0 \tag{5}$$

and

$$\sum_{i=1}^{n} [y_i - (a + bx_i)]x_i = 0 \tag{6}$$

Simplifying (5) and (6), we get

$$na + b\sum_{i=1}^{n} x_i = \sum y_i$$

$$a\sum x_i + b\sum_{i=1}^{n} x_i^2 = \sum x_{i2}y_i$$

Since $\bar{x} = \frac{1}{n}\sum_{i=1}^{n} x_i$, these equations reduces to

$$na + nb\bar{x} = n\bar{y} \tag{7}$$

$$na\bar{x} + b\sum_{i=1}^{n} x_i^2 = \sum_{i=1}^{n} x_i y_i \tag{8}$$

Since

$$\text{Cov}(X, Y) = \frac{1}{n}\sum_{i=1}^{n} x_i y_i - \bar{x}\,\bar{y},$$

We get

$$\sum x_i y_i = n\text{Cov}(X, Y) + n\,\bar{x}\,\bar{y}.$$

Also

$$\sigma_x^2 = \frac{1}{n}\sum_{i=1}^{n} (x_i - \bar{x})^2 = \frac{1}{n}\left[\sum_{i=1}^{n} x_i^2 - n(\bar{x})^2\right],$$

which yields

$$\sum_{i=1}^{n} x_i^2 = n\,\sigma_x^2 + n(\bar{x})^2 = n[\sigma_x^2 + (\bar{x})^2].$$

Substituting these values in (8), we get

$$n\,a\,\bar{x} + n\,b\,[\sigma_x^2 + (\bar{x})^2] = n\,[\text{Cov}(X, Y) + \bar{x}\,\bar{y}],$$

that is,

$$a\,\bar{x} + b[\sigma_x^2 + (x)^2] = \text{Cov}(X, Y) + \bar{x}\,\bar{y}. \qquad (9)$$

Multiplying (7) by x and subtracting from (9), we get

$$b\sigma_x^2 = \text{Cov}(X, Y)$$

and so

$$b = \frac{\text{Cov}(X, Y)}{\sigma_x^2}.$$

Then (7) yields

$$a = \bar{y} - \frac{\bar{x}\,\text{Cov}(X, Y)}{\sigma_x^2}.$$

Hence, the line of regression of Y on X is

$$y = a + bx = \bar{y} - \frac{\bar{x}\,\text{Cov}(X, Y)}{\sigma_x^2} + \frac{\text{Cov}(X, Y)}{\sigma_x^2}x$$

or

$$y - \bar{y} = \frac{\text{Cov}(X, Y)}{\sigma_x^2}(x - \bar{x})$$
$$= b_{yx}(x - \bar{x}), \qquad (10)$$

where

$$b_{yx} = \frac{\text{Cov}(X, Y)}{\sigma_x^2} = \frac{\sum_{i=1}^{n} (x_i - \bar{x})(y_i - \bar{y})}{\sum_{i=1}^{n} (x_i - \bar{x})^2} \qquad (11)$$

$$= \frac{n\sum_{i=1}^{n} x_i y_i - \sum_{i=1}^{n} x_i \sum_{i=1}^{n} y_i}{n\sum_{i=1}^{n} x_i^2 - \left(\sum_{i=1}^{n} x_i\right)^2}$$

is called *regression coefficient* of Y on X Since $\sigma_x^2 > 0$, the sign of b_{yx} is the same as that of Cov(X, Y) or of $\rho(X, Y)$.

Similarly, the regression line of X on Y is

$$x - \bar{x} = b_{xy}(y - \bar{y}), \qquad (12)$$

where

$$b_{xy} = \frac{\text{Cov}(X, Y)}{\sigma_y^2}$$
$$= \frac{n\sum_{i=1}^{n} x_i y_i - \sum_{i=1}^{n} x_i \sum_{i=1}^{n} y_i}{n\sum_{i=1}^{n} y_i^2 - \left(\sum_{i=1}^{n} y_i\right)^2} \qquad (13)$$

is the *regression coefficient of X on Y*.
We observe that

$$b_{xy}\,.\,b_{yx} = \frac{[\text{Cov}(X, Y)]^2}{\sigma_x^2\sigma_y^2} = \left[\frac{\text{Cov}(X, Y)}{\sigma_x\sigma_y}\right]^2$$
$$= [\rho(X, Y)]^2.$$

Hence, *the correlation coefficient is the geometric mean of the regression coefficients.* Since $-1 \le \rho(X, Y) \le 1$, it follows that

$$b_{xy}\,b_{yx} \le 1.$$

Remarks 13.2.

(i) The point of intersection of the two lines of regression obtained above is $(\bar{x}, \bar{y})$.
(ii) The regression coefficients are *independent of change of origin but not of scale.*

13.9 ANGLE BETWEEN THE REGRESSION LINES

The regression line of Y on X is

$$y - \bar{y} = b_{yx}(x - \bar{x}).$$

The slope of this line is

$$b_{yx} = \frac{\text{Cov}(X, Y)}{\sigma_x^2} = \frac{\text{Cov}(X, Y)}{\sigma_x\sigma_y}\cdot\frac{\sigma_y}{\sigma_x}$$
$$= \frac{\rho(X, Y)}{\sigma_x}\sigma_y.$$

The regression line of X on Y is

$$x - \bar{x} = b_{xy}(y - \bar{y}),$$

whose slop is

$$\frac{1}{b_{xy}} = \frac{\sigma_y^2}{\text{Cov}(X, Y)} = \frac{\sigma_y}{\rho(X, Y)\,\sigma_x}.$$

Hence the angle θ between the lines of regression is given by

$$\tan\theta = \pm \frac{\frac{1}{b_{xy}} - b_{yx}}{1 + \frac{1}{b_{xy}} b_{yx}} = \frac{\frac{\sigma_y}{\rho\sigma_x} - \frac{\rho\sigma_y}{\sigma_x}}{1 + \frac{\sigma_y}{\rho\sigma_x}\cdot\frac{\rho\sigma_y}{\sigma_x}}$$

$$= \pm \frac{(1-\rho^2)\sigma_x\sigma_y}{\rho(\sigma_x^2 + \sigma_y^2)} \quad (14)$$

The angle θ is usually taken as the acute angle, that is, $\tan\theta$ is taken as positive.

Deductions. If follows from (14) that

(i) if $\rho(X, Y) = \pm 1$, then $\tan\theta = 0$ and so $\theta = 0$ or π. Hence the two lines of regression *coincides*.

(ii) if $\rho(X, Y) = 0$, then $\tan\theta = \infty$ which implies $\theta = 90°$. Hence the lines are *perpendicular* in this case. The lines of regression in this case are $x = \bar{x}$ and $y = \bar{y}$, that is, they are parallel to the axes.

Least Square Error

(i) Least square error of prediction of Y on X is

$$\sum [y_i - (a + bx_i)]^2$$

which on simplification equals to

$$n\sigma_y^2\{1 - [\rho(X, Y)]^2\}.$$

(ii) Least square error of prediction of X on Y is similarly

$$n\sigma_x^2\{1 - [\rho(X, Y)]^2\}.$$

Clearly, if $\rho(X, Y) = \pm 1$, then the sum of least square of deviation (least square error) from either line of regression is 0. Hence each deviation is 0 and all the points lie on both lines of regression and so the lines coincide.

EXAMPLE 13.17

Find the regression of Y on X for the following data:

$$\sum x = \sum y = 15, \sum x^2 = \sum y^2 = 49,$$

$$\sum xy = 44, n = 5.$$

Solution. The regression of Y on X is given by

$$b_{yx} = \frac{n\sum xy - \sum x \sum y}{n\sum x^2 - \left(\sum x\right)^2}$$

$$= \frac{5(44) - 15(15)}{5(49) - (15)^2} = \frac{-5}{20} = -\frac{1}{4}.$$

Hence the regression line is

$$y - \bar{y} = b_{yx}(x - \bar{x})$$

or

$$y - \frac{15}{5} = -\frac{1}{4}\left(x - \frac{15}{5}\right)$$

or

$$y - 3 = -\frac{1}{4}(x - 3).$$

EXAMPLE 13.18

Find the equation of the lines of regression based on the following data:

x:	4	2	3	4	2
y:	2	3	2	4	4

Solution. For the given data, we have the following table:

x	y	xy	x^2	y^2
4	2	8	16	4
2	3	6	4	9
3	2	6	9	4
4	4	16	16	16
2	4	8	4	16
15	15	44	49	49

Since $n = 5$, we have

$$\bar{x} = \frac{\sum x}{5} = \frac{15}{5} = 3 \text{ and } \bar{y} = \frac{\sum x}{5} = \frac{15}{5} = 3.$$

As in Example 13.17, the regression of Y on X is

$$y - 3 = -\frac{1}{4}(x - 3) \text{ or } x + 4y = 15.$$

For the line of regression of X on Y, we have

$$b_{xy} = \frac{n\sum xy - \sum x \sum y}{n\sum y^2 - \left(\sum y\right)^2} = \frac{5(44) - 15(15)}{5(49) - (15)^2} = -\frac{1}{4}$$

Hence the regression of X on Y is given by

$$x - 3 = -\frac{1}{4}(y - 3) \text{ or } 4x + y = 15.$$

Hence the lines of regression are

$$x + 4y = 15 \text{ and } 4x + y = 15.$$

EXAMPLE 13.19

Out of the following two regression lines, find the regression line of Y on X.

$$x + 4y = 3 \text{ and } y + 3x = 15.$$

Solution. The line of regression of Y on X is

$$y = y + b_{yx}(x - \bar{x})$$

and the line of regression of X on Y is

$$x = x + b_{xy}(y - \bar{y}).$$

Suppose that the line of regression of Y on X is $x + 4y = 3$, that is, $y = \frac{1}{4}x + \frac{3}{4}$. The other line is $x = \frac{1}{3}y + 5$. Hence $b_{yx} = -\frac{1}{4}$ and $b_{xy} = \frac{1}{3}$. Therefore,

$$\rho^2 = b_{yx}(b_{xy}) = \frac{1}{12} < 1.$$

Hence the required line of regression of Y on X is $x + 4y = 3$.

Remark 13.3. If we begin taking $y + 3x = 15$ as the line of regression of Y on X, then

$$y = 3x + 5.$$

The other line is

$$x = 4y + 3.$$

Thus

$$b_{yx} = 3, b_{xy} = 4$$

and so

$$\rho^2 = b_{yx}(b_{xy}) = 12,$$

which is absurd, since $\rho^2 \leq 1$. Hence the line of regression is $x + 4y = 3$.

EXAMPLE 13.20

Two random variables have the regression lines with equation $3x + 2y = 26$ and $6x + y = 31$. Find the mean values and the correlation coefficient between x and y. Also find the angle between these lines.

Solution. Since the point of intersection of the regression lines is $(\bar{x}, \bar{y})$, the mean $\bar{x}$ and $\bar{y}$ lie on the two regression lines. Thus we have

$$3\bar{x} + 2\bar{y} = 26 \text{ and } 6\bar{x} + \bar{y} = 31.$$

Solving these equations, we get $\bar{x} = 4, \bar{y} = 7$.

As in the above example, we can verify that $3x + 2y = 26$ is the line of regression of Y on X and $6x + y = 31$ is the line of regression of X on Y. These lines can be written as

$$y = -\frac{3}{2}x + 13 \text{ and } x = -\frac{1}{6}y + \frac{31}{6}.$$

Therefore the regression coefficients are $b_{yx} = -\frac{3}{2}$ and $b_{xy} = -\frac{1}{6}$. Since $\rho^2 = b_{yx}, b_{xy}$, it follows that $\rho(x, y)$ is the geometric mean of these two regression coefficients. Hence

$$\rho(x, y) = \sqrt{b_{yx}.b_{xy}} = \sqrt{-\frac{1}{4}} = -0.5,$$

the minus sign is taken because both of the regression coefficients b_{yx} and b_{xy} are negative.

The angle between the regression lines is given by

$$\tan\theta = \pm\frac{\frac{1}{b_{xy}} - b_{yx}}{1 + \frac{b_{yx}}{b_{xy}}} = \mp\frac{8}{15}.$$

Taking positive value, we get $\theta = \tan^{-1}(\frac{8}{15})$.

EXAMPLE 13.21

In partially destroyed laboratory record of an analysis of correlation data, the following regression results are legible:

Variance of $x = 9$,
Regression lines

$$8x - 10y + 66 = 0 \qquad (15)$$

$$40x - 18y - 214 = 0. \qquad (16)$$

What were
(i) The mean value of x and y
(ii) The standard deviation of y
(iii) The coefficient of correlation between x and y.

Solution. Since the point of intersection of the regression lines is $(\bar{x}, \bar{y})$, the mean $\bar{x}$ and $\bar{y}$ lie on the two regression lines. Thus we have

$$8\bar{x} - 10\bar{y} + 66 = 0$$
$$40\bar{x} - 18\bar{y} - 214 = 0.$$

Solving these equations, we get the means

$$x = 13, y = 17.$$

It can be seen that (15) is the line of regression of Y on X and (16) in the line of regression of X and Y. From (15) and (16) we have respectively

$$y = \frac{4}{2}x + 6.6,$$

$$x = \frac{9}{20}y + \frac{214}{40}.$$

Thus the regression coefficients are

$$b_{yx} = \frac{4}{5} \text{ and } b_{xy} = \frac{9}{20}.$$

Therefore coefficient of correlation is given by

$$[\rho(X, Y)]^2 = b_{yx} b_{xy} = \frac{4(9)}{5(20)} = \frac{9}{25}$$

or

$$\rho(X, Y) = \frac{3}{5}.$$

Also,

$$\sigma_x^2 = 9 \text{(given)}$$

Further

$$b_{yx} = \frac{\text{cov}(X, Y)}{\sigma_x^2}$$

or

$$\text{cov}(X, Y) = b_{yx} \cdot \sigma_x^2$$
$$= \frac{4}{5}(9) = \frac{36}{5}.$$

Now

$$b_{xy} = \frac{\text{cov}(X, Y)}{\sigma_y^2}$$

so that

$$\sigma_y^2 = \frac{\text{cov}(X, Y)}{b_{xy}} = 16.$$

Hence

$$\sigma_y = 4.$$

EXAMPLE 13.22

The regression lines of Y on X and X on Y are respectively $y = ax + b$ and $x = cy + d$. Show that the means are

$$\bar{x} = \frac{bc + d}{1 - ac} \text{ and } \bar{y} = \frac{ad + b}{1 - ac}$$

and correlation coefficient between x and y is $\sqrt{ac}$. Also show that the ratio of the standard deviations of y and x is $\sqrt{\frac{a}{c}}$.

Solution. The regression lines of Y on X and of X on Y are respectively

$$y = ax + b$$

and

$$x + cy + d.$$

Since the point of intersection of the regression lines is $(\bar{x}, \bar{y})$, the means $\bar{x}$ and $\bar{y}$ lie on the two regression lines. Thus, we have

$$a\bar{x} - \bar{y} + b = 0 \text{ and } \bar{x} - c\bar{y} - d = 0.$$

Solving these equations, we get the means

$$\bar{x} = \frac{bc + d}{1 - ac} \text{ and } y = \frac{ad + b}{1 - ac}.$$

Further, the given equations simply

$$b_{yx} = a \text{ and } b_{xy} = c$$

We know that

$$b_{yx} = \frac{\text{cov}(X, Y)}{\sigma_x^2} \text{ and } b_{xy} = \frac{\text{cov}(X, Y)}{\sigma_y^2}$$

or

$$a = \frac{\text{cov}(X, Y)}{\sigma_x^2} \text{ and } c = \frac{\text{cov}(X, Y)}{\sigma_y^2}.$$

Hence

$$\frac{a}{c} = \frac{\sigma_y^2}{\sigma_x^2}$$

ort

$$\frac{\sigma_y}{\sigma_x} = \sqrt{\frac{a}{c}}.$$

Also

$$[\rho(X, Y)]^2 = b_{yx} b_{xy}$$

or

$$\rho(X, Y) = \sqrt{b_{yx} b_{xy}} = \sqrt{ac}.$$

13.10 PROBABILITY

Probability theory was developed in the seventeenth century to analyse games and so directly involved counting. It is a mathematical modelling of the phenomenon of chance or randomness. The measure of *chance* or *likelihood* for a statement to be true is called the *probability* of the statement. Thus, probability is an expression of an outcome of which we are not certain. For example, if we toss a coin, we cannot predict in advance whether a head or tail will show up. Similarly, if a dice (die) is thrown, then any one of the six faces can turn up. We cannot predict in advance which number (face) is going to turn up. Similarly, if we consider a pack of 52 playing cards, in which there are two colours, black and red, and four suits namely spades, hearts, diamonds, and clubs. Each suit has 13 cards. If we shuffle the pack of cards and draw a card from it, we are not sure to get a desired card.

An *experiment* is a process that yields an outcome. A *random experiment* or *experiment of chance* is an experiment in which (i) all the outcomes of the experiment are known in advance and (ii) the exact outcome of any specific performance of the experiment is not known in advance. For example, tossing of a fair coin is a random experiment. The possible outcomes of the experiment are head and tail. But we do not know in advance what the outcome will be on any performance of experiment.

The set of all the possible outcomes of a random experiment is called the *sample space* of that random experiment. It is denoted by S. An element of a sample space is called a *sample point*. An *event* is a subset of a sample space. An event may not contain any element. Such event is represented by ϕ and is called *impossible event*. An event may include the whole sample space S. Such event is called *sure* (*certain*) *event*. An event containing exactly one element is called a *simple event*.

For example, if we toss a fair coin, the sample space is

$$S_1 = \{T, H\},$$

where T stands for tail and H stands for head. Thus S_1 consists of $2^1 = 2$ sample points.

If the same coin is tossed twice, then

$$S_2 = \{TT, TH, HT, HH\}$$

consists of $2^2 = 4$ sample points.

Thus, in case of n toss, the sample space S_n shall have 2^n sample points.

The sample space of a random experiment can also be determined with the help of a *tree diagram*. For example, if a fair coin is tossed thrice, then the tree diagram for the sample space is as given below:

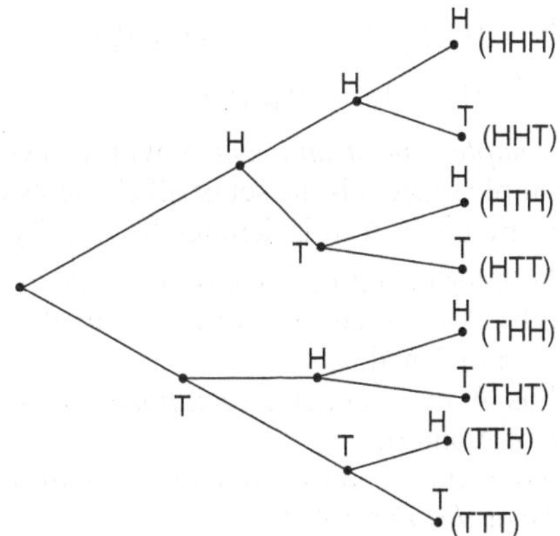

Thus,
S_3 = {HHH, HHT, HTH, HTT, THH, THT, TTH, TTT}.

Similarly, if an unbiased cubical dice is thrown, then

$$S_1 = \{1, 2, 3, 4, 5, 6\}.$$

If it is thrown again, then S_2 shall consists of $6^2 = 36$ sample points. These points can be determinet in the following way:

	⚀	⚁	⚂	⚃	⚄	⚅
⚀	(1,1)	(1,2)	(1,3)	(1,4)	(1,5)	(1,6)
⚁	(2,1)	(2,2)	(2,3)	(2,4)	(2,5)	(2,6)
⚂	(3,1)	(3,2)	(3,3)	(3,4)	(3,5)	(3,6)
⚃	(4,1)	(4,2)	(4,3)	(4,4)	(4,5)	(4,6)
⚄	(5,1)	(5,2)	(5,3)	(5,4)	(5,5)	(5,6)
⚅	(6,1)	(6,2)	(6,3)	(6,4)	(6,5)	(6,6)

If two coins are tossed simultaneously, then the first coin may show up either H or T and the second coin may also show up either H or T. Therefore, the outcomes of the experiment are

$$S = \{HH, HT, TH, TT\}.$$

In general, when two random experiment having m outcomes $e_1, e_2, \ldots, e_m$ and n outcomes $p_1, p_2, \ldots, p_n$, respectively, are performed simultaneously, the sample space consists of mn sample points and so

$$S = \{(e_1, p_1), (e_1, p_2), \ldots, (e_1, p_n), \ldots, (e_m, p_1), \ldots, (e_m, p_n)\}.$$

The *complement of an event* A with respect to the sample space S is the set of all elements of S which are not in A. It is denoted by $\overline{A}$ or by A′.

The intersection of two events A and B, denoted by $A \cap B$, consists of all points that are common to A and B.

Thus $A \cap B$ denotes *simultaneous occurrence* of A and B.

Two events A and B are called *mutually exclusive* or *disjoint* if $A \cap B = \phi$.

The *union* of the two events A and B, denoted by $A \cup B$, is the event containing all the elements that belong to A or to B or to both.

EXAMPLE 13.23

Let A be the event that a "sum of 6" appears on the dice when it is rolled *twice* and B denote the event that a "sum of 8" appears on the dice when rolled twice. Then

$$A = \{(1, 5), (2, 4), (3, 3), (4, 2), (5, 1)\},$$

$$B = \{(2, 6), (3, 5), (4, 4), (5, 3), (6, 2)\}.$$

We observe that $A \cap B = \phi$. Therefore, A and B cannot occur simultaneously and are mutually exclusive (disjoint).

The following combinations of events are usually needed in probability theory:

Combination	Meaning
$A \cup B$	Either A or B or both
$A \cap B$	Both A and B
$\overline{A}$ or A^c or A'	Not A
$A \cap B = \varphi$	Mutually exclusive events A and B
$A' \cap B'$ or $(A \cup B)'$	Neither A nor B
$A \cap B'$	Only A
$A' \cap B$	Only B
$(A \cap B') \cup (A' \cap B)$	Exactly one of A and B
$A \cup B \cap C$	Atleast one of A, B and C
$A \cap B \cap C$	All the three A, B and C

A collection of events $E_1, E_2, \ldots, E_n$ of a given-sample space S is said to be *mutually exclusive* and *exhaustive system* of events if

(i) $E_i \cap E_j = \phi, \quad i \neq j;\ i, j = 1, 2, \ldots, n$

(ii) $E_1 \cup E_2 \cup \ldots \cup E_n = S.$

A collection of events is said to be *equally likely* if all the outcomes of the sample space have the-same chance of occurring.

If an event E_1 can occur in m ways and an event E_2 can occur in n ways, then E_1 or E_2 canoccur in $m+n$ ways. This rule is called *Addition Rule*.

If an operation (task) is performed in 2 steps such that the first step can be performed in n_1 ways and the second step can be performed in n_2 ways (regardless of how the first step was performed), then the entire operation can be performed in $n_1\, n_2$ ways. This rule is called *Multiplication Rule*. The rule can be extended to k steps.

EXAMPLE 13.24

A coin is tossed thrice. If the event E denotes the "number of heads is odd" and event F denotes the "number of tails is odd", determine the cases favourable to $E \cap F$.

Solution. The coin is tossed thrice, therefore, the sample space is

$$S = \{HHH, HHT, HTH, HTT, THT, THH, TTH, TTT\}$$

The events E and F are

$$E = \{ HHH, HTT, THT, TTH\} \text{ and}$$
$$F = \{HHT, HTH, THH, TTT\}$$

We note that $E \cap F = \phi$.

EXAMPLE 13.25

From a group of 2 men and 3 women, two persons are to be selected. Describe the sample space of the experiment. If E is the event in which a man and one woman are selected, determine the favourable cases to E.

Solution. Let M_1, M_2 and W_1, W_2, and W_3 be the men and women in the group. Then number of ways selecting two persons is equal to

$$\binom{5}{2} = \frac{5!}{3!2!} = 10.$$

The sample space is

$$S = \{M_1 M_2, W_1 W_2, W_2 W_3, W_1 W_3, M_1 W_1, M_1 W_2, M_1 W_3, M_2 W_1, M_2 W_2, M_2 W_3\}.$$

If E is the event where one man and one woman is selected, then

$$E = \{M_1 W_1, M_1 W_2, M_1 W_3, M_2 W_1, M_2 W_2, M_2 W_3\}$$

Thus, there are six favorable cases to the event E.

If S is a finite sample space having n mutually exclusive, equally likely and exhative outcomes out of which m are favourable to the occurrence of an event E, then the *probability* of occurrence of E, denoted by P(E), is

$$P(E) = \frac{\text{The number of favourable outcomes in E}}{\text{The total number of outcomes in S}}$$

$$= \frac{|E|}{|S|} = \frac{m}{n}.$$

It follows from the definition that

1. The probability of the sure event is 1, that is, $P(S) = 1$
2. The probability of the impossible event is 0, that is, $P(\phi) = 0$.
3. Since $0 \le m \le n$, we have

   $$0 \le \frac{m}{n} \le 1 \text{ or } 0 \le P(E) \le 1$$

   This relation is called the *axiom of calculus of probability*.
4. The cases favourable to non-occurrence of event E is $n - m$. Therefore,

   $$P(\text{not } E) = \frac{n-m}{n} = 1 - \frac{m}{n} = 1 - P(E),$$

   that is,

   $$P(\overline{E}) = 1 - P(E) \text{ or } P(E) + P(\overline{E}) = 1.$$

EXAMPLE 13.26

Three coins are tossed simultaneously. What is the probability that at least two tails are obtained?

Solution. The sample space consists of $2^3 = 8$ outcomes and

$$S = \{HHH, HHT, HTH, THH, HTT, THT, TTH, TTT\}.$$

Let E be the event obtaining at least 2 tails. Then

$$E = \{ HTT, THT, TTH, TTT\} .$$

Thus, there are four favourable cases to the event E. Hence $P(E) = \frac{4}{8} = \frac{1}{2}$.

EXAMPLE 13.27

In a single throw of two distinct dice, what is the probability of obtaining

(i) a total of 7?

(ii) a total of 13?

(iii) a total as even number?

Solution. The sample space shall consist of $6^2 = 36$ points. We list the total number of out comes as given below:

(1, 1)	(1, 2)	(1, 3)	(1, 4)	(1, 5)	(1, 6)
(2, 1)	(2, 2)	(2, 3)	(2, 4)	(2, 5)	(2, 6)
(3, 1)	(3, 2)	(3, 3)	(3, 4)	(3, 5)	(3, 6)
(4, 1)	(4, 2)	(4, 3)	(4, 4)	(4, 5)	(4, 6)
(5, 1)	(5, 2)	(5, 3)	(5, 4)	(5, 5)	(5, 6)
(6, 1)	(6, 2)	(6, 3)	(6, 4)	(6, 5)	(6, 6)

(i) Let E_1 be the event in which a total of sev-enis obtained. Then

$$E_1 = \{(6, 1), (5, 2), (4, 3), (3, 4), (2, 5), (1, 6)\}$$

and so number of favourable outcomes to the event E_1 is 6. Hence

$$P(E_1) = \frac{6}{36} = \frac{1}{6}.$$

(ii) Since the sum of outcomes on the two dices cannot exceed 6 + 6 = 12, there is no favourable outcome to an event E_2 having sum 13. Hence

$$P(E_2) = \frac{0}{36} = 0$$

(iii) Let E_3 be the event in which we get even number as the sum. Then

$$E3 = \{(1, 1), (1, 3), (1, 5), (2, 2), (2, 4), (2, 6), (3, 1), (3, 3), (3, 5), (4, 2), (4, 4), (4, 6), (5, 1), (5, 3), (5, 5), (6, 2), (6, 4), (6, 6)\}.$$

Thus, number of favourable outcomes to the event E_3 is 18. Hence

$$P(E_3) = \frac{18}{36} = \frac{1}{2}$$

EXAMPLE 13.28

What is the probability that
(i) a non-leap year will have 53 Sunday?
(ii) a leap year will have 53 Sunday?

Solution.

(i) A non-leap year contains 365 days. So it has $\frac{365}{7} = 52$ complete weeks and one extra day. The extra day can be any one of seven days—Sunday, Monday, Tuesday, Wednesday, Thursday, Friday, Saturday. Out of these seven possibilities, the first one is the only favourable to the event "53 Sundays". Therefore,

$$P(\text{53 Sunday}) = \frac{1}{7}.$$

(ii) A leap year contains 366 days. So, it has 52 complete weeks and 2 extra days. These days can be any one of the following seven combinations

[Sunday and Monday,]	Monday and Tuesday
Tuesday and Wednesday	Wednesday and Thursday
Thursday and Friday	Friday and Saturday
[Saturday and Sunday.]	

Out of these seven possibilities only two possibilities (enclosed in boxes) are favourable to the event "53 Sunday". Hence

$$P(\text{53 Sunday in a leap year}) = \frac{2}{7}.$$

EXAMPLE 13.29

Ten persons among whom are A and B, sit down at random at a round table. Find the probability that there are three persons between A and B.

Solution. Let A occupy any seat at the round table. Then there are nine seats available to B. If there are three persons between A and B, then B has only two ways to sit as shown in the diagram below:

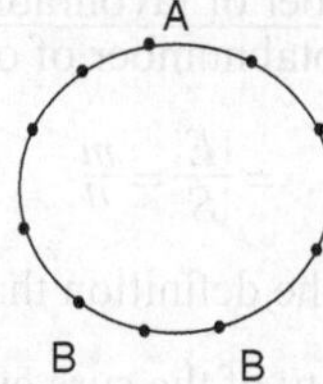

Thus, the probability of the required event is $\frac{2}{9}$.

EXAMPLE 13.30

Four microprocessors are randomly selected from a lot of 20 microprocessor among which five are defective. Find the probability of obtaining no defective microprocessor.

Solution. The sample space will consist of $\binom{20}{4}$ sample points since there are $20C_4$ ways to select 4 microprocessors out of 20 microprocessors. Further, since five microprocessors are defective,the number of favourable outcomes to the event "no defective microprocessor is obtained" is $\binom{15}{4}$. Hence

P(no defective microprocessor)

$$= \frac{\binom{15}{4}}{\binom{20}{4}} = \frac{15.14.13.12}{20.19.18.17} = \frac{32760}{116280} = 0.2817337$$

EXAMPLE 13.31

A bag contains 5 distinct white and 10 distinct-black balls. Random samples of three balls are taken out without replacement. Find the probability that the sample contains

(i) exactly one white ball

(ii) no white ball.

Solution. The total number of ways of chooing 3 balls out of 15 balls is 15 C_3. Thus, the sample space consists of 15 C_3 points. Now

(i) The number of ways of choosing one white ball out of five white balls is 5 C_1. Similarly the number of ways of choosing 2 black balls out of 10 is 10 C_2. Therefore, by multiplication rule, the total number of outcomes for the event "sample consists exactly one white ball" is $5C_1$. 10 C_2. Hence

$$P\text{ (exactly one white ball)} = \frac{5C_1.10C_2}{15C_3}$$

$$= \frac{5.10.9.3.2}{2.15.14.13}$$

$$= \frac{45}{91}.$$

(ii) The event "no white ball" means that all balls selected should be black. So we have to choose 3 balls out of 10 black balls. Hence the number of favourable outcomes to the event is $10C_3$. Therefore,

$$P\text{ (no white ball)} = \frac{10C_3}{15C_3} = \frac{24}{91}.$$

EXAMPLE 13.32

Given a group of four persons, find the proba-ity that

(i) No two of them have their birthday on the same day

(ii) All of them have birthday on the same day.

Solution. Each of the four persons can have his birthday on any of 365 days. Thus, the sample space consists of $(365)^4$ points. Now

(i) Since no two persons have their birthday on the same day, the number of favourable out comes to this event is

$$365 . 364 . 363 . 362$$

Hence

$$P\text{ (distinct birthday)} = \frac{365.364.363.362}{365^4}$$

$$= \frac{364.336.362}{365^3}$$

$$= \frac{364P_3}{365^3}$$

(ii) If all the four persons have their birthday on the same day, then we have to choose just 1 day out of 365. Thus the number of favourable outcomes to the event is 365.

Hence

$$P\text{ (birthday on the same day)} = \frac{365}{365^4}$$

$$= \frac{1}{365^3}$$

EXAMPLE 13.33

A bag contains n distinct white and n distinct red balls. Pair of balls are drawn *without replacement* until the bag is empty. Show that the probability that each pair consists of one white and one red ball is $\frac{2^n}{2nC_n}$.

Solution. The bag contains $2n$ distinct balls. Since the pairs are drawn without replacement, the total number of outcomes in the sample space is

$$\binom{2n}{2}\cdot\binom{2n-2}{2}\cdots\binom{4}{2}\cdot\binom{2}{2}$$

$$=\frac{(2n)!}{2!(2n-2)}\cdot\frac{(2n-2)!}{2!(2n-4)!}\cdots\frac{4!}{2!2!}=\frac{(2n)!}{2^n}.$$

Now, suppose that E is the event in which a pair of balls drawn consists of one white ball and one red ball. Then the first pair can be chosen in $n.n$ ways. Since there is no replacement, the second pair can be selected in $(n-1).(n-1)$ ways, and so on. Therefore, the number of favourable outcomes to the event is

$$n^2(n-1)^2(n-2)^2\ldots 2^2.1^2$$

$$=[n(n-1)(n-2)\ldots 2.1]=(n!)^2.$$

Hence

$$P(E)=\frac{(n!)^2}{(2n)!}\cdot 2^n=\frac{2^n}{\frac{(2n)!}{n!n!}}=\frac{2^n}{\binom{2n}{n}}.$$

Theorem 13.3. If E and F are two mutually exclusive events of a random experiment, then

$$\text{P}(E \text{ or } F)=\text{P}(E\cup F)=\text{P}(E)+\text{P}(F).$$

Thus, the probability that at least one of the mutually exclusive event E or F occurs is the sum of their individual probabilities.

Proof: Suppose that a random experiment results in n mutually exclusive, equally likely, and exhaustive outcomes of which m_1 are favourable to the occurrence of the event E and m_2 to the occurrence of the event F. Then,

$$P(E)=\frac{m_1}{n} \text{ and } P(F)=\frac{m_2}{n}.$$

Since E and F are mutually exclusive, by addition rule, the number of favourable outcomes to the occurrence of E or F is m_1+m_2. Hence

$$\text{P}(E \text{ or } F)=\text{P}(E\cup F)=\frac{m_1+m_2}{n}=\frac{m_1}{n}+\frac{m_2}{n}$$

$$=\text{P}(E)+\text{P}(F).$$

Corollary (1). If $E_1, E_2, \ldots, E_n$ are n manully exclusive, then

$$\text{P}(E_1\cup E_2\cup\ldots E_n)=\text{P}(E_1)+\text{P}(E_2)+\ldots+\text{P}(E_n):$$

Proof: We shall prove the result by mathematical induction on n. By the above theorem

$$\text{P}(E_1\cup E_2)=\text{P}(E_1)+\text{P}(E_2)$$

Let the result be true for $n=k$, that is,

$$\text{P}(E_1\cup E_2\cup\ldots\cup E_k)$$

$$=\text{P}(E_1)+\text{P}(E_2)+\ldots+\text{P}(E_k) \tag{17}$$

We put $E=E_1\cup\ldots\ldots\cup E_k$. Then

$$\text{P}(E_1\cup E_2\cup\ldots\cup E_{k+1})$$

$$=\text{P}(E\cup E_{k+1})=\text{P}(\text{E}_1)$$

$$+\text{P}(E_2)+\ldots+\text{P}(E_k)+\text{P}(E_k+1) \text{ using (17).}$$

Hence, the result holds by mathematical induction.

Corollary (2). If $E_1, E_2, \ldots, E_n$ are n mutually exclusive and exhaustive events, then

$$\text{P}(E_1)+\text{P}(E_2)+\ldots+\text{P}(E_n)=1.$$

Proof: Since $E_1, E_2, \ldots, E_n$ are mutually exclusive and exhaustive,

$$E_1\cup E_2\cup\ldots\cup E_n=\text{S (sample space).}$$

simple P(S) = 1, we have

$$1=\text{P(S)}=\text{P}(E_1\cup E_2\cup\ldots\cup E_n)$$

$$=\text{P}(E_1)+\text{P}(E_2)+\ldots+\text{P}(E_n).$$

Corollary (3). If E and F are two events, then

$$\text{P}(\text{E}\cap\text{F})=\text{P(E)}-\text{P}(\text{E}\cap\text{F}).$$

Proof: The events $\text{E}\cap\overline{\text{F}}$ and $\text{E}\cap\text{F}$ are mutually exclusive. Also

$$(\text{E}\cap\overline{\text{F}})\cap(\text{E}\cap\text{F})=\text{E}.$$

Hence, by the above theorem

$$\text{P(E)}=\text{P}(\text{E}\cap\overline{\text{F}})+\text{P}(\text{E}\cap\text{F}) \text{ or}$$

$$\text{P}(\text{E}\cap\overline{\text{F}})=\text{P(E)}-\text{P}(\text{E}\cap\text{F}).$$

Corollary (4). If E and F are two events such that $E \subseteq F$, then $P(E) \leq P(F)$.

Proof: Since $E \subseteq F$, we have $F = E \cup (F - E)$. Also $E \cap (F - E) = \varphi$. Hence, by Theorem 13.3, we have

$$P(F) = P(E) + P(F - E), \qquad (18)$$

Since $P(F \backslash E) \geq 0$, it follows from (18) that $P(F) \geq P(E)$.

Theorem 13.4. (Addition Rule or Law of Addition of Probability). If *E* and *F* are any arbitrary events associated with a random experiment, then

$$P(E \text{ or } F) = P(E \cup F) = P(E) + P(F) - P(A \cap B).$$

Proof: The events $E \cap \bar{F}$ and F are two mutually exclusive events and

$$(E \cap \bar{F}) \cup F = E \cup F.$$

Hence

$$P(E \cap \bar{F}) + P(F) = P(E \cup F) \qquad (19)$$

But $E \cap F$ and $E \cap \bar{F}$ are mutually exclusive, that is,

$$(E \cap F) \cup (E \cap \bar{F}) = E$$

and so

$$P(E \cap \bar{F}) + P(E \cap F) = P(E)$$

$$\text{or } P(E \cap \bar{F}) = P(E) - P(E \cap F) \qquad (20)$$

From (19) and (20), it follows that

$$P(E \cup F) = P(E) + P(F) - P(E \cap F).$$

Remark 13.4. If E and F are mutually exclusive, then $E \cap F = \phi$ and $P(\phi) = 0$, and so the above result reduces to

$$P(E \cup F) = P(E) + P(F),$$

a result proved already.

EXAMPLE 13.34

Two fair dices are rolled. Find the probability of getting doubles (two dices showing the same numbers) or the sum of 7.

Solution. The sample space S is given by

(1,1) (1,2) (1,3) (1,4) (1,5) (1,6)
(2,1) (2,2) (2,3) (2,4) (2,5) (2,6)
(3,1) (3,2) (3,3) (3,4) (3,5) (3,6)
(4,1) (4,2) (4,3) (4,4) (4,5) (4,6)
(5,1) (5,2) (5,3) (5,4) (5,5) (5,6)
(6,1) (6,2) (6,3) (6,4) (6,5) (6,6)

The total number of outcomes in S is 36. Let E_1 be the event "get doubles" and E_2 is the event "sum of 7". Then

$E_1 = \{(1,1), (2,2), (3,3), (4,4), (5,5), (6,6)\}$ and

$E_2 = \{(1,6), (2,5), (3,4), (4,3), (5,2), (6,1)\}$.

We notice that E_1 and E_2 are mutually exclusive. Therefore,

$$P(E_1 \text{ or } E_2) = P(E_1 \cup E_2) = P(E_1) + P(E_2).$$

But

$$P(E_1) = \frac{\text{The number of favourable outcome in } E_1}{\text{Number of outcomes in S}} = \frac{6}{36} = \frac{1}{6}.$$

Similarly,

$$P(E_2) = \frac{6}{36} = \frac{1}{6}.$$

Hence

$$P(E_1 \text{ or } E_2) = \frac{1}{6} + \frac{1}{6} = \frac{1}{3}.$$

EXAMPLE 13.35

Two fair dices are thrown simultaneously. Find the probability of getting doubles or a multiple of 3 as the sum.

Solution. The sample space S consists of the points.

(1, 1,) (1, 2) (1, 3) (1, 4) (1, 5) (1, 6)
(2, 1) (2, 2) (2, 3) (2, 4) (2, 5) (2, 6)
---- ----- ----- ----- ---- ----
---- ----- ----- ----- ----- ----
(6, 1) (6, 2) (6, 3) (6, 4) (6, 5) (6, 6)

Thus S consists of 36 outcomes. Let E_1 be the event of getting doubles. Then

$$E_1 = \{(1, 1), (2, 2), (3, 3), (4, 4), (5, 5), (6, 6)\}$$

and so the number of favourable outcomes to the event E_1 is 6. So

$$P(E_1) = \frac{6}{36} = \frac{1}{6}.$$

Let E_2 be the event of getting a multiple of 3 as the sum. Then

$$E_2 = \{(1,2), (1,5), (2,1), (2,4), (3,3), (3,6), (4,2), (4,5), (5,1), (5,4), (6,3), (6,6)\}$$

and so the number of favourable outcomes to the event E_2 is 12. Thus

$$P(E_2) = \frac{12}{36} = \frac{1}{3}.$$

Further,

$$E_1 \cap E_2 = \{(3,3), (6,6)\}.$$

Thus

$$P(E_1 \cap E_2) = \frac{2}{36} = \frac{1}{18}.$$

Hence

$$P(E_1 \cup E_2) = P(E_1) + P(E_2) - P(E_1 \cap E_2)$$

$$= \frac{1}{6} + \frac{1}{3} - \frac{1}{18} = \frac{4}{9}.$$

EXAMPLE 13.36

A bag contains five white, seven black, and eight red balls. A ball is drawn at random. What is the probability that it is a red ball or a white ball?

Solution. The number of outcomes in the sample space is

$$20\ C_1 = 20.$$

Let E_1 be the event where red ball is obtained and E_2 be the event where white ball is obtained. Then

$$P(E_1) = \frac{8C_1}{20} = \frac{8}{20} = \frac{2}{5} \text{ and}$$

$$P(E_2) = \frac{5C_1}{20} = \frac{5}{20} = \frac{1}{4}.$$

Also the events are mutually exclusive. Therefore,

$$P(E_1 \text{ or } E_2) = P(E_1) + P(E_2)$$

$$= \frac{2}{5} + \frac{1}{4} = \frac{13}{20}$$

EXAMPLE 13.37

Let A and B be two mutually exclusive events of an experiment. If P(not A) = 0.65, $P(A \cup B) = 0.65$ and $P(B) = p$, find p.

Solution. We have

$$P(\text{not } A) = P(\overline{A}) = 0.65.$$

But

$$P(A) + P(\overline{A}) = 1 \text{ and so}$$

$$P(A) = 1 - P(\overline{A}) = 1 - 0.65 = 0.65 = 0.35.$$

Further, since A and B are mutually exclusive,

$$P(A \cup B) = P(A) + P(B) = P(A) + p$$

and so

$$p = P(A \cup B) - P(A) = 0.65 - 0.35 = 0.30$$

13.11 CONDITIONAL PROBABILITY

Let E and F be events and let $P(F) > 0$. Then the conditional probability of E, given F, is defined as

$$P(E\backslash F) = \frac{P(E \cap F)}{P(F)}.$$

EXAMPLE 13.38

Let two fair dice be rolled. If the sum of 7 is obtained, find the probability that at least one of the dice shows 2.

Solution. Let E be the event "sum of 7 is obtained". Thus

$$E = \{(1,6), (2,5), (3,4), (4,3), (5,2), (6,1)\}.$$

Let F be the event "at least one dice shows 2". Then

$$F = \{(1, 2), (2, 2), (3, 2), (4, 2), (5, 2), (6, 2), (2, 1), (2, 3), (2, 4), (2, 5), (2, 6)\}.$$

Since $E \cap F = \{(2,5), (5,2)\}$, by the definition of conditional probability, we have

$$P(F\backslash E) = \frac{P(E\cap F)}{P(E)} = \frac{\frac{2}{36}}{\frac{6}{36}} = \frac{1}{3}.$$

EXAMPLE 13.39

Weather records show that the probability of-high barometric pressure is 0.82 and the probability of rain and high barometric pressure is 0.20. Find the probability of rain, given high barometric pressure?

Solution. Let E denote the event "rain" and F denote the event "high barometric pressure."Then

$$P(E\backslash F) = \frac{P(E\cap F)}{P(F)} = \frac{0.20}{0.82} = 0.2446.$$

Theorem 13.5. (Multiplication Law of Probability). Let $P(A\backslash B)$ denote the conditional probability of A when B has occurred. Then

$$P(A \cap B) = P(B)\,P(A\backslash B) = P(A)\,P(B\backslash A).$$

Proof: We know that

$$P(A\backslash B) = \frac{P(A\cap B)}{P(B)} \qquad (21)$$

and

$$P(B\backslash A) = \frac{(A\cap B)}{P(A)}. \qquad (22)$$

From (21) and (22), we have

$$P(A \cap B) = P(B)\,P(A\backslash B) = P(A)\,P(B\backslash A).$$

EXAMPLE 13.40

A fair coin is tossed four times. Find the probabilitythat they are all heads if the first two tosses results in head.

Solution. The sample space consists of $2^4 = 16$ outcomes. Let A be the event "all heads." Then A = {HHHH}. Let B be the event "first two heads". Then

$$B = \{HHHH, HHHT, HHTH, HHTT\}.$$

We notice that

$$A \cap B = \{HHHH\}.$$

Therefore

$$P(B) = \frac{4}{16} = \frac{1}{4} \text{ and } P(A \cap B) = \frac{1}{16}$$

and so

$$P(A\backslash B) = \frac{P(A\cap B)}{P(B)} = \frac{1/16}{1/4} = \frac{1}{4}.$$

13.12 INDEPENDENT EVENTS

Two events A and B are said to be *independent* if the occurrence or non-occurrence of one event does not affect the probability of the occurrence or non-occurrence of the other event. Mathematically, A and B are independent if and only if any one of the following conditions is satisfied.

$$P(A\backslash B) = P(A),\ P(\overline{A}\backslash B) = P(\overline{A}),$$

$$P(A\backslash \overline{B}) = P(A),\ P(\overline{A}\backslash \overline{B}) = P(\overline{A}),$$

$$P(B\backslash A) = P(B),\ P(\overline{B}\backslash A) = P(\overline{B}),$$

$$P(B\backslash \overline{A}) = P(B),\ P(\overline{B}\backslash \overline{A}) = P(\overline{B}),$$

Thus, if A and B are independent events, then

$$P(A) = P(A\backslash B) = \frac{P(A\cap B)}{P(B)}$$

or

$$P(A \cap B) = P(A)\ P(B).$$

This relation is called *multiplication rule for independent events*.

Hence, we may also define independence of events as follows:

Events A and B are called independent if $P(A \cap B) = P(A)\,P(B)$.

EXAMPLE 13.41

A married couple (husband and wife) appear for an interview for two vacancies against the same post. The probability of husband's selection is $\frac{1}{6}$ and the probability of wife's selection is $\frac{2}{5}$. What is the probability that

(i) both of them will be selected
(ii) only one of them will be selected
(iii) none of them will be selected
(iv) at least one of them will be selected?

Solution. Let E be the event "husband is selected" and F denote the event "wife is seleced". We are given that

$$P(E) = \frac{1}{6} \text{ and } P(E) = \frac{2}{5}.$$

Since there are two vacancies, selection of one does not affect the other. Hence E and F are independent events. Then

(i) P(both of them are selected) = $P(E \cap F)$

$= P(E)\, P(F)$ since E and F are independent

$= \frac{1}{6}\frac{2}{5} = \frac{1}{15}.$

(ii) Since $E \cap \overline{F}$ and $\overline{E} \cap F$ are exclusive, we have

P(only one of them is selected)

$= P(E \cap \overline{F}) \cup (\overline{E} \cap F)$

$= P(E \cap \overline{F}) + (\overline{E} \cap F)$ (exclusive events)

$= P(E)\, P(\overline{F}) + P(\overline{E})\, P(F),$

since E and F are independent

$= P(E)\,(1 - P(F)) + (1 - P(E))\, P(F)$

$= \frac{1}{6}\left(1 - \frac{2}{5}\right) + \left(1 - \frac{1}{6}\right)\frac{2}{5} = \frac{1}{10} + \frac{1}{3} = \frac{13}{30}.$

(iii) We have

P(none of them is selected)=P(not E and not F)

$= P(\overline{E} \cap \overline{F}) = P(\overline{E})P(\overline{F})$

since E and F are independent

$= (1 - P(E))\,(1 - P(F))$

$= \left(1 - \frac{1}{6}\right)\left(1 - \frac{2}{5}\right) = \frac{5}{6} \cdot \frac{3}{5} = \frac{1}{2}.$

(iv) We have

P(at least one of them gets selected)

$= P(E \text{ or } F) = P(E \cup F)$

$= P(E) + P(F) - P(E \cap F)$

$= \frac{1}{6} + \frac{2}{5} - \frac{1}{15}$, using(i)

$= \frac{1}{2}.$

Second Method: $P(E \cup F) = 1 - P(\overline{E \cup F})$

$= 1 - P(\overline{E} \cap \overline{F}) = 1 - \frac{1}{2} = \frac{1}{2}$

EXAMPLE 13.42

If $P(B) \neq 1$, show that

$$P(\overline{A} \backslash \overline{B}) = \frac{1 - P(A \cup B)}{P(\overline{B})}.$$

Solution. We have

$$P(\overline{A} \backslash \overline{B}) = \frac{P(\overline{A} \cap \overline{B})}{P(\overline{B})} = \frac{P(\overline{A \cup B})}{P(\overline{B})}$$

$$= \frac{1 - P(A \cup B)}{P(\overline{B})}.$$

EXAMPLE 13.43

A problem in mathematics is given to three students whose chances of solving the problem are $\frac{1}{2}, \frac{1}{3}, \frac{1}{4}$. What is the probability that the problem is solved?

Solution. Let

A be the event "first student solves the problem"
B be the event "second student solves the problem" and C be the event "third student solves the problem"

It is given that

$$P(A) = \frac{1}{2},\ P(B) = \frac{1}{3},\ P(C) = \frac{1}{4}$$

and so

$$P(\overline{A}) = 1 - \frac{1}{2} = \frac{1}{2},\ P(\overline{B}) = 1 - \frac{1}{3} = \frac{2}{3},$$

$$P(\overline{C}) = 1 - \frac{1}{4} = \frac{3}{4}.$$

Hence

P(the problem is solved) = P(A or B or C)

$= P(A \cup B \cup C) = 1 - P[\overline{(A \cup B \cup C)}]$

$= 1 - P(\overline{A} \cap \overline{B} \cap \overline{C})$

$= 1 - P(\overline{A})P(\overline{B})P(\overline{C})$

since A, B, and C are independent

$= 1 - \frac{1}{2} \cdot \frac{2}{3} \cdot \frac{3}{4} = \frac{3}{4}.$

Theorem 13.6. (Baye's Theorem). Let $A_1, A_2, \ldots, A_m$ be pairwise mutually exclusive and exhaustive random events, where $P(A_i) \geq 0$,

$i = 1, 2, \ldots, m$. Then for any arbitrary event B of the random experiment,

$$P(A_i \backslash B) = \frac{P(A_i)P(B\backslash A_i)}{\sum_{i=1}^{m} P(A_i)P(B\backslash A_i)}.$$

Proof: Let S be the sample space of the random experiment. Since the events $A_1, A_2, \ldots, A_m$ are pairwise exclusive and exhaustive, we have

$$S = A_1 \cup A_2 \cup \ldots\ldots \cup A_m.$$

Therefore, we have

$$B = S \cap B = (A_1 \cup A_2 \cup \ldots\ldots \cup A_m) \cap B$$
$$= (A_1 \cap B) \cup (A_2 \cap B) \cup \ldots\ldots \cup (A_m \cap B)$$

Since $A_1 \cap B, A_2 \cap B, \ldots\ldots, A_m \cap B$ are mutually exclusive, it follows by addition law that

$$P(B) = P(A_1 \cap B) + P(A_2 \cap B) + \ldots + \ldots + P(A_m \backslash B)$$
$$= P(B \backslash A_1)\, P(A_1) + P(B \backslash A_2)\, P(A_2) + \ldots + P(B \backslash A_m)\, P(A_m).$$

This relation is called the "*theorem on total probability*." Using this relation, we have

$$P(A_i \backslash B) = \frac{P(A_i \cap B)}{P(B)} = \frac{P(B\backslash A_i)P(A_i)}{P(B)}$$

$$= \frac{P(A_i)P(B\backslash A_i)}{\sum_{i=1}^{m} P(A_i)P(B\backslash A_i)}.$$

EXAMPLE 13.44

A university purchased computers from three firms. The percentage of computer purchased and percentage of defective computers is shown in the table below:

	Firm		
	HCL	WIPRO	IBM
Percent purchase	45	25	30
Percent defective	2	3	1

Let A be the event "computer purchased from HCL"

B be the event "computer purchased from WIPRO"

C be the event "computer purchased from IBM"

D be the event "computer was defective".

Find P(A), P(B), P(C), P(D\A), P(D\B), P(D\C) and P(D).

Solution. We note that

$$P(A) = \frac{45}{45+25+30} = 0.45,\ P(B) = \frac{25}{100} = 0.25,$$

$$P(C) = \frac{30}{100} = 0.30,\ P(D \backslash A) = \frac{2}{100} = 0.02,$$

$$P(D \backslash B) = \frac{3}{100} = 0.03,\ P(D/C) = \frac{1}{100} = 0.01$$

$$P(D) = P(D\backslash A)\, P(A) + P(D\backslash B)\, P(B) + P(D\backslash C)\, P(C)$$
$$= (0.02)\,(0.45) + (0.03)(0.25) + (0.01)(0.30)$$
$$= 0.0090 + 0.0075 + 0.0030 = 0.0195.$$

EXAMPLE 13.45

In a test, an examinee either guesses, or copies or knows the answer to multiple choice questions with four choices. The probability that he makes a guess is $\frac{1}{3}$ and the probability that he copies the answer is $\frac{1}{6}$. The probability that his answer is correct, given that he copied it is $\frac{1}{8}$. Find the probability that he knew the answer to the question given that he correctly answered.

Solution. Let us consider the following events:

A: the examinee guesses the answer

B: the examinee copies the answer

C: the examinee knows the answer

D: the examinee answers correctly.

It is given that

$$P\,(A) = \frac{1}{3},\quad P\,(B) = \frac{1}{6} \text{ and } P\,(D\backslash B) = \frac{1}{8}.$$

Also, the hypothesis that examinee either guesses or copies or knows the answer implies that

$$P\,(C) = 1 - P\,(A) - P\,(B) = 1 - \frac{1}{3} - \frac{1}{6} = \frac{1}{2}.$$

Further,

$P\,(D\backslash C) = 1$ since he knows the answer correctly.

$P\,(D\backslash A) = \frac{1}{4}$ (since if he guesses, he can tick any one of the four choices):

Then, by Baye's law,

P(C\D)

$$= \frac{P(D\backslash C)P(C)}{P(D\backslash A)P(A) + P(D\backslash B)P(B) + P(D\backslash C)P(C)}$$

$$= \frac{1.\frac{1}{2}}{\frac{1}{4}\cdot\frac{1}{3} + \frac{1}{8}\cdot\frac{1}{6} + 1\cdot\frac{1}{2}} = \frac{24}{29}.$$

EXAMPLE 13.46

The following observations were made at a clinic where HIV virus test was performed.

(i) 15% of the patients at the clinic have HIV virus

(ii) among those who have HIV virus, 95% test positive on the ELISA test

(iii) among those that do not have HIV virus, 2% test positive on the ELISA test.

Find the probability that a patient has the HIV virus if the ELISA test is positive.

Solution. We consider the following events:

A: "has the HIV virus"

B: "does not have the HIV virus"

C: "test positive".

We are given that

$$P(A) = \frac{15}{100} = 0.15.$$

Therefore,

$$P(B) = P(\bar{A}) = 1 - P(A)$$

$$= 1 - 0.15 = 0.85$$

$$P(C\backslash A) = \frac{95}{100} = 0.95 \text{ and}$$

$$P(C\backslash B) = \frac{2}{100} = 0.02.$$

We want to find P (A\C). By Baye's theorem, we have

$$P(A\backslash C) = \frac{P(C\backslash A)\,P(A)}{P(C\backslash A)\,P(A) + P(C\backslash B)\,P(B)}$$

$$= \frac{(0.95)\,(0.15)}{(0.95)\,(0.15) + (0.2)\,(.085)}$$

$$= 0.89.$$

EXAMPLE 13.47

An item is manufactured by three factories F_1, F_2, and F_3. The number of units of the item produced by F_1, F_2, and F_3 are $2x$, x, and x, respectively. It is known that 2% of the items produced by F_1 and F_2 are defective and 4% of the items produced by F_3 are defective. All units produced by these factories are put together in one stockpile and one unit is chosen at random. It is found that this item is defective. What is the probability that this defective unit came from (i) factory F_1, (ii) factory F_2, or (iii) factory F_3?

Solution. Consider the events:

A: "the unit is defective"

B: "the defective unit came from F_1"

C: "the defective unit came from F_2"

D: "the defective unit came from F_3"

We have then, as per given hypothesis,

$$P(B) = \frac{2x}{4x} = 12, \quad P(C) = \frac{x}{4x} = \frac{1}{4},$$

$$P(D) = \frac{x}{4x} = \frac{1}{4}$$

$$P(A\backslash B) = \frac{2}{100} = 0.02, \; P(A\backslash C) = \frac{2}{100} = 0.02,$$

$$P(A\backslash D) = \frac{4}{100} = 0.04.$$

Then the theorem on total probability implies that

$$P(A) = P(A\backslash B)\,P(B) + P(A\backslash C)\,P(C) + P(A\backslash D)\,P(D)$$

$$= (0.02)\left(\frac{1}{2}\right) + (0.02)\left(\frac{1}{4}\right) + (0.04)\left(\frac{1}{4}\right)$$

$$= 0.025.$$

We then have, by Baye's theorem,

$$P(B\backslash A) = \frac{P(A\backslash B)P(B)}{P(A)} = \frac{(0.02)\left(\frac{1}{2}\right)}{0.025} = 0.4,$$

$$P(C\backslash A) = \frac{P(A\backslash C)P(C)}{P(A)} = \frac{(0.02)\left(\frac{1}{4}\right)}{0.025} = 0.2,$$

$$P(D\backslash A) = \frac{P(A\backslash D)P(D)}{P(A)} = \frac{(0.04)\left(\frac{1}{4}\right)}{0.0025} = 0.4.$$

13.13 PROBABILITY DISTRIBUTION

Let S be a sample space of an random experiment. *A random variable X* is a function of the possible events of S which assigns a numerical value of each outcome in S. A random variable is also called a *Variate.*

Let a random variable X assume the values $x_1, x_2, \ldots, x_n$ corresponding to various outcomes of a random experiment. If the probability of x_i is $P(x_i) = p_i$, $1 \le i \le n$ such that $p_1 + p_2 + \ldots + p_n = 1$, then the function $P(X)$ is called the *probability* function of the random variable X and the set $\{P(x_i)\}$ is called the *probability distribution* of X. Since random variable X takes a finite set of values, it is called discrete variate and $\{P(x_i)\}$ is called the *discrete probability distribution.*

The probability distribution of X is denoted by the table:

X	x_1	x_2	$x_3 \ldots x_n$
$P(X)$	p_1	p_2	$p_3 \ldots p_n$

If x is an integer, then the function F defined by

$$F(X) = P(X \le x) = \sum_{i=1}^{x} p(x_i)$$

is called the *distribution function* or *cumulative distribution function* of the discrete variate X.

If a variate X takes every value in an interval, the number of events is infinitely large and so the probability for an event to occur is practically zero. In such a case, the probability of x falling in a small interval is determined. The function f defined by

$$P\left(x - \frac{1}{2}\,dx \le x \le x + \frac{1}{2}\,dx\right) = f(x)dx$$

is called the *probability density function* and the continuous curve $y = f(x)$ is called the probability curve.

If the range of x is finite, we may consider it as infinite by supposing the density function f to be zero outside the given range. Thus if $f(x) = \phi(x)$ be the density function for x in $[a, b]$ then we take

$$f(x) = \begin{cases} 0, & x < a \\ \phi(x), & x \in [a, b] \\ 0, & x > b. \end{cases}$$

Further, the density function f is always positive and $\int_{-\infty}^{\infty} f(x)dx = 1$, that is, the total area under the probability curve and the x-axis is unity, This fulfills the requirement that the total probability of the occurrence of an event is 1.

If X is continuous variate, then the function F defined by

$$F(x) = P(X \le x) = \int_{-\infty}^{\infty} f(x)dx$$

is called the *cumulative distribution function of the continuous variate X.*

The cumulative distribution function F has the following important properties:

(i) $F(x) = f(x) \ge 0$ and so F *is* nondecreasing function.

(ii) $F(-\infty) = 0$ *and* $F(\infty) = 1$.

(iii) $P(a \le x \le b) = \int_a^b f(x)\,dx$

$$= \int_{-\infty}^{b} f(x)\,dx - \int_{-\infty}^{a} f(x)\,dx$$

$$= F(b) - F(a).$$

13.14 MEAN AND VARIANCE OF A RANDOM VARIABLE

Let X be a random variable which takes the values $x_1, x_2, \ldots, x_m$ with corresponding probabilities $p_1, p_2, \ldots, p_m$. Then the *mean* (also called *expectation*) and *variance* of the random variables are defined by

$$\text{Mean: } \propto = \frac{\sum_{i=1}^{m} p_i x_i}{\sum_{i=1}^{m} p_i}$$

$$= \sum_{i=1}^{m} p_i x_i \text{ since } \sum_{i=1}^{m} p_i = 1,$$

$$\text{Variance: } \sigma^2 = \sum_{i=1}^{m} (x_i - \mu)^2 p_i$$

$$= \sum_{i=1}^{m} (x_i^2 - 2 \propto x_i + \propto^2) p_i$$

$$= \sum_{i=1}^{m} p_i x_i^2 - 2\propto \sum_{i=1}^{m} p_i x_i + \propto^2 \sum_{i=1}^{m} p_i$$

$$= \sum_{i=1}^{m} p_i x_i^2 - 2\propto^2 + \propto^2, \text{ since}$$

$$\sum_{i=1}^{m} p_i x_i = \propto \text{ and } \sum_{i=1}^{m} p_i = 1$$

$$= \sum_{i=1}^{m} p_i x_i^2 - \propto^2$$

where σ is the *standard deviation of the distribution.*

In case of continuous probability distribution, the mean (expected value) and variations are defined by

$$\propto = \int_{-\infty}^{\infty} x f(x) dx,$$

$$\sigma^2 = \int_{-\infty}^{\infty} (x-\mu)^2 f(x) dx.$$

For the discrete probability distribution, the rth moment about the mean $\propto$, denoted by $\propto_r$, is defined by

$$\propto_r = \frac{1}{N} \sum (x_i - \propto)^r f(x_i)$$

$$= \sum (x_i - \propto)^r f(x_i), \text{ since } N = \sum_{i=1}^{n} f(x_i) = 1$$

Putting $r = 0, 1, 2, 3$, and 4, we get

$$\propto_0 = \sum f(x_i)$$

$$\propto_1 = \sum (x_i - \propto) f(x_i)$$

$$\propto_2 = \sum (x_i - \propto)^2 f(x_i)$$

$$\propto_3 = \sum (x_i - \propto)^3 f(x_i)$$

$$\propto_4 = \sum (x_i - \propto)^4 f(x_i).$$

The *four Pearson's β and γ coefficients* are

$$\beta_1 = \frac{\mu_3^2}{\mu_2^3}, \beta_2 = \frac{\mu_4}{\mu_2^2},$$

$$\gamma_1 = \sqrt{\beta_1}, \gamma_2 = \beta_2 - 3,$$

The coefficient β_1 gives *the measure of departure from symmetry* or the *measure of skewness.*

Similarly, β_2 gives the *measure of kurtosis.* Consider the function

$$M_a(t) = \sum p_i e^{t(x_i - a)}.$$

This function is a function of the parameter t and is nothing but mean (expected value) of the probability distribution of $e^{t(x_i - a)}$. Expanding the exponential, we get

$$M_a(t) = \sum p_i [1 + t(x_i - a) + \frac{t^2}{2}(x_i - a)^2$$
$$+ \ldots + \frac{t^r}{r!}(x_i - a)^r + \ldots]$$

$$= \sum p_i + t \sum p_i (x_i - a) + \frac{t^2}{2!} \sum p_i (x_i - a)^2$$
$$+ \ldots + \frac{t^r}{r!} \sum p_i (x_i - a)^r + \ldots$$

$$= 1 + t\mu_1 + \frac{t^2}{2!}\mu_2 + \ldots + \frac{t^r}{r!}\mu_r + \ldots \quad (23)$$

where μ_r is the moment of order r about a. Thus $M_a(t)$ generates moments and is, therefore, called the *moment generating function* of the discrete probability distribution of the variate X about the value $x = a$. Thus, the moment generating function of the discrete probability distribution of the variate X about $x = a$ is defined as the expected value of the function $e^{t(x-a)}$.

We observe that μ_r, the rth moment, is equal to the coefficient of $\frac{t^r}{r!}$ in the expansion of the moment generating function $M_a(t)$.

Alternately, μ_r can be obtained by differentiating (23) r times with respect to t and then putting $t = 0$. Thus

$$\mu_r = \left[\frac{d^r}{dt^r} M_a(t)\right]_{t=0}$$

Also

$$M_a(t) = \sum p_i e^{t(x_i - a)} = e^{-at} \sum p_i e^{tx_i} = e^{-at} M_o(t).$$

Hence moment generating function about the value a is e^{-at} times the moment generating function about the origin.

For continuous distribution of variate X, the moment generating function about $x = a$ is defined by

$$M_a(t) = \int_{-\infty}^{\infty} e^{t(x-a)} f(x)dx.$$

EXAMPLE 13.48

A random variable x has the following probability function:

x:	−2	−1	0	1	2	3
$p(x)$:	0.1	k	0.2	$2k$	0.3	k

Find the value of k and calculate the mean and variance.

Solution. Since $\sum p_i = 1$, we have

$$0.1 + k + 0.2 + 2k + 0.3 + k = 1,$$

which yields $k = 0.1$. Further,

$$\text{Mean: } \mu = \sum_{i=1}^{n} p_i x_i = \sum_{i=1}^{6} p_i x_i$$

$$= -2(0.1) + (-1)(0.1) + 0(0.2)$$

$$+ 2(0.1) + 2(0.3) + 3(0.1) = 0.8$$

and

$$\text{Variance: } \sigma^2 = \sum_{i=1}^{n} (x_i - \mu)^2 p_i = \sum_{i=1}^{n} p_i x_i^2 - \mu^2$$

$$= 0.4 + 0.1 + 0 + 0.2 + 1.2$$

$$+ 0.9 - 0.64 = 2.16.$$

EXAMPLE 13.49

A die is tossed thrice. A success is "getting 1 or 6"on a toss. Find the mean and variance of the number of successes.

Solution. We have $n = 3$. Let X denote the number of success. Then

$$\text{Probability of success} = \frac{2}{6} = \frac{1}{3},$$

$$\text{Probability of failure} = 1 - \frac{1}{3} = \frac{2}{3}$$

and

$$P(X = 0) = P\text{ (no success)} = P\text{ (all failures)}$$

$$= \frac{2}{3} \cdot \frac{2}{3} \cdot \frac{2}{3} = \frac{8}{27}$$

$$P(X = 1) = P\text{ (1 success and 2 failures)}$$

$$= \left({}^3C_1 \cdot \frac{1}{3}\right)\frac{2}{3} \cdot \frac{2}{3} = \frac{4}{9},$$

$$P(X = 2) = P\text{ (2 success and 1 failure)}$$

$$= \left({}^3C_2 \cdot \frac{1}{3}\right)\frac{1}{3} \cdot \frac{2}{3} = \frac{2}{9},$$

$$P(X = 3) = P\text{ (3 success)} = \frac{1}{3} \cdot \frac{1}{3} \cdot \frac{1}{3} = \frac{1}{27}$$

Therefore, the probability distribution is

X:	0	1	2	3
$P(X)$:	$\frac{8}{27}$	$\frac{4}{9}$	$\frac{2}{9}$	$\frac{1}{27}$

Now

$$\text{Mean}(\mu) = \sum p_i x_i = \frac{4}{9} + \frac{4}{9} + \frac{1}{9} = 1,$$

$$\text{Variance } (\sigma^2) = \sum p_i x_i^2 - \mu^2 = \frac{4}{9} + \frac{8}{9} + \frac{9}{27} - 1 = \frac{2}{3}.$$

EXAMPLE 13.50

Find the standard deviation for the following discrete distribution:

x:	8	12	16	20	24
$p(x)$:	$\frac{1}{8}$	$\frac{1}{6}$	$\frac{3}{8}$	$\frac{1}{4}$	$\frac{1}{12}$

Solution. For the given discrete distribution, $n = 5$ and the mean

$$\mu = \sum_{i=1}^{5} p_i x_i = 1 + 2 + 6 + 5 + 2 = 16.$$

Thus the variance is

$$\sigma^2 = \sum_{i=1}^{5} p x_i^2 - \mu^2$$

$$= 8 + 24 + 96 + 100 + 48 - 256 = 20.$$

Hence, the standard deviation is

$$\sigma = \sqrt{20} = 2\sqrt{5}.$$

EXAMPLE 13.51

The diameter x of an electric cable is assumed to be a continuous variate with possible probability density function $f(x) = 6x(1 - x)$, $0 \le x \le 1$. Verify whether f is a probability density function. Also find the mean and variance.

Solution. The given function is non-negative and

$\int_0^1 f(x)dx = \int_0^1 6x dx - \int_0^1 6x^2 dx = 1$. Hence f is a probability density function. Further

$$\text{Mean}(\mu) = \int_0^1 x f(x)dx = \int_0^1 6x^2 dx - \int_0^1 6x^3 dx$$

$$= 6\left[\frac{x^3}{3}\right]_0^1 - 6\left[\frac{x^4}{4}\right]_0^1 = 2 - \frac{3}{2} = \frac{1}{2},$$

$$\text{Variance}(\sigma^2) = \int_0^1 (x - \mu)^2 f(x)dx$$

$$= \int_0^1 \left(x - \frac{1}{2}\right)^2 [6x(1 - x)]dx$$

$$= \int_0^1 \left(-6x^4 + 12x^3 - \frac{15}{2}x^2 + \frac{3}{2}x\right)dx$$

$$= -\frac{6}{5} + 3 - \frac{15}{6} + \frac{3}{4} = \frac{1}{20}.$$

EXAMPLE 13.52

The probability density $p(x)$ of a continuous random variable is given by

$$p(x) = y_0 e^{-|x|}, -\infty < x < \infty.$$

Prove that $y_0 = \frac{1}{2}$. Find the mean and variance of distribution.

Solution. Since $e^{-|x|}$ is an even function of a, we have

$$\int_{-\infty}^{\infty} p(x)dx = y_0 \int_{-\infty}^{\infty} e^{-|x|} dx = 2y_0 \int_0^{\infty} e^{-|x|} dx$$

$$= 2y_0 \int_0^{\infty} e^{-x} dx = 2y_0 \left[\frac{e^{-x}}{-1}\right]_0^{\infty} = 2y_0.$$

But, $p(x)$ being probability density function, we have

$$\int_{-\infty}^{\infty} p(x)dx = 1.$$

Therefore $2y_0 = 1$, which yields $y_0 = \frac{1}{2}$. Further since $xe^{-|x|}$ is an odd function, we have

$$\text{Mean}(\mu) = \int_{-\infty}^{\infty} xp(x)dx = \frac{1}{2}\int_{-\infty}^{\infty} xe^{-[x]}dx = 0.$$

Since $x^2 e^{-|x|}$ is an even function, we have

$$\text{Variance}(\sigma^2) = \int_{-\infty}^{\infty} (x-\mu)^2 p(x)dx$$

$$= \frac{1}{2}\int_{-\infty}^{\infty} x^2 e^{-|x|} dx$$

$$= \frac{2}{2}\int_0^{\infty} x^2 e^{-|x|} dx$$

$$= \int_0^{\infty} x^2 e^{-x} dx = \Gamma(3) = 2! = 2:$$

EXAMPLE 13.53

Show that the function f defined by

$$f(x) = \begin{cases} \dfrac{3 + 2x}{18}, & 2 \le x \le 4 \\ 0 & \text{otherwise} \end{cases}$$

is a density function. Find mean, variance, standard deviation, and mean deviation from the mean of the distribution.

Solution. The function f is non-negative and

$$\int_{-\infty}^{\infty} f(x)dx = \frac{1}{18}\int_{2}^{4}(3+2x)dx$$

$$= \frac{1}{18}\left|3x + \frac{x^2}{2}\right|_2^4 = \frac{1}{18}(28-10) = 1$$

Hence f is a density function. Also

$$\text{Mean } \mu) = \int_{-\infty}^{\infty} xf(x)dx$$

$$= \frac{1}{18}\int_{2}^{4}(3x+2x^2)dx$$

$$= \frac{1}{18}\left[\frac{3x^2}{2} + 2\frac{x^3}{3}\right]_2^4 = \frac{83}{27}$$

$$\text{Variance } (\sigma^2) = \int_{-\infty}^{\infty}(x-\mu)^2 f(x)dx$$

$$= \frac{1}{18}\int_{2}^{4}\left(x - \frac{83}{27}\right)^2(3+2x)dx$$

$$= \frac{239}{729}.$$

Therefore,

$$\text{Standard deviation} = \sqrt{\sigma^2} = \sqrt{\frac{239}{729}} = 0.57.$$

$$\text{Mean deviation} = \int_{-\infty}^{\infty}|x-\mu| f(x)dx$$

$$= \int_{2}^{4}\left|x - \frac{83}{27}\right|\left(\frac{3+2x}{18}\right)dx$$

$$= \int_{2}^{\frac{83}{27}}\left(\frac{83}{27} - x\right)\left(\frac{3+2x}{18}\right)dx$$

$$+ \int_{\frac{83}{27}}^{4}\left(x - \frac{83}{27}\right)\left(\frac{3+2x}{18}\right)dx = 0.49.$$

EXAMPLE 13.54

Two cards are drawn successively *with replacement* from a well-shuffled pack of 52 playing cards. Find the probability distribution of the number of aces.

Solution. Let X be the random variable that is the number of aces obtained in the draw of two cards. There are three possibilities: (i) there is no ace, (ii) there is one ace, and (iii) there are two aces. Thus, the random variable takes the values 0, 1, 2. Then

$$\text{P(no ace is drawn)} = P(X=0) = \frac{48}{52}\cdot\frac{48}{52} = \frac{144}{169}$$

P(one ace is drawn) = $(X = 1)$

= P(one ace is drawn in the first draw and no ace is drawn in the second draw)

+ P(no ace is drawn in the first draw and one ace is drawn in the second draw)

$$= \frac{4}{52}\cdot\frac{48}{52} + \frac{48}{52}\cdot\frac{4}{52} = \frac{24}{169}$$

P(two aces are drawn)

$$= P(X=2) = \frac{4}{52}\cdot\frac{4}{52} = \frac{1}{169}.$$

Hence the probability distribution is

X:	0	1	2
P(X):	$\frac{144}{169}$	$\frac{24}{169}$	$\frac{1}{169}$

EXAMPLE 13.55

Find the probability distribution of the number of green balls drawn when three balls are drawn one by one *without replacement* from a bag containing three greens and five white balls.

Solution. Let X be the random variable which is the number of green balls drawn when three balls are drawn without replacement. The random variable takes the values 0, 1, 2, 3.

We represent green ball by G and white ball by W. Then we have

P(no green ball is drawn) = P(X = 0)

$= \text{P(WWW)} = \frac{5}{8} \cdot \frac{4}{7} \cdot \frac{3}{6} = \frac{5}{28}$

P(one green ball is drawn) = P(X = 1)

= P(GWW) + P(WGW) + P(WWG)

$= \frac{3}{8} \cdot \frac{5}{7} \cdot \frac{4}{6} + \frac{5}{8} \cdot \frac{3}{7} \cdot \frac{4}{6} + \frac{5}{8} \cdot \frac{4}{7} \cdot \frac{3}{6} = \frac{15}{28}.$

P(two green balls are drawn) = P(X = 2)

= P(GGW) + P(GWG) + P(WGG)

$= \frac{3}{8} \cdot \frac{2}{7} \cdot \frac{5}{6} + \frac{3}{8} \cdot \frac{5}{7} \cdot \frac{2}{6} + \frac{5}{8} \cdot \frac{3}{7} \cdot \frac{2}{6} = \frac{15}{56}.$

P(three green balls are drawn) = P(GGG)

$= \frac{3}{8} \cdot \frac{2}{7} \cdot \frac{1}{6} = \frac{1}{56}.$

Therefore, the probability distribution is

X:	0	1	2	3
P(X):	$\frac{5}{28}$	$\frac{15}{28}$	$\frac{15}{56}$	$\frac{1}{56}$

13.15 BINOMIAL DISTRIBUTION

Let S be a sample space for a random experiment. Let A be an event associated with a subset of S and let P(A) = p, then we know that P($\overline{A}$) = $1-p$. If we denote P($\overline{A}$) = q, then $p + q = 1$.

If we call the occurrence of the event A as "success" and non-occurrence of the event A as a "failure", then

P(failure) = 1 − P(success) and so

P(failure) + P(success) = 1.

Suppose that X is a random variable on the sample space as the "number of success." Then the probability distribution associated with the above random experiment is

X:	0	1
$p(X)$:	q	p

If the experiment is conducted two times, then the possible outcomes are success success, success failure, failure success, and failure failure.

Since the trials are independent, we have

P(success success) = P(both success)

= P(success) P(success)

$= p.p = p^2,$

P(success failure) = P(success) P(failure)= pq,

P(failure success) = P(failure) P(success) = qp,

P(failure failure) = P(failure) P(failure) = q^2.

Thus, in term of random variable, we have

P(X = 0) = P(failure, failure) = q^2,

P(X = 1) = $pq + qp = 2pq$,

P(X = 2) = p(success, success) = p^2.

Also we note that

$P(X = 0) + P(X = 1) + P(X = 2)$

$= p^2 + q^2 + 2pq = (p + q)^2 = (1)^2 = 1.$

Thus, the probability distribution associated with the two experiments is

X:	0	1	2
P(X):	q^2	$2pq$	p^2

The term of P(X) are the terms in the binomial expansion of $(q+p)^2$.

Similarly, the probability distribution associated with the three experiments is

X:	0	1	2	3
$P(X)$:	q^3	$3q^2p$	$3qp^2$	p^3

Thus probabilities are the terms in the binomial expansions of $(q+p)^3$. If the experiment is repeated n times, then the probability distribution is

X:	0	1	2	r........	n
$P(X)$:	q^n	$n_{c_1} q^{n-1} p$	$n_{c_2} q^{n-2} p^2$	$n_{c_r} q^{n-r} p^r$	p^n

Clearly the probabilities are terms in the binomial expansion of $(q+p)^n$.

This probability distribution is called the *binomial distribution* and X is called a *binomial random variable.*

Further, *mean of the binomial distribution* is given by

Mean:

$$\mu = \sum_{r=0}^{n} rP(r) = P(1) + 2P(2) + \ldots\ldots + nP(n)$$

$$= n_{c_1} q^{n-1} p + 2n_{c_2} q^{n-2} p^2 + \ldots\ldots + n.n_{c_n} p^n$$

$$= npq^{n-1} + \frac{2n(n-1)}{2!} p^2 q^{n-2} + \ldots\ldots + np^n$$

$$= np[q^{n-1} + (n-1)pq^{n-2} + \ldots\ldots + p^{n-1}]$$

$$= np[(q+p)^{n-1}] = np, \text{ since } q + p = 1.$$

The *variance of the binomial distribution* is

$$\text{Variance:} \sigma^2 = \sum_{r=0}^{n} r^2 P(r) - \mu^2. \quad (24)$$

Now

$$\sum_{r=0}^{n} r^2 P(r) = \sum_{r=0}^{n} [r + r(r-1)]P(r)$$

$$= \sum_{r=0}^{n} rP(r) + \sum_{r=0}^{n} r(r-1)P(r)$$

$$= \mu + \sum_{r=2}^{n} r(r-1)P(r)$$

$$= np + \sum_{r=2}^{n} r(r-1)P(r) \text{ since } \mu = np.$$

$$= np + \sum_{r=2}^{n} r(r-1)n_{C_r} p^r q^{n-r}$$

$$= np + n(n-1)\, p^2 (q+p)^{n-2}$$

$$= np + n(n-1)p^2, \text{ since } (q+p) = 1$$

Hence, the expression (24) reduces to

$$\sigma^2 = np + n(n-1)p^2 - \mu^2$$

$$= np + n(n-1)p^2 - n^2p^2, \text{ since } \mu = np$$

$$= np + n^2p^2 - np^2 - n^2p^2$$

$$= np(1-p) = npq.$$

Thus the variance of the binomial distribution is

$$\sigma^2 = npq$$

and the standard deviation of the binomial distribution is

$$\sigma = \sqrt{npq}.$$

To derive a recurrence formula for the binomial distribution, we note that

$$P(r) = {}^nC_r q^{n-r} p^r = \frac{n!}{r!(n-r)!} q^{n-r} p^r$$

and so

$$P(r+1) = {}^nC_{r+1} q^{n-(r+1)} p^{r+1}$$

$$= \frac{n!}{(r+1)!(n-r-1)!} q^{n-r-1} p^{r+1}.$$

Then

$$\frac{P(r+1)}{P(r)} = \frac{n-r}{r+1} \cdot \frac{p}{q}.$$

Hence

$$P(r+1) = \frac{n-r}{r+1} \cdot \frac{p}{q} P(r),$$

which is the required recurrence formula. Thus, if $P(0)$ is known, we can determine $P(1)$, $P(2)$, $P(3)$, ...

13.16 PEARSON'S CONSTANTS FOR BINOMIAL DISTRIBUTION

We know that moment generating function about the origin is

$$M_0(t) = \sum p_i e^{t(x_i - 0)} = \sum p_i e^{tx_i}.$$

Thus, for binomial distribution,

$$M_0(t) = \sum_{i=0}^{n} {}^nC_i q^{n-i} p^i e^{ti}$$

$$= {}^nC_i (pe^t)^i q^{n-i} = (q + pe^t)^n. \quad (25)$$

Differentiating with respect to t and then putting $t = 0$, we get

$$\left[\frac{d}{dt}(q + pe^t)^n\right]_{t=0} = [n(q + pe^t)^{n-1} . pe^t]_{t=0}$$

$$= n(q+p)^{n-1} . p$$

$$= np, \text{ since } p + q = 1.$$

Thus the mean $(\mu) = np$. Further

$$M_a(t) = e^{-at} M_0(t)$$

or

$$1 + t\mu_1 + \frac{t^2}{2!}\mu^2 + \frac{t^3}{3!}\mu^3$$

$$+ \frac{t^4}{4!}\mu^4 + \ldots = e^{-at} M_0(t) \quad (26)$$

If we take $a = \mu = np$, then (26) reduces to

$$1+t\mu_1+\frac{t^2}{2!}\mu_2+\frac{t^3}{3!}\mu_3+\frac{t^4}{4!}\mu_4+\ldots = e^{-npt}M_0(t)$$

$$= e^{-npt}(q+pe^t)^n, \text{ using}(25)$$

$$= (qe^{-pt}+pe^{(1-p)t})^n = (qe^{-pt}+pe^{qt})^n$$

$$=\left[1+pq\frac{t^2}{2!}+pq(q^2-p^2)\frac{t^3}{3!}+pq(q^3+p^3)\frac{t^4}{4!}+\ldots\right]^n$$

$$=1+npq\frac{t^2}{2!}+npq(q-p)\frac{t^3}{4!}$$

$$+npq[1+3(n-2)pq]\frac{t^4}{4!}+\ldots$$

Comparing the coefficients of the power of t on both sides, we get

$$\text{Variance } (\mu_2) = npq, \mu_3 = npq(q-p).$$

$$\mu_4 = npq\,[1+3(n-2)pq].$$

Therefore, Pearson's constants for binomial distributions are

$$\beta_1 = \frac{\mu_3^2}{\mu_2^3} = \frac{(q-p)^2}{npq} = \frac{(1-2p)^2}{npq},$$

$$\beta_2 = \frac{\mu_4}{\mu_2^2} = 3+\frac{1-6pq}{npq},$$

$$\gamma_1 = \sqrt{\beta_1} = \frac{q-p}{\sqrt{npq}} = \frac{1-2p}{\sqrt{npq}},$$

$$\gamma_2 = \beta_2 - 3 = \frac{1-6pq}{\sqrt{npq}}.$$

Hence, for a binomial distribution,

Mean $(\mu) = np$,

Variance $(\sigma^2) = npq$,

Standard deviation $(\sigma) = \sqrt{npq}$,

Skewness $\sqrt{\beta_1} = \dfrac{1-2p}{\sqrt{npq}}$,

Kurtosis $(\beta_2) = 3 + \dfrac{1-6pq}{npq}$.

We observe that

(i) skewness of the binomial distribution is 0 for $p = \frac{1}{2}$,

(ii) skewness is positive for $p < \frac{1}{2}$,

(iii) skewness is negative for $p > \frac{1}{2}$.

EXAMPLE 13.56

The incidence of occupational disease in an industry is such that the workers have a 20% chance of suffering from it. What is the probability that out of six workers chosen at random, four or more will suffer from the disease?

Solution. We are given that $p = \frac{20}{100} = \frac{1}{5}$. Therefore, $q = 1 - p = \frac{4}{5}$. Let $P(X > 3)$ denote the probability that out of six workers chosen four or more will suffer from the disease. Then

$$P(X > 3) = P(X = 4) + P(X = 5) + P(X = 6)$$

$$= {}^6C_4 q^{6-4}p^4 + {}^6C_5 q^{6-5}p^5 + {}^6C_6 q^0 p^6$$

$$= {}^6C_4 q^2 p^4 + {}^6q\,p^5 + p^6$$

$$= \frac{15\times 16}{25\times 625} + \frac{6\times 4}{5\times 3125} + \frac{1}{25\times 625}$$

$$= \frac{240+24+1}{25\times 625} = \frac{265}{25\times 625} = \frac{53}{3125}.$$

EXAMPLE 13.57

The probability that a bomb dropped from a plane will strike the target is $\frac{1}{5}$. If six bombs are dropped, find the probability that (i) exactly two will strike the target and (ii) at least two will strike the target.

Solution. The probability to strike the target is $p = \frac{1}{5}$. Therefore, $q = 1 - \frac{1}{p} = \frac{4}{5}$. Then

$$\text{(i) } P(X = 2) = {}^6C_2 q^{6-2}p^2 = {}^6C_2 q^4 p^2$$

$$= \frac{15\times 256\times 1}{625\times 25} = 0.24576$$

$$\text{(ii) } P(X \geq 2) = P(X = 2) + P(X = 3) + P(X = 4) + P(X = 5) + P(X = 6)$$

$$= 1 - [P(X = 0) + P(X = 1)]$$

$$= 1 - [q^6 + 6q^5 p]$$

$$= 1 - \left[\frac{4096}{15625} - \frac{6144}{15625}\right]$$

$$= 0.34478.$$

EXAMPLE 13.58

The probability that a pen manufactured by a company will be defective is $\frac{1}{10}$. If 12 such pens are manufactured, find the probability that

(i) exactly two pens will be defective
(ii) at least two pens will be defective
(iii) none will be defective.

Solution. We have $p = \frac{1}{10}$ and so, $q = 1 - \frac{1}{10} = \frac{9}{10}$. Then since $n = 12$, we have

(i) $P(X = 2) = {}^{12}C_2 q^{10} p^2 = 66(0.1)^2 (0.9)^{10}$
$= 0.2301.$

(ii) $P(X \geq 2) = 1 - [P(X = 0) + P(X = 1)]$
$= 1 - [q^{12} + {}^{12}C_1 q^{11} p]$
$= 1 - [(0.9)^{12} + 12(0.9)^{11}(0.1)]$
$= 0.3412.$

(iii) $P(X = 0) = q^{12} = (0.9)^{12} = 0.2833.$

EXAMPLE 13.59

Out of 800 families with 5 children each, how many families would be expected to have

(i) Three boys and two girls
(ii) Two boys and three girls
(iii) No girl
(iv) At the most two girls, under the assumption that probabilities for boys and girls are equal.

Solution. We have $n = 5$. Further

p = probability to have a boy $= \frac{1}{2}$

q = probability to have a girl $= \frac{1}{2}$:

Then

(i) The expected number of families to have three boys and two girls is

$$800\,[{}^5C_3 q^{5-3} p^3] = 800\left[10\left(\frac{1}{2}\right)^2\left(\frac{1}{2}\right)^3\right]$$
$$= 250.$$

(ii) The expected number of families to have two boys and three girls is

$$800[{}^5C_2 q^{5-2} p^2] = 800\left[10\left(\frac{1}{2}\right)^3\left(\frac{1}{2}\right)^2\right]$$
$$= 250.$$

(iii) The expected number of families to have no girls that is to have five boys is

$$800\,[{}^5C_5 q^0 p^5] = 800\left(\frac{1}{2}\right)^5 = \frac{800}{32} = 25.$$

(iv) The expected number of families to have at the most two girls, that is, at least three boys is

$$800[P(X = 3) + P(X = 4) + P(X = 5)]$$
$$= 800[{}^5C_3 q^{5-3} p^3 + {}^5C_4 q^{5-4} p^4 + {}^5C_5 q^0 p^5]$$
$$= 800\left[10\left(\frac{1}{2}\right)^2\left(\frac{1}{2}\right)^3 + 5\left(\frac{1}{2}\right)\left(\frac{1}{2}\right)^4 + \left(\frac{1}{2}\right)^5\right]$$
$$= 800\left[\frac{10}{32} + \frac{5}{32} + \frac{1}{32}\right] = 400.$$

EXAMPLE 13.60

The following data shows the number of seeds germinating out of 10 on damp filter for 80 sets of seeds. Fit a binomial distribution to this data:

x:	0	1	2	3	4	5	6	7	8	9	10
f:	6	20	28	12	8	6	0	0	0	0	0

Solution. We note that $n = 10$ and $\sum f_i = 6 + 20 + 28 + 12 + 8 + 6 = 80$.
Therefore, the mean of the binomial distribution μ is given by

$$\mu = \frac{\Sigma\ f_i x_i}{\Sigma\ f_i} = \frac{20 + 56 + 36 + 32 + 30}{80} = \frac{174}{80}$$
$$= 2.175.$$

But $\mu = np$. Therefore,

$$p = \frac{\mu}{n} = \frac{2.175}{10} = 0.2175 \text{ and}$$

$$q = 1 - p = 0.7825.$$

Therefore, the probability distribution is

x:	0	1	2	3	......	9	10
$p(x)$:	q^{10}	${}^{10}C_1 q^9 p$	${}^{10}C_2 q^8 p^2$	${}^{10}C_3 q^7 p^3$	......	${}^{10}C_9 q p^9$	p^{10}

Hence the frequencies are given by $f = 80p(x)$. Putting the values of p and q, we get

x:	0	1	2	3	4	5	6	7	8	9	10
f:	6.9	19.1	24.0	17.8	8.6	2.9	0.7	0.1	0	0	0

EXAMPLE 13.61

If the chance that one of the 10 telephone lines is busy at an instant is 0.2, then (i) what is the chance that five of the lines are busy? and (ii) what is the probability, that all lines are busy?

Solution. Here $n = 10$, $p = 0.2$, and so $q = 1 - 0.2 = 0.8$. Then

(i) Probability of five lines to be busy is

$$P(X = 5) = {}^{10}C_5 q^{10-5} p^5$$

$$= {}^{10}C_5 q^5 p^5 = 252(0.8)^5(0.2)^5$$

$$= 252(0.32768)(0.00032) = 0.0264.$$

(ii) Probability that all the lines are busy is

$$P(X = 10) = {}^{10}C_{10} p^{10} = (0.2)^{10}$$

$$= 1024 \times 10^{-10}.$$

EXAMPLE 13.62

In sampling a large number of parts manufactured by a machine, the mean number of defective parts in a sample of 20 is 2. Out of 1000 such samples, how many would be expected to contain at least 3 defective parts.

Solution. We are given that $n = 20$ and $m = np = 2$ and so

$$p = \frac{2}{n} = \frac{2}{20} = \frac{1}{10}, q = 1 - \frac{1}{10} = \frac{9}{10}.$$

Then

$$P(X > 2) = 1 - [P(X = 0) + P(X = 1) + P(X = 2)]$$

$$= 1 - \left[{}^{20}C_0 q^{20} + {}^{20}C_1 q^{19} p + {}^{20}C_2 q^{18} p^2\right]$$

$$= 1 - [(0.9)^{20} + 20(0.9)^{19}(0.1)]$$

$$+ 190(0.9)^{18}\ (0.9)^2 = 0.323.$$

Hence the number of sample having atleast three defective parts out of the 1000 samples is $1000 \times 0.323 = 323$.

EXAMPLE 13.63

Fit a binomial distribution to the following data and compare the theoretical frequencies with the actual ones

x:	0	1	2	3	4	5
f:	2	14	20	34	22	8

Solution. We have $n = 5$, $\Sigma f_i = 100$. Therefore,

$$\mu = \frac{\Sigma f_i x_i}{\Sigma f_i} = \frac{14 + 40 + 102 + 88 + 40}{100} = 2.84.$$

But for binomial distribution, $\mu = np$. Therefore, $p = \frac{\mu}{n} = \frac{2.84}{5} = 0.568$ and $q = 1 - p = 0.432$.
Therefore, the probability distribution is

x:	0	1	2	3	4	5
P(x):	q^5	${}^5C_1 q^4 p$	${}^5C_2 q^3 p^2$	${}^5C_3 q^2 p^3$	${}^5C_4 q p^4$	p^5

Therefore the expected (theoretical) frequencies are

$100(0.432)^5$, $500(0.432)^4(0.568)$, $10^3(0.432)^3\ (0.568)^2$, $10^3(0.432)^2(0.568)^3$, $500(0.432)(0.568)^4$, $100(0.568)^5$.

After computation, we get the theoretical frequencies as

1.504, 9.891, 26.010, 34.199, 22.483, 5.918.

EXAMPLE 13.64

Find the probability of number 4 turning up at least once in *two* tosses of a fair dice.

Solution. Let X denote the number of times the number 4 turn up. We note that

P(4 turns up) $= p = \frac{1}{6}$ and so $q = 1 - p = 1 - \frac{1}{6} = \frac{5}{6}$.

Thus the probability distribution is

X:	0	1	2
P(X):	q^2	$2pq$	p^2

Hence

P(4 turns up at least once)

$$= P(X = 1) + P(X = 2) = 2pq + p^2$$

$$= 2.\frac{1}{6}.\frac{5}{6} + \left(\frac{1}{6}\right)^2 = \frac{11}{36}.$$

EXAMPLE 13.65

A coin is tossed five times. What is the probability of getting at least three heads?

Solution. Let X denote the "number of heads obtained". We know that

$$p = \text{P(head obtained)} = \frac{1}{2}.$$

Therefore,

$$q = 1 - p = 1 - \frac{1}{2} = \frac{1}{2}.$$

The random variable X takes the values 0, 1, 2, 3, 4, 5, and $n = 5$. Hence

$$\text{P(at least three heads)} = \text{P}(X \geq 3)$$

$$= \text{P}(X = 3) + \text{P}(X = 4) + \text{P}(X = 5)$$

$$= {}^5C_3 p^3 q^2 + {}^5C_4 p^4 q + {}^5C_5 p^5$$

$$= 10\left(\frac{1}{2}\right)^3\left(\frac{1}{2}\right)^2 + 5\left(\frac{1}{2}\right)^4\left(\frac{1}{2}\right) + \left(\frac{1}{2}\right)^5$$

$$= \frac{10}{32} + \frac{5}{32} + \frac{1}{32} = \frac{1}{2}.$$

EXAMPLE 13.66

The mean and variance of a binomial variable X are 2 and 1, respectively. Find the probability that X takes a value greater than 1.

Solution. Suppose n is the number of independent trials. Since X is a binomial variate, we have

$$\text{Mean} = np = 2 \text{ (given)} \quad (25)$$

$$\text{Variance} = np\,q = 1 \text{ (given)} \quad (26)$$

Dividing (26) by (25), we get $q = \frac{1}{2}$, which yields $p = 1 - q = \frac{1}{2}$. Also then (25) gives $n = 4$. Hence

$$\text{P}(X > 1) = 1 - [\text{P}(X = 0) + \text{P}(X = 1)]$$

$$= 1 - \left[{}^4C_0 q^4 + {}^4C_1 q^3 p\right]$$

$$= 1 - [q^4 - 4p\,q^3] = 1 - \left[\left(\frac{1}{2}\right)^4 - 4\left(\frac{1}{2}\right)\left(\frac{1}{2}\right)^3\right]$$

$$= 1 - \frac{5}{16} = \frac{11}{16}.$$

13.17 POISSON DISTRIBUTION

The Poisson distribution is a limiting case of binomial distribution when n is very large and p is very small in such a way that mean np remains constant. To derive Poisson distribution, we assume that when n is large and p is very small, then $np = \lambda$ (constant). In the binomial distribution,the probability of r successes is given by

$$P(r) = {}^nC_r q^{n-r} p^r$$

$$= \frac{n(n-1)(n-2)\ldots(n-r+1)}{r!}(1-p)^{n-r}p^r$$

$$= \frac{n(n-1)(n-2)\ldots(n-r+1)}{r!}\left(1-\frac{\lambda}{n}\right)^{n-r}\left(\frac{\lambda}{n}\right)^r$$

$$= \frac{\lambda^r}{r!}\cdot\frac{n(n-1)(n-2)\ldots(n-r+1)}{n^r}\frac{\left(1-\frac{\lambda}{n}\right)^n}{\left(1-\frac{\lambda}{n}\right)^r}$$

$$= \frac{\lambda^r}{r!}\left(1-\frac{1}{n}\right)\left(1-\frac{2}{n}\right)\cdots\left(1-\frac{r-1}{n}\right)\frac{\left[\left(1-\frac{\lambda}{n}\right)^{-\frac{n}{\lambda}}\right]^{-\lambda}}{\left(1-\frac{\lambda}{n}\right)^r}.$$

Therefore, the Poisson distribution is given by

$$\lim_{n\to\infty} P(r) = \frac{\lambda^r}{r!}e^{-\lambda}\ (r = 0, 1, 2, 3, \ldots),$$

where $\lambda = np$ is called *parameter of the Poisson distribution.* Thus, the probabilities of 0, 1, 2,..., r,... of successes in a Poisson distribution are

$$e^{-\lambda}, \lambda e^{-\lambda}, \frac{\lambda^2}{2!}e^{-\lambda}, \ldots, \frac{\lambda^r}{r!}e^{-\lambda}, \ldots$$

The sum of the probabilities $P(r)$, $r = 0, 1, 2,\ldots$ is

$$e^{-\lambda} + \lambda e^{-\lambda} + \frac{\lambda^2}{2!}e^{-\lambda} + \ldots + \frac{\lambda^r}{r!}e^{-\lambda} + \ldots$$

$$= e^{-\lambda}\left(1 + \lambda + \frac{\lambda^2}{2!} + \ldots + \frac{\lambda^r}{r!} + \ldots\right) = e^{-\lambda}.e^{\lambda} = 1.$$

Further, for the Poisson distribution,

$$\frac{P(r+1)}{P(r)} = \frac{\lambda^{r+1}e^{-\lambda}}{(r+1)!}\cdot\frac{r!}{\lambda^r e^{-\lambda}} = \frac{\lambda}{r+1}$$

and so

$$P(r+1) = \frac{\lambda}{r+1}P(r),$$

which is the *recurrence formula* for the Poisson distribution. Some examples of Poisson distribution are

(i) The number of defective screws per box of 100 screws
(ii) The number of fragments from a shell hitting a target
(iii) Number of typographical error per page in typed material
(iv) Mortality rate per thousand.

13.18 CONSTANTS OF THE POISSON DISTRIBUTION

The constants of the Poisson distribution can be derived from the corresponding constants of the binomial distribution by letting $n \to \infty$ and $p \to 0$. Since $q = 1 - p$, $p \to 0$ if $q \to 1$. Therefore, mean (μ), variation (σ^2), standard deviation σ, skewness ($\sqrt{\beta_1}$),and kurtosis (β_2) are given by

$$\mu = \lim_{\substack{n \to \infty \\ p \to 0}} np = \lambda, \text{ since } np = \lambda \text{ (constant)},$$

$$\sigma^2 = \mu_2 = \lim_{\substack{n \to \infty \\ p \to 0}} npq = \lim_{q \to 1} \lambda q = \lambda,$$

$$\sigma = \sqrt{\lambda},\ \mu_3 = \lambda,\ \mu_4 = 3\lambda^2 + \lambda,$$

$$\text{Skewness } \left(\sqrt{\beta_1}\right) = \sqrt{\frac{\mu_3^2}{\mu_2^3}} = \sqrt{\frac{\lambda^2}{\lambda^3}} = \sqrt{\frac{1}{\lambda}}$$

$$\text{Kurtosis } (\beta_2) = \frac{\mu_4}{\mu_2^2} = 3 + \frac{1}{\lambda}.$$

EXAMPLE 13.67

A certain screw making machine produces an average of 2 defective screws out of 100 and pack them in boxes of 500. Find the probability that a box contains 15 defective screws.

Solution. The probability of occurrence is $\frac{2}{100}$ = 0.02 and, therefore, it follows a Poisson distribution. Now $n = 500$, and $p = 0.02$. Therefore,

$$\lambda = \text{mean} = np = 10.$$

Now the probability that a box contains 15 defective screws is

$$\frac{\lambda^{15}}{15!} e^{-\lambda} = \frac{10^{15}}{15!} e^{-10} = 0.035.$$

EXAMPLE 13.68

A book of 520 pages has 390 typographical errors. Assuming Poisson law for the number of errors per page, find the probability that a random sample of five pages will contain no error.

Solution. The average number of typographical error per page is given by

$$\lambda = \frac{390}{520} = 0.75.$$

Therefore, probability of zero error per page is

$$P(X = 0) = e^{-\lambda} = e^{-0.75}.$$

Hence, required probability that a random sample of five pages contains no error is

$$[P(X = 0)]^5 = (e^{-0.75})^5 = e^{-3.75}.$$

EXAMPLE 13.69

Fit a Poisson distribution to the following:

x:	0	1	2	3	4
y:	46	38	22	9	1

Solution. The mean of the Poisson distribution is

$$\lambda = \frac{\sum f_i x_i}{\sum f_i} = \frac{0 + 38 + 44 + 27 + 4}{116} = \frac{113}{116}$$
$$= 0.974.$$

Therefore, frequencies are

$$116e^{-\lambda}, 116\lambda e^{-\lambda}, 116\frac{\lambda^2}{2}e^{-\lambda}, 116\frac{\lambda^3}{3!}e^{-\lambda}, 116\frac{\lambda^4}{4!}e^{-\lambda}.$$

Since $e^{-\lambda} = e^{-0.974} = 0.3776$, the required Poisson distribution is

x:	0	1	2	3	4
y:	44	43	21	7	1

EXAMPLE 13.70

An insurance company insures 6,000 people against death by tuberculosis (TB). Based on the previous data, the rates were computed on the assumption that 5 persons in 10,000 die due to TB each year. What is the probability that more than two of the insured policy will get refund in a given year?

Solution. Here $n = 6000$ is large and the probability p of death due to TB is $\frac{5}{10000} = 0.0005$ (small). Therefore, the data follows Poisson distribution. The parameter of the distribution is

$$\lambda = np = 6000 \times 0.0005 = 3.0.$$

The required probability that more than two of the insured policies will get refunded is

$$P(X>2) = 1 - [P(X=0) + P(X=1) + P(X=2)]$$

$$= 1 - \left[e^{-\lambda} + \lambda e^{-\lambda} + \frac{\lambda^2}{2!}e^{-\lambda}\right] = 1 - e^{-\lambda}\left[1 + \lambda + \frac{\lambda^2}{2}\right]$$

$$= 1 - 0.04979\left[1 + 3 + \frac{9}{2}\right], \text{ since } e^{-\lambda} = 0.4979$$

$$= 1 - 0.4232 = 0.5768.$$

EXAMPLE 13.71

Fit a Poisson distribution to the following data

x:	0	1	2	3	4
f:	122	60	15	2	1

Solution. If the above distribution is approximated by a Poisson distribution, then the parameter of the Poisson distribution is given by

$$\lambda = \text{Mean} = \frac{\Sigma f_i x_i}{\Sigma f_i} = \frac{0 + 60 + 30 + 6 + 4}{200}$$

$$= 0.5.$$

$$Ne^{-\lambda}, N\lambda e^{-\lambda}, N\frac{\lambda^2}{2!}e^{-\lambda}, N\frac{\lambda^3}{3!}e^{-\lambda}, N\frac{\lambda^4}{4!}e^{-\lambda},$$

where $N = 200$. Also $e^{-\lambda} = e^{-0.5} = 0.6065$.

Therefore, the required Poisson distribution is

x:	0	1	2	3	4
f:	121	61	15	2	0

EXAMPLE 13.72

A car hire firm has two cars which it hires out day by day. The number of demands for a car on each day is distributed as a Poisson distribution with mean 1.5. Calculate the proportion of days (i) on which there is no demand and (ii) on which demand is refused:

Solution. We are given that $\lambda = 1.5$. When there is no demand, the probability is

$$e^{-\lambda} = e^{-1.5} = 0.2231.$$

When demand is refused, then the probability of number of demands exceeds 2. Therefore, the probability for this event is

$$1 - [P(X=0) + P(X=1) + P(X=2)]$$

$$= 1 - \left[e^{-\lambda} + \lambda e^{-\lambda} + \frac{\lambda^2}{2!}e^{-\lambda}\right]$$

$$= 1 - e^{-\lambda}\left[1 + \lambda + \frac{\lambda^2}{2!}\right]$$

$$1 - 0.2231\left[1 + 1.5 + \frac{(1.5)^2}{2}\right]$$

$$= 1 - 0.80874 = 0.19126.$$

EXAMPLE 13.73

The mortality rate for a certain disease is 6 per 1000. What is the probability for just four deaths from that disease in a group of 400?

Solution. The parameter of the Poisson distribution is given by

$$\lambda = np = 400p.$$

But

$$p = \frac{6}{1000} = 0.0006.$$

Therefore,

$$\lambda = 400 \times 0.006 = 2.4$$

and so

$$P(X=4) = \frac{\lambda^4}{4!}e^{-\lambda} = \frac{(2.4)^4}{4!}e^{-2.4}$$

$$= \frac{(2.4)^4}{4!}(0.09072) = 0.1254.$$

EXAMPLE 13.74

Find the probability that at most 5 defective diodes will be found in a pack of 600 diodes if previous data shows that 1% of such diodes are defective.

Solution. Here $n = 600, p = 0.01$. Therefore, parameter of Poisson distribution is

$$\lambda = np = 600(0.01) = 6.$$

Therefore,

$$P(X \le 5) = e^{-\lambda} + \lambda e^{-\lambda} + \frac{\lambda^2}{2!}e^{-\lambda} + \frac{\lambda^3}{3!}e^{-\lambda} + \frac{\lambda^4}{4!}e^{-\lambda} + \frac{\lambda^5}{5!}e^{-\lambda}$$

$$= e^{-\lambda}\left[1 + \lambda + \frac{\lambda^2}{2!} + \frac{\lambda^3}{3!} + \frac{\lambda^4}{4!} + \frac{\lambda^5}{5!}\right]$$

$$= e^{-6}\left[1 + 6 + \frac{6^2}{2} + \frac{6^3}{6} + \frac{6^4}{24} + \frac{6^5}{120}\right]$$

$$= 0.00248[179.8] = 0.4459.$$

13.19 NORMAL DISTRIBUTION

The normal distribution is a continuous distribution, which can be regarded as the limiting form of the Binomial distribution when n, the number of trials, is very large but neither p nor q is very small. The limit approach more rapidly if p and q are nearly equal, that is, if p and q are close to $\frac{1}{2}$. In fact, using Stirling's formula, the following theorem can be proved:

A binomial probability density function

$$P(x) = {}^nC_x q^{n-x} p^x,$$

in which n becomes infinitely large, approaches as a limit to the so-called normal probability density function

$$f(x) = \frac{1}{\sqrt{2\pi npq}}\, e^{-\frac{(x-np)^2}{2npq}}.$$

Since for a Binomial distribution, the mean and standard deviations are given by

$$\mu = np \text{ and } \sigma = \sqrt{npq},$$

the normal frequency function becomes

$$f(x) = \frac{1}{\sigma\sqrt{2\pi}}\, e^{-\frac{1}{2}\left(\frac{x-\mu}{\sigma}\right)^2},$$

where the variable x can assume all values from $-\infty$ to ∞.

The graph of the normal frequency function is called the *normal curve*. The normal curve is bell-shaped and is symmetrical about the mean μ. This curve is unimodal and its mode coincide with its mean μ. The two tails of the curve extend to $+\infty$ and $-\infty$, respectively, towards the positive and negative directions of x-axis, approaching the x-axis without ever meeting it. Thus the curve is asymptotic to the x-axis. Since the curve is symmetrical about $x = \mu$, its mean, median, and mode are the same. Its points of inflexion are found to be $x = \mu \pm \sigma$, that is, the points are equidistant from the mean on either side. As we shall prove, the total area under the normal curve above the x-axis is unity. Thus the graph of the normal frequency curve is as shown in the Figure 13.1.

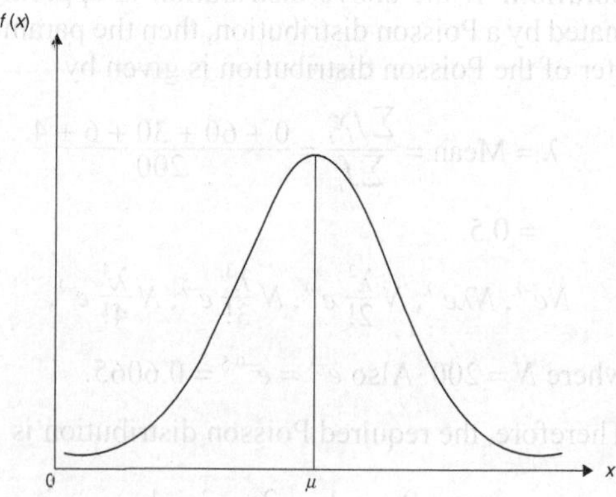

Figure 13.1

The parameters μ and σ determines the position and relative proportions of the normal curve. If two populations defined by normal frequency functions have different means μ_1 and μ_2 but identical standard deviations $\sigma_1 = \sigma_2$, then their graphs appear as shown in the Figure 13.2.

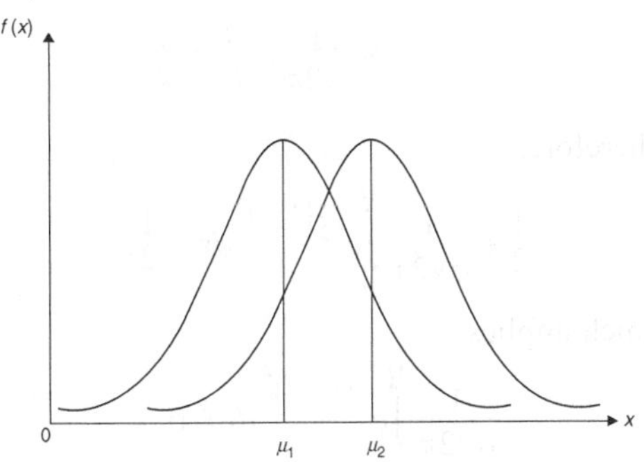

Figure 13.2

On the other hand, if the two populations have identical means $\mu_1 = \mu_2$ and different standard deviations σ_1 and σ_2, then their graphs would appear as shown in the Figure 13.3

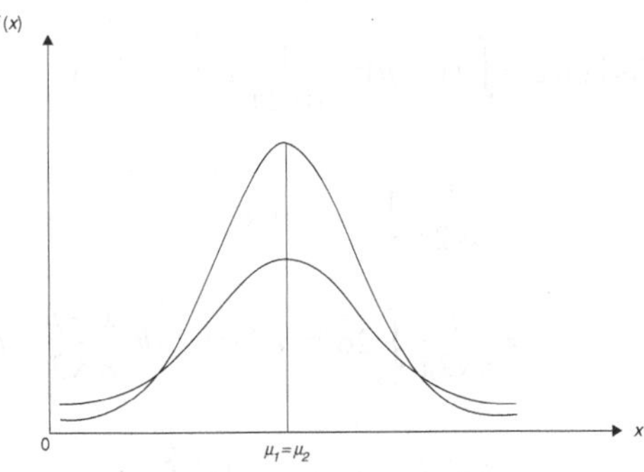

Figure 13.3

13.20 CHARACTERISTICS OF THE NORMAL DISTRIBUTION

The normal distribution has the following properties:

1. **Normal distribution is a continuous distribution:** The probability density function of the normal distribution is

$$f(x) = \frac{1}{\sigma\sqrt{2\pi}} e^{-\frac{1}{2}\left(\frac{x-\mu}{\sigma}\right)^2}$$

Therefore, area under the normal curve is equal to

$$\int_{-\infty}^{\infty} f(x)dx = \frac{1}{\sigma\sqrt{2\pi}} \int_{-\infty}^{\infty} e^{-\frac{1}{2}\left(\frac{x-\mu}{\sigma}\right)^2} dx.$$

Putting $\frac{x-\mu}{\sigma\sqrt{2}} = t$, we have $dx = \sigma\sqrt{2}\, dt$ and so

$$\int_{-\infty}^{\infty} f(x)dx = \frac{1}{\sigma\sqrt{2\pi}} \int_{-\infty}^{\infty} e^{-t^2}\sigma\sqrt{2}\, dt = \frac{1}{\sqrt{\pi}}\int_{-\infty}^{\infty} e^{-t^2} dt$$

$$= \frac{1}{\sqrt{\pi}}\sqrt{\pi} = 1.$$

Thus $f(x) \geq 0$ and $\int_{-\infty}^{\infty} f(x)\, dx = 1$. Hence f is a continuous distribution.

2. **Mean, mode, and median of the normal distribution coincide. Hence the distribution is symmetrical**

(i) **Mean:** *The general form of the normal* curve is

$$y = f(x) = \frac{N}{\sigma\sqrt{2}} e^{-\frac{1}{2}\left(\frac{x-\mu}{\sigma}\right)^2}$$

Therefore,

$$\text{Mean} = \frac{1}{N}\int_{-\infty}^{\infty} yx dx = \frac{1}{\sigma\sqrt{2\pi}} \int_{-\infty}^{\infty} x\, e^{-\frac{1}{2}\left(\frac{x-\mu}{\sigma}\right)^2} dx$$

Putting $\frac{x-\mu}{\sigma\sqrt{2}} = t$, so that $x = \mu + t\sigma\sqrt{2}$ and $dx = \sigma\sqrt{2}\, dt$. Hence

$$\text{Mean} = \frac{1}{\sigma\sqrt{2\pi}} \int_{-\infty}^{\infty} (\mu + t\sigma\sqrt{2})\, e^{-t^2} .\sigma\sqrt{2}\, dt$$

$$= \frac{1}{\sqrt{\pi}} \int_{-\infty}^{\infty} (\mu + t\sigma\sqrt{2})\, e^{-t^2} dt$$

$$= \frac{\mu}{\sqrt{\pi}} \int_{-\infty}^{\infty} e^{-t^2} + \frac{\sigma\sqrt{2}}{\sqrt{\pi}} \int_{-\infty}^{\infty} te^{-t^2} dt$$

$$= \frac{\mu}{\sqrt{\pi}}\sqrt{\pi} = \mu,$$

because $\int_{-\infty}^{\infty} e^{-t^2}\, dt = \sqrt{\pi}$ and $\int_{-\infty}^{\infty} te^{-t^2}\, dt = 0$ due to oddness of te^{-t^2}.

(ii) Mode: Mode is the value of x for which f is maximum. In other words, mode is the solution of $f'(x) = 0$ and $f''(x) < 0$.

For normal distribution, we have

$$f(x) = \frac{1}{\sigma\sqrt{2\pi}} e^{-\frac{1}{2}\left(\frac{x-\mu}{\sigma}\right)^2}$$

Taking log, we get

$$\log f(x) = \log\left(\frac{1}{\sigma\sqrt{2\pi}}\right) - \frac{1}{2\sigma^2}(x-\mu)^2.$$

Differentiating with respect to x, we get

$$\frac{f'(x)}{f(x)} = -\frac{1}{\sigma^2}(x-\mu)$$

and so $f'(x) = \frac{-1}{\sigma^2}(x-\mu) f(x)$. Then

$$f''(x) = -\frac{1}{\sigma^2}[f(x) + (x-\mu) f'(x)]$$

$$= -\frac{f(x)}{\sigma^2}\left[1 + \frac{(x-\mu)}{\sigma^2}\right]$$

New $f'(x) = 0$ implies $x = \mu$. Also at $x = \mu$, we have

$$f''(\mu) = -\frac{1}{\sigma^2} \cdot \frac{1}{\sigma\sqrt{2\pi}} < 0.$$

Hence $x = \mu$ *is mode of the normal distribution.*

(iii) Median: We know that

$$\int_{-\infty}^{\infty} f(x)dx = 1.$$

Therefore, if M is the median of the normal distribution, we must have

$$\int_{-\infty}^{M} f(x)dx = \frac{1}{2}.$$

Therefore,

$$\frac{1}{\sigma\sqrt{2\pi}} \int_{-\infty}^{M} e^{-\frac{1}{2}\left(\frac{x-\mu}{\sigma}\right)^2} dx = \frac{1}{2}$$

or

$$\frac{1}{\sigma\sqrt{2\pi}} \int_{-\infty}^{\mu} e^{-\frac{1}{2}\left(\frac{x-\mu}{\sigma}\right)^2} dx + \frac{1}{\sigma\sqrt{2\pi}} \int_{\mu}^{M} e^{-\frac{1}{2}\left(\frac{x-\mu}{\sigma}\right)^2} dx = \frac{1}{2}.$$

But

$$\frac{1}{\sigma\sqrt{2\pi}} \int_{-\infty}^{\mu} e^{-\frac{1}{2}\left(\frac{x-\mu}{\sigma}\right)^2} dx = \frac{1}{\sqrt{2\pi}} \int_{-\infty}^{0} e^{-\frac{t^2}{2}} dx$$

$$= \frac{1}{\sqrt{2\pi}} \cdot \frac{\sqrt{\pi}}{\sqrt{2}} = \frac{1}{2}.$$

Therefore,

$$\frac{1}{2} + \frac{1}{\sigma\sqrt{2\pi}} \int_{\mu}^{M} e^{-\frac{1}{2}\left(\frac{x-\mu}{\sigma}\right)^2} dx = \frac{1}{2},$$

which implies

$$\frac{1}{\sigma\sqrt{2\pi}} \int_{\mu}^{M} e^{-\frac{1}{2}\left(\frac{x-\mu}{\sigma}\right)^2} dx = 0.$$

Consequently, $M = \mu$.

Thus, mean, mode, and median coincide for the normal distribution. Hence the normal curve is symmetrical.

3. The variance of the normal distribution is σ^2 and so the standard deviation is σ: In fact, we have

$$\text{Variance} = \int_{-\infty}^{\infty} (x-\mu)^2 \cdot \frac{1}{\sigma\sqrt{2\pi}} e^{-\frac{1}{2}\left(\frac{x-\mu}{\sigma}\right)^2} dx$$

$$= \frac{1}{\sigma\sqrt{2\pi}} \int_{-\infty}^{\infty} (x-\mu)^2 e^{-\frac{1}{2}\left(\frac{x-\mu}{\sigma}\right)^2} dx$$

$$= \frac{1}{\sigma\sqrt{2\pi}} \int_{-\infty}^{\infty} 2\sigma^2 t^2 . e^{-t^2} \sigma\sqrt{2}\, dt, \quad \frac{x-\mu}{\sigma\sqrt{2}} = t,$$

$$= \frac{2\sigma^2}{\sqrt{2}} \int_{-\infty}^{\infty} t^2 e^{-t^2} dt = \frac{4\sigma^2}{\sqrt{\pi}} \int_{0}^{\infty} t^2 e^{-t^2} dt$$

$$= \frac{4\sigma^2}{\sqrt{\pi}} \int_{0}^{\infty} z e^{-z} \frac{dz}{2\sqrt{z}}, \quad t^2 = z$$

$$= \frac{2\sigma^2}{\sqrt{\pi}} \int_{0}^{\infty} z^{\frac{1}{2}} e^{-z} dz = \frac{2\sigma^2}{\sqrt{\pi}} \int_{0}^{\infty} z^{\left(\frac{3}{2}-1\right)} e^{-z} dz$$

$$= \frac{2\sigma^2}{\sqrt{\pi}} \Gamma\left(\frac{3}{2}\right)$$

$$= \frac{2\sigma^2}{\sqrt{\pi}} \cdot \frac{1}{2}\sqrt{\pi} = \sigma^2.$$

Therefore,

Standard deviation = $\sqrt{\text{variance}} = \sigma$.

4. Points of inflexion of normal curve: At the point of inflexion of the normal curve, we should have $f''(x) = 0$ and $f'''(x) \neq 0$. As we have seen, for normal distribution

$$f''(x) = -\frac{f(x)}{\sigma^2}\left[1 - \frac{(x-\mu)^2}{\sigma^2}\right].$$

Therefore, $f''(x) = 0$ yields

$$1 - \frac{(x-\mu)^2}{\sigma^2} = 0 \quad \text{or} \quad (x-\mu)^2 = \sigma^2.$$

Hence $x = \mu \pm \sigma$. Further, at $x = \mu \pm \sigma$, we have $f'''(x) \neq 0$. Thus the normal curve has two points of inflexion given by $x = \mu - \sigma$ and $x = \mu + \sigma$. Clearly, the points of inflexions are equidistant (at a distance σ) from the mean.

5. Mean deviation about the mean: The mean deviation from the mean is given by

Mean deviation (about mean)

$$= \int_{-\infty}^{\infty} |x-\mu| f(x)\, dx$$

$$= \frac{1}{\sigma\sqrt{2\pi}} \int_{-\infty}^{\infty} |x-\mu|\, e^{-\frac{1}{2}\left(\frac{x-\mu}{\sigma}\right)^2} dx$$

$$= \frac{\sigma}{\sqrt{2\pi}} \int_{-\infty}^{\infty} |t|\, e^{\frac{-t^2}{2}} dt, \left(\frac{x-\mu}{\sigma}\right) = t$$

$$= \frac{2\sigma}{\sqrt{2\pi}} \int_0^{\infty} |t|\, e^{\frac{-t_2}{2}} dt,$$

since the integral is an even function of z. Since $|t| = t$ for $t \in [0, \infty]$, we have

$$\text{Mean deviation} = \sqrt{\frac{2}{\pi}}\, \sigma \int_0^{\infty} t\, e^{\frac{-t^2}{2}} dt$$

$$= \sqrt{\frac{2}{\pi}}\sigma \int_0^{\infty} t\, e^{-z} dt, \frac{t^2}{2} = z$$

$$= \sqrt{\frac{2}{\pi}}\sigma \left[\frac{e^{-z}}{-1}\right]_0^{\infty} = \sqrt{\frac{2}{\pi}}\sigma$$

$$= \frac{4}{5}\sigma, \text{ approximately.}$$

Thus for the normal distribution, the mean deviation is approximately $\frac{4}{5}$ times the standard deviation.

13.21 NORMAL PROBABILITY INTEGRAL

If X is a normal random variable with mean μ and variance σ^2, then the probability that random value of X will lie between $X = \mu$ and $X = x_1$ is given by

$$P(\mu < X < x_1)$$

$$= \int_{\mu}^{x_1} f(x)\, dx = \frac{1}{\sigma\sqrt{2\pi}} \int_{\mu}^{x_1} e^{-\frac{1}{2}\left(\frac{x-\mu}{\sigma}\right)^2} dx$$

Put $\frac{X-\mu}{\sigma} = Z$. Then $X - \mu = \sigma Z$. Therefore, when $X = \mu$, we have $Z = 0$ and when $X = x_1$, $Z = \frac{x_1 - \mu}{\sigma} = z_1$, say. Therefore,

$P(\mu < X < x_1) = P(0 < Z < z_1)$

$$= \frac{1}{\sqrt{2\pi}} \int_0^{z_1} e^{-\frac{z^2}{2}} dz = \int_0^{z_1} \phi(z) dz,$$

where $\phi(z) = \frac{1}{2\sqrt{\pi}} e^{-\frac{z^2}{2}}$ is the probability function of the *standard normal variate* $Z = \frac{X-\mu}{\sigma}$.

The definite integral $\int_0^{z_1} \phi(z) dz$ is called the *normal probability integral* which gives the area under the standard normal curve (Figure 13.4) between the ordinates at $z = 0$ and $z = z_1$.

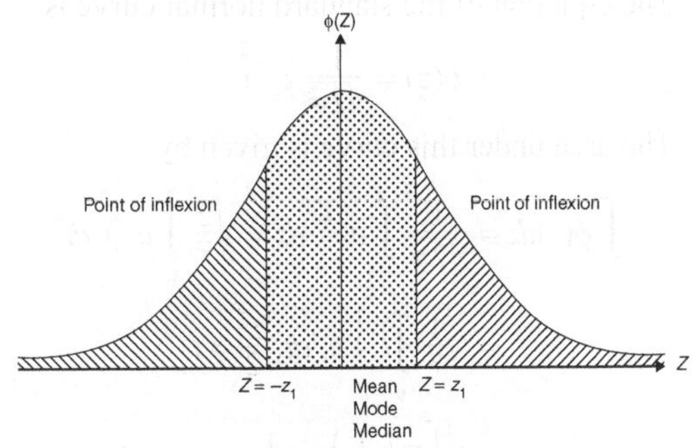

Figure 13.4

The standard normal curve

$$\phi(z) = \frac{1}{\sqrt{2\neq}} e^{-\frac{z^2}{2}}$$

is symmetrical with respect to $\phi(z)$ – axis since $\phi(z)$ remain unchanged if z is replaced by $-z$. Thus arithmetic mean and the median of a normal frequency distribution coincide at the centre of it. The exponent of e in $\phi(z)$ is negative, $-\frac{z^2}{2}$. Hence $\phi(z)$ is maximum when $z = 0$. All other values of z make $\phi(z)$ smaller since $e^{-\frac{z^2}{2}} = \frac{1}{e^{\frac{z^2}{2}}}$.

Thus the maximum value of $\phi(z)$ is

$$\phi(0) = \frac{1}{\sqrt{2\pi}} = 0.3989.$$

As z increases numerically, $e^{-\frac{z^2}{2}}$ decreases and approaches zero when z becomes infinite. Thus the standard normal curve is asymptotic to the z-axis in both the positive and negative directions. Differentiating $\phi(z)$ with respect to z, we get

$$\phi'(z) = z\phi(z), \text{ and}$$

$$\phi''(z) = -\phi(z) - z\phi'(z) = -\phi(z) + z^2\phi(z)$$

$$= (z^2 - 1)\phi(z).$$

Therefore, $\phi''(z) = 0$ implies $z = \pm 1$. Thus the points of inflexion (at which the curve changes from concave downward to concave upward) are situated at a unit distance from the $\phi(z)$– axis.

13.22 AREAS UNDER THE STANDARD NORMAL CURVE

The equation of the standard normal curve is

$$\phi(z) = \frac{1}{\sqrt{2\neq}} e^{-\frac{z^2}{2}}.$$

The area under this curve is given by

$$\int_{-\infty}^{\infty} \phi(z)dz = \frac{1}{\sqrt{2\neq}} \int_{-\infty}^{\infty} e^{-\frac{z^2}{2}} dz = \sqrt{\frac{2}{\neq}} \int_{0}^{\infty} e^{-\frac{z^2}{2}} dz$$

$$= \sqrt{\frac{1}{\neq}} \int_{0}^{\infty} t^{-\frac{1}{2}} e^{-t} dt, \quad \frac{z^2}{2} = t$$

$$= \sqrt{\frac{1}{\pi}} \Gamma\left(\frac{1}{2}\right) = \frac{1}{\sqrt{\neq}} \cdot \sqrt{\neq} = 1.$$

If follows, therefore, that the *area under* $\phi(z)$ from $z = z_1$ to $z = z_2$, that is, $\int_{z_1}^{z_2} \phi(z)dz$ is *always less than* 1, where z_1 and z_2 are finite.

Further, because of the symmetry of the curve with respect to $\phi(z)$-axis, the area from any $z = z_1$ to $+^\circ$ is equal to the area from $-^\circ$ to $-z_1$. Thus

$$\int_{z_1}^{\infty} \phi(z)dz = \int_{-\infty}^{-z_1} \phi(z)dz.$$

Since the area under the curve from $z = 1$ to $z = ^\circ$ is 0.1587, we have

$$\int_{-1}^{1} \phi(z)dz = \int_{-\infty}^{\infty} \phi(z)dz - \int_{-\infty}^{-1} \phi(z)dz - \int_{1}^{\circ} \phi(z)dz$$

$$= \int_{-^\circ}^{^\circ} \phi(z)dz - 2\int_{1}^{^\circ} \phi(z)dz$$

$$= 1 - 2(0.1587) = 0.6826.$$

In term of statistics, this means that 68% of the normal variates deviate from their mean by less than one standard deviation. Similarly,

$$\int_{-2}^{2} \phi(z)dz = 0.9544, \qquad \int_{-3}^{3} \phi(z)dz = 0.9974.$$

Thus, over 95% of the area is included between the limits –2 and 2 and over 99% of the area is included between –3 and 3 as shown in the Figure 13.5.

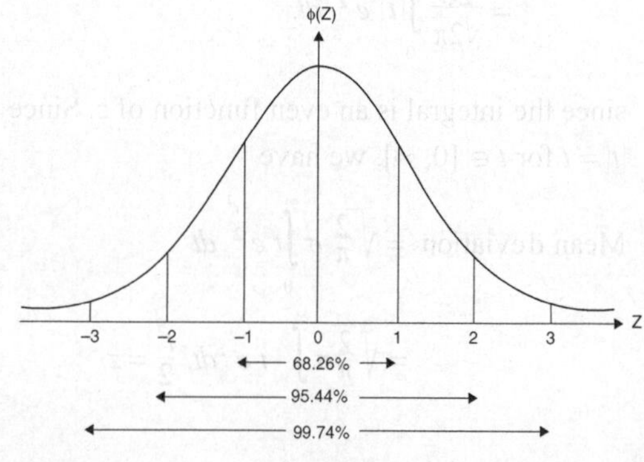

Figure 13.5

13.23 FITTING OF NORMAL DISTRIBUTION TO A GIVEN DATA

The equation of the normal curve fitted to a given data is

$$y = f(x) = \frac{1}{\sigma\sqrt{2\neq}} e^{-\frac{1}{2}\left(\frac{x-\mu}{\sigma}\right)^2}, -\infty < x < \infty.$$

Therefore, first calculate the mean μ and the standard deviation σ. Then find the standard normal variate $Z = \frac{X-\mu}{\sigma}$ corresponding to the lower limits of each of the class interval, that is, determine $z_1 = \frac{x_1-\mu}{\sigma}$, where x_1 is the lower limit of the ith class. The third step is to calculate the area under the normal curve to the left of the ordinate $Z = z_1$, say $\Phi(z_i)$, from the tables. Then areas for the successive class intervals are obtained by subtraction, *viz*, $\Phi(z_i + 1) - \Phi(z_i)$, $i = 1, 2, 3\ldots$ Then

Expected frequency $= N[\Phi(z_i+1) - \Phi(z_i)]$.

EXAMPLE 13.75

The scores in a competitive examination is normally distributed with mean 400 and standard deviation 80. Out of 10,000 candidates appeared in the examination, it is desired to pass 350 candidates. What should be the lowest score permitted for passing the examination?

Solution. The fraction of the passing candidate is $\frac{350}{10000} = 0.035$. Thus the fraction of the failing candidates is 0.965. The passing fraction is shown in the right—tail area of the Figure 13.6 of normal curve.

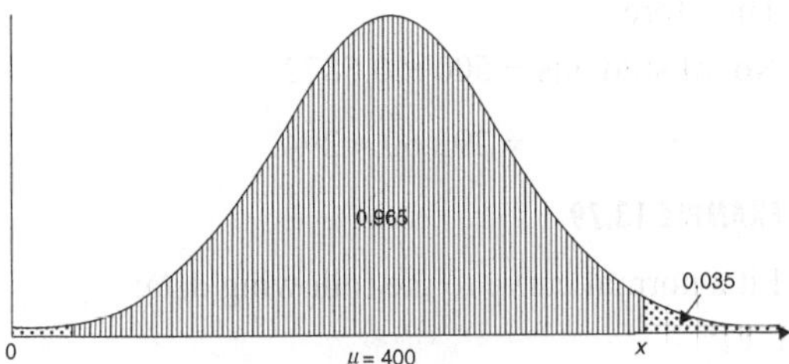

Figure 13.6

Thus the area of the standard normal curve is

$$\int_z^{\infty} \phi(z)\, dz = 0.035$$

as shown in the Figure 13.7.

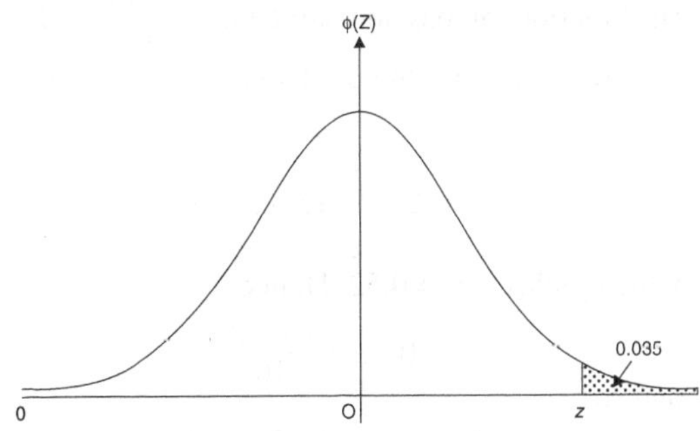

Figure 13.7

Consulting the table for area under the normal curve, we have $z = 1.81$. Therefore, the relation $Z = \frac{x-\mu}{\sigma}$ yields

$$1.81 = \frac{x-400}{80} \text{ or } x = 400 = 80(1.81) = 545.$$

Thus the candidates having scores of 545 or above will be declared pass.

EXAMPLE 13.76

In an examination taken by 500 candidates the average and standard deviation of marks obtained (normally distributed) are 40% and 10%. Find approximately

(i) How many will pass if 50 is fixed as a minimum?

(ii) What should be minimum score if 350 candidates are to be declared as pass?

(iii) How many candidates have scored marks above 60%?

Solution. We are given that

$$N = 500, \mu = 40, \text{ and } \sigma = 10.$$

Then

(i) $Z = \frac{50-40}{10} = 1.$

Therefore, consulting table for standard normal curve, we have

$$P(X \geq 50) = P(Z \geq 1) = 0.1587.$$

Hence the number of candidates passed, if 50 is fixed as minimum, is

$$N \times 0.1587 = 500 \times 0.1587 = 79.35 \approx 79$$

(ii) Fraction of passing students $= \dfrac{350}{500} = 0.7$

Fraction of failing students $= 1 - 0.7 = 0.3$

Thus

$$\int_{-\infty}^{z_1} \phi(z)\, dz = \int_{-z_1}^{0} \phi(z)\, dz = 0.3$$

which yields $-z_1 = 0.52$. Hence

$$-0.52 = \frac{x - 40}{10}$$

and so

$$x = 40 - 5.2 = 34.8 \approx 35\%.$$

(iii) We have

$$Z = \frac{60 - 40}{10} = 2.$$

Therefore, from the standard normal curve table

$$P(X \geq 60) = P(Z \geq 2) = 0.0288.$$

Hence number of candidates scoring more than 60% $= 500 \times 0.0288 \approx 11$.

EXAMPLE 13.77

For a normally distributed variate X with mean 1 and standard deviation 3, find out the probability that

(i) $3.43 \leq x \leq 6.19$

(ii) $-1.43 \leq x \leq 6.19$.

Solution. We have $\mu = 1$ and $\sigma = 3$.

(i) When $x = 3.43$,

$$Z = \frac{x - \mu}{\sigma} = \frac{3.43 - 1}{3} = 0.81$$

and when $x = 6.19$,

$$Z = \frac{6.19 - 1}{3} = 1.73.$$

Therefore,

$$P(3.43 \leq x \leq 6.19) = P(0.81 \leq Z \leq 1.73)$$
$$= P(Z \geq 0.81) - P(Z \geq 1.73)$$
$$= 0.2090 - 0.0418 = 0.1672.$$

(ii) When $x = -1.43$,

$$Z = \frac{-1.43 - 1}{3} = -\frac{2.43}{3} = -0.81$$

and when $x = 6.19$,

$$Z = \frac{6.19 - 1}{3} = 1.73.$$

Therefore

$$P(-1.43 \leq x \leq 6.19) = P(-0.81 \leq Z \leq 1.73)$$
$$= P(-0.81 \leq Z \leq 0) + P(0 \leq Z \leq 1.73)$$
$$= P(0 \leq Z \leq 0.81) + P(0 \leq Z \leq 1.73) \quad \text{(by symmetry)}$$
$$= 0.2910 + 0.4582 = 0.7492.$$

EXAMPLE 13.78

The mean height of 500 students is 151 cm and the standard deviation is 15 cm. Assuming that the heights are normally distributed, find the number of students whose heights lie between 120 and 155 cm.

Solution. We have $N = 500$, $\mu = 151$, $\sigma = 15$.

If $x = 120$, then

$$Z = \frac{120 - \mu}{\sigma} = \frac{120 - 151}{15} = -\frac{31}{15} = -2.07.$$

If $x = 155$, then

$$Z = \frac{155 - 151}{15} = \frac{4}{15} = 0.27.$$

Therefore,

$$P(120 \leq x \leq 155) = P(-2.07 \leq Z \leq 0.27)$$
$$= P(-2.07 \leq Z \leq 0) + P(0 \leq Z \leq 0.27)$$
$$= P(0 \leq Z \leq 2.07) + P(0 \leq Z \leq 0.27) \quad \text{(By symmetry)}$$
$$= 0.4808 + 0.1064 = 0.5872$$

Therefore,

$$\text{No. of students} = 500 \times 0.5872$$
$$= 293.60 \approx 294.$$

EXAMPLE 13.79

Fit a normal curve to the following data:

Class:	1–3	3–5	5–7	7–9	9–11
Frequency:	1	4	6	4	1

Also obtain the expected normal frequency.

Solution. The class marks (mid-values) are 2, 4, 6,8, 10. Therefore, for the given data, we have

$$\text{Mean}(\mu) = \frac{\sum fx}{\sum f}$$

$$= \frac{2\times1+4\times4+6\times6+8\times4+10\times1}{1+4+6+4+1}$$

$$= \frac{96}{16} = 6,$$

Standard deviation (σ)

$$= \sqrt{\frac{\sum fx^2}{\sum f} - \mu^2} = \sqrt{40-36} = 2.$$

Hence the equation of the normal curve fitted to the given data is

$$f(x) = \frac{1}{\sigma\sqrt{2\neq}}\, e^{-\frac{1}{2}\left(\frac{x-\mu}{\sigma}\right)^2} = \frac{1}{2\sqrt{2\neq}}\, e^{-\frac{1}{8}(x-6)^2}.$$

To calculate the expected frequency, we note that the area under $f(x)$ in (z_1, z_2) is

$$\Delta\,\Phi(z) = \frac{1}{\sqrt{2\neq}}\int_0^{z_2} e^{-\frac{z^2}{2}}\,dz - \frac{1}{\sqrt{2\neq}}\int_0^{z_1} e^{-\frac{z^2}{2}}\,dz,$$

$$z = \frac{x-\mu}{\sigma} = \frac{x-6}{2}.$$

Thus, the theoretic normal frequencies $N\Phi(z)$ are given by the following table:

Class interval	mid-value	(z_1, z_2)	$\Delta\Phi(z) = \Phi(z+1) - \Phi(z)$	Expected frequency
1–3	2	(−2.5, −1.5)	0.4938 − 0.4332 = 0.0606	16(0.0606) = 0.97 ≈ 1
3–5	4	(−1.5, −0.5)	0.4332 − 0.1915 = 0.2417	16(0.2417) = 3.9 ≈ 4
5–7	6	(−0.5, 0.5)	0.1915 + 0.1915 = 0.383	16(0.383) = 6.1 ≈ 6
7–9	8	(0.5, 1.5)	0.4332 − 0.1915 = 0.2417	16(0.2417) = 3.9 ≈ 4
9–11	10	(1.5, 2.5)	0.4938 − 0.4332 = 0.0606	16(0.0606) = 0.97 ≈ 1

Thus, the expected frequencies agree with the observed frequencies. Hence the normal curve obtained above is a proper fit to the given data.

EXAMPLE 13.80

In a normal distribution, 31% of the items are under 45 and 8% are over 64. Find the means and the standard deviation of the distribution.

Solution. When $x = 45$, we have

$$z_1 = \frac{x-\mu}{\sigma} = \frac{45-\mu}{\sigma}.$$

When $x = 64$, we have

$$z_2 = \frac{64-\mu}{\sigma}.$$

Further

$$\int_{-\infty}^{z_1} \phi(z)dz = 0.31 \text{ and } \int_{z_2}^{\infty} \phi(z)dz = 0.08,$$

that is,

$$\int_{-z_1}^{0} \phi(z)dz = 0.31 \text{ and } \int_{z_2}^{0} \phi(z)dz = 0.08.$$

Hence $-z_1 = 0.5$ or $z_1 = -0.5$ and $z_2 = 1.4$. Thus

$$45 - \mu = -0.5\sigma \text{ and } 64 - \mu = 1.45.$$

Solving these questions, we get $\sigma = 10$ and $\mu = 50$.

EXAMPLE 13.81

The marks obtained by the number of students for a certain subject are assumed to be approximately distributed with mean value 65 and with a standard deviation of 5. If three students are taken at random from this set of students, what is the probability that exactly two of them will have marks over 70?

Solution. We are given that $\mu = 65$ and $\sigma = 5$. If $x = 70$, we have

$$Z = \frac{x-\mu}{\sigma} = \frac{70-65}{5} = 1.$$

Thus

$$P(X > 70) = P(Z > 1)$$
$$= 0.1587 \text{ (using the table).}$$

Since this probability is the same for each student, the required probability that out of three students selected at random, exactly two will get marks over 70 is

$${}^3C_2 p^2 q, \text{ where } p = 0.1587 \text{ and}$$
$$q = 1 - p = 0.8413,$$

which is equal to $3(0.1587)^2(0.8413) = 0.06357$.

13.24 SAMPLING

A *population* or *universe* is an aggregate of objects, animate, or inanimate, under study. More precisely, a population consists of numerical values connected with these objects. A population containing a finite number of objects is called a *finite population,* while a population with infinite number of objects is called an *infinite population.*

For any statistical investigation, complete enumeration of the infinite population is not practicable. For example, to calculate average per capita income of the people of a country, we have to enumerate all the earning individuals in the country, which is a very difficult task. So we take the help of sampling in such a case.

A *sample* is a finite subset of statistical individual of a population. The number of individual in a sample is called the *sample size.* A sample is said to be *large* if the number of objects in the sample is at least 30, otherwise it is called *small.* The process of selecting a sample from a population is called *sampling.*

A sampling in which the objects are chosen in such a manner that one object has as good chance of being selected as another is called a *random sampling*. Thus, sample obtained in a random sampling is called a *random sample.* The error involved in approximation by sampling technique is known as *sampling error* and is inherent and unavoidable in any and every sampling scheme. But sampling results in considerable gains, especially in time and cost.

The statistical constants of the population, namely, mean, variance etc., are denoted by, μ, σ^2, etc., respectively, and are called *parameters* whereas the statistical measures computed from the sample observations alone, namely, mean, variance, etc., are denoted by $\bar{x}$, s^2, etc., and are called *statistics.*

Suppose that we draw possible samples of size *n* from a population at random. For each sample, we compute the mean. The means of the samples are not identical. The frequency distribution obtained by grouping the different means according to their frequencies is called *sampling distribution of the mean.* Similarly, the frequency distribution obtained by grouping different variances according to their frequency is called *sampling distribution of the variance.*

The sampling of large samples is assumed to be normal. The standard deviation of the sampling distribution of a statistics is called *standard error of that statistics*. The standard error of the sampling distribution of means is called *standard error of means*. Similarly, standard error of the sampling distribution of variances is called *standard error of the variances*. The standard error is used to assess the difference between the expected and observed values. The reciprocal of the standard error is called *precision.*

Certain assumptions about the population are made to reach decisions about populations based on sample information. Such assumptions, true or false, are called *statistical hypothesis*.

A hypothesis which is a definite statement about the population parameter is called *null hypothesis* and is denoted by H_0. In fact, the *null hypothesis is that which is tested for possible rejection under the assumption that it is true*. For example, let us take the hypothesis that a coin is unbiased (true). Thus H_0 is that $p = \frac{1}{2}$, where *p* is probability for head. We toss this coin 10 times and observe the number of times a head appears. If head appears too often or too seldom, we shall reject the hypothesis H_0 and, thus, decide that the coin is biased, otherwise we shall decide that the penny is a fair one.

A hypothesis which is complementary to the null hypothesis is called the *alternative hypothesis*, which is denoted by H_1. For example, if $H_0 : p = \frac{1}{2}$, then the alternative hypothesis H_1 can be

(i) $H_1 . p \neq \frac{1}{2}$,

(ii) $H_1 . p > \frac{1}{2}$,

(iii) $H_1 . p < \frac{1}{2}$.

The alternative hypothesis in (i) is called a *twotailed alternative*, in (ii) it is called *right tailed alternative*, and in (iii) it is known as *left-tailed alternative* If a hypothesis is rejected while it should have been accepted, we say that a *type I error* is committed. If a hypothesis is accepted while it should have been rejected, we say that the *type II error* has been committed.

13.25 LEVEL OF SIGNIFICANCE AND CRITICAL REGION

The probability level, below which we reject the hypothesis, is called the *level of significance*. A region in the sample space where hypothesis is rejected is called the *critical region or region of rejection*. The levels of significance, usually employed in testing of hypothesis, are 5% and 1%. We know that for large n,

$$Z = \frac{x - np}{\sqrt{npq}}$$

is distributed as a standard normal variate. Thus, the shaded area is the standard normal curve shown in Figure 13.8 corresponds to 5% level of significance.

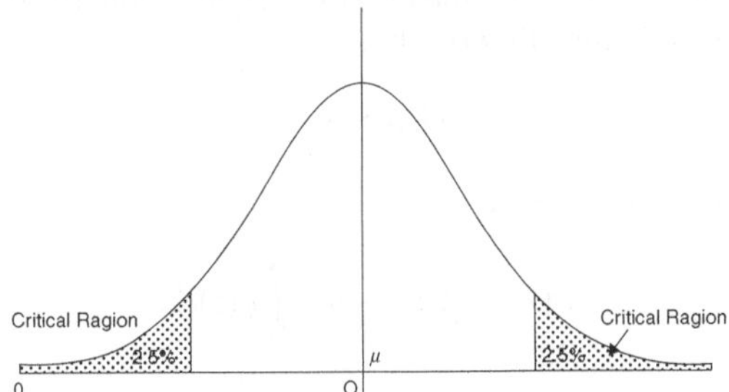

Figure 13.8

The probability of the value of the variate falling in the critical region is the level of significance.

We use a single-tail test or double-tail test to estimate for the significance of a result. In a double-tail test, the areas of both the tails of the curve representing the sampling distribution are taken into account whereas in the single-tail test, only the area on the right of an ordinate is taken into account. For example, we should use double-tail test to test whether a coin is biased or not because a biased coin gives either more number of heads than tails (right tail) or more number of tails than heads (left-tail).

The procedure which enables us to decide whether to accept or reject a hypothesis is called the *test of significance*. The procedure usually consists in assuming or accepting the hypothesis as correct and then calculating the probability of getting the observed or more extreme sample. If this probability is less than a certain pre-assigned value, the hypothesis is rejected, since samples with small probabilities should be rare and we assume that a rare event has not happened.

13.26 TEST OF SIGNIFICANCE FOR LARGE SAMPLES

We know that for large number of trials, the binomial and Poisson distributions are very closely approximated by normal distribution. Therefore, for large samples we apply the *normal test*, which is based on the area property of normal probability curve. In standard normal curve, the standard normal variate Z is given by

$$Z = \frac{X - \mu}{\sigma}.$$

Then

$$P(-3 \le Z \le 3) = P(-3 \le Z \le 0) + P(0 \le Z \le 3)$$

$$= P(0 \le Z \le 3) + P(0 \le Z \le 3) \text{ (by symmetry)}$$

$$= 2P(0 \le Z \le 3) = 2(0.4987) = 0{:}9974$$

and so

$$P(|Z| > 3) = 1 - 0.9974 = 0.0026.$$

If follows therefore that, in all probability, we should expect a *standard normal variate to lie between –3 and 3*. Further,

$$P(-1.96 \le Z \le 1.96) = P(-1.96 \le Z \le 0) + P(0 \le Z \le 1.96)$$

$$= 2P(0 \le Z \le 1.96)$$

$$= 2(0.4750) = 0.9500$$

and so

$$P(|Z| > 1.96) = 1 - 0.95 = 0.05.$$

If follows that the *significant value of Z at 5% level of significance for a two-tailed test is 1.96.*

Also, we note that

$$P(-2.58 \leq Z \leq 2.58) = P(-2.58 \leq Z \leq 0) + P(0 \leq Z \leq 2.58)$$

$$= 2P(0 \leq Z \leq 2.58)$$

$$= 2(0.4951) = 0.9902$$

and so

$$P(|Z| > 2.58) = 0.01.$$

Hence the *significant value of Z at 1% level of significance for a two-tailed test is 2.58.*

Now we find value of Z for single-tail test. From normal probability tables, we note that

$$P(Z > 1.645) = 0.5 - P(0 \leq Z \leq 1.645)$$

$$= 0.5 - 045 = 0.05$$

$$P(Z > 2.33) = 0.5 - P(0 \leq Z \leq 2.33)$$

$$= 0.5 - 0.49 = 0.01.$$

Hence, *significant value of Z at 5% level of significance of a single-tail test is 1.645, whereas the significant value of Z at 1% level of significance is 2.33.*

As a consequence of the above discussion, the steps to be used in the normal test are:

(i) Compute the test statistic Z under the null hypothesis

(ii) If $|Z| > 3$, H_0 is always rejected

(iii) If $|Z| \leq 3$, we test its level of significance at 5% or 1% level.

(iv) For a two-tailed test, if $|Z| > 1.96$, H_0 is rejected at 5% level of significance. If $|Z| > 2.58$, H_0 is rejected at 1% level of significance and if $|Z| \leq 2.58$, H_0 may be accepted at 1% level of significance

(v) For a single-tailed test, if $|Z| > 1.645$, then H_0 is rejected at 5% level and if $|Z| > 2.33$, then H_0 is rejected at 1% level of significance.

The following theorem of statistics helps us to determine sample mean $\bar{x}$ and sample variance S^2 in terms of population mean μ and population variance σ^2.

Theorem 13.7. (The Central Limit Theorem). The mean $\bar{x}$ of a sample of size N drawn from any population (continuous or discrete) with mean μ and finite variance σ^2 will have a distribution that approaches the normal distribution as $N \to \infty$, with mean μ and variance σ^2/N.

The quantity $\frac{\sigma}{\sqrt{N}}$ *is called the standard error of the mean*

13.27 CONFIDENCE INTERVAL FOR THE MEAN

For the standard normal distribution, let Z_a be a point on the z-axis for which the area under the density function $\phi(z)$ to its right is equal to a (see Figure 13.9a). Thus

$$P(Z > z_a) = a,$$

or equivalently,

$$P(Z < z_a) = 1 - a = \int_{-\infty}^{z_a} \phi(z)dz.$$

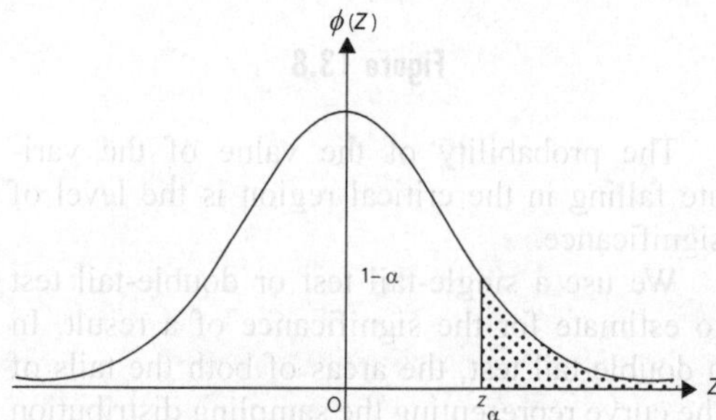

Figure 13.9 (a)

Since standard normal curve is symmetrical about $\phi(z)$-axis, we have

$$P(-z_{\frac{a}{2}} < Z < z_{\frac{a}{2}}) = 1 - a \text{ (see Figure 13.9b).}$$

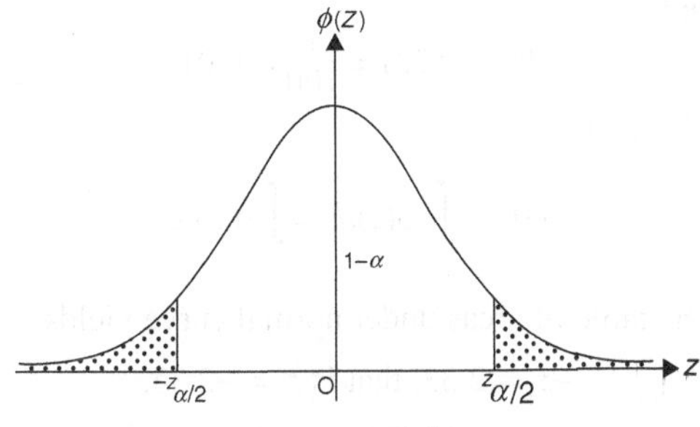

Figure 13.9 (b)

But, assuming normality of the sample average, the Central Limit theorem yields

$$Z = \frac{\bar{x}-\mu}{\frac{\sigma}{\sqrt{n}}}.$$

Therefore,

$$P\left(-z_{\frac{\alpha}{2}} < \frac{\bar{x}-\mu}{\frac{\sigma}{\sqrt{n}}} < z_{\frac{\alpha}{2}}\right) = 1 - \alpha$$

and so cross multiplication and change of sign yields

$$P\left(\bar{x} - z_{\frac{\alpha}{2}}\frac{\sigma}{\sqrt{n}} < \mu < \bar{x} + z_{\frac{\alpha}{2}}\frac{\sigma}{\sqrt{n}}\right)1 - \alpha.$$

The interval defined by

$$\left(\bar{x} - z_{\frac{\alpha}{2}}\frac{\sigma}{\sqrt{n}}, x + z_{\frac{\alpha}{2}}\frac{\sigma}{\sqrt{n}}\right)$$

is called a $100(1 - \alpha)\%$ *confidence interval for the mean* with variance (σ) known. Thus if a is specified, the upper and lower limit of this interval can be calculated from the sample average.

We know that if $\alpha = 0.05$, then $z_{0.05} = 1.645$ and then the 95% confidence interval for singletailed test is

$$\left(\bar{x} - 1.645\frac{\sigma}{\sqrt{n}}, \bar{x} + 1.645\frac{\sigma}{\sqrt{n}}\right).$$

EXAMPLE 13.82

The temperature, in degree celsius, at 12 points chosen at random in New Delhi is measured. The observations at these points are:

25°	23°	22.5°	26.5°	27°	27.5°
23.5°	22.5°	26°	24°	24.5°	25.5°

The past experience shows that the standard deviation of temperature in Delhi is 1°C. Find a 95% confidence interval for the mean temper ture in the city.

Solution. We note that

$$\mu = \frac{297.5}{12} = 24.792.$$

Since $z_{0.05} = 1.645$, the 95% confidence interval is

$$\left(24.79 \pm 1.645\left(\frac{1}{\sqrt{12}}\right)\right) = (24.79 \pm 0.4748)$$
$$= (24.32, 25.26).$$

EXAMPLE 13.83

For all children taking an examination, the mean mark was 60% with a standard deviation of 8%. A particular class of 30 children achieved an average of 63%. Is this unusual?

Solution. Let H_0 be null hypothesis that the achievement is usual. We have

$$N = 30, \mu = 60, \sigma = 8.$$

Since the sample is large, the distribution tends to normal distribution. The standard normal variate is given by

$$Z = \frac{\bar{x}-\mu}{\frac{\sigma}{\sqrt{N}}} = \frac{63-60}{\frac{8}{\sqrt{30}}} = 2.0539.$$

Since $|Z| < 2.58$, H_0 is accepted at 1% level of significance and rejected at 5% level of significance since $|Z| > 1.96$.

EXAMPLE 13.84

A coin was tossed 400 times and the head turned up 216 times. Test the hypothesis that the coin is unbiased.

Solution. The null hypothesis is

H_0: The coin is unbiased, that is, p (head) $= \frac{1}{2}$.

The number of trial $(n) = 400$. Therefore,

$$\text{Expected number of success} = np = \frac{1}{2} \cdot 400 = 200.$$

Observed number of success = 216.

Further $p = \frac{1}{2}$ implies $q = 1 - p = \frac{1}{2}$. Therefore,

$$\sigma = \sqrt{npq} = \sqrt{400 \cdot \frac{1}{2} \cdot \frac{1}{2}} = 10.$$

Hence, standard normal variate is

$$Z = \frac{\bar{x} - np}{\sigma} = \frac{216 - 200}{10} = 1.6.$$

Since $|Z| = 1.6 < 1.96$, H_0, is accepted at 5% level of significance. We conclude that the coin is unbiased.

EXAMPLE 13.85

In IIT joint entrance test, the score showed $\mu = 64$ and $\sigma = 8$. How large a sample of candidates appearing in the test must be taken in order that there be a 10% chance that its mean score is less than 62%?

Solution. We are given that $\mu = 64$, $\sigma = 8$, $\bar{x} = 62$ and

$$P(\bar{x} < 62) = \frac{10}{100} = 0.1.$$

Therefore,

$$0.1 = \int_{-\infty}^{z_1} \phi(z)dz = \int_{z_1}^{\infty} \phi(z)dz.$$

The table of areas under the normal curve yield

$$-z_1 = 1.28 \text{ and so } z_1 = 1.28.$$

Hence

$$-1.28 = \frac{\bar{x} - \mu}{\frac{\sigma}{\sqrt{N}}} = \frac{(\bar{x} - \mu)\sqrt{N}}{\sigma} = \frac{(62 - 64)}{8}\sqrt{N},$$

which yields $N = 26.21$. Hence, we must take a sample of size 26.

EXAMPLE 13.86

If the mean breaking strength of copper wire is 575 kg with a standard deviation of 8.3 kg, how large a sample must be used so that there be one chance in 100 that the mean breaking strength of the sample is less than 572 kg?

Solution. We are given that

$$\mu = 575 \text{ kg}, \sigma = 8.3 \text{ kg}, \text{ and } \bar{x} = 572$$

and

$$P(\bar{x} < 572) = \frac{1}{100} = 0.01.$$

Therefore,

$$0.01 = \int_{-\infty}^{z_1} \phi(z)dz = \int_{-z_1}^{\infty} \phi(z)dz.$$

The table of areas under normal curve yields

$$-z_1 = 2.33, \text{ that is } z_1 = -2.33,$$

Therefore,

$$-2.33 = \frac{\bar{x} - \mu}{\frac{\sigma}{\sqrt{N}}} = \frac{572 - 575}{8.3}\sqrt{N},$$

which gives $N = 41.602$. Hence, we must take a sample of size 42.

EXAMPLE 13.87

A normal population has a mean of 6.8 and standard deviation of 1.5. A sample of 400 members gave a mean of 6.75. Is the difference between the means significant?

Solution. Let the null and alternative hypothesis be

H_0: there is no significant difference between $\bar{x}$ and μ,

H_1: there is significant difference between $\bar{x}$ and μ.

It is given that $\mu = 6.8$, $\sigma = 1.5$, $N = 400$, and $\bar{x} = 6.75$. Therefore, the standard normal variate is given by

$$Z = \frac{\bar{x} - \mu}{\frac{\sigma}{\sqrt{N}}} = \frac{6.75 - 6.8}{1.5}\sqrt{400}$$

$$= -0.6660 \approx -0.67.$$

Since $|Z| = 0.67 < 1.96$, H_0 is accepted at 5% level of significance and so there is no significant difference between $\bar{x}$ and μ.

EXAMPLE 13.88

A research worker wishes to estimate mean of a population by using sufficiently large sample. The probability is 95% that sample mean will not differ from the true mean by more than 25% of the standard deviation. How large a sample should be taken?

Solution. We are given that

$$P(|\bar{x}-\mu| < 0.25\sigma) = 0.95$$

Also

$$P(|Z| \le 1.96) = 0.95,$$

that is,

$$P\left(\frac{\bar{x}-\mu}{\sigma}\sqrt{n} \le 1.96\right) = 0.95$$

or

$$P\left(|\bar{x}-\mu| \le 1.96\left(\frac{\sigma}{\sqrt{n}}\right)\right) = 0.95.$$

Therefore,

$$1.96\left(\frac{\sigma}{\sqrt{n}}\right) < 0.25$$

or

$$n > \left(\frac{1.96}{0.25}\right)^2 = (7.84)^2 = 61.47.$$

Therefore, the sample should be of the size *62*.

EXAMPLE 13.89

As an application of Central Limit theorem, show that if E is such that $P(|\bar{x}-\mu| < E) > 0.95$, then the minimum sample size n is given by $n = \frac{(1.96)^2\sigma^2}{E^2}$, where μ and σ^2 are the mean and variance, respectively, of the population and $\bar{x}$ is the mean of the random variable.

Solution. By Central Limit theorem $Z = \frac{\bar{x}-\mu}{\sigma}\sqrt{n}$ is a standard normal variate and, therefore, $P(|Z| \le 1.96) = 0.95$ implies

$$P\left\{\left|\frac{\bar{x}-\mu}{\sigma}\sqrt{n}\right| \le 1.96\right\} = 0.95$$

or

$$P\left\{|\bar{x}-\mu| \le 1.96\frac{\sigma}{\sqrt{n}}\right\} = 0.95.$$

Also, it is given that

$$P\{|\bar{x}-\mu| < E\} > 0.95.$$

Thus

$$E > \frac{1.96}{\sqrt{n}}\sigma \quad \text{or} \quad n > \frac{(1.96)^2\sigma^2}{E^2} = \frac{3.84\sigma^2}{E^2}.$$

Hence, minimum sample size is given by $n = \frac{3.84\sigma^2}{E^2}$.

13.28 TEST OF SIGNIFICANCE FOR SINGLE PROPORTION

Let X be the number of successes in n independent trials with probability p of success for each trial. Then

$$E(X) = np, \text{ variation } (X) = npq,\ q = 1 - p.$$

Let $P = \frac{X}{n}$ be called the observed proportion of success. Then

$$\text{Variance }(P) = \text{Variation}\left(\frac{X}{n}\right)$$

$$= n\cdot\frac{p}{n}\cdot\frac{q}{n} = \frac{pq}{n}$$

$$\text{Standard error (P)} = \sqrt{\frac{pq}{n}}$$

and

$$Z = \frac{P - E(P)}{S.E(P)} = \frac{P - p}{\sqrt{\frac{pq}{n}}},$$

where Z is test statistics used to test the significant difference of sample and population proportion. Further, the limit for p at the level of significance is given by $P \pm z_{\alpha}\sqrt{\frac{pq}{n}}$. In particular,

95% confidence limits for p are given by

$$P \pm 1.96\sqrt{\frac{pq}{n}}$$

99% confidence limits for p are given by

$$P \pm 2.58\sqrt{\frac{pq}{n}}.$$

EXAMPLE 13.90

Solve Example 13.84 using significance for single proportion.

Solution. Let the null and alternative hypothesis be

H_0: The coin is unbiased, that is, $p = \frac{1}{2} = 0.5$.

H_1: The coin is biased, that is, $p \ne 0.5$.

We are given that $n = 400$ and number of successes $(X) = 216$. Therefore,

Proportion of success in the sample $(P) = \frac{X}{n}$

$$= \frac{216}{400} = 0.54.$$

Further, population proportion $= p = 0.5$ and so $q = 1 - p = 1 - 0.5 = 0.5$. Hence

$$\text{Test statistics } (Z) = \frac{P-p}{\sqrt{\frac{pq}{n}}} = \frac{0.54 - 0.50}{\sqrt{0.25}}\sqrt{400}$$

$$= \frac{0.04 \times 20}{0.5} = 1.6.$$

Since $|Z| = 1.6 < 1.96$, H_0 is accepted at 5% level of significance. Hence the coin is unbiased.

EXAMPLE 13.91

In an opinion poll conducted with a sample of 2,000 people chosen at random, 40% people told that they support a certain political party. Find a 95% confidence interval for the actual proportion of the population who support this party.

Solution. The required 95% confidence interval is

$$0.4 \pm 1.96\sqrt{\frac{(0.4)(0.6)}{2000}} = 0.4 \pm 1.96(0.01095)$$

$$= 0.4 \pm 0.214 = (0.3786, 0.4214).$$

This shows that a variation of about 4% either way is expected when conducting opinion poll with sample size of this order.

EXAMPLE 13.92

A randomsample of 400 mangoes was taken from a large consignment out of which 80 were found to be rotten. Obtain 99% confidence limits for the percentage of rotten mangoes in the consignment.

Solution. We have $n = 400$ and proportion of rotten mangoes in the sample $(P) = \frac{80}{400} = 0.2.$ Since significant value of Z at 99% confidence coefficient (level of significance 1%) is 2.58, the 99% confidence limits are

$$P \pm 2.58\sqrt{\frac{PQ}{n}} = 0.2 \pm 2.58\sqrt{\frac{0.2-0.8}{400}}$$

$$= 0.2 \pm 2.58\sqrt{\frac{0.16}{400}}$$

$$= 0.2 \pm 2.58(0.02)$$

$$= (0.148, 0.252).$$

Hence, 99% confidence limits for percentage of rotten mangoes in the consignments are (14.8, 25.2).

EXAMPLE 13.93

A die was thrown 9,000 times and a throw of 3 or 4 was observed 3,240 times. Show that the die cannot be regarded as an unbiased one.

Solution. Let the null and alternative hypothesis be

H_0: die is unbiased,

H_1: die is biased.

Further,

p = probability of success (getting 3 or 4)

$$= \frac{1}{6} + \frac{1}{6} = \frac{1}{3} = 0.333$$

q = probability of failure $= 1 - \frac{1}{3} = \frac{2}{3}.$

P = proportion of success in the sample

$$= \frac{3240}{9000} = 0.360.$$

Therefore, the test statistics Z is given by

$$Z = \frac{P-p}{\sqrt{\frac{pq}{n}}} = \frac{0.360 - 0.333}{\sqrt{\frac{2}{9}}}\sqrt{9000} = 5.37.$$

Since $|Z| = 5.37 > 3$, the null hypothesis is rejected. So, we conclude that the die is almost certainly biased.

EXAMPLE 13.94

Out of 650 truck drivers, 40 were found to have consumed alcohol more than the legal limit. Find 95% confidence interval for the true proportion of drivers who were over the limit during the time of the tests.

Solution. The observed proportion of the sample is

$$P = \frac{40}{650} = \frac{4}{65}.$$

Therefore,

$$Q = 1 - P = 1 - \frac{4}{65} = \frac{61}{65}.$$

The 95% confidence interval of the proportion is

$$P \pm 1.96\sqrt{\frac{PQ}{n}} = \frac{4}{65} \pm 1.96\sqrt{\frac{\frac{4}{65}\cdot\frac{61}{65}}{650}}$$

$$= \frac{4}{65} \pm 1.96(0.009426)$$

$$= 0.00615 \pm 0.01847$$

$$= (0.0431, 0.0800).$$

This mean, 4% to 8% of the drivers were over the limit during the tests.

13.29 TEST OF SIGNIFICANCE FOR DIFFERENCE OF PROPORTION

Let X_1 and X_2 be the number of persons possessing the given attribute A in random samples of sizes n_1 and n_2 from two populations, respectively. The sample proportions are given by.

$$P_1 = \frac{X_1}{n_1},\ P_2 = \frac{X_2}{n_2}.$$

Then

$$E(P_1) = E\left(\frac{X_1}{n_1}\right) = \frac{1}{n_1}E(X_1) = \frac{1}{n_1}(n_1p_1) = p_1$$

$$V(P_1) = \frac{p_1q_1}{n_1},\ V(P_2) = \frac{p_2q_2}{n_2}.$$

Since for large samples P_1 and P_2 (the probability of success) are independent and normally distributed, $P_1 - P_2$ is also normally distributed. Therefore, the standard normal variate corresponding to the difference $P_1 - P_2$ is given by

$$Z = \frac{(P_1 - P_2) - E(P_1 - P_2)}{\sqrt{V(P_1 - P_2)}}.$$

Let H_0: $P_1 = P_2$, that is, the population are similar be the null hypothesis. Then

$$E(P_1 - P_2) = E(P_1) - E(P_2) = p_1 - p_2 = 0.$$

Also $V(P_1 - P_2) = V(P_1) + V(P_2)$

$$= \frac{p_1q_1}{n_1} + \frac{p_2q_2}{n_2} = pq\left(\frac{1}{n_1} + \frac{1}{n_2}\right),$$

because under H_0, $p_1 = p_2 = p$, say. Therefore,

$$Z = \frac{P_1 - P_2}{\sqrt{pq\left(\frac{1}{n_1} + \frac{1}{n_2}\right)}},$$

where an unbiased pooled estimate of proportion is taken as

$$p = \frac{n_1P_1 + n_2P_2}{n_1 + n_2} = \frac{X_1 + X_2}{n_1 + n_2}.$$

If $|Z| > 1.96$, H_0 is rejected at 5% level of significance. If $|Z| < 2.58$, H_0 is accepted at 1% level of significance.

EXAMPLE 13.95

In a sample of 600 men from a certain city, 450 are found to be smokers. In another sample of 900 men from another city, 450 are smokers. Does the data indicate the habit of smoking among men?

Solution. We have

$$\text{Proportion } P_1 = \frac{450}{600} = \frac{3}{4},$$

$$\text{Proportion } P_2 = \frac{450}{900} = \frac{1}{2}.$$

Then the test statistics is given by

$$Z = \frac{P_1 - P_2}{\sqrt{pq\left(\frac{1}{n_1} + \frac{1}{n_2}\right)}}$$

where

$$p = \frac{X_1 + X_2}{n_1 + n_2} = \frac{450 + 450}{900 + 600} = \frac{900}{1500} = \frac{3}{5},$$

$$q = 1 - p = \frac{2}{5}.$$

Therefore,

$$Z = \frac{\frac{3}{4} - \frac{1}{2}}{\sqrt{\frac{6}{25}\left(\frac{1}{600} + \frac{1}{900}\right)}} = \frac{1}{4(0.0258)} = 9.68.$$

and hence the cites are significantly different.

EXAMPLE 13.96

A drug manufacturer claims that the proportion of patients exhibiting side effects to their new antiarthritis drug is at least 8% lower than for the standard brand X. In a controlled experiment, 31 out of 100 patients receiving the new drug exhibited side effects, as did 74 out of 150 patients receiving brand X. Test the manufacturer's claim using 95% confidence for the true proportion.

Solution. We have $n_1 = 100$, $n_2 = 150$, and

$$\text{Proportion } (P_1) \text{ for new drug} = \frac{31}{100},$$

$$\text{Proportion } (P_2) \text{ for the standard drug } X = \frac{74}{150}.$$

The test statistics is

$$Z = \frac{P_1 - P_2}{\sqrt{pq\left(\frac{1}{n_1} + \frac{1}{n_2}\right)}},$$

where

$$p = \frac{X_1 + X_2}{n_1 + n_2} = \frac{31 + 74}{100 + 150} = \frac{21}{50},$$

$$q = 1 - p = 1 - \frac{21}{50} = \frac{29}{50}.$$

Therefore,

$$Z = \frac{\frac{31}{100} - \frac{74}{150}}{\sqrt{\frac{609}{2500}\left(\frac{5}{300}\right)}} = \frac{-11}{60\sqrt{\frac{203}{150000}}} = -4.984.$$

Thus $|Z| = 4.984 > 1.96$. Thus the difference between the two brands are significant at 5% level of significance.

Also, the 95% confidence interval is

$$P_1 - P_2 \pm 1.96\sqrt{\frac{P_1Q_1}{n_1} + \frac{P_2Q_2}{n_2}}$$

$$= \frac{-11}{60} \pm 1.96\sqrt{\frac{31 \times 69}{(100)^3} + \frac{74 \times 26}{(150)^3}}$$

$$= -0.1833 \pm 0.1020 = (-0.2853, -0.0813).$$

Since 0 does not lie within the interval, the difference is significant. Further, the claim of the manufacturer is accepted as it lies within the confidence interval.

EXAMPLE 13.97

In two large populations, there are 30% and 25%, respectively, of fair-haired people. Is this difference likely to be hidden in samples of 1,200 and 900, respectively, from the two populations?

Solution. Let

P_1 = proportion of fair – haired people in first population

$$= \frac{30}{100} = 0.30$$

P_2 = proportion of fair – haired people in second population

$$= \frac{25}{100} = 0.25.$$

Accordingly,

$$Q_1 = 1 - 0.3 = 0.7, Q_2 = 1 - 0.25 = 0.75.$$

Let the null and alternative hypothesis be

H_0: Sample proportion are equal, that is, $P_1 = P_2$.

H_1: $P_1 \neq P_2$.

Then the test statistics is given by

$$Z = \frac{P_1 - P_2}{\sqrt{pq\left(\frac{1}{n_1} + \frac{1}{n_2}\right)}},$$

where pooled estimate of proportion is

$$p = \frac{X_1 + X_2}{n_1 + n_2} = \frac{1200(0.3) + 900(0.25)}{1200 + 900}$$

$$= \frac{360 + 225}{2100} = 0.2786.$$

Therefore,

$$q = 1 - p = 0.7214.$$

Hence

$$Z = \frac{0.3 - 0.25}{\sqrt{(0.2786)(0.7214)\left(\frac{1}{1200} + \frac{1}{900}\right)}}$$

$$= \frac{0.05}{0.019768}$$

$$= 2.53.$$

Since $|Z| = 2.53 > 1.96$, the proportions are si nificantly different and so H_0 is rejected. The differences are unlikely to be hidden.

EXAMPLE 13.98

Random samples of 400 men and 600 women were asked whether they would like to have a flyover near their residence. Two hundred men and 325 women were in favour of the proposal. Test the hypothesis that proportions of men and women in favour of the proposal are same against that they are not, at 5% level.

Solution. The null hypothesis is

H_0: $P_1 = P_2$, that is, no significant difference between the opinion of men and women as far as the proposal of flyover is concerned.

We have

$$n_1 = 400, X_1 = 200,$$
$$n_2 = 600, X_2 = 325,$$
$$P_1 = \frac{200}{400} = 0.5 \text{ and } P_2 = \frac{325}{600} = 0.541.$$

Therefore, the test statistics is

$$Z = \frac{P_1 - P_2}{\sqrt{pq\left(\frac{1}{n_1} + \frac{1}{n_2}\right)}},$$

where

$$p = \frac{X_1 + X_2}{n_1 + n_2} = \frac{200 + 325}{400 + 600} = \frac{525}{1000} = 0.525$$

and

$$q = 1 - p = 0.475.$$

Hence

$$Z = \frac{0.5 - 0.541}{\sqrt{(0.525)(0.475)\left(\frac{1}{400} + \frac{1}{600}\right)}} = \frac{-0.041}{0.3223}$$
$$= -1.272.$$

Since $|z| = 1.272 < 1.96$, H0 may be accepted at 5% level of significance, that is, men and women do not differ significantly in their opinions.

EXAMPLE 13.99

In a referendumsubmitted to the student body at a university, 850 men and 560 women voted. Out of these 500 men and 320 women voted "yes". Does this indicate a significant difference of opinion between men and women on the matter at 1% level of significance?

Solution. We have $n_1 = 850$, $n_2 = 560$, $X_1 = 500$, $X_2 = 320$.

Let the null hypothesis be

H_0: there is no significant difference in voting pattern, that is, $P_1 = P_2$,

where

$$\text{Proportion } (P_1) = \frac{500}{850} = \frac{10}{17} = 0.588,$$
$$\text{Proportion } (P_2) = \frac{320}{560} = \frac{4}{7} = 0.571.$$

Then the test statistics is

$$Z = \frac{P_1 - P_2}{\sqrt{pq\left(\frac{1}{n_1} + \frac{1}{n_2}\right)}},$$

where

$$p = \frac{X_1 + X_2}{n_1 + n_2} = \frac{820}{1410} = 0.582 \text{ and}$$
$$q = 1 - p = 0.418.$$

Therefore,

$$Z = \frac{0.588 - 0.57}{\sqrt{(0.582)(0.418)(0.1765 + 0.1786)}}$$
$$= \frac{0.017}{0.294} = 0.578.$$

Since $|Z| = 0.578 < 2.58$, the hypothesis H_0 is accepted at 1% level of significance.

EXAMPLE 13.100

Suppose that 10 years ago 500 people were working in a factory, and 180 of them were exposed to a material which is now suspected as being carcinogenic. Of these 180, 30 have developed cancer, whereas 32 of the other workers, who were not exposed, have also developed cancer. Obtain 95% confidence interval for the difference between the proportions with cancer among those exposed and not exposed, and assess whether the material should be considered carcinogenic on this evidence.

Solution. According to the given data

Total No. of workers = 500,

n_1 = No. of people exposed to materials = 180,

n_2 = No. of people not exposed to materials = 320,

X_1 = No. of people out of n_1, who suffered with cancer = 30,

X_2 = No. of people out of n_2, who suffered withcancer = 32,

$$\text{Proportion } (P_1) = \frac{30}{180} = 0.167,$$
$$\text{Proportion } (P_2) = \frac{32}{320} = 0.100.$$

Therefore, a 95% confidence interval for the difference between the true proportions is

$$P_1 - P_2 \pm 1.96\sqrt{\frac{P_1Q_1}{n_1} + \frac{P_2Q_2}{n_2}}$$
$$= 0.067 \pm 1.96(0.0325)$$
$$= 0.07 \pm 0.0637 = (0.033, 0.131).$$

On the other hand, the test statistics is given by

$$Z = \frac{P_1 - P_2}{\sqrt{pq\left(\frac{1}{n_1} + \frac{1}{n_2}\right)}},$$

where

$$p = \frac{X_1 + X_2}{n_1 + n_2} = \frac{62}{500} = 0.124 \text{ and}$$
$$q = 1 - p = 0.876.$$

Therefore,

$$Z = \frac{0.167 - 0.100}{\sqrt{(0.124)(0.876)\left(\frac{1}{180} + \frac{1}{320}\right)}}$$
$$= \frac{0.067}{\sqrt{0.000942}}$$
$$= \frac{0.067}{0.031} = 2.16.$$

Since $|Z| > 1.96$, the difference is significant at 5% level and so the material should be considered carcinogenic on this evidence.

13.30 TEST OF SIGNIFICANCE FOR DIFFERENCE OF MEANS

Let $\bar{x}_1$ be the mean of a sample of size n_1 from a population with mean μ_1 and variance σ_1^2 and let $\bar{x}_2$ be the mean of sample of size n_2 from another population with mean μ_2 and variance σ_2^2. Then $\bar{x}_1$ and $\bar{x}_2$ are two independent normal variates. Therefore, $\bar{x}_1 - \bar{x}_2$ is also a normal variate. The value of the standard normal variate Z corresponding to $\bar{x}_1 - \bar{x}_2$ is given by

$$Z = \frac{(\bar{x}_1 - \bar{x}_2) - E(\bar{x}_1 - \bar{x}_2)}{\text{Standard error of } (\bar{x}_1 - \bar{x}_2)}.$$

If $H_0: \mu_1 = \mu_2$, that is, there is no significant difference between the sample means is the null hypothesis, then

$$E(\bar{x}_1 - \bar{x}_2) = E(\bar{x}_1) - E(\bar{x}_2) = \mu_1 - \mu_2 = 0$$

$$V(\bar{x}_1 - \bar{x}_2) = V(\bar{x}_1) + V(\bar{x}_2) = \frac{\sigma_1^2}{n_1} + \frac{\sigma_2^2}{n_2},$$

since the covariance term vanishes due to independence of $\bar{x}_1$ and $\bar{x}_2$. Therefore, under the null hypothesis H_0, the test statistics Z is given by

$$Z = \frac{\bar{x}_1 - \bar{x}_2}{\sqrt{\frac{\sigma_1^2}{n_1} + \frac{\sigma_2^2}{n_2}}}.$$

If $\sigma_1^2 = \sigma_2^2$, that is, if the samples have been drawn from the same population, then the test statistics reduces to

$$Z = \frac{\bar{x}_1 - \bar{x}_2}{\sigma\sqrt{\frac{1}{n_1} + \frac{1}{n_2}}}.$$

If σ is not known, then its estimate $\hat{\sigma}$ based on sample variance is used and

$$\hat{\sigma}^2 = \frac{(n_1 - 1)S_1^2 + (n_2 - 1)S_2^2}{(n_1 + n_2 - 2)}$$

If σ_1, σ_2, are known and $\sigma_1^2 \neq \sigma_2^2$ then they are estimated from sample values and we have

$$Z = \frac{\bar{x}_1 - \bar{x}_2}{\sqrt{\frac{S_1^2}{n_1} + \frac{S_2^2}{n_2}}}.$$

EXAMPLE 13.101

A sample of 100 electric bulbs produced by a manufacture A showed a mean life time of 1,190 hours and a standard deviation of 90 hours. A sample of 75 bulbs produced by manufacturer B showed a mean life time of 1,230 hours with a standard deviation of 120 hours. Is there a difference between the mean life time of the two brands at significance levels of 5% and 1%?

Solution. We have

$$n_1 = 100, \bar{x}_1 = 1190, \sigma_1 = 90$$
$$n_2 = 75, \bar{x}_2 = 1230, \sigma_2 = 120.$$

Therefore, the test statistics is

$$Z = \frac{\bar{x}_1 - \bar{x}_2}{\sqrt{\frac{\sigma_1^2}{n_1} + \frac{\sigma_2^2}{n_2}}} = \frac{1190 - 1230}{\sqrt{\frac{90^2}{100} + \frac{120^2}{75}}}$$
$$= -\frac{40}{\sqrt{81 + 192}} = -\frac{40}{16.523} = -2.42.$$

Since $|Z| = 2.42 > 1.96$, there is a difference between the mean life time of the two brands at a significant level of 5%.

On the other hand $|Z| = 2.42 < 2.58$, therefore, there is no difference between the mean life time of the two brands at a significant level of 1%.

EXAMPLE 13.102

The means of simple samples of sizes 1,000 and 2,000 are 67.5 and 68.0, respectively. Can the samples be regarded as drawn from the same population of standard deviation 2.5?

Solution. We are given that

$n_1 = 1000, \bar{x}_1 = 67.5, n_2 = 2000, \bar{x}_2 = 68.0, \sigma = 2.5.$

Therefore, the test statistics is

$$Z = \frac{\bar{x}_1 - \bar{x}_2}{\sigma\sqrt{\frac{1}{n_1} + \frac{1}{n_2}}} = \frac{67.5 - 68.0}{2{:}5\sqrt{\frac{1}{1000} + \frac{1}{2000}}}$$

$$= -\frac{0.5}{2.5(0.03873)} = 5.16$$

Since $|Z| = 5.16 > 1.96$, the difference between the mean is very significant. Therefore, the samples cannot be regarded drawn from the same population.

EXAMPLE 13.103

The mean height of 50 male students who showed above average participation in college athletics was 68.2 inches with a standard deviation of 2.5 inches, whereas 50 male students who showed no interest in such participation had a mean height of 67.5 inches with a standard deviation of 2.8 inches. Test the hypothesis that male students who participiple in college athletics are taller than other male students.

Solution. It is given that

$$n_1 = 50, \bar{x}_1 = 68.2, s_1 = 2.5,$$
$$n_2 = 50, \bar{x}_2 = 67.5, s_2 = 2.8.$$

Let the null and alternative hypothesis be

Null hypothesis : $\mu_1 = \mu_2$,

Alternative hypothesis : $\mu_1 > \mu_2$ (right tailed).

The test statistics is

$$Z = \frac{\bar{x}_1 - \bar{x}_2}{\sqrt{\frac{s_1^2}{n_1} + \frac{s_2^2}{n_2}}} = \frac{68.2 - 67.5}{\sqrt{\frac{(2.5)^2}{50} + \frac{(2.8)^2}{50}}}$$

$$= \frac{0.7}{\sqrt{0.282}} = \frac{0.7}{0.53} = 1.32.$$

Since $Z = 1.32 < 1.645$ (critical value of Z at 5% level of significance). Therefore, it is not significant at 5% level of significance. Hence, the null hypothesis is accepted. Hence the students who participate in college athletics are not taller than other students.

13.31 TEST OF SIGNIFICANCE FOR THE DIFFERENCE OF STANDARD DEVIATIONS

Let s_1 and s_2 be the standard deviations of two independent samples of size n_1 and n_2, respetively. Let the null hypothesis be that the sample standard deviation does not differ significantly. Then the statistics of the hypothesis is

$$Z = \frac{s_1 - s_2}{S.E(s_1 - s_2)}.$$

For large samples,

$$S.E(s_1 - s_2) = \sqrt{\frac{\sigma_1^2}{2n_1} + \frac{\sigma_2^2}{2n_2}}$$

and so

$$Z = \frac{s_1 - s_2}{\sqrt{\frac{\sigma_1^2}{2n_1} + \frac{\sigma_2^2}{2n_2}}}.$$

If σ_1^2 and σ_2^2 are unknown, then s_1^2 and s_2^2 are used in place of them. Hence, in that case, we have

$$Z = \frac{s_1 - s_2}{\sqrt{\frac{s_1^2}{2n_1} + \frac{s_2^2}{2n_2}}}.$$

EXAMPLE 13.104

The yield of wheat in a random sample of 1,000 farms in a certain area has a standard deviation of 192 kg. Another random sample of 1,000 farms gives a standard deviation of 224 kg. Are the standard deviations significantly different?

Solution. We are given that

$n_1 = 1000, s_1 = 192, n_2 = 1000, s_2 = 224.$

Therefore, the test statistics for the null hypothesis that standard deviations are same is

$$Z = \frac{s_1 - s_2}{\sqrt{\frac{s_1^2}{n_1} + \frac{s_2^2}{n_2}}} = \frac{192 - 224}{\sqrt{\frac{(192)^2}{1000} + \frac{(224)^2}{1000}}}$$

$$= \frac{-32}{\sqrt{36.864 + 50.176}} = \frac{-32}{9.33} = -3.43.$$

Since $|Z| = 3.43 > 1.96$. Hence the null hypothesis is rejected and so the standard deviations are significantly different.

13.32 SAMPLING WITH SMALL SAMPLES

In large sample theory, the sampling distribution approaches a normal distribution. But in case of small size, the distributions of the various statistics like $Z = \frac{\bar{x}-\mu}{\sigma}\sqrt{n}$ are far from normality and as such normal test cannot be applied to such samples.

The problem of testing the significance of the deviation of a sample mean from a given population mean when sample size is small and only the sample variance is known was first solved by W. S. Gosset, who wrote under the pen-name "student." Later on R.A. Fisher modified the method given by Gosset. The test discovered by them is known as *Students Fisher t-test.*

Let $x_1, x_2, \ldots, x_n$ be a random small sample of size n drawn from a normal population with mean μ and variance σ. The statistics t is defined as

$$t = \frac{\bar{x} - \mu}{S}\sqrt{n},$$

where $\bar{x} = \frac{1}{n}\sum_{i=1}^{n} x_i$ is the sample mean and $S^2 = \frac{1}{n-1}\sum_{i=1}^{n}(x_i - \bar{x})^2$ is an unbiased estimate of the population variance σ^2. If we calculate t for each sample, we obtain a distribution for t, known as *Student Fisher t-distribution*, defined by

$$y = f(t) = C\,(1 + t^2)^{-\frac{v+1}{2}},$$

where the parameter $v = n - 1$ is called the *number of degrees of freedom* and C is a constant, depending upon v, such that the area under the curve is unity.

The curve $y = f(t)$ is symmetrical about y-axis like the normal curve. But it is more peaked than the normal curve with the same standard deviation. Further, this curve (Figure 13.10) approaches the horizontal t-axis less rapidly than the normal curve. If attains its maximum value at $t = 0$ and so its mode coincides with the mean.

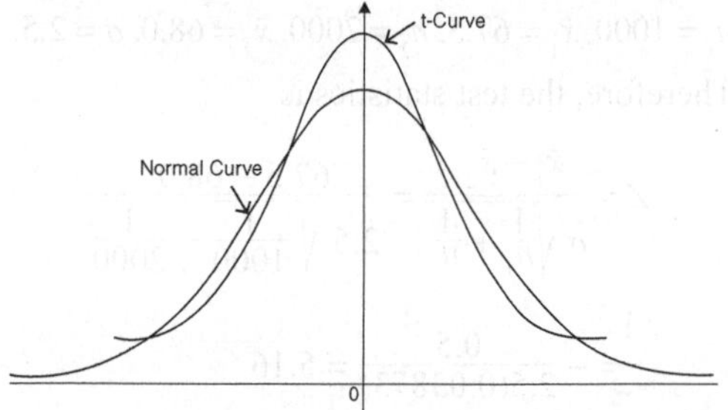

Figure 13.10

If $v \to \infty$, we have

$$y = Ce^{-\frac{t^2}{2}},$$

which is a normal curve. Hence t is normally distributed for large samples.

The probability p that the value of t will exceed t_0 is given by

$$P = \int_{t_0}^{\infty} y dx.$$

Fisher tabulated the values of t corresponding to various levels of significance for different values of v. For example for $v = 10$ and $p = 0.022$ we note that $t = 2.76$. Thus

$$P(t > 2.76) = P(t < -2.76) = 0.02$$

or

$$P(|t| > 2.76) = 0.02.$$

If the calculated value of t is greater than $t_{0.05}$ (the tabulated value), then the difference between $\bar{x}$ and μ is said to be significant at 5% level of significance. Similarly if $t > t_{0.01}$, then the difference between $\bar{x}$ and μ is said to be significant at 1% level of significance.

Since the probability P that $t > t_{0.05}$ is 0.95 the 95% confidence limits for μ are given by

$$\left|\frac{\bar{x}-\mu}{S}\sqrt{n}\right| \le t_{0.05}$$

or

$$|\bar{x}-\mu| \le \frac{S}{\sqrt{n}}t_{0.05}.$$

Thus 95% confidence interval for μ is

$$\left(\bar{x}-\frac{S}{\sqrt{n}}t_{0.05,}\ \bar{x}+\frac{S}{\sqrt{n}}t_{0.05}\right)$$

EXAMPLE 13.105

A randomsample of 10 boys had the following IQ: 70, 120, 110, 101, 88, 83, 95, 98, 107, 100.

Do these data support the assumption of population mean IQ of 100 at 5% level of significance?

Solution. The statistics t is defined by

$$t = \frac{\bar{x}-\mu}{\frac{S}{\sqrt{n}}}$$

So, we first find $\bar{x}$ and S. We have

$$\bar{x} = \frac{1}{n}\sum_{i=1}^{n} x_i = \frac{972}{10} = 97.2.$$

To calculate S, we use the following table:

x	70	120	110	101	88
$x-\bar{x}$	−27.2	22.8	12.80	3.80	−9.2
$(x-\bar{x})^2$	739.84	519.84	163.84	14.44	84.64
x	83	95	98	107	100
$x-\bar{x}$	−14.2	− 2.2	0.8	9.80	2.80
$(x-\bar{x})^2$	201.64	4.84	0.64	96.04	7.84

We have

$$S^2 = \frac{1}{n-1}\sum_{i=1}^{10}(x-x_i)^2 = \frac{1}{9}(1833.96) = 203.773.$$

Therefiore, $S = 14.275$.Then

$$t = \frac{97.2-100}{14.275}\sqrt{10} = -\frac{2.80}{14.275}(3.1623) = -0.620.$$

But

$$t_{0.05} = 2.26 \text{ for } v = 10-1 = 9.$$

Since $|t| = 0.62 < 2.26$, the value of t is not significant at 5% level of significance. Therefore, the data supports the population mean 100. Further, 95% confidence interval is

$$\bar{x} \pm t_{0.05}\frac{S}{\sqrt{n}} = 97.2 \pm 2.26\left(\frac{14.275}{\sqrt{10}}\right)$$

$$= 97.2 \pm 10.20 = (87, 107.4).$$

Since 100 lies within this interval, the data support the population mean.

EXAMPLE 13.106

A certain stimulus administered to each of 12 patients resulted in the following change in blood pressure:

5, 2, 8, −1, 3, 0, −2, 1, 5, 0, 4, 6.

Can it be concluded that the stimulus will increase the blood pressure?

Solution. The mean of sample is

$$\bar{x} = \frac{1}{12}\sum_{i=1}^{12} x_i = \frac{31}{12} = 2.583.$$

Therefore,

$$S^2 = \frac{1}{n-1}\sum_{i=1}^{12}(x_i-\bar{x})^2$$

$$= \frac{1}{11}[(5-2.583)^2 + (2-2.583)^2$$

$$+ (8-2.583)^2 + (-1-12.583)^2$$

$$+ (3-2.583)^2 + (0-2.583)^2$$

$$+ (-2-2.583)^2 + (1-2.583)^2$$

$$+ (5-2.583)^2 + (0-2.583)^2$$

$$+ (4-2.583)^2 + (6-2.583)^2$$

$$= \frac{1}{11}[5.842 + 0.340 + 29.344$$

$$+ 12.838 + 0.174 + 6.672 + 21.004 + 2.506$$

$$+ 5.842 + 6.672 + 2.008 + 11.676]$$

$$= \frac{104.918}{11} = 9.538.$$

Therefore, $S = 3.088$ and so the 95% confidence interval for the mean is

$$\left(\overline{x} - \frac{S}{\sqrt{n}} t_{0.05}, \overline{x} + \frac{S}{\sqrt{n}} t_{0.05}\right)$$

$$= \left(2.583 - \frac{3.088(2.2)}{\sqrt{12}}, 2.583 + \frac{3.088(2.2)}{\sqrt{12}}\right)$$

$$= (2.583 - 1.910, 2.583 + 1.910)$$

$$= (0.673, 4.493).$$

Since the average change in blood pressure of the population (μ) is positive, the stimulus will increase the blood pressure.

EXAMPLE 13.107

The measured lifetime of a sample of 15 electronic components gave an average of 750 hours with a sample standard deviation of 85 hours. Find a 95% confidence interval for the mean life time of the population and test the hypothesis that the mean is 810 hours.

Solution. We have

$$n = 15, \overline{x} = 750, S = 85.$$

The table value $t_{0.05}$ for $v = 14$ is 2.14. Therefore, 95% confidence interval is

$$750 \pm \frac{2.14(85)}{\sqrt{15}} = (750 - 46.97, 750 + 46.97)$$

$$= (703.03, 796.97).$$

Since 810 is not included in this interval, the hypothesis that the mean is 810 hours is rejected at 5% significance level. The same conclusion is reached by evaluating the test statistics:

$$Z = \frac{\overline{x} - \mu}{S}\sqrt{n} = \frac{750 - 810}{85}\sqrt{15} = -\frac{60}{85}(3.873)$$

$$= -2.73.$$

Since $|Z| = 2.73 > t_{0.05}$ (for $v = 14$), the difference is significant at 5% level of significance.

13.33 SIGNIFICANCE TEST OF DIFFERENCE BETWEEN SAMPLE MEANS

Let $x_1, x_2, \ldots, x_{n_1}$ and $y_1, y_2, \ldots, y_{n_2}$ be two independent samples with means $\overline{x}$ and $\overline{y}$ and standard deviation S_1 and S_2, respectively, from a normal population with the same variance. The test hypothesis is that the means are the same.

The test statistics is

$$t = \frac{\overline{x} - \overline{y}}{S\sqrt{\frac{1}{n_1} + \frac{1}{n_2}}},$$

where

$$\overline{x} = \frac{1}{n_1}\sum_{i=1}^{n_1} x_i, \; \overline{y} = \frac{1}{n_2}\sum_{i=1}^{n_2} y_i.$$

$$S^2 = \frac{1}{n_1+n_2-2}[(n_1-1)S_1^2 + (n_2-1)S_2^2]$$

$$= \frac{1}{n_1+n_2-2}\left[\sum_{i=1}^{n_1}(x_i-\overline{x})^2 + \sum_{i=1}^{n_2}(y_1-\overline{y})^2\right].$$

The variate t defined above follow the t-distribution with $n_1 + n_2 - 2$ degree of freedom.

If $t > t_{0.05}$, the difference between the sample means is significant at 5% level of significance. If $t < t_{0.05}$, the data is consistent with the hypothesis that the means are the same.

Similarly if $t > t_{0.01}$, the difference between the sample means is significant at 1% level of significance. If $t < t_{0.01}$, the data is consistent with the hypothesis that the means are the same.

If $n_1 = n_2$, that is, if the samples are of the same size and the data are paired, then the test statistics is given by

$$t = \frac{\overline{d}}{S}\sqrt{n},$$

where

$$S^2 = \frac{1}{n-1}\sum_{i=1}^{n}(d_i - \overline{d})^2, \; d_i = x_i - y_i, \; \overline{d} = \frac{\sum_{i=1}^{n} d_i}{n},$$

No. of degree of freedom = $n - 1$

EXAMPLE 13.108

A group of 10 boys fed on a diet A and another group of 8 boys fed on a different diet B, recorded the following increase in weights (in kg):

Diet A:	5	6	8	1	12	4	3	9	6	10
Diet B:	2	3	6	8	10	1	2	8		

Does it show the superiority of diet A over that of B?

Solution. We have

$$\bar{x} = \frac{1}{n_1}\sum_{i=1}^{n_1} x_i = \frac{1}{10}(64) = 6.4,$$

$$\bar{y} = \frac{1}{n_2}\sum_{i=1}^{n_2} y_2 = \frac{1}{8}(40) = 5.0,$$

$$S^2 = \frac{1}{n_1+n_2-2}\left[\sum_{i=1}^{n_1}(x_i-\bar{x})^2 + \sum_{i=1}^{n_2}(y_i-y)^2\right]$$

$$= \frac{1}{16}\left[\sum_{i=1}^{10}(x_i-\bar{x})^2 + \sum_{i=1}^{8}(y_i-\bar{y})^2\right]$$

$$= \frac{1}{16}[(1.4)^2 + (0.4)^2 + (1.6)^2 + (5.4)^2 + (5.6)^2 + (2.4)^2 + (3.4)^2 + (2.6)^2 + (0.4)^2 + (3.6)^2 + 3^2 + 2^2 + 1^2 + 3^2 + 5^2 + 4^2 + 3^2 + 3^2]$$

$$= \frac{1}{16}[1.96 + 0.16 + 2.56 + 2.32 + 3.14 + 5.76 + 11.56 + 6.76 + 0.16 + 12.96 + 9 + 4 + 1 + 9 + 25 + 16 + 9 + 9]$$

$$= \frac{129.34}{16},$$

which yields $S = 2.843$. Then the test statistics is

$$t = \frac{\bar{x}-\bar{y}}{S\left(\frac{1}{n_1}+\frac{1}{n_2}\right)^{\frac{1}{2}}} = \frac{6.4-5.0}{2.843\sqrt{\left(\frac{1}{10}+\frac{1}{8}\right)}}$$

$$= \frac{1.4}{2.843(0.474)} = 1.038.$$

From the table, $t_{0.05}$ for $v = n_1 + n_2 - 2 = 16$ is 2.12. Since calculated t is less than $t_{0.05}$, we conclude that the difference between sample mean is not significant. Hence, there is no superiority of diet A over the diet B.

EXAMPLE 13.109

Eleven school boys were given a test in drawing. They were given a month's further tuition and a second test of equal difficulty was held at the end of the month. Do the marks give evidence that the students have been benefited by extra coaching?

Marks in 1st test	23	20	19	21	18	
Marks in 2nd test	24	19	22	18	20	
Marks in 1st test	20	18	17	23	16	19
Marks in 2nd test	22	20	20	23	20	17

Solution. We have $n_1 = n_2 = 11$. Representing marks in second test by x_i and that of first test by y_i, we have the differences $d_i = x_i - y_i$ as

1, −1, 3, −3, 2, 2, 2, 3, 0, 4, −2.

Therefore,

$$\bar{d} = \frac{\sum d_i}{n} = \frac{\sum(x_i - y_i)}{11} = \frac{11}{11} = 1,$$

$$S^2 = \frac{1}{n-1}\sum_{i=1}^{11}(d_i-\bar{d})^2$$

$$= \frac{1}{10}[0^2 + (-2)^2 + 2^2 + (-4)^2 + 1^2 + 1^2 + 1^2 + 2^2 + (-1)^2 + 3^2 + (-3)^2 = 5$$

and so $S = \sqrt{5} = 2.24$.

The test statistics for equal sample means is

$$t = \frac{\bar{d}}{\frac{S}{\sqrt{n}}} = \frac{1}{2.24}\sqrt{11} = 1.481.$$

The tabular value of $t_{0.05}$ for $v = 10$ is 2.228. Thus the calculated value of t is less than $t_{0.05}$.

Therefore, the hypothesis that the mean are same, is accepted. Hence, the data provides no evidence that the students have benefited by extra coaching.

EXAMPLE 13.110

A group of boys and girls were given an intelligent test. The mean score, standard deviations, and number in each group are as follows:

	Boys	Girls
Mean	124	121
S.D	12	10
N	18	14

Is the mean score of boys significantly different from that of girls?

Solution. We have

$$n_1 = 18, \quad n_2 = 14,$$
$$S_1 = 12, \quad S_2 = 10,$$
$$\bar{x} = 124, \quad \bar{y} = 121.$$

Therefore,

$$S^2 = \frac{1}{n_1 + n_2 - 2}\left[(n_1 - 1)S_1^2 + (n_2 - 2)S_2^2\right]$$

$$= \frac{1}{20}[17(144) + 13(100) = 187.40$$

and so $S = 13.69$. Therefore, the test statistics for the hypothesis that the mean are same is

$$t = \frac{\bar{x} - \bar{y}}{S\sqrt{\frac{1}{18} + \frac{1}{14}}}$$

$$= \frac{124 - 121}{13.69(0.350)} = \frac{3}{4.792} = 0.626.$$

From the table, $t_{0.05}$ for $v = n_1 + n_2 - 2 = 20$ is 2.09. Since calculated value of t is less than the tubular value of $t_{0.05}$ for $v = 20$, the difference in mean is not significant.

EXAMPLE 13.111

A manufacturer claims that the lifetime of a particular electronic component is unaffected by temperature variation within the range 0–60° *C*. Two samples of these components were tested and their measured lifetimes are (in hours) recorded as follows:

0°*C*	7050	6970	7370	7910
60°*C*	7030	7270	6510	6700
0°*C*	6790	6850	7280	7830
60°*C*	7350	6770	6220	7230

Solution. The sample sizes are equal, that is, $n_1 = n_2 = 8$. Representing the lifetimes at 0°*C* by x_i and the lifetimes at 60°*C* by y_i, we get the differences $d_i = x_i - y_i$ as

20, –300, 860, 1210, –560, 80, 1060, 600.

Therefore,

$$\bar{d} = \frac{\sum d_i}{n} = \frac{2970}{8} = 371.25.$$

$$S_2 = \frac{1}{n-1}\sum_{i=1}^{8}(d_i - \bar{d})^2$$

$$= \frac{1}{7}[123376.56 + 450576.56 + 238876.56$$
$$+ 703501.56 + 867226.56 + 84826.56$$
$$+ 474376.56 + 52326.56 = 427869.56$$

and so $S = 654.12$. The test statistics for equal sample mean is

$$t = \frac{\bar{d}}{S}\sqrt{n} = \frac{371.25}{654.12}\sqrt{8} = 1.61.$$

The tabular value of $t_{0.05}$ for $v = 7$ is 2.36. Since the calculated values of t is less than $t_{0.05}$, the difference in the mean is not significant at 5% level of significance. Hence, the manufacture claims is accepted at 5% level of significance.

If we calculate the 95% confidence interval, we get

$$\bar{d} \pm 2.36\left(\frac{S}{\sqrt{8}}\right) = \bar{d} \pm \frac{2.36(654.12)}{\sqrt{8}}$$
$$= (371.25 - 545.87, 371.25 + 545.82)$$
$$= (-174.62, 917.12).$$

Since zero lies within the 95% confidence interval, the difference in mean is not significant and so the manufacturer's claim is accepted.

EXAMPLE 13.112

Two kinds of photographic films were tested for sharpness of definition in the same camera under varying conditions. Each pair of readings given below was produced under the same conditions except for difference of film. Is there any unusual difference between the sharpness of the definition of the two films?

Film X:	27	30	30	32
Film Y:	25	28	30	30
Film X:	24	26	40	35
Film Y:	27	28	37	28.

Solution. The sample sizes are $n_1 = n_2 = 8$. The null and alternative hypotheses are

H_0: Mean μ of the population of difference is zero

H_1: Mean μ of the population of difference is not zero.

We shall test under 5% level of significance. We have

$$\bar{d} = \frac{\sum d_i}{n} = \frac{\sum (x_i - y_i)}{n}$$

$$= \frac{2+2+0+2-3-2+3+7}{8} = 1.38$$

and

$$S_2 = \frac{\sum d_i^2}{n} - (\bar{d})^2$$

$$= \frac{4+4+0+4+9+4+9+49}{8} - (1.38)^2 = 8.47.$$

Thus $S = 2.910$. Therefore, test statistics is given by

$$t = \frac{\bar{d} - \mu}{S}\sqrt{n} = \frac{1.38}{2.91}\sqrt{8} = 1.34.$$

From the table, for $v = 7$, we have $t_{0.005} = 2.36$. Thus, the calculated value of t is less than the tabulated $t_{0.005}$. Therefore, the difference is not significant at 5% level of confidence. Hence H_0 is accepted and consequently there is no unusual difference between the sharpness of definitions of the two films.

13.34 CHI-SQUARE DISTRIBUTION

Let f_{o_i} and f_{e_i} be the observed and expected frequencies of a class interval, then χ^2 is defined by the relation

$$\chi^2 = \frac{\sum_{i=1}^{n} (f_{o_i} - f_{e_i})^2}{f_{e_i}},$$

where summation extends to all class intervals.

Note that χ^2 describes the magnitude of discrepancy between the observed and expected frequencies.

For large sample sizes, the sampling distribution of χ^2 can be closely approximated by a continuous curve known as χ^2-distribution. Thus χ^2-distribution is defined by means of the function

$$y = C e^{-\frac{\chi^2}{2}} (\chi^2)^{\frac{v-1}{2}},$$

where v is the degree of freedom and C is a constant. In the case of binomial distribution, the degree of freedomis $n - 1$. In case of Poisson distribution, the degree of freedom is $n - 2$, whereas in case of normal distribution, the degree of freedom is $n - 3$ In fact, if we have $s \times t$ contingency table, then the degree of freedom is $(s - 1)(t - 1)$. If $v = 1$, the χ^2-curve reduces to $y = C e^{-\frac{x^2}{2}}$, which is right half of a normal curve as shown in Figure 13.14.

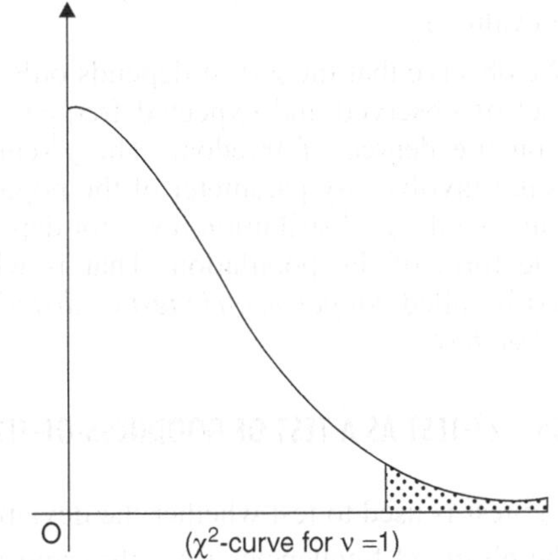

Figure 13.14

If $v > 1$, the χ^2-curve is tangential to the x-axis at the origin, as shown in Figure 13.15.

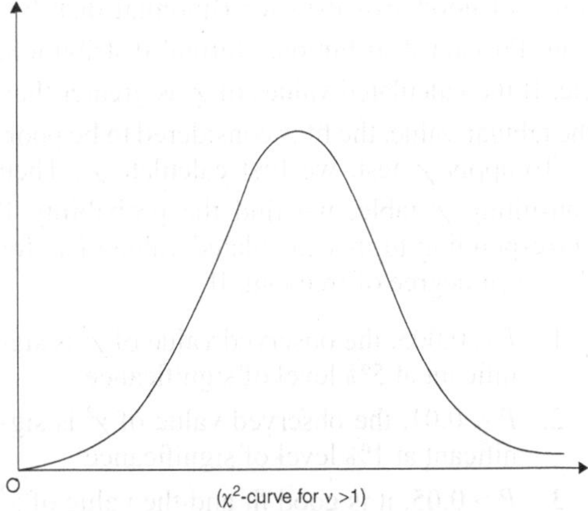

Figure 13.15

As v increases, the curve becomes more symmetrical. If $v > 30$, the χ^2-curve approximates to the normal curve and in such case the sample is of large size and we should refer to normal distribution table.

The probability P that the value of χ^2 from a random sample will exceed χ_0^2 is given by

$$P = \int_{\chi_0^2}^{\infty} y\,dx.$$

The values of χ^2 for degree of freedom from $v = 1$ to $v = 30$ have been tabulated for various convenient probability values. The table yields the values for the probability P that χ^2 exceeds a given value, χ_0^2.

We observe that the χ^2-test depends only on the set of observed and expected frequencies and on the degree of freedom. The χ^2-curve does not involve any parameter of the population and so the χ^2-distribution does not depend on the form of the population. That is why, χ^2-test is called *nonparametric test* or *distribution-free test.*

13.35 χ^2-TEST AS A TEST OF GOODNESS-OF-FIT

The χ^2-test is used to test whether the deviation of the observed frequencies from the expected (theoretical) frequencies are significant or not. Thus, this test tells us how a set of observations fits a given distribution. Hence χ^2-test provides a test of goodness-of-fit for Binomial distribution, Poisson distribution, Normal distribution, etc. If the calculated values of χ^2 is greater than the tabular value, the fit is considered to be poor.

To apply χ^2-test, we first calculate χ^2. Then consulting χ^2-table, we find the probability P corresponding to this calculated value of χ^2 for the given degree of freedom. If

1. $P < 0.005$, the observed value of χ^2 is significant at 5% level of significance
2. $P < 0.01$, the observed value of χ^2 is significant at 1% level of significance
3. $P > 0.05$, it is good fit and the value of χ^2 is not significant.

This mean that we accept the hypothesis if calculated χ^2 is less than the tabulated value, otherwise reject it.

Conditions for the validity of χ^2-test: In 13.34, we pointed out that χ^2-test is used for large sample size. For the validity of χ^2-test as a test of goodness-of-fit regarding significance of the deviation of the observed frequencies from the expected (theoretical) frequencies, the following conditions must be satisfied:

1. The sample observations should be independent.
2. The total frequency (the sum of the observed frequencies or the sum of expected frequency) should be larger than 50.
3. No theoretical frequency should be less than 5 because χ^2-distribution cannot maintain continuity character if frequency is less than 5.
4. Constraints on the frequencies, if any, should be linear.

EXAMPLE 13.113

Fit a binomial distribution to the data

x:	0	1	2	3	4	5
y:	38	144	342	287	164	25

and test for goodness-of-fit at the level of significance 0.05.

Solution. We have $n = 5$, $\sum f_i = 1000$. Therefore,

$$\mu = \frac{\sum f_i x_i}{\sum f_i}$$

$$= \frac{0 + 144 + 684 + 861 + 656 + 125}{1000} = 2.470.$$

But, for a binomial distribution, $\mu = np$ and so

$$p = \frac{\mu}{n} = \frac{2.470}{5} = 0.494,\; q = 1 - p = 0.506.$$

Therefore, the binomial distribution to be fitted is

$$1000(0.506 + 0.494)^5 = 1000\,[{}^5C_0(0.506)^5$$

$$+\,{}^5C_1(0.506)^4(0.494)$$

$$+\,{}^5C_2(0.506)^3(0.494)^2 + {}^5C_3(0.506)^2(0.494)^3$$

$$+\,{}^5C_4(0.506)(0.494)^4({}^5C_5(0.494)^5$$

$= 1000\,[0.0332 + 0.1619 + 0.3161 + 0.3086$

$+ 0.1507 + 0.02942$

$= 33.2 + 161.9 + 316.1 + 308.6 + 150.7 + 29.42.$

Thus the theoretical frequencies are

x:	0	1	2	3	4	5
y:	33.2	161.9	316.1	308.6	150.7	29.42

Therefore,

$$\chi^2 = \frac{(38-33.2)^2}{33.2} + \frac{(144-161.9)^2}{161.9}$$

$$+ \frac{(342-316.1)^2}{316.1} + \frac{(287-308.6)^2}{308.6}$$

$$+ \frac{(164-150.7)^2}{150.7} + \frac{(25-29.42)^2}{29.42}$$

$$= 0.6940 + 1.9791 + 2.1222 + 1.5119$$

$$+ 1.1738 + 0.6640 = 8.145.$$

The number of degree of freedom is 6 – 1 = 5. For $v = 5$, $\chi^2_{0.05} = 11.07$. Thus the calculated value of χ^2 is less than $\chi^2_{0.05}$ and so the binomial distribution gives a good fit at 5% level of significance.

EXAMPLE 13.114

The following table gives the frequency of occupancy of digits 0, 1, 2,..., 9 in the last place in four logarithms of numbers 10–99. Examine if there is any peculiarity.

Digits:	0	1	2	3	4
Frequency:	6	16	15	10	12

Digits:	5	6	7	8	9
Frequency:	12	3	2	9	5

Solution. Let the null hypothesis be H_0: frequency of occupance of digits is equal, that is, there is no significant difference between the observed and the expected frequency.

Therefore under the null hypothesis, the expected frequency is $f_e = \frac{90}{10} = 9$. Then

$$\chi^2 = \frac{\sum (f_{oi} - f_{ei})^2}{f_{ei}}$$

$$= \frac{9 + 49 + 36 + 1 + 9 + 9 + 36 + 49 + 0 + 16}{9}$$

$$= 23.777.$$

Number of degree of freedom is 10 –1 = 9. The tabulated value of $\chi^2_{0.05}$ for $v = 9$ is 16.92. Since the calculated value of χ^2 is greater than the tabulated value of $\chi^2_{0.05}$, the hypothesis is rejected and so there is a significant difference between the observed and expected frequency.

EXAMPLE 13.115

In a locality, 100 persons were randomly selected and asked about their academic qualifications. The results are as given below:

Education

Sex	Middle standard	High school	Graduation	Total
Male	10	15	25	50
Female	25	10	15	50
Total	35	25	40	100

Can you say that education depends on sex?

Solution. Let the null hypothesis be

H_0: Education does not depend on sex.

On this hypothesis the expected frequencies are

$$\left(\text{using}\ \frac{\text{row total} \times \text{column total}}{\text{grand total}}\right).$$

Sex	Middle standard	High school	Graduation	Total
Male:	17.5	12.5	20	50
Female:	17.5	12.5	20	50
Total	35	25	40	100

Therefore,

$$\chi^2 = \frac{(10-17.5)^2}{17.5} + \frac{(15-12.5)^2}{12.5} + \frac{(25-20)^2}{20}$$

$$+ \frac{(25-17.5)^2}{17.5} + \frac{(10-12.5)^2}{12.5} + \frac{(15-20)^2}{20}$$

$$= 9.93.$$

Further, the number of degree of freedom (v) = $(s-1)(t-1) = (3-1)(2-1) = 2$.

From χ^2-table, $\chi^2_{0.05}$ for $v = 2$ is 5.99. Thus the calculated value of χ^2 is greater than the tabulated value of χ^2. Hence H_0 is rejected and so the education depends on sex.

EXAMPLE 13.116

Fit a Poisson distribution to the following data and test for its goodness-of-fit at 5% level of significance.

x:	0	1	2	3	4
y:	419	352	154	56	19

Solution. If the given distribution is approximated by a Poisson distribution, then the parameter of the Poisson distribution is given by

$$\lambda = \frac{\sum f_i x_i}{\sum f_i} = \frac{0+352+308+168+76}{1000}$$

$$= 0.904.$$

Therefore, the theoretical frequencies are

$$1000\, e^{-\lambda},\ 1000\lambda e^{-\lambda},\ 1000\frac{\lambda^2}{2}e^{-\lambda},$$

$$1000\frac{\lambda^3}{3\,!}e^{-\lambda}, 1000\frac{\lambda^4}{4\,!}e^{-\lambda}.$$

Also $e^{-\lambda} = e^{-0.904} = 0.4049$. Therefore, the theoretical frequencies are

x:	0	1	2	3	4	Total
f:	404.9	366	165.4	49.8	11.3	997.4
	406				12.8	

To make the total of frequencies 1000, we take the first frequency as 406 and the last frequency as 12.8. Then

$$\chi^2 = \frac{(419-406)^2}{406} + \frac{(352-366)^2}{366}$$

$$+ \frac{(154-165.4)^2}{165.4} + \frac{(56-49.8)^2}{49.8} + \frac{(19-12.8)^2}{12.8}$$

$$= 0.416 + 0.536 + 0.786 + 0.772 + 3.003 = 5.513.$$

The number of degree of freedom in case of Poisson distribution is $n-2 = 5-2 = 3$. Therefore, the tabular value of χ^2 for $v = 3$ is 7.82. Thus the calculated value of χ^2 is less than the tabulated value of $\chi^2_{0.05}$. Therefore, the Poisson distribution provides a good fit to the data.

EXAMPLE 13.117

Obtain the equation of the normal curve that may be fitted to the data given below and test the goodness- of-fit.

x:	4	6	8	10	12
y:	1	7	15	22	35

x:	14	16	18	20	22	24
y:	43	38	20	13	5	1

Solution. For the given data, we have

x	x^2	f	fx	fx^2
4	16	1	4	16
6	36	7	42	252
8	64	15	120	960
10	100	22	220	220
12	144	35	42	5040
14	196	43	602	8428
16	256	38	608	9728
18	324	20	360	6480
20	400	13	260	5200
22	484	5	110	2420
24	576	1	24	576
		200	**2770**	**41300**

Therefore,

$$\text{Mean } (\mu) = \frac{\sum fx}{\sum f} = \frac{2770}{200} = 13.85.$$

Standard deviation (σ) = $\sqrt{\frac{\sum fx^2}{\sum f} - \mu^2}$ = $\sqrt{\frac{41300}{200} - (13.85)^2} = 14.678 = 3.8311.$

Hence, the equation of the normal curve fitted to the given data is

$$f(x) = \frac{1}{\sigma\sqrt{2\pi}} e^{-\frac{1}{2}\left(\frac{x-\mu}{\sigma}\right)^2}$$

$$= \frac{1}{13.85\sqrt{2\neq}} e^{-\frac{1}{29.36}(x-13.85)^2}.$$

To calculate the theoretical normal frequencies,we note that the area under $f(x)$ in (z_1, z_2) is

$$\Delta\Phi(z) = \frac{1}{\sqrt{2\neq}}\int_0^{z_2} e^{-\frac{z^2}{2}} dz - \frac{1}{\sqrt{2\neq}}\int_0^{z_1} e^{-\frac{z^2}{2}} dz,$$

where $z = \frac{\overline{x}-\mu}{\sigma} = \frac{x-13.85}{3.83}$. Thus, the expected normal frequencies are given by

Class interval	Mid-value	(z_1, z_2)	$\Delta\Phi(z)$	Expected frequency $N\Delta\Phi(z)$
3–5	4	(−2.83, −2.31)	0.4977 − 0.4896 = 0.0081	200(0.0081) = 1.62
5–7	6	(−2.83, −1.79)	0.4896 − 0.4633 = 0.0263	200(0.0263) = 5.26
7–9	8	(−1.79, −1.27)	0.4633 − 0.3980 = 0.0653	200(0.0653) = 13.06
9–11	10	(−1.27, −0.74)	0.3980 − 0.2704 = 0.1276	200(0.1276) = 25.52
11–13	12	(−0.74, −0.22)	0.2704 − 0.0871 = 0.1833	200(0.1833) = 36.66
13–15	14	(−0.22, 0.30)	0.1179 + 0.0871 = 0.2050	200(0.2050) = 41.00
15–17	16	(0.30, 0.82)	0.2939 − 0.1179 = 0.1760	200(0.1760) = 35.20
17–19	18	(0.82, 1.34)	0.4099 − 0.2939 = 0.1160	200(0.1160) = 23.20
19–21	20	(1.34, 1.86)	0.4686 − 0.4099 = 0.0587	200(0.0587) = 11.74
21–23	22	(1.86, 2.38)	0.4913 − 0.4686 = 0.0227	200(0.0227) = 4.54
23–25	24	(2.38, 2.91)	0.4982 − 0.4913 = 0.0069	200(0,0069) = 1.38

Therefore,

$$\chi^2 = \frac{(1-1.62)^2}{1.62} + \frac{(7-5.26)^2}{5.26} + \frac{(15-13.06)^2}{13.06}$$

$$+ \frac{(22-25.52)^2}{25.52} + \frac{(35-36.66)^2}{36.66}$$

$$+ \frac{(43-41)^2}{41} + \frac{(38-35.20)^2}{35.20} + \frac{(20-23.20)^2}{23.20}$$

$$+ \frac{(13-11.74)^2}{11.74} + \frac{(5-4.54)^2}{4.54} + \frac{(1-1.38)^2}{1.38}$$

$= 0.0912 + 0.5756 + 0.2882 + 0.4855 + 0.0752$

$+ 0.0976 + 0.2227 + 0.4414 + 0.1352$

$+ 0.0466 + 0.1046 = 2.56.$

The number of degree of freedomis $n - 3 = 11 - 3 = 8$ and $\chi^2_{0.005}$ at $v = 8$ is 15.51. Therefore, the normal distribution provides a good fit.

13.36 SNEDECOR'S F-DISTRIBUTION

Let $x_1, x_2, \ldots x_{n_1}$ and $y_1, y_2, \ldots y_{n_2}$ be the values of two independent random samples drawn from two normal populations with equal variance σ^2. Let $\overline{x}$ and $\overline{y}$ be the sample means and let

$$S_1^2 = \frac{1}{n_1 - 1}\sum_{i=1}^{n_1}(x_i - \overline{x})^2,$$

$$S_2^2 = \frac{1}{n_2 - 1}\sum_{i=1}^{n_2}(y_i - \overline{y})^2.$$

Then we define the statistics F by the relation

$$F = \frac{S_1^2}{S_2^2}.$$

The Snedecor's F-distribution is defined by the function

$$y = C\,F^{\frac{v_1-2}{2}}\left(1+\frac{v_1}{v_2}F\right)^{-\frac{v_1+v_2}{2}},$$

where the constant C depends on v_1 and v_2 and is so chosen that area under the curve is unity. The F-distribution is independent of the population variance σ^2 and depends only on v_1 and v_2, the numbers of degree of freedom of the samples. The F-curve is bell-shaped for $v_1 > 2$, as shown in Figure 13.16.

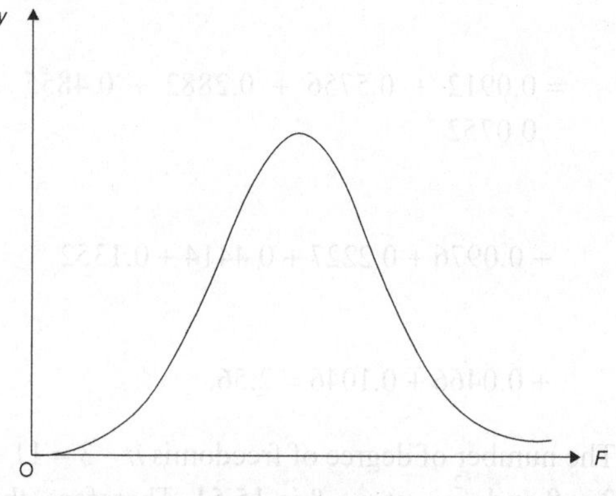

Figure 13.16

Significant test is performed by means of Snedecor's F-tables which provides 5% and 1% of points of significance for F. Five percent points of F means that area under the F-curve, to the right of the ordinate at a value of F is 0.05. Further the F-tables give only single tail test. However, if we are testing the hypothesis that the population variances are same, then we should use both tail areas under the F-curve and in that case F-table will provide 10% and 2% levels of significance.

13.37 FISHER'S Z-DISTRIBUTION

Putting $F = e^{2z}$ in the F-distribution, we get

$$y = Ce^{v_1 z}(v_1 e^{2z} + v_2),$$

which is called the *Fisher's z-distribution,* where C is a constant depending upon v_1 and v_2 such that area under the curve is unity. The curve forthis distribution is more symmetrical than F-distribution.

Significance test are performed from the z-table in a similar way as in the case of Fdistribution.

EXAMPLE 13.118

In testing for percent of ash content, 17 tests from one shipment of coal shows $S^2 = 7.08$ percent and 21 tests from a second shipment shows $S^2 = 20.70$. Can these samples be regarded as drawn from the same shipment?

Solution. We have $n_1 = 21, n_2 = 17, S_1^2 = 20.70$ and $S_2^2 = 7.08$. Therefore, the test statistics is

$$F(v_1, v_2) = F(20, 16) = \frac{20.70}{7.08} = 2.92.$$

From the F-table, we have $F_{0.05}(20, 16) = 2.18$. Since $F(v_1, v_2)$ is greater than $F_{0.05}$, the population variances are significantly different.

EXAMPLE 13.119

Two independent samples of sizes 7 and 6 havethe following values:

Sample A:	28	30	32	33	33	29	34
Sample B:	29	30	30	24	27	29	

Examine whether the samples have been drawn from normal populations having the same variance.

Solution. The means for the sample A and B are, respectively

$$\bar{x} = \frac{219}{7} = 31.285 \text{ and } \bar{y} = \frac{169}{6} = 28.166.$$

Then

$$S_1^2 = \frac{1}{n_1 - 1}\sum (x_i - \bar{x})^2$$

$$= \frac{1}{6}\left[(28 + 31.285)^2 + (30 - 31.285)^2\right.$$

$$+ (32 - 31.285)^2 + (33 - 31.285)^2$$

$$+ (33 - 31.285)^2 + (29 - 31.285)^2$$

$$+ (34 - 31:285)^2]$$

$$= \frac{1}{6}[10.791 + 1.651 + 0.511 + 2.941$$

$$+ 2.941 + 5.221 + 7.371] = 5.238$$

and

$$S_2^2 = \frac{1}{n_2 - 1}\sum (y_i - \bar{y})^2$$

$$= \frac{1}{5}[29 - 28.166)^2 + (30 - 28.166)^2$$

$$+ (30 - 28.166)^2 + (24 - 28.16)^2$$

$$+ (27 - 28.166)^2 + (29 - 28.166)^2]$$

$$= \frac{1}{5}[0.695 + 3.364 + 17.355$$

$$+ 1.359 + 0.695] = 5.366.$$

Therefore, the teststatistics is given by

$$F = \frac{S_1^2}{S_2^2} = \frac{5.238}{5.366} = 0.976.$$

Further, since numbers of degree of freedom are 6 and 5, we have

$$F_{0.05}(6, 5) = 4.95.$$

Thus, the calculated value of F is less than the tabular value. Hence the samples have been drawn from normal population having the same variance.

EXAMPLE 13.120

Two samples of sizes 9 and 8 give the sum of squares of deviations from their respective means equal to 160 and 91, respectively. Examine, whether the samples have been drawn from normal population having the same variance.

Solution. We have

$$\sum_{i=1}^{9} (x_i - \bar{x})^2 = 160 \quad \text{and} \quad \sum_{i=1}^{8} (y_i - \bar{y}) = 91.$$

Therefore, their variances are

$$S_1^2 = \frac{1}{8}(160) = 20, \quad \text{and} \quad S_2^2 = \frac{1}{7}(91) = 13.$$

Their test statistics for F-test is

$$F = \frac{S_1^2}{S_2^2} = \frac{20}{13} = 1.54.$$

From the F-table, we have

$$F_{0.05}(8,7) = 3.73.$$

Since the calculated value of F is less than $F_{0.05}$ (8, 7), the population variances are not significantly different. So the samples can be regarded as drawn from the populations having the same variance.

EXAMPLE 13.121

The nicotine content (in mg) of two samples of tobacco were found to be as follows:

Sample A:	24	27	26	21	25	
Sample B:	27	30	28	31	22	36

Can it be said that the two samples came from the same population?

Solution. Suppose that $\bar{x}$ be the sample mean for the sample B and $\bar{y}$ be the sample mean of the sample A. Then

$$\bar{x} = \frac{174}{6} = 29 \text{ and } \bar{y} = \frac{123}{5} = 24.6$$

$$S_1^2 = \frac{1}{n - 1}\sum (x_1 - \bar{x})^2$$

$$= \frac{1}{5}[(27-29)^2 + (30-29)^2 + (28-29)^2$$

$$+ (31-29)^2 + (22-29)^2 + (36-29)^2]$$

$$= \frac{1}{5}[4 + 1 + 1 + 4 + 49 + 49] = 21.6,$$

$$S_2^2 = \frac{1}{n-1}\sum (y_i - \bar{y})^2$$

$$= \frac{1}{4}[(24-24.6)^2 + (27-24.6)^2 + (26-24.6)^2$$

$$+ (21-24.6)^2 + (25-24.6)^2$$

$$= \frac{1}{4}[0.36 + 5.76 + 1.96 + 12.96 + 0.16] = 5.3.$$

Therefore, the statistics for F-test is

$$F = \frac{S_1^2}{S_2^2} = \frac{21.6}{5.3} = 4.08.$$

But tabular value of $F_{0.05}$ (5, 4) is 6.26. The calculated value of F is less than the tabular value. So there is no significant difference. Hence the two samples may be considered to come from the same population.

13.38 MISCELLANEOUS EXAMPLES

EXAMPLE 13.122

An insurance company insured 2,000 scooter drivers, 4,000 car drivers and 6,000 truck drivers. The probability of accident is 0.01, 0.03 and 0.15 respectively. One of the insured persons meets an accident. What is the probability that he is a scooter driver?

Solution. Consider the events

A: Accident takes place
B: Person met with accident is a scooter driver
C: Person met with accident is a car driver
D: Person met with accident is a truck driver.

Then

$$P(B) = \frac{2000}{12000} = \frac{1}{6}, \; P(C) = \frac{4000}{12000} = \frac{1}{3},$$

$$P(D) = \frac{6000}{12000} = \frac{1}{2}.$$

Further, it is given that

$$P(A\backslash B) = 0.01, \; P(A\backslash C) = 0.03, \; P(A\backslash D) = 0.15.$$

Then, by theorem on total probability, we have

$$P(A) = P(A\backslash B)P(B) + P(A\backslash C) = 0.03,$$

$$P(A\backslash D) = 0.15.$$

$$= 0.01\left(\frac{1}{6}\right) + 0.03\left(\frac{1}{3}\right) + 0.15\left(\frac{1}{2}\right)$$

$$= \frac{1}{6}[0.01 + 0.06 + 0.45] = \frac{0.52}{6}.$$

Now, Baye's Theorem implies

$$P(B\backslash A) = \frac{P(A\backslash B)P(B)}{P(B)} = \frac{0.01\left(\frac{1}{6}\right)}{0.\,\frac{52}{6}} = \frac{1}{52}.$$

EXAMPLE 13.123

If A, B, C are mutually exclusive and exhaustive events associated with a random experiment and $P(B) = 0.6P(A)$ and $P(C) = 0.2P(A)$. Then find $P(A)$.

Solution. Since A, B and C are mutually exclusive and exhaustive events, we have

$$P(A) + P(B) + P(C) = 1.$$

Since $P(B) = 0.6\,P(A)$ and $P(C) = 0.2\,P(A)$, we get

$$(1 + 0.6 + 0.2)\,P(A) = 1$$

or

$$P(A) = \frac{1}{1.8} = \frac{5}{9}.$$

EXAMPLE 13.124

A factory is manufacturing electric bulbs, there is a chance of 1/500 for any bulb to be defective. The bulbs are packed in packets of 50. Calculate the approximate number of packets containing no defective, one, two and three defective bulbs in a consignment of 10,000 packets.

Solution. We are given that $p = \frac{1}{150}$ and $n = 50$. Therefore parameter λ of the Poisson distribution is

$$\lambda = np = 0.1.$$

Then

$$p(r) = \frac{e^{-\lambda}\lambda^r}{r!} = \frac{e^{-0.1}(0.1)^r}{r!}.$$

Thus

(i) $p(\text{no defective}) = p\ (x = 0) = \frac{e^{-0.1}(0.1)^0}{0!} = e^{-0.1} = 0.90483$. Therefore number of packets containing no defective bulb is $10000 \times 0.90483 \approx 9048$.

(ii) $P(\text{one defective}) = P(x = 1) = \frac{e^{-0.1}(0.1)}{1} = (0.9048)\ (0.1)$

$$\approx 0.090483.$$

Therefore the number of packets containing one defective bulb is $10000 \times 0.090483 \approx 904$

(iii) $P\ (\text{two defective}) = p(x = 2)$

$$= \frac{e^{-0.1}(0.1)^2}{2} = \frac{(0.90483)(0.01)}{2}$$

$\approx 0.00452.$

Therefore the number of packets containing two defective bulbs is $10000 \times 0.00452 \approx 45$

(iv) $P(\text{three defective}) = P(x = 3)$

$$= \frac{e^{-0.1}(0.01)^3}{2} = \frac{(0.90483)(0.001)}{6} \approx 0.00015.$$

Therefore the number of packets containing three defective bulbs is $10000 \times 0.00015 \approx 2$.

EXAMPLE 13.125

Two random samples have the following values:

Sample 1	15	22	28	26	18	17	29	21	24
Sample 2	8	12	9	16	15	10			

Test the difference of the estimates of the population variances at 5% level of significance (Given that $F_{0.05}$ for $v_1 = 8$ and $v_2 = 5$ is 4.82).

Solution. The means for samples 1 and 2 are respectively

$$\bar{x} = 22.22 \text{ and } \bar{y} = 11.66.$$

Then for $n_1 = 9$, $n_2 = 6$, we have

$$S_1^2 = \frac{1}{n_2 - 1}\ \Sigma(x_i - \bar{x})^2$$

$$= \frac{171.10364}{8} = 21.387955,$$

$$S_2^2 = \frac{1}{n_2 - 1}\ \Sigma(y_i - \bar{y})^2$$

$$= \frac{53.3336}{5} = 10.66672.$$

Therefore the test statistics is given by

$$F = \frac{S_1^2}{S_2^2} = \frac{21.387955}{10.66672} = 2.005.$$

Further, the numbers of degree of freedom are 8 and 5.

Therefore, we have

$$F_{0.05}(8,5) = 4.82 \text{ (given)}.$$

Thus the calculated value of F is less than the tabulated value. Hence the samples have been drawn from normal population having the same variance, that is, there is no significant difference between the population variances.

EXERCISES

1. Find the mean, median, and mode of the following data relating to weight of 120 articles.

Weight in gm:	0–10	10–20	20–30	30–40	40–50	50–60
No. of articles	14	17	22	26	23	18

Ans. Mean: 32.58, Median: 32.6 Mode: 35.1

2. Determine the mean and standard deviation for the following data

Size of item:	6	7	8	9	10	11	12
Frequency:	3	6	9	13	18	5	4

Ans. Mean: 9, S.D:1.61

3. Find (i) mean $\bar{x}$ and $\bar{y}$ (ii) regression coefficients b_{yx}, and b_{xy} (iii) coefficient of correlation between x and y for the two regression lines $2x + 3y - 10 = 0$ and $4x + y - 5 = 0$

Ans. $\bar{x} = \frac{1}{2}, y = 3, b_{yx} = -\frac{2}{3}, b_{xy} = \frac{1}{4},$

$\rho = -\frac{1}{\sqrt{6}}$

4. Out of the following two regression lines, find the regression line of Y on X:

$3x + 12y = 9,\ 3x + 9x = 46.$

Ans. $3x + 12y = 9$

5. Calculate the coefficient of correlation between X and Y from the following data:

x:	43	44	46	40	44	42	45	50
y:	29	31	19	18	19	27	27	22

Ans. −0.057

6. Using the following data, calculate the Karl Pearson's coefficient of correlation between the experience and sales performance of five sales personnel.

Sales Personal	Experience in years	Sale performance in Rs Lakhs
1	2	20
2	4	12
3	6	18
4	8	10
5	10	40

Ans. 0.504.

7. Calculate the coefficient of correlation between the values of x and y form the following data:

x	1	3	5	7	8	10
y	8	12	15	17	18	29

Ans. 0.988.
(High positive correlation).

8. Find the coefficient of correlation between ages and marks of five students from the following data:

marks \ age	0–4	4–8	8–12	12–16	Total
0-5	7	–	–	–	7
5-10	6	8	–	–	14
10-15	–	5	3	–	8
15-20	–	7	2	–	9
20-25	–	–	–	9	9
Total	13	20	5	9	47

Ans. 0.87.

9. Calculate the rank correlation coefficient from the following data of marks obtained by 10 students:

Marks in Physics	Marks in Mathematics
78	84
36	51
98	91
25	60
75	68
82	62
90	86
62	58
65	63
39	47

Ans. 0.908.

10. Two sales person gave the following rank to seven different types of mobile phones:

2	1	4	3	5	7	6
1	3	2	4	5	6	7

Find Spearman's Rank Correlation Coefficient.

Ans. 0.786.

11. Find the value of coefficient of correlation if regression coefficients are 0.2 and 0.8.

Ans. 0.4.

12. Find the line of regression of Y on X for the following data:

x:	1.53	1.78	2.60	2.95	3.42
y:	33.50	36.30	40.00	45.80	53.50

Ans. $y = 9.72x + 17.931$.

13. Find the linear regression coefficients for the following data:

x:	1	2	3	4	5	6	7	8
y:	3	7	10	12	14	17	20	24

Ans. $b_{yx} = 2.798$, $b_{xy} = 0.354$

14. Find the lines of regression for price (x) and supply (y) from the following data. Also estimate the supply when price is 16 units.

$$\sum x = 130, \sum y = 220, \sum x^2 = 2288$$

$$\sum y^2 = 5506 \text{ and } \sum xy = 3467.$$

Hints: $b_{yx} = 1.015$, $b_{xy} = 0.9114$.
Therefore regression lines are
$y = 1.015 + 8.805$ and $x = 0.9114y - 7.0508$.
Putting $x = 16$ in the first line, we have $y = 25.045$ units.

15. In a single throw of two distinct dice, what is the probability of getting a total of 11?

Ans. $\frac{1}{18}$

16. Find the probability that a randomly chosen three-digit integer is divisible by 5.

Ans. $\frac{1}{5}$

17. Show that the number of distinguishable words that can be formed from the letters of MISSISSIPPI is 34650.

18. A certain defective dice is tossed. The probabilities of getting the faces 1 to 6 are respectively

$$p_1 = \frac{2}{18}, p_2 = \frac{3}{18}, p_3 = \frac{4}{18}, p_4 = \frac{3}{18},$$
$$p_5 = \frac{4}{18}, p_6 = \frac{2}{18}.$$

What is the probability that a prime number is on the top?

Ans. $\frac{11}{18}$

19. Let A and B be two events such that $P(A) = 0.4$, $P(B) = p$ and $P(A \cup B) = 6$. Find p so that A and B are independent.

Ans. $\frac{1}{3}$

20. A bag contains 3 red and 5 black balls and a second bag contains 6 red and 4 black balls. A ball is drawn from each bag. Find the probability that one ball is red and the other is black.

Ans. $\frac{21}{40}$

21. The probability of a man hitting a target is 1 3. If he fires six times, what is the probability that he hits the target

(i) at least twice
(ii) at most twice

Ans. $\frac{473}{729}, \frac{496}{729}$

22. A candidate takes on 20 questions, each with four multiple choices. One of the choice in every question is incorrect. The candidate makes guess of the remaining choices. Find the expected number of correct answers and the standard deviation.

Ans. $\frac{20}{3}, \sqrt{\frac{40}{9}}$

23. A random variable X has the following probability function:

x:	0	1	2	3	4	5	6	7
y:	0	k	$2k$	$3k$	k^2	$2k^2$	$2k^2$	$7k^2+k$

Find k, evaluate $P(X < 6)$, $P(X < 6)$, $P(3 < X \le 6)$ and find the minimum value of x so that $P(X \le x) > \frac{1}{2}$.

Ans. $k = 10$, $P(X < 6) = \frac{81}{100}$, $P(X \ge 6) = \frac{19}{100}$ $P(3 < X6) = \frac{33}{100}$, $x = 4$.

24. A die is tossed twice. Getting a number greater than 4 is considered a success. Find the variance of the probability distribution of the number of successes.

Ans. $\frac{4}{9}$

25. The frequency distribution of a measurable characteristic varying between 0 and 2 is as follows:

$$f(x) = \begin{cases} x^3, 0 \le x \le 1 \\ (2-x)^3, 1 \le x \le 2. \end{cases}$$

Calculate the standard deviation and the mean deviation about the mean.

Hint:

$$\mu = \frac{1}{2}\left[\int_0^2 xf(x)dx\right]$$

$$= \frac{1}{2}\left[\int_0^1 x^4dx + \int_0^1 x(2-x)^3dx\right] = 1,$$

$$\sigma^2 = \frac{1}{2}\left[\int_0^2 (x-1)^2 f(x)dx\right]$$

$$= \frac{1}{15} \text{ and so } \sigma = \frac{1}{\sqrt{15}}$$

Mean deviation for the mean

$$= \frac{1}{2}\left[\int_0^2 |x-\mu| f(x)dx\right] = \frac{1}{5}.$$

26. The diameter X of an electric cable is a sumed to be a continuous random variable with probability density function $f(x) = 6x(1-x)$, $0 \le x \le 1$. Determine a number k such that $P(X < k) = P(x > k)$.

Hint:

$$P(X < k) = P(X > k) \Rightarrow \int_0^k f(x)dx = \int_k^1 f(x)$$

$$\Rightarrow 6\int_0^k x(1-x)dx = 6\int_k^1 x(1-x)dx$$

$$\Rightarrow 3k^2 - 2k^3 = 1 - 3k^2 + 2k^3$$

$$\Rightarrow 4k^3 - 6k^2 + 1 = 0 \Rightarrow k = \frac{1 \pm \sqrt{3}}{2} \text{ and } k = \frac{1}{2}.$$

$k = \frac{1}{2}$ lies between 0 and 1

Ans. $\frac{1}{2}$

27. In a precision bombing attack there is a 50% chance that any bomb will strike the target. Two direct hits are required to destroy the target completely. How many bombs must be dropped to give a 99% chance or better to completely destroy the target?

Hint:

$$p = 1/2,\ q = 1 - \frac{1}{2} = 1/2$$

$$P(X = r) = \binom{n}{r}\left(\frac{1}{2}\right)^r\left(\frac{1}{2}\right)^{n-r} = \binom{n}{r}\left(\frac{1}{2}\right)^n$$

We should have $P(X \ge 2) \ge 0.99$ or $[1 - p(X \le 1)] \ge 0.99$

or $[1 - p(0) - p(1) \ge 0.99$

$$\text{or } \left[1 - \left\{\binom{n}{0} + \binom{n}{1}\right\}\left(\frac{1}{2}\right)^n\right] \ge 0.99$$

or $0.01 \ge \frac{1+n}{2^n}$ or $2n \ge 100 + 100n$.

Note that $n = 11$ satisfies this equation.

28. If, on an average 1 vessel in every 10 is wrecked, find the probability that out of 5 vessel's expected to arrive, at least 4 will arrive safely.

Hint: $p = \frac{1}{10}$, $q = \frac{9}{10}$. P (at the most one will be wrecked).

Therefore,

$$P(X \le 1) = p(0) + p(1) = nc_0 q^n + nc_1 q^{n-1} p.$$

$$= \left(\frac{9}{10}\right)^5 + 5\left(\frac{9}{10}\right)^4\left(\frac{1}{10}\right)$$

$$= \left(\frac{9}{10}\right)^4\left[\frac{9}{10} + \frac{5}{10}\right] = \frac{9^4(7)}{10^5} = \frac{45927}{50000}.$$

29. Fit a binomial distribution to the following frequency distribution:

x:	0	1	2	3	4	5	6
y:	13	25	52	58	32	16	4

Ans. $200(0.554 + 0.446)^6$

30. Six dice are thrown 729 times. How many times do you expect at least three dice to show a five or six?

Hint: Calculate $P(X \ge 3)$.

Ans. 233

Poisson's Distribution

31. In a certain factory turning razor blades, there is a small chance of 0.002 for any blade to be defective. The blades are supplied in packets, of 10. Use Poisson's distribution to calculate the approximate number of packets containing no defective, one defective, and two defective blades, respectively, in a consignment of 10,000 blades.

Ans. 9802,196,2

32. Show that in a Poisson distribution with unit mean, mean deviation about mean is $\frac{2}{e}$ times the standard deviation.

Hint: $P(X = x) = e^{-\lambda}\lambda^x/x$, Here $\lambda = 1$.

Therefore, $P(X = x) = \dfrac{e^{-1}}{x!}$

Mean deviation about mean 1 is

$$\sum |x-1| P(X = x)$$

$$= e^{-1}\left(1 + \frac{1}{2!} + \frac{2}{3!} + \ldots\right)$$

$$= e^{-1}\left\{1 + \left(1 - \frac{1}{2!}\right) + \left(\frac{1}{2!} - \frac{1}{3!}\right) + \left(\frac{1}{3!} - \frac{1}{4!}\right) + \ldots\right\}$$

$$= e^{-1}(1+1) = \frac{2}{e} \times 1 = \frac{2}{e} \times \textit{standard deviation}.$$

33. Fit a Poisson distribution to the following data:

x:	0	1	2	3	4
y:	419	352	154	56	19

Ans.

0	1	2	3	4
404.9	366	165.4	49.8	11.3

34. If the probability of a bad reaction from a certain injection is 0.001, determine the chance that out of 2,000 individuals more than 2 will get a bad reaction.

Hint: $\lambda = np = 2000(0.001)$, Probability

$$= 1 - \left(e^{-\lambda} + \frac{\lambda e^{-\lambda}}{1} + \frac{\lambda^2 e^{-\lambda}}{2\,!}\right) = 0.32.$$

35. If a random variable has a Poisson distribution such that $P(1) = P(2)$, find (i) mean of the distribution (ii) $P(4)$

Hint:

$$P(1) = P(2) \Rightarrow \lambda e^{-1} = \lambda^2 e^{-1}/2 \Rightarrow \lambda = 2$$

$$P(4) = \lambda^4 e^{-\lambda}/4\,! = 2^4 e^{-2}/4\,! = 2/3\ e^{-2}.$$

36. Fit a Poisson distribution to the following data:

x:	0	1	2	3	4
y:	192	100	24	3	1

Hint: $\lambda = \dfrac{\sum f_i x_i}{\sum f_i} = 0.503$, then the frequencies are $320\left[\dfrac{e^{0.503}(0.503)^r}{r!}\right]$.

37. The incidence of occupational disease in an industry is such that the workmen have a 10% chance of suffering from it. What is probability that in a group of 7, five, or more will suffer from the disease?

Ans. 0.0008

Normal Distribution

38. The mean yield of a crop for one-acre plot is 662 kg with a standard deviation 32 kg. Assuming normal distribution how many one-acre plots in a batch of 1,000 plots would you expect to have yield over 700 kg?

Hint:

$\mu = 662, \sigma = 32, z = \dfrac{x-\mu}{32} = 1.19$

$P(z > 1.19) = 0.1170.$

No. of plots $= 1000 \times 0.117 = 117.$

39. The mean and standard deviation of the marks obtained by 1,000 students in an examination are respectively, 34.4 and 16.5. Assuming the normality of the distribution, find the approximate number of students expected to obtain marks between 30 and 60.

Hint:

$$z_1 = \frac{30-34.4}{16.5} = -0.266,$$

$$z_2 = \frac{60-34.4}{16.5} = 1.552$$

$$P(-0.266 \le z \le 1.552)$$
$$= P(-0.27 \le z \le 0) + P(0 \le z \le 1.56)$$
$$= P(0 \le z \le 0.27) + P(0 \le z \le 1.56)$$
$$= 0.1064 + 0.4406 = 0.5470.$$

Therefore, number of students $= 1000 \times 0.5470 = 547.$

40. Fit a normal curve to the following data:

x:	0	1	2	3	4	5
frequency:	13	23	24	15	11	4

Hint:

$$\mu = \frac{\sum fx}{\sum f} = \frac{23+68+45+44+20}{100} = 2$$

$$\sigma = \sqrt{\frac{\sum fx}{\sum f} - \mu^2} = \sqrt{5.70-4} = 1.304$$

Normal curve is $y = \dfrac{100}{\sigma\sqrt{2\neq}} e^{-\frac{1}{2}\left(\frac{x-\mu}{\sigma}\right)^2} = \dfrac{100}{2\sqrt{2\neq}} e^{-\frac{(x-2)^2}{3.4}}$

41. If is known from past-experience that the number of telephone calls made daily in a certain community between 3 pm and 4 pm have a mean of 352 and a standard deviation of 31. What percentage of the time will there be for more than 400 telephone calls made in this community between 3 pm and 4 pm?

Ans. 6% approx.

42. If X is a normal variate with mean 30 and standard deviation 5, find the probability that $|X-5| > 5$.

Hint:

$$P(|X-5| \le 5) = P(25 \le X \le 35)$$
$$= P(-1 \le z \le 1)$$
$$= 2P(0 \le z \le 1)$$
$$= 2(0.3413) = 0.6826$$

Therefore, $P(|x-5| > 5) = 1 - 0.6826 = 0.3174.$

43. In a normal distribution, 10.03% of the items are under 25 kg weight and 89.97% of the items are under 70 kg weight. Find the mean and standard deviation of the distribution.

Ans. $\mu = 47.5kg, \sigma = 17.578kg$

Significance for Means

44. A sample of 900 members has a mean 3.4 cm and standard deviation 2.61 cm. Is this a sample from a large population of mean 3.25 and standard deviation 2.61 cm?

Ans. $z = \dfrac{\bar{x}-\mu}{\sigma}\sqrt{n} = 1.73$. Also 95% confidence interval: (3.2295, 3.5705). The mean 3.25 lies in the interval.

45. A sample of 30 pieces of a semi-conduction metrical gave an average of resistivity of 73.2 units with a sample standard deviation of 5.4 units. Obtain a 95% confidence interval for the resistivity of the material and test the hypothesis that this is 75 units.

Hint: $\bar{x} \pm 1.96 \frac{5}{\sqrt{n}}$.

Ans. (71.2, 75.2), accepted.

46. The mean of a certain normal population is equal to the standard error of the mean of the samples of 100 from that distribution. Find the probability that the mean of the sample of 25 from the distribution will be negative.

Hint:

$$\mu = \frac{\sigma}{\sqrt{100}} = \frac{\sigma}{10}, z = \frac{\bar{x} - \mu}{\sigma}\sqrt{n} = \frac{\bar{x} - \frac{\sigma}{10}}{\frac{\sigma}{\sqrt{n}}}$$

$$= \frac{5\bar{x}}{\sigma} - \frac{1}{2}$$

Since $\bar{x}$ *is* –ve, $z < -\frac{1}{2}$. Therefore,

$$P\left(z < \frac{1}{2}\right) = -\frac{1}{\sqrt{2\neq}} \int_{-\infty}^{-\frac{1}{2}} e^{-\frac{z^2}{2}} dz$$

$$= \frac{1}{\sqrt{2\pi}} \int_{\frac{1}{2}}^{\infty} e^{-\frac{z^2}{2}} dz = 0.3085$$

47. A sample of height of 6,400 soldiers has a mean of 67.85 inches and a standard deviation of 2.56 inches whereas a simple sample of heights of 1,600 sailors has a mean of 68.55 inches and a standard deviation of 2.52 inches. Do the data indicate that the sailors are on the average taller than soldiers?

Ans. Yes

48. A sample of 400 individuals is found to have a mean height of 67.47 inches. Can it be reasonably regarded as a sample from a large population with mean height of 67.39 inches and standard deviation 1.30 inches?

Hint: $\mu = 69.39$, $\sigma = 1.30$, $\bar{x} = 67.47$, $n = 400$, $z = 1.23$, Yes.

49. If 60 new entrants in a given university are found to have a mean height of 68.60 inches and 50 seniors a mean height of 69.51 inches, can we conclude that the mean height of the senior is greater than that of new entrants. Assume the standard deviation of height to be 2.48 inches.

Ans. No

50. Two kinds of a new plastic material are to be compared for strength. From tensile strength, measurement of 10 similar pieces of each type, the sample average and standard deviations were found as follows: $\bar{x}_1 = 78.3$, $S_1 = 5.6$, $x_2 = 84.2$, $S_2 = 6.3$ compare the mean strength, assuming normal data.

Hint: is not known, so calculate

$$\sigma^2 = \frac{n_1 S_1^2 + n_2 S_2^2}{n_1 + n_2}$$

$$= \frac{10}{20}[(5.6)^2 + (6.3)^2] = 35.525$$

$$\therefore \sigma = 5.96$$

(pooled estimate of standard deviation)

Then

$$z = \frac{\bar{x}_1 - \bar{x}_2}{\sigma\left(\frac{1}{n_1} + \frac{1}{n_2}\right)^{\frac{1}{2}}} = \frac{78.3 - 84.2}{\frac{5.96}{\sqrt{5}}} = -2.21$$

$|z| = 2.21 > 1.96$ implies that the difference is significant. Also 95% confidence interval

$$\text{is } \bar{x}_1 - \bar{x}_2 \pm 1.96\left[\sigma\sqrt{\left(\frac{1}{n_1} + \frac{1}{n_2}\right)}\right] = 78.3$$

$$-84.2 \pm 1.96\left(\frac{5.96}{\sqrt{5}}\right) = -5.90 \pm 4.95$$

$$= (-10.85, -0.95)$$

Since 0 does not lie within the interval, the difference is significant.

51. An examination was given to 50 students of a college A and to 60 students of college B. For A, the mean grade was 75 with standard deviation of 9 and for B, the mean grade was 79 with standard deviation of 7. Is there any significant difference between the performance of the students of college A and those of college B?

Ans. No

52. The mean yield and standard deviation of a set of 40 plots are 1258 kg and 34 kg whereas mean yield and standard deviation of another set of 60 plots are 1243 kg and 28 kg. Is the difference in the mean yields of two sets of plots significant?

Ans. $z = 2.3$, Yes at 5% level of confidence

Significance for single proportion

53. A random sample of 500 apples was taken from a large consignment and 60 were found to be bad. Obtain 98% confidence limits far the percentage of bad apples in the consignment.

Ans. (0.086, 0.154), that is, 8.6% to 15.4%

54. A bag contains defective articles, the exact number of which is not known. A sample of 100 from the bag gives 10 defective articles. Find the limits for the proportion of defective articles in the bag.

Ans. $0.1 \pm 1.96\sqrt{\dfrac{0.1(0.9)}{100}} = (0.0412, 0.1589)$

55. A sample of 1,000 days is taken from meteorological records of a certain district and 120 of them are found to be foggy. What are the probable limits to percentage of foggy days in the district?

Ans. 8.91% to 15.07%

Significance for Difference of Proportion

56. Before an increase in excise duty on tea, 800 persons out of a sample of 1,000 persons were found to be tea drinkers. After an increase in duty, 800 people were the drinkers in a sample of 1,200 people. Using standard error of proportion, state whether there is a significant decrease in the consumption of tea after the increase in excise duty.

Ans. $z = 6.84$, significant decrease

57. One type of aircraft is found to develop engine trouble in 5 flights out of a total of 100 and another type in 7 flights out of 200 flights. Is there a significant difference in the two types of aircrafts so far as defects are concerned?

Ans. Difference is not significant

58. In a random sample of 400 students of the university teaching departments, it was found that 300 students failed in the examination. On another random sample of 500 students of the affiliated colleges, the number of failures in the same examination was found to be 300. Find out whether the proportion of failures in the university teaching departments is significantly greater than the proportion of failures in the university teaching departments and affiliated colleges taken together.

Ans. $z = 4.08$

59. In a random sample of 100 men taken from village A, 60 were found to be consuming alcohol. In another sample of 200 men taken from village B, 100 were found to be consuming alcohol. Do the two villages differ significantly of the proportion of men who consume alcohol?

Ans. $z = 1.64$

60. 500 articles from a factory are examined and found to be 2% defective. 800 similar articles from another factory are found to be only 1.5% defective. Can we conclude that the products of the first factory are inferior to those of the second?

Ans. $z = 0.68$, No

Significance for Difference of Standard Deviations

61. Random samples drawn from two countries A and B gave the following data regarding the heights (in inches) of the adult males

	Country A	Country B
Mean height	67.42	67.25
Standard deviation	2.58	2.50
Number in sample	1000	1200

Is the difference between the standard deviations significant?

Ans. $z = \dfrac{s_1 - s_2}{\sqrt{\dfrac{S_1^2}{2n_1} + \dfrac{S_2^2}{2n_2}}}$

62. In Exercise 10, examine whether the difference in the variability in yields is significant.

    Ans. $z = 1.31$, Difference not significant at 5% level of significance.

t-distribution

63. A random sample of eight envelops is taken from letter box of a post office. The weights in grams are found to be 12.1, 11.9, 12.4, 12.3, 11.9, 12.1, 12.4, and 12.1. Find 99% confidence limits for the mean weight of the envelopes received at the post office.

    Hint:

$$\bar{x} = 12.15, S = 0.2, \bar{x} \pm t_{0.05}\cdot \frac{S}{\sqrt{n}}$$

$$= 12.15 \pm 2.35 \frac{0.2}{\sqrt{8}}$$

$$= (11.984, 12.316).$$

64. The nine items of a sample have the following values: 45, 47, 50, 52, 48, 47, 49, 53, 51. Does the mean of these differ significantly from the assumed mean of 47.5?

    Ans. Not significant at 5% level of significances

65. Two horses A and B were tested according to the time (in seconds) to run a particular track with the following results:

Horse A:	28	30	32	33	33	29	34
Horse B:	29	30	30	24	27	29	

Test whether the two horses have the same running capacity (use t-test)

Ans. t = 2.5, Yes

66. A sample of 10 measurements of the diameter of a sphere gave a mean of 12 cm and a standard deviation of 0.15 cm. Find 95% confidence limit for the actual diameter.

    Ans. (11.887, 12.113)

67. For a random sample of 10 pigs fed on diet A, the increase in weight in a certain period were 16, 6, 16, 17, 13, 12, 8, 14, 15, 9 kg. For another random sample of 12 pigs fed on diet B, the increases in the same period were 7, 13, 22, 15, 12, 14, 18, 8, 21, 23, 10, 17 kg. Are these two samples significantly different regarding the effect of diet?

    Ans. $t = 1.51$, Sample mean do not differ significantly

68. A car company has to decide between two brands A and B of tyre for its car. A trial is conducted using 12 of each brand, run until they wear out. The sample average and standard deviations of running distance (in km) are, respectively, 36,300 and 5,000 for A, and 39100 and 6100 for B. Obtain a 95% confidence interval for the difference in means assuming the distribution to be normal and test the hypothesis that brand B tyres outrun brand A tyres.

    Hint:

$$S^2 = \frac{1}{n_1 + n_2 - 2}[(n_1 - 1)S_1^2 + (n_2 - 1)S_2^2].$$

Here $n_1 = n_2 = 12$.

Degree of freedom $= n_1 + n_2 - 2 = 24 - 2 = 22$, $t_{0.05}$ at $v = 22$ is 1.71.

95% confidence interval is 36300 - 39100 ±

$$1.71\, S\sqrt{\frac{1}{n_1} + \frac{1}{n_2}}.$$

Also $t = \dfrac{36300 - 39100}{S\sqrt{\frac{1}{n_1} + \frac{1}{n_2}}}$. Find t and compare with $t_{0.05}$

x^2–Distribution

69. The following figures show the distribution of digits in numbers chosen at random from a telephone directory:

Digits:	Frequency:
0	1026
1	1107
2	997
3	966

Digits:	Frequency:
4	1075
5	933
6	1107
7	972
8	964
9	853
Total	10000

Test whether the digits may be taken to occur equally and frequently in the directory.

Ans. $x^2 = 58.542$, $x^2_{0.05}(9) = 16.92$

70. A set of five similar coins is tossed 320 times and the result is

N.of heads:	0	1	2	3	4	5
Frequency:	6	27	72	112	71	32

Test the hypothesis that the data follow a binomial distribution.

Ans. $x^2 = 78.68$;,$x^2_{0.05}(5) = 11.07$, hypothesis rejected

71. Fit a normal distribution to the data given below and test the goodness-of-fit.

x:	50	55	60	65	70	75	80	85	90	95	100
f:	2	3	5	9	10	12	7	2	3	1	0

Ans. good fit.

72. The following table gives the number of aircraft accidents occurred during the days of the week. Find whether the accidents are uniformly distributed over the week.

Days:	No.of accidents:
Monday	14
Tuesday	18
Wednesday	12
Thursday	11
Friday	15
Saturday	14

Ans. $x^2 = 2.14$

73. During proof reading 392 pages of a book of 1,200 pages were read. The distribution of printing mistakes were found to be as follows:

No.of mistakes in pages(x)	0	1	2	3	4	5	6
No.of page(f)	275	72	30	7	5	2	1

Fit a Poisson distribution to the above data and test the goodness-of-fit.

Hints: The expected (theoretical) frequencies are 242.1, 116.7, 28.1, 4.5, 0.5, 0.1, 0. Further,

$x^2 = 40.937$, $x^2_{0.05}(2) = 5.99$. Not a good fit.

74. A survey of 800 families with four children were taken. Each revealed the following distribution:

No.of boys:	0	1	2	3	4
No of girls:	4	3	2	1	0
No.of families:	32	178	290	236	64

Test the hypothesis that male and female births are equally possible.

Hint: Probability for boy's birth $(p) = \frac{1}{2}$, so $q = \frac{1}{2}$. Fit binomial distribution to male birth, which is 50, 200, 300, 200, 50. Then proceed to find x^2, which is 19.63 and $x^2_{0.05}(4) = 9.488$. Hypothesis rejected.

F-distribution

75. Two samples of sizes 8 and 10, respectively, give the sum of the squares of deviations from their respective means equals to 84.4 and 102.6, respectively. Examine whether the samples have been drawn from normal population having the same variance.

Ans. $F = 1.057$, $F_{0.05}(7,9) = 3.29$ Hypothesis accepted at 5% level of significance.

76. Two random samples from two normal populations are given below. Do the estimates of population variance differ significantly?

Sample A:	16	26	27	23	24	22
Sample B:	33	42	35	32	28	31

Ans. $F = 1.49$, Do not differ significantly

77. The following are the values in thousands of an inch obtained by two engineers in 10 successive measurements with the same micrometer. Is one engineer significantly more consistent than the other?

Engineer A:	Engineer B:
503	502
505	497
497	492
505	498
495	499
502	495
499	497
493	496
510	498
501	

Ans. $F = 2.4$ $F_{0.05}(9, 8) = 5.47$ Equally consistent.

14 Linear Programming

Linear programming is an extremely efficient algorithm, developed by Danzig in the 1940s, to optimize (maximizing or minimizing) a real valued linear function of several real variables subject to a number of constraints expressed in the form of linear inequalities or linear equations. In engineering, our aim is always to get the best out of a system. We desire to obtain maximum amount of product with minimum cost of the process involved. Such problems of optimization occur in expansive areas in engieering fields such as steel industries, chemical industries, and space industries. Linear programming provides satisfactory solutions to such problems.

14.1 LINEAR PROGRAMMING PROBLEMS

A problem involving linear programming in its solution is called a *linear programming* problem, generally written as LPP.

A general LPP with n variables and m costraints can be expressed in the following way:

Optimize

$$z = a_1x_1 + a_2x_2 + \ldots + a_nx_n \quad (1)$$

subject to the constraints

$$\left.\begin{aligned} &b_{11}x_1 + b_{12}x_2 + \ldots + b_{1n}x_n (\leq, =, \geq) c_1 \\ &b_{21}x_1 + b_{22}x_2 + \ldots + b_{2n}x_n (\leq, =, \geq) c_2 \\ &\ldots\ldots\ldots\ldots\ldots\ldots \\ &\ldots\ldots\ldots\ldots\ldots\ldots \\ &b_{m1}x_1 + b_{m2}x_2 + \ldots + b_{mn}x_n (\leq, =, \geq) c_m \end{aligned}\right\} \quad (2)$$

and

$$x_1, x_2, x_3, \ldots, x_n \geq 0. \quad (3)$$

The linear function z, which is to be optimized, is called the *objective function* of the LPP. The variables $x_1, x_2, \ldots, x_n$ involved in LPP are called *decision* or *structural variables*. The equations or inequalities (2), with one of the signs $\leq$, $=$ or $\geq$ are called the constraints of the LPP. The inequalities (3) represent the set of *non-negative restrictions* of the LPP. The constants $a_1, a_2, \ldots, a_n$ represent the contribution to the objective function by $x_1, x_2, \ldots, x_n$, respectively. The constants $c_1, c_2, \ldots, c_m$ are the constants representing the availability of the constraints. The coefficients b_{ij}, $i = 1, 2, \ldots, m; j = 1, 2, \ldots, n$ are called *technological constants.*

If the inequalities/equalities in an LPP are plotted as a graph, then the area for which all the inequalities/equalities are satisfied is called the *feasible region.*

A region or a set of points is said to be *convex* if the line segment joining any two points of the region (set) lies entirely in the region (set). For example, the regions

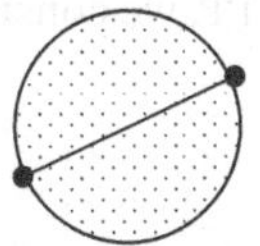

and

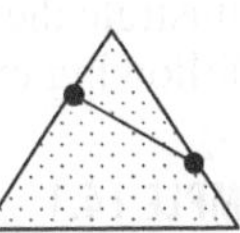

are convex, whereas the regions

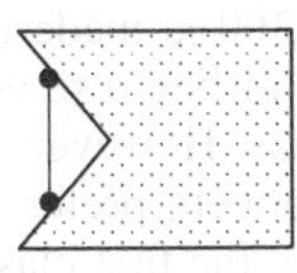

and

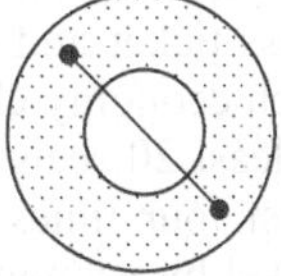

are not convex.

The feasible region of an LPP *is always a convex region whose boundary consists of line segments.*

A set of values of the decision variables which satisfy all the constraints of an LPP is called a *solution* to that LPP.

A solution of an LPP that also satisfies the non-negativity restrictions of the problem is called a *feasible solution* to that problem.

Any feasible solution which optimizes (maximize or minimize) the objective function of an LPP is called an *optimal* (or *optimum*) solution of the LPP. The value of the objective function at an optimal solution in called an *optimal value.*

14.2 FORMULATION OF AN LPP

To solve a given LPP, we first define the variables involved, establish relationship between these variables, identify the objective function, and then express the constraints as linear inequations/equations.

Thus formulation of LPP involves the following steps:

1. Identification of the decision variables x_1, $x_2, x_3 \ldots$
2. Identification of the objective function z and its expression as a linear function of the decision variables $x_1, x_2, \ldots$
3. Identification of constraints and then representation as linear equations/inequations in terms of the decision variables $x_1, x_2, x_3, \ldots$
4. Addition of non-negativity restriction on the decision variables, that is, $x_1 \geq 0$, $x_2 \geq 0$, $\ldots x_n \geq 0$. This is done since negative value of the decision variables have no meaning.

To illustrate the formulation of LPP, we consider the following examples.

EXAMPLE 14.1

An aeroplane can carry a maximum of 200 passengers. A profit of Rs 400 is made on each first class ticket and a profit of Rs 300 is made on each economy class ticket. The airline reserves at least 20 seats for the first class. However, at least four times as many passengers prefer to travel by economy class than by the first class. How many tickets of each class must be sold to maximize profit for the airline? Formulate the problem as linear programming (LP) model.

Solution. Let x and y denote the number of passengers travelling by first class and economy class, respectively. Thus the decision variables are x and y. The profit on first class ticket is Rs 400 and the profit on the economy class ticket is Rs 300. Therefore, the objective function is

$$z = 400x + 300y.$$

Since the maximum number of passengers carried is 200, we have the restriction

$$x + y \leq 200.$$

Since at least 20 seats are reserved for first-class category, we have $x \geq 20$. Since at least four times as many passengers prefer to travel by economy class than by the first class, we have $y \geq 4x$. Further, the non-negative restrictions are $x \geq 0$, $y \geq 0$. Hence, the mathematical formulation of the given LPP is

Maximize $z = 400x + 300y$

subject to the constraints

$$x + y \leq 200$$

$$x \geq 20,\ y \geq 4x$$

and $x \geq 0$, $y \geq 0$.

EXAMPLE 14.2

A factory manufactures two types of cylinder, C_1 and C_2. Three materials M_1, M_2, and M_3 are required for the manufacture of each cylinder. The quantities of materials required and available are as follows:

Quantities of materials required

	M_1	M_2	M_3
C_1	1	1	2
C_2	5	2	2

Quantities of materials available

M_1	M_2	M_3
45	21	24

Rs 50 profit is earned on one C_1 and Rs 40 profit is earned on one C_2. How many of each cylinder should the factory manufacture to maximize the profit? Formulate the problem as an LP model.

Solution. Suppose

No. of C_1 cylinders manufactured $= x$

No. of C_2 cylinders manufactured $= y$

The profit earned on each C_1 cylinder is Rs 50, whereas the profit earned on each C_2 cylinder is Rs 40. Therefore, the objective function is

$$z = 50x + 40y.$$

Each C_1 requires 1 unit of M_1 whereas each C_2 requires 5 units of M_1 and since the quantity of M_1 available is 45, we have

$$x + 5y \leq 45.$$

Similarly, for the material M_2 and M_3, we have

$$x + 2y \leq 21 \text{ and } 2x + 2y \leq 24.$$

Further, the non-negative restriction is $x \geq 0$, $y \geq 0$.

Hence, the mathematical model of the given LPP is

Maximize

$$z = 50x + 40y$$

subject to the constraints

$$x + 5y \leq 45$$
$$x + 2y \leq 21$$
$$2x + 2y \leq 24$$
$$\text{and } x \geq 0, y \geq 0.$$

EXAMPLE 14.3

A pineapple firm produces two products: canned pineapples and canned juice. The specific amounts of material, labour, and equipments required to produce each product and availability of each of the resources are shown below:

	Canned juice	Canned pine apple	Available resource
Labour (man hours)	3	2.0	12.0
Equipments (machine hours)	1	2.3	6.9
Material (units)	1	1.4	4.9

Profit margin on 1 unit each of canned juice and canned pineapple is Rs 2 and Rs 1, respectively. How many of each product should the firm produce to maximize the profit? Formulate the problem as LP model.

Solution. Let the firm should produce x units of canned juice and y units of canned pineapple so as to get maximum profit. Taking profit margin on each unit into account, the objective function is

$$z = 2x + y.$$

Now considering the requirement of material, labour, equipments and the corresponding availability of these, we have from the given table,

$$3x + 2y \leq 12$$
$$x + 2.3y \leq 6.9$$
$$x + 1.4y \leq 4.9.$$

Also, the productions cannot be negative. Therefore $x \geq 0$ and $y \geq 0$. Hence the mathematical formulation of the given problem is

Maximize

$$z = 2x + y$$

subject to the constraints

$$x + 2y \leq 12$$
$$x + 2.3y \leq 6.9$$
$$x + 1.4y \leq 4.9$$
$$\text{and } x \geq 0, y \geq 0.$$

14.3 GRAPHICAL METHOD TO SOLVE LPP

This method is used to solve LPPs involving only two variables. In this method, we plot the given constraints as equations in the co-ordinate plane and find the convex region (feasible region) formed by them. The points lying within the feasible region satisfies all the constraints. The value of objective function at each vertex of the feasible region is determined. The vertex which gives the optimal (maximum or minimum) value of the objective function gives the required solution to the LPP. Since the solution occurs at an extreme point (corner) of the feasible region, this method is also called *corner point method.*

EXAMPLE 14.4

A factory manufactures nails and screws. The profit earned is Rs 2/kg nails and Rs 3/kg screws. Three units of labours are required to manufacture 1 kg nails and 6 units to make 1 kg screws. Twenty four units of labour are available. Two units of raw materials are needed to make 1 kg nails and 1 unit for 1 kg screws. Determine the manufacturing policy that yields maximum profit from 10 units of raw materials.

Solution. The formulation of LPP yields

Maximize

$$z = 2x + 3y$$

subject to the constraints

$$3x + 6y \leq 24$$

$$2x + y \leq 10$$

$$\text{and } x \geq 0; y \geq 0$$

Consider a rectangular set of axes (Ox, Oy) with O as the origin. Since $x \geq 0$ and $y \geq 0$, it follows that the feasible region lies in the first quadrant. Converting the inequalities of the constraints into equations, we get respectively

$$3x + 6y = 24,$$

$$2x + y = 10,$$

which are straight lines. The convex region bounded by these lines and $x \geq 0, y \geq 0$ is shown as shaded area in the Figure 14.1.

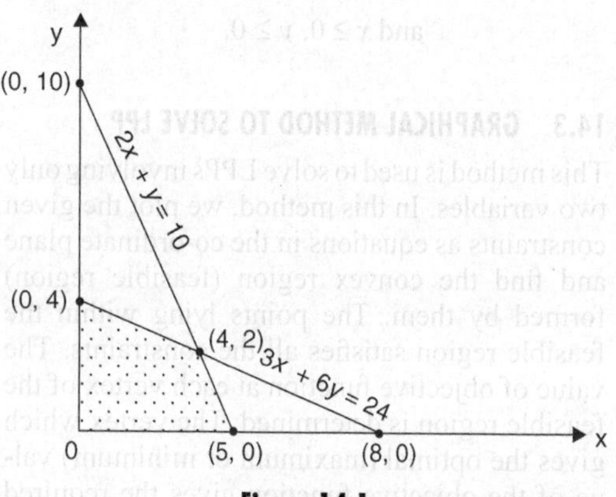

Figure 14.1

The vertices (corners) of the feasible (convex) region are (0,0), (5,0), (0,4), and (4,2). The values of the objective function $z = 2x + 3y$ at these points are 0, 10, 12, and 14, respectively. Since the maximum value of z is 14 at the vertex (4,2), the optimal solution to give LPP is

$$x = 4, y = 2, \text{ and Max. } z = 14.$$

EXAMPLE 14.5

Using Corner method, solve the following LPP:

Minimize

$$z = 8x_1 + 12x_2$$

subject to the constraints

$$60x_1 + 30x_2 \geq 240$$

$$30x_1 + 60x_2 \geq 300$$

$$30x_1 + 180x_2 \geq 540$$

$$\text{and } x_1, x_2 \geq 0.$$

Solution. Since $x_1, x_2 \geq 0$, the feasible region lies in the first quadrant. Plot each of the constraints treating them as linear equations. Using the inequality condition of each constraint, the feasible region of the given LPP is as shown by shaded area in the Figure 14.2.

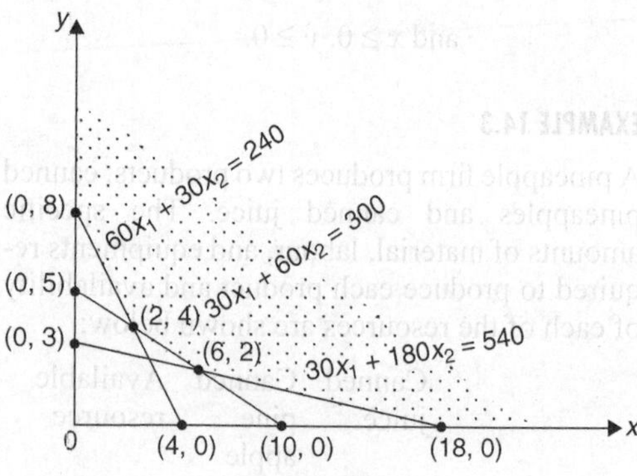

Figure 14.2

The corners of the feasible regions are (0,8), (2,4), (6,2), and (18,0). The values of the objective function $z = 8x_1 + 12x_2$ at these vertices are 96, 64, 72, and 144, respectively. Since the minimum value of z is at (2, 4), the optimal solution of the given LPP is $x_1 = 2, x_2 = 4$, and $z = 64$.

EXAMPLE 14.6

Using graphical method, find the maximum value of $z = 2x + 3y$ subject to the constraints $x + y \le 30, y \ge 3, 0 \le y \le 12, 0 \le x \le 20, x - y \ge 0,$ and $x, y \ge 0$.

Solution. The non-negativity of constraints $x, y \ge 0$ shows that the feasible region lies in the first quadrant only. Treating each constraint as linear equation, we plot each of them in xy plane. Then using the inequality condition of each constraint, the feasible region of the given LPP is shown by shaded area in the Figure 14.3.

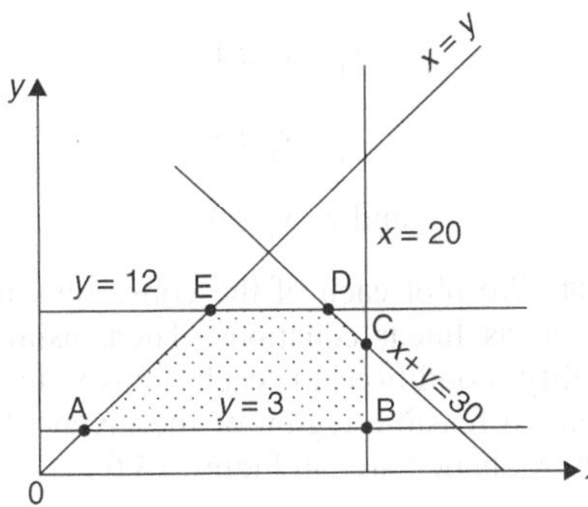

Figure 14.3

The co-ordinates of the vertices (corners) of the feasible region are

$$A(3, 3), B(20, 3), C(20, 10), D(18, 12), \text{ and } E(12, 12).$$

The values of the objective function $z = 2x + 3y$ at these corners are

$$z(A) = 15, z(B) = 49, z(C) = 70, z(D) = 72, \text{ and } z(E) = 60.$$

The maximum value of the objective function is 72 and it occurs at the corner $D(18, 12)$ Hence, the solution of the given LPP is

$$x = 18, y = 12, \text{ and Max. } z = 72.$$

EXAMPLE 14.7

Using graphical method, solve the LPP

$$\text{Max. } z = 5x_1 + 3x_2$$

subject to the constraints

$$3x_1 + 5x_2 \le 15$$

$$5x_1 + 2x_2 \le 10$$

$$\text{and } x_1, x_2 \ge 0.$$

Solution. The convex region OABCO of the LPP is shown by the shaded area in the Figure 14.4.

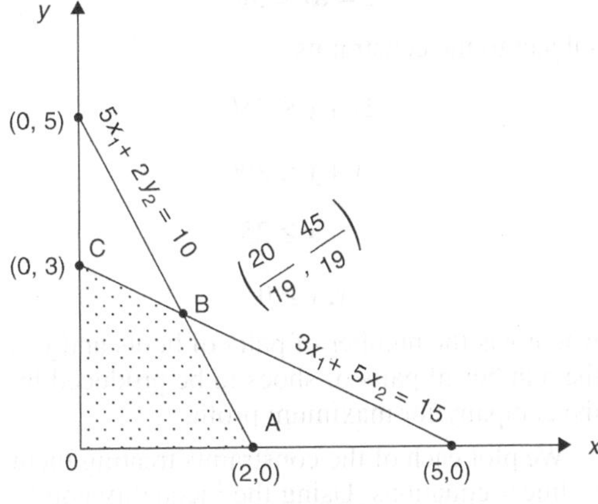

Figure 14.4

The co-ordinates of the vertices (corners) are

$$A(2, 0), B\left(\frac{20}{19}, \frac{45}{19}\right), C(0, 3), \text{ and } O(0, 0).$$

The values of the objective function $z = 5x_1 + 3x_2$ at these corners are

$$z(A) = 10, z(B) = \frac{235}{19}, z(C) = 9, \text{ and } z(0) = 0.$$

The maximum value of z is at $\left(\frac{20}{19}, \frac{45}{19}\right)$. Therefore, the solution to the given LPP is

$$x_1 = \frac{20}{19}, x_2 = \frac{45}{19}, \text{ and Max. } z = \frac{235}{19}.$$

EXAMPLE 14.8

A footwear company produces boots and shoes. If no boots are made, the company can produce a maximum of 250 pairs of shoes in a day. Each pair of boots takes twice as long to make as each pair of shoes. The maximum sales of boots and shoes daily are 200, but 25 pairs of boots must be produced to satisfy an important customer. The profit per pair of boots and shoes are Rs 8 and Rs 5, respectively. Determine the daily production plan to maximize the profits.

Solution. The formulation of the LPP yields

Maximize

$$z = 8x + 5y$$

subject to the constraints

$$2x + y \le 250$$

$$x + y \le 200$$

$$x \ge 25$$

$$x, y \ge 0,$$

where x is the number of pairs of boots and y is the number of pairs of shoes to be produced by the company for maximum profit.

We plot each of the constraints treating them as linear equations. Using the inequality condition of each constraint, the feasible region of the given LPP is shown in the Figure 14.5

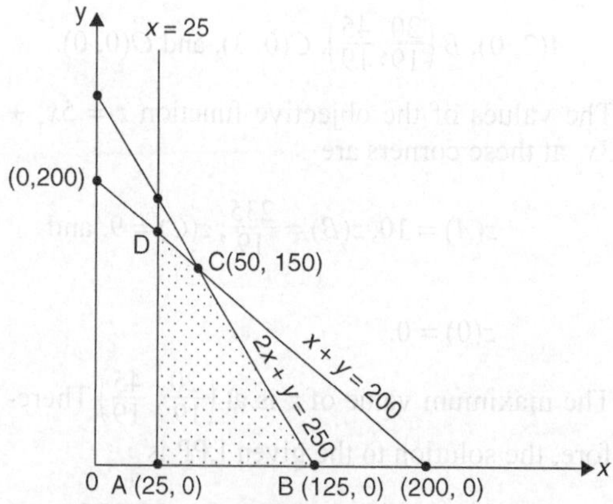

Figure 14.5

The feasible region is ABCD. The co-ordinates of A, B, C, and D are

$$A(25, 0), B(125, 0), C(50, 150), D(25, 175).$$

The values of the objective function at these vertices are

$$z(A) = 200, z(B) = 1000, z(C) = 1150,$$

$$z(D) = 1075.$$

Hence the solution is

$$x = 50, y = 150, \text{ and Max. } z = 1150.$$

EXAMPLE 14.9

Use graphical method to solve

Maximize

$$z = 3x_1 + 2x_2$$

subject to the constraints

$$x_1 - x_2 \ge 1$$

$$x_1 + x_2 \ge 3$$

$$\text{and } x_1, x_2 \ge 0.$$

Solution. We plot each of the constraints treating them as linear equations. Then, using the inequality condition of each constraint, the unbounded feasible region of the given LPP is shown by shaded area in Figure 14.6.

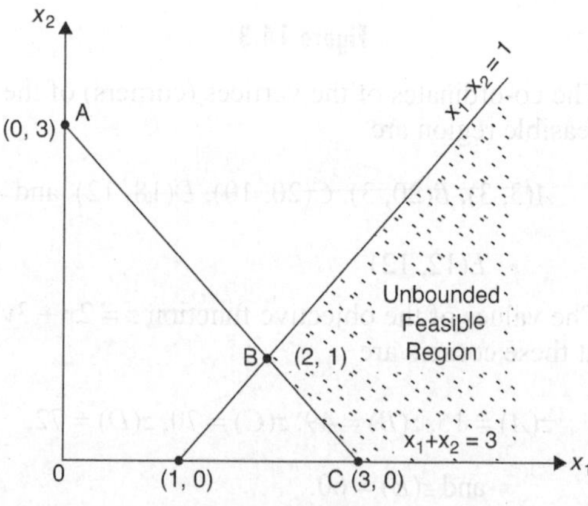

Figure 14.6

The convex region is unbounded. The two corner points are C(3, 0) and B(2, 1). The values of the objective function $z = 3x_1 + 2x_2$ at these points are 9 and 8, respectively. But there are other points, e.g., (3, 1), at which the value is more than 9. In fact, in this region the value of z becomes as large as we please. Hence the given LPP has an *unbounded solution.*

EXAMPLE 14.10

Does the following LPP has a feasible solution?

Maximize

$$z = x + y$$

subject to the constraints

$$x - y \geq 0$$
$$3x - y \leq -3$$
$$\text{and } x, y \geq 0.$$

Solution. Plot the graph of constraints treating them as linear equations. Use the inequality condition of both constraints to get the region shown in Figure 14.7.

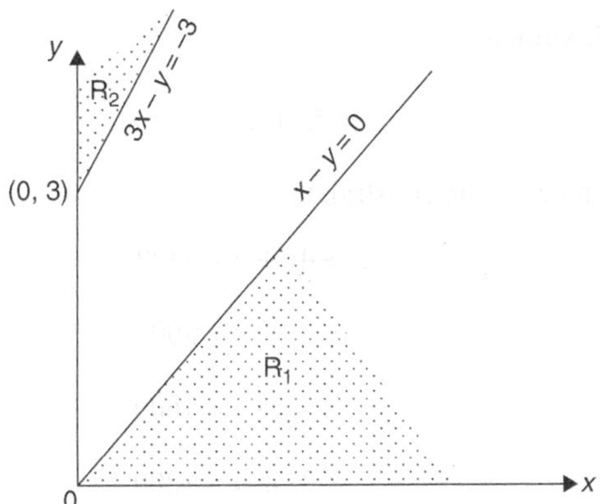

Figure 14.7

It follows from the graph that the feasible region R is the region common to the regions R_1 and R_2, that is, $R = R_1 \cap R_2$. But $R_1 \cap R_2 = \phi$ and so $R = \phi$. This means that there is no set of values of the variables x and y that satisfies both the constraints. Hence there is *no feasible solution* to the given LPP.

14.4 CANONICAL AND STANDARD FORMS OF LPP

Before solving an LPP, it should be presented in a suitable form. In fact, an LPP is presented in one of the following form:

1. **Canonical form:** Applying some elementary transformations, the LPP can be expressed as Maximize

$$z = a_1x_1 + a_2x_2 + \ldots + a_nx_n$$

subject to the constraints

$$b_{11}x_1 + b_{12}x_2 + \ldots + b_{1n}x_n \leq c_1$$
$$b_{21}x_1 + b_{22}x_2 + \ldots + b_{22}n_{xn} \leq c_2$$
$$\vdots$$
$$b_{m1}x_1 + b_{m2}x_2 + \ldots + b_{mn}x_n \leq c_m,$$
$$x_1, x_2, \ldots, x_n \geq 0.$$

This form of the LPP is called *canonical form.* We observe that, *in canonical form of an LPP, the objective function is of maximum type, the constraints are of "≤" type, and the variables x_i are non-negative.*

2. **Standard form:** By adding or subtracting some variables to each constraint, LPP can be expressed as

Maximize

$$z = a_1x_1 + a_2x_2 + \ldots + a_nx_n$$

subject to the constraints

$$b_{11}x_1 + b_{12}x_2 + \ldots + b_{1n}x_n + s_1 = c_1$$
$$b_{21}x_1 + b_{22}x_2 + \ldots + b_{22}x_n + s_2 = c_2$$
$$b_{m1}x_1 + b_{m2}x_2 + \ldots + b_{mn}x_n + s_m = c_m$$
$$\vdots$$
$$x_1, x_2, \ldots, x_n \geq 0.$$

This form of the LPP is called its *standard form.* Thus, *in standard form of an LPP, the objective function is of maximize type, all constraints are in the form of equations, the right-hand side of each constraint and all variables are nonnegative.*

The non-negative variables added to the lefthand side of the constraints

$$\sum_{j=1}^{n} b_{ij}x_i \leq c_i (i = 1, 2 \ldots, m)$$

to convert the constraints into equalities are called *slack variables.* Thus if s_i are slack variables, then

$$\sum_{j=1}^{n} b_{ij}x_i + s_i = c_i (i = 1, 2, \ldots, m).$$

The non-negative variables subtracted from the left-hand sides of the constraints

$$\sum_{j=1}^{n} b_{ij}x_i \geq c_i, \; i = 1, 2, \ldots, m$$

to convert the constraints into equalities are called *surplus variables.*

Obviously, an LPP can be expressed in standard form using slack/surplus variables.

14.5 BASIC FEASIBLE SOLUTION OF AN LPP

Suppose in an LPP, there are m constraints and $m + n$ variables. Then the starting solution, if it exits, of the LPP is obtained by setting n variable equal to zero and then solving the remaining m equations. The n zero variables are called *non-basic variables* whereas the remaining m variables are called *basic variables.* The solution so obtained is called *basic solution.* Obviously, the number of basic solution is ${}^{m+n}C_m$.

The solution in which each basic variable is non-negative is called *basic feasible solution.*

If one or more of the basic variables in a basic feasible solution are zero, then the solution is called a *degenerate solution.*

If all the basic variables in a basic feasible solution are positive, then the solution is called a *non-degenerate solution.*

EXAMPLE 14.11

Express the following LPP in standard form:

Maximize

$$z = 5x + 9y$$

subject to the constraints

$$3x + 4y \leq 2400$$

$$x + 2y \leq 900$$

$$2x + 3y \leq 1600$$

$$2x \leq 1200$$

$$\text{and } x \geq 0, y \geq 0$$

and hence obtain its basic feasible solutions and optimal basic feasible solution.

Solution. Introducing the slack variables r, s, t, and u, the standard form of the given LPP is

Maximize

$$z = 5x + 9y$$

subject to the constraints

$$3x + 4y + r = 2400$$

$$x + 2y + s = 900$$

$$2x + 3y + t = 1600$$

$$2x + u = 1200$$

$$x, y, r, s, t, u \geq 0.$$

We have four constraint equations and six variables. Thus for a basic solution, we put 6 – 4 = 2 variables equal to zero and solve the constraint equations for the remaining four variables. There are ${}^6C_4 = 15$ various possibilities of putting two variables equal to zero. Thus we have the following table:

Non-basic variables	*Basic variables*	*Basic solution*	*Feasibility*	*Value of Z*
$x = y = 0$	r, s, t, u	$r = 2400, s = 900$ $t = 1600, u = 1200$	Feasible	0
$x = r = 0$	y, s, t, u	$y = 600, s = -300$ $t = -200, u = 1200$	Non-feasible	–

Continued

Non-basic variables	*Basic variables*	*Basic solution*	*Feasibility*	*Value of Z*
$x = s = 0$	y, r, t, u	$y = 450, s = 600$ $t = 250, u = 1200$	Feasible	4050
$x = t = 0$	y, r, s, u	$y = \frac{1600}{3}, r = \frac{800}{3}$ $s = -\frac{500}{3}, u = 1200$	Non-feasible	–
$x = u = 0$		The equation $2x + u = 1200$ is not satisfied		
$y = r = 0$	x, s, t, u	$x = 800, s = 100$ $t = 0, u = -400$	Non-feasible	–
$y = s = 0$	x, r, t, u	$x = 900, r = 0$ $t = -200, u = -600$	Non-feasible	–
$y = t = 0$	x, r, s, u	$x = 800, r = 0$ $s = 100, u = -400$	Non-feasible	–
$y = u = 0$	x, r, s, t	$x = 600, r = 600$ $s = 300, t = 400$	Feasible	3000
$r = s = 0$	x, y, t, u	$x = 600, y = 150$ $t = -50, u = 0$	Non-feasible	–
$r = t = 0$	x, y, s, u	$x = 800, y = 0$ $s = 100, u = -400$	Non-feasible	–
$r = u = 0$	x, y, s, t	$x = 600, y = 150$ $s = 0, t = -50$	Non-feasible	–
$s = t = 0$	x, y, r, u	$x = 500, y = 200$ $r = 100, u = 200$	Feasible	4300
$s = u = 0$	x, y, r, t	$x = 600, y = 150$ $r = 0, t = -50$	Non-feasible	–
$t = u = 0$	x, y, r, s	$x = 600, y = \frac{400}{3}$ $r = \frac{200}{3}, s = \frac{100}{3}$	Feasible	4200

Thus the basic feasible solutions are

(i) $x = y = 0$,

(ii) $x = 0, y = 450$,

(iii) $x = 600, y = 0$,

(iv) $x = 500, y = 200$, and

(v) $x = 600, y = \frac{400}{3}$.

The optimal basic feasible solution is

$$x = 500, y = 200, \text{ and Max. } z = 4300.$$

EXAMPLE 14.12

Express the following LPP in standard form and determine the vertices algebraically.

Maximize

$$u = 4x + 3y$$

subject to constraints

$$x + y \leq 4$$

$$-x + y \leq 2$$

$$\text{and } x, y \geq 0.$$

Solution. Introducing the slack variables r and t, the standard form (slack form) of the given LPP is

Maximize

$$u = 4x + 3y$$

subject to constraints

$$x + y + r = 4$$

$$-x + y + t = 2$$

$$\text{and } x, y, r, t \geq 0.$$

We have four variables and two constraint equations. Thus for a basic solution, we put 4 – 2 = 2 variables equal to zero and solve algebraically the constraint equations for the remaining two variables. There are ${}^4C_2 = 6$ various possibilities of putting two variables equal to zero. Thus, we have the following table:

Non-basic variables	Basic variables	Basic solution	Feasibility
$x = y = 0$	r, t	$r = 4$, $t = 2$	Feasible
$x = r = 0$	y, t	$y = 4$, $t = -2$	Non-feasible
$x = t = 0$	y, r	$y = 2$, $r = 2$	Feasible
$y = r = 0$	x, t	$x = 4$, $t = 6$	Feasible
$y = t = 0$	x, r	$x = -2$, $r = 6$	Non-feasible
$r = t = 0$	x, y	$x = 1$, $y = 3$	Feasible

The feasible solution yields the following four vertices:

$$(0, 0), (0, 2), (4, 0), (1, 3).$$

EXAMPLE 14.13

Determine the basic solution to the following LPP:

Maximize

$$u = x + 3y + 3z$$

subject to the constraints

$$x + 2y + 3z = 4$$

$$2x + 3y + 5z = 7$$

$$\text{and } x, y, z \geq 0.$$

Also point out the degenerate basic feasible solution and optimal basic feasible solution.

Solution. In the given LPP, there are three variables x, y, z and two constraints equations. Thus to get solutions, we put 3 – 2 = 1 variable equal to zero and solve for the other two variables. The total number of basic solution is ${}^3C_2 = 3$. The various possibilities of choosing basic and nonbasic variables are shown in the following table.

Non-basic variables	Basic variables	Basic solution	Solution feasible or not	Solution degenerate or not
z	x, y	$x = 2$, $y = 1$	Yes	No
y	x, z	$x = 1$, $z = 1$	Yes	No
x	y, z	$y = -1$, $z = 2$	No	No

The values of the objective function z for these solutions are, respectively, 5, 4, and 3. Hence the optimal basic feasible solution is

$$x = 2, y = 1, z = 0, \text{ and Max. } u = 5.$$

EXAMPLE 14.14

Show that the following system of linear equations has two degenerate feasible basic solutions and the non-degenerate basic solution is not feasible:

$$2x_1 + x_2 - x_3 = 2, 3x_1 - 2x_2 + x_3 = 3.$$

Solution. There are three variables and only two equations. Thus to get basic solution, we put $3 - 2 = 1$ variable equal to zero and solve for the other two variables. The number of basic solutions in this case ${}^3C_2 = 3$. The possibilities of selecting basic and non-basic variables are shown in the table follows.

Non-basic variable	Basic variables	Basic solution	Feasibility	Degenerate or not
$x_1 = 0$	x_2, x_3	$x_2 = \frac{5}{3}$, $x_3 = -\frac{1}{3}$	Non-feasible	Non-degenerate
$x_2 = 0$	x_1, x_3	$x_1 = 1$, $x_3 = 0$	Feasible	Degenerate
$x_3 = 0$	x_1, x_2	$x_1 = 1$, $x_2 = 0$	Feasible	Degenerate

Thus the given system of linear equation has two degenerate basic solutions. Further, the nondegenerates basic solution

$$x_1 = 0, x_2 = \frac{5}{3}, x_3 = -\frac{1}{3} \text{ is not feasible:}$$

14.6 SIMPLEX METHOD

The graphical method to solve a LPP will only work if the LPP has two variables because the treatment for three variables is very difficult and complicated. Further, in the graphical method, the feasible region was found to be convex and bounded by vertices. Then the optimal solution occurred at some corner of the region. Thus it is only necessary to inspect the corners of the feasible region. In many-dimensional problem, the most popular method for finding the vertex at which optimal solution exists is the *simplex method* developed by George B. Danzig.

In simplex method, we select a starting corner, choose the neighbouring corner that increases the objective function the most and then repeat the process until no improvement is possible. *The algebraic equivalent of moving to a neighbouring corner is to increase one of the non-basic variables in the standard form of the LPP from zero to its largest possible value.*

We now describe *essentials of the simplex method* with the help of an example. Consider the following LPP.
Maximize

$$z = 2x + 3y$$

subject to the constraints

$$3x + 6y \leq 24$$

$$2x + y \leq 10$$

$$\text{and } x \geq 0, y \geq 0,$$

which was already solved graphically in Example 14.4.

Introducing slack variables r and s, the constraints convert into

$$3x + 6y + r = 24 \quad (1)$$

$$2x + y + s = 10 \quad (2)$$

$$\text{and } x, y, r, s \geq 0.$$

We now have more variables than the number of equations. To construct a feasible basic solution,we have $x = y = 0$ (non-basic variables) and $r = 24$, $s = 10$ (basic variables) with two *basic variables* (which are non-zero and equal to the number of constraints) and two of *non-basic variables* (the remainder of the variables and which are zero). The objective function is

$$z = 2x + 3y.$$

With the above selection of basic and non-basic variables, the value of the objective function z is zero. To move to a neighbouring vertex, we increase one of the non-basic variables from zero to its largest possible value. Since the coefficient of y is larger, it will increase z the most. So, we keep $x = 0$ and increase y to its maximum value. There are two possibilities:

(a) Change y to 4 and reduce r to zero.

(b) Change y to 10 and reduce s to zero.

If we change y to 10, then (4) gives $r = 24 - 60 = -36$ (negative) which violates the condition that r is positive. Thus we choose $y = 4$. We interchange r and y between the set of basic and non-basic variables and rewrite (4) and (5). This is done by solving for y from (4) and substituting in (5). We get

$$3x + 6y + r = 24 \tag{6}$$

$$\frac{3}{2}x - \frac{r}{6} + s = 6 \tag{7}$$

and the objective function

$$z = \frac{1}{2}x - \frac{1}{2}r + 12. \tag{8}$$

The problem has reduced again to the standard form of LPP and so the above procedure can be repeated. Now we have

$x = r = 0$ (non-basic variables) and $y = 4$, $s = 6$ (basic variables) and the value of z is 12.

Since the objective function is (8), the increase in r would decrease z. Therefore, only x can be increased. Again there are two possibilities:

(c) Change x to 8 and reduce y to zero.

(d) Change x to 4 and reduce s to zero.

The choice (c) is ruled out because taking $x = 8$, the equation (7) yields s as negative which contradicts the fact that basic variables are positive. So we apply (d). From (7), we have $x = \frac{2}{3}\left(6 + \frac{r}{6} - s\right)$. We interchange x and s in the set of basic and non-basic variables.

Then putting $x = \frac{2}{3}\left(6 + \frac{r}{6} - s\right)$ in (6), the new constraint equations are

$$\frac{4}{3}r - s + 6y = 12$$

and

$$\frac{3}{2}x - \frac{r}{6} + s = 6.$$

The basic and non-basic variables are

$$r = s = 0 \text{ and } x = 4, y = 2.$$

The objective function (8) takes the form

$$z = \frac{1}{2}\left[\frac{2}{3}\left(6 + \frac{r}{6} - s\right)\right] - \frac{r}{2} + 12$$

$$= 14 - \frac{5r}{9} - \frac{s}{3}.$$

Since increasing r and s now would decrease z, we have reached the solution. Hence the solution of the given LPP is

$$x = 4, y = 2, \text{ and Max. } z = 14,$$

which is in total agreement with the graphical solution.

14.7 TABULAR FORM OF THE SOLUTION

The objective function z is placed in the first row with minus inserted because it ensure that z remains positive in subsequent steps. The basic variables are written in the left-hand column. The coefficients in the constraint equations are placed in appropriate array elements. Thus the standard display to start the simplex method (for the above example) is as shown below.

		Non-basic variables		Basic variables		Solution
		x	y	r	s	
Objective function	z	−2	−3	0	0	0
Basic variables	r	3	6	1	0	24
	s	2	1	0	1	10

The *current solution* is $z = 0$. The basic variables are $r = 24$, $s = 10$. The non-basic variables are $x = y = 0$. In the basic variables columns, a 2x2 unit matrix occurs in the table with zeros occurring above in z-row.

The simplex algorithm is now performed in a series of the following steps:

Step 1. Choose the most negative entry in the z-row and mark the column in which this entry lies (the y-column in the present example) with an arrow at the bottom.

Step 2. Calculate the ratios of the solution and *positive entries* in the marked column (y-column in the present case). Choose the smallest ratio and mark the row in which this ratio lies (the r-row in the present case) with an arrow to the right. Encircle the element where the marked row and marked column intersect. The table, in view of Step1 and Step2, reduces to

	x	y	r	s	Solution	Ratios	
z	−2	−3	0	0	0		
r	3	(6)	1	0	24	4	←
s	2	1	0	1	10	10	
		↑					

The element 6 is the *pivot element* and the r-row is the *pivot row.*

Step 3. Change the marked basic variable in the left-hand column to the marked non-basic variables in the top row (in the present example, replace r by y).

Step 4. Make the pivot equal to 1 by dividing through (in the present case, we divide the element of r-row by 6). The table corresponding to Step 3 and Step 4 becomes

	x	y	r	s	Solution
z	−2	−3	0	0	0
y	$\frac{1}{2}$	(1)	$\frac{1}{6}$	0	4
s	2	1	0	1	10

Step 5. Use Gaussian elimination method to annihilate the elements in the y column using pivot and pivot row. This leads us to the table given below:

	x	y	r	s	Solution
z	$-\frac{1}{2}$	0	$\frac{1}{2}$	0	12
y	$\frac{1}{2}$	1	$\frac{1}{6}$	0	4
s	$\frac{3}{2}$	0	$-\frac{1}{6}$	1	6

This table is again in the standard form and so the above five steps can be applied again. Steps 1 and 2 yield the following table:

	x	y	r	s	Solution	Ratios	
z	$-\frac{1}{2}$	0	$\frac{1}{2}$	0	12		
y	$\frac{1}{2}$	1	$\frac{1}{6}$	0	4	8	
s	($\frac{3}{2}$)	0	$-\frac{1}{6}$	1	6	4	←
	↑						

The applications of Steps 3, 4, and 5 yields the following tables:

	x	y	r	s	Solution
z	0	0	$\frac{4}{9}$	1/3	14
y	0	1	$\frac{1}{9}$	-1/3	2
x	1	0	$-\frac{1}{9}$	2/3	4

All the entries in the z-row are now non-negative. Hence the optimum is achieved. From the table, the required solution to the given LPP is

$$x = 4, y = 2, \text{ and Max. } z = 14.$$

14.8 GENERALIZATION OF SIMPLEX ALGORITHM

Consider the LPP:

Maximize

$$z = a_1x_1 + a_2x_2 + \ldots + a_nx_n$$

subject to the constraints

$$b_{11}x_1 + b_{12}x_2 + \ldots + b_{1n}x_n \le c_1$$

$$b_{21}x_1 + b_{22}x_2 + \ldots + b_{2n}x_n \le c_2$$

$$\ldots\ldots\ldots\ldots$$

$$\ldots\ldots\ldots\ldots$$

$$b_{m1}x_1 + b_{m2}x_2 + \ldots + b_{mn}x_n \le c_m$$

$$x_1, x_2, \ldots x_n \ge 0 \text{ and } c_1, c_2, \ldots c_n \ge 0.$$

If $x_{n+1}, x_{n+2}, \ldots, x_{n+m}$ are slack variables, then the standard table for this problem is

	x_1	x_2	...	x_n	x_{n+1}	x_{n+2}	...	x_{n+m}	Solution
z	$-a_1$	$-a_2$	...	$-a_n$	0	0	...	0	0
x_{n+1}	b_{11}	b_{12}	...	b_{1n}	1	0	...	0	c_1
x_{n+2}	b_{21}	b_{22}	...	b_{2n}	0	1	...	0	c_2
.	.	.	...	.	.	.	...	.	.
.	.	.	...	.	.	.	...	.	.
.	.	.	...	.	.	.	...	.	.
x_{n+m}	b_{m1}	b_{m2}	...	b_{mn}	0	0	...	1	c_m

Then the simplex algorithm is

1. *Identify optimal column i:* Choose the most negative value in the z-row, say $-a_i$. In case all entries are non-negative, then the maximum has been achieved.
2. *Identify pivot row j:* Determine ratios $\frac{c_1}{b_{1i}}, \frac{c_2}{b_{2i}}, \ldots, \frac{c_m}{b_{mi}}$. Choose the minimum ratio, say $\frac{c_j}{b_{ji}}$. Then j-row is the pivot row and b_{ji} is the pivot element.
3. *Change the basic variables:* Replace the basic variable x_{n+j} in the left-hand column by x_i.
4. *Reduce pivot to 1:* In row j replace b_{jk} by $\frac{b_{jk}}{b_{ji}}$ for $k = 1, 2, \ldots, n + m + 1$.
5. *Gaussian elimination:* Using Gaussian elimination, annihilate the column i except the pivot. The algorithm is repeated until at Step 1, the maximum is achieved.

Remarks:

1. In case we proceed without taking negative of the coefficient a_i in objective function in the initial table, then we have to take negative of the optimal solution in z-row. The process is stopped when all entries in z-row become non-positive.
2. One exception occurs at Step 2 above, when all the $b_{1i}, b_{2i}, \ldots, b_{mi}$ in the optimal column i are *zero* or *negative* and it becomes impossible to identify the row j to continue the method. The feasible region in this case is *unbounded* and so the solution is *unbounded.*

EXAMPLE 14.15

Find the maximum of

$$z = 4x_1 + 10x_2$$

subject to

$$2x_1 + x_2 \le 50$$
$$2x_1 + 5x_2 \le 100$$
$$2x_1 + 3x_2 \le 90$$
$$\text{and } x_1, x_2 \ge 0$$

Solution. Introducing the slack variables r, s, and t, the basic feasible solution is

$$x_1 = x_2 = 0, r = 50, s = 100 \text{ and } t = 90.$$

Therefore, the standard (initial) tableau for this LPP is

	x_1	x_2	r	s	t	Solution	
z	−4	−10	0	0	0	0	Ratios
r	2	1	1	0	0	50	$\frac{50}{1} = 50$
s	2	(5)	0	1	0	100	$\frac{100}{5} = 20 \leftarrow$
t	2	3	0	0	1	90	$\frac{90}{3} = 30$
		↑					

The most negative value in the z-row is -10. Therefore, we have marked the column containing this element. Since minimum ratio lies is the s-row, the pivot element is 5. Making the pivot equal to 1 and replacing s by x_2, we get the following table:

	x_1	x_2	r	s	t	Solution
z	-4	-10	0	0	0	0
r	2	1	1	0	0	50
x_2	$\frac{2}{5}$	(1)	0	$\frac{1}{5}$	0	20
t	2	3	0	0	1	90

Using Gauss elimination, this table transforms to

	x_1	x_2	r	s	t	Solution
z	0	0	0	2	0	200
r	$\frac{8}{5}$	0	1	$-\frac{1}{5}$	0	30
x_2	$\frac{2}{5}$	1	0	$\frac{1}{5}$	0	20
t	$\frac{4}{5}$	0	0	$-\frac{3}{5}$	1	30

Since all the entries in z-row are non-negative, the required solution is

$$x_1 = 0, x_2 = 20, \text{ and Max. } z = 200.$$

EXAMPLE 14.16

Use the simplex method to solve the problem Maximize

$$u = 2x + 3y$$

subject to
$$-2x + 3y \leq 2$$
$$3x + 2y \leq 5$$

$$\text{and } x, y \geq 0.$$

Solution. Introducing the slack variables, the standard form of the given LPP is

Maximize

$$u = 2x + 3y$$

subject to

$$-2x + 3y + r = 2$$

$$3x + 2y + t = 5$$

$$\text{and } x, y, r, t \geq 0.$$

The basic feasible solution is $x = y = 0$, $r = 2$, $t = 5$. Therefore, the initial basic feasible solution table is

	x	y	r	t	Solution	Ratios
u	-2	-3	0	0	0	
r	-2	(3)	1	0	2	$\frac{2}{3}$ ←
t	3	2	0	1	5	$\frac{5}{2}$
		↑				

The most negative value in the u-row is -3 which lies in y-column. Thus the key column is y-column. The minimum positive ratio lies in r-row. Therefore, the pivot element is 3 and the pivotal row is r-row. We divide the pivotal row throughout by 3 so that pivot becomes 1. We replace r by y in the left column and perform Gauss elimination. Thus we get the following table:

	x	y	r	t	Solution	Ratios
u	-4	0	1	0	2	
y	$-\frac{2}{3}$	1	$\frac{1}{3}$	0	$\frac{2}{3}$	$-$ve
t	($\frac{13}{3}$)	0	$-\frac{2}{3}$	1	$\frac{11}{3}$	$\frac{11}{13}$ ←
	↑					

Now the most negative value in the u-row is -4. Therefore the key column is x-column. The minimum positive ratio lies in t-row. Therefore the pivot element is $\frac{13}{3}$. We divide the pivot row throughout by $\frac{13}{3}$ so that pivot becomes 1. Then replace t by x in the left column and perform Gauss elimination to get the following table:

	x	y	r	t	Solution
u	0	0	$\frac{5}{13}$	$\frac{12}{13}$	$\frac{70}{13}$
y	0	1	$\frac{9}{39}$	$\frac{6}{39}$	$\frac{48}{39}$
x	1	0	$-\frac{2}{13}$	$\frac{3}{13}$	$\frac{11}{13}$

Since all entries in u-row are non-negative, we have achieved the solution. Hence the solution is

$$x = \frac{11}{13}, y = \frac{48}{39} \text{ and Max. } u = \frac{70}{13}.$$

EXAMPLE 14.17

A publisher has three books available for printing. B_1, B_2, and B_3. The paper requirement of the books, total paper supplies, and profit per thousand copies are given in the table below:

	B_1	B_2	B_3	Total units available
Units of paper required per 1,000 copies	3	2	1	60
Profit per 1,000 copies	Rs 900	Rs 800	Rs 300	

Books B_1 and B_2 are similar in contents and total combined market for these two books is estimated to be at most 15,000 copies. Determine how many copies of each book should be printed to maximize overall profit?

Solution. Considering 1,000 copies as one unit, let x, y, and z be the units of books to be published. Then the LPP is

Maximize

$$p = 900x + 800y + 300z$$

subject to

$$3x + 2y + z \le 60$$

$$x + y \le 15$$

$$\text{and } x, y \ge 0.$$

Since B_1 and B_2 are similar in contents and profit, in term of unit of paper required, is Rs 300 per unit of paper for B_1 and Rs 400 per unit of paper for B_2, the publisher should not publish B_1. Thus, $x = 0$ in the above set of equations/inequalities. Hence the problem reduces to Maximize

$$p = 800y + 300z$$

subject to

$$2y + z \le 60$$

$$y \le 15$$

$$\text{and } y, z \ge 0$$

Introducing slack variables, we have

$$p = 800y + 300z + 0r + 0s$$

$$2y + z + r = 60$$

$$y + s = 15$$

$$y, z, r, s \ge 0.$$

Therefore, the initial table corresponding to $y = z = 0$, $r = 60$, $s = 15$, $p = 0$ is

	y	z	r	s	Solution	
p	−800	−300	0	0	0	Ratios
r	2	1	1	0	60	$\frac{60}{2} = 30$
s	(1)	0	0	1	15	$\frac{15}{1} = 15 \leftarrow$
	↑					

The Gauss elimination yields the following table:

	y	z	r	s	Solution	
p	0	−300	0	800	12,000	Ratios
r	0	(1)	1	−2	30	$\frac{30}{1} = 30 \leftarrow$
y	1	0	0	1	15	$\frac{15}{0} = \infty$
		↑				

Now the pivot element is 1 in r-row. Thus Gauss elimination yields

	y	z	r	s	Solution
p	0	0	300	200	21,000
z	0	1	1	−2	30
y	1	0	0	1	15

Since all the entries in p-row are non-negative, the optimal solution has been achieved. The solution is

$$x = 0 \text{ unit}, y = 15 \text{ unit}, z = 30 \text{ units, and}$$

$$p = \text{Rs } 21000.$$

Thus,
$B_1 = 0$ and $B_2 = 15,000$ books, $B_3 = 30,000$ books with the profit of Rs 21,000.

EXAMPLE 14.18

Use simplex method to solve the following LPP:

Maximize

$$z = x_1 + x_2 + 3x_3$$

subject to the constraints

$$3x_1 + 2x_2 + x_3 \leq 3$$

$$2x_1 + x_2 + 2x_3 \leq 2$$

$$\text{and } x_1, x_2, x_3 \geq 0.$$

Solution. Introducing the slack variables, the standard form of the LPP is

Maximize

$$z = x_1 + x_2 + 3x_3 + 0r + 0s$$

subject to conditions

$$3x_1 + 2x_2 + x_3 + r + 0s = 3$$

$$2x_1 + x_2 + 2x_3 + 0r + s = 2$$

$$x_1, x_2, x_3, r, s \geq 0.$$

Therefore, the basic feasible solution is

$$x_1 = x_2 = x_3 = 0, r = 3, s = 2.$$

Thus, the initial basic feasible solution is shown by the following table:

	x_1	x_2	x_3	r	s	Solution	
z	−1	−1	−3	0	0	0	Ratios
r	3	2	1	1	0	3	$\frac{3}{1} = 3$
s	2	1	(2)	0	1	2	$\frac{2}{2} = 1 \leftarrow$
			↑				

The most negative value of z is −3. The minimum ratio is 1. Therefore, pivot element is 2. Making pivot equal to 1, replacing s by x_3 and annihilating the x_3-column using Gaussian elimination, we get the following table:

	x_1	x_2	x_3	r	s	Solution
z	2	$\frac{1}{2}$	0	0	$\frac{3}{2}$	3
r	2	$\frac{3}{2}$	0	1	$-\frac{1}{2}$	2
x_3	1	$\frac{1}{2}$	1	0	$\frac{1}{2}$	1

Since all the entries in the z-row are non-negative, the optimal solution has been achieved. The optimal solution is

$$x_1 = x_2 = 0, x_3 = 1, \text{ and Max. } z = 3.$$

EXAMPLE 14.19

Use simplex method to solve the following LPP:

Minimize

$$z = x_1 - 3x_2 + 2x_3$$

subject to the constraints

$$3x_1 - x_2 + 2x_3 \leq 7$$

$$-2x_1 + 4x_2 \leq 12$$

$$-4x_1 + 3x_2 + 8x_3 \leq 10$$

$$\text{and } x_1, x_2, x_3 \geq 0.$$

Solution. We first convert the given problem to the maximization problem by taking

Maximize

$$z' = -z = -x_1 + 3x_2 - 2x_3$$

subject to conditions

$$3x_1 - x_2 + 2x_3 \le 7$$

$$-2x_1 + 4x_2 \le 12$$

$$-4x_1 + 3x_2 + 8x_3 \le 10.$$

Introducing slack variables, the standard form of the LPP in question is

Maximize

$$z' = -x_1 + 3x_2 - 2x_3 + 0r + 0s + 0t$$

subject to constraints

$$3x_1 - x_2 + 2x_3 + r + 0s + 0t = 7$$

$$-2x_1 + 4x_2 + 0x_3 + 0r + s + 0t = 12$$

$$-4x_1 + 3x_2 + 8x_3 + 0r + 0s + t = 10$$

$$x_1, x_2, x_3, r, s, t \ge 0.$$

The basic feasible solution is

$$x_1 = x_2 = x_3 = 0, r = 7, s = 12, t = 10.$$

Thus, the initial basic feasible solution is shown-by the table:

	x_1	x_2	x_3	r	s	t	Solution	
z'	1	−3	2	0	0	0	0	Ratios
r	3	−1	2	1	0	0	7	
s	−2	(4)	0	0	1	0	12	3 ←
t	−4	3	8	0	0	1	10	$\frac{10}{3}$
		↑						

The most negative value of z' is in x_2-column and the minimum ratio is in s-row. Therefore, pivot element is 4. Making the pivot element 1 by dividing s-row throughout by 4, replacing s by x_2 and applying Gauss elimination to annihilate the element in x_2-column, we get the following table:

	x_1	x_2	x_3	r	t	s	Solution	
z'	$-\frac{1}{2}$	0	2	0	$\frac{3}{4}$	0	9	Ratio
r	$\left(\frac{5}{2}\right)$	0	2	1	$\frac{1}{4}$	0	10	4 ←
x_2	$-\frac{1}{2}$	1	0	0	$\frac{1}{4}$	0	3	
t	$-\frac{5}{2}$	0	8	0	$-\frac{3}{4}$	1	1	
	↑							

Now, the pivot element is $\frac{5}{2}$. Making it equal to 1 by dividing throughout by $\frac{5}{2}$, replacing r by x_1 and using Gauss elimination method, we get the table:

	x_1	x_2	x_3	r	s	r	Solution
z'	0	0	$\frac{12}{5}$	$\frac{1}{5}$	$\frac{4}{5}$	0	11
x_1	1	0	$\frac{4}{5}$	$\frac{2}{5}$	$\frac{1}{10}$	0	4
x_2	0	1	$\frac{2}{5}$	$\frac{1}{5}$	$\frac{1}{20}$	3	5
t	0	0	10	$\frac{1}{2}$	$-\frac{1}{2}$	1	11

Since all the entries in z'-row are non-negative, the solution to the problem is achieved. Therefore, the solution to the problem is

$$x_1 = 4, x_2 = 5, x_3 = 0 \text{ and } z = -z' = -11.$$

EXAMPLE 14.20

Use simplex method to solve the following LPP:

Maximize

$$p = x + 4y - z$$

subject to the constraints

$$-5x + 6y - 2z \le 30$$

$$-x + 3y + 6z \le 12$$

$$\text{and } x, y, z \ge 0.$$

Solution. Introducing the slack variables, the standard form of the given problem is

$$p = x + 4y - z + 0r + 0s$$

subject to the constraints

$$-5x + 6y - 2z + r = 30$$

$$-x + 3y + 6z + s = 12$$

$$x, y, z, r, s \ge 0.$$

The basic feasible solution is

$$x = y = z = 0, r = 30, s = 12.$$

Thus the initial basic feasible solution is shown in the following table:

	x	y	z	r	s	Solution	
p	−1	−4	1	0	0	0	Ratios
r	−5	6	−2	1	0	30	5
s	−1	(3)	6	0	1	12	4 ←
		↑					

The most negative value of p is in y-column and then the minimum ratio is in s-row. Therefore, the pivot element is 3. Making the pivot element 1 by dividing s-row throughout by 3, replacing s by y and applying Gauss elimination to annihilate the element in y-column, we get the following table:

	x	y	z	r	s	Solution
p	$-\frac{7}{3}$	0	9	0	$\frac{4}{3}$	16
r	−3	0	−14	1	−2	6
y	$-\frac{1}{3}$	1	2	0	$\frac{1}{3}$	4

The most negative value of p is in x-column but the elements −3 and $-\frac{1}{3}$ in the optimal column are both negative and it becomes impossible to identify a row to continue the method. The region in this case is *unbounded*. Hence the problem has *unbounded solution*.

EXAMPLE 14.21

Solve the following LPP by simplex method.
Maximize

$$u = x + y$$

subject to

$$-x + y \le 1$$

$$x - 2y \le 4$$

$$\text{and } x, y \ge 0.$$

Solution. Introducing the slack variables, the standard form of the given LPP is
Maximize

$$u = x + y + 0r + 0t$$

subject to

$$-x + y + r = 1$$

$$x - 2y + t = 4$$

$$x, y, r, t \ge 0.$$

The basic feasible solution is

$$x = y = 0, r = 1, t = 4$$

Therefore, the initial basic solution table is

	x	y	r	t	Solution	
u	−1	−1	0	0	0	Ratios
r	−1	(1)	1	0	1	1 ←
t	1	−2	0	1	4	−2
		↑				

Since the coefficient of u in x-column and y-column are equal, we may choose any of these columns as the key column. Let us choose y-column as the key column. Then positive ratio lies with r-row. Therefore the pivot is 1 and pivotal row is r-row. Replacing r by y and using Gaussian elimination, we get the following table:

	x	y	r	t	Solution
u	−2	0	1	0	1
y	−1	1	1	0	1
t	−1	0	2	1	6

The most negative value of u is now in x-column but the coefficients in x-column are both negative. Thus it is impossible to identify a row to continue the process. The region in this case is *unbounded.* Hence the LPP has unbounded solution.

14.9 TWO-PHASE METHOD

While dealing with maximizing linear programming so far, we observed that the constraints contained only '≤' sign. The question arises "What happens if some constraints contains '≥' sign? In such a case, we do not have obvious initial feasible solution. Such problems are solved by *two-phase method.* The first phase of the method is concerned only to get initial feasible solution. After obtaining initial feasible solution, the *phase 2 is simply the standard simplex method* discussed already. As an illustration, we consider the following example:

Maximize

$$p = 3x + 2y + 3z$$

subject to constraints

$$2x + y + z \le 2$$

$$3x + 4y + 2z \ge 8$$

$$\text{and } x, y, z \ge 0.$$

Using slack variable r and surplus variable t, the constraints yield

$$2x + y + z + r = 2$$

$$3x + 4y + 2z - t = 8$$

$$x, y, z, r, t \ge 0.$$

Then the solution $x = y = z = 0, r = 2, t = -8$ does not satisfy the condition that all variables are non-negative. Thus origin is not in the feasible region. The last equation is forced into standard form by adding one more variable u, called an *artificial variable*, to give

$$3x + 4y + 2z - t + u = 8.$$

Now we have a feasible solution $x = y = z = t = 0, r = 2, u = 8$. Of course, this is not feasible solution to the original problem. Our aim is to get rid of the artificial variable u so that the problem reduces to the original one. Thus our task in phase 1 is to make u equal to zero. So we take new objective (cost) function

$$p' = -u.$$

Obviously, if we maximize p', it will happen at $u = 0$. As soon as $u = 0$, the task of phase 1 shall be over. Thus, the simplex table for phase 1 is of the form:

	x	y	z	r	t	u	Solution
p	−3	−2	−3	0	0	0	0
p'	0	0	0	0	0	1	0
r	2	1	1	1	0	0	2
u	3	4	2	0	−1	1	8

Here p has been included for the elimination and so does not enter the optimization.

Since (p', u) entry is 1 and not zero, the above table is not of standard form. We convert it to standard form by subtracting u-row from the p'-row. Thus the standard initial table for phase 1 becomes

	x	y	z	r	t	u	Solution	Ratio
p	−3	−2	−3	0	0	0	0	
p'	−3	−4	−2	0	1	0	−8	
r	2	1	1	1	0	0	2	2
u	3	(4)	2	0	−1	1	8	2 ←
		↑						

The most negative value of p' is −4 and that lies in y-column. Thus optimal column is the y-column. Since the ratios are equal, it is a tie case. Dividing the coefficients of r and u column by respective elements in the y-column, the ratios are $\frac{1}{1}, \frac{0}{1}$ under the r-column and $\frac{0}{1}$ and $\frac{1}{4}$ in the u column. Therefore, u-row is the pivot row and 4 is the pivot. Making the pivot equal to 1 by dividing throughout by 4, replacing u by y and applying Gauss elimination, we get the following table:

	x	y	z	r	t	u	Solution
p	$-\frac{3}{2}$	0	−2	0	$-\frac{1}{2}$	$\frac{1}{2}$	4
p'	0	0	0	0	0	1	0
r	$\frac{5}{4}$	0	$\frac{1}{2}$	1	$\frac{1}{4}$	$-\frac{1}{4}$	0
y	$\frac{3}{4}$	1	$\frac{1}{2}$	0	$-\frac{1}{4}$	$\frac{1}{4}$	2

The table shows that p' is zero and $u = 0$. Thus the phase 1 is over.

To enter into the phase 2, we delete p'-row and the u-column from the above table to get the initial table given below:

	x	y	z	r	t	Solution	Ratios
p	$-\frac{3}{2}$	0	−2	0	$-\frac{1}{2}$	4	
r	$\frac{5}{4}$	0	$\boxed{\frac{1}{2}}$	1	$\frac{1}{4}$	0	0 ←
y	$\frac{3}{4}$	1	$\frac{1}{2}$ ↑	0	$-\frac{1}{4}$	2	4

The most negative value in p-row is −2. Therefore, the optimal column is z-column. Further minimum ratio lies in r-row. Therefore the pivot is $\frac{1}{2}$. Making pivot equal to 1, replacing r by z and using Gauss elimination, we get the following table:

	x	y	z	r	t	Solution
p	$\frac{7}{2}$	0	0	4	$\frac{1}{2}$	4
z	$\frac{5}{2}$	0	1	2	$\frac{1}{2}$	0
y	$-\frac{1}{2}$	1	0	−1	$-\frac{1}{2}$	2

Since all entries in p-row are non-negative, we have arrived at the solution. The solution of the given LPP is

$$x = 0, y = 2, z = 0, \text{ and Max. } p = 4.$$

Thus the steps for two-phase method are:

Phase 1:

1. Introduce slack and surplus variables.
2. *Introduce artificial variables in the constraints where surplus variables are used.*
3. If $u_1, u_2, \ldots, u_n$ are artificial variables, then write the artificial objective function.
$$z' = -u_1 - u_2 - \ldots - u_n$$
4. Subtract rows $u_1, u_2, \ldots, u_n$ form the objective function row z' to ensure that there are zeros in the entries in the z'-row corresponding to the basic variables.
5. Maximize z' until $z' = 0$ and $u_1, u_2, \ldots, u_n = 0$.

Phase 2:

1. Eliminates the z'-row and artificial columns $u_1, u_2, \ldots, u_n$ from the table reached at in phase1.
2. Maximize z using standard simplex method.

EXAMPLE 14.22

Solve the following LPP:

Maximize

$$p = 8x + 5y$$

subject to the constraints

$$2x + y \leq 250$$

$$x + y \leq 200$$

$$x \geq 25.$$

$$\text{and } x, y \geq 0.$$

Solution. We have solved this example graphically in Example 14.8. We now solve it using two-phase method. Introducing slack variables r, t, surplus variable u, and artificial variable w, the given LPP reduces to

Maximize

$$p = 8x + 5y$$

$$p' = -w$$

subject to the constraints

$$2x + y + r = 250$$

$$x + y + t = 200$$

$$x - u + w = 25.$$

The initial simplex table for the problem is

	x	y	r	t	u	w	Solution
p	–8	–5	0	0	0	0	0
p'	0	0	0	0	0	1	0
r	2	1	1	0	0	0	250
t	1	1	0	1	0	0	200
w	1	0	0	0	–1	1	25

Since (p', w) entry is 1, we make it 0 by subtracting w-row from p'-row. Thus we have the following table:

	x	y	r	t	u	w	Solution	
p	–8	–5	0	0	0	0	0	
p'	–1	0	0	0	1	0	–25	Ratios
r	2	1	1	0	0	0	250	125
t	1	1	0	1	0	0	200	200
w	(1)	0	0	0	–1	1	25	25 ←
	↑							

The most negative value of p' is –1. The x-column is the optimal column. Since the minimum ratio is 25, the pivot row is w-row and the pivot is 1. Using Gauss elimination to annihilate the entries in x-column and replacing w by x in the left column, we get

	x	y	r	t	u	w	Solution
p	0	–5	0	0	–8	8	200
p'	0	0	0	0	0	1	0
r	0	1	1	0	2	–2	200
t	0	1	0	1	1	–1	175
x	1	0	0	0	–1	1	25

Since $p' = 0$ and $w = 0$, the first phase of the two-phase method is over. So we delete p'-row and w-column from the above table to get the following table.

	x	y	r	t	u	Solution	
p	0	–5	0	0	–8	200	Ratios
r	0	1	1	0	(2)	200	100 ←
t	0	1	0	1	1	175	175
x	1	0	0	0	–1	25	
					↑		

The most negative value of p is –8 and that lies in u-column. The u-column is the current optimal column. The minimum ratio is 100 and so r-row is the pivot row and 2 in the pivot. Dividing r-row throughout by 2, we make the pivot element equal to 1. Thus the table becomes

	x	y	r	t	u	Solution	
p	0	–5	0	0	–8	200	
r	0	$\frac{1}{2}$	$\frac{1}{2}$	0	(1)	100	←
t	0	1	0	1	1	175	
x	1	0	0	0	–1	25	
					↑		

Replacing r in the left column by u and using Gauss elimination, we get

	x	y	r	t	u	Solution	
p	0	–1	4	0	0	1,000	Ratios
u	0	$\frac{1}{2}$	$\frac{1}{2}$	0	1	100	200
t	0	($\frac{1}{2}$)	$-\frac{1}{2}$	1	0	75	150 ←
x	1	$\frac{1}{2}$	$\frac{1}{2}$	0	0	125	250
		↑					

Multiplying t-row throughout by 2, we make the pivot element equal to 1. Thus we get the table shown below:

	x	y	r	t	u	Solution
p	0	−1	4	0	0	1000
u	0	$\frac{1}{2}$	$\frac{1}{2}$	0	1	100
t	0	(1)	−1	2	0	150 ←
x	1	$\frac{1}{2}$	$\frac{1}{2}$	0	0	125
		↑				

Using Gaussian elimination and replacing t in the left column by y, we get

	x	y	r	t	u	Solution
p	0	0	3	0	0	1150
u	0	0	0	−1	1	25
y	0	1	−1	2	0	150
x	1	0	1	−1	0	50

Since all entries in the p-row are non-negative, the solution has been achieved. Reading from the table, the solution to the given LPP is

$$x = 50, y = 150, \text{ and Max. } p = 1150.$$

EXAMPLE 14.23

Solve the following LPP:

Minimize

$$z = 3x_1 + x_2$$

subject to

$$x_1 + x_2 \geq 1$$
$$2x_1 + 3x_2 \geq 2$$
$$\text{and } x_1, x_2 \geq 0.$$

Solution. Converting the problem to maximizing problem, introducing the surplus variables r, t, and artificial variables u and w, and writing the artificial objective function p^*, we have

Maximize

$$z' = -3x_1 - x_2$$
$$p^* = -u - w$$

subject to

$$x_1 + x_2 - r + u = 1$$
$$2x_1 + 3x_2 - t + w = 2$$
$$x_1, x_2, r, t, u, w \geq 0.$$

The initial simplex table for the problem is

	x_1	x_2	r	t	u	w	Solution
z'	3	1	0	0	0	0	0
p^*	0	0	0	0	1	1	0
u	1	1	−1	0	1	0	1
w	2	3	0	−1	0	1	2

We make (p^*, u) and (p^*, w) equal to zero by subtracting u-row and w-row from p^*-row. Thus we get the following table:

	x_1	x_2	r	t	u	w	Solution	
z	3	1	0	0	0	0	0	
p^*	−3	−4	1	1	0	0	−3	Ratio
u	1	1	−1	0	1	0	1	1
w	2	(3)	3	−1	0	1	2	$\frac{2}{3}$ ←
		↑						

The pivot is 3. We make the pivot equal to 1 by dividing throughout the w-row by 3. Then replacing w by x_2 and applying Gaussian elimination, we get the following table:

	x_1	x_2	r	t	u	w	Solution	
z'	$\frac{7}{3}$	0	0	$\frac{1}{3}$	0	$-\frac{1}{3}$	$-\frac{2}{3}$	
p^*	$\frac{1}{3}$	0	1	$-\frac{1}{3}$	0	$\frac{4}{3}$	$\frac{2}{3}$	Ratio
u	$\frac{1}{3}$	0	−1	$\left(\frac{1}{3}\right)$	1	$-\frac{1}{3}$	$\frac{1}{3}$	1 ←
x_2	$\frac{2}{3}$	1	0	$-\frac{1}{3}$	0	$\frac{1}{3}$	$\frac{2}{3}$	
				↑				

Now the pivot is $\frac{1}{3}$. Making it equal to 1, and repeating the process, we get

	x_1	x_2	r	t	u	w	Solution
z'	2	0	1	0	−1	0	−1
p^*	$\frac{2}{3}$	0	0	0	1	1	0
t	1	0	−3	1	3	−1	1
x_2	1	1	−1	0	1	0	1

Since the entries in p^*-row are non-negative, the first phase is over. Leaving aside the p^*-row, u-column, and w-column, the table for the second phase is

	x_1	x_2	r	t	Solution
z'	2	0	1	0	−1
t	1	0	−3	1	1
x_2	1	1	−1	0	1

Since the variables in z'-row are non-negative, the optimal solution has been achieved. The solution is

$x_1 = 0, x_2 = 1$, Max. $z' = -1$, that is, Min. $z = 1$.

EXAMPLE 14.24

Solve the following LPP by two phase method:

Maximize

$$z = 6x_1 + 4x_2$$

subject to the constraints

$$x_1 + x_2 \leq 5$$
$$x_2 \geq 8$$
$$\text{and } x_1, x_2 \geq 0.$$

Solution. We shall use two-phase method to solve this problem. By adding slack variable r, surplus variable s, and artificial variable u, the problem reduces to

Maximize

$$z = 6x_1 + 4x_2$$
$$p = -u$$

subject to the constraints

$$x_1 + x_2 + r = 5$$
$$0x_1 + x_2 - s + u = 8$$
$$x_1, x_2, r, s, u \geq 0.$$

The initial simplex table for the problem is

	x_1	x_2	r	s	u	Solution
z	−6	−4	0	0	0	0
p	0	0	0	0	1	0
r	1	1	1	0	0	5
u	0	1	0	−1	1	8

The entry $(p, u) = 1$. To make this 0, we subtract u-row from p-row. So the table reduces to

	x_1	x_2	r	s	u	Solution	Ratios
z	−6	−4	0	0	0	0	
p	0	−1	0	1	0	−8	
r	1	(1)	1	0	0	5	5 ←
u	0	1	0	−1	1	8	8
		↑					

The most negative value in p-row is −1. The minimum ratio is in r-row and so 1 is the pivot element. Replacing r in the left column by x_2 and using Gaussian elimination, we get the following table:

	x_1	x_2	r	s	u	Solution
z	−2	0	4	0	0	20
p	1	0	1	0	0	−3
x_2	1	1	1	0	0	5
u	−1	0	−1	−1	1	3

Since the entries in p-row are non-negative, the first phase is over. We remove p-row and u-column and get the table given below:

	x_1	x_2	r	s	Solution	Ratio
z	−2	0	4	0	20	
x_2	1	1	1	0	5	5 ←
u	−1	0	−1	−1	3	

Since $u = 3$ in this table, this shows that artificial variable is not equal to zero. The process cannot be carried out further. Further, the solution attained, that is, $x_1 = 0, x_2 = 5, z = 20$, is not feasible because it violates the constraints $x_2 \geq 8$.

14.10 DUALITY PROPERTY

Every LPP can be analysed in two different manners without changing its data. For example, if an LPP is concerning profit maximization, it can be viewed as cost minimization problem. Similarly, a cost minimization problem can be thought of profit maximization problem. Thus to every LPP, there exits another LPP. These two problems are called *duals* of each other. The original problem is called *primal* and the associated problem is called its *dual*. In fact, either of the duals can be considered as primal and the other as its dual. The benefit to convert a given linear programming into its dual is that it reduces the computational work considerably. In fact, *the duality concept reduces the number of constraints.*

Characteristics of Duality

1. If the primal problem is maximization problem, than its dual shall be a minimization problem and vice-versa.
2. The "$\le$" type constraints in the primal becomes "$\ge$" type constraints in the dual problem and vice-versa.
3. The coefficients a_1, a_2,..., a_n in the objective function of the primal become c_1, c_2,..., c_m in the objective function of the dual and vice-versa.
4. If the primal has n variables and m constraints, then the dual will have m variables and n constraints. Therefore, the body *matrix of the dual is the transpose of the body matrix of the primal and vice-versa.*
5. A new set of variables appears in the dual problem.
6. The variables in both the primal and its dual are non-negative.
7. The dual of the dual problem is the original (primal) problem.
8. If the primal variable corresponds to a slack starting variable in the dual problem, then the optimal value of the primal variable is given by the coefficient of the slack variable in the optimal solution, *z-row* of the dual simplex table.
9. In case the primal variable corresponds to an artificial variable in the dual problem, then its optimal value is equal to the coefficient of the artificial variable, in the *z-row* of the dual simplex table.

To illustrate the construction of the dual problem, consider the general form of a LPP:

Maximize

$$z = a_1x_1 + a_2x_2 + \ldots + a_nx_n$$

subject to the constraints

$$b_{11}x_1 + b_{12}x_2 + \ldots + b_{1n}x_n \le c_1$$
$$b_{21}x_1 + b_{22}x_2 + \ldots + b_{2n}x_n \le c_2$$
$$\ldots\ldots\ldots\ldots\ldots\ldots\ldots\ldots$$
$$\ldots\ldots\ldots\ldots\ldots\ldots\ldots\ldots$$
$$b_{m1}x_1 + b_{m2}x_2 + \ldots + b_{mn}x_n \le c_m$$
$$x_1, x_2, \ldots, x_n \ge 0$$

In view of the above-mentioned characteristic, the dual problem of the above LPP is

Minimize

$$z^* = c_1y_1 + c_2y_2 + \ldots + c_my_m$$

subject to the constraints

$$b_{11}y_1 + b_{21}y_2 + \ldots + b_{m1}y_m \ge a_1$$
$$b_{12}y_1 + b_{22}y_2 + \ldots + b_{m2}y_m \ge a_2$$
$$\ldots\ldots\ldots\ldots\ldots\ldots\ldots\ldots$$
$$\ldots\ldots\ldots\ldots\ldots\ldots\ldots\ldots$$
$$b_{1n}y_1 + b_{2n}y_2 + \ldots + b_{mn}y_m \ge a_n$$
$$y_1, y_2, \ldots, y_m \ge 0$$

EXAMPLE 14.25

With the help of the following problem, show that dual of the dual problem is the primal problem:

Maximize

$$z = 2x_1 + 5x_2 + 6x_3$$

subject to the constraints

$$5x_1 + 6x_2 - x_3 \le 3$$
$$-2x_1 + x_2 + 4x_3 \le 4$$
$$x_1 - 5x_2 + 3x3 \le 1$$
$$3x_1 - 3x_2 + 7x_3 \le 6$$
$$x_1, x_2, x_3 \ge 0.$$

Solution. The dual of the given LPP is

Minimize

$$z^* = 3y_1 + 4y_2 + y_3 + 6y_4$$

subject to the constraints

$$5y_1 - 2y_2 + y_3 - 3y_4 \geq 2$$
$$6y_1 + y_2 - 5y_3 - 3y_4 \geq 5$$
$$-y_1 + 4y_2 + 3y_3 + 7y_4 \geq 6$$
$$y_1, y_2, y_3, y_4 \geq 0.$$

Taking dual again, we obtain

Maximize

$$Z^* = 2z_1 + 5z_2 + 6z_3$$

subject to the constraints

$$5z_1 + 6z_2 - z_3 \leq 3$$
$$-2z_1 + z_2 + 4z_3 \leq 4$$
$$z_1 - 5z_2 + 3z_2 \leq 1$$
$$-3z_1 - 3z_2 + 7z_3 \leq 6$$
$$z_1, z_2, z_3 \geq 0.$$

which is nothing but original primal problem.

EXAMPLE 14.26

Write the dual of the following LPP in standard form.

Minimize

$$z = 2x_1 + 3x_2 + 4x_3$$

subject to the constraints

$$2x_1 + 3x_2 + 5x_3 \geq 2$$
$$3x_1 + x_2 + 7x_3 = 3$$
$$x_1 + 4x_2 + 6x_3 \leq 5$$
$$x_1, x_2 \geq 0,\ x_3 \text{ unrestricted.}$$

Solution. Changing the third constraints, the given LPP is

Minimize

$$z = 2x_1 + 3x_2 + 4x_3$$

subject to the constraints

$$2x_1 + 3x_2 + 5x_3 \geq 2$$
$$3x_1 + x_2 + 7x_3 = 3$$
$$-x_1 - 4x_2 - 6x_3 \geq -5$$
$$x_1, x_2 \geq 0,\ x_3 \text{ unrestricted.}$$

The dual of this problem is

Maximize

$$z^* = 2y_1 + 3y_2 - 5y_3$$

subject to the constraints

$$2y_1 + 3y_2 - y_3 \leq 2$$
$$3y_1 + y_2 - 4y_3 = 3$$
$$5y_1 + 7y_2 - 6y_3 \leq 4$$
$$y_1, y_2 \geq 0,\ y_3 \text{ unrestricted.}$$

The standard form of the dual is

Maximize

$$z^* = 2y_1 + 3y_2 - 5(y'_3 - y_3'').$$

subject to the constraints

$$2y_1 + 3y_2 - (y_3' - y_3'') \leq 2$$
$$3y_1 + y_2 - 4(y_3' - y_3'') = 3$$
$$5y_1 + 7y_2 - 6(y_3' - y_3'') \leq 4$$
$$y_1, y_2, y_3', y_3'' >= 0$$

where y_3, being unrestricted, is equal to $y_3' - y_3''$.

EXAMPLE 14.27

Construct the dual of the following primal LPP.

Maximize

$$z = x_1 - 2x_2 + 3x_3$$

subject to the constraints

$$-2x_1 + x_2 + 3x_3 = 2$$
$$2x_1 + 3x_2 + 4x_3 = 1$$
$$x_1, x_2, x_3 \geq 0.$$

Solution. We note that both the primal constraints are equality. Therefore, the corresponding dual variables y_1 and y_2 will be unrestricted in sign. The dual of the given problem is

Minimize

$$z^* = 2y_1 + y_2$$

subject to

$$-2y_1 + 2y_2 \geq 1$$

$$y_1 + 3y_2 \geq 2$$

$$3y_1 + 4y_2 \geq 3$$

y_1, y_2 unrestricted in sign.

EXAMPLE 14.28

Using duality concept, solve the following problem:

Minimize

$$z = 2x_1 + x_2$$

subject to the constraints

$$3x_1 + x_2 \geq 3$$

$$4x_1 + 3x_2 \geq 6$$

$$x_1 + 2x_2 \leq 3$$

$$x_1, x_2 \geq 0.$$

Solution. Changing the third constraint in the given LPP, the constraints become

$$3x_1 + x_2 \geq 3$$

$$4x_1 + 3x_2 \geq 6$$

$$-x_1 - 2x_2 \geq -3$$

$$x_1, x_2 \geq 0.$$

Therefore, the dual of the problem is

Maximize

$$z^* = 3y_1 + 6y_2 - 3y_3$$

subject to the constraints

$$3y_1 + 4y_2 - y_3 \leq 2$$

$$y_1 + 3y_2 - 2y_3 \leq 1.$$

Using slack variables, we have

Maximize

$$z^* = 3y_1 + 6y_2 - 3y_3$$

subject to the condition

$$3y_1 + 4y_2 - y_3 + r = 2$$

$$y_1 + 3y_2 - 2y_3 + t = 1$$

$$y_1, y_2, y_3, r, t \geq 0.$$

The basic feasible solution is

$$y_1 = y_2 = y_3 = 0, r = 2, t = 1.$$

Thus initial simplex table for the problem is

	y_1	y_2	y_3	r	t	Solution		
z^*	-3	-6	3	0	0	0	Ratios	
r	3	4	-1	1	0	2	$\frac{1}{2}$	
t	1	$\textcircled{3}$	-2	0	1	1	$\frac{1}{3}$	←
		↑						

Most negative value in z^*-row is -6, which lies in y_2-column. Thus the y_2-column is the optimal column. The minimum ratio is $\frac{1}{3}$, which lies in the t-row. Therefore, t-row is the pivot row and 3 is the pivot element. Dividing throughout the t-row by 3, we make the pivot equal to 1.

Thus the table reduces to the new table given below:

	y_1	y_2	y_3	r	t	Solution	
z^*	-3	-6	3	0	0	0	
r	3	4	-1	1	0	2	
t	$\frac{1}{3}$	$\textcircled{1}$	$-\frac{2}{3}$	0	$\frac{1}{3}$	$\frac{1}{3}$	←
		↑					

Replacing t in the left column by y_2 and using Gauss elimination to annihilate the entries in y_2-column, we get

	y_1	y_2	y_3	r	t	Solution		
z^*	-1	0	-1	0	2	2	Ratio	
r	$\frac{5}{3}$	0	$\textcircled{\frac{5}{3}}$	1	$-\frac{4}{3}$	$\frac{2}{3}$	$\frac{2}{5}$	←
t	$\frac{1}{3}$	1	$-\frac{2}{3}$	0	$\frac{1}{3}$	$\frac{1}{3}$		
			↑					

Now the pivot is $\frac{5}{3}$. Making it equal to 1 by dividing throughout, we get

	y_1	y_2	y_3	r	t	Solution
z^*	-1	0	-1	0	2	2
r	1	0	(1)	$\frac{3}{5}$	$-\frac{4}{5}$	$\frac{2}{5}$ ←
y_2	$\frac{1}{3}$	1	$-\frac{2}{3}$ ↑	0	$\frac{1}{3}$	$\frac{1}{3}$

Gauss elimination now yields

	y_1	y_2	y_3	r	t	Solution
z^*	0	0	0	$\frac{3}{5}$	$\frac{6}{5}$	$\frac{12}{5}$
y_3	1	0	1	$\frac{3}{5}$	$-\frac{4}{5}$	$\frac{2}{5}$
y_2	1	1	0	$\frac{2}{5}$	$-\frac{1}{5}$	$\frac{3}{5}$

Since all the entries in the z^*-row are nonnegative, we have achieved the solution of the dual problem. From the last table, the solution is

$$y_1 = 0, y_2 = \frac{3}{5}, y_3 = \frac{2}{5},$$

$$\text{and Max. } z^* = \frac{12}{5}.$$

Hence, by duality principle, the solution to the primal is

$$x_1 = \frac{3}{5} \text{ (coefficient of } r \text{ in } z^*\text{-row)}$$

$$x_2 = \frac{6}{5} \text{ (coefficient of } t \text{ in } z^*\text{-row)}$$

$$\text{Max } z^* = \text{Min. } z = \frac{12}{5}$$

EXAMPLE 14.29

Using duality concept, solve the following LPP.

Maximize

$$z = 3x_1 + 2x_2$$

subject to the constraints

$$x_1 + x_2 \geq 1$$

$$x_1 + x_2 \leq 7$$

$$x_1 + 2x_2 \leq 10$$

$$x_2 \leq 3$$

$$x_1, x_2 \geq 0.$$

Solution. Changing the sign of first constraint, we have

$$-x_1 - x_2 \leq -1.$$

So, the dual LPP is

Minimize

$$z^* = -y_1 + 7y_2 + 10y_3 + 3y_4$$

subject to the constraints

$$-y_1 + y_2 + y_3 \geq 3$$

$$-y_1 + y_2 + 2y_3 + y_4 \geq 2$$

$$y_1, y_2, y_3, y_4 \geq 0.$$

Introducing the surplus and artificial variables, the dual problem in standard form is

Maximize

$$z^{**} = y_1 - 7y_2 - 10y_3 - 3y_4$$

Subject to

$$-y_1 + y_2 + y_3 - r + u = 3$$

$$-y_1 + y_2 + 2y_3 + y_4 - t + w = 2.$$

The artificial objective function is

$$p = -u - w.$$

The initial table for the two-phase method is

	y_1	y_2	y_3	y_4	r	t	u	w	Solution
z^{**}	-1	7	10	3	0	0	0	0	0
p	0	0	0	0	0	0	1	1	0
u	-1	1	1	0	-1	0	1	0	3
w	-1	1	2	1	0	-1	0	1	2

Since (p, u) =1 and (p, w) =1, we subtract u-row and w-row from from the p-row. Thus we get the following table:

	y_1	y_2	y_3	y_4	r	t	u	w	Solution	
z^{**}	−1	7	10	3	0	0	0	0	0	
p	2	−2	−3	−1	1	1	0	0	−5	Ratio
u	−1	1	1	0	−1	0	1	0	3	3
w	−1	1	(2)	1	0	−1	0	1	2	1 ←
			↑							

The next tables are:

	y_1	y_2	y_3	y_4	r	t	u	w	Solution
z^{**}	4	2	0	−2	0	5	0	−5	−10
p	$\frac{1}{2}$	$-\frac{1}{2}$	0	$\frac{1}{2}$	1	$-\frac{1}{2}$	0	$\frac{3}{2}$	−2
u	$-\frac{1}{2}$	$\frac{1}{2}$	0	$-\frac{1}{2}$	−1	$\frac{1}{2}$	1	$-\frac{1}{2}$	2
y_3	$-\frac{1}{2}$	($\frac{1}{2}$)	1	$\frac{1}{2}$	0	$-\frac{1}{2}$	0	$\frac{1}{2}$	1 ←
		↑							

	y_1	y_2	y_3	y_4	r	t	u	w	Solution
z^{**}	6	0	4	−4	0	7	0	7	−14
p	0	0	1	1	1	−1	0	2	−1
u	0	0	−1	−1	−1	(1)	1	−1	1 ←
y_2	−1	1	2	1	0	−1	0	1	2
						↑			

and

	y_1	y_2	y_3	y_4	r				Solution
z^{**}	6	0	3	3	7	0	−7	0	−21
p	0	0	0	0	0	0	1	1	0
t	0	0	−1	−1	−1	1	(1)	−1	1 ←
y_2	−1	1	1	0	−1	0	1	0	3
							↑		

Since elements in p-row are non-negative, phase one is over. We remove p-row, u-column, and w-column from the table and get the following table:

	y_1	y_2	y_3	y_4	r	t	Solution
z^{**}	6	0	3	3	7	0	−21
t	0	0	−1	−1	−1	1	1
y_2	−1	1	1	0	−1	0	3

Since entries in z^{**} are also non-negative, the solution has been achieved. Hence $x_1 = 7, x_2 = 0$, and Max. z = Min. $z^* = -$Max. $z^{**} = 21$.

14.11 DUAL SIMPLEX METHOD

We have seen that if in a maximizing LPP, someconstraints contains "≥" sign, then two-phase method is applicable to find the optimal solution to the problem. Based on the primal-dual relationship, we have another method, called the *dual simplex method,* to solve the above-mentioned problems. The difference between the simplex method and the dual simplex method is that in simplex method we begin with an initial basic feasible solution and find an optimal solution, whereas in case of dual simplex method, we start with a basic unfeasible but optimal solution and find feasibility of that solution. In the dual simplex method, we first identify the optimal row and then identify the column with the help of ratios.

The *algorithm for dual simplex method* is given below:

1. Convert the problem to maximization form if it is not given in that form.
2. Convert "≥" sign, if any, to "≤" sign in the constraints by multiplying such constraints throughout by −1.
3. Express the LPP in standard form by introducing slack variables.
4. Determine the initial basic solution and its simplex table.
5. If all entries in the solution column are positive, then there is no need to apply dual simplex method and the optimal solution can be determined by ordinary simplex method.

6. If there exits a row in which the solution value is negative, choose the *key row* in which longest negative solution exists.
7. Determine the minimum ratio only for those columns which have negative elements in the key row.
8. Choose the *key column* as the column in which the ratio is minimum.
9. The pivot element lies at the intersection of key row and key column.
10. Use Gauss elimination method to annihilate the elements in the key column.
11. Repeat the above steps till all elements in solution column are turned into greater than or equal to zero.

EXAMPLE 14.30

Use simplex method to solve the following:
Maximize

$$z = 5x_1 + 2x_2$$

subject to

$$6x_1 + x_2 \geq 6$$

$$4x_1 + 3x_2 \geq 12$$

$$x_1 + 2x_2 \geq 4$$

$$\text{and } x_1, x_2 \geq 0.$$

Solution. Making all the constraints of the type by multiplying throughout by −1 and introducing slack variables, the problem reduces to
Maximize

$$z = 5x_1 + 2x_2$$

subject to

$$-6x_1 - x_2 + r = -6$$

$$-4x_1 - 3x_2 + s = -12$$

$$-x_1 - 2x_2 + t = -4$$

$$x_1, x_2, r, s, t \geq 0.$$

An initial basic solution is obtained by setting $x_1 = x_2 = 0$. This gives the solution values as $r = -6$, $s = -12$, $t = -4$, and Max. $z = 0$.

Thus the initial solution table is

	x_1	x_2	r	s	t	Solution	
z	−5	−2	0	0	0	0	
r	−6	−1	1	0	0	−6	
s	−4	(−3)	0	1	0	−12	←
t	−1	−2	0	0	1	−4	
Ratios	$\frac{5}{4}$	$\frac{2}{3}$ ↑					

The largest negative value in the solution column is −12 and so the s-row is the key row. Further, minimum ratio is in x_2 − *column*. Thus the pivot is −3. Make the pivot element equal to 1 by dividing s-row by −3. Replace s by x_2 in the left column and use Gauss elimination to get the table given below:

	x_1	x_2	r	s	t	Solution	
z	$-\frac{7}{3}$	0	0	$-\frac{2}{3}$	0	8	
r	($-\frac{14}{3}$)	0	1	$-\frac{1}{3}$	0	−2	←
x_2	$\frac{4}{3}$	1	0	$-\frac{1}{3}$	0	4	
t	$\frac{5}{3}$	0	0	2	1	4	
Ratio	$\frac{1}{2}$ ↑						

The largest negative value in the solution column is −2 and so the key row is r-row. The minimum ratio $\frac{1}{2}$ lies in the x_1-column. Thus the pivot is $-\frac{14}{3}$. Making pivot element equal to 1, replacing r by x_1 in the left column and using Gaussian elimination, we get the table given below:

	x_1	x_2	r	s	t	Solution
z	0	0	$-\frac{7}{6}$	0	0	9
x_1	1	0	$-\frac{3}{14}$	$\frac{1}{14}$	0	$\frac{3}{7}$
x_2	0	1	$\frac{2}{7}$	$-\frac{3}{7}$	0	$\frac{24}{7}$
t	0	0	$\frac{5}{14}$	$\frac{79}{42}$	1	$\frac{23}{7}$

Since all the entries in the solution column are non-negative, the optimal and feasible solution have been achieved. Hence the solution is

$$x_1 = \frac{3}{7},\ x_2 = \frac{24}{7}, \text{ and Max. } z = 9.$$

EXAMPLE 14.31

Use dual simplex method to solve the following LPP:

Minimize

$$z = 2x_1 + x_2$$

subject to

$$3x_1 + x_2 \geq 3$$
$$4x_1 + 3x_2 \geq 6$$
$$x_1 + 2x_2 \leq 3$$
$$x_1, x_2 \geq 0.$$

Solution. Writing the problem in maximizing form, converting the constraints with "≥" into the constraints with "≤" and introducing the slack variables, the given problem reduces to Maximize

$$z' = -2x_1 - x_2$$

subject to

$$-3x_1 - x_2 + r = -3$$
$$-4x_1 - 3x_2 + s = -6$$
$$x_1 + 2x_2 + t = 3$$
$$x_1, x_2, r, s, t \geq 0.$$

A basic solution is

$$x_1 = x_2 = 0,\ r = -3,\ s = -6,\ t = 3,\ \text{Max.}z' = 0.$$

Thus the initial table is

	x_1	x_2	r	s	t	Solution	
$-z'$	−2	−1	0	0	0	0	
r	−3	−1	1	0	0	− 3	
s	−4	(−3)	0	1	0	− 6	←
t	1	2	0	0	1	3	
Ratios	$\frac{1}{2}$	$\frac{1}{3}$ ↑					

The largest negative value in the solution column is −6. So s-row is the key row. Further, minimum ratio lies in x_2 − column. Therefore, x_2 − column is the key-column and −3 is the pivot.

We make the pivot equal to 1 by dividing s–row throughout by −3. Replace s by x_2 in the left column and apply Gaussian elimination to get the table.

	x_1	x_2	r	s	t	Solution	
$-z'$	$-\frac{2}{3}$	0	0	$-\frac{1}{3}$	0	2	
r	$-\frac{5}{3}$	0	1	$-\frac{1}{3}$	0	−1	
x_2	$\frac{4}{3}$	1	0	$-\frac{1}{3}$	0	2	
t	($-\frac{5}{3}$)	0	0	$-\frac{2}{3}$	1	−1	←
Ratio	$\frac{2}{5}$ ↑						

Now the pivot is $-\frac{5}{3}$. Making it equal to 1 by dividing t-row throughout by $-\frac{5}{3}$, replacing t by x_1, and using Gaussian elimination, we get the following table:

	x_1	x_2	r	s	t	Solution
$-z'$	0	0	0	$-\frac{3}{5}$	$-\frac{2}{5}$	$\frac{12}{5}$
r	0	0	1	− 1	− 1	0
x_2	0	1	0	$\frac{1}{5}$	$\frac{4}{5}$	$\frac{6}{5}$
x_1	1	0	0	$-\frac{2}{5}$	$-\frac{3}{5}$	$\frac{3}{5}$

Since all the entries in the solution column are non-negative, the solution has been achieved. Reading from the table, we have

$$x_1 = \frac{3}{5},\ x_2 = \frac{6}{5},\ \text{Max. } z' = -\frac{12}{5} \text{ and so Min.} z = \frac{12}{5}.$$

EXAMPLE 14.32

Use the dual-simplex method to solve the following LPP:

Minimize

$$z = 3x_1 + x_2$$

subject to the constraints

$$x_1 + x_2 \geq 1$$

$$2x_1 + 3x_2 \geq 2.$$

Solution. First convert the given problem into maximization problem and then convert constraints with ≥ sign into constraints with ≤ sign multiplying throughout by −1.

Thus the given LPP reduces to

Maximize

$$z' = -3x_1 - x_2, z' = -z$$

subject to

$$-x_1 - x_2 \leq -1$$

$$-2x_1 - 3x_2 \leq -2$$

$$x_1, x_2 \geq 0.$$

Converting the problem into standard form by adding slack variables r and t, we get

Maximize

$$z' = -3x_1 - x_2$$

subject to

$$-x_1 - x_2 + r = -1$$

$$-2x_1 - 3x_2 + t = -2$$

$$x_1, x_2 \geq 0.$$

An initial solution is

$$x_1 = x_2 = 0, r = -1, t = -2, \text{Max. } z' = 0.$$

The initial solution table is

	x_1	x_2	r	t	Solution	
$-z'$	−3	−1	0	0	0	
r	−1	−1	1	0	−1	
t	−2	(−3)	0	1	−2	←
Ratios	$\frac{3}{2}$	$\frac{1}{3}$ ↑				

The key row is t-row and key column is x_2 – *column*. Making the pivot element equal to 1 and using Gauss-elimination, we get the following table:

	x_1	x_2	r	t	Solution	
$-z'$	$-\frac{7}{3}$	0	0	$-\frac{1}{3}$	$\frac{2}{3}$	
r	$\frac{1}{3}$	0	1	$\left(-\frac{1}{3}\right)$	$-\frac{1}{3}$	←
x_2	$\frac{2}{3}$	1	0	$-\frac{1}{3}$ ↑	$\frac{2}{3}$	

The next table is

	x_1	x_2	r	t	Solution
$-z'$	−2	0	−1	0	1
t	1	0	−3	1	1
x_2	1	1	−1	0	1

Since solution values are now positive, the solution has been achieved and is

$$x_1 = 0, x_2 = 1, \text{Max. } z' = -1, \text{ or Min. } z = 1$$

EXAMPLE 14.33

Use duality to solve the following LPP:

Maximize

$$z = 2x_1 + x_2$$

subject to the constraints

$$x_1 + 2x_2 \leq 10$$

$$x_1 + x_2 \leq 6$$

$$x_1 - x_2 \leq 2$$

$$x_1 - 2x_2 \leq 1$$

$$x_1, x_2 \geq 0.$$

Solution. The dual of the given problem is

Minimize

$$z^* = 10y_1 + 6y_2 + 2y_3 + y_4$$

subject to the constraints

$$y_1 + y_2 + y_3 + y_4 \geq 2$$

$$2y_1 + y_2 - y_3 - 2y_4 \geq 1$$

$$y_1, y_2, y_3, y_4 \geq 0.$$

Converting the dual to maximizing problem and converting $\geq$ sign into $\leq$ sign in the constraints, we have

Maximize

$$z^{**} = -10y_1 - 6y_2 - 2y_3 - y_4$$

subject to

$$-y_1 - y_2 - y_3 - y_4 \leq -2$$

$$-2y_1 - y_2 - y_3 + 2y_4 \leq 1.$$

We shall use dual simplex method to solve it. Introducing slack variables in the constraints, we have

$$-y_1 - y_2 - y_3 - y_4 + r = -2$$

$$-2y_1 - y_2 + y_3 + 2y_4 + t = -1.$$

A basic solution is

$$y_1 = y_2 = y_3 = y_4 = 0, r = -2,$$

$$t = -1, \text{ Max. } z^{**} = 0.$$

Thus the initial table is

	y_1	y_2	y_3	y_4	r	t	Solution	
$-z^{**}$	-10	-6	-2	-1	0	0	0	
r	-1	-1	-1	(−1)	1	0	-2	←
t	-2	-1	1	2	0	1	-1	
Ratios	10	6	2	1 ↑				

The pivot is -1. Therefore using Gausselimination, we get the table

	y_1	y_2	y_3	y_4	r	t	Solution	
$-z^{**}$	-9	-5	-1	0	-1	0	2	
y_4	1	1	1	1	-1	0	2	
t	-4	-3	(−1)	0	2	1	-5	←
Ratios	$\frac{9}{4}$	$\frac{5}{3}$	1 ↑					

The next tables are

	y_1	y_2	y_3	y_4	r	t	Solution	
$-z'$	-5	-2	0	0	-3	-1	7	
y_4	-3	(−2)	0	1	1	1	-3	←
t	4	3	1	0	-2	-1	5	
Ratios	$\frac{5}{3}$	1 ↑						

and

	y_1	y_2	y_3	y_4	r	t	Solution
$-z^{**}$	-2	0	0	-1	-4	-2	10
y_4	$\frac{3}{2}$	1	0	$-\frac{1}{2}$	$-\frac{1}{2}$	$-\frac{1}{2}$	$\frac{3}{2}$
t	$-\frac{1}{2}$	0	1	$\frac{3}{2}$	$-\frac{1}{2}$	$-\frac{1}{2}$	$\frac{1}{2}$

Since x_1 and x_2 corresponds to the slack variables r and t, respectively, the solution to the problem is

$$x_1 = 4, x_2 = 2 \text{ and Min. } z = 10.$$

14.12 TRANSPORTATION PROBLEMS

A special class of LPPs, in which our aim is to transport a single product from various production units, called the origins, to different locations, called *destinations*, at a minimum cost is called a *transportation problem*. Thus, in a transportation problem, we wish to determine transporting schedule which minimize the total cost of transportation.

Suppose that there are m origins and n destinations. Let a_i be the quantity of the product available at the origin i and let b_j the quantity of the product required at the destination j. Suppose that the cost of transportation of unit of the product from origin i to the destination j be c_{ij}. If x_{ij} is the quantity (in units) transported from the origin i to the destination j, then our aim is to find $x_{ij} \geq 0$, which satisfy the $m + n$ constraints

$$\sum_{j=1}^{n} x_{ij} = a_i, \; i = 1, 2, \ldots, m$$

$$\sum_{i=1}^{m} x_{ij} = b_j, \; j = 1, 2, \ldots, n$$

and for which

$$\sum_{i=1}^{m} \sum_{j=1}^{n} c_{ij} x_{ij}$$

is minimum.

In general transportation problems, we assume that

$$\sum_{i=1}^{m} a_i = \sum_{j=1}^{n} b_j,$$

which means that the *total quantity of the product available at the origins is equal to the total quantity required at the destinations.* Problems satisfying these conditions are called *balanced transportation problems.* A transportation problem will have a feasible solution under this condition, known as *consistency condition for the constraints.* In fact, it is a *necessary and sufficient condition* for the existence of a feasible solution to the constraints.

Thus, the general transportation problem can be expressed as
Minimize

$$z = \sum_{i=1}^{m} \sum_{j=1}^{n} c_{ij} x_{ij}$$

subject to the constraints

$$\sum_{j=1}^{n} x_{ij} = a_i,\ i = 1, 2, \ldots, m$$

$$\sum_{i=1}^{m} x_{ij} = b_j,\ j = 1, 2, \ldots, n$$

$$x_{ij} \geq 0.$$

We observe that the *coefficients of all x_{ij} in the constraints are unity.*

14.13 MATRIX FORM OF THE TRANSPORTATION PROBLEM

The constraints are

$$\begin{aligned}
x_{11} + x_{12} + \ldots + x_{1n} &= a_1\\
x_{21} + x_{22} + \ldots + x_{2n} &= a_2\\
&\ldots\ldots\\
&\ldots\ldots\\
x_{m1} + x_{m2} + \ldots + x_{mn} &= a_m\\
x_{11} + x_{21} + \ldots + x_{m1} &= b_1\\
x_{12} + x_{22} + \ldots + x_{m2} &= b_2\\
&\ldots\ldots\\
&\ldots\ldots\\
x_{1n} + x_{2n} + \ldots + x_{mn} &= b_n
\end{aligned}$$

We write these constraints as

$$\begin{array}{llll}
x_{11} + \ldots + x_{1n} & & & = a_1\\
 & x_{21} + \ldots + x_{2n} & & = a_2\\
 & \vdots & & \\
 & & x_{m1} + \ldots + x_{mn} & = a_m\\
x_{11} & + x_{21} + \ldots + x_{m1} & & = b_1\\
\quad x_{12} & + x_{22} + \ldots + x_{m2} & & = b_2\\
 & \vdots & & \\
\quad x_{1n} & + x_{2n} + \ldots + x_{mn} & & = b_n
\end{array}$$

Then the standard matrix form of the transportation problem is

$$AX = B,$$

where

$$A = \begin{pmatrix}
1_n & 0 & 0 & \ldots & \ldots & 0\\
0 & 1_n & 0 & \ldots & \ldots & 0\\
0 & 0 & 1_n & \ldots & \ldots & 0\\
\ldots & \ldots & \ldots & \ldots & \ldots & \ldots\\
0 & 0 & 0 & 0 & 0 & 1_n\\
I_n & I_n & I_n & I_n & I_n & I_n
\end{pmatrix}
\begin{array}{l}\left.\vphantom{\begin{matrix}1\\1\\1\\1\\1\end{matrix}}\right\} m \text{ rows}\\ \} n \text{ rows}\end{array}$$

is an $(m + n) \times (mn)$ matrix. Here 1_n is the sum vector having n components and I_n is a unit matrix of order n. Also

$$X = \begin{pmatrix} x_{11}\\ x_{12}\\ \vdots\\ x_{1n}\\ x_{21}\\ x_{22}\\ \vdots\\ x_{1n}\\ \vdots\\ x_{mn} \end{pmatrix}, \quad B = \begin{pmatrix} a_1\\ a_2\\ \vdots\\ a_m\\ b_1\\ b_2\\ \vdots\\ b_n \end{pmatrix}$$

The simplex method can now be applied to solve the problem. But the number of variables being large (sometimes in thousands), there will be too many calculations. However, the simple structure of the matrix A (consisting of the entries as 0 or 1) allow us to develop algorithm for solving transportation problems.

The rank of the matrix A is $m+n-1$. Therefore, it follows that "an optimal solution to a transportation problem with m origins and n destinations contains at most $m+n-1$ of the x_{ij} different from zero."

One more important property of the matrix A used in developing algorithm for solving the transportation problem is that "every minor of the matrix A has the value ± 1 or 0."

14.14 TRANSPORTATION PROBLEM TABLE

The table for transportation problem involves *m* rows and *n* columns. We use O_i as the heading for row *i* to indicate that this row pertains to the origin *i*. Similarly, we use D_j as a heading for the columns *j* to indicate that this column pertains to destination *j*. The mn squares formed by these *m* rows and *n* columns are called cells. The per unit cost c_{ij} of the transporting from the ith origin to the jth destination is shown in the lower right side of the $(i,j)^{th}$ cell. Any feasible solution x_{ij} is shown encircled inside the $(i,j)^{th}$ cell. The availabilities a_i are shown as a column on the right-hand side of *mn* cells, while the requirements b_j have been shown as a row below the *mn* cells. The a_i and b_j are called *rim requirements*. The consistency condition is shown on lower right corner of the table. Thus the table for the transportation problem is as shown below:

	D_1	D_2	D_j	D_n	a_i
O_1	c_{11}	c_{12}	c_{1j}	c_{1n}	a_1
O_2	c_{21}	c_{22}	c_{2j}	c_{2n}	a_2
O_i	c_{i1}	c_{i2}	x_{ij} c_{ij}	c_{in}	a_i
O_m	c_{m1}	c_{m2}	c_{mj}	x_{mn} c_{mn}	a_m
b_j	b_1	b_2	b_j	b_n	$\sum a_i = \sum b_j$

14.15 BASIC INITIAL FEASIBLE SOLUTION OF TRANSPORTATION PROBLEM

The initial basic feasible solution of a transportation problem is determined by any of the fol lowing method:

1. North-west corner method
2. Column minima method
3. Row minima method
4. Matrix minima method (or least cost method)
5. Vogel's approximation method (or Vogel's penalty method).

It has not been established that which one of these methods is better than the others. We shall discuss the working procedures for north-west corner method, matrix minima, and Vogel's approximation method (VAM). To get optimal solution, generally VAM followed by MODI method is used.

A. North-west Corner Method

The following steps are involved in this method:

1. Begin with the cell (1, 1) at the upper left (north-west) corner of the transportation matrix. Set $x_{11} = \min(a_1, b_1)$ Thus at this step, we satisfy either an origin or a destination requirement.
2. (a) If allocation made in Step 1 is equal to the availability of the first origin, that is, $b_1 > a_1$, then we move to cell (2, 1) in second row and first column and allocate $x_{21} = \min(b_1 - a_1, a_2)$ to the cell (2, 1).

   (b) If allocation made in Step 1 is equal to the requirement of the first destination, that is, if $a_1 > b_1$, then we move to the cell (1, 2) in first row and second column and allocate $x_{12} = \min(a_1 - b_1, b_2)$ to the cell (1, 2).

   (c) If $a_1 = b_1$ then we allocate $x_{11} = a_1$ or b_1 in the first step and move diagonally to the cell (2, 2).
3. Continue the process satisfying, at the kth step, either an origin or a destination requirement.

B. Matrix Minima or Least Cost Method

The steps involved in the least cost method are as follows:

1. Choose the cell $(O_i, D_j) \equiv (i, j)$ with smallest unit cost in the transportation table and allocate maximum possible to this cell. Eliminate the row i or column j in which either availability or demand is exhausted. In case both row and column are satisfied simultaneously, only one of these will be eliminated. If the smallest unit cost cell is not unique, then choose the cell to which the maximum allocation can be made.
2. Adjust the supply and demand for all remaining rows and columns and repeat the process with the smallest unit cost among the remaining rows and columns of the transportation table, allocating maximum possible to the cell and then eliminating the row or column in which either supply or demand is exhausted.
3. Continue with the procedure till supply at various origins and demand at various destinations are satisfied.

C. VAM or Vogel's Penalty Method

The steps involved in this method are:

1. Find the difference, called *penalty*, between the lowest cost and the next lowest cost in each row and display it in bracket ( ) to the right of the row. In a similar way, display in bracket the difference between the lowest cost and the next lowest cost in each column below that column.
2. Choose the largest of these $m + n$ differences. Suppose that the largest of these differences was associated with the difference in column j and let (i,j) be the cell containing the lowest cost in column j. Allocate $x_{ij} = \min (a_i, b_j)$ to the cell (i, j). Adjust the supply and demand and cross out the row i or column j depending on the requirement whish is satisfied.

   Any row or column with zero supply or demand should not be used in computing. If the maximum difference is not unique, arbitrary choice can be made.

   However, in such a case, it is better to allocate to the cell with the lower cost.
3. Repeat the whole process for the resulting table after Step 2.
4. Continue the process till all the rim requirements are satisfied.

14.16 TEST FOR THE OPTIMALITY OF BASIC FEASIBLE SOLUTION

To test the optimality of basic feasible solution, we shall use *modified distribution method* (MODI). The steps involved in this method are:

1. Mark the numbers $u_i (i = 1, 2, \ldots, m)$ and $v_j (j = 1, 2, \ldots, n)$ along the left and top of the cost matrix, respectively, such that their sum equals the original cost c_{ij} of the *occupied cell* (i, j). Thus $u_i + v_j = c_{ij}$. Starting initially with some u_i or v_j equal to zero, solve these equations for u_i and v_j.
2. For unoccupied cells, calculate the *net evaluations* $d_{ij} = u_i + v_j - c_{ij}$.
3. Examine the sign of each d_{ij}. If
   (i) Each $d_{ij} < 0$, the current basic feasible solution is optimal.
   (ii) $d_{ij} = 0$, the current basic feasible solution will remain unaffected but an alternative solution exists.
   (iii) At least one $d_{ij} > 0$ for some i, j, then the solution is not optimal. A better solution exists in such a case. For this purpose, choose an unoccupied cell with the largest positive d_{ij} and mark θ inside that cell. Call this cell a θ-cell.
4. In case of 3(iii), construct a loop (closed path) consisting of horizontal and vertical lines beginning and ending at the θ-cell and having its other corners at the occupied cells. Trace a path along the rows (or columns) to an occupied cell, mark the corner with $- \theta$ and continue down the column (or row) to an occupied cell and mark the column with θ and $- \theta$, alternately.
5. Assign a value to θ so that one basic variable in the loop becomes zero and the other basic variables remain nonnegative. The basic cell whose allocation has been reduced to zero leaves the basis.

6. Repeat steps 1, 2, 3 and continue the process till an optimal basic feasible solution is attained.

EXAMPLE 14.34

Find initial basic feasible solution of the following transportation problem by

1. North-west corner rule
2. Least cost method
3. VAM

and hence find its optimal solution.

	D_1	D_2	D_3	D_4	Availability
O_1	21	16	25	13	11
O_2	17	18	14	23	13
O_3	32	27	18	41	19
Requirement	6	10	12	15	43

Solution.

1. North-west Corner Rule

(i) Compare a_1 and b_1. Since $b_1 < a_1$, allocate 6 to (1,1)-cell. Thus $x_{11} = 6$. This exhausted the demand at D_1.

(ii) Move to the cell (1,2) and assign $x_{12} = \min(11 - 6, 10) = \min(5, 10) = 5$.

(iii) Since 10 > 5, we move down to the cell (2,2) and allocate $x_{22} = \min(5, 13) = 5$. This exhausted the demand at D_2.

(iv) Move to the cell (2,3) and assign $x_{23} = \min(13 - 5, 12) = 8$.

(v) We move down the cell (3,3) and allocate $x_{33} = \min(4, 19) = 4$ and then allocate $x_{34} = 19 - 4 = 15$.

Thus the transportation table reduces to

(6) 21	(5) 16	25	13	11
17	(5) 18	(8) 14	23	13
32	27	(4) 18	(15) 41	19
6	10	12	15	43

We note that the number of allocated cell is 6 which is equal to $m + n - 1 = 3 + 4 - 1 = 6$. Hence the solution is non-degenerate. Also transportation cost $= 21 \times 6 + 16 \times 5 + 18 \times 5 + 14 \times 8 + 18 \times 4 + 41 \times 15 = 1095$.

2. Least Cost Method

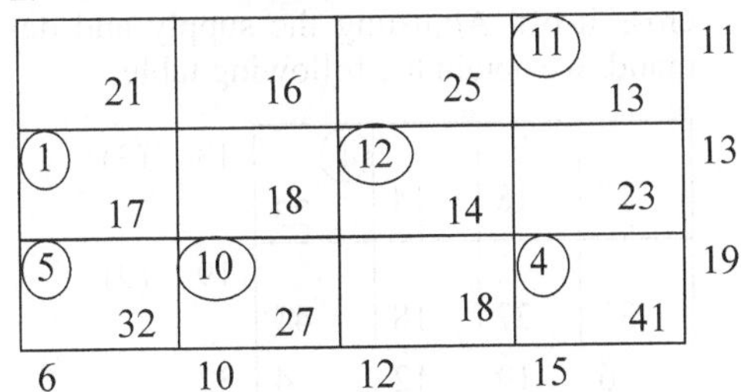

21	16	25	(11) 13	11
(1) 17	18	(12) 14	23	13
(5) 32	(10) 27	18	(4) 41	19
6	10	12	15	

(i) The lowest unit cost is 13 in cell(1,4). Therefore, we allocate min(11,15) = 11 to this cell. Thus $x_{14} = 11$. This exhausted the first row and, therefore, we omit it.

(ii) The next lowest unit cost is 14 in cell (2,3). Therefore, we allocate min (12,13) = 12 to this cell. Thus $x_{23} = 12$. This exhausted the third column and so we omit it.

(iii) The next lowest unit cost is 17 in cell (2,1). Therefore, we allocate $x_{21} = \min(6, 1) = 1$. This exhausted second row also.

(iv) Since the demand at D_1, D_2, D_3 are now 5,10, and 4 and their sum is equal to the availability 19 of origin O_3, we allocate $x_{31} = 5$, $x_{32} = 10$, and $x_{34} = 4$.

Since the number of allocation is equal to $m + n - 1$, the solution is non-degenerate. The transportation cost in this case is

$$13 \times 11 + 17 \times 1 + 14 \times 12$$

$$+ 32 \times 5 + 27 \times 10 + 41 \times 4 = 922.$$

3. VAM

					Row Penalty
21	16	25	(11) 13	11	(3)
17	18	14	23	13	(3)
32	27	18	41	19	(9)
6	10	12	15	43	
Column penalty (4)	(2)	(4)	(10)		

(i) The largest penalty is 10 which corresponds to the fourth column. The smallest cost in fourth column is $c_{14} = 13$. So we allocate $x_{14} = \min(11,15) = 11$. The availability in the first row is, thus, exhausted. Hence we cross it off. Adjusting the supply and demand, we obtain the following table

17	18	14	(4) 23	13	(3)
32	27	18	41	19	(9)
6	10	12	4	32	
(15)	(9)	(4)	(18)		

(ii) The largest penalty is 18 associated with fourth column and the smallest cost in this column is $c_{24} = 23$. Therefore, we allocate $x_{24} = \min(4,13) = 4$. This exhausted the demand at column four and so we cross fourth column. Adjusting again the supply and demand and using VAM, we obtain the following table:

(6) 17	18	14	9 (3)
32	27	18	19 (9)
6	10	12	28
(15)	(9)	(4)	

(iii) Using VAM again, we get the following tables:

(3) 18	14	3 (4)
27	18	19 (9)
10	12	22
(9)	(4)	

and

(7) 27	(12) 18	19 (9)
7	12	

Therefore, the initial basic feasible solution is shown in the following table:

	v_1	v_2	v_3	v_4
u_1	21	16	25	(11) 13
u_2	(6) 17	(3) 18	14	(4) 23
u_3	32	(7) 27	(12) 18	41

Since the number of occupied cell = 6 = $m + n - 1$, the initial solution is non-degenerate.

To get the optimal solution, we introduce u_i and v_i according to MODI and have

$$u_1 + v_4 = 13,\ u_2 + v_1 = 17,\ u_2 + v_2 = 18$$

$$u_2 + v_4 = 23,\ u_3 + v_2 = 27,\ u_3 + v_3 = 18.$$

Taking $u_2 = 0$ and then solving the above equations, we get

$$u_1 = -10,\ u_3 = 9,\ v_1 = 17,\ v_2 = 18,\ v_3 = 9,$$

$$v_4 = 23.$$

The net evaluations $d_{ij} = (u_i + v_j) - c_{ij}$ for the empty cells are

$$d_{11} = u_1 + v_1 - 21 = -14$$

$$d_{12} = u_1 + v_2 - 16 = -8$$

$$d_{13} = u_1 + v_3 - 25 = -26$$

$$d_{23} = u_2 + v_3 - 14 = -5$$

$$d_{31} = u_3 + v_1 - 32 = -6$$

$$d_{34} = u_3 + v_4 - 41 = -9.$$

Since all the net evaluations are negative, the solution obtained is optimal. Thus the optimal solution is

$$x_{14} = 11, x_{21} = 6, x_{22} = 3,$$

$$x_{24} = 4, x_{32} = 7, x_{33} = 12,$$

and minimum transportation cost is

$$13 \times 11 + 17 \times 6 + 18 \times 3 + 23 \times 4 + 27 \times 7 + 18 \times 12 = 796.$$

EXAMPLE 14.35

Consider four bases of operations B_i and three targets T_j. The tons of bombs per aircraft from any base that can be delivered to any target are given in the following table:

	T_1	T_2	T_3
B_1	8	6	5
B_2	6	6	6
B_3	10	8	4
B_4	8	6	4

The daily sortie capability of each of the four bases is 150 sorties per day. The daily requirement in sorties over each target is 200. Find the allocation of sorties from each base to each target which maximizes the total tonnage over all the three targets.

Solution. Since total capability is equal to total requirement, the problem is balanced. Using VAM, the initial basic feasible solution can be obtained through the following tables:

Table 1

				Row Penalty
	8	6	5	150 (1)
	6	6	6	150 (0)
	10	8	(150) 4	150 (4)
	8	6	4	150 (2)
	200	200	200	
Column Penalty	(2)	(2)	(1)	

Table 2

8	6	5	150 (1)
(150) 6	6	6	150 (0)
8	6	4	150 (2)
200	200	50	
(2)	(0)	(1)	

Table 3

8	6	5	150 (1)
8	6	(50) 4	150 (2)
50	200	50	
(0)	(0)	(1)	

Table 4

8	(150) 6	150 (2)
8	6	100 (2)
50	200	
(0)	(0)	

Table 5

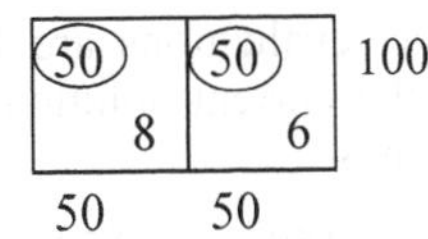

(50) 8	(50) 6	100
50	50	

Thus the table for initial basic feasible solution is

Table 6

	v_1	v_2	v_3	
u_1	8	(150) 6	5	150
u_2	(150) 6	6	6	150
u_3	10	8	(150) 4	150
u_4	(50) 8	(50) 6	(50) 4	150
	200	200	200	

To check the optimality of the current solution, we have

$$u_1 + v_2 = 6, \quad u_2 + v_1 = 6$$

$$u_3 + v_3 = 4, \quad u_4 + v_1 = 8$$

$$u_4 + v_2 = 6, \quad u_4 + v_3 = 4.$$

Putting $u_4 = 0$, the solution to these equations is

$$u_1 = 0,\ u_2 = -2,\ u_3 = 0,\ u_4 = 0$$

$$v_1 = 8,\ v_2 = 6,\ v_3 = 4.$$

Then the net evaluations for the empty cells are

$$d_{11} = u_1 + v_1 - 8 = 0 + 8 - 8 = 0$$

$$d_{13} = u_1 + v_3 - 5 = 0 + 0 - 5 = -5(-ve)$$

$$d_{22} = u_2 + v_2 - 6 = -2 + 6 - 6 = -2(-ve)$$

$$d_{23} = u_2 + v_3 - 6 = -2 + 4 - 6 = -4(-ve)$$

$$d_{31} = u_3 + v_1 - 10 = 0 + 8 - 10 = -2(-ve)$$

$$d_{32} = u_3 + v_2 - 8 = 0 + 6 - 8 = -2(-ve).$$

Since all the net evaluations are less than or equal to zero, the current solution is optimal. Thus the solution is

$$x_{12} = 150,\ x_{21} = 150,\ x_{33} = 150$$

$$x_{41} = 50,\ x_{42} = 50,\ x_{43} = 50,$$

and also the cost is

$$6 \times 150 + 6 \times 150 + 4 \times 150 + 8 \times 50 + 6 \times 50 + 4 \times 50 = 3300.$$

Further, since $d_{ij} = 0$ for $i, j = 1$, *there exists an alternate solution* also. In fact, in Table 4, we could have allocated 100 to lower right corner to get **Table 4'**

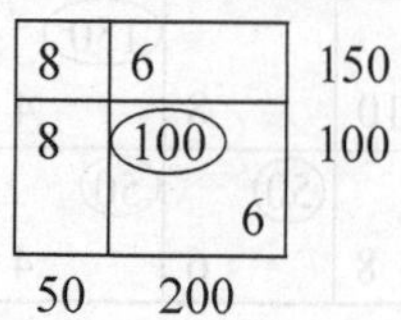

8	6	150
8	(100) 6	100
50	200	

and then we have

Table 5'

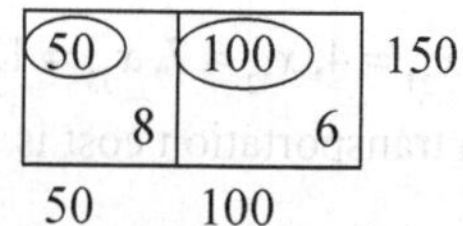

(50) 8	(100) 6	150
50	100	

Thus an alternate solution is

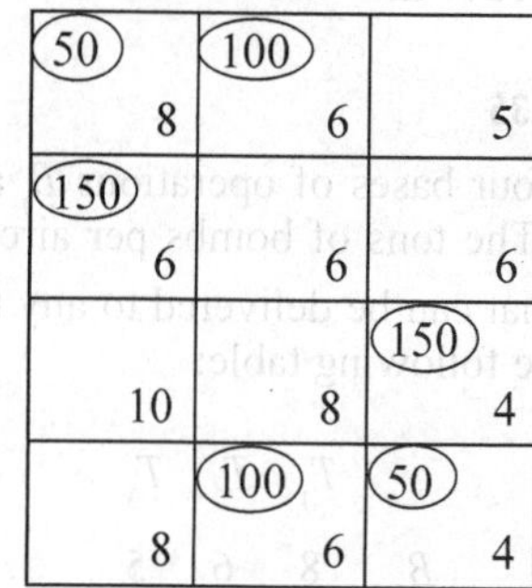

(50) 8	(100) 6	5
(150) 6	6	6
10	8	(150) 4
8	(100) 6	(50) 4

Hence

$$x_{11} = 50,\ x_{12} = 100,\ x_{21} = 150$$

$$x_{33} = 150,\ x_{42} = 100,\ x_{43} = 50.$$

EXAMPLE 14.36

Express the following transportation problem as an LPP. Find its initial basic solution by VAM.

	D_1	D_2	D_3	D_4	Supply a_i
O_1	2	3	11	7	6
O_2	1	0	6	1	1
O_3	5	8	15	9	10
Demand b_j	7	5	3	2	

Solution. Let x_{ij}, $i = 1, 2, 3, j = 1, 2, 3, 4$ represent the quantity of the product to be transported from origin O_i to the distribution centre (destination) D_j. Then the linear programming expression of this problem is

Minimize

$$z = 2x_{11} + 3x_{12} + 11x_{13} + 7x_{14} + x_{21} + 0x_{22} + 6x_{23} + x_{24} + 5x_{31} + 8x_{32} + 15x_{33} + 9x_{34}$$

subject to the constraints

$$\left.\begin{aligned} x_{11}+x_{12}+x_{13}+x_{14}&=6\\ x_{21}+x_{22}+x_{23}+x_{24}&=1\\ x_{31}+x_{32}+x_{33}+x_{34}&=10 \end{aligned}\right\} \text{availability constraints}$$

$$\left.\begin{aligned} x_{11}+x_{21}+x_{31}&=7\\ x_{12}+x_{22}+x_{32}&=3\\ x_{13}+x_{23}+x_{33}&=3\\ x_{14}+x_{24}+x_{34}&=2 \end{aligned}\right\} \text{demand constraints}$$

$x_{ij} \geq 0$ for all i and j (non-negativity).

Since total demand is equal to total supply, the consistency condition $\Sigma a_i = \Sigma b_j$ is satisfied. Hence the given transportation problem is balanced. Further, the rank of the matrix of the given problem is $m+n-1=6$. Applying VAM, we obtain the following table:

Table 1

					a_j	Row Penalty
	2	3	11	7	6	(1)
	1	0	6	(1) 1	(1)	(1)
	5	8	15	9	(10)	(3)
b_{ij}	7	5	3	2		
Column Penalty	(1)	(3)	(5)	(6)		

The largest penalty is 6 which lie in the fourth column. Since $c_{24} = 1$ is the minimum cost in the fourth column, we allocate $x_{24} = \min(a_2, b_4) = \min(1,2) = 1$ and enter 1 in the cell (2,4). The availability of the second row is thus exhausted and so we omit the second row. Adjusting the supply and demand, we again apply VAM to obtain the second table as

Table 2

2	(5) 3	11	7	6	(1)
5	8	15	9	10	(3)
7	5	3	1	16	
(3)	(5)	(4)	(2)		

Since demand in second column has been exhausted, we omit this column and get the following table:

Table 3

(1) 2	11	7	1	(5)
5	15	9	10	(4)
7	3	1	11	
(3)	(4)	(2)		

Similarly, the next tables are

Table 4

5	(3) 15	9	10	(4)
6	3	1	10	
(5)	(15)	(9)		

and

Table 5

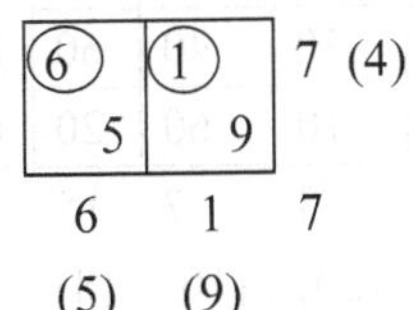

(6) 5	(1) 9	7	(4)
6	1	7	
(5)	(9)		

Thus the initial basic feasible solution is shown in the table below:

Table 6

(1) 2	(5) 3	11	7
1	0	6	(1) 1
(6) 5	8	(3) 15	(1) 9

The number of allocation is $6 = m+n-1$ and so the solution is non-degenerate. Thus the solution is

$$x_{11}=1, x_{12}=5, x_{24}=1, x_{31}=5, x_{33}=3, x_{34}=1.$$

The minimum transportation cost is

$$2\times1+3\times5+1\times1+5\times6+15\times3+9\times1$$

$$=102.$$

EXAMPLE 14.37

Obtain an optimal solution to the following transportation problem using MODI method.

	D_1	D_2	D_3	D_4	Supply
O_1	19	30	50	12	7
O_2	70	30	40	60	10
O_3	0	10	60	20	18
Demand	5	8	7	15	35

Solution. Since total demand is equal to total supply, the consistency condition is satisfied. Hence the given transportation problem is balanced. Applying Vogel's Approximation method, we get the following tables:

Table 1

				Row Penalty
(5) 19	30	50	12	7 (7)
70	30	40	60	10 (10)
40	10	60	20	18 (10)
5	8	7	15	35
Column penalty (21)	(20)	(10)	(8)	

Table 2

30	50	12	2	(18)
30	40	60	10	(10)
(8) 10	60	20	18	(10)
8	7	15	30	
(20)	(10)	(8)		

Table 3

50	12	2 (38)
40	60	10 (20)
60	(10) 20	10 (40)
7	15	22
(10)	(8)	

Table 4

50	(2) 12	2 (38)
40	60	10 (20)
7	5	22
(10)	(38)	

Table 5

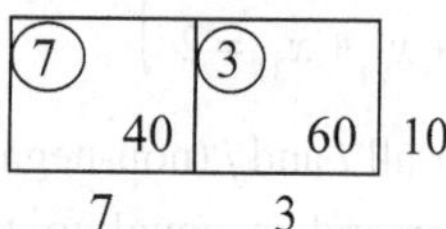

(7) 40	(3) 60	10
7	3	

Thus the initial basic feasible solution is shown in the table given below:

Table 6

	v_1	v_2	v_3	v_4
u_1	(5) 19	30	50	(2) 12
u_2	70	$\theta =$ (3) 30	(7) 40	(3) $-\theta$ 60
u_3	40	$-\theta$ (8) 10	60	(10) $+\theta$ 20

Since the number of occupied cells is $6 = m + n - 1$, which is the rank of the cost matrix, the current initial solution is non-degenerated. Thus an optimal solution can be obtained. The transportation cost for this solution is: $19 \times 5 + 12 \times 2 + 40 \times 7 + 60 \times 3 + 10 \times 8 + 20 \times 10 = 859$.

We find the values of u_i and v_i from the equations

$$u_1 + v_1 = 19,\ u_1 + v_4 = 12,\ u_2 + v_4 = 60$$

$$u_2 + v_3 = 40,\ u_3 + v_2 = 10,\ u_3 + v_4 = 20.$$

We arbitrarily assign $v_4 = 0$. Then the above equations yields

$$u_1 = 12,\ u_2 = 60,\ u_3 = 20,\ v_1 = 7,$$

$$v_2 = -10,\ v_3 = -20,\ v_4 = 0.$$

Therefore, the net evaluations for the non-occupied cells are

$$d_{12} = u_1 + v_2 - 30 = 12 - 10 - 30 = -28$$

$$d_{13} = u_1 + v_3 - 50 = 12 - 20 - 50 = -58$$

$$d_{21} = u_2 + v_1 - 70 = 60 + 7 - 70 = -3$$

$$d_{22} = u_2 + v_2 - 30 = 60 - 10 - 30 = 20(+\text{ve})$$

$$d_{31} = u_3 + v_1 - 40 = 20 + 7 - 40 = -13$$

$$d_{33} = u_3 + v_3 - 60 = 20 - 20 - 60 = -60.$$

Since one of the net evaluation (d_{22}) is +ve, the current solution is not optimal. Also it follows that the cost can be reduced in a multiple of 20. We mark θ in the (2,2) cell. Draw the closed path beginning and ending at the cell (2,2). Take $\theta = 3$ to annihilate x_{24}. The effect of this step to other value of x_{ij} in other nodes of the loop is shown in Table 6. Hence the new solution is

$$x_{11} = 5, x_{14} = 2$$

$$x_{22} = 3, x_{23} = 7$$

$$x_{32} = 5, x_{34} = 13$$

and the solution table is

Table 7

	v_1	v_2	v_3	v_4
u_1	(5) 19	30	50	(2) 12
u_2	70	(3) 30	(7) 40	60
u_3	40	(5) 10	60	(13) 20

For the optimal test, we have

$$u_1 + v_1 = 19, u_1 + v_4 = 12$$

$$u_2 + v_2 = 30, u_2 + v_3 = 40$$

$$u_3 + v_2 = 10, u_3 + v_4 = 20.$$

We have seven variables and six equations. Thus degree of the freedom is 1. So, taking $v_4 = 0$ and solving the above equations, we get

$$u_1 = 12, u_2 = 40, u_3 = 20$$

$$v_1 = 7, v_2 = -10, v_3 = 0, v_4 = 0.$$

Therefore, the net evaluations for the unoccupied cells are

$$d_{12} = u_1 + v_2 - 30 = 12 - 10 - 30 = -28$$

$$d_{13} = u_1 + v_3 - 50 = 12 - 0 - 50 = -38$$

$$d_{21} = u_2 + v_1 - 70 = 40 + 7 - 70 = -23$$

$$d_{24} = u_2 + v_4 - 60 = 40 + 0 - 60 = -20$$

$$d_{31} = u_3 + v_1 - 40 = 20 + 7 - 40 = -13$$

$$d_{33} = u_3 + v_3 - 60 = 20 - 0 - 60 = -40.$$

Since all d_{ij} are negative, the current solution is optimal. Hence the minimum transportation cost is

$$19 \times 5 + 12 \times 2 + 30 \times 3 + 40 \times 7 + 10 \times 5 + 20 \times 13 = 799.$$

EXAMPLE 14.38

Determine an initial basic feasible solution to the following transportation problem using

(a) Least cost method

(b) Vogel's approximation method.

	D_1	D_2	D_3	D_4	Supply
S_1	1	2	1	4	30
S_2	3	3	2	1	50
S_3	4	2	5	9	20
Demand	20	40	30	10	

Solution. Since total demand is equal to total supply, the given transportation problem is *balanced.*

(a) Using least cost method, the initial basic feasible solution is shown in the following table:

(20) 1	2	(10) 1	4	30
3	(20) 3	(20) 2	(10) 1	50
4	(20) 2	5	9	20
20	40	30	10	

Therefore, the initial basic feasible solution is

$$x_{11} = 20, x_{13} = 10, x_{22} = 20,$$

$$x_{23} = 20, x_{24} = 10, x_{32} = 20.$$

$$\text{Total cost} = 1 \times 20 + 1 \times 10 + 3 \times 20$$

$$+ 2 \times 20 + 1 \times 10 + 2 \times 20 = 180.$$

(b) Using VAM, we obtain the following tables:

Table 1

(20) 1	2	1	4	30 (1)
3	3	2	1	50 (1)
4	2	5	9	20 (2)
20	40	30	10	
(2)	(1)	(1)	(3)	

Table 2

2	1	4	10 (1)
3	2	(10) 1	50 (1)
2	5	9	20 (3)
40	30	10	
(1)	(1)	(3)	

Table 3

2	1	10 (1)
3	2	40 (1)
(20) 2	5	20 (3)
40	30	
(1)	(1)	

Table 4

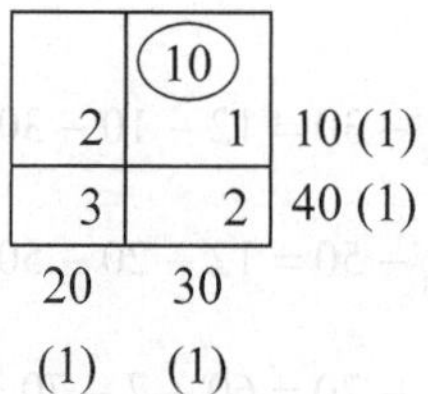

2	(10) 1	10 (1)
3	2	40 (1)
20	30	
(1)	(1)	

Table 5

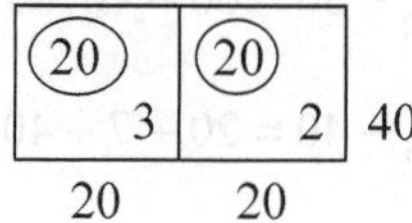

(20) 3	(20) 2	40
20	20	

Hence the table for the initial basic feasible solution is as shown below:

Table 6

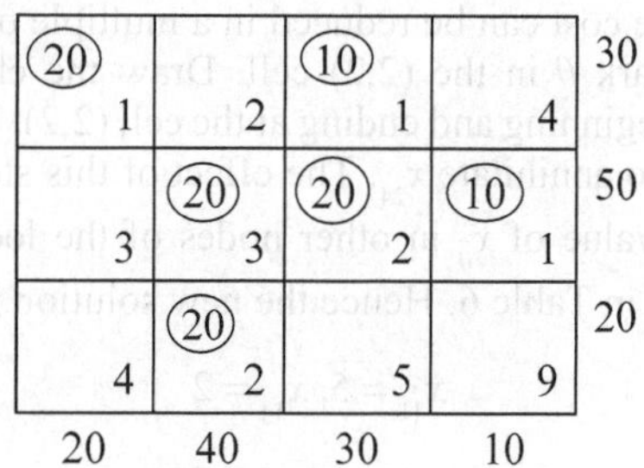

(20) 1	2	(10) 1	4	30
3	(20) 3	(20) 2	(10) 1	50
4	(20) 2	5	9	20
20	40	30	10	

Hence the solution is

$$x_{11} = 20, x_{13} = 10, x_{22} = 20, x_{23} = 20,$$

$$x_{24} = 10, x_{32} = 20,$$

and the total cost is

$$1 \times 20 + 1 \times 10 + 3 \times 20 + 2 \times 20 + 1 \times 10$$

$$+ 2 \times 20 = 180.$$

EXAMPLE 14.39

Using north-west corner rule, find the initial basic feasible solution to the following transportation problem:

	D_1	D_2	D_3	D_4	Supply
O_1	6	4	1	5	14
O_2	8	9	2	7	16
O_3	4	3	6	2	5
Demand	6	10	15	4	

Solution. Since total demand is equal to total supply, the given transportation problem is balanced. Using north-west corner rule, the table showing the initial basic feasible solution is

(6) 6	(8) 4	1	5	14
8	(2) 9	(14) 2	7	16
4	3	(1) 6	(4) 2	5
6	10	15	4	

Starting from north-west corner, we allocated min (6, 14) = 6 to (1, 1) cell. Thus demand column for D_1 is exhausted. So, we move to the cell (1, 2) and allocate min (10, 8) = 8 to it. This exhausts the availability of O_1. So, we move down to the cell (2, 2) and allocate min (2, 16) to this cell. This exhausts the demand column of D_2. Therefore, we move to the cell (2, 3) and allocate min (15, 14) = 14 to this cell. This exhausts the availability of origin O_2. Hence we move down to the cell (3, 3) and allocate min (1, 5) = 1 to it. This exhausts the demand column of the destination D_3. So, we move to the right to enter the cell (3, 4) and allocate min (4, 5) = 4 to it. Hence the initial basic feasible solution is

$$x_{11} = 6, x_{12} = 8, x_{22} = 2,$$
$$x_{23} = 14, x_{33} = 1, x_{34} = 4$$

and the total cost is

$$6 \times 6 + 4 \times 8 + 9 \times 2 + 2 \times 14 + 6 \times 1 + 2 \times 4$$
$$= 128.$$

EXAMPLE 14.40

Find an optimum basic feasible solution to the following transportation problem:

	D_1	D_2	D_3	Availability
O_1	7	3	4	2
O_2	2	1	3	3
O_3	3	4	6	5
Demand	4	1	5	10

Solution. Since total demand and total supply are equal, the given transportation problem is balanced. So an initial basic feasible solution exists. To find an initial basic feasible solution, we use VAM in the form of the following tables:

Table 1

7	3	4	2 (1)
2	(1) 1	3	3 (1)
3	4	6	5 (1)
4	1	5	
(1)	(2)	(1)	

Table 2

7	4	2	(3)
2	3	2	(1)
(4) 3	6	5	(3)
4	5		
(1)	(1)		

Table 3

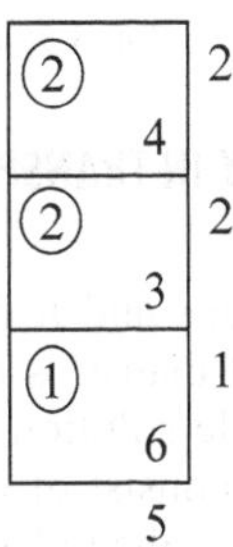

(2) 4	2
(2) 3	2
(1) 6	1
5	

Thus the solution table is

	v_1	v_2	v_3	
u_1	7	3	(2) 4	2
u_2	2	(1) 1	(2) 3	3
u_3	(4) 3	4	(1) 6	5
	4	1	5	

Thus the number of occupied cells is $5 = m + n - 1 = 3 + 3 - 1$. Hence the solution is nondegenerate. To check the optimality, we have

$$u_1 + v_3 = 4,\ u_2 + v_2 = 1,\ u_2 + v_3 = 3,$$

$$u_3 + v_1 = 3,\ u_3 + v_3 = 6,$$

Setting $v_3 = 0$, we get

$$u_1 = 4,\ u_2 = 3,\ u_3 = 6,\ v_1 = -3,$$

$$v_2 = -2,\ v_3 = 0.$$

Therefore, the net evaluations are

$$d_{11} = u_1 + v_1 - 7 = -6(-ve)$$

$$d_{12} = u_1 + v_2 - 3 = -1(-ve)$$

$$d_{21} = u_2 + v_1 - 2 = -2(-ve)$$

$$d_{32} = u_3 + v_2 - 4 = 0.$$

Since none of the net evaluation is positive, it follows that the current solution is optimal. Hence the optimal solution is

$$x_{13} = 2, x_{22} = 1, x_{23} = 2, x_{31} = 4, x_{33} = 1$$

and the total cost is

$$4 \times 2 + 1 \times 1 + 3 \times 2 + 3 \times 4 + 6 \times 1 = 33.$$

14.17 DEGENERACY IN TRANSPORTATION PROBLEM

If there are m origins and n destinations, then the rank of the cost matrix is $m + n - 1$. Therefore, a basic feasible solution for a transportation problem must consist of exactly $m + n - 1$ positive allocations in the independent positions in the transportation table. A solution is called degenerate if the number of occupied cells is less than the rank $m + n - 1$ of the cost matrix. In such a case, the current solution cannot be improved because in such a situation, we are not able to draw loop.

To remove degeneracy in the initial basic solution, we allocate a very small quantity ε which is very close to zero to one or more unoccupied cells so as to get $m + n - 1$ number of occupied cells. We treat these additional cells like other basic cells and solve the problem in the usual way.

EXAMPLE 14.41

Solve the following transportation problem:

	D_1	D_2	D_3	D_4	Supply
O_1	1	2	3	4	6
O_2	4	3	2	0	10
O_3	0	2	2	1	8
Demand	4	6	8	6	

Solution. Since total demand is equal to total supply, the problem is balanced. Using VAM, we have the following tables:

Table 1

	D_1	D_2	D_3	D_4	
O_1	1	2	3	4	6 (1)
O_2	4	3	2	(6) 0	10 (2)
O_3	0	2	2	1	8 (1)
	4	6	8	6	
	(1)	(1)	(1)	(3)	

Table 2

1	2	3	6 (1)
4	3	2	4 (1)
(4) 0	2	2	8 (2)
4	6	8	
(1)	(1)	(1)	

Table 3

(6) 2	3	6 (1)
3	2	4 (1)
2	2	4 (0)
6	8	
(1)	(1)	

Table 4

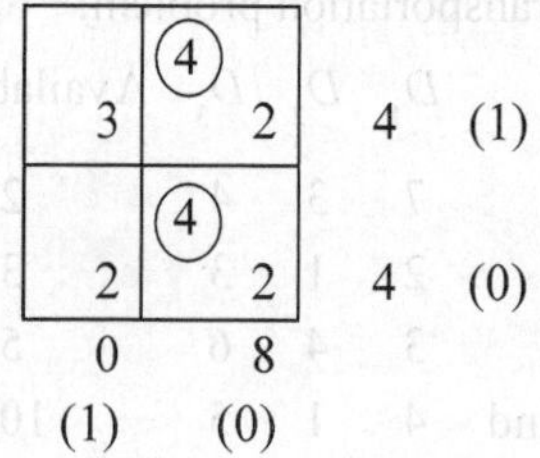

3	(4) 2	4	(1)
2	(4) 2	4	(0)
0	8		
(1)	(0)		

Thus the initial basic feasible solution is shown in the table below.

	v_1	v_2	v_3	v_4	
u_1	1	(6) 2	(ε) 3	4	6
u_2	4	3	(4) 2	(6) 0	10
u_3	(4) 0	2	(4) 2	1	8
	4	6	8	6	

(If we use least cost method, the same initial basic solution is obtained.)

Since number of occupied cells = 5 ≠ $m + n - 1 = 6$, there is degeneracy in the problem. So we assign ε to an unoccupied cell (1, 3) say. To check the optimality, we have

$$u_1 + v_2 = 2,\ u_1 + v_3 = 3,$$

$$u_2 + v_3 = 2,\ u_2 + v_4 = 0,$$

$$u_3 + v_1 = 0,\ u_3 + v_3 = 2.$$

Putting $v_3 = 0$, we have

$$u_1 = 3,\ u_2 = 2,\ u_3 = 2$$

$$v_1 = -2,\ v_2 = -1,\ v_3 = 0,\ v_4 = -2.$$

Therefore, the net evaluations are

$$d_{11} = u_1 + v_1 - 1 = 3 - 2 - 1 = 0$$

$$d_{14} = u_1 + v_4 - 4 = 3 - 2 - 4 = -3(-ve)$$

$$d_{21} = u_2 + v_1 - 4 = 2 - 2 - 4 = -4(-ve)$$

$$d_{22} = u_2 + v_2 - 3 = 2 - 1 - 3 = -2(-ve)$$

$$d_{32} = u_3 + v_2 - 2 = 2 - 1 - 2 = -1(-ve)$$

$$d_{34} = u_3 + v_4 - 1 = 2 - 2 - 1 = -1(-ve).$$

Since none of the net value is + ve, the current solution is optimal. Hence the solution is

$$x_{12} = 6,\ x_{23} = 4,\ x_{24} = 6,\ x_{31} = 4,\ x_{33} = 4$$

and the total cost is

$$2 \times 6 + 2 \times 4 + 0 \times 4 + 2 \times 4 = 28.$$

EXAMPLE 14.42

Solve the following transportation problem:

		D_1	D_2	D_3	Supply
	O_1	50	30	220	1
Source	O_2	90	45	170	3
	O_3	250	200	50	4
	Demand	4	2	2	8

Solution. The given transportation problem is balanced. Using VAM, we get the following table:

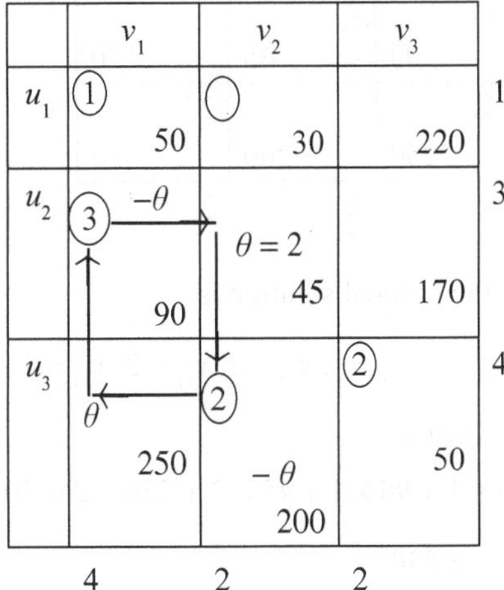

Since number of occupied cells is not equal to $m + n - 1 = 5$, the solution is *degenerate*. So we allocate an arbitrarily small number ε to the cell (1, 2). To check the optimality of solution, we have

$$u_1 + v_1 = 50,\ u_1 + v_2 = 30,\ u_2 + v_1 = 90,$$

$$u_3 + v_2 = 200 \text{ and } u_3 + v_3 = 50.$$

Putting $v_2 = 0$ and solving the above equations, we get

$$u_1 = 30,\ u_2 = 70,\ u_3 = 200,$$

$$v_1 = 20,\ v_2 = 0,\ v_3 = -150.$$

Then the net evaluations are

$$d_{13} = u_1 + v_3 - 220 = -340(-ve)$$

$$d_{22} = u_2 + v_2 - 45 = 25(+ve)$$

$$d_{23} = u_2 + v_3 - 170 = -250(-ve)$$

$$d_{31} = u_3 + v_1 - 250 = -30(-ve).$$

Since d_{22} is positive, the current solution is not optimal. Drawing the loop beginning and ending at (2, 2) cell as shown in the table. Taking $\theta = 2$, we get the following table:

① 50	30	220	1
① 90	② 45	170	3
② 250	200	② 50	4
4	2	2	

Hence the optimal solution is

$$x_{11} = 1, x_{21} = 1, x_{22} = 2, x_{31} = 2, x_{33} = 2$$

and the cost is

$$50 \times 1 + 90 \times 1 + 45 \times 2 + 250 \times 2 + 50 \times 2$$

$$= 830.$$

EXAMPLE 14.43

Solve the following degenerate transportation problem:

	D_1	D_2	D_3	D_4	D_5	D_6	Supply
O_1	9	12	9	6	9	10	5
O_2	7	3	7	7	5	5	6
O_3	6	5	9	11	3	11	2
O_4	6	8	11	2	2	10	9
Demand	4	4	6	2	4	2	22

Solution. Since total demand and total supply are equal, the given transportation problem is balanced. So, an initial basic feasible solution exists. To find an initial basic feasible solution, we use VAM and get the following tables:

Table 1

9	12	9	6	9	10	5 (3)
7	3	7	7	5	② 5	6 (2)
6	5	9	11	3	11	2 (2)
6	8	11	2	2	10	9 (4)
4	4	6	2	4	2	
(1)	(2)	(2)	(4)	(1)	(5)	

Table 2

9	12	9	6	9	5 (3)
7	3	7	7	5	4 (2)
6	5	9	11	3	2 (2)
6	8	11	② 2	2	9 (4)
4	4	6	2	4	
(1)	(2)	(2)	(4)	(1)	

Table 3

9	12	9	9	5 (3)
7	3	7	5	4 (2)
6	5	9	3	2 (2)
6	8	11	④ 2	7 (4)
4	4	6	4	
(1)	(2)	(2)	(1)	

Table 4

9	12	9	5 (3)
7	④ 3	7	4 (4)
6	5	9	2 (1)
6	8	11	3 (2)
4	4	6	
(1)	(2)	(2)	

Table 5

9	9	5 (0)
6	9	2 (3)
(3) 6	11	3 (5)
4	6	
(3)	(2)	

Table 6

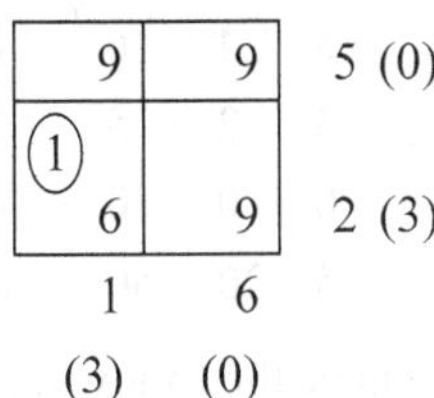

9	9	5 (0)
(1) 6	9	2 (3)
1	6	
(3)	(0)	

Table 7

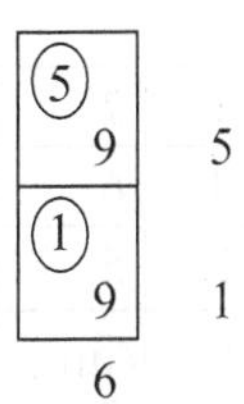

(5) 9	5
(1) 9	1
6	

Hence, the table for initial basic feasible solution is

	v_1	v_2	v_3	v_4	v_5	v_6	
u_1	9	12	(5) 9	6	9	10	5
u_2	7	(4) 3	(ε) 7	7	5	(2) 5	6+ε
u_3	(1) 6	5	(1) 9	11	3	11	2
u_4	(3) 6	8	11	(2) 2	(4) 2	10	9
	4	4	6+ε	2	4	2	

Note that number of occupied cells is 8 which is not equal to $m + n - 1 = 4 + 6 - 1 = 9$. Therefore, the current basic solution is degenerate. To remove the degeneracy we allocate a very small positive quantity to unoccupied cell (2, 3).

To check the optimality of the current solution, we note that

$$u_1 + v_3 = 9,\ u_2 + v_2 = 3,\ u_2 + v_3 = 7$$

$$u_2 + v_6 = 5,\ u_3 + v_1 = 6,\ u_3 + v_3 = 9$$

$$u_4 + v_1 = 6,\ u_4 + v_4 = 2,\ u_4 + v_5 = 2.$$

Putting $v_3 = 0$ and solving the above equations, we get

$$u_1 = 9,\ u_2 = 7,\ u_3 = 9,\ u_4 = 9$$

$$v_1 = -3,\ v_2 = -4,\ v_3 = 0,\ v_4 = -7,$$

$$v_5 = -7,\ v_6 = -2.$$

Then the net evaluations are

$$d_{11} = u_1 + v_1 - 9 = 9 - 3 - 9 = -3(-ve)$$

$$d_{12} = u_1 + v_2 - 12 = 9 - 4 - 12 = -7(-ve)$$

$$d_{14} = u_1 + v_4 - 6 = 9 - 7 - 6 = -4(-ve)$$

$$d_{15} = u_1 + v_5 - 9 = 9 - 7 - 9 = -7(-ve)$$

$$d_{16} = u_1 + v_6 - 10 = 9 - 2 - 10 = -3(-ve)$$

$$d_{21} = u_2 + v_1 - 7 = 7 - 3 - 7 = -3(-ve)$$

$$d_{24} = u_2 + v_4 - 7 = 7 - 7 - 7 = -7(-ve)$$

$$d_{25} = u_2 + v_5 - 5 = 7 - 7 - 5 = -5(-ve)$$

$$d_{32} = u_3 + v_2 - 5 = 9 - 4 - 5 = 0$$

$$d_{34} = u_3 + v_4 - 11 = 9 - 7 - 11 = -9(-ve)$$

$$d_{35} = u_3 + v_5 - 3 = 9 - 7 - 3 = -1(-ve)$$

$$d_{36} = u_3 + v_6 - 11 = 9 - 2 - 11 = -4(-ve)$$

$$d_{42} = u_4 + v_2 - 8 = 9 - 4 - 8 = -3(-ve)$$

$$d_{43} = u_4 + v_3 - 11 = 9 + 0 - 11 = -2(-ve)$$

$$d_{46} = u_4 + v_6 - 10 = 9 - 2 - 10 = -3(-ve).$$

Since none of the net evaluation is positive, it follows that the current solution is optimal. Thus the optimal solution is

$x_{13} = 5, x_{22} = 4, x_{23} = \varepsilon$ (however small),

$x_{26} = 2, x_{31} = 1, x_{33} = 1, x_{41} = 3,$

$x_{44} = 2, x_{45} = 4.$

The minimum transport cost is

$$9 \times 5 + 3 \times 4 + 7 \times \varepsilon + 5 \times 2 + 6 \times 1 + 9 \times 1$$
$$+ 6 \times 3 + 2 \times 2 + 2 \times 4 = 112 + 7\varepsilon = 112,$$

since ε is arbitrarily small.

14.18 UNBALANCED TRANSPORTATION PROBLEMS

A transport problem is called unbalanced if either the total demand exceeds the total supply or the total supply exceeds the total demand. No feasible solution exists for such problems. To convert an unbalanced problem into a balanced problem, we proceed as follows:

Suppose that total demand is m units and the total supply is *n* units. Then

1. If *m* exceeds *n*, then the requirement of *m* – *n* units is handled by adding a dummy plant O_{excess} with a capacity of *m* –*n* units. We use zero unit transportation costs to the dummy plant (origin). Thus one row with 0 as the entries is added to the cost matrix of the problem.
2. If *n* exceeds *m*, then the supply of *n* – m units is handled by a dummy destination D_{excess} to absorb the excess supply. The associated cost entries in dummy destination are taken as zero and this surplus quantity remains lying in the factories (source or origin). Thus one column with 0 entries is added to the cost matrix of the problem.

EXAMPLE 14.44

Solve the following transportation problem under the condition that there are penalty costs for every unsatisfied demand units which are 5, 3, and 2 for destinations D_1, D_2, and D_3, respectively.

	D_1	D_2	D_3	Supply
O_1	5	1	7	10
O_2	6	4	6	80
O_3	3	2	5	15
Demand	75	20	50	

Solution. We note that

Total demand = 145 units

Total supply = 105 units.

Therefore, the problem is *unbalanced.* So we add a dummy plant O_4 with 0 entries in the cost matrix. Thus the modified transportation table is

				Supply
O_1	5	1	7	10
O_2	6	4	6	80
O_3	3	2	5	15
O_4	0	0	0	40
Demand	75	20	50	145

Using VAM, we have the following tables:

Table 1

			Row Penalty
5	1	7	10 (4)
6	4	6	80 (2)
3	2	5	15 (1)
0	0	(40) 0	40 (0)
75	20	50	
Column penalty (3)	(1)	(5)	

Table 2

5	(10) 1	7	10 (4)
6	4	6	80 (2)
3	2	5	15 (1)
75	20	10	
(2)	(1)	(1)	

Table 3

6	4	6	80 (2)
(15) 3	2	5	15 (1)
75	10	10	
(3)	(2)	(1)	

Table 4

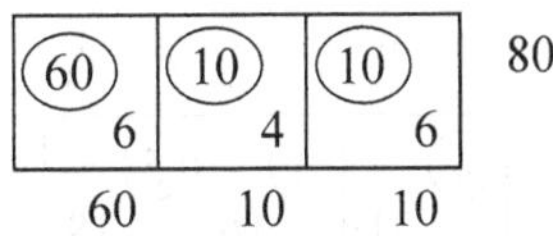

(60) 6	(10) 4	(10) 6	80
60	10	10	

Thus the initial basic solution is shown the following table:

	v_1	v_2	v_3
u_1	5	(10) 1	7
u_2	(60) 6	(10) 4	(10) 6
u_3	(15) 3	2	5
u_4	0	0	(40) 0

To check the optimality of the solution, we have

$$u_1 + v_2 = 1,\ u_2 + v_1 = 6,\ u_2 + v_2 = 4$$

$$u_2 + v_3 = 6,\ u_3 + v_1 = 3,\ u_4 + v_3 = 0.$$

Put $v_2 = 0$ and solve the above equations to get

$$u_1 = 1,\ u_2 = 4,\ u_3 = 1,\ u_4 = -2$$

$$v_1 = 2,\ v_2 = 0,\ v_3 = 2.$$

Then the net evaluations are

$$d_{11} = u_1 + v_1 - 5 = 1 + 2 - 5 = -2(-ve)$$

$$d_{13} = u_1 + v_3 - 7 = 1 + 2 - 7 = -4(-ve)$$

$$d_{32} = u_3 + v_2 - 2 = 1 + 0 - 2 = -1(-ve)$$

$$d_{33} = u_3 + v_3 - 5 = 1 + 2 - 5 = -2(-ve)$$

$$d_{41} = u_4 + v_1 - 0 = -2 + 2 - 0 = 0$$

$$d_{42} = u_4 + v_2 - 0 = -2 + 0 - 0 = -2(-ve).$$

Since none of the net evaluations is positive, it follows that the current solution is optimal. Thus the solution is

$$x_{12} = 10,\ x_{21} = 60,\ x_{22} = 10,\ x_{23} = 10,$$

$$x_{31} = 15,\ x_{43} = 40.$$

So the transportation cost is

$$1 \times 10 + 6 \times 60 + 4 \times 10 + 6 \times 10 + 3 \times 15$$

$$= 515.$$

Further penalty for the less supply to the destination D_3 is $40 \times 2 = 80$. Thus the total cost of transportation is

$$515 + 80 = 595.$$

EXAMPLE 14.45

Solve the following unbalanced transportation problem.

	D_1	D_2	D_3	D_4	Supply
O_1	20	21	16	18	10
O_2	17	28	14	16	9
O_3	29	23	19	20	7
Demand	6	10	4	5	

Does there exists an alternative solution?

Solution. Since total supply exceeds total demands, the given problem is unbalanced. So we add dummy column D_{excess} with demand 1. Thus we have the following table:

D_1	D_2	D_3	D_4	D_{excess}	
20	21	16	18	0	10
17	28	14	16	0	9
29	23	19	20	0	7
6	10	4	5	1	

Using VAM, we get the following tables:

Table 1

20	21	16	18	0	10 (16)
17	28	14	16	0	9 (14)
29	23	19	20	(1) 0	7 (19)
6	10	4	5	1	
(3)	(2)	(2)	(2)	(0)	

Table 2

20	21	16	18	10	(2)
⑥ 17	28	14	16	9	(2)
29	23	19	20	6	(1)
6	10	4	5		
(3)	(2)	(2)	(2)		

Table 3

21	16	18	10 (2)
28	③ 14	16	3 (2)
23	19	20	6 (1)
10	4	5	
(2)	(2)	(2)	

Table 4

21	① 16	18	10 (2)
23	19	20	6 (1)
10	1	5	
(2)	(3)	(2)	

Table 5

21	⑤ 18	9 (3)
23	20	6 (3)
10	5	
(2)	(2)	

Table 6

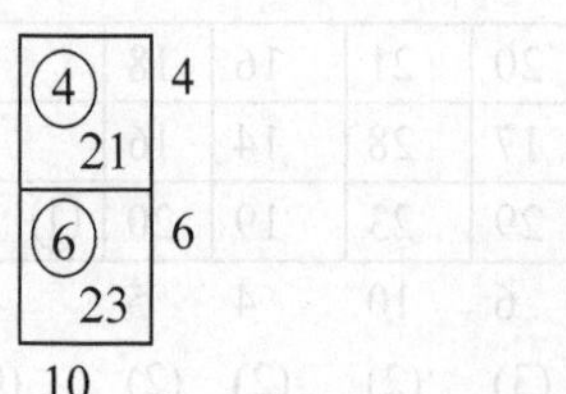

④ 21	4
⑥ 23	6
10	

Thus the table for initial feasible solution is

	v_1	v_2	v_3	v_4	v_5	
u_1	20	④ 21	① 16	⑤ 18	0	10
u_2	⑥ 17	28	③ 14	16	0	9
u_3	29	⑥ 23	19	20	① 0	7
	6	10	4	5	1	

The number occupied cell is equal to $m + n - 1$. Hence the solution is non-degenerate.

To check whether the solution is optimal, we have

$$u_1 + v_2 = 21,\ u_1 + v_3 = 16,\ u_1 + v_4 = 18,$$

$$u_2 + v_1 = 17,\ u_2 + v_3 = 14,\ u_3 + v_2 = 23,$$

$$u_3 + v_5 = 0.$$

Putting $v_3 = 0$, we have

$$u_1 = 16,\ u_2 = 14,\ u_3 = 18$$

$$v_1 = 3,\ v_2 = 5,\ v_3 = 0,\ v_4 = 2,\ v_5 = -18.$$

Then the net evaluations are

$$d_{11} = u_1 + v_1 - 20 = 16 + 3 - 20 = -1(-ve)$$

$$d_{15} = u_1 + v_5 - 0 = 16 - 18 - 0 = -2(-ve)$$

$$d_{22} = u_2 + v_2 - 28 = 14 + 5 - 28 = -9(-ve)$$

$$d_{24} = u_2 + v_4 - 16 = 14 + 2 - 16 = 0$$

$$d_{25} = u_2 + v_5 - 0 = 14 - 18 - 0 = -4(-ve)$$

$$d_{31} = u_3 + v_1 - 29 = 18 + 3 - 29 = -8(-ve)$$

$$d_{33} = u_3 + v_3 - 19 = 18 + 0 - 19 = -1(-ve)$$

$$d_{34} = u_3 + v_4 - 20 = 18 + 2 - 20 = 0.$$

Since none of the net evaluations is positive, the current solution is optimal. Hence the optimal solution is

$$x_{12} = 4,\ x_{13} = 1,\ x_{14} = 5$$

$$x_{21} = 6,\ x_{23} = 3,\ x_{32} = 6,\ x_{35} = 1.$$

$$\text{Cost} = 21 \times 4 + 16 \times 1 + 18 \times 5 + 17 \times 6$$
$$+ 14 \times 3 + 23 \times 6 + 1 \times 0 = 472.$$

Alternative solution exists because net evaluations in cells (2, 4) and (3,4) are zeros. In fact, if in Table 5 we would have allocated 5 to the cell having 20 as the cost, then the Tables 4 and 5 would have been

Table 4

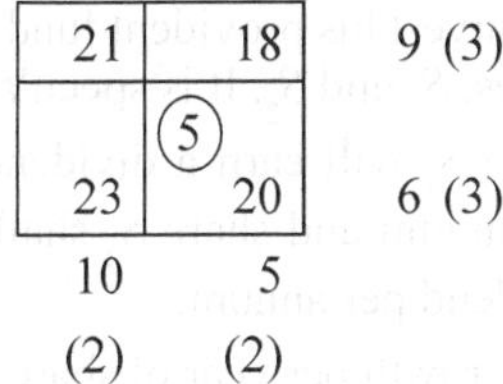

21	18	9 (3)
23	(5) 20	6 (3)
10	5	
(2)	(2)	

Table 5

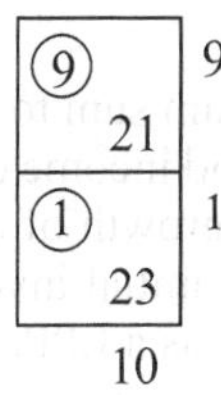

(9) 21	9
(1) 23	1
10	

Thus the solution table becomes

21	(9) 21	(1) 16	18	0
(6) 17	28	(3) 14	16	0
29	(1) 23	19	(5) 20	(1) 0
6	10	4	5	1

and so the solution is

$$x_{12} = 9, x_{13} = 1, x_{21} = 6, x_{23} = 3,$$
$$x_{32} = 1, x_{34} = 5, x_{35} = 1.$$

The cost of transportation is

$$21 \times 9 + 16 \times 1 + 17 \times 6 + 14 \times 3$$
$$+ 23 \times 1 + 20 \times 5 = 472.$$

EXERCISES

Foundation of LPP

1. A person invest in four firms F_1, F_2, F_3, and F_4. He can spare only Rs 50,000. He invests not more than 50% of the total investment in firms F_2 and F_3. Further he prefers to invest at least Rs 2 in F_1 for every Rs 5 invested in the firm F_4. The anticipated return on investment are as follows:

Firms	F_1	F_2	F_3	F_4
% Anticipatory return	10	13	14	16

The investor wants to know how much to invest in each firm to maximize the total return. Formulate this problem as an LPP.

Ans. If x_1, x_2, x_3, x_4 is %age of total amount to be invested in F_1, F_2, F_3, and F_4, respectively, then the LPP is

Max. $z = 0.10x_1 + 0.13x_2 + 0.14x_3 + 0.16x_4$
subject to

$$x_1 + x_2 + x_3 + x_4 \leq 50{,}000$$
$$x_2 + x_3 = 0.50(x_1 + x_2 + x_3 + x_4)$$
$$x_1 \geq \frac{2}{5}x_4$$
$$x_1, x_2, x_3, x_4 \geq 0.$$

2. A manufacturing company manufacturers two circuit boards C_1 and C_2 constructed as follows:

C_1 Consists of 3 registers, 1 capacitor, 2 transistors, and 2 inductances

C_2 Consists of 4 registers, 2 capacitors, and 3 transistors

The available stocks for daily production are 2,400 resistors, 900 capacitors, 1,600 transistor, and 1,200 inductances. How many C_1 and C_2 should be produced to maximize the profit if the profit on C_1 is Rs 5 and on C_2 board is Rs 9?

Ans. Max. $z = 5x + 9y$

subject to

$$3x + 4y \le 2400,\ x + 2y \le 900,$$
$$2x + 3y \le 1600,\ 2x \le 1200$$
$$x, y \ge 0.$$

3. A firm manufactures three items A, B, and C and earns profit of Rs 3, Rs 2, and Rs 4 on these items, respectively. The items are being manufactured by two machines M_1 and M_2. The required processing time in minutes for each machine on each product is shown below:

Machine	Items		
	A	B	C
M_1	4	3	5
M_2	2	2	4

Time limits for M_1 and M_2 are 2,000 and 5,000 minutes, respectively. The firm must manufacture 100A's, 200B's, and 50C's but not more than 150A's. Formulate an LPP to maximize the profit of the firm.

Ans. If x_1, x_2, and x_3 are the numbers of A's, B's, and C's, then the LPP is

$$\text{Max. } z = 3x_1 + 2x_2 + 4x_3$$

subject to

$$4x_1 + 3x_2 + 5x_3 \le 2000$$
$$2x_1 + 2x_2 + 4x_3 \le 5000$$
$$100 \le x_1 \le 150$$
$$x_2 \ge 200,\ x_3 \ge 50$$
$$x_1, x_2, x_3 \ge 0.$$

4. A factory manufactures nuts and bolts and earns a profit of Rs 5/kg on nuts and Rs 8/kg on bolts. Four units of labour are required to manufacture 1 kg of nuts and 6 units to manufacture 1 kg of bolts. Twenty four units of labour are available. Further, 3 units of raw materials are required to produce 1 kg of nuts and 1 unit for 1 kg of bolts. Twelve units of raw materials are available. What should be the manufacturing policy to earn maximum profit?

Ans. If x and y kg of nuts and bolts are produced, then the LPP is

$$\text{Max. } z = 5x + 8y$$

subject to

$$4x + 6y \le 24$$
$$3x + y \le 12$$
$$x, y \ge 0.$$

5. A person after retirement from services wants to invest his provident fund money in two shares, S_1 and S_2. It is speculated that
   (a) Share S_1 will earn a dividend of 12% per annum and share S_2 shall earn 4% dividend per annum.
   (b) The growth per year of share S_1 will be 10 paise per rupee invested while that of share S_2 it will be 40 paise per rupee invested.

Find the maximum sum to be invested if he desires a dividend income of at least Rs 600 per annum and growth of at least Rs 1,000 in 1 year on the initial investment. Formulate the problem as a LPP.

Ans. Let x and y be the number of shares.

Then the LPP is

$$\text{Min. } z = x + y$$

subject to

$$0.12x + 0.04y \ge 600$$
$$0.10x + 0.40y \ge 1000$$
$$x, y \ge 0.$$

Graphical Method

6. Solve the following LPP graphically:

Maximize

$$z = 3x_1 + 5x_2$$

subject to

$$x_1 + 2x_2 \le 2000$$
$$x_1 + x_2 \le 1500$$
$$x_2 \le 600$$
$$x_1, x_2 \ge 0.$$

Ans. $x_1 = 1,000$, $x_2 = 500$, Max. $z = 5,500$.

7. Minimize

$$z = 3x + 2y$$

subject to

$$5x + y \geq 10$$
$$x + y \geq 6$$
$$x + 4y \geq 12$$
$$x, y \geq 0.$$

Ans. $x = 1, y = 5$, Min. $z = 13$

8. Minimize

$$z = 20x_1 + 10x_2$$

subject to

$$x_1 + 2x_2 \leq 40$$
$$3x_1 + x_2 \geq 30$$
$$4x_1 + 3x_2 \geq 60$$
$$x_1, x_2 \geq 0.$$

Ans. $x_1 = 6, x_2 = 12$, Min. $z = 240$

9. A diet for a sick person must contain at least 4,000 units of vitamins, 50 units of minerals, and 1,400 calories. Two foods A and B are available at a cost of Rs 4 and Rs 3 per unit, respectively. If 1 unit of food A contains 200 units of vitamin, 1 unit of mineral, and 40 calories whereas 1 unit of B contains 100 units of vitamins, 2 units of minerals, and 40 units of calories, determine graphically what combination of these foods be used to have least cost?

Ans. 5 units of food A and 30 units of food B worth Rs 110.

10. Maximize

$$z = 200x_1 + 120x_2$$

subject to

$$40x_1 + 80x_2 \leq 800$$
$$10x_1 + 4x_2 \leq 80$$
$$x_1 \leq 6, x_2 \leq 9$$
$$x_1, x_2 \geq 0.$$

Ans. $x_1 = 5, x_2 = 7.5$, Max. $z = 1900$

11. Maximize

$$z = 9x_1 + 10x_2$$

subject to

$$2x_1 + 4x_2 \geq 50$$
$$4x_1 + 3x_2 \geq 24$$
$$3x_1 + 2x_2 \geq 60$$
$$x_1, x_2 \geq 0$$

Ans. $x_1 = \frac{35}{2}, x_2 = \frac{15}{4}$, Max. $z = 195$.

12. Maximize

$$z = 5x_1 + 7x_2$$

subject to

$$x_1 + x_2 \leq 4$$
$$5x_1 + 8x_2 \leq 24$$
$$10x_1 + 7x_2 \leq 35$$
$$x_1, x_2 \geq 0.$$

Ans. $x_1 = \frac{8}{15}, x_2 = \frac{12}{5}$, Max. $z = 24.8$.

13. Maximize

$$z = 4x + 5y$$

subject to

$$3x + 7y \leq 10$$
$$2x + y \leq 3$$
$$x, y \geq 0.$$

Ans. $x = 1, y = 1$, Max. $z = 9$.

14. Maximize

$$z = 60x + 15y$$

subject to

$$x + y \leq 50$$
$$3x + y \leq 90$$
$$x, y \geq 0.$$

Ans. $x = 30, y = 0$, Max. $z = 1800$.

15. Maximize

$$z = 30x_1 + 40x_2$$

subject to

$$4x_1 + 2x_2 \leq 100$$
$$4x_1 + 6x_2 \leq 180$$

$$x_1 \le 20, x_2 \ge 10$$
$$x_1, x_2 \ge 0.$$

Ans. $x_1 = 15, x_2 = 20$, Max. $z = 1250$

Simplex Method

16. Maximize
$$z = 4x_1 + 3x_2 + 4x_3 + 6x_4$$
subject to the constraints
$$x_1 + 2x_2 + 2x_3 + 4x_4 \le 80$$
$$2x_1 + 2x_3 + x_4 \le 60$$
$$3x_1 + 3x_2 + x_3 + x_4 \le 80$$
$$x_1, x_2, x_3, x_4 \ge 0.$$

Ans. $x_1 = \frac{280}{13}, x_2 = 0, x_3 = \frac{20}{13}, x_4 = \frac{180}{13}$, Max. $z = \frac{2280}{13}$

17. Maximize
$$z = 5x_1 + 4x_2 + 6x_2$$
subject to the constraints
$$4x_1 + x_2 + x_3 \le 19$$
$$3x_1 + 4x_2 + 6x_3 \le 30$$
$$2x_1 + 4x_2 + x_3 \le 25$$
$$x_1 + x_2 + 2x_3 \le 15$$
$$x_1, x_2, x_3 \ge 0$$

Ans. $x_1 = 4, x_2 = 0, x_3 = 3$, Max. $z = 38$

18. Maximize
$$z = 10x_1 + x_2 + 2x_3$$
subject to the constraints
$$x_1 + x_2 - 2x_3 \le 10$$
$$4x_1 + x_2 + x_3 \le 20$$
$$x_1, x_2, x_3 \ge 0$$

Ans. $x_1 = 5, x_2 = x_3 = 0$, Max. $z = 50$

19. Write the following LPP in standard form and hence solve it using simplex method.
Maximize
$$z = 5x_1 + 3x_2$$
subject to the constraints
$$x_1 + x_2 \le 2$$
$$5x_1 + 2x_2 \le 10$$
$$3x_1 + 8x_2 \le 12$$
$$x_1, x_2 \ge 0.$$

Ans. $x_1 = 2, x_2 = 0$, Max. $z = 10$.

20. Maximize
$$z = x_1 + 3x_2$$
subject to
$$x_1 + 2x_2 \le 10$$
$$x_1 \le 5, x_2 \le 4$$
$$x_1, x_2 \ge 0$$

Ans. $x_1 = 2, x_2 = 4$, Max. $z = 14$

21. Maximize
$$z = 3x_1 + 5x_2 + 4x_3$$
subject to the constraints
$$2x_1 + 3x_2 \le 8$$
$$2x_2 + 5x_3 \le 10$$
$$3x_1 + 2x_2 + 4x_3 \le 15$$
$$x_1, x_2, x_3 \ge 0$$

Ans. $x_1 = \frac{89}{41}, x_2 = \frac{50}{41}, x_3 = \frac{62}{41}$, Max. $z = \frac{765}{41}$

22. Maximize
$$z = 2x_1 + x_2$$
subject to the constraints
$$3x_1 + 4x_2 \le 6$$
$$6x_1 + x_2 \le 3$$
$$x_1, x_2 \ge 0$$

Ans. $x_1 = \frac{2}{7}, x_2 = \frac{9}{7}$, Max. $z = \frac{13}{7}$

23. Maximize
$$z = 3x_1 + 5x_2$$
subject to the constraints
$$3x_1 + 2x_2 \le 18$$
$$x_1 \le 4, x_2 \le 6$$
$$x_1, x_2 \ge 0$$

Ans. $x_1 = 2, x_2 = 6$, Max. $z = 36$

24. Maximize

$$z = 6x_1 - 2x_2 + 3x_3$$

subject to the constraints

$$2x_1 - x_2 + 2x_3 \le 2$$
$$x_1 + 4x_3 \le 4$$
$$x_1, x_2 \ge 0$$

Ans. $x_1 = 4, x_2 = 6, x_3 = 0$, Max. $z = 12$

25. Maximize

$$z = 3x_1 + 2x_2 + 5x_3$$

subject to the constraints

$$x_1 + 2x_2 + x_3 \le 430$$
$$3x_1 + 2x_3 \le 460$$
$$x_1 + 4x_2 \le 420$$
$$x_1, x_2, x_3 \ge 0$$

Ans. $x_1 = 0, x_2 = 100, x_3 = 230$, Max. $z = 1350$

26. Minimize

$$z = x_1 - 3x_2 + 3x_3$$

subject to

$$3x_1 - x_2 + 2x_3 \le 7$$
$$2x_1 + 4x_2 \ge -12$$
$$-4x_1 + 3x_2 + 8x_3 \le 10$$
$$x_1, x_2, x_3 \ge 0$$

Ans. $x_1 = \frac{31}{5}, x_2 = \frac{58}{5}, x_3 = 0$,

Min. $z = -\frac{143}{5}$

Two-Phase Method

27. Maximize

$$z = x_1 + 2x_2 + 3x_3 - x_4$$

subject to

$$x_1 + 2x_2 + 3x_3 = 15$$
$$2x_1 + x_2 + 5x_3 = 20$$
$$x_1 + 2x_2 + x_3 + x_4 = 10$$
$$x_1, x_2, x_3, x_4 \ge 0$$

Ans. $x_1 = x_2 = x_3 = \frac{5}{2}, x_4 = 0$, Max. $z = 15$

28. Minimize

$$z = x_1 - 3x_2 + 2x_3$$

subject to

$$3x_1 - x_2 + 2x_3 \le 7$$
$$-2x_1 + 4x_2 \le 12$$
$$-4x_1 + 3x_2 + 8x_3 \le 10$$
$$x_1, x_2, x_3 \ge 0.$$

Ans. $x_1 = 4, x_2 = 5$, Min. $z = -11$

29. Maximize

$$z = 2x_1 + x_2 + 3x_3$$

subject to

$$x_1 + x_2 + 2x_3 \le 5$$
$$2x_1 + 3x_2 + 4x_3 = 12$$
$$x_1, x_2, x_3 \ge 0$$

Ans. $x_1 = 3, x_2 = 2, x_3 = 0$, Max. $z = 8$.

30. Maximize

$$z = x_1 + 2x_2 + 3x_3 - x_4$$

subject to

$$x_1 + 2x_2 + 3x_3 = 15$$
$$2x_1 + x_2 + 5x_3 \ge 20$$
$$x_1 + 2x_2 + x_3 + x_4 \ge 10$$
$$x_1, x_2, x_3, x_4 \ge 0$$

Ans. $x_1 = 0, x_2 = \frac{15}{7}, x_3 = \frac{25}{7}, x_4 = 0$,

Max. $z = 15$

31. Maximize

$$z = 22x + 30y + 25z$$

subject to

$$2x + 2y \le 100$$
$$2x + y + z \ge 100$$
$$x + 2y + 2z \le 100$$
$$x, y, z \ge 0$$

Ans. $x = \frac{100}{3}, y = \frac{50}{3}, z = \frac{50}{3}$,
Max. $z = 1650$.

Duality

32. Write the dual of the following LPP.
Maximize

$$z = x_1 - x_2 + 3x_3$$

subject to

$$x_1 + x_2 + x_3 \le 10$$
$$2x_1 - x_3 \le 2$$
$$2x_1 - 2x_2 - 3x_3 \le 6$$
$$x_1, x_2, x_3 \ge 0$$

Ans.

Minimize $z^* = 10y_1 + 2y_2 + 6y_3$

subject to

$$y_1 + 2y_2 + 2y_3 \ge 1$$
$$y_1 - 2y_3 \ge -1$$
$$y_1 - y_2 - 3y_3 \ge 3$$
$$y_1, y_2, y_3 \ge 0.$$

33. Write the dual to the following LPP:
Minimize

$$z = x_1 + x_2 + x_3$$

subject to

$$x_1 - 3x_2 + 4x_3 = 5$$
$$x_1 - 2x_2 \le 3$$
$$2x_2 - x_3 \ge 4$$
$$x_1, x_2 \ge 0$$

x_3 is unrestricted

Ans.

Maximize $z^* = -5y_1 - 3y_2 + 4y_3$

subject to

$$-y_1 - y_2 \le 1$$
$$3y_1 + 2y_2 + 2y_3 \le 1$$
$$-4y_1 - y_3 = 1$$
$$y_2, y_3 \ge 0$$

y_1 is unrestricted

34. Write the dual to the following LPP:
Minimize

$$z = 2x_1 + 3x_2 + 4x_3$$

subject to

$$2x_1 + 3x_2 + 5x_3 \ge 2$$
$$3x_1 + x_2 + 7x_3 = 3$$
$$x_1 + 4x_2 + 6x_3 \le 5$$
$$x_1, x_2 \ge 0$$

and x_3 is unrestricted.

Ans. Maximize $z^* = 2y_1 + 3y_2 - 5y_3$

subject to

$$2y_1 + 3y_2 - y_3 \le 2$$
$$3y_1 + y_2 - 4y_3 \le 3$$
$$5y_1 + 7y_2 - 6y_3 = 4$$
$$y_1, y_2 \ge 0$$

and y_3 is unrestricted.

35. Using duality solve the following LPP:
Maximize

$$z = 3x_1 + 2x_2 + 5x_3$$

subject to the constraints

$$x_1 + 2x_2 + x_3 \le 430$$
$$3x_1 + 2x_3 \le 460$$
$$x_1 + 4x_2 \le 420$$
$$x_1, x_2, x_3 \ge 0.$$

Ans. $x_1 = 0, x_2 = 100, x_3 = 230,$
Max. $z = 1350$

36. Use duality concept to solve the following LPP.
Maximize

$$z = 7x_1 + 5x_2$$

subject to the constraints

$$3x_1 + x_2 \le 48$$
$$2x_1 + x_2 \le 40$$
$$x_1, x_2 \ge 0$$

Ans. Dual is Min $z^* = 48y_1 + 40y_2$

subject to

$$3y_1 + 2y_2 \ge 7$$
$$y_1 + y_2 \ge 5$$
$$y_1, y_2 \ge 0.$$

Solution is $y_1 = 0, y_2 = 5$, Min. $z^* = 200$.
Solution of the primal is $x_1 = 0, x_2 = 40$,
Min. $z = 200$.

37. Write the dual of the following LPP and find the solution of the both dual and primal problem:

Maximize

$$z = 40x_1 + 25x_2 + 50x_3$$

subject to the constraints

$$x_1 + x_2 + x_3 \le 36$$
$$2x_1 + x_2 + 4x_3 \le 60$$
$$2x_1 + 5x_2 + x_3 \le 45$$
$$x_1, x_2, x_3 \ge 0.$$

Ans. The dual is

Minimize $z^* = 36y_1 + 60y_2 + 45y_3$

subject to

$$y_1 + 2y_2 + 2y_3 \ge 40$$
$$y_1 + y_2 + 5y_3 \ge 25$$
$$y_1 + 4y_2 + y_3 \ge 50$$
$$y_1, y_2, y_3 \ge 0$$

Solution is $y_1 = y_2 = 0$, $y_3 = 10$, Min. $z^* = 1050$. Solution of the primal is

$x_1 = 20, x_2 = 0, x_3 = 5$, Max. $z = 1050$.

38. Solve the dual of the following LPP and, hence, find max z:

Maximize

$$z = 20x_1 + 30x_2$$

subject to the constraints

$$3x_1 + 3x_2 \le 36$$
$$5x_1 + 2x_2 \le 50$$
$$2x_1 + 6x_2 \le 60$$
$$x_1, x_2 \ge 0.$$

Ans. Dual's solution is

$y_1 = 5, y_2 = 0, y_3 = \frac{5}{2}$, Min. $z^* = 330$.

Solution of the primal is

$x_1 = 3, x_2 = 9$, Max. $z = 330$.

39. Use duality concept to solve the following LPP.

Maximize

$$z = 2x_1 + x_2$$

subject to the constraints

$$-x_1 + 2x_2 \le 2$$
$$x_1 + x_2 \le 4$$
$$x_1 \le 3$$
$$x_1, x_2 \ge 0.$$

Ans. Dual's solution is

$y_1 = 0, y_2 = 1$, Min. $z^* = 7$

Primal solution is

$x_1 = 3, x_2 = 1$, Max. $z = 7$.

Dual Simplex Method

40. Use dual simplex method to solve the following LPP.

Minimize

$$z = 6x_1 + 7x_2 + 3x_3 + 5x_4$$

subject to the constraints

$$5x_1 + 6x_2 - 3x_3 + 4x_4 \ge 12$$
$$x_2 + 5x_3 - 6x_4 \ge 10$$
$$2x_1 + 5x_2 + x_3 + x_4 \ge 8$$
$$x_1, x_2, x_3, x_4 \ge 0$$

Ans. $x_1 = 0, x_2 = \frac{30}{11}, x_3 = \frac{16}{11}, x_4 = 0,$

Min. $z = \frac{258}{11}$.

41. Use dual simplex method to solve the following LPP.

Minimize

$$z = 3x_1 + 2x_2 + x_3 + 4x_4$$

subject to the constraints

$$2x_1 + 4x_2 + 5x_3 + x_4 \ge 10$$
$$3x_1 - x_2 + 7x_3 - 2x_4 \ge 2$$
$$5x_1 + 2x_2 + x_3 + 6x_4 \ge 15$$
$$x_1, x_2, x_3, x_4 \ge 0.$$

Ans. $x_1 = \frac{65}{23}, x_2 = 0, x_3 = \frac{20}{23}, x_4 = 0,$

Min. $z = \frac{215}{23}$

42. Using dual simplex method, solve the following LPP:

Maximize

$$z = -2x_1 - 2x_2 - 4x_3$$

subject to the constraints

$$2x_1 + 3x_2 + 5x_3 \geq 2$$
$$3x_1 + x_2 + 7x_3 \leq 3$$
$$x_1 + 4x_2 + 6x_3 \leq 5$$
$$x_1, x_2, x_3 \geq 0.$$

Ans. $x_1 = 0, x_2 = \frac{2}{3}, x_3 = 0$, Max. $z = -\frac{4}{3}$

43. Apply dual simplex method to solve the following problem:

Maximize

$$z = -3x_1 - 2x_2$$

subject to the constraints

$$x_1 + x_2 \geq 1$$
$$x_1 + x_2 \leq 7$$
$$x_1 + 2x_2 \geq 10$$
$$x_2 \geq 3$$
$$x_1, x_2 \geq 0.$$

Ans. $x_1 = 4, x_2 = 3, x_3 = 0$, Max. $z = -18$.

44. Using dual simplex method, solve the following problem:

Maximize

$$z = -2x_1 - x_3$$

subject to the constraints

$$x_1 + x_2 - x_3 \geq 5$$
$$x_1 - 2x_2 + 4x_3 \geq 8$$
$$x_1, x_2, x_3 \geq 0.$$

Ans. $x_1 = 0, x_2 = 14, x_3 = 9$ Max. $z = -9$.

Transportation problems

45. Solve the following transportation problem:

	D_1	D_2	D_3	
O_1	2	7	4	5
O_2	3	3	7	8
O_3	5	4	1	7
O_4	1	6	2	14
	7	9	18	34

Ans. $x_{11} = 5, x_{22} = 8, x_{32} = 1, x_{33} = 6,$ $x_{41} = 2, x_{43} = 12$, total cost is 70.

46. Solve the following transportation problem:

	D_1	D_2	D_3	D_4	D_5	
O_1	4	1	3	4	4	60
O_2	2	3	2	2	3	35
O_3	3	5	2	4	4	40
	22	45	20	18	30	

Ans. $x_{12} = 45, x_{15} = 15,$

$x_{21} = 17, x_{24} = 18, x_{31} = 5, x_{35} = 15,$

total cost is 290.

47. Solve the following transportation problem:

	D_1	D_2	D_3	D_4	D_5	
O_1	4	2	3	2	6	8
O_2	5	4	5	2	1	12
O_3	6	5	4	7	7	14
	4	4	6	8	8	

Ans. $x_{12} = 4, x_{14} = 4,$

$x_{24} = 2, x_{25} = 8, x_{31} = 4, x_{33} = 4$

Total cost is 80.

48. Solve the following transportation problem:

	D_1	D_2	D_3	D_4	D_5	
O_1	275	350	425	225	150	300
O_2	300	325	450	175	100	250
O_3	250	350	475	200	125	150
O_4	325	275	400	250	175	200
	150	100	75	250	200	

Ans. (unbalanced problem)

$x_{12} = 25, x_{14} = 50, x_{15} = 200, x_{21} = 150,$

$x_{32} = 75, x_{33} = 75, x_{44} = 200.$

49. Solve the following transportation problem:

	D_1	D_2	D_3	D_4	
O_1	3	2	7	6	5,000
O_2	7	5	2	3	6,000
O_3	2	5	4	5	2,500
	6,000	4,000	2,000	1,500	

Ans. $x_{11} = 3,500$, $x_{12} = 1,500$, $x_{22} = 2,500$, $x_{23} = 2,000$, $x_{24} = 1,500$, $x_{31} = 2,500$
Total cost is 39,500.

50. Solve the following transportation problem:

	D_1	D_2	D_3	D_4	D_5	
O_1	5	8	6	6	3	800
O_2	4	7	7	6	6	500
O_3	8	4	6	6	3	900
	400	400	500	400	800	

Ans. $x_{13} = 0$, $x_{15} = 800$, $x_{21} = 400$, $x_{24} = 100$, $x_{32} = 400$, $x_{33} = 200$, $x_{34} = 300$, $x_{43} = 300$, cost: 9200.

15 Basic Numerical Methods

Numerical analysis is a branch of mathematics in which we analyse and solve the problems which require calculations. The methods (techniques) used for this purpose are called *numerical methods* (techniques). These techniques are used to solve algebraic or transcendental equations, an ordinary or partial differential equations, integral equations, and to obtain functional value for an argument in some given interval where some values of the function are given. In numerical analysis, we do not strive for exactness but try to device a method which will yield an approximate solution differing from the exact solution by less than a specified tolerance. The approximate calculation is one which involves approximate data, approximate methods, or both. The error in the computed result may be due to errors in the given data and errors of calculation. There is no remedy to the error in the given data but the second kind of error can usually be made as small as we please. The calculations are carried out in such a way as to make the error of calculation negligible.

15.1 APPROXIMATE NUMBERS AND SIGNIFICANT FIGURES

The numbers of the type 3, 6, 2, $\frac{5}{4}$, 7.35 are called *exact numbers* because there is no approximation associated with them. On the other hand, numbers like $\sqrt{2}$, π are exact numbers but cannot be expressed exactly by a finite number of digits when expressed in digital form. Such numbers are approximated by numbers having finite number of digits. An *approximate number* is a number which is used as an approximation to an exact number and differ only slightly from the exact number for which it stands. For example,

(1) 1.4142 is an approximate number for $\sqrt{2}$

(2) 3.1416 is an approximate number for π

(3) 2.061 is an approximate number for $\frac{27}{13.1}$.

A *significant figure* is any of the digits 1, 2, ..., 9, and 0 is a significant figure except when it is used to fix the decimal point or to fill the places of unknown or discarded digits. For example, 1.4142 contains five significant figures, whereas 0.0034 has only two significant figures, 3 and 4.

If we attempt to divide 22 by 7, we get

$$\frac{22}{7} = 3.142857\ldots.$$

In practical computation, we must cut it down to a manageable form such as 3.14, 3.143. The process of cutting off superfluous digits and retaining the desired is called *rounding off*. Thus to round off a number, we retain a certain number of digits, counted from the left, and drop the others. However, the numbers are rounded off so as to cause the least possible error.

To round off a number to n significant figures.

1. Discard all digits to the right of the nth digit.
2. (a) if the discarded number is less than half a unit in the nth place, leave the nth digit unchanged.

   (b) if the discarded number is greater than half a unit in the nth place, increase the nth digit by 1.

   (c) if the discarded number is exactly half a unit in the nth place, increase the nth digit by 1 if it is odd, otherwise leave the nth digit unaltered. Thus, in this case, the nth digit shall be an even number. The reason for this step is that even numbers are more exactly divisible by many more numbers than are odd numbers and so there will be fewer left–over errors in computation when the rounded numbers are left even.

When a given number has been rounded off according to the above rules, it is said to be *correct to n significant figures.*

EXAMPLE 15.1

Round off the following numbers correctly to *four significant figures:*

81.9773, 48.365, 21.385, 12.865, 27.553.

Solution. After rounding off,

81.9773 becomes 81.98,
48.365 becomes 48.36,
21.385 becomes 21.38,
12.865 becomes 12.86,
27.553 becomes 27.55.

15.2 CLASSICAL THEOREMS USED IN NUMERICAL METHODS

The following theorems will be used in the derivation of some of the numerical methods and in the study of error analysis of the numerical methods.

Theorem 15.1. (Rolle's Theorem). Let f be a function such that

1. f is continuous in $[a, b]$
2. f is derivable in (a, b)
3. $f(a) = f(b)$

Then, there exists at least one $\xi \in (a, b)$ such that $f'(\xi) = 0$.

Theorem 15.2. (Generalized Rolle's Theorem). Let f be n times differentiable function in $[a, b]$. If f vanishes at $(n + 1)$ distinct points $x_0, x_1, x_2, \ldots x_n$ in (a, b), then there exists a number $\xi \in (a, b)$ such that $f^{(n)}(\xi) = 0$.

It follows from Theorem 15.2 that between any two zeroes of a polynomial $f(x)$ of degree ≥ 2, there lies at least one zero of the polynomial $f'(x)$.

Theorem 15.3. (Intermediate Value Theorem). Let f be continuous in $[a, b]$ and $f(a) < k < f(b)$. then there exists a number $\xi \in (a, b)$ such that $f(\xi) = k$.

Theorem 15.4. (Mean Value Theorem). If

1. f is continuous in $[a, b]$.
2. f is derivable in (a, b).

then there exists at least one $\xi \in (a, b)$ such that

$$\frac{f(b) - f(a)}{b - a} = f'(\xi), \ a < \xi < b.$$

Theorem 15.5. If f is continuous in $[a, b]$ and if $f(a)$ and $f(b)$ are of opposite signs, then there exists at least one $\xi \in (a, b)$ such that $f(\xi) = 0$.

Theorem 15.6. (Taylor's Theorem). Let f be continuous and possess continuous derivatives of order n in $[a, b]$. If $x_0 \in [a, b]$ is a fixed point, then for every $x \in [a, b]$, there exists a number ξ lying between x_0 and x such that

$$f(x) = f(x_0) + (x - x_0) f'(x_0) + \frac{(x - x_0)^2}{2!} f''(x_0)$$

$$+ \ldots + \frac{(x - x_0)^{n-1}}{(n - 1)!} f^{(n-1)}(x_0) + R_n(x),$$

where

$$R_n(x) = \frac{(x - x_0)^n}{n!} f^{(n)}(\xi), \ x_0 < \xi < x.$$

If $x = x_0 + h$, then we get

$$f(x_0 + h) = f(x_0) + hf'(x_0) + \frac{h^2}{2!} f''(x_0) + \ldots$$

$$+ \frac{h^{n-1}}{(n-1)!} f^{(n-1)}(x_0) + \frac{h^n}{n!} f^{(n)}(\xi)$$

$$= f(x_0) + hf'(x_0) + \frac{h^2}{2!} f''(x_0) + \ldots$$

$$+ \frac{h^{(n-1)}}{(n-1)!} f^{(n-1)}(x_0) + O(h^n).$$

As a corollary to Taylor's theorem, we have

$$f(x) = f(0) + xf'(0) + \frac{x^2}{2!} f''(0) + \ldots$$

$$+ \frac{x^n}{n!} f^{(n)}(0) + \ldots,$$

which is called *Maclaurin's expansion for the function f.*

Theorem 15.7. (Taylor's Theorem for Function of Several Variables). If $f(x, y)$ and all its partial derivatives of order n are finite and continuous for all points (x, y) in the domain $a \le x \le a + h$, $b \le y \le b + k$, then

$$f(a+h, b+k) = f(a, b) + df(a, b) + \frac{1}{2!} d^2 f(a, b)$$

$$+ \ldots + \frac{1}{(n-1)!} d^{n-1} f(a, b) + R_n,$$

where $d = h\frac{\partial}{\partial x} + k\frac{\partial}{\partial y}$ and

$$R_n = \frac{1}{n!} d^n f(a + \theta h, b + \theta k), 0 < \theta < 1.$$

Putting $a = b = 0$, $h = x$, $k = y$, we get

$$f(x, y) = f(0, 0) + df(0, 0) + \frac{1}{2!} d^2 f(0, 0) + \ldots$$

$$+ \frac{1}{(n-1)!} d^{n-1} f(0, 0) + R_n,$$

where

$$R_n = \frac{1}{n!} d^n f(\theta x, \theta y), 0 < \theta < 1.$$

This result is called *Maclaurin's* theorem for functions of several variables.

Theorem 15.8. (Fundamental Theorem of Integral Calculus). If f is continuous over $[a, b]$, then there exists a function F, called the anti-derivative of f such that

$$\int_a^b f(x)\,dx = F(b) - F(a),$$

where $F'(x) = f(x)$.

The second version of the above theorem is as given below:

Theorem 15.9. If f is continuous over $[a, b]$ and $a < x < b$, then

$$\frac{d}{dx}\int_a^x f(t)\,dt = f(x) \text{ or } F'(x) = f(x),$$

where

$$F(x) = \int_a^x f(t)\,dt.$$

15.3 TYPES OF ERRORS

In numerical computation, the quantity [true value–approximate value] is called the *error.*

We come across the following types of errors in numerical computation.

1. *Inherent error (initial error).* Inherent error is the quantity which is already present in the statement (data) of the problem before its solution. This type of error arises due to the use of approximate value in the given data because there are limitations of the mathematical tables and calculators. This type of error can be there due to mistakes by human. For example, one can write, by mistake, 67 instead of 76. The error in this case is called *transposing error.*
2. *Round-off error.* This error arises due to rounding off the numbers during computation and occurs due to the limitation of computing aids. However, this type of error can be minimized by
   - (i) Avoiding the subtraction of nearly equal numbers or division by a small number.
   - (ii) Retaining at least one more significant figure at each step of calculation.
3. *Truncation error.* It is the error caused by using approximate formulas during computation such as the one that arise when a function $f(x)$ is evaluated form an infinite series for x after truncating it at certain stage.

For example, we will see that in Newton–Raphson method for finding the roots of an equation, if x is the true value of the root of $f(x) = 0$ and x_0 and h are approximate value and correction, respectively, then by Taylor's theorem,

$$f(x_0 + h) = f(x_0) + hf'(x_0) + \frac{h^2}{2!}$$
$$f''(x_0) + \ldots + = 0.$$

To find the correction h, we truncate the series just after the first derivative. Therefore, some error occurs due to this truncation.

4. *Absolute error.* If x is the true value of a quantity and x_0 is the approximate value, then $|x - x_0|$ is called the *absolute error.*
5. *Relative error.* If x is the true value of a quantity and x_0 is the approximate value, then $\left(\frac{x - x_0}{x}\right)$ is called the *relative error.*
6. *Percentage error.* If x is the true value of quantity and x_0 is the approximate value, then $\left(\frac{x - x_0}{x}\right) \times 100$ is called the *percentage error.* Thus, percentage error is 100 times the relative error.

15.4 GENERAL FORMULA FOR ERRORS

Let

$$u = f(u_1, u_2, \ldots, u_n) \qquad (1)$$

be a function of $u_1, u_2, \ldots, u_n$ which are subject to the errors $\Delta u_1, \Delta u_2, \ldots, \Delta u_n$, respectively. Let Δu be the error in u caused by the errors $\Delta u_1, \Delta u_2, \ldots, \Delta u_n$ in $u_1, u_2, \ldots, u_n$, respectively. Then

$$u + \Delta u = f(u_1 + \Delta u_1, u_2 + \Delta u_2, \ldots, u_n + \Delta u_n). \quad (2)$$

Expanding the right-hand side of (2) by Taylor's theorem for a function of several variables, we have

$$u + \Delta u = f(u_1, u_2, \ldots, u_n)$$
$$+ \left(\Delta u_1 \frac{\partial}{\partial u_1} + \ldots + \Delta u_n \frac{\partial}{\partial u_n}\right) f$$
$$+ \frac{1}{2}\left(\Delta u_1 \frac{\partial}{\partial u_1} + \ldots + \Delta u_n \frac{\partial}{\partial u_n}\right)^2 f + \ldots$$

Since the errors are relatively small, we neglect the squares, products, and higher powers and have

$$u + \Delta u = f(u_1, u_2, \ldots, u_n)$$
$$+ \left(\Delta u_1 \frac{\partial}{\partial u_1} + \ldots + \Delta u_n \frac{\partial}{\partial u_n}\right) f. \quad (3)$$

Subtracting (1) from (3), we have

$$\Delta u = \frac{\partial f}{\partial u_1}\Delta u_1 + \frac{\partial f}{\partial u_2}\Delta u_2 + \ldots + \frac{\partial f}{\partial u_n}\Delta u_n$$

or

$$\Delta u = \frac{\partial u}{\partial u_1}\Delta u_1 + \frac{\partial u}{\partial u_2}\Delta u_2 + \ldots + \frac{\partial u}{\partial u_n}\Delta u_n,$$

which is known as *general formula for error.* We note that the right-hand side is simply the total derivative of the function u.

For a relative error E_r of the function u, we have

$$E_r = \frac{\Delta u}{u}$$
$$= \frac{\partial u}{\partial u_1}\frac{\Delta u_1}{u} + \frac{\partial u}{\partial u_2}\frac{\Delta u_2}{u} + \ldots + \frac{\partial u}{\partial u_n}\frac{\Delta u_n}{u}.$$

EXAMPLE 15.2

If $u = \frac{5xy^2}{z^3}$ and errors in x, y, and z are 0.001, compute the relative maximum error $(E_r)_{\max}$ in u when $x = y = z = 1$.

Solution. We have $u = \frac{5xy^2}{z^3}$. Therefore,

$$\frac{\partial u}{\partial x} = \frac{5y^2}{z^3}, \frac{\partial u}{\partial y} = \frac{10xy}{z^3}, \frac{\partial u}{\partial z} = -\frac{15xy^2}{z^4}$$

and so

$$\Delta u = \frac{5y^2}{z^3}\Delta x + \frac{10xy}{z^3}\Delta y - \frac{15xy^2}{z^4}\Delta z.$$

But it is given that $\Delta x = \Delta y = \Delta z = 0.001$ and $x = y = z = 1$. Therefore,

$$(\Delta u)_{\max} \approx \left|\frac{5y^2}{z^3}\Delta x\right| + \left|\frac{10xy}{z^3}\Delta y\right| + \left|\frac{15xy^2}{z^4}\Delta z\right|$$

$$= 5(0.001) + 10(0.001) + 15(0.001) = 0.03.$$

Thus the relative maximum error $(E_r)_{\max}$ is given by

$$(E_r)_{\max} = \frac{(\Delta u)_{\max}}{u} = \frac{0.03}{u} = \frac{0.03}{5} = 0.006.$$

15.5 SOLUTION OF NON-LINEAR EQUATIONS

The aim of this section is to discuss the most useful methods for finding the roots of any equation having numerical coefficients. Polynomial equations of degree ≤ 4 can be solved by standard algebraic methods. But no general method exists for finding the roots of the equations of the type a log x + $bx = c$ or $ae^{-x} + b \tan x = 4$ etc in terms of their coefficients. These equations are called *transcendental equations*. Therefore, we take help of numerical methods to solve such type of equations.

Let f be a continuous function. Any number ξ for which $f(\xi) = 0$ is called a *root* of the equation $f(x) = 0$. Also, ξ is called a *zero* of function $f(x)$.

A zero ξ is called of *multiplicity p*, if we can write

$$f(x) = (x - \xi)^p\, g(x),$$

where $g(x)$ is bounded at ξ and $g(\xi) \neq 0$. If $p = 1$, then ξ is said to be *simple zero* and if $p > 1$, then ξ is called a *multiple zero.*

We shall discuss bisection method, Regula-Falsi method, Newton–Raphson method, and iteration method to solve non-linear equations.

1. Bisection Method (Bolzano Method)

Suppose that we want to find a zero of a continuous function f. We start with an initial interval $[a_0, b_0]$, where $f(a_0)$ and $f(b_0)$ have opposite signs. Since f is continuous, the graph of f will cross the x axis at a root $x = \xi$ lying in $[a_0, b_0]$. This is shown in Figure 15.1

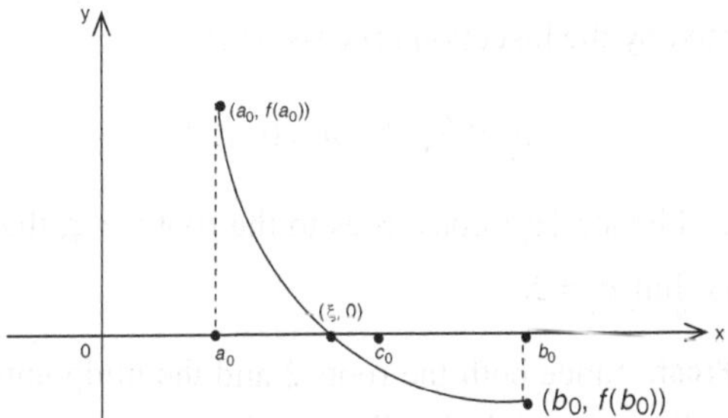

Figure 15.1

The bisection method systematically moves the end points of the interval closer until we obtain an interval of arbitrary small width that contains the root. We choose the midpoint $c_0 = \dfrac{a_0 + b_0}{2}$ and then consider the following possibilities:

(i) If $f(a_0)$ and $f(c_0)$ have opposite signs, then a root lies in $[a_0, c_0]$.

(ii) If $f(c_0)$ and $f(b_0)$ have opposite signs, then a root lies in $[c_0, b_0]$.

(iii) If $f(c_0) = 0$ then $x = c_0$ is a root.

If (iii) happens, then there is nothing to proceed as c_0 is the root in that case. If anyone one of (i) or (ii) happens, then let $[a_1, b_1]$ be the interval (representing $[a_0, c_0]$ or $[c_0, b_0]$) containing the root, where $f(a_1)$ and $f(b_1)$ have opposite signs. Let $c_1 = \dfrac{a_1 + b_1}{2}$ and $[a_2, b_2]$ represents $[a_1, c_1]$ or $[c_1, b_1]$ such that $f(a_2)$ and $f(b_2)$ have opposite signs. Then the root lies between a_2 and b_2. Continue with the process to construct an interval $[a_{n+1}, b_{n+1}]$, which contains the root and its width is half that of $[a_n, b_n]$. In this case

$$[a_{n+1}, b_{n+1}] = [a_n, c_n] \text{ or } [c_n, b_n] \text{ for all } n.$$

Theorem 15.10. Let f be a continuous function on $[a, b]$ and $\xi \in [a, b]$ be a root of $f(x) = 0$. If $f(a)$ and $f(b)$ have opposite signs and $\{c_n\}$

represents the sequence of the midpoints generated by the bisection process, then

$$|\xi - c_n| \le \frac{b-a}{2^{n+1}},\ n = 0, 1, 2, \ldots$$

and hence $\{c_n\}$ converges to the root $x = \xi$, that is, $\lim_{n\to\infty} c_n = \xi$.

Proof: Since both the roots ξ and the midpoints c_n lie in $[a_n, b_n]$, the distance from c_n to ξ cannot be greater than half the width of $[a_n, b_n]$ as shown in the Figure 15.2.

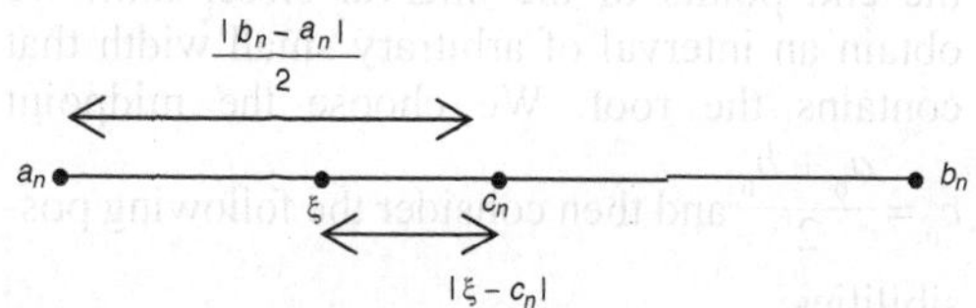

Figure 15.2

Thus

$$|\xi - c_n| \le \frac{|b_n - a_n|}{2} \text{ for all } n.$$

But, we note that

$$|b_1 - a_1| = \frac{|b_0 - a_0|}{2},$$

$$|b_2 - a_2| = \frac{|b_1 - a_1|}{2} = \frac{|b_0 - a_0|}{2^2},$$

$$|b_3 - a_3| = \frac{|b_2 - a_2|}{2} = \frac{|b_0 - a_0|}{2^3}$$

--- --- --- ---

--- --- --- ---

$$|b_n - a_n| = \frac{|b_{n-1} - a_{n-1}|}{2} = \frac{|b_0 - a_0|}{2^n}.$$

Hence

$$|\xi - c_n| \le \frac{|b_0 - a_0|}{2^{n+1}} \text{ for, all } n.$$

and so $\lim_{n\to\infty} |\xi - c_n| = 0$ or $\lim_{n\to\infty} c_n = \xi$.

EXAMPLE 15.3

Find a real root of the equation $x^3 + x^2 - 1 = 0$ using bisection method.

Solution. Let

$$f(x) = x^3 + x^2 - 1.$$

Then $f(0) = -1, f(1) = 1$. Thus a real root of $f(x) = 0$ lies between 0 and 1. Therefore, we take $x_0 = 0.5$. then $f(0.5) = (0.5)^3 + (0.5)^2 - 1 = 0.125 + 0.25 - 1 = -0.625$.

This shows that the root lies between 0.5 and 1 and we get

$$x_1 = \frac{1 + 0.5}{2} = 0.75.$$

Then $f(x_1) = (0.75)^3 + (0.75)^2 - 1 = 0.421875 + 0.5625 - 1 = -0.015625$.

Hence the root lies between 0.75 and 1. Thus we take

$$x_2 = \frac{1 + 0.75}{2} = 0.875$$

and then

$$f(x_2) = 0.66992 + 0.5625 - 1 = 0.23242 \text{ (+ve)}.$$

If follows that the root lies between 0.75 and 0.875. We take

$$x_3 = \frac{0.75 + 0.875}{2} = 0.8125,$$

and then

$$f(x_3) = 0.53638 + 0.66015 - 1 = 0.19653(+ve).$$

Therefore, the root lies between 0.75 and 0.8125. So, let

$$x_4 = \frac{0.75 + 0.8125}{2} = 0.781,$$

which yields

$$f(x_4) = (0.781)^3 + (0.781)^2 - 1 = 0.086(+ve).$$

Thus the root lies between 0.75 and 0.781. We take

$$x_5 = \frac{0.750 + 0.781}{2} = 0.765$$

and note that

$$f(0.765) = 0.0335(+ve).$$

Hence the root lies between 0.75 and 0.765. So, let

$$x_6 = \frac{0.750 + 0.765}{2} = 0.7575$$

and then

$$f(0.7575) = 0.4346 + 0.5738 - 1$$

$$= 0.0084\,(+ve).$$

Therefore, the root lies between 0.75 and 0.7575. Proceeding in this way, the next approximations shall be

$$x_7 = 0.7538, \quad x_8 = 0.7556, \quad x_9 = 0.7547,$$

$$x_{10} = 0.7551, \quad x_{11} = 0.7549, \quad x_{12} = 0.75486.$$

and so on.

EXAMPLE 15.4

Find a root of the equation $x^3 - 3x - 5 = 0$ by bisection method.

Solution. Let $f(x) = x^3 - 3x - 5$ Then we observe that $f(2) = -3$ and $f(3) = 13$. Thus a root of the given equation lies between 2 and 3. Let $x_0 = 2.5$. Then

$$f(2.5) = f(2.5)^3 - 3(2.5) - 5 = 3.125(+ve).$$

Thus, the root lies between 2.0 and 2.5. Then

$$x_1 = \frac{2 + 2.5}{2} = 2.25.$$

We note that $f(2.25) = -0.359375$ (−ve). Therefore, the root lies between 2.25 and 2.5. Then we take

$$x_2 = \frac{2.25 + 2.5}{2} = 2.375$$

and observe that $f(2.375) = 1.2715$(+ve). Hence the root lies between 2.25 and 2.375. therefore, we take

$$x_3 = \frac{2.25 + 2.375}{2} = 2.3125.$$

Now $f(2.3125) = 0.4289$ (+ve). Hence the root lies between 2.25 and 2.3125. We take

$$x_4 = \frac{2.25 + 2.3125}{2} = 2.28125$$

Now

$$f(2.28125) = 0.0281(+ve).$$

We observe that the root lies very near to 2.28125. Let us try 2.280. Then

$$f(2.280) = 0.0124.$$

Thus the root is 2.280 approximately.

2. Regula-Falsi Method

The *Regula-Falsi method,* also known as *method of false position, chord method,* or *secant method,* is the oldest method for finding the real roots of a numerical equation. We know that the root of the equation $f(x) = 0$ corresponds to abscissa of the point of intersection of the curve $y = f(x)$ with the x-axis. In Regula Falsi method, we replace the curve by a chord in the interval, which contains a root of the equation $f(x) = 0$. We take the point of intersection of the chord with the x-axis as an approximation to the root.

Suppose that a root $x = \xi$ lies in the interval (x_{n-1}, x_n) and that the corresponding ordinates $f(x_{n-1})$ and $f(x_n)$ have opposite signs. Then the equation of the straight line through the points $P(x_n, f(x_n))$ and $Q(x_{n-1}, f(x_{n-1}))$ is

$$\frac{f(x) - f(x_n)}{f(x_{n-1}) - f(x_n)} = \frac{x - x_n}{x_{n-1} - x_n}. \tag{4}$$

Let this straight line cut the x-axis at x_{n+1}. Since $f(x) = 0$ where the line (4) cut the x-axis, we have, $f(x_{n+1}) = 0$ and so

$$x_{n+1} = x_n - \frac{x_{n+1} - x_n}{f(x_{n-1}) - f(x_n)} f(x_n). \tag{5}$$

Now $f(x_{n-1})$ and $f(x_{n+1})$ have opposite signs. Therefore, it is possible to apply the approximation again to determine a line through the points Q and P_1. Proceeding in this way we find that as the points approach ξ, the curve becomes more nearly a straight line. The equation (5) can also be written in the form

$$x_{n+1} = \frac{x_n f(x_{n-1}) - x_{n-1} f(x_n)}{f(x_{n-1}) - f(x_n)},\ n = 1, 2, \ldots, \quad (6)$$

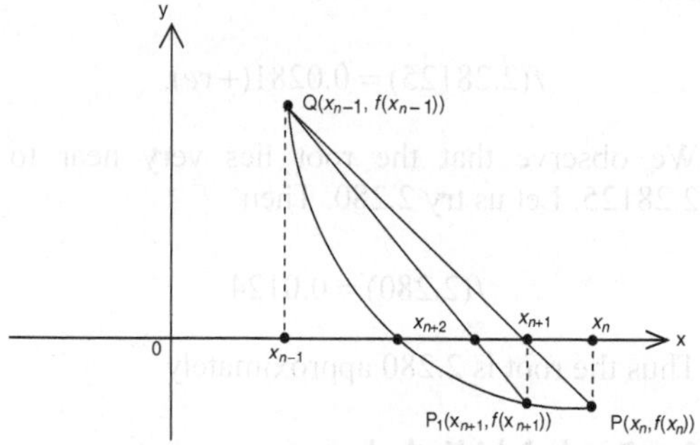

Figure 15.3

(a) Convergence of Regula-Falsi Method

Let ξ be the actual root of the equation $f(x) = 0$. Thus $f(\xi) = 0$. Let $x_n = \xi + \varepsilon_n$, where ε_n is the error involved at the nth step while determining the root. Using

$$x_{n+1} = \frac{x_n f(x_{n-1}) - x_{n-1} f(x_n)}{f(x_{n-1}) - f(x_n)},\ n = 1, 2, \ldots,$$

we get

$$\xi + \varepsilon_{n+1} = \frac{(\xi+\varepsilon_n) f(\xi+\varepsilon_{n-1}) - (\xi+\varepsilon_{n-1}) f(\xi+\varepsilon_n)}{f(\xi+\varepsilon_{n-1}) - f(\xi+\varepsilon_n)}$$

and

$$\varepsilon_{n+1} = \frac{(\xi+\varepsilon_n) f(\xi+\varepsilon_{n-1}) - (\xi+\varepsilon_{n-1}) f(\xi+\varepsilon_n)}{f(\xi+\varepsilon_{n-1}) - f(\xi+\varepsilon_n)} - \xi$$

$$= \frac{\varepsilon_n f(\xi+\varepsilon_{n-1}) - \varepsilon_{n-1} f(\xi+\varepsilon_n)}{f(\xi+\varepsilon_{n-1}) - f(\xi+\varepsilon_n)}$$

Expanding the right-hand side by Taylor's series, we get

$$\varepsilon_{n+1} = \frac{\left\{\begin{array}{l}\varepsilon_n[f(\xi) + \varepsilon_{n-1} f'(\xi) + \frac{1}{2}\varepsilon_{n-1}^2 f''(\xi) + \ldots] \\ -\varepsilon_{n-1}[f(\xi) + \varepsilon_n f'(\xi) + \frac{1}{2}\varepsilon_n^2 f''(\xi) + \ldots]\end{array}\right\}}{\begin{array}{l}[f(\xi) + \varepsilon_{n-1} f'(\xi) + \frac{1}{2}\varepsilon_{n-1}^2 f''(\xi) + \ldots \\ -f(\xi) - \varepsilon_n f'(\xi) - \frac{1}{2}\varepsilon_n^2 f''(\xi) - \ldots]\end{array}}$$

That is,

$$\varepsilon_{n+1} = k\,\varepsilon_{n-1}\,\varepsilon_n + O\left(\varepsilon_n^2\right), \quad (7)$$

where

$$k = \frac{1}{2}\frac{f''(\xi)}{f'(\xi)}.$$

We now try to determine some number in m such that

$$\varepsilon_{n+1} = A\varepsilon_n^m \quad (8)$$

and

$$\varepsilon_n = A\varepsilon_{n-1}^m \text{ or } \varepsilon_{n-1} = A^{-\frac{1}{m}}\varepsilon_n^{\frac{1}{m}}$$

From (7) and (8), we get

$$\varepsilon_{n+1} = k\,\varepsilon_{n-1}\,\varepsilon_n = kA^{-\frac{1}{m}}\varepsilon_n^{\frac{1}{m}}\varepsilon_n$$

and so

$$A\varepsilon_n^m = kA^{-\frac{1}{m}}\varepsilon_n^{\frac{1}{m}}\varepsilon_n = kA^{-\frac{1}{m}}\varepsilon_n^{1+\frac{1}{m}}.$$

Equating powers of ε_n on both sides, we get

$$m = m + \frac{1}{m} \text{ or } m^2 - m - 1 = 0,$$

Which yields $m = \frac{1 \pm \sqrt{5}}{2} = 1.618$ (+ve value). Hence

$$\varepsilon_{n+1} = A\varepsilon_n^{1.618}.$$

Thus *Regula Falsi method is of order 1.618.*

EXAMPLE 15.5

Find a real root of the equation $x^3 - 5x - 7 = 0$ using Regula Falsi method.

Solution. Let $f(x) = x^3 - 5x - 7 = 0$. We note that $f(2) = -9$ and $f(3) = 5$. Therefore, one root of the given equation lies between 2 and 3. By Regula Falsi method, we have

$$x_{n+1} = \frac{x_n f(x_{n-1}) - x_{n-1} f(x_n)}{f(x_{n-1}) - f(x_n)},\ n = 1, 2, 3, \ldots.$$

We start with $x_0 = 2$ and $x_1 = 3$. Then

$$x_2 = \frac{x_1 f(x_0) - x_0 f(x_1)}{f(x_0) - f(x_1)} = \frac{3(-9) - 2(5)}{-9 - 5} = \frac{37}{14} \approx 2.6.$$

But $f(2.6) = -2.424$ and $f(3) = 5$. Therefore,

$$x_3 = \frac{x_2 f(x_1) - x_1 f(x_2)}{f(x_1) - f(x_2)} = \frac{(2.6)5 + 3(2.424)}{5 + 2.424} = 2.73.$$

Now $f(2.73) = -0.30583$. Since we are getting close to the root, we calculate $f(2.75)$, which is found to be 0.046875. Thus the next approximation is

$$x_4 = \frac{2.75f(2.73) - (2.73)f(2.75)}{f(2.73) - f(2.75)}$$

$$= \frac{2.75(-0.303583) - 2.73(0.0468675)}{-0.303583 - 0.0468675}$$

$$= 2.7473.$$

Now $f(2.747) = -0.0062$. Therefore,

$$x_5 = \frac{2.75f(2.747) - 2.747f(2.75)}{f(2.747) - f(2.75)}$$

$$= \frac{2.75(-0.0062) - 2.747(0.046875)}{-0.0062 - 0.046875}$$

$$= 2.74724.$$

Thus the root is 2.747 correct up to three places of decimal.

EXAMPLE 15.6

Solve $x \log_{10} x = 1.2$ by Regula Falsi method.

Solution. We have $f(x) = x \log_{10} x - 1.2 = 0$. Then $f(2) = -0.060$ and $f(3) = 0.23$. therefore, the root lies between 2 and 3. Then

$$x_2 = \frac{x_1 f(x_0) - x_0 f(x_1)}{f(x_0) - f(x_1)}$$

$$= \frac{3(-0.6) - 2(0.23)}{-0.6 - 0.23} = 2.723.$$

Now $f(2.72) = 2.72 \log(2.72) - 1.2 = -0.01797$. Since we are getting closer to the root, we calculate $f(2.75)$ and have

$$f(2.75) = 2.75 \log(2.75) - 1.2$$

$$= 2.75(0.4393) - 1.2 = 0.00816.$$

Therefore,

$$x_3 = \frac{2.75(-0.01797) - 2.72(0.00816)}{-0.01797 - 0.00816}$$

$$= \frac{-0.04942 - 0.02219}{-0.02613} = 2.7405.$$

Now $f(2.74) = 2.74 \log (2.74) - 1.2 = 2.74\,(0.43775) - 1.2 = -0.00056$.

Thus the root lies between 2.74 and 2.75 and it is more closer to 2.74. Therefore,

$$x_4 = \frac{2.75(-0.00056) - 2.74(0.00816)}{-0.00056 - 0.00816} = 2.7408.$$

This value of the root is correct up to three decimal places.

3. Newton–Raphson Method

If the derivative of a function f can be easily found and if it is a simple expression, then the real roots of the equation $f(x) = 0$ can be computed rapidly by Newton–Raphson method.

Let x_0 denote the approximate value of the desired root and let h be the correction which must be applied to x_0 to give the exact value of the root x. Thus $x = x_0 + h$ and so the equatioin $f(x) = 0$ reduces to $f(x_0 + h) = 0$. Expanding by Taylor's theorem, we have

$$f(x_0 + h) = f(x_0) + h f'(x_0) + \frac{h^2}{2!} f''(x_0 + \theta h), \quad 0 < \theta < 1.$$

Hence

$$f(x_0) + h f'(x_0) + \frac{h^2}{2} f''(x_0 + \theta h) = 0.$$

If h is relatively small, we may neglect the term containing h^2 and have

$$f(x_0) + h f'(x_0) = 0.$$

Hence

$$h = -\frac{f(x_0)}{f'(x_0)}$$

and the improved value of the root becomes

$$x_1 = x_0 + h = x_0 - \frac{f(x_0)}{f'(x_0)}.$$

If we use x_1 as the approximate value, then the next approximation to the root is

$$x_2 = x_1 - \frac{f(x_1)}{f'(x_1)}.$$

In general, the $(n+1)$th approximation is

$$x_{n+1} = x_n - \frac{f(x_n)}{f'(x_n)}, \quad n = 0, 1, 2, 3, \ldots \quad (9)$$

Formula (9) in called *Newton–Raphson method.*

The expression $h = -\frac{f(x_0)}{f'(x_0)}$ is the *fundamental formula* in Newton–Raphson method. This formula tells us that larger the derivative, smaller is the correction to be applied to get the correct value of the root. This means when the graph of f is nearly vertical, where it crosses the x-axis, the correct value of the root can be found very rapidly with very little labour. On the other hand, if the value of $f'(x)$ is small in the neighbourhood of the root, the value of h given by the fundamental formula would be large and, therefore, the computation of the root shall be a slow process. Thus, Newton–Raphson method should not be used when the graph of f is nearly horizontal where it crosses the x-axis. Further, the method fails if $f'(x) = 0$ in the neighbourhood of the root.

EXAMPLE 15.7

Find the smallest positive root of $x^3 - 5x + 3 = 0$.

Solution. We observe that there is a root between -2 and -3, a root between 1 and 2, and a (smallest) root between 0 and 1. We have

$$f(x) = x^3 - 5x + 3, \quad f'(x) = 3x^2 - 5.$$

Then taking $x_0 = 1$, we have

$$x_1 = x_0 - \frac{f(x_0)}{f'(x_0)} = 1 - \frac{f(1)}{f'(1)} = 1 - \frac{(-1)}{-2} = 0.5,$$

$$x_2 = x_1 - \frac{f(x_1)}{f'(x_1)} = 0.5 + \frac{5}{34} = 0.64,$$

$$x_3 = 0.64 + \frac{0.062144}{3.7712} = 0.6565,$$

$$x_4 = 0.6565 + \frac{0.000446412125}{3.70702325} = 0.656620,$$

$$x_5 = 0.656620 + \frac{0.00000115976975}{3.70655053}$$

$$= 0.656620431.$$

We observe that the convergence is very rapid even though x_0 was not very near to the root.

EXAMPLE 15.8

Find the positive root of the equation $x^4 - 3x^3 + 2x^2 + 2x - 7 = 0$ by Newton–Raphson method.

Solution. We have $f(0) = -7$, $f(1) = -5$, $f(2) = -3$, $f(3) = 17$. Thus the positive root lies between 2 and 3. The Newton–Raphson formula becomes

$$x_{n+1} = x_n - \frac{x_n^4 - 3x_n^3 + 2x_n^2 + 2x_n - 7}{4x_n^3 - 9x_n^2 + 4x_n + 2}.$$

Taking $x_0 = 2.1$, the improved approximations are

$$x_1 = 2.39854269, x_2 = 2.33168543,$$
$$x_3 = 2.32674082, x_4 = 2.32671518,$$
$$x_5 = 2.32671518.$$

Since $x_4 = x_5$, the Newton–Raphson formula gives no new values of x and the approximate root is correct to eight decimals.

EXAMPLE 15.9

Use Newton–Raphson method to solve the transcendental equation $e^x = 5x$.

Solution. Let $f(x) = e^x - 5x = 0$. Then $f'(x) = e^x - 5$. The Newton–Raphson formula becomes

$$x_{n+1} = x_n - \frac{e^{x_n} - 5x_n}{e^{x_n} - 5}, \; n = 0, 1, 2, 3, \ldots.$$

The successive approximations are

$$x_0 = 0.4, x_1 = 0.2551454079, x_2 = 0.2591682786$$
$$x_3 = 0.2591711018, x_4 = 0.2591711018.$$

Thus the value of the root is correct to 10 decimals places.

(a) Square Root of a Number Using Newton-Raphson Method

Suppose that we want to find the square root of N. Let

$$x = \sqrt{N} \text{ or } x^2 = N.$$

We have

$$f(x) = x^2 - N = 0.$$

Then, Newton–Raphson method yields

$$x_{n+1} = x_n - \frac{f(x_n)}{f'(x_n)} = x_n - \frac{x_n^2 - N}{2x_n}$$

$$= \frac{1}{2}\left[x_n + \frac{N}{x_n}\right], \quad n = 0, 1, 2, 3, \ldots$$

For example, if N = 10, taking $x_0 = 3$ as an initial approximation, the successive approximations are

$$x_1 = 3.166666667, x_2 = 3.162280702,$$
$$x_3 = 3.162277660, x_4 = 3.162277660$$

correct up to nine decimal places.

However, if we take $f(x) = x^3 - Nx$ so that if $f(x) = 0$, then $x = \sqrt{N}$. Now $f'(x) = 3x^2 - N$ and so the Newton–Raphson method gives

$$x_{n+1} = x_n - \frac{f(x_n)}{f'(x_n)} = x_n - \frac{x_n^3 - Nx_n}{3x_n^2 - N} = \frac{2x_n^3}{3x_n^2 - N}.$$

Taking $x_0 = 3$, the successive approximations to $\sqrt{10}$ are

$$x_1 = 3.176, \quad x_2 = 3.1623,$$
$$x_3 = 3.16227, x_4 = 3.16227$$

correct up to five decimal places.

Suppose that we want to find the pth root of N. Then consider $f(x) = x^p - N$. The Newton–Raphson formula yields

$$x_{n+1} = x_n - \frac{f(x_n)}{f'(x_n)} = x_n - \frac{x_n^p - N}{px_n^{p-1}}$$
$$= \frac{(p-1)x_n^p + N}{px_n^{p-1}}, \quad n = 0, 1, 2, 3, \ldots$$

For $p = 3$, the formula reduces to

$$x_{n+1} = \frac{2x_n^3 + N}{3x_n^2} = \frac{1}{3}\left(2x_n + \frac{N}{x_n^2}\right).$$

If $N = 10$ and we start with the approximation $x_0 = 2$. Then

$$x_1 = \frac{1}{3}\left(4 + \frac{10}{4}\right) = 2.16666, x_2 = 2.154503616,$$

$$x_3 = 2.154434692, x_4 = 2.154434690,$$

$$x_5 = 2.154434690$$

correct up to eight decimal places.

(b) Order of Convergence of Newton–Raphson Method

Suppose $f(x) = 0$ has a simple root at $x = \xi$ and let ε_n be the error in the approximation. Then $x_n = \xi + \varepsilon_n$. Applying Taylor's expansion of $f(x_n)$ and $f'(x_n)$ about the root ξ, we have

$$f(x_n) = \sum_{r=1}^{\infty} a_r \varepsilon_n^r$$

and

$$f'(x_n) = \sum_{r=1}^{\infty} r a_r \varepsilon_n^{r-1}, \text{ where } a_r = \frac{f^{(r)}(\xi)}{r!}.$$

Then

$$\frac{f(x_n)}{f'(x_n)} = \varepsilon_n - \frac{a_2}{a_1}\varepsilon_n^2 + O\left(\varepsilon_n^3\right).$$

Therefore, Newton–Raphson formula

$$x_{n+1} = x_n - \frac{f(x_n)}{f'(x_n)}$$

gives

$$\xi + \varepsilon_{n+1} = \xi + \varepsilon_n - \left[\varepsilon_n - \frac{a_2}{a_1}\varepsilon_n^2 + O(\varepsilon_n^3)\right]$$

and so

$$\varepsilon_{n+1} = \frac{a_2}{a_1}\varepsilon_n^2 = \frac{1}{2}\frac{f''(\xi)}{f'(\xi)}\varepsilon_n^2.$$

If

$$\frac{1}{2}\frac{f''(\xi)}{f'(\xi)} < 1,$$

then

$$\varepsilon_{n+1} < \varepsilon_n^2. \qquad (10)$$

It follows, therefore, that Newton–Raphson method has a *quadratic convergence* (*or second–order convergence*) *if* $\frac{1}{2}\frac{f''(\xi)}{f'(\xi)} < 1$.

The inequality (10) implies that if the correction term $\frac{f(x_n)}{f'(x_n)}$ begins with n zeros, then the result is correct to about 2^n decimals. Thus, in

Newton–Raphson method the number of correct decimal roughly doubles at each stage.

4. Fixed Point Iteration

Let f be a real–valued function $f: \Re \to \Re$. Then a point $x \in \Re$ is said to be a *fixed point* of f if $f(x) = x$.

For example, let $I: \Re \to \Re$ be an identity mapping. Then all points of $\Re$ are fixed points for I since $I(x) = x$ for all $x \in \Re$. Similarly, a constant map of $\Re$ into $\Re$ has a unique fixed point.

Consider the equation

$$f(x) = 0. \tag{11}$$

The fixed point iteration approach to the solution of (11) is that it is rewritten in the form of an equivalent relation

$$x = \phi(x) \tag{12}$$

Then any solution of (11) is a fixed point of the iteration function ϕ. Thus the task of solving the equation is reduced to find the fixed points of the iteration function ϕ.

Let x_0 be an initial solution [approximate value of the root of (11)] obtained from the graph of f or otherwise. We substitute this value of x_0 in the right-hand side of (12) and obtain a better approximation x_1 given by

$$x_1 = \phi(x_0).$$

Then the successive approximations are

$$x_2 = \phi(x_1),$$

$$x_3 = \phi(x_2),$$

$$\ldots \ldots \ldots$$

$$\ldots \ldots \ldots$$

$$x_{n+1} = \phi(x_n),\ n = 0, 1, 2, 3, \ldots$$

$$\ldots$$

The iteration

$$x_{n+1} = \phi(x_n),\ n = 0, 1, 2, 3, \ldots$$

is called *fixed point iteration*.

Obviously, Regula Falsi method and Newton–Raphson methods are iteration processes.

(a) Convergence of Iteration Method

We are interested in determining the condition under which the iteration method converges, that is, for which x_{n+1} converges to the solution of $x = \phi(x)$ as $n \to \infty$. Thus, if $x_{n+1} = x_n$ up to the number of significant figures considered, then x_n is a solution to that degree of approximation. Let ξ be the true solution of $x = \phi(x)$, that is,

$$\xi = \phi(\xi) \tag{13}$$

The first approximation is

$$x_1 = \phi(x_0) \tag{14}$$

Subtracting (14) from (13), we get

$$\begin{aligned} \xi - x_1 &= \phi(\xi) - \phi(x_0) \\ &= (\xi - x_0)\,\phi'(\xi_0),\ x_0 < \xi_0 < \xi, \end{aligned}$$

by mean value theorem. Similar equations hold for successive approximations so that

$$\xi - x_2 = (\xi - x_1)\phi'(\xi_1), x_1 < \xi_1 < \xi.$$
$$\xi - x_3 = (\xi - x_2)\phi'(\xi_2), x_2 < \xi_2 < \xi$$
$$\ldots\ldots\ldots\ldots$$
$$\ldots\ldots\ldots\ldots$$
$$\xi - x_{n+1} = (\xi - x_n)\phi'(\xi_n), x_n < \xi_n < \xi.$$

Multiplying together all the equations, we get

$$\xi - x_{n+1} = (\xi - x_0)\phi'(\xi_0)\phi'(\xi_1)\ldots\phi'(\xi_n)$$

and so

$$|\xi - x_{n+1}| = |\xi - x_0||\phi'(\xi_0)|\ldots|\phi'(\xi_n)|.$$

If each of $|\phi'(\xi_0)|, \ldots |\phi'(\xi_n)|$ is less than or equal to $k < 1$, then

$$|\xi - x_{n+1}| \le |\xi - x_0|\, k^{n+1} \to 0 \text{ as } n \to \infty.$$

Hence the error $\xi - x_{n+1}$ can be made as small as we please by repeating the process a sufficient number of times. Thus the *condition for convergence is*

$$|\phi'(x)| < 1$$

in the neighbourhood of the desired root.

Consider the iteration formula $x_{n+1} = \phi(x_n)$, $n = 0, 1, 2, ---$. If ξ is the true solution of $x = \phi(x)$, then $\xi = \phi(\xi)$. Therefore,

$$\xi - x_{n+1} = \phi(\xi) - \phi(x_n) = (\xi - x_n)\phi'(\xi)$$

$$= (\xi - x_n)\, k\,,\ |\phi'(\xi)| \le k < 1,$$

which shows that the *iteration method has a linear convergence.* This slow rate of convergence can be accelerated in the following way. We write

$$\xi - x_{n+1} = (\xi - x_n)k$$

$$\xi - x_{n+2} = (\xi - x_{n+1})k.$$

Dividing we get

$$\frac{\xi - x_{n+1}}{\xi - x_{n+2}} = \frac{\xi - x_n}{\xi - x_{n+1}}$$

or

$$(\xi - x_{n+1})^2 = (\xi - x_{n+2})(\xi - x_n)$$

or

$$\xi = x_{n+2} - \frac{(x_{n+2} - x_{n+1})^2}{x_{n+2} - 2x_{n+1} + x_n}$$

$$= x_{n+2} - \frac{(\Delta x_{n+1})^2}{\Delta^2 x_n}. \tag{15}$$

Formula (15) is called the *Aitken's* Δ^2–method.

(b) Square Root of a Number Using Iteration Method

Suppose that we want to find square root of a number, say N. This is equivalent to say that we want to find x such that $x^2 = N$, that is, $x = \frac{N}{x}$ or $x + x = x + \frac{N}{x}$. Thus

$$x = \frac{x + \frac{N}{x}}{2}.$$

Thus, if x_0 is the initial approximation to the square root, then

$$x_{n+1} = \frac{x_n + \frac{N}{x_n}}{2},\ n = 0, 1, 2, 3, \ldots$$

Suppose $N = 13$. We begin with the initial approximation of $\sqrt{13}$ found by bisection method. The solution lies between 3.5625 and 3.625. we start with $x_0 = \frac{3.5625 + 3.6250}{2} = 3.59375$. Then, using the above iteration formula, we have

$$x_1 = 3.6055705,\ x_2 = 3.6055513,$$

$$x_3 = 3.6055513$$

correct up to seven decimal places.

(c) Sufficient Condition for the Convergence of Newton–Raphson Method

We know that an iteration method $x_{n+1} = \phi(x_n)$ converges if $|\phi'(x)| < 1$. Since Newton–Raphson method is an iteration method, where $\phi(x) = x - \frac{f(x)}{f'(x)}$, it converges if $|\phi'(x)| < 1$, that is, if

$$\left| 1 - \frac{(f'(x))^2 - f(x)f''(x)}{(f'(x))^2} \right| < 1,$$

that is, if

$$|f(x)f''(x)| < (f'(x))^2,$$

which is the required sufficient condition for the convergence of Newton–Raphson method.

EXAMPLE 15.10

Derive an iteration formula to solve $f(x) = x^3 + x^2 - 1 = 0$ and solve the equation.

Solution. Since $f(0)$ and $f(1)$ are of opposite signs, there is a root between 0 and 1. We write the equation in the form $x^3 + x^2 = 1$, that is

$$x^2(x + 1) = 1, \text{ or } x^2 = \frac{1}{x+1} \text{ or equivalently}$$

$$x = \frac{1}{\sqrt{1+x}}.$$

Then

$$x=\phi(x)=\frac{1}{\sqrt{1+x}},\quad \phi'(x)=-\frac{1}{2(1+x)^{\frac{3}{2}}}$$

so that

$$|\phi'(x)|<1 \text{ for } x<1.$$

Hence this iteration method is applicable. We start with $x_0 = 0.75$ and obtain the next approximations to the root as

$$x_1=\phi(x_0)=\frac{1}{\sqrt{1+x_0}}\approx 0.7559,$$

$$x_2=\phi(x_1)\approx 0.7546578\ ,\ x_3\approx 0.7549249$$

$$x_4\approx 0.7548674,\quad x_5\approx 0.754880,$$

$$x_6\approx 0.7548772,\quad x_7\approx 0.75487767$$

correct up to six decimal places.

EXAMPLE 15.11

Find, by the method of iteration, a root of the equation $2x - \log_{10} x = 7$.

Solution. The fixed point form of the given equation is

$$x=\frac{1}{2}(\log_{10} x+7).$$

From the intersection of the graphs $y_1 = 2x - 7$ and $y_2 = \log_{10} x$, we find that the approximate value of the root is 3.8. Therefore,

$$x_0 = 3.8,$$

$$x_1=\frac{1}{2}(\log 3.8+7)\approx 3.78989,$$

$$x_2=\frac{1}{2}(\log 3.78989+7)\approx 3.789313,$$

$$x_3=\frac{1}{2}(\log 3.789313+7)\approx 3.78928026,$$

$$x_4\approx 3.789278,\quad x_5\approx 3.789278,$$

correct up to six decimal places.

EXAMPLE 15.12

Use iteration method to solve the equation $e^x = 5x$.

Solution. The iteration formula for the given problem is

$$x_{n+1}=\frac{1}{5}e^{x_n}.$$

We start with $x_0 = 0.3$ and get the successive approximations as

$$x_1=\frac{1}{5}(1.34985881)=0.269972,\ x_2=0.26198555$$

$$x_3=0.25990155,\ x_4=0.259360482,$$

$$x_5=0.259220188\ \ x_6=0.259183824,$$

$$x_7=0.259174399,\ x_8=0.259171956$$

$$x_9=0.259171323,\ x_{10}=0.259171159$$

correct up to six decimal places.

If we use Aitken's Δ^2–method, then

$$x_3=x_2-\frac{(\Delta x_1)^2}{\Delta^2 x_0}=x_2-\frac{(x_2-x_1)^2}{x_2-2x_1+x_0}$$

$$=0.26198555-\frac{0.000063783}{0.02204155}=0.259091$$

and so on.

5. Newton's Method for Finding Multiple Roots

If ξ is a multiple root of an equation $f(x) = 0$, then $f(\xi) = f'(\xi) = 0$ and, therefore, the Newton–Raphson method fails. However, in case of multiple roots, we proceed as follows:

Let ξ be a root of multiplicity m. Then

$$f(x)=(x-\xi)^m A(x). \tag{16}$$

We make use of a localized approach that in the immediate vicinity (neighbourhood) of $x = \xi$, the relation (16) can be written as

$$f(\xi) = A(x - \xi)^m,$$

where $A = A(\xi)$ is effectively constant. Then

$$f'(x) = mA(x - \xi)^{m-1}$$

$$f'(x) = m(m - 1)\, A(x - \xi)^{m-2}, \text{ and so on.}$$

We thus obtain

$$\frac{f'(x)}{f(x)} = \frac{m}{x - \xi}$$

or

$$\xi = x - \frac{m\, f(x)}{f'(x)},$$

where x is close to ξ, which is a modification of Newton's rule for a multiple root. Thus if x_1 is in the neighbourhood of a root ξ of multiplicity m of an equation $f(x) = 0$, then

$$x_2 = x_1 - m\frac{f(x_1)}{f'(x_1)}$$

is an even more close approximation to ξ. Hence in general, we have

$$x_{n+1} = x_n - m\frac{f(x_n)}{f'(x_n)} \qquad (17)$$

Remark 15.1 (i). The case $m = 1$ of (17) yields Newton–Raphson method

(ii) If two roots are close to a number, say x, then

$$f(x + \varepsilon) = 0 \text{ and } f(x - \varepsilon) = 0,$$

that is,

$$f(x) + \varepsilon f'(x) + \frac{\varepsilon^2}{2!} f''(x) + - - - - = 0,$$

$$f(x) - \varepsilon f'(x) + \frac{\varepsilon^2}{2!} f''(x) - - - - - = 0.$$

Since ε is small, adding the above expressions, we get

$$0 = 2f(x) + \varepsilon^2 f''(x) = 0$$

or

$$\varepsilon^2 = -2\frac{f(x)}{f''(x)}$$

or

$$\varepsilon = \pm\sqrt{\frac{-2f(x)}{f''(x)}}.$$

So in this case, we take two approximations as $x + \varepsilon$ and $x - \varepsilon$ and then apply Newton–Raphson method.

For example, consider the equation $x^4 - 5x^3 - 12x^2 + 76x - 79 = 0$ having two roots close to $x = 2$. We have

$$f(x) = x^4 - 5x^3 - 12x^2 + 76x - 79$$
$$f'(x) = 4x^3 - 15x^2 - 24x + 76$$
$$f''(x) = 12x^2 - 30x - 24.$$

Thus

$$f(2) = 16 - 40 - 48 + 152 - 79 = 1$$

$$f''(2) = 48 - 60 - 24 = -36.$$

Therefore,

$$\varepsilon = \pm\sqrt{\frac{-2f(2)}{f''(2)}} = \pm\sqrt{\frac{-2}{-36}} = \pm 0.2357$$

Thus the initial approximations to the roots are

$$x_0 = 2.2357 \text{ and } y_0 = 1.7643.$$

The application of Newton–Raphson method yields

$$x_1 = x_0 - \frac{f(x_0)}{f'(x_0)} = 2.2357 + 0.00083 = 2.0365.$$

$$x_2 = x_1 - \frac{f(x_1)}{f'(x_1)} = 2.2365 + 0.00459 = 2.24109.$$

$$x_3 = x_2 - \frac{f(x_2)}{f'(x_2)} = 2.24109 - 0.0009 = 2.2410.$$

Thus, one root, correct to four decimal places is 2.2410. similarly, the second root correct to four decimal places will be found to be 1.7684.

Similarly, consider the equation $x^3 - 5x^2 + 8x - 4 = 0$, which has double root near 1.8. we have

$$f(x) = x^3 - 5x^2 + 8x - 4,$$

$$f'(x) = 3x^2 - 10x + 8.$$

and $x_0 = 1.8$ therefore,

$$f(x_0) = f(1.8) = 5.832 - 16.2 + 14.4 - 4 = 0.032,$$

$$f'(x_0) = 9.72 - 18 + 8 = -0.28.$$

Hence

$$x_1 = x_0 - 2\frac{f(x_0)}{f'(x_0)} = 1.8 - 2\frac{f(1.8)}{f'(1.8)}$$

$$= 1.8 - 2\frac{0.032}{-0.28} = 2.02857.$$

We take $x_1 = 2.028$. Then

$$f(x_1) = 8.3407 - 20.5639 + 16.224 - 4 = 0.0008$$

$$f'(x_1) = 12.3384 - 20.28 + 8 = 0.0584.$$

Therefore,

$$x_2 = x_1 - 2\frac{f(x_1)}{f'(x_1)}$$

$$= 2.028 - \frac{2(0.0008)}{0.0584} = 2.0006,$$

which is quite close to the actual double root 2.

15.6 LINEAR SYSTEM OF EQUATIONS

In this section, we shall study *direct and iterative* methods to solve linear system of equations. Among the direct methods, we shall study Gaussian elimination method and its modification by Jordan. Among the iterative methods, we shall study Jacobi and Gauss – Seidel methods.

1. Gauss's Elimination Method

This is the simplest method of step-by-step elimination and it reduces the system of equations to an equivalent *upper triangular system,* which can be solved by back substitution.

Let the system of equations be

$$\begin{aligned}
a_{11}x_1 + a_{12}x_2 + \ldots + a_{1n}x_n &= b_1 \\
a_{21}x_1 + a_{22}x_2 + \ldots + a_{2n}x_n &= b_2 \\
\ldots \quad \ldots \quad \ldots \quad \ldots & \\
\ldots \quad \ldots \quad \ldots \quad \ldots & \\
a_{n1}x_1 + a_{n2}x_2 + \ldots + a_{nn}x_n &= b_n.
\end{aligned}$$

The matrix form of this system is

$$AX = B.$$

where

$$A = \begin{bmatrix} a_{11} & a_{12} & \ldots & \ldots & a_{1n} \\ a_{21} & a_{22} & \ldots & \ldots & a_{2n} \\ \ldots & \ldots & \ldots & \ldots & \ldots \\ \ldots & \ldots & \ldots & \ldots & \ldots \\ a_{n1} & a_{n2} & \ldots & \ldots & a_{nn} \end{bmatrix},$$

$$X = \begin{bmatrix} x_1 \\ x_2 \\ \ldots \\ x_n \end{bmatrix}, \quad B = \begin{bmatrix} b_1 \\ b_2 \\ \ldots \\ b_n \end{bmatrix}.$$

The augmented matrix is

$$[A:B] = \left[\begin{array}{ccccc|c} a_{11} & a_{12} & \ldots & \ldots & a_{1n} & b_1 \\ a_{21} & a_{22} & \ldots & \ldots & a_{2n} & b_2 \\ \ldots & \ldots & \ldots & \ldots & \ldots & \ldots \\ \ldots & \ldots & \ldots & \ldots & \ldots & \ldots \\ a_{n1} & a_{n2} & \ldots & \ldots & a_{nn} & b_n \end{array}\right].$$

The number a_{rr} in position (r, r), that is used to eliminate x_r in rows $r + 1, r + 2, \ldots, n$ is called the rth *pivotal element and* the rth row is called the *pivotal row*. Thus the augmented matrix can be written as

$$\begin{array}{r} \text{pivot}\rightarrow \\ m_{2,1}=a_{21}/a_{11} \\ \\ \\ m_{n,1}=a_{n1}/a_{11} \end{array} \left[\begin{array}{cccccc|c} \underline{a_{11}} & a_{12} & \cdots & \cdots & a_{1n} & b_1 \\ a_{21} & a_{22} & \cdots & \cdots & a_{2n} & b_2 \\ \cdots & \cdots & \cdots & \cdots & \cdots & \cdots \\ \cdots & \cdots & \cdots & \cdots & \cdots & \cdots \\ a_{n1} & a_{n2} & \cdots & \cdots & a_{nn} & b_n \end{array}\right] \begin{array}{l} \leftarrow\text{pivotal} \\ \text{row} \\ \\ \\ \\ \end{array}$$

The first row is used to eliminate elements in the first column below the diagonal. In the first step, the element a_{11} is pivotal element and the first row is pivotal row. The values $m_{k,1}$ are the *multiples of row* 1 that are to be subtracted from row k for $k = 2, 3, 4, \ldots, n$. The result after elimination becomes

$$\begin{array}{r} \\ \text{pivot}\rightarrow \\ m_{3,2}=c_{32}/c_{22} \\ \\ m_{n,2}=c_{n2}/c_{22} \end{array} \left[\begin{array}{ccccc|c} a_{11} & a_{12} & \cdots & \cdots & a_{1n} & b_1 \\ & \underline{c_{22}} & \cdots & \cdots & c_{2n} & d_2 \\ & c_{32} & \cdots & \cdots & c_{3n} & d_3 \\ & \cdots & \cdots & \cdots & \cdots & \cdots \\ & c_{n2} & \cdots & \cdots & c_{nn} & d_n \end{array}\right] \begin{array}{l} \\ \leftarrow\text{pivotal} \\ \text{row} \\ \\ \\ \end{array}.$$

The second row (now pivotal row) is used to eliminate elements in the second column that lie below the diagonal. The elements $m_{k,2}$ are the multiples of row 2 that are to be subtracted from row k for $k = 3, 4, \ldots, n$.

Continuing this process, we arrive at the matrix:

$$\left[\begin{array}{cccccc|c} a_{11} & a_{12} & \cdots & \cdots & a_{1n} & b_1 \\ & c_{22} & \cdots & \cdots & c_{2n} & d_2 \\ & \cdots & \cdots & \cdots & \cdots & \cdots \\ & \cdots & \cdots & \cdots & \cdots & \cdots \\ & & & & h_{nn} & p_n \end{array}\right].$$

Hence the given system of equations reduces to

$$\begin{aligned} a_{11}x_1 + a_{12}x_2 + \ldots + a_{1n}x_n &= b_1 \\ c_{22}x_2 + \ldots + c_{2n}x_n &= d_2 \\ \cdots \quad \cdots \quad & \cdots \\ \cdots \quad \cdots \quad & \cdots \\ h_{nn}x_n &= p_n, \end{aligned}$$

which can be solved by back substitution.

Remark 15.2. It may occur that the pivot element, even if it is different from zero, is very small and gives rise to large errors. The reason is that the small coefficient usually has been formed as the difference between two almost equal numbers. This difficulty is overcome by suitable permutations of the given equations. It is recommended, therefore, that the *pivotal equation should be the selected equation which has the largest leading coefficient.*

EXAMPLE 15.13

Express the following system in augmented matrix form and find an equivalent upper–triangular system and the solution;

$$2x_1 + 4x_2 - 6x_3 = -4$$

$$x_1 + 5x_2 + 3x_3 = 10$$

$$x_1 + 3x_2 + 2x_3 = 5.$$

Solution. The augmented matrix for the system is

$$\begin{array}{r} \text{pivot} \rightarrow \\ m_{2,1} = 0.5 \\ m_{3,1} = 0.5 \end{array} \left[\begin{array}{ccc|c} \underline{2} & 4 & -6 & -4 \\ 1 & 5 & 3 & 10 \\ 1 & 3 & 2 & 5 \end{array}\right] \begin{array}{l} \leftarrow \text{pivotal row} \\ \\ \\ \end{array}$$

The result after first elimination is

$$\begin{array}{r} \\ \text{pivot} \rightarrow \\ m_{3,2} = 1/3 \end{array} \left[\begin{array}{ccc|c} 2 & 4 & -6 & -4 \\ 0 & \underline{3} & 6 & 12 \\ 0 & 1 & 5 & 7 \end{array}\right] \begin{array}{l} \\ \leftarrow \text{pivotal row} \\ \\ \end{array}$$

The result after second elimination is

$$\left[\begin{array}{ccc|c} 2 & 4 & -6 & -4 \\ 0 & 3 & 6 & 12 \\ 0 & 0 & 3 & 3 \end{array}\right].$$

Therefore, back substitution yields

$$3x_3 = 3 \text{ and so } x_3 = 1,$$

$$3x_2 + 6x_3 = 12 \text{ and so } x_2 = 2,$$

$$2x_1 + 4x_2 - 6x_3 = -4 \text{ and so } x_1 = -3.$$

Hence the solution is $x_1 = -3$, $x_2 = 2$ and $x_3 = 1$.

EXAMPLE 15.14

Solve by Gaussian elimination method:

$$10x - 7y + 3z + 5u = 6$$

$$-6x + 8y - z - 4u = 5$$

$$3x + y + 4z + 11u = 2$$

$$5x - 9y - 2z + 4u = 7.$$

Solution. The augmented matrix for the given system is

$$\begin{matrix} \text{pivot} \rightarrow \\ m_{2,1} = -0.6 \\ m_{3,1} = 0.3 \\ m_{4,1} = 0.5 \end{matrix} \left[\begin{array}{cccc|c} \underline{10} & -7 & 3 & 5 & 6 \\ -6 & 8 & -1 & -4 & 5 \\ 3 & 1 & 4 & 11 & 2 \\ 5 & -9 & -2 & 4 & 7 \end{array}\right] \leftarrow \text{pivotal row}$$

The first elimination yields

$$\begin{matrix} \\ \text{pivot} \rightarrow \\ m_{3,2} = 0.81579 \\ m_{4,2} = -1.4474 \end{matrix} \left[\begin{array}{cccc|c} 10 & -7 & 3 & 5 & 6 \\ 0 & \underline{3.8} & 0.8 & -1 & 8.6 \\ 0 & 3.1 & 3.1 & 9.5 & 0.2 \\ 0 & -5.5 & -3.5 & 1.5 & 4 \end{array}\right] \leftarrow \text{pivotal row}$$

The result after second elimination is

$$\begin{matrix} \\ \\ \text{pivot} \rightarrow \\ m_{4,3} = 0.957 \end{matrix} \left[\begin{array}{cccc|c} 10 & -7 & 3 & 5 & 6 \\ 0 & 3.8 & 0.8 & -1 & 8.6 \\ 0 & 0 & \underline{2.4474} & 10.3158 & -6.8158 \\ 0 & 0 & -2.3421 & 0.0526 & 16.44764 \end{array}\right] \leftarrow \text{pivotal row}$$

The result after third elimination is

$$\left[\begin{array}{cccc|c} 10 & -7 & 3 & 5 & 6 \\ 0 & 3.8 & 0.8 & -1 & 8.6 \\ 0 & 0 & 2.4474 & 10.3158 & -6.8158 \\ 0 & 0 & 0 & 9.9248 & 9.9249 \end{array}\right]$$

Therefore, back substitution yields

$$9.9248u = 9.9249 \text{ and so } u \approx 1$$

$$2.4472z + 10.3158u = -6.8158 \text{ and so}$$

$$z = -6.9999 \approx -7$$

$$3.8y + 0.8z - u = 8.6 \text{ and so } y = 4$$

$$10x - 7y + 3z + 5u = 6 \text{ and so } x = 5.$$

Hence the solution of the given system is $x = 5$, $y = 4$, $z = -7$ and $u = 1$.

2. Jordan's Modification to Gauss's Method

Jordan modification means that the elimination is performed not only in the equation below but also in the equation above the pivotal row so that we get a diagonal matrix. In this way we have the solution without further computation.

Comparing the methods of Gauss and Jordan, we find that the number of operations is essentially $\frac{n^3}{3}$ for Gauss's method and $\frac{n^3}{2}$ for Jordan's method where n is the order of the coefficient matrix. Hence the Gauss's method should usually be preferred over Jordan's method.

To illustrate this modification we reconsider Example 15.13. The result of first elimination is unchanged and we have

$$\begin{matrix} m_{1,2} = 4/3 \\ \text{pivot} \rightarrow \\ m_{3,2} = 1/3 \end{matrix} \left[\begin{array}{ccc|c} 2 & 4 & -6 & -4 \\ 0 & \underline{3} & 6 & 12 \\ 0 & 1 & 5 & 7 \end{array}\right] \leftarrow \text{pivotal row}$$

Now, the second elimination as per Jordan's modification yields

$$\begin{matrix} m_{1,3} = -14/3 \\ m_{2,3} = 2 \\ \text{pivot} \rightarrow \end{matrix} \left[\begin{array}{ccc|c} 2 & 0 & -14 & -20 \\ 0 & 3 & 6 & 12 \\ 0 & 0 & \underline{3} & 3 \end{array}\right] \leftarrow \text{pivotal row}$$

The third elimination as per Jordan's modification yields

$$\left[\begin{array}{ccc|c} 2 & 0 & 0 & -6 \\ 0 & 3 & 0 & 6 \\ 0 & 0 & 3 & 3 \end{array}\right].$$

Hence

$$2x_1 = -6 \text{ and so } x_1 = -3,$$

$$3x_2 = 6 \text{ and so } x_2 = 2,$$

$$3x_3 = 3 \text{ and so } x_3 = 1.$$

EXAMPLE 15.15

Solve

$$x + 2y + z = 8$$

$$2x + 3y + 4z = 20$$

$$4x + 3y + 2z = 16$$

by Gauss–Jordan method.

Solution. The augmented matrix for the given system of equations is

$$\begin{matrix} \text{pivot} \rightarrow \\ m_{2,1} = 2 \\ m_{3,1} = 4 \end{matrix} \left[\begin{array}{ccc|c} \underline{1} & 2 & 1 & 8 \\ 2 & 3 & 4 & 20 \\ 4 & 3 & 2 & 16 \end{array}\right] \begin{matrix} \leftarrow \text{pivotal row} \\ \\ \\ \end{matrix}$$

The result of first elimination is

$$\begin{matrix} m_{1,2} = -2 \\ \text{pivot} \rightarrow \\ m_{3,2} = 5 \end{matrix} \left[\begin{array}{ccc|c} 1 & 2 & 1 & 8 \\ 0 & \underline{-1} & 2 & 4 \\ 0 & -5 & -2 & -16 \end{array}\right] \begin{matrix} \\ \leftarrow \text{pivotal row} \\ \\ \end{matrix}$$

The second Gauss–Jordan elimination yields

$$\begin{matrix} m_{1,3} = -5/12 \\ m_{2,3} = -1/6 \\ \text{pivot} \rightarrow \end{matrix} \left[\begin{array}{ccc|c} 1 & 0 & 5 & 16 \\ 0 & -1 & 2 & 4 \\ 0 & 0 & \underline{-12} & -36 \end{array}\right] \begin{matrix} \\ \\ \leftarrow \text{pivotal row} \end{matrix}$$

The third Gauss–Jordan elimination yields

$$\left[\begin{array}{ccc|c} 1 & 0 & 0 & 1 \\ 0 & -1 & 0 & -2 \\ 0 & 0 & -12 & -36 \end{array}\right].$$

Therefore $x = 1$, $y = 2$, and $z = 3$ is the required solution.

EXAMPLE 15.16

Solve

$$10x + y + z = 12$$

$$x + 10y + z = 12$$

$$x + y + 10z = 12$$

by Gauss–Jordan method.

Solution. The augmented matrix for the given system is

$$\begin{matrix} \text{pivot} \rightarrow \\ m_{2,1} = 1/10 \\ m_{3,1} = 1/10 \end{matrix} \left[\begin{array}{ccc|c} \underline{10} & 1 & 1 & 12 \\ 1 & 10 & 1 & 12 \\ 1 & 1 & 10 & 12 \end{array}\right] \begin{matrix} \leftarrow \text{pivotal row} \\ \\ \\ \end{matrix}$$

The first Gauss – Jordan elimination yields

$$\begin{matrix} m_{1,2} = 10/99 \\ \text{pivot} \rightarrow \\ m_{3,2} = 1/11 \end{matrix} \left[\begin{array}{ccc|c} 10 & 1 & 1 & 12 \\ 0 & \underline{99/10} & 9/10 & 108/10 \\ 0 & 9/10 & 99/10 & 108/10 \end{array}\right] \begin{matrix} \\ \leftarrow \text{pivotal row} \\ \\ \end{matrix}$$

Now the Gauss–Jordan elimination gives

$$\begin{matrix} m_{1,3} = 10/108 \\ m_{2,3} = 11/120 \\ \text{pivot} \rightarrow \end{matrix} \left[\begin{array}{ccc|c} 10 & 0 & 10/11 & 120/11 \\ 0 & 99/10 & 9/10 & 108/10 \\ 0 & 0 & \underline{108/11} & 108/11 \end{array}\right] \begin{matrix} \\ \\ \leftarrow \text{pivotal row} \end{matrix}$$

The next Gauss–Jordan elimination yields

$$\left[\begin{array}{ccc|c} 10 & 0 & 0 & 10 \\ 0 & 99/10 & 0 & 99/10 \\ 0 & 0 & 108/11 & 108/11 \end{array}\right].$$

Hence the solution of the given system is $x = 1$, $y = 1, z = 1$.

3. Iterative Methods for Linear systems

We have seen that direct methods for the solution of simultaneous linear equations yield the solution after an amount of computation that is known in advance. On the other hand, in case of *iterative or indirect* methods, we start from an approximation to the true solution and, if convergent, we form a sequence of closer approximations repeated till the required accuracy is obtained. The difference between direct and iterative method is that in *direct method, the amount of computation is fixed* whereas *in iterative method, the amount of computation depends upon the accuracy required.*

In general, we *prefer a direct method for solving system of linear equations.* But, in case of *matrices with a large number of zero elements, it is economical to use iterative methods.*

(a) Jacobi Iteration Method

Consider the system

$$\left.\begin{aligned} a_{11}x_1 + a_{12}x_2 + \ldots + a_{1n}x_n &= b_1 \\ a_{21}x_1 + a_{22}x_2 + \ldots + a_{2n}x_n &= b_2 \\ a_{31}x_1 + a_{32}x_2 + \ldots + a_{3n}x_n &= b_3 \\ \ldots \quad \ldots \quad \ldots \quad \ldots & \\ \ldots \quad \ldots \quad \ldots \quad \ldots & \\ a_{n1}x_1 + a_{n2}x_2 + \ldots + a_{nn}x_n &= b_n \end{aligned}\right\} \qquad (18)$$

in which the diagonal coefficients a_{ii} do not vanish. If this is not the case, the equations should be rearranged so that this condition is satisfied. The equation (18) can be written as

$$\left.\begin{aligned} x_1 &= \tfrac{b_1}{a_{11}} - \tfrac{a_{12}}{a_{11}}x_2 - \ldots - \tfrac{a_{1n}}{a_{11}}x_n \\ x_2 &= \tfrac{b_2}{a_{22}} - \tfrac{a_{21}}{a_{22}}x_1 - \ldots - \tfrac{a_{2n}}{a_{22}}x_n \end{aligned}\right\}$$

$$\ldots \quad \ldots \quad \ldots \quad \ldots$$

$$\ldots \quad \ldots \quad \ldots \quad \ldots$$

$$x_n = \frac{b_n}{a_{nn}} - \frac{a_{n1}}{a_{nn}}x_1 - \ldots - \frac{a_{n,n-1}}{a_{nn}}x_{n-1}. \qquad (19)$$

Suppose $x_1^{(1)}, x_2^{(1)}, x_3^{(1)}, \ldots, x_n^{(1)}$ are first approximation to the unknowns $x_1, x_2, \ldots, x_n$. Substituting in the right side of (19), we find a system of second approximations:

$$x_1^{(2)} = \frac{b_1}{a_{11}} - \frac{a_{12}}{a_{11}}x_2^{(1)} - \ldots - \frac{a_{1n}}{a_{11}}x_n^{(1)}$$

$$x_2^{(2)} = \frac{b_2}{a_{22}} - \frac{a_{21}}{a_{22}}x_1^{(1)} - \ldots - \frac{a_{2n}}{a_{22}}x_n^{(1)}$$

$$\ldots \quad \ldots \quad \ldots \quad \ldots$$

$$\ldots \quad \ldots \quad \ldots \quad \ldots$$

$$x_n^{(2)} = \frac{b_n}{a_{nn}} - \frac{a_{n1}}{a_{nn}}x_1^{(1)} - \ldots - \frac{a_{n,n-1}}{a_{nn}}x_{n-1}^{(1)}.$$

In general, if $x_1^{(n)}, x_2^{(n)}, \ldots, x_n^{(n)}$ is system of nth approximations, then the next approximations, then the next approximation is given by the formula

$$x_1^{(n+1)} = \frac{b_1}{a_{11}} - \frac{a_{12}}{a_{11}}x_2^{(n)} - \ldots - \frac{a_{1n}}{a_{11}}x_n^{(n)}$$

$$x_2^{(n+1)} = \frac{b_2}{a_{22}} - \frac{a_{21}}{a_{22}}x_1^{(n)} - \ldots - \frac{a_{2n}}{a_{22}}x_n^{(n)}$$

$$\ldots \quad \ldots \quad \ldots \quad \ldots$$

$$x_n^{(n+1)} = \frac{b_n}{a_{nn}} - \frac{a_{n1}}{a_{nn}}x_1^{(n)} - \ldots - \frac{a_{n,n-1}}{a_{nn}}x_n^{(n)}.$$

This method, due to Jacobi, is called the *method of simultaneous displacements* or *Jacobi method.*

(b) Gauss–Seidel Method

A simple modification of Jacobi method yields faster convergence. Let $x_1^{(1)}, x_2^{(1)}, \ldots, x_n^{(1)}$ be the first approximation to the unknowns $x_1, x_2, \ldots, x_n$. Then the second approximations are given by

$$x_1^{(2)} = \frac{b_1}{a_{11}} - \frac{a_{12}}{a_{11}}x_2^{(1)} - \ldots - \frac{a_{1n}}{a_{11}}x_n^{(1)}$$

$$x_2^{(2)} = \frac{b_2}{a_{22}} - \frac{a_{21}}{a_{22}}x_1^{(2)} - \frac{a_{23}}{a_{22}}x_3^{(1)} - \ldots - \frac{a_{2n}}{a_{22}}x_n^{(1)}$$

$$x_3^{(2)} = \frac{b_3}{a_{33}} - \frac{a_{31}}{a_{33}}x_1^{(2)} - \frac{a_{32}}{a_{33}}x_2^{(2)} - \ldots - \frac{a_{3n}}{a_{33}}x_n^{(1)}$$

$$\ldots \quad \ldots \quad \ldots \quad \ldots$$

$$\ldots \quad \ldots \quad \ldots \quad \ldots$$

$$x_n^{(2)} = \frac{b_n}{a_{nn}} - \frac{a_{n1}}{a_{nn}} x_1^{(2)} - \frac{a_{n2}}{a_{nn}} x_2^{(2)} - \ldots - \frac{a_{n,n-1}}{a_{nn}} x_{n-1}^{(2)}.$$

The entire process is repeted till the values of x_1, $x_2, \ldots, x_n$ are obtained to the accuracy required. Thus this method uses an improved component as soon as available and so it is called *the method of successive displacements or Gauss–Seidel Method.*

It may be mentioned that Gauss–Seidel method converges twice as fast as the Jacobi method.

EXAMPLE 15.17

Starting with $(x_0, y_0, z_0) = (0, 0, 0)$ and using Jacobi method, find the next five iterations for the system:

$$5x - y + z = 10$$

$$2x + 8y - z = 11$$

$$-x + y + 4z = 3.$$

Solution. The given equations can be written in the form

$$x = \frac{y - z + 10}{5},$$

$$y = \frac{-2x + z + 11}{8} \text{ and}$$

$$z = \frac{x - y + 3}{4}.$$

Therefore, starting with $(x_0, y_0, z_0) = (0, 0, 0)$, we get

$$x_1 = \frac{y_0 - z_0 + 10}{5} = 2,$$

$$y_1 = \frac{-2x_0 + z_0 + 11}{8} = 1.375$$

$$z_1 = \frac{x_0 - y_0 + 3}{4} = 0.75.$$

The second iteration gives

$$x_2 = \frac{y_1 - z_1 + 10}{5} = \frac{1.375 - 0.75 + 10}{5} = 2.125,$$

$$y_2 = \frac{-2x_1 + z_1 + 11}{8} = \frac{-4 + 0.75 + 11}{8} = 0.96875,$$

$$z_2 = \frac{x_1 - y_1 + 3}{4} = \frac{2 - 1.375 + 3}{4} = 0.90625.$$

The third iteration gives

$$x_3 = \frac{y_2 - z_2 + 10}{5}$$

$$= \frac{0.96875 - 0.90625 + 10}{5} = 2.0125,$$

$$y_3 = \frac{-2x_2 + z_2 + 11}{8}$$

$$= \frac{-4.250 + 0.90625 + 11}{8} = 0.95703125,$$

$$z_3 = \frac{x_2 - y_2 + 3}{4}$$

$$= \frac{2.125 - 0.96875 + 3}{4} = 1.0390625.$$

The fourth iteration yields

$$x_4 = \frac{y_3 - z_3 + 10}{5}$$

$$= \frac{0.95703125 - 1.0390625 + 10}{5} = 1.98359375,$$

$$y_4 = \frac{-2x_3 + z_3 + 11}{8}$$

$$= \frac{-4.0250 + 1.0390625 + 11}{4} = 0.8767578,$$

$$z_4 = \frac{x_3 - y_3 + 3}{4}$$

$$= \frac{2.0125 - 0.95703125 + 3}{4} = 1.0138672,$$

whereas the fifth iteration gives

$$x_5 = \frac{y_4 - z_4 + 10}{5} = 1.9725781,$$

$$y_5 = \frac{-2x_4 + z_4 + 11}{8}$$

$$= \frac{-3.9671875 + 1.0138672 + 11}{8} = 1.005834963,$$

$$z_5 = \frac{x_4 - y_4 + 3}{4}$$

$$= \frac{1.98359375 - 0.8767578 + 3}{4} = 1.027670898.$$

We find that the iterations converge to (2, 1, 1).

EXAMPLE 15.18

Using Gauss–Seidel iteration and the first iteration as (0, 0, 0), calculate the next three iterations for the solution of the system of equations given in Example 15.17.

Solution. The first iteration is (0, 0,0). The next iteration is

$$x_1 = \frac{y_0 - z_0 + 10}{5} = 2,$$

$$y_1 = \frac{-2x_1 + z_0 + 11}{8} = \frac{-4 + 0 + 11}{8} = 0.875,$$

$$z_1 = \frac{x_1 - y_1 + 3}{4} = \frac{2 - 0.875 + 3}{4} = 1.03125.$$

Then

$$x_2 = \frac{y_1 - z_1 + 10}{5}$$

$$= \frac{0.875 - 1.03125 + 10}{5} = 1.96875,$$

$$y_2 = \frac{-2x_2 + z_1 + 11}{8}$$

$$= \frac{-3.9375 + 1.03125 + 11}{8} = 1.01171875,$$

$$z_2 = \frac{x_2 - y_2 + 3}{4}$$

$$= \frac{1.96875 - 1.01171875 + 3}{4} = 0.989257812.$$

Further,

$$x_3 = \frac{y_2 - z_2 + 10}{5}$$

$$= \frac{1.01171875 - 0.989257812 + 10}{5}$$

$$= 2.004492188,$$

$$y_3 = \frac{-2x_3 + z_2 + 11}{8}$$

$$= \frac{-4.008984376 + 0.989257812 + 11}{8}$$

$$= 0.997534179,$$

$$z_3 = \frac{x_3 - y_3 + 3}{4}$$

$$= \frac{2.004492188 - 0.997534179 + 3}{4}$$

$$= 1.001739502$$

We find that the iterations converge to (2,1,1).

Remark 15.3. If follows from Examples 15.17 and 15.18 that *Gauss–seidel method converges rapidly in comparison to Jacobi's method.*

EXAMPLE 15.19

Solve

$$54x + y + z = 110$$

$$2x + 15y + 6z = 72$$

$$-x + 6y + 27z = 85$$

by Gauss–Seidel method.

Solution. From the given equations, we have

$$x = \frac{110 - y - z}{54}$$

$$y = \frac{72 - 2x - 6z}{15}$$

$$z = \frac{85 + x - 6y}{27}.$$

We take the initial approximation as $x_0 = y_0 = z_0 = 0$. Then the first approximation is given by

$$x_1 = \frac{110}{54} = 2.3070,$$

$$y_1 = \frac{72 - 2x_1 - 6z_0}{15} = 4.5284,$$

$$z_1 = \frac{85 + x_1 - 6y_1}{27} = 2.2173.$$

The second approximation is given by

$$x_2 = \frac{110 - y_1 - z_1}{54} = 1.9122,$$

$$y_2 = \frac{72 - 2x_2 - 6z_1}{15} = 3.6581,$$

$$z_2 = \frac{85 + x_2 - 6y_2}{27} = 2.4061.$$

The third approximation is

$$x_3 = \frac{110 - y_2 - z_2}{54} = 1.9247,$$

$$y_3 = \frac{72 - 2x_3 - 6z_2}{15} = 3.5809,$$

$$z_3 = \frac{85 + x_3 - 6y_3}{27} = 2.4237.$$

The fourth approximation is

$$x_4 = \frac{110 - y_3 - z_3}{54} = 1.9258,$$

$$y_4 = \frac{72 - 2x_4 - 6z_3}{15} = 3.5738,$$

$$z_4 = \frac{85 + x_4 - 6y_4}{27} = 2.4253.$$

The fifth approximation is

$$x_5 = \frac{110 - y_4 - z_4}{54} = 1.9259,$$

$$y_5 = \frac{72 - 2x_5 - 6z_4}{15} = 3.5732,$$

$$z_5 = \frac{85 + x_5 - 6y_5}{27} = 2.4254.$$

Thus the required solution, correct to three decimal places, is

$$x = 1.916.\ y = 3.573,\ z = 2.425.$$

EXAMPLE 15.20

Solve

$$28x + 4y - z = 32$$

$$2x + 17y + 4z = 35$$

$$x + 3y + 10z = 24$$

By Gauss–Seidel method.

Solution. From the given equations, we have

$$x = \frac{32 - 4y + z}{28}$$

$$y = \frac{35 - 2x - 4z}{17}$$

$$z = \frac{24 - x - 3y}{10}.$$

Taking first approximation as $x_0 = y_0 = z_0 = 0$, we have the next approximations as

$x_1 = 1.1428571,\ y_1 = 1.9243697,\ z_1 = 1.7084034$

$x_2 = 0.9289615,\ y_2 = 1.5475567,\ z_2 = 1.8428368$

$x_3 = 0.9875932,\ y_3 = 1.5090274,\ z_3 = 1.8485325$

$x_4 = 0.9933008,\ y_4 = 1.5070158,\ z_4 = 1.8485652$

$x_5 = 0.9935893,\ y_5 = 1.5069741,\ z_5 = 1.8485488$

$x_6 = 0.9935947,\ y_6 = 1.5069774,\ z_6 = 1.8485473.$

Hence the solution, correct to four decimal places, is

$$x = 0.9935,\ y = 1.5069,\ z = 1.8485.$$

15.7 FINITE DIFFERENCES

Suppose that a function $y = f(x)$ is tabulated for the equally spaced arguments x_0, $x_0 + h$, $x_0 + 2h, \ldots, x_0 + nh$ giving the functional values y_0, $y_1, y_2, \ldots, y_n$. The constant difference between two consecutive values of x is called the *interval of differencing* and is denoted by h.

The operator Δ defined by

$$\Delta y_0 = y_1 - y_0,$$

$$\Delta y_1 = y_2 - y_1,$$

$$\ldots\ldots\ldots\ldots$$

$$\ldots\ldots\ldots\ldots$$

$$\Delta y_{n-1} = y_n - y_{n-1},$$

is called the *Newton–forward difference operator.* We note that the first difference $\Delta y_n = y_{n+1} - y_n$ is itself a function of x. consequently, we can repeat the operation of differencing to obtain

$$\Delta^2 y_0 = \Delta(\Delta y_0) = \Delta(y_1 - y_0) = \Delta y_1 - \Delta y_0$$

$$= y_2 - y_1 - (y_1 - y_0) = y_2 - 2y_1 + y_0,$$

which is called the *second forward difference.* In general, the nth difference of f is defined by

$$\Delta^n y_r = \Delta^{n-1} y_{r+1} - \Delta^{n-1} y_r.$$

For example, let

$$f(x) = x^3 - 3x^2 + 5x + 7.$$

Taking the arguments as 0, 2, 4, 6, 8, 10, we have $h = 2$ and

$$\Delta f(x) = (x+2)^3 - 3(x+2)^2 + 5(x+2) + 7$$

$$- (x^3 - 3x^2 + 5x + 7) = 6x^2 + 6,$$

$$\Delta^2 f(x) = \Delta(\Delta f(x)) = \Delta(6x^2 + 6) = 6(x+2)^2 + 6$$

$$- (6x^2 + 6) = 24x + 24,$$

$$\Delta^3 f(x) = 24(x+2) + 24 - (24x + 24) = 48,$$

$$\Delta^4 f(x) = \Delta^5 f(x) = \ldots = 0.$$

In tabular form, we have

Difference Table

x	$f(x)$	$\Delta f(x)$	$\Delta^2 f(x)$	$\Delta^3 f(x)$	$\Delta^4 f(x)$	$\Delta^5 f(x)$
0	7					
		6				
2	13		24			
		30		48		
4	43		72		0	
		102		48		0
6	145		120		0	
		222		48		
8	367		168			
		390				
10	757					

Theorem 15.11. If $f(x)$ is a polynomial of degree n, that is,

$$f(x) = \sum_{i=0}^{n} a_i x^i,$$

Then $\Delta^n f(x)$ is constant and is equal to $n!\, a_n h^n$

Proof: We shall prove the theorem by induction on n. if $n = 1$, then $f(x) = a_1 x + a_0$ and $\Delta f(x) = f(x + h) - f(x) = a_1 h$ and so the theorem holds for $n = 1$. Assume now that the result is true for all degrees $1, 2, \ldots, n - 1$. Consider

$$f(x) = \sum_{i=0}^{n} a_i x^i.$$

Then by the linearity of the operator Δ, we have

$$\Delta^n f(x) = \sum_{i=0}^{n} a_i \Delta^n x^i.$$

For $i < n$, $\Delta^n x^i$ is the nth difference of a polynomial of degree less than n and hence must vanish, by induction hypothesis. Thus

$$\Delta^n f(x) = a_n \Delta^n x^n = a_n \Delta^{n-1} (\Delta x^n)$$

$$= a_n \Delta^{n-1} [(x+h)^n - x^n]$$

$$= a_n \Delta^{n-1} [nhx^{n-1} + g(x)],$$

where $g(x)$ is a polynomial of degree less than $n-1$. Hence, by induction hypothesis,

$$\Delta^n f(x) = a^n \Delta^{n-1} (nhx^{n-1}) = a^n (hn)(n-1)!\ H^{n-1}$$

$$= a_n n!\ h^n.$$

Hence, by induction, the theorem holds.

Let $y_0, y_1, \ldots, y_n$ be the functional values of a function f for the arguments $x_0, x_0+h, x_0+2h, \ldots, x_0+nh$. Then the operator ∇ defined by

$$\nabla y_r = y_r - y_{r-1}$$

is called the *Newton-backward difference operator*. The higher oder backward differences are:

$$\nabla^2 y_r = \nabla y_r - \nabla y_{r-1}$$

$$\nabla^3 y_r = \nabla^2 y_r - \nabla^2 y_{r-1}$$

....

$$\nabla^n y_r = \nabla^{n-1} y_r - \nabla^{n-1} y_{r-1}.$$

Thus the backward difference table becomes:

x	y	1st Diff.	2nd Diff.	3rd Diff.
x_0	y_0			
		∇y_1		
x_1	y_1		$\nabla^2 y_2$	
		∇y_2		$\nabla^3 y_3$
x_2	y_2		$\nabla^2 y_3$	
		∇y_3		
x_3	y_3			

EXAMPLE 15.21

Form the table of backward differences for the function

$$f(x) = x^3 - 3x^2 + 5x - 7$$

for $x = -1, 0, 1, 2, 3, 4,$ and 5.

Solution.

x	y	1st Diff.	2nd Diff.	3rd Diff.	4th Diff.
–1	–16				
		9			
0	–7		–6		
		3		6	
1	–4		0		0
		3		6	
2	–1		6		0
		9		6	
3	8		12		0
		21		6	
4	29		18		
		39			
5	68				

An operator E, known as *enlargement operator, displacement operator,* or *shifting operator,* is defined by

$$Ey_r = y_{r+1}.$$

Thus, shifting operator moves the functional value $f(x)$ to the next higher value $f(x+h)$. Further,

$$E^2 y_r = E(Ey_r) = E(y_{r+1}) = y_{r+2}$$

$$E^3 y_r = E(E^2 y_r) = E(y_{r+2}) = y_{r+3}$$

....

....

$$E^n y_r = y_{r+n}.$$

Relations between Δ, ∇ and E

We know that

$$\Delta y_r = y_{r+1} - y_r = Ey_r - y_r = (E - I)y_r,$$

where I is the identity operator. Hence

$$\Delta = E - I \text{ or } E = I + \Delta. \tag{20}$$

Also, by definition,

$$\nabla y_r = y_r - y_{r-1} = y_r - E^{-1} y_r = y_r (I - E^{-1}),$$

and so

$$\nabla = I - E^{-1} \text{ or } E^{-1} = I - \nabla$$

or

$$E = \frac{I}{I - \nabla}. \tag{21}$$

From (20) and (21), we have

$$I + \Delta = \frac{I}{I - \nabla} \tag{22}$$

or

$$\Delta = \frac{I}{I - \nabla} - I = \frac{\nabla}{I - \nabla}. \tag{23}$$

From (22) and (23)

$$\nabla = I - \frac{I}{I + \Delta} = \frac{\Delta}{1 + \Delta}. \tag{24}$$

Theorem 15.12.

$$f_{x+nh} = \sum_{k=0}^{\infty} \binom{n}{k} \Delta^k f_x$$

Proof: We shall prove our result by mathematical induction. For $n = 1$, the theorem reduces to $f_{x+h} = f_x + \Delta f_x$ which is true. Assume now that the theorem is true for $n - 1$. Then

$$f_{x+nh} = E^n f_x = E(E^{n-1} f_x)$$

$$= E \sum_{i=0}^{\infty} \binom{n-1}{i} \Delta^i f_x \text{ by induction hypothesis.}$$

But $E = I + \Delta$. So

$$E^n f_x = (I + \Delta) E^{n-1} f_x = E^{n-1} f_x + \Delta E^{n-1} f_x$$

$$= \sum_{i=0}^{\infty} \binom{n-1}{i} \Delta^i f_x + \sum_{i=0}^{\infty} \binom{n-1}{i} \Delta^{i+1} f_x$$

$$= \sum_{i=0}^{\infty} \binom{n-1}{i} \Delta^i f_x + \sum_{j=1}^{\infty} \binom{n-1}{j-1} \Delta^j f_x.$$

The coefficient of $\Delta^k f_x$ ($k = 0, 1, 2, \ldots, n$) is given by

$$\binom{n-1}{k} + \binom{n-1}{k-1} = \binom{n}{k}.$$

Hence

$$f_{x+nh} = E^n f_x = \sum_{k=0}^{\infty} \binom{n}{k} \Delta^k f_x,$$

which completes the proof of the theorem. As a special case of this theorem, we get

$$f_x = E^x f_0 = \sum_{k=0}^{\infty} \binom{x}{k} \Delta^k f_0,$$

which is known as *Newton's advancing difference formula* and expresses the general functional value f_x in term of f_0 and its differences.

Let h be the interval of differencing. Then the operator δ defined by

$$\delta f_x = f_{x+\frac{h}{2}} - f_{x-\frac{h}{2}}$$

is called the *central difference operator.*

We note that

$$\delta f_x = f_{x+\frac{h}{2}} - f_{x-\frac{h}{2}} = E^{\frac{1}{2}} f^x - E^{-\frac{1}{2}} f_x$$

$$= (E^{\frac{1}{2}} - E^{-\frac{1}{2}}) f_x.$$

Hence

$$\delta = E^{\frac{1}{2}} - E^{-\frac{1}{2}}. \tag{25}$$

Multiplying both sides by $E^{\frac{1}{2}}$, we get

$$E - \delta E^{\frac{1}{2}} - I = 0 \text{ or } \left(E^{\frac{1}{2}} - \frac{\delta}{2}\right)^2 - \frac{\delta^2}{4} - I = 0$$

or

$$E^{\frac{1}{2}} - \frac{\delta}{2} = \sqrt{I + \frac{\delta^2}{4}} \text{ or } E^{\frac{1}{2}} = \frac{\delta}{2} + \sqrt{I + \frac{\delta^2}{4}}$$

or

$$E = \frac{\delta^2}{4} + I + \frac{\delta^2}{4} + \delta\left(1 + \frac{\delta^2}{4}\right)^{\frac{1}{2}}$$

$$= I + \frac{\delta^2}{2} + \delta\sqrt{I + \frac{\delta^2}{4}}. \tag{26}$$

Also, using (26), we note that

$$\Delta = E - I = \frac{\delta^2}{2} + \delta\sqrt{I + \frac{\delta^2}{4}}, \tag{27}$$

$$\nabla = I - \frac{I}{E} = I - \left(I + \frac{\delta^2}{2} + \delta\sqrt{I + \frac{\delta^2}{4}}\right)^{-1} \tag{28}$$

$$= -\frac{\delta^2}{2} + \delta\sqrt{I + \frac{\delta^2}{4}}.$$

Conversely,

$$\delta = E^{\frac{1}{2}} - E^{-\frac{1}{2}} = (I + \Delta)^{\frac{1}{2}} - \frac{I}{(I + \Delta)^{1/2}} \tag{29}$$

$$= \frac{I + \Delta - I}{\sqrt{I + \Delta}} = \frac{\Delta}{\sqrt{I + \Delta}}$$

and

$$\delta = E^{\frac{1}{2}} - E^{-\frac{1}{2}} = \frac{I}{\sqrt{1 - \nabla}} - \sqrt{1 - \nabla}$$

$$= \frac{\nabla}{\sqrt{I - \nabla}} \tag{30}$$

Let h be the interval of differencing. Then the operator μ defined by

$$\mu f_x = \frac{1}{2}\left[f_{x+\frac{h}{2}} + f_{x-\frac{h}{2}}\right]$$

is called the *mean value operator or averaging operator*. We have

$$\mu f_x = \frac{1}{2}\left[f_{x+\frac{h}{2}} + f_{x-\frac{h}{2}}\right] = \frac{1}{2}\left[E^{\frac{1}{2}} f_x + E^{-\frac{1}{2}} f_x\right].$$

Hence

$$\mu = \frac{1}{2}\left[E^{\frac{1}{2}} + E^{-\frac{1}{2}}\right] \tag{31}$$

or

$$2\mu = E^{\frac{1}{2}} + E^{-\frac{1}{2}}. \tag{32}$$

Also, we know that

$$\delta = E^{\frac{1}{2}} - E^{-\frac{1}{2}}. \tag{33}$$

Adding (32) and (33), we get

$$2\mu + \delta = 2E^{\frac{1}{2}} \text{ or } E^{\frac{1}{2}} = \mu + \frac{\delta}{2}.$$

Also,

$$E^{\frac{1}{2}} = \frac{\delta}{2} + \sqrt{I + \frac{\delta^2}{4}}.$$

Hence

$$\mu + \frac{\delta}{2} = \frac{\delta}{2} + \sqrt{I + \frac{\delta^2}{4}} \text{ or } \mu = \sqrt{I + \frac{\delta^2}{4}} \tag{35}$$

The relation (35) yields

$$I + \frac{\delta^2}{4} = \mu^2 \text{ or } \delta = 2\sqrt{\mu^2 - I}. \tag{36}$$

Multiplying (32) throughout by $E^{\frac{1}{2}}$, we get

$$E + I = 2\mu E^{\frac{1}{2}} \text{ or } E - 2\mu E^{\frac{1}{2}} + I = 0$$

or

$$(E^{\frac{1}{2}} - \mu)^2 - \mu^2 + I = 0 \text{ or } E^{\frac{1}{2}} - \mu = \sqrt{\mu^2 - I}$$

or

$$E^{\frac{1}{2}} = \mu + \sqrt{\mu^2 - I}$$

or

$$E = 2\mu^2 - I + 2\mu\sqrt{\mu^2 - I}. \tag{37}$$

Then

$$\Delta = E - I = 2\mu^2 - 2I + 2\mu\sqrt{\mu^2 - I} \tag{38}$$

and

$$\nabla = I - \frac{I}{E} = I - (2\mu^2 + 2\mu\sqrt{\mu^2 - I} - I)^{-1}$$

$$= \frac{2\mu(\mu + \sqrt{\mu^2 - 1}) - 2I}{2\mu^2 + 2\mu\sqrt{\mu^2 - I} - I}. \tag{39}$$

The *differential operator* D *is defined by*

$$Df(x) = f'(x).$$

By Taylor's theorem, we have

$$f(x+h) = f(x) + hf'(x) + \frac{h^2}{2!}f''(x) + \ldots$$

$$= f(x) + hDf(x) + \frac{h^2}{2!}D^2f(x) + \ldots$$

$$= (1 + hD + \frac{h^2}{2!}D^2 + \ldots)f(x)$$

and so

$$Ef(x) = f(x+h) = (1 + h\,D + \frac{h^2}{2!}D^2 + \ldots)f(x).$$

Hence

$$E = 1 + hD + \frac{h^2}{2!}D^2 + \ldots = e^{hD}$$

$$= e^U, \text{ where } U = hD. \quad (40)$$

Then

$$\Delta = E - I = e^U - I \text{ and } \nabla = I - e^{-U}.$$

We note that

$$\delta = E^{\frac{1}{2}} - E^{-\frac{1}{2}} = e^{\frac{U}{2}} - e^{-\frac{U}{2}} = 2\sinh\frac{U}{2} \quad (41)$$

$$\mu = \frac{1}{2}(E^{\frac{1}{2}} + E^{-\frac{1}{2}}) = \frac{1}{2}(e^{\frac{U}{2}} + e^{-\frac{U}{2}}) = \cosh\frac{U}{2} \quad (42)$$

Conversely

$$e^{\frac{U}{2}} + e^{-\frac{U}{2}} = 2\mu \text{ or}$$

$$e^U + 1 = 2\mu e^{\frac{U}{2}} \text{ (quadratic in } e^{U/2}).$$

Thus

$$e^{\frac{U}{2}} = \mu + \sqrt{\mu^2 - I} \text{ or}$$

$$U = \log(2\mu^2 + I + 2\mu\sqrt{\mu^2 - I}). \quad (43)$$

Since, by (41), $\delta = 2\sinh\frac{U}{2}$, it follows that

$$U = 2\sinh^{-1}\frac{\delta}{2}. \quad (44)$$

From the above discussion, we obtain the following table for the relations among the finite difference operators.

	Δ	∇	δ	E	$U = hD$
Δ	Δ	$(I-\nabla)^{-1} - I$	$\frac{\delta^2}{2} + \delta\sqrt{I + \frac{\delta^2}{4}}$	$E - I$	$e^U - I$
∇	$I - \frac{I}{\Delta + I}$	∇	$-\frac{\delta^2}{2} + \delta\sqrt{I + \frac{\delta^2}{4}}$	$I - \frac{1}{E}$	$I - e^{-U}$
δ	$\frac{\Delta}{\sqrt{I+\Delta}}$	$\frac{\nabla}{\sqrt{I-\nabla}}$	δ	$E^{\frac{1}{2}} - E^{-\frac{1}{2}}$	$2\sinh\frac{U}{2}$
E	$I + \Delta$	$\frac{I}{I - \nabla}$	$I + \frac{\delta^2}{2} + \delta\sqrt{I + \frac{\delta^2}{4}}$	E	e^U
$U = hD$	$\text{Log}(I+\Delta)$	$\log\frac{I}{I-\nabla}$	$2\sinh^{-1}\frac{\delta}{2}$	log E	U

EXAMPLE 15.22

Find the cubic polynomial $f(x)$ which takes on the values $f_0 = -5, f_1 = 1, f_2 = 9, f_3 = 25, f_4 = 55, f_5 = 105$.

Solution. The difference table for the given function is given below:

x	$f(x)$	$\Delta f(x)$	$\Delta^2 f(x)$	$\Delta^3 f(x)$	$\Delta^4 f(x)$
0	–5				
		6			
1	1		2		
		8		6	
2	9		8		0
		16		6	
3	25		14		0
		30		6	
4	55		20		
		50			
5	105				

Now

$$f_x = E^x f_0 = (I + \Delta)^x f_0$$

$$= \left[1 + x\Delta + \frac{x(x-1)}{2!}\Delta^2 + \frac{x(x-1)(x-2)}{3!}\Delta^3\right]f_0$$

$$= f_0 + x\Delta f_0 + \frac{x^2 - x}{2}\Delta^2 f_0 + \frac{x^3 - 3x^2 + 2x}{6}\Delta^3 f_0$$

$$= -5 + 6x + \frac{x^2 - x}{2}(2) + \frac{x^3 - 3x^2 + 2x}{6}(6)$$

$$= x^3 - 2x^2 + 7x - 5,$$

which is the required cubic polynomial.

15.8 ERROR PROPAGATION

Let $y_0, y_1, y_2, y_3, y_4, y_5, y_6, y_7, y_8$ be the values of the function f at the arguments $x_0, x_1, x_2, x_3, x_4, x_5, x_6, x_7, x_8$, respectively. Suppose an error E is committed in y_4 during tabulation. To study the error propagation, we use the difference table. For the sake of convenience, we construct difference table up to fourth difference only. If the error in y_4 is ε, then the value of the function f at x_4 is $y_4 + \varepsilon$. the difference table of the data is as shown below. We note that

(i) Error propagates in a triangular pattern (shown by fan lines) and grows quickly with the order of difference.

(ii) The coefficients of the error ε in any column are the binomial coefficients of $(1 - \varepsilon)^n$ with the alternating signs. Thus errors in the third column are ε, −3ε, 3ε, −ε

(iii) The algebraic sum of the errors in any difference column is zero.

x	y	Δy	$\Delta^2 y$	$\Delta^3 y$	$\Delta^4 y$
x_0	y_0				
		Δy_0			
x_1	y_1		$\Delta^2 y_0$		
		Δy_1		$\Delta^3 y_0$	
x_2	y_2		$\Delta^2 y_1$		$\Delta^4 y_0+\varepsilon$
		Δy_2		$\Delta^3 y_1+\varepsilon$	
x_3	y_3		$\Delta^2 y_2+\varepsilon$		$\Delta^4 y_1-4\varepsilon$
		$\Delta y_3+\varepsilon$		$\Delta^3 y_2-3\varepsilon$	
x_4	$y_4+\varepsilon$		$\Delta^2 y_3-2\varepsilon$		$\Delta^4 y_2+6\varepsilon$
		$\Delta y_4-\varepsilon$		$\Delta^3 y_3+3\varepsilon$	
x_5	y_5		$\Delta^2 y_4+\varepsilon$		$\Delta^4 y_3-4\varepsilon$
		Δy_5		$\Delta^3 y_4-\varepsilon$	
x_6	y_6		$\Delta^2 y_5$		$\Delta^4 y_4+\varepsilon$
		Δy_6		$\Delta^3 y_5$	
x_7	y_7		$\Delta^2 y_6$		
		Δy_7			
x_8	y_8				

(iv) If the difference table has even differences, then the maximum error lies on the same horizontal line on which the tabular value in error lies.

EXAMPLE 15.23

One entry in the following table of a polynomial of degree 4 is incorrect. Correct the entry by locating it.

x	1.0	1.1	1.2	1.3	1.4
y	1.0000	1.5191	2.0736	2.6611	3.2816

x	1.5	1.6	1.7	1.8	1.9	2.0
y	3.9375	4.6363	5.3771	6.1776	7.0471	8.0

Solution. The difference table for the given data is shown below. Since the degree of the polynomial is four, the fourth difference must be constant. But we note that the fourth differences are oscillating for the larger values of x. The largest numerical fourth difference 0.0186 is at $x = 1.6$. This suggest that the error in the value of f is at $x = 1.6$. Draw the fan lines as shown in the difference table.

x	y	Δy	$\Delta^2 y$	$\Delta^3 y$	$\Delta^4 y$
1.0	1.0000				
		0.5191			
1.1	1.5191		0.0354		
		0.5545		−0.0024	
1.2	2.0736		0.0330		0.0024
		0.5875		0	
1.3	2.6611		0.0330		0.0024
		0.6205		0.0024	
1.4	3.2816		0.0354		0.0051 Fan line
		0.6559		0.0075	
1.5	3.9375		0.0429		−0.0084
		0.6988		−0.0009	
1.6	4.6363		0.0420		0.0186
		0.7408		0.0177	
1.7	5.3771		0.0597		−0.0084
		0.8005		0.0093	
1.8	6.1776		0.0690		0.0051
		0.8695		0.0144	
1.9	7.0471		0.0834		
		0.9529			
2.0	8.0000				

Then taking 1.6 as x_0, we have

$$\Delta^4 f_{-4} + \varepsilon = 0.0051,$$

$$\Delta^4 f_{-3} - 4\varepsilon = -0.0084,$$

$$\Delta^4 f_{-2} + 6\varepsilon = 0.0186,$$

$$\Delta^4 f_{-1} - 4\varepsilon = -0.0084, \Delta^4 f_0 + \varepsilon = 0.0051.$$

We want all fourth differences to be alike. Eliminating $\Delta^4 f$ between any two of the compatible equations and solving for ε will serve our purpose. For example, subtracting the second equation from the first, we get

$$5\varepsilon = 0.0135 \text{ and so } \varepsilon = 0.0027.$$

Putting this value of E in the above equations, we note that all the fourth differences become 24. Further,

$$f(1.6) + \varepsilon = 4.6363$$

which yields

$$f(1.6) = 4.6363 - \varepsilon = 4.6363 - 0.0027 = 4.6336.$$

Thus the error was a transposing error, that is, writing 63 instead of 36 while tabulation.

EXAMPLE 15.24

Find and correct the error, by means of differences, in the given data:

x	0	1	2	3	4	5	6	7	8	9	10
y	2	5	8	17	38	75	140	233	362	533	752

Solution. The difference table for the given data is shown below. The largest numerical fourth difference of −12 is at $x = 5$. So there is some error in the value $f(5)$. The fan lines are drawn and we note from the table that

$$\Delta^4 f_{-4} + \varepsilon = -2, \Delta^4 f_{-3} - 4\varepsilon = 8,$$

$$\Delta^4 f_{-2} + 6\varepsilon = -12, \Delta^4 f_{-1} - 4\varepsilon = 8,$$

$$\Delta^4 f_0 + \varepsilon = -2,$$

and

$$\Delta^3 f_{-3} + \varepsilon = 4, \Delta^3 f_{-2} - 3e = 12,$$

$$\Delta^3 f_{-1} + 3\varepsilon = 0, \Delta^3 f_0 - \varepsilon = 8.$$

Subtracting second equation from the first (for both sets shown above), we get $5\varepsilon = -10$ (for the first set) and $4\varepsilon = -8$ (for the second set). Hence $\varepsilon = -2$.

Difference Table

x	y	Δ	Δ^2	Δ^3	Δ^4
0	2				
		3			
1	5		0		
		3		6	
2	8		6		0
		9		6	
3	17		12		−2 Fan line
		21		4	
4	38		16		8
		37		12	
5	75		28		−12
		65		0	
6	140		28		8
		93		8	
7	233		36		−2
		129		6	
8	362		42		0
		171		6	
9	533		48		
		219			
10	752				

We now have

$$f(5) + \varepsilon = 75 \text{ and so } f(5) = 75 - \varepsilon = 75 - (-2) = 77.$$

Therefore, the true value of $f(5)$ is 77.

15.9 INTERPOLATION

Interpolation is the process of finding the value of a function for any value of argument (independent variable) within an interval for which some values are given.

Thus, *interpolation is the art of reading between the lines in a given table.*

Extrapolation is the process of finding the value of a function outside an interval for which some values are given.

We now discuss interpolation processes for equal spacing.

(a) Newton's Forward Difference Formula

Let $\ldots, f_{-2}, f_{-1}, f_{-0}, f_1, f_2, \ldots$ be the values of a function for $\ldots, x_0 - 2h, x_0 - h, x_0, x_0 + h, x_0 + 2h, \ldots$ Suppose that want to compute the functional value f_p for $x = x_0 + ph$, where in general $-1 < p < 1$. We have

$$f_p = f(x_0 + ph) \text{ and } p = \frac{x - x_0}{h},$$

where h is the interval of differencing. Then using shift operator and binomial theorem, we have

$$f_x = E^p f_0 = (I + \Delta)^p f_0$$

$$= \left[I + p\Delta + \frac{p(p-1)}{2!}\Delta^2 + \frac{(p(p-1)(p-2)}{3!}\Delta^3 + \ldots\right] f_0$$

$$= f_0 + \binom{p}{1}\Delta f_0 + \binom{p}{2}\Delta^2 f_0 + \binom{p}{3}\Delta^3 f_0 + \cdots \quad (45)$$

The expression (45) is called *Newton's forward difference formula for interpolation.*

(b) Newton's Backward Difference Formula

Let $\ldots, f_{-2}, f_{-1}, f_0, f_1, f_2, \ldots$ be the values of a function for $\ldots, x_0 - 2h, x_0 - h, x_0, x_0 + h, x_0 + 2h, \ldots$ Suppose that we want to compute the functional value f_p for $x = x_0 + ph$, $-1 < p < 1$. We have

$$f_p = f(x_0 + ph),\ p = \frac{x - x_0}{h}.$$

Using Newton's backward differences, we have

$$f_x = E^p f_0 = (I - \nabla)^{-p} f_0$$

$$= \left[I + p\nabla + \frac{p(p+1)}{2!}\nabla^2 + \frac{p(p+1)(p+2)}{3!}\nabla^3 + \ldots\right] f_0$$

$$= f_0 + p\nabla f_0 + \frac{p(p+1)}{2!}\nabla^2 f_0 + \frac{p(p+1)(p+2)}{3!}\nabla^3 f_0 + \ldots,$$

which is known as *Newton's backward difference formula for interpolation.*

Remark 15.4. It is clear from the differences used that

(i) Newton's forward difference formula is used for interpolating the values of the function near the beginning of a set of tabulated values.

(ii) Newton's backward difference formula is used for interpolating the values of the function near the end of a set of tabulated values.

EXAMPLE 15.25

Calculate the approximate value of sin x for $x = 0.54$ and $x = 1.36$ using the following table:

x:	0.5	0.7	0.9	1.1	1.3	1.5
sin:	0.47943	0.64422	0.78333	0.89121	0.96356	0.99749

Solution. The difference table for the given data is

x	sin x	1st diff.	2nd diff.	3rd diff.	4th diff.	5th diff.
0.5	0.47943					
		0.16479				
0.7	0.64422		−0.02568			
		0.13911		−0.00555		
0.9	0.78333		−0.03123		0.00125	
		0.10788		−0.00430		0.00016
1.1	0.89121		−0.03553		0.00141	
		0.07235		−0.00289		
1.3	0.96356		−0.03842			
		0.03393				
1.5	0.99749					

We take

$$x_0 = 0.50,\ x_p = 0.54 \text{ and } p = \frac{0.54 - 0.50}{0.2} = 0.2.$$

Using Newton's forward-difference method, we have

$$f_p = f_0 + p\Delta f_0 + \frac{p(p-1)}{2!}\Delta^2 f_0 + \frac{p(p-1)(p-2)}{3!}\Delta^3 f_0$$
$$+\frac{p(p-1)(p-2)(p-3)}{4!}\Delta^4 f_0$$
$$+\frac{p(p-1)(p-2)(p-3)(p-4)}{5!}\Delta^5 f_0$$
$$=0.47943+0.2(0.16479)+\frac{0.2(0.2-1)}{2}(-0.0268)$$
$$+\frac{0.2(0.2-1)(0.2-2)}{6}(-0.00555)$$
$$+\frac{0.2(0.2-1)(0.2-2)(0.2-3)}{4!}(0.00125)$$
$$+\frac{0.2(0.2-1)(0.2-2)(0.2-3)(0.2-4)}{5!}(0.00016)$$
$$\approx 0.51386.$$

Further, the point $x = 1.36$ lies towards the end of the tabulated values. Therefore, to find the value of the function at $x = 1.36$, we use Newton's backward differences method. We have

$$x_p = 1.36, x_0 = 1.3 \text{ and } p = \frac{1.36-1.30}{0.2} = 0.3,$$

$$f_p = f_0 + p\nabla f_0 + \frac{p(p+1)}{2!}\nabla^2 f_0$$
$$+\frac{p(p+1)(p+2)}{3!}\nabla^3 f_0 + \frac{p(p+1)(p+2)(p+3)}{4!}\nabla^4 f_0$$
$$=0.96356+0.3(0.07235)$$
$$+\frac{0.3(0.3+1)}{2}(-0.03553)$$
$$+\frac{0.3(0.3+1)(0.3+2)}{6}(-0.00430)$$
$$+\frac{0.3(0.3+1)(0.3+2)(0.3+3)}{24}(0.00125)$$
$$=0.96356+0.021705-0.006128-0.000642$$
$$+0.000154$$
$$\approx 0.977849.$$

EXAMPLE 15.26

Find the cubic polynomial $f(x)$ which takes on the values $f(0) = -4, f(1) = -1, f(2) = 2, f(3) = 11, f(4) = 32, f(5) = 71$. Find $f(6)$ and $f(2.5)$.

Solution. The difference table for the given data is

x	$f(x)$	$\Delta f(x)$	$\Delta^2 f(x)$	$\Delta^3 f(x)$
0	−4			
		3		
1	−1		0	
		3		6
2	2		6	
		9		6
3	11		12	
		21		6
4	32		18	
		39		
5	71			

Using Newton's forward difference formula, we have

$$f_x = f_0 + x\Delta f_0 + \frac{x(x-1)}{2!}\Delta^2 f_0 + \frac{x(x-1)(x-2)}{3!}\Delta^3 f_0$$

$$= -4 + x(3) + \frac{x^2-x}{2}(0) + \frac{x^3-3x^2+2x}{6}(6)$$

$$= x^3 - 3x^2 + 2x + 3x - 4$$

$$= x^3 - 3x^2 + 5x - 4,$$

which is the required cubic polynomial. Therefore,

$$f(6) = 6^3 - 3(6^2) + 5(6) - 4$$
$$= 216 - 108 + 30 - 4 = 134.$$

On the other hand, if we calculate $f(6)$ using Newton's forward difference formula, then take $x_0 = 0, p = \frac{x-x_0}{h} = \frac{6-0}{1} = 6$ and have

$$f(6) = f_6 = f_0 + 6\Delta f_0 + \frac{(6)(5)}{2}\Delta^2 f_0 + \frac{(6)(54)}{6}\Delta^3 f_0$$

$$= -4 + 6(3) + 15(0) + 20(6)$$

$$= 134 \text{ [exact value of } f(6)].$$

Again taking $x_0 = 2$, we have $p = \frac{x-x_0}{h} = 2.5 - 2.0 = 0.5$.

Therefore

$$f(2.5) = f_0 + p\Delta f_0 + \frac{p(p-1)}{2!}\Delta^2 f_0 + \frac{p(p-1)(p-2)}{6}\Delta^3 f_0$$

$$= 2 + 0.5(9) + \frac{(0.5)(0.5-1)}{2}(12) + \frac{0.5(05-1)(0.5-2)}{6}(6)$$

$$= 2 + 4.50 - 1.50 + 0.375$$

$$= 6.875 - 1.500$$

$$= 5.375 \text{ [exact value of } f\ (2.5)].$$

Remark 15.5. We note (in the above example) that if a tabulated function is a polynomial, then interpolation and extrapolation would give exact values.

15.10 INTERPOLATION WITH UNEQUAL SPACED POINTS

The classical polynomial interpolating formulae discussed so far are limited to the case in which intervals of independent variables were equally spaced. We shall now discuss interpolation formulate with unequally spaced values of the argument.

(a) Divided Differences

Let $f(x_0), f(x_1), \ldots, f(x_n)$ be the values of a function f corresponding to the arguments $x_0, x_1, \ldots, x_n$, where the intervals $x_1 - x_0, x_2 - x_1, \ldots, x_n - x_{n-1}$ are not necessarily equally spaced. Then the *first divided differences* of f for the arguments $x_0, x_1, x_2, \ldots$ are defined by

$$f(x_0, x_1) = \frac{f(x_1) - f(x_0)}{x_1 - x_0},$$

$$f(x_1, x_2) = \frac{f(x_2) - f(x_1)}{x_2 - x_1},$$

and so on. The *second divided difference* (divided difference of order 2) of f for three arguments x_0, x_1, and x_2 is defined by

$$f(x_0, x_1, x_2) = \frac{f(x_1, x_2) - f(x_0, x_1)}{x_2 - x_0}$$

and similarly, the divided difference of order n is defined by

$$f(x_0, x_1, \ldots, x_n) = \frac{f(x_1, x_2, \ldots, x_n) - f(x_0, x_1, \ldots, x_{n-1})}{x_n - x_0}.$$

Remark 15.6. Even if the arguments are equal, the divided difference may still have a meaning. For example, if we set $x_1 = x_0 + \varepsilon$, then

$$f(x_0, x_1) = f(x_0, x_0 + \varepsilon) = \frac{f(x_0 + \varepsilon) - f(x_0)}{\varepsilon}$$

and in the limit when $\varepsilon \to 0$, we have

$$f(x_0, x_0) = f'(x_0) \text{ if } f \text{ is derivable.}$$

Similarly,

$f(x_0, x_0, \ldots\ldots, x_0) = \dfrac{f^{(r)}(x_0)}{r!}$ for $r + 1$ equal arguments x_0.

Further, we observe that

$$f(x_0, x_1) = \frac{f(x_1) - f(x_0)}{x_1 - x_0} = \frac{f(x_0) - f(x_1)}{x_0 - x_1} = f(x_1, x_0),$$

$$f(x_0, x_1, x_2) = \frac{f(x_1, x_2) - f(x_0, x_1)}{x_2 - x_0}$$

$$= \frac{1}{x_2 - x_0}\left[\frac{f(x_2) - f(x_1)}{x_2 - x_1} - \frac{f(x_1) - f(x_0)}{x_1 - x_0}\right]$$

$$= \frac{f(x_0)}{(x_0 - x_1)(x_0 - x_2)} + \frac{f(x_1)}{(x_1 - x_0)(x_1 - x_2)} + \frac{f(x_2)}{(x_2 - x_0)(x_2 - x_1)}$$

and in general,

$$f(x_0, x_1, \ldots, x_n) = \frac{f(x_0)}{(x_0 - x_1)\ldots(x_0 - x_n)} + \frac{f(x_1)}{(x_1 - x_0)(x_1 - x_2)\ldots(x_1 - x_n)} + \ldots + \frac{f(x_n)}{(x_n - x_0)(x_n - x_1)\ldots(x_n - x_{n-1})}.$$

Hence the divided differences are symmetrical in their arguments. It follows, therefore, that for any function f, the value of the divided difference remains unaltered when any of the arguments involved are interchanged. Thus the value of the divided difference depends only on the value of the arguments involved and not on the order in which they are taken. Thus

$$f(x_0, x_1) = f(x_1, x_0)$$

$$f(x_0, x_1, x_2) = f(x_2, x_1, x_0) = f(x_1, x_0, x_2).$$

Theorem 15.13. The nth divided differences of a polynomial of the nth degree are constant

Proof: Consider the function $f(x) = x^n$. The first divided difference

$$f(x_r,\, x_{r+1}) = \frac{f(x_{r+1}) - f(x_r)}{x_{r+1} - x_r} = \frac{x_{r+1}^n - x_r^n}{x_{r+1} - x_r}$$
$$= x_{r+1}^{n-1} + x_r x_{r+1}^{n-2} + \ldots + x_r^{n-2} x_{r+1} \ldots + x_r^{n-1}$$

is a homogeneous polynomial of degree $n - 1$ in x_r, x_{r+1}. Similarly, it can be shown that second divided differences are homogeneous polynomial of degree $n - 2$. Proceeding by mathematical induction, it can be shown that *divided difference of nth order is a polynomial of degree $n - n = 0$ and so is a constant.* For a polynomial of the nth degree with leading term $a_0 x^n$, the nth divided difference of all terms except the leading term are zero. So the nth divided differences of this polynomial are constant and of value a_0.

Remark 15.7. Let the arguments be equally spaced so that $x_1 - x_0 = x_2 - x_1 = \ldots = x_n - x_{n-1} = h$. Then

$$f(x_0,\, x_1) = \frac{f(x_1) - f(x_0)}{h} = \frac{\Delta f_0}{h}$$

$$f(x_0,\, x_1,\, x_2) = \frac{f(x_1, x_2) - f(x_0, x_1)}{x_2 - x_0}$$
$$= \frac{1}{2h}\left[\frac{\Delta f_1}{h} - \frac{\Delta f_0}{h}\right] = \frac{1}{2h^2}\Delta^2 f_0$$
$$= \frac{1}{2!}\frac{1}{h^2}\Delta^2 f_0$$

and, in general,

$$f(x_0, x_1, \ldots, x_n) = \frac{1}{n!} \cdot \frac{1}{h^n}\,\Delta^n f_0.$$

If the tabulated function is a polynomial of nth degree, then $\Delta^n f_0$ would be constant and hence the nth divided difference would also be a constant.

15.11 NEWTON'S FUNDAMENTAL (DIVIDED DIFFERENCE) FORMULA

Let $f(x_0), f(x_1), f(x_2), \ldots\ldots, f(x_n)$ be the values of a function f corresponding to the arguments $x_0, x_1, \ldots, x_n$, where the intervals $x_1 - x_0$, $x_2 - x_1, \ldots, x_n - x_{n-1}$ are not necessarily equally spaced. By the definition of divided differences, we have

$$f(x, x_0) = \frac{f(x) - f(x_0)}{x - x_0}$$

and so

$$f(x) = f(x_0) + (x - x_0) f(x, x_0). \tag{46}$$

Further,

$$f(x, x_0, x_1) = \frac{f(x, x_0) - f(x_0, x_1)}{x - x_1},$$

which yields

$$f(x, x_0) = f(x_0, x_1) + (x - x_1) f(x, x_0, x_1). \tag{47}$$

Similarly,

$$f(x, x_0, x_1) = f(x_0, x_1, x_2) + (x - x_2) f(x, x_0, x_1, x_2) \tag{48}$$

....

....

$$f(x, x_0, \ldots, x_{n-1}) = f(x_0, x_1, \ldots, x_n) + (x - x_n) f(x, x_0, x_1, \ldots, x_n)$$

Multiplying (47) by $(x - x_0)$, (48) by $(x - x_0)(x - x_1)$ and so on, and finally the last term (49) by $(x - x_0), (x - x_1) \ldots (x - x_{n-1})$ and adding we obtain

$$f(x) = f(x_0) + (x - x_0) f(x_0, x_1)$$
$$+ (x - x_0)(x - x_1) f(x_0, x_1, x_2)$$
$$+ \ldots + (x - x_0)(x - x_1) \ldots$$
$$(x - x_{n-1}) f(x_0, x_1, \ldots, x_n) + R,$$

where

$$R = (x - x_0)(x - x_1) \ldots (x - x_n) f(x, x_0, \ldots x_n).$$

This formula is called *Newton's divided difference formula.* The last term R is the remainder term after $(n + 1)$ terms.

Remark 15.8. If we consider the case of equal spacing, then we have

$$f(x_0, x_1, \ldots, x_n) = \frac{1}{h^n\ n!}\Delta^n f_0$$

and so

$$f(x)=f(x_0)+\frac{x-x_0}{h}\Delta f_0+\frac{(x-x_0)(x-x_1)}{h^2\ 2!}\Delta^2 f_0+\ldots$$
$$=f_0+\frac{x_0+ph-x_0}{h}\Delta f_0$$
$$+\frac{(x_0+ph-x_0)(x_0+ph-x_1)}{h^2\ 2!}\Delta^2 f_0+\ldots$$
$$=f_0+p\Delta f_0+\frac{p(p-1)}{2!}\Delta^2 f_0+\ldots,$$

which is nothing but Newton's forward difference formula.

EXAMPLE 15.27

Find a polynomial satisfied by (–4, 1245), (–1, 33), (0, 5), (2, 9), and (5, 1335).

Solution. The divided difference table based on the given nodes is shown below:

x	y				
–4	1,245				
		–404			
–1	33		94		
		–28		–14	
0	5		10		3
		2		13	
2	9		88		
		442			
5	1,335				

In fact,

$$f(x_0,x_1)=\frac{f(x_0)-f(x_1)}{x_0-x_1}=\frac{1245-33}{-3}=-404,$$
$$f(x_1,x_2)=\frac{f(x_1)-f(x_2)}{x_1-x_2}=\frac{28}{-1}=-28,$$
$$f(x_2,x_3)=\frac{f(x_2)-f(x_3)}{x_2-x_3}=\frac{5-9}{-2}=2,$$
$$f(x_3,x_4)=\frac{f(x_3)-f(x_4)}{x_3-x_4}=\frac{9-1335}{-3}=442,$$
$$f(x_0,x_1,x_2)=\frac{f(x_0,x_1)-f(x_1,x_2)}{x_0-x_2}=\frac{-404+28}{-4}=94,$$
$$f(x_1,x_2,x_3)=\frac{f(x_1,x_2)-f(x_2,x_3)}{x_1-x_3}=\frac{-28-2}{-3}=10,$$
$$f(x_2,x_3,x_4)=\frac{f(x_2,x_3)-f(x_3,x_4)}{x_2-x_4}=\frac{2-442}{-5}=88$$
$$f(x_0,x_1,x_2,x_3)=\frac{f(x_0,x_1,x_2)-f(x_1,x_2,x_3)}{x_0-x_3}$$
$$=\frac{94-10}{-6}=-14,$$
$$f(x_1,x_2,x_3,x_4)=\frac{f(x_1,x_2,x_3)-f(x_2,x_3,x_4)}{x_1-x_4}$$
$$=\frac{10-88}{-6}=13,$$
$$f(x_0,x_1,x_2,x_3,x_4)=\frac{f(x_0,x_1,x_2,x_3)-f(x_1,x_2,x_3,x_4)}{x_0-x_4}$$
$$=\frac{-14-13}{-9}=3.$$

Putting these values in Newton's fundamental formula, we have

$$f(x)=f(x_0)+(x-x_0)f(x_0,x_1)+(x-x_0)(x-x_1)$$
$$f(x_0,x_1,x_2)+(x-x_0)(x-x_1)(x-x_2)f(x_0,x_1,x_2,x_3)$$
$$+(x-x_0)(x-x_1)(x-x_2)(x-x_3)f(x_0,x_1,x_2,x_3,x_4)$$
$$=1245-404(x+4)+94(x+4)(x+1)-14(x+4)(x+1)x$$
$$+3(x+4)(x+1)x(x-2)=3x^4-5x^3+6x^2-14x+5.$$

EXAMPLE 15.28

Using the table given below, find $f(x)$ as a polynomial in x.

x	−1	0	3	6	7
$f(x)$	3	−6	39	822	1611.

Solution. The divided difference table for the given data is shown below:

	x	f(x)				
x_0	–1	3				
			–9			
x_1	0	–6		6		
			15		5	
x_2	3	39		41		1
			261		13	
x_3	6	822		132		
			789			
x_4	7	1611				

Putting these values in the Newton's divided difference formula, we have

$$f(x)=f(x_0)+(x-x_0)f(x_0,x_1)+(x-x_0)(x-x_1)$$
$$f(x_0,x_1,x_2)+(x-x_0)(x-x_1)(x-x_2)f(x_0,x_1,x_2,x_3)$$
$$+(x-x_0)(x-x_1)(x-x_2)(x-x_3)f(x_0,x_1,x_2,x_3,x_4)$$
$$=3+(-9)(x+1)+6(x+1)x+5(x+1)x(x-3)$$
$$+1(x+1)x(x-3)(x-6)=x^4-3x^3+5x^2-6.$$

EXAMPLE 15.29

By means of Newton's divided difference formula, find the value of $f(8)$ and $f(15)$ from the following table:

x	4	5	7	10	11	13
$f(x)$	48	100	294	900	1210	2028

Solution. The divided difference table is

	x	$f(x)$					
x_0	4	48					
			52				
x_1	5	100		15			
			97		1		
x_2	7	294		21		0	
			202		1		0
x_3	10	900		27		0	
			310		1		
x_4	11	1,210		33			
			409				
x_5	13	2,028					

Using the formula

$$f(x) = f(x_0) + (x - x_0)f(x_0, x_1) + (x - x_0)(x - x_1)f(x_0, x_1, x_2) + (x - x_0)(x - x_1)(x - x_2)f(x_0, x_1, x_2, x_3),$$

we obtain

$$f(8) = 48 + (8-4)(52) + (8-4)(8-5)15 + (8-4)(8-5)(8-7)(1) = 448,$$

and

$$f(15) = 48 + (15-4)(52) + (15-4)(15-5)(15) + (15-4)(15-5)(15-7)(1) = 3150.$$

15.12 LAGRANGE'S INTERPOLATION FORMULA

Let f be continuous and differentiable (n+1) times in an interval (a, b) and let $f_0, f_1, f_2, \ldots, f_n$ be the values of at $x_0, x_1, x_2, \ldots, x_n$ where $x_0, x_1, x_2, \ldots, x_n$ are not necessarily equally spaced. We wish to find a polynomial of degree n, say $P_n(x)$ such that

$$P_n(x_i) = f(x_i) = f_i, \; i = 0, 1, \ldots\ldots, n. \quad (50)$$

Let

$$P_n(x) = a_0 + a_1x + a_2x^2 + \ldots + a_nx^n, \quad (51)$$

be the desired polynomial. Substituting the condition (50) in (51), we obtain the following system of equations

$$\left.\begin{aligned} f_0 &= a_0 + a_1x_0 + a_2x_0^2 + \ldots + a_nx_0^n \\ f_1 &= a_0 + a_1x_1 + a_2x_1^2 + \ldots + a_nx_1^n \\ f_2 &= a_0 + a_1x_2 + a_2x_2^2 + \ldots + a_nx_2^n \\ &\ldots\ldots\ldots\ldots\ldots\ldots\ldots\ldots \\ &\ldots\ldots\ldots\ldots\ldots\ldots\ldots\ldots \end{aligned}\right\} \quad (52)$$

$$f_n = a_0 + a_1x_n + a_2x_n^2 + \ldots + a_nx_n^n.$$

This set of equation will have a solution if the determinant

$$\begin{vmatrix} 1 & x_0 & x_0^2 \ldots\ldots\ldots x_0^n \\ 1 & x_1 & x_1^2 \ldots\ldots\ldots x_1^n \\ 1 & x_2 & x_2^2 \ldots\ldots\ldots x_2^n \\ \ldots & \ldots & \ldots\ldots \\ \ldots & \ldots & \ldots\ldots \\ 1 & x_n & x_n^2 \ldots\ldots\ldots x_n^n \end{vmatrix} \neq 0.$$

The value of this determinant, called *Vandermonde's determinant*, is $(x_0 - x_1)(x_0 - x_2) \ldots (x_0 - x_n)(x_1 - x_2)(x_1 - x_3) \ldots\ldots (x_1 - x_n) \ldots\ldots (x_{n-1} - x_n)$.

Eliminating $a_0, a_1, \ldots, a_n$ from equations (51) and (52), we obtain

$$\begin{vmatrix} P_n(x) & 1 & x & x^2 \ldots\ldots x^n \\ f_0 & 1 & x_0 & x_0^2 \ldots\ldots x_0^n \\ f_1 & 1 & x_1 & x_1^2 \ldots\ldots x_1^n \\ \ldots & \ldots & \ldots & \ldots \\ \ldots & \ldots & \ldots & \ldots \\ f_n & 1 & x_n & x_n^2 \ldots\ldots x_n^n \end{vmatrix} = 0, \quad (53)$$

which shows that $P_n(x)$ is a linear combination of $f_0, f_1, f_2, \ldots, f_n$. Hence, we write

$$P_n(x) = \sum_{i=0}^{n} L_i(x) f_i, \quad (54)$$

where $L_i(x)$ are polynomials in x of degree n. But $P_n(x_j) = f_j$ for $j = 0,1,2,3,\ldots, n$. Therefore, equation (54) yields

$$L_i(x_j) = 0 \text{ for } i \neq j \quad \text{for all } j. \quad (55)$$

$$L_i(x_j) = 1 \text{ for } i = j$$

Hence we may take $L_i(x)$ as

$$L_i(x) = \frac{(x-x_0)(x-x_1)\ldots(x-x_{i-1})(x-x_{i+1})\ldots(x-x_n)}{(x_i-x_0)(x_i-x_1)\ldots(x_i-x_{i-1})(x_i-x_{i+1})\ldots(x_i-x_n)} \quad (56)$$

which clearly satisfies the condition (55). Let

$$\prod(x) = (x-x_0)(x-x_1)\ldots (x-x_{i-1})(x-x_i)(x-x_{i+1})\ldots(x-x_n). \quad (57)$$

Then

$$\prod{}'(x_i) = \left[\frac{d}{dx}\prod(x)\right]_{x=x_i} = (x_i-x_0)(x_i-x_1)\ldots(x_i-x_{i-1})(x_i-x_{i+1})\ldots(x_i-x_n)$$

and so (56) becomes

$$L_i(x) = \frac{\prod(x)}{(x-x_i)\prod{}'(x_i)}. \quad (58)$$

Hence (54) becomes

$$P_n(x) = \sum_{i=0}^{n} \frac{\prod(x)}{(x-x_i)\prod{}'(x_i)} f_i, \quad (58)$$

which is called *Lagrange's interpolation formula.* The coefficients $L_i(x)$ defined in (56) are called *Lagrange's interpolation coefficients.*

Interchanging x and y in (58), we get the formula

$$P_n(y) = \sum_{i=0}^{n} \frac{\prod(y)}{(y-y_i)\prod{}'(y_i)} x_i, \quad (59)$$

which is useful for inverse interpolation.

Second Method Let $f(x_0), f(x_1), \ldots, f(x_n)$ be the values of the function f corresponding to the arguments $x_0, x_1, \ldots, x_n$, not necessarily equally spaced. We wish to find a polynomial $P_n(x)$ in x of degree n such that

$$P_n(x_0) = f(x_0), P_n(x_1) = f(x_1), \ldots, P_n(x_n) = f(x_n).$$

Suppose that

$$\begin{aligned} P_n(x) =& A_0(x-x_1)(x-x_2)\ldots(x-x_n) \\ &+ A_1(x-x_0)(x-x_2)\ldots(x-x_n) \\ &+ A_2(x-x_0)(x-x_1)(x-x_3)..(x-x_n) + \ldots \\ &+ A_n(x-x_0)(x-x_1)\ldots(x-x_{n-1}), \end{aligned} \quad (60)$$

where $A_0, A_1, A_2, \ldots, A_n$ are the constants to be determined.

To determine A_0, we put $x = x_0$ and $P_n(x_0) = f(x_0)$ and have

$$f(x_0) = A_0(x_0-x_1)(x_0-x_2)\ldots(x_0-x_n)$$

and so

$$A_0 = \frac{f(x_0)}{(x_0-x_1)(x_0-x_2)\ldots(x_0-x_n)}.$$

Similarly, putting $x = x_1, x_2, \ldots, x_n$, we get

$$A_1 = \frac{f(x_1)}{(x_1-x_0)(x_1-x_2)(x_1-x_3)\ldots(x_1-x_n)}$$

$$A_2 = \frac{f(x_2)}{(x_2-x_0)(x_2-x_1)(x_2-x_3)\ldots(x_2-x_n)},$$

....

....

$$A_n = \frac{f(x_n)}{(x_1-x_0)(x_n-x_1)\ldots(x_n-x_{n-1})}.$$

Substituting these values in(60), we get

$$P_n(x) = \sum_{i=0}^{n} L_i(x) f(x_i),$$

where

$$L_i(x) = \frac{(x-x_0)(x-x_1)\ldots(x-x_{i-1})(x-x_{i+1})\ldots(x-x_n)}{(x_i-x_0)(x_i-x_1)\ldots(x_i-x_{i-1})(x_i-x_{i+1})\ldots(x_i-x_n)},$$

which is *Lagrange's interpolation formula.* Clearly,

$$L_i(x_j) = 0 \text{ for } i \neq j,$$
$$L_i(x_j) = 1 \text{ for } i = j.$$

Remark 15.9. If f takes same value, say k, at each of the points $x_0, x_1, x_2, \ldots, x_n$, we have

$$P_n(x) = \sum_{i=0}^{n} L_i(x)k = k\sum_{i=0}^{n} L_i(x).$$

This yields,

$$\sum_{i=0}^{n} L_i(x) = 1,$$

which is an *important check during calculations.* Further, dividing both sides of Lagrange's interpolation formula by $(x-x_0)(x-x_1)\ldots(x-x_n)$, we obtain

$$\frac{P_n(x)}{(x-x_0)(x-x_1)\ldots(x-x_n)}$$

$$=\frac{f(x_0)}{(x_0-x_1)(x_0-x_2)\ldots(x_0-x_n)}\cdot\frac{1}{x-x_0}$$

$$+\frac{f(x_1)}{(x_1-x_0)(x_1-x_2)\ldots(x_1-x_n)}\cdot\frac{1}{x-x_1}$$

$$+\ldots.+\frac{f(x_n)}{(x_n-x_0)(x_n-x_1)\ldots(x_n-x_{n-1})}\frac{1}{x-x_n}$$

Thus $\frac{P_n(x)}{(x-x_0)(x-x_1)\ldots(x-x_n)}$ has been expressed as the sum of partial fractions.

EXAMPLE 15.30

Use Lagrange's formula to express the function $\frac{x^2+6x-1}{(x-1)(x+1)(x-4)(x-6)}$ as a sum of partial fractions.

Solution. We have

and so
$$P_n(x) = x^2+6x-1,$$
$$P_n(x_0) = f(x_0) = f(1) = 6$$
$$P_n(x_1) = f(x_1) = f(-1) = -6$$
$$P_n(x_2) = f(x_2) = f(4) = 39$$
$$P_n(x_3) = f(x_3) = f(6) = 71.$$

Therefore,

$$\frac{x^2+6x-1}{(x-1)(x+1)(x-4)(x-6)}$$

$$=\frac{6}{(x-1)(2)(-3)(-5)}$$

$$+\frac{-6}{(x+1)(-2)(-5)(-7)}+\frac{39}{(x-4)(3)(5)(-2)}$$

$$+\frac{71}{(x-6)(5)(7)(2)}$$

$$=\frac{1}{5(x-1)}+\frac{3}{35(x+1)}-\frac{13}{10(x-4)}+\frac{71}{70(x-6)}.$$

EXAMPLE 15.31

Use Lagrange's interpolation formula to express the function

$$\frac{x^2+x-3}{x^3-2x^2-x+2}$$

as sum of partial functions.

Solution. We have

$$\frac{x^2+x-3}{x^3-2x^2-x+2}=\frac{x^2+x-3}{(x-1)(x+1)(x-2)}.$$

Let

$$P_n(x) = x^2+x-3,$$

and let $x_0 = 1$, $x_1 = -1$, $x_2 = 2$. Then

$$P_n(x_0) = f(x_0) = f(1) = -1$$

$$P_n(x_1) = f(x_1) = f(-1) = -3$$

$$P_n(x_2) = f(x_2) = f(2) = 3.$$

Therefore,

$$\frac{x^2+x-3}{(x-1)(x+1)(x-2)}$$

$$=\frac{f(x_0)}{(x_0-x_1)(x_0-x_2)}\cdot\frac{1}{x-x_0}$$

$$+\frac{f(x_1)}{(x_1-x_0)(x_1-x_2)}\cdot\frac{1}{x-x_1}$$

$$+\frac{f(x_2)}{(x_2-x_0)(x_2-x_1)}\cdot\frac{1}{x-x_2}$$

$$=\frac{-1}{(x-1)(2)(-1)}+\frac{-3}{(x+1)(-2)(-3)}+\frac{3}{(x-2)(1)(3)}$$

$$=\frac{1}{2(x-1)}-\frac{1}{2(x+1)}+\frac{1}{(x-2)}.$$

EXAMPLE 15.32

Using Lagrange's interpolation formula, prove that $32f(1) = -3f(-4) + 10f(-2) + 30f(2) - 5f(4)$.

Solution. We have

$$x_0 = -4,\ x_1 = -2,\ x_3 = 4 \text{ and } x = 1$$

Then

$$L_0(x) = \frac{(x-x_1)(x-x_2)(x-x_3)}{(x_0-x_1)(x_0-x_2)(x_0-x_3)}$$

$$= \frac{(1+2)(1-2)(1-4)}{(-4+2)(-4-2)(-4-4)} = -\frac{3}{32}.$$

Similarly

$$L_1(x) = \frac{5}{16},\ L_2(x) = \frac{15}{16},\ L_3(x) = -\frac{5}{32}.$$

We observe that $\sum_{i=0}^{3} L_i(x) = 1$. Therefore,

$$f(x) = \sum_{i=0}^{3} L_i(x) f(x_i)$$

or

$$f(1) = \frac{5}{16} f_1 + \frac{15}{16} f_2 - \frac{5}{32} f_3 - \frac{3}{32} f_0$$

or

$$32 f(1) = -3f(-4) + 10f(-2) + 30f(2) - 5f(4).$$

EXAMPLE 15.33

The function $y = f(x)$ is given in the points (7,3), (8,1), (9,1), and (10,9). Find the value of y for $x = 9.5$ using Lagrange's interpolation formula.

Solution. We have

	x	$y=f(x)$
x_0	7	3
x_1	8	1
x_2	9	1
x_3	10	9

By Lagrange's formula, we have

$$f(x) \approx P_n(x) = \sum_{i=0}^{n} L_i(x) f(x_i),$$

where

$$L_i(x) = \frac{(x-x_0)(x-x_1)\ldots(x-x_{i-1})(x-x_{i+1})\ldots(x-x_n)}{(x_i-x_0)(x_i-x_1)\ldots(x_i-x_{i-1})(x_i-x_{i+1})\ldots(x_i-x_n)}.$$

In the present problem, $x = 9.5$ and we have

$$L_0(x) = \frac{(x-x_1)(x-x_2)(x-x_3)}{(x_0-x_1)(x_0-x_2)(x_0-x_3)}$$

$$= \frac{(9.5-8)(9.5-9)(9.5-10)}{(7-8)(7-9)(7-10)}$$

$$= \frac{0.375}{6} = 0.06250,$$

$$L_1(x) = \frac{(x-x_0)(x-x_2)(x-x_3)}{(x_1-x_0)(x_1-x_2)(x_1-x_3)}$$

$$= \frac{(9.5-7)(9.5-9)(9.5-10)}{(8-7)(8-9)(8-10)}$$

$$= -\frac{0.625}{2} = -0.3125,$$

$$L_2(x) = \frac{(x-x_0)(x-x_1)(x-x_3)}{(x_2-x_0)(x_2-x_1)(x_2-x_3)}$$

$$= \frac{(9.5-7)(9.5-8)(9.5-10)}{(9-7)(9-8)(9-10)}$$

$$= \frac{1.875}{2} = 0.9375,$$

$$L_3(x) = \frac{(x-x_0)(x-x_1)(x-x_2)}{(x_3-x_0)(x_3-x_1)(x_3-x_2)}$$

$$= \frac{(9.5-7)(9.5-8)(9.5-9)}{(10-7)(10-8)(10-9)}$$

$$= \frac{1.875}{6} = 0.3125.$$

We observe that $L_0(x) + L_1(x) + L_2(x) + L_3(x) = 1$ and, therefore, so far, our calculations are correct.

Hence

$$P(x) = P(9.5) = \sum_{i=0}^{3} L_i(x) f(x_i)$$

$$= L_0 f_0 + L_1 f_1 + L_2 f_2 + L_3 f_3$$

$$= (0.06250)\,(3) - 0.3125(1) + 0.9395(1) + 0.3125(9)$$

$$= 0.1875 - 0.3125 + 0.9375 + 2.8125 = 3.625$$

EXAMPLE 15.34

Find the interpolating polynomial for (0,2), (1,3), (2, 12), and (5, 147).

Solution. The given data is

x	0	1	2	5
$f(x)$	2	3	12	147

By Lagrange's formula reads

$$f(x) = \sum_{i=0}^{3} L_i(x) f(x_i)$$

where

$$L_i(x) = \frac{(x-x_0)(x-x_1)\dots(x-x_{i-1})(x-x_{i+1})\dots(x-x_n)}{(x_i-x_0)(x_i-x_1)\dots(x_i-x_{i-1})(x_i-x_{i+1})\dots(x_i-x_n)}.$$

Thus

$$L_0(x) = \frac{(x-x_1)(x-x_2)(x-x_3)}{(x_0-x_1)(x_0-x_2)(x_0-x_3)} = \frac{(x-1)(x-2)(x-5)}{(0-1)(0-2)(0-5)}$$
$$= -\frac{1}{10}(x^3 - 8x^2 + 17x - 10),$$
$$L_1(x) = \frac{(x-x_0)(x-x_2)(x-x_3)}{(x_1-x_0)(x_1-x_2)(x_1-x_3)} = \frac{(x-0)(x-2)(x-5)}{(1-0)(1-2)(1-5)}$$
$$= \frac{1}{4}(x^3 - 7x^2 + 10x),$$
$$L_2(x) = \frac{(x-x_0)(x-x_1)(x-x_3)}{(x_2-x)(x_2-x_1)(x_2-x_3)} = \frac{(x-0)(x-1)(x-5)}{(2-0)(2-1)(2-5)}$$
$$= -\frac{1}{6}(x^3 - 6x^2 + 5x),$$
$$L_3(x) = \frac{(x-x_0)(x-x_1)(x-x_2)}{(x_3-x_0)(x_3-x_1)(x_3-x_2)} = \frac{(x-0)(x-1)(x-2)}{(5-0)(5-1)(5-2)}$$
$$= \frac{1}{60}(x^3 - 3x^2 + 2x).$$

Putting these values in Lagrange's formula, we have

$$P(x) = \sum_{i=0}^{3} L_i(x) f(x_i)$$
$$= -\frac{2}{10}(x^3 - 8x^2 + 17x - 10) + \frac{3}{4}(x^3 - 7x^2 + 10x)$$
$$- \frac{12}{6}(x^3 - 6x^2 + 5x) + \frac{147}{60}(x^3 - 3x^2 + 2x)$$
$$= x^3 + x^2 - x + 2.$$

15.13 CURVE FITTING

So far, we have considered the construction of a polynomial, which approximates a given function and takes the same values as the function at certain given points. This is called the *method of collocation* and the conditions are satisfied by the approximate Lagrange's interpolation polynomial. When the given points are equally spaced, we can form difference table and find the polynomial using Newton's forward difference formula. For example, the polynomial $4x - 4x^2$ agree with the function $\sin \pi x$ for $x = 0, \frac{1}{2}, 1$, but this approximation is not very satisfactory because the polynomial $4x - 4x^2$ is larger than $\sin \pi x$ in the range (0,1) except at the point $x = \frac{1}{2}$. Similarly, the Lagrangian interpolation polynomials constructed for the function $\frac{1}{x^2+1}$ in the interval [−5, 5] with uniformly distributed nodes give rise to arbitrary large deviations for increasing degree n.

If the functional values at the given points (nodes) are the result of experiments or if they are rounded values or if the nodes are subject to error, then the advantages of the method of collocation are, to some extent, lost. In such a case, Weierstrass's approximation theorem is of remarkable utility.

Theorem 15.14. (Weierstrass's Approximation Theorem). If f is a continuous function in the interval $[a,b]$, then to each $\varepsilon > 0$ there exists a polynomial $p(x)$ such that

$$|f(x) - p(x)| < \varepsilon \text{ for all } x \in [a, b].$$

Weierstrass's theorem allows us to consider other methods of approximations. We will discuss these methods one-by-one.

(b) Least Square Line Approximation

Suppose that we have an empirical data in the form of n pairs of values $(x_1, y_1), (x_2, y_2), \dots, (x_n, y_n)$, where the experimental errors are associated with the functional values $y_1, y_2, \dots, y_n$ only. Then we seek a linear function

$$y = f(x) = a + bx \qquad (61)$$

fitting the given points as much as possible. The equation (61) will not in general be satisfied by any of the n pairs. Substituting in (61) each of the n pairs of values in turn, we get

$$\left.\begin{aligned} e_1 &= y_1 - a - bx_1 \\ e_2 &= y_2 - a - bx_2 \\ &\dots\dots\dots \\ &\dots\dots\dots \\ e_n &= y_n - a - bx_n \end{aligned}\right\}, \qquad (62)$$

where e_k, $k = 1, \dots, n$ are measurement errors, called *residuals or deviations*. To know how far the curve $y = f(x)$ lies from the given data, the following errors are considered:

1. Maximum error

$$e_{\infty}(f) = \max_{1 \le k \le n} \{|y_k - a - bx_k|\}$$

2. Average error

$$e_A(f) = \frac{1}{n}\sum_{k=1}^{n} | y_k - a - bx_k |$$

3. Root mean square (RMS) error

$$e_{rms}(f) = \left[\frac{e_1^2 + \ldots + e_n^2}{n}\right]^{1/2}.$$

The least square line $y = f(x) = a+bx$ is the line that minimize the RMS error $e_{rms}(f)$. But the quantity $e_{rms}(f)$ is minimum if and only if $\sum_{k=1}^{n} (y_k - a - bx_k)^2 = \sum_{k=1}^{n} e_k^2$ is minimum. Thus, in case of least – square line we are looking for a linear function $a + bx$ as an approximation to a function $y=f(x)$ when we are given the value of y at the points $x_1, \ldots, x_n$. We aim at minimizing the sum of the squared errors:

$$e(a, b) = \sum_{i=1}^{n} (y_i - a - bx_i)^2. \tag{63}$$

Geometrically, if d_i is the vertical distance from the data point (x_i, y_i) to the point $(x_i, a+bx_i)$ on the line, then $d_i = y_i - a - bx_i$ (see Figure 26.4). We must minimize the sum of the squares of the vertical distances d_i, that is, the sum $\sum_{i=1}^{n} d_i^2$.

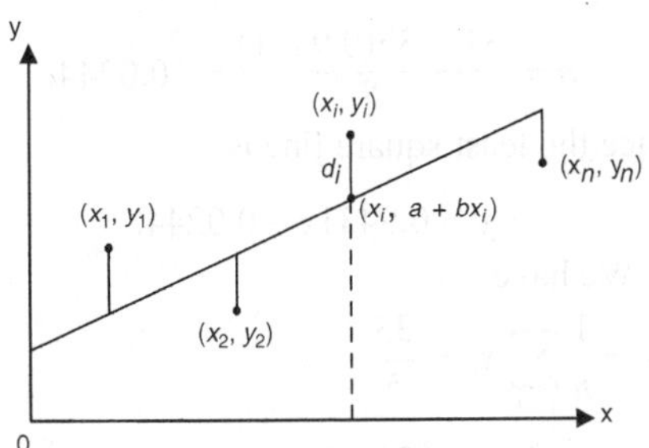

Figure 15.4

To minimize $e(a,b)$, we equate to zero the partial derivatives of (63) with respect to 'a' and 'b'. Thus

$$\frac{\partial e(a, b)}{\partial a} = \sum_{i=1}^{n} 2(y_i - a - bx_i) = 0$$

and

$$\frac{\partial e(a, b)}{\partial b} = \sum_{i=1}^{n} 2x_i(y_i - a - bx_i) = 0,$$

which are known as *normal equations.* We write these equations in the form

$$na + b\sum_{i=1}^{n} x_i = \sum_{i=1}^{n} y_i \tag{64}$$

and

$$a\sum_{i=1}^{n} x_i + b\sum_{i=1}^{n} x_i^2 = \sum_{i=1}^{n} x_i y_i. \tag{65}$$

The normal equations (64) and (65) can the solved for 'a' and 'b' using Cramer's rule or by some other method.

EXAMPLE 15.35

Show that, according to the principle of least squares, the best fitting linear function for the points (x_i, y_i), $i = 1,2,\ldots,n$ may the expressed in the form

$$\begin{vmatrix} x & y & 1 \\ \sum_{i=1}^{n} x_i & \sum_{i=1}^{n} y_i & n \\ \sum_{i=1}^{n} x_i^2 & \sum_{i=1}^{n} y_i^2 & \sum_{i=1}^{n} x_i \end{vmatrix} = 0.$$

Solution. Eliminating 'a' and 'b' from (64), (65), and $y = a + bx$, we get the required result.

We have supposed in the above derivation that the errors in x values can be neglected compared with the errors in the y values. Now we suppose that the x values as well as the y values are subject to errors of about the same order of magnitude. Now we minimize the sum of the squares of the perpendicular distances to the line. Thus, if $y = a + bx$ is the equation of the line, then

$$e(a, b) = \frac{1}{1+b^2} \sum_{i=1}^{n} (y_i - a - bx_i)^2.$$

For minimum, partial derivatives with respect to 'a' and 'b' should vanish. Thus

$$\frac{\partial\, e\,(a,\, b)}{\partial a} = \frac{2}{1+b^2}\sum_{i=1}^{n}(y_i - a - bx_i) = 0$$

and

$$\frac{\partial\, e\,(a,\, b)}{\partial b} = -2(1+b^2)\sum_{i=1}^{n}(y_i - a - bx_i)x_i$$

$$-\,2b\sum_{i=1}^{n}(y_i - a - bx_i)^2 = 0,$$

that is,

$$\sum_{i=1}^{n}(y_i - a - bx_i) = 0 \qquad (66)$$

and

$$(1+b^2)\sum_{i=1}^{n}(y_i - a - bx_i)x_i$$

$$= b\sum_{i=1}^{n}(y_i - a - bx_i)^2. \qquad (67)$$

From (66), we get

$$a = y_0 - bx_0, \qquad (68)$$

where

$$x_0 = \frac{1}{n}\sum_{i=1}^{n} x_i \quad \text{and} \quad y_0 = \frac{1}{n}\sum_{i=1}^{n} y_i.$$

After simplification, (67) yields

$$b^2 + \frac{A-C}{B}\,b\;1 = 0, \qquad (69)$$

where

$$A = \sum_{i=1}^{n} x_i^2 - n\,x_0^2,$$

$$B = \sum_{i=1}^{n} x_i y_i - n\,x_0 y_0,$$

$$C = \sum_{i=1}^{n} y_i^2 - n\,y_0^2.$$

Finding the value of 'b' from (69), we obtain the corresponding value of 'a' from (68).

EXAMPLE 15.36

The points (2,2), (5,4), (6,6), (9,9), and (11,10) should be approximated by a straight line. Perform this assuming

(i) the error in the x values can be neglected.

(ii) that the errors in x and y values are of the same order of magnitude.

Solution. (a). The sum table for the given problem is

n	x	x^2	y	xy	y^2
1	2	4	2	4	4
1	5	25	4	20	16
1	6	36	6	36	36
1	9	81	9	81	81
1	11	121	10	110	100
5	**33**	**267**	**31**	**251**	**237**

Let the least square line be $y = a + bx$. Therefore, the normal equations are

$$5a + 33b = 31 \qquad (70)$$

$$33a + 267b = 251. \qquad (71)$$

Multiplying (70) by 33 and (71) by 5, we obtain

$$165\,a + 1089\,b = 1023$$

$$165a + 1335b = 1255.$$

Subtracting, we get

$$246b = 232 \text{ and so } b = \frac{116}{123} = 0.9431.$$

Then (70) yields

$$a = \frac{31 - 33(0.9431)}{5} = -0.0244.$$

Hence the least square line is

$$y = 0.9431x - 0.0244.$$

(b) We have

$$x_0 = \frac{1}{n}\sum_{i=1}^{n} x_i = \frac{33}{5},$$

$$y_0 = \frac{1}{n}\sum_{i=1}^{n} y_i = \frac{31}{5}.$$

$$A = \sum_{i=1}^{n} x_i^2 - nx_0^2 = 267 - 5\left(\frac{33}{5}\right)^2 = \frac{246}{5} = 49.2$$

$$B = \sum_{i=1}^{n} x_i y_i - nx_0 y_0 = 251 - \frac{5(33)(31)}{25} = 46.4$$

$$C = \sum_{i=1}^{n} y_i^2 - ny_0^2 = 237 - 5\left(\frac{31}{5}\right)^2 = 44.8.$$

Therefore, equation $b^2 + \frac{A-C}{B} b - 1 = 0$ becomes

$$b^2 + \frac{4.4}{46.4} b - 1 = 0 \quad \text{or} \quad b^2 + 0.0948\ b - 1 = 0.$$

Hence

$$b = \frac{-0.948 \pm \sqrt{4.0089}}{2} = 0.9537\ (+\text{ve}).$$

Then $a = y_0 - bx_0$ yields $a = -0.0944$.

Hence

$$y = 0.9537x - 0.0944$$

is the required least square line.

EXAMPLE 15.37

In the following data, x and y are subject to error of the same order of magnitude:

x	1	2	3	4	5	6	7	8
y	3	3	4	5	5	6	6	7

Find a straight-line approximation using the least square method.

Solution. The sum table for the given problem is

n	x	x^2	y	xy	y^2
1	1	1	3	3	9
1	2	4	3	6	9
1	3	9	4	12	16
1	4	16	5	20	25
1	5	25	5	25	25
1	6	36	6	36	36
1	7	49	6	42	36
1	8	64	7	56	49
8	36	204	39	200	205

Let the equation be $y = a + bx$. Then

$$a = y_0 - bx_0, \tag{72}$$

where

$$x_0 = \frac{1}{n}\sum_{i=1}^{n} x_i = \frac{36}{8},$$

$$y_0 = \frac{1}{n}\sum_{i=1}^{n} y_i = \frac{39}{8}.$$

Further

$$A = \sum_{i=1}^{n} x_i^2 - nx_0^2 = 204 - 8\left(\frac{36}{8}\right)^2$$

$$= \frac{408 - 324}{2} = 42,$$

$$B = \sum_{i=1}^{n} x_i y_i - nx_0 y_0 = 200 - 8\frac{(36)(39)}{8^2}$$

$$= \frac{400 - 351}{2} = \frac{49}{2} = 24.5,$$

$$C = \sum_{i=1}^{n} y_i^2 - ny_0^2 = 205 - 8\left(\frac{39}{8}\right)^2$$

$$= \frac{119}{8} = 14.87$$

Then the value of b is given by

$$b^2 + \frac{A - C}{B}\ b - 1 = 0$$

or

$$b^2 + \frac{42.0 - 14.87}{24.5}\ b - 1 = 0$$

or

$$b^2 + 1.107\ b - 1 = 0,$$

which yields

$$b = \frac{-1.107 \pm \sqrt{5.225}}{2} = 0.5895 \quad (+\text{ve}).$$

Then (72) gives a=2.225. Hence the least square line is

$$y = 0.59x + 2.22.$$

(b) The Power Fit $y = ax^m$

Suppose we require ax^m as an approximation to a function y, where m is a known constant. We must find the value of 'a' such that the equation.

$$y = ax^m \tag{73}$$

is satisfied as nearly as possible by each of the n pairs of observed values $(x_1, y_1), (x_2, y_2), \ldots, (x_n, y_n)$. Using least square technique, we should minimize the error function

$$e\ (a) = \sum_{i=1}^{n} (ax_i^m - y_i)^2. \tag{74}$$

For this purpose, partial derivative of (74) with respect to 'a' must vanish. So, we have

$$0 = 2 \sum_{i=1}^{n} (ax_i^m - y_i)(x_i^m)$$

and so

$$0 = a \sum_{i=1}^{n} x_i^{2m} - \sum_{i=1}^{n} x_i^m y_i,$$

which yields

$$a = \frac{\sum_{i=1}^{n} x_i^m y_i}{\sum_{i=1}^{n} x_i^{2m}}.$$

Putting the value of 'a' in (73), we get the required equation.

Second Method. Taking logarithms of both sides of (73) yields

$$\log y = \log a + m \log x,$$

which is of the form Y = A + BX, where Y = log y, A = log a, B = m, and X = log x. Now the least square line can be found. Then 'a' and 'm' are found.

EXAMPLE 15.38

Find the gravitational constant g using the given data below and the relation $h = \frac{1}{2} g t^2$, where h is the distance in metres and t the time in seconds.

t	0.200	0.400	0.600	0.800	1.000
h	0.1960	0.7850	1.7665	3.1405	4.9075.

Solution. The sum table for the given problem is

t	h	$t^{2m} (m = 2)$	ht^2
0.200	0.1960	0.0016	0.00784
0.400	0.7850	0.0256	0.12560
0.600	1.7665	0.1296	0.63594
0.800	3.1405	0.4096	2.00992
1.000	4.9075	1.0000	4.90750
		1.5664	**7.68680**

Then using the formula $y = ax^m$ for power fit, we have

$$\frac{1}{2} g = \frac{\sum_{k=1}^{n} h_k t_k^m}{\sum_{k=1}^{n} t_k^{2m}} = \frac{7.68680}{1.5664} = 4.9073$$

and so the gravitational constant $g = 9.8146$ m/sec^2.

EXAMPLE 15.39

Find the power fits $y = ax^2$ and $y = bx^3$ for the data given below and determine which curve fits best:

x	2.0	2.3	2.6	2.9	3.2
y	5.1	7.5	10.6	14.4	19.0

Solution. The sum table for the given problem is

x	x^2	x^3	x^4	x^6
2	4	8	16	64
2.3	5.29	12.167	27.984	148.035
2.6	6.76	17.576	45.698	308.918
2.9	8.41	24.389	70.729	594.831
3.2	10.24	32.768	104.858	1073.746
			265.269	**2189.530**

y	yx^2	yx^3
5.1	20.4	40.8
7.5	39.675	91.252
10.6	71.656	186.306
14.4	121.104	351.202
19.0	194.560	622.592
	447.395	**1292.152**

Then for $y = ax^2$, we have

$$a = \frac{\sum y_i x_i^2}{\sum x_i^4} = \frac{447.395}{265.269} = 1.6866.$$

Hence the power fit is

$$y = 1.6866x^2.$$

On the other hand, for $y = bx^3$, we have

$$b = \frac{\sum y_i x_i^3}{\sum x_i^6} = \frac{1292.152}{2189.530} = 0.5902.$$

Hence the power fit is

$$y = 0.5902x^3.$$

To know which of these is best fit, we calculate the corresponding errors. For the first power fit, we have

$$e_{rms} = \left[\frac{1}{5}\left\{(ax_1^2 - y_1)^2 + (ax_2^2 - y_2)^2 + (ax_3^2 - y_3)^2 + (ax_4^2 - y_4)^2 + (ax_5^2 - y_5)^2\right\}\right]^{1/2}$$

$$= \left[\frac{1}{5}\left\{(1.646)^2 + (1.4330)^2 + (0.8014)^2 + (0.2157)^2 + (-1.7292)^2\right\}\right]^{1/2}$$

$$= \left[\frac{1}{5}(2.704 + 2.053 + 0.642 + 0.046 + 2.990\right]^{1/2} \approx 1.3.$$

Similarly for the second curve, we have

$$e_{rms} \approx 0.29.$$

Hence the power fit curve $y = 0.5902\, x^3$ is the best.

(C) Least Square Parabola (Parabola of Best Fit)

Suppose that we want to approximate a given function $y = f(x)$ by a quadratic $a+bx+cx^2$. We must find the values of 'a', 'b', and 'c' such that the equation

$$y = a+bx+cx^2 \tag{75}$$

is satisfied as nearly as possible by each of the n pairs of observed values $(x_1, y_1), (x_2, y_2), \ldots,$ (x_n, y_n). The equation will not in general be satisfied exactly by any of the n pairs. Substituting in (75) each of the pairs of values in turn, we get the following residual equations:

$$e_1 = a + bx_1 + cx_1^2 - y_1$$
$$e_2 = a + bx_2 + cx_2^2 - y_2$$
$$-\ -\ -\ -\ -\ -\ -\ -\ -\ -$$
$$-\ -\ -\ -\ -\ -\ -\ -\ -\ -$$
$$e_n = a + bx_n + cx_n^2 - y_n.$$

The principle of least square says that the best values of the unknown constants 'a', 'b', and 'c' are those which make the sum of the squares of the residuals a minimum, that is,

$$\sum_{i=1}^{n} e_i^2 = e_1^2 + e_2^2 + \ldots. + e_n^2$$

must be minimum. Thus

$$e\ (a,\ b,\ c) = \sum_{i=1}^{n} (a + bx_i + cx_i^2 - y_i)^2$$

is to be minimum. For this, the partial derivatives of e (a, b, c) with respect to 'a', 'b', and 'c' should be zero. We, therefore, have

$$\frac{\partial\, e\ (a,b,c)}{\partial\, a} = 2(a + bx_1 + cx_1^2 - y_1) + 2(a + bx_2 + cx_2^2 - y_2) + \ldots + 2(a + bx_n + cx_n^2 - y_n) = 0,$$

$$\frac{\partial\, e\ (a,\ b,\ c)}{\partial\, b} = 2(a + bx_1 + cx_1^2 - y_1)x_1 + 2(a + bx_2 + cx_2^2 - y_2)x_2 + \ldots + 2(a + bx_n + cx_n^2 - y_n)x_n = 0,$$

$$\frac{\partial\, e\ (a,\ b,\ c)}{\partial\, c} = 2(a + bx_1 + cx_1^2 - y_1)x_1^2 + 2(a + bx_2 + cx_2^2 - y_2)x_2^2 + \ldots + 2(a + bx_n + cx_n^2 - y_n)x_n^2 = 0.$$

Hence the normal equations are

$$(a + bx_1 + cx_1^2 - y_1) + (a + bx_2 + cx_2^2 - y_2) + \ldots + (a + bx_n + cx_n^2 - y_n) = 0,$$

$$(a + bx_1 + cx_1^2 - y_1)x_1 + (a + bx_2 + cx_2^2 - y_2)x_2 + \ldots + (a + bx_n + cx_n^2 - y_n)x_n = 0, \quad \text{and}$$

$$(a + bx_1 + cx_1^2 - y_1)x_1^2 + (a + bx_2 + cx_2^2 - y_2)x_2^2 + \ldots + (a + bx_n + cx_n^2 - y_n)x_n^2 = 0.$$

These equations can further be written as

$$na + b(x_1 + x_2 + \ldots + x_n) + c(x_1^2 + x_2^2 + \ldots + x_n^2) = y_1 + y_2 + \ldots + y_n,$$

$$a\,(x_1 + x_2 + \ldots + x_n) + b(x_1^2 + x_2^2 + \ldots + x_n^2) + c(x_1^3 + x_2^3 + \ldots + x_n^3) = x_1y_1 + x_2y_2 + x_3y_3, + \ldots + x_ny_n,$$

$$a(x_1^2 + x_2^2 + \ldots + x_n^2) + b(x_1^3 + x_2^3 + \ldots\ x_n^3) + c(x_1^4 + x_2^4 + \ldots + x_n^4) = x_1^2y_1 + x_2^2y_2 + \ldots + x_n^2y_n.$$

The above normal equations are solved by ordinary methods of algebra for solving simultaneous equations of first degree in two or more unknowns.

Remark 15.10. The number of normal equations is always the same as the number of unknown constants, whereas the number of residual equations is equal to the number of observations. The number of observations must always be greater than the number of undetermined constants if the method of least square is to be useful in the solution.

EXAMPLE 15.40

Find the parabola of best fit (with equation of the form $a+bx + cx^2$) for the data in the following table:

x	0	1	2	3	4
y	−2.1	−0.4	2.1	3.6	9.9.

Solution. We establish the following sum table:

n	x	x^2	x^3	x^4	y	xy	x^2y
1	0	0	0	0	−2.1	0	0
1	1	1	1	1	−0.4	−0.4	−0.4
1	2	4	8	16	2.1	4.2	8.4
1	3	9	27	81	3.6	10.8	32.4
1	4	16	64	256	9.9	39.6	158.4
5	10	30	100	354	13.1	54.2	198.8

The normal equations are

$$5a + 10b + 30c = 13.1 \tag{76}$$

$$10a + 30b + 100c = 54.2 \tag{77}$$

$$30a + 100b + 354c = 198.8. \tag{78}$$

Multiplying (76) by 2 and then subtracting from (77), we get

$$10b + 40c = 28 \tag{79}$$

Multiplying (77) by 3 and then subtracting from (78), we get

$$10b + 54\,c = 36.2. \tag{80}$$

Subtracting (79) from (80), we get $c = 0.58571$. Then (79) yields $b = 0.45716$ and then (76) yields $a = -1.80858$. Hence the parabola of best fit is

$$y = -1.80858 + 0.45716\,x + 0.58571x^2.$$

EXAMPLE 15.41

Find the least square polynomial of degree two for the following data:

x	0.78	1.56	2.34	3.12	3.81
y	2.50	1.20	1.12	2.25	4.28

Solution. Let the required polynomial be $a + bx + cx^2$. To make the calculations simple, we use the substitution

$$X = \frac{x - 2.34}{0.78}$$

making use of the equal spacing of the arguments. The sum table then becomes

n	X	X^2	X^3	X^4
1	−2	4	−8	16
1	−1	1	−1	1
1	0	0	0	0
1	1	1	1	1
1	1.88	3.53	6.64	12.49
5	−0.12	9.53	−1.36	30.49

n	y	Xy	X^2y
1	2.50	−5.00	10.00
1	1.20	−1.20	1.20
1	1.12	0	0
1	2.25	2.25	2.25
1	4.28	8.05	15.13
5	11.35	4.10	28.58

The normal equations are

$$5a - 0.12b + 9.53\,c = 11.35$$

$$-0.12a + 9.53b - 1.36\,c = 4.10$$

$$9.53a - 1.36b + 30.49c = 28.58.$$

Solving these equations by Cramer's rule, we get

$$a = 1.1155021, \quad b = 0.5316061,$$

$$c = 0.612401.$$

Hence the parabola of best fit is

$$y = 1.1155 + 0.5316\,X + 0.6124\,X^2,$$

where $X = \dfrac{x - 2.34}{0.78}$.

EXAMPLE 15.42

Find the least square fit $y = a + bx + cx^2$ for the data

x	−3	−1	1	3
y	15	5	1	5.

Solution. The sum table for the given problem is

n	x	x^2	x^3	x^4	y	xy	x^2y
1	−3	9	−27	81	15	−45	135
1	−1	1	−1	1	5	−5	5
1	1	1	1	1	1	1	1
1	3	9	27	81	5	15	45
4	0	20	0	164	26	−34	186

The normal equations are

$$4a + 20\,c = 26$$

$$20b = -34$$

$$20a + 164\,c = 186.$$

Solving these equations, we have

$$b = -\frac{34}{20} = -1.70, \quad c = 0.875, \quad a = 2.125.$$

Hence the least square parabola is

$$y = 2.125 - 1.700\,x + 0.875\,x^2.$$

EXAMPLE 15.43

Fit a parabola to the following data

x:	1	2	3	4
y:	0.30	0.64	1.32	5.40

Solution. The sum table for the given problem is

n	x	x^2	x^3	x^4	y	xy	x^2y
1	1	1	1	1	0.30	0.30	0.30
1	2	4	8	16	0.64	1.28	2.56
1	3	9	27	81	1.32	3.96	11.88
1	4	16	64	256	5.40	21.60	86.40
4	10	30	100	354	7.66	27.14	101.14

The normal equations are

$$4a + 10b + 30\,c = 7.66,$$

$$10a + 30b + 100\,c = 27.14,$$

$$30a + 100b + 354\,c = 101.14.$$

Solving these equations by Gauss elimination method or Cramer's rule, we get

$$a = -1.09, \quad b = 0.458, \quad c = 0.248.$$

Hence the parabola of fit is

$$y = -1.09 + 0.458\,x + 0.248\,x^2.$$

15.14 NUMERICAL QUADRATURE

Numerical integration is the process of computing the approximate value of a definite integral using a set of numerical values of the integrand. If the integrand is a function of single variable, the process is called *mechanical quadrature.* If the integrand is a function of two independent variables, the process of computing double integral is called *mechanical cubature.*

The numerical integration is performed by representing the integrand by an interpolation formula and then integrating the interpolation formula between the given limits. Thus to find $\int_a^b f(x)dx$, we replace the function f by an interpolation formula involving differences and then integrate this formula between the limits a and b.

In equidistant interpolation formulas, the relation between x and p is

$$x = x_0 + ph, \qquad (81)$$

where h is the equidistance between the given nodes. Then

$$dx = h\,dp. \qquad (82)$$

We integrate Newton's forward difference formula over n equidistant intervals of width h. Let the limit of integration for x be x_0 and $x_0 + nh$. Then (81) yields the corresponding limits of p as 0 and n. Therefore, integration of Newton's forward difference formula

$$f(x) = f_0 + p\Delta f_0 + \frac{p(p-1)}{2!}\Delta^2 f_0 + \frac{p(p-1)(p-2)}{3!}\Delta^3 f_0 + \frac{p(p-1)(p-2)(p-3)}{4!}\Delta^4 f_0 + \ldots$$

yields

$$\int_{x_0}^{x_0+nh} f(x)dx = h\int_0^n \left[f_0 + p\Delta f_0 + \binom{p}{2}\Delta^2 f_0 + \binom{p}{3}\Delta^3 f_0 + \binom{p}{4}\Delta^4 f_0 + \ldots\right] dp$$

$$= h\left[nf_0 + \frac{n^2}{2}\Delta f_0 + \left(\frac{n^3}{3} - \frac{n^2}{2}\right)\frac{\Delta^2 f_0}{2} + \left(\frac{n^4}{4} - n^3 + n^2\right)\frac{\Delta^3 f_0}{3!} + \ldots\right]. \qquad (83)$$

From this general formula, we obtain district quadrature formulas by putting $n = 1, 2, 3, \ldots$.

(a) Trapezoidal rule: Setting $n = 1$ in the general formula (83), we get the differences $\Delta^2, \Delta^3, \ldots$ to be zero and, therefore, for the interval $[x_0, x_1]$, we have

$$\int_{x_0}^{x_1} f(x)\,dx = h\left[f_0 + \tfrac{1}{2}\Delta f_0\right] = h\left[f_0 + \tfrac{1}{2}(f_1 - f_0)\right] = \frac{h}{2}(f_0 + f_1),$$

which is called *trapezoidal rule.*

For the next intervals $[x_1, x_2], [x_2, x_3], \ldots [x_{n-1}, x_n]$, we have

$$\int_{x_1}^{x_2} f(x)dx = \frac{h}{2}(f_1 + f_2)$$

$$\ldots\ldots\ldots\ldots\ldots\ldots$$

$$\ldots\ldots\ldots\ldots\ldots\ldots$$

$$\int_{x_{n-1}}^{x_n} f(x)dx = \frac{h}{2}(f_{n-1} + f_n).$$

Adding all these expressions, we get

$$\int_{x_0}^{x_n} f(x)dx = \frac{h}{2}[f_0 + 2(f_1 + f_2 + \ldots + f_{n-1}) + f_n],$$

which is known as the *composite trapezoidal rule.*

(b) Simpson's one-third rule: Setting $n = 2$ in the general formula (83), the differences $\Delta^3, \Delta^4, \ldots$ are all zero. The interval of integration is from x_0 to $x_0 + 2h$ and the functional values available to us are f_0, f_1, and f_2. Thus we have, from general formula (83),

$$\int_{x_0}^{x_0+2h} f(x)\,dx = h\left[2f_0 + 2\Delta f_0 + \left(\frac{8}{3} - 2\right)\frac{\Delta^2 f_0}{2}\right]$$

$$= h\left[2f_0 + 2(f_1 - f_0) + \tfrac{1}{3}(f_2 - 2f_1 + f_0)\right]$$

$$= \frac{h}{3}[f_0 + 4f_1 + f_2],$$

which is known as *Simpson's one-third rule.*

Similarly,

$$\int_{x_2}^{x_4} f(x)dx = \frac{h}{3}[f_2 + 4f_3 + f_4]$$

$$\int_{x_4}^{x_6} f(x)dx = \frac{h}{3}[f_4 + 4f_5 + f_6]$$

....

....

$$\int_{x_{n-2}}^{x_n} f(x)dx = \frac{h}{3}[f_{n-2} + 4f_{n-1} + f_n].$$

Thus for even n, adding the above expressions gives

$$\int_{x_0}^{x_{0+nh}} f(x)dx = \frac{h}{3}[(f_0+f_n) + 4(f_1+f_3+\ldots+f_{n-1}) + 2(f_2+f_4+\ldots+f_{n-2})]$$

which is known as *composite Simpson's rule* or *parabolic rule* and is probably the most useful formula for mechanical quadrature. Obviously, to use this formula, we divide the interval of integration into an even number of sub-intervals of width h. The geometric significance of *Simpson's rule* is that we replace the graph of the given function by $\frac{n}{2}$ arcs of the second degree polynomials or parabolas with vertical axis.

(c) Simpson's three-eight rule: If we put $n = 3$ in the general formula (83), then the values available are f_0, f_1, f_2, f_3, and so the differences Δ^4, Δ^5, ... are all zero. Then we shall obtain

$$\int_{x_0}^{x_3} f(x)dx = \int_{x_0}^{x_0+3h} f(x)dx = \frac{3h}{8}[f_0 + 3f_1 + 3f_2 + f_3],$$

$$\int_{x_3}^{x_6} f(x)dx = \frac{3h}{8}[f_3 + 3f_4 + 3f_5 + f_6],$$

....

....

$$\int_{x_{n-3}}^{x_n} f(x)dx = \frac{3h}{8}[f_{n-3} + 3f_{n-2} + 3f_{n-1} + f_n].$$

Thus if n is a multiple of 3, then adding the above expressions, we get

$$\int_{x_0}^{x_0+nh} f(x)dx = \frac{3h}{8}[(f_0+f_n) + 3(f_1+f_2+f_4+f_5+\ldots + f_{n-1}) + 2(f_3+f_6+\ldots+f_{n-3})],$$

which is called *Simpson's three–eight rule.* Thus, in this method, we divide the interval of integration into multiple of 3 subintervals.

(d) Boole's rule: If $n = 4$, the available values of f are f_0, f_1, f_2, f_3, f_4 and, therefore, Δ^5, Δ^6, ... are zero. So, putting $n = 4$ in the general quadrature formula, we get

$$\int_{x_0}^{x_0+4h} f(x)dx = \int_{x_0}^{x_4} f(x)dx$$

$$= h\left[4f_0 + 8\Delta f_0 + \frac{20}{3}\Delta^2 f_0 + \frac{8}{3}\Delta^3 f_0 + \frac{28}{90}\Delta^4 f_0\right]$$

$$= \frac{2h}{45}[7f_0 + 32f_1 + 12f_2 + 32f_3 + 7f_4],$$

$$\int_{x_4}^{x_8} f(x)dx = \frac{2h}{45}[7f_4 + 32f_5 + 12f_6 + 32f_7 + 7f_8],$$

....

....

Adding these integrals, we get

$$\int_{x_0}^{x_0+nh} f(x)dx = \frac{2h}{45}[7f_0 + 32f_1 + 12f_2 + 32f_3 + 14f_4 + 32f_5 + 12f_6 + 32f_7 + 14f_8 + \ldots],$$

where n is a multiple of 4. This formula is known as *Boole's rule.*

(e) Weddle's rule: If $n = 6$, then Δ^7, Δ^8, ... are zero and we have

$$\int_{x_0}^{x_0+6h} f(x)dx = h\left[6f_0 + 18\Delta f_0 + 27\Delta^2 f_0 + 24\Delta^3 f_0 + \frac{123}{10}\Delta^4 f + \frac{33}{10}\Delta^5 f_0 + \frac{41}{140}\Delta^6 f_0\right].$$

The coefficient of $\Delta^6 f_0$ differs from $\frac{3}{10}$ by a small function $\frac{1}{140}$. Therefore, if we replace this coefficient by $\frac{3}{10}$, we commit an error of only $\frac{h}{140}\Delta^6 f_0$.

For small values of h, this error is negligible. Making this change, we get

$$\int_{x_0}^{x_0+6h} f(x)dx = \frac{3h}{10}[f_0 + 5f_1 + f_2 + 6f_3 + f_4 + 5f_5 + f_6].$$

Similarly,

$$\int_{x_0+6h}^{x_0+12h} f(x)dx = \frac{3h}{10}[f_6 + 5f_7 + f_8 + 6f_9 + f_{10} + 5f_{11} + f_{12}],$$

....

....

So, if n is a multiple of 6, adding all such above expressions, we get

$$\int_{x_0}^{x_0+nh} f(x)dx = \frac{3h}{10}[f_0 + 5f_1 + f_2 + 6f_3 + f_4 + 5f_5 + 2f_6 + 5f_7 + \ldots + 5f_{n-1} + f_n]$$

$$= \frac{3h}{10}\sum_{i=0}^{n} Kf_i,$$

where

$$K = 1, 5, 1, 6, 1, 5, 2, 5, 16, 15, 2 \text{ etc.}$$

This formula is known as *Weddle's rule.* It is more accurate, in general, than Simpson's rule but requires at least seven consecutive values of the function. The geometric meaning of Weddle's rule is that we replace the graph of the given function by $\frac{n}{6}$ arcs of sixth degree polynomials.

If the integrate *Newton's backward difference formula*

$$f_p = f_0 + p\nabla f_0 + \frac{p(p+1)}{2!}\nabla^2 f_0 + \frac{p(p+1)(p+2)}{3!}\nabla^3 f_0 + \ldots,$$

then we get

$$\int_{x_0}^{x_1} f(x)dx = \int_{x_0}^{x_0+h} f(x)dx = h[f_0 + \tfrac{1}{2}\nabla f_0 + \tfrac{5}{12}\nabla^2 f_0 + \tfrac{3}{8}\nabla^3 f_0 + \tfrac{251}{720}\nabla^4 f_0 + \ldots] \quad (84)$$

If we multiply the right-hand side of (84) by the identity operator $(I - \nabla)E$, we get

$$\int_{x_0}^{x_1} f(x)dx = h[f_1 - \tfrac{1}{2}\nabla f_1 - \tfrac{1}{12}\nabla^2 f_1 - \tfrac{1}{24}\nabla^3 f_1 - \tfrac{19}{720}\nabla^4 f_1 - \ldots \quad (85)$$

The above two formulas are used for the numerical solution of differential equations. Formula (84) is an extrapolation formula because it uses the ordinates at $x_0, x_{-1}, x_{-2}, \ldots$ to find the integral up to x_1. For this reason, it is called a *predictor,* whereas (85) is called *corrector* and is more accurate as its coefficients are smaller which make it more rapidly convergent than the predictor.

EXAMPLE 15.44

The prime number theorem states that the number of primes in the interval $a < x < b$ is approximately $\int_a^b \frac{dx}{\log x}$. Use this for $a = 100$ and $b = 200$ and compare with the exact value.

Solution. We know that

$$\log_e x = \log_{10} x \log_e 10 = (2.302585)\log_{10} x.$$

Therefore,

$$\int_{100}^{200} \frac{dx}{\log x} = \int_{100}^{200} \frac{dx}{(2.3025)\log_{10} x}.$$

We have the following table for the integrand values:

x	100	150	200
f	$\frac{1}{2(2.302585)}$	$\frac{1}{2.1760(2.302585)}$	$\frac{1}{2.3010(2.302585)}$

Here $h = 50$. We use Simpson's rule and get

$$\int_{100}^{200} \frac{dx}{\log x} = \frac{h}{3}(f_0 + 4f_1 + f_2)$$

$$= \frac{50}{3}\left(\frac{1}{4.60517} + \frac{4}{5.0104} + \frac{1}{5.282}\right)$$

$$= 16.6667\,(0.2171 + 0.7983 + 0.1887)$$
$$= 20.068.$$

If $h = 25$, then the table is

x	100	125	150	175	200
f	0.2171	0.2071	0.1996	0.1936	0.1887

and, therefore, Simpson's formula now yields

$$\int_{100}^{200} \frac{dx}{\log x} = \frac{h}{3}[(f_0+f_4)+4(f_1+f_3)+2f_2]$$

$$= \frac{25}{3}[0.4058 + 4(0.4007) + 2(0.1996)]$$

$$= 20.065.$$

The exact number of primes between 100 and 200 is 21.

EXAMPLE 15.45

Evaluate $\int_0^1 \frac{dx}{1+x^2}$ using

1. Trapezoidal rule taking $h = \frac{1}{4}$
2. Simpson's $\frac{1}{3}$ rule taking $h = \frac{1}{4}$
3. Simpson's $\frac{3}{8}$ rule taking $h = \frac{1}{6}$
4. Weddle's rule taking $h = \frac{1}{6}$.

Solution. The value of $f(x) = \frac{1}{1+x^2}$ for first two cases are

x:	0	$\frac{1}{4}$	$\frac{1}{2}$	$\frac{3}{4}$	1
$f(x)$:	1	0.9412	0.8000	0.6400	0.5000

Case (i): By Trapezoidal rule, we have

$$\int_0^1 \frac{dx}{1+x^2} = \frac{h}{2}[f_0 + 2(f_1+f_2+f_3)+f_4]$$

$$= \frac{1}{8}[1 + 2(0.9412 + 0.8000 + 0.6400) + 0.5000] = 0.7828.$$

Case (ii): Using Simpson's one-third rule, we have

$$\int_0^1 \frac{dx}{1+x^2} = \frac{h}{3}[f_0 + 4(f_1+f_3) + 2f_2 + f_4]$$

$$= \frac{1}{12}[1 + 4(0.9412 + 0.6400) + 2(0.8000) + 0.5000] = 0.7854.$$

The values of $f(x)$ for the cases (iii) and (iv) are:

x	0	$\frac{1}{6}$	$\frac{1}{3}$	$\frac{1}{2}$	$\frac{2}{3}$	$\frac{5}{6}$	1
$f(x)$	1	0.9730	0.9000	0.8000	0.6923	0.5902	0.5000

Case (iii): By Simpson's $\frac{3}{8}$ rule, we have

$$\int_0^1 \frac{dx}{1+x^2} = \frac{3h}{8}[(f_0+f_6)+3(f_1+f_2+f_4+f_5)+2f_3]$$

$$= \frac{1}{16}[(1+0.5)+3(0.9730+0.9000+0.6923+0.5902)+2(0.8)] = 0.78541.$$

Case(iv): By Weddle's rule we have

$$\int_0^1 \frac{dx}{1+x^2} = \frac{3h}{10}[f_0 + 5f_1 + f_2 + 6f_3 + f_4 + 5f_5 + f_6]$$

$$= \frac{1}{20}[1+5(0.9730) + 0.9000 + 6(0.8) + 0.6923 + 5(0.5902) + 0.5000] = 0.78542.$$

15.15 ORDINARY DIFFERENTIAL EQUATIONS

An *ordinary differential equation* is an equation containing one independent variable and one dependent variable and at least one of its derivatives with respect to the independent variable. We know that a differential equation of nth order has n independent arbitrary constants in its general solution. Therefore, we need n conditions to compute the numerical solution of a nth order differential equation.

Problems in which all the initial conditions are specified only at the initial points are called *initial value problems or marching problems.* Thus, in an initial value problem, all the auxiliary conditions are specified at a point, for example value of $y, y', \ldots, y^{(n-1)}$ at the point x_0.

As an illustration, we note that the equation

$$y' = x - y^2, y(0) = 1$$

is an initial value problem.

Problems involving second and higher order differential equations in which auxiliary conditions are specified at two or more points are called *boundary value problems or jury problems.*

As an illustration, we note that the equation

$$y'' = xy, y(0) = 0, y(2) = 1$$

is a boundary value problem.

Classification of Methods of Solution

Consider first order differential equation $y' = f(x, y)$. Let $x_n = x_0 + nh$ and let y_n be the corresponding value of y obtained from a particular method. If the value y_{n+1} appears as a function of just one y-value y_n, then the method is called a *Single step method.* On the other hand, if the value y_{n+1} appears as a function of several values $y_n, y_{n-1}, \ldots, y_{n-p}$, then the method is called a *multi-step method.* Thus, a single-step method is a method which requires only one preceeding value of y whereas a multi-step method requires two or more preceeding values of y.

1. Taylor Series Method

Let $f(x, y)$ be a function that is differentiable for sufficient number of times and let

$$\frac{dy}{dx} = y' = f(x, y), y(x_0) = y_0 \tag{86}$$

be the initial value theorem. We expand $y(x)$ into Taylor series about the point x_0. Thus

$$y(x_0 + h) = y_0 + h\, y'_0 + \frac{h^2}{2!} y''_0 + \ldots + \frac{h^p}{p!} y_0^{(p)} + \frac{h^{p+1}}{(p+1)!} y^{(p+1)}(\xi), \tag{87}$$

where ξ is a point in $[x_0, x]$. Since the solution is not known, the derivatives in the expansion are not known. However, they can be obtained by taking total derivative of the differential equation (86). Therefore,

$$y' = f(x, y),$$

$$y'' = f_x + f_y y' = f_x + f f_y,$$

$$y''' = f_{xx} + f_{xy} f + f_{yx} f + f_{yy} f^2 + f_y f_x + f_y^2 f$$

$$= f_{xx} + 2 f_{xy} f + f^2 f_{yy} + f_y^2 f,$$

and so on.

The number of terms to be included in (87) is fixed by permissible error. If the permissible error is ε and the series in (87) is truncated after the term in $y^{(p)}$, then we have

$$\frac{h^{p+1}}{(p+1)!} |y^{(p+1)}(\xi)| < \varepsilon$$

or

$$\frac{h^{p+1}}{(p+1)!} |f^p(\xi)| < \varepsilon.$$

For a given h, we can find p and obtain an upper bound on h. For computational purposes $|f^{(p)}(\xi)|$ is replaced by max. $|f^p(\xi_n)|$ in $[x_0, x_n]$.

Advantages:

1. A large interval can be used by increasing the number of terms.
2. No special starting procedure is required.
3. The values computed can be checked by applying Taylor's expansion equally on either side of the point x_n. Thus corresponding to y_{n+1}, we may also compute y_{n-1} from the series

$$y_{n+1} = y_n + h y'_n + \frac{h^2}{2} y''_n + \frac{h^3}{3!} y'''_n + \ldots,$$

$$y_{n-1} = y_n - h y'_n + \frac{h^2}{2!} y''_n - \frac{h^3}{3!} y'''_n + \ldots$$

Disadvantages:

1. The necessity of calculating the higher derivatives makes this method completely unsuitable on high speed computers.
2. The method is laborious and so is not recommended except for a few equations.

EXAMPLE 15.46

Solve by Taylor series method:

$$y' = y - \frac{2x}{y}, y(0) = 1 \text{ for } x = 0.1 \text{ and } -0.1.$$

Solution. The given equation is

$$y' = y - \frac{2x}{y}, y(0) = 1.$$

Therefore,

$$y'' = \frac{y(2yy' - 2) - (y^2 - 2x)y'}{y^2} = \frac{2yy' - 2 - y'^2}{y},$$

$$y''' = \frac{2yy'' - 3y'y'' + 2y'^2}{y},$$

....

so that

$$y'(0) = y(0) - \frac{2(0)}{y(0)} = y(0) = 1,$$

$$y''(0) = \frac{2y(0)y'(0) - y'^2(0) - 2}{y(0)} = \frac{2-1-2}{1} = -1,$$

$$y'''(0) = \frac{2y(0)y''(0) - 3y'(0)y''(0) + 2y'^2(0)}{y(0)},$$

$$= \frac{2(1)(-1) - 3(1)(-1) + 2}{1} = 3,$$

....

....

Therefore,

$$y(0.1) = y(0) + (0.1)y'(0) + \frac{(0.1)^2}{2\,!}y''(0)$$

$$+ \frac{(0.1)^3}{3\,!}y'''(0) + \ldots,$$

$$= 1 + 0.1 + \frac{0.01}{2}(-1) + \frac{0.001}{3\,!}(3)$$

$$+ \ldots = 1.0955.$$

Similarly,

$$y(-0.1) = y(0) - (0.1)y'(0) + \frac{(0.1)^2}{2\,!}y''(0)$$

$$- \frac{(0.1)^3}{3\,!}y'''(0) + \ldots,$$

$$= 1 - 0.1 + \frac{0.01}{2}(-1) + \frac{0.001}{6}(3)$$

$$+ \ldots = 0.8955.$$

EXAMPLE 15.47

Solve the differential equation $y' = x - y^2$, $y(0) = 1$ by series expansion, for $x = 0.2, 0.4, 0.6, 0.8$ and 1 taking step size $h = 0.2$.

Solution. We have

$$y' = x - y^2,$$

$$y'' = 1 - 2yy' = 1 - 2y(x - y^2) = 1 - 2xy + 2y^3,$$

$$y''' = -2yy'' - 2y'^2 = -2(y - 4xy^2 + 3y^4 + x^2),$$

$$y^{iv} = -2yy''' - 2y'y'' - 4y'y'' = -2yy''' - 6y'y'',$$

....

....

Using the initial condition $y(0) = 1$, we get

$$y'(0) = 0 - (y(0))^2 = -1,$$

$$y''(0) = 1 - 2y(0)y'(0) = 1 - 2(1)(-1) = 3,$$

$$y'''(0) = -2y(0)y''(0) - 2y'^2 = 2(1)(3) - 2(-1)^2 = -8,$$

$$y^{iv}(0) = -2y(0)y'''(0) - 6y'(0)y''(0)$$

$$= -2(1)(-8) - 6(-1)(3) = 34.$$

Therefore,

$$y_1 = y(0.2) \approx y(0) + 0.2y'(0) + \frac{(0.2)^2}{2!}y''(0)$$

$$+ \frac{(0.2)^3}{3!}y'''(0) + \frac{(0.2)^4}{4!}y^{iv}(0) + \ldots,$$

$$= 1 - 0.2 + 0.06 - 0.01066 + 0.002266 = 0.8516.$$

Now

$$y_2 = y(0.4) = y_1 + 0.2y_1' + \frac{(0.2)^2}{2!}y_1''$$

$$+ \frac{(0.2)^3}{3!}y_1''' + \frac{(0.2)^4}{4!}y_1^{iv} + \ldots$$

But

$$y_1' = x_1 - y_1^2 = 0.2 - (0.8516)^2 = -0.5252,$$

$$y_1'' = 1 - 2y_1y_1' = 1 - 2(0.8516)(-0.5252) = 1.8945,$$

$$y_1''' = -2y_1y_1'' - 2y_1'^2$$

$$= -2(0.8516)(1.8945) - 2(0.5252)^2$$

$$= -3.2267 - 0.5517 = -3.7784,$$

$$y_1^{(iv)} = -2y_1y_1''' - 6y'y_1''$$

$$= -2(0.8516)(-3.7784) - 6(-0.5252)(1.8945)$$

$$= 6.43537 + 5.96995 = 12.40532.$$

Therefore,

$$y_2 \approx 0.8516 + 0.2(-0.5252) + \frac{0.04}{2}(1.8945)$$

$$+ \frac{0.008}{6}(-3.7784) + \frac{.0016}{24}(12.40532)$$

$$= 0.8516 - 0.10504 + 0.03789$$
$$- 0.00504 + 0.000827 = 0.7802.$$

Similarly, we can calculate $y(0.6)$, $y(.8)$, and $y(1)$.

EXAMPLE 15.48

Solve the differential equation $y'' = xy$ for $x = 0.5$ and $x = 1$ by Taylor series method. Initial values: $x = 0, y = 0, y' = 1$.

Solution. We have

$$y'' = xy,$$
$$y''' = xy' + y,$$
$$y^{iv} = xy'' + y' + y' = xy'' + 2y'$$
$$y^{(v)} = xy''' + y'' + 2y'' = xy''' + 3y''.$$

Initial conditions are $y(0) = 0, y'(0) = 1$. Further,

$$y''(0) = 0,$$
$$y'''(0) = 0 + y(0) = 0,$$
$$y^{iv}(0) = 0 + 2y'(0) = 2(1) = 2,$$
$$y^{(v)}(0) = 0 + 3y''(0) = 0.$$

Hence

$$y_1 = y(0.5) = y(0) + 0.5y'(0) + \frac{(0.5)^2}{2!}y''(0)$$
$$+ \frac{(0.5)^3}{3!}y'''(0) + \frac{(0.5)^4}{4!}y^{iv}(0) + \frac{(0.5)^5}{5!}y^{(v)}(0) + \dots$$
$$= 0 + 0.5(1) + \frac{.0625}{24}(2) = 0.5 + 0.00521$$
$$= 0.50521.$$

Now we find $y_2 = y(1)$. We have

$$y_2 = y_1 + 0.5y_1' + y_1''\frac{(0.5)^2}{2!}$$
$$+ \frac{(0.5)^3}{3!}y_1''' + \frac{(0.5)^4}{4!}y_1^{(iv)} + \dots .$$

But

$$y_1' = y_0' + hy_0'' + \frac{h^2}{2!}y_0''' + \frac{h^3}{3!}y_0^{(iv)} + \dots$$
$$= 1 + 0.5(0) + \frac{0.25}{2}(0) + \frac{0.125}{6}(2)$$
$$+ \dots = 1.04167.$$
$$y_1'' = x_1y_1 = 0.5(0.50521) = 0.25261$$
$$y_1''' = x_1\,y_1' + y_1 = 0.5(1.04167)$$
$$+ 0.50521 = 1.02604.$$
$$y_1^{(iv)} = x_1y_1'' + 2y_1' = 0.5(0.25261)$$
$$+ 2(1.04167) = 2.2096.$$

Hence

$$y_2 = y(1) = 0.50521 + 0.5(1.04167) + \frac{0.25}{2}(0.25261)$$
$$+ \frac{0.125}{6}(1.02604) + \frac{0.0625}{24}(2.2096) + \dots$$
$$\approx 0.50521 + 0.52084 + 0.03157$$
$$+ 0.021376 + 0.00575 = 1.08475.$$

2. Euler's Method

Consider the initial value problem

$$y' = \frac{dy}{dx} = f(x, y), \; y(x_0) = y_0. \qquad (88)$$

The Euler's method is based on the property that in a small interval, a curve is nearly a straight line. Thus if $x \in [x_0, x_1]$, a small interval, we approximate the curve by the tangent at the point (x_0, y_0). But the equation of the tangent at (x_0, y_0) is

$$y - y_0 = \left(\frac{dy}{dx}\right)_{(x_0,\, y_0)} (x - x_0)$$
$$= f(x_0, y_0)(x - x_0), \text{ using } (88)$$

or

$$y = y_0 + (x - x_0) f(x_0, y_0).$$

Therefore, the value of y corresponding to x_1 is

$$y_1 = y_0 + (x_1 - x_0) f(x_0, y_0).$$

If $x_n = x_0 + nh$, then we get

$$y_1 = y_0 + hf(x_0, y_0).$$

Similarly, approximating the curve by the tangent in $[x_1, x_2]$ at the point (x_1, y_1) with slope $f(x_1, y_1)$, we have

$$y_2 = y_1 + hf(x_1, y_1),$$

and so, in general

$$y_{n+1} = y_n + hf(x_n, y_n). \qquad (89)$$

The *Euler's method is very slow.* We have to take h very small to obtain accuracy.

Geometric Interpretation: The Euler method has a very simple geometric interpretation. In the interval $x_n \le x \le x_{n+1}$, the solution is assumed to follow the line tangent to $y(x)$ at (x_n, y_n). When this method is applied repeatedly across several intervals in sequence, the numerical solutions traces a *polygon segment* with sides of slope $f(x_n, y_n)$, $n = 0,1,2,\ldots$ That is why, this method is also called *polygon method.*

Error Analysis of Euler's Method: Let $y(x_n)$ be exact value of y at $x = x_n$ and let y_{n+1} be the computed value of y at $x = x_{n+1}$. Then the *truncation error* after one step, called the local *truncation error,* is given by

$$\begin{aligned} T_{n+1} &= y_{n+1} - y(x_{n+1}) \\ &= y_n + hy'(x_n) - y(x_{n+1}) \quad \text{(by Euler' formula)} \\ &= y_n + hy'(x_n) - [y_n + hy'_n(x_n) + \frac{h^2}{2}y''_n(\xi)] \quad \xi \in [x_n, x_{n+1}] \\ &= -\frac{h^2}{2}y''_n(\xi). \end{aligned}$$

Hence the local truncation error is $O(h^2)$.

The total truncation error is

$$e_n = y_n - y(x_n).$$

We assume that

(i) y_0 is exact so that $e_0 = 0$ and y_i are the values of y computed by Euler's method

(ii) Lipschitz condition

$$|f(x, y) - f(x, y^*)| \le L|y - y^*|$$

is satisfied and

(iii) $|y''(\xi)| \le M$ in the given interval. By Euler's method, we have

$$y_{n+1} = y_n + hf(x_n, y_n) \tag{90}$$

and by Taylor's expansion, we have

$$y(x_{n+1}) = y(x_n) + hf(x_n, y(x_n)) + \frac{h^2}{2!}y''(\xi). \tag{91}$$

Subtracting (91) from (90), we have

$$e_{n+1} = e_n + h[f(x_n, y_n) - f(x_n, y(x_n)] - \frac{h^2}{2!}y''(\xi).$$

Hence

$$|e_{n+1}| \le |e_n| + hL|y_n - y(x_n)| + \frac{h^2}{2}M$$

or

$$|e_{n+1}| \le (1 + hL)|e_n| + \frac{h^2}{2}M.$$

Putting $1 + hL = A$ and $\frac{h^2}{2}M = B$, we get

$$|e_{n+1}| \le A|e_n| + B, n = 0, 1, 2, \ldots, N-1.$$

Thus

$$\begin{aligned} |e_1| &\le A|e_0| + B \\ |e_2| &\le A|e_1| + B \le A[A|e_0| + B] \\ &= A^2|e_0| + (A+1)B = \frac{A^2-1}{A-1}B + A^2|e_0|, \\ |e_3| &\le A|e_2| + B = A^3|e_0| + \frac{A^3-1}{A-1}B, \end{aligned}$$

....

$$|e_N| \le A^N|e_0| + \frac{A^N-1}{A-1}B.$$

But $e_0 = 0$ and

$$A^N = (1 + hL)^N \le e^{NhL} = e^{L(x_N - x_0)}.$$

Hence

$$|e_N| \le \frac{1}{2}hM\frac{e^{L(x_N-x_0)} - 1}{L} = O(h).$$

The error tends to zero as $h \to 0$ in such a way that $nh = x_n - x_0$ remains constant. From this computation it follows that the *Euler method is convergent.*

Improved Euler's Method: In this method, the curve in the interval $[x_0, x_1]$ is approximated by a line through (x_0, y_0) whose slope is the *average of the slopes at* (x_0, y_0) and $(x_1, y_1^{(1)})$ such that

$$y_1^{(1)} = y_0 + hf(x_0, y_0).$$

Thus the equation of the line becomes

$$y - y_0 = (x - x_0)\left[\frac{1}{2}\{f(x_0, y_0) + f(x_1, y_1^{(1)})\}\right]$$

and so the line through (x_0, y_0) and (x_1, y_1) is

$$y_1 - y_0 = (x_1 - x_0)\frac{1}{2}[f(x_0, y_0) + f(x_1, y_1^{(1)})]$$

or

$$y_1 = y_0 + \frac{h}{2}[f(x_0, y_0) + f(x_1, y_1^{(1)})]$$
$$= y_0 + \frac{h}{2}[f(x_0, y_0) + f(x_0 + h, y_0 + h f(x_0, y_0))].$$

Hence the general formula becomes

$$y_{n+1} = y_n + \frac{h}{2}[f(x_n, y_n) + f(x_n + h, y_n + hf(x_n, y_n))],$$

where $x_n - x_{n-1} = h$.

Modified Euler's Method: In this method, the curve in the interval $[x_0, x_1]$ is approximated by the line through (x_0, y_0) with slope $f(x_0 + \frac{h}{2}, y_0 + \frac{h}{2} f(x_0, y_0))$, that is, the slope at the mid-point whose abscissa is the average of x_0 and x_1, that is, the slope at $x_0 + \frac{h}{2}$. Thus, the equation of the line is

$$y - y_0 = (x - x_0)\left\{f(x_0 + \frac{h}{2},\ y_0 + \frac{h}{2} f(x_0, y_0))\right\}.$$

Taking $x = x_1$, we have

$$y_1 = y_0 + h\left\{f\left(x_0 + \frac{h}{2},\ y_0 + \frac{h}{2} f(x_0, y_0)\right)\right\}.$$

Hence the general formula becomes

$$y_{n+1} = y_n + h\left\{f\left(x_n + \frac{h}{2},\ y_n + \frac{h}{2} f(x_n, y_n)\right)\right\}.$$

EXAMPLE 15.49

Solve, by Euler's method, the initial value problem

$$\frac{dy}{dx} = \frac{x - y}{2},\ y(0) = 1$$

over [0,3], using step size $\frac{1}{2}$.

Solution. By Euler's method,

$$y_{n+1} = y_n + hf(x_n, y_n).$$

We are given that $h = \frac{1}{2}$ and $f(x, y) = \frac{x - y}{2}$. Therefore,

$$y_{n+1} = y_n + 0.5\left(\frac{x_n - y_n}{2}\right) = 0.25\ x_n + 0.75\ y_n.$$

Thus

$$y_1 = 0.25x_0 + 0.75y_0 = 0.25(0) + 0.75(1) = 0.75,$$
$$y_2 = 0.25x_1 + 0.75y_1 = 0.25(0.5) + 0.75(0.75) = 0.125 + 0.5625 = 0.6875,$$
$$y_3 = 0.25(1) + 0.75(0.6875) = 0.25 + 0.515625 = 0.765625,$$
$$y_4 = 0.25(1.5) + 0.75(0.765625) = 0.375 + 0.57421875 = 0.94921875,$$
$$y_5 = 0.25(2) + 0.75(0.94921875) = 0.50 + 0.711914062 = 1.211914063,$$
$$y_6 = 0.25(2.5) + 0.75(1.211914063) = 0.625 + 0.908935546 = 1.533935547 \approx 1.533936.$$

EXAMPLE 15.50

Solve the initial value problem

$$\frac{dy}{dx} = \frac{y - x}{y + x},\ y(0) = 1$$

for $x = 0.1$ by Euler's method.

Solution. By Euler's method

$$y_{n+1} = y_n + hf(x_n, y_n).$$

We take $h = 0.02$. Therefore

$$y_{n+1} = y_n + 0.02\left(\frac{y_n - x_n}{y_n + x_n}\right)$$

and so

$$y_1 = y_0 + 0.02\left(\frac{y_0 - x_0}{y_0 + x_0}\right) = 1 + 0.02\left(\frac{1 - 0}{1 + 0}\right) = 1.02,$$
$$y_2 = y_1 + 0.02\left(\frac{y_1 - x_1}{y_1 + x_1}\right) = 1.02 + 0.02\left(\frac{1.02 - 0.02}{1.02 + 0.02}\right) = 1.0392,$$
$$y_3 = y_2 + 0.02\left(\frac{y_2 - x_2}{y_2 + x_2}\right)$$

$$= 1.0392 + 0.02\left(\frac{1.0392 - 0.04}{1.0392 + 0.04}\right) = 1.05918,$$

$$y_4 = y_3 + 0.02\left(\frac{y_3 - x_3}{y_3 + x_3}\right)$$

$$= 1.05918 + 0.02\left(\frac{1.05918 - 0.06}{1.05918 + 0.06}\right) = 1.07917,$$

$$y_5 = y_4 + 0.02\left(\frac{y_4 - x_4}{y_4 + x_4}\right)$$

$$= 1.07917 + 0.02\left(\frac{1.07917 - 0.08}{1.07917 + 0.08}\right) = 1.09916.$$

Hence, the required solution is 1.09916.

EXAMPLE 15.51

Use Euler's method and its modified form to obtain $y(0.2)$, $y(0.4)$, and $y(0.6)$ correct to three decimal places given that $y' = y - x^2$ with initial condition $y(0) = 1$.

Solution. By Euler's method,

$$y_{n+1} = y_n + hf(x_n, y_n).$$

Here $f(x, y) = y - x^2$ and $h = 0.2$. Therefore,

$$y_{n+1} = y_n + 0.2\,(y_n - x_n^2) = 1.2\,y_n - 0.2x_n^2.$$

Thus

$$y_1 = 1.2y_0 - 0.2\,x_0^2 = 1.2(1) = 1.2,$$

$$y_2 = 1.2y_1 - (0.2)x_1^2 = (1.2)^2 - (0.2)^3$$

$$= 1.44 - 0.008 = 1.4320,$$

$$y_3 = 1.2y_2 - (0.2)x_2^2 = (1.2)(1.432)$$

$$- (0.2)(0.4)^2 = 1.6864.$$

Modified Euler's formula is

$$y_{n+1} = y_n + h\,f\left(x_n + \frac{h}{2},\ y_n + \frac{h}{2}f(x_n, y_n)\right).$$

Taking $h = 0.2$, we have

$$y_1 = y_0 + 0.2\left[y_0 + \frac{0.2}{2}(y_0 - x_0^2) - \left(x_0 + \frac{0.2}{2}\right)^2\right]$$

$$= 1 + 0.2[1 + 0.1(1 - 0) - (0 + 0.1)^2]$$

$$= 1 + 0.2(1 + 0.1 - 0.01) = 1.218,$$

$$y_2 = y_1 + 0.2[y_1 + 0.1(y_1 - x_1^2) - (x_1 + 0.1)^2]$$

$$= 1.218 + 0.2[1.218 + 0.1(1.218 + (0.2)^2)$$

$$- (0.2 + 0.1)^2]$$

$$= 1.218 + 0.2[1.218 + 0.1178 - 0.09] = 1.4672,$$

$$y_3 = y_2 + 0.2[y_2 + 0.1(y_2 - x_2^2) - (x_2 + 0.1)^2]$$

$$= 1.4672 + 0.2[1.4672 + 0.1(1.4672 - (0.4)^2)$$

$$- (0.4 + 0.1)^2]$$

$$= 1.4672 + 0.2[1.4672 + 0.13072 - 0.25]$$

$$= 1.7368.$$

3. Picard's Method of Successive Integration

Consider the initial value problem $y'(x) = f(x, y(x))$ over $[a, b]$ with $y(x_0) = y_0$.

Using fundamental theorem of calculus, we have

$$\int_{x_0}^{x_1} f(x, y(x))dx = \int_{x_0}^{x_1} y'(x)dx = y(x_1) - y(x_0).$$

Thus

$$y(x_1) = y(x_0) + \int_{x_0}^{x_1} f(x, y(x))dx.$$

Thus, if we start with the approximation $y(x_0)$, then

$$y_1 = y_0 + \int_{x_0}^{x_1} f(x, y_0)\, dx,$$

$$y_2 = y_0 + \int_{x_0}^{x_1} f(x, y_1)\, dx,$$

$$\cdots \quad \cdots \quad \cdots$$

$$\cdots \quad \cdots \quad \cdots$$

$$y_{n+1} = y_0 + \int_{x_0}^{x_1} f(x, y_n)dx.$$

We stop the process when $y_{n+1} = y_n$ up to the desired decimal places.

The *Picard's method of successive integration fails if the function is not easily integrable.*

EXAMPLE 15.52

Using Picard's method, solve

$$\frac{dy}{dx} = x^2 - y,\ y(0) = 1$$

for $x = 0.2$.

Solution. We start with the approximation $y(0) = 1$. Then

$$y_1 = y_0 + \int_0^{0.2} (x^2 - y_0)dx = 1 + \int_0^{0.2} (x^2 - 1)\, dx$$

$$= 1 + \left[\frac{x^3}{3} - x\right]_0^{0.2} = 1 + \left[\frac{(0.2)^3}{3} - 0.2\right] = 0.8027,$$

$$y_2 = 1 + \int_0^{0.2} (x^2 - y_1)\, dx = 1 + \int_0^{0.2} (x^2 - 0.8027)dx$$

$$= 1 + \left[\frac{x^3}{3} - 0.8027x\right]_0^{0.2}$$
$$= 1 + [.00267 - 0.16054] = 0.8421,$$

$$y_3 = 1 + \int_0^{0.2} (x^2 - y_2)dx = 1 + \left[\frac{x^3}{3} - y_2 x\right]_0^{0.2}$$

$$= 1 + [0.00267 - (0.8421)(0.2)] = 0.8342,$$
$$y_4 = 1 + [0.00267 - (0.8342)(0.2)] = 0.8358,$$
$$y_5 = 1 + [0.00267 - (0.8358)(0.2)]= 0.8355.$$

Hence $y(0.2) = 0.835$ up to three decimal places.

EXAMPLE 15.53

Solve

$$y' = x^2 + 2xy, y(0) = 0.$$

Solution. We take first approximation to be $y(0) = 0$. Then

$$y_1 = y_0 + \int_0^x (x^2 + 2xy(0))dx = 0 + \int_0^x x^2 dx = \frac{x^3}{3},$$

$$y_2 = 0 + \int_0^x \left(x^2 + 2x\left(\frac{x^3}{3}\right)\right) dx = \frac{x^3}{3} + \frac{2x^5}{15},$$

$$y_3 = 0 + \int_0^x \left[x^2 + 2x\left(\frac{x^3}{3} + \frac{2x^5}{15}\right)\right]dx$$
$$= \frac{x^3}{3} + \frac{2x^5}{3(5)} + \frac{4x^7}{3(5)(7)},$$

$$y_4 = 0 + \int_0^x \left[x^2 + 2x\left(\frac{x^3}{3} + \frac{2x^5}{3(5)} + \frac{4x^7}{3(5)(7)}\right)\right]dx$$

$$= \frac{x^3}{3} + \frac{2x^5}{3(5)} + \frac{4x^7}{3(5)(7)} + \frac{8x^9}{3(5)(7)(9)}.$$

EXAMPLE 15.54

Solve by Picard's method,

$$\frac{dy}{dx} = 1 + xy, y(0) = 1$$

for $x = 0.1$.

Solution. We take first approximation to be $y(0) = 1$. Then

$$y_1 = y_0 + \int_0^x f(x, y(0))dx$$
$$= 1 + \int_0^x (1 + x)dx = 1 + x + \frac{x^2}{2},$$

$$y_2 = 1 + \int_0^x \left[1 + x\left(1 + x + \frac{x^2}{2}\right)\right]dx$$
$$= 1 + x + \frac{x^2}{2} + \frac{x^3}{3} + \frac{x^4}{8},$$

$$y_3 = 1 + \int_0^x \left[1 + x\left(1 + x + \frac{x^2}{2} + \frac{x^3}{3} + \frac{x^4}{8}\right)\right]dx$$
$$= 1 + x + \frac{x^2}{2} + \frac{x^3}{3} + \frac{x^4}{8} + \frac{x^5}{15} + \frac{x^6}{48}.$$

Thus

$$y_3(0.1) = 1 + 0.1 + \frac{(0.1)^2}{2} + \frac{(0.1)^3}{3} + \frac{(0.1)^4}{8} + \frac{(0.1)^5}{15} + \frac{(0.1)^6}{48}$$
$$= 1 + 0.1 + \frac{0.01}{2} + \frac{0.001}{3} + \frac{0.0001}{8} + \frac{0.00001}{15} + \frac{0.000001}{48} = 1.105346.$$

Further

$$y_4 = 1 + \int_0^x \left[1 + x\left(1 + x + \frac{x^2}{2} + \frac{x^3}{3} + \frac{x^4}{8} + \frac{x^5}{15} + \frac{x^6}{48}\right)\right] dx$$
$$= 1 + x + \frac{x^2}{2} + \frac{x^3}{3} + \frac{x^4}{8} + \frac{x^5}{15} + \frac{x^6}{48} + \frac{x^7}{105} + \frac{x^8}{384}.$$

Thus

$$y_4(0.1) = 1 + 0.1 + \frac{0.01}{2} + \frac{0.001}{3} + \frac{0.0001}{8} + \frac{0.00001}{15} + \frac{0.000001}{48} + \frac{0.0000001}{105} + \frac{0.00000001}{384} = 1.1053465.$$

Hence

$$y(0.1) = 1.1053465.$$

4. Fourth Order Runge-Kutta Method

Consider the initial value problem

$$y'(x) = f(x, y), y(x_0) = y_0.$$

We define

$$\left.\begin{aligned} K_1 &= h\,f(x_r, y_r) \\ K_2 &= h\,f(x_r + mh, y_r + mK_1) \\ K_3 &= h\,f(x_r + nh, y_r + nK_2) \end{aligned}\right\} \quad (92)$$
$$K_4 = h\,f(x_r + ph, y_r + pK_3)$$

We wish to obtain a formula of the type

$$y_{r+1} = y_r + aK_1 + bK_2 + cK_3 + dK_4. \quad (93)$$

Let

$$F_1 = f_x + ff_y$$
$$F_2 = f_{xx} + 2ff_{xy} + f^2 f_{yy}$$
$$F_3 = f_{xxx} + 3ff_{xxy} + 3f^2 f_{xyy} + f^3 f_{yyy}.$$

Expanding y_{r+1} in series, we obtain

$$y_{r+1} = y_r + hf + \frac{h^2}{2}F_1 + \frac{h^3}{3!}(F_2 + F_1 f_y) + \frac{h^4}{4!}(F_3 + F_1 f_y^2 + 3F_1(f_{xy} + f_{yy}f)) + O(h^5). \quad (94)$$

Further, using Taylor's theorem for two variables, we have

$$K_1 = h\,f(x_r, y_r),$$
$$K_2 = h[f(x_r, y_r) + mh\,F_1 + \frac{m^2h^2}{2}F_2 + \frac{m^3h^3}{3\,!}F_3 + \ldots],$$
$$K_3 = h[f(x_r, y_r) + nh\,F_1 + \frac{h^2}{2}(n^2F_2 + 2mnF_1 f_y) + \frac{h^3}{6}(n^3F_3 + 3m^2nF_2 f_y + 6mn^2F_1 f_y') + \ldots],$$
$$K_4 = h[f(x_r, y_r) + phF_1 + \frac{h^2}{2}(p^2F_2 + 2npF_1 f_y) + \frac{h^3}{6}(p^3F_3 + 3n^2pF_2 f_y + 6np^2F_1 f_y' + 6mnpF_1 f_y^2) + \ldots].$$

Putting these values of K_1, K_2, K_3, and K_4 in (93) and equating the like powers of h in the corresponding expressions for y_{r+1}, we obtain

$$\left.\begin{aligned} &a + b + c + d = 1, & &cmn + dnp = \tfrac{1}{6} \\ &bm + cn + dp = \tfrac{1}{2}, & &cmn^2 + dnp^2 = \tfrac{1}{8} \\ &bm^2 + cn^2 + dp^2 = \tfrac{1}{3}, & &cm^2n + dn^2p = \tfrac{1}{12} \\ &bm^3 + cn^3 + dp^3 = \tfrac{1}{4}, & &dmnp = \tfrac{1}{24}. \end{aligned}\right\} \quad (95)$$

Any solution of (94) will serve our purpose. Let us take $m = n = \frac{1}{2}, p = 1, a = d = \frac{1}{6}, b = c = \frac{1}{3}$. Then

$$K_1 = hf(x_r, y_r),$$
$$K_2 = hf(x_r + \frac{h}{2}, y_r + \frac{K_1}{2}),$$
$$K_3 = hf(x_r + \frac{h}{2}, y_r + \frac{K_2}{2}),$$
$$K_4 = hf(x_r + h, y_r + K_3)$$

and

$$y_{r+1} = y_r + \frac{1}{6}(K_1 + 2K_2 + 2K_3 + K_4),$$

which is the required *fourth order Runge–Kutta method.*

Remark 15.11. Whenever we mention only Runge–Kutta method, we mean the Runge–Kutta method of order 4.

EXAMPLE 15.55

Use Runge–Kutta method to solve $y' = x + y$, $y(0) = 1$, for $x = 0.1$.

Solution. Taking $h = 0.1$, we obtain

$$K_1 = hf(x_0, y_0) = 0.1(x_0 + y_0) = 0.1(0 + 1) = 0.1,$$
$$K_2 = hf\left(x_0 + \frac{h}{2},\ y_0 + \frac{K_1}{2}\right) = 0.1\left(x_0 + \frac{h}{2} + y_0 + \frac{K_1}{2}\right) = 0.1(0 + 0.05 + 1 + 0.05) = 0.11,$$
$$K_3 = hf\left(x_0 + \frac{h}{2},\ y_0 + \frac{K_2}{2}\right) = 0.1\left(0 + 0.05 + 1 + \frac{0.11}{2}\right) = 0.1105,$$
$$K_4 = hf(x_0 + h, y_0 + K_3) = 0.1(0 + 0.1 + 1 + 0.1105) = 0.12105.$$

Therefore,

$$y_1 = y(0.1) = y_0 + \frac{1}{6}(K_1 + 2K_2 + 2K_3 + K_4) = 1 + \frac{1}{6}(0.1 + 0.22 + 0.2210 + 0.12105) = 0.11034167.$$

EXAMPLE 15.56

Apply fourth order Runge–Kutta method to

$$\frac{dy}{dx} = 3x + \frac{1}{2}y, y(0) = 1$$

to determine $y(0.1)$ and $y(0.2)$ correct to four decimal places.

Solution. Taking $h = 0.1$, we have

$$K_1 = hf(x_0, y_0) = 0.1\left(0 + \frac{1}{2}\right) = 0.05,$$

$$K_2 = hf\left(x_0 + \frac{h}{2},\ y_0 + \frac{K_1}{2}\right)$$
$$= 0.1\left[3(0 + 0.05) + \frac{1}{2}(1 + 0.025)\right] = 0.06625,$$

$$K_3 = hf\left(x_0 + \frac{h}{2},\ y_0 + \frac{K_2}{2}\right)$$
$$= 0.1\left[3(0 + 0.05) + \frac{1}{2}\left(1 + \frac{0.06625}{2}\right)\right]$$
$$= 0.06665625,$$

$$K_4 = hf(x_0 + h, y_0 + K_3) = 0.1[3(0 + 0.1) + \frac{1}{2}(1 + 0.06665625)]$$
$$= 0.1[0.3 + 0.533328125] = 0.0833328125.$$

Hence

$$y_1 = y_0 + \frac{1}{6}[K_1 + 2K_2 + 2K_3 + K_4]$$
$$= 1 + \frac{1}{6}[0.05 + 2(0.0625) + 2(0.06665625) + 0.0833328125]$$
$$= 1.06652421875 \approx 1.0665.$$

To find $y(0.2)$, we note that

$$K_1 = hf(x_1, y_1) = 0.1[3(0.1) + \frac{1}{2}(1.066524)]$$
$$= 0.0833262,$$

$$K_2 = hf\left(x_1 + \frac{h}{2}, y_1 + \frac{K_1}{2}\right)$$
$$= 0.1\left[3(0.1 + 0.05) + \frac{1}{2}\left(1.066524 + \frac{0.0833262}{2}\right)\right]$$
$$= 0.100409515,$$

$$K_3 = hf\left(x_1 + \frac{h}{2}, y_1 + \frac{K_2}{2}\right)$$
$$= 0.1\left[3(0.1 + 0.05) + \frac{1}{2}\left(1.066524 + \frac{0.100409515}{2}\right)\right]$$
$$= 0.100836437,$$

$$K_4 = hf(x_1 + h, y_1 + K_3)$$
$$= 0.1\left[3(0.1 + 0.1) + \frac{1}{2}\left(1.066524 + \frac{0.100836437}{2}\right)\right]$$
$$= 0.11584711.$$

Hence

$$y(0.2) = y_1 + \frac{1}{6}(K_1 + 2K_2 + 2K_3 + K_4)$$
$$= 1.06652422 + \frac{1}{6}[0.0833262 + 2(0.100409515) + 2(0.100836437) + 0.11584711]$$
$$= 1.166801756 \approx 1.1668.$$

EXAMPLE 15.57

Apply the fourth order Runge-Kutta method to solve

$$\frac{dy}{dx} = x^2 + y^2,\ y(0) = 1.$$

Take step size $h = 0.1$ and determine approximations to $y(0.1)$ and $y(0.2)$ correct to four decimal places.

Solution. Taking $h = 0.1$, we have

$$K_1 = hf(x_0, y_0) = 0.1(0 + 1) = 0.1,$$

$$K_2 = hf\left(x_0 + \frac{h}{2},\ y_0 + \frac{K_1}{2}\right)$$
$$= 0.1\left[(0.05)^2 + \left(1 + \frac{0.1}{2}\right)^2\right]$$
$$= 0.1105,$$

$$K_3 = hf\left(x_0 + \frac{h}{2},\ y_0 + \frac{K_2}{2}\right)$$
$$= 0.1\left[(0.05)^2 + \left(1 + \frac{0.1105}{2}\right)^2\right]$$
$$= 0.111605256,$$

$$K_4 = hf(x_0 + h, y_0 + K_3) = 0.1[(0.1)^2 + (1 + 0.111605256)^2]$$
$$= 0.124566624.$$

Therefore,

$$y_1 = y(0.1) = y_0 + \frac{1}{6}(K_1 + 2K_2 + 2K_3 + K_4)$$
$$= 1 + \frac{1}{6}(0.1 + 2(0.1105) + 2(0.111605256) + 0.124566624)$$
$$= 1.111462856 \approx 1.11146.$$

To find $y(0.2)$, we have

$$K_1 = hf(x_1, y_1) = 0.1[x_1^2 + y_1^2]$$
$$= 0.1[(0.1)^2 + (1.1114628)^2] = 0.124534956,$$
$$K_2 = hf\left(x_1 + \frac{h}{2},\ y_1 + \frac{K_1}{2}\right)$$
$$= 0.1\left[\left(0.1 + \frac{0.1}{2}\right)^2 + \left(1.1114628 + \frac{0.124534956}{2}\right)^2\right]$$
$$= 0.1400142,$$
$$K_3 = hf\left(x_1 + \frac{h}{2}, y_1 + \frac{K_2}{2}\right)$$
$$= 0.1\left[0.0225 + \left(1.1114628 + \frac{0.1400142}{2}\right)^2\right]$$
$$= 0.1418371125,$$
$$K_4 = hf(x_1 + h,\ y_1 + K_3)$$
$$= 0.1[(0.2)^2 + (1.1114628 + 0.141837112)^2]$$
$$= 0.161076063.$$

Hence

$$y_2 = y(0.2) = y_1 + \frac{1}{6}(K_1 + 2K_2 + 2K_3 + K_4)$$
$$= 1.11142856 + \frac{1}{6}[0.124534956 + 2(0.1400142)$$
$$+ 2(0.1418371125) + 0.161076063]$$
$$= 1.2529808 \approx 1.2530.$$

15.16 NUMERICAL SOLUTION OF PARTIAL DIFFERENTIAL EQUATIONS

Partial differential equations appear in the description of physical processes in applied sciences and engineering. A differential equation which involves more than one independent variable is called a *partial differential equation.* We restrict ourselves to second order partial differential equations. The general second order linear partial differential equation is of the form

$$Au_{xx} + Bu_{xy} + Cu_{yy} + Du_x + Eu_y + Fu = G,$$

where A, B, C, D, E, F, and G are all functions of x and y. Equations of the above form can be classified into three types:

(i) If $B^2 - 4AC < 0$ at a point in the (x, y) plane, then the equation is called *elliptic.* For example, the equation $u_{xx} + u_{yy} = 0$, known as *Laplace equation,* is elliptic.

(ii) If $B^2 - 4\,AC = 0$ at a point in the (x, y) plane, then the equation is called *parabolic.* For example, the equation $u_{xx} - u_t = 0$, called the *heat conduction equation,* is parabolic.

(iii) If $B^2 - 4AC > 0$ at a point in the (x, y) plane, then the equation is called *hyperbolic.* For example, the equation $u_{xx} - \frac{1}{c^2} u_{tt} = 0$, known as the *wave equation,* is hyperbolic.

The most popular method for solving partial differential equation is *finite-difference method.* This method is based on formulas for approximating the first and second derivatives of a function.

15.16.1 Formation of Difference Equation

To get finite difference analogue of partial differential equation, we replace the derivatives in the equation by their corresponding difference approximations.

We first derive difference formula for approximating u_x. By Taylor's series about the point (x_0, y_0), we have

$$u(x_0 + h, y_0) = u(x_0, y_0) + hu_x(x_0, y_0) + \frac{h^2}{2}u_{xx}(\xi, y_0), x_0 \le \xi \le x_0 + h.$$

and so

$$u_x(x_0, y_0) = \frac{u(x_0 + h, y_0) - u(x_0, y_0)}{h} + O(h) \text{ (error)}$$

Thus, the finite difference formula for the first derivative is

$$u_x(x_0, y_0) = \frac{u(x_0 + h, y_0) - u(x_0, y_0)}{h} + O(h).$$

Dropping the term $O(h)$ and using $u_{i,j}$ for $u(x_i, y_j)$, $i = 0,1,2, \ldots$, we get

$$u_x \approx \frac{u_{i+1,j} - u_{i,j}}{h}, \tag{96}$$

which is called *forward difference approximation* to u_x. Similarly, expanding $u(x_0 - h, y_0)$ by Taylor series, we get

$$u_x(x_0, y_0) \approx \frac{u(x_0, y_0) - u(x_0 - h, y_0)}{h}$$

or

$$u_x \approx \frac{u_{i,j} - u_{i-1,j}}{h}, \qquad (97)$$

which is known as *backward difference approximation* to u_x.

Now using formular (97), we get by Taylor's series,

$$u_{xx} \approx \frac{u(x_0+h,y_0) - 2u(x_0,y_0) + u(x_0-h,y_0)}{h^2}$$

or

$$u_{xx} \approx \frac{u_{i+1,j} - 2u_{i,j} + u_{i-1,j}}{h^2} \qquad (98)$$

as the difference approximation to u_{xx}.

Similarly, we have the approximation with $y = i\,k$, $k = 0,1,2,\ldots,$

$$u_y \approx \frac{u_{i,j+1} - u_{i,j}}{k}, \qquad (99)$$

$$u_y \approx \frac{u_{i,j} - u_{i,j-1}}{k} \qquad (100)$$

and

$$u_{yy} \approx \frac{u(x_0,y_0+k) - 2u(x_0,y_0) + u(x_0,y_0-k)}{k^2}$$

or

$$u_{yy} \approx \frac{u_{i,j+1} - 2u_{i,j} + u_{i,j-1}}{k^2}.$$

15.16.2 Geometric Representation of Partial Difference Quotients

Let (x, y) plane be partitioned into a network of rectangles of sides $\Delta x = h$ and $\Delta y = k$ by drawing the sets of lines

$$x = ih,\ i = 0, 1, 2, 3, \ldots,$$

$$y = jk,\ j = 0, 1,2,3, \ldots.$$

The points of intersection of these families of lines are called *mesh points, grid points,* or *lattice points.* Thus the points (x, y), $(x + h, y)$, $(x + 2h, y)$, $(x - h, y)$, $(x - 2h, y)$ . . . , $(x, y + k)$, $(x, y + 2h)$, . . . are the grid points as shown in Figure 15.5.

If we represent (x, y) by (i, j), then $(x + h, y) = (i + 1, j)$, $(x + 2h, y) = (i + 2, j)$, and so on.

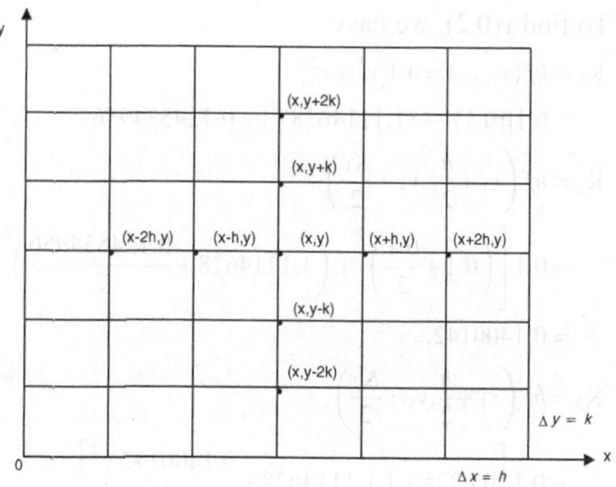

Figure 15.5

15.16.3 Standard Five Point Formula and Diagonal Five-Point Formula

Consider the Laplace equation in two dimensions,

$$u_{xx} + u_{yy} = 0.$$

Its finite difference analogue is

$$\frac{u_{i+1,j} - 2u_{i,j} + u_{i-1,j}}{h^2} + \frac{u_{i,j+1} - 2u_{i,j} + u_{i,j-1}}{k^2} = 0. \qquad (101)$$

If we consider square mess, that's is, $h = k$, then the equation (101) yields

$$u_{i,j} = \frac{1}{4}(u_{i+1,j} + u_{i-1,j} + u_{i,j+1} + u_{i,j-1}). \qquad (102)$$

The equation (102) shows that the *value of u at any point is the mean of its values at the four neighbouring points* as shown in Figure 15.6a.

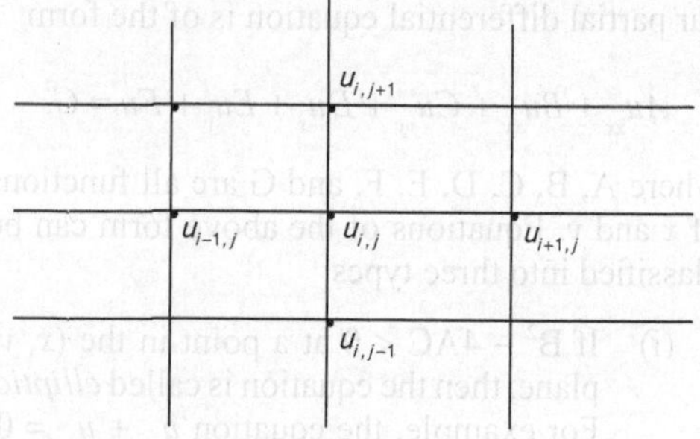

Figure 15.6a Standard five-point formula

The formula (102) is called the *standard five -point formula.*

If we rotate the co-ordinate axes through 45°, then the Laplace equation remains invariant. In fact, if $X = x\cos\theta + y\sin\theta$, $Y = x\sin\theta - y\cos\theta$, where $\theta = 45°$, then $u_{xx} + u_{yy} = 0$. Therefore, we may use the function values at the diagonal points (Fig. 15.6b.) in place of the neighbouring points. Then we may use the formula

$$u_{i,j} = \frac{1}{4}(u_{i-1,j-1} + u_{i+1,j-1} + u_{i+1,j+1} + u_{i-1,j+1})$$

in place of (102). This formula is called the *diagonal five-point formula.*

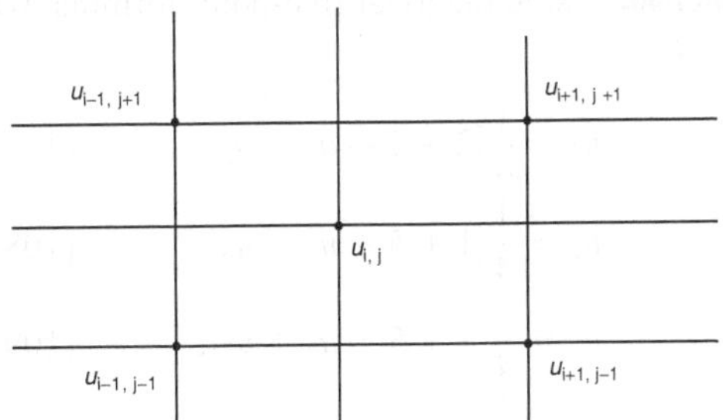

Figure 15.6b Diagonal five-point formula

15.16.4 Point Jacobi's Method

Let $u_{i,j}^{(n)}$ be the nth iterative value of $u_{i,j}$. Then the iterative procedure to solve (102) is

$$u_{i,j}^{(n+1)} = \frac{1}{4}(u_{i-1,j}^{(n)} + u_{i+1,j}^{(n)} + u_{i,j-1}^{(n)} + u_{i,j+1}^{(n)})$$

for the interior mesh points. This procedure is called the *point Jacobi method.*

15.16.5 Gauss-Seidel Method

This method uses the latest iterative values available and scans the mesh points systematically from left to right along successive rows. The formula is

$$u_{i,j}^{(n+1)} = \frac{1}{4}\left[u_{i-1,j}^{(n+1)} + u_{i+1,j}^{(n)} + u_{i,j-1}^{(n+1)} + u_{i,j+1}^{(n)}\right].$$

It can be shown that the *Gauss-Seidel method converges twice as fast as the Jacobi's method.*

EXAMPLE 15.58

Solve Laplace equation $u_{xx} + u_{yy} = 0$ for the following square meshes with boundary conditions exhibited in Figure 15.7

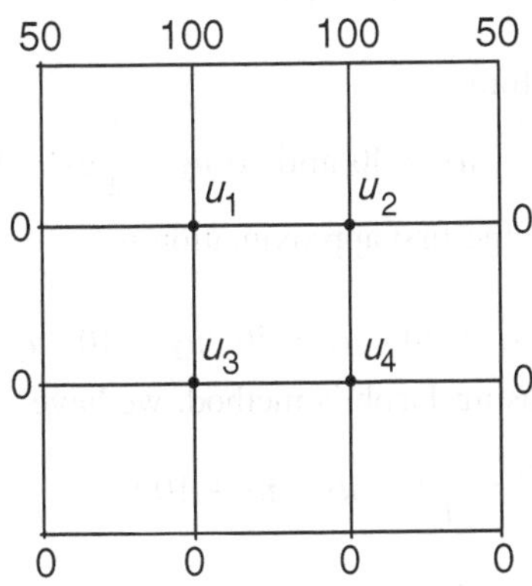

Figure 15.7

Solution. Using diagonal five-point formula, we have

$$u_1 = \frac{1}{4}[0 + 100 + 50 + u_4] \qquad (103)$$

$$u_2 = \frac{1}{4}[100 + 0 + u_3 + 50] \qquad (104)$$

$$u_3 = \frac{1}{4}[0 + u_2 + 0 + 0] \qquad (105)$$

$$u_4 = \frac{1}{4}[0 + 0 + 0 + u_1]. \qquad (106)$$

From (105) and (106), we have

$$u_3 = \frac{1}{4}u_2,$$

$$u_4 = \frac{1}{4}u_1.$$

Then (103) yields

$$u_1 = \frac{1}{4}\left(150 + \frac{1}{4}u_1\right),$$

or

$$\frac{15}{16}u_1 = \frac{150}{4}$$

or

$$u_1 = 40 \text{ and so } u_4 = 10.$$

Similarly (104) yields

$$u_2 = \frac{1}{4}[150 + u_3] = \frac{1}{4}\left(150 + \frac{1}{4}u_2\right)$$
$$= \frac{150}{4} + \frac{1}{16}u_2.$$

Therefore,

$$u_2 = 40 \text{ and so } u_3 = \frac{1}{4}u_2 = 10.$$

Hence the first approximation is

$$u_1 = 40,\ u_2 = 40,\ u_3 = 10,\ u_4 = 10.$$

Now using Jacobi's method, we have

$$u_1^{(1)} = \frac{1}{4}[0 + u_2 + u_3 + 100]$$
$$= \frac{1}{4}[0 + 40 + 10 + 100] = \frac{150}{4} = 37.5,$$
$$u_2^{(1)} = \frac{1}{4}[40 + 10 + 0 + 100] = \frac{150}{4} = 37.5,$$
$$u_3^{(1)} = \frac{1}{4}[0 + 0 + 40 + 10] = \frac{50}{4} = 12.5,$$
$$u_4^{(1)} = \frac{1}{4}[10 + 0 + 0 + 40] = \frac{50}{4} = 12.5.$$

The next approximation is

$$u_1^{(2)} = \frac{1}{4}[0 + 12.5 + 37.5 + 100] = \frac{150}{4} = 37.5,$$
$$u_2^{(2)} = \frac{1}{4}[37.5 + 0 + 12.5 + 100] = \frac{150}{4} = 37.5,$$
$$u_3^{(2)} = \frac{1}{4}[0 + 0 + 12.5 + 37.5] = \frac{50}{4} = 12.5,$$
$$u_4^{(2)} = \frac{1}{4}[0 + 0 + 12.5 + 37.5] = \frac{50}{4} = 12.5.$$

Hence the solution is

$$u_1 = 37.5,\ u_2 = 37.5,\ u_3 = 12.5,\ u_4 = 12.5.$$

EXAMPLE 15.59

Solve Laplace equation

$$u_{xx} + u_{yy} = 0$$

for the squares meshes with the boundary values shown in the Figure 15.8

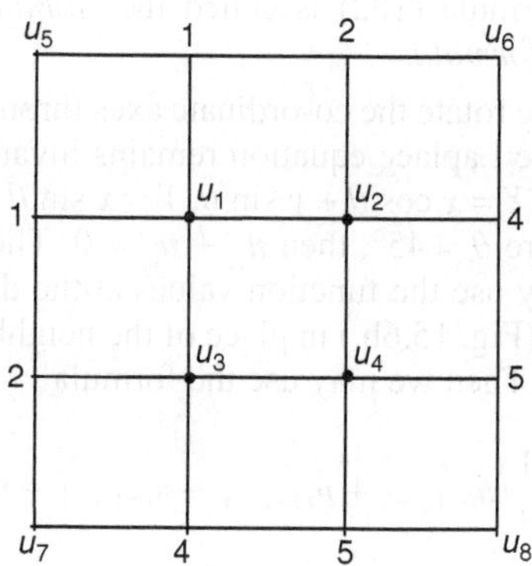

Figure 15.8

Solution. Using diagonal five-point formula, we have

$$u_1 = \frac{1}{4}[2 + 2 + u_4 + u_5], \quad (107)$$
$$u_2 = \frac{1}{4}[1 + 5 + u_3 + u_6], \quad (108)$$
$$u_3 = \frac{1}{4}[1 + 5 + u_2 + u_7], \quad (109)$$
$$u_4 = \frac{1}{4}[4 + 4 + u_1 + u_8]. \quad (110)$$

If we use standard five-point formula, we have

$$u_1 = \frac{1}{4}[1 + u_2 + u_3 + 1], \quad (111)$$
$$u_2 = \frac{1}{4}[4 + 2 + u_1 + u_4], \quad (112)$$
$$u_3 = \frac{1}{4}[2 + 4 + u_4 + u_1], \quad (113)$$
$$u_4 = \frac{1}{4}[5 + 5 + u_2 + u_3]. \quad (114)$$

The expressions (112) and (113) shows that $u_2 = u_3$. Therefore, (111), (112), (113), and (114) reduces to

$$u_1 = \frac{1}{4}[2 + 2u_2],$$
$$u_2 = \frac{1}{4}[6 + u_1 + u_4],$$
$$u_3 = \frac{1}{4}[6 + u_1 + u_4],$$
$$u_4 = \frac{1}{4}[10 + 2u_2].$$

If we start with the approximation $u_2 = 0$, then

$$u_1 = \frac{1}{2} \cdot u_2 = 0,\ u_3 = 0,\ u_4 = \frac{5}{2}.$$

Then, by Gauss-Seidel's method, we have

$$u_1^{(1)} = \frac{1}{4}[1 + 0 + 1 + 0] = \frac{1}{2} = 0.5,$$

$$u_2^{(1)} = \frac{1}{4}\left[\frac{1}{2} + \frac{5}{2} + 4 + 2\right] = \frac{9}{4} = 2.25,$$

$$u_3^{(1)} = \frac{1}{4}\left[2 + 4 + \frac{5}{2} + \frac{1}{2}\right] = \frac{9}{4} = 2.25,$$

$$u_4^{(1)} = \frac{1}{4}\left[5 + 5 + \frac{9}{4} + \frac{9}{4}\right] = \frac{5}{2} = 2.50.$$

$$u_1^{(2)} = \frac{1}{4}[1 + 2.25 + 1 + 2.25] = 1.625,$$

$$u_2^{(2)} = \frac{1}{4}[1.625 + 4 + 2.50 + 2] = 2.53125,$$

$$u_3^{(2)} = \frac{1}{4}[2 + 4 + 1.625 + 2.50] = 2.53125,$$

$$u_4^{(2)} = \frac{1}{4}[5 + 5 + 2.53125 + 2.53125] = 3.765625.$$

$$u_1^{(3)} = \frac{1}{4}[1 + 1 + 2.53125 + 2.53125] = 1.765625,$$

$$u_2^{(3)} = \frac{1}{4}[4 + 2 + 1.765625 + 3.765625] = 2.8828125,$$

$$u_3^{(3)} = 2.8828125 \text{ since } u_2 = u_3,$$

$$u_4^{(3)} = \frac{1}{4}[5 + 5 + 2.8828125 + 2.8828125] = 3.9414031.$$

$$u_1^{(4)} = \frac{1}{4}[1 + 1 + 2.8828125 + 2.8828125] = 1.94140625,$$

$$u_2^{(4)} = \frac{1}{4}[2 + 4 + 1.94140625 + 3.9414031] = 2.970702338,$$

$$u_3^{(4)} = 2.970702338, \text{ since } u_2 = u_3,$$

$$u_4^{(4)} = \frac{1}{4}[5 + 5 + 2.970702338 + 2.970702338] = 3.985351169.$$

$$u_1^{(5)} = \frac{1}{4}[1 + 1 + 2.970702338 + 2.970702338] = 1.985351169,$$

$$u_2^{(5)} = \frac{1}{4}[2 + 4 + 1.985351169 + 3.985331169] = 2.992675585,$$

$$u_3^{(5)} = 2.992675585, \text{ since } u_2 = u_3,$$

$$u_4^{(5)} = \frac{1}{4}[5 + 5 + 2.992675585 + 2.992675585] = 3.996337793.$$

$$u_1^{(6)} = \frac{1}{4}[1 + 1 + 2.992675585 + 2.992675585] = 1.996337793,$$

$$u_2^{(6)} = \frac{1}{4}[2 + 4 + 1.996337793 + 3.996337793] = 2.998168897,$$

$$u_3^{(6)} = 2.998168897, \text{ since } u_2 = u_3,$$

$$u_4^{(6)} = \frac{1}{4}[5 + 5 + 2.998168897 + 2.998168897] = 3.999084449.$$

$$u_1^{(7)} = \frac{1}{4}[1 + 1 + 2.998168897 + 2.998168897] = 1.999084449,$$

$$u_2^{(7)} = \frac{1}{4}[2 + 4 + 1.999084449 + 3.999084449] = 2.999542224,$$

$$u_3^{(7)} = 2.999542224, \text{ since } u_2 = u_3,$$

$$u_4^{(7)} = \frac{1}{4}[5 + 5 + 2.999542224 + 2.999542224] = 3.999771112.$$

We observe that the values of sixth and seventh iterations agree up to two decimal places. Hence

$$u_1 = 1.99,\ u_2 = 2.99,$$

$$u_3 = 2.99,\ u_4 = 3.99.$$

EXAMPLE 15.60

Using the given boundary values, solve the Laplace equation $\nabla^2 u = 0$ at the nodal points of the square grid shown in Figure 15.9

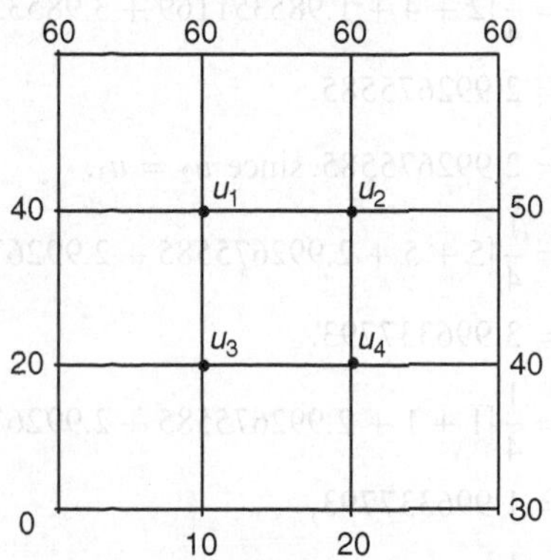

Figure 15.9

Solution. We assume that $u_4 = 0$. Then the initial approximation is

$$u_1 = \frac{1}{4}[20+60+60+0]$$
$$= 35 \text{ (diagonal five-point formula)},$$
$$u_2 = \frac{1}{4}[35+60+50+0]$$
$$= 36.25 \text{ (standard five-point formula)},$$
$$u_3 = \frac{1}{4}[35+20+10+0]$$
$$= 16.25 \text{ (standard five-point formula)},$$
$$u_4 = \frac{1}{4}[36.25+16.25+20+40]$$
$$= 28.125 \text{ (standard five-point formula)}.$$

Now using Gauss-Seidel's method, we have

$$u_1^{(1)} = \frac{1}{4}[60+40+u_2+u_3]$$
$$= \frac{1}{4}[100+36.25+16.25] = 38.125,$$
$$u_2^{(1)} = \frac{1}{4}[60+50+u_1^{(1)}+u_4]$$
$$= \frac{1}{4}[110+38.125+28.125] = 44.0625,$$
$$u_3^{(1)} = \frac{1}{4}[20+10+u_1^{(1)}+u_4]$$
$$= \frac{1}{4}[30+38.125+28.125] = 24.0625,$$
$$u_4^{(1)} = \frac{1}{4}[40+20+u_2^{(1)}+u_3^{(1)}]$$
$$= \frac{1}{4}[60+44.0625+24.0625] = 32.03125.$$

$$u_1^{(2)} = \frac{1}{4}[100+44.0625+24.0625] = 42.03125,$$
$$u_2^{(2)} = \frac{1}{4}[110+42.03125+32.03125] = 46.015625,$$
$$u_3^{(2)} = \frac{1}{4}[30+42.03125+32.03125] = 26.015625,$$
$$u_4^{(2)} = \frac{1}{4}[60+46.015625+26.015625]$$
$$= 33.0078125.$$
$$u_1^{(3)} = \frac{1}{4}[100+46.015625+26.015625]$$
$$= 43.0078125,$$
$$u_2^{(3)} = \frac{1}{4}[110+43.0078125+33.0078125]$$
$$= 46.50390625,$$
$$u_3^{(3)} = \frac{1}{4}[30+43.0078125+33.0078125]$$
$$= 26.50390625,$$

$$u_4^{(3)} = \frac{1}{4}[60+46.50390625+26.50390625]$$
$$= 33.25195311.$$
$$u_1^{(4)} = \frac{1}{4}[100+46.50390625+26.50390625]$$
$$= 43.25195311,$$
$$u_2^{(4)} = \frac{1}{4}[110+43.25195311+33.25195311]$$
$$= 46.62597655,$$
$$u_3^{(4)} = \frac{1}{4}[30+43.25195311+33.25195311]$$
$$= 26.625997656,$$
$$u_4^{(4)} = \frac{1}{4}[60+46.62597655+26.62597656]$$
$$= 33.31298827.$$
$$u_1^{(5)} = \frac{1}{4}[100+46.62597655+26.62597656]$$
$$= 43.31298827,$$
$$u_2^{(5)} = \frac{1}{4}[110+43.31298827+33.31298827]$$
$$= 46.65649412,$$
$$u_3^{(5)} = \frac{1}{4}[30+43.31298827+33.31298827]$$
$$= 26.65649414,$$

$$u_4^{(5)} = \frac{1}{4}[60 + 46.65649412 + 26.65649414]$$
$$= 33.32824706$$

Hence $u_1 = 43.313$, $u_2 = 46.656$, $u_3 = 26.60$, $u_4 = 33.328$

EXAMPLE 15.61

Solve the Laplace equation $u_{xx} + u_{yy} = 0$ for the square mesh with boundary values shown in the Figure 15.10.

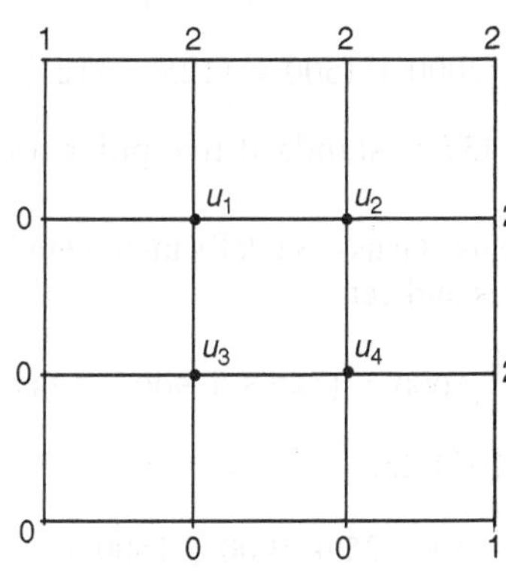

Figure 15.10

Solution. We assume that $u_4 = 0$. Then the first approximation is

$$u_1 = \frac{1}{4}[1 + 2 + +0 + 0]$$
$$= 0.75 \text{(diagonal five-point formula)},$$
$$u_2 = \frac{1}{4}[0.75 + 2 + 2 + 0]$$
$$= 1.1875 \text{ (standard five-point formula)},$$
$$u_3 = \frac{1}{4}[0 + 0 + 0.75 + 0]$$
$$= 0.1875 \text{ (standard five-point formula)},$$
$$u_4 = \frac{1}{4}[0.1875 + 2 + 0 + 1.1875]$$
$$= 0.84375 \text{ (standard five-point formula)}.$$

Now using Gauss–Seidel's method, we have

$$u_1^{(1)} = \frac{1}{4}[0 + 1.1875 + 0.1875 + 2]$$
$$= 0.84375,$$
$$u_2^{(1)} = \frac{1}{4}[0.84375 + 2 + 2 + 0.84375]$$
$$= 1.421875,$$
$$u_3^{(1)} = \frac{1}{4}[0 + 0.84375 + 0 + 0.84375]$$
$$= 0.421875,$$
$$u_4^{(1)} = \frac{1}{4}[0.421875 + 2 + 0 + 1.421875]$$
$$= 0.9609375.$$
$$u_1^{(2)} = \frac{1}{4}[0 + 2 + 1.421875 + 0.421875]$$
$$= 0.9609375,$$
$$u_2^{(2)} = \frac{1}{4}[0.9609375 + 2 + 2 + 0.9609375]$$
$$= 1.48046875,$$
$$u_3^{(2)} = \frac{1}{4}[0 + 0.9609375 + 0 + 0.9609375]$$
$$= 0.48046875,$$
$$u_4^{(2)} = \frac{1}{4}[0.4804675 + 2 + 0 + 1.48046875]$$
$$= 0.990234375.$$
$$u_1^{(3)} = \frac{1}{4}[0 + 1.48046875 + 0.48046875 + 2]$$
$$= 0.990234375,$$
$$u_2^{(3)} = \frac{1}{4}[0.990234375 + 2 + 0.990234375 + 2]$$
$$= 1.495117188,$$
$$u_3^{(3)} = \frac{1}{4}[0 + 0.990234375 + 0 + 0.990234375]$$
$$= 0.495117188,$$
$$u_4^{(3)} = \frac{1}{4}[0.495117187 + 2 + 0 + 1.495117187]$$
$$= 0.997558593.$$
$$u_1^{(4)} = \frac{1}{4}[0 + 1.495117188 + 2 + 0.495117188]$$
$$= 0.997558594,$$
$$u_2^{(4)} = \frac{1}{4}[0.997558594 + 2 + 2 + 0.997558593]$$
$$= 1.498779297,$$

$$u_3^{(4)} = \frac{1}{4}[0 + 0.997558593 + 0 + 0.997558594]$$
$$= 0.498779296,$$
$$u_4^{(4)} = \frac{1}{4}[0.498779296 + 2 + 0 + 1.498779297]$$
$$= 0.999389648.$$

Hence up to two decimal places, we have

$$u_1 = 0.99, u_2 = 1.49, u_3 = 0.49, u_4 = 0.99.$$

EXAMPLE 15.62

Solve the elliptic equation $u_{xx} + u_{yy} = 0$ for the square mesh with boundary values shown in the Figure 15.11.

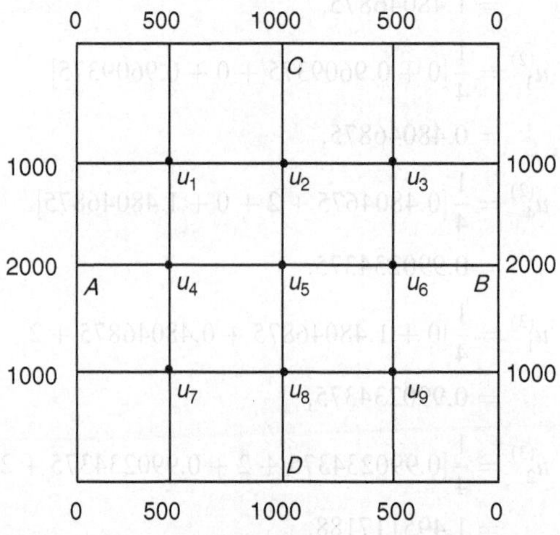

Figure 15.11

Solution. We observe that the figure is symmetrical about AB and so

$$u_1 = u_7, u_2 = u_8, u_3 = u_9$$

Similarly, the figure's symmetry about CD yields

$$u_1 = u_3, u_4 = u_6, u_7 = u_9.$$

Thus

$$u_1 = u_3 = u_7 = u_9, u_2 = u_8, u_4 = u_6.$$

Hence it is sufficient to find u_1, u_2, u_4, and u_5. To get initial values, we have

$$u_5 = \frac{1}{4}[2000 + 2000 + 1000 + 1000]$$
$$= 1500 \text{ (standard five-point formula)},$$
$$u_1 = \frac{1}{4}[0 + 1000 + 2000 + 1500]$$
$$= 1125 \text{ (diagonal five-point formula)},$$
$$u_2 = \frac{1}{4}[1125 + 1125 + 1000 + 1500]$$
$$= 1187.5 \text{ (standard five-point formula)},$$
$$u_4 = \frac{1}{4}[2000 + 1500 + 1125 + 1125]$$
$$= 1437.5 \text{ (standard five-point formula)}.$$

Now we use Gauss–Seidel's method to improve the values and get

$$u_1^{(1)} = \frac{1}{4}[1000 + 1187.5 + 500 + 1437.5]$$
$$= 1031.25,$$
$$u_2^{(1)} = \frac{1}{4}[1031.25 + 1000 + 1500 + 1031.25]$$
$$= 1140.625,$$
$$u_4^{(1)} = \frac{1}{4}[2000 + 1500 + 1031.25 + 1031.25]$$
$$= 1390.625$$
$$u_5^{(1)} = \frac{1}{4}[1390.625 + 1390.625 + 1140.625 + 1140.625] = 1265.625.$$

After nine iterations, we shall obtain

$$u_1 = u_3 = u_7 = u_9 = 938.05$$
$$u_2 = u_8 = 1000.55$$
$$u_4 = u_6 = 1250.55$$
$$u_5 = 1125.55.$$

EXAMPLE 15.63

Determine the system of four equations in four unknowns p_1, p_2, p_3, and p_4 for computing approximation for the harmonic function $u(x, y)$ in the rectangle $R = \{(x, y): 0 \le x \le 3, 0 \le y \le 3\}$ shown in the Figure 15.12, under the conditions

$$u(x,0) = 10, u(x,3) = 90 \text{ for } 0 < x < 3$$
$$u(0,y) = 70, u(3,y) = 0 \text{ for } 0 < y < 3.$$

Hence find p_1, p_2, p_3, and p_4

Solution. We want to solve $u_{xx} + u_{yy} = 0$ under the given conditions in the problem. Taking $h = k = 1$, the square mess is shown in the Figure 15.12.

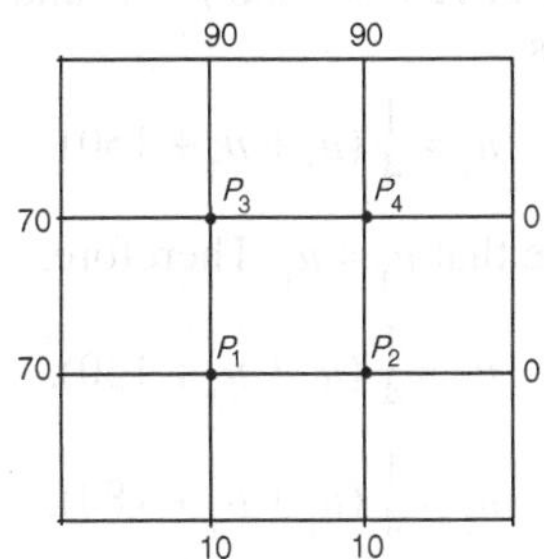

Figure 15.12

Therefore, using standard five–point formula, we have

$$-4p_1 + p_2 + p_3 + 0p_4 = -80,$$

$$p_1 - 4p_2 + 0p_3 + p_4 = -10,$$

$$p_1 - 0p_2 - 4P_3 + p_4 = -160,$$

$$0p_1 + p_2 + p_3 - 4p_4 = -90.$$

We solve this system by Gaussian elimination method. The augmented matrix is

$$\begin{matrix}\text{Pivot} \rightarrow \\ m_{21} = -1/4 \\ m_{31} = -1/4 \\ \\ \end{matrix}\left[\begin{array}{cccc|c} -4 & 1 & 1 & 0 & -80 \\ 1 & -4 & 0 & 1 & -10 \\ 1 & 0 & -4 & 1 & -160 \\ 0 & 1 & 1 & -4 & -90 \end{array}\right]$$

The result after first elimination is

$$\begin{matrix} \\ \text{Pivot} \rightarrow \\ m_{32} = -1/15 \\ m_{42} = -4/15 \end{matrix}\left[\begin{array}{cccc|c} -4 & 1 & 1 & 0 & -80 \\ 0 & -15/4 & 1/4 & 1 & -30 \\ 1 & 1/4 & -15/4 & 1 & -180 \\ 0 & 1 & 1 & -4 & -90 \end{array}\right]$$

The second elimination yields

$$\begin{matrix} \\ \\ \text{Pivot} \rightarrow \\ m_{43} = -2/7 \end{matrix}\left[\begin{array}{cccc|c} -4 & 1 & 1 & 0 & -80 \\ 0 & -15/4 & 1/4 & 1 & -30 \\ 0 & 0 & -56/15 & 16/15 & -182 \\ 0 & 0 & 16/15 & -56/15 & -98 \end{array}\right]$$

The third elimination yields

$$\left[\begin{array}{cccc|c} -4 & 1 & 1 & 0 & -80 \\ 0 & -15/4 & 1/4 & 1 & -30 \\ 0 & 0 & -56/15 & 16/15 & -182 \\ 0 & 0 & 0 & -24/7 & -1050/7 \end{array}\right].$$

Back substitution yields

$$p_4 = \frac{1050}{24} = 43.75, \; -\frac{56}{15}p_3 + \frac{16}{15}p_4 = -182$$

and so $p_3 = 61.16$,

$$-\frac{15}{4}p_2 + \frac{1}{4}p_3 + p_4 = -30$$

and so $sp_2 = 23.61$,

and

$$-4p_1 + p_2 + p_3 = -80$$

and so $p_1 = 41.19$.

15.16.6 Poisson's Equation

The elliptic partial differential equation

$$u_{xx} + u_{yy} = f(x, y), \tag{115}$$

where $f(x,y)$ is a given function of x and y, is called the *Poisson's equation.*

The Poisson's equation is solved numerically by replacing the derivatives by difference expressions at the points $x = ih$ and $y = jh$. Thus we have

$$\frac{u_{i-1,j} - 2u_{i,j} + u_{i+j}}{h^2} + \frac{u_{i,j-1} - 2u_{i,j} + u_{i,j+1}}{h^2}$$
$$= f(ih, jh)$$

or

$$u_{i-1,j} - 4u_{i,j} + u_{i+1,j} + u_{i,j-1} + u_{i,j+1}$$

$$= h^2 f(ih, jh). \qquad (116)$$

The error involved in (116) is $O(h^2)$.

EXAMPLE 15.64

Solve the Poisson's equation

$$u_{xx} + u_{yy} = -10(x^2 + y^2 + 10)$$

over the square with sides $x = 0 = y$, $x = 3 = y$ with $u = 0$ on the boundary and mesh length 1.

Solution. Since mess length is 1 and side of the square is 3, the Figure 15.13 of the problem is

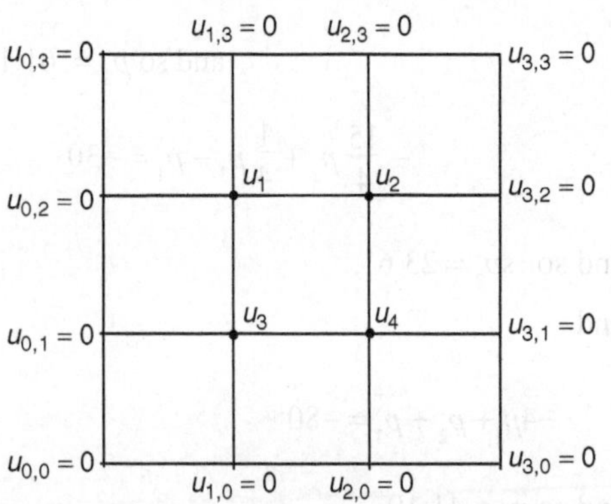

Figure 15.13

By standard formula (116), we have

$$u_{i-1,j} - 4u_{i,j} + u_{i+1,j} + u_{i,j-1} + u_{i,j+1} = h^2 f(ih, jh).$$

For u_1, we have $i = 1$, and $j = 2$ and so the formula gives

$$u_{0,2} - 4u_{12} + u_{2,2} + u_{1,1} + u_{1,3} = h^2 f(h, 2h)$$

or

$$0 - 4u_1 + u_2 + u_3 + 0 = f(1,2)$$

$$= -10(1 + 4 + 10)$$

or

$$u_1 = \frac{1}{4}(u_2 + u_3 + 150).$$

Now, for u_2, we have $i = 2$ and $j = 2$. Therefore, the formula yields

$$u_2 = \frac{1}{4}(u_1 + u_4 + 180).$$

For u_3, we have $i = 1$ and $j = 1$ and so the formula yields

$$u_3 = \frac{1}{4}(u_1 + u_4 + 120).$$

For u_4, we have $i = 2$ and $j = 1$ and so the formula yields

$$u_4 = \frac{1}{4}(u_2 + u_3 + 150).$$

We observe that $u_1 = u_4$. Therefore,

$$u_1 = \frac{1}{4}(u_2 + u_3 + 150),$$

$$u_2 = \frac{1}{4}(u_1 + u_4 + 180),$$

$$u_3 = \frac{1}{4}(u_1 + u_4 + 120).$$

We start with $u_2 = u_3 = 0$ and use Gauss–Seidel's method to improve the values. We have

$$u_1^{(1)} = \frac{1}{4}(0 + 0 + 150) = 37.5,$$

$$u_2^{(1)} = \frac{1}{4}[2(37.5) + 180] = 63.75,$$

$$u_3^{(1)} = \frac{1}{4}[2(37.5) + 120] = 48.75.$$

$$u_1^{(2)} = \frac{1}{4}[63.75 + 48.75 + 150] = 65.625,$$

$$u_2^{(2)} = \frac{1}{4}[2(65.625) + 180] = 77.8125,$$

$$u_3^{(2)} = \frac{1}{4}[2(65.625) + 120] = 62.8125.$$

$$u_1^{(3)} = \frac{1}{4}[77.8125 + 62.8125 + 150] = 72.65625,$$

$$u_2^{(3)} = \frac{1}{4}[2(72.65625) + 180] = 81.328125,$$

$$u_3^{(3)} = \frac{1}{4}[2(72.65625) + 120] = 66.328125.$$

$$u_1^{(4)} = \frac{1}{4}[81.328125 + 66.328125 + 150] = 74.4140625,$$

$$u_2^{(4)} = \frac{1}{4}[\,2(74.4140625) + 180] = 82.20703125,$$

$$u_3^{(4)} = \frac{1}{4}[2(74.4140625) + 120] = 67.20703125.$$

$$u_1^{(5)} = \frac{1}{4}[82.2070125 + 67.20703125 + 150] = 74.8535,$$

$$u_2^{(5)} = \frac{1}{4}[2(74.8535) + 180] = 82.4268,$$

$$u_3^{(5)} = \frac{1}{4}[\,2(74.8535) + 120] = 67.4268.$$

$$u_1^{(6)} = \frac{1}{4}[\,82.4268 + 67.4268 + 150] = 74.9634,$$

$$u_2^{(6)} = \frac{1}{4}[2(74.9634) + 180] = 82.4817,$$

$$u_3^{(6)} = \frac{1}{4}[2(74.9634) + 120] = 67.4817.$$

The values obtained by the fifth and sixth iteration are nearly equal and so the solution is

$$u_1 \approx 74.9,\ u_2 \approx 82.5,\ u_3 \approx 67.5,\ u_4 = u_1 = 74.9.$$

EXAMPLE 15.65

The function ϕ satisfies the equation

$$\frac{\partial^2\phi}{\partial x^2} + \frac{\partial^2\phi}{\partial y^2} + 2 = 0$$

at every point inside the square bounded by the straight lines $x = \pm 1$, $y = \pm 1$, and is zero on the boundary. Calculate a finite difference solution using a square mesh of side $\frac{1}{2}$. Assuming that the error is $O(h^2)$, calculate the improved value of ϕ at (0,0).

(The example is the non–dimensional form of the torsion problem for a solid elastic cylinder with a square cross–section).

Solution. The mesh points and the boundary values are shown in Figure 15.14.

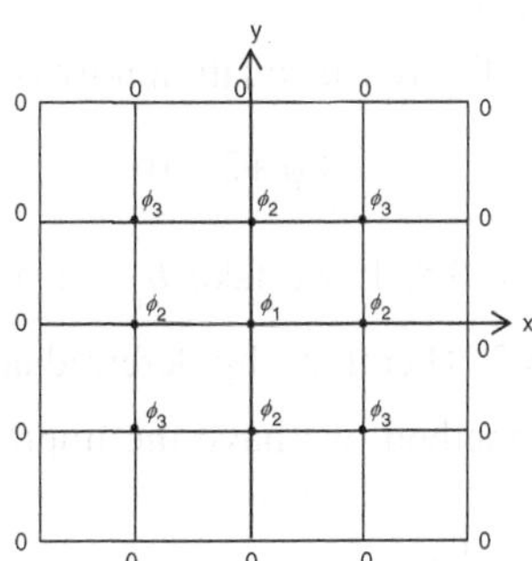

Figure 15.14

Because of the symmetry with respect to x–axis, y–axis, and the diagonals, there are only three unknowns, ϕ_1 at (0,0), ϕ_2 at $\left(\frac{1}{2}, 0\right)$ and ϕ_3 at $\left(\frac{1}{2}, \frac{1}{2}\right)$. The difference equation for the given problem is

$$\frac{\phi(x_0 - h, y_0) - 2\,\phi(x_0, y_0) + \phi(x_0 + h, y_0)}{h^2} + \frac{\phi(x_0, y_0 - h) - 2\,\phi(x_0, y_0) + \phi(x_0, y_0 + h)}{h_2} + 2 = 0$$

or

$$\phi(x_0 - h, y_0) - 4\phi(x_0, y_0) + \phi(x_0 + h, y_0) + \phi(x_0, y_0 - h) + \phi(x_0, y_0 + h) + 2h^2 = 0.$$

Taking $h = \frac{1}{2}$, the above formula yields

$$2\phi_2 - 8\,\phi_1 + 2\,\phi_2 + 2\,\phi_2 + 1 = 0,$$

$$2\phi_1 - 8\,\phi_2 + 0 + 2\,\phi_3 + 2\,\phi_3 + 1 = 0,$$

$$2\phi_2 - 8\,\phi_3 + 0 + 2\,\phi_2 + 0 + 1 = 0.$$

Thus, we have

$$8\phi_2 - 8\,\phi_1 + 1 = 0,$$

$$4\phi_3 + 2\phi_1 - 8\phi_2 + 1 = 0,$$

$$4\phi_2 - 8\phi_3 + 1 = 0.$$

Solving these three equations, we get

$$\phi_1 = 0.562,\ \phi_2 = 0.438,\ \phi_3 = 0.344.$$

On the other hand, if we use coarse mesh of side $h = 1$, then the figure becomes as shown below (Figure.15.15).

The finite difference equation now is

$$-4\phi + 2 = 0$$

and so $\phi = 0.5$. If we take $h_1 = 1$ and $h_2 = \frac{1}{2}$, then $\frac{h_1}{h_2} = 2$. Therefore, by deferred approach to the limit method, we have the improved value of ϕ as

$$\phi^* = \phi_1 + \frac{1}{3}(\phi_1 - \phi)$$

$$= 0.562 + \frac{1}{3}(0.562 - 0.500) = 0.583,$$

which is very close to the exact value 0.589 of ϕ at (0,0).

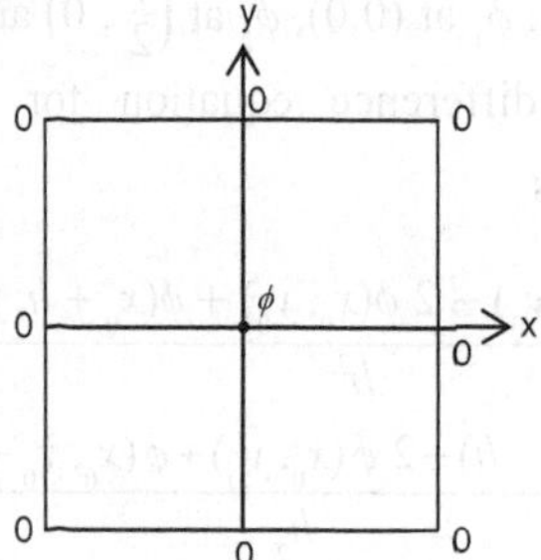

Figure 15.15

15.16.7 Parabolic Equations

The simplest example of parabolic equation is one dimensional heat equation

$$\frac{\partial u}{\partial t} = c^2 \frac{\partial^2 u}{\partial x^2}. \qquad (117)$$

Its solution gives the temperature u at a distance x units of length from one end of a thermally insulated bar after t seconds of heat conduction. In this problem, the temperatures at the ends of a bar of length L are often known for all time. Thus the boundary conditions are known. Also the temperature distribution along the bar is known at some particular instant. This instant is usually taken as zero time and the temperature distribution is called the initial condition. The solution gives u for all values of x between 0 and L and values of t from 0 to ∞.

Let the (x, t) plane be divided into smaller rectangles with sides $\Delta x = h$ and $\Delta t = k$. Our aim is to develop a difference formula for the solution of the problem. The difference formulas used for $u_t(x, t)$ and $u_{xx}(x, t)$ are

$$u_t(x, t) = \frac{u(x, t+k) - u(x, t)}{k} + O(k) \qquad (118)$$

and

$$u_{xx}(x, t) = \frac{u(x-h, t) - 2u(x,t) + u(x+h, t)}{h^2} + O(h^2) \qquad (119)$$

Since grid spacing in uniform, we have

$$x_{i+1} = x_i + h \text{ and } t_{j+1} = t_j + k.$$

Dropping the terms $O(k)$ and $O(h^2)$ and using $u_{i,j}$ for $u(x_i, t_j)$, and putting the values from (118) and (119) in (117), we get

$$\frac{u_{i,j+1} - u_{i,j}}{k} = c^2 \frac{u_{i-1,j} - 2u_{i,j} + u_{i+1,j}}{h^2}$$

Putting $r = \frac{c^2 k}{h^2}$, we get

$$u_{i,j+1} = u_{i,j} + r[u_{i-1,j} - 2u_{i,j} + u_{i+1,j}]. \qquad (120)$$

Equation (120) creates the $(j+1)^{\text{th}}$ row across the grid assuming that approximations in the jth row are known. This formula is called the *explicit formula*. However, it can be shown that this formula is valid only for $0 < r \leq \frac{1}{2}$.

For $r = \frac{1}{2}$, the formula (120) reduces to

$$u_{i,j+1} = \frac{u_{i-1,j} + u_{i+1,j}}{2},$$

which is called *Bender–Schmidt method*.

Crank–Nicholson Method. This formula is based on numerical approximations for the solution of the equation (117) at the point $(x, t+\frac{k}{2})$, which lies between the rows in the grid. The approximation used for $u_t\left(x, t+\frac{k}{2}\right)$ is obtained from the central difference formula

$$u_t\left(x, t+\frac{k}{2}\right) = \frac{u(x, t+k) - u(x, t)}{k} + O(k^2) \qquad (121)$$

The approximation for $u_{xx}(x, t+\frac{k}{2})$ is the average of $u_{xx}(x, t)$ and $u_{xx}(x, t+k)$. Thus

$$u_{xx}(x, t+\frac{k}{2}) = \frac{1}{2}\left[\frac{u_{i-1,j}-2u_{i,j}+u_{i+1,j}}{h^2} + \frac{u_{i-1,j+1}-2u_{i,j+1}+u_{i+1,j+1}}{h^2}\right] + O(h^2) \quad (122)$$

Thus using (121) and (122), the difference equation for the heat equation (117) becomes

$$\frac{u_{i,j+1}-u_{i,j}}{k} = \frac{c^2}{2h^2}\left[u_{i-1,j}-2u_{i,j} + u_{i+1,j}+u_{i-1,j+1}-2u_{i,j+1}+u_{i+1,j+1}\right].$$

Putting $\frac{c^2 k}{h^2} = r$, we get

$$-ru_{i-1,j+1} + (2+2r)u_{i,j+1} - ru_{i+1,j+1} = ru_{i-1,j} + (2-2r)u_{i,j} + ru_{i+1,j} \quad (123)$$

On the left–hand side of (123), we have three unknown quantities and on the right–hand side all the three quantities are known. The implicit formula (123) is called *Crank–Nicolson formula* which is convergent for all finite values of r. If we have m internal mesh points on each row, then Crank–Nicolson formula gives m simultaneous equations in m unknowns in term of the given boundary values. Thus the solution at each interval point on all rows can be obtained.

EXAMPLE 15.66

Solve the heat equation

$$\frac{\partial u}{\partial t} = \frac{\partial^2 u}{\partial x^2},$$

subject to the conditions $u(x,0) = 0$, $u(0,t) = 0$, and $u(1, t) = t$

Solution. The given equation is $\frac{\partial u}{\partial t} = \frac{\partial^2 u}{\partial x^2}$.

(i) Here $c^2 = 1$. We first choose $k = \frac{1}{8}$ and $h = \frac{1}{2}$ so that $r = \frac{c^2 k}{h^2} = \frac{1}{2}$. The Crank–Nicolson formula becomes

$$-u_{i-1,j+1} + 6u_{i,j+1} - u_{i+1,j+1} = u_{i-1,j} - 2u_{i,j} + u_{i+1,j} \quad (124)$$

The grid for the solution is shown in Figure 15.16.

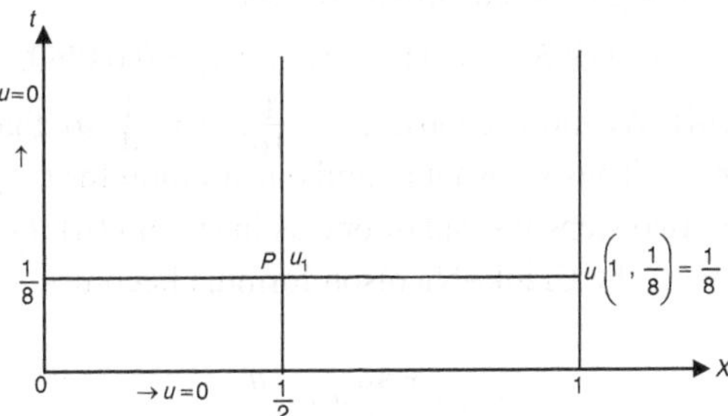

Figure 15.16

Suppose that u_1 is the value of u at the mess point $P\left(\frac{1}{2}, \frac{1}{8}\right)$. Then formula (124) yields $0 + 6u_1 - \frac{1}{8} = 0$ and so $u_1 = \frac{1}{48} = 0.02083$.

(ii) We now choose $k = \frac{1}{8}$, $h = \frac{1}{4}$ so that $r = 2$. For this value of r, the Crank–Nicolson formula takes the form

$$-u_{i-1,j+1} + 3u_{i,j+1} - u_{i+1,j+1} = u_{i-1,j} - u_{i,j} + u_{i+1,j} \quad (125)$$

The grid for the solution is now as shown in Figure 15.17.

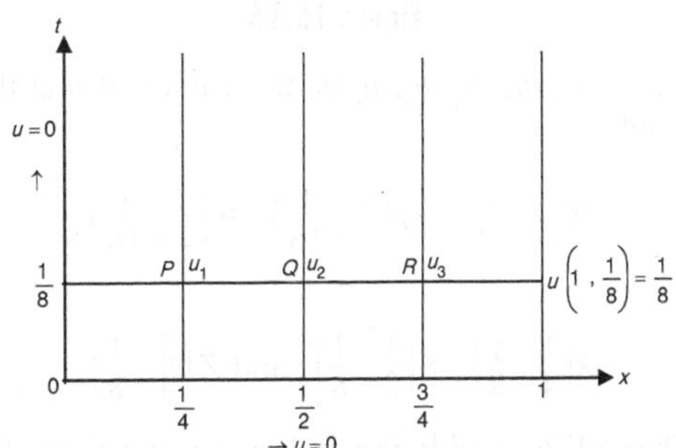

Figure 15.17

Let u_1, u_2, and u_3 be the value of u at $P\left(\frac{1}{4}, \frac{1}{8}\right)$, $Q\left(\frac{1}{2}, \frac{1}{8}\right)$, and $R\left(\frac{3}{4}, \frac{1}{8}\right)$. Then (125) yields

$$0 + 3u_1 - u_2 = 0,$$

$$-u_1 + 3u_2 - u_3 = 0, \text{ and}$$

$$-u_2 + 3u_3 - \frac{1}{8} = 0.$$

Solving these equations, we get

$u_1 = 0.00595$, $u_2 = 0.01785$, and $u_3 = 0.04760$.

(iii) We now choose $k = \frac{1}{16}$, $h = \frac{1}{4}$ so that $r = 1$. Thus we want to find our solution for $t = \frac{1}{8}$ in two steps instead of one as in (i) and (ii). For $r = 1$, the Crank–Nicolson formula becomes

$$-u_{i-1,j+1} + 4u_{i,j+1} - u_{i+1,j+1}$$

$$= u_{i-1,j} + 0 + u_{i+1,j} \quad (126)$$

The grid for the solution in this case is shown in the Figure 15.18.

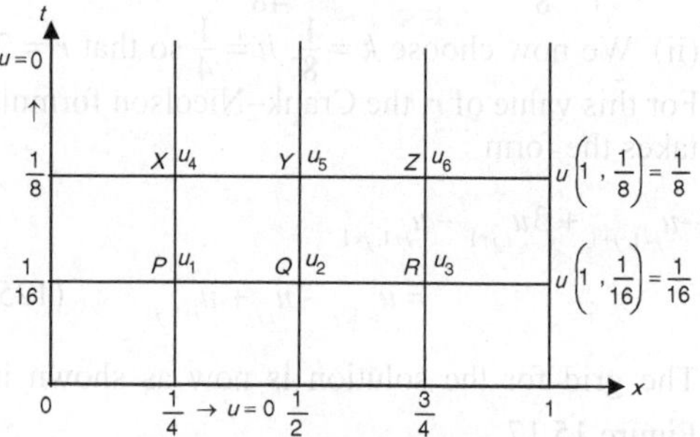

Figure 15.18

Let u_1, u_2, u_3, u_4, u_5, u_6 be the values of u at the points

$$P\left(\frac{1}{4}, \frac{1}{16}\right), Q\left(\frac{1}{2}, \frac{1}{16}\right), R\left(\frac{3}{4}, \frac{1}{16}\right),$$

$$X\left(\frac{1}{4}, \frac{1}{8}\right), Y\left(\frac{1}{2}, \frac{1}{8}\right), \text{ and } Z\left(\frac{3}{4}, \frac{1}{8}\right).$$

Then (126) yields the following equations for u_1, u_2, u_3.

$$4u_1 - u_2 = 0, \quad (127)$$

$$-u_1 + 4u_2 - u_3 = 0, \text{ and} \quad (128)$$

$$-u_2 + 4u_3 - \frac{1}{16} = 0 \quad (129)$$

Solving these equations, we get

$$u_1 = \frac{1}{56(16)}, \; u_2 = \frac{1}{56(4)}, \; u_3 = \frac{15}{56(16)}.$$

Also (126) yields the following equations for u_4, u_5, u_6.

$$4u_4 - u_5 = \frac{1}{4(56)},$$

$$-u_4 + 4u_5 - u_6 = \frac{1}{56}, \text{ and}$$

$$-u_5 + 4u_6 - \frac{1}{8} = \frac{1}{4(56)} + \frac{1}{16}.$$

Solving these equations, we get

$$u_4 = 0.005899, \; u_5 = 0.019132, \; u_6 = 0.052771.$$

The exact solution of the problem by Fourier series method is

$$u(x, t) = \frac{1}{6}(x^3 - x + 6xt)$$

$$+ \frac{2}{\pi^3} \sum_{n=1}^{\infty} \frac{(-1)^{n+1}}{n^3} e^{-n^2\pi^2 t} \sin n\pi x,$$

which yields

$$u\left(\frac{1}{4}, \frac{1}{8}\right) = 0.00541, \; u\left(\frac{1}{2}, \frac{1}{8}\right) = 0.01878, \text{ and}$$

$$u\left(\frac{3}{4}, \frac{1}{8}\right) = 0.5240.$$

EXAMPLE 15.67

Solve, by Bender–Schmidt method, the parabolic equation

$$\frac{\partial u}{\partial t} = \frac{1}{2} \frac{\partial^2 u}{\partial x^2}$$

subject to the condition $u(0,t) = u(4,t) = 0$ and $u(x,0) = x(4 - x)$.

Solution. We have $c^2 = \frac{1}{2}$. We first choose $k = 1$ and $h = 1$. Then $r = \frac{c^2 k}{h^2} = \frac{1}{2}$. The Bender–Schmidt method is applicable and we have

$$u_{i,j+1} = \frac{u_{i-1,j} + u_{i+1,j}}{2}. \quad (130)$$

The grid for the solution is shown in Figure 15.19

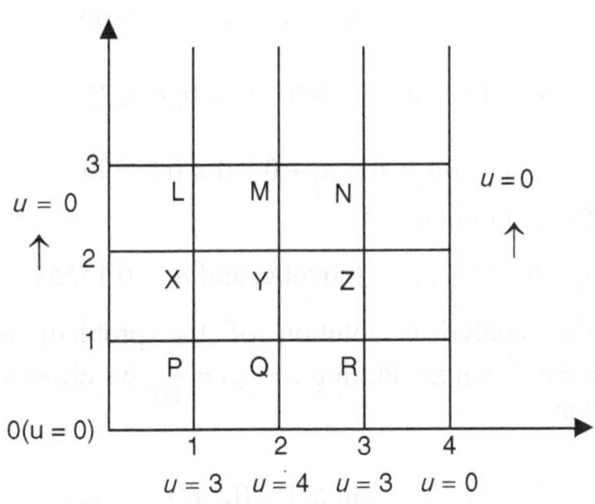

Figure 15.19

We have

$$u(1,0) = 1(4-1) = 3,\ u(2,0) = 2(4-2) = 4$$

and

$$u(3,0) = 3(4-3) = 3.$$

Let u_1, u_2, and u_3 be the values of u at $P(1,1)$, $Q(2,1)$ and $R(3,1)$, respectively. Then (130) yields

$$u_1 = \frac{0+4}{2} = 2,$$

$$u_2 = \frac{3+3}{2} = 3, \text{ and}$$

$$u_3 = \frac{4+0}{2} = 2.$$

Similarly, it u_4, u_5, and u_6 are the values of u at $X(1,2)$, $Y(2,3)$, and $Z(3,2)$, respectively, then

$$u_4 = \frac{0+u_2}{2} = \frac{0+3}{2} = 1.5,$$

$$u_5 = \frac{u_1+u_3}{2} = \frac{2+2}{2} = 2, \text{ and}$$

$$u_6 = \frac{u_2+0}{2} = \frac{3+0}{2} = 1.5.$$

Similarly, the values u_7, u_8, and u_9 at L, M, and N are, respectively,

$$u_7 = \frac{0+u_5}{2} = \frac{0+2}{2} = 1,$$

$$u_8 = \frac{u_4+u_6}{2} = \frac{1.5+1.5}{2} = 1.5, \text{ and}$$

$$u_9 = \frac{u_5+0}{2} = \frac{2+0}{2} = 1.$$

EXAMPLE 15.68

Use *Crank–Nicolson* method to solve

$$\frac{\partial u}{\partial t} = \frac{\partial^2 u}{\partial x^2}$$

subject to the conditions

$u(x,0) = \sin \pi x,\ 0 \le x \le 1,\ u(0,t) = u(1,t) = 0.$

Solution. We first take $k = \frac{1}{8}$ and $h = \frac{1}{4}$ so that $r = 2$. The Crank–Nicolson scheme corresponding to $r = 2$ is given by

$$-u_{i-1,\,j+1} + 3u_{i,\,j+1} - u_{i+1,\,j+1}$$
$$= u_{i-1,\,j} - u_{i,\,j} + u_{i+1,\,j}.$$

The grid corresponding to these values of h, k, and r is shown in the Figure 15.20

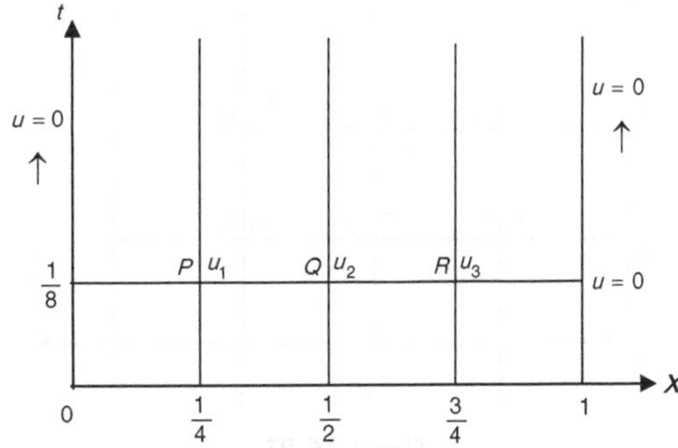

Figure 15.20

Let u_1, u_2, and u_3 be the values at $P\left(\frac{1}{4}, \frac{1}{8}\right)$, $Q\left(\frac{1}{2}, \frac{1}{8}\right)$, and $R\left(\frac{3}{4}, \frac{1}{8}\right)$, respectively. Then the above difference equation yields

$$0 + 3u_1 - u^2 = 0 - \sin\frac{\pi}{4} + \sin\frac{\pi}{2}$$
$$= -0.7071 + 1 = 0.2929,$$

$$-u_1 + 3u_2 - u_3 = \sin\frac{\pi}{4} - \sin\frac{\pi}{2} + \sin\frac{3\pi}{4}$$
$$= 0.7071 - 1 + 0.7071 = 0.4142,$$

and

$$-u_2 + 3u_3 - 0 = \sin\frac{\pi}{2} - \sin\frac{3\pi}{4} + \sin\pi$$

$$= 1 + 0.7071 = 1.7071.$$

Solving these equations, we obtain

$$u_1 = 0.252042,\ u_2 = 0.463228, \text{ and}$$

$$u_3 = 0.7234426.$$

Now we choose $k = \frac{1}{16}$ and $h = \frac{1}{4}$ so that $r = 1$. The Crank–Nicolson scheme corresponding to this value of r is

$$-u_{i-1,\,j+1} + 4u_{i,\,j+1} - u_{i+1,\,j+1}$$

$$= u_{i-1,\,j} + 0 + u_{i+1,\,j}. \qquad (131)$$

The grid corresponding to these values of h, k, and r is shown in Figure 15.21.

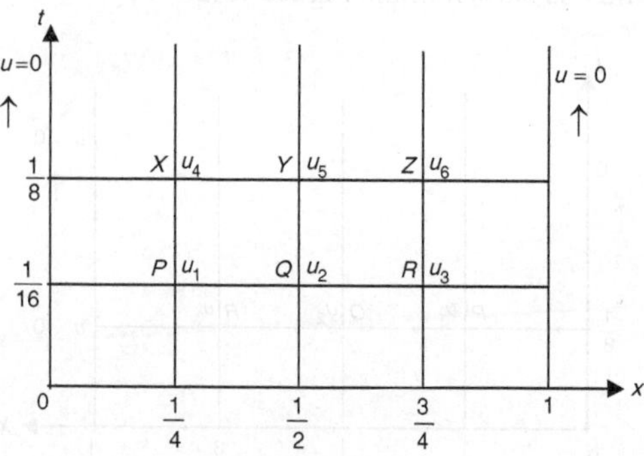

Figure 15.21

Applying (131) at the grid points P, Q, and R, we get

$$0 + 4u_1 - u_2 = 0 + 0 + \sin\frac{\pi}{2} = 1,$$

$$-u_1 + 4u_2 - u_3 = \sin\frac{\pi}{4} + \sin\frac{3\pi}{4} + 0 = 1.4142,$$

$$-u_2 + 4u_3 + 0 = \sin\frac{\pi}{2} + 0 + \sin\pi - 0.7071.$$

Solving these equations, we get

$u_1 = 0.381497$, $u_2 = 0.52599$, and $u_3 = 0.30827$. Now using the scheme (131) at each of the points X, Y, and Z, we have

$$4u_4 - u_5 = u_2 = 0.52599,$$

$$-u_4 + 4u_5 - u_6 = u_1 + 0 - u_3 = 0.073227,$$

$$-u_5 + 4u_6 - 0 = u_2 + 0 - 0 = 0.52599.$$

The solution is

$u_4 = 0.15551$, $u_5 = 0.09606$, and $u_6 = 0.15551$.

The analytical solution of the problem is $u = e^{-\pi^2 t}\sin\pi x$. Putting $x = \frac{1}{4}$, $t = \frac{1}{16}$, we observe that

$$u_1 = u\left(\frac{1}{4}, \frac{1}{16}\right) = \frac{\sin\pi/4}{e\pi^2/16} = \frac{0.7071}{1.853} = 0.38159,$$

which is in good agreement with the calculated value.

15.16.8 Hyperbolic Equations

A simple example of an hyperbolic partial differential equation is the *wave equation*

$$u_{tt}(x,t) = c^2 u_{xx}(x,t). \qquad (132)$$

We divide the (x,t) plane into grids consisting of small rectangles with sides $\Delta x = h$ and $\Delta t = k$. We shall use a difference equation method to compute approximations $\{u_{i,j} : i = 1,2,\ldots,n\}$ in successive rows for $j = 2,3,\ldots$. The true solution at the grid point (x_i, t_i) is $u(x_i, t_i)$.

We know that the central–difference formulas for approximating $u_{tt}(x,t)$ and $u_{xx}(x,t)$ are

$$u_{tt}(x,t) = \frac{u(x,t+k) - 2u(x,t) + u(x,t-k)}{k^2} + O(k^2), \qquad (133)$$

$$u_{xx}(x,t) = \frac{u(x+h,t) - 2u(x,t) + u(x-h,t)}{h^2} + O(h^2). \qquad (134)$$

Since grid spacing is uniform, we have

$$x_{i+1} = x_i + h,\ x_{i-1} = x_i - h,$$

$$t_{j+1} = t_j + k,\ t_{j-1} = t_j - k.$$

Dropping the terms $O(h^2)$, $O(k^2)$, using $u_{i,j}$ for $u(x_i, t_j)$, and putting the values from (133) and (134) in (132), we get

$$\frac{u_{i,j-1} - 2u_{i,j} + u_{i,j+1}}{k^2} = \frac{c^2\, u_{i-1,j} - 2u_{i,j} + u_{i+1,j}}{h^2}.$$

Putting $r = \frac{ck}{h}$, this equation reduces to

$$u_{i,j-1} - 2u_{i,j} + u_{i,j+1} = r^2\,[\,u_{i-1,j} - 2u_{i,j} + u_{i+1,j}]. \tag{135}$$

Thus, using (135), we can find row $j + 1$ across the grid assuming that approximation in both rows j and $j - 1$ are known. Therefore,

$$u_{i,j+1} = (2 - 2r^2)\, u_{i,j} + r^2\,(u_{i+1,j} + u_{i-1,j}) - u_{i,j-1} \tag{136}$$

for $i = 2,3, \ldots .\, n - 1$. We observe that the four known values on the right–hand side of (136), which are used to find $u_{i,j+1}$ can be shown as in Figure 15.22.

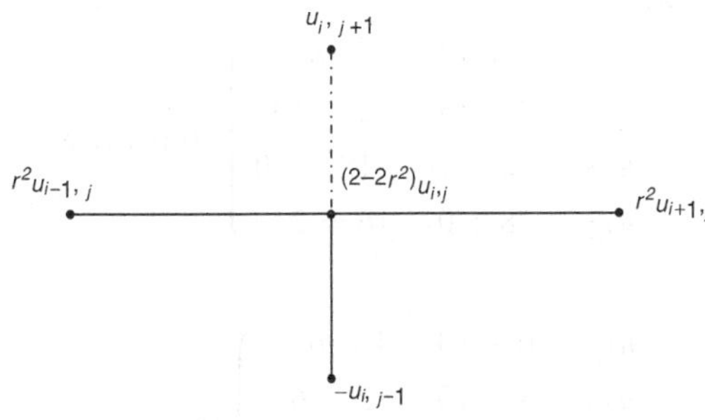

Figure 15.22

EXAMPLE 15.69

Use the finite difference method to solve the wave equation for a vibrating string

$u_{tt}(x,t) = 4u_{xx}(x,t)$ for $0 \le x \le 4$ and $0 \le t \le 2$ with the boundary conditions

$$u(0,t) = u(4,t) = 0,\ t > 0$$

and the initial conditions

$$\left.\begin{aligned} u(x,0) &= x(4 - x) \\ u_t(x,0) &= 0 \end{aligned}\right\} \qquad 0 \le x \le 4.$$

Solution. We have $c^2 = 4$. We take $h = 1$ and $k = 0.5$. Then $r^2 = \frac{c^2 k^2}{h^2} = 1$. Therefore, the difference equation for the problem is

$$u_{i,j+1} = u_{i+1,j} + u_{i-1,j} - u_{i,j-1}. \tag{137}$$

Since $u(t,0) = u(4,t) = 0$ we have

$$u_{0,j} = 0,\ u_{4,j} = 0.$$

Further, since $u(x,0) = x(4 - x)$ we have

$$\begin{aligned} u_{i,0} &= i(4 - i) \text{ for } t = 0 \\ &= 3, 4, 3 \text{ for } i = 1, 2, 3 \text{ at } t = 0, \end{aligned}$$

which are entities for the first row.

Finally, since $u_t(x,0) = 0,\ 0 \le x \le 4,\ t = 0,$ we have

$$\frac{u_{i,j+1} - u_{i,j}}{k} = 0 \text{ for } t = 0 \text{ and } j = 0$$

and so

$$u_{i,1} = u_{i,0},$$

which show that the entries in the second row are same as those of the first row. Thus, the grid for the solution is as shown in Figure 15.23.

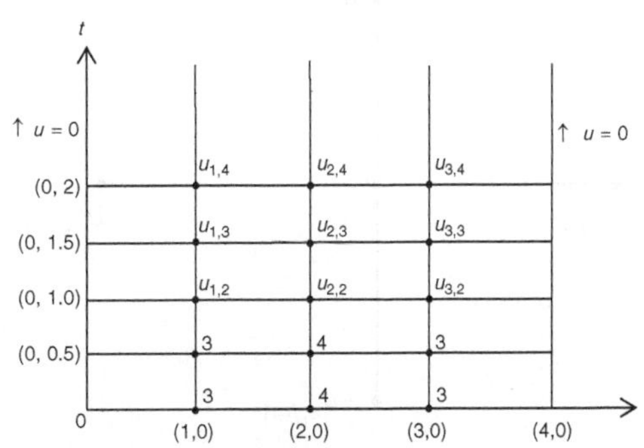

Figure 15.23

Using formula (137), we have

$$\left.\begin{aligned} u_{1,2} &= 4 + 0 - 3 = 1 \\ u_{2,2} &= 3 + 3 - 4 = 2 \\ u_{3,2} &= 4 + 0 - 3 = 1 \end{aligned}\right\} \text{ third row}$$

$$\left.\begin{aligned} u_{1,3} &= 0+2-3=-1 \\ u_{2,3} &= 1+1-4=-2 \\ u_{3,3} &= 2+0-3=-1 \end{aligned}\right\} \text{fourth row}$$

$$\left.\begin{aligned} u_{1,4} &= 0-2-1=-3 \\ u_{2,4} &= -1-1-2=-4 \\ u_{3,4} &= -2+0-1=-3 \end{aligned}\right\} \text{fifth row.}$$

EXAMPLE 15.70

Solve the wave equation

$u_{tt}(x,t) = 16\, u_{xx}(x,t), \quad 0 \le x \le 5, \quad 0 \le t \le 1.25$

subject to the conditions

$$u\,(0,t) = u\,(5,t) = 0, \; t > 0$$

$$\left.\begin{aligned} u(x,0) &= x^2(5-x) \\ u_t\,(x,0) &= 0 \end{aligned}\right\} 0 \le x \le 5.$$

Solution. In the given problem, we have $c^2 = 16$. We take $h = 1$, $k = 0.25$. Then $r^2 = \frac{c^2\,k^2}{h^2} = 16(0.0625) = 1$. Therefore, the difference equation for the problem becomes

$$u_{i,j+1} = u_{i+1,j} + u_{i-1,j} - u_{i,j-1} \tag{138}$$

and the scheme for calculation becomes as shown in Figure 15.24.

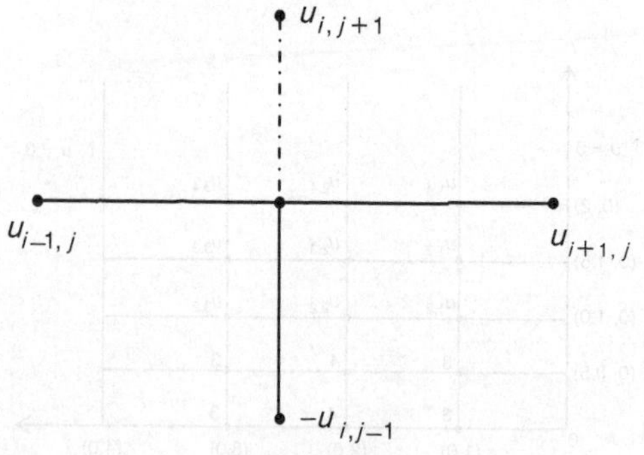

Figure 15.24

Since $u(0,t) = u(5,t) = 0$, we have

$$u_{0,j} = 0 \text{ and } u_{5,j} = 0 \text{ for all } j.$$

Hence the entries in the first and last columns are zero. Since $u(x,0) = x^2\,(5-x)$, we have

$$u_{i,0} = i^2\,(5-i),$$

which yields 4,12,18, and 16 for $i = 1,2,3,4$ and $t = 0$. Thus values of u on the first row are 0,4,12,18,16, and 0. Also $u_t(x,0) = 0$, and so

$$\frac{u_{i,j+1} - u_{i,j}}{k} = 0 \text{ for } j = 0.$$

This implies $u_{i,1} = u_{i,0}$ and so the entries in the second row are the same as those of first row. Thus the grid of the solution of the given equation is given in Figure 15.25

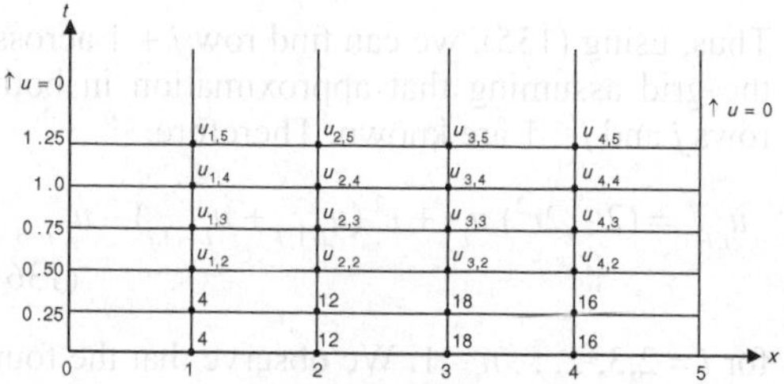

Figure 15.25

Using (138) and the scheme given above, we have

$$\left.\begin{aligned} u_{1,2} &= 0+12-4=8 \\ u_{2,2} &= 4+18-12=10 \\ u_{3,2} &= 12+16-18=10 \\ u_{4,2} &= 18+0-16=2 \end{aligned}\right\} \text{(third row)}$$

$$\left.\begin{aligned} u_{1,3} &= 0+10-4=6 \\ u_{2,3} &= 8+10-12=6 \\ u_{3,3} &= 10+2-18=-6 \\ u_{4,3} &= 10+0-16=-6 \end{aligned}\right\} \text{(fourth row)}$$

$$\left.\begin{aligned} u_{1,4} &= 0+6-8=-2 \\ u_{2,4} &= 6-6-10=-10 \\ u_{3,4} &= 6-6-10=-10 \\ u_{4,4} &= -6+0-2=-8 \end{aligned}\right\} \text{(fifth row)}$$

$$\left.\begin{aligned} u_{1,5} &= 0-10-6=-16 \\ u_{2,5} &= -2-10-6=-18 \\ u_{3,5} &= -10-8+6=-12 \\ u_{4,5} &= -10+0+6=-4 \end{aligned}\right\} \text{(sixth row).}$$

EXERCISES

1. Round off the following numbers to three decimal places

   (i) 498.5561 (ii) 52.2756 (iii) 0.70035

   (iv) 48.21416.

   Ans. (i) 498.556 (ii) 52.276 (iii) 0.700 (iv) 48.214

2. Round off to four significant figures

   (i) 19.235101 (ii) 49.8556 (iii) 0.0022218

   Ans. (i) 19.24 (ii) 49.860 (iii) 0.002222

3. Find the number of term of the exponential series such that their sum gives the value of e^x, correct to eight decimal places at $x = 1$

$$e^x = 1 + x + \frac{x^2}{2!} + \frac{x^3}{3!} + \ldots + \frac{x^{n-1}}{(n-1)!} + \frac{x^n}{n!}e^{\xi}, 0 < \xi < x.$$

   Hint:

   The Maximum absolute error at $\xi = x$ is equal to $\frac{x^n}{n!}$ and so

$$\text{Maximum relative error} = \left(\frac{x^n e^x}{n!}\right) e^x$$

   $= \frac{x^n}{n!} = \frac{1}{n!}$, since x = 1. For, an 8 decimal accuracy at $x = 1$, use have

$$\frac{1}{n!} < \frac{1}{2}10^{-8},$$

   which yields $n = 12$.

4. If $n = 10x^3 y^2 z^2$ and error in x,y, z are respectively, 0.03, 0.01, 0.02 at $x = 3$, $y = 1$, $z = 2$. Calculate the absolute error and percentage relative error in the calculation of it.

   Ans. 140.4, 13%

5. Find the root of the equation $x - \cos x = 0$ by bisection method.

   Ans. 0.739

6. Find a positive root of equation $xe^x = 1$ lying between 0 and 1 using bisection method.

   Ans. 0.567

7. Solve $x^3 - 4x - 9 = 0$ by Bolzano method.

   Ans. 2.706

8. Use Regula Falsi method to solve $x^3 + 2x^2 + 10x - 20 = 0$

   Ans. 1.3688

9. Use the method of false position to obtain a root of the equation, $x^3 + 2x^2 + 10x - 20 = 0$

   Ans. 1.796

10. Solve $e^x \sin x = 1$ by Regula Falsi method.

    Ans. 0.5885

11. Use Newton – Raphson method to obtain a root of the equation $x \log_{10} x = 1.2$.

    Ans. 2.7406

12. Use Newton–Raphson method to obtain a root of $x - \cos x = 0$.

    Ans. 0.739

13. Solve $\sin x = 1 + x^3$ by Newton–Raphson method

    Ans. – 1.24905

14. Find the real root of the equation $3x = \cos x + 1$ using Newton–Raphson method

    Ans. 0.6071

15. Derive the formula $x_{i+1} = \frac{1}{2}\left(x_i + \frac{N}{x_i}\right)$ to determine square root of N. Hence calculate the square root of 2.

    Ans. 1.414214

16. Find a real root of the equation $\cos x = 3x - 1$ correct to there decimal places using iteration method.

    Hint: Iteration formula is $x_n = \frac{1}{3}(1 + \cos x_n)$.

    Ans. 0.607

17. Using iteration method, find a root of the equation $x^3 + x^2 - 100 = 0$.

    Ans. 4.3311

18. Find the double root of the equation $x^3 - x^2 - x + 1 = 0$ near 0.9

Ans. 1.0001

19. Solve the system

$$2x + y + z = 10$$
$$3x + 2y + 3z = 18$$
$$x + 4y + 9z = 16$$

by Gauss elimination method.

Ans. $x = 7, y = -9, z = 5$

20. Solve the following system of equations by Gauss elimination method:

$$x_1 + 2x_2 - x_3 = 3$$
$$3x_1 - x_2 + 2x_3 = 1$$
$$2x_1 - 2x_2 + 3x_3 = 2$$

Ans. $x_1 = -1, x_2 = 4, x_3 = 4$

21. Solve the following system of equations by Gauss elimination method:

$$2x + 2y + z = 12$$
$$3x + 2y + 2z = 8$$
$$5x + 10y - 8z = 10.$$

Ans. $x = -12.75, y = 14.375, z = 8.75$

22. Solve the following system of equations by Gauss – Jordan method:

$$5x - 2y + z = 4$$
$$7x + y - 5z = 8$$
$$3x + 7y + 4z = 10.$$

Ans. $x = 11.1927, y = 0.8685, z = 0.1407$

23. Solve by Gauss – Jordan method:

$$2x_1 + x_2 + 5x_3 + x_4 = 5$$
$$x_1 + x_2 - 3x_3 + 4x_4 = -1$$
$$3x_1 + 6x_2 - 2x_3 + x_4 = 8$$
$$2x_1 + 2x_2 + 2x_3 - 3x_4 = 2.$$

Ans. $x_1 = 2, x_2 = \frac{1}{5}, x_3 = 0, x_4 = \frac{4}{5}$

24. Solve by Gauss–Jordan method:

$$x + y + z = 9$$
$$2x - 3y + 4z = 13$$
$$3x + 4y + 5z = 40.$$

Ans. $x = 1, y = 3, z = 5$

25. solve the following using Gauss–Jordan method:

$$2x - 3y + z = -1$$
$$x + 4y + 5z = 25$$
$$3x - 4y + z = 2.$$

Ans. $x = 8.7, y = 5.7, z = -1.3$

26. Use Jacobi's iteration method to solve the following

$$5x + 2y + z = 12$$
$$x + 4y + 2z = 15$$
$$x + 2y + 5z = 20.$$

Ans. $x = 1.08, y = 1.95, z = 3.16$

27. Solve by Jacobi's iteration method

$$10x + 2y + z = 9$$
$$2x + 20y - 2z = -44$$
$$-2z + 3y + 10z = 22.$$

Ans. $x = 1, y = -2, z = 3$

28. Solve by Jacobi's method

$$5x - y + z = 10$$
$$2x + 4y = 12$$
$$x + y + 5z = -1$$

Ans. $x = 2.556, y = 1.722, z = -1.055$

29. Use Gauss–Seidel method to solve

$$54x + y + z = 110$$
$$2x + 15y + 6z = 72$$
$$x + 6y + 27z = 85$$

Ans. $x = 1.926, y = 3.573, z = 2.425$

30. Find the solution, to three decimal places, of the system

$$83x + 11y - 4z = 95$$
$$7x + 52y + 13z = 104$$
$$3x + 8y + 29z = 71$$

using Gauss–Seidel method.

Ans. $x = 1.052, y = 1.369, z = 1.962$

31. Evaluate
(i) $\Delta^2 \cos 2x$ (ii) $\Delta^n\left(\frac{1}{x}\right)$

Ans. (i) $-4 \sin^2 h \cos (2x + 2h)$

(ii) $\dfrac{(-1)^n n!}{x(x+1)(x+2)\ldots(x+n)}$

32. Show that $\Delta + \nabla = \frac{\Delta}{\nabla} - \frac{\nabla}{\Delta}$.

33. Show that $\Delta^3 y_i = y_{i+3} - 3y_{i+2} + 3y_{i+1} - y_i$.

34. Find the function whose first difference is $9x^2 + 11x + 5$.

Ans. $3x^3 + x^2 + x + k$

35. Find the missing values in the following data:

x:	45	50	55	60	65
y:	3.0	–	2.0	–	−2.4

Ans. $f(50) = 2.925, f(60) = 0.225$

36. Form a difference table to fourth differences

x:	1	2	3	4
$f(x)$:	7.93	10.05	12.66	15.79

x:	5	6	7	8
$f(x)$:	19.47	23.73	28.60	34.11

Repeat the procedure for the same table when $f_5 = 19.47 + \varepsilon$, where ε represents an error. How many $\Delta^n f_x$ are affected?

37. If $f(x)$ is a cubic polynomial, use the difference table to locate and correct the error in the data:

x:	0	1	2	3	4	5	6	7
$f(x)$:	25	21	18	18	27	45	76	123

Ans. $f(3)$ is in error, true value is 19

38. If $f(x)$ is a polynomial of degree 4, locate and correct the error in the given table

x:	1	2	3	4
y:	3010	3424	3802	4105

x:	5	6	7	8
y:	4472	4771	5051	5315

39. Evaluate $f(3.75)$ from the table

x:	2.5	3.0	3.5	4.0	4.5	5.0
y:	24.145	22.043	20.225	18.644	17.262	16.047

(use Gauss forward formula)

Ans. 19.40746093

40. Using Newton's divided difference formula, find $f(x)$ as a polynomial in x for the table:

x	0	1	2	4	5	6
y	1	14	15	5	6	19

Ans. $x^3 - 9x^2 + 21x + 1$

41. Let $f(x) = x^3 - 4x$. Construct the divided difference table based on the nodes $x_0 = 1, x_1 = 2, \ldots, x_5 = 6$ and find the Newton's polynomial $P_3(x)$ based on x_0, x_1, x_2, x_3.

Ans. $P_3(x) = -3 + 3(x-1) + 6(x-1)(x-2) + (x-1)(x-2)(x-3)$

42. Using Lagranges interpolation formula, find the value of t for $A = 85$ using the table

t:	2	5	8	14
A:	94.8	87.9	81.3	68.7

Ans. 6.5928

43. Use Lagrange's interpolation formula to find the value of y for $x = 10$ using the table given below:

x:	5	6	9	11
y:	12	13	14	16

Ans. 14.3

44. Find the Lagrange's interpolating polynomial for (−1, 3), (3, 9), (4, 30), and (6, 132).

Ans. $x^3 - 3x^2 + 5x - 6$

45. Using Lagrange's interpolation formula, express the function

$$\frac{3x^2 + x + 1}{(x-1)(x-2)(x-3)}$$

as a sum of partial function.

Ans. $\dfrac{5}{2(x-1)} - \dfrac{15}{x-2} + \dfrac{31}{2(x-3)}$

46. Find the least square line for the data given below:

x	−1	0	1	2	3	4	5	6
y	10	9	7	5	4	3	0	−1

Ans. $y = -1.60714\,x + 8.64286$

47. The result of measurement of electric resistance R of a copper wire at various temperatures is listed below:

t	19	25	30	36	40	45	50
R	76	77	79	80	82	83	85

Using the method of least square, find the straight line $R = a + bt$ that fits best in the data.

Ans. $R = 70.052 + 0.290t$

48. The points (1, 14), (2, 27), (3, 40), (4, 55), and (5, 68) should be approximated by a straight line. Find the line assuming that the error in the x values can be neglected.

Ans. $y = 13.6x$

49. Find the power fit $y = ax$ (straight line through the origin) for the data

x	1	2	3	4	5
y	1.6	2.8	4.7	6.4	8.0

Ans. $y = 1.58x,\ e_2(f) = 0.1720$

50. Find the power fit $y = ax^m$ for the data

x	1	2	3	4	5
y	0.5	2	4.5	8	12.5

Hint: Taking log we have $\log y = \log a + m \log x$ that is, $Y = A + BX$, where $Y = \log y$, $A = \log a$, and $X = \log x$. Form table in X and Y and find A and B. Then take anti-logarithm to find 'a' and 'm'.

Ans. $y = 0.5012\, x^{1.998}$

51. Find the least square parabolic fit $y = a + bx + cx^2$ for the data

x	1	2	3	4
y	1.7	1.8	2.3	3.2

Ans. $y = 1.53 + 0.063x + 0.074x^2$

52. Using Simpson's rule, find the volume of the solid of revolution formed by rotating about x- axis, the area between the x axis, the lines $x = 0$ and $x = 1$ and a curve through the points (0, 1), (0.25, 0.9896), (0.50, 0.9589), (0.75, 0.9089), and (1, 0.8415).

Hint: $\text{Volume} = \int_0^1 \pi y^2 dx = \pi \int_0^1 y^2 dx$

$$= \pi \frac{h}{3}[y_0^2 + 4(y_1^2 + y_3^2) + 2y_2^2 + y_4^2]$$

Ans. 2.8192

53. Find the approximate value of

$$\int_0^{\pi/2} \sqrt{\cos\theta}\, d\theta$$

by dividing the interval into six parts.

Ans. 1.1873

54. Evaluate

$$\int_1^2 \frac{dx}{x}$$

by Simpson's rule and compare the approximate value obtained with the exact solution

Ans. 0.6932

Exact value: $\log_2 2 = 0.693147$

55. Evaluate

$$\int_0^{\pi/2} \sin x\, dx$$

by Simpson's $\frac{1}{3}$ rule using 11 ordinates.

Ans. 0.9985

56. The velocity v of a particle at distance s from a point on its path is given by the table:

s ft.:	0	10	20	30	40	50	60
v ft/sec.:	47	58	64	65	61	52	38

Using Simpson's $\frac{1}{3}$ rule, determine the time taken by the particle to travel 60 ft.
Hint: $v = \frac{ds}{dt}$ and so $dt = \frac{1}{v}\, ds$. So find $\int_0^{60} \frac{1}{v} ds$.

Ans. 1.063 Sec

57. For the case of six known ordinates, show that

$$\int_0^5 f(x)dx = \frac{5}{288} [19(f_0+f_5) + 75(f_1+f_4) + 50(f_2+f_3)]$$

58. The velocity v km/min of a moped started from rest is given at fixed intervals of time t (minutes) as follows:

t:	2	4	6	8	10	12	14	16	18	20
v:	10	18	25	29	32	20	11	5	2	0

Using Simpson's rule, find the distance covered in 20 minutes.
Hint: $v = \frac{ds}{dt}$ and so $ds = v\,dt$. So find $\int_0^{20} v\,dt$. Take interval length equal to 2 and use Simpson's formula

Ans. 309.33 km

59. Obtain an estimate of the number of subintervals that should be chosen so as to guarantee that the error committed in evaluating $\int_1^2 \frac{1}{x}dx$ by trapezoidal rule is less than 0.001.
Hint: $E_n(x) \le -\frac{n h^3}{12} f''(\xi)$,

Ans. n = 8

60. Compute the value of
$$\int_0^1 \frac{dx}{1+x^2}$$
using trapezoidal rule with $h = 0.5, 0.25$, and 0.125.

Ans. 0.77500, 0.78279 0.78475, 0.7854

61. Calculate by Simpson's rule an approximate value of $\int_{-3}^{3} x^4 dx$ by taking seven equidistant ordinates. Compare it with exact value and the estimate obtained by using trapezoidal rule.

Ans. by Simpson's rule: 98
Exact value: 97.2
By trapezoidal rule: 115
So Simpson's rule yields better results

62. Calculate $\int_2^{10} \frac{dx}{1+x}$ by dividing the range into eight equal parts.

Ans. 1.299

63. If $e^0 = 1$, $e^1 = 2.72$, $e^2 = 7.39$, $e^3 = 20.09$, $e^4 = 54.60$, find $\int_0^4 e^x$ by Simpson's rule.

Ans. 2.97049

64. A river is 80 feet wide. The depth d (in feet) of the river at a distance x from one bank is given by the following table:

x:	0	10	20	30	40	50	60	70	80
d:	0	4	7	9	12	15	14	8	3

Find approximately the area of the cross-section of the river
Hint: Since $A = \int y\,dx$ and $h = 10$ we have by Simpson's rule,
$$A = \frac{10}{3}\,[(0+3) + 4(4+9+15+8) + (7+12+14)] = 710 \text{ sq.ft}$$

65. Show that
$$\int_{-1}^{1} f(x)dx = \frac{13}{12}[f(1) + f(-1) - f(3) - f(-3)].$$

66. Solve $\frac{dy}{dx} = 1 - 2xy$, $y(0) = 0$ by Taylor's series method for $x = 0.2$.

Ans. 0.1947

67. Using Taylor's series method, obtain the values of y at $x = 0.1, 0.2, 0.3$ if y satisfies the equation $\frac{d^2y}{dx^2} + xy = 0$ and $y(0) = 1$, $y'(0) = 0.5$.

Ans. $y(0.1) = 1.050$, $y(0.2) = 1.099$, $y(0.3) = 1.145$

68. Solve $\frac{dy}{dx} = -xy$, $y(0) = 1$ over [0, 0.1] with $h = 0.05$ using Taylor's series method.

Ans. $y(0.05) = 0.9987508$, $y(0.1) = 0.9950125$

69. Solve $\frac{dy}{dx} = 1 - y$, $y(0) = 0$ in [0, 0.3] by modified Euler's method taking $h = 0.1$.

Ans. $y(0.1) = 0.095$, $y(0.2) = 0.180975$, $y(0.3) = 0.2587823$

70. Solve $\frac{dy}{dx} = x + y^2$, $y(0) = 1$ for $x = 0.5$ by modified Euler's method.

Ans. 2.2352

71. Solve $\frac{dy}{dx} = y - \frac{2x}{y}$, $y(0) = 1$ in [0, 0.2] using Euler's method and taking $h = 0.1$.

Ans. $y(0.1) = 1.095909$, $y(0.2) = 1.184097$

72. Use Picard's method to solve $\frac{dy}{dx} = x - y^2$, $y(0) = 1$.

Ans. 0.9138

73. Use Picard's method to solve $y'' + 2xy' + y = 0$, $y(0) = 0.5$, $y'(0) = 0.1$ for $x = 0.1$.

Ans. 0.5075

74. Use Picard's method to solve $\frac{dy}{dx} = x^2 + y^2$, $y(0) = 0$ for $x = 0.4$.

Ans. 0.0214

75. Solve for $x = 0.1$, the equation $\frac{dy}{dx} = 3x + y^2$, $y(0) = 1$ by Picard's method.

Ans. $y(0.1) = 1.127$

76. Use Runge–Kutta method of order four to solve the differential equation $\frac{dy}{dx} = \frac{y^2 - x^2}{y^2 + x^2}$, $y(0) = 1$ at $x = 0.2$.

Ans. $y(0.2) = 1.196$

77. Use fourth order Runge–Kutta method to find $y(0.2)$ for the equation $\frac{dy}{dx} = \frac{y - x}{y + x}$, $y(0) = 1$.

Ans. $y(0.2) = 1.1749$

78. Solve $y' = \frac{y^2 - 2x}{y^2 + x}$, $y(0) = 1$ for $x = 0.1$ and $x = 0.2$ using Runge–Kutta method.

Ans. $y(0.1) = 1.091, y(0.2) = 1.168$

79. Using Runge–Kutta method, solve $y' = x + y$, $y(0) = 1$ for $x = 0.2$.

Ans. $y(0.2) = 0.2428$

80. Use Runge–Kutta method to solve $y' = -xy$, $y(0) = 1$ for $x = 0.2$.

Ans. 0.9801987

81. Solve the elliptic equation in the square region $0 \le x \le 4$, $0 \le y \le 4$ subject to the conditions

$$u(0,y) = 0, \quad u(4,y) = 8 + 2y$$

$$u(x,0) = \frac{x^2}{2}, \quad u(x,y) = x^2$$

and taking $h = k = 1$.

Hint: Use standard five-point formula and diagonal five-point formula to find initial value at the mess point and then use iteration method.

Ans. $u_1 \approx 1.99, u_2 \approx 4.91, u_3 \approx 8.99,$
$u_4 \approx 2.06, u_5 \approx 4.69, u_6 \approx 8.06,$
$u_7 \approx 1.57, u_8 \approx 3.71, u_9 \approx 6.57.$

82. Solve Laplace equation $u_{xx} + u_{yy} = 0$ in the domain of Figure 15.26

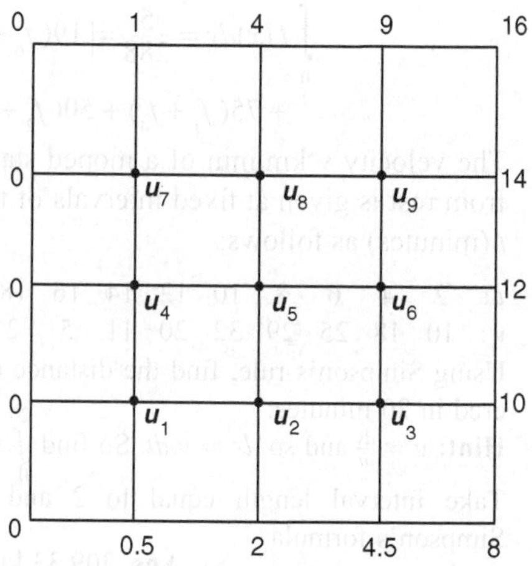

Figure 15.26

Ans. $u_1 \approx 1.57$, $u_2 \approx 3.71$, $u_3 \approx 6.57$,
$u_4 \approx 2.06$, $u_5 \approx 4.69$, $u_6 \approx 8.06$,
$u_7 \approx -2$, $u_8 \approx -2$

83. Solve Laplace equation $u_{xx} + u_{yy} = 0$ at the internal mesh points of the square region with the boundary values shown in Figure 15.27

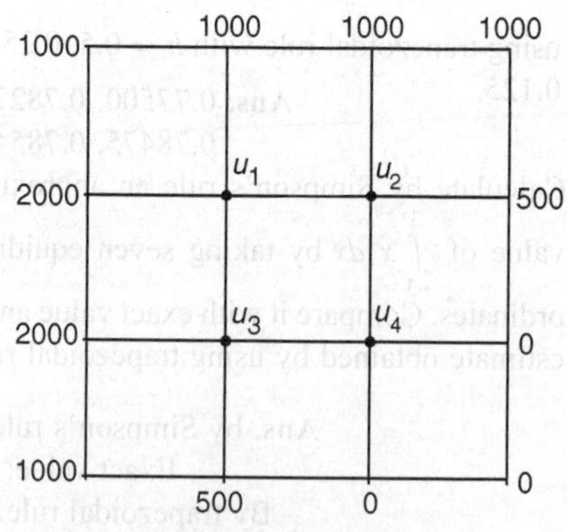

Figure 15.27

Ans. $u_1 = 1208.3$, $u_2 = 791.7$,
$u_3 = 1041.7$, $u_4 = 458.4$

84. Solve the Laplace equation $\nabla^2 u = 0$ in the square region with mesh points and boundary conditions shown in the Figure 15.28

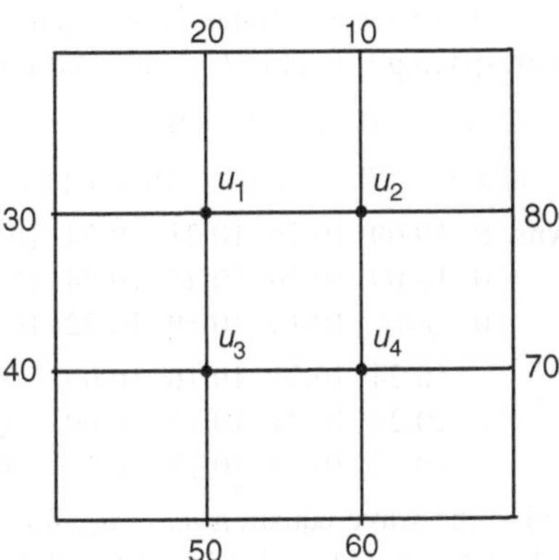

Figure 15.28

Ans. $u_1 = 34.99,\ u_2 = 44.99,$
$u_3 = 44.99,\ u_4 = 55.0$

85. Solve the Poisson's equation $u_{xx} + u_{yy} - 8x^2 y^2 = 0$ for the square mesh shown in Figure 15.29 under the condition that $u = 0$ on the boundary and that mesh length $h = k = 1$.

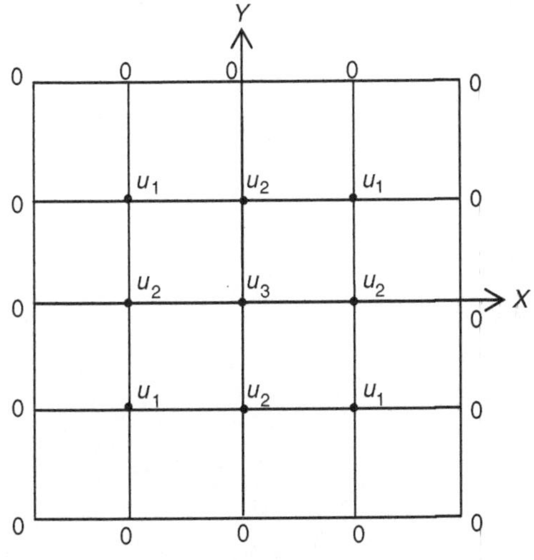

Figure 15.29

Hint: use symmetry first, so there are only three values to be determined

Ans. $u_1 = -3,\ u_2 = -2,\ u_3 = -2$

86. Determine the system of four equation in the four unknowns P_1, P_2, P_3, and P_4 for solving the Laplace equation $\nabla^2 u = 0$ on the 4×4 grid shown in Figure 15.30.

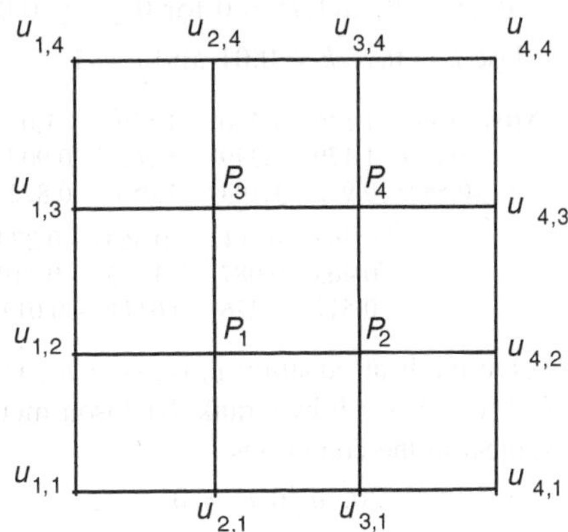

Ans.
$$-4p_1 + u_{1,2} + p_2 + p_3 + u_{2,1} = 0$$
$$p_1 - 4p_2 + p_4 + u_{4,2} + u_{3,1} = 0$$
$$p_1 + u_{1,3} - 4p_3 + p_4 + u_{2,4} = 0$$
$$p_2 + p_3 - 4p_4 + u_{4,3} + u_{3,4} = 0$$

87. Solve the heat equation $\dfrac{\partial u}{\partial t} = \dfrac{\partial^2 u}{\partial x^2}$ subject to the conditions

$$u(0,t) = 0,\ u(x,0) = x(1-x),\ u(1,t) = 0.$$

Take $h = 0.1$ and $t = 0,\ 1, 2.$

Ans.

0	0.09	0.16	0.21	0.24	0.25
0	0.08	0.15	0.20	0.23	0.24
0	0.24	0.21	0.16	0.09	0
0	0.23	0.20	0.15	0.08	0

88. Solve the heat equation

$$\frac{\partial u}{\partial t} = \frac{\partial^2 u}{\partial x^2} \text{ for } 0 < x < 1;\ 0 < t < 0{:}04$$

subject to the conditions

$$u(x,0) = 4x - 4x^2,\ 0 \le x \le 1$$
$$u(0,t) = 0,\ 0 \le t \le 0.04$$
$$u(1,t) = 0 \text{ for } 0 \le t \le 0.04$$

Ans.

0	0.64	0.96	0.96	0.64	0
0	0.48	0.80	0.80	0.48	0
0	0.40	0.64	0.64	0.40	0

89. Use Crank–Nicolson method for solving the heat equation $u_t = u_{xx}$ for $0 < x < 1$ and $0 < t < 0.03$ subject to the conditions

$$u(x,0) = \sin(\pi x) + \sin(2\pi x),\ 0 \le x \le 1$$

$$u(x,t) = 0,\ u(1,t) = 0 \text{ for } 0 \le t \le 0.03$$

Use $h = 0.1,\ k = 0.01$ and $r = 1$

Ans.

0.897	1.539	1.760	1.539	1.0
0.679	1.179	1.379	1.262	0.907
0.5525	0.922	1.104	1.053	0.822

0.363	−0.142	−0.363	−0.279
0.463	0.087	−0.113	−0.119
0.511	0.226	0.0444	−0.017.

90. Solve the heat equation $u_t\ (x, t) = u_{xx}\ (x, t)$, $0 < x < 5,\ t > 0$ by Crank–Nicloson method subject to the conditions

$$u\ (x,0) = 20,\ u\ (0,t) = 0,$$

$$u\ (5,t) = 100 \text{ and taking } h = 1,\ k = 1$$

Ans.

0	20	20	20	20	100
0	9.80	20.19	30.72	59.92	100.

91. Solve the wave equation $u_{tt} = u_{xx}$ up to $t = 0.2$ with spacing 0.1 subject to the conditions

$$u\ (0,t) = 0,\ u\ (1,t) = 0$$

$$u_t(x,0) = 0,\ u\ (x,0) = 10 + x\ (1 - x).$$

Ans.

0	10.09	10.16	10.21	10.24	10.25
0	10.09	10.16	10.21	10.24	10.25
0	0.07	10.14	10.19	10.22	10.23

10.24	10.21	10.16	10.09	0
10.24	10.21	10.16	10.09	0
10.22	10.19	10.17	0.7	0

92. Solve the wave equation $u_{tt} = u_{xx}$ for $x = 0$, 0.1, 0.2, 0.3, 0.4, and 0.5 and $t = 0, 0.1, 0.2$ subject to the conditions

$$u\ (x,0) = \frac{1}{8}\sin \pi x,\ u_t(x,0) = 0,\ 0 \le x \le 1$$

$$u\ (0,t) = u\ (1,t) = 0,\ t \le 0$$

Ans.

0	0.037	0.070	0.096	0.113	0.119
0	0.031	0.059	0.082	0.096	0.101
0	0.023	0.043	0.059	0.07	0.074.

Statistical Tables

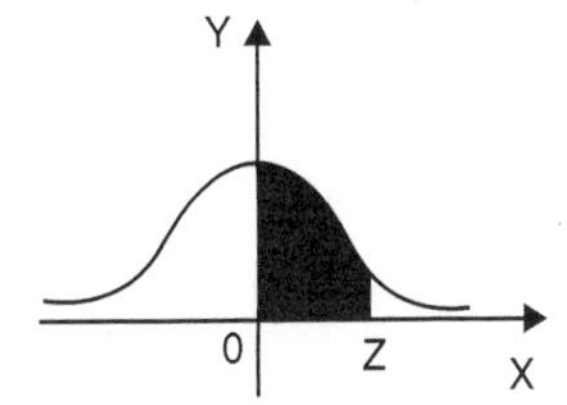

Area Under Normal Curve from 0 to z

z	.00	.01	.02	.03	.04	.05	.06	.07	.08	.09
0.0	0.0000	0.004	0.0080	0.0120	0.0160	0.0199	0.0239	0.0279	0.0319	0.0359
0.1	0.0398	0.0438	0.0479	0.0517	0.0557	0.0596	0.0636	0.0675	0.0714	0.0754
0.2	0.0793	0.0832	0.0871	0.0910	0.0948	0.0987	0.1026	0.1064	0.1103	0.1141
0.3	0.1179	0.1217	0.1255	0.1293	0.1331	0.1368	0.1406	0.1443	0.1480	0.1517
0.4	0.1554	0.1591	0.1628	0.1664	0.1700	0.1736	0.1772	0.1808	0.1844	0.1879
0.5	0.1915	0.1950	0.1985	0.2019	0.2054	0.2088	0.2123	0.2157	0.2190	0.2224
0.6	0.2258	0.2291	0.2324	0.2357	0.2389	0.2422	0.2454	0.2486	0.2518	0.2549
0.7	0.2580	0.2612	0.2642	0.2673	0.2704	0.2734	0.2764	0.2794	0.2823	0.2852
0.8	0.2881	0.2910	0.2939	0.2967	0.2996	0.3023	0.3051	0.3078	0.3106	0.3133
0.9	0.3159	0.3186	0.3212	0.3238	0.3264	0.3289	0.3315	0.3340	0.3365	0.3389
1.0	0.3413	0.3438	0.3461	0.3485	0.3508	0.3531	0.3554	0.3577	0.3599	0.3621
1.1	0.3643	0.3665	0.3686	0.3708	0.3729	0.3749	0.3770	0.3790	0.3810	0.3830
1.2	0.3849	0.3869	0.3888	0.3907	0.3925	0.3944	0.3962	0.3980	0.3997	0.4015
1.3	0.4032	0.4049	0.4066	0.4082	0.4099	0.4115	0.4131	0.4147	0.4162	0.4177
1.4	0.4192	0.4207	0.4222	0.4236	0.4251	0.4265	0.4279	0.4292	0.4306	0.4319
1.5	0.4332	0.4345	0.4357	0.4370	0.4382	0.4394	0.4406	0.4418	0.4429	0.4441
1.6	0.4452	0.4463	0.4474	0.4484	0.4495	0.4505	0.4515	0.4525	0.4535	0.4545
1.7	0.4554	0.4564	0.4573	0.4582	0.4591	0.4599	0.4608	0.4616	0.4625	0.4633
1.8	0.4641	0.4649	0.4656	0.4664	0.4671	0.4678	0.4686	0.4693	0.4699	0.4706
1.9	0.4713	0.4719	0.4726	0.4732	0.4738	0.4744	0.4750	0.4756	0.4761	0.4767
2.0	0.4772	0.4778	0.4783	0.4788	0.4793	0.4798	0.4803	0.4808	0.4812	0.4817
2.1	0.4821	0.4826	0.4830	0.4834	0.4838	0.4842	0.4846	0.4850	0.4854	0.4857
2.2	0.4861	0.4864	0.4868	0.4871	0.4875	0.4878	0.4881	0.4884	0.4887	0.4890
2.3	0.4893	0.4896	0.4898	0.4901	0.4904	0.4906	0.4909	0.4911	0.4913	0.4916
2.4	0.4918	0.4920	0.4922	0.4925	0.4927	0.4929	0.4931	0.4932	0.4934	0.4936
2.5	0.4938	0.4940	0.4941	0.4943	0.4945	0.4946	0.4948	0.4949	0.4951	0.4952
2.6	0.4953	0.4955	0.4956	0.4957	0.4959	0.4960	0.4961	0.4962	0.4963	0.4964
2.7	0.4965	0.4966	0.4967	0.4968	0.4969	0.4970	0.4971	0.4972	0.4973	0.4974
2.8	0.4974	0.4975	0.4976	0.4977	0.4977	0.4978	0.4979	0.4979	0.4980	0.4981
2.9	0.4981	0.4982	0.4982	0.4983	0.4984	0.4984	0.4985	0.4985	0.4986	0.4986
3.0	0.4987	0.4987	0.4987	0.4988	0.4988	0.4989	0.4989	0.4989	0.4990	0.4990
3.1	0.4990	0.4991	0.4991	0.4991	0.4992	0.4992	0.4992	0.4992	0.4993	0.4993

Examples: (i) $P(0 \leq z \leq 0.27) = 0.1064$ (ii) $P(z \geq 0.81) = 0.5 - P(0 \leq z \leq 0.81) = 0.5 - 0.2910 = 0.2090$
(iii) $P(-3 \leq z \leq 3) = P(-3 \leq z \leq 0) + P(0 \leq z \leq 3) = P(0 \leq z \leq 3) + P(0 \leq z \leq 3) = 2\,P(0 \leq z \leq 3)$
$= 2(0.4987) = 0.9974.$

Area under the Normal Curve from z to ∞

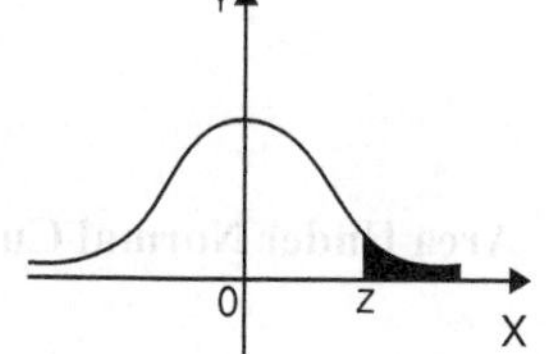

z	.00	.01	.02	.03	.04	.05	.06	.07	.08	.09
0.0	0.5000	0.4960	0.4920	0.4880	0.4840	0.4801	0.4761	0.4721	0.4681	0.4641
0.1	0.4602	0.4562	0.4521	0.4483	0.4443	0.4404	0.4364	0.4325	0.4286	0.4246
0.2	0.4207	0.4168	0.4129	0.4090	0.4052	0.4013	0.3974	0.3936	0.3897	0.3859
0.3	0.3821	0.3783	0.3745	0.3707	0.3669	0.3632	0.3594	0.3557	0.3520	0.3483
0.4	0.3446	0.3409	0.3372	0.3336	0.3300	0.3264	0.3228	0.3192	0.3156	0.3121
0.5	0.3085	0.3050	0.3015	0.2981	0.2946	0.2912	0.2877	0.2843	0.2810	0.2776
0.6	0.2742	0.2709	0.2676	0.2643	0.2611	0.2578	0.2546	0.2514	0.2482	0.2451
0.7	0.2420	0.2388	0.2358	0.2327	0.2296	0.2266	0.2236	0.2206	0.2177	0.2148
0.8	0.2119	0.2090	0.2061	0.2033	0.2004	0.1977	0.1949	0.1922	0.1894	0.1867
0.9	0.1841	0.1814	0.1788	0.1762	0.1736	0.1711	0.1685	0.1660	0.1635	0.1611
1.0	0.1587	0.1562	0.1539	0.1515	0.1492	0.1469	0.1446	0.1423	0.1401	0.1379
1.1	0.1357	0.1335	0.1314	0.1292	0.1271	0.1251	0.1230	0.1210	0.1190	0.1170
1.2	0.1151	0.1131	0.1112	0.1093	0.1075	0.1056	0.1038	0.1020	0.1003	0.0985
1.3	0.0968	0.0951	0.0934	0.0918	0.0901	0.0885	0.0869	0.0853	0.0838	0.0823
1.4	0.0808	0.0793	0.0778	0.0764	0.0749	0.0735	0.0721	0.0708	0.0694	0.0681
1.5	0.0668	0.0655	0.0643	0.0630	0.0618	0.0606	0.0594	0.0582	0.0571	0.0559
1.6	0.0548	0.0537	0.0526	0.0516	0.0505	0.0495	0.0485	0.0475	0.0465	0.0455
1.7	0.0446	0.0436	0.0427	0.0418	0.0409	0.0401	0.0392	0.0384	0.0375	0.0367
1.8	0.0359	0.0351	0.0344	0.0336	0.0329	0.0322	0.0314	0.0307	0.0301	0.0294
1.9	0.0287	0.0281	0.0274	0.0268	0.0262	0.0256	0.0250	0.0244	0.0239	0.0233
2.0	0.0228	0.0222	0.0217	0.0212	0.0207	0.0202	0.0197	0.0192	0.0188	0.0183
2.1	0.0179	0.0174	0.0170	0.0166	0.0162	0.0158	0.0154	0.0150	0.0146	0.0143
2.2	0.0139	0.0136	0.0132	0.0129	0.0125	0.0122	0.0119	0.0116	0.0113	0.0110
2.3	0.0107	0.0104	0.0102	0.0099	0.0096	0.0094	0.0091	0.0089	0.0087	0.0084
2.4	0.0082	0.0080	0.0078	0.0075	0.0073	0.0071	0.0069	0.0068	0.0066	0.0064
2.5	0.0062	0.0060	0.0059	0.0057	0.0055	0.0054	0.0052	0.0051	0.0049	0.0048
2.6	0.0047	0.0045	0.0044	0.0043	0.0041	0.0040	0.0039	0.0038	0.0037	0.0036
2.7	0.0035	0.0034	0.0033	0.0032	0.0031	0.0030	0.0029	0.0028	0.0027	0.0026
2.8	0.0026	0.0025	0.0024	0.0023	0.0023	0.0022	0.0021	0.0021	0.0020	0.0019
2.9	0.0019	0.0018	0.0018	0.0017	0.0016	0.0016	0.0015	0.0015	0.0014	0.0014
3.0	0.0013	0.0013	0.0013	0.0012	0.0012	0.0011	0.0011	0.0011	0.0010	0.0010
3.1	0.0010	0.0009	0.0009	0.0009	0.0008	0.0008	0.0008	0.0008	0.0007	0.0007

Examples: (i) $P(z \geq 0.81) = 0.2090$ (ii) $P(0 \leq z \leq 0.27) = 0.5 - P(z \geq 0.27) = 0.5 - 0.3936 = 0.1064$.

Values of |*t*| with probability P and degree of freedom *v*

v	*P* = 0.50	*P* = 0.10	*P* = 0.05	*P* = 0.02	*P* = 0.01
1	1.000	6.340	12.710	31.820	63.660
2	0.816	2.920	4.300	6.960	9.920
3	0.765	2.350	3.180	4.540	5.840
4	0.741	2.130	2.780	3.750	4.600
5	0.727	2.020	2.570	3.360	4.030
6	0.718	1.940	2.450	3.140	3.710
7	0.711	1.900	2.360	3.000	3.500
8	0.706	1.860	2.310	2.900	3.360
9	0.703	1.830	2.260	2.820	3.250
10	0.700	1.810	2.230	2.760	3.170
11	0.697	1.800	2.200	2.720	3.110
12	0.695	1.780	2.180	2.680	3.060
13	0.694	1.770	2.160	2.650	3.010
14	0.692	1.760	2.140	2.620	2.980
15	0.691	1.750	2.130	2.600	2.950
16	0.690	1.750	2.120	2.580	2.920
17	0.689	1.740	2.110	2.570	2.900
18	0.688	1.730	2.100	2.550	2.880
19	0.688	1.730	2.090	2.540	2.860
20	0.687	1.720	2.090	2.530	2.840
21	0.686	1.720	2.080	2.520	2.830
22	0.686	1.720	2.070	2.510	2.820
23	0.685	1.710	2.070	2.500	2.810
24	0.685	1.710	2.060	2.490	2.800
25	0.684	1.710	2.060	2.480	2.790
26	0.684	1.710	2.060	2.480	2.780
27	0.684	1.700	2.050	2.470	2.770
28	0.683	1.700	2.050	2.470	2.760
29	0.683	1.700	2.040	2.460	2.760
30	0.683	1.700	2.04	2.460	2.750

Examples: (i) $t_{0.05} = 2.26$ for $v = 9$ (ii) $t_{0.05} = 2.14$ for $v = 14$ (iii) $t_{0.01} = 3.50$ for $v = 7$.

Values of χ^2 with probability *P* and degree of freedom *v*

v	$P = 0.99$	$P = 0.95$	$P = 0.50$	$P = 0.30$	$P = 0.20$	$P = 0.10$	$P = 0.05$	$P = 0.01$
1	0.0002	0.004	0.460	1.070	1.640	2.710	3.840	6.640
2	0.020	0.103	1.390	2.410	3.220	4.600	5.990	9.210
3	0.115	0.350	2.370	3.660	4.640	6.250	7.820	11.340
4	0.300	0.710	3.360	4.880	5.990	7.780	9.490	13.280
5	0.550	1.140	4.350	6.060	7.290	9.240	11.070	15.090
6	0.870	1.640	5.350	7.230	8.560	10.64	12.590	16.810
7	1.240	2.170	6.350	8.380	9.800	12.020	14.070	18.480
8	1.650	2.730	7.340	9.520	11.030	13.360	15.510	20.090
9	2.090	3.320	8.340	10.660	12.240	14.680	16.920	21.670
10	2.560	3.940	9.340	11.780	13.440	15.990	18.310	23.210
11	3.050	4.580	10.340	12.900	14.630	17.280	19.680	24.720
12	3.570	5.230	11.340	14.010	15.810	18.550	21.030	26.220
13	4.110	5.890	12.340	15.120	16.980	19.810	22.360	27.690
14	4.660	6.570	13.340	16.220	18.150	21.060	23.680	29.140
15	5.230	7.260	14.340	17.320	19.310	22.310	25.000	30.580
16	5.810	7.960	15.340	18.420	20.460	23.540	26.300	32.000
17	6.410	8.670	16.340	19.510	21.620	24.770	27.590	33.410
18	7.020	9.390	17.340	20.600	22.760	25.990	28.870	34.800
19	7.630	10.120	18.340	21.690	23.900	27.200	30.140	36.190
20	8.260	10.850	19.340	22.780	25.040	28.410	31.410	37.570
21	8.900	11.590	20.340	23.860	26.170	29.620	32.670	38.930
22	9.540	12.340	21.340	24.940	27.300	30.810	33.920	40.290
23	10.200	13.090	22.340	26.020	28.430	32.010	35.170	41.640
24	10.860	13.850	23.340	27.100	29.550	33.200	36.420	42.980
25	11.520	14.610	24.340	28.170	30.680	34.680	37.650	44.310
26	12.200	15.380	25.340	29.250	31.800	35.560	38.880	45.640
27	12.880	16.150	26.340	30.320	32.910	36.740	40.110	46.960
28	13.560	16.930	27.340	31.390	34.030	37.920	41.340	48.280
29	14.260	17.710	28.340	32.460	35.140	39.090	42.560	49.590
30	14.950	18.490	29.340	33.530	36.250	40.260	43.770	50.890

Examples: (i) $\chi^2_{0.05} = 11.07$ for $v = 5$.

F–Distribution $F_{0.05}(n_1, n_2)$

$n_1 \backslash n_2$	1	2	3	4	5	6	7	8	9	10	12
1	161.45	199.50	215.71	224.58	230.16	233.99	236.77	238.88	240.54	241.88	243.91
2	18.513	19.000	19.164	19.247	19.296	19.330	19.353	19.371	19.385	19.396	19.413
3	10.128	9.5521	9.2766	9.1172	9.0135	8.9406	8.8868	8.8452	8.8123	8.7855	8.7446
4	7.7086	6.9443	6.5914	6.3883	6.2560	6.1631	6.0942	6.0410	5.9988	5.9644	5.9117
5	6.6079	5.7861	5.4095	5.1922	5.0503	4.9503	4.8753	4.8183	4.7725	4.7351	4.6777
6	5.9874	5.1433	4.7571	4.5337	4.3874	4.2839	4.2066	4.1468	4.0990	4.0600	3.9999
7	5.5914	4.7374	4.3468	4.1203	3.9715	3.8660	3.7870	3.7257	3.6767	3.6365	3.5747
8	5.3177	4.4590	4.0662	3.8378	3.6875	3.5806	3.5005	3.4381	3.3881	3.3472	3.2840
9	5.1174	4.2565	3.8626	3.6331	3.4817	3.3738	3.2927	3.2296	3.1789	3.1373	3.0729
10	4.9646	4.1028	3.7083	3.4780	3.3258	3.2172	3.1355	3.0717	3.0204	2.9782	2.9130
11	4.8443	3.9823	3.5874	3.3567	3.2039	3.0946	3.0123	2.9480	2.8962	2.8536	2.7876
12	4.7272	3.8853	3.4903	3.2502	3.1059	2.9961	2.9134	2.8486	2.7964	2.7534	2.6866
13	4.6672	3.8056	3.4105	3.1791	3.0254	2.9153	2.8321	2.7669	2.7144	2.6710	2.6037
14	4.6001	3.7389	3.3439	3.1122	2.9582	2.8477	2.7642	2.6987	2.6458	2.6021	2.5342
15	4.5431	3.6823	3.2874	3.0556	2.9013	2.7905	2.7066	2.6408	2.5876	2.5437	2.4753
16	4.4940	3.6337	3.2389	3.0069	2.8524	2.7413	2.6572	2.5911	2.5377	2.4935	2.4247
17	4.4513	3.5915	3.1968	2.9647	2.8100	2.6987	2.6143	2.5480	2.4943	2.4499	2.3807
18	4.4139	3.5546	3.1599	2.9277	2.7729	2.6613	2.5767	2.5102	2.4563	2.4117	2.3421
19	4.3808	3.5219	3.1274	2.8951	2.7401	2.6283	2.5435	2.4768	2.4227	2.3779	2.3080
20	4.3513	3.4928	3.0984	2.8661	2.7100	2.5990	2.5140	2.4471	2.3928	2.3479	2.2776
21	4.3248	3.4668	3.0725	2.8401	2.6848	2.5727	2.4876	2.4205	2.3661	2.3210	2.2504
22	4.3009	3.4434	3.0491	2.8167	2.6613	2.5491	2.4638	2.3965	2.3419	2.2967	2.2258
23	4.2793	3.4221	3.0280	2.7955	2.6500	2.5277	2.4422	2.3748	2.3201	2.2747	2.2036
24	4.2597	3.4028	3.0088	2.7763	2.6207	2.5082	2.4226	2.3551	2.3002	2.2547	2.1834
25	4.2417	3.3852	2.9912	2.7587	2.6030	2.4904	2.4047	2.3371	2.2821	2.2365	2.1649
26	4.2252	3.3690	2.9751	2.7426	2.5868	2.4741	2.3883	2.3205	2.2655	2.2197	2.1479
27	4.2100	3.3541	2.9604	2.7278	2.5719	2.4591	2.3732	2.3053	2.2501	2.2043	2.1323
28	4.1960	3.3404	2.9467	2.7141	2.5581	2.4453	2.3593	2.2913	2.2360	2.1900	2.1179
29	4.1830	3.3277	2.9340	2.7014	2.5454	2.4324	2.3463	2.2782	2.2229	2.1768	2.1045
30	4.1709	3.3158	2.9223	2.6896	2.5336	2.4205	2.3343	2.2662	2.2107	2.1646	2.0921

Examples: (i) $F_{0.05}(3.8) = 4.0662$ (ii) $F_{0.05}(2.9) = 4.2565$ (iii) $F_{0.05}(5.24) = 2.6207$.

Index